Theory of Machines

Kinematics and Dynamics

Third Edition

Sadhu Singh
Ex-Director (Colleges)
Punjab Technical University, Jalandhar

PEARSON

Delhi • Chennai • Chandigarh

Assistant Editor–Acquisitions: Gauravjeet Singh Reen
Associate Editor–Production: M. R. Ramesh

ISBN 978-81-317-6069-7

First Impression

Published by Dorling Kindersley (India) Pvt. Ltd, licensees of Pearson Education in South Asia.

Head Office: 7th Floor, Knowledge Boulevard, A-8(A), Sector 62, Noida 201 309, UP, India
Registered Office: 11 Community Centre, Panchsheel Park, New Delhi 110 017, India

Laser typeset by Nittany Creative Solutions Ltd
Printed at HT Media Ltd., Noida

To my wife,
Smt. Manjit Kaur

About the Author

Dr Sadhu Singh retired as Professor and Head of the Department of Mechanical Engineering, Govind Ballabh Pant University of Agriculture and Technology, Pantnagar, Uttarakhand. An honours graduate in mechanical engineering from Punjab University, Chandigarh, he did his M.Sc. (Mechanical Design and Production Engineering) and received his Ph.D. from Kurukshetra University. He was Director (Colleges), Punjab Technical University, Jalandhar, and has taught at Regional Engineering College (now National Institute of Technology), Kurukshetra, for 15 years and later at Pantnagar University for 19 years.

Dr Singh has published more than 50 research papers in journals of national and international repute and has authored many textbooks in mechanical engineering. He is a renowned academician, researcher, writer and an able administrator.

Brief Contents

Contents

Preface to the Third Edition

This third enlarged edition of the book is presented after thorough revision, incorporating the valuable suggestions sent by readers on enhancing the contents of the second edition. The book is ideal for students of B.Tech./B.E. students of mechanical, production, industrial, aeronautical, agricultural, civil, chemical, metallurgical and automobile engineering of all Indian and foreign universities. I am thankful to the readers who have adopted this book as a text or reference book.

The chapter on kinematics has been split into two chapters: Velocity in Mechanisms and Acceleration in Mechanisms. Three new chapters on Kinematic Synthesis of Planar Mechanisms, Mechanical Vibrations and Automatic Control have been added. Ample multiple-choice questions from IES, IAS and GATE examinations, with explanatory notes, have been added for the benefit of students who aspire to appear for competitive examinations. Each chapter has been enriched with solved numerical examples for the benefit of the readers. An added feature of the book is the inclusion of Summary for Quick Revision at the end of each chapter to enable quick recapitulation of the discussed concepts. I hope that readers will like the book even more in its present form.

Utmost care has been taken while proofreading to make the text error-free. However, it is possible that a few errors might have been left unnoticed inadvertently. Suggestions from readers are invited for improving the book further.

I appreciate the continued support received from my family while preparing the manuscript for this book.

Sadhu Singh

Preface

A machine is a device which transforms energy available in one form to another to do certain type of useful work. A machine consists of many mechanisms and they in turn comprise of many links. The subject of **Theory of Machines** is concerned with the kinematics and kinetics of the various links of the machine. The elasticity of the links and the effect of clearances in the joints of the links in transmitting motion is neglected, as this is beyond the scope of this book.

An undergraduate course on this subject is studied by most of the students of engineering and technology. The course material in this book has been limited to undergraduate level. The author has taught this course for over thirty years and his rich experience has been reflected in the book. The reading material has been presented in a simple manner which can be easily assimilated by the readers.

The book contains twelve chapters. The first chapter deals with various types of mechanisms, their velocity and acceleration diagrams. This is the fundamental chapter and more stress must be laid in understanding the various concepts. Second chapter deals with the lower pairs to generate intermittent, approximate and accurate straight line motion. The working of Oldham's coupling, automobile steering gears, parallel linkages and engine pressure indicators have been explained in this chapter. The gyroscopic and precessional motion is described in chapter three and laws of friction in chapter four. Power transmission methods by belts, chains and ropes are explained in chapter five. Brakes, clutches, and dynamometers are dealt with in chapter six, governors in chapter seven and cams in chapter eight. The effect of inertia forces and turning moment is described in chapter nine. Chapter ten is concerned with the problems of balancing of rotating and reciprocating parts. The various types of gears and gear trains are described in chapter eleven and twelve respectively.

All chapters contain a large number of solved and unsolved problems. The author has taken utmost care in solving the numerical problems. However, errors if still left, may be brought to his notice.

An added feature of the book is the inclusion of Machine Theory Laboratory Practice, Glossary of terms and Multiple-choice Questions as appendices. These shall be highly useful to the readers.

It is hoped that the book in its present form shall be liked by the readers. Suggestions for the improvement of the book shall be welcomed.

I express my sincere thanks to all my family members for their patience and support during the writing of this book. The financial support provided by the publishers to partially meet the expenses to prepare the manuscript of the book is highly acknowledged.

Sadhu Singh

1

MECHANISMS

Chapter Outline

1.1 INTRODUCTION

A mechanism is a set of machine elements or components or parts arranged in a specific order to produce a specified motion. The machine elements or components are considered rigid or resistant bodies that do not deform under the action of forces. Resistant bodies are bodies that do not suffer appreciable distortion or change in physical form due to forces acting on them, e.g. springs, belts, and fluids. Elastic bodies are also resistant bodies. They are capable of transmitting the required forces with negligible deformation. Rigid bodies are bodies that do not deform under the action of forces. All resistant bodies are considered rigid bodies for the purpose of transmitting motion. In this chapter, we shall study the different ways of connecting rigid (resistant) bodies to obtain various types of mechanisms.

Kinematics is a subject that deals with the study of relative motion of parts constituting a machine, neglecting forces producing the motion. A structure is an assemblage of a number of resistant bodies meant to take up loads or subjected to forces having straining actions, but having no relative motion between its members. Frame is a structure that supports the moving parts of a machine.

1.2 KINEMATIC JOINT

A kinematic joint is the connection between two links by a pin. There is clearance between the pin and the hole in the ends of the links being connected so that there is free motion of the links.

1.2.1 Type of Kinematic Joints

The type of kinematic joints generally used in mechanisms are:

1. *Binary joint*: In a binary joint, two links are connected at the same joint by a pin, as shown in Fig.1.1(a).
2. *Ternary joint*: In a ternary joint, three links are connected at the same joint by a pin. It is equivalent to two binary joints. In Fig.1.1(b), joints B and C are ternary joints and others are binary joints.
3. *Quaternary joint*: When four links are connected at the same joint by a pin, it is called a quaternary joint. One quaternary joint is equivalent to four binary joints. In Fig.1.1(c), joint B is a quaternary joint; A, C, E, F are ternary joints; and D is a binary joint.

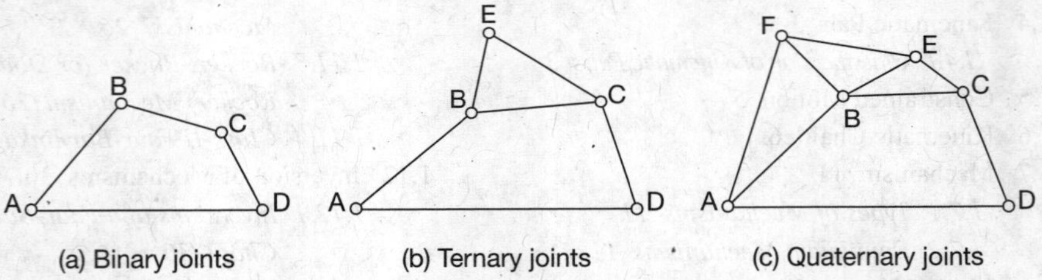

Fig.1.1 Type of kinematic joints

1.3 ELEMENTS OR LINKS

A link (or element or kinematic link) is a resistant body (or assembly of resistant bodies) constituting a part (or parts) of the machine, connecting other parts, which have motion, relative to it. A slider crank mechanism of an internal combustion engine, shown in Fig.1.2, consists of four links, namely, (1) frame, (2) crank, (3) connecting rod and (4) slider.

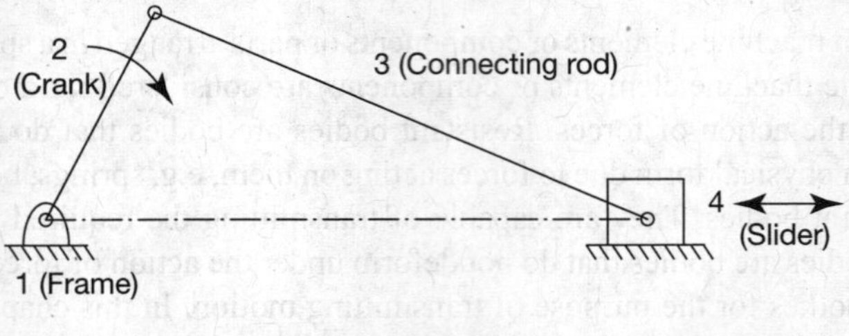

Fig.1.2 Kinematic links of a slider crank mechanism

1.3.1 Classification of Links

Links can be classified as binary, ternary, or quaternary depending upon the ends on which revolute or turning pairs can be placed, as shown in Fig.1.3. A binary link has two vertices, a ternary has three vertices, and a quaternary link has four vertices, and so on.

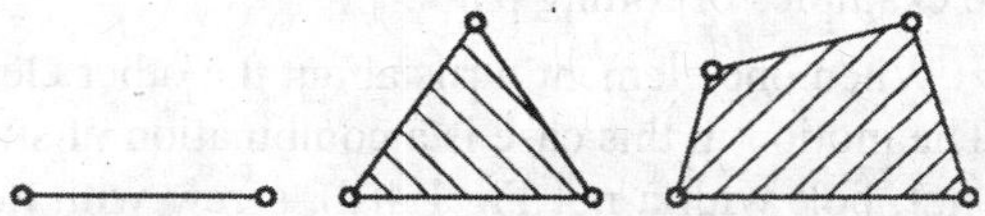

Fig.1.3 Types of links

There are four types of links: rigid, flexible, fluid, and floating links.

- *Rigid link*: A rigid link does not undergo any deformation while transmitting motion. Links in general are elastic in nature. They are considered rigid if they do not undergo appreciable deformation while transmitting motion, e.g. connecting rod, crank, tappet rod, etc.
- *Flexible link*: A flexible link is one which while transmitting motion is partly deformed in a manner not to affect the transmission of motion, e.g. belts, ropes, chains, springs, etc.
- *Fluid link*: A fluid link is deformed by having fluid in a closed vessel and the motion is transmitted through the fluid by pressure, as in the case of a hydraulic press, hydraulic jack, and fluid brake.
- *Floating link*: It is a link which is not connected to the frame.

1.4 KINEMATIC PAIR

The two links of a machine, when in contact with each other, are said to form a pair. A kinematic pair consists of two links that have relative motion between them. In Fig.1.2, links 1 and 2, 2 and 3, 3 and 4, and 4 and 1 constitute kinematic pairs.

1.4.1 Classification of Kinematic Pairs

Kinematic pairs may be classified according to the following considerations:

- Type of relative motion
- Type of contact
- Type of mechanical constraint.

1. Kinematic Pairs According to the Relative Motion

- *Sliding pair*: It consists of two elements connected in such a manner that one is constrained to have sliding motion relative to another. For example, a rectangular bar in a rectangular hole (Fig.1.4(a)), piston and cylinder of an engine, cross-head and guides of a steam engine, ram and its guides in a shaper, tailstock on the lathe bed, etc. all constitute sliding pairs.
- *Turning (revolute) pair*: It consists of two elements connected in such a manner that one is constrained to turn or revolve about a fixed axis of another element. For example, a shaft with

collar at both ends revolving in a circular hole (Fig.1.4(b)) crankshaft turning in a bearing, cycle wheels revolving over their axles, etc. all constitute turning pairs.

- *Rolling pairs*: When two elements are so connected that one is constrained to roll on another element which is fixed, forms a rolling pair. Ball and roller bearings, a wheel rolling on a flat surface (Fig.1.4(c)) are examples of rolling pairs.
- *Screw (or helical) pair*: When one element turns about the other element by means of threads, it forms a screw pair. The motion in this case is a combination of sliding and turning. The lead screw of a lathe with nut, bolt with a nut Fig.1.4(d), screw with nut of a jack, etc. are some examples of screw pairs.
- *Spherical pair*: When one element in the form of a sphere turns about the other fixed element, it forms a spherical pair. The ball and socket joint Fig.1.4(e), pen stand, the mirror attachment of vehicles, etc. are some examples of spherical pair.

2. Kinematic Pairs According to the Type of Contact
   - *Lower pair*: When the two elements have surface (or area) contact while in motion and the relative motion is purely turning or sliding, they are called a lower pair. All sliding pairs, turning pairs, and screw pairs form lower pairs. For example, nut turning on a screw, shaft rotating in a bearing, universal joint, all pairs of a slider crank mechanism, pantograph etc., are lower pairs.

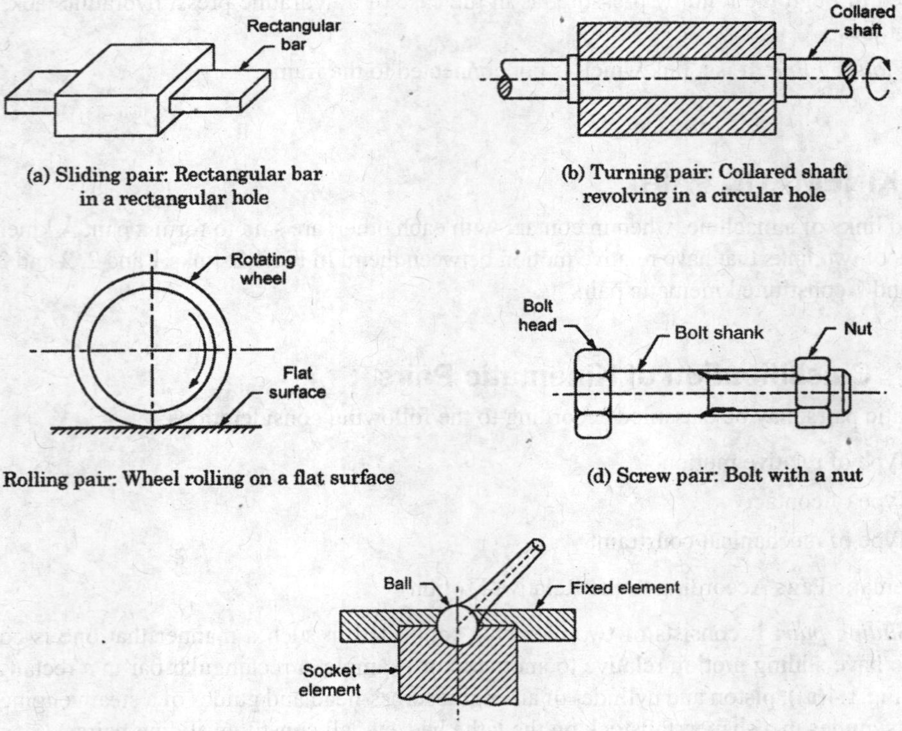

(a) Sliding pair: Rectangular bar in a rectangular hole

(b) Turning pair: Collared shaft revolving in a circular hole

(c) Rolling pair: Wheel rolling on a flat surface

(d) Screw pair: Bolt with a nut

(e) Spherical pair: Ball and socket joint

Fig.1.4 Types of kinematic pairs according to the type of relative motion

- *Higher pairs*: When the two elements have point or line contact while in motion and the relative motion being the combination of sliding and turning, then the pair is known as a higher pair. Belts, ropes, and chains drive, gears, cam and follower, ball and roller bearings, wheel rolling on a surface, etc., all form higher pairs.

3. Kinematic Pairs According to the Type of Mechanical Constraint

- *Closed pair*: When the two elements of a pair are held together mechanically in such a manner that only the required type of relative motion occurs, they are called a closed pair. All lower pairs and some higher pairs (e.g. enclosed cam and follower) are closed pairs (Fig.1.5(a)).
- *Unclosed pair*: When the two elements of a pair are not held mechanically and are held in contact by the action of external forces, are called unclosed pair, e.g. cam and spring loaded follower pair (Fig.1.5(b)).

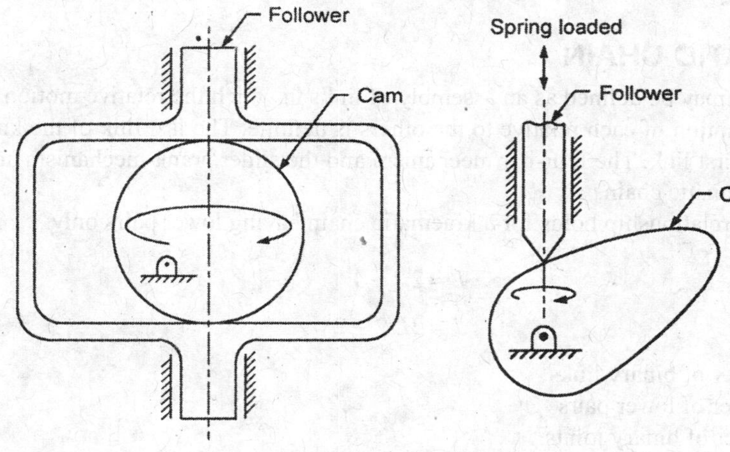

(a) Closed pair (b) Unclosed pair

Fig.1.5 Closed and unclosed pairs

1.5 CONSTRAINED MOTION

The three types of constrained motion are as follows:

- *Completely constrained motion*: When the motion between a pair takes place in a definite direction irrespective of the direction of force applied, then the motion is said to be a completely constrained motion. For example, a square bar in a square hole, a shaft with collars at each end in a circular hole, a piston in the cylinder of an internal combustion engine, have all completely constrained motion.
- *Partially (or successfully) constrained motion*: When the constrained motion between a pair is not completed by itself but by some other means, it is said to be successfully constrained motion. For example, the motion of a shaft in a footstep bearing becomes successfully constrained motion when compressive load is applied to the shaft (Fig.1.6(a)).
- *Incompletely constrained motion*: When the motion between a pair can take place in more than one direction, it is said to be incompletely constrained motion, e.g. a circular shaft in a circular hole. (Fig.1.6(b)).

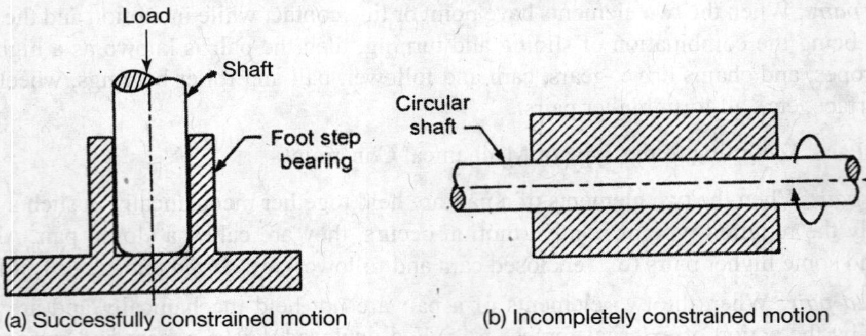

Fig.1.6 Types of constrained motion

1.6 KINEMATIC CHAIN

A kinematic chain may be defined as an assembly of links in which the relative motion of the links is possible and the motion of each relative to the others is definite. The last link of the kinematic chain is attached to the first link. The four-bar mechanism and the slider crank mechanism are some of the examples of a kinematic chain.

The following relationship holds for a kinematic chain having lower pairs only:

$$L = 2P - 4 \tag{1.1a}$$

$$J = 3L/2 - 2 \tag{1.1b}$$

where L = number of binary links
P = number of lower pairs
J = number of binary joints.

If LHS > RHS, then chain is called locked chain or redundant chain.
LHS = RHS, then chain is constrained
LHS < RHS, then chain is unconstrained

For a kinematic chain having higher pairs, each higher pair is taken equivalent to two lower pairs and an additional link. In that case,

$$J + \frac{H}{2} = \frac{3}{2}L - 2 \tag{1.1c}$$

where H = number of higher pairs.

Example 1.1

A chain with three links is shown in Fig.1.7. Prove that the chain is locked.

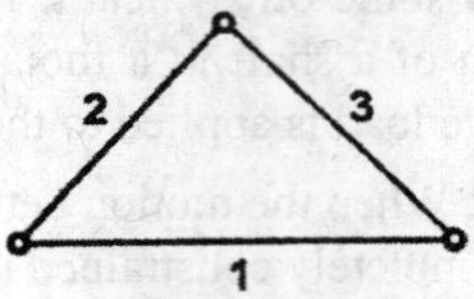

Fig.1.7 Three-bar chain

■ **Solution**

Number of binary joints, $J = 3$
Number of binary links, $L = 3$
Number of lower pairs, $P = 3$

Now
$$L = 2P - 4$$
$$3 = 2 \times 3 - 4 = 2$$
$\therefore$
$$\text{LHS} > \text{RHS}$$

Also
$$J = \left(\frac{3}{2}\right) L - 2$$
$$3 = \left(\frac{3}{2}\right) \times 3 - 2 = 2.$$
$\therefore$
$$\text{LHS} > \text{RHS}$$

Therefore, it is a locked chain.

Example 1.2

A four-bar chain is shown in Fig.1.8. Prove that it is a constrained chain.

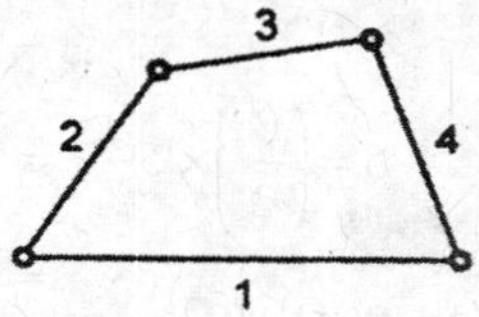

Fig.1.8 Four-bar chain

■ **Solution**

Number of binary joints, $J = 4$
Number of binary links, $L = 4$
Number of lower pairs, $P = 4$

Now
$$L = 2P - 4$$
$$4 = 2 \times 4 - 4 = 4$$
$\therefore$
$$\text{LHS} = \text{RHS}$$

Also
$$J = \left(\frac{3}{2}\right) L - 2$$
$$4 = \left(\frac{3}{2}\right) \times 4 - 2 = 4$$
$\therefore$
$$\text{LHS} = \text{RHS}$$

Therefore, it is a constrained chain.

Example 1.3

A five-bar chain is shown in Fig.1.9. Prove that it is an unconstrained chain.

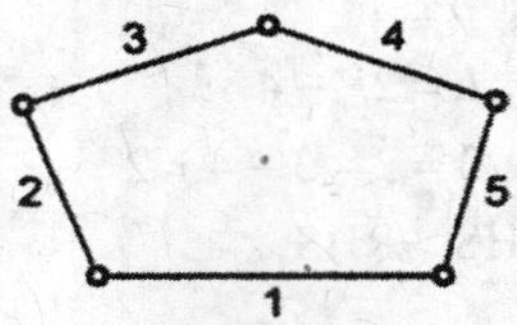

Fig.1.9 Five-bar chain

■ Solution

Number of binary joints, $J = 5$
Number of binary links, $L = 5$
Number of lower pairs, $P = 5$

Now
$$L = 2P - 4$$
$$5 = 2 \times 5 - 4 = 6$$

∴ LHS < RHS

Also
$$J = \left(\frac{3}{2}\right) L - 2$$
$$5 = \left(\frac{3}{2}\right) \times 5 - 2 = 5.5$$

∴ LHS < RHS

Therefore, it is an un-constrained chain.

Example 1.4

Show that the chain shown in Fig.1.10 is an unconstrained kinematic chain.

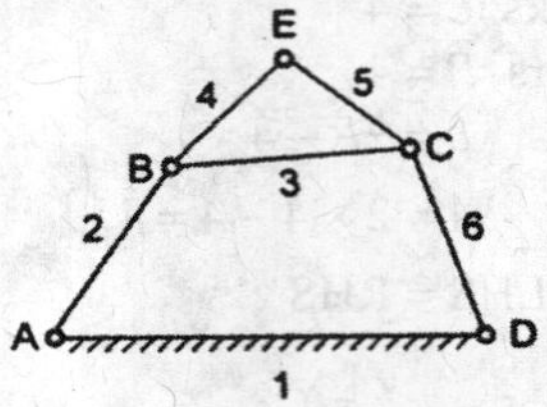

Fig.1.10 Six-bar chain

■ Solution

Number of binary joints, $J = 7$ ($A = 1, B = 2, C = 2, D = 1, E = 1$)
Number of binary links, $L = 6$
Number of lower pairs, $P = 9$ ($1 - 2, 2 - 3, 3 - 4, 2 - 4, 4 - 5, 5 - 3, 5 - 6, 3 - 6, 1 - 6$)

Now
$$L = 2P - 4$$
$$6 = 2 \times 9 - 4 = 14$$
$$\therefore \quad \text{LHS} < \text{RHS}$$

Also
$$J = \left(\frac{3}{2}\right) L - 2$$
$$7 = \left(\frac{3}{2}\right) \times 6 - 2 = 7$$
$$\therefore \quad \text{LHS} < \text{RHS}$$

Therefore, it is an un-constrained chain.

Example 1.5

Show that the chain shown in Fig.1.11 is not a kinematic chain.

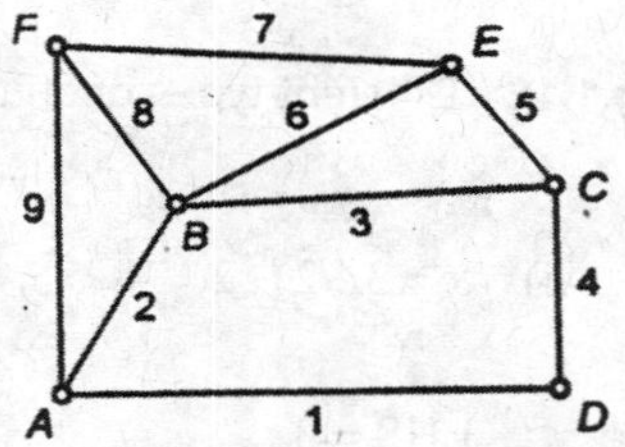

Fig.1.11 Nine-bar chain

■ Solution

Number of binary joints, $J = 13$ ($A = 2, B = 4, C = 2, D = 1, E = 2, F = 2$)
Number of binary links, $L = 9$
Number of lower pairs, $P = 13$ ($D = 1$, $A, C, E, F = 2$ each, $B = 4$)

Now
$$L = 2P - 4$$
$$9 = 2 \times 13 - 4 = 22$$
$$\text{LHS} < \text{RHS}$$

Also
$$J = \left(\frac{3}{2}\right) L - 2$$
$$13 = \left(\frac{3}{2}\right) \times 9 - 2 = 11.5$$
$$\text{LHS} > \text{RHS}$$

Therefore, it is not a kinematic chain. It is a locked chain or a frame.

Example 1.6

Determine the type of chain in Fig.1.12(a)–(e).

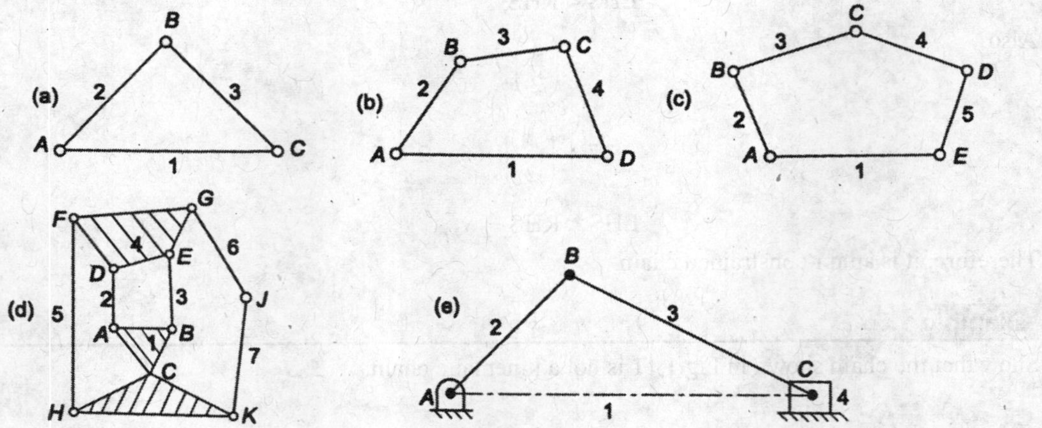

Fig.1.12 Different types of chains

■ Solution

(a) (i) $L = 2P - 4$

$L = 3, P = 3, J = 3$

LHS = 3

RHS = $2 \times 3 - 4 = 2$

LHS > RHS

(ii) $J = 3L/2 - 2$

LHS = 3

RHS = $3 \times 3/2 - 2 = 2.5$

LHS > RHS

It is a locked chain and not a kinematic chain.

(b) $L = 4, P = 4, J = 4$

LHS = 4	LHS = 4
RHS = $2 \times 4 - 4 = 4$	RHS = $3 \times 4/2 - 2 = 4$
LHS = RHS	LHS = RHS

It is a constrained kinematic chain.

(c) $L = 5, P = 5, J = 5$

LHS = 5	LHS = 5
RHS = $2 \times 5 - 4 = 6$	RHS = $3 \times 5/2 - 2 = 5.5$
LHS < RHS	LHS < RHS

It is an unconstrained chain and not a kinematic chain.

(d) $L = 6, P = 5, J = 7$

LHS = 6	LHS = 7
RHS = $2 \times 5 - 4 = 6$	RHS = $3 \times 6/2 - 2 = 7$
LHS = RHS	LHS = RHS

It is a constrained kinematic chain.

(e) $L = 4, P = 4, J = 4$

LHS = 4	LHS = 4
RHS = $2 \times 4 - 4 = 4$	RHS = $3 \times 4/2 - 2 = 4$
LHS = RHS	LHS = RHS

It is a constrained kinematic chain.

1.7 MECHANISM

When one of the links of a kinematic chain is fixed, the chain is called a mechanism.

1.7.1 Types of Mechanisms

The mechanisms are of the following types:

- *Simple mechanism*: A mechanism which has four links.
- *Compound mechanism*: A mechanism which has more than four links.
- *Complex mechanism*: It is formed by the inclusion of ternary or higher order floating link to a simple mechanism.
- *Planar mechanism*: When all the links of the mechanism lie in the same plane.
- *Spatial mechanism*: When the links of the mechanism lie in different planes.

1.7.2 Equivalent Mechanisms

Turning pairs of plane mechanisms may be replaced by other types of pairs such as sliding pairs or cam pairs. The new mechanism thus obtained having the same number of degrees of freedom as the original mechanism is called the equivalent mechanism. The equivalent mechanism will have same degrees of freedom and shall be kinematically similar.

The following rules may be used to obtain the equivalent mechanism:

1. A sliding pair is equivalent to a turning pair, as shown in Fig.1.13(a).
2. A spring can be replaced by two binary links, as shown in Fig.1.13(b).
3. A cam pair can be replaced by one binary link together with two turning pairs at each end, as shown in Fig.1.13(c).

1.8 MECHANISM AND MACHINES

A machine is a device that transforms energy available in one form to another to do certain type of desired useful work. The parts of the machine move relative to one another. Its links may transmit both power and motion. On the other hand, a mechanism is a combination of rigid or restraining bodies, which are so shaped and connected that they move upon each other with definite relative motion. A mechanism is obtained when one of the links of the kinematic chain is fixed. A machine is a combination of two or more mechanisms arranged in such a way so as to obtain the required motion and transfer the energy to some desired point by the application of energy at some other convenient point. A machine is not able to move itself and must get the motive power from some source.

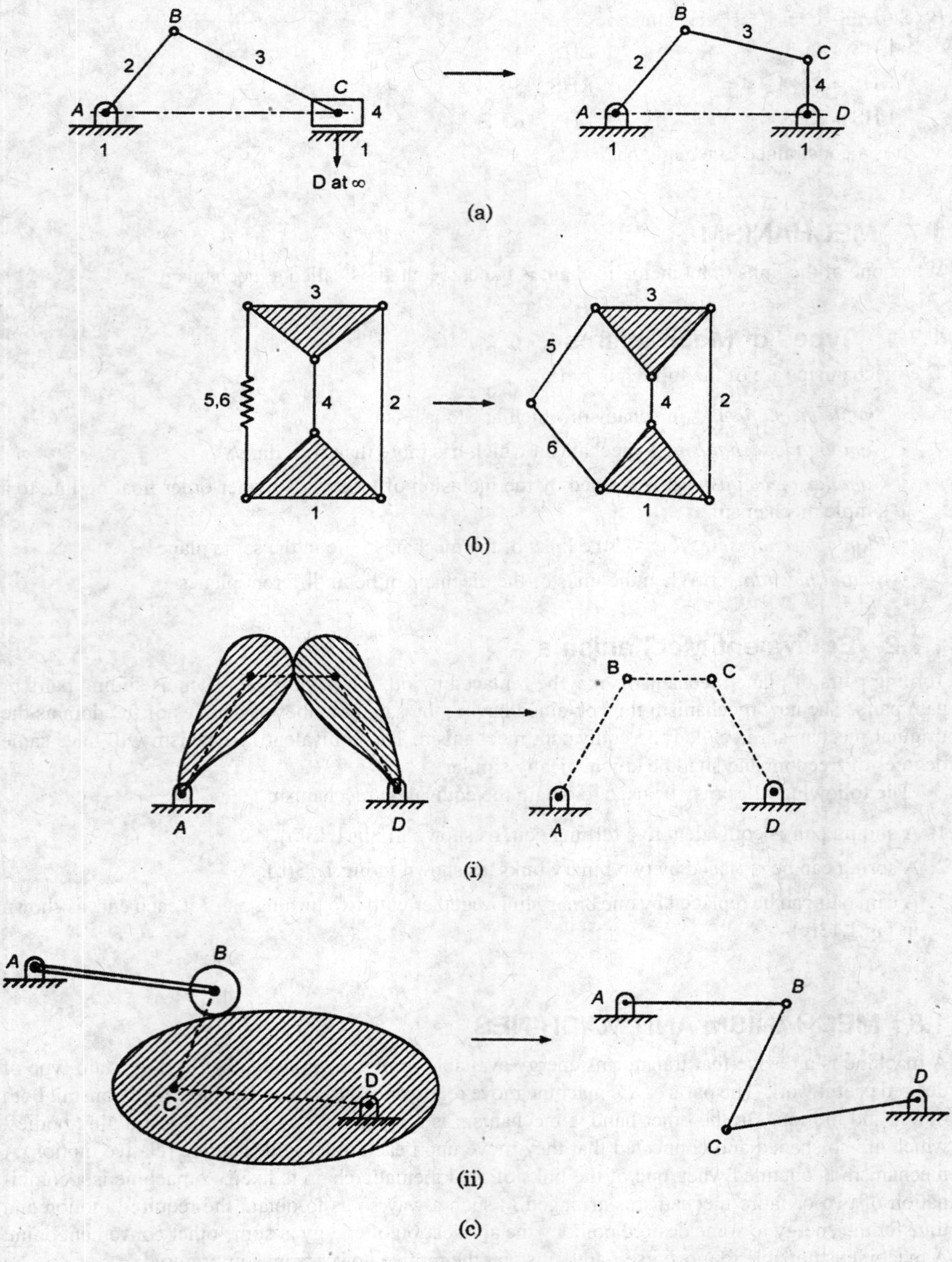

Fig.1.13 Equivalent mechanisms

Some examples of mechanisms are: slider crank, typewriter, clocks, watches, spring toys, etc. Steam engine, internal combustion engine, lathe, milling machine, drilling machine, etc. are some examples of machines.

1.8.1 Classification of Machines

The machines may be classified as the following:

- *Simple machine*: In a simple machine, there is one point of application for the effort and one point for the load to be lifted. Some examples of simple machines are lever, screw jack, inclined plane, bicycle, etc.
- *Compound machine*: In a compound machine, there are more than one point of application for the effort and the load. It may be thought of as a combination of many simple machines. Some examples of compound machines are lathe machine, grinding machine, milling machine, printing machine, etc.

1.9 DEGREES OF FREEDOM

An unconstrained rigid body moving in space can have three translations and three rotational motions (i.e. six motions) about the three mutually perpendicular axes. Degrees of freedom of a kinematic pair is defined as the number of independent relative motions, both translational and rotational, a kinematic pair can have.

$$\text{Degrees of freedom} = 6 - \text{number of restraints} \tag{1.2}$$

The degrees of freedom of some of the systems are as follows:

- A rigid body has 6 degrees of freedom.
- A rectangular bar sliding in a rectangular hole has one degree of freedom as the motion can be expressed by the linear displacement only.
- The position of the crank of a slider crank mechanism can be expressed by the angle turned through and thus has one degree of freedom.
- A circular shaft rotating in a hole and also translating parallel to its axis has two degrees of freedom, i.e. angle turned through and displacement.
- A ball and a socket joint has three degrees of freedom.

1.9.1 Degrees of Freedom of Planar Mechanisms

- *Mobility of a mechanism*: The mobility of a mechanism is defined as the number of degrees of freedom it possesses. An equivalent definition of mobility is the minimum number of independent parameters required to specify the location of every link within a mechanism.
- *Kutzbach criterion*: The Kutzbach criterion for determining the number of degrees of freedom of a planar mechanism is:

$$F = 3(n - 1) - 2p - h \tag{1.3}$$

where F = degrees of freedom

n = total number of links in a mechanism out of which one is a fixed link.

$n - 1$ = number of movable links
p = number of simple joints or lower pairs having one degree of freedom
h = number of higher pairs having two degrees of freedom and so on.

When two links are joined by a hinge, two degrees of freedom are lost. Hence for each joint two degrees of freedom are lost. Therefore, for p number of joints the number of degrees of freedom lost are $2p$. When a kinematic chain is made up of different type of links, then the number of lower pairs p is computed as follows:

$$p = (1/2)\ [2n_2 + 3n_3 + 4n_4 + \cdots\cdots] \quad (1.4)$$

where n_2 = number of binary links
n_3 = number of ternary links, and so on.

To determine the degrees of freedom of a mechanism, the presence of a redundant link or redundant pair may also be considered.

(i) A mechanism may have one or more links which do not introduce any extra constraint. Such links are called redundant links (n_r) and should not be taken into account. Similarly redundant joints (p_r) should also not be taken into account.

In Fig.1.14(a), links 3 and 4 are parallel and are termed as redundant links, as none of them produces extra constraint. By removing one of the two links, the motion remains the same. So one of the two links is considered for calculating the degrees of freedom.

The corresponding kinematic pairs either between links 4 and 2, and 4 and 5; or 3 and 2, and 3 and 5 are considered as redundant pair. Therefore, either of the two links and the corresponding kinematic pair should be considered while calculating the degrees of freedom.

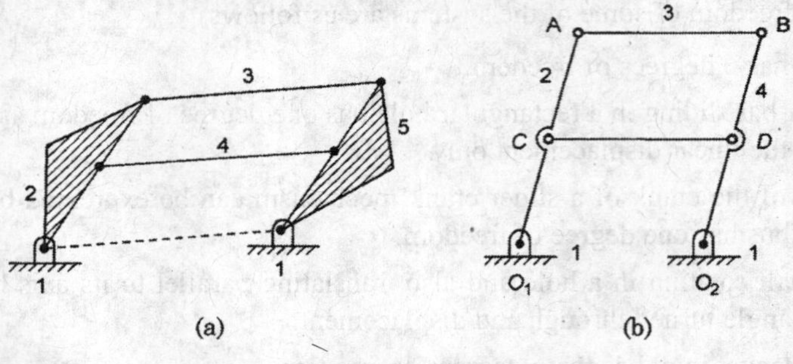

Fig.1.14

In Fig.1.14(b), links AB and CD are identical and each leads to same constraint.

(ii) Sometimes one or more links of a mechanism may have redundant degrees of freedom. If a link can be moved without causing any movement in the rest of the mechanism then the link is said to have redundant degree of freedom (F_r).

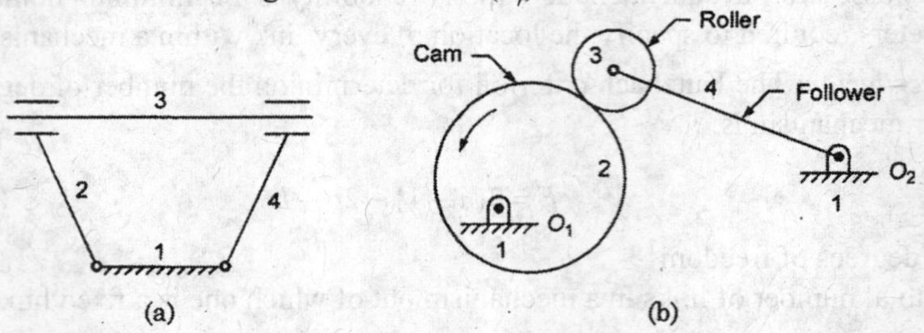

Fig.1.15

In Fig.1.15(a), the link 3 can slide without causing any movement to the mechanism. Thus link 3 represents one redundant degree of freedom. In Fig.1.15(b), roller can rotate without causing any movement in the rest of the mechanism.

Thus Eq. (1) can be modified as:

$$F = 3\,(n - n_r - 1) - 2(p - p_r) - h - F_r \tag{1.5}$$

1.9.2 Planar Mechanisms with Lower Pairs Only

For linkages with lower pairs only, $h = 0$, and

$$F = 3(n - 1) - 2p \tag{1.6}$$

A joint connecting k links at a single joint must be counted as $(k - 1)$ joints. Only four types of joints are commonly found in planar mechanisms. These are the revolute, the prismatic, the rolling contact joints (each having one degree of freedom), and the cam or gear joint (each having two degrees of freedom). These joints are depicted in Fig.1.16. The following definitions apply to the actual degrees of freedom of a device.

$F \geq 1$: the device is a mechanism with F degrees of freedom.
$F = 0$: the device is a statically determinate structure.
$F < -1$: the device is a statically indeterminate structure.

Joint type (Symbol)	Physical form	Schematic representation	Degrees of freedom
Revolute (R)	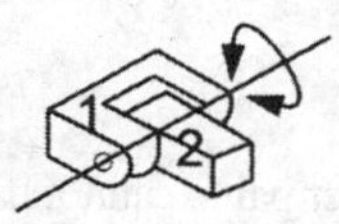	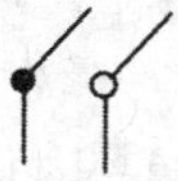	1 (Pure rotation)
Prismatic (P)	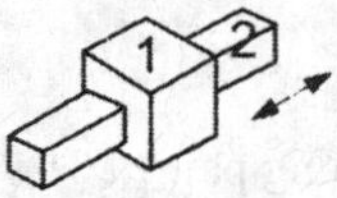		1 (Pure sliding)
Cam or gear	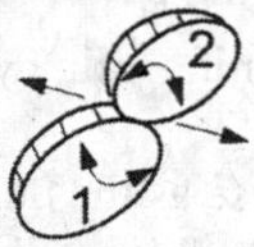	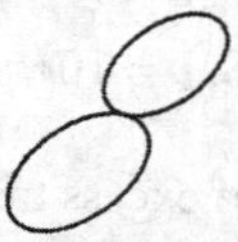	2 (Rolling and sliding)
Rolling contact		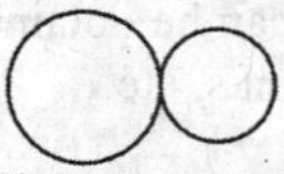	1 (Rolling without sliding)

Fig.1.16 Common types of joints found in planar mechanisms

The degrees of freedom of some of the planar mechanisms have been listed in Table 1.1.

Table 1.1 Degrees of Freedom of Planar Mechanisms

Mechanism	n	p	h	$F = 3(n-1) - 2p - h$
1. Three – bar	3	3	0	0
2. Four – bar	4	4	0	1
3. Five – bar	5	5	0	2
4. Five – bar	5	6	0	0
5. Six – bar	6	8	0	–1
6. Four – bar	4	5	0	–1
7. Three – bar	3	2	1	1
8. Four – bar	4	3	1	2
9. Five – bar	5	11/2	0	1
10. Six – bar	6	7	0	1

Gruebler's criterion: For a constrained motion, $F = 1$, so that

$$1 = 3(n-1) - 2p - h$$

or

$$2p + h - 3n + 4 = 0 \tag{1.7}$$

Eq. (1.7) represents the Gruebler's criterion.

If $h = 0$, then

$$p = 3n/2 - 2 \tag{1.8}$$

Therefore, a planar mechanism with $F = 1$ and having only lower pairs, cannot have odd number of links.

Eq. (1.8) is similar to Eq. (1.1 b) with $p = J$ and $n = L$. As p and n are to be whole numbers, the relation can be satisfied only if n is even.

For possible linkages made of binary links only,

$n = 4, p = 4$ No excess turning pair

$n = 6, p = 7$ One excess turning pair

$n = 8, p = 10$ Two excess turning pair

and so on.

Thus, we find that the number of excess turning pairs increase as the number of links increase. To get the required number of turning pairs from the same number of binary links is not possible. Therefore, the additional pairs or joints can be obtained only from the links having more than two joining points, i.e., ternary or quaternary links, etc.

1.10 FOUR-BAR CHAIN

A four-bar chain has been shown in Fig.1.17. It consists of four binary links. Link AD is fixed (called frame), AB is the crank (or driver link), BC is the coupler (or connecting rod), and CD the lever (or rocker or follower link). θ is the input angle and ϕ the angle of transmission. The coupler BC may be a ternary link. The number of degrees of freedom of the four-bar chain is one.

A link that makes complete revolutions is the *crank,* the link opposite to the fixed link is the *coupler,* and the fourth link a *lever* or *rocker*, if it oscillates or another *crank*, if it rotates.

The four-bar mechanism with all its pairs as turning pairs is called the "quadric cycle chain." When one of these turning pairs is replaced by a slider pair, the chain becomes "single slider chain." When two turning pairs are replaced by slider pairs, it is called a "double slider chain" or a "crossed double slider chain," depending on whether the two slider pairs are adjacent or crossed.

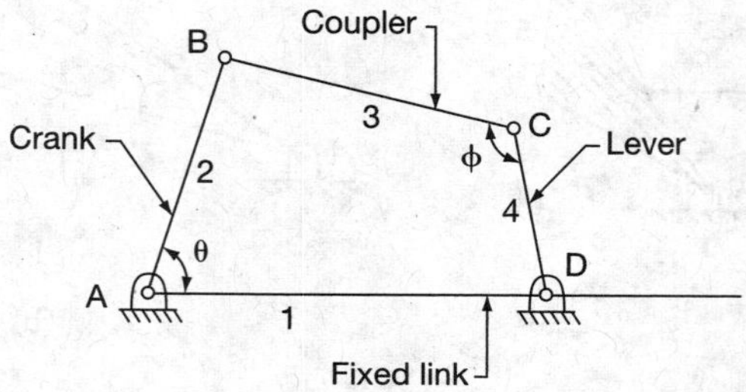

Fig.1.17 Four-bar chain

Example 1.7

Calculate the number of degrees of freedom of the linkages shown in Fig.1.18(a) and (b).

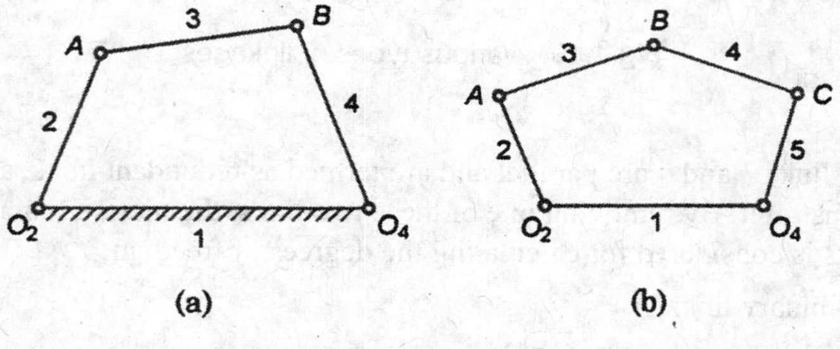

Fig.1.18 Four- and five-bar chains

■ Solution

(a) Number of binary links, $n = 4$

Number of lower pairs, $p = 4$

Degrees of freedom, $F = 3(n - 1) - 2p$

$= 3(4 - 1) - 2 \times 4 = 1$

(b) Number of binary links, $n = 5$

Number of lower pairs, $p = 5$

Degrees of freedom, $F = 3(n - 1) - 2p$

$= 3(5 - 1) - 2 \times 5 = 2$

Example 1.8

Calculate the number of degrees of freedom of the linkages shown in Fig.1.19(a) to (c).

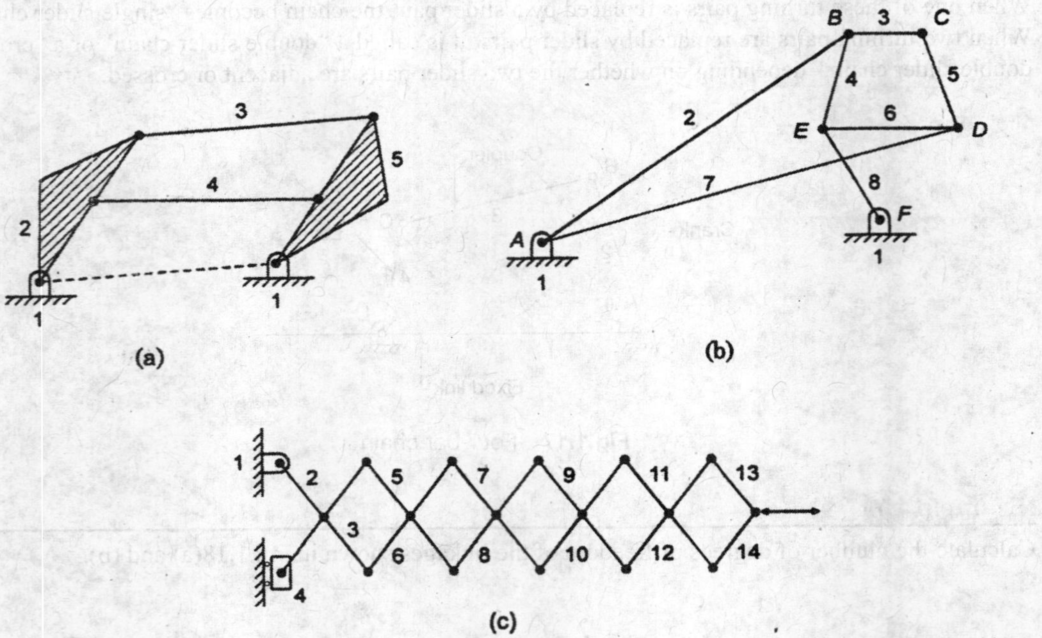

Fig.1.19 Various types of linkages

■ Solution

(a) In Fig.1.19(a), links 3 and 4 are parallel and are termed as redundant links, as one of them produces extra constraint. By removing one of these two links, the motion remains same. So one of the two links is considered for calculating the degrees of freedom.

Number of binary links, $n = 4$

Number of lower pairs, $p = 4$

Degrees of freedom, $F = 3\,(n - 1) - 2p$

$= 3\,(4 - 1) - 2 \times 4$

$= 9 - 8 = 1$

(b) Number of binary links, $n = 8$

Number of simple joints:

Binary joints at C and $F = 2$

Ternary joints at A, B, D and $E = 4$

$P = 2 + 2 \times 4 = 10$

Degrees of freedom, $F = 3\,(n - 1) - 2p$

$= 3\,(8 - 1) - 2 \times 10$

$= 21 - 20 = 1$

(c) Number of binary links, $n = 14$

Number of lower pairs, $p = 18$

Number of higher pairs, $h = 1$
(Slider can rotate and slide)

Degrees of freedom, $F = 3\,(n - 1) - 2p - h$

$= 3\,(14 - 1) - 2 \times 18 - 1$

$= 39 - 36 - 1 = 2$

Example 1.9

Determine the number of degrees of freedom of the mechanism shown in Fig.1.20(a) to (f).

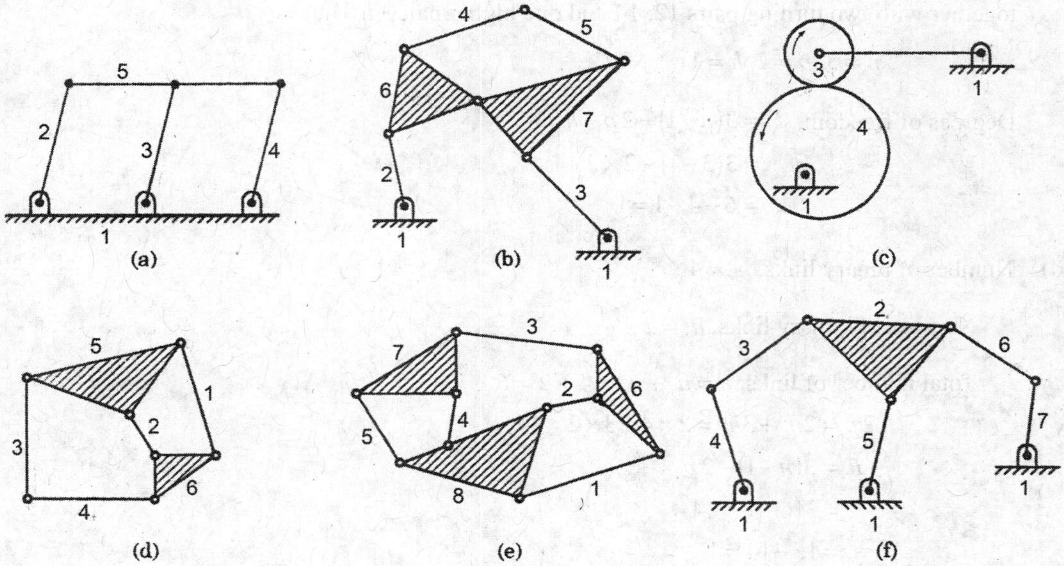

Fig.1.20 Various types of mechanisms

■ Solution

(a) Number of binary links, $n_2 = 3$

Number of ternary links, $n_3 = 2$

Total number of links, $n = n_2 + n_3 = 3 + 2 = 5$

$$2p = 2n_2 + 3n_3 = 2 \times 3 + 3 \times 2 = 12$$

Degrees of freedom, $F = 3(n-1) - 2p$

$$= 3(5-1) - 12$$

$$= 12 - 12 = 0$$

It is a structure.

(b) Number of binary links, $n_2 = 5$

Number of ternary links, $n_3 = 2$

Total number of links, $n = n_2 + n_3 = 5 + 2 = 7$

$$2p = 2n_2 + 3n_3 = 2 \times 5 + 3 \times 2 = 16$$

Degrees of freedom, $F = 3(n-1) - 2p$

$$= 3(7-1) - 16$$

$$= 18 - 16 = 2$$

(c) The roller 3 carried at the end of the output link 2 can be rotated without causing any motion in the rest of the mechanism. Thus roller 3 is a link with redundant degree of freedom. The roller can thus be considered welded to the output link. Hence, there are three binary links 1, 2 and 4 together with two turning pairs 12, 14 and one higher pair 34. Thus

$$n = 3, p = 2, h = 1$$

Degrees of freedom, $F = 3(n-1) - 2p - h$

$$= 3(3-1) - 2\times 2 - 1$$
$$= 6 - 4 - 1 = 1$$

(d) Number of binary links, $n_2 = 4$

Number of ternary links, $n_3 = 2$

Total number of links, $n = n_2 + n_3 = 4 + 2 = 6$

$$2p = 2n_2 + 3n_3 = 2\times 4 + 3\times 2 = 14$$
$$F = 3(n-1) - 2p$$
$$= 3(6-1) - 14$$
$$= 15 - 14 = 1$$

(e) Number of binary links, $n_2 = 5$

Number of ternary links, $n_3 = 2$

Number of quaternary links, $n_4 = 1$

Total number of links, $n = n_2 + n_3 + n_4 = 5 + 2 + 1 = 8$

$$2p = 2n_2 + 3n_3 + 4n_4 = 2\times 5 + 3\times 2 + 4\times 1 = 20$$
$$F = 3(n-1) - 2p$$
$$= 3(8-1) - 20$$
$$= 21 - 20 = 1$$

(f) Number of binary links, $n_2 = 5$

Number of ternary links, $n_3 = 2$

Total number of links, $n = n_2 + n_3 = 5 + 2 = 7$

$$2p = 2n_2 + 3n_3 = 2\times 5 + 3\times 2 = 16$$
$$F = 3(n-1) - 2p$$
$$= 3(7-1) - 16$$
$$= 18 - 16 = 2$$

Example 1.10

Determine the number of degrees of freedom of the mechanism shown in Fig.1.21(a) to (f).

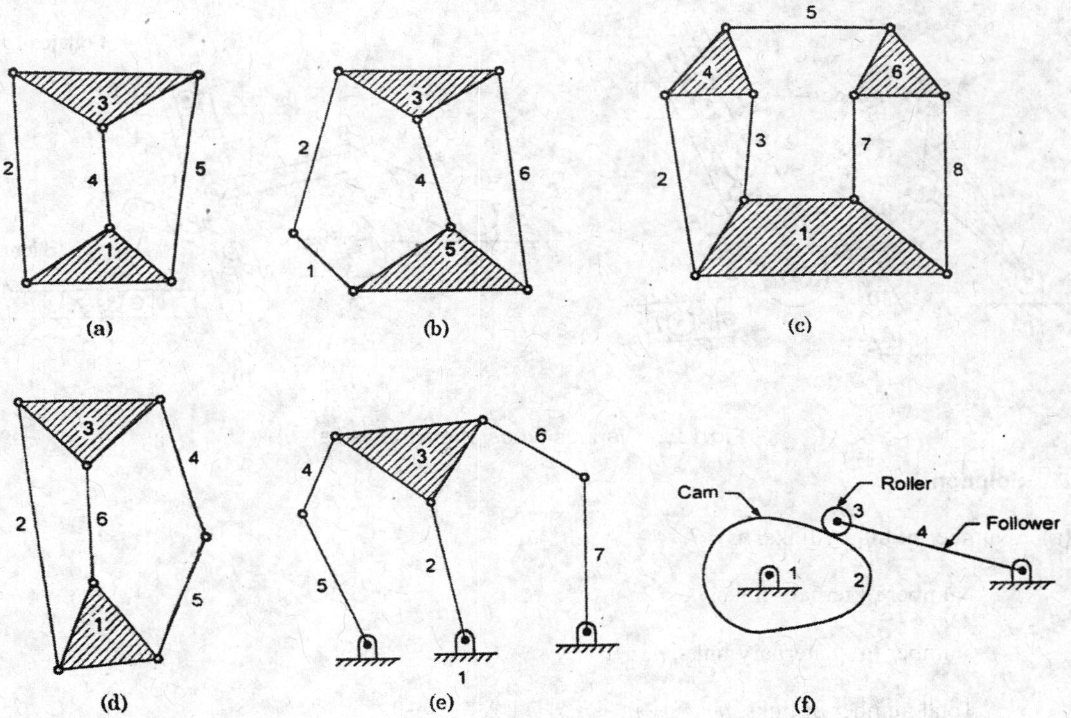

Fig.1.21 Various types of mechanisms

■ Solution

(a) $n_2 = 3, n_3 = 2, n = 5$

$2p = 2 \times 3 + 3 \times 2 = 12$

$F = 3(5 - 1) - 12 = 0$

(b) $n_2 = 4, n_3 = 2, n = 6$

$2p = 2 \times 4 + 3 \times 2 = 14$

$F = 3(6 - 1) - 14 = 1$

(c) $n_2 = 5, n_3 = 2, n_4 = 1, n = 8$

$2p = 2 \times 5 + 3 \times 2 + 4 + 1 = 20$

$F = 3(8 - 1) - 20 = 1$

(d) $n_2 = 4, n_3 = 2, n = 6$

$2p = 2 \times 4 + 3 \times 2 = 14$

$F = 3(6 - 1) - 14 = 1$

(e) $n_2 = 5, n_3 = 2, n = 5 + 2 = 7$

$2p = 2n_2 + 3n_3 = 2\times5+3\times2 = 16$

$F = 3(n-1) - 2p$

$= 3(7-1) - 16$

$= 18 - 16 = 2$

(f) The solution has been given in Example 1.9(c).

Example 1.11

Determine the number of degrees of freedom of the mechanism shown in Fig.1.20 (a)–(h).

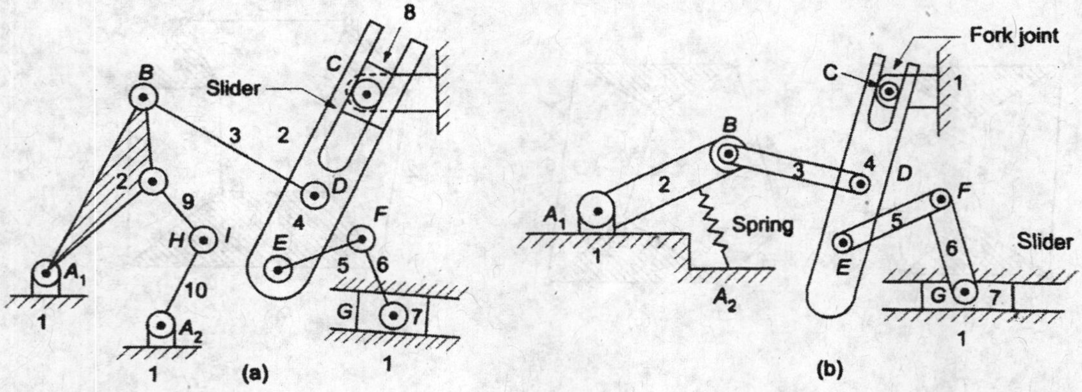

Fig.1.22 Various types of mechanisms

Solution

(a) Number of binary links, $n_2 = 7$

Number of ternary links, $n_3 = 2$

Number of quaternary links, $n_4 = 1$

Total number of links, $n = n_2 + n_3 + n_4 = 7 + 2 + 1 = 10$

$$2p = 2n_2 + 3n_3 + 4n_4 = 2\times7+3\times2+4\times1 = 24$$
$$F = 3(n-1) - 2p$$
$$= 3(10-1) - 24$$
$$= 27 - 24 = 3$$

(b) $n = 7,\ p = 7,\ h = 1$

$$F = 3(n-1) - 2p - h$$
$$= 3(7-1) - 2\times7 - 1$$
$$= 18 - 14 - 1 = 3$$

Example 1.12

Find the equivalent mechanisms with turning pairs for the mechanisms shown in Figs.1.23(a) to (d).

Solution

(a) A spring can be replaced by two binary links. Therefore, the equivalent mechanism is as shown in Fig.1.23a(ii).

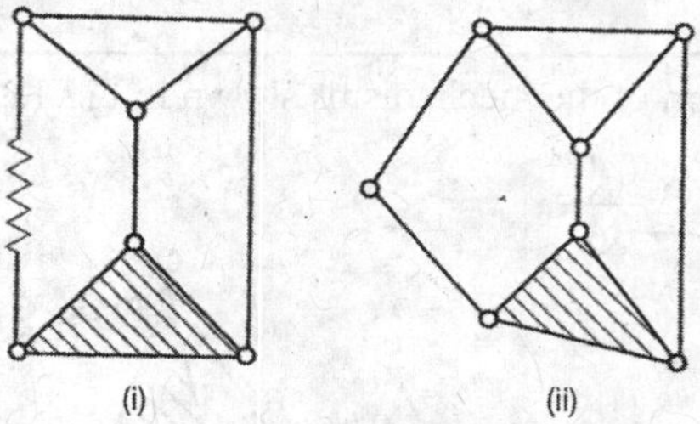

(a) Equivalent mechanism for a spring

(b) A cam pair can be replaced by one binary link with two turning pairs at each end. Therefore, the equivalent mechanism is shown in Fig.1.23b(ii). The centres of curvature at the point of contact *E* lie at *B* and *C*, respectively.

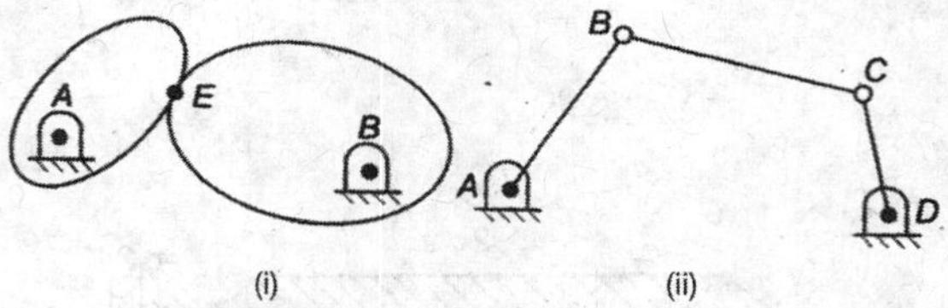

(b) Equivalent mechanism for a cam pair

(c) The equivalent mechanism is shown in Fig.1.23c(ii) as explained in (b) above.

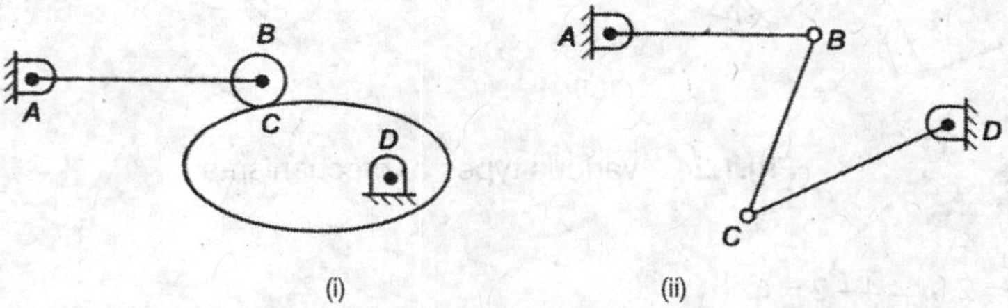

(c) Equivalent mechanism for a cam with roller follower

(d) A spring is equivalent to two binary links connected by a turning pair. A cam follower is equivalent to one binary link with turning pairs at each end. The equivalent chain with turning pairs is shown in Fig.1.23d(ii).

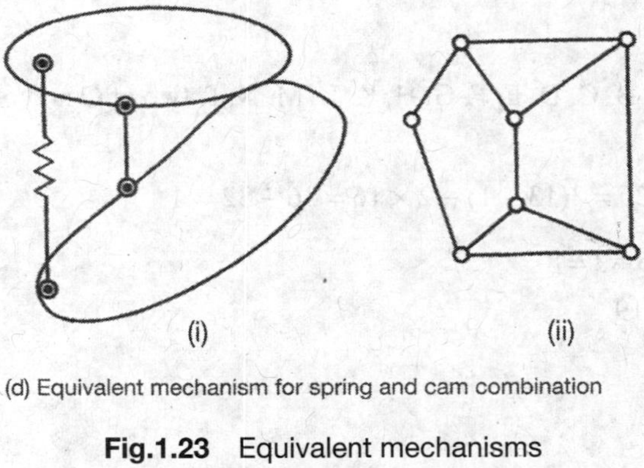

(d) Equivalent mechanism for spring and cam combination

Fig.1.23 Equivalent mechanisms

Example 1.13

Calculate the degrees of freedom of the mechanisms shown in Fig.1.24(a–e).

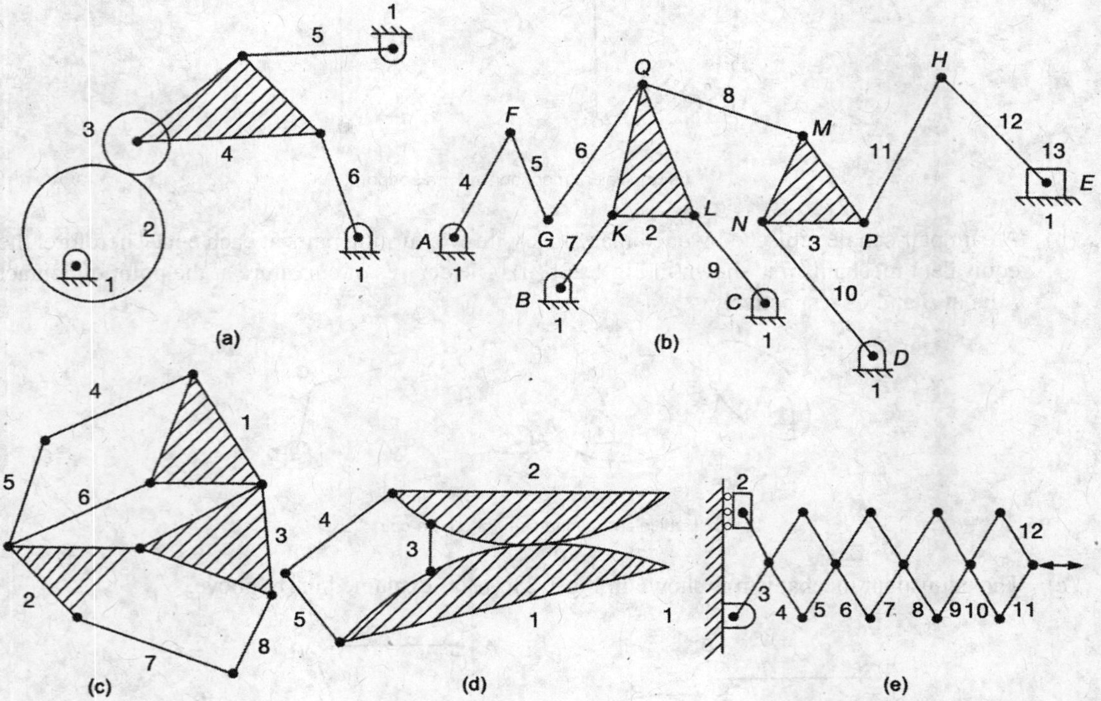

Fig.1.24 Various types of mechanisms

■ Solution

(a) $n_2 = 4, n_3 = 2, n = 4 + 2 = 6$

$$2p = 2n_2 + 3n_3 = 2 \times 4 + 3 \times 2 = 14$$

$$F = 3(n-1) - 2p - h$$

$$= 3(6-1) - 14 - 0$$

$$= 15 - 14 = 1$$

(b) $n = 13$

Binary pairs: A, B, C, D, E, F, G, H, K, L, M, N, P, two at Q and one slider.

$P = 16$

$F = 3\ (n - 1) - 2p = 3(13 - 1) - 2 \times 16 = 36 - 32 = 4$

(c) $n_2 = 5, n_3 = 3, n = 5 + 3 = 8$

$$2p = 2 \times 5 + 3 \times 3 = 19$$

$$F = 3(n-1) - 2p$$

$$= 3(8-1) - 19$$

$$= 21 - 19 = 2$$

(d) $n = 5, p = 5, h = 1$ (cam pair)

$$F = 3(n-1) - 2p - h$$
$$= 3(5-1) - 2\times5 - 1$$
$$= 12 - 10 - 1 = 1$$

(e) $n = 12, p = 15, h = 1$

$$F = 3(n-1) - 2p - h$$
$$= 3(12-1) - 2\times15 - 1$$
$$= 33 - 30 - 1 = 2$$

1.11 GRASHOF'S LAW

This law states that for a four-bar mechanism the sum of the lengths of the largest and the shortest links should be less than or equal to the sum of the lenghts of the other links, that is,

$$(l + s) \le (p + q) \tag{1.9}$$

where l, s = lengths of the longest and the shortest links, respectively.
p, q = lengths of the other two links.

Consider the four-bar chain shown in Fig.1.25. Let the length of fixed link $O_2O_4 = l_1$, crank $O_2A = l_2$, coupler $AB = l_3$, and lever $BO_4 = l_4$. The following types of mechanisms are obtained by adjusting the lengths of various links:

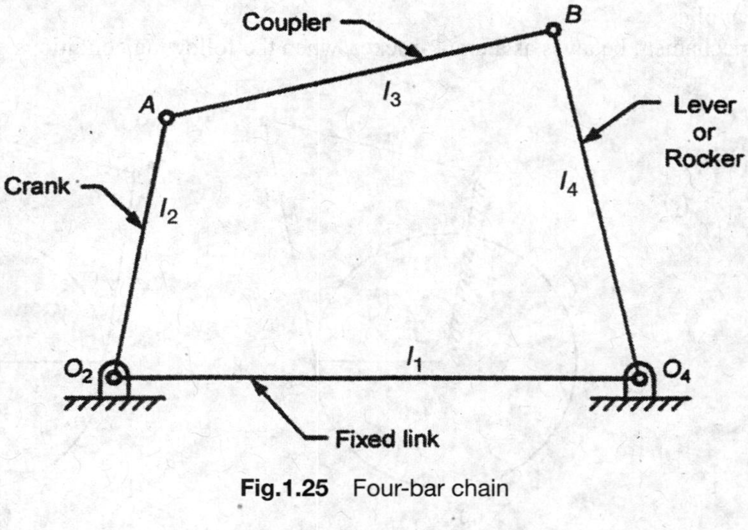

Fig.1.25 Four-bar chain

1.11.1 Crank–Crank (or Double Crank) Mechanism

When the shortest link is fixed and $(l + s) \le (p + q)$, crank-crank or double crank mechanism is obtained, as shown in Fig.1.26. Links O_2A or O_4B may be the inputs links or cranks. They are able to make complete rotations about points O_2 and O_4 respectively. Shortest link $O_2O_4 = l_1$ is fixed. A four-bar mechanism behaves as a crank-crank when the following conditions exist:

$l_1 < l_2$ or l_3 or l_4
and $l_1 < (l_3 + l_4 - l_2)$
and $l_1 < (|l_3 - l_4| + l_2)$
where | | stands for absolute value.

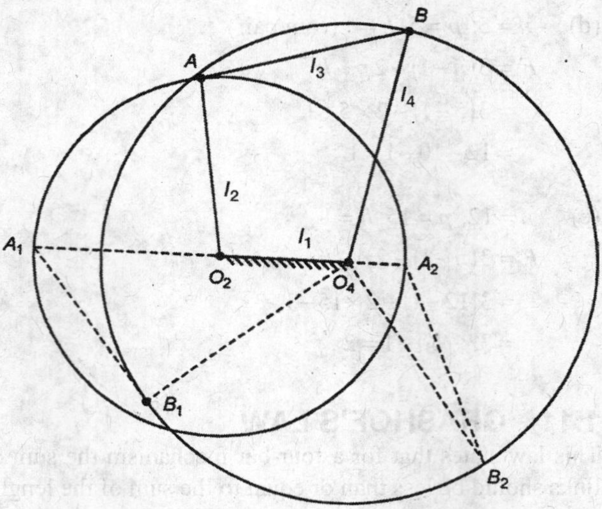

Fig.1.26 Crank–crank mechanism

1.11.2 Crank–Rocker (or Lever) Mechanism

If any of the adjacent links of shortest link l_1 i.e., 'l_2' or l_4 is fixed then l_1 can have full revolution and the link l_3 opposite to it oscillates. In Fig.1.27(a), 'l_2' is fixed, l_1 is the crank to rotate about O_2 and l_3 oscillates, whereas in Fig.1.27(b), l_4 is fixed, l_1 is the crank to rotate about O_4 and l_3 oscillates. Again $(l + s) \le (p + q)$ applies.

A four-bar mechanism behaves as a crank-rocker, when the following conditions exist:

$l_2 < l_1$ or l_3 or l_4
and $l_1 < (l_3 + l_4 - l_2)$
and $l_1 > (|l_3 - l_4| + l_2)$

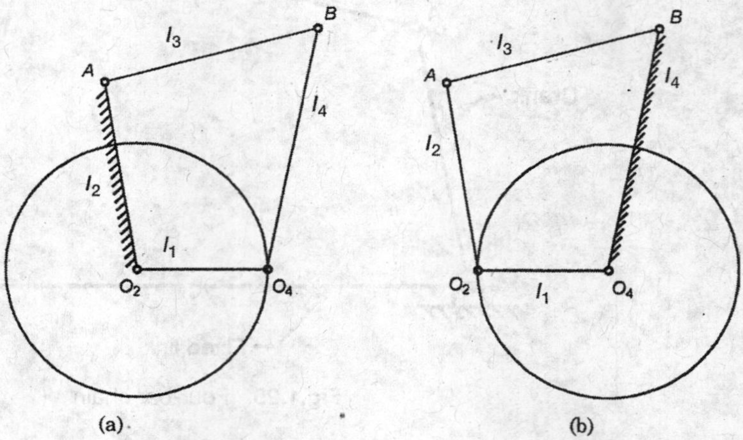

Fig.1.27 Crank–rocker mechanism

1.11.3 Rocker–Rocker (or Double Rocker) Mechanism

If the link l_3 opposite to the shortest link l_1 is fixed and shortest link l_1 is made the coupler, the other two links 'l_2' and l_4 would oscillate, as shown in Fig.1.28(a). When $(l + s) \le (p + q)$, the linkage is known as a class-*I* four-bar linkage. A four-bar mechanism behaves as a double rocker mechanism when the following condition is met:

$$l_3 < l_1 \text{ or } l_2 \text{ or } l_4$$

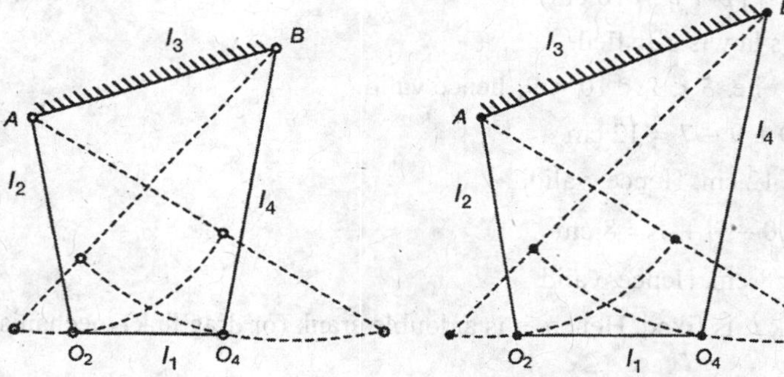

(a) $(l + s) < (p + q)$ (b) $(l + s) > (p + q)$

Fig.1.28 Rocker–rocker mechanism

1.11.4 Class-II Four-Bar Linkage

When $(l + s) > (p + q)$, the linkage is known as a class-II four-bar linkage. In such a mechanism, fixing of any of the links always results in rocker-rocker mechanism, as shown in Fig.1.28(b).

Example 1.14

Identify the nature of each mechanism shown in Fig.1.29(a) to (d).

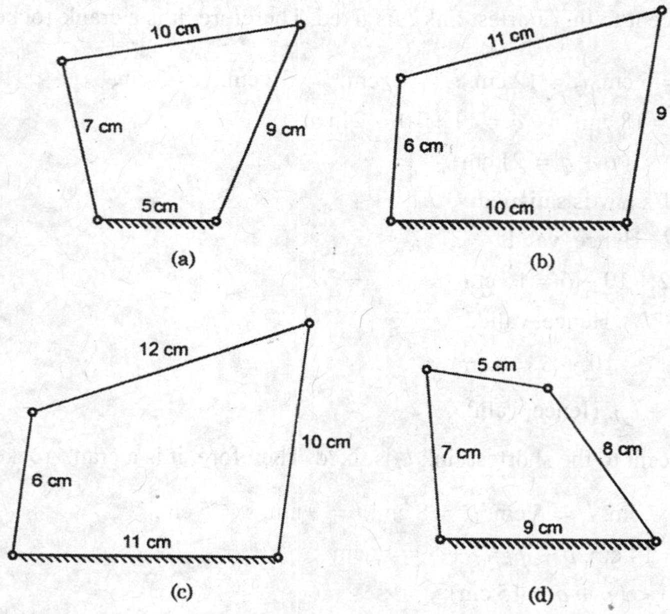

Fig.1.29 Four-bar mechanisms

Solution

(a) $l_1 = 5$ cm, $l_2 = 7$ cm, $l_3 = 10$ cm, $l_4 = 9$ cm; $l = 10$ cm, $s = 5$ cm, $p + q = 7 + 9 = 16$ cm

$l + s = 10 + 5 = 15$ cm

$(l + s = 15 \text{ cm}) < (p + q = 16 \text{ cm})$

Hence Grashof's law is satisfied.

$l_1 < l_2 < l_3 < l_4$, i.e. $5 < 7 < 10 < 9$, hence valid.

$l_3 + l_4 - l_2 = 10 + 9 - 7 = 12$ cm

$\therefore (l_1 = 5 \text{ cm}) < 12$ cm. Hence, valid.

$|l_3 - l_4| + l_2 = |10 - 9| + 7 = 8$ cm

$\therefore (l_1 = 5 \text{ cm}) < 8$ cm. Hence, valid.

The shortest link l_1 is fixed. Hence, it is a double crank (or drag link) mechanism.

(b) $l_1 = 10$ cm, $l_2 = 6$ cm, $l_3 = 11$ cm, $l_4 = 9$ cm; $l = 11$ cm, $s = 6$ cm, $l + s = 11 + 6 = 17$ cm,

$p + q = 10 + 9 = 19$ cm

$(l + s = 17 \text{ cm}) < (p + q = 19 \text{ cm})$

Hence Grashof's law is satisfied.

$l_2 < l_1$ or l_3 or l_4. Hence, valid.

$l_3 + l_4 - l_2 = 11 + 9 - 6 = 14$ cm

$(l_1 = 10 \text{ cm}) < (l_3 + l_4 - l_2 = 14 \text{ cm})$. Hence, valid.

$|l_3 - l_4| + l_2 = |11 - 9| + 6 = 8$ cm

$(l_1 = 10 \text{ cm}) > (l_3 - l_4) + l_2 = 8$ cm. Hence, valid.

The link l_1 adjacent to the shortest link l_2 is fixed. Therefore, it is a crank-rocker mechanism.

(c) $l_1 = 11$ cm, $l_2 = 6$ cm, $l_3 = 12$ cm, $l_4 = 10$ cm; $l = 12$ cm, $s = 6$ cm,

$l + s = 12 + 6 = 18$ cm, $p + q = 11 + 10 = 21$ cm

$(l + s = 18 \text{ cm}) < (p + q = 21 \text{ cm})$

Hence, Grashof's law is satisfied.

$l_2 < l_1$ or l_3 or l_4. Hence, valid.

$l_3 + l_4 - l_2 = 12 + 10 - 6 = 16$ cm

$\therefore l_1 < (l_3 + l_4 - l_2)$. Hence, valid.

$|l_3 - l_4| + l_2 = |12 - 10| + 6 = 8$ cm

$\therefore l_1 > (|l_3 - l_4| + l_2)$. Hence, valid.

The link l_1 adjacent to the shortest link l_2 is fixed. Therefore, it is a crank-rocker mechanism.

(d) $l_1 = 9$ cm, $l_2 = 7$ cm, $l_3 = 5$ cm, $l_4 = 8$ cm; $l = 9$ cm, $s = 5$ cm,

$l + s = 9 + 5 = 14$ cm, $p + q = 7 + 8 = 15$ cm

$(l + s = 14 \text{ cm}) < (p + q = 15 \text{ cm})$

Hence, Grashof's law is satisfied.

$l_3 < l_1 < l_2 < l_4$. Hence, valid.

The link l_1 opposite to the shortest link l_3 (coupler) is fixed. Therefore, it is a rocker-rocker mechanism.

Example 1.15

Identify the nature of the mechanisms shown in Fig.1.30(a) to (d).

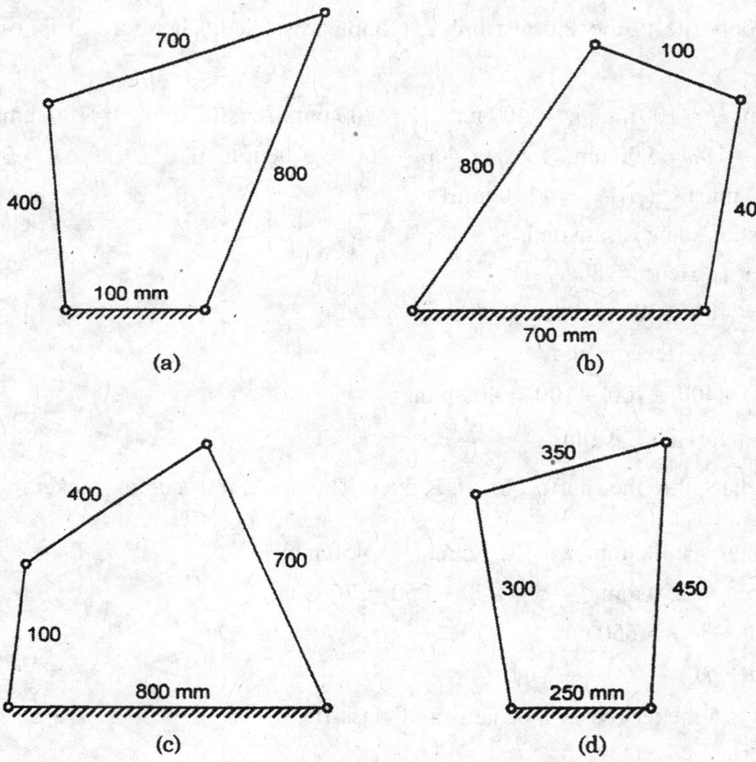

Fig.1.30 Four-bar mechanisms

■ Solution

(a) $l_1 = 100$ mm, $l_2 = 400$ mm, $l_3 = 700$ mm, $l_4 = 800$ mm; $l = 800$ mm, $s = 100$ mm,

$l + s = 800 + 100 = 900$ mm, $p + q = 400 + 700 = 1100$ mm

$(l + s = 900 \text{ mm}) < (p + q = 1100 \text{ mm})$

Hence, Grashof's law is satisfied.

$l_1 < l_2 < l_3 < l_4$, i.e. $100 < 400 < 700 < 800$. Hence, valid.

$l_3 + l_4 - l_2 = 700 + 800 - 400 = 1100$ mm

$l_1 < (l_3 + l_4 - l_2)$. Hence, valid.

$|l_3 - l_4| + l_2 = |700 - 800| + 400 = 500$ mm

$l_1 < |l_3 - l_4| + l_2$

The shortest link is fixed. Therefore, it is a crank-crank mechanism.

(b) $l_1 = 700$ mm, $l_2 = 800$ mm, $l_3 = 100$ mm, $l_4 = 400$ mm; $l = 800$ mm, $s = 100$ mm,

$l + s = 800 + 100 = 900$ mm, $p + q = 700 + 400 = 1100$ mm

$(l + s = 900\,\text{mm}) < (p + q = 1100\text{ mm})$

Hence, Grashof's law is valid.

$l_3 < l_1 < l_2 < l_4$. Hence, valid.

The link l_1 opposite to the shortest link l_3 (coupler) is fixed. Therefore, it is a rocker-rocker mechanism.

(c) $l_1 = 800$ mm, $l_2 = 100$ mm, $l_3 = 400$ mm, $l_4 = 700$ mm; $l = 800$ mm, $s = 100$ mm,

$l + s = 800 + 100 = 900$ mm, $p + q = 400 + 700 = 1100$ mm

$(l + s = 900\text{ mm}) < (p + q = 1100\text{ mm})$

Hence, Grashof's law is satisfied.

$l_2 < l_1 < l_3 < l_4$. Hence, valid.

$l_3 + l_4 - l_2 = 400 + 700 - 100 = 1000$ mm

$l_1 < (l_3 + l_4 - l_2)$. Hence, valid.

$|l_3 - l_4| + l_2 = |400 - 700| + 100 = 400$ mm

$l_1 > |l_3 - l_4| + l_2$. Hence, valid.

The link l_1 adjacent to the shortest link l_2 is fixed. Therefore, it is a crank-rocker mechanism.

(d) $l_1 = 250$ mm, $l_2 = 300$ mm, $l_3 = 350$ mm, $l_4 = 450$ mm;

$l = 450$ mm, $s = 250$ mm, $l + s = 450 + 250 = 700$ mm,

$p + q = 300 + 350 = 650$ mm

$(l + s) > (p + q)$

Therefore, it is a rocker-rocker mechanism of class-II.

1.12 INVERSION OF MECHANISMS

A kinematic chain becomes a mechanism when one of its links is fixed. Therefore, as many number of mechanisms can be obtained as many are the links in the kinematic chain. This method of obtaining different mechanisms by fixing different links of a kinematic chain is called inversion of the mechanism. The relative motion between the various links is not altered as a result of inversion, but their absolute motion with respect to the fixed link may alter drastically.

1.12.1 Inversions of a Four-Bar Chain

Some of the important inversions of a four-bar chain are:

1. Beam engine
2. Coupled wheel of locomotive
3. Watt's indicator mechanism
4. Slider-crank chain.

1. *Beam engine:* The beam engine mechanism is shown in Fig.1.31. It consists of four links. When the crank *AB* rotates about the fixed centre *A*, the lever oscillates about a fixed centre *D*. The end *E* of the lever *CDE* is connected to a piston rod which moves the piston up and down in the cylinder. This is also called crank and lever mechanism.

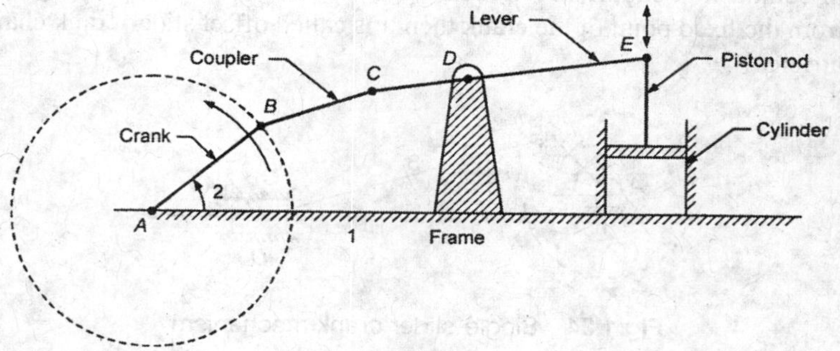

Fig.1.31 Beam engine

2. *Coupled wheel of a locomotive:* In this mechanism, as shown in Fig.1.32, the links *AB* and *CD* are of equal lengths and act as cranks. These cranks are connected to the respective wheels. The link *BC* acts as the connecting rod. The link *AD* is fixed to maintain constant distance between the wheels. This mechanism is used to transmit rotary motion from one wheel to the other. This is also called the double crank mechanism.

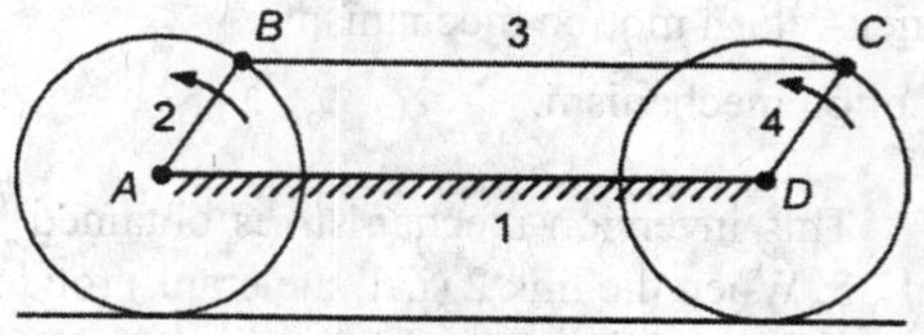

Fig.1.32 Coupled wheel of a locomotive

3. *Watt's indicator mechanism:* This mechanism is shown in Fig.1.33. It consists of four links: a fixed link at *A*, link *AC*, link *CE*, and link *BFD*. The links *CE* and *BFD* act as levers. The displacement of link *BFD* is directly proportional to the pressure in the indicator cylinder. The point *E* on link *CE* traces out an approximate straight line. It is also called double lever mechanism.

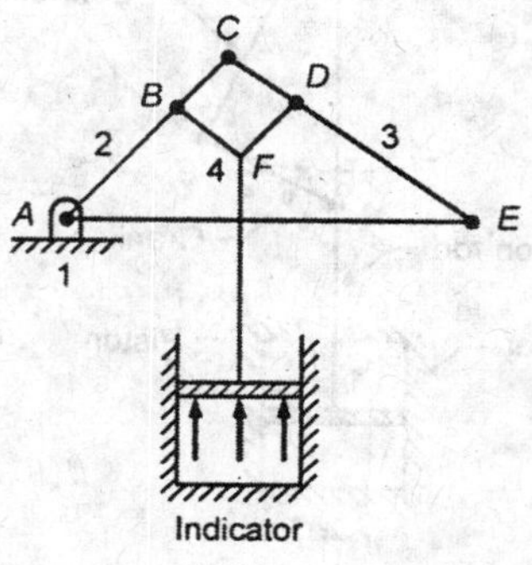

Fig.1.33 Watt's indicator mechanism

4. *Single slider-crank chain:* The single slider-crank chain shown in Fig.1.34 consists of three turning pairs and one sliding pair. Link 1 corresponds to the frame of the mechanism, which is fixed. Link 2 is the crank and link 3 the connecting rod. The link 4 is the slider. It is used to convert rotary motion into reciprocating motion and vice-versa. Its important applications are in steam engines, internal combustion engines, reciprocating compressors, etc. If the straight line path of the slider is offset from the fixed point of the crank then it is called offset slider-crank chain. The offset is called the eccentricity.

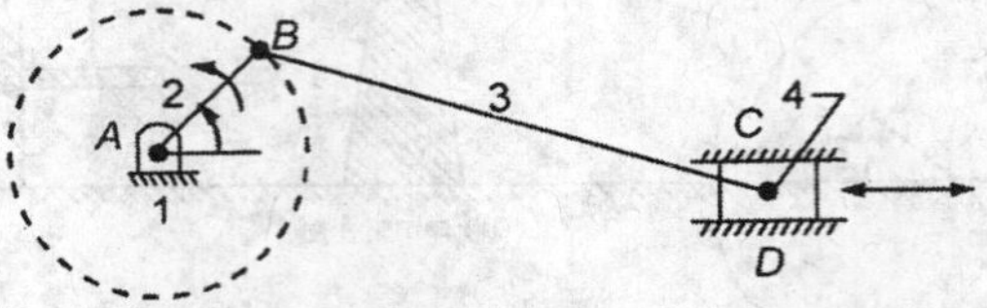

Fig.1.34 Single-slider crank mechanism

1.12.2 Inversions of a Single-Slider Crank Chain

The inversions of a single slider crank chain are as follows:

1. Pendulum pump
2. Oscillating cylinder engine
3. Rotary internal combustion engine
4. Crank and slotted lever quick-return motion mechanism
5. Whitworth quick-return motion mechanism.

1. *Pendulum pump:* This inversion mechanism is obtained by fixing the link 4, i.e., the sliding pair, as shown in Fig.1.35. When the link 2 (i.e., the crank) rotates, the link 3 (i.e., connecting rod) oscillates about a pin pivoted to fixed link 4 at C and the piston attached to the piston rod (link 1) reciprocates in the cylinder. It is used to supply feed water to a boiler.

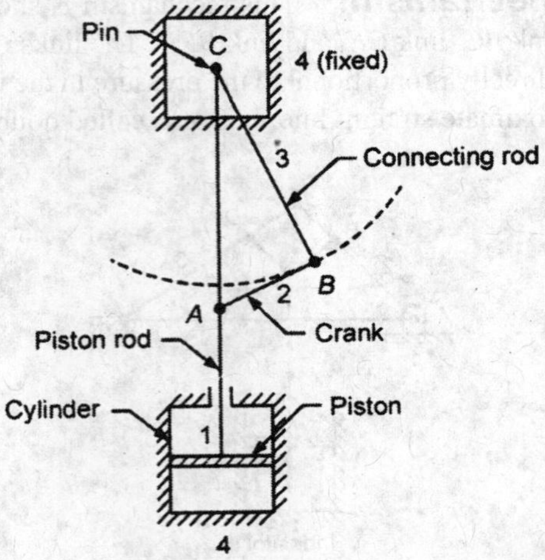

Fig.1.35 Pendulum pump

2. *Oscillating cylinder engine:* In this mechanism, as shown in Fig.1.36, link 3 is fixed. When the crank (link 2) rotates, the piston attached to piston rod (link 1) reciprocates and the cylinder (link 4) oscillates about a pin pivoted to the fixed link at *A*.

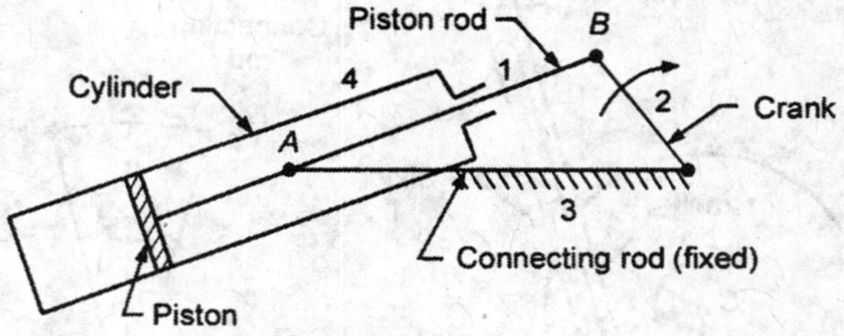

Fig.1.36 Oscillating cylinder engine

3. *Rotary internal combustion engine (Gnome engine):* It consists of several cylinders in one plane and all revolve about fixed centre *O*, as shown in Fig.1.37. The crank (link 2) is fixed. When the connecting rod (link 4) rotates, the piston (link 3) reciprocates inside the cylinder forming link 1.

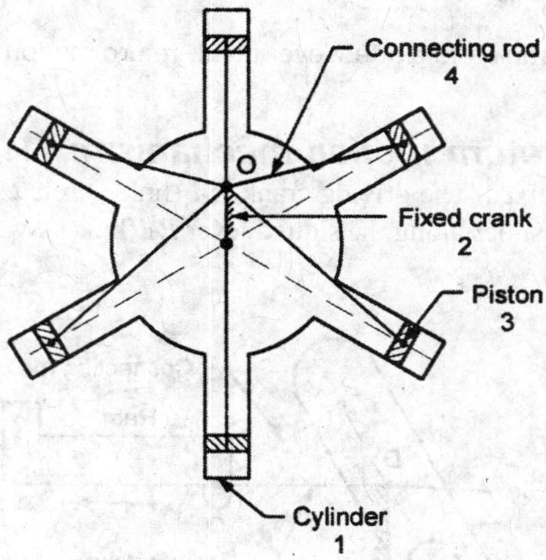

Fig.1.37 Rotary internal combustion engine

4. *Crank and slotted lever quick-return motion mechanism:* In this mechanism, as shown in Fig.1.38, the link *AC* (link 3) corresponding to the connecting rod is fixed. The driving crank *CB* (link 2) revolves about centre *C*. A slider (link 1) attached to the crank pin at *B* slides along the slotted lever *AP* (link 4) and make the slotted lever oscillate about the pivoted point *A*. A short link *PQ* transmits the motion from *AP* to the arm which reciprocates with the tool along the line of stroke. The line of stroke is perpendicular to *AC* produced. This mechanism is mostly used in shaping machines, slotting machine, and rotary internal combustion engines.

$$\text{Time of cutting stroke/Time of return stroke} = \frac{\alpha}{\beta} \tag{1.10}$$

$$\text{Length of stroke} = 2\,AP\,(CB/AC) \tag{1.11}$$

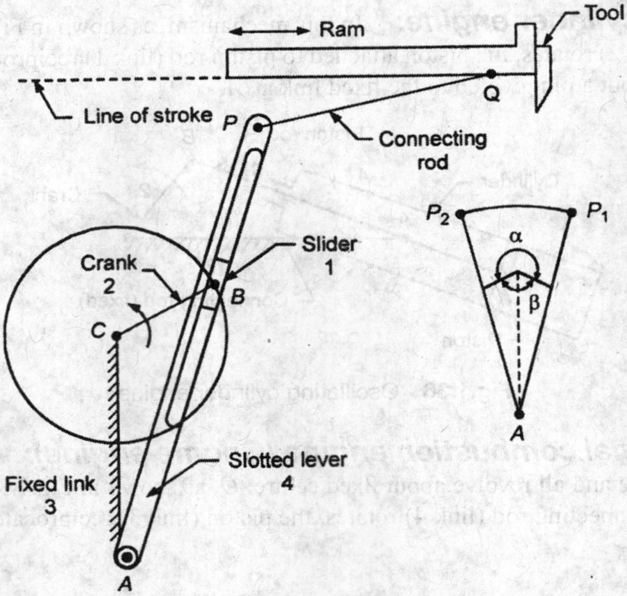

Fig.1.38 Crank and slotted lever quick-return motion mechanism

5. *Whitworth quick-return motion mechanism:* In this mechanism, as shown in Fig.1.39, link *CD* (link 2) is fixed. The driving crank *CA* (link 3) rotates about *C*. The slider (link 4) attached to the crank pin at *A* slides along the slotted lever *PA* (link 1), which oscillates about pivot *D*.

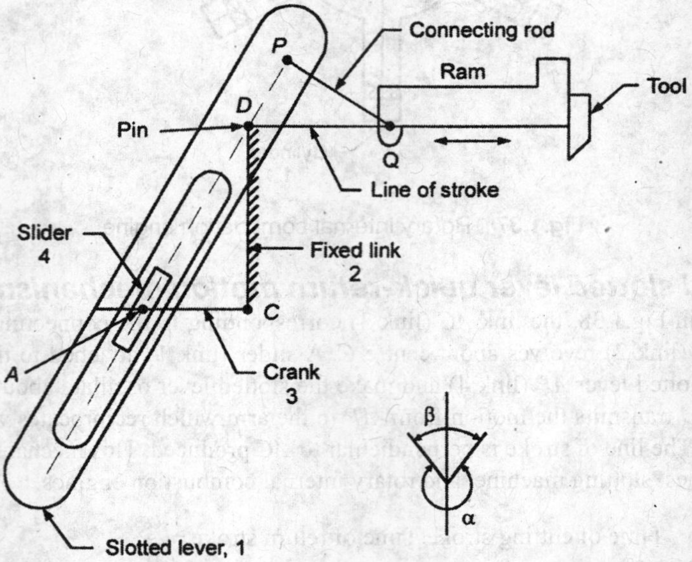

Fig.1.39 Whitworth quick-return motion mechanism

The connecting rod PQ carries the ram at Q with cutting tool. The ram reciprocates along the line of stroke. It is used in shaping and slotting machines.

$$\text{Time of cutting stroke / Time of return stroke} = \frac{\alpha}{\beta} \tag{1.12}$$

$$\text{Length of stroke} = 2\,PD \tag{1.13}$$

6. *Toggle mechanism:* This mechanism has many applications where it is necessary to overcome a large resistance with a small driving force. Fig.1.40 shows the toggle mechanism; links 4 and 5 are of equal length. As the angles α decrease and links 4 and 5 approach being collinear, the force F required to overcome a given resistance P decreases as:

$$F = 2P \tan \alpha \tag{1.14}$$

If α approaches zero, for a given F, P approaches infinity. A stone crusher utilizes this mechanism to overcome a large resistance with a small force. It can be used in numerous toggle clamping devices, for holding work pieces.

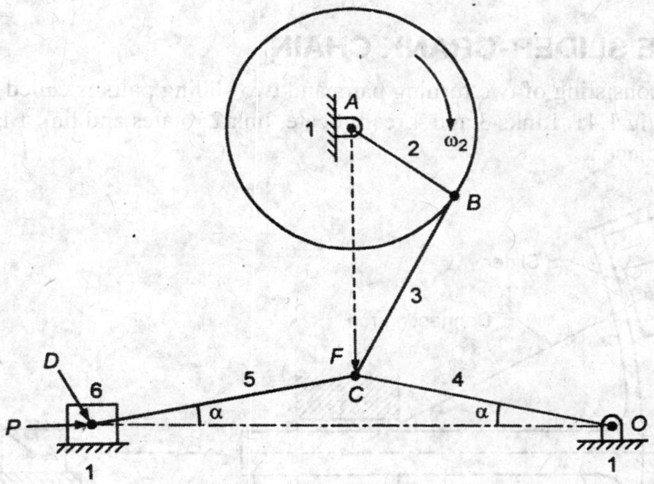

Fig.1.40 Toggle mechanism

The summary of single slider crank chain and its inversions is given in Table 1.2.

Table 1.2 Summary Of Single Slider Crank Chain And Its Inversions

Mechanism	Links			
	Fixed	Rotates	Oscillates	Reciprocates
Single slider-crank chain	1	2	3	4
Inversions:				
Pendulum pump	4	2	3	1
Oscillating cylinder engine	3	2	4	1
Crank and slotted lever	3	2	4	1
Whitworth mechanism	2	3	1	4
Gnome engine	2	3	1	4

1.13 DOUBLE SLIDER-CRANK CHAIN

A kinematic chain consisting of two turning pairs and two sliding pairs is called double slider–crank chain, as shown in Fig.1.41. Links 3 and 4 reciprocate, link 2 rotates and link 1 is fixed. Two pairs of the same kind are adjacent.

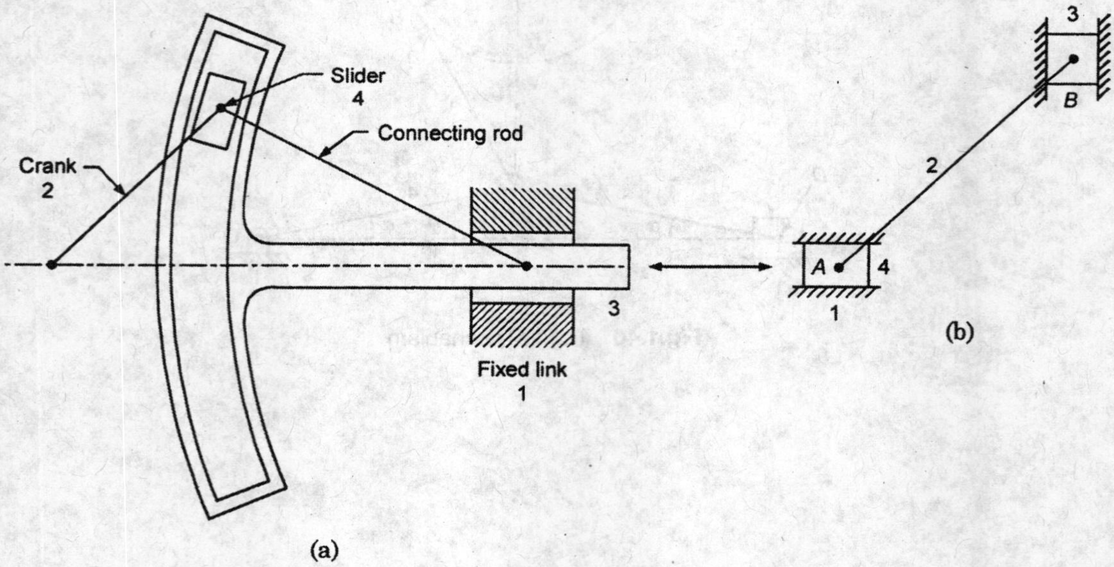

Fig.1.41 Double slider–crank chain

1.13.1 Inversions of Double Slider–Crank Chain

The inversions of double slider-crank chain are as follows:

1. Donkey pump
2. Oldham's coupling
3. Elliptical trammel
4. Scotch yoke.

1. *Donkey Pump:* Figure 1.42 shows a donkey pump, in which link 2 (crank) rotates about point A. One end of the crank is connected to the piston, through the piston rod, which reciprocates vertically in the pump cylinder. This cylinder together with the body of the pump represents the fixed link 1. The other end of the crank is connected to the slider (link 3) which reciprocates horizontally in the cylinder.

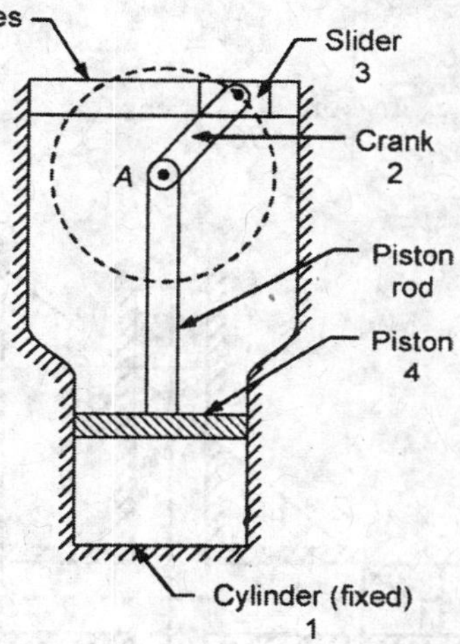

Fig.1.42 Donkey pump

2. *Oldham's coupling:* The Oldham's coupling shown in Fig.1.43, is used to connect two parallel shafts, the distance between whose axes is small and variable. The shafts connected by the coupling rotate at the same speed. The shafts have flanges at the ends, in which slots are cut. These

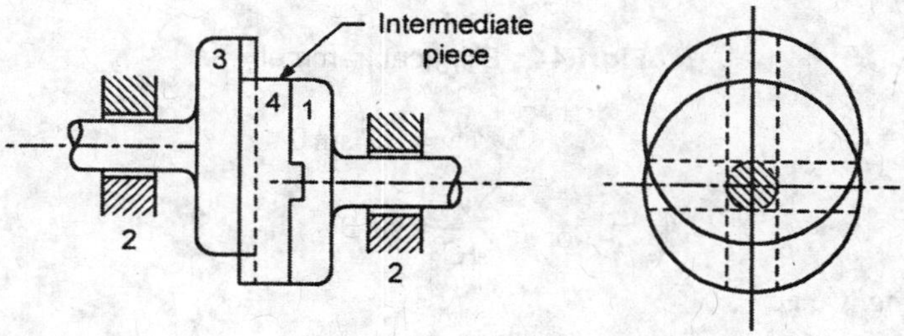

Fig.1.43 Oldham's coupling

These form links 1 and 3. An intermediate piece circular in shape and having tongues at right angles on opposite sides, is fitted between the flanges of the two shafts in such a way that the tongues of the intermediate piece get fitted in the slots of the flanges. The intermediate piece forms link 4, which slides or reciprocates in links 1 and 3. The link 2 is fixed.

Maximum sliding speed of each tongue along its slot
= Distance between the axes of the shafts × angular velocity of each shaft (1.15)

3. *Elliptical trammel:* It is a device to draw ellipses. Fig.1.44 shows an elliptical trammel in which two grooves are cut at right angles in a plate that is fixed. The plate forms the fixed link 4. Two sliding blocks are fitted into the grooves. The slides form two sliding links 1 and 3. The link joining slides form the link 2. Any point on link 2 or on its extension traces out an ellipse on the fixed plate, when relative motion occurs.

$$x = BC \cos \theta$$

$$\text{or} \quad \frac{x}{BC} = \cos \theta$$

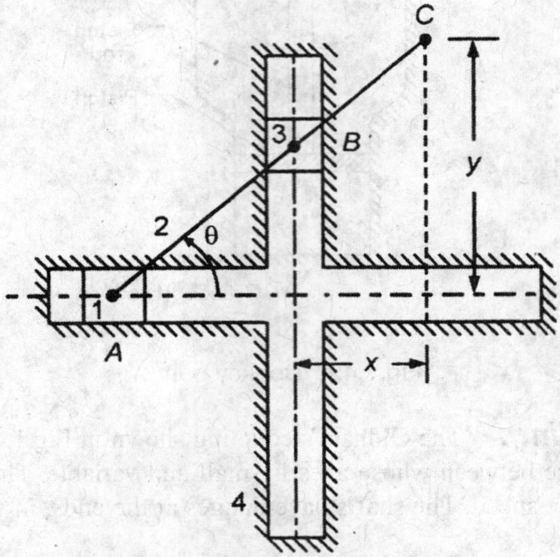

Fig.1.44 Elliptical trammel

$$y = AC \sin \theta$$

$$\text{or} \quad \frac{y}{AC} = \sin \theta$$

Squaring and adding, we get

$$\frac{x^2}{BC^2} + \frac{y^2}{AC^2} = 1 \tag{1.16}$$

which is the equation of an ellipse.

4. *Scotch yoke:* This mechanism gives simple harmonic motion. Its early application was on steam pumps, but it is now used as a mechanism on a test machine to produce vibrations. It is also used as a sine-cosine generator for computing elements. Fig.1.45 shows a sketch of scotch yoke mechanism.

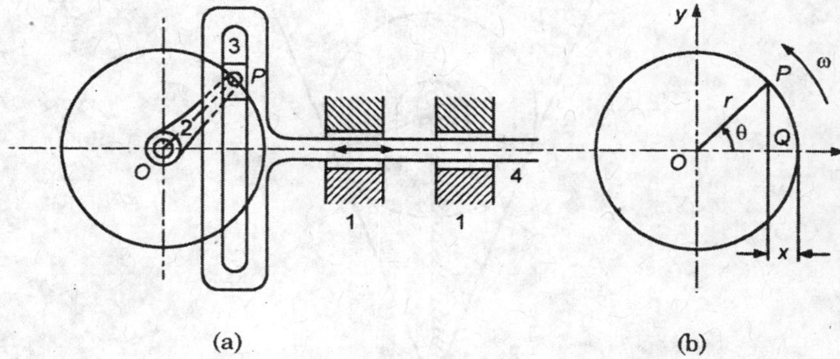

Fig.1.45 Scotch yoke

$$x = r - r\cos\theta$$

$$= r(1 - \cos\omega t) \quad (1.17)$$

$$v = \frac{dx}{dt} = r\omega\sin\omega t \quad (1.18)$$

$$a = \frac{d^2x}{dt^2} = r\omega^2\cos\omega t \quad (1.19)$$

Example 1.16

In a crank and slotted lever quick-return mechanism shown in Fig.1.46, the distance between the fixed centres is 300 mm and the length of the driving crank is 150 mm. Find the inclination of the slotted lever with the vertical in the extreme position and the ratio of time of cutting stroke to return stroke.

■ Solution

Given $AC = 300$ mm, $B_1C = 150$ mm

$$\cos(\beta/2) = B_1C/AC$$

$$= \frac{150}{300} = 0.50$$

$$\beta = 120^\circ$$

$$\alpha = 360^\circ - 120^\circ = 240^\circ$$

$$\text{Time of cutting stroke/Time of return stroke} = \frac{\alpha}{\beta}$$

$$= \frac{240^\circ}{120^\circ} = 2$$

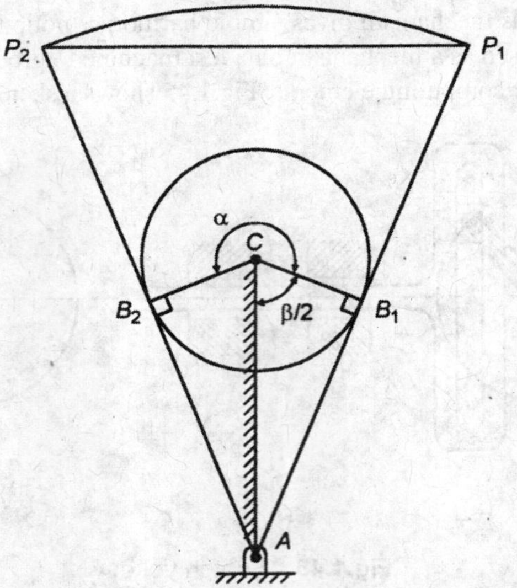

Fig.1.46 Crank and slotted lever mechanism

$$\text{Inclination of slotted lever with the vertical} = 90^\circ - \frac{\beta}{2} = 90^\circ - \frac{120^\circ}{2}$$
$$= 30^\circ$$

Example 1.17

In a Whitworth quick return motion mechanism, as shown in Fig.1.47, the distance between the fixed centres is 80 mm and the length of the driving crank is 100 mm. The length of the slotted lever is 180 mm and the length of the connecting rod is 150 mm. Calculate the ratio of the time of cutting to return strokes.

Solution

Given: $CD = 80$ mm, $CA = 100$ mm, $PA = 180$ mm, $PQ = 150$ mm

$$\cos\frac{\beta}{2} = \frac{CD}{CA_2} = \frac{60}{100} = 0.6$$
$$\beta = 106.26^\circ$$

$$\text{Time of cutting stroke / Time of return stroke} = \left(\frac{360^\circ - \beta}{\beta}\right)$$
$$= \frac{253.74^\circ}{106.26^\circ}$$
$$= 2.388$$

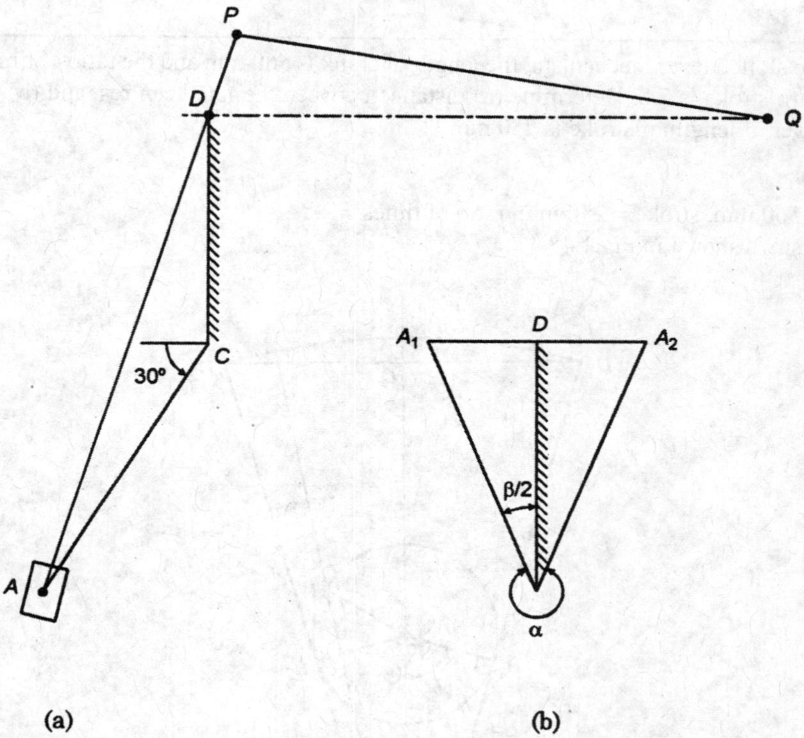

Fig.1.47 Whitworth mechanism

Example 1.18

The distance between two parallel shafts connected by Oldham's coupling is 25 mm. The driving shaft revolves at 240 rpm. Determine the maximum speed of sliding of the tongue of the intermediate piece along its groove.

■ Solution

$$d = 25 \text{ mm}, n = 240 \text{ rpm}$$

$$\omega = 2\pi \times \frac{240}{60} = 25.133 \text{ rad/s}$$

Maximum velocity of sliding $= \omega \times d = 25.133 \times 0.025 = 0.628$ m/s

Example 1.19

In a crank and slotted lever mechanism, the length of crank is 560 mm and the ratio of time of working stroke to return stroke is 2.8. Determine (a) distance between the fixed centres, and (b) the length of the slotted lever, if length of stroke is 250 mm.

■ Solution

Given: $AB = 560$ mm, stroke = 250 mm, ratio of times = 2.8
The mechanism is shown in Fig.1.48.

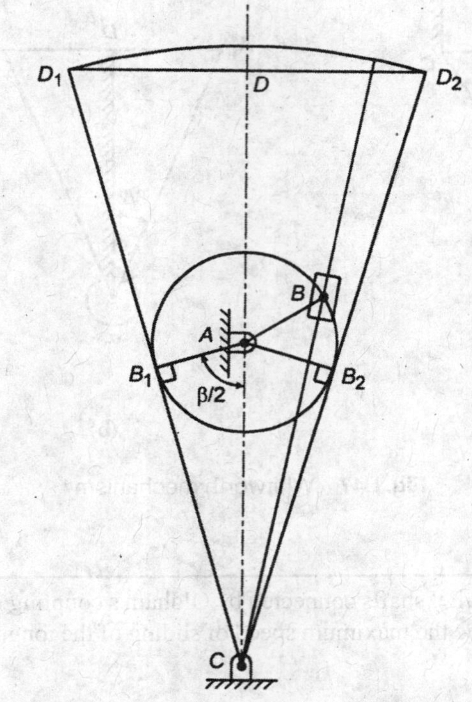

Fig. 1.48 Crank and slotted lever mechanism

$$\frac{\text{Time of working stroke}}{\text{Time of return stroke}} = 2.8$$

$$\frac{360^o - \beta}{\beta} = 2.8$$

$$\beta = 94.737^o \text{ Now}$$

$$\frac{AB_1}{AC} = \cos\left(\frac{\beta}{2}\right)$$

$$AC = \frac{560}{\cos 47.368^\circ} = 826.83 \text{ mm}$$

$$\text{Length of stroke} = D_1D_2 = 2D_1D$$

$$250 = 2CD_1 \sin\left(90^\circ - \frac{\beta}{2}\right)$$

$$= 2CD_1 \sin 47.368^\circ$$

Length of slotted lever. $CD_1 = 169.9$ mm.

Example 1.20

The configuration of a drag link mechanism is shown in Fig.1.49. Determine the time ratio and the length of stroke. The crank O_2A rotates clockwise.

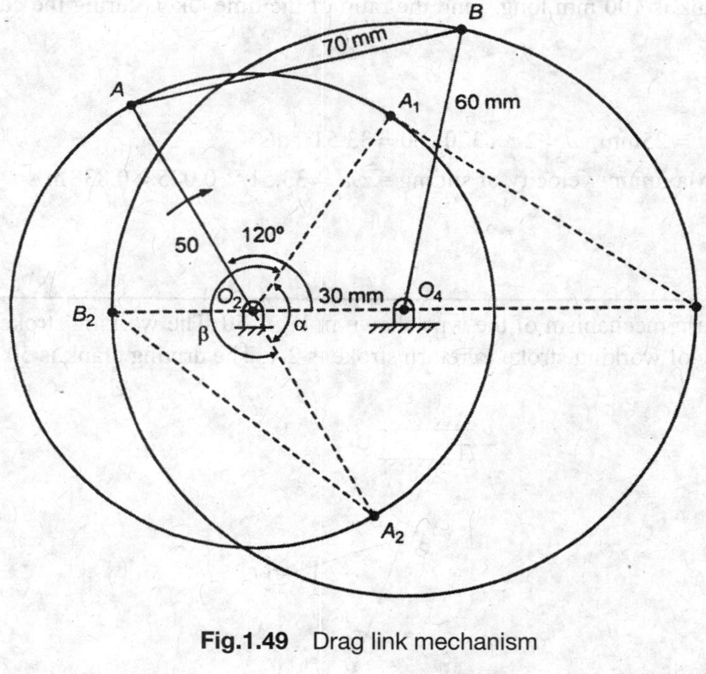

Fig.1.49 Drag link mechanism

■ Solution

The drag link mechanism has been drawn as O_2ABO_4 to scale of 1 cm = 10 mm. The extreme positions of B are B_1 and B_2. The length of stroke is $2 \times O_4B = B_1B_2 = 2 \times 60 = 120$ mm.

When B is at B_1, then A is at A_1 and when B is at B_2, then A is at A_2. By measurement, we have

$$\alpha = 110^\circ \text{ and } \beta = 250^\circ$$

$$\text{Ratio of times} = \frac{\beta}{\alpha} = \frac{250}{110} = 2.27$$

Example 1.21

The distance between the axes of parallel shafts connected by Oldham's coupling is 25 mm. The speed of rotation of the shafts is 320 rpm. Determine the maximum velocity of sliding of each tongue in its slot.

■ Solution

$$\cos(\beta/2) = 100/150 = 2/3, \ \beta = 96.4^\circ$$

$$\text{Time of cutting stroke / Time of return stroke} = (360^\circ - \beta)/\beta$$

$$= 263.6/96.4 = 2.735$$

Example 1.22

In crank and slotted lever quick return mechanism, the distance between the fixed centres is 150 mm and the driving crank is 100 mm long. Find the ratio of the time taken during the cutting and return strokes.

■ Solution

$$d = 25\text{mm}, \ \omega = 2\pi \times 320/60 = 33.51 \text{ rad/s}$$

$$\text{Maximum velocity of sliding} = \omega d = 33.51 \times 0.025 = 0.838 \text{ m/s}$$

Example 1.23

Design a quick return mechanism of the type shown in Fig.1.50. The working stroke is 200 mm and the ratio of the time of working stroke to return stroke is 2:1. The driving crank is 50 mm long.

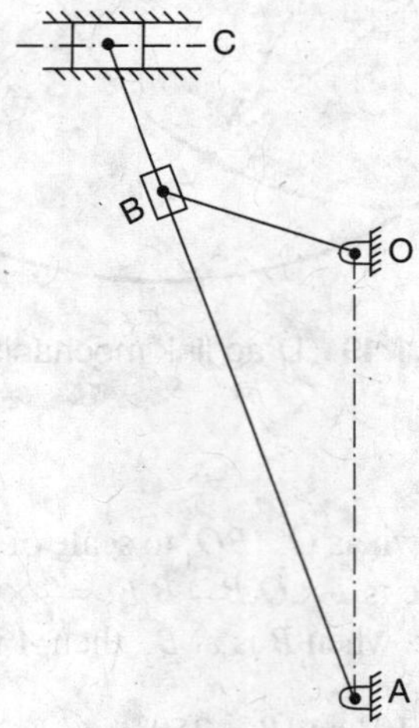

Fig.1.50 Quick-return mechanism

■ **Solution**

$$(360^\circ - \beta)/\beta = 2, \; \beta = 120^\circ$$
$$\cos(\beta/2) = OB/OA$$
$$OA = 50/\cos 60^\circ = 100 \text{ mm}$$
$$\text{Working stroke} = C_1C_2 = 2C_1C$$
$$200 = 2AC_1 \sin 30^\circ$$
$$AC_1 = 200 \text{ mm}$$
$$\text{Length of lever, } AC = 200 \text{ mm}$$

Example 1.24

Design a Whitworth quick return motion mechanism shown in Fig.1.51 to have the following particulars:
Return stroke = 200 mm
Time ratio of working to return stroke = 2
Length of driving crank = 50 mm.

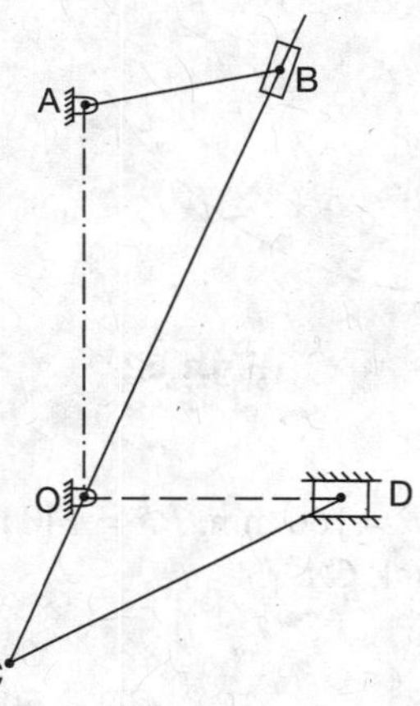

Fig.1.51 Whitworth mechanism

■ **Solution**

$$(360^\circ - \beta)/\beta = 2$$
$$\beta = 120^\circ$$
$$OA = \frac{AB}{\cos(\beta/2)}$$
$$= 50/\cos 60^\circ = 100 \text{ mm}$$
$$\text{Return stroke} = 2 \times OC$$
$$OC = 200/2 = 100 \text{ mm}$$
$$CD > OC$$

Example 1.25

In a Whitworth quick return motion mechanism, as shown in Fig.1.52, the distance between the fixed centres is 60 mm and the length of the driving crank is 80 mm. The length of the slotted lever is 160 mm and length of the connecting nod is 140 mm. Find the ratio of the time of cutting stroke to the time of return stroke and also the effective stroke.

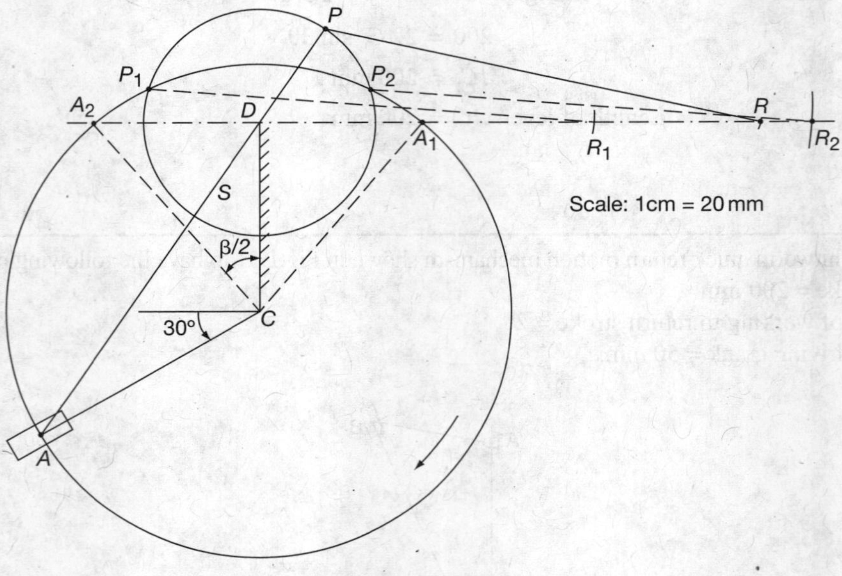

Fig.1.52

Solution

Given: $CD = 60$ mm, $CA = 80$ mm, $AP = 160$ mm, $PR = 140$ mm.
Draw extreme positions of driving crank CA.

$$\cos\frac{\beta}{2} = \frac{CD}{CA_2} = \frac{60}{80} = 0.75$$

$$\frac{\beta}{2} = 41.41^\circ$$

$$\beta = 82.82^\circ$$

$$\frac{\text{Time of cutting stroke}}{\text{Time of return stroke}} = \frac{360^\circ - \beta}{\beta}$$

$$= \frac{360^\circ - 82.82^\circ}{82.82^\circ} = 3.347$$

Draw a circle with centre D and radius equal to DP to intersect the crank circle at P_1 and P_2. Mark $P_1R_1 = P_2R_2 = PR$.
Length of effective stroke $= R_1R_2 = 3.5$ cm or 70 mm

Summary for Quick Revision

1 Kinematics is a subject that deals with the study of relative motion of parts constituting a machine, neglecting forces producing the motion.

2 Dynamics is a subject that deals with the study of relative motion of parts constituting a machine, considering the forces producing the motion.

3 A mechanism is a set of machine elements or components or parts, arranged in a specific order to produce a specified motion.

4 A kinematic joint is the connection between two links by a pin. Kinematic joints can be binary, ternary, quaternary and so on. A ternary joint is equivalent to two binary joints and a quaternary joint is equivalent to four binary joints.

5 A link (or element or kinematic link) is a resistant body (or assembly of resistant bodies) which constitute part (or parts) of a machine, connecting other parts, which have motion, relative to it.

6 Links can be classified as: binary, ternary, quaternary, etc. depending upon its ends on which revolute or turning pairs can be placed. A binary link has two vertices, a ternary link has three vertices, and a quaternary link has four vertices, and so on. Links can be rigid, flexible, fluid, and floating.

7 The two links of a machine, when in contact with each other, are said to form a pair. A kinematic pair consists of two links that have relative motion between them. Kinematic pairs may be classified according to the type of relative motion, contact or mechanical constraint. The kinematic pairs could be of the sliding, turning, rolling, screw or spherical types.

8 Lower pairs are those which have surface (or area) contact while in motion and the relative motion being purely turning or sliding. Higher pairs have point or line contact while in motion and the relative motion being the combination of sliding and turning.

9 Closed pair consists of two elements held together mechanically in such a manner that only required type of relative motion occurs. Unclosed pair consists of two elements not held mechanically and are held in contact by the action of external forces.

10 A kinematic chain may be defined as an assembly of links in which the relative motion of the links is possible and the motion of each relative to the others is definite. The last link of the kinematic chain is attached to the first link.

11 When one of the links of a kinematic chain is fixed, the chain is called a mechanism. The mechanisms can be simple, compound, complex, planar, or spatial.

12 A machine is a device which transforms energy available in one form to another to do certain type of desired useful work.

13 A four-bar chain consists of four binary links. The four-bar mechanism with all its pairs as turning pairs is called the "quadric cycle chain." When one of these turning pairs is replaced by a slider pair, the chain becomes "single slider chain." When two turning pairs are replaced by slider pairs, it is called a "double slider chain" or a "crossed double slider chain," depending on whether the two slider pairs are adjacent or crossed.

14 A kinematic chain becomes a mechanism when one of its links is fixed.

15 For a kinematic chain: $L = 2P - 4$ and $J = (3/2)\,L - 2$, where L = number of binary links, P = number of lower pairs, J = number of joints.

For a locked chain, LHS > RHS; for a constrained chain, LHS = RHS; and for an unconstrained chain, LHS < RHS.

For a kinematic chain having higher pairs, $J + (H/2) = (3/2)L - 2$, where H = number of higher pairs.

16 Degrees of freedom of a kinematic pair is defined as the number of independent relative motions, both translational and rotational, a kinematic pair can have.

17 Mobility of a mechanism is defined as the number of degrees of freedom it possesses.

18 According to Kutzbach criterion for degrees of freedom of planar mechanisms, $F = 3(n-1) - 2p - h$, where n = total number of links in a mechanism, p = number of simple joints or lower pairs, h = number of higher pairs. For a kinematic chain having different type of links, $p = (1/2)[2n_2 + 3n_3 + 4n_4 + \cdots]$, where n_2, n_3, etc. are binary, ternary and so on links.

19 Gruebler's criterion: For a constrained motion, $F = 1$, so that $2p + h - 3n + 4 = 0$.

20 A new mechanism obtained by replacing sliding, spring or cam pairs by turning having the same number of degrees of freedom as the original mechanism is called an equivalent mechanism.

21 Grashof's law: For a kinematic chain. $(l + s) \leq (p + q)$.

Double crank mechanism: when the shortest link is fixed.

Crank–rocker mechanism: when any of the adjacent links of shortest link are fixed.

Rocker–rocker mechanism: when link opposite to the shortest link is fixed and shortest link is made the coupler.

22 The method of obtaining different mechanisms by fixing different links of a kinematic chain is called inversion of the mechanism.

Inversions of four-bar chain: Beam engine. Coupled wheel of locomotive. Watt's indicator mechanism. Slider–crank chain

Inversions of single slider crank chain: Pendulum pump, Oscillating cylinder engine, Rotary internal combustion engine, Crank and slotted lever quick return motion mechanism, Whitworth quick return motion mechanism, Toggle mechanism.

Inversions of double slider crank chain: Donkey pump, Oldham's coupling, Elliptical trammel, Scotch yoke.

23 Quick return motion mechanisms are used in shaping and slotting machines. Scotch yoke is used as a sine – cosine function generator. Oldham's coupling is used to connect two shafts when the distance between their axes is small and variable.

24 In a quick–return motion mechanism,

Time of cutting stroke/Time of return stroke = angle covered by crank during cutting stroke/angle covered by crank during return stroke.

25 Maximum sliding speed of each tongue along its slot in the Oldham's coupling = Distance between the axes of shafts × angular speed of each shaft.

Multiple Choice Questions

1 The purpose of a link is to

(a) transmit motion
(b) guide other links
(c) act as a support
(d) all of the above.

2 A kinematic chain requires at least
(a) 2 links and 3 turning pairs
(b) 3 links and 4 turning pairs
(c) 4 links and 4 turning pairs
(d) 5 links and 4 turning pairs.

3 Which of the following is a lower pair?
(a) ball and socket
(b) piston and cylinder
(c) cam and follower
(d) (a) and (b) above.

4 Quick return motion mechanism is used in
(a) milling machine
(b) broaching machine
(c) grinding machine
(d) slotter.

5 A kinematic chain becomes a mechanism when
(a) first link is fixed
(b) any one link is fixed
(c) all links are fixed
(d) none of the links are fixed.

6 A slider crank mechanism consists of the following number of turning and sliding pairs
(a) 1, 3
(b) 2, 2
(c) 3, 1
(d) 4, 0.

7 A typewriter constitutes a
(a) machine
(b) structure
(c) mechanism
(d) inversion.

8 The lead screw of a lathe with nut is a
(a) rolling pair
(b) screw pair
(c) turning pair
(d) sliding pair.

9 In kinematic chain, a ternary joint is equivalent to
(a) two binary joints
(b) three binary joints
(c) four binary joints
(d) single binary joint.

10 A four-bar mechanism satisfying Grashof's criteria will act as a drag-crank mechanism, if
(a) the longest link is fixed
(b) the shortest link is fixed
(c) any link adjacent to shortest link is fixed.

11 In a four-bar mechanism, $(l + s) < (p + q)$. It will act as a crank-rocker mechanism, if
(a) the link opposite to the shortest link is fixed
(b) the shortest link is fixed
(c) any link adjacent to shortest link is fixed.

12 In a four-bar mechanism, the sum of the lengths of shortest and longest links is less than the sum of other two links. It will act as a rocker-rocker mechanism, if
(a) the link opposite to the shortest link is fixed
(b) the shortest link is fixed
(c) any link adjacent to shortest link is fixed.

13 Which of the following is an inversion of single-slider crank chain?
(a) elliptical trammel
(b) hand pump
(c) Scotch yoke
(d) Oldham's coupling.

14 Which one of the following is an inversion of double-slider crank chain?
(a) Whitworth quick return motion mechanism
(b) reciprocating compressor
(c) Scotch yoke
(d) rotary engine.

15 The number of degrees of freedom of the mechanism shown below is:

(a) 0 (b) 1
(c) 2 (d) 3

16 The number of degrees of freedom of the mechanism shown below is:

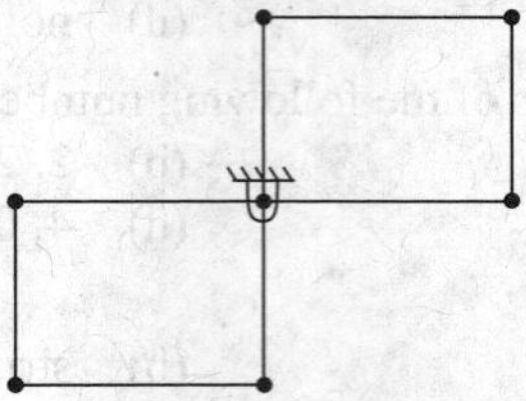

(a) 0 (b) 1
(c) 2 (d) 3

17 The number of equivalent lower pairs of a higher pairs are
(a) 2 (b) 3
(c) 4 (d) 1

18 The equivalent number of binary joints of a quaternary joint are:
(a) 2 (b) 3
(c) 4 (d) 5

19 For a kinematic chain, if $L = 2P - 4$, then
$J = CL - 2$
where C is equal to

(a) $\frac{1}{2}$ (b) 1

(c) $\frac{3}{2}$ (d) 2

20 In the equation, $L = 2P - 4$, the kinematic chain is called locked, when
(a) LHS < RHS (b) LHS = RHS
(c) LHS > RHS

21 The equivalent number of binary links of a spring are:
(a) 1 (b) 2
(c) 3 (d) 4

Answers

1. (d) 2. (c) 3. (d) 4. (d) 5. (b) 6. (c) 7. (c) 8. (b) 9. (a) 10. (b) 11. (c)
12. (a) 13. (b) 14. (c) 15. (b) 16. (b) 17. (a) 18. (c) 19. (c) 20. (c) 21. (b)

Review Questions

1 Differentiate between a mechanism and a machine.

2 Define kinematic link, kinematic pair and a kinematic chain.

3 What is a resistant body?

4 How do you classify kinematic pairs? Illustrate with examples.

5 Differentiate between (a) lower and higher pairs, (b) turning and screw pairs, (c) rolling and spherical pairs, and (d) closed and unclosed pairs.

6 Define degrees of freedom of a mechanism. How this is determined?

7 Explain Gruebler's criterion for degrees of freedom for planar mechanisms.

8 What do you understand by an equivalent mechanism?

9 Discuss various types of constrained motions.

10 What is mobility (or movability) of a mechanism?

11 What is a kinematic chain? How this can be ascertained?

12 What do you understand by inversion of a mechanism? List various inversions of a four-bar chain.

13 Describe various inversions of a single slider–crank chain.

14 List the various inversions of a double slider–crank chain and give their applications.

Exercises

1.1 Calculate the number of degrees of freedom of the mechanisms shown in Fig.1.53.

[Ans. 1, 0, 3]

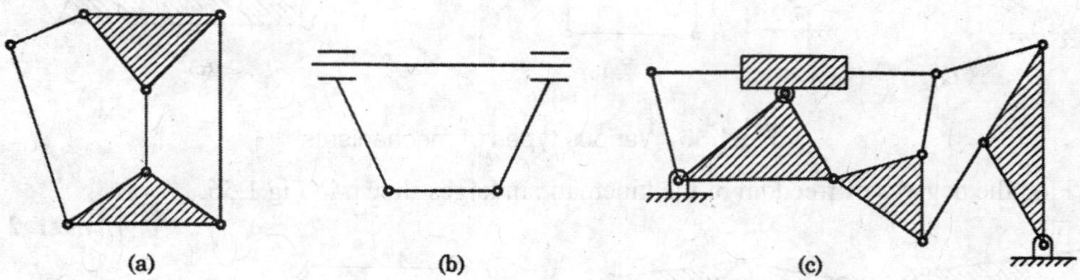

Fig.1.53 Various types of mechanisms

1.2 Determine the number of degrees of freedom of all the devices shown in Fig.1.54.

[Ans. 0, 1, 2, 2]

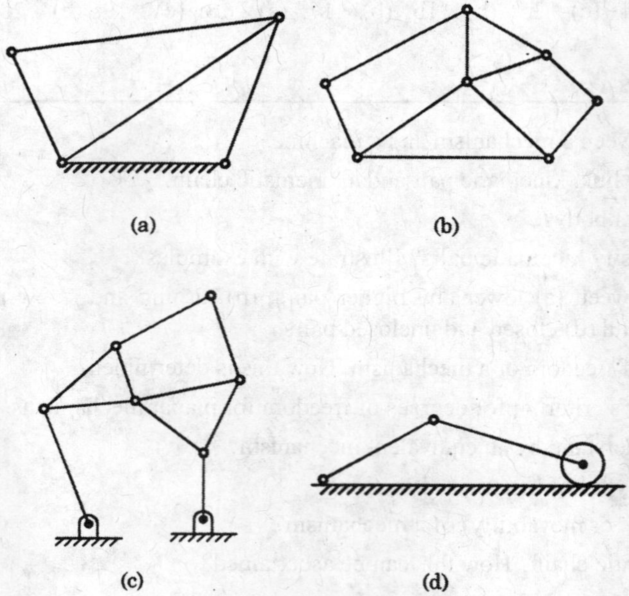

Fig.1.54 Various types of devices

1.3 Determine the degrees of freedom of the mechanisms shown in Fig.1.55.

[Ans. 1, 1, 1]

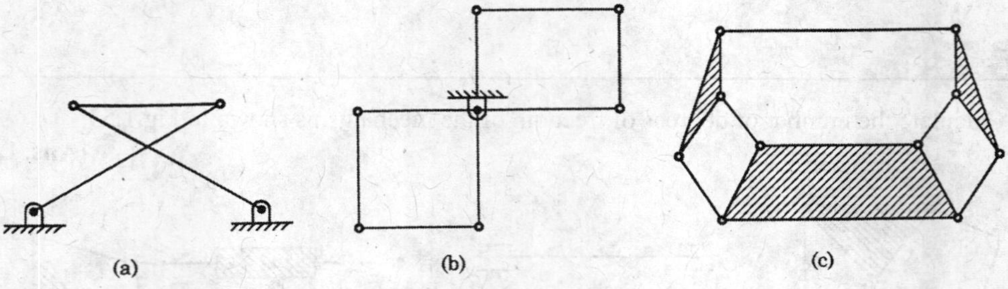

Fig.1.55 Various types of mechanisms

1.4 Find the degrees of freedom of the kinematic linkages shown in Fig.1.56.

[Ans. 1, 2, 1, 0]

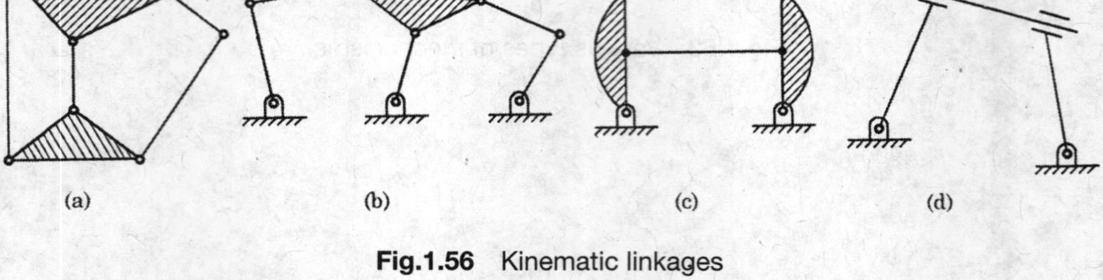

Fig.1.56 Kinematic linkages

1.5 Draw the equivalent mechanisms of the following and find their degrees of freedom, as shown in Fig.1.57.

[Ans. 1, 1, 2]

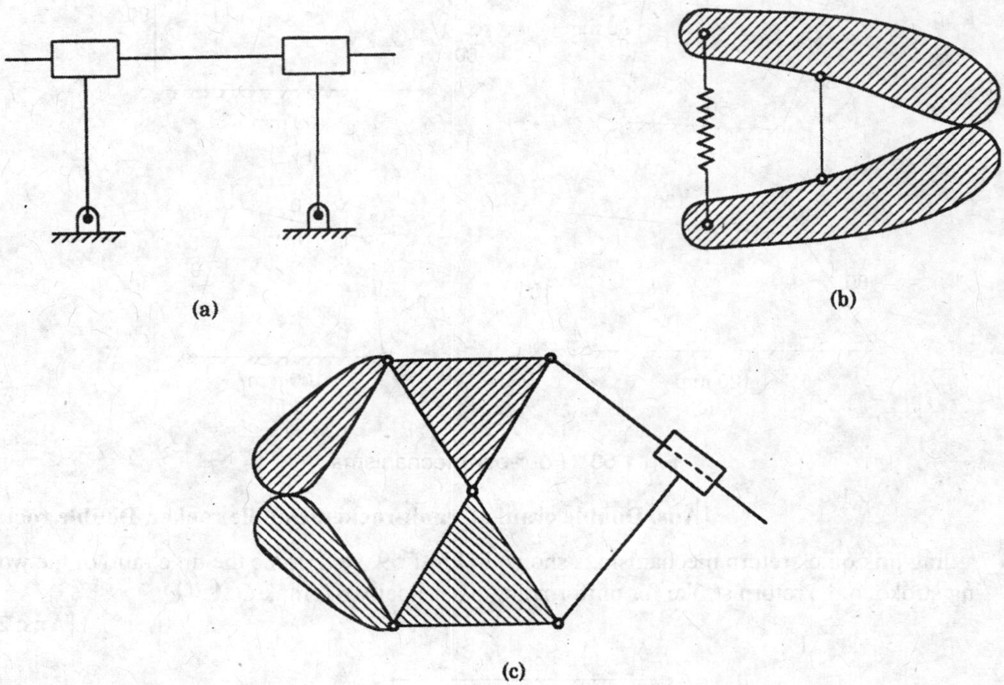

Fig.1.57 Various types of mechanisms

1.6 In a quick-return motion mechanism of the oscillating link type, the distance between the fixed centres is 80mm and the length of the driving crank is 20 mm. Determine the time ratio of the working stroke to the return stroke.

[Ans. 1.38]

1.7 In an off-set slider crank mechanism, the eccentricity is 50 mm, length of crank is 300 mm, and length of connecting rod is 500 mm, determine the quick return ratio.

[Ans. 1.293]

1.8 In a quick-return motion mechanism of the crank and slotted lever type, the ratio of maximum velocities is 2. If the length of stroke is 250 mm, find (a) the length of the slotted lever, (b) the ratio of times of cutting and return strokes, and (c) the maximum cutting velocity in m/s if the crank rotates at 30 rpm.

[Ans. 375 mm, 1.552, 0.294 m/s]

1.9 The distance between the parallel shafts of an Oldham's coupling is 15 mm. The driving shaft revolves at 160 rpm. Calculate the maximum speed of sliding of the tongue of the intermediate piece along its groove.

[Ans. 0.251 m/s]

1.10 Determine the type of chains shown in Fig.1.58.

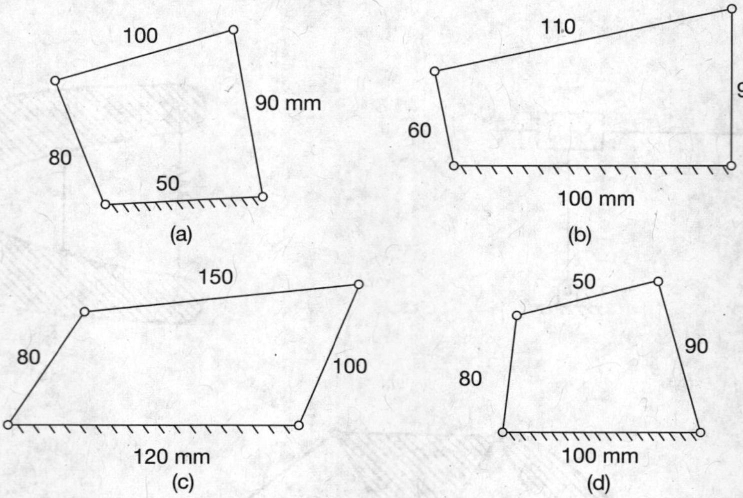

Fig.1.58 Four-bar mechanisms

[Ans. Double crank, Crank-rocker, Double rocker, Double rocker]

1.11 A drag link quick return mechanism is shown in Fig.1.59. Determine the time ratio of the working stroke to the return stroke for uniform angular velocity of crank O_1A.

[Ans. 2.3]

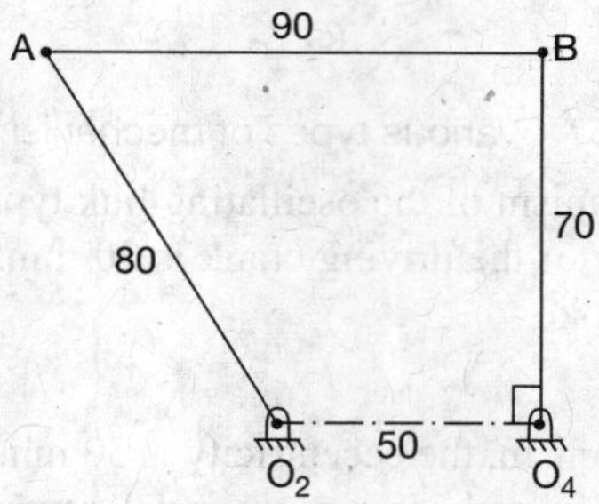

Fig.1.59 Drag link mechanism

1.12 Classify the motion of the four-bar mechanism shown in Fig.1.60, $l_1 = 80$ cm, $l_2 = 30$ cm, $l_3 = 75$ cm, $l_4 = 65$ cm

[Ans. Crank-rocker]

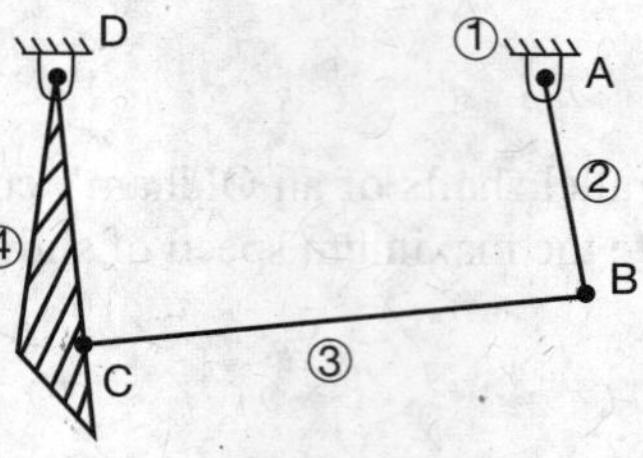

Fig.1.60 Four-bar mechanism

1.13 Find the number of degrees of freedom of the mechanism shown in Fig.1.61.

[Ans. 1]

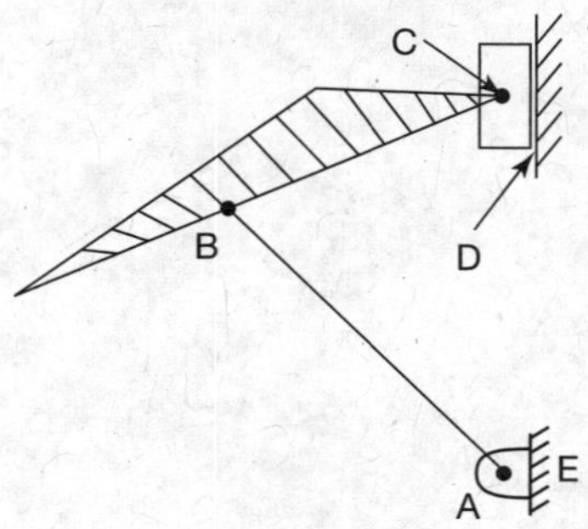

Fig.1.61 Four-bar mechanism

1.14 Classify the four-bar mechanism based on its possible motion, when the lengths of the links are: $l_1 = 30$ cm, $l_2 = 12.5$ cm, $l_3 = 30$ cm, and $l_4 = 10$ cm.

[Ans. Crank-rocker]

2 VELOCITY IN MECHANISMS

Chapter Outline

2.1 INTRODUCTION

Kinematics deals with the study of relative motion between the various links of a machine ignoring the forces involved in producing such motion. Thus, kinematics is the study to determine the displacement velocity and acceleration of the various links of the mechanism. A machine is a mechanism or a combination of mechanisms that not only imparts definite motion to the various links but also transmits and modifies the available mechanical energy into some kind of useful energy. In this chapter, we shall study the various methods of determination of velocity in mechanisms.

2.2 VELOCITY DIAGRAMS

Displacement: The displacement of a body is its change of position with reference to a certain fixed point.

Velocity: Velocity is the state of change of displacement of a body with respect to time. It is a vector quantity.

Linear velocity: It is the rate of change of velocity of a body along a straight line with respect to time. Its units are m/s.

Angular velocity: It is the rate of change of angular position of a body with respect to time. Its units are rad/s.

The relationship between velocity v and angular velocity ω is:

$$v = r\,\omega \tag{2.1}$$

where r = distance of point undergoing displacement from the centre of rotation.

Relative velocity: The relative velocity of a body A with respect to a body B is obtained by adding to the velocity of A the reversed velocity of B. If $v_a > v_b$, then

$$v_{ab} = \vec{v_a} - \vec{v_b}$$

or

$$ba = oa - ob$$

Similarly,

$$v_{ba} = \vec{v_b} - \vec{v_a}$$

or

$$ab = ob - oa$$

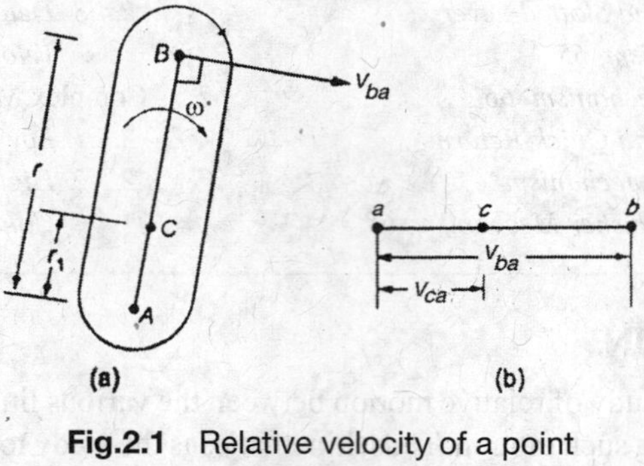

Fig.2.1 Relative velocity of a point

Consider two points A and B on a rigid link rotating clockwise about A as shown in Fig.2.1(a). There can be no relative motion between A and B as long as the distance between A and B remains the same. Therefore, the relative motion of B with respect to A must be perpendicular to AB. Hence, the direction of relative velocity of two points in a rigid link is always along the perpendicular to the straight line joining these points. Let relative velocity of B with respect to A be represented by $v_{ba} = \omega \cdot AB$, then ab is drawn perpendicular to AB to a convenient scale, as shown in Fig.2.1(b). Similarly, the linear velocity of any other point C on AB with respect to A is $v_{ca} = \omega \cdot AC$ and is represented by vector ac. Hence,

$$v_{ba}/v_{ca} = (\omega \cdot AB)/(\omega \cdot AC) = AB/AC$$

or

$$ab/ac = AB/AC = r/r_1 \qquad (2.2)$$

Hence the point C divides the vector ab in the same ratio as the point C divides the link AB.

2.3 DETERMINATION OF LINK VELOCITIES

There are two methods to determine the velocities of links of mechanisms.

1. Relative velocity method
2. Instantaneous centre method

2.3.1 Relative Velocity Method

Consider a rigid link AB, as shown in Fig.2.2(a), such that the velocity of A (v_a) is vertical and velocity of B (v_b) is horizontal. To construct the velocity diagram, take a point o. Draw oa representing the magnitude and direction of velocity of A. Draw ob along the direction of v_b. From point 'a' draw a line ab perpendicular to AB, meeting ob in b. Then oab is the velocity triangle, as shown in Fig.2.2(b). $Ob = v_b$; $ab = v_{ba}$, i.e. the velocity of B with respect to A. Vector ab is called the *velocity image* of link AB. The velocity of any point C in AB with respect to A is given by,

$$v_{ca} = ac = v_{ba} \cdot \left(\frac{AC}{AB}\right)$$

$$v_c = v_a + \left(\frac{AC}{AB}\right) v_{ba}$$

$$= oa + \left(\frac{ac}{ab}\right) \cdot ab$$

$$= oa + ac$$

$$= oc$$

Hence vector oc represents the velocity of point C.

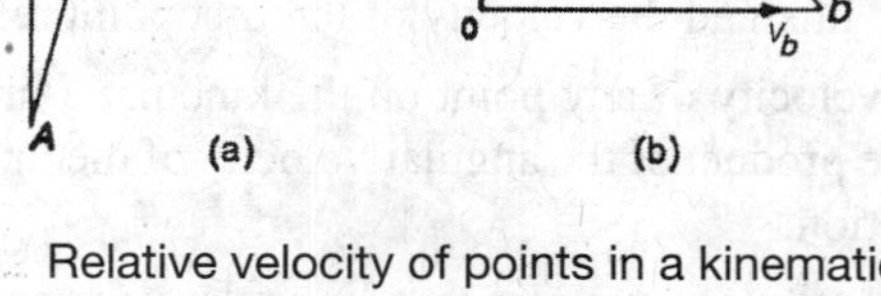

Fig.2.2 Relative velocity of points in a kinematic link

2.3.2 Relative Velocity of Points in a Kinematic Link

Consider a link A_1B_1 first undergoing rotation by an amount $\Delta\theta$ to A_1B_1' and then undergoing translation by an amount Δs_A to occupy the new position A_2B_2, as shown in Fig.2.3(a). Then

$$\Delta s_B = \Delta s_{BA} + \Delta s_A \tag{a}$$

Now let the link A_1B_1 first undergo linear translation Δs_A to A_2B_1'' and then angular rotation of $\Delta\theta$ to A_2B_2, as shown in Fig.2.3(b). Then

$$\Delta s_B = \Delta s_A + \Delta s_{BA} \tag{b}$$

Eqs. (a) and (b) are same. Dividing by Δt, we get

$$\frac{\Delta s_B}{\Delta t} = \frac{\Delta s_A}{\Delta t} + \frac{\Delta s_{BA}}{\Delta t}$$

or

$$v_b = v_a + v_{ba}$$

Therefore, the velocity of point B is obtained by adding vectorially the relative velocity of point B w.r.t. point A to the velocity of point A.

Now

$$\Delta s_{BA} = A_1B_1 \cdot \Delta\theta$$

or

$$\frac{\Delta s_{BA}}{\Delta t} = A_1B_1 \cdot \frac{\Delta\theta}{\Delta t}$$

$$v_{ba} = A_1B_1 \cdot \omega$$

Also

$$\angle\phi = 90^\circ$$

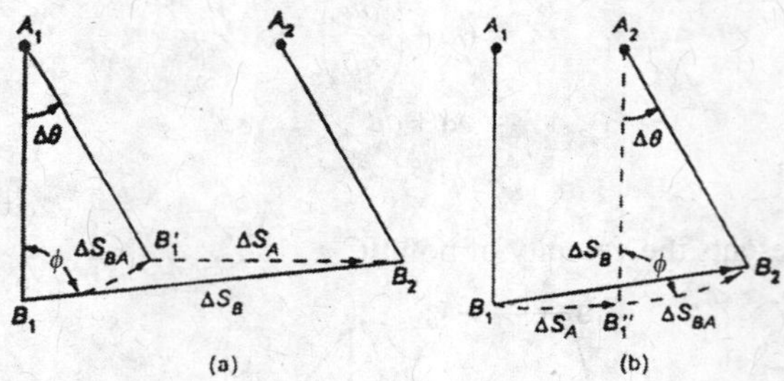

Fig.2.3 Relative velocity of points in a kinematic link

The following conclusions may be drawn from the above analysis:

- The velocity of any point on the kinematic link is given by the vector sum of the velocity of some other point in the link and the velocity of the first point relative to the other.
- The magnitude of the velocity of any point on the kinematic link relative to the other point in the kinematic link is the product of the angular velocity of the link and distance between the two points under consideration.
- The direction of the velocity of any point on a link relative to any other point on the link is perpendicular to the line joining the two points.

2.3.3 Relative Angular Velocities

Consider two links OA and OB connected by a pin joint at O, as shown in Fig.2.4. Let ω_1 and ω_2 be the angular velocities of the links OA and OB, respectively. Relative angular velocity of OA with respect to OB is,

$$\omega_{12} = \omega_1 - \omega_2$$

and relative angular velocity of OB with respect to OA is,

$$\omega_{21} = \omega_2 - \omega_1$$
$$= -\omega_{12}$$

If r = radius of the pin at joint O, then rubbing (or sliding) velocity at the pin joint O,

$$= (\omega_1 - \omega_2)r, \text{ when the links move in the same direction} \quad (2.3a)$$
$$= (\omega_1 + \omega_2)r, \text{ when the links move in the opposite direction} \quad (2.3b)$$

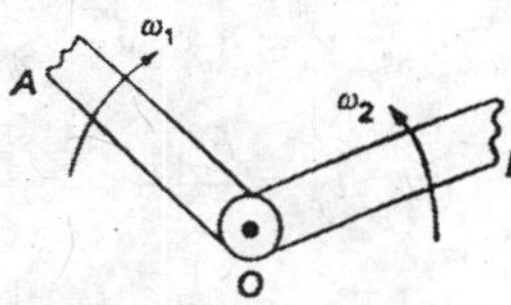

Fig.2.4 Relative angular velocities

2.3.4 Relative Velocity of Points on the Same Link

Consider a ternary link ABC, as shown in Fig.2.5(a), such that C is any point on the link. Let v_a and v_b be the velocities of points A and B, respectively. Then

$$v_b = v_a + v_{ba}$$

$$\omega_{ab} = \frac{v_{ba}}{AB} = \frac{ab}{AB}$$

$$v_c = v_a + v_{ca}$$

or

$$= v_b + v_{cb}$$

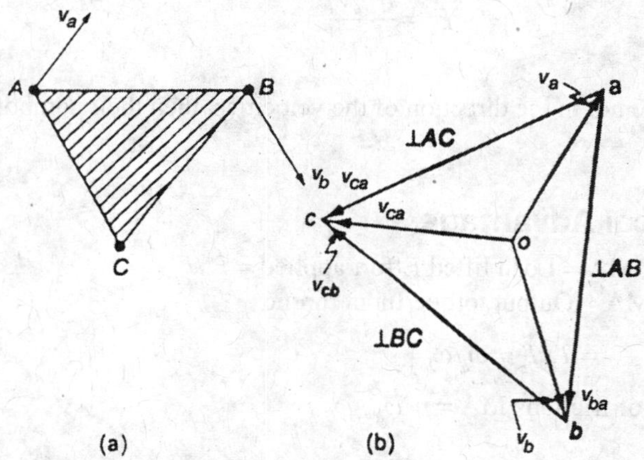

Fig.2.5 Relative velocity of points on the same link

The velocity diagram is shown in Fig.2.5(b).
Angular velocity of link ABC,

$$\omega_{abc} = \frac{v_{ba}}{AB} = \frac{v_{cb}}{BC} = \frac{v_{ca}}{AC}$$

$$= \frac{ab}{AB} = \frac{bc}{BC} = \frac{ac}{AC} \tag{2.4}$$

2.3.5 Forces in a Mechanism

Consider a link AB subjected to the action of forces and velocities, as shown in Fig.2.6. Let A be the driving end and B the driven end. When the direction of the forces and velocities is the same, then

$$\text{Energy at } A = \text{Energy at } B$$

$$F_a \times v_a = F_b \times v_b$$

or $$F_b = \frac{F_a v_a}{v_b} \tag{2.5a}$$

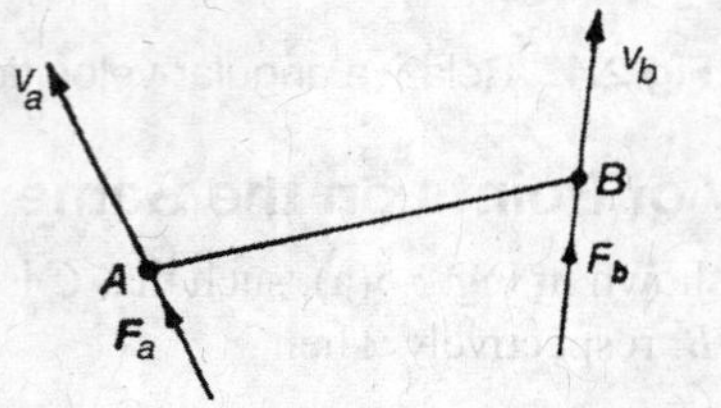

Fig.2.6 Force and velocity diagram

Considering the effect of friction, the efficiency of transmission,

$$\eta = \frac{\text{output}}{\text{input}} = \frac{(F_b v_b)}{(F_a v_a)}$$

or $$F_b = \frac{\eta F_a v_a}{v_b} \tag{2.5b}$$

When the forces are not in the direction of the velocities, then their components along the velocities should be taken.

2.3.6 Mechanical Advantage

Mechanical advantage, MA = Load lifted/Effort applied = F_b/F_a

For a mecahnism, MA = Output torque/Input torque

$$= T_b/T_a = \omega_a/\omega_b \tag{2.6a}$$

Considering the effect of friction, MA = $\eta\ \omega_a / \omega_b$ (2.6b)

2.3.7 Four-Bar Mechanism

(a) Consider the four-bar mechanism, as shown in Fig.2.7(a), in which the crank O_1A is rotating clockwise with uniform angular speed ω. The linear velocity of point A is $v_a = \omega \times O_1A$ and it is perpendicular to O_1A. Therefore, draw $o_1a \perp O_1A$ to a convenient scale in Fig.2.7(b). The velocity of point B is perpendicular to O_2B. Therefore, at point o_1, draw $o_1b \perp O_2B$. The relative velocity of B with respect to A is perpendicular to AB. Therefore, draw $ab \perp AB$ meeting the line drawn perpendicular to O_2B at b. Then $v_b = o_1b$, and $v_{ba} = ab$. To find the velocity of joint C, draw $ac \perp AC$ and $bc \perp BC$ to meet at c. Join o_1c. Then $v_c = o_1c$.

Now

$$v_b = v_a + v_{ba}$$
$$= o_1a + ab = o_1b$$

$$v_c = v_b + v_{cb}$$
$$= o_1b + bc = o_1c$$

and also,

$$v_c = v_a + v_{ca}$$
$$= o_1a + ac = o_1c$$

To find the velocity of any point D in AB, we have

$$\frac{BD}{BA} = \frac{bd}{ba}$$

or

$$bd = \left(\frac{BD}{BA}\right) \cdot ba$$

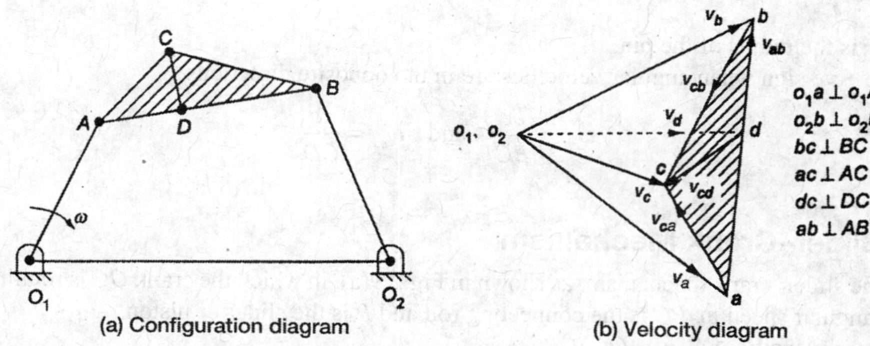

Fig.2.7 Four-bar mechanism with ternary link

Thus, locate point d in ab and join o_1d. Then $o_1d = v_d$. To find the relative velocity of C w. r. t. D, join cd. Then $v_{cd} = dc$. The velocity image of link ABC is abc.

(b) Now consider the four-bar mechanism, as shown in Fig.2.8(a), in which the crank AB is rotating at angular velocity ω in the anticlockwise direction. The absolute linear velocity of B is $\omega \cdot AB$ and is perpendicular to AB. Draw $ab \perp AB$ to a convenient scale to represent v_{ba}, as shown in Fig.2.8(b). From b, draw a line perpendicular to BC and from 'a' another line perpendicular to CD to meet each other at point c. Then, $ac = v_{ca}$ and $cb = v_{bc}$. To find the velocity of any point E in BC, we have

$$\frac{CE}{CB} = \frac{ce}{cb}$$

or

$$ce = \left(\frac{CE}{CB}\right) \cdot cb$$

Thus, locate point e in cb and join ae. Then $ae = v_{ea}$.

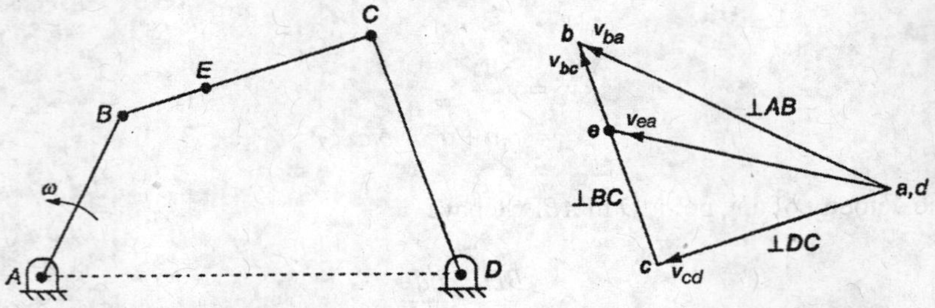

(a) Configuration diagram (b) Velocity diagram

Fig.2.8 Four-bar mechanism

The rubbing velocities at pins of joints are:

A : $\omega_{ab} \cdot r_a$
B : $(\omega_{ab} \pm \omega_{cb})\, r_b$
C : $(\omega_{bc} \pm \omega_{dc})\, r_c$
D : $\omega_{cd} \cdot r_d$

where r is the radius of the pin.
Use the + ve sign when angular velocities are in the opposite directions.

$$\omega_{cb} = \frac{bc}{BC} \quad \text{and} \quad \omega_{cd} = \frac{cd}{CD}$$

2.3.8 Slider–Crank Mechanism

Consider the slider–crank mechanism, as shown in Fig.2.9(a), in which the crank OC is rotating clockwise with angular speed ω. PC is the connecting rod and P is the slider or piston.

The linear velocity of C, $v_c = OC \cdot$

To draw the velocity diagram, draw a line $oc = v_c$ from point o, as shown in Fig.2.9(b), representing the velocity of point C to a convenient scale. From point c draw a line perpendicular to CP. The velocity of slider P is horizontal. Therefore, from point o draw a line parallel to OP to intersect the

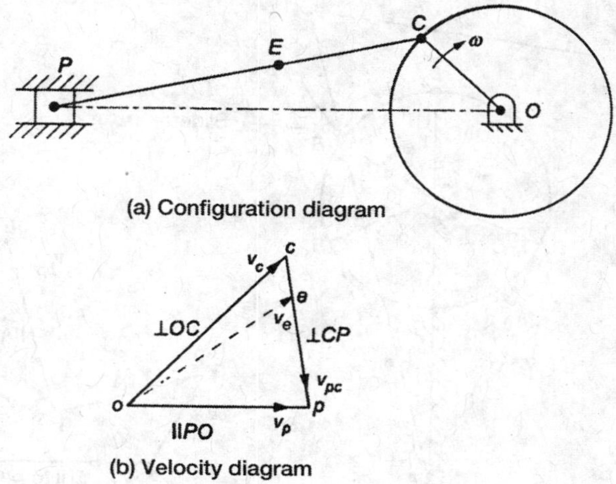

Fig.2.9 Slider-crank mechanism

line drawn perpendicular to *CP* at *p*. Then the velocity of the piston, $v_p = op$ and the velocity of piston *P* relative to crank pin *C* is v_{pc} = cp. To find the velocity of any point *E* in *CP*, we have

$$CE/CP = ce/cp$$

$$\text{or} \quad ce = (CE/CP) \cdot cp$$

Thus locate point *e* in *cp*, join *oe*. Then $v_e = oe$

Rubbing velocities are:

$$O : \omega_{oc} \cdot r_o$$

$$P : \omega_{cp} \cdot r_p$$

$$C : \left(\omega_{oc} + \omega_{cp}\right) \cdot r_c$$

2.3.9 Crank and Slotted Lever Mechanism

Consider the crank and slotted lever mechanism, as shown in Fig.2.10(a). The crank *OB* is rotating at a uniform angular speed ω. Let *OB* = *r*, *AC* = *l*, and *OA* = *d*. Linear velocity of *B*, $v_b = r\omega$. Draw $ob = v_b$ and perpendicular to *OB* to a convenient scale, as shown in Fig.2.10(b). v_b is the velocity of point *B* on the crank *OB*. The velocity of the slotted lever is perpendicular to *AC*. The velocity of the slider is along the slotted lever. Hence draw a line from *b* parallel to *AC* to meet the line perpendicular to *AC* at *p*. Now, $ap = v_{pa}$, and

$$ac = v_c = \left(\frac{AC}{AP}\right) . v_{pa}$$

From point *c*, draw a line perpendicular to *CD* and from point o draw a line parallel to the tool motion, to intersect at point *d*.

Velocity of cutting tool, $v_d = ad$

The component of the velocity of the crank perpendicular to the slotted lever is zero at positions B_1 and B_2. Thus for these positions of the crank, the slotted lever reverses its direction of motion.

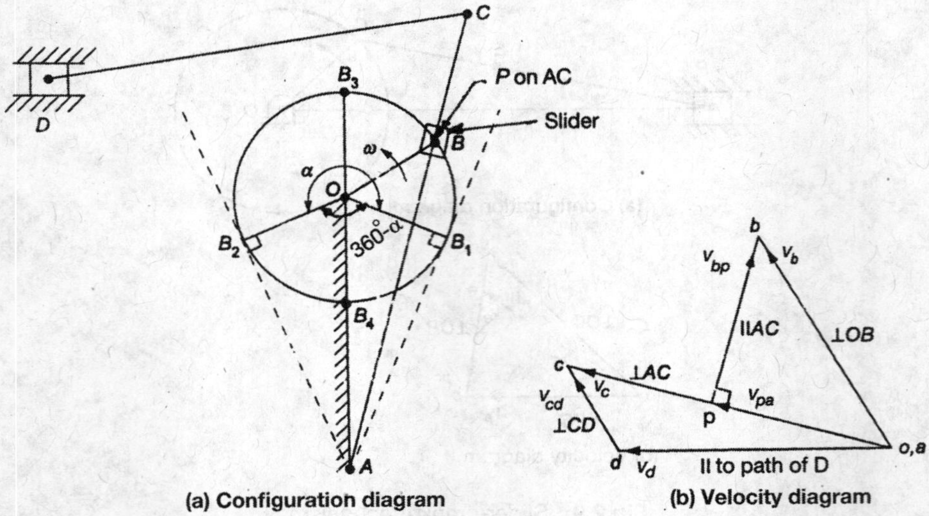

(a) Configuration diagram (b) Velocity diagram

Fig.2.10 Crank and slotted lever mechanism

$$\text{Time of cutting stroke/Time of return stroke} = \frac{\alpha}{(360^\circ - \alpha)}$$

At positions B_3 and B_4, the component of velocity along the lever is zero, i.e. the velocity of the slider at the crank pin is zero. Thus the velocity of the lever at crankpin is equal to the velocity of crankpin, i.e. $r\omega$. The velocities of lever and tool at these points are minimum. The maximum cutting velocity occurs at B_3 and maximum return velocity occurs at B_4.

$$\text{Maximum cutting velocity} = (OB \cdot \omega)\left(\frac{AC}{AB_3}\right)$$

$$= (r\omega)\left[\frac{l}{d+r}\right] \tag{2.7a}$$

$$\text{Maximum return velocity} = (OB \cdot \omega)\left(\frac{AC}{AB_4}\right)$$

$$= (r\omega)\left[\frac{l}{d-r}\right] \tag{2.7b}$$

$$\text{Maximum cutting velocity/Maximum return velocity} = \left(\frac{d-r}{d+r}\right) \tag{2.8}$$

2.3.10 Drag Mechanism

The drag mechanism is shown in Fig.2.11(a). Link 2 rotates at constant angular speed ω. Link 4 rotates at a nonuniform velocity. Ram 6 will move with nearly constant velocity over most of the upward stroke to give a slow upward stroke and a quick downward stroke when link 2 rotates clockwise.

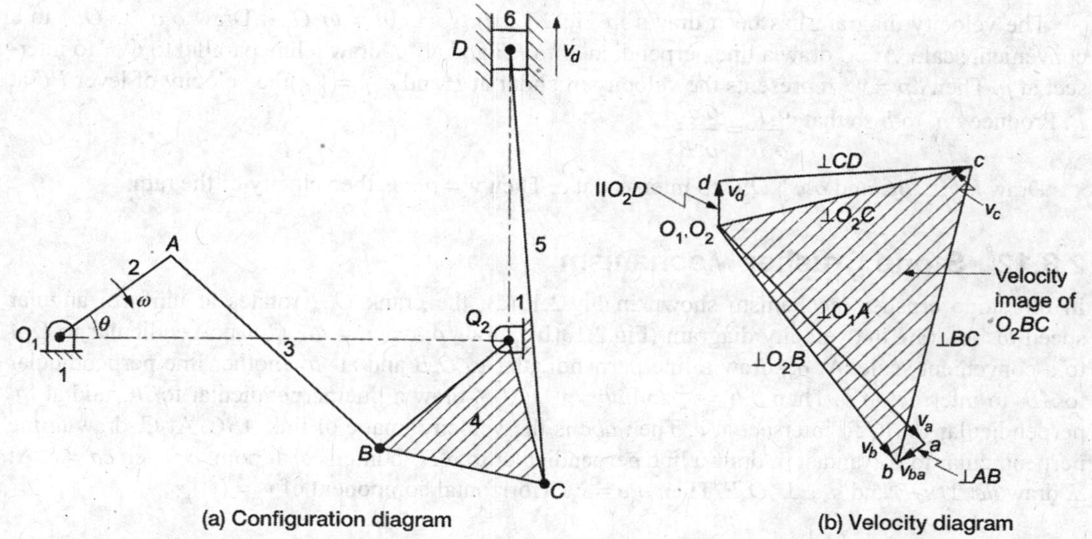

Fig.2.11 Drag mechanism

To determine the velocity diagram, draw $o_1a = \omega \cdot O_1A$, perpendicular to O_1A, as shown in Fig.2.11(b). From 'a' draw $ab \perp AB$ and from o_1 draw $o_1b \perp O_2B$ to intersect at b. Then, $o_1b = v_b$ and $ab = v_{ba}$. At b draw $bc \perp BC$ and at o_1 draw $o_1c \perp O_2C$, to intersect at c. Then $o_1c = v_c$.

Now at o_2 draw $o_2d \parallel O_2D$ and from c draw a line perpendicular to CD to intersect at d. Then $o_2d = v_d$, the velocity of ram 6. o_2bc is the velocity image of O_2BC.

2.3.11 Whitworth Quick-Return Motion Mechanism

In the Whitworth mechanism, as shown in Fig.2.12(a), the crank O_1A rotates with constant angular speed ω. The link PB oscillates about pin O_2. The ram C reciprocates on the guides.

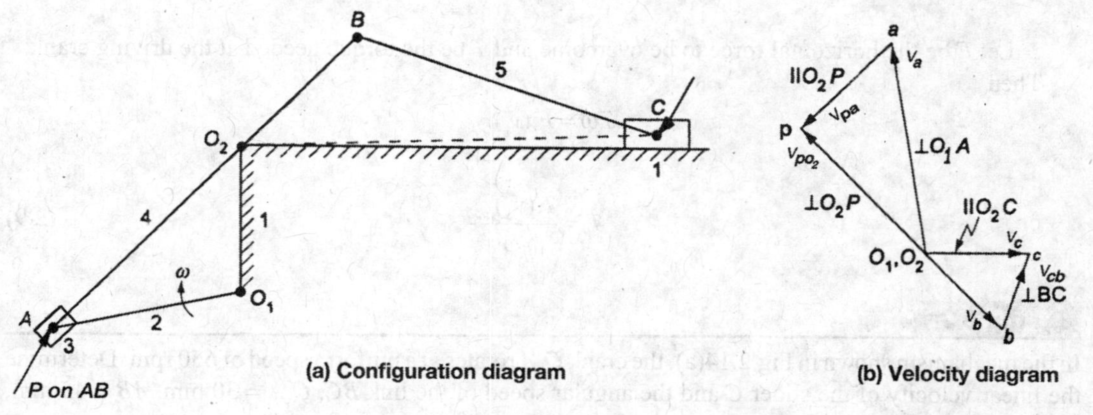

Fig.2.12 Whitworth quick-return motion mechanism

The velocity diagram has been drawn in Fig.2.12(b). $v_a = o_1a = \omega \cdot O_1A$. Draw $o_1a \perp O_1A$ to a convenient scale. At o_2, draw a line perpendicular to PB and at 'a' draw a line parallel to O_2P to intersect at p. Then $ap = v_{pa}$ represents the velocity of slider at P and $o_2p = v_{po_2}$ the velocity of lever PB at P. Produce po_2 to b so that $\dfrac{o_2p}{o_2b} = \dfrac{o_2P}{o_2B}$.

Draw $bc \perp BC$ and $o_2c \parallel O_2C$ to intersect at c. Then $v_c = o_2c$ is the velocity of the ram.

2.3.12 Stone Crusher Mechanism

In the stone crusher mechanism shown in Fig.2.13(a), the crank O_1A rotates at uniform angular speed ω. To draw the velocity diagram (Fig.2.13(b)), draw $o_1a = v_a = \omega \cdot O_1A$ perpendicular to O_1A to a convenient scale. At o_2, draw a line perpendicular to O_2B and at 'a' another line perpendicular to AB to intersect at b. Then $o_2b = v_b$ and $ab = v_{ba}$. At b, draw a line perpendicular to BC and at 'a' perpendicular to AC to intersect at c. Then abc is the velocity image of link ABC. At C, draw a line perpendicular to CD and at o_3 draw a line perpendicular to o_3D to intersect at point d. Then $cd = v_d$. At d, draw $de \perp DE$ and $o_3e \perp O_3E$. Then $o_3e = v_e$. Horizontal component of $v_e = (v_e)_{hor}$.

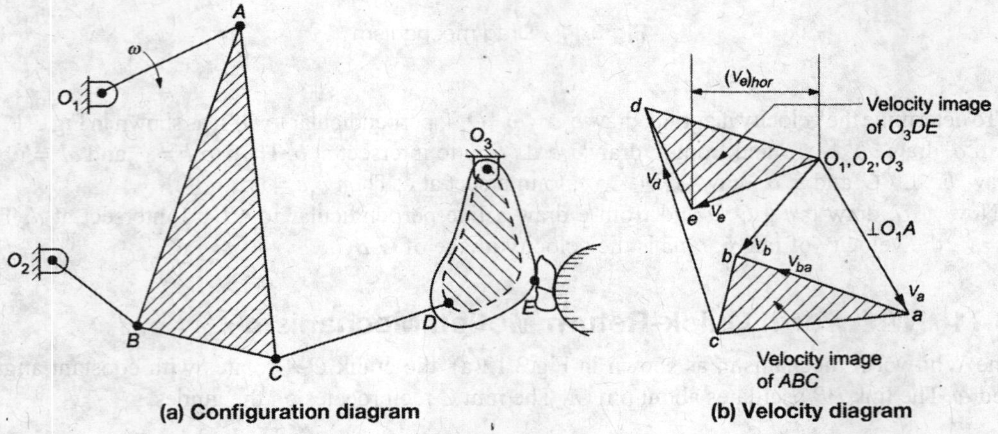

Fig.2.13 Stone crusher mechanism

Let F be the horizontal force to be overcome and T be the torque needed at the driving crank. Then

$$T\omega = F\,(v_e)_{hor}$$

or

$$T = \frac{F(v_e)_{hor}}{\omega} \tag{2.9}$$

Example 2.1

In the mechanism shown in Fig.2.14(a), the crank O_1A rotates at a uniform speed of 650 rpm. Determine the linear velocity of the slider C and the angular speed of the link BC. $O_1A = 30$ mm, $AB = 45$ mm, $BC = 50$ mm, $O_2B = 65$ mm, and $O_1O_2 = 70$ mm.

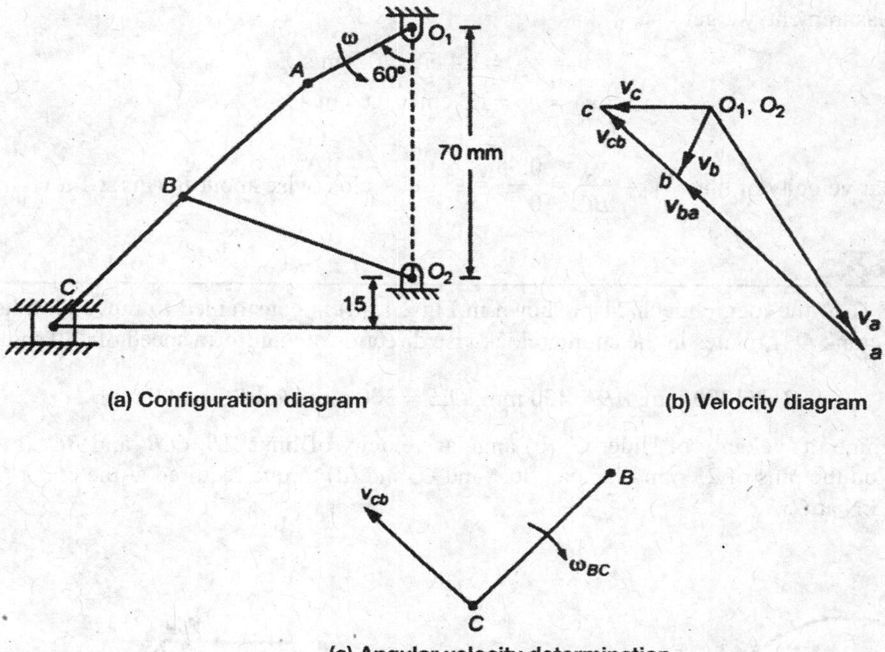

Fig.2.14 Diagram for Example 2.1

■ Solution

Given: $N = 650$ rpm, lengths of links.

Procedure:

1. Draw the coniguration diagram to a convenient scale as shown in Fig.2.14(a).
2. Calculate the angular velocity of crank O_1A.

$$\omega = 2\pi \times \frac{650}{60} = 68.07 \text{ rad/s}$$

3. Calculate the linear velocity of point A,

$$v_a = \omega \cdot O_1A = 68.07 \times 0.03 = 2.04 \text{ m/s}$$

4. Draw the velocity diagram as shown in Fig.2.14(b) to a scale of 1 cm = 0.5 m/s.
5. Draw $v_a = o_1a \perp O_1A$ such that $o_1a = 4.08$ cm.
6. Draw a line from o_2 perpendicular to O_2B and another line from 'a' perpendicular to AB to intersect at b. Then

$$ab = v_{ba} \text{ and } o_2b = v_b$$

7. From b, draw a line perpendicular to BC and another line from o_2 parallel to the path of motion of the slider to intersect at c. Then

   Velocity of the sider C, $v_c = o_2c$

By measurement, we get

$$v_c = o_2c = 1.4 \text{ cm} = 0.7 \text{ m/s}$$

$$v_{cb} = bc = 1.3 \text{ cm} = 0.65 \text{ m/s}$$

$$\text{Angular velocity of link } BC = \frac{v_{cb}}{BC} = \frac{0.65}{0.05} = 13 \text{ rad/s, clockwise about } B. \text{ (Fig.2.14c)}$$

Example 2.2

The slider C of the toggle mechanism shown in Fig.2.15(a) is constrained to move on a horizontal path. The crank O_1A rotates in the counterclockwise direction at a uniform speed of 180 rpm.

$$O_1A = 200 \text{ mm}, AB = 400 \text{ mm}, O_2B = 300 \text{ mm, and } BC = 600 \text{ mm.}$$

Determine (a) velocity of slider C, (b) angular velocity of links AB, O_2B, and BC, (c) rubbing velocities on the pins of 25 mm diameter at A and C, and (d) torque required at the crank O_1A for a force of 2 kN at C.

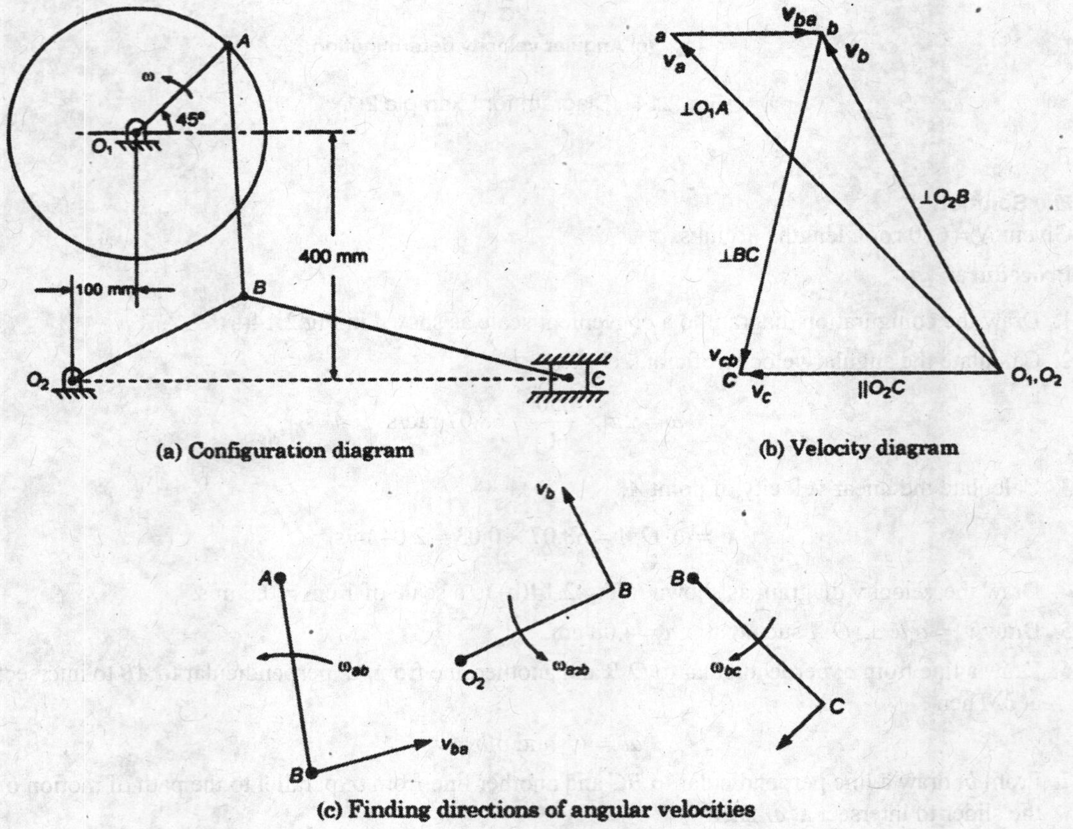

(a) Configuration diagram

(b) Velocity diagram

(c) Finding directions of angular velocities

Fig.2.15 Toggle mechanism

Solution

Given: $N = 180$ rpm, radius of pin, $r = 12.5$ mm, lengths of links.

Procedure:

1. Draw the configuration diagram to a convenient scale as shown in Fig.2.15(a).
2. Calculate the angular velocity of crank O_1A.

$$\omega = 2\pi \times \frac{180}{60} = 18.85 \text{ rad/s}$$

3. Calculate the linear velocity of point A.

$$v_a = \omega \times O_1A = 18.85 \times 0.2 = 3.77 \text{ m/s}$$

4. Draw the velocity diagram as shown in Fig.2.15(b) to a scale of 1 cm = 0.5 m/s.
5. Draw $v_a = o_1a \perp O_1A$ such that $o_1a = 7.54$ cm.
6. From 'a' draw a line perpendicular to AB and another line from o_2 perpendicular to o_2B to intersect at b. Then $o_2b = v_b$ and ab $= v_{ba}$.
7. From b, draw a line perpendicular to BC and another line from o_2 parallel to the path of motion of the slider C to meet at c. Then $o_2c = v_c$. By measurement, we get

(a) Velocity of the slider, $v_c = o_2c = 4.2 \text{ cm} = 2.1 \text{ m/s}$

(b) Velocity of link AB $AB = v_{ba} = ab = 2.4 \text{ cm} = 1.2 \text{ m/s}$

Velocity of point B, $v_b = o_2b = 6.2 \text{ cm} = 3.1 \text{ m/s}$

Velocity of link BC, $v_{cb} = bc = 5.7 \text{ cm} = 2.85 \text{ m/s}$

Angular velocity of B about A, $\omega_{ab} = \dfrac{v_{ba}}{AB} = \dfrac{1.2}{0.4} = 3$ rad/s (ccw about A)

Angular velocity of B about O_2, $\omega_b = \dfrac{v_b}{O_2B} = \dfrac{3.1}{0.3} = 10.33$ rad/s (ccw about O_2)

Angular velocity of C about B, $\omega_{bc} = \dfrac{v_{cb}}{BC} = \dfrac{2.85}{0.6} = 4.75$ rad/s (cw about B)

For direction of angular velocities, refer to Fig.2.15(c).

(c) Radius of pin, $r = \dfrac{25}{2} = 12.5$ mm

Rubbing velocity on pin at $C = \omega_{cb} \cdot r = 4.75 \times 0.0125 = 0.0594$ m/s

Relative angular velocity at $A = \omega_b - \omega_{ba} + \omega_{cb}$

$$= 10.33 - 3 + 4.75 = 12.08 \text{ rad/s}$$

Rubbing velocity on pin at $A = 12.08 \times 0.0125 = 0.151$ m/s

(d) Let T be the torque required at the crank O_1A.

$$T\omega = F_c v_c$$

$$18.85\ T = 2000 \times 2.1$$

$$T = 222.81 \text{ Nm}$$

Example 2.3

Determine the mechanical advantage of the toggle mechanism shown in Fig.2.16(a).

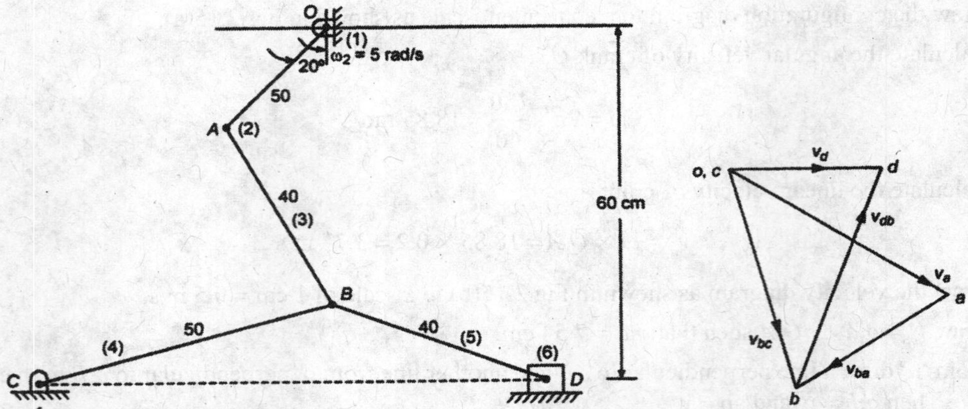

Scale: 1 cm = 10 cm
(a) Configuration diagram

Scale: 1 cm = 20 cm/s
(b) Velocity diagram

Fig.2.16 Toggle mechanism

Solution

Given: $\omega = 5$ rad/s, lengths of links.

Procedure:

1. Calculate the linear velocity of point *A*.

$$v_a = \omega_2 \times OA = 5 \times 20 = 100 \text{ cm/s}$$

2. Draw the configuration diagram to a convenient scale as shown in Fig.2.16(a).
3. Draw the velocity diagram to a scale of 1 cm = 20 cm/s, as shown in Fig.2.16(b).
4. Draw $v_a = oa \perp OA$, such that $oa = 5$ cm.
5. Draw $ab \perp AB$ and $cb \perp BC$ to meet at *b*.
6. Draw $bd \perp BD$ and $cd \parallel CD$ to meet at *d*.

By measurement, $cd = 2.5$ cm

Let T_a = torque input at *A*

Force, $$F_a = \frac{T_a}{OA}$$

$$\text{Work input} = F_a \times v_a$$

$$F_d = \text{force at } D$$

$$\text{Work output} = F_d \times v_d$$

Let

$$\text{Mechanical advantage} = \frac{F_d}{F_a} = \frac{v_a}{v_d} = \frac{oa}{cd} = \frac{5}{2.5} = 2$$

Example 2.4

Determine the angular velocity of the follower 3 and the velocity of sliding at the point of contact in Fig.2.17(a). The speed of driver link 2 is 3 rad/s.

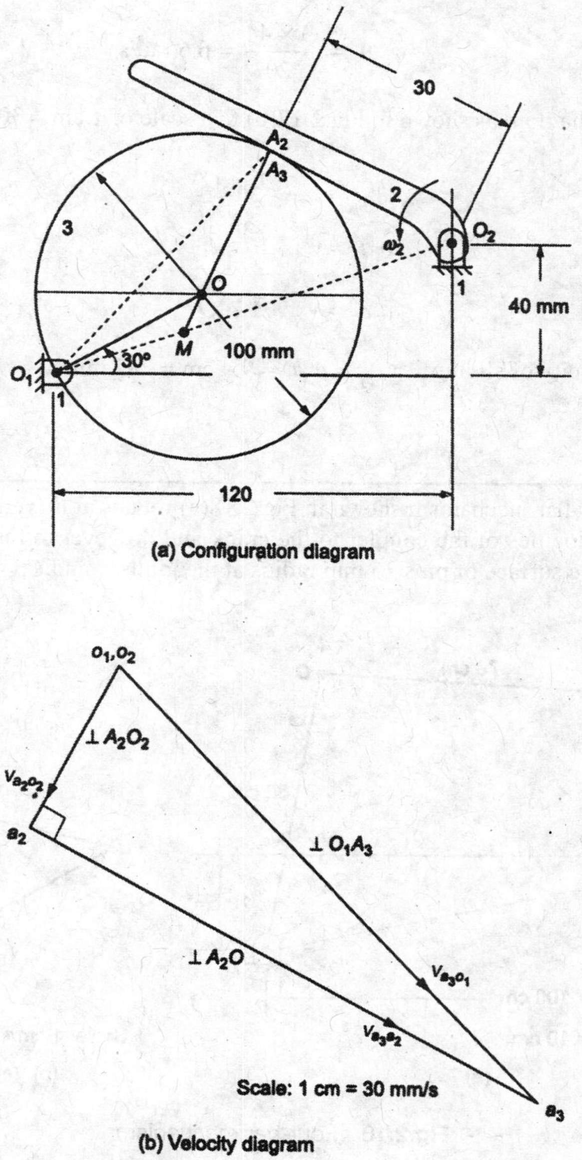

Fig.2.17 Diagram for Example 2.4

■ Solution

Join A_2O, O_1O, and O_1O_2. Produce A_2O to meet O_1O_2 at M.

$$v_{a_2o_2} = \omega_2 \times O_2A_2 = 3 \times 30 = 90 \text{ mm/s}$$

Now $MA_2 \perp O_2A_2$. Therefore,

$$\frac{\omega_3}{\omega_2} = \frac{O_2M}{O_1M}$$

$$\omega_3 = \frac{3 \times 44}{20} = 6.6 \text{ rad/s}$$

Draw the velocity diagram as shown in Fig.2.17(b) to a scale of 1 cm = 30 mm/s.

$$o_1a_2 = v_{a_2o_2} \perp A_2O_2$$

$$o_1a_3 = v_{a_3o_1} \perp O_1A_3$$

$$a_2a_3 = v_{a_3a_2} \perp a_2O$$

Velocity of sliding, $v_{a_3a_2} = a_2a_3 = 9.4 \text{ cm} = 282 \text{ mm/s}$

Example 2.5

The crank *AB* of a four-bar mechanism shown in Fig.2.18(a) rotates at 60 rpm clockwise. Determine the relative angular velocities of the coupler to the crank and the lever to the coupler. Find also the rubbing velocities at the surface of pins 25 mm radius at the joints *B* and *C*.

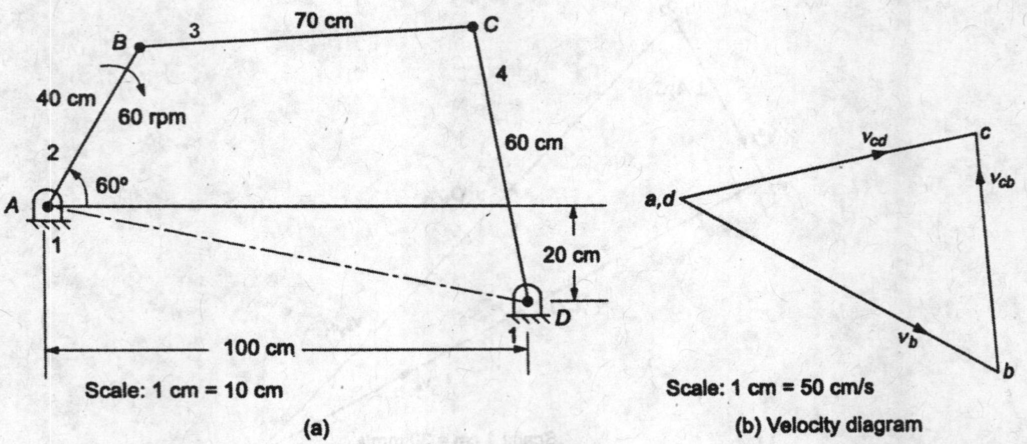

Fig.2.18 Four-bar mechanism

■ Solution

$$\omega_2 = 2\pi \times \frac{60}{60} = 6.28 \text{ rad/s}$$

$$v_b = \omega_2 \times AB = 6.28 \times 40 = 251.3 \text{ cm/s}$$

Draw the velocity diagram as shown in Fig.2.18(b), to a scale of 1 cm = 50 cm/s.

$$v_b = ab \perp AB$$

$$bc \perp BC$$

$$dc \perp CD$$

$$v_{cb} = bc = 2.3 \text{ cm} = 165 \text{ cm/s}$$

$$v_{cd} = dc = 4.2 \text{ cm} = 210 \text{ cm/s}$$

$$\omega_3 = \frac{bc}{BC} = \frac{165}{70} = 2.36 \text{ rad/s (ccw)}$$

$$\omega_{23} = \omega_2 - \omega_3 = 6.28 - (-2.36) = 8.64 \text{ rad/s (cw)}$$

Rubbing velocity of pin $B = \omega_{23} \times r_p = 8.64 \times 2.5 = 21.6$ cm/s

$$\omega_4 = \frac{cd}{CD} = \frac{210}{60} = 2.5 \text{ rad/s (cw)}$$

$$\omega_{34} = \omega_3 - \omega_4 = 2.36 - (-2.5) = 5.86 \text{ rad/s (ccw)}$$

Rubbing velocity of pin $C = \omega_{34} \times r_p = 5.86 \times 2.5 = 14.65$ cm/s

Example 2.6

Wheel 2 in Fig.2.19(a) rotates at 1500 rpm and is driving the wheel 7 pivoted at O_2. Determine the linear velocity of slider and angular velocities of links 3, 4 and 6.

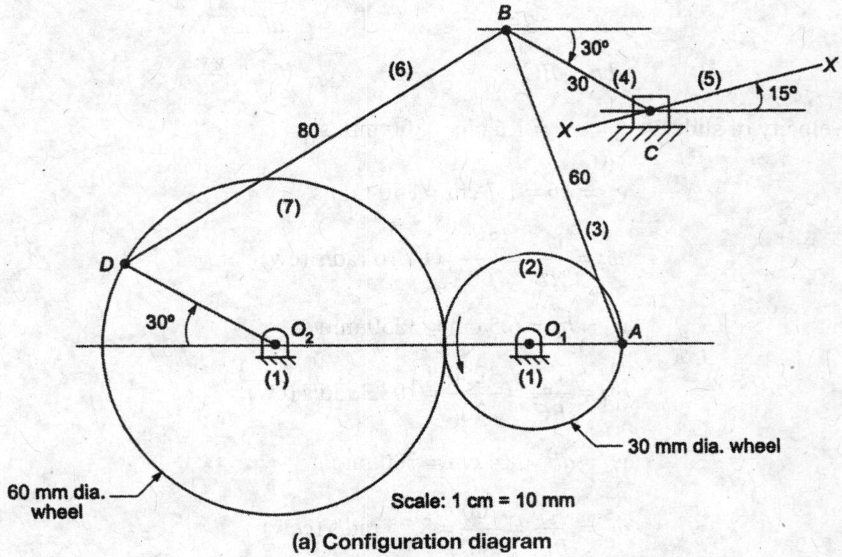

(a) Configuration diagram

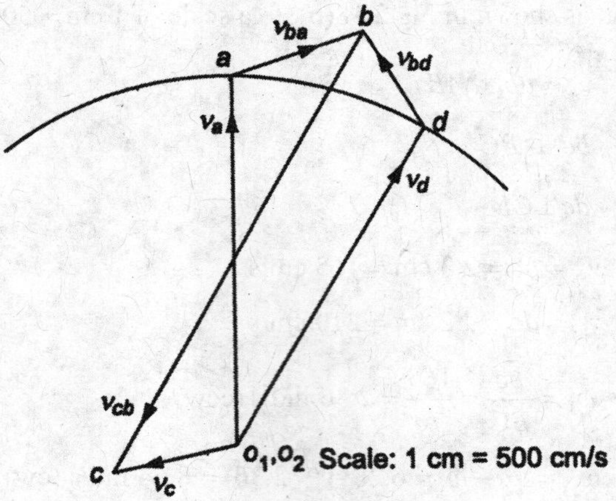

(b) Velocity diagram

Fig.2.19 Diagram for Example 2.6

■ Solution

$$\omega_2 = 2\pi \times \frac{1500}{60} = 157.08 \text{ rad/s}$$

$$v_a = \omega_2 \times O_1A = 157.08 \times 15 = 2356.2 \text{ mm/s}$$

Draw the velocity diagram as shown in Fig.2.19(b) to a scale of 1 cm = 500 mm/s.

$$v_a = o_1a \perp O_1A \qquad v_d = v_a \perp O_2D$$

$$ab \perp AB \qquad db \perp DB$$

$$bc \perp BC \qquad o_1c \parallel XX$$

Linear velocity of slider $= v_c = o_1c = 1.6 \text{ cm} = 800 \text{ mm/s}$

$$v_{ba} = ab = 1.7 \text{ cm} = 850 \text{ mm/s}$$

$$\omega_3 = \frac{v_{ba}}{AB} = \frac{850}{60} = 14.16 \text{ rad/s (cw)}$$

$$v_{cb} = bc = 6.5 \text{ cm} = 3250 \text{ mm/s}$$

$$\omega_4 = \frac{v_{cb}}{BC} = \frac{3250}{30} = 108.3 \text{ rad/s (cw)}$$

$$v_{bd} = db = 1.4 \text{ cm} = 700 \text{ mm/s}$$

$$\omega_6 = \frac{v_{bd}}{BD} = \frac{700}{80} = 8.75 \text{ rad/s (ccw)}$$

Example 2.7

The dimensions of the mechanism for hydraulic riveter, as shown in Fig.2.20(a), are: $OA = 200$ mm, $AB = 210$ mm, $AD = 550$ mm and $BC = 330$ mm.

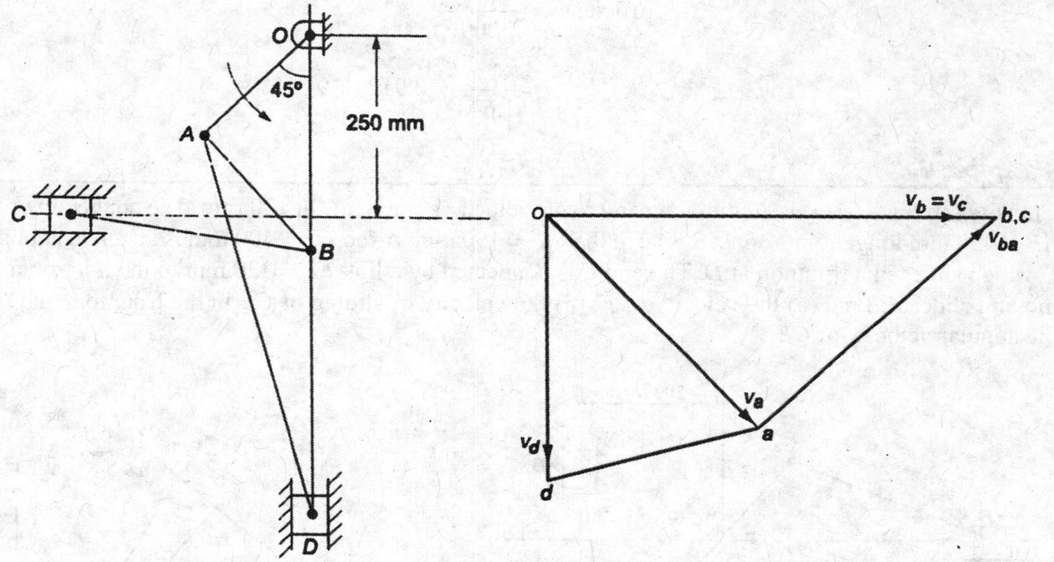

(a) Configuration diagram (b) Velocity diagram

Fig.2.20 Hydraulic riveter

Determine the velocity ratio between the piston *C* and ram *D*. Also calculate the efficiency of the machine if a load of 3 kN on piston *C* causes a thrust of 4.5 kN at ram *D*.

■ Solution

Let *N* be the speed of the crank *OA* in rpm. Then $\omega = \dfrac{2\pi N}{60}$ rad/s.

$$v_a = \omega \cdot OA = 2\pi N \times \frac{0.2}{60} = 0.0209\ N \text{ m/s}$$

Since *N* is unknown, let $v_a = 25$ mm. Draw $oa \perp OA$ to represent v_a (Fig.2.20b). From '*a*' draw a vector $ab \perp AB$, to represent v_{ba}, and from *o* draw $ob \perp OB$, to represent v_b, to intersect it at *b*.

From point *b*, draw a line perpendicular to *BC* and from *o* draw a line parallel to the path of motion of the piston *C*. These lines intersect at *c* so that *b* and *c* coincide. Thus $v_b = v_c$.

From point '*a*' draw a line perpendicular to *AD* and from *o* draw another line parallel to the path of motion of ram *D*, to intersect at *d*. Then $od = v_d$, the velocity of ram *D*. By measurement, we get

$$v_c = oc = 74 \text{ mm}$$
$$v_d = od = 45 \text{ mm}$$
$$\text{Velocity ratio} = \frac{v_c}{v_d} = \frac{74}{45} = 1.644$$

$$\text{Work done on the piston } C = \text{Load on the piston} \times v_c = 3000v_c$$

$$\text{Work done by ram } D = 4500\ v_d$$

$$\text{Efficiency} = \frac{4500v_d}{3000v_c} = \frac{1.5}{1.644} = 0.912 \text{ or } 91.2\%$$

Example 2.8

Crank OA in Fig.2.21(a) is 80 mm long and rotates clockwise about O at 120 rpm. The connecting rod AB is 450 mm long. Point C on AB is such that $AC = 150$ mm. A rod $CE = 400$ mm is attached at point C which slides in a trunnion at D. The end E is connected by a link $EF = 320$ mm to the horizontally moving slider F. Find (a) the velocity of F, (b) the velocity of sliding of CE in the trunnion, and (c) the angular velocity of CE.

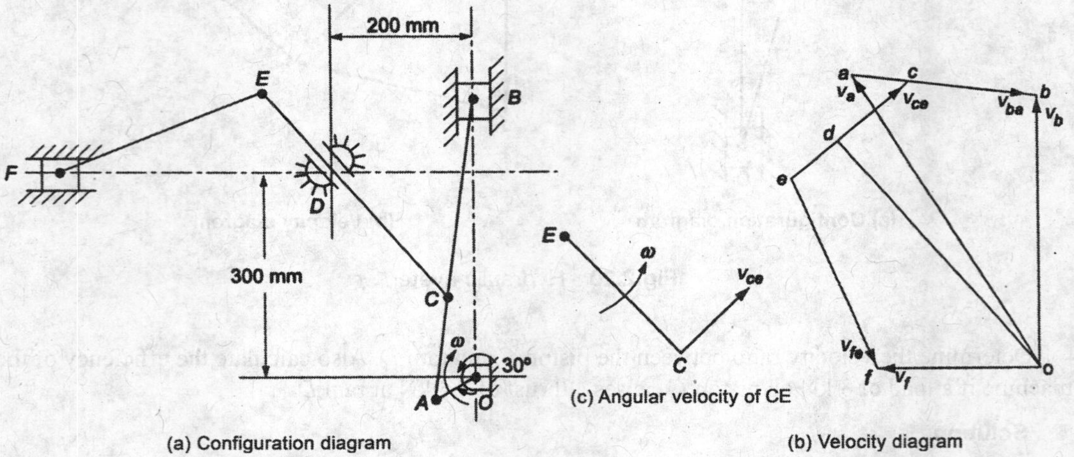

Fig.2.21 Swivelling pin mechanism

■ Solution

Angular speed of OA, $$\omega = 2\pi \times \frac{120}{60} = 12.57 \text{ rad/s}$$

$$v_a = \omega \cdot OA = 12.57 \times 0.08 = 1.005 \text{ m/s}$$

(a) Draw $oa \perp OA$ to represent v_a to a scale of 1 cm = 0.2 m/s, as shown in Fig.2.21(b). From 'a' draw a line perpendicular to AB and from o draw another line parallel to the line of motion of the slider to intersect at b. Locate point c on ab such that, $ac/ab = AC/AB$. From point c draw a line perpendicular to CD to represent v_{dc} and at point o draw another line parallel to the motion of CD to meet at d, which moves along CD only, to represent v_d. Locate point e on cd such that, $cd/ce = CD/CE$. From point e draw a line perpendicular to EF and from point o draw another line parallel to the path of motion of the slider F, to meet at point f. Then $of = v_f$. By measurement, we have

$$v_f = 2.3 \text{ cm} = 0.46 \text{ m/s}$$

(b) Velocity of sliding of CE in trunnion D is, $v_d = od = 4.3$ cm $= 0.86$ m/s

(c) $v_{ce} = ec = 2.4$ cm $= 0.48$ m/s

$$\text{Angular velocity of } CE = \frac{v_{ce}}{CE} = \frac{0.48}{0.4} = 1.2 \text{ rad/s ccw about } E. \text{ (Fig.2.21c)}$$

Example 2.9

An engine crankshaft drives a reciprocating pump through a mechanism, as shown in Fig.2.22(a). The crank OA rotates in the counter-clockwise direction at 150 rpm. The diameter of the pump piston at F is 180 mm and $OA = 175$ mm, $AB = 650$ mm, $CD = 160$ mm, and $DE = 600$ mm.

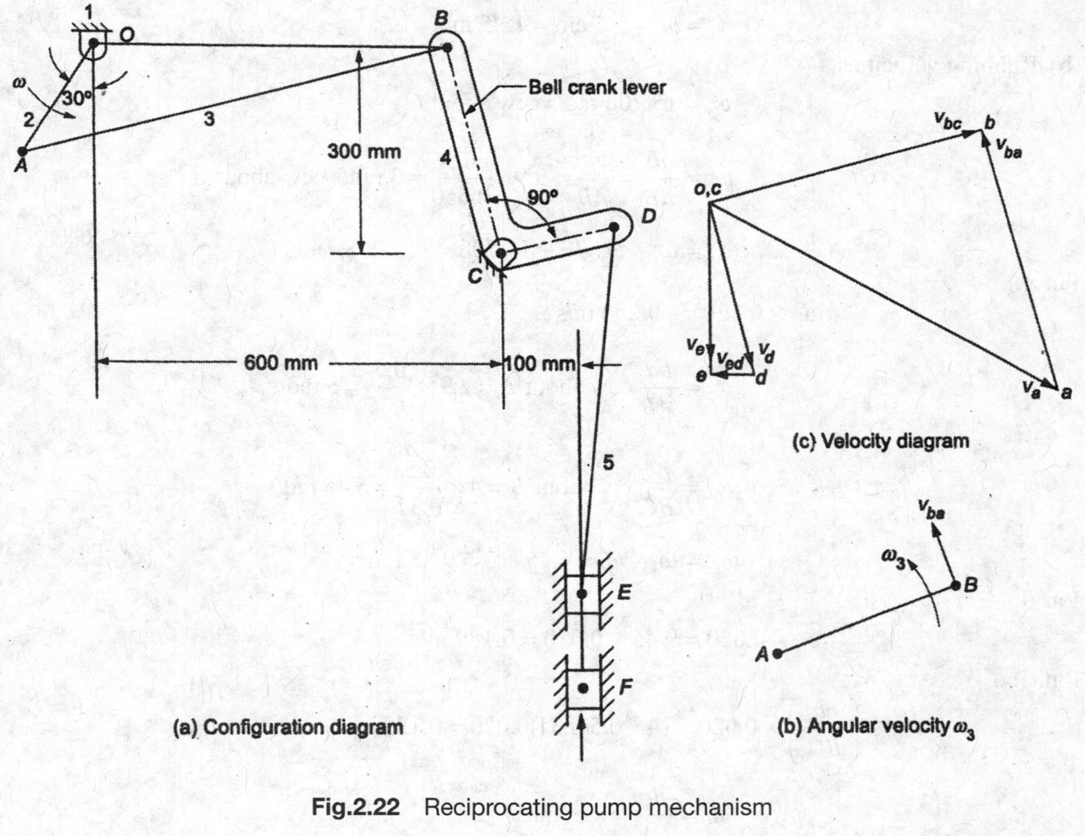

Fig.2.22 Reciprocating pump mechanism

Determine (a) the velocity of cross head E, (b) the rubbing velocities at pins A, B, C, and D having diameters of 40 mm each, and (c) the torque required at the crank to overcome a pressure of 0.35 MPa at the pump piston at F.

■ **Solution**

$$\omega = 2\pi \times \frac{150}{60} = 15.708 \text{ rad/s}$$

$$v_a = 15.708 \times 0.175 = 2.75 \text{ m/s}$$

Draw $oa \perp OA$ to represent v_a to a scale of 1 cm = 0.5 m/s, as shown in Fig.2.22(b). From 'a' draw a line perpendicular to AB and from c draw another line perpendicular to BC to intersect at b. Then $ab = v_{ba}$ and $cb = v_{bc}$.

$$cd = cb \cdot \left(\frac{CD}{BC}\right) = 5 \times \frac{160}{310} = 2.58 \text{ cm}$$

Draw a line perpendicular to CD from c and cut it equal to cd. From d draw a line perpendicular to DE and from o draw another line parallel to the path of motion of the slider E, to meet at e. Then $de = v_{ed}$ and $oe = v_e$. By measurements, we have

(a) Velocity of cross-head E,

$$v_e = oe = 2.5 \text{ cm} = 1.25 \text{ m/s}$$

(b) Rubbing velocities,

$$\omega_2 = 15.708 \text{ rad/s ccw about } O$$

$$\omega_3 = \frac{ab}{AB} = \frac{v_{ba}}{AB} = 2.9 \times \frac{0.5}{0.650} = 3 \text{ rad/s ccw about } A$$

$$\omega_{32} = \omega_3 - \omega_2 = 3 - 15.708 = -12.708 \text{ rad/s}$$

Pin A:

$$\omega_{32} \times 0.020 = -0.254 \text{ m/s}$$

$$\omega_3 = \frac{ba}{AB} \text{ ccw about } B = 2.9 \times \frac{0.5}{0.65} = 3 \text{ rad/s}$$

$$\omega_4 = \frac{bc}{BC} \text{ cw about } B = 4 \times \frac{0.5}{0.31} = 6.45 \text{ rad/s}$$

$$\omega_{43} = \omega_4 - \omega_3 = 6.45 + 3 = 9.45 \text{ rad/s}$$

Pin B:

$$\omega_{43} \times 0.020 = 9.45 \times 0.020 = 0.189 \text{ m/s}$$

Pin C:

$$\left(\frac{cb}{BC}\right) \times 0.020 = (4 \times 0.5/0.31)0.020 = 0.129 \text{ m/s}$$

$$\omega_5 = \frac{de}{DE} \text{ cw about } D$$

$$= 0.6 \times \frac{0.5}{0.6} = 0.5 \text{ rad/s}$$

$$\omega_4 = \frac{dc}{DC} \text{ cw about } D$$

$$= 2.6 \times \frac{0.5}{0.160} = 8.125 \text{ m/s}$$

$$\omega_{54} = \omega_5 - \omega_4 = 0.5 - 8.125 = -7.625 \text{ rad/s}$$

Pin D:

$$\omega_{54} \times 0.020 = 7.625 \times 0.020 = 0.1525 \text{ m/s}$$

(c) Velocity of the piston, $v_f = v_e = 2.5 \times 0.5 = 1.25$ m/s

Let T be the torque required at the crank OA. Then,

$$T\omega = Fv_f$$

$$15.708T = \left(\frac{\pi}{4}\right)(180)^2 \times 0.35 \times 10^6 \times 1.25$$

$$T = 708.75 \text{ Nm}$$

Example 2.10

The various dimensions of the mechanism, as shown in Fig.2.23(a), are $OA = 120$ mm, $AB = 500$ mm, $BC = 120$ mm, $CD = 300$ mm, and $DE = 150$ mm. The crank OA rotates at 150 rpm. The bell crank lever is DE. Determine the absolute velocity of point E.

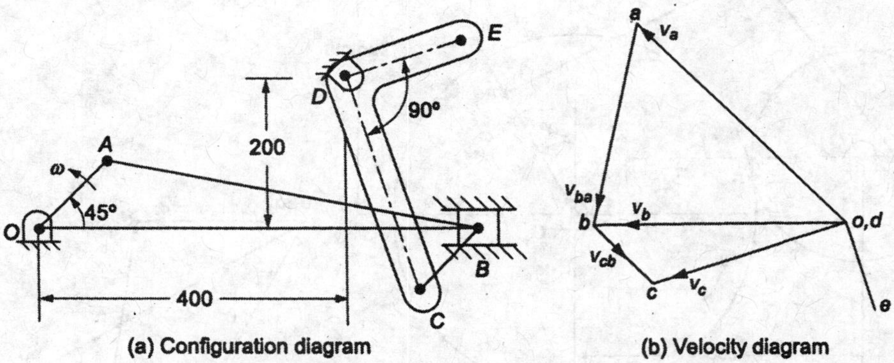

Fig.2.23 Diagram for Example 2.10

■ Solution

$$\omega = 2\pi \times \frac{150}{60} = 15.708 \text{ rad/s}$$

$$v_a = 15.708 \times 0.120 = 1.885 \text{ m/s}$$

Draw $oa \perp OA$ to represent the velocity v_a of point A to a scale of 1 cm = 0.5 m/s, as shown in Fig.2.23(b). At 'a' draw a line perpendicular to AB and at o draw another line parallel to the path of motion of the slider at B to intersect at b. Then $ob = v_b$ and $ab = v_{ba}$.

At b draw a line perpendicular to BC and at d draw another line perpendicular to DC to intersect at c. Then $bc = v_{cb}$ and $dc = v_c$. Now draw $de \perp dc$ such that

$$\frac{de}{dc} = \frac{DE}{DC}$$

or

$$de = 2.5 \times \frac{150}{300} = 1.25 \text{ cm}$$

Then $$v_e = de = 1.25 \text{ cm} = 0.625 \text{ m/s}$$

Example 2.11

In the mechanism shown in Fig.2.24(a), the crank O_1A and O_2B are 100 and 50 mm, respectively. The diameters of wheels with centres O_1 and O_2 are 260 and 150 mm, respectively. $BC = AC = 200$ mm, $CD = 250$ mm. The wheels roll on each other. The crank O_1A rotates at 120 rpm. Determine (a) the velocity of the slider D, (b) the angular velocities of links BC and CD and (c) the torque at O_2B when the force required at D is 4 kN.

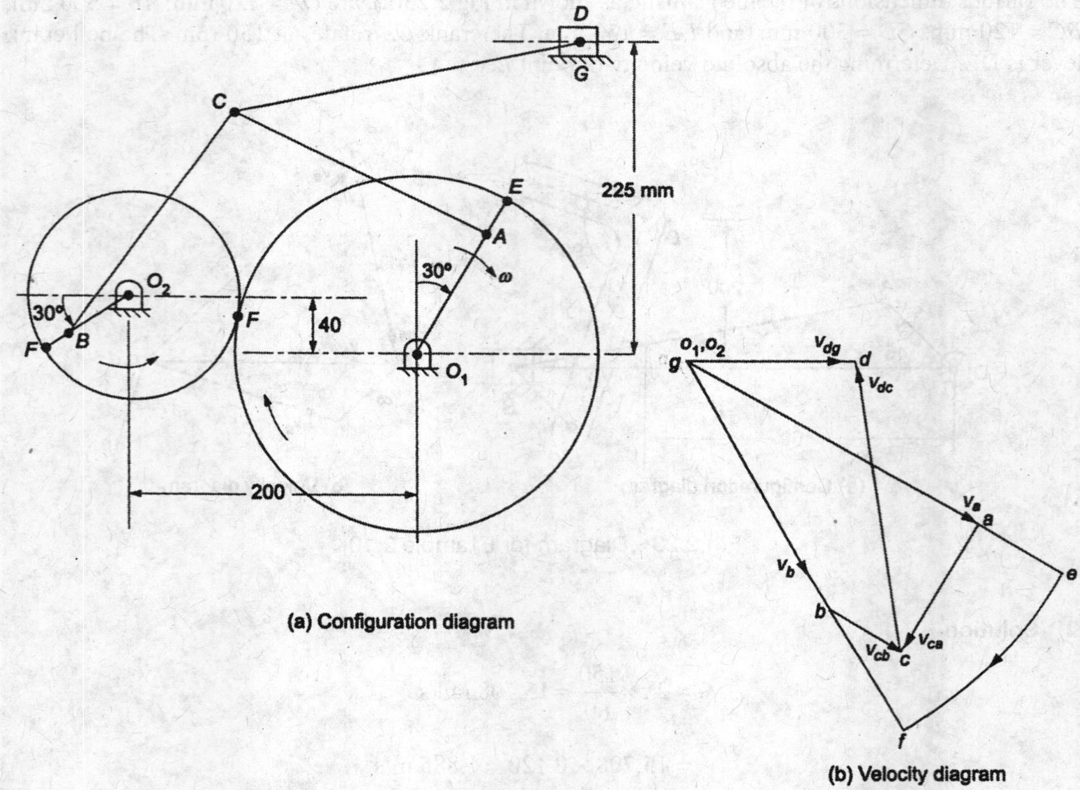

Fig.2.24 Diagram for Example 2.11

Solution

$$\omega = 2\pi \times \frac{120}{60} = 12.57 \text{ rad/s}$$

$$v_a = 12.57 \times 0.1 = 1.257 \text{ m/s}$$

$$v_e = v_a \times \frac{O_1E}{O_1A} = 1.257 \times \frac{130}{100} = 1.634 \text{ m/s}$$

$$v_f = v_e$$

$$v_b = v_f \times \frac{O_2B}{O_2F} = 1.634 \times \frac{50}{60} = 1.362 \text{ m/s}$$

Draw $o_1a \perp O_1A$ to represent the velocity v_a of point A to a scale of 1 cm = 0.25 m/s, as shown in Fig.2.24(b). Extend o_1a to e such that

$$o_1e = o_1a \times \frac{O_1E}{O_1A}$$

$$= 5 \times \frac{130}{100} = 6.5 \text{ cm}$$

From o_2 draw a line perpendicular to O_2B to intersect o_1e rotated about o_1 at f. Locate point b on o_2f such that

$$o_2b = o_2f \times \frac{O_2B}{O_2F}$$

$$= 6.5 \times \frac{50}{75} = 4.33 \text{ cm} \qquad (\because o_2f = o_2e)$$

Draw a line perpendicular to BC at b and draw another line perpendicular to AC at 'a' to intersect at c. Now draw a line perpendicular to CD at c and draw another line from g parallel to the path of motion of slider D to intersect at d. Then,

(a) Velocity of slider, $v_{dg} = gd = 2.4 \text{ cm} = 0.6 \text{ m/s}$

(b) $v_{bc} = cb = 1.4 \text{ cm} = 0.35 \text{ m/s}$

$$\omega_{bc} = \frac{v_{bc}}{BC} = \frac{0.35}{0.2} = 1.75 \text{ rad/s (cw)}$$

$$v_{dc} = cd = 4.6 \text{ cm} = 1.15 \text{ m/s}$$

$$\omega_{cd} = \frac{v_{dc}}{DC} = \frac{1.15}{0.25} = 4.6 \text{ rad/s (ccw)}$$

(c) $T\omega = F_d v_{dg}$

$$T = 4000 \times \frac{0.6}{12.57} = 190.93 \text{ Nm}$$

Example 2.12

In the mechanism shown in Fig.2.25(a), $v_a = 120$ m/s. Determine the angular velocities ω_4, ω_5 of the two gears and the velocity v_d on gear 5. $O_2A = 50$ mm, $AB = 200$ mm and $O_6C = 150$ mm.

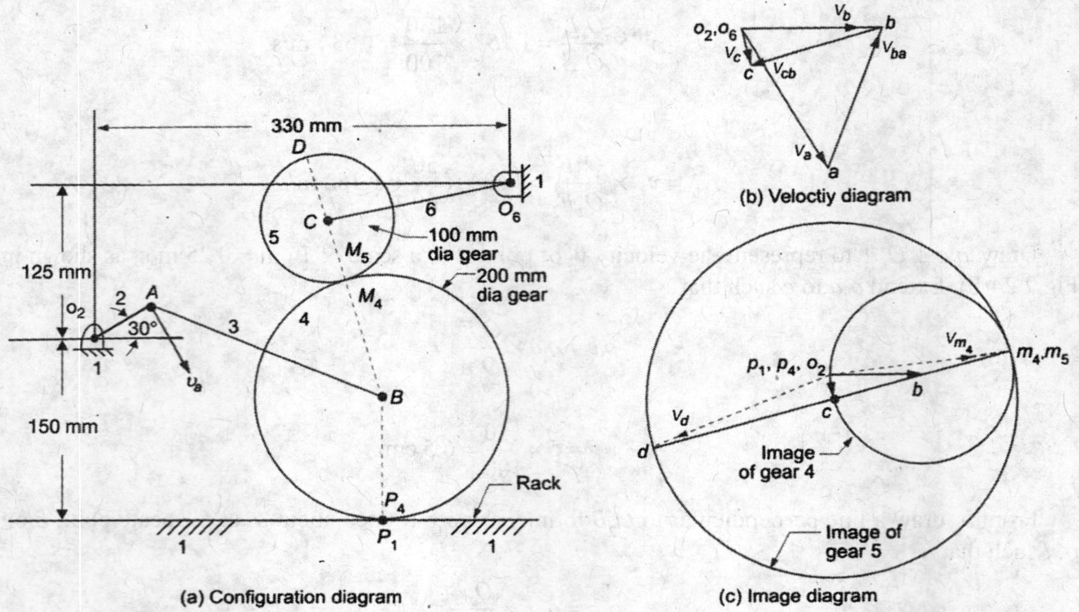

Fig.2.25 Diagram for Example 2.12

■ Solution

Draw $o_2 a \perp O_2 A$ to represent $v_a = 120$ m/s to a scale of 1 cm = 40 m/s, as shown in Fig.2.25(b). Draw $ab \perp AB$ and $o_2 b$ parallel to pitch line of rack. Then $o_2 b = v_b$ and $ab = v_{ba}$. Draw $bc \perp BC$ and $o_6 c \perp O_6 C$. Then $o_6 c = v_c$ and $bc = v_{cb}$. By measurement,

$$v_b = o_2 b = 2.6 \text{ cm} = 104 \text{ m/s}$$

$$v_c = o_6 c = 0.7 \text{ cm} = 28 \text{ m/s}$$

$$v_{ba} = ab = 2.9 \text{ cm} = 116 \text{ m/s}$$

$$v_{cb} = bc = 2.6 \text{ cm} = 104 \text{ m/s}$$

$$\omega_4 = \frac{(v_{bp4} = v_b)}{BP} = \frac{104}{0.1} = 1040 \text{ rad/s (cw)}$$

$$\omega_5 = \frac{v_{cm5}}{CM} = 5.2 \times \frac{40}{0.05} = 4160 \text{ rad/s (ccw)}$$

$$v_{m4} = v_{m5} = 5.2 \text{ cm} = 2.08 \text{ m/s}$$

$$v_d = 5.3 \text{ cm} = 212 \text{ m/s}$$

Because $v_{p1} = 0$ and $v_{p4} = v_{p1}$, the images of the points P_1 and P_4 both are at the pole point o_2. Draw velocity image of gear 4 with b as centre and radius bp_4. Produce cb to meet the circle at m_4, m_5 since $v_{m4} = v_{m5}$. The velocity image of gear 5 is drawn with c as centre and cm_5 as radius. Point d is located on the circle opposite to m_5.

Example 2.13

The dimensions of various links for the mechanism shown in Fig.2.26(a) are: $OA = 0.5$ m, $AB = 1.5$ m, $AC = CD = 0.9$ m. The crank OA has uniform angular speed of 180 rpm. Determine the velocities of sliders *B* and *D*. Also calculate the turning moment at *O* if a force of 500 N acts on *B* and a force of 800 N acts on *D*, as shown.

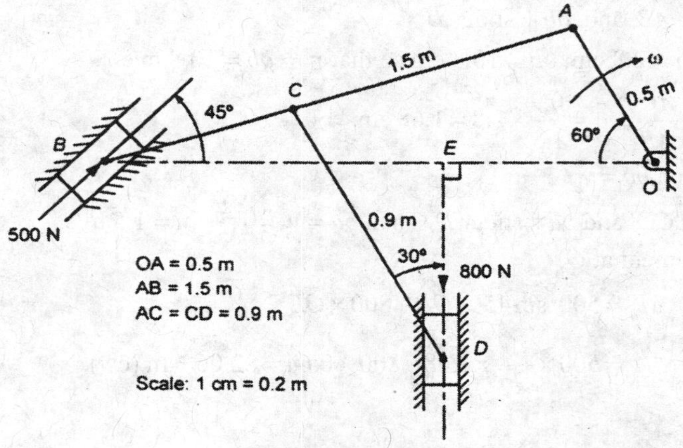

(a) Configuration diagram

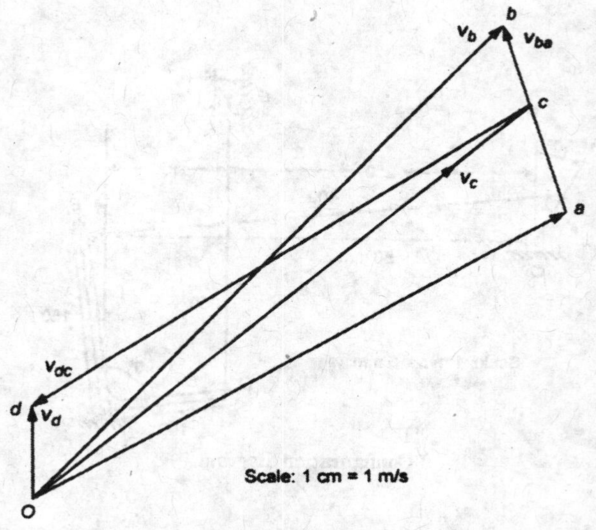

(b) Velocity diagram

Fig.2.26 Diagram for Example 2.13

■ Solution

Procedure:

1. Draw configuration diagram as shown in Fig.2.26(a) to a scale of 1 cm = 0.2 m, OB = 1.68 m, OE = 0.64 m.

2. Angular speed of OA, $\omega = \dfrac{2\pi \times 180}{60} = 18.85$ rad/s

$$v_a = \omega \times OA = 18.85 \times 0.5 = 9.425 \text{ m/s}.$$

3. Draw the velocity diagram as shown in Fig.2.26(b) to a scale of 1 cm = 1 m/s.
   (i) Draw $oa \perp OA, oa = 9.425$ cm.
   (ii) Draw $ab \perp AB$ and $ob \parallel$ slider B.
   (iii) Measure $ab = 2.8$ cm, $ob = 10.8$ cm so that $v_b = ob = 10.8$ m/s
   (iv) Now $\frac{ac}{ab} = \frac{AC}{AB}, ac = \frac{0.9}{1.5} \times 2.8 = 1.68$ cm.
   (v) Join oc. Then $oc = v_c$.
   (vi) Draw $cd \perp CD$ and $od \parallel$ slider D. Then $od = v_d = 1.5$ cm = 1.5 m/s.
   (vii) Turning moment at O,

$$M_o = 500 \sin 45^\circ \times OB - 800 \times OE$$

$$= 500 \times \frac{1}{\sqrt{2}} \times 1.68 - 800 \times 0.64 = 82.06 \text{ Nm (cw)}.$$

Example 2.14

The dimensions of the various links of the mechanism shown in Fig.2.27(a) are: OA = 50 mm, AB = 400 mm, BC = 150 mm, CD = 100 mm, and DE = 250 mm. The crank OA rotates at 60 rpm. Find velocity of slider E.

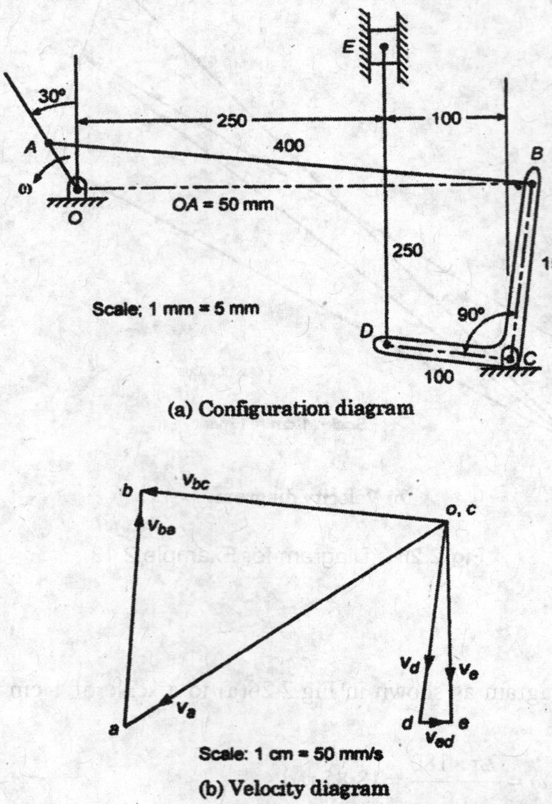

Fig.2.27 Diagram for Example 2.14

Solution

Procedure:

1. Angular speed of crank OA, $\omega = 2\pi \times \frac{60}{60} = 6.28$ rad/s

$$v_a = \omega \times OA = 6.28 \times 50 = 314.16 \text{ mm/s}$$

2. Draw the configuration diagram to a scale of 1 mm = 5 mm, as shown in Fig.2.27(a).
3. Draw the velocity diagram as shown in Fig.2.27(b) to a scale of 1 cm = 50 mm/s.
   (i) Draw $v_a = oa \perp OA, oa = 6.28$ cm.
   (ii) Draw $ab \perp AB$ and $cb \perp BC$ meeting at b.
   (iii) Measure $cd = v_{bc}$
   (iv) Now $\frac{cd}{cb} = \frac{CD}{BC}$, $cd = \frac{100}{150} \times 5 = 3.33$ cm.
   (v) Draw $cd \perp CD = 3.33$ cm.
   (vi) Draw $de \perp DE$ and $oe \perp OB$ Then $v_e = oe = 3.4 \text{ cm} = 117$ mm/s

   Velocity of slider $E = 117$ mm/s.

Example 2.15

The dimensions of the various links of the mechanism shown in Fig.2.28(a) are OA = 30 mm, $AB = 75$ mm, $BD = 100$ mm. The crank OA rotates at 120 rpm. Determine the velocity of the slider D and angular speed of links AB, BC, and BD.

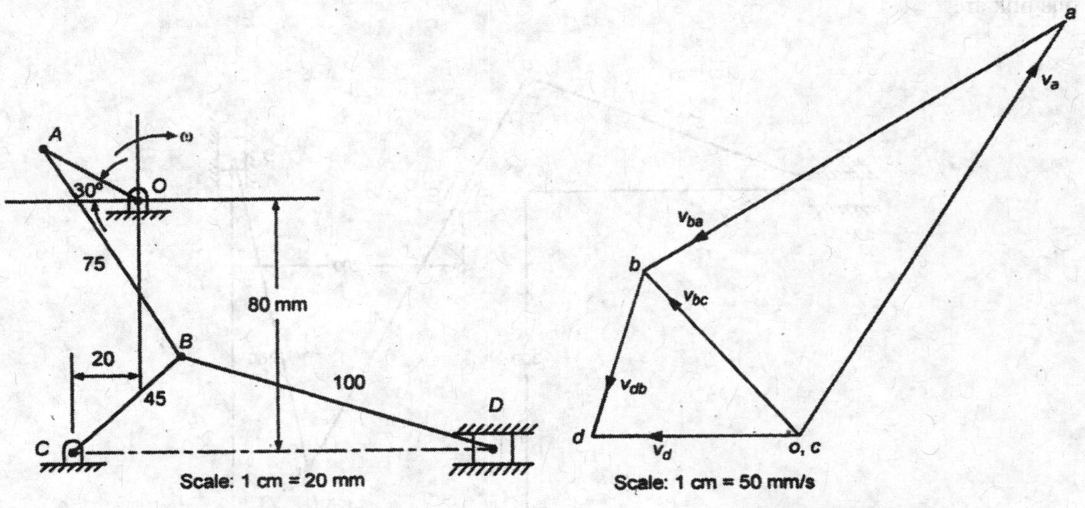

(a) Configuration diagram (b) Velocity diagram

Fig.2.28 Toggle mechanism

Solution

Procedure:

1. Draw configuration diagram as shown in Fig.2.28(a) to a scale of 1 cm = 20 mm.

2. Angular speed of the crank OA, $\omega = 2\pi \times \frac{120}{60} = 12.57$ rad/s

$$v_a = \omega \times OA = 12.57 \times 30 = 377 \text{ mm/s.}$$

3. Draw the velocity diagram as shown in Fig.2.28(b) to a scale of 1 cm = 50 mm/s.
   (i) Draw $v_a = oa \perp OA$, $oa = 7.54$ cm.
   (ii) Draw $ab \perp AB$ and $cb \perp BC$, intersecting at point b.
   (iii) Draw $bd \perp BD$ and $cd \parallel CD$, intersecting at point d.
   (iv) Then $v_{ba} = ab = 7.4 \text{ cm} = 370 \text{ mm/s}$

$$v_d = cd = 3.2 \text{ cm} = 160 \text{ mm/s}$$
$$v_{bc} = cb = 3.5 \text{ cm} = 175 \text{ mm/s}$$
$$v_{db} = bd = 2.8 \text{ cm} = 140 \text{ mm/s}$$

   (v) $\omega_{AB} = \frac{v_{ba}}{AB} = \frac{370}{75} = 4.93$ rad/s, cw

$$\omega_{BC} = \frac{v_{bc}}{BC} = \frac{175}{45} = 3.89 \text{ rad/s, cw}$$

$$\omega_{BD} = \frac{v_{bd}}{BD} = \frac{140}{100} = 1.4 \text{ rad/s, cw.}$$

Example 2.16

The crank OA of the mechanism shown in Fig.2.29(a) rotates at 120 rpm. The dimensions of the various link are:

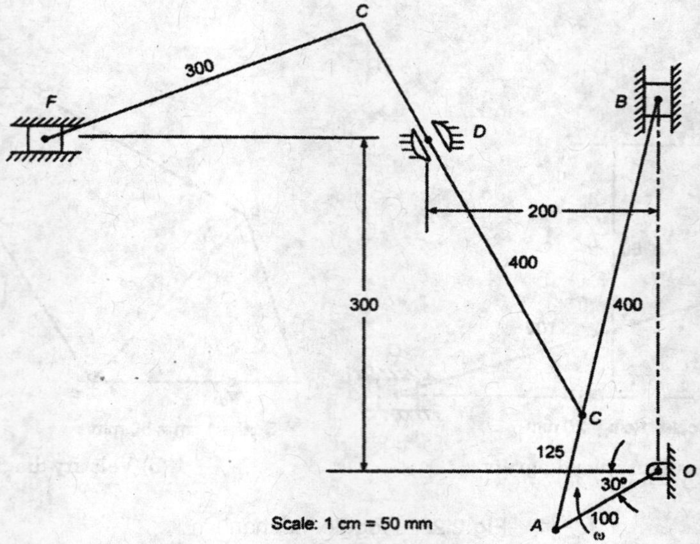

(a) Configuration diagram

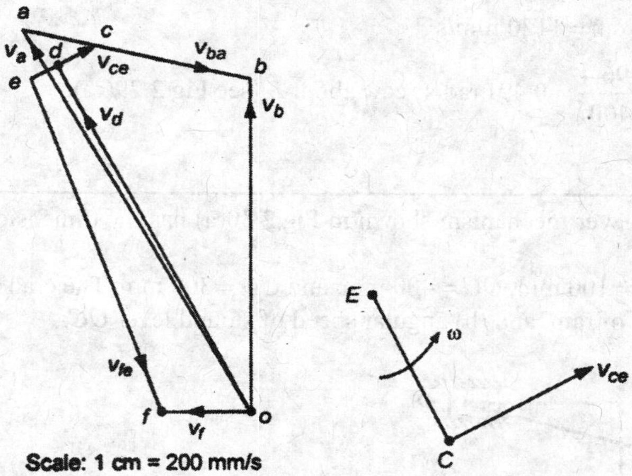

(b) Velocity diagram (c)

Fig.2.29 Swivelling pin mechanism

OA = 100 mm, AB = CE = 400 mm, AC = 125 mm, and EF = 300 mm. The rod CE slides in a slot in trunnion at D. Determine (a) velocity of F, (b) velocity of sliding of CE in D, and (c) angular velocity of CE.

■ **Solution**

Procedure:

1. Draw the configuration diagram shown in Fig.2.29(a) to a scale of 1 cm = 50 mm.

2. Angular speed of crank OA, $\omega = 2\pi \times \dfrac{120}{60} = 12.567$ rad/s

$$v_a = \omega \times oA = 12.567 \times 100 = 1256.7 \text{ mm/s}$$

3. Draw the velocity diagram as shown in Fig.2.29(b) to a scale of 1 cm = 200 mm/s.
   (i) Draw $v_a = oa \perp OA$, $oa = 6.28$ cm.
   (ii) Draw $ab \perp AB$ and $ob \parallel OB$, intersecting at point b. Measure $ab = 3.2$ cm.
   (iii) Now $\dfrac{ac}{ab} = \dfrac{AC}{AB}$, $ac = 3.2 \times \dfrac{125}{400} = 1$ cm.
   (iv) Draw $cd \perp CD$ and $od \parallel CD$, intersecting at d. Measure $cd = 0.7$ cm.
   (v) Now $\dfrac{ce}{cd} = \dfrac{CE}{CD}$, $ce = 0.7 \times \dfrac{400}{285} = 0.982$ cm, $v_{ce} = 196.4$ mm/s. Produce cd to e so that $ce = 0.982$ cm.
   (vi) Draw $ef \perp EF$ and $of \parallel FD$, intersecting at point f.
   Then $v_f = of = 1.2$ cm = 240 mm/s.

Thus, velocity of F, $v_f = 240$ mm/s

(b) $v_d = od = 5.6 \text{ cm} = 1120 \text{ mm/s}$

(c) $\omega_{ce} = \dfrac{v_{ce}}{CE} = \dfrac{196.4}{400} = 0.491 \text{ rad/s}$, ccw about E (see Fig.2.29(c)).

Example 2.17

The crank and slotted lever mechanism shown in Fig.2.30(a) has the dimensions of its various links as follows:

$OA = 200$ mm, $AB = 100$ mm, $OC = 400$ mm, and $CD = 300$ mm. The crank AB rotates at 75 rpm. Determine (a) velocity of ram, and (b) angular speed of slotted lever OC.

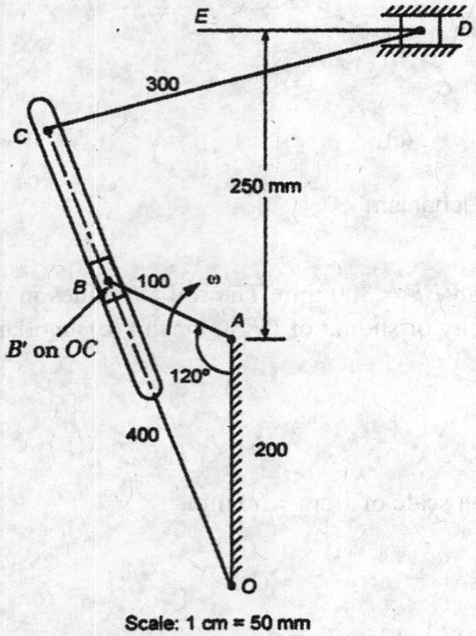

(a) Configuration diagram

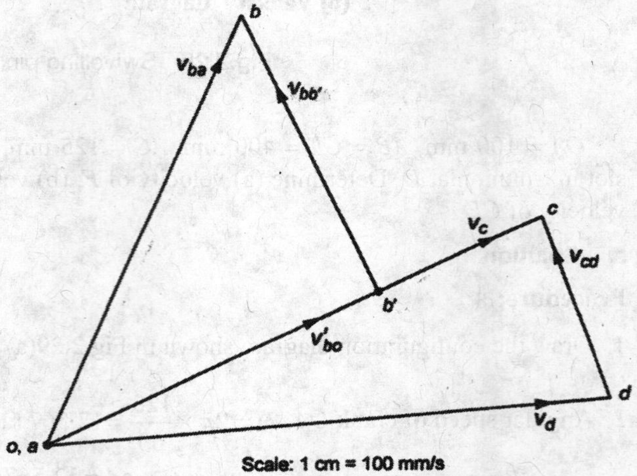

(b) Velocity diagram

Fig.2.30 Crank and slotted lever mechanism

■ Solution

Procedure:

1. Draw the configuration diagram shown in Fig.2.30(a) to a scale of 1 cm = 50 mm.

2. Angular speed of crank AB, $\omega = 2\pi \times \dfrac{75}{60} = 7.854$ rad/s

   $v_b = \omega \times AB = 7.854 \times 100 = 785.4$ mm/s

3. (a) Draw the velocity diagram as shown in Fig.2.30(b) to a scale of 1 cm = 100 mm/s

   (i) Draw $v_b = ab \perp AB$, $ab = 7.85$ cm.

   (ii) Draw $bb' \parallel OB$ and $ob' \perp OB$, intersecting at point b'. Measure $ob' = 5.8$ cm.

(iii) $oc = ob' \times \dfrac{OC}{OB} = 5.8 \times \dfrac{400}{265} = 8.75$ cm. Produce ob' to c so that $oc = 8.75$ cm.

(iv) Draw $cd \perp CD$ and $od \parallel DE$, intersecting at point d.

(v) Then $v_d = od = 9 \text{ cm} = 900 \text{ mm/s}$.

(b) Angular speed of slotted lever OC, (by measurement, $OB' = 265$ mm)

$$\omega_{oc} = \frac{ob'}{OB'} = \frac{5.8 \times 100}{265} = 2.19 \text{ rad/s, } cw.$$

Example 2.18

For the mechanism shown in Fig.2.31(a), determine the velocities of points C, E and F. Also calculate the angular velocities of the links BC, CDE, and EF. Crank AB rotates at 120 rpm. The dimensions of various links are: $AB = 60$ mm, $BC = 120$ mm, $AD = 50$ mm, $CD = 100$ mm, $DE = 120$ mm, $CE = 50$ mm, and $EF = 150$ mm.

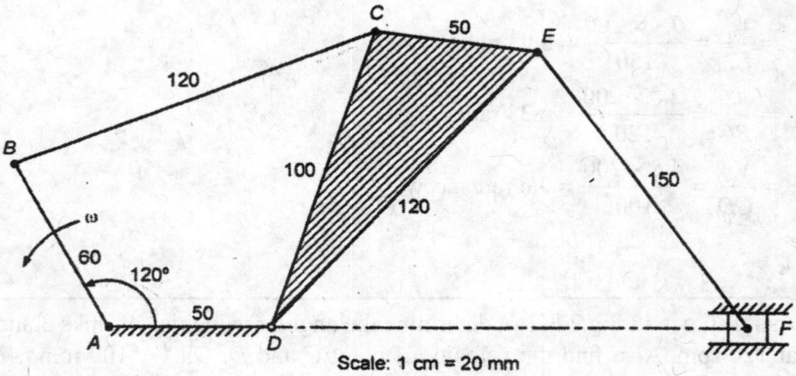

(a) Configuration diagram

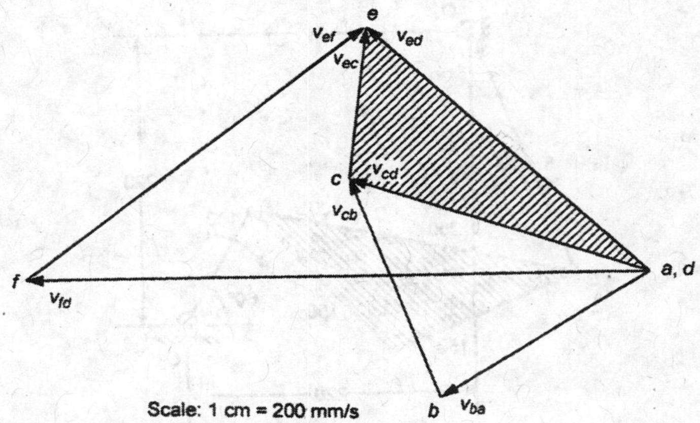

(b) Velocity diagram

Fig.2.31 For Example 2.18

■ Solution

Procedure:

1. Draw the configuration diagram shown in Fig.2.31(a) to a scale of 1 cm = 20 mm.
2. Angular speed of crank AB, $\omega = 2\pi \times \dfrac{120}{160} = 12.57$ rad/s

   $v_b = \omega \times AB = 12.57 \times 60 = 752.98$ mm/s.
3. Draw the velocity diagram as shown in Fig.2.31(b) to a scale of 1 cm = 200 mm/s.
   (i) Draw $v_b = ab \perp AB$, $ab = 3.77$ cm.
   (ii) Draw $bc \perp BC$ and $dc \perp DC$ to intersect at point c.
   (iii) Measure $v_{cd} = dc = 4.8$ cm $= 960$ mm/s to get the velocity of point c.
   (iv) Draw $ce \perp CE$ and $de \perp DE$ to intersect at point e. Then $v_{ed} = de \doteq 5.8$ cm $= 1160$ mm/s, the velocity of point E.
   (v) Draw $ef \perp EF$ and $df \perp DF$ to intersect at point f. Then $v_{fd} = df = 9.5$ cm $= 1900$ mm/s, the velocity of point F.
   (vi) $\omega_{EF} = \dfrac{v_{ef}}{EF} = \dfrac{6.7 \times 200}{150} = 8.93$ rad/s, ccw

   $\omega_{BC} = \dfrac{v_{bc}}{BC} = \dfrac{2.8 \times 200}{120} = 6.33$ rad/s, ccw

   $\omega_{CDE} = \dfrac{v_{cd}}{CD} = \dfrac{4.8 \times 200}{100} = 9.6$ rad/s, ccw.

Example 2.19

For the mechanism shown in Fig.2.32(a), determine the angular velocities of links 3 and 4 when link 2 is rotating at 120 rpm. Also find the velocity of point C and D. $O_2A = 100$ mm, $AB = 250$ mm, $AC = 150$ mm, $BC = 150$ mm, $O_4B = 100$ mm, and $AD = 100$ mm.

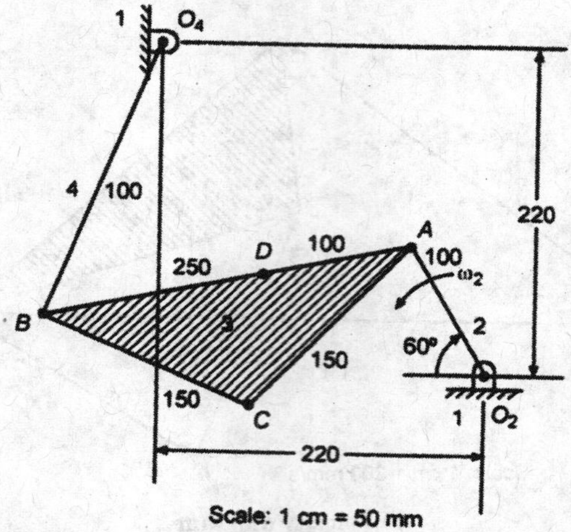

(a) Configuration diagram

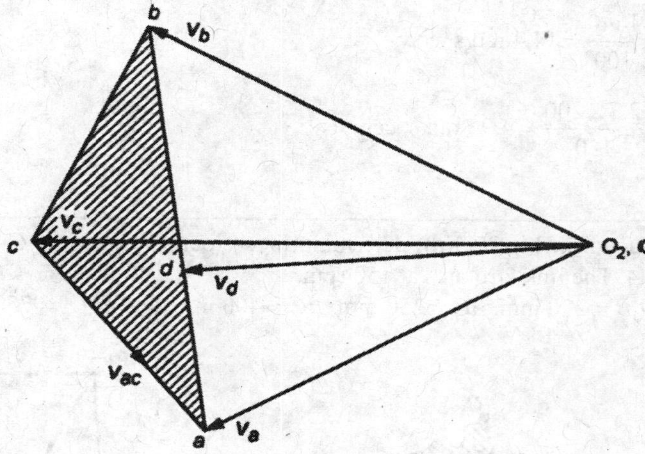

(b) Velocity diagram

Fig.2.32 Diagram for Example 2.19

■ Solution

Procedure:

1. Draw the configuration diagram shown in Fig.2.32(a), to a scale of 1 cm = 50 mm.
2. Angular speed of link 2,

$$\omega_2 = 2\pi \times \frac{120}{60} = 12.57 \text{ rad/s}$$

$$v_a = \omega_2 \times O_2A = 12.57 \times 100 = 1257 \text{ mm/s}$$

3. Draw the velocity diagram shown in Fig.2.31(b) to a scale of 1 cm = 200 mm/s.
   (i) Draw $v_a = o_2a \perp O_2A$, $o_2a = 6.20$ cm.

   (ii) Draw $o_4b \perp O_4B$ and $ab \perp AB$ to intersect at point b. Also draw $ac \perp AC$ and $bc \perp BC$ to intersect at point c.

   (iii) Measure $ab = 6$ cm.

Now,
$$\frac{ad}{ab} = \frac{AD}{AB}$$

∴
$$ad = 6 \times \frac{100}{250} = 2.4 \text{ cm}$$

Join o_2d ·Measure $o_2d = 6$ cm

$$v_d = o_2d = 1200 \text{ mm/s}$$

$$v_c = o_4c = 8 \text{ cm} = 1600 \text{ mm/s}$$

$$v_b = o_4b = 7.2 \text{ cm} = 1440 \text{ mm/s}$$

(iv) $\omega_4 = \dfrac{v_b}{O_4B} = \dfrac{1440}{100} = 14.4$ rad/s, ccw

$$\omega_3 = \frac{v_{ac}}{AC} = \frac{2.7 \times 200}{150} = 4.93 \text{ rad/s, ccw.}$$

Example 2.20

For the mechanism shown in Fig.2.33(a), determine the velocities of points C and A and angular speedy of links 3 and 4. The link 2 rotates at 150 rpm.

$O_2A = 380$ mm, $O_4B = 250$ mm, $AC = 250$ mm, $BC = 400$ mm, and $O_2O_4 = 750$ mm.

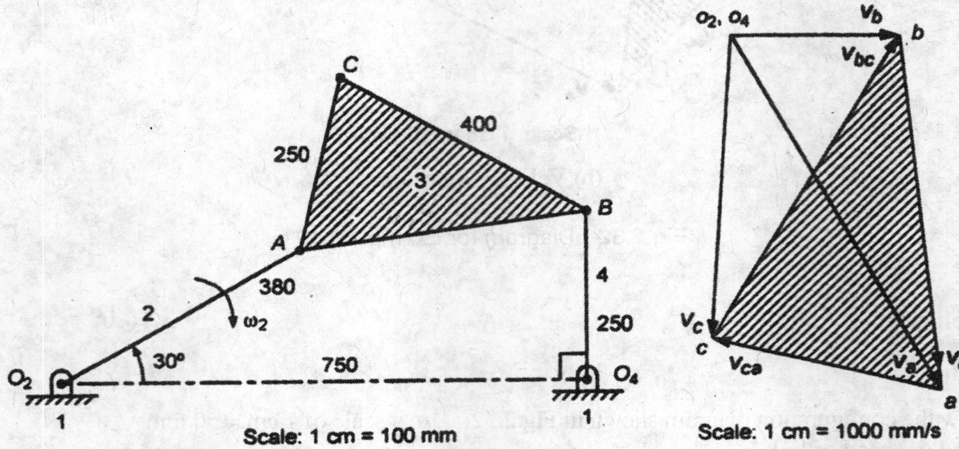

(a) Configuration diagram (b) Velocity diagram

Fig.2.33 Four-bar mechanism having ternary link

■ **Solution**

Procedure :

1. Draw the configuration diagram in Fig.2.33(a) to a scale of 1 cm = 100 mm.
2. Angular speed of link 2,

$$\omega_2 = 2\pi \times \frac{150}{60} = 15.7 \text{ rad/s}$$

$$v_a = \omega_2 \times O_2A = 15.7 \times 380 = 5969 \text{ mm/s.}$$

3. Draw the velocity diagram shown in Fig.2.33(b) to a scale of 1 cm = 1000 mm/s.
   (i) Draw $o_2a \perp O_2A$, $o_2a = 5.97$ cm.
   (ii) Draw $o_4b \perp O_4B$ and $ab \perp AB$ to intersect at b.
   (iii) Draw $ac \perp AC$ and $bc \perp BC$ to intersect at point c. Join o_2c.
   (iv) Measure $v_c = o_2c = 4.5$ cm $= 4500$ mm/s, $v_b = o_2b = 2.3$ cm $= 2300$ mm/s, and $v_{ab} = ba = 5.2$ cm $= 5200$ mm/s.

(v) $\omega_4 = \dfrac{v_b}{OB_4} = \dfrac{2300}{250} = 9.2$ rad/s, cw

$\omega_3 = \dfrac{v_{ab}}{AB} = \dfrac{5200}{400} = 13$ rad/s, cw.

Example 2.21

For the mechanism shown in Fig.2.34(a), determine the velocity of the slider C. The link 2 rotates at 180 rpm.

$O_2A = 50$ mm, $AB = 100$ mm, $AC = 200$ mm, $BD = 100$ mm, $O_2D = 100$ mm, and $O_2O_6 = 100$ mm.

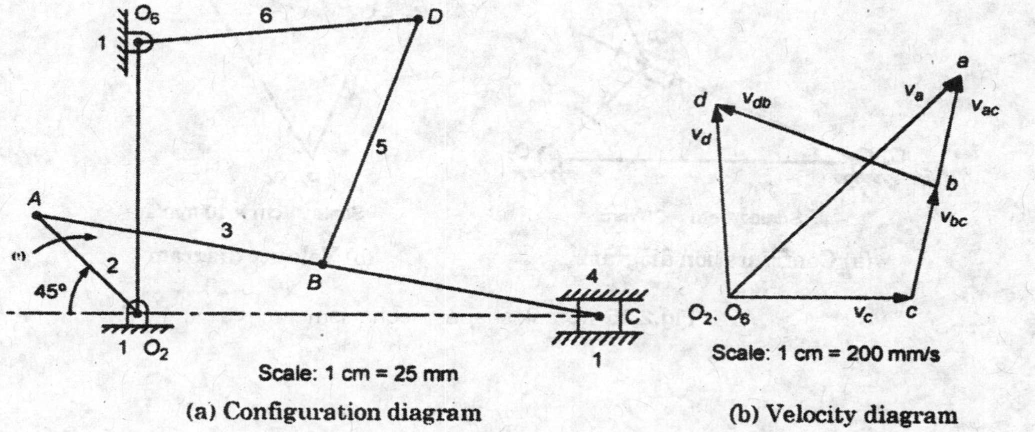

(a) Configuration diagram (b) Velocity diagram

Fig.2.34 Diagram for Example 2.21

■ Solution

Procedure:

1. Draw the configuration diagram shown in Fig.2.34(a) to a scale of 1 cm = 25 cm.
2. Angular speed of link 2,

$$\omega_2 = 2\pi \times \frac{180}{60} = 18.85 \text{ rad/s}$$

$$v_a = \omega_2 \times o_2A = 18.85 \times 50 = 942.5 \text{ mm/s.}$$

3. Draw the velocity diagram in Fig.2.34(b) to a scale of 1 cm = 200 mm/s.
   (i) Draw $v_a = o_2a \perp O_2A$, $o_2a = 4.71$ cm.
   (ii) Draw $ac \perp AC$ and $o_2c \parallel O_2C$ to intersect at c. Measure $ac = 3.5$ cm.
   (iii) Now $\dfrac{ab}{ac} = \dfrac{AB}{AC}$, $ab = 3.5 \times \dfrac{100}{200} = 1.75$ cm.
   (iv) Draw $o_6d \perp O_6D$ and $bd \perp BD$ to intersect at d.
   (v) Velocity of $c = v_c = o_2c = 2.6$ cm $= 520$ mm/s.

Example 2.22

In the crank–shaper mechanism shown in Fig.2.35(a), the link 2 rotates at a constant angular speed of 1 rad/s. Determine the angular speed of link 4, v_{A4}, v_{A2A3}, and v_{A3A4}, $O_2A = 50$ mm, $O_4A = 80$ mm, and $O_2O_4 = 120$ mm.

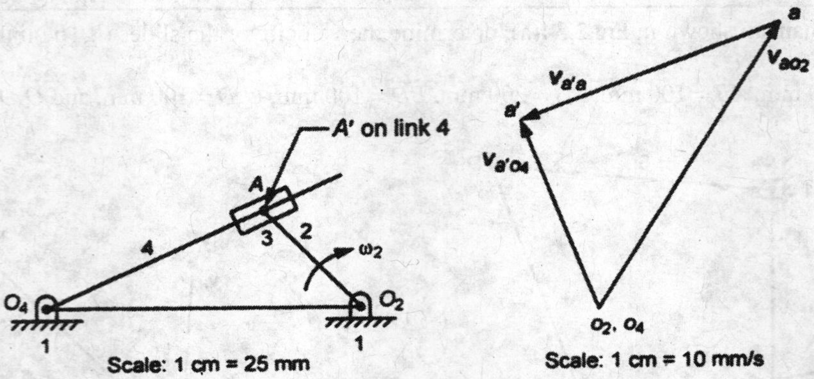

(a) Configuration diagram (b) Velocity diagram

Fig.2.35 Crank shaper mechanism

Solution

Procedure:

1. Draw the configuration diagram shown in Fig.2.35(a) to a scale of 1 cm = 25 mm.
2. $\omega_2 = 1$ rad/s (given),

$$v_a = \omega_2 \times o_2 A$$
$$= 1 \times 50 = 50 \text{ mm/s}$$

3. Draw the velocity diagram shown in Fig.2.35 (b) to a scale of 1 cm = 10 mm/s.

(i) $v_a = o_2 a \perp O_2 A,\ o_2 a = 5$ cm.

(ii) Draw $o_4 a' \perp O_4 A'$ and $a'a \parallel O_4 A'$ to intersect at a'. Then

$$v_{A4} = v_{a'o_4} = o_4 a' = 3 \text{ cm} = 30 \text{ mm/s}$$

$$\omega_4 = \frac{v_{a'o_4}}{o_4 A} = \frac{30}{80} = 0.375 \text{ rad/s, ccw}$$

$$v_{A3A4} = aa' = 4 \text{ cm} = 40 \text{ mm/s}$$
$$v_{A2A3} = v_a = 50 \text{ mm/s.}$$

Example 2.23

For the mechanism shown in Fig.2.36(a), the link 2 rotates at 160 rad/s. Determine v_b, ω_4 and v_{ba}. O_2A = 150 mm, AB = 200 mm, and O_4B = 150 mm.

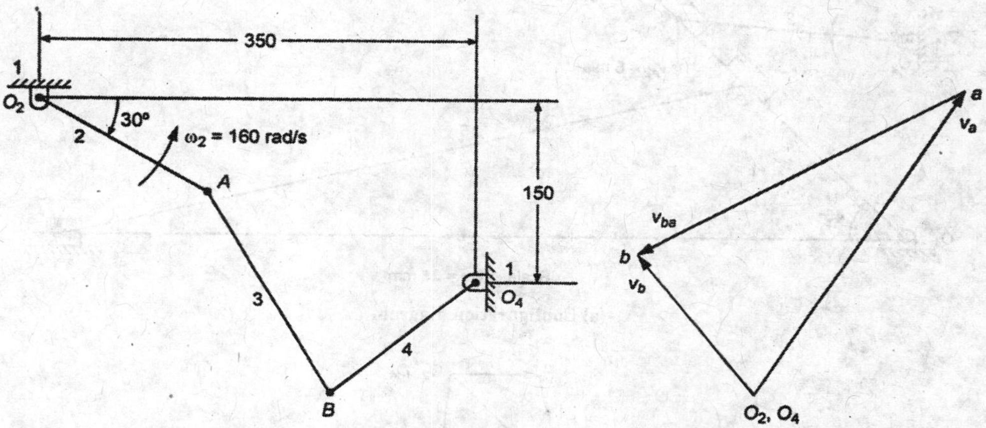

Scale: 1 cm = 50 mm

(a) Configuration diagram

Scale: 1 cm = 4000 mm/s

(b) Velocity diagram

Fig.2.36 Four-bar mechanism

■ Solution

Procedure:

1. Draw the configuration diagram shown in Fig.2.36(a) to a scale of 1 cm = 50 mm.
2. ω_2 = 160 rad/s (given)
   $v_a = \omega_2 \times o_2A = 160 \times 150 = 24{,}000$ mm/s.
3. Draw the velocity of diagram shown in Fig.2.36(b) to a scale of 1 cm = 4000 mm/s.
   (i) Draw $v_a = o_2a \perp O_2A, o_2a = 6$ cm.
   (ii) Draw $o_4b \perp O_4B$ and $ab \perp AB$ to intersect at *b*.
   (iii) Measure $v_b = o_4b = 3$ cm = 12,000 mm/s.
   (iv) $\omega_2 = \dfrac{v_b}{O_4B} = \dfrac{12{,}000}{150} = 80$ rad/s, ccw

   $v_{ba} = ab = 5.8$ cm = 23,200 mm/s.

Example 2.24

The driving link 2 of the Whitworth quick-return motion mechanism shown in Fig.2.37(a) rotates at a constant speed of 6 m/s. Determine the velocity of tool holder. $O_1A = 100$ mm, $O_4B = 100$ mm, BC = 300 mm, $O_2O_4 = 90$ mm.

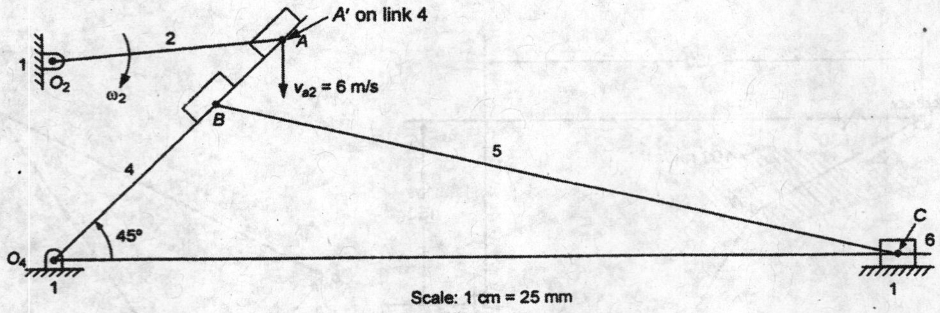

(a) Configuration diagram

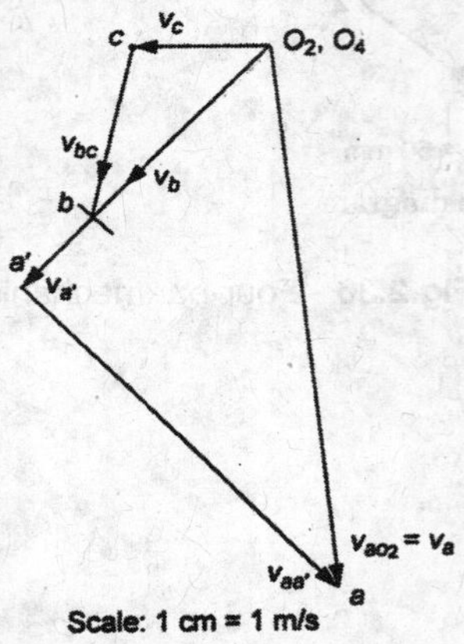

(b) Velocity diagram

Fig.2.37 Whitworth quick-return motion mechanism

■ Solution

Procedure:

1. Draw the configuration diagram shown in Fig.2.37(a) to a scale of 1 cm = 25 mm and velocity diagram shown in Fig.2.37(b) to a scale of 1 cm = 1 m/s.
2. Draw $v_{ao_2} = o_2a = v_{a2} \perp O_2A, o_2a = 6$ m/s.
3. Draw $o_4a' \parallel O_4A$ and $a'a \perp O_4A$ to intersect at a'.
4. Measure $o_4a' = 3.8$ cm, $O_4A = 5.7$ cm, $O_4B = 4$ cm
5. Now $\dfrac{o_4b}{o_4a'} = \dfrac{O_4B}{O_4A}, o_4b = 3.8 \times \dfrac{4}{5.7} = 2.67$ cm.
6. Draw $bc \perp BC$ and $o_2c \parallel O_4C$ to intersect at point C.
7. Velocity of tool holder, $v_c = o_2c = 1.5$ cm $= 1.5$ m/s.

Example 2.25

For the mechanism shown in Fig.2.38(a), determine velocity of slider and ω_3, ω_4, and ω_5.

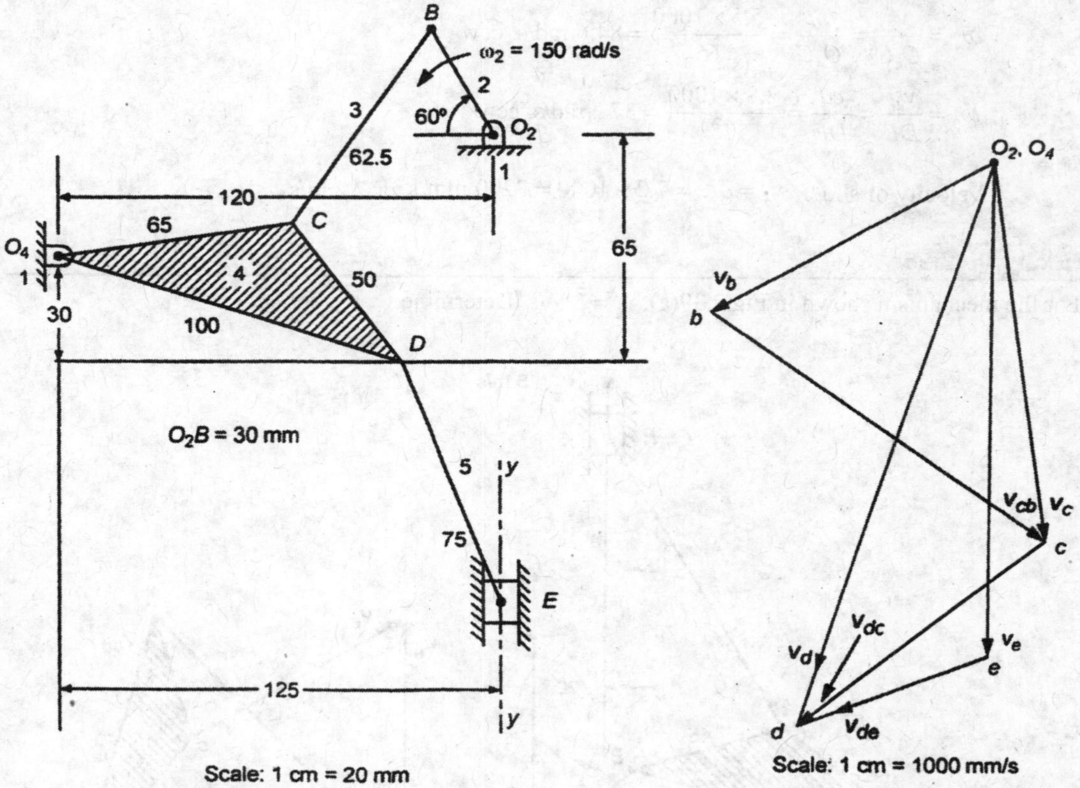

(a) Configuration diagram (b) Velocity diagram

Fig.2.38 Diagram for Example 2.25

■ Solution

Procedure:

1. Draw the configuration diagram shown in Fig.2.38(a) to a scale of 1 cm = 20 mm.
2. Angular speed of link 2,

$$\omega_2 = 150 \ rad/s$$

$$v_b = \omega_2 \times O_2B = 150 \times 30 = 4500 \ mm/s.$$

3. Draw the velocity diagram shown in Fig.2.38(b) to a scale of 1 cm = 1000 mm/s.
   (i) Draw $v_b = o_2b \perp O_2B$, $o_2b = 4.5$ cm.
   (ii) Draw $bc \perp BC$ and $o_4c \perp O_4C$ to intersect at point c.
   (iii) Draw $cd \perp CD$ and $o_4d \perp O_4D$ to intersect at d.
   (iv) Draw $de \perp DE$ and $o_2e \parallel yy$ to intersect at e.

(v) $\omega_3 = \dfrac{v_{cb}}{BC} = \dfrac{bc}{BC} = \dfrac{5.8 \times 1000}{62.5} = 92.8$ rad/s, ccw

$$\omega_4 = \frac{v_c}{O_4C} = \frac{o_4c}{O_4C} = \frac{5.5 \times 1000}{65} = 84.6 \text{ rad/s, ccw}$$

$$\omega_5 = \frac{v_{de}}{DE} = \frac{ed}{DE} = \frac{2.8 \times 1000}{75} = 37.3 \text{ rad/s, ccw}$$

Velocity of slider, $v_e = o_4e = 7.2 \times 1000 = 7200$ mm/s or 7.2 m/s.

Example 2.26

For the mechanism shown in Fig.2.39(a), $v_E = 5$ m/s. Determine v_d, ω_3, and ω_5.

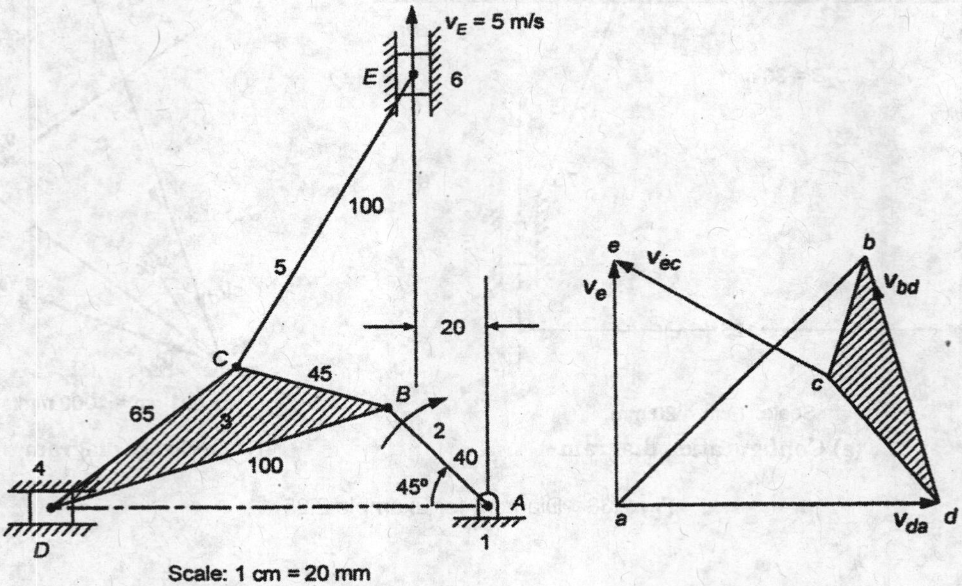

(a) Configuration diagram (b) Velocity diagram

Fig.2.39 Diagram for Example 2.26

■ Solution

Procedure:

1. Draw the configuration diagram shown in Fig.2.39(a) to a scale of 1 cm = 20 mm.
2. Draw the velocity diagram shown in Fig.2.39(b) by assuming $v_b = 5$ cm and crank AB rotating clockwise.
3. Draw $v_b = ab \perp Ab, ab = 5$ cm.
4. Draw $bd \perp BD$, $ad \parallel AD$ to meet at point d.
5. Draw $bc \perp BC$ and $cd \perp CD$ to meet at point c.
6. Draw $ae \parallel v_E$ and $ce \perp CE$ to meet at e. Then $ae = v_E = 5$ m/s = 3.5 cm. Thus scale is: 1 cm = 1.428 m/s.

7. $v_{da} = ad = 4.5 \text{ cm} = 6.45 \text{ m/s}$

$v_{bd} = db = 2.7 \text{ cm} = 5.28 \text{ m/s}$

$$\omega_3 = \frac{v_{bd}}{BD} = \frac{bd}{BD} = \frac{5.28}{0.1} = 52.8 \text{ rad/s, ccw}$$

$v_{ec} = ce = 2.4 \text{ cm} = 4.85 \text{ m/s}$

$$\omega_5 = \frac{v_{ec}}{EC} = \frac{4.85}{0.10} = 48.5 \text{ rad/s, cw}$$

Example 2.27

The crank O_1A of the four-bar linkage shown in Fig.2.40(a) is rotating at a uniform angular velocity of 30 rad/s. Draw the velocity polygon and determine the velocity of point B, the angular velocities of the links 3 and 4. O_1A =100 mm, AB = 200 mm, O_2B = 75 mm, AC = 100 mm, BC = 150 mm, and AD = 50 mm. Also determine the velocity images of all links.

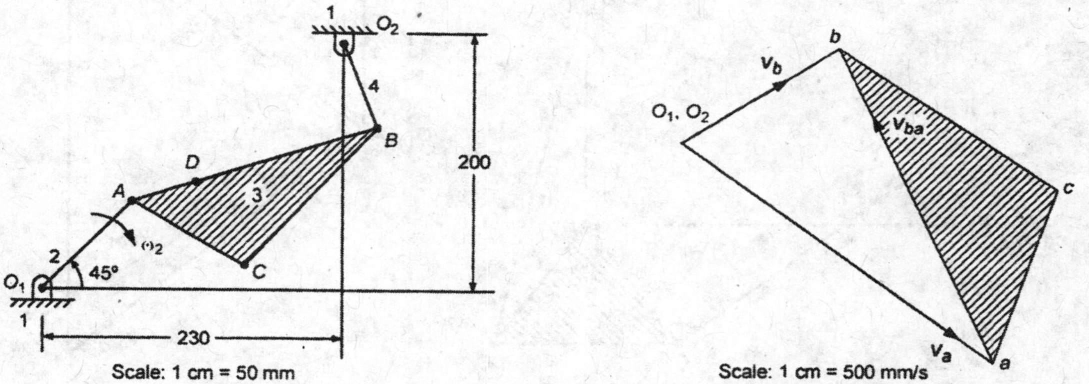

(a) Configuration diagram (b) Velocity diagram

Fig.2.40 Four-bar mechanism with ternary link

■ Solution

Procedure:

1. Draw the configuration diagram shown in Fig.2.40(a) to a scale of 1 cm = 50 mm.
2. $\omega_2 = 30$ rad/s, $v_a = \omega_2 \times O_1A = 30 \times 100 = 3000$ mm/s
3. Draw the velocity polygon as shown in Fig.2.40(b) to a scale of 1 cm = 500 mm/s.
   (i) Draw $v_a = o_1a \perp O_1A,\ o_1a = 6$ cm.
   (ii) Draw $ab \perp AB$ and $o_2b \perp O_2B$ to intersect at point b.
   (iii) Draw $ac \perp AC$ and $bc \perp BC$ to intersect at point c.
   (iv) $v_{ba} = ab = 5.6 \text{ cm} = 2800 \text{ mm/s}$
   $v_b = o_1b = 2.9 \text{ cm} = 1450 \text{ mm/s}.$

(v) $\omega_3 = \frac{v_{ba}}{AB} = \frac{ab}{AB} = \frac{2800}{200} = 14\,\text{rads/s, cw.}$

$\omega_4 = \frac{v_b}{O_2B} = \frac{o_2b}{O_2B} = \frac{1450}{75} = 19.3\,\text{rads/s, cw.}$

Example 2.28

In the mechanism shown in Fig.2.41(a), the piston D moves in the vertical direction upwards with a velocity of 5 m/s.

$O_1A = 7.5$ cm, $O_1O_2 = 30$ cm, $AB = 25$ cm, $O_2C = O_2B = 10$ cm, $CD = 25$ cm, and $BC = 12.5$ cm. Find the speed in rpm and direction of rotation of crank O_1A.

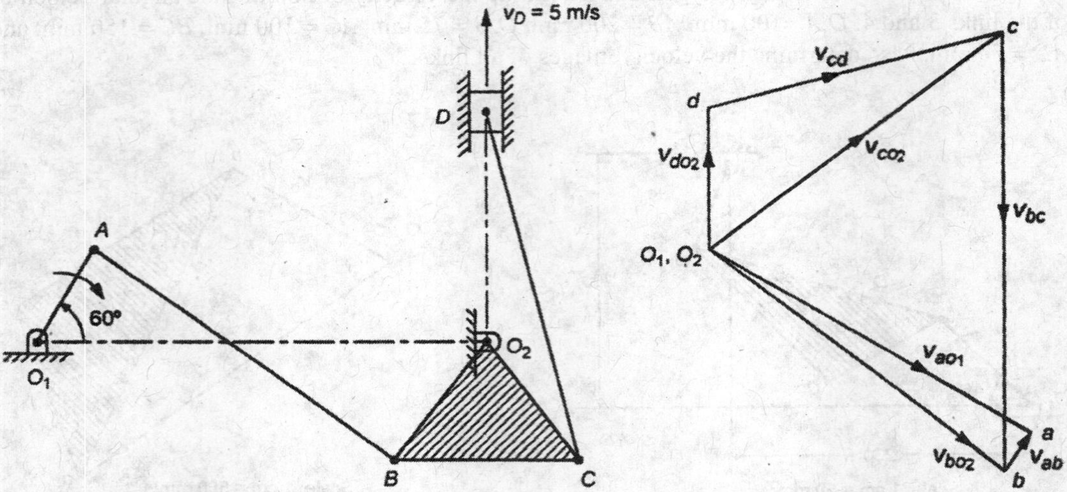

(a) Configuration diagram (b) Velocity diagram

Fig.2.41 Diagram for Example 2.28

Solution

Procedure:

Draw the configuration diagram shown in Fig.2.41(a) to a scale of 1 cm = 5 cm. Let us assume that crank O_1A is rotation clockwise. Assume that $v_a = 5$ cm to some scale. Then draw the velocity diagram shown in Fig.2.41(b) as follows:

1. Draw $v_a = o_1a \perp O_1A, o_1a = 5\,\text{cm}$.
2. Draw $ab \perp AB$ and $o_2b \perp O_2B$ meeting at point b.
3. Draw $bc \perp BC$ and $o_2c \perp O_2C$ meeting at point c.
4. Draw $o_2d \parallel O_2D$ and $cd \perp CD$ meeting at point d.

   Then $v_D = o_2d = 2$ cm = 5 m/s

   Hence scale: 1 cm = 2.5 m/s

   and $v_a = 5 \times 2.5 = 12.5$ m/s

$$\omega = \frac{v_a}{O_1A} = \frac{12.5 \times 100}{7.5} = 166.67 \text{ rad/s}$$

$$\frac{2\pi N}{60} = 166.67$$

$$N = 1591.5 \text{ rpm}$$

Example 2.29

The quick-return motion mechanism of the crank and slotted lever type shaping machine is shown in Fig.2.42(a).

$O_{1O2} = 800$ mm, $O_1B = 300$ mm, $O_2D = 1300$ mm, $DR = 400$ mm.

The crank O_1B makes and angle of 45° with the vertical and rotates at 40 rpm in the counter clock-wise direction. Find: (a) velocity of ram R, and (b) angular velocity of link O_2D.

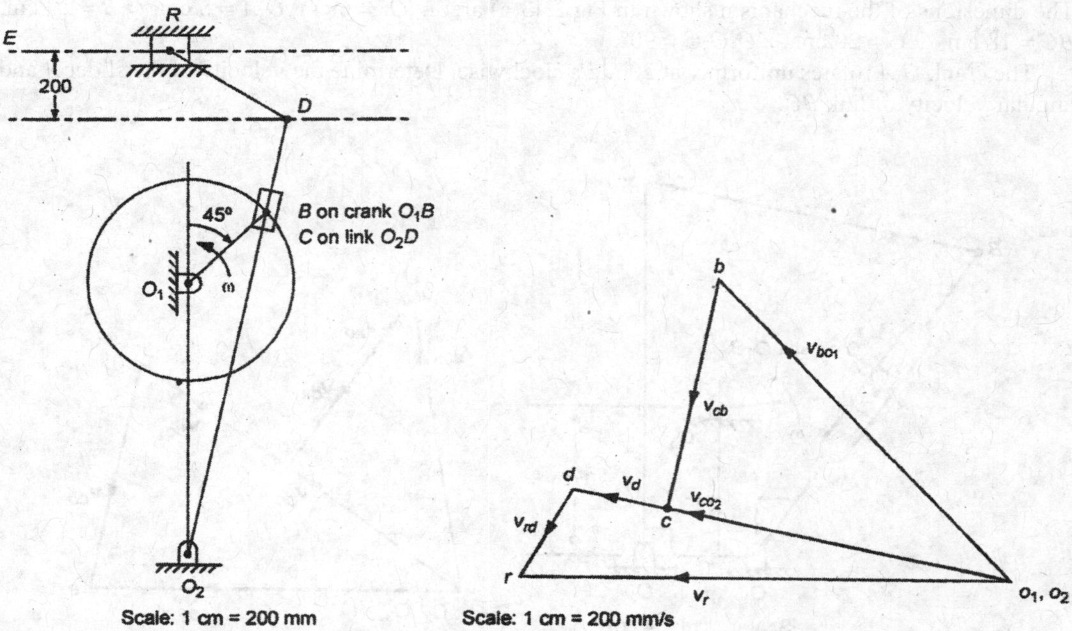

(a) Configuration diagram (b) Velocity diagram

Fig.2.42 Shaping machine mechanism

■ Solution

Procedure:

1. Draw the configuration diagram to a scale of 1 cm = 200 mm.
2. Angular speed of crank O_1B,

$$\omega = \frac{2\pi \times 40}{60} = 4.189 \text{ rad/s}$$

$$v_b = \omega \times O_1B = 4.189 \times 300 = 1256.6 \text{ mm/s}$$

3. Draw the velocity diagram as shown in Fig.2.42(b) to a scale of 1 cm = 200 mm/s.
4. Draw $o_1b \perp O_1B, o_1b = 6.28$ cm. Also draw $bc \parallel O_2D$ and $o_2d \perp O_2D$
5. Measure $o_2c = 5.1$ cm.
6. Now $\dfrac{o_2d}{o_2c} = \dfrac{O_2D}{O_2C}$, $o_2d = \dfrac{5.1 \times 6.5}{5.1} = 6.5$ cm.
7. Produce o_2c to d so that $o_2d = 6.5$ cm.
8. Draw $dr \perp DR$ and $o_2r \parallel RE$ to meet at r.
9. Then $v_r = o_1r = 7.1$ cm = 1420 mm/s is the velocity of ram.
10. $v_d = o_2d = 6.5\,\text{cm} = 1300\,\text{mm/s}$, $\omega_{O_2D} = \dfrac{v_d}{O_2D} = \dfrac{1300}{1300} = 1\,\text{rad/s}$ ccw about O_2.

Example 2.30

The dimesions of the mechanism shown in Fig.2.43(a) are: $O_1O_2 = 6$ cm, $O_3A = 8$ cm, $O_1B = 20$ cm, $BC = 18$ cm, $O_2C = 20$ cm, $\angle O_1O_2C = 90°$.

The crank O_3A rotates uniformly at 20 rad/s clockwise. Determine the velocity of the slider A and angular velocity of link BC.

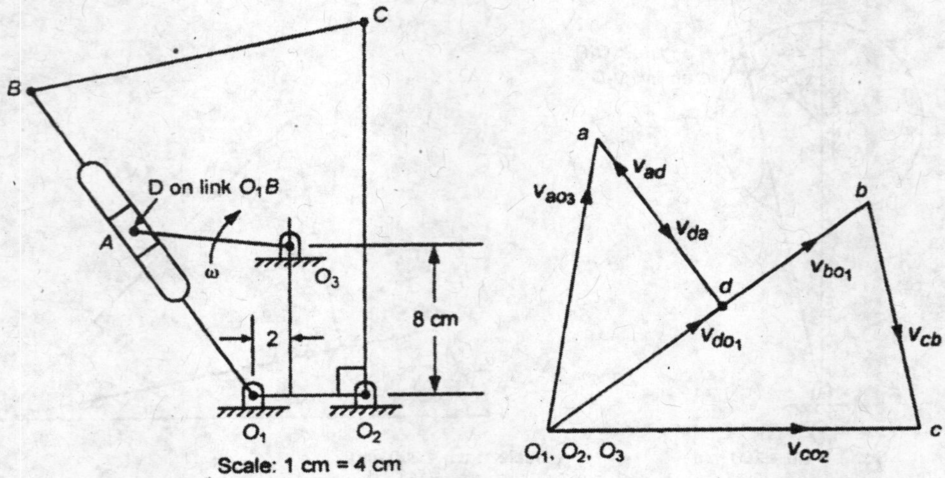

(a) Configuration diagram (b) Velocity diagram

Fig.2.43 Diagram for Example 2.30

Procedure:

1. Angular speed of link $O_3A = 20$ rad/s

   $v_{ao_3} = \omega \times O_3A = 20 \times 8 = 160$ cm/s
2. Draw velocity diagram as shown in Fig.2.43(b)
3. Draw $v_{ao_3} = o_3a \perp O_3A$ to a scale of 1 cm = 40 cm/s so that $o_3a = 4$ cm.

4. Draw $ad \parallel BO_1$ and $o_1b \perp O_1B$. Measure $o_1d = 2.9$ cm and $O_1B = 2.8 \times 4 = 11.2$ cm.

   Now $\dfrac{o_1d}{o_1b} = \dfrac{O_1D}{O_1B}$, $o_1b = 2.9 \times \dfrac{20}{11.2} = 5.2$ cm

   Produce o_1d to b so that $o_1b = 5.2$ cm

5. Draw $bc \perp BC$ and $o_2c \perp O_2C$ to meet at c.

   Velocity of the slider A, $v_{da} = 2.8 \times 40 = 112$ cm/s

   Angular velocity of link $BC = \omega_{BC} = \dfrac{v_{cb}}{BC} = \dfrac{3.1 \times 40}{18} = 6.9$ rad/s, cw.

Example 2.31

Using relative velocity method find the absolute velocity of the slider E in the mechanism shown in Fig.2.44(a). The crank O_1A rotates at 60 rpm. $O_1A = 50$ mm, $O_2B = 120$ mm, $AB = 270$ mm $O_2D = 90$ mm, and $DE = 180$ mm.

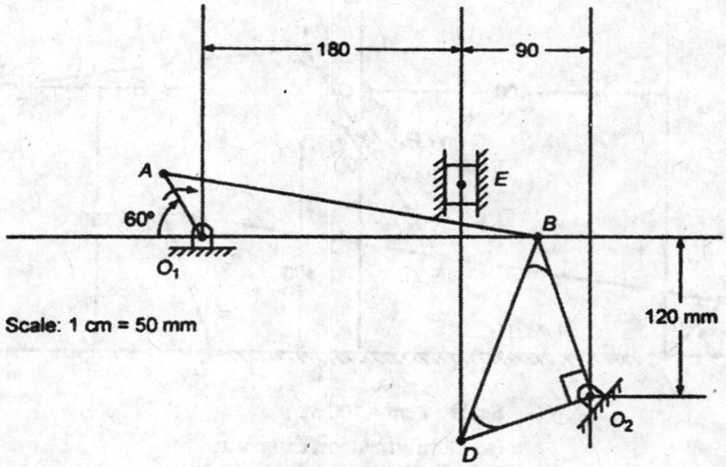

(a) Configuration diagram

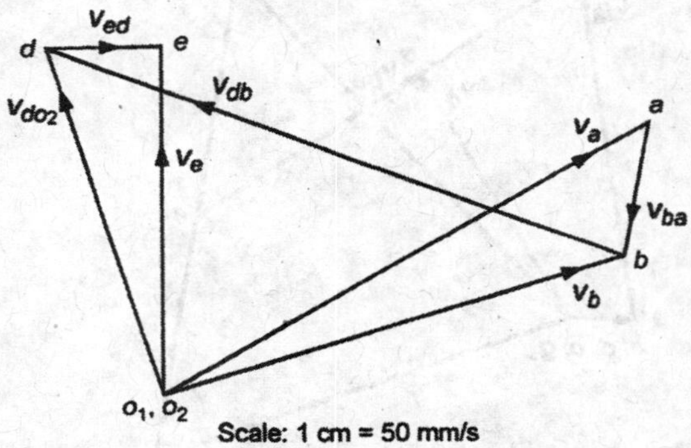

(b) Velocity diagram

Fig.2.44 Diagram for Example 2.31

■ **Solution**

$$\omega = \frac{2\pi N}{60} = \frac{2\pi \times 60}{60} = 6.28 \text{ rad/s}$$

$$v_a = \omega \times O_1A = 6.28 \times 50 = 314.16 \text{ mm/s}$$

1. Draw $v_a = o_1a \perp O_1A, o_1a = 6.28$ cm (Fig.2.44(b)).
2. Draw $ab \perp AB$ and $o_2b \perp O_2B$ to meet at point *b*.
3. Draw $bd \perp BD$ and $o_2d \perp O_2D$ to meet at point *d*.
4. Draw $de \perp DE$ and $o_1e \perp O_1B$ to meet at point *e*.

Absolute velocity of the slider $v_e = o_1e = 4$ cm $= 200$ mm/s.

Example 2.32

Figure 2.45(a) shows a swivelling joint mechanism in which *AB* is the driving crank which rotates at 240 rpm clockwise.

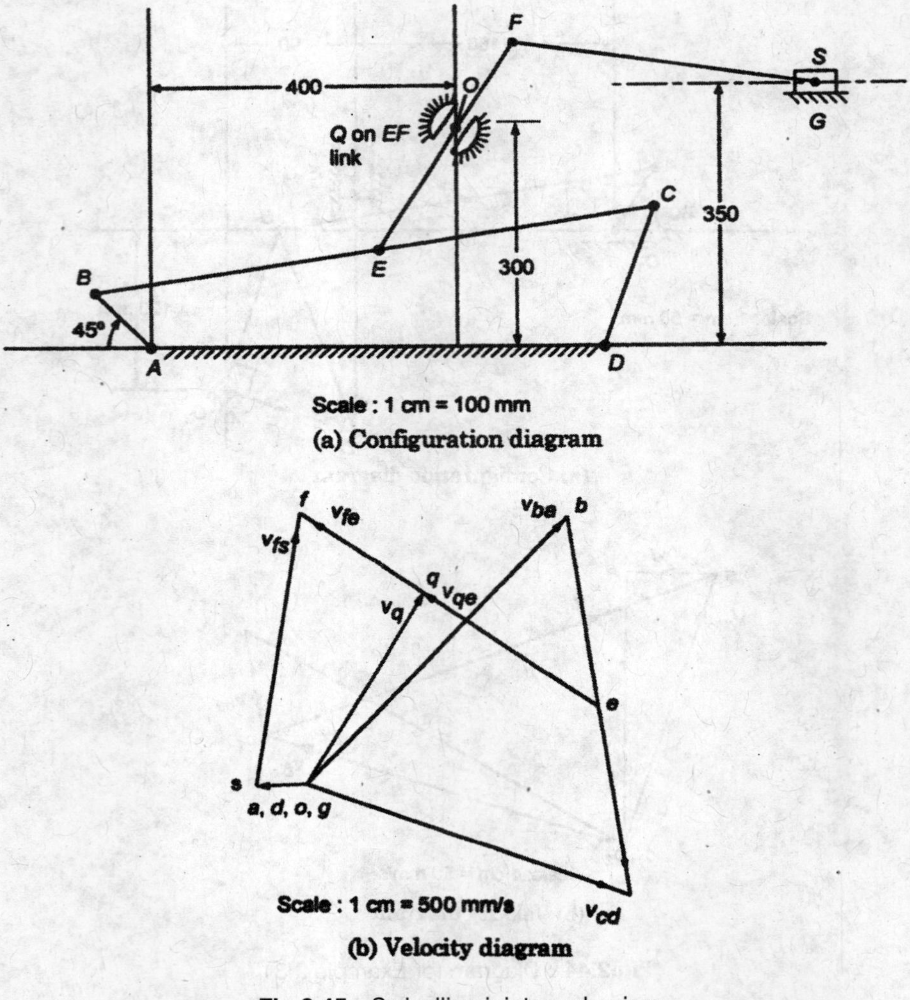

(a) Configuration diagram

(b) Velocity diagram

Fig.2.45 Swivelling joint mechanism

$AB = 100$ mm, $BC = 750$ mm, $CD = 200$ mm, $AD = 600$ mm, $BE = EC$, $EF = 340$ mm, $FG = 400$ mm.
Determine (a) the velocity of slider G, (b) angular velocity of link EF, and (c) the velocity of link EF in the swivel block.

■ Solution

$$\omega = \frac{2\pi \times 240}{60} = 25.13 \text{ rad/s}$$

$$v_b = \omega \times AB = 25.13 \times 100 = 2513 \text{ mm/s}$$

1. Draw $v_{ba} = ab \perp AB, ab = 5.02$ cm (Fig.2.45(b)).
2. Draw $bc \perp BC$ and $cd \perp CD$ to meet at c.
3. Since $BE = EC$, therefore $ce = eb$. Locate point e.
4. Let Q be a point on link EF at joint O. Draw $eq \perp EF$ and $oq \parallel EF$ to locate point q.
5. Now $\dfrac{ef}{eq} = \dfrac{EF}{EQ}$, $ef = 2.8 \times \dfrac{340}{200} = 4.76$ cm. Extend eq to f so that $ef = 4.76$ cm.
6. Draw $fs \perp FS$ and $gs \parallel$ line of stroke of slider to meet at point s.
   (a) Velocity of slider $s = gs = 0.6$ cm $= 300$ mm/s.
   (b) Angular velocity of link $EF = \dfrac{v_{fe}}{EF} = \dfrac{4.76 \times 500}{340} = 7$ red/s, ccw.
   (c) Velocity of link EF in swivel block $= oq = 3$ cm $= 1500$ mm/s.

Example 2.33

The angular velocity of crank OA shown in Fig.2.46(a) is 600 rpm. Determine the linear velocity of slider D and the angular velocity of link BD when the crank is inclined at an angle of 75° to the vertical. The dimensions of the various links are: $OA = 28$ mm, $AB = 44$ mm, $BC = 49$ mm, and $BD = 46$ mm. The centre distance between the centres of rotation O and C is 65 mm. The path of travel of the slider is 11 mm below the fixed point C. The slider moves along a horizontal path and OC is vertical.

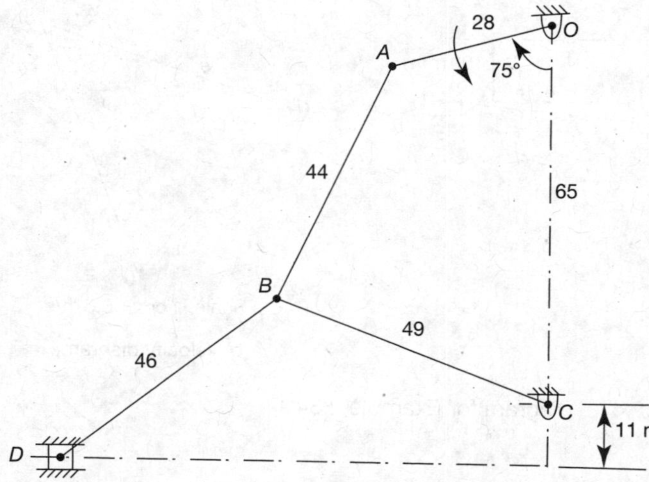

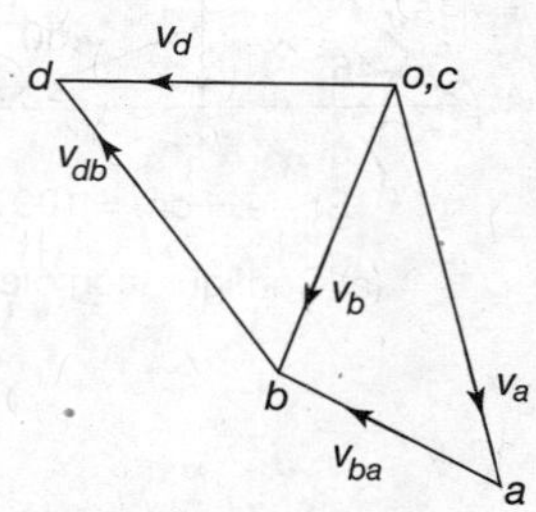

Fig.2.46 Diagram for Example 2.33

■ **Solution**

Angular velocity of OA, $\omega = \dfrac{2\pi N}{60} = \dfrac{2\pi \times 600}{60} = 62.83$ rad/s

Linear velocity of A, $v_a = \omega \times OA = 62.83 \times 28 \times 10^{-3} = 1.76$ m/s

1. Draw configuration diagram as shown in Fig.2.46(a).
2. Draw velocity diagram as shown in Fig.2.46(b).
3. Draw oa perpendicular to OA such that $oa = v_a = 3.52$ cm.
4. Draw $ab \perp AB$ and $ob \perp BC$ to intersect at b.
5. Draw $bd \perp BD$ and od parallel to the path of slider D to intersect at d.

Linear velocity of slider D, $v_d = od = 2.8$ cm $= 1.4$ m/s

Linear velocity of BD, $v_{db} = 3.2$ cm $= 1.6$ m/s

Angular velocity of BD, $\omega_{bd} = \dfrac{v_{db}}{BD} = \dfrac{1.6}{0.046} = 34.78$ rad/s (clockwise about B)

Example 2.34

The mechanism shown in Fig.2.47 has the dimensions of various links as follows:

$$AB = DE = 150 \text{ mm}, BC = CD = 450 \text{ mm}, EF = 375 \text{ mm}$$

The crank AB makes an angle of 45° with the horizontal and rotates about A in the clockwise direction at a uniform speed of 120 rpm. The lever DC oscillates about the fixed point D, which is connected to AB by the coupler BC.

The block F moves in the horizontal guides, being driven by the link EF. Determine (a) linear velocity of block F, (b) angular velocity of DC, and (c) rubbing speed at pin C, which is 50 mm diameter.

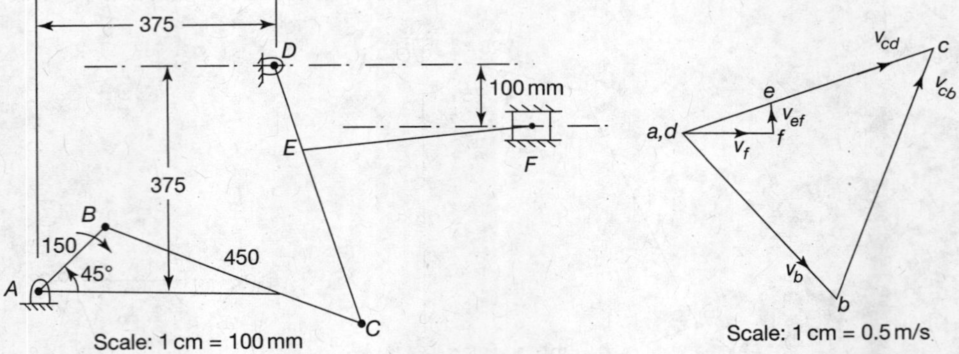

(a) Configuration diagram (b) Velocity diagram

Fig.2.47 Diagram for Example 2.34

■ **Solution**

Angular velocity of AB, $\omega = \dfrac{2\pi N}{60} = \dfrac{2\pi \times 120}{60} = 12.57$ rad/s

Linear velocity of B, $v_b = \omega \times AB = 12.57 \times 0.15 = 1.885$ m/s

1. Draw the configuration diagram as shown in Fig.2.47(a).
2. Draw the velocity diagram as shown in Fig.2.47(b).
3. Draw $ab \parallel AB$ such that $ab = v_b = 3.77$ cm.
4. Draw $bc \perp BC$ and $dc \perp DC$ to intersect at c.
5. Measure $dc = 4.4$ cm.
6. $de = \dfrac{DE}{CD} \times dc = \dfrac{150}{450} \times 4.4 = 1.47$ cm.
7. Draw $ef \perp EF$ and af parallel to the line of stroke of F to meet at f.
8. Linear velocity of F, $v_f = af = 1.5$ cm $= 0.75$ m/s.
9. $v_{cd} = dc = 4.4$ cm $= 2.2$ m/s.
10. $\omega_{cd} = v_{cd}/CD = 2.2/0.45 = 4.9$ rad/s (ccw about D)
11. $v_{cb} = bc = 4.4$ cm $= 2.2$ m/s
12. $\omega_{cb} = \dfrac{v_{cb}}{BC} = \dfrac{2.2}{0.45} = 4.9$ rad/s (ccw about B)
13. Rubbing velocity at pin C $v_r = (\omega_{cb} - \omega_{cd})\, r_c = (4.9 - 4.9) \times 0.025 = 0$

Example 2.35

The dimensions of a four-bar mechanism are:

$AD = 300$ mm, $BC = AB = 360$ mm, $CD = 600$ mm. The link CD is fixed and $\angle ADC = 60°$. The driving link AD rotates uniformly at a speed of 120 rpm clockwise and the constant driving torque is 60 Nm. Determine (a) the velocity of the point B and angular velocity of BC, (b) actual mechanical advantage and the resisting torque if efficiency of the mechanism is 75 percent.

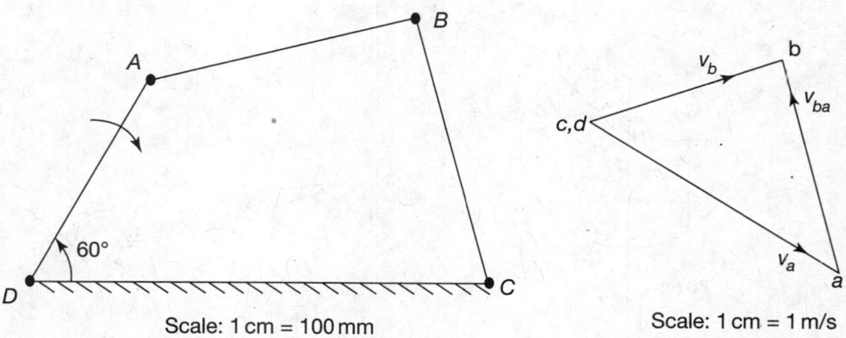

Fig.2.48 Four-bar mechanism

■ **Solution**

Angular velocity of link AD, $\omega = \dfrac{2\pi N}{60} = \dfrac{2\pi \times 120}{60} = 12.57$ rad/s

Linear velocity of A, $v_b = \omega \times AB = 12.57 \times 0.3 = 3.77$ m/s

1. Draw configuration diagram as shown in Fig.2.48(a).
2. Draw velocity diagram as shown in Fig.2.48(b).
3. Draw $da \perp DA$ such that $da = v_a = 3.77$ cm.
4. Draw $ab \perp AB$ and $cb \perp CB$ to intersect at b.
5. $v_b = cb = 2.6$ cm $= 2.6$ m/s
6. $\omega_{bc} = \dfrac{v_b}{BC} = \dfrac{2.6}{0.36} = 7.22$ rad/s (cw about c)

$$\text{Mechanical advantage} = \eta \frac{\omega_a}{\omega_b} = 0.75 \times \frac{12.57}{7.22} = 1.30$$

$$\text{Now Efficiency, } \eta = \frac{T_b \cdot \omega_b}{T_a \cdot \omega_a}$$

$$T_b = 0.75 \times 60 \times \frac{12.57}{7.22}$$

$$= 78.345 \text{ Nm}$$

Example 2.36

A quick-return mechanism of a shaper is shown in Fig.2.49(a). The crank O_1A rotates in the counter-clockwise direction. Determine the linear velocity of the cutting tool when the crank O_1A is at 45° with the horizontal. All dimensions are given in the figure.

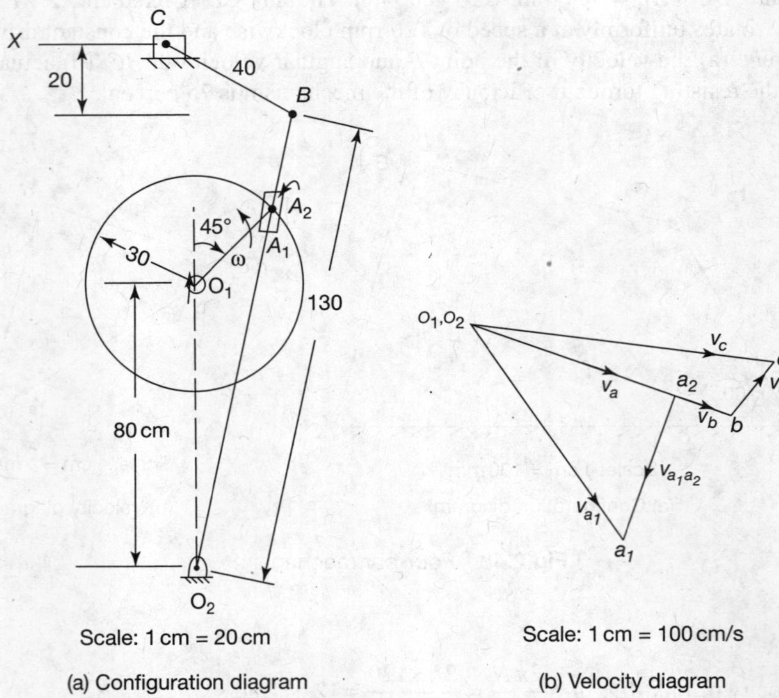

Fig.2.49 Quick-return mechanism

■ **Solution**

Draw configuration diagram shown in Fig.2.49(a).

$\omega = 2\pi \times 120/60 = 12.57$ rad/s

$v_{a1} = \omega \times O_1A_1 = 12.57 \times 30 = 377$ cm/s

Draw the velocity diagram as shown in Fig.2.49(b).

$v_{a1} = o_1a_1 \perp O_1A_1$

$a_1a_2 \parallel O_2B$

$o_2a_2 \perp O_2B$

$o_2a_2 = 3$ cm

$o_2b = o_2a_2 \times O_2B/O_2A_2 = 3 \times 6.5/5.1 = 2.82$ cm

$bc \perp BC$

$o_2c \parallel XC$

Velocity of cutting tool, $v_c = o_1c = 4.30$ cm/s

2.4 INSTANTANEOUS CENTRE METHOD

A link as a whole may be considered to be rotating about an imaginary centre or about a given centre at a given instant. Such a centre has zero velocity, i.e. the link is at rest at this point. This is known as the instantaneous centre or centre of rotation. This centre varies from instant to instant for different positions of the link. The locus of these centres is termed the *centrode*.

2.4.1 Velocity of a Point on a Link

Consider two points *A* and *B* on a rigid link, having velocities v_a and v_b, respectively, as shown in Fig.2.50(a). From *A* and *B* draw lines perpendicular to the directions of motion and let them meet at *I*. Then *I* is the instantaneous centre of rotation of the link *AB* for its give position.

If ω is the instantaneous angular velocity of the link *AB*, then $v_a = \omega \cdot IA$ and $v_b = \omega \cdot IB$. Thus

$$v_b = \left(\frac{IB}{IA}\right) \cdot v_a \tag{2.10}$$

The velocity diagram for the link *AB* has been drawn in Fig.2.50(b). Triangles IAB and oab are similar. Hence

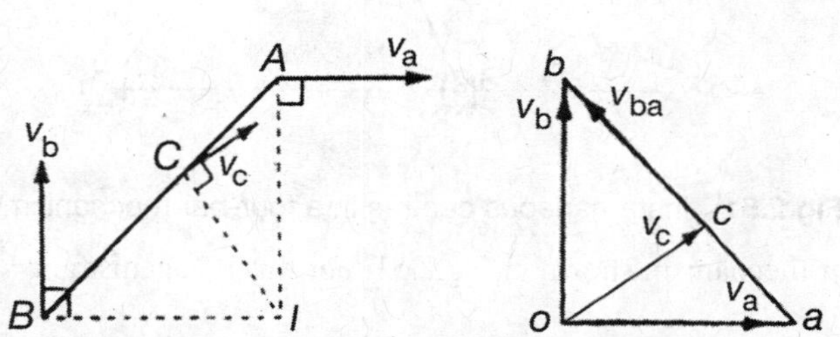

(a) Instantaneous centre (b) Velocity diagram

Fig.2.50 Concept of instantaneous centre

$$\frac{oa}{IA} = \frac{ob}{IB} = \frac{ab}{AB} \tag{2.11}$$

or $$\frac{v_a}{IA} = \frac{v_b}{IB} = \frac{v_{ba}}{AB} = \frac{v_c}{IC} = \omega$$

where C is any point on the link AB.

2.4.2 Properties of Instantaneous Centre

The properties of the instantaneous centre are as follows:

1. At the instantaneous centre of rotation, one rigid link rotates instantaneously relative to another for the configuration of the mechanism considered.
2. The two rigid links have no linear velocities relative to each other at the instantaneous centre.
3. The two rigid links have the same linear velocity relative to the third rigid link, or any other link.

2.4.3 Number of Instantaneous Centres

The number of instantaneous centres in a mechanism is equal to the number of possible combinations of two links. The number of instantaneous centres,

$$N = \frac{n(n-1)}{2} \tag{2.12}$$

where n = number of links.

2.4.4 Types of Instantaneous Centres

The instantaneous centres for a mechanism are of the following types:

1. Fixed instantaneous centres.
2. Permanent instantaneous centres.
3. Neither fixed nor permanent instantaneous centres.

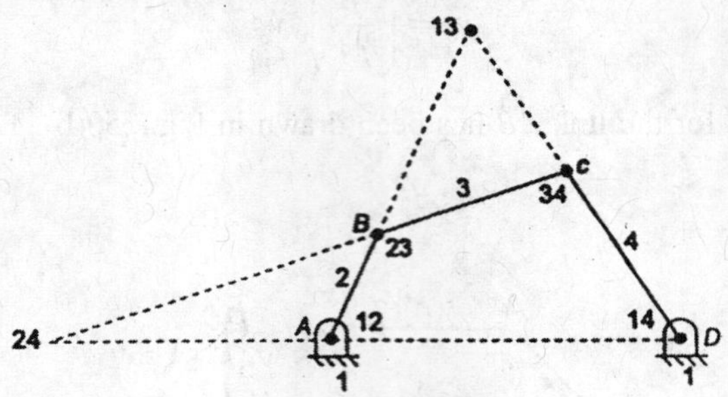

Fig.2.51 Instantaneous centres in a four-bar mechanism

Consider a four-bar mechanism shown in Fig.2.51. For this mechanism, $n = 4$. Hence

$$N = 4(4 - 1)/2 = 6$$

The instantaneous centres are:

12, 13, 14, 23, 24, 34.

The instantaneous centres 12 and 14 remain at the same place for the configuration of the mechanism, and are therefore called fixed instantaneous centres. The instantaneous centres 23 and 34 move when the mechanism moves. But the joints are permanent, therefore, they are called permanent instantaneous centres. The instantaneous centres 13 and 24 vary with the configuration of the mechanism and are neither fixed nor permanent instantaneous centres.

2.4.5 Location of Instantaneous Centres

The following observations are quite helpful in locating the instantaneous centres:

1. For a pivoted or pin joint, the instantaneous centre for the two links lies on the centre of the pin (see Fig.2.52(a)).
2. In a pure rolling contact of the two links, the instantaneous centre lies at their point of contact (see Fig.2.52(b)). This is because the relative velocity between the two links at the point of contact is zero.
3. In a sliding motion, the instantaneous centre lies at infinity in a direction perpendicular to the path of motion of the slider. This is because the sliding motion is equivalent to a rotary motion of the links with radius of curvature equal to infinity (see Fig.2.52(c)). If the slider (link 2) moves on a curved surface (link 1), then the instantaneous centre lies at the centre of curvature of the curved surface (see Fig.2.52(d) and (e)).

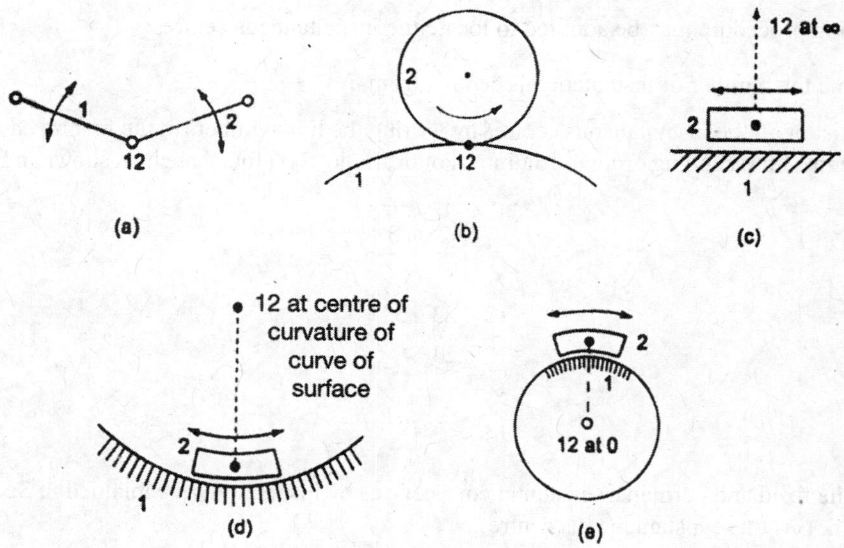

Fig.2.52 Locating instantaneous centres

2.4.6 Arnold–Kennedy Theorem

This theorem states that if three-plane bodies have relative motion among themselves their three instantaneous centres must lie on a straight line.

Consider three rigid links 1, 2, and 3; 1 being a fixed link. 12 and 13 are the instantaneous centres of links 1, 2 and 1, 3 respectively. Let 23 be the instantaneous centre of links 2, 3, lying outside the line joining 12 and 13, as shown in Fig.2.53. The links 2 and 3 are moving relative to link 1. Therefore, the motion of their instantaneous centre 23 is to be the same whether it is considered in body 2 or 3. If the point 23 is considered on link 2, then its velocity v_2 is perpendicular to the line joining 12 and 23. If the point 23 lies on link 3, then its velocity v_3 must be perpendicular to the line joining 13 and 23. The velocities v_2 and v_3 of point 23 are in different directions, which is not possible. The velocities v_2 and v_3 of instantaneous centre 23 will be equal only if it lies on the line joining 12 and 13. Hence all the three instantaneous centres 12, 13 and 23 must lie on a straight line.

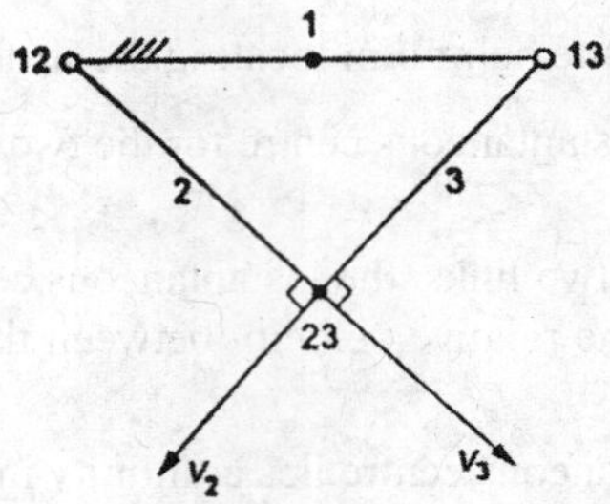

Fig.2.53 Proving three-centres theorem

2.4.7 Method of Locating Instantaneous Centres

The following procedure may be adopted to locate the instantaneous centres:

1. Determine the number of instantaneous centres from $N = \dfrac{n(n-1)}{2}$.
2. Make a list of all the instantaneous centres by writing the link numbers in the first row and instantaneous centres in ascending order in columns. For example, for a four-bar chain shown in Fig.2.54(a), we have

1	2	3	4
✓	✓	✓	
12	23	34	
(13)	(24)		
✓			
14			

3. Locate the fixed and permanent instantaneous centres by inspection, as explained in Section 2.4.5. Tick mark (✓) these instantaneous centres.
4. Locate the remaining neither fixed nor permanent instantaneous centres (circled) by using Arnold–Kennedy's theorem. This is done by a circle diagram, as shown in Fig.2.54(b). Mark points on a circle equal to the number of links in the mechanism. Join the points by solid lines for which instantaneous centres are known by inspection. Now join the points forming the other instantaneous centres by dotted lines. The instantaneous centre shall lie at the intersection of the lines joining the instantaneous centres of the two adjacent triangles of the dotted line. For example, in Fig.2.54(b), the centre 13 is located at the intersection of lines (produced) joining the instantaneous centres 12, 23, and 14, 34. Similarly, the centre 24 is located at the intersection of the lines (produced) joining the centres 23, 34 and 12, 14.

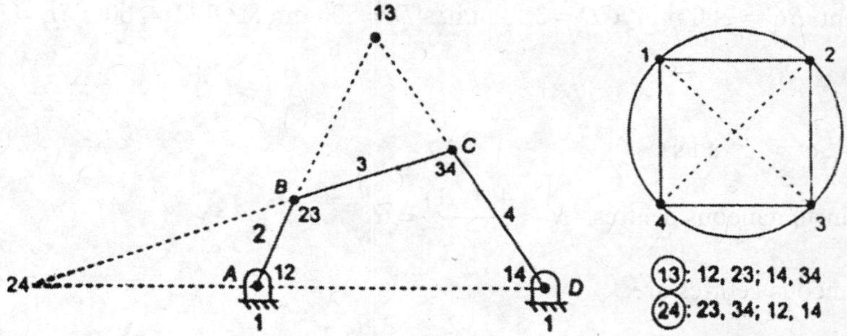

(a) Locating instantaneous centres (b) Circle diagram

Fig.2.54 Four-bar mechanism

2.4.8 Determination of Angular Velocity of a Link

The angular velocities of two links vary inversely as the distances from their common instantaneous centre to their respective centres of rotation relative to the frame. For example, for the four-bar mechanism shown in Fig.2.54(a), if ω_2 is the angular velocity of link 2, then angular velocity ω_4 of link 4 will be given by the following relationship:

$$\frac{\omega_4}{\omega_2} = \frac{(24-12)}{(24-14)}$$

If the respective centres of rotation are on the same side of the common instantaneous centre, then the direction of angular velocities will be same. However, if the respective centres of rotation are on opposite sides, then the direction of angular velocities will be opposite.

Similarly,
$$\frac{\omega_3}{\omega_2} = \frac{(23-12)}{(23-13)}$$

Example 2.37

In the four-bar mechanism shown in Fig.2.55(a), link 2 is rotating at angular velocity of 15 rad/s $c\omega$. Locate all the instantaneous centres of the mechanism and find (a) the angular speeds of links 3 and 4, (b) the linear velocities of links 3 and 4, and (c) the linear velocities of points E and F.

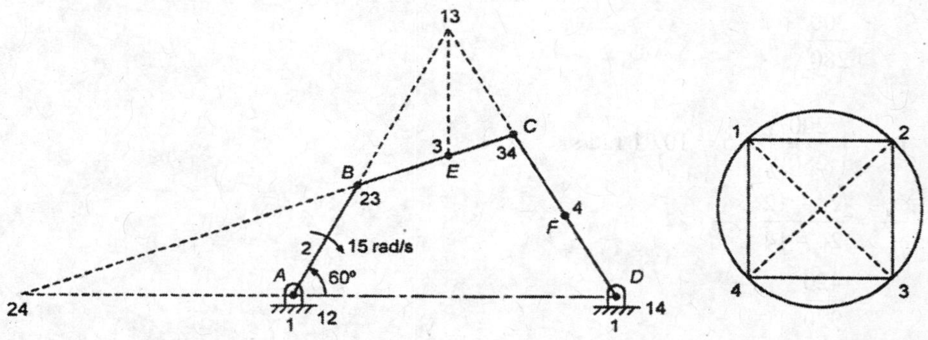

(a) Locating instantanous centres (b) Circle diagram

Fig.2.55 Four-bar mechanism

$AB = 200$ mm, $BC = 250$ mm, $CD = 300$ mm, $AD = 500$ mm, $\angle BAD = 60°$, $BE = FD = 150$ mm.

Solution

Given: $n = 4$, $\omega = 15$ rad/s

Number of instantaneous centres, $N = \frac{4(4-1)}{2} = 6$

The instantaneous centres are:

1	2	3	4
✓	✓	✓	
12	23	34	
(13)	(24)		
✓			
14			

1. Draw the configuration diagram of the four-bar mechanism to a convenient scale, as shown in Fig.2.55(a).
2. Locate the fixed and permanent instantaneous centres 12, 14, 23, and 34.
3. Locate the centres 13 and 24 by using Arnold–Kennedy's theorem of three centres from Fig.2.55(b).
   (13): 12, 23; 14, 34
   (24): 12, 14; 23, 34
4. By measurement, we have
   23 – 13 = 280 mm
   24 – 12 = 420 mm
   24 – 14 = 920 mm
   13 – 34 = 190 mm
   13 – E = 200 mm

(a) $$\frac{\omega_3}{\omega_2} = \frac{23-12}{23-13} = \frac{200}{280}$$

or $$\omega_3 = \left(\frac{200}{280}\right)\times 15 = 10.71 \text{ rad/s}$$

$$\frac{\omega_4}{\omega_2} = \frac{24-12}{24-14} = \frac{420}{920}$$

or $$\omega_4 = 420\times\frac{15}{920} = 6.848 \text{ rad/s}$$

(b) $v_b = \omega_2 \cdot AB = 15 \times 0.2 = 3 \text{ m/s}$

$$v_c = v_b \left[\frac{13-34}{13-23}\right]$$

$$= 3 \times \frac{190}{280} = 2.04 \text{ m/s}$$

$$v_{bc} = v_b \left[\frac{BC}{13-23}\right]$$

$$= 3 \times \frac{250}{280} = 2.678 \text{ m/s}$$

(c) $v_e = \omega_3 \cdot (13 - E)$

$$= 10.71 \times 0.2 = 2.142 \text{ m/s}$$

$$v_f = v_c \left(\frac{FD}{CD}\right)$$

$$= 2.04 \times \frac{150}{300} = 1.02 \text{ m/s}$$

Example 2.38

Locate the instantaneous centres of the slider crank mechanism shown in Fig.2.56(a). Find the velocity of the slider. $OA = 160$ mm, $AB = 470$ mm, and $OB = 600$ mm, $\omega_2 = 12$ rad/s cw.

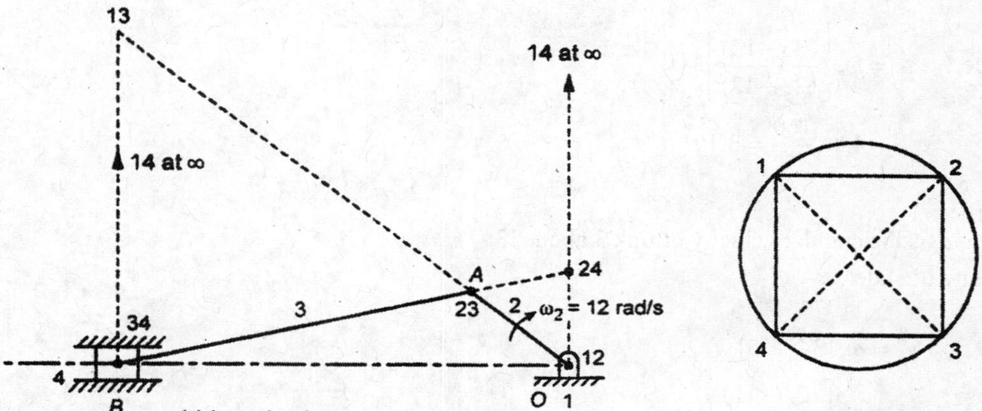

(a) Locating instantaneous centres (b) Circle diagram

Fig.2.56 Slides-crank mechanism

■ Solution

Given: $n = 4$, $\omega_2 = 12$ rad/s cw, lengths of links.

Number of instantaneous centres, $N = \dfrac{4(4-1)}{2} = 6$

1	2	3	4
✓	✓	✓	
12	23	34	
⑬	㉔		
✓			
14			

1. Draw the configuration diagram to a convenient scale as shown in Fig.2.56(a).
2. Locate fixed centres 12 and 34.
3. Locate permanent centre 23. Instantaneous centre 14 is at infinity.
4. Locate neither fixed nor permanent centres 13 and 24 by using Arnold–Kennedy's three centres theorem.

   As shown in Fig.2.56(b), we have

   ⑬: 12, 23; 14, 34

   ㉔: 12, 14; 23, 34
5. By measurements, we have

$$13 - 23 = 550 \text{ mm}$$

$$13 - 34 = 390 \text{ mm}$$

$$v_a = \omega_2 \cdot OA = 12 \times 0.16 = 1.92 \text{ m/s}$$
$$= \omega_3 \cdot (23 - 13)$$

Hence, $$\omega_3 = \omega_2 \left[\frac{(23-12)}{(23-13)}\right]$$

$$= 12 \times \frac{160}{550} = 3.191 \text{ m/s}$$

When ω_3 is the angular velocity of link 3 about 13.

Velocity of slider,

$$v_b = \omega_3 (34 - 13)$$

$$= \omega_2 \left[\frac{(23-12)}{(23-13)}\right](34-13)$$

$$= \omega_2 \cdot OA \left[\frac{(34-13)}{(23-13)}\right]$$

$$= 12 \times 0.16 \times \frac{390}{550} = 1.36 \text{ m/s}$$

Example 2.39

Locate the instantaneous centres of the mechanisms shown in Fig.2.57(a).

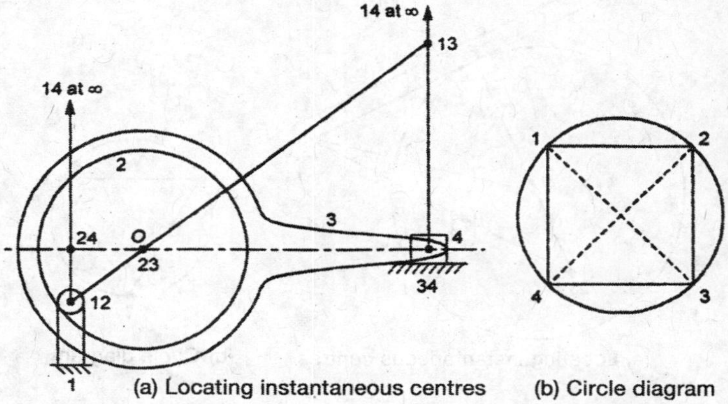

(a) Locating instantaneous centres (b) Circle diagram

Fig.2.57 Four-bar mechanism

■ Solution

(a) Here $n = 4$

Number of instantaneous centres, $N = 4 \times \frac{3}{2} = 6$

1	2	3	4
✓	✓	✓	✓
12	23	34	
(13)	(24)		
✓			
14			

Instantaneous centres 12, 14, 23, and 34 are located by inspection and centres 13, 24 are located by using Arnold–Kennedy's theorem as follows:

(13): 12, 23; 14, 34

(24): 23, 34; 12, 14

Example 2.40

Locate the instantaneous centres of the mechanism shown in Fig.2.58.

■ Solution

Here $n = 3$; Number of instantaneous centres $= \frac{3 \times 2}{2} = 3$

1	2	3
✓		
12	(23)	
✓		
13		

Instantaneous centres 12, 13 are located by inspection. (23) : 12, 13

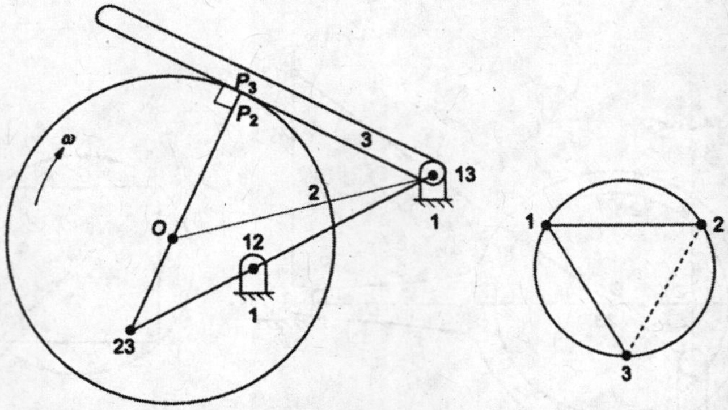

(a) Locating instantaneous centres (b) Circle diagram

Fig.2.58 Three-bar mechanism

Example 2.41

In the toggle mechanism, shown in Fig.2.59, crank O_1A rotates at 30 rpm clockwise. O_1A = 40 mm, AB = 140 mm, BC = 100 mm, BD = 80 mm, and DE = 80 mm. Neglecting friction and inertia effects, calculate the torque required to overcome a resistance of 500 N at D. Use the instantaneous centre method.

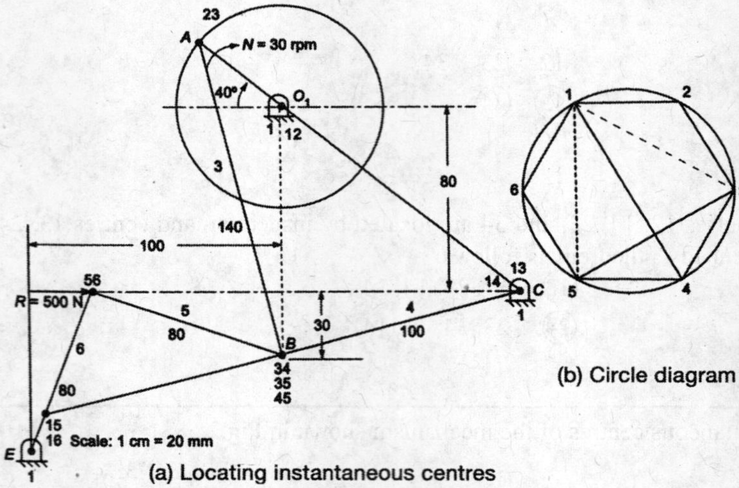

(a) Locating instantaneous centres (b) Circle diagram

Fig.2.59 Toggle mechanism

■ Solution

Here n = 6; Number of instantaneous centres $= \dfrac{n(n-1)}{2} = \dfrac{6\times 5}{2} = 15$

1	2	3	4	5	6
✓	✓	✓	✓	✓	
12	23	34	45	56	
		✓	✓		
(13)	(24)	35	46		
✓	✓	✓			
14	25	36			
(15)	26				
✓					
16					

(13): 12, 23; 14, 34

(15): 16, 56; 14, 45

(24): 23, 34; 12, 14

$$\omega_2 = 2\pi \times \frac{30}{60} = 2.14 \text{ rad/s}; \qquad v_a = \omega_2 \times O_1A = 2.14 \times 40 = 125.7 \text{ mm/s}$$

$$\omega_3 = \frac{v_a}{(13-23)} = \frac{125.7}{(8.4 \times 20)} = 0.748 \text{ rad/s}; \qquad v_b = \omega_3\,(13-34) = 0.748 \times 5.3 \times 20 = 79.29 \text{ mm/s}$$

$$\omega_5 = \frac{v_b}{(15-35)} = \frac{79.29}{(4.8 \times 20)} = 0.826 \text{ rad/s}; \qquad v_d = \omega_5\,(15-56) = 0.826 \times 2.1 \times 20 = 51.21 \text{ mm/s}$$

$$\omega_6 = \frac{v_d}{(16-56)} = \frac{51.21}{80} = 0.64 \text{ rad/s}$$

$$\text{Torque} \times \text{Angular velocity} = \text{Resistance force} \times v_d$$

$$T \times 2.14 = 500 \times 51.21 \times 10^{-3}; \; T = 8.15 \text{ Nm}$$

Example 2.42

Determine all the instantaneous centres of the double slider–crank mechanism shown in Fig.2.60(a).

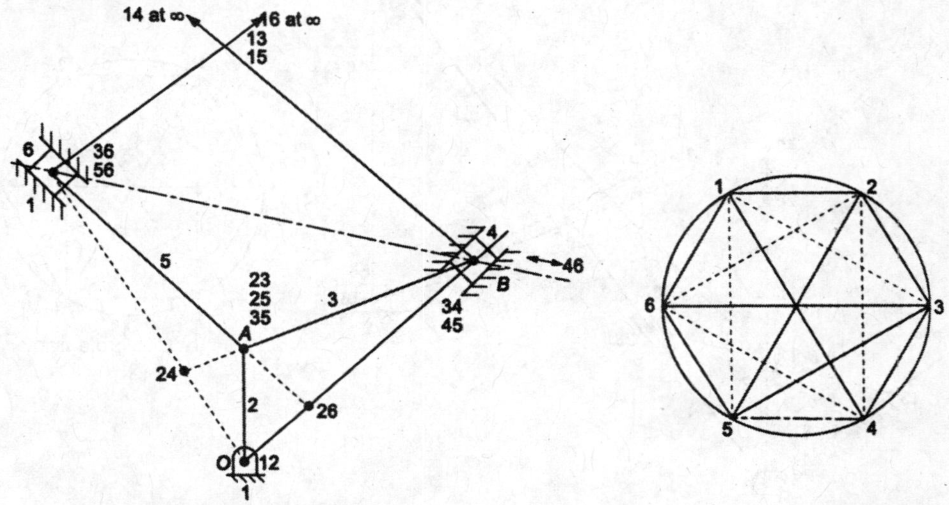

(a) Locating instantaneous centres

(b) Circle diagram

Fig.2.60 Double slider-crank mechanism

■ Solution

Here $n = 6$; so that $N = \dfrac{6(6-1)}{2} = 15$

1	2	3	4	5	6
✓	✓	✓		✓	
12	23	34	(45)	56	
(13)	(24)	(35)	(46)		
✓	✓				
14	25	(36)			
(15)	(26)				
✓					
16					

As shown in Fig.2.60(b), we have

(13): 12, 23; 14, 34

(15): 16, 56; 12, 25

(24): 23, 34; 12, 14

(26): 12, 16; 25, 56

(36): 13, 16; 35, 56

(45): 14, 15; 34, 35

(46): 45, 56; 34, 36

Example 2.43

Locate all the instantaneous centres of the Whitworth mechanism shown in Fig.2.61(a).

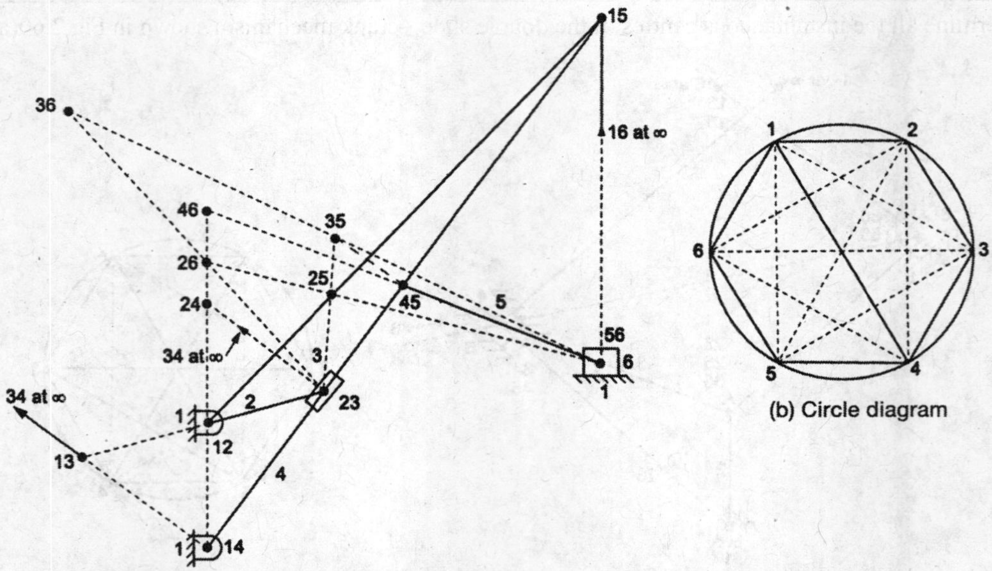

(a) Locating instantaneous centres

(b) Circle diagram

Fig.2.61 Whitworth mechanism

■ Solution

$$N = \frac{6(6-1)}{2} = 15$$

```
1   2   3   4   5   6
✓   ✓   ✓   ✓   ✓

12  23  34  45  56
(13) (24) (35) (46)
✓

14  (25) (36)
(15) (26)
✓

16
```

As shown in Fig.2.61(b), we have

(13): 12, 23; 14, 34
(15): 14, 45; 16, 56
(24): 23, 34; 12, 14
(25): 24, 45; 12, 15
(26): 25, 56; 12, 16
(35): 34, 45; 23, 25
(36): 35, 56; 23, 26
(46): 45, 46; 14, 16

Example 2.44

Determine all the instantaneous centres of the mechanism shown in Fig.2.62(a). Calculate the velocities of the slider E and the joints B and D when the crank OA is rotating at 120 rpm. Also find ω_{AB}, ω_{BD}, and ω_{DE}. $OA = 200$ mm, $AB = 500$ mm, $BC = 300$ mm, $BD = 600$ mm, and $DE = 450$ mm.

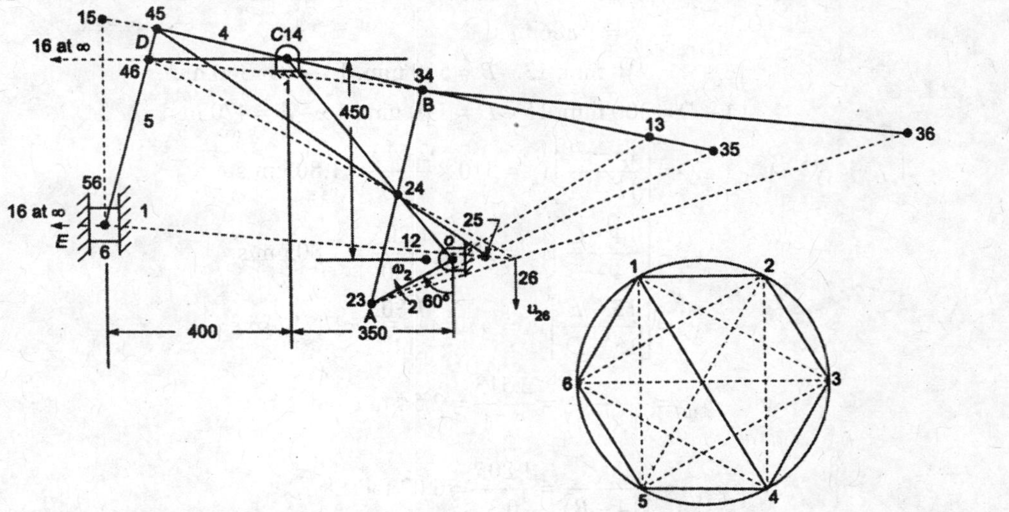

(a) Locating instantaneous centres (b) Circle diagram

Fig.2.62 Mechanism for Example 2.44

■ Solution

$$\omega_2 = 2\pi \times \frac{120}{60} = 12.57 \text{ rad/s}$$

$$v_a = \omega_2 \cdot OA = 12.57 \times 0.2 = 2.513 \text{ m/s}$$

$$N = \frac{6(6-1)}{2} = 15$$

1	2	3	4	5	6
✓	✓	✓	✓	✓	
12	23	34	45	56	
⑬	㉔	㉟	㊻		
✓					
14	㉕	㊱			
⑮	㉖				
✓					
16					

As shown in Fig.2.62(b), we have

⑬: 12, 23; 14, 34
⑮: 14, 45; 16, 56
㉔: 23, 34; 12, 14
㉕: 24, 45; 12, 15
㉖: 25, 56; 12, 16
㉟: 34, 45; 23, 25
㊱: 35, 56; 23, 26
㊻: 45, 46; 14, 16

$$v_{26} = \omega_2(12 - 26) = 12.57 \times 0.15 = 1.886 \text{ m/s}$$

$$v_e = v_{26} = 1.886 \text{ m/s}$$

$$13 - A = 710 \text{ mm},\ 13 - B = 510 \text{ mm},\ 14 - B = 300 \text{ mm},$$
$$14 - D = 300 \text{ mm},\ 15 - D = 110 \text{ mm},\ 15 - E = 470 \text{ mm}$$

$$v_b = \left[\frac{13 - B}{13 - A}\right] v_a = 510 \times \frac{2.513}{710} = 1.805 \text{ m/s}$$

$$v_d = \left[\frac{14 - C}{14 - B}\right] v_b = 300 \times \frac{1.805}{300} = 1.805 \text{ m/s}$$

$$v_e = \left[\frac{15 - E}{15 - D}\right] v_d = 470 \times \frac{1.805}{110} = 7.712 \text{ m/s}$$

$$\omega_{AB} = \frac{v_a}{(13 - A)} = \frac{2.513}{0.71} = 2.54 \text{ rad/s}$$

$$\omega_{BD} = \frac{v_b}{(14 - B)} = \frac{1.805}{0.3} = 6.02 \text{ rad/s}$$

$$\omega_{DE} = \frac{v_d}{(15 - D)} = \frac{1.805}{0.11} = 16.41 \text{ rad/s}$$

Example 2.45

A wrapping mechanism is shown in Fig.2.63(a). The crank O_1A rotates at a uniform speed of 1200 rpm. Determine the velocity of point E on the bell crank lever.

$$O_1A = 300 \text{ mm}, AC = 650 \text{ mm}, BC = 100 \text{ mm}, O_3C = 400 \text{ mm},$$
$$O_2E = 400 \text{ mm}, O_2D = 200 \text{ mm, and } BD = 200 \text{ mm}.$$

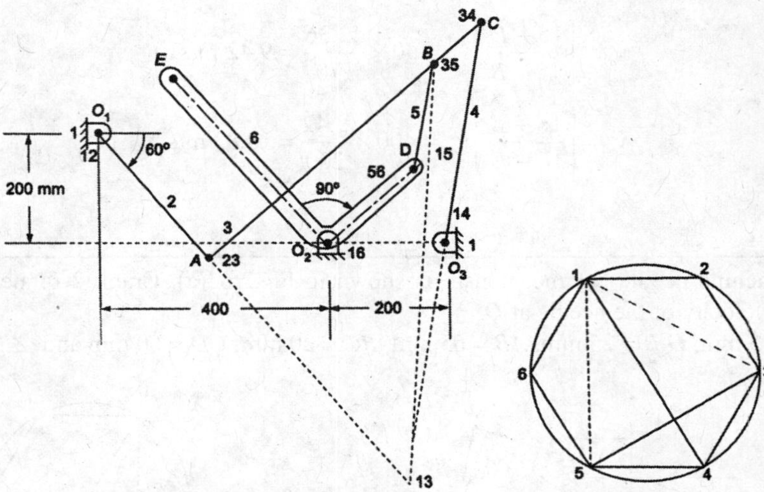

(a) Locating instantaneous centres

(b) Circle diagram

Fig.2.63 Wrapping mechanism

■ Solution

$$\omega = 2\pi \times \frac{1200}{60} = 125.7 \text{ rad/s}$$

$$v_a = 125.7 \times 0.3 = 37.71 \text{ m/s}$$

$$N = \frac{6(6-1)}{2} = 15$$

As shown in Fig.2.63(b), we have

1	2	3	4	5	6
✓	✓	✓	✓	✓	
12	23	34	45	56	
		✓			
(13)	24	35	46		
✓					
14	25	36			
(15)	26				
✓					
16					

(13): 12, 23; 14, 34

(15): 14, 45; 16, 56

13 – A = 530 mm, 13 – B = 750 mm, 15 – B = 170 mm,
15 – D = 30 mm, 16 – D = 200 mm, 16 – E = 400 mm

$$v_b = \left[\frac{13-B}{13-A}\right] v_a = 750 \times \frac{37.71}{530} = 52.36 \text{ m/s}$$

$$v_d = \left[\frac{15-D}{15-B}\right] v_b = 30 \times \frac{52.36}{170} = 9.42 \text{ m/s}$$

$$v_e = \left[\frac{16-E}{16-D}\right] v_d = 400 \times \frac{9.42}{200} = 18.83 \text{ m/s}$$

Example 2.46

The sewing machine needle bar mechanism is shown in Fig.2.64(a). Crank 2 rotates at 450 rpm. Determine the velocity of the needle at D.

O_1A = 15 mm, O_2B = 25 mm, AB = 65 mm, BC = 20 mm, CD = 60 mm and $\angle O_2BC = 90°$.

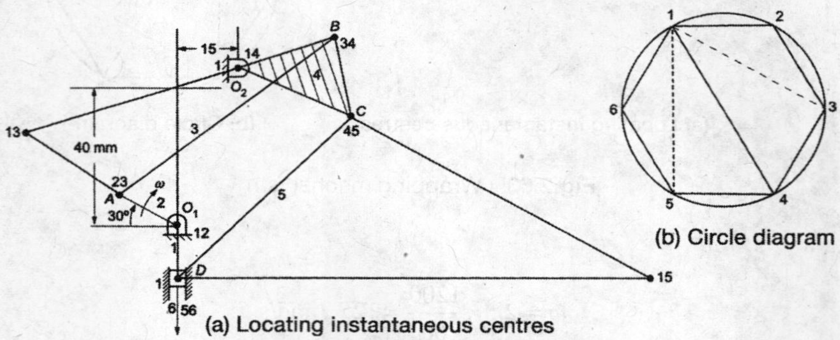

(a) Locating instantaneous centres

(b) Circle diagram

Fig.2.64 Sewing machine needle bar mechanism

■ Solution

Here n = 6 and N = 15

As shown in Fig.2.64(b), we have

1	2	3	4	5	6
✓	✓	✓	✓	✓	
12	23	34	45	56	
(13)	24	35	46		
✓					
14	25	36			
(15)	26				
✓					
16					

(13): 12, 23; 14, 34
(15): 14, 45; 16, 56

13 – A = 32 mm, 13 – B = 84 mm, 14 – B = 25 mm,
14 – C = 30 mm, 15 – C = 98 mm, 15 – D = 132 mm

$$\omega = 2\pi \times \frac{450}{60} = 47.124 \text{ rad/s}$$

$$v_a = 47.124 \times 0.015 = 0.71 \text{ m/s}$$

$$v_b = \left[\frac{13 - B}{13 - A}\right] v_a = 84 \times \frac{0.71}{32} = 1.864 \text{ m/s}$$

$$v_c = \left[\frac{14 - C}{14 - B}\right] v_b = 30 \times \frac{1.864}{25} = 2.236 \text{ m/s}$$

$$v_d = \left[\frac{15 - D}{15 - C}\right] v_c = 132 \times \frac{2.236}{98} = 3.01 \text{ m/s}$$

Example 2.47

Figure 2.65 shows the Whitworth mechanism. The crank O_1A rotates at 120 rpm. O_1O_2 = 100 mm, O_1A = 200, O_2C = 150 mm, and CD = 500 mm. Locate all the instantaneous centres and find the velocity of ram D.

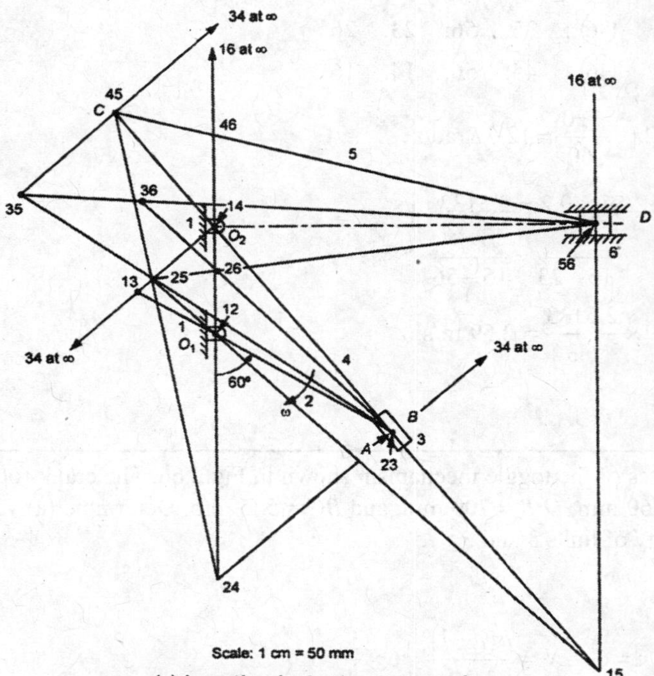

(a) Locating instantaneous centres

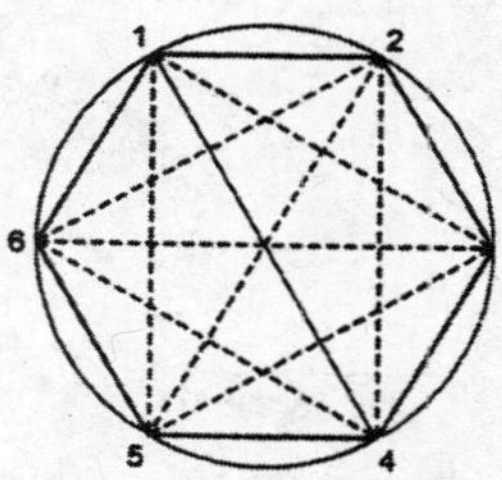

(b) Circle diagram

Fig.2.65 Whitworth mechanism

■ Solution

$$n = 6, \quad N = \frac{n(n-1)}{2} = 6 \times \frac{5}{2} = 15$$

Locate the centres as shown in Fig.2.65(a) and using circle diagram shown in Fig.2.65(b).

1	2	3	4	5	6
✓	✓	✓	✓	✓	
12	23	34	45	56	
(13)	(24)	(35)	(46)		
14	(25)	(36)			
(15)	(26)				
✓					
16					

(13) : 12, 23; 14, 34

(15) : 14, 45; 16, 56

(24) : 23, 34; 12, 14

(25) : 24, 45; 12, 15

(26) : 25, 56; 12, 16

(35) : 34, 45; 23, 25

(36) : 35, 56; 23, 26

(46) : 45, 56; 14, 16

$$\omega = 2\pi \times \frac{120}{60} = 12.57 \text{ rad/s}$$

$$v_a = 12.57 \times 0.2 = 2.5133 \text{ m/s}$$

$$\frac{v_a}{15-23} = \frac{v_d}{15-56}$$

$$v_d = 9 \times \frac{2.5133}{6.3} = 2.59 \text{ m/s.}$$

Example 2.48

Locate all the instantaneous centres of the toggle mechanism shown in Fig.2.66. The crank rotates at 240 rpm. $O_1A = 200$ mm, $AB = 360$ mm, $O_2B = 200$ mm, and $BC = 525$ mm. Determine (a) velocity of slider C and (b) angular velocity of links 3 and 5.

■ Solution

$$n = 6, \quad N = \frac{n(n-1)}{2} = 6 \times \frac{5}{2} = 15$$

Locate the centres as shown in Fig.2.66.

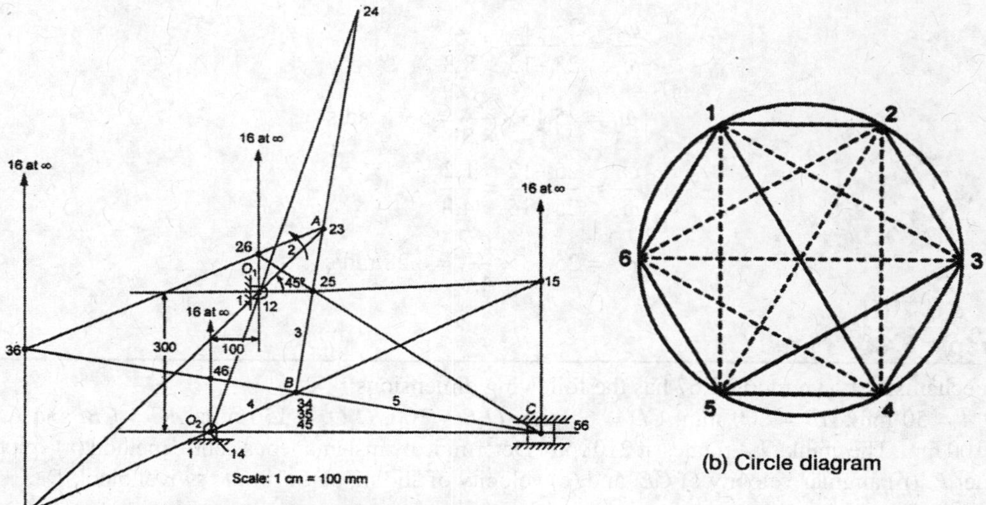

(a) Locating instantaneous centres

(b) Circle diagram

Fig.2.66 Toggle mechanism

1	2	3	4	5	6
✓	✓	✓	✓	✓	
12	23	34	45	56	
⑬	㉔	㉟	㊻		
14	㉕	㊱			
⑮	㉖				
✓					
16					

⑬: 12, 23; 14, 34

⑮: 14, 45; 16, 56

㉔: 23, 34; 12, 14

㉕: 24, 45; 12, 15

㉖: 25, 56; 12, 16

㉟: 34, 45; 23, 25

㊱: 35, 56; 23, 26

㊻: 45, 56; 14, 16

$$\omega_2 = 2\pi \times \frac{240}{60} = 25.13 \text{ rad/s}$$

$$v_a = \omega_2 \times O_1A = 25.13 \times 0.2 = 5.026 \text{ m/s}$$

$$v_c = v_{26} = \omega_2(12 - 26) = 25.13 \times 0.8 \times 100 \times 10^{-3} = 2.01 \text{ m/s}$$

$$\frac{\omega_3}{\omega_2} = \frac{23-12}{23-13} = \frac{2}{8.8}$$

$$\omega_3 = 25.13 \times \frac{2}{8.8} = 5.71 \text{ rad/s}$$

$$\frac{\omega_5}{\omega_2} = \frac{25-12}{25-15} = \frac{1.2}{4.8}$$

$$\omega_5 = 25.13 \times \frac{1.2}{4.8} = 6.28 \text{ rad/s}$$

Example 2.49

The mechanism shown in Fig.2.67 has the following dimensions:

O_1A = 50 mm, AB = 200 mm, CD = 60 mm, O_2B = 100, O_1O_2 = 150 mm, $AC = CB$, and $CE = EF$ = 100 mm. The crank O_1A rotates at 210 rpm. Determine by instantaneous centre method (a) velocity of slider F, (b) angular velocity of CE, and (c) velocity of sliding of CE in the swivel block D.

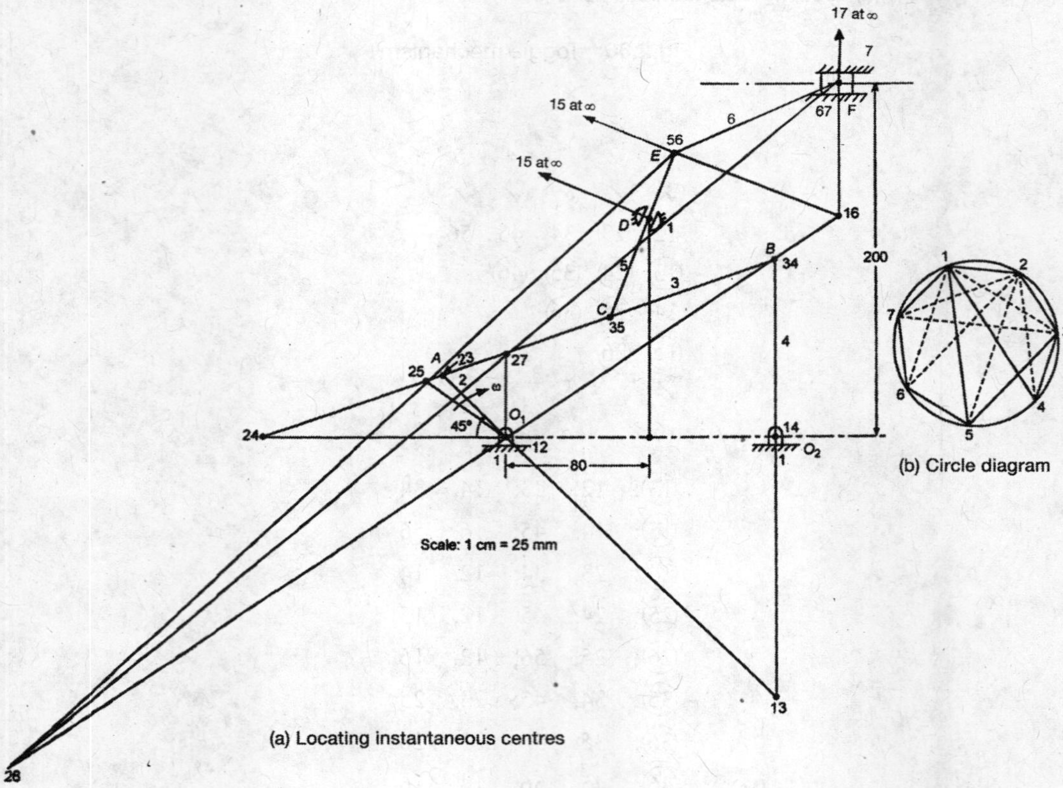

Fig.2.67 Diagram for Example 2.49

■ **Solution**

$$n = 7, \quad N = \frac{n(n-1)}{2} = 7 \times \frac{6}{2} = 21$$

Locate the centres as shown in Fig.2.67, which are required for the solution.

1	2	3	4	5	6	7
✓	✓	✓	✓	✓	✓	
12	23	34	45	56	67	
(13)	(24)	35	46	57		
✓						
14	(25)	36	47			
✓						
15	(26)	37				
(16)	(27)					
17						

(13): 12, 23; 14, 34

(16): 17, 67; 15, 56

(24): 23, 34; 12, 14

(25): 25, 56; 12, 16

(27): 12, 17, 26, 67

$$\omega_2 = 2\pi \times \frac{210}{60} = 22 \text{ rad/s}$$

$$v_f = v_{27} = \omega_2(12 - 27) = 22 \times 1.8 \times 25 \times 10^{-3} = 0.99 \text{ mm/s}$$

$$v_d = v_{25} = \omega_2\,(12 - 25) = 22 \times 2.2 \times 25 \times 10^{-3} = 1.21 \text{ mm/s}$$

$$\frac{\omega_5}{\omega_2} = \frac{25-12}{25-35} = \frac{2.2}{4.4} = \frac{1}{2}$$

$$\omega_5 = 22 \times \frac{1}{2} = 11 \text{ rad/s}$$

Example 2.50

A mechanism shown in Fig.2.68 has the following dimensions:

OA = 200 mm, AB = 1500 mm, BC = 600 mm, CD = 500 mm and BE = 40 mm. Locate all the instantaneous centres.

If the crank OA rotates uniformly at 120 rpm clockwise, determine (a) the velocity of B, C and D, and (b) the angular velocity of links AB, BC, and CD.

■ Solution

Given: N = 120 rpm, OA = 200 mm = 0.2 m

Angular speed of OA, $\omega_{OA} = \frac{2\pi N}{60} = \frac{2\pi \times 120}{60} = 12.57$ rad/s

$$v_{OA} = \omega_{OA} \times OA = 12.57 \times 0.2 = 2.51 \text{ m/s}$$

(a) Number of instantaneous centres:

$$N = \frac{n(n-1)}{2}$$

Here $n = 6$, $\quad N = \frac{6 \times 5}{2} = 15$

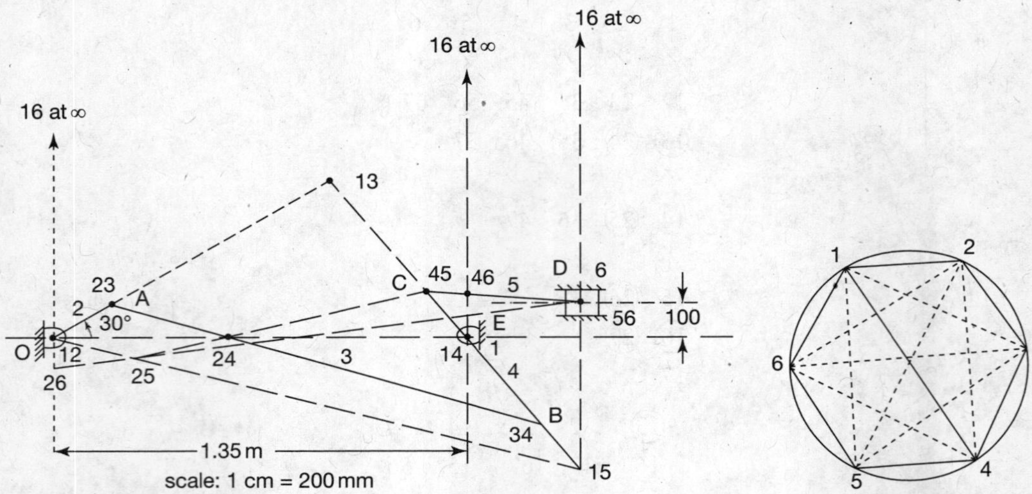

(a) Locating instantaneous centres (b) Circle diagram

Fig.2.68 Diagram for Example 2.50

1	2	3	4	5	6
✓	✓	✓	✓	✓	
12	23	34	45	56	
(13)	(24)	(35)	(46)		
14	(25)	(36)			
(15)	(26)				
✓					
16					

(13): 12, 23; 14, 34
(15): 16, 56; 14, 45
(24): 23, 34; 12, 14
(25): 24, 45; 12, 15
(26): 12, 16; 25, 56
(35): 34, 45; 23, 25
(36): 13, 16; 23, 26
(46): 45, 56; 14, 16

$$\frac{v_A}{v_B} = \frac{13-23}{13-34} = \frac{43}{53}$$

$$v_B = \frac{53}{43} \times 2.51 = 3.1 \text{ m/s}$$

$$\frac{v_c}{v_B} = \frac{14-45}{14-34} = \frac{10}{20} = 0.5$$

$$v_c = 3.1 \times 0.5 = 1.55 \text{ m/s}$$

$$\frac{v_D}{v_c} = \frac{15-56}{15-45} = \frac{29}{40}$$

$$v_D = 1.55 \times \frac{29}{40} = 1.12 \text{ m/s}$$

$$\omega_{AB} = \frac{v_A}{13-23} = \frac{2.51}{43 \times 20 \times 10^{-3}} = 2.92 \text{ rad/s}$$

$$\omega_{BC} = \frac{v_B}{14-34} = \frac{3.1}{20 \times 20 \times 10^{-3}} = 7.75 \text{ rad/s}$$

$$\omega_{CD} = \frac{v_c}{15-45} = \frac{1.55}{40 \times 20 \times 10^{-3}} = 1.94 \text{ rad/s}$$

Example 2.51

For the two rolling wheels shown in Fig.2.69(a), determine the angular velocity of links 3 and 4 and v_{eb}.

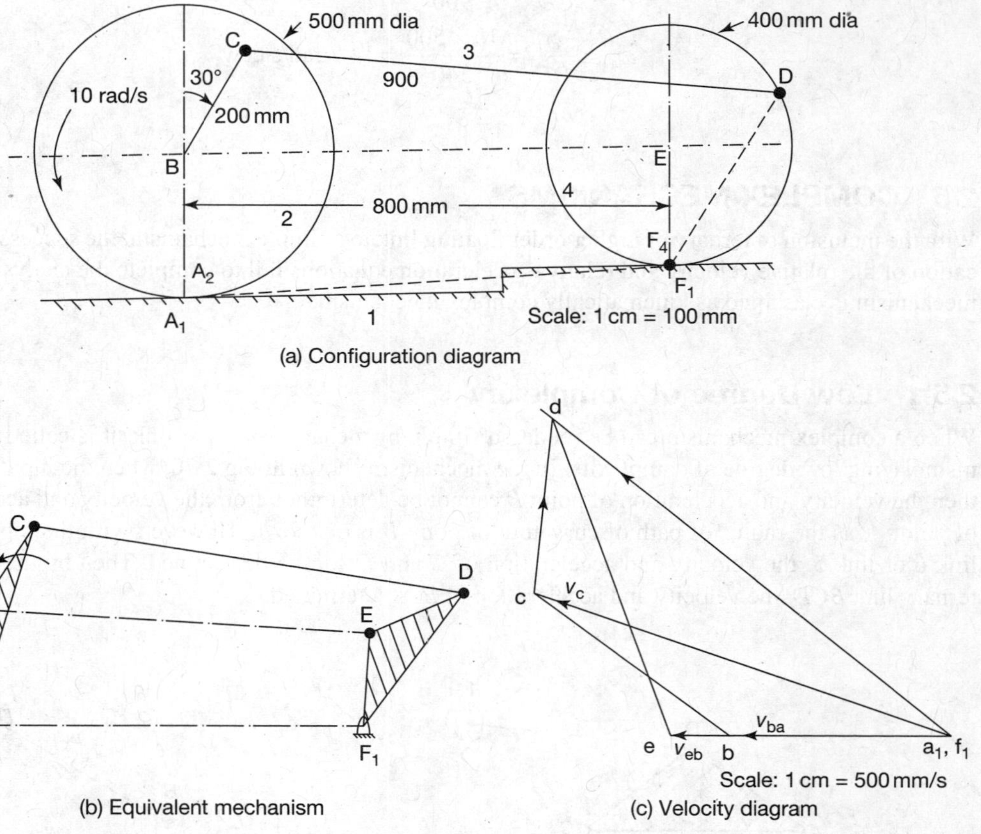

Fig.2.69 Two rolling wheels mechanism

■ Solution

Draw the equivalent mechanism as shown in Fig.2.69(b).

$$A_1C = 430 \text{ mm}, F_1D = 340 \text{ mm}, \omega = 10 \text{ rad/s}$$

$$v_c = \omega \times A_1C = 10 \times 430 = 4300 \text{ mm/s}$$

Draw the velocity diagram as shown in Fig.2.69(c).

$$v_c = a_1c \perp A_1C$$

$$cd \perp CD$$

$$f_1d \perp F_1D$$
$$cb \perp CB$$
$$a_1b \perp A_1B$$
$$ed \perp ED$$
$$f_1e \perp F_1E$$
$$v_{eb} = be = 1 \text{ cm} = 500 \text{ mm/s}$$
$$\omega_3 = \frac{v_{dc}}{CD} = \frac{3.4 \times 500}{900} = 1.9 \text{ rad/s ccw}$$
$$\omega_4 = \frac{v_{df1}}{DF_1} = \frac{10 \times 500}{340} = 14.7 \text{ rad/s ccw}$$

2.5 COMPLEX MECHANISMS

With the inclusion of ternary or higher order floating link to a simple mechanism, the successive application of the relative velocity and relative acceleration equations fail to complete the analysis. Such a mechanism is classified as kinematically complex mechanism.

2.5.1 Low Degree of Complexity

When a complex mechanism can be rendered simple by a change of input link, it is called a mechanism having low degree of complexity. In the mechanism shown in Fig.2.70, when the input link is 2, then the velocity and acceleration of point *B* cannot be determined from the velocity and acceleration of point *A,* as the radius of path of curvature of point *B* is unknown. However, with the input link as link 6 or link 5, the velocity and acceleration of *C* and *D* can be determined. Then by the image of ternary link *BCD,* the velocity and acceleration of *B* is determined.

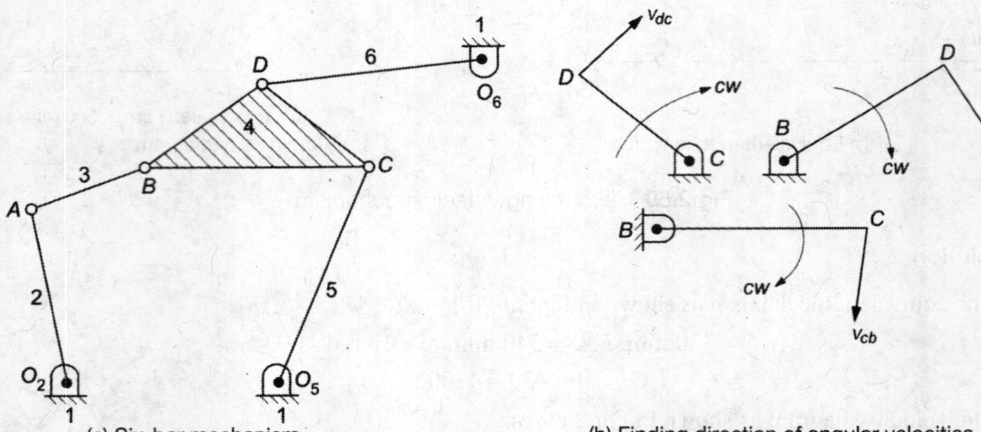

Fig.2.70 Complex mechanism having low degree of complexity

2.5.2 High Degree of Complexity

When a complex mechanism cannot be rendered simple by a change of input link, it is called a mechansim with high degree of complexity. In such a mechanism, the radii of the paths of curvature of two or more motion transfer points of a floating link are not known. The mechanisms shown in Figs. 2.71(a) and (b) have high degree of complexity.

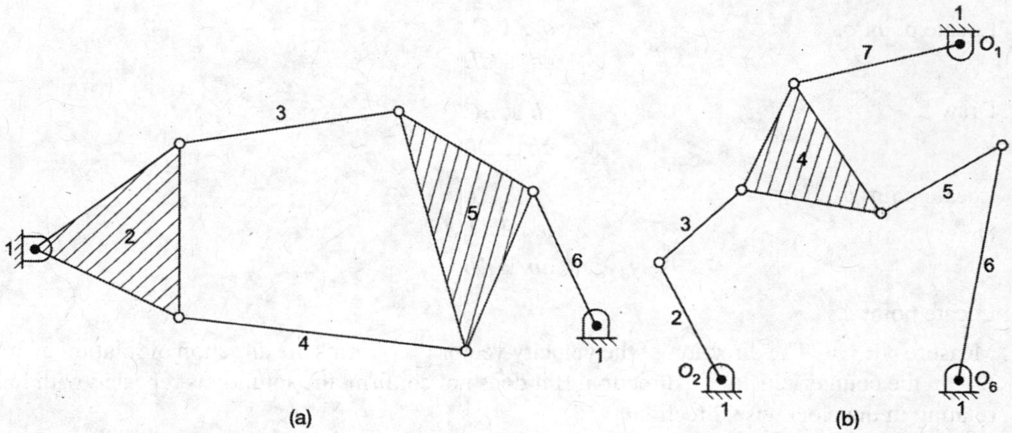

Fig.2.71 High degree of complexity mechanisms

Example 2.52

For the mechanism shown in Fig.2.72(a), determine ω_4 and ω_6.

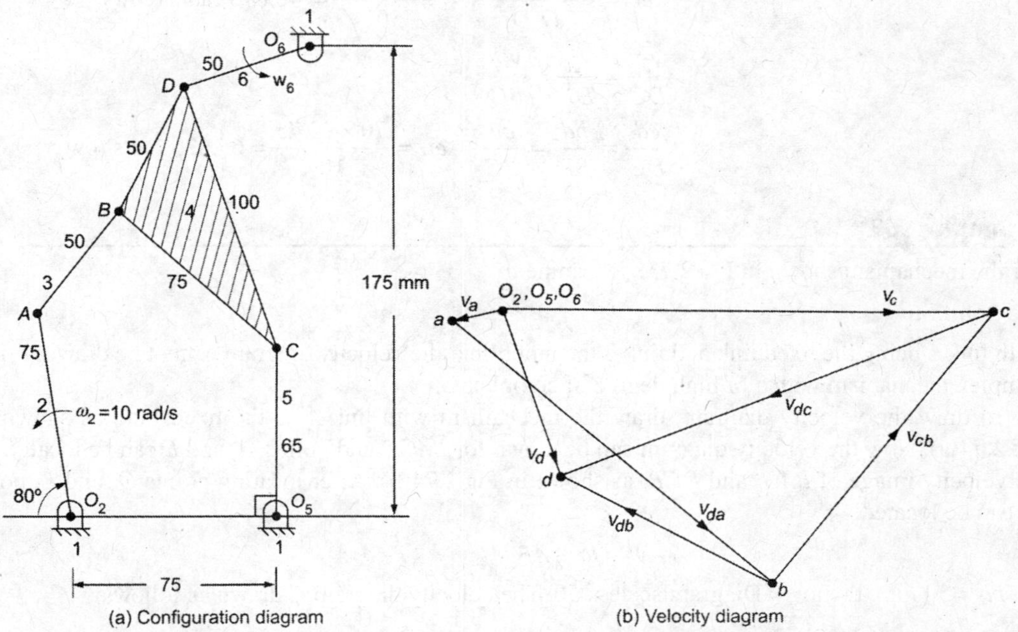

Fig.2.72 Low degree of complexity mechanism

■ **Solution**

With link 2 as the input link, the mechanism is complex. Make link 6 the input link. Then it becomes a simple mechanism.

Let $\omega_6 \cdot O_6D = 25\text{ mm} = v_d$.

(a) Draw $o_6d \perp O_6D = v_d = 25$ mm, as shown in Fig.2.58(b).

Locate point c.
$$o_5c \perp O_5C$$
$$cd \perp CD$$

(b) Draw
$$cb \perp BC$$
$$db \perp BD$$

Locate point b.

(c) Draw
$$o_2a \perp O_2A$$
$$ab \perp AB$$

Locate point a.

(d) Measure $o_2a = v_a$. The direction of the velocity vector v_a confirms the direction of rotation of crank O_2A in the counter-clockwise direction. If it does not confirm, the solution is repeated with link 6 rotating in the clockwise direction.

By measurement,
$$o_2a = 8\text{ mm}.$$
$$\omega_2 \cdot O_2A = 8 \times \text{scale}$$

or
$$\text{scale, 1 mm} = 10 \times \frac{75}{8} = 92.75\text{ mm/s}$$
$$\omega_6 = \frac{o_2d \times \text{scale}}{O_2D} = \frac{25 \times 92.75}{50} = 46.875\text{ rad/s (ccw)}$$
$$\frac{v_{dc}}{DC} = \frac{v_{db}}{DB} = \frac{v_{cb}}{BC}$$
$$\frac{cd}{DC} = \frac{bd}{BD} = \frac{bc}{BC}; \quad \omega_4 = \frac{70 \times 92.75}{10} = 65.625\text{ rad/s (cw)}$$

Example 2.53

For the mechanism shown in Fig.2.73, determine ω_6.

■ **Solution**

With link 1 being the fixed link and link 2 the input link, the velocity diagram cannot be drawn. It is a complex mechanism with a of high degree of complexity.

To draw the velocity diagram, draw the mechanism with link 4 as the fixed link, as shown in Fig.2.74(a). Now the velocity diagram can be drawn for $BACD$ and points O_2 and E can be located on the velocity image of BAO_2 and DCE, as shown in Fig.2.74(b). After locating points O_2 and E, point O_6 can be located.

$$v_a = \omega_{24} \cdot AB$$

Let $ba = 50$ mm, then v_a = Diagram scale × 50. The velocity diagram is drawn as follows:

(a) Draw $ab \perp AB = 50$ mm.

(b) Draw $ac \perp AC$ and $dc \perp CD$ to locate c.

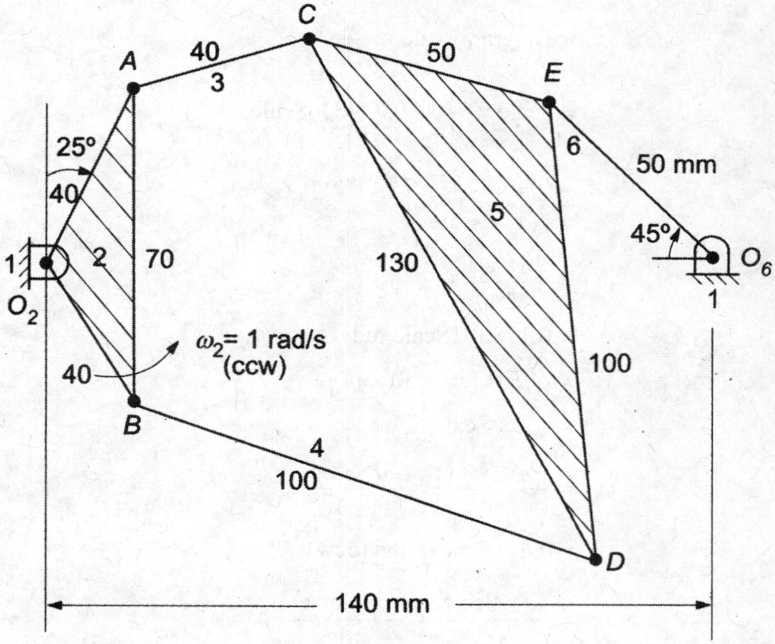

Fig.2.73 Complex mechanism

(c) Draw $ce \perp CE$ and $de \perp DE$ to locate *e*.

(d) Draw $ao_2 \perp AO_2$ and $bo_2 \perp BO_2$ to locate o_2.

(e) Draw $eo_6 \perp EO_6$ and $o_2o_6 \perp O_2O_6$ to locate o_6.

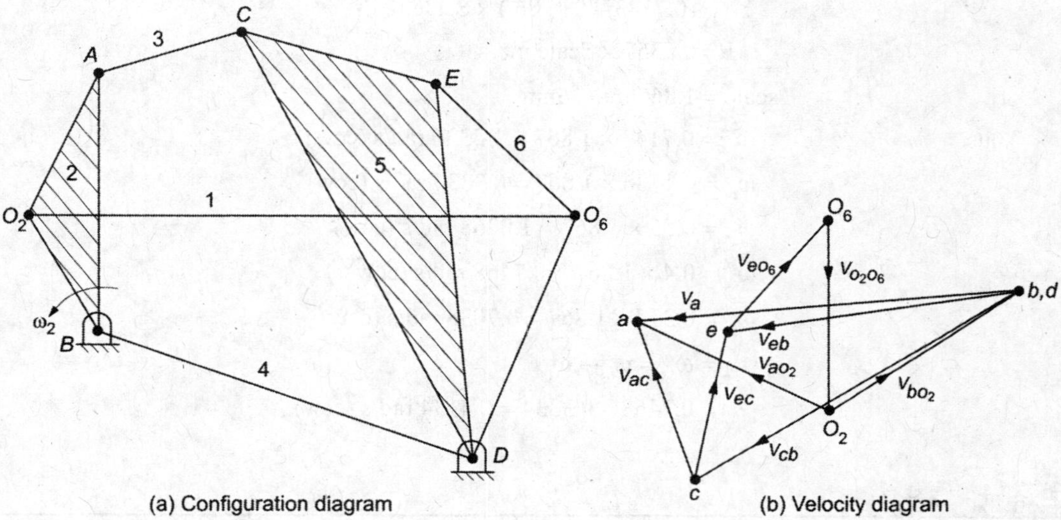

Fig.2.74 High degree of complexity mechanism

$$\omega_{24} = \text{Diagram scale} \times \frac{ba}{AB}$$

$$= \text{Scale} \times \frac{50}{70} = 0.7143 \times \text{Scale (ccw)}$$

$$v_{o6o2} = o_2o_6 \times \omega_{14} = 140 \times \omega_{14}$$

$$= o_2o_6 \times \text{scale}$$

or $$\omega_{14} = \frac{25 \times \text{scale}}{140}$$

$$= 0.1786 \times \text{Scale rad/s}$$

$$v_{o6e} = O_6E \times \omega_{64} = 50 \times \omega_{64}$$

$$= eo_6 \times \text{Scale}$$

or $$\omega_{64} = \frac{eo_6 \times \text{scale}}{50} = \frac{20 \times \text{scale}}{50}$$

$$= 0.4 \times \text{Scale rad/s (ccw)}$$

Similarly, $$\omega_{34} = \frac{ac \times \text{Scale}}{AC} = \frac{22 \times \text{Scale}}{40}$$

$$= 0.55 \times \text{Scale rad/s (ccw)}$$

$$\omega_{54} = \frac{dc \times \text{Scale}}{DC} = \frac{49 \times \text{Scale}}{130}$$

$$= 0.377 \times \text{Scale rad/s (ccw)}$$

But, $$\omega_2 = \omega_{21} = 1 \text{ rad/s (ccw)}$$

Also, $$\omega_{21} = \omega_{24} - \omega_{14+}$$

$$= (0.7143 - 0.1786) \times \text{Scale rad/s}$$

$$1 = 0.5357 \times \text{Scale rad/s}$$

or $$\text{scale} = 1.867 \text{ mm/s/mm}$$

Thus, $$\omega_{24} = 0.7143 \times 1.867 = 1.3336 \text{ rad/s (ccw)}$$

$$\omega_{14} = 0.1786 \times 1.867 = 0.3334 \text{ rad/s (ccw)}$$

$$\omega_{34} = 0.55 \times 1.867 = 1.0268 \text{ rad/s (ccw)}$$

$$\omega_{64} = 0.4 \times 1.867 = 0.7468 \text{ rad/s (ccw)}$$

$$\omega_{54} = 0.377 \times 1.867 = 0.7038 \text{ rad/s (ccw)}$$

$$\omega_6 = \omega_{61} = \omega_{64} - \omega_{14}$$

$$= 0.7468 - 0.3334 = 0.4134 \text{ rad/s (ccw)}$$

Example 2.54

For the mechanism, shown in Fig.2.75(a), the linear velocity of point E is 2.3 m/s. Link 2 rotates at uniform angular velocity in the counter-clockwise direction. Determine ω_2, ω_3, and ω_4.

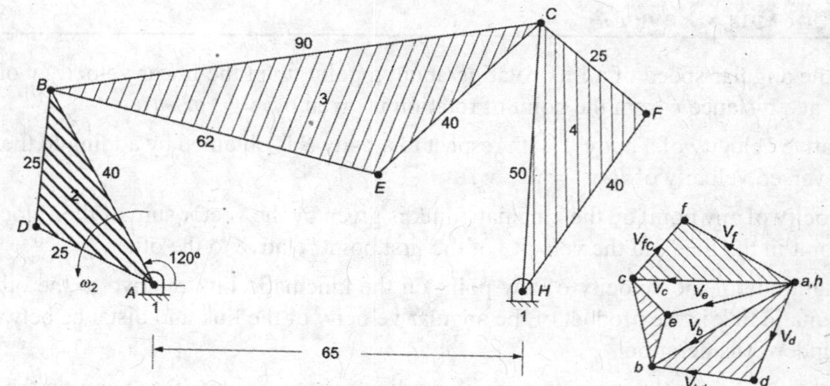

(a) Configuration diagram (a) Velocity diagram

Fig.2.75 Complex mechanism

■ Solution

The magnitude of ω_2 is unknown. Let us take ab = 30 mm (arbitrarily).

Draw the velocity diagram as shown in Fig.2.61(b).

$$ab \perp AB = 30 \text{ mm}$$

$$bd \perp BD$$

$$ad \perp AD$$

Locate d.

$$bc \perp BC$$

$$hc \perp HC$$

Locate c.

$$be \perp BE$$

$$ce \perp CE$$

Locate e. Join ae.

$$cf \perp CF$$

$$hf \perp HF$$

Locate f.

$$v_e \text{ Scale of diagram} \times ae$$

$$2300 = \text{Scale} \times 23$$

or

$$\text{scale} = \frac{100 \text{ mm/s}}{\text{mm}}$$

$$\omega_2 = \frac{v_b}{AB} = \frac{ab \times \text{scale}}{AB} = \frac{30 \times 100}{40}$$

$$= 75 \text{ rad/s (ccw)}$$

$$\omega_3 = \frac{v_{cb}}{CB} = \frac{bc \times \text{scale}}{CB} = \frac{17 \times 100}{90}$$

$$= 18.9 \text{ rad/s (cw)}$$

$$\omega_4 = \frac{v_c}{HC} = \frac{hc \times \text{scale}}{HC} = \frac{28 \times 100}{50}$$

$$= 56 \text{ rad/s (ccw)}$$

Summary for Quick Revision

1 If ω is the angular speed of a link rotating about a point, then the linear velocity v of a point on the link at a distance r from the point of rotation is given by: $v = r\omega$.

2 The relative velocity of a body A with respect to a body B is obtained by adding to the velocity of A the reversed velocity of B. $v_{ba} = v_b - v_a$.

3 The velocity of any point on the kinematic link is given by the vector sum of the velocity of some other point in the link and the velocity of the first point relative to the other.

4 The magnitude of the velocity of any point on the kinematic link relative to the other point in the kinematic link is the product of the angular velocity of the link and distance between the two points under consideration.

5 The direction of the velocity of any point on a link relative to any other point on the link is perpendicular to the line joining the two points.

6 If C is any point on a link AB, then the corresponding point c in the velocity diagram will divide the vector ab in the same ratio as the point C divides the link AB. i.e.

$$ac/ab = AC/AB$$

7 Relative angular velocity of 1.w.r.t. link 2,

$$\omega_{12} = \omega_1 - \omega_2$$

and relative angular velocity of 2 w.r.t. 1,

$$\omega_{21} = \omega_2 - \omega_1 = -\omega_{12}$$

8 If r = radius of the pin at joint O, then

Rubbing velocity at the pin joint O,

$v_r = (\omega_1 - \omega_2)r$, when the links move in the same direction

$= (\omega_1 + \omega_2)r$, when the links move in the opposite direction.

9 Forces in a mechanism

$$F_a \times v_a = F_b \times v_b$$

Considering the effect of friction, the efficiency of transmission,

$$\eta = \text{output/input} = (F_b\, v_b)/(F_a v_a)$$

$$F_b = \eta F_a v_a / v_b$$

10 Mechanical advantage, $MA = F_b/F_a$

For a mechanism, MA = Output torque/Input torque

$$= T_b/T_a = \omega_a/\omega_b$$

Considering the effect of friction, $MA = \eta\omega_a/\omega_b$

11 The number of instantaneous centers in a mechanism,

$$N = n\,(n-1)/2$$

Where n = number of links.

12 The instantaneous centre is a point about which a link as a whole may be considered to be rotating at a given instant. The velocity of the link is zero at the instantaneous centre.

13 The angular velocities of two links vary linearly as the distances from their common instantaneous centre to their respective centres of rotation relative to the frame.

14 Arnold–Kennedy's theorem states that if three plane bodies have relative motion among themselves, their three centres must lie on a straight line.

15 Instaneous centres may be fixed, permanent, and neither fixed nor permanent.

16 For a pin joint, the instantaneous centre lies at the centre of the pin.

17 In a pure rolling contact two links, the instantaneous centre lies at the point of contact of two links.

18 In a sliding motion, the instantaneous centre lies at infinity in a direction perpendicular to the path of motion of the slider.

19 When the slider moves on a curved surface then the instantaneous centre lies at the centre of curvature of the curved surface.

20 Instantaneous centres can be located by the circular diagram.

Multiple Choice Questions

1 The total number of instantaneous centres for a mechanism of n links are
(a) $n(n-1)/2$ (b) n (c) $n-1$ (d) $n/2$.

2 A mechanism has 7 links with all binary pairs except one which is a ternary pair. The number of instantaneous centres of this mechanism are
(a) 14 (b) 21 (c) 28 (d) 42.

3 The direction of the linear velocity of any point on the kinematic link relative to any other point on the same kinematic link is
(a) parallel to the line joining the points
(b) perpendicular to the line joining the points
(c) at 45° to the line joining the points
(d) dependent on the angular speed of rotation of the link.

4 Two kinematic links have absolute angular velocities of ω_1 (clockwise) and ω_2 (anti-clockwise). The angular velocity of link 1 relative to link 2 , is
(a) $\omega_1 + \omega_2$ (b) $\omega_1 - \omega_2$ (c) $\omega_2 - \omega_1$ (d) $\omega_1\omega_2$.

5 The linear velocity of a point B on a link rotating at an angular velocity ω relative to another point A on the same link is
(a) $\omega \times AB$ (b) $\dfrac{\omega}{AB}$ (c) $\omega^2 \times AB$ (d) $\omega \times (AB)^2$

6 According to Kennedy's theorem, the instantaneous centres of three bodies having relative motion lie on a
(a) straight line (b) point (c) curved path (d) circle

7 The instantaneous centre of a slider moving in a linear guide lies at
(a) their point of contact (b) infinity perpendicular to the path of guide
(c) the pin point (d) infinity parallel to the path of guide.

8 The instantaneous centre of a slider moving on a curved surface lies
(a) infinity (b) their point of contact
(c) the centre of curvature of curved surface (d) the pin point.

9 The instantaneous centre of rotation of a circular disc rolling on a straight path lies at
(a) the centre of the disc (b) their point of contact
(c) the centre of gravity of the disc (d) infinity.

10 The number of types of instantaneous centres are:
(a) 2 (b) 3 (c) 4 (d) 6.

Answers:

1. (a) 2. (c) 3. (b) 4. (a) 5. (a) 6. (a) 7. (b) 8. (c) 9. (b) 10. (b)

Review Questions

1 Define relative velocity.
2 Define linear velocity and angular velocity.
3 What is relative angular velocity? How this is determined?
4 What is mechanical advantage for a mechanism?
5 What is a velocity diagram? What are its uses.
6 What is a velocity image? Give its uses.
7 What is rubbing velocity? How this is determined?
8 State angular velocity theorem.
9 Define instantaneous centre of a link.
10 What are the various types of instantaneous centres?
11 How number of instantaneous centres are determined?
12 State Arnold–Kennedy theorem of three centres.
13 What are the properties of instantaneous centre?
14 How instantaneous centres are located?

Exercises

2.1 The dimensions of a four-bar chain shown in Fig.2.76 are: $AD = BE = 120$ mm, $AB = 30$ mm and $CD = 60$ mm. The crank AB rotates at 100 rpm. Determine the angular speed of link CD.

[Ans. 3.9 rad/s, cw]

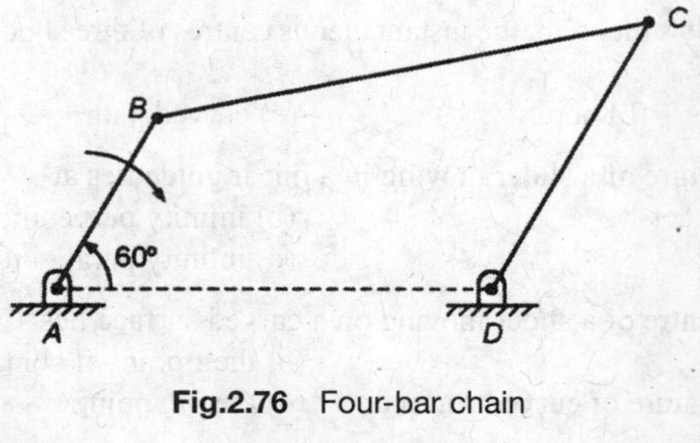

Fig.2.76 Four-bar chain

2.2 In the mechanism shown in Fig.2.77, $O_1O_2 = 210$ mm, $O_1B = 300$ mm and $O_2A = 60$ mm. The crank O_2A rotates at 300 rpm in the ccw direction. Find (a) angular speed of link O_1A, and (b) velocity of slider.

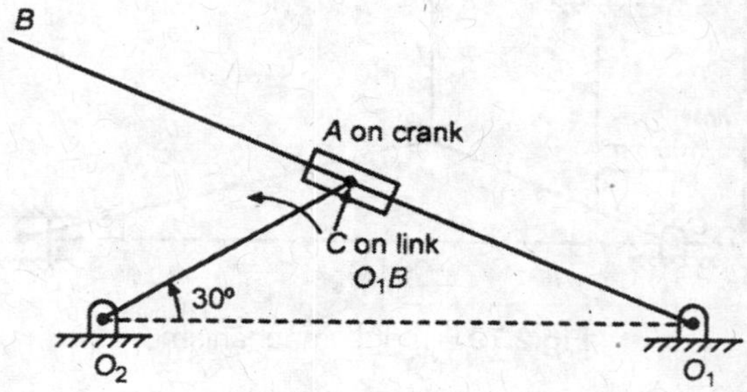

Fig.2.77 Four-bar chain

[Ans. 8.9 rad/s ccw, 1.425 m/s]

2.3 The dimensions of the various links of the mechanism shown in Fig.2.78 are: $AD = DE = 150$ mm, $BC = CD = 450$ mm, $EF = 375$ mm.

The crank *AB* rotates at 120 rpm. The lever *DC* oscillates about the fixed point *D*. Determine (a) velocity of slider *F*, and (b) angular speed of *CD*.

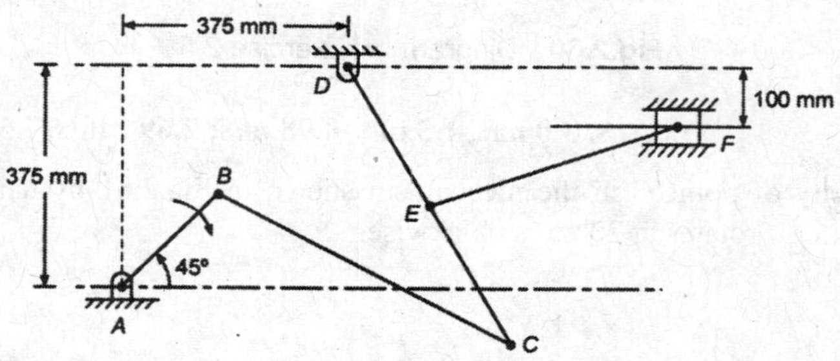

Fig.2.78 Six-bar chain

[Ans. 0.7 m/s, 4.9 rad/s ccw]

2.4 In the toggle mechanism shown in Fig.2.79, the crank *OA* rotates at 180 rpm and the slider is constrained to move on a horizontal path. $OA = 180$ mm, $BC = 240$ mm, $AB = 360$ mm, and $BD = 540$ mm

Find (a) velocity of slider *D*, (b) angular speed of links *AB*, *BC* and *BD*, (c) velocity of rubbing on the pins of diameter 30 mm at *A* and *D*, and (d) torque applied to crank *OA* for a force of 2 kN at *D*.

[Ans. 1.75 m/s, 3.33 rad/s ccw, 11.04 rad/s ccw, 4.53 rad/s cw; 0.1836 m/s, 0.0679 m/s; 185.68 Nm]

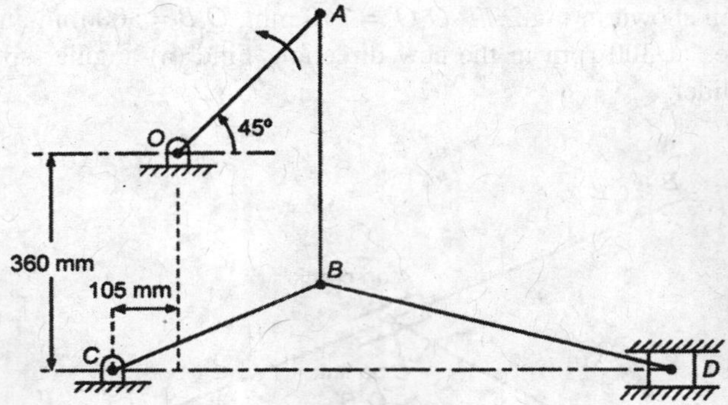

Fig.2.79 Toggle mechanism

2.5 The crank OA of the mechanism shown in Fig.2.80 rotates at 100 rpm clockwise. Using instantaneous centre method determine the linear velocities of points B, C and D, and angular speeds of links AB, BC and CD.

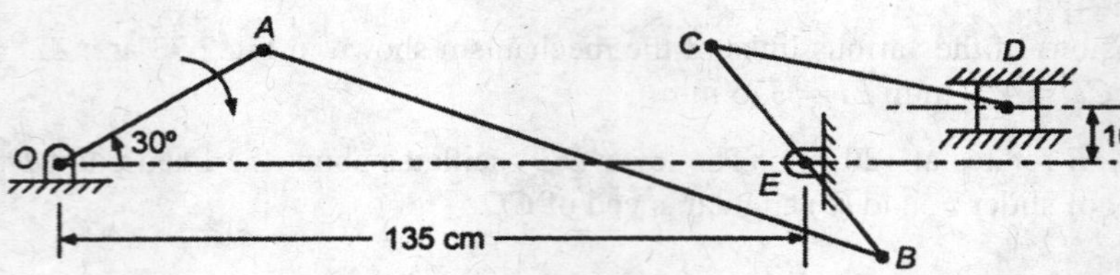

Fig.2.80 Diagram for Exercise 2.5

[Ans. 3.0 m/s, 1.5 m/s, 0.98 m/s; 2.69 rad/s, 7.50 rad/s, 2.08 rad/s]

2.6 Find the velocity of point C in the mechanism shown in Fig.2.81 by using relative velocity method. Crank O_2A rotates at 20 rad/s clockwise.

[Ans. 1.7 m/s]

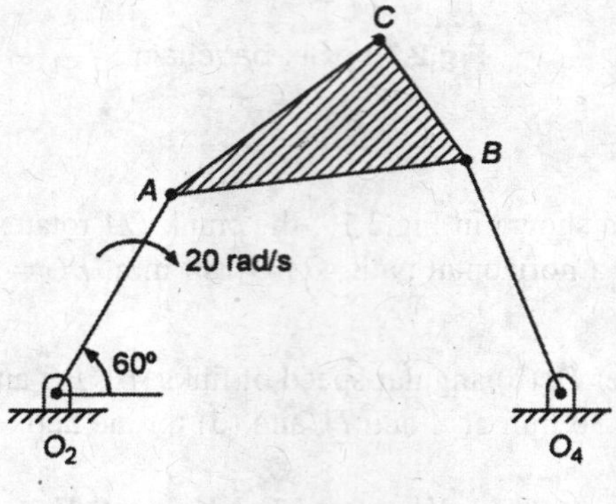

Fig.2.81 Four-bar mechanism

3 ACCELERATION IN MECHANISMS

Chapter Outline

3.1 INTRODUCTION

In the Chapter 2, we have defined the concept of velocity. In this chapter, we shall study the concept of acceleration. The acceleration may be defined as the rate of change of velocity of a body with respect to time. The acceleration can be linear or angular. Linear acceleration is the rate of change of linear velocity of a body with respect to time. Angular acceleration is the rate of change of angular velocity of a body with respect to time.

Acceleration diagram. It is the graphical representation of the accelerations of the various links of a mechanism drawn on a suitable scale. It helps us to determine the acceleration of various links of the mechanism.

The determination of acceleration of various links is important from the point of view of calculating the forces and torques in the various links to carry out the dynamic analysis.

3.2 ACCELERATION OF A BODY MOVING IN A CIRCULAR PATH

Consider a body moving in a circular path of radius r with angular speed ω, as shown in Fig.3.1(a). The body is initially at point A and in time δt moves to point B. Let the velocity change from v to $v+\delta v$ at B and the angle covered be $\delta\theta$.

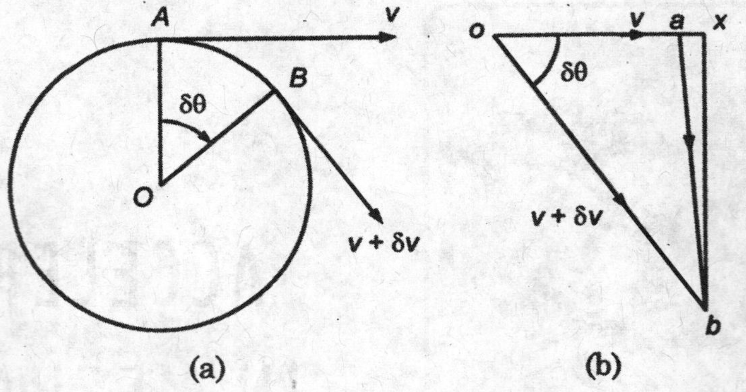

Fig.3.1 Acceleration for a link

The change in velocity can be determined by drawing the velocity diagram as shown in Fig.3.1(b). In this diagram, $oa = v$, $ob = v + \delta v$, and ab = change in velocity during time δt. The vector ab is resolved into two components ax and xb, parallel and perpendicular to oa, respectively.

Now $$ax = ox - oa = ob \cos \delta\theta - oa = (v + \delta v) \cos \theta - v$$

$$xb = ob \sin \delta\theta = (v + \delta v) \sin \delta\theta$$

The rate of change of velocity is defined as the acceleration. It has two components: tangential and normal. The tangential component of acceleration f^t is the acceleration in the tangential direction. It is defined by,

$$f^t = \text{change of velocity in the tangential direction per unit time}$$

$$= \frac{ax}{\delta t}$$

$$= \frac{(v + \delta v)\cos\theta - v}{\delta t}$$

$$\approx \frac{(v + \delta v) - v}{\delta t} \qquad \text{[Because for small angle } \cos \delta\theta \approx 1]$$

$$= \frac{\delta v}{\delta t} = \frac{\mathrm{d}v}{\mathrm{d}t}$$

$$= \frac{\mathrm{d}(\omega r)}{\mathrm{d}t} = r\left(\frac{\mathrm{d}\omega}{\mathrm{d}t}\right) = ra \qquad (3.1)$$

where α = angular acceleration.

The normal component of acceleration f^n is the acceleration in the direction normal to the tangent at that instant. This component is directed towards the centre of the circular path. It is also called the radial or centripetal acceleration. It is defined by,

$$f^n = \text{change of velocity component in a direction normal to tangent per unit time}$$

$$= \frac{xb}{\delta t}$$

$$= \frac{(v+\delta v)\ \sin\theta}{\delta t}$$

$$\approx \frac{(v+\delta v)\,\delta\theta}{\delta t} \qquad \text{[Because for small angle } \sin\delta\theta \sim \delta\theta]$$

$$= \frac{v\delta\theta}{\delta t} + \frac{\delta v\delta\theta}{\delta t} \approx \frac{v\delta\theta}{\delta t} \qquad \text{[Neglecting second term, being small]}$$

$$= \frac{v\,\mathrm{d}\theta}{\mathrm{d}t} = v\omega = r\omega^2 = \frac{v^2}{r} \tag{3.2}$$

Two cases arise regarding the motion of the body.

(i) When the body is rotating with uniform angular velocity, then $d\omega/dt = 0$, so that $f^t = 0$. The body will have only normal acceleration, $f^n = r\,\omega^2$.

(ii) When the body is moving on a straight path, r will be infinitely large and $\frac{v^2}{r}$ will tend to zero, so that $f^n = 0$. The body will have only tangential acceleration, $f^t = r\alpha$.

3.3 ACCELERATION DIAGRAMS

3.3.1 Total Acceleration of a Link

Consider two points A and B on a rigid link, as shown in Fig.3.2(a), such that the point B moves relative to point A with an angular velocity ω and angular acceleration α.

Centripetal (or normal or radial) acceleration of point B with respect to point A is,

$$f_{ba}^n = \omega^2 \cdot AB = \frac{v_{ba}^2}{AB} \tag{3.3}$$

Tangential acceleration of point B with respect to point A,

$$f_{ba}^t = \alpha \cdot AB \tag{3.4}$$

Total acceleration of B w.r.t. A,

$$f_{ba} = f_{ba}^n + f_{ba}^t \quad \text{(vector sum)}$$

$$= \omega^2 \times AB + \alpha \times AB$$

$$= (\omega^2 + \alpha) \cdot AB \tag{3.5}$$

$$\tan\beta = \frac{f_{ba}^t}{f_{ba}^n} = \frac{\alpha}{\omega^2} \tag{3.6}$$

The acceleration diagram has been represented in Fig.3.2(b).

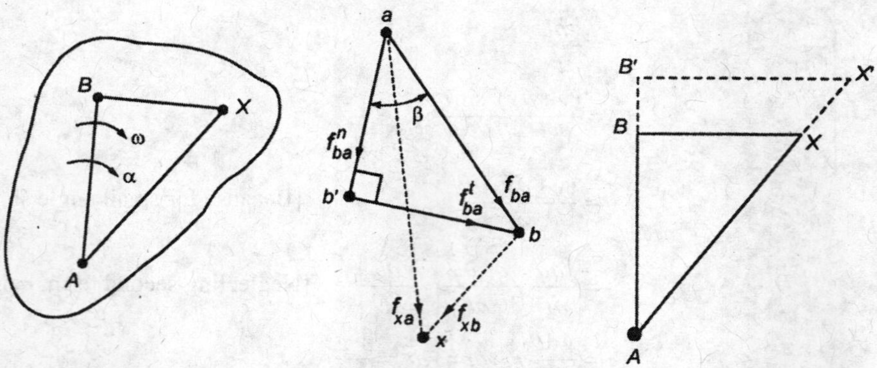

(a) Link subjected to acceleration (b) Acceleration diagram (c) Acceleration image

Fig.3.2 Total acceleration of a link

3.3.2 Acceleration of a Point on a Link

The accelerations of any point X on the rigid link w.r.t. A as shown in Fig.3.2(a) are:

$$f^n_{xa} = \omega^2 \cdot AX \tag{3.7}$$

$$f^t_{xa} = \alpha \cdot AX \tag{3.8}$$

Total acceleration, $$f_{xa} = f^n_{xa} + f^t_{xa} \quad \text{(vector sum)} \tag{3.9}$$

Therefore, f_{xa} Denoted by ax in the acceleration diagram shown in Fig.3.2(b) is inclined to XA at the same angle β. Triangles abx and ABX are similar. Thus, point x can be fixed on the *acceleration image*, corresponding to point X on the link. Total acceleration of X relative to A,

$$f_{xa} = ax$$

Total acceleration of X relative to B,

$$f_{xb} = bx$$

Acceleration Image The concept of velocity image was explained in Section 2.3.7(b) in Chapter 2. It was stated that the velocity images are useful in finding velocities of offset points of links. In the same way, acceleration images are also helpful to find the accelerations of offset points of the links. The acceleration image is obtained in the same manner as a velocity image.

An easier method of making Δabx similar to ΔABX is by making AB' on AB equal to ab and drawing a line parallel to BX, meeting AX in X'. $AB'X'$ is the exact size of the triangle to be made on ab.

Take $ax = AX'$ and $bx = B'X'$

Thus the point x is located.

The method is illustrated in Fig.3.2(c).

3.3.3 Absolute Acceleration for a Link

Consider the rigid link AB such that point B is rotating about A with angular velocity ω and angular acceleration α, as shown in Fig.3.3(a). The point A itself has acceleration f_a. The acceleration diagram is shown in Fig.3.3(b). Absolute acceleration of B,

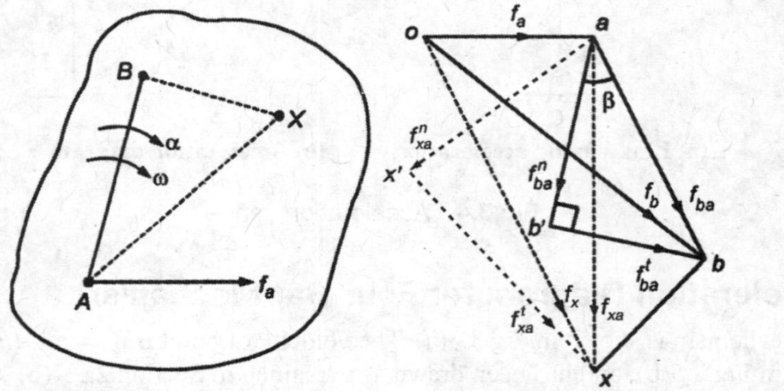

(a) Rigid link subjected to acceleration

(b) Acceleration diagram

Fig.3.3 Absolute acceleration of a rigid link

$$f_b = f_a + f^n_{xa} + f^t_{xa} \quad \text{(vector sum)}$$

$$= f_a + \omega^2 \cdot AB + \alpha \cdot AB \tag{3.10}$$

Similarly for any other point X,

$$f_x = f_a + f^n_{xa} + f^t xa$$

$$= f_a + \omega^2 \cdot AX + \alpha \cdot AX \tag{3.11}$$

3.3.4 Acceleration Centre

Consider a rigid link AB whose ends A and B have accelerations f_a and f_b respectively, as shown in Fig.3.4(a). The acceleration diagram is shown in Fig.3.4(b). If we select a point O on the link such that triangles aob and AOB are similar, then the acceleration of point O relative to a fixed link or fixed point O is zero. The point O is called the instantaneous centre of acceleration of link AB or *acceleration centre*.

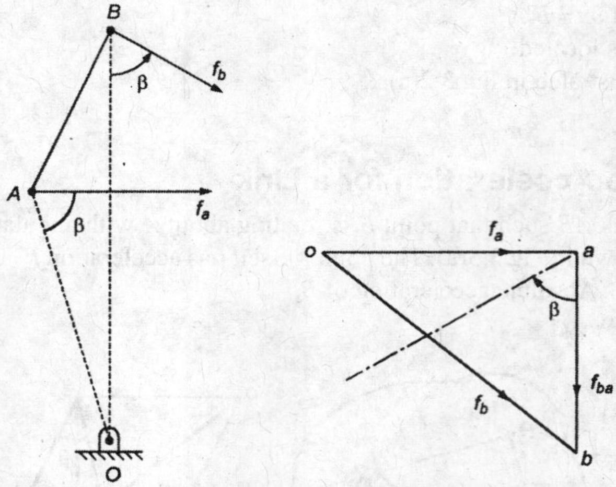

(a) Link having acceleration (b) Acceleration diagram

Fig.3.4 Acceleration centre

3.3.5 Acceleration Diagram for Four-Bar Mechanism

The four-bar mechanism is shown in Fig.3.5(a). The velocity of point B, $v_b = \omega \cdot AB$. The velocity diagram is shown in Fig.3.5(b), and has been drawn as explained in Section 2.3.7(b). In this diagram,

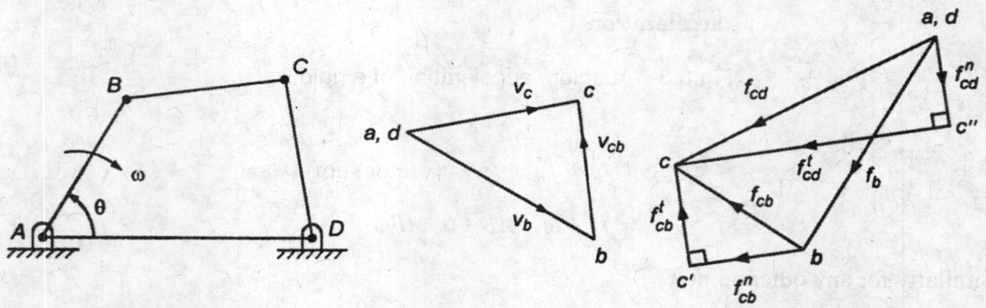

(a) Configuration diagram (b) Velocity diagram (c) Acceleration diagram

Fig.3.5 Acceleration diagram for four-bar mechanism

$$v_b = ab \perp AB$$

$$dc \perp DC$$

and $$bc \perp BC$$

Then Velocity of C relative to D, $v_{cd} = v_c = dc$

Velocity of C relative to B, $v_{cb} = bc$

Now we calculate the accelerations of various points and links.

$$f_{ba}^n = \frac{v_b^2}{AB} = ab$$

$$f_{cb}^n = \frac{v_{cb}^2}{CB} = bc'$$

$$f_{cd}^n = \frac{v_c^2}{CD} = dc''$$

$$f_{cb} = f_{cb}^n + f_{cb}^t$$

$$bc = bc' + c'c$$

$$f_{cd} = f_{cd}^n + f_{cd}^t$$

$$dc = dc'' + c''c$$

The acceleration diagram is shown in Fig.3.5(c) to a suitable scale. To construct the acceleration diagram, proceed as follows:

1. Draw $ab = f_b$ parallel to AB, which is known in magnitude and direction.
2. Draw $bc' = f_{cb}^n$ parallel to BC, which is known in magnitude and direction.
3. Draw cc', representing f_{cb}^n perpendicular to bc', which is known in direction only.
4. Now draw $dc'' = f_{cd}^n$ parallel to CD, which is known in magnitude and direction.
5. Draw $c''c$, representing f_{cd}^t, perpendicular to dc'' to intersect $c'c$ at point c.
6. Join dc and bc. Then

   Acceleration of C relative to $B, f_{cb} = bc$

   Acceleration of C relative to $D, f_{cd} = f_c = dc$

3.3.6 Four-Bar Mechanism with Ternary Link

The four-bar mechanism with a ternary link is shown in Fig.3.6(a), in which the driving crank has angular velocity ω and angular acceleration $\alpha \cdot v_a = \omega \times O_1A$. The velocity diagram is shown in Fig.3.6(b), and has been drawn as explained in Section 2.3.7 (a). In this diagram,

(i) Velocity diagram

$$v_a = o_1a \perp O_1A$$

$$ab \perp AB$$

$$o_2b \perp O_2B$$

Then velocity of B relative to A, $v_{ba} = ab$
Velocity of B relative to O_2, $v_{bo_2} = v_b = o_2b$
Now $bc \perp BC$
and $ac \perp AC$
which locates point c.
Velocity of C relative to O_1, $v_c = o_1c$
Then Δabc is the velocity image of ternary link ABC.

Now $$\frac{ad}{ab} = \frac{AD}{AB}$$

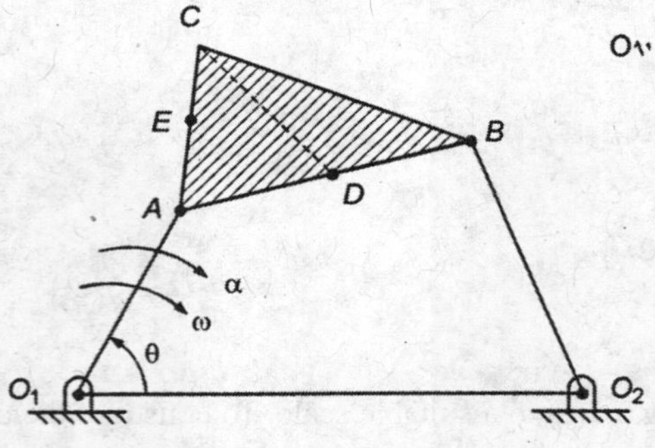

(a) Configuration diagram

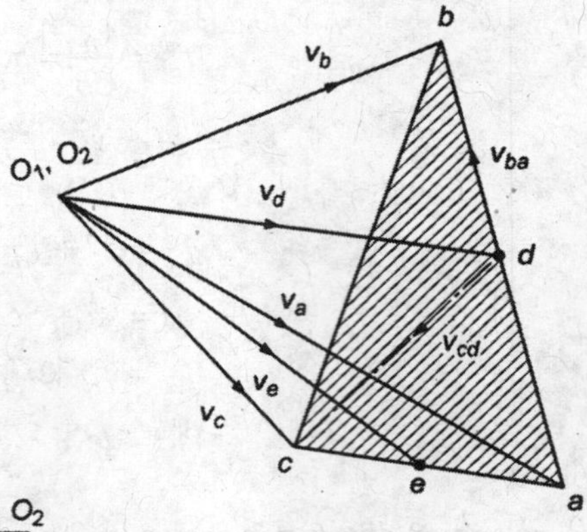

(b) Velocity diagram

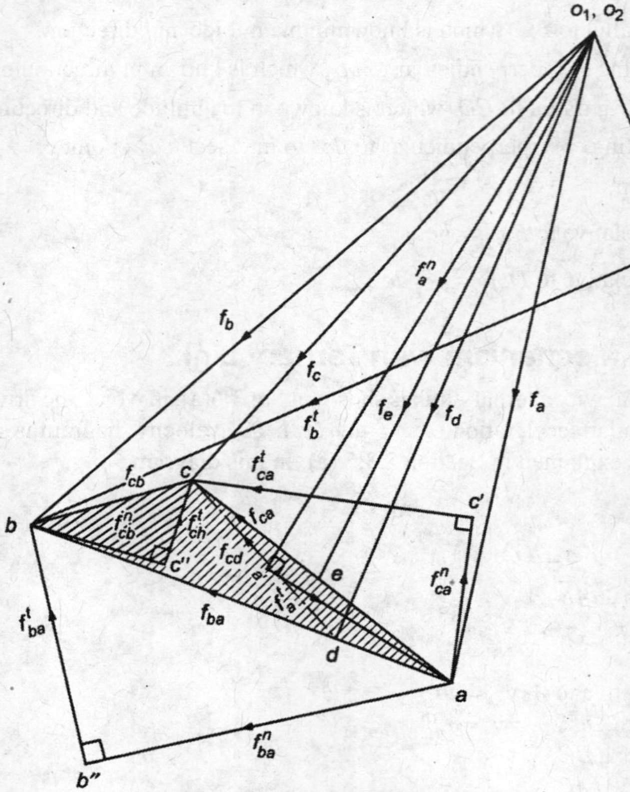

(c) Acceleration diagram

Fig.3.6 Acceleration diagram for four-bar chain with ternary link

Locate point d in ab. Join o_1d and cd. Then $v_d = o_1d$ and $v_{cd} = dc$

Now $$\frac{ae}{ac} = \frac{AE}{AC}$$

Locate point e in ac. Join o_1e. Then velocity of point E, $v_e = o_1e$
This completes the construction of velocity diagram.

(ii) Acceleration diagram

Now we calculate the accelerations of various points and links.

$$f_a^n = f_{ao1}^n = \frac{v_a^2}{O_1A} = o_1a'$$

$$f_a^t = f_{ao1}^t = \alpha \times O_1A = a'a$$

Total acceleration of A, $f_a = f_{ao1} = f_{ao1}^n + f_{ao1}^t = o_1a$ (vector sum)

$$f_b^n = f_{bo2}^n = \frac{v_b^2}{O_2B} = o_2b'$$

$$f_{ba}^n = \frac{v_{ba}^2}{AB} = ab''$$

$$bb'' = f_{ba}^t \perp ab''$$

$$f_b^t = f_{bo2}^t \perp o_2b' = b'b$$

$$f_b = f_{bo2} = f_{ao1} + f_{ba} \quad \text{(vector sum)}$$

$$= f_{ao1}^n + f_{ao1}^t + f_{ba}^n + f_{ba}^t \quad \text{(vector sum)}$$

The following steps may be adopted to draw the acceleration diagram as shown in Fig.3.6(c).

1. Draw $f_a^n = o_1a' \parallel O_1A$, to a convenient scale, to represent the normal acceleration of A, which is known in magnitude and direction.
2. Draw $f_a^t = a'a \perp O_1A$ at a' to represent the tangential acceleration of A, which is known in magnitude and direction. Join o_1a to get the total acceleration of A, $f_a = o_1a$.
3. Draw $f_{ba}^n = ab'' \parallel AB$ to represent the normal acceleration of AB, which is known in magnitude and direction.
4. Draw $f_{ba}^t \perp AB$ at b'' to represent the tangential acceleration of AB, which is not known in magnitude.
5. Draw $f_b^n = o_2b' \parallel$ to O_2B to represent the normal acceleration of B, which is known in magnitude and direction.
6. Draw $f_b^t \perp O_2B$ at b' to represent the tangential acceleration of B, which is not known in magnitude.
7. The lines of f_{ba}^t and f_b^t meet at point b. Join ab and o_2b. Then

   Total acceleration of B relative to A, $f_{ba} = ab$

   Total acceleration of B relative to O_2, $f_b = o_2b$

(iii) Acceleration of Intermediate Points

8. To calculate the acceleration of point D in AB, we have

$$\frac{ad}{ab} = \frac{AD}{AB}$$

This locates point d in ab. Join o_1d. Then acceleration of point $D, f_d = o_1d$

Similarly acceleration of any point in link O_1A and O_2B can be determined.

(iv) Acceleration of Offset Points

9. Draw a line ac' parallel to AC to represent f^n_{ac}, which is known in magnitude and direction. Draw another line perpendicular to AC at c' to represent f^t_{ac}, which is unknown in magnitude.

10. Draw a line bc'' parallel to BC to represent f^n_{cb} which is known in magnitude and direction. Draw another line perpendicular to BC at c'' to represent f^t_{cb}, which is unknown in magnitude. These two lines meet at c. Join ac and bc.

    The triangles abc and ABC are similar. Then Δabc is the *acceleration image* of ternary link ABC. To find the acceleration of offset point C, join o_1c. Then

    Total acceleration of $C, f_c = o_1c$

11. To find the acceleration of point E in AC, we have

$$\frac{ae}{ac} = \frac{AE}{AC}$$

Thus locate point e in ac. Join o_1e. Then

Total acceleration of $E, f_e = o_1e$

(v) Angular Acceleration of Links

It may be observed that the tangential component of acceleration occurs due to the angular acceleration of a link. Thus, angular acceleration can be determined if the tangential acceleration is known.

In Fig.3.6(c), tangential acceleration of B relative to A, $f^t_{ba} = b''b$

i.e., tangential acceleration of B relative to A is in a direction b'' to b or in a counter-clockwise direction about A.

Now $\quad f^t_{ba} = \alpha_{ba} \times AB$

$$\therefore \quad \alpha_{ba} = \frac{f^t_{ba}}{AB}$$

Tangential acceleration of A relative to B, $f^t_{ab} = bb''$

i.e., tangential acceleration A relative to B is in a direction b to b'' or in a counter-clockwise direction about B with magnitude $\alpha_{ab} = \dfrac{f^t_{ab}}{AB}$, which is the same as α_{ba}.

Thus, angular acceleration of a link about one end of the link is the same in magnitude and direction as the angular acceleration about the other end

Tangential acceleration of B relative to O_2, $f^t_{bo2} = b'b = f^t_b$

i.e., B relative to O_2 moves in a direction from b' to b or B moves in the counter-clockwise direction about O_2.

$$\alpha_b = \alpha_{bo2} = \frac{f^t_b}{O_2B} \text{ ccw}$$

3.3.7 Acceleration Diagram for Slider-Crank Mechanism

(i) Velocity diagram

For the slider crank mechanism as shown in Fig.3.7(a). $v_\alpha = \omega \cdot OA$. The velocity diagram is shown in Fig.3.7(b) and has been drawn as explained in Section 2.3.8. In this diagram,

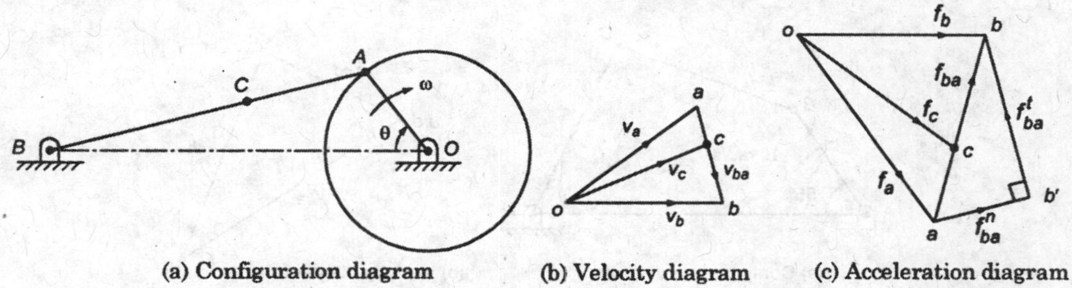

Fig.3.7 Acceleration diagram of slider-crank mechanism

$$v_a = oa \perp OA$$
$$ab \perp AB$$
$$ob \parallel OB$$

Then
Velocity of slider, $v_b = ob$
Velocity of connecting rod, $v_{ba} = ab$

(ii) Acceleration Diagram

The acceleration diagram is shown in Fig.3.7(c), in which

$$f_a = \omega^2 \times OA = \frac{v_a^2}{OA} = oa$$

$$f_{ba}^n = \frac{v_{ba}^2}{AB} = ab'$$

The following steps may be adopted to draw the acceleration diagram:

1. Draw $oa = f_a$ parallel to OA on a convenient scale to represent the normal acceleration of the crank OA.
2. Draw $ab' = f_{ba}^n$ parallel to AB to represent the normal acceleration of the connecting rod, which is known in magnitude and direction.
3. Draw a line perpendicular to AB at b' to represent the tangential acceleration f_{ba}^t of AB, which is unknown in magnitude.
4. Draw a line parallel to OB at o to represent the acceleration f_b of the slider B. Let these two lines intersect at b. Join ab. Then

   Linear acceleration of slider $B, f_b = ob$
5. To find the acceleration of any point C in AB, we have

$$\frac{ac}{ab} = \frac{AC}{AB}$$

Locate point c in ab. Join oc. Then
Acceleration of point $C, f_c = oc$

Example 3.1

In the four-bar mechanism shown in Fig.3.8, the lengths of the various links are: $AB = 190$ mm, $BC = CD = 280$ mm, $AD = 500$ mm, $\angle BAD = 55°$. The crank AB rotates at 10 rad/s in the clockwise direction. Determine (a) the acceleration of the links BC and CD, and (b) angular accelerations of BC and CD.

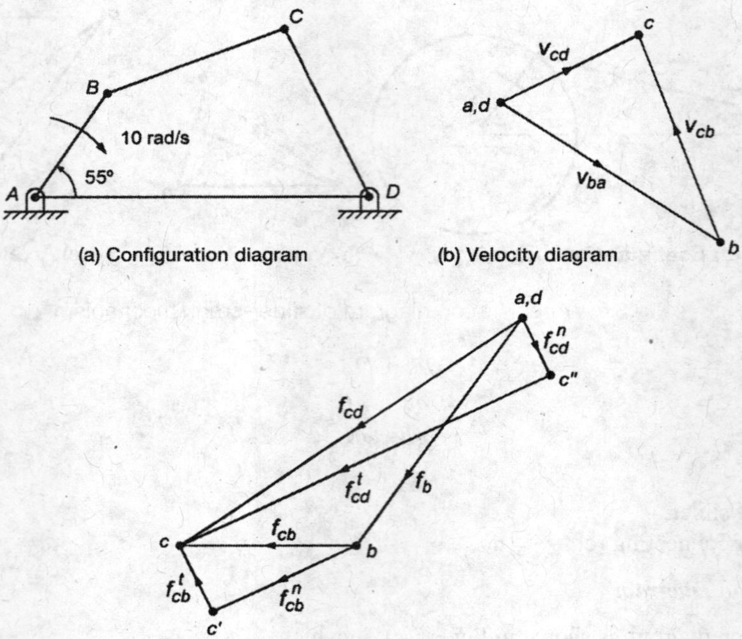

(a) Configuration diagram (b) Velocity diagram

(c) Acceleration diagram

Fig.3.8 Four-bar mechanism

■ Solution

Linear velocity of point B, $v_b = v_{ba} = AB \times \omega = 0.19 \times 10 = 1.9$ m/s

1. Draw the configuration diagram to a scale of 1 cm = 100 mm, as shown in Fig.3.8(a).
2. Draw the velocity diagram shown in Fig.3.8(b), as explained in Section 3.3.5, to a scale of 1 cm = 0.5 m/s.
3. Measure the velocity of C relative to B, $v_{cb} = bc = 3.3$ cm = 1.65 m/s, and the velocity of C relative to D, $v_{cd} = dc = 2.2$ cm = 1.1 m/s.
4. Calculate the normal accelerations:

$$f^n_b = f^n_{ba} = \frac{v^2_{ba}}{AB} = \frac{(1.9)^2}{0.9} = 19 \text{ m/s}^2$$

$$f^n_{cb} = \frac{v^2_{cb}}{BC} = \frac{(1.65)^2}{0.28} = 9.72 \text{ m/s}^2$$

$$f^n_{cd} = \frac{v^2_{cd}}{CD} = \frac{(1.1)^2}{0.28} = 4.32 \text{ m/s}^2$$

5. Draw the acceleration diagram to a scale of 1 cm = 5 m/s², as shown in Fig.3.8(c) and as explained in Section 3.3.5.

6. Measure the acceleration of C relative to $B, f_{cb} = bc = 2.1$ cm $= 10.5$ m/s^2.
   Acceleration of C relative to $D, f_{cd} = dc = 5.1$ cm $= 25.5$ m/s^2
7. Tangential acceleration of BC, $f^t_{cb} = cc' = 0.9\ cm = 4.5$ m/s^2

   Angular acceleration of BC, $\alpha_{bc} = \dfrac{f^t_{bc}}{BC} = \dfrac{4.5}{0.28} = 16.07$ rad/s^{2}cw

   Tangential acceleration of CD, $f^t_{cd} = cc'' = 5.2$ cm $= 26$ m/s^2

   Angular acceleration of CD, $\alpha_{cd} = \dfrac{f^t_{cd}}{CD} = \dfrac{26}{0.28} = 92.857$ rad/s^2 ccw

Example 3.2

A four-bar mechanism with ternary link is shown in Fig.3.9(a). The lengths of various links is given as below:

$$O_1O_2 = 600 \text{ mm}, O_1A = 300 \text{ mm}, AB = 400 \text{ mm}, O_2B = 450 \text{ mm},$$
$$AC = 300 \text{ mm}, BC = 250 \text{ mm}, AD = 100 \text{ mm, and } \angle AO_1O_2 = 75°.$$

Angular velocity of crank $O_1A = 20$ rad/s
Angular acceleration of crank $O_1A = 100$ rad/s^2
Determine (a) acceleration of coupler AB, (b) acceleration of lever O_2B, (c) acceleration of points C and D, and (d) angular acceleration of ternary link.

Solution

Linear velocity of A, $v_a = \omega \times O_1A = 20 \times 0.3 = 6$ m/s

1. Draw the configuration diagram as shown in Fig.3.9(a) to a scale of 1 cm = 100 mm.
2. Draw the velocity diagram as shown in Fig.3.9(b) to a scale of 1 cm = 1 m/s, and as explained in Section 3.3.6.
3. Measure $ab = v_{ba} = 3.5$ cm $= 3.5$ m/s; $o_2b = v_b = 4.7$ cm $= 4.7$m/s; $o_1c = v_c = 3.5$ cm $= 3.5$ m/s, $v_{ca} = ac = 2.7$ cm $= 2.7$ m/s; $v_{cb} = bc = 2.1$ cm $= 2.1$ m/s
4. Now $\dfrac{ad}{ab} = \dfrac{AD}{AB}$, $ad = \dfrac{3.5 \times 100}{400} = 0.875$ cm. Locate point d in ab and join o_1d. Then $v_d = o_1d = 5.5$ cm $= 5.5$ m/s.
5. Calculate the accelerations of various points as follows:

$$f^n_a = \frac{v^2_a}{O_1A} = \frac{6^2}{0.3} = 120 \text{ m/s}^2$$

$$f^t_a = a \times O_1A = 100 \times 0.3 = 30 \text{ m/s}^2$$

$$f^n_{ba} = \frac{v^2_{ba}}{AB} = \frac{(3.5)^2}{0.4} = 30.625 \text{ m/s}^2$$

$$f^n_b = \frac{v^2_b}{O_2B} = \frac{(4.7)^2}{0.45} = 49.1 \text{ m/s}^2$$

$$f^n_{ac} = \frac{v^2_{ac}}{AC} = \frac{(2.7)^2}{0.3} = 24.3 \text{ m/s}^2$$

$$f^n_{cb} = \frac{v^2_{cb}}{BC} = \frac{(2.1)^2}{0.25} = 17.64 \text{ m/s}^2$$

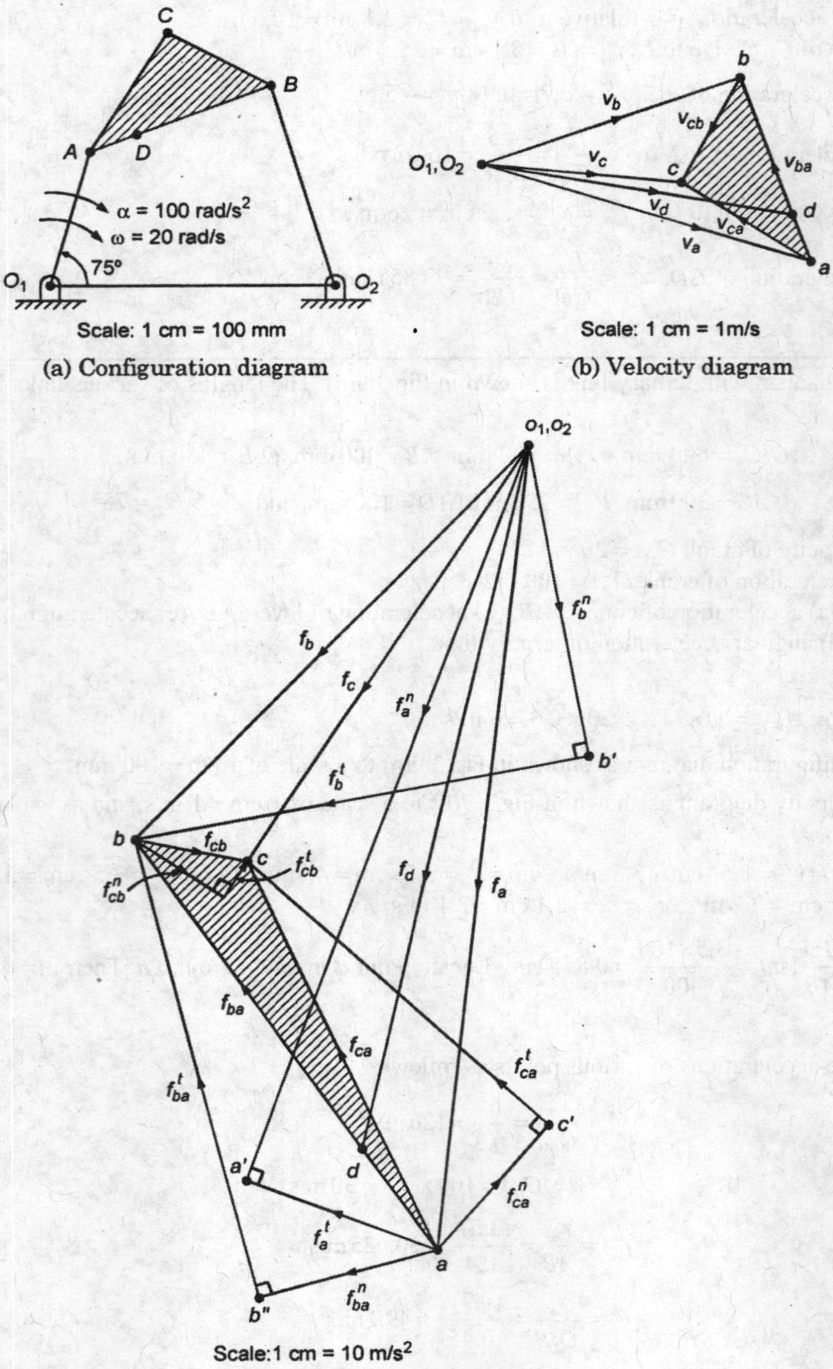

Fig.3.9 Four-bar mechanism with ternary link

6. Draw the acceleration diagram on a scale of 1 cm = 10 m/s², as shown in Fig.3.9(c), and as explained in section. 3.3.6.
7. Measure the various required lengths on the acceleration diagram.
   (a) Acceleration of coupler AB, $f_{ba} = ab = 7.8$ cm $= 78$ m/s²
   (b) Acceleration of lever O_2B, $f_b = o_2b = 8.6$ cm $= 86$ m/s²
   (c) Acceleration of point C, $f_c = o_1c = 7.9$ cm $= 79$ m/s²
8. Locate point d in ab from the relation, $ad = \dfrac{ab \times AD}{AB} = 1.95$ cm. Then
   Acceleration of point D, $f_d = o_1d = 11$ cm $= 110$ m/s².
9. Tangential acceleration of AB, $f^t_{ba} = b''b = 7.2$ cm $= 72$ m/s²

   Angular acceleration of AB, $\alpha_{ab} = \dfrac{f^t_{ba}}{AB} = \dfrac{72}{0.4} = 180$ rad/s²cw.

Example 3.3

In the slider-crank shown in Fig.3.10(a), the lengths of the various links are:

$$OA = AC = 200 \text{ mm}, AB = 600 \text{ mm}, \angle AOB = 30°.$$

The crank rotates at 10 rad/s. Determine (a) the acceleration of the connecting rod AB, (b) acceleration of slider B, and (c) acceleration of a point C in AB.

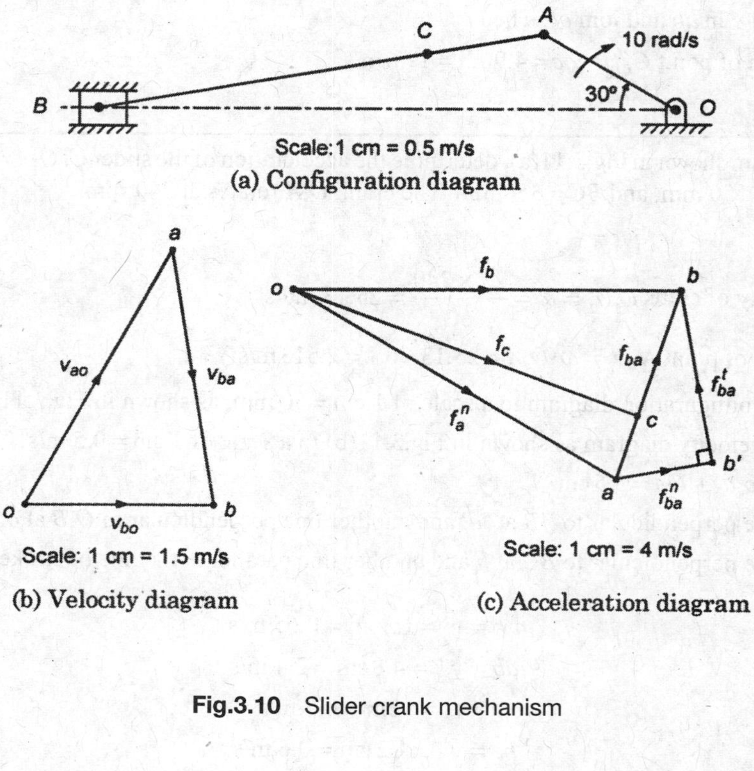

(a) Configuration diagram

(b) Velocity diagram

(c) Acceleration diagram

Fig.3.10 Slider crank mechanism

Solution

Linear velocity of point A, $v_a = \omega \times OA = 10 \times 0.2 = 2$ m/s

1. Draw the configuration diagram to a scale of 1 cm = 100 mm, as shown in Fig.3.10(a).
2. Draw the velocity diagram to a scale of 1 cm = 0.5 m/s, as shown in Fig.3.10(b), and following the procedure explained in Section 3.3.7.
3. Measure the velocity of B relative to A, $v_{ba} = ab = 3.5\text{ cm} = 1.75$ m/s.
4. Calculate the normal accelerations:

$$f_a^n = \frac{v_a^2}{OA} = \frac{2^2}{0.2} = 20\text{ m/s}^2$$

$$f_{ba}^n = \frac{v_{ba}^2}{AB} = \frac{(1.75)^2}{0.6} = 5.1\text{ m/s}^2$$

5. Draw the acceleration diagram as shown in Fig.3.10(c) to a scale of 1 cm = 4 m/s², and following the procedure as explained in Section 3.3.7.
6. Measure the acceleration of connecting rod, $f_{ba} = ab = 2.7\text{ cm} = 10.8\text{ m/s}^2$

   Acceleration of slider B, $f_b = ob = 5.2\text{ cm} = 20.8\text{ m/s}^2$
7. Now $\dfrac{ac}{ab} = \dfrac{AC}{AB}$

$$ac = \frac{2.7 \times 200}{600} = 0.9\text{ cm}$$

8. Locate point c in ab and join oc. Then

   Acceleration of point C, $f_c = oc = 4.9\text{ cm} = 19.6\text{ m/s}^2$.

Example 3.4

In the mechanism shown in Fig.3.11(a), determine the acceleration of the slider C. $O_1A = 100$ mm, $AB = 120$ mm, $O_2B = 150$ mm, and $BC = 350$ mm. The crank O_1A rotates at 240 rpm.

Solution

$$\text{Angular velocity of crank } O_1A_2 = \omega = \frac{2\pi \times 240}{60} = 25.13\text{ rad/s}$$

Linear velocity of point A, $v_a = \omega \cdot O_1A = 25.13 \times 0.1 = 2.513$ m/s

1. Draw the configuration diagram to a scale of 1 cm = 50 mm, as shown in Fig.3.11(a).
2. Draw the velocity diagram as shown in Fig.3.11(b) to a scale of 1 cm = 0.5 m/s.
3. Draw $v_a = o_1a \perp OA = 5.5$ cm.
4. Draw a line perpendicular to AB at 'a' and another line perpendicular to O_2B at o_2 to meet at b.
5. Draw a line perpendicular to BC at b and another line parallel to the line of stroke of the slider C at o_1. Then

$$o_2b = v_b = 1.3\text{ cm} = 0.65\text{ m/s}$$

$$ab = v_{ba} = 4.8\text{ cm} = 2.4\text{ m/s}$$

$$o_1c = v_c = 1\text{ cm} = 0.5\text{ m/s}$$

$$bc = v_{cb} = 1.2\text{ cm} = 0.6\text{ m/s}$$

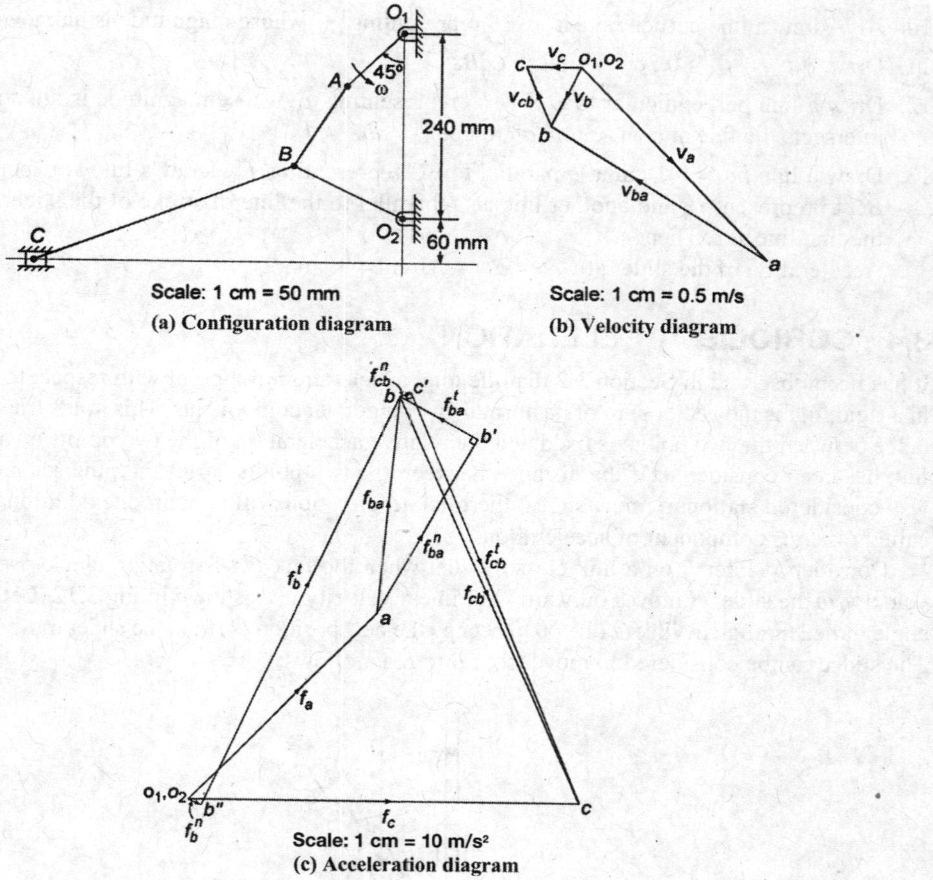

Fig.3.11 Six-link mechanism

6. Calculate the accelerations as follows:

$$f^n_{ao1} = \frac{v_a^2}{O_1A} = \frac{(2.513)^2}{0.1} = 63.15\,\text{m/s}^2$$

$$f^n_{ba} = \frac{f_{ba}^2}{AB} = \frac{(2.4)^2}{0.12} = 48\,\text{m/s}^2$$

$$f^n_{bo2} = \frac{v_b^2}{O_2B} = \frac{(0.65)^2}{0.15} = 2.81\,\text{m/s}^2$$

$$f^n_{cb} = \frac{v_{cb}^2}{BC} = \frac{(0.6)^2}{0.35} = 1.03\,\text{m/s}^2$$

7. Draw acceleration diagram as shown in Fig.3.11(c) to a scale of 1 cm = 10 m/s².
8. Draw $f_a^n = o_1a = 6.3$ cm parallel to O_1A.
9. Draw $ab' = f^n_{ba} = 4.8$ cm at 'a' parallel to AB.

10. At b' draw a line perpendicular to AB representing f^t_{ba} whose magnitude is unknown.
11. Draw $o_2b'' = f^n_b = 0.28$ parallel to O_2B.
12. Draw a line perpendicular to O_2B at b'' representing f^t_b whose magnitude is unknown. This line intersects the line at b' at b. Join ab to get $f_{ba} = ab$.
13. Draw a line $bc' = 0.1$ cm at b parallel to BC representing f^n_{cb}. Draw a line at c' perpendicular to BC to represent f^t_{cb} and another line at o_1 parallel to the line of stroke of the slider at C meeting the first line at c. Then
Acceleration of the slider at $C, f_c = o_1c = 9.1$ cm $= 91$ m/s^2

3.4 CORIOLIS ACCELERATION

It has been observed in Section 3.2 that the total acceleration of a point with respect to another point in a rigid link is the vector sum of its normal and tangential components. This holds true when the distance between the two points is fixed and the relative acceleration of the two points on a moving rigid link has been considered. If the distance between the two points varies, i.e., the second point which was considered stationary, now slides, the total acceleration will contain one additional component, called *Coriolis* component of acceleration.

Consider a slider B on a link OA such that when the link OA is rotating clockwise with angular velocity ω the slider B moves outward with linear velocity v, as shown in Fig.3.12. Let in time δt the angle turned through by link OA be $\delta\theta$ to occupy the new position OA' and the slider moves to position E. The slider can be considered to move from B to E as follows:

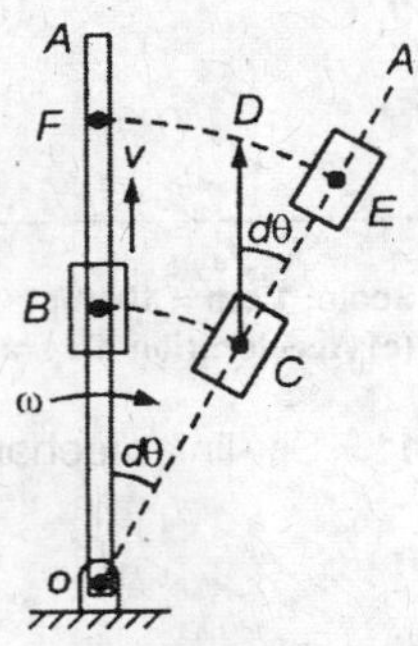

Fig.3.12 Concept of coriolis acceleration

1. From B to C due to angular velocity ω of link OA.
2. C to D due to outward velocity v of the slider.
3. D to E due to acceleration perpendicular to the rod, i.e., due to Coriolis acceleration.Now

$$\text{arc } DE = \text{arc } EF - \text{arc } FD$$
$$= \text{arc } EF - \text{arc } BC$$
$$= OF \cdot d\theta - OB \cdot d\theta$$
$$= (OF - OB)\, d\theta$$
$$= BF \cdot d\theta$$
$$= CD \cdot d\theta$$

Now $\quad CD = v \cdot dt$

and $\quad d\theta = \omega \cdot dt$

Hence arc $\quad DE = (v \cdot dt) \cdot (\omega \cdot dt)$

Now arc DE = chord DE, as $d\theta$ is very small.

Therefore, $DE = v\,\omega\,(dt)^2$ (1)

But $DE = \frac{1}{2} \cdot f^{cr}\,(dt)^2$ (2)

Where f^{cr} is the constant Coriolis acceleration of the particle. Hence from (1) and (2), we get

$$f^{cr} = 2\,v\omega \tag{3.12}$$

The direction of the Coriolis acceleration component is such so as to rotate the sliding velocity vector in the same sense as the angular velocity of OB. This is achieved by turning the sliding velocity vector through 90° in a manner that the velocity of this vector is the same as that of angular velocity of OA. The method of finding the direction of Coriolis acceleration is illustrated in Fig.3.13.

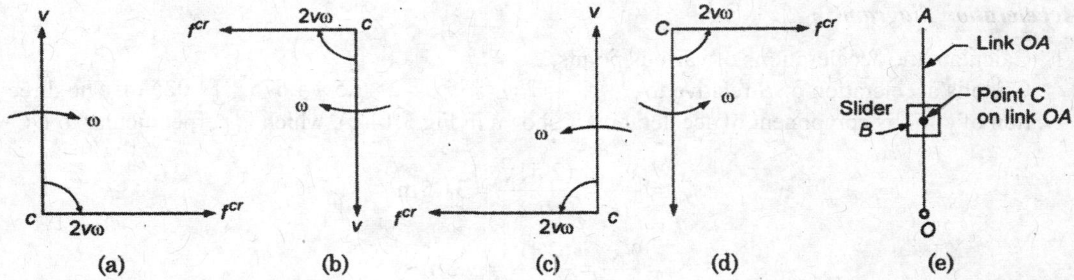

Fig.3.13 Finding direction of Coriolis acceleration

Example 3.5

In the crank and slotted lever type quick return motion mechanism shown in Fig.3.14(a), the crank AB rotates at 120 rpm. Determine (a) velocity of ram at D, (b) magnitude of Coriolis acceleration component, and (c) acceleration of ram at $D \cdot AB = 200$ mm, $OC = 800$ mm, $CD = 600$ mm, $OA = 300$ mm.

■ **Solution**

Velocity diagram

$$\omega = 2\pi \times \frac{120}{60} = 12.57 \text{ rad/s}$$

$$v_b = v_{ba} = \omega \cdot AB = 12.57 \times 0.2 = 2.51 \text{ m/s}$$

1. Draw the configuration diagram to a scale of 1 cm = 100 mm, as shown in Fig.3.14(a). By measurement, $OP = 450$ mm.
2. Draw the velocity diagram as shown in Fig.3.14(b) to a scale of 1 cm = 0.5 m/s, by adopting the following steps.
3. Draw $ba \perp AB$ = 5.02 cm to represent the linear velocity of point B on link AB.
4. Let the coincident point of B on the slotted lever OC be P. Draw a line at o perpendicular to OC and another line at b parallel to OC to meet at point p.
5. By measurement, $op = 4.2$ cm. Then $oc = \frac{op \times OC}{OP} = \frac{4.2 \times 800}{600} = 7.47$ cm.

6. Produce *op* to *oc*.
7. Draw a line parallel to *XX*, the line of stroke of the ram *D*, and a line at *c* perpendicular to *CD* to meet the previous line at *d*. Then

   Velocity of the slider *D*, $v_d = od = 7.8$ cm $= 3.9$ m/s

   Velocity of point *C*, $vc = oc = 7.47$ cm $= 3.74$ m/s

   Angular velocity of slotted lever *OC*, $\omega_{oc} = \frac{v_c}{OC} = \frac{3.74}{0.8} = 4.675$ rad/s cw

$$v_{po} = op = 4.2\,\text{cm} = 2.1\text{ m/s}$$
$$v_{bp} = pb = 3\,\text{cm} = 1.5\text{ m/s}$$
$$v_{cd} = cd = 3.2\,\text{cm} = 1.6\text{ m/s}$$

Acceleration diagram

1. Calculate the accelerations of various points.
   Coriolis acceleration of *B* relative to *P*, $f_{bp}^{cr} = 2v_{bp} \times \omega_{oc} = 2 \times 1.5 \times 4.675 = 14.025$ m/ The direction of coriolis component of acceleration is shown in Fig.3.14(d), which is perpendicular to *OC*.

$$f_{ba}^{n} = \frac{v_{ba}^2}{AB} = \frac{(2.51)^2}{0.2} = 31.5\,\text{m/s}^2$$
$$f_{po}^{n} = \frac{v_{po}^2}{OP} = \frac{(2.1)^2}{0.45} = 9.8\,\text{m/s}^2$$
$$f_{dc}^{n} = \frac{v_{dc}^2}{CD} = \frac{(1.6)^2}{0.6} = 4.27\,\text{m/s}^2$$

2. Draw the acceleration diagram as shown in Fig.3.14(c) to a scale of 1 cm = 2 m/s², by adopting the following steps.
3. Draw $ab = f_{ba}^{n} = 15.75$ cm parallel to *AB*.
4. Draw $p''b = f_{bp}^{cr} = 7.0$ cm perpendicular to *OC* at point *b*, and another line perpendicular to *OC* representing f_{bp}^{s}, the sliding acceleration of point *B* relative to point *P*. $f_{bp} = pb$ is the total acceleration of *B* relative to *P*.
5. Draw $op' = f_{po}^{n} = 4.9$ cm parallel to *OP* and another line *p'p* perpendicular to it to meet the previous line at point *p*.
6. Join *bp* and *op*. By measurement, $op = 5.3$ cm.
7. Now $oc = \frac{op \times OC}{OP} = \frac{5.3 \times 800}{450} = 9.4$ cm.
8. Extend *op* to point *c* so that $oc = 9.4$ cm.
9. Draw $cd' = f_{dc}^{n} = 2.13$ cm parallel to *CD* and draw a line perpendicular to *CD* at *d'* to represent the tangential acceleration of *CD*.
10. Draw a line at *o* parallel to the line of action of the ram at *D* meeting previous line at *d*. Join *cd*. Then

    Acceleration of the ram at *D*, $f_d = od = 2.5$ cm $= 5$ m/s²

    (a) Velocity of ram = 3.9 m/s

    (b) Coriolis acceleration = 14.025 m/s²

    (c) Acceleration of ram = 2.5 m/s²

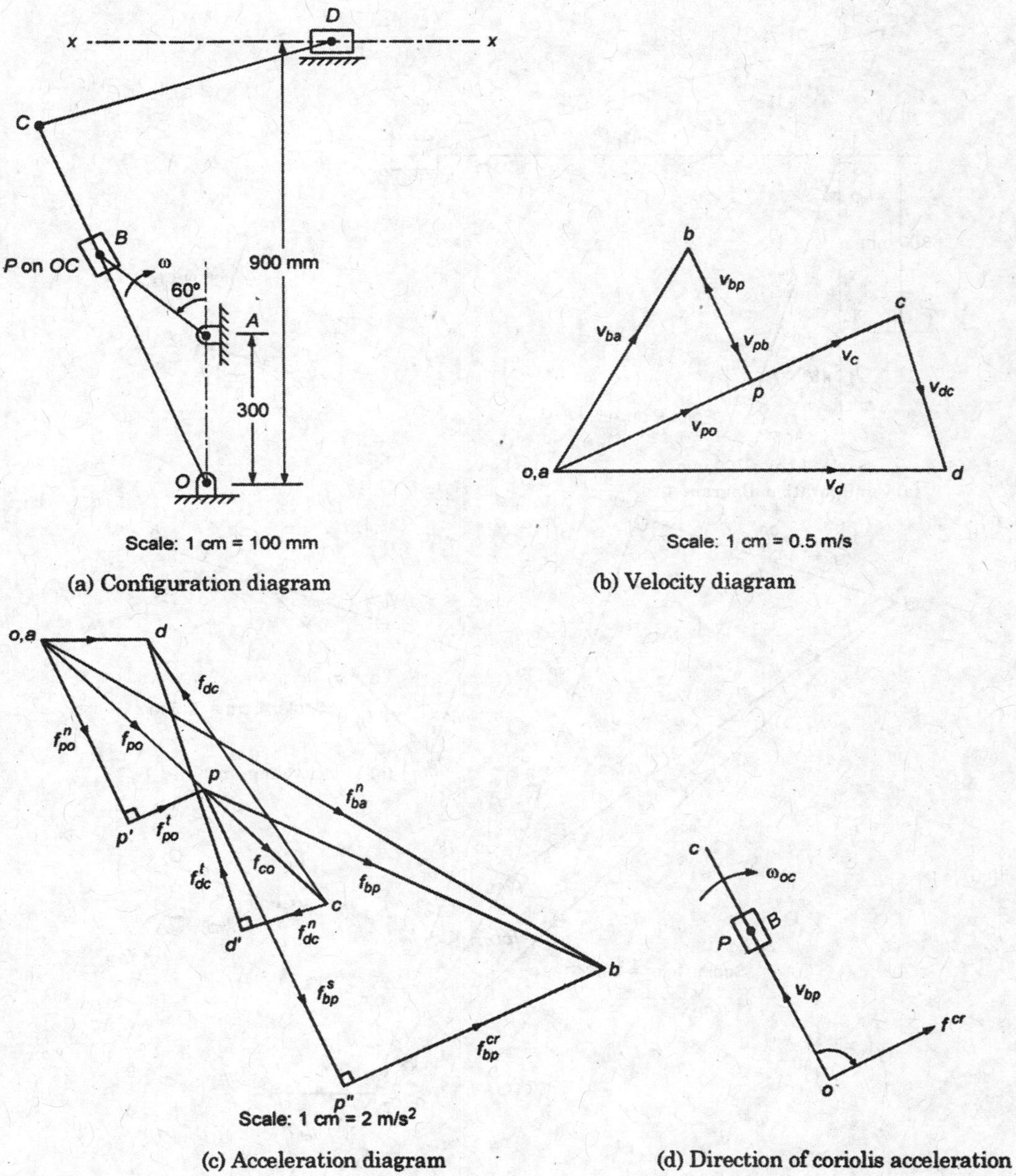

Fig.3.14 Crank and slotted lever mechanism

Example 3.6

Draw the acceleration diagram for the Whitworth mechanism shown in Fig.3.15(a).

$$O_1O_2 = 300 \text{ mm}, \ O_1A = 200 \text{ mm},$$
$$AB = 700 \text{ mm}, \ BC = 800 \text{ mm},$$
$$\angle AO_1O_2 = 45°.$$

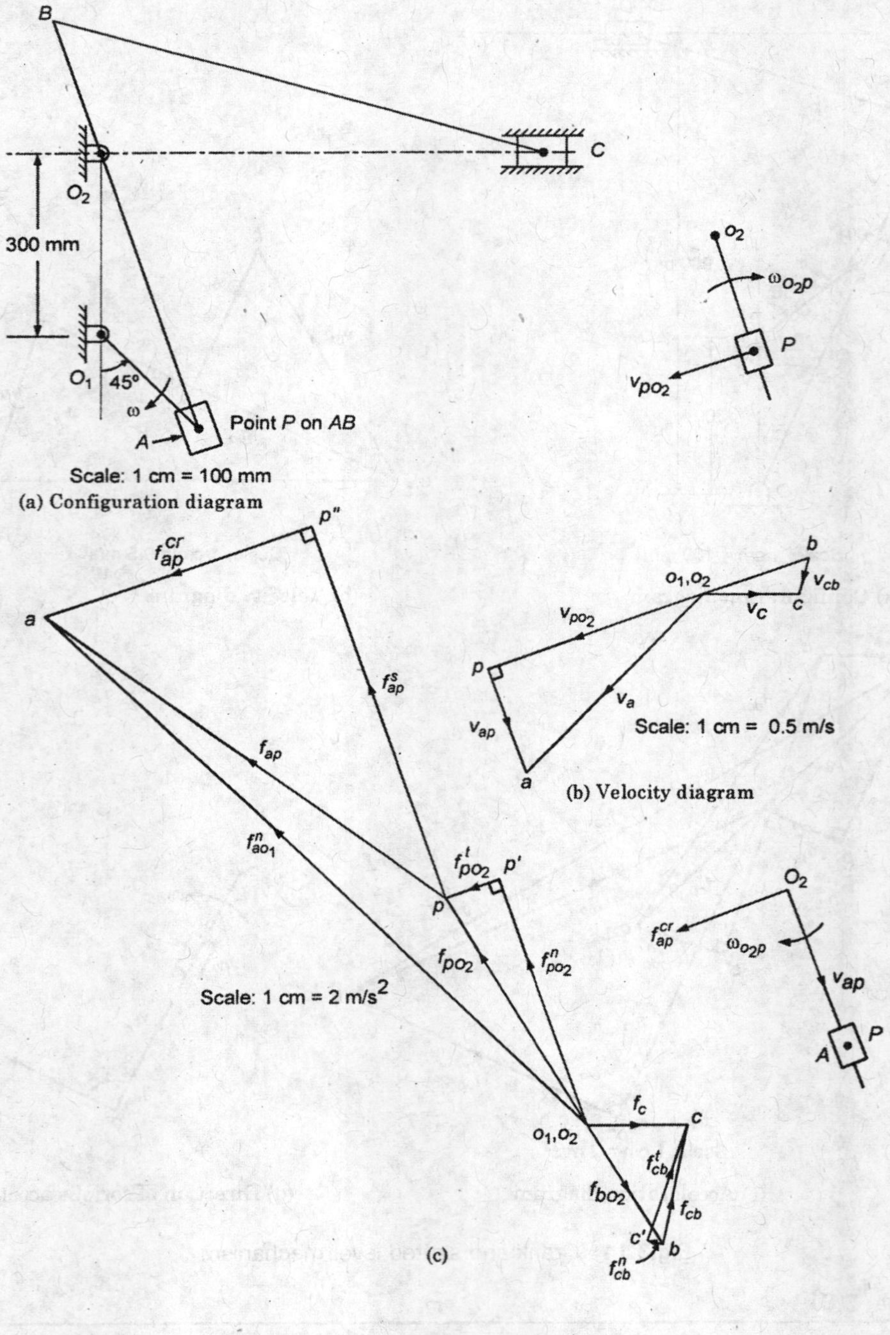

Fig.3.15 Whitworth mechanism

The crank O_1A rotates at 120 rpm clockwise. Determine (a) Velocity of ram C, (b) Coriolis acceleration, (c) angular acceleration of AB, (d) angular acceleration of BC, and (e) acceleration of ram C.

■ Solution

$$\omega = 2\pi \times \frac{120}{60} = 12.57 \text{ rad/s}$$

$$v_a = 12.57 \times 0.2 = 2.51 \text{ m/s}$$

Velocity diagram

1. Draw the configuration diagram to a scale of 1 cm = 100 mm, as shown in Fig.3.15(a). By measurement, $O_2P = 460$ mm. Point P is the coincident point of A on the link AB, whereas point A is on the crank O_1A.
2. Draw the velocity diagram to a scale of 1 cm = 0.5 m/s, as shown in Fig.3.15(b), by adopting the following steps.
3. Draw $o_1a = v_a = 5.02$ cm $\perp O_1A$.
4. Draw a line at 'a' parallel to PB and another line at o_2 perpendicular to PB to meet at point p. $v_{ap} = pa$ represents the velocity of sliding between A and P.
5. By measurement, $o_2p = 4.5$ cm.

Now $$pb = o_2p \times \frac{PB}{O_2P} = \frac{4.5 \times 700}{460} = 6.85 \text{ cm}$$

6. Extend line po_2 to b such that $pb = 6.85$ cm.
7. Draw a line at b perpendicular to BC and another line parallel to the path of the ram at C to meet at point c. Then

$$v_{ap} = pa = 2.3 \text{ cm} = 1.15 \text{ m/s}$$

$$v_{po2} = o_2p = 4.5 \text{ cm} = 2.25 \text{ m/s}$$

$$v_{cb} = bc = 0.8 \text{ cm} = 0.4 \text{ m/s}$$

Velocity of ram C, $vc = o_1c = 1.9$ cm $= 0.85$ m/s

Angular velocity of link O_2P, $\omega_{o2p} = \dfrac{v_{po2}}{O_2P} = \dfrac{2.25}{0.46} = 4.89$ rad/s cw (see Fig.3.15(d))

Angular velocity of BC, $\omega_{bc} = \dfrac{v_{bc}}{BC} = \dfrac{0.4}{0.8} = 0.5$ rad/s

Acceleration diagram

Coriolis acceleration, $f_{ap}^{cr} = 2v_{ap}\omega_{o2p} = 2 \times 1.15 \times 4.89 = 11.247$ m/s

Refer to figure for direction of Coriolis acceleration.

$$f_a^n = \frac{v_a^2}{O_1A} = \frac{(2.51)^2}{0.2} = 31.5 \text{ m/s}^2$$

$$f_{po2}^n = \frac{v_{po2}^2}{O_2P} = \frac{(2.25)^2}{0.46} = 11 \text{ m/s}^2$$

$$f_{cb}^n = \frac{v_{cb}^2}{BC} = \frac{(0.4)^2}{0.8} = 0.2 \text{ m/s}^2$$

1. Draw the acceleration diagram as shown in Fig.3.15(c) to a scale of 1 cm = 2 m/s², by adopting the following procedure:
2. Draw $o_1 a = f_a^n = 15.75$ cm parallel to O_1A.
3. At point 'a' draw a line $ap'' = f_{ap}^{cr} = 5.6$ cm perpendicular to PB. At point p' draw another line perpendicular to PB representing the acceleration of sliding f_{ap}^s between A and P. $f_{ap} = pa$ is the total acceleration of A relative to P.
4. Draw $o_2 p' = f_{po2}^n = 5.5$ cm parallel to O_2P. At p' draw another line perpendicular to O_2P representing f_{o2p}^t to meet the line f_{ap}^s at point p. Join ap and o_2p. Then

$$f_{ap} = pa \text{ and } f_{po2} = o_2 p$$

5. By measurement $po_2 = 5.7$ cm $= 11.4$ m/s²
6. Now $pb = \dfrac{po_2 \times O_2B}{O_2P} = \dfrac{5.7 \times 700}{460} = 8.67$ cm.
7. Produce po_2 to b so that $pb = 8.67$ cm.
8. At point b draw $bc' = f_{cb}^n = 0.1$cm parallel to BC and draw a line at c' perpendicular to BC representing f_{cb}^t.
9. Draw a line at o_2 parallel to the path of ram at C meeting the line drawn at c' at c. Then Acceleration of ram $C, f_c = o_2c = 2.2$ cm $= 4.4$ m/s²
10. Calculate the angular accelerations as follows:

Tangential acceleration of $O_2P, f_{po2}^t = p'p = 1.2\,\text{cm} = 4.4\text{m/s}^2$

$$\text{Angular acceleration of } O_2P = \frac{f_{po2}^t}{O_2P} = \frac{2.4}{0.46} = 5.21 \text{ rad/s}^2$$

Tangential acceleration of $BC, f_{cb} = c'c = 2.7$ cm $= 5.4$ m/s²

$$\text{Angular acceleration of } BC = \frac{f_{cb}^t}{BC} = \frac{5.4}{0.8} = 6.75 \text{ rad/s}^2$$

(a) 0.85 m/s (b) 11.247 m/s² (c) 5.21 rad/s² (d) 6.75 rad/s² (e) 4.4 m/s²

3.5 LINK SLIDING IN A SWIVELLING PIN

Consider a link AB sliding through a swivelling pin O, as shown in Fig.3.16(a). C is a point on link AB. The point A moves up with velocity v_a and acceleration f_a.

Velocity diagram

1. Draw $oa = v_a$ parallel to the path of motion of point A, as shown in Fig.3.16(b).
2. From 'a' draw a line perpendicular to AB to represent the velocity of A relative to B, v_{ba}.
3. From o draw a line parallel to AB to represent the sliding velocity between C and O, v_{co}, to meet the above line at c.
4. Produce ac to b such that $\dfrac{ac}{ab} = \dfrac{AC}{AB}$. Join o to b. Then

$$v_b = ob$$
$$v_{co} = oc$$
$$v_{ca} = ac$$
$$v_{bc} = cb$$

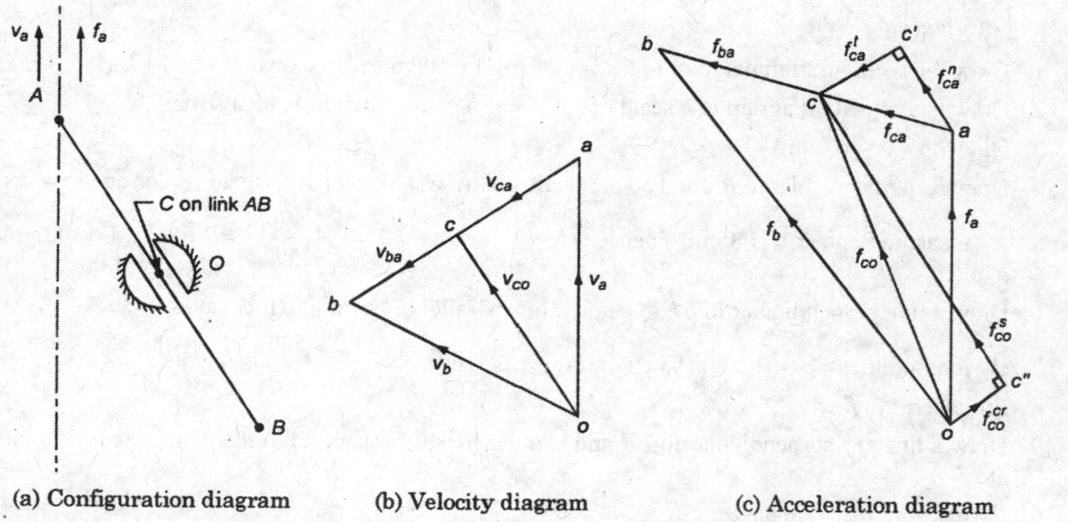

Fig.3.16 Link sliding in a swivelling pin

Acceleration diagram

$$f_{co}^{cr} = 2\, v_{CD} \times \omega_{ab}$$

$$f_{ca}^{n} = \frac{v_{ca}^2}{AC}$$

1. Draw $oa = f_a$ parallel to the path of A, as shown in Fig.3.16(c).
2. Draw $ac' = f_{ca}^n = \frac{v_{ca}^2}{CA} = \frac{ac^2}{CA}$ parallel to AB and a line perpendicular to AB at c' to represent the tangential acceleration f_{ca}^t.
3. Draw the coriolis acceleration of CO, $f_{co}^{cr} = oc''$ perpendicular to AB at o.
4. Draw the sliding acceleration f_{ca}^s between C and O parallel AB at c'' to meet the f_{ca}^t line at c.
5. Join ac and oc.
6. Measure ac. Then $\frac{ab}{ac} = \frac{AB}{AC}$.
7. Produce ac to b.
8. Join ob. Then $f_b = ob$.

Example 3.7

In the swivelling joint mechanism shown in Fig.3.17(a), $AB = 300$ mm, $BC = 800$ mm, $CD = 400$ mm, $AD = 500$ mm, $BE = 400$ mm, $EF = 500$ mm, $ES = 250$ mm, and $FR = 600$ mm. The crank AB rotates at 20 rads/s and 200 rad/s². Determine (a) the coriolis acceleration, and (b) the sliding acceleration of link EF in the trunnion.

■ Solution

Given: $\omega = 20$ rad/s, $\alpha = 200$ rad/s²

$$v_{ba} = 20 \times 0.3 = 6 \text{ m/s}$$

Velocity diagram

1. Draw the configuration diagram to a scale of 1 cm = 100 mm, as shown in Fig.3.17(a).
2. Draw the velocity diagram to a scale of 1 cm = 1 m/s, as shown in Fig.3.17(b).
3. Draw $ab = v_{ba} = 6$ cm.
4. Draw bc perpendicular to BC and dc perpendicular to CD to meet at c. Then $v_{cb} = bc$ and $v_{cd} = dc$.
5. By measurement, $bc = 5.3$ cm. Then $\frac{be}{bc} = \frac{BE}{BC}$, so that $be = \frac{5.3 \times 400}{259} = 2.65$ cm. Locate point e in bc.
6. Draw a line perpendicular to EF at e and a line parallel to EF at g to meet at s.
7. By measurement, $es = 4.2$ cm. Then $\frac{ef}{es} = \frac{EF}{ES}$, so that $ef = \frac{4.2 \times 500}{259} = 8.4$ cm.
8. Produce es to f.
9. Draw a line at f perpendicular to FR and another line at 'a' parallel to the path of ram R to meet at r.
10. By measurement, we have

$$v_{se} = es = 4.2 \text{ cm} = 4.2 \text{ mm/s}$$

$$v_{sg} = gs = 3.6 \text{ cm} = 3.6 \text{ mm/s}$$

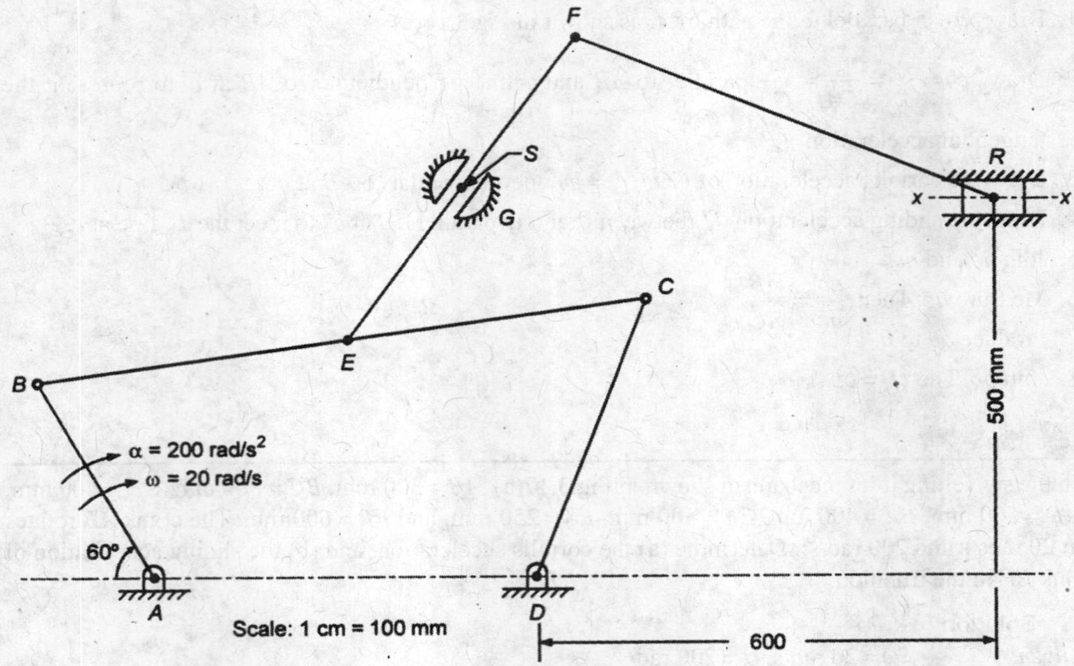

(a) Configuration diagram

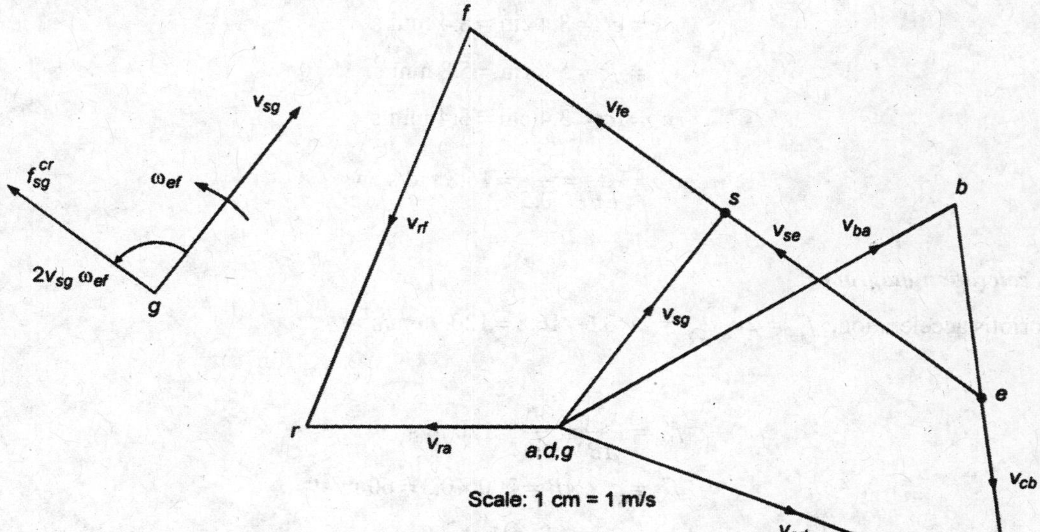

(b) Velocity diagram

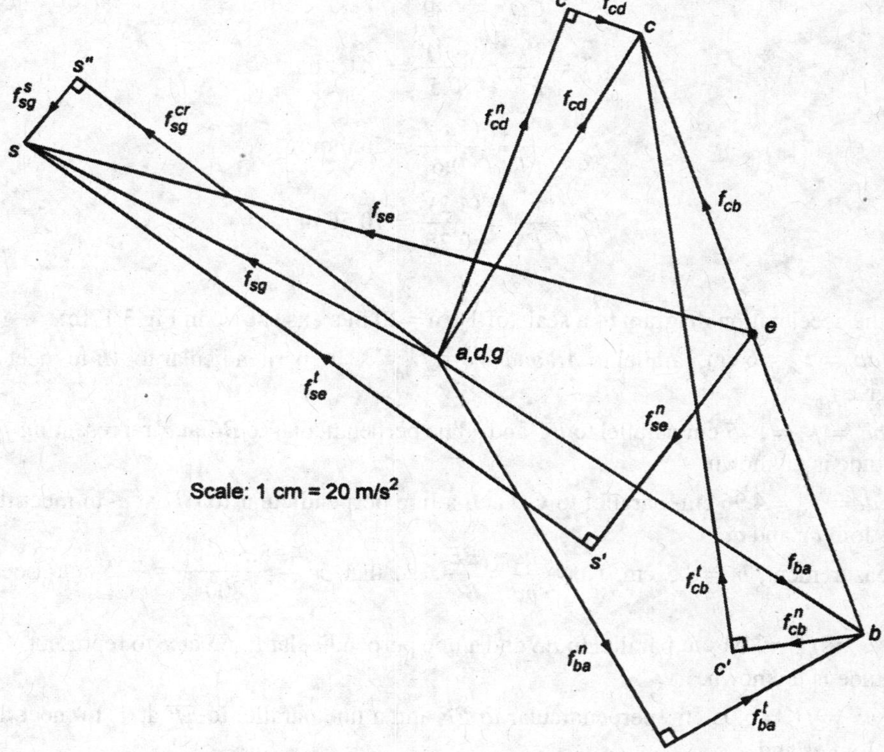

(c) Acceleration diagram

Fig.3.17 Swivelling joint mechanism

$$v_{fe} = ef = 8.4 \text{ cm} = 8.4 \text{ mm/s}$$

$$v_{rf} = fr = 5.8 \text{ cm} = 5.8 \text{ mm/s}$$

$$v_r = ar = 3.4 \text{ cm} = 3.4 \text{ mm/s}$$

$$\omega_{ef} = \frac{v_{fe}}{EF} = \frac{8.4}{0.5} = 16.8 \text{ rad/s cw}$$

Acceleration diagram

Coriolis acceleration, $f_{sg}^{cr} = 2v_{sg}\omega_{ef} = 2 \times 3.6 \times 16.8 = 120.96 \text{ m/s}^2$

$$f_{ba}^{cr} = \frac{v_{ba}^2}{AB} = \frac{36}{0.3} = 120 \text{ m/s}^2$$

$$f_{ba}^{t} = a \times AB = 200 \times 0.3 = 60 \text{ m/s}^2$$

$$f_{cb}^{n} = \frac{v_{cb}^2}{BC} = \frac{(5.3)^2}{0.8} = 35.1 \text{ m/s}^2$$

$$f_{cd}^{n} = \frac{v_{cd}^2}{CD} = \frac{(6.3)^2}{0.4} = 99.2 \text{ m/s}^2$$

$$f_{fe}^{n} = \frac{v_{fe}^2}{EF} = \frac{(8.4)^2}{0.5} = 141.1 \text{ m/s}^2$$

$$f_{rf}^{n} = \frac{v_{rf}^2}{FR} = \frac{(5.8)^2}{0.6} = 56.0 \text{ m/s}^2$$

$$f_{se}^{n} = \frac{v_{se}^2}{ES} = \frac{(4.2)^2}{0.25} = 70.56 \text{ m/s}^2$$

1. Draw the acceleration diagram to a scale of 1 cm = 20 m/s², as shown in Fig.3.17(c).
2. Draw $ab' = f_{ba}^{n} = 6$ cm parallel to *AB* and $b'b = f_{ba}^{t} = 3$ cm perpendicular to *AB* to meet at *b*. Join *ab* to give f_{ba}.
3. Draw $bc' = f_{cb}^{n} = 1.75$ cm parallel to *BC* and a line perpendicular to *BC* at *c′* representing f_{bc}^{t}, whose magnitude is unknown.
4. Draw $dc = f_{cd}^{n} = 4.96$ cm parallel to *CD* and a line perpendicular to *CD* at *c′* to meet the line at *c′* at *c*. Join *bc* and *dc*.
5. By measurement, *bc* = 8.5 cm. Now $\frac{be}{bc} = \frac{BE}{BC}$, so that $be = \frac{8.5 \times 400}{800} = 4.25$ cm. Locate point *e* in *bc*.
6. Draw $es' = f_{se}^{n} = 3.51$ cm parallel to *ES* and a line perpendicular to *ES* at *s′* to represent f_{se}^{t}, whose magnitude is unknown.
7. Draw $as'' = f_{sg}^{cr} = 6.05$ cm perpendicular to *EF* and a line parallel to *EF* at *s″* to meet the line at *s′* at *s*. Join *es* and *gs*.
8. Sliding acceleration of link *EF* in the trunnions is, $f_{sg}^{s} = s''s = 1\text{cm} = 20 \text{ m/s}^2$.

Example 3.8

For the mechanism shown in Fig.3.18(a), find the angular accelerations of the links AB, BO_2 and the linear accelerations of points C, D, and E. $\omega = 10$ rad/s, $\alpha = 200$ rad/s^2.

$$O_1O_2 = 100 \text{ mm}, O_1A = 50 \text{ mm}, AB = 45 \text{ mm}, AD = 30 \text{ mm}, AC = 45 \text{ mm},$$
$$BC = 30 \text{ mm}, O_2B = 55 \text{ mm}, O_2E = 45 \text{ mm, and } BE = 25 \text{ mm}.$$

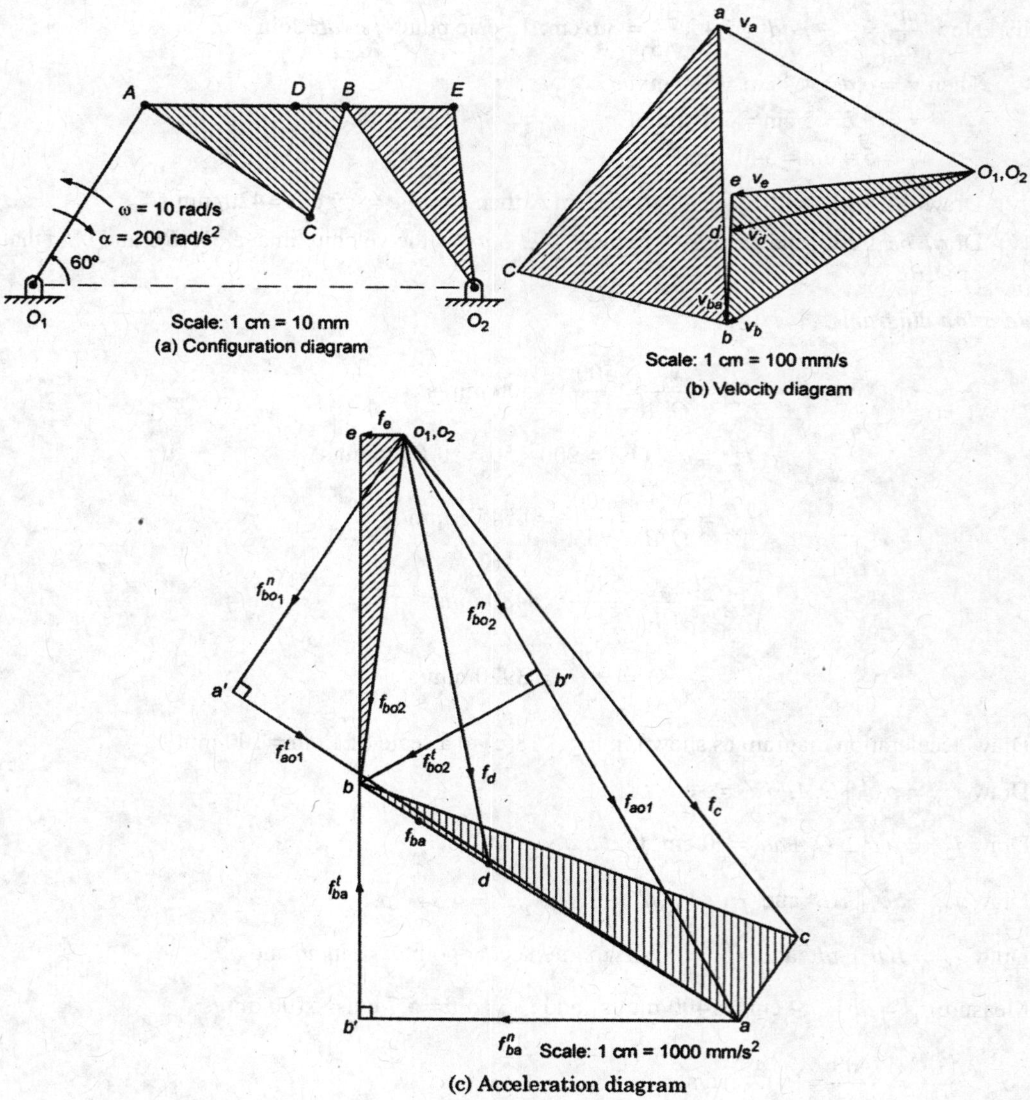

Fig.3.18 Four-bar mechanism with two ternary links

Solution

Velocity diagram

1. Draw the configuration diagram shown in Fig.3.18(a) to a scale of 1 cm = 10 mm.
2. Given: $\omega = 10$ rad/s, $v_a = \omega \times O_1A = 10 \times 50 = 500$ mm/s.
3. Draw the velocity diagram shown in Fig.3.18(b) to a scale of 1 cm = 100 mm/s.
   (i) Draw $v_a = o_1a \perp O_1A, o_1a = 5$ cm.
   (ii) Draw $o_2b \perp O_2B$ and $ab \perp AB$ to intersect at point b. Measure $ab = 5.4$ cm.
   (iii) Now $\dfrac{ad}{ab} = \dfrac{AD}{AB}$, $ad = 5.4 \times \dfrac{30}{45} = 3.6$ cm. Locate point d in ab. Join o_1d.
   Then $v_d = o_1d = 4.4$ cm = 440 mm/s
   $v_b = o_2b = 5$ cm = 500 mm/s
   $v_{ba} = 5.4$ cm = 540 mm/s
   (iv) Draw $be \perp BE$ and $o_2e \perp O_2$ to meet at e. then $v_e = o_2\,e = 4.2$ cm = 420 mm/s
   (v) Draw $bc \perp BC$ and $ac \perp AC$ to meet at c. Δabc is the velocity image of ABC and o_2be that of O_2BE.

Acceleration diagram

$$f^n_{ao1} = \frac{v_a^2}{O_1A} = \frac{500^2}{50} = 500 \text{ mm/s}^2$$

$$f^t_{ao1} = a \times O_1A = 200 \times 50 = 10{,}000 \text{ mm/s}^2$$

$$f^n_{bo2} = \frac{v_b^2}{O_2B} = \frac{500^2}{55} = 4545.5 \text{ mm/s}^2$$

$$f^n_{ba} = \frac{v_{ba}^2}{AB} = \frac{540^2}{45} = 6480 \text{ mm/s}^2$$

$$f^n_{eo2} = \frac{v_e^2}{O_2E} = \frac{420^2}{45} = 3920 \text{ mm/s}^2$$

1. Draw acceleration diagram as shown in Fig.3.18(c) to a scale of 1 cm = 100 mm/s^2.
2. Draw $f^n_{ao1} = o_1a' \parallel O_1A,\ o_1a' = 5$ cm.
3. Draw $f^t_{ao1} = a'a \perp O_1A\ aa' = 10$ cm. Join o_1a. $o_1a = f_{aol}$.
4. Draw $f^n_{ba} = ab' \parallel AB$ and $f^n_{bo2} = o_2b'' \parallel O_2B,\ o_2b'' = 4.54$ cm.
5. Draw $f^t_{ba} = b'b \perp ab'$ and $f^t_{bo2} = b''b \perp$ to intersect at point b. Join ab and o_2b.
6. Measure $f_{ba} = ab = 7.4$ cm = 7400 mm/s^2 and $f_{bo2} = o_2b = 5.7$ cm = 5700 mm/s^2.
7. $\alpha_{ba} = \dfrac{f_{ba}}{AB} = \dfrac{7400}{45} = 164.4 \text{ rad/s}^2$

   $\alpha_{bo2} = \dfrac{f_{bo2}}{O_2B} = \dfrac{5700}{55} = 103.64 \text{ rad/s}^2.$

8. Now $\frac{ad}{ab} = \frac{AD}{AB}$, $ad = 7.4 \times \frac{30}{45} = 4.86$ cm. Locate point d in ab. Join o_1d. Then $f_d = o_1d$.

9. Draw $ac \perp AC$ and $be \perp BC$. Join o_1c. $f_c = o_1c = 10.6$ cm $= 10600$ mm/s².

10. Draw $be \perp BC$ and $o_2e \perp O_2E$ to meet at e. Join o_2e. Then $f_e = o_2e = 0.4$ cm $= 400$ mm/s².

Example 3.9

For the slider-crank mechanism shown in Fig.3.19(a), determind (a) acceleration of slider B, (b), acceleration of point C, and (c) acceleration of link AB. The crank OA rotates at 180 rpm. $OA = 500$ mm, $AB = 1500$ mm, and $AC = 250$ mm.

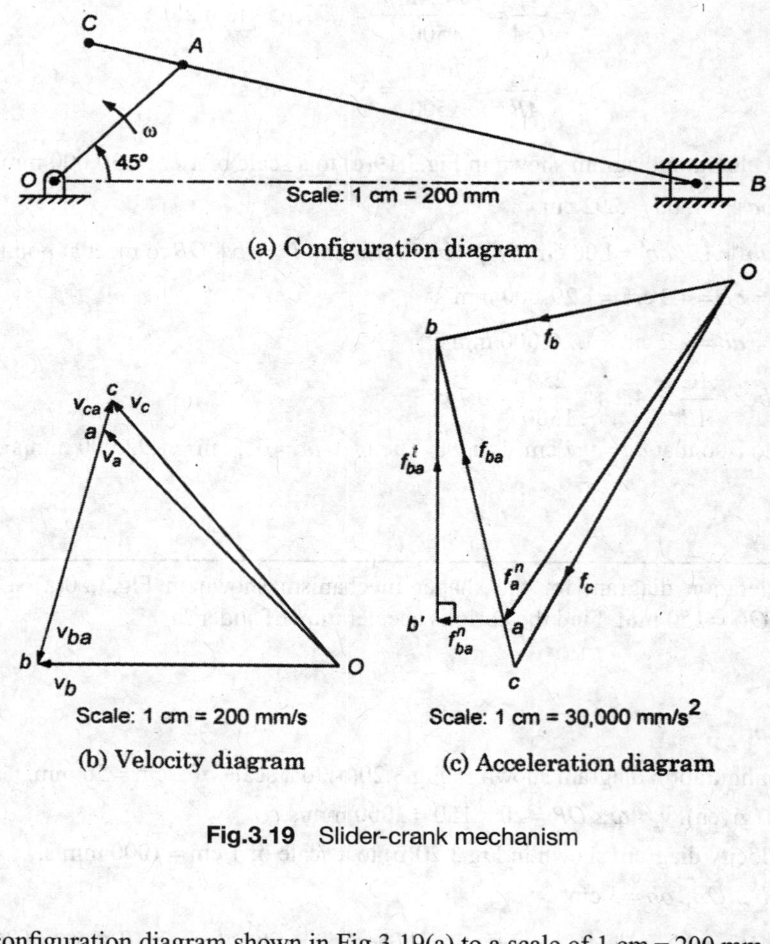

Fig.3.19 Slider-crank mechanism

Solution

1. Draw the configuration diagram shown in Fig.3.19(a) to a scale of 1 cm = 200 mm.

2. Angular speed of crank OA, $\omega = 2\pi \times \frac{180}{60} = 18.85$ rad/s

$$v_a = \omega \times OA = 18.85 \times 500 = 4424.8 \text{ mm/s.}$$

3. Draw the velocity diagram shown in Fig.3.19(b) to a scale of 1 cm = 200 mm/s.

4. Draw $v_a = oa \perp OA$, $oa = 4.71$ cm.
5. Draw $ab \perp AB$ and $ob \parallel OB$ to meet at point b. $v_b = ab = 4.5$ cm $= 8400$ mm/s

$$v_{ba} = ab = 3.5 \text{ cm} = 7000 \text{ mm/s}$$

6. Now $\dfrac{ac}{ab} = \dfrac{AC}{AB}, ac = 3.5 \times \dfrac{250}{1500} = 0.583$ cm. Produce ba to c so that $ac = 0.583$ cm. Join oc. Then $v_c = oc = 5$ cm $= 10{,}000$ mm/s.

Acceleration diagram

$$f_{oa}^n = \frac{v_a^2}{OA} = \frac{(9424.8)^2}{500} = 1{,}77{,}654 \text{ mm/s}^2$$

$$f_{ba}^n = \frac{v_{ba}^2}{AB} = \frac{7000^2}{1500} = 32667 \text{ mm/s}^2$$

1. Draw the acceleration diagram shown in Fig.3.19(c) to a scale of 1 cm = 30,000 mm/s².
2. Draw $f_{oa}^n = oa \parallel OA$, $oa = 5.92$ cm.
3. Draw $f_{ba}^n = ab' \parallel AB$, $ab' = 1.08$ cm. $f_{ba}^t = b'b \perp AB$ and $f_b = ob \parallel OB$ to meet at point b. Join ab.
4. Measure $f_b = ob = 4.3 \text{ cm} = 1{,}29{,}000 \text{ mm/s}^2$

   $f_{ba} = ab = 4.2 \text{ cm} = 1{,}26{,}000 \text{ mm/s}^2$
5. Now $ac = ab \times \dfrac{AC}{AB} = 4.2 \times \dfrac{250}{1500} = 0.7$ cm.
6. Produce ba to c so that $ac = 0.7$ cm. Join oc. Then $f_c = oc = 6.4$ cm $= 1{,}92{,}000$ mm/s².

Example 3.10

Draw the acceleration diagram for the shaper mechanism shown in Fig.3.20(a). $OB = 150$ mm, $CB = 225$ mm, $OC = 150$ mm. Find the coriolis acceleration of slider B.

■ Solution

Velocity diagram

1. Draw the configuration diagram shown in Fig.3.20(a) to a scale of 1 cm = 50 mm.
2. $\omega = 20$ rad/s (given). $v_b = \omega \times OB = 20 \times 150 = 3000$ mm/s.
3. Draw the velocity diagram shown in Fig.3.20(b) to a scale of 1 cm = 1000 mm/s.
4. Draw $v_b = ob \perp OB$. $ob = 3$ cm.
5. Draw $cd \perp AC$ and $db \parallel AC$ to meet at d. Then, velocity of slider B along ABC, $v_{bd} = db = 2$ cm $= 2000$ mm/s.

$$v_{dc} = cd = 1.5 \text{ cm} = 1.5 \text{ m/s}$$

6. Angular speed of ABC link, $\omega_{AC} = \dfrac{v_{dc}^2}{CD} = \dfrac{1.5 \times 100}{225} = 6.67$ rad/s.

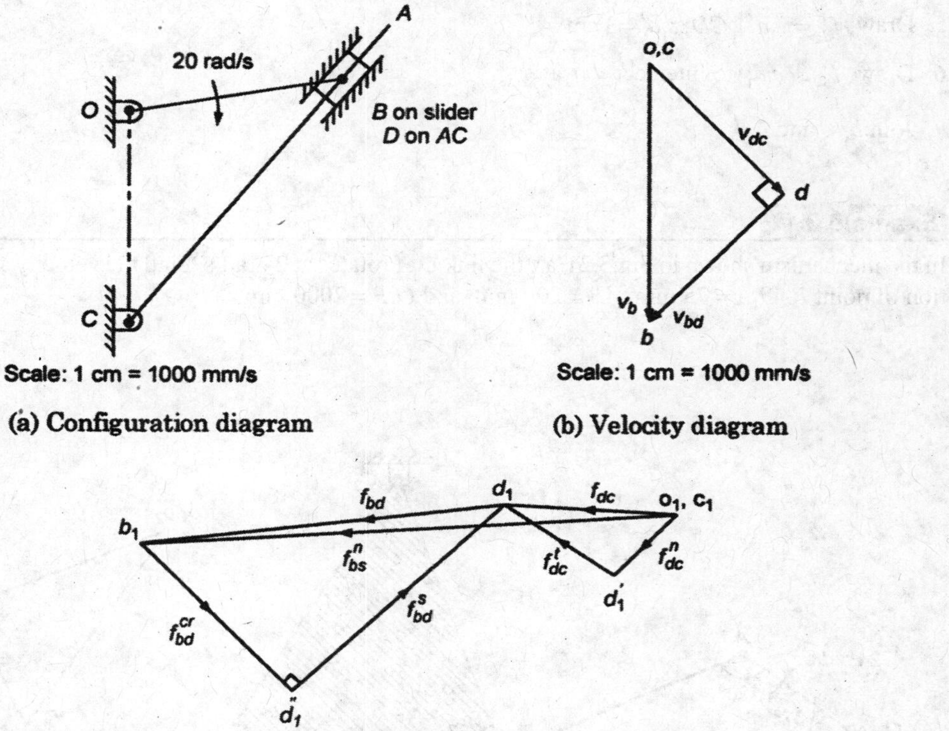

(a) Configuration diagram

(b) Velocity diagram

(c) Acceleration diagram

Fig.3.20 Shaper mechanism

Acceleration diagram

Coriolis acceleration of slider *B*, $f_{bd}^{cr} = 2\, v_{bd}\, \omega_{AC} = 2 \times 2000 \times 6.67 = 26680$ mm/s^2 and is perpendicular to *CB*.

$$f_{bo}^n = \frac{v_b^2}{OB} = \frac{3000^2}{150} = 60{,}000 \text{ mm/s}^2 \text{ along } BO.$$

$$f_{dc}^n = \frac{v_{dc}^2}{CD} = \frac{1500^2}{225} = 10000 \text{ mm/s}^2 \text{ along } CD.$$

1. Draw acceleration diagram shown in Fig.3.20(c) to a scale of 1 cm = 10,000 mm/s^2.
2. Draw $f_{bn}^n = o_1 b_1 \parallel OB,\ o_1 b_1 = 6$ cm.
3. Draw $f_{bd}^{cr} = b_1 d_1'' \perp CD,\ b_1 d_1'' = 2.67$ cm.
4. Draw $f_{bd}^s \perp b_1 d_1''$.

5. Draw $f^n_{dc} = c_1 d_1' \parallel CD, c_1 d_1' = 1\,\text{cm}$

6. Draw $f^t_{bc} \perp c_1 d_1'$ to intersect f^s_{bd} at d_1.

7. Join $b_1 d_1$ and $c_1 d_1$.

Example 3.11

In the mechanism shown in Fig.3.21(a), the link O_1A rotates at 24 rad/s. Find the velocity and acceleration of point B. $O_1A = 75$ mm, $AB = 200$ mm, and $O_2B = 2000$ mm.

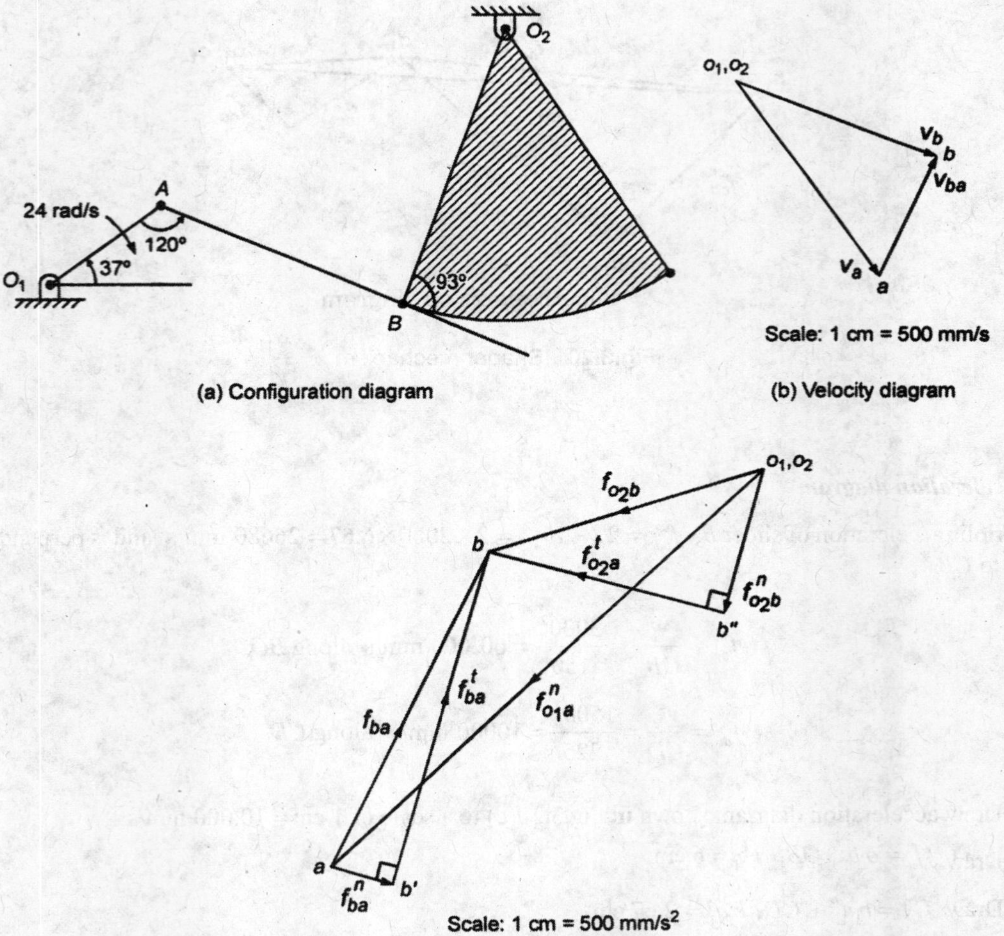

Fig.3.21 Four-bar mechanism with a sector type ternary link

Solution

Velocity diagram

1. Draw the configuration diagram shown in Fig.3.21(a) to a scale of 1 cm = 50 mm.
2. $\omega = 24$ red/s, $v_a = \omega \times O_1A = 24 \times 75 = 1800$ mm/s.
3. Draw the velocity diagram shown in Fig.3.21(b) to a scale of 1 cm = 500 mm/s.
   (i) Draw $v_a = o_1a \perp O_1A$, $o1\alpha = 3.6$ cm.
   (ii) Draw $ab \perp AB$ and $o_2b \perp O_2B$ to meet at point b. $v_b = o2b = 2$ cm 1500 mm/s.
   (iii) $v_{ba} = ab = 2$ cm = 1000 mm/s.

Acceleration diagram

$$f_{o1a}^{n} = \frac{v_a^2}{O_1A} = \frac{1800^2}{75} = 43,200 \text{ mm/s}^2 \text{ along } AO_1.$$

$$f_{ba}^{n} = \frac{v_{ba}^2}{AB} = \frac{1000^2}{200} = 500 \text{ mm/s}^2 \text{ along AB.}$$

$$f_{o2b}^{n} = \frac{v_b^2}{O_2B} = \frac{1500^2}{200} = 11,250 \text{ mm/s}^2 \text{ along } BO_2.$$

1. Draw the acceleration diagram shown in Fig.3.21(c) to a scale of 1 cm = 500 mm/s².
2. Draw $f_{o1a}^{n} = o_1a \parallel O_1A,\ o_1a = 8.64$ cm

   $f_{ba}^{n} = ab' \parallel AB,\ ab' = 1$ cm
3. Draw $b'b \perp AB, f_{o2b}^{n} = o_2b'' \parallel O_2B, o_2b'' = 2.25$ cm
4. Draw $b''b \perp o_2B$ to meet $b'b$ line at point b. Join ba and bo_2. Then $f_{o2b} = o_2b = 4.2$ cm = 21,000 mm/s².

Example 3.12

A double slider-crank mechanism is shown in Fig.3.22(a). Crank 2 rotates at constant angular speed $\omega_2 = 10$ rad/s. Determine the velocity and acceleration of each slider. $O_2A = 1000$ mm, $AB = 200$ mm, and $AC = 200$ mm.

Solution

Velocity diagram

1. Draw the configuration diagram shown in Fig.3.22(a) to a scale of 1 cm = 25 mm.
2. $\omega_2 = 10$ rad/s, $v_a = \omega_2 \times O_2A = 10 \times 100 = 1000$ mm/s.
3. Draw the velocity diagram as shown in Fig.3.22(b) to a scale of 1 cm = 200 mm/s.
4. Draw $v_a = o_{2a} \perp O_2A,\ o_{2a} = 5$ cm.

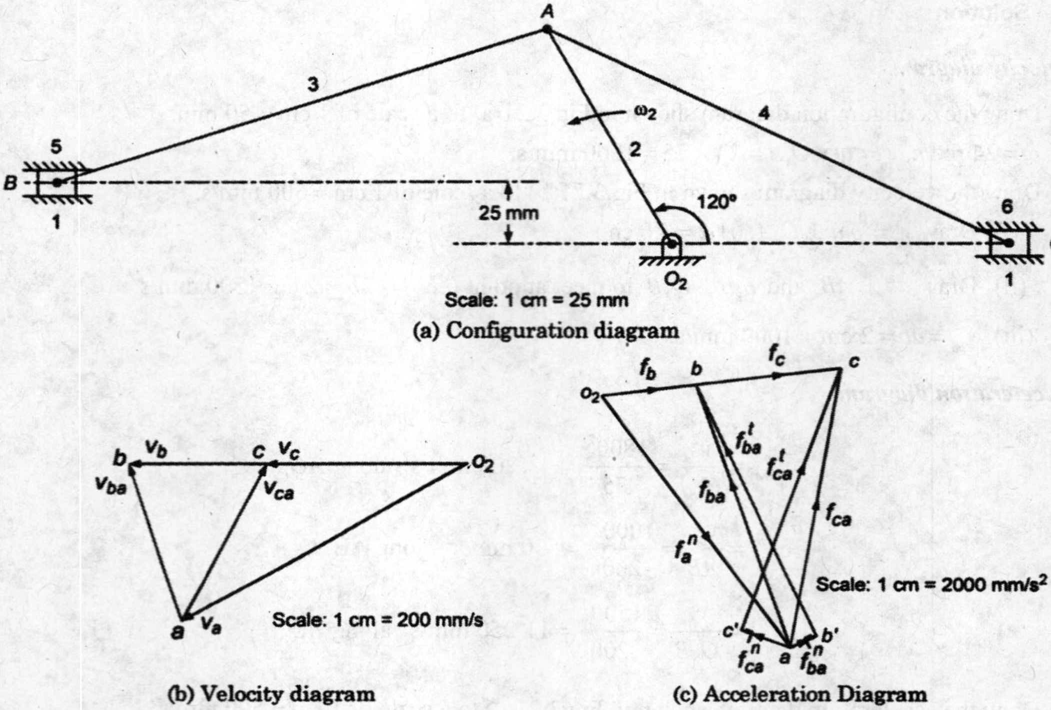

Fig.3.22 Double slider-crank mechanism

5. Draw $ab \perp AB$ and $ac \perp AC$, $o_2cb \parallel o_2c$ to meet ab and ac lines at b and c respectively.
6. Measure $v_c = o_2c = 3.1 \text{ cm} = 620 \text{ mm/s}$
   $v_b = o_2b = 5.2 \text{ cm} = 1040 \text{ mm/s}$
   $v_{ba} = ab = 2.2 \text{ cm} = 440 \text{ mm/s}$
   $v_{ca} = ac = 2.8 \text{ cm} = 560 \text{ mm/s}$

Acceleration diagram

$$f^n_{o2a} = \frac{v_a^2}{O_2A} = \frac{1000^2}{100} = 10{,}000 \text{ mm/s}^2 \text{ along } AO_2.$$

$$f^n_{ba} = \frac{v_{ba}^2}{AB} = \frac{440}{200} = 968 \text{ mm/s}^2 \text{ along } AB.$$

$$f^n_{ca} = \frac{v_{ca}^2}{AC} = \frac{500^2}{200} = 1568 \text{ mm/s}^2 \text{ along } AC.$$

1. Draw the acceleration diagram as shown in Fig.3.22(c) to a scale of 1 cm = 2000.
2. Draw $f^n_{o2a} = f^n_a = o_2a \parallel O_2A, o_2a = 5 \text{ cm}$
   $f^n_{ba} = ab \parallel AB, ab' = 0.484 \text{ cm}$

$b'b \perp AB$ and $f^n_{o2b} = o_2b \parallel O_2C$ to meet $b'b$ at point b

$f_b = o_2b = 1.4 \times 200 = 2800$ mm/s^2

3. Draw $f_{ca} = ac' \parallel AC, ac' = 0.784$ cm.

$c'c \perp AC$ and $o_2c \parallel O_2C$ to meet at c.

Then $f_c = o_2c = 3.7$ cm $= 7400$ mm/s^2.

Example 3.13

A part of a mechanism which operates a horizontally sliding block *E* is shown in Fig.3.23(a). In the given configuration, the lever *OB* swings about *O* in the clockwise direction with an angular velocity of 11.2 rad/s and an angular acceleration of 56.5 rad/s^2. The reciprocating mass at *E* is 70 kg. Determine the value of force *P* necessary to ensure this motion, neglecting the inertia of the links *OB* and *CE*.

■ Solution

Velocity diagram

1. Draw configuration diagram shown in Fig.3.23(a) to a scale of 1 cm = 50 mm.
2. $\omega = 11.2$ rad/s,

$$v_a = \omega \times OA = 11.2 \times 150 = 1680 \text{ mm/s}$$
$$v_b = \omega \times OB = 11.2 \times 250 = 2800 \text{ mm/s}$$

3. Draw the velocity diagram as shown in Fig.3.23(b) to a scale of 1 cm = 500 mm/s
4. Draw $v_b = ob \perp AB$, $ob = 5.6$ cm. $v_a = oa = 3.36$ cm
5. Draw $ae \perp AE$ and $oe \parallel XE$ to meet at point *e*. Then

$$v_e = oe = 2.9 \text{ cm} = 1450 \text{ mm/s}$$
$$v_{ea} = ae = 0.8 \text{ cm} = 400 \text{ mm/s}$$

Acceleration diagram

$$f^n_a = \frac{v_a^2}{OA} = \frac{1680^2}{150} = 18{,}816 \text{ mm/s}^2 \text{ along } AO.$$

$$f^t_a = a \cdot OA = 56.5 \times 150 = 8475 \text{ mm/s}^2 \perp AO.$$

$$f^n_{ea} = \frac{v_{ea}^2}{AE} = \frac{400^2}{250} = 640 \text{ mm/s}^2 \text{ along } AE.$$

1. Draw the acceleration diagram as shown in Fig.3.23(c) to a scale of 1 cm = 2000 mm/s^2.
2. Draw $f^n_a = oa' \parallel OA, oa' = 9.4$ cm.
3. Draw $f^t_a = a'a \perp o'a, a'a = 4.2$ cm. Join *oa*. f_a = oa
4. Draw $f^n_{ea} = ae' \parallel AE, ae' = 0.32$ cm:
5. Draw $e'e \perp AE$ and $oe \parallel XE$ to meet at point *e*. Then

$$f_e = oe = 8.7 \text{ cm} = 1\,7400 \text{ mm/s}^2$$

$$P \cos [90 - (45° + 15°)] = 70 \times f_e$$

$$P = \frac{70 \times 17{,}400 \times 10^{-3}}{\cos 30°} = 1406.5 \text{ N}$$

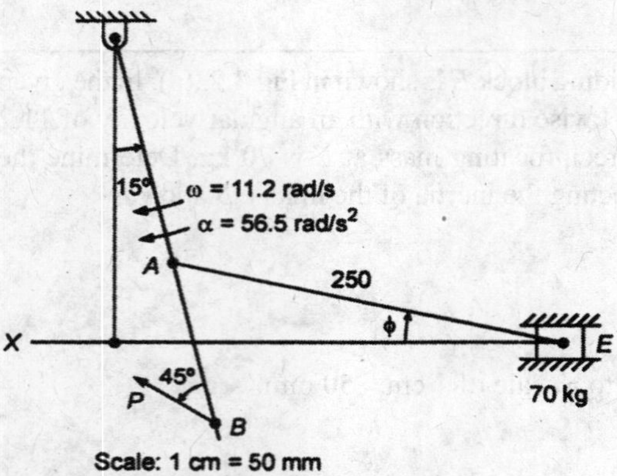

(a) Configuration diagram

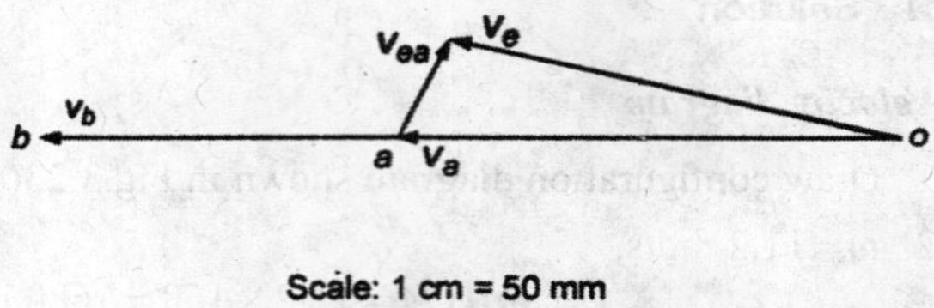

(b) Velocity diagram

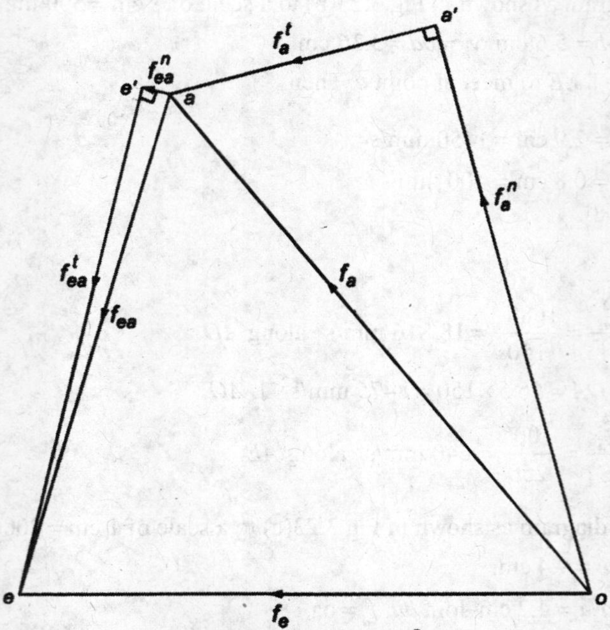

(c) Acceleration diagram

Fig.3.23 Diagram for Example 3.13

3.6 KLEIN'S CONSTRUCTION

Klein's construction is used to draw the velocity and acceleration diagrams for a single slider-crank mechanism on its configuration diagram. The line that represents the crank in the configuration diagram also represents the velocity and the acceleration of its moving end to some scale. The following steps may be adopted to draw the velocity and acceleration diagrams, as shown in Fig.3.24:

1. Draw the configuration diagram *OAB* of the slider-crank mechanism to a convenient scale.
2. Draw $OI \perp OB$ and produce *BA* to meet *OI* at *C*. Then *OAC* represents the velocity diagram. $OA = \omega r$ to some scale from which the scale of velocity diagram is determined. $OC = v_{bp}$, $CA = v_{ba}$, $OA = v_{ao}$.
3. With *AC* as the radius and *A* as the centre, draw a circle.
4. Locate the mid-point *D* of *AB*. With *D* as centre and *DA* as radius, draw the circle to intersect the previously drawn circle at *E* and *F*. Join *EF* intersecting *AB* at *G*.
5. *EF* meets *OB* at *H*. If it does not meet *OB* then produce *EF* to meet *OB* at *H*. Join *AH*.

   Then *OAGH* is the acceleration diagram.

$$OA = f^n_{ao}$$

$$AG = f^n_{ba}$$

Acceleration of piston (slider *B*), $f_b = \omega^2 \cdot OH$

$$f^n_{ba} = \omega^2 \cdot AG$$

$$f^t_{ba} = \omega^2 \cdot GH.$$

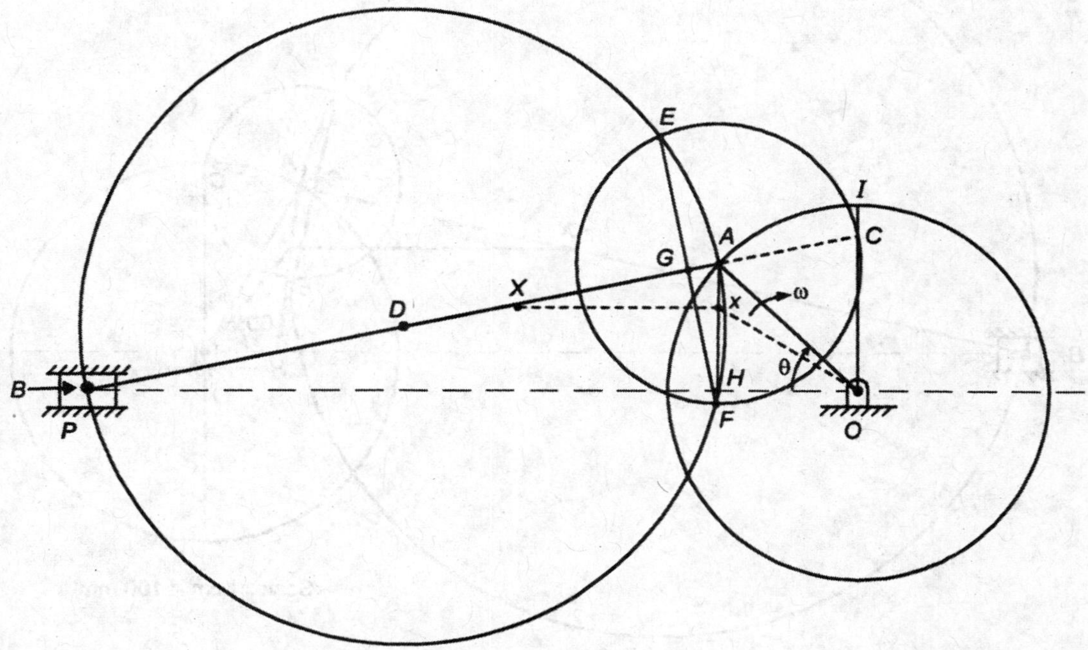

Fig.3.24 Klein's construction

6. To find the acceleration of any point X in AB, draw a line $XX \parallel OB$ to intersect HA at X. Join OX. Then

$$f_x = \omega^2 \cdot OX$$

$$\alpha_{AB} = \frac{f_{ba}^t}{AB} = \frac{\omega^2 \cdot GH}{AB} \text{ (ccw)}$$

Example 3.14

The crank of an engine 300 mm long rotates at a uniform speed of 300 rpm. The ratio of connecting rod length to crank radius is 4. Determine (a) acceleration of the piston. (b) angular acceleration of the rod, and (c) acceleration of a point X on the connecting rod at 400 mm from crank pin. The crank position is 60° from inner dead centre.

■ Solution

The acceleration diagram using Klein's construction has been drawn in Fig.3.25, following the steps as explained in Section. 3.6.

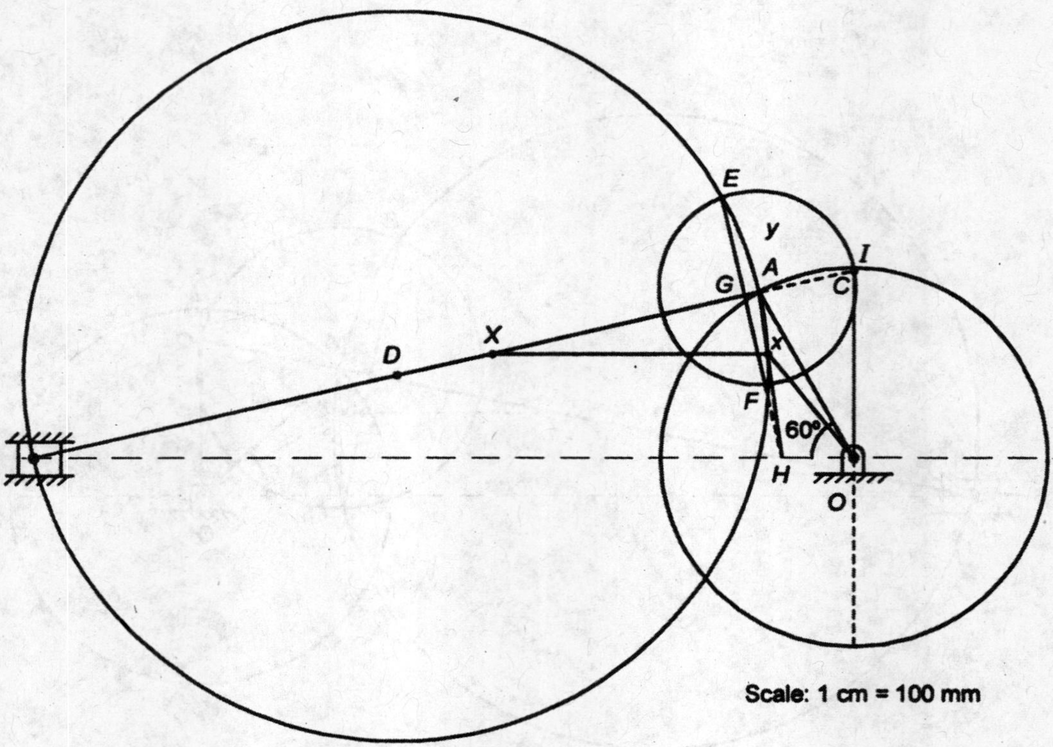

Fig.3.25 Klein's construction for Example 3.14

OAGH is the acceleration diagram.

$$f_{ba}^{n} = \omega^2 \cdot AG$$

$$f_{ba}^{t} = \omega^2 \cdot GH$$

$$f_{ba} = \omega^2 \cdot AH$$

$$\omega^2 = \left(\frac{2\pi \times 300}{60}\right)^2 = 986.96 \text{ (rad/s)}^2$$

Locate point X in AB, such that $AX = 400$ mm. Draw Xx parallel to BO to meet AH at x. Join Ox. Then

$$Ox \propto f_x$$

$$f_b = \omega^2 OH$$

(a) Acceleration of piston, $f_b = \omega^2 \times OH = \dfrac{986.96 \times 1.1 \times 100}{1000} = 108.56 \text{ m/s}^2$

(b) Tangential acceleration of rod, $f_{ba}^{t} = \omega^2 \times GH = \dfrac{986.96 \times 2.6 \times 100}{1000} = 256.61 \text{ m/s}^2$

Angular acceleration of rod, $\alpha_{AB} = \dfrac{f_{ba}^{t}}{AB} = \dfrac{256.61}{1.2} = 213.84 \text{ rad/s}^2$

(c) Acceleration of point X on rod $= \omega^2 \times Ox = \dfrac{986.96 \times 2.2 \times 100}{1000} = 217.13 \text{ m/s}^2$

3.7 ANALYTICAL ANALYSIS OF SLIDER-CRANK MECHANISM

Consider the slider-crank mechanism shown in Fig.3.26. Let θ be the angle turned through by the crank $OA = r$ when the slider B has moved by an amount x to the right, and ϕ the angle, which the connecting rod $AB = l$ makes with the line of stroke.

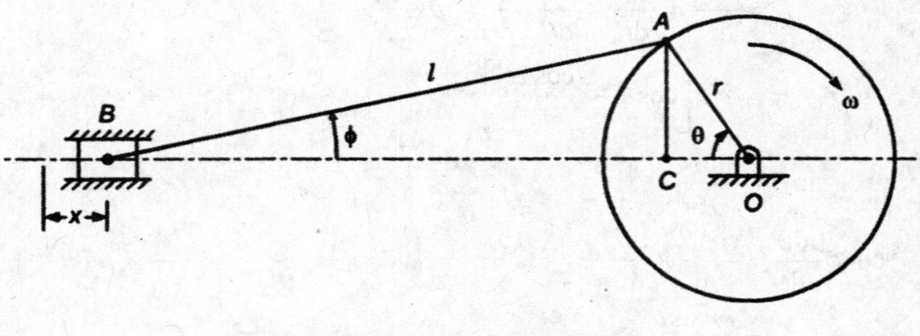

Fig.3.26 Slider crank mechanism

$$x = r + l - (OC + BC)$$
$$= r + l - (r\cos\theta + l\cos\phi)$$
$$= r(1 - \cos\theta) + l(1 - \cos\phi)$$

Now $$AC = r\sin\theta = l\sin\phi$$

$$\sin\phi = \frac{r}{l}\sin\theta$$

Let $$n = \frac{l}{r}$$

Then $$\cos\phi = (1 - \sin^2\phi)^{0.5}$$

$$= \left[1 - \frac{\sin^2\theta}{n^2}\right]^{0.5}$$

$$x = r(1 - \cos\theta) + l\left\{1 - \left(1 - \frac{\sin^2\theta}{n^2}\right)^{0.5}\right\}$$

$$= r\left[(1 - \cos\theta) + n\left\{1 - \left(1 - \frac{\sin^2\theta}{n^2}\right)^{0.5}\right\}\right]$$

$$= r\left[(1 - \cos\theta) + n\left\{1 - \left(1 - \frac{\sin^2\theta}{2n^2} - \ldots\right)\right\}\right]$$

$$\approx r\left[(1 - \cos\theta) + \frac{\sin^2\theta}{2n}\right]$$

$$\approx r\left[(1 - \cos\theta) + \frac{(1 - \cos 2\theta)}{4n}\right] \tag{3.13}$$

Velocity of slider, $$v_b = \frac{dx}{dt} = \frac{dx}{d\theta}\cdot\frac{d\theta}{dt} = \frac{dx}{d\theta}\cdot\omega$$

$$= \omega r\left[\sin\theta + \frac{\sin 2\theta}{2n}\right] \tag{3.14}$$

Acceleration of slider,

$$f_b = \frac{d^2x}{dt^2} = \frac{dv_b}{dt} = \frac{dv_b}{d\theta}\cdot\frac{d\theta}{dt} = \omega\cdot\frac{dv_b}{d\theta}$$

$$= \omega^2 r\left[\cos\theta + \frac{\cos 2\theta}{n}\right]$$

Now $$\sin\phi = \frac{\sin\theta}{n}$$

$$\cos\phi\cdot\frac{d\phi}{dt} = \frac{\cos\theta}{n}\times\frac{d\theta}{dt}$$

$$= \frac{\omega\cos\theta}{n}$$

Angular velocity of connecting rod,

$$\omega_{ba} = \frac{d\phi}{dt} = \frac{\omega}{n} \cdot \frac{\cos\theta}{\cos\phi}$$

$$= \frac{\omega\cos\theta}{(n^2 - \sin^2\theta)^{0.5}}$$

$$\approx \left(\frac{\omega}{n}\right)\cos\theta \tag{3.15}$$

Angular acceleration of connecting rod,

$$\alpha_{ba} = \frac{d\omega_{ba}}{dt} = -\omega^2 \sin\theta \, \frac{(n^2 - 1)}{(n^2 - \sin^2\theta)^{3/2}}$$

$$\approx -\frac{\omega^2}{n} \cdot \sin\theta \tag{3.16}$$

Example 3.15

In Example 3.14, calculate analytically, the acceleration of the piston and angular acceleration of the rod.

■ **Solution**

Given:

$$n = \frac{l}{r} = 4, \ \theta = 60°, \ r = 300 \text{ mm}, \ N = 300 \text{ rpm}$$

$$\omega^2 = \left(\frac{2\pi \times 300}{60}\right)^2 = 986.96 \text{ (rad/s)}^2$$

Acceleration of piston,

$$f_b = \omega^2 r\left[\cos\theta + \frac{\cos 2\theta}{n}\right]$$

$$= 986.96 \times 0.3\left[\cos 60° + \frac{\cos 120°}{4}\right]$$

$$= 111.0 \text{ m/s}^2$$

Angular acceleration of rod $\cong \dfrac{\omega^2}{n}\sin\theta$

$$= \frac{986.96 \times \sin 60°}{4}$$

$$= 213.68 \text{ rad/s}^2$$

Example 3.16

For the Scotch yoke mechanism shown in Fig.3.27, find the velocity and acceleration of point B. $\omega_2 = 5$ rad/s, and $O_2A = 100$ mm.

Solution

Draw configuration diagram as shown in Fig.3.27(a).

$$\omega_2 = 5 \text{ rad/s}$$
$$v_a = \omega_2 \times O_2A = 5 \times 100 = 500 \text{ mm/s}$$

Draw the velocity diagram as shown in Fig.3.27(b).

$$v_a = o_2a \perp O_2A$$
$$ab \parallel AB$$
$$o_2b \parallel O_2B$$
$$v_b = o_2b = 2.55 \text{ cm} = 355 \text{ mm/s}$$
$$v_{ba} = ab = 2.55 \text{ cm} = 355 \text{ mm/s}$$

The accelerations of various links are;

$$f^n_a = v^2_a / O_2A = (500)^2/100 = 2500 \text{ mm/s}^2 \text{ along } AO_2.$$
$$f^n_{ba} = v^2ba/AB = (355)^2/70.72 = 1782 \text{ mm/s}^2 \text{ along } AB$$
$$f^n_b = v^2_b/O_2B = (355)^2/70.72 = 1782 \text{ mm/s}^2 \text{ along } BO_2.$$

Draw acceleration diagram as shown in Fig.3.27(c).

$$o_2a = f^n_a \parallel O_2A$$
$$ab = f^n_{ba}$$
$$ab \parallel AB$$
$$o_2b = f^n_b \parallel O_2B$$
$$f_b = o_2b = 2.55 \text{ cm} = 1775 \text{ mm/s}^2$$

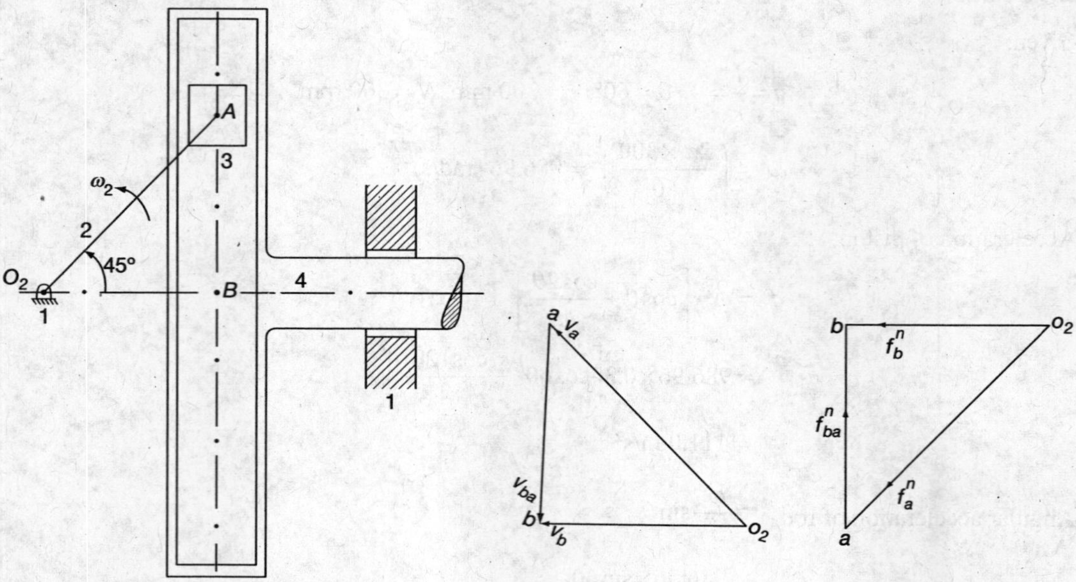

Scale: 1 cm = 25 mm
(a) Configuration diagram

Scale: 1 cm = 100 mm/s
(b) Velocity diagram

Scale: 1 cm = 500 mm/s²
(c) Acceleration diagram

Fig.3.27 Scotch yoke mechanism

Example 3.17

Bar AB is connected by pin C to slider D that slides along the fixed vertical rod EF as shown in Fig.3.28. Find the velocity and acceleration of the slider D if the bar AB rotates at a constant angular velocity of 10 rad/s in counter-clockwise direction.

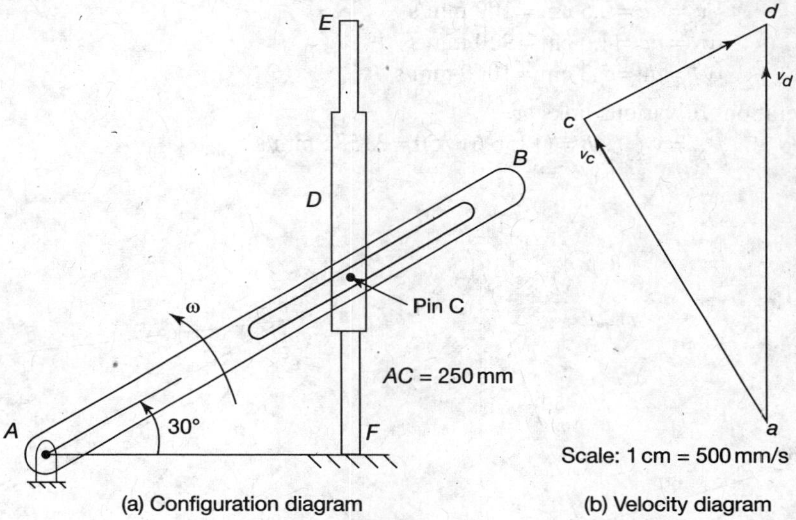

Fig.3.28 Diagram for Example 3.17

■ Solution

Draw configuration diagram as shown in Fig.3.28(a).

$$\omega = 10 \text{ rad/s}$$

$$v_c = \omega \times AC = 10 \times 250 = 2500 \text{ mm/s},$$

Draw the velocity diagram as shown in Fig.3.28(b).

$$v_d = 5.6 \text{ cm} = 2800 \text{ mm/s} \uparrow$$

$$f^{cr}_d = 2v_d\omega = 2 \times 2800 \times 10 = 56{,}000 \text{ mm/s}^2 \leftarrow$$

Example 3.18

In the mechanism shown in Fig.3.29, link AB rotates clockwise at a speed of 240 rpm. At the instant shown, find the velocity and acceleration of slider C as well as those of slider E. $AB = 50$ mm, $BC = 120$ mm, $BD = DC = 60$ mm, $DE = 80$ mm.

■ Solution

Draw configuration diagram shown in Fig.3.29(a).

$$\omega = 2\pi \times 240 / 60 = 25.13 \text{ rad/s}$$

$$v_d = \omega \times AB = 25.13 \times 50 = 1256.6 \text{ mm/s}$$

Draw the velocity diagram as shown in Fig.3.29(b).

$$v_b = ab \perp AB$$

$$bc \perp BC$$

$ac \parallel AC$

$ac = v_c = 5.7 \text{ cm} = 1140 \text{ mm/s}$

As $CD = DB$, hence $cd = db = 2.3$ cm

$de \perp DE$

$ae \parallel YE$

$v_e = ae = 0.5 \text{ cm} = 100 \text{ mm/s}$

$v_{cb} = bc = 4.6 \text{ cm} = 920 \text{ mm/s}$

$v_{ed} = de = 5.3 \text{ cm} = 1060 \text{ mm/s}$

The accelerations of various links are;

$$f^n_{ba} = v^2_{ba}/AB = (1256.6)^2/50 = 31581 \text{ mm/s}^2$$

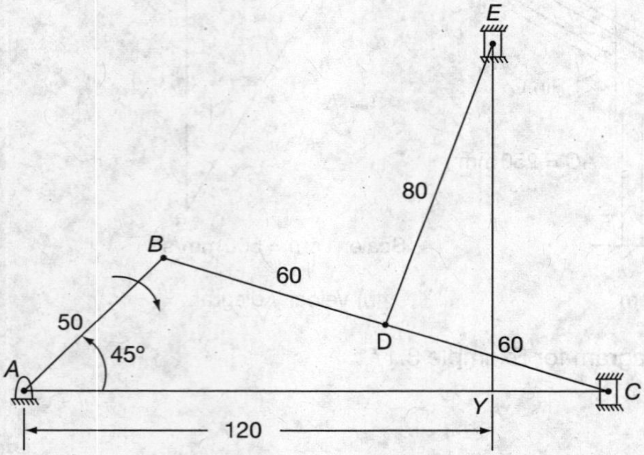

Scale: 1 cm = 20 mm

(a) Configuration diagram

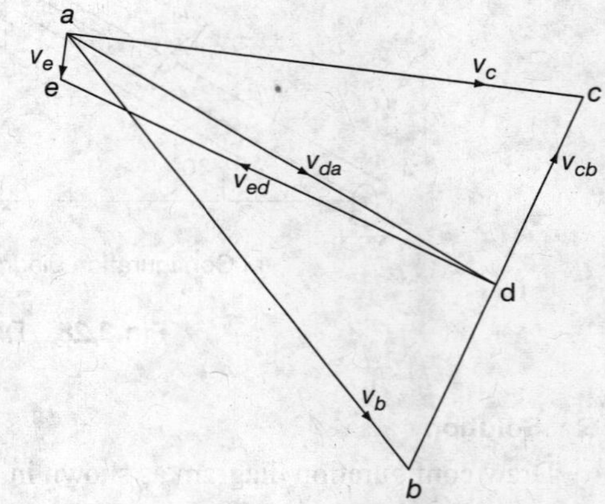

Scale: 1 cm = 200 mm/s

(b) Velocity diagram

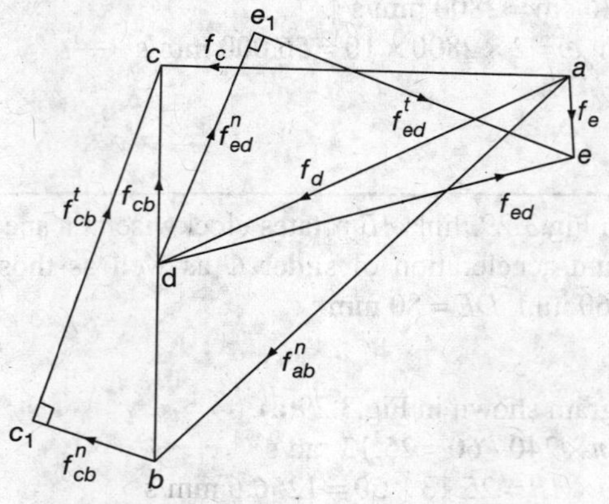

Scale: 1 cm = 5000 mm/s²

(c) Acceleration diagram

Fig.3.29 Diagram for Example 3.18

$f^n{}_{cb} = v^2{}_{cb}/BC = (920)^2/120 = 7052.3\ \text{mm/s}^2$
$f^n{}_{ed} = v^2{}_{ed}/DE = (1060)^2/80 = 14045\ \text{mm/s}^2$

Draw acceleration diagram as shown in Fig.3.29(c).

$f^n{}_{ba} = ab \parallel AB$
$f^n{}_{cb} = bc_1 \parallel BC$
$f^t{}_{cb} = c_{1c} \perp BC$
$ac \parallel AC$

Join bc.

$f_c = ac = 4.4\ \text{cm} = 22{,}000\ \text{mm/s}^2$
$cd = bd = 2.2\ \text{cm}$
$f^n{}_{ed} = de_1 \parallel DE$
$ee_1 \perp de_1$
$ae \parallel EY$

Join de.

$f_e = ae = 0.9\ \text{cm} = 4500\ \text{mm/s}^2$.

Example 3.19

Fig.3.30 depicts the structure of Whitworth quick return mechanism used in reciprocating machine tools. The various dimensions of the mechanism for a specified stroke of the tool are:

$OQ = 120$ mm, $OP = 240$ mm, $RQ = 180$ mm, and $RS = 600$ mm.

Crank OP makes an angle of 60° with the vertical. Determine the velocity of the slider S (cutting tool) when crank rotates at 120 rpm clockwise. Find also the angular velocity of the link RS and the velocity of the sliding block on the slotted lever QT.

■ Solution

Draw configuration diagram shown in Fig.3.30(a).

$\omega = 2\pi \times 120/60 = 12.57$ rad/s
$v_{p1} = \omega \times OP = 12.57 \times 240 = 3015.93$ mm/s

Draw the velocity diagram as shown in Fig.3.30(b).

$v_{p1} = op_1 \perp OP_1$
$p_1p_2 \parallel P_2Q$
$pb \perp PQ$
$v_{p2} = qp_2 = 2.8\ \text{cm} = 2800$ mm/s
$\omega_{p2q} = v_{p2}/P_2Q = 2800/(5.2 \times 60) = 8.97$ rad/s cw
$p_2q = 2.8$ cm, $p_2r = p_2q \times P_2R/P_2Q = 2.8 \times 8.2/5.2 = 4.41$ cm,
$v_s = qs = 0.9\ \text{cm} = 900$ mm/s
$v_{sr} = rs = 1.1\ \text{cm} = 1100$ mm/s.

Velocity of sliding block on QT, $v_{p1p2} = p_2p_1 = 1\ \text{cm} = 1000$ mm/s
Angular velocity of link RS, $\omega_{rs} = v_{sr}/RS = 1100/600 = 1.83$ rad/s cw
The accelerations of various links are

$f^n{}_{p1o} = v^2{}_{p1}/OP_1 = (3015.93)^2/240 = 37899\ \text{mm/s}^2$
$f^n{}_{p2q} = v^2{}_{p2}/P_2Q = (2800)^2/312 = 25128\ \text{mm/s}^2$

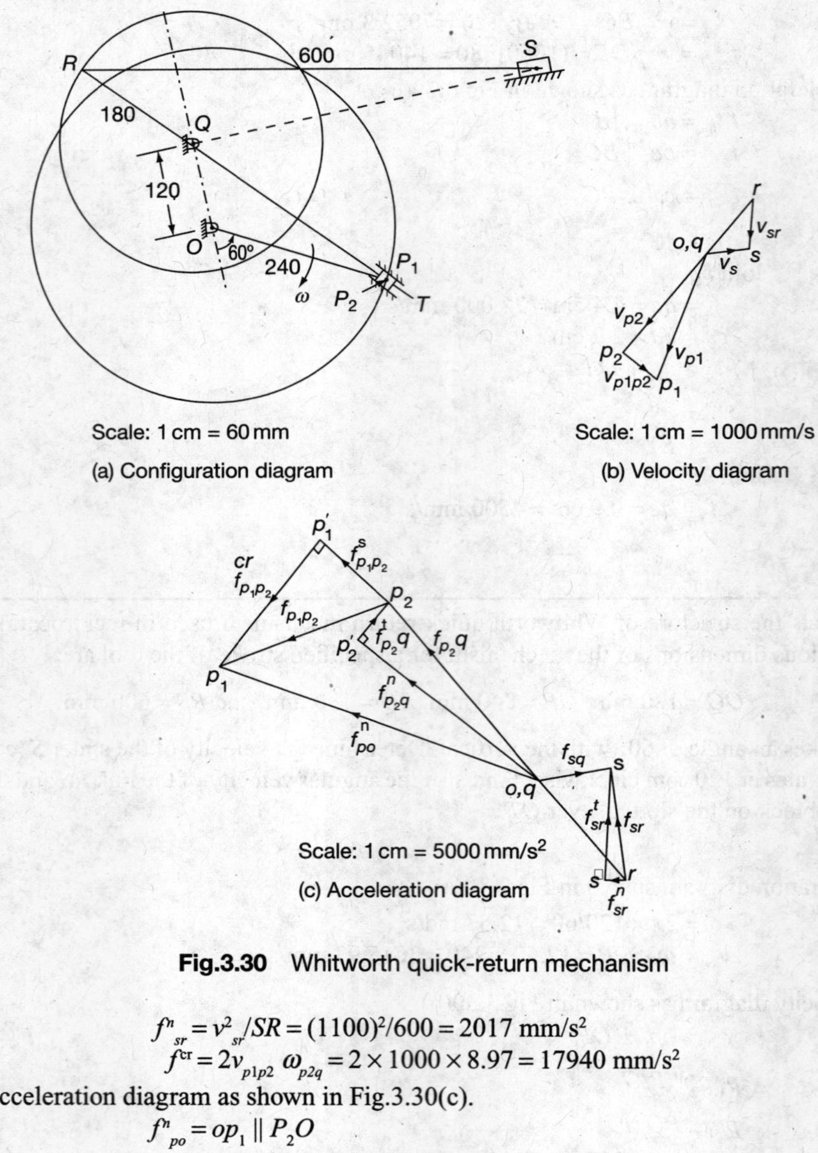

(a) Configuration diagram (b) Velocity diagram (c) Acceleration diagram

Fig.3.30 Whitworth quick-return mechanism

$$f^{n}_{sr} = v^{2}_{sr}/SR = (1100)^2/600 = 2017 \text{ mm/s}^2$$
$$f^{cr} = 2v_{p1p2}\ \omega_{p2q} = 2 \times 1000 \times 8.97 = 17940 \text{ mm/s}^2$$

Draw acceleration diagram as shown in Fig.3.30(c).

$$f^{n}_{po} = op_1 \parallel P_2O$$
$$p_1p_2' = f^{cr}_{p1p2} \perp TQ$$
$$p_1'p_2 \perp p_1p_1'$$
$$f^{n}_{p2q} = qp'_2 \parallel P_2Q,\ f_{p2q} = p_2'p_2 \perp p_2'q.$$

Join p_1p_2 and $qp_2 \cdot p_2q = 5.2$ cm

$$p_2r = p_2q \times P_2R/P_2Q = 5.2 \times 8.2/5.2 = 8.2 \text{ cm}$$
$$f^{n}_{sr} = rs' \parallel RS$$
$$s's \perp s'r \text{ and } qs \parallel QS.$$

Join rs.

$$QS = f_{sq} = 1.6 \text{ cm} = 8000 \text{ mm/s}^2.$$

Example 3.20

In the swivelling point mechanism shown in Fig.3.31(a), $OA = 25$ m… $AD = DE$, $DE = 150$ mm, $EF = 100$ mm, $BC = 60$ mm, $DS = 40$ mm, and $OC = 15$… …otates at 200 rpm. Determine the acceleration of sliding link DE in the trunnion.

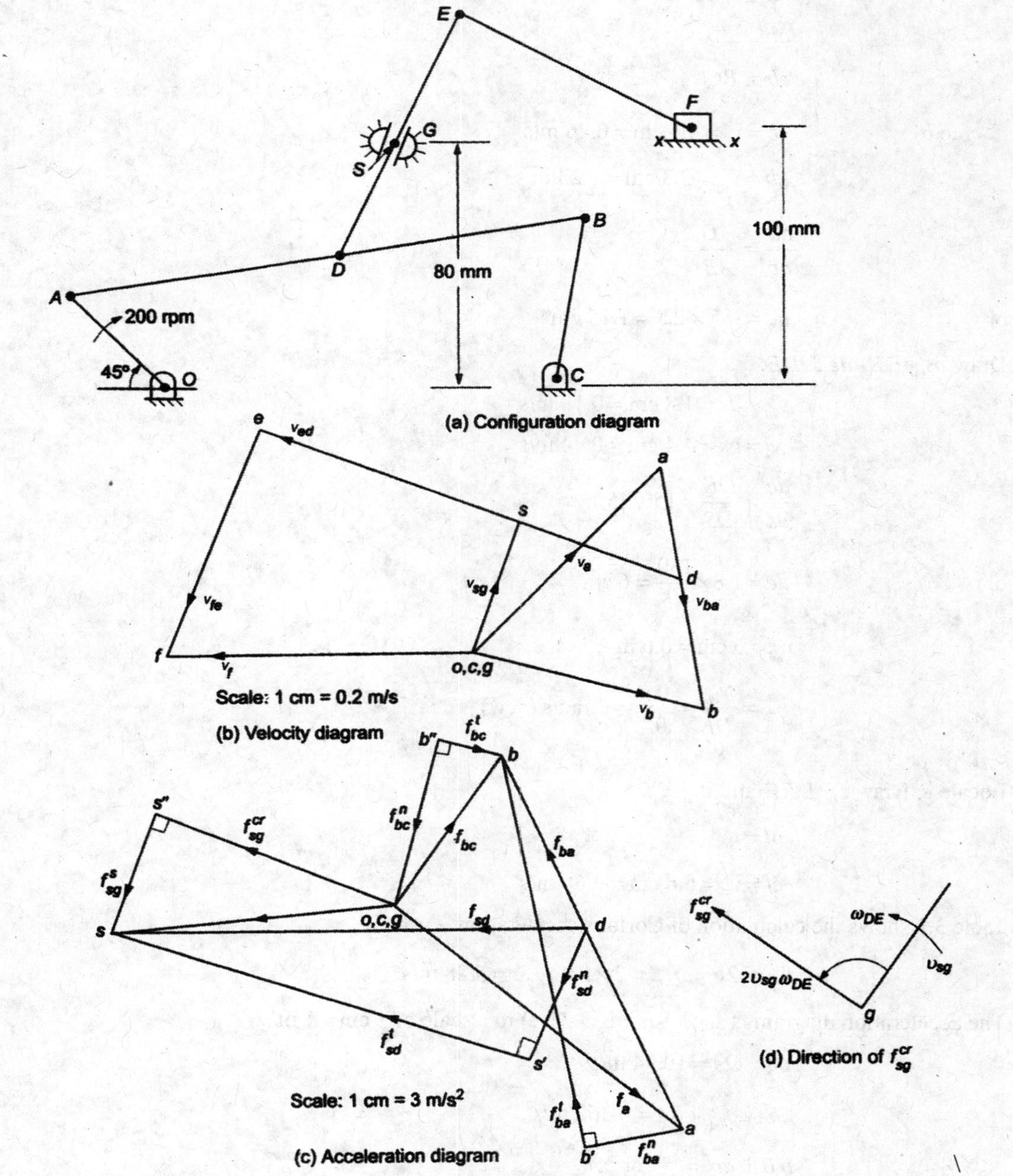

Fig.3.31 Swivelling point mechanism

■ Solution

$$\omega = 2\pi \times 200/60 = 20.944 \text{ rad/s}$$
$$v_a = 20.944 \times 0.025 = 0.523 \text{ m/s}$$

The velocity diagram is shown in Fig.2.31(b) to a scale of 1 cm = 0.1 m/s, in which

$$oa = v_a \perp OA$$

$$ab \perp AB$$

$$cb \perp BC$$

~ate b. $\quad cb = v_b = 4.6 \text{ cm} = 0.46 \text{ m/s}$

$$ab = v_{ba} = 2.3 \text{ cm} = 0.23 \text{ m/s}$$

$$\frac{ad}{ab} = \frac{AD}{AB} = \frac{1}{2}$$

or $\quad ad = 0.5 \times 2.3 = 1.15 \text{ cm}$

Draw $os \parallel DE$, $de \perp DE$:

$$v_{sd} = ds = 1.8 \text{ cm} = 0.18 \text{ m/s}$$
$$v_s = gs = 4.4 \text{ cm} = 0.44 \text{ m/s}$$
$$\frac{de}{ds} = \frac{DE}{DS} = 150/45$$

or $\quad de = 1.8 \times \dfrac{150}{45} = 6 \text{ cm}$

$$v_{ed} = 6 \text{ cm} = 0.6 \text{ m/s}$$

$$\omega_{DE} = \frac{v_{de}}{DE} = \frac{0.6}{0.1} = 6 \text{ rad/s (ccw)}$$

Locate e. Draw $ef \perp EF$ and if $\parallel XX$.

$$of = v_f$$

$$ef = v_{fe} = 6.6 \text{ cm} = 0.66 \text{ m/s}$$

Table 3.1 shows the calculation of Coriolis acceleration.

$$f_{sg}^{cr} = 2v_{sg}\ \omega_{DE} = 2 \times 0.44 \times 6 = 5.28 \text{ m/s}^2$$

The acceleration diagram is shown in Fig.3.31(c) to a scale of 1 cm = 1 m/s^2.

$$oa \parallel OA = 10.94 \text{ m/s}^2$$

$$ab' = f_{ba}^n \parallel BA = 0.352 \text{ m/s}^2$$

$$b'b \perp ab' = f_{ba}^n \text{ or } BA$$

Table 3.1

Link	Length r m	Velocity v m/s	Normal acceleration $f^n = v^2/r$ m/s²	Tangential acceleration $f^t = \alpha \cdot r$ m/s²	Coriolis acceleration $f^{cr} = 2v\omega$ m/s²
OA	0.025	0.523	10.94	–	–
AB	0.150	0.23	0.352	–	–
BC	0.06	0.46	2.52	–	–
DE	0.15	0.6	2.4	–	–
EF	0.10	0.66	4.356	–	–
DS	0.045	0.18	0.72	–	–
S fixed	–	0.44	–	–	5.28
S on link					

$$cb'' = f_{bc}^n \parallel BC = 3.52 \text{ m/s}^2$$

$$b''b \perp cb'' = f_{bc}^t$$

Locate point *b*. Join *ob* and *ab*.

$$ab = f_{ba};\ ob = f_b$$

$$\frac{ad}{ab} = \frac{AD}{AB}$$

or $$ad = 0.5 \times 6.5 = 3.25 \text{ cm}$$

$$ds' \parallel DS = f_{sd}^n = 0.72 \text{ m/s}^2$$

$$s's \perp ds'$$

Draw $$gs' = f_{sg}^{cr} = 5.28 \text{ m/s}^2 \perp SD.$$

$$s''s \parallel SD$$

Locate *s*. Join *os* and *sd*.

The acceleration of sliding in the trunnion, $s''s = f_{sg}^s = 0.3 \text{ cm} = 0.3 \text{ m/s}^2$

Example 3.21

The crank of an engine 250 mm long rotates at a uniform speed of 240 rpm. The ratio of connecting rod length to crank radius is 4. Determine (a) the acceleration of the piston, (b) the angular acceleration of the rod, and (c) the acceleration of a point *X* on the connecting rod at 1/3rd length from crank pin. The crank position is 30° from inner dead centre.

■ Solution

The acceleration diagram using Klein's construction has been drawn in Fig.3.32.
Draw $OM \perp BO$. The acceleration diagram is shown by $OAPQ$.

$$AP = f_{ba}^{n}$$
$$PQ = f_{ba}^{t}$$
$$AQ = f_{ba}$$
$$\omega = 2\times\frac{240}{60} = 25.13 \text{ rad/s}$$

Draw Xx parallel to BO. Join AQ and Ox. Then,

$$Ox \propto f_x$$
$$f_b = \omega^2 \cdot OQ$$

(a) Acceleration of the piston $= \omega^2 \times OQ = (25.13)^2 \times 2.5 \times 10 = 15788 \text{ cm/s}^2$

(b) Tangential acceleration of the rod $= \omega^2 \times PQ = (25.13)^2 \times 1.25 \times 10 = 7894 \text{ cm/s}^2$

Angular acceleration of the rod $= \frac{7894}{100} = 78.94 \text{ rad/s}^2$

(c) Acceleration of the point X on the rod $= \omega^2 \times Ox = (25.13)^2 \times 2.4 \times 10 = 15156 \text{ cm/s}^2$.

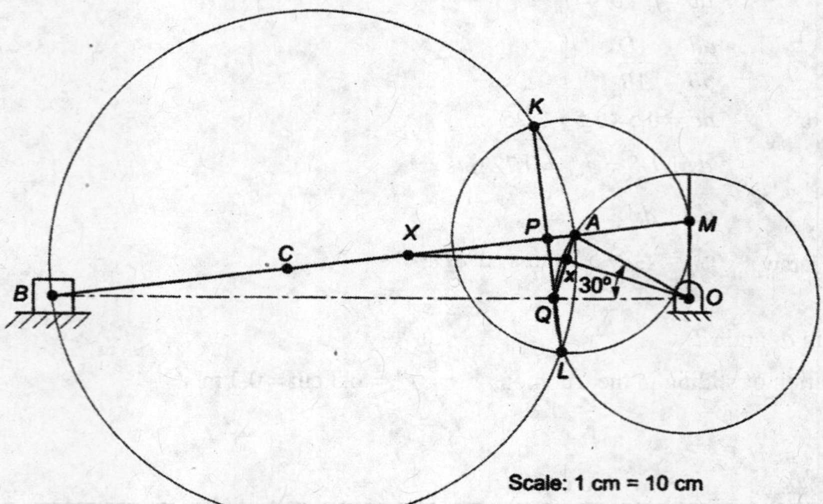

Fig.3.32 Klein's construction for Example 3.21

Example 3.22

In the mechanism shown in Fig.3.33, the crank OA rotates at 20 rpm anti-clockwise and gives motion to the sliding blocks B and D. The dimensions of the various links are: $OA = 300$ mm, $AB = 1200$ mm,

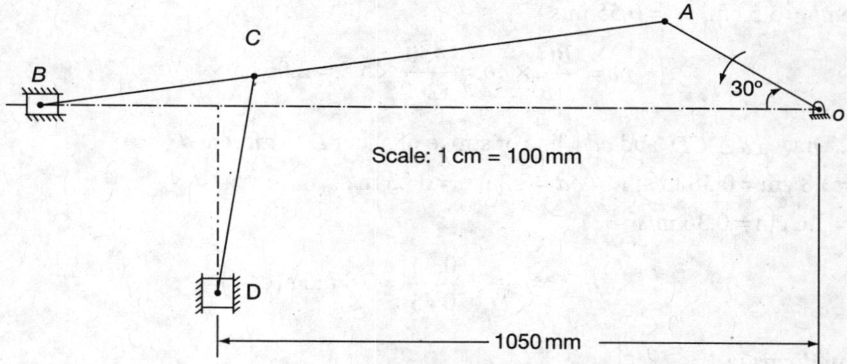

(a) Configuration diagram

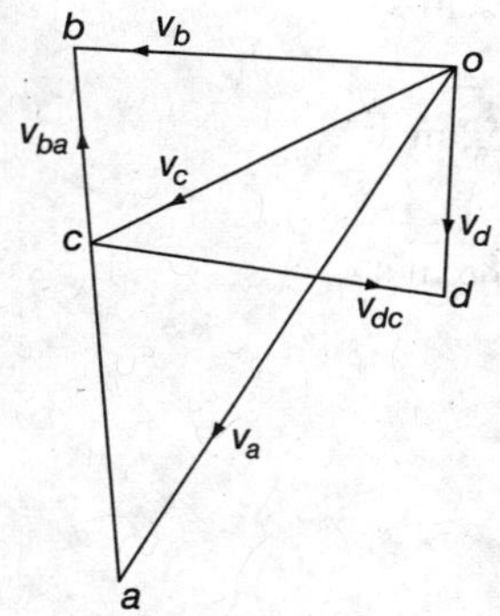

Scale: 1 cm = 0.1 m/s

(c) Velocity diagram

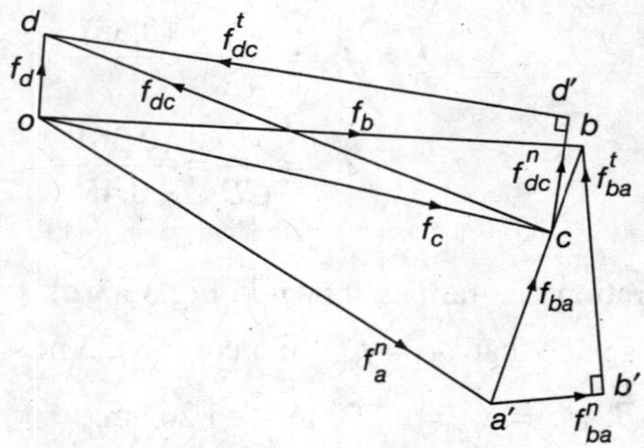

Scale: 1 cm = 0.2 m/s

(c) Acceleration diagram

Fig.3.33 Double slider mechanism

$BC = 450$ mm, and $CD = 450$ mm. Determine (a) velocities of sliding at B and D, (b) angular velocity of CD, (c) linear acceleration of D, and (d) angular acceleration of CD.

■ Solution

Angular velocity of OA, $\omega = \dfrac{2\pi\ N}{60} = \dfrac{2\pi \times 20}{60} = 2.1$ rad/s

Linear velocity of A, $v_a = \omega \times OA = 2.1 \times 0.3 = 0.63$ m/s

Velocity diagram

1. Draw configuration diagram as shown in Fig.3.33(a).
2. Draw velocity diagram as shown in Fig.3.33(b)
3. Draw $oa \perp OA$ such that $oa = 6.3$ cm.

   $ab \perp AB$ and $ob \parallel OB$ to intersect at b.

4. Measure $ab = 5.5$ cm, $v_{ba} = 0.55$ m/s

$$bc = \frac{BC}{AB} \times ab = \frac{450}{1200} \times 5.5 = 2.06 \text{ cm}$$

5. Join *OC*. Draw $cd \perp CD$ and $od \parallel$ line of stroke of slider *D* to meet at *d*.
6. $v_b = ob = 3.8$ cm $= 0.38$ m/s; $v_d = od = 2.3$ cm $= 0.23$ m/s.
7. $v_{cd} = cd = 3.6$ cm $= 0.36$ m/s

$$\omega_{cd} = \frac{v_{cd}}{CD} = \frac{0.36}{0.45} = 0.8 \text{ rad/s(ccw)}$$

Acceleration diagram

$$f_a^n = \frac{v_a^2}{OA} = \frac{(0.63)^2}{0.3} = 1.323 \text{ m/s}^2$$

$$f_{ba}^n = \frac{v_{ba}^2}{AB} = \frac{(0.55)^2}{1.2} = 0.252 \text{ m/s}^2$$

$$f_{cd}^n = \frac{v_{dc}^2}{CD} = \frac{(0.36)^2}{0.45} = 0.288 \text{ m/s}^2$$

1. Draw acceleration diagram as shown in Fig.3.33(c).
2. Draw $oa \parallel OA$ such that $oa = f_a^n = 6.6$ cm $= 1.323$ m/s^2.
3. Draw $ab' \parallel AB$, $ab' = f_{ba}^n = 0.252$ m/s^2=1.26 cm.
4. Draw $b'b \perp AB$ and $ob \parallel OB$ to intersect at *b*.

   Joint *ab*. Measure $ab = 3.3$ cm.
5. $bc = \frac{BC}{AB} \times ab = \frac{450}{1200} \times 3.3 = 1.237$ cm. Join *oc*.
6. Draw $cd'' \parallel CD$, $cd' = f_{dc}^n = 0.288$ m/s^2 $= 1.44$ cm.
7. Draw $d'd \perp CD$ and $od \parallel$ the line of stroke of slider *D* to intersect of *d*. Join *cd*.
8. Linear acceleration of *D*, $f_d = od = 1$ cm $= 0.2$ m/s^2.
9. Linear tangential acceleration of *CD*, $f_{dc}^t = d'd = 6.5$ cm $= 1.3$ m/s^2

   Angular acceleration of *CD*, $\propto_{cd} = \frac{f_{dc}^t}{CD} = \frac{1.3}{0.45} = 2.9$ rad/s^2 (cw).

Example 3.23

The kinematic diagram of one of the cylinders of a rotary engine is shown in Fig.3.34(a). The crank *OA* which is vertical and fixed, is 50 mm long. The length of connecting rod *AB* is 125 mm. The line of stroke *OB* is inclined at 50° to the vertical. The cylinder is rotating at a uniform speed of 300 rpm, in a clockwise direction, about the fixed centre *O*. Calculate (a) acceleration of the piston inside the cylinder, and (b) angular acceleration of connecting rod.

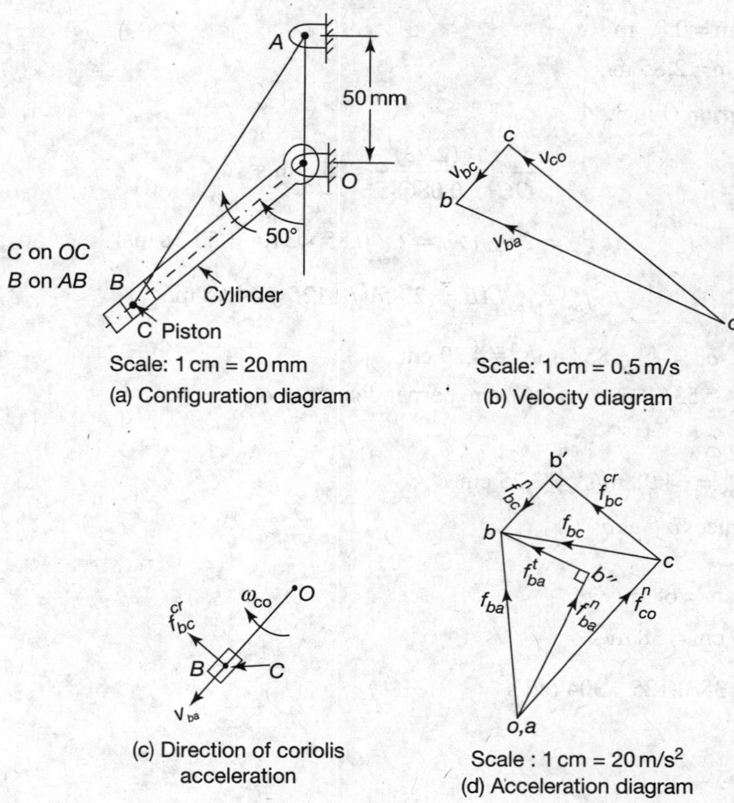

Fig.3.34 Rotary engine

■ Solution

Angular velocity of OC, $\omega_{CO} = \dfrac{2\pi N}{60} = \dfrac{2\pi \times 300}{60} = 31.4$ rad/s

Linear velocity of C, $v_c = \omega_{CD} \times OC$

$$AB^2 = OB^2 + OA^2 - 2\ OB \times OA \times \cos 130°$$
$$125^2 = \mathrm{OB}^2 + 50^2 - 2\ OB \times 50 \times \cos 130°$$
$$OB^2 + 64.28 \times OB - 13125 = 0$$
$$OB = 86.85 \text{ mm} = OC$$
$$v_c = 31.4 \times 86.85 \times 10^{-3} = 2.73 \text{ m/s}$$

Velocity diagram

1. Draw velocity diagram as shown in Fig.3.34(b).
2. Draw $oc \perp OC$, $oc = 5.46$ cm $= 2.73$ m/s.

$$ab \perp AB$$
$$cb \parallel OC$$

3. $v_{bc} = cb = 1.7\ \text{cm} = 0.85\ \text{m/s}$

   $v_{ba} = ab = 5.7\ \text{cm} = 2.85\ \text{m/s}$

Acceleration diagram (Fig.3.34d):

$$f_{co}^{n} = \frac{v_{co}^2}{OC} = \frac{(2.73)^2}{0.08685} = 85.8\ \text{m/s}^2$$

$$f_{bc}^{cr} = 2\, v_{bc}\, \omega_{oc} = 2 \times 0.85 \times 31.4 = 53.38\ \text{m/s}^2$$

$$f_{ba}^{n} = v_{ba}^2 / AB = (2.85)^2 / 0.125 = 64.98\ \text{m/s}^2$$

1. Draw $oc \parallel OC$, $oc = f_{co}^{n} = 85.8\ \text{m/s}^2 = 4.29\ \text{cm}$.
2. Draw $cb' = f_{bc}^{cr} = 53.38\ \text{m/s}^2 = 2.67\ \text{cm}$ perpendicular to oc.

   $b'b \perp cb'$
3. Draw $ab'' = f_{ba}^{n} = 64.98\ \text{m/s}^2 = 3.25\ \text{cm}$

   $b''b \perp ab''$ to meet $b'b$ at b.
4. Join cb and ab.

   $f_{bc} = cb = 3.4\ \text{cm} = 68\ \text{m/s}^2$

   $f_{ba}^{t} = b''b = 1.9\ \text{cm} = 38\ \text{m/s}^2$

   $\alpha_{ba} = f_{ba}^{t} / AB = 38/0.125 = 304\ \text{rad/s}^2$

Example 3.24

In the mechanism shown in Fig.3.35(a), the link 2 rotates with angular velocity of 30 rad/s and an angular acceleration of 240 rad/s². Determine (a) the acceleration of points B and C, (b) the angular accelerations of link 3 and 4 and (c) the relative acceleration α_{34}. $O_2A = 100$ mm, $AB = 200$ mm, $AC = 100$ mm and $BC = 150$ mm.

■ Solution

$$v_a = 30 \times 0.1 = 3\ \text{m/s}$$

The velocity diagram is shown in Fig.3.35(b) to a scale of 1 cm = 1 m/s.

$$v_b = v_a + v_{ba}$$

$$v_c = v_a + v_{ca} = v_b + v_{cb}$$

$$v_a = o_2 a \quad \text{and} \quad \perp O_2 A$$

$$v_b = o_2 b \quad \text{and} \quad \perp o_4 b \perp = 2.6\ \text{cm} = 2.6\ \text{m/s}$$

$$v_{ba} = ab \quad \text{and} \quad \perp AB = 2.3\ \text{cm} = 2.3\ \text{m/s}$$

$$v_{cb} = bc \quad \text{and} \quad \perp BC = 1.7\ \text{cm} = 1.7\ \text{m/s}$$

$$v_{ca} = o_2 b \text{ and } \perp o_4 b = 1.2\ \text{cm} = 1.2\ \text{m/s}$$

$$v_c = o_2 c$$

The acceleration diagram is shown in Fig.3.35(c) to a scale of 1 cm = 20 m/s².

$$f_b^n = \frac{v_b^2}{O_4B} = \frac{2.6^2}{0.2} = 64.8 \text{ m/s}^2 \quad (\text{from } B \text{ to } O_4)$$

$$f_b^t \perp f_b^n$$

$$f_a^n = \frac{v_a^2}{O_2A} = \frac{3^2}{0.1} = 90 \text{ m/s}^2 \quad (\text{from } A \text{ to } O_2)$$

$$f_{ba}^n = \frac{v_{ba}^2}{AB} = \frac{2.3^2}{0.2} = 26.45 \text{ m/s}^2 \quad (\text{from } B \text{ to } A)$$

$$f_{ba}^t \perp f_{ba}^n$$

$$f_b = f_a + f_{ba}$$

$$f_b^n + f_b^t = f_a^n + f_a^t + f_{ba}^n + f_{ba}^t$$

$$f_b = o_2b = 2.5 \text{ cm} = 70 \text{ m/s}^2$$

$$f_b^t = b'b = 1.4 \text{ cm} = 28 \text{ m/s}^2$$

$$f_{ba}^t = b''b = 6.2 \text{ cm} = 124 \text{ m/s}^2$$

$$\alpha_3 = \frac{f_{ba}^t}{BA} = \frac{124}{0.2} = 620 \text{ rad/s}^2 \text{ (ccw)}$$

$$\alpha_4 = \frac{f_b^t}{O_4B} = \frac{28}{0.2} = 140 \text{ rad/s}^2 \text{ (cw)}$$

$$\alpha_{34} = \alpha_3 - \alpha_4 = 620 + 140 = 760 \text{ rad/s}^2 \text{ (ccw)}$$

$$f_c = f_a + f_{ca}^n + f_{ca}^t$$

$$= f_b + f_{cb}^n + f_{cb}^t$$

$$f_{ca}^n = \frac{v_{ca}^2}{CA} = \frac{1.2^2}{0.1} = 14.4 \text{ m/s}^2$$

$$f_{ca}^t \perp f_{ca}^n$$

$$f_{cb}^n = \frac{v_{cb}^2}{CB} = \frac{1.7^2}{0.150} = 19.27 \text{ m/s}^2$$

$$f_{cb}^t \perp f_{cb}^n$$

$$f_c = o_2c = 5 \text{ cm} = 100 \text{ m/s}^2$$

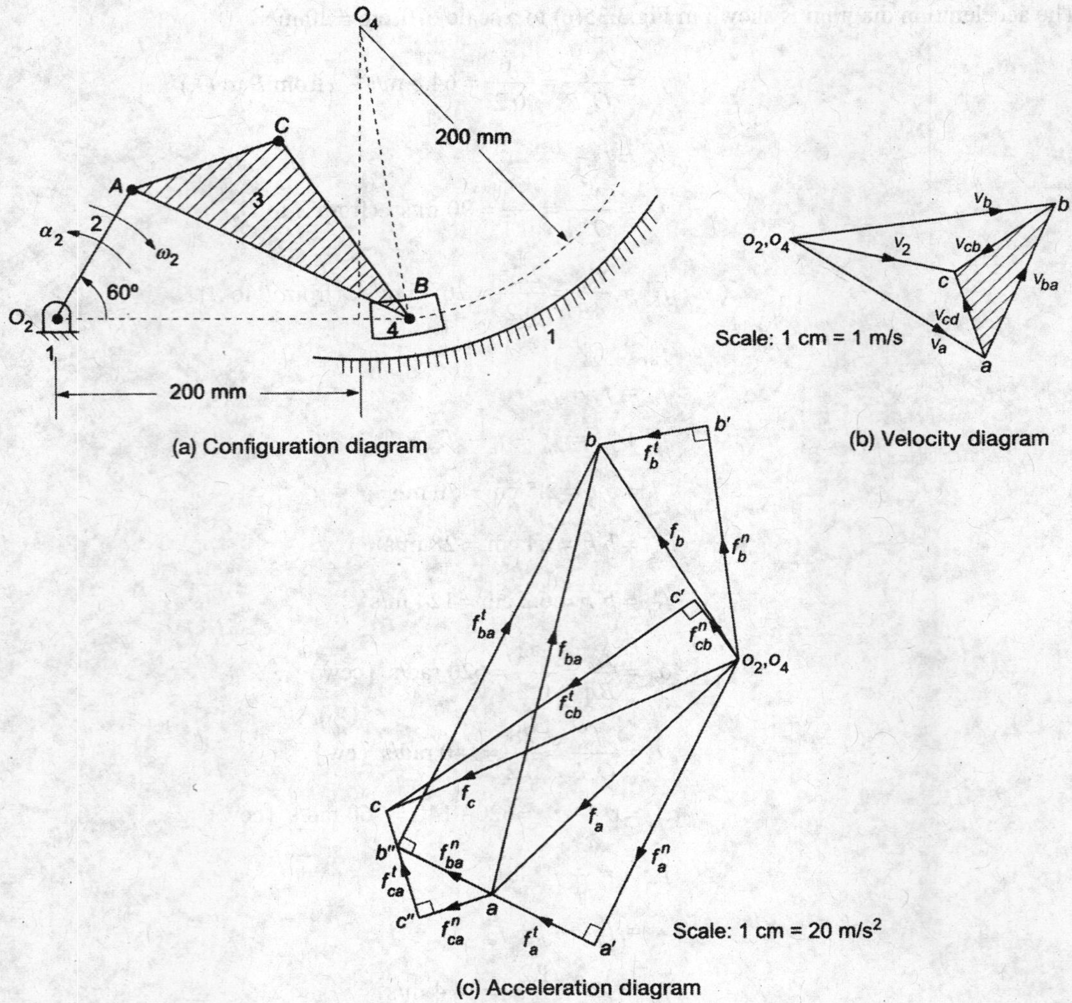

Fig.3.35 Diagram for Example 3.24

Summary for Quick Revision

1 A particle moving in a circular path has two components of accelerations, normal and tangential, which are perpendicular to each other.
Normal (or radial or centripetal) component of acceleration, $f^n = \omega^2 r = v^2/r$
Tangential component of acceleration, $f = \alpha\, r$
where ω = angular velocity, α = angular acceleration, v = linear velocity, and r = radius

2 Total acceleration of a point B w.r.t. point A on the same link (vector sum),

$$f_{ba} = f_{ba}^n + f_{ba}^t = (\omega^2 + \alpha)\ \text{AE}$$

$$\tan \beta = \alpha/\omega^2$$

3 Acceleration centre is a point in a link about which the acceleration is zero.

4 Coriolis components of acceleration acts when the second point which was considered stationary, now slides, e.g. a slider sliding along a rotating link, as in the case of a crank and slotted lever or Whitworth quick-return motion mechanisms.
Coriolis acceleration, $f^{cr} = 2v\omega$
Where v = linear velocity of the slider along the link
ω = angular speed of the rotating link

5 The direction of the coriolis component of acceleration is such as to rotate the sliding velocity vector through 90° in the same sense as the angular velocity of the rotating link.

6 Klein's construction is used to draw the velocity and acceleration diagrams for a single slider-crank mechanism.

7 In a slider–crank mechanism,
Velocity of slider $= \omega r\,[\sin\theta + \sin 2\theta/(2n)]$
Acceleration of slider $= \omega^2 r\,[\cos\theta + \cos 2\theta/n]$
Angular velocity of connecting rod $= \omega\cos\theta\,/\,(n^2 - \sin^2\theta)^{0.5}$
$\approx (\omega/n)\cos\theta$
Angular acceleration of connecting rod $= -\omega^2 \sin\theta\,(n^2 - 1)/(n^2 - \sin^2\theta)^{3/2}$
$\approx (\omega^2/n)\cdot\sin\theta$

Multiple Choice Questions

1 The C.G. of a link in any mechanism would experience
(a) zero acceleration
(b) linear acceleration
(c) angular acceleration
(d) both angular and linear accelerations.

2 Coriolis components of acceleration is encountered in
(a) four bar mechanism
(b) lower pairs
(c) higher pairs
(d) Whitworth quick return motion.

3 Coriolis components of acceleration exists whenever a point moves along a path that has
(a) linear displacement
(b) rotational motion
(c) tangential acceleration
(d) centripetal acceleration.

4 The total acceleration of B relative to A is
(a) $[(\omega^2_{AB} \times AB)^2 + (\alpha \times AB)2]^{1/2}$
(b) $\omega^2_{AB} \times AB + \alpha \times AB$
(c) $[\omega_{AB} \times AB^2 + \alpha \times AB^2]^{1/2}$
(d) $[(\omega_{AB} \times AB)^2 + (\alpha \times AB)^2]^{1/2}$.

5 The total acceleration of B relative to A is inclined to AB at an angle tangent of which is given by
(a) α/ω^2
(b) α/ω
(c) ω^2/α
(d) ω/α

6 The velocity of slider-crank mechanism is given by
(a) $(\omega L/2)\,[\sin\theta + (L/4l)\sin 2\theta]$
(b) $(\omega^2 L/2)\,[\sin\theta + (L/2l)\sin 2\theta]$
(c) $(\omega L/2)\,[\cos\theta + (L/4l)\cos 2\theta]$
(d) $(\omega^2 L/2)\,[\cos\theta + (L/2l)\cos 2\theta]$
where L = length of stroke, I = length of connecting rod.
ω = angular speed of crank, θ = crank angle.

7 The linear acceleration of piston of a reciprocating engine is
(a) $\omega^2 r\,[\cos\theta + \cos 2\theta/(2n)]$
(b) $\omega^2 r\,[\cos\theta + \cos 2\theta/n]$
(c) $\omega^2 r\,[\sin\theta + \sin 2\theta/(2n)]$
(d) $\omega^2 r\,[\sin\theta + \sin 2\theta/n]$
where $n = l/r$.

8 The angular velocity of the connecting rod of a reciprocating engine is approximately given by
(a) $\omega \cos \theta/n$ (b) $\omega \sin \theta/n$
(c) $\omega \cos \theta/(2n)$ (d) $\omega \sin \theta/(2n)$.

9 The angular acceleration of the connecting rod of a reciprocating engine is approximately given by
(a) $\omega^2 \cos \theta/n$ (b) $\omega^2 \sin \theta/n$
(c) $\omega^2 \cos \theta/(2n)$ (d) $\omega^2 \sin \theta/(2n)$.

10 Klein's construction is used mainly to determine the
(a) linear velocity of the piston (b) linear acceleration of the piston
(c) linear displacement of the piston (d) all of the above.

11 A slider slides along a straight link with uniform velocity v and the link rotates about a point with uniform angular speed ω. The Coriolis components of acceleration of a point on the slider at a distance r from the centre of rotation is
(a) v^2/r parallel to link (b) ωr perpendicular to link
(c) $2\,\omega v$ perpendicular to link (d) $v\,\omega$ parallel to link.

12 The Coriolis component of acceleration occurs in
(a) slider-crank mechanism (b) Scotch-yoke mechanism
(c) oscillating cylinder mechanism (d) four-bar chain.

13 The direction of the Coriolis component of acceleration is
(a) along the surface of sliding
(b) perpendicular to the surface of sliding in the direction of angular speed
(c) perpendicular to the surface of sliding in the direction opposite to the direction of angular speed
(d) inclined to the surface of sliding depending on the magnitude of normal and tangential accelerations.

Answers

1. (d) 2. (d) 3. (b) 4. (a) 5. (a) 6. (a) 7. (b) 8. (a) 9. (b) 10. (b) 11. (c) 12. (c) 13. (b)

Review Questions

1 Define normal and tangential components of acceleration.

2 How do you determine normal and tangential components of acceleration?

3 What is acceleration image? How it is helpful in determining the acceleration of offset points in a mechanism?

4 Define Coriolis component of acceleration. When it occurs?

5 How do you determine coriolis components of acceleration?

6 What is the use of Klein's construction?

Exercises

3.1 A link *AB* of a four-bar mechanism *ABCD* revolves uniformly at 120 rpm in a clockwise direction. Find the angular acceleration of links *BC* and *CD* and acceleration of pont *E* in link *BC*. Given: $AB = 75$ mm, $BC = 175$ mm, $EC = 50$ mm, $CD = 150$ mm, $AD = 100$ mm, and $\angle BAD = 90°$.

[Ans. 12.57 rad/s², 12.0 rad/s², 10.8 cm/s²]

3.2 In the steam engine mechanism shown in Fig.3.36, the crank AB rotates at 200 rpm clockwise. Find the velocities of C, D, E, F, and G, and the acceleration of slider at C. $AB = 120$ mm, $BC = 4780$ mm, $CD = 180$ mm, $DE = 360$ mm, $EF = 120$ mm, and $FG = 360$ mm.

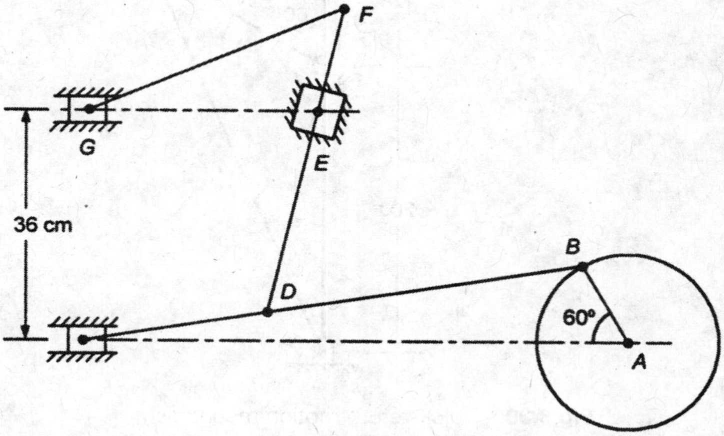

Fig.3.36 Steam engine mechanism

[Ans. 2.55 m/s, 2.5 m/s, 1.5 m/s, 1.65 m/s, 0.7 m/s; 0.85 m/s²]

3.3 In the Fig.3.37, the washer is sliding outward on the rod with a velocity of 1.2 m/s when distance from point O is 0.6 m. Its velocity along the rod is increasing at the rate of 0.9 m/s². The angular velocity of rod is 5 rad/s counter-clockwise and its angular acceleration is 10 rad/s² clockwise. Determine the absolute acceleration of a point on the washer.

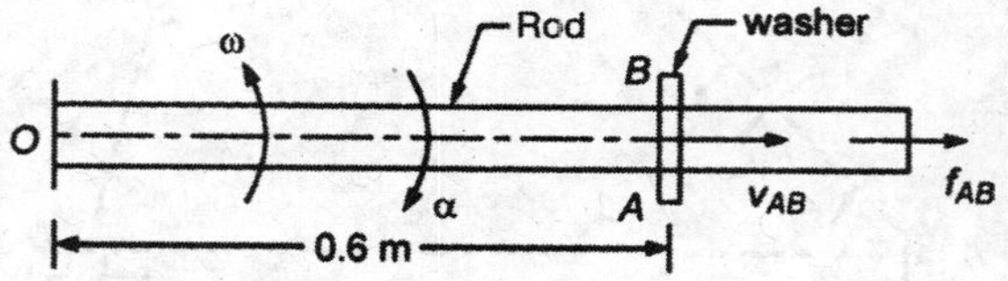

Fig.3.37 Washer sliding on a rod

[Ans. 15.32 m/s²]

3.4 The driving crank O_1A of the quick-return motion mechanism shown in Fig.3.38 resolves at a uniform speed of 200 rpm clockwise. Find the velocity and acceleration of tool post C when the crank makes an angle of 60° with the vertical line of centres O_1O_2. What is the acceleration of sliding of block A along the slotted lever O_2B?

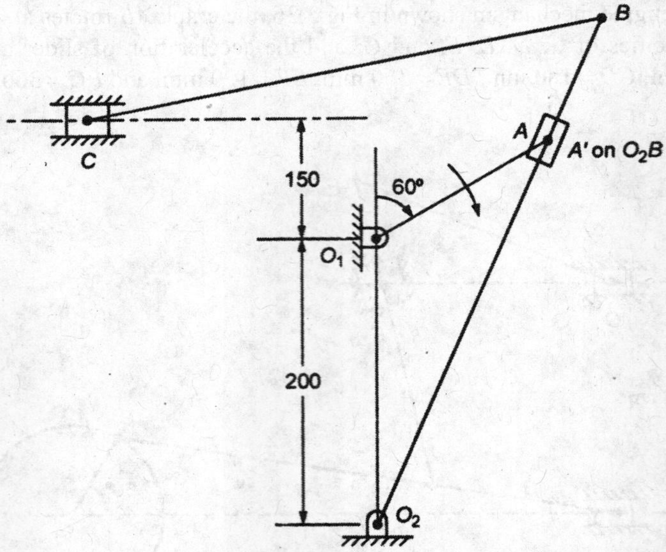

Fig.3.38 Quick-return motion mechanism

[Ans. 1.6 m/s, 20.5 m/s², 22 m/s²]

3.5 The crank *AB* of a four-bar mechanism shown in Fig.3.39 rotates with an angular speed of 100 rad/s and an angular acceleration of 4400 rad/s² when the crank makes an angle of 53° to the horizontal. Determine (a) the angular acceleration of *BC*, and (b) linear acceleration of point *E*. Take *AB* = 75 mm, *BC* = 80 mm, *AD* = 125 mm, and *BE* = 28 mm.

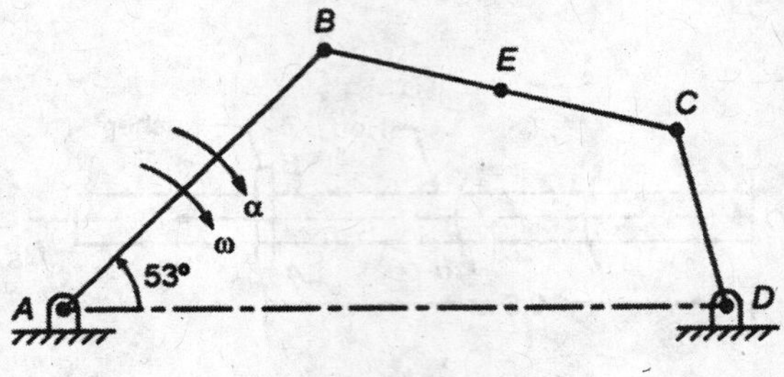

Fig.3.39 Four-bar mechanism

[Ans. 10,750 rad/s², 1160 m/s²]

3.6 Fig.3.40 shows a mechanism in which the hydraulic actuator O_2A is expanding at a constant rate of 10 cm/s. Determine the directions and magnitudes of the angular velocity and acceleration of link O_4A.

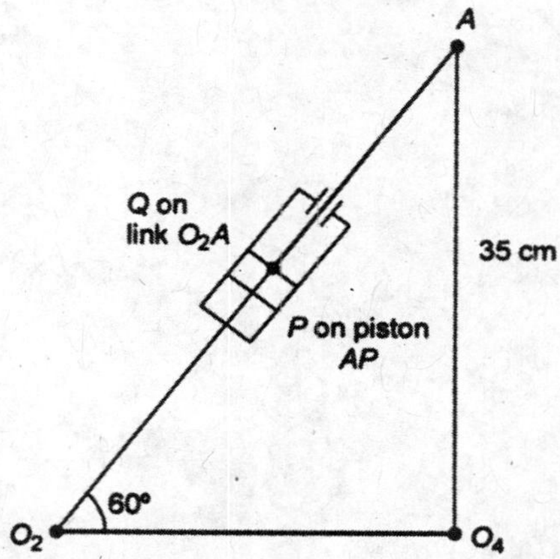

Fig.3.40 Mechanism with hydraulic actuator

[Ans. 0.566 rad/s]

3.7 For the mechanism shown in Fig.3.41, find the acceleration of point B in link 3. Given: $\omega_2 = 30$ rad/s, $O_2A = 200$ mm, $AB = BC = 175$ mm, $AC = 600$ mm.

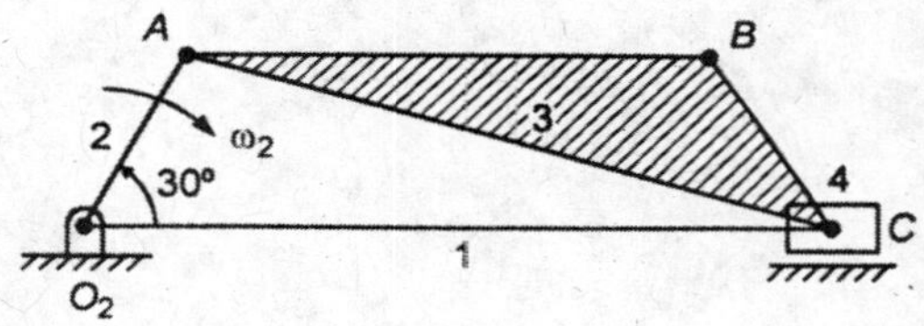

Fig.3.41 Four-bar mechanism

[Ans. 196 m/s²]

3.8 Determine the angular acceleration of links 3 and 4 and the absolute acceleration of point C on link 3 in the mechanism shown in Fig.3.42. $O_2A = 45$ mm, $AB = 130$ mm, $O_2O_4 = 90$ mm, $O_4B = 60$ mm, $AC = 55$ mm, $BC = 100$ mm.

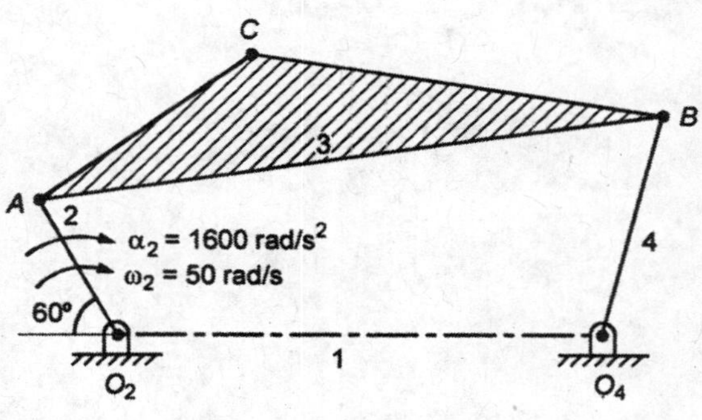

Fig.3.42 Four-bar mechanism with ternary link

[Ans. 492.3 rad/s², 2600 rad/s, 160 m/s²]

4

MECHANISMS WITH LOWER PAIRS

Chapter Outline

4.1 INTRODUCTION

The lower pairs are those which have surface contact between them. We study, in this chapter, the various mechanisms for the generation of straight line motion, both exact and approximate. These mechanisms have a vital role to play in generating the configurations for machines. The steering gears and Hooke's joint are other lower pairs that are discussed.

4.2 PANTOGRAPH

It is a mechanism to produce the path traced out by a point on enlarged or reduced scale. Fig.4.1(a) shows the line diagram of a pantograph in which $AB = CD$, $BC = AD$ and $ABCD$ is always a parallelogram. OQP is a straight line. P describes a path similar to that described by Q. It is used as a copying mechanism.

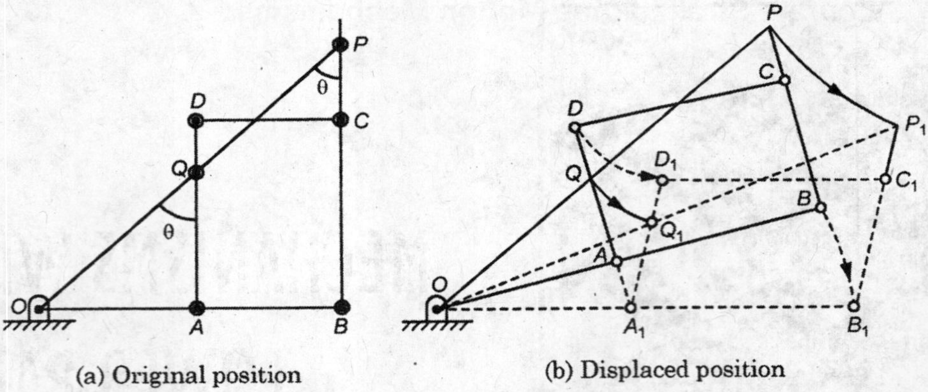

Fig.4.1 Pantograph

Proof: To prove that the path decribed by P is similar to that described by Q, consider Δs OAQ and OBP which are similar, because $\angle BOP$ is common.

$$\angle AQO = \angle BPO \text{ being corresponding angles as } AQ \parallel BP.$$

Hence $$\frac{OA}{OB} = \frac{OQ}{OP} = \frac{AQ}{BP} \quad (4.1)$$

In the displaced position shown in Fig.4.1(b), as all inks are rigid,

$$B_1O = BO,\ D_1A_1 = DA,\ A_1O = AO$$

and $$P_1B_1 = PB,\ B_1A_1 = BA,\ A_1Q_1 = AQ$$

Hence $$\frac{OA_1}{OB_1} = \frac{A_1Q_1}{B_1P_1}$$

As $A_1B_1C_1D_1$ is a parallelogram, $A_1D_1 \parallel B_1C_1$, i.e., $A_1Q_1 \parallel B_1P_1$.
OQ_1P_1 is again a straight line so that Δs QA_1Q_1 and OB_1P_1 are similar.

$$\frac{OA_1}{OB_1} = \frac{OQ_1}{OP_1} \quad (4.2)$$

From Eqs. (4.1) and (4.2), we get

$$\frac{OQ}{OP} = \frac{OQ_1}{OP_1} \quad \text{because } OA = OA_1 \text{ and } OB = OB_1.$$

Hence QQ_1 is similar to PP_1 or they are parallel. Thus path traced by P is similar to that of Q.

The pantograph is used in geometrical instruments, manufacture of irregular objects, to guide cutting tools, and as indicator rig for cross head.

4.3 STRAIGHT LINE MOTION MECHANISMS

Now we will look at lower pairs generating straight line motion. Straight line motion can be generated by either sliding pairs or turning pairs. Sliding pairs are bulky and gets worn out rapidly. Therefore, turning pairs are preferred over sliding pairs for generating straight line motion. Straight line motion can be generated either accurately or approximately.

4.3.1 Accurate Straight Line Motion Mechanisms

The mechanisms for the generation of accurate straight line motion are as follows:

1. Peaucellier mechanism
2. Hart mechanism
3. Scott-Russel mechanism.

1. *Peaucellier mechanism:* A line diagram of the Peaucellier mechanism is shown in Fig.4.2.

$$OR = OS \quad \text{and} \quad OO_1 = O_1Q.$$

Point P describes a straight line perpendicular to OO_1 produced.

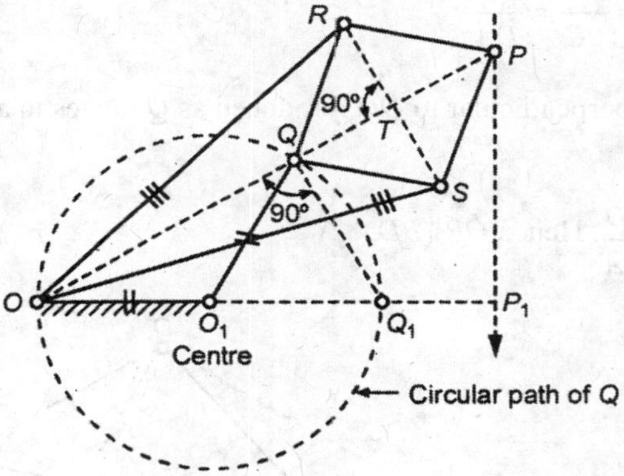

Fig.4.2 Peaucellier straight line mechanism

Proof: Δs ORQ and OSQ are congruent, because $OR = OS$, $OR = QS$ and OQ is common.

$$\angle ROQ = \angle SOQ. \text{ Also } \angle OQR = \angle OQS \text{ and } \Delta s\ PRO = \Delta s\ PSQ.$$

$$\angle OQR + \angle RQP = \angle OQS + \angle SQP$$

But OQ is a straight line.

$$\angle OQR + \angle RQP = \angle OQS + \angle SQP = 180°$$

Hence OQP is a straight line.

Now
$$OR^2 = OT^2 + RT^2$$
$$RP^2 = RT^2 + TP^2$$
$$OR^2 - RP^2 = OT^2 - TP^2$$
$$= (OT + TP)(OT - TP) = OP \cdot OQ$$

But OR and RP are always constant.

Hence
$$OP \cdot OQ = \text{constant.}$$

Draw $PP_1 \perp OO_1$ produced and join QQ_1. Δs OQQ_1 and OPP_1 are similar, because $\angle OQQ_1 = \angle OP_1P = 90°$ and $\angle QOQ_1$ is common.

Hence $\dfrac{OQ}{OQ_1} = \dfrac{OP_1}{OP}$

or $OQ \cdot OP = OQ_1 \cdot OP_1 = \text{constant}$

Now $OQ_1 = 2OO_1 = \text{constant}$

Hence $OP_1 = \text{constant}$

or point P moves in a straight line.

2. *Hart mechanism*: The Hart mechanism is shown in Fig.4.3. OO_1 is a fixed link and O_1Q is the rotating link. The point Q moves in a circle with centre O_1 and radius O_1Q. $ABCD$ is a trapezium so that $AB = CD$, $BC = AD$ and $BD \parallel AC$.

$$\frac{BO}{BA} = \frac{BQ}{BC} = \frac{DP}{DA}$$

P describes a straight line perpendicular to OO_1 produced as Q moves in a circle with centre O_1.

Proof: In ΔABD, $\dfrac{BO}{BA} = \dfrac{PD}{DA}$. Hence $OP \parallel BD$.

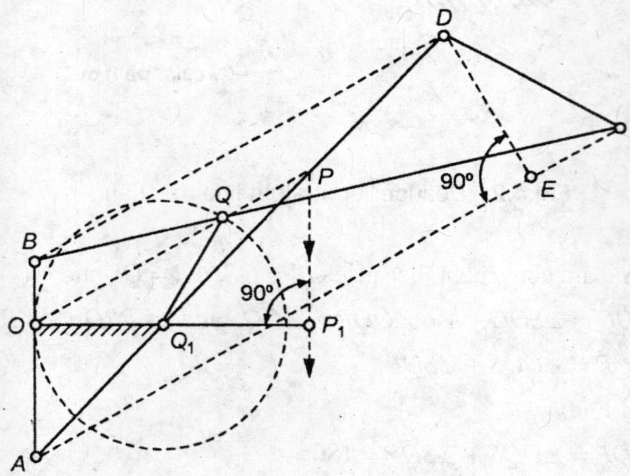

Fig.4.3 Hart straight line mechanism

Δs ABD and AOP are similar, because $\angle AOP = \angle ABD$ being corresponding angles and $\angle DAB$ is common.

$$\therefore \quad \frac{AB}{AO} = \frac{BD}{OP}$$

or
$$OP = \frac{BD \cdot AO}{AB} \tag{4.3}$$

Similarly as $\dfrac{BO}{AB} = \dfrac{BQ}{BC}$.

Hence $OQ \parallel AC$.

Now $AC \parallel BD$, $OQ \parallel BD$, $OP \parallel BD$. Hence $OP \parallel OQ$.
Since O is a common point, therefore OQP is a straight line.
Now Δs BOQ and BAC are similar.

$$\therefore \quad \frac{AC}{OQ} = \frac{AB}{OB}$$

or
$$OQ = \frac{OB \cdot AC}{AB} \tag{4.4}$$

From Eqs. (4.3) and (4.4), we get

$$OP \cdot OQ = \left(\frac{BD \cdot AO}{AB}\right) \cdot \left(\frac{OB \cdot AC}{AB}\right)$$
$$= \left(\frac{AO \cdot OB}{AB^2}\right)(BD \cdot AC) \tag{4.5}$$

Draw $DE \perp AC$.

$$BD = AC - 2EC$$
$$AC = AE + EC$$
$$BD \cdot AC = (AC - 2EC)(AE + EC)$$
$$= (AE + EC - 2EC)(AE + EC)$$
$$= (AE - EC)(AE + EC)$$
$$= AE^2 - EC^2 \tag{4.6}$$

From ΔAED, $\quad AE^2 = AD^2 - DE^2$
From ΔGED, $\quad EC^2 = CD^2 - DE^2$

$$AE^2 - EC^2 = AD^2 - CD^2 \tag{4.7}$$

From Eqs. (4.6) and (4.7), we get

$$BD \cdot AC = AD^2 - CD^2 \tag{4.8}$$

$$OP \cdot OQ = \left(\frac{AO \cdot BO}{AB^2}\right)(AD^2 - CD^2$$

= constant, as AO, $BO \cdot AB$, AD, and CD are fixed.

Hence P describes a straight line perpendicular to OO_1 produced as point Q moves in a circle with centre O_1.

3. *Scott-Russel mechanism*: The Scott-Russel mechanism is shown in Fig.4.4. It consists of a sliding pair and turning pairs. It can be used to generate approximate and accurate straight lines.

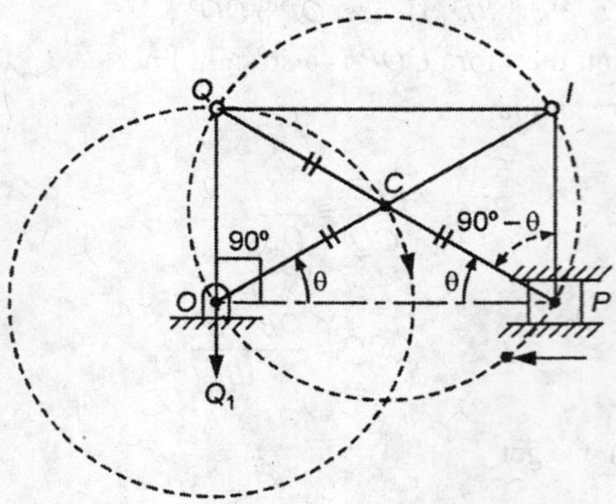

Fig.4.4 Scott-Russel straight line mechanism (CP = CQ)

(a) When $OC = CP = CQ$, Q describes a straight line $QO \perp OP$ provided P moves in a straight line along OP.

(b) If $CP \neq CQ$ then Q describes an approximate straight line perpendicular to OP provided P moves along a straight line OP such that $OC = \dfrac{CP^2}{CQ}$.

Proof:

(a) As $OC = CP = CQ$, hence $\angle POQ = 90°$ and $OQ \perp OP$. If P moves along OP then Q moves along a line perpendicular to OP.

Hence $\quad OP = OC\cos\theta + CP\cos\theta = PQ\cos\theta$

$\angle CPI = 90° - \theta$

Now $\quad \angle ICP = \angle COP + \angle CPO = 2\theta$

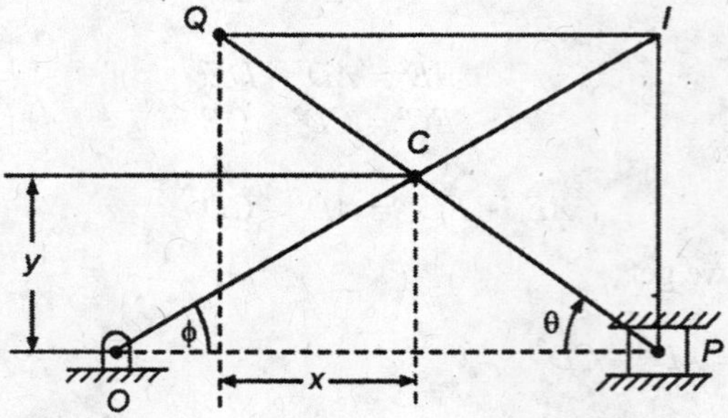

Fig.4.5 Scott-Russel mechanism when CP ≠ CQ

Then $\quad \angle CIP = 180° - \angle CPI - \angle ICP$

$= 180° - 90° + \theta - 2\theta$

$= 90° - \theta$

$\angle CIP = \angle CPI$

or $$CI = CP$$
∴ $$CP = CQ = CO = CI \quad \text{and} \quad \angle POQ = 90°$$

Hence $OPIQ$ is a square. As the path of Q is $\perp OI$, I is the instantaneous centre. Q will move along a line perpendicular to OP.

(b) When $CP \neq CQ$, it will form an elliptical trammel, as shown in Fig.4.5.

$$\frac{x^2}{CQ^2} + \frac{y^2}{CP^2} = 1$$

Because $x = CQ \cos \theta$ and $y = PC \sin \theta$

$$OC = \frac{\text{Semi-minor axis}^2}{\text{Semi-major axis}} = \frac{CP^2}{CQ}$$

As point C moves in a circle, for Q to move along an approximate straight line, $OC = \dfrac{CP^2}{CQ}$.

Limitations: When $OC \perp OP$, P coincides with O and $OQ = 2OC$. Here a small displacement of P shall cause a large displacement of Q, requiring a relatively small displacement of P to give displacement to Q. This requires highly accurate sliding surfaces.

4.3.2 Approximate Straight Line Motion Mechanisms

The mechanisms for the generation of approximate straight line motion are:

1. Grasshopper mechanism
2. Watt mechanism
3. Tchebicheff mechanism

1. *Grasshopper mechanism*: The Grasshopper mechanism is shown in Fig.4.6. The crank OC rotates about a fixed point $O \cdot O_1$ is a fixed pivot for link O_1P. For small angular displacements of O_1P, point Q on link PCQ will trace approximately a straight line path perpendicular to OP if

$$OC = \frac{CP^2}{CQ}$$

2. *Watt mechanism*: The Watt mechanism is shown in Fig.4.7(a).

Links OA and O_1B oscillates about O and O_1 respectively. AB is a connecting link. P will trace an approximate straight line if $\dfrac{PA}{PB} \cong \dfrac{O_1B}{OA}$. In Fig.4.7(b), θ and ϕ are the amplitudes of oscillation and I is the instantaneous centre of $A'B'$. P' is the point which lies on the approximate straight line described by P.

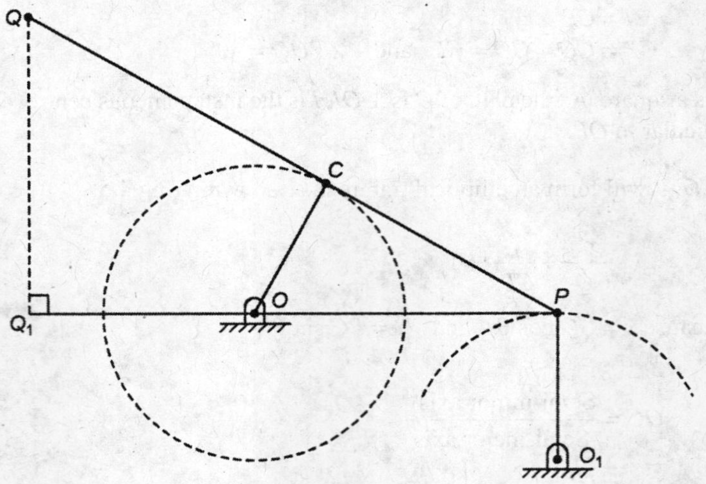

Fig.4.6 Grasshopper mechanism

Proof:

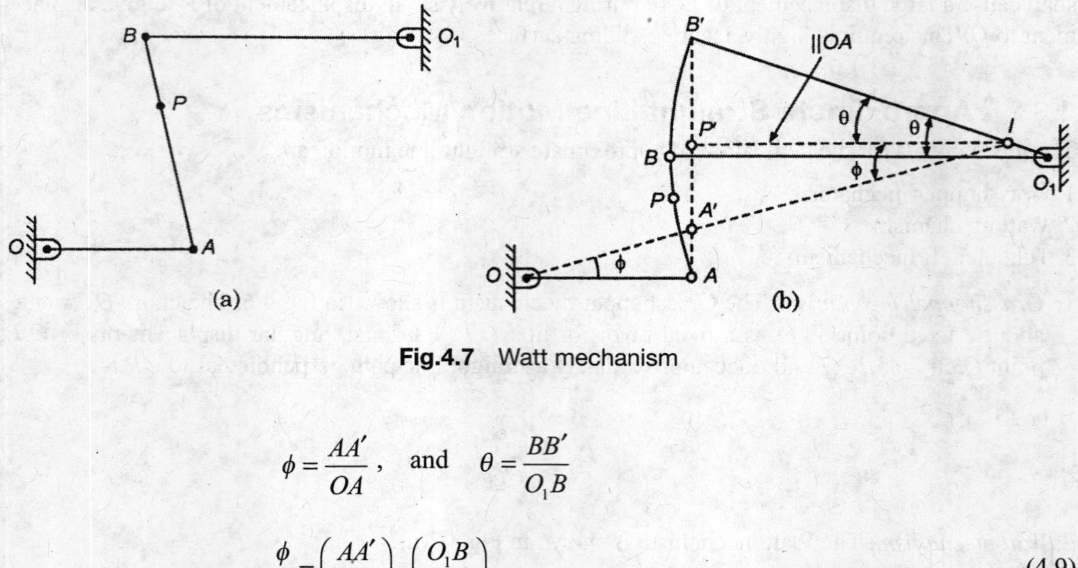

Fig.4.7 Watt mechanism

$$\phi = \frac{AA'}{OA}, \quad \text{and} \quad \theta = \frac{BB'}{O_1B}$$

$$\frac{\phi}{\theta} = \left(\frac{AA'}{OA}\right) \cdot \left(\frac{O_1B}{BB'}\right) \tag{4.9}$$

If $\angle B'P'I = 90°$, then $\sin\theta = \frac{B'P'}{B'I}$

or $\theta \approx \frac{B'P'}{B'I'}$ and $\phi \approx \frac{A'P'}{A'I}$

$$\frac{\phi}{\theta} \approx \left(\frac{A'P'}{A'I}\right) \cdot \left(\frac{B'I}{B'P}\right) \tag{4.10}$$

From Eqs. (4.9) and (4.10), we get

$$\left(\frac{AA'}{OA}\right) \cdot \left(\frac{O_1B}{BB'}\right) \approx \frac{A'P'}{A'I'} \cdot \frac{B'I}{B'P}$$

Now $$\frac{AA'}{BB'} = \frac{B'I}{A'I}$$

Δs OAA' and $IP'A$ are approximately similar, as well as Δs O_1BB' and $B'P'I$ are approximately similar.

$$\frac{O_1B}{OA} \approx \frac{A'P'}{B'P'} \approx \frac{AP}{BP}$$

Therefore, P divides the coupler AB in the ratio of the lengths of oscillating links. Hence P will describe an approximate straight line for a certain position of its path.

3. *Tchebicheff mechanism*: The Tchebicheff mechanism is shown in Fig.4.8. In this mechanism, $OA = O_1B$ and $AP = PB$. P is the tracing point. Let $AB = 1$, $OA = O_1B = x$, and $OO_1 = y$.

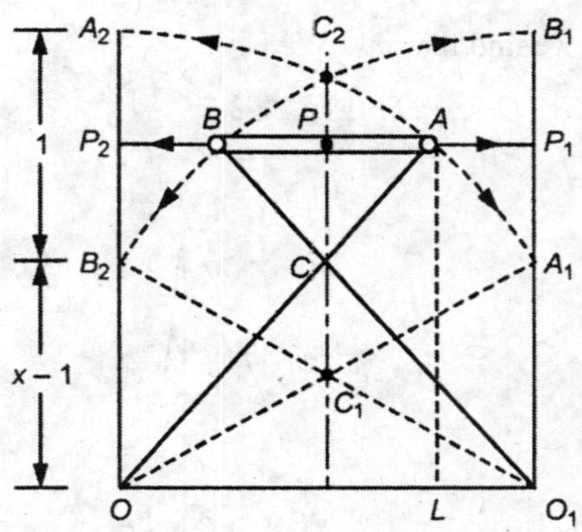

Fig.4.8 Tchebicheff mechanism

Now $$OB_2^2 = B_2O_1^2 - OO_1^2$$

$$(x-1)^2 = x^2 - y^2$$

or $$x = \frac{y^2+1}{2} \tag{4.11}$$

Draw $AL \perp OO_1$, then $OL = OO_1 - O_1L$

Also $$O_1L = AP_1$$

and $$AP_1 = \left(\frac{P_2P_1 - BA}{2}\right) = \left(\frac{y-1}{2}\right)$$

Again $$OL = OO_1 - O_1L$$

$$= y - \frac{y-1}{2} = \frac{y+1}{2}$$

Further $$OA^2 = AL^2 + OL^2$$

$$x^2 = \left(x - \frac{1}{2}\right)^2 + \left[\frac{y+1}{2}\right]^2$$

or $$x = \frac{y^2}{4} + \frac{1}{2} + \frac{y}{2} \tag{4.12}$$

From Eqs. (4.11) and (4.12), we get

$$\left(\frac{y^2+1}{2}\right) = \frac{y^2}{4} + \frac{1}{2} + \frac{y}{2}$$

or $$y = 2$$

Substituting in Eq. (4.11), we find that, $x = 3.5$

i.e., $$AB : OO_1, : OA :: 1 : 2 : 3.5$$

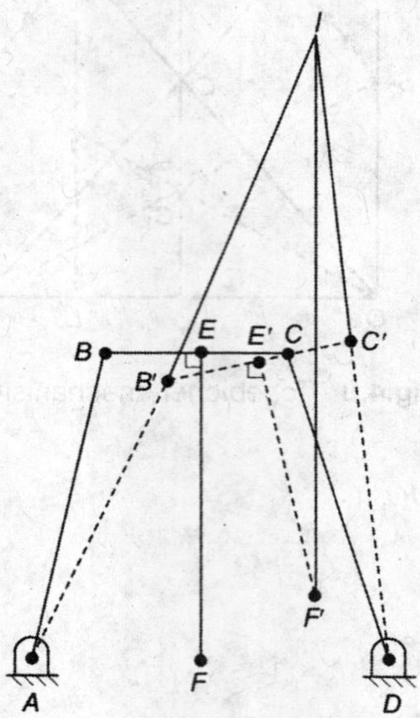

Fig.4.9 Robert's mechanism

P moves horizontally because instantaneous centre of AB will lie on point C which is the intersection of OA and BO_1. Thus C lies just below P.

4. *Robert's mechanism*: This mechanism is shown in Fig.4.9. The lengths of links are such that $AB = CD$, $BE = EC$ and EF is perpendicular to BC. Here, F is the tracing point. When the mechanism is displaced (shown by dotted lines), the point F will approximately trace a straight line parallel to BC. Produce AB' and DC' to meet at I, the instantaneous centre of link BC. From I drop a vertical line to intersect $E'F'$ at F'. The velocity v_f of point F' is perpendicular to the line joining I and F' and is thus a horizontal line.

4.4 INTERMITTENT MOTION MECHANISMS

These mechanisms are used to convert continuous motion into intermittent motion. The mechanisms used for this purpose are the Geneva wheel and the ratchet mechanism.

1. *Geneva wheel*: The Geneva wheel as shown in Fig.4.10, consists of a plate 1, which rotates continuously and contains a driving pin P that engages in a slot in the driven member 3. Member 2 is turned $\frac{1}{4}$th of a revolution for each revolution of plate 1. The slot in member 2 must be tangential to the path of pin upon engagement in order to reduce shock. The angle β is half the angle turned through by member 2 during the indexing period. The locking plate serves to lock member 2 when it is not being indexed. Cut the locking plate back to provide clearance for member 2 as it swings through the indexing angle. The clearance arc in the locking plate will be equal to twice the angle α.
2. *Ratchet mechanism*: This mechanism is used to produce intermittent circular motion from an oscillating or reciprocating member. Fig.4.11 shows the details of a ratchet mechanism. Wheel 4 is given intermittent circular motion by means of arm 2 and driving pawl 3. A second pawl 5 prevents 4 from turning backward when 2 is rotated clockwise in preparation for another stroke. The line of action PN of the driving pawl and tooth must pass between centers O and A in order to have the pawl 3 remain in contact with the tooth. This mechanism is used particularly in counting devices.

4.5 PARALLEL LINKAGES

These are the four-bar linkages in which the opposite links are equal in length and always form a parallelogram. There are three types of parallel linkages: parallel rules, universal drafting machine and lazy tongs. They are used for producing parallel motion.

1. *Parallel rules*: A parallel rule is shown in Fig.4.12, in which $AB = CD = EF = GH = IJ$ and $AC = BD$, $CE = DF$, $EG = FH$, $GI = HJ$. Here AB, C, EF, GH and IJ will always be parallel to each other.
2. *Universal drafting machine*: A universal drafting machine is shown in Fig.4.13, in which $AB = CD$; $AC = BD$; $EF = GH$; and $EG = FH$. Position of points A and B are fixed. Similarly the positions of points E and F are fixed with respect to C and D. The positions of scales I and II are fixed with respect to points G and H. Then $ABDC$ is a parallelogram. Line CD will always be parallel to AB so that the direction of CD is fixed. Therefore, the direction of EF is fixed. Further $EFHG$ is a parallelogram, so GH is always parallel to EF such that the direction of GH is also fixed, whatever their actual position may be.

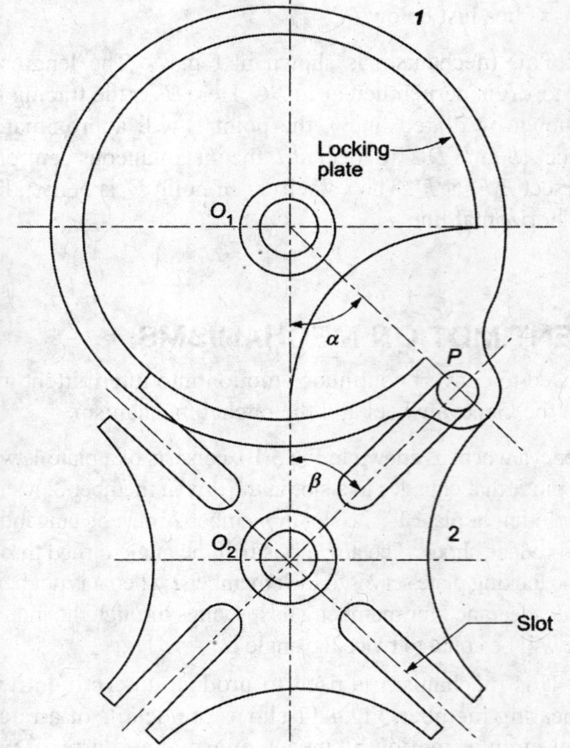

Fig.4.10 Geneva wheel

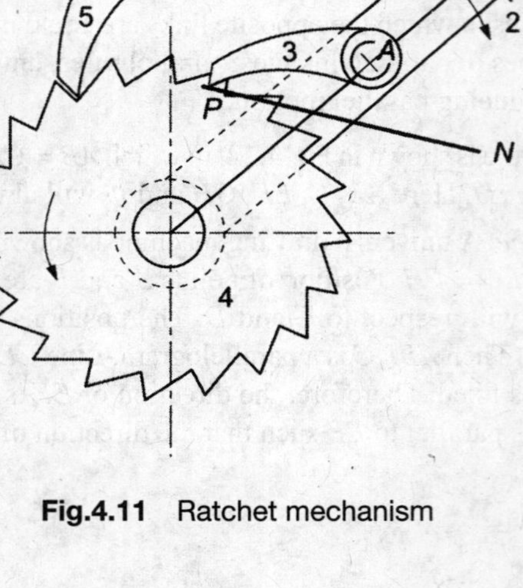

Fig.4.11 Ratchet mechanism

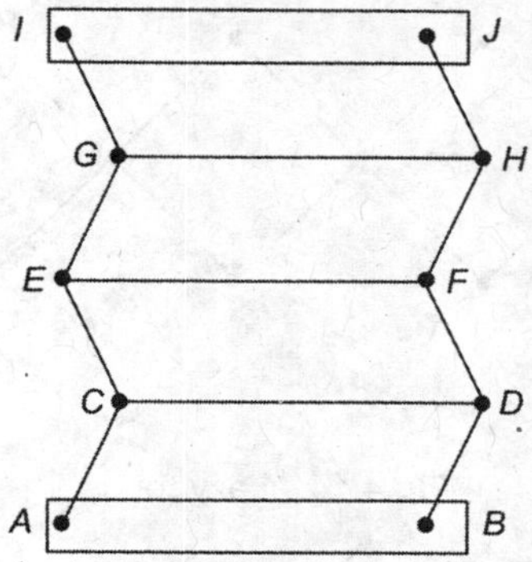

Fig.4.12 Parallel rule

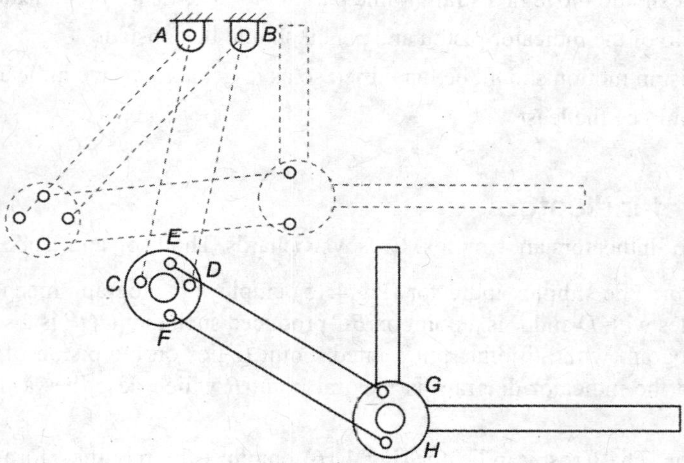

Fig.4.13 Universal drafting machine

3. *Lazy tongs*: Lazy tongs, as shown in Fig.4.14, consists of pin joint A attached to a fixed point. Point B moves on a roller. All other joints are pin jointed. The vertical movement of the roller affects the movement of the end C to move in a horizontal direction. It is used to support a telephone or a bulb at point C for horizontal movements.

4.6 ENGINE PRESSURE INDICATORS

A pressure indicator is an instrument used to obtain a graphical record of the pressure–volume diagram of a reciprocating engine. It consists of a cylinder with a piston, a straight line motion mechanism with a pencil and a drum with paper wrapped around it. The indicator cylinder is connected to the engine cylinder, which causes the movement of the indicator piston with change of pressure in the engine cylinder. The piston motion is constrained by a spring so that the piston displacement is a direct measure of the working fluid pressure acting upon it. The displacement and hence volume is then recorded by the pencil with the help of straight line mechanism on the drum paper. The requirements of the straight line mechanism are as follows:

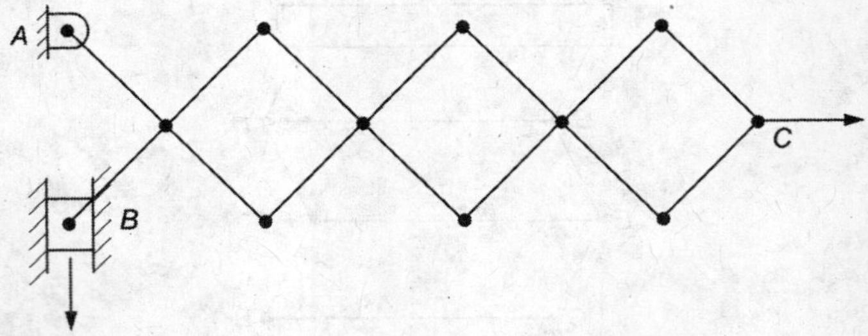

Fig.4.14 Lazy tongs

1. The pencil point should move in a straight line parallel to the axis of the indicator piston.
2. The velocity ratio of the indicator piston and pencil should be constant.
3. The indicator piston motion should be magnified so as to get a good size indicator diagram.
4. The friction should be the least.

4.6.1 Types of Indicators

The various types of indicators are simplex, Crosby, Richards, Thomson and Dobbie–McInnes.

1. *Simplex indicator*: The simplex indicator (Fig.4.15) employs pantograph mechanism. Point Q on link AD coincides with D and P is a point on BC produced such that OQP is a straight line. $ABCD$ form a parallelogram with all joints pin jointed. Point Q lies on the piston of the indicator. The pencil to record the indicator diagram is fixed at point P which describes a path similar to that of Q.
2. *Crosby indicator*: The Crosby indicator (Fig.4.16) employs the modified form of pantograph to generate motion of pencil point P similar to that of point Q lying on the indicator piston. The following conditions are to satisfied by this mechanism:

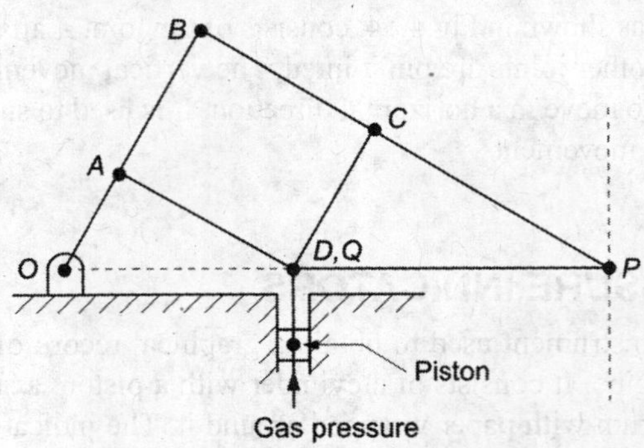

Fig.4.15 Simplex indicator

1. Velocity ratio between P and Q is constant, i.e. $\frac{v_p}{v_q}$ = constant.
2. Point P travels along a straight line.

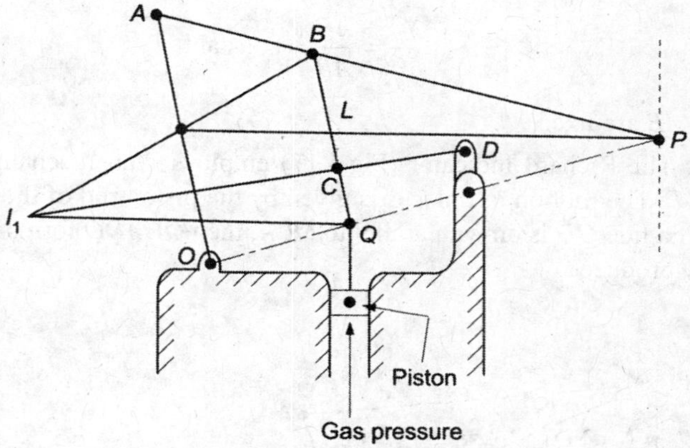

Fig.4.16 Crosby indicator

Proof:
Draw instantaneous centres I_1 and I_2 of links QB and AP, respectively. I_1 is obtained by drawing a horizontal line from Q which meets the line DC produced in I_1. For I_2, draw a horizontal line from P meeting OA in I_3. The line I_2P cuts the link QB in I_1. The point I_2 will lie in I_1B.

Now $$\frac{v_b}{v_p} = \frac{I_1B}{I_1Q}$$

but $$\frac{I_1B}{I_1Q} = \frac{I_2B}{I_2L}$$

∴ $$\frac{v_b}{v_q} = \frac{I_2B}{I_2L} \qquad (4.13)$$

Also $$\frac{v_p}{v_b} = \frac{I_2P}{I_2B} \qquad (4.14)$$

Multiplying (4.13) and (4.14), we get

$$\frac{v_p}{v_q} = \frac{I_2P}{I_2L}$$

As OA is parallel to QB or I_2A is parallel to BL, Δs PI_2A and PLB are similar.

∴ $$\frac{I_2P}{I_2L} = \frac{AP}{AB}$$

$$\frac{v_p}{v_q} = \frac{AP}{AB} = \text{constant}$$

Since lengths AP and AB are fixed.

3. *Richard indicator*: The Richard indicator (Fig.4.17) employs Watt mechanism $OABO_1$ to guide the pencil point at P. The motion to link AC is given by the piston rod of the indicator piston at C through link QC. If a line QD is drawn parallel to OCA, then $OCADQ$ forms a pantograph, having point P on link AD produced.

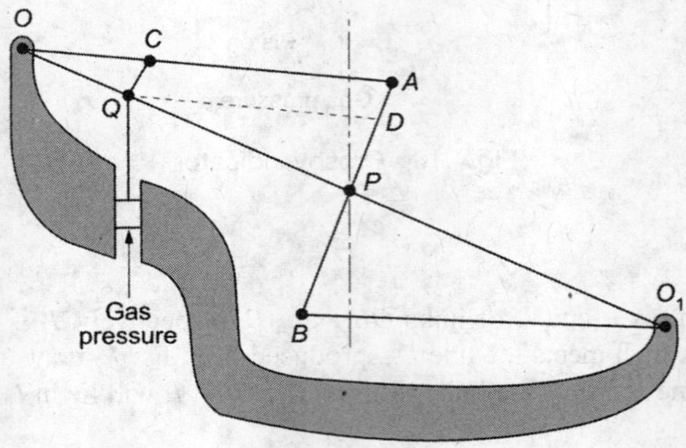

Fig.4.17 Richard indicator

4. *Thomson indicator*: The Thomson indicator (Fig.4.18) employs Grasshopper mechanism $OABO_1$, the tracing point P lying on link AB produced. Link AB gets the motion from the piston rod of the indicator at C which is connected by link QC at Q to the end of indicator piston rod. Link QC is parallel to link OA.

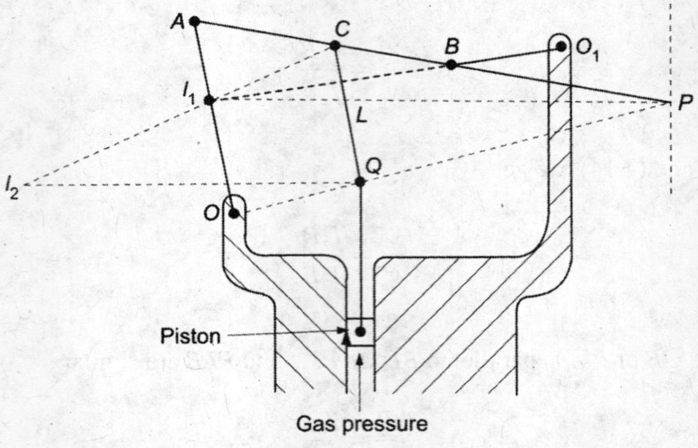

Fig.4.18 Thomson indicator

The velocity ratio $\frac{v_p}{v_q}$ constant. To prove this, draw the instantaneous centres I_1 and I_2 of the links AB and QC respectively. I_1P cuts CQ in L.

Proof:

$$\frac{v_c}{v_q} = \frac{I_2C}{I_2Q}$$

Now triangles $I_1\,CL$ and I_2CQ are similar. Therefore

$$\frac{I_2C}{I_2Q} = \frac{I_1C}{I_1L}$$

Also
$$\frac{v_p}{v_c} = \frac{I_1P}{I_1C}$$

$\therefore$
$$\frac{v_p}{v_q} = \frac{I_1P}{I_1L}$$

Δs PAI_1 and PCL are similar. Hence

$$\frac{I_1P}{I_1L} = \frac{AP}{AC}$$

$\therefore$
$$\frac{v_p}{v_q} = \frac{AP}{AC} = \text{constant}$$

Since the lengths AP and AC are fixed.

5. *Dobbie–McInnes indicator*: In the Dobbie–McInnes indicator (Fig.4.19), the motion is given to link O_1B by the link QC connected to the indicator piston and I_1 and I_2 are the instantaneous centres of AB and QC, respectively. The line I_1P cuts QC in L. Draw $BM \perp I_1P$ from point B. OQP is a straight line. The ratio $\frac{v_p}{v_q} = \text{constant}$.

Proof:

$$\frac{v_c}{v_q} = \frac{I_2C}{I_2Q}$$

Triangles $I_1\,BM$, I_1CL and I_1CQ are similar.

$$\frac{I_2C}{I_2Q} = \frac{I_1C}{I_1L} = \frac{I_1B}{I_1M}$$

$$\frac{v_c}{v_q} = \frac{I_1B}{I_1M}$$

Now
$$\frac{v_b}{v_c} = \frac{O_1B}{O_1C}$$

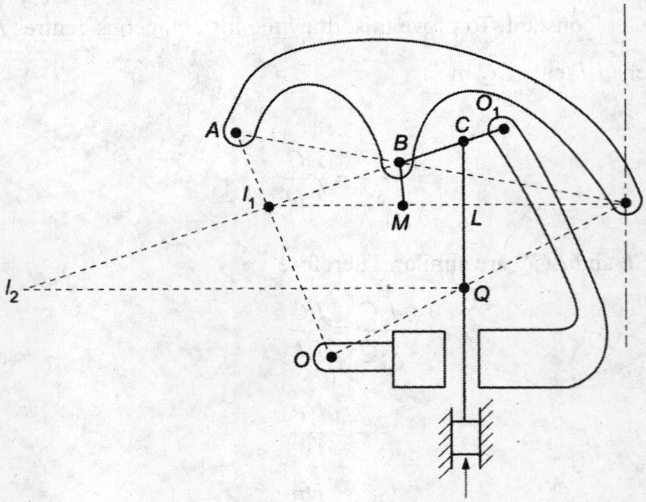

Fig.4.19 Dobbie-McInnes indicator

Also

$$\frac{v_p}{v_b} = \frac{I_1P}{I_1B}$$

$$\frac{v_p}{v_q} = \left(\frac{I_1P}{I_1M}\right)\cdot\left(\frac{O_1B}{O_1C}\right)$$

In similar traingles PBM and PAI_1.

$$\frac{I_1P}{I_1M} = \frac{AP}{AB}$$

$$\frac{v_p}{v_q} = \frac{PA}{AB}\cdot\frac{O_1B}{O_1C} = \text{constant}$$

Since the lengths of all four links PA, AB, O_1B and O_1C are fixed

Example 4.1

A circle with AB as diameter has a point C on its circumference. D is a point on AC produced such that if C turns about A, the product $AC \times AD$ is constant. Prove that the point D moves in a straight line perpendicular to AB produced.

Solution

Let D_1D be perpendicular to AB produced, as shown in Fig.4.20.

Now $\angle ACB = 90°$, being angle in a semicircle. Also $\angle AD_1D = 90°$. Therefore, Δs ACB and AD_1D are similar, as $\angle CAB$ is common.

$$\frac{AC}{AB} = \frac{AD_1}{AD}$$

or $$AB \times AD_1 = AC \times AD$$

or $$AD_1 = \frac{AC \times AD}{AB} = \text{constant}$$

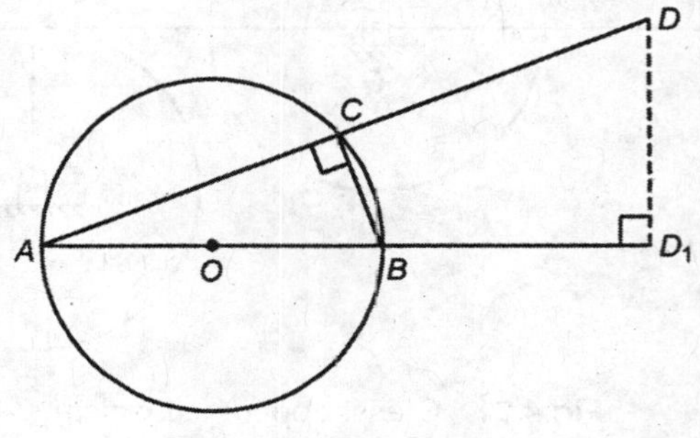

Fig.4.20 For Example 4.1

Because AB is fixed and $AC \times AD$ = constant.

Thus AD_1 will be constant for all positions of C. Therefore, the location of D_1, is fixed, which means that D moves in a straight line perpendicular to AB produced.

Example 4.2

In a Grasshopper mechanism shown in Fig.4.21, $OC = 100$ mm, $PC = 150$ mm, and $PQ = 375$ mm. Determine the magnitude of the vertical force at Q necessary to resist a torque of 200 N m applied to the link OC when it makes an angle of 30° with the horizontal.

■ Solution

$$QC = PQ - PC = 375 - 150 = 225 \text{ mm}$$

$$\frac{OC}{PC} = \frac{PC}{QC}$$

or $$\frac{100}{150} = \frac{150}{225}$$

Thus the condition for the dimensions of a Grasshopper mechanism is satisfied, and point Q will trace an approximate straight line perpendicular to OP.

Now $$F_q \cdot v_q = T_c \cdot v_c$$

or $$F_q = \frac{T_c \cdot v_c}{v_q \cdot OC}$$

The instantaneous centre of PQ is at I. Let $\angle QIC = \theta$. Also Δs QIC and OCP are similar.

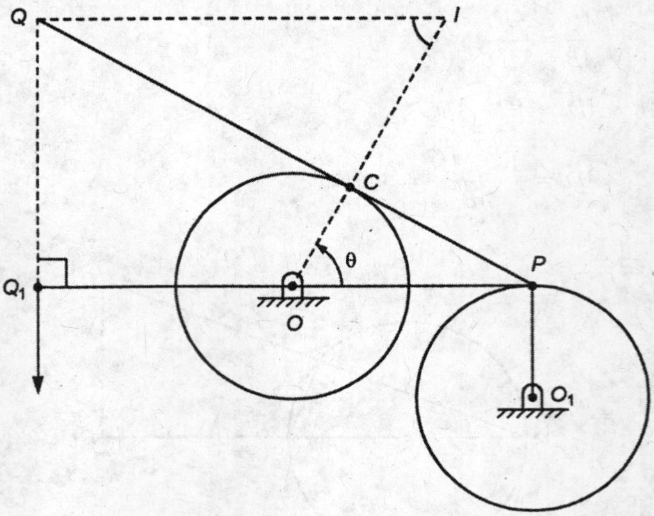

Fig.4.21 Grasshopper mechanism

$$\frac{v_c}{v_q} = \frac{IC}{IQ} = \frac{OC}{OP}$$

Thus $$F_q = \frac{T_c}{OP}$$

When $$\theta = 30°,$$

$$OP = 100 \cos 30° + \left[(150)^2 - (100 \sin 30°)^2\right]^{0.5}$$
$$= 228 \text{ mm}$$

$$F_q = \frac{200}{0.228} = 877.2\text{N}$$

4.7 AUTOMOBILE STEERING GEAR MECHANISMS

The mechanisms that are used for changing the direction of motion of an automobile are called steering gears. Steering is done by changing the direction of motion of front wheels only as the rear wheels have a common axis which is fixed and moves in a straight line only.

4.7.1 Fundamental Equation for Correct Steering

When a vehicle takes a turn all the wheels should roll on the road smoothly preventing excessive tyre wear. This is achieved by mounting the two front wheels on two short axles, called stub axles. The stub axles are pin-jointed with the main front axle which is rigidly attached to the rear axle, on which the back wheels are attached. Steering is done by front wheels only.

Figure 4.22 shows an automobile taking a right turn. When the vehicle takes a turn, the front wheels along with stub axles turn about their respective pin-joints. The inner front wheel turns through a greater angle as compared to that of outer front wheel. In order to avoid skidding, for correct steering, the two front wheels must turn about the same instantaneous centre I, which lies on the axis of the rear wheels. Thus, the condition for correct steering is that all the four wheels must turn about the same instantaneous centre.

Let $AE = l$ = wheel base

$CD = a$ = wheel track

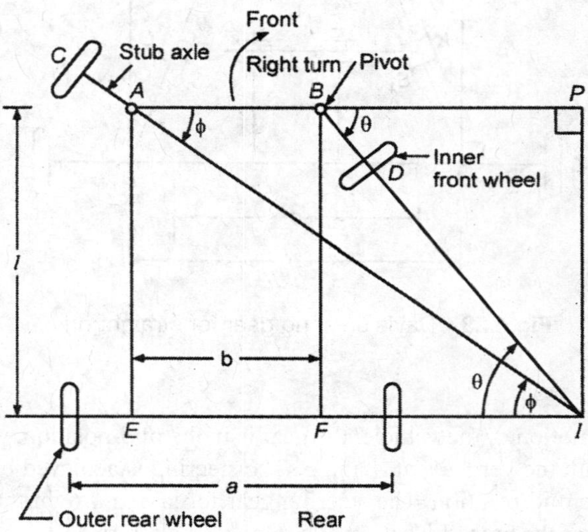

Fig.4.22 Automobile steering gear

$AB = b$ = distance between the pivots of front axles

I = common instantaneous centre of all wheels

θ, ϕ = angles turned through by stub-axles of the inner and outer front wheels respectively

Now
$$b = AP - BP$$
$$= l(\cot\phi - \cot\theta)$$

or
$$\cot\phi - \cot\theta = \frac{b}{l} \tag{4.12}$$

Eq. (4.12) is called the fundamental equation for correct steering.

4.7.2 Steering Gears

A steering gear is a mechanism for automatically adjusting values of θ, ϕ for correct steering. The following steering gears are commonly used in automoiles:

1. Davis steering gear
2. Ackermann steering gear

1. Davis steering gear

Davis steering gear is shown in Fig.4.23. This type of gear has only sliding pairs. Two arms *AG* and *BH* are fixed to the stub axles *AC* and *BD* respectively. *CAG* and *DBH* form two similar bell-crank levers pivoted at *A* and *B* respectively. *KL* is a cross-link which is constrained to slide parallel to *AB*. The ends of the cross-link *KL* are pin-jointed to two sliders S_1 and S_2 as shown. These sliders are free to slide on links *AG* and *BH* respectively. The whole mechanism is in front of the front wheels.

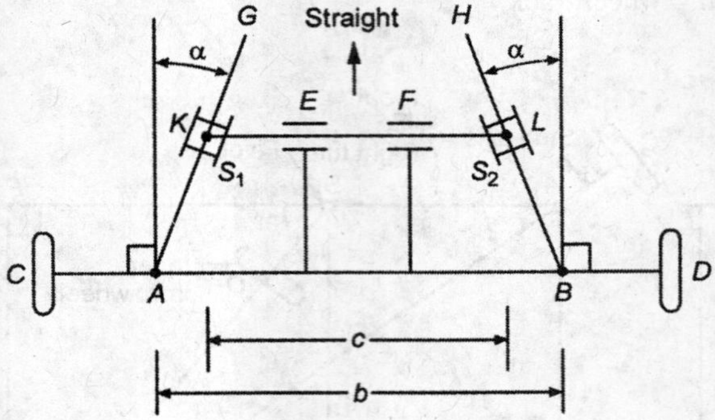

Fig.4.23 Davis steering gear for straight drive

During the straight motion of the vehicle, the gear is in the mid-position, with equal inclination of the arms *AG* and *BH* with the verticals at *A* and *B*. The steering is achieved by moving cross-link *KL* to the right or left of the mid-position. The steering gear for taking a right turn is shown in Fig.4.24. *K'L'* shows the position of the cross-link *KL* while taking a right turn.

Determination of angle α:

Let x = distance moved by *KL* from mid-position

$= KK' - LL'$

y = horizontal distance of points *K* and *L* from *A* and *B* respectively

h = vertical distance between *AB* and *KL*

α = angle of inclination of track arms *AG* and *BH* with the vertical in mid-position

θ, ϕ = angles turned by stub axles

From Fig.4.24, we have

$$y = AK \sin\alpha = \text{BL} \sin\alpha$$

$$\tan(\alpha - \theta) = \frac{y - x}{h}$$

or

$$\frac{\tan a - \tan\theta}{1 + \tan a \tan\theta} = \frac{y - x}{h}$$

Now

$$\tan\alpha = \frac{y}{h}$$

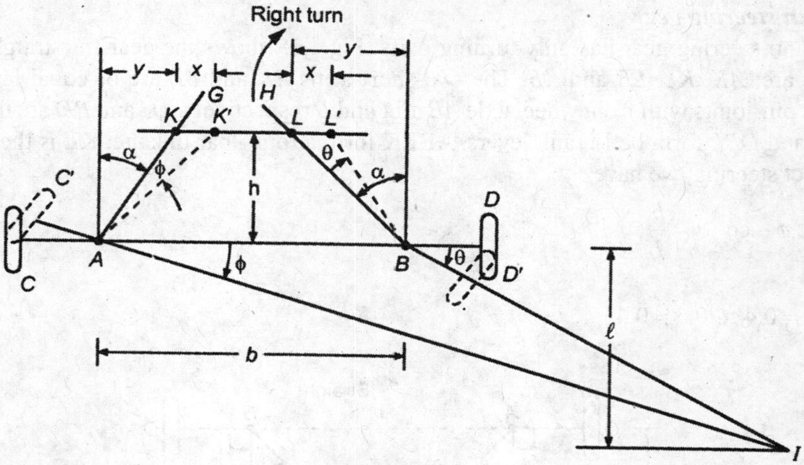

Fig.4.24 Davis steering gear taking a right turn

$$\therefore \qquad \frac{\frac{y}{h} - \tan\theta}{1 + \frac{y}{h}\tan\theta} = \frac{y - x}{h}$$

or
$$\tan\theta = \frac{xh}{y^2 - xy + h^2} \tag{4.16}$$

Similarly
$$\tan(\alpha + \phi) = \frac{y + x}{h}$$

So that
$$\tan\phi = \frac{xh}{y^2 + xy + h^2} \tag{4.17}$$

From Eqs. (4.16) and (4.17), we get

$$\cot\phi - \cot\theta = \frac{2y}{h} = \frac{b}{l}$$

or
$$\tan\alpha = \frac{y}{h} = \frac{b}{2l}$$

Generally, $\frac{b}{l} =$ 0.4 to 0.5, so that $\alpha = 11.3°$ to $14.1°$. There will be friction and more wear due to sliding pairs in the Davis steering gear. Therefore, it becomes inaccurate after some use. On account of these reasons, it is not used in practice.

2. Ackermann steering gear

The Ackermann steering gear has only turning pairs. Fig.4.25 shows the gear for straight drive. The turning pairs are: *AK*, *KL*, *LB* and *AB*. The two short arms *AK* and *BL* are of equal length and are connected by pin-joints with front wheel axle *AB* at *A* and *B* respectively. *AC* and *BD* are the stub-axles so that *CAK* and *DBL* form bell-crank levers. ABLK form a four - bar linkage. KL is the track rod.

For correct steering, we have

$$\cot \phi - \cot \theta = \frac{b}{l}$$

Generally, $\frac{b}{l} = 0.4 \text{ to } 0.5 \approx 0.455.$

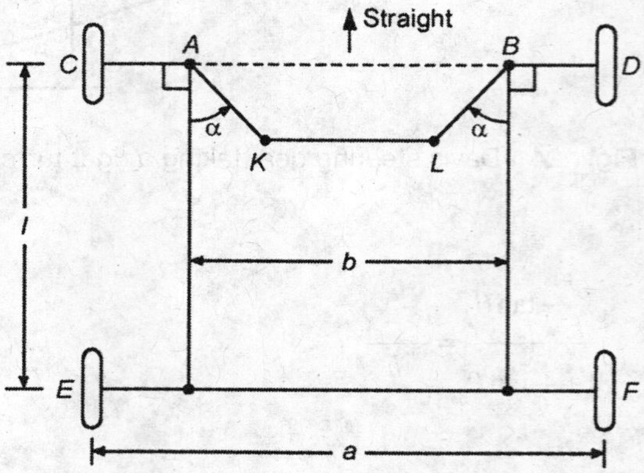

Fig.4.25 Ackermann steering gear for straight drive

Determination of angle α

1. Analytical method: Consider the Ackermann steering gear, as shown in Fig.4.26(a), taking a right turn. The instantaneous centre *I* lies on *a* line parallel to the rear axis at a distance of approximately $0.3l$ above the rear axis. It may be seen that the whole mechanism of the Ackermann steering gear is on the back of the front whells. From Fig.4.26(b), we have

Projection of arc *K'K* on *AB* = Projection of arc *L'L* on *AB*

or
$$K_1K_1' = L_1L_1'$$

$$AK\,[\sin \alpha - \sin (\alpha - \phi) = BL\,[\sin (\alpha + \theta) - \sin \alpha]$$

Now
$$AK = BL$$

$$\sin \alpha - \sin (\alpha - \phi) = \sin (\alpha + \theta) - \sin \alpha$$

or
$$\sin \alpha - \sin \alpha \cos \phi + \cos \alpha \sin \phi = \sin \alpha \cos \theta + \cos \alpha \sin \theta - \sin \alpha$$

$$\sin \alpha\,(2 - \cos \phi - \cos \theta) = \cos \alpha\,(\sin \theta - \sin \phi)$$

$$\tan \alpha = \frac{\sin\theta - \sin\phi}{2 - \cos\phi - \cos\theta} \tag{4.18}$$

The values of θ and ϕ are known for correct steering. Hence, the value of α can be determined from, Eq. (4.18).

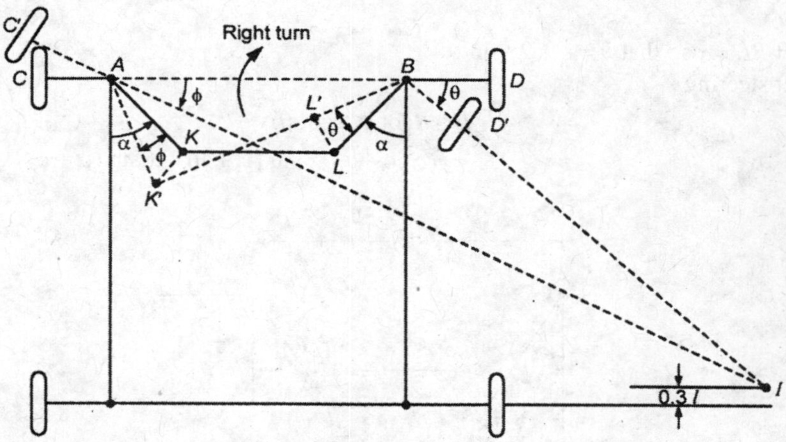

(a) Ackermann steering gear taking a right turn

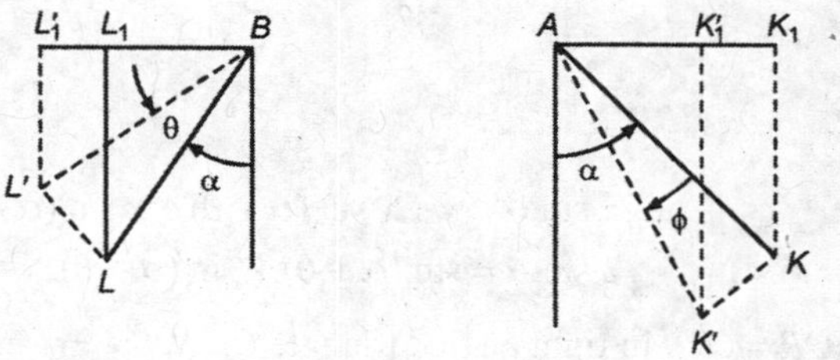

(b) Determination of angle α

Fig.4.26 Ackermann steering gear in displaced position

2. Graphical method to determine angle α: Draw a horizontal line OX (Fig.4.27). Make $\angle XOQ = \theta$ and $\angle XOP = \phi$. With any radius, draw arc PRQ. Join PR and produce it so that $PR = RM$. Join MQ and produce. Draw $ON \perp NQM$. Then $\alpha = 90° - \angle XON$.

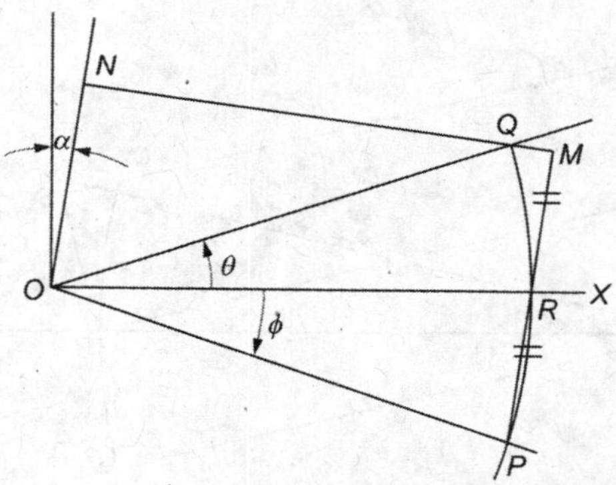

Fig.4.27 Graphical method to determine angle α

Proof

In Fig.4.28, let $BL = x$ so that $b = c + 2x \sin \alpha$.

For correct steering,

$$b = l(\cot \phi - \cot \theta)$$
$$= c \cos \psi + x \sin (\alpha + \theta) + x \sin (\alpha - \phi)$$

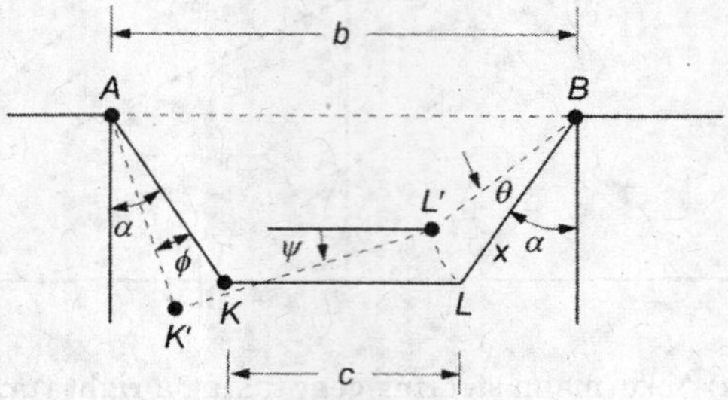

Fig.4.28 Graphical method to determine angle α

If ψ is small, c $\cos \psi = c$.

$$c + 2x \sin \alpha = c + x \sin (\alpha + \theta) + x \sin (\alpha - \phi)$$

or
$$2 \sin \alpha = \sin (\alpha + \theta) + \sin (\alpha + \theta)$$

In Fig.4.29, let r = $OP = OR = OQ$. In right angled triangle OQN,

$$\angle NOQ = 90° - \alpha - \theta = 90 - (\alpha - \theta)$$
$$ON = r \cos[90° - (\alpha + \theta)] = r \sin(\alpha + \theta)$$

Draw $RL \perp ON$ then $RL \parallel SN$.

i.e.
$$\angle SRQ = \angle \alpha$$

Also
$$\angle ORL = \angle \alpha$$
$$OL = r \sin \alpha$$

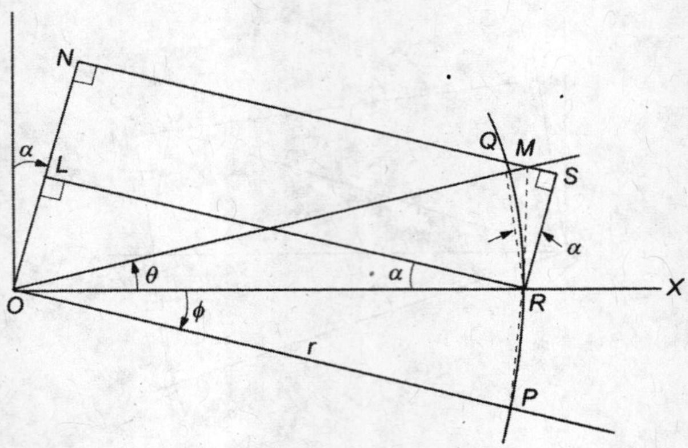

Fig.4.29 Graphical method to determine angle α

Now $NL = RS$. In ΔOPR,

$$\frac{PR}{\sin\phi} = \frac{r}{\sin(90° - 0.5\phi)}$$

or

$$PR = \frac{r\sin\phi}{\cos(0.5\phi)}$$

$$= 2r\sin(0.5\phi)$$

From triangle *RSM* in Fig.4.30, we have

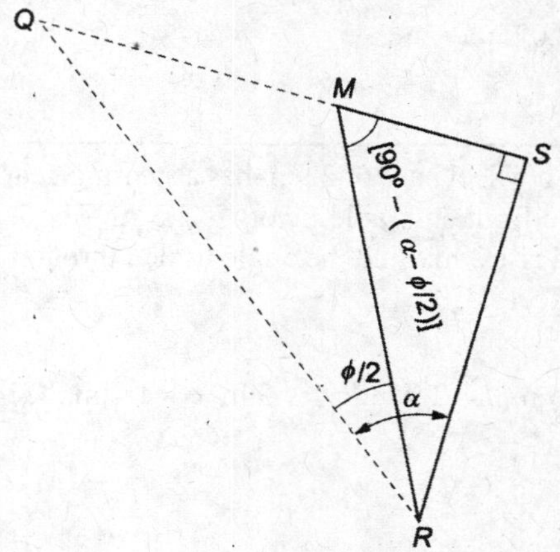

Fig.4.30 Graphical method to determine angle α

$$\angle MRS = \alpha - 0.5\phi$$
$$\angle SMR = 90° - (\alpha - 0.5\phi)$$

Also

$$\frac{RS}{RM} = \cos(\alpha - 0.5\phi)$$

Now $RM = PR$.

$$\frac{RS}{PR} = \cos(\alpha - 0.5\phi)$$

or

$$RS = 2r\sin(0.5\phi)\cos(\alpha - 0.5\phi)$$
$$= r\sin\alpha - r\sin(\alpha - \phi)$$
$$ON = OL + LN = OL + RS$$
$$r\sin(\alpha + \theta) = r\sin\alpha + r\sin\alpha - r\sin(\alpha - \phi)$$

or

$$2\sin\alpha = \sin(\alpha + \theta) + \sin(\alpha - \phi) \tag{4.19a}$$

or

$$\tan\alpha = \frac{\sin\theta - \sin\phi}{2 - \cos\theta - \cos\phi} \tag{4.19b}$$

Hence proved.

Example 4.3

In a Davis steering gear, the distance between the pivots of the front axle is 1.2 m and the wheel base is 2.8 m. When the automobile is moving along a straight path, find the inclination of the track arms to the longitudinal axis of the automobile?

■ Solution

Given: $b = 1.2$ m, $l = 2.8$ m

Now
$$\tan \alpha = \frac{b}{2l}$$
$$= \frac{1.2}{5.6} = 0.2143$$
$$\alpha = 12.095°$$

Example 4.4

A car with a track of 1.5 m and a wheel base of 2.9 m has a steering gear mechanism of the Ackermann type. The distance between the front stub axle pivots is 1.3 m. The length of each track arm is 150 mm and the length of track rod is 1.2 m. Find the angle turned through by the outer wheel if the angle turned through by the inner wheel is 30°.

■ Solution

Given: $a = 1.5$ m, $l = 2.9$ m, $b = 1.3$ m, $c = 1.2$ m, $AK = 0.15$ m.

For correct steering, we have

$$\cot\phi - \cot\theta = \frac{b}{l}$$
$$= \frac{1.3}{2.9} = 0.44827$$
$$\cot \phi - \cot 30° = 0.44827$$
$$\cot \phi - 1.73205 = 0.44827$$
$$\cot \phi = 2.18032$$
$$\phi = 24.64°$$

4.8 HOOKE'S JOINT OR UNIVERSAL COUPLING

It is a device to connect two shafts whose axes are neither coaxial nor parallel but intersect at a point. This is used to transmit power from the engine to the rear axle of an automobile and similar other applications. It is also called universal coupling.

The Hooke's joint, as shown in Fig.4.31, consists of two forks connected by a centre piece, having the shape of a cross or square carrying four trunnions. The ends of the two shafts to be connected together are fitted to the forks.

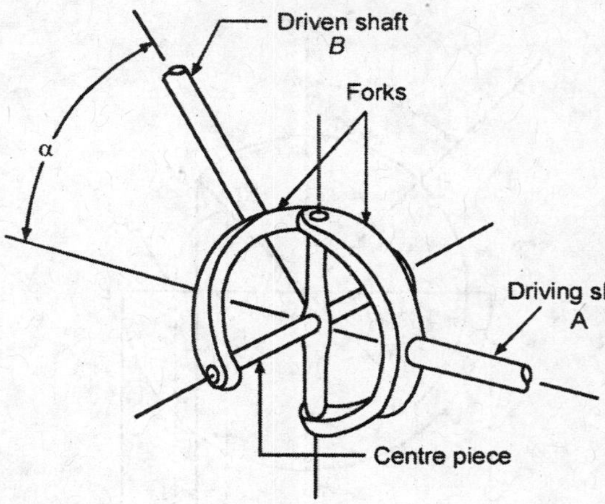

Fig.4.31 Hooke's joint

4.8.1 Velocities of Shafts

Let the driving shaft A and the driven shaft B be connected by a joint having two forks KL and MN. The two forks are connected by a cross $KLMN$ intersecting at O, as shown in Fig.4.32. Let the angle between the axes of the shafts be α. Let fork KL move through an angle θ in a circle to the position K_1L_1 in the front view. The fork MN will also move through the same angle θ. MN being not in the same plane shall move in an ellipse in the front view to its new position M_1N_1. To find the true angle, project M_1 to the top view, which cuts the horizontal axis in R and fork MN in R_1. Rotate R_1 to R_2 on the horizontal axis with centre O. Project R_2 back in the front view cutting the circle in M_2. Join OM_2 Measure angle MOM_2, which is the true angle ϕ. Thus when the driving shaft A revolves through an angle θ, the driven shaft B will revolve through an angle ϕ.

Now $$\tan\phi = \frac{OR_2'}{R_2'M_2}$$

and $$\tan\theta = \frac{OR'}{R'M_1}$$

$$\frac{\tan\phi}{\tan\theta} = \frac{OR_2'}{R_2'M_2} = \frac{R'M_1}{OR'}$$

But $$R'M_1 = R_2'M_2$$

$\therefore$ $$\frac{\tan\phi}{\tan\theta} = \frac{OR_2'}{OR'} = \frac{OR_1}{OR}$$

Now $$OR = OR_1 \cos\alpha$$

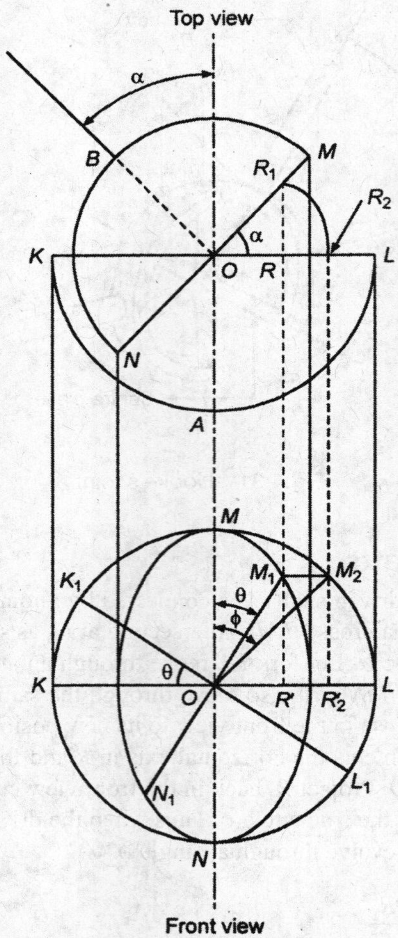

Fig.4.32 Hooke's coupling forks in displaced position

$$\frac{\tan\phi}{\tan\theta} = \frac{1}{\cos\alpha}$$

or $$\tan\theta = \cos\alpha \tan\phi \tag{4.20}$$

Angular velocity of driven shaft, $\omega_A = \dfrac{d\theta}{dt}$

Angular velocity of driven shaft, $\omega_B = \dfrac{d\phi}{dt}$

Differentiating Eq. (4.20), we get

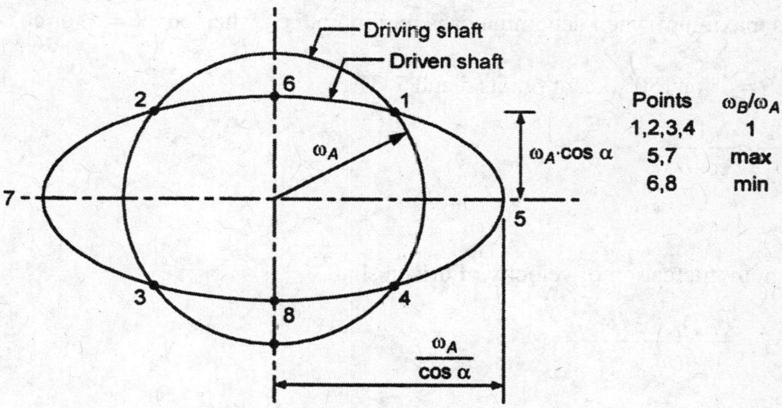

Fig.4.33 Speed polar diagram

$$\sec^2\theta.\ \frac{d\theta}{dt} = \cos\alpha \sec^2\phi.\ \frac{d\phi}{dt}$$

$$\frac{\frac{d\phi}{dt}}{\frac{d\theta}{dt}} = \frac{\cos\alpha}{1-\sin^2\alpha\cos^2\theta}$$

or $$\frac{\omega_B}{\omega_A} = \frac{\cos\alpha}{1-\sin^2\alpha\cos^2\theta} \tag{4.21}$$

Case 1: $$\frac{\omega_B}{\omega_A} = 1$$

$$\cos\alpha = 1 - \sin^2\alpha\cos^2\theta$$

or $$\cos\theta = \pm\sqrt{\frac{1}{1+\cos\alpha}} \tag{4.22}$$

This condition occurs once in each quadrant, as shown in Fig.4.33 by points 1, 2, 3 and 4.

Case 2: $\frac{\omega_B}{\omega_A}$ is minimum when the denominator is maximum, i.e., $\sin^2\alpha\cos^2\theta$ must be minimum, or $\cos\theta = 0°$. Thus $\frac{\omega_B}{\omega_A}$ is minimum at $\theta = 90°, 270°$, i.e. at points 6 and 8. Then

$$\frac{\omega_B}{\omega_A} = \cos\alpha \tag{4.23}$$

Case 3: $\frac{\omega_B}{\omega_A}$ is maximum when denominator is minimum, i.e., when $\cos^2\theta = 1$, or $\cos\theta = \pm 1$,

$\theta = 0°$ or $180°$, i.e. at points 5 and 7. Then

$$\frac{\omega_B}{\omega_A} = \frac{1}{\cos\alpha} \tag{4.24}$$

Case 4: Maximum fluctuation of velocity of driven shaft

$$= \frac{(\omega_B)_{\max} - (\omega_B)_{\min}}{(\omega_B)_{mean}}$$

Now $\quad (\omega_B)_{\text{mean}} = \omega_A$

$$\text{Maximum speed fluctuation} = \frac{\frac{\omega_A}{\cos\alpha} - \omega_A\cos\alpha}{\omega_A}$$

$$= \frac{1-\cos^2\alpha}{\cos\alpha}$$

$$= \tan\alpha\ \sin\alpha \tag{4.25}$$

For α to be small, $\tan\alpha \approx \alpha$ and $\sin\alpha \approx \alpha$

Maximum speed fluctuation $\approx \alpha^2$ (4.26)

4.8.2 Angular Acceleration of Driven Shaft

$$\frac{\omega_B}{\omega_A} = \frac{\cos\alpha}{1-\sin^2\alpha\cos^2\theta}$$

$$\frac{d\omega_B}{dt} = \omega_A \cdot \frac{d\theta}{dt} \cdot \cos a \left[\frac{2\sin^2 a \cdot \cos\theta \cdot \sin\theta}{(1-\sin^2 a\cos^2\theta)^2}\right]$$

Angular acceleration of driven shaft,

$$\alpha_B = \frac{\omega_A^2 \cos\alpha \sin^2\alpha \sin 2\theta}{(1-\sin^2\alpha\cos^2\theta)^2} \tag{4.27}$$

For acceleration to be maximum or minimum $\frac{d\alpha_B}{d\theta} = 0$

$$\frac{d\alpha_B}{d\theta} = \omega_A^2 \cos\alpha \sin^2\alpha \left[\frac{2\sin 2\theta(1-\sin^2\alpha\cos^2\theta)\sin^2\alpha\sin 2\theta - (1-\sin^2\alpha\cos^2\theta)^2 .2\cos 2\theta}{(1-\sin^2\alpha\cos^2\theta)^4}\right]$$

$$= \cos^2 2\theta + \left(\frac{2}{\sin^2\alpha} - 1\right)\cos 2\theta - 2$$

$$= 0$$

or $$\cos 2\theta \approx \frac{2\sin^2\alpha}{2 - \sin^2\alpha} \tag{4.28}$$

4.9 DOUBLE HOOKE'S JOINT

The double Hooke's joint, as shown in Fig.4.34, is used to maintain the speed of driven shaft equal to the driving shaft at every instant. To achieve this, the driving and the driven shafts should make equal angles with the intermediate shaft and the forks of the intermediate shaft should lie in the same plane. Let γ be the angle turned by the intermediate shaft while the angle turned by the driving shaft and the driven shaft be θ and ϕ respectively.

Then $\tan\theta = \cos\alpha \tan\gamma$

and $\tan\phi = \cos\alpha \tan\gamma$

Therefore, $\theta = \phi$

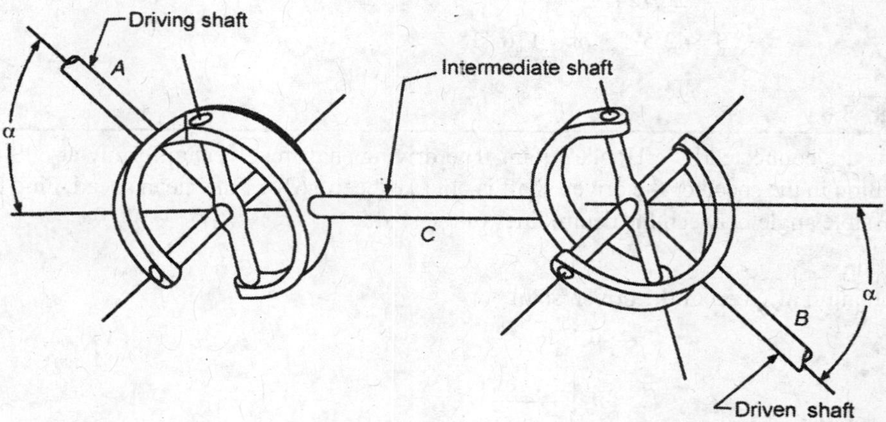

Fig.4.34 Double Hooke's joint

However, if the forks of the intermediate shafts lie in perpendicular planes to each other, then the variation of speed of the driven shaft will be there.

$$\left(\frac{\omega_c}{\omega_A}\right)_{\min} = \cos\alpha$$

$$\left(\frac{\omega_B}{\omega_C}\right)_{\min} = \cos\alpha$$

$$\left(\frac{\omega_B}{\omega_A}\right)_{min} = \cos^2 \alpha \tag{4.29}$$

Similarly,
$$\left(\frac{\omega_B}{\omega_A}\right)_{max} = \frac{1}{\cos^2 \alpha} \tag{4.30}$$

Example 4.5

The angle between the axes of two shafts connected by Hooke's joint is 25°. Determine the angle turned through by the driving shaft when the velocity ratio is unity.

Solution

For velocity ratio to be unity,

$$\cos\theta = \pm\left[\frac{1}{1+\cos\alpha}\right]^{0.5}$$

$$= \pm\left[\frac{1}{1+\cos 25^\circ}\right]^{0.5}$$

$$= \pm 0.72427$$

$$\theta = 43.59^\circ \quad \text{or} \quad 136.41^\circ$$

Example 4.6

Two shafts are connected by a Hooke's joint. The driving shaft revolves uniformly at 600 rpm. The total variation in the speed of the driven shaft is not to exceed ±6% of the mean speed. Find the greatest permissible angle between the centre lines of the shafts.

Solution

Total fluctuation in speed of the driven shaft

$$= 0.12\, \omega_m$$

$$= \omega_m \left[\frac{1-\cos^2\alpha}{\cos\alpha}\right]$$

or $\cos^2 \alpha + 0.12 \cos \alpha - 1 = 0$

$$\cos a = \left[\frac{-0.12 \pm \left\{(0.12)^2 + 4\right\}^{0.5}}{2}\right]$$

$$= 0.9418$$

$$a = 19.64^\circ$$

Example 4.7

Two shafts are connected by the Hooke's joint. The driving shaft rotates at a uniform speed of 1000 rpm. The angle between the shafts is 20°. Calculate the maximum and minimum speeds of the driven shaft, when the acceleration of the driven shaft is maximum?

■ **Solution**

$$\text{Maximum speed, } N_{max} = \frac{N}{\cos\alpha} = \frac{1000}{\cos 20°} = 1064.2 \text{ rpm}$$

$$\text{Minimum speed, } N_{min} = N\cos\alpha = 1000\cos 20° = 939.7 \text{ rpm}$$

For acceleration to be maximum,

$$\cos 2\theta \approx \frac{2\sin^2\alpha}{2-\sin^2\alpha} \approx \frac{2\sin^2 20°}{2-\sin^2 20°} = \frac{2\times 0.11698}{2-2\times 0.11698} = 0.13248$$

$$\theta = 41.19°,\ 138.61°,\ 221.19°, \text{ and } 318.81°$$

Example 4.8

The driving shaft of a Hooke's joint runs at a speed of 400 rpm. The angle between the shafts is 25°. The driven shaft with attached masses has a mass of 50 kg at a radius of gyration of 200 mm. If a steady torque of 900 N.m resists rotation of the driven shaft, find the torque required at the driving shaft, when angle turned through by the driving shaft is 45°.

■ **Solution**

Given: $N_A = 400$ rpm, $\alpha = 25°$, $m = 50$ kg, $K = 0.2$ m, $T = 900$ N m, $\theta = 45°$

Moment of inertia of the driven shaft, $I = mK^2$

$$= 50 \times (0.2)^2 = 2 \text{ kg m}^2$$

Angular acceleration of the driven shaft

$$a_B = -\omega_A^2\left[\frac{\cos a\,\sin 2\theta \sin^2 a}{(1-\cos^2\theta\sin^2 a)^2}\right]$$

$$= -\left(\frac{2\pi\times 400}{60}\right)^2\left[\frac{\cos 25°\sin 90°\sin^2 25°}{(1-\cos^2 45°\sin^2 25°)^2}\right]$$

$$= -1754.6\left[\frac{0.16187}{0.910697}\right]$$

$$= -311.868 \text{ rad/s}^2$$

Torque required to accelerate the driven shaft $= I\alpha_B$

$$= -2 \times 311.868$$
$$= -623.73 \text{ N.m}$$

Total torque required on the driven shaft, $T_B = 900 - 623.73$
$$= 276.26 \text{ N.m}$$

Torque required on the driving shaft, $T_A = \dfrac{T_B \omega_B}{\omega_A}$

$$= T_B \left[\frac{\cos\alpha}{1 - \cos^2\theta \sin^2\alpha} \right]$$

$$= 276.26 \left[\frac{\cos 25°}{1 - \cos^2 45° \sin^2 25°} \right]$$

$$= 274.93 \text{ N.m}$$

Example 4.9

In a double Hooke's coupling, the driving and the driven shafts are parallel and the angle between each and the intermediate shaft is 30°. Find the maximum and minimum velocities of the driven shaft if the axis of the driving pin carried by the intermediate shaft has inadvertently been placed 90° in advance of the correct position. The driving shaft rotates uniformly at 350 rpm.

■ Solution

$$(N_B)_{\max} = \frac{N_A}{\cos^2\alpha}$$

$$= \frac{350}{\cos^2 30°}$$

$$= 466.67 \text{ rpm}$$

$$(N_B)_{\min} = N_A \cos^2\alpha$$

$$= 350 \cos^2 305°$$

$$= 262.5 \text{ rpm}$$

Example 4.10

A torque of 100 N·m is applied to the link OC of a Grasshopper mechanism shown in Fig.4.35. Link OC makes an angle of 20° with the horizontal. Find the magnitude of the vertical force exerted at Q to overcome this torque. $OC = 80$ mm, $PC = 120$ mm, and $PQ = 300$ mm. Also calculate the force required if the link makes an angle of 0°.

■ Solution

$$QC = QP - CP = 300 - 120 = 180 \text{ mm}$$

$$\frac{OC}{PC} = \frac{PC}{QC}$$

$$\therefore \quad \frac{80}{120} = \frac{120}{180} = \frac{2}{3}$$

Let $\quad F_q$ = vertical force exerted at Q

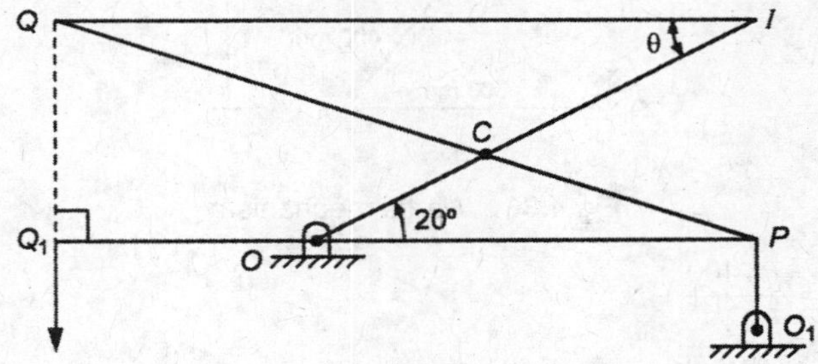

Fig.4.35 Grasshopper mechanism

Then $\quad F_q \times v_q = T_c \times v_c$

$$F_q = \frac{T_c \times v_c}{v_q \times OC}$$

$$\frac{v_c}{v_q} = \frac{IC}{IQ} = \frac{OC}{OP} = \frac{80}{OP}$$

$$F_q = \frac{T_c \times 80}{80 \times OP} = \frac{T_c}{OP}$$

$$OP = OC \cos 20° = 80 \cos 20° = 75.17 \text{ mm}$$

$$F_q = \frac{100}{75.17 \times 10^{-3}} = 1330N$$

When $\theta = 0°$, then

$$OP = OC = 80 \text{ mm}$$

$$F_q = \frac{100}{80 \times 10^{-3}} = 1250\text{N}$$

Example 4.11

The distance between the fixed centres of a Watt's straight line mechanism shown in Fig.4.36 is 320 mm. The lengths of links are: $OA = 300$ mm, $AB = 400$ mm, and $BO_1 = 250$ mm. Locate the position of a point P on AB which will trace approximately straight line.

■ Solution

For the Watt's mechanism, we have

$$\frac{AP}{BP} = \frac{O_1B}{OA} = \frac{250}{300} = \frac{5}{6}$$

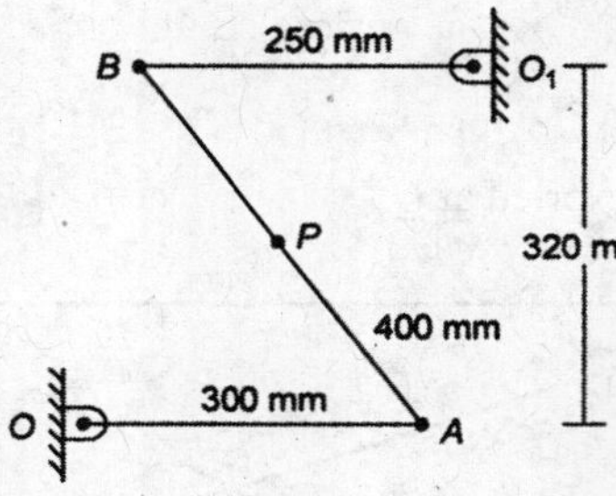

Fig.4.36 Watt's mechanism

$$AP = \frac{5 \times 400}{11}$$
$$= 181.8 \text{ mm}$$
$$BP = 213.2 \text{ mm}$$

Example 4.12

In a Robert mechanism shown in Fig.4.37(a), $AB = BC = CD = \frac{AD}{2}$. Locate the point P on the central vertical arm that approximately describes a straight line.

Solution

Draw the mechanism with $BC = 30$ mm (assumed) in its displaced position $AB'C'D$, as shown in Fig.4.37(b). Produce AB' and DC' to meet at I the instantaneous centre. Draw a vertical line from I to intersect the perpendicular on $B'C'$ at E' at P'. Measure $E'P'$. Then

$$E'P' = 3.6 \text{ cm} = 1.2\ BC$$

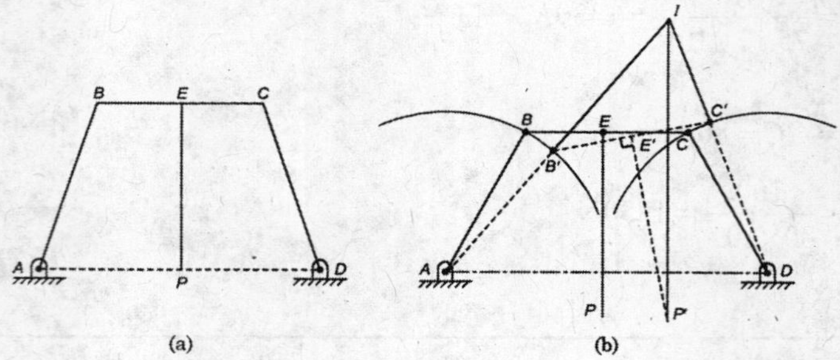

Fig.4.37 Robert mechanism

Example 4.13

In the Watt's mechanism shown in Fig.4.38(a), plot the path of point P and mark and measure the straight line segment of the path of P.

Solution

1. Draw the Watt's mechanism $OABO_1$.
2. Draw straight lines at 10° interval at points O and O_1 on both sides of lines OA and O_1B, as shown in Fig.4.38(b).
3. Cut off AB on arc with centre O_1 from the points on arc with centre O.

4. Join the mid-points P of lines AB to get the desired straight line.
5. It is observed that beyond 20°, the line becomes curved.

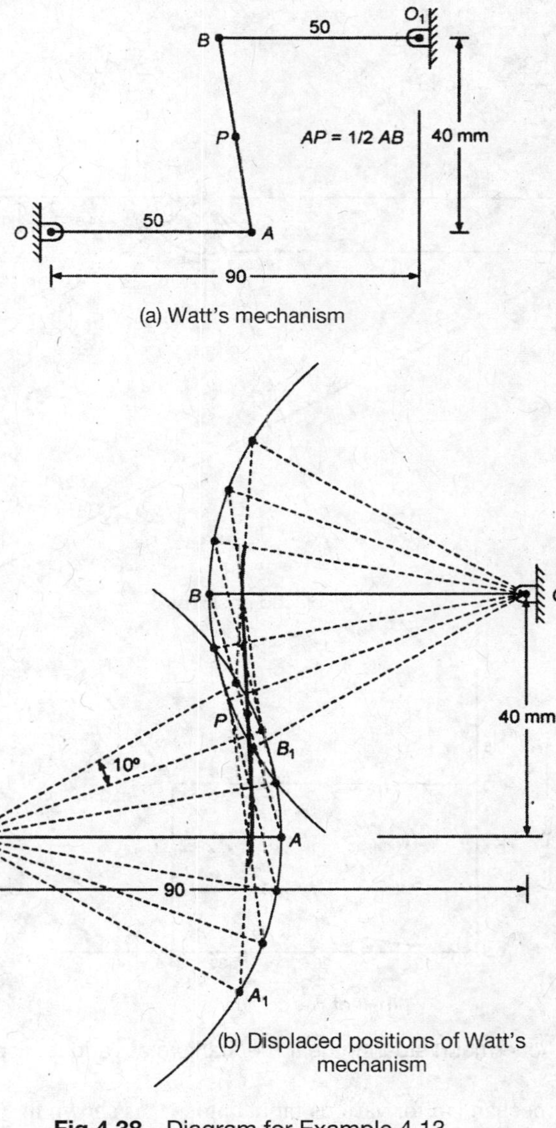

Fig.4.38 Diagram for Example 4.13

Example 4.14

A car using Ackermann steering gear has a wheel base of 2.8 m and a track of 1.5 m. The track rod is 1.2 m and each track arm is 150 mm long. The distance between the pivots of front stub axles is 1.3 m. If the car is turning to the right find the radius of curvature of the path followed by the inner front wheel for correct steering.

■ Solution

Given: $l = 2.8$ m, $a = 1.5$ m, $b = 1.3$ m, $KL = 1.2$ m, $AK = BL = 0.15$ m

$$AC = BD = 0.5\,(a - b) = 0.5\,(150 - 130) = 10 \text{ cm}$$

For correct steering, we have

$$\cot\phi - \cot\theta = \frac{b}{l}$$

$$= \frac{1.3}{2.8} = 0.4643$$

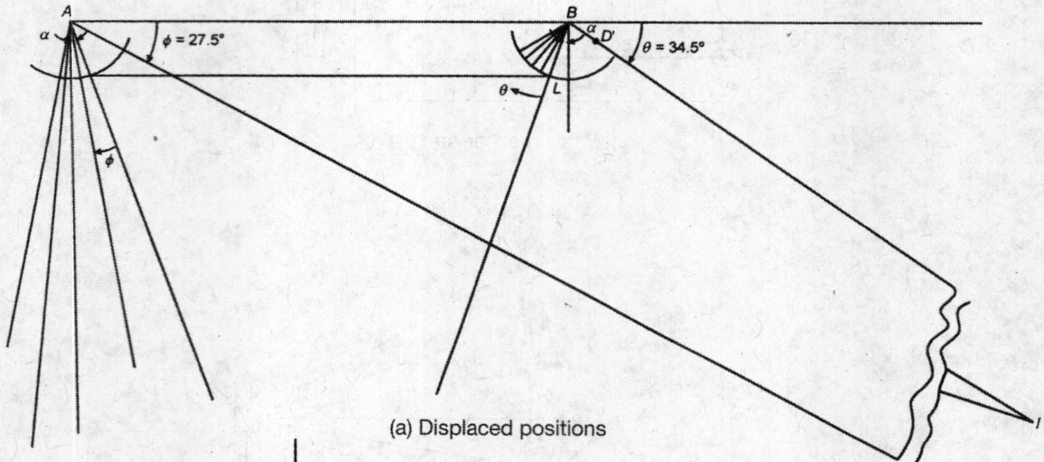

(a) Displaced positions

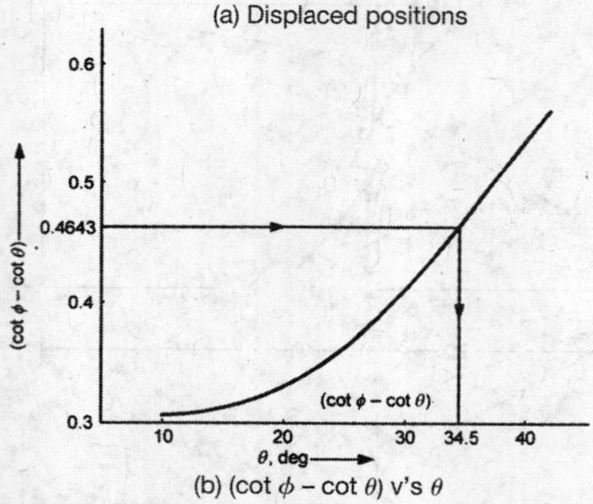

(b) (cot ϕ – cot θ) v's θ

Fig.4.39 Ackermann steering gear mechanism at various displaced positions

Draw the steering mechanism for various input angles θ as shown in Fig.4.39(a), and measure the output angles ϕ. Complete the following table:

θ, deg	10	20	30	40
cot θ	5.6728	2.74747	1.73205	1.19175
ϕ, deg	9.5	18.0	25.0	30.0
cot ϕ	5.97576	3.07768	2.14450	1.73205
cot ϕ – cot θ	0.304	0.330	–0.412	0.540

$$\sin\alpha = \frac{AB-KL}{2BL} = \frac{1.3-1.2}{2\times0.15} = 0.3333$$

$$\alpha = 19.47°$$

Plot ($\cot\phi - \cot\theta$) v's θ as shown in Fig.4.39(b). Determine value of θ corresponding to $\cot\phi - \cot\theta = 0.4643$. We find that $\theta_C = 34.5°$ and $\phi = 27.5°$. Locate the centre of curvature by drawing these angles. We find that,

$$BI = 548 \text{ cm}, \quad BD' = 10 \text{ cm}$$

Radius of curvature of path, $ID' = 538$ cm.

Example 4.15

A universal joint is used to connect two shafts which are inclined at 20° and the speed of the driving shaft is 1000 rpm. Find the extreme angular velocities of the driven shaft and its maximum acceleration.

■ **Solution**

Angular speed of driving shaft, $\omega_1 = \dfrac{2\pi\times1000}{60} = 104.72$ rad/s

Maximum speed of driven shaft, $(\omega_2)_{max} = \dfrac{\omega_1}{\cos\alpha} = \dfrac{104.72}{\cos 20°} = 111.44$ rad/s

Minimum speed of driven shaft, $(\omega_2)_{min} = \omega_1 \cos\alpha = 104.72 \cos 20° = 98.40$ rad/s

For acceleration of driven shaft to be maximum,

$$\cos 2\theta = \frac{2\sin^2\alpha}{2-\sin^2\alpha} = \frac{2\sin^2 20°}{2-\sin^2 20°} = 0.12424$$

$$2\theta = 82.86° \text{ or } 277.14°$$

$$\therefore \quad \theta = 41.43° \text{ or } 138.57°$$

Maximum angular acceleration of driven shaft,

$$(\alpha_2)_{max} = \frac{\omega_1^2 \cos\alpha \sin^2\alpha \sin^2\theta}{(1-\cos^2\theta\sin^2\alpha)^2}$$

$$= \frac{(104.72)^2 \times \cos 20° \times \sin^2 20° \times \sin 82.86°}{(1-\cos^2 41.43° \sin^2 20°)^2} = \frac{1196.1}{0.87281} = 1370.4 \text{ rad/s}^2$$

Example 4.16

The distance between the pivots of the front stub axles of a car is 1.35 m. The length of track rod is 1.25 m. The wheel track is 1.5 m and the wheel base is 2.8 m. What should be the length of the track arm if the gear is to be given a correct steering, when rounding a corner of radius 5 m.

Solution

$$b = 1.35 \text{ m}, \ a = 1.5 \text{ m}, \ \ell = 2.8 \text{ m}, \ KL = 1.25 \text{ m}$$

Draw the steering mechanism as shown in Fig.4.24.

$$a = CD = CA + AB + BD = 2CA + AB$$

$$CA = BD = 0.5\,(CD - AB) = 0.5\,(a - b) = 0.5\,(1.5 - 1.35) = 0.075 \text{ m}$$

Now $ID = 5$ m

$$IB = ID' - D/B' = 5 + 0.075 = 5.075 \text{ m}$$

$$\sin\theta = IE/IB = \ell/IB = 2.8/5.075 = 0.55172$$

$$\theta = 33.485°$$

For correct steering,

$$\cot\phi - \cot\theta = b/\ell$$

$$\cot\phi - \cot 33.485° = 1.35/2.8$$

$$\cot\phi = 0.48214 + 1.51169 = 1.99383$$

$$\phi = 63.364°$$

$$\tan a = b/(2\ell) = 1.35/5.6 = 0.24107$$

$$a = 13.554°$$

$$\sin a = (b - KL)/(2 \times AK)$$

$$AK = (1.35 - 1.25)/(2 \times \sin 13.554°)$$

$$= 0.213 \text{ m}$$

Example 4.17

A Hooke's coupling is used to connect two shafts whose axes are inclined at 30°. The driving shaft rotates uniformly at 600 rpm. What are the extreme angular velocities of the driven shaft? Find the maximum value of retardation or acceleration and state the angle where both will occur.

Solution

$$\alpha = 30°, \ N_A = 600 \text{ rpm}$$

$$\omega_B/\omega_A = \cos\alpha\,/\,(1 - \sin^2\alpha\,\cos^2\theta)$$

$$(\omega_B/\omega_A)_{\max} = 1/\cos\alpha \text{ at } \theta = 0° \text{ and } 180°$$

$$\omega_A = 2\pi \times 600/60 = 62.83 \text{ rad/s}$$

$$(\omega_B)_{\max} = \omega_A/\cos\alpha = 62.83/\cos 30° = 72.552 \text{ rad/s}$$

$$(\omega_B/\omega_A)_{\min} = \cos\alpha \text{ at } \theta = 90° \text{ and } 270°$$

$$(\omega_B)_{\min} = \omega_A \cos\alpha = 62.83 \times \cos 30° = 54.41 \text{ rad/s}$$

For acceleration to be maximum,

$$\cos 2\theta = 2\sin^2 a\,/\,(2 - \sin^2 a)$$

$$= 2\sin^2 30°\,/\,(2 - \sin^2 30°)$$

$$= 0.28571$$

$$\theta = 36.7°$$

$$a_B = \omega_A^2\,[\,(\cos a\,\sin^2 a\,\sin 2\theta)\,/\,(1 - \sin^2 a\,\cos^2\theta)^2]$$

$$= (20\pi)^2\,[(\cos 30°\,\sin^2 30°\,\sin 73.4°)\,/\,(1 - \sin^2 30°\,\cos^2 36.7°)^2]$$

$$= 1007 \text{ rad/s}^2$$

Example 4.18

Two shafts are to be joined by a Hooke's coupling. The driving shaft rotates at a uniform speed of 600 rpm and the speed of the driven shaft must lie between 500 and 550 rpm. Determine the maximum permissible angle between the shafts.

Solution

$$N_1 = 600 \text{ rpm},\ N_2 = 500 \text{ to } 550 \text{ rpm},\ \alpha = ?$$

$$\text{Maximum variation of speed} = \left[(N_2)_{max} - (N_2)_{min}\right]/N_1$$

$$= (550 - 500)/600 = 1/12$$

$$= (1 - \cos^2\alpha)/\cos\alpha$$

$$12\cos^2\alpha + \cos\alpha - 12 = 0$$

$$\cos\alpha = 0.9592$$

$$\alpha = 16.42°$$

Example 4.19

A Hooke's coupling connects a shaft running at a uniform speed of 900 rpm to a second shaft. The angle between their axes being 20°. Find the velocity and acceleration of the driven shaft at the instant when the fork of the driving shaft has turned through an angle of 15° from the plane containing the shaft axes.

Solution

$$N_1 = 900 \text{ rpm},\ a = 20°,\ \theta = 15°$$

$$N_2/N_1 = \cos a/(1 - \sin^2 a\ \cos^2\theta)$$

$$= \cos 20°/(1 - \sin^2 20° \cos^2 15°)$$

$$= 1.0548$$

$$N_2 = 949.3 \text{rpm}$$

$$\omega_1 = 2\pi \times 900/60 = 94.248 \text{ rad/s}$$

$$a_2 = \omega^2_1\ [(\cos a\ \sin^2 a\ \sin 2\theta)/(1 - \sin^2 a\ \cos^2\theta)^2]$$

$$= (94.248)^2\ [(\cos 20°\ \sin^2 20°\ \sin 30°)/(1 - \sin^2 20°\ \cos^2 15°)^2]$$

$$= 586.6 \text{ rad/s}^2$$

Example 4.20

The angle between the axes of two horizontal shafts to be connected by Hooke's joint is 150°. The speed of the driving shaft is 150 rpm. The driven shaft carries a flywheel of mass 15 kg and having a radius of gyration of 100 mm. If the forked end of the driving shaft rotates 30° from the horizontal plane, find the torque required to drive the shaft to overcome the inertia of the flywheel.

Solution

$$a = 150°,\ \theta = 30°,\ N_1 = 150 \text{ rpm}$$

$$\omega_1 = 2\pi \times 150/60 = 15.7 \text{ rad/s}$$

$$a_2 = \omega^2_1\ [(\cos a\ \sin^2 a\ \sin 2\theta)/(1 - \sin^2 a\ \cos^2\theta)^2]$$

$$= (15.7)^2\ [(\cos 150°\ \sin^2 300°\ \sin 60°)/(1 - \sin^2 150°\ \cos^2 30°)^2]$$

$$= -70.076 \text{ rad/s}^2$$

Moment of inertia of flywheel on driven shaft,

$$I_2 = MK^2 = 15\times(100\times10^{-3})^2 = 0.15 \text{ kg}\cdot\text{m}^2$$
$$T_2 = I_2\alpha_2 = 0.15\times70.076 = 10.51 \text{ N}\cdot\text{m}$$

Example 4.21

The driving shaft of a double Hooke's joint rotates at 500 rpm. The angle of the driving and driven shafts with the intermediate shaft is 25°. Determine the maximum and minimum velocities of the driven shaft.

Solution

$$N_A = 500\text{rpm},\ \alpha = 25°$$
$$(N_B)_{\max} = N_A/\cos^2\alpha = 500/\cos^2 25° = 603.72 \text{ rpm}$$
$$(N_B)_{\min} = N_A\cos^2\alpha = 500\times\cos^2 25° = 410.7 \text{ rpm}$$

Example 4.22

The driving shaft of a Hooke's joint runs at a uniform speed of 240 rpm and the angle between the shafts is 25°. The driven shaft with attached masses has a mass of 50 kg at a radius of gyration of 150 mm.

(a) If a steady torque of 200 N·m resists rotation of the driven shaft, find the torque required at the driving shaft when angle turned through by the driving shaft is 45°.
(b) At what angle between the shafts will the total fluctuation of speed of the driven shaft be limited to 20 rpm ?

Solution

Given: $N_1 = 240$ rpm, $\alpha = 25°$, $M = 50$ kg, $K = 150$ mm, $T_1 = 200$ N m, $\theta = 45°$, $N_2 - N_1 = 20$ rpm

$\omega_1 = 2\pi N_1/60 = 2\pi\times 240/60 = 25.14$ rad/s

(a) $I_2 = MK^2 = 50\times(0.15)^2 = 1.125 \text{ kg}\cdot\text{m}^2$

Angular acceleration of driven shaft,

$$\alpha_2 = \frac{-\omega_1^2\cos\alpha\sin^2\alpha\sin 2\theta}{(1-\sin^2\alpha\cos^2\theta)^2} = \frac{-(25.14)^2\cos 25°\times\sin^2 25°\times\sin 90°}{(1-\sin^2 25°\cos^2 45°)^2}$$
$$= \frac{-102.3}{0.8294} = -123.35 \text{ rad/s}^2$$
$$T_2 = I_2\alpha_2 = 1.125\times(-123.35) = -138.76 \text{ N}\cdot\text{m}$$

Total torque required on driven shaft, $T = T_1 + T_2 = 200 - 138.76 = 61.24$ N·m

Now $$T_1\,\omega_1 = T_2\,\omega_2$$

$$T_1 = \frac{\omega_2}{\omega_1} \times T_2 = \left(\frac{\cos\alpha}{1-\sin^2\alpha\cos^2\theta}\right)T_2 = \frac{61.24\cos 25^\circ}{1-\sin^2 25^\circ \cos^2 45^\circ} = 60.94\,\text{N.m}$$

(b) Total fluctuation of speed = $N_1 \tan\alpha \sin\alpha$

$$20 = \frac{240(1-\cos^2\alpha)}{\cos\alpha}$$

$$12\,(1-\cos^2\alpha) = \cos\alpha$$

$$1-\cos^2\alpha = 0.0833\,\cos\alpha$$

$$\cos^2\alpha + 0.0833\,\cos\alpha - 1 = 0$$

$$\cos\alpha = \frac{-0.0833 \pm \sqrt{(0.0833)^2 + 4}}{2} = 0.9592$$

$$\alpha = 16.42^\circ$$

Example 4.23

A double universal joint is used to connect two shafts in the same plane. The intermediate shaft is inclined at an angle of 20° to the driving shaft as well as the driven shaft. Find the maximum and minimum speed of the intermediate shaft and the driven shaft if the driving shaft has a constant speed of 600 rpm.

Solution

Given: $\alpha = 20^\circ,\ N_A = 600$ rpm

Let N_A, N_B and N_C be the speed of driving, intermediate and driven shaft respectively.

$$(N_B)_{max} = \frac{N_A}{\cos\alpha} = \frac{600}{\cos 20^\circ} = 638.5 \text{ rpms}$$

$$(N_B)_{min} = N_A \cos\alpha = 600\cos 20^\circ = 563.8 \text{ rpm}$$

$$(N_C)_{max} = \frac{(N_B)_{max}}{\cos\alpha} = \frac{638.5}{\cos 20^\circ} = 679.5 \text{ rpm}$$

$$(N_C)_{min} = (N_B)_{min}\cos\alpha = 563.8\cos 20^\circ = 529.8 \text{ rpm}$$

Summary for Quick Revision

1. Pantograph is a mechanism to produce the path traced out by a point on enlarged or reduced scale
2. Mechanisms for accurate straight line motion are : Peaucellier, Hart, and Scott-Russel.
3. Mechanisms for approximate straight line motion are: Grasshopper. Watt. Tchebicheff and Robert.
4. Mechanisms for intermittent motion are: Geneva wheel and Ratchet.

5 Parallel linkages are: Parallel rules, universal drafting machine, and lazy tongs.

6 Engine pressure indicators are: Simplex, Crosby, Richards, Thomson, and Dobbie–McInnes.

7 Steering gear mechanism is used for changing the direction of motion of an automobile.

8 Automobile steering gear mechanisms are: Davis, and Ackermann.

9 Davis steering gear lies at the front of front axle.

10 To avoid skidding, the two front wheels must turn about the same instantaneous centre.

11 The fundamental equation for correct steering is:

$$\cot \phi - \cot \theta = b/l$$

where θ, ϕ = angles turned through by the stub – axles of inner and outer front wheels respectively

b = distance between the pivots of front axles.
ℓ = wheel base

12 For the Davis steering gear, $\tan \alpha = b/(2l)$

13 Davis steering gear has all sliding pairs.

14 Ackermann steering gear lies at the back of the front axle.

15 Ackermann steering gear has all turning pairs.

16 For Ackermann steering gear,
$\tan \alpha = (\sin \theta - \sin \phi)/(2 - \cos \phi - \cos \theta)$
Generally, $b/l = 0.4$ to $0.5 \approx 0.455$.

17 The instantaneous centre for Davis steering gear lies on the axis of rear wheels.

18 The instantaneous centre for Ackermann steering gear lies approximately at $0.3\,\ell$ above the rear wheels axis.

19 Hook's coupling is a device to connect two shafts whose axes are neither coaxial nor parallel but intersect at a point. This is used to transmit power from the engine to the rear axle of an automobile and similar other applications.
$\tan \theta = \cos \alpha \tan \phi$
Where θ, ϕ = angles turned through by driving and driven shafts respectively
α = angle of inclination of driven shaft with driving shaft.

20 Ratio of angular speeds of driven and driving shafts is,
$\omega_B/\omega_A = \cos \alpha/(1 - \sin^2 \alpha \cos^2 \theta)$

21 For $\omega_B/\omega_A = 1$, $\cos \theta = \pm \sqrt{1/(1 + \cos a)}$

22 $(\omega_B/\omega_A)_{max} = 1/\cos \alpha$

23 $(\omega_B/\omega_A)_{min} = \cos \alpha$

24 Maximum variation of velocity of driven shaft = $[(\omega_B)_{max} - (\omega_B)_{min}]/(\omega_B)_{mean}$
$= \tan \alpha \sin \alpha$
$\approx \alpha^2$

25 Angular acceleration of driven shaft,
$\alpha_B = -\omega_A^2 \cos \alpha \sin^2 \alpha \sin 2\theta/(1 - \sin^2 \alpha \cos^2 \theta)^2$

26 For acceleration of driven shaft to be maximum or minimum,
$\cos 2\theta \approx 2 \sin^2 \alpha/(2 - \sin^2 \alpha)$

27 Double Hooke's joint is used to maintain the speed of driven shaft equal to the driving shaft at every instant. To achieve this, the driving and the driven shafts should make equal angles with the intermediate shaft and the forks of the intermediate shaft should lie in the same plane.

28 For a double Hooke's joint,

$\tan\theta = \cos\alpha \tan\gamma$

$\tan\phi = \cos\alpha \tan\gamma$

$\therefore \theta = \phi$

where γ = angle turned by the intermediate shaft

θ, ϕ = angles turned through by the driving and driven shafts respectively.

29 For a double Hooke's joint, the speed ratios are:

$(\omega_B/\omega_A)_{min} = \cos^2\alpha$

$(\omega_B/\omega_A)_{max} = 1/\cos^2\alpha$

Multiple Choice Questions

1 The number of links in a pantograph mechanism is equal to
(a) 2 (b) 3 (c) 4 (d) 5

2 Automobile steering gear is an example of
(a) higher pair (b) sliding pair (c) turning pair (d) lower pair.

3 In automobiles, the power is transmitted from gear box to differential through
(a) bevel gears (b) knuckle joint (c) Hooke's joint (d) Cotter joint.

4 Scott-Russel mechanism for generating straight line has
(a) four lower kinematic turning pairs
(b) two lower kinematic turning pairs and one lower kinematic sliding pairs
(c) one lower kinematic turning pair and two lower kinematic sliding pairs
(d) two lower kinematic turning pairs and two lower kinematic sliding pairs.

5 Watt mechanism is capable of generating
(a) approximate straight line (b) exact straight line
(c) approximate circular path (d) exact circular path.

6 Which of the following mechanisms generetes approximate straight line?
(a) Hart mechanism (b) Watt mechanism
(c) Peaucellier mechanism (d) Scott-Russel mechanism

7 Which of the following mechanisms generates accurate straight line?
(a) Scott-Russel mechanism (b) Grasshopper mechanism
(c) Watt mechamism (d) Tehebicheff mechanism

8 Geneva wheel is used to generete
(a) circular motion (b) intermittent motion
(c) continuous motion (d) parcbolic motion

9 Lazy tongs generete
(a) straight line motion (b) circular motion
(c) simple harmonic motion (d) uniformly acceleretedmotion.

10 Davis steering gear has
(a) only turning pairs (b) only sliding pairs
(c) both sliding and turning pairs (d) rolling pairs

11 Ackermann steering gear has
(a) only sliding pairs (b) only turning pairs
(c) both sliding and turning pairs (d) spherical pairs

Answers

1. (c) 2. (d) 3. (c) 4. (b) 5. (a) 6.(b) 7. (a) 8. (b) 9. (a) 10. (b) 11. (c)

Review Questions

1 What is a pantograph? What are its uses?
2 Name the exact straight line mechanisms.
3 List the approximate straight line mechanisms.
4 Define a steering gear.
5 What is the function of a steering gear?
6 Name the steering gears used in automobiles.
7 Compare Davis and Ackermann steering gears.
8 Write the fundamental equation for correct steering.
9 What is a Hooke's joint? Where it is used?
10 Write the expression for ratio of angular velocities of shafts for a Hooke's joint.
11 Draw the polar velocity diagram of a Hooke's joint and mark its salient features.
12 Write the expressions for maximum and minimum speeds of driven shaft in a Hooke's joint.
13 What is a double Hooke's joint? What is its use?
14 Write the condition for the speeds of two shafts to be same in a Hooke's joint.

Exercises

4.1 Two shafts connected by a Hooke's joint have their axes inclined at 20°. The driving shaft rotates at 1440 rpm and the driven shaft carries a flywheel of mass 20 kg. The radius of gyration is 10 cm. Find the maximum torque in the driven shaft.

[Ans. 568.36 N · m]

4.2 The axes of two shafts connected by a Hooke's joint are inclined at 20°. At what positions of the driving shaft, the velocities of two shafts are equal? State whether the accelerations are positive or negative at these positions.

If the driving shaft rotates at 1800 rpm, determine acceleration of the driven shaft at an the of the above positions.

[Ans. 44.1°, 135.9°, 224.1°, 315.9°; +ve at 135.9°, 315.9 and –ve at 44.1°, 224.1°; –4421.23 rad/s^2 at 44.1°]

4.3 The driving shaft of a Hooke's joint is rotating at a uniform speed of 600 rpm. The speed of the driven shaft must be 575 and 625 rpm. Determine the maximum permissible angle between the shafts.

[Ans. 16.54°]

4.4 A shaft running at 1200 rpm is connected to a second shaft by a Hooke's joint. The angle between the axes of the shafts is 15°. Determine the velocity and acceleration of the driven shaft when it has turned through an angle of 10° from the plane containing the shaft axes.

[Ans. 1239.6 rpm, –400 rad/s²]

4.5 The intermediate shaft of a double Hooke's joint is inclined at 15° to each. The input and output forks on the intermediate shaft have been assembled by mistake at 90° to one another. Determine the maximum and minimum speeds of output shaft if the speed of input shaft is 600 rpm. Also calculate the coefficient of fluctuation in speed.

[Ans. 643 rpm, 559.8 rpm; 13.87%]

4.6 Two shafts are connected by a Hooke's joint. The driving shaft rotates uniformly at 600 rpm. If the total permissible variation in speed of the driven shaft is not to exceed ±5% of the mean speed, find the greatest permissible angle between the centre lines of the shafts.

[Ans. 17.97°]

4.7 The moment of inertia of the driven shaft in a Hooke's joint is 35 kg. m². The driven shaft is inclined at 30° to the axis of the driving shaft. The driving shaft rotates at 2700 rpm with a steady torque of 275 N.m. Determine the maximum fluctuation of output torque.

[Ans. 96.91% of input torque]

4.8 In a Davis steering gear, the distance between the pivots of the front axle is 1.2 m and the wheel base is 2.8 m. Find the inclination of each arm to the longitudinal axis of the car, when it is moving along a straight path.

[Ans. 12.1°]

4.9 The distance between the pivots of the front stub axles of a car is 1.35 m. The length of track rod is 1.25 m. The wheel track is 1.5 m and the wheel base is 2.8 m. What should be the length of the track arm if the gear is to be given a correct steering, when rounding a corner of radius 5 m.

[Ans. 0.213 m]

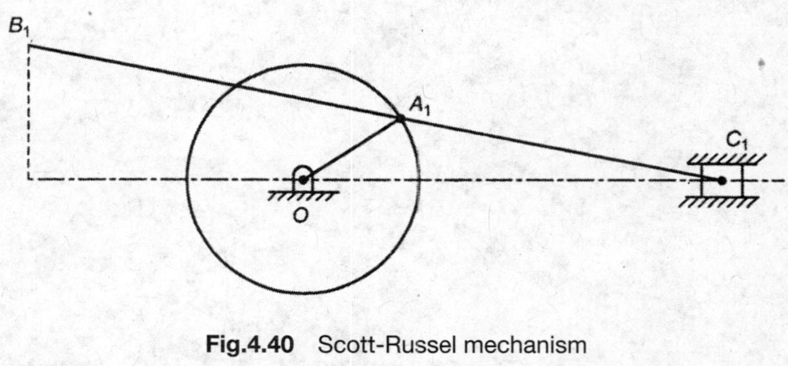

Fig.4.40 Scott-Russel mechanism

4.10 In a Scott-Russel mechanism shown in Fig.4.40, $OA1$ = 30 cm, $A1C1$ = 40 cm. Find the length of the extended link A_1B_1 so that point B_1 is generating an approximate straight line motion.

[Ans. 53.3 cm]

5

FRICTION

Chapter Outline

5.1 INTRODUCTION

When a body moves or tends to move on another body, the property of the two bodies by virtue of which a force is developed between the two bodies, which opposes the motion, is called *friction*, and the opposing force is called the *force of friction*. The force of friction acts opposite to the impending motion.

Friction: A Blessing or a Curse

Friction is a blessing when it is necessary to increase the force of friction between two contacting bodies when power is being transmitted through them, as in the case of a belt and pulley and friction clutches. In the case of brakes also, it is necessity to have friction between contacting surfaces to reduce the speed of the moving member or stop it altogether.

In the case of lathe slides, journal bearings, etc., it is necessary to reduce the force of friction in order to decrease the power lost due to friction. In such situations, friction is a curse. To decrease the friction, proper lubrication has to be provided.

5.2 TYPES OF FRICTION

Friction is of the following types:

Static friction: It is the friction experienced by two bodies in contact, while at rest.

Dynamic (or kinetic) friction: It is the friction experienced by two bodies in contact, while in motion. It is less than the static friction.

Sliding friction: It is due to sliding of two bodies on each other.

Rolling friction: It is due to rolling of two bodies on each other.

Pivot friction: It is the friction experienced by two bodies due to the rotation of one body around the other, as in the case of a foot step bearing.

Dry (or solid) friction: It is due to two dry and unlubricated surfaces in contact.

Boundary (or skin or greasy) friction: It is the friction experienced by two bodies separated by a very thin layer of a lubricant.

Fluid (or film) friction: It is the friction experienced by two bodies in contact when separated by a thick film of a lubricant.

5.3 LAWS OF FRICTION

a. Static friction
   1. The force of friction always acts in a direction opposite to the impending motion.
   2. The limiting force of friction is directly proportional to the normal reaction.
   3. The force of friction is independent of the area of contact between the two surfaces.
   4. The force of friction depends upon the roughness of the surfaces of two materials.

b. Kinetic friction
   1. The force of friction always acts in a direction opposite to that in which the body is moving.
   2. The magnitude of kinetic friction bears a constant ratio to the normal reaction between the two surfaces.
   3. The force of friction is independent of the relative velocity between the two surfaces in contact, but it decreases slightly with increase in velocity.
   4. The force of friction increases with reversal of motion.
   5. The coefficient of friction changes slightly due to temperature changes.

c. Fluid friction
   1. The force of friction is almost independent of the load.
   2. The force of friction reduces with the increase of temperature of the lubricant.
   3. The force of friction depends upon the type and viscosity of the lubricant.
   4. The force of friction is independent of the nature of surfaces.
   5. The frictional force increases with the increase in the relative velocity of the frictional surfaces.

5.4 DEFINITIONS

a. Coefficient of friction

Consider a block of weight W resting on a plane rough surface, as shown in Fig.5.1. The normal reaction because of the weight of the block is R. Let a force P be applied to the block towards the right.

The frictional force F will set up between the block and the rough surface. The direction of F will be towards the left. The *coefficient of friction* (μ) is the ratio of the force of friction (F) to the normal reaction (R). Thus

$$\mu = \frac{F}{R} \tag{5.1}$$

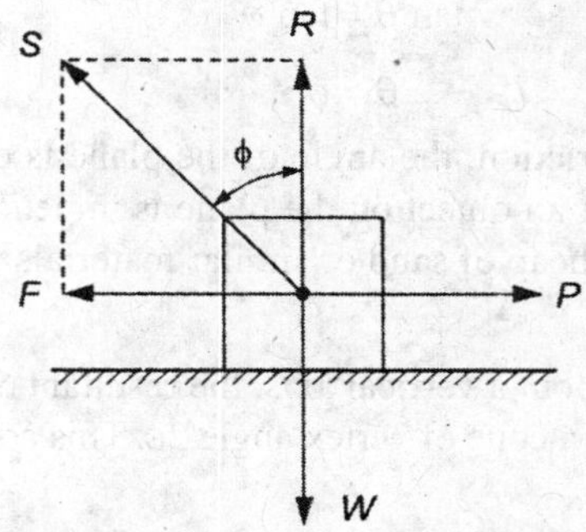

Fig.5.1 Block resting on rough horizontal plane

b. Angle of friction

Let S be the resultant of F and R making an angle of ϕ with R that is called the angle of friction, such that

$$\tan\phi = \frac{F}{R} \tag{5.2}$$

From Eqs. (5.1) and (5.2), we find that

$$\mu = \tan\phi \tag{5.3}$$

Therefore, in the case of limiting friction, the coefficient of friction is equal to the tangent of the angle of friction.

Now consider the block resting on rough inclined plane, as shown in Fig.5.2. By resolving the forces along and perpendicular to the plane, we have

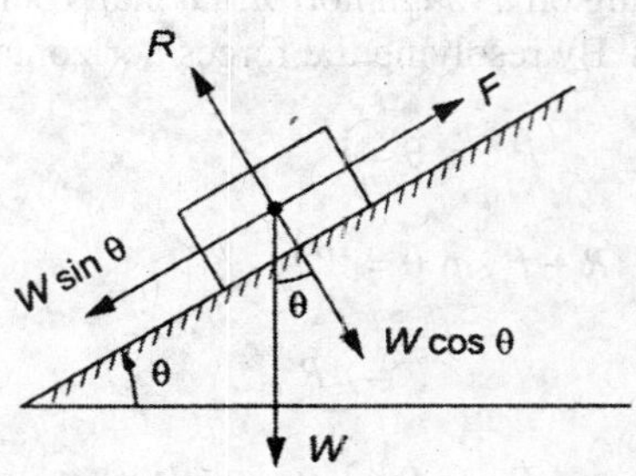

Fig.5.2 Block resting on rough inclined plane

$$F = W \sin \theta$$
$$R = W \cos \theta$$

or
$$\frac{F}{R} = \tan\theta$$

Comparing with Eq. (5.2), we find that

$$\tan \theta = \tan \phi$$

or
$$\theta = \phi \tag{5.4}$$

Thus, in the case of limiting friction, the angle of the plane is equal to the angle of friction. The angle of the plane when motion of an object on the plane is impending is called the *angle of repose*. This is the maximum angle that a heap of sand or similar materials will make with the horizontal.

c. Cone of friction

If the force P is made to revolve about a vertical axis, the resultant S will also revolve about a vertical axis. As S revolves, it will generate a cone of vertex angle 2ϕ. This cone is called the *cone of friction*, as shown in Fig.5.3.

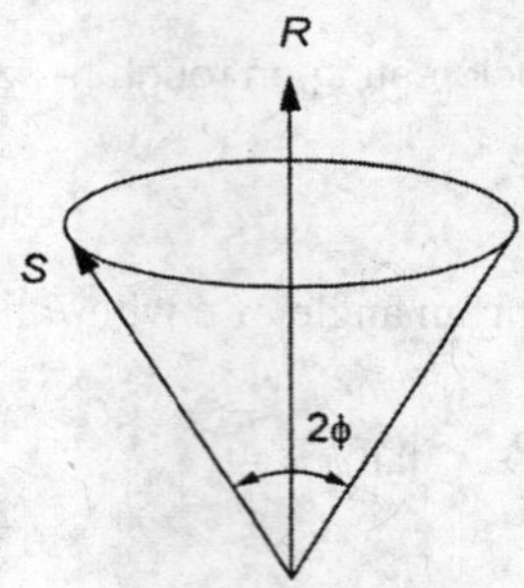

Fig.5.3 Cone of friction

5.5 FORCE ANALYSIS OF A SLIDING BODY

A rough body resting on another rough body may be made to slide by a force of the pull or push type. Now we study the force required to slide the body under various situations.

a. Body resting on a horizontal plane

Consider a body of weight W resting on a rough horizontal plane being pulled by a force P inclined at an angle θ, as shown in Fig.5.4(a). By resolving the forces horizontally and vertically, we have

$$P \cos \theta = F$$

$$R + P \sin \theta = W$$

Now,
$$F = \mu R$$

Therefore,
$$P \cos \theta = \mu R$$

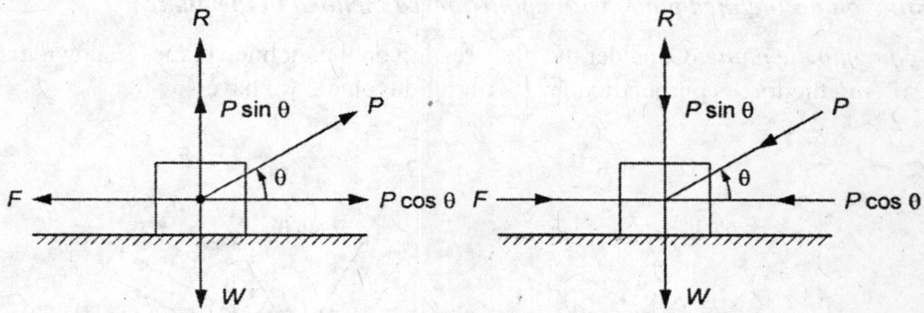

(a) *Inclined pull* (b) *Inclined push*

Fig.5.4 Force analysis of a sliding body on horizontal plane

$$\frac{P\cos\theta}{\mu}+P\sin\theta=W$$

$$P\left(\frac{\cos\theta}{\mu}+\sin\theta\right)=W$$

$$P\,(\cos\theta\cos\phi+\sin\phi\sin\theta)=W\sin\phi$$

or $$P\cos(\theta-\phi)=W\sin\phi$$

or $$P=\frac{W\sin\phi}{\cos(\theta-\phi)} \qquad (5.5)$$

For P to be minimum, $\cos(\theta-\phi)$ should be maximum, that is,

$$\cos(\theta-\phi)=1$$

or $$\theta=\phi$$

and $$P_{min}=W\sin\theta \qquad (5.6)$$

If the force P is of the push type, as shown in Fig.5.4(b), then

$$P\cos\theta=F=\mu R$$

$$P\sin\theta+W=R$$

$$=P\frac{\cos\theta}{\mu}$$

$$P\,(\cos\theta\cos\phi-\sin\theta\sin\phi)=W\sin\phi$$

$$P\cos(\theta+\phi)=W\sin\phi$$

or $$P=\frac{W\sin\phi}{\cos(\theta+\phi)} \qquad (5.7)$$

b. Body resting on an inclined plane with applied force inclined to the plane

1. *Body going up the plane:* Consider the body resting on the inclined plane as shown in Fig.5.5(a). By resolving the forces perpendicular and along the plane, we have

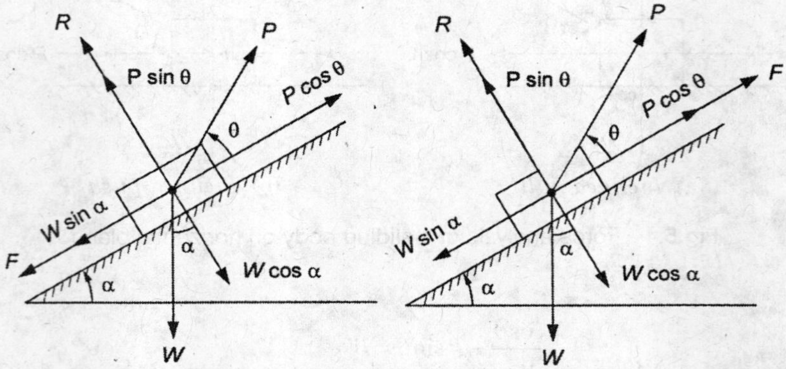

(a) *Body going up the plane* (b) *Body going down the plane*

Fig.5.5 Inclined pull acting on body resting on inclined plane

$$P \cos \theta = W \sin \alpha + \mu R$$

$$W \cos \alpha = P \sin \theta + R$$

or
$$P \cos \theta = W \sin \alpha + \tan \phi \,(W \cos \alpha - P \sin \theta)$$

or
$$P (\cos \theta + \sin \theta \tan \phi) = W (\sin \alpha + \cos \alpha \tan \phi)$$

or
$$P (\cos \theta \cos \phi + \sin \theta \sin \phi) = W (\sin \alpha \cos \phi + \cos \alpha \sin \phi)$$

$$P \cos (\theta - \phi) = W \sin (\alpha + \phi)$$

$$P = \frac{W \sin(\alpha + \phi)}{\cos(\theta - \phi)} \tag{5.8}$$

For P to be minimum, $\cos (\theta - \phi) = 1$, or $\theta = \phi$. In that case,

$$P_{\min} = W \sin (\alpha + \theta) \tag{5.9}$$

2. *Body going down the plane*: When the body is going down the plane, as shown in Fig.5.5(b), then, we get

$$P = \frac{W \sin(a + \phi)}{\cos(\theta + \phi)} \tag{5.10}$$

For P to be minimum, $\cos (\theta + \phi) = 1$, or $\phi = -\theta$. In that case

$$P_{\min} = W \sin (\alpha - \theta). \tag{5.11}$$

c. Body resting on an inclined plane with horizontal applied force

1. *Body going up the plane:* Consider the body as shown in Fig.5.6(a). By resolving the forces perpendicular and along the plane, we have

$$P \cos \alpha = W \sin \alpha + \mu R$$

$$P \sin \alpha + W \cos \alpha = R$$

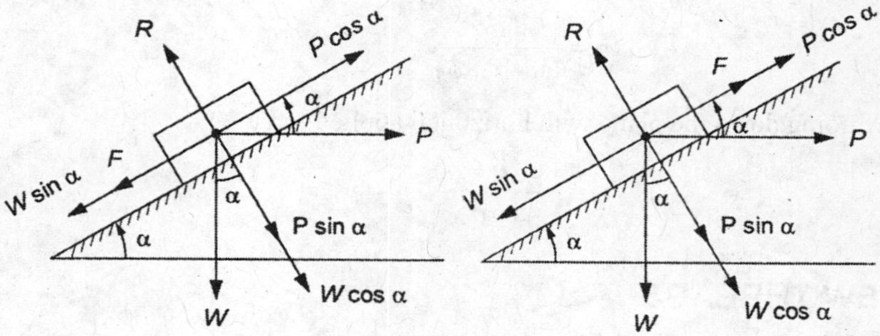

(a) *Body going up the plane* (b) *Body going down the plane*

Fig.5.6 Horizontal pull acting on body resting on inclined plane

or
$$P \cos \alpha = W \sin \alpha + \tan \phi\, (P \sin \theta + W \cos \alpha)$$

$$P (\cos \alpha \cos \phi - \sin \alpha \sin \phi) = W (\sin \alpha \cos \phi + \cos \alpha \sin \phi)$$

$$P \cos (\alpha + \phi) = W \sin (\alpha + \phi)$$

$$P = W \tan (\alpha + \phi) \tag{5.12}$$

2. *Body going down the plane:* For the body shown in Fig.5.6(b), we get

$$P = W \tan (\phi - \alpha) \tag{5.13}$$

d. *Efficiency of the inclined plane*

The efficiency of the inclined plane is defined as the ratio of the effort required P_o without friction and P with friction.

$$\eta = \frac{P_o}{P}$$

1. For body going up the plane with inclined applied force,

$$\eta = \frac{\dfrac{\sin \alpha}{\cos \theta}}{\dfrac{\sin(\alpha + \phi)}{\cos(\theta - \phi)}}$$

$$= \frac{1 + \mu \tan \theta}{1 + \mu \cot \alpha} \tag{5.14}$$

2. For body going down the plane with inclined applied force

$$\eta = \left(-\frac{\sin \alpha}{\cos \theta}\right) \times \left[\frac{\cos(\theta + \phi)}{\sin(\phi - \alpha)}\right]$$

$$= \frac{1-\mu\tan\theta}{1-\mu\cot\alpha} \tag{5.15}$$

3. For body going up the plane with horizontal applied force

$$\eta = \frac{\tan\alpha}{\tan(\alpha+\phi)} \tag{5.16}$$

4. For body going down the plane with horizontal applied force

$$\eta = \frac{\tan(\alpha-\phi)}{\tan\alpha} \tag{5.17}$$

5.6 SCREW THREADS

Screws are used for fastening, load lifting, and power transmission purposes. The screws may have single start or multi-start threads. Lead is the product of pitch and number of starts. A screw thread is obtained when the hypotenuse of a right-angled triangle is wrapped round the circumference of a cylinder. Figure 5.7(a) shows the development of a helix of diameter d and lead L or pitch p.

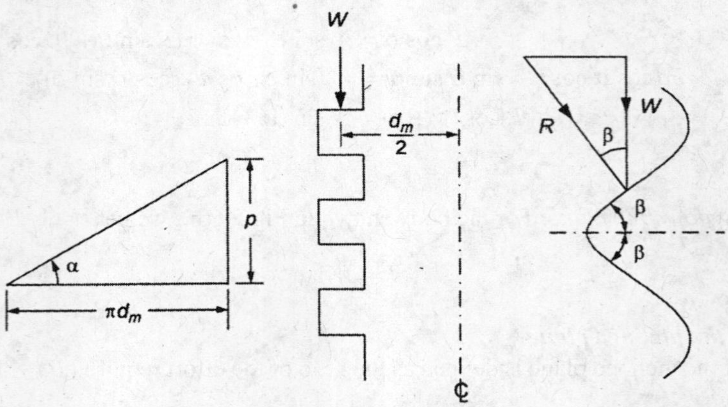

(a) *Helix angle* (b) *Square threads* (c) *V-threads*

Fig.5.7 Square and *V*-threads

Let p = pitch of the threads
L = lead of the threads
d_m = mean pitch circle diameter of the screw
α = helix angle, that is, angle of inclination of the threads with a line perpendicular to the axis of the screw.

Then $$\tan\alpha = \frac{p}{\pi d_m} \quad \text{or} \quad \frac{L}{\pi d_m} \tag{5.18}$$

(a) Square threads

In the case of square threads as shown in Fig.5.7(b), the faces of the threads are normal to the axis of the spindle. The force P acting horizontally required at the screw threads to slide the load W up the inclined plane is given by Eq. (5.12), that is

$$P = W\tan(\alpha+\phi)$$

(b) V-threads

For a screw having V-threads, as shown in Fig.5.7(c), let 2β be the angle of the threads. Then normal reaction on the threads is given by:

$$R = \frac{W}{\cos\beta}$$

Frictional force,

$$F = \mu R = \left(\frac{\mu}{\cos\beta}\right)\cdot W$$
$$= \mu_e W$$

Where $\mu_e = \mu/\cos\beta$ is called the virtual or equivalent coefficient of friction.

Therefore, for a screw having *V*-threads, the virtual coefficient of friction should be used to calculate the torque required to lift the load.

5.7 SCREW JACK

A screw jack is a device for lifting of loads. The principle of working of a screw jack is similar to that of an inclined plane. Fig.5.8 shows the common form of the screw jack.

Let P = effort applied at the circumference of the screw to lift the load
W = load to be lifted
μ = coefficient of friction between the screw and the nut
$= \tan\phi$

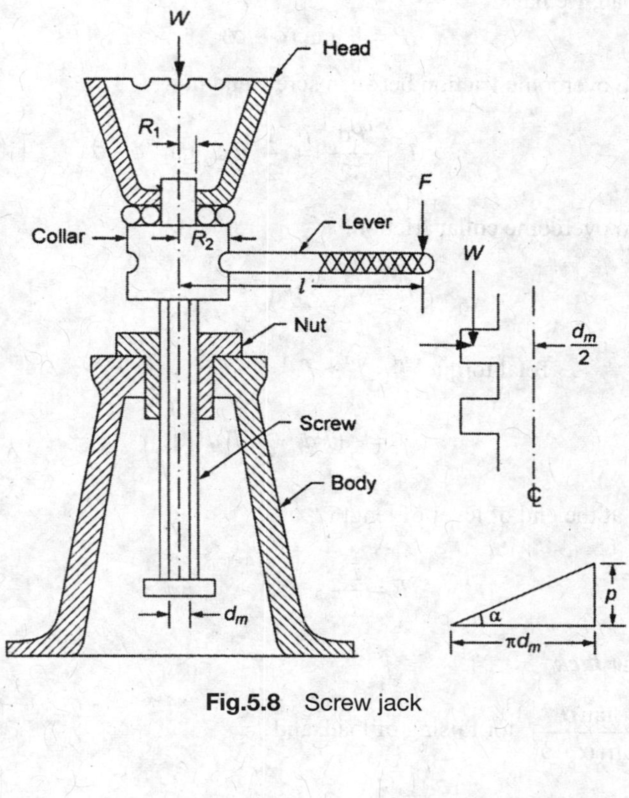

Fig.5.8 Screw jack

1. ***Raising of load***

As derived in Section 5.5, for raising the load, we have

$$P = W\tan(\alpha + \phi)$$

Torque required to overcome friction between the screw and nut is

$$T_1 = \frac{P \times d_m}{2} = \frac{Wd_m \tan(\alpha + \phi)}{2}$$

Torque required to overcome friction at the collar,

$$T_2 = \frac{\mu_c W d_c}{2}$$

where μ_c = collar coefficient of friction

d_c = mean diameter of the collar = $R_1 + R_2$

Total torque, $$T = T_1 + T_2 = \frac{1}{2}(Pd_m + \mu_c W d_c) \tag{5.19}$$

Let F = effort applied at the end of a lever of length l

Then $T = Fl$

2. ***Lowering of load***

For lowering of load, we have

$$P = W\tan(\phi - \alpha)$$

Torque required to overcome friction between screw and nut,

$$T_1 = \frac{Pd_m}{2} = \frac{1}{2}\,Wd_m \tan(\phi - \alpha)$$

Torque required to overcome collar friction,

$$T_2 = \frac{1}{2}\,\mu_c W d_c$$

Total torque, $T = T_1 + T_2$

$$= \frac{1}{2}(Pd_m + \mu_c W d_c)$$

Effort F required at the end of lever of length l,

$$F = \frac{T}{l}$$

3. ***Efficiency of Screw Jack***

Efficiency, $\eta = \dfrac{\tan\alpha}{\tan(\alpha + \phi)}$, for raising of load, and (5.20a)

$$= \frac{\tan(\phi - \alpha)}{\tan \alpha}, \text{ for lowering of load} \tag{5.20b}$$

For efficiency to be maximum, $\frac{d\eta}{d\alpha} = 0$

Now
$$\eta = \frac{\tan\alpha(1 - \tan\alpha \times \tan\phi)}{\tan\alpha + \tan\phi}$$

$$= \left(\frac{\sin\alpha}{\cos\alpha}\right)\frac{\left[1 - \left(\frac{\sin\alpha}{\cos\alpha}\right)\left(\frac{\sin\phi}{\cos\phi}\right)\right]}{\left(\frac{\sin\alpha}{\cos\alpha}\right) + \left(\frac{\sin\phi}{\cos\phi}\right)}$$

$$= \frac{\sin\alpha[\cos\alpha\cos\phi - \sin\alpha\sin\phi]}{\cos\alpha[\sin\alpha\cos\phi + \sin\phi\cos\alpha]}$$

$$= \frac{\sin\alpha\cos(\alpha + \phi)}{\cos\alpha\sin(\alpha + \phi)} = \frac{\sin(2\alpha + \phi) - \sin\phi}{\sin(2\alpha + \phi) + \sin\phi}$$

For efficiency to be maximum, $\sin(2\alpha + \phi) = 0$

or
$$2\alpha + \phi = 90°$$

or
$$\alpha = 45° - \frac{\phi}{2} \tag{5.21}$$

$$\eta_{max} = \frac{\sin 90° - \sin\phi}{\sin 90° + \sin\phi} = \frac{1 - \sin\phi}{1 + \sin\phi} \tag{5.22}$$

4. *Self-locking and over-hauling screw*

The effort required for lowering of load is,

$$P = W\tan(\phi - \alpha)$$

Torque,
$$T = \frac{Pd_m}{2} = W\left(\frac{d_m}{2}\right)\tan(\phi - \alpha)$$

Torque will remain positive if $\phi > \alpha$. Such a screw is called *self-locking screw*. If $\phi < \alpha$, then the torque will become negative. In other words, the screw will lower of its own. Such a screw is called *over-hauling screw*.

For $\alpha = \phi$, the efficiency of the screw for raising of load becomes,

$$\eta = \frac{\sin\phi\,\cos 2\phi}{\cos\phi\,\sin 2\phi} = \frac{\tan\phi}{\tan 2\phi}$$

For a self-locking screw,

$$\eta \le \frac{\tan\phi}{\tan 2\phi} \le \frac{\tan\phi(1 - \tan^2\phi)}{2\tan\phi} \le \left[\frac{1}{2} - \frac{\tan^2\phi}{2}\right]$$

$$\eta \le 50\% \tag{5.23}$$

5.8 FRICTION IN BEARINGS

When a shaft is rotating in a bearing, a pivot or a collar is provided on the shaft to take up the axial thrust. Common examples are steam turbines, hydraulic turbines, propeller shaft of a ship, etc.

5.8.1 Flat Pivot Bearing

When the axial force is taken by the end of a shaft that is inserted in a recess to take up the thrust, it is called a pivot bearing or foot step bearing.

A flat pivot bearing like the foot step bearing is shown in Fig.5.9(a).

Let W = load on the bearing

r = radius of bearing surface

p = intensity of pressure between rubbing surfaces

(a) Uniform pressure

When the pressure is uniformly distributed over the bearing surface area, then

$$p = \frac{W}{\pi r^2} \tag{5.24}$$

Consider a ring of radius x and thickness dx of the bearing area, as shown in the figure.

Area of bearing surface, $\delta A = 2\pi x \cdot \mathrm{d}x$

Load transmitted to the ring, $\delta W = p \cdot \delta A$

Frictional resistance to sliding at radius r,

$$F_r = \mu \cdot \delta W = 2\pi\mu \cdot px \, \mathrm{d}x$$

Frictional torque on the ring, $\mathrm{d}T_f = F_r x = 2\pi\mu \cdot px^2 \mathrm{d}x$

Total frictional torque, $T_f = \int_0^r 2\pi\mu \, px^2 \mathrm{d}x = 2\pi\mu p \int_0^r x^2 \mathrm{d}x = 2\pi\mu p \left| \frac{x^3}{3} \right|_0^r$

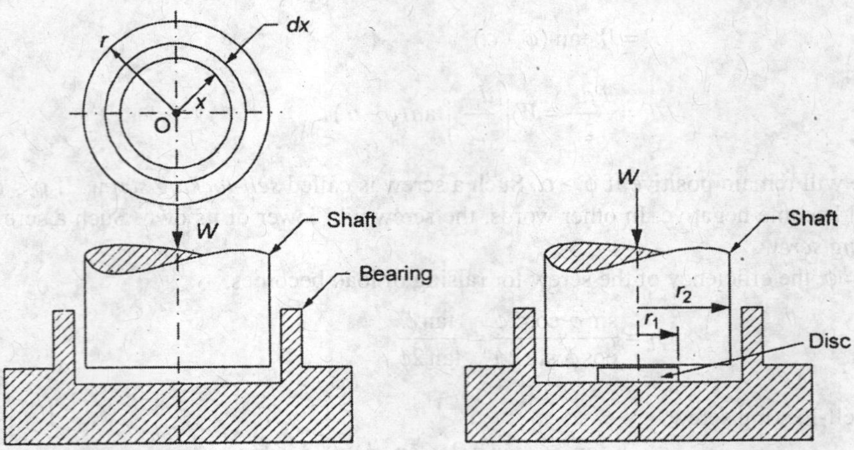

(a) Flat pivot

(b) Flat pivot with a disc

Fig.5.9 Flat Pivot bearing

$$= \frac{2\pi\mu\, pr^3}{3}$$

$$= \frac{2\mu Wr}{3} = \mu W r_m \quad (5.25)$$

where $$r_m = \frac{2}{3} r \quad (5.26)$$

Power lost in friction, $P = T_f \cdot \omega$ (5.27)

Where $\omega = \frac{2\pi N}{60}$ rad/s and N is the speed of the shaft in rpm.

If the shaft of radius r_2 is resting on a disc of radius r_1 as shown in Fig.5.9(b), then

$$p = \frac{W}{\pi(r_2^2 - r_1^2)} \quad (5.28)$$

$$T_f = 2\pi\mu p \int_{r_1}^{r_2} x^2 \mathrm{d}x = 2\pi\mu p \left| \frac{x^3}{3} \right|_{r_1}^{r_2}$$

$$= 2\pi\mu p \frac{(r_2^3 - r_1^3)}{3} = 2\mu W \left(\frac{r_2^3 - r_1^3}{r_2^2 - r_1^2} \right) = \mu W r_m \quad (5.29)$$

where $$r_m = \frac{2}{3} \left(\frac{r_2^3 - r_1^3}{r_2^2 - r_1^2} \right) = \text{mean radius} \quad (5.30)$$

(b) Uniform wear

The rate of wear depends upon the intensity of pressure and the rubbing velocity. The rubbing velocity increases with the increase in radius. Therefore, the wear rate is proportional to the pressure p and radius x. For uniform wear,

$$px = C$$

or $$p = \frac{C}{x}$$

where C is a constant.

Load transmitted to the ring, $\delta W = p \cdot 2\pi x \, \mathrm{d}x$

$$= 2\pi C \cdot \mathrm{d}x$$

Total load transmitted to the bearing, $W = 2\pi C \int_0^r \mathrm{d}x = 2\pi C \, |x|_0^r$

$$= 2\pi C r$$

or $$C = \frac{W}{2\pi r}$$

Frictional torque acting on the ring,

$$\mathrm{d}T_f = 2\pi\mu p x^2 \mathrm{d}x = 2\pi\mu C x \, \mathrm{d}x$$

Total frictional torque on the bearing,

$$T_f = 2\pi\mu C \int_0^r x\mathrm{d}x = 2\pi\mu C \left| \frac{x^2}{2} \right|_0^r = \pi\mu C r^2 = \frac{\pi W r}{2} \quad (5.31)$$

If the shaft of radius r_2 is resting on a disc of inner radius r_1, then

$$T_f = 2\pi\,\mu C \int_{r_1}^{r_2} x\,\mathrm{d}x = 2\pi\,\mu C \left|\frac{x^2}{2}\right|_{r_1}^{r_2} = \pi\,\mu C\,(r_2^2 - r_1^2) = \frac{\mu W (r_2 + r_1)}{2}$$

$$T_f = W r_m \tag{5.32}$$

Where $r_m = \frac{1}{2}(r_1 + r_2)$ is the mean radius. (5.33)

5.8.2 Conical Pivot Bearing

Consider a conical pivot bearing as shown in Fig.5.10.

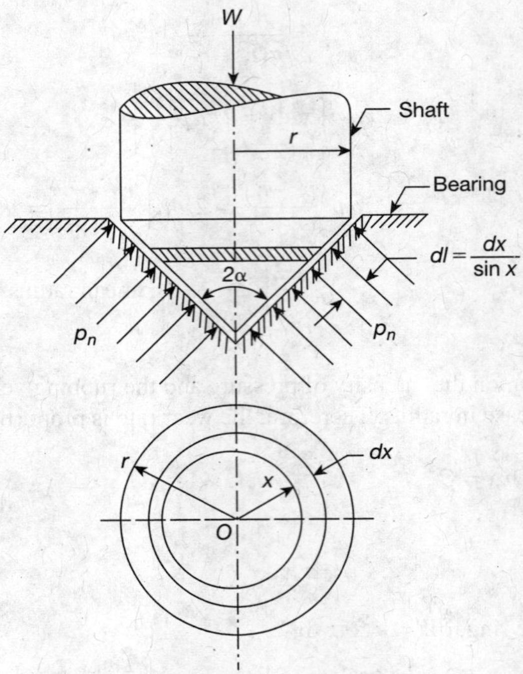

Fig.5.10 Conical pivot bearing

Let p_n = intensity of normal pressure on the cone
α = semicone angle
μ = coefficient of friction between the shaft and the bearing
r = radius of the shaft

Consider a small ring of radius x and thickness dx. Let dl be the length of the ring along the cone, so that

$$\mathrm{d}l = \mathrm{d}x \cdot \operatorname{cosec}\alpha$$

Area of the ring, $\mathrm{d}A = 2\pi x \cdot \mathrm{d}l = 2\pi x \cdot \mathrm{d}x\,\operatorname{cosec}\alpha$

(a) Uniform pressure

Normal load acting on the ring, $\delta W_n = 2\pi x \times p_n \times dx \text{ cosec } \alpha$

Vertical load acting on the ring, $\delta W = \delta W_n \sin \alpha$

Total vertical load transmitted to the bearing,

$$W = \int_0^r \delta W = 2\pi\, p_n \int_0^r x\, dx = \pi\, p_n\, |x^2|_0^r$$

$$W = \pi r^2 p_n$$

or
$$p_n = \frac{W}{\pi r^2}$$

Frictional force acting on the ring tangentially at radius x, $F_f = \mu \times \delta W_n$

Frictional torque acting on the ring,

$$dT_f = F_f, x = 2\pi\mu \cdot p_n \cdot \text{cosec } \alpha \cdot x^2 \cdot dx$$

Total frictional torque,

$$T_f = 2\pi\mu \cdot p_n \cdot \text{cosec}\,\alpha \int_0^r x^2 dx = \frac{2}{3}\pi\mu\, p_n\, \text{cosec}\,\alpha\, |x^3|_0^r$$

$$= \left(\frac{2}{3}\right) \cdot \pi r^3 \cdot \mu\, p_n \cdot \text{cosec}\,\alpha$$

$$T_f = \left(\frac{2}{3}\right) \cdot \mu W r \text{ cosec}\,\alpha \qquad (5.34)$$

(b) Uniform wear

In the case of uniform wear, the intensity of normal pressure varies inversely with the distance. Therefore,

$$p_n \cdot x = C, \text{ where } C \text{ is a constant}$$

Load transmitted to the ring, $\delta W = p_n \cdot 2\pi x \cdot dx = 2\pi C \cdot dx$

Total load transmitted to the bearing, $W = \int_0^r \delta W = 2\pi C \int_0^r dx = 2\pi C \cdot r$

or
$$C = \frac{W}{2\pi r}$$

Frictional torque acting on the ring,

$$dT_f = 2\pi\mu \cdot p_n \cdot \text{cosec } \alpha \cdot x^2 \cdot dx$$
$$= 2\pi\mu \cdot C \cdot \text{cosec } \alpha \cdot x \cdot dx$$

Total frictional torque acting on the bearing,

$$T_f = 2\pi\mu \cdot C \cdot \text{cosec}\,\alpha \int_0^r x \cdot dx$$

$$= \pi\mu \cdot C \cdot r^2 \cdot \text{cosec } \alpha$$

$$T_f = \left(\frac{1}{2}\right) \cdot \mu W r \text{ cosec}\,\alpha \qquad (5.35)$$

5.8.3 Truncated Conical Pivot Bearing

Consider a truncated conical pivot bearing as shown in Fig.5.11. Then

Intensity of uniform pressure is given by: $p_n = \dfrac{W}{\pi(r_2^2 - r_1^2)}$

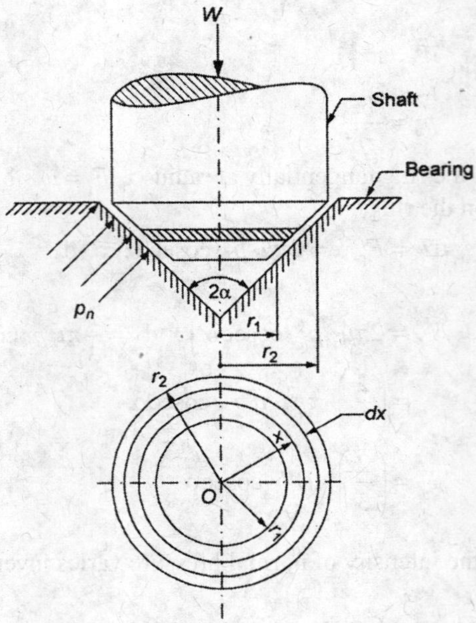

Fig.5.11 Truncated pivot bearing

(a) Uniform pressure

Total frictional torque acting on the bearing,

$$T_f = 2\pi\mu p_n \operatorname{cosec}\alpha \int_{r_1}^{r_2} x^2 dx = 2\pi\mu p_n \operatorname{cosec}\alpha \frac{(r_2^3 - r_1^3)}{3}$$

$$= 2\pi\mu\left[\frac{W}{\pi(r_2^2 - r_1^2)}\right]\frac{\operatorname{cosec}\alpha(r_2^3 - r_1^3)}{3}$$

$$T_f = \left(\frac{2}{3}\right)\cdot \mu W \cdot \operatorname{cosec}\alpha\left(\frac{r_2^3 - r_1^3}{r_2^2 - r_1^2}\right)$$

$$T_f = \mu W r_m \operatorname{cosec}\alpha \tag{5.36}$$

Where $$r_m = \frac{2}{3}\left(\frac{r_2^3 - r_1^3}{r_2^2 - r_1^2}\right) \tag{5.37}$$

(b) Uniform wear

Total frictional torque acting on the bearing,

$$T_f = 2\pi\mu C \operatorname{cosec}\alpha \int_{r_1}^{r_2} x dx = \pi\mu C \operatorname{cosec}\alpha(r_2^2 - r_1^2)$$

$$= \pi\mu\left[\frac{W}{2\pi(r_2 - r_1)}\right]\operatorname{cosec}\alpha(r_2^2 - r_1^2) = \mu W \operatorname{cosec}\alpha\,\frac{(r_2 + r_1)}{2}$$

$$T_f = \mu W r_m \operatorname{cosec}\alpha \tag{5.38}$$

where $$r_m = \frac{1}{2}(r_1 + r_2) \tag{5.39}$$

5.8.4 Flat Collar Bearing

A single collar bearing is shown in Fig.5.12(a) and the multiple-collar bearing in Fig.5.12(b).

Let r_1 and r_2 = inner and outer radii of the bearing, respectively.

Area of the bearing surface, $A = \pi(r_2^2 - r_1^2)$

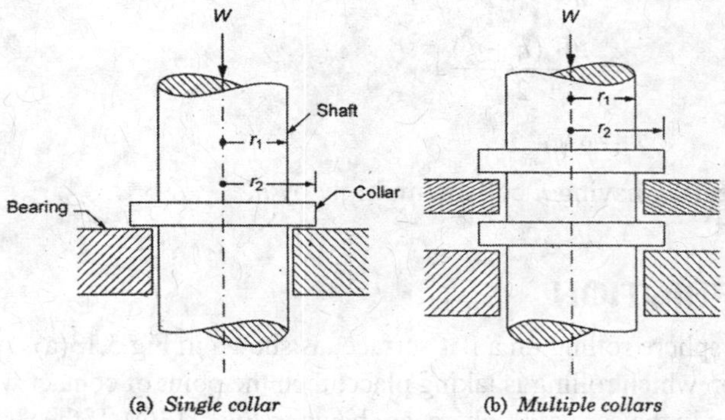

Fig.5.12 Flat collar bearings

(a) Uniform pressure

Intensity of pressure, $p = \dfrac{W}{A} = \dfrac{W}{\pi(r_2^2 - r_1^2)}$

Frictional torque on the ring of radius x and thickness dx,

$$dT_f = 2\,\pi\mu \cdot p \cdot x^2 \cdot dx$$

Total frictional torque, $T_f = 2\pi\mu p \int_{r_1}^{r_2} x^2 \cdot dx$

$$= \left(\frac{2}{3}\right)\cdot \mu p\left(r_2^3 - r_1^3\right)$$

$$= \left(\frac{2}{3}\right)\cdot \mu W\left(\frac{r_2^3 - r_1^3}{r_2^2 - r_1^2}\right) \tag{5.40a}$$

$$= \mu W r_m \tag{5.40b}$$

(b) Uniform wear

For uniform wear, the load transmitted on the ring,

$$\delta W = p_n \cdot 2\,\pi x \cdot dx$$

$$= 2\pi\, C \cdot dx$$

Total load transmitted to the collar,

$$W = 2\pi C\int_{r_1}^{r_2} dx = 2\pi C(r_2 - r_1)$$

or
$$C = \frac{W}{2\pi(r_2 - r_1)}$$

Frictional torque on the ring, $dT_f = \mu \cdot \delta W \cdot r = 2\pi \cdot \mu \cdot C \cdot x \cdot dx$

Total frictional torque on the bearing,

$$T_f = 2\pi\mu \cdot C\int_{r_1}^{r_2} x \cdot dx$$

$$= \pi\mu \cdot C\left(r_2^2 - r_1^2\right)$$

$$= \frac{\mu W(r_1 + r_2)}{2} \tag{5.41a}$$

$$T_f = \mu W r_m \tag{5.41b}$$

For a multi-collared bearing having n collars, multiply by n.

5.9 ROLLING FRICTION

Consider a cylinder or sphere rolling on a flat surface, as shown in Fig.5.13(a). When there is no deformation of the surface on which rolling is taking place, then the point of contact will be a line in the case of a cylinder and a point in the case of a sphere, as shown in Fig.5.13(a). If the surface deforms, then the shape of the surface will be as shown in Fig.5.13(b). Let the distance between the point of contact B and the point A through which the load W passes be b and F be the force required for rolling. Then rolling moment is equal to $F \cdot h$ and the resisting moment is $W \cdot b$. For the equilibrium of forces, we have

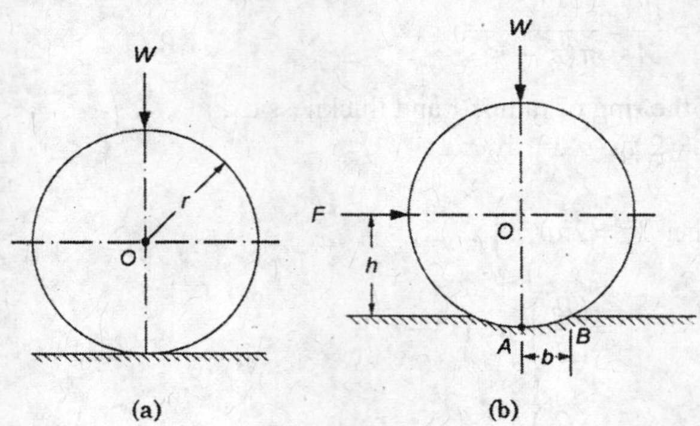

Fig.5.13 Rolling of a cylinder or sphere

$$F \cdot h = W \cdot b$$

b is known as the coefficient of rolling friction, and has linear dimension.

Let F_r = force applied to the body for rolling

F_s = force applied to the body for sliding

Then $F_r = \left(\frac{b}{h}\right) \cdot W$

and $F_s = \mu \cdot W$

The body rolls without sliding, if $F_r < F_s$ or $\mu > b/h$. The body will slide if $F_s < F_r$ or $\mu < b/h$. The body will either roll or slide if $\mu = b/h$.

5.10 ANTI-FRICTION BEARINGS

In the case of anti-friction bearings, the point of contact between the journal and the bearing elements is either a point (as in the case of ball bearings) or a line (as in the case of roller bearings).

The ball bearing consists of a number of hardened balls mounted between two hardened races. The inner race is fitted on the shaft and the outer race is a tight fit into the bearing housing. The balls are kept at a fixed distance from one another in a brass cage. The distortion of balls is very little and the rolling friction is very low. They are mainly used to carry radial loads.

In roller bearings, either right cylindrical or tapered rollers are present. They are used to carry heavy loads, both radial and thrust.

5.11 FRICTION CIRCLE

Consider a journal bearing, as shown in Fig.5.14. When the journal is at stand still, then the point of contact is at A. The load W is balanced by the reaction R. When the journal starts rotating in the clockwise direction, then the point of contact shifts from A to B. The resultant of normal reaction R and force of friction $F = \mu R$ is S, as shown in Fig.5.14(b). For the equilibrium of the journal, we have

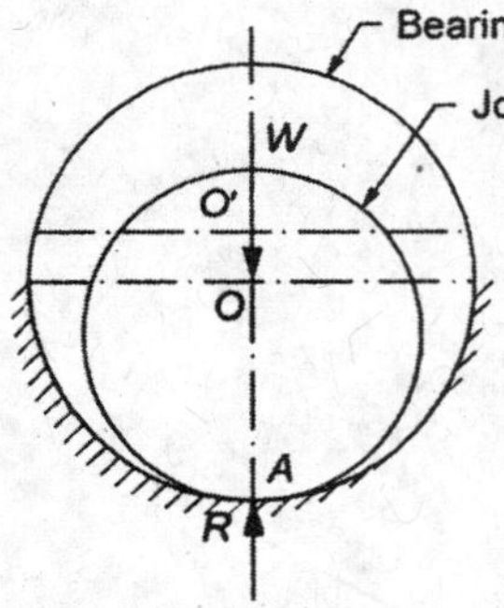

(a) *Journal bearing*

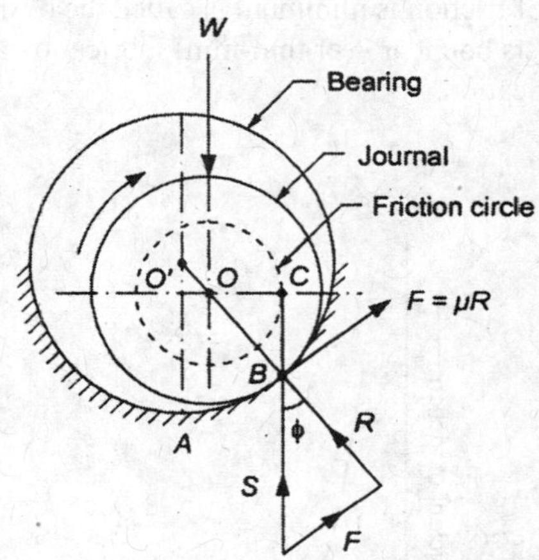

(b) *Forces in journal bearing contact*

Fig.5.14 Friction circle

$$W = S$$

$$\text{Torque, } T = W \cdot OC$$

$$= W \cdot OB \sin \phi$$

$$= W \cdot r \tan \phi \text{ (angle } \phi \text{ being small)}$$

$$= W \cdot r\mu \tag{5.42}$$

A circle drawn with centre O and radius $OC = r \sin \phi = r\mu$ is called the *friction circle*.

5.12 FILM FRICTION

Thin-film lubrication: Thin-film lubrication is defined as a condition of lubrication in which the lubricant film is relatively thin and there is partial metal-to-metal contact. This mode of lubrication is seen in door hinges and machine tool slides. It is also called boundary lubrication. The conditions resulting in thin-film lubrication are excessive load, insufficient surface area or oil supply, low speed, and misalignment.

Thick-film lubrication: When the bearing and the journal are completely separated from each other by the lubricant film, then it is called thick-film lubrication. Since there is no contact between the surfaces, properties of surface, like surface finish, have little or no influence on the performance of the bearing. The resistance to relative motion between the journal and the bearing arises from the viscous resistance of the lubricant. Therefore, the performance of the bearing is affected by the viscosity of the lubricant. Thick-film lubrication could be hydrodynamic or hydrostatic.

In a journal bearing working under thick-film lubrication regime, the frictional resistance depends upon the following parameters:

1. Dynamic viscosity of the lubricant.
2. Speed of the journal.
3. Unit pressure of the lubricant, and
4. Radial clearance ratio.

The coefficient of friction depends upon the factor ($\mu n/p$), where μ = dynamic viscosity of lubricant, n = speed of journal, and p = unit pressure of lubricant. The value of this factor at which the coefficient of friction is minimum is called the bearing modulus, as shown in Fig.5.15. Below this value, there exists boundary (or thin-film) lubrication and above this value there shall be full-film (or thick-film) lubrication.

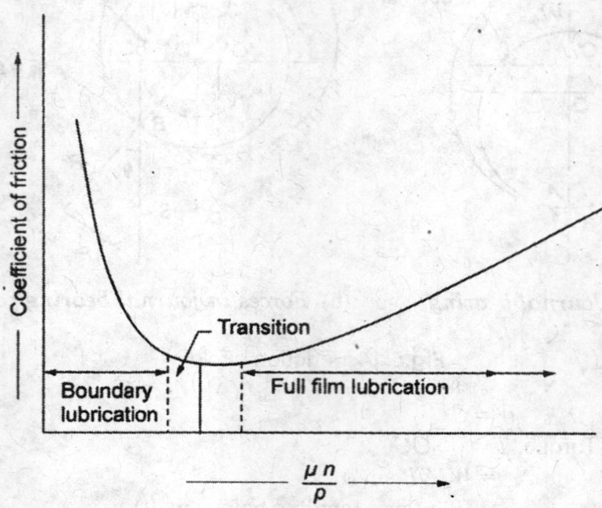

Fig.5.15 Variation of coefficient of friction with $\mu n/p$

5.13 MITCHELL (OR TILTING PAD) THRUST BEARING

To maintain thick-film lubrication in the case of rotating members, a wedge-shaped converging space should be available for the lubricant between the journal and the bearing. For flat surfaces, the same purpose can be served if one surface is made slightly inclined. A Mitchell thrust bearing consists of a series of metallic pads arranged around a rotating collar fixed to the shaft, as shown in Fig.5.16. Each pad is held by the housing of the bearing to prevent rotation but free to tilt about its stepped edge. The oil carried by the moving collar is dragged around the pad. Thus, an oil film of wedge shape is formed and a considerable pressure is built up to carry the axial load.

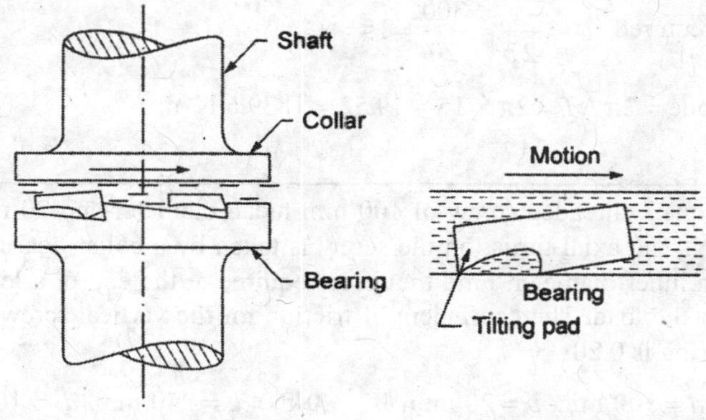

(a) *Bearing* (b) *Pad*

Fig.5.16 Tilting pad thrust bearing

Example 5.1

A square threaded screw of mean diameter 30 mm and pitch of threads 5 mm is used to lift a load of 15 kN by a horizontal force applied at the circumference of the screw. Find the force required if the coefficient of friction between screw and nut is 0.02.

■ **Solution** Given: $d_m = 30$ mm, $p = 5$ mm, $W = 15$kN, $\mu = 0.02$

$$\phi = \tan^{-1}\mu = \tan^{-1} 0.02 = 1.145°$$

$$\tan\alpha = \frac{p}{\pi d_m} = \frac{5}{\pi \times 30} = 0.05305, \quad \alpha = 3.037°$$

$$P = W\tan(\alpha + \phi) = 15 \tan 4.182° = 1.0967 \text{ kN}$$

Example 5.2

A turnbuckle with right- and left-hand single-start square threads is used to couple two railway coaches. The pitch of thread is 10 mm over a mean diameter of 30 mm. The coefficient of friction is 0.15. Find the work to be done in drawing the coaches together a distance of 300 mm against a steady load of 2.5 kN.

■ **Solution** Given $p = 10$ mm, $d_m = 30$ mm, $W = 2.5$ kN, $\mu = 0.15$, $s = 300$ mm

$$\phi = \tan^{-1}\mu = \tan^{-1} 0.15 = 8.530°$$

$$\tan\alpha = \frac{p'}{\pi d_m} = \frac{10}{\pi \times 30} = 0.10610, \quad \alpha = 6.057°$$

$$P = W \tan(\alpha + \phi) = 2500 \tan 14.587° = 650.6 \text{ N}$$

$$\text{Torque on each rod} = \frac{pd_m}{2} = 650.6 \times 0.015 = 9.76 \text{ N}\cdot\text{m}$$

Total torque required on the coupling nut, $T = 2 \times 9.76 = 19.52$ N·m

For one revolution of rod, coaches are drawn nearer by a distance = $2p = 20$ mm

$$\text{Number of turns required, } N = \frac{s}{2p} = \frac{300}{20} = 15$$

Work done by torque = $2\pi NT = 2\pi \times 15 \times 19.52 = 1839.6$ N·m

Example 5.3

A vertical two-start square threaded screw of 100 mm mean diameter and 20 mm pitch supports a vertical load of 20 kN. The axial thrust on the screw is taken by a collar bearing of 250 mm outer diameter and 100 mm inner diameter. Find the force required at the end of a lever 450 mm long in order to lift and lower the load. The coefficient of friction for the vertical screw and nut is 0.15 and that for the collar bearing is 0.20.

■ **Solution** Given: $d = 100$ mm, $p = 20$ mm, $W = 20$ kN, $d_2 = 250$ mm, $d_1 = 100$ mm, $l = 450$ mm, $\mu = 0.15$, and $\mu_c = 0.20$.

$$\phi = \tan^{-1}\mu = \tan^{-1} 0.15 = 8.530°$$

$$\tan a = \frac{L}{\pi d_m} = \frac{2\times 20}{\pi \times 100} = 0.12732, \; a = 7.256°$$

$$P = W \tan(\alpha + \phi) = 20 \tan 15.786° = 5.6544 \text{ kN}$$

$$d_c = 0.5(100 + 250) = 175 \text{ mm}$$

$$Fl = 0.5\,(Pd + \mu_c W d_c)$$

$$F = \frac{0.5(5.6544\times 100 + 0.20\times 20\times 175)}{450} = 2.812 \text{ kN}$$

Example 5.4

The spindle of a screw jack has single-start square threads with an outer diameter of 50 mm and a pitch of 10 mm. The spindle moves in a fixed nut. The load is carried on a swivel head but is not free to rotate. The bearing surface of the swivel head has a mean diameter of 60 mm. The coefficient of friction between the nut and the screw is 0.12 and that between the swivel head and the spindle is 0.10. Calculate the load that can be raised by efforts of 125 N, each applied at the end of two levers with effective length 400 mm. Also find the efficiency of the lifting system.

■ **Solution** Given: $d_o = 50$, $p = 10$ mm, $W = ?\,N$, $d_c = 60$ mm, $l = 400$ mm, $\mu = 0.12$, $\mu_c = 0.10$, and $F = 125$ N.

$$\phi = \tan^{-1}\mu = \tan^{-1} 0.12 = 6.843°$$

$$d_m = d_o - 0.5p = 50 - 5 = 45 \text{ mm}$$

$$\tan a = \frac{p}{\pi d_m} = \frac{10}{\pi \times 45} = 0.0707, \quad a = 4.046°$$

$$P = W\tan(a+\phi) = W\tan 10.889° = 0.19237\,W$$

$$2Fl = 0.5(Pd_m + \mu_c W d_c)$$

$$2 \times 125 \times 400 = 0.5(0.19237\,W \times 45 + 0.10 \times W \times 60)$$

$$100{,}000 = 7.3283\,W$$

$$W = 13645.7\,\text{N}$$

$$\text{Efficiency, } \eta = \frac{\tan a}{\tan(a+\phi)} = \frac{\tan 4.046°}{\tan 10.889°} = \frac{0.0707 \times 100}{0.19237} = 36.75\%$$

Example 5.5

A flat foot step bearing of 300 mm diameter supports a load of 8 kN. If the coefficient of friction is 0.10 and speed of the shaft is 80 rpm, find the power lost in friction, assuming (*a*) uniform pressure, and (*b*) uniform wear.

■ **Solution** Given: $r = 150$ mm, $W = 8$ kN, $\mu = 0.1$, $N = 80$ rpm

(a) Uniform pressure

$$T_f = \frac{2}{3}\mu W r = \frac{2}{3} \times 0.1 \times 8000 \times 150 \times 10^{-3} = 80\,\text{N}\cdot\text{m}$$

$$\omega = \frac{2\pi \times 80}{60} = 8.3776\,\text{rad/s}$$

$$\text{Power lost, } P = T_f\omega = 80 \times 8.3776 = 670.2\,\text{W}$$

(b) Uniform wear

$$T_f = 0.5\mu W r = 0.5 \times 0.1 \times 8000 \times 150 \times 10^{-3} = 60\,\text{N}\cdot\text{m}$$

$$\text{Power lost, } P = T_f\omega = 60 \times 8.3776 = 502.6\,\text{W}$$

Example 5.6

A vertical pivot bearing 200 mm diameter has a cone angle of 150°. If the shaft supports an axial load of 25 kN, and the coefficient of friction is 0.25, find the power lost in friction when the shaft rotates at 300 rpm, assuming (a) uniform pressure, and (b) uniform wear.

■ **Solution** Given: $r = 100$ mm, $W = 25$ kN, $\mu = 0.25$, $\alpha = 75°$, $N = 300$ rpm

(a) Uniform pressure

$$T_f = \frac{2}{3}\mu W r\,\text{cosec}\,\alpha = \left(\frac{2}{3}\right) \times 0.25 \times 25000 \times 0.1 \times \text{cosec}\,75° = 431.365\,\text{N}\cdot\text{m}$$

$$\omega = \frac{2\pi \times 300}{60} = 31.416\,\text{rad/s}$$

$$\text{Power lost, } P = T_f\omega = 431.365 \times 31.416 = 13.552\,\text{W}$$

(b) Uniform wear

$$T_f = 0.5\ \mu\ Wr\ \text{cosec}\ \alpha = 0.5 \times 0.25 \times 25000 \times 0.1 \times \text{cosec}\ 75° = 323.5\ \text{N}\cdot\text{m}$$

Power lost, $P = T_f\omega = 323.5 \times 8.3776 = 10.164$ W

Example 5.7

A vertical shaft supports a load of 30 kN in a conical pivot bearing. The external radius of the cone is three times the internal radius and the cone angle is 120°. Assuming uniform intensity of pressure of 0.40 MPa, determine the radii of the bearing. If the coefficient of friction between the shaft and bearing is 0.05 and the shaft rotates at 150 rpm, find the power lost in friction.

■ **Solution** Given: $W = 30$ kN, $\alpha = 60°$, $r_2 = 3r_1$, $p_n = 0.40$ MPa, $\mu = 0.05$, $N = 150$ rpm

$$\omega = \frac{2\pi \times 150}{60} = 15.708 \text{ rad/s}$$

$$p_n = \frac{W}{\pi(r_2^2 - r_1^2)}$$

$$0.40 \times 10^6 = \frac{30 \times 10^3}{\pi(9-1)r_1^2}$$

$$r_1^2 = 2985.15 \times 10^{-6}$$

$$r_1 = 54.627 \text{ mm}$$

$$r_2 = 163.882 \text{ mm}$$

$$T_f = \frac{2}{3}\mu W \text{ cosec } \alpha \left(\frac{r_2^3 - r_1^3}{r_2^2 - r_1^2}\right)$$

$$= \left(\frac{2}{3} \times 0.05 \times 30{,}000 \times \text{cosec } 60° \times 54.627\right)\left(\frac{27-1}{9-1}\right)$$

$$= 205 \text{ N}\cdot\text{m}$$

Example 5.8

The thrust on the propeller shaft of a marine engine is taken by eight collars whose outer and inner diameters are 650 and 400 mm, respectively. The thrust pressure is 0.5 MPa and may be assumed uniform. The coefficient of friction between the shaft and collars is 0.05. If the shaft rotates at 120 rpm, find (a) total thrust on the collars and (b) power absorbed by friction at the bearing.

■ **Solution** Given: $n = 8$, $d_o = 650$ mm, $d_i = 400$ mm, $p = 0.5$ MPa, $\mu = 0.04$, $N = 120$ rpm

(a)

$$p = \frac{W}{A} = \frac{4W}{\pi\left(d_o^2 - d_i^2\right)}$$

$$W = \frac{0.5 \times \pi(650^2 - 400^2)}{4} = 1{,}03{,}083 \text{ N}$$

(b) Uniform pressure

$$T_f = \frac{2}{3} n\mu W\left(\frac{r_o^3 - r_i^3}{r_o^2 - r_i^2}\right)$$

$$= \left(\frac{2}{3}\times 8\times 0.04\times 103083\right)\left(\frac{325^3 - 200^3}{325^2 - 200^2}\right)\times 10^{-3}$$

$$= 8822.6 \text{ N}\cdot\text{m}$$

$$\omega = \frac{2\pi\times 120}{60} = 12.566 \text{ rad/s}$$

$$P = T_f\omega = 8822.6\times 12.566\times 10^{-3} = 110.87 \text{ kW}$$

Example 5.9

The movable jaw of a bench vice is at the upper end of a hinged arm 0.5 m long, the centre line of the screw being 400 mm above the hinge. The screw has outer diameter of 25 mm and pitch of 6 mm. The mean radius of the thrust collar is 30 mm. Find the tangential force to be applied to the screw at a radius of 300 mm to produce a force of 6 kN at the jaw. Also find the mechanical efficiency of the vice. Assume thread and collar coefficient of friction to be 0.1 and 0.15, respectively.

■ **Solution** Given: $d_o = 25$ mm, $p = 6$ mm, $d_c = 60$ mm, $l = 300$ mm, $W = 6$ kN, $\mu = 0.10$, $F_c = 0.15$

$$d_m = d_o - 0.5p = 25 - 3 = 22 \text{ mm}$$

$$\tan\alpha = \frac{p}{\pi d_m} = \frac{6}{\pi\times 22} = 0.08681, \quad a = 4.96°$$

$$\phi = \tan^{-1}\mu = \tan^{-1} 0.10 = 5.71°$$

$$P = W\tan(\alpha+\phi) = 6000\tan 10.67° = 1130.6 \text{ N}$$

$$T = 0.5(Pd_m + \mu_c W d_c)$$

$$300\,F = 0.5(1130.6\times 22 + 0.15\times 6000\times 60)$$

$$F = 131.46 \text{ N}$$

$$\text{Efficiency}, \eta = \frac{\tan\alpha}{\tan(\alpha+\phi)} = \frac{0.08681\times 100}{0.18841} = 46.06\%$$

Example 5.10

A pivot bearing of a shaft consists of a frustrum of a cone. The diameters of the frustrum are 200 mm and 400 mm and its semi-cone angle is 60° . The shaft carries a load of 40 kN and rotates at 240 rpm. The coefficient of friction is 0.02. Assuming the intensity of pressure to be uniform, determine (a) the magnitude of pressure, and (b) the power lost in friction.

■ **Solution** Given: $d_2 = 400$ mm, $d_1 = 200$ mm, $W = 40$ kN, $\alpha = 30°$, $\mu = 0.02$, and $N = 240$ rpm

$$\omega = \frac{2\pi \times 240}{60} = 25.133 \text{ rad/s}$$

$$p = \frac{4W}{\pi(d_2^2 - d_1^2)}$$

$$= \frac{4 \times 40 \times 10^3}{\pi(400^2 - 200^2) \times 10^{-6}}$$

$$= 0.4244 \text{ MPa}$$

$$T_f = \frac{2}{3}\mu W \text{ cosec } \alpha \left(\frac{r_2^3 - r_1^3}{r_2^2 - r_1^2}\right)$$

$$= \left[\frac{2}{3} \times 0.02 \times 40{,}000 \times \text{cosec } 30° \times 10^{-3}\right]\left(\frac{200^3 - 100^3}{200^2 - 100^3}\right)$$

$$= 248.9 \text{ N} \cdot \text{m}$$

$$P = T_f\omega = 248.9 \times 25.133 = 6255.54 \text{ W}$$

Example 5.11

What force will be required at a radius of 80 mm to raise and lower an 11 kN crossbar of a planer that is raised and lowered by two 38 mm single-start square thread screw having a pitch of 7 mm? The outer and inner diameters of the collar are 76 and 38 mm respectively. Assume the coefficient of friction at the threads as 0.11 and at the collar as 0.13.

■ **Solution** Given: $d_o = 76$ mm, $d_i = 38$ mm, $p = 7$ mm, $d_m = 38$ mm, $l = 80$ mm, $F = W = 11$ kN, $\mu = 0.11$, $\mu_c = 0.13$

$$d_c = 0.5\,(d_o + d_i) = 0.5\,(76 + 38) = 57 \text{ mm}$$

$$\tan \alpha = \frac{p}{\pi d_m} = \frac{7}{\pi \times 38} = 0.05864, \quad \alpha = 3.356°$$

$$\phi = \tan^{-1}\mu = \tan^{-1} 0.11 = 6.277°$$

$$P = W \tan(\alpha + \phi) = 11{,}000 \tan 9.633° = 1867 \text{ N}$$

$$T = 0.5(Pd_m + \mu_c W d_c)$$

$$2F \times 80 = 0.5(1867 \times 38 + 0.13 \times 11{,}000 \times 57)$$

$$F = 476.43 \text{ N}$$

Example 5.12

A square threaded bolt of root diameter 20 mm and pitch 5 mm is tightened by screwing a nut whose mean diameter of bearing surface is 50 mm. The coefficient of friction for nut and bolt is 0.10 and for nut and bearing surface is 0.15. Find the force required at the end of a 450 mm long spanner when the load on the bolt is 10 kN.

■ **Solution** Given: $d_c = 20$ mm, $p = 5$ mm, $R_m = 25$ mm, $\mu = 0.10$, $\mu_c = 0.15$, $\ell = 450$ mm, $W = 10$ kN,

$$\text{Mean diameter of screw, } d_m = d_c + \frac{p}{2} = 20 + 2.5 = 22.5 \text{ mm}$$

$$\tan a = \frac{p}{\pi d_m} = \frac{5}{\pi \times 22.5} = 0.07073$$

$$a = 4.05°$$

$$\phi = \tan^{-1} 0.10 = 5.71°$$

$$P = W \tan(a + \phi)$$

$$= 10 \times 10^3 \times \tan(4.05 + 5.71)° = 1720.1 \text{ N}$$

$$T = \frac{Pd_m}{2} + \mu_c W R_m$$

$$= 1720.1 \times \frac{22.5}{2} + 0.15 \times 10^4 \times 25$$

$$= 19351.12 + 37500 = 56851.12 \text{ Nmm}$$

$$Fl = T$$

$$F \times 450 = 56851.12$$

$$F = 126.33 \text{ N}$$

Example 5.13

The mean diameter of a bolt having *V*-threads is 25 mm. The pitch of the thread is 5 mm and the angle of threads is 55°. The bolt is tightened by screwing a nut whose mean radius of bearing surface is 25 mm. The coefficient of friction for nut and bolt is 0.10 and for nut and bearing is 0.15. Find the force required at the end of a 0.5 m long lever when the load on the bolt is 15 kN.

■ **Solution** Given: $d_m = 25$ mm, $p = 5$ mm, $2\beta = 55°$, $R_m = 25$ mm, $\mu = 0.10$, $\mu_c = 0.15$, $\ell = 0.5$ m, $W = 15$ kN,

$$a = \tan^{-1}\left(\frac{p}{\pi d_m}\right) = \tan^{-1}\left(\frac{5}{25\pi}\right) = 3.64°$$

$$\phi = \tan^{-1}\left(\frac{\mu}{\cos\beta}\right) = \tan^{-1}\left(\frac{0.1}{\cos 27.5°}\right) = 6.43°$$

$$P = W \tan(a + \phi)$$

$$= 15 \times 10^3 \times \tan(3.64 + 6.43)° = 2663.8 \text{ N}$$

$$T = \frac{Pd_m}{2} + \mu_c W R_m$$

$$= 2663.8 \times 12.5 + 0.15 \times 15000 \times 25$$

$$= 33297.5 + 56250 = 89547.5 \text{ N}$$

$$Fl = T$$

$$F = \frac{89547.5}{500} = 179.1 \text{ N}$$

Example 5.14

Two tie rods are connected by a turnbuckle having right- and left-hand metric threads of V-type. The pitch of the threads is 5 mm on a mean diameter of 30 mm and a thread angle of 60°. Assuming coefficient of friction of 0.12, find the torque required to produce a pull of 40 kN.

■ **Solution** Given: $P = 5$ mm, $d_m = 30$ mm, $2\beta = 60°$, $\mu = 0.12$, $W = 40$ kN,

$$\mu_e = \frac{\mu}{\cos\beta} = \frac{0.12}{\cos 30°} = 0.13856$$

$$\phi = \tan^{-1}\mu_e = 7.89°$$

$$a = \tan^{-1}\left(\frac{p}{\pi d_m}\right) = \frac{5}{\pi \times 30} = 3.04°$$

(a) When rods are tightened,

$$P = W \tan(a + \phi)$$
$$= 40 \times 10^3 \times \tan(3.04 + 7.89)°$$
$$= 7724.5\text{ N}$$

Torque, $$T = \frac{Pd_m}{2}$$
$$= 7724.5 \times 15 \times 10^{-3} = 115.86\text{ Nm}$$

(b) When the rods are slackened

$$P = W \tan(\phi - a)$$
$$= 40 \times 10^3 \times \tan(7.89 - 3.04)°$$
$$= 3394\text{ N}$$
$$T = 3394 \times 15 \times 10^{-3}$$
$$= 50.91\text{ Nm}$$

Example 5.15

A vertical shaft 140 mm in diameter rotating at 120 rpm rests on a flat end foot step bearing. The shaft carries a vertical load of 30 kN. The coefficient of friction is 0.06. Estimate the power lost in friction, assuming (a) uniform pressure and (b) uniform wear.

■ **Solution** Given: $R = 70$ mm, $N = 120$ rpm, $W = 30$ kN, $\mu = 0.06$

(a) Uniform pressure

$$T = \left(\frac{2}{3}\right) \cdot \mu W R$$
$$= \left(\frac{2}{3}\right) \times 0.06 \times 30 \times 10^3 \times 0.07$$
$$= 84\text{ Nm}$$

$$\text{Power lost in friction} = T\omega$$
$$= 84 \times \left(2\pi \times \frac{150}{60}\right) \times 10^{-3}$$
$$= 1.056\text{ kW}$$

(b) Uniform wear

$$T = \mu W \frac{R}{2}$$

$$= 0.06 \times 30 \times 10^3 \times \frac{0.07}{2}$$

$$= 63 \text{ Nm}$$

$$\text{Power lost in friction} = 63 \times \left(2\pi \times \frac{120}{60}\right) \times 10^{-3}$$

$$= 0.79 \text{ kW}$$

Example 5.16

A conical pivot supports a load of 25 kN, the cone angle being 120°, and the intensity of normal pressure does not exceed 0.25 MPa. The external radius is twice the internal radius. Find the outer and inner radii of the bearing surface. If the shaft rotates at 180 rpm and the coefficient of friction is 0.15, find the power lost in friction, assuming uniform pressure.

■ **Solution** Given: $W = 25$ kN, $2a = 120°$, $p_n = 0.25$ mpa, $r_2 = 2r_1$, $N = 180$ rpm, $\mu = 0.15$

Intensity of normal pressure, $$p_n = \frac{W}{\pi(r_2^2 - r_1^2)}$$

$$0.25 = \frac{25 \times 10^3}{\pi\left[(2r_1)^2 - r_1^2\right]}$$

$$r_1 = 103 \text{ mm}$$

$$r_2 = 206 \text{ mm}$$

Frictional torque, $$T_f = \left(\frac{2}{3}\right) \cdot \mu W \operatorname{cosec} a \left[\frac{r_2^3 - r_1^3}{r_2^2 - r_1^2}\right]$$

$$= \left(\frac{2}{3}\right) \times 0.15 \times 25 \times 10^3 \times \operatorname{cosec} 60° \left[\frac{206^3 - 103^3}{206^2 - 103^2}\right]$$

$$= 693.8 \text{ Nm}$$

Power lost in friction $$= \frac{T_f \omega}{1000}$$

$$= 693.8 \times \frac{2\pi(\frac{180}{60})}{1000} = 13.08 \text{ kW}$$

Example 5.17

A thrust shaft of a ship has six collars of 600 mm outer diameter and 300 mm inner diameter. The total thrust from the propeller is 120 kN. If the coefficient of friction is 0.15 and speed of the engine 100 rpm, find the power lost in friction at the thrust block, assuming (a) uniform pressure, and (b) uniform wear.

■ **Solution** Given: $n = 6$, $r_2 = 300$ mm, $r_1 = 150$ mm, $W = 120$ kN, $\mu = 0.15$, $N = 100$ rpm

(a) Uniform pressure

Friction torque,
$$T_f = \left(\frac{2}{3}\right)\cdot n\mu W\left[\frac{r_2^3 - r_1^3}{r_2^2 - r_1^2}\right]$$
$$= \left(\frac{2}{3}\right)\times 6\times 0.15\times 120\times 10^3\times\left[\frac{300^3 - 150^3}{300^2 - 150^2}\right]\times 10^{-3}$$
$$= 25{,}200 \text{ Nm}$$

Power lost in friction,
$$= 25{,}200\times\frac{(2\pi\times 100/60)}{1000} = 263.9 \text{ kW}$$

(b) Uniform wear

Friction torque,
$$T_f = n\mu W r_m$$
$$= 6\times 0.15\times 120\times 10^3\times\frac{(300+150)10^{-3}}{2} = 24{,}300 \text{ Nm}$$

Power lost in friction,
$$= 24{,}300\times\frac{(2\pi\times 100/60)}{1000} = 254.47 \text{ kW}$$

Example 5.18

A force of 250 N is required to pull a body resting on a horizontal plane with a force applied at 30° to the horizontal. If the direction of the force is reversed to push the body, the force required is 300 N. determine the mass of the body and the coefficient of friction between the body and the surface. Also calculate the minimum force required to pull the body.

■ **Solution**

Given: $\theta = 30°$, $P_1 = 250$ N, $P_2 = 300$ N

To pull the body, $P_1 = W \sin\phi/\cos(\theta - \phi) = 250$

To push the body, $P_2 = W \sin\phi/\cos(\theta + \phi) = 300$

$$\cos(\theta+\phi)/\cos(\theta-\phi) = 250/300$$

$1.2\,[\cos 30°\cos\phi - \sin 30°\sin\phi] = \cos 30°\cos\phi + \sin 30°\sin\phi$

$1.2\,[0.866\cos\phi - 0.5\sin\phi] = 0.866\cos\phi + 0.5\sin\phi$

$1.2\,[0.866 - 0.5\tan\phi] = 0.866 + 0.5\tan\phi$

$1.0392 - 0.6\,\mu = 0.866 + 0.5\,\mu$

$1.1\,\mu = 0.1732$

$\mu = 0.1575$, $\phi = 8.948°$

$W = 250\cos(30° - 8.948°)/\sin 8.948° = 1500$ N

Minimum force $= W\sin\phi = 1500\sin 8.948° = 233.3$ N

Example 5.19

A bolt with square threads has outer diameter 30 mm and pitch of 5 mm. It is tightened by screwing a nut whose mean diameter of bearing surface is 50 mm. The coefficient of friction for nut and bolt is 0.12 and for nut and bearing is 0.15. The nut is tightened by a spanner whose effective lever arm is 450 mm. The load on the bolt is 12 kN. Determine the effort required at the end of the spanner.

■ **Solution** Given: $d_0 = 30$ mm, $p = 5$ mm, $d_b = 50$ mm, $\mu = 0.12$, $\mu_c = 0.15$, $\ell = 450$ mm, $W = 12$ kN

Mean diameter of screw, $d_m = d_o - p/2 = 30 - 2.5 = 27.5$ mm

Helix angle, $\alpha = \tan^{-1}[p/(\pi d_m)] = \tan^{-1}[5/(\pi \times 27.5)] = 3.3123°$

Friction angle, $\phi = \tan^{-1}\mu = \tan^{-1}0.12 = 6.8428°$

Force required on the screw, $P = W\tan(\alpha + \phi)$

$= 12\tan(3.3123° + 6.8428°) = 2.149$ kN

Total frictional torque, $T_f = Pd_m/2 + \mu_c Wd_b/2$

$= 2.149 \times 27.5/2 + 0.15 \times 12 \times 50/2$

$= 74.555$ N·m

If F is the effort applied at the end of the lever, then

$450F = 74555$

$F = 165.7$ N

Example 5.20

A turnbuckle is used to couple two railway coaches. It has single-start square threads with pitch of 12 mm on a mean diameter of 38 mm. Calculate the work to be done in drawing the coaches together by a distance of 200 mm, if the coefficient of friction is 0.12, when (a) the steady load is 2500 N, (b) the load increases from 2500 N to 8000 N, and (c) efficiency of the turnbuckle.

■ **Solution** Given: $p = 12$ mm, $d_m = 38$ mm, $s = 200$ mm, $\mu = 0.12$, $W_1 = 2500$ N, $W_2 = 8000$ N

(a) Helix angle, $a = \tan^{-1}[p/(\pi d_m)] = \tan^{-1}[12/(\pi \times 38)] = 5.74°$

Friction angle, $\phi = \tan^{-1}\mu = \tan^{-1}0.12 = 6.8428°$

Force required on the screw, $P = W\tan(a + \phi)$

$= 2500\tan(5.74° + 6.8428°) = 558\text{ N}$

Torque on each rod, $T_r = Pd_m/2 = 558 \times 38/2 = 10602\text{ N·mm}$

Total torque required on the coupling nut, $T = 2T_r = 21{,}204\text{ N·mm}$

Number of turns required to bring the coaches nearer by 200 mm, $N = s/2p$

$= 200/24 = 8.33$

Work done $= 2\pi NT = 2\pi \times 8.33 \times 21204 \times 10^{-3} = 1110.24\text{ N·m}$

(b) Work done $= 1110.24\,(8000 - 2500)/2500 = 2442.528$ N·m

(c) Efficiency $= \tan\alpha/\tan(\alpha + \phi) = \tan 5.74°/\tan(5.74° + 6.8428°) = 0.45$ or 45%

Example 5.21

In a thrust bearing the outer and inner radii of the contact surface are 200 mm and 150 mm, respectively. Total axial load is 80 kN and the coefficient of friction is 0.05. The shaft is rotating at 420 rpm. Intensity of pressure is not to exceed 0.35 MPa. Assuming uniform pressure, calculate (a) power lost in overcoming friction, and (b) the number of collars required for the thrust bearing.

■ **Solution** Given: $r_2 = 200$ mm, $r_1 = 150$ mm, $W = 80$ kN, $\mu = 0.05$, $N = 420$ rpm, $p = 0.35$ mpa

(a) Frictional torque, $T_f = (2/3)\cdot\mu W[(r_2^3 - r_2^3)/(r_2^2 - r_1^2)]$

$= (2/3) \times 0.05 \times 80{,}000[(200^3 - 150^3)/(200^2 - 150^2)] \times 10^{-3}$

$= 704.762\text{ N·m per collar}$

Power lost in friction $= 2\pi NT_f/(60 \times 10^3)$

$= 2\pi \times 420 \times 704.762/(60 \times 10^3) = 31\text{ kW per collar}$

(b) Number of collars required, $n = W/\left[\pi(r_2^2 - r_1^2)p\right]$

$$= 80{,}000/[\pi(200^2 - 150^2) \times 0.35] = 4.16 \approx 4$$

$$\text{Total power lost} = 4 \times 31.997$$

$$= 124 \text{ kW}$$

Example 5.22

The mean diameter of a screw jack having pitch of 10 mm is 50 mm. A load of 25 kN is lifted through a distance of 180 mm. Find the work done in lifting the load and efficiency of the screw when

(a) The load rotates with the screw.

(b) The load rests on the loose head that does not rotate with the screw. The outer and inner diameters of the bearing surface of the loose head are 60 mm and 10 mm, respectively. The coefficient of friction for the screw as well the bearing surface may be taken as 0.10.

Solution

Given: $p = 10$ mm, $d_m = 50$ mm, $W = 25$ kN, $h = 180$ mm, $\mu = \mu_c = 0.10$, $R_2 = 30$ mm, $R_1 = 5$ mm

$$\text{Helix angle of screw, } a = \tan^{-1}\left(\frac{p}{\pi d_m}\right) = \tan^{-1}\left(\frac{10}{\pi \times 50}\right) = 3.6426°$$

$$\text{Angel of friction, } \phi = \tan^{-1}\mu = \tan^{-1} 0.10 = 5.7106°$$

Force required at the circumference of screw to lift the load,

$$P = W\tan(\alpha + \phi) = 25\tan(3.6426° + 5.7106°) = 4.118\,\text{kN}$$

$$\text{Frictional torque of screw, } T_f = P \times \frac{d_m}{2} = 4.118 \times \frac{50}{2} = 102.94\,\text{N}\cdot\text{m}$$

(a) Number of rotations made by screw to lift the load through 180 mm,

$$N = \frac{h}{p} = \frac{180}{10} = 18$$

$$\text{Work done in lifting the load} = 2\pi N T_f = 2\pi \times 18 \times 102.94 = 11642.6\,\text{N}\cdot\text{m}$$

$$\text{Efficiency, } \eta = \frac{\tan\alpha}{\tan(\alpha+\phi)} = \frac{\tan 3.6426°}{\tan(3.6426 + 5.7106°)} = 0.3865 \text{ or } 38.65\%$$

(b) Mean radius of bearing surface, $R_m = \frac{1}{2}(R_1 + R_2) = \frac{1}{2}(5 + 30) = 17.5$ mm

Total frictional torque required at the screw and collar,

$$T_f = P \times \frac{d_m}{2} + \mu_C W R_m$$

$$= 4.118 \times \frac{50}{2} + 0.10 \times 25 \times 17.5 = 146.7\,\text{N}\cdot\text{m}$$

Work done in lifting the load = $2\pi NT_f = 2\pi \times 18 \times 146.7 = 16591.4$ N · m
Torque required to lift the load without friction,

$$T_o = W \tan\alpha \times \frac{d_m}{2}$$

$$= 25 \tan 3.6426° \times \frac{50}{2} = 39.79 \text{ N}\cdot\text{m}$$

Efficiency of screw jack, $\eta = \frac{T_o}{T_f} = \frac{39.79}{146.7} = 0.2712$ or 27.12%

Example 5.23

Two co-axial roads are connected by a turnbuckle having pitch diameter of 20 mm and pitch of 3 mm threads. The included angle of the threads is 60°. Calculate the torque required on the nut to produce a pull of 50 kN. The coefficient of friction is 0.15.

■ **Solution** Given: $d_m = 20$ mm, $p = 3$ mm, $2\beta = 60°$, $W = 50$ kN, $\mu = 0.15$

$$\tan\alpha = \frac{p}{\pi d_m} = \frac{3}{\pi \times 20} = 0.04775$$

$$\alpha = 2.7336°$$

Virtual coefficient of friction, $\mu_e = \frac{\mu}{\cos\beta} = \frac{0.15}{\cos 30°} = 0.1732$

$$\phi = \tan^{-1}\mu_e = \tan^{-1} 0.1732 = 9.8264°$$

Force required at the circumference of screw,

$$P = W \tan(a + \phi) = 50 \tan(2.7336° + 9.8264°) = 11.14 \text{ kN}$$

Torque on one rod, $T = P \times \frac{d_m}{2} = 11.14 \times \frac{20}{2} = 111.4$ N · m

Torque required on the nut = $2T = 2 \times 111.4 = 222.8$ N · m

Summary for Quick Revision

1. Friction is the property of the two bodies moving or tending to move on one another by virtue of which a force is developed between them. Force of friction is the force that opposes the motion of one body over another.
2. Friction is a blessing in some cases but a necessary evil in other cases.
3. *Static friction:* It is the friction experienced by two bodies in contact, while at rest.
   Dynamic (or kinetic) friction: It is the friction experienced by two bodies in contact, while in motion. It is less than the static friction.
   Sliding friction: It is due to sliding of two bodies on each other.
   Rolling friction: It is due to rolling of two bodies on each other.

Pivot friction: It is the friction experienced by two bodies due to rotation of one body around other, as in the case of foot step bearing.
Dry (or solid) friction: It is due to two dry and unlubricated surfaces in contact.
Boundary (or skin or greasy) friction: It is the friction experienced by two bodies in contact when separated by a thin film of a lubricant.

4 Laws of static friction
(i) The force of friction always acts in a direction opposite to the impending motion.
(ii) The limiting force of friction is directly proportional to the normal reaction.
(iii) The force of friction is independent of the area contact between the two surfaces.
(iv) The force of friction depends upon the roughness of the surfaces of two materials.

5 The coefficient of friction (μ) is the ratio of the force of friction (F) to the normal reaction (R), that is, $\mu = F/R$. Angle of friction is the angle that the resultant of F and R makes with R. In the case of limiting friction, the coefficient of friction is equal to the tangent of the angle of friction, that is, $\mu = \tan\phi$. In the case of limiting friction, the angle of the inclined plane is equal to the angle of friction. The angle of the plane when motion of an object on the plane is impending is called the angle of repose. Cone of friction is the cone generated by the resultant of F and R when it is rotated about the vertical axis.

6 Forces on sliding bodies:
(a) Body resting on a horizontal plane subjected to inclined force:

$$P = W\sin\phi/\cos(\theta - \phi) \text{ for pull type force}$$
$$= W\sin\phi/\cos(\theta + \phi) \text{ for push type force}$$

Note that pull type force is smaller than push type force.
(b) Body resting on an inclined plane with force inclined to the plane:

$$P = W\sin(\alpha + \phi)/\cos(\theta - \phi) \text{ for body going up the plane}$$
$$= W\sin(\phi - \alpha)/\cos(\theta + \phi) \text{ for body going down the plane}$$

(c) Body resting on an inclined plane with horizontal force:

$$P = W\tan(\alpha + \phi) \text{ for body going up the plane}$$
$$P = W\tan(\phi - \phi) \text{ for body going down the plane}$$

7 Efficiency of the inclined plane:
(i) For body going up the plane with inclined force:

$$\eta = (1 + \mu\tan\theta)/(1 + \mu\cot\alpha)$$

(ii) For body going down the plane with inclined force:

$$\eta = (1 - \mu\tan\theta)/(1 - \mu\cot\alpha)$$

(iii) For body going up the plane with horizontal force:

$$\eta = \tan\alpha/\tan(\alpha + \phi)$$

(iv) For body going down the plane with horizontal force:

$$\eta = \tan(\alpha - \phi)/\tan\alpha$$

8 Screw threads:
(a) Square threads:

$$P = W\tan(\alpha + \phi)$$

where $\tan a = p/(\pi d_m)$
(b) V-threads:
Frictional force, $F = (\mu/\cos\beta)\,W$

9 Screw jack:

A screw jack is a device for lifting of load.

(i) Horizontal effort required:

$= W \tan(\alpha + \phi)$ for raising a load

$= W \tan(\phi - \alpha)$ for lowering a load

Torque required to overcome friction between the screw and the nut:

$T_1 = W d_m \tan(\alpha + \phi)/2$

Torque required to overcome friction at the collar:

$T_2 = \mu_c W d_c/2$

where μ_c = collar coefficient of friction and

d_c = mean diameter of the collar

Total torque, $T = T_1 + T_2 = (P d_m + \mu_c W d_c)/2$

F = effort applied at the end of a lever of length $\ell = T/\ell$

(ii) Efficiency:

$\eta = \tan\alpha/\tan(\alpha + \phi)$, for raising of load

$= \tan(\phi - \alpha)/\tan\alpha$, for lowering of load

For efficiency to be maximum, $\alpha = 45° - \phi/2$

$\eta_{max} = (1 - \sin\phi)/(1 + \sin\phi)$

(iii) Self locking and over-hauling screw:

If $\phi > \alpha$, then the screw is called *self-locking*. If $\phi < \alpha$, then the screw will lower of its own. Such a screw is called *over-hauling*. For a self-locking screw, $\eta \leq 50\%$.

10 Friction in bearings:

(i) Flat pivot bearing.

(a) Uniform pressure:

Total frictional torque, $T_f = 2\,\mu\,W R/3 = 2\pi\mu\,p\,(R_2 - R_1)/3$

$p = W/[\pi(R_2 - R_1)]$

Power lost in friction, $P = T_f\,\omega$

(b) Uniform wear

Total frictional torque on the bearing, $T_f = \dfrac{1}{2}\mu\,W(R_2 + R_1) = \mu\,W R_m$

(ii) Conical pivot bearing

(a) Uniform pressure

Total frictional torque, $T_f = (2/3)\,\mu\,W R \operatorname{cosec}\alpha$

(b) Uniform wear

Total frictional torque on the bearing, $T_f = 1/2 \times \mu\,W R \operatorname{cosec}\alpha$

(iii) Truncated conical pivot bearing.

(a) Uniform pressure:

Total frictional torque, $T_f = (2/3)\,\mu W \operatorname{cosec}\alpha\,[(r_2^3 - r_1^3)/(r_2^2 - r_1^2)]$

(b) Uniform wear:

Total frictional torque, $T_f = \mu W r_m \operatorname{cosec}\alpha$

(iv) Flat collar bearing.

(a) Uniform pressure:

Intensity of pressure, $p = W/[\pi(r_2^2 - r_1^2)]$

Frictional torque, $T_f = (2/3)\mu W[(r_2^3 - r_1^3)/(r_2^2 - r_1^2)]$

(b) Uniform wear:

Total frictional torque on the bearing, $T_f = \mu\,W r_m$

11 Rolling friction:
Coefficient of rolling friction, $b = Fh/W$
Let F_r = force applied to the body for rolling = bW/h
F_s = force applied to the body for sliding = μW
The body rolls without sliding, if $F_r < F_s$ or $\mu > b/h$. The body will slide if $F_s < F_r$ or $\mu < b/h$. The body will either roll or slide if $\mu = b/h$.

12 Anti-friction bearings:
In the case of anti-friction bearings, the point of contact between the journal and the bearing elements is either a point (as in the case of ball bearings) or a line (as in the case of roller bearings).

13 Friction circle:
A cricle drawn with centre O and radius OC = $r \sin \phi = r \mu$ is called the friction circle. r is the radius of the journal.

14 Mitchell bearing is a tilting pad thrust bearing.

Multiple Choice Questions

1 Friction is a curse in case of
(a) Belt drive (b) Lathe slide (c) Brakes (d) Clutches

2 Force of friction does not depend upon
(a) Area of contacting surfaces (b) Material of contacting surfaces
(c) Velocity of sliding (d) Temperature

3 If μ = coefficient of friction, and ϕ = angle of friction, then μ is equal to
(a) $\sin \phi$ (b) $\cos \phi$ (c) $\tan \phi$ (d) $\cot \phi$

4 Angle of cone of friction is equal to (ϕ = angle of friction)
(a) ϕ (b) 2ϕ (c) 1.5ϕ (d) 3ϕ

5 The efficiency of screw jack for raising the load is
(a) $\dfrac{\tan\alpha}{\tan(\alpha-\phi)}$ (b) $\dfrac{\tan\alpha}{\tan(\alpha+\phi)}$ (c) $\dfrac{\tan(\alpha-\phi)}{\tan\alpha}$ (d) $\dfrac{\tan(\alpha+\phi)}{\tan\alpha}$

6 The maximum efficiency of a screw jack is
(a) $\dfrac{1-\sin\phi}{1+\sin\phi}$ (b) $\dfrac{1+\sin\phi}{1-\sin\phi}$ (c) $\dfrac{1-\cos\phi}{1+\cos\phi}$ (d) $\dfrac{1-\sin\phi}{1+\cos\phi}$

7 The efficiency of a screw jack is maximum when
(a) $\alpha = \dfrac{\pi}{4} - \dfrac{\phi}{2}$ (b) $\alpha = \dfrac{\pi}{4} + \dfrac{\phi}{2}$ (c) $\dfrac{\pi}{2} - \dfrac{\phi}{2}$ (d) $\dfrac{\pi}{2} + \dfrac{\phi}{2}$

8 For a self-locking screw, efficiency should be
(a) <50% (b) >50% (c) =50% (d) ≥50%

9 For a flat pivot bearing of radii r_1 and r_2, the mean radius with uniform pressure is
(a) $\dfrac{1}{2}(r_1 + r_2)$ (b) $\dfrac{2}{3}\left[\dfrac{r_2^3 - r_1^3}{r_2^2 - r_1^2}\right]$ (c) $\dfrac{1}{2}(r_1 - r_2)$ (d) $\dfrac{2}{3}\left[\dfrac{r_2^3 + r_1^3}{r_2^2 + r_1^2}\right]$

10 The effective coefficient of friction for V-threads of thread angle 2β is

(a) $\mu \sin \beta$ (b) $\dfrac{\mu}{\sin \beta}$ (c) $\mu \cos \beta$ (d) $\dfrac{\mu}{\cos \beta}$

Answers

1. (b) 2. (a) 3. (c) 4. (b) 5. (b) 6. (a) 7. (a) 8. (a) 9. (b) 10. (b)

Review Questions

1 What is friction? Is it necessary evil or blessing?

2 Describe various types of friction.

3 State the laws of friction.

4 Define coefficient of friction, angle of friction, and angle of repose.

5 When a screw jack is self-locking and self-hauling?

6 What are the limits on the efficiency of a screw jack to be of the self-locking type?

7 Explain uniform pressure and uniform wear theories for a clutch.

8 What are thin-film and thick-film lubrications?

9 What are anti-friction bearings?

10 Explain friction circle.

Exercises

5.1 An effort of 1 kN is required to move a certain body up an inclined plane of angle 15° when the force is acting parallel to the plane. If the angle of plane is increased to 20°, then the effort required is 1.25 kN. Find the weight of the body and coefficient of friction.

[Ans. 3072.5N, 0.069]

5.2 A shaft has a number of collars integral with it. The outer diameter of the collars is 450 mm and the shaft diameter is 300 mm. If the uniform intensity of pressure is 0.3 MPa and the coefficient of friction is 0.05, determine (a) power absorbed in overcoming friction when the shaft is running at 120 rpm and carries a load of 150 kN, and (b) the number of collars required.

[Ans. 19.03kW, 6]

5.3 A thrust shaft of a ship has six collars of 600 mm outer diameter and 300 mm inner diameter. The total thrust from the propeller is 120 kN. If the coefficient of friction is 0.15 and the speed of the engine is 100 rpm, determine the power absorbed in friction when (a) pressure is uniform, and (b) wear is uniform.

[Ans. 43.98kW, 42.41kW]

5.4 The mean diameter of a bolt having *V*-threads is 25 mm. The pitch of the threads is 5 mm and the angle of threads is 55°. The bolt is tightened by screwing a nut whose mean radius of the bearing surface is 25 mm. If the coefficient of friction for nut and bolt is 0.10 and for the nut and bearing surface is 0.15, find the force required at the end of a 0.5 m long spanner when the load on the bolt is 10 kN.

[Ans. 119.4N]

5.5 A vertical screw with single-start square threads 50 mm mean diameter and 10 mm pitch is raised against a load of 6 kN by means of a hand wheel, the boss of which is threaded to act as a nut. The axial load is taken by a thrust collar that supports the wheel boss and has a mean diameter of 70 mm. If the coefficient of friction is 0.15 for the screw and 0.20 for the collar, find the suitable diameter of hand wheel. Assume that a person can apply a force of 150 N by each hand.

[Ans. 1m]

5.6 The pitch of a 50 mm mean diameter threaded screw of a jack is 12.5 mm. The coefficient of friction between the screw and the nut is 0.15. Determine the torque required on the screw to raise a load of 20 kN. Assume that the load does not rotate with the screw. Calculate the ratio of torque required for raising and lowering the load. Also calculate the efficiency of the screw in both the cases.

[Ans. 3.34, 34.25%, 87.43%]

6 BELTS, CHAINS AND ROPES

Chapter Outline

6.1 INTRODUCTION

Belts, chains and ropes are used for power transmission. Belts and chains are used for short centre drives whereas ropes are used for long centre distances. Belts are of two types: flat belts and V-belts. Chains give a more positive drive than belts. There are problems of slip and creep in belts. Belts can be either of the open type or cross type. In this chapter, we shall study belts, chains and ropes from the point of view of power transmission.

6.2 FLAT BELT DRIVE

6.2.1 Angular Velocity Ratio

Consider the open belt drive shown in Fig.6.1. Let the smaller pulley be the driver and the bigger pulley the driven (or follower). When the driver is rotating anticlockwise, the top side of the belt will be the tight side and the bottom side the slack side.

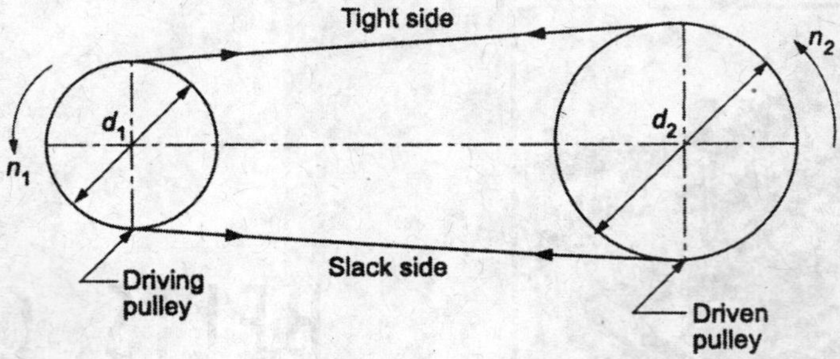

Fig.6.1 Flat belt drive

Let d_1 = diameter of the driving pulley
d_2 = diameter of the driven pulley
n_1 = rpm of the driving pulley
n_2 = rpm of the driven pulley

Then angular velocity of driver, $\omega_1 = \dfrac{2\pi n_1}{60}$ rad/s

and angular velocity of follower, $\omega_2 = \dfrac{2\pi n_2}{60}$ rad/s

Angular velocity ratio, $\dfrac{\omega_1}{\omega_2} = \dfrac{n_1}{n_2}$

Linear velocity of driver, $v_1 = \pi d_1 n_1$

Linear velocity of follower, $v_2 = \pi d_2 n_2$

Assuming there is no slip between the belt and the pulleys, $v_1 = v_2$

$$\frac{n_1}{n_2} = \frac{d_2}{d_1} \tag{6.1}$$

If t is the thickness of the belt, then

$$n_1/n_2 = (d_2 + t)/(d_1 + t) \tag{6.2}$$

6.2.2 Effect of Slip

Slip is the difference between the speed of the driver and the belt on the driver side and the belt and the follower on the driven side.

Let s_1 = percentage slip between the driver and the belt
s_2 = percentage slip between the belt and the follower

Linear speed of belt on driver $= v_1\left(1 - \dfrac{s_1}{100}\right)$

Linear speed of follower, $v_2 = v_1\left(1 - \dfrac{s_1}{100}\right)\left(1 - \dfrac{s_2}{100}\right)$

$$= v_1\left[1-\frac{(s_1+s_2-0.01s_1s_2)}{100}\right]= v_1(1-s)$$

where $s=\dfrac{(s_1+s_2-0.01s_1s_2)}{100}$ is the total fractional slip.

Hence
$$\frac{n_2}{n_1}=(1-s)\cdot\frac{d_1+t}{d_2+t} \tag{6.3}$$

6.2.3 Law of Belting

The law of belting states that the centre line of the belt as it approaches the pulley must lie in a plane perpendicular to the axis of that pulley, or must lie in the plane of the pulley, otherwise the belt will run off the pulley.

6.2.4 Length of Open Belt

Consider the open flat belt drive shown in Fig.6.2. The length of open belt,

$$L_o = \text{arc } AEB + \text{arc } CFD + AC + BD$$

$$\text{arc } AEB = \frac{d_1(\pi-2\alpha)}{2}$$

$$\text{arc } CFD = \frac{d_2(\pi+2\alpha)}{2}$$

$$AC = BD = O_1G = C\cos\alpha$$

$$L_o = \frac{d_1(\pi-2\alpha)}{2}+\frac{d_2(\pi+2\alpha)}{2}+2C\cos\alpha$$

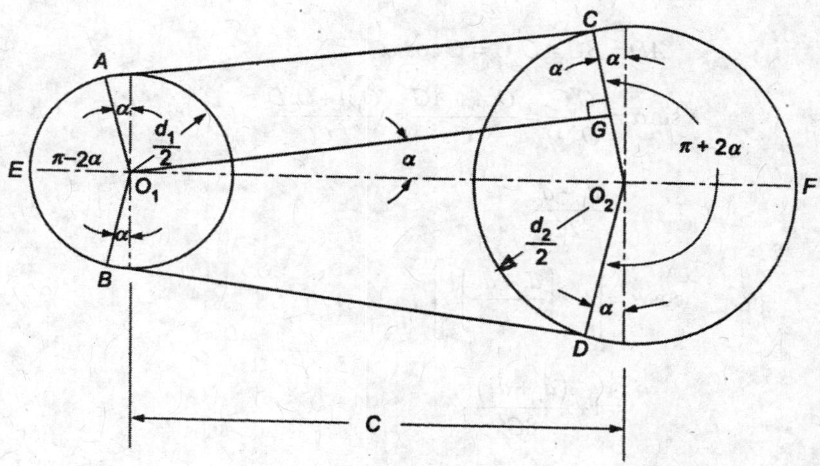

Fig.6.2 Open flat belt drive

Now $$\sin\alpha = \frac{O_2G}{O_1O_2} = \frac{(O_2C - CG)}{C} = \frac{(O_2C - AO_1)}{C}$$

$$\alpha \approx \sin\alpha = \frac{d_2 - d_1}{2C}$$

$$\cos a = \left[1 - \sin^2 a\right]^{0.5} = \left[1 - \left\{\frac{d_2 - d_1}{2C}\right\}^2\right]^{0.5}$$

$$\approx 1 - \frac{(d_2 - d_1)^2}{8C^2}$$

$$L_o = \frac{\pi(d_1 + d_2)}{2} + \alpha(d_2 - d_1) + 2C\left[1 - \frac{(d_2 - d_1)^2}{8C^2}\right]$$

$$L_o = \frac{\pi(d_1 + d_2)}{2} + \frac{(d_2 - d_1)^2}{2C} + 2C\left[\frac{1 - (d_2 - d_1)^2}{8C^2}\right]$$

$$= 2C + \frac{(d_2 - d_1)^2}{4C} + \frac{\pi(d_1 + d_2)}{2} \qquad (6.4)$$

6.2.5 Length of Cross Belt

Consider the cross belt shown in Fig.6.3. The length of cross flat belt,

$$L_c = \text{arc } AEB + \text{arc } CEB + AD + BC$$

$$\text{arc } AEB = (\pi + 2\alpha)\frac{d_1}{2}$$

$$\text{arc } CEB = (\pi + 2\alpha)\frac{d_2}{2}$$

$$AD = BC = O_2G = C\cos\alpha$$

Now $$\sin\alpha = \frac{O_1G}{O_1O_2} = \frac{O_1A + AG}{C} = \frac{O_1A + O_2D}{C}$$

$$= \frac{d_1 + d_2}{2C}$$

$$\cos\alpha = \left[1 - \left\{\frac{d_1 + d_2}{2C}\right\}^2\right]^{0.5}$$

$$= 1 - \frac{(d_1 + d_2)^2}{8C^2}$$

$$L_c = 2C\left[\frac{1-(d_1+d_2)^2}{8C^2}\right]+\frac{\pi(d_1+d_2)}{2}+\frac{(d_1+d_2)^2}{2C}$$

$$= 2C+\frac{\pi(d_1+d_2)}{2}+\frac{(d_1+d_2)^2}{4C} \tag{6.5}$$

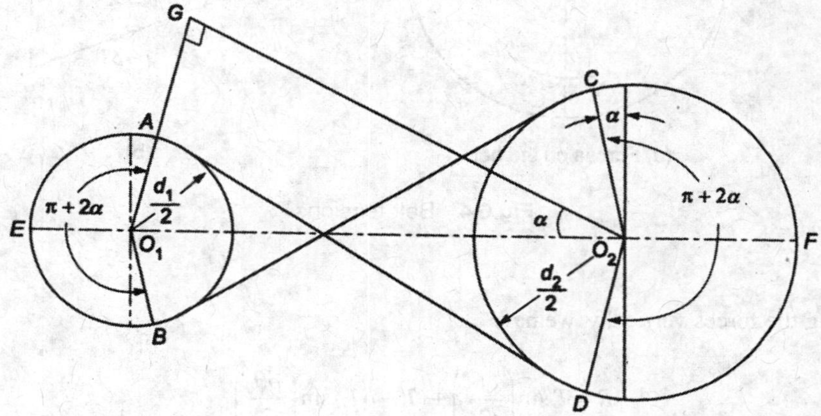

Fig.6.3 Cross flat belt drive

6.2.6 Angle of Arc of Contact

1. *Open belt* The angle of arc of contact on the smaller pulley,

$$\theta = \pi - 2\alpha$$

$$= \pi - 2\sin^{-1}\left[\frac{d_2-d_1}{2C}\right] \tag{6.6a}$$

The possibility of slip is more on the smaller pulley due to smaller angle of arc of contact.

2. *Cross belt* The angle of arc of contact on either pulley,

$$\theta = \pi + 2\alpha$$

$$= \pi + 2\sin^{-1}\left[\frac{d_1+d_2}{2C}\right] \tag{6.6b}$$

6.2.7 Ratio of Belt Tensions

Let T_1 and T_2 be the belt tensions on the tight and slack sides respectively, as shown in Fig.6.4(a). Let θ be the angle of contact on the pulley. Consider a portion AB of the belt on angular arc $\delta\theta$. Let the tension change from T to $T+\delta T$ in going from A to B. Let R be the normal reaction on the pulley and μR the force of friction. The forces acting on the belt-pulley system are shown in Fig.6.4(b).

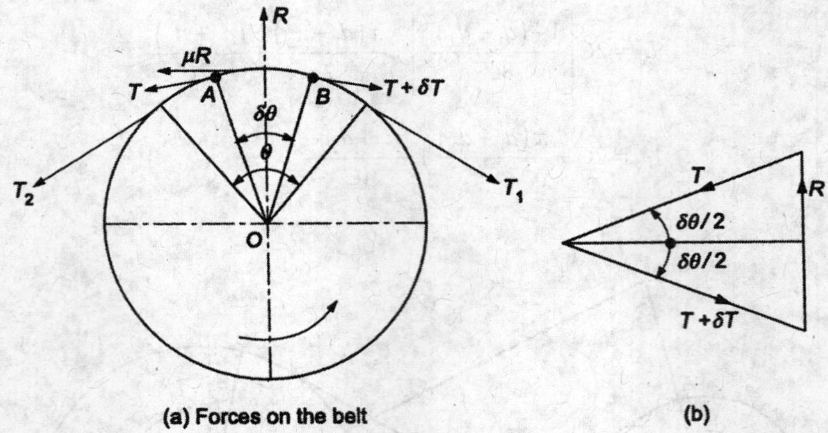

Fig.6.4 Belt tensions

Resolving the forces vertically, we have

$$R = T\sin\left(\frac{\delta\theta}{2}\right) + (T+\delta T)\sin\left(\frac{\delta\theta}{2}\right)$$

$$\approx T\frac{\delta\theta}{2} + (T+\delta T)\frac{\delta\theta}{2}$$

$$\approx T\delta\theta$$

Resolving the forces horizontally, we have

$$\mu \mathrm{R} = (T+\delta T)\cos\left(\frac{\delta\theta}{2}\right) - T\cos\left(\frac{\delta\theta}{2}\right) \approx \delta T$$

or $$\mu T\delta\theta = \delta T$$

or $$\frac{\delta T}{T} = \mu \cdot \delta\theta$$

$$\int_{T_2}^{T_1} \frac{\mathrm{d}T}{T} = \int_0^{\theta} \mu \cdot \mathrm{d}\theta$$

$$\ln\left(\frac{T_1}{T_2}\right) = \mu\theta$$

or $$\frac{T_1}{T_2} = \exp(\mu\theta) \tag{6.7}$$

The ratio of belt tensions is given by Eq. (6.7).

6.2.8 Power Transmitted

Power transmitted, $$P = \frac{(T_1 - T_2)v}{1000} \text{ kW} \tag{6.8a}$$

$$= T_1 v \times \frac{[1 - \exp(-\mu\theta)]}{10^3} \text{kW} \tag{6.8b}$$

6.2.9 Centrifugal Tension

The centrifugal tension is introduced in the belt due to its mass. Let T_c be the centrifugal tension in the belt. Consider the length of the belt over an angular arc $\delta\theta$, as shown in Fig.6.5. The force acting on the belt due to belt tension,

$$= 2T_c \sin\left(\frac{\delta\theta}{2}\right)$$

$$\approx T_c\,\delta\theta$$

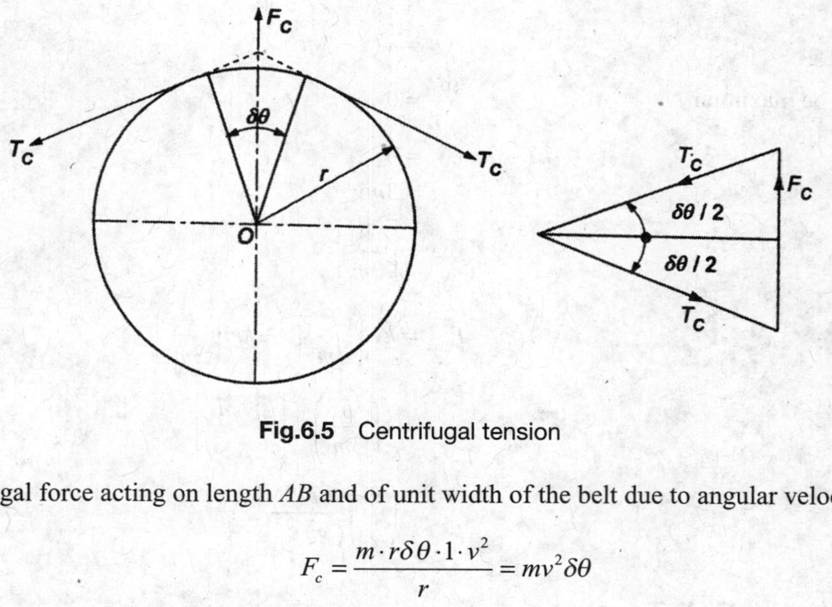

Fig.6.5 Centrifugal tension

Centrifugal force acting on length *AB* and of unit width of the belt due to angular velocity,

$$F_c = \frac{m \cdot r\delta\theta \cdot 1 \cdot v^2}{r} = mv^2\delta\theta$$

where m = mass of the belt per unit length

r = radius of the pulley

v = speed of belt

For equilibrium of the belt,

$$T_c\delta\theta = mv^2\,\delta\theta$$

$$T_c = mv^2 \tag{6.9}$$

Centrifugal stress, $$\sigma_c = \frac{T_c}{bt} = \frac{mv^2}{bt} \quad (6.10)$$

where b = width of the belt and
t = thickness of the belt

Effective tension on tight side = $T_1 + T_c$
Effective tension on slack side = $T_2 + T_c$

6.2.10 Condition for Maximum Power Transmission

Power transmitted,
$$P = (T_1 - T_2)\,v$$
$$= T_1[1 - \exp(-\mu\theta)]\,v$$

Let $1 - \exp(-\mu\theta) = k$, so that
$$P = kT_1 v$$

Now maximum belt tension,
$$T = T_1 + T_c$$
or
$$T_1 = T - T_c$$
$\therefore$
$$P = k(T - T_c)\,v$$
$$= kTv - kT_c v$$
$$= kTv - kmv^3$$

For P to be maximum,
$$\frac{dP}{dv} = 0$$
or
$$kT - 3kmv^2 = 0$$
or
$$T = 3mv^2$$
or
$$v = \left[\frac{T}{3m}\right]^{0.5} \quad (6.11)$$

$$P_{max} = kT\left(\frac{T}{3m}\right)^{0.5} - km\left(\frac{T}{3m}\right)^{1.5}$$
$$= k\left(\frac{T}{3m}\right)^{0.5}\left[T - \frac{T}{3}\right]$$
$$= \left(\frac{2}{3}\right)\frac{kT^{1.5}}{(3m)^{0.5}} \quad (6.12)$$

6.2.11 Initial Belt Tension

Let T_o = initial tension in belt
Resultant tension on tight side = $T_1 - T_o$
Resultant tension on slack side = $T_o - T_2$
Since the belt length remains constant, therefore
$$T_1 - T_o = T_o - T_2$$
or
$$T_o = \frac{T_1 + T_2}{2} \quad (6.13a)$$

Considering centrifugal tension,
$$T_o = \frac{T_1 + T_2 + 2T_c}{2} \quad (6.13b)$$

6.2.12 Effect of Initial Tension on Power Transmission

Ratio of tensions, $$\frac{T_1}{T_2} = \exp(\mu\theta) = c$$

Initial tensions, $$T_o = \frac{T_1 + T_2 + 2T_c}{2}$$

or $$T_1 + T_2 = 2(T_o - T_c)$$

$$T_1\left(1 + \frac{1}{c}\right) = 2(T_o - T_c)$$

$$T_1 = \left[\frac{2c}{1+c}\right](T_o - T_c)$$

and $$T_2 = \left[\frac{2}{1+c}\right](T_o - T_c)$$

$$T_1 - T_2 = \left[\frac{2(c-1)}{c+1}\right](T_o - T_c)$$

Power transmitted, $$P = (T_1 - T_2)\, v$$

$$= \left[\frac{2(c-1)}{c+1}\right](T_o - T_c)v$$

$$= \left[\frac{2(c-1)}{c+1}\right](T_o - mv^2)v$$

For P to be maximum, $$\frac{dP}{dv} = 0$$

$$\frac{2(c-1)}{(c+1)}(T_o - 3mv^2) = 0$$

$$T_o = 3mv^2$$

$$= 3T_c \qquad (6.14)$$

$$v = \left[\frac{T_o}{3m}\right]^{0.5} \qquad (6.15)$$

$$P_{max} = \left[\frac{2(c-1)}{c+1}\right]\left(\frac{T_o - T_o}{3}\right)\left[\frac{T_o}{3m}\right]^{0.5}$$

$$= \left(\frac{4}{3}\right)\left[\frac{c-1}{c+1}\right]\frac{T_o^{1.5}}{(3m)^{0.5}} \qquad (6.16)$$

While starting, $v = 0$ and $T_c = 0$. Hence

$$T_1 = \left[\frac{2c}{c+1}\right]T_o$$

While running, $$T_1 = \left[\frac{2c}{c+1}\right](T_o - T_c)$$

Hence maximum belt tension, $$T_1 = \left[\frac{2c}{c+1}\right]T_o \tag{6.17}$$

6.2.13 Belt Creep

The tension on the tight side is more than the tension on the slack side. As a result of this, the belt is stretched more on the tight side as compared to the slack side. Therefore, the driver pulley receives more length of the belt and delivers less. Hence, the belt creeps forward. The reverse occurs on the follower pulley. The follower pulley receives less length of the belt and delivers more. As a result of it, the belt creeps backward. This phenomenon is called *creeping of the belt.*

$$\text{Creep} = \frac{T_1 - T_2}{bt\,E} \tag{6.18}$$

where E is the modulus of elasticity of belt material.

The velocity ratio becomes,

$$\frac{n_2}{n_1} = \left(\frac{d_1}{d_2}\right)\left[\frac{E+\sqrt{\sigma_2}}{E+\sqrt{\sigma_1}}\right] \tag{6.19}$$

6.2.14 Crowning of Pulleys

The pulley face is given a convex curvature and is never kept flat. This is called crowning of the pulleys. This helps in running the belt in the centre of the pulley width. Crowning prevents any tendency of the belt to fall off the pulley face.

6.2.15 Cone Pulleys

Consider the cone pulley block shown in Fig.6.6. Let N be the speed of the driver block, and R_1, R_2, R_3, etc. the radii of its pulleys. Let the radii of the driven pulleys be r_1, r_2, r_3, etc. and speeds n_1, n_2, n_3, etc. The centre distance between the pulleys is C.

Now $$R_1 N = r_1 n_1$$

or $$\frac{N}{n_1} = \frac{r_1}{R_1}$$

Similarly $$\frac{N}{n_2} = \frac{r_2}{R_2}$$

$$\frac{N}{n_3} = \frac{r_3}{R_3} \quad \text{and so on.}$$

Let $$n_2 = kn_1$$

Therefore $$\frac{n_2}{N} = \frac{kn_1}{N} = k\left(\frac{R_1}{r_1}\right)$$

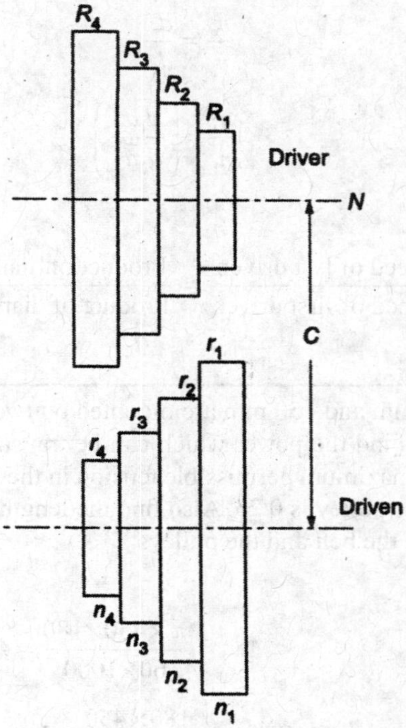

Fig.6.6 Cone pulley system

Similarly
$$\frac{n_3}{N} = k^2\left(\frac{R_1}{r_1}\right)$$

or in general,
$$\frac{n_i}{N} = k^{(i-1)}\left(\frac{R_1}{r_1}\right) = \frac{R_i}{r_i} \quad (6.20)$$

and
$$\frac{n_i}{n_1} = k^{(i-1)} \quad (6.21)$$

The length of the belt is same for all the cones, therefore
$$R_i + r_i = \text{constant}, \quad i = 1, 2, 3 \text{ etc.} \quad (6.22)$$

6.2.16 Compound Belt Drive

In compound belt drive, the driven pulley of the first set is mounted on the same shaft on which the driver of the second set is mounted. Let pulley 1 be the driver for the first set and pulley 2 its follower. The driver of the second set, pulley 3, is mounted on the same shaft on which pulley 2 is mounted. The follower of second set is pulley 4.

$$\frac{n_2}{n_1} = \frac{d_1}{d_2}$$

$$\frac{n_4}{n_3} = \frac{d_3}{d_4}$$

Hence
$$\frac{n_4}{n_1} = \left(\frac{d_1 d_3}{d_2 d_4}\right)$$

or in general,

$$\frac{\text{Speed of last driven}}{\text{Speed of first driver}} = \frac{\text{Product of diameters of drivers}}{\text{Product of diameters of drivens}} \tag{6.23}$$

Example 6.1

Two pulleys of diameters 450 mm and 150 mm are mounted on two parallel shafts 2 m apart and are connected by a flat belt drive. Find the power which can be transmitted by the belt when the larger pulley rotates at 180 rpm. The maximum permissible tension in the belt is 1 kN, and the coefficient of friction between the belt and the pulley is 0.25. Also find the length of the cross belt required and the angle of arc of contact between the belt and the pulleys.

■ **Solution**

$$v_2 = \frac{\pi \times 450 \times 180}{60 \times 1000} = 4.24 \text{ m/s}$$

$$n_1 = \frac{180 \times 450}{150} = 540 \text{ rpm}$$

Now
$$\sin\alpha = \frac{d_1 + d_2}{2C} = \frac{150 + 450}{4000} = 0.15$$

$$\alpha = 8.627°$$

Angle of contact,
$$\theta = \pi + 2\alpha = 197.254° \text{ or } 3.443 \text{ rad}$$

$$\frac{T_1}{T_2} = \exp(\mu\theta)$$

$$= \exp(0.25 \times 3.443) = 2.365$$

$$T_2 = \frac{T_1}{2.365} = \frac{1000}{2.365} = 422.83 \text{ N}$$

Power transmitted,
$$P = (T_1 - T_2)\frac{v_2}{1000}$$

$$= (1000 - 422.83) \times \frac{4.24}{1000} = 2.447 \text{ kW}$$

Length of cross belt,
$$L_c = 2C + \frac{\pi(d_1 + d_2)}{2} + \frac{(d_1 + d_2)^2}{4C}$$

$$= 4 + \frac{\pi(0.450 + 0.150)}{2} + \frac{(0.450 + 0.150)^2}{8}$$

$$= 4.987 \text{ m}$$

Example 6.2

A shaft running at 200 rpm drives another shaft at 400 rpm, and transmits 7.5 kW through an open belt. The belt is 80 mm wide and 10 mm thick. The centre distance is 4 m. The smaller pulley is of 500 mm diameter, and the coefficient of friction between the belt and pulley is 0.30. Calculate the stress in the belt.

■ Solution

$$d_2 = \frac{n_1 d_1}{n_2} = 400 \times \frac{500}{200} = 1000 \text{ mm}$$

$$v_1 = \frac{\pi \times 500 \times 400}{60 \times 1000} = 10.472 \text{ m/s}$$

$$\sin\alpha = \frac{d_2 - d_1}{2C} = \frac{1000 - 500}{8000} = 0.0625$$

$$\alpha = 3.583°$$

Angle of arc of contact, $\theta = 180° - 2\alpha = 172.83° = 3.0165 \text{ rad}$

$$\frac{T_1}{T_2} = \exp(\mu\theta) = \exp(0.3 \times 3.0165) = 2.472 \qquad (1)$$

Power transmitted, $P = (T_1 - T_2)\dfrac{v}{1000} \text{ kW}$

$$(T_1 - T_2) = \frac{7.5 \times 1000}{10.472} = 716.2 \text{ N} \qquad (2)$$

From (1) and (2), we get

$$T_1 = 1202.7 \text{ N} \quad \text{and} \quad T_2 = 486.5 \text{ N}$$

Maximum stress in the belt $= \dfrac{T_1}{bt}$

$$= \frac{1202.7}{80 \times 10} = 1.503 \text{ N/mm}^2$$

Example 6.3

A leather belt is required to transmit 8 kW from a pulley 1.5 m diameter running at 240 rpm. The angle of contact is 160° and the coefficient of friction between belt and pulley is 0.25. The safe working stress for leather is 1.5 MPa and density of leather is 1000 kg/m³. Determine the width of the belt if its thickness is 10 mm. Take into account the effect of centrifugal tension.

■ Solution

Velocity of the belt, $v = \dfrac{\pi \times 1.5 \times 240}{60} = 18.85 \text{ m/s}$

Power transmitted, $P = (T_1 - T_2)\dfrac{v}{1000}$

$$T_1 - T_2 = 8\times\frac{1000}{18.85} = 424.4 \text{ N}$$

$$\frac{T_1}{T_2} = \exp(\mu\theta)$$

$$= \exp\left(0.25\times 160\times\frac{\pi}{180}\right) = 2.01$$

$$T_1 = 844.6 \text{ N and } T_2 = 420.2 \text{ N}$$

Mass of the belt per metre length, $m = b \times 0.01 \times 1 \times 1000 = 10b\ kg$

Centrifugal tension, $T_c = mv^2 = 10b(18.85)^2 = 3553.2 \times b$ N

Maximum tension in the belt, $T = \sigma \cdot bt$

$$= 1.5 \times 10^6 \times b \times 0.01 = 15000 \times b \text{ N}$$

$$T = T_1 + T_c$$

or $15000b = 844.6 + 3553.2b$

$$b = 0.0738 \text{ m or } 73.8 \text{ mm}$$

Example 6.4

A pulley is driven by a flat belt 100 mm wide and 6 mm thick. The density of belt material is 1000 kg/m³. The angle of lap is 120° and the coefficient of friction 0.3. The maximum stress in the belt does not exceed 2 MPa. Find the maximum power that can be transmitted and the corresponding speed of the belt.

■ Solution

Maximum tension in the belt, $T = 2 \times 10^6 \times 0.1 \times 0.006 = 1200$ N

Mass of the belt per metre length, $m = 0.1 \times 0.006 \times 1 \times 1000 = 0.6$ kg/m

Speed of the belt for maximum power, $v = \left[\frac{T}{3m}\right]^{0.5}$

$$= \left[\frac{1200}{3\times 0.6}\right]^{0.5}$$

$$= 25.82 \text{ m/s}$$

For maximum power to be transmitted, the centrifugal tension,

$$T_c = \frac{T}{3} = \frac{1200}{3} = 400 \text{ N}$$

$$\frac{T_1}{T_2} = \exp(0.3\times 120\times \pi/180) = 1.874$$

$$T_1 = T - T_c = 1200 - 400 = 800 \text{ N}$$

$$T_2 = \frac{T_1}{1.874} = \frac{800}{1.874} = 426.8 \text{ N}$$

$$\text{Maximum power transmitted} = (T_1 - T_2)\frac{v}{1000}$$

$$= (800 - 426.8)\times\frac{25.82}{1000}$$

$$= 9.636 \text{ kW}$$

Example 6.5

An open belt drive is used to connect two parallel shafts 4 m apart. The diameter of bigger pulley is 1.5 m and that of the smaller pulley 0.5 m. The mass of the belt is 1 kg/m length. The maximum tension is not to exceed 1500 N. The coefficient of friction is 0.25. The bigger pulley, which is the driver, runs at 250 rpm. Due to slip, the speed of the driven pulley is 725 rpm. Calculate the power transmitted, power lost in friction, and the efficiency of the drive.

Solution

$$v_1 = \pi \times 1.5 \times \frac{250}{60} = 19.635 \text{ m/s}$$

$$v_2 = \pi \times 0.5 \times \frac{725}{60} = 18.98 \text{ m/s}$$

$$T_c = mv_1^2 = 1 \times (19.635)^2 = 385.53 \text{ N}$$

$$T_1 = T - T_c = 1500 - 385.53 = 1114.47 \text{ N}$$

$$\sin\alpha = \frac{d_2 - d_1}{2C} = \frac{1.5 - 0.5}{8} = 0.125$$

$$\alpha = 7.18°$$

$$\theta = 180° - 2\alpha = 165.64° = 2.89 \text{ rad}$$

$$\frac{T_1}{T_2} = \exp(\mu\theta) = \exp(0.25 \times 2.89) = 2.06$$

$$T_2 = \frac{T_1}{2.06} = \frac{1114.47}{2.06} = 541.0 \text{ N}$$

$$T_1 - T_2 = 1114.47 - 541.00 = 573.47 \text{ N}$$

Torque on bigger pulley, $M_1 = 573.47 \times 0.75 = 430.1$ Nm

Torque on smaller pulley, $M_2 = 573.47 \times 0.25 = 143.37$ Nm

Power transmitted, $P = (T_1 - T_2)\dfrac{v_1}{1000} = 573.47 \times \dfrac{19.635}{1000} = 11.26$ kW

Input power, $P_1 = M_1\omega_1 = 430.1 \times \dfrac{\left(2\pi \times \dfrac{250}{60}\right)}{1000} = 11.26$ kW

Output power, $P_2 = M_2\omega_2 = 143.37 \times \dfrac{\left(2\pi \times \dfrac{725}{60}\right)}{1000} = 10.885$ kW

Power lost in friction, $P_f = 11.26 - 10.885 = 0.375$ kW

Efficiency of the drive $= \dfrac{10.885}{11.26} = 96.67\%$

Example 6.6

Two parallel shafts 5 m apart are connected by open flat belt drive. The diameter of the bigger pulley is 1.5 m and that of the smaller pulley 0.75 m. The initial tension in the belt is 2.5 kN. The mass of the belt is 1.25 kg/m length and coefficient of friction is 0.25. Taking centrifugal tension into account, find the power transmitted, when the smaller pulley rotates at 450 rpm.

■ **Solution**

$$v = \frac{\pi \times 0.75 \times 450}{60} = 17.67 \text{ m/s}$$

$$T_c = mv_2^2 = 1.25 \times (17.67)^2 = 390.28 \text{ N}$$

$$T_o = \frac{T_1 + T_2 + 2T_c}{2}$$

$$2500 = \frac{T_1 + T_2 + 2 \times 390.28}{2}$$

$$T_1 + T_2 = 4219.44 \text{ N}$$

$$\alpha = \sin^{-1}\left[\frac{1.5 - 0.75}{10}\right] = 4.3°$$

$$\theta = 180° - 2\alpha = 171.4° = 2.991 \text{ rad}$$

$$T_1/T_2 = \exp(\mu\theta) = \exp(0.25 \times 2.991) = 2.1125$$

$$T_2 = 1355.65 \text{ N and } T_1 = 2863.81 \text{ N}$$

Power transmitted,
$$P = (T_1 - T_2)\frac{v}{1000}$$

$$= (2863.81 - 1355.65) \times \frac{17.67}{1000} = 26.65 \text{ kW}$$

6.3 V-BELT DRIVE

The V-belt drive is more positive than the flat belt drive. It is a short centre drive and is preferred for power transmission from the prime mover. The belt touches the sides of the grooved pulley only. The V-belts are classified as *A*, *B*, *C*, *D*, *E*.

6.3.1 Ratio of Belt Tensions

A V-belt in a grooved pulley is shown in Fig.6.7.

Let 2β = pulley groove angle;

R = total reaction on the pulley;

R_n = normal reaction between the belt and the sides of the groove and

μ = coefficient of friction between the belt and the groove sides.

Then $R = 2R_n \sin\beta$

or
$$R_n = \frac{R}{2\sin\beta}$$

Force of friction, $F = 2\mu R_n = \left(\dfrac{\mu}{\sin\beta}\right) R = \mu_e R$

where μ_e is called the virtual, apparent or equivalent coefficient of friction.

Ratio of tensions, $\dfrac{T_1}{T_2} = \exp(\mu_e \theta)$ (6.24)

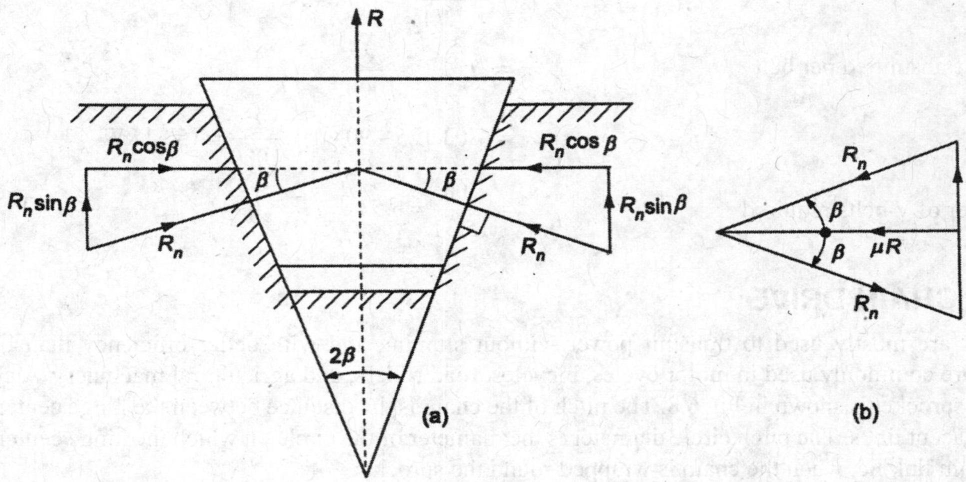

Fig.6.7 Forces on V-belt

Example 6.7

A compressor requires 100 kW to run at 240 rpm from an electric motor of speed 750 rpm, by means of a V-belt drive. The diameter of the compressor shaft pulley should not be more than 1 m while the centre distance between the shafts is 2 m. The belt speed should not exceed 25 m/s.

Determine the number of V-belts required to transmit the power if each belt has a cross-sectional area of 375 mm², density 1000 kg/m³, and an allowable tensile stress of 2.5 MPa. The pulley groove angle is 40 and coefficient of friction between the belt and the pulley sides is 0.25.

Solution

Diameter of motor pulley, $d_1 = \dfrac{240 \times 1}{750} = 0.32$ m

Mass of belt per metre length, $m = 375 \times 10^{-6} \times 1 \times 1000 = 0.375$ kg/m

Velocity of belt, $v = 25$ m/s

Centrifugal tension, $T_c = mv^2 = 0.375 \times 625 = 234.375$ N

Maximum tension in the belt, $T = \sigma A = 2.5 \times 10^6 \times 375 \times 10^{-6}$

$= 937.5$ N

Tight side tension, $T_1 = T - T_c = 937.5 - 234.375 = 703.125$ N

$$\sin\alpha = \frac{d_2 - d_1}{2C} = \frac{1 - 0.32}{4} = 0.17$$

$$\alpha = 9.78°$$

$$\theta = 180° - 2\alpha = 160.44° = 2.8 \text{ rad}$$

$$T_1/T_2 = \exp(\mu_e\,\theta)$$

$$= \exp\left[\left(\frac{0.25}{\sin 20°}\right)\times 2.8\right]$$

$$= \exp(2.0466) = 7.74$$

$$T_2 = \frac{703.125}{7.74} = 90.84\text{ N}$$

Power transmitted per belt $= (T_1 - T_2)\dfrac{v}{1000}$

$$= (703.125 - 90.84)\times\frac{25}{1000} = 15.31\text{ kW}$$

Number of V-belts required $= \dfrac{100}{15.31} = 6.53 \approx 7$

6.4 CHAIN DRIVE

Chains are mostly used to transmit power without slipping and with better efficiency than belts. They are commonly used in motor cycles, bicycles, road rollers, and agricultural machinery. A chain on the sprocket is shown in Fig.6.8. The pitch of the chain is the distance between the hinge centers of the adjacent links. The pitch circle diameter is the diameter of the circle on which the hinge centers of the chain link lie, when the chain is wrapped round the sprocket.

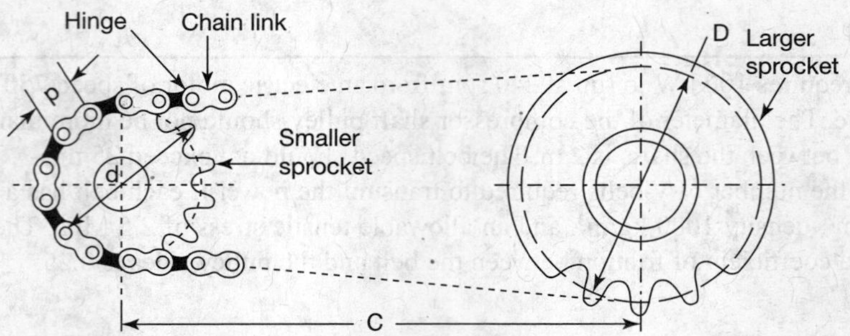

Fig.6.8 Chain drive

6.4.1 Chain Pitch

As shown in Fig.6.9, the chain pitch is,

$$p = \frac{2\times D}{2}\times\sin\left(\frac{\theta}{2}\right) = D\sin\left(\frac{\theta}{2}\right) \tag{6.25a}$$

where $\theta = \dfrac{360°}{z}$ and z = Number of teeth

Hence,

$$p = D\sin\left(\frac{180°}{z}\right) \tag{6.25b}$$

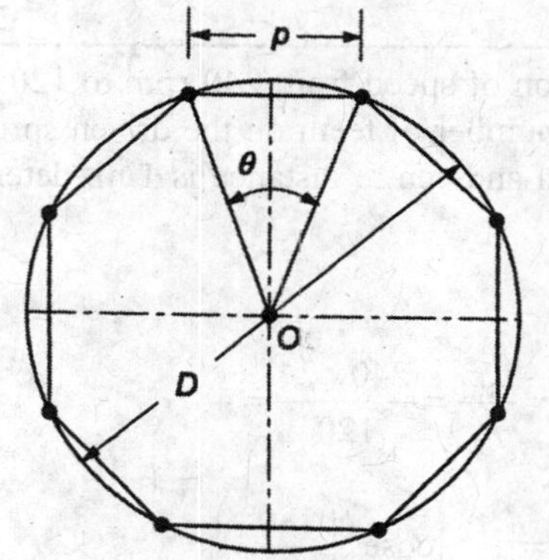

Fig.6.9 Chain pitch

6.4.2 Chain Length

As shown in Fig.6.10, the chain length

$$L = 2C + \pi(R_1 + R_2) + \frac{(R_1 - R_2)^2}{C} \tag{6.26a}$$

$$\pi(R_1 + R_2) = \frac{p(z_1 + z_2)}{2}$$

$$R_1 = \left(\frac{p}{2}\right)\operatorname{cosec}\left(\frac{180°}{z_1}\right)$$

$$R_2 = \left(\frac{p}{2}\right)\operatorname{cosec}\left(\frac{180°}{z_2}\right)$$

$$L = 2C + \frac{p(z_1 + z_2)}{2} + \frac{p^2\left[\operatorname{cosec}\left(\frac{180°}{z_1}\right) - \operatorname{cosec}\left(\frac{180°}{z_2}\right)\right]^2}{4C} \tag{6.26b}$$

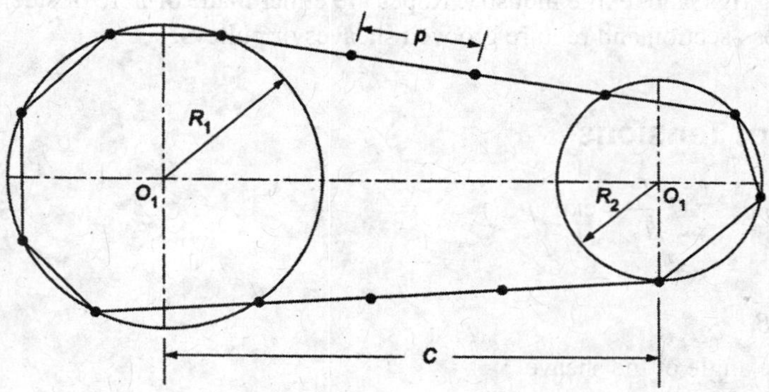

Fig.6.10 Chain length

Example 6.8

A chain drive is used for reduction of speed from 240 rpm to 120 rpm. The number of teeth on the driving sprocket is 24. Find the number of teeth on the driven sprocket. If the pitch circle diameter of the driven sprocket is 600 mm and centre distance is 1 m, determine the pitch and length of the chain.

Solution

$$z_2 = \frac{n_1 z_1}{n_2} = \frac{240 \times 24}{120} = 48$$

$$R_2 = \left(\frac{p}{2}\right) \text{cosec}\left(\frac{180°}{z_2}\right)$$

$$0.3 = \left(\frac{p}{2}\right) \text{cosec}\left(\frac{180°}{48}\right)$$

$$p = 0.0392 \text{ m or } 39.2 \text{ mm}$$

$$L = 2C + \frac{p(z_1 + z_2)}{2} + \left(\frac{p^2}{4C}\right)\left[\text{cosec}\left(\frac{180°}{z_1}\right) - \text{cosec}\left(\frac{180°}{z_2}\right)\right]^2$$

$$= 2 \times 1 + \left(\frac{0.0392}{2}\right)(24 + 48)$$

$$+ \left(\frac{0.03922}{4}\right)\left[\text{cosec}\left(\frac{180°}{24}\right)\text{cosec}\left(\frac{180°}{48}\right)\right]^2$$

$$= 2 + 2.8224 + 0.0223 = 4.8447 \text{ m}$$

6.5 ROPE DRIVE

Ropes are used for power transmission over long distances. They are commonly used in hoisting equipment, drilling rigs, and textile industry. Ropes are either made of fibre or steel. Ropes are generally of circular cross-section and require grooved sheaves or pulleys.

6.5.1 Ratio of Tensions

The ratio of tensions, $\frac{T_1}{T_2} = \exp(\mu_e \theta)$ (6.27)

where $\mu_e = \frac{\mu}{\sin \beta}$

β = semi-groove angle of the sheave
θ = angle of contact.

Example 6.9

A pulley of groove angle 45°, diameter 4 m and having 15 grooves is used to transmit power. The angle of contact is 170° and the coefficient of friction between the ropes and the groove sides is 0.30. The maximum possible tension in the ropes is 1 kN and the mass of the rope is 1.5 kg/m length. Determine the speed of the pulley for maximum power conditions.

■ Solution

For maxim power to be transmitted, $v = \left[\dfrac{T}{3m}\right]^{0.5} = \left[\dfrac{1000}{3\times1.5}\right]^{0.5} = 14.91\text{ m/s}$

Speed of the pulley, $n = \dfrac{60\times14.91}{4\pi} = 71.2\text{ rpm}$

For maximum power to be transmitted, $T_c = \dfrac{T}{3} = \dfrac{1000}{3} = 333.3\text{ N}$

Tension on the side, $T_1 = T - T_c = 1000 - 333.33 = 666.67\text{ N}$

$$\frac{T_1}{T_2} = \exp(\mu_e\theta)$$

$$= \exp\left[0.3\times\text{cosec}\,(22.5°)\times170\times\frac{\pi}{180}\right]$$

$$= \exp(2.326) = 10.23$$

$$T_2 = \frac{T_1}{10.23} = \frac{666.67}{10.23} = 65.12\text{ N}$$

Power transmitted per rope $= (T_1 - T_2)\dfrac{v}{1000}$

$$= (666.67 - 65.12)\times\frac{14.91}{1000} = 8.97\text{ kW}$$

Total power transmitted, $P = 8.97\times15 = 134.5\text{ kW}$

Example 6.10

A prime mover running at 400 rpm drives a generator at 600 rpm by a belt drive. Diameter of prime mover pulley 600 mm. Assuming a slip of 2%, determine the diameter of the generator pulley, if the belt thickness is 6 mm.

■ Solution

Given: $n_1 = 400$ rpm, $n_2 = 600$ rpm, $d_1 = 600$ mm, $s = 2\%$, $t = 6$ mm.

$$n_2/n_1 = [(d_1 + t)/(d_2 + t)]\,(1 - s/100)$$

$$600/400 = [(600 + 6)/(d_2 + 6)]\,(1 - 0.02)$$

$$d_2 = (606\times400/600)\,(0.98) - 6 = 389.92\text{ mm}$$

Example 6.11

A pulley 200 mm in diameter is mounted on a motor shaft running at 900 rpm and drives a main shaft at 300 rpm by means of a flat belt. The belt is 6 mm thick and 20 mm wide. Determine the size of the pulley on the main shaft if there is a slip of 3 % on each pulley.

■ **Solution**

Given: $d_1 = 200$ mm, $n_1 = 900$ rpm, $n_2 = 300$ rpm, $t = 6$ mm, $b = 20$ mm, $s_1 = s_2 = 3\%$
Now linear speed of belt, $v_1 = v_2$
The slip will occur on smaller pulley only.

$$\pi (d_1 + t)\, n_1 (1 - s_1/100)(1 - s_2/100) = \pi (d_2 + t)\, n_2$$

$$d_2 = (206 \times 900 \times 0.97 \times 0.97/300) - 6 = 575.5 \text{ mm}$$

Example 6.12

Two parallel shaft at 6 m centre distance are connected by a cross belt have pulley diameters of 300 mm and 600 mm, respectively. Calculate the length of the belt required. If the same belt is to be used for the open belt drive, what is the remedy?

■ **Solution**

Given: $d_1 = 300$ mm, $d_2 = 600$ mm, $C = 6$ m

Length of cross belt, $L_c = 2C \cos\alpha + (d_1 + d_2)(\pi/2 + \alpha)$

$$\sin\alpha = (d_1 + d_2)/(2C) = 900/12000 = 0.075$$

$$\alpha = 4.3° = 0.075 \text{ rad}$$

$$L_c = 2 \times 600° \times \cos 4.3 + 900 \times (\pi/2 + 0.075) = 13447 \text{ mm}$$

Length of open belt, $L_o = 2C + \pi (d_1 + d_2)/2 + (d_2 - d_1)^2/(4C)$

$$= 12000 + 450 \times \pi + 3600/24000 = 13413.87 \text{ mm}$$

The remedy is that the length of the belt be shortened by $(L_c - L_o) = 13447 - 13413.87 = 33.13$ mm

Example 6.13

A shaft running at 120 rpm is to drive another shaft at 240 rpm and transmits 10 kW. The belt is 120 mm wide and 10 mm thick. The coefficient of friction between belt and pulley is 0.25. The distance between the shafts is 3 m and the smaller pulley is of 600 mm diameter. Calculate the stress in the belt, assuming the drive to be of the open belt type.

■ **Solution**

Given: $n_1 = 120$ rpm, $n_2 = 240$ rpm, $d_1 = 600$ mm, $b = 120$ mm, $t = 10$ mm, $\mu = 0.25$, $P = 10$ kW, $C = 3$ m

$$d_2 = d_1 n_1/n_2 = 600 \times 120/240 = 300 \text{ mm}$$

$$v_m = \pi (d_1 + t)\, n_1/60 = \pi \times 610 \times 10^{-3} \times 120/60 = 3.833 \text{ m/s}$$

Power transmitted, $P = (T_1 - T_2)\, v_m/10^3$

$$10 = (T_1 - T_2) \times 3.833/10^3$$

$$T_1 - T_2 = 10 \times 10^3/3.833 = 2609.1 \text{ N} \quad (1)$$

For open belt drive, $\cos(\theta/2) = (d_1 - d_2)/(2C) = (600 - 300)/6000 = 0.05$

$$\theta/2 = 87.13°$$

$$\theta = 174.27° = 3.0416 \text{ rad}$$

Now $T_1/T_2 = \exp(\mu\theta) = \exp(0.25 \times 3.0416) = 2.1391 \quad (2)$

Substituting in Eq. (1), we get

$$T_2 (2.1391 - 1) = 2609.1$$

$$T_2 = 1875.7 \text{ N}$$

$$T_1 = 4012.3 \text{ N}$$

Maximum stress in the belt, $\sigma_{max} = T_1/(bt) = 4012.3/(120 \times 10) = 3.343 \text{ N/mm}^2$

Example 6.14

A pulley of 600 mm diameter mounted on the driving shaft rotates at 120 rpm. A countershaft is to be driven at 300 rpm by an open belt drive, the centre distance being 2.5 m. The belt is to transmit 3 kW. The coefficient of friction between the belt and the pulley is 0.3. Determine the width of the belt if safe tension in the belt material in tension is 15 N/mm.

■ Solution

Given: $d_1 = 600$ mm, $n_1 = 120$ rpm, $n_2 = 300$ rpm, $C=2.5$ m, $P=2.5$ kW, $\mu=0.3$, $\sigma_w = 15$ N/mm

$$d_2 = d_1 n_1/n_2 = 600 \times 120/300 = 240 \text{ mm}$$

$$\sin \alpha = (d_1 - d_2)/(2C) = (600 - 240)/5000 = 0.072$$

$$\alpha = 4.12887° = 0.072 \text{ rad}$$

Angle of contact, $\theta = \pi - 2\alpha = 2.99746$ rad

$$T_1/T_2 = \exp(\mu\theta) = \exp(0.3 \times 2.99746) = 2.45774$$

$$P = (T_1 - T_2)\, v/10^3$$

$$3 = (T_1 - T_2)\, \pi \times 600 \times 120/(60 \times 10^6)$$

$$T_1 - T_2 = 795.78 \text{ N}$$

$$T_2\,(2.45774 - 1) = 795.78$$

$$T_2 = 545.896 \text{ N}$$

$$T_1 = 1341.67 \text{ N}$$

Width of belt, $b = T_1/\sigma_w = 1341.67/15 = 89.45$ mm

Example 6.15

In a belt drive, the ratio of belt tensions is 2.2, when the effect of centrifugal tension is not considered. Linear velocity of belt is 15 m/s and the safe stress for the belt material is 1.5 MPa. Determine the width of the belt to transmit 9 kW. Density of belt material is 1000 kg/m³ and the thickness of belt is 10 mm.

■ Solution

Power, $P = (T_1 - T_2)\, v/10^3$ kW

$$T_1 - T_2 = 9 \times 10^3/15 = 600 \text{ N}$$

$$T_1/T_2 = 2.2$$

$$T_2 = 600/1.2 = 500 \text{ N}$$

$$T_1 = 1100 \text{ N}$$

Mass of the belt per metre length, $m = \rho bt = 1000 \times b \times 10 \times 10^{-6} = 10^{-2}\,b$ kg/m length

Centrifugal tension, $T_c = mv^2 = 10^{-2}\, b \times 225 = 2.25\,b$ N

Maximum tension, $T_{max} = T_1 + T_c = 1100 + 2.25\,b$ N

Maximum stress in belt $= T_{max}/bt$

$$= (1100 + 2.25\,b)/(10\,b) = 1.5$$

or $1100 + 2.25\,b = 15\,b$

$$b = 1100/12.75 = 86.27 \text{ mm}$$

Example 6.16

A leather belting of mass 1000 kg/m³ has a maximum permissible tension of 2.1 MPa. Determine the maximum power that can be transmitted by the belt 200 mm × 10 mm, if the ratio of belt tensions is 2.

Solution

Mass of belt per metre length, $m = bt\rho = 200 \times 10 \times 10^{-6} \times 1000 = 2$ kg/m length

Maximum permissible tension in belt, $T_{max} = bt\sigma_w = 200 \times 10 \times 2.1 = 4200$ N

For maximum power to be transmitted, $T_c = T_{max}/3 = 4200/3 = 1400$ N

$$T_c = mv^2 = 1400$$
$$v = (1400/2)^{0.5} = 26.45 \text{ m/s}$$
$$T_1 = 2T_{max}/3 = 2 \times 4200/3 = 2800 \text{ N}$$
$$T_2 = T_1/2 = 1400 \text{ N}$$
$$P_{max} = (T_1 - T_2)v/10^3 = (2800 - 1400) \times 26.45/10^3 = 37 \text{ kW}$$

Example 6.17

An open belt drive transmits power from a 300 mm diameter pulley running at 240 rpm to a pulley 450 mm diameter. Angle of lap on smaller pulley is 160°. The belt is on the point of slipping when 3 kW is being transmitted. The coefficient of friction between belt and pulley is 0.3.

It is desired to increase the power transmitted. State which of the following methods suggested would be more effective ?

(a) Initial tension in the belt is increased by 10%.

(b) Suitable dressing is given to the belt surface to increase the coefficient of friction by 10%. Assume that initial tension is kept the same.

Solution

Linear velocity of belt, $v = \pi d_1 n_1/(60 \times 10^3) = \pi \times 300 \times 240/(60 \times 10^3) = 3.77$ m/s

Power transmitted, $P = (T_1 - T_2)\, v/10^3$

$$3 = (T_1 - T_2) \times 3.77/10^3$$
$$T_1 - T_2 = 3 \times 10^3/3.77 = 795.76 \text{ N}$$
$$T_1/T_2 = \exp(\mu\theta) = \exp(0.3 \times 160 \times \pi/180) = 2.3112$$
$$T_2(2.3112 - 1) = 795.76$$
$$T_2 = 606.89 \text{ N}$$
$$T_1 = 1402.65 \text{ N}$$

(a) Initial tension, $T_i = (T_1 + T_2)/2 = (1402.65 + 606.89)/2 = 1004.77$ N

When the initial tension is increased by 10%,

$$T_1 + T_2 = 2 \times 1.1 \times 1004.77 = 2210.494 \text{ N}$$
$$T_2\,(2.3112 + 1) = 2210.494$$
$$T_2 = 667.58 \text{ N}$$
$$T_1 = 1542.91 \text{ N}$$
$$P = (1542.91 - 667.58) \times 3.77/10^3 = 3.30 \text{ kW}$$

(b) When coefficient of friction is increased by 10%, then $\mu = 1.1 \times 0.3 = 0.33$

$$T_1/T_2 = \exp(\mu\theta) = \exp(0.33 \times 160 \times \pi/180) = 2.51314$$

When the initial tension is kept constant.

$$T_1 + T_2 = 2 \times 1004.77 = 2009.54 \text{ N}$$
$$T_2\,(2.51314 + 1) = 2009.54.494$$
$$T_2 = 572 \text{ N}$$
$$T_1 = 1437.5 \text{ N}$$
$$P = (1437.5 - 572) \times 3.77/10^3 = 3.26 \text{ kW}$$

Example 6.18

A belt 200 mm wide and 8 mm thick embraces the smaller pulley by 160°. The density of belt material is 1000 kg/m³. Determine the maximum power that can be transmitted by the belt if maximum permissible stress in belt is 2 MPa and coefficient of friction is 0.25.

Solution

Maximum allowable tension in belt, $T_{max} = bt\sigma_w = 200 \times 8 \times 2 = 3200$ N

Mass of belt per metre length, $m = bt\rho = 200 \times 8 \times 10^{-6} \times 1000 = 1.6$ kg/m

Speed for maximum power tranmission, $v = [T/(3m)]^{0.5} = [3200/(3 \times 1.6)]^{0.5} = 25.82$ m/s

Centrifugal tension, $T_c = mv^2 = 1.6 \times (25.82)^2 = 1066.67$ N

$$T_1/T_2 = \exp\,(0.25 \times \pi \times 160/180) = 2.01$$

$$T_{max} = T_1 + T_c,\ T_1 = 3200 - 1066.67 = 2133.33 \text{ N}$$

$$T_2 = 2133.33/2.01 = 1061.36 \text{ N}$$

Power transmitted, $P = (T_1 - T_2)\, v/10^3 = (2133.33 - 1061.36) \times 25.82/10^3 = 27.678$ kW

Example 6.19

An open belt 100 mm × 10 mm connects two pulleys 1500 mm and 600 mm diameters on parallel shafts 4 m apart. The material density of the belt is 985 kg/m³. The maximum tension is not to exceed 2 kN and the coefficient of friction is 0.3. The larger driver pulley runs at 240 rpm. Due to slip the larger pulley runs at 590 rpm. Determine (a) the torque exerted on each of the shafts, (b) the power transmitted, (c) power lost in friction, and (d) the efficiency of the drive.

Solution

Linear speed of belt, $v = \pi d_1 n_1/(10^3 \times 60) = \pi \times 1500 \times 240/(10^3 \times 60) = 18.85$ m/s

Mass of belt per metre length, $m = bt\rho = 100 \times 10 \times 10^{-6} \times 985 = 0.985$ kg/m

Centrifugal tension, $T_c = mv^2 = 0.985 \times (18.85)^2 = 350$ N

$$T_1 = T_{max} - T_c = 2000 - 350 = 1650 \text{ N}$$

Angle of contact on smaller pulley, $\theta = 180° - 2\sin^{-1}[(d_1 - d_2)/2C]$

$$= 180° - 2\sin^{-1}[(1500 - 600)/8000]$$

$$= 167.1° \text{ or } 2.916 \text{ rad}$$

$$T_1/T_2 = \exp\,(\mu\theta) = \exp\,(0.3 \times 2.916) = 2.3984$$

$$T_2 = 1650/2.3984 = 687.96 \text{ N}$$

(a) Torque on larger pulley shaft $= (T_1 - T_2)\, d_1/2 = (1650 - 687.96) \times 0.750 = 721.53$ Nm

Torque on smaller pulley shaft $= (T_1 - T_2)\, d_2/2 = (1650 - 687.96) \times 0.3 = 288.612$ Nm

(b) Power transmitted by larger pulley, $P_1 = (T_1 - T_2)\, v_1/10^3 = (1650 - 687.96) \times \pi \times 1.5 \times 240/(10^3 \times 60) = 18.134$ kW

Power transmitted by smaller pulley, $P_2 = (T_1 - T_2)\, v_2/10^3 = (1650 - 687.96) \times \pi \times 0.6 \times 590/(10^3 \times 60) = 17.832$ kW

(c) Power lost in friction, $P_f = P_1 - P_2 = 18.134 - 17.832 = 0.302$ kW

(d) Efficiency of drive $= P_2/P_1 = 17.832 \times 100/18.134 = 98.33\%$

Example 6.20

An open belt drive connects two parallel shafts 4 m apart. The diameter of larger pulley is 1.5 m and that of the smaller pulley is 1 m. The initial tension in the belt before starting is 2 kN. The belt is 120 mm wide and 15 mm thick. Its density is 1100 kg/m^3 and the coefficient of friction is 0.34. Taking centrifugal tension into account, determine the maximum power transmitted.

■ Solution

Mass of belt per metre length, $m = bt\rho = 120 \times 15 \times 10^{-6} \times 1100 = 1.98$ kg/m

For maximum power transmission, $v = [T_i/(3m)]^{0.5} = [2000/(3 \times 1.98)]^{0.5} = 18.35$ m/s

Centrifugal tension, $T_c = mv^2 = 1.98 \times (18.35)^2 = 666.7$ N

Also $T_i = [T_1 + T_2 + 2T_c]/2$

$$T_1 + T_2 = 2 \times 2000 - 2 \times 666.7 = 2666.6 \text{ N}$$

Angle of contact, $\theta = 180° - 2 \sin^{-1} [(d_1 - d_2)/2C]$

$$= 180° - 2 \sin^{-1} [(1.5 - 1)/8] = 172.83° \text{ or } 3.0165 \text{ rad}$$

$$T_1/T_2 = \exp (0.34 \times 3.0165) = 2.7888$$

$$T_2 (2.7888 + 1) = 2666.6$$

$$T_2 = 703.8 \text{ N}$$

$$T_1 = 1962.8 \text{ N}$$

Power transmitted, $P = (T_1 - T_2)\, v/10^3 = (1962.8 - 703.8) \times 18.35/10^3 = 23.1$ kW

Example 6.21

A V-belt of 6.5 cm^2 cross-section has a groove angle of 40° and angle of lap of 165°. The mass of the belt is 1.2 kg/m run. The maximum safe stress is 8 N/mm^2. Calculate the power that can be transmitted at 20 m/s speed, if coefficient of friction is 0.15.

■ Solution

Centrifugal tension, $T_c = mv^2 = 1.2 \times 20^2 = 480$ N

Maximum tension, $T_{max} = \sigma_w \times A = 8 \times 500 = 4000$ N

Apparent coefficient of friction, $\mu_e = \mu/\sin \beta = 0.15/\sin 20° = 0.4386$

$$T_1/T_2 = \exp (\mu_e \theta) = \exp (0.4386 \times 165 \times \pi/180) = 3.536$$

$$T_{max} = T_1 + T_c,\ T_1 = 4000 - 480 = 3520 \text{ N}$$

$$T_2 = 3520/3.536 = 995.5 \text{ N}$$

Power transmitted, $P = (T_1 - T_2)\, v/10^3 = (3520 - 995.5) \times 20/10^3 = 50.49$ kW

Example 6.22

A rope drive transmits 120 kW at 220 rpm by ropes 25 mm in diameter and 0.6 kg/m length mass. The maximum rope tension is 1360 N and it is designed for maximum power transmission conditions. The angle of contact is 160° and the coefficient of friction is 0.25. The sheave groove angle is 45°. Determine the diameter of the sheave and the number of ropes.

■ Solution

For maximum power transmission, $T_{max} = 3T_c$

$$1360 = 3mv^2$$

$$v^2 = 1360/(3 \times 0.6) = 755.56$$

$$v = 27.487 \text{ m/s}$$

Diameter of sheave $= 60\,v/\pi N = 60 \times 27.487/(\pi \times 220) = 2.386$ m

$$T_c = mv^2 = 0.6 \times (27.487)^2 = 453.32 \text{ N}$$
$$T_1 = 1360 - 453.32 = 906.68 \text{ N}$$
$$T_1/T_2 = \exp\,[(0.25/\sin 22.5°)\,(\pi \times 160/180)] = 6.1985$$
$$T_2 = 906.68/6.1985 = 146.27 \text{ N}$$

Power transmitted, $P = n\,(906.68 - 146.27) \times 27.487/10^3 = 120$

Number of ropes, $n = 5.74 \approx 6$

Example 6.23

A chain drive is used for reduction of speed from 300 to 150 rpm. The number of teeth on the driving sprocket is 25. The pitch circle diameter of the driven sprocket is 500 mm. Determine the number of teeth on the driven sprocket and the pitch.

■ **Solution**

(a) From $n_1 z_1 = n_2 z_2$, $z_2 = 300 \times 25/150 = 50$

(b) $d_2 = p \text{ cosec } (180°/z_2)$

$$500 = p \text{ cosec } (180°/50) = 15.926\,p$$
$$p = 31.4 \text{ mm}$$

Example 6.24

A flat belt is required to transmit 20 kW from a pulley 1.5 m diameter running at 300 rpm. The angle of contact between the belt and the pulley is 160° and the coefficient of friction is 0.25. The safe working stress for the belt material is 3 MPa. The thickness of belt is 6 mm and its density is 1100 kg/m³. Find the width of the belt required.

■ **Solution**

Given: $P = 20$ kW, $d = 1.5$ m, $n = 300$ rpm, $\theta = 160°$, $\mu = 0.25$, $\sigma = 3$ MPa, $t = 6$ mm, $\rho = 1100$ kg/m³, $b = ?$

$$v = \pi d\, n/60 = \pi \times 1.5 \times 300/60 = 23.562 \text{ m/s}$$
$$m = \rho b\, t = 1100 \times b \times 0.006 = 6.6b \text{ kg}$$
$$T_c = mv^2 = 6.6b \times (23.562)^2 = 3664.1\, b \text{ N}$$
$$(T_1 + T_c)/(T_2 + T_c) = \exp\,(\mu\theta) = \exp\,(0.25 \times \pi \times 160/180) = \exp\,(0.698) = 2$$
$$T_1 - 2T_2 = T_c$$
$$(T_1 + T_c)/(bt) = 3 \times 10^6$$
$$T_1 + T_c = 3 \times 10^6 \times b \times 6 \times 10^3 = 18 \times 10^3 \times b$$
$$T_1 + 3664.1\, b = 18 \times 10^3 \times b$$
$$T_1 = 14335.9 \times b$$
$$T_2 = 5335.9 \times b$$
$$P = (T_1 - T_2)\, v/10^3$$
$$20 \times 10^3 = (14335.9 - 5335.9)\, b \times 23.562$$
$$b = 0.0943 \text{ m or } 94.3 \text{ mm}$$

Example 6.25

A shaft running at 200 rpm carries a pulley 1.25 m diameter which drives a dynamo at 1200 rpm by means of a belt 6 mm thick. Allowing for the thickness of the belt and a slip of 4%, find the size of the dynamo pulley and width of belt. Find also the power required if ratio of belt tensions is 2.5. Maximum tension in the belt is not to exceed 3 MPa.

■ Solution

Given: $n_1 = 200$ rpm, $d_1 = 1.25$ m, $n_2 = 1200$ rpm, $t = 6$ mm, $s = 4\%$, $T_1/T_2 = 2.5$, $P = 15$ kW, $\sigma_{max} = 3$ MPa, $b = ?$, $d_2 = ?$

$$n_2/n_1 = (d_1 + t)(100 - s)/[(d_2 + t) \times 100]$$
$$1200/200 = (1250 + 6)(100 - 4)/[100(d_2 + 6)]$$
$$d_2 = 195 \text{ mm}$$
$$v = \pi d_1 n_1/60 = \pi \times 1.25 \times 200/60 = 13.09 \text{ m/s}$$
$$\sigma_{max} = T_1/(bt)$$
$$T_1 = 3 \times 10^6 \times 6 \times 10^{-3} \times b = 18 \times 10^3 \times b$$
$$T_2 = T_1/2.5 = 7.2 \times 10^3 \times b$$
$$P = (T_1 - T_2)\, v/10^3$$
$$15 = (18 - 7.2)\, b \times 13.09$$
$$b = 0.1061 \text{ m or } 106.1 \text{ mm}$$

Example 6.26

A belt is required to transmit 40 kW from a pulley 1.5 m diameter running at 300 rpm. The angle of contact is spread over 11/24th of the circumference of the pulley, and the coefficient of friction is 0.3. Determine the width of the belt required, if thickness of belt is 10 mm, safe working stress for belt meterial is 2.5 MPa, and density of belt material is 1100 kg/m³.

■ Solution

Given: $P = 40$ kW, $d = 1.5$ m, $n = 300$ rpm, $\theta = 11 \times 360/24 = 165°$, $\mu = 0.3$, $t = 10$ mm, $\sigma = 2.5$ MPa, $\rho = 1100$ kg/m³

$$v = \pi dn/60 = \pi \times 1.5 \times 300/60 = 23.562 \text{ m/s}$$
$$m = \rho bt = 1100 \times b \times 0.01 = 11\, b \text{ kg}$$
$$T_c = mv^2 = 11\, b \times (23.562)^2 = 6106.82 \times b \text{ N}$$
$$\exp(\mu\theta) = \exp(0.3 \times \pi \times 165/180) = \exp(0.86394) = 2.3725$$
$$(T_1 + T_c)/(T_2 + T_c) = 2.3725$$
$$T_1 - 2.3725\, T_2 = 1.3725 T_c$$
$$(T_1 + T_c)/(bt) = 2.5 \times 10^6$$
$$T_1 + T_c = 2.5 \times 10^6 \times b \times 10 \times 10^3 = 25 \times 10^3 \times b$$
$$T_1 + 6106.82 \times b = 25 \times 10^3 \times b$$
$$T_1 = 18893.18 \times b$$
$$T_2 = 4430.6 \times b$$
$$P = (T_1 - T_2)\, v/10^3$$
$$40 \times 10^3 = (18893.18 - 4430.6)\, b \times 23.562$$
$$b = 0.1174 \text{ m or } 117.4 \text{ mm}$$

Example 6.27

A V-belt having a lap angle of 180° has a cross-sectional area of 250 mm², and runs in a groove of included angle 40°. The density of the belt material is 1500 kg/m³ and maximum stress is limited to 4 MPa. The coefficient of friction is 0.15.

Find the maximum power that can be transmitted, if the wheel has a mean diameter of 300 mm and runs at 900 rpm.

■ **Solution**

Given: $\theta = 180°$, $a = 250\text{ mm}^2$, $\beta = 20°$, $\rho = 1500\text{ kg/m}^3$, $\sigma = 4\text{ MPa}$,

$\mu = 0.15$, $d = 300\text{ mm}$, $n = 900\text{ rpm}$, $P_{max} = ?$

$$\mu_e = \mu/\sin\beta = 0.15/\sin 20° = 0.43857$$

$$\exp(\mu_e\theta) = \exp(0.43857 \times \pi) = 3.9662$$

$$m = \rho A = 1500 \times 250 \times 10^{-6} = 0.375\text{ kg}$$

$$v = \pi d\, n/60 = \pi \times 0.3 \times 900/60 = 14.137\text{ m/s}$$

$$T_c = mv^2 = 0.375 \times (14.137)^2 = 74.95\text{ N}$$

For maximum power transmission, $\sigma = (T_1 + T_c)/A$

$$(T_1 + 74.95)/(250 \times 10^{-6}) = 4 \times 10^6$$

$$T_1 = 925.05\text{ N}$$

$$T_2 = T_1/3.9662 = 233.23\text{ N}$$

$$P_{max} = (T_1 - T_2)\, v/10^3$$

$$= (925.05 - 233.23) \times 14.137/10^3 = 9.78\text{ kW}$$

Example 6.28

A machine which is to rotate at 400 rpm is run by an engine turning at 1500 rpm, through a silent chain, having a pitch of 15 mm. The number of teeth on a sprocket should be from 18 to 105. The linear velocity of chain drive is not to exceed 10 m/s. Find the suitable number of teeth for both the sprockets.

■ **Solution**

Given: $n_1 = 1500\text{ rpm}$, $n_2 = 400\text{ rpm}$, $p = 15\text{ mm}$, $v = 10\text{ m/s}$

$$v = \pi D_1\, n_1/60$$

$$10 = \pi D_1 \times 1500/60$$

$$D_1 = 0.127\text{ m}$$

$$p = d\sin(180°/z)$$

$$15 \times 10^{-3} = 0.127\sin(180°/z_1)$$

$$z_1 = 26.5 \cong 28$$

$$z_2/z_1 = n_1/n_2$$

$$z_2 = 28 \times 1500/400 = 105$$

Example 6.29

A rope drive transmits 120 kW at 225 rpm by ropes, each 25 mm diameter and density 6800 kg/m³. The maximum rope tension is 1.5 kN and it is designed for maximum power conditions. The angle of contact is 160° and coefficient of friction is 0.25. Determine the diameter of pulley and number of ropes, if groove angle is 45°.

■ **Solution**

Given: $P_t = 120\text{ kW}$, $n = 225\text{ rpm}$, $d_r = 25\text{ mm}$, $\rho = 6800\text{ kg/m}^3$,

$T_{max} = 1.5\text{ kN}$, $\theta = 160°$, $\mu = 0.25$, $2\beta = 45°$, $d\,?$, $i = ?$

$$\mu_e = \mu/\sin\beta = 0.25/\sin 22.5° = 0.65328$$

$$m = \pi d^2_r\, \rho/4 = \pi \times 625 \times 10^{-6} \times 6800/4 = 3.338\text{ kg}$$

For maximum power transmission, $v = [T_{max}/(3m)]^{0.5}$

$$= [1500/(3 \times 3.338)]^{0.5} = 12.239 \text{ m/s}$$

$$T_c = T_{max}/3 = 1500/3 = 500 \text{ N}$$

$$T_1 = T_{max} - T_c = 1500 - 500 = 1000 \text{ N}$$

$$\exp(\mu_e \theta) = \exp(0.65328 \times \pi \times 160/180) = 6.1985$$

$$T_1/T_2 = 6.1985$$

$$T_2 = 1000/6.9815 = 161.33 \text{ N}$$

$$P_{max} = (T_1 - T_2)\, v/10^3$$

$$= (1000 - 161.33) \times 12.239/10^3 = 10.264 \text{ kW}$$

Number of ropes, $i = P_t/P = 120/10.264 = 11.69 \cong 12$

$$v = \pi\, d\, n/60$$

$$12.239 = \pi \times d \times 225/60$$

$$d = 1.039 \text{ m}$$

Example 6.30

An open belt 100 mm wide connects two pulleys mounted on parallel shafts with their centres 2.5 m apart. The pulleys are of 500 mm and 250 mm diameters. The coefficient of friction between the belt and the pulleys is 0.3. The maximum stress in the belt is limited to 15 N/mm width. Find the maximum power which can be transmitted if the larger pulley rotates at 120 rpm.

■ Solution

Given: $b = 100$ mm, $C = 2.5$ m, $d_1 = 250$ mm, $d_2 = 500$ mm, $\mu = 0.3$, $\sigma_{max} = 15$ N/mm width, $n_2 = 120$ rpm, $P_{max} = ?$

$$\sin\alpha = (d_2 - d_1)/(2C) = (500 - 250)/(2 \times 2500) = 0.05$$

$$\alpha = 2.866°$$

$$\theta = 180° - 2\,\alpha = 180° - 5.732° = 174.268°$$

$$\exp(\mu\,\theta) = \exp(0.3 \times \pi \times 174.268/180) = 2.49$$

$$v = \pi\, d_2 n_2/60 = \pi \times 500 \times 10^{-3} \times 120/60 = 3.14 \text{ m/s}$$

$$T_{max} = \sigma_{max} \times b = 15 \times 100 = 1500 \text{ N},\ T_c = T_{max}/3 = 500 \text{ N}$$

$$T_1 = T_{max} - T_c = 1500 - 500 = 1000 \text{ N}$$

$$T_2 = T_1/2.49 = 401.6 \text{ N}$$

$$P = (T_1 - T_2)\, v/10^3$$

$$= (1000 - 401.6) \times 3.14/10^3 = 1.879 \text{ kW}$$

Example 6.31

A leather belt 120 mm wide and 6 mm thick transmits power from a pulley 800 mm diameter which rotates at 450 rpm. The angle of lap is 160° and coefficient of friction is 0.3. The mass of the belt is 1000 kg/m^3 and the stress is not to exceed 2.5 MPa. Find the maximum power that can be transmitted.

■ Solution

Given: $b = 120$ mm, $t = 6$ mm, $d_1 = 800$ mm, $n_1 = 450$ rpm, $\theta = 160°$, $\mu = 0.3$, $\rho = 100$ kg/m^3, $\sigma_{max} = 2.5$ MPa, $P_{max} = ?$

$$m = bt\rho = 120 \times 6 \times 10^{-6} \times 1000 = 0.72 \text{ kg}$$

$$v = \pi\, d_1\, n_1/60 = \pi \times 800 \times 10^{-3} \times 450/60 = 18.85 \text{ m/s}$$

$$\exp(\mu\,\theta) = \exp(0.3 \times \pi \times 160/180) = 2.3112$$

For maximum power transmission, $T_c = T_{max}/3$

$$= \sigma_{max}\, b\, t/3$$
$$= 2.5 \times 120 \times 6/3 = 600 \text{ N}$$
$$T_1 = T_{max} - T_c = 1800 - 600 = 1200 \text{ N}$$
$$T_1/T_2 = 2.3112$$
$$T_2 = 1200/2.3112 = 519.21 \text{ N}$$
$$P_{max} = (T_1 - T_2)\, v/10^3$$
$$= (1200 - 519.21) \times 18.85/10^3 = 12.83 \text{ kW}$$

Example 6.32

An open belt drive connects two pulleys 1.5 and 0.5 m diameter on parallel shafts 3.5 m apart. The belt has a mass of 1 kg/m length and the maximum tension in the belt is not to exceed 2 kN. The 1.5 m pulley, which is the driver, runs at 250 rpm. Due to belt slip, the velocity of the driven shaft is only 730 rpm. If the coefficient of friction between the belt and the pulley is 0.25, find (a) torque on each shaft, (b) power transmitted, (c) power lost in friction, and (d) efficiency of the drive.

■ Solution

Given: $d_1 = 1.5$ m, $d_2 = 0.5$ m, $C = 3.5$ m, $m = 1$ kg/m, $T_{\max} = 2$ kN, $n_1 = 250$ rpm, $n_2 = 730$ rpm, $\mu = 0.25$

$$v_1 = \pi d_1 n_1/60 = \pi \times 1.5 \times 250/60 = 19.635 \text{ m/s}$$
$$v_2 = \pi d_2 n_2/60 = \pi \times 0.5 \times 730/60 = 19.111 \text{ m/s}$$
$$T_c = m v_1^2 = 1 \times (19.635)^2 = 385.53 \text{ N}$$
$$T_1 = T_{\max} - T_c = 2000 - 385.53 = 1614.467 \text{ N}$$
$$\sin\alpha = (d_1 - d_2)/(2C) = (1.5 - 0.5)/(2 \times 3.5) = 0.14286$$
$$\alpha = 8.213°$$
$$\theta = 180° - 2\alpha = 180° - 16.426° = 163.574°$$
$$\exp(\mu\theta) = \exp(0.25 \times \pi \times 163.574/180) = 2.0416$$
$$T_2 = T_1/2.0416 = 1614.467/2.0416 = 790.792 \text{ N}$$
$$T_1 - T_2 = 1614.467 - 790.792 = 823.675 \text{ N.}$$

Torque on bigger pulley, $M_1 = (T_1 - T_2)\, d_1/2 = 823.675 \times 0.75 = 617.76$ Nm

Torque on bigger pulley, $M_2 = (T_1 - T_2)\, d_2/2 = 823.675 \times 0.25 = 205.92$ Nm

Input power, $P_1 = (T_1 - T_2)\, v_1/10^3 = 823.675 \times 19.635/10^3 = 16.173$ kW

Output power, $P_2 = (T_1 - T_2)\, v_2/10^3 = 823.675 \times 19.111/10^3 = 15.741$ kW

Power lost in friction, $P_f = P_1 - P_2 = 16.173 - 15.741 = 0.431$ kW

Efficiency = 15.741/16.173 = 0.973 or 97.3%

Example 6.33

The power transmitted between two shafts 4 m apart by a cross belt drive is 7.5 kW. The pulleys are 600 and 300 mm diameters, bigger pulley being the driver, and running at 225 rpm. The permissible load on the belt is 25 N/mm width of the belt, which is 5 mm thick. The coefficient of friction is 0.35.

Determine (a) length of the belt, (b) width of the belt, and (c) initial tension in the belt.

■ Solution

Given: $C = 4$ m, $P = 7.5$ kW, $n_1 = 225$ rpm, $d_1 = 600$ mm, $d_2 = 300$ mm, $T = 25$ N/mm of belt width, $t = 5$ mm, $\mu = 0.35$, $L_c = ?$, $b = ?$, $T_o = ?$

$$v_1 = \pi d_1 n_1/60 = \pi \times 0.6 \times 225/60 = 7.068 \text{ m/s}$$
$$\sin \alpha = (d_1 + d_2)/(2C) = (0.6 + 0.3)/(2 \times 4) = 0.1125$$
$$\alpha = 6.46°$$
$$\theta = 180° + 2\alpha = 180° + 12.92° = 192.92°$$
$$\exp(\mu\,\theta) = \exp(0.35 \times \pi \times 192.92/180) = 3.249$$
$$L_c = 2\,C + \pi(d_1 + d_2)/2 + (d_1 + d_2)^2/(4C)$$
$$= 2 \times 4 + \pi \times 0.9/2 + (0.9)^2/(4 \times 4)$$
$$= 9.4643 \text{ m}$$
$$T_1 = 25\,b$$
$$T_2 = 25\,b/3.247 = 7.6947\,b$$
$$P = (T_1 - T_2)\,v_1/10^3$$
$$7.5 \times 10^3 = (25 - 7.6947) \times b \times 7.068/10^3$$
$$b = 61.32 \text{ mm}$$
$$\text{Initial tension, } T_o = (T_1 + T_2)/2 = (25 + 7.6947) \times 61.32/2 = 1002.42 \text{ N}$$

Example 6.34

A V-belt drive consists of three belts in parallel on grooved pulleys of the same size. The angle of groove is 40°, and the coefficient of friction 0.15. The cross-sectional area of each belt is 800 mm² and the permissible stress in the belt material is 3 MPa. Calculate the power that can be transmitted between two pulleys 400 mm in diameter rotating at 960 rpm.

■ Solution

Given: $\beta = 40°$, $\mu = 0.15$, $A = 800$ mm², $\sigma_p = 3$ MPa,
$d_1 = d_2 = 400$ mm, $n_1 = n_2 = 960$ rpm, $P = ?$, $i = 3$

$$\mu_e = \mu/\sin\beta = 0.15/\sin 20° = 0.43857$$
$$\exp(\mu_e\theta) = \exp(0.43857 \times \pi) = 3.9662$$
$$v = \pi d\,n/60 = \pi \times 0.4 \times 960/60 = 20.1 \text{ m/s}$$
$$T_1 = \sigma_p A = 3 \times 800 = 2400 \text{ N}$$
$$T_1/T_2 = 3.9662,\ T_2 = 2400/3.9662 = 605.1 \text{ N}$$
$$P = i(T_1 - T_2)\,v/10^3$$
$$= 3\,(2400 - 605.1) \times 20.1/10^3 = 108.23 \text{ kW}$$

Example 6.35

A rope drive is required to transmit 250 kW from a sheave of 1 m diameter running at 450 rpm. The safe pull in each rope is 800 N and the mass of the rope is 0.46 kg/m length. The angle of lap is 160° and the groove angle is 45°. If the coefficient of friction is between the rope and the sheave is 0.3, find the number of ropes required.

■ Solution

Given: $P_t = 250$ kW, $d = 1$ m, $n = 450$ rpm, $m = 0.46$ kg/m length of belt,
$T_{max} = 800$ N, $\theta = 160°$ $2\beta = 45°$, $\mu = 0.3$, $i = ?$.

$$\mu_e = \mu/\sin\beta = 0.3/\sin 22.5° = 0.784$$
$$\exp(\mu_e\theta) = \exp(0.784 \times \pi \times 160/180) = 8.9278$$
$$v = \pi d\,n/60 = \pi \times 1 \times 450/60 = 23.562 \text{ m/s}$$

$T_c = m\,v^2 = 0.46 \times (23.562)^2 = 255.4$ N
$T_1 = 800 - 255.4 = 544.6$ N, $T_2 = 544.6/8.9278 = 61$ N
$P = (T_1 - T_2)\,v/10^3$
$= (544.6 - 61) \times 23.562/10^3 = 11.4$ kW
$i = P_t/P = 250/11.4 = 21.93 \cong 22$

Example 6.36

The reduction of speed from 360 to 120 rpm is desired by the use of a chain drive. The driving sprocket has 18 teeth. Find the number of teeth on the driven sprocket and the pitch length of the chain, if the pitch radius of the driven sprocket is 250 mm and the centre distance between the two sprockets is 400 mm.

■ Solution

Given: $n_1 = 360$ rpm, $n_2 = 120$ rpm, $z_1 = 18$, $d_2 = 250$ mm, $C = 400$ mm, $z_2 = ?$, $L_p = ?$

$z_2 = n_1 z_1/n_2 = 360 \times 18/120 = 54$

$p = d_2 \sin(180°/z_2) = 250 \sin(180°/54) = 14.54$ mm

$L_p = 2C + p(z_1 + z_2)/2 + \{p^2/(4C)\}\,[\text{cosec}\,(180°/z_1) - \text{cosec}\,(180°/z_2)]^2$

$= 2\times400 + 14.54\,(18+54)/2 + \{(14.54)^2/(4\times400)\}[\text{cosec}\,(180°/18) - \text{cosec}\,(180°/54)]^2$

$= 1340.8$ mm

Example 6.37

A leather belt 150 mm wide, 6 mm thick, and weighing 6 N/m connects two pulleys each 1 m in diameter and on parallel shafts. The belt is found to slip when the moment of resistance is 600 Nm and the speed is 500 rpm. If the coefficient of friction between the belt and the pulleys is 0.24, find the largest tension in the belt. [IAS, 1982]

■ Solution

Given: $b = 150$ mm, $t = 6$ mm, $w = 6$ N/m, $d_1 = d_2 = 1$ m, $\mu = 0.24$, $\theta = 180°$, $M = 600$ N.m, $n = 500$ rpm, $T_{max} = ?$.

$\omega = 2\,\pi\,n/60 = 2\,\pi \times 500/60 = 52.36$ rad/s

$v = r\,\omega = 0.5 \times 52.36 = 26.18$ m/s

$P = M\omega/10^3 = 600 \times 52.36/10^3 = 31.416$ kW

$m = 6/9.81 = 0.61162$ kg/m

$T_c = mv^2 = 0.61162 \times (26.18)^2 = 419.2$ N

$\exp(\mu\,\theta) = \exp(0.24 \times \pi) = 2.1254$

$T_{max} = T_1 + T_c$

$(T_1 + T_c)/(T_2 + T_c) = 2.1254$

$T_1 + 419.2 = 2.1254\,T_2 + 890.97$

$T_1 - 2.1254,\ T_2 = 471.77$ (1)

$P = (T_1 - T_2)\,v/10^3$

$31.416 \times 10^3 = (T_1 - T_2) \times 26.18$

$T_1 - T_2 = 1200$ (2)

Solving Eqs. (1) and (2), we get

$T_1 = 1847.1$ N, $T_2 = 647.1$ N, and $T_{max} = 2266.3$ N

Example 6.38

What is the effect of centrifugal force on the transmission of power in a belt drive? A prime mover running at 300 rpm drives a D.C. generator at 500 rpm by a belt drive. Diameter of the pulley on the output shaft of the prime mover is 600 mm. Assuming a slip of 3%, determine the diameter of the generator pulley if the belt running over it is 6 mm thick. [IAS, 1997]

Solution

Given: $n_1 = 300$ rpm, $n_2 = 500$ rpm, $d_1 = 600$ mm, $s = 3\%$, $t = 6$ mm, $d_2 = ?$,

$$n_2/n_1 = (1 - s/100)\,[(d_1 + t)/(d_2 + t)]$$
$$500/300 = (1 - 0.03)\,[(600 + 6)/(d_2 + 6)]$$
$$d_2 = 346.7 \text{ mm}$$

Example 6.39

An 8 mm thick leather open belt connects two pulleys. The smaller pulley is 300 mm diameter and runs at 200 rpm. The angle of lap of this pulley is 160°, and the coefficient of friction between the belt and the pulley is 0.25. The belt is on the point of slipping when 3 kW is transmitted. Safe working stress in the belt material is 1.6 N/mm². Determine the required width of the belt for 20% overload capacity. Initial tension may be taken equal to mean of the driving tensions.

It is proposed to increase the power transmitting capacity of the drive by adopting one of the following alternatives:

(a) by increasing initial tension by 10%.

(b) by increasing the coefficient of friction to 0.3 by applying a dressing to the belt.

Examine the two alternatives and recommend the one which you think will be more effective. How much power would the drive transmit adopting either of the alternatives? [IES, 1976]

Solution

Given: $t = 8$ mm, $d_1 = 300$ mm, $n_1 = 200$ rpm, $\theta = 160°$, $\mu = 0.25$, $P_{max} = 3$ kW, $\sigma = 1.6$ N/m², overload = 20%, $b = ?$, $T_o = ?$.

$$v = \pi d n/60 = \pi \times 0.3 \times 200/60 = 3.1416 \text{ m/s}$$
$$\exp(\mu\theta) = \exp(0.25 \times \pi \times 160/180) = 2$$
$$T = \sigma b t = 1.6 \times b \times 8 = 12.8\,b$$
$$T_1 = T = 12.8b, \quad T_2 = T_1/2 = 6.4b$$
$$1.2\,P = (T_1 - T_2)\,v/10^3$$
$$1.2 \times 3 \times 10^3 = (12.8 - 6.4)\,b \times 3.1416$$
$$b = 179 \text{ mm}$$
$$T_1 = 2309.1 \text{ N},\ T_2 = 1154.6 \text{ N}$$
$$T_o = (T_1 + T_2)/2 = 1731.85 \text{ N}$$

(a)
$$T_o = 1.1 \times 1731.85 = 1905 \text{ N}$$
$$T_1 + T_2 = 3810 \text{ N}$$
$$T_1/T_2 = 2,\ 3T_2 = 3810,\ T_2 = 1270 \text{ N},\ T_1 = 2540 \text{ N}$$
$$P = (2540 - 1270) \times 3.1416/10^3 = 3.99 \text{ kW}$$

(b)
$$\exp(\mu\theta) = \exp(0.3 \times \pi \times 160/180) = 2.311$$
$$T_1 = 2309.1 \text{ N},\ T_2 = 999.2 \text{ N}$$
$$P = (2309 - 999.2) \times 3.1416/10^3 = 4.11 \text{ kW}$$

Hence, increasing coefficient of friction gives better results.

Example 6.40

A rope drive transmits 600 kW from a pulley of effective diameter 4 m, which runs at a speed of 90 rpm. The angle of lap is 160°, the angle of groove 45°, the coefficient of friction 0.28, the weight of the rope 15 N/m, and the allowable tension in each rope 2.4 kN. Find the number of ropes required. [IES, 1980]

■ Solution

Given: $P_t = 600$ kW, $d = 4$ m, $n = 90$ rpm, $\theta = 160°$, $2\beta = 45°$, $\mu = 0.28$, $w = 15$ N/m, $T = 2.4$ kN, $i = ?$

$$\mu_e = \mu/\sin\beta = 0.28/\sin 22.5° = 0.73167$$

$$\exp(\mu_e\theta) = \exp(0.73167 \times \pi \times 160/180) = 7.715$$

$$v = \pi dn/60 = \pi \times 4 \times 90/60 = 18.85 \text{ m/s}$$

$$m = w/g = 15/9.81 = 1.529 \text{ kg/m}$$

$$T_c = mv^2 = 1.529 \times (18.85)^2 = 543.3 \text{ N}$$

$$T_1 = 2400 - 543.3 = 1856.7 \text{ N}, \quad T_2 = 1856.7/7.715 = 240.7 \text{ N}$$

$$P = (T_1 - T_2)\, v/10^3$$

$$= (1856.7 - 240.7) \times 18.85/10^3 = 30.462 \text{ kW}$$

$$i = P_t/P = 600/30.462 = 19.69 \cong 20$$

Example 6.41

An electric motor is to drive a compressor by a belt drive.
Power to be transmitted = 7.5 kW
Diameter of motor pulley = 200 mm
Centre distance between pulleys = 1 m
Motor speed = 750 rpm
Compressor speed = 250 rpm
Direction of rotation of both the pulleys is same. Find the width of the belt required if the permissible belt tension is 16 N/mm belt width. Coefficient of friction between the belt and the pulleys is 0.3. Neglect the effect of centrifugal tension. [IES, 1982]

■ Solution

Given: $P = 7.5$ kW, $d_1 = 200$ mm, $C = 1$ m, $n_1 = 750$ rpm, $n_2 = 250$ rpm, $T = 16b$, $\mu = 0.3$, $b = ?$

$$\sin\alpha = (d_2 - d_1)/(2C) = (600 - 200)/(2 \times 1000) = 0.2$$

$$\alpha = 11.537°$$

$$\theta = 180° - 2\,\alpha = 180° - 23.074° = 156.93°$$

$$\exp(\mu\,\theta) = \exp(0.3 \times \pi \times 156.93/180) = 2.274$$

$$v = \pi\, d_1\, n_1/60 = \pi \times 0.2 \times 750/60 = 7.854 \text{ m/s}$$

$$T_1 = T = 16\,b,\ T_2 = 16b/2.274 = 7.035\,b$$

$$P = (T_1 - T_2)\, v/10^3$$

$$7.5 \times 10^3 = (16 - 7.035)\, b \times 7.854$$

$$b = 106.52 \text{ mm}$$

Example 6.42

A blower is driven by an electric motor through a belt drive. The motor runs at 750 rpm. For this power transmission, a flat belt of 8 mm thickness and 250 mm width is used. The diameter of the motor pulley is 350 mm and that of the blower pulley 1350 mm. The centre distance between these pulleys is

1350 mm and an open belt configuration is adopted. The pulleys are made out of cast iron. Frictional coefficient between belt and pulley is 0.35 and the permissible stress for the belt meterial can be taken as 2.5 N/mm^2 with sufficient factor of safety. Belt Weighs 20 N/m length. What is the maximum power transmitted without belt slipping on any one of the pulleys? [IES, 1983]

■ **Solution**

Given: $n_1 = 750$ rpm, $t = 8$ mm, $b = 250$ mm, $d_1 = 350$ mm, $d_2 = 1350$ mm,
$C = 1350$ mm, $\mu = 0.35$, $\sigma = 2.5$ N/mm^2, $w = 20$ N/m, $P_{max} = ?$.

$$\sin \alpha = (d_2 - d_1)/(2C) = (1350 - 350)/(2 \times 1350) = 0.3704$$
$$\alpha = 21.74°$$
$$\theta = 180° - 2\alpha = 180° - 43.48° = 136.52°$$
$$\exp(\mu\,\theta) = \exp(0.35 \times \pi \times 136.52/180) = 2.3024$$
$$v = \pi\, d_1\, n_1/60 = \pi \times 0.35 \times 750/60 = 13.74 \text{ m/s}$$
$$m = 20/9.81 = 2.0387 \text{ kg/m}$$
$$T_c = mv^2 = 2.0387 \times (13.74)^2 = 385.13 \text{ N}$$
$$T = T_1 + T_c,\ T_1 = 5000 - 385.37 = 4614.87 \text{ N}$$
$$T_2 = 4614.87/2.3024 = 2004.37 \text{ N}$$
$$P = (T_1 - T_2)\, v/10^3$$
$$= (4614.87 - 2004.37) \times 13.74/10^3 = 35.87 \text{ kW}$$

Example 6.43

Determine the width of a 9.75 mm thick leather belt required to transmit 15 kW from a motor running at 900 rpm. Diameter of the driving pulley of the motor is 300 mm. The driven pulley runs at 300 rpm and the distance between the centre of the two pulleys is 3 m. The weight of the leather is 0.1 × 10^{-3} N/mm^2. Maximum allowable stress in the leather is 2.5 N/mm^2. Coefficient of friction between leather and pulley is 0.3. Assume open belt drive and neglect the sag and slip of the belt. [IES, 1989]

■ **Solution**

Given: $t = 9.75$ mm, $P = 15$ kW, $n_1 = 900$ rpm, $d_1 = 300$ mm, $n_2 = 300$ rpm,
$C = 3$ m, $w = 0.1 \times 10^{-3}$ N/mm^2, $\sigma = 2.5$ n/mm^2, $\mu = 0.3$

$$d_2 = d_1 n_1/n_2 = 300 \times 900/300 = 900 \text{ mm}$$
$$\sin \alpha = (d_2 - d_1)/(2C) = (900 - 300)/(2 \times 3000) = 0.1$$
$$\alpha = 5.739°$$
$$\theta = 180° - 2\alpha = 180° - 11.478° = 168.52°$$
$$\exp(\mu\theta) = \exp(0.3 \times \pi \times 168.52/180) = 2.41662$$
$$v = \pi d_1 n_1/60 = \pi \times 0.3 \times 900/60 = 14.137 \text{ m/s}$$
$$m = (0.1 \times 10^{-3} \times b \times 9.75)/9.81 = 0.0994 \times 10^{-3} \times b \text{ kg/m}$$
$$T_c = mv^2 = 0.0994 \times 10^{-3} \times b \times (14.137)^2 = 19.866 \times 10^{-3} \times b$$
$$T = \sigma bt = 2.5 \times 9.75 \times b = 24.375\, b$$
$$T = T_1 + T_c,\ T_1 = (24.375 - 19.866 \times 10^{-3})\, b = 24.355\, b$$
$$T_2 = 24.355\, b/2.41662 = 10.078\, b$$
$$P = (T_1 - T_2)\, v/10^3$$
$$15 \times 10^3 = (24.355 - 10.078)\, b \times 14.137$$
$$b = 74.32 \text{ mm}$$

Example 6.44

A prime mover running at 300 rpm, drives a DC generator at 500 rpm by a belt drive. Diameter of the pulley on the output shaft of the prime mover is 600 mm. Assuming a slip of 3%, determine the diameter of the generator pulley if the belt running over it is 6 mm thick. [IES, 1994]

Solution

Given: $n_1 = 300$ rpm, $n_2 = 500$ rpm, $d_1 = 600$ mm, $s = 3\%$, $t = 6$ mm, $d_2 = ?$

$$n_2/n_1 = [(d_1 + t)/(d_2 + t)]\,(1 - s/100)$$
$$600/500 = [(600 + 6)/(d_2 + 6)]\,(1 - 0.03)$$
$$d_2 = (606 \times 500/600)\,(0.97) - 6 = 483.85 \text{ mm}$$

Example 6.45

A V-belt of 6 cm² cross-section has a groove angle of 40° and angle of lap of 150°, $\mu = 0.1$. The mass of belt per metre run is 1.2 kg. The maximum allowable stress in the belt is 850 N/cm². Calculate the power that can be transmitted at a belt speed of 30 m/s. [IES, 1998]

Solution

Given: $A = 6$ cm², $2\beta = 40°$, $\theta = 150°$, $\mu = 0.1$, $m = 1.2$ kg/m run, $\sigma = 850$ N/cm², $v = 30$ m/s, $P = ?$

$$\mu_e = \mu/\sin\beta = 0.1/\sin 20° = 0.2924$$
$$\exp(\mu_e\,\theta) = \exp(0.2924 \times \pi \times 150/180) = 2.15$$
$$T_c = m\,v^2 = 1.2 \times (30)^2 = 1080 \text{ N}$$
$$T = \sigma a = 850 \times 6 = 5100 \text{ N}$$
$$T_1 = 5100 - 1080 = 4020 \text{ N},\ T_2 = 4020/2.15 = 1869.77 \text{ N}$$
$$P = (T_1 - T_2)\,v/10^3$$
$$= (4020 - 1869.77) \times 30/10^3 = 64.5 \text{ kW}$$

Example 6.46

A shaft which rotates at a constant speed of 160 rpm is connected by belting to a parallel shaft 720 mm apart, which has to run at 60, 80 and 100 rpm. The smallest pulley on the driving shaft is 40 mm in radius. Determine the remaining radii of the two stepped pulleys for (a) an open belt, and (b) a crossed belt.

Solution

Refer to Fig.6.11.

Given: $n_1 = n_3 = n_5 = 160$ rpm, $C = 720$ mm, $n_2 = 60$ rpm, $n_4 = 80$ rpm, $n_6 = 100$ rpm, $r_1 = 40$ mm

(a) Open belt

$$r_2 = \frac{r_1\,n_1}{n_2} = \frac{40 \times 160}{60} = 106.7 \text{ mm}$$

$$r_4 = \frac{r_3\,n_3}{n_4} = r_3 \times \frac{160}{80} = 2r_3$$

$$L_o = \pi(r_1 + r_2) + \frac{(r_2 - r_1)^2}{C} + 2C$$

$$= \pi(40 + 106.7) + \frac{(106.7 - 40)^2}{720} + 2 \times 720 = 1907 \text{ mm}$$

$$= \pi(r_3 + r_4) + \frac{(r_4 - r_3)^2}{C} + 2C$$

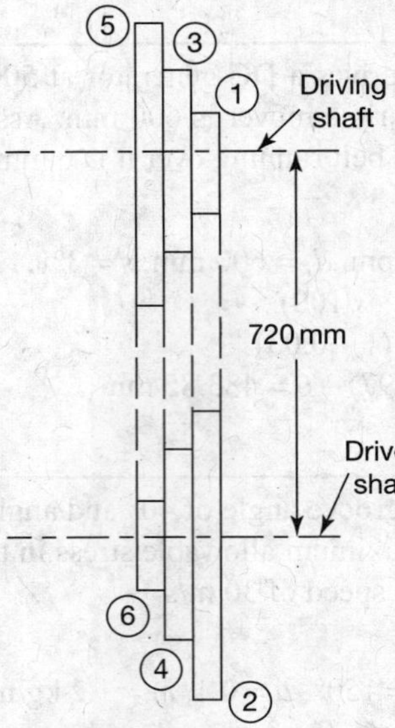

Fig.6.11 Cone pulley drive

$$= \pi\left(r_3 + 2r_3\right) + \frac{\left(2r_3 - r_3\right)^2}{720} + 2\times 720$$

$$1907 = 9.426\ r_3 + 0.0014\ r_3^2 + 1440$$

$$r_3^2 + 6732.86\ r_3 - 333571 = 0$$

$$r_3 = \frac{1}{2}\left(-6732.86 + \sqrt{(6732.86)^2 + 4\times 333571}\right) = 49.2 \text{ mm}$$

$$r_4 = 2\times 49.2 = 98.4 \text{ mm}$$

$$r_6 = \frac{r_5\ n_5}{n_6} = r_5 \times \frac{160}{100} = 1.6\ r_5$$

$$L_o = \pi\left(r_5 + r_6\right) + \frac{\left(r_6 - r_5\right)^2}{C} + 2C$$

$$= \pi\left(r_5 + 1.6\ r_5\right) + \frac{\left(1.6\ r_5 - r_5\right)^2}{720} + 2\times 720 = 1907$$

$$8.168\ r_5 + 5\times 10^{-4} r_5^2 = 467$$

$$r_5^2 + 16336\ r_5 - 934000 = 0$$

$$r_5 = \frac{1}{2}\left(-16336 + \sqrt{(16336)^2 + 4\times 934000}\right) = 57 \text{ mm}$$

$$r_6 = 1.6\times 57 = 91.2 \text{ mm}$$

(b)

$$r_2 = \frac{r_1\ n_1}{n_2} = 40\times\frac{160}{60} = 106.7 \text{ mm}$$

$$r_4 = \frac{r_3\ n_3}{n_4} = r_3 \times \frac{160}{80} = 2r_3$$

$$\text{Now } r_1 + r_2 = r_3 + r_4 = r_5 + r_6 = 40 + 106.7 = 146.7 \text{ mm}$$

$$r_3 + 2r_3 = 146.7$$

$$r_3 = 48.9 \text{ mm}$$

$$r_4 = 97.8 \text{ mm}$$

$$r_6 = \frac{r_5\, n_5}{n_6} = r_5 \times \frac{160}{100} = 1.6\, r_5$$

$$r_5 + 1.6\, r_5 = 146.7$$

$$r_5 = 56.4 \text{ mm}$$

$$r_6 = 1.6 \times 56.4 = 90.2 \text{ mm}$$

Example 6.47

A leather belt is required to transmit 7.5 kW from a pulley 1.5 m in diameter, running at 240 rpm. The angle of wrap is 165° and the coefficient of friction between belt and pulley is 0.35. If the safe working stress for leather belt is 1.5 MPa, density of leather 1000 kg/m³ and thickness of belt 10 mm, determine the width of belt taking centrifugal tension into account.

■ **Solution**

Given: $P = 7.5$ kW, $d = 1.5$ m, $n = 240$ rpm, $\theta = 165°$, $\mu = 0.35$, $\sigma_\omega = 1.5$ MPa, $\rho = 1000$ kg/m³, $t = 10$ mm

$$\text{Velocity of belt, } v = \frac{\pi d n}{60} = \frac{\pi \times 1.5 \times 240}{60} = 18.85 \text{ m/s}$$

$$\text{Power transmitted, } P = \frac{(T_1 - T_2)v}{10^3}$$

$$T_1 - T_2 = \frac{7.5 \times 10^3}{18.85} = 397.88 \text{ N} \qquad (1)$$

$$\frac{T_1}{T_2} = e^{\mu\theta} = e^{0.35 \times \frac{\pi}{180} \times 165} = 2.74 \qquad (2)$$

From Eqs. (1) and (2), we have

$T_2(2.74 - 1) = 397.88$

$T_2 = 228.67\ N$

$T_1 = 2.74 \times 228.67 = 626.55\ N$

Mass of belt per metre length, $m = b \times 0.01 \times 1 \times 1000 = 10\, b$ kg

Centrifugal tension, $T_c = mv^2 = 10\,b \times (18.85)^2 = 3553.2\, b$ N

Maximum tension in belt, $T_{max} = \sigma_\omega bt = 1.5 \times 10^6 \times b \times 0.01 = 15{,}000 \times b$ N

Now $\quad T_{max} = T_1 + T_c$

$15{,}000\, b = 626.55 + 3553.2\, b$

$b = 0.0547$ m or 54.7 mm

Example 6.48

A pulley is driven by a flat belt, the angle of lap being 160°. The belt is 100 mm wide and 6 mm thick. The density of belt material is 1000 kg/m³. The coefficient of friction is 0.32 and the maximum stress in the belt is not to exceed 2 MPa. Determine the greatest power that the belt can transmit and corresponding speed of belt.

■ **Solution**

Given: $\sigma = 160°$, $b = 100$ mm, $t = 6$ mm, $\rho = 1000$ kg/m^3, $\mu = 0.32$, $\sigma_\omega = 2$ MPa

Maximum tension in belt, $T_{max} = \sigma_\omega bt = 2 \times 10^6 \times 0.1 \times 0.006 = 1200\ N$

Mass of belt per metre length, $m = bt\,l\,\rho = 0.1 \times 0.006 \times 1 \times 1000 = 0.6$ kg/m

Speed of belt for greatest power transmitted, $v = \sqrt{\dfrac{T_{max}}{3m}} = \sqrt{\dfrac{1200}{3\times0.6}} = 25.82$ m/s

For maximum power to be transmitted, centrifugal tension,

$$T_c = \frac{T_{max}}{3} = \frac{1200}{3} = 400 \text{ N}$$

Tension on tight side, $T_1 = T_{max} - T_c = 1200 - 400 = 800$ N

$$\frac{T_1}{T_2} = e^{\mu\theta} = e^{0.32\times\frac{\pi}{180}\times160} = 2.444$$

$$T_2 = \frac{800}{2.444} = 327.34 \text{ N}$$

$$\text{Power transmitted, } P = \frac{(T_1 - T_2)v}{10^3} = \frac{(800 - 327.34)\times25.82}{10^3} = 12.2 \text{ kW}$$

Example 6.49

A belt drive consists of two V-belts in parallel on grooved pulley having groove angle of 40°. The cross-sectional area of each belt is 750 mm^2 and coefficient of friction is 0.15. The density of belt material is 1200 kg/m^3 and safe stress is 7 MPa. Calculate the power that can be transmitted between pulleys 300 mm diameter rotating at 1500 rpm. Find also the shaft speed in rpm at which power transmitted would be maximum.

■ **Solution**

Given: $2\beta = 40°$, $A = 750$ mm^2, $\mu = 0.15$, $\rho = 1200$ kg/m^3, $\sigma_\omega = 7$ mpa,
$d = 300$ mm, $n = 1500$ rpm, $\theta = \pi$, $i = 2$

$$\text{Velocity of belt, } v = \frac{\pi dn}{10^3\times60} = \frac{\pi\times300\times1500}{10^3\times60} = 23.56 \text{ m/s}$$

Mass of belt per metre length, $m = Al\rho = 750\times10^{-6}\times1\times1200 = 0.9$ kg/m

Centrifugal tension, $T_c = mv^2 = 0.9\times(23.56)^2 = 500$ N

Maximum tension in belt, $T_{max} = \sigma_\omega \times A = 7\times750 = 5250$ N

Tension in belt on tight side, $T_1 = T_{max} - T_c = 5250 - 500 = 4750$ N

$$\text{Equivalent coefficient of friction, } \mu_e = \frac{\mu}{\sin\beta} = \frac{0.15}{\sin 20°} = 0.4386$$

$$\text{Now } \frac{T_1}{T_2} = e^{\mu_e\theta} = e^{0.4386\times\pi} = 3.9662$$

$$T_2 = \frac{4750}{3.9662} = 1197.6 \text{ N}$$

$$\text{Power transmitted} = \frac{i(T_1 - T_2)v}{10^3} = \frac{2(4750-1197.6)\times 23.56}{10^3} = 167.4 \text{ kW}$$

For maximum power transmitted, $T_c = \frac{T_{max}}{3} = \frac{5250}{3} = 1750 \text{ N}$

$$T_c = mv^2$$

$$v = \sqrt{\frac{T_c}{m}} = \sqrt{\frac{1750}{0.9}} = 44.1 \text{ m/s}$$

$$= \frac{pdn}{60}$$

$$n = \frac{60\times 44.1}{p\times 0.3} = 2807 \text{ rpm}$$

Summary for Quick Revision

(a) Flat Belt Drive

1 Belts are used for short centre drive to transmit motion on parallel shafts.

2 Angular velocity ratio:

$n_2/n_1 = d_1/d_2 = (d_1 + t)/(d_2 + t) = (1 - s/100)\cdot(d_1 + t)/(d_2 + t)$

where $s = (s_1 + s_2 - 0.01\, s_1 s_2)$ is the total percentage slip.

t = thickness of belt

3 Law of Belting

The law of belting states that the centre line of the belt as it approaches the pulley must lie in a plane perpendicular to the axis of that pulley, or must lie in the plane of the pulley, otherwise the belt will run off the pulley.

4 Length of open belt,

$L_o = 2C + (d_2 - d_1)^2/4C + \pi(d_1 + d_2)/2$

5 Length of cross belt,

$L_c = 2C + \pi(d_1 + d_2)/2 + (d_1 + d_2)^2/4C$

6 Angle of arc of contact

$\theta = \pi - 2\sin^{-1}[(d_2 - d_1)/2C]$ for open type belt

$= \pi + 2\sin^{-1}[(d_1 + d_2)/2C]$ for cross type belt

7 Ratio of belt tensions

$T_1/T_2 = \exp(\mu\,\theta)$

8 Power transmitted

$P = (T_1 - T_2)\,v/10^3 = T_1\, v\,[1 - \exp(-\mu\,\theta)]/10^3$ kW

9 Centrifugal tension

$T_c = m\,v^2$

10 Effective tension on tight side = $T_1 + T_c$

Effective tension on slack side = $T_2 + T_c$

11 Condition for maximum power transmission

$v = [T/(3\text{ m})]^{0.5}$

12 Initial belt tension

$T_o = (T_1 + T_2)/2$

$= (T_1 + T_2 + 2T_c)/2$ considering centrifugal tension.

13 The jumping of the belt forward and backward on the driving and driven pulleys respectively due to unequal tension on the two sides of the belt, is called creep.

14 Creep $= (T_1 - T_2)/(btE)$.

15 Velocity ratio after accounting for creep,

$(n_2/n_1) = (d_1/d_2)\ [(E + \sqrt{\sigma_2})/(E + \sqrt{\sigma_1})]$

16 The convex curvature given to the pulley rim is called crowning.

17 Crowning helps in running the belt in the centre of the pulley width.

18 Cone pulleys are used to obtain a range of speeds between the connected shafts.

19 Fast pulley is keyed to the shaft and loose pulley is free on the shaft.

(b) V-Belt Drive

20 Virtual (or apparent) coefficient of friction, $\mu_e = \mu/\sin\beta$
where β = semi-groove angle of pulley.

21 Ratio of belt tensions, $T_1/T_2 = \exp(\mu_e\theta)$.

(c) Chain Drive

22 Chains are used for short centre drive.

23 Chain pitch, $p = D\sin(180°/z)$.

24 Chain length, $L = 2C + \pi(R_1 + R_2) + (R_1 - R_2)^2/C$

$R = (p/2)\ \text{cosec}\ (180°/z)$.

(d) Rope Drive

25 Ropes are used for long distance drive.

26 Ratio of rope tensions, $T_1/T_2 = \exp(\mu_e\ \theta)$, $\mu_e = \mu/\sin\beta$, β = semi-groove angle of sheave.

Multiple Choice Questions

1 The centrifugal tension in belts
(a) reduces power transmission
(b) increases power transmission
(c) does not affect power transmission
(d) increases or decreases power transmission depending on speed.

2 In case of a flat belt drive with T as the maximum permissible tension, v as linear speed of belt, w as weight per metre length of belt, the maximum permissible speed is given by
(a) $T = wv^2/g$ (b) $T = 2\ wv^2/g$ (c) $T = 3wv^2/2g$ (d) $T = 3\sqrt{\ } wv^2/2g$.

3 With the same set of pulleys, belt and centre distance, the maximum power transmitted by
(a) cross belt is more than open belt
(b) cross belt is less than open belt
(c) cross and open belts is same
(d) cross and open belts depends upon pulley diameters.

4 The ratio of tensions in the tight and slack sides of a belt drive is
(a) $\mu \theta$ (b) exp $(\mu\theta)$ (c) $1/\mu \theta$ (d) exp $(1/\mu \theta)$.

5 If the percentage slip is same on both the driving and driven pulleys, then the speed ratio will
(a) increase (b) decrease (c) remain same (d) unpredictable.

6 The crowning of pulleys is done to
(a) make the belt run in the centre of the pulley face width
(b) strengthen the pulley
(c) give better shape to pulley
(d) decrease slip.

7 Considering centrifugal tension in a belt, the maximum linear velocity of belt is proportional to
(a) cube root of maximum tension (b) square root of maximum tension
(c) maximum tension (d) reciprocal of maximum tension.

8 If the initial tension in the belt is increased then the power transmitted by the belt
(a) reduces (b) increases (c) remains same (d) depends on speed.

9 The initial tension in the belt due to centrifugal tension, for the same power to be transmitted
(a) increases (b) decreases (c) remains same (d) depends on speed.

10 The maximum tension in the belt, for limiting friction conditions, occurs at
(a) starting (b) stopping (c) maximum power speed (d) specified speed.

11 The apparent coefficient of friction for V-belts is
(a) $\mu/\cos \beta$ (b) $\mu \cos \beta$ (c) $\mu \sin \beta$ (d) $\mu/\sin \beta$.
where β = semi-angle of pulley groove.

12 For maximum power to be transmitted by belt drive, the ratio of centrifugal tension to permissible tension is
(a) 1/2 (b) 1/3 (c) 2/3 (d) 1/4.

13 For maximum power to be transmitted by belt drive, the ratio of centrifugal tension to effective tight side tension is
(a) 1/2 (b) 1/3 (c) 2/3 (d) 1/4.

14 If the ratio of the tensions on tight and slack sides of a belt drive is increased by 20%, the power is
(a) increased by 20% (b) decreased by 20% (c) unaffected (d) unpredictable.

15 The net effect of creep in belts is to
(a) increase the speed of driven pulley (b) decrease the speed of driven pulley
(c) increase the power output (d) decrease the power output.

Answers:

1. (a) 2. (c) 3. (a) 4. (b) 5. (a) 6. (a) 7. (b) 8. (b) 9. (b) 10. (a) 11. (d) 12. (b) 13. (a) 14. (a) 15. (b)

Review Questions

1 State the law of belting.

2 What is the effect of belt thickness and slip on speed ratio?

3 What is the effect of centrifugal tension on power transmission?

4 What is the role of initial tension in flat belt drive?

5 Write the expression for ratio of belt tensions.

6 What is centrifugal tension in a belt?

7 What is the condition for maximum power transmission by a belt?

8 What is belt creep?

9 Why crowning of pulley is done?

10 Differentiate between a fast and loose pulley.

11 What is virtual coefficient of friction in a V-belt drive?

12 Define chain pitch.

13 When rope drive is preferred?

14 When chain drive is preferred?

15 When V-belt drive is preferred?

Exercises

6.1 A flat belt drive is required to transmit 20 kW at 300 rpm of 2 m diameter pulley. The angle of contact is 170° and coefficient of friction between belt and pulley is 0.30. The permissible stress for belt material is 3 MPa. Thickness of belt is 8 mm and density of its material is 0.95 kg/m^3. Find the width of belt required taking centrifugal tension into account.

[Ans. 66 mm]

6.2 A shaft running at 500 rpm carries a pulley 1 m diameter and drives another pulley by means of ropes with a speed ratio of 2:1. The drive transmits 200 kW. Angle of groove is 40° and distance between pulley centres is 2 m. The coefficient of friction between rope and pulley is 0.20. The mass of rope is 0.12 kg/m and has a safe stress of 1.75 MPa. The initial tension in the rope should not exceed 800 N. Calculate the number of ropes required and rope diameter.

[Ans. 8, 30 mm]

6.3 If the difference between tight and slack side tensions for a leather belt does not exceed 100 N/cm of width for a belt 5 mm thick, find the maximum stress in the belt. Assume the following data:

Angle of lap = 170°, coefficient of friction = 0.25, density of leather = 10^{-3} kg/cm^3, belt speed = 1000 m/min.

[Ans. 3.82 MPa]

6.4 The initial tension in a flat belt drive is 1800 N and angle of lap on the smaller pulley is 170°. The coefficient of friction between belt and pulley surface is 0.25. The pulley diameter is 0.9 m and runs at 540 rpm. Neglecting centrifugal tension, determine the power that can be transmitted.

[Ans. 32.5 kW]

6.5 It is required to reduce speed from 360 to 120 rpm by the use of chain drive. The driving sprocket has 10 teeth. Calculate: (a) the number of teeth on the follower, (b) the pitch of chain if pitch circle diameter of follower is 0.5 m, (c) the pitch circle diameter of driver, and (d) the length of chain, if centre distance is 0.4 m.

[Ans. 30, 52.26 mm, 169 mm, 1.92 m]

6.6 A rope pulley having a mean diameter of 1.5 m rotates at 90 rpm; angle of lap of ropes = 170°; angle of groove = 45°, safe tension per rope = 750 N, and coefficient of friction between the ropes and sides of groove = 0.25. Calculate the number of ropes for transmitting 50 kW.

[Ans. 11]

6.7 A belt of density 1000 kg/m^3 has a maximum permissible stress of 2.5 MPa. Calculate the maximum power that can be transmitted by a belt of 200 mm × 12 mm size if the ratio of tensions is 2.

[Ans. 57.734 kW]

6.8 Determine the maximum power that can be transmitted by a belt of 100 mm × 10 mm size with an angle of lap of 160°. The belt density is 1000 kg/m^3 and coefficient of friction is 0.25. The tension in the belt should not exceed 1.5 MPa.

[Ans. 11.18 kW]

6.9 The included angle of a V-grooved pulley is 30°. The belt is 20 mm deep and maximum width is 20 mm. The mass of belt is 0.35 kg per metre length and maximum allowable stress is 1.4 MPa. Determine the maximum power that can be transmitted if angle of lap is 140° and coefficient of friction is 0.15.

[Ans. 4.09 kW]

6.10 A pulley used to transmit power by means of ropes has a diameter of 3.6 m and has 15 grooves of 45° angle. The angle of contact is 170° and coefficient of friction between ropes and groove side is 0.28. The maximum possible tension in the ropes is 960 N and rope mass is 1.5 kg per metre length. Calculate the pulley speed in rpm and maximum power transmitted.

[Ans. 77.45 rpm, 124.16 kW]

7 BRAKES, CLUTCHES, AND DYNAMOMETERS

Chapter Outline

7.1 INTRODUCTION

Brakes are the devices that reduce the speed of a moving machine component by absorbing energy. The energy thus absorbed is converted into heat and released into the atmosphere or absorbed in another medium. Clutches are the devices that are used to engage or disengage two rotating machine components as and when desired. Dynamometers, on the other hand, are the devices that measure the power developed by a prime mover. In this chapter, we shall study these devices from the point of view of machine theory.

7.2 BRAKES

Brakes can be classified as follows:

1. Block or shoe brake
2. Band brake
3. Band and block brake
4. Internal expanding shoe brake

7.2.1 Block or Shoe Brake

These brakes may be classified as:

1. Single-block or shoe brake
2. Pivoted-block or shoe brake
3. Double-block or shoe brake

1. Single-block or Shoe Brake A single-block brake, shown in Fig.7.1(a), comes in to play when the force of friction passes through the fulcrum of the lever. When the angle of contact of the block on the brake drum is small (<60°), then the normal pressure between the block and the drum can be assumed to be uniform.

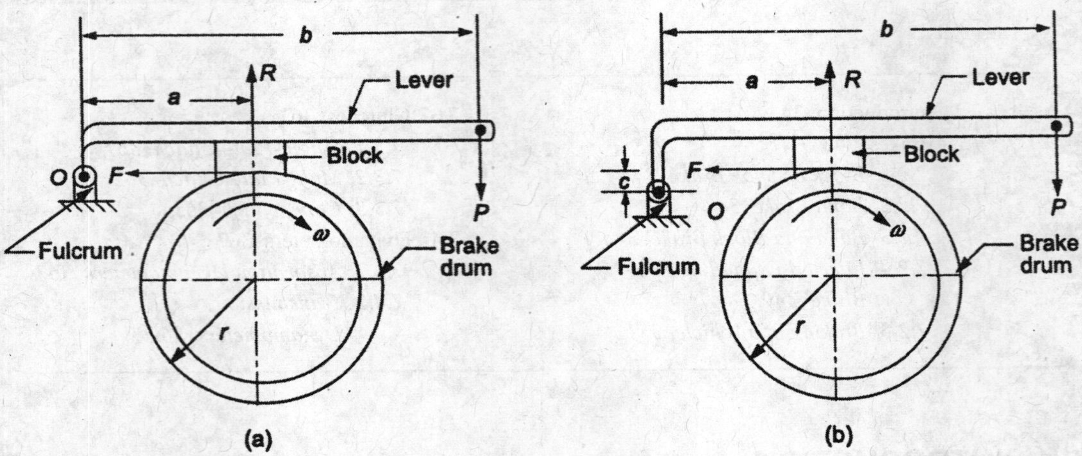

Fig.7.1 Block brake

Normal force between the block and drum, $R = \dfrac{Pb}{a}$

Tangential braking or frictional force on the drum, $F = \mu R = \dfrac{\mu Pb}{a}$

Braking torque, $$T_b = Fr = \frac{\mu Pbr}{a} \tag{7.1}$$

where r is the radius of the drum.

When the brake drum is moving on the rails or road with speed v and the braking distance is s, then

Work done against friction $= F \cdot s = \dfrac{\mu Pb \cdot s}{a}$

Kinetic energy lost $= \dfrac{1}{2} \cdot mv^2 + \dfrac{1}{2} \cdot I\omega^2$

where m = mass of brake drum,
I = moment of inertia of brake drum
ω = angular speed of drum.

For the conservation of energy, we have

$$\frac{\mu Pb\cdot s}{a}=\frac{1}{2}\cdot mv^2+\frac{1}{2}\cdot I\omega^2 \tag{7.2}$$

1. If the frictional force F is above the lever fulcrum by a distance c, as shown in Fig.7.1(b), then

$$\begin{aligned}P\cdot b &= R\cdot a+F\cdot c\\ &= R\cdot a+\mu R\cdot c\\ &= R(a+\mu c)\end{aligned}$$

or
$$R=\frac{Pb}{a+\mu c} \tag{7.3}$$

We find that the frictional force helps in applying the brake. Such a brake is called *self-energizing* brake.

2. If the fulcrum is above the frictional force F by an amount c, then

$$P\cdot b+F\cdot c=R\cdot a$$
$$P\cdot b=R(a-\mu c)$$

or
$$R=\frac{P\cdot b}{a-\mu c} \tag{7.4}$$

If $a \le \mu c$, then P will be zero or negative, that is no external force will be required to apply the brake. Such a brake is called *self-locking* type of brake.

2. Pivoted-block or Shoe Brake When the angle subtended by the shoe at the drum centre is more than 60°, then the normal pressure is lesser at the sides than at the centre. In such a case (Fig.7.2), consider an element of the block between ϕ and ϕ + dϕ.

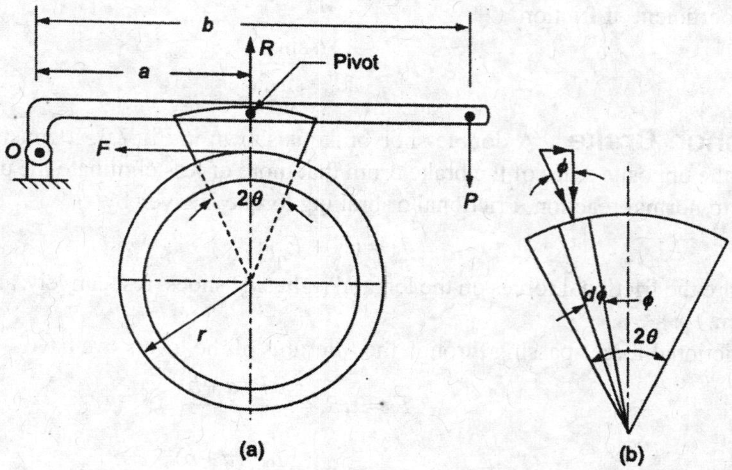

Fig.7.2 Pivoted shoe brake

Let b = width of the drum
r = radius of the drum
p = normal pressure between the block and the drum

Area of the drum element, $\quad dA = br\, d\phi$

Normal force on the drum, $\quad dR = pbr\, d\phi$

Vertical component of normal force $\quad = pbr\, d\phi \cos\phi$

Total normal force, $\quad R = br\int_{-\theta}^{+\theta} p\cos\phi\, d\phi$

Frictional force on element of the block $\quad = \mu pbr\, d\phi$

Resisting torque on the drum, $\quad dT = \mu pbr^2\, d\phi$

Total torque, $\quad T = \mu br^2 \int_{-\theta}^{+\theta} p\, d\phi$

Normal wear is proportional to the product of normal pressure and rubbing velocity. The component of wear in the direction of applied force P is proportional to $\cos\phi$. Hence, normal pressure is also proportional to $\cos\phi$.

Let $\quad p = k\cos\phi$

where k is the constant of proportionality. Then

$$R = brk\int_{-\theta}^{+\theta} \cos^2\phi\, d\phi$$

$$= \frac{brk(2\theta + \sin 2\theta)}{4}$$

$$T = \mu bkr^2 \int_{-\theta}^{+\theta} \cos\phi\, d\phi$$

$$= 2\mu bkr^2 \sin\theta$$

Eliminating k, we get

$$T = \frac{4\mu Rr\sin\theta}{2\theta + \sin 2\theta} \tag{7.5}$$

Equivalent coefficient of friction,

$$\mu_e = \frac{4\mu\sin\theta}{2\theta + \sin 2\theta}$$

3. Double-shoe Brake A double-shoe brake is shown in Fig.7.3. It consists of two brake shoes applied at the opposite sides of the brake drum that more or less eliminate the unbalanced force on the shaft due to normal reaction. Frictional or braking torque is given by,

$$T_b = (F_l + F_r)r \tag{7.6}$$

where F_l and F_r are the frictional forces on the left and right side shoes, respectively, and r is the radius of the brake drum.

Assuming frictional forces passing through the fulcrums of the levers, we have

$$F_l = \mu R_l = \frac{\mu P_l(a+b)}{a}$$

$$F_r = \mu R_r = \frac{\mu P_r(a+b)}{a}$$

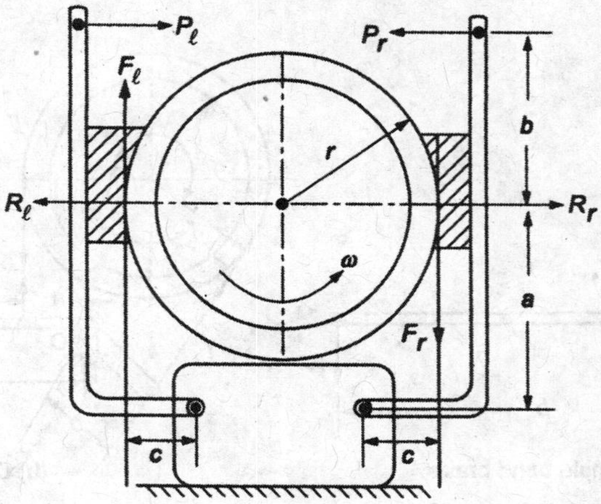

Fig.7.3 Double-shoe brake

When the frictional force is not passing through the fulcrums, then

$$P_l(a+b) + F_l c = R_l a$$

or
$$P_l(a+b) + \mu R_l c = R_l a$$

or
$$R_l = \frac{P_l(a+b)}{a-\mu c}$$

$$F_l = \frac{\mu P_l(a+b)}{a-\mu c}$$

and
$$P_r(a+b) = R_r a + F_r c = R_r a + \mu c$$

or
$$R_r = \frac{P_r(a+b)}{a+\mu c}$$

$$F_r = \frac{\mu P_r(a+b)}{a+\mu c}$$

7.2.2 Band Brake

The band brakes may be classified as: (a) simple-band brake, and (b) differential-band brake.

(1) Simple-band Brake A simple band brake is shown in Fig.7.4(a). Let T_1 and T_2 be the tensions on the tight and slack sides, respectively. Taking moments about the fulcrum, we have

$$Pb = T_1 a, \text{ for counter-clockwise rotation of the drum}$$
$$= T_2 a, \text{ for clockwise rotation of the drum}$$

Also
$$\frac{T_1}{T_2} = \exp(\mu\theta)$$

Braking torque on the drum, $$T_b = (T_1 - T_2)r \tag{7.7}$$

(2) Differential-band Brake The differential band brake is shown in Fig.7.4(b). Taking moments about the fulcrum, we have

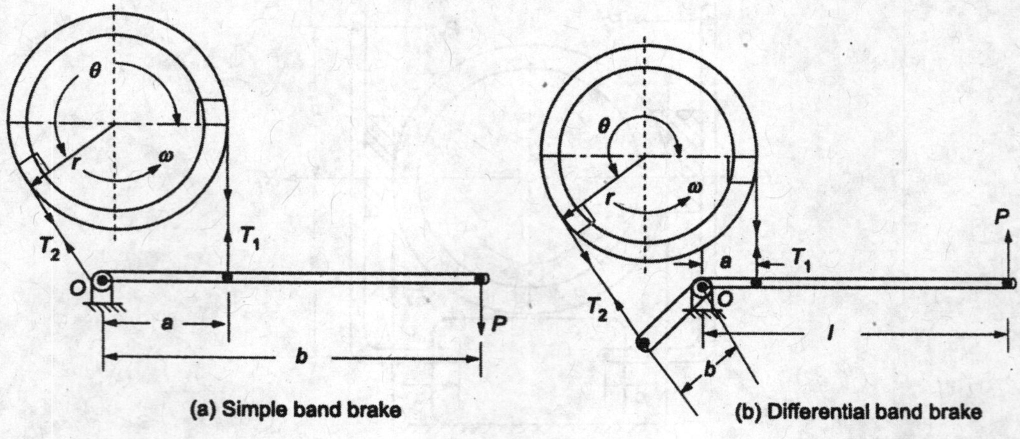

Fig.7.4 Band brake

$$Pl + T_1a = T_2b, \text{ for counter-clockwise rotation of the drum}$$

or $$Pl = T_2b - T_1a \tag{7.8a}$$

$$Pl + T_2a = T_1b, \text{ for clockwise rotation of the drum}$$

or $$Pl = T_1b - T_2a \tag{7.8b}$$

For a self-locking brake, $P \leq 0$, therefore, for counter-clockwise rotation,

$$\frac{T_1}{T_2} \geq \frac{b}{a} \tag{7.9a}$$

and for clockwise rotation,

$$\frac{T_1}{T_2} \leq \frac{a}{b} \tag{7.9b}$$

$$\frac{T_1}{T_2} = \exp(\mu\theta)$$

7.2.3 Band and Block Brake

The band and block brake is shown in Fig.7.5(a).

Let T_0 = tight side tension in the band on the first block
T_1 = slack side tension in the band on the first block
T_n = slack side tension in the band on the nth block
2θ = angle subtended by the block at the drum centre

The forces acting on the first block are shown in Fig.7.5(b). Resolving the forces horizontally and vertically, we get

$$(T_1 - T_0)\cos\theta = \mu R$$
$$(T_1 + T_0)\sin\theta = R$$

or $$\frac{T_1 - T_0}{T_1 + T_0}\cot\theta = \mu$$

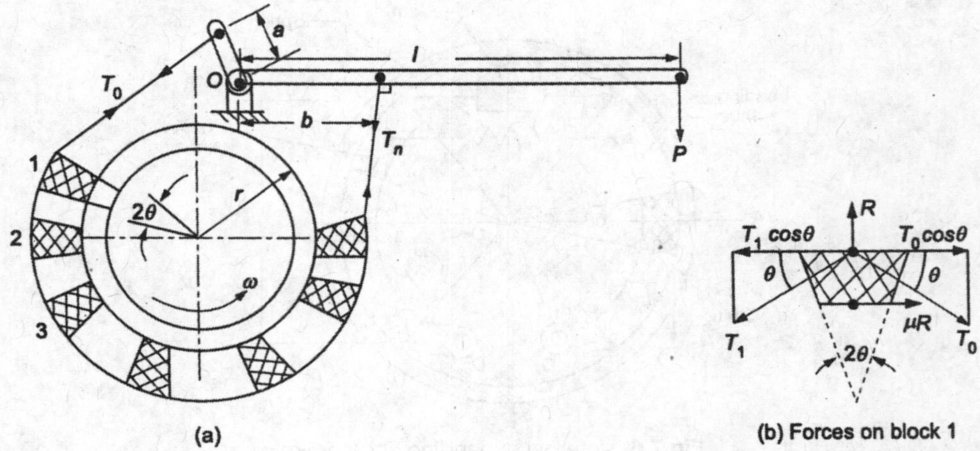

Fig.7.5 Block and band brake

or $$\frac{T_1}{T_0} = \frac{1+\mu\tan\theta}{1-\mu\tan\theta}$$

Similarly $$\frac{T_2}{T_1} = \frac{1+\mu\tan\theta}{1-\mu\tan\theta}$$

$$\frac{T_n}{T_{n-1}} = \frac{1+\mu\tan\theta}{1-\mu\tan\theta}$$

Hence $$\frac{T_n}{T_0} = \left[\frac{1+\mu\tan\theta}{1-\mu\tan\theta}\right]^n \tag{7.10}$$

Braking torque, $$T_b = (T_0 - T_n)r \tag{7.11}$$

7.2.4 Internal Expanding Shoe Brake

The internal expanding shoe brake is shown in Fig.7.6. The shoes are pinned at O_1 and O_2. The shoes are kept in non-braking position by the spring.

The brakes are applied when the cam is pressed down, in the case of a mechanically operated brake, or when the shoes are pressed on the brake drum, in the case of a hydraulically operated brake.

Consider a small element BC of the shoe between θ and $\theta + d\theta$.

Let r = radius of the brake drum
b = width of the brake lining
p = normal pressure between the shoe and the drum
p_{max} = maximum intensity of normal pressure
F_1 = force exerted by the cam on the leading shoe
F_2 = force exerted by the cam on the trailing shoe
a = distance between the fulcrum O_1 and O.

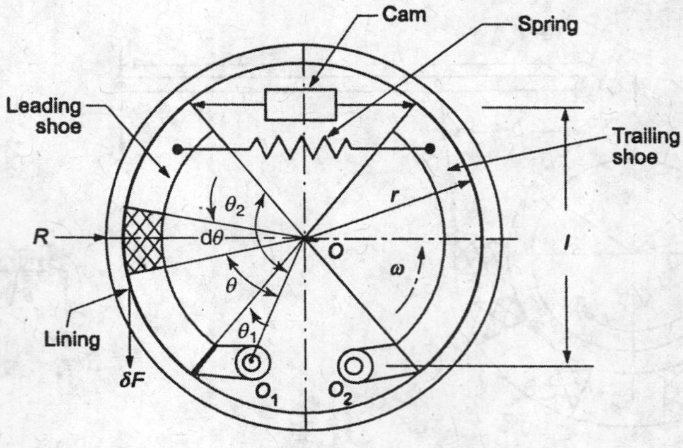

Fig.7.6 Internal expanding shoe brake

It is assumed that the pressure distribution on the shoe is nearly uniform. The shoe turns about point O_1. Therefore, the rate of wear of the shoe lining at B will be proportional to the radial displacement of that point. The rate of wear of the shoe lining varies directly as the perpendicular distance from O_1 to OB, that is, O_1A.

Now $$O_1A = OO_1 \sin\theta = a \sin\theta$$

Normal pressure at B,

$$p \propto \sin\theta$$

or $$p = p_{\max} \sin\theta$$

Normal force acting on the element,

$$\delta R = \text{Normal pressure} \times \text{Area of the element}$$
$$= p \cdot br\delta\theta = p_{\max} \sin\theta \cdot br\delta\theta$$

Friction force on the element,

$$\delta F = \mu \cdot \delta R = \mu \cdot p_{\max} \cdot \sin\theta \cdot br\delta\theta$$

Braking torque due to the element about O,

$$\delta T_b = \delta F \cdot r = \mu p_{\max} br^2 \sin\theta \cdot \delta\theta$$

Total braking torque about O,

$$(T_b = \mu p_{\max}\, br^2 \int_{\theta_1}^{\theta_2} \sin\theta\, d\theta)$$
$$= \mu p_{\max}\, br^2 (\cos\theta_1 - \cos\theta_2)$$

Moment of the normal force δR about the fulcrum O_1,

$$\delta M_n = \delta R \cdot O_1A = \delta R \cdot OO_1 \sin\theta = \delta R \cdot a \sin\theta$$

Total moment of normal force about the fulcrum O_1,

$$M_n = p_{\max} \cdot b \cdot r \cdot a \int_{\theta_1}^{\theta_2} \sin^2\theta\, d\theta$$
$$= p_{\max} \cdot b \cdot r \cdot a \int_{\theta_1}^{\theta_2} \left[\frac{1 - \cos 2\theta}{2}\right] d\theta$$
$$= \frac{(p_{\max} \cdot b \cdot r \cdot a)}{2} \cdot \left[\frac{\theta - \sin 2\theta}{2}\right]_{\theta_1}^{\theta_2}$$

$$= \frac{(p_{\max} \cdot b \cdot r \cdot a)}{2} \cdot \left[(\theta_2 - \theta_1) + \left(\frac{1}{2}\right) \cdot (\sin 2\theta_1 - \sin 2\theta_2) \right] \tag{7.12}$$

Moment of frictional force δF about the fulcrum O_1,

$$\delta M_f = \delta F \cdot AB = \delta F (r - a\cos\theta)$$
$$= \mu p_{\max} \sin\theta \cdot br\,\delta\theta \cdot (r - a\cos\theta)$$
$$= \mu p_{\max} \cdot br(r\sin\theta - a\sin\theta\cos\theta)\,\delta\theta$$
$$= \mu p_{\max} \cdot br\left(r\sin\theta - \frac{a\sin 2\theta}{2}\right)\delta\theta$$

Total moment of the frictional force about the fulcrum O_1,

$$M_f = \mu p_{\max} \cdot br \int_{\theta_1}^{\theta_2} \left[r\sin\theta - \frac{a\sin 2\theta}{2} \right] d\theta$$
$$= \mu p_{\max} \cdot br \left[-r\cos\theta + \frac{a\cos 2\theta}{4} \right]_{\theta_1}^{\theta_2}$$
$$= \mu p_{\max} \cdot br \left[r(\cos\theta_1 - \cos\theta_2) + \left(\frac{a}{4}\right) (\cos 2\theta_2 - \cos 2\theta_1) \right] \tag{7.13}$$

For the leading shoe, taking moments about the fulcrum O_1,

$$F_1 \cdot l = M_n - M_f \tag{7.14}$$

and for the trailing shoe, taking moments about the fulcrum O_2,

$$F_2 \cdot l = M_n + M_f \tag{7.15}$$

7.2.5 Braking of a Vehicle

Consider a vehicle going up an inclined plane with acceleration 'a', as shown in Fig.7.7. To stop the vehicle, let brakes be applied to all the four wheels.

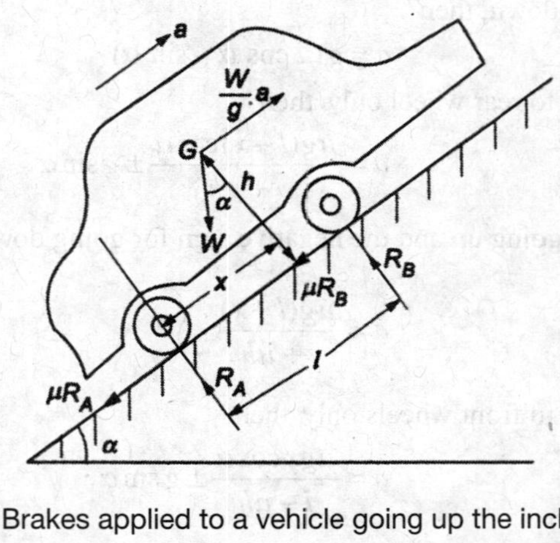

Fig.7.7 Brakes applied to a vehicle going up the inclined plane

Let $F_A = \mu R_A$ = braking force applied at the rear wheels
$F_B = \mu R_B$ = braking force applied at the front wheels
R_A, R_B = normal reactions at A and B
W = weight of the vehicle
h = height of the C.G. of the vehicle from the ground level
l = wheel base

Resolving the forces parallel to the plane, we have

$$\mu(R_A + R_B) + W \sin\alpha = \frac{W}{g}\cdot a \tag{7.16}$$

Resolving the forces perpendicular to the plane, we have

$$R_A + R_B = W \cos\alpha \tag{7.17}$$

From (7.17), we get

$$R_B = W \cos\alpha - R_A \tag{7.18}$$

Taking moments about G, we have

$$\mu(R_A + R_B)h + R_A \cdot x = (W\cos\alpha - R_A)(l - x)$$

$$\mu Wh\cos\alpha + R_A(x + l - x) = W(l - x)\cos\alpha$$

$$R_A = W\cos\alpha\left[\frac{l - x - \mu h}{l}\right]$$

$$R_B = W\cos\alpha\left[\frac{l - (l - x - \mu h)}{l}\right]$$

$$= W\cos\alpha\left[\frac{x + \mu h}{l}\right]$$

From (7.16), we have

$$\mu W\cos\alpha + W\sin\alpha = \frac{W}{g}\cdot a$$

or

$$a = g(\mu\cos\alpha + \sin\alpha) \tag{7.19}$$

(a) When the vehicle moves on a level track, then $\alpha = 0$, and

$$a = \mu g \tag{7.20}$$

(b) When the vehicle moves down, then

$$a = g(\mu\cos\alpha - \sin\alpha) \tag{7.21}$$

(c) When brakes are applied to rear wheel only, then

$$a = \frac{\mu g(l - x)\cos\alpha}{l + \mu h} \pm g\sin\alpha \tag{7.22}$$

Use the positive sign for going up and the negative sign for going down the plane.
On a level track,

$$a = \frac{\mu g(l - x)}{l + \mu h} \tag{7.23}$$

(d) When brakes are applied to front wheels only, then

$$a = \frac{\mu g x\cos\alpha}{l - \mu h} \pm g\sin\alpha \tag{7.24}$$

Use the positive sign for going up the plane and the negative sign for going down.
On a level track,

$$a = \frac{\mu g x}{l - \mu h} \tag{7.25}$$

Example 7.1

A bicycle and rider of mass 120 kg are travelling at a speed of 15 km/h on a level road. The rider applies brake to the rear wheel that is 0.9 m in diameter. How far will the bicycle travel before it comes to rest? The pressure applied on the brake is 100 N and coefficient of friction between the brake and the cycle rim is 0.05. Assume that no other resistance is acting on the bicycle.

Solution

Frictional for, $F = \mu R = 0.05 \times 100 = 5$ N

Let s = be the distance travelled after the bicycle comes to rest in m

$$\text{Work done} = F \times s = 5 \cdot s \text{ Nm}$$

Kinetic energy of the wheel, neglecting rotational energy,

$$= \frac{1}{2} m v^2$$

$$= \frac{1}{2} \times 120 \times \left(\frac{15 \times 1000}{3600}\right)^2 = 1041.67 \text{ Nm}$$

Hence,

$$5s = 1041.67$$

$$s = 208.33 \text{ m}$$

Example 7.2

A double-shoe brake (Fig.7.8) is capable of absorbing a torque of 1500 Nm. The diameter of the brake drum is 300 mm and the angle of contact for each shoe is 90°. The coefficient of friction between the brake drum and the lining is 0.35. Find (a) the spring force necessary to set the brake and (b) width of the brake shoes. The bearing pressure on the lining material is not to exceed 0.25 MPa.

Solution

(a) Let P be the spring force to set the brake. Since the angle of contact is greater than 60°, therefore, equivalent coefficient of friction,

$$\mu_e = \frac{4\mu \sin\theta}{2\theta + \sin 2\theta}$$

$$= \frac{4 \times 0.35 \times \sin 45°}{2 \times \frac{\pi}{4} + \sin\frac{\pi}{2}}$$

$$= 0.385$$

Taking moments about the fulcrum O_1, we have

$$P_r \times 480 = R_r \times 220 + F_r \times (150 - 30)$$

$$= \left(\frac{F_r}{0.385}\right) \times 220 + F_r \times 120 = 691.428 F_r$$

or

$$F_r = 0.694\, P_r$$

Now taking moments about O_2, we have

$$P_l \times 480 + F_l \times (150-30) = R_l \times 220 = \left(\frac{F_l}{0.385}\right) \times 220$$

$$F_l = 1.0633\, P_l$$

Let the spring force, $P_l = P_r = P$

Torque capacity of the brake, $T_b = (F_l + F_r)r$

$$1500 = (1.0633 + 0.694)P \times 0.150$$

$$P = 5690.5 \text{ N}$$

(b) Let b be the width of the brake shoes in mm.

Projected bearing area for one shoe, $A_b = b \cdot 2r \sin\theta$

$$= b\,(2 \times 150 \sin 45°) = 212.1\, b \text{ mm}^2$$

$$R_r = \frac{F_r}{0.385} = \frac{0.694 \times 5690.5}{0.385} = 10257.7 \text{ N}$$

$$R_l = \frac{F_l}{0.385} = \frac{1.0633 \times 5690.5}{0.385} = 15716.1 \text{ N}$$

The normal force is maximum on the left-hand side shoe.

$$0.25 = \frac{15716.1}{212.1\, b}$$

or $b = 296.4$ mm

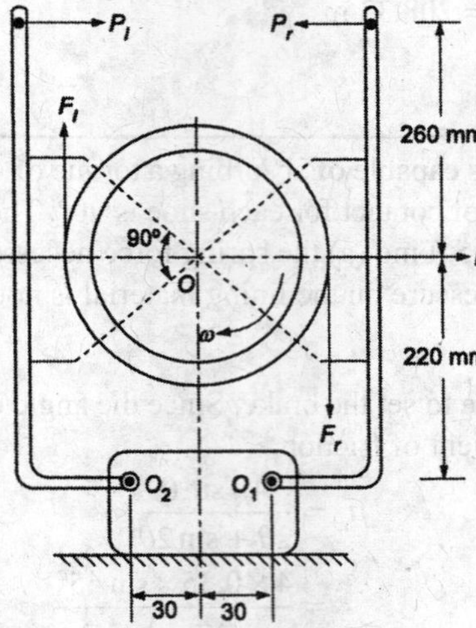

Fig.7.8 Double-shoe brake mechanism

Example 7.3

A simple band brake, as shown in Fig.7.9, is used on a shaft carrying a flywheel of mass 450 kg. The radius of gyration of the flywheel is 500 mm and runs at 320 rpm. The coefficient of friction is 0.2 and the brake drum diameter is 250 mm. Find (a) torque applied due to a hand load of 150 N, (b) the number of turns of the wheel before it is brought to rest, and (c) the time required to bring it to rest from the moment of application of the brake.

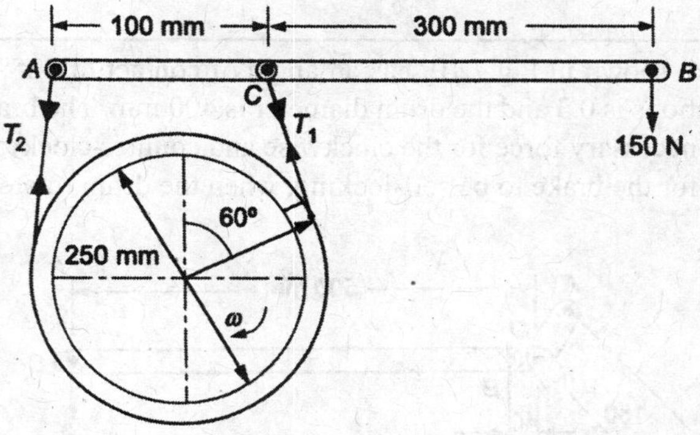

Fig.7.9 Simple band brake mechanism

■ Solution

(a) Angle of contact, $\theta = 210° = 3.6652$ rad

$$\frac{T_1}{T_2} = \exp(\mu\theta) = \exp(0.2 \times 3.6652) = 2.08$$

Taking moments about the fulcrum C, we have

$$T_2 \times 100 = 150 \times 300$$

$$T_2 = 450 \text{ N}$$

$$T_1 = 936 \text{ N}$$

Torque applied, $T_b = (T_1 - T_2)\, r = (936 - 450) \times 0.125$
$= 60.75$ Nm

(b) Rotational kinetic energy of the wheel $= \frac{1}{2} \cdot I\omega^2$

$$= \frac{1}{2} \cdot mK^2 \cdot \omega^2$$

$$= \frac{1}{2} \times 450 \times 0.5^2 \times \left(2\pi \times \frac{320}{60}\right)^2$$

$$= 63165.5 \text{ Nm}$$

Energy used to overcome the braking torque $= 2\pi n T_b$

$$= 2\pi \times n \times 60.75$$

$$= 381.7n \text{ Nm}$$

$$= 63165.5$$

or $n = 165$

(c) Time required to bring the wheel to rest $= \dfrac{n}{\left(\frac{N}{2}\right)}$

$$= \frac{2 \times 165}{320} = 1.031 \text{ min or } 61.87 \text{ s}$$

Example 7.4

A differential band brake shown in Fig.7.10, has an angle of contact of 225°. The band has a lining whose coefficient of frictions is 0.3 and the drum diameter is 400 mm. The brake is to sustain a torque of 375 Nm. Find (a) the necessary force for the clockwise and counter-clockwise rotation of the drum and (b) the value of *OA* for the brake to be self-locking, when the drum rotates clockwise.

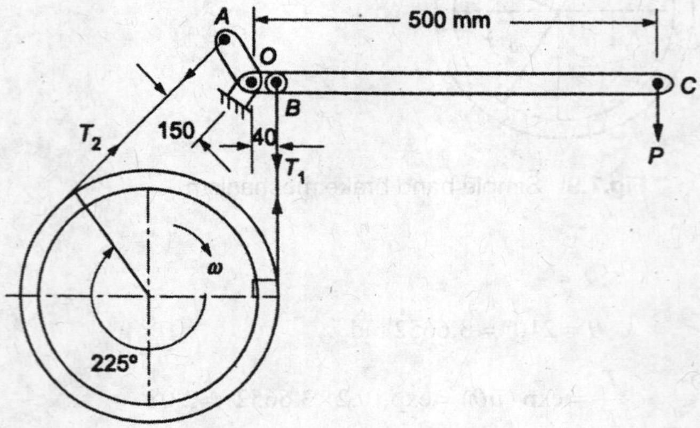

Fig.7.10 Differential band brake mechanism

Solution

1. Force required

(a) Clockwise rotation of the drum

$$\frac{T_1}{T_2} = \exp(\mu\theta) = \exp\left(\frac{0.3 \times \pi \times 225}{180}\right) = 3.248$$

$$T_b = (T_1 - T_2)r$$

$$375 = (T_1 - T_2) \times 0.2$$

$$T_1 - T_2 = 1875 \text{ N}$$

$$T_1 = 2709 \text{ N}, \quad T_2 = 834 \text{ N}$$

Taking moments about the fulcrum *O*, we have

$$P \times 500 + T_1 \times 40 = T_2 \times 150$$

$$P \times 500 + 2709 \times 40 = 834 \times 150, \quad P = 46.7 \text{ N}$$

(b) Counter-clockwise rotation of the drum

Taking moments about *O*, we have

$$P \times 500 + T_2 \times 40 = T_1 \times 150$$

$$P \times 500 + 834 \times 40 = 2790 \times 150, \quad P = 903.7 \text{ N}$$

2. For the brake to be self-locking, $P = 0$. For clockwise rotation of the drum,

$$T_1 \times 40 = T_2 \times OA$$

$$OA = \frac{2709 \times 40}{834} = 129.92 \text{ mm}$$

Example 7.5

A band and block brake, with 15 blocks, each of which subtends an angle of 15°, is applied to a drum of 1 m diameter, as shown in Fig.7.11. The drum and the flywheel mounted on the same shaft has a mass of 1500 kg and a combined radius of gyration of 500 mm. Find (a) maximum braking torque, (b) angular retardation of the drum, and (c) time taken by the system to come to rest from the rated speed of 380 rpm. The coefficient of friction between the drum and the blocks can be taken as 0.25.

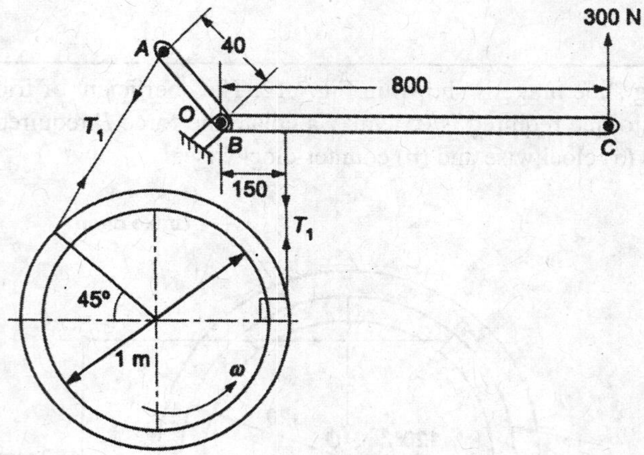

Fig.7.11 Band and block brake mechanism

■ **Solution**

(a) The braking torque will be maximum when the drum rotates counter-clockwise and the force P is upwards. Taking moments about O, we have

$$300 \times 800 + T_1 \times 40 = T_2 \times 150$$
$$15T_2 - 4T_1 = 24000$$
$$\frac{T_1}{T_2} = \left[\frac{1+\mu\tan\theta}{1-\mu\tan\theta}\right]^n$$
$$= \left[\frac{1+0.25\times\tan 7.5°}{1-0.25\times\tan 7.5°}\right]^{15} = 2.685$$
$$15T_2 - 4\times 2.685T_2 = 24000$$
$$4.26T_2 = 24000$$
$$T_2 = 5633.8 \text{ N}$$
$$T_1 = 15126.76 \text{ N}$$

Braking torque, $T_b = (T_1 - T_2)\,r = (15126.76 - 5633.8) \times 0.5$

$$= 4744 \text{ Nm}$$

(b) Let α be the angular retardation of the drum

$$T_b = I\alpha = m\,K^2 \cdot \alpha$$
$$4744 = 1500 \times (0.5)^2 \cdot \alpha$$
$$\alpha = 12.65 \text{ rad/s}^2$$

(c) Let t be the time taken to come to rest

Initial angular speed, $\omega_1 = \dfrac{2\pi \times 380}{60} = 39.793$ rad/s

Final angular speed, $\omega_2 = 0$

$$\omega_2 = \omega_1 - \alpha t$$

$$t = \frac{39.793}{12.65} = 3.15 \text{ s}$$

Example 7.6

An external expanding shoe brake is shown in Fig.7.12. The coefficient of friction may be taken as 0.35, and the braking torque required is 25 Nm. Calculate the force P required to operate the brake when the drum rotates (a) clockwise and (b) counter-clockwise.

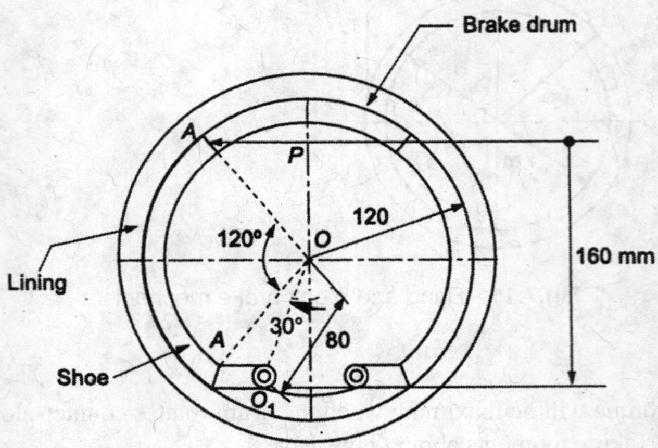

Fig.7.12 External expanding shoe brake mechanism

■ Solution

(a) When drum rotates clockwise,

Total braking torque, $T_b = \mu P_{max} \cdot b \cdot r^2 (\cos\theta_1 - \cos\theta_2)$

$$25 = 0.35 \times P_{max} \times b \times 0.12^2 (\cos 30° - \cos 150°)$$

$$P_{max} b = 2863.84$$

Total moment of the normal force about the fulcrum O_1,

$$M_n = 0.5\, P_{max}\, bra\, [(\theta_2 - \theta_1) + 0.5 (\sin 2\theta_1 - \sin 2\theta_2)]$$

$$= 0.5 \times 2863.84 \times 0.12 \times 0.08 \left[\frac{(150-30)\times\pi}{180} + 0.5(\sin 60° - \sin 300°)\right]$$

$$= 40.695 \text{ Nm}$$

Total moment of friction force about the fulcrum O_1,

$$M_f = \mu P_{max} br\, [r (\cos\theta_1 - \cos\theta_2) + 0.25\, a (\cos 2\theta_2 - \cos 2\theta_1)]$$

$$= 0.35 \times 2863.84 \times 0.12\, [0.12(\cos 30° - \cos 150°) + 0.25 \times 0.08(\cos 300° - \cos 60°)]$$

$$= 25 \text{ Nm}$$

Total moment $= M_n + M_f = 65.695$ Nm

Taking moments about O_1, we have

$$P \times 0.16 = 65.695$$
$$P = 410.6 \text{ N}$$

(b) When drum rotates counter-clockwise, taking moments about O_1, we have

$$P \times 0.16 = M_n - M_f = 15.695$$
$$P = 98.1 \text{ N}$$

Example 7.7

A vehicle moving on a rough plane inclined at 15° with the horizontal at a speed of 40 km/h has a wheel base of 1.8 m. The centre of gravity of the vehicle is 0.8 m from the rear wheels and 0.9 m above the inclined plane. Find the distance travelled by the vehicle before coming to rest and the time taken to do so when the vehicle is moving (a) up the plane and (b) down the plane. The brakes are applied to all the four wheels and the coefficient of friction is 0.45.

Solution

(a) For the vehicle moving up the plane,

$$a = g(\mu \cos \alpha + \sin \alpha)$$
$$= 9.81(0.45 \times \cos 15° + \sin 15°)$$
$$= 6.8 \text{ m/s}^2$$

Distance travelled, $$s = \frac{u^2}{2a} = \frac{\left(\frac{40 \times 1000}{3600}\right)^2}{2 \times 6.8}$$
$$= 9.078 \text{ m}$$

Final velocity of vehicle, $v = u + at$

$$0 = 11.11 - 6.8t$$
$$t = 1.63 \text{ s}$$

(b) For the vehicle going down the plane,

$$a = g(\mu \cos \alpha - \sin \alpha)$$
$$= 9.81(0.45 \times \cos 15° - \sin 15°)$$
$$= 1.725 \text{ m/s}^2$$

$$s = \frac{u^2}{2a} = \frac{(11.11)^2}{2 \times 1.725} = 35.78 \text{ m}$$

$$t = \frac{u}{a} = \frac{11.11}{1.725} = 6.44 \text{ s}$$

Example 7.8

The wheel base of a car is 3 m and its centre of gravity is 1.2 m ahead of the rear axle and 0.75 m above the ground level. The coefficient of friction between the wheels and the road is 0.5. Determine the maximum deceleration of the car when it moves on a level if the braking force on all the four wheels is the same and no wheel slip occurs.

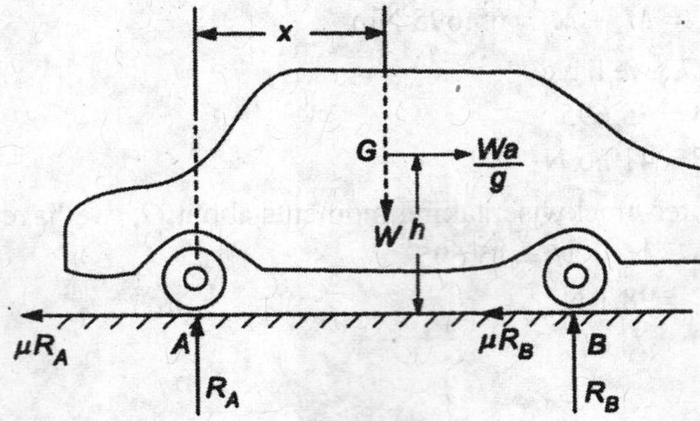

Fig.7.13 Forces on the vehicle

■ Solution

(a) When slipping is imminent on the rear wheels (Fig.7.13),

$$R_A = W\left[\frac{l - \mu h - x}{l}\right]$$

$$= W\left[\frac{3 - 0.5\times 0.75 - 1.2}{3}\right] = 0.475\,W\,\text{N}$$

$$F_A + F_B = W\frac{a}{g}$$

Now $\quad F_A = F_B \text{ and } F_A = \mu R_A$

$$2\mu R_A = W$$

or $\quad 2\times 0.5\times 0.475\,W = W\frac{a}{g}$

$$a = 4.66\text{ m/s}^2$$

(b) When slipping is imminent on the front wheels,

$$R_B = W\left[\frac{\mu h + x}{l}\right]$$

$$= W\left[\frac{0.5\times 0.75 + 1.2}{3}\right] = 0.525\,W\,\text{N}$$

$$F_A + F_B = \frac{Wa}{g}$$

Now $\quad F_A = F_B \text{ and } F_B = \mu R_B$

$$2\mu R_B = W$$

or $\quad 2\times 0.5\times 0.525\,W = \frac{Wa}{g}$

$$a = 5.15\text{ m/s}^2$$

7.3 CLUTCHES

The friction clutches may be classified as follows:

1. Plate (or disc) clutches
   (a) Single-plate clutches
   (b) Multiple-plate clutches
2. Cone clutch

7.3.1 Single-Plate Clutch

Consider a single-plate clutch as shown in Fig.7.14.

Let r_1, r_2 = inner and outer radii of the plate, respectively
p = intensity of axial pressure on the plate
W = axial load on the clutch
μ = coefficient of friction between the friction surfaces and the plate
T = torque transmitted by the clutch

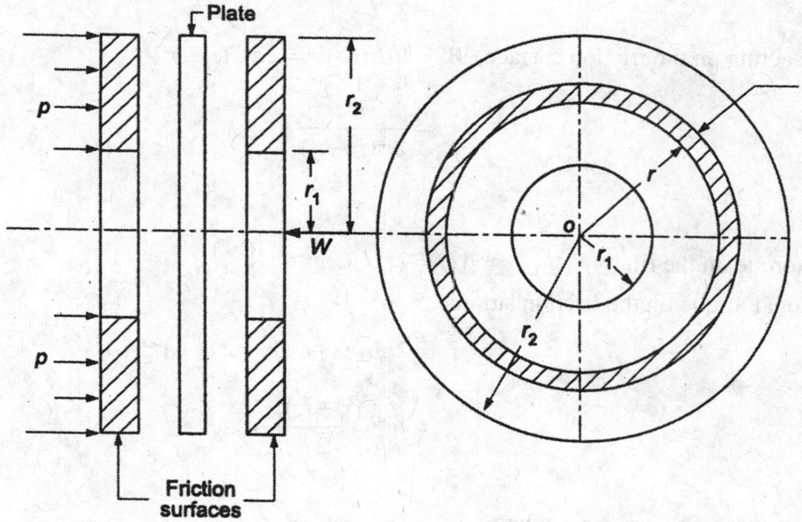

Fig.7.14 Single-plate clutch

Consider an elementary ring of the friction surface of radius r and thickness dr.

Contact area of the friction surface, $\mathrm{d}A = 2\pi r \cdot \mathrm{d}r$
Arial force on the ring, $\mathrm{d}W = p \cdot \mathrm{d}A$
Frictional force on the ring, $\mathrm{d}F = \mu \cdot \mathrm{d}W$
Frictional torque on the ring, $\mathrm{d}T_f = \mathrm{d}F \cdot r = 2\pi\mu \cdot p \cdot r^2 \cdot \mathrm{d}r$

(a) Uniform pressure
When the pressure is uniform over the entire area of friction surface, then the intensity of pressure,

$$p = \frac{W}{\pi(r_2^2 - r_1^2)}$$

Total frictional torque on the frictional surface,

$$T_f = 2\pi\mu\pi \int_{r_1}^{r_2} r^2 \mathrm{d}r = \frac{2}{3}\pi\mu p(r_2^3 - r_1^3)$$

$$= \left(\frac{2}{3}\right)\cdot \mu W\left[\frac{r_2^3 - r_1^3}{r_2^2 - r_1^2}\right] \quad (7.26)$$

Mean radius of friction surface, $r_m = \left(\frac{2}{3}\right)\cdot\left[\frac{r_2^3 - r_1^3}{r_2^2 - r_1^2}\right]$ (7.27)

(b) Uniform wear

For uniform wear, the intensity of pressure varies inversely with the distance, therefore

$$p\cdot r = \mathrm{C}$$

or

$$p = \frac{C}{r}$$

At $r = r_1, p = p_{\max}$ and at $r = r_2, p = p_{\min}$

Force acting on the ring, $\mathrm{d}W = \frac{C}{r}\cdot 2\pi r\cdot \mathrm{d}r = 2\pi C\cdot \mathrm{d}r$

Total force acting on the friction surface, $W = \int_{r_1}^{r_2} 2\pi C\cdot \mathrm{d}r = 2\pi C(r_2 - r_1)$

or

$$C = \frac{W}{2\pi(r_2 - r_1)}$$

$$W = p_{\max}\times r_1 \times 2\pi(r_2 - r_1)$$

Frictional torque on the ring, $\mathrm{d}T_f = 2\pi\mu\cdot C\cdot r\cdot \mathrm{d}r$ (7.28)

Total frictional torque on the friction suface,

$$T_f = 2\pi\mu\cdot C\int_{r_1}^{r_2} r\cdot \mathrm{d}r = \pi\mu\cdot C(r_2^2 - r_1^2)$$

$$= \mu W\frac{(r_1 + r_2)}{2} \quad (7.29)$$

where $r_m = \frac{r_1 + r_2}{2}$

7.3.2 Multi-Plate Clutch

A multi-plate friction clutch is shown in Fig.7.15. Multi-plate friction clutches are used where space is a limitation, as in the case of two wheelers, scooters, etc.

Let n_1 = number of plates on the driving shaft

n_2 = number of plates on the driven shaft

Then, number of pairs of contact surfaces, $n = n_1 + n_2 - 1$

Total frictional torque transmitted, $T = n\mu W r_m$ (7.30)

Dry and Wet Clutches In case of single-plate clutches, the contacting surfaces are either one or two. Due to large surface area available, heating is not a problem and the clutch is of the dry type. In case of multi-plate clutches, the work done during engagement and disengagement is

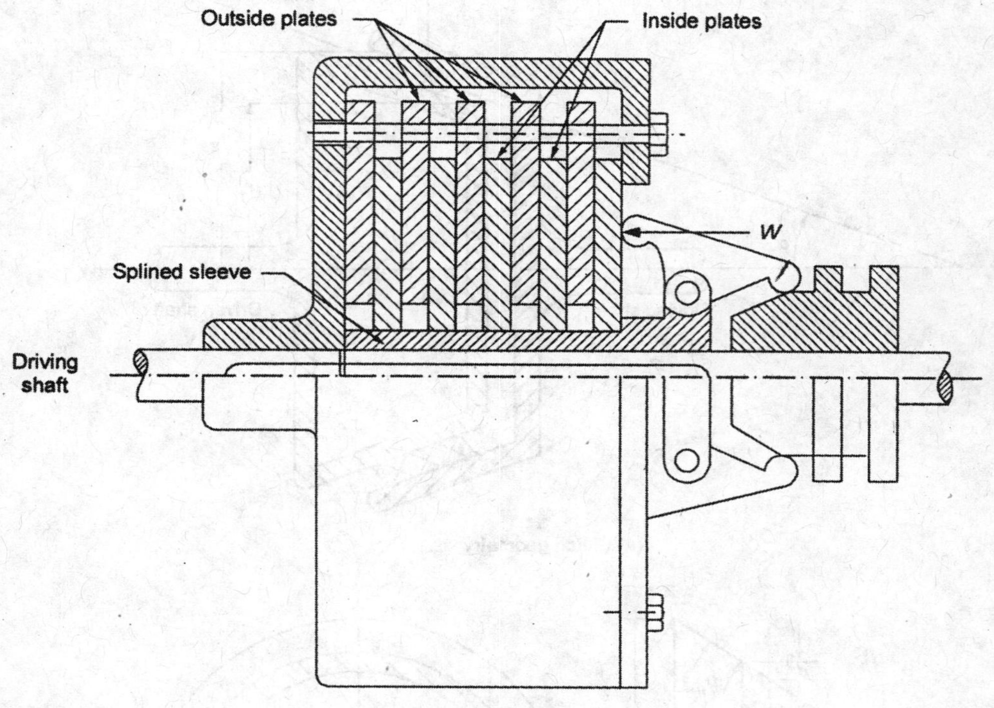

Fig.7.15 Multi-plate friction clutch

converted into heat. Heat dissipation is a serious problem in multi-plate clutches because of the large number of friction surfaces. Cooling oil is used to dissipate this heat. Therefore, these clutches are wet clutches.

7.3.3 Cone Clutch

Consider a cone clutch as shown in Fig.7.16(a).

Let r_1 and r_2 = inner and outer radii of the frictional conical surface, respectively
p_n = normal pressure between the contact surfaces
b = width of the conical surfaces
μ = coefficient of friction between contact surfaces
α = semi-cone angle.

Consider an elementary ring of the conical surface of radii r and $r + \mathrm{d}r$ and of length $\mathrm{d}l$, as shown in Fig.7.16(b).

$$\mathrm{d}l = \mathrm{d}r \cdot \operatorname{cosec} \alpha$$

Area of the ring,
$$\mathrm{d}A = 2\pi r \cdot \mathrm{d}l = 2\pi r \cdot \mathrm{d}r \cdot \operatorname{cosec} \alpha$$

(a) Uniform pressure

Normal load acting on the ring, $\mathrm{d}Wn = p_n \cdot \mathrm{d}A$

Axial load acting on the ring, $\mathrm{d}W = \mathrm{d}W_n \cdot \sin \alpha = 2\pi p_n \cdot r \cdot \mathrm{d}r$

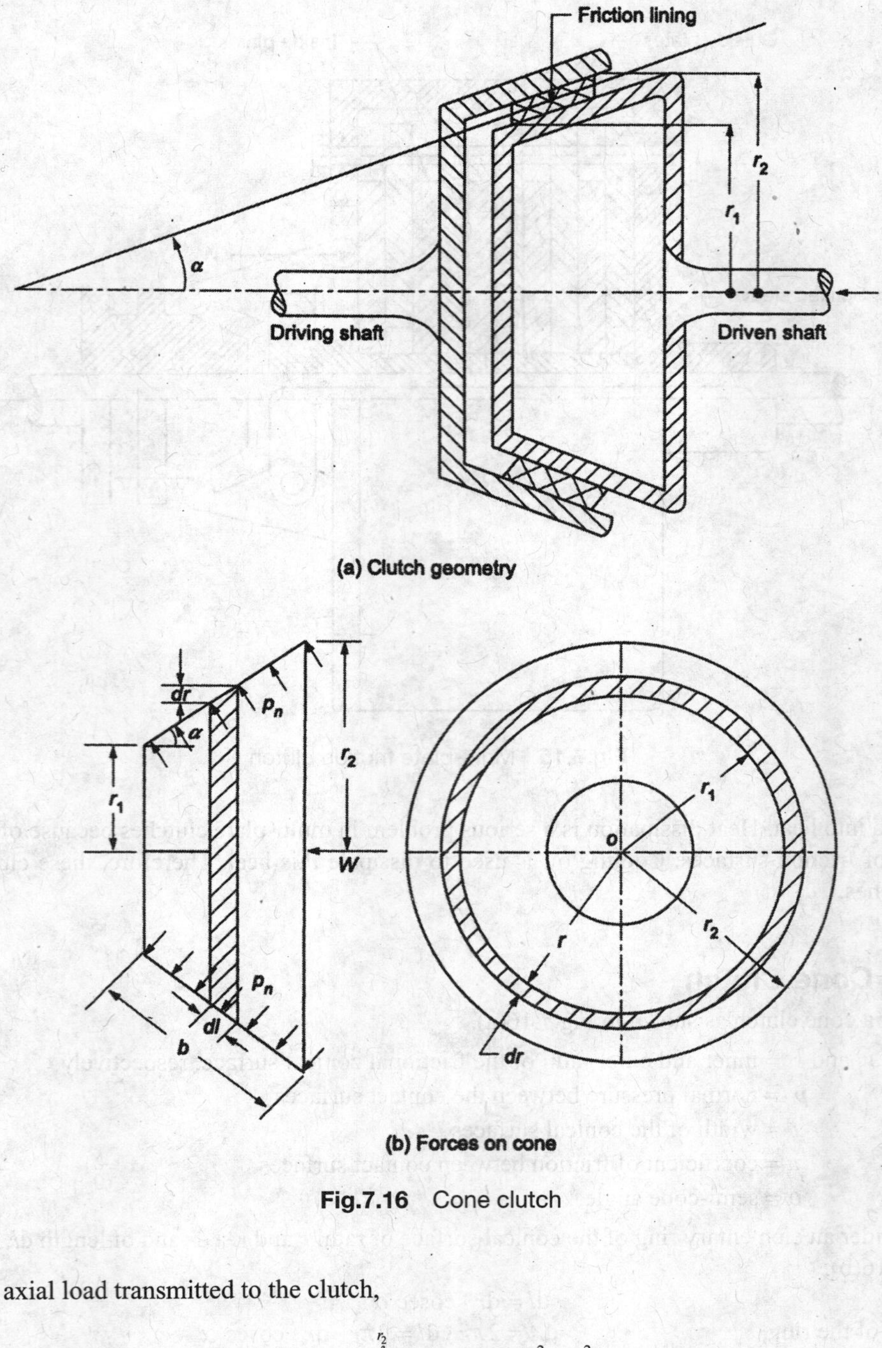

Fig.7.16 Cone clutch

Total axial load transmitted to the clutch,

$$W = 2\pi\, p_n \int_{r_1}^{r_2} r \cdot \mathrm{d}r = \pi\, p_n (r_2^2 - r_1^2)$$

or
$$p_n = \frac{W}{\pi (r_2^2 - r_1^2)}$$

Frictional force on the ring, $dF = \mu \cdot dW_n$

Frictional torque, $dT_f = dF \cdot r = 2\pi\mu \cdot p_n \cdot \text{cosec}\ \alpha \cdot r^2\, dr$

Total frictional torque, $T_f = 2\pi\mu \cdot p_n \cdot \text{cosec} \cdot \alpha \int_{r_1}^{r_2} r^2 dr$

$$= \left(\frac{2}{3}\right) \cdot \pi\mu \cdot p_n \cdot \text{cosec}\,\alpha (r_2^3 - r_1^3)$$

$$= \left(\frac{2}{3}\right) \cdot \mu W \cdot \text{cosec}\ \alpha \left[\frac{r_2^3 - r_1^3}{r_2^2 - r_1^2}\right] \quad (7.31a)$$

$$= \mu W \cdot r_m\ \text{cosec}\ \alpha \quad (7.31b)$$

(b) Uniform wear

For uniform wear, $p_n \cdot r = C$

or $p_n = \dfrac{C}{r}$

Normal load on the ring, $dW_n = p_n \cdot dA$

Axial load on the ring, $dW = dW_n \cdot \sin\alpha = p_n \cdot 2\pi r \cdot dr = 2\pi C \cdot dr$

Total axial load on the clutch, $W = 2\pi C \int_{r_1}^{r_2} dr = 2\pi C(r_2 - r_1)$

or $C = \dfrac{W}{2\pi(r_2 - r_1)}$

Frictional force on the ring, $dF = \mu \cdot dW_n$

Frictional torque on the ring, $dT_f = dF \cdot r = 2\pi\,\mu p_n \cdot \text{cosec}\ \alpha \cdot r_2 \cdot dr$

$$= 2\pi\mu\, C \cdot \text{cosec}\ \alpha \cdot r \cdot dr$$

Frictional torque on the clutch, $T_f = 2\pi\mu C \cdot \text{cosec}\ \alpha \int_{r_1}^{r_2} r \cdot dr$

$$= \pi\mu C \cdot \text{cosec}\alpha (r_2^2 - r_1^2)$$

$$= \mu W \text{cosec}\alpha \frac{(r_1 + r_2)}{2}$$

$$= \mu W r_m\ \text{cosec}\ \alpha \quad (7.32)$$

where $r_m = \dfrac{r_1 + r_2}{2}$ is the mean radius.

The frictional torque can also be written as,

$$T_f = 2\pi\mu \cdot p_n \cdot b \cdot r^2 m \quad (7.33)$$

Axial force required to engage the clutch,

$$W_e = W + \mu W_n \cos\alpha$$

$$= W_n \sin\alpha + \mu W_n \cos\alpha$$

$$= W_n(\sin\alpha + \mu\cos\alpha) \quad (7.34)$$

Example 7.9

A single-plate clutch, with both sides effective, has inner and outer diameters of friction surface 250 mm and 350 mm, respectively. The maximum intensity of pressure is not to exceed 0.15 MPa. The coefficient of friction is 0.3. Determine the power transmitted by the clutch at a speed of 2400 rpm for (a) uniform wear and (b) uniform pressure.

■ **Solution**

(a) For uniform wear, $T_f = n \cdot \mu W \cdot r_m$

$$r_m = \frac{250+350}{4} = 150 \text{ mm} = 0.15 \text{ m}$$

$$n = 2$$

$$C = pr_1 = 0.15 \times 125 = 18.75 \text{ N/mm}$$

$$W = 2\pi C(r_2 - r_1) = 2\pi \times 18.75(175 - 125) = 5890.48 \text{ N}$$

$$T_f = 2 \times 0.3 \times 5890.48 \times 0.15 = 530.14 \text{ Nm}$$

Power transmitted, $P = \dfrac{T_f\,\omega}{1000} \text{kW} = 530.14 \dfrac{\left(2\pi \times \frac{2400}{60}\right)}{1000}$

$= 133.24$ kW

(b) For uniform pressure, $r_m = \left(\dfrac{2}{3}\right) \cdot \left[\dfrac{r_2^3 - r_1^3}{r_2^2 - r_1^2}\right] = \left(\dfrac{2}{3}\right) \cdot \left[\dfrac{175^3 - 125^3}{175^2 - 125^2}\right]$

$= 151.4$ mm

$$W = \pi\left(r_2^2 - r_1^2\right)p$$

$$= \pi(175^2 - 125^2) \times 0.15 = 7068.58 \text{ N}$$

$$T_f = n\mu W r_m$$

$$= 2 \times 0.3 \times 7068.58 \times 151.4 \times 10^{-3}$$

$$= 642.1 \text{ Nm}$$

Power transmitted, $P = 642.1 \times \dfrac{\left(2\pi \times \frac{2400}{60}\right)}{1000}$

$= 161.38$ kW

Example 7.10

A plate clutch has three discs on the driving shaft and two discs on the driven shaft, providing four pairs of contact surfaces. The inside and outside diameters of the friction surfaces are 125 mm and 250 mm, respectively. Assuming uniform pressure and coefficient of friction equal to 0.3, find the total spring load pressing the plates together to transmit 30 kW at 1500 rpm.

■ **Solution**

Power transmitted, $P = \dfrac{T_f\omega}{1000} \text{kW}$

$$30 = \frac{T_f \times \left(2\pi \times \frac{1500}{60}\right)}{1000}$$

$$T_f = \frac{(30 \times 1000 \times 60)}{(2\pi \times 1500)} = 190.986 \text{ Nm}$$

$$r_m = \left(\frac{2}{3}\right) \cdot \left[\frac{r_2^3 - r_1^3}{r_2^2 - r_1^2}\right]$$

$$= \left(\frac{2}{3}\right) \cdot \left[\frac{125^3 - 62.5^3}{125^2 - 62.5^2}\right] = 97.2 \text{ mm}$$

$$T_f = n\mu W r_m$$
$$190.986 = 4 \times 0.3 \times W \times 0.0972$$
$$W = 1637.4 \text{ N}$$

Example 7.11

An engine developing 50 kW at 1200 rpm is fitted with a cone clutch. The cone angle is 12° and a maximum mean diameter of 500 mm. The coefficient of friction is 0.25. The normal pressure on the clutch face is not to exceed 0.1 MPa. Determine (a) the axial spring force to engage the clutch, and (b) the face width required.

■ Solution

(a)
$$P = \frac{T_f \omega}{1000} \text{ kW}$$
$$50 = \frac{T_f\left(2\pi \times \frac{1200}{60}\right)}{1000}$$
$$T_f = 397.9 \text{ Nm}$$
$$= \mu W_n r_m = 0.25 \times W_n \times 0.25$$
$$W_n = 6366.4 \text{ N}$$

Axial spring force required to engage the clutch,
$$W_e = W_n(\sin\alpha + \mu\cos\alpha)$$
$$= 6366.4(\sin 12° + 0.25\cos 12°)$$
$$= 2880.5 \text{ N}$$

(b)
$$W_n = p_n \cdot 2\pi r_m b$$

Face width,
$$b = \frac{2880.5}{0.1 \times 2\pi \times 250} = 18.3 \text{ mm}$$

7.4 DYNAMOMETERS

The two types of dynamometers are: (1) absorption dynamometers, and (2) transmission dynamometers.

7.4.1 Absorption Dynamometers

In these type of dynamometers, the entire power produced by the prime mover is absorbed by the frictional resistance of the brake and is transformed into heat, during the process of measurement. The absorption type of dynamometers can be classified as: (a) Prony brake dynamometer and (b) Rope brake dynamometer.

1. Prony Brake Dynamometer The prony brake dynamometers is shown in Fig.7.17. It consists of two wooden blocks placed around a pulley fixed to the shaft of the prime mover, whose power is to be measured. The blocks are clamped by means of two bolts and nuts. A helical spring is provided between the nut and the upper block to adjust the pressure on the pulley to control its speed. The upper block has a long lever attached to it and carries a weight W at its other end. A counter weight is placed at the other end of the lever that balances the brake when unloaded. Two stops S are provided to limit the motion of the lever.

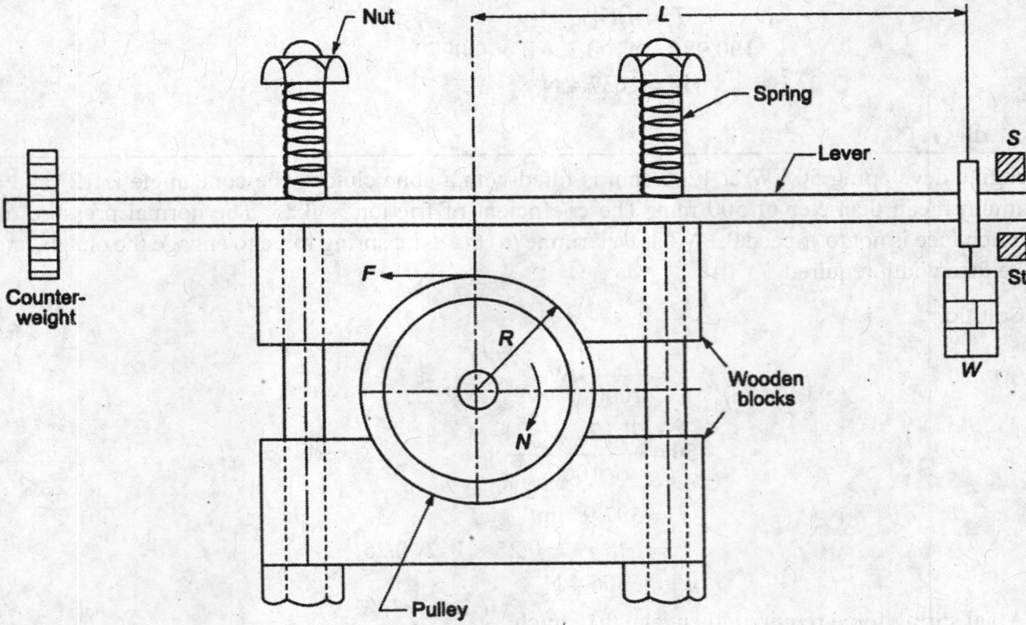

Fig.7.17 Prony brake dynamometer

When the brake is to be put in operation, the long end of the lever is loaded with suitable weight W and the nuts are tightened until the prime mover shaft runs at a constant speed and the lever is in horizontal position. Under these conditions, the moment due to the weight W must balance the moment of the frictional resistance between blocks and pulley.

Moment of the frictional resistance, $T_f = WL = FR$

Work done per minute $= 2\pi NT_f$

Brake power,
$$BP = \frac{2\pi NT_f}{60\times 1000} = \frac{2\pi NWL}{60\times 1000}\text{ kW} \tag{7.35}$$

2. Rope brake dynamometers The rope brake dynamometers is shown in Fig.7.18. It consists of one or more ropes wound around the flywheel or rim of the pulley, fixed rigidly to the shaft of the prime mover. The upper end of the ropes is attached to a spring balance while the lower end is kept in position by applying a dead weight. In order to prevent the slipping of the ropes over the flywheel, wooden blocks are placed at intervals around the circumference of the flywheel.

During operation of the brake, the prime mover is made to run at a constant speed. The frictional torque due to the ropes must be equal to the torque being transmitted by the prime mover.

Let W = dead load on the rope
S = spring balance reading
D = diameter of the pulley
d = diameter of the rope
N = speed of the pulley

Work done per minute $= (W - S)\,\pi\,(D + d)N$

Brake power,
$$BP = \frac{(W-S)\pi(D+d)N}{60\times 1000}\text{ kW} \tag{7.36}$$

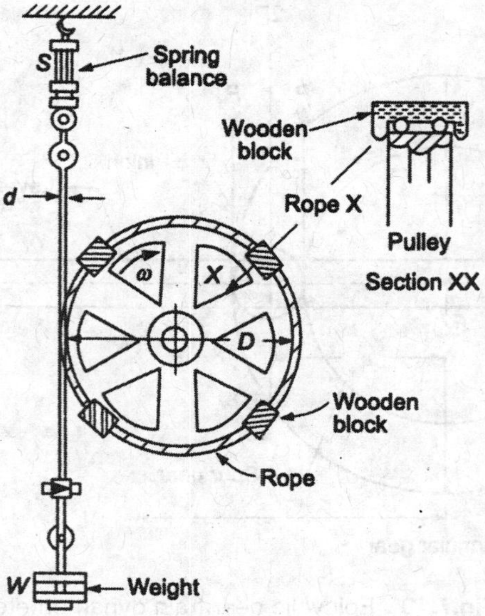

Fig.7.18 Rope brake dynamometer

7.4.2 Transmission Dynamometers

In these type of dynamometers, the energy is used for doing work. The power developed by the prime mover is transmitted through the dynamometer to some other machine where the power is suitably measured. This type of dynamometer can be classified as follows:

1. Epicyclic train dynamometer
2. Belt transmission dynamometer
3. Torsion dynamometer.

1. Epicyclic Train Dynamometer The epicyclic train dynamometer is shown in Fig.7.19. It consists of a simple epicyclic train of gears: a spur gear, an annular gear, and a pinion. The spur gear is keyed to the engine shaft and rotates in counter-clockwise direction. The annular gear is also keyed to the driving shaft and rotates in clockwise direction: The pinion meshes with both the spur and annular gears. The pinion revolves freely on a lever that is pivoted to the common axis of the driving and driven shafts. A weight w is placed at the smaller end of the lever in order to keep it in position.

Let P be the force between the pinion, the spur gear, and the annular gear. Then the total upward force on the lever through the axis of the pinion is $2P$. This force is balanced by a dead weight W at the other end of the lever. The stops control the movement of the lever.

For the equilibrium of the lever, taking moments about the fulcrum, we have

$$2P \cdot a = WL$$

or

$$P = \frac{WL}{2a}$$

Torque transmitted, $T = PR$

where R = pitch circle radius of the spur gear.

Power transmitted,

$$BP = \frac{2\pi N T}{60\times1000} = \frac{2\pi N P R}{60\times1000}\text{kW}$$

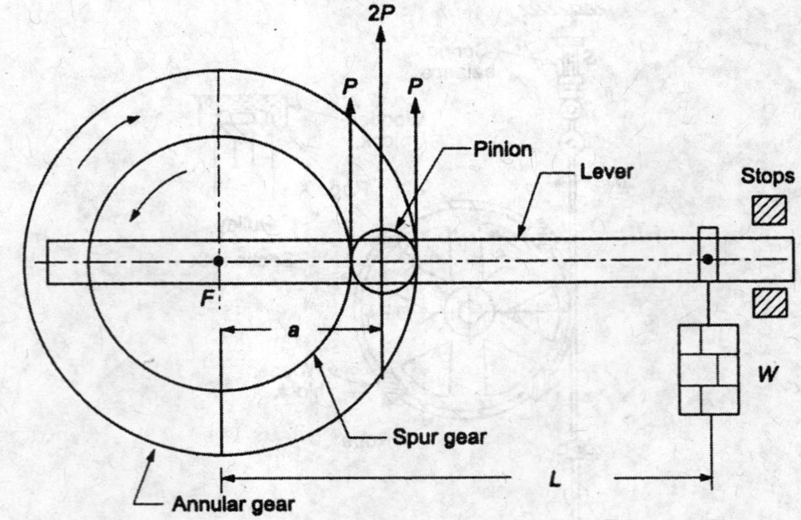

Fig.7.19 Epicyclic gear train dynamometer

2. Belt Transmission Dynamometer A belt transmission dynamometer, as shown in Fig.7.20, consists of a driving pulley A, rigidly fixed to the shaft of the prime mover. There is another driven pulley B mounted on another shaft, to which the power from pulley A is transmitted. The pulleys A and B are connected by means of a continuous belt passing round the two loose pulleys C and D, which are mounted on the lever, pivoted at E. The lever carries a dead weight W at the one end, and a balancing weight is attached at the other end. The total force acting on the pulley D is $2T_1$ and on the pulley, C is $2T_2$. Taking moments about the fulcrum E, we have

$$2T_1 L = 2T_2 a + W L$$

or
$$T_1 - T_2 = \frac{W L}{2a}$$

Brake power developed,

$$BP = \frac{\pi D N (T_1 - T_2)}{60 \times 1000} \text{ kW} \tag{7.37}$$

3. Torsion Dynamometer A torsion dynamometer is used to measure large power developed by a turbine or marine engines. The torque developed by a shaft of diameter d, length l, and modulus of rigidity G is

$$T = \left(\frac{GJ}{l}\right)\theta = k\theta$$

where θ = angle of twist of the shaft.

Therefore, the torque acting on the shaft is proportional to the angle of twist.

By measuring the angle of twist, the power developed by the machine can be measured.

Power developed,
$$P = \frac{2\pi N T}{60 \times 1000} \text{ kW} \tag{7.38}$$

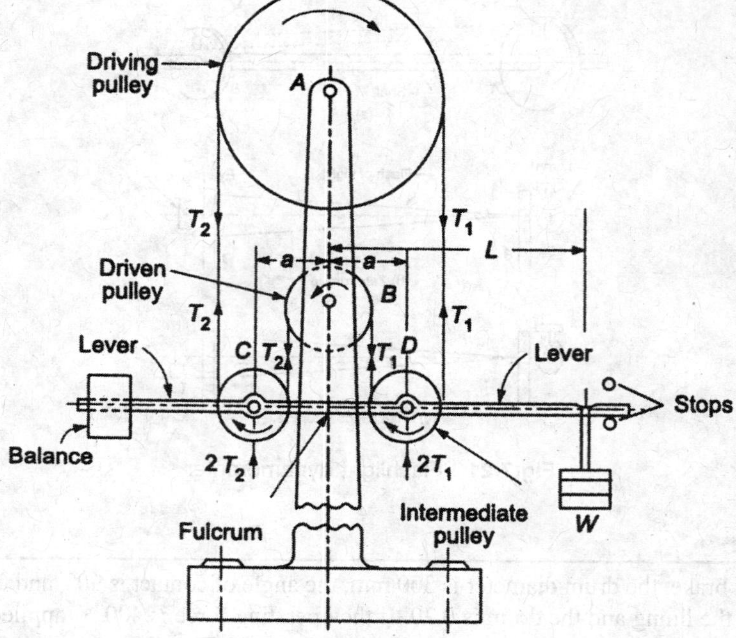

Fig.7.20 Belt transmission dynamometer

A large number of torsion dynamometers are used to measure the angle of twist. We describe below the flashlight dynamometer.

4. FlashLight Dynamometer The principle of the flashlight dynamometer is shown in Fig.7.21. It consists of two discs *A* and *B* fixed on a shaft at a convenient distance apart. Each disc has a small radial slot and these two slots are in the same line when no power is transmitted and there is no torque on the shaft. A bright electric lamp behind the disc *A* is fixed on the bearing of the shaft. This lamp is masked having a slot directly opposite to the slot of disc *A*. At every revolution of the shaft, a flash of light is projected through the slot in the disc *A* towards the disc *B* in a direction parallel to the shaft. An eye piece is fitted behind the disc *B* on the shaft bearing and is capable of slight circumferential adjustment.

When the shaft does not transmit any torque, a flash of light may be seen after every revolution of the shaft, as the positions of the slot do not change relative to one another, as shown in Fig.7.21(b). When the torque is transmitted, the shaft twists and the slot in the disc *B* changes its position, though the slots in the lamp, disc *A*, and eye piece are still in line. Because of this, the light does not reach the eye piece, as shown in Fig.7.21(c). If the eye piece is now moved round by an amount equal to the lag of the disc *B*, then the slot in the eye piece will be opposite to the slot in disc *B*, and hence the eye piece will receive flash of light. The eye piece is moved by operating a micrometer spindle and by means of scale and Vernier, the angle of twist may be measured upto $1/100^{th}$ of a degree. For the measurement of variable torque, the discs *A* and *B* should be perforated with slots arranged in the form of a spiral.

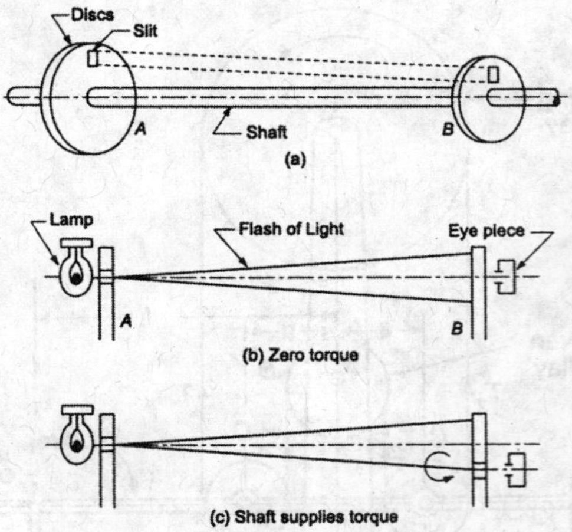

Fig.7.21 Flashlight dynamometer

Example 7.12

In a single-block brake, the drum diameter is 300 mm, the angle of contact is 90°, and the coefficient of friction between the lining and the drum is 0.30. If the operating force is 400 N, applied at the end of a lever 400 mm long, determine the torque transmitted by the brake. The distance of the fulcrum from the centre of the brake drum is 200 mm and assume that the force of friction passes through the fulcrum.

Solution

Given: $d = 300$ mm, $2\theta = 90°$, $\mu = 0.3$, $P = 400$ N, $a = 200$ mm, $b = 400$ mm.

$$R = P\,b/a = 400 \times 400/200 = 800 \text{ N}$$

$$T = 4\mu R\, r \sin\theta/(2\theta + \sin 2\theta)$$

$$= 4 \times 0.35 \times 800 \times 0.15 \times \sin 45°/(\pi/2 + \sin 90°)$$

$$= 39.614 \text{ Nm}$$

Example 7.13

In a double-shoe brake, the diameter of the brake drum is 350 mm and the contact angle for each shoe is 120°. The coefficient of friction for the brake lining and drum is 0.35. Find the necessary spring force to transmit a torque of 40 Nm. The distance of the centre of brake drum from the fulcrum and spring is 250 and 300 mm, respectively.

Solution

Given: $d = 350$ mm, $2\theta = 120°$, $\mu = 0.35$, $T = 40$ Nm, $a = 250$ mm, $b = 300$ mm

$$R = P\,b/a = 400 \times 400/200 = 800 \text{ N}$$

$$\mu_e = 4\mu \sin\theta/(2\theta + \sin 2\theta)$$

$$= 4 \times 0.35 \times \sin 60°/(2\pi/3 + \sin 120°)$$

$$= 1.55$$

Let P_ℓ = spring force on left side

P_r = spring force on right side

$$P_\ell(a + b) = R_\ell a$$

$$R_\ell = P_\ell(a + b)/a$$

$$= P_\ell \times 550/250 = 2.2\ P_L$$

$$P_r\,(a+b) = R_r\,a$$

$$R_r = P_r\,(a+b)/a = 2.2\ P_R$$

$$T = \mu_e\,(R_\ell + R_r)\,r$$

For $P_L = P_R$

$$40 = 1.55\ (2 \times 2.2)\ P_r \times 0.35/2$$

Spring force, $P_r = 33.51$ N

Example 7.14

A simple band brake is operated by a lever of length 450 mm. The brake drum has a diameter of 600 mm and the brake band embraces fifth-eigth of the circumference. One end of the band is attached to the fulcrum of the lever, whereas the other end is attached to a pin on the lever 120 mm from the fulcrum. The effort applied to the end of the lever is 2 kN, and the coefficient of friction is 0.30. Find the maximum braking torque on the drum.

■ Solution

Given: $b = 450$ mm, $d = 600$ mm, $\theta = 5 \times 360/8 = 225°$, $\mu = 0.3$, $a = 120$ mm, $P = 2$ kN, $T_b = ?$.

$$T_1/T_2 = \exp\,(\mu\theta) = \exp\,(0.3 \times \pi \times 225/180) = 3.248$$

$$T_1 = Pb/a = 2000 \times 450/120 = 7500 \text{ N}$$

$$T_2 = 7500/3.248 = 2309 \text{ N}$$

Braking torque for ccw of drum, $T_b = (T_1 - T_2)\,d/2 = (7500 - 2309) \times 0.3 = 1557.3$ Nm

Example 7.15

In the differential band brake, the diameter of the drum is 900 mm, and the coefficient of friction between the drum and the band is 0.3. The angle of contact is 240°. When a force of 650 N is applied at the free end of the lever, find the maximum and the minimum force in the band and the torque that can be applied by the brake. Take $a = 120$ mm and $b = 100$ mm.

■ Solution

Given: $a = 120$ mm, $b = 100$ mm, $d = 900$ mm, $\theta = 240°$, $\mu = 0.3$, $P = 650$ N, $\ell = 500$ mm.

$$T_1/T_2 = \exp\,(\mu\theta) = \exp\,(0.3 \times \pi \times 240/180) = 3.5136$$

$$T_1 a + P\ell = T_2 b$$

$$650 \times 500 = T_2 \times 100 - T_1 \times 120$$

$$T_2 - 1.2\ T_1 = 3250$$

$$T_2 - 4.216\ T_2 = 3250$$

$$T_2 = -1010.5 \text{ N}$$

$$T_1 = -3.5136 \times 1010.5 = -3550.4 \text{ N}$$

Braking torque for ccw of drum, $T_b = (T_1 - T_2)\,d/2$

$$= (3550.4 - 1010.5) \times 0.45 = 1143 \text{ Nm}$$

Example 7.16

A vehicle is moving on a level track at a speed of 40 km/h. Its centre of gravity lies at a distance of 0.6 m from the ground level. The wheel base is 2.4 m and the distance of the CG from the rear is 1 m. Find the distance travelled by the vehicle before coming to rest when brakes are applied (a) to the rear wheels, (b) to the front wheels, and (c) to all the four wheels.

The coefficient of friction between the tyres and the road surface is 0.40.

■ **Solution**

Given: $h = 0.6$ m, wheel base, $\ell = 2.4$ m, $x = 1$ m, $\mu = 0.4$

$v = 40 \times 1000/3600 = 11.11$ m/s

(a)

$$R_A + R_B = W,\ M_G = 0 \text{ gives,}$$
$$R_A x + \mu R_A h = R_B (\ell - x)$$
$$R_A \times 1 + 0.4 \times R_a \times 0.6 = R_B (2.4 - 1)$$
$$1.24\ R_A = 1.4\ (W - R_A)$$
$$R_A = 0.53\ W,\ R_B = 0.47\ W$$
$$F_A = -Wa/g$$
$$\mu R_A = -Wa/g$$
$$a = -0.4 \times 0.53\ W \times 9.81/W = -2.08 \text{ m/s}^2$$
$$v^2 - u^2 = 2\ a\ s$$
$$0 - (11.11)^2 = -2 \times 2.08 \times s$$
$$s = 29.67 \text{ m}$$

(b)

$$\mu R_B = -Wa/g$$
$$a = -0.4 \times 0.47\ W \times 9.81/W = -1.844 \text{ m/s}^2$$
$$v^2 - u^2 = 2\ a\ s$$
$$0 - (11.11)^2 = -2 \times 1.844 \times s$$
$$s = 33.47 \text{ m}$$

(c)

$$\mu\ (R_A + R_B) = -Wa/g = 0.4\ W$$
$$a = -0.4 \times W \times 9.81/W = -3.924 \text{ m/s}^2$$
$$v^2 - u^2 = 2\ a\ s$$
$$0 - (11.11)^2 = -2 \times 3.924 \times s$$
$$s = 15.72 \text{ m}$$

Example 7.17

A single-plate clutch having both sides effective is required to transmit 30 kW at 1500 rpm. The outer diameter of the plate is limited to 300 mm and the intensity of pressure between the plates is not to exceed 0.07 MPa. Assuming uniform wear and a coefficient of friction 0.35, determine the inside diameter of the plate.

■ **Solution**

Given: $P = 30$ kW, $n = 1500$ rpm, $d_2 = 0.3$ m, $p = 0.07$ MPa, $i = 2$, $\mu = 0.35$, $d_1 = ?$.

$$r_m = 0.25\ (d_1 + d_2) = 0.25\ (d_1 + 0.3)$$

For uniform wear,

$$W = \pi d_1 (d_2 - d_1)\ p/2 = \pi d_1 (0.3 - d_1) \times 0.07 \times 10^6/2 = 109956\ d_1 (0.3 - d_1) \text{ N}$$
$$\omega = 2\pi \times 1500/60 = 157.08 \text{ rad/s}$$
$$P = T_f \omega$$
$$30 \times 10^3 = T_f \times 157.08$$
$$T_f = 191 \text{ Nm}$$
$$T_f = \mu W r_m i$$
$$191 = 0.35 \times 109956\ d_1 (0.3 - d_1) \times 0.25\ (d_1 + 0.3) \times 2$$
$$9.926 \times 10^{-3} = d_1 (0.09 - d_1^2)$$

We determine the value of d_1 by hit and trial.

d_1 (m)	RHS
0.10	8×10^{-3}
0.12	9.072×10^{-3}
0.14	9.856×10^{-3}
0.145	9.917×10^{-3}

Hence, we adopt $d_1 = 0.145$ m or 145 mm

Example 7.18

A multi-plate clutch has three pairs of contact surfaces. The outer and inner radii of the contact surfaces are 150 and 80 mm, respectively. The maximum axial spring force is limited to 03 kN and the coefficient of friction is 0.3. Assuming uniform uniform wear find the power transmitted by the clutch at 1500 rpm.

■ Solution

Given: $n = 3$, $r_1 = 80$ mm, $r_2 = 150$ mm, $\mu = 0.3$, N = 1500 rpm, $W = 0.9$ kN

$$r_m = 0.5\,(r_1 + r_2) = 0.5\,(80 + 150) = 115 \text{ mm}$$

$$T_f = n\mu W r_m = 3 \times 0.3 \times 0.9 \times 115 = 93.15 \text{ Nm}$$

$$\omega = 2\,\pi \times 1500/60 = 157.08 \text{ rad/s}$$

$$P = T_f\omega = 93.15 \times 157.08/10^3 = 14.632 \text{ kW}$$

Example 7.19

A cone clutch with cone angle 25° is to transmit 8 kW at 750 rpm. The normal intensity of pressure between the contact faces is not to exceed 0.15 MPa. The coefficient of friction is 0.25. If face width is one-fifth of mean diameter, find (a) the main dimensions of the clutch and (b) axial force required while running.

■ Solution

Given: $2\alpha = 25°$, $P = 8$ kW, $N = 750$ rpm, $p_n = 0.15$ MPa, $\mu = 0.25$, $b = d_m/5$

$$r_m = 0.25\,(d_1 + d_2)$$

For uniform wear,

$$\omega = 2\,\pi \times 750/60 = 78.54 \text{ rad/s}$$

$$P = T_f\omega$$

$$8000 = T_f \times 78.54$$

$$T_f = 101.86 \text{ Nm}$$

$$T_f = 2\pi\mu p_n b r_m^2$$

$$101.86 = 2\,\pi \times 0.25 \times 0.15 \times 10^6 \times 0.2 \times 0.5\,(d_1 + d_2) \times (d_1 + d_2)^2/16$$

$$(d_1 + d_2)^3 = 69168.6 \times 10^{-6}$$

$$d_1 + d_2 = 410.5 \text{ mm}$$

Let $d_2/d_1 = 2$, then $d_1 = 136.8$ mm, $d_2 = 273.6$ mm,

$$b = 0.2 \times 0.5 \times 410.5 = 41.05 \text{ mm}$$

Example 7.20

A torsion dynamometer is fitted on a turbine shaft to measure the angle of twist. It is observed that the shaft twists 2° in a length of 5 m at 600 rpm. The shaft is solid and has a diameter of 250 mm. If the modulus of rigidity is 84 GPa, find the power transmitted by the turbine.

■ **Solution**

Given: $\theta = 2°$, $\ell = 5$ m, $N = 600$ rpm, $d = 250$ mm, $G = 84$ GPa, $P = ?$

$$T = GJ\theta/\ell = 84 \times 10^9 \times \pi \times (0.25)^4 \times \pi \times 2/(5 \times 32 \times 180)$$
$$= 224893 \text{ Nm}$$
$$P = 2\pi NT/(60 \times 10^3)$$
$$= 2\pi \times 600 \times 224893/(60 \times 10^3) = 14130 \text{ kW}$$

Example 7.21

A single-plate clutch is required to transmit 22 kW at 6000 rpm. The clutch facings available provide a coefficient of friction of 0.25 and the average pressure is to be limited to 75 kN/m². Determine the dimensions of the working surface of the clutch plate if its maximum dimension is not to exceed 260 mm due to space restrictions. [IAS, 1983]

■ **Solution**

Given: $P = 22$ kW, $N = 6000$ rpm, $\mu = 0.25$, $p = 75$ kN/m², $d_2 = 0.26$ m, $i = 1$, $d_1 = ?$

For uniform wear, $r_m = 0.25\,(d_1 + d_2) = 0.25\,(d_1 + 0.26)$

$$W = \frac{\pi d_1 (d_2 - d_1) p_{max}}{2}$$
$$= \frac{\pi d_1 (0.16 - d_1) \times 75 \times 10^3}{2} = 117809.7\, d_1 (0.16 - d_1) \text{ N}$$
$$\omega = 2\,\pi \times 6000/60 = 628.3 \text{ rad/s}$$
$$P = T_f \omega$$
$$22 \times 10^3 = T_f \times 628.3$$
$$T_f = 35.015 \text{ Nm}$$
$$T_f = \mu W r_m i$$
$$35.015 = 0.25 \times 117809.7\, d_1 (0.26 - d_1) \times 0.25\,(d_1 + 0.26) \times 1$$
$$\text{or } 4.755 \times 10^{-3} = d_1 (0.0676 - d_1^2)$$

We shall solve this by hit and trial method to find d_1.

d_1, (m)	RHS
0.100	5.76×10^{-3}
0.080	4.896×10^{-3}
0.078	4.798×10^{-3}
0.077	4.749×10^{-3}

Hence, we adopt $d_1 = 77$ mm

Example 7.22

In a winch, the rope supports a load W and is wound round a barrel of 450 mm diameter. A differential band brake acts on a drum 800 mm diameter that is keyed to the same shaft as the barrel. Two ends of the band are attached to pins on opposite sides of the fulcrum of the brake lever and at a distance of 25 and 100 mm, respectively, from the fulcrum. The angle of lap of the brake band is 250° and coefficient of friction is 0.25. What is the maximum load W that can be supported by the brake when a force of 750 N is applied to the lever at a distance of 3 m from the fulcrum? [IAS, 1987]

■ Solution

Given: $d_b = 450$ mm, $d = 800$ mm, $a = 100$ mm, $b = 25$ mm, $\theta = 250°$, $\mu = 0.25$, $P = 750$ N, $\ell = 3$ m.

$$P\ell + T_1 a = T_2 b$$
$$750 \times 3 + 0.1\, T_1 = 0.025\, T_2$$
$$T_2 - 4\, T_1 = 90{,}000$$
$$T_1/T_2 = \exp(\mu\theta) = \exp(0.25 \times \pi \times 250/180) = 2.9767$$
$$T_2 - 4 \times 2.9767\, T_2 = 90{,}000$$
$$T_2 = -8251.6 \text{ N},\ T_2 = -24562.5 \text{ N}$$
$$W d_b/2 = (T_1 - T_2)\, d/2$$
$$W \times 450 = (24562.5 - 8251.6) \times 800$$
$$W = 28997 \text{ N}$$

Example 7.23

A differential band brake under certain conditions can provide self-locking. Where this facility finds applications?

A differential band brake has a force of 220 N applied at the end of a pedal as shown in Fig.7.22. The coefficient of friction between the band and the drum is 0.4. Angle of lap is 180°.

(a) What is the maximum torque the brake may sustain, for a counter-clockwise rotation, when the force applied at the pedal is 220 N?

(b) If a clockwise torque of 450 Nm is applied to the drum, determine the maximum and minimum force in the band. [IES, 1984]

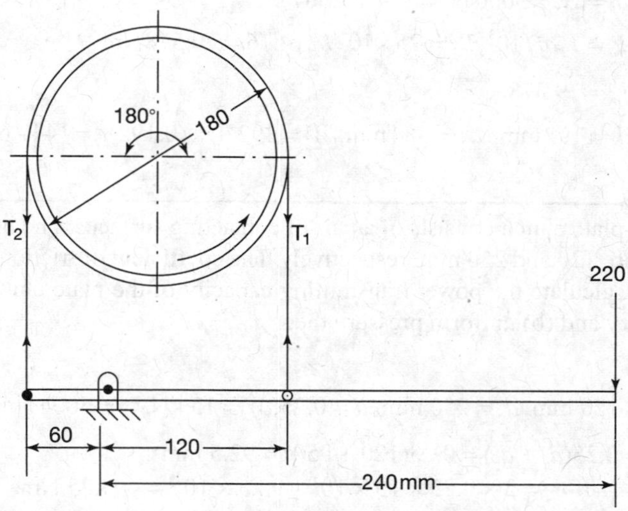

Fig.7.22 Differential band brake

■ **Solution**

Given: $P = 220$ N, $\ell = 240$ mm, $\mu = 0.4$, $\theta = 180°$, $b = 60$ mm, $a = 120$ mm, $d = 180$ mm

$$P\ell + T_1 a = T_2 b$$
$$220 \times 240 + 60\ T_2 = 120 T_1$$
$$2T_1 - T_2 = 880$$
$$T_1/T_2 = \exp(\mu\theta) = \exp(0.4 \times \pi) = 3.5136$$
$$2 \times 3.5136\ T_2 - T_2 = 880$$
$$T_2 = 146 \text{ N}, T_1 = 513 \text{ N}$$

(a) Braking torque $= (T_1 - T_2)\ d/2 = (513 - 146) \times 0.09 = 33.03$ Nm

(b) If $T_b = 450$ Nmcw, then $T_1 - T_2 = \dfrac{450 \times 10^3}{90} = 5000$ N

$$T_2(3.5136 - 1) = 5000$$
$$T_2 = 1989.2 \text{ N}$$
$$T_1 = 6989.2 \text{ N}$$

Example 7.24

A single-plate clutch, effective on both sides, is required to transmit 25 kW at 3000 rpm. Determine the outer and inner diameters of frictional surface if the coefficient of friction is 0.255, ratio of diameters is 1.25, and the maximum pressure is not to exceed 0.1 N/mm². Also determine the axial thrust to be provided by springs, assuming the theory of uniform wear. [IES, 1984]

■ **Solution**

Given: $P = 25$ kW, $N = 3000$ rpm, $\mu = 0.255$, $p = 0.1\text{N}/\text{mm}^2$, $d_2/d_1 = 1.25$, $i = 2$, $W = ?$, d_1, $d_2 = ?$.

$$W = \pi d_1(d_2 - d_1) \times p_{\max}/2 = \pi d_1^2(1.25 - 1) \times 0.1 \times 10^6/2 = 39.270\ d_1^2$$
$$r_m = 0.25(d1 - d_2) = 0.25 \times d_1(1 + 1.25) = 0.5625 d_1$$
$$T_f = i\mu W r_m = 2 \times 0.255 \times 39270 d_1^2 \times 0.5625 d_1 = 11265.6 \times d_1^3$$
$$\omega = 2\pi \times 3000/60 = 314.16 \text{ rad/s}$$
$$P = T_f\omega/10^3, T_f = 25 \times 10^3/314.16 = 79.578 \text{Nm}$$
$$11265.6\ d_1^3 = 79.578$$
$$d_1 = 192 \text{mm},\ d_2 = 384 \text{mm},\ W = 39270 \times (0.192)^2 = 1447.6 \text{ N}$$

Example 7.25

An automobile single-plate clutch consists of a pair of contacting surfaces. The inner and outer diameters of friction plate are 120 and 250 mm, respectively. The coefficient of friction is 0.25 and the total axial force is 15 kN. Calculate the power transmitting capacity of the plate clutch at 500 rpm using (a) uniform wear theory and (b) uniform pressure theory. [IES, 1996]

■ **Solution**

Given: $d_1 = 120$ mm, $d_2 = 250$ mm, $\mu = 0.25$, $W = 15$ kN, $N = 500$ rpm

(a)
$$r_m = 0.25(d_1 + d_2) = 0.25(120 + 250) = 92.5 \text{ mm}$$
$$T_f = i\mu W r_m = 2 \times 0.25 \times 15 \times 10^3 \times 92.5 \times 10^{-3} = 693.75 \text{ Nm}$$
$$\omega = 2\pi \times 500/60 = 52.36 \text{ rad/s}$$
$$P = T_f\omega/10^3,\ T_f = 693.75 \times 52.36/10^3 = 36.32 \text{ kW}$$

(b) $r_m = (2/3)\,[(125^3 - 60^3)/(125^2 - 60^2)] = 96.3$ mm

$T_f = i\mu W r_m = 2 \times 0.25 \times 15 \times 10^3 \times 96.3 \times 10^{-3} = 722.25$ Nm

$P = T_f\omega/10^3,\ T_f = 722.25 \times 52.36/10^3 = 37.82$ kW

Example 7.26

The semi-cone angle of a cone clutch is 12.5° and the contact surfaces have a mean diameter of 80 mm. The coefficient of friction is 0.32.
(a) What is the maximum torque required to produce slipping of the clutch for an axial force of 200 N?
(b) What is the time needed to attain the full speed?
(c) What is the total energy supplied during slipping? Motor speed is 900 rpm and the moment of inertia of the flywheel is 0.4 kg m².
(d) What are the considerations in the selection of plate clutches and cone clutches? [IES, 1998]

■ Solution

Given: $\alpha = 12.5°,\ d_m = 80$ mm, $\mu = 0.32$, $W = 200$ N, $I = 0.4$ kg m²

(a) $T_f = \mu W r_m$ cosec $\alpha = 0.32 \times 200 \times 40 \times 10^{-3} \times$ cosec $12.5° = 11.83$ Nm

$T_f = \mathrm{I} \cdot \alpha$

Angular acceleration, $\alpha = 11.83/0.4 = 29.575$ rad/s²

$\omega = 2\pi \times 900/60 = 94.248$ rad/s

(b) Time taken to obtain full speed, $t = \omega/\alpha = 94.248/29.575 = 3.18$ s Angle turned through by the driving shaft during slipping period, $\theta_1 = \omega\, t = 94.248 \times 3.18 = 299.7$ rad

Angle turned by the driven shaft, $\theta_2 = 0.5\,\alpha\, t^2$

$= 0.5 \times 29.575 \times (3.18)^2 = 149.54$ rad

Energy lost in friction due to clutch slip $= T_f(\theta_1 - \theta_2)$

$= 11.83 \times (299.7 - 149.54)$

$= 1776.4$ Nm

(c) Kinetic energy of driven shaft $= 0.5\ \mathrm{I}\ \omega^2$

$= 0.5 \times 0.4 \times (94.248)^2 = 1776.53$ Nm

Total energy supplied during slipping $= 1776.53 + 1776.4 = 3552.93$ Nm

Example 7.27

A single-plate friction clutch has the following data:
Power developed = 30 kW
Speed = 2400 rpm
Axial load = 1500 N
Outside diameter = 300 mm
Coefficient of friction = 0.32
Overload = 10%

(a) Determine the inside diameter of the clutch plate for its both sides effective and assuming uniform wear.

(b) The moment of inertia of rotating parts attached to driven shaft is 2.5 kg m² and average torque is 80% of maximum torque. Determine the time lapse before the engine attains the full speed, if the clutch is suddenly applied.

■ Solution

$$T_{max} = 30 \times 10^3 \times 60/(2\,\pi \times 2400) = 119.37 \text{ Nm}$$
$$T_f = 1.1\,T_{max} = 131.3 \text{ Nm}$$
$$r_m = 0.25\,(d_1 + 0.3)$$
$$T_f = 2\,\mu W r_m$$
$$131.3 = 2 \times 0.32 \times 1500 \times 0.25\,(d_1 + 0.3)$$
$$d_1 = 247 \text{ mm}$$
$$T_{avg} = 0.8\,T_{max} = 95.496 \text{ Nm}$$
$$T_{avg} = I\,\omega/t$$
$$t = 2.5 \times 2\pi \times 2400/(60 \times 95.496) = 6.57 \text{ s}$$

Example 7.28

A machine is driven from a constant speed shaft rotating at 270 rpm by a disc friction clutch having both sides effective. The moment of inertia of rotating parts of the machine is 4.5 kg · m². The inner and outer diameters of friction plate are 250 and 150 mm, respectively, and the axial pressure applied is 0.075 MPa. Assuming uniform pressure and coefficient of friction to be 0.25, determine the time required for the machine to attain full speed when the clutch is suddenly engaged. Also determine the energy supplied during clutch slip.

■ Solution

Given: $d_1 = 150$ mm, $d_2 = 250$ mm, $\mu = 0.25$, $p = 0.075$ MPa, $I = 4.5$ kg·m², $N = 270$ rpm

(a) $W = \pi\,(250^2 - 150^2) \times 0.075/4 = 2356$ N

$r_m = (2/3)\,[(125^3 - 75^3)/(125^2 - 75^2)] = 102.1$ mm

$T_f = 2\mu W r_m = 2 \times 0.25 \times 2356 \times 102.1 \times 10^{-3} = 120.284$ Nm

$T_f = I\alpha$

Angular acceleration, $\alpha = 120.284/4.5 = 26.73$ rad/s²

$\omega = 2\pi \times 270/60 = 28.27$ rad/s

(b) Time taken to obtain full speed, $t = \omega/\alpha = 28.27/26.73 = 1.06$ s Angle turned through by the driving shaft during slipping period, $\theta_1 = \omega\,t = 28.27 \times 1.06 = 29.9$ rad

Angle turned by the driven shaft, $\theta_2 = 0.5\,\alpha t^2$
$= 0.5 \times 26.73 \times (1.06)^2 = 15.017$ rad

Energy lost in friction due to clutch slip $= T_f(\theta_1 - \theta_2)$
$= 120.284 \times (29.9 - 15.017)$
$= 1790.2$ Nm

(c) Kinetic energy of driven shaft $= 0.5\,I\omega^2$
$= 0.5 \times 4.5 \times (28.27)^2 = 1798.2$ Nm

Total energy supplied during slipping $= 1790.2 + 1798.2 = 3588.4$ Nm

Example 7.29

A crane is required to hold a load of 100 kN. This load is attached to a rope wound round the crane barrel that is 450 mm in diameter. The brake drum that is fixed to the barrel shaft has diameter of 600 mm. The band embraces three-fourth of the circumference of the drum and the coefficient of friction between the band and the drum is 0.35. The brake is to be applied by a hand lever above the drum and the operating force acting vertically downwards must not exceed 500 N. Find suitable length of lever on both sides of fulcrum, assuming that one of the bands is attached to the fulcrum pin directly.

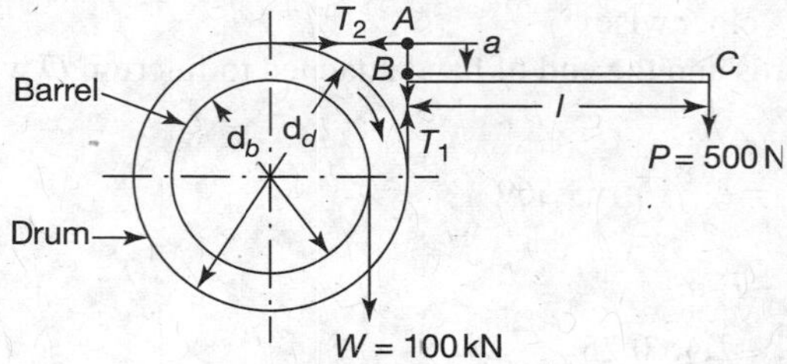

Fig.7.23 Rope brake

■ Solution

Refer to Fig.7.23.

Given: $W = 100$ kN, $d_b = 450$ mm, $d_d = 600$ mm, $\theta = 1.5\ \pi$ rad, $\mu = 0.35$

$$T_1/T_2 = \exp\ (0.35 \times 1.5\pi) = 5.2$$

$$(T_1 - T_2)\ d_d/2 = Wd_b/2$$

$$T_1 - T_2 = 100 \times 450/600 = 75 \text{ kN}$$

$$T_2 = 17857 \text{ N},\ T_1 = 92857 \text{ N}$$

$$Pl = T_2 a$$

$$l/a = 17857/500 = 35.714$$

If the rotation of drum is ccw, then T_2 will become, T_1. In that case,

$$l/a = 92857/500 = 185.714$$

Example 7.30

The power of a turbine is to be determined by observing the angle of twist of a 6 m long shaft at 520 rpm; the angle was 2°. The solid shaft has a diameter of 200 mm and modulus of rigidity 84 GPa. Neglecting end thrust, determine the power of the turbine.

■ Solution

$$\theta = 2\ \pi/180 = 0.0349 \text{ rad},\ l = 6 \text{ m},\ G = 84 \text{ GPa}$$

$$J = \pi \times (0.2)^4/32 = 1.57 \times 10^{-4} \text{ m}^{-4}$$

$$T = GJ\theta/l = 84 \times 10^9 \times 1.57 \times 10^{-4} \times 0.0349/6 = 76.749 \text{ Nm}$$

$$\omega = 2\ \pi \times 520/60 = 54.\ 45 \text{ rad/s}$$

$$P = T\omega/10^3 = 76{,}749 \times 54.45/10^3 = 4179 \text{ kW}$$

Example 7.31

A band brake acts on three-fourth of circumference of a drum of 500 mm diameter that is the keyed to the shaft. The band brake provides a breaking torque of 250 Nm. One end of the band is attached to a fulcrum pin of the lever and other end to a pin 100 mm from the fulcrum. If the operating force is applied at 550 mm from the fulcrum and the coefficient of friction is 0.27, find the operating force when the drum rotates in (a) clockwise direction and (b) anti-clockwise direction.

■ Solution

Refer to Fig.7.4(a).

Given: $d = 500$ mm, $\theta = \dfrac{3}{4} \times 360 = 270°$, $T_b = 250$ Nm, $a = 100$ mm, $b = 550$ mm, $\mu = 0.27$

(a) When drum rotates clockwise
P will act downwards and the end of band attached to fulcrum O will be tight with tension T_1.

$$\frac{T_1}{T_2} = e^{\mu_\theta} = e^{0.27\times\frac{3\pi}{2}} = 3.569$$

$$T_b = (T_1 - T_2)r$$

$$250 = (T_1 - T_2)\times 0.25$$

$$T_1 - T_2 = 1000\,\text{N}$$

$$T_2(3.569 - 1) = 1000$$

$$T_2 = 389.25\,\text{N}, \quad T_1 = 1389.25\,\text{N}$$

Taking moments about O, we have

$$Pb = T_2\,a$$

$$P = 389.25\times\frac{100}{550} = 70.8\,\text{N}$$

(b) When drum rotates anticlockwise
Right hand side band will be tight.

$$Pb = T_1\,a$$

$$P = 1389.25\times\frac{100}{550} = 252.6\,\text{N}$$

Example 7.32

In the internal expanding shoe brake shown in Fig.7.24, the distance $OO_1 = 80$ mm. The internal radius of the brake drum is 105 mm. Angle $BOO_1 = 45°$ and $AOO_1 = 135°$. The brake is applied by means of a force at C perpendicular to the line O_1C, the distance O_1C being 150 mm. The coefficient of friction may be taken as 0.4 and the braking torque required is 25 Nm. Calculate the force required at C to operate the brake when (a) the drum rotates clockwise, and (b) the drum rotates anticlockwise.

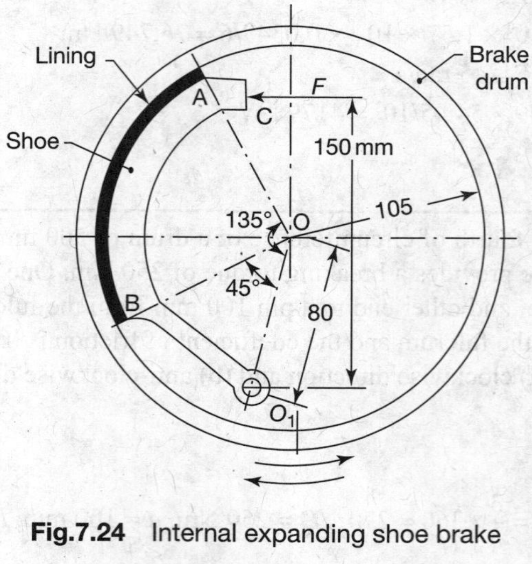

Fig.7.24 Internal expanding shoe brake

■ **Solution**

Given: $OO_1 = 80$ mm, $r = 105$ mm, $\theta_1 = 45° = \frac{\pi}{4}, \theta_2 = 135° = \frac{3\pi}{4}$, $l = 150$ mm, $\mu = 0.4$, $T_b = 25$ Nm.
Let F = force applied at C.

(a) When drum rotates clockwise.

$$T_b = \mu p_1 b r^2 (\cos\theta_1 - \cos\theta_2)$$

$$25\times10^3 = 0.4\times p_1 b\times105^2\left(\cos\frac{\pi}{4}-\cos\frac{3\pi}{4}\right)$$

$$= 6236.68\times p_1 b$$

$$p_1 Pb = 4$$

Total moment of normal forces about the fulcrum O_1,

$$M_r = \frac{1}{2}(p_1\, b)r\times OO_1[(\theta_2-\theta_1)+\frac{1}{2}(\sin2\theta_1-\sin2\theta_2)]$$

$$= \frac{1}{2}\times4\times105\times80\left[\left(\frac{3\pi}{4}-\frac{\pi}{4}\right)+\frac{1}{2}(\sin90°-\sin270°)\right]$$

$$= 43189 \text{ N mm}$$

Total moment of friction force about the fulcrum O_1,

$$M_f = \mu\, p_1\, br\left[r(\cos\theta_1-\cos\theta_2)+\frac{OO_1}{4}(\cos2\theta_2-\cos2\theta_1)\right]$$

$$= 0.4\times4\times105\left[105(\cos45°-\cos135°)+\frac{80}{4}(\cos270°-\cos90°)\right]$$

$$= 24947 \text{ N mm}$$

Taking moments about O_1,

$$Fl = M_n + M_f$$

$$F\times150 = 43189 + 24947 = 68136$$

$$F = 454.24 \text{ N}$$

(b) When drum rotates anticlockwise.

$$Fl = M_n - M_f$$

$$F\times150 = 43189 - 24947 = 18242$$

$$F = 121.6 \text{ N}$$

Example 7.33

A car moving on a level road at a speed of 60 km/h has a wheel base 2.8 m, distance of CG from ground level is 0.6 m, and the distance of CG from rear wheel is 1.2 m. Find the distance travelled by the car before coming to rest when brakes are applied (a) to the rear wheels, (b) to the front wheels, and (c) to all the four wheels.

The coefficient of friction between types and road may be taken as 0.5.

■ **Solution**

Given: $u = \frac{60\times10^3}{3600} = 16.67$ m/s, $l = 2.8$ m, $h = 0.6$ m, $x = 1.2$ m, $\mu = 0.5$

(a) Brakes applied to rear wheels

Retardation, $a = \dfrac{\mu g(l-x)}{l+\mu h} = \dfrac{0.5\times 9.81(2.8-1.2)}{2.8+0.5\times 0.6} = 2.53\,\text{m/s}^2$

Distance travelled, $s = \dfrac{u^2}{2a} = \dfrac{(16.67)^2}{2\times 2.53} = 54.9\,\text{m}$

(b) Brakes applied to front wheels

$$a = \frac{\mu g x}{l-\mu h} = \frac{0.5\times 9.81\times 1.2}{2.8-0.5\times 0.6} = 2.354\text{ m/s}^2$$

$$s = \frac{u^2}{2a} = \frac{(16.67)^2}{2\times 2.354} = 59\text{ m}$$

(c) Brakes applied to all four wheels

$$a = g\mu = 9.81\times 0.5 = 4.905\text{ m/s}^2$$

$$s = \frac{u^2}{2a} = \frac{(16.67)^2}{2\times 4.905} = 28.3\text{ m}$$

Example 7.34

A rotor is driven by a coaxial motor through a single-plate clutch, both sides of the plate being effective. The external and internal diameters of the plate are 240 mm and 180 mm, respectively, and total spring load pressing the plates together is 600 *N*. The motor armature and shaft has a mass of 800 kg with an effective radius of gyration of 200 mm. The rotor has a mass of 1250 kg with an effective radius of gyration of 180 mm. The coefficient of friction for the clutch is 0.36.

The driving motor is brought up to a speed of 1250 rpm when the current is switched off and the clutch is suddenly engaged. Determine:

(a) the final speed of motor and rotor, (b) the time to reach this speed, and (c) the kinetic energy lost during the period of slipping.

How long would slipping continue if it is assumed that a constant resisting torque of 60 Nm were present? If instead of a resisting torque, it is assumed that a constant driving torque of 60 Nm is maintained on the armature shaft, what would then be slipping time?

Solution

Given: $r_1 = 90$ mm, $r_2 = 120$ mm, $W = 600$ N, $m_m = 800$ kg, $K_m = 200$ mm,
$m_r = 1250$ kg, $K_r = 180$ mm, $\mu = 0.36$, $N_1 = 1250$ rpm

$$\omega_1 = \frac{2\pi N_1}{60} = \frac{2\pi\times 1250}{60} = 130.9\,\text{rad/s}$$

(a) $I_m = m_m K_m^2 = 800\times(0.2)^2 = 32\text{kg}\cdot\text{m}$

$I_r = m_r K_r^2\ 1250\times(0.18)^2 = 40.5\text{ kg}\cdot\text{m}^2$

Let ω_3 = final speed of motor and rotor

For conservation of angular momentum,

$I_m\omega_1 + I_r\omega_2 = (I_m + I_r)\,\omega_3$

$32\times 130.9 + 40.5\times 0 = (32 + 40.5)\,\omega_3$

$\omega_3 = 57.78$ rad/s or 551.7 rpm

(b) Let t = time in seconds to reach ω_3

Mean radius, $r_m = \frac{1}{2}(r_1 + r_2) = \frac{1}{2}(90+120) = 105\,\text{mm}$

Total frictional torque, $T = n\mu W r_m = 2 \times 0.36 \times 600 \times 0.105 = 45.36$ Nm

Angular acceleration of rotor, $\alpha_2 = \frac{T}{I_r} = \frac{45.36}{40.5} = 1.12\,\text{rad/s}^2$

$$t = \frac{\omega_3 - \omega_1}{\alpha_2} = \frac{57.78-0}{1.12} = 51.59\,\text{s}$$

(c) Angular kinetic energy before impact,

$$E_1 = \frac{1}{2}I_m\omega_1^2 + \frac{1}{2}I_r\omega_2^2 = \frac{1}{2}I_m\omega_1^2$$

$$= \frac{1}{2}\times 32 \times (130.9)^2 = 274157\ \text{Nm}$$

After impact, $E_2 = \frac{1}{2}(I_m + I_r)\omega_3^2 = \frac{1}{2}(32+40.5)\times(57.78)^2$

$= 121022$ Nm

Kinetic energy lost during slipping $= E_1 - E_2 = 274157 - 121022$

$= 153135$ Nm

Let t_1 = time of slipping with constant resisting torque

Torque on armature shaft, $T_1 = -60 - 45.36 = -105.36$ Nm

Torque on rotor shaft, $T_2 = T = 45.36$ Nm

For armature shaft,

$$\omega_3 = \omega_1 + \alpha_1 t_1 = \omega_1 + \frac{T_1}{I_m}t_1$$

$$= 130.9 - \frac{105.36}{32}\times t_1$$

$$= 130.9 - 3.2925t_1 \qquad (1)$$

For rotor shaft,

$$\omega_3 = \alpha_2 t_1 = \frac{T_2}{I_r}t_1 = \frac{45.36}{40.5}t_1 = 1.12t_1 \qquad (2)$$

From Eqs. (1) and (2),

$$1.12\,t_1 = 130.9 - 3.2925\,t_1$$

$$t_1 = 29.7\ \text{s}$$

With constant driving torque,

$$T_1 = 60 - 45.36 = 14.64\ \text{Nm}$$

Now

$$\omega_1 + \frac{T_1}{I_m}t_1 = \frac{T_2}{I_r}t_1$$

$$130.9 + \frac{14.64}{32}t_1 = \frac{45.36}{40.5}t_1$$

$$130.9 + 0.4575t_1 = 1.12t_1$$

$$t_1 = 197.6\ \text{s}$$

Example 7.35

The contact surfaces in a cone clutch have an effective diameter of 80 mm. The semi-angle of cone is 15° and coefficient of friction is 0.32. Find the torque required to produce slipping of the clutch if an axial force applied is 200 N.

This clutch is employed to connect an electric motor running uniformly at 1200 rpm with a flywheel, which is initially stationary. The flywheel has a mass of 15 kg and its radius of gyration is 150 mm. Calculate the time required for the flywheel to attain full speed and also the energy lost in the slipping of the clutch.

■ Solution

Given: $R = 40$ mm, $\alpha = 15°$, $\mu = 0.32$, $W = 200$ N, $N_f = 1200$ rpm, $m_f = 15$ kg, $K_f = 150$ mm

Torque required to produce slipping, $T = \mu W r \operatorname{cosec} \alpha$

$= 0.32 \times 200 \times 0.04 \times \operatorname{cosec} 15° = 9.891$ Nm

Mass moment of inertia of flywheel, $I_f = m_f K_f^2 = 15 \times (0.15)^2 = 0.3375 \text{ kg}\cdot\text{m}^2$

Now $T = I_f \times \alpha_f$

Angular acceleration of flywheel, $\alpha_f = \dfrac{T}{I_f} = \dfrac{9.891}{0.3375} = 29.3 \text{ rad/s}^2$

Angular speed of flywheel, $\omega_f = \dfrac{2\pi N_f}{60} = \dfrac{2\pi \times 1200}{60} = 125.664 \text{ rad/s}$

Now $\omega_f = \alpha_f \times t_f$

∴ $t_f = \dfrac{125.664}{29.3} = 4.29 \text{ s}$

Angle turned through by the clutch in 4.29 s from rest,

θ = average angular velocity × time

$= \dfrac{1}{2} \times \omega_f \times t_f = \dfrac{1}{2} \times 125.664 \times 4.29 = 269.55 \text{ rad}$

Energy lost in slipping of clutch = $T\theta = 8.891 \times 269.55 = 2396.6$ Nm.

Example 7.36

The arrangement of an internal expanding shoe brake is shown in Fig.7.25. The width of the brake lining is 40 mm and the intensity of pressure at any point A is $4 \times 10^5 \sin \theta$ MPa, where θ is measured as shown from either pivot. The coefficient of friction is 0.35. Determine the braking torque and magnitude of forces F_1 and F_2.

■ Solution

Given: $\theta_1 = 30°$, $\theta_2 = 120°$, $\mu = 0.35$, $b = 40$ mm, $r = 150$ mm, $p_n = 4 \times 10^5 \sin \theta$ MPa, $\theta_0 = 25°$

Maximum intensity of pressure, $p_{max} = 4 \times 10^5$ MPa

Distance of force F_1 from fulcrum O_1, $l = 200$ mm

Distance of force F_2 from fulcrum O_2, $l = 200$ mm

Braking torque, $T_b = \mu p_{max} b r^2 (\cos \theta_1 - \cos \theta_2)$

$= 0.35 \times 4 \times 10^5 \times 0.04 \times (0.15)^2 (\cos 30° - \cos 120°)$

$= 172.12$ Nm

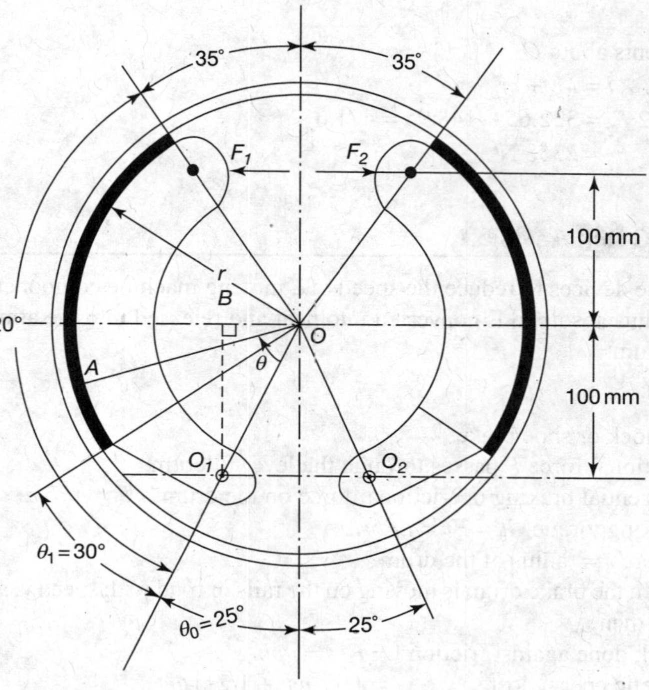

Fig.7.25 Internal expanding shoe brake

Total braking torque for two shoes, $T_b = 2 \times 172.12 = 344.24$ Nm

Now $$OO_1 = \frac{O_1B}{\cos\theta_0} = \frac{100}{\cos 25°} = 110.34 \text{ mm}$$

Total moment of normal forces about fulcrum O_1,

$$M_n = \frac{1}{2} p_{\max} br \times OO_1[(\theta_2 - \theta_1) + \frac{1}{2}(\sin 2\theta_1 - \sin 2\theta_2)]$$

$$= \frac{1}{2} \times 4 \times 10^5 \times 0.04 \times 0.15 \times 0.11034 \times \left[\frac{\pi}{180}(120° - 30°) + \frac{1}{2}(\sin 60° - \sin 240°)\right]$$

$$= 322.65 \text{ Nm}$$

Moment of friction forces about fulcrum O_1,

$$M_f = \mu p_{\max} br[r(\cos\theta_1 - \cos\theta_2) + \frac{OO_1}{4}(\cos 2\theta_2 - \cos 2\theta_1)]$$

$$= 0.35 \times 4 \times 10^5 \times 0.04 \times 0.15 \left[0.15(\cos 30° - \cos 120°) + \frac{0.11034}{4}(\cos 240° - \cos 60°)\right]$$

$$= 148.95 \text{ Nm}$$

Leading shoe:

Taking moments about O_1,

$$F_1 \times l = m_n - m_f$$

$$0.2\, F_1 = 322.65 - 148.95 = 173.7$$

$$F_1 = 868.5 \text{ N}$$

Trailing shoe:

Taking moments about O_2,

$$F_2 \times l = M_n + m_f$$
$$0.2\,F_2 = 322.65 + 148.95 = 471.6$$
$$F_2 = 2358\text{ N}$$

Summary for Quick Revision

1 Brakes are the devices to reduce the speed of a moving machine component by absorbing energy. The energy thus absorbed is converted into heat and released into the atmosphere or absorbed in another medium.

2 Block brakes:

(a) Single-block or shoe brake.

(i) Frictional force F passes through the lever fulcrum.

Tangential braking or frictional force on the drum, $F = \mu Pb/a$

Braking torque, $T_b = Fr = \mu Pbr/a$

Where r = radius of the drum

When the brake drum is moving on the rails or road with speed v, and the braking distance is s, then

Work done against friction $= F \cdot s = \mu pb \cdot s/a$

Kinetic energy lost $= 1/2 \cdot mv^2 + 1/2 \cdot 1\omega^2$

where m = mass of brake drum

I = moment of intertia of brake drum

ω = angular speed of drum

For the conservation of energy, we have

$\mu Pb \cdot s/a = 1/2 \cdot mv^2 + 1/2 \cdot I\omega^2$

(ii) Frictional force F is above the lever fulcrum by a distance c.

$R = P \cdot b/(a + \mu c)$

In this case, the frictional force is helping to apply the brake. Such a brake is called *self-energizing brake*.

(iii) Frictional force F is below the lever fulcrum by a distance c.

$R = P \cdot b/(a - \mu c)$

If $a \le \mu c$, then P will be zero or negative, that is, no external force will be required to apply the brake. Such a brake is called *self-locking* type of brake.

(b) Pivoted shoe brake

Frictional torque, $T = 4\,\mu Rr \sin\theta/(2\theta + \sin 2\theta)$

Equivalent coefficient of friction,

$\mu_e = 4\,\mu \sin\theta/(2\theta + \sin 2\theta)$

(c) Double-shoe brake

(i) Frictional forces passing through the fulcrums of the levers.

$R_l = P_l\,(a + b)/(a - \mu c)$

$F_l = \mu P_l\,(a + b)/(a - \mu c)$

$R_r = P_r\,(a + b)/(a + \mu c)$

$F_r = \mu P_r\,(a + b)/(a + \mu c)$

3 Band brakes

(a) Simple band brake

Ratio of tensions, $F_1/F_2 = \exp(\mu\theta)$

Braking torque on the drum, $T_b = (F_1 - F_2)\, r$

(b) Differential band brake

$PI + F_1 a = F_2 b$, for anti-clockwise rotation of drum

$PI + F_2 a = F_1 b$, for clockwise rotation of drum

For a self-locking brake, $P \le 0$, therefore, for anti-clockwise rotation,

$F_1/F_2 \ge b/a$

and for clockwise rotation, $F_1/F_2 \le a/b$

$F_1/F_2 = \exp(\mu\theta)$

4 Band and block brake.

$$F_n/F_o = [(1 + \mu \tan\theta)/(1 - \mu \tan\theta)]^n$$

Braking torque, $T_b = (F_o - F_n)\, r$

5 Internal expanding shoe brake

Total moment of normal force about the fulcrum,

$$M_n = (p_{max} \cdot b \cdot r \cdot a)/2 \cdot [(\theta_2 - \theta_1) + (1/2) \cdot (\sin 2\theta_1 - \sin 2\theta_2)]$$

Total moment of the frictional force about the fulcrum O_1,

$$M_f = \mu\, (p_{max} \cdot br\, [r\, (\cos\theta_1 - \cos\theta_2) + (a/4)(\cos 2\theta_2 - \cos 2\theta_1)]$$

For the leading shoe, taking moments about the fulcrum,

$I = M_n - M_f$

and for the trailing shoe, taking moments about the fulcrum,

$F_2 I = M_n + M_f$

6 Braking of a vehicle.

Acceleration, $f = g\, (\mu \cos\alpha + \sin\alpha)$

(a) Vehicle moves on a level track, $f = \mu g$

(b) Vehicle moves down, $f = g\, (\mu \cos\alpha - \sin\alpha)$

(c) Brakes are applied to rear wheel only, $f = \{\mu g\, (I - x) \cos\alpha\}/(I + \mu h) \pm g \sin\alpha$

Use positive sign for going up and negative sign for going down the plane.

On a level track, $f = \mu g\, (I - x)/(I + \mu h)$

(d) Brakes are applied to front wheels only: $f = (\mu g \times \cos\alpha)/(I - \mu h) \pm g \sin\alpha$

Use positive sign for going up the plane and negative sign for going down.

On a level track, $f = \mu g \times /(I - \mu h)$

7 A clutch is a device used to transmit the rotary motion of one shaft to another when desired.

8 Friction clutches:

(i) Single-plate clutch

(a) Uniform pressure

Intensity of pressure, $p = W/\pi(r_2^2 - r_1^2)$

Total frictional torque, $T_f = (2/3) \cdot \mu W[(r_2^3 - r_1^3)/(r_2^2 - r_1^2)]$

Mean radius of friction surface, $r_m = (2/3) \cdot [(r_2^3 - r_1^3)/(r_2^2 - r_1^2)]$

(b) Uniform wear

Total frictional torque, $T_f = \mu W\, (r_1 + r_2)/2$

$r_m = (r_1 + r_2)/2$

(ii) Multi-plate clutch
n_1 = number of plates on the driving shaft
n_2 = number of plates on the driven shaft
Number of pairs of contact surfaces, $n = n_1 + n_2 - 1$
Total frictional torque transmitted, $T = n\mu W r_m$

(iii) Cone clutch
(a) Uniform pressure
Total frictional torque, $T_f = \mu W r_m \operatorname{cosec} \alpha$
(b) Uniform wear
Total frictional torque, $T_f = \mu W r_m \operatorname{cosec} \alpha$
where $r_m = (r_1 + r_2/2)$
Also $T_f = 2\pi \cdot \mu p_n \cdot b \cdot r_m^2$
Axial force required to engage the clutch,

$$W_e = W_n (\sin \alpha + \mu \cos \alpha)$$

9 Dynamometers are the devices to measure the power developed by a prime mover.

10 Dynamometers.

(a) In absorption dynamometers, the entire power produced by the prime mover is absorbed by the frictional resistance of the brake and is transformed into heat, during the process of measurement. The absorption type of dynamometer are: Prony brake dynamometer, and Rope brake dynamometer.

(i) Prony brake dynamometer:
Brake power, $BP = 2\pi NWL/(60 \times 10^3)$ kW

(ii) Rope brake dynamometer:
Brake power, $BP = (W - S) \times \pi (D + d) N/(60 \times 10^3)$ kW

(b) In transmission dynamometers, the energy is used for doing work. The power developed by the prime mover is transmitted through the dynamometer to some other machine where the power is suitably measured. These type of dynamometers are: Epicyclic train dynamometer, belt transmission dynamometer, and Torsion dynamometer.

(i) Epicyclic gear train dynamometer:
Brake power, $BP = 2\pi NPR/(60 \times 10^3)$ kW, $P = WL/(2a)$

(ii) Belt transmission dynamometer:
Brake power, $BP = \pi DNWL/(2a \times 60 \times 10^3)$ kW

(iii) Torsion dynamometer:
Brake power, $BP = 2\pi NT/(60 \times 10^3)$ kW, $T = (GJ/l)\,\theta$

Multiple Choice Questions

1 The following dynamometer is widely used for the absorption of wide range of power at wide range of speed
(a) Hydraulic (b) Belt transmission (c) Rope brake (d) Electric generator.

2 The equivalent coefficient of friction for a block brake is
(a) $4\mu \sin\theta/(\sin\theta + \theta)$
(b) $4\mu \sin(\theta/2)/[\sin(\theta/2) + \theta/2]$
(c) $4\mu \sin(\theta/2)/(\sin\theta + \theta)$
(d) $\mu \sin\theta/(\sin\theta + \theta)$.

3 The equivalent radius of a block brake is
(a) $4\,r \sin\theta/(\sin\theta + \theta)$ (b) $4\,r \sin(\theta/2)/(\sin(\theta/2) + \theta/2)$
(c) $4\,r \sin(\theta/2)/(\sin\theta + \theta)$ (d) $r \sin\theta/(\sin\theta + \theta)$
where r = drum radius.

4 The ratio of tensions on the tight side to slack side of multi-block brake is
(a) $(1 - n\mu \tan\theta)/(1 + n\mu \tan\theta)$ (b) $(1 + n\mu \tan\theta)/(1 - n\mu \tan\theta)$
(c) $[(1 + \mu \tan\theta)/(1 - \mu \tan\theta)]^n$ (d) $[(1 - \mu \tan\theta)/(1 + \mu \tan\theta)]^n$.

5 The stopping distance for a vehicle by applying brakes when all the four wheels are sliding as compared with when all the four wheels are in a limiting state of sliding is
(a) More (b) Less (c) Same (d) Unpredictable.

6 The stopping distance for a four-wheel vehicle is
(a) Unaltered by an increase in weight of vehicle
(b) Decreases with increase of coefficient of friction
(c) Directly proportional to square of velocity of vehicle
(d) All of the above.

7 Dynamometer is a device used on a prime mover for measuring
(a) Torque developed (b) Power developed
(c) Power absorbed (d) All of the above.

8 Which of the following is an absorption dynamometer?
(a) Prony brake dynamometer (b) Rope brake dynamometer
(c) Froude's hydraulic dynamometer (d) All of the above.

9 Which of the following is a transmission dynamometer?
(a) Torsion dynamometer (b) Belt dynamometer
(c) Hydraulic dynamometer (d) Prony brake dynamometer.

10 Which type of brake is commonly used in cars?
(a) Band brake (b) Shoe brake
(c) Band and block brake (d) Internal expanding shoe brake.

11 In a self-locking brake, the force required to apply the brake is
(a) Zero (b) Minimum (c) Maximum (d) Average.

12 When the frictional force helps the applied force in applying the brake, the brake is called
(a) Automatic (b) Self-locking (c) Self-energizing.

Answers:

1. (a) 2. (c) 3. (c) 4. (c) 5. (a) 6. (d) 7. (d) 8. (d) 9. (b) 10. (d) 11. (a) 12. (c)

Review Questions

1 What is a brake?

2 What are the various types of brakes?

3 Differentiate between a self-locking and self-energizing brake.

4 What are the advantages of internal expanding shoe brake?

5 What is the effect of applying brakes only to the front wheels of a vehicle?

6 What is the advantage of a pivoted shoe brake?

7 What is the difference between a shoe brake and band brake?

8 How do the internal expanding shoe brakes become self-locking?

9 What is a dynamometer?

10 What are the various types of dynamometers?

11 What is the principle of working of an absorption dynamometer?

12 What is a transmission dynamometer?

13 Why the pulley of a rope brake dynamometer water cooled?

14 What is a clutch? State its different types.

15 Differentiate between dry and wet clutches.

16 Where do we use multi-plate clutches?

Exercises

7.1 The drum diameter of a single-block brake is 1 m. It sustains 240 Nm of torque at 400 rpm. The coefficient of friction is 0.32. The distance of the fulcrum from the vertical centre line of the drum is 150 mm and length of lever is 800 mm. The frictional force acts above the fulcrum by a distance of 25 mm. Determine the required force to be applied when the drum rotates clockwise and angle of contact is 100°.

[Ans. 265.7 N]

7.2 A bicycle and rider of mass 90 kg are travelling at a speed of 15 km/h on a level road. A brake is applied to the rear wheel which is 0.7 m in diameter. How far will the bicycle travel? The pressure applied on the brake is 100 N and coefficient of friction is 0.06.

[Ans. 130.2 m]

7.3 A car moving on a level road at a speed of 45 km/h has a wheel base 2.8 m, distance of CG from ground level 0.6 m, and the distance of CG from rear wheels 1.1 m. Find the distance travelled by the car before coming to rest when the brakes are applied to (a) the rear wheels, (b) to the front wheels, and (c) all the four wheels.

The coefficient of friction between road and types may be taken as 0.5.

[Ans. 29 m, 36.2 m, 15.93 m]

7.4 In a belt transmission dynamometer, the distance between the centre of driving pulley and dead weights is 1 m. Find the value of dead weights required to keep the lever in horizontal position if power transmitted is 7.5 kW and the diameter of each of the driving as well as the intermediate pulleys is equal to 0.4 m. The driving pulley runs at 400 rpm.

[Ans. 716.2 N]

7.5 In a Prony brake dynamometer, the spring balance reading is 200 N, radius of brake drum is 0.3 m, and distance between the drum axis and hinge of the blocks is 0.6 m. Determine the pressure exerted on the drum by tightening the screw, tangential force acting on the brake drum, and the output power of the prime mover if the record speed is 300 rpm. Take coefficient of friction equal to 0.25.

[Ans. 3048 N, 762 N, 14.36 kW]

7.6 A single-plate clutch transmits 20 kW at 1000 rpm. The maximum pressure intensity between the plates is 0.09 MPa. The outer diameter of the plate is 350 mm and both the sides of the plate are effective. The coefficient of friction is 0.25. Determine (a) inner diameter of the plate and (b) axial force required to engage the clutch.

[Ans. 0.147 m, 2327.5 N]

7.7 A clutch in a motor car is of single-plate type having both sides of the plate effective. It is required to transmit 35 kW at 1500 rpm. The axial thrust is 0.075 MPa. The ratio between the external and internal radii of the plate is 1.5 and coefficient of friction is 0.25. Assuming uniform pressure, determine the dimensions of the clutch.

[Ans. 106 mm, 159 mm]

7.8 A multi-plate clutch of alternate bronze and steel plates having effective diameters of 175 mm and 72.5 mm has to transmit 25 kW at 2000 rpm. The end thrust is 1600 N and coefficient of friction is 0.1. Calculate the number of plates necessary assuming uniform pressure.

[Ans. 12]

7.9 A cone clutch has a radii of 130 mm and 150 mm. The semi-cone angle is 20°. If coefficient of friction is 0.25 and uniform normal pressure is 0.15 MPa, find (a) necessary axial load and (b) power that can be transmitted at 1000 rpm.

[Ans. 902.57 N, 28.327 kW]

7.10 The external and internal radii of a friction plate of a clutch are 120 mm and 60 mm respectively. The total axial thrust is 1500 N. For uniform wear, find the maximum, minimum and average pressure on the contact surfaces.

[Ans. 0.066 MPa, 0.033 MPa, 0.044 MPa]

7.11 A power of 60 kW is transmitted by a multi-plate clutch at 1500 rpm. The axial intensity of pressure is not to exceed 15 MPa. The coefficient of friction for the friction surfaces is 0.15. The external radius of friction surface is 120 mm and internal radius is 100 mm. Find the number of plates required to transmit the power of 60 kW.

[Ans. 12]

7.12 A cone clutch of semi-cone angle 15° is used to transmit 30 kW at 800 rpm. The mean frictional surface radius is 150 mm and normal intensity of pressure of the mean radius is not to exceed 0.15 MPa. The coefficient of friction is 0,2. Assuming uniform wear, determine:

(a) Width of contact surface, and

(b) Axial force required to engage the clutch

[Ans. 84 mm, 3089.4 N]

8 CAMS

Chapter Outline

8.1 INTRODUCTION

A cam is a rotating or a reciprocating element of a mechanism, which imparts rotating, reciprocating or oscillating motion to another element, called the follower. There is line contact between the cam and the follower, and thereby forms a higher pair. Cams are used in clocks, printing machines, automatic screw cutting machines, internal combustion engines for operating the valves, and shoe making machines etc. In this chapter, we shall study the various types of planar cams from the point of view of drawing their profile and motion analysis.

The three essential components of a cam mechanism are as follows:

(i) the cam,

(ii) the follower, and

(iii) the frame.

The cam revolves at a constant speed and drives the follower. The motion of the follower depends upon the profile of the cam. The frame supports and guides the follower and the cam.

8.2 CLASSIFICATION OF CAMS

The planar cams can be classified according to their shape as follows:

1. *Wedge and flat cams*: Such a cam has a wedge *A* to which translational motion is given to actuate the follower *B* in order to either reciprocate or oscillate it, as shown in Fig.8.1(a) and (b). The follower is guided in the guides *C*. In Fig.8.1(c), the cam is stationary and the follower guide *C* causes the relative motion of the cam *A* and follower *B*. Fig.8.1(d) shows a flat plate with a groove in which the follower is held to obtain the desired motion.

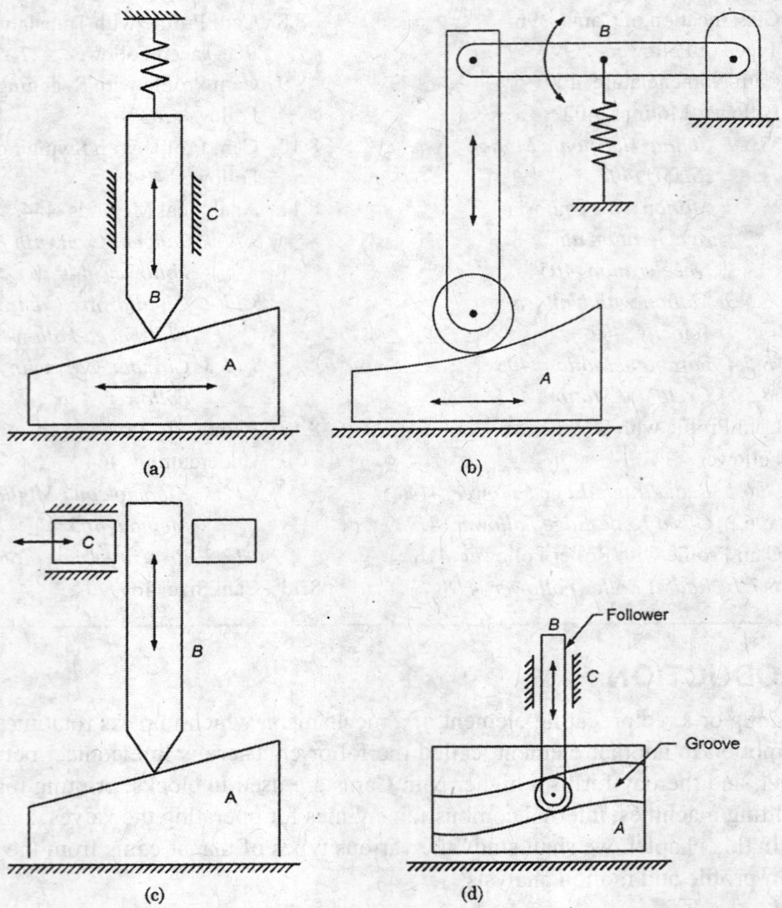

Fig.8.1 Wedge and flat cams

2. *Radial and offset cams*: A cam in which the follower moves radially from the centre of rotation of the cam is called a radial (or disc or plate) cam, as shown in Fig.8.2(a) and (b). In such a cam, the axis of the follower passes through the axis of the cam.
   If the axis of the follower does not pass through the axis of the cam, it is called an offset cam, as shown in Fig.8.2(c).
3. *Cylindrical cams*: In a cylindrical cam, a cylinder which has a circumferential groove cut in the surface, rotates about its axis. The follower motion can be either oscillatory or reciprocating type, as shown in Fig.8.3(a) and (b). They are also called barrel or drum cams.

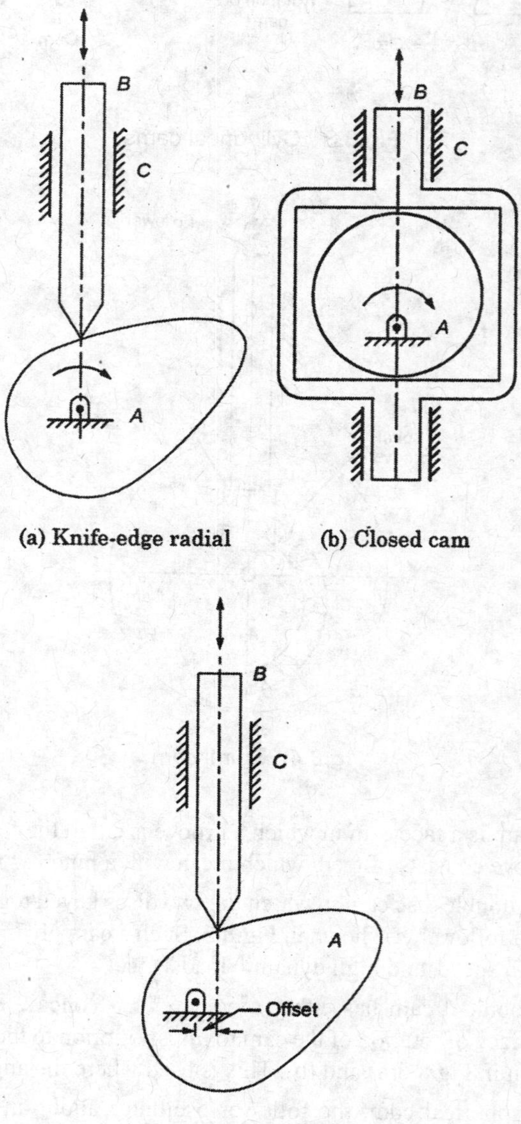

(a) Knife-edge radial

(b) Closed cam

(c) Offset cam

Fig.8.2 Radial and offset cams

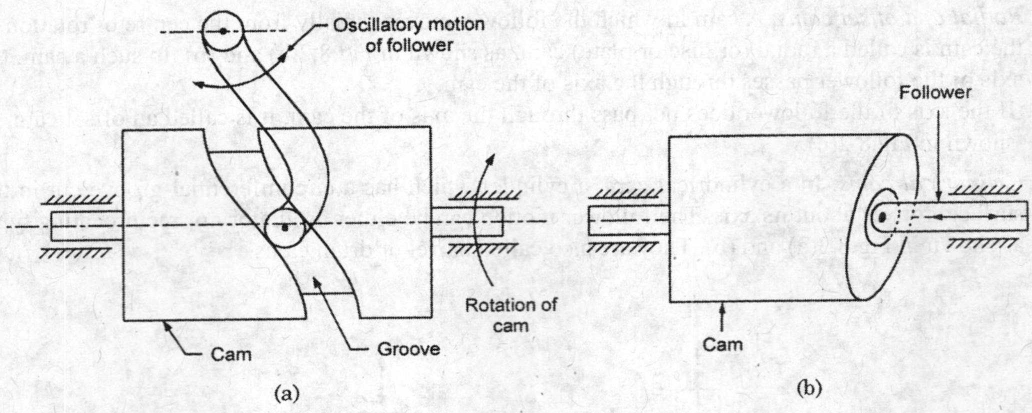

Fig.8.3 Cylindrical cams

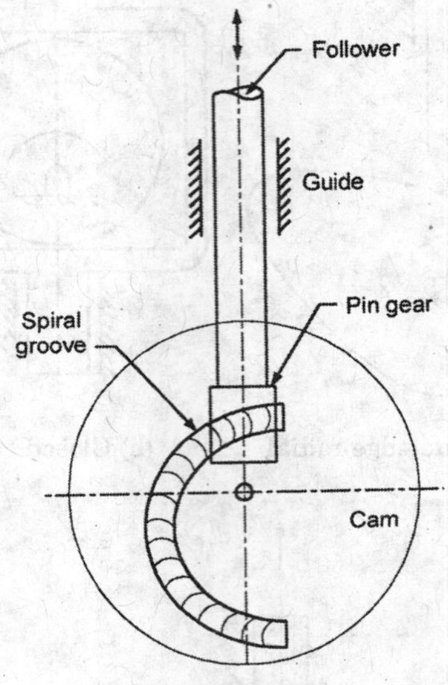

Fig.8.4 Spiral cam

4. *Spiral cams*: A spiral cam is a face cam in which a groove is cut in the form of a spiral, as shown in Fig.8.4. The spiral groove consists of teeth which mesh with a pin gear follower.
5. *Conjugate cams*: It is a double disc cam in which the two discs keyed together are in constant touch with the two rollers of a follower, as shown in Fig.8.5. Such a cam gives low wear, low noise, better control of follower, high speed, and high dynamic loads, etc.
6. *Globoidal cams*: A globoidal cam has either a convex or a concave surface. A circumferential groove is cut on the surface of rotation of the cam to impart motion to the follower, which has oscillatory motion, as shown in Fig.8.6(a) and (b). This is used where the angle of oscillation is large.
7. *Spherical cams*: In a spherical cam, the follower oscillates about an axis perpendicular to the axis of rotation of the cam. The spherical cam is in the form of a spherical surface, as shown in Fig.8.7.

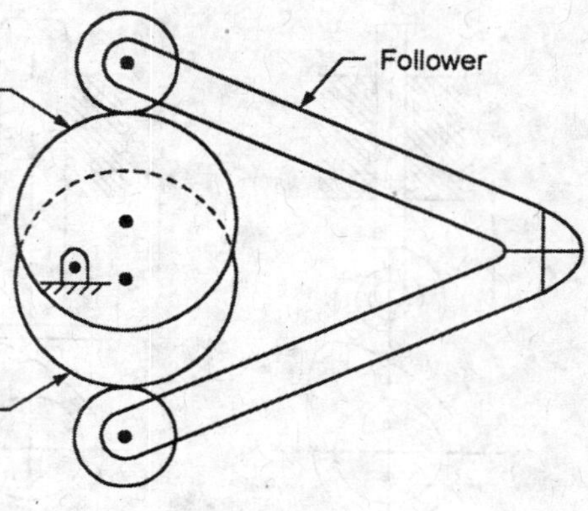

Fig.8.5 Conjugate cam

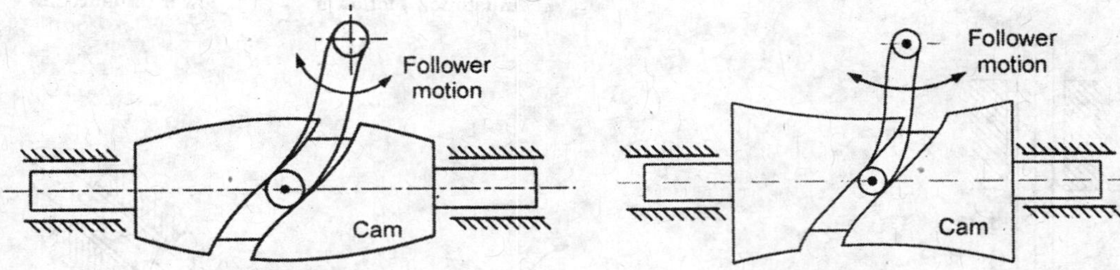

Fig.8.6 Globoidal cam

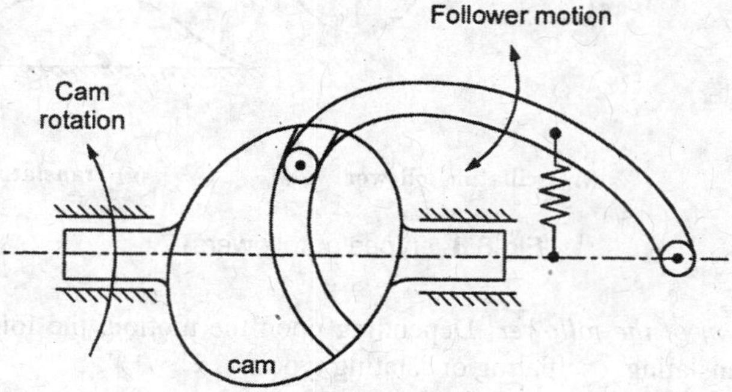

Fig.8.7 Spherical cam

8.3 TYPES OF FOLLOWERS

The followers can be of the following types:

1. *Based on the surface in contact*: The followers based on the type of surface in contact can be classified as: knife edge, roller, flat faced (or mush room), and spherical-faced follower, as shown in Fig.8.8(a) to (g).

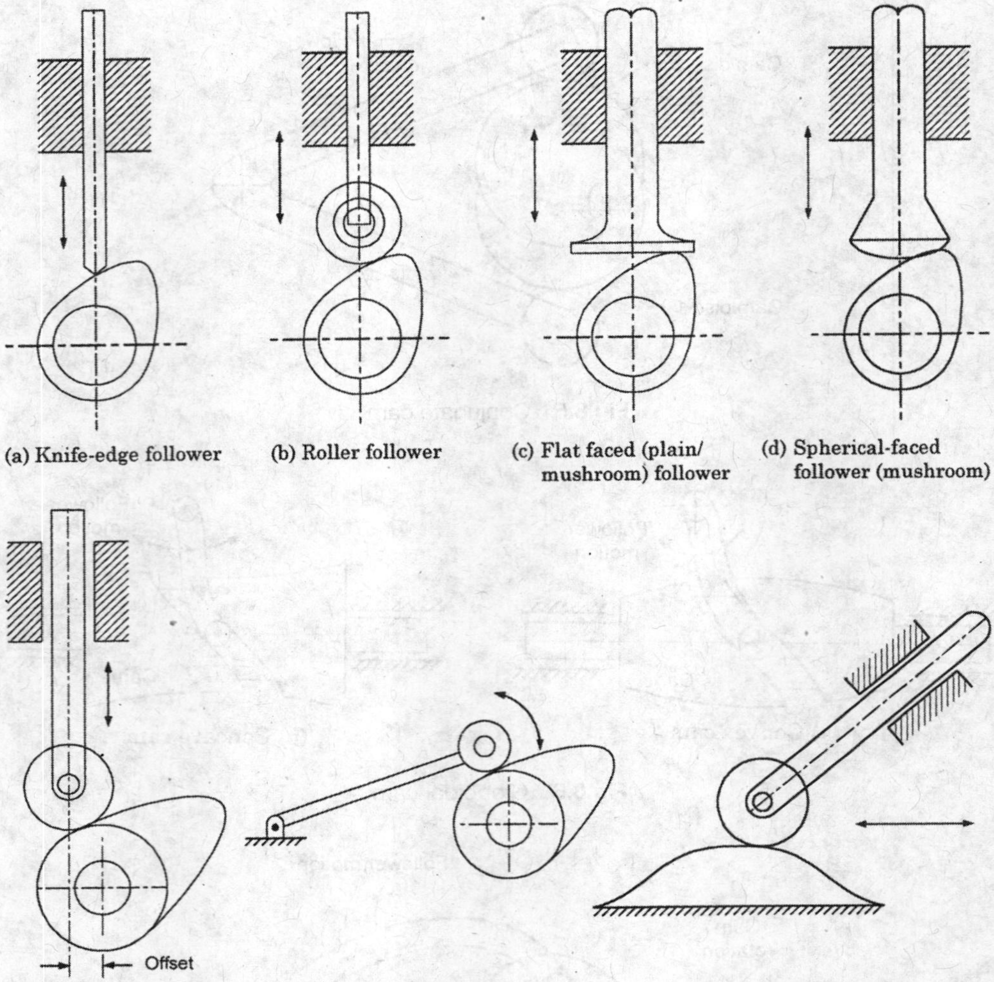

(a) Knife-edge follower (b) Roller follower (c) Flat faced (plain/ mushroom) follower (d) Spherical-faced follower (mushroom)

(e) Offset follower (f) Oscillating follower (g) Translating follower

Fig.8.8 Types of followers

2. *Based on the motion of the follower*: Depending upon the motion, the follower could be of the reciprocating or translating, oscillating or rotating types.
3. *Based on the path of motion of follower:* When the axis of the follower passes through the centre of rotation of the cam, it is called a radial follower, and when the axis of the follower does not pass through the axis of the cam, it is called an offset follower.

8.4 CAM NOMENCLATURE

A radial cam with reciprocating roller follower is shown in Fig.8.9. The following nomenclature is used in reference to planar cam mechanisms:

Base circle: It is the smallest circle that can be drawn to the cam profile from the centre of rotation.

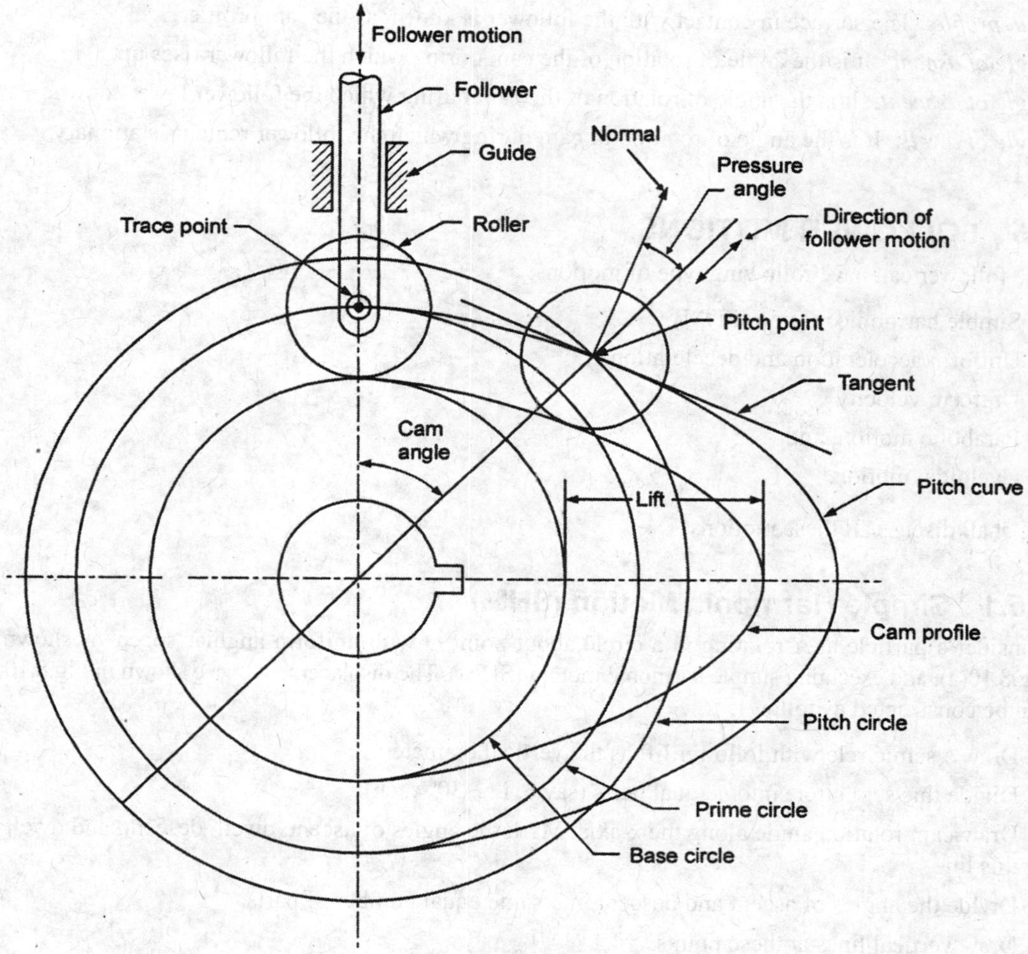

Fig.8.9 Cam nomenclature

Prime circle: It is the smallest circle drawn to the pitch curve from the centre of rotation of the cam.

Pitch point: It is a point on the pitch curve having the maximum pressure angle.

Pitch circle: It is the circle drawn through the centre and pitch point.

Trace point: It is a reference point on the follower and is used to generate the pitch curve. In the case of a knife edge follower, it is the knife edge, and in the case of a roller follower, it is the centre of the roller.

Pitch curve: It is the curve generated by the trace point as the follower moves relative to the cam.

Cam angle: It is the angle turned through by the cam from the initial position.

Pressure angle: It is the angle between the direction of the follower motion and a normal to the pitch curve.

Lift: It is the maximum travel of the follower from the lowest position to the topmost position. It is also called throw or stroke of the cam.

Cam profile: The surface in contact with the follower is known as the cam profile.

Angle of ascent: It is the angle of rotation of the cam during which the follower rises up.

Angle of descent: It is the angle of rotation of the cam during which the follower lowers down.

Angle of dwell: It is the angle of rotation of cam during which the follower remains stationary.

8.5 FOLLOWER MOTIONS

The follower can have following type of motions:

1. Simple harmonic motion (SHM)
2. Uniform acceleration and deceleration
3. Uniform velocity
4. Parabolic motion, and
5. Cycloidal motion.

We shall discuss all these motions.

8.5.1 Simple Harmonic Motion (SHM)

Consider a particle at A rotating in a circle about point O with uniform angular speed, as shown in Fig.8.10(a), and executing simple harmonic motion (SHM). The displacement curve shown in Fig.8.10(b) can be constructed as follows:

1. Draw a semicircle with follower lift as the vertical diameter.
2. Divide this semicircle into n equal parts (say 6, i.e., 30° each).
3. Draw cam rotation angle along the x-axis. Mark the angles of ascent, dwell, descent, and dwell on this line.
4. Divide the angles of ascent and descent into same equal number of parts.
5. Draw vertical lines at these points.
6. Draw horizontal lines from the points on the circumference of the semicircle to intersect the vertical lines.
7. Mark the points of intersection and join by a smooth curve to obtain the displacement diagram.

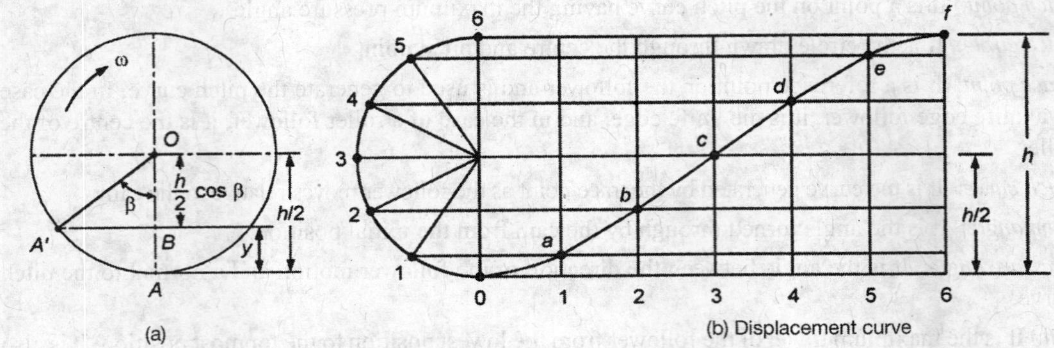

Fig.8.10 Simple harmonic motion of follower

Motion Analysis

Let y = displacement of the follower
h = lift of the follower (maximum displacement or rise)
θ = angle turned through by the crank OA from given datum
ϕ = cam rotation angle for the maximum follower displacement
β = angle on harmonic circle

Then $y = OA - OB$

$$= \left(\frac{h}{2}\right) \times (1 - \cos\beta) \tag{8.1a}$$

For the ascent or descent h of the follower displacement, the cam is rotated through an angle ϕ, whereas a point on the harmonic semicircle traverses an angle π radians. Thus, the cam rotation is proportional to the angle turned through by the point on the harmonic semicircle, i.e.,

$$\beta = \frac{\pi\theta}{\phi}$$

Thus Eq. (8.1a) becomes,

$$y = \left(\frac{h}{2}\right)\left[1 - \cos\left(\frac{\pi\theta}{\phi}\right)\right] \tag{8.1b}$$

Now $\theta = \omega t$

$$y = \left(\frac{h}{2}\right)\left[1 - \cos\left(\frac{\pi\omega t}{\phi}\right)\right] \tag{8.1c}$$

Velocity, $v = \dfrac{dy}{dt}$

Differentiating Eq. (8.1c), we get

$$v = \left(\frac{h}{2}\right)\left(\frac{\pi\omega}{\phi}\right)\sin\left(\frac{\pi\omega t}{\phi}\right)$$

$$= \left(\frac{h}{2}\right)\left(\frac{\pi\omega}{\phi}\right)\sin\left(\frac{\pi\theta}{\phi}\right) \tag{8.2}$$

Let θ_1 = angle of ascent
θ_2 = angle of dwell
θ_3 = angle of descent

Then, Velocity during ascent, $v_a = \left(\frac{h}{2}\right)\left(\frac{\pi\omega}{\theta_1}\right)\sin\left(\frac{\pi\theta}{\theta_1}\right)$ (8.3a)

Velocity during descent, $v_d = \left(\frac{h}{2}\right)\left(\frac{\pi\omega}{\theta_3}\right)\sin\left(\frac{\pi\theta}{\theta_3}\right)$ (8.3b)

The velocity is maximum when $\theta = \dfrac{\phi}{2}$.

$$v_{max} = \left(\frac{h}{2}\right)\left(\frac{\pi\omega}{\phi}\right) \tag{8.4}$$

Maximum velocity of follower during ascent $= \left(\frac{h}{2}\right)\left(\frac{\pi\omega}{\theta_1}\right)$ (8.5a)

Maximum velocity of follower during descent $= \left(\frac{h}{2}\right)\left(\frac{\pi\omega}{\theta_3}\right)$ (8.5b)

Acceleration, $$f = \frac{dv}{dt}$$

Differentiating Eq. (8.2), we get $$f = \left(\frac{h}{2}\right)\left(\frac{\pi\omega}{\phi}\right)^2 \cos\left(\frac{\pi\theta}{\phi}\right) \quad (8.6)$$

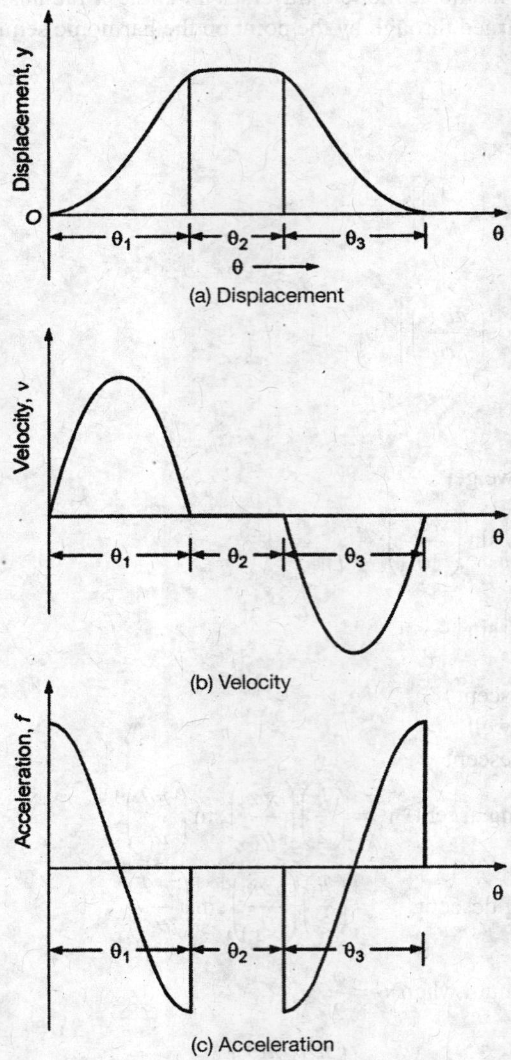

Fig.8.11 Displacement, velocity and acceleration distribution in SHM of follower

Acceleration during ascent, $$f_a = \left(\frac{h}{2}\right)\left(\frac{\pi\omega}{\theta_1}\right)^2 \cos\left(\frac{\pi\theta}{\theta_1}\right) \quad (8.7a)$$

Acceleration during descent, $$f_d = \left(\frac{h}{2}\right)\left(\frac{\pi\omega}{\theta_3}\right)^2 \cos\left(\frac{\pi\theta}{\theta_3}\right) \quad (8.7b)$$

The acceleration is maximum when $\theta = 0°$.

$$f_{\max} = \left(\frac{h}{2}\right)\left(\frac{\pi\omega}{\phi}\right)^2 \quad (8.8)$$

Maximum acceleration of follower during ascent $$= \left(\frac{h}{2}\right)\left(\frac{\pi\omega}{\theta_1}\right)^2 \quad (8.9a)$$

Maximum acceleration of follower during descent $$= \left(\frac{h}{2}\right)\left(\frac{\pi\omega}{\theta_3}\right)^2 \quad (8.9b)$$

The motion of the follower is shown in Fig.8.11.

It can be observed from Fig.8.11 that there is an abrupt change of acceleration from zero to maximum at the beginning of the follower motion and also from maximum (negative) to zero at the end of the follower motion. The same pattern is repeated during descent. This leads to jerk, vibration and noise etc. Therefore, SHM should be adopted only for low and moderate cam speeds.

8.5.2 Motion with Uniform Acceleration and Deceleration

In such a motion, there is acceleration during the first half of the follower motion and deceleration during the later half. The magnitude of both acceleration and deceleration is the same in the two halves.

The displacement diagram, as shown in Fig.8.12, can be constructed as follows:

1. Draw the cam rotation angle along the x-axis and the follower lift along the y-axis.
2. Mark the angles of ascent, dwell, descent and dwell on the horizontal line and the lift on the vertical line at the origin.
3. Divide the angles of ascent and descent into equal number of parts (say 6). Also divide the lift line into same equal number of parts.
4. Draw horizontal and vertical lines at these points.

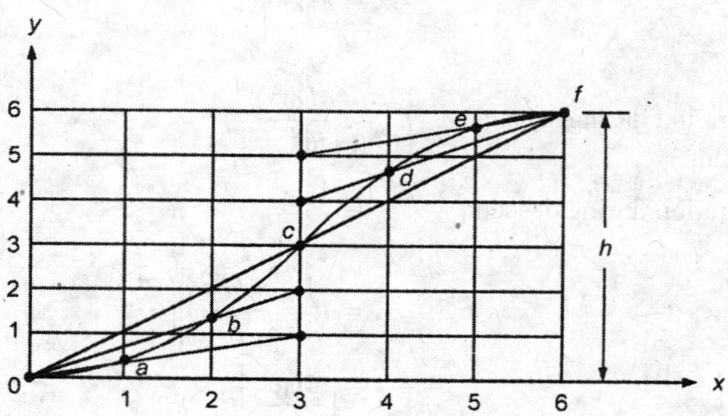

Fig.8.12 Displacement diagram for uniform acceleration and deceleration of follower

5. Join the origin with the points of intersection on the middle line of ascent upto half the lift. Also join the points on the other half of the middle line of lift with the topmost point of the last line.
6. Mark the points of intersection of these lines with the vertical lines.
7. Join these points with a smooth curve to get the displacement diagram.

Motion analysis

Let f = uniform acceleration or deceleration of the follower.
y = displacement of follower

Then
$$\frac{dv}{dt} = f$$

Integrating, we have
$$v = ft + C_1$$

where C_1 is a constant of integration
If at $t = 0$, $v = 0$, then $C_1 = 0$. Hence
$v = ft$

Now
$$v = \frac{dy}{dt} = ft$$

Integrating again, we have
$$y = \frac{ft^2}{2} + C_2$$

where C_2 is another constant of integration.
As $y = 0$ at $t = 0$, therefore $C_2 = 0$. Hence
$$y = \frac{ft^2}{2}$$
$$f = \frac{2y}{t^2} = \text{const.} \tag{8.10 a}$$

Considering the follower at midway, we have
$$y = \frac{h}{2} \quad \text{and} \quad t = \frac{(\phi/2)}{\omega}$$
$$f = 2\left(\frac{h}{2}\right) / \left[\phi^2 (4\omega^2)\right]$$
$$= \frac{4h\omega^2}{\phi^2} \tag{8.10 b}$$

Maximum acceleration during ascent, $$f_{max} = \frac{4h\omega^2}{\theta_1^2} \tag{8.11 a}$$

Maximum acceleration during descent, $$f_{max} = \frac{4h\omega^2}{\theta_3^2} \tag{8.11 b}$$

Velocity, $$v = ft = \left[\frac{4h\omega^2}{\phi^2}\right]\left(\frac{\theta}{\omega}\right)$$
$$= \left[\frac{4h\omega}{\phi^2}\right]\theta \tag{8.12}$$

The velocity is maximum when the follower is at midway position, i.e., $\theta = \frac{\phi}{2}$

Maximum velocity, $$v_{max} = \left(\frac{4h\omega}{\phi^2}\right)\left(\frac{\phi}{2}\right)$$

$$= \frac{2h\omega}{\phi} \qquad (8.13)$$

Maximum velocity during ascent $$= \frac{2h\omega}{\theta_1} \qquad (8.14\ a)$$

Maximum velocity during descent $$= \frac{2h\omega}{\theta_3} \qquad (8.14\ b)$$

The motion of the follower is shown in Fig.8.13.

It may be observed from Fig.8.13 that there are abrupt changes in the acceleration at the beginning, midway and at the end of the follower motion. At the midway, an infinite jerk is produced. Therefore, this motion is adopted for moderate speeds only.

8.5.3 Motion with Uniform Velocity

In this case, the displacement of the follower is proportional to the angle of cam rotation. Therefore, slope of the displacement curve is constant.

Let $y = c\theta$

Where c = constant of proportionality, and θ = angle of cam rotation.

If h = follower rise

ϕ = angle through which the cam is to rotate to lift the follower by h.

Then $h = c\phi$

so that $c = \frac{h}{\phi}$

$$\therefore \quad y = \frac{h\theta}{\phi} \qquad (8.15)$$

Velocity, $$v = \left(\frac{h}{\phi}\right)\cdot\left(\frac{d\theta}{dt}\right)$$

$$= \frac{h\omega}{\phi} \qquad (8.16)$$

Acceleration, $$f = \left(\frac{h}{\phi}\right)\cdot\left(\frac{d\omega}{dt}\right) = 0 \qquad (8.17)$$

The variation of displacement, velocity and acceleration are shown in Fig.8.14. It may be observed that although the acceleration is zero during ascent or descent of the follower, it is infinite at the beginning and end of the motion. There are abrupt changes in velocity at these points. This results in infinite inertia forces and is therefore unsuitable from practical point of view.

This can be avoided by rounding the sharp corners of the displacement curve so that the velocity changes are gradual at the beginning and end of the follower motion. During these periods the acceleration may be assumed to be constant and of finite values. A modified uniform velocity motion is shown in Fig.8.15.

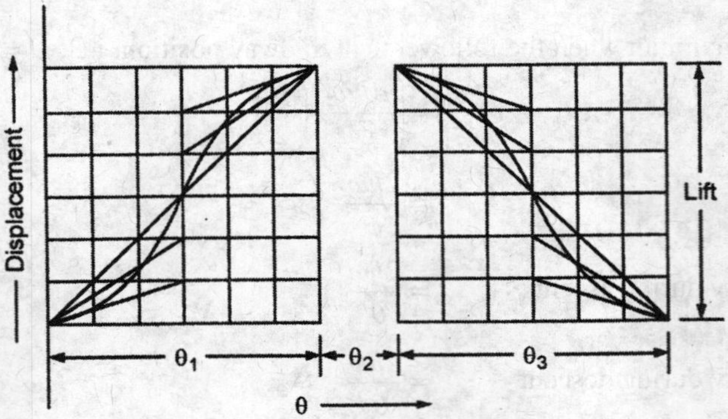

(a) Displacement

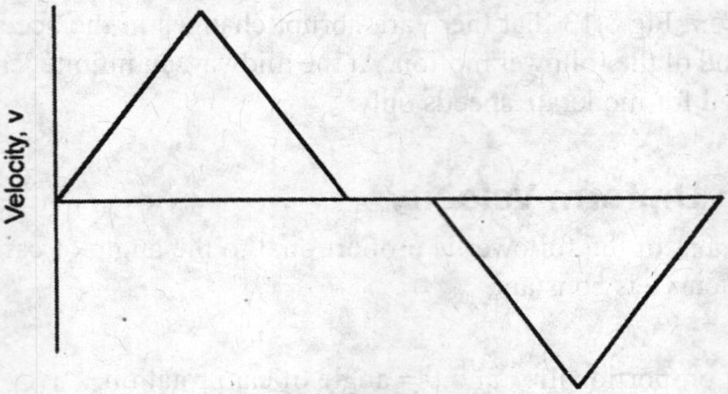

(b) Velocity

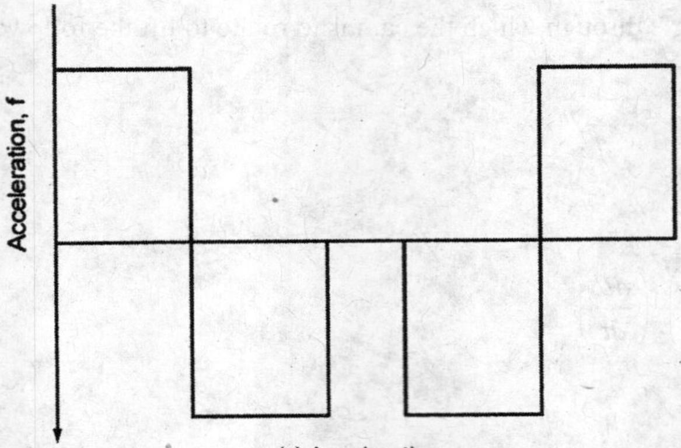

(c) Acceleration

Fig.8.13 Motion with uniform acceleration and deceleration

8.5.4 Parabolic Motion

In the parabolic motion, the displacement of the follower is proportional to the square of the angle of cam rotation.

Let the parabolic motion, for the first half, be represented by

$$y = c\theta^2$$

where $c = a$ constant of proportionality

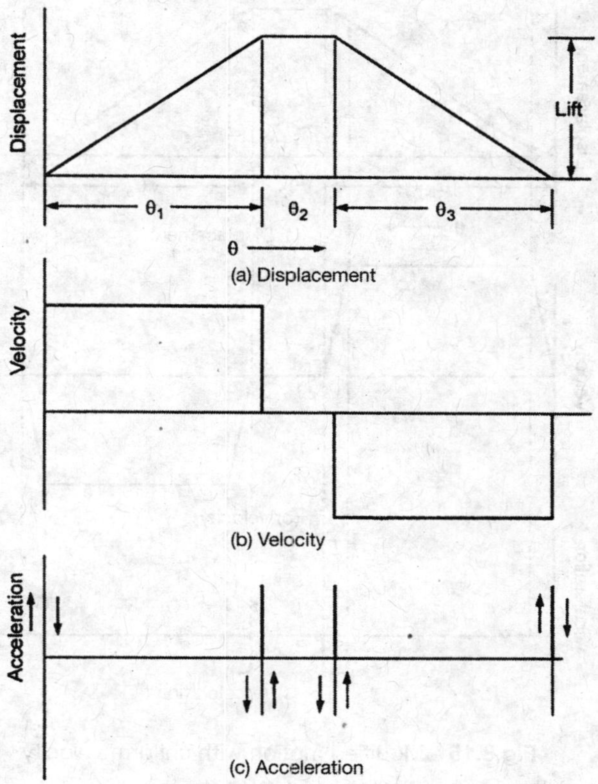

Fig.8.14 Motion with uniform velocity

For $$y = \frac{h}{2},\ \theta = \frac{\phi}{2}$$

∴ $$\frac{h}{2} = \frac{c\phi^2}{4}$$

or $$c = \frac{2h}{\phi^2}$$

Hence, $$y = 2h\left(\frac{\theta}{\phi}\right)^2 \tag{8.18}$$

Velocity, $$v = \frac{4h\omega\theta}{\phi^2} \tag{8.19}$$

Acceleration, $$f = 4h\left(\frac{\omega}{\phi}\right)^2 = \text{const.} \tag{8.20}$$

For velocity to be maximum, $\theta = \dfrac{\phi}{2}$, and

$$v_{\max} = \frac{2h\omega}{\phi} \tag{8.21}$$

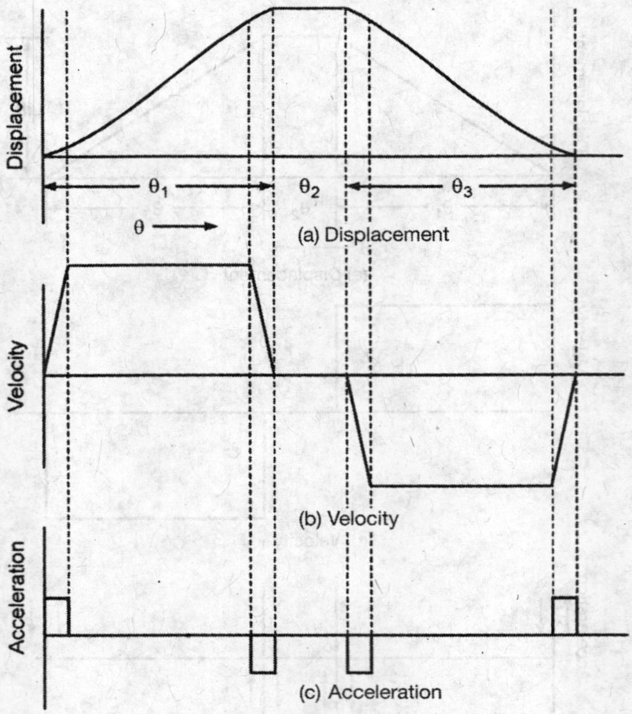

Fig.8.15 Modified motion with uniform velocity

For the second half, we have

$$y = c_1 + c_2\theta + c_3\theta^2$$

For $\theta = \beta, y = h$ and $v = 0$. Also for $\theta = \dfrac{\phi}{2}, v = v_{max}$.

Therefore, $h = c_1 + c_2\phi + c_3\phi^2$

$$0 = c_2\omega + 2c_3\,\omega\,\phi$$

or $$0 = c_2 + 2c_3\,\phi$$

Hence, $$c_1 = -h, c_2 = \frac{4h}{\phi}, c_3 = \frac{-2h}{\phi^2}$$

$$y = h\left[1 - 2\left(1 - \frac{\theta}{\phi}\right)^2\right] \tag{8.22}$$

$$v = \left(\frac{4h\omega}{\phi}\right)\left(1 - \frac{\theta}{\phi}\right) \tag{8.23}$$

$$f = -4h\left(\frac{\omega}{\phi}\right)^2 \tag{8.24}$$

The variation of displacement, velocity and acceleration are shown in Fig.8.16.

In this case, the displacement of the follower is proportional to the angle of cam rotation.

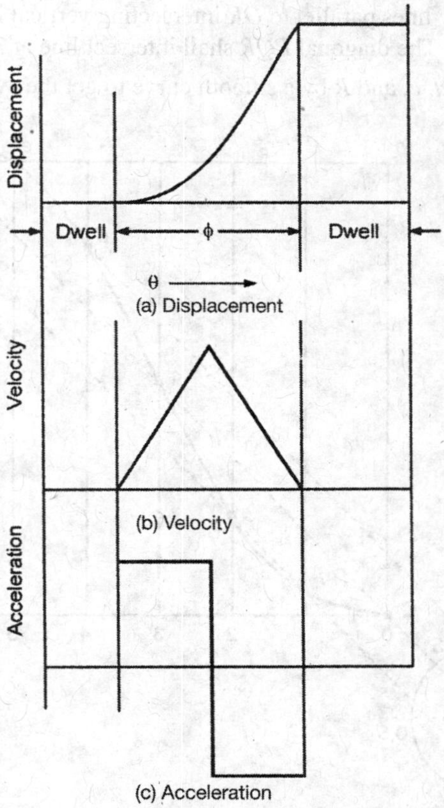

Fig.8.16 Parabolic motion

8.5.5 Cycloidal Motion

A cycloid is the locus of a point on a circle rolling on a straight line. The cycloidal curve, as shown in Fig.8.17, can be constructed by adopting the following steps:

1. Draw a horizontal line to a convenient scale equal to the angle of ascent of the cam.
2. At the origin, draw another line perpendicular to the previous line, on a convenient scale, equal to the lift of the cam.
3. Divide the angle of ascent into equal number of parts (say 6), and number them from 0 to 6.
4. Erect vertical lines at these points.
5. Draw the diagonal QR passing through the origin and produce it backward.
6. Calculate the radius of the circle, $r = \dfrac{h}{2\pi}$, where h = lift.
7. Select a convenient point P on the diagonal produced backwards and draw a circle with radius equal to r.
8. Divide the circle into six equal parts and number the ends of the diameters from 0 to 6.
9. Join points 1–2 and 4–5, intersecting the vertical diameter at points m and n, respectively.

10. From points *m* and *n*, draw lines parallel to *QR* intersecting vertical lines at 1 and 2 at *a* and *b* and lines at 4 and 5 at *d* and *e*. The diagonal *PQR* shall intersect line at 3 at *c*.
11. Join the points *Q*, *a*, *b*, *c*, *d*, *e*, and *R* by a smooth curve to get the cycloidal curve.

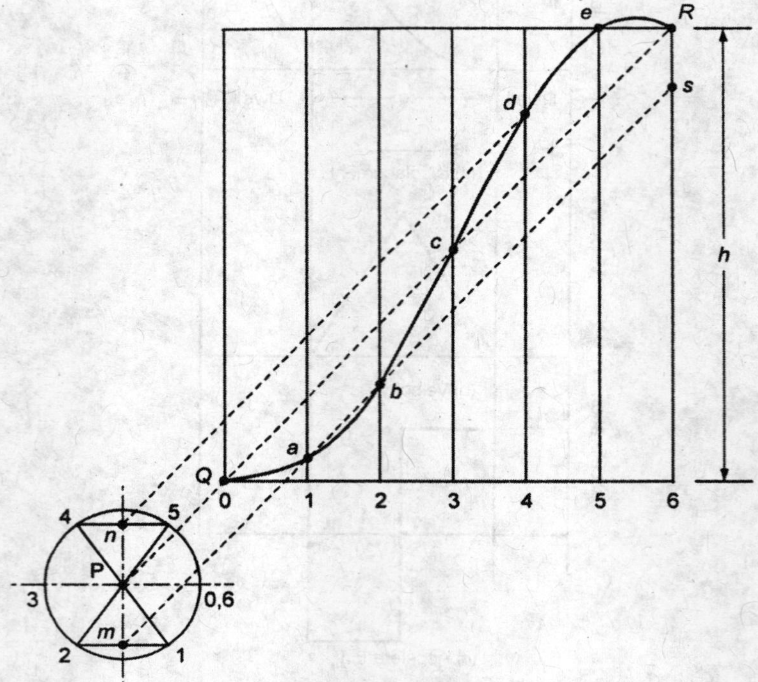

Fig.8.17 Cycloidal curve

Motion Analysis

A cycloid is expressed by,

$$y = \left(\frac{h}{\pi}\right)\left[\frac{\pi\theta}{\phi} - \left(\frac{1}{2}\right)\sin\left(\frac{2\pi\theta}{\phi}\right)\right] \tag{8.25}$$

$$v = \frac{dy}{dt} = \left(\frac{dy}{d\theta}\right)\left(\frac{d\theta}{dt}\right)$$

$$= \left[\frac{h}{\phi} - \left\{\frac{h}{2\pi}\right\}\left(\frac{2\pi}{\phi}\right)\cos\left(\frac{2\pi\theta}{\phi}\right)\right]\omega$$

$$= \frac{h\omega}{\phi} - \left(\frac{h\omega}{\phi}\right)\cos\left(\frac{2\pi\theta}{\phi}\right)$$

$$= \left(\frac{h\omega}{\phi}\right)\left[1 - \cos\left(\frac{2\pi\theta}{\phi}\right)\right] \tag{8.26}$$

$$v_{max} = \frac{2h\omega}{\phi} \quad \text{at } \theta = \frac{\phi}{2} \tag{8.27}$$

$$f = \frac{dv}{dt} = \left(\frac{dv}{d\phi}\right)\frac{d\theta}{dt}$$

$$= \left(\frac{h\omega}{\phi}\right)\left(\frac{2\omega}{\phi}\right) \times \left[\sin\left(\frac{2\pi\theta}{\phi}\right)\right]\omega$$

$$= (2h\pi) \times \left(\frac{\omega}{\phi}\right)^2 \sin\left(\frac{2\pi\theta}{\phi}\right) \tag{8.28}$$

$$f_{\max} = (2h\pi) \times \left(\frac{\omega}{\phi}\right)^2 \quad \text{at } \theta = \frac{\phi}{4} \tag{8.29}$$

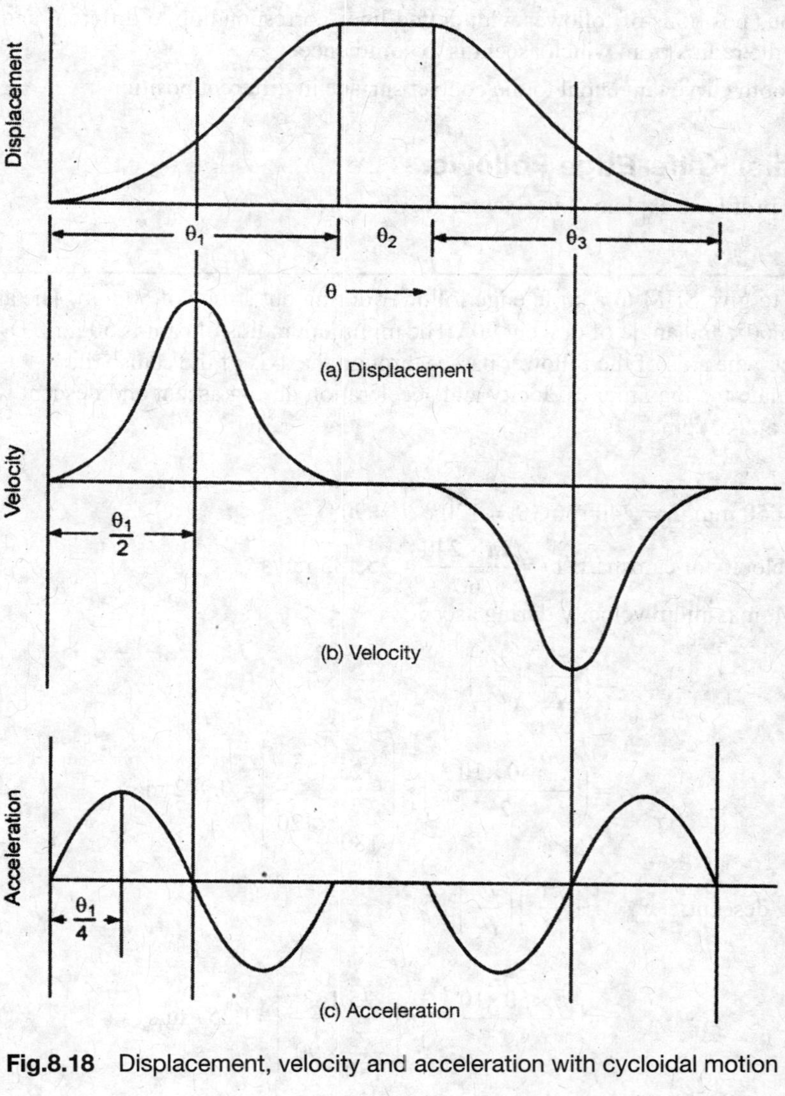

Fig.8.18 Displacement, velocity and acceleration with cycloidal motion

The variation of cycloidal motion is shown in Fig.8.18. It may be observed that there are no abrupt changes in the velocity and acceleration at any stage of the motion. Therefore, cycloidal motion is the most ideal for high speed follower motion.

8.6 CAM PROFILE WITH KNIFE–EDGE FOLLOWER

The following procedure may be adopted to draw the cam profile with knife edge follower:

1. Draw the displacement diagram for follower motion.
2. Consider that cam remains stationary and that the follower moves round it in a direction opposite to the direction of cam rotation.
3. Draw the cam base circle and divide its circumference into equal number of divisions depending upon the divisions used in the displacement diagram.
4. Draw various positions of follower with dotted lines corresponding to different angular displacement from the radius from which ascent is to commence.
5. Draw a smooth curve tangential to the contact surface in different positions.

8.6.1 Radial Knife-Edge Follower

Draw the cam profile as explained in Example 8.1.

Example 8.1

A disc cam is to give SHM to a knife edge follower during out stroke of 50 mm. The angle of ascent is 120° , dwell 60°, and angle of descent 90°. The minimum radius of cam is 50 mm. Draw the profile of the cam when the axis of the follower passes through the axis of the camshaft.

Also calculate the maximum velocity and acceleration during ascent and descent when the camshaft revolves at 240 rpm.

Solution

Given: $h = 50$ mm, $N = 240$ rpm, $\theta_1 = 120°$, $\theta_3 = 90°$

Angular velocity of camshaft, $\omega = \dfrac{2\pi \times 240}{60} = 25.133$ rad/s

With SHM, maximum velocity during ascent,

$$v_{max} = \left(\frac{\pi h}{2}\right)\left(\frac{\omega}{\theta_1}\right)$$

$$= \left(\frac{\pi \times 50 \times 10^{-3}}{2}\right) \times \left(\frac{25.133}{\frac{\pi}{180} \times 120}\right) = 0.942 \text{ m/s}$$

and during descent, $$v_{max} = \left(\frac{\pi h}{2}\right)\left(\frac{\omega}{\theta_3}\right)$$

$$= \left(\frac{\pi \times 50 \times 10^{-3}}{2}\right) \times \left(\frac{25.133}{\frac{\pi}{180} \times 90}\right) = 1.257 \text{ m/s}$$

Maximum acceleration during ascent,

$$f_{max} = \left(\frac{\pi\omega}{\theta_1}\right)^2 \times \left(\frac{h}{2}\right)$$

$$= \left[\frac{\pi \times 25.133}{\frac{\pi}{180} \times 120}\right]^2 \times \left(\frac{50 \times 10^{-3}}{2}\right) = 35.53 \text{ m/s}^2$$

and during descent,

$$f_{max} = \left(\frac{\pi\omega}{\theta_3}\right)^2 \times \left(\frac{h}{2}\right)$$

$$= \left[\frac{\pi \times 25.133}{\frac{\pi}{180} \times 90}\right]^2 \times \left(\frac{50 \times 10^{-3}}{2}\right) = 63.17 \text{ m/s}^2$$

Displacement diagram

The displacement diagram shown in Fig.8.19(a) may be drawn as follows:

1. Draw a vertical line 0–6 equal to the lift of 50 mm.
2. Draw a semicircle on this line and divide the semicircle into six equal parts of 30° each.
3. Draw a horizontal line 00′ perpendicular to 0–6 line representing cam rotation angle θ on a scale of 1 cm = 20°.
4. Divide the ascent angle of 120° into six equal parts and also the descent angle of 90° into six equal parts.
5. Erect perpendiculars at points 0 to 6 and 6′ to 0′.
6. Draw horizontal lines from points 1 to 5 on the semicircle to intersect vertical lines drawn previously, as shown in the figure.
7. Join the points of intersection with a smooth curve to get the displacement diagram.

Cam profile

The cam profile shown in Fig.8.19(b) may be drawn as follows:

1. Draw a circle of radius equal to the base circle radius of 50 mm with center *O*.
2. Draw angles of ascent, dwell and descent of 120°, 60° and 90°, respectively. Divide the angles of ascent and descent into six equal parts. Draw radial lines for these angles.
3. Mark points 0 to 6 and 6′ to 0′ on the base circle in the angles of ascent and descent, respectively.
4. Measure distance 1*a*, 2*b*, 3*c*, 4*d*, 5*e*, and 6*f* from the displacement diagram for ascent and cut-off corresponding distances on the radial lines drawn in the cam profile. Repeat the same process during descent.
5. Join the points so obtained by a smooth curve to get the cam profile.

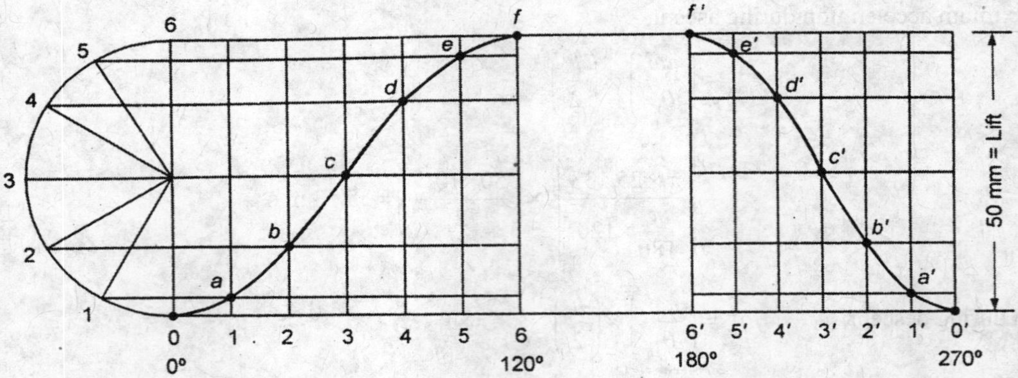

(a) Displacement diagram

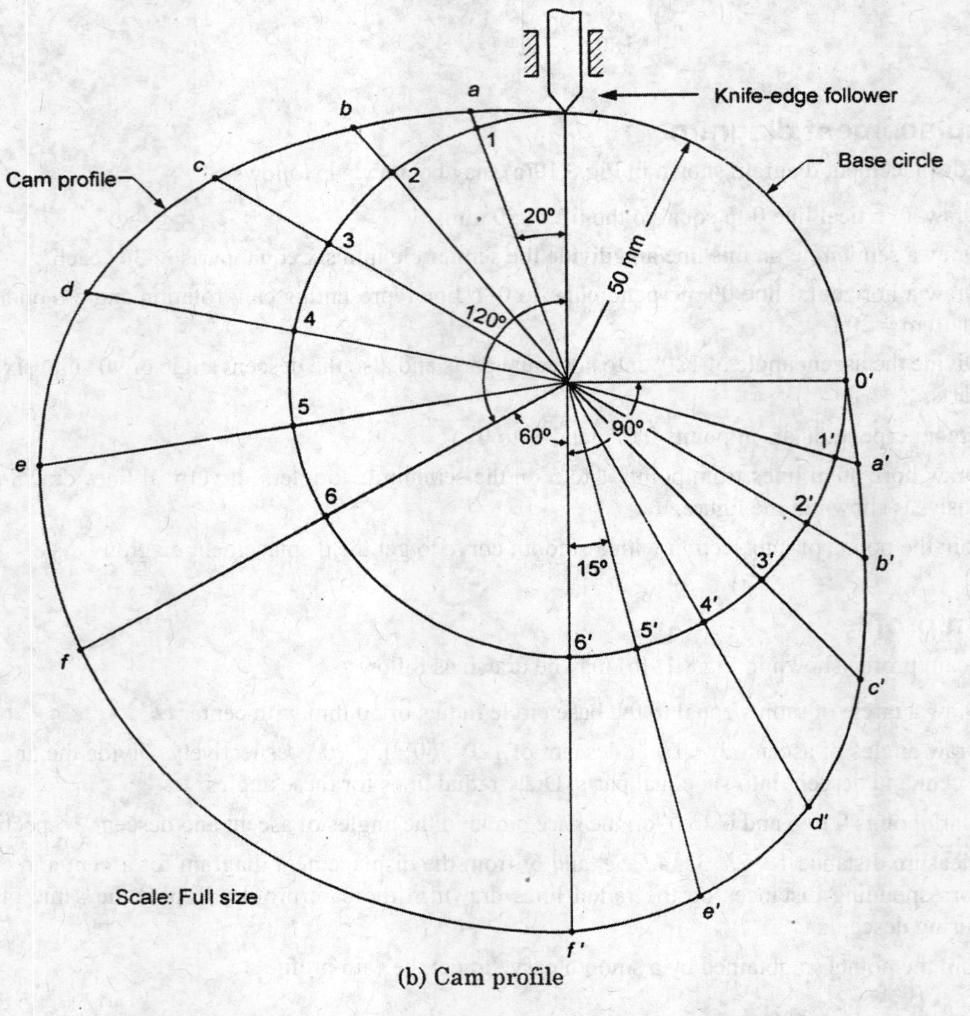

(b) Cam profile

Fig.8.19 Cam profile with radial knife-edge follower having SHM

8.6.2 Offset Knife-Edge Follower

Draw the cam profile as explained in Example 8.2.

Example 8.2

Draw the cam profile for the data given in Example 8.1 when the knife-edge follower is offset by 20 mm.

■ Solution

Displacement diagram remains the same as in Example 8.1.

The cam profile as shown in Fig.8.20 can be drawn as follows:

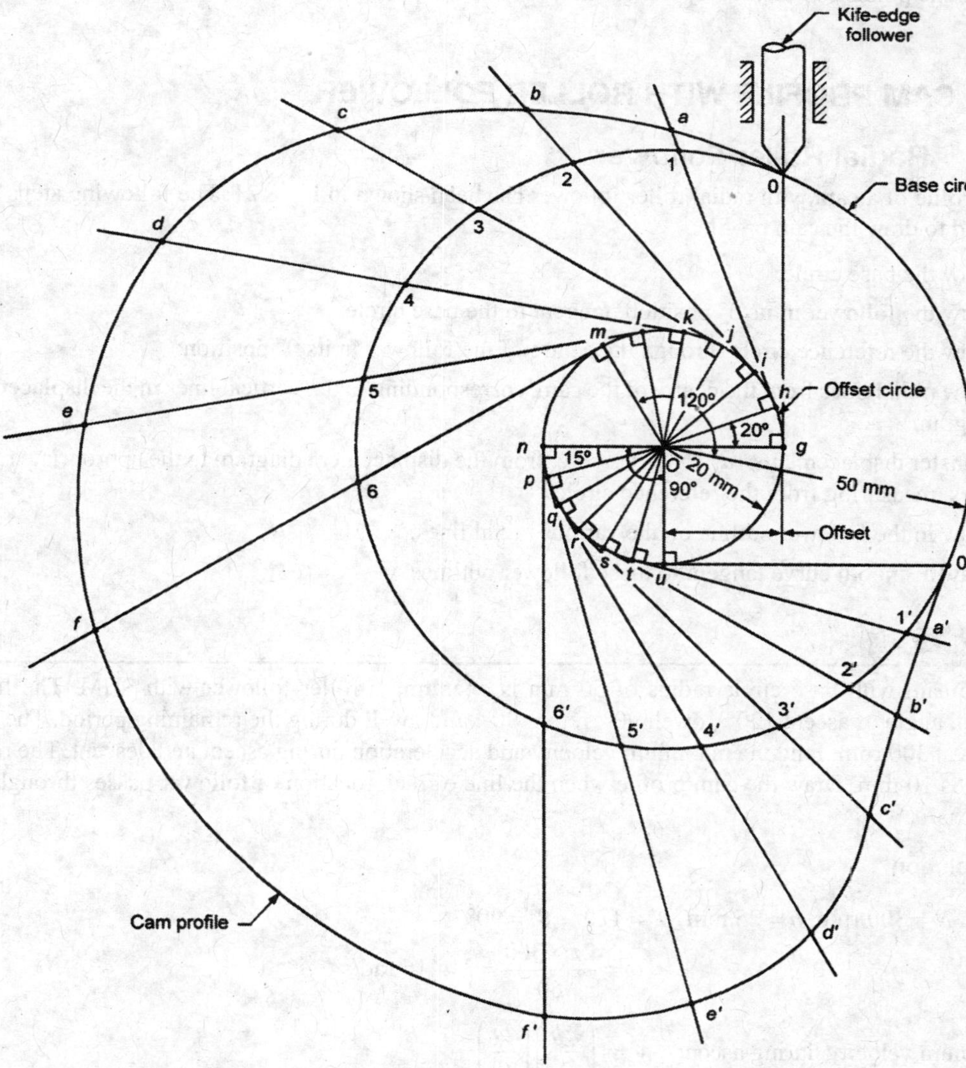

Fig.8.20 Cam profile with offset knife-edge follower having SHM

1. Draw the offset circle with radius equal to the offset of 20 mm and the base circle with radius 50 mm at centre *O*.
2. Divide the angles of ascent and descent in the offset circle into six equal parts. Draw radial lines intersecting the offset circle from *g* to *m* and *n* to *u* in the angle of ascent and descent respectively.
3. Draw tangents at points *g* to *u* intersecting the base circle at 0 to 6 and 6′ to 0′ in the angle of ascent and descent, respectively.
4. Measure distances 1*a*, 2*b*, 3*c*, 4*d*, 5*e*, and 6*f* on the displacement diagram for ascent and cut-off corresponding distances on the tangential lines drawn in the cam profile. Repeat the same process for the angle of descent.
5. Join the points so obtained by a smooth curve to get the cam profile.

8.7 CAM PROFILE WITH ROLLER FOLLOWER

8.7.1 Radial Roller Follower

The profile of a cam with radial roller follower has been shown in Fig.8.21. The following steps may be used to draw the cam profile:

1. Draw the base circle.
2. Draw the follower in its 0° position, tangent to the base circle.
3. Draw the reference circle through the centre of the follower in its 0° position.
4. Draw radial lines from the centre of the cam, corresponding to the vertical lines in the displacement diagram.
5. Transfer displacements 1*a*, 2*b*, 3*c*, …, etc. from the displacement diagram to the appropriate radial lines, measuring from the reference circle.
6. Draw in the follower outline on the various radial lines.
7. Draw a smooth curve tangent to these follower outlines.

Example 8.3

A disc cam with base circle radius of 50 mm is operating a roller follower with SHM. The lift is 25 mm, angle of ascent 120° , dwell 90°, return 90°, and dwell during the remaining period. The cam rotates at 300 rpm. Find the maximum velocity and acceleration during ascent and descent. The roller radius is 10 mm. Draw the cam profile when the line of reciprocation of follower passes through the cam axis.

Solution

Given: $N = 300$ rpm, $h = 25$ mm, $\theta_1 = 120°$, $\theta_3 = 90°$

$$\omega = \frac{2\pi \times 300}{60} = 31.416 \text{ rad/s}$$

Maximum velocity during ascent, $v_a = \left(\frac{\pi h}{2}\right)\left(\frac{\omega}{\theta_1}\right)$

$$= \frac{\pi \times 0.025}{2} \times \frac{31.416 \times 180}{\pi \times 120}$$

$$= 0.589 \text{ m/s}$$

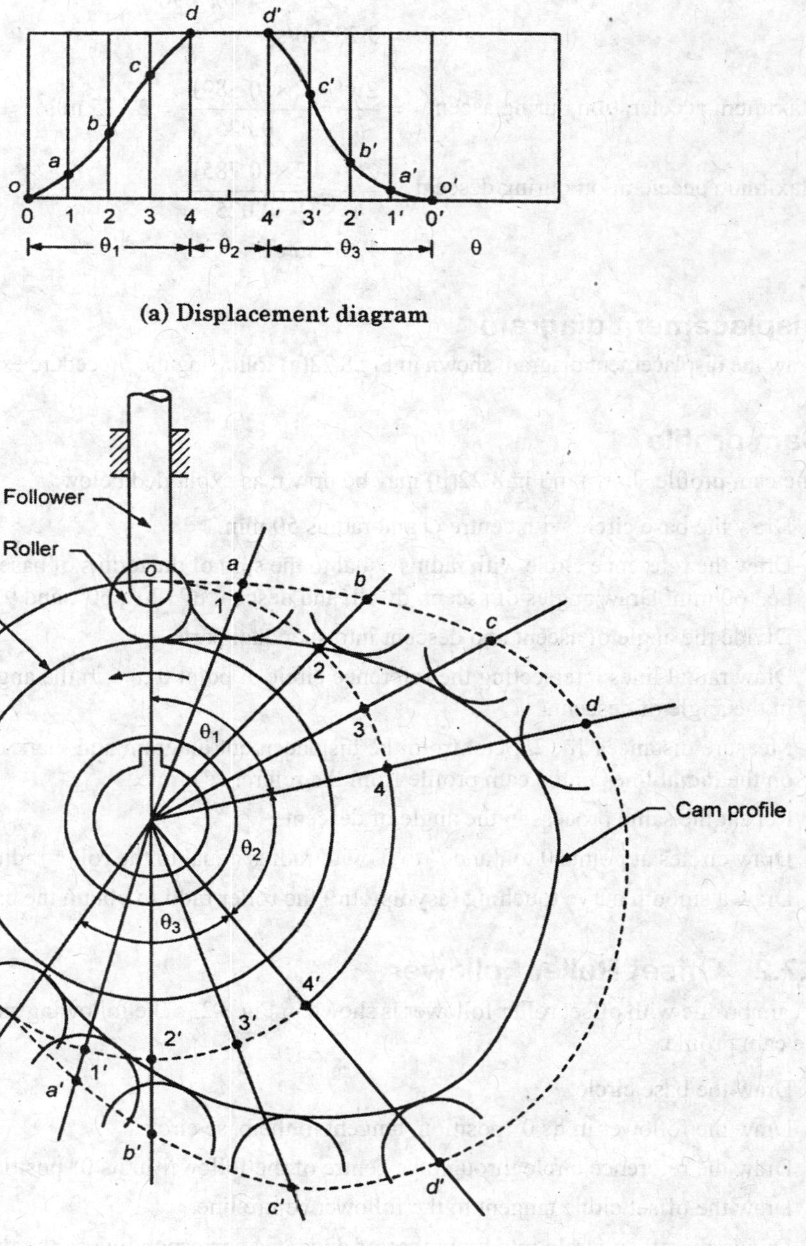

(b) Cam profile

Fig.8.21 Cam with radial roller follower

Maximum velocity during descent, $v_d = \left(\frac{\pi h}{2}\right)\left(\frac{\omega}{\theta_3}\right)$

$$= \frac{\pi \times 0.025}{2} \times \frac{31.416 \times 180}{\pi \times 90}$$

$$= 0.785 \text{ m/s}$$

Maximum acceleration during ascent $= \frac{2v_a^2}{h} = \frac{2 \times (0.589)^2}{0.025} = 27.75 \text{ m/s}^2$

Maximum acceleration during descent $= \frac{2v_d^2}{h} = \frac{2 \times (0.785)^2}{0.025}$

$$= 49.35 \text{ m/s}^2$$

Displacement diagram

Draw the displacement diagram shown in Fig.8.22(a) following the procedure explained in Example 8.1.

Cam profile

The cam profile shown in Fig.8.22(b) may be drawn as explained below:

1. Draw the base circle with centre *O* and radius 50 mm.
2. Draw the reference circle with radius equal to the sum of the radius of base circle and roller radius, i.e., 60 mm. Draw angles of ascent, dwell, and descent of 120°, 60°, and 90°, respectively.
3. Divide the angle of ascent and descent into six equal parts.
4. Draw radial lines intersecting the reference circle at point 0 to 6 in the angle of ascent and 6′ to 0′ in the angle of descent.
5. Measure distances 1*a*, 2*b*, etc. from the displacement diagram and mark corresponding distances on the radial lines in the cam profile from the reference circle.
6. Repeat the same process in the angle of descent.
7. Draw circles at points 0 to f and *f*′ to 0′ with radius equal to the roller radius of 10 mm.
8. Draw a smooth curve touching (asymptotic) the roller radii to obtain the cam profile.

8.7.2 Offset Roller Follower

A cam profile with offset roller follower is shown in Fig.8.23. The following steps may be used to draw the cam profile:

1. Draw the base circle.
2. Draw the follower in its 0° position, tangent to the base circle.
3. Draw the reference circle through the centre of the follower in its 0° position.
4. Draw the offset circle tangent to the follower centre line.
5. Divide the offset circle into a number of divisions corresponding to the divisions in the displacement diagram.
6. Draw tangents to the offset circle at each number.

7. Lay off various displacements 1*a*, 2*b*, 3*c*,..., etc. along the appropriate tangent lines, measuring from the reference circle.
8. Draw in the follower outlines on the various tangent lines.
9. Draw a smooth curve to these follower outlines.

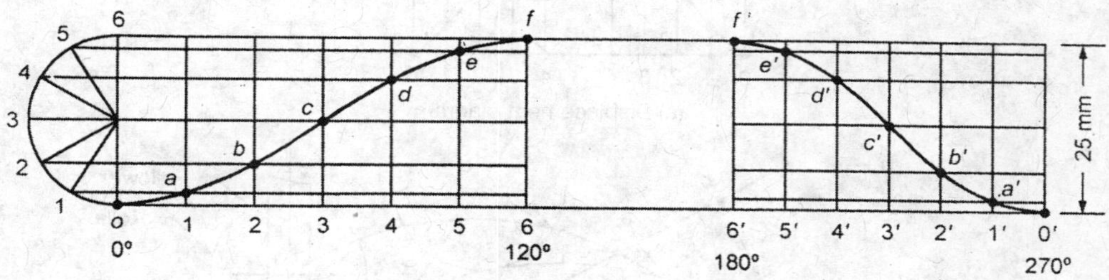

(a) Displacement diagram

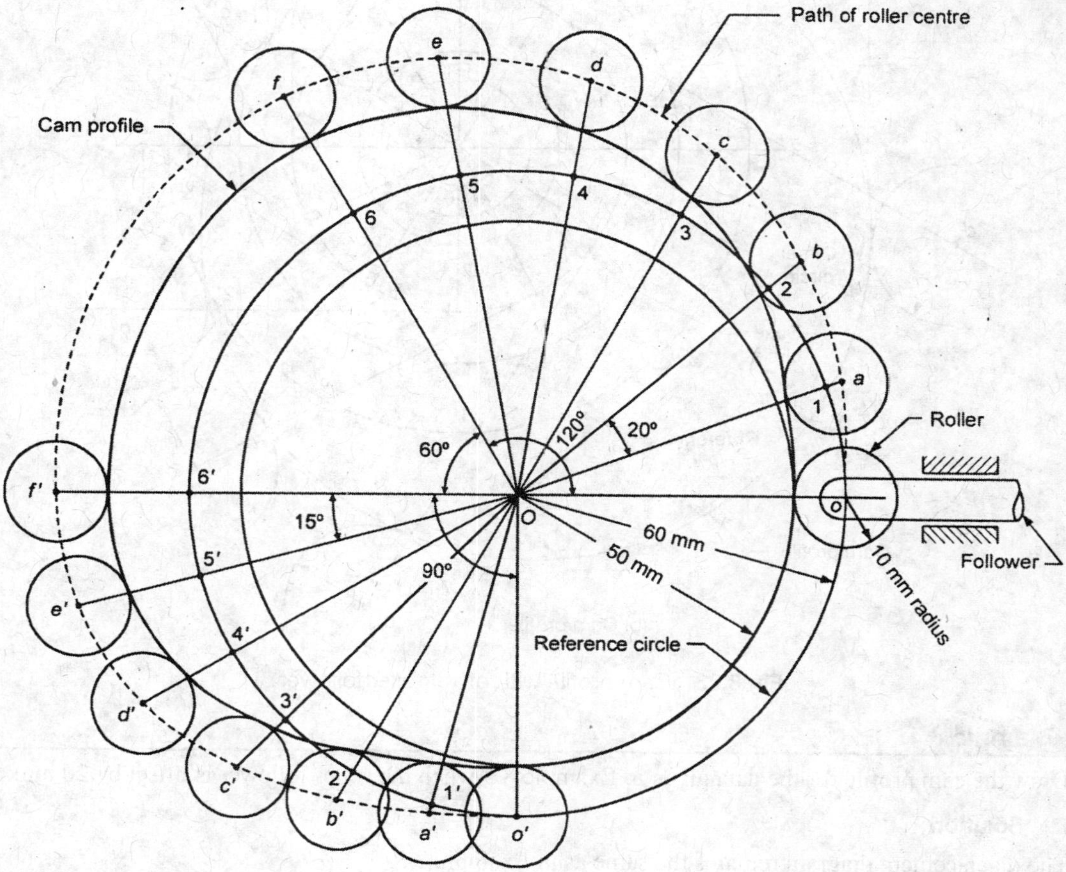

(b) Cam profile

Fig.8.22 Cam profile with radial roller follower having SHM

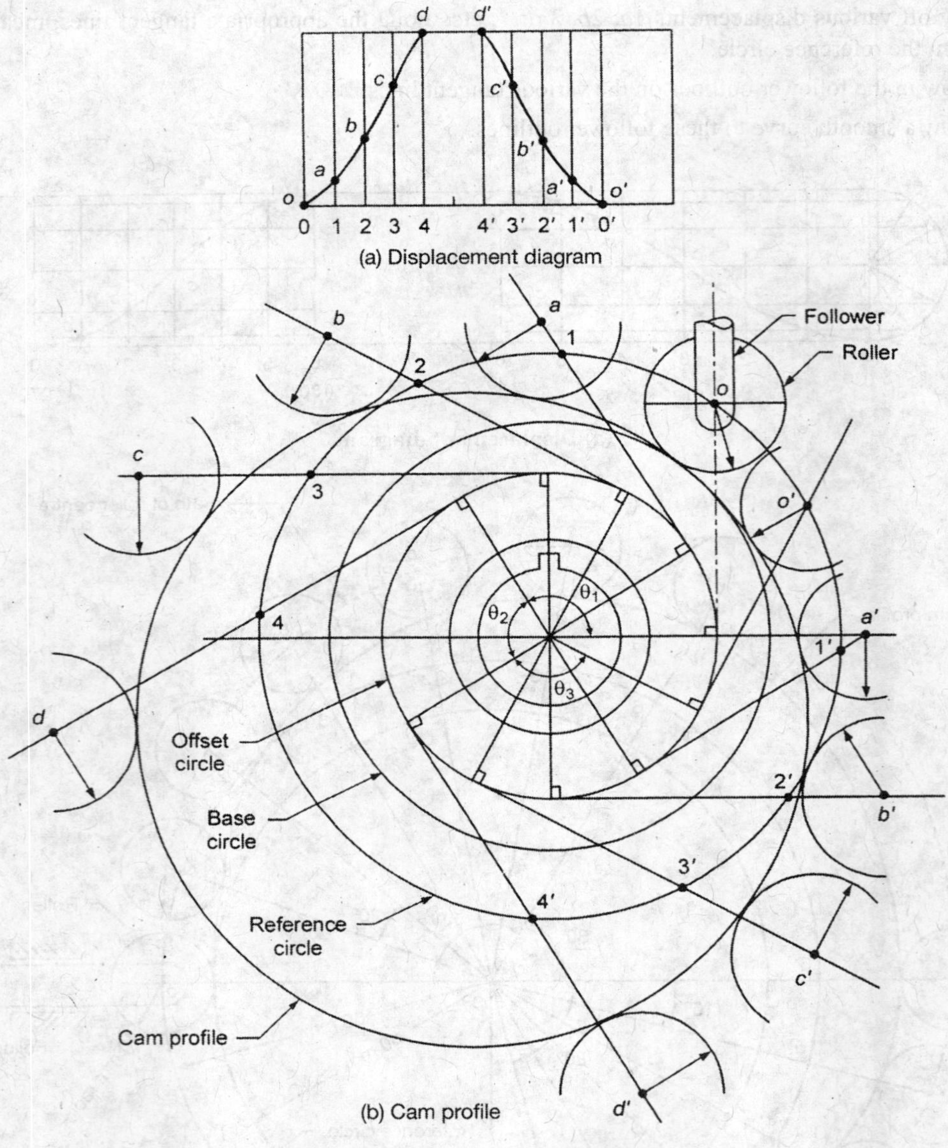

Fig.8.23 Cam profile with offset roller follower

Example 8.4

Draw the cam profile for the data given in Example 8.3 when the roller follower is offset by 20 mm.

■ Solution

The displacement diagram remains the same as in Example 8.3.

The cam profile shown in Fig.8.24 may be drawn as explained below:

1. Draw the base circle at centre *O* with radius 50 mm and reference circle with radius 60 mm.
2. Draw the offset circle with radius 20 mm.

3. Draw angles of ascent, dwell, and descent of 120°, 60°, and 90° respectively in the offset circle.
4. Divide the angle of ascent and descent into six equal parts intersecting the offset circle at points *g* to *m* and *n* to *u*, respectively.
5. Draw tangents at points *g* to *u* intersecting the reference circle at points 0 to 6 and 6′ to 0′ in the angle of ascent and descent, respectively.
6. Measure off distances 1*a*, 2*b*, etc. from the displacement diagram in the angle of ascent and measure the corresponding distances in the cam profile. Repeat the same process for the angle of descent.
7. Draw circles with radius of roller at points 0 to *f* and *f′* to 0′.
8. Draw a smooth curve touching these circles to get the cam profile.

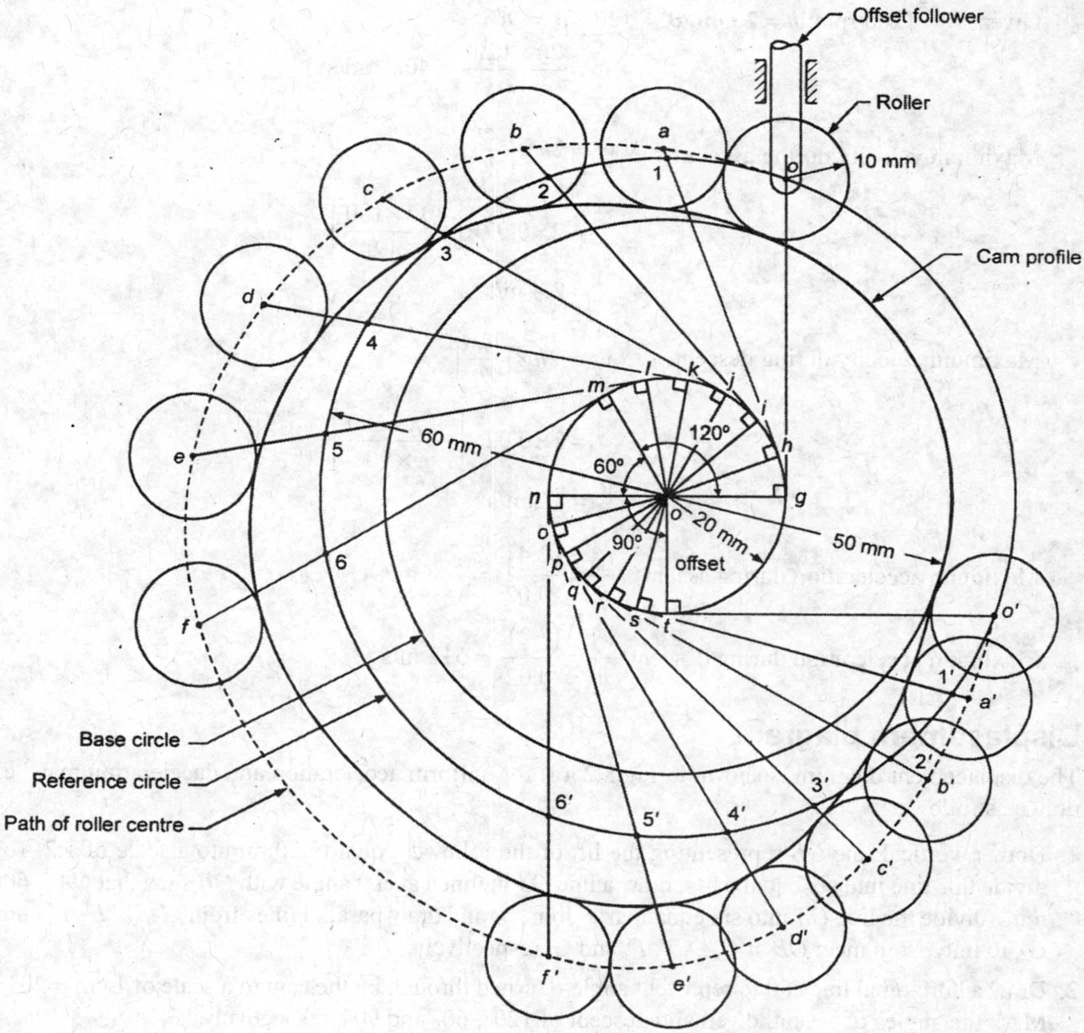

Fig.8.24 Cam profile with off-set roller follower having SHM

Example 8.5

A cam with 30 mm as minimum diameter and 20 mm lift is rotating clockwise at a uniform speed of 1200 rpm and has to given the following motion to a roller follower 10 mm in diameter:

1. Outward stroke during 120° with equal uniform acceleration and deceleration.
2. Dwell for 60°.
3. Return during 90°.
4. Dwell during the remaining period.

Draw the cam profile if the cam axis coincides with the follower axis. Also calculate the maximum velocity and acceleration during ascent and return strokes.

Solution

Given: $N = 1200$ rpm, $h = 2$ mm, $\theta_1 = 120°$, $\theta_3 = 90°$.

$$\omega = \frac{2\pi \times 1200}{60} = 40\pi \text{ rad/s}$$

Maximum velocity during ascent,

$$v_a = 2h \times \left(\frac{\omega}{\theta_1}\right)$$

$$= 2 \times 0.02 \times \left[\frac{40\pi \times 180}{\pi \times 120}\right]$$

$$= 2.4 \text{ m/s}$$

Maximum velocity during descent,

$$v_d = 2h \times \left(\frac{\omega}{\theta_3}\right)$$

$$= 2 \times 0.02 \times \left[\frac{40\pi \times 180}{\pi \times 90}\right]$$

$$= 3.2 \text{ m/s}$$

Maximum acceleration during ascent $\dfrac{v_a^2}{h} = \dfrac{(2.4)^2}{0.02} = 288 \text{ m/s}^2$

Maximum acceleration during descent $\dfrac{v_d^2}{h} = \dfrac{(3.2)^2}{0.02} = 512 \text{ m/s}^2$

Displacement diagram

The displacement diagram is shown in Fig.8.25(a) for uniform acceleration and deceleration may be drawn as follows:

1. Draw a vertical line OB representing the lift of the follower equal to 20 mm to a scale of 1:2. To divide this line into six equal parts, draw a line OA inclined at any angle with OB such that $OA = 60$ mm. Divide the line OA into six equal parts. Join AB and draw parallel lines from C', D', E', F' and G', to line AB to meet OB at C, D, E, F, and G, respectively.
2. Draw a horizontal line at 0 to represent angle θ turned through by the cam to a scale of 1 cm = 20°. Mark the angles of ascent, dwell and descent of 120°, 60° and 90°, respectively.
3. Divide the angles of ascent and descent into six equal parts and erect perpendiculars at these points to intersect the horizontal lines from points B to G.

4. Join the points *o* and *f* with the points of intersection on the middle vertical line in the angle of ascent. Join the points of intersection of these lines with the vertical lines by a smooth curve *o a b c d e f*. Repeat the similar process for the angle of descent to get the curve *f′ e′ d′ c′ b′ a′o′*.

Cam profile

The cam profile has been drawn in Fig.8.25(b) to a scale of 1:2. The procedure described in Example 8.3 may be followed to draw the profile.

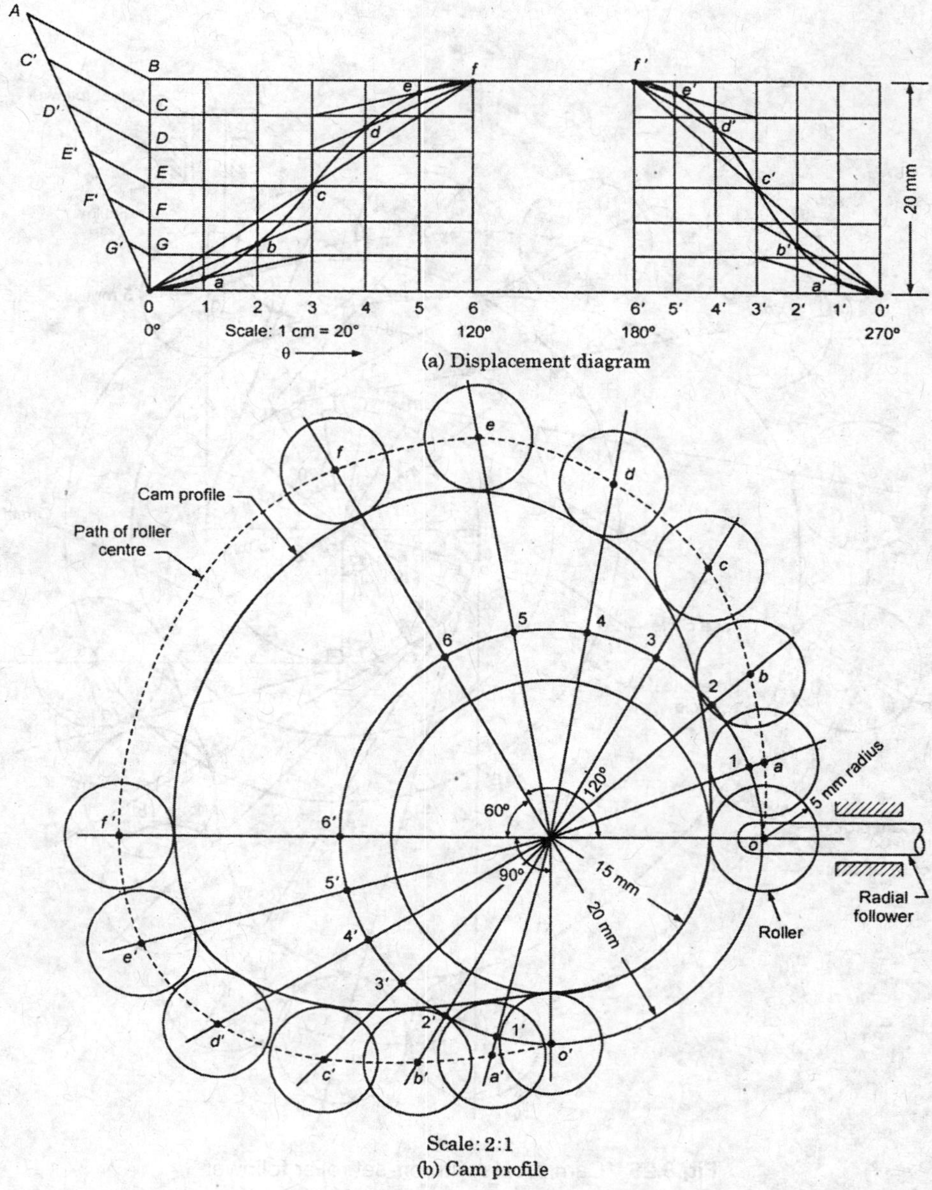

Fig.8.25 Cam profile with radial roller follower having uniform acceleration and deceleration

Example 8.6

Draw the cam profile for the data given in Example 8.5 when the roller follower is offset by 10 mm from the axis of the cam.

■ Solution

The displacement diagram remains the same as drawn in Example 8.5. The cam profile has been drawn in Fig.8.26 following the same procedure as explained in Example 8.4.

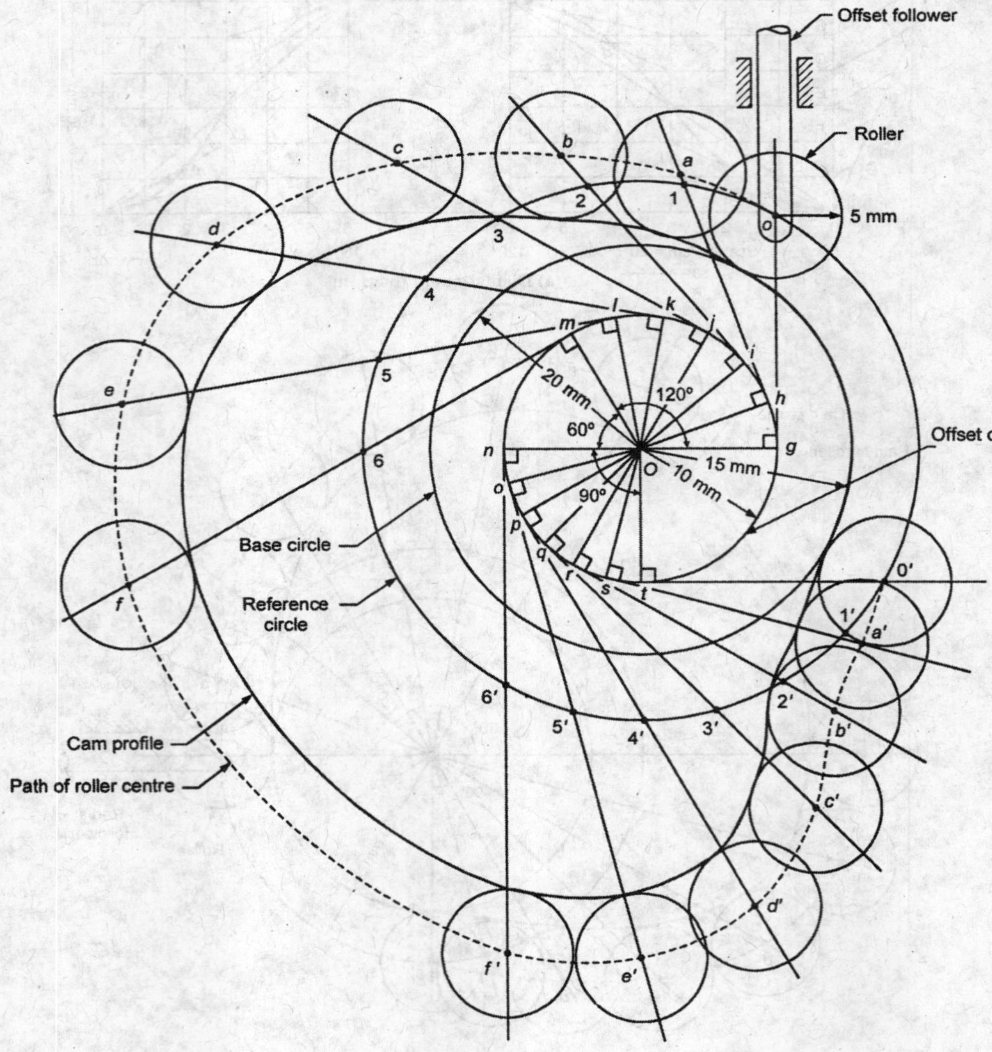

Fig.8.26 Cam profile with off-set roller follower

8.8 CAM PROFILE WITH TRANSLATIONAL FLAT-FACED FOLLOWER

A cam with flat-faced follower is shown in Fig.8.28. The following steps may be adopted to draw the cam profile:

1. Draw the base circle, which also serves the reference circle in this case.
2. Draw the follower in the home position, tangent to the base circle.
3. Draw radial lines corresponding to the divisions in the displacement diagram, and number accordingly.
4. Draw in the follower outline on the various radial lines by laying off the appropriate displacements and drawing lines perpendicular to the radial lines.
5. Draw a smooth curve tangent to the follower lines.

Example 8.7

A cam is to operate a flat faced follower having uniform acceleration and deceleration during ascent and descent. The least radius of the cam is 50 mm. During descent, the deceleration period is half of the acceleration period. The ascent lift is 37.5 mm. The ascent is for 1/4th period, dwell for 1/4th, descent for 1/3rd, and dwell for the remaining 1/6th period. The cam rotates at 600 rpm. Find the maximum velocity and acceleration during ascent and descent. Draw the cam profile.

Solution

$$\text{Acceleration period during descent} = \frac{1}{3}\times 360\times\frac{2}{3} = 80°$$

$$\text{Deceleration period during descent} = 40°$$

$$\text{Distance moved during acceleration period} = \frac{80}{120}\times 37.5 = 25 \text{ mm}$$

$$\text{Distance moved during deceleration period} = 12.5 \text{ mm}$$

$$\text{Angular velocity of cam, } \omega = \frac{2\pi\times 600}{60} = 20\pi \text{ rad/s}$$

$$\text{Maximum velocity during ascent, } v_{max} = \frac{2\omega h}{\theta_1}$$

$$= \frac{2\times 20\pi\times 30.5\times 10^{-3}}{\frac{\pi}{2}} = 3 \text{ m/s}$$

$$\text{Maximum velocity during descent, } v_{max} = \frac{2h}{t}$$

$$= \frac{2\times 37.5\times 180}{80\times\pi} = 3.375 \text{ m/s}$$

$$\text{Maximum acceleration during ascent} = \frac{4\omega^2 h}{\theta_1^2}$$

$$= \frac{4\times(20\pi)^2\times 37.5\times 10^{-3}}{\left(\frac{\pi}{2}\right)^2}$$

$$= 240 \text{ m/s}^2$$

$$\text{Maximum acceleration during descent} = \frac{v_{max}}{t} = \frac{3.3750\times360\times600}{80\times60} = 151.875 \text{ m/s}^2$$

$$\text{Maximum deceleration during descent} = \frac{v_{max}}{t} = \frac{3.375\times360\times600}{40\times60} = 303.75 \text{ m/s}^2$$

The velocity and acceleration diagrams are shown in Fig.8.27.

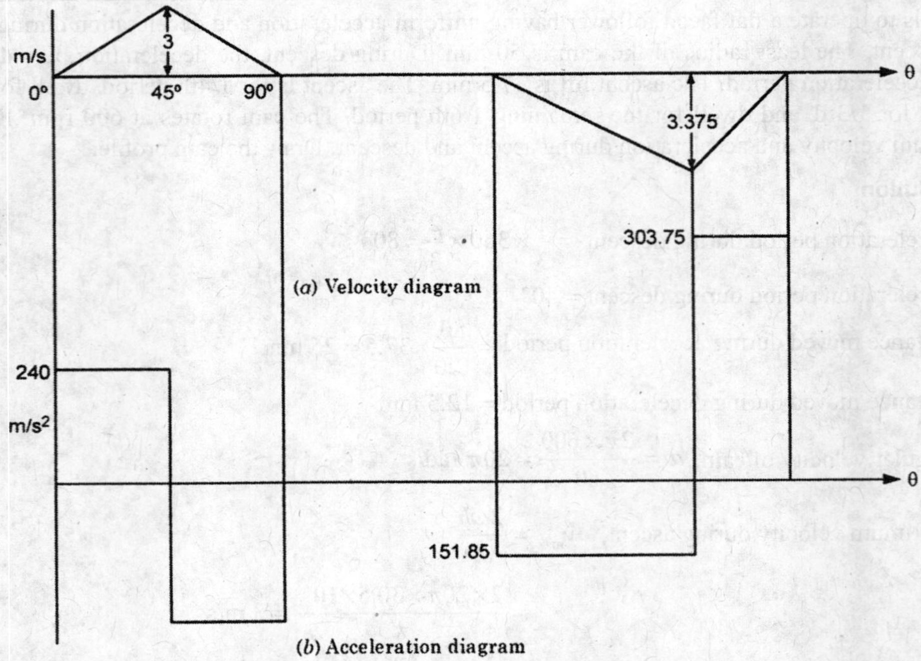

Fig.8.27 Velocity and acceleration diagram for a flat-faced follower having uniform acceleration and deceleration

Displacement diagram

1. Draw a vertical line equal to the lift of 37.5 mm and a horizontal line perpendicular to it representing the cam angle. Mark angle of ascent = 90°, dwell = 90° and descent = 120°.
2. Divide the angle of ascent and lift into six equal parts. Join point *O* with three points of intersection of vertical and horizontal lines at point 3, and the other three points with point *f*.

   Join the points of intersection of these inclined lines with the vertical lines with a smooth curve.
3. Divide the angle of descent into two parts of 80° and 40°. Divide these angles and the lift into four equal parts. Join the intersection of vertical and horizontal lines with a smooth curve. The displacement diagram has been shown in Fig.8.28(a).

Cam profile

1. Draw the base circle with 50 mm radius.
2. From the vertical line, measure angle of ascent = 90°, dwell = 90°, and angle of descent = 120°, in the anti-clockwise direction.
3. Divide the angle of ascent into six equal parts. Divide the angle of descent into 80° and 40°. Divide 80° and 40° angles into four equal parts.
4. From the circumference of the base circle, mark distances 1*a*, 2*b*, 3*c*, ..., etc. along the radial lines at 1, 2, 3, ..., etc. Draw perpendiculars at these points to the radial lines. Draw a smooth curve tangential to these perpendiculars to get the cam profile.
5. Repeat this procedure for the descent.

The cam profile has been shown in Fig.8.28(b).

Example 8.8

A flat-faced reciprocating follower has the following motion with uniform acceleration and retardation: Ascent for 80°, dwell for 80°, and return for 120°. The base circle diameter of the cam is 60 mm and the stroke of the follower is 20 mm. The line of motion of the follower passes through the axis of the cam. Draw the cam profile.

■ **Solution**

Displacement diagram: Draw the displacement diagram shown in Fig.8.29(a) according to the procedure explained in Example 8.5.

Cam profile

1. Draw the base circle with radius 30 mm and centre O as shown in Fig.8.29(b).
2. Lay off angles of ascent, dwell and descent of 80°, 80° and 120°, respectively.
3. Divide the angles of ascent and descent into six equal parts and draw radial lines intersecting the base circle at points 0 to 6 and 6′ to 0′ in the angles of ascent and descent, respectively.
4. Measure distances 1*a*, 2*b*, 3*c*, 4*d*, 5*e*, and 6*f* from the displacement diagram in the angle of ascent. Cut off the corresponding distances on the redial lines in the cam profile diagram.
5. Draw perpendicular on the radial lines at these points '*a*' to *f*.

   Repeat the same process in the angle of descent.
6. Draw a smooth curve touching the perpendicular lines to get the cam profile.

Example 8.9

Draw the profile of a cam to raise a valve with harmonic motion through 50 mm in 1/3 of a revolution, keep it fully raised through 1/12 of a revolution, and to lower it with harmonic motion 1/6 of a revolution. The valve remains closed during the rest of the revolution. The diameter of the roller is 20 mm and the minimum radius of the cam is to be 25 mm. The diameter of the cam shaft is 25 mm. The axis of the valve rod passes through the axis of the cam shaft. Assume the camshaft to rotate with a uniform velocity.

■ **Solution**

Given: Life = 50 mm, angle of ascent $= \frac{1}{3} \times 360° = 120°$,

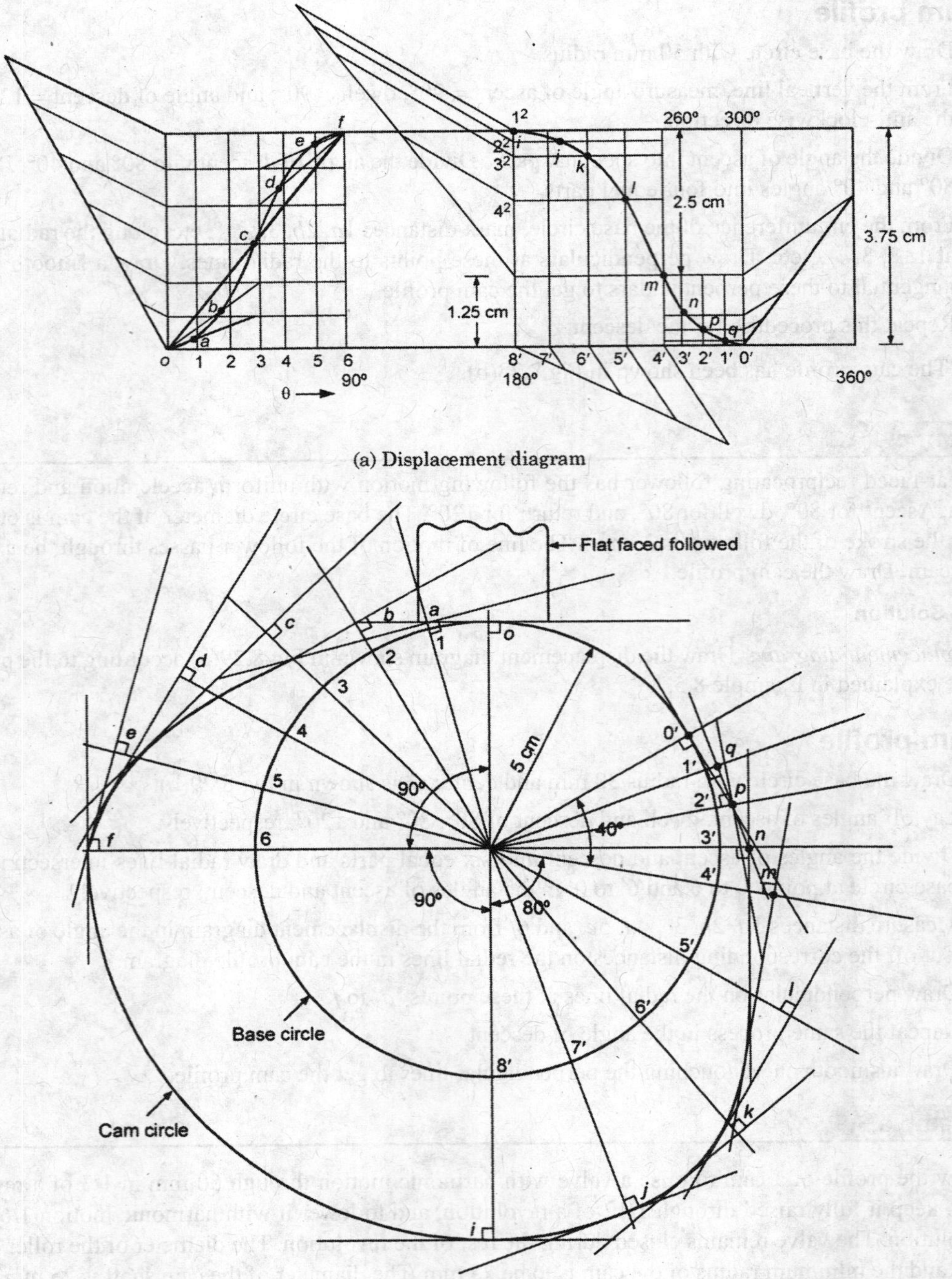

(a) Displacement diagram

(b) Cam profile

Fig.8.28 Cam profile with a flat-faced follower

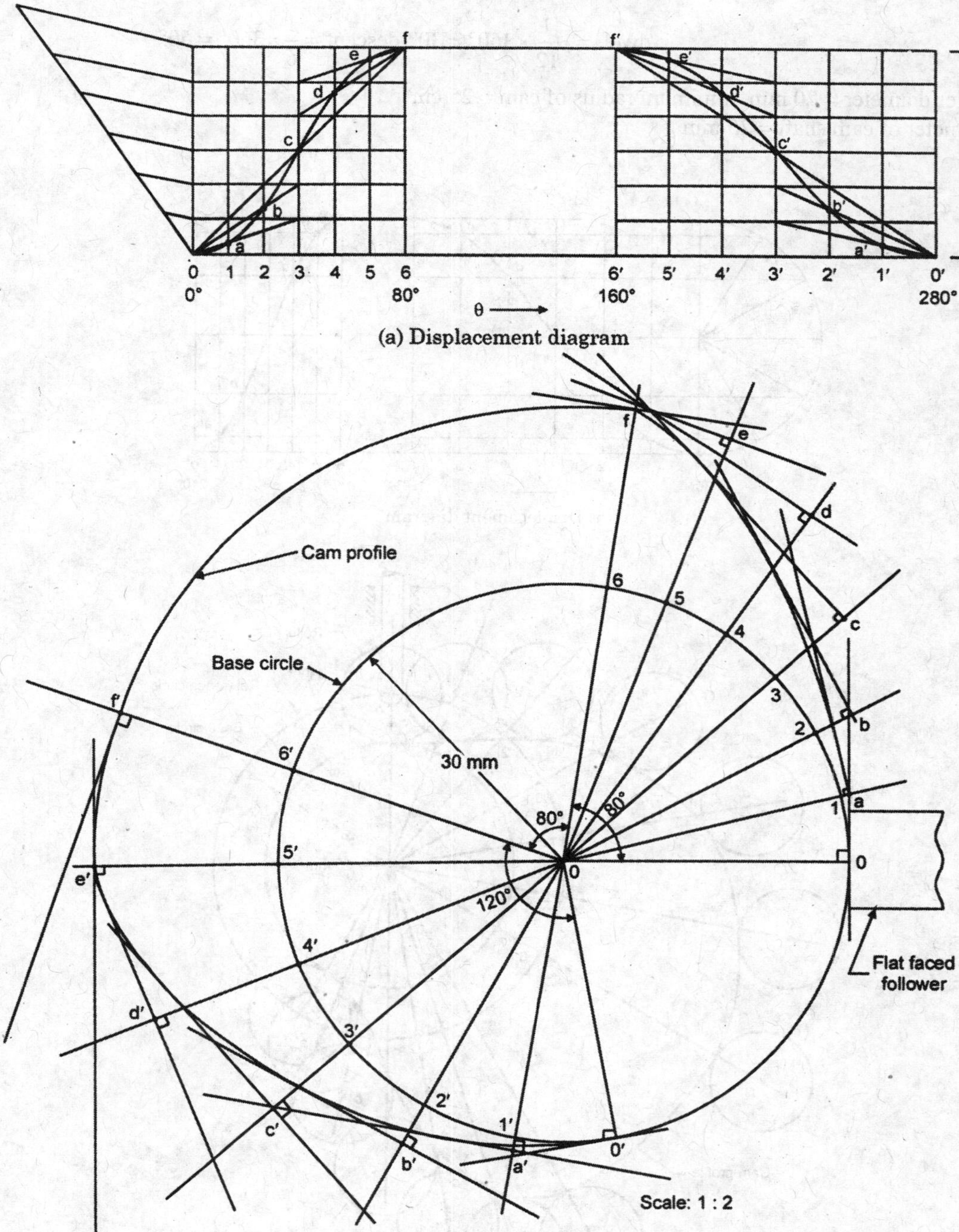

Fig.8.29 Cam profile for a flat faced follower having uniform acceleration and retardation

$$\text{dwell} = \frac{1}{12}\times 360° = 30°, \text{descent} = \frac{1}{6}\times 360° = 60°$$

Roller diameter = 20 mm, minimum radius of cam = 25 cm
Diameter of cam shaft = 25 mm

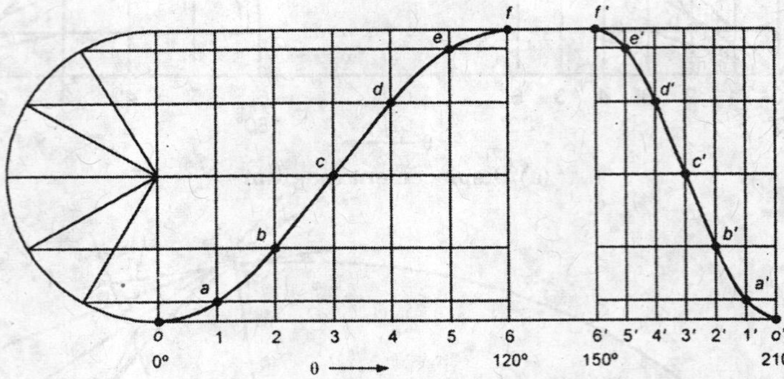

(a) Displacement diagram

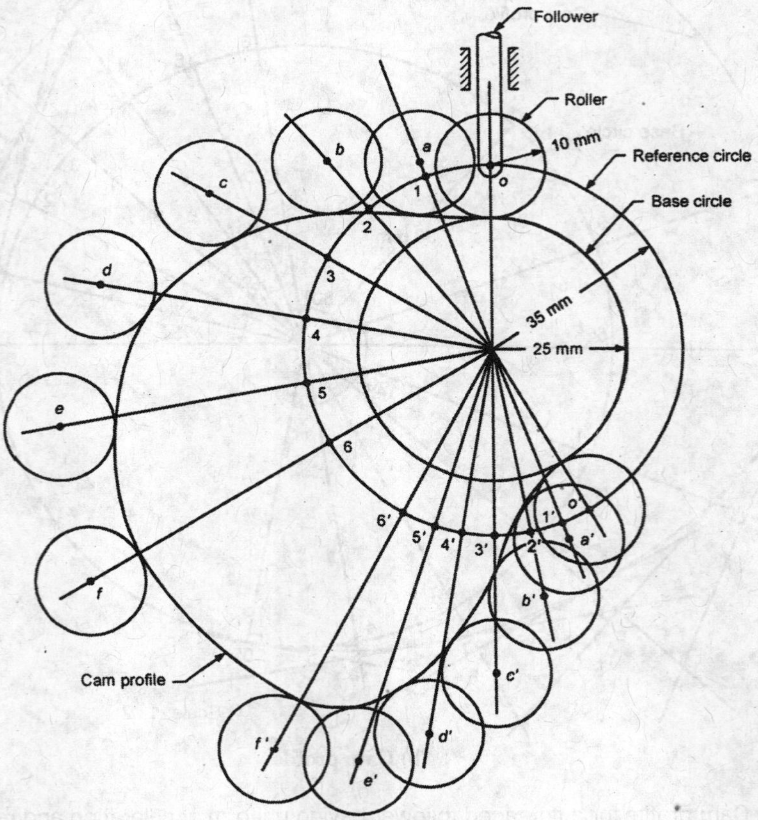

(b) Cam profile

Fig.8.30 Cam profile with roller follower having SHM

Displacement diagram: Draw the displacement diagram shown in Fig.8.30 (a) according to the procedure explained in Example 8.1.

Cam profile: Draw the cam profile shown in Fig.8.30(b) according to the procedure described in Example 8.3.

8.9 CAM PROFILE WITH SWINGING ROLLER FOLLOWER

The cam with swinging roller follower is shown in Fig.8.31. The following steps may be used to draw the cam profile:

1. Draw the base circle.
2. Draw the follower in its 0° position, tangent to the base circle.
3. Draw the reference circle through the centre of the follower.
4. Locate points around the reference circle corresponding to the divisions in the displacement diagram and number them accordingly.
5. Draw a pivot circle through the follower point.
6. Locate the pivot points around the pivot circle corresponding to each point on the reference circle, and number them accordingly.
7. From each of the pivot points, draw an arc where radius is equal to the length of the follower arm.

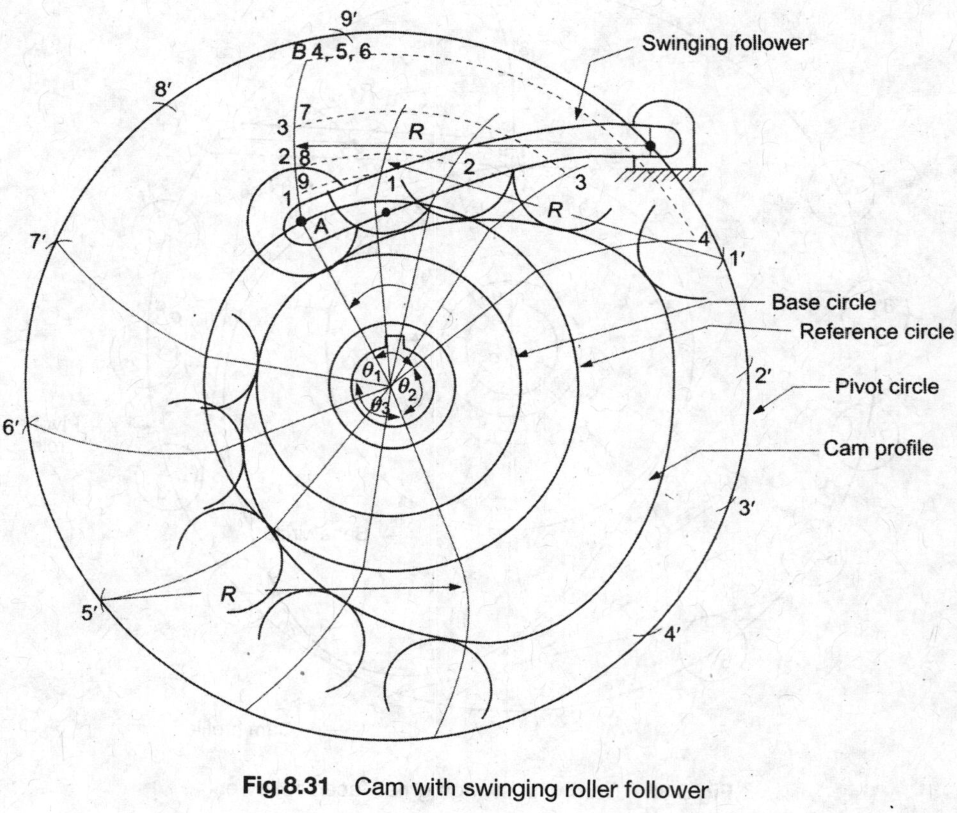

Fig.8.31 Cam with swinging roller follower

8. At the zero position, draw the two extreme positions of the follower lever by laying off the arc *AB* equal to the maximum displacement.
9. Lay off the various displacements 1*a*, 2*b*, 3*c*, …, along this arc.
10. Rotate each of the points on arc *AB* to its proper position around the cam profile.
11. Draw the follower outline at each of the points just located.
12. Draw a smooth curve tangent to the follower outlines.

8.10 CAM PROFILE WITH SWINGING FLAT-FACED FOLLOWER

A cam with swinging flat-faced follower is shown in Fig.8.32. The following steps may be followed to draw the cam profile:

1. Draw the base circle, which in this case also serves the reference circle.
2. Draw the follower in its home position, tangent to the base circle.
3. Draw radial lines corresponding to the divisions in the displacement diagram, and number accordingly.
4. Draw the pivot circle through the follower pivot.

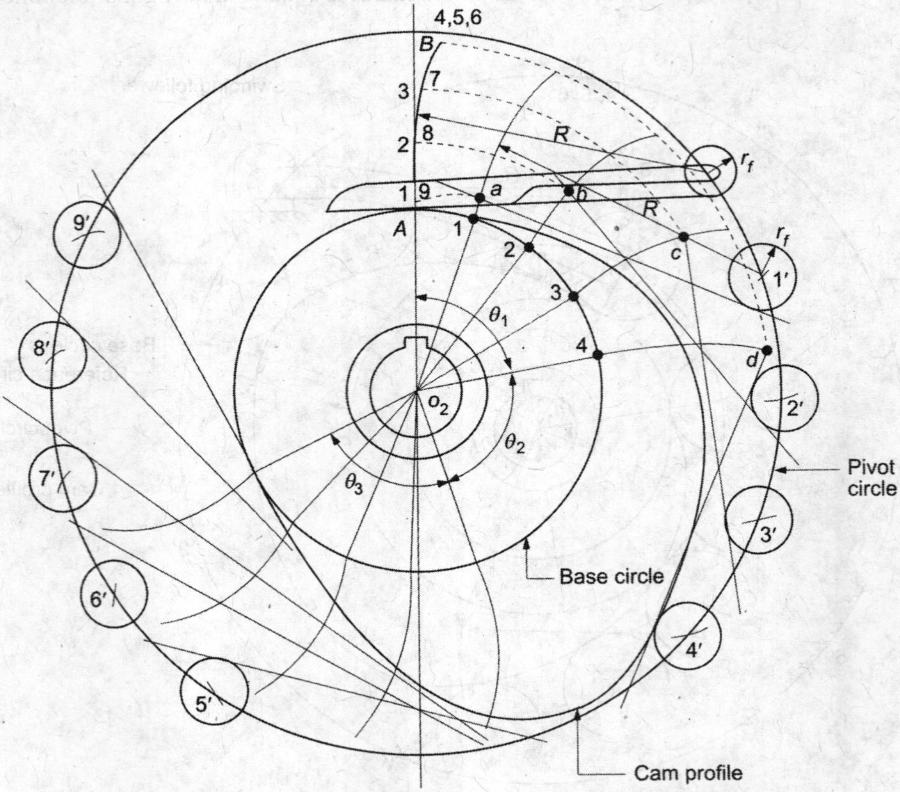

Fig.8.32 Cam with swinging flat-faced follower

5. Locate the pivot points around the pivot circle.
6. Locate the trace point on the flat face at a radius *r* from the pivoted point at zero position.
7. At the zero position, draw the two extreme positions of the follower lever laying off the arc *AB* equal to the maximum displacement.
8. Lay off the various displacements, 1*a*, 2*b*, 3*c*, ... along this arc.
9. Locate the trace point *a* relative to the cam at the intersection of the arc of radius *R*, centred at 1′ and the arc 1*a* with centre O_2. Establish points *b*, *c*, *d*, ... similarly.
10. The first position of the flat follower, relative to the cam is the straight line through point *a* that is tangent to the circle of radius r_f, centred at 1′. Construct the successive positions of follower face in a similar manner.
11. Draw a smooth curve tangent to the family of straight lines representing the follower face.

Example 8.10

Draw the cam profile for the data given below:

Base circle radius of cam = 50 mm
Lift = 40 mm
Angle of ascent = 60°
Angle of dwell = 40°
Angle of descent = 90°
Speed of cam = 300 rpm
Motion of follower = SHM
Type of follower = knife-edge

Also calculate the maximum velocity and acceleration during ascent and descent.

■ Solution

Angular velocity, $\omega = 2\pi \times 300/60 = 31.416$ rad/s

Maximum velocity, $v_{max} = (\pi h/2)\cdot(\omega/\theta_1)$

$$= \left(\pi \times 40 \times 10^{-3}/2\right)\cdot\left[10\pi/(\pi/3)\right]$$

$= 1.885$ m/s during ascent

$$= (\pi h/2)\cdot(\omega/\theta_3)$$

$$= \left(\pi \times 40 \times 10^{-3}/2\right)\cdot\left[10\pi/(\pi/2)\right]$$

$= 1.257$ m/s during descent

Maximum acceleration, $f_{max} = (\pi\omega/\theta_1)\cdot(h/2)$

$$= (\pi \times 30\pi/\pi)^2 \times 0.02$$

$= 174.65$ m/s^2 during ascent

$$= (\pi\omega/\theta_3)^2\cdot(h/2)$$

$$= (\pi \times 20\pi/\pi)^2 \times 0.02$$

$= 78.957$ m/s^2 during descent.

Displacement diagram

1. Draw a vertical line 0–6 equal to the lift of 40 mm.
2. Draw a semicircle on this line and divide the semicircle into six equal parts of 60° each.
3. Choosing a scale of 1 mm = 2°, mark of the angles of ascent, dwell, descent, and dwell of 60°, 40°, 90°, and 170° respectively.
4. Divide the angles of ascent and descent into six equal parts and draw vertical lines at those points parallel to 0–6.
5. Draw horizontal lines from the points 0 to 6 on the semicircle to intersect the vertical lines.
6. Join the points of intersection with a smooth curve to get the displacement diagram.

   The displacement diagram is shown in Fig.8.33(a).

Cam profile

1. Draw a circle of radius equal to the base circle radius of 50 mm.
2. Draw angles of ascent, dwell, and descent equal to 60°, 40°, and 90°, respectively. Divide the angles of ascent and descent into six equal parts and draw radial lines.
3. Mark points 0 to 6 and 6′ to 0′ on the base circle in the angles of ascent and descent respectively.
4. Measure distances 1*a*, 2*b*, 3*c*, 4*d*, 5*e* and 6f from the displacement diagram and cut off corresponding distances on the radial lines. Repeat the same procedure during the descent.
5. Join the points so obtained by a smooth curve to get the cam profile.

   The cam profile has been drawn in Fig.8.33(b).

Example 8.11

A cam of base circle 50 mm is to operate a roller follower of 20 mm diameter. The follower is to have SHM. The angular speed of the cam is 360 rpm. Draw the cam profile for the cam lift of 40 mm. Angle of ascent = 60°, angle of dwell = 40°, and angle of descent = 90°, followed by dwell again. Also calculate the maximum velocity and acceleration during ascent and descent.

■ Solution

Angular velocity of cam, $\omega = 2\pi \times 360/60 = 12\pi \text{ rad/s}$

Maximum velocity, $v_{max} = (\pi h/2)\cdot(\omega/\theta_1)$

$$= \left(\pi \times 40 \times 10^{-3}/2\right)\cdot\left[12\pi/(\pi/3)\right]$$

$$= 2.262 \text{ m/s during ascent}$$

$$= \left(\pi h/2\right)\cdot\left(\omega/\theta_3\right)$$

$$= \left(\pi \times 40 \times 10^{-3}/2\right)\cdot\left[12\pi/(\pi/2)\right]$$

$$= 1.507 \text{ m/s during descent}$$

Maximum acceleration, $f_{max} = (\pi\omega/\theta_1)^2. (h/2)$

$$= \left(\pi \times 36\pi/\pi\right)^2 \times 0.02$$

$$= 255.82 \text{ m/s}^2 \text{ during ascent}$$

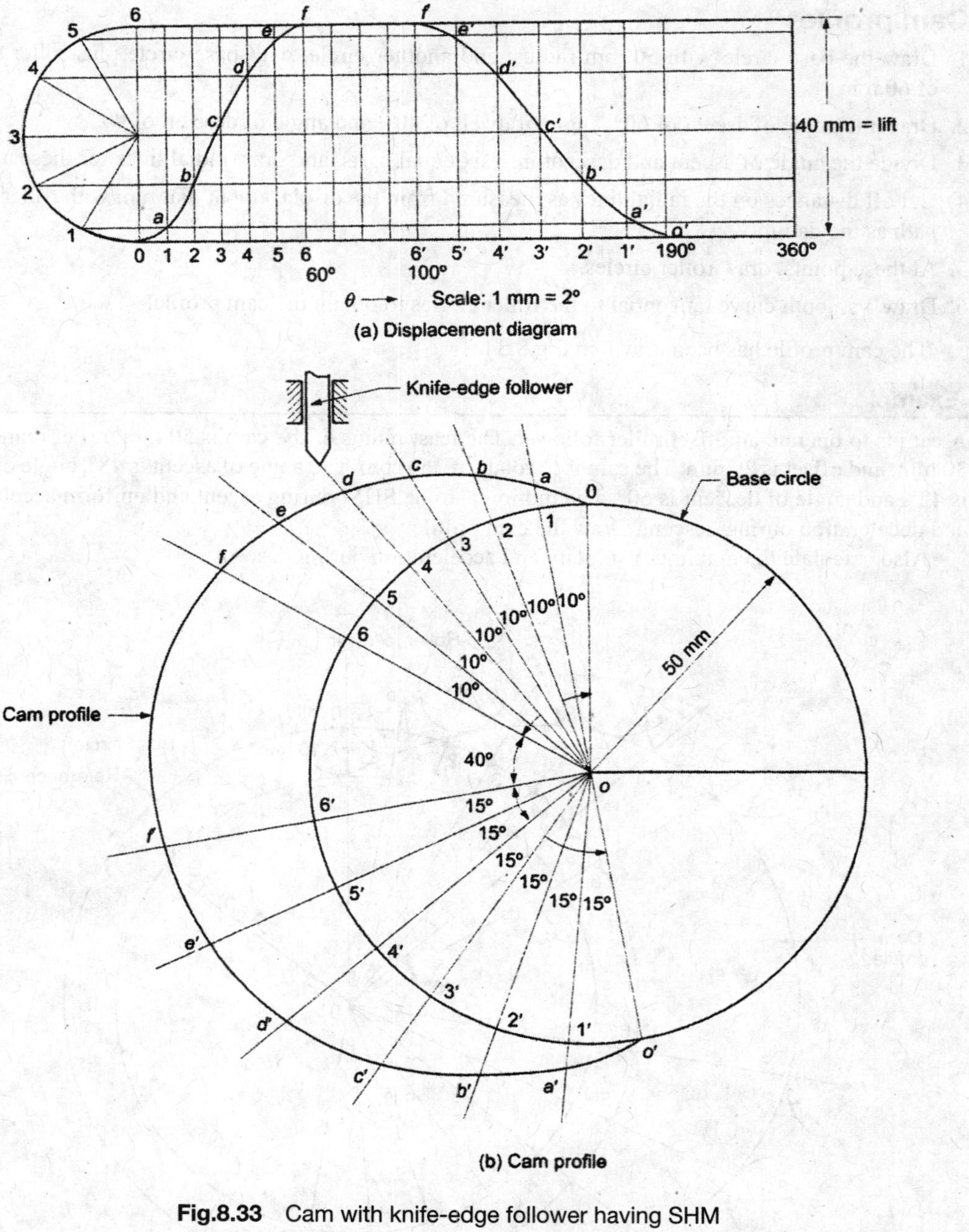

Fig.8.33 Cam with knife-edge follower having SHM

$$= (\pi\omega/\theta_3)^2 \cdot (h/2)$$

$$= (\pi \times 24\pi/\pi)^2 \times 0.02$$

$$= 113.7 \text{ m/s}^2 \text{ during descent}$$

Draw the displacement diagram as explained in Example 8.10.

Cam profile

1. Draw the base circle with 50 mm radius, and another circle, with base circle plus roller radius, of 60 mm.
2. Draw the angle of ascent of 60°, angle of dwell of 40°, and angle of descent of 90°.
3. Divide the angle of ascent and descent into six equal parts, and draw radial lines for these angles.
4. Cut off distances on the radial lines as measured from the displacement diagram with roller centre path as the datum.
5. At these points, draw roller circles.
6. Draw a smooth curve tangential to the roller circles to obtain the cam profile.

The cam profile has been drawn in Fig.8.34.

Example 8.12

A cam is to operate an offset roller follower. The least radius of the cam is 50 mm, roller diameter is 30 mm, and offset is 20 mm. The cam is to rotate at 360 rpm. The angle of ascent is 48°, angle of dwell is 42°, and angle of descent is 60°. The motion is to be SHM during ascent and uniform acceleration and deceleration during descent. Draw the cam profile.

Also calculate the maximum velocity and acceleration during descent.

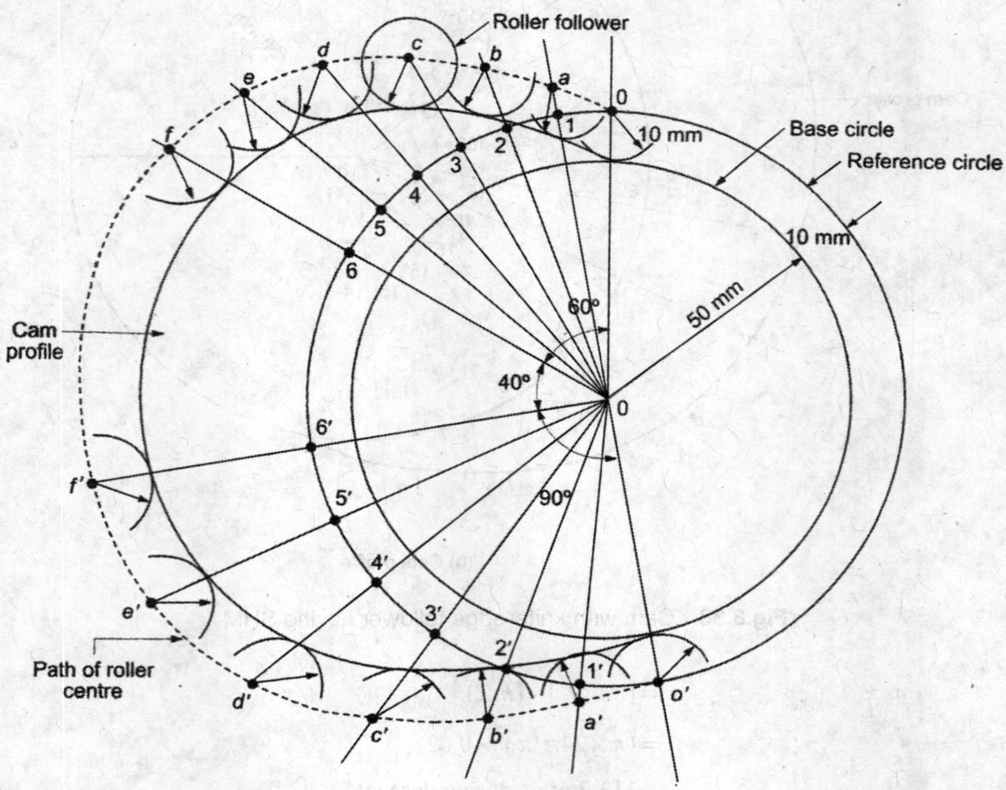

Fig.8.34 Cam with roller follower having SHM

■ **Solution**

$$\omega = 2\pi \times 360/60 = 12\pi \text{rad/s}$$

Maximum velocity during descent with uniform acceleration and deceleration,

$$v_{max} = 2h\,\omega/\theta_3 \text{ during descent}$$
$$= 2\times 40\times 36\pi/\pi$$
$$= 2.88 \text{ m/s}$$

Maximum acceleration, $f_{max} = 4w^2h/\theta_3^2$ during descent

$$= 4\times(12\pi)^2\, 40\times 10^{-3}/(\pi/3)^2$$
$$= 204.36 \text{ m/s}^2$$

Cam profile

1. Draw the displacement diagram as shown in Fig.8.35(a).
2. Draw the base circle with 50 mm radius. Draw another circle with 65 mm radius to represent the path traced out by the roller centre.
3. Draw a circle 20 mm radius equal to the offset.
4. Divide the angle of ascent and the angle of descent into six equal parts.
5. Draw tangents at the circumference of the offset circle at the above points.
6. From the circle of the path of the roller centre, measure off distances from the displacement diagram along the tangential lines.
7. Draw circles at these points equal to the roller radius.
8. Draw a smooth curve tangential to these circles to get the cam profile.

The cam profile has been shown in Fig.8.35(b).

Example 8.13

A cam is operating an oscillating roller follower having SHM, as shown in Fig.8.36. Draw the cam profile for the data given below:

Roller centre from cam centre at beginning of ascent = 60 mm
Angle of ascent = 60°
Minimum radius of cam = 40 mm
Dwell = 45°
Angle of descent = 90°
Angle of oscillation of arm during ascent or descent = 15°

■ **Solution**

Length of circular arc during movement of follower arm

$$= 80\times 15\times \pi/180 = 20.94 \text{ mm}$$

Length of chord joining the roller centre in extreme position

$$= 2\times 80 \sin 4.5° = 20.88 \text{ mm}$$

For small angle, ≈ arc ≈ chord. Therefore, we take displacement = 20.88 mm

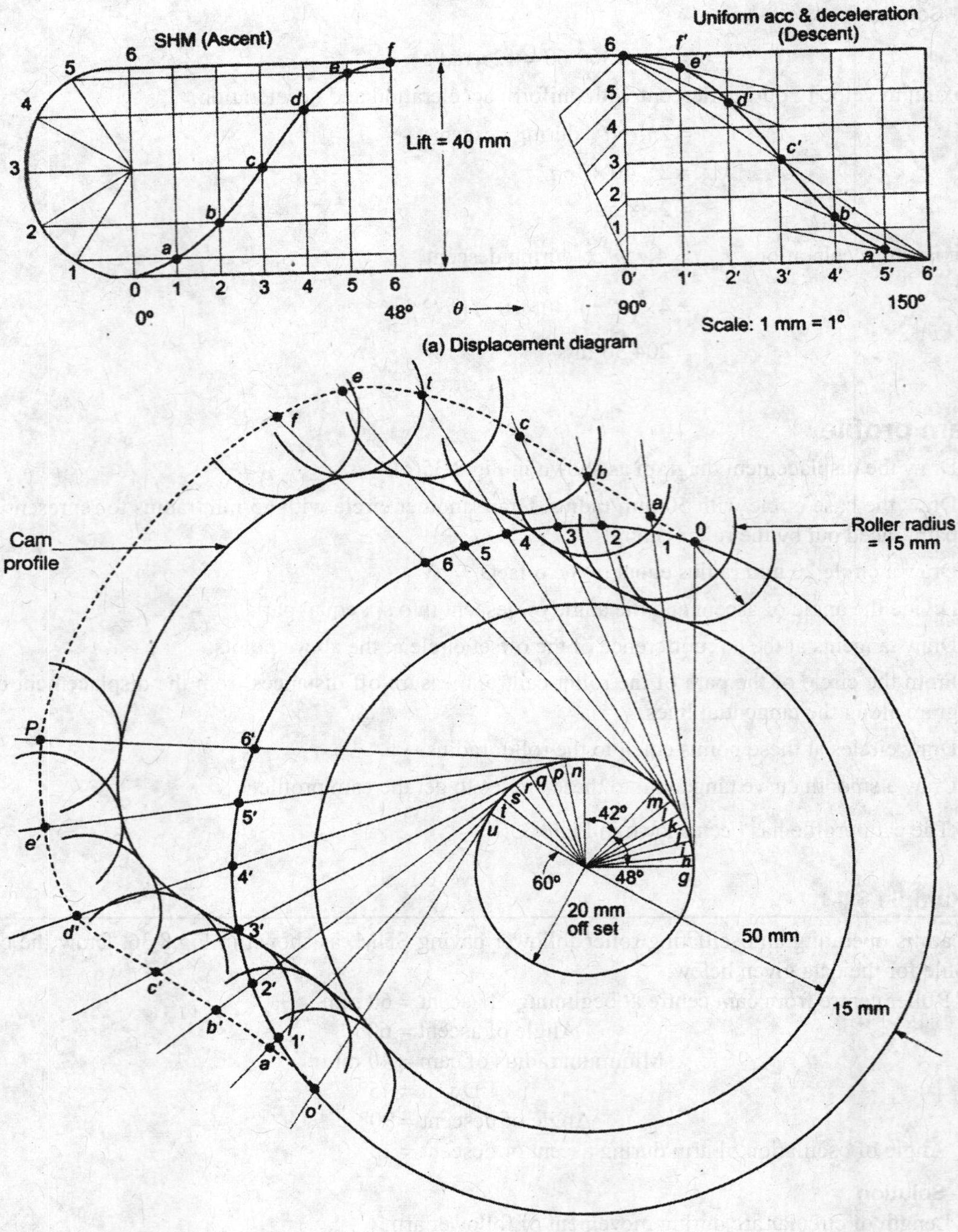

Fig.8.35 Cam with offset roller follower having SHM during ascent uniform acceleration and retardation during descent

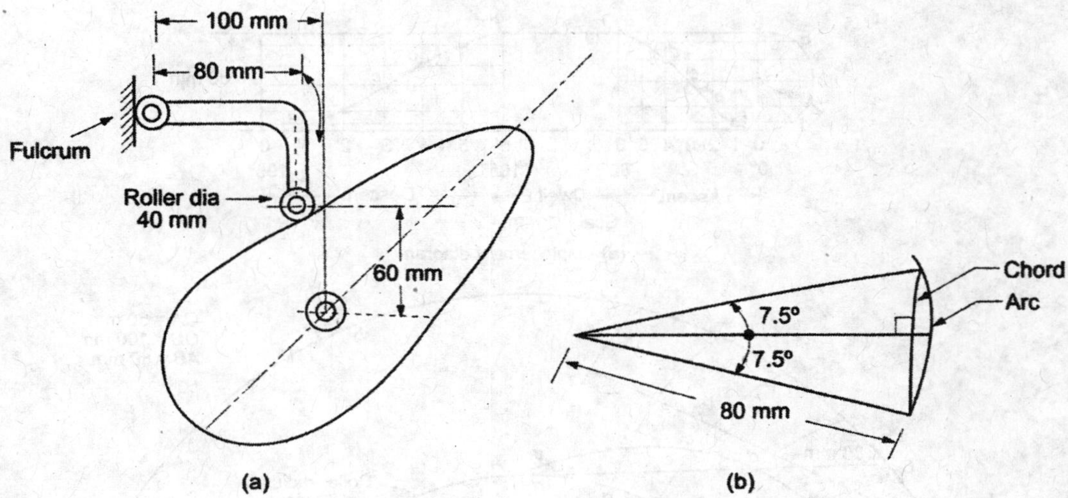

Fig.8.36 Cam with oscillating roller follower

Cam profile

1. Taking lift of cam to be equal to 20.88 mm, draw the displacement diagram, as shown in Fig.8.37(a) and explained in Example 8.1.
2. Draw cam base circle with 40 mm radius, roller centre path with 60 mm radius, and another circle with 100 mm radius, which is equal to the distance between cam centre and the fulcrum.
3. Now $OA = 60$ mm, and let it be a vertical line. With O and A as centres, and radii equal to 100 mm and 80 mm, respectively, draw arcs to meet at B. Point B represents the fulcrum position before follower begins to ascend.
4. Mark off $\angle BOC = 60°$, $\angle COD = 45°$, and $\angle DOE = 90°$. Divide the angles of ascent and descent into six equal parts.
5. With $a_1, b_1, c_1, \ldots$, etc. As centers, draw arcs of 80 mm radius. From the points of intersection of these arcs with the 60 mm radius circle, mark along the chords of the displacement arcs, distances pa, qb, rc, …, etc, equal to $1a$, $2b$, $3c$, …, etc., of the displacement diagram.
6. Repeat the above procedure for the angle of descent.
7. Join the points a, b, c, …, etc. To obtain the pitch profile of the cam.
8. From the above pitch points, draw the roller radius circles.
9. The envelope of the arcs of roller radius gives the cam profile.

The cam profile has been shown in Fig.8.37(b).

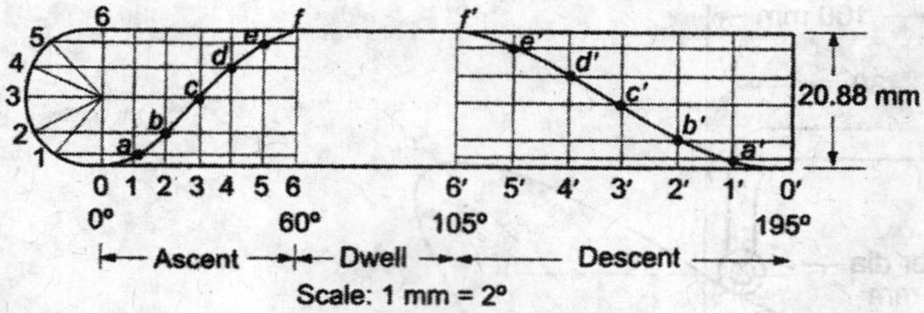

(a) Displacement diagram

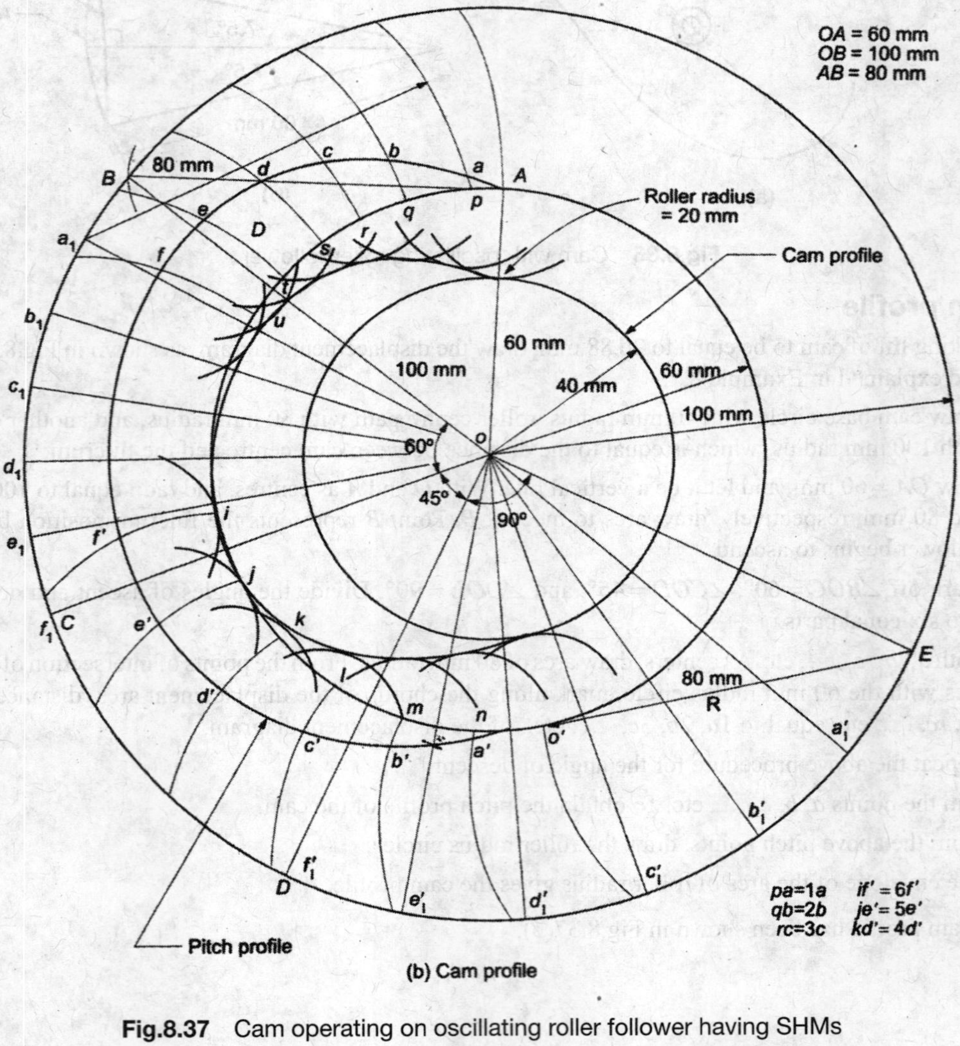

(b) Cam profile

Fig.8.37 Cam operating on oscillating roller follower having SHMs

Example 8.14

It is required to set out the profile of a cam to give motion to the follower in such a way that it rises through 31.4 mm during 180° of cam rotation with cycloidal motion and returns with cycloidal motion during 180° of cam rotation. Determine the maximum velocity and acceleration of the follower during the outstroke when the cam rotates at 1800 rpm clockwise. The base circle diameter of the cam is 25 mm and roller diameter of the follower is 10 mm. The axis of the follower passes through the cam centre.

■ **Solution**

Stroke, $h = 2\pi R$

Radius of circle generating the cycloid, $R = 31.4/2\pi = 5$ mm

The displacement diagram is shown in Fig.8.38(a).

$$\omega = 2\pi \times 1800/60 = 188.5 \text{ rad/s}$$

$$v_{\max} = 2\omega h/\beta = 2 \times 188.5 \times 31.4/(\pi \times 1000) = 3.77 \text{ m/s}$$

$$a_{\max} = 2\pi\omega^2 h/\beta^2 = 2\pi \times (188.5)^2 \times 31.4/(\pi^2 \times 1000) = 710.3 \text{ m/s}^2$$

The cam profile has been drawn in Fig.8.38(b).

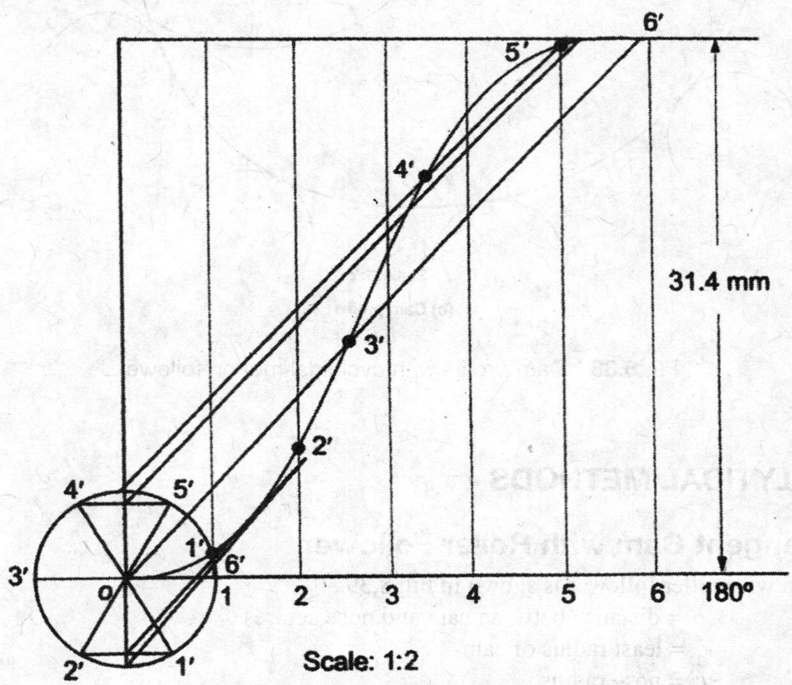

(a) Displacement diagram

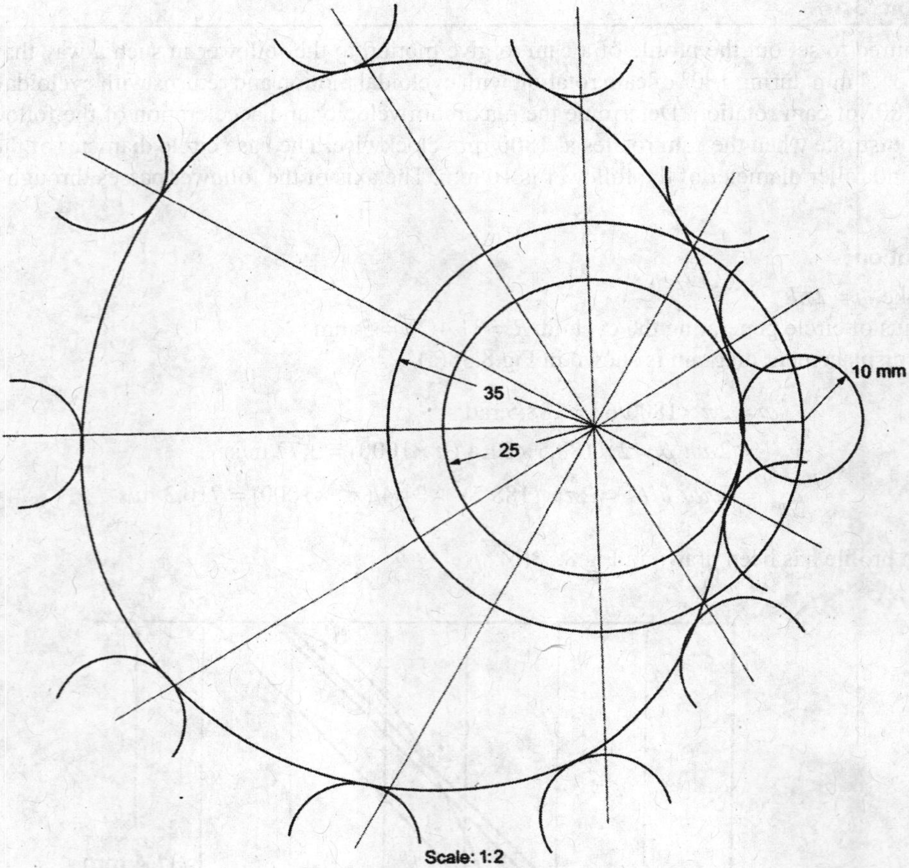

(b) Cam profile

Fig.8.38 Cam profile with cycloidal motion follower

8.11 ANALYTICAL METHODS

8.11.1 Tangent Cam with Roller Follower

A tangent cam with roller follower is shown in Fig.8.39.

Let r = distance between cam and nose centres

r_1 = least radius of cam

r_2 = nose radius

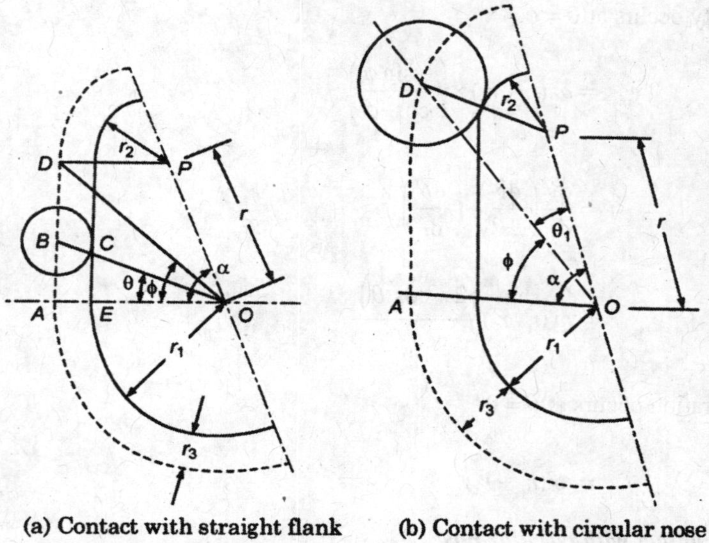

Fig.8.39 Tangent cam with roller follower

r_3 = roller radius

$l = r_2 + r_3$

α = angle of ascent

ϕ = angle of contact of cam with straight flank

(a) Roller in contact with straight flank

At position B, let θ be the angle turned through by the cam, as shown in Fig.8.39(a). Then lift,

$$x = OB - OA$$

$$= \frac{OA}{\cos\theta} - OA$$

$$= \frac{OA(1-\cos\theta)}{\cos\theta}$$

$$= \frac{(r_1 + r_3)(1-\cos\theta)}{\cos\theta} \tag{8.30}$$

Velocity,
$$v = \frac{dx}{dt} = \left(\frac{dx}{d\theta}\right)\left(\frac{d\theta}{dt}\right)$$

$$= \omega\,(r_1 + r_3)\left(\frac{-\sin\theta}{\cos^2\theta}\right) \tag{8.31}$$

Maximum velocity occurs at $\theta = \phi$.

$$v_{max} = \omega\,(r_1 + r_3) \times \left(\frac{-\sin\phi}{\cos^2\phi}\right) \tag{8.32}$$

Acceleration, $$f = \frac{dv}{dt}\left(\frac{dv}{d\theta}\right)\cdot\left(\frac{d\theta}{dt}\right)$$

$$= \omega\,(r_1 + r_3)\frac{\left(2-\cos^2\theta\right)}{\cos^3\theta} \tag{8.33}$$

Minimum acceleration occurs at $\theta = 0°$

$$f_{min} = \omega^2(r_1 + r_3) \tag{8.34}$$

(b) Follower in contact with circular nose

Fig.8.39(b) shows the follower in contact with circular nose.

Let $OP = r = \text{const.}$

$PD = r_2 + r_3 = l = \text{const.}$

OPD is a slider crank chain in which *OP* is the crank, *PD* the connecting rod, and *D* the slider.

Let $\theta_1 = \alpha - \theta$

For a slider crank chain, the displacement from top dead centre is given by,

$$x = r\left[(1-\cos\theta) + n - \left(n^2 - \sin^2\theta\right)^{0.5}\right]$$

where $n = l/r$

For the cam mechanism shown in Fig.8.41, we have

$$x = r\left[(1-\cos\theta_1) + (r_2 + r_3)/r - \left[\left\{(r_2 + r_3)/r\right\}^2 - \sin^2\theta_1\right]^{0.5}\right]$$

$$= r(1-\cos\theta_1) + l - (l^2 - r^2\sin^2\theta_1)^{0.5} \tag{8.35}$$

$$v = \omega\frac{dx}{d\theta}$$

$$= \omega r\left[r\sin\theta_1 + \frac{r^2\sin 2\theta_1}{2\left(l^2 - r^2\sin^2\theta_1\right)^{0.5}}\right] \tag{8.36}$$

$$f = \omega \frac{dv}{d\theta}$$

$$= \omega^2 r \left[\cos\theta_1 + \frac{\left(l^2 r \cos 2\theta_1 + r^3 \sin^4 \theta_1\right)}{\left(l^2 - r^2 \sin^2 \theta_1\right)^{3/2}} \right] \quad (8.37)$$

Example 8.15

A symmetrical tangent cam operating a roller follower has the following particulars:

Radius of base circle of cam = 40 mm
Roller radius = 20 mm
Angle of ascent = 75°
Total lift = 20 mm
Speed of cam shaft = 300 rpm

Determine (a) the principal dimensions of the cam, (b) the equation of the displacement curve, when the follower is in contact with the straight flank, and (c) the acceleration of the follower, when it is in contact with the straight flank where it merges into the circular nose.

Solution

Given: $r_1 = 40$ mm, $r_3 = 20$ mm, $h = 20$ mm, $N = 300$ rpm, $\alpha = 75°$

(a) With reference to Fig.8.40, we have

$$OP + r_2 = r_1 + h = 40 + 20$$

$$OP = 60 - r_2$$

$$OQ + r_2 = r_1 = 40$$

$$OQ = 40 - r_2$$

$$\cos\alpha = \frac{OQ}{OP}$$

$$\cos 75° = \frac{40 - r_2}{60 - r_2}$$

$$0.25882 = \frac{40 - r_2}{60 - r_2}$$

$$15.52914 - 0.25882\, r_2 = 40 - r_2$$

Nose radius, $$r_2 = \frac{24.47085}{0.74118} = 33 \text{ mm}$$

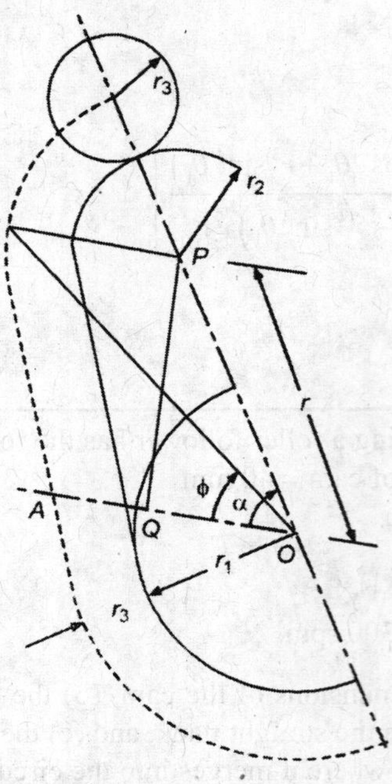

Fig.8.40 Tangent cam operating roller follower

Distance between cam and nose centre,

$$r = OP = 60 - 33 = 27 \text{ mm}$$

$$\tan\phi = \frac{RA}{OA} = \frac{PQ}{OA}$$

$$= \frac{OP\sin\alpha}{OA}$$

$$= \frac{27\sin 75^\circ}{60}$$

$$= 0.43467$$

$$\phi = 23.49^\circ$$

(b) Equation of displacement curve

1. When contact is with straight flank

$$x = (r_1 + r_3)\left(\frac{1}{\cos\theta} - 1\right)$$

$$= (40 + 20)\left(\frac{1}{\cos\theta} - 1\right)$$

$$= 60\left(\frac{1}{\cos\theta} - 1\right)\text{mm}$$

2. When contact is with circular nose

$$x = r(1-\cos\theta_1)+(r_2+r_3)-(l^2-r^2\sin^2\theta_1)^{0.5}$$
$$= 27(1-\cos\theta_1)+(33+20)-(53^2-27^2\sin^2\theta_1)^{0.5}$$
$$= 27(1-\cos\theta_1)+53-(2809-729\sin^2\theta_1)^{0.5}$$

where θ_1 is measured from apex position,

(c) Acceleration of follower

When in contact with straight flank,

$$f = \omega^2(r_1+r_3)\left[\frac{2-\cos^2\theta}{\cos^3\theta}\right]$$

$$\omega = \frac{2\pi\times 300}{60} = 31.416 \text{ rad/s}$$

When $\theta = 0°$, $$f = (31.416)^2(0.040+0.020)\left[\frac{2-1}{1}\right]$$

$$= 59.22 \text{ m/s}^2$$

When in contact with straight flank, $\theta = \phi$.

$$f = \omega^2(r_1+r_3)\times\left[\frac{2-\cos^2\phi}{\cos^3\phi}\right]$$
$$= \frac{(31.416)^2(0.04+0.02)(2-\cos^2 23.493°)}{\cos^3 23.439°}$$
$$= 52.94 \text{ m/s}^2$$

When contact is on circular nose,

$$\theta_1 = \alpha - \phi = 75 - 23.493 = 51.507°$$

$$f = \omega^2 r\times\left[\cos\theta_1\frac{\left(l^2 r\cos 2\theta_1 + r^3\sin^4\theta_1\right)}{\left(l^2-r^2\sin^2\theta_1\right)^{1.5}}\right]$$
$$= (31.416)^2\times 0.027\,[\cos 51.507° + (0.060^2\times 0.027\cos 103.014°$$
$$+ 0.027^3\sin^4 51.503°)/(0.060^2 - 0.027^2\sin^2 51.507°)^{1.5}]$$
$$= 13.33 \text{ m/s}^2$$

When at apex, $\theta_1 = 0°$

$$f = \omega^2 r \times \left[1 + \frac{r}{l}\right]$$

$$= (31.416)^2 \times 0.027 \left[1 + \frac{0.027}{0.060}\right] = 38.64 \text{ m/s}^2$$

Example 8.16

The follower of a tangent cam is operated through a roller of 50 mm diameter and its line of stroke passes through the axis of the cam. The minimum radius of the cam is 40 mm and the nose radius 15 mm. The lift is 25 mm. If the speed of the camshaft is 600 rpm, calculate the velocity and acceleration of the follower at the instant when the cam is (a) in full lift position, and (b) 20° from full lift position.

Solution

Given: $r_3 = 25$ mm, $r_1 = 40$ mm, $r_2 = 15$ mm, $h = 25$ mm, $N = 600$ rpm

(a)
$$\cos\alpha = \frac{r_1 - r_3}{r_1 + h - r_2} = \frac{40-15}{40+25-15} = \frac{25}{50} = 0.5$$

$$\alpha = 60°$$

$$\omega = \frac{2\pi \times 600}{60} = 62.83 \text{ rad/s}$$

At full lift position, $\theta = \alpha$.

$$v = \omega\,(r_1 + r_3)\frac{\sin a}{\cos^2 a} = \frac{62.83 \times 65 \times \sin 60°}{\cos^2 60} = 14147 \text{ mm/s}$$

$$f = \omega^2 (r_1 + r_3)\left[\frac{2 - \cos^2 a}{\cos^3 a}\right]$$

$$= (62.83)^2 \times 65 \times \left[\frac{2 - \cos^2 60°}{\cos^3 60°}\right] = 35{,}92{,}324 \text{ mm/s}^2$$

(b) $\theta = 60° = 20° = 40°$

$$v = \omega \times (r_1 + r_3)\frac{\sin a}{\cos^2 a} = \frac{62.83 \times 65 \times \sin 40°}{\cos^2 40°} = 4473.4 \text{ mm/s}$$

$$f = \omega^2 \times (r_1 + r_3)\left[\frac{2 - \cos^2 a}{\cos^3 a}\right]$$

$$= (62.83)^2 \times 65 \times \left[\frac{2 - \cos^2 40°}{\cos^3 40°}\right] = 8{,}06{,}649 \text{ mm/s}^2$$

Example 8.17

A cam profile consists of two circular arcs of radii 30 mm and 15 mm, joined by straight lines, giving the follower a lift of 15 mm. The follower is a roller of 25 mm radius and its line of action is a straight line passing through the cam shaft axis. When the cam shaft has a uniform speed of 600 rpm, find the maximum velocity and acceleration of the follower while in contact with the straight flank of the cam.

■ **Solution**

Given: $r_1 = 30$ mm, $r_2 = 15$ mm, $r_3 = 25$ mm, $h = 15$ mm, $N = 600$ rpm

$$OP = r_1 + h - r_2 = 30 + 15 - 15 = 30 \text{ mm}$$

$$OQ = r_1 - r_2 = 30 - 15 = 15 \text{ mm}$$

$$\cos\alpha = \frac{OQ}{OP} = \frac{15}{30} = 0.5$$

$$\alpha = 60°$$

$$\tan\phi = \frac{OP\sin\alpha}{OA} = \frac{30\sin 60°}{55} = 0.27272$$

$$\phi = 15.25°$$

$$\omega = \frac{2\pi\times 600}{60} = 62.83 \text{ rad/s}$$

Contact with straight flank:

$$v_{max} = \omega\,(r_1 + r_3)\left(\frac{-\sin\phi}{\cos^2\phi}\right)$$

$$= 62.83\times 55\times\left(\frac{-\sin 15.25°}{\cos^2 15.25°}\right) = 976.4 \text{ mm/s}$$

$$f = \omega^2(r_1 + r_3)\left[\frac{2-\cos^2\theta}{\cos^3\theta}\right]$$

At $\theta = 0°$

$$f = (62.83)^2\times 0.055 = 217.12 \text{ m/s}^2$$

At $\theta = \phi$

$$= (62.83)^2\times 0.055\times\frac{(2-\cos^2 15.25°)}{\cos^3 15.25°} = 258.5 \text{ m/s}^2$$

8.11.2 Circular Arc Cam Operating Flat-Faced Follower

(a) Follower in contact with circular flank

The follower in contact with circular flank is shown in Fig.8.41(a).

Let r = distance between cam and nose centers

$r_1 = OB$ = least circle radius

r_2 = nose circle radius
$r_3 = QD$ = flank circle radius
α = angle of ascent
ϕ = angle of contact on circular flank

Displacement of follower, $x = OC - OB$

$$= DE - r_1$$
$$= (QD - QE) - r_1$$
$$= (r_3 - OQ \cos\theta) - r_1$$
$$= r_3 - (r_3 - r_1)\cos\theta - r_1 \quad (8.38)$$

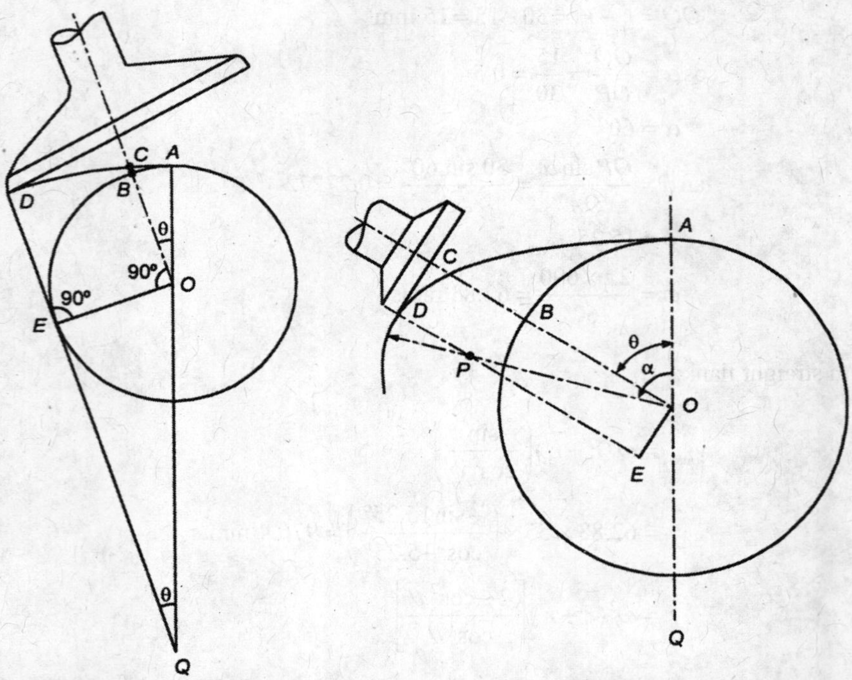

(a) Contact with circular flank

(b) Contact with circular nose

Fig.8.41 Circular arc cam operating flat-faced follower

Velocity,
$$v = \frac{dx}{dt} = \left(\frac{dx}{d\theta}\right)\cdot\left(\frac{d\theta}{dt}\right)$$

$$= (r_3 - r_1)\,\omega \sin\phi \quad (8.39)$$

Maximum velocity occurs at $\theta = \phi$.

$$v_{max} = (r_3 - r_1)\,\omega \sin\phi \quad (8.40)$$

Acceleration, $f = \frac{dv}{dt} = \left(\frac{dv}{d\theta}\right) \cdot \left(\frac{d\theta}{dt}\right)$

$$= (r_3 - r_1)\, \omega^2 \cos \theta \qquad (8.41)$$

Maximum acceleration occurs at $\theta = 0°$.

$$f_{max} = (r_3 - r_1)\omega^2 \qquad (8.42)$$

Minimum acceleration occurs at $\theta = \phi$.

$$f_{min} = (r_3 - r_1)\omega^2 \cos \phi \qquad (8.43)$$

(b) Follower in contact with circular nose

The follower in contact with circular nose is shown is Fig.8.41(b).

$$x = OC - OB$$

$$= DE - r_1$$

$$= DP + PE - r_1$$

$$= r_2 + OP \cos(\alpha - \theta) - r_1 \qquad (8.44)$$

$$= r_2 + r\cos(a - \theta) - r_1$$

$$v = \omega \cdot \frac{dx}{d\theta}$$

$$= \omega r \sin(\alpha - \theta) \qquad (8.45)$$

Velocity is maximum when $\alpha - \theta$ is maximum. This happens when contact changes from circular flank to circular nose. Minimum velocity occurs when $\alpha - \theta = 0°$, i.e., at the apex of the circular nose.

$$v_{min} = 0 \qquad (8.46)$$

Acceleration, $f = \omega \cdot \frac{dv}{d\theta}$

$$= -\omega^2 r \cos(\alpha - \theta) \qquad (8.47)$$

Maximum acceleration occurs when $\alpha - \theta = 0°$, i.e., apex of circular nose.

$$f_{max} = -\omega^2 r \qquad (8.48)$$

Minimum acceleration occurs when $\alpha - \theta$ is maximum, i.e., when contact changes from circular flank to circular nose.

Example 8.18

The following data refers to a circular arc cam working with a flat-faced reciprocating follower:

Minimum radius of cam = 30 mm, total angle of cam action = 120° , radius of circular arc = 80 mm and nose radius = 10 mm.

Find (a) the distance of the centre of the nose circle from the cam axis, (b) the angle through which the cam turns when the point of contact moves from the junction of minimum radius arc and circular arc of the junction of nose arc and circular arc, and (c) velocity and acceleration of the follower when the cam has turned through an angle of 20° . The angular velocity of the cam is 10 rad/s.

■ Solution

Given: $r_1 = 30$ mm, $r_2 = 10$ mm, $r_3 = 80$ mm, $2\alpha = 120°$, $\omega = 10$ rad/s, lift = 10 mm

(a) $OP + r_2 = r_1 + \text{lift}$

$OP = 30 - 10 + \text{lift} = 20 + \text{lift}$

$x = (r_3 - r_1)(1 - \cos\theta)$

At $\theta = \alpha = 60°$, Lift, $x = (80 - 30)(1 - \cos 60°) = 25$ mm

$OP = 20 + 25 = 45$ mm

(b) $$\tan\phi = PM/CM = \frac{OP\sin\alpha}{OC + OM} = \frac{OP\sin\alpha}{OC + OP\sin\alpha}$$

$$= \frac{45\sin 60°}{80 + 45\cos 60°} = 0.38$$

$\phi = 20.8°$ (see Fig.8.42)

(c) Velocity, $v = \omega(r_3 - r_1)\sin\theta = 10 \times (80 - 30)\sin 20° = 171$ mm/s

Acceleration, $f = \omega^2 (r_3 - r_1)\cos\theta = 100 \times (80 - 30)\cos 20° = 4698.5$ mm/s^2

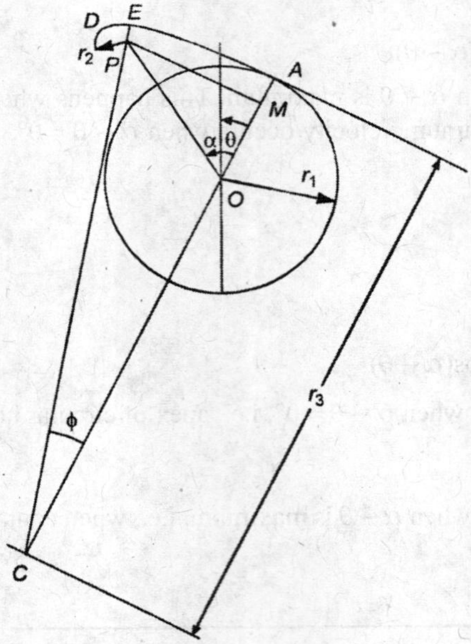

Fig.8.42 Circular arc cam operating a flat-faced follower

Example 8.19

The suction valve of a four stroke petrol engine is operated by a circular arc cam with a flat-faced follower. The lift of the follower is 10 mm; base circle diameter of the cam is 40 mm and the nose radius is 2.5 mm. The crank angle when suction valve opens is 4° after top dead centre and when the suction valve closes, the crank angle is 50° after bottom dead centre. If the cam shaft rotates at 600 rpm, determine (a) maximum velocity of the valve, and (b) maximum acceleration and retardation of the valve.

■ Solution

Given: $h = 10$ mm, $r_1 = 20$ mm, $r_2 = 2.5$ mm, $N = 600$ rpm,

$2\alpha = (180° - 4° + 50°)/2 = 226/2 = 113°$, $\alpha = 56.5°$

In Fig.8.44, $r_3 = CP + r_2 = OC + r_1$

$$CP = OC + 20 - 2.5 = OC + 17.5$$

$$OP + r_2 = r_1 + h$$

$$OP = r = 20 + 10 - 2.5 = 27.5 \text{ mm}$$

In Δ *COP*, we have

$$CP^2 = OC^2 + OP^2 - 2 \cdot OC \cdot OP \cos(180° - \alpha)$$

$$(OC + 17.5)^2 = OC^2 + (27.5)^2 - 2 \cdot OC \times 27.5 \times \cos 123.5°$$

$$OC = 96.9 \text{ mm}$$

$$CP = 114.4 \text{ mm}$$

$$r_3 = OC + r_1 = 96.9 + 20 = 116.9 \text{mm}$$

$$\frac{OP}{\sin\phi} = \frac{CP}{\sin(180° - \alpha)}$$

$$\sin\phi = (27.5/114.4)\sin 123.5° = 0.20045$$

$$\phi = 11.56°$$

$$\omega = \frac{2\pi \times 600}{60} = 62.83 \text{ rad/s}$$

Maximum velocity, $v_{max} = \omega(r_3 - r_1)\sin\phi = 62.83 \times (116.9 - 20)\sin 11.56° = 1.22$ m/s

$$f_{max} = -\omega^2(r_3 - r_1) = -(62.83)^2 \times 96.9 = -382.52 \text{ m/s}^2$$

Example 8.20

A cam having a lift of 10 mm operates the suction valve of a four stroke SI engine. The least radius of the cam is 20 mm and nose radius is 2.5 mm. The crank angle for the engine when suction valve opens is 4° after TDC and it is 50° after BDC when the suction valve closes. The crank shaft speed is 2000 rpm. The cam is of circular type with circular nose and flanks. It is integral with cam shaft and operates a flat faced-follower. Estimate (a) the maximum velocity of the valve, (b) the maximum acceleration and retardation of the valve, and (c) the minimum force to be exerted by the spring to overcome inertia of the valve parts which weigh 2 *N*.

■ Solution

Given: $r_1 = 20$ mm, $r_2 = 2.5$ mm, $h = 10$ mm, $N = 2000$ rpm,

$2\alpha = (180° - 4° + 50°)/2 = 226/2 = 113°$, $\alpha = 56.5°$

In Fig.8.42, $r_3 = CP + r_2 = OC + r_1$

$$CP = OC + 20 - 2.5 = OC + 17.5$$
$$OP + r_2 = r_1 + h$$
$$OP = r = 20 + 10 - 2.5 = 27.5 \text{ mm}$$

In ΔCOP, we have

$$CP^2 = OC^2 + OP^2 - 2 \cdot OC \cdot OP \cos (180° - a)$$
$$(OC + 17.5)^2 = OC^2 + (27.5)^2 - 2 \times OC \times 27.5 \times \cos 123.5°$$
$$OC = 96.98 \text{ mm}$$
$$CP = 114.48 \text{ mm}$$
$$r_3 = CP + r_2 = 114.48 + 2.5 = 116.98 \text{ mm}$$
$$\frac{OP}{\sin\phi} = \frac{CP}{\sin(180° - a)}$$
$$\sin\phi = (27.5/114.48)\sin 123.5° = 0.2003$$
$$\phi = 11.55°$$
$$\omega = \frac{2À\times 2000}{60} = 209.44 \text{ rad/s}$$
$$v_{max} = \omega(r_3 - r_1)\sin\phi = 209.44 \times (116.9 - 20)\sin 11.55° = 4.070 \text{ m/s}$$
$$f_{max} = \omega^2(r_3 - r_1) = -(209.44)^2 \times 96.98 = -4254 \text{ m/s}^2$$

Inertia force = spring force = 2 × 4254/9.81 = 867.3 N.

8.11.3 Circular Cam with Roller Follower

The circular arc cam with roller follower is shown in Fig.8.43.

Let $R = CF$ = radius of flank
r_1 = base circle radius of cam
r_2 = nose radius of cam
r_3 = roller follower radius
h = total lift or stroke
x = lift at the instant the cam has rotated by an angle θ
α = semi-angle of action of cam
ϕ = angle of action of cam from the beginning of rise to the point it leaves the flank
$\beta = \angle ODC$

(a) Roller Follower on Flank

The roller follower on flank is shown in Fig.8.43(a).

Let $R - r_1 = A$
$R + r_3 = B$
$CG = CD \sin\beta = OC \sin\theta$

or $(R + r_3)\sin\beta = (R - r_1)\sin\theta$

or $B\sin\beta = A\sin\theta$

$$\sin\beta = \left(\frac{A}{B}\right)\sin\theta$$

$$\cos\beta = \left[1 - \left(\frac{A}{B}\right)^2 \sin^2\theta\right]^{0.5}$$

Lift, $x = AB = OB - OA = OB - OF$

Cams

$$= (OD - BD) - OF = (DG - OG) - BD - OF$$

$$= (CD \cos\beta - r_3) - r_1$$

$$= (R + r_3)\cos\beta - (R - r_1)\cos\theta - (r_3 + r_1)$$

$$= B\cos\beta - A\cos\theta - (B - A)$$

$$= [B^2 - A^2\sin^2\theta]^{0.5} - A\cos\theta - (B - A) \tag{8.49}$$

Velocity, $v = \frac{dx}{dt} = \left(\frac{dx}{d\theta}\right)\cdot\left(\frac{d\theta}{dt}\right)$

$$= \omega\,[0.5\,(B^2 - A^2\sin^2\theta)^{-0.5}\cdot(-2A^2\sin\theta\cos\theta) + A\sin\theta]$$

$$= \omega A\left[\sin\theta - \frac{A\sin 2\theta}{2(B^2 - A^2\sin^2\theta)^{0.5}}\right] \tag{8.50}$$

Acceleration, $f = \frac{dv}{dt} = \left(\frac{dv}{d\theta}\right)\cdot\left(\frac{d\theta}{dt}\right)$

$$= \omega^2 A\left[\cos\theta - \frac{A\cos 2\theta}{\left(B^2 - A^2\sin^2\theta\right)^{0.5}} - \frac{A^3\sin^2 2\theta}{4\left(B^2 - A^2\sin^2\theta\right)^{0.5}}\right] \tag{8.51}$$

(b) Roller Follower on Nose

The roller follower on circular nose has been shown in Fig.8.43(b).

Let $PQ = r_2 + r_3 = l,\ OP = r$

$\angle PQO = \beta,\ \angle POQ = (\alpha - \theta) = \theta_1$

$$v = \omega\left[l\sin\theta_1 + \left\{l^2\sin 2\theta_1\right\}/\left\{l^2 - r^2\sin^2\theta_1\right\}\right]v = \omega\left[l\sin\theta_1 + \frac{l^2\sin\theta_1}{l^2 - r^2\sin^2\theta_1}\right] \tag{8.52}$$

$$f = \omega^2\left[-l\cos\theta_1 - \frac{l^2\cos 2\theta_1}{\left\{l^2 - r^2\sin^2\theta_1\right\}^{0.5}} - \frac{l^4\sin^2 2\theta_1}{4\left(l^2 - r^2\sin^2\theta_1\right)^{1.5}}\right] \tag{8.53}$$

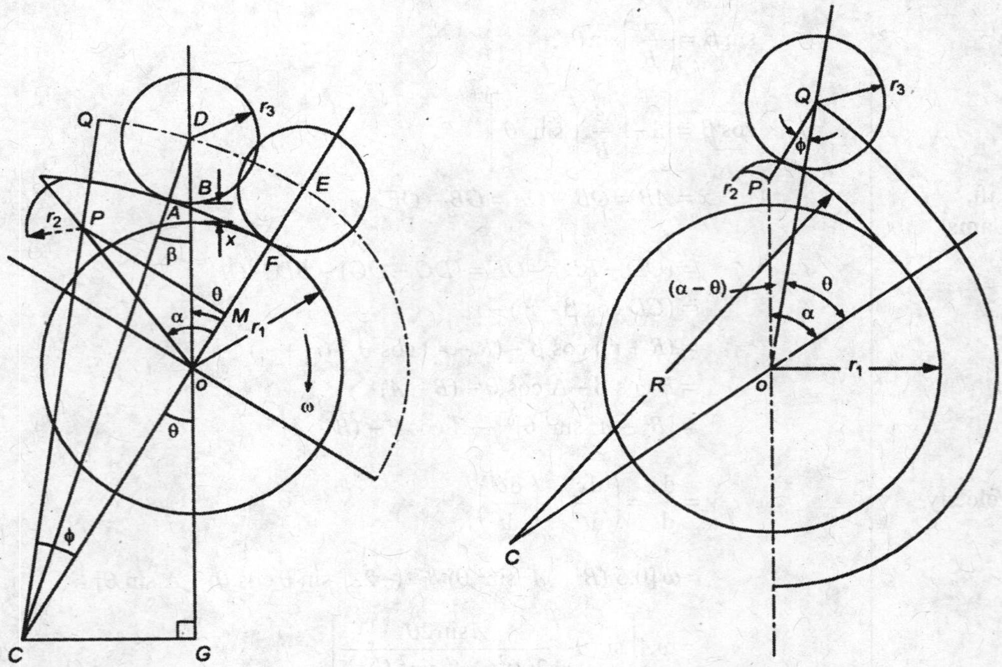

(a) Roller follower on flank

(b) Roller follower on circular nose

Fig.8.43 Circular arc cam operating roller follower

Example 8.21

A cam of circular arc type is to operate a flat-faced follower of a four-stroke engine. The exhaust valve opens 50° before top dead centre and closes 15° after bottom dead centre. The valve lift is 10 mm, base circle radius of cam 20 mm and nose radius 3 mm. Calculate the maximum velocity, acceleration and retardation, if cam shaft rotates at 1800 rpm. Also calculate the minimum force required, which must be exerted by the spring in order to overcome the inertia of moving parts of mass 0.25 kg.

Solution

Given:

$$2\alpha = \frac{50° + 180° + 15°}{2} = 122.5°$$

$$\alpha = 61.25°$$

$$\omega = \frac{2\pi \times 1800}{60} = 188.5 \text{ rad/s}$$

In Fig.8.44, we have

$$CP = CE - EP = (OC + OG) - EP = OC + 20 - 3 = OC + 17$$

$$OP = OD - PD = OG + R - r_2 = r_1 + h - r_2 = 20 + 10 - 3 = 27 \text{ mm}$$

$$CP^2 = OC^2 + OP^2 - 2\, OC \times OP \cos 118.75°$$

$$(OC + 17)^2 = OC^2 + 27^2 - 2 \times OC \times 27 \times \cos 118.75°$$

$$OC = 55 \text{ mm}$$

$$\frac{OC}{\sin 61.25^\circ} = \frac{OP}{\sin \beta} = \frac{CP}{\sin 118.75^\circ}$$

$$\frac{27}{\sin \beta} = \frac{OC + 17}{\sin 118.75^\circ}$$

$$\sin \beta = 27 \times \sin 118.75^\circ / 72 = 0.32877$$

$$\beta = 19.19^\circ$$

$$R = OC + r_1 = 55 + 20 = 75 \text{ mm}$$

$$v_{max} = \omega (R - r_1) \sin \beta = 188.5 \times 0.055 \times \sin 19.19^\circ = 3.4 \text{ m/s}$$

$$f_{max} = \omega^2 (R - r_1) = (188.5)^2\, 0.055 = 1954.3 \text{ m/s}^2$$

$$\text{Maximum retardation} = \omega^2 OP = (188.5)^2 \times 0.027 = 959.4 \text{ m/s}^2$$

$$\text{Force, } F = 0.25 \times 959.4 = 239.84 \text{ N}$$

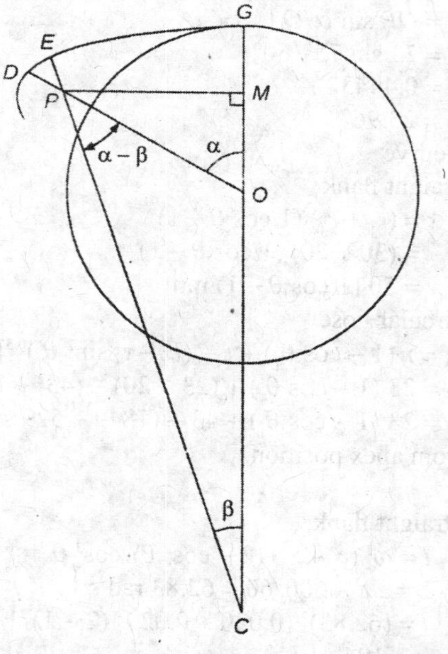

Fig.8.44 Circular arc cam operssating a flat-faced follower

Example 8.22

The particulars of a symmetrical tangent cam operating a roller follower are given below:

Least radius of cam = 30 mm; roller radius = 20 mm; angle of ascent = 75°; total lift = 20 mm; speed of cam shaft = 600 rpm. Calculate the (a) principal dimensions of the cam, (b) the equation of the displacement curve when the follower is in contact with straight flank and circular nose and (c) acceleration of the follower at the follower at the beginning of lift, where straight flank merges into the circular nose and at the apex of the circular nose. Assume that there is no dwell between ascent and return.

■ **Solution**

Here $r_1 = 30$ mm, $r_3 = 20$ mm, $\alpha = 75^\circ$, lift = 20 mm and $N = 600$ rpm.

(a) With reference to Fig 8.45, we have

$$OP + r_2 = 30 + 20$$
$$OP = 50 - r_2$$
$$OQ + r_2 = 30$$
$$OQ = 30 - r_2$$
$$\cos\alpha = OQ/OP$$
$$\cos 75° = (30 - r_2)/(50 - r_2)$$
$$0.25882 = (30 - r_2)/(50 - r_2)$$
$$12.941 - 0.25882\, r_2 = 30 - r_2$$
$$\text{Nose radius, } r_2 = 14.059/0.7412 = 23 \text{ mm}$$

Distance between cam and nose centre,

$$r = OP = 50 - 23 = 27 \text{ mm}$$
$$\tan\phi = RA/OA = PQ/OA$$
$$= OP\sin\alpha/OA$$
$$= 23\sin 75°/50°$$
$$= 0.44432$$
$$\phi = 23.96°$$

(b) Equation of displacement curve

1. When contact is with straight flank

$$x = (r_1 + r_3)(1/\cos\theta - 1)$$
$$= (30 + 20)(1/\cos\theta - 1)$$
$$= 50(1/\cos\theta - 1) \text{ mm}$$

2. When contact is with circular nose

$$x = r(1 - \cos\theta_1) + \ell - (\ell^2 - r^2\sin^2\theta_1)^{0.5}]$$
$$= 23(1 - \cos\theta_1) + (23 + 20) - (43^2 - 23^2\sin^2\theta_1)^{0.5}]$$
$$= 23(1 - \cos\theta_1) + 43 - (1849 - 529\sin^2\theta_1)^{0.5}]$$

where θ_1 is measured from apex position.

(c) Acceleration of follower

When in contact with straight flank,

$$f = \omega^2(r_1 + r_3)(2 - \cos^2\theta)/\cos^3\theta$$
$$\omega = 2\pi \times 600/60 = 62.83 \text{ rad/s}$$

When $\theta = 0°$,

$$f = (62.83)^2(0.030 + 0.020)(2 - 1)/1$$
$$= 194.4 \text{ m/s}^2$$

When in contact with straight flank, $\theta = \phi$.

$$f = \omega^2(r_1 + r_3)(2 - \cos^2\phi)/\cos^3\theta$$
$$= (62.83)^2(0.03 + 0.02)(2 - \cos^2 23.96°)/\cos^3 23.96°$$
$$= 301.3 \text{ m/s}^2$$

When contact is on circular nose,

$$\theta_1 = \alpha - \phi = 75 - 23.96 = 51.04°$$
$$f = \omega^2 r[\cos\theta_1 + (l^2 r\cos 2\theta_1 + r^3\sin^4\theta_1)/(l^2 - r^2\sin^2\theta_1)^{1.5}]$$
$$= (62.83)^2 \times 0.023[\cos 51.04° + (0.043^2 \times 0.023\cos 102.08°$$
$$+ 0.023^3\sin^4 51.04°)/(0.043^2 - 0.023^2\sin^2 51.04°)^{1.5}]$$
$$= 50.33 \text{ m/s}^2$$

When at apex, $\theta_1 = 0°$

$$f = \omega^2 r[1 + r/\ell]$$
$$= (62.83)^2 \times 0.023[1 + 0.023/0.043]$$
$$= 139.36 \text{ m/s}^2$$

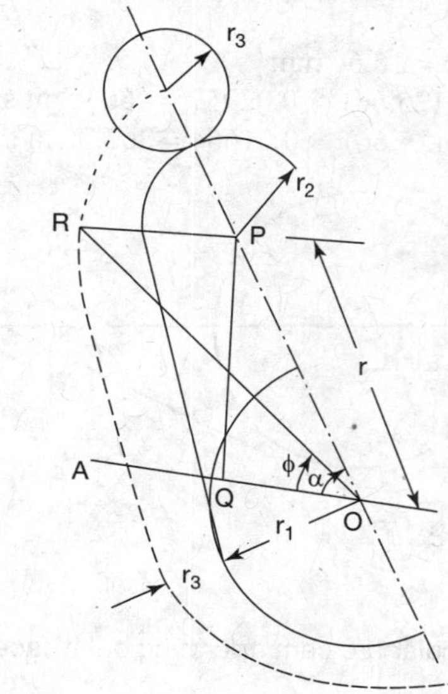

Fig.8.45 Tangent cam operating a roller follower

Example 8.23

A flat-faced mushroom follower is operated by a symmetrical cam with circular arc flank and nose profile. The axis of tappet passed through the cam axis. Total angle of action is 162°, lift 10 mm and base circle diameter 40 mm. Period of acceleration is half the period of retardation during the lift. The cam rotates at 1200 rpm. Determine (a) nose and flank radii and (b) maximum acceleration and retardation during lift.

■ Solution

$$\alpha = 162/2 = 81°,\ h = 10 \text{ mm},\ r_1 = 40/2 = 20 \text{ mm},\ N = 1200 \text{ rpm}$$

Acceleration takes place on flank and retardation on nose.

Angle of action on flank, $\phi = 81/3 = 27°$

Angle of action on nose, $\alpha = 2 \times 81/3 = 54°$

$$\omega = 2\pi \times 1200/60 = 125.66 \text{ rad/s}$$

In *COP* (Fig.8.46), we have

$$CP = CE - EP = OC + OG - EP = OC + 20 - r_2$$

$$OP = OD - PD = OG + h - r_2 = 20 + 10 - r_2 = 30 - r_2$$

$$\frac{OC}{\sin 54°} = \frac{OP}{\sin 27°} = \frac{CP}{\sin 99°}$$

$$\frac{OC}{0.809} = \frac{30 - r_2}{0.454} = \frac{OC + 20 - r_2}{0.9877}$$

$$OC = 0.819\,(OC + 20 - r_2)$$

$$OC = 90.5 - 4.525\, r_2 \qquad \text{(a)}$$

Also $\quad 30 - r_2 = 0.45965\,(OC + 20 - r_2)$

or $\quad OC = 45.267 - 1.1756\, r_2 \qquad$ (b)

From Eqs. (a) and (b), we get

$$r_2 = 13.36 \text{ mm}$$

$$OC = 29.57 \text{ mm}$$
$$OP = 30 - 13.36 = 16.64 \text{ mm}$$

Maximum acceleration $= \omega^2\, OC = (125.66)^2 \times 0.02957 = 466.95 \text{ m/s}^2$

Maximum retardation $= \omega^2\, OP = (125.66)^2 \times 0.01664 = 262.75 \text{ m/s}^2$

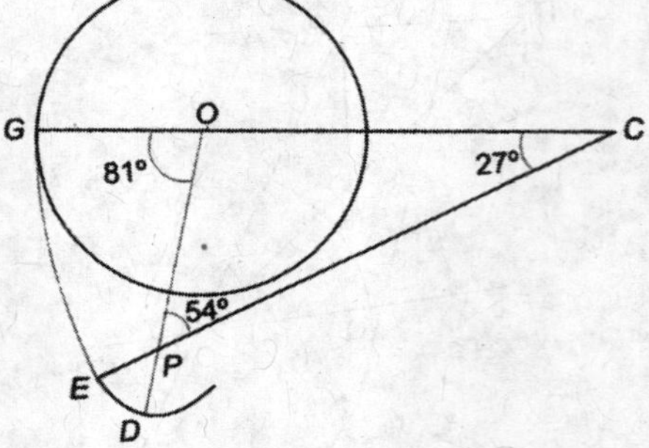

Fig.8.46 Circular arc cam operating a flat-faced follower

Example 8.24

For a flat-faced follower $\alpha = 75°$, $\phi = 40°$ and the cam is of circular arc type with $r_1 = 30$ mm, $r_2 = 10$ mm and $r = 40$ mm, $N = 600$ rpm. Combined mass of follower and valve is 5 kg. Calculate the spring force required to close the valve.

■ Solution

$$\omega = 2\pi \times 600/60 = 62.83 \text{ rad/s}$$

In ΔCOP (Fig.8.49), we have

$$CP = OC + OG - r_2 = OC + 30 - 10 = OC + 20$$
$$OP = r = 40 \text{ mm}$$
$$CP^2 = OC^2 + OP^2 - 2OC \cdot OP\cos(180° - \alpha)$$
$$(OC + 20)^2 = OC^2 + 40^2 - 2 \times OC \times 40 \times \cos 105°$$
$$OC = 62.17 \text{ mm}$$

At the start of the nose, acceleration, $f_{sn} = -\omega^2 \times r\cos(\alpha - \phi)$

$$= -(62.83)^2\, 0.04 \times \cos(75° - 40°)$$
$$= -129.35 \text{ m/s}^2$$

At the top of the nose, acceleration, $f_{tn} = -\omega^2 \times r$

$$= -(62.83)^2 - 0.04 = -157.9 \text{ m/s}^2$$

Spring force required to maintain contact between the follower and cam,

$$F = m \cdot f_{tn} = 5 \times 157.9 = 789.5 \text{ N}$$

Acceleration at beginning of the flank,

$$f_{sf} = \omega^2 \times OC\cos\phi$$
$$= (62.83)^2 \times 0.06217 \times \cos 0°$$
$$= 245 \text{ m/s}^2$$

Acceleration at the end of the flank, $f_{ef} = \omega^2\, OC\cos\phi$

$$= (62.83)^2 \times 0.06217 \times \cos 40°$$
$$= 188 \text{ m/s}^2$$

Force required to stop the acceleration of 188 m/s² and decelerate at 129.35 m/s²,

$$F = m\,(f_{ef} - f_{sn}) = 5(188 + 129.35) = 1586.75 \text{ N}$$

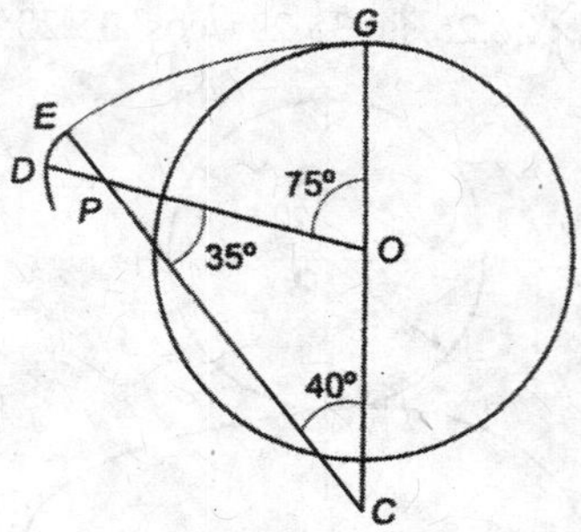

Fig.8.47 Circular arc cam operating a flat-faced follower

Example 8.25

A reciprocating roller follower has cycloidal motion and its stroke of 30 mm is completed in 90° of the cam rotation. The follower is offset against the direction of rotation by 6.25 mm and the radius of the roller is 12.5 mm. Determine the base circle radius which would limit the pressure angle to 30°. Also discuss the conditions that would lead to undercutting in the present case.

■ Solution

$h = 30$ mm, $\beta = 90°$, $e = 6.25$ mm, $R_r = 12.5$ mm, $\alpha_{max} = 30°$, $r_1 = ?$

For cycloidal motion, $y = f(\theta) = h\,[\theta/\beta - (1/2)\sin(2\pi\,\theta/\beta)]$

$=30[2\theta/-(1/2)\sin 4\theta]$

$= (15/\pi)\,[4\theta - \sin 4\theta]$

$df/d\theta = (60/\pi)\,(1 \cos 4\theta)$

$d^2f/d\theta^2 = (240/\pi)\sin 4\theta$

$\tan\alpha_{max} = d^2f/d\theta^2/df/d\theta = (240/\pi)\sin 4\theta/(60/\pi)\,(1-\cos 4\theta)$

$\tan 30° = 4\cot 2\theta$

$\theta = 40.89°$

$\tan\alpha = [df/d\theta - e]/[f(\theta) + (r_1^2 - e^2)^{0.5}]$

$= [(60/\pi)\,(1 - \cos 4\theta) - e]/[(15/\pi)(4\theta - \sin 4\theta) + (r_1^2 - e^2)^{0.5}]$

$\tan 30° = [(60/\pi)(1 - \cos 163.56°) - 6.25]/[(15/\pi)(2.8547 - \sin 163.56°) + (r_1^2 - 39.06)^{0.5}]$

$r_1 = 42.22$ mm

Example 8.26

The cam shown in Fig 8.48 rotates about O at an uniform speed of 500 rpm and operates a follower attached to the roller with centre Q. The path of Q is a straight line passing through O. Draw the time-lift diagram of the roller centre on a base of 10 mm equaling 0.005 s and to a vertical scale of four times full size for a movement of 180° from the position shown. Find also the cam angle for which the velocity of the follower is a maximum and determine the magnitude of the velocity.

■ Solution

$N = 500$ rpm, $r_3 = 11.25$ mm, $r_1 = 15$ mm, $r_2 = 7.5$ mm, $r = 20$ mm

$\omega = 2\pi \times 500/60 = 52.36$ rad/s

For maximum velocity, $\theta = \phi$

$\cos\alpha = (r_1 - r_2)/r = 7.5/20 = 0.375$

$\alpha = 67.97°$, $2\alpha = 135.94°$

$\tan\phi = r\sin\alpha/(r_1 + r_3) = 20\sin 67.97°/26.25 = 0.706$

$\phi = 25.23°$

$v = \omega (r_1 + r_3) \sin \phi / \cos^2 \phi = 52.36 \times 26.25 \sin 35.23° / \cos^2 35.23° = 1188.5$ mm/s

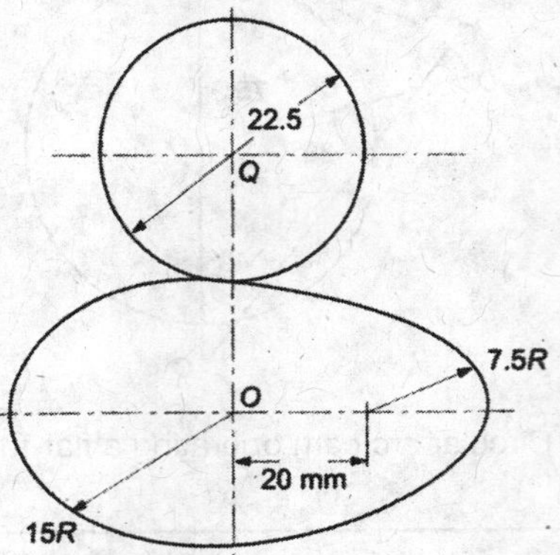

Fig.8.48

8.12 RADIUS OF CURVATURE AND UNDERCUTTING

There is no restriction on the radius of curvature of the cam profile with a knife-edge follower. The cam profile must be convex everywhere for a flat-faced follower. In the case of a roller follower, the concave portion of the cam profile must have a radius of curvature greater than that of the roller to ensure proper contact along the cam profile.

To determine the pitch surface of a disc cam with radial roller follower, let the displacement R of the centre of the follower from the centre of the cam be given by (Fig.8.49):

$$R = R_0 + f(\theta) \tag{8.54}$$

where R_0 = minimum radius of the pitch surface of the cam.

$f(\theta)$ = radial motion of the follower as a function of cam angle.

Once the value of R_0 is known, the polar coordinates of the centres of the roller follower can be generated.

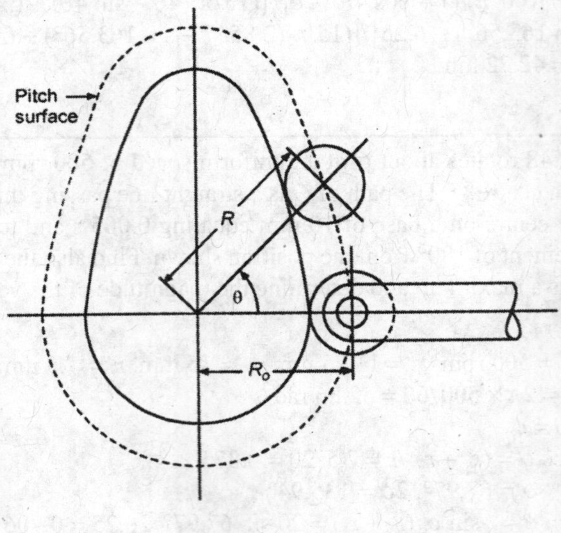

Fig.8.49 Disc cam operating roller follower

8.12.1 Kloomok and Muffley Method

Let ρ = radius of curvature of the pitch surface

R_r = radius of roller

These values are shown in Fig.8.50 together with the radius of curvature ρ_c of the cam surface, ρ is held constant and R_r is increased so that ρ_c decreases. If this continued until R_r equals ρ, then ρ_c will be zero and the cam becomes pointed as shown in Fig.8.51(a). As R_r is further increased, the cam becomes undercut as shown in Fig.8.51(b), and the motion of the follower will not be as prescribed. Therefore, to prevent a point or an undercut from occurring on the cam profile, R_r must be less than ρ_{min}, where ρ_{min} is the minimum value of ρ over the particular segment of profile being considered. It is impossible to undercut a concave portion of a cam.

The radius of curvature at a point on a curve can be expressed as:

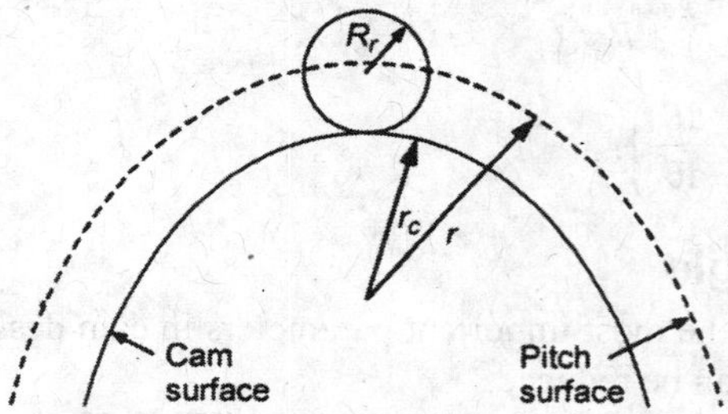

Fig.8.50 Cam and pitch surfaces

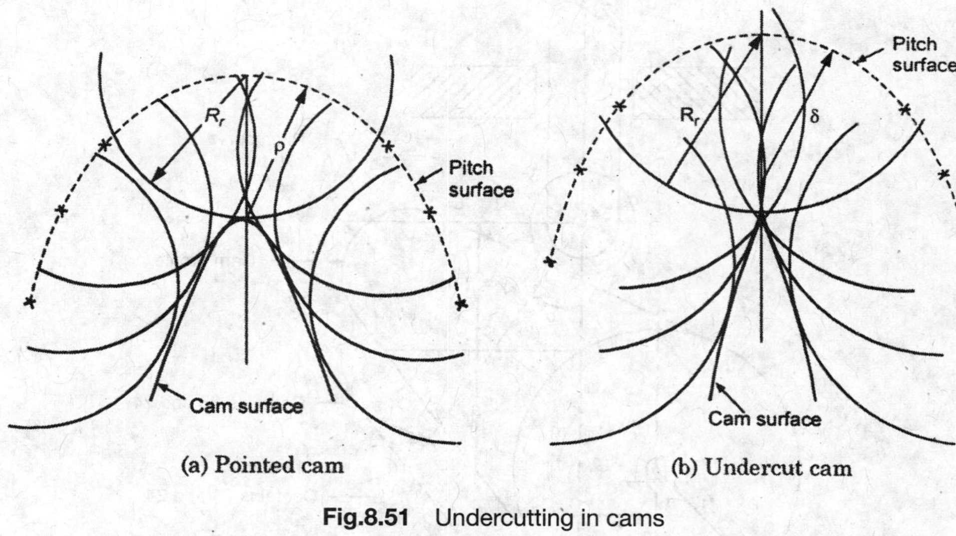

Fig.8.51 Undercutting in cams

$$\rho = \frac{\left[R^2 + \left(\frac{dR}{d\phi}\right)^2\right]^{3/2}}{\left[R^2 + 2\left(\frac{dR}{d\phi}\right)^2 - r\left(\frac{d^2R}{d\phi^2}\right)\right]}$$

Where $R = f(\phi)$. Here $f(\theta) = f(\phi)$.

$$\frac{dR}{d\theta} = f'(\theta)$$

$$\frac{d^2 R}{d\theta^2} = f''(\theta)$$

Hence
$$\rho = \frac{[R^2 + \{f'(\theta)^2\}]^{3/2}}{[R^2 + 2\{f''(\theta)\}^2 - R\{f''(\theta)\}]} \quad (8.55)$$

For a convex cam surface, $R_r \le \rho_{min}$
For concave cam surface, $\rho + R_r \ge R_{cutter}$

Pressure angle, $\alpha = \tan^{-1}\left(\frac{1}{R}\cdot\frac{dR}{d\theta}\right)$ (8.56)

8.12.2 Pressure Angle

The pressure angle is one of the most important parameters in cam design. By increasing the size of the cam, the pressure angle can be reduced.

Consider a cam with offset roller follower, as shown in Fig.8.52

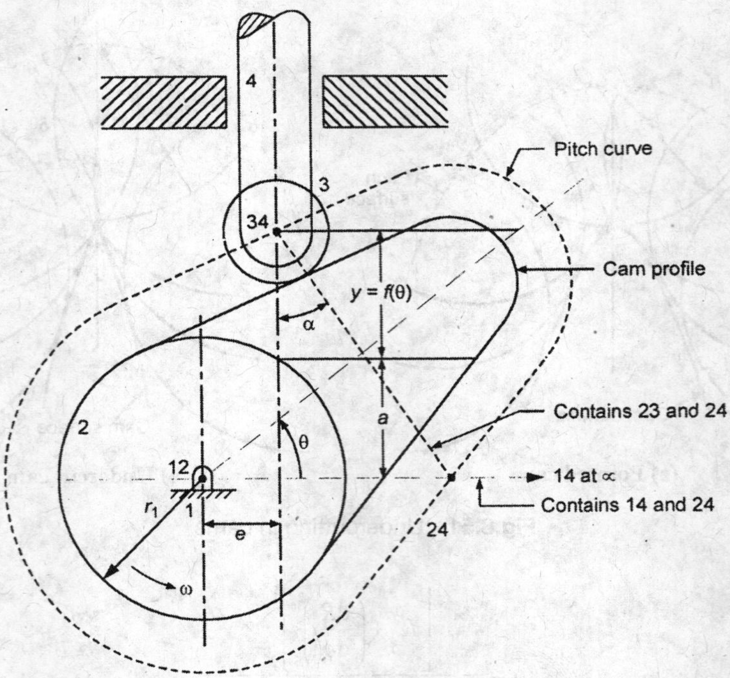

Fig.8.52 Cam with offset roller follower

Let α = pressure angle
r_1 = prime circle radius
e = offset
$y = f(\theta)$
θ = angle of cam rotation
ω = angular velocity of cam

The cam mechanism has four elements, namely, the fixed link 1, cam 2, roller 3, and follower rod 4. The instantaneous centres are :

12 : at O
34 : at roller centre
14 : at infinity
23 : lies on the common normal at the point of contact of the roller and cam surface
24 : at the intersection of common normal at the point of contact and horizontal axis on which 14 lies.

As the movement of the follower rod is pure translation, all points on it have the same velocity. Thus, the velocity of follower during rise is:

$$\frac{\mathrm{d}y}{\mathrm{d}t} = v_{24} = \omega(12-24) = \omega\,[e+(a+y)\tan a] \tag{8.57}$$

where $a = (r^2_1 - e^2)^{0.5}$

$$\tan\alpha = \frac{\left[\dfrac{\mathrm{d}y}{\mathrm{d}t} - e\right]}{[f(\theta)+(r^2_1 - e^2)^{0.5}]}$$

$$= \frac{\left[\dfrac{\mathrm{d}f}{\mathrm{d}\theta} - e\right]}{[f(\theta)+(r^2_1 - e^2)^{0.5}]} \tag{8.58}$$

During the return, we get

$$\tan\alpha = \frac{\left[\left|\dfrac{\mathrm{d}f}{\mathrm{d}\theta}\right| + e\right]}{\left[f(\theta)+(r_1^2 - e^2)^{0.5}\right]} \tag{8.59}$$

For maximum pressure angle, using rise, we have

$$\frac{\mathrm{d}(\tan\alpha)}{\mathrm{d}\theta} = [f(\theta)+(r^2_1 - e^2)^{0.5}]\left(\frac{\mathrm{d}^2 f}{\mathrm{d}\theta^2}\right) - \left(\frac{\mathrm{d}f}{\mathrm{d}\theta} - e\right)\frac{\mathrm{d}f}{\mathrm{d}\theta} = 0$$

$$\frac{\left(\dfrac{\mathrm{d}f}{\mathrm{d}\theta} - e\right)}{\left[f(\theta)+(r^2_1 - e^2)^{0.5}\right]} = \frac{\left(\dfrac{\mathrm{d}^2 f}{\mathrm{d}\theta^2}\right)}{\left(\dfrac{\mathrm{d}f}{\mathrm{d}\theta}\right)} = 0 \tag{8.60}$$

Solve Eq. (8.60) to get the value of θ_0. Then

$$\alpha_{max} = \tan^{-1}\left[\frac{\left(\dfrac{d^2 f}{d\theta^2}\right)}{\left(\dfrac{df}{d\theta}\right)}\right]_{\theta=\theta_0} \tag{8.61}$$

Example 8.27

An off-set translating roller follower is driven by a SHM cam rotating at 450 rpm. The lift of the follower is 25 mm during 120° cam rotation. The base circle radius of cam is 40 mm. The off-set is 10 mm. Calculate (a) the pressure angle at a cam angle of 60°, and (b) the pressure angle for radial follower.

■ Solution

Given: $N = 450$ rpm, $h = 25$ mm, $\theta_1 = 120°$, $\theta = 60°$, $e = 10$ mm, $r_1 = 40$ mm

$$\omega = \frac{2\pi \times 450}{60} = 47.124 \text{ rad/s}$$

For SHM,

$$x = \frac{h}{2}\left[1 - \cos\left(\frac{\pi\theta}{\theta_1}\right)\right]$$

$$= \frac{25}{2}\left[1 - \cos\left(\frac{\pi \times 60}{120}\right)\right] = 12.5 \text{ mm}$$

Velocity,

$$v = \frac{dx}{dt} = \frac{h}{2} \times \frac{\pi}{\theta_1} \times \omega \sin\left(\frac{\pi\theta}{\theta_1}\right)$$

$$= \frac{25}{2} \times \frac{\pi \times 180}{\pi \times 120} \times 47.124 \times \sin\left(\frac{\pi \times 60}{120}\right) = 883.6 \text{ mm/s}$$

Now

$$\tan a = \frac{\dfrac{v}{\omega} - e}{\sqrt{r_1^2 - e^2} + x}$$

$$= \frac{\dfrac{883.6}{47.124} - 10}{\sqrt{40^2 - 10^2} + 12.5} = 0.171$$

$$\alpha = 9.693°$$

For a radial follower, $e = 0$

$$\tan a = \frac{v}{\omega(r_1 + x)} = \frac{883.6}{47.124(40 + 12.5)} = 0.357$$

$$a = 19.654°$$

8.13 CAM SIZE

The cam size is defined by the following parameters:

1. Pressure angle
2. Radius of curvature of cam profile
3. Hub size.

The following methods may be used to reduce the pressure angle:

1. Increase the diameter of the base circle.
2. Increase the angle of rotation of the cam, thereby lengthening the pitch curve for the specified follower displacement. The cam profile becomes flatter and the pressure angle becomes smaller.
3. Select the motion curve for a smaller pressure angle.
4. By changing the offset of the follower.

Summary for Quick Revision

1 A cam is a mechanical member used to impart desired motion to a follower by direct contact.

2 The essential components of a cam mechanism are: cam, follower and frame.

3 Cams can be classified as: wedge and flat cams, radial and offset cams, cylindrical cams, spiral cams, conjugate cams, spheroidal cams and spherical cams.

4 The followers can be classified as: knife edge, roller, flat faced and spherical faced; translating, oscillating and rotating; radial and offset followers.

5 Pressure angle of a cam is the angle between the direction of follower motion and a normal to the pitch curve.

6 Lift is the maximum travel of the follower from the lowest position to the topmost position.

7 Follower motions can be: simple harmonic, uniform acceleration and deceleration, uniform velocity, parabolic motion and cycloidal motion.

(a) Simple Harmonic Motion

Displacement, $y = (h/2)\,[1 - \cos(\pi\theta/\phi)]$

Velocity, $v = (h/2)\,(\pi\,\omega/\phi)\sin(\pi\theta/\phi)$

Acceleration, $f = (h/2)\,(\pi\omega/\phi)^2 \cos(\pi\theta/\phi)$

(b) Motion with Uniform Acceleration and Deceleration

Displacement, $y = ft2/2$

Velocity, $v = 2hw/\theta$

Acceleration, $f = 4\,\omega^2\,h/\theta^2$

(c) Motion with Uniform Velocity

Displacement, $y = h\,\theta/\phi$

Velocity, $v = h\,\omega/\phi$

Acceleration, $f = 0$

(d) Parabolic Motion

For the first half, $y = 2\,h\,(\theta/\phi)^2$

$v = 4\,h\,\omega\theta/\phi^2$

$f = 4\,h\,(\omega/\phi)^2 = \text{const}$

For the second half, $y = h\,[1 - 2\,(1 - \theta/\phi)^2]$

$v = (4h\omega/\beta)\,(1 - \theta/\phi)$

$f = -\,4h(\omega/\phi)^2$

(e) Cycloidal Motion

$y = (h/\pi)\,[\pi\theta/\phi - (1/2)\sin(2\,\pi\,\theta/\phi)]$

$v = (h\,\omega/\phi)\,[1 - \cos(2\pi\theta/\phi)]$

$f = (2h\pi)\,(\omega/\phi)^2 \sin(2\pi\theta/\phi)$

Cycloidal motion is the most ideal for high speed follower motion.

8 Tangent cam with roller follower

Let r = distance between cam and nose centres

r_1 = least radius of cam

r_2 = nose radius

r_3 = roller radius

$l = r_2 + r_3$

α = angle of ascent

ϕ = angle of contact of cam with straight flank

(a) Roller in contact with straight flank

Lift, $x = (r_1 + r_3)\,(1 - \cos\theta)/\cos\theta$

Velocity, $v = \omega(r_1 + r_3)\,(-\sin\theta/\cos^2\theta)$

Maximum velocity occurs at $\theta = \phi$

$v_{max} = \omega(r_1 + r_2)\,(-\sin\phi/\cos^2\phi)$

Acceleration, $f = \omega^2(r_1 + r_2)\,(2 - \cos^2\theta)/\cos^3\theta$

Minimum acceleration occurs at $\theta = 0°$.

$f_{min} = \omega^2\,(r_1 + r_3)$

(b) Follower in contact with circular nose

Let $\theta_1 = \alpha - \theta$

For a slider crank chain the displacement from top dead centre is given by,

$x = r\,[\,(1 - \cos\theta) + n - (n^2 - \sin^2\theta)^{0.5}]$

where $n = r/l$.

For the cam mechanism,

$x = r\,(1 - \cos\theta_1) + l - (l^2 - r^2\sin^2\theta_1)^{0.5}]$

$v = \omega\,[r\sin\theta_1 + r^2\sin 2\theta_1/2\,(l^2 - r^2\sin^2\theta_1)^{0.5}]$

$f = \omega^2 r\,[\cos\theta_1 + (l^2 r\cos 2\theta_1 + r^3\sin^4\theta_1)/(l^2 - r^2\sin^2\theta_1)^{1.5}]$

9 Circular arc cam operating flat-faced follower

(a) Follower in contact with circular flank

Let r = distance between cam and nose centers

r_1 = least circle radius

r_2 = nose circle radius

r_3 = flank circle radius

α = angle of ascent

ϕ = angle of contact on circular flank

Displacement of follower, $x = (r_3 - r_1)\,(1 - \cos\theta)$

Velocity, $v = (r_3 - r_1)\,\omega\sin\theta$

Maximum velocity occurs at $\theta = \phi$.

$v_{max} = (r_3 - r_1)\,\omega\sin\phi$

Acceleration, $f = (r_3 - r_1)\,\omega^2\cos\theta$

Maximum acceleration occurs at $\theta = 0°$.

$f_{max} = (r_3 - r_1)\,\omega^2$

Minimum acceleration occurs at $\theta = \phi$.

$f_{min} = (r_3 - r_1)\,\omega^2\cos\phi$

(b) Follower in contact with circular nose

$x = r_2 + r \cos(\alpha - \theta) - r_1$

$v = \omega r \sin(\alpha - \theta)$

Velocity is maximum when $\alpha - \theta$ is maximum. This happens when contact changes from circular flank to circular nose. Minimum velocity occurs when $\alpha - \theta = 0°$, i.e. at the apex of circular nose.

$v_{min} = 0$

Acceleration, $f = -\omega^2 r \cos(\alpha - \theta)$

Maximum acceleration occurs when $\alpha - \theta = 0°$, i.e. apex of circular nose.

$f_{max} = -\omega^2 r$

Minimum acceleration occurs when $\alpha - \theta$ is maximum, i.e. when contact changes from circular flank to circular nose.

10 Circular arc cam with roller follower

Let R = radius of flank

r_1 = base circle radius of cam

r_2 = nose radius of cam

r_3 = roller follower radius

h = total lift or stroke

x = lift at the instant the cam has rotated by an angle θ

α = semi – angle of action of cam

ϕ = angle of action of cam from the beginning of rise to the point it leaves the flank

$\beta = \angle ODC$

(a) Roller Follower on Flank

Let $R - r_1 = A$

$R + r_3 = B$

Lift, $x = [B^2 - A^2 \sin^2 \theta]^{0.5} - A \cos \theta = (B - A)$

Velocity, $v = \omega A [\sin \theta - \{(A \sin 2\theta) / 2 (B^2 - A^2 \sin^2 \theta)^{0.5}\}]$

Acceleration, $f = \omega^2 A[\cos \theta - (A \cos 2\theta)/(B^2 - A^2 \sin^2 \theta)^{0.5} - (A^3 \sin^2 2\theta)/\{4(B^2 - A^2 \sin^2 \theta)^{0.5}\}]$

(b) Roller Follower on Nose

Let $r_2 + r_3 = \phi$, $(\alpha - \theta) = \theta_1$

$v = \omega [l \sin \theta_1 + \{l^2 \sin 2\, \theta_1\} / \{l^2 - r^2 \sin^2 \theta_1\}]$

$f = \omega^2 [-l \cos \theta_1 - \{l^2 \cos 2\theta_1\} / \{l^2 - r^2 \sin^2 \theta_1\}^{0.5} - \{l^4 \sin^2 2\theta_1\} / 4\{l^2 - r^2 \sin^2 \theta_1\}^{1.5}]$

11 Radius of curvature

The cam profile must be convex everywhere for a flat-face follower. In the case of a roller follower the concave portion of the cam profile must have a radius of curvature greater than that of the roller to ensure proper contact along the cam profile.

12 Pressure angle

Pressure angle is the angle between the direction of the follower motion and a normal to the pitch curve. The following methods may be used to reduce the pressure angle:

1. Increase the diameter of the base circle.
2. Increase the angle of rotation of the cam, thereby lengthening the pitch curve for the specified follower displacement. The cam profile becomes flatter and the pressure angle becomes smaller.
3. Select the motion curve for a smaller pressure angle.
4. By changing the offset of the follower.

13 Cam Size

The cam size is defined by the following parameters:

1. Pressure angle
2. Radius of curvature of cam profile
3. Hub size.

Multiple Choice Questions

1 The pitch point on a cam is
(a) any point on the pitch curve
(b) the point on cam pitch curve having the maximum pressure angle
(c) any point on pitch circle
(d) a point at a distance equal to pitch circle radius from the centre.

2 In its simplest form, a cam mechanism consists of following number of links
(a) 1 (b) 2 (c) 3 (d) 4.

3 The type of follower used in automobiles is
(a) knife edge (b) roller
(c) mushroom with flat face (d) mushroom with spherical face.

4 The minimum radius circle drawn to the cam profile is called
(a) prime circle (b) base circle (c) pitch circle (d) pitch curve.

5 The reference point on the follower for the purpose of laying the cam profile is known as
(a) pitch point (b) trace point (c) roller centre (d) cam centre.

6 The pressure angle of a cam is defined as the angle between the line of motion of the follower and the
(a) tangent on the pitch curve (b) normal on the pitch curve
(c) tangent on the cam profile (d) normal on the cam profile.

7 The cam profile and pitch curves are same for
(a) roller follower (b) knife edge follower
(c) mushroom follower (d) flat-faced follower.

8 The size of the cam depends upon
(a) base circle (b) prime circle (c) pitch circle (d) pitch curve.

9 The point on the cam with maximum pressure angle is called the
(a) pitch point (b) trace point (c) cam centre (d) roller centre.

10 Pressure angle of a cam is directly proportional to
(a) base circle diameter (b) pitch circle diameter
(c) prime circle diameter (d) lift of cam.

11 The throw of a cam is the maximum lift of the follower from
(a) base circle (b) pitch circle (c) prime circle (d) pitch curve.

12 For S.H.M cam and follower, the maximum velocity is
(a) $\pi wh/(2\theta)$ (b) $\pi\omega h/\theta$ (c) $\pi\omega h/4\theta$ (d) $\pi\omega h/(3\theta)$
where h = lift.

13 For S.H.M cam and follower, the maximum acceleration is
(a) $(\pi\omega/\theta)^2(h/2)$ (b) $(\pi\omega/\theta)^2h$ (c) $(\pi\omega/\theta)^2(h/4)$ (d) $2h(\pi/\omega)^2$.

14 For a uniformly accelerated cam, the maximum velocity is
(a) $2(\omega h/\theta)$ (b) $0.5(\omega h/\theta)$ (c) $(\omega h/\theta)$ (d) $4(\omega h/\theta)$.

15 For a uniformly accelerated cam, the maximum acceleration is
(a) $2(\omega/\theta)^2h$ (b) $0.5(\omega/\theta)^2h$ (c) $4(\omega/\theta)^2h$ (d) $0.25(\omega/\theta)^2h$.

Answers

1. (b) 2. (c) 3. (d) 4. (b) 5. (b) 6. (b) 7. (b) 8. (a) 9. (a) 10. (a) 11. (a) 12. (a)
13. (a) 14. (a) 15. (c)

Review Questions

1 What is a cam? What is its use?

2 How cams can be classified?

3 What are the various types of followers?

4 Name the different motions that a follower can have.

5 Differentiate between (a) base circle and prime circle and (b) cam angle and pressure angle.

6 Compare the knife-edge follower with roller follower.

7 What is a tangent cam?

8 Compare various types of follower motions.

9 Define pressure angle of a cam.

10 Differentiate between trace point and pitch point.

11 How the cam size is defined?

12 What are the methods for reducing pressure angle of a cam?

13 How undercutting can be avoided in cams?

14 What is an offset follower?

Exercises

8.1 Draw the profile of a cam operating a knife-edge follower when the axis of the follower passes through the axis of the cam shaft. The following data is given: Lift = 40 mm, angle of ascent = 60°, dwell = 45°, angle of descent = 90° and dwell for the remaining period of cam rotation.

The motion of the cam is simple harmonic during both ascent and descent. The least radius of cam is 50 mm. If the cam rotates at 300 rpm, determine the maximum velocity and acceleration of the follower during ascent and descent.

[Ans. 1.88 m/s, 1.26m/s; 177.47 m/s², 78.87 m/s²]

8.2 If the follower in Exercise 8.1 is offset by 25 mm, then draw the cam profile.

8.3 A cam with 30 mm as minimum diameter is rotating clockwise at a uniform speed of 1200 rpm and operates a roller follower of 10 mm diameter as given below:

(i) Outward stroke of 30 mm during 120° of can rotation with equal uniform acceleration and retardation.

(ii) Follower is to dwell for 50° of cam rotation.

(iii) Inward stroke during 90° of cam rotation with equal uniform acceleration and retardation.

(iv) Follower is to dwell for the remaining period of cam rotation.

Draw the cam profile if the axis of follower passes through the axis of the cam. Determine the maximum velocity and acceleration during outward and inward strokes.

[Ans. 3.6 m/s, 4.8 m/s; 432 m/s², 768 m/s²]

8.4 If the follower in Exercise 8.3 is offset by 20 mm, then draw the cam profile.

8.5 Draw the cam profile from the following data if the radial follower moves with simple harmonic motion during ascent and uniform acceleration and deceleration during descent: Lift = 40 mm, Least radius of cam = 60 mm, Angle of ascent = 54°, Dwell = 40°, Angle of descent = 72°, Roller diameter = 20 mm.

8.6 If the roller follower is offset by 20 mm in Exercise 8.5, then draw the cam profile.

8.7 Draw the profile of a cam to give reciprocating motion to a flat-faced follower for the following data:
Lift = 25 mm, Ascent = 120°, Dwell = 30°, Descent = 120°, Dwell = 90°
Minimum radius of cam = 25 mm
The ascent and descent is to take place with SHM. The line of movement of follower passes through the cam centre.

8.8 The following data refers to a symmetrical circular arc cam operating a flat-faced follower: Least radius of cam = 30 mm, Lift = 12.5 mm, Angle of lift = 55°, Nose radius = 3 mm, Speed of cam = 600 rpm. Calculate (a) distance between cam and nose centres, (b) radius of circular flank, and (c) angle of contact on the circular flank.

[Ans. 39.5 mm, 125.6 mm, 15.3°]

8.9 The following data refers to a circular arc cam operating a flat-faced reciprocating follower: Minimum radius of cam = 30 mm, Total angle of cam = 120°, Radius of circular arc = 100 mm, Nose radius = 10 mm, angular velocity of cam = 10 rad/s
Determine the velocity and acceleration of the follower when the cam has turned through 20°.

[Ans. 239.4 mm/s, 6578 mm/s²]

8.10 A symmetrical tangent cam with least radius of 30 mm operates a roller follower of 10 mm radius.
The angle of ascent is 60° and lift is 20 mm. The speed of cam is 450 rpm. Calculate (a) distance between cam and nose centres, (b) nose radius, (c) angle of contract of cam with straight flank, and (d) acceleration of follower:
(i) at the beginning of lift,
(ii) where the roller just touches the nose, and
(iii) at the apex of circular nose.

[Ans. 40 mm, 10 mm; 88.83 m/s², 293.7 m/s², –266.5 m/s²]

8.11 The following data refers to a tangent cam operating a radial roller follower: Minimum radius of cam = 45 mm, Lift = 15 mm, Nose radius = 18 mm, Radius of roller = 20 mm, Semi-angle of cam action = 70°, Angular velocity of cam = 10 rad/s. Draw the displacement, velocity and acceleration diagrams for one rotation of cam.

8.12 The following data refers to a cam with circular nose and flanks operating the suction valve of a four-stroke petrol engine:
Lift = 10 mm, Least radius of cam = 20 mm, Nose radius = 2.5 mm, Crank angle when suction valve opens after TDC = 4°, Crank angle when suction valve closes after BDC = 50°, camshaft speed = 1000 rpm.
The follower is of flat-faced type. Determine (a) the maximum velocity of valve, (b) the maximum acceleration and retardation of valve, and (c) maximum force to be exerted by the spring to overcome inertia of the valve parts of weight 0.2 kg.

[Ans. 2.03 m/s, 1063.73 m/s², 301.57 m/s², 60.3 N]

9

GOVERNORS

Chapter Outline

9.1 INTRODUCTION

A governor is a device to maintain, as closely as possible, a constant mean speed of rotation of the crankshaft over long periods during which the load on the engine may vary. The governor meets the varying demand for power by regulating the supply of working fluid.

9.2 TYPES OF GOVERNORS

There are basically two types of governors.

1. Centrifugal governors, and
2. Inertia governor.

In centrifugal governors, the centrifugal force is balanced by the controlling force. These types of governors are used extensively. In inertia type of governors, the inertia force is balanced by the controlling force. They are not used popularly.

The centrifugal governors can be further classified as follows:

1. Pendulum type—Watt governor
2. Loaded type

(a) Dead weight type
   (i) Porter governor
   (ii) Proell governor
(b) Spring loaded type
   (i) Hartnell governor
   (ii) Hartung governor
   (iii) Wilson–Hartnell governor
   (iv) Pickering governor

9.3 TERMINOLOGY

1. *Height of governor* (h): It is the vertical distance from the centre of the ball to a point on the spindle axis where the axes of upper arms intersect.
2. *Centrifugal force* (F_c): It is the radially outward force acting on the balls due to the rotational speed. $F_c = m r \omega^2$, where r = radius of rotation of balls, m = mass of ball, ω = angular speed.
3. *Controlling force*: An equal and opposite force to the centrifugal force is called the controlling force.
4. *Equilibrium speed*: It is the speed of the governor at which the sleeve does not move upward or downward on the spindle.
5. *Radius of rotation*: It is the horizontal distance between centre of ball and the axis of rotation.
6. *Mean equilibrium speed*: It is the average of the maximum and minimum speeds of rotation.
7. *Sleeve lift*: It is the vertical distance travelled by the sleeve on the spindle due to change in equilibrium speed.

9.4 CENTRIFUGAL GOVERNORS

These governors work on the principle of centrifugal action. They have two balls which rotate along with the sleeve. As the centrifugal forcer is directly proportional to the radius of rotation of the mass, therefore, when the speed of a device increases, the balls rotate at a larger radius. As a result of it, the sleeve slides upwards on the spindle and with the help of lever, the throttle is closed to the required extent. With the decrease in speed, the balls rotate at smaller radius of rotation, compelling the sleeve to move down on the spindle. The downward movement of the sleeve opens the throttle to the required extent to admit more fuel into the prime movers. By this process, the speed of the prime mover and in turn that of the driven device is maintained constant.

9.4.1 Simple Watt Governor

A simple Watt governor is shown in Fig.9.1(a). It consists of two balls attached to the spindle through four arms. The upper two arms meet at the pivot, which may be on the spindle axis or offset from the spindle axis. The arms may be of the open type or crossed type. The lower arms are connected to a sleeve by pin joints. The movement of the sleeve is restricted by means of two stops.

(a) Neglecting weight of the arms.

Let w = weight of the balls
r = radius of the balls
h = height of the governor
T = tension in the arms
ω = angular speed of rotation

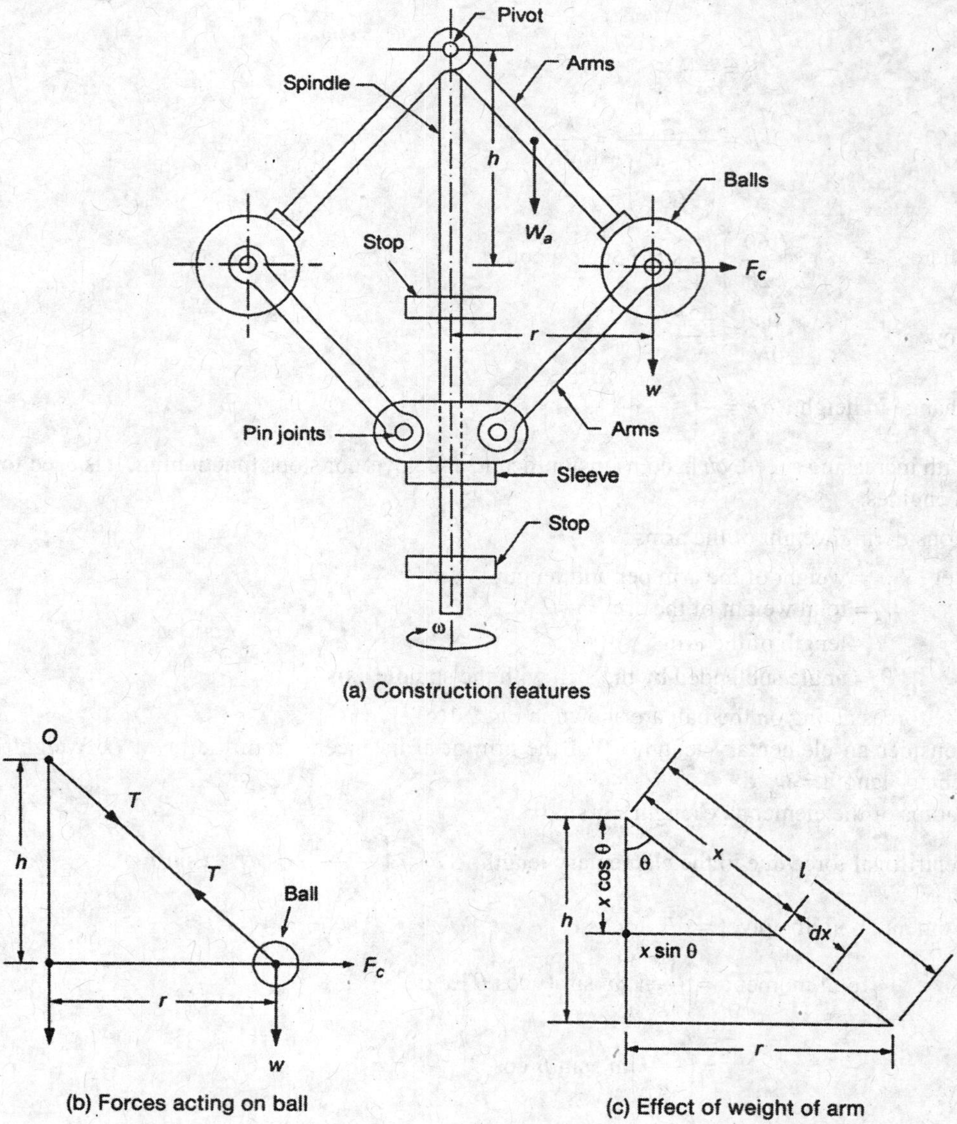

Fig.9.1 Simple Watt governor

The forces acting on the ball are shown in Fig.9.1(b).

Centrifugal force due to the balls, $F_c = \left(\frac{w}{g}\right) \cdot \omega^2 r$

Taking moments about the pivot O, we have

$$F_c h = wr$$

or $$\left(\frac{w}{g}\right) \cdot \omega^2 rh = wr$$

or $$h = \frac{g}{\omega^2}$$

or $$h = \frac{g}{\left(\frac{2\pi N}{60}\right)^2} = \frac{k}{N^2} \qquad (9.1)$$

where $k = 9.81 \times \left(\frac{60}{2\pi}\right)^2 = 894.565$ is a constant.

Now $$\frac{dh}{dN} = \frac{-2k}{N^3}$$

Change in height, $$\delta h = -\left(\frac{2k}{N^2}\right) \cdot \left(\frac{\delta N}{N}\right) \qquad (9.2)$$

With increasing speed, δh becomes insignificant, and governor stops functioning. It is used for slow speed engines.

(b) Considering weight of the arms.

Let w_a = weight of the arm per unit length
W_a = total weight of the arm = $w_a l$
l = length of the arm
θ = angle subtended by the arm with the spindle axis

The forces acting on the ball are shown in Fig.9.1(c).

Consider an elementary length dx of the arm at a distance x from the pivot O. Weight of the elementary length = $w_a \cdot dx$

Radius of the elementary length = $x \sin \theta$

Centrifugal force due to the elementary length, $dF_a = \left(\frac{w_a \cdot dx}{g}\right) \cdot \omega^2 \cdot x \sin\theta$

Moment about the pivot = $dF_a \cdot x \cos \theta$

$$\text{Total moment} = \left(\frac{w_a}{g}\right)\omega^2 \sin\theta \cos\theta \int_0^l x^2 dx$$

$$= \left(\frac{w_a}{g}\right)\omega^2 \sin\theta \cos\theta \left(\frac{l^3}{3}\right)$$

$$= \left(\frac{w_a l}{3g}\right)\omega^2 \cdot l \sin\theta \cdot l \cos\theta = \left(\frac{1}{3}\right) \cdot \left(\frac{W_a}{g}\right)\omega^2 rh$$

where $r = l \sin \theta$ and $h = l \cos \theta$

Therefore, the effect of the weight of the arm is equivalent to that produced by a weight $\frac{W_a}{3}$ placed at the centre of the ball.

Taking moments about the pivot O, we have

$$\left(\frac{W_a}{3g}\right)\omega^2 rh + \left(\frac{W}{g}\right)\omega^2 rh = \left(\frac{W_a}{2}\right) l \sin\theta + wr$$

$$\left[\frac{\left(w + \frac{W_a}{3}\right)}{g}\right]\omega^2 rh = \left(\frac{w + W_a}{2}\right) r$$

or $$h = \left(\frac{g}{\omega^2}\right)\left[\frac{w + \frac{W_a}{2}}{w + \frac{W_a}{3}}\right] \tag{9.3}$$

9.4.2 Gravity-Loaded Type Governors

In the gravity-loaded governors, a central load is attached to the sleeve, which slides on the spindle. There is a force of friction between the loaded sleeve and the spindle. The frictional force acts downwards when the sleeve moves up and acts upwards when the sleeve moves down. Thus, the height of the governor increases or decreases from normal value.

(a) *Porter governor*

The Porter governor is a modification of the Watt governor in which a central mass is attached to the sleeve. The Porter governor is shown in Fig.9.2(a). The forces acting on the governor are shown in Fig.9.2(b).

Let W = dead weight of sleeve
w = weight of ball
T_1 = tension in upper arm
T_2 = tension in lower arm
r = radius of the balls
$r_o = r - c$
c = distance of hinge B from axis of rotation
ω = angular speed of rotation

The forces acting at the hinge B are:

1. Half of the central load W.
2. Tension T_2 in the lower arm.
3. Reaction of the hinge.

Resolving the forces vertically, we have

$$T_2 \cos\beta = \frac{W}{2}$$

or $$T_2 = \frac{W}{2\cos\beta}$$

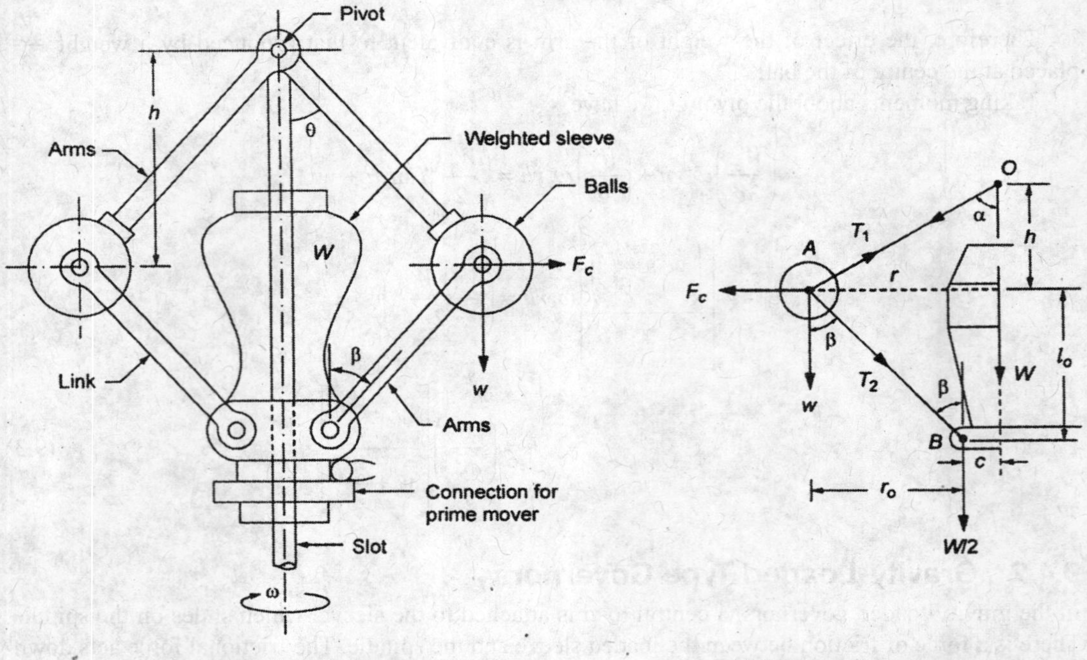

(a) Construction features (b) Forces acting on governor

Fig.9.2 Porter governor

The ball is in equilibrium under the following forces:

1. Centrifugal forces, F_c.
2. Weight of the ball, w.
3. Tension in upper arm, T_1.

Resolving the forces horizontally, we have

$$F_c = T_1 \sin \alpha + T_2 \sin \beta$$
$$= T_1 \sin \alpha + \left(\frac{W}{2}\right) \tan \beta$$

Resolving the forces vertically, we have

$$T_1 \cos \alpha = w + T_2 \cos \beta = w + \frac{W}{2}$$

Therefore,
$$F_c = \left(\frac{W}{2}\right) \tan \beta + \left(w + \frac{W}{2}\right) \tan \alpha$$
$$= \left(\frac{w}{g}\right) \omega^2 r$$

If $\tan\alpha = \dfrac{r}{h}$ and $\tan\beta = \dfrac{r_o}{l_o}$, then

$$\left(\frac{w}{g}\right)\omega^2 r = \left(\frac{W}{2}\right)\tan\beta + \left(w + \frac{W}{2}\right)\tan\alpha$$

$$= \left[\left(w + \frac{W}{2}\right) + \left(\frac{W}{2}\right)\left(\frac{\tan\beta}{\tan\alpha}\right)\right]\left(\frac{r}{h}\right)$$

$$= \left[w + \left(\frac{W}{2}\right)(1+k)\right]\left(\frac{r}{h}\right)$$

or

$$h = \left(\frac{g}{\omega^2}\right)\left[1 + \left(\frac{W}{2w}\right)(1+k)\right] \tag{9.4}$$

where $k = \dfrac{\tan\beta}{\tan\alpha}$.

If $\alpha = \beta$, then

$$\left(\frac{w}{g}\right)\omega^2 r = (w + W)\cdot\left(\frac{r}{h}\right)$$

$$h = \left(\frac{g}{\omega^2}\right)\left(1 + \frac{W}{w}\right) \tag{9.5}$$

If F is the frictional force acting on the sleeve, then

$$h = \left(\frac{g}{\omega^2}\right)\left(\frac{w + W \pm F}{w}\right) \tag{9.6}$$

Take +ve sign when the sleeve moves upwards or the governor speed increases and –ve sign when the sleeve moves downwards or the governor speed decreases.

(b) *Proell governor*

The Proell governor is shown in Fig.9.3(a) in which the balls are fixed at C and D to the extension of links EB and FA. The forces acting on the governor are shown in Fig.9.3(b).

Considering the equilibrium of forces at point E, we have

$$T_2 \cos\beta = \frac{W}{2}$$

or

$$T_2 = \frac{W}{2\cos\beta}$$

At point B, we have

$$T_1 \cos\alpha = T_2 \cos\beta + w = \frac{W}{2} + w$$

Taking moments about E, we have

$$F_c \times CF + w \times FE - T_1 \cos\alpha \times FE - T_1 \sin\alpha \times BF = 0$$

$$F_c \times CF + w \times BF \tan\beta - \left(w + \frac{W}{2}\right) \times BF \tan\beta - \left(w + \frac{W}{2}\right) \times BF \tan\alpha = 0$$

$$F_c = \left(\frac{BF}{CF}\right)\left[\left(w + \frac{W}{2}\right)\tan\alpha + \left(\frac{W}{2}\right)\tan\beta\right]$$

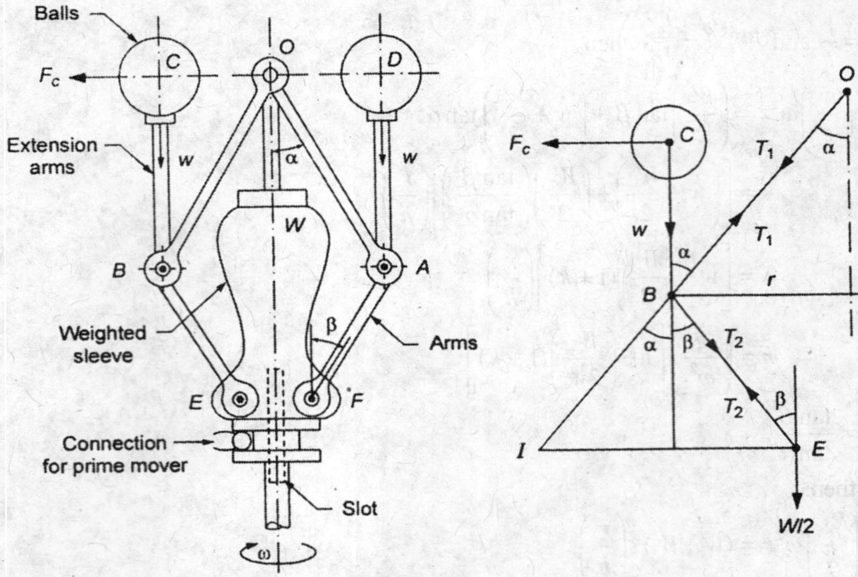

(a) Construction features (b) Forces acting on governor

Fig.9.3 Proell governor

Now $\tan\alpha = \dfrac{r}{h}$, and let $k = \dfrac{\tan\beta}{\tan\alpha}$.

$$\left(\frac{w}{g}\right)\omega^2 r = \left(\frac{BF}{CF}\right)\left[\left(\frac{W}{2}\right)(1+k) + w\right]\cdot\left(\frac{r}{h}\right)$$

$$\omega^2 = \left(\frac{BF}{CF}\right)\left[1 + \left(\frac{W}{2w}\right)(1+k)\right]\cdot\left(\frac{g}{h}\right) \tag{9.7}$$

If $\alpha = \beta$, *i.e.* $k = 1$, then

$$\omega^2 = \left(\frac{BF}{CF}\right)\left(1 + \frac{W}{w}\right)\left(\frac{g}{h}\right) \tag{9.8}$$

9.4.3 Spring-Loaded Governors

In a spring-loaded governor, a compressed spring is placed on the sleeve so that it may exert some force on it. When speed increases, balls move outwards compelling the sleeve to slide on the spindle upwards against the spring force. If the speed decreases, the sleeve moves downwards.

(a) Hartnell governor

The Hartnell governor is of the spring loaded type, and is shown in Fig.9.4(a). It consists of two bell crank levers pivoted at point A to the frame. The frame is attached to the governor spindle and rotates with it. Each lever carries a ball at the end of the vertical arm AB and a roller at the other end of the horizontal arm. A helical compression spring provides equal downwards force on the two rollers through the sleeve. The spring forces may be adjusted by the nut.

Let w = weight of the ball
W = weight of the sleeve
r_1, r_2 = maximum and minimum radii of rotation of the ball
ω_1, ω_2 = maximum and minimum angular speeds of rotation
S_1, S_2 = maximum and minimum spring forces
F_{c1}, F_{c2} = centrifugal forces at speeds ω_1 and ω_2 respectively
k = stiffness of the spring
a, b = vertical and horizontal length of arms of bell crank lever
r = radius of the ball

The forces acting on half of the governor are shown in Fig.9.4(b). The forces acting on half of the governor at maximum and minimum speeds are shown in Fig.9.4(c) and (d), respectively.

From Fig.9.4(b), taking moments about A, we have

$$F_c \times a = (W + S)\frac{b}{2}$$

$$W + S = \frac{2F_c a}{b} \tag{9.9}$$

From Fig.9.4(c), taking moments about A, we have

$$F_{c1}a_1 + wc_1 = (W + S_1)\frac{b_1}{2} \tag{9.10}$$

From Fig.9.4(d), taking moments about A, we have

$$F_{c2}a_2 - wc_2 = (W + S_2)\frac{b_2}{2}$$

Now $$c_1 + c_2 = r_1 - r_2 \tag{9.11}$$

Neglecting obliquity effect of arms, we have

$$a_1 = a_2 = a \quad \text{and} \quad b_1 = b_2 = b$$

$$W + S_1 = \frac{2F_{c1}a}{b}$$

and $$W + S_2 = \frac{2F_{c2}a}{b}$$

$$S_1 - S_2 = \frac{2a(F_{c1} - F_{c2})}{b} \tag{9.12}$$

Now $$\sin\theta = \frac{c_1}{a} = \frac{h_1}{b}$$

or $$h_1 = \frac{c_1 b}{a}$$

Similarly $$h_2 = \frac{c_2 b}{a}$$

$$h_1 + h_2 = \frac{b(c_1 + c_2)}{a}$$

Lift of sleeve, $h = h_1 + h_2 = \dfrac{b(r_1 - r_2)}{a}$

$$S_1 - S_2 = kh = \frac{kb(r_1 - r_2)}{a} \tag{9.13}$$

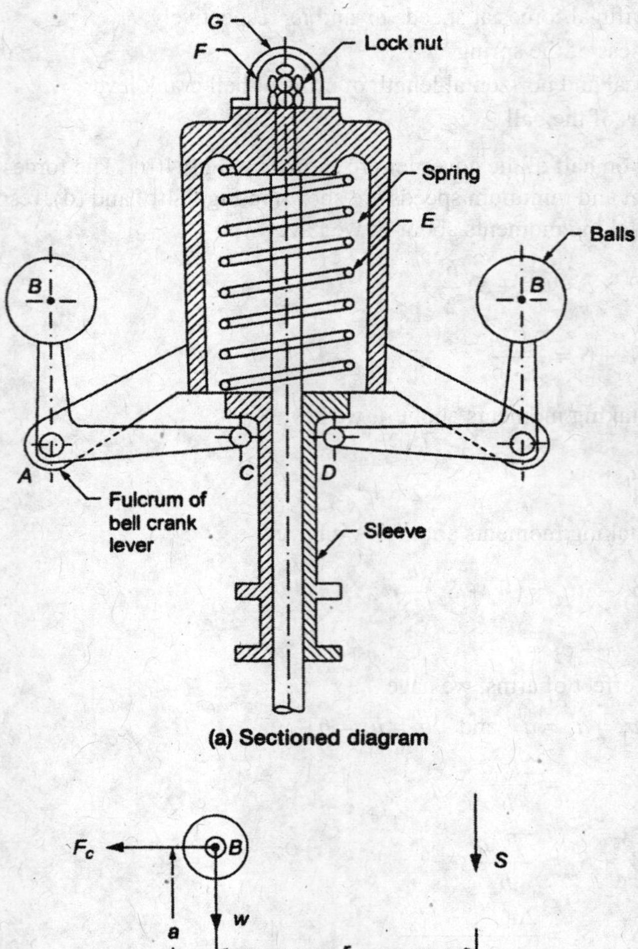

(a) Sectioned diagram

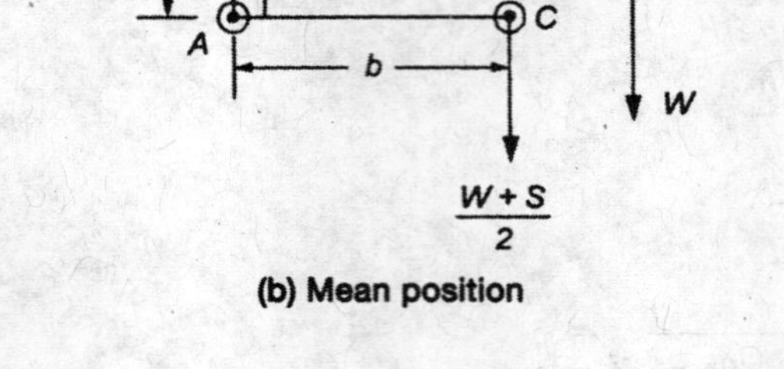

(b) Mean position

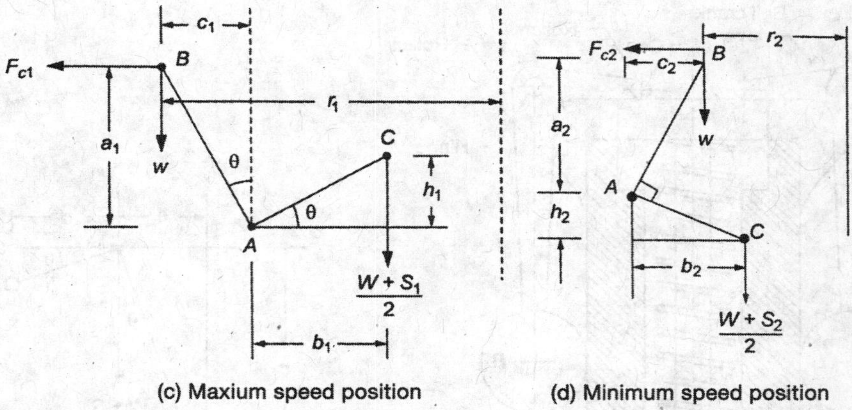

Fig.9.4 Hartnell governor

Comparing Eqs. (9.12) and (9.13), we get

$$k = 2\left(\frac{a}{b}\right)^2 \frac{(F_{c1} - F_{c2})}{r_1 - r_2} \tag{9.14}$$

(b) *Gravity and spring controlled governor*

The gravity and spring controlled governor is shown in Fig.9.5(a). This type of governor has the pivots for the bell crank levers on the moving sleeve. The spring is compressed between the sleeve and the cap, which is fixed to the end of the governor shaft. As the rollers of the bell crank lever press on the cap, the sleeve is lifted against the spring compression. The forces acting on half the governor are shown in Fig.9.5(b). Taking moments about D, we have

$$F_c c = w\,(d + e) + (W + S)\frac{d}{2}$$

$$W + S = \frac{2[F_c c - w(d + e)]}{d} \tag{9.15}$$

(c) *Wilson–Hartnell governor*

In this governor, the balls are connected by a spring in tension, as shown in Fig.9.6(a). An auxiliary spring is attached to the sleeve mechanism through a lever by means of which the equilibrium speed for a given radius may be adjusted. The forces acting on the governor are shown in Fig.9.6(b).

Let w = weight of each ball
W = weight of sleeve
P = combined pull of the ball springs
S = pull of the auxiliary spring
k_a = stiffness of auxiliary spring
k_b = stiffness of each ball spring
r = radius of the balls
F_c = centrifugal force of each ball
a, b = lengths of the arms of the bell crank lever
x, y = distance of hinge O for the lever from M and N, respectively.

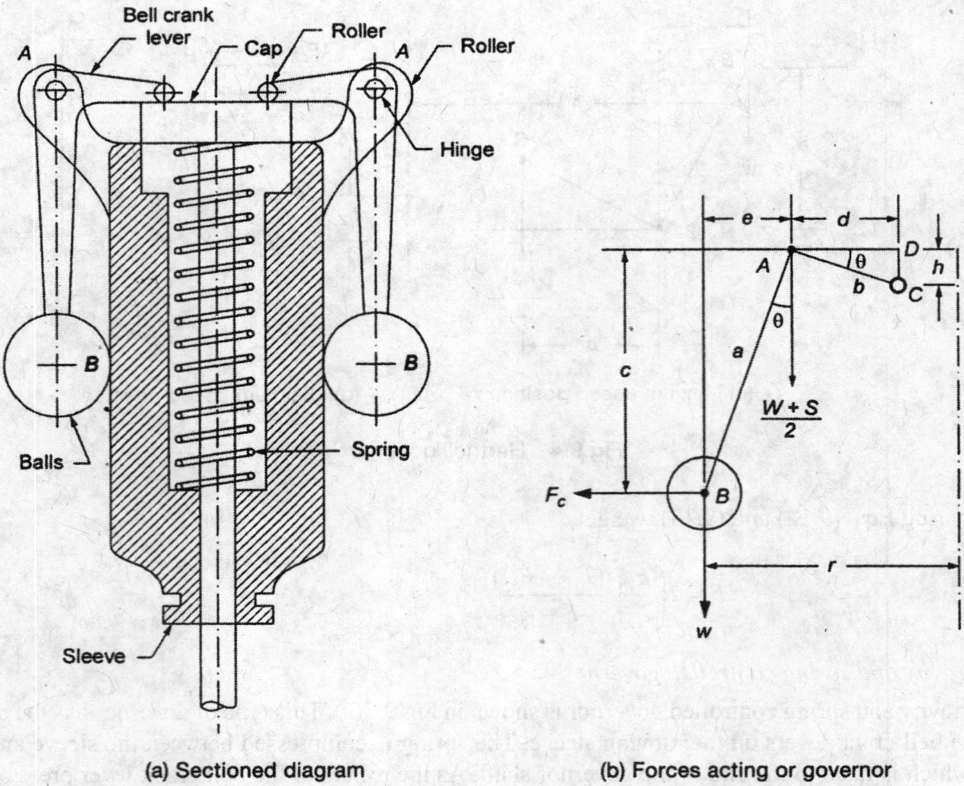

Fig.9.5 Governor with gravity and spring control

$$\text{Total downward force on the sleeve} = W + \frac{S\,y}{x}$$

Taking moments about the fulcrum A of the bell crank lever, and neglecting the pull of gravity on the balls, we have

$$(F_c - P)\,a = \left(W + \frac{S\,y}{x}\right)\frac{b}{2}$$

If suffices 1 and 2 refer to maximum and minimum equilibrium speeds respectively, then

$$(F_{c1} - P_1)\,a = \left(W + \frac{S_1\,y}{x}\right)\frac{b}{2}$$

$$(F_{c2} - P_2)\,a = \left(W + \frac{S_2\,y}{x}\right)\frac{b}{2}$$

Subtracting, we get

$$[(F_{c1} - F_{c2}) + (P_2 - P_1)]\,a = \frac{(S_1 - S_2)\,y\,b}{2x}$$

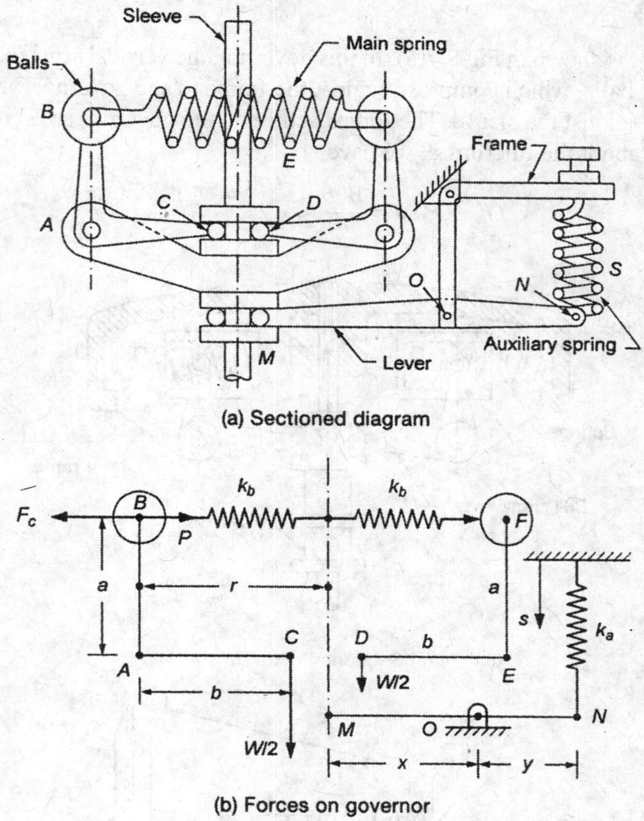

Fig.9.6 Wilson–Hartnell governor

If the radius increases from r_2 to r_1, the ball springs extend by the amount $2(r_1 - r_2)$ and the auxiliary spring extends by the amount $\dfrac{(r_1 - r_2)\,b\,y}{ax}$.

$$P_1 - P_2 = 4k_b(r_1 - r_2)$$

and
$$S_1 - S_2 = \frac{k_a b\,y\,(r_1 - r_2)}{a\,x}$$

$$F_{c1} - F_{c2} = 4k_b\left(r_1 - r_2\right) + k_a\left(\frac{by}{ax}\right)^2 \frac{(r_1 - r_2)}{2}$$

or
$$4k_b + \frac{k_a\left(\dfrac{by}{ax}\right)^2}{2} = \frac{(F_{c1} - F_{c2})}{(r_1 - r_2)} \tag{9.16}$$

If $k_a = 0$, then

$$k_b = \frac{F_{c1} - F_{c2}}{4(r_1 - r_2)} \tag{9.17}$$

(d) *Hartung governor*

The Hartung governor is shown in Fig.9.7(a). In this governor, the vertical arms of the bell crank levers are fitted with spring balls, which compress against the frame of the governor when the rollers at the horizontal arm press against the sleeve. The forces acting on the governor are shown in Fig.9.7(b).

Taking moments about the fulcrum A, we have

$$(F_c - P)\, a = \frac{W\, b}{2} \tag{9.18}$$

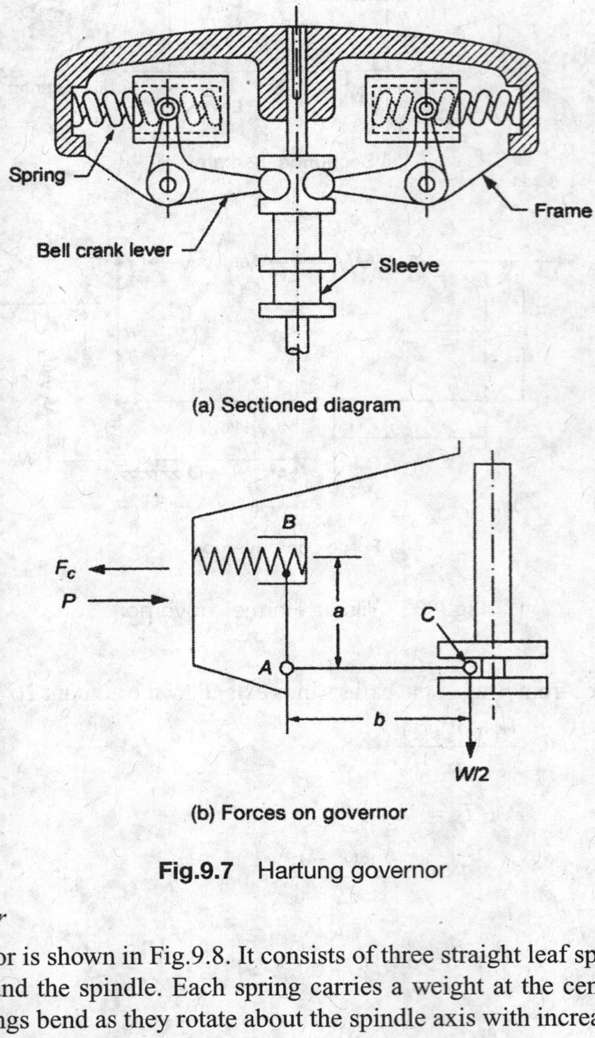

Fig.9.7 Hartung governor

(e) *Pickering governor*

The Pickering governor is shown in Fig.9.8. It consists of three straight leaf springs arranged at equal angular intervals around the spindle. Each spring carries a weight at the centre. The weights move outwards and the springs bend as they rotate about the spindle axis with increasing speed.

It is mostly used for driving gramophones.

Let m = mass attached at the centre of the leaf spring

a = distance from the spindle axis to the centre of gravity of the mass when the governor is at rest

δ = deflection of the centre of the leaf spring

ω = angular speed of the spindle

h = lift of the sleeve

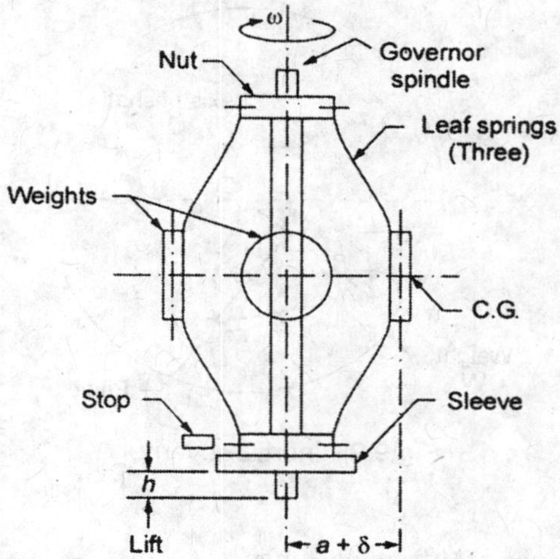

Fig.9.8 Pickering governor

The deflection of a leaf spring with both ends fixed and carrying a central load W, is given by

$$\delta = \frac{WL^3}{192EI}$$

where L = length of the spring
E = modulus of elasticity
I = moment of inertia of the spring cross-section about the neutral axis

$$= \frac{bt^3}{12}$$

b = width of the spring leaf
t = thickness of the spring leaf

In the Pickering governor, the central load is the centrifugal force.

$$W = F_c = m(a+\delta)\omega^2$$

Hence $$\delta = \frac{m(a+\delta)\omega^2 L^3}{192EI} \tag{9.19}$$

The lift of the sleeve, $h \approx 2.4\ \delta^2/L$ (9.20)

9.4.4 Inertia Governor

The principle of the inertia governor is depicted in Fig.9.9. A mass of weight W, whose centre of gravity is at G is attached to an arm, the other end of which is pivoted at a point A on the rotating disc. The point A is selected such that points O, A and G are not collinear. The end of the arm is connected to an eccentric, which operates the fuel supply valve to the prime mover.

Let v = velocity of G
r = radial distance of G from the centre of disc

Centrifugal force due to the rotating weight W, $F_c = \left(\frac{W}{g}\right)\cdot\frac{v^2}{r}$

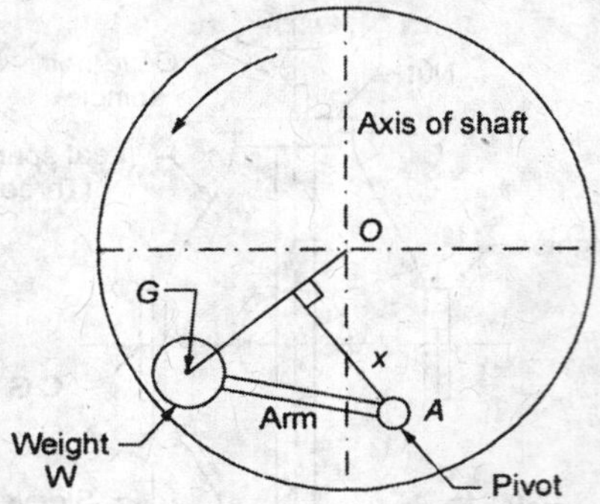

Fig.9.9 Inertia governor

If x = perpendicular distance of A from OG, then
Moment of F_c about $A = F_c \cdot x$

The inertia force acting on the ball perpendicular to $OG = \left(\frac{W}{g}\right) \cdot \frac{dv}{dt}$

For a governor to be rapid in action, the arm should be arranged such that as the mass moves outwards, the arm rotates in a direction opposite to that of the rotation of shaft.

9.5 PERFORMANCE OF GOVERNORS

9.5.1 Definitions

1. *Sensitiveness*: For maintaining constant speed of rotation, the movement of sleeve of governor should be as large as possible and the corresponding change of equilibrium speed as small as possible. The bigger the displacement of the sleeve for a given fractional change of speed, the more sensitive is the governor. Sensitiveness is more correctly defined as the ratio of the difference between the maximum and minimum equilibrium speeds to the mean equilibrium speed.

If N_{max} = maximum equilibrium speed
N_{min} = minimum equilibrium speed
N_{mean} = mean equilibrium speed

$$= \frac{N_{max} + N_{min}}{2}$$

Range of speed $= N_{max} - N_{min}$

$$\text{Then, Sensitiveness} = \frac{N_{max} - N_{min}}{N_{mean}} = \frac{2(N_{max} - N_{min})}{(N_{max} + N_{min})} \tag{9.21}$$

A too sensitive governor changes the fuel supply by a large amount when a small change in the speed of rotation takes place. This causes wide fluctuations in the engine speed, resulting in hunting of the governor.

2. *Stability*: A governor is said to be stable, when for each speed within the working range, there is only one radius of the governor balls at which the governor is in equilibrium. For a stable governor, if the equilibrium speed increases, the radius of the governor balls must also increase.

3. *Isochronism*: A governor is said to be isochronous, when the equilibrium speed is constant for all radii of rotation of the balls, within the working range. An isochronous governor will be infinitely sensitive. For a Porter governor,

$$\omega_1^2 = \left(\frac{g}{wh_1}\right)\left[w + \left(\frac{W}{2}\right)\cdot\left(1+\frac{h_1}{\ell}\right)\right]$$

$$\omega_2^2 = \left(\frac{g}{wh_2}\right)\left[w + \left(\frac{W}{2}\right)\cdot\left(1+\frac{h_2}{\ell}\right)\right]$$

For the governor to be isochronous, $N_{max} - N_{min} = 0$, or $N_{max} = N_{min}$. Therefore, $h_1 = h_2$, which is not possible. Hence, a Porter governor cannot be isochronous.
For a Hartnell governor, we have

$$W + S_1 = \frac{2F_{c1}a}{b} = 2\left(\frac{w}{g}\right)\frac{(2\pi N_{min})^2 r_1 a}{b}$$

and

$$W + S_2 = \frac{2F_{c2}a}{b} = 2\left(\frac{w}{g}\right)\frac{(2\pi N_{max})^2 r_2 a}{b}$$

For isochronism, $N_{max} = N_{min}$. Hence

$$\frac{W + S_1}{W + S_2} = \frac{r_1}{r_2}$$

Therefore, a Hartnell governor can be isochronous.

An isochronous governor is not of much practical use, as the sleeve will move to one of its extreme positions immediately when the speed deviates from its isochronous speed.

4. *Hunting*: It is a condition in which the speed of the engine controlled by the governor fluctuates continuously above and below the mean speed. It is caused by a governor which is too sensitive.
5. *Governor effort*: The effort of a governor is the force it can exert at the sleeve on the mechanism, which controls the supply of fuel to the engine. The mean force exerted during the given change of speed is termed as effort. Generally efforts are defined for 1% change of speed.
6. *Power*: The power of a governor is defined as the work done at the sleeve for a given percentage change of speed.

$$\text{Power} = \text{effort} \times \text{displacement of sleeve}$$

9.5.2 Effort and Power of a Porter Governor

For the Porter governor, shown in Fig.9.2, let

N = equilibrium speed
c = a factor by which speed increases

For $\alpha = \beta$, we have

$$h = \left(\frac{w + W}{w}\right)\left(\frac{g}{\omega^2}\right) = \left(\frac{w + W}{w}\right)\left(\frac{3600g}{4\pi^2 N^2}\right) \tag{9.22}$$

If the speed increases to $(1 + c)N$, and the height remains the same, a downward force has to be exerted on the sleeve, then we have

$$h = \left(\frac{w + W_1}{w}\right)\left[\frac{3600g}{4\pi^2(1+c)^2 N^2}\right] \tag{9.23}$$

where W_1 is the required sleeve load.
From Eqs. (9.22) and (9.23) we have

$$W_1 + w = (W + w)(1 + c)^2$$

or $$W_1 = (W + w)(1 + c)^2 - w$$

and $$W_1 - W = (W + w)[(1 + c)^2 - 1]$$

Let $P = W_1 - W$ = downward force which must be applied in order to prevent the sleeve from rising when increase of speed takes place.

and $$Q = \frac{W_1 - W}{2}$$

$$= \frac{P}{2} = \text{mean force exerted by the sleeve during the change of speed from } N \text{ to } (1 + c)N.$$

Now $(1 + c)^2 \approx 1 + 2c$

Therefore, $P \approx 2c(W + w)$

Governor effort, $$Q = \frac{P}{2} \approx c(W + w) \tag{9.24}$$

If $x = 2(h - h_1)$ = lift of the sleeve

Where h_1 = height corresponding to the increased speed $(1 + c)N$

$$= \frac{h}{(1+c)^2}$$

Then $$x = 2h\left[1 - \frac{1}{(1+c)^2}\right]$$

$$\approx \frac{4hc}{1 + 2c}$$

Governor power $= Qx$

$$\approx \left[\frac{4c^2}{1+2c}\right](w + W)h \tag{9.25}$$

When $\alpha \neq \beta$, then

$$Q \approx c\left[W + \frac{2w}{1+k}\right]$$

where $$k = \frac{\tan\beta}{\tan\alpha}$$

and $$x \approx (1 + k)(h - h_1)$$

where $$h_1 = \frac{h}{(1+c)^2}$$

so that $$x \approx (1 + k)h\left[1 - \frac{1}{(1+c)^2}\right]$$

$$\approx (1+k)h\left(\frac{2c}{1+2c}\right)$$

Governor power $\approx \left[\frac{2c^2}{1+2c}\right][W(1+k)+2w]h$

$$\approx \left[\frac{4c^2}{1+2c}\right]\left[\frac{W(1+k)}{2}+w\right]h \tag{9.26}$$

9.5.3 Quality of a Governor

The quality of a governor is ascertained by the following:

1. Sensitiveness,
2. Stability, and
3. Effort and power.

9.5.4 Controlling Force

When the speed of rotation is uniform, each ball of the governor is subjected either directly or indirectly to an inward pull, which is equal and opposite to the outward centrifugal reaction. This inward pull is termed the controlling force. A curve drawn to show how the pull varies with the radius of rotation of the ball is called a controlling force curve, as shown in Fig.9.10.

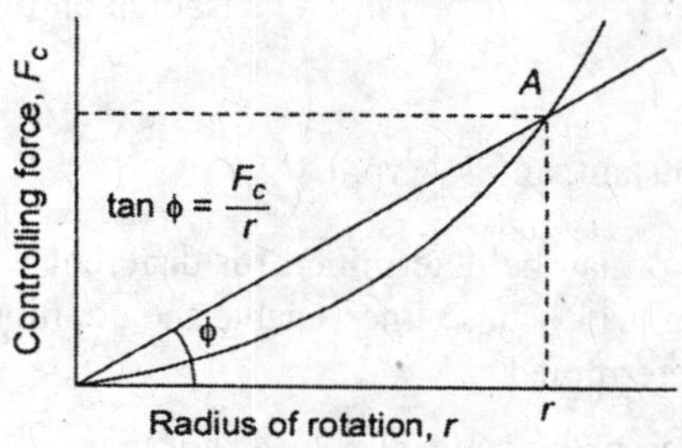

Fig.9.10 Controlling force diagram

(a) *Controlling force diagram for a Porter governor*

Controlling force, $F_c = \left(\frac{w}{g}\right)\omega^2 r$

or $\omega = \left[\frac{gF_c}{wr}\right]^{0.5}$

$$= \left[\left(\frac{g}{w}\right)\cdot \tan\phi\right]^{0.5} \tag{9.27}$$

If controlling force curve is a straight line, then

$$\tan\phi = \left(\frac{w}{g}\right)\omega^2$$

$$= kN^2 \tag{9.28}$$

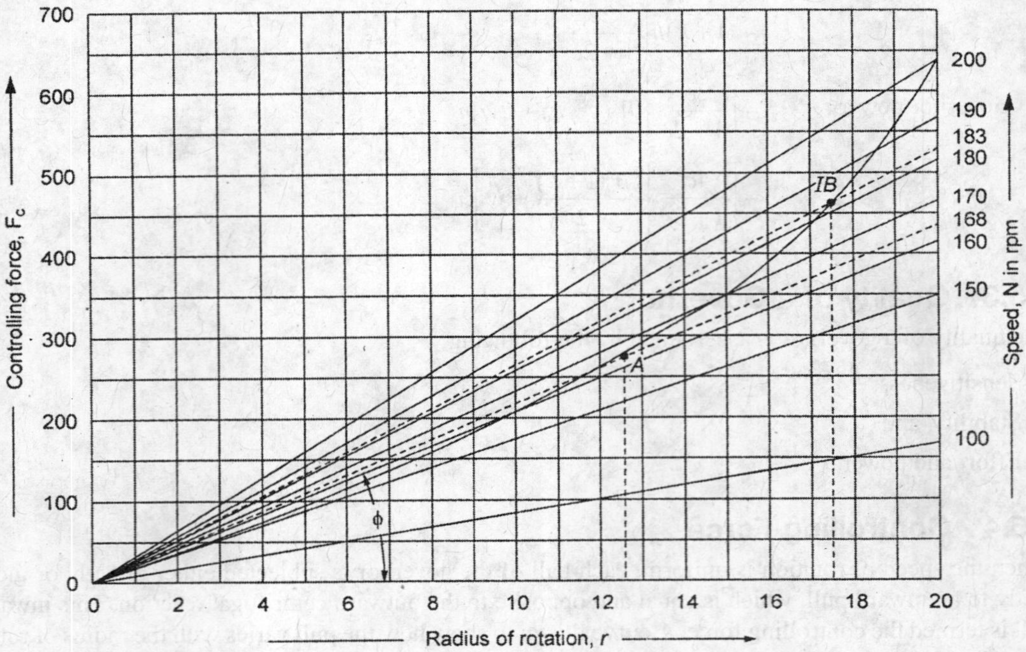

Fig.9.11 Controlling force vs Radius of rotation

where $k = \left(\frac{w}{g}\right)\left(\frac{2\pi}{60}\right)^2$ = a constant, and N = rpm.

Using Eq. (9.28), the angle ϕ may be determined for different values of N, and lines are drawn from the origin, as shown in Fig.9.11. These lines enable the equilibrium speed, corresponding to a given radius of rotation, to be determined.

(b) *Controlling force diagram for spring controlled governors*

The controlling force diagram for spring controlled governors is a straight line, as shown in Fig.9.12.

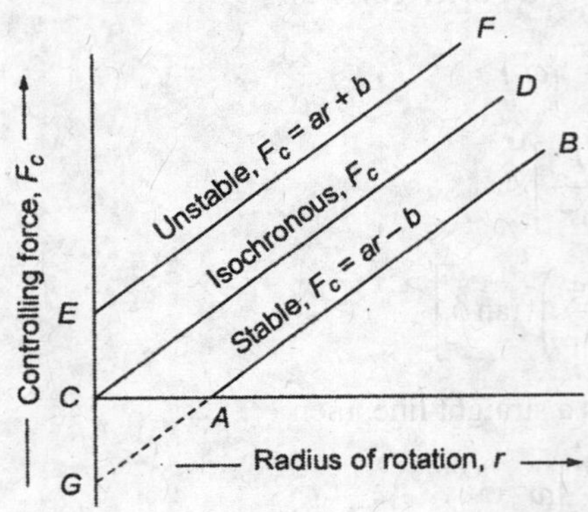

Fig.9.12 Controlling force for a spring loaded governor

Controlling force, $F_c = \left(\frac{w}{g}\right)\omega^2 r$

or $\frac{F_c}{r} = \left(\frac{w}{g}\right)\omega^2$

The stability of a spring controlled governor can be ascertained as follows:

1. For a stable governor, the controlling force must increase as radius increases, i.e. F_c/r must increase as r increases. Therefore, the controlling force line AB when produced must intersect the controlling force axis below the origin, as shown in Fig.9.12. The equation of line AB is given by,

$$F_c = ar - b \tag{9.29}$$

where a and b are constants.

2. When $b = 0$, the controlling force line CD passes through the origin, and the governor becomes isochronous, because F_c/r will remain constant for all radii of rotation. The equation of line CD is

$$F_c = ar \tag{9.30}$$

3. If b is positive, then F_c/r decreases as r increases, so that the equilibrium speed of the governor decreases with an increase of the radius of rotation of the balls, which is not possible. Such a governor is said to be unstable. The equation of line EF is,

$$F_c = ar + b \tag{9.31}$$

9.5.5 Coefficient of Insensitiveness

There is always friction in the joints and operating mechanism of the governor. The friction force opposes the motion of the sleeve. To account for the forces, let

F_s = force required at the sleeve to overcome friction
F_b = corresponding radial force required at the balls
F_c = controlling force on each ball
W = total load on the sleeve
For decrease in speed, sleeve load, $W_1 = W - F_s$
For increase in speed, sleeve load, $W_2 = W + F_s$
For decrease in speed, controlling force, $F_{c1} = F_c - F_b$
For increase in speed, controlling force, $F_{c2} = F_c + F_b$

Now $$F_c = \left(\frac{w}{g}\right)\omega^2 r = KN^2$$

$$F_{c1} = K N_1^2$$

$$F_{c2} = K N_2^2$$

Similarly, $$F_c + F_b = K N_1^2$$

$$F_c - F_b = K N_2^2$$

Subtracting, we get

$$\frac{2F_b}{F_c} = \frac{N_1^2 - N_2^2}{N^2}$$

$$= \frac{(N_1 + N_2)(N_1 - N_2)}{N^2}$$

Now $\qquad N_1 + N_2 \approx 2N$

$$\frac{F_b}{F_c} = \frac{N_1 - N_2}{N} \qquad (9.32)$$

The coefficient of insensitiveness = F_b/F_c, and is defined as the ratio of the difference of speed at ascent and descent for same radius of rotation to the steady speed at same radius of rotation.

For a Porter governor, as shown in Fig.9.2, we have

$$F_c = \tan\alpha \left[w + \left(\frac{W}{2}\right)(1+k) \right]$$

and

$$F_c \pm F_b = \tan\alpha \left[w + \left(\frac{(W \pm F_s)}{2}\right)(1+k) \right] \qquad (9.33)$$

and

$$F_b = \frac{\tan\alpha \cdot F_s(1+k)}{2} \qquad (9.34)$$

Similarly, for a spring loaded governor of Hartnell type (Fig.9.4), neglecting obliquity of the arms, we have

$$F_c a = \frac{Pb}{2}$$

and

$$(F_c \pm F_b)\, a = \frac{(P \pm F_s)b}{2}$$

or

$$F_b = \frac{F_s b}{2a} \qquad (9.35)$$

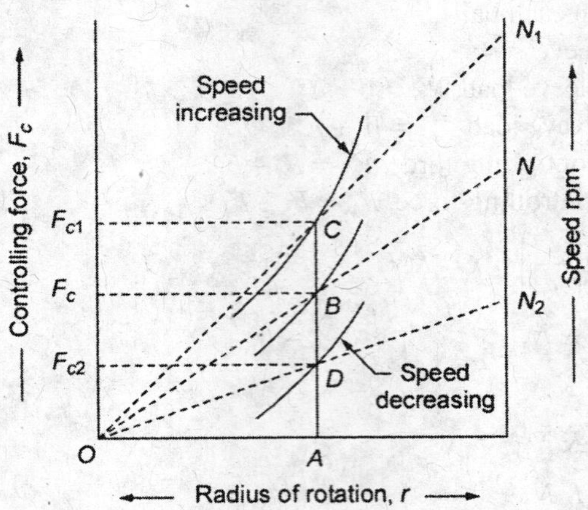

Fig.9.13 Effect of friction on the controlling force diagram

Figure 9.13 shows the effect of friction on the controlling force diagram. We see that for one value of the radius of rotation, there are three values of controlling force, as discussed below:

1. For decreasing speed, the controlling force reduces to F_{c2}, and the corresponding speed is N_2.
2. For increasing speed, the controlling force increases to F_{c1}, and the corresponding speed is N_1.
3. For friction neglected, the controlling force is F_c, and the corresponding speed is N.

Example 9.1

A simple Watt governor rotates at 75 rpm. Calculate its vertical height and the change if the speed increases to 80 rpm. Also calculate the height at 75 rpm if the weight of the ball is 20 N and that of the arm 5 N.

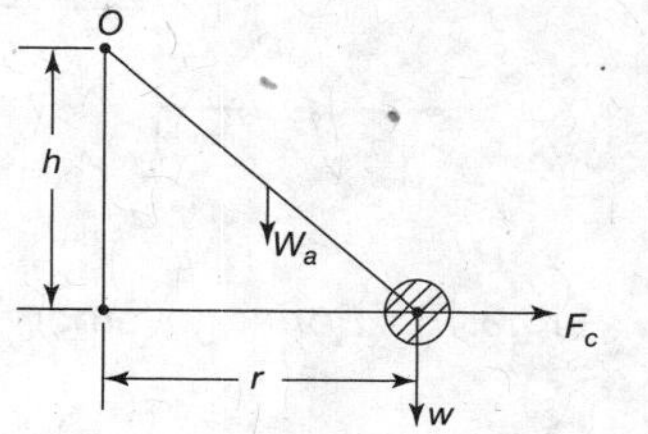

Fig.9.14 Simple Watt governor

■ Solution

Refer to Fig.9.14. $N_1 = 75$ rpm, $N_2 = 80$ rpm, $w = 20$ N, $W_a = 5$ N

$$h_1 = g/(4\,\pi^2 N^2)$$
$$= (9.81 \times 3600)/(4\,\pi^2 \times 75^2)$$
$$= 0.159 \text{ m}$$
$$h_2 = 0.159\,(75/80)^2$$
$$= 0.14 \text{ m}$$

Change in height $= h_1 - h_2$

$$= 0.159 - 0.14 = 0.019 \text{ m or } 19 \text{ mm}$$

Height of the governor considering the weight of the arm,

$$h = (g/\omega^2)[(w + W_a/2)/(w + W_a/3)]$$
$$= \left[\frac{9.81 \times 3600}{4\pi^2 \times 75^2}\right]\left[\frac{20 + \frac{5}{2}}{20 + \frac{5}{3}}\right]$$
$$= 0.165 \text{ m}$$

Example 9.2

The arms of a Porter governor are each 200 mm long. The weight of each ball is 40 N and that of the sleeve is 200 N. The radius of rotation of the balls is 125 mm when the sleeve begins to rise and reaches a value of 150 mm for maximum speed. Determine the speed range of the governor. If the friction at the sleeve is equivalent to 20 N of load at the sleeve, determine how the speed range is modified.

■ Solution

Refer to Fig.9.15. $w = 40$ N, $W = 200$ N, $F = 20$ N, $l = 200$ mm

$$h_1 = [200^2 - 125^2]^{0.5} = 156.12 \text{ mm}$$
$$\omega_1^2 = (1 + W/w)\,(g/h_1)$$
$$= (1 + 200/40)\left[\frac{9.81}{156.12 \times 10^{-3}}\right]$$
$$= 378.02$$

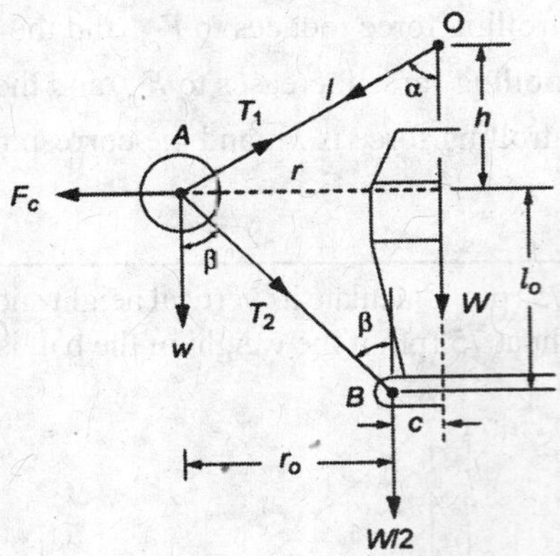

Fig.9.15 Porter governor

$\omega_1 = 19.417$ rad/s
$N_1 = 185.4$ rpm
$h_2 = [200^2 - 150^2]^{0.5} = 132.28$ mm

$$\omega_2^2 = (1 + 200/40)\left[\frac{9.81}{132.28 \times 10^{-3}}\right]$$
$$= 444.96$$

$\omega_2 = 21.09$ rad/s
$N_2 = 201.4$ rpm

Range of speed $= N_2 - N_1 = 201.4 - 185.4 = 16$ rpm

When the sleeve moves downwards, the force of friction acts upwards, therefore

$$\omega_1^2 = [(w + W - F)/w]\,(g/h_1)$$
$$= [(40 + 200 - 20)/40]\left[\frac{9.81}{156.12 \times 10^{-3}}\right]$$
$$= 345.6$$

$\omega_1 = 18.59$
$N_1 = 178.52$ rpm

When the sleeve moves upwards, the force of friction acts downwards, therefore

$$\omega_2^2 = [(w + W + F)/w]\,(g/h_2)$$
$$= [(40 + 200 + 20)/40]\left[\frac{9.81}{132.28 \times 10^{-3}}\right]$$
$$= 482.04$$

$\omega_2 = 21.96$ rad/s
$N_2 = 209.66$ rpm

Range of speed $= N_2 - N_1 = 209.66 - 178.52 = 32.14$ rpm

Example 9.3

The arms of a Porter governor are 250 mm long. The upper arms are pivoted on the axis of revolution, but the lower arms are attached to a sleeve at a distance of 50 mm from the axis of rotation. The weight on the sleeve is 600 N and the weight of each ball is 80 N. Determine the equilibrium speed when the radius of rotation of the balls is 150 mm. If the friction is equivalent to a load of 25 N at the sleeve, determine the range of speed for this position.

■ Solution

Refer to Fig.9.2. $l = 250$ mm, $c = 50$ mm, $W = 600$ n, $w = 80$ N, $F = 25$ N

$$h = [250^2 - 150^2]^{0.5} = 200 \text{ mm} = 0.2 \text{ m}$$
$$r_o = r - c = 150 - 50 = 100 \text{ mm}$$
$$l_o = [250^2 - 100^2]^{0.5} = 229.13 \text{ mm} = 0.22913 \text{ m}$$
$$\tan\alpha = r/h = 150/200 = 0.75$$
$$\tan\beta = r_o/l_o = 100/229.13 = 0.43643$$
$$k = \tan\beta/\tan\alpha = 0.43643/0.75 = 0.582$$
$$\omega^2 = [1 + \{W/(2w)\}(1 + k)]\,(g/h)$$
$$= [1 + (600/160)\,(1 + 0.582)]\,(9.81/0.2)$$
$$= 340.04$$
$$\omega = 18.44 \text{ rad/s}$$
$$N = 176 \text{ rpm}$$

Maximum equilibrium speed shall occur when the sleeve is going upwards.

$$\omega_2^2 = [1 + \{(W + F)/(2w)\}(1 + k)]\,(g/h)$$
$$= [1 + \{(600 + 25)/160\}\,(1 + 0.582)]\,(9.81/0.2)$$
$$= 352.16$$
$$\omega_{max} = 18.766 \text{ rad/s}$$
$$N_{max} = 179.2 \text{ rpm}$$
$$\omega_1^2 = [1 + \{(W - F)/(2w)\}(1 + k)]\,(g/h)$$
$$= [1 + \{(600 - 25)/160\}\,(1 + 0.582)]\,(9.81/0.2)$$
$$= 328.91$$
$$\omega_{min} = 18.11 \text{ rad/s}$$
$$N_{min} = 172.9 \text{ rpm}$$

Range of speed = 179.2 – 172.9 = 6.3 rpm

Example 9.4

The arms of a Proell governor are 300 mm long. The pivots of the upper and lower arms are 30 mm from the axis. The load on the sleeve is 250 N and the weight of each ball is 30 N. When the governor sleeve is in mid-position, the extension link of the lower arm is vertical and the radius of rotation of the balls is 160 mm. The vertical height of the governor is 200 mm.

If the speed of the governor is 150 rpm when in mid-position, find (a) length of the extension link, and (b) tension in the upper arm.

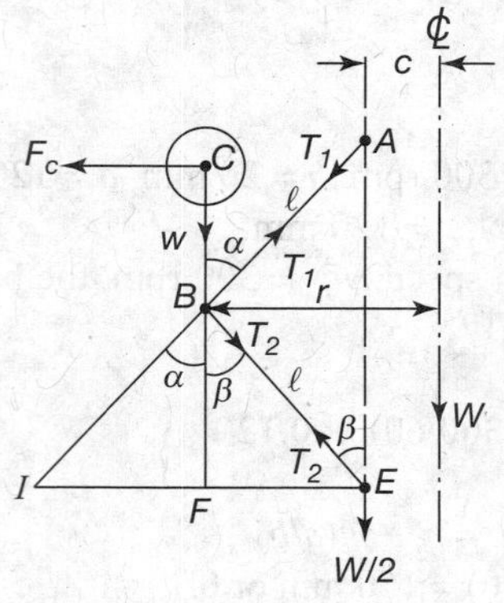

Fig.9.16 Proell governor

■ Solution

Given: $l = 300$ mm, $c = 30$ mm, $W = 250$ N, $w = 30$ N, $r = 160$ mm

(a) Let BC = length of the extension link (Fig.9.16)

In mid-position, $\alpha = \beta$, or $k = 1$.

$$w^2 = (BF/CF)\,[1 + W/w](g/h)$$

$$BF = [300^2 - 130^2]^{0.5} = 270.37 \text{ mm}$$

$$(2\pi \times 150/60)^2 = (270.37/CF)\,[1 + 250/30]\,(9.81/0.2)$$

$$246.74 = 123775.4/CF$$

or $$CF = 501.64 \text{ mm}$$

$$BC = CF - BF = 501.64 - 270.37 = 231.27 \text{ mm}$$

(b) $$\cos\alpha = 200/300 = 0.667$$

$$T_1 \cos\alpha = w + W/2$$

$$= 30 + 250/2 = 155$$

$$T_1 = 232.5 \text{ N}$$

Example 9.5

A Hartnell governor moves between 300 rpm and 320 rpm for a sleeve lift of 20 mm. The sleeve arms and the ball arms are 80 mm and 120 mm, respectively. The levers are pivoted at 120 mm from the governor axis. The weight of each ball is 25 N. The ball arms are parallel to the governor axis at the lowest equilibrium speed. Determine (a) loads on the spring at the minimum and the maximum speeds, and (b) stiffness of the spring.

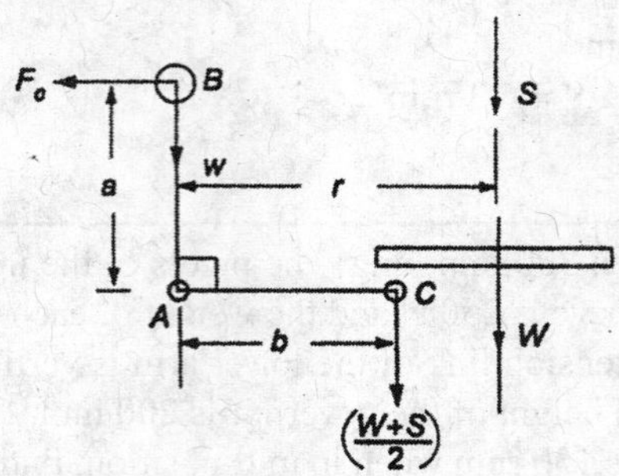

Fig.9.17 Hartnell governor

■ Solution

Refer to Fig.9.17.

(a) Here $N_{max} = 320$ rpm, $N_{min} = 300$ rpm, $h = 20$ mm, $a = 120$ mm, $b = 80$ mm, $w = 25$ N, $r = 120$ mm.

At the minimum equilibrium speed $N_{min} = 300$ rpm, the ball arms are parallel to the governor axis. Therefore $r = r_2 = 120$ mm.

$$F_{c2} = (w/g)\,\omega^2_{min} r_2$$

$$= (25/9.81)\,(2\pi \times 300/60)^2 \times 0.12$$

$$= 301.82 \text{ N}$$

Radius at the maximum speed, $r_1 = r_2 + h\,(a/b)$

$$= 120 + 20\,(120/80) = 150 \text{ mm or } 0.15 \text{ m}$$

$$F_{c1} = (25/9.81)\,(2\pi \times 320/60)^2 \times 0.15$$

$$= 429.26 \text{ N}$$

For lowest position,

$$W + S_2 = 2\,F_{c2} \times a/b$$
$$0 + S_2 = 2 \times 301.82 \times 120/80$$
$$S_2 = 905.46 \text{ N}$$

For highest position,

$$W + S_1 = 2\,F_{c1} \times a/b$$
$$0 + S_1 = 2 \times 429.26 \times 120/80$$
$$S_1 = 1288.78 \text{ N}$$

(b) Stiffness of spring, $k = (S_1 - S_2)/h$

$$= (1288.78 - 905.46)/20$$
$$= 19.116 \text{ N/mm}$$

Example 9.6

In a gravity and spring controlled governor, the weight of each ball is 15 N, and the weight of the sleeve is 80 N. The two arms of the bell crank lever are at right angles and their lengths are $AB = 100$ mm and $AC = 50$ mm. The distance of the fulcrum A of each lever from the axis of rotation is 50 mm. The minimum radius of rotation of the governor balls is also 50 mm, with a corresponding equilibrium speed of 250 rpm. The sleeve lift is 15 mm for an increase in speed of 5%. Find the stiffness and initial compression of the spring.

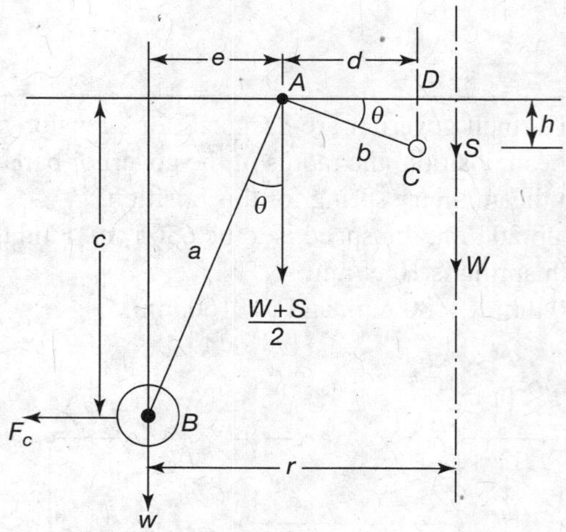

Fig.9.18 Gravity and spring-controlled governor

■ Solution

Here $w = 15$ N, $W = 80$ N, $a = 100$ mm, $b = 50$ mm, $r = 50$ mm, $r_2 = 50$ mm, $N_2 = 250$ rpm, $h = 15$ mm, $N_1 = 1.05$ rpm.

From Fig.9.18, we have

$$F_{c2} = (w/g)\,\omega_2^2 r_2$$
$$= (15/9.81)\,(2\pi \times 250/60)^2 \times 0.05$$
$$= 52.4 \text{ N}$$
$$F_{c2} \times a = [w + (W + S_2)/2]\,b$$
$$52.4 \times 100 = [15 + (80 + S_2)/2] \times 50$$

$$5240 = 750 + 2000 + 25\, S_2$$
$$S_2 = 99.6 \text{ N}$$
$$r_1 = r_2 + h\, a/b$$
$$= 50 + 15 \times 100/50 = 80 \text{ mm}$$
$$N_1 = 1.05 \times 250 = 262.5 \text{ rpm}$$
$$F_{c1} = (15/9.81)(2\pi \times 262.5/60)^2 \times 0.08$$
$$= 92.43 \text{ N}$$
$$W + S_1 = 2\,[\,F_{c1}c - w\,(d + e)]/d$$
$$d = b \cos \theta,\ e = a \sin \theta, \quad \text{and} \quad c = a \cos \theta$$
$$\sin \theta = h/b = 15/50 = 0.3, \quad \text{and} \quad \cos \theta = 0.954$$
$$d = 50 \times 0.954 = 48.7 \text{ mm}$$
$$e = 100 \times 0.3 = 30 \text{ mm}$$
$$c = 100 \times 0.954 = 95.4 \text{ mm}$$
$$80 + S_1 = 2\,[92.43 \times 95.4 - 15\,(48.7 + 30)]/48.7$$
$$= 320.85$$
$$S_1 = 240.85 \text{ N}$$
$$\text{Stiffness of spring, } k = (S_1 - S_2)/h$$
$$= (240.85 - 99.6)/15$$
$$= 9.41 \text{ N/mm}$$

Example 9.7

Two springs of the Wilson–Hartnell governor are designed for a tension of 1 kN in each. The weight of each ball is 75 N. In the mean position, the radius of the governor balls is 125 mm and the speed is 600 rpm. Find the tension in the auxiliary spring for this position.

When the sleeve moves up 20 mm, the speed is to be 650 rpm. Find the stiffness of the auxiliary spring, if the stiffness of each spring is 10 N/mm.

Take $a = 100$ mm, $b = 90$ mm, $x = 80$ mm, and $y = 160$ mm.

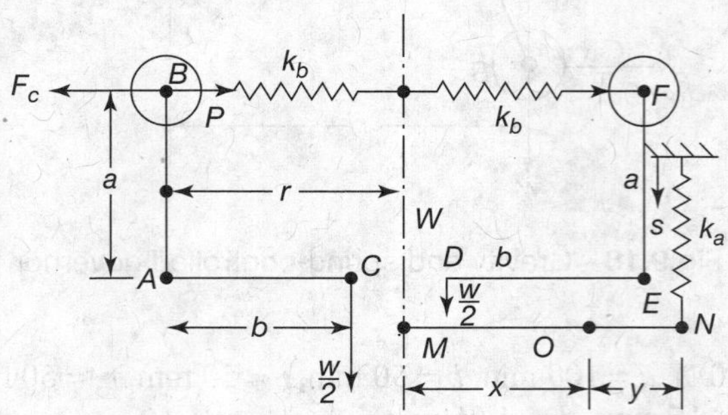

Fig.9.19 Forces on Wilson–Hartnell governor

■ **Solution**

Given: $w = 75$ N, $r_2 = 125$ mm, $N_2 = 600$ rpm

At minimum speed, from Fig.9.19 we have

$$F_{c2} = (w/g)\,\omega_2^2 r_2$$
$$= (75/9.81)\,(2\pi \times 600/60)^2 \times 0.125$$
$$= 3772.78 \text{ N}$$

Total pull of ball springs, $P = 2$ kN

Taking moments about the fulcrum A, we have

$$F_{c2} \times a = P \times a + W \times b/2$$

$$3772.78 \times 100 = 2000 \times 100 + W \times 90/2$$

or $$W = 3939.5 \text{ N}$$

Taking moments about point O, we have

$$S_2 \times y = W \times x$$

$$S_2 = 3939.5 \times 80/160 = 1969.75 \text{ N}$$

Let k_a = stiffness of auxiliary spring

$$h = (r_1 - r_2)\, b/a$$

or $$r_1 = r_2 + h\, a/b$$

$$= 125 + 20 \times 100/90 = 148.22 \text{ mm}$$

$$F_{c1} = (w/g)\, \omega_1^2 r_1$$

$$= (75/9.81)\,(2\pi \times 650/60)^2 \times 148.22 \times 10^{-3}$$

$$= 5214.86 \text{ N}$$

Extension of the spring = $2\,(r_1 - r_2) \times$ number of springs

$$= 2\,(148.22 - 125) \times 2$$

$$= 88.88 \text{ mm}$$

Total spring force = $2000 + 88.88 \times 10$

$$= 2888.8 \text{ N}$$

Taking moments about A, neglecting the obliquity of arms, we have

$$F_{c1} \times 100 = 2888.8 \times 100 + (W/2) \times 90$$

$$W = [5214.86 \times 100 - 288880]/45$$

$$= 5169 \text{ N}$$

Now taking moments about O, we have

$$S_1 \times 160 = 5169 \times 80$$

$$S_1 = 2584.5 \text{ N}$$

Extension of auxiliary spring = $20 \times 80/160$

$$= 10 \text{ mm}$$

Stiffness of auxiliary spring = $(S_1 - S_2)$/extension

$$= (2584.5 - 1969.75)/10$$

$$= 61.47 \text{ N/mm}$$

Example 9.8

In a Hartung governor, the length of the ball and sleeve arms are 80 mm and 120 mm, respectively. The total travel of the sleeve is 25 mm. In the mid-position, each spring is compressed by 50 mm and the radius of rotation of the balls is 140 mm. The weight of each ball is 40 N and the spring has a stiffness of 10 N/mm. The equivalent weight of the governor gear at the sleeve is 160 N. Neglecting the moment due to the revolving masses when the arms are inclined, determine the ratio of range of speed to the mean speed of the governor. Also find the speed in the mid position.

■ Solution

Refer to Fig.9.20.

Here $a = 80$ mm, $b = 120$ mm, $h = 25$ mm, $r = 140$ mm, $w = 40$ N,

$W = 160$ N, $k = 10$ N/mm, initial compression = 50 mm.

$$F_c = (w/g)\, \omega^2 r = (40/9.81)\, \omega^2 \times 0.14 = 0.57\, \omega^2 \text{ N}$$

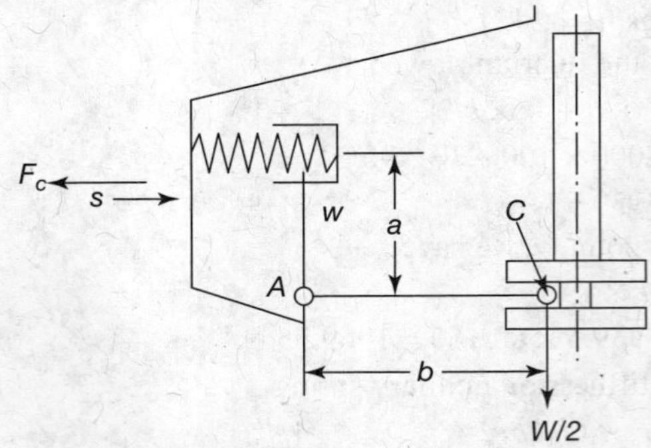

Fig.9.20 Hartung governor

Spring force, $S = 10 \times 50 = 500$ N

Taking moments about the fulcrum A of the lever (Fig.9.20), we have

$$F_c \times a = S \times a + W \times b/2$$
$$0.57\,\omega^2 \times 80 = 500 \times 80 + 160 \times 120/2$$
$$45.6\,\omega^2 = 49600$$
$$\omega^2 = 1088.72$$
$$\omega = 32.98 \text{ rad/s}$$

Mean speed of the governor, $N = 314.9$ rpm

At maximum position:

$$(r_1 - r)/h_1 = a/b$$

or

$$r_1 = r + h_1 a/b$$
$$= 140 + (25 \times 80)/(2 \times 120)$$
$$= 148.3 \text{ mm}$$
$$F_{c1} = (w/g)\,\omega_1^2 r_1$$
$$= (40/9.81)\,\omega_1^2 \times 0.1483$$
$$= 0.6047\,\omega_1^2 \text{ N}$$

Spring force, S_1 = [initial compression + $(r_1 - r)$] × stiffness

$$= [50 + (148.3 - 140)] \times 10$$
$$= 583 \text{ N}$$

Taking moments about the fulcrum of the lever, we have

$$F_{c1} a = S_1 a + W\,b/2$$
$$0.6047\,\omega_1^2 \times 80 = 583 \times 80 + 160 \times 120/2$$
$$48.376\,\omega_1^2 = 56240$$
$$= 1162.56$$
$$\omega_1 = 34.1 \text{ rad/s}$$
$$N_1 = 325.63 \text{ rpm}$$

At minimum position:

$$(r - r_2)/h_2 = a/b$$

or

$$r_2 = r - h_2 a/b$$
$$= 140 - (25 \times 80)/(2 \times 120)$$
$$= 131.67 \text{ mm}$$
$$F_{c2} = (w/g)\,\omega_1^2 r_1$$
$$= (40/9.81)\,\omega_1^2 \times 0.13167$$

$= 0.537\ \omega_1^2$ N

S_2 = [initial compression – $(r - r_2)$] × stiffness

= [50 – (140 – 131.67)] × 10

= 416.7 N

Taking moments about the fulcrum of the lever, we have

$F_{c2}a = S_2a + W\,b/2$

$0.537\ \omega_2^2 \times 80 = 416.7 \times 80 + 160 \times 120/2$

$42.96\ \omega_2^2 = 42936$

$\omega_2^2 = 999.44$

$\omega^2 = 31.614$ rad/s

$N_2 = 301.9$ rpm

Range of speed = $N_1 - N_2$ = 325.63 – 301.9 = 23.73 rpm

Ratio of range of speed = $(N_1 - N_2)/N$ = 23.73/314.9 = 0.07536 or 8.536%

Example 9.9

A gramophone is driven by a Pickering governor. The mass of each disc attached to the centre of a leaf spring is 20 g. Each spring is 5 mm wide and 0.125 mm thick. The effective length of each spring is 40 mm. The distance from the spindle axis to the centre of gravity of the mass when the governor is at rest, is 10 mm. Find the speed of the turntable when the sleeve has risen 1 mm and the ratio of the governor speed to the turntable speed is 10. Take E = 210 GPa.

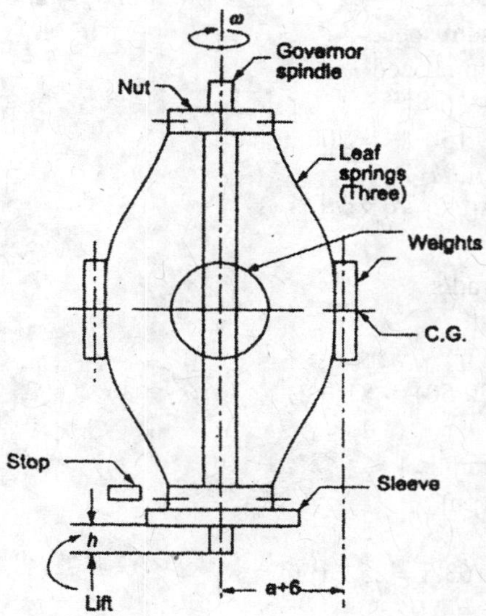

Fig.9.21 Pickering governor

■ Solution

Refer to Fig.9.21.

Moment of inertia of the spring about the neutral axis,

$$I = b\,t^3/12 = 5\,(0.125)^3/12 = 0.000814 \text{ mm}^4$$

Length of spring between fixed ends, L = 40 – 1 = 39 mm

Lift of the sleeve, $h = 2.4\ \delta^2/L$

$$1 = 2.4\ \delta^2/39 = 0.06154\ \delta^2$$
$$\delta = 4.03 \text{ mm}$$

Let N = speed of the governor

N_t = speed of the turntable

$$N/N_t = 10$$
$$\delta = m\omega^2 (a + \delta) L^3/(192 \text{ EI})$$
$$4.03 = 0.02\ \omega^2 (10 + 4.03)\times 39^3/(192 \times 210 \times 10^3 \times 0.000814)$$
$$\omega^2 = 8.946$$
$$\omega = 2.819 \text{ rad/s}$$
$$N = 26.92 \text{ rpm}$$
$$N_t = 2.692 \text{ rpm}$$

Example 9.10

A Porter governor has equal arms of 250 mm length each and pivoted on the axis of rotation. Each ball is of 50 N weight and the weight of the central load is 250 N. The radius of rotation of the ball is 150 mm when the governor begins to rise and 200 mm when the governor is at maximum speed. Find the range of speed, sleeve lift, governor effort, and power of the governor, when the friction at the sleeve is neglected.

■ Solution

Given: $l = 240$ mm, $w = 50$ N, $W = 250$ N, $r_1 = 150$ mm, $r_2 = 200$ mm

Let N_1 = maximum speed

N_2 = minimum speed

$$h_1 = [250^2 - 200^2]^{0.5} = 150 \text{ mm}$$
$$h_2 = [250^2 - 150^2]^{0.5} = 200 \text{ mm}$$
$$\omega_1^2 = (1 + W/w)(g/h_1)$$
$$= (1 + 250/50)(9.81/0.15)$$
$$= 392.4$$
$$\omega_1 = 19.81 \text{ rad/s}$$
$$N_1 = 189 \text{ rpm}$$

and
$$\omega_2^2 = (1 + W/w)(g/h_2)$$
$$= (1 + 250/50)(9.81/0.2)$$
$$= 294.3$$
$$\omega_2 = 18.15 \text{ rad/s}$$
$$N_2 = 163.8 \text{ rpm}$$

Range of speed $= N_1 - N_2$

$$= 189 - 163.8 = 25.2 \text{ rpm}$$

Sleeve lift, $x = 2(h_2 - h_1) = 2(r_2 - r_1)$

$$= 2(200 - 150) = 100 \text{ mm}$$

Let c = percentage increase in speed

$$= (N_1 - N_2)/N_2$$
$$= 25/163.8 = 0.1526$$

Governor effort, $P = c(w + W)$

$$= 0.1526(50 + 250)$$
$$= 45.78 \text{ N}$$

Power of the governor $= P \cdot x$

$$= 45.78 \times 0.1 = 4.578 \text{ Nm}$$

Example 9.11

The radius of rotation of the balls of a Hartnell governor is 100 mm at the minimum speed of 300 rpm. Neglecting gravity effects, determine the speed after the sleeve has lifted by 50 mm. Also determine the initial compression of the spring, governor effort, and the power. Take length of ball arm of lever = 150 mm, length of sleeve arm = 100 mm, weight of each ball = 40 N, and stiffness of spring = 25 N/mm.

Solution

Given: $r_2 = 100$ mm, $N_2 = 300$ rpm, $h = 50$ mm, $a = 150$ mm, $b = 100$ mm, $w = 40$ N, $k = 25$ N/mm

For maximum speed:

$$h = (r_1 - r_2)\, b/a$$
$$r_1 = r_2 + h\, a/b$$
$$= 100 + 50 \times 150/100 = 175 \text{ mm}$$
$$F_{c1} = (w/g)\, \omega_1^2 r_1$$
$$= (40/9.81)\,(2\pi N_1/60)^2 \times 0.175$$
$$= 0.007825\, N_1^2 \text{ N}$$
$$F_{c2} = (w/g)\, \omega_2^2 r_2$$
$$= (40/9.81)\,(2\pi \times 300/60)^2 \times 0.1$$
$$= 402.43 \text{ N}$$

Taking moments about the fulcrum of the lever, we have

$$F_{c2}\, a = (W + S_2)\, b/2$$
$$402.43 \times 150 = (0 + S_2) \times 100/2$$
$$S_2 = 1208.3 \text{ N}$$
$$S_1 - S_2 = h\, k$$
$$S_1 = 1208.3 + 50 \times 25 = 2458.3 \text{ N}$$
$$F_{c1}\, a = (W + S_1)\, b/2$$
$$0.007825\, N_1^2 \times 150 = (0 + 2458.3) \times 100/2$$
$$N_1 = 323.54 \text{ rpm}$$

Initial compression of spring = $S_2/k = 1208.3/25 = 48.29$ mm

Governor effort, $P = (S_1 - S_2)/2$

$$= (2458.3 - 1208.3)/2 = 625 \text{ N}$$

Governor power = Ph

$$= 625 \times 0.05 = 31.25 \text{ Nm}$$

Example 9.12

A Porter governor has equal arms 250 mm long pivoted on the axis of rotation. The weight of each ball is 25 N and the weight of the sleeve is 150 N. The ball path is 120 mm when the governor begins to lift and 150 mm at the maximum speed. Determine the range of speed. If the friction at the sleeve is equivalent to a force of 15 N, find the coefficient of insensitiveness.

Solution

Given: $l = 250$ mm, $w = 25$ N, $W = 150$ N, $r_1 = 120$ mm, $r_2 = 150$ mm, $F = 15$ N

At maximum speed, $h_1 = [250^2 - 150^2]^{0.5} = 200$ mm

At minimum speed, $h_2 = [250^2 - 120^2]^{0.5} = 219.32$ mm

$$\omega_1^2 = (1 + W/w)\,(g/h_1)$$
$$= (1 + 150/25)\,(9.81/0.2)$$
$$= 343.35$$
$$\omega_1 = 18.53 \text{ rad/s}$$
$$N_1 = 176.95 \text{ rpm}$$
$$\omega_2^2 = (1 + W/w)\,(g/h_2)$$
$$= (1 + 150/25)\,(9.81/0.21932)$$
$$= 313.1$$
$$\omega_2 = 18.69 \text{ rad/s}$$
$$N_2 = 168.97 \text{ rpm}$$

Range of speed $= N_1 - N_2$

$$= 176.95 - 168.97 = 8.98 \text{ rpm}$$

Coefficient of insensitiveness $= F/(w + W)$

$$= 15/(25 + 150) = 0.0857 \quad \text{or } 8.57\%$$

Example 9.13

In a spring-controlled governor, the curve of controlling force is a straight line. When balls are 400 mm apart, the controlling force is 1200 N, and when 200 mm apart, 450 N. At what speed will the governor run when the balls are 250 mm apart. What initial tension on the spring would be required for isochronism and what would then be the speed? The weight of each ball is 100 N.

■ Solution

Given: $F_{c1} = 1200$ N, $r_1 = 400$ mm
$F_{c2} = 450$ N, $r_2 = 200$ mm
$w = 100$ N

For the stability of the spring controlled governor, we have

$$F_c = a\,r - b$$
$$1200 = 0.2\,a - b$$
$$450 = 0.1\,a - b$$

Solving for a and b, we get

$$a = 7500,\ b = 300$$
$$F_c = 7500\,r - 300$$

For $r = 125$ mm, $F_c = 7500 \times 0.125 - 300 = 638.5$ N

$$F_c = (w/g)\,(2\pi\,N/60)^2 r$$
$$638.5 = (100/9.81)\,(2\pi \times N/60)^2 \times 0.125$$
$$N = 213.6 \text{ rpm}$$

For an isochronous governor, $b = 0$. Therefore, the initial tension required is 300 N.

$$F_c = a\,r$$
$$(w/g)\,\omega^2 r = a\,r$$

or
$$\omega^2 = a\,g/w = 7500 \times 9.81/100$$
$$= 735.75$$
$$\omega = 28.125 \text{ rad/s}$$
$$N = 259 \text{ rpm}$$

Example 9.14

A Porter governor has two balls of 25 N weight each and a central load 150 N. The arms are 200 mm long, pivoted on the axis. If the maximum and minimum radii of rotation of the balls are 150 mm and 120 mm, respectively, find the range of speed.

■ Solution

Given: $w = 25$ N, $W = 150$ N, $r_1 = 120$ mm, $r_2 = 150$ mm, $l = 200$ mm

$$h_1 = [l^2 - r_1^2]^{0.5} = [200^2 - 120^2]^{0.5} = 160 \text{ mm}$$
$$h_2 = [l^2 - r_2^2]^{0.5} = [200^2 - 150^2]^{0.5} = 132.3 \text{ mm}$$
$$\omega_1^2 = (1 + W/w)(g/h_1) = (1 + 150/25)(9.81/0.16) = 429.2$$
$$\omega_1 = 20.717 \text{ rad/s}$$
$$N_1 = 60 \times 20.717/(2\pi) = 197.83 \text{ rpm}$$
$$\omega_2^2 = (1 + W/w)(g/h_2)(1 + 150/25)(9.81/0.1323) = 519.09$$
$$\omega_2 = 22.78 \text{ rad/s}$$
$$N_2 = 60 \times 22.78/(2\pi) = 217.56 \text{ rpm}$$

Range of speed $= N_2 - N_1 = 217.56 - 197.83 = 19.73$ rpm

Example 9.15

A loaded governor of the Porter type has equal arms and links each 300 mm long. The weight of each ball is 20 N and the central weight is 120 N. When the ball radius is 150 mm, the valve is fully open and when the radius is 180 mm, the valve is closed. Find the maximum speed and the range of speed. If the maximum speed is to be increased 25% by an addition of weight to the central load, find its value.

■ Solution

Given: $w = 20$ N, $W = 120$ N, $r_1 = 150$ mm, $r_2 = 180$ mm, $l = 300$ mm

$$h_1 = [l^2 - r_1^2]^{0.5} = [300^2 - 150^2]^{0.5} = 259.8 \text{ mm}$$
$$h_2 = [l^2 - r_2^2]^{0.5} = [300^2 - 180^2]^{0.5} = 240 \text{ mm}$$
$$\omega_1^2 = (1 + W/w)(g/h_1) = (1 + 120/20)(9.81/0.2598) = 264.32$$
$$\omega_1 = 16.258 \text{ rad/s}$$
$$N_1 = 60 \times 16.258/(2\pi) = 155.25 \text{ rpm}$$
$$\omega_2^2 = (1 + W/w)(g/h_2)(1 + 120/20)(9.81/0.240) = 286.125$$
$$\omega_2 = 16.91 \text{ rad/s}$$
$$N_2 = 60 \times 16.91/(2\pi) = 161.53 \text{ rpm}$$

Range of speed $= N_2 - N_1 = 161.53 - 155.25 = 6.28$ rpm

$$N_{max} = 161.53 \text{ rpm}$$

with 25% increase in maximum speed, $N_2' = 161.53 \times 1.25 = 201.91$ rpm

$$\omega_2' = 2\pi \times 201.91/60 = 21.144 \text{ rad/s}$$
$$(21.144)^2 = (1 + W/20)(9.81/0.24)$$
$$W' = 198.75 \text{ kg}$$

Example 9.16

The arms of a Porter governor are 250 mm long. The upper arms are pivoted on the axis of rotation and the lower arms are attached to the sleeve at a distance of 40 mm from the axis of rotation. The load on the sleeve is 525 N and the weight of each ball is 75 N. Determine the equilibrium speed when the radius of the balls is 200 mm. What will be the range of speed for this position, if the frictional resistance to the motion of the sleeve are equivalent to a force of 30 N?

■ Solution

Given: $l = 250$ mm, $c = 40$ mm, $r = 200$ mm, $W = 525$ N, $w = 75$ N, $F = 30$ N

$h = [l^2 - r^2]^{0.5} = [250^2 - 200^2]^{0.5} = 150$ mm

$r_o = r - c = 200 - 40 = 160$ mm

$l_o = [l^2 - r_o^2]^{0.5} = [250^2 - 160^2]^{0.5} = 192.1$ mm

$\tan \alpha = r/h = 200/150 = 4/3$

$\tan \beta = r_o/l_o = 160/192.1 = 0.833$

$k = \tan \beta/\tan \alpha = (0.833 \times 3)/4 = 0.6247$

$h = (g/\omega^2)\,[1 + (1 + k)\,\{W/(2w)\}]$

$0.15 = (9.81/\omega^2)\,[1 + 1.6247 \times 525/150]$

$\omega^2 = 437.29$

$\omega = 20.91$ rad/s

$N = 60 \times 20.91/(2\pi) = 199.69$ rpm

(i) Sleeve going upwards:

$\omega_2^2 = [1 + (1 + \text{k})\,\{(W + F)/(2w)\}]\,(g/h)$

$= [1 + (1 + 0.6247)\,\{(525 + 30)/(2 \times 75)\}]\,(9.81/0.15)$

$= 458.545$

$\omega^2 = 21.41$ rad/s

$N^2 = 21.48 \times 60/(2\pi) = 204.48$ rpm

(ii) Sleeve going downwards:

$\omega_1^2 = [1 + (1 + k)\,\{(\text{W} - \text{F})/(2w)\}]\,(g/h)$

$= [1 + (1 + 0.6247)\,\{(525 - 30)/(2 \times 75)\}]\,(9.81/0.15)$

$= 416.04$

$\omega_1 = 20.397$ rad/s

$N_1 = 20.397 \times 60/(2\pi) = 194.78$ rpm

Range of speed $= N_2 - N_1 = 204.48 - 194.78 = 9.70$ rpm

Example 9.17

A Proell governor has all the four arms of length 300 mm. The upper and lower ends of the arms are pivoted on the axis of rotation of the governor. The extension arms of the lower links are each 100 mm long and parallel to the axis when the radius of the ball path is 150 mm. The weight of each ball is 45 N and the weight of the central load is 350 N. Determine the equilibrium speed of the governor.

■ Solution

Refer to Fig.9.22.

Given: $w = 45$ N, $W = 350$ N, r = 150 mm, $l_o = 100$ mm, $l = 300$ mm

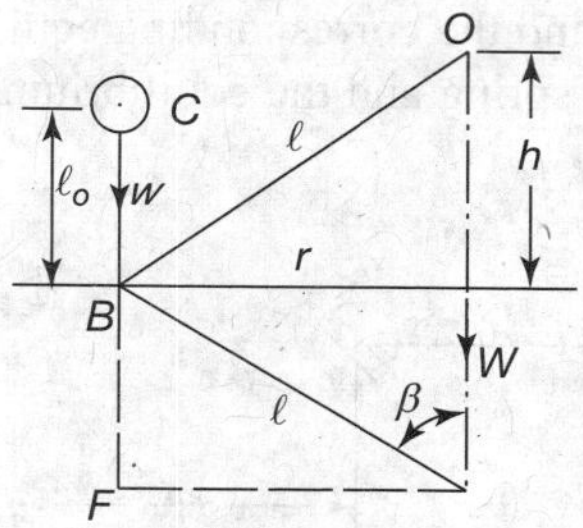

Fig.9.22 Proell governor

$$h = BF = [l^2 - r^2]^{0.5} = [300^2 - 150^2]^{0.5} = 259.8 \text{ mm}$$
$$CF = CB + BF = 100 + 259.8 = 359.8 \text{ mm}$$
$$\omega^2 = (BF/CF)\,(1 + W/w)\,(g/h)$$
$$= (259.8/359.8)\,(1+350/45)\,(9.81/0.2598) = 240.07$$
$$\omega = 15.494 \text{ rad/s}$$
$$N = 60 \times 15.494/(2\pi) = 147.96 \text{ rpm}$$

Example 9.18

A spring controlled governor of the Hartnell type with a central spring under compression has balls of weight 20 N. The ball and sleeve arms of the bell crank lever are, respectively, 120 mm and 80 mm long and are right angles. In the lower position of the sleeve, the radius of rotation of the balls is 80 mm and the ball arms are parallel to the governor axis. Find the initial load on the spring in order that the sleeve may begin to lift at 325 rpm. If the stiffness of the spring is 25 N/mm, what is the equilibrium speed corresponding to a sleeve lift of 15 mm?

Solution

Given: $w = 20$ N, $a = 120$ mm, $b = 80$ mm, $r_2 = 80$ mm, $S = ?$, $N_2 = 325$ rpm, $k = 25$ N/mm, $h = 15$ mm, $N_1 = ?$

For ball arms parallel to the governor axis, $r = r_2 = 80$ mm

$$F_{c2} = (w/g)\ \omega_2^2 r_2 = (20/9.81)\,(2\pi \times 325/60)^2 \times 0.08 = 188.92 \text{ N}$$
$$F_{c2} \times a = S_2 \times b/2$$
$$S_2 = 188.92 \times 120 \times 2/80 = 566.76 \text{ N}$$
$$r_1 = r_2 + h \times a/b = 80 + 15 \times 120/80 = 102.5 \text{ mm}$$
$$F_{c1} = (w/g)\ \omega_1^2 r_1 = (20/9.81)\,(2\pi \times N_1/60)^2 \times 0.1025 = 2.2916 \times 10^{-3} \times N_1^2$$
$$r_1 - r_2 = ha/b = 15 \times 120/80 = 22.5 \text{ mm}$$
$$k = 2\,(a/b)^2\,[\,(F_{c1} - F_{c2})/(r_1 - r_2)]$$
$$25 = 2 \times (120/80)^2\,[(2.2916 \times 10^{-3} \times N_1^2 - 188.92)/22.5]$$
$$N_1 = 290.9 \text{ rpm}$$

Example 9.19

A governor of the Hartnell type has each ball of weight 15 N and the lengths of vertical and horizontal arms of the bell crank lever are 120 mm and 60 mm, respectively. The fulcrum of the bell crank lever is at a distance of 100 mm from the axis of rotation. The maximum and minimum radii of rotation of the balls are 120 mm and 80 mm and the corresponding equilibrium speeds are 325 and 300 rpm, respectively. Find the stiffness of the spring and the equilibrium speed when the radius of rotation is 100 mm.

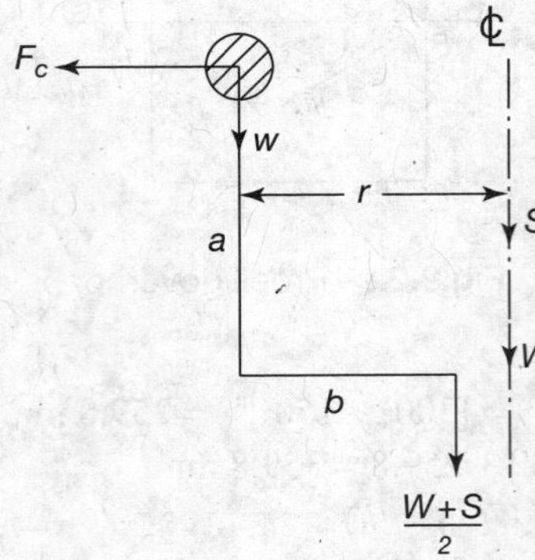

Fig.9.23 Hartnell governor

Solution

Refer to Fig.9.23.

Given: $w = 15$ N, $a = 120$ mm, $b = 60$ mm, $r = 100$ mm, $r_1 = 120$ mm, $r_2 = 80$ mm, $N_1 = 325$, $N_2 = 300$ rpm, $k = ?$.

$$F_{c1} = (w/g)\,\omega_1^2 r_1 = (20/9.81)\,(2\pi \times 325/60)^2 \times 0.08 = 188.92 \text{ N}$$

$$F_{c2} \times a = S_2 \times b/2$$

$$S_2 = 188.92 \times 120 \times 2/80 = 566.76 \text{ N}$$

$$r_1 = r_2 + h \times a/b = 80 + 15 \times 120/80 = 102.5 \text{ mm}$$

$$F_{c1} = (w/g)\,\omega_1^2 r = (20/9.81)\,(2\pi \times N_1/60)^2 \times 0.1025 = 2.2916 \times 10^{-3} \times N_1^2$$

$$r_1 - r_2 = ha/b = 15 \times 120/80 = 22.5 \text{ mm}$$

$$k = 2\,(a/b)^2\,[(F_{c1} - F_{c2})/(r_1 - r_2)]$$

$$25 = 2 \times (120/80)^2\,[\,(2.2916 \times 10^{-3} \times N_1^2 - 188.92)/22.5]$$

$$N_1 = 290.9 \text{ rpm}$$

Example 9.20

The Hartnell governor balls are of 30 N weight each. The balls radius is 120 mm in the mean position when the ball arms are vertical and the speed is 150 rpm, with the sleeve rising. The length of the ball arms is 150 mm and the sleeve arm 100 mm. The stiffness of the spring is 8 N/mm and the total sleeve movement is 15 mm from the mean position. Allowing for a constant frictional force of 15 N acting at the sleeve, determine the speed range of the governor in the lowest and highest sleeve positions. Neglect the obliquity of the ball arms.

Solution

Refer to Fig.9.24.

Given: $w = 30$ N, $a = 150$ mm, $b = 100$ mm, $r = 120$ mm, $N = 150$ rpm, $h = 15$ mm, $F = 15$ N, $k = 8$ N/mm

$$r_1 = r + h\,a/b = 120 + 15 \times 150/100 = 142.5 \text{ mm}$$

$$r_2 = r - h\,a/b = 120 - 15 \times 150/100 = 97.5 \text{ mm}$$
$$F_{c1} = (w/g)\,\omega_1^2 r_1 = (30/9.81) \times \omega_1^2 \times 0.1425 = 0.43578\,\omega_1^2$$
$$F_{c2} = (w/g)\,\omega_2^2 r_2 = (30/9.81) \times \omega_2^2 \times 0.0975 = 0.29816\,\omega_2^2$$

Total compression of spring = $2h = 30$ mm

$$S_1 = S_2 = 2kh = 8 \times 30 = 240 \text{ N}$$
$$F_{c1} \times a = (S_1 + F)\,b/2$$
$$0.43578\,\omega_1^2 \times 0.15 = (240 + 15) \times 0.1/2$$
$$\omega_1^2 = 195.05,\ \omega_1 = 13.966 \text{ rad/s},\ N_1 = 133.4 \text{ rpm}$$
$$F_{c2} \times a = (S_2 - F)\,b/2$$
$$0.29816\,\omega_2^2 \times 0.15 = (240 - 15) \times 0.1/2$$
$$\omega_2^2 = 251.54,\ \omega_2 = 15.86 \text{ rad/s},\ N_2 = 151.45 \text{ rpm}$$

Range of speed at the highest position = $N - N_1 = 150 - 133.4 = 16.6$ rpm

Range of speed at the lowest position = $N_2 - N = 151.45 - 150 = 1.45$ rpm

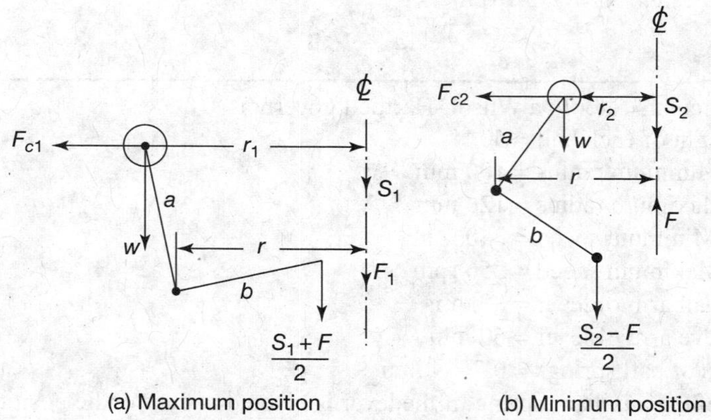

Fig.9.24 Maximum and minimum positions of Hartnell governor

Example 9.21

In a spring controlled governor of the Hartung type, the lengths of the horizontal and vertical arms of the bell crank lever are 120 mm and 90 mm, respectively. The fulcrum of the bell crank lever is at a distance of 120 mm from the axis of rotation of the governor. Each revolving weight is 100 N. The stiffness of the spring is 25 N/mm. If the length of each spring is 120 mm when the radius of rotation is 75 mm and the equilibrium speed is 350 rpm, find the free length of the spring. If the radius of rotation increases to 120 mm, what will be the corresponding percentage increase in speed?

■ Solution

Given: $a = 90$ mm, $b = 120$ mm, $w = 100$ N, $r = 120$ mm, $k = 25$ N/mm

(i) When radius of rotation, $r = 75$ mm

$$\omega = 2\pi \times 350/60 = 36.652 \text{ rad/s}$$
$$F_c = (w/g)\,\omega^2 r = (100/9.81) \times (36.652)^2 \times 0.075 = 1027 \text{ N}$$

Distance of weight of ball from the fulcrum towards the spindle,

$c = 120 - 75 = 45$ mm

Taking moments about the fulcrum, we have

$F_c \times (a^2 - c^2)^{0.5} = w \times c + S \times a$

$1027 \times (90^2 - 45^2)^{0.5} = 100 \times 45 + S \times 90$

$S = 839.4$ N

Compression of spring = 839.4/25 = 33.57 mm

Length of each spring = 120 mm

Free length of each spring = 120 + 33.57 = 153.57 mm

(ii) When radius is 120 mm,

Additional compression of spring = 120 – 75 = 45 mm

Force exerted by spring, $S_1 = 839.4 + 25 \times 45 = 1964.4$ N

At this position the bell crank lever arms are horizontal and vertical.

$F_{c1} = S_1$

$(100/9.81) \times \omega_1^2 \times 0.12 = 1964.4$

$\omega_1 = 40.07$ rad/s, $N_1 = 382.7$ rpm

Example 9.22

The following particulars refer to a Wilson-Hartnell governor:

Weight of each ball = 50 N

Minimum radius = 100 mm

Maximum radius = 120 mm

Minimum speed = 240 rpm

Maximum speed = 256 rpm

Length of ball arm of lever = 80 mm

Length of sleeve arm of lever = 60 mm

Combined stiffness of ball springs = 0.75 N/mm

Find the stiffness of the auxiliary spring, if the lever is pivoted at the middle position.

Solution

Given: $w = 50$ N, $r_1 = 100$ mm, $r_2 = 120$ mm, $N_1 = 240$ rpm, $N_2 = 256$ rpm,

$a = 80$ mm, $b = 60$ mm, $k_b = 0.75$ N/mm, $x = y$, $k_a = ?$

$F_{c1} = (w/g)\, \omega_1^2 r_1 = (50/9.81)\, (2\pi \times 240/60)^2 \times 0.1 = 321.94$ N

$F_{c2} = (w/g)\, \omega_2^2 r_2 = (20/9.81)\, (2\pi \times 256/60)^2 \times 0.12 = 439.56$ N

$4k_b + k_a \times 0.5\, (by/ax)^2 = (F_{c1} - F_{c2})/(r_1 - r_2)$

$4 \times 0.75 + k_a \times 0.5\, (60/80) = (321.94 - 439.56)/(100 - 120)$

$k_a = 10.24$ N/mm

Example 9.23

A Porter governor has arms 250 mm long. The upper arms are pivoted on the axis of rotation and the lower arms are attached to a sleeve at a distance of 30 mm from the axis. Each ball has a weight of 20 N and the weight of the load on the sleeve is 200 N. If the radius of rotation of the balls at a speed of 260 rpm is 100 mm, find the speed of the governor after the sleeve has lifted 50 mm. Also determine the effort and power of the governor.

■ **Solution**

Given: $l = 250$ mm, $c = 30$ mm, $w = 20$ N, $W = 200$ N, $N = 260$ rpm, $r = 100$ mm, $h = 50$ mm.

$$r_o = r - c = 100 - 30 = 70 \text{ mm}$$
$$\omega_1 = 2\pi \times 260/60 = 27.227 \text{ rad/s}$$
$$h_1 = [l^2 - r^2]^{0.5} = [250^2 - 100^2]^{0.5} = 229.13 \text{ mm}$$
$$l_o = [l^2 - r_0^2]^{0.5} = [250^2 - 70^2]^{0.5} = 240 \text{ mm}$$
$$h_2 = h_1 - 50 = 229.13 - 50 = 179.13 \text{ mm}$$
$$r_2 = [l^2 - h_2^2]^{0.5} = [250^2 - (179.13)^2]^{0.5} = 174.4 \text{ mm}$$
$$\tan\alpha = r_2/h_2 = 174.4/179.13 = 0.9736$$
$$r_{o2} = r_2 - c = 174.4 - 30 = 144.4 \text{ mm}$$
$$l_{o2} = [l^2 - r_{o2}^2]^{0.5} = [250^2 - (144.4)^2]^{0.5} = 204 \text{ mm}$$
$$\tan\beta = r_{02}/l_{o2} = 144.4/204 = 0.7078$$
$$k = \tan\beta/\tan\alpha = 0.7078/0.9736 = 0.727$$
$$h_2 = (g/\omega^2)\,[1 + (1 + k)\{W/(2w)\}]$$
$$0.17913 = (9.81/\omega_2^2)\,[1 + 1.727 \times 200/20]$$
$$\omega_2^2 = 527.65$$
$$\omega_2 = 22.97 \text{ rad/s}$$
$$N = 60 \times 22.97/(2\pi) = 219.35 \text{ rpm}$$

Percentage decrease in speed, $s = (N_1 - N_2)/N_1$

$= (260 - 219.35)/260 = 0.1563$ or 15.63%

Effort, $P = s\,(w + W) = 0.1563\,(20 + 200) = 34.392$ N

Total lift of sleeve, $x = 2h = 100$ mm

Power $= Px = 34.392 \times 0.1 = 3.4392$ N m

Example 9.24

The upper arms of a Porter governor are pivoted on the axis of rotation and the lower arms are pivoted to the sleeve at a distance of 40 mm from the axis of rotation. The length of each arm is 350 mm and the weight of each ball is 50 N. If the equilibrium speed is 240 rpm, when the radius of rotation is 200 mm, find the weight of the sleeve. If the friction is equivalent to a force of 50 N at the sleeve, find the coefficient of insensitiveness at 200 mm radius.

■ **Solution**

Given: $c = 40$ mm, $l = 350$ mm, $w = 50$ N, $N_1 = 240$ rpm, $r = 200$ mm, $W = ?$, $F = 50$ N.

$$h = [l^2 - r^2]^{0.5} = [350^2 - 200^2]^{0.5} = 287.23 \text{ mm}$$
$$r_o = r - c = 200 - 40 = 160 \text{ mm}$$
$$l_o = [l^2 - r_o^2]^{0.5} = [350^2 - 160^2]^{0.5} = 311.29 \text{ mm}$$
$$\tan\alpha = r/h = 200/287.23 = 0.6963$$
$$\tan\beta = r_o/l_o = 160/311.69 = 0.5133$$
$$k = \tan\beta/\tan\alpha = 0.5133/0.6963 = 0.7372$$
$$h = (g/\omega_1^2)\,[1 + (1 + k)\{(W/(2w)\}]$$
$$0.28723 = [9.81/(25.13)^2]\,[1 + 1.7372 \times W/100]$$

$$W = 1007 \text{ N}$$
$$F_b = \tan\alpha \times F \times (1 + k)/2$$
$$= 0.6963 \times 25 \times 1.7372 = 30.24 \text{ N}$$

Considering friction:

$$h = (g/\omega_2^2)\,[1 + (1 + k)\,\{(W + F)/(2w)\}]$$
$$0.28723 = [\,9.81/\omega_2^2]\,[1 + 1.7372 \times 1057/100]$$
$$\omega_2^2 = 661.293,\ \omega_2 = 25.716 \text{ rad/s},\ N_2 = 245.57 \text{ rpm}$$
$$N_{mean} = (N_1 + N_2)/2 = (240 + 245.57)/2 = 242.783 \text{ rpm} \cong 242.8 \text{ rpm}$$

Coefficient of insensitiveness $= (N_1 - N_2)/N_{mean}$

$$= (245.57 - 240)/242.8 = 0.0229$$

Example 9.25

A Pickering governor employed in a gramophone consists of three leaf springs each 40 mm long, 5 mm wide and 1.5 mm thick. Each of the springs has disc of mass 20 g, attached to the centre. The distance between the axis of the governor spindle and the centre of gravity of the disc when at rest is 10 mm. Find the equilibrium speed of the turntable to which this governor is fixed, if the ratio of the governor speed to the speed of the turntable is 10 and the lift of the sleeve is 0.75 mm. Take modulus of elasticity of leaves as 210 GPa.

■ Solution

Given: $L = 40$ mm, $b = 5$ mm, $t = 1.5$ mm, $m = 20$ gram, $a = 10$ mm, $N = ?$, $h = 0.75$ mm, $E = 210$ GPa, $i = 10$.

$$h = 2.4\,\delta^2/L$$
$$0.75 = 2.4\,\delta^2/40,\ \delta = 3.535 \text{ mm}$$
$$I = bt^3/12 = 5 \times (1.5)^3/12 = 1.406 \text{ mm}^4$$
$$\delta = m\,(a + \delta)\,\omega^2 L^3/(192\text{ EI})$$
$$3.535 = 20 \times 10^{-3} \times 13.535 \times \omega^2 \times 40^3/(192 \times 210 \times 10^3 \times 1.406)$$
$$\omega^2 = 11567,\ \omega = 107.55 \text{ rad/s},\ N = 1027 \text{ rpm}$$

Speed of turntable $= N/10 = 102.7$ rpm

Example 9.26

In an inertia governor the disc has two weights, each weighing 50 N, attached to the ends of a 300 mm long rod. The rod is pivoted at its centre to the disc such that the pivot falls on the vertical axis of the disc. The rod is horizontal in its neutral position and is 50 mm from the horizontal axis of the disc. Each of the weights is circular in shape of 50 mm radius. Find the torque required about the pivot of the rod if the disc revolves at 300 rpm and receives an angular acceleration of 1 rad/s^2.

■ Solution

Refer to Fig.9.25.

Given: $W = 50$ N, $N = 300$ rpm, $\alpha = 1$ rad/s^2.

$$AG = 150 \text{ mm},\ OA = 50 \text{ mm}$$
$$r = OG = \sqrt{(AG)^2 + (OA)^2} = [150^2 + 50^2]^{0.5} = 158.1 \text{ mm}$$
$$\tan\theta = AG/OA = 150/50 = 3$$
$$\theta = 71.565°$$

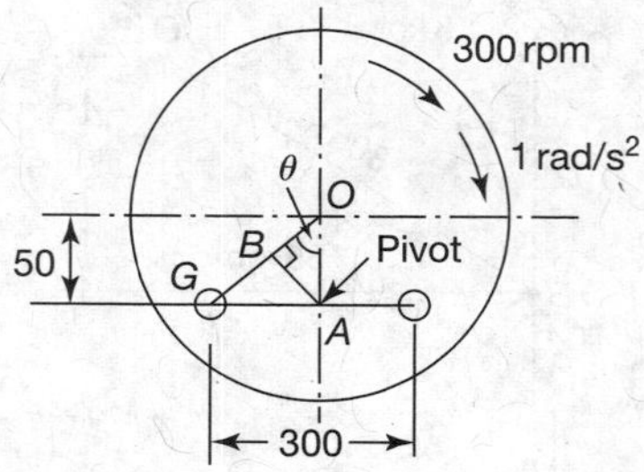

Fig.9.25 Inertia governor

$$\sin\theta = AB/OA = 0.94868$$
$$AB = 50 \times 0.94868 = 47.43 \text{ mm}$$
$$\omega = 2\pi \times 300/60 = 31.416 \text{ rad/s}$$

Centrifugal force, $F_c = (w/g)\,\omega^2 r = (50/9.81)\,(31.416)^2 \times 0.1581 = 795.3$ N

Moment of F_c about $A = F_c \times AB = 795.3 \times 0.04743 = 37.72$ N m

Inertia force acting on the ball perpendicular to $OG = (W/g)\, r\, \alpha$

$$= (50/9.81) \times 0.1581 \times 1 = 0.8058 \text{ N}$$

Example 9.27

The controlling force F_c for a spring loaded governor is given by $F_c = 300r - 80$ N where r is in mm. The weight of each ball is 50 N and the extreme radii of rotation of the balls are 100 mm and 175 mm, respectively. Find the maximum and minimum equilibrium speeds. If the friction of the governor mechanism is equivalent to a radial force of 5 N at each ball, find the extent to which the equilibrium speeds are affected at the extreme radii of rotation.

■ Solution

Given: $w = 50$ N, $r_1 = 100$ mm, $r_2 = 175$ mm

$F_c = 300r - 80$

At $r = 100$ mm $\quad (50/9.81) \times \omega_1^2 \times 0.1 = 300 \times 100 - 80$

$\omega_1^2 = 58703,\ \omega_1 = 242.287$ rad/s, $N_1 = 2313.7$ rpm

At $r = 175$ mm, $\quad (50/9.81) \times \omega_2^2 \times 0.1 = 300 \times 175 - 80$

$\omega_2^2 = 58770,\ \omega_2 = 242.42$ rad/s, $N_2 = 2315$ rpm

Considering friction force, $F = 5$ N

$F_c - F = (w/g) \times \omega^2 \times r$

At $r = 100$ mm, $\quad (50/9.81) \times \omega_1^2 \times 0.1 = 300 \times 100 - 80 - 5$

$\omega_1^2 = 58693,\ \omega_1 = 242.27$ rad/s, $N_1 = 2313.48$ rpm

At $r = 175$ mm, $\quad (50/9.81) \times \omega_2^2 \times 0.1 = 300 \times 175 - 80 - 5$

$\omega_2^2 = 58764.7,\ \omega_2 = 242.414$ rad/s, $N_2 = 2314.88$ rpm

Example 9.28

A Proell governor has arms of 305 mm length. The upper arms are hinged on the axis of rotation, whereas the lower arms are pivoted at a distance of 38 mm from the axis of rotation. The extension of lower arms to which the balls are attached are 102 mm long. Each ball mass is 4.8 kg and the load on the sleeve is 54 kg. At minimum radius of rotation of 165 mm, the extensions are parallel to the governor axis, determine the equilibrium speed at radii 165 mm and 216 mm.

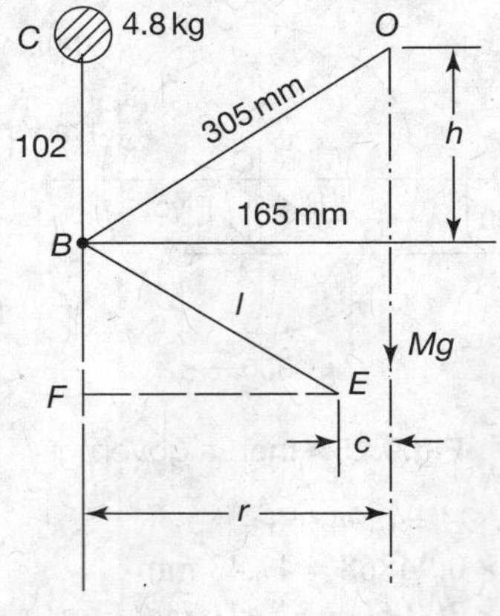

Fig.9.26 Proell governor

■ Solution

Refer to Fig.9.26.

Given: $l = 305$ mm, $c = 38$ mm, $m = 4.8$ kg, $M = 54$ kg, $r_1 = 165$ mm, $r_2 = 216$ mm

$FE = r - c = 165 - 38 = 127$ mm

$BF = [l^2 - FE^2]^{0.5} = [305^2 - 127^2]^{0.5} = 277.3$ mm

$h_1 = [l^2 - r_1^2]^{0.5} = [305^2 - 165^2]^{0.5} = 256.5$ mm

$\omega_1^2 = (BF/CF)\,[1 + M/m\}]\,(g/h_1)$

$= (277.3/379.3)\,[1 + 54/4.8]\,(9.81/0.2565)$

$= 342.52$

$\omega_1 = 18.5$ rad/s

$N_1 = 18.5 \times 60/(2\pi) = 176.73$ rpm

For $r_2 = 216$ mm, $FE = 216 - 38 = 178$ mm

$BF = [l^2 - FE^2]^{0.5} = [305^2 - 178^2]^{0.5} = 247.67$ mm

$h_2 = [l^2 - r_2^2]^{0.5} = [305^2 - 216^2]^{0.5} = 215.33$ mm

$\omega_2^2 = (BF/CF)\,[1 + M/m\}]\,(g/h_1)$

$= (247.67/376.67)\,[1 + 54/4.8]\,(9.81/0.21533)$

$= 375.937$

$\omega_2 = 19.389$ rad/s

$N_2 = 19.389 \times 60/(2\pi) = 185.15$ rpm

Example 9.29

A centrifugal governor shown in Fig.9.27 has two masses each of weight w connected by a helical spring. The arms carrying the weights are parallel to the axis of rotation at the speed of 900 rpm. If the speed is increased by 10%, it requires a force of 30 N to maintain the sleeve at the same position. Determine (a) the value of the masses, (b) the stiffness of the spring and its initial extensions if the sleeve moves by 10 mm for a change of speed of 250 rpm.

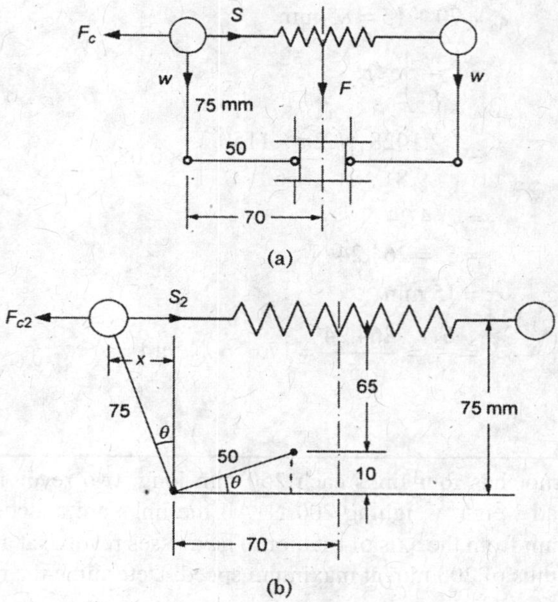

Fig.9.27 Centrifugal governor

Solution

Refer to Fig.9.27.

Given: $N = 900$ rpm, $r = 70$ mm, $a = 75$ mm, $b = 50$ mm

(a) $$\omega = \frac{2\pi \times 900}{60} = 94.248 \text{ rad/s}$$

Let S = spring force. Taking moments about fulcrum, we have

$$\left(\frac{w}{g}\right) \times \omega^2 \times r \times 75 = S \times 75$$

$$\left(\frac{w}{g}\right) \times \omega^2 \times r = S \qquad (1)$$

When speed is increased by 10%, then

$$\left(\frac{w}{g}\right) \times (1.1\omega)^2 \times r \times 75 = S \times 75 + 30 \times 70 \qquad (2)$$

Solving Eqs. (1) and (2), we get

$$w = 2.1028 \text{ N}$$

(b) $$N_2 = 900 + 250 = 1150 \text{ rpm}$$

$$\sin\theta = \frac{10}{50} = 0.2$$

$$\theta = 11.537°$$

$$x = 75 \sin\theta = 15 \text{ mm}$$

$$r_2 = 70 + 15 = 85 \text{ mm}$$

$$F_{c2} = \left(\frac{w}{g}\right)\omega_2^2 r_2$$

$$= \left(\frac{2.1028}{9.81}\right)\left(\frac{2\pi \times 1150}{60}\right)^2 \times 0.085$$

$$= 264.24 \text{ N}$$

$$F_{c2} = S_2 = 264.24 \text{ N}$$

Extension of spring, $x = 15$ mm

Stiffness of spring, $k = \frac{S_2}{x} = \frac{264.24}{15} = 17.616 \text{ N/mm}$

Example 9.30

A loaded Porter governor has four links each 250 mm long, two revolving masses each weighing 30 N and a central dead weight weighing 200 N. All the links are attached to respective sleeves at radial distances of 40 mm from the axis of rotation. The masses revolve at a radius of 150 mm at minimum speed and at a radius of 200 mm at maximum speed. Determine the range of speed.

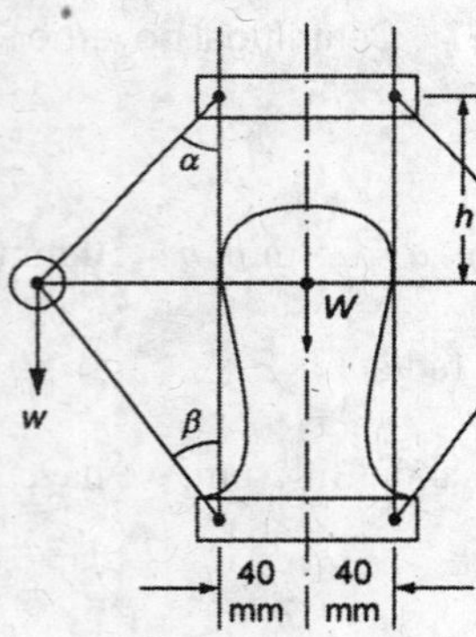

Fig.9.28 Loaded Porter governor

■ Solution

Refer to Fig.9.28.

Given: $l = 250$ mm, $w = 30$ N, $W = 200$ N, $c = 40$ mm, $r_1 = 150$ mm, $r_2 = 200$ mm

For $r = r_1$ $\quad r_o = r_1 - c = 150 - 40 = 110$ mm

$$h_1 = (l^2 - r_o^2)^{0.5} = (250^2 - 110^2)^{0.5} = 224.5 \text{ mm}$$

$$h_1 = \left(\frac{g}{\omega_1^2}\right)\left(1 + \frac{W}{w}\right)$$

$$0.2245 = \left(\frac{9.81}{\omega_1^2}\right)\left(1 + \frac{200}{30}\right)$$

$$\omega_1^2 = 335.01, \ \omega_1 = 18.3 \text{ rad/s}, \ N_1 = 174.78 \text{ rpm}$$

For $r = r_2$, $\quad r_o = r_2 - c = 200 - 40 = 160$ mm

$$h_2 = (l^2 - r_o^2)^{0.5} = (250^2 - 160^2)^{0.5} = 192.1 \text{ mm}$$

$$h_2 = \left(\frac{g}{\omega_2^2}\right)\left(1 + \frac{W}{w}\right)$$

$$0.1921 = \left(\frac{9.81}{\omega_2^2}\right)\left(1 + \frac{200}{30}\right)$$

$$\omega_2^2 = 391.51, \omega_2 = 19.786 \text{ rad/s}, N_2 = 188.95 \text{ rpm}$$

Range of speed = 192.1 – 188.95 = 3.15 rpm

Example 9.31

In a loaded Proell governor shown in Fig.9.29 each ball weighs 3 kg and the central sleeve weighs 25 kg. The arms are of 200 mm length and pivoted about axis displaced from the central axis of rotation by 38.5 mm, $y = 238$ mm, $x = 303.5$ mm, $CE = 85$ mm, $MD = 142.5$ mm. Determine the equilibrium speed.

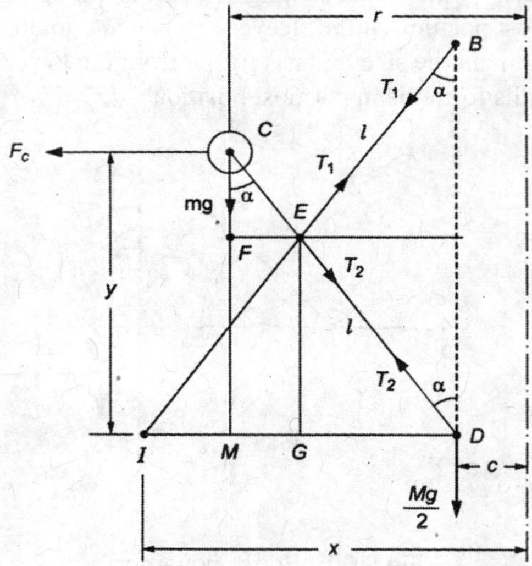

Fig.9.29 Loaded Proell governor

■ **Solution**

Refer to Fig.9.29.

Given: $\quad m = 3$ kg, $M = 25$ kg, $l = 200$ mm, $c = 37.5$ mm, $x = 303.5$ mm, $y = 238$ mm, $CE = 85$ mm, $MD = 142.5$ mm

$$\tan\alpha = \frac{MD}{CM} = \frac{MD}{y} = \frac{142.5}{238} = 0.59874$$

$$\alpha = 30.91°$$

$$\cos\alpha = 0.858$$

$$FE = CE \sin\alpha = 43.66 \text{ mm}$$

$$GD = MD - FE = 98.84 \text{ mm}$$
$$EG = FM = CM - CF = y - CE \cos\alpha = 165.07 \text{ mm}$$
$$r = c + MD = 180 \text{ mm}$$

At point D, $T_2 \cos\alpha = \dfrac{Mg}{2}$ or $T_2 = 142.92$ N

At point E, $T_1 \cos\alpha = T_2 \cos\alpha$ or $T_1 = T_2 = 142.92$ N

$$F_c = mr\omega^2 = 3 \times 0.18 \times \omega^2 = 0.54\,\omega^2$$

Taking moments about D, we have

$$F_c \times y + mg \times MD - T_1 \cos\alpha \times GD - T_1 \cos\alpha \times EG = 0$$
$$0.54\,\omega^2 \times 238 + 3 \times 9.81 \times 142.5 - 142.92 \times 0.858 \times (98.84 + 165.07) = 0$$
$$\omega^2 = 219.17,\ \omega = 14.8 \text{ rad/s},\ N = 141.37 \text{ rpm}$$

Example 9.32

The arms of a Porter governor are pivoted on the governor axis and are each 250 mm long. The mass of each ball is 0.5 kg and mass of the sleeve is 2 kg. The arms are inclined at an angle of 30° to the governor axis in the lowermost position of the sleeve. Lift is equal to 50 mm. Determine the force of friction if the speed at the moment the sleeve starts lifting from the lowermost position is the same as the speed at the moment it falls form the uppermost position.

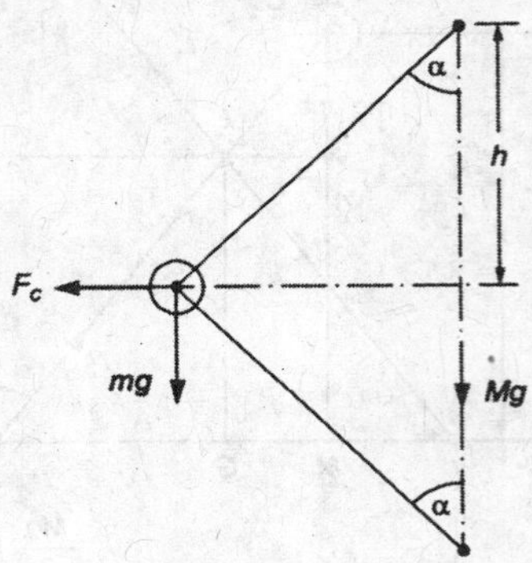

Fig.9.30 Porter governor

Solution

Refer to Fig.9.30.

Given: $l = 250$ mm, $m = 0.5$ kg, $M = 2$ kg, $\alpha = \beta = 30°$, lift, $h = 50$ mm

$$h_1 = l\cos\alpha = 250\cos 30° = 217.5 \text{ mm}$$
$$\text{Lift} = 2(h_1 - h_2)$$
$$25 = 2\,(217.5 - h_2)$$
$$h_2 = 204 \text{ mm}$$

When the sleeve starts lifting from lower position,

$$h_1 = \left(\frac{g}{\omega^2}\right)\left(1 + \frac{Mg + F}{mg}\right)$$

$$0.2165 = \left(\frac{9.81}{\omega^2}\right)\left(1 + \frac{2 \times 9.82 + F}{0.5 \times 9.81}\right)$$

$$\omega^2 = 227.559 + 9.238\,F \quad (1)$$

When the sleeve starts lowering down,

$$h_2 = \left(\frac{g}{\omega^2}\right)\left(1 + \frac{Mg - F}{mg}\right)$$

$$0.204 = \left(\frac{9.81}{\omega^2}\right)\left(1 + \frac{2 \times 9.82 - F}{0.5 \times 9.81}\right)$$

$$\omega^2 = 240.49 - 9.804\,F \quad (2)$$

Solving Eqs. (1) and (2), we get

$$F = 0.729 \text{ N}$$

Example 9.33

In a spring-controlled governor of the Hartung type, the length of the horizontal and vertical arms of the bell crank levers are 100 mm and 80 mm, respectively. The fulcrum of the bell crank lever is at a distance of 120 mm from the axis of the governor. Each revolving mass is 8 kg. The stiffness of the spring is 20 kN/m. If the length of each spring is 120 mm, when the radius of rotation is 70 mm and the equilibrium speed is 380 rpm, find the free length of the spring. If the radius of rotation increases to 120 mm, what will be the corresponding percentage increase in speed? Ignore sleeve mass.

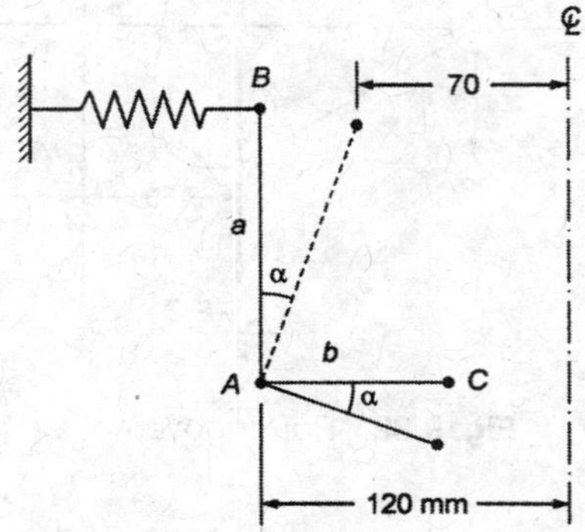

Fig.9.31 Harting governor

■ Solution

Refer to Fig.9.31.

Given: $a = 80$ mm, $b = 100$ mm, $m = 8$ kg, $k = 20$ kN/m, $r = 70$ mm, $N = 380$ rpm

$$\omega = \frac{2\pi \times 380}{60} = 39.79 \text{ rad/s}$$

Let δ = extensions of spring, *m*

Spring force, $S = 20\delta$ N

$$F_c = mr\,\omega^2 = 8 \times 0.70 \times (39.79)^2 = 887.77 \text{ N}$$

$$20\,\delta = 887.7, \quad \text{or} \quad \delta = 44.34 \text{ mm}$$

$$\text{Free length of spring} = 120 - 44.34 = 75.66 \text{ mm}$$

When r increase to 120 mm, $F_c = 8 \times 0.12 \times \omega_1^2 = 0.96\ \omega_1^2$ N

Spring force, $S = 20 \times [(r_1 - r_2) + \delta]$

$$= 20 \times [(120 - 70) + 44.34] = 1887.8 \text{ N}$$

For $F_c = S$, we have

$$0.96\,\omega_1^2 = 1887.8$$

$$\omega_1^2 = 1965.4, \omega_1 = 44.33 \text{ rad/s}, N_1 = 423.35 \text{ rpm}$$

$$\text{Percentage increases in speed} = \frac{423.35 - 380}{380} = 11.4\%$$

Example 9.34

The length of the arms of a Porter governor is 300 mm long. The upper and lower arms are pivoted to links at 50 mm and 60 mm, respectively, from the axis of rotation. The mass of each ball is 5 kg and the sleeve is of mass 60 kg. The frictional force on the sleeve is 35 N. Determine the range of speed for extreme radii of rotation of 120 mm and 150 mm.

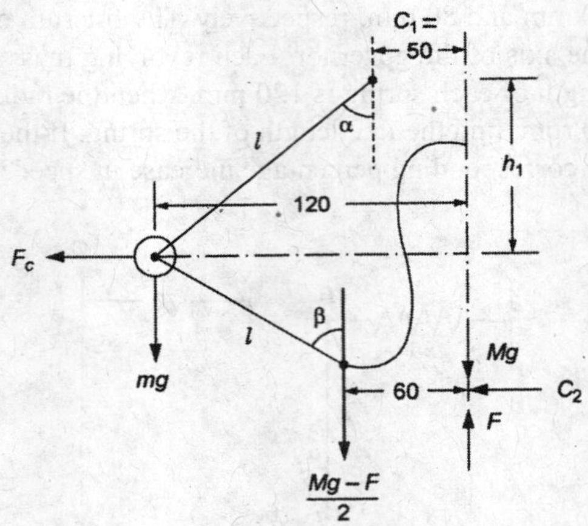

Fig. 9.32 Porter governor

■ **Solution**

Refer to Fig.9.32.

Given: $l = 300$ mm, $c_1 = 50$ mm, $c_2 = 60$ mm, $r_1 = 120$ mm, $r_2 = 150$ mm, $m = 5$ kg, $M = 60$ kg, $F = 35$ N

$$r_o = r_1 - c_1 = 120 - 50 = 70 \text{ mm}$$

$$h_1 = (l^2 - r_o^2)^{0.5} = (300^2 - 70^2)^{0.5} = 291.72 \text{ mm}$$

$$\sin\alpha = \frac{r_o}{l} = \frac{70}{300} = 0.233$$

$$\alpha = 13.4934°, \tan\alpha = 0.24$$

$$\sin\beta = \frac{r_1 - c_2}{l} = \frac{60}{300} = 0.2$$

$$\beta = 11.537°,\ \tan\beta = 0.204$$

$$k = \frac{\tan\beta}{\tan\alpha} = \frac{0.204}{0.24} = 0.8505$$

For $r = r_1$, the sleeve lowers and the force of friction acts upwards.

$$h_1 = \left(\frac{g}{\omega_1^2}\right)\left[1 + (1+k)\left\{\frac{W - F}{2w}\right\}\right]$$

$$0.29172 = \left(\frac{9.81}{\omega_1^2}\right)\left(1 + 1.8505 \times \left\{\frac{60 \times 9.81 - 35}{2 \times 5 \times 9.81}\right.\right.$$

$$\omega_1^2 = 384.8,\ \omega_1 = 19.616 \text{ rad/s},\ N_1 = 187.3 \text{ rpm}$$

For $r = r_2$, the sleeve rises and F acts downwards.

$$r_o = r_2 - c_1 = 150 - 50 = 100 \text{ mm}$$

$$h_2 = (l^2 - r_o^2)^{0.5} = (300^2 - 100^2)^{0.5} = 282.84 \text{ mm}$$

$$\sin\alpha = \frac{r_o}{l} = \frac{100}{300} = 0.333$$

$$\alpha = 19.47°,\ \tan\alpha = 0.3535$$

$$\sin\beta = \frac{r_2 - c_2}{l} = \frac{90}{300} = 0.3$$

$$\beta = 17.4576°,\ \tan\beta = 0.3144$$

$$k = \frac{\tan\beta}{\tan\alpha} = \frac{0.3144}{0.3535} = 0.8896$$

For $r = r_1$, the sleeve lowers and the force of friction acts upwards.

$$h_2 = \left(\frac{g}{\omega_2^2}\right)\left[1 + (1+k)\left\{\frac{W + F}{2w}\right\}\right]$$

$$0.28284 = \left(\frac{9.81}{\omega_2^2}\right)\left(1 + 1.8896 \times \frac{60 \times 9.81 + 35}{2 \times 5 \times 9.81}\right)$$

$$\omega_2^2 = 451.3,\ \omega_2 = 21.244 \text{ rad/s},\ N_2 = 202.86 \text{ rpm}$$

Range of speed = 202.86 – 187.3 = 15.56 rpm

Example 9.35

In a Porter governor, all four arms are of equal length of 250 mm and are hinged on the spindle axis. Mass of each ball is 2.5 kg and sleeve mass is 25 kg. The force of friction at the sleeve is 30 N. The inclination of arms to spindle axis is 30° and 45° in the lowest and highest position respectively. Calculate (a) the sleeve lift, (b) speeds at the bottom, middle and top of the sleeve by neglecting and considering friction.

■ Solution

Given: $l = 250$ mm, $m = 2.5$, $M = 25$ kg, $F = 30$ N, $\alpha = 30°$ and $45°$

(a) $h_1 = l\cos 30° = 250\cos 30° = 217.5$ mm
$h_2 = l\cos 45° = 250\cos 45° = 177.78$ mm
The arms being of equal length, form a rhombus. Hence,
Lift of sleeve = 2 (217.5 − 177.78) = 79.45 mm

(b) Neglecting friction:

At the bottom position,
$$\omega_1^2 = \left(\frac{g}{h_1}\right)\left(1+\frac{M}{m}\right) = \left(\frac{9.81}{0.2165}\right)\left(1+\frac{25}{2.5}\right) = 498.43$$

$$\omega_1 = 22.32 \text{ rad/s}, \quad N_1 = 213.2 \text{ rpm}$$

At the top position,
$$N_2 = N_1\left(\frac{h_1}{h_2}\right)^{0.5} = 213.2\left(\frac{217.5}{177.78}\right)^{0.5} = 235.94 \text{ rpm}$$

At the middle position, $h = 0.5(h_1 + h_2) = 0.5\,(217.5 + 177.78) = 197.64$ mm

$$N_m = N_1\left(\frac{h_1}{h}\right)^{0.5} = 213.2\left(\frac{217.5}{197.64}\right)^{0.5} = 223.65 \text{ rpm}$$

(c) Considering friction:
At the lowest position, (i) when the sleeve is lowering

$$\omega_1^2 = \left(\frac{g}{h_1}\right)\left(1+\frac{Mg-F}{mg}\right) = \left(\frac{9.81}{0.2165}\right)\left(1+\frac{25\times 9.81+30}{2.5\times 9.81}\right) = 443$$

$$\omega_1 = 21.05 \text{ rad/s}, \quad N_1 = 201 \text{ rpm}$$

(i) When sleeve is rising

$$\omega_1^2 = \left(\frac{g}{h_1}\right)\left(1+\frac{Mg+F}{mg}\right) = \left(\frac{9.81}{0.2165}\right)\left(1+\frac{25\times 9.81+30}{2.5\times 9.81}\right) = 553.86$$

$$\omega_1 = 23.53 \text{ rad/s}, \quad N_1 = 224.7 \text{ rpm}$$

At the highest position, (i) when the sleeve is lowering

$$\omega_2^2 = \left(\frac{g}{h_2}\right)\left(1 + \frac{Mg - F}{mg}\right)$$

$$= \left(\frac{9.81}{0.17678}\right)\left(1 + \frac{25 \times 9.81 - 30}{2.5 \times 9.81}\right)$$

$$= 542.54$$

$$\omega_2 = 23.29 \text{ rad/s}, \quad N_1 = 222.4 \text{ rpm}$$

(ii) When sleeve is rising

$$\omega_2^2 = \left(\frac{g}{h_2}\right)\left(1 + \frac{Mg + F}{mg}\right)$$

$$= \left(\frac{9.81}{0.2165}\right)\left(1 + \frac{25 \times 9.81 + 30}{2.5 \times 9.81}\right)$$

$$= 680.56$$

$$\omega_2 = 27.08 \text{ rad/s}, \quad N_1 = 249.1 \text{ rpm}$$

Example 9.36

The following data refer to a Proell governor:
Mass of each ball = 5 kg
Mass of sleeve = 60 kg
Length of each arm = 250 mm
Distance of pivots of lower arms from axis of rotation = 30 mm
Length of extensions of lower arms = 100 mm

The extensions arms are parallel to the axis of the governor at the minimum radius. Determine the equilibrium speeds corresponding to extreme radii of 160 mm and 220 mm.

■ Solution

Refer to Fig.9.33.

Given: $m = 5$ kg, $M = 60$ kg, $l = 250$ mm, $c = 30$ mm, $BC = 100$ mm, $r_1 = 160$ mm, $r_2 = 220$ mm

$$AF = r_o = r_1 - c = 160 - 30 = 130 \text{ mm}$$

$$BF = l_o = (l^2 - r_o^2)^{0.5} = \left(250^2 - 130^2\right)^{0.5} = 213.54 \text{ mm}$$

$$FC = e = BF + BC = 213.54 + 100 = 313.54 \text{ mm}$$

$$\sin\beta = \frac{r_o}{l} = \frac{130}{250} = 0.52$$

$$\beta = 31.33°$$

$$AC = (e^2 + AF^2)^{0.5} = [(313.54)^2 + 130^2]^{1/2}\ 339.4 \text{ mm}$$

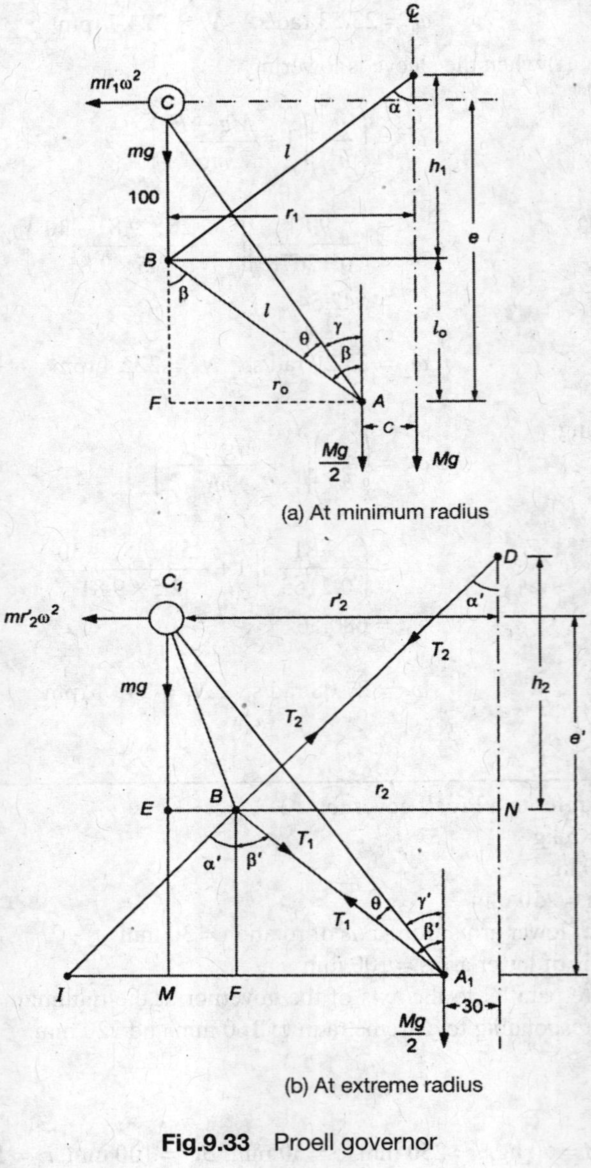

Fig.9.33 Proell governor

$$\tan\gamma = \frac{AF}{FC} = \frac{130}{313.54} = 0.41462$$

$$\gamma = 22.52°$$

$$\theta = \beta - \gamma = 31.33 - 22.528.8°$$

$$h_1 = (l^2 - r_1^2)^{0.5} = (250^2 - 160^2)^{0.5} = 192.1 \text{ mm}$$

$$\tan\alpha = \frac{r}{h_1} = \frac{160}{192.1} = 0.8329$$

$$\tan\beta = \frac{r_o}{l_o} = \frac{130}{213.54} = 0.6087$$

$$k = \frac{\tan\beta}{\tan\alpha} = \frac{0.6087}{0.8329} = 0.7309$$

$$CF = CB + BF = 100 + 213.54 = 313.54 \text{ mm}$$

$$\omega_1^2 = \left(\frac{BF}{CF}\right)\left(\frac{g}{h_1}\right)\left[1 + (1+k)\left(\frac{M}{2m}\right)\right]$$

$$= \left(\frac{213.54}{313.54}\right)\left(\frac{9.81}{0.1921}\right)\left(1 + \frac{1.7309 \times 60}{10}\right)$$

$$= 395.983$$

$$\omega_1 = 19.9 \text{ rad/s}, \quad N_1 = 190 \text{ rpm}$$

$$h_2 = (l^2 - r_2^2)^{0.5} = \left(250^2 - 220^2\right)^{0.5} = 118.74 \text{ mm}$$

$$\tan\alpha' = \frac{r_2}{h_2} = \frac{220}{118.74} = 1.853$$

$$\alpha' = 61.64°$$

$$r_o = r_2 - c = 220 - 30 = 190 \text{ mm}$$

$$BF = (l^2 - r_o^2)^{0.5} = \left(250^2 - 190^2\right)^{0.5} = 162.48 \text{ mm}$$

$$\tan\beta' = \frac{r_o}{BF} = \frac{190}{162.48} = 1.169$$

$$\beta' = 49.46°$$

At point A, $$T_1\cos\beta' = \frac{Mg}{2}$$

$$T_1 = \frac{60 \times 9.81}{2 \times 0.65} = 452.78 \text{ N}$$

At point B, $$T_1\cos\beta' = T_2\cos\alpha'$$

$$T_2 = \frac{60 \times 9.81}{2 \times 0.475} = 619.56 \text{ N}$$

$$A_1F = A_1B \sin\beta' = 250 \sin\beta' = 190 \text{ mm}$$

$$IF = BF \tan\alpha' = 162.48 \times 1.853 = 301 \text{ mm}$$

$$A_1F = A_1F + IF = 491 \text{ mm}$$

$$\gamma' = \beta' - \theta = 49.46° - 8.8° = 40.66°$$

$$AC \equiv A_1C_1$$

$$e' = A_1C_1 \cos\gamma' = AC\cos\gamma' = 339.4 \cos 40.66° = 287.46 \text{ mm}$$

$$r_2' = 30 + AC \sin\gamma' = 251.14 \text{ mm}$$

$$IM = (IF + A_1F + 30) - r_2' = 269.86 \text{ mm}$$

Taking moments about the lowest point, we have

$$mr_2'\omega_2^2 e = mg \times IM = \frac{MG}{2} \times IA_1$$

$$5 \times 0.25114 \times \omega_2^2 \times 0.25746 = 5 \times 9.81 \times 0.26986 + 60 \times 8.81 \times 0.491$$

$$\omega_2^2 = 487.9, \quad \omega_2 = 22.09 \text{ rad/s}, \quad N_2 = 210.93 \text{ rpm}$$

Example 9.37

In a spring-controlled governor, the controlling force curve is a straight line. The balls are 450 mm apart when the controlling force is 1450 N and 250 mm when it is 750 N. The mass of each ball is 8 kg. Determine the speed at which the governor runs when the balls are 300 mm apart. By how much should the initial tension be increased to make the governor isochronous? Also find the isochronous speed.

■ Solution

(i) $F_c = ar + b$

$$\text{For } r = 225 \text{ mm, } 1450 = 0.225\, a + b$$

$$\text{For } r = 125 \text{ mm, } 750 = 0.125\, r + b$$

Solving for a and b, we get

$$a = 7000, \quad b = -125$$

$$F_c = mr\omega^2 = 7000r - 125$$

When $r = 150$ mm, and $m = 8$ kg, we have

$$8 \times 0.15 \times \omega^2 = 7000 \times 0.15 - 125$$

$$\omega^2 = 770.83, \quad \omega = 27.764 \text{ rad/s}, \quad N = 265.1 \text{ rpm}$$

(ii) For governor to be isochronous, $b = 0$. This can be done by increasing initial tension by 125 N.

(iii) $8 \times r \times \omega^2 = 7000 \times r$

$$\omega^2 = 875, \quad \omega = 29.58 \text{ rad/s}, \quad N = 282.47 \text{ rpm}$$

Example 9.38

In a Porter governor, each arm is 250 mm long and is pivoted at the axis of rotation. The mass of each ball is 4.5 kg and the load on the sleeve is 25 kg. The extreme radii of rotation are 100 mm and 150 mm. Plot a graph of controlling force's radius of rotation and set off a speed scale along the ordinate corresponding to a radius of 150 mm.

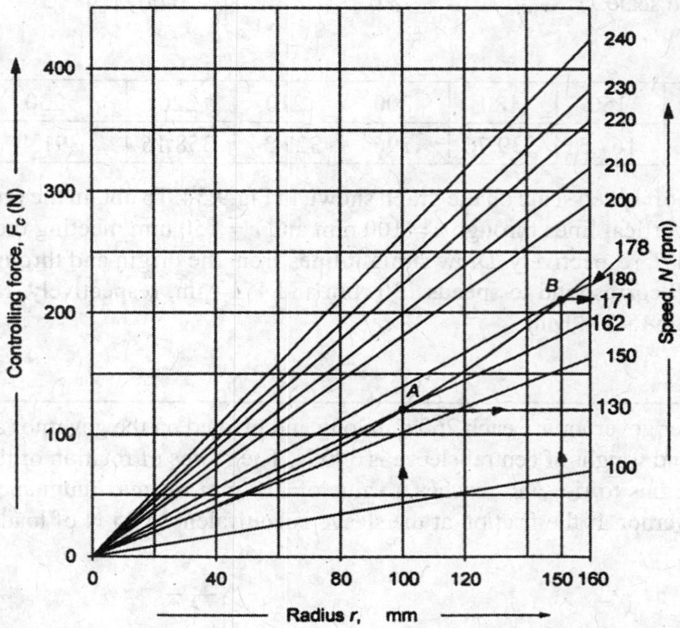

Fig.9.34 Controlling force curves for Porter governor

■ Solution

Controlling force for a Porter governor is,

$$F_c = \tan\alpha \left[mg + (1+k)\left\{\frac{Mg \pm F}{2}\right\}\right]$$

where

$$\tan\alpha = \frac{r}{h}, \quad k = \frac{\tan\beta}{\tan\alpha}$$

Here $k = 1$ and $F = 0$

$$F_c = \left(\frac{r}{h}\right)(m+M)g = \frac{(m+M)gr}{(l^2 - r^2)^{0.5}}$$

Here $l = 250$ mm, $m = 4.5$ kg, $M = 25$ kg

$$F_c = \frac{(4.5+25)\times 9.81\times r}{(250^2 - r^2)^{0.5}}$$

$$= \frac{289.395r}{(250^2 - r^2)^{0.5}}$$

r, mm	20	40	60	80	100	120	140	150	160
F_c, N	23.22	47.91	71.54	97.75	127.30	158.34	195.61	217.05	241.04

The graph has been plotted in Fig.9.34.

To set off speed scale, $F_c = mr\omega^2 = 4.5 \times 0.15 \times \left(\dfrac{2\pi N}{60}\right)^2 = 0.0074\ \text{N}^2$

N, rpm	100	150	180	200	210	220	230	240
F_c, N	74	167.5	239.76	296	327.34	358.16	391.46	427.24

Now mark off speed scale on the graph shown in Fig.9.34. To obtain the range of equilibrium speeds, draw vertical lines through r = 100 mm and r = 150 mm meeting the controlling force curve at A and B, respectively. Draw straight lines from the origin and through points A and B. Points A and B correspond to speeds 130 rpm and 171 rpm, respectively. The range of speed is from 162 rpm to 178 rpm.

Example 9.39

The arms of a Porter governor are each 260 mm long and pivoted on the governor axis. The weight of each ball is 50 N and weight of central sleeve is 300 N. The radius of rotation of the balls is 140 mm when the sleeve begins to rise and reaches a value of 210 mm for maximum speed. Determine the speed range of governor. If the friction at the sleeve is equivalent to 25 N of load at the sleeve, find the range of speed.

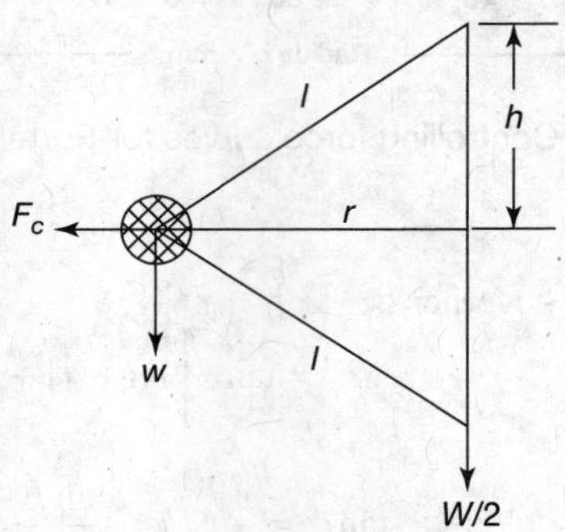

Fig.9.35 Porter governor

■ Solution

Refer to Fig.9.35.

Given: $l = 260$ mm, $w = 50$ N, $W = 300$ N, $r_1 = 140$ mm, $r_2 = 210$ mm, $F = 25$ N

$$h_1 = \sqrt{l^2 - r_1^2} = \sqrt{260^2 - 140^2} = 219.1\ \text{mm}$$

$$h_1 = \frac{g}{\omega_1^2}\left[1 + \frac{W}{w}\right]$$

$$\omega_1^2 = \frac{9.81}{0.2191}\left[1+\frac{300}{50}\right] = 313.42$$

$$\omega_1 = 17.7 \text{ rad/s}$$

$$N_1 = \frac{60\times 17.7}{2\pi} = 169 \text{ rpm}$$

$$h_2 = \sqrt{l^2 - r_2^2} = \sqrt{260^2 - 210^2} = 153.3 \text{ mm}$$

$$\omega_2^2 = \frac{9.81}{0.1533}\left[1+\frac{300}{50}\right] = 447.94$$

$$\omega_2 = 21.165 \text{ rad/s}$$

$$N_2 = \frac{60\times 21.165}{2\pi} = 202 \text{ rpm}$$

Range of speed $= N_2 - N_1 = 202 - 169 = 33$ rpm

With friction on the sleeve,

$$h = \frac{g}{\omega^2}\left[\frac{w + W \pm F}{w}\right]$$

When sleeve is moving downwards, F is –ve.

$$\omega_1^2 = \frac{9.81}{0.2191}\left[\frac{50+300-25}{50}\right] = 291.03$$

$$\omega_1 = 17.06 \text{ rad/s}$$

$$N_1 = 162.9 \text{ rpm}$$

When sleeve is moving upwards, F is +ve.

$$\omega_2^2 = \frac{9.81}{0.1533}\left[\frac{50+300+25}{50}\right] = 479.94$$

$$\omega_2 = 21.91 \text{ rad/s}$$

$$N_2 = 209.2 \text{ rpm}$$

Range of speed $= 209.2 - 162.9 = 46.3$ rpm

Example 9.40

A Porter governor has all four arms 240 mm long. The upper arms are attached on the axis of rotation and the lower arms are attached to the sleeve at a distance of 25 mm from the axis. The weight of each ball is 50 N and the sleeve weighs 500 N. The extreme radii of rotation are 160 mm and 220 mm. Determine the range of speed of the governor.

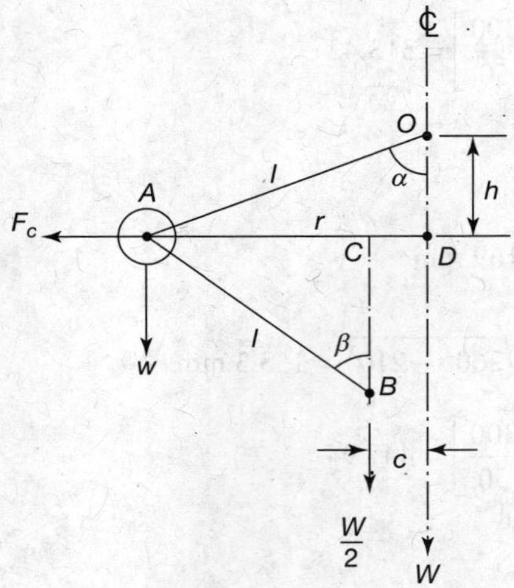

Fig.9.36 Porter governor

■ Solution

Refer to Fig.9.36.

Given: $l = 240$ mm, $c = 25$ mm, $w = 50$ N, $W = 500$ N, $r_1 = 160$ mm, $r_2 = 220$ mm

$$h_1 = \sqrt{l^2 - r_1^2} = \sqrt{240^2 - 160^2} = 178.9 \text{ mm}$$

$$\tan\alpha_1 = \frac{AD}{OD} = \frac{r_1}{h_1} = \frac{160}{178.9} = 0.894$$

$$AC = AD - CD = r_1 - c = 160 - 25 = 135 \text{ mm}$$

$$BC = \sqrt{AB^2 - AC^2} = \sqrt{240^2 - 135^2} = 198.43 \text{ mm}$$

$$\tan\beta_1 = \frac{AC}{BC} = \frac{135}{198.43} = 0.680$$

$$k_1 = \frac{\tan\beta_1}{\tan\alpha_1} = \frac{0.680}{0.894} = 0.761$$

$$h_1 = \frac{g}{\omega_1^2}\left[1 + \frac{W}{2w}(1 + k_1)\right]$$

$$\omega_1^2 = \frac{9.81}{0.1789}\left[1 + \frac{500}{2 \times 50}(1 + 0.761)\right] = 537.66$$

$$\omega_1 = 23.187 \text{ rad/s}$$

$$N_1 = \frac{60 \times 23.187}{2\pi} = 221.4 \text{ rpm}$$

$$h_2 = \sqrt{l^2 - r_2^2} = \sqrt{240^2 - 220^2} = 95.92 \text{ mm}$$

$$\tan\alpha_2 = \frac{r_2}{h_2} = \frac{220}{95.92} = 2.29$$

$$AC = r_2 - c = 220 - 25 = 195 \text{ mm}$$

$$BC = \sqrt{240^2 - 195^2} = 139.91$$

$$\tan\beta_2 = \frac{195}{139.91} = 1.39$$

$$k_2 = \frac{\tan\beta_2}{\tan\alpha_2} = \frac{1.39}{2.29} = 0.61$$

$$h_2 = \frac{g}{\omega_2^2}\left[1 + \frac{W}{2w}(1 + k_2)\right]$$

$$\omega_2^2 = \frac{9.81}{0.09592}\left[1 + \frac{500}{2\times 50}(1 + 0.61)\right] = 925.68$$

$$\omega_2 = 30.42 \text{ rad/s}$$

$$N_2 = \frac{60\times 30.42}{2\pi} = 290.5 \text{ rpm}$$

Range of speed $= N_2 - N_1 = 290.5 - 221.4 = 69.1$ rpm

Example 9.41

The following particulars refer to a Proell governor with open arms:

Length of all arms = 220 mm, distance of pivot of arms from axis of rotation = 50 mm, length of extension of lower arms to which each ball is attached = 100 mm, weight of each ball = 60 N, and weight of central load = 1500 N.

If the radius of rotation of the balls is 180 mm when the arms are inclined at an angle of 30° to the axis of rotation, find the equilibrium speed for the above configuration.

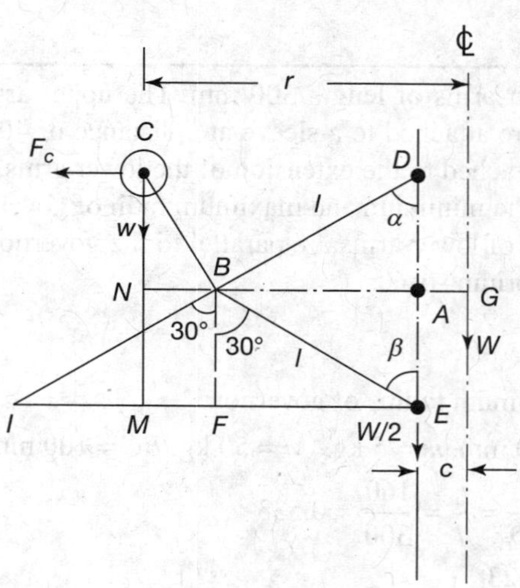

Fig.9.37 Proell governor

■ **Solution**

Refer to Fig.9.37.

$l = 220$ mm, $c = 50$ mm, $BC = 100$ mm, $w = 60$ N, $W = 1500$ N, $r = 180$ mm, $\alpha = \beta = 30°$

$$AD = BD\cos\alpha = l\cos\alpha = 220\cos 30° = 190.5 \text{ mm}$$

$$AB = BD\sin\alpha = l\sin\alpha = 220\sin 30° = 110 \text{ mm}$$

$$NB = NG - AG - AB = r - c - AB$$

$$= 180 - 50 - 110 = 20 \text{ mm}$$

$$CN = \sqrt{BC^2 - NB^2} = \sqrt{100^2 - 20^2} = 97.98 \text{ mm}$$

$$CM = CN + NM = CN + AE = CN + AD = 97.98 + 190.5 = 288.48 \text{ mm}$$

$$IM = IF - MF = FE - MF = AB - NB = 110 - 20 = 90 \text{ mm}$$

$$IE = IF + FE = 2 \times AB = 2 \times 110 = 220 \text{ mm}$$

Taking moments about the instantaneous centre I, we have

$$F_c \times CM = w \times IM + \frac{W}{2} \times IE$$

$$F_c \times 288.48 = 60 \times 90 + \frac{1500}{2} \times 220$$

$$F_c = 590.68\ N$$

$$F_c = \frac{w}{g}\omega^2 r$$

$$590.68 = \frac{60}{9.81} \times \left(\frac{2\pi N}{60}\right)^2 \times 0.18$$

$$N = 221.2 \text{ rpm}$$

Example 9.42

A Proell governor has all four arms of length 300 mm. The upper arms are pivoted on the axis of rotation and the lower arms are attached to a sleeve at a distance of 40 mm from the axis. The mass of each ball is 5 kg and are attached to the extension of the lower arms, which are 100 mm long. The mass on the sleeve is 50 kg. The minimum and maximum radii of governor are 160 mm and 210 mm. Assuming that the extensions of lower arms are parallel to the governor axis at the minimum radius, find the corresponding equilibrium speeds.

■ **Solution**

Refer to Fig.9.38 for minimum radius of governor.

Given: $l = 300$ mm, $c = 40$ mm, $m = 5$ kg, $M = 50$ kg, $BC = 100$ mm, $r_1 = 160$ mm

$$\sin\alpha = \frac{BG}{BD} = \frac{r_1}{l} = \frac{160}{300} = 0.533$$

$$\alpha = 32.23°$$

$$AB = BG - AG = r_1 - c = 160 - 40 = 120 \text{ mm}$$

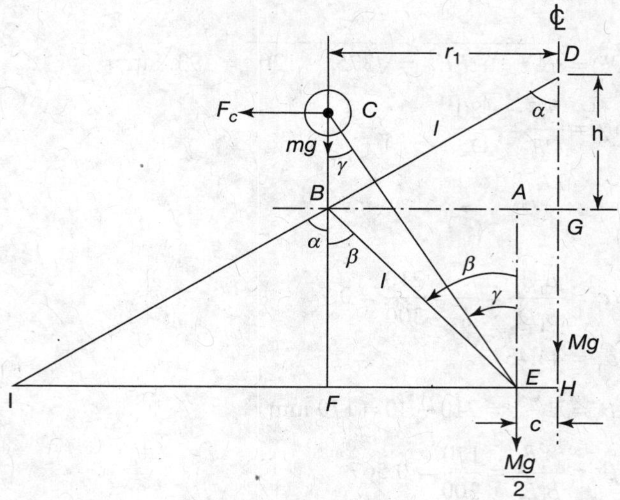

Fig.9.38 Proell governor

$$\sin\beta = \frac{AB}{BE} = \frac{120}{300} = 0.4$$

$$\beta = 23.58°$$

$$\tan\alpha = \tan 32.23° = 0.6305$$

$$\tan\beta = \tan 23.58° = 0.4365$$

$$k = \frac{\tan\beta}{\tan\alpha} = \frac{0.4365}{0.6305} = 0.6923$$

$$h = DG = l\cos\alpha = 300\cos 32.23° = 253.8 \text{ mm}$$

$$FE = r_1 - c = AB = 120 \text{ mm}$$

$$BF = l\cos\beta = 300\cos 23.58° = 275 \text{ mm}$$

$$CF = BC + BF = 100 + 275 = 375 \text{ mm}$$

$$\omega_1^2 = \left(\frac{BF}{CF}\right)\left[1 + \left(\frac{M}{2m}\right)(1+k)\right]\left(\frac{g}{h}\right)$$

$$= \left(\frac{275}{375}\right)\left[1 + \frac{50}{2\times 5}(1+0.6923)\right]\left(\frac{9.81}{0.2538}\right)$$

$$= 268.188$$

$$\omega_1 = 16.376 \text{ rad/s}$$

$$N_1 = \frac{60\times 16.376}{2\pi} = 156.4 \text{ rpm}$$

The configuration of governor at maximum radius is shown in Fig.9.39. *CBE* is a one rigid link. Therefore C_1E_1 and $\angle C_1E_1B_1$ does not change.

$$r_2 = 210 \text{ mm}$$

From Fig.9.38,

$$CE = \sqrt{CF^2 + FE^2} = \sqrt{375^2 + 120^2} = 393.7 \text{ mm}$$

$$\tan\gamma = \frac{FE}{CF} = \frac{120}{375} = 0.32$$

$$\gamma = 17.7°$$

From Fig.9.39,

$$\sin\alpha_1 = \frac{B_1G_1}{B_1D_1} = \frac{r_2}{l} = \frac{210}{300} = 0.7$$

$$\alpha_1 = 44.43°$$

$$A_1B_1 = r_2 - c = 210 - 40 = 170 \text{ mm}$$

$$\sin\beta_1 = \frac{A_1B_1}{B_1E_1} = \frac{170}{300} = 0.567$$

$$\beta_1 = 34.52°$$

$$C_1E_1 = CE = 393.7 \text{ mm}$$

$$\beta_1 - \gamma_1 = \beta - \gamma$$

$$\gamma_1 = \beta_1 - \beta + \gamma = 34.52 - 23.58 + 17.7 = 28.64°$$

Radius of rotation, $r_2' = L_1E_1 + E_1H_1 = C_1E_1 \sin\gamma_1 + c$

$$= 393.7 \sin 28.64° + 40 = 228.7 \text{ mm}$$

$$C_1L_1 = C_1E_1 \cos\gamma_1 = 393.7 \cos 28.64° = 345.53 \text{ mm}$$

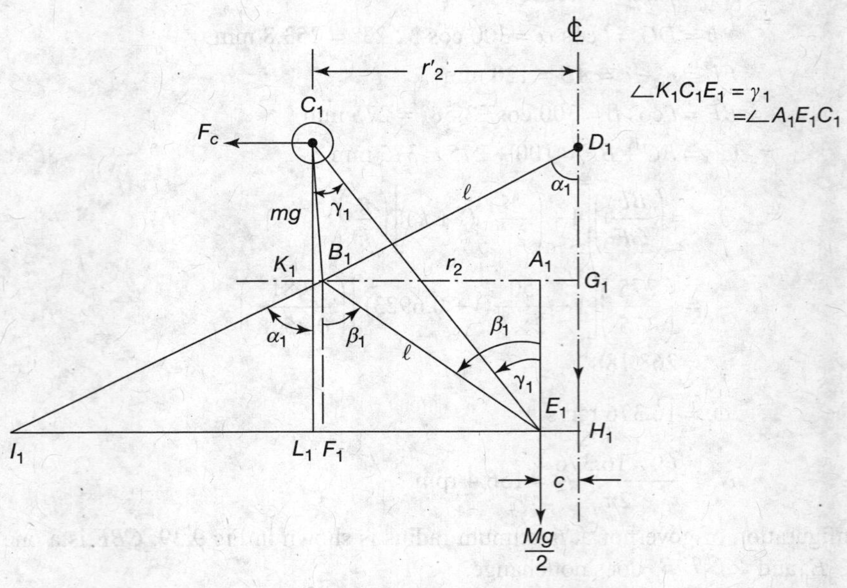

Fig.9.39 Proell governor

$$B_1F_1 = l\cos\beta_1 = 300\cos 34.52° = 247.18 \text{ mm}$$
$$I_1F_1 = B_1F_1 \tan\alpha_1 = 247.18 \tan 44.43° = 242.9 \text{ mm}$$
$$F_1E_1 = l\sin\beta_1 = 300\sin 34.52° = 170 \text{ mm}$$
$$I_1E_1 = I_1F_1 + F_1E_1 = 242.9 + 170 = 412.9 \text{ mm}$$
$$L_1E_1 = C_1E_1 \sin\gamma_1 = 393.7 \sin 28.64° = 188.7 \text{ mm}$$
$$I_1L_1 = I_1E_1 - L_1E_1 = 412.9 - 188.7 = 224.2 \text{ mm}$$
$$F_c = m\,\omega_2^2\, r_2' = 5\times\left(\frac{2\pi N_2}{60}\right)^2 \times 0.2287 = 0.01254\, N^2 \text{ N}$$

Taking moments about I_1,

$$F_c \times C_1L_1 = mg \times I_1L_1 + \frac{Mg}{2}\times I_1E_1$$

$$0.01254\,\text{N}^2 \times 345.53 = 5\times 9.81\times 224.2 + \frac{50\times 9.81}{2}\times 412.9$$

$$\text{N}^2 = 25908.6$$
$$\text{N} = 161 \text{ rpm}$$

Example 9.43

In a spring-loaded governor of the Hartnell type, the mass of each ball is 5 kg and lift of sleeve is 50 mm. The governor begins to float at 240 rpm when the ball path is 120 mm. The mean working speed of the governor is 20 times the range of speed when friction is neglected. The lengths of ball and roller arm of bell crank lever are 120 mm and 90 mm respectively. The distance between the centre of pivot of bell crank lever and axis of governor spindle is 140 mm. Determine the initial compression of spring taking into account the obliquity of arms.

If friction is equivalent to a force of 30 N at the sleeve, find the total alteration in speed before the sleeve beings to move from mid-position.

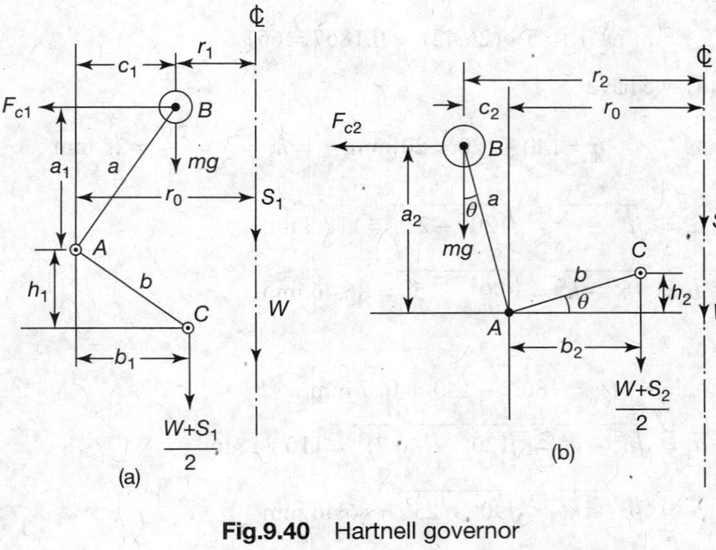

Fig.9.40 Hartnell governor

Solution

Given: $m = 5$ kg, $h = 50$ mm, $N_1 = 240$ rpm, $r_1 = 120$ mm, $a = 120$ mm, $b = 90$ mm, $r_o = 140$ mm, $F = 30$ N

Mean working speed, $\omega_m = \dfrac{\omega_1 + \omega_2}{2}$

Range of speed, neglecting friction, $\omega_r = \omega_2 - \omega_1$

$$\omega_m = 20\,\omega_r$$

$$\frac{\omega_1 + \omega_2}{2} = 20(\omega_1 - \omega_2)$$

$$39\,\omega_2 = 41\,\omega_1$$

$$\omega_2 = \frac{41}{39}\,\omega_1$$

$$\omega_1 = \frac{2\pi N_1}{60} = \frac{2\pi \times 240}{60} = 25.13 \text{ rad/s}$$

$$\omega_2 = \frac{41}{39} \times 25.13 = 26.42 \text{ rad/s}$$

Lift of sleeve, $h = (r_2 - r_1)\dfrac{b}{a}$

$$r_2 = r_1 + \frac{ha}{b} = 120 + 50 \times \frac{120}{90} = 186.7 \text{ mm}$$

At minimum speed,

$$F_{c1} = m\,\omega_1^2\,r_1 = 5 \times (25.13)^2 \times 0.12 = 378.9 \text{ N}$$

At maximum speed,

$$F_{c2} = m\,\omega_2^2\,r_2 = 5 \times (26.42)^2 \times 0.1867 = 651.6 \text{ N}$$

From Fig.9.40(a), we have

$$c_1 = r_0 - r_1 = 140 - 120 = 20 \text{ mm}, \; h_1 = h_2 = \frac{h}{2} = \frac{50}{2} = 25 \text{ mm}$$

$$a_1 = \sqrt{a^2 - c_1^2} = \sqrt{120^2 - 20^2} = 118.3 \text{ mm}$$

$$b_1 = \sqrt{b^2 - h_1^2} = \sqrt{90^2 - 25^2} = 86.46 \text{ mm}$$

From Fig.9.40(b),

$$c_2 = r_2 - r_0 = 186.7 - 140 = 46.7 \text{ mm}$$

$$a_2 = \sqrt{a^2 - c_2^2} = \sqrt{120^2 - (46.7)^2} = 110.54 \text{ mm}$$

$$b_2 = \sqrt{b^2 - h_2^2} = \sqrt{90^2 - 25^2} = 86.46 \text{ mm}$$

Taking moments about fulcrum A of bell crank lever, (Here $W = 0$)

$$\frac{S_1}{2} \times b_1 = F_{c1} \times a_1 - \text{mg} \times c_1$$

$$S_1 \times \frac{86.46}{2} = 378.9 \times 118.3 - 5 \times 9.81 \times 20$$

$$S_1 = 1014.2 \text{ N}$$

$$\frac{S_2}{2} \times b_2 = F_{c2} \times a_2 + mg \times c_2$$

$$S_2 \times \frac{86.46}{2} = 651.6 \times 110.54 + 5 \times 9.81 \times 46.7$$

$$S_2 = 1719.14 \text{ N}$$

Spring stiffness, $k = \dfrac{S_2 - S_1}{h} = \dfrac{1719.14 - 1014.2}{50} = 14.1$ N/mm

Initial compression of spring $= \dfrac{S_1}{k} = \dfrac{1014.2}{14.1} = 71.93$ mm

Taking friction into account:

Spring force in mid-position, $S = S_1 + h_1 k = 1014.2 + 25 \times 14.1 = 1366.7$ N

$$\omega_m = \frac{25.13 + 26.42}{2} = 25.775 \text{ rad/s}$$

$$N_m = \frac{60 \times 25.775}{2\pi} = 246 \text{ rpm}$$

Speed when the sleeve begins to move downwards from mean position,

$$N' = N_m \sqrt{\frac{S - F}{S}} = 246 \sqrt{\frac{1366.7 - 30}{1366.7}} = 243.3 \text{ rpm}$$

and when begins moving upwards,

$$N'' = N_m \sqrt{\frac{S + F}{S}} = 246 \sqrt{\frac{1366.7 + 30}{1366.7}} = 248.7 \text{ rpm}$$

Alteration in speed $= N'' - N' = 248.7 - 243.3 = 5.4$ rpm

Example 9.44

In a spring-controlled governor of Hartung type, the length of the ball and sleeve arms are 80 mm and 120 mm respectively. The total travel of sleeve is 30 mm. In the mid-position, each spring is compressed by 50 mm and radius of rotation of mass centres is 150 mm. Each ball has a mass of 5 kg and the spring has a stiffness of 10 N/mm. The equivalent mass of governor gear at the sleeve is 15 kg. Neglecting the moment due to revolving masses when the arms are inclined, determine the ratio of range of speed to mean speed of the governor. Also find the speed in mid-position.

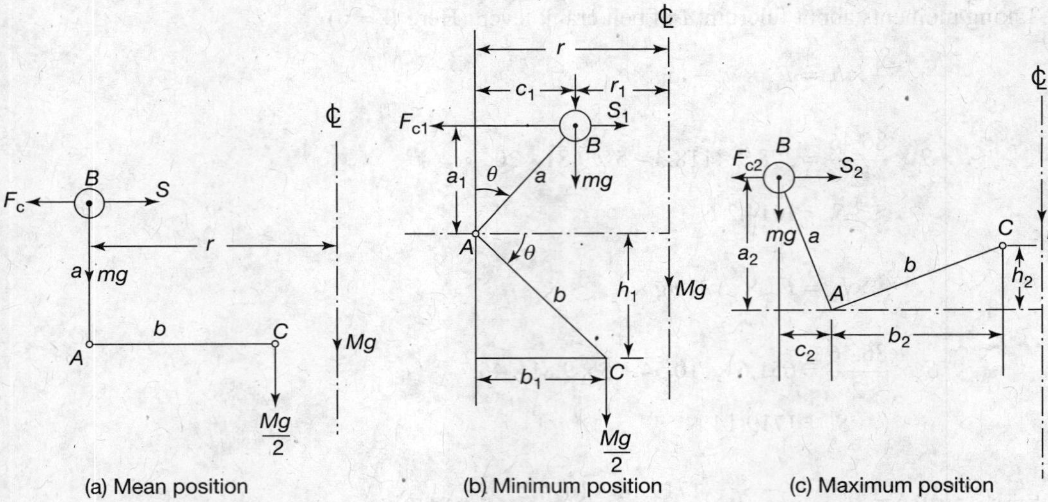

Fig.9.41 Harting governor

■ Solution

Given: $a = 80$ mm, $b = 120$ mm, $h = 30$ mm, $r = 150$ mm, $m = 5$ kg,

$k = 10$ N/mm, $M = 15$ kg, $\delta_i = 50$ mm

Refer to Fig.9.41. Let ω_m = mean angular speed

Mean position:

$$F_c = m\omega_m^2\, r = 4 \times \omega_m^2 \times 0.15 = 0.6\omega_m^2 \text{ N}$$

Spring force, $S = k\delta_i = 10 \times 50 = 500$ N

Taking moments about fulcrum A,

$$F_c \times a = S \times a + \frac{Mg}{2} \times b$$

$$0.6\,\omega_m^2 \times 80 = 500 \times 80 + \frac{15 \times 9.81}{2} \times 120$$

$$\omega_m^2 = 1017.27$$

$$\omega_m = 31.895 \text{ rad/s}$$

$$N_m = \frac{60 \times 31.895}{2\pi} = 304.6 \text{ rpm}$$

From Fig.9.41(b), we have

$$\frac{r - r_1}{h_1} = \frac{a}{b}, \quad h_1 = \frac{h}{2} = \frac{30}{2} = 15 \text{ mm} = h_2$$

$$r_1 = r - h_1 \frac{a}{b} = 150 - \frac{30}{2} \times \frac{80}{120} = 140 \text{ mm}$$

$$F_{c1} = m\omega_1^2 r_1 = 5 \times \omega_1^2 \times 0.14 = 0.7\omega_1^2$$

$$S_1 = [\delta_i - (r - r_1)]\, k$$

$$= [50 - (150 - 140)] \times 10 = 400 \text{ N}$$

Taking moments about A,

$$F_{c1} \times a = S_1 \times a + \frac{Mg}{2} \times b$$

$$0.7\omega_1^2 \times 80 = 400 \times 80 + \frac{15 \times 9.81}{2} \times 120$$

$$\omega_1^2 = 729.09$$

$$\omega_1 = 27 \text{ rad/s}$$

$$N_1 = \frac{60 \times 27}{2\pi} = 257.8 \text{ rpm}$$

From Fig.9.41(c),

$$\frac{r_2 - r}{h_2} = \frac{a}{b}$$

$$r_2 = r + \frac{a}{b} h_2$$

$$= 150 + \frac{80}{120} \times 15 = 160 \text{ mm}$$

$$F_{c2} = m\omega_2^2 r_2 = 5\ \omega_2^2 \times 0.16 = 0.8\ \omega_2^2$$

$$S_2 = [\delta_i + (r_2 - r)]\, k$$

$$= [50 + (160 - 150)] \times 10 = 600 \text{ N}$$

Taking moments about A,

$$F_{c2} \times a = S_2 \times a + \frac{Mg}{2} \times l$$

$$0.8\ \omega_2^2 \times 80 = 600 \times 80 + \frac{15 \times 9.81}{2} \times 120$$

$$\omega_2^2 = 887.95$$

$$\omega_2 = 29.8 \text{ rad/s}$$

$$N_2 = \frac{60 \times 29.8}{2\pi} = 284.5 \text{ rpm}$$

Range of speed $= N_2 - N_1 = 284.5 - 257.8 = 26.7$ rpm

$$\frac{N_2 - N_1}{N_m} = \frac{26.7}{304.6} = 0.0876 \quad \text{or} \quad 8.76\%$$

Example 9.45

The two balls of a Wilson-Hartnell governor, each of mass 6 kg, are connected across by two springs. An auxiliary spring provides an additional force at the sleeve through the medium of a lever which pivots about a fixed centre as shown in Fig.9.42. In the mean position, the radius of the governor balls is 120 mm and the speed is 600 rpm. Find the tension in the auxiliary spring in this position when tension in each spring is 800 N.

$x = 80$ mm, $y = 180$ mm, $a = b = 100$ mm, $r = r_1 = 120$ mm, $k_b = 10$ N/mm

When the sleeve moves up by 20 mm, the speed is to be 650 rpm. Find the stiffness k_a.

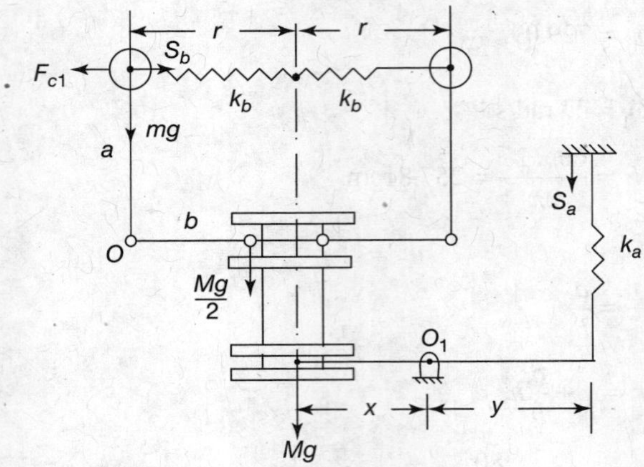

Fig.9.42 Wilson–Hartnell governor

■ **Solution**

$$\omega_1 = 2\pi \times 600/60 = 62.83 \text{ rad/s}$$

$$F_{c1} = m\,\omega_1^2 r_1 = 6 \times (62.83)^2 \times 0.12 = 2842.3 \text{ N}$$

$$S_{b1} = 2 \times 800 = 1600 \text{ N}$$

Taking moments about the fulcrum O of bell crank lever,

$$F_{c1} \times a = S_{b1} \times a + \frac{Mg}{2} \times b$$

$$2842.3 \times 100 = 1600 \times 100 + \frac{Mg}{2} \times 100$$

$$Mg = 2484.6 \text{ N}$$

Taking moments about O_1,

$$Mg \times x = s_a \times y$$

$$S_a = \frac{2484.6 \times 80}{180} = 1104.25 \text{ N}$$

Now $h = 20$ mm, $N_2 = 650$ rpm

$$\omega_2 = \frac{2\pi N_2}{60} = \frac{2\pi \times 650}{60} = 68.1 \text{ rad/s}$$

$$k_b = 10 \text{ N/mm}$$

$$h = (r_2 - r_1)\frac{a}{b}$$

$$r_2 = \frac{a}{b}h + r_1 = \frac{100}{100} \times 20 + 120 = 140 \text{ mm}$$

$$F_{c2} = m\omega_2^2 r_2 = 6 \times (68.1)^2 \times 0.14 = 3892 \text{ N}$$

Extension of springs of balls, $S_b = 2\,(r_2 - r_1) \times$ number of springs

$$= 2(140 - 120) \times 2 = 80 \text{ mm}$$

Total spring force, $S_{b2} = S_{b1} + \delta_b \times k_b$

$$= 1600 + 80 \times 10 = 2400 \text{ N}$$

Taking moments about O,

$$F_{c2} \times 100 = S_{b2} \times 100 + \frac{Mg}{2} \times 100$$

$$3892 = S_{b2} + \frac{Mg}{2} = 2400 + \frac{Mg}{2}$$

$$Mg = 2984 \text{ N}$$

Taking moments about O_1

$$Mg \times x = S_{a2} \times y$$

$$S_{a2} = 2984 \times \frac{80}{100} = 1326.2 \text{ N}$$

Extension of auxiliary spring, $\quad S_a = h\frac{y}{x} = 20 \times \frac{180}{80} = 45 \text{ mm}$

Stiffness of auxiliary spring, $\quad k_a = \frac{S_{a2} - S_{a1}}{\delta_a}$

$$= \frac{1326.2 - 1104.25}{45} = 4.93 \text{ N/mm}$$

Summary for Quick Review

1 A governor is a device to maintain, as closely as possible, a constant mean speed of rotation of the crankshaft over long periods during which the load on the engine may vary. The governor meets the varying demand for power by regulating the supply of working fluid.

2 There are basically two types of governors: centrifugal governors and inertia governor. In centrifugal governors, the centrifugal force is balanced by the controlling force. In inertia type of governors, the inertia force is balanced by the controlling force. The centrifugal governors can be either of pendulum type or dead weight and spring loaded type.

3 Dead weight type of governors are: Porter governor and Proell governor.

4 Spring loaded type governors are: Hartnell governor, Hartung governor, Wilson – Hartnell governor, and Pickering governor.

5 Simple Watt governor.

(a) Neglecting weight of the arms.

$$h = g/\omega^2$$

With increasing speed, governor stops functioning. It is used for slow speed engines.

(b) Considering weight of the arms.

$$h = (g/\omega^2)[(w + W_a/2)/(w + W_a/3)]$$

6 Porter Governor

$$h = (g/\omega^2)\,[1 + (W/2w)\,(1 + k)]$$

where $k = \tan\beta/\tan\alpha$.

If $\alpha = \beta$, then

$$h = (g/\omega^2)\,[1 + W/w]$$

If F is the frictional force acting on the sleeve, then

$$h = (g/\omega^2)\,[(w + W \pm F)/w]$$

Take +ve sign when the sleeve moves upwards or the governor speed increases and –ve sign when the sleeve moves downwards or the governor speed decreases.

7 Proell governor.

$$\omega^2 = (BF/CF)\,[1 + \{(W/(2w)\}(1 + k)] \cdot (g/h)$$

Where $\tan\alpha = r/h$, and $k = \tan\beta/\tan\alpha$.

If $\alpha = \beta$, i.e. $k = 1$, then

$$\omega^2 = (BF/CF)\,[\,1 + W/w]\,(g/h)$$

8 Hartnell governor

$$k = 2\,(a/b)^2\,[(F_{c1} - F_{c2})/(r_1 - r_2)]$$

9 Gravity and spring controlled governor

$$W + S = 2\,[F_c c - w\,(d + e)]/d$$

10 Wilson-Hartnell governor

$$4\,k_b + k_a\,(by/ax)^2/2 = (F_{c1} - F_{c2})/(r_1 - r_2)$$

If $k_a = 0$, then

$$k_b = (F_{c1} - F_{c2})/[4(r_1 - r_2)]$$

11 Hartung governor

$$(F_c - P)\,a = W\,b/2$$

12 Pickering governor

h = lift of the sleeve

The deflection of a leaf spring with both ends fixed and carrying a central load W, is given by

$$\delta = m\,(a + \delta)\,\omega^2L^3/(192\,EI)$$

The lift of the sleeve, $h \approx 2.4\,\delta^2/L$

13 Inertia governor

For a governor to be rapid in action, the arm should be arranged such that as the mass moves outwards, the arm rotates in a direction opposite to that of the rotation of shaft.

14 Performance of governors.

(a) *Sensitiveness*: It is defined as the ratio of the difference between the maximum and minimum equilibrium speeds to the mean equilibrium speed.

(b) *Stability*: A governor is said to be stable, when for each speed within the working range, there is only one radius of the governor balls at which the governor is in equilibrium.

(c) *Isochronism*: A governor is said to be isochronous, when the equilibrium speed is constant for all radii of rotation of the balls, within the working range. An isochronous governor will be infinitely sensitive.

(d) *Hunting*: It is a condition in which the speed of the engine controlled by the governor fluctuates continuously above and below the mean speed. It is caused by a governor which is too sensitive.

(e) *Governor effort*: The effort of a governor is the force which it can exert at the sleeve on the mechanism which controls the supply of fuel to the engine. The mean force exerted during the given change of speed is termed the effort. Generally effort is defined for 1% change of speed.

(f) *Power*: The power of a governor is defined as the work done at the sleeve for a given percentage change of speed.

$$\text{Power} = \text{Effort} \times \text{Displacement of sleeve}$$

15 Quality of a governor

The quality of a governor is ascertained by 1. Sensitiveness, 2. Stability, and 3. Effort and power.

16 Controlling force

When the speed of rotation is uniform, each ball of the governor is subjected either directly or indirectly to an inward pull, which is equal and opposite to the outward centrifugal reaction. This inward pull is termed the controlling force. A curve drawn to show how the pull varies with the radius of rotation of the ball is called a controlling force curve.

17 The stability of a spring controlled governor can be ascertained as follows:

(a) For a stable governor, the controlling force must increase as radius of rotation increases, i.e. F_c/r must increase as r increases. Therefore, the controlling force line, $F_c = a\,r - b$, produced must intersect the controlling force axis below the origin

(b) When $b = 0$, the controlling force line, $F_c = ar$, passes through the origin, and the governor becomes isochronous, because F_c/r will remain constant for all radii of rotation.

(c) If b is positive, then F_c/r decreases as r increases, so that the equilibrium speed of the governor decreases with an increase of the radius of rotation of the balls, which is not possible. Such a governor is said to be unstable. The equation of line is, $F_c = a\,r + b$.

18 Coefficient of insensitiveness

The coefficient of insensitiveness is defined as the ratio of the difference of speed at ascent and descent for same radius of rotation to the steady speed at the same radius of rotation.

Multiple Choice Questions

1 The height of Watt's governor is proportional to
(a) speed (N) (b) N^2 (c) 1/N (d) $1/N^2$.

2 In a Hartnell governor, if the stiffness of spring is increased, the governor will
(a) become more sensitive (b) become less sensitive
(c) remain unaffected (d) start hunting.

3 The function of a governor is to
(a) reduce the speed fluctuations during a cycle
(b) maintain the prime mover speed within prescribed limits
(c) not to influence the speed of the prime mover
(d) not to control the variation in load on the prime mover.

4 The following governor is spring loaded
(a) Watt governor (b) Porter governor
(c) Proell governor (d) Hartnell governor.

5 The gravity controlled governor is
(a) Hartnell governor (b) Pickering governor
(c) Hartung governor (d) Proell governor.

6 The height of a Watt's governor is
(a) ω^2/g (b) g/ω^2 (c) $g\omega^2$ (d) $2\,g\omega^2$

7 The Proell governor as compared to Porter governor, at same speed
(a) is more sensitive (b) requires smaller size
(c) has less lift (d) all of the above.

8 The sensitivity of a governor due to frictional resistance at the sleeve
(a) increases (b) decreases (c) remains same (d) depends on speed.

9 The spring loaded governors as compared to gravity controlled governors
(a) can operate at higher speeds (b) are more compact and smaller in size
(c) are capable of being fixed at any inclination (d) all of the above.

10 If the ball masses of a governor occupy a definite specified position for each speed, it is said to be
(a) stable (b) hunting (c) isochronous (d) sensitive.

11 If the ball masses of a governor have same equilibrium speed for all the radii of rotation, it is said to be
(a) stable (b) hunting (c) isochronous (d) sensitive.

12 Isochronous governor is
(a) more sensitive (b) less stable (c) less sensitive (d) less stable.

13 Governor effort is defined as the force applied for
(a) 1% change in speed (b) 2% change in speed
(c) 5% change in speed (d) the total range of speed.

14 Governor which is hunting is
(a) more sensitive (b) less sensitive (c) more stable (d) less stable.

15 Governor power is defined as the product of governor effort and
(a) sleeve lift (b) reciprocal of sleeve lift
(c) difference of radii of rotation for maximum and minimum speeds
(d) square of sleeve lift.

Answers

1. (d) 2. (b) 3. (b) 4. (d) 5. (d) 6. (b) 7. (d) 8. (d) 9. (d) 10. (a) 11. (c) 12. (c)
13. (a) 14. (d) 15. (a)

Review Questions

1. What is the main function of a governor? How does it differ from that of a flywheel?
2. What are the various types of governors?
3. How does centrifugal governor differ from an inertia governor?
4. What is the effect of friction on the functioning of a Porter governor?
5. Why an auxiliary spring is used along with main springs in a Wilson-Hartnell governor?
6. Which type of governor is used in a gramophone?
7. Explain the meaning of sensitiveness, hunting and stability of a governor?
8. What is the condition of isochronism in case of a Hartnell governor?
9. Define effort and power of a governor.
10. What is the controlling force of a governor? How does the controlling force curve help in establishing the stability or instability of a governor?

Exercises

9.1 A Porter governor has all the four arms of 250 mm length each. All the upper arms and the sleeve arms are pivoted on the axis of rotation. The mass of each governor ball is 0.9 kg. The mass on the sleeve is 15 kg. Find the speed of rotation when the balls rotate at a radius of 160 mm.

9.2 The lengths of upper and lower arms of a Porter governor are 220 mm and 260 mm respectively. All the arms are pivoted on the axis of rotation. The central load is 125 N and weight of each ball is 15 N. The friction of sleeve together with the resistance of the operating gear is equivalent to a force of 25 N at the sleeve. If the limiting inclinations of the upper and lower arms to the vertical are 30° and 40°, respectively, determine the range of speed of the governor.

9.3 A Porter governor has all four arms 200 mm long. The upper arms are attached on the axis of rotation and the lower arms are attached to the sleeve at a distance of 25 mm from the axis. The mass of each ball is 5 kg and the mass of sleeve is 40 kg. The extreme radii of rotation are 150 mm and 200 mm. Determine the range of speed of governor.

9.4 A loaded Porter governor has four arms each 240 mm long, two revolving masses each weighing 25 N and a central dead weight of 200 N. All the arms are attached to respective sleeves at radial distances of 40 mm from the axis of rotation. The masses revolve at a radius of 150 mm at minimum speed and 200 mm at maximum speed. Determine the range of speed.

9.5 In a Proell governor, the mass of each ball is 5 kg and central load on sleeve is 50 kg. The length of each upper and lower arms is 250 mm. The minimum and maximum radii are 120 mm and 180 mm respectively. The distances of point of suspension of upper and lower links from governor axis are 20 mm and 30 mm, respectively. The length of links to which balls are attached is 75 mm. Assuming that the links to which balls are attached are parallel to governor axis at the minimum radius, determine the equilibrium speeds corresponding to extreme radii.

9.6 The arms of a Proell governor are 250 mm long. The upper arms are pivoted on the axis of rotation, while the lower arms are pivoted at a radius of 35 mm. Mass of each ball is 5 kg and is fixed to the extension 100 mm long of the lower arms. The central sleeve load is 100 N. At minimum radius of 160 mm the extension to which the balls are attached are parallel to the governor axis. Determine the equilibrium speeds for radii of 160 and 200 mm.

9.7 The fly balls of spring loaded governor of Hartnell type running at 600 rpm have a radius of rotation of 80 mm with sleeve in mid-position and ball arms vertical. The ball arms and sleeve arms are of equal length. The maximum sleeve movement is 20 mm with ±5% variation in speed. The mass of sleeve is 5 kg and friction may be assumed to be equivalent to an additional load of 20 N at the sleeve. The governor effort is sufficient to overcome the friction at the sleeve by 1% change of speed at mid-point. Determine (a) mass of each ball, (b) spring rate, (c) initial compression of spring, (d) governor effort for 1% change of speed, and (e) governor power.

9.8 In a Hartnell type governor, the two masses are 5 kg each and the load on the sleeve is 45 kg. With the mass arms vertical, the path radius is 80 mm and the equilibrium speed, neglecting friction, is 450 rpm. Determine (a) the corresponding compressive force in the spring, and (b) the friction force at the sleeve, which can be overcome in this position for an increase in speed of 1%.

9.9 In Hartnell governor, the lengths of ball and sleeve arms of a bell crank lever are 120 mm and 100 mm respectively. The distance of the fulcrum of the bell crank lever from the governor axis is 150 mm. Each governor ball has a mass of 4 kg. The governor runs at a mean speed of 300 rpm with the ball arms vertical and sleeve arms horizontal. For an increase of speed of 4%, the sleeve moves 10 mm upwards. Neglecting friction, determine (a) minimum equilibrium speed if the total sleeve movement is limited to 20 mm, (b) spring stiffness, (c) sensitiveness of governor, and (d) spring stiffness if the governor is to be isochronous at 300 rpm.

9.10 In a Wilson-Hartnell spring loaded governor, the two balls are of 5 kg each, which are connected by two springs. The speed of the governor is 600 rpm in its mean position. The radius of the governor ball is 150 mm. The tension in each spring is 1200 N. Find the tension in the other spring and its stiffness, if the speed is 650 rpm when the sleeve moves up by 20 mm from mean position and stiffness of each spring is 10 N/mm.

9.11 In a spring-controlled governor, the curve of controlling force is a straight line. When balls of 8 kg mass each are 400 mm apart, the controlling force is 1 kN and when 200 mm apart is 560 N. At what speed will the governor run when the balls are 250 mm apart? What initial tension on the spring would be required for isochronisms and what would then be the speed?

9.12 The controlling force F in N and radius of rotation r in cm for a spring loaded governor are related by the expression: $F = 3r - 8$. The mass of each ball is 5 kg and extreme radii of rotation of the balls are 100 mm and 175 mm, respectively. Find the maximum and minimum speeds. If the friction of governor mechanism is equivalent to a radial force of 5 N at each ball, find the extent to which the equilibrium speeds are affected at the extreme radii of rotation.

9.13 The controlling force curve of a spring controlled governor is a straight line. The weight of each governor ball is 40 N and the extreme radii of rotation are 120 and 200 mm. The values of the controlling force at the above radii are respectively 200 N and 400 N. The friction of the mechanism is equivalent to 3 N at each ball. Determine (a) the extreme equilibrium speeds of the governor, and (b) the equilibrium speed and the coefficient of insensitiveness at a radius of 150 mm.

9.14 The following data refer to a Hartnell governor:

Length of ball arm of bell crank lever = 150 mm; length of sleeve arm = 100 mm;

Mass of ball = 5 kg; spring stiffness = 250 N/mm.

At minimum speed of 300 rpm the radius of rotation of ball is 80 mm. Neglecting gravity effect, determine (a) the speed after the sleeve has lifted by 50 mm, (b) the initial compression of the spring, (c) governor effort, and (d) power.

10 INERTIA FORCE AND TURNING MOMENT

Chapter Outline

10.1 INTRODUCTION

The inertia force arises due to the mass and acceleration of the reciprocating parts whereas turning moment arises due to the crank effort. Inertia force is equal to the acceleration force in magnitude and opposite in direction. Dynamic forces arise due to eccentricity of the centre of mass from the axis of rotation or the geometric centre. It causes vibrations in the system, which are undesirable.

10.2 MOTION ANALYSIS OF RECIPROCATING MECHANISM

Consider the slider crank mechanism, as shown in Fig.10.1. *OC* is the crank, *BC* the connecting rod, *C* the crank pin, and *B* the gudgeon pin or the cross head. The crank is rotating clockwise with angular speed ω.

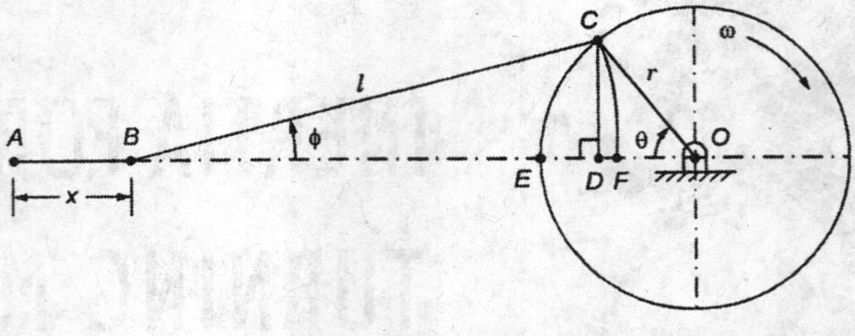

Fig.10.1 Slider-crank mechanism

10.2.1 Velocity and Acceleration of Piston

Displacement of the piston from top dead centre,

$$\begin{aligned} x &= AB = EF \\ &= ED + DF \\ &= (OE - OD) + (BF - BD) \\ &= (r - r\cos\theta) + (l - l\cos\phi) \\ &= r(1 - \cos\theta) + l(1 - \cos\phi) \end{aligned}$$

Now $$CD = r\sin\theta = l\sin\phi$$

or $$\sin\phi = \left(\frac{r}{l}\right)\sin\theta = \frac{\sin\theta}{n}$$

where $\frac{l}{r} = n$, is the ratio of length of connecting rod to that of crank.

$$\begin{aligned} \cos\phi &= [1 - \sin^2\phi]^{0.5} \\ &= \frac{\left(n^2 - \sin^2\theta\right)^{0.5}}{n} \end{aligned}$$

Therefore, $$x = r(1 - \cos\theta) + l\left(1 - \frac{\left(n^2 - \sin^2\theta\right)^{0.5}}{n}\right) \tag{10.1}$$

$$= r[(1 - \cos\theta) + \{n - (n^2 - \sin^2\theta)^{0.5}\}]$$

Velocity of piston, $$v_p = \frac{dx}{dt}$$

$$= \frac{dx}{d\theta}\cdot\frac{d\theta}{dt}$$

$$= \omega \cdot \frac{dx}{d\theta}$$

$$= \omega r \left(\sin\theta + \left(\frac{1}{2}\right)\left(n^2 - \sin^2\theta\right)^{-0.5} 2\sin\theta\cos\theta \right)$$

$$= \omega r \left(\sin\theta + \frac{\sin 2\theta}{2\left(n^2 - \sin^2\theta\right)^{-0.5}} \right)$$

$$\approx \omega r \left(\sin\theta + \frac{\sin 2\theta}{2n} \right) \tag{10.2}$$

Acceleration of piston, $f_p = \dfrac{dv}{dt}$

$$= \frac{dv}{d\theta} \cdot \frac{d\theta}{dt}$$

$$= \omega \cdot \frac{dv}{d\theta}$$

$$= \omega^2 r \left[\cos\theta + \frac{0.5\left\{\sin 2\theta \times 0.5\left(n^2 - \sin^2\theta\right)^{-0.5}(-2\sin\theta\cos\theta) - \left(n^2 - \sin^2\theta\right)^{0.5} \cdot 2\cos 2\theta\right\}}{\left(n^2 - \sin^2\theta\right)} \right]$$

$$= \omega^2 r \left[\cos\theta - \frac{\left\{0.5\sin^2 2\theta + \cos 2\theta (n^2 - \sin^2\theta)\right\}}{\left(n^2 - \sin^2\theta\right)^{1.5}} \right]$$

$$f_p \approx \omega^2 r \left[\cos\theta + \frac{\cos 2\theta}{n} \right] \tag{10.3}$$

10.2.2 Angular Velocity and Acceleration of Connecting Rod

Now

$$\sin\phi = \frac{\sin\theta}{n}$$

$$\cos\phi \cdot \frac{d\phi}{dt} = \frac{\cos\theta}{n} \cdot \frac{d\theta}{dt}$$

Angular velocity of connecting rod,

$$\omega_c = \frac{d\phi}{dt}$$

$$= \frac{\cos\theta}{\cos\phi} \cdot \frac{\omega}{n}$$

$$= \frac{\omega \cos\theta}{\left(n^2 - \sin^2\theta\right)^{0.5}} \tag{10.4}$$

Angular acceleration of connecting rod,

$$\alpha_c = \frac{\mathrm{d}^2\phi}{\mathrm{d}t^2}$$

$$= \frac{\mathrm{d}}{\mathrm{d}\theta} \cdot \frac{\mathrm{d}\phi}{\mathrm{d}t} \cdot \frac{\mathrm{d}\theta}{\mathrm{d}t}$$

$$= \omega \cdot \frac{\mathrm{d}}{\mathrm{d}\theta} \frac{\mathrm{d}\phi}{\mathrm{d}t}$$

$$= \omega^2 \left[\frac{-\cos\theta \times 0.5\left(n^2 - \sin^2\theta\right)^{-0.5} (-2\sin\theta\cos\theta) - \left(n^2 - \sin^2\theta\right)\sin\theta}{\left(n^2 - \sin^2\theta\right)^{1.5}} \right]$$

$$= \omega^2 \left[\frac{-\sin\theta \times \cos^2\theta + \sin\theta\left(n^2 - \sin^2\theta\right)}{\left(n^2 - \sin^2\theta\right)^{1.5}} \right]$$

$$= -\omega^2 \sin\theta \left[\frac{-\cos^2\theta + n^2 - \sin^2\theta}{\left(n^2 - \sin^2\theta\right)^{1.5}} \right]$$

$$= -\omega^2 \sin\theta \left[\frac{n^2 - 1}{\left(n^2 - \sin^2\theta\right)^{1.5}} \right] \tag{10.5a}$$

$$\approx -\omega^2 \frac{\sin\theta}{n} \tag{10.5b}$$

Example 10.1

The length of the crank of a reciprocating engine is 120 mm and its connecting rod is 600 mm long. It rotates at 360 rpm and at a particular instant makes an angle of 50° with the inner dead centre. Find the (a) velocity and acceleration of the piston, (b) velocity and acceleration of the midpoint of the connecting rod, and (c) angular velocity and angular acceleration of the connecting rod.

Solution

Given: $r = 120$ mm, $l = 600$ mm, $N = 360$ rpm, $\theta = 50°$

$$n = \frac{l}{r} = \frac{600}{120} = 5$$

$$\omega = \frac{2\pi \times 360}{60} = 37.7 \text{ rad/s}$$

The slider-crank mechanism is shown in Fig.10.2(a).

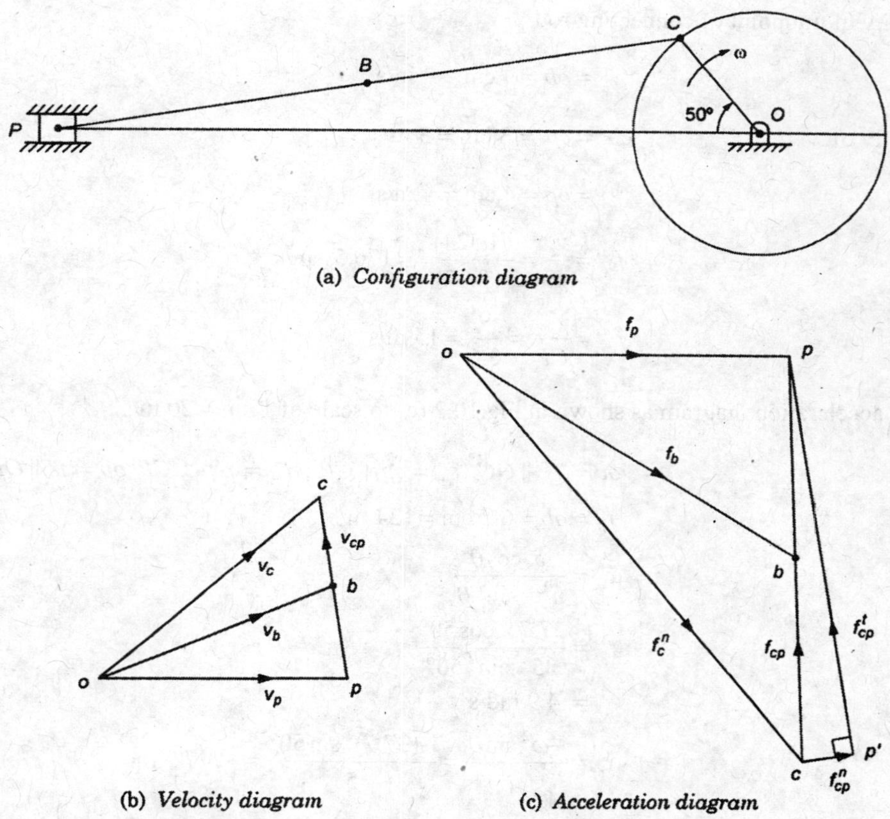

Fig.10.2 Slider-crank mechanism

(a) $$v_p = \omega r\left[\sin\theta + \frac{\sin 2\theta}{2n}\right] = 37.7 \times 0.12\left[\sin 50° + \frac{\sin 100°}{10}\right]$$

$$= 3.91 \text{ m/s}$$

$$f_p = \omega^2 r\left[\cos\theta + \frac{\cos 2\theta}{n}\right]$$

$$= (37.7)^2 \times 0.12\left[\cos 50° + \frac{\cos 100°}{5}\right]$$

$$= 103.7 \text{ m/s}^2$$

(b) $$v_c = \omega r = 37.7 \times 0.12 = 4.524 \text{ m/s}$$

Draw velocity polygon as shown in Fig 10.2(b) to a scale of 1 cm = 1 m/s.

$$v_c = oc \perp OC,\ c_p \perp CP,\ op \parallel OP.$$

Velocity of midpoint of connecting rod,

$$v_b = ob = 4 \text{ cm} = 4 \text{ m/s}$$

Velocity of $CP = cp = 3 \text{ cm} = 3 \text{ m/s}$

$$v_p = op = 4 \text{ cm} = 4 \text{ m/s}$$

$$f_c^n = \frac{v_c^2}{r} = \frac{(4.524)^2}{0.12} = 170.55 \text{ m/s}^2$$

$$f_{cp}^n = \frac{v_{cp}^2}{CP} = \frac{9}{0.6} = 15 \text{ m/s}^2$$

Draw acceleration diagram as shown in Fig.10.2 (c) to scale of 1 cm = 20 m/s².

$$oc = f_c^n \parallel OC,\ f_{cp}^n = cp' \parallel CP',\ f_{cp}^t = p'p \perp CP, op = f_p \parallel OP.$$

$$f_b = ob = 6.2 \text{ cm} = 124 \text{ m/s}^2$$

(c)

$$\omega_r = \frac{\omega \cos\theta}{n^2 - \sin^2\theta}$$

$$= \frac{37.7 \times \cos 50°}{25 - \sin^2 50°}$$

$$= 4.9 \text{ rad/s}$$

$$\alpha_r = \frac{-\omega^2 \sin\theta}{n} = \frac{-(37.7)^2 \sin 50°}{5} = -217.7 \text{ rad/s}^2$$

10.3 INERTIA FORCES IN THE RECIPROCATING ENGINE

The piston reciprocates along the line of stroke. On account of the acceleration and mass of the piston pin, inertia force is generated along the line of stroke. On the other hand, the connecting rod oscillates between its two extreme positions. Some mass of the connecting rod may be considered as reciprocating with the piston and remaining mass rotating with the crank pin. In this article, we shall study the inertia force generated due to the reciprocating parts.

10.3.1 Analytical Method

Consider the slider crank chain, as shown in Fig.10.3.

Let W_c = weight of the connecting rod

M_r = mass of reciprocating parts

K = radius of gyration of connecting rod about its centre of gravity and perpendicular to the plane of rotation

l_1 = distance of centre of gravity G of connecting rod from the gudgeon (or piston) pin (point P)

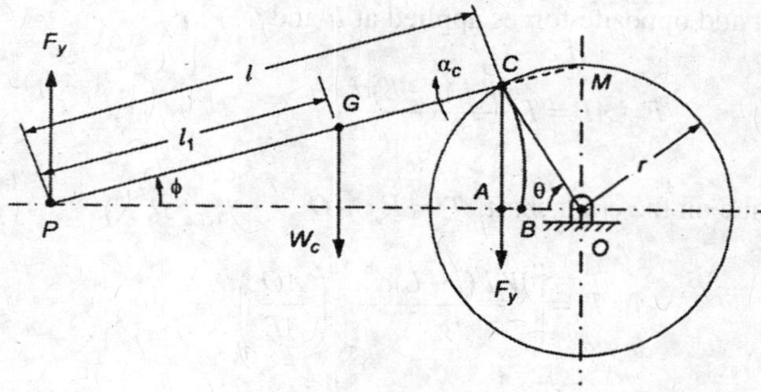

Fig.10.3 Inertia forces in reciprocating engine

l = length of connecting rod

$$L = l_1 + \frac{K^2}{l_1}$$

r = radius of crank

Total equivalent reciprocating mass, $M_{re} = M_r + (l - l_1)\frac{W_c}{gl}$

The inertia force due to M_{re},

$$F_i = -M_{re} \times f_p$$

where f_p = acceleration of reciprocating parts

$$\approx \omega^2 r\left(\cos\theta + \frac{\cos 2\theta}{n}\right)$$

$$n = \frac{l}{r}$$

Torque exerted on the crankshaft due to inertia force,

$$T_i = F_i \times OM$$

where $OM = \dfrac{r\sin(\theta+\phi)}{\cos\phi}$ and $\sin\phi = \dfrac{\sin\theta}{n}$

Correction couple, $T_o = W_c l_1 \dfrac{(l-L)\alpha_c}{g}$ (10.6)

where α_c = angular acceleration of connecting rod

$$= -\omega^2 \frac{(n^2-1)\sin\theta}{(n^2 - \sin^2 2\theta)^{1.5}}$$

$$\approx -\omega^2 \frac{\sin\theta}{n}$$

Let F_y be two equal and opposite forces applied at P and C.

Then $$F_y \cdot AP = T_o \frac{W_c l_1 (l-L)\alpha_c}{g}$$

Corresponding torque on the crankshaft, $T_c = F_y \cdot AO$

or $$T_c = \left[\frac{W_c l_1 (l-L)\alpha_c}{g}\right] \cdot \left(\frac{AO}{AP}\right)$$

Now $$AO = OC\cos\theta \quad \text{and} \quad AP = CP\cos\phi$$

Also $$\cos\phi = \left[1-\sin^2\phi\right]^{0.5} = \frac{(n^2-\sin^2\theta)^{0.5}}{n}$$

Where $$n = \frac{CP}{OC} = \frac{l}{r}$$

$$\therefore \quad T_c = \left[\frac{W_c l_1 (l-L)\alpha_c}{g}\right] \cdot \left[\frac{\cos\theta}{(n^2-\sin^2\theta)^{0.5}}\right]$$

or $$= -\left[\frac{W_c l_1 (l-L)}{g}\right] \cdot \left[\frac{\omega^2 (n^2-1)\sin 2\theta}{2\left(n^2-\sin^2\theta\right)2}\right]$$

$$\approx \left[\frac{W_c l_1 (l-L)}{g}\right] \cdot \left[\frac{\omega^2 \sin 2\theta}{2n^2}\right] \tag{10.7}$$

Vertical force through $C = W_c \cdot \dfrac{PG}{PC} = \dfrac{W_c l_1}{l}$

Torque exerted on crankshaft by gravity,

$$T_g = -\left(\frac{W_c l_1}{l}\right) \cdot AO$$

Now $$AO = OC\cos\theta = r\cos\theta$$

$$T_g = -\left(\frac{W_c l_1}{n}\right) \cdot \cos\theta \tag{10.8}$$

Total torque exerted on the crankshaft by the inertia of moving parts

$$\therefore \quad = T_i + T_c + T_g \tag{10.9}$$

10.3.2 Graphical Method

The graphical construction for calculating the inertia forces in a reciprocating engine is shown in Fig.10.4. The following procedure may be adopted for this purpose:

1. Draw the acceleration diagram *OCQN* by Klein's construction. The acceleration of the piston *P* with respect to the crank centre *O* is,

$$f_p = \omega^2 \cdot NO$$

and is acting in the direction from *N* to *O*. Therefore the inertia force F_i shall act in the opposite direction from *O* to *N*.

2. Replace the connecting rod by dynamically equivalent system of two masses as explained in Section 10.8. Let one of the masses be placed at *P*. To obtain the position of the other mass, draw *GZ* perpendicular to *CP* such that *GZ* = *K*, the radius of gyration of the connecting rod. Join *PZ* and from *Z* draw perpendicular to *DZ* which intersects *CP* at *D*. Now *D* is the position of the second mass. Otherwise, GP · GD = K^2.

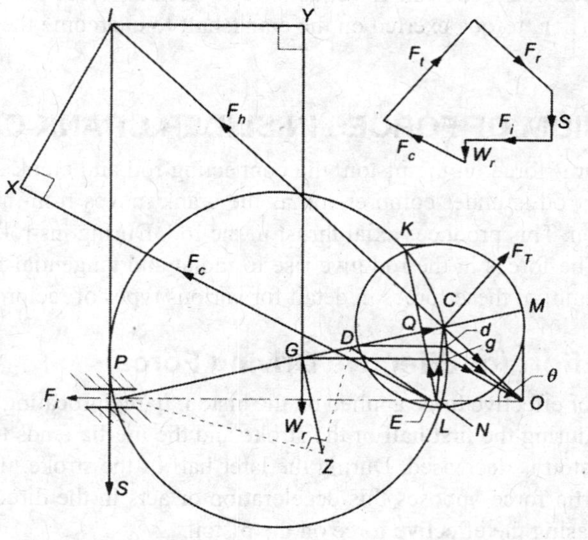

Fig.10.4 Graphical method to determine inertia forces in reciprocating engine

3. Locate the points *g* and *d* on *NC*, the acceleration image of the rod, by drawing parallel lines from *G* and *D* to the line of stroke, Join *gO* and *dO*. Then

$$f_G = \omega^2 \cdot gO \text{ and } f_D = \omega^2 \cdot dO$$

4. From *D*, draw *DE* parallel to *dO* to interect the line of stroke at *E*. The inertia force of the rod F_r acts through *E* and in the opposite direction.

$$F_r = m_r\omega^2 \cdot gO$$

where m_r is the mass of the rod.

The forces acting on the connecting rod are:

(i) Inertia force of the reciprocating parts F_i acting along the line of stroke PO.

(ii) The side thrust between the cross-head and the guide bars S acting at P and right angles to the line of stroke.

(iii) The weight of the connecting rod, $W_r = m_r g$.

(iv) Inertia force of the connecting rod F_r.

(v) The radial force F_r acting through O and parallel to the crank OC.

(vi) The force F_t acting perpendicular to the crank OC.

5. Now produce the line of action of F_r and S to intersect at a point I. From I draw IX and IY perpendicular to the lines of action of F_r and W_r. Taking moments about I, we have

$$F_t \cdot IC = F_i \cdot IP + F_r \cdot IX + W_r \cdot IY$$

The value of F_t may be obtained from this equation and from the force polygon, the forces S and F_r may be calculated. Then, torque exerted on the crankshaft to overcome the inertia of moving parts is $= F_t \cdot OC$.

10.4 EQUILIBRIUM OF FORCES IN SLIDER CRANK CHAIN

Due to the gas (or steam) force on the piston, the connecting rod and crank are subjected to various forces. The connecting rod is under compression as the crank moves from inner dead centre during outstroke, i.e. expansion. This produces axial thrust in the rod. During instroke, i.e. compression, the rod is under tension. The forces in the rod give rise to radial and tangential forces on the crank. We shall study the equilibrium of these forces in detail for various types of reciprocating engines.

10.4.1 Piston Effort (or Effective Driving Force)

Piston effort is the net or effective force applied on the piston. In reciprocating engines, the reciprocating masses accelerate during the first half of the stroke and the inertia tends to resist the same. Thus, the net force on the piston is decreased. During the later half of the stroke, the reciprocating masses decelerate and the inertia force opposes this deceleration or acts in the direction of the applied gas pressure, thereby increasing the effective force on the piston.

In a vertical engine, the weight of the reciprocating masses assists the piston during the outstroke (i.e. downstroke), thus increasing the piston effort. During the instroke (i.e. upstroke), piston effort is decreased by the same amount.

The various forces acting on the links of the slider crank chain during outstroke are shown in Fig.10.5(a).

Let F = piston effort

F_c = thrust in the connecting rod

F_t = crank pin effort

F_r = force in the crank or bearing reaction

F_n = reaction of guide bars

In ΔPCD, $\dfrac{F}{\sin(90° - \phi)} = \dfrac{F_c}{\sin 90°} = \dfrac{F_n}{\sin \phi}$

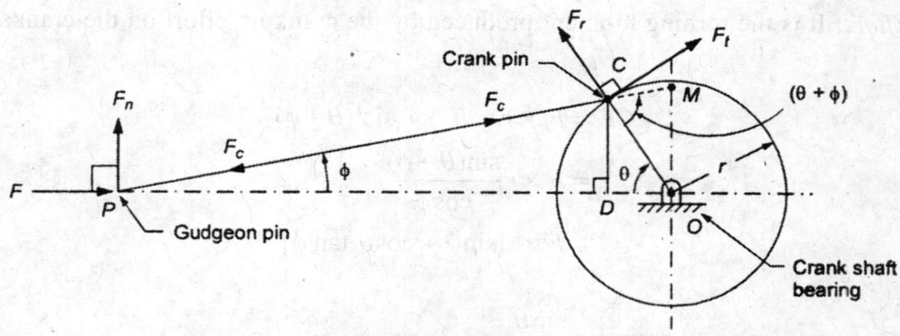

(a) *Connecting rod under compression*

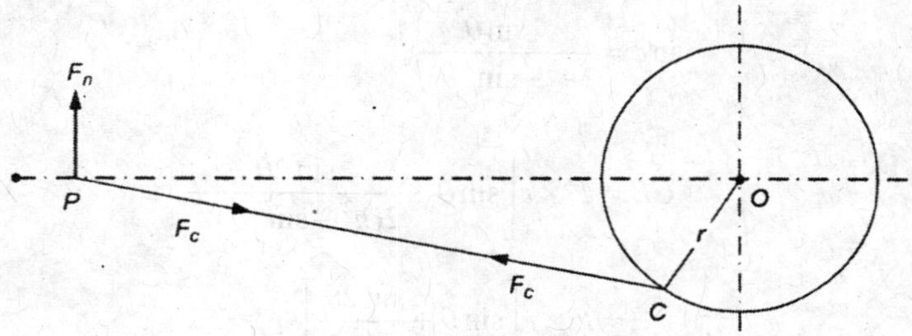

(b) *Connecting rod under tension*

Fig.10.5 Equilibrium forces in slider-crank mechanism

or
$$\frac{F}{\cos\phi} = \frac{F_c}{1} = \frac{F_n}{\sin\phi}$$

Thrust along the connecting rod, $F_c = \dfrac{F}{\cos\phi}$ (10.10)

(a) *Thrust on the sides of the cylinder:* The piston effort produces thrust on the sides of the cylinder. This results into reaction of the guide bars.

$$F_n = F\tan\phi \tag{10.11}$$

The connecting rod is in compression during outstroke.

(b) *Crank pin effort:* It is the net force applied at the crankpin perpendicular to the crank.

$$F_t = F_c \sin(\theta+\phi) = \frac{F\sin(\theta+\phi)}{\cos\phi} \tag{10.12}$$

(c) *Thrust on the bearings:* The component of the thrust in the connecting rod along the crank in the radial direction F_r produces thrust on the crankshaft bearings.

$$F_r = F_c \cos(\theta+\phi) = \frac{F\cos(\theta+\phi)}{\cos\phi} \tag{10.13}$$

(d) *Crank Effort:* It is the turning moment produced by the crankpin effort on the crankshaft. Crank effort,

$$CE = F_t \times r = F_c \times r \sin(\theta+\phi)$$

$$= F \times r \frac{\sin(\theta+\phi)}{\cos\phi}$$

$$= F \times r[\sin\theta + \cos\theta \tan\phi]$$

Now $$\sin\phi = \frac{\sin\theta}{n}$$

$$\tan\phi = \frac{\sin\theta}{(n^2-\sin^2\theta)^{0.5}}$$

$$CE = F \times r\left[\sin\theta + \frac{\sin 2\theta}{2(n^2-\sin^2\theta)^{0.5}}\right] \quad (10.14a)$$

$$\approx F \times r\left[\sin\theta + \frac{\sin 2\theta}{2n}\right] \quad (10.14b)$$

$$= F \times OM \quad (10.14c)$$

10.4.2 Piston Effort for Various Types of Engines

(a) *Double Acting Horizontal Steam Engine:* The net force acting on the cross-head pin along the line of stroke is called piston effort.

Let D = diameter of piston
d = diameter of piston rod
p_1, p_4 = steam pressure on cylinder end during outstroke and instroke respectively
p_1, p_3 = steam pressure on crank end during outstroke and instroke respectively
M_r = mass of reciprocating parts

(i) *Outstroke*

Force on piston due to steam pressure during outstroke, $F_{po} = \frac{\pi}{4}\left[D^2 p_1 - (D^2-d^2)p_2\right]$

Accelerating force due to mass of reciprocating parts, $F_i = M_r \cdot \omega^2 r\left(\cos\theta + \frac{\cos 2\theta}{n}\right)$

Piston effort during outstroke,

$$\text{PEO, } F = F_{po} - F_i \text{, neglecting frictional resistance} \quad (10.15a)$$

$$= F_{po} - F_i - F_f \text{, considering frictional resistance} \quad (10.15b)$$

where F_f = frictional resistance

(ii) *Instroke*

Force on piston due to steam pressure during instroke, $F_{pi} = \frac{\pi}{4}\left[D^2 p_3 - (D^2 - d^2) p_4\right]$

Net force on piston during instroke, PEI, $F = F_{pi} - F_i$ neglecting frictional resistance (10.16a)

$$= F_{pi} - F_i - F_f \text{, considering frictional resistance} \quad (10.16b)$$

(b) *Double Acting Vertical Steam Engine*

(i) *Downstroke*

Piston effort during downstroke for vertical steam engine,

PED, $$F = F_{pd} + M_r g - F_i - F_f \quad (10.17)$$

(ii) *Upstroke*

Piston effort during upstroke for vertical steam engine,

PEU, $$F = F_{pu} + M_r g - F_i - F_f \quad (10.18)$$

(c) *Four Stroke Horizontal Internal Combustion Engine*

Substitute $p_2, p_3 = p_a$ = atmospheric pressure, and d = 0 in Eqs. (10.14) and (10.15) to obtain the piston effort during outstroke and instroke, respectively.

PED, $$F = \left(\frac{\pi}{4}\right) D^2 (p_1 - p_a) - M_r \cdot \omega^2 r \left[\cos\theta + \frac{\cos 2\theta}{n}\right] \quad (10.19)$$

PEI, $$F = \left(\frac{\pi}{4}\right) D^2 (p_a - p_4) - M_r \cdot \omega^2 r \left[\cos\theta + \frac{\cos 2\theta}{n}\right] \quad (10.20)$$

(d) *Four Stroke Vertical Internal Combustion Engine*

PED, $$F = \left(\frac{\pi}{4}\right) D^2 (p_1 - p_a) - M_r \cdot \omega^2 r \left[\cos\theta + \frac{\cos 2\theta}{n}\right] + M_r g \quad (10.21)$$

PEU, $$F = \left(\frac{\pi}{4}\right) D^2 (p_a - p_4) - M_r \cdot \omega^2 r \left[\cos\theta + \frac{\cos 2\theta}{n}\right] - M_r g \quad (10.22)$$

10.5 CRANK EFFORT (OR TURNING MOMENT) DIAGRAMS

The diagrams obtained on plotting crank effort for various positions of crank are known as crank effort diagrams, From Eq. (10.1), we have

Displacement, $$x = r\left[(1 - \cos\theta) + \left\{n - (n^2 - \sin^2\theta)^{0.5}\right\}\right]$$

Length of stroke, $$L = 2r$$

$$\therefore \qquad \frac{x}{L} = 0.5\left[(1-\cos\theta) + \left\{n-(n^2-\sin^2\theta)^{0.5}\right\}\right] \qquad (10.23)$$

$\frac{x}{L}$ is called the displacement constant.

10.5.1 Procedure for Determination of Turning Moment Diagram

The following steps may be followed to determine the turning moment diagram:

1. Determine the pressure of the working fluid on both sides of the piston either from the indicator diagram or from theoretical calculations.
2. Calculate the load on the piston due to steam or gas pressure, F_p.
3. Calculate the acceleration of the piston, f_p.
4. Calculate the inertia force due to reciprocating parts, $F_i = M_r \times f_p$.
5. Calculate the piston effort, $PE = F = F_p \pm M_r g - M_r \times f_p - F_f$

   Remember that second term is zero for a horizontal engine. For a vertical engine, take positive sign for downstroke and negative sign for upstroke before $M_r g$.
6. Calculate the crank effort, $CE = T = F \times r\left(\sin\theta + \frac{\sin 2\theta}{2n}\right)$

10.5.2 Turning Moment Diagram for a Vertical Steam Engine

The turning moment diagram for a vertical steam engine is shown in Fig.10.6. The hatched area below the mean torque shows deficient energy, and the area above the mean torque is the surplus energy. One cycle is completed during 360° of the crank rotation.

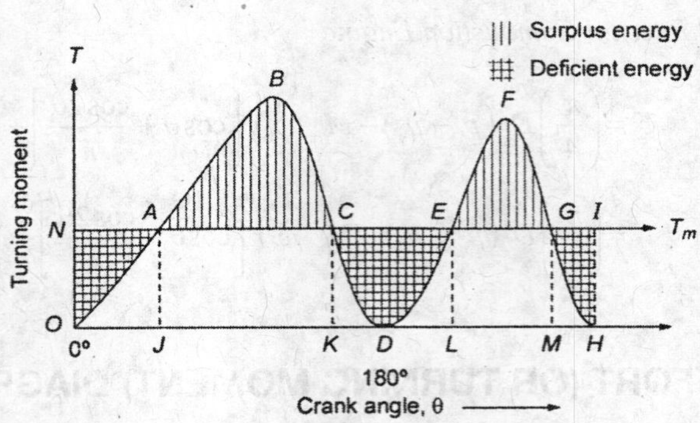

Fig.10.6 Turning moment diagram for a vertical steam engine

10.5.3 Turning Moment Diagram for a Four Stoke I.C. Engine

The turning moment diagram for a four stroke I.C. engine is shown in Fig.10.6. Here one cycle is completed during 720° of the crank rotation. Energy is supplied mainly during the expansion stroke. This diagram is more nonuniform as compared to Fig.10.6.

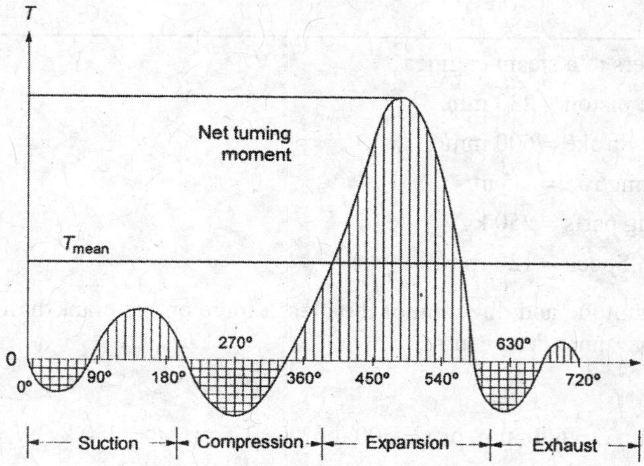

Fig.10.7 Turning moment diagram for a four stroke I.C. engine

10.5.4 Turning Moment Diagram for a Multicylinder Engine

The turning moment diagram for a three-cylinder engine is shown in Fig.10.8. The diagram becomes more uniform above and below the mean torque.

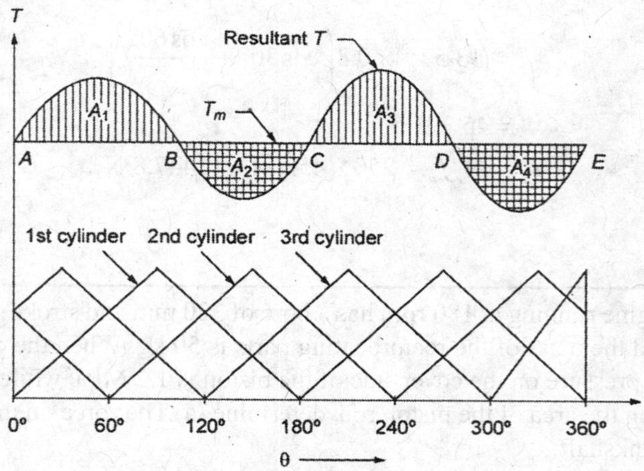

Fig.10.8 Turning moment diagram for a multi-cylinder engine

10.5.5 Uses of Turning Moment Diagram

The uses of turning moment diagram are as follows:

1. The area under the turning moment diagram represents work done per cycle. This area multiplied by number of cycles per second gives the power developed by the engine in Watts.
2. Dividing the area of the turning moment diagram with the length of the base gives the mean turning moment. This enables to find the fluctuation of energy.
3. The maximum ordinate of the turning moment diagram gives the maximum torque to which the crankshaft is subjected to. This enables to find the diameter of the crankshaft.

Example 10.2

The following data refers to a steam engine:

Diameter of the piston = 230 mm

Stroke = 600 mm

Length of connecting rod = 1.5 m

Mass of reciprocating parts = 250 kg

Speed = 120 rpm

Determine the magnitude and direction of the inertia force on the crankshaft when the crank has turned through 30° from inner dead centre.

Solution

Given: $d_p = 230$ mm, $L = 600$ mm, or $r = 300$ mm, $l = 1.5$ m, $M_r = 250$ kg, $N = 120$ rpm, $\theta = 30°$

$$n = \frac{l}{r} = \frac{1.5}{0.3} = 5$$

$$\omega = \frac{2\pi \times 120}{60} = 13.57 \text{ rad/s}$$

$$f_p = \omega^2 r\left(\cos\theta + \frac{\cos 2\theta}{n}\right)$$

$$= (13.57)^2 \times 0.3\left(\cos 30° + \frac{\cos 60°}{5}\right)$$

$$= 45.791 \text{ m/s}^2$$

$$F_i = M_r f_p = 250 \times 45.791 = 11447.8 \text{ N}$$

Example 10.3

A horizontal steam engine running at 180 rpm has a bore of 320 mm and stroke 560 mm. The connecting rod is 1 m long and the mass of the reciprocating parts is 50 kg. When the crank is 50° past inner dead centre, the steam pressure on the cover side of the piston is 1.2 MPa, while that on the crank side is 0.15 MPa. Neglecting the area of the piston rod, determine (a) The force on the piston, and (b) turning moment on the crankshaft.

Solution

Given: $N = 180$ rpm, $d_p = 320$ mm, $L = 560$ mm or $r = 280$ mm, $l = 1$m, $M_r = 50$ kg, $\theta = 50°$

$$p_1 = 1.2 \text{ MPa}, p_2 = 0.15 \text{ MPa}.$$

$$n = \frac{l}{r} = \frac{1}{0.28} = 3.571$$

$$\omega = \frac{2\pi \times 180}{60} = 18.85 \text{ rad/s}$$

(a) Force on piston, $F_p = \dfrac{\pi d_p^2(p_1 - p_2)}{4} = \dfrac{\pi \times 320^2(1.2-0.15)}{4} = 84446 \text{ N}$

(b) $$f_p = \omega^2 r\left(\cos\theta + \frac{\cos 2\theta}{n}\right)$$

$$= (18.85)^2 \times 0.28\,[\cos 50° + 0.28 \times \cos 100°]$$

$$= 59.11 \text{ m/s}^2$$

Inertia force due to reciprocating parts, $F_i = M_r f_p = 50 \times 59.11 = 2955.66 \text{ N}$

Piston effort, $F = F_p - F_i = 84446 - 2955.66 = 81490.34 \text{ N}$

Crank effort, $CE = F \times r\left(\sin\theta + \dfrac{\sin 2\theta}{2n}\right)$

$$= 81490.34 \times 0.28\left(\sin 50° + \frac{\sin 100°}{2 \times 3.571}\right)$$

$$= 20625.3 \text{ N m}$$

Example 10.4

A vertical single cylinder engine has a cylinder diameter of 240 mm and a stroke of 420 mm. The mass of the reciprocating parts is 200 kg. The connecting rod is 4.25 times the crank radius and the speed is 340 rpm. When the crank has turned through 40° from the top dead centre, the net pressure on the piston is 1 MPa. Calculate the effective moment on the crankshaft for this position.

■ Solution

Given: $N = 340$ rpm, $d_p = 240$ mm, $L = 420$ mm or $r = 210$ mm, $n = 4.25$, $M_r = 200$ kg, $\theta = 40°$, $p = 1$ mpa

$$\omega = \frac{2\pi \times 340}{60} = 35.6 \text{ rad/s}$$

(a) Force on piston, $F_p = \dfrac{\pi d_p^2 p}{4} = \dfrac{\pi \times 240^2 \times 1}{4} = 45238.93 \text{ N}$

(b) $$f_p = \omega^2 r\left(\cos\theta + \frac{\cos 2\theta}{n}\right)$$

$$= (35.6)^2 \times 0.21\left[\cos 40° + \frac{\cos 80°}{4.25}\right]$$

$$= 214.75 \text{ m/s}^2$$

Inertia force due to reciprocating parts, $F_i = M_r f_p = 200 \times 214.75 = 42950 \text{ N}$

Piston effort, $F = F_p + M_r g - F_i = 45238.93 + 200 \times 9.81 - 42950 = 4250.93 \text{ N}$

Crank effort, $$CE = F \times r\left(\sin\theta + \frac{\sin 2\theta}{2n}\right)$$

$$= 4250.93 \times 0.21\left(\sin 40° + \frac{\sin 80°}{2 \times 4.25}\right)$$

$$= 677.24 \text{ N m}$$

Example 10.5

A petrol engine 100 mm in diameter and 120 mm stroke has a connecting rod 250 mm long. The piston has a mass of 1 kg and the speed is 1800 rpm. The gas pressure is 0.5 MPa at 30° from top dead centre during the explosion stroke. Find (a) the resultant load in the gudgeon pin, (b) the thrust on the cylinder walls, (c) the speed above which the gudgeon pin load will be reversed, and (d) the crank effort at this position.

Solution

Given: $N = 1800$ rpm, $d_p = 100$ mm, $L = 120$ mm or $r = 60$ mm, $l = 250$ mm, $M_r = 1$ kg, $\theta = 30°$, $p = 0.5$ MPa.

$$n = \frac{l}{r} = \frac{250}{50} = 5$$

$$\omega = \frac{2\pi \times 1800}{60} = 188.5 \text{ rad/s}$$

Force on piston due to gas pressure, $$F_p = \frac{\pi d_p^2 p}{4} = \frac{\pi \times 100^2 \times 0.5}{4} = 3927 \text{ N}$$

Acceleration of reciprocating parts, $$f_p = \omega^2 r\left(\cos\theta + \frac{\cos 2\theta}{n}\right)$$

$$= (188.5)^2 \times 0.05\left[\cos 30° + \frac{\cos 60°}{5}\right]$$

$$= 1716.25 \text{ m/s}^2$$

Inertia force due to reciprocating parts, $F_i = M_r f_p = 1 \times 1716.26 = 1716.25$ N

(a) Piston effort, $$F = F_p - F_i = 3927 - 1716.25 = 2210.75 \text{ N}$$

(b) $$\sin\phi = \frac{\sin\theta}{n} = \frac{\sin 30°}{5} = 0.1, \phi = 5.74°$$

Thrust on cylinder walls, $$F_n = F \tan\phi = 2210.75 \ \tan 5.74° = 222.22 \text{ N}$$

Resultant load on gudgeon pin $= \left(F^2 + F_n^2\right)^{0.5} = [(2210.75)^2 + (222.22)^2]^{0.5} = 2221.9$ N

(c) $$f_p = \omega^2 r\left(\cos\theta + \frac{\cos 2\theta}{n}\right)$$

$$= \omega^2 \times 0.05\left(\cos 30° + \frac{\cos 60°}{5}\right)$$

$$= 0.0483\ \omega^2$$

$$F_i = M_r f_p = 1 \times 0.0483 \times \omega^2 = 0.0483\ \omega^2$$

For $$F_p - F_i = 0$$

$$0.0483\ \omega^2 = 3927$$

$$\omega^2 = 81304$$

$$\omega = 285.14 \text{ rad/s}$$

$$N = \frac{285.14 \times 60}{2\pi} = 2723.8 \text{ rpm}$$

(d) Crank effort, $$CE = F \times r\left(\sin\theta + \frac{\sin 2\theta}{2n}\right)$$

$$= 2210.75 \times 0.05\left(\sin 30° + \frac{\sin 60°}{2 \times 5}\right)$$

$$= 64.84 \text{ N m}$$

Example 10.6

The following data refers to a horizontal reciprocating engine:

Length of crank = 300 mm
Length of connecting rod = 1.5 m
Speed = 120 rpm, cw
Mass of reciprocating parts of engine = 290 kg
Mass of connecting rod = 250 kg
Centre of gravity of connecting rod from crankpin centre = 475 mm
Radius of gyration of connecting rod about an axis through centre of gravity = 625 mm
Find the inertia torque on the crankshaft when $\theta = 40°$.

■ Solution

Given: $r = 300$ mm, $l = 1.5$ m, $N = 120$ rpm, $M_r = 290$ kg, $M_c = 250$ kg, $l_2 = 475$ mm, $K = 625$ mm, $\theta = 40°$

$$n = \frac{l}{r} = \frac{1.5}{0.3} = 5$$

$$\omega = \frac{2\pi \times 120}{60} = 13.57 \text{ rad/s}$$

Distance of centre of gravity of connecting rod from gudgeon pin: $l_1 = l - l_2 = 1500 - 475 = 1025$ mm

Let m_1 = mass of connecting rod placed at gudgeon pin

$$= \frac{(l - l_1)M_c}{l} = \frac{475 \times 250}{1500} = 79.17 \text{ kg}$$

Total equivalent reciprocating mass, $M_{re} = M_r + m_1 = 290 + 79.17 = 369.17$ kg

Acceleration of piston, $f_p = r\omega^2 \left(\cos\theta + \frac{\cos 2\theta}{n} \right)$

$$= 0.3 \times (13.57)^2 \left(\cos 40° + \frac{\cos 80°}{5} \right)$$

$$= 37.96 \text{ m/s}^2$$

Total equivalent inertia force, $F_i = -M_{re} \,.\, f_p = -369.17 \times 37.96 = -14013.9$ N

$$\sin\phi = \frac{\sin\theta}{n} = \frac{\sin 40°}{5} = 0.12855$$

$$\phi = 7.386°, \ \theta + \phi = 47.386°, \ \beta = 180° - (50° + 47.386°) = 83.614°$$

$$OC = \frac{OA \sin(\theta + \phi)}{\sin\beta} = \frac{0.3 \times \sin 47.386°}{\sin 83.614°} = 0.2226 \text{ m}$$

Approximate inertia torque exerted on crankshaft $= F_i \times OC$

$$T_i = -14013.9 \times 0.2226 = -3119.6 \text{ N m}$$

Correction couple, $T_c = M_c l_1 (l - L_e) \alpha_c$

where $L = \frac{K^2 + l_1^2}{l_1} = \frac{625^2 + 1025^2}{1025} = 1406.1$ mm

$$\alpha_c = \frac{-\omega^2 \sin\theta}{n} = \frac{-(13.57)^2 \sin 40°}{5} = -20.31 \text{ rad/s}$$

$$T_c = -250 \times 1.025 \times (1.5 - 1.4061) \times 20.31 = -488.7 \text{ N m}$$

Torque on crankshaft due to correction couple, $T_c^* = \frac{T_c \cos\theta}{n}$

$$= \frac{-4888.7 \cos 40°}{5} = -74.87 \text{ N m}$$

Mass of connecting rod at crankpin, $m_2 = M_c - m_1 = 250 - 79.17 = 170.83$ kg

Weight of connecting rod at crankpin, $W_2 = m_2 g = 170.83 \times 9.81 = 1675.87$ N

Torque exerted by W_2 on crankshaft, $T_w = -W_2 r \cos\theta = -1675.87 \times 0.3\cos 40° = -385.14$ N m

Total torque exerted on the crankshaft due to inertia of moving parts

$$= T_i + T_c^* + T_w = -3119.6 - 74.87 - 385.14 = -3579.6 \text{ N m}$$

Example 10.7

A single cylinder vertical engine has a bore of 300 mm and a stroke of 400 mm. The connecting rod is 1 m long. The mass of the reciprocating parts is 150 kg. The gas pressure is 0.7 MPa with the crank at 30° from the top dead centre during expansion stroke. The speed of crank is 250 rpm. Determine (a) net force acting on the piston, (b) resultant load on the gudgeon pin, (c) thrust on the cylinder wall, and (d) the speed above which, other things remaining same, the gudgeon pin load would be reversed in direction.

Solution

Given: $D = 300$ mm, $L = 400$ mm or $r = 200$ mm, $l = 1$ m, $M_r = 150$ kg, $p = 0.7$ MPa, $\theta = 30°$, $N = 250$ rpm, $n = \frac{l}{r} = \frac{1}{0.2} = 5$

$$\omega = \frac{2\pi \times 250}{60} = 26.18 \text{ rad/s}$$

Inertia force due to reciprocating parts, $F_i = -M_r f_p$

$$= -M_r\, r\, \omega^2 \left[\cos\theta + \frac{\cos 2\theta}{n}\right]$$

$$= -150 \times 0.2 \times (26.18)^2 \left[\cos 30° + \frac{\cos 60°}{5}\right]$$

$$= -19863.2 \text{ N}$$

Force due to gas pressure, $F_p = \frac{\pi D^2 p}{4} = \frac{\pi \times (300)^2 \times 0.7}{4} = 49480$ N

Weight of reciprocating parts, $W_r = M_r g = 150 \times 9.81 = 1471.5$ N

(a) Net force acting on the piston, $F = F_p - F_i + W_r = 49480 - 19863.2 + 1471.5 = 31088.3$ N

(b) $\sin\phi = \frac{\sin\theta}{n} = \frac{\sin 30°}{5} = 0.1$, $\phi = 5.739°$

Resultant load on gudgeon pin $= \frac{F}{\cos\phi} = \frac{31088.3}{\cos 5.739°} = 31249.9$ N

(c) Thrust on cylinder wall $= F \tan\phi = 31088.3 \tan 5.739° = 3124.4$ N

(d) The gudgeon pin load would be reversed if $F_i > (F_p + W_r)$

$$M_r r\omega^2 \left(\cos\theta + \frac{\cos 2\theta}{n}\right) > (F_p + W_r)$$

$$150 \times 0.2 \times \left(\frac{2\pi}{60}\right)^2 N^2 \left(\cos 30° + \frac{\cos 60°}{5}\right) > (49480 + 1471.5)$$

$$N^2 > 160320.8$$

$$N > 400.4 \text{ rpm}$$

Example 10.8

A vertical double-acting steam engine has a cylinder 300 mm diameter and 450 mm stroke and runs at 200 rpm. The reciprocating parts weigh 2250 N and piston rod is 50 mm diameter and the connecting rod is 1.2 m long. When the crank has turned through 125° from the top dead centre, the steam pressure above the piston is 0.3 MPa and below the piston is 0.015 MPa gauge. Calculate the effective turning moment on the crankshaft.

Solution

Given: $N = 200$ rpm, $d_p = 50$ mm, $r = 225$ mm, $l = 1.2$ m, $W_r = 2250$ N, $\theta = 125°$, $d_c = 300$ mm, $p_1 = 0.3$ MPa, $p_2 = 0.015$ MPa.

$$n = \frac{l}{r} = \frac{1200}{225} = 5.33$$

$$\omega = \frac{2\pi \times 200}{60} = 20.944 \text{ rad/s}$$

(a) Force on piston, $F_p = \left(\frac{\pi}{4}\right)[d_c^2 p_1 - (d_c^2 - d_p^2) p_2]$

$$= \left(\frac{\pi}{4}\right)[300^2 \times 0.3 - (300^2 - 50^2) \times 0.015] = 20175 \text{ N}$$

(b)

$$f_p = \omega^2 r \left(\cos\theta + \frac{\cos 2\theta}{n}\right)$$

$$= (20.944)^2 \times 0.255 \left(\cos 125° + \frac{\cos 250°}{5.33}\right)$$

$$= -63.94 \text{ m/s}^2$$

Inertia force due to reciprocating parts, $F_i = M_r f_p = \frac{-2250 \times 63.94}{9.81} = -14436$ N

Piston effort, $F = F_p - F_i = 20175 - 14436 = 5739$ N

$$\sin\phi = \frac{\sin\theta}{n} = \frac{\sin 125°}{5.33} = 0.1537, \ \phi = 8.84°$$

Force in connecting rod, $F_c = \frac{F}{\cos\phi} = \frac{5739}{\cos 8.84°} = 5808$ N

$$F_t = F_c \sin(\theta + \phi) = 5808 \sin(125° + 8.84°) = 4189 \text{ N}$$

Turning moment on crankshaft, $T = F_t \times r = 4189 \times 0.225 = 943.5$ N m

10.6 FLUCTUATION OF ENERGY

Fluctuation of Energy (E_f): It is the excess energy developed by the engine between two crank positions.

$$E_f = C_e E \tag{10.24}$$

where $E = \left(\frac{1}{2}\right) \cdot I\omega_m^2$, I = moment of inertia of the flywheel, and ω_m its mean angular speed.

Coefficient of Fluctuation of Energy (C_e): It is the ratio of the maximum fluctuation of energy to the indicated work done by the engine during one revolution of crank.

$$C_e = \frac{E_{max} - E_{min}}{T_m \cdot \theta} \tag{10.25}$$

$\theta = 4\pi$ for steam engines and 4π for four stroke I.C. engines

$$\text{Mean torque, } T_m = \frac{\text{Power developed}}{\omega_m}$$

Coefficient of Fluctuation of Speed (C_s): It is defined as the ratio of the difference between the maximum and minimum angular velocities of the crankshaft to its mean angular velocity.

$$C_s = \frac{\omega_{max} - \omega_{min}}{\omega_m}$$

$$= \frac{N_{max} - N_{min}}{N_m} \tag{10.26}$$

where $N_m = \dfrac{N_{max} + N_{min}}{2}$ = mean speed

10.6.1 Determination of Maximum Fluctuation of Energy

As shown in Fig.10.9, let E be the energy at point A. Then

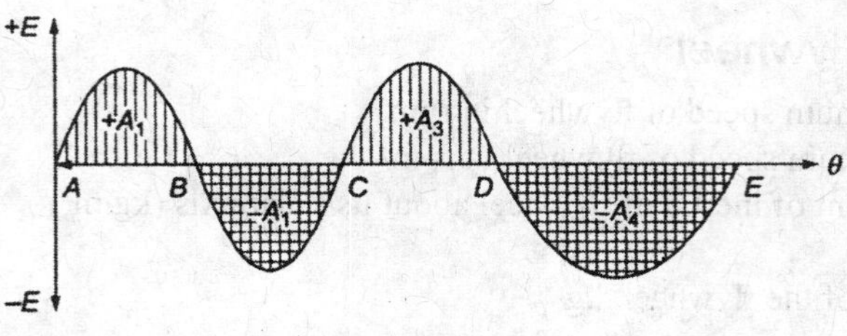

Fig.10.9 Fluctuation of energy diagram

Energy at $B = E + A_1$

Energy at $C = E + A_1 - A_2$

Energy at $D = E + A_1 - A_2 + A_3$

Energy at $E = E + A_1 - A_2 + A_3 - A_4$

= Energy at A, i.e., E

Let the maximum energy be at point B and minimum at point D. Then

$E_{max} = E + A_1$

$E_{min} = E + A_1 - A_2 + A_3$

Maximum fluctuation of energy $= E_{max} - E_{min}$

$= A_2 - A_3$

10.7 FLYWHEEL

A flywheel is a device that serves as a reservoir to store energy when the supply of energy is more than the requirement, and releases energy when the requirement is more than the supply. Thereby, it controls the fluctuation of speed of the prime mover during each cycle of operation. The differences between the functions of a flywheel and governor are given in Table 10.1.

Table 10.1 Differences between the Functions of a Flywheel and a Governor

Flywheel	Governor
1. It decreases the variation of speed of prime mover during each cycle of operation	1. It regulates the speed of the prime mover from cycle to cycle.
2. It decreases the fluctuation of speed due to difference in output and input.	2. It decreases the fluctuation of speed by adjusting the output of the prime mover
3. A flywheel controls $\frac{dN}{dt}$.	3. A governor controls dN.
4. It stores energy and gives up when required.	4. It regulates speed by regulating the quantity of working medium of the prime mover.
5. It has no control over the quality of the working medium.	5. It takes care of the quality of the working medium.
6. It is not an essential part of every prime mover.	6. It is an essential part of every prime mover.

10.7.1 Size of Flywheel

Let N_{max} = maximum speed of flywheel in rpm

N_{min} = Minimum speed of flywheel in rpm

I = moment of inertia of flywheel about its polar axis (kg m²)

$= mK^2$

m = mass of the flywheel, kg

$= \pi D b t \rho$

D = mean diameter of flywheel (m)

b = width of rim (m)

t = thickness of rim (m)

ρ = density of rim material (kg/m³)

K = radius of gyration of the flywheel (m)

$K^2 = \frac{D^2}{8}$, for a solid flywheel

Maximum kinetic energy of the flywheel, $E_{max} = \left(\frac{1}{2}\right) \cdot I\omega^2$

$$= \left(\frac{1}{2}\right) \cdot mK^2 \left(\frac{2\pi N_{max}}{60}\right)^2$$

Minimum kinetic energy of the flywheel, $E_{min} = \left(\frac{1}{2}\right) \cdot I\omega_{min}^2 = \left(\frac{1}{2}\right) \cdot mK^2 \left(\frac{2\pi N_{min}}{60}\right)^2$

Fluctuation of energy, $E_f = E_{max} - E_{min}$

$$= \left(\frac{1}{2}\right) \cdot mK^2 \left(\frac{2\pi}{60}\right)^2 \left(N_{max}^2 - N_{min}^2\right)$$

$$= \left(\frac{\pi^2}{1800}\right) \cdot mK^2 \cdot (N_{max} + N_{min})(N_{max} - N_{min})$$

$$= \left(\frac{\pi^2}{1800}\right) \cdot mK^2 \cdot (2N_m \times N_m C_s)$$

or $$m = \frac{900 \times E_f}{\pi^2 \times K^2 \times N_m^2 \times C_s} kg \quad (10.27)$$

Also $$E_f = \left(\frac{1}{2}\right) I\left(\omega_{max}^2 - \omega_{min}^2\right) = \left(\frac{1}{2}\right) I(\omega_{max} + \omega_{min})(\omega_{max} - \omega_{min})$$

$$= \left(\frac{1}{2}\right) I \times 2\omega_m \times C_s\omega_m$$

$$= \left(\frac{1}{2}\right) I \times \omega_m^2 \times 2C_s$$

$$= I\omega_m^2 C_s = 2C_s E \quad (10.28)$$

There are two types of flywheels: Disc type and arm type. In the arm type of flywheels, the weight of the flywheel is mainly located in the rim and the arms and boss do not contribute much in storing the energy. A rimmed type flywheel is shown in Fig.10.10.

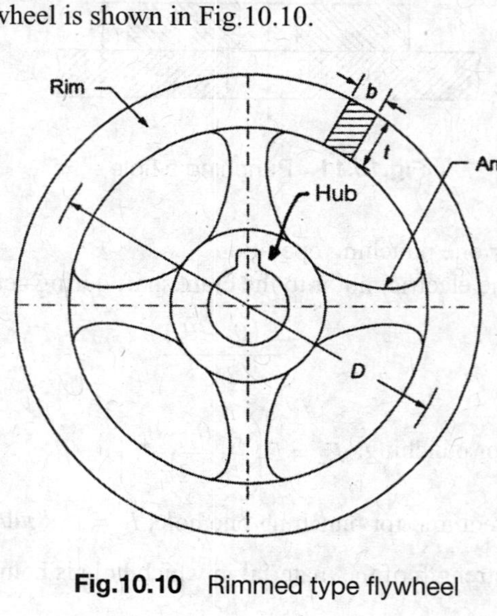

Fig.10.10 Rimmed type flywheel

Mass of the rim, $m_r \approx 0.9$ m, m = total mass of flywheel

The hoop stress in the flywheel can be determined by assuming it as a ring.

Hoop stress, $\sigma_\theta = \rho v^2$ (10.29)

Where ρ is the density of the rim, and v is its peripheral speed.

10.7.2 Flywheel for a Punching Press

In a punching press, the crank is driven at a constant torque by an electric motor. The load is maximum when operation of punching takes place and it is zero during the rest of the cycle. Thus, there are variations in load which causes fluctuations in speed of the press.

Consider the punching press shown in Fig.10.11. Punching operation is performed during the period when the crank rotates from θ_1 to θ_2. There is no load on the crank during the remaining angle of rotation of the crank.

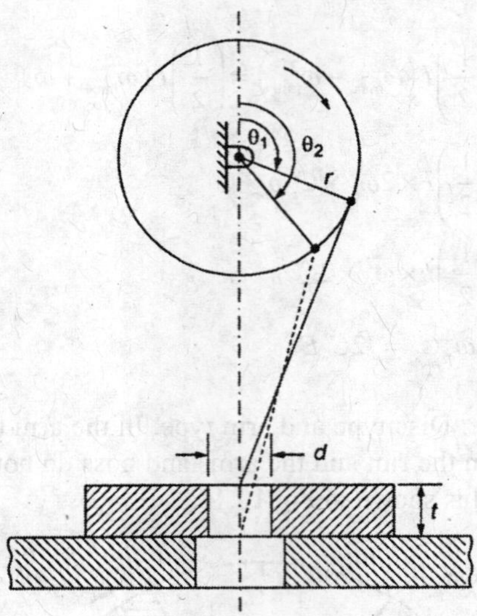

Fig.10.11 Punching a hole

Let E = energy required for one punching operation

Then energy supplied by the electric motor to the crankshaft during actual punching operation,

$$E_a = \frac{E(\theta_2 - \theta_1)}{2\pi}$$

Balance energy available for punching, $E_b = E\left(1 - \frac{\theta_2 - \theta_1}{2\pi}\right)$ (10.30)

Maximum shearing force required for punching one hole, $F_s = \tau_u \times \pi d t$ (10.31)

where τ_u = ultimate shear strength of the material in which hole is being punched, N/mm^2

d = diameter of hole punched, mm
t = thickness of plate, mm

Average energy consumed for punching one hole, $E = \frac{1}{2} F_s \times t$ (10.32)

Eq. (10.30) represents the fluctuation of energy, which can be written as:

$$\left(\frac{1}{2}\right) I \left(\omega_{\max}^2 - \omega_{\min}^2\right) = E \left(1 - \frac{\theta_2 - \theta_1}{2\pi}\right)$$

For approximation, $\frac{\theta_2 - \theta_1}{2\pi} \approx \frac{t}{2L} = \frac{t}{4r}$ (10.33)

where L = stroke of punch
r = crank radius

$$\left(\frac{1}{2}\right) I \left(\omega_{\max}^2 - \omega_{\min}^2\right) = E \left(1 - \frac{t}{4r}\right) \tag{10.34}$$

The mechanism used for punching of holes in plates is shown in Fig.10.12.

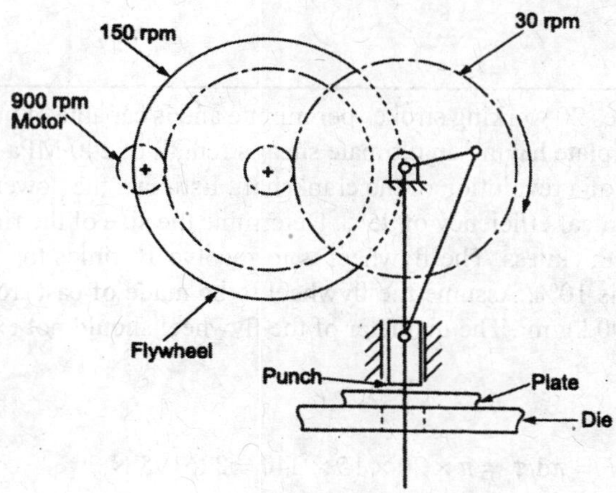

Fig.10.12 Punching press

It consists of a driving motor whose speed is reduced to the desired value by a set of gears to drive the slider–crank mechanism. The slider of the slider–crank is used as the punch to punch a hole in the plate placed on the die. The punching takes place for a very short interval of the angle of rotation of the crank. The punching operation requires huge amount of energy for a short time. For the remaining period, the device remains idle. Therefore, the flywheel is used to store energy during the idle period and supply the desired energy during the working period. The distribution of force on the punch during the punching operation is shown in Fig.10.13.

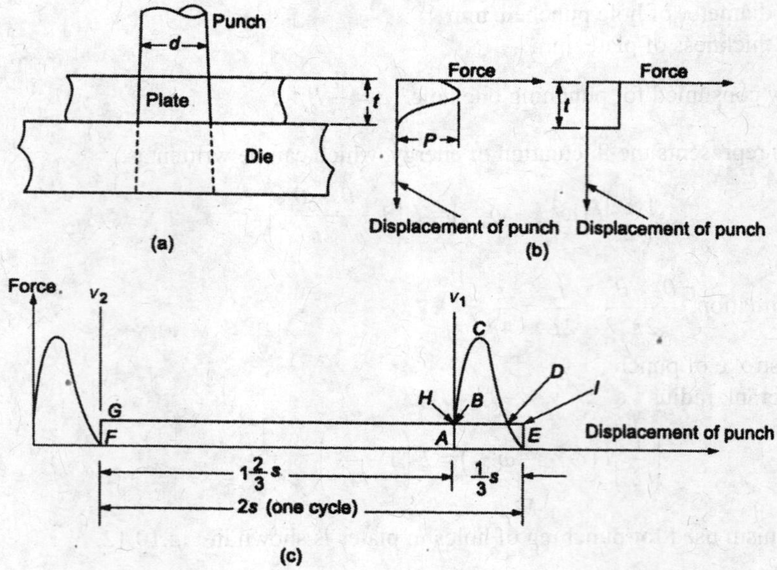

Fig.10.13 Distribution of forces during punching

Example 10.9

A punching machine makes 20 working strokes per minute and is capable of punching 20 mm diameter hole in 15 mm thick steel plate having an ultimate shear strength of 240 MPa. The punching operation takes place during 1/10th of a revolution of the crankshaft. Estimate the power required for the driving motor, assuming a mechanical efficiency of 95%. Determine the size of the rim of the flywheel having width equal to twice the thickness. The flywheel is to revolve 10 times the speed of the crankshaft. The fluctuation of speed is 10%. Assume the flywheel to be made of cast iron having working stress of 6 MPa and density 7300 kg/m³. The diameter of the flywheel should not exceed 1.5 m. Neglect the effect of arms and hub.

Solution

Punching force required, $F = \pi d t \tau_u = \pi \times 20 \times 15 \times 240 = 226195$ N

Work done in punching the hole, $E_1 = 0.5\,Ft = 0.5 \times 226195 \times 0.015 = 1696.5$ Nm

Time between punching operations $= \dfrac{60}{20} = 3\,\text{s}$

Punching time $= \dfrac{3}{10} = 0.3\,\text{s}$

Average power required without flywheel, $E_2 = \dfrac{1696.5}{0.3} = 5654.87$ W

Instantaneous power required $= 2 \times 5654.87 = 11309.74$ W

Motor power required in punching the hole $= \dfrac{1696.5}{3} = 565.5$ W

Work supplied by the motor during 0.3s of punching time = 565.5 × 0.3 = 169.5 N m

Energy to be taken from flywheel = 1696.5 – 169.5 = 1527 Nm

Taking motor efficiency = 95%

Power of motor required = $\frac{565.5}{0.95} = 595.2$ W

Flywheel:

$$\sigma_\theta = \rho v^2$$
$$6 \times 10^6 = 7300\, v_m^2$$
$$v_m = 28.67 \text{ m/s}$$

Mass of flywheel, $m = \frac{E_f}{C_s v_m^2}$

$$= \frac{1527}{0.1 \times 821.97} = 18.57 \text{ kg}$$

$$C_s = \frac{v_{max} - v_{min}}{v_m}, \; v_{max} - v_{min} = 28.67 \times 0.1 = 3.867 \text{ m/s}$$

$$v_{max} + v_{min} = 2v_m = 57.34 \text{ m/s}$$

$$m = \rho \times \pi d \times bt$$

$$18.57 = 7300 \times \pi \times 1.5 \times 2t^2$$

$$t = 16.4 \text{ mm}, \; b = 33.8 \text{ mm}$$

Example 10.10

A vertical double acting steam engine develops 80 kW at 240 rpm. The maximum fluctuation of energy is 25% of the work done per stroke. The maximum and minimum speeds are not to vary more than ±1% of the mean speed. Find the mass of the flywheel, if the radius of gyration is 0.65 m.

■ Solution

Given: $P = 80$ kW, $N_m = 240$ rpm, $C_e = 25\%$, $C_s = \pm 1\%$, $K = 0.65$ m

$$m = \frac{900\, E_f}{\pi^2 K^2 N_m^2 C_s}, E_f = C_e\, E$$

$$\omega_m = \frac{2\pi \times 240}{60} = 25.13 \text{ rad/s}$$

$$\text{P} = \frac{T\omega_m}{10^3}$$

$$T = \frac{80 \times 10^3}{25.13} = 3183 \text{ N m} = E$$

$$m = \frac{900 \times 3183 \times 0.25}{\pi^2 \times 0.65^2 \times 240^2 \times 0.01} = 298 \text{ kg}$$

Example 10.11

The turning moment diagram for a multicylinder engine shown in Fig.10.14 has been drawn to a scale of 1 mm = 5000 N m vertically and 1 mm = 3.5° horizontally. The areas between output torque curve and mean resistance line taken in order from one end are: –340, +21, –245, +300, –118, +230, –225, +377 mm², when the engine is running at 180 rpm. If the mass of the flywheel is 1000 kg and the total fluctuation of speed is not to exceed 3% of the mean speed, find the minimum value of the radius of gyration.

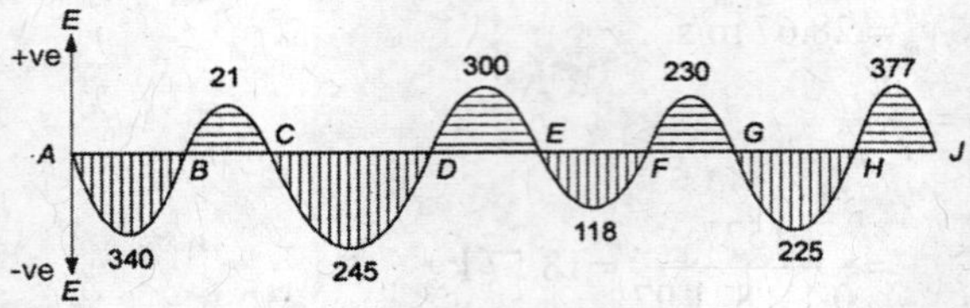

Fig.10.14 Turning moment diagram for a multi-cylinder engine

Point	Energy (mm²)
A	*E*
B	*E* – 340
C	*E* – 319
D	*E* – 564
E	*E* – 264
F	*E* – 382
G	*E* – 152
H	*E* – 377
J	*E*

■ Solution

Let E = energy at A

$$E_{max} = E - 152$$

$$E_{min} = E - 564$$

$$E_f = [(E - 152) - (E - 564)] \times \text{Horizontal scale} \times \text{Vertical scale}$$

$$= \frac{412 \times 5000 \times \pi \times 3.5}{180} = 89884 \text{ N m}$$

$$m = \frac{900 E_f}{\pi^2 K^2 N_m^2 C_s}$$

$$K^2 = \frac{900 \times 89884}{\pi^2 \times 180^2 \times 0.03 \times 1000} = 8.4326 \text{ m}^2$$

Radius of gyration, $K = 2.9$ m

Example 10.12

The torque exerted on the crankshaft of a two stroke engine is given by

$$T = 15000 + 2500 \sin\theta - 2000 \cos 2\theta$$

where θ is the crank angle measured from the inner dead centre. Assuming the resulting torque to be constant, determine (a) the power of the engine when the speed is 180 rpm, (b) the moment of inertia of the flywheel if the speed variation is not to exceed ±0.5% of the mean speed, and (c) the angular acceleration of the flywheel when the crank has turned through 30° from inner dead centre.

Solution

(a) For a two-stroke engine, cycle is repeated after every crank rotation of 180°.

$$T_m = \left(\frac{1}{\pi}\right)\int_0^{\pi} T\mathrm{d}\theta = \left(\frac{1}{\pi}\right)\int_0^{\pi}(15000 + 2500\sin 2\theta - 2000\cos 2\theta)\mathrm{d}\theta$$

$$= \frac{1}{\pi}\left|15000\,\theta - 1250\cos 2\theta - 1000\sin 2\theta\right|_0^{\pi} = 15000 \text{ N m}$$

$$\text{Power} = \frac{2\pi \times 180 \times 15000}{60 \times 10^3} = 283.74 \text{ kW}$$

(b) Change in torque, $\Delta T = T - T_m = (2500 \sin 2\theta - 2000 \cos 2\theta)$

For $\Delta T = 0$, $\tan 2\theta = 0.8$, $2\theta = 38.66°$ and $218.66°$

The T-θ diagram is shown in Fig.10.15.

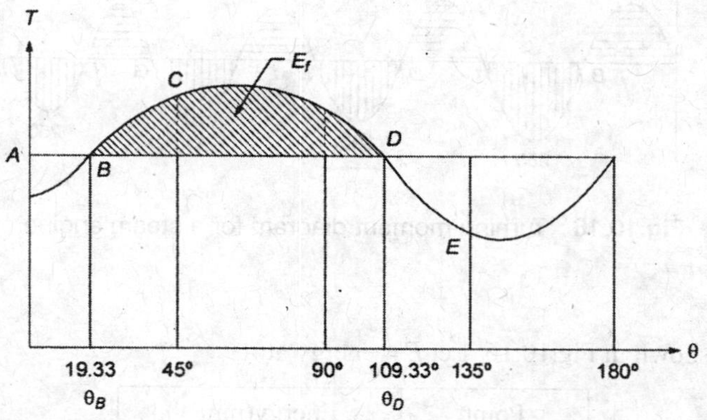

Fig.10.15 T-θ diagram

$$\text{Fluctuation of energy, } E_f = \int_{38.66°}^{218.66°} (2500 \sin 2\theta - 2000 \cos 2\theta)\mathrm{d}\theta$$

$$= \left|-1250\cos 2\theta - 1000\sin 2\theta\right|_{38.66°}^{218.66°}$$

$$= -1250\cos 218.66° - 1000\sin 218.66° + 1250\cos 38.66° + 1000\sin 38.66°$$

$$= 3201.6 \text{ N m}$$

$$I = \frac{900 E_f}{\pi^2 N_m^2 C_s}$$

(c) $$= \frac{900 \times 3201.6}{\pi^2 \times 180^2 \times 0.005} = 1802 \text{ kg m}^2$$

$$T_{excess} = 2500 \sin 2\theta - 2000 \cos 2\theta$$

At $\theta = 30°$, $T_{excess} = 1165$ N m

$$T_{excess} = I \cdot a$$

Angular acceleration, $$a = \frac{1165}{1802} = 0.6465 \text{ rad/s}^2$$

Example 10.13

A steam engine runs at 150 rpm. Its turning moment diagram gave the following area measurements in mm² taken in order above and below the mean torque line:

500, –250, 270, –390, 190, –340, 270, –250

The scale for the turning moment is 1 mm = 500 N m and for crank angle is 1 mm = 5°. If the fluctuation of speed is not to exceed ±1.5% of the mean speed, determine the cross-section of the rim of the flywheel assuming rectangular with axial dimension equal to 1.5 times the radial dimension. The hoop stress is limited to 3.5 MPa and the flywheel is 7470 kg/m³.

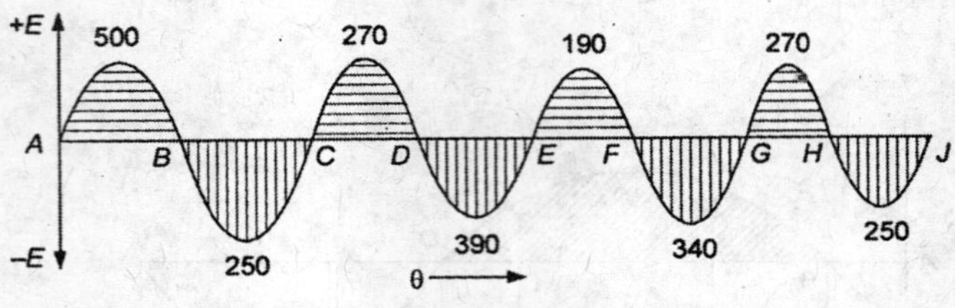

Fig.10.16 Turning moment diagram for a steam engine

■ Solution

The T–θ diagram is shown in Fig.10.16. Let E = energy at A.

Point	Energy(mm³)
A	E
B	$E + 500$
C	$E + 250$
D	$E + 520$
E	$E + 130$
F	$E + 320$
G	$E - 20$
H	$E + 250$
J	E

$$E_{max} = E + 520, E_{min} = E - 20$$

$$E_f = [(E+520)-(E-20)] \times \text{Horizontal scale} \times \text{Vertical scale}$$

$$= \frac{540 \times 500 \times \pi \times 5}{180} = 23562 \text{ N m}$$

$$\sigma_\theta = \rho v^2,\ 3.5 \times 10^6 = 7470\ v^2,\ v = 18.3 \text{ m/s}$$

$$\omega = \frac{2\pi \times 150}{60} = 15.7 \text{ rad/s}$$

$$v = \frac{\omega D}{2}, D = \frac{18.3 \times 2}{15.7} = 3.33 \text{ m}$$

$$m = \rho \pi D b t = 7470 \times \pi \times 3.33 \times 1.5 \times t^2 = 82019\ t^2$$

Considering flywheel rim as a ring, $K^2 = \frac{D^2}{8} = \frac{3.33^2}{8} = 0.6786 \text{ m}^2$

$$m = \frac{900 E_f}{\pi^2 K^2 N_m^2 C_s}$$

$$82019\ t^2 = \frac{900 \times 23562}{\pi^2 \times 0.6786 \times 150^2 \times 0.015}$$

$$t^2 = 0.11438$$

$$t = 0.338 \text{ m or } 338 \text{ mm}, b = 1.5\,t = 507 \text{ mm}$$

Example 10.14

A cast iron flywheel used for a four stroke I.C. engine is developing 150 kW at 240 rpm. The hoop stress developed in the flywheel is 5 MPa. The total fluctuation of speed is to be limited to 3% of the mean speed. If the work done during the power stroke is 1/3 times more than the average work done during the whole cycle, find (a) mean diameter of the flywheel, (b) mass of the flywheel, and (c) cross-sectional dimensions of the rim when the width is twice the thickness. The density of cast iron may be taken as 7300 kg/m³.

Solution

Power, $P = \frac{2\pi \times 240 \times T_m}{60} = 150$

$$T_m = 5968.31 \text{ N m}$$

During power stroke, $T_{max} = \frac{4T_m}{3} = 7957.75 \text{ N m}$

$$E_f = 7957.75 - 5968.31 = 1989.4 \text{ N m}$$

$$\sigma_\theta = \rho v^2$$

$$5 \times 10^6 = 7300\ v^2$$

$$v = 26.17 \text{ m/s}$$

$$\omega = \frac{2\pi \times 240}{60} = 25.13 \text{ rad/s}$$

$$v = \frac{\omega D}{2}$$

$$D = \frac{26.17 \times 2}{25.13} = 3.08 \text{ m}$$

$$m = \rho \pi D b t = 7300 \times \pi \times 3.08 \times 2 \times t^2 = 95404\, t^2$$

Considering flywheel rim as a ring, $K^2 = \frac{D^2}{8} = \frac{3.08^2}{8} = 0.542 \text{ m}^2$

$$m = \frac{900 E_f}{\pi^2 K^2 N_m^2 C_s}$$

$$95404\, t^2 = \frac{900 \times 1989.4}{\pi^2 \times 0.542 \times 240^2 \times 0.03}$$

$$t^2 = 0.3.03 \times 10^{-3}$$

$$t = 0.045 \text{ m or } 45 \text{ mm}$$

$$b = 2t = 90 \text{ mm}$$

Example 10.15

A machine has to carry out punching operations at the rate of 10 holes per minute. It does 6 kN m of work per mm² of the sheared area in cutting 25 mm diameter holes in 20 mm thick plates. A flywheel is fitted to the machine shaft, which is driven by a constant torque. The fluctuation of speed is between 180 and 200 rpm. The actual punching operation takes 2 *s*. The frictional losses are equivalent to 1/6th of the work done during punching. Find (a) power required to drive the punching machine, and (b) mass of the flywheel, if the radius of gyration of the wheel is 0.5 m.

■ Solution

Given: $d = 25$ mm, $t = 20$ mm, $C_e = 25\%$, $N_{min} = 180$ rpm, $N_{max} = 200$ rpm, $K = 0.5$ m

Time taken for punching a hole $= \frac{60}{10} = 6\,s$

Energy required $= \pi \times 25 \times 20 \times 6 = 9424.8$ N m

$$N_m = \frac{180 + 200}{2} = 190 \text{ rpm}$$

$$N_{max} - N_{min} = 200 - 180 = 20 \text{ rpm}$$

$$C_s = \frac{20}{190} = 0.105$$

Energy required during punching operation $= \frac{9424.8 \times 2}{6} = 1341.6$ N m

Energy stored by flywheel $= 9424.8 - 1343.6 = 6283.2$ N m

Energy lost in friction $= \frac{1341.6}{6} = 22.36$ N m

Total energy required $= 9424.8 + 223.6 = 9648.4$ N m

(a) Power required $= \dfrac{9648.4}{6\times10^3} = 1.6$ kW

(b) $m = \dfrac{900E_f}{\pi^2K^2N_m^2C_s} = \dfrac{900\times6283.2}{\pi^2\times0.5^2\times190^2\times0.105} = 604.6$ kg

Example 10.16

A 17 mm diameter hole is to be punched in a steel plate 19 mm thick at the rate of 20 holes per minute. The actual punching takes place in 1/5th the interval between punches. The driving motor runs at 1200 rpm and is geared to a countershaft, which runs at 160 rpm and upon which the flywheel is mounted. The countershaft in turn is geared to the crankshaft of the press. The resistance to shear for the plate may be taken as 310 MPa. Find (a) the power required for the motor if no flywheel is used, (b) the power required for the motor assuming a flywheel is used, and (c) the mass of the flywheel rim required assuming that 90% of the effective mass at the rim is due to the rim alone. The average speed at the rim diameter is 20 m/s, and the coefficient of speed fluctuation is 0.10.

Solution

Time taken for punching a hole $= \dfrac{60}{20} = 3\,s$

Actual punching time $= \dfrac{3}{5} = 0.6\,s$

Energy required for punching $= \pi\times17\times19\times310 = 314567$ N m

$$v = 20 \text{ m/s},\ C_s = 0.1,\ \text{N}_m = 160 \text{ rpm}$$

(a) Motor power without flywheel $= \dfrac{314567}{3\times10^3} = 104.85$ kW

(b) Energy required during punching operation $= \dfrac{314567\times0.6}{3} = 62913.4$ N m

Energy stored by flywheel $= 314567 - 62913.4 = 251653.6$ N m

Motor power with flywheel $= \dfrac{251653.6}{3\times10^3} = 83.88$ kW

(c) $$\omega = \frac{2\pi\times160}{60} = 16.75 \text{ rad/s}$$

$$D = \frac{2v}{\omega} = \frac{40}{16.75} = 3.388 \text{ m}$$

$$K^2 = \frac{D^2}{8} = 0.713 \text{ m}^2$$

Mass of flywheel, $m = \dfrac{900E_f}{\pi^2K^2N_m^2C_s} = \dfrac{900\times251653.6}{\pi^2\times0.713\times160^2\times0.1} = 12572$ kg

Example 10.17

The crankshaft of a punching machine runs at a mean speed of 300 rpm. During punching of 10 mm diameter holes in mild steel sheets, the torque required by the machine increases uniformly from 1 kN m to 4 kN.m, while the shaft turns through 40°, remains constant for the next 100°, decreases uniformly to 1 kN m for the next 180°. This cycle is repeated during each revolution. The power is supplied by a constant torque motor and the fluctuation of the speed is to be limited to ±3% of the mean speed. Find the power of the motor and the moment of inertia of the flywheel fitted to the machine.

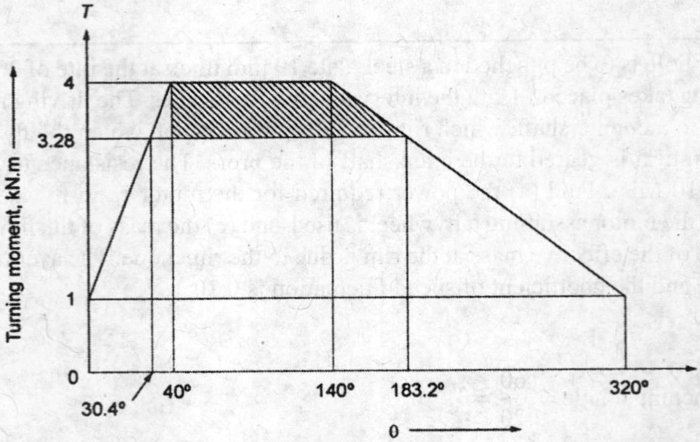

Fig.10.17 T-θ diagram for a punching machine

Solution

Given: $N_m = 300$ rpm, $C_s = \pm 3\%$, $d = 10$ mm

The T–θ diagram is shown in Fig.10.17.

$$\text{Mean torque, } T_m = \frac{0.5\times3\times40+4\times100+0.5\times3\times180+1\times320}{320}$$

$$= 3.28 \text{ kN m}$$

$$\text{Surplus torque} = \frac{[0.5\times0.72\times9.6+0.72\times100+0.5\times0.72\times43.2]\times\pi}{180}$$

$$= 1.5884 \text{ kN m}$$

$$E_f = 1.5884 \text{ kN m}, C_s = \pm3\%, N_m = 300 \text{ rpm}$$

$$I = mK^2 = \frac{900E_f}{\pi^2 N_m^2 C_s} = \frac{900\times1.5884\times10^3}{\pi^2\times300^2\times0.03} = 53.65 \text{ kg m}^2$$

$$\text{Motor power} = \frac{2\pi\times300\times3.28\times10^3}{60\times10^3} = 103 \text{ kW}$$

Example 10.18

A single-cylinder double acting pump is driven through gearing at 50 rpm. The resisting torque of pump shaft may be assumed to follow a sine curve in half revolution with a maximum value of 6 kN m at 90° and 270°. Find the weight of the flywheel required to be mounted on a pump shaft to keep the speed within 1.5% of the mean speed, if the radius of gyration of the flywheel is 1.5 m. The effect of motor armature and gear wheel is equivalent to a flywheel of 4.5 kN with a radius of gyration of 1 m on the pump shaft.

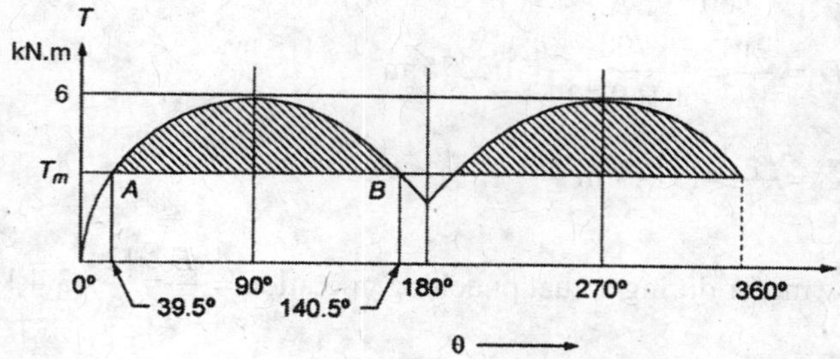

Fig.10.18 T-θ diagram for a single cylinder double acting pump

■ Solution

Given: $N_m = 50$ rpm, $C_s = 1.5\%$, $K = 1.5$ m

The T–θ diagram is shown in Fig.10.18.

Moment of inertia of motor armature and gear wheel

$$= \frac{4.5\times10^3\times1^2}{9.81} = 458.7 \text{ kg m}^2$$

$$T_m = \left(\frac{1}{\pi}\right)\int_0^{\pi} 6\sin\theta\, d\theta = \frac{6}{\pi}\left|-\cos\theta\right|_0^{\pi} = \frac{6}{\pi}\times 2 = 3.82 \text{ kN m}$$

At points A and B, $6 \sin \theta = 3.82$

$$\sin\theta = 0.06367,\ \theta = 39.5° \text{ and } 140.5°$$

$$E_f = \int_{39.5°}^{140.5°} (6\sin\theta - 3.82)d\theta$$

$$= 6\left|-\cos\theta - 3.82\theta\right|_{39.5°}^{140.5°}$$

$$= 6(-\cos 140.5° + \cos 39.5°) - 3.82(140.5 - 39.5)\times\frac{\pi}{180}$$

$$= 9.2595 - 6.7338$$

$$= 2.5257 \text{ kN.m}$$

$$\text{Mass of flywheel, } m = \frac{900E_f}{\pi^2K^2N_m^2C_s} = \frac{900\times2.5257\times10^3}{\pi^2\times1.5^2\times50^2\times0.015} = 2729.7 \text{ kg}$$

Work done per stroke $= 3.82 \times \pi = 12$ kN m

$$\text{Power} = \frac{2\times50\times12\times10^3}{60\times10^3} = 20 \text{ kW}$$

Example 10.19

A cast iron flywheel is fitted to a punch press to run at 90 rpm and must supply 12 kN m of energy during 1/5th revolution and allow 15% change of speed. The rim speed is limited to 350 m/min. Find the mean diameter and weight of the flywheel and the motor power. Assume overall efficiency as 80%.

■ **Solution**

Given: $N_m = 90$ rpm, $E = 12$ kN m, $C_s = 0.15$, $v = 350$ m/min, $D = ?$, $W = ?$

$$\omega = \frac{2\pi \times 90}{60} = 9.427 \text{ rds/s}$$

$$D = \frac{2v}{\omega} = \frac{700}{60 \times 9.427} = 1.2375 \text{ m}$$

$$K^2 = \frac{D^2}{8} = 19145 \text{ m}^2$$

Energy supplied by motor during actual punching operation $= \frac{E}{5} = \frac{12}{5} = 3.4$ kN m

Balance energy supplied by flywheel $= E\left(1 - \frac{1}{5}\right) = \frac{12 \times 4}{5} = 9.6$ kN m

Mass of flywheel, $m = \frac{900 E_f}{\pi^2 K^2 N_m^2 C_s}$

$$= \frac{900 \times 9600}{\pi^2 \times 0.19145 \times 90^2 \times 0.15} = 3763 \text{ kg}$$

Time taken to punch one hole $= \frac{60}{90 \times 5} = \frac{2}{15} s$

Motor power required $= \frac{12 \times 10^3 \times 15}{2 \times 10^3 \times 0.8} = 113.5$ kW

Example 10.20

A machine punching 38 mm diameter holes in a 32 mm thick plate, does 6 N m of work per square mm of sheared area? The punch has a stroke of 102 mm and punches 6 holes per minute. The maximum speed of the flywheel at its radius of gyration is 27.5 m/s. Find the weight of the flywheel so that its speed at the same radius does not fall below 24.5 m/s. Also determine the power of the motor driving the machine.

■ **Solution**

Given: $d = 38$ mm, $t = 32$ mm, $L = 102$ mm

Energy supplied in punching a hole $= \pi \times 38 \times 32 \times 6 = 22{,}921$ N m

Time taken for punching one hole $= \frac{60}{6} = 10 s$

Power required $= \frac{22921}{10} = 2293.1$ N m/s or 3.2921 kW

Speed of punching $= \frac{2 \times 102}{10} = 20.4$ mm/s

$$C_s = \frac{v_{max} - v_{min}}{v_m} = \frac{27.5 - 24.5}{26} = 0.1154$$

Time to punch a hole in 32 mm thick plate $= \frac{32}{20.4} = 1.568 s$

Energy supplied by motor in 1.568 $s = \dfrac{22921 \times 1.568}{10} = 3595$ N m

Energy supplied by flywheel, $E_f = 22921 - 3595 = 19326$ N m

$$v_m = K\omega_m = K\left(\frac{2\pi K_m}{60}\right),\ KN_m = \frac{26 \times 60}{2\pi} = 248.3$$

Mass of flywheel, $m = \dfrac{900E_f}{\pi^2 K^2 N_m^2 C_s}$

$$= \frac{900 \times 19326}{\pi^2 \times (248.3)^2 \times 0.1154} = 247.7 \text{ kg}$$

Example 10.21

An engine runs at a constant load at a speed of 480 rpm. The crank effort diagram to a scale of 1 cm = 2 kN m torque and 1 cm = 36° crank angle. The areas of the diagram above and below the mean torque line are measured in square cm and are in the following order:

$$+1.1, -1.32, +1.53, -1.66, +1.97, -1.62$$

Design the flywheel if the total fluctuation of speed is not to exceed 10 rpm and the centrifugal stress in the rim is not to exceed 5 N/mm³. You may assume that the rim breadth is approximately 3.5 times the rim thickness and 90% of the moment of inertia is due to the rim. The density of the material of the flywheel is 7250 kg/m³.

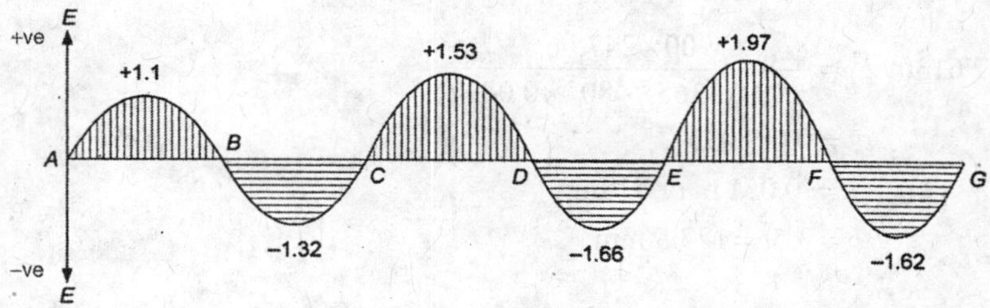

Fig.10.19 Crank effort diagram for an engine

■ Solution

Let E = energy at A (Fig.10.19)

Point	Energy, mm³
A	E
B	$E + 1.1$
C	$E - 0.22$
D	$E + 1.31$
E	$E - 0.35$
F	$E + 1.62$
G	E

$$E_{max} = E + 1.62,\ E_{min} = E - 0.35$$

$$E_f = [(E + 1.62) - (E - 0.35)] \times \text{Horizontal scale} \times \text{Vertical scale}$$

$$= \frac{1.97 \times 2000 \times \pi \times 36}{180} = 2475.6 \text{ N m}$$

$$C_s = \frac{490 - 470}{480} = 0.0417$$

$$\sigma_\theta = \rho v^2$$

$$5 \times 10^6 = 7250\ v^2$$

$$v = 26.26 \text{ m/s}$$

$$\omega_m \frac{2\pi \times 480}{60} = 50.26 \text{ rad/s}$$

$$v = \omega_m \frac{D}{2}$$

$$D = \frac{26.26 \times 2}{50.26} = 1.045 \text{ m}$$

$$m = \rho \pi D b t = 7470 \times \pi \times 1.045 \times 3.5 \times t^2 = 61309\ t^2$$

Considering flywheel rim as a ring, $K^2 = \frac{D^2}{8} = \frac{1.045^2}{8} = 0.1365 \text{ m}^2$

Mass of flywheel, $m = \frac{900 E_f}{\pi^2 K^2 N_m^2 C_s}$

$$61309\ t^2 = \frac{900 \times 2475.6}{\pi^2 \times 0.1365 \times 480^2 \times 0.0417}$$

$$t^2 = 0.006035$$

$$t = 0.051 \text{ m or } 51 \text{ mm}$$

$$b = 3.5t = 127.5 \text{ mm}$$

Example 10.22

The equation of the turning moment curve of a three crank engine is (5 + 1.5 sin 3θ) kN m, where θ is the crank angle. The moment of inertia of the flywheel is 1000 kg m² and the mean engine speed is 300 rpm. Calculate (a) power of the engine, (b) the maximum fluctuation of the speed of the flywheel in percentage, (i) when the resisting torque is constant, and (ii) when the resisting torque is (5 + 0.6 sin θ) kN m.

■ Solution

$$T = (5 + 1.5 \sin 3\theta) \text{ kN m}$$

Work done per crank per cycle = $\int_{0°}^{120°} (5 + 1.5 \sin 3\theta)\, d\theta = 10.472$ kN m

(a) Power of engine $= \dfrac{3\times300\times10.472\times10^3}{60\times10^3} = 157$ kW

$$T_m = \frac{10.472\times180}{120\times\pi} = 5 \text{ kN m}$$

For $T_m = T$, $5 + 1.5 \sin 3\theta = 5$

$\sin 3\theta = 0$, $3\theta = 0°, 180°, 360°$ or $\theta = 0°, 60°, 120°$

Excess torque $= 5 + 1.5 \sin 3\theta - 5 = 1.5 \sin 3\theta$

$$E_f = \int_{0°}^{60°} 1.5 \sin 3\theta \, d\theta = 1 \text{ kN m}$$

$$m = \frac{900E_f}{\pi^2 K^2 N_m^2 C_s}$$

$$C_s = \frac{900\times1000}{\pi^2\times1000\times300^2} = 0.001 \text{ or } 0.1\%$$

(ii) Change in torque $= (5 + 1.5 \sin 3\theta) - (5 + 0.6 \sin \theta)$

$= 1.5 \sin 3\theta - 0.6 \sin \theta$

For change in torque to be $= 0$, $\theta = 53.72°, 126.28°$

The T–θ diagram is shown in Fig.10.20.

$$E_f = \int_{126.28°}^{180°} (1.5 \sin 3\theta - 0.6 \sin \theta) \, d\theta = 0.7282 \text{ kN m}$$

$C_s = 1.01 \times 10^{-3} \times 0.7282 = 0.7355 \times 10^{-3}$ or 0.07355%.

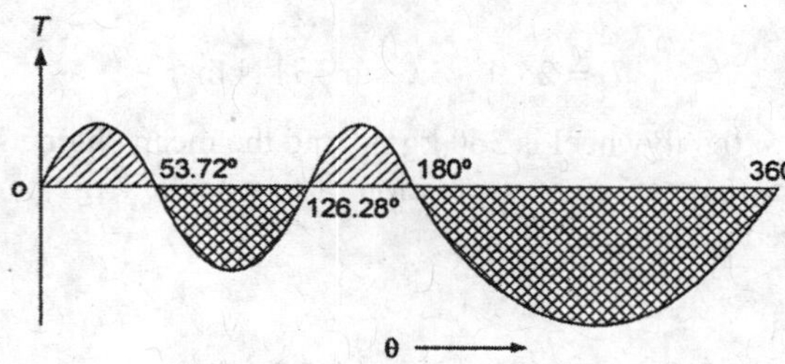

Fig.10.20 Turning moment diagram for a three-crank engine

Example 10.23

A punch press is fitted with a flywheel capable of furnishing 3 kN m of energy during quarter of a revolution near the bottom dead centre while blanking a hole on sheet metal. The maximum speed of the flywheel during the operation is 200 rpm and the speed decreases by 10% during the cutting stroke. The mean radius of the rim is 900 mm. Calculate the approximate weight of the flywheel rim assuming that it contributes 90% of the energy requirements.

■ **Solution**

$$N_{min} = 200 \times 0.9 = 180 \text{ rpm}$$

$$N_m = \frac{200+180}{2} = 190 \text{ rpm}$$

$$C_s = \frac{20}{190} = \frac{2}{19}$$

$$R_m = 900 \text{ mm}$$

$$K^2 = \frac{R_m^2}{2} = 0.405 \text{ m}^2$$

Work required for one punch = 3 kN m

$$\text{Energy supplied by motor } = \frac{300}{4} = 750 \text{ kN m}$$

Energy to be supplied by flywheel, $E_f = 3000 - 750 = 2250$ kN m

$$m = \frac{900E_f}{\pi^2 K^2 N_m^2 C_s}$$

$$= \frac{900 \times 2250 \times 0.9 \times 19}{\pi^2 \times 0.405 \times 190^2 \times 2} = 120 \text{ kg}$$

Example 10.24

The equation of turning moment for a three-crank engine is:

$$T_c = 25.0 - 7.5 \sin\theta \text{ kN m}$$

where θ is the crank angle measured from inner dead centre. The resisting torque exerted by the driven machine is given by:

$$T_r = 25.0 + 3.6 \sin 3\theta \text{ kN m}$$

The moment of inertia of the flywheel is 360 kg m^2 and the mean engine speed is 450 rpm. Calculate (a) power of the engine, (b) maximum fluctuation of flywheel energy per cycle, and (c) the coefficient of fluctuation of speed [IES, 1996]

■ **Solution**

Change in torque = $T_r - T_c = (25 + 3.6 \sin\theta) - (25 - 7.5 \sin 3\theta)$

$= 3.6 \sin\theta + 7.5 \sin 3\theta$

$$\text{Work done per crank per cycle} = \int_{0°}^{120°} (3.6 \sin\theta + 7.5 \sin 3\theta)\, d\theta = 5.4 \text{ kN m}$$

$$\text{(a) Power of engine} = \frac{3 \times 450 \times 5.4 \times 10^3}{60 \times 10^3} = 121.5 \text{ kW}$$

$$T_m = \frac{5.4 \times 180}{120 \times \pi} = 3.578 \text{ kN m}$$

For $T_m = T$, $3.6 \sin \theta + 7.5 \sin 3\theta = 0$

$\sin \theta = 0$, $\theta = 0°, 180°, 360°$, and $68.866°, 111.134°$

The T–θ diagram is shown in Fig.10.21.

(b) $$E_f = \int_{111.134°}^{180°} (3.6 \sin \theta = 7.5\ 3\theta)\, d\theta = 7\,039\ \text{kN m}$$

$$m = \frac{900 E_f}{\pi^2 K^2 N_m^2 C_s}$$

(c) $$C_s = \frac{900 \times 7039}{\pi^2 \times 360 \times 450^2} = 0.0088 \text{ or } 0.88\%$$

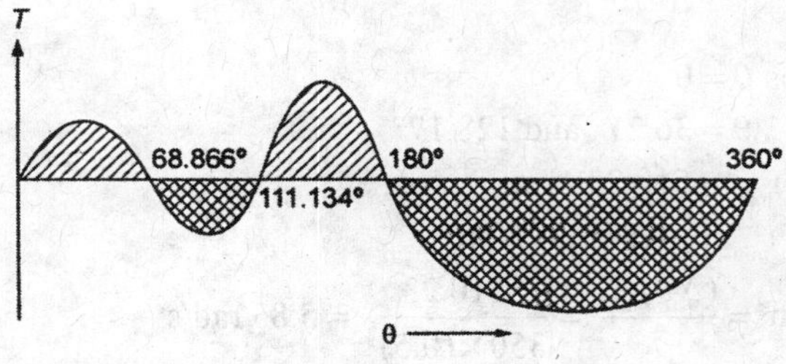

Fig.10.21 Turning moment diagram for a three-crank engine

Example 10.25

An engine coupled to a machine produces a torque given by $T_c = 10 + \sin 2\theta$ kN m where θ is the angle of rotation of shaft. The resisting torque of machine is $T_r = 10 + 0.75 \sin \theta$ kNm. The engine runs at a mean speed of 240 rpm and has a flywheel of mass 350 kg and radius of gyration 0.5 m fixed to it. Determine (a) fluctuation of energy, (b) fluctuation of speed, and (c) maximum and minimum acceleration of flywheel.

Solution

Change in torque, $\Delta T = T_r - T_c = (10 + \sin 2\theta) - (10 + 0.75 \sin \theta)$
$= \sin 2\theta - 0.75 \sin \theta$

For $\Delta T = 0$, $\sin 2\theta - 0.75 \sin \theta = 0$

$\sin \theta = 0$, $\theta = 0°, 180°, 360°$, and $67.98°, 293.02°$

The T–θ diagram is shown in Fig.10.22.

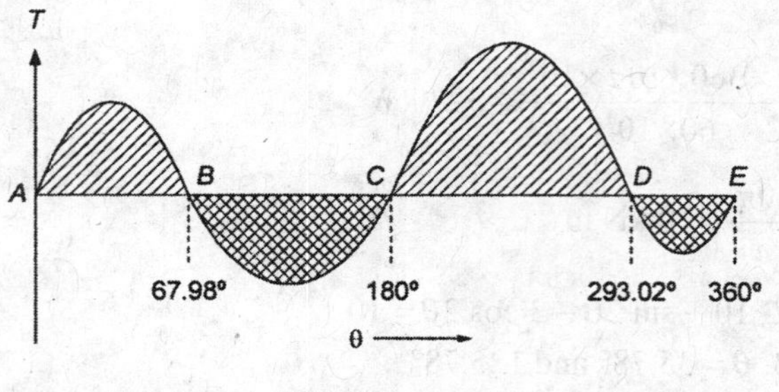

Fig.10.22 T-θ diagram for an engine

(a) Maximum fluctuation of energy,

$$E_f = \int_{180^\circ}^{293.02^\circ} (\sin 2\theta - 0.75 \sin\theta)\, d\theta = 1890.62 \text{ N m}$$

(b)

$$m = \frac{900 E_f}{\pi^2 K^2 N_m^2 C_s}$$

$$C_s = \frac{900 \times 1890.62}{\pi^2 \times 350 \times 0.5^2 \times 240^2}$$

$$= 0.0342 \text{ or } 3.42\%$$

(c) For acceleration to be maximum or minimum, $\frac{d(\Delta T)}{d\theta} = 0$

$$2 \cos 2\theta - 0.75 \cos \theta = 0$$

$$\theta = 36.27^\circ \text{ and } 128.17^\circ$$

$$(\Delta T)_{max} = 510.23 \text{ N m}$$

$$(\Delta T)_{min} = -1561.34 \text{ N m}$$

$$\text{Maximum acceleration} = \frac{(\Delta T)_{max}}{I} = \frac{510.23}{350 \times (0.5)^2} = 5.83 \text{ rad/s}^2$$

$$\text{Minimum acceleration} = \frac{(\Delta T)_{min}}{I} = \frac{1561.34}{350 \times (0.5)^2} = 17.84 \text{ rad/s}^2$$

Example 10.26

The turning moment exerted by two stroke engine at crankshaft is given by:

$$T = 10 + \sin 2\theta - 3 \cos 2\theta \text{ kN m}$$

where θ = inclination of crank to inner dead centre.

The mass of the flywheel is 600 kg and its radius of gyration 0.8 m. The engine speed is 360 rpm. Assuming external resistance as constant, determine (a) power developed, (b) fluctuation of speed, and (c) maximum angular retardation of flywheel.

■ Solution

$$T = (10 + \sin 2\theta - 3 \cos 2\theta) \text{ kN m}$$

(a) Work done per stroke $= \int_{0^\circ}^{\pi} (10 + \sin 2\theta - 3 \cos 2\theta)\, d\theta = 10\pi$ kN m

$$\text{Power of engine} = \frac{2 \times 360 \times 10\pi \times 10^3}{60 \times 10^3} = 377 \text{ kW}$$

$$T_m = \frac{10\pi}{\pi} = 10 \text{ kN m}$$

For $T_m = T$, $10 + \sin 2\theta - 3 \cos 2\theta = 10$

$\tan 2\theta = 3$, $\theta = 35.78^\circ$ and 125.78°

The T–θ diagram is shown in Fig.10.23.

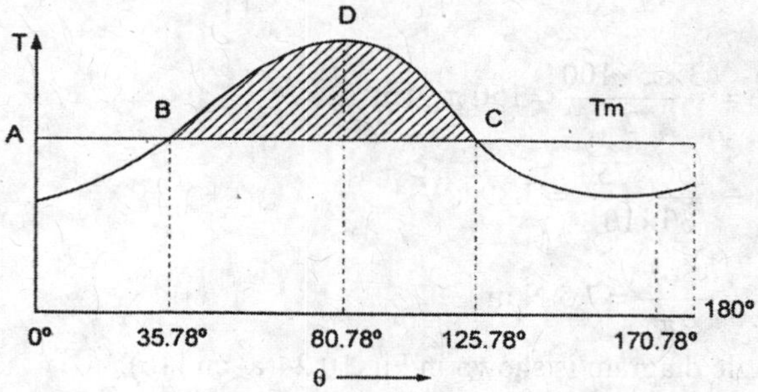

Fig.10.23 Turning moment diagram for a two-stroke engine

Excess torque $= T - T_m = \sin 2\theta - 3\cos 2\theta$

$$E_f = \int_{35.78^\circ}^{125.78^\circ} (\sin 2\theta - 3\cos 2\theta)\, d\theta = 3.1621 \text{ kN m}$$

$$m = \frac{900 E_f}{\pi^2 K^2 N_m^2 C_s}$$

$$C_s = \frac{900 \times 3163.1}{\pi^2 \times 600 \times 0.8^2 \times 360^2} = 0.0058 \text{ or } 0.58\%$$

(b) For torque to be maximum or minimum, $\dfrac{d(\Delta T)}{d\theta} = 0$

$$2\cos 2\theta + 6\sin 2\theta = 0$$

$$\tan 2\theta = -\frac{1}{3}$$

$$2\theta = 161.56^\circ \text{ and } 341.56^\circ$$

$$\text{or } \theta = 80.78^\circ \text{ and } 170.78^\circ$$

$$T_{max} = 10 + \sin 161.56^\circ - 3\cos 161.56^\circ = 13.1623 \text{ kN m}$$

$$T_{min} = 6.8377 \text{ kN m}$$

$$\text{Maximum acceleration} = \frac{(T_m - T_{min})}{I}$$

$$= \frac{(10 - 6.8377) \times 10^3}{600 \times (0.8)^2} = 8.235 \text{ rad/s}^2$$

Example 10.27

A three-cylinder single-acting engine has its cranks set equally at 120° and runs at 750 rpm. The torque-crank angle diagram for each cylinder is a triangle for the power with maximum torque 100 N m at 60° after dead centre of the corresponding crank. The torque on return stroke is zero. Determine (a) the power developed, (b) the coefficient of fluctuation of speed if the mass of the flywheel is 10 kg and the radius of gyration of 100 mm, (c) coefficient of fluctuation of energy, and (d) maximum angular acceleration of flywheel.

Solution

$$\text{Work done per cycle} = \frac{3\times\pi\times100}{2} = 150\,\pi\ \text{N m}$$

$$\text{(a) Power developed} = \frac{150\times750}{60\times10^3} = 5.89\ \text{kW}$$

$$\text{(b) Mean torque, } T_m = \frac{150\pi}{2\pi} = 7.5\ \text{N m}$$

The torque-crank angle diagram is shown in Fig.10.24(a) and (b).

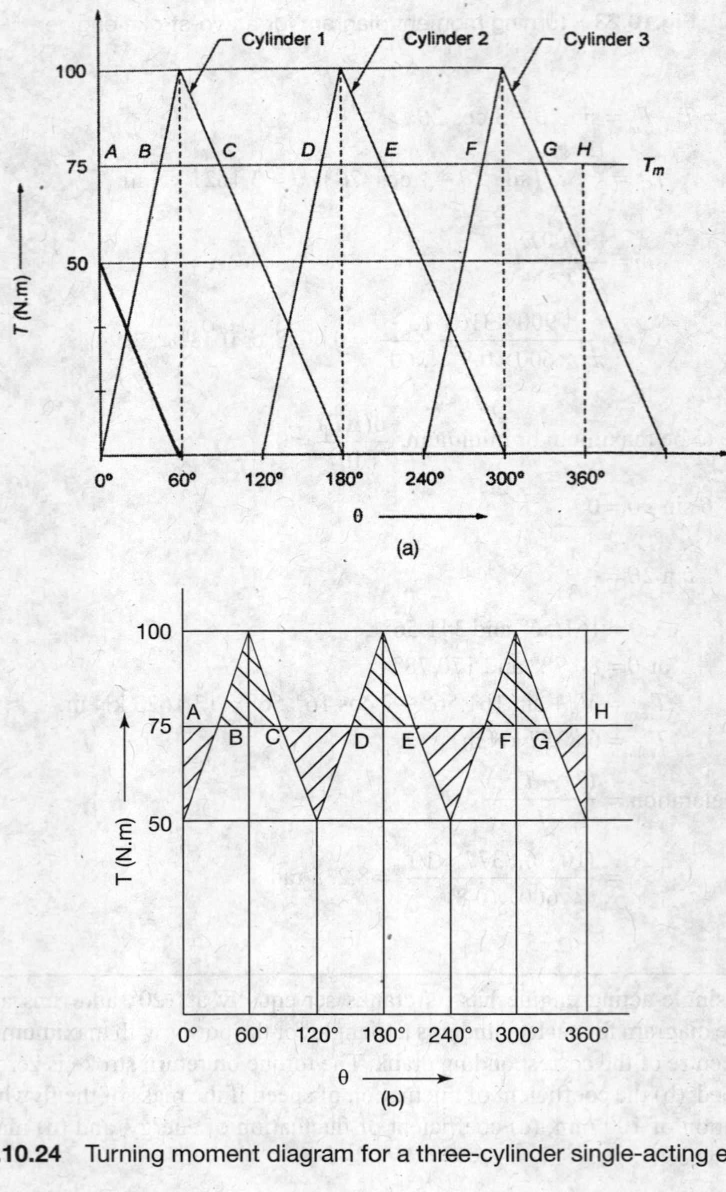

Fig.10.24 Turning moment diagram for a three-cylinder single-acting engine

Let total energy at $A = E$

Location	Energy
B	$E - \dfrac{0.5\times\pi\times 25}{6} = E - 6.545$
C	$E + 6.545$
D	$E - 6.545$
E	$E + 6.545$
F	$E - 6.545$
G	$E + 6.545$
H	E

$$E_f = E_{max} - E_{min} = (E + 6.545) - (E - 6.545) = 13.09 \text{ N m}$$

$$m = \frac{900E_f}{\pi^2 K^2 N_m^2 C_s}$$

$$C_s = \frac{900\times 13.09}{\pi^2\times 10\times 0.1^2\times 750^2} = 0.0212 \text{ or } 3.12\%$$

(c) $$C_e = \frac{E_f}{E} = \frac{13.09}{150\pi} = 0.0278 \text{ or } 3.78\%$$

(d) $$I\alpha = T_{max} - T_m = 100 - 75 = 25 \text{ N m}$$

$$\therefore \alpha = \frac{25}{10\times(0.1)^2} = 250 \text{ rad/s}^2$$

10.8 EQUIVALENT DYNAMICAL SYSTEM

A continuous body may be replaced by a body by two masses assumed to be concentrated at two points and connected rigidly together. Such a system of two masses is termed an equivalent dynamical system. The conditions to be satisfied by an equivalent dynamical system are as follows:

1. The total mass must be equal to that of the rigid body.
2. The centre of gravity must coincide with that of the rigid body.
3. The total moment of inertia about an axis through centre of gravity must be equal to that of the rigid body.

Consider a rigid body as shown in Fig.10.25.

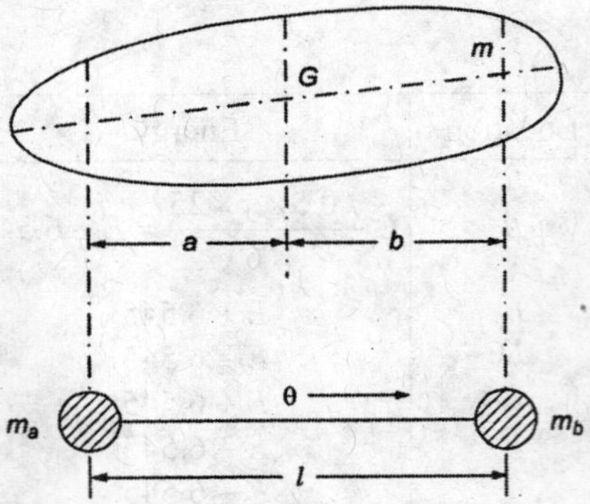

Fig.10.25 Truely dynamical system

Let m = mass of the rigid body

K = radius of gyration about an axis through G

m_a, m_b = two masses for equivalent dynamical system

a, b = distance of m_a, m_b from G, respectively.

Then
$$m_a + m_b = m \tag{10.34}$$
$$m_a \cdot a = m_b \,.\, b \tag{10.35}$$
$$m_a \cdot a^2 + m_b \cdot b^2 = m K^2 \tag{10.36}$$

Solving Eqs. (10.35) and (10.36), we have

$$m_a \,.\, a^2 + m_a \,.\, ab = mK^2$$

or
$$m_a = \frac{mK^2}{a(a+b)} \tag{10.37}$$

From Eqs. (10.34) and (10.35), we get

$$m_a = \frac{mb}{(a+b)} \tag{10.38}$$

Comparing Eqs. (10.37) and (10.38), we get

$$K^2 = ab$$

Let L = length of a simple pendulum which has the same period of oscillations as the body with length equal to $(a + b)$.

Therefore the second mass is situated at the centre of percussion of the body.

A compound pendulum equivalent to the rigid body is shown in Fig.10.26. The centre of oscillation is A and the centre of percussion is at A'.

Therefore
$$L = a + \frac{K^2}{a} = a + b$$

or
$$b = \frac{K^2}{a}$$

or
$$K^2 = ab$$

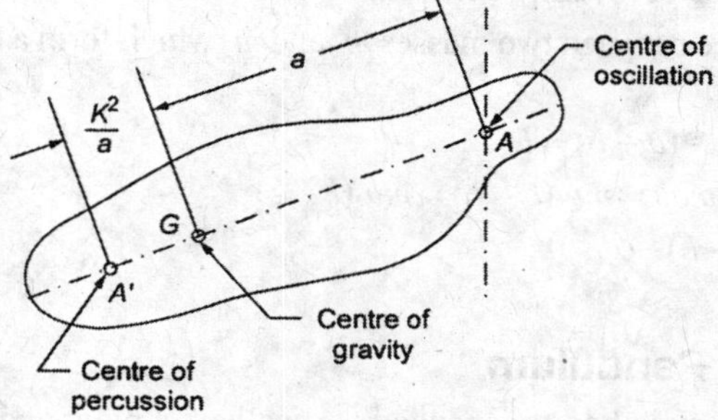

Fig.10.26 Compound pendulum

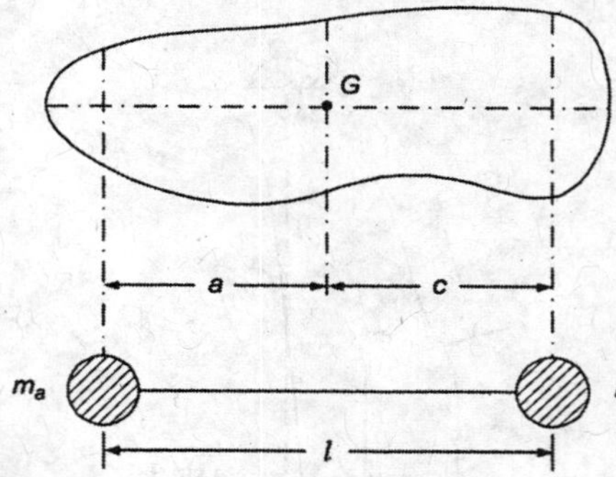

Fig.10.27 Approximate dynamical system

In an approximate dynamical system, as shown in Fig.10.27, the distances 'a' and 'c' are fixed arbitrarily, then

$$m_a = \left[\frac{c}{a+c}\right] m$$

$$m_c = \left[\frac{a}{a+c}\right] m$$

Mass moment of inertia of m_a, and m_c about G is

$$I_1 = \left[\frac{m}{a+c}\right] a^2 c + c^2 a = m \cdot ac = m_a a^2 + m_c c^2$$

Let $\quad K_1$ = radius of gyration of two-mass system.

Then $\quad I_1 = mK_1^2$

Therefore $K_1^2 = ac$ (10.39)

Difference in mass moment of inertia $= I_1 - I$

$$= m\,(K_1^2 - K^2)$$

Let $\quad \alpha$ = angular acceleration of the body

Difference in torque or correction couple,

$$T_0 = m\,(K_1^2 - K^2) \cdot \alpha$$

Let l = Distance between two masses m_a and m_c fixed arbitrarily

L = distance between two masses m_a and m_b which form a true dynamically equivalent system.

Then $c - b = (a + c) - (a + b) = l - L$

Now $I_1 - I = mac - mab = ma\,(c - b) = ma\,(l - L)$

and $T_o = ma\,(l - L) \cdot \alpha$ (10.40)

10.8.1 Compound Pendulum

A compound pendulum is a rigid body suspended vertically at a point and oscillating with very small amplitude under the action of gravitational force. Consider a compound pendulum shown in Fig.10.28 suspended from a point *A*. *G* is its centre of gravity at which its weight acts.

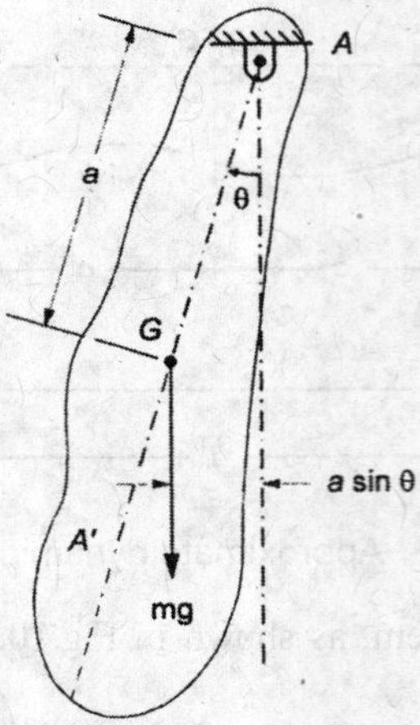

Fig.10.28 Compound pendulum

Let m = mass of the pendulum

K = radius of gyration about an axis passing through centre of gravity perpendicular to the plane of rotation.

a = Distance of the point of suspension from centre of gravity

If the pendulum is displaced through a small angle θ from its mean position, then restoring couple T acting on it is,

$T = mga \sin\theta \approx mga\,\theta$ (for θ to be small, $\sin\theta \approx \theta$)

Mass moment of inertia of the pendulum, $I = m(K^2 + a^2)$

$$\text{Angular acceleration, } \alpha = \frac{T}{I} = \frac{mga\cdot\theta}{m(K^2 + a^2)}$$

$$= \left(\frac{ga}{K^2 + a^2}\right)\theta$$

$$= \text{const.} \times \theta$$

Thus, angular speed, $\omega_n = \left(\frac{ga}{K^2 + a^2}\right)^{1/2}$ rad/s (10.41)

Equivalent length of a simple pendulum, $l_e = \frac{K^2}{a} + a$ (10.42)

The point A' on the other side of centre of gravity at a distance of $\left(\frac{K^2}{a}\right)$ is called the centre of percussion.

Example 10.28

The length of the connecting rod of an engine is 600 mm and its mass is 20 kg. The centre of gravity is 150 mm from the crankpin centre and the crank radius is 120 mm. Determine the dynamically equivalent system keeping one mass at the small end. The frequency of oscillations of the rod when suspended from the centre of the small end is 40 vibrations per minute.

Solution

Given: $l = 600$ mm, $m = 20$ kg, $a = 150$ mm, $r = 120$ mm, $f_n = 40$ cycles/min

$$f_n = \frac{1}{2\pi}\left(\frac{gl_1}{K^2 + l_1^2}\right)^{0.5}$$

Here $\quad l_1 = a = 150$ mm

$$\frac{40}{60} = \frac{1}{2\pi}\left(\frac{9.81 \times 0.15}{K^2 + 0.15^2}\right)^{0.5}$$

$$K^2 = 0.06136$$

$$K = 0.2477 \text{ m}$$

$$K^2 = ab$$

$$b = \frac{(0.2477)^2}{0.15} = 0.409 \text{ m}$$

$$m_a = \frac{bm}{a+b} = \frac{0.409 \times 20}{0.15 + 0.409} = 14.63 \text{ kg}$$

$$m_b = 20 - 14.63 = 5.37 \text{ kg}$$

Example 10.29

A connecting rod 240 mm long has a mass of 2 kg and a moment of inertia of 0.02 kg m^2 about the centre of gravity. The centre of gravity is located at a distance of 150 mm from the small end centre. Determine the dynamically equivalent two mass system when one mass is located at the small end centre. If the connecting rod is replaced by two mass system located at the two centres, find the correction couple that must be applied for complete dynamical equivalence of the system when the angular acceleration of the connecting rod is 20,000 rad/s^2 ccw.

■ Solution

Given: $l = 240$ mm, $m = 2$ kg, $I = 0.02$ kg m², $a = 150$ mm

$$I = mK^2, K^2 = \frac{I}{m} = \frac{0.02}{2} = 0.01 \text{ m}^2$$

$$K^2 = ab, b = \frac{(0.01)^2}{0.15} = 66.67 \text{ mm}$$

$$m_a = \frac{bm}{a+b} = \frac{66.67 \times 2}{150 + 66.67} = 0.615 \text{ kg}$$

$$m_b = 2 - 0.615 = 1.385 \text{ kg}$$

Correction couple, $T_o = ma\,(l - L) \cdot \alpha$

$$l = a + c = 240 \text{ mm}, L = a + b = 216.67 \text{ mm}$$

$$T_o = 2 \times 0.15\,(240 - 216.67) \times 10^{-3} \times 20{,}000$$

$$= 139.98 \text{ Nm}$$

Example 10.30

The connecting rod of an oil engine weighs 600 N and the distance between the bearing centers is 1 m. The diameter of the big end bearing is 120 mm and of the small end bearing is 75 mm. When suspended vertically with a knife edge through the small end, it makes 100 oscillations in 190 s, and with knife edge through the big end it makes 100 oscillations in 165s. Find the moment of inertia of the rod in kg m² and distance of the centre of gravity from the small end centre.

■ Solution

Given: $m = \dfrac{600}{9.81} = 61.16$ kg, $l = 1$ m, $D = 120$ mm, d = 75 mm

Refer to Fig.10.29.

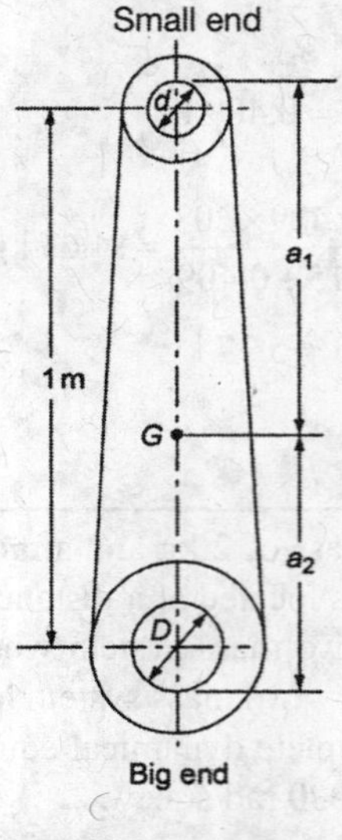

Fig.10.29 Connecting rod

$$f_{n1} = \frac{100}{190} \text{ cps}$$

$$f_{n2} = \frac{100}{165} \text{ cps}$$

Let a_1 = distance of centre of gravity from the top of small end bearing

a_2 = distance of centre of gravity from the top of big end bearing

For a simple pendulum, natural frequency of rod when suspended from small end bearing is given by,

$$f_{n1} = \frac{1}{2\pi}\left(\frac{g}{l_{e1}}\right)$$

$$\frac{100}{190} = \frac{1}{2\pi}\sqrt{\frac{9.81}{l_{e1}}}$$

$$l_{e1} = 0.897 \text{ m}$$

$$= \frac{K^2 + a_1^2}{a_1}$$

$$K^2 = a_1\,(0.897 - a_1) \tag{10.43}$$

When suspended from the big end,

$$f_{n2} = \frac{1}{2\pi}\left(\frac{g}{l_{e2}}\right)$$

$$\frac{100}{165} = \frac{1}{2\pi}\sqrt{\frac{9.81}{l_{e2}}}$$

$$l_{e2} = 0.676 \text{ m}$$

$$= \frac{K^2 + a_2^2}{a_2}$$

or $$K^2 = a_2\,(0.676 - a_2) \tag{10.44}$$

From Eqs. (10.43) and (10.44), we have

$$a_1\,(0.897 - a_1) = a_2\,(0.676 - a_2) \tag{10.45}$$

Also $$a_1 + a_2 = l + \frac{1}{2}\,(D + \text{d})$$

$$= 1.0 + \frac{1}{2}\,(0.120 + 0.075)$$

$$= 1.0975 \text{ m}$$

or $$a_2 = 1.0975 - a_1 \tag{10.46}$$

Substituting in Eq. (10.45), we have

$$0.897\,a_1 - a_1^2 = 0.676\,(1.0975 - a_1) - (1.0975 - a_1)^2$$

$$= 0.742 - 0.676\,a_1 - 1.204 - a_1^2 + 2.195\,a_1$$

$$0622\, a_1 = 0.462$$

$$a_1 = 0.743 \text{ m}$$

$$K^2 = 0.743\,(0.897 - 0.743) = 0.181 \text{ m}^2$$

$$I = mK^2 = 61.16 \times 0.181 = 11.1 \text{ kg m}^2$$

Distance of centre of gravity from small end centre

$$= a_1 - \frac{d}{2}$$

$$= 743 - 37.5 = 705.5 \text{ mm}$$

Example 10.31

The length of connecting rod of a gas engine running at 340 rpm is 600 mm and the crank is 120 mm long. When the piston has moved 1/4th stroke during outstroke, determine (a) the angular position of crank, (b) the angular speed of connecting rod and (c) the acceleration of the piston.

■ Solution

(a)

$$n = \frac{\ell}{r} = \frac{600}{120} = 5$$

$$\omega = 2\pi \times \frac{340}{60} = 35.6 \text{ rad/s}$$

$$r = \frac{L}{2}, \text{ where } L \text{ is the stroke length}$$

$$x = r\left[(1-\cos\theta) + \left\{n - \left(n^2 - \sin^2\theta\right)^{0.5}\right\}\right]$$

$$\frac{x}{L} = \frac{x}{2r} = 0.5\left[(1-\cos\theta) + \left\{n - \left(n^2 - \sin^2\theta\right)^{0.5}\right\}\right]$$

$$\frac{1}{4} = 0.5\left[(1-\cos\theta) + \left\{n - \left(n^2 - \sin^2\theta\right)^{0.5}\right\}\right]$$

$$0.5 = 1 - \cos\theta + 5 - \left(25 - \sin^2\theta\right)^{0.5}$$

$$5.5 = \cos\theta + 5 - \left(25 - \sin^2\theta\right)^{0.5}$$

or $$(5.5 - \cos\theta)^2 = 25 - \sin^2\theta$$

$$30.25 + \cos^2\theta - 11\cos\theta = 25 - \sin^2\theta$$

$$6.25 - 11\cos\theta = 0$$

$$\theta = 55.37°$$

Angular speed of the connecting rod,

$$= \frac{\omega\cos\theta}{(n^2 - \sin^2\theta)^{0.5}}$$

$$= \frac{35.6 \times \cos 55.37°}{(25 - \sin^2 55.37°)^{0.5}} = 4.1 \text{ rad/s}$$

Acceleration of the piston,

$$\approx \omega^2 r\left[\cos\theta + \frac{\cos 2\theta}{n}\right]$$

$$= (35.6)^2 \times 0.12\left[\cos 55.37° + \frac{\cos 110.74°}{5}\right] = 75.65 \text{ m/s}^2$$

Example 10.32

In a vertical double-acting steam engine running at 360 rpm, the cylinder diameter is 0.3 m, piston rod diameter is 40 mm and length of connecting rod is 0.7 m. When the crank has moved 120° from top dead centre, the pressure of steam at the cover end is 0.35 N/mm² and that at the crank end is 0.03 N/mm². If the weight of reciprocating parts is 500 N and length of stroke is 300 mm, find (a) piston effort and (b) turning moment on the crankshaft for the given crank position.

■ Solution

(a) Net force on the piston, $F_p = p_1A_1 - p_2A_2$

$$= \left(\frac{\pi}{4}\right)\left[0.35\times300^2 - 0.03\times40^2\right] = 24702 \text{ N}$$

$$n = \frac{\ell}{r} = \frac{700}{150} = 4.67$$

Acceleration of the piston, $f_p = \omega^2 r\left[\cos\theta + \frac{\cos 2\theta}{n}\right]$

$$= \left(\frac{2\pi\times360}{60}\right)^2 \times 0.150\times\left[\cos120° + \frac{\cos 240°}{4.67}\right]$$

$$= -129.4 \text{ m/s}^2$$

Piston effort, $PE = F_p + R - \frac{Rf_p}{g}$

$$= 24702 + 500 + \frac{500\times129.4}{9.81} = 31797 \text{ N}$$

(b) Turning moment on the crankshaft

$$= PE\cdot r\left[\sin\theta + \frac{\sin 2\theta}{2(n^2 - \sin^2\theta)^{0.5}}\right]$$

$$= 31797\times0.150\left[\sin120° + \frac{\sin 240°}{2(4.672 - \sin^2 120°)^{0.5}}\right]$$

$$= 3680.5 \text{ N m}$$

Example 10.33

The radius of crank of a horizontal engine is 300 mm. The mass of the reciprocating parts is 200 kg. The difference between the driving and the back pressures is 0.4 N/mm^2 when the crank has travelled 60° from I.D.C. The length of connecting rod is 1.2 m and the cylinder bore is 0.5 m. The engine runs at 240 rpm. Neglecting the effect of the piston rod, find (a) pressure on the slide bar, (b) thrust in the connecting rod, (c) tangential force, and (d) turning moment on the crankshaft.

Solution

Given : $r = 300$ mm, $M_r = 200$ kg, $l = 1.2$ m, $D = 0.5$ m, $N = 240$ rpm, $p_1 - p_2 = 0.4$ MPa

$\omega = 2\pi \times 240/60 = 25.13$ rad/s

$n = l/r = 1.2/0.3 = 4$

Force on the piston, $F_p = (p_1 - p_2) \times \pi D^2/4$

$= 0.40 \times \pi \times 500^2/4$

$= 78540$ N

Inertia force due to reciprocating parts, $F_i = M_r\, \omega^2 r\, [\cos\theta + \cos 2\theta/.n]$

$= 200 \times (25.13)^2 \times 0.3\, [\cos 60° + \cos 120°/4]$

$= 14209$ N

Piston effort, PE $= F_p - F_i = 78540 - 14209 = 64331$ N

(a) Pressure on the slide bar:

$\sin\phi = \sin\theta/\text{n} = \sin 60°/4 = 0.2165$

$\phi = 12.5°$

Pressure on the slide bars, $F_n =$ PE $\tan\phi$

$= 64331 \tan 12.5° = 14266$ N

(b) Thrust in the connecting rod, $F_n =$ PE$/\cos\phi$

$= 64331/\cos 12.5° = 65893$ N

(c) Tangential force on the crankpin, $F_t = F_c \sin(\theta + \phi)$

$= 65893 \sin 72.5° = 62843$ N

(d) Turning moment on the crankshaft $= F_t\, r$

$= 62843 \times 0.3 = 18853$ N m

Example 10.34

The turning moment diagram for one revolution of a multicylinder engine is shown in Fig.10.30. The vertical and horizontal scales are:
1 mm = 600 N m and 2.5°, respectively. The fluctuation of speed is limited to ±1.5 % of mean speed, which is 250 rpm. The hoop stress in rim material is limited to 5.6 N/mm^2. Neglecting effect of boss and arms, determine the suitable diameter and cross-section of flywheel rim. Take density of rim material as 7200 kg/m^3, and width to be four times the thickness.

Solution

Hoop stress, $\sigma_\theta = \rho v^2$

$v = \pi\, \text{d} N_m K_s/60$

$= 13.286$ d m/s

$5.6 \times 10^6 = 7200 \times (13.286\ \text{d})^2$

d = 2.1 m

Let E = energy at A

Point	Energy (mm^2)
B	$E - 30$
C	$E - 30 + 410 = E + 380$
D	$E + 380 - 275 = E + 105$
E	$E + 105 + 340 = E + 445$
F	$E + 445 - 320 = E + 125$
G	$E + 125 + 245 = E + 370$
H	$E + 370 - 385 = E - 15$
J	$E - 15 + 276 = E + 261$
K	$E + 261 - 261 = E$

Fluctuation of energy, $E_f = E_{max} - E_{min}$

$$= (E + 445) - (E - 30)$$
$$= 475 \text{ mm}^2$$
$$= 475 \times 10^{-6} \times (2.5 \times \pi \times 10^3/180) \times 600 \times 10^3$$
$$= 12435.47 \text{ N m}$$
$$W = (900 \times g \times E_f)/(\pi^2 \times K^2 \times N^2_m \times C_s)$$
$$= (900 \times 9.81 \times 12435.47)/(\pi^2 \times 1.05^2 \times 250^2 \times 0.03)$$
$$= 5381.4 \text{ N}$$

Now $\quad W = \pi\, d\, b\, t\, \rho\, g$

$$5381.4 = \pi \times 2.1 \times 4\, t \times t \times 7200 \times 9.81$$
$$t = 53.73 \text{ mm} \quad \text{and} \quad b = 214.92 \text{ mm}$$

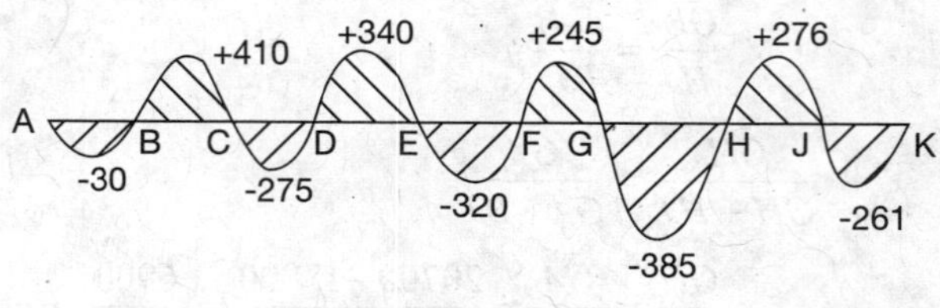

Fig.10.30 Turning moment diagram for a multi-cyclinder engine

Example 10.35

The variation of torque for an intermittent operation of a machine is shown in Fig.10.31. The machine is directly coupled to a motor, which exerts a constant torque at a mean speed of 200 rpm. The flywheel has a moment of inertia of 2000 kg m^2. Determine (a) the mean power of the motor, and (b) total fluctuation of speed of machine shaft.

Solution

Area of turning moment diagram = Area $OAEF$ + Area $ABCD$

$$= 13800\left[6\pi + \frac{4\pi + 2\pi}{2}\right] = 13800 \times 9\pi$$

$$= T_m \times 6\pi$$

or $$T_m = 20700 \text{ N m}$$

Power developed, $$P = \frac{2\pi N_m T_m}{60\times 1000} \text{ kW}$$

$$= \frac{2\pi\times 200\times 20700}{60\times 1000} = 433.54 \text{ kW}$$

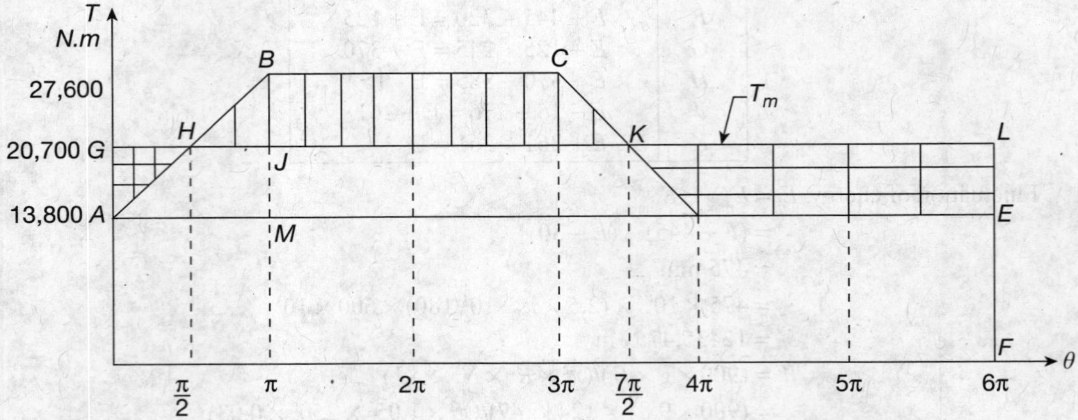

Fig.10.31 Variation of torque for an intermittent operation

Surplus energy is represented by the area *HBCK* and deficient energy by the areas *AGH* and *KLED*.

Now $$\frac{GH}{HJ} = \frac{GA}{BJ}$$

or $$\frac{GH}{GH+HJ} = \frac{GA}{GA+BJ}$$

$$\frac{GH}{GJ} = \frac{GA}{BM} = \frac{20700-13800}{27800-13800} = \frac{6900}{13800}$$

$$GH = \frac{\pi\times 6900}{13800} = \frac{\pi}{2}$$

Therefore, $$\theta_H = \frac{\pi}{2}$$

Similarly $$\theta_K = \frac{7\pi}{2}$$

Surplus energy $= \text{Area } HBCK$

$$= \left[\frac{3\pi + 2\pi}{2}\right](27600 - 20700) = 54192 \text{ N m}$$

$$= \left(\frac{\pi^2}{1800}\right)\cdot I\cdot (N_{max} + N_{min})\,(N_{max} - N_{min})$$

$$= \left(\frac{\pi^2}{1800}\right) \times 200 \times (200 \times 2)(N_{max} - N_{min})$$

$$N_{max} - N_{min} = 12.35 \text{ rpm}$$

Percentage fluctuation $= \dfrac{12.35 \times 100}{200} = 6.18\%$

Example 10.36

A certain machine requires a torque of (500 + 50 sin θ) N m to drive it, where θ is the angle of rotation of shaft measured from a certain datum. The machine is directly coupled to an engine which produces a torque of (500 + 60 sin 2θ) N m. The flywheel and the other rotating parts attached to the engine weigh 500 N and have a radius of gyration of 0.4 m. The mean speed is 180 rpm. Determine (a) The fluctuation of energy, (b) the percentage fluctuation of speed, and (c) the maximum and minimum angular acceleration of the flywheel and corresponding shaft positions.

■ **Solution**

(a) Change in torque $= (500 + 60 \sin 2\theta) - (500 + 50 \sin \theta)$

$$= 60 \sin 2\theta - 50 \sin \theta$$

$$= 120 \sin \theta \cos \theta - 50 \sin \theta$$

$$= \sin \theta\,(120 \cos \theta - 50)$$

For change in torque to be zero, $\sin \theta = 0$, or $\theta = 0°$ 180° and 360°.

Also, $\cos\theta = \dfrac{50}{120} = 0.4167$ or $\theta = 65.4°$ and 294.6°

The variation of T against θ is shown in Fig.10.32.

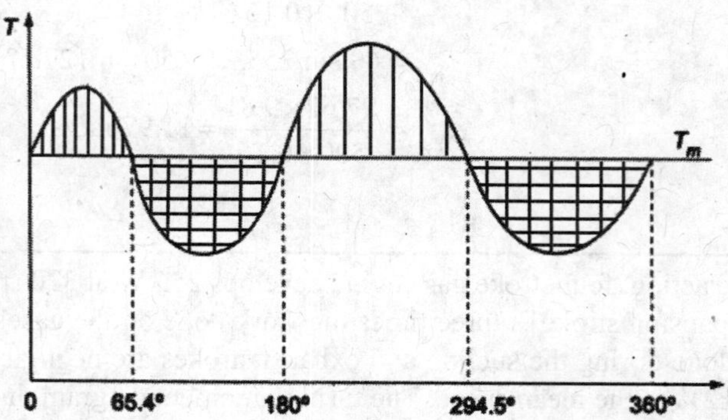

Fig.10.32 Variation of T against θ for a machine

Fluctuation of energy, $E_f = \displaystyle\int_{180°}^{294.6°} (60 \sin 2\theta - \sin\theta)\,d\theta$

$$= \left[-30 \cos 2\theta + 50 \cos\theta\right]_{180°}^{294.6°} = 121 \text{ N m}$$

(b)
$$E_f = \left(\frac{\pi^2}{1800}\right)\cdot\left(\frac{WK^2}{g}\right)\cdot(N_{max} + N_{min})(N_{max} - N_{min})$$

$$121 = \left(\frac{\pi^2}{1800}\right)\cdot\left(\frac{500\times 0.4^2}{9.81}\right)\cdot(180\times 2)(N_{max} - N_{min})$$

$$N_{max} - N_{min} = 7.45 \text{ rpm}$$

Percentage fluctuation of speed $= \dfrac{7.45\times 100}{180} = 4.14\%$.

(c) $T = 60 \sin 2\theta - 50 \sin \theta$

For maximum or minimum value of T,

$$\frac{dT}{d\theta} = 120 \cos 2\theta - 50 \cos \theta = 0$$

or $120(2 \cos^2 \theta - 1) - 50 \cos \theta = 0$

or $240\cos^2\theta - 5 \cos \theta - 120 = 0$

or $24 \cos^2\theta - 5 \cos \theta - 12 = 0$

$$\cos \theta = \frac{5 \pm (25 + 1152)^{0.5}}{48}$$

$$= 0.8189 \quad \text{and} \quad -0.61057$$

$$\theta = 35^\circ \quad \text{and} \quad 127.6^\circ$$

$$T_{max} = 60 \sin 70^\circ - 50 \sin 35^\circ = 27.7 \text{ Nm}$$

Maximum acceleration, $\alpha_{max} = \dfrac{T_{max}\, g}{W K^2}$

$$= \frac{27.7\times 9.81}{500\times 0.16} = 3.397 \text{ rad/s}^2$$

$$T_{min} = 60 \sin 255.2^\circ - 50 \sin 127.6^\circ = -97.62 \text{ Nm}$$

Minimum acceleration, $\alpha_{min} = \dfrac{97.62\times 9.81}{500\times 0.16} = 11.97 \text{ rad/s}^2$

Example 10.37

A single cylinder, single acting, four stroke gas engine develops 25 kW at 320 rpm. The work done by the gases during the expansion stroke is three times the work done on the gases during the compression stroke. The work done during the suction and exhaust strokes are negligible. The fluctuation of speed is not to exceed ±2% of the mean speed. The turning moment diagram during compression and expansion is assumed to be triangular in shape. Find the weight of the flywheel if its radius of gyration is 0.5 m.

■ Solution

Coefficient of fluctuation of speed, $K_s = 4\%$

Work done per cycle $= \dfrac{60P}{n}$

$$= \frac{60 \times 25000}{160} = 9375 \text{ N m}$$

Net work done per cycle = Work done during expansion – Work done during compression

$$= W_e - W_c = \frac{W_e - W_e}{3}$$

$$= \frac{2W_e}{3} = 9375$$

or $$W_e = 14062.5 \text{ N m}$$

The T–θ diagram is shown in Fig.10.33.

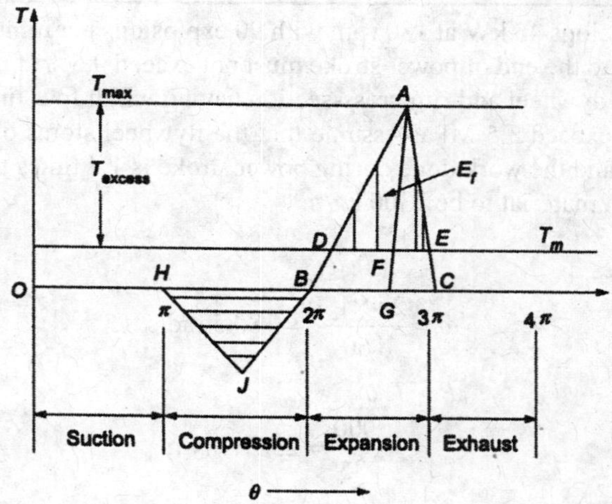

Fig.10.33 Variation of T against θ for a single cyclinder, single acting, four-stroke engine

Work done during expansion stroke = area $ABC = 0.5 \times BC \times AG$

$$14062.5 = 0.5 \times \pi \times AG$$

or $$AG = 8952.5 \text{ N m} = T_{max}$$

Mean turning moment, $$T_m = FG = \frac{9375}{4\pi} = 746 \text{ N m}$$

Excess turning moment, $$T_{excess} = AF = AG - FG$$

$$= 8952.5 - 746 = 8206.5 \text{ N m}$$

From similar triangles ADE and ABC, we have

$$\frac{DE}{BC} = \frac{AF}{AG}$$

or $$DE = \frac{AF}{AG} \cdot BC = \left(\frac{8206.5}{8952.5}\right)\pi = 2.88 \text{ rad}$$

Maximum fluctuation of energy, E_f = area $ADE = 0.5 \cdot DE \cdot AF$

$$= 0.5 \times 2.88 \times 8206.5 = 11817.4 \text{ N m}$$

Now $$E_f = K_s \left(\frac{W}{g}\right) K^2 \omega^2$$

$$11817.4 = 0.04 \left(\frac{W}{9.81}\right) \times (0.5)^2 \left(\frac{2\pi \times 320}{60}\right)^2$$

$$W = 10324 \text{ N}$$

Example 10.38

An Otto cycle engine develops 45 kW at 180 rpm with 90 explosions per minute. The change of speed from the commencement to the end of power stroke must not exceed 0.5% of mean on either side. Find the mean diameter of the flywheel and rim cross-section having width four times the thickness so that the hoop stress does not exceed 3.5 MPa. Assume that the flywheel stores 6% more energy than the energy stored by the rim and the work done during power stroke is 1.4 times the work done during the cycle. Take density of rim material to be 7300 kg/m³.

Solution

$$\omega = \frac{2\pi \times 180}{60} = 18.85 \text{ rad/s}$$

Power developed, $P = T_m \omega$

or $$T_m = \frac{45000}{18.85} = 2387.3 \text{ N m}$$

Since the number of explosions are half of the rpm, therefore, it is a four-stroke engine. The turning moment diagram is shown in Fig.10.34.

Work done per cycle $= T_m \theta = 2387.3 \times 4\pi = 30000$ N m.
Work done during the power stroke $= 1.4 \times 30000 = 42000$ N m.
Triangle ABC shows the work done during the power stroke.
Work done during the working stroke = area ABC $= 0.5 \cdot AC \cdot BF$

$$42000 = 0.5 T_{max} \times \pi$$

or $$T_{max} = 26738 \text{ Nm}$$

Excess torque, $T_{excess} = BG = BF - GF = T_{max} - T_m$

$$= 26738 - 2387.3 = 24350.7 \text{ N m}$$

From similar triangles ABC and BDE, we have

$$\frac{DE}{AC} = \frac{BG}{BF}$$

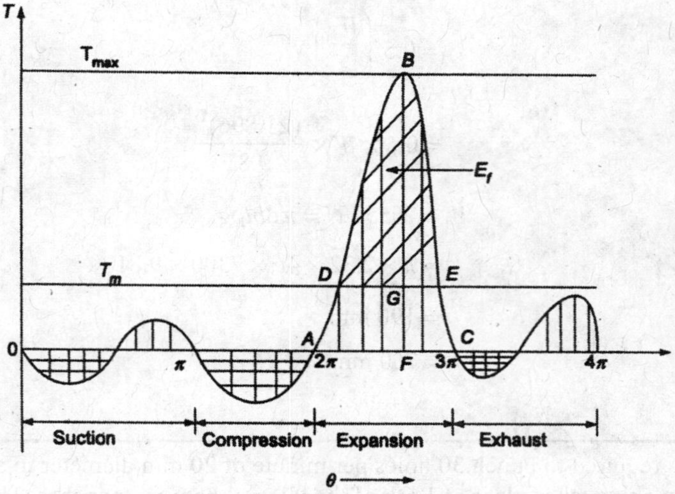

Fig.10.34 Variation of T against θ for a four-stroke petrol engine

or
$$DE = \left(\frac{BG}{BF}\right) \cdot AC = \frac{T_{excess}}{T_{max}} \times \pi$$
$$= \left(\frac{24350.7}{26738}\right)\pi = 0.9107\pi$$

Maximum fluctuation of energy,
$$E_f = \text{area } BDE = 0.5DE \cdot BG$$
$$= 0.5 \times 0.9107\pi \times 24350.7$$
$$= 34834 \text{ N m}$$

Hoop stress in flywheel rim $= \rho v^2$
$$3.5 \times 10^6 = 7300\, v^2$$
or
$$v = 21.896 \text{ m/s}$$
Let d be the diameter of flywheel.
$$21.896 = \pi \text{d} \times \frac{180}{60}$$
or
$$\text{d} = 2.32 \text{ m}$$
Fluctuation of speed = 1%

Coefficient of fluctuation of speed, $K_s = 0.01$

Now
$$E_f = 2\, EK_s$$
$$34834 = 2 \times 0.01 \times E$$
$$E = 1741.7 \times 10^3 \text{ N m}$$
Energy stored by the flywheel $= 1.06\, E = 1846.2 \times 10^3$ N m

$$= 0.5\left(\frac{W}{g}\right)v^2$$

$$= 0.5 \times W \times \frac{(21.896)^2}{9.81}$$

or $$W = 75552 \text{ N} = \pi dbt\rho g$$

$$= \pi \times 2.32 \times 4t^2 \times 7300 \times 9.81$$

or $$t = 190 \text{ mm}$$

$$b = 760 \text{ mm}$$

Example 10.39

A punching press is required to punch 30 holes per minute of 20 mm diameter in a steel plate 13 mm thick. The actual punching takes place at 1/6th of the interval between punches. The shear strength of the plate is 310 N/mm². The driving motor runs at 900 rpm with a reduction in the velocity through gears to give the desired speed of 30 punching operations per minute. Find the mass of the flywheel required if its mean diameter is 900 mm. Take $K_s = 10\%$.

■ Solution

Required punching force, $$F = \pi dt\tau_u$$

$$= \pi \times 20 \times 13 \times 310 = 253212 \text{ N}$$

The force versus displacement diagram is shown in Fig.10.13. The area under the curve can be approximated as a triangle, so that the work done in punching the hole is,

$$= 0.5\,Ft$$

$$= 0.5 \times 253212 \times 0.013 = 1645.9 \text{ N m}$$

Time between punching operations $$= \frac{60}{30} = 2\,s$$

Punching time $= \frac{2}{6} = \frac{1}{3}\,s$

Average power required without the flywheel $= \frac{1645.9}{(1/3)} = 4937.6$ W

Since F in Fig.10.13(a) is twice as large as in Fig.10.13(b), therefore the instantaneous power required will be 9875 W.

When a flywheel is used, the force-displacement curve is shown in Fig.10.13(c). The work required to punch the hole is represented by the area *ABCDE*. The same amount of energy to be supplied by the flywheel is represented by the area *FGIE*. Therefore, 1645.9 N m of energy is to be supplied in 2 *s*, that is, a 822.95 W motor is required. During the 1/3 second punching interval, the motor supplies the energy represented by the area *AHIE,* which is 822.95/3 = 274.3 N m. But the energy required is 1645.9 N m. Therefore, the energy to be taken from the flywheel is, $E_f = 1645.9 - 274.3 = 1371.6$ N m.

Mean velocity, $$v_m = \pi \times 0.9 \times \frac{150}{60} = 7.07 \text{ m/s}$$

$$v_1 + v_2 = 2v_m = 14.14$$

$$v_1 - v_2 = K_s v_m = 0.1 \times 7.07 = 0.707$$

$$v_1 = 7.424 \text{ m/s} \quad \text{and} \quad v_2 = 6.717 \text{ m/s}$$

Mass of flywheel required, $M = \dfrac{E_f}{K_s \cdot v_m^2} = \dfrac{1371.6}{0.1 \times 7.07^2} = 274.4$ kg

Example 10.40

A punching machine is required to punch 5 holes per minute of 50 mm diameter in 40 mm thick plate. The ultimate shear strength of plate material is 225 MPa. The punch has a stroke of 100 mm. Find the power of motor required if mean speed of flywheel is 18 m/s. If coefficient of fluctuation of energy is 4%, find the mass of the flywheel.

■ Solution

Punching force, $F = \pi \text{d} \text{t} \tau_u = \pi \times 50 \times 40 \times 225 = 1413717$ N

Punching time per hole $= \dfrac{60}{5} = 12\ s$

Energy required in punching one hole, $E_1 = 0.5\ FK_e$

$$= 0.5 \times 1413717 \times 0.04 = 28274 \text{ N m}$$

Power required = E_1/punching time $= \dfrac{28274}{12} = 2356$ W or 2.356 kW

The punch travels a total distance of $2 \times 100 = 200$ mm (upstroke + downstroke) in 12 s.

Time required to punch a hole in 40 mm thick plate $= \dfrac{12 \times 40}{200} = 2.4\ s$

Energy required to be supplied by motor in 12 s = 28,274 N m

Energy supplied by the motor in 2.4 $s = 28274 \times \dfrac{2.4}{12} = 5654.8$ N m

Energy supplied by flywheel, $E_f = 28274 - 5654.8 = 22619.8$ N m

$$E = \frac{E_f}{2K_e} = \frac{22619.2}{2 \times 0.04} = 282740 \text{ N m}$$

If M is the mass of the flywheel, then

$$0.5Mv^2 = E$$

$$M = 2 \times \frac{282740}{(18)^2} = 1745 \text{ kg}$$

Example 10.41

A punching press makes 25 holes of 20 mm diameter per minute in a plate 15 mm thick. This causes variation in the speed of flywheel attached to press from 240 to 220 rpm. The punching operation takes 2 seconds per hole. Assuming 6 N m of work is required to shear 1 mm² of the area and frictional losses account for 15% of the work supplied for punching, determine (a) the power required to operate the punching press, and (b) the mass of flywheel with radius of gyration of 0.5 m.

■ Solution

Work required for punching one hole = Area of shear in mm² × Work per mm²

$$= \pi dt \times 6 = \pi \times 20 \times 15 \times 6$$

$$= 5654.86 \text{ N m}$$

Accounting 15% for frictional losses, the actual work supplied

$$= \frac{5654.86}{0.85} = 6652.78 \text{ N m}$$

Total work required per minute for drilling 25 holes

$$= 6652.78 \times 25 = 166,319 \text{ N m}$$

(a) Power required $= \dfrac{166319}{60 \times 10^3} = 2.772$ kW

Energy supplied during the punching operation $= 2.772 \times 1000 \times 2 = 5544$ N m

Energy supplied by the flywheel, $E_f = 6652.78 - 5544 = 1108.78$ N m

$$E_f = 0.5I\left(\omega_1^2 - \omega_2^2\right)$$

$$1108.78 = 0.5M \times (0.5)^2 \times \left(\frac{2\pi}{60}\right)^2 \left[240^2 - 220^2\right]$$

$$M = 87.92 \text{ kg}$$

Example 10.42

An electric motor drives a punching press to which a flywheel of radius of gyration 0.5 m is fitted. The flywheel runs at 240 rpm. The press is capable of punching 600 holes per hour with each punching operation taking 2 seconds and requiring 15 kN m of work. Determine (a) the rating of the motor, and (b) mass of the flywheel if its speed does not drop below 220 rpm.

■ Solution

(a) Total work required per hour = Work per hole × Number of holes per hour

$$= 15 \times 600 = 9 \times 10^6 \text{ Nm}$$

Motor power $= \dfrac{9 \times 10^6}{10^3 \times 3600} 0 = 2.5$ kW

(b) Energy delivered by motor during the punching operation,

$$E_2 = 2.5 \times 1000 \times 2 = 5000 \text{ N m}$$

Energy required per punch operation, $E_1 = 15{,}000$ N m

Fluctuation of energy, $E_f = E_1 - E_2 = 15{,}000 - 5000 = 10{,}000$ N m

$$E_f = 0.5I\ (\omega_1^2 - \omega_2^2)$$

$$10000 = 0.5M \times (0.5)^2 \times \left(\frac{2\pi}{60}\right)^2 \left[240^2 - 220^2\right]$$

$$M = 792.95 \text{ kg}$$

Example 10.43

A 5 kW induction motor running at 750 rpm operates a rivetting machine. A flywheel of mass 80 kg and radius of gyration 0.45 m is fitted to it. Each rivetting takes 1 s and requires 10 kW. Determine (a) number of rivets closed per hour and (b) fall in speed of the flywheel after the riveting operation.

Solution

Mean speed, $\omega_m = 2\pi \times \dfrac{750}{60} = 78.54$ rad/s

Energy supplied by the motor in one hour $= 5 \times 10^3 \times 3600 = 18 \times 10^6$ N m

Energy required for one riveting operation $= 10 \times 10^3 \times 1 = 10^4$ N m

Number of rivets closed per hour $= \dfrac{18 \times 10^6}{10^4} = 1800$

Energy supplied by the motor in 1 $s = 5 \times 10^3 \times 1 = 5000$ N m

Energy to be supplied by the flywheel = 10000 – 5000 = 5000 N m

$$E_f = E_{max} - E_{min} = 0.5I\left(\omega_{max}^2 - \omega_{min}^2\right)$$

Now $\omega_{max} = \omega_m$

$$5000 = 0.5 \times 80 \times (0.45)^2 \left[(78.540^2 - \omega_{min}^2\right]$$

$$\omega_{min} = 74.5 \text{ rad/s}$$

or $N_{min} = 711.5$ rpm

Fall in speed = 750 – 711.5 = 38.5 rpm

Example 10.44

A connecting rod of an internal combustion engine has a mass of 1.5 kg and the length of the rod is 250 mm. The centre of gravity of the rod is located at a distance of 100 mm from the gudgeon pin. The radius of gyration about an axis through the centre of gravity perpendicular to the plane of rotation is 110 mm. Find the equivalent dynamical system if only one of the masses is located at the gudgeon pin.

If the connecting rod is replaced by two masses, one at the gudgeon pin and the other at the crank pin and the angular acceleration of the rod is 24000 rad/s² clockwise, determine the correction couple applied to the system to reduce it to a dynamically equivalent system.

■ **Solution**

Here $m = 1.5$ kg, $L = 250$ mm, $a = 100$ mm, $K = 110$ mm

Now $$ab = K^2$$

$$b = \frac{110^2}{100} = 121 \text{ mm}$$

Let m_a = mass at the gudgeon pin
m_b = mass at the crank pin

Then $$m_a = \frac{m_b}{L} = \frac{1.5 \times 121}{250} = 0.72 \text{ kg}$$

$$m_b = 1.5 - 0.72 = 0.78 \text{ kg}$$

Correction couple

Now $$a = 100 \text{ mm}, c = 150 \text{ mm}$$

$$K_1^2 = ac = 100 \times 150 = 15000$$

$$K_1 = 122.47 \text{ mm}$$

Correction couple, $$T_o = m\left(K_1^2 - K^2\right)\alpha$$

$$= 1.5(15000 - 12100) \times 24000 \times 10^{-6}$$

$$= 104.4 \text{ N m}$$

Example 10.45

A vertical engine running at 1200 rpm with a stroke of 120 mm, has a connecting rod 300 mm long and of 1.5 kg mass. The mass centre of the rod is 100 mm from the big end centre. When the rod is suspended from the gudgeon pin as a pendulum, it makes 20 complete oscillations in 20 seconds.

(a) Calculate the radius of gyration of the rod about an axis through the mass centre. (b) When the crank is at 35° from the top dead centre and the piston is moving downwards, find the acceleration of the piston and the angular acceleration of the rod. Hence, find the inertia torque exerted on the crankshaft.

■ **Solution**

Angular speed, $$\omega = \frac{2\pi \times 1200}{60} = 126.66 \text{ rad/s}$$

$$L = 120 \text{ mm or r} = \frac{L}{2} = 60 \text{ mn}$$

$$l = 300 \text{ mm}, \; m = 1.5 \text{ kg}, \; \theta = 35°, \; n = \frac{l}{r} = \frac{300}{600} = 5$$

(a) Radius of gyration of the rod

Distance of the centre of gravity of rod from the point of suspension,

$$l_1 = 300 - 100 = 200 \text{ mm}$$

Frequency of oscillation of a compound pendulum,

$$f_n = \left(\frac{1}{2\pi}\right)\left[\frac{gl_1}{K^2 + l_1^2}\right]^{0.5}$$

$$\frac{20}{20} = \left(\frac{1}{2\pi}\right)\left[\frac{9.81 \times 200 \times 10^3}{K^2 + 200^2}\right]^{0.5}$$

$$4\pi^2 = \frac{9.81 \times 200 \times 10^3}{K^2 + 200^2}$$

or $$K^2 + 40000 = 49698$$

or $$K^2 = 9698$$

or $$K = 98.47 \text{ mm}$$

(b) Acceleration of the connecting rod:

Angular acceleration of the rod, $$f_r = -\omega^2 \frac{\sin\theta}{n}$$

$$= -(126.66)^2 \frac{\sin 35°}{5}$$

$$= -1840.35 \text{ rad/s}^2$$

Acceleration of the piston, $$f_p = \omega^2 r\left[\cos\theta + \frac{\cos 2\theta}{n}\right]$$

$$= (126.66)^2 \times 0.06\left[\cos 35° + \frac{\cos 70°}{5}\right]$$

$$= 854.33 \text{ m/s}^2$$

Inertia torque exerted on the crankshaft

Mass of the rod at the gudgeon pin, $m_g = \dfrac{m(l - l_1)}{l} = \dfrac{1.5(300 - 200)}{300} = 0.5$ kg

Vertical inertia force due to m_g, $F_i = m_g f_p = 0.5 \times 854.33 = 427.16$ N

Now $$\sin\phi = \frac{\sin\theta}{n} = \frac{\sin 35°}{5} = 0.1147$$

$$\phi = 6.587°$$

$$\frac{OM}{\sin(\theta + \phi)} = \frac{r}{\cos\phi}$$

$$OM = \frac{60\sin 41.587°}{\cos 6.587°} = 40.1 \text{ mm}$$

Torque due to F_i

$$T = -F_i \cdot OM = -427.16 \times 0.0401$$

$$= -17.125 \text{ Nm or } 17.125 \text{ Nm (counter-clockwise)}$$

Equivalent length of a pendulum, $l_e = \frac{K^2 + l_1^2}{l_1} = \frac{9698 + 200^2}{200} = 248.49 \text{ mm}$

Correction couple,

$$T_o = -ml_1(l - l_e)\alpha_r$$

$$= -1.5 \times 200(300 - 248.49)10^{-6} \times 1840.35$$

$$= -28.44 \text{ N m}$$

Corresponding torque on the crankshaft, $T_{cs} = \frac{T_o \cos\theta}{n} = \frac{-28.44 \cos 35°}{5} = -4.66 \text{ Nm}$

$$= 4.66 \text{ N m (counter-anticlockwise)}$$

Torque due to the mass at the gudgeon pin,

$$T_g = m_g g \cdot OM = 0.5 \times 9.81 \times 0.0401$$

$$= 0.1967 \text{ N m (clockwise)}$$

Equivalent mass of the rod at the crank pin, $m_c = \frac{1.5 \times 200}{300} = 1 \text{ kg}$

Torque due to this mass,

$$T_c = m_c g \cdot l \sin\phi = 1 \times 9.81 \times 0.3 \times \sin 6.587°$$

$$= 0.3376 \text{ N m (clockwise)}$$

Inertia torque exerted on the crankshaft $= T + T_{cs} - T_g - T_c$

$$= 17.125 + 4.66 - 0.1967 - 0.3376$$

$$= 21.25 \text{ N m (counter-clockwise)}$$

Example 10.46

The connecting rod of an oil engine weighs 800 N, the distance between the bearing centers is 1 m. The diameter of the big end bearing is 120 mm and of the small end bearing is 75 mm. When suspended vertically with a knife edge through the small end it makes 100 oscillations is 200 *s*, and with knife edge through the big end it makes 100 oscillations in 170 *s*. Find the moment of inertia of the rod in kg m² and distance of the center of gravity from the small end centre.

Solution:

Given: $m = \frac{800}{9.81} = 81.55$ kg, $l = 1$m, $D = 120$ mm, d = 75 mm

$$f_{n1} = \frac{100}{200} = 0.5 \text{ cps}$$

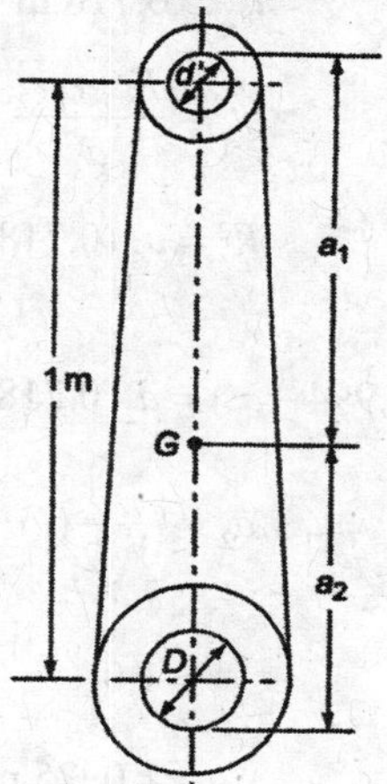

Fig.10.35 Connecting rod

$$f_{n2} = \frac{100}{170} \text{ cps}$$

Let a_1 = distance of centre of gravity from the top of small end bearing

a_2 = distance of centre of gravity from the top of big end bearing

For a simple pendulum, natural frequency of rod when suspended from small end bearing is given by,

$$f_{n1} = \frac{1}{2\pi}\left(\frac{g}{l_{e1}}\right)$$

$$0.5 = \frac{1}{2\pi}\sqrt{\frac{9.81}{l_{e1}}}$$

$$l_{e1} = 0.994 \text{ m}$$

$$= \frac{K^2 + a_1^2}{a_1}$$

$$K^2 = a_1(0.994 - a_1) \quad (1)$$

When suspended from the big end,

$$f_{n2} = \frac{1}{2\pi}\left(\frac{g}{l_{e2}}\right)$$

$$\frac{100}{170} = \frac{1}{2\pi}\sqrt{\frac{9.81}{l_{e2}}}$$

$$l_{e2} = 0.718 \text{ m}$$

$$= \frac{K^2 + a_2^2}{a^2}$$

or $$K^2 = a_2 (0.718 - a_2) \quad (2)$$

From Eqs. (1) and (2), we have

$$a_1(0.994 - a_1) = a_2(0.718 - a_2) \quad (3)$$

Also $$a_1 + a_2 = l + \frac{1}{2}(D + \text{d}) \quad \text{(see Fig.10.35)}$$

$$= 1.0 + \frac{1}{2}(0.120 + 0.075)$$

$$= 1.0975 \text{ m}$$

or $$a_2 = 1.0975 - a_1 \quad (10.50)$$

Substituting in Eq. (3), we have

$$0.994a_1 - a_1^2 = 0.718(1.0975 - a_1) - (1.0975 - a_1)^2$$

$$= 0.788 - 0.718\, a_1 - 1.204 - a_1^2 + 2.195\, a_1$$

$$0.483a_1 = 0.416$$

$$a_1 = 0.861 \text{ m}$$

$$K^2 = 0.861(0.994 - 0.861) = 0.1145 \text{ m}^2$$

$$\text{I} = \text{mK}^2 = 81.55 \times 0.1145 = 9.34 \text{ kg m}^2$$

Distance of centre of gravity from small end centre

$$= a_1 - \frac{d}{2}$$

$$= 861 - 37.5 = 823.5 \text{ mm}$$

Example 10.47

A punching press is required to punch 40 mm diameter holes in a plate of 15 mm thickness at a rate of 30 holes per minute. It requires 6 N m of energy per mm² of sheared area. Determine the moment of inertia of the flywheel if the punching takes one-tenth of a second and the speed of the flywheel varies from 160 to 140 rpm.

Solution

Given: $d = 40$ mm, $t = 15$ mm

Energy supplied in punching a hole = $\pi \times 40 \times 15 \times 6 = 11309.7$ N m

Time taken for punching one hole = 60/30 = 2 s

Power required = 11309.7/2 = 5654.87 N m/s or 5.655 kW

$$C_s = (v_{max} - v_{min})/v_m = (27.5 - 24.5)/26 = 0.1154$$

Time to punch a hole in 15 mm thick plate = 1/10 *s*

Energy supplied by motor in 0.1 *s* = 11309.7 × 0.1/2 = 565.5 N m

Energy supplied by flywheel, E_f = 11309.7 – 565.5 = 10744.2 N m

$$N_m = (160 + 140)/2 = 150 \text{ rpm}, \; C_s = 20/150 = 2/15$$

$$W = 900\, gE_f/(\pi^2 K^2 N^2_m C_s)$$

$$I = WK^2/g = 900 \times 10744.2 \times 15/[\pi^2 \times (150)^2 \times 2] = 326.6 \text{ kg m}^2$$

Example 10.48

A machine is required to punch 4 holes of 40 mm diameter in a plate of 25 m thickness per minute. The work required is 6 N m per square mm of sheared area. The stroke of punch is 100 mm and maximum speed of flywheel at its radius of gyration is 30 m/s. Find the mass of the flywheel so that the speed does not fall below 27 m/s at the radius of gyration. Also determine the motor power required.

■ Solution

Work required per punch, $E = \pi \times 40 \times 25 \times 6 = 18849.6$ N m

Now $(\theta_2 - \theta_1)/2\pi = t/2L = 25/200 = 1/8$

$E_f = (1 - t/2L)\,E = (1 - 1/8) \times 18849.6 = 16493.4$

$= 0.5\, MK^2\,(\omega_1^2 - \omega_2^2)$

$= 0.5\, M(K^2\omega_1^2 - K^2\omega_2^2)$

$= 0.5\, M(30^2 - 27^2)$

$= 85.5\, M$

$M = 192.9$ kg

Energy supplied per minute = 18849.6 ×4 = 75398.4 N m

Motor power = 75398.4/(60 ×1000) = 1.257 kW

Example 10.49

A punching press is required to punch 30 mm dia holes in a plate of 20 mm thickness at the rate of 20 holes per minute. It requires 6 N m of energy per mm² of sheared area. If punching takes place 1/10th of a second and the speed of the flywheel varies from 160 to 140 rpm, determine the mass of the flywheel having radius of gyration of 1 m.

■ Solution

Work required per punch, $E_1 = \pi \times 30 \times 20 \times 6 = 11309.7$ N m

Time required to punch a hole = 60/20 = 3 *s*

Energy required for punching per second = 11309.7/3 = 3769.91 N m/s

Energy supplied by motor in 1/10th second , E_2 = 3769.91/10 = 376.991 N m

$E_f = E_1 - E_2 = 11309.7 - 376.991 = 10932.71$ N m

$= 0.5 \times MK^2 \times (2\pi/60)^2(N^2_1 - N_2^2)$

$= 0.5M \times 1 \times (2\pi/60)^2(160^2 - 140^2)$

$M = 332.3$ kg

Example 10.50

The turning moment diagram for a four-stroke gas engine may be assumed for simplicity to be represented by four triangles, the areas of which from the line of zero pressure are as follows: suction stroke = 0.45×10^{-3} m^2, compression stroke = 1.7×10^{-3} m^2, expansion stroke = 6.8×10^{-3} m^2, exhaust stroke = 0.65×10^{-3} m^3. Each m^2 of area represents 3 MN m of energy.

Assuming the resisting torque to be uniform, find the mass of rim of flywheel required to keep the speed between 202 and 198 rpm. The mean radius of the rim is 1.2 m.

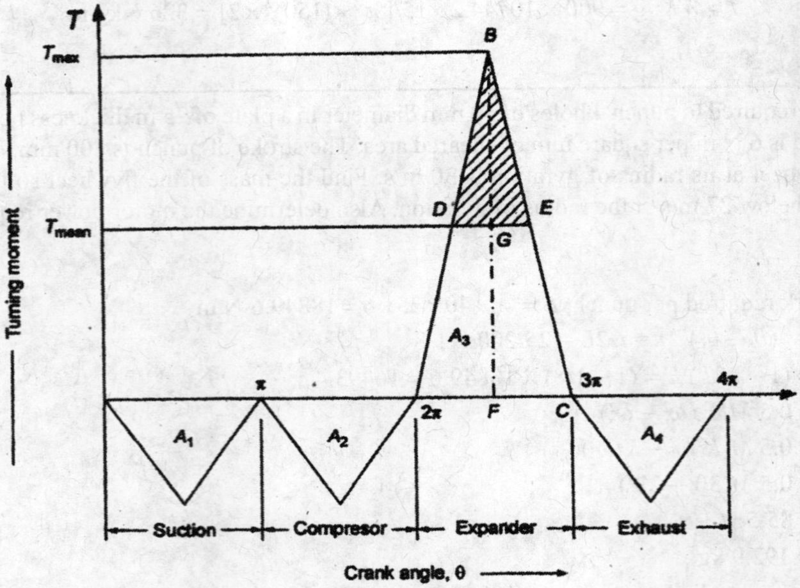

Fig.10.36 Turning moment diagram for a four-stroke engine

Solution

Given: $N_{max} = 202$ rpm, $N_{min} = 198$ rpm, $R = 1.2$ m.

$N_m = (N_{max} + N_{min})/2 = (202 + 198)/2 = 200$ rpm

Coefficient of fluctuation of speed, $C_s = (N_{max} - N_{min})/N_m$

$= (202 - 198)/200 = 4/200 = 0.02$

$\omega = 2\pi \times 200/60 = 20.944$ rad/s

The turning moment diagram is shown in Fig.10.36.

$$\text{Net area, } A = A_3 - (A_1 + A_2 + A_4) = [6.8 - (0.45 + 1.7 + 0.65)] \times 10^{-3}$$

$$= 4 \times 10^{-3} \text{ m}^2$$

Net work done = $A \times$ Scale = $4 \times 10^{-3} \times 3 \times 10^6 = 12{,}000$ N m

Work done per cycle = $T_{mean} \times 4\pi$ N m

$$T_{mean} = \frac{12{,}000}{4\pi} = 954.93 \text{ N m}$$

Work done during expansion stroke $= A_3 \times \text{Scale} = 6.8 \times 10^{-3} \times 3 \times 10^6 = 20.400 \text{ N m}$

$$= \text{Area } \Delta ABC$$

$$= \left(\frac{1}{2}\right) \times AC \times BF = \left(\frac{1}{2}\right) \times \pi \times BF$$

$$BF = \frac{20400 \times 2}{\pi} = 12987 \text{ N m}$$

Excess torque, $BG = BF - FG = 12987 - 954.93 = 12033.07$ N m

Now Δs ABC and DBE are similar. Thus

$$\frac{DE}{AC} = \frac{BG}{BF}$$

$$DE = \left(\frac{12033.07}{12987}\right)\pi = 3.91 \text{ radian}$$

Maximum fluctuation of energy, $E_f = \text{Area } \Delta\ BDE = \left(\frac{1}{2}\right) DE \times BG$

$$= \left(\frac{1}{2}\right) \times 3.91 \times 12033.07 = 17506.7 \text{ N m}$$

Mass of flywheel, $$m = \frac{E_f}{R^2 \omega^2 C_s}$$

$$= \frac{17506.7}{(1.2)^2 \times (20.944)^2 \times 0.02} = 1385.8 \text{ kg}$$

Example 10.51

A vertical single cylinder engine has a cylinder diameter 300 mm, stroke length 500 mm, and connecting rod length 4.5 times the crank length. Engine runs at 180 rpm. The mass of reciprocating parts is 280 kg, compression ratio is 14, and the pressure remains constant during the injection of oil for 1/10th of the stroke. If the compression and expansion follow the law $pV^{1.35}$ = constant, find:

(i) Crank pin effort, (ii) Thrust on the bearings, (iii) Turning moment on the crankshaft when the crank displacement is 45° from the I.D.C. position during expansion stroke. Suction pressure may be taken as 0.1 N/mm³

■ Solution

Given: $D = 300$ mm, $L = 500$ mm, or $r = 250$ mm, $n = \frac{l}{r} = 4.5, N = 180$ rpm, $M_r = 280$ kg

$$r_c = \frac{V_1}{V_2} = 14,\ V_3 - V_2 = 0.1\ V_s = 0.1\ (V_1 - V_2),\ p_1 = 0.1 \text{ N/mm}^2,\ \theta = 45°$$

The p-V diagram for the diesel engine is shown in Fig.10.37.

Angular speed of engine, $\omega = \frac{2\pi N}{60} = \frac{2\pi \times 180}{60} = 18.85$ rad/s

Process 1–2: $p_1V_1^{1.35} = p_2V_2^{1.35}$

$$p_2 = p_1\left(\frac{V_1}{V_2}\right)^{1.35} = 0.1(14)^{1.35} = 3.526\ \text{N/mm}^2$$

Swept volume, $V_s = \left(\frac{\pi}{4}\right)D^2L = \frac{\pi}{4}\times 300^2\times 500\times 10^{-9} = 0.035343\ \text{m}^3$

Now $$\frac{V_1}{V_2} = \frac{V_2+V_s}{V_2}$$

$$14 = \frac{1+0.035343}{V_2}$$

Also $p_2 = p_3$ and $V_3 = V_2 + 0.1V_s = 0.00272 + 0.0035343 = 0.00625\ \text{m}^3$

The displacement of the piston for 45° crank rotation from I.D.C. during expansion stroke is indicated by point 5 in Fig.10.37.

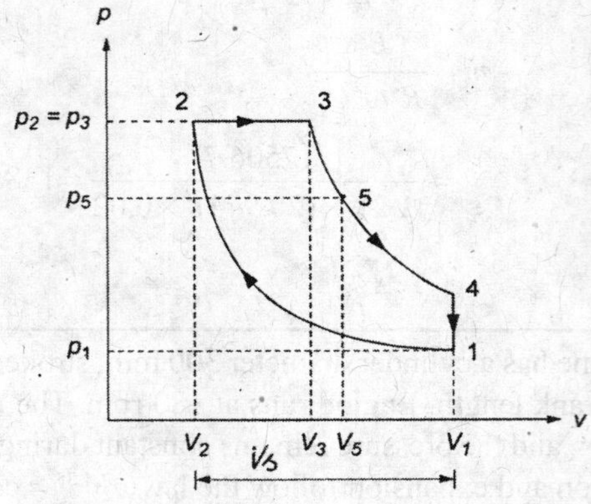

Fig.10.37 *p-V diagram for a diesel engine*

$$x = r\left[(1-\cos\theta)+\left\{n-(n^2-\sin^2\theta)^{0.5}\right\}\right]$$

$$= 250\left[(1-\cos 45°)+\left\{4.5-(4.5^2-\sin^2 45°)^{0.5}\right\}\right]$$

$$= 87.2\ \text{mm or }\ 0.0872\ \text{m}$$

$$V_5 = V_2 + \left(\frac{\pi}{4}\right)D^2x = 0.00272 + \left(\frac{\pi}{4}\right)\times 300^2\times 87.2\times 10^{-9}$$

$$= 0.0089\ \text{m}^3$$

Process 3–5:

$$p_3V_3^{1.35} = p_5V_5^{1.35}$$

$$p_5 = p_3\left(\frac{V_3}{V_5}\right)^{1.35} = 3.526\left(\frac{0.00625}{0.0089}\right)^{1.35} = 3.188 \text{ N/mm}^2$$

Difference of pressure on two sides of the piston, $\Delta p = p_5 - p_1 = 3.188 - 0.1 = 3.088 \text{ N/mm}^2$

Force on the piston due to gas pressure during outstroke,

$$F_p = \left(\frac{\pi}{4}\right)D^2 \times \Delta p = \left(\frac{\pi}{4}\right)\times 300^2 \times 3.088 = 147592 \text{ N}$$

Inertia force due to the reciprocating parts, $F_i = M_r\omega^2 r\left(\cos\theta + \frac{\cos 2\theta}{n}\right)$

$$= 280 \times (18.85)^2 \times 0.25 \times \left(\cos 45° + \frac{\cos 90°}{4.5}\right)$$

$$= 17587.6 \text{ N}$$

Piston effort during down stroke of a vertical four-stroke diesel engine.

$$F = F_p - F_i + M_r g = 147592 - 17587.6 + 280\times 9.81 = 132751.2 \text{ N}$$

(i) Crank pin effort.

Angle of inclination of the connecting rod with the line of stroke,

$$\sin\phi = \frac{\sin\theta}{n} = \frac{\sin 45°}{4.5} = 0.15713$$

$$\phi = 9.04°$$

$$\text{CPE, } F_t = F \times \frac{\sin(\theta+\phi)}{\cos\phi}$$

$$= 132751.2 \times \frac{\sin(45° + 9.04°)}{\cos 9.04°}$$

$$= 108803.9 \text{ N}$$

(ii) Thrust on bearings,

$$F_r = F \times \frac{\cos(\theta+\phi)}{\cos\phi}$$

$$= 132751.2 \times \frac{(45° + 9.04°)}{\cos 0.94°}$$

$$= 78934.65 \text{ N}$$

(iii) Turning moment on the crankshaft or crank effort,

$$T = F_t \times r = 108803.9 \times 0.25 = 27200.975 \text{ N m}$$

Example 10.52

The following data relate to a connecting rod of a reciprocating engine:

Mass = 60 kg, distance between bearing centres = 900 mm, diameter of small end bearing = 80 mm, diameter of big end bearing = 100 mm, time of oscillation when the connecting rod is suspended from small end = 1.85 *s*, time of oscillation when the connecting rod is suspended from big end = 1.70 *s*. Determine (a) the radius of gyration of the rod about an axis passing through the centre of gravity and perpendicular to the plane of oscillation, (b) the moment of inertia of the rod about the same axis, and (c) the dynamically equivalent system for the connecting rod, consisting of two masses, one of which is situated at the small end centre.

■ **Solution**

Refer to Fig 10.38.

Given: $m = 60$ kg, $l = 900$ mm, $d_1 = 80$ mm, $d_2 = 100$ mm, $T_1 = 1.85$ s, $T_2 = 1.70$ s

(a) For an equivalent simple pendulum,

$$T_1 = 2\pi\sqrt{\frac{l_{e1}}{g}}$$

$$1.85 = 2\pi\sqrt{\frac{l_{e1}}{9.81}}$$

$$l_{e1} = 0.85 \text{ m}$$

$$= \frac{K^2 + a_1^2}{a_1}$$

$$K^2 = a_1(0.85 - a_1) \quad (1)$$

$$T_2 = 2\pi\sqrt{\frac{l_{e2}}{g}}$$

$$1.70 = 2\pi\sqrt{\frac{l_{e2}}{9.81}}$$

$$l_{e2} = 0.718 \text{ m}$$

$$= \frac{K^2 + a_2^2}{a_2}$$

$$K^2 = a_2\,(0.718 - a_2) \quad (2)$$

From Eqs. (1) and (2)

$$a_1\,(0.85 - a_1) = a_2\,(0.718 - a_2)$$

Now $$a_1 + a_2 = 0.900 + \frac{1}{2}(80 + 100)\,10^{-3} = 0.990 \text{ m}$$

or $$a_2 = 0.990 - a_1$$

$\therefore$ $$a_1\,(0.85 - a_1) = (0.99 - a_1)\,(0.718 - 0.99 + a_1)$$

$$0.85\,a_1 - a_1^2 = (0.99 - a_1)\,(a_1 - 0.272)$$

$$= 0.99\,a_1 - a_1^2 - 0.26928 + 0.272\,a_1$$

$$= 1.262\, a_1 - a_1^2 - 0.26928$$
$$0.412\, a_1 = 0.26928$$
$$a_1 = 0.6536 \text{ m}$$
$$K^2 = 0.6536\,(0.85 - 0.6536) = 0.1284 \text{ m}^2$$
$$K = 0.3583 \text{ m}$$

(b) $$I = mK^2 = 60 \times 0.1284 = 7.704 \text{ kg m}^2$$

(c) Distance of centre of gravity from small end centre, $a = a_1 - \dfrac{d_1}{2}$

$$= 0.6536 - \frac{0.08}{2} = 0.6136 \text{ m}$$

Let m_a = mass placed at small end centre
m_b = second mass
b = distance of m_b from centre of gravity

For dynamically equivalent system,

$$ab = K^2$$

$$b = \frac{K^2}{a} = \frac{0.1284}{0.6136} = 0.2092 \text{ m}$$

Now $$m_a = \frac{bm}{a+b} = \frac{0.2092 \times 60}{0.6136 + 0.2092} = 15.25 \text{ kg}$$

$$m_b = m - m_a = 60 - 15.25 = 44.75 \text{ kg}$$

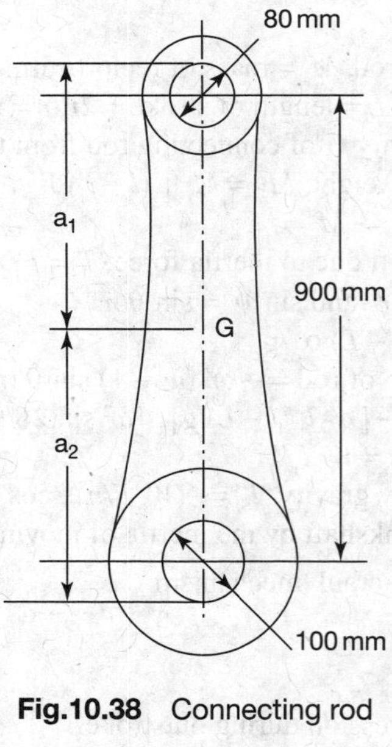

Fig.10.38 Connecting rod

Summary for Quick Revision

1 The inertia force arises due to the mass and acceleration of the reciprocating parts.

2 Piston effort is the net force applied on the piston.

3 Crank effort is the net force applied to the crank pin perpendicular to the crank which gives the required turning moment on the crankshaft.

4 The plot of torque T v's. crank angle θ is called the turning moment diagram.

5 The area under the turning moment diagram represents the work done per cycle.

6 Mean turning moment is obtained by dividing the area of turning moment diagram by the length of base.

7 The maximum ordinate of the turning moment diagram gives the maximum torque to which the crankshaft is subjected to.

8 Displacement, Velocity and Acceleration of Piston:
Displacement of the piston from top dead centre,
$x = r\,[(1 - \cos\,\theta) + \{n - (n^2 - \sin^2\theta)^{0.5}\}]$
Velocity of piston, $v_p = \omega\, r\,[\,\sin\,\theta + \sin\,2\theta/\{2\,(n^2 - \sin^2\theta\,)^{1/2}\}]$
$\approx \omega r\,[\,\sin\,\theta + \sin\,2\theta/(2n)]$
Acceleration of piston, $f_p \approx \omega^2 r\,[\,\cos\,\theta + (\cos\,2\theta)/n]$
where $n = \ell/r$

9 Angular velocity and acceleration of connecting rod
Angular velocity of connecting rod, $\omega_c = \omega\cos\,\theta/(n^2 - \sin^2\theta)^{0.5}$
Angular acceleration of connecting rod
$f_c = -\omega^2 \sin\,\theta\,[(n^2 - 1)/(n^2 - \sin^2\theta)^{1.5}]$
$\approx -\omega^2 \sin\,\theta/n$

10 Inertia forces in reciprocating engines.
(a) Slider crank chain.
W_c = weight of the connecting rod, M_r = mass of reciprocating parts,
l = length of connecting rod, L = length of stroke = $2r$, r = radius of crank,
l_1 = distance of centre of gravity G of connecting rod from the gudgeon pin
Total equivalent reciprocating weight, $M_{re} = M_r + (l - l_1)\,W_c\,/\,(gl)$
The inertia force due to M_{re}, $F_i = -M_{re}\cdot f_p$
Torque exerted on the crankshaft due to inertia force, $T_i = F_i \times OM$
where $OM = r\sin\,(\theta + \phi\,)/\cos\,\phi$ and $\sin\,\phi = \sin\,\theta/n$.
Correction couple, $T_o = W_c\,l_1\,(l - L)\,\alpha_c\,/g$
where α_c = angular acceleration of rod = $-\,\omega^2\,(n^2 - 1)\sin\,\theta/(n^2 - \sin^2\,2\theta)^{1.5} \approx -\omega^2 \sin\,\theta/n$
Torque on the crankshaft, $T_c \approx -\,[W_c\,l_1\,(l - L)/g]\,.\,[\omega^2 \sin\,2\theta/(2n^2)]$
Vertical force through crankpin = $W_c\,l_1/1$
Torque exerted on crankshaft by gravity, $T_o = -(W_c\,l_1/n)\,.\cos\,\theta$
Total torque exerted on the crankshaft by the inertia of moving parts = $T_i + T_c + T_g$

11 Equilibrium of forces in slider–crank mechanism
(a) Outstroke

$F_c = F/\cos\,\phi$, $F_n = F\tan\,\phi$

The connecting rod is in compression during outstroke.

Crank pin effort, $F_t = F_c \sin(\theta + \phi) = F \sin(\theta + \phi)/\cos\phi$

Thrust on the bearings, $F_b = F_c \cos(\theta + \phi) = F \cos(\theta + \phi)/\cos\phi$

(b) In stroke

The connecting rod is under tension now.

(c) Crank effort

Crank effort , $CE = Fr\,[\sin\theta + \sin 2\theta/\{2(n^2 - \sin^2\theta)^{0.5}\}]$

$= F \times OM$

12 Piston effort:

(a) Double acting horizontal steam engine

Force on piston due to steam pressure during outstroke, $F_p = (\pi/4)\,[D^2 p_1 - (D^2 - d^2)\,p_2]$

Inertia force due to mass of reciprocating parts, $F_i = M_r \cdot \omega^2 r\,[\cos\theta + \cos 2\theta/n]$

Piston effort during outstroke (PEO), $F = F_p - F_i$

Piston effort during instroke, PEI, $F = (\pi/4)\,[D^2 p_3 - (D^2 - d^2)\,p_4] - F_i$

(b) Double acting vertical steam engine

$PED = F_2 - F_i + M_r g$

$PEU = F_p - F_i - M_r g$

(c) Four-stroke horizontal internal combustion engine

$PEO = (\pi/4)\,D^2\,(p_1 - p_a) - F_i$

$PEI = (\pi/4)D^2\,(p_a - p_4) - F_i$

(d) Four-stroke vertical Internal Combustion Engine

$PED = (\pi/4)\,D^2\,(p_1 - p_a) - F_i + M_r g$

$PEU = (\pi/4)D^2\,(p_a - p_4) - F_i - M_r g$

13 Fluctuation of energy (E_f): It is the excess energy developed by the engine between two crank positions.

$E_f = C_e E$

where $E = 1/2 \cdot I\,\omega_m^2$, I = moment of inertia of the flywheel, and ω_m its mean angular speed.

14 Coefficient of fluctuation of energy (C_e): It is the ratio of the maximum fluctuation of energy to the indicated work done by the engine during one revolution of crank.

$C_e = (E_{max} - E_{min})/(T_m \cdot \theta)$

$\theta = 4\pi$ for steam engines and four stroke I.C. engines.

Mean torque, T_m = power developed/ω_m.

15 Coefficient of fluctuation of speed (C_s): It is defined as the ratio of the difference between the maximum and minimum angular velocities of the crankshaft to its mean angular velocity.

$C_s = (\omega_{max} - \omega_{min})/\omega_m = (N_{max} - N_{min})/N_m$

where $N_m = (N_{max} + N_{min})/2$

16 A flywheel is a device which serves as a reservoir to store energy when the supply of energy is more than the requirement, and releases energy when the requirement is more than the supply.

17 A flywheel controls the fluctuation of speed pf the prime mover during each cycle.

18 Size of flywheel:

Mass, $m = (900 \times E_f)/(\pi^2 \times K^2 \times N_m^2 \times C_s)$ kg

$E_f = 2C_s E$

Hoop stress in rim of flywheel, $\sigma_\theta = \rho v^2$
For a punching press, $(1/2) \mid (\omega^2_{max} - \omega^2_{min}) = E\,[1 - t/(4r)\,]$

19 Equivalent dynamical system
A continuous body may be replaced by a body two masses assumed to be concentrated at two points and connected rigidly together. Such a system of two masses is termed an equivalent dynamical system. The conditions to be satisfied by an equivalent dynamical system are as follows:

1. The total mass must be equal to that of the rigid body.
2. The centre of gravity must coincide with that of the rigid body.
3. The total moment of inertia about an axis through centre of gravity must be equal to that of the rigid body.

$$m_a + m_b = m$$
$$m_a \cdot a = m_b \cdot b$$
$$m_a \cdot a^2 + m_b \cdot b^2 = mK^2$$
$$K^2 = ab$$

20 In an approximate dynamical system, the distances 'a' and 'c' are fixed arbitrarily, then

$$m_a = [c/(a + c)\,]\,m$$
$$m_c = [\,a/(a + c)]\,m$$

Correction couple,

$$T_o = ma\,(\mathrm{I} - L) \cdot \alpha$$

where I = distance between two masses m_a and m_c fixed arbitrarily
L = distance between two masses m_a and m_b, which form a true dynamically equivalent system.

21 Compound Pendulum
Natural frequency, $\omega_n = [\,ga/(K^2 + a^2\,)\,]^{1/2}$ rad/s
Equivalent length of a simple pendulum, $I_e = K^2/a + a$

Multiple Choice Questions

1 The flywheel influences the
(a) variation of load demand on prime mover
(b) mean speed of the prime mover
(c) cyclic variation in speed of the prime mover
(d) mean torque developed by the prime mover.

2 If mean speed of the prime mover is increased then the coefficient of fluctuation of speed will
(a) increase (b) decrease (c) remains same (d) unpredictable.

3 The maximum fluctuation of energy of flywheel is directly proportional to
(a) coefficient of fluctuation of speed
(b) square of angular speed of flywheel
(c) moment of inertia of flywheel
(d) all of the above.

4 The acceleration of piston of a reciprocating engine is:

(a) $\omega^2 r\left[\sin\theta + \frac{\cos 2\theta}{n}\right]$ (b) $\omega^2 r\left[\cos\theta + \frac{\cos 2\theta}{n}\right]$

(c) $\omega^2 r\left[\sin\theta + \frac{\sin 2\theta}{n}\right]$ (d) $\omega^2 r\left[\cos\theta + \frac{\sin 2\theta}{n}\right]$

5 Crank pin effort in a reciprocating engine is:

(a) $\frac{F\sin(\theta+\phi)}{\cos\phi}$ (b) $\frac{F\sin(\theta+\phi)}{\sin\phi}$

(c) $\frac{F\cos(\theta+\phi)}{\cos\phi}$ (d) $\frac{F\cos(\theta+\phi)}{\sin\phi}$

where F = piston effort

6 Crank effort in a reciprocating engine is

(a) $F_r\left[\sin\theta + \frac{\sin 2\theta}{2n}\right]$ (b) $F_r\left[\cos\theta + \frac{\cos 2\theta}{2n}\right]$

(c) $F_r\left[\sin\theta + \frac{\cos 2\theta}{2n}\right]$ (d) $F_r\left[\cos\theta + \frac{\sin 2\theta}{2n}\right]$

7 Coefficient of fluctuation of speed is given by:

(a) $\frac{\omega_{max} + \omega_{min}}{\omega_m}$ (b) $\frac{\omega_{max} - \omega_{min}}{\omega_m}$

(c) $\frac{\omega_{max} - \omega_{min}}{2\omega_m}$ (d) $\frac{\omega_{max} - \omega_m}{\omega_m}$

8 Fluctuation of energy of a flywheel is

(a) $C_s E$ (b) $2C_s E$ (c) $\frac{1}{2}C_s E$ (d) $4C_s E$

Answers

1. (c) 2. (c) 3. (d) 4. (b) 5. (a) 6. (a) 7. (b) 8. (b)

Review Questions

1 Define piston effort and crank effort.

2 What do you mean by dynamically equivalent system?

3 How do you account for the inertia of the connecting rod?

4 What is a turning moment diagram? What are its advantages?

5 Define coefficient of fluctuation of energy and coefficient fluctuation of speed.

6 What is the main function of a flywheel?

7 Write the relationship between coefficient of fluctuation of speed, maximum fluctuation of energy and kinetic energy of flywheel.

8 Write the procedure for determining the turning moment diagram.

9 What is a compound pendulum?

Exercises

10.1 A horizontal steam engine running at 250 rpm has a bore of 210 mm and a stroke of 350 mm. The piston rod is 20 mm in diameter and connecting rod length is 1050 mm. The mass of the reciprocating parts is 6 kg and frictional resistance is equivalent to a force of 500 N. Determine the following when the crank is at 120° form IDC, the mean pressure is 5 kN/m^2 on the cover side and 0.1 kN/m^2 on the crank side:

(a) Thrust in the connecting rod, (b) thrust on the cylinder walls, (c) load on the bearings, and (d) turning moment on the crankshaft.

10.2 A single cylinder vertical engine has a bore of 250 mm and a stroke of 500 mm. The connecting rod is 1000 mm long. The mass of the reciprocating parts is 150 kg. The gas pressure is 0.75 MPa during expansion stroke when the crank is at 30° form TDC. The speed of the engine is 240 rpm. Determine: (a) net force acting on the piston, (b) resultant load on the gudgeon pin, (c) thrust on the cylinder walls, and (d) the speed above which, other things remaining same, the gudgeon pin load would be reversed in direction.

10.3 In a reciprocating engine, the length of stroke is 250 mm and connecting rod is 500 mm long between centres. Determine: (a) the angular position of the crank, (b) velocity and acceleration of the piston, and (c) angular velocity of the connecting rod if the engine speed is 240 rpm.

10.4 The following data refer to a horizontal reciprocating engine:

Mass of reciprocating parts = 125 kg, crank length = 100 mm, length of connecting rod between centres = 500 mm, engine speed = 600 rpm, mass of connecting rod = 100 kg, distance of mass centre of connecting rod from the small end centre = 200 mm, radius of gyration about mass centre axis = 160 mm.

Determine the magnitude and direction of the inertia torque on the crankshaft when the crank has turned 30° from IDC.

10.5 The following data refer to a connecting rod of a reciprocating engine:

Mass = 60 kg; Distance between bearing centres = 900 mm; Diameter of small end bearing = 80 mm; Diameter of big end bearing = 120 mm; Time of oscillation when the connecting rod is suspended from small end = 1.85 s; time of oscillation when the connecting rod if suspended from big end = 1.70 *s*.

Determine:

(a) The radius of gyration of the rod about an axis passing through the centre of gravity and perpendicular to the plane of oscillation, (b) the moment of inertia of the rod about the same axis, and (c) the dynamically equivalent system for the connecting rod, consisting of two masses, one of which is situated at the small end centre.

10.6 A connecting rod of length 400 mm between centres has a mass of 4.5 kg. The centre of gravity is 260 mm from the small end and its radius of gyration about an axis through the centre of gravity perpendicular to the plane of rotation is 125 mm. Determine a dynamically equivalent system having one mass at the centre of small end and the other at a point somewhere in between the centre of big end and the centre of gravity.

10.7 The connecting rod of a vertical reciprocating engine is 2 m long between centres and its mass is 250 kg. The mass centre is 750 mm from the big end bearing. When suspended as a pendulum from the gudgeon pin axis, it makes 10 complete oscillations in 25 seconds. Calculate the radius of gyration of the rod about an axis through its mass centre. The crank is 400 mm long and rotates at 250 rpm.
When the crank has turned through 35° from the TDC and the piston is moving downwards, find the inertia torque exerted on the crankshaft.

10.8 A single cylinder vertical engine has a bore of 300 mm, a stroke of 350 mm and a connecting rod of length 700 mm. The weight of the reciprocating parts is 125 kg. When the piston is at quarter-stroke from TDC and is moving downwards, the net pressure on it is 0.55 MPa. If the speed of the engine is 240 rpm, calculate the turning moment on the crankshaft.

10.9 The turning moment curve for one revolution of a multicylinder engine above and below the line of mean resisting torque are given by:
–0.32, +4.06, –3.71, +3.29, –3.16, +3.32, –3.74, +3.71, and –3.45 sq. cm.
The vertical and horizontal scales are: 1 cm = 60000 kg cm and 1 cm = 24°, respectively. The fluctuation of speed is limited to ±1.5% of mean speed, which is 250 rpm. The hoop stress in the rim material is limited to 5.5 N/mm^2. Neglecting effect of boss and arms, determine the suitable diameter and cross-section of flywheel rim. The density of rim material is 7200 kg/m^3. Assume width of rim equal to four times its thickness.

10.10 A constant torque 3 kW motor drives a punching machine. The mass of the moving parts including the flywheel is 130 kg at 750 mm radius. One punching operation absorbs 1 kg of energy and takes 1 *s*. Speed of the flywheel is 240 rpm before punching: Determine (a) the number of punches per hour and (b) reduction in speed after the punching operation.

11

STATIC AND DYNAMIC FORCE ANALYSIS

Chapter Outline

11.1 INTRODUCTION

Forces in mechanisms arise form various sources, e.g. forces of gravity, forces of assembly, forces from applied loads, forces from energy transmission, frictional forces, spring forces, impact forces, and forces due to change of temperature. All these forces must be considered in the final design of a machine for its successful operation.

In the design of mechanisms, the following forces are generally considered:

1. Applied forces
2. Inertia forces, and
3. Frictional forces.

The applied forces act from outside on the mechanism. The inertia forces arise due to the mass of the links of the mechanism and their acceleration. Frictional forces is the outcome of friction in the joints. A pair of action and reaction forces acting on a body are called constraint forces.

11.2 STATIC FORCE ANALYSIS

In the analysis of static forces, the inertia forces are not taken into account. Often the gravity forces are also small, and are neglected as compared to other forces.

11.2.1 Static Equilibrium

A body is in static equilibrium if it remains in its state of rest or of motion. The conditions for static equilibrium are the following:

1. The vector sum of all the forces acting on the body is zero.
2. The vector sum of all the moments about any arbitrary point is zero.

Mathematically, this can be stated as:

$$\Sigma F = 0 \tag{11.1}$$

$$\Sigma M = 0 \tag{11.2}$$

In a planar mechanism, forces can be described by two-dimensional vectors. Thus

$$\Sigma F_x = 0 \tag{11.3}$$

$$\Sigma F_y = 0 \tag{11.4}$$

$$\Sigma M_x = 0 \tag{11.5}$$

11.2.2 Equilibrium of Members

(a) *Two-force member*

A member under the action of two forces, as shown in Fig.11.1, shall be in equilibrium, if:

1. The forces are of the same magnitude,
2. The forces are collinear i.e. act along the same line, and
3. The forces act in opposite directions.

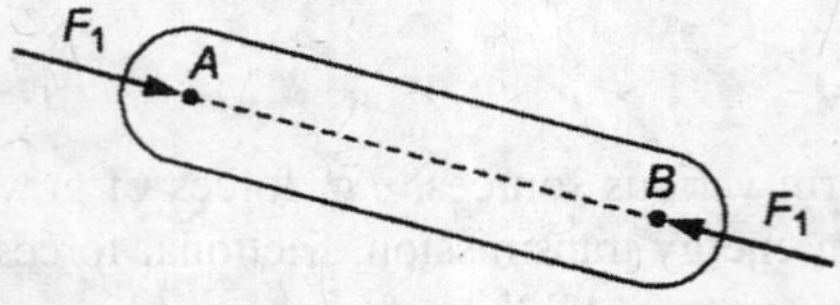

Fig.11.1 Two-force system

(b) *Three-force member*

A member under the action of three forces shall be in equilibrium, if:

1. The resultant of the forces is zero, and
2. The forces are concurrent, i.e. the line of action of the forces intersect at the same point.

Fig.11.2(a) shows a member acted upon by three forces F_1, F_2, and F_3 such that the lines of action of these forces intersect at point O and their resultant is zero. The resultant of three forces shall be zero if the triangle of forces is closed, as shown in Fig.11.2(b).

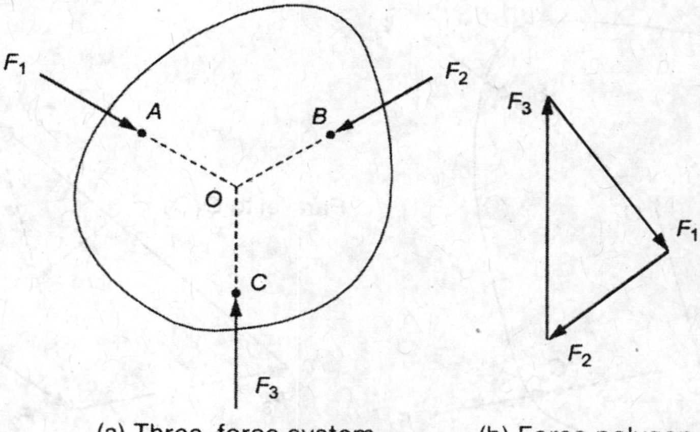

(a) Three-force system (b) Force polygon

Fig.11.2 Three-force system

(c) *Two-forces and torque*

A member under the action of two forces and an applied torque shall be in equilibrium, if:

1. the forces are equal in magnitude, parallel and opposite in direction, and
2. the forces form a couple, which is equal and opposite to the applied torque.

Fig.11.3 shows a member acted upon by two equal and opposite forces F_1 and F_2 and an applied torque T. For equilibrium, we have

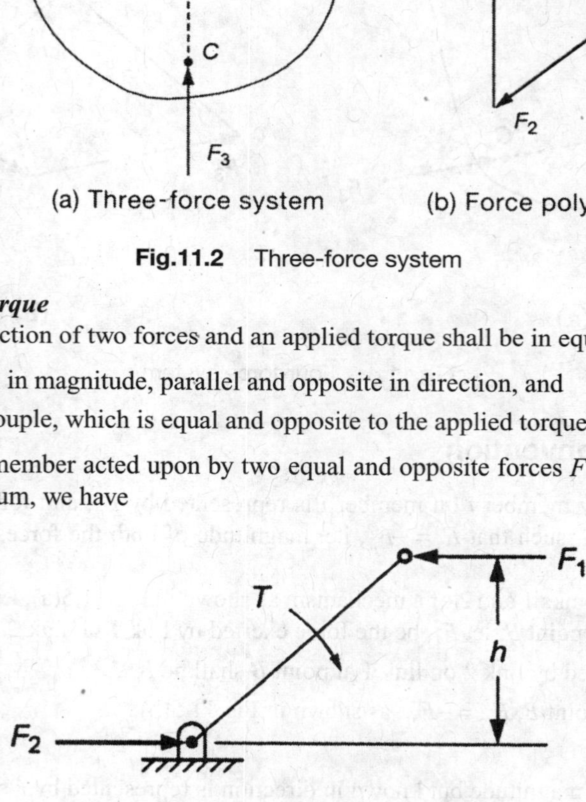

Fig.11.3 Two-forces and torque system

$$T = F_1 \times h = F_2 \times h \tag{11.6}$$

We may remember that a couple can be balanced by a couple only of opposite sense.

(d) *Four-force members*

A four-force member is completely solvable if one force is known completely in magnitude and direction along with lines of action of the other three forces. The conditions of equilibrium as stated above are sufficient.

Consider a system of four non-parallel forces as shown in Fig.11.4(a). Let O_1 be the point of intersection of the lines of action of F_1 and F_2. Similarly, O_2 is the point of intersection of the lines of action of the forces F_3 and F_4. Join O_1O_2. The resultant of F_1 and F_2 and that of F_3 and F_4 is parallel to O_1O_2. The force polygon for the four forces can be drawn as shown in Fig.11.4(b) and the forces F_3 and F_4 can be known completely.

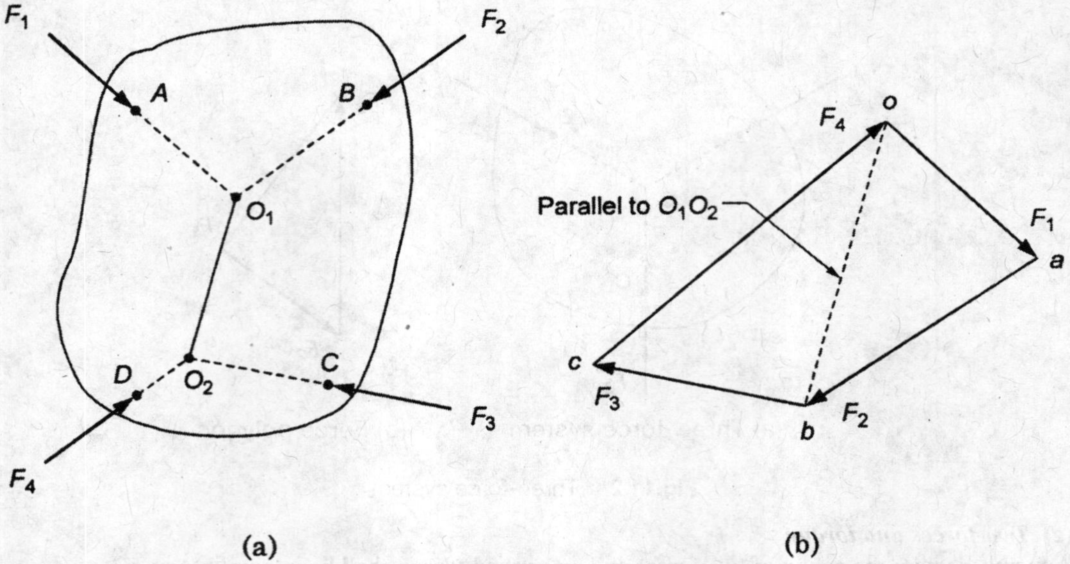

Fig.11.4 Four-force system

11.2.3 Force Convention

1. The force exerted by member *i* on member *j* is represented by F_{ij}, and force exerted by member *j* on member *i* by F_{ji}, such that $F_{ij} = -F_{ji}$, i.e. magnitude of both the forces is same but direction is opposite.

   Consider two links 1 and 2 of a mechanism as shown in Fig.11.5(a). Link 1 rotates clockwise about point *A*. At point *B*, let F_{12} be the force exerted by link 1 on link 2.

   Then the force exerted by link 2 on link 1 at point *B* shall be F_{21}.

   For equilibrium of point *B*, $F_{12} = -F_{21}$, as shown in Fig.11.5(b).

   In general, $F_{ij} = -F_{ji}$.
2. A force unknown in magnitude but known in direction is represented by a solid straight line without arrowhead, e.g. ____________.
3. A force unknown in magnitude and direction is represented by a wavy line, e.g. ∿∿∿∿∿.

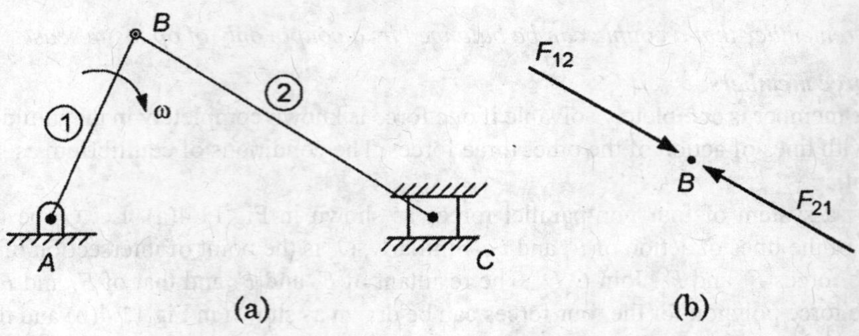

Fig.11.5 Force convention

11.2.4 Free Body Diagrams

A free body diagram is the diagram of a link isolated from the mechanism showing all the active and reactive forces acting on the link in order to determine the nature of forces acting on the link. Refer to Section 11.2.6 to understand the concept of free-body diagrams.

11.2.5 Principle of Superposition

This principle states that if number of forces act on a system, the net effect is equal to the sum of the individual effects of the forces taken one at a time. In a linear system, the output force is directly proportional to the input force.

11.2.6 Static Force Analysis of Four-Bar Mechanism

(a) *One known force*

Consider a four-bar mechanism subjected to a force F applied to link 4, as shown in Fig.11.6(a). The free body diagrams of the forces acting on the various members are shown in Figs.11.6(b) to (d):

1. Member 2 is subjected to two forces F_{12}, F_{32} and a torque T_2 (Fig.11.6(b)).
2. Member 3 is subjected to two forces F_{23} and F_{43} (Fig.11.6(c))
3. Member 4 is subjected to three forces F, F_{34} and F_{14} (Fig.11.6(d))

 (i) We observe that member 3 is a two-force member. For its equilibrium, F_{23} and F_{43} must act along BC. Their magnitudes are not known at this stage.

 (ii) Draw the force polygon for member 4, as shown in Fig.11.6(e). At point C draw a line parallel to BC to represent F_{34} to intersect the line of action of F at G. Then the line of action of F_{14} shall also pass through G. Now draw the triangle of forces as shown in Fig.11.6(f) to know the forces F_{34} and F_{14} completely in magnitude and direction. From the triangle of forces, we have

$$F_{34} = -F_{43} = F_{23} = -F_{32}$$

 (iii) Member 2 shall be in equilibrium, as shown in Fig.11.6(g), if F_{12} is equal, parallel and opposite to F_{32} and,

$$T = -F_{32} \times h$$

Input torque, $$T_2 = -T$$

(b) *Two known forces*

Consider a four-bar mechanism as shown in Fig.11.7(a), subjected to two forces, P and Q. A moment T_2 must be applied to link 2 to maintain equilibrium. The free body diagram of the various links is shown in Fig.11.7(b). The unknowns for the various links are: five (magnitude + direction of F_{12} and F_{32} and magnitude of T_2) for link 2, four (magnitude + direction of F_{23} and F_{43}) for link 3, and four (magnitude + direction of F_{34} and F_{14}) for link 4. Therefore, these links cannot be solved by the equilibrium equations. If we consider links 3 and 4 together then there are six unknowns, because $F_{ij} = F_{ji}$. Since there are six equations of equilibrium, three for each link, we can obtain a solution. The forces on links 3 and 4 are shown in Fig.11.7(c). The force F_{34} is broken into components F_{34}^n and F_{34}^t, which are parallel and perpendicular respectively, to O_4C. The magnitude of F_{34}^t is found by taking moments about O_4, i.e.

$$F_{34}^t = \frac{Pa}{O_4C}$$

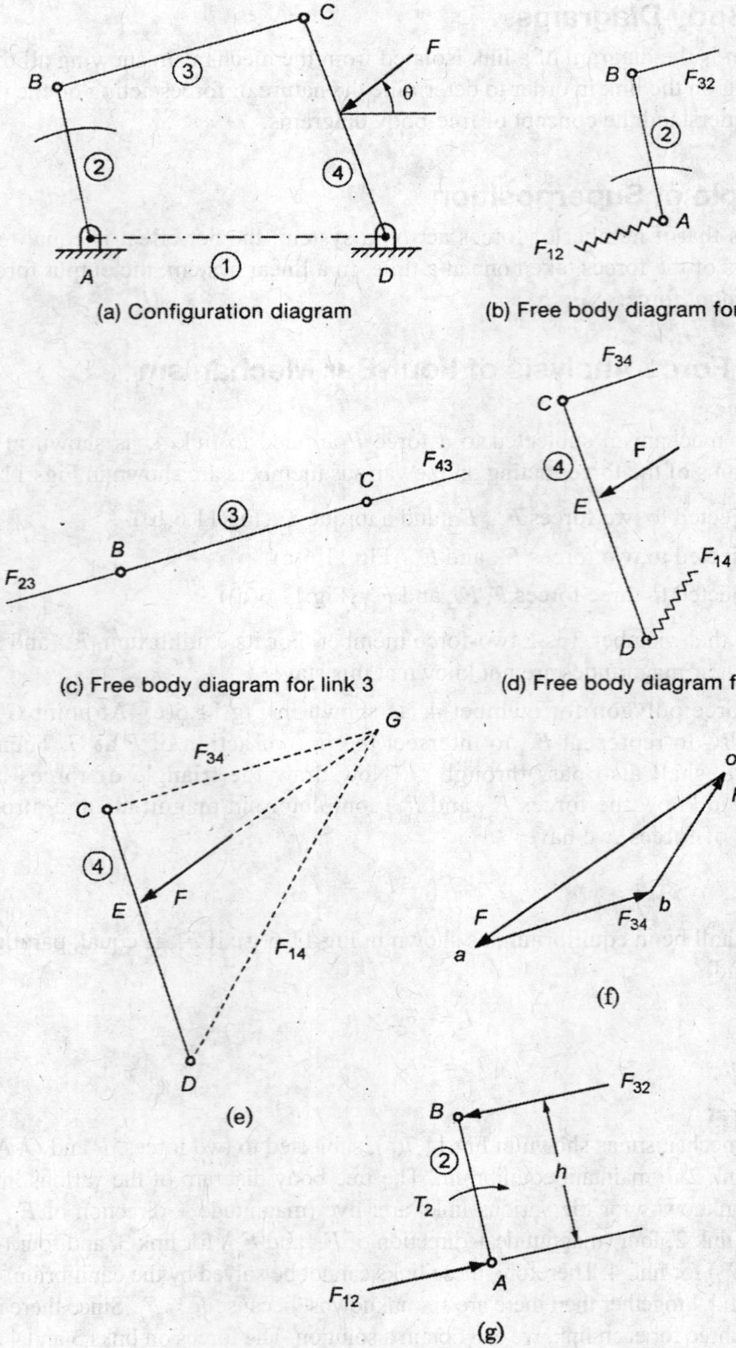

Fig.11.6 Static force analysis of four-bar mechanism: one known force

On link 3 the reactions at C are equal and opposite to those at C on link 4. The magnitude of $F_{43}^t = -F_{34}^t$ On link 3, there are three unknowns: magnitude and direction of F_{23} and magnitude of F_{34}^n. The magnitude of F_{43}^n can be found by taking moments about point B:

$$Qb - F_{43}^t\, d + F_{43}^n\, e = 0$$

or

$$F_{43}^n = \frac{F_{43}^t d - Qb}{e}$$

Next we draw the force polygon for link 3, as shown in Fig.11.7(d), to obtain the magnitude and direction of F_{23}. In Fig.11.7(e), $F_{32} = F_{23}$. Then $F_{12} = F_{32}$. Taking moments about O_2, we obtain T_2, the torque which the shaft at O_2 exerts on link 2.

$$T_2 = F_{32}\, h$$

F_{14} is obtained from the force polygon for bodies 2, 3 and 4, taken as a whole system as shown in Fig.11.7(f).

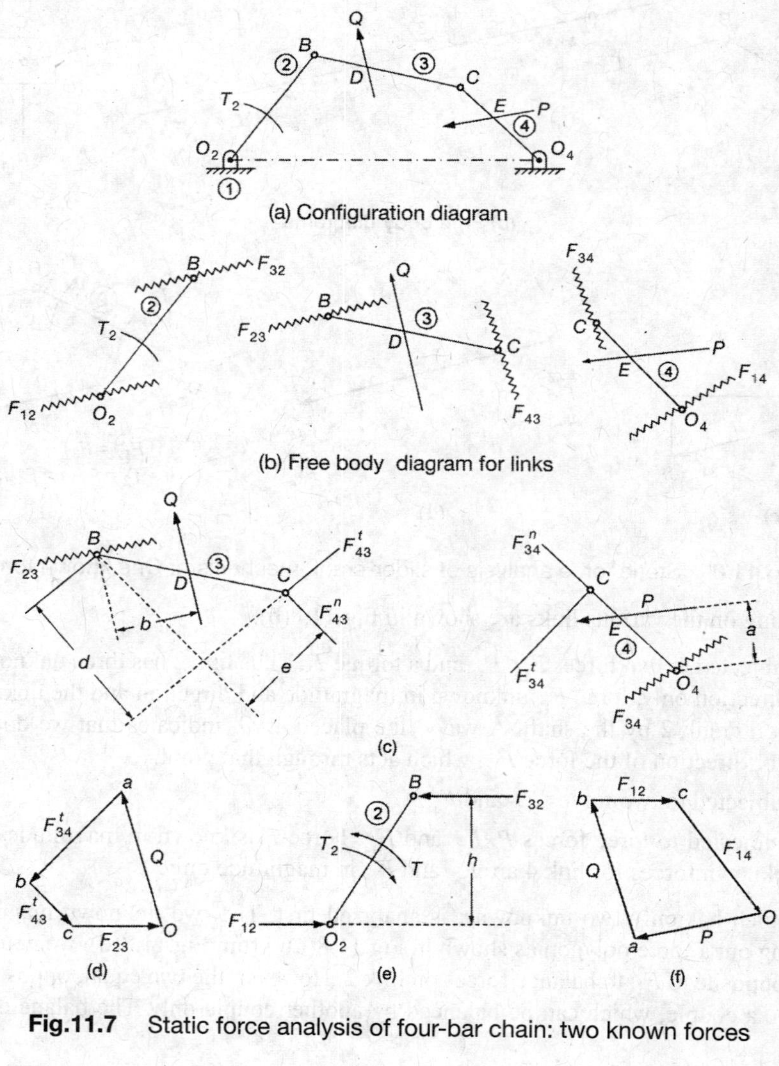

Fig.11.7 Static force analysis of four-bar chain: two known forces

11.2.7 Static Force Analysis of Slider–Crank Mechanism

(a) *One known force*

Consider the slider crank mechanism with one known force P only due to the gas force on the piston, as shown in Fig.11.8(a). The system is kept in equilibrium by applying a couple T_2 to crank link 2 through the shaft at O_2. It is required to find the forces in all the links and the couple applied to link 2.

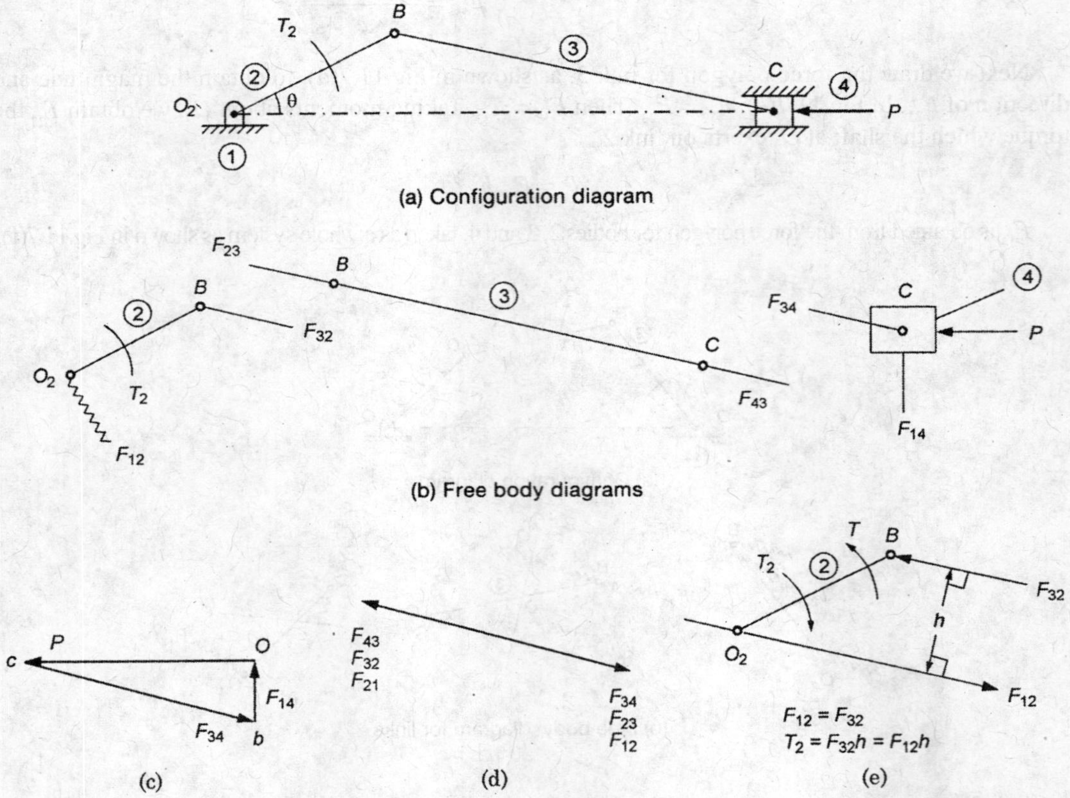

Fig.11.8 Static force analysis of slider-crank mechanism: One known force

The forces acting on the various links are shown in Fig.11.8(b).

1. Link 2 is subjected to two forces F_{12}, F_{32} and a torque T_2. Thus link 2 has three unknowns: force F_{32} known in direction only, force F_{12} unknown in magnitude and direction and the unknown moment T_2 exerted on crank 2 by the shaft. A wavy line placed at O_2 indicates that we do not know the magnitude or direction of the force F_{12}, which acts through that point.
2. Link 3 is subjected to two forces F_{23} and F_{43}.
3. Link 4 is subjected to three forces P, F_{34} and F_{14}. Force F is known in magnitude and direction. The two unknown forces for link 4 are F_{34} and F_{14} in magnitude only.

Link 4, which has only two unknowns, is analyzed first. The two unknown magnitudes can be found by laying out a force polygon as shown in Fig.11.8(c). From Fig.11.8(d), we note that F_{12} must be equal and opposite to F_{32} to balance forces on link 2. However, the two equal, opposite and parallel forces produce a couple, which can be balanced by another couple only. The balancing couple T_2 is equal to $F_{32} \times h$, where h is the perpendicular distance between F_{32} and F_{12}. It is clockwise and is the torque, which the shaft exerts on the crank 2. From the triangle of forces, we have

$$F_{34} = -F_{43} = F_{23} = -F_{23} = cb$$
$$F_{14} = bo$$

Member 2 shall be in equilibrium, as shows in Fig.11.8(e), if F_{12} is equal, paraller and opposite to F_{23}.

$$T = -F_{32} \times h$$

Input torque, $$T_2 = -T$$

(b) *Two known forces*

Figure 11.9(a) shows a slider crank mechanism with two known forces P and Q. A force P is applied to the piston due to gas pressure and force Q is applied to link 3. It is required to determine the forces in the links and the torque T_2. The free body diagrams are shown in Fig.11.9(b).

Link 3: Link 3 is a three-force member, F_{23}, F_{43}, and Q. Let F_{43} be broken into its normal and tangential components F_{43}^n and F_{43}^t, respectively. The normal component is along the link 3 and tangential component is perpendicular to the link 3. Now consider the equilibrium of link 3 by taking moments about point B, as shown in Fig.11.9(c):

$$F_{43}^t \times CB = Q \sin \alpha \times DB$$

or $$F_{43}^t = Q \sin \theta \times \left(\frac{DB}{CB} \right)$$

Then $$F_{43}^t = -F_{43}^t$$

Link 4: The forces acting on link 4 are shown in Fig.11.9(d). F_{43}^t and P are known completely. Their resultant is found as shown in Fig.11.9(e). F_{34}^n is perpendicular to F_{34}^t and F_{14} is perpendicular to the path of the slider link 4 and passes through O. $F_{34} = -F_{43}$ can be determined from Fig.11.9(e).

Again consider the equilibrium of link 3. There are three forces F_{43} known fully, Q known fully and F_{23}, as shown in Fig.11.9(f). By polygon of forces, the direction and magnitude of F_{23} is determined as shown in Fig.11.9(g).

Link 2: $F_{23} = -F_{32} = F_{21} = -F_{12}$. With the help of Fig.11.9(h), the couple T_2 is given by,

$$T_2 = F_{32} \times h \text{ (cw)}$$

Example 11.1

A four-bar mechanism shown in Fig.11.10(a) is acted upon by a force $P = 100\ \angle 120°$ N on link CD. The dimensions of the various links are:

$$AB = 40 \text{ mm}, BC = 60 \text{ mm}, CD = 50 \text{ mm}, AD = 30 \text{ mm}, DE = 20 \text{ mm}$$

Determine the input torque on link AB for the static equilibrium of the mechanism.

■ Solution

Draw the configuration diagram to a scale of 1 cm = 10 mm as shown in Fig.11.10(a). The forces acting on the various members are as follows:

1. Member 2 is subjected to two forces F_{12}, F_{32}, and a torque T_2 (Fig.11.10(b)).
2. Member 3 is subjected to two forces F_{23} and F_{43} (Fig.11.10(c)).
3. Member 4 is subjected to three forces F, F_{34}, and F_{14} (Fig.11.10(d)).

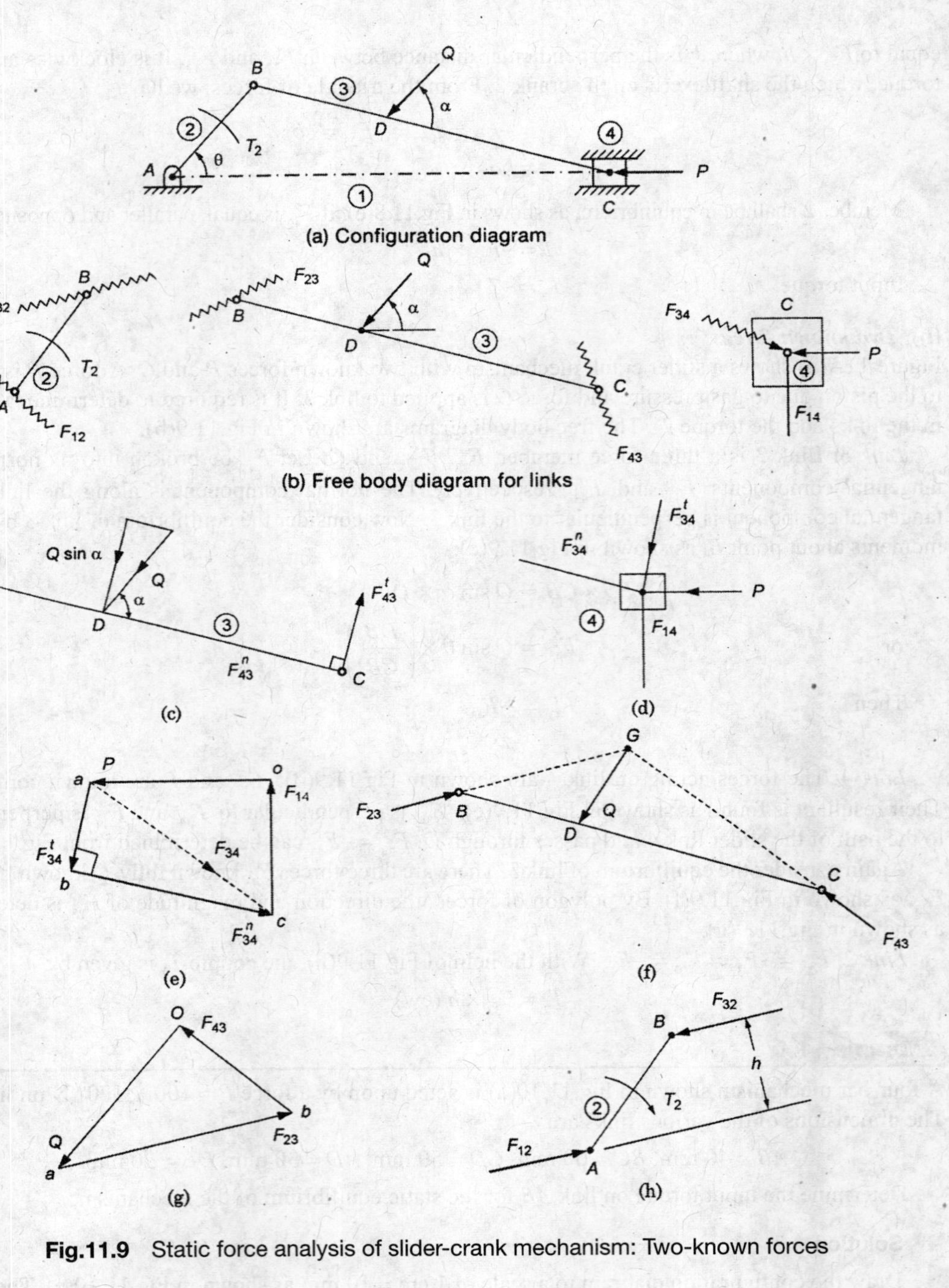

Fig.11.9 Static force analysis of slider-crank mechanism: Two-known forces

(i) We observe that member 3 is a two-force member, as shown in Fig.11.10(c). For its equilibrium, F_{23} and F_{43} must act along BC. Their magnitudes and direction are not known at this stage.

(ii) Draw free body diagram for member 4, as shown in Fig.11.10(d). At point C, draw a line parallel to BC to intersect the line of action of P at G. Then the line of action of F_{14} shall also pass through G. Now draw the triangle of forces to a scale of 1 cm = 20 N, as shown in Fig.11.10(e) to know the forces F_{34} and F_{14} completely in magnitude and direction. From the triangle of forces, we have

$$F_{34} = -F_{43} = F_{23} = -F_{32} = ab = 2.4 \text{ cm} = 28 \text{ N}$$
$$F_{14} = bo = 4.9 \text{ cm} = 98 \text{ N}$$

(iii) Member 2 shall be in equilibrium, as shown in Fig.11.10(b), if F_{12} is equal, parallel and opposite to F_{32}. By measurement, $h = 3.9$ cm = 39 mm.

$$T = -F_{32} \times h = -28 \times 39 = -1092 \text{ N mm (ccw)}$$

Input torque, $T_2 = -T = 1092$ N mm (cw)

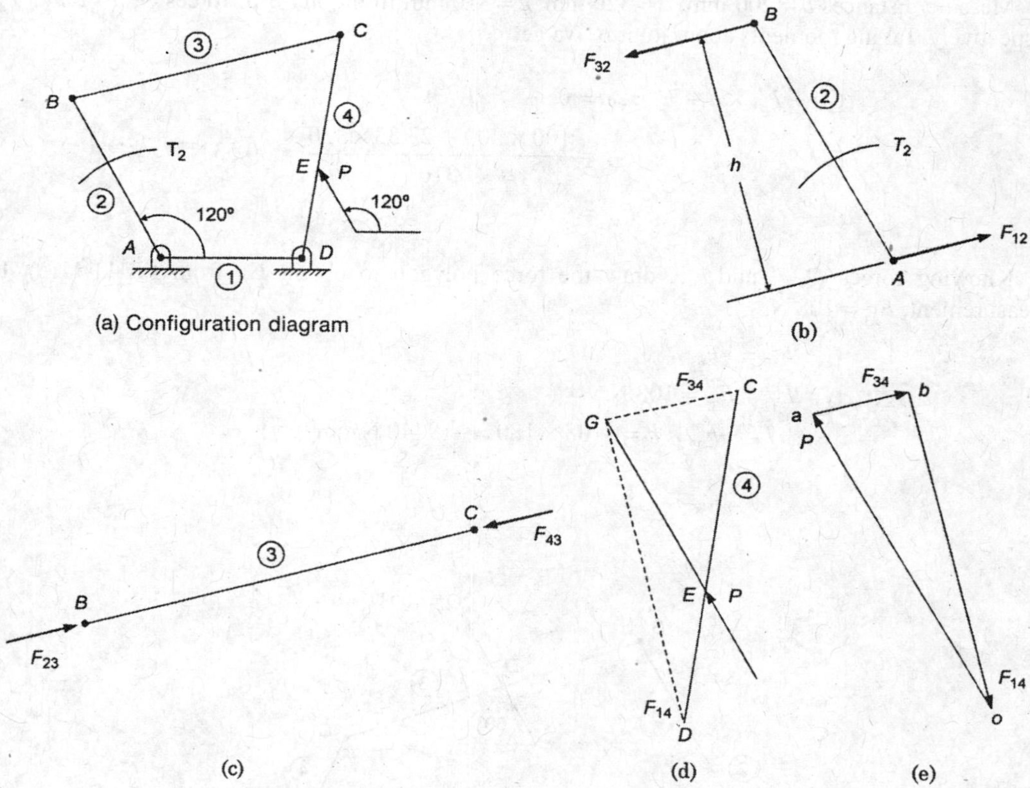

Fig.11.10 Four-bar mechanism static force analysis

Example 11.2

The links 3 and 4 of a four-bar mechanism are subjected to forces of $Q = 100 \angle 60°$ N and $P = 50 \angle 45°$ N. The dimensions of various links are:

$O_2O_4 = 800$ mm, $O_2B = 500$ mm, $BC = 450$ mm, $O_4C = 300$ mm, $BD = 200$ mm, $O_4E = 150$ mm.

Calculate the shaft torque T_2 on the link 2 for static equilibrium of the mechanism. Also find the forces in the joints.

■ Solution

The mechanism has been drawn in Fig.11.11(a) and forces in various links have been shown in Fig.11.11(b).

Let F^t_{34} and F^n_{34} be the forces at joint C on link 4, perpendicular and parallel to the link O_4C. Draw a line at O_4 parallel to force $P = 50$ N $\angle 60°$. The perpendicular distance between these two lines is 'a' = 140 mm. Taking moments about O_4, we get

$$F^t_{34} = \frac{Pa}{O_4C} = \frac{50 \times 140}{300} = 23.33 \text{ N}$$

$$F^t_{43} = -F^t_{34} = -23.33 \text{ N}$$

Measure distances $b = 200$ mm, $d = 320$ mm, $e = 310$ mm from joint B of forces Q, F^t_{43}, and F^n_{43}, respectively. Taking moments about joint B, we get

$$Qb + F^t_{43} \times d - F^n_{43} \times e = 0$$

$$F^n_{43} = \frac{100 \times 200 + 23.33 \times 320}{310} = 30.53 \text{ N}$$

$$F^n_{34} = -F^n_{43}$$

Knowing forces Q, F^t_{43} and F^n_{43}, draw the force polygon to obtain F_{23} from Fig.11.11(c). By Measurement, $F_{23} = 108$ N.

$$F_{32} = -F_{23}$$

$$F_{12} = F_{32} = 108 \text{ N}$$

$$T = -F_{32} \times h = -108 \times 180 = -19440 \text{ N mm (cw)}$$

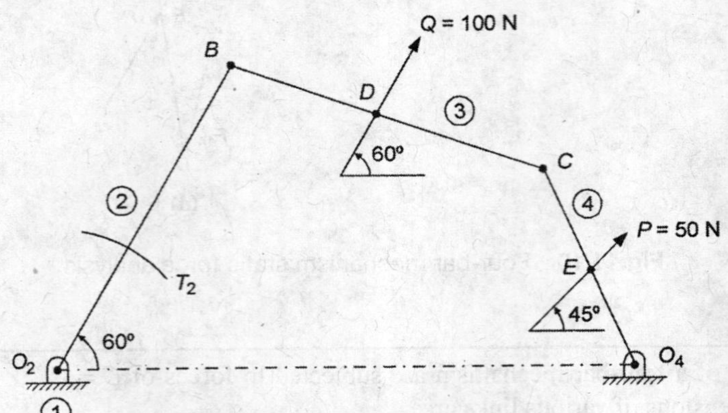

(a) Configuration diagram

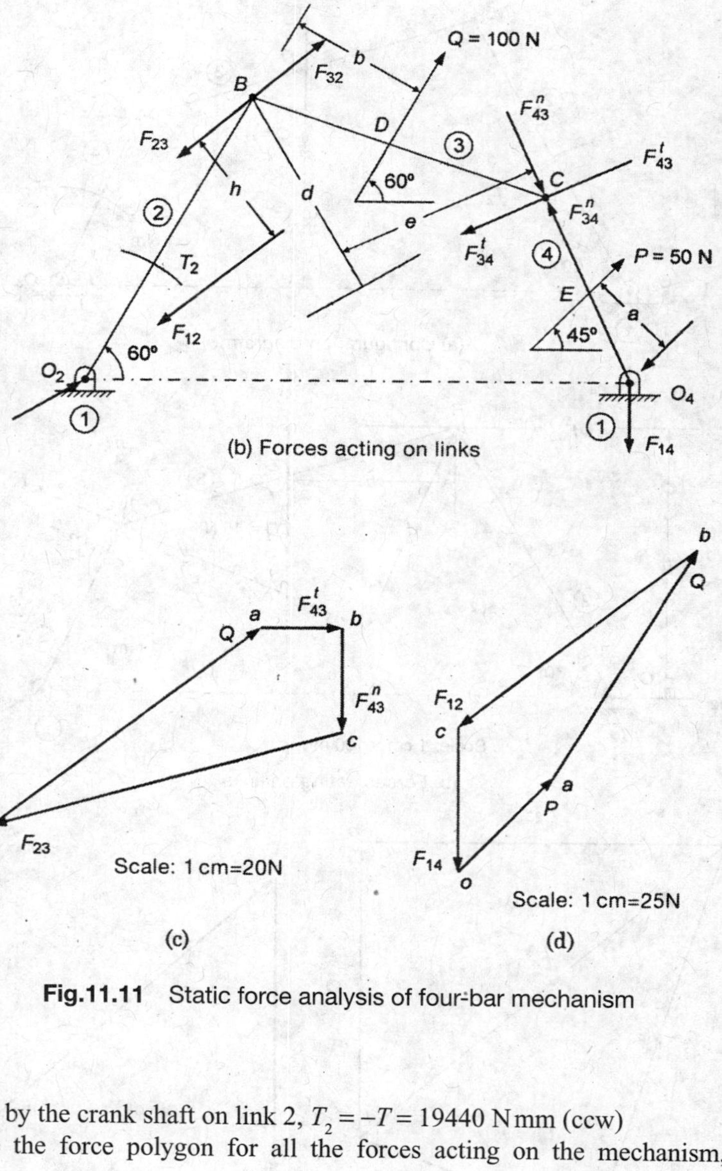

Fig.11.11 Static force analysis of four-bar mechanism

Torque exerted by the crank shaft on link 2, $T_2 = -T = 19440$ N mm (ccw)

Now complete the force polygon for all the forces acting on the mechanism, as shown in Fig.11.11(d).

$$F_{14} = 52.5 \text{ N}$$

Example 11.3

Link O_4C of a four-bar mechanism is subjected to a torque $T_4 = 1$ N m (ccw). The link BC is subjected to a force $Q = 45\angle 90°$ N downwards. Determine the torque T_2 on link O_2B and the reactions at O_2 and O_4. The lengths of the various links are as follows:

$$O_2O_4 = 90 \text{ mm}, O_2B = 50 \text{ mm}, BC = 55 \text{ mm}, O_4C = 30 \text{ mm}, BD = BC = 27.5 \text{ m}.$$

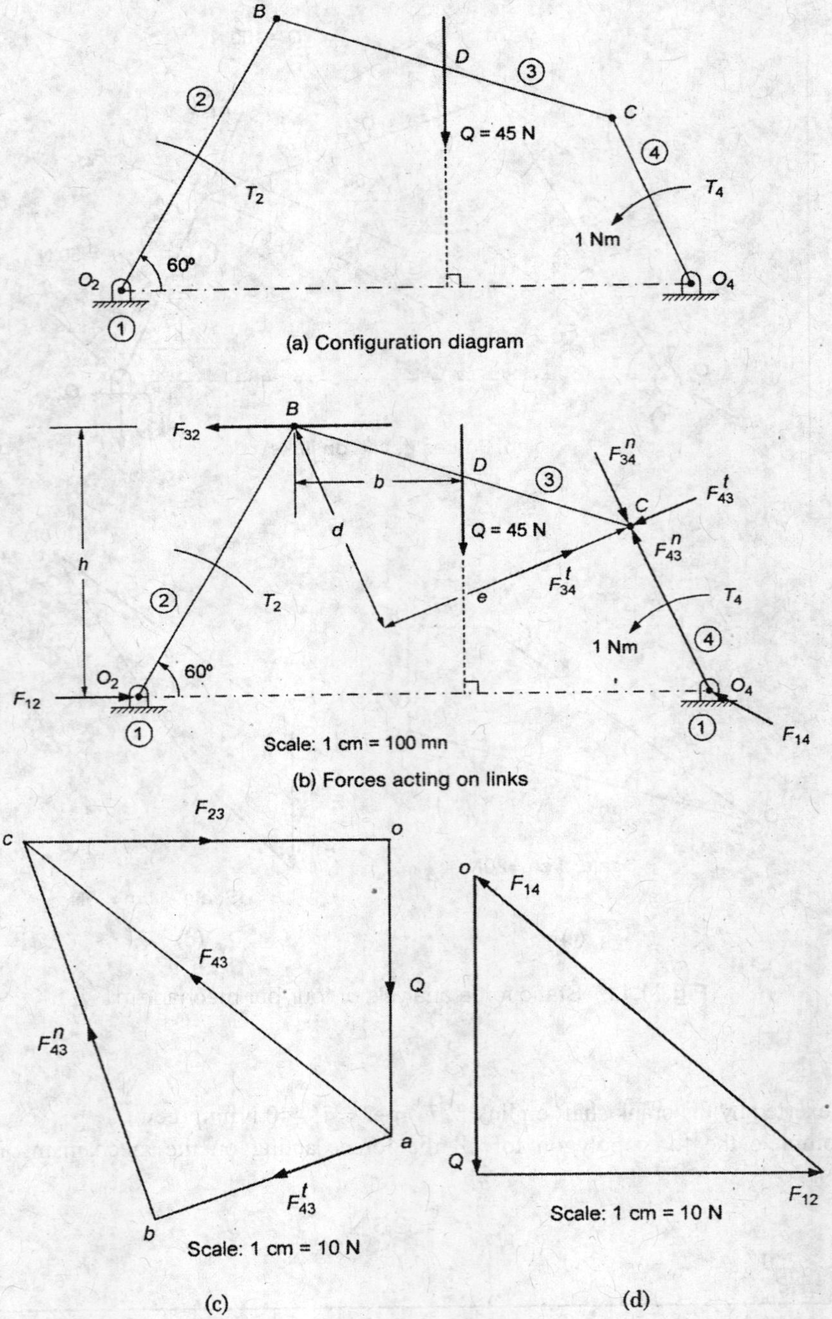

Fig.11.12 Static force analysis of four-bar mechanism

■ Solution

The mechanism has been drawn in Fig.11.12(a) and the forces have been shown in various links in Fig.11.12(b). Taking moments about O_4, we have

$$F_{34}^t \times O_4C = T_4 = 1$$

$$F_{34}^t = \frac{1}{0.03} = 33.3 \text{ N}$$

$$F_{34}^t = F_{43}^n$$

Measure perpendicular distances $b = 26$ mm, $d = 36$ mm, and $e = 40$ mm, as shown in Fig.11.12(b).

Taking moments about joint B, we have

$$Qb - F_{43}^n \times e = F_{43}^t \times d = 0$$

$$F_{43}^n = \frac{45 \times 26 + 33.3 \times 36}{40}$$

$$= 59.22 \text{ N}$$

$$F_{34}^n = -F_{43}^n$$

Draw the force polygon for link BC, as shown in Fig.11.12(c) $F_{23} = 53\text{ N} = -F_{32}$. Also $F_{12} = F_{32}$. Now draw the force polygon for Q, F_{12} and F_{14}, as shown in Fig.11.12(d), $F_{14} = 63$ N and $h = 42$ mm. Then

$$T = F_{32} \times h$$

$$= -53 \times 0.042 = -2.226 \text{ Nm (ccw)}$$

Torque on link O_2B, $T_2 = -T = 2.226$ Nm (cw)

Example 11.4

For the four-bar mechanism shown in Fig.11.13(a), T_3 on link BC is 30 Nm clockwise and T_4 on CD is 20 Nm counter-clockwise. Find the torque exerted by crankshaft on AB. $AD = 800$ mm, $AB = 300$ mm, $BC = 700$ mm, $CD = 400$ mm.

■ Solution

The forces in various links have been shown in Fig.11.13(b).
Taking moments about joint D, we get

$$F_{34}^t \times CD = T_4 = 20$$

$$F_{34}^t = \frac{20}{0.4} = 50 \text{ N}$$

$$F_{43}^t = -F_{34}^t$$

$$a = 670 \text{ mm}, \; b = 200 \text{ mm}$$

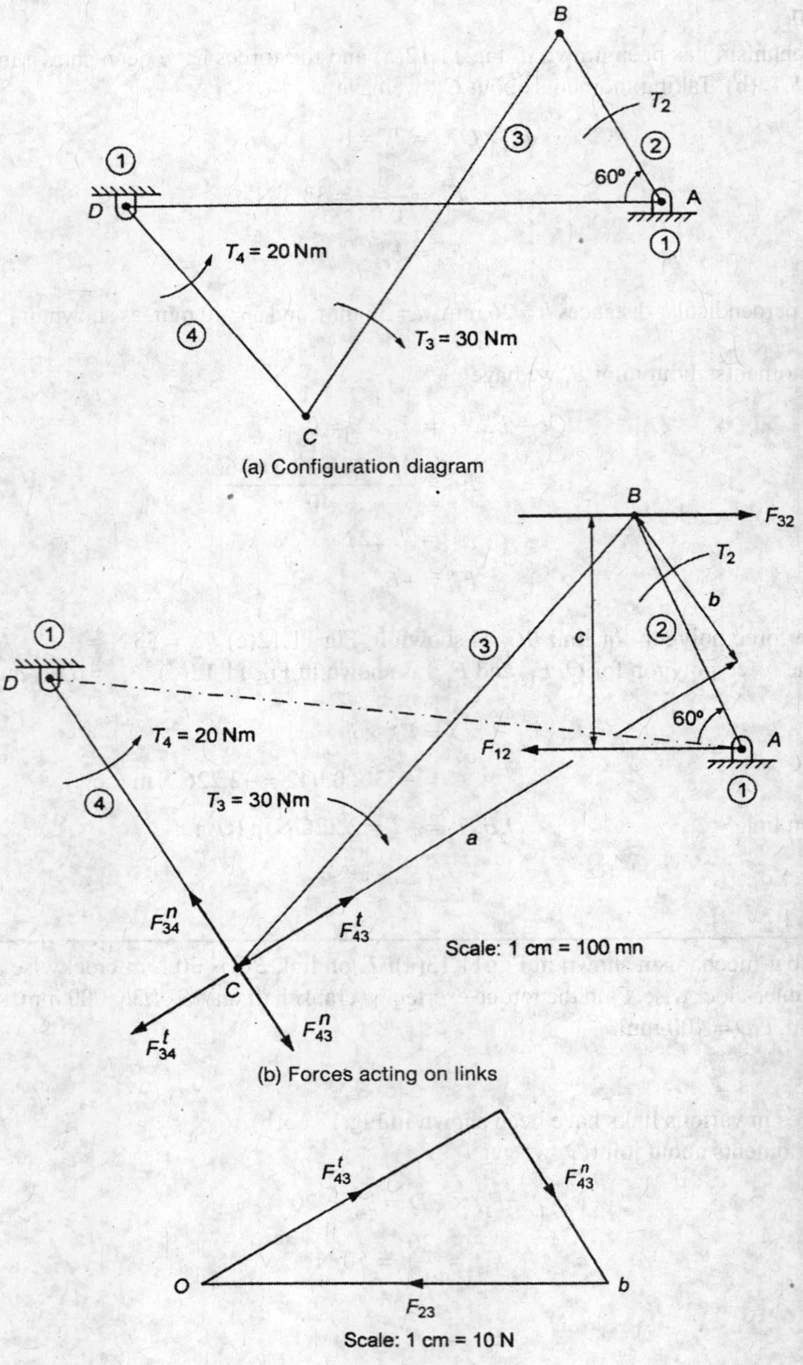

(a) Configuration diagram

(b) Forces acting on links

(c)

Fig.11.13 Static force analysis of four-bar mechanism

Taking moments about joint B, we get

$$F_{43}^{t} \times b + F_{43}^{n} \times a = T_3$$

$$50 \times 0.2 + F_{43}^{n} \times 0.67 = 30$$

$$F_{43}^{n} = \frac{20}{0.67} = 29.85 \text{ N}$$

$$F_{34}^{n} = -F_{43}^{n}$$

Draw force polygon for link BC, as shown in Fig.11.13(c). $F_{23} = 59$ N. $F_{32} = -F_{23}$, $F_{12} = F_{32}$. $c = 280$ mm.

$$T = F_{32} \times c = -59 \times 0.28 = -16.52 \text{ N m (cw)}$$

Torque exerted by crankshaft on crank, $T_2 = -T = 16.52$ N m (ccw)

Example 11.5

In the slider crank mechanism shown in Fig.11.14(a), the value of force applied to slider 4 is 2 kN. The dimensions of the various links are:

$$AB = 80 \text{ mm}, BC = 240 \text{ mm}, \theta = 60°$$

Determine the forces on various links and the driving torque T_2.

Solution

Draw the configuration diagram to a scale of 1 cm = 40 mm, as shown in Fig.11.14(a). The free body diagrams for the links 2, 3 and 4 are shown in Fig.11.14(b).

On link 4 there are three forces: $F = 2$ kN to the right; F_{14} in unknown in magnitude, perpendicular to F but direction is unknown; F_{34} whose magnitude and direction is unknown but acts along BC. To determine the forces and couple, the following procedure may be adopted:

1. Draw the force polygon for link 4, as shown in Fig.11.14(c).
   (a) Draw a line $oa = 5$ cm and parallel to F to a scale of 1 cm = 400 N.
   (b) From 'a' draw a line perpendicular to oa representing F_{14}.
   (c) From 'o' draw a line parallel to BC representing F_{34} to intersect the pervious line at b.
   (d) They by measurement, we have

$$F_{14} = ab = 1.5 \text{ cm} = 600 \text{ N, and } F_{34} = ob = 5.3 \text{ cm} = 2120 \text{ N}$$

2. Now $F_{34} = -F_{43} = F_{32} = -F_{23} = F_{21} = -F_{12}$
3. The forces on link 2 are shown in Fig.11.14(d). F_{12} is parallel to F_{32}, By measurement, $h = 2.98$ cm = 79.2 mm

   Couple $$T = F_{32} \times h = -2120 \times 79.2 = -167.9 \text{ N m (ccw)}$$

   Torque on link AB, $T_2 = -T = 167.9$ N m (cw)

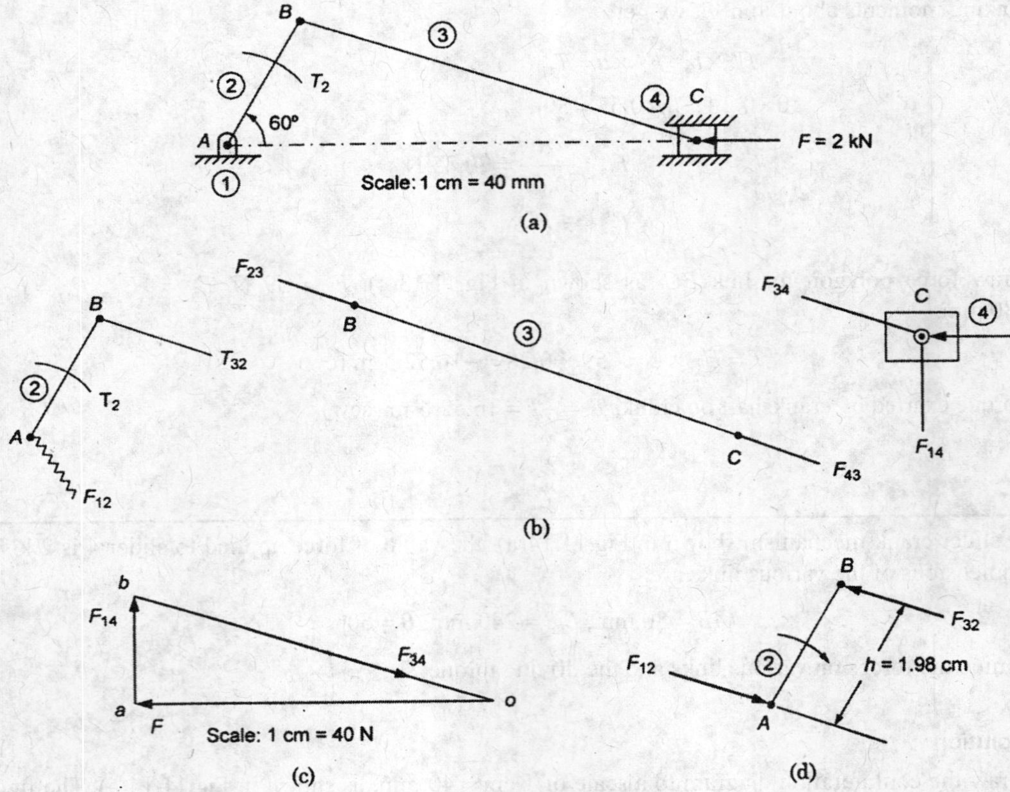

Fig.11.14 Static force analysis of slider-crank mechanism

Example 11.6

A slider crank mechanism shown in Fig.11.15(a) is subjected to two forces: $P = 3$ kN and $Q = 1000\angle 60°$. The dimensions of various links are:

$$AB = 250 \text{ mm}, BC = 600 \text{ mm}, BD = 250 \text{ mm}, \theta = 45°$$

Determine the torque T_2 applied to link 2.

Solution

Draw the configuration diagram to a scale of 1 cm = 100 mm, as shown in Fig.11.15(a). The forces on the various links are shown in Fig.11.15(b), (c), and (d).

1. Consider the equilibrium of link 3 shown in Fig.11.15(c).
   Taking moments about point B, we have

$$Q \sin 60° \times BD = F_{43}^t \times BC$$

$$F_{43}^t = \frac{Q \sin 60° \times BD}{BC} = \frac{1000 \times 0.866 \times 250}{600} = 360.8 \text{ N}$$

2. Now consider the equilibrium of link 4 shown in Fig.11.15(b). The direction of F_{14} is perpendicular to force P. F^t_{34} is completely known. Draw the force polygon as shown in Fig.11.15(d).

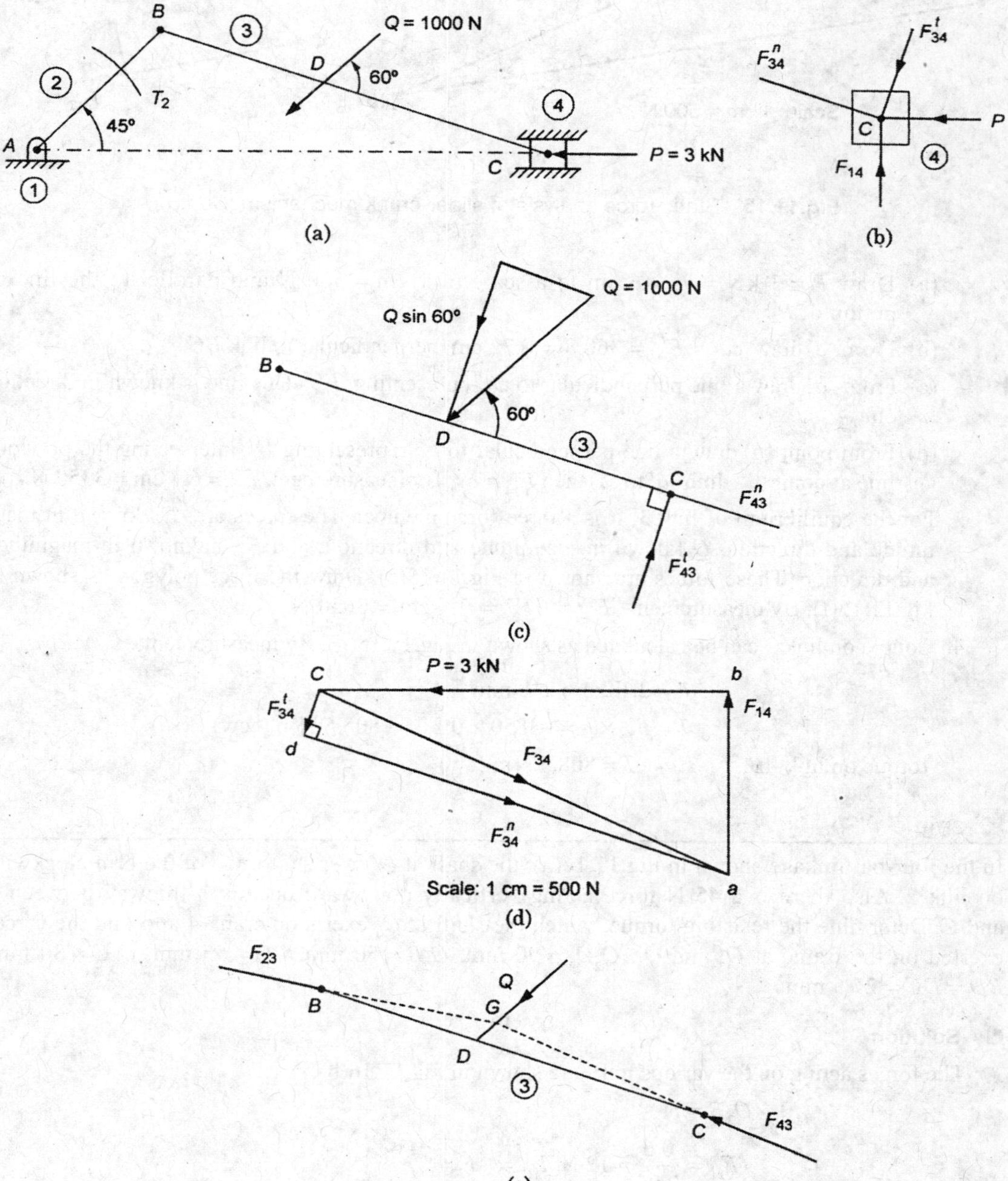

Fig.11.15 Static force analysis of slider-crank mechanism

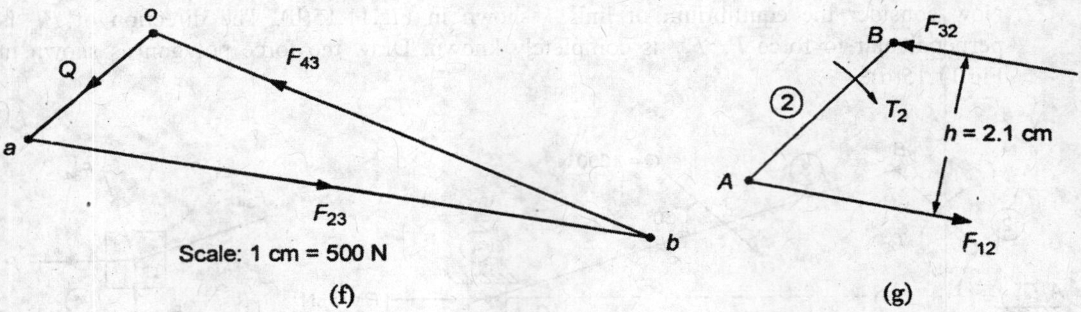

Fig.11.15 Static force analysis of slider-crank mechanism (*Contd.*)

(a) Draw $P = 3$ kN $= bc = 6$ cm to a scale of 1 cm = 500 N and parallel to the line of action of P.

(b) From 'c' draw $cd = F_{34}^t = 360.8 = 0.72$ cm perpendicular to link BC.

(c) From 'd' draw a line perpendicular to cd representing F_{34}^n. This line is known in direction only.

(d) From point 'b' draw a line perpendicular to bc representing F_{14} intersecting the previous line at point 'a'. Join 'a' to 'c' then $F_{34} = ca$. By measurement, $F_{34} = 6.7$ cm $= 3350$ N.

3. For the equilibrium of link 3, it is a three-force member. The forces are: F_{43} known in magnitude and direction, Q known in magnitude and direction, and F_{23} unknown in magnitude and direction. These forces are shown in Fig.11.15(e). Draw the force polygon as shown in Fig.11.15(f). By measurement, $F_{23} = -F_{23} = 7.7$ cm $= 3850$ N.

4. Couple on link 2 can be calculated as shown in Fig.11.15(g). By measurement,

$$h = 2.1 \text{ cm} = 210 \text{ mm}$$

$$T = F_{32} \times h = -3850 \times 0.21 = -808.5 \text{ N m (ccw)}$$

Torque on link AB, $\quad T_2 = -T = 808.5$ N m (cw)

Example 11.7

In the four-bar linkage shown in Fig.11.16(a), the shaft at O_2 exerts a torque of 0.6 N m clockwise on link 2. Also there is a 45 N force acting vertically downward on link 3 midway between B and C. Determine the resisting torque, which the shaft at O_4 exerts on crank 4 and find the forces exerted on the frame at O_2 and O_4. $O_2O_4 = 90$ mm, $O_2B = 50$ mm, $BC = 55$ mm, $O_4C = 30$ mm, $BD = DC = 27.5$ mm.

Solution

The forces acting on the various links are shown in Fig.11.16(b).

$$F_{21}^t \times O_2B = T_2$$

$$F_{21}^t = \frac{0.6}{0.05} = 12 \text{ N}, \; F_{21}^t = -F_{12}^t$$

$$b = 25 \text{ mm}, \; d = 14 \text{ mm}, \; e = 51 \text{ mm}, \; h = 7 \text{ mm}$$

$$\Sigma M_c = 0 \text{ gives}$$

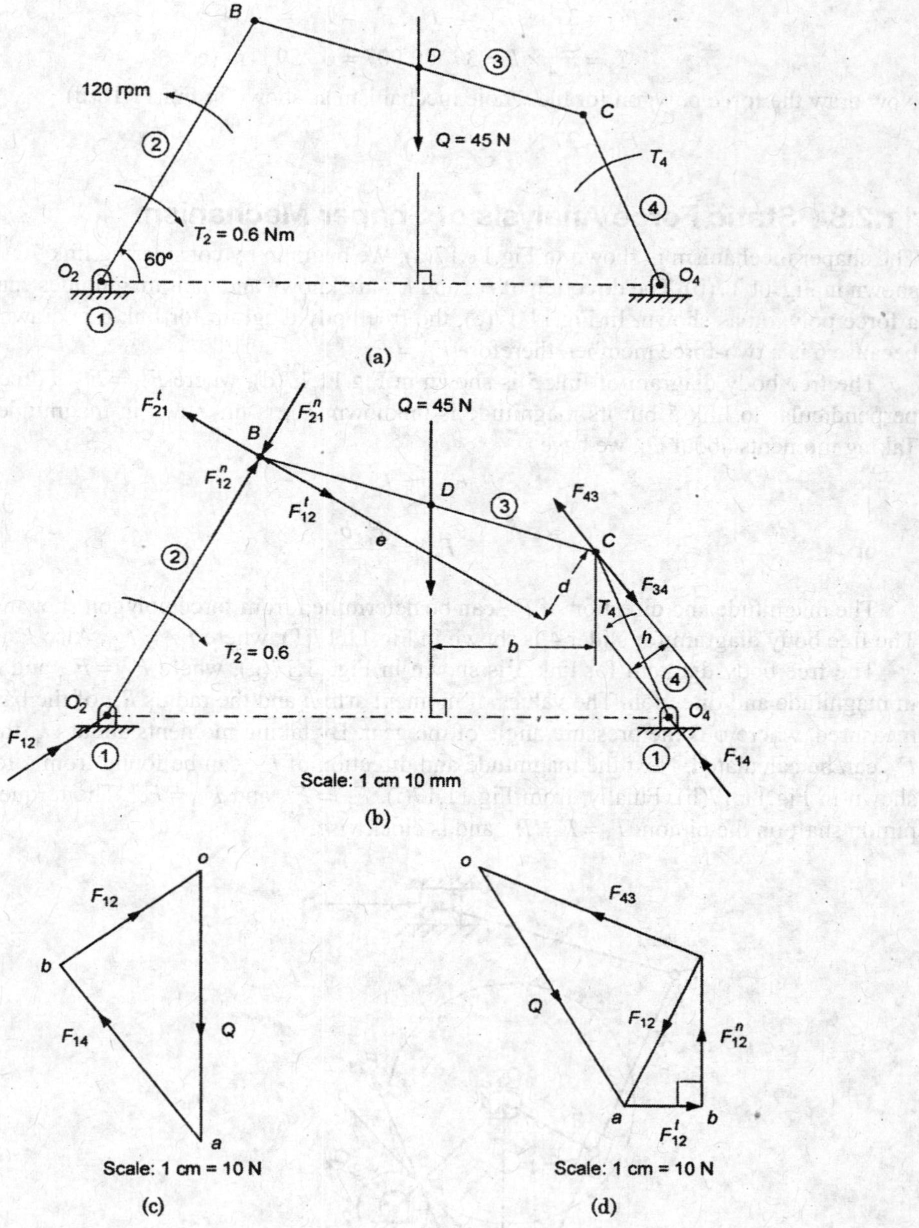

Fig.11.16 Static force analysis of four-bar linkage

$$Q \times b + F_{12}^t \times d = F_{12}^n \times e$$

$$F_{12}^n = \frac{45 \times 25 + 12 \times 14}{53} = 24.4 \text{ N}$$

Draw the force polygon as shown in Fig.11.16(c). From the figure, we have

$$F_{43} = 37 \text{ N}, F_{34} = -F_{43}, F_{14} = F_{34} = 37 \text{ N}$$

$$T_4 = F_{34} \times h = 37 \times 0.007 = 0.259 \text{ Nm (ccw)}$$

Now draw the force polygon for the whole mechanism as shown in Fig.11.16(d).

$$F_{12} = 27 \text{ N}$$

11.2.8 Static Force Analysis of Shaper Mechanism

The shaper mechanism is shown in Fig.11.17(a). We begin in by considering link 7 as a free body, as shown in Fig.11.17(b). The direction of F_{67} and F_{17} are known and their magnitudes can be found from a force polygon as shown. In Fig.11.17(c), the free body diagram for link 6 is drawn. $F_{76} = F_{67}$, and because 6 is a two-force member, therefore $F_{56} = F_{76}$.

The free body diagram of link 5 is shown in Fig.11.17(d), where $F_{65} = F_{56}$. Force F_{45} is directed perpendicular to link 5 but its magnitude is unknown. F_{15} is unknown in magnitude and direction. Taking moments about O_5, we have

$$F_{65}\, b = F_{45}\, e$$

or

$$F_{45} = \frac{F_{65}\, b}{e}$$

The magnitude and direction of F_{15} can be determined from force polygon shown in Fig.11.17(e). The free body diagram for slider 4 is shown in Fig.11.17(f), where $F_{54} = F_{45}$. Also $F_{34} = F_{54}$.

The free body diagram for link 3 is shown in Fig.11.17(g), where $F_{45} = F_{34}$ and F_{13} is unknown in magnitude and direction. The values of moment arm h and the radius R_{b3} of the base circle can be measured, where ϕ is the pressure angle of the gear. By taking moments about O_3, the magnitude of F_{23} can be calculated. Next the magnitude and direction of F_{13} can be found from a force polygon as shown in Fig.11.17(h). Finally, from Fig.11.17(i), $F_{32} = F_{23}$ and $F_{12} = F_{32}$. The torque exerted by the pinion shaft on the pinion, $T_2 = F_{32} \cdot R_{b2}$ and is clockwise.

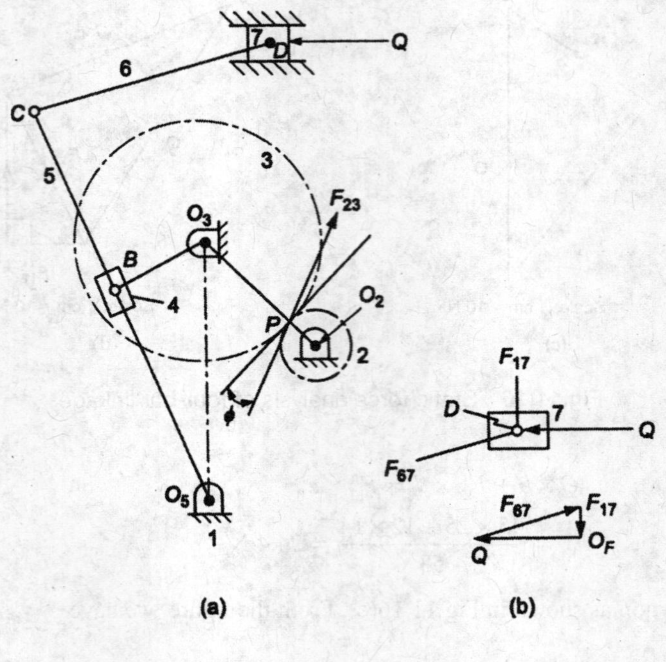

(a) (b)

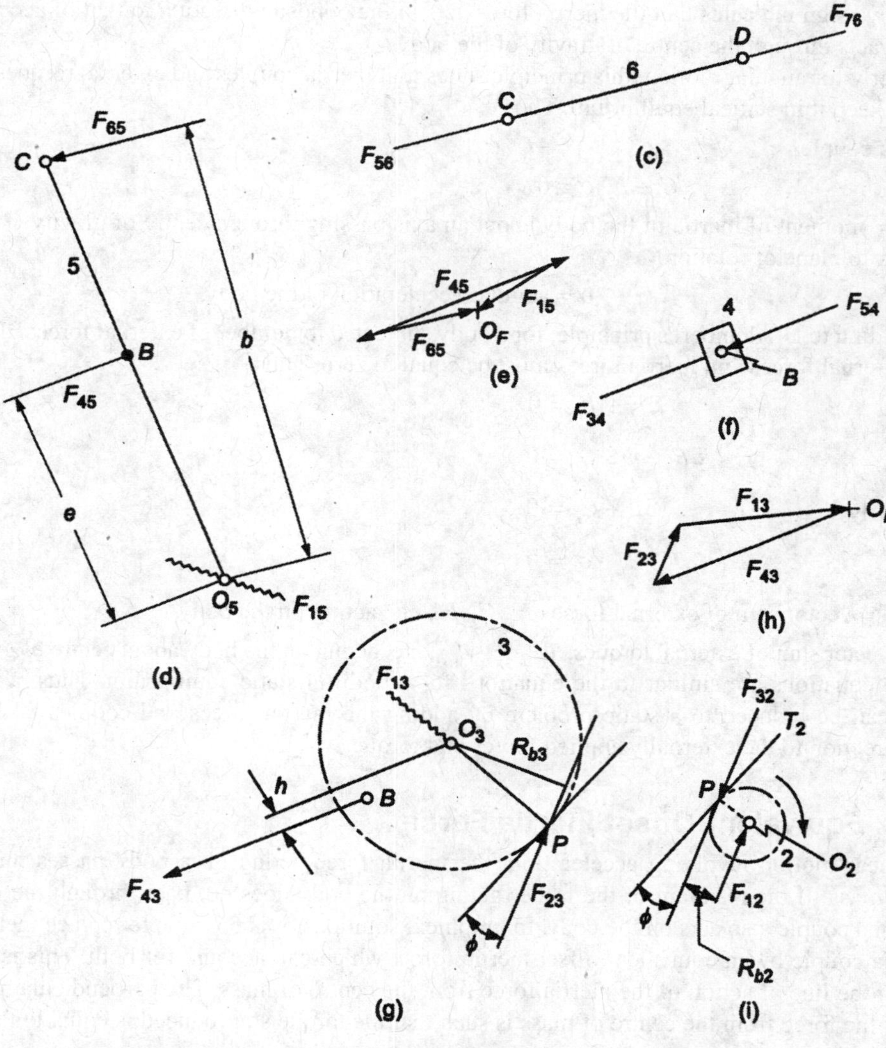

Fig.11.17 Static force analysis of shaper mechanism

11.3 DYNAMIC FORCE ANALYSIS

Dynamic forces in mechanisms arise due to mass of the links and their accelerations. Dynamic analysis has to be carried out when the dynamic forces are comparable with the externally applied forces.

11.3.1 D' Alembert's Principle

For rectilinear motion, this principle states that inertia forces and external forces acting on a body taken together give statical equilibrium. Thus

Inertia force, $$F_i = -m \cdot f_G \tag{11.7a}$$

or $$F_i + m \cdot f_G = 0 \tag{11.7b}$$

where m = mass of the body, and f_G = acceleration of centre of gravity or (mass) of the body.

Negative sign indicates that the inertia force acts in the opposite direction to that of acceleration. The force acts through the centre of gravity of the body.

Similarly for angular motion, this principle states that inertia couples and external torques applied to a body keep it in statical equilibrium. Thus,

Inertia couple, $$C_i = -I_G \cdot \alpha \tag{11.8a}$$

or $$C_i + I_G \cdot \alpha = 0 \tag{11.8b}$$

where I_G = moment of inertia of the body about an axis passing through centre of gravity G and perpendicular to plane of rotation.

α = angular acceleration of the body.

According to D'Alembert's principle, for a body subjected to number of external forces, the vector sum of external forces and inertia forces must be equal to zero. Thus,

$$\Sigma F + F_i = 0 \tag{11.9a}$$

$$\Sigma F - m \cdot f_G = 0 \tag{11.9b}$$

Similarly $$\Sigma T + C_i = 0 \tag{11.10a}$$

$$\Sigma T - I_G \cdot \alpha = 0 \tag{11.10b}$$

where ΣF = vector sum of external forces F_1, F_2, F_3, etc. acting on the body.

ΣT = vector sum of external torques, T_{G1}, T_{G2}, T_{G3}, etc. acting on the body about centre of gravity.

These equations are similar to the equations for a body in static equilibrium. Thus, a dynamic problem can be reduced to a static problem by adding the inertia forces and couples taken in the reverse direction to the externally applied forces and torques.

11.3.2 Equivalent Offset Inertia Force

In rectilinear motion involving acceleration, the inertia force acting on a body passes through its centre of mass. If the resultant of the forces acting on the body does not pass through the centre of mass, then a couple also acts on the body. In graphical solutions, it is possible to replace inertia force and inertia couple by an equivalent offset inertia force, which can account for both. This is done by displacing the line of action of the inertia force from the centre of mass. The perpendicular displacement h of the force from the centre of mass is such that the torque so produced is equal to the inertia couple acting on the body.

Consider a body whose centre of mass is G, its linear acceleration f_G, and angular acceleration α, as shown in Fig.11.18(a). Let a force $F_i = m \cdot f_G$ be applied at G, from left to right upwards. This force can be replaced by another force F_i acting at a distance h (Fig.11.18(b)) together with a torque $T_i = I_G \cdot \alpha$, where I_G is the moment of inertia of the body about an axis passing through G and perpendicular to the plane of rotation. Hence

$$T_i = C_i$$

$$F_i \times h = I_G \cdot \alpha$$

or $$h = \frac{I_G \cdot \alpha}{F_i} = \frac{mK^2 \cdot \alpha}{m \cdot f_G} = \frac{K^2 \alpha}{f_G} \tag{11.11}$$

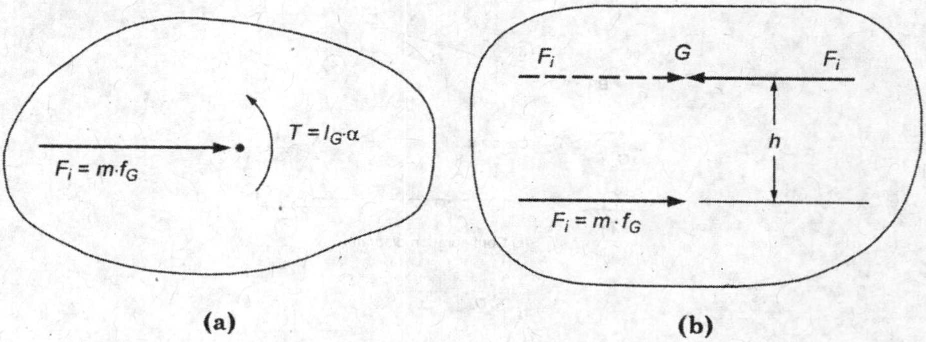

Fig.11.18 Equivalent offset inertial force

The radius of gyration K is taken in such a way that the force produces a moment about the centre of mass, which is opposite in sense to angular acceleration α.

11.3.3 Dynamic Force Analysis of Four-Bar Mechanism

Consider a four-bar mechanism shown in Fig.11.19(a), where the magnitude of ω_2 is assumed known and constant. Points G_2, G_3, and G_4 denote the centres of mass of the links 2, 3, and 4 respectively. We are interested to determine the torque which the shaft at O_2 must exert on crank 2 to give the desired motion.

To determine the linear acceleration of the points G_2, G_3, and G_4, we construct the acceleration polygon. From the magnitude and sense of the tangential components of acceleration, the magnitude and sense of α_3 and α_4 can be determined.

Link 2 is shown in Fig.11.19(c), where f_{G2} is the acceleration of the centre of mass G_2. The resultant force $F_2 = m_2 \cdot f_{G2}$, where m_2 is the mass of the link 2, has the same sense and line of action as f_{G2}. The inertia force $f_2 = -F_2$.

Link 3 is shown in Fig.11.19(d) with the acceleration of the centre of mass G_3 indicated as f_{G3}. The resultant force $F_3 = m_3 \cdot f_{G3}$, where m_3 is the mass of the link 3, has the same sense and line of action as f_{G3}. $f_3 = -F_3$ is the inertia force. In order to produce α_3, there must be a resultant torque $T_3 = I_3 \cdot \alpha_3$ having the same sense as α_3. Inertia torque $t_3 = -T_3$. Link 3 is again shown in Fig.11.19(e), where the inertia force f_3 and inertia torque t_3 have been replaced by a single force f_3. The direction and sense of f_3 is the same as in Fig.11.19(d), but the line of action is displaced from G_3 by an amount h_3, such that

$$f_3 h_3 = t_3 \quad \text{or} \quad h_3 = \frac{t_3}{f_3} = \frac{I_3 \cdot \alpha_3}{m_3 \cdot f_{G3}}$$

In Fig.11.19(e), f_3 can be located by drawing a circle of radius h_3 with its centre at G_3. f_3 is drawn tangent to the left side of the circle rather than the right side because f_3 must produce a torque about G_3 in the same sense as t_3.

Link 4 is shown in Fig.11.19(f) where $f_4 = -F_4$ and $T_4 = I_4 \cdot \alpha_4$. Inertia torque $t_4 = -T_4$. Link 4 appears again in Fig.11.19(g), where the inertia force f_4 and inertia torque t_4 have been replaced by a single force f_4. Since $f_4 \cdot h_4$ must equal t_4, therefore

$$h_4 = \frac{t_4}{f_4} = \frac{I_4 \cdot \alpha_4}{m_4 \cdot f_{G4}}$$

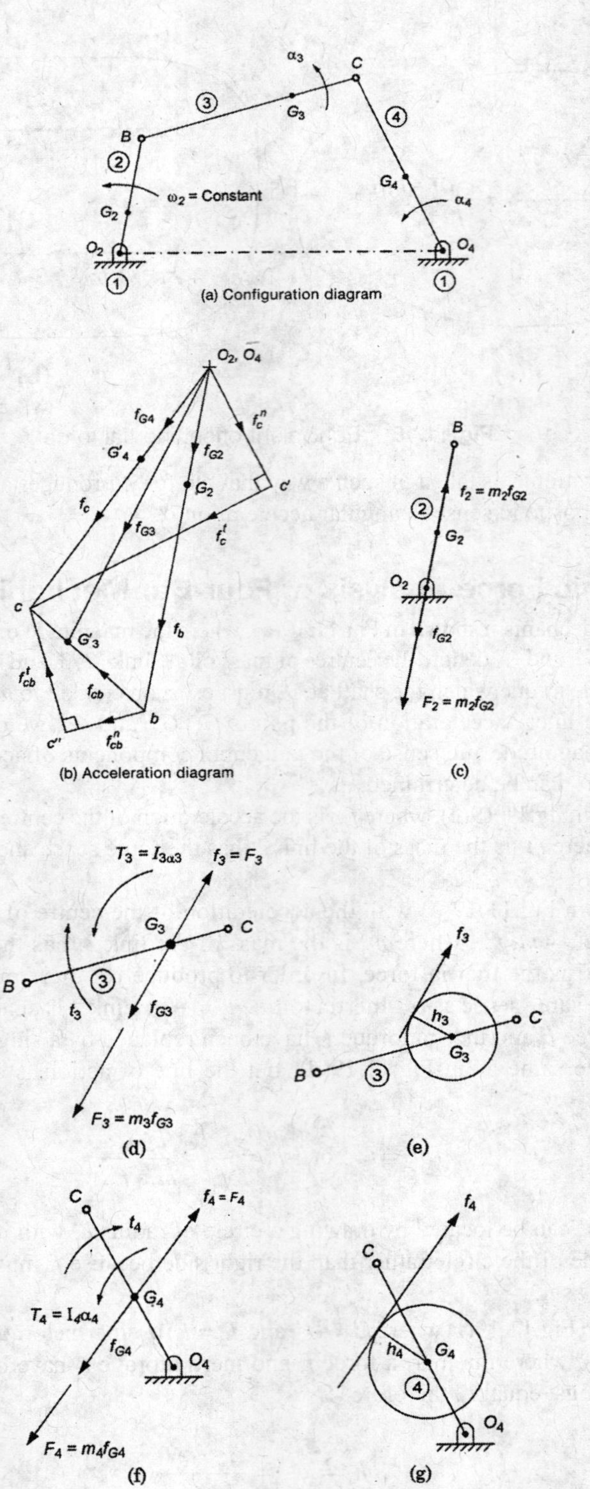

Fig.11.19 Inertia forces in a four-bar mechanism

To find the forces at each pin connection and the torque which the shaft exerts on crank 2, we draw the free body diagrams of links 2, 3, and 4 as shown in Fig.11.20(a) to (c). For the known inertia forces in each link, the forces in each pin can be determined by using the equilibrium equations. Starting with link 4, we take the moments about point O_4 and determine F^t_{34}. Then on link 3, $F^t_{43} = -F^t_{34}$ For Equilibrium of link 3, the sum of the moments about B equal zero. This determines F^n_{43}. The force polygon for link 3 is shown in Fig.11.20(d) to determine F_{23}.

Link 2 appears in Fig.11.20(e). Here $F_{32} = -F_{23}$. Then $F_{12} = -(f_2 +\!\!\!\rightarrow f_{32})$. Taking moments about O_2, we obtain T_2, as

$$T_2 = (f_2 +\!\!\!\rightarrow F_{32})\,a$$

where $+\!\!\!\rightarrow$ represents vector sum.

Force F_{14} obtained from the force polygon for bodies 2, 3, and 4 taken as a whole system as shown in Fig.11.20(f).

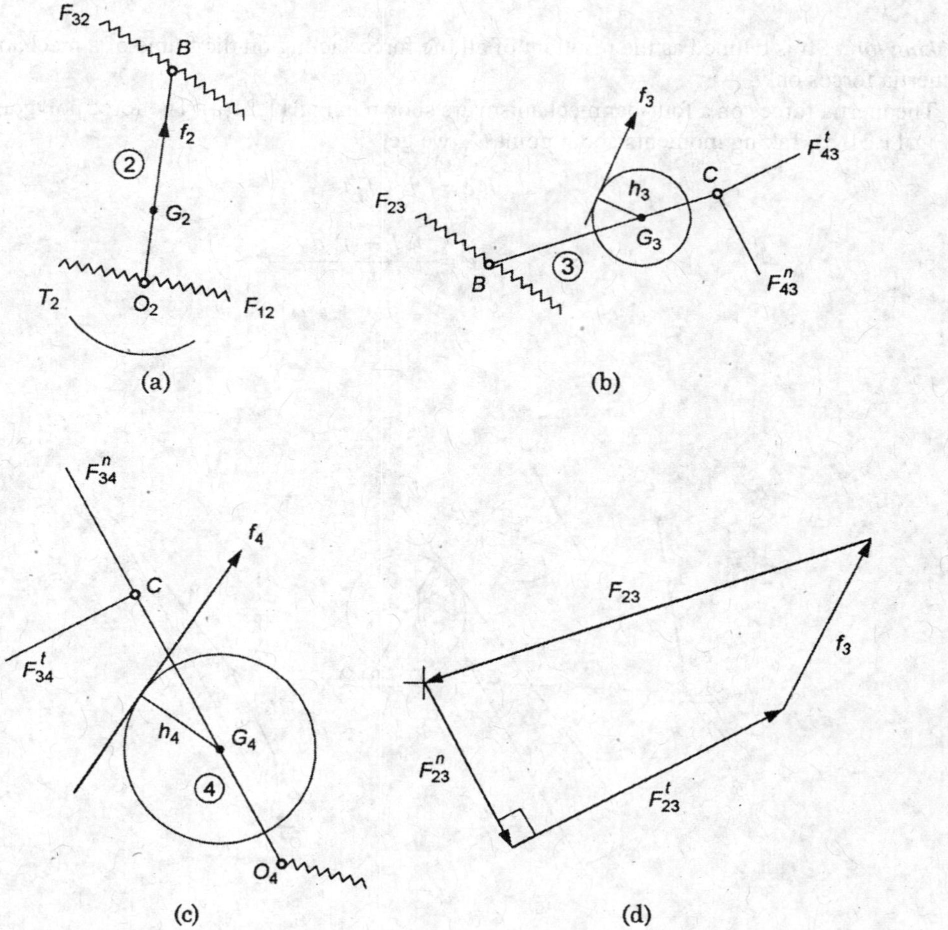

Fig.11.20 Force polygons

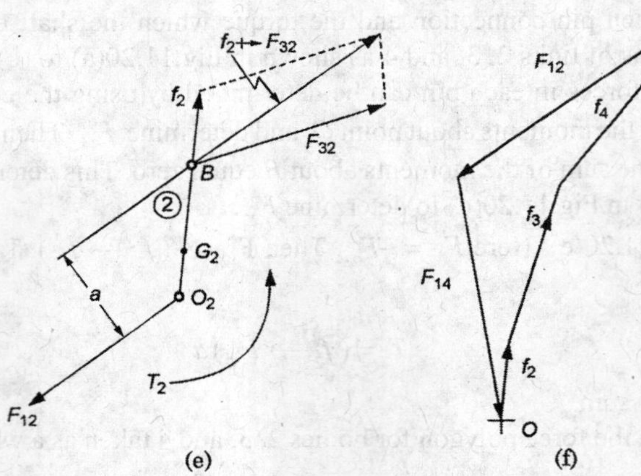

Fig.11.20 Force polygons (*Contd.*)

Shaking force: It is defined as the resultant of all the forces acting on the frame of a mechanism due to inertia forces only.

The inertia forces on a four-bar mechanism are shown in Fig.11.21(a). The force polygon is shown in Fig.11.21(b). Taking moments about point O_2, we get

$$F_s e = f_3 b + f_4 d$$

or

$$e = \frac{f_3\, b + f_4\, d}{F_s}$$

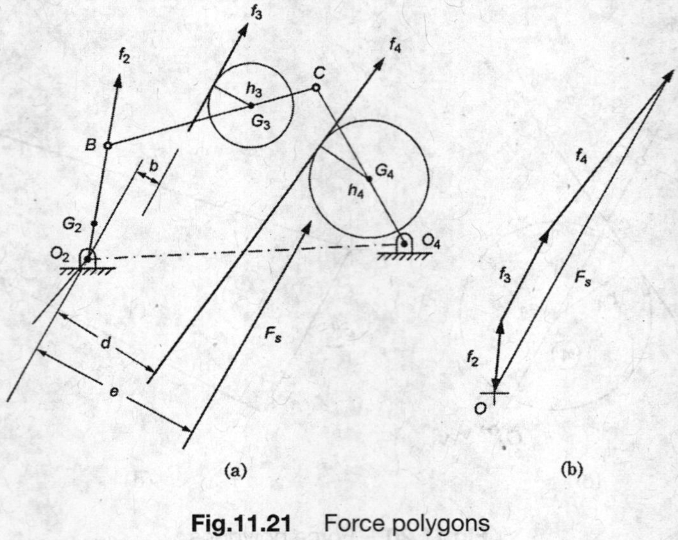

Fig.11.21 Force polygons

11.3.4 Dynamic Force Analysis of Slider–Crank Mechanism

The slider crank mechanism is shown in Fig.11.22(a). Let P be the force on the piston due to gas pressure and ω_2 the angular velocity of link 2, be known. Points G_2, G_3, and G_4 are the centres of mass of links 2, 3, and 4. We are interested to find the torque T_2, which the crank 2 exerts on the crankshaft and the shaking force.

The velocity and acceleration polygons are constructed first, as shown in Fig.11.22(b) and (c), respectively. Link 3 and 4 combined as a free body are shown in Fig 11.22(d). The inertia force f_3, its moment about G_3 and f_4 are determined as explained in Section 11.3.3. The unknowns are the magnitudes of F_{23} and F_{14}. By taking moments about B, we have

$$F_{14}a + f_3 b + f_4 d - Pd = 0$$

or

$$F_{14} = \frac{Pd - f_3 b - f_4 d}{a}$$

Force F_{23} can then be found by a summation of forces on bodies 3 and 4 together as a free body. The force polygon is shown in Fig.11.22(e).

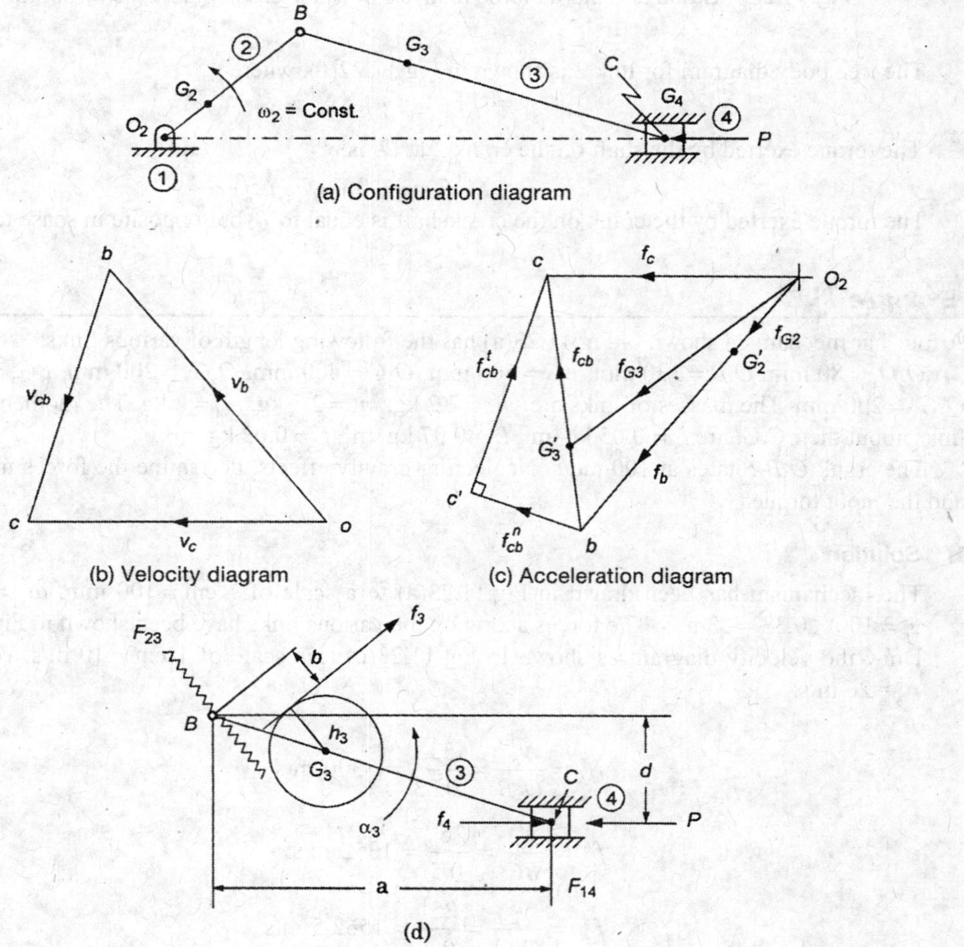

Fig.11.22 Static and inertia force analysis of slider-crank mechanism

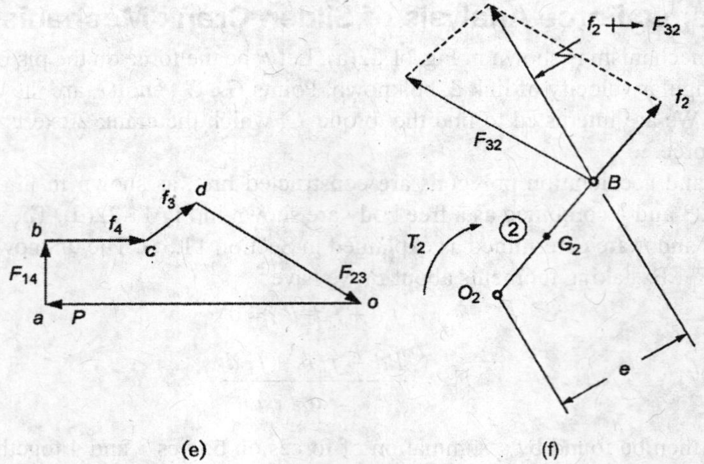

Fig.11.22 Static and inertia force analysis of slider-crank mechanism (*Contd.*)

The free body diagram for link 2 is shown in Fig.11.22(f), where

$$F_{12} = -(f_2 \mapsto F_{32})$$

The torque exerted by the shaft on the crank 2 at O_2 is,

$$T_2 = -(f_2 \mapsto F_{32})\, e$$

The torque exerted by the crank on the crankshaft is equal to T_2 but opposite in sense to T.

Example 11.8

A four-bar mechanism shown in Fig.11.23(a) has the following length of various links:

$O_2O_4 = 80$ mm, $O_2B = 330$ mm, $BC = 500$ mm, $O_4C = 400$ mm, $O_2G_2 = 200$ mm, $BG_3 = 250$ mm, $O_4G_4 = 200$ mm. The masses of links are: $m_2 = 2.2$ kg, $m_3 = 2.5$ kg, $m_4 = 2$ kg. The moment of inertia links about their C.G. are $l_2 = 0.05\ \text{kg}\cdot\text{m}^2$, $l_3 = 0.07\ \text{kg}\cdot\text{m}^2$. $l_4 = 0.02\ \text{kg}\cdot\text{m}^2$.

The crank O_2B rotates at 100 rad/s. Neglecting gravity effects, determine the forces in the joints and the input torque.

■ Solution

The mechanism has been drawn in Fig.11.23(a) to a scale of 1 cm = 100 mm. $\omega_2 = 100$ rad/s. $v_b = 100 \times 0.33 = 33$ m/s. The forces acting on the various links have been shown in Fig.11.23(b). Draw the velocity diagram as shown in Fig.11.23(c) to a scale of 1 cm = 10 m/s. $v_c = 25$ m/s. $v_{cb} = 26$ m/s.

$$f_b^n = \frac{v_b^2}{O_2B} = \frac{33^2}{0.33} = 3300 \text{ m/s}^2$$

$$f_{cb}^n = \frac{v_{cb}^2}{BC} = \frac{26^2}{0.5} = 1352 \text{ m/s}^2$$

$$f_{cb}^n = \frac{v_c^2}{O_4C} = \frac{25^2}{0.4} = 1562.5 \text{ m/s}^2$$

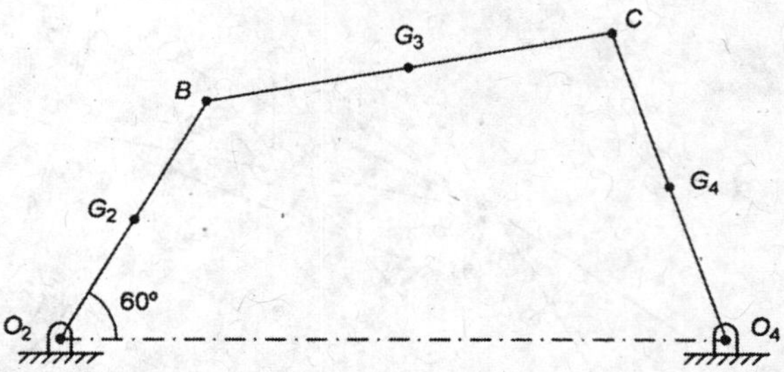

(a) Configuration diagram

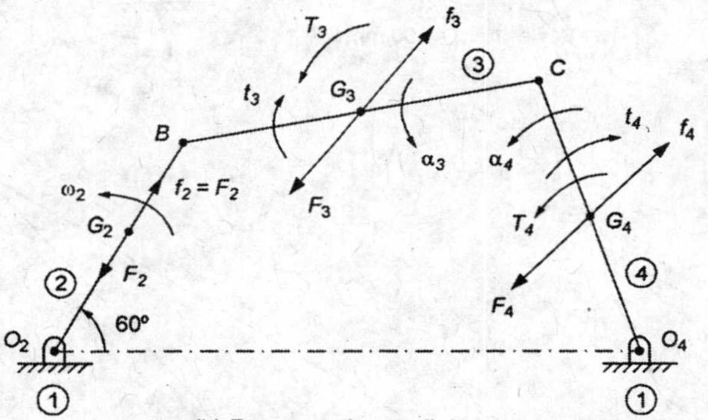

(b) Forces acting on links

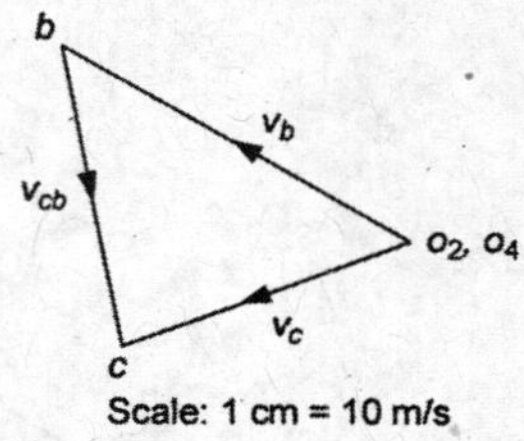

(c) Velocity diagram

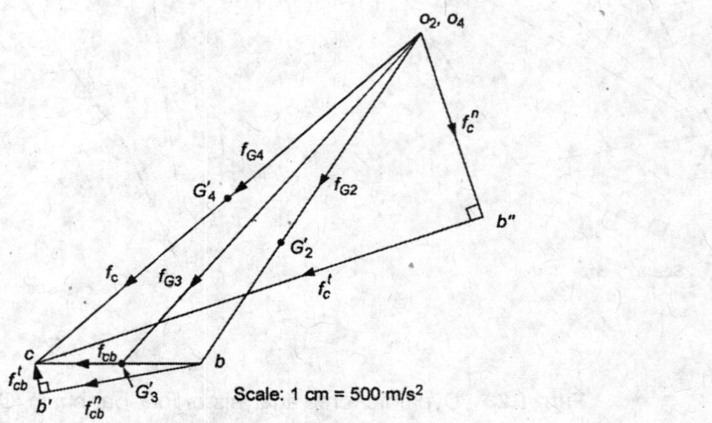

(d) Acceleration diagram

Fig.11.23 Dynamic force analysis of four-bar chain

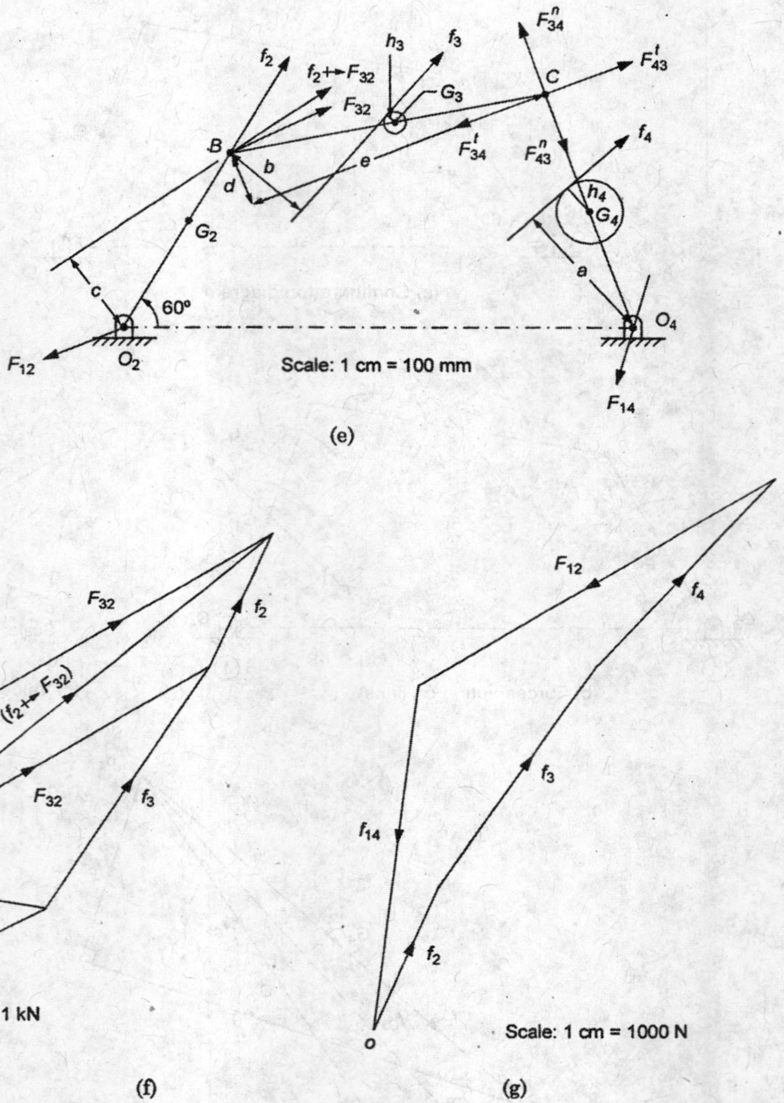

Fig.11.23 Dynamic force analysis of four-bar chain (*Contd.*)

Draw the acceleration diagram, as shown in Fig.11.23 (d to a scale of 1 cm = 500 m/s²). $o_2b = 6.6$ cm, $bc = 2.8$ cm, $O_4C = 8.4$ cm.

$$o_2G_2' = \frac{O_2G_2 \times o_2b}{O_2B} = \frac{200 \times 6.6}{330} = 4 \text{ cm}$$

$$bG_3' = \frac{BG_3 \times bc}{BC} = \frac{250 \times 2.8}{500} = 2.4 \text{ cm}$$

$$o_4G_4' = \frac{O_4G_4 \times o_4c}{O_4C} = \frac{200 \times 8.4}{400} = 4.2 \text{ cm}$$

$$\text{Acceleration of } G_2,\ f_{G2} = o_2G_2' = 4 \times 500 = 2000 \text{ m/s}^2$$

$$\text{Acceleration of } G_3,\ f_{G3} = o_2G_3' = 7.4 \times 500 = 3700 \text{ m/s}^2$$

$$\text{Acceleration of } G_4,\ f_{G4} = o_2G_4' = 4.2 \times 500 = 2100 \text{ m/s}^2$$

$$f_{cb}^t = b'c = 0.5 \text{ cm} = 250 \text{ m/s}^2$$

$$f_c^t = b''c = 7.8 \text{ cm} = 3900 \text{ m/s}^2$$

$$\alpha_4 = \frac{f_{cb}^t}{BC} = \frac{250}{0.5} = 500 \text{ rad/s}^2 \text{ (ccw)}$$

$$\alpha_4 = \frac{f_c^t}{O_4C} = \frac{3900}{0.4} = 9750 \text{ rad/s}^2 \text{ (ccw)}$$

$$F_2 = m_2 f_{G2} = 2.2 \times 2000 = 2400 \text{ N}$$

Inertia force, $f_2 = -F_2$

$$F_3 = m_3 f_{G3} = 2.5 \times 3700 = 4550 \text{ N}$$

Inertia force, $f_3 = -F_3$

$$T_3 = I_3\, \alpha_3 = 0.07 \times 500 = 35 \text{ N m (ccw)}$$

Inertia torque, $t_3 = -T_3 = 35 \text{ N m (cw)}$

$$F_4 = m_4 f_{G4} = 2 \times 2100 = 4200 \text{ N}$$

Inertia force, $f_4 = -F_4$

$$T_4 = I_4\, \alpha_4 = 0.02 \times 9750 = 195 \text{ N m (ccw)}$$

Inertia torque, $t_4 = -T_4 = 195 \text{ N m (cw)}$

$$h_3 = \frac{t_3}{f_3} = \frac{35}{4550} = 7.7 \text{ mm}$$

$$h_4 = \frac{t_4}{f_4} = \frac{195}{4200} = 46 \text{ mm}$$

The forces and perpendicular distances are shown in Fig.11.23(e).

$$a = 230 \text{ mm},\ b = 150 \text{ mm},\ d = 90 \text{ mm},\ e = 490 \text{ mm}$$

$$F_{43}^t = \frac{f_4 a}{O_4C} = \frac{4200 \times 230}{400} = 2415 \text{ N}$$

$$F_{43}^t = -F_{34}^t = -2415 \text{ N}$$

Taking moments about B, we have

$$f_3 b + f_{43}^t d - F_{43}^n e = 0$$

$$4550 \times 150 + 2415 \times 90 = F_{43}^n \times 490$$

$$F_{43}^n = 1836.4 \text{ N}$$

Draw the force polygon for link BC, as shown in Fig.11.23(f).

$$F_{23} = 6500 \text{ N}$$
$$F_{32} = -F_{23}$$
$$F_{12} = F_{32}$$

The resultant of f_2 and F_{32} has been obtained in Fig.11.23(f) and is equal to $f_2 \leftrightarrow F_{32} = 8600$ N.

$$T = -(f_2 \leftrightarrow F_{32})\,c$$
$$= -8600 \times 0.13 = -1118 \text{ N m (cw)}$$

Torque exerted by the crankshaft on crank O_2B, $T_2 = -T = 1118$ N m (ccw)

Now draw the force polygon for the whole mechanism, as shown in Fig.11.23(g).

$$F_{14} = 5600 \text{ N}$$

Example 11.9

The slider crank mechanism of a single cylinder diesel engine is shown in Fig.11.24(a). A gas force $P = 17800$ N acts to the left through piston pin C. The crank rotates counter-clockwise at a constant speed of 1800 rpm. Determine (a) the force F_{14} and F_{12} and the torque T_2 exerted by the crank-shaft on the crank for equilibrium and (b) the magnitude and direction of the shaking force and its location from point O_2. $O_2B = 75$ mm, $O_2G_2 = 50$ mm, $BC = 280$ mm, $BG_3 = 125$ mm, $m_2 = 2.25$ kg, $m_3 = 3.65$ kg, $m_4 = 2.75$ kg, $I_2 = 0.0055$ kg·m², $I_3 = 0.041$ kg·m².

Solution

The forces acting on the various links have been shown in Fig.11.24(b).

$$\omega = \frac{2\pi \times 1800}{60} = 188.5 \text{ rad/s}$$

$$v_b = O_2B \times \omega = 0.075 \times 188.5 = 14.14 \text{ m/s}$$

Draw the velocity diagram to a scale of 1 cm = 2 m/s, as shown in Fig.11.24(c).

$$v_{cb} = bc = 3.7 \text{ cm} = 7.4 \text{ m/s}$$
$$v_c = o_2c = 0.7 \text{ cm} = 2.4 \text{ m/s}$$

$$f_b^n = \frac{v_b^2}{O_2B} = \frac{(14.14)^2}{0.075} = 2665.8 \text{ m/s}^2$$

$$f_{cb}^n = \frac{v_{cb}^2}{BC} = \frac{(7.4)^2}{0.28} = 195.6 \text{ m/s}^2$$

Draw the acceleration diagram to a scale of 1 cm = 500 m/s², as shown in Fig.11.24(d).

$$f_{cb}^t = b'c = 4.6 \text{ cm} = 2300 \text{ m/s}^2,\ f_{cb} = bc = 4.7 \text{ cm} = 2350 \text{ m/s}^2$$

$$f_c = o_2c = 2.1 \text{ cm} = 1050 \text{ m/s}^2$$

$$\alpha_3 = \frac{f_{cb}^t}{BC} = \frac{2300}{0.28} = 8214.3 \text{ rad/s}^2$$

$$o_2G_2' = \frac{O_2G_2 \times o_2b}{O_2B} = \frac{50 \times 5.3}{75} = 3.55 \text{ cm}$$

$$bG_3' = \frac{BG_3 \times bc}{BC} = \frac{125 \times 4.7}{280} = 2.1 \text{ cm}$$

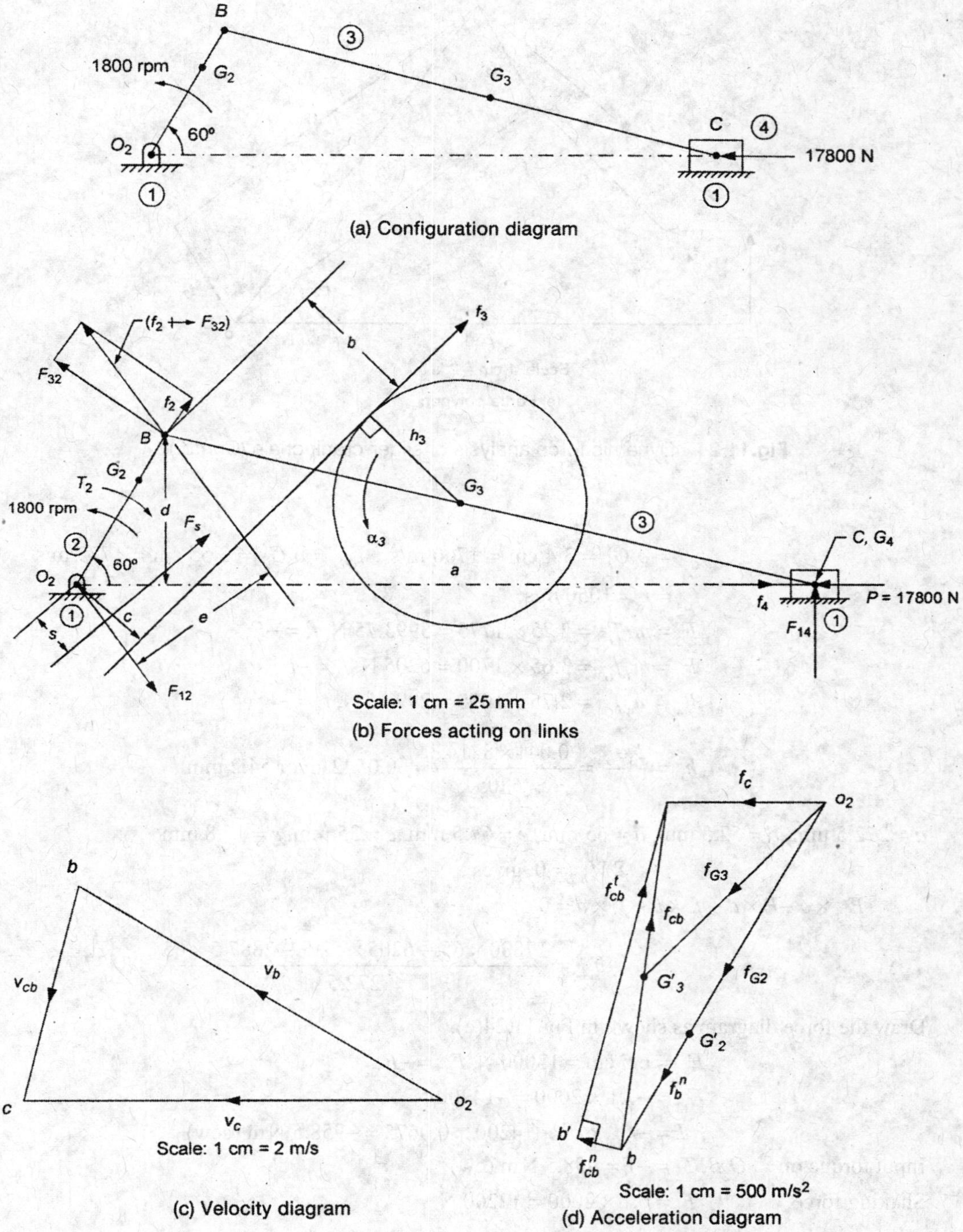

Fig.11.24 Dynamic force analysis of slider-crank chain

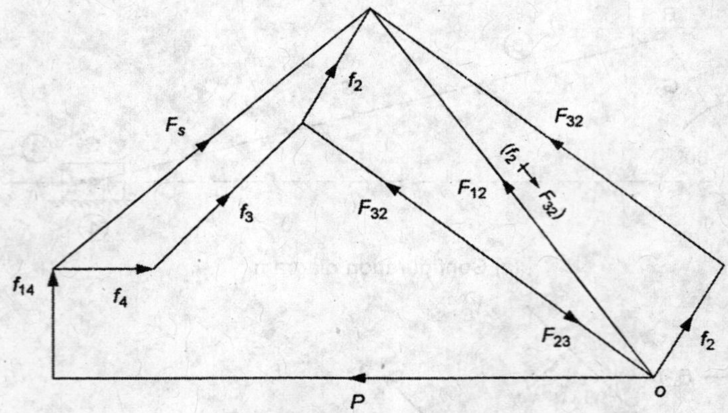

(e) Force polygon

Fig.11.24 Dynamic force analysis of slider-crank chain (*Contd.*)

$$f_{G3} = o_2G_3' = 3.4 \text{ cm} = 1700 \text{ m/s}^2,\ f_{G2} = o_2G_2' = 3.55 \text{ cm} = 1775 \text{ m/s}^2$$

$$f_{G4} = f_c = 1050 \text{ m/s}^2$$

$$F_2 = m_2 f_{G2} = 2.25 \times 1775 = 3993.75 \text{ N}, f_2 = -F_2$$

$$F_3 = m_3 f_{G3} = 3.65 \times 1700 = 6205 \text{ N}, f_3 = -F_3$$

$$F_4 = m_4 f_{G4} = 2.75 \times 1050 = 2997.5 \text{ N}, f_4 = -F_4$$

$$h_3 = \frac{I_3\alpha_3}{f_3} = \frac{0.041 \times 8214.3}{6205} = 0.0542 \text{ m or } 54.2 \text{ mm}$$

$a = 272.5$ mm, $b = 52.5$ mm, $d = 65$ mm, $e = 67.5$ mm, $c = 35$ mm, $s = 17.8$ mm

$$\Sigma M_B = 0 \text{ gives}$$

$$F_{14} \times a - P \times d + f_3 \times b + f_4 \times d = 0$$

$$F_{14} = \frac{17800 \times 65 - 6205 \times 52.5 - 2887.5 \times 65}{272.5} = 3437.7 \text{ N}$$

Draw the force diagram as shown in Fig.11.24(e).

$$F_{23} = 6.5 \text{ cm} = 13000 \text{ N}, F_{32} = -F_{23}$$

$$F_{12} = -7.1 \times 2000 = -14200 \text{ N}$$

$$T = -F_{12} \times e = -14200 \times 0.0675 = -958.5 \text{ Nm (ccw)}$$

Input torque on O_2B, $T_2 = -T = 958.5$ Nm (cw)

Shaking force, $F_s = 6.1 \times 2000 = 12200$ N

$$F_s \times s = f_3 \times c$$

$$s = \frac{6205 \times 32}{12200} = 17.8 \text{ mm.}$$

Example 11.10

For the reciprocating engine mechanism in Figure 11.25(a), the following data is given:

Length of crank is 7.5 cm, Length of connecting rod is 28 cm, Distance of centre of gravity (C.G.) of link 2 from main bearings is 5 cm, Distance of centre of gravity of link 3 from crank pin is 12 cm, Crank angle from line of stroke is 60°, Crank speed is 2000 rpm counter-clockwise, Mass of link 2 is 2.5 kg, Mass of link 3 is 4 kg, Mass of link 4 is 3 kg, Mass moment of inertia of link 2 is 60 kg·cm² and mass moment of inertia of link 3 is 500 kg·cm². Make a complete inertia force analysis.

Solution

Given: $r = 7.5$ cm, $l = 28$ cm, $O_2G_2 = 5$ cm, $BG_3 = 12$ cm, $\theta = 60°$, $N_2 = 2000$ rpm ccw, $m_2 = 2.5$ kg, $m_3 = 4$ kg, $m_4 = 3$ kg, $I_2 = 60\ \text{kg}\cdot\text{cm}^2$, $I_3 = 500\ \text{kg}\cdot\text{cm}^2$

Let force on the piston, $P = 20$ kN to the left (not given in the data)

The forces acting on the various links have been shown in Fig 11.25(b).

$$\omega = \frac{2\pi \times 2000}{60} = 209.44 \text{ rad/s}$$

$$v_b = O_2B \times \omega = 0.075 \times 209.44 = 15.71 \text{ m/s}$$

Draw the velocity diagram to a scale of 1 cm = 2 m/s, as shown in Fig.11.25(c).

$$v_{cb} = bc = 4.1 \text{ cm} = 8.2 \text{ m/s}$$

$$v_c = o_2c = 7.8 \text{ cm} = 15.6 \text{ m/s}$$

$$f_b^n = \frac{v_b^2}{O_2B} = \frac{(15.71)^2}{0.075} = 3290.7 \text{ m/s}$$

$$f_{cb}^n = \frac{v_{cb}^2}{BC} = \frac{(8.2)^2}{0.28} = 240.14 \text{ m/s}^2$$

Draw the acceleration diagram to a scale of 1 cm = 300 m/s², as shown in Fig.11.25(d).

$$f_{cb}^t = b'c = 9.5 \text{ cm} = 2850 \text{ m/s}^2,\ f_{cb} = bc = 9.6 \text{ cm} = 2880 \text{ m/s}^2$$

$$f_c = o_2c = 4.3 \text{ cm} = 1290 \text{ m/s}^2$$

$$\alpha_3 = \frac{f_{cb}^t}{BC} = \frac{2880}{0.28} = 10178.6 \text{ rad/s}^2$$

$$o_2G_2' = \frac{O_2G_2 \times o_2b}{O_2B} = \frac{5 \times 11}{7.5} = 7.3 \text{ cm}$$

$$bG_3' = \frac{BG_3 \times bc}{BC} = \frac{12 \times 9.6}{28} = 4.11 \text{ m/s}^2$$

$$f_{G3} = o_2G_3' = 7.2 \text{ cm} = 2160 \text{ m/s}^2, f_{G2} = o_2G_2 = 7.3 \text{ cm}$$
$$= 2190 \text{ m/s}^2$$

$$f_{G4} = f_c = 4.3 \text{ cm} = 1290 \text{ m/s}^2$$

$$F_2 = m_2 f_{G2} = 2.5 \times 2190 = 5475 \text{ N}, f_2 = -F_2$$
$$F_3 = m_3 f_{G3} = 4 \times 2160 = 8640 \text{ N}, f_3 = -F_3$$
$$F_4 = m_4 f_{G4} = 3 \times 1290 = 3870 \text{ N}, f_4 = -F_4$$
$$h_3 = \frac{I_3 \alpha_3}{f_3} = \frac{0.05 \times 10178.6}{8640} = 0.058 \text{ m or } 58 \text{ mm}$$
$$a = AC = 275 \text{ mm}, b = 45 \text{ mm}, d = 65 \text{ m}, e = 66.25 \text{ mm}$$

$\Sigma M_B = 0$ gives

$$F_{14} \times a - P \times d + f_3 \times b + f_4 \times d = 0$$
$$F_{14} = \frac{20000 \times 65 - 8640 \times 45 - 3870 \times 65}{275} = 2398.7 \text{ N}$$

Draw the force diagram as shown in Fig.11.25(e).

$$F_{23} = 6.9 \text{ cm} = 13800 \text{ N}, F_{32} = -F_{23}$$
$$F_{12} = -7.9 \times 2000 = -15800 \text{ N}$$
$$T = F_{12} \times e = -15800 \times 0.06625 = -1046.75 \text{ N m (ccw)}$$

Torque on link O_2B, $T_2 = -T = 1046.75$ N m (cw)

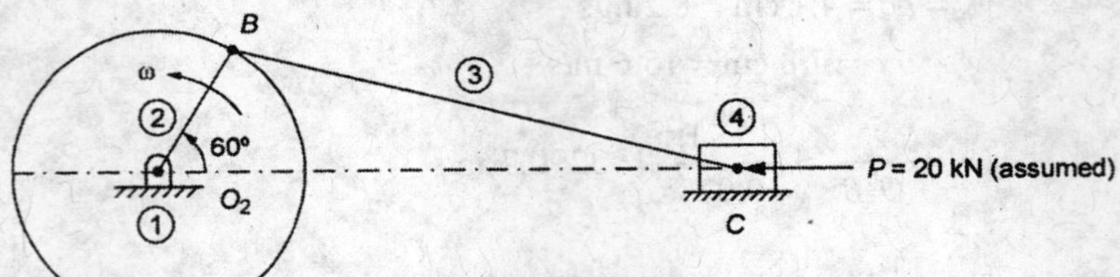

(a) Configuration diagram

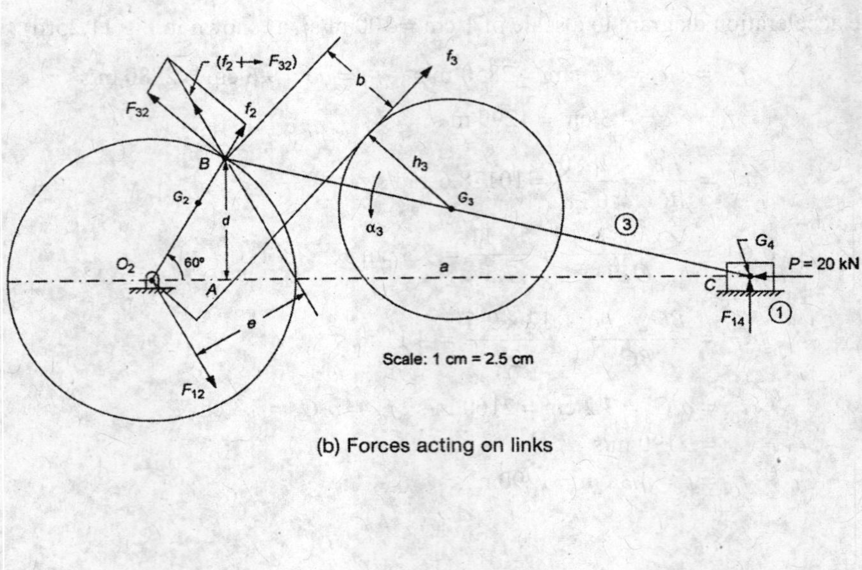

(b) Forces acting on links

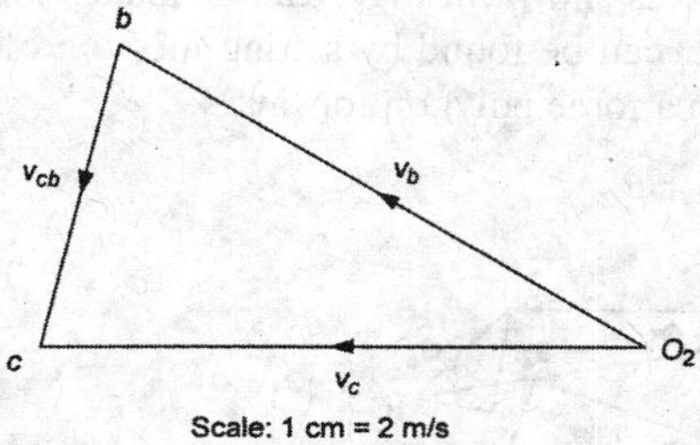

(c) Velocity diagram

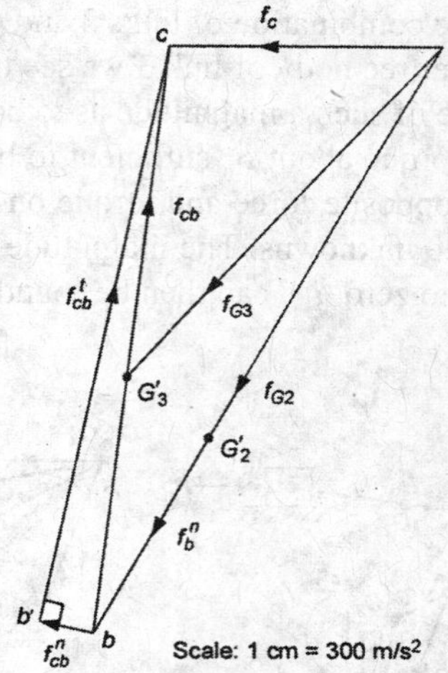

(d) Acceleration diagram

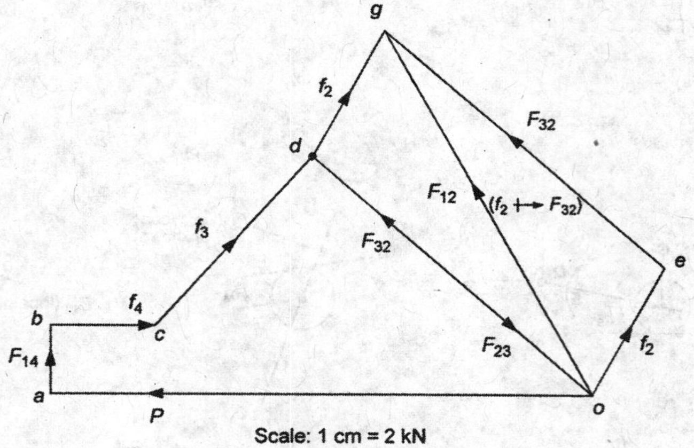

(e) Force polygen

Fig.11.25 Dynamic force analysis of slider-crank machanism

11.3.5 Static and Inertia Force Analysis of Shaper Mechanism

Consider the shaper mechanism shown in Fig.11.26(a), where the link 2 rotates with constant velocity ω_2 and the force P is known on the slider 6. We are interested to determine the forces at all the points and the torque T_2 exerted by the shaft to drive the crank. The acceleration diagram is shown in Fig.11.26(b).

The force analysis is started with link 6, shown in Fig.11.26(c). The unknowns are: magnitude of F_{16}, magnitude and direction of F_{56}. The horizontal component of F_{56} is F_{56}^h and its magnitude can be found from the summation of horizontal forces on link 6.

In Fig.11.26(d), $F_{65}^h = -F_{56}^h$. The magnitude of F_{65}^v can be found by summation of moments about C. Then from a force polygon for link 5, the magnitude and direction of F_{45} are found. Next, in Fig.11.26(e), $F_{54} = -F_{45}$ is known. There are four unknowns in Fig.11.26(e): the magnitude and direction on F_{14} and the magnitude and location of F_{34}. For link 3 shown in Fig.11.26(f), there are also four unknowns: F_{23} in magnitude and direction and F_{43} in magnitude, which is perpendicular to link 4.

However, for the combination of links 3 and 4, there are six unknowns that can be analyzed in combination. From the free body of link 3 we see that F_{23} causes no torque about the centre of mass B_3 and thus F_{43} must be of such a magnitude as to balance the forces and its line of action must be displaced from B_3 and a torque about B_3 sufficient to balance the inertia torque. This is shown in Fig.11.26(g). The equal and opposite force and torque on link 4 as shown in Fig.11.26(h) makes the free body of link 4 with three unknowns. The magnitude of F_{34} can be found by setting the sum of the moments about O_4 equal to zero. F_{14} can then be found from a force polygon for link 4.

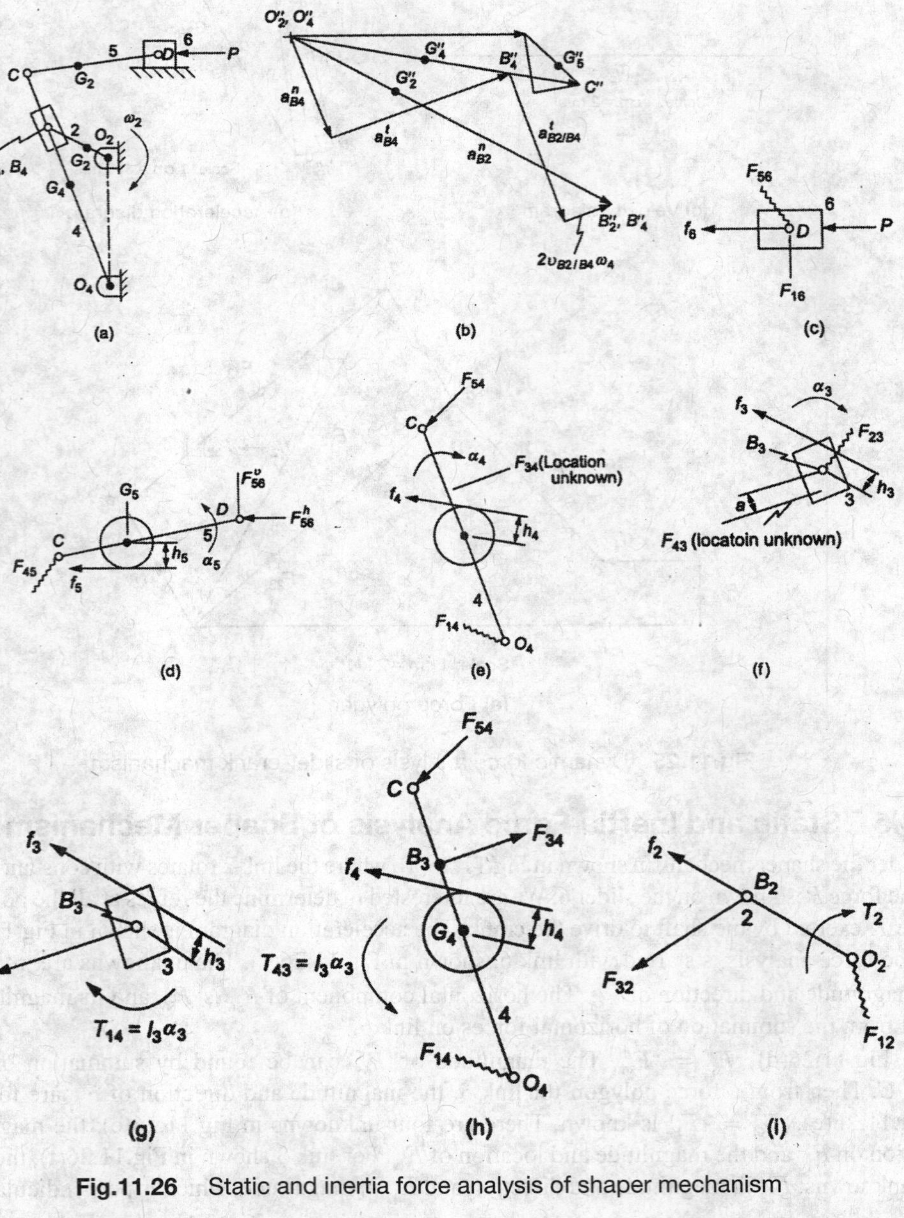

Fig.11.26 Static and inertia force analysis of shaper mechanism

We replaced F_{43} in Fig.11.26(f) with the force F_{43} and T_{43}, which are shown in Fig.11.26(g). Thus in Fig.11.26(f),

$$F_{43}a = T_{43} = I_3\alpha_3$$

or
$$a = \frac{I_3\alpha_3}{F_{43}}$$

F_{23} can now be determined from a force polygon for link 3. The free body diagram of link 2 is shown in Fig.11.26(i) and $F_{32} = -F_{23}$. F_{12} can be determined from a force polygon on link 2. Finally, by summing moments about O_2, the torque T_2 can be determined.

Example 11.11

The value of force applied to slider in a four-bar mechanism shown in Fig.11.27(a) is 4 kN. Determine the forces in various links and driving torque T_2.

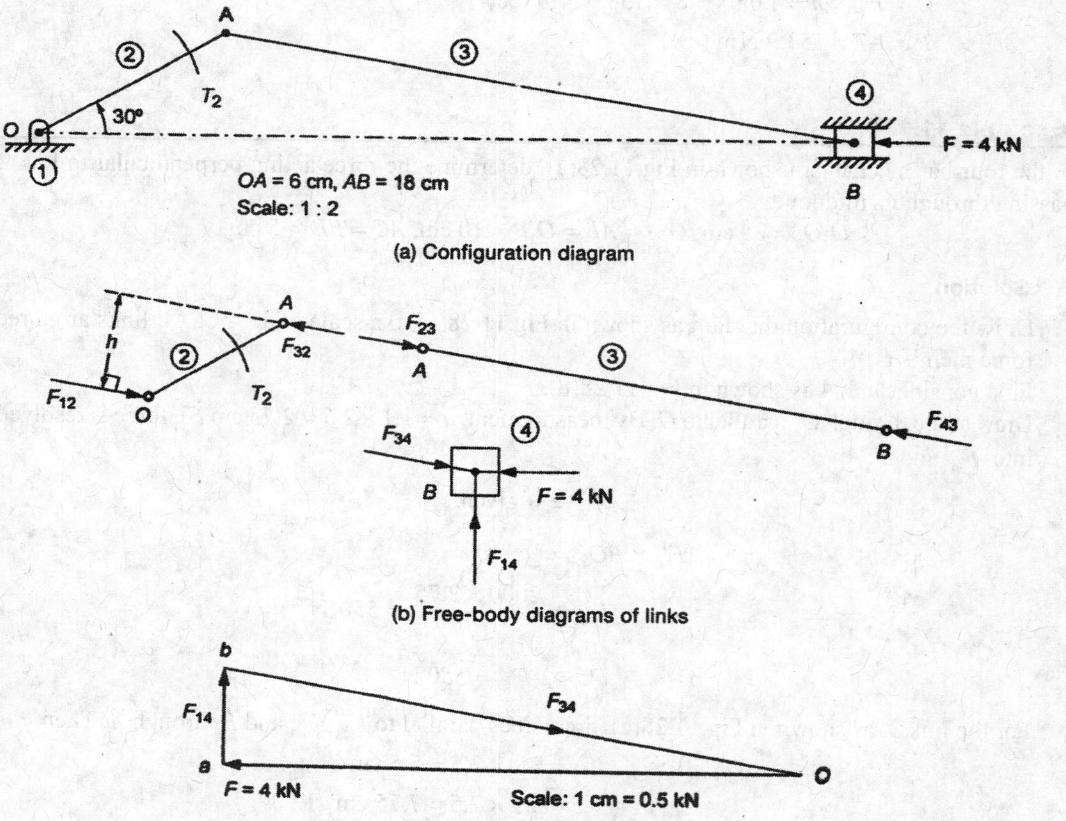

Fig.11.27 Static force analysis of a slider-crank mechanism

■ **Solution**

The free-body diagrams of links are shown in Fig.11.27(b).
There are three forces acting on link 4.

(i) $F = 4$ kN towards left, which is known as magnitude and direction.

(ii) F_{34} acting along AB whose magnitude is unknown.

(iii) F_{14} acting perpendicular to OB whose magnitude is unknown.

Draw the force polygon for member 4 as shown in Fig.11.27(c).

(i) Draw $Oa = F = 4$ kN to a scale of 1 cm = 0.5 kN

(ii) Draw ab perpendicular to OB and ob parallel to ab to intersect at b.

Then $F_{14} = ab = 1.3 \text{ cm} = 0.65 \text{ kN}$

$F_{34} = bO = 8.1 \text{ cm} = 4.05 \text{ kN}$

$h = 3.8 \text{ cm}$

Now $F_{34} = F_{43} = F_{23} = F_{32} = F_{21} = F_{12}$

$T = F_{32} \times h = 4.05 \times 3.8 = 153.9 \text{ Nm (ccw)}$

$T_2 = -T = 153.9 \text{ Nm (cw)}$

Example 11.12

In the four-bar mechanism shown in Fig.11.28(a), determine the force acting perpendicular to link 2 passing through its midpoint.

$$O_2O_4 = 25 \text{ cm}, O_2A = AB = O_4B = 10 \text{ cm}, AC = BD = 5 \text{ cm}.$$

■ **Solution**

Draw the configuration diagram as shown in Fig.11.28(a) to a scale of 1 : 2.5. All links are three force members.
First consider link 4 as shown in Fig.11.28(b).
Draw a line through O_4 parallel to Q. By measurement, $a = 1.1 \times 2.5 = 2.75$ cm F_{34} force is resolved into F_{34}^t and F_{34}^n

$$\Sigma M_{04} = 0 \text{ given}$$

$$F_{34}^t \times BO_4 = Q \times a$$

$$F_{34}^t = \frac{2000 \times 2.75}{10} = 550 \text{ N}$$

$$F^t_{34} = -F^t_{34} = 550 \text{ N}$$

For the link 3, as shown in Fig.11.28(c), draw lines parallel to F_{43}^t, F_{43}^t and P through A. Then

$$b = 2.2 \times 2.5 = 5.5 \text{ cm}$$

$$c = 3.1 \times 2.5 = 7.75 \text{ cm}$$

$$d = 1.5 \times 2.5 = 3.75 \text{ cm}$$

$$\Sigma M_A = 0 \text{ gives}$$

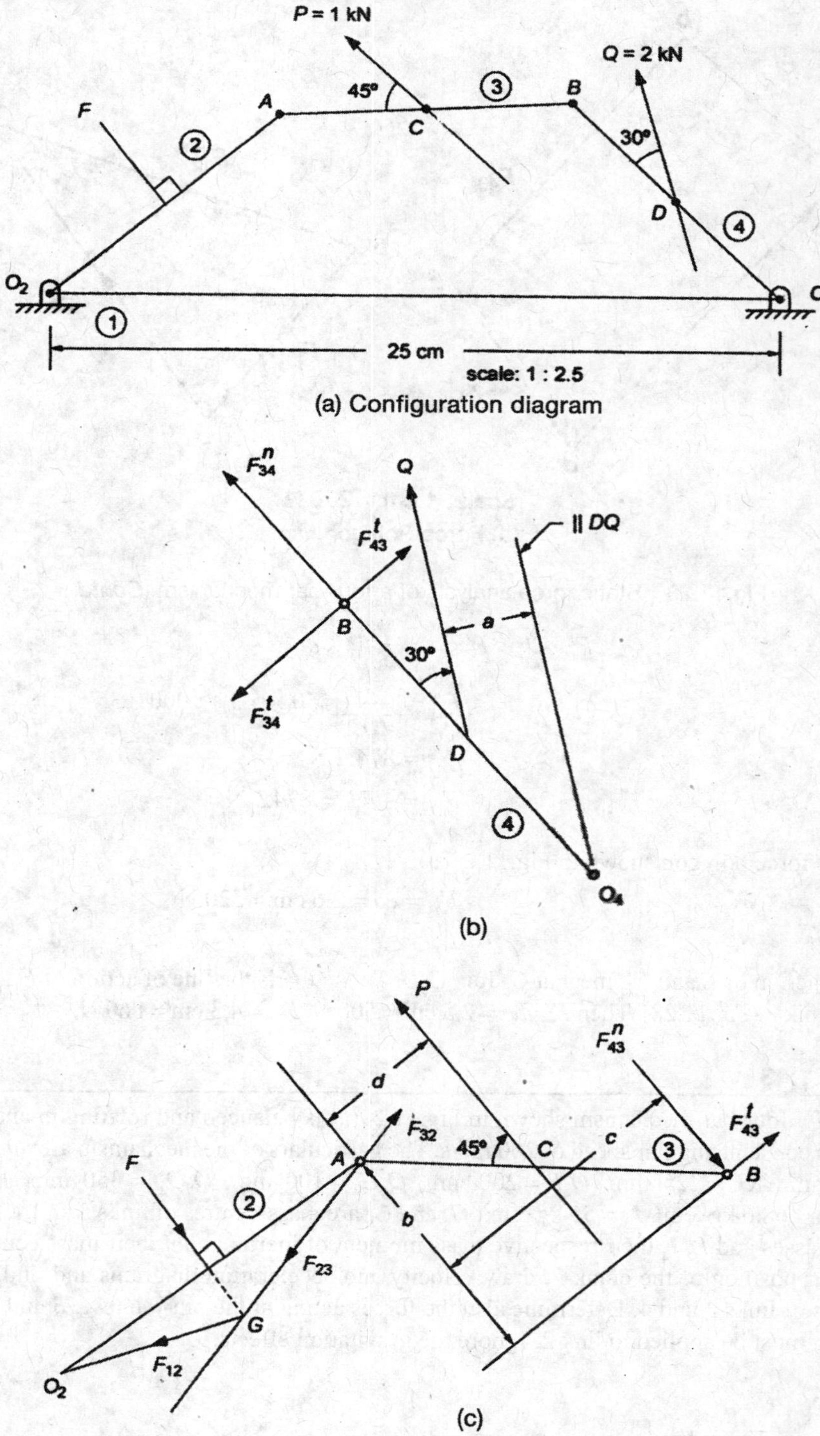

Fig.11.28 Static force analysis of a four-bar mechanism

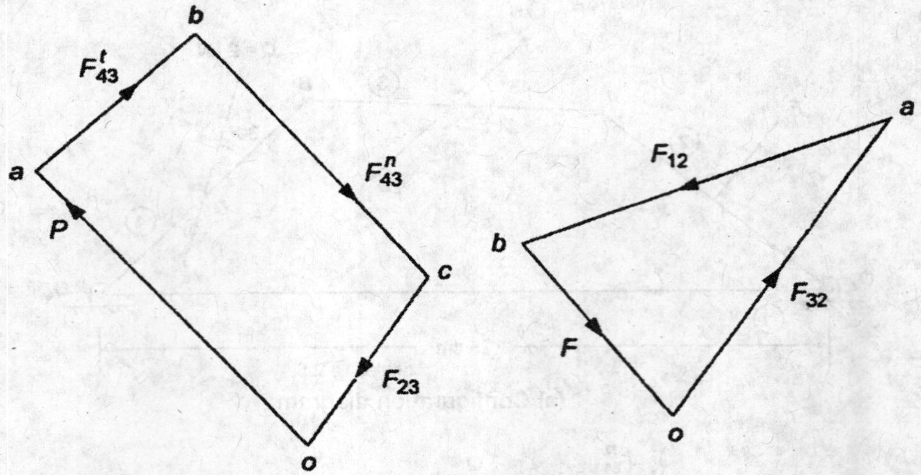

Scale: 1 cm = 200 N

(d) Force Polygon

Fig.11.28 Static force analysis of a four-bar mechanism (*Contd.*)

$$F^t_{43} \times b + P \times d = F^n{}_{43} \times c$$

$$F^n_{43} = \frac{1}{7.75}[550 \times 5.5 + 1000 \times 3.75]$$

$$= 874.2 \text{ N}$$

$$F^t_{43} = -F^n_{43} = 874.5 \text{ N}$$

Draw the force polygon shown in Fig.11.28(d).

$$F_{23} = co = 2.6 \text{ cm} = 520 \text{ N}$$

$$F_{32} = -F_{23}$$

Lines of action of F and F_{32} meet at G. Join O_2G. Then O_2G is the line of action of F_{12}. Draw force polygon for link 2 (Fig.11.28). Then $F = bo = 2.5$ cm = 500 N, $F_{12} = 4.3$ cm = 860 N.

Example 11.13

The crank of a four-bar mechanism shown in Fig.11.29(a) is balanced and rotating in anti-clockwise direction at a constant angular speed of 200 rad/s. The particulars of the mechanism are: $O_2A = 50$ mm, $AB = 450$ mm, $AG_3 = 225$ mm, $O_4B = 200$ mm, $O_4G_4 = 100$ mm, $O_2O_4 = 350$ mm, $W_3 = 1.2$ kg, $W_4 = 3$ kg, $I_3 = 68.6$ kg·cm², $I_4 = 55$ kg·cm². G_3 and G_4 are mass-centres of links 3 and 4, W_3, W_4 their respective masses and I_3, I_4 their respective mass moment of inertia about their mass centres. For the given angular position of the crank 2, draw velocity and acceleration diagrams and find the angular accelerations of links 3 and 4. Determine also the forces acting at the pin-joints A, B and the external torque which must be applied to link 2. Ignore the gravitation effects.

■ **Solution**

$$\omega = 200 \text{ rad/s}$$

$$v_a = O_2A \times \omega = 0.05 \times 200 = 10 \text{ m/s}$$

Draw the velocity diagram as shown in Fig.11.29(b).

$$v_{ba} = ab = 7.3 \text{ cm} = 14.6 \text{ m/s}, \; v_b = o_4b = 7.6 \text{ cm} = 15.2 \text{ m/s}$$

$$f_a^n = v_a^2/O_2A = (10)^2/0.05 = 2000 \text{ m/s}^2, f_{ba}^n = v_{ba}^2/AB \; (14.6)^2/0.45 = 473.7 \text{ m/s}^2$$

$$f_b^n = v_b^2/O_4B = (15.2)^2/0.2 = 1155.2 \text{ m/s}^2$$

Draw the acceleration diagram as shown in Fig.11.29(c).

$$f_{ba}^t = 0, \text{ Hence, } f_{ba}^n = f_{ba}$$

$$f_b = o_2b = 9.7 \text{ cm} = 1940 \text{ m/s}^2, f_b^t = 7.2 \text{ cm} = 1440 \text{ m/s}^2$$

$$\alpha_3 = f_{ba}^t/AB = 0, \; \alpha_4 = f_b^t/O_4B = 1440/0.2 = 7200 \text{ rad/s}^2 \text{ (cw)}$$

$$o_4G'_4 = O_4G_4 \times o_4b/O_4B = 100 \times 9.7/200 = 4.85 \text{ cm}$$

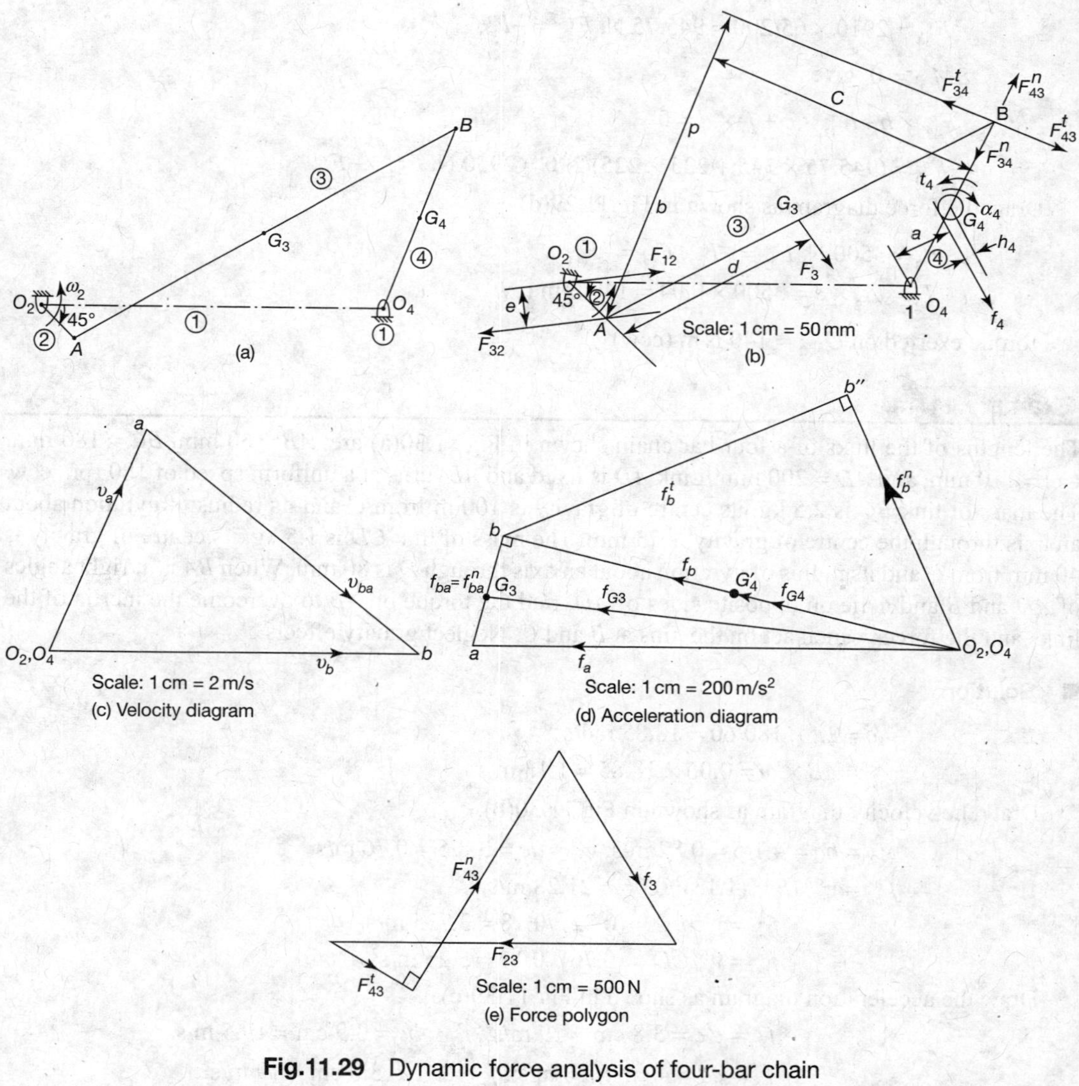

Fig.11.29 Dynamic force analysis of four-bar chain

$aG_3' = AG_3 \times ab/AB = 225 \times 2.368/450 = 1.184$ cm

$f_{G3} = o_2G_3' = 9.8$ cm $= 1960$ m/s^2, $f_{G4} = o_4G_4' = 4.85$ cm $= 970$ m/s^2

$F_3 = m_3 f_{G3} = 1.2 \times 1960 = 2352$ N, $f_3 = -F_3$

$F_4 = m_4 f_{G4} = 3 \times 970 = 2910$ N, $f_4 = -F_4$

$T_3 = I_3\, \alpha_3 = 0,\ t_3 = -T_3 = 0$

$T_4 = I_4\, \alpha_4 = 55 \times 10^{-4} \times 7200 = 39.6$ N m (cw), $t_4 = -T_4$ ccw

$h_3 = t_3/f_3 = 0,\ h_4 = t_4/f_4 = 39.6/2910 = 13.6$ mm

$a = 65$ mm, $b = 345$ mm, $c = 290$ mm, $d = 225$ mm, $e = 40$ mm

$\Sigma M_{O4} = 0$ gives

$F_{34}^t \times BO_4 = f_4 \times a$

$F_{34}^t = 2910 \times 65/200 = 945.75$ N, $F_{43}^t = -F_{34}^t$

$\Sigma M_A = 0$ gives

$F_{43}^t \times b - F_{43}^n \times c + f_3 \times d = 0$

$F_{34}^n = (945.75 \times 345 + 235 \times 225)/290 = 2950$ N, $F_{34}^n = -F_{43}^n$

Draw the force diagram as shown in Fig.11.29(d).

$F_{23} = 3500$ N, $F_{32} = -F_{23}$, $F_{12} = -F_{32}$

$T_2 = F_{32} \times e = 3500 \times 0.04 = 140$ N m (cw)

Torque exerted on $O_2A = 140$ N m (ccw)

Example 11.14

The lengths of the links of a four-bar chain shown in Fig.11.30(a) are: $AB = 60$ mm, $BC = 180$ mm, $CD = 110$ mm, and $AD = 200$ mm. Link AD is fixed and AB turns at a uniform speed of 180 rpm ccw. The mass of link BC is 2.5 kg, its centre of gravity is 100 m from C and its radius of gyration about an axis through the centre of gravity is 75 mm. The mass of link CD is 1.5 kg, its centre of gravity is 40 mm from C and its radius of gyration about an axis through D is 80 mm. When BA is at right angles of AD and B and C lie on opposite sides of AD, find the torque on AB to overcome the inertia of the links and the forces which act on the pins at B and C. Neglect gravity effects.

■ Solution

$\omega = 2\pi \times 180/60 = 18.85$ rad/s

$v_{ba} = AB \times \omega = 0.06 \times 18.85 = 1.13$ m/s

Draw the velocity diagram as shown in Fig.11.30(b).

$v_{cb} = bc = 4.1/5 = 0.82$ m/s, $v_{cd} = dc = 3.8/5 = 0.76$ m/s

$f_{ba}^n = v_{ba}^2/AB = (1.13)^2/0.06 = 21.28$ m/s ^{2}

$f_{cb}^n = v_{cd}^2/BC = (0.82)^2/0.18 = 3.735$ m/s^2

$f_{cd}^n = v_{cd}^2/CD = (0.76)^2/0.11 = 5.25$ m/s^2

Draw the acceleration diagram as shown in Fig.11.30(c).

$f_{cb}^t = c'c = 3.8$ cm $= 19$ m/s^2, $f_{cb} = bc = 3.9$ cm $= 19.5$ m/s^2

$f_{cd} = cd = 4$ cm $= 20$ m/s^2, $f_{cb}^t = c''c = 3.8$ cm $= 19$ m/s^2

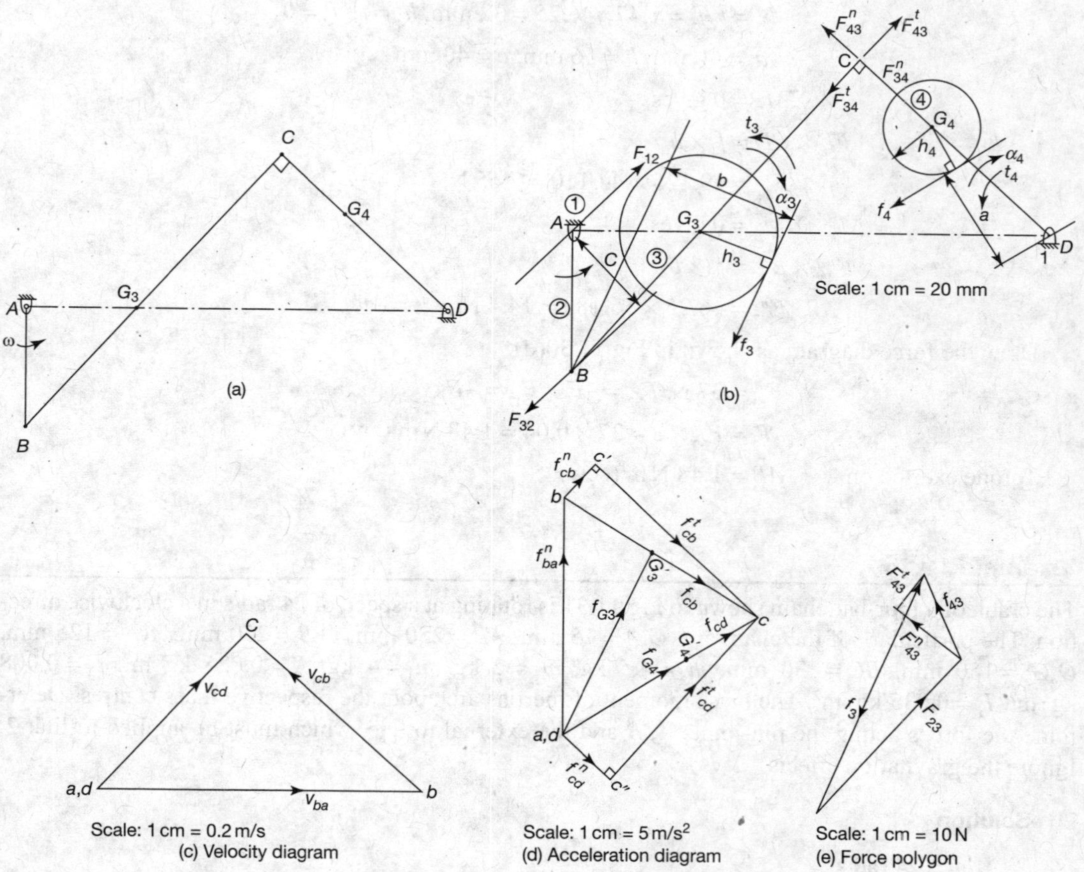

Fig.11.30 Dynamic force analysis of a four-bar chain

$$\alpha_3 = f^t_{cb}/BC = 19/0.18 = 105.5 \text{ rad/s}^2 \text{ (cw)},$$

$$\alpha_4 = f^t_{cd}/CD = 19/0.11 = 172.7 \text{ rad/s}^2 \text{ (cw)}$$

$$cG'_4 = CG_4 \times cd/CD = 40 \times 4/110 = 1.45 \text{ cm}$$

$$bG'_3 = BG_3 \times bc/BC = 80 \times 3.9/180 = 1.73 \text{ cm}$$

$$dG'_4 = 4 - 1.45 = 2.55 \text{ cm}$$

$$f_{G3} = aG'_3 = 3.7 \text{ cm} = 18.5 \text{ m/s}^2, f_{G4} = dG'_2 = 2.55 \text{ cm} = 12.75 \text{ m/s}^2$$

$$F_3 = m_3 f_{G3} = 2.5 \times 18.5 = 46.25 \text{ N}, f_3 = -F_3$$

$$F_4 = m_4 f_{G4} = 1.5 \times 12.75 = 19.125 \text{ N}, f_4 = -F_4$$

$$I_3 = m_3 K^2_3 = 2.5 \times (0.075)^2 = 0.014 \text{ kg}\cdot\text{m}^2$$

$$I_4 = m_4 K^2_4 = 1.5 \times (0.08^2 - 0.07^2) = 0.00225 \text{ kg}\cdot\text{m}^2$$

$$T_3 = I_3\, \alpha_3 = 0.014 \times 105.5 = 1.477 \text{ Nm (cw)}, t_3 = -T_3 \text{ ccw}$$

$$T_4 = I_4\, \alpha_4 = 0.00225 \times 172.7 = 0.388 \text{ Nm (cw)}, t_4 = -T_4 \text{ ccw}$$

$$h_3 = t_3/f_3 = 1.477/46.25 = 32 \text{ mm}, h_4 = t_4/f_4 = 0$$
$$a = 44 \text{ mm}, b = 56 \text{ mm}, c = 40 \text{ mm}$$
$$\Sigma M_D = 0 \text{ gives}$$
$$F^t_{34} \times CD = f_4 \times a$$
$$F^t_{34} = 19.125 \times 44/110 = 7.65 \text{ N}, F^t_{34} = -F^t_{34}$$
$$\Sigma M_B = 0 \text{ gives}$$
$$F^n_{43} \times BC = f_3 \times b$$
$$F^n_{43} = 46.25 \times 56/180 = 14.4 \text{ N}, F^n_{34} = -F^n_{43}$$

Draw the force diagram as shown in Fig 11.30(d).

$$F_{23} = 37 \text{ N}, F_{32} = -F_{23}, F_{12} = -F_{32}$$
$$T_2 = F_{32} \times c = 37 \times 0.04 = 1.48 \text{ N m (cw)}$$

Torque exerted on $AB = 1.48$ N m (ccw)

Example 11.15

The crank of a four-bar chain shown in Fig.11.31 is rotating at a speed of 24 rad/s in a clockwise direction. The particulars of the chain are: $O_2A = 75$ mm, $AB = 250$ mm, $O_4B = 250$ mm, $AG_3 = 125$ mm, $O_2G_4 = 150$ mm, $BC = 130$ mm, $m_2 = 4.5$ kg, $m_3 = 2$ kg, $m_4 = 4$ kg, $I_2 = 0.025$ kg·m², $I_3 = 0.008$ kg·m², $I_4 = 0.035$ kg·m². The mass moment of inertias are about the respective mass centres. Determine the forces acting the pin-joints A, B and the external torque, which must be applied to link 2. Ignore the gravitation effects.

Solution

$$\omega_2 = 24 \text{ rad/s}$$
$$v_a = AB \times \omega_2 = 0.075 \times 24 = 1.8 \text{ m/s}$$

Draw the velocity diagram as shown in Fig.11.31(b)

$$v_{cb} = bc = 1.6 \text{ cm} = 0.8 \text{ m/s}, v_c = o_4c = 3.1 \text{ cm} = 1.55 \text{ m/s},$$
$$v_{ba} = ab = 1.9 \text{ cm} = 0.95 \text{ m/s}, v_b = o_4b = 3.1 \text{ cm} = 1.55 \text{ m/s}$$
$$v_{G4} = o_4G_4 = 1.8 \text{ cm} = 0.9 \text{ m/s}$$

Draw the acceleration diagram as shown in Fig.11.31(c).

$$f^n_a = v^2_a/O_2A = (1.8)^2/0.075 = 43.2 \text{ m/s}^2, f^n_{ba} = v^2_{ba}/AB = (0.95)^2/0.25 = 3.61 \text{ m/s}^2$$
$$f^n_b = v^2_b/O_4B = (1.55)^2/0.25 = 9.61 \text{ m/s}^2, f^n_c = v^2_c/O_4C = (1.55)^2/0.25 = 9.61 \text{ m/s}^2$$
$$f^n_{cb} = v^2_{cb}/BC = (0.8)^2/0.13 = 4.92 \text{ m/s}^2, f^n_{G4} = v^2_{G4}/O_4G_4 = (0.9)^2/0.15 = 5.4 \text{ m/s}^2$$
$$o_4G'_4 = O_4G_4 \times o_4d/O_4D = 150 \times 2.5/250 = 1.5 \text{ cm}, f_{G4} = 1.5 \times 10 = 15 \text{ m/s}^2$$
$$f^t_{ba} = b'b = 4.6 \text{ cm} = 46 \text{ m/s}^2, f_{ba} = ab = 4.7 \text{ cm} = 47 \text{ m/s}^2$$
$$f^t_b = b''b = 2.4 \text{ cm} = 24 \text{ m/s}^2, f_b = o_4b = 2.6 \text{ cm} = 26 \text{ m/s}^2$$
$$\alpha_3 = f^t_{ba}/AB = 46/0.25 = 18.4 \text{ rad/s}^2 \text{ (ccw)}, \alpha_4 = f^t_b/O_4B = 24/0.25 = 96 \text{ rad/s}^2 \text{ (cw)}$$
$$aG'_3 = AG_3 \times ab/AB = 125 \times 4.7/250 = 2.35 \text{ cm}, f_{G3} = o_2G'_3 = 2.5 \text{ cm} = 25 \text{ m/s}^2$$
$$F_2 = 0$$

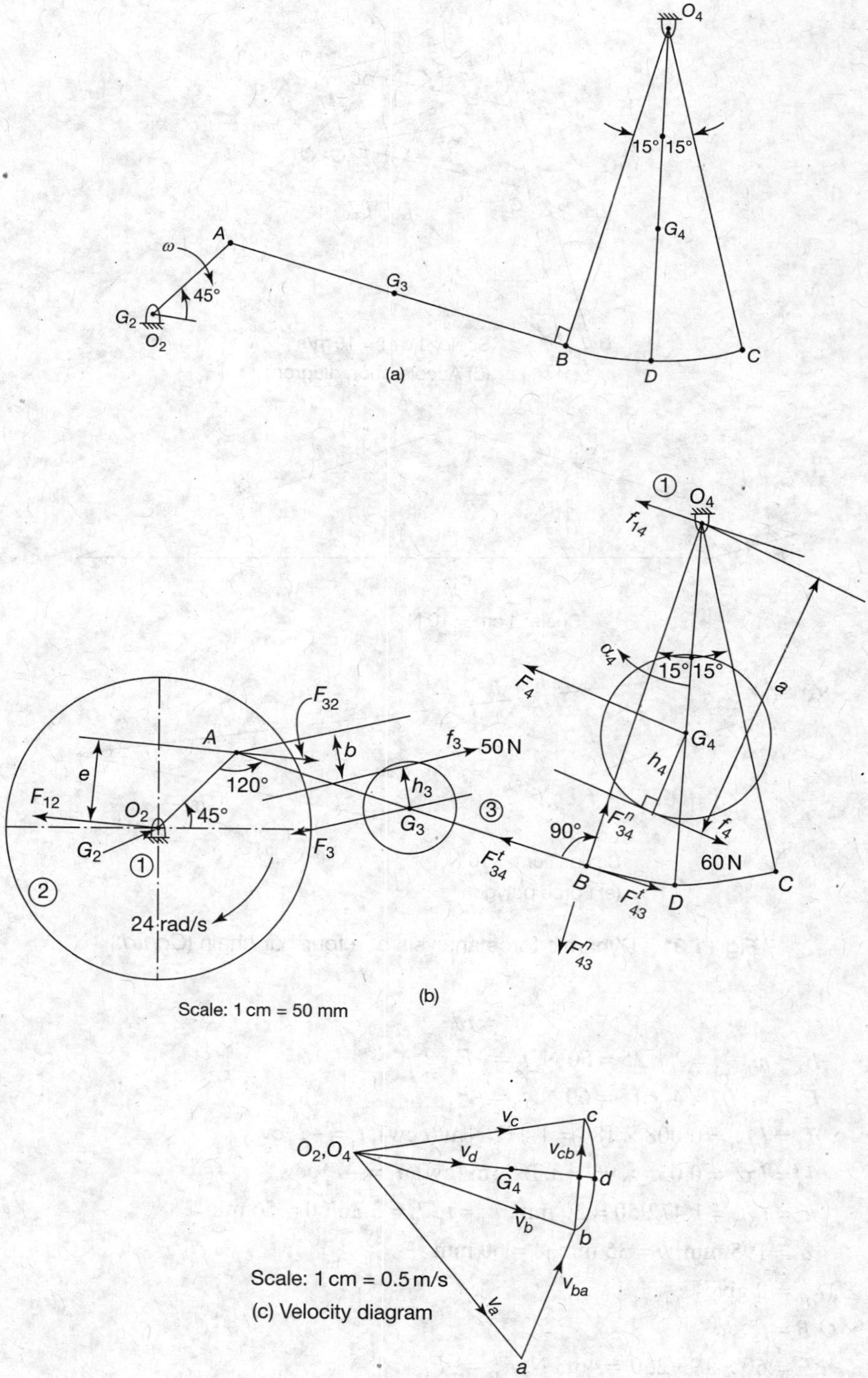

Fig.11.31 Dynamic force analysis of a four-bar chain

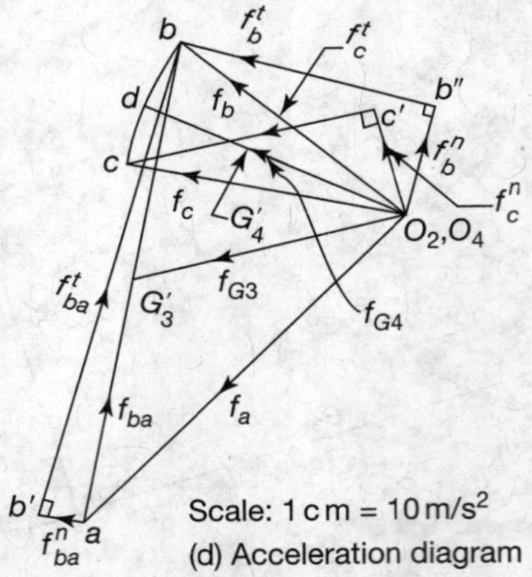

Scale: 1 cm = 10 m/s²

(d) Acceleration diagram

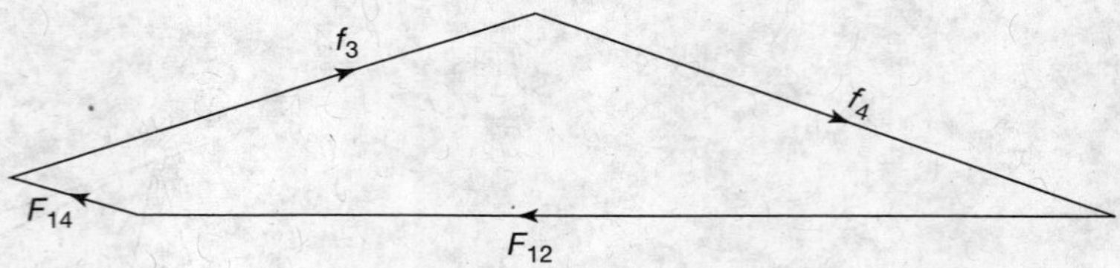

Scale: 1 cm = 10 N

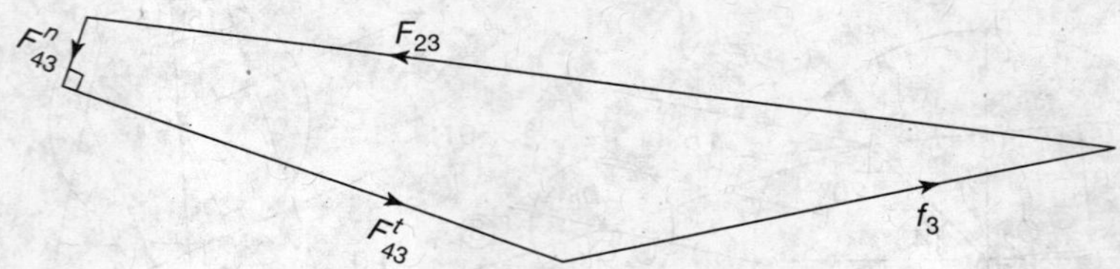

Scale: 1 cm = 10 N

(e) Force polygon

Fig.11.31 Dynamic force analysis of a four-bar chain (*Contd.*)

$$F_3 = m_3 f_{G3} = 2 \times 25 = 50 \text{ N}, f_3 = -F_3$$
$$F_4 = m_4 f_{G4} = 4 \times 15 = 60 \text{ N}, f_4 = -F_4$$
$$T_3 = I_3 \alpha_3 = 0.008 \times 184 = 1.472 \text{ N m (ccw)}, t_3 = -T_3 \text{ cw}$$
$$T_4 = I_4 \alpha_4 = 0.035 \times 96 = 3.36 \text{ N m (cw)}, t_4 = -T_4 \text{ ccw}$$
$$h_3 = t_3/f_3 = 1.472/50 = 29 \text{ mm}, h_4 = t_4/f_4 = 3.36/60 = 56 \text{ mm}$$
$$a = 195 \text{ mm}, b = 35 \text{ mm}, e = 60 \text{ mm}$$
$$\Sigma M_{O4} = 0 \text{ gives}$$
$$F_{34}^t \times O_4B = f_4 \times a$$
$$F_{34}^t = 60 \times 195/250 = 46.8 \text{ N}, F_{43}^t = -F_{34}^t$$
$$\Sigma M_a = 0 \text{ gives}$$

$F^n_{43} \times BC = f_3 \times b$

$F^n_{43} = 50 \times 35/250 = 7$ N, $F^n_{34} = -F^n_{43}$

Draw the force diagram as shown in Fig.11.31(d).

$F_{23} = 93$ N, $F_{32} = -F_{23}$, $F_{12} = F_{32}$

$T_2 = F_{32} \times e = 93 \times 0.06 = 5.58$ Nm (cw)

Torque exerted on $AB = 5.58$ Nm (ccw)

Draw the force polygon for the mechanism as shown in Fig.11.31(e).

$F_{14} = 12$ N

Summary for Quick Revision

1. Forces acting on the system from outside are called applied forces.
2. A member under the action of two forces shall be in equilibrium if the forces are of the same magnitude, act along the same line, and are in equilibrium.
3. A member under the action of three forces shall be in equilibrium if the resultant of the forces is zero and the lines of action of the forces intersect at a point, i.e. the forces are concurrent.
4. A member under the action of two forces and an applied torque shall be in equilibrium if the forces are equal in magnitude, parallel having different lines of action and opposite in sense, i.e. form a couple which is equal and opposite to the applied torque.
5. The force exerted by a member i on another member j is represented by F_{ij}.
6. A force unknown in magnitude and known in direction is represented by a solid line without an arrow.
7. A force unknown in magnitude and direction is represented by a wavy line.
8. A free body diagram is a diagram of a link isolated from the mechanism showing both active and reactive forces acting on it.
9. The principle of superposition states that if a number of forces act on a system the net effect is equal to the superposition of the effects of the individual forces taken one at a time.
10. A linear system is one in which the output force is directly proportional to the input force. The principle of superposition holds good for a linear system.
11. D'Alembert's principle can be used to convert a dynamic system into an equivalent static system.
12. Equivalent offset inertia force accounts for both inertia force and inertia couple. This is obtained by displacing the line of action of the inertia force from the centre of mass.

Multiple Choice Questions

1. The forces generally considered in the design of mechanisms are:
   (a) applied forces (b) inertia forces (c) frictional forces (d) all of them.
2. A pair of action and reaction forces acting on a body are called
   (a) applied forces (b) inertia forces (c) frictional forces (d) constraint forces.
3. For the static equilibrium of planar mechanisms
   (a) $\Sigma F_x = 0$ b) $\Sigma F_y = 0$ (c) $\Sigma M_O = 0$ (d) all of the above.

4 If the resultant of forces acting on a body does not pass through the centre of mass, then the inertia force and inertia couple is replaced by
(a) Equivalent inertia force (b) equivalent inertia couple
(b) Equivalent offset inertia force (c) equivalent offset inertia couple

Answers

1. (d) 2. (d) 3. (d) 4. (c)

Review Questions

1 State the conditions of static equilibrium.
2 State the principle of superposition.
3 State the D'Alembert's principle.
4 What is equivalent offset inertia force?
5 What is inertia force?
6 Which do you mean by a static force?
7 What is an applied force?
8 State the conditions for the equilibrium of a body under the following system of loading. (a) Two forces, (b) three forces, and (c) two forces and a torque.

Exercises

11.1 A four-bar mechanism shown in Fig.11.32 is subjected to a force of 2 $\angle 60°$ kN of link CD. The dimensions of the various links are:

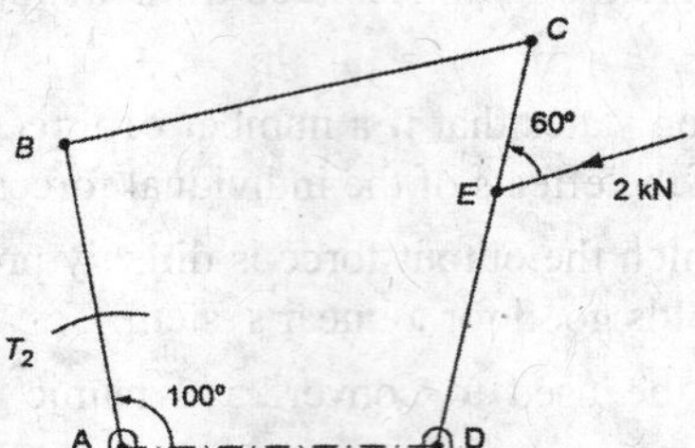

Fig.11.32 Four-bar mechanism

$AB = AD = 300$ mm, $BC = 450$ mm, $CD = 400$ mm, $CE = 150$ mm.

Calculate the required value of torque to be applied to link AB for static equilibrium of the mechanism.

11.2 A four-bar mechanism shown in Fig.11.33 is subjected to a force as shown. The dimensions of the various links are:

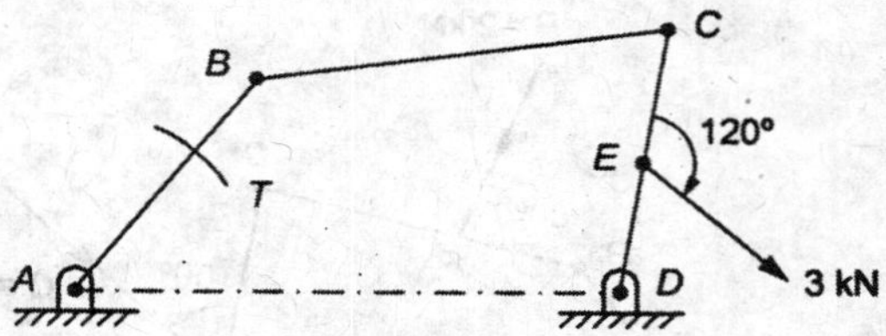

Fig.11.33 Four-bar mechanism

$AB = CD = 200$ mm, $BC = 300$ mm, $AD = 400$ mm, $CE = 100$ mm.

Calculate the input torque to link AB for the static equilibrium of the mechanism.

11.3 A four-bar mechanism shown in Fig.11.34 is subjected to torques $T_3 = 50$ N m and $T_4 = 60$ N m. The dimensions of the various links are:

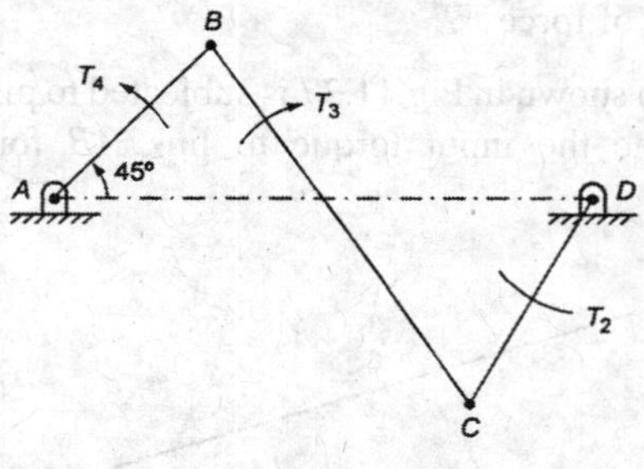

Fig.11.34 Four-bar mechanism

$AB = CD = 400$ mm, $BC = 800$ mm, $AD = 1000$ mm

Calculate the input torque to link CD for the static equilibrium of the mechanism.

11.4 A slider crank mechanism is loaded as shown in Fig.11.35. $AB = 400$ mm, $BC = 600$ mm, $AD = 200$ mm, $CE = 300$ mm. Calculate the input torque for the static equilibrium of the mechanism.

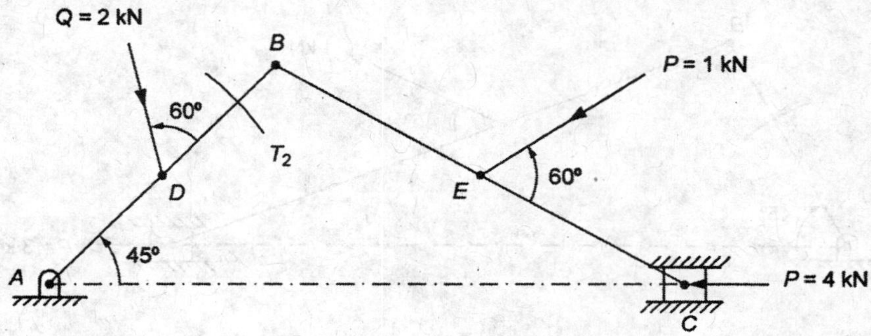

Fig.11.35 Slider-crank mechanism

11.5 The lengths of crank and connecting rod of a slider crank mechanism are 40 mm and 100 mm, respectively. It is subjected to piston force of 2000 N. Determine the required input torque on the crank for the static equilibrium.

11.6 A four-bar mechanism is loaded as shown in Fig.11.36. $AB = CD = 300$ mm. $BC = 250$ mm, $AE = CG = 150$ mm, $BF = 100$ mm, $AD = 500$ mm

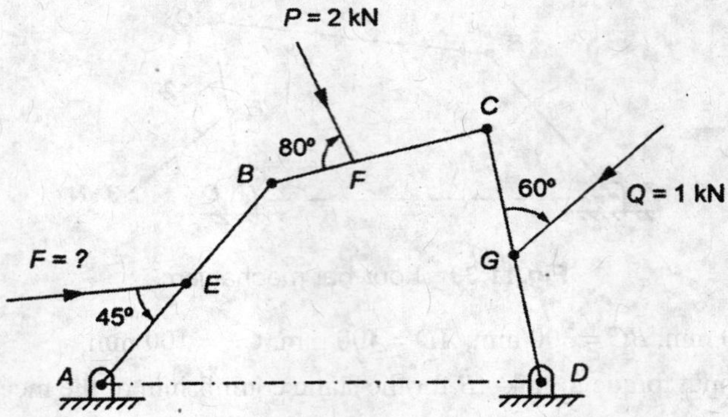

Fig.11.36 Four-bar mechanism

Determine the magnitude of force F.

11.7 A slider-crank mechanism shown in Fig.11.37 is subjected to piston load of 1 kN, $AB = 250$ mm, $BC = 600$ mm. Determine the input torque to link AB for the static equilibrium of the mechanism.

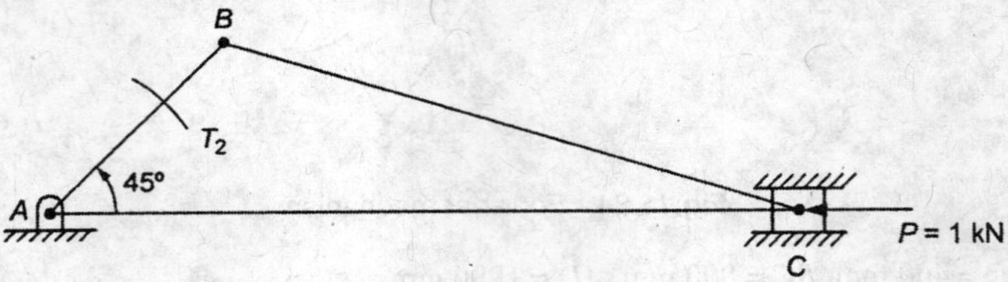

Fig.11.37 Slider-crank mechanism

11.8 A slider-crank mechanism shown in Fig.11.38 is subjected to piston load of 3 kN and a force $1 \angle 45°$ kN or link BC.

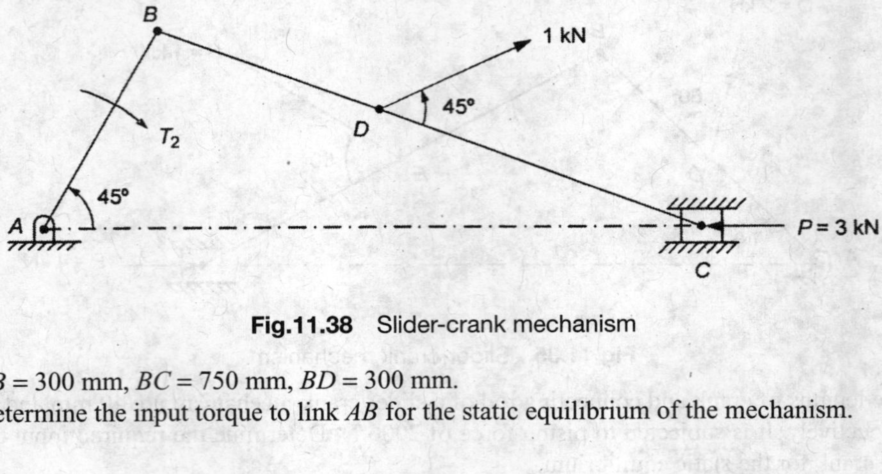

Fig.11.38 Slider-crank mechanism

$AB = 300$ mm, $BC = 750$ mm, $BD = 300$ mm.
Determine the input torque to link AB for the static equilibrium of the mechanism.

12

BALANCING

Chapter Outline

12.1 INTRODUCTION

The high-speed engines and other machines are being used frequently. Unbalance in these machines arises either due to eccentric rotating or reciprocating masses or geometric centre not coinciding with the mass centre of the rotating components. These masses give rise to dynamic forces that increase the bearing loads and introduce severe stresses in the machine components.

The eccentric rotating or reciprocating mass is called the disturbing mass.

The various causes of unbalance are:

1. Eccentric rotating or reciprocating masses.
2. Unsymmetry caused during production process.

3. Non-homogeneity of materials.
4. Elastic deformations during running.
5. Faulty mounting resulting in eccentricity.
6. Misalignment of bearings.
7. Plastic deformations.

Unbalance introduces severe stresses and result in undersirable vibrations in the machines.

By balancing we mean to eliminate either partially or completely the effects due to unbalanced resultant inertia forces and couples to avoid vibration of a machine or device.

12.2 BALANCING OF ROTATING MASSES

A system of rotating masses are in static balance if the combined mass centre of the system of masses lies on the axis of rotation. For dynamic balance of a system of masses, there does not exist any resultant centrifugal force as well as resultant couple.

12.2.1 Single Rotating Mass

(a) Balance mass in the same plane as the disturbing mass: Consider a single mass M rotating with angular speed ω at a radius r, as shown in Fig.12.1. The centrifugal force due to this mass is

$$F_m = Mr\omega^2$$

If a balancing mass B is placed on this rotating machine component in the same plane at a radius b and in line with the mass M at 180°, then the centrifugal force due to mass B will be

$$F_b = Bb\omega^2$$

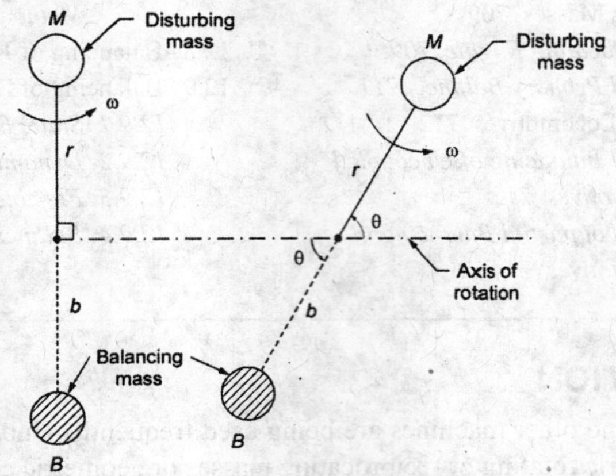

Fig.12.1 Single rotating mass

For the equilibrium of the system, we have

$$F_m = F_b$$

or

$$Mr = Bb \tag{12.1}$$

(b) Two balance masses in different planes: If the balance mass cannot be placed in the same plane as the rotating mass, then two parallel masses in different planes may be used to balance the rotating mass. A force in a plane can be replaced by a force in a parallel plane having the direction of the original force along with a couple in the reference plane formed by the product of the force and the perpendicular distance (arm of the couple) between the parallel planes. A plane passing through a point on the parallel plane perpendicular to the axis of the shaft is called the *reference plane* (*RP*).

The balance masses may be either on the same side of the unbalance mass or on opposite sides. The equilibrium equations would require that the resultant sum of their moments about any point in the same plane must be zero.

1. *Balance masses on the same side of the disturbing mass*

Consider a mass M at A rotating at a radius r and two balance masses B_1 and B_2 at B and C, parallel to M, rotating at radii b_1 and b_2 respectively, as shown in Fig.12.2(a). Let l_1 and l_2 be the distances of these masses from M.

Taking moments about B, we have

$$Mrl_1 = B_2b_2(l_2 - l_1)$$

or
$$B_2b_2 = Mr\left(\frac{l_1}{l_2 - l_1}\right) \tag{12.2a}$$

Taking moments about C, we have

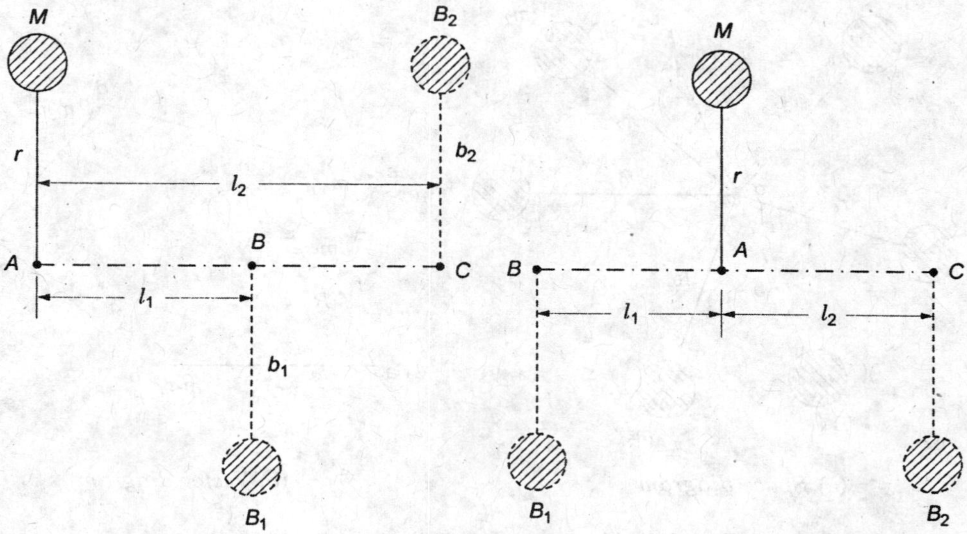

(a) Balance masses on one side of disturbing mass

(a) Balance masses on both sides of disturbing mass

Fig.12.2 Two balance masses

$$Mrl_2 = B_1b_1(l_2 - l_1)$$

or
$$B_1b_1 = Mr\left(\frac{l_2}{l_2 - l_1}\right) \tag{12.2b}$$

2. *Balance masses on the opposite sides of the disturbing mass*

Consider now that the two balance masses B_1 and B_2 are on the opposite sides of the disturbing mass M, as shown in Fig.12.2(b). Taking moments about B, we have

$$Mrl_1 = B_2 b_2 (l_1 + l_2)$$

or $$B_2 b_2 = Mr\left(\frac{l_1}{l_1 + l_2}\right) \tag{12.3a}$$

Now taking moments about C, we have

$$Mrl_2 = B_1 b_1 (l_1 + l_2)$$

or $$B_1 b_1 = Mr\left(\frac{l_2}{l_1 + l_2}\right) \tag{12.3b}$$

12.2.2 Many Masses Rotating in the Same Plane

Let there be M_i, $i = 1$ to n, masses rotating in the same plane with radii r_i, $i = 1$ to n and with same angular speed ω, as shown in Fig.12.3(a), so that the centrifugal force due to each mass is,

$$F_i = M_i r_i \omega^2$$

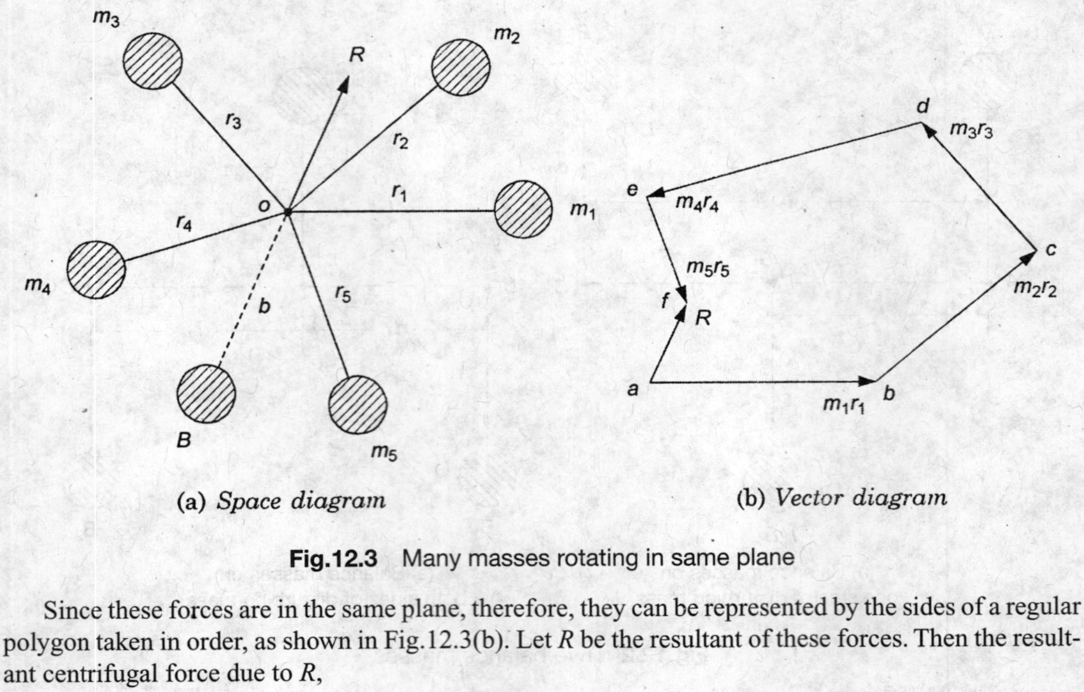

(a) *Space diagram* (b) *Vector diagram*

Fig.12.3 Many masses rotating in same plane

Since these forces are in the same plane, therefore, they can be represented by the sides of a regular polygon taken in order, as shown in Fig.12.3(b). Let R be the resultant of these forces. Then the resultant centrifugal force due to R,

$$R = \Sigma M_i r_i \omega^2$$

If a balancing mass B is placed at a radius b at 180° with R, then the centrifugal force due to B is

$$F_b = Bb\omega^2$$

For the equilibrium of the system, we have

$$R = F_b$$

or

$$\Sigma M_i r_i = Bb \tag{12.4}$$

From Eq. (12.4), it may be seen that the force polygon may be drawn for $M_i\, r_i$ instead of $M_i\, r_i\, \omega^2$.

Example 12.1

Four masses 150, 250, 200 and 300 kg are rotating in the same plane at radii of 0.25, 0.2, 0.3 and 0.35 m, respectively. Their angular location is 40°, 120° and 250° from mass 150 kg, respectively, measured in counter-clockwise direction. Find the position and magnitude of the balance mass required, if its radius of rotation is 0.25 m.

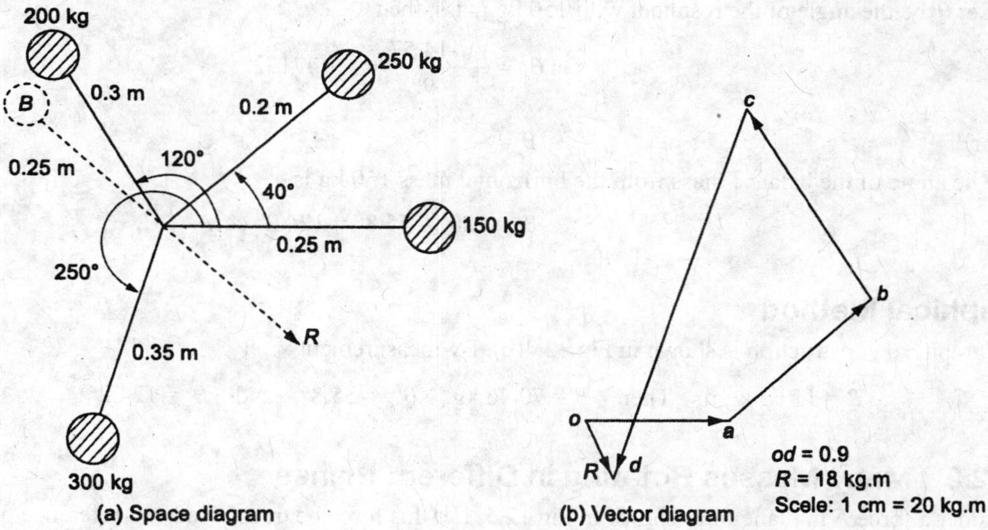

Fig.12.4 Balancing of four masses in the same plane

Solution

The mass space diagram is shown in Fig.12.4(a). The problem can be solved either analytically or graphically.

Analytical Method

Table 12.1

M kg	r (m)	Mr (kgm)	θ (deg)	$H = Mr\cos\theta$ (kgm)	$V = Mr\sin\theta$ (kgm)
150	0.25	37.5	0	37.5	0
250	0.20	50.0	40	38.3	32.14
200	0.30	60.0	120	–30.0	51.96
300	0.35	105.0	250	–35.9	–98.67

From Table 12.1,

$$\Sigma H = 9.9 \quad \Sigma V = -14.57$$

Resultant,

$$R = \left[(\Sigma H)^2 + (\Sigma V)^2\right]^{0.5}$$

$$= \left[(9.9)^2 + (-14.57)^2\right]^{0.5} = 17.61 \text{ kgm}$$

Let B be the balancing mass, then

$$0.25\,B = 17.61$$

or

$$B = 70.46 \text{ kg}$$

Let θ_r be the angle of the resultant with 150 kg mass, then

$$\tan\theta_r = -\frac{-14.57}{-9.9} = -1.47172$$

or

$$\theta_r = -55.8°$$

The angle of the balance mass from the horizontal mass 150 kg is

$$\theta_b = 180° - 55.8° = 124.2° \text{ ccw}$$

Graphical Method

The graphical construction is shown in Fig.12.4(b). By measurement

$R = 17.61$ kgm. Then $B = 70.46$ kg, $\theta_r = -55.8°$ and $\theta_b = 124.2°$

12.2.3 Many Masses Rotating in Different Planes

Consider a force F in plane B, as shown in Fig.12.5. Let this force be transferred to a reference plane A at a distance 'a'. The effect of transferring a force F from plane B to plane A is:

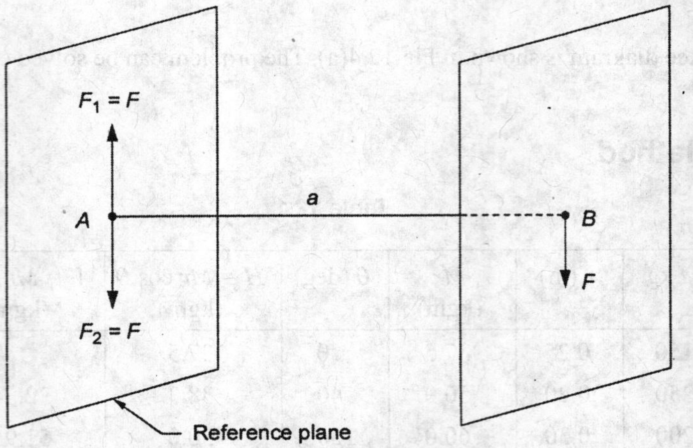

Fig.12.5 Equivalence of a single force into a force and a couple

(i) an unbalance force $F_2 = F$ on plane A, and

(ii) an unbalanced couple, $C = F_a$.

The couple is represented by a vector at right angles to the plane of the couple and the arrow head points in the direction in which a right hand screw would move if acted upon by the couple. In practice, the phase of the couple diagram is rotated through 90° ccw.

This leads to the balancing equations, in general

$$\Sigma M_i r_i = 0$$

$$\Sigma M_i r_i a = 0$$

Let us consider the mass system as shown in Fig.12.6(a). The orientation of the forces is shown in Fig.12.6(b). The couples acting on the system are:

$$C_1 = M_1 r_1 l_1$$

$$C_2 = M_2 r_2 l_2$$

$$C_3 = M_3 r_3 l_3$$

$$C_4 = M_4 r_4 l_4$$

The couples are shown in Fig.12.6(c), and when turned through 90°, are shown in Fig.12.6(d). The couple vectors may be fixed in their correct relative positions by drawing them radially outwards along the corresponding radii for all masses, which lie on one side for the fixed point A. Similarly all masses which lie on the other side of the fixed point A can be represented radially inwards along the corresponding radii. The fixed point A is taken as the point of intersection of the plane of rotation of one of the balancing masses B_1 and the axis of rotation, in order to eliminate the couple due to the mass in this plane. The plane at A is known as the *reference plane*. The couple polygon has been drawn in Fig.12.5(e). The resultant couple is represented by C_B, the closing side of the couple polygon, as shown

Now $$C_B = B_2 b_2 d$$

or $$B_2 = \frac{C_B}{b_2 d} \quad (12.5)$$

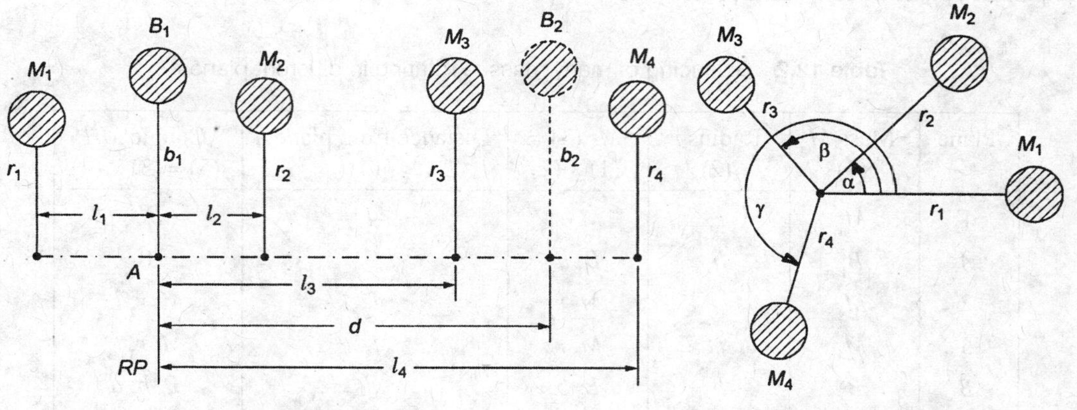

(a) *Position of planes* (b) *Position of masses*

Fig.12.6 Several masses rotating in different planes

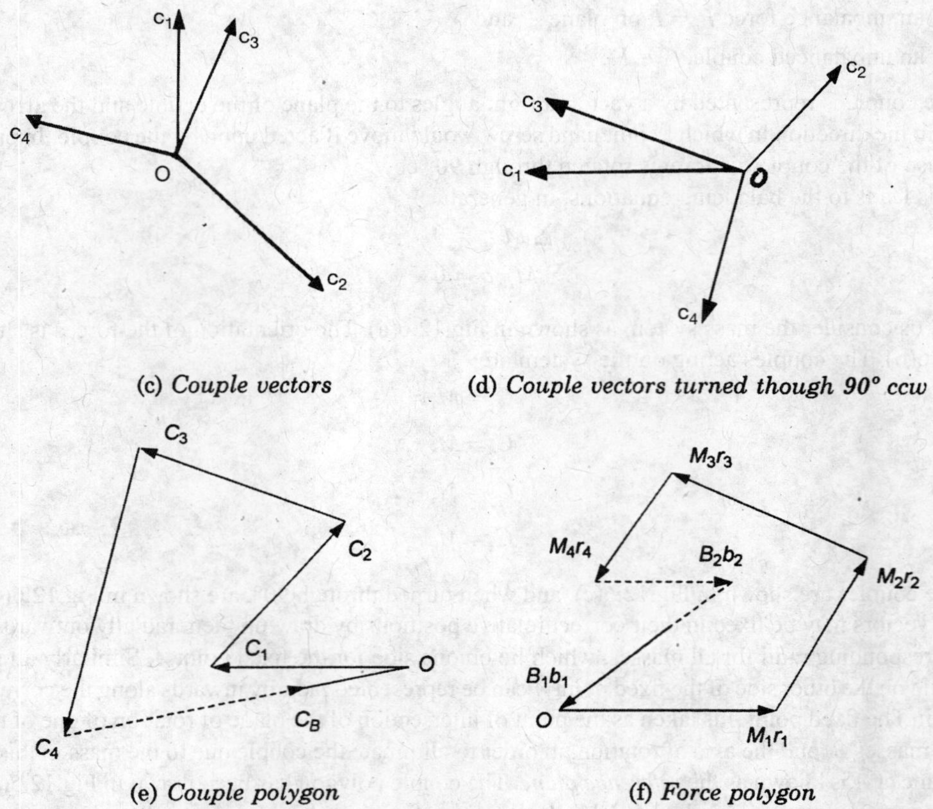

(c) *Couple vectors* (d) *Couple vectors turned though 90° ccw*

(e) *Couple polygon* (f) *Force polygon*

Fig.12.6 Several masses rotating in different planes (*Contd.*)

Knowing B_2 and its direction, draw the force polygon, as shown in Fig.12.6(f). The closing side of the force polygon will represent the magnitude and direction of the force due to the balancing mass required in plane A. The whole process can be represented as shown in Table 12.2.

Table 12.2 Balancing of many masses rotating in different planes

Plane	Mass M (1)	Radius r (2)	Mr (3) = (1) × (2)	Distance from plane A l (4)	Couple Mrl (5) = (3) × (4)
1	M_1	r_1	$M_1 r_1$	$-l_1$	$-M_1 r_1 l_1$
A	B_1	b_1	$B_1 b_1$	0	0
2	M_2	r_2	$M_2 r_2$	l_2	$M_2 r_2 l_2$
3	M_3	r_3	$M_3 r_3$	l_3	$M_3 r_3 l_3$
B	B_2	b_2	$B_2 b_2$	d	$B_2 b_2 d$
4	M_4	r_4	$M_4 r_4$	l_4	$M_4 r_4 l_4$

12.2.4 Analytical Method for Balancing of Rotating Masses

(a) ***Several masses in the same plane***

Let M_i = number or masses, $i = 1, 2, 3, \ldots 1$
r_i = radius of mass M_i
θ_i = angle of mass M_i with x-axis measured ccw
B = Balancing mass
b = radius of balancing mass
θ_b = angle of mass B with x-axis measured ccw.

Considering forces along the x- and y-axis, we have

$$\Sigma M_i r_i \cos\theta_i + Bb\cos\theta_b = 0$$

$$\Sigma M_i r_i \sin\theta_i + Bb\sin\theta_b = 0$$

$$\left[(\Sigma M_i r_i \cos\theta_i)^2 + (\Sigma M_i r_i \sin\theta_b)^2\right]^{0.5} = Bb$$

or

$$B = \frac{\left[(\Sigma M_i r_i \cos\theta_i)^2 + (\Sigma M_i r_i \sin\theta_b)^2\right]^{0.5}}{b} \quad (12.6)$$

$$\tan\theta_b = \frac{\Sigma M_i r_i \sin\theta_i}{\Sigma M_i r_i \cos\theta_i} \quad (12.7)$$

(b) ***Several masses in different planes***

If M_L and M_M be the balance forces at radii r_L and r_M respectively, then for the balance of couples about plane L, we have

$$\left[(\Sigma M_i r_i l_i \cos\theta_i)^2 + (\Sigma M_i r_i \sin\theta_i)^2\right]^{0.5} = M_M r_M l_M \quad (12.8)$$

$$\tan\theta_M = \frac{-\Sigma M_i r_i l_i \sin\theta_i}{-\Sigma M_i r_i l_i \cos\theta_i} \quad (12.9)$$

For the balance of forces, we have

$$\left[(\Sigma M_i r_i \cos\theta_i)^2 + (\Sigma M_i r_i \sin\theta_i)^2\right]^{0.5} = \left[(M_L r_L \cos\theta_L)^2 + (M_M r_M \sin\theta_M)^2\right]^{0.5}$$

$$M_L r_L = \left[(\Sigma M_i r_i \cos\theta_i + M_M r_M \cos\theta_M)^2 + (\Sigma M_i r_i \sin\theta_i + M_M r_M \sin\theta_M)^2\right]^{0.5} \quad (12.10)$$

$$\tan\theta_L = \frac{-(\Sigma M_i r_i \sin\theta_i + M_M r_M \sin\theta_M)}{-(\Sigma M_i r_i \cos\theta_i + M_M r_M \cos\theta_M)} \quad (12.11)$$

Example 12.2

A shaft carries four masses as shown in Fig.12.7(a) and (b). The balancing masses are to be placed in planes L and M. If the balancing masses revolve at a radius of 100 mm, find their magnitude and *angular positions*.

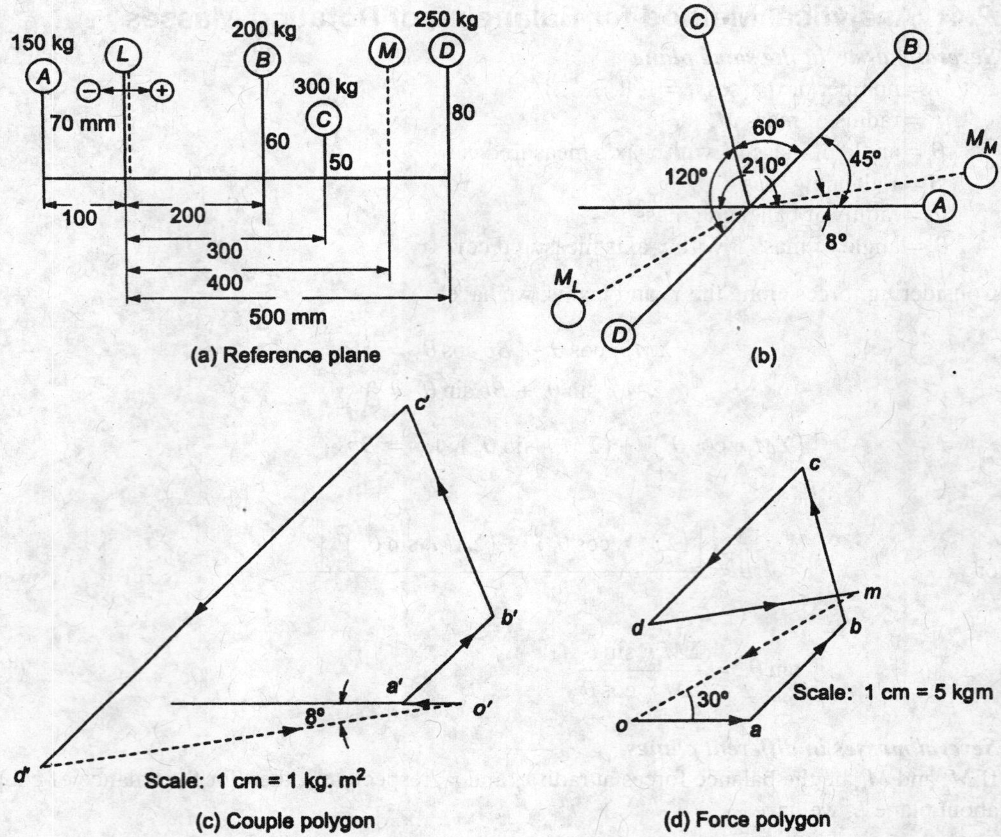

Fig.12.7 Shaft carrying four masses in different planes

■ Solution

Assume the plane L as the reference plane.

Table 12.3

Plane	Mass, M (kg)	Radius, r (m)	Mr (kgm)	Distance from plane L, l (m)	$Mr\,l$ (kgm^2)
A	150	0.07	10.5	−0.1	−1.05
L	M_L	0.10	0.1 M_L	0	0
B	200	0.06	12.0	0.2	2.40
C	300	0.05	15.0	0.3	4.50
M	M_M	0.10	0.1 M_M	0.4	0.04 M_M
D	250	0.18	20.0	0.5	10.0

1. Draw the couple polygon from the data in the last column of Table 12.3 as shown in Fig.12.7(c). By measurement,

$$0.04\ M_M = \text{vector } d'\ o' = 7.7 \text{ cm}$$

or

$$M_M = 192.5 \text{ kg}$$

The angular position of M_M is obtained by drawing OB parallel to $d'o'$ in Fig.12.7(b). $\theta_M = 8°$.

2. Now draw the force polygon from the data in column 4 of the table, as shown in Fig.12.7(d). The vector mo represents the balance force. By measurement

$$0.1\ M_L = \text{vector } mo = 4.7 \times 5$$

or

$$M_L = 235 \text{ kg}$$

The angular position of M_L is obtained by drawing a line parallel to mo in Fig.12.7(b). $\theta_L = 30° + 180° = 210°$.

Analytical Method

Reference plane L

From Table 12.4, we have

$$\Sigma M_i r_i l_i \cos\theta_i = 0.961$$

$$\Sigma M_i r_i \sin\theta_i = 8.832$$

$$\Sigma M_i r_i l_i \cos\theta_i = -7.589$$

$$\Sigma M_i r_i l_i \sin\theta_i = -1.028$$

$$\left[(\Sigma M_i r_i l_i \cos\theta_i)^2 + (\Sigma M_i r_i l_i \sin\theta_i)^2\right]^{0.5} = M_M r_M l_M$$

$$\left[(-7.589)^2 + (-1.028)^2\right]^{0.5} = 0.04\ M_M$$

$$M_M = \frac{7.658}{0.04} = 191.45 \text{ kg}$$

$$\tan\theta_M = \frac{-\Sigma M_i r_i l_i \sin\theta_i}{-\Sigma M_i r_i l_i \cos\theta_i}$$

$$= \frac{-(-1.028)}{-(-7.589)} = 0.13546$$

$$\theta_M = 7.71°$$

Table 12.4

Plane	M(kg)	r (m)	Mr (kg.m)	θ deg	Mr × cos θ (kg.m)	Mr × sinθ (kg.m)	l (m)	Mrl (kg. m²)	Mrl × cos θ (kg.m²)	Mrl × sin θ (kg.m²)
A	150	0.07	10.5	0	10.5	0	–0.1	–1.05	–1.05	0
B	200	0.06	12	45	8.485	8.485	0.2	2.40	1.697	1.697
C	300	0.05	15	105	–3.882	14.489	0.3	4.50	–1.165	4.346
D	250	0.08	20	225	–	–	0.5	10.0	–7.071	–7.071
L	M_L	0.10	$0.1M_L$	θ_L	$0.1\,M_L \cos\theta_L$	$0.1\,M_L \sin\theta_L$	0	0	0	0
M	M_M	0.10	$0.1\,M_M$	θ_M	$0.1\,M_M \cos\theta_M$	$0.1\,M_M \sin\theta_M$	0.4	$0.04\,M_M$	$0.04\,M_M \times \cos\theta_M$	$0.04\,M_M \times \sin\theta_M$

Since the numerator and denominator are both positive, therefore θ_M lies in the first quadrant.

$$M_L r_L = \left[\left(\Sigma M_i r_i \cos\theta_i\ M_M r_M \cos\theta_M\right)^2 + \left(\Sigma M_i r_i \sin\theta_i + M_M\ r_M \sin\theta_M\right)^2\right]^{0.5}$$

$$0.1\,ML = \left[(0.961 + 0.1 \times 191.45 \times \cos 7.71°)^2 + (8.832 + 0.1 \times 191.45 \times \sin 7.71°)^2\right]^{0.5}$$

$$= \left[(19.933)^2 + (11.400)^2\right]^{0.5} = 22.963$$

$$M_L = 229.63 \text{ kg}$$

$$\tan\theta L = \frac{-\left(\Sigma M_i r_i \sin\theta_i + m_m r_m \sin\theta_M\right)}{-\left(\Sigma M_i r_i \cos\theta_i + m_m r_m \cos\theta_M\right)}$$

$$= \frac{-11.400}{-19.933} = 0.5719; \quad \theta L = 29.76° + 180° = 209.76°$$

Since the numerator and denominator are both negative, therefore θ_L lies in the third quadrant.

Example 12.3

A shaft has three eccentrics of mass 1 kg each. The central plane of the eccentrics is 50 mm apart. The distances of the centers from the axis of rotation are 20, 30 and 20 mm and their angular positions are 120° apart. Find the amount of out-of-balance force and couple at 600 rpm. If the shaft is balanced by adding two masses at a radius of 70 mm and at a distance of 100 mm from the central plane of the middle eccentric, find the amount of the masses and their angular positions.

■ **Solution**

Analytical Method

Let L and M be the planes at which the balancing masses are to be placed, as shown in Fig.12.8(a). Take L as the reference plane.

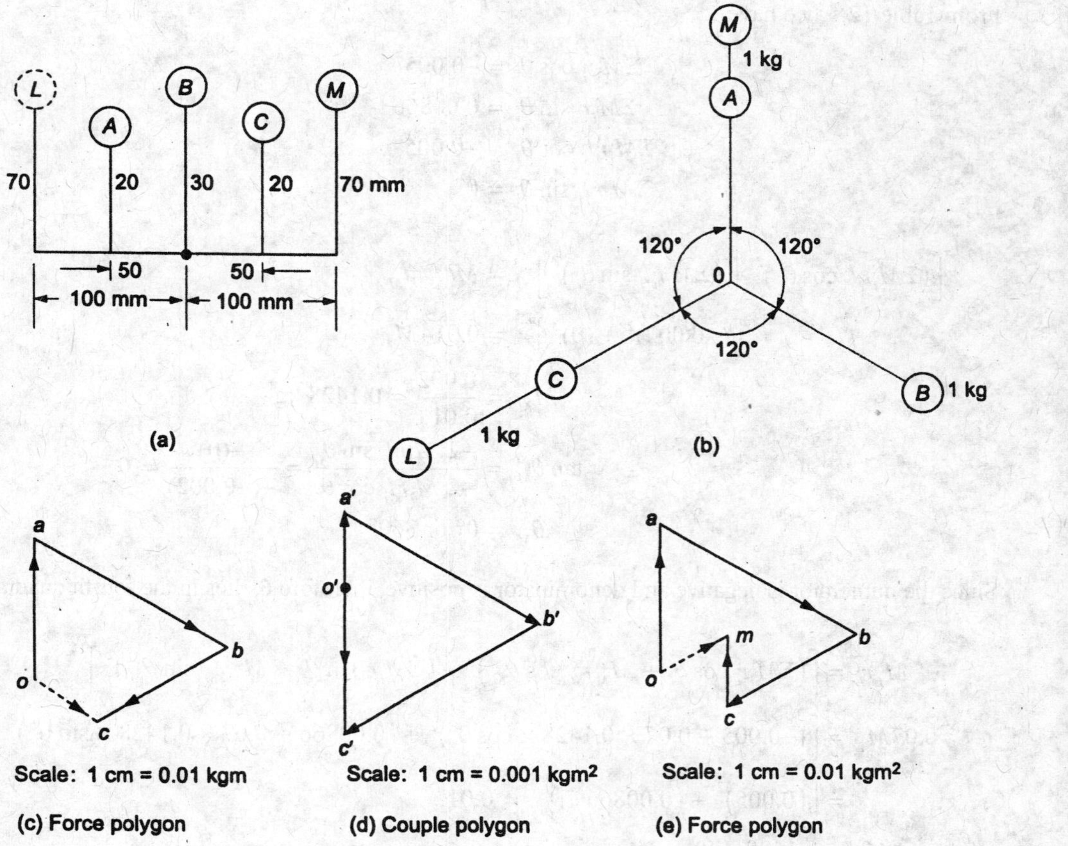

Fig.12.8 Three rotating masses in different planes

Table 12.5

Plane	M (kg)	r (m)	Mr (kg.m)	θ deg	$Mr \times \cos\theta$	$Mr \times \sin\theta$	l (m)	$Mr\,l$ (kg m²)	$Mrl \times \cos\theta$	$Mrl \times \sin\theta$
A	1	0.02	0.02	0	0.02	0	0.05	0.001	0.001	0
B	1	0.03	0.03	120	–0.015	0.0259	0.10	0.003	–	0.0026
C	1	0.02	0.02	240	–0.01	8	0.15	0.003	0.0015	–0.0026
L	M_L	0.07	$0.07 \times M_L$	θ_L	$0.07\,M_L \times \cos\theta_L$	–0.01732 0.07	0	0	–0.0015	0
M	M_M	0.07	$0.07 \times M_M$	θ_M	$0.07\,M_M \times \cos\theta_M$	$0.07\,M_M \times \sin\theta_M$	0.2	$0.014 \times M_M$	$0.014\,M_M \times \cos\theta_M$	$0.014\,M_M \times \sin\theta_M$

From Table 12.5, we have

$$\Sigma M_i r_i \cos\theta_i = -0.005$$
$$\Sigma M_i r_i \sin\theta_i = 0.00866$$
$$\Sigma M_i r_i l_i \cos\theta_i = -0.005$$
$$\Sigma M_i r_i l_i \sin\theta_i = 0$$

$$\left[(\Sigma M_i r_i l_i \cos\theta_i)^2 + (\Sigma M_i r_i l_i \sin\theta_i)^2\right]^{0.5} = MMr_M l_M$$
$$\left[(-0.002)^2 + (0)^2\right]^{0.5} = 0.014 M_M$$
$$M_M = \frac{0.002}{0.014} = 0.1428 \text{ kg}$$
$$\tan\theta_M = \frac{-\Sigma M_i r_i l_i \sin\theta_i}{-\Sigma M_i r_i l_i \cos\theta_i} = \frac{-(0)}{-(-0.002)} = 0$$
$$\theta_M = 0° \text{ or } 360°$$

Since the numerator is negative and denominator is positive, therefore θ_M lies in the fourth quadrant.

$$M_L r_L = \left[\left(\Sigma M_i r_i \cos\theta_i + M_M r_M \cos\theta_M\right)^2 + \left(\Sigma M_i r_i \sin\theta_i + M_M r_M \sin\theta_M\right)^2\right]^{0.5}$$
$$0.07 M_L = \left[(-0.005 + 0.07 \times 0.1428 \times \cos 0°)^2 + \left(0.00866 + 0.07 \times 0.1428 \times \sin 0°\right)^2\right]^{0.5}$$
$$= \left[(0.005)^2 + (0.00866)^2\right]^{0.5} = 0.01$$
$$M_L = 0.1428 \text{ kg}$$
$$\tan\theta_L = \frac{-\left(\Sigma M_i r_i \sin\theta_i + M_M r_M \sin\theta_M\right)}{-\left(\Sigma M_i r_i \cos\theta_i + M_M r_M \cos\theta_M\right)}$$
$$= \frac{-0.00866}{-0.005} = 1.732, \quad \theta_L = 240°$$

Since the numerator and denominator are both negative, therefore θ_L lies in the third quadrant.

Graphical Method

Out-of-balance force: The out-of-balance force is obtained by drawing the force polygon, as shown in Fig.12.8(c), drawn from the data in column 4 of the Table 12.5.

The resultant oc represents the out-of-balance force.

$$\text{Out-of-balance force} = \text{vector } oc \times \omega^2 = 0.01 \times \left(2\pi \times \frac{600}{60}\right)^2$$
$$= 39.48 \text{ N}$$

Out-of-balance couple: The out-of-balance couple is obtained by drawing the couple polygon from the data in column 9, as shown in Fig.12.8(d).

$$\text{Out-of-balance couple} = o'c' \times \omega^2 = 0.002\left(2\pi \times \frac{600}{60}\right)^2$$
$$= 7.9 \text{ Nm}$$

Balancing masses: The vector $c'o'$ from c' to o', as shown in Fig.12.8(d), represents the balancing couple.

$$0.014\ M_M = \text{vector } c'o' = 0.002$$

or
$$M_M = 0.1428 \text{ kg}$$

Draw OM parallel to $c'o'$ in Fig.12.8(b). We find that the angular position of mass M is from mass A.

To find the balancing mass M_L, draw the force polygon, as shown in Fig.12.8(e). The closing side do of the polygon represents the balancing force.

$$0.07\ M_L = \text{Vector } om = 0.01$$

or
$$M_L = 0.1428 \text{ kg}$$

Now draw O_L in Fig.12.8(b), parallel to om. We find that the angular position of M_L is 120° from mass A.

Example 12.4

Three masses $M_1 = 3$ kg, $M_2 = 4$ kg, and $M_3 = 3$ kg are rotating in different planes as shown in Fig.12.9. Two balancing masses are B_1 and B_2 are placed at 100 mm from each end at 80 mm radius. Find the magnitude and angular location of the balancing masses.

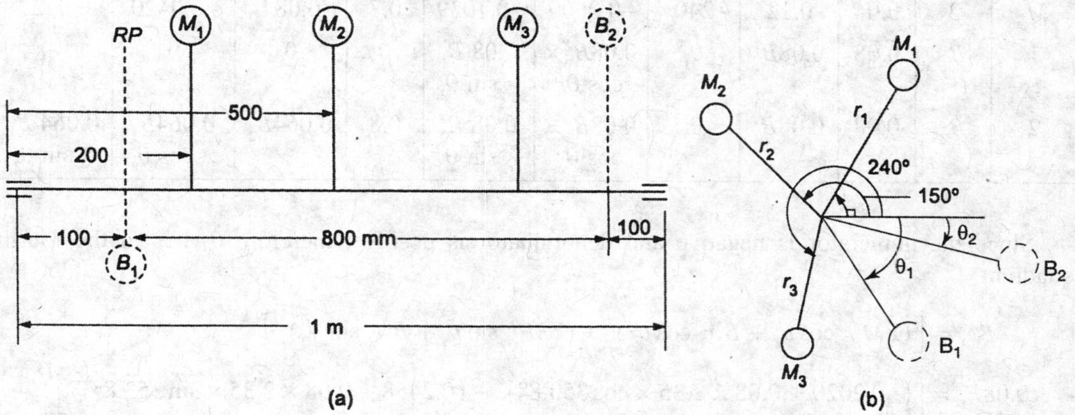

Fig.12.9 Three masses rotating in different planes

■ Solution

Reference plane 1.
From Table 12.6, we have

$$\Sigma M_i r_i \cos\theta_i = -0.2020$$
$$\Sigma M_i r_i \sin\theta_i = 0.2458$$
$$\Sigma M_i r_i l_i \cos\theta_i = -0.14973$$
$$\Sigma M_i r_i l_i \sin\theta_i = 0.01627$$

$$\left[(\Sigma M_i r_i l_i \cos\theta_i)^2 + (\Sigma M_i r_i l_i \sin\theta_i)^2\right]^{0.5} = B_2 r_2 l_2$$

$$\left[(-0.14973)^2 + (0.01627)^2\right]^{0.5} = 0.064 B_2$$

$$B_2 = \frac{0.15061}{0.064} = 2.35 \text{ kg}$$

$$\tan\theta_2 = \frac{-\Sigma M_i r_i l_i \sin\theta_i}{-\Sigma M_i r_i l_i \cos\theta_i}$$

$$= \frac{-(0.01627)}{-(-0.14973)} = -0.10866$$

$$\theta_2 = -6.2° \text{ or } 353.8° \text{ ccw}$$

Table 12.6

Plane	M (kg)	r (m)	Mr (kg.m)	θ (deg)	$Mr \times$ cos θ	$Mr \times$ sin θ	l (m)	Mrl (kg m²)	$Mrl \times$ cos θ	$Mrl \times$ sin θ
M_1	3	0.08	0.24	45	0.1697	0.1697	0.1	0.024	0.01697	0.01697
M_2	4	0.09	0.36	150	0.3117	0.1800	0.4	0.144	0.1247	0.0720
M_3	3	0.04	0.12	240	–0.0600	–0.1039	0.7	0.084	–0.0420	0.0727
1	B_1	0.08	$0.08B_1$	θ_1	$0.08B_1 \times \cos\theta_1$	$0.08\,B_1 \times \sin\theta_1$	0	0	0	–0
2	B_2	0.08	$0.08B_2$	θ_2	$0.08B_2 \times \cos\theta_2$	$0.08\,B_2 \times \sin\theta_2$	0.8	$0.064B_2$	$0.064B_2 \times \cos\theta_2$	$0.064\,B_2 \times \sin\theta_2$

Since the numerator is negative and denominator is positive, therefore θ_2 lies in the fourth quadrant.

$$B_1 r_1 = \left[(\Sigma M_i r_i \cos\theta_i + B_2 r_2 \cos\theta_2)^2 + (\Sigma M_i r_i \sin\theta_i + B_2 r_2 \sin\theta_2)^2\right]^{0.5}$$

$$0.08B_1 = \left[(-0.2020 + 0.08 \times 2.35 \times \cos 353.8°)^2 + (0.2458 + 0.08 \times 2.35 \times \sin 353.8°)^2\right]^{0.5}$$

$$= \left[(-0.0151)^2 + (0.22549)^2\right]^{0.5} = 0.226$$

$$B_1 = 2.825 \text{ kg}$$

$$\tan\theta_1 = -(\Sigma M_i r_i \sin\theta_i + B_2 r_2 \sin\theta_2)/-(\Sigma M_i r_i \cos\theta_i + B_2 r_2 \cos\theta_2)$$

$$= -0.22549/(-0.0151) = -14.93311$$

$$\theta_1 = -86.169° \text{ or } 273.831°$$

Since the numerator is negative and denominator is positive, therefore θ_1 lies in the fourth quadrant.

12.3 RECIPROCATING MASSES

Reciprocating masses occur in internal combustion engines and steam engines. The reciprocating masses are due to the mass of the piston, piston pin and part of the mass of connecting rod considered as reciprocating. For static balancing or reciprocating masses, the resultant of all centrifugal forces should be equal to zero. For dynamic balancing, the resultant of centrifugal forces and resultant of couples must be equal to zero.

12.3.1 Reciprocating Engine

Consider the reciprocating engine mechanism shown in Fig.12.10.

Acceleration of the piston, $f_p = f_c\left(\cos\theta + \dfrac{\cos 2\theta}{n}\right)$

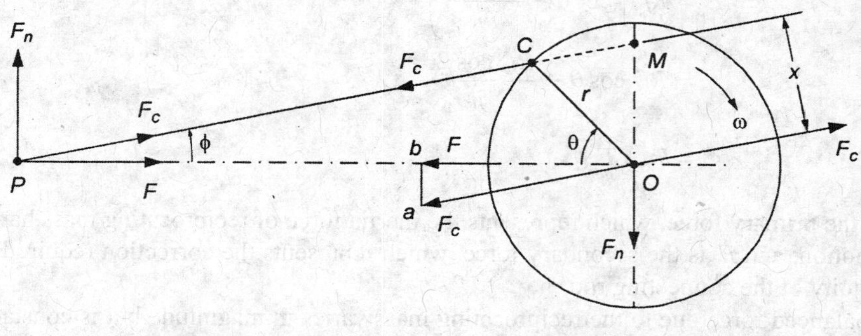

Fig.12.10 Forces in reciprocating engine mechanism

where f_c = acceleration of the crankpin

$= \omega^2 r$

$n = \dfrac{\text{length of connecting rod}}{\text{radius of crank}} = \dfrac{l}{r}$

θ = inclination of crank to inner dead centre

F_c = force in the connecting rod

F_n = thrust on the piston or guide bars

x = perpendicular distance between the connecting rod and the crankpin

Let R = mass of the reciprocating parts

Accelerating force, $F = Rf_p$

where f_p = acceleration of piston or reciprocating parts.

The force F_c at C is equivalent to a force F_c at O and a couple $F_c x$ tending to retard the rotation of the crankshaft.

$$\text{Thrust couple} = F_n \cdot OP$$

Now triangles Oba and POM are similar. Therefore

$$\frac{ba}{Ob} = \frac{F_n}{F} = \frac{OM}{OP}$$

or

$$F_n \cdot OP = F \cdot OM$$

Also

$$\frac{ba}{Oa} = \frac{F_n}{F_c} = \frac{OM}{PM}$$

and $$\frac{x}{OM} = \cos\phi = \frac{OP}{PM}$$

So that $$\frac{OM}{PM} = \frac{x}{OP}$$

Hence $$\frac{F_n}{F_c} = \frac{x}{OP}$$

and $$F_n \cdot OP = F \cdot OM = F_c \cdot x$$

The full effect on the engine frame of the inertia of the reciprocating mass is equivalent to the force F along the line of stroke at O and the clockwise thrust couple of magnitude $F_n \cdot OP$.

Now $$F = Rf_c\left(\cos\theta + \frac{\cos 2\theta}{n}\right)$$

$$= Rf_c \cos\theta + \frac{Rf_c \cos 2\theta}{n}$$

$$= F_p + F_s \tag{12.12}$$

where F_p is the primary force, which represents the inertia force of reciprocating mass having simple harmonic motion, and F_s is the secondary force, which represents the correction required to account for the obliquity of the connecting rod.

The unbalanced force due to the reciprocating mass varies in magnitude but is constant in direction. A single revolving mass can neither be used to balance a reciprocating mass, nor vice-versa.

The graphical representation of the various forces is shown in Fig.12.11.

where oa = primary disturbing force

ob = centrifugal force due to the revolving balance mass

oc = residual unbalanced force parallel to the line of stroke

oe = unbalanced force at right angles to the line of stroke

of = resultant unbalanced force on engine frame

Let $od = oa/2$, for 50% balancing of reciprocating parts

Then $oc = od$ and $\angle cof = \theta$.

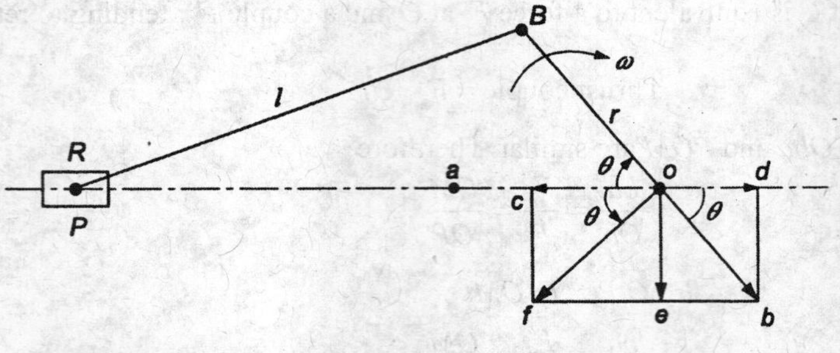

Fig.12.11 Graphical representation of forces in reciprocating engine mechanism

12.3.2 Partial Primary Balance

Consider the reciprocating engine mechanism shown in Fig.12.12. The primary unbalanced force,

$$F_p = Rf_c \cos\theta$$
$$= R\omega^2 r \cos\theta$$

= component parallel to the line of stroke of the centrifugal force produced by an equal mass attached to and revolving with the crankpin.

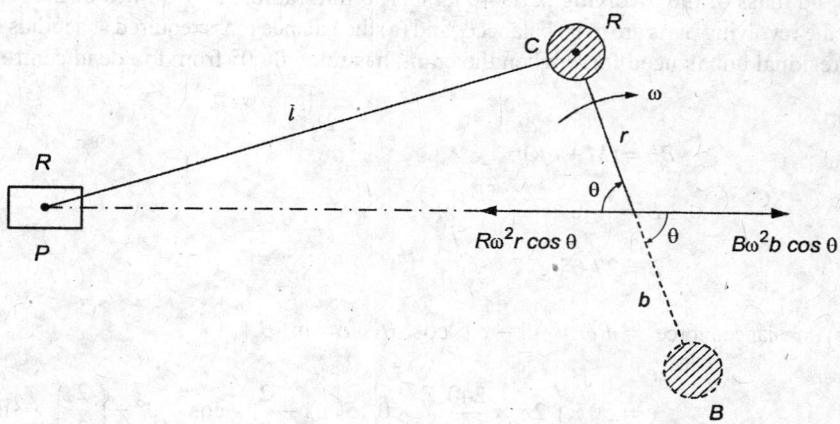

Fig.12.12 Primary balancing of reciprocating mass

Let a balance mass B be placed along the line of crank at a radius b opposite to the crankpin. Comparing the horizontal components of forces, we have

$$R\omega^2 r \cos\theta = B\,\omega^2\, b \cos\theta$$

or

$$Rr = Bb$$

Component of revolving balance mass perpendicular to the line of stroke = $B\,\omega^2\, b \sin\theta$.

This is the component of the balancing force which remains unbalanced. It is usually preferable to make $Bb = cRr$.

where c is a fraction less than one.

Reduced value of unbalanced force parallel to line of stroke,

$$F_h = (1 - c)\, R\, \omega^2 r \cos\theta \qquad (12.13)$$

Unbalanced force perpendicular to line of stroke, $F_v = c\, R\omega^2\, r \sin\theta$ (12.14)

Resultant unbalanced force, $F = R\omega^2 r\,[(1 - c)^2 \cos^2\theta + c^2 \sin^2\theta]^{0.5}$ (12.15)

For F to be minimum or maximum, $\dfrac{dF}{dc} = 0$

$$2(1 - c)(-1) + 2c = 0$$

$$c = 0.5$$

Thus for unbalanced force to be least, $c = 0.5$

If the balance mass B has to balance the revolving parts M as well as give a partial balance of the reciprocating parts R, then

$$Bb = Mr + cRr = (M + cR)\,r \tag{12.16}$$

In practice, two balance masses, each equal to $B/2$, would be attached to the crank webs.

Example 12.5

A single cylinder reciprocating engine has speed 240 rpm, stroke 300 mm, mass of the reciprocating parts 50 kg and mass of the revolving parts 40 kg at 150 mm radius. If two-third of the reciprocating parts and all the revolving parts are to be balanced, find (a) the balance mass required at a radius of 400 mm, and (b) the residual unbalanced force when the crank has rotated 60° from top dead centre.

Solution

(a)
$$Bb = (M + cR)r$$

$$0.4\,B = \left(40 + \frac{2 \times 50}{3}\right)0.15$$

or
$$B = 27.5 \text{ kg}$$

(b) Residual unbalanced force
$$= R\omega^2 r\left[(1-c)^2 \cos^2\theta + c^2 \sin^2\theta\right]^{0.5}$$

$$= 50 \times \left(2\pi \times \frac{240}{60}\right)^2 \times 0.15\left[\left(1-\frac{2}{3}\right)^2 \cos^2 60° + \left(\frac{2}{3}\right)^2 \times \sin^2 60°\right]^{0.5}$$

$$= 2846.9 \text{ N}$$

12.4 BALANCING OF LOCOMOTIVES

Locomotives are of two types, coupled or uncoupled. If two or more pairs of wheels are coupled together to increase the adhesive force between the wheels and the track, it is called a coupled locomotive. Otherwise, it is an uncoupled locomotive. Locomotives usually have two cylinders. If the cylinders are mounted between the wheels, it is called an inside cylinder locomotive and if the cylinders are outside the wheels, it is an outside cylinder locomotive. The cranks of the two cylinders are set at 90° to each other so that the engine can be started easily after stopping in any position. Balance masses are placed on the wheels in both types.

Wheels are coupled by connecting their crank pins with coupling rods in coupled locomotives. In uncoupled locomotives, there are four planes, two of the cylinders and two of the driving wheels, for consideration. In coupled locomotives, there are six planes, two each for cylinders, coupling rods and driving wheels. Coupled locomotives result in reduced hammer blow as the mass of reciprocating parts is distributed among all the coupled wheels.

12.4.1 Partial Balancing of Uncoupled Locomotives

In an uncoupled locomotive, two cylinders are placed symmetrically either inside or outside the frames. The two cranks are at right angles to each other, as shown in Fig.12.12(a). In an uncoupled locomotive, the effort is transmitted to one pair of wheels only, whereas in a coupled locomotive, the driving wheels are connected to the leading and trailing wheels by an outside coupling rod. $c \approx 2/3$ to $3/4$ with two pairs of coupled wheels, and $c = 2/5$ for four cylinder locomotives.

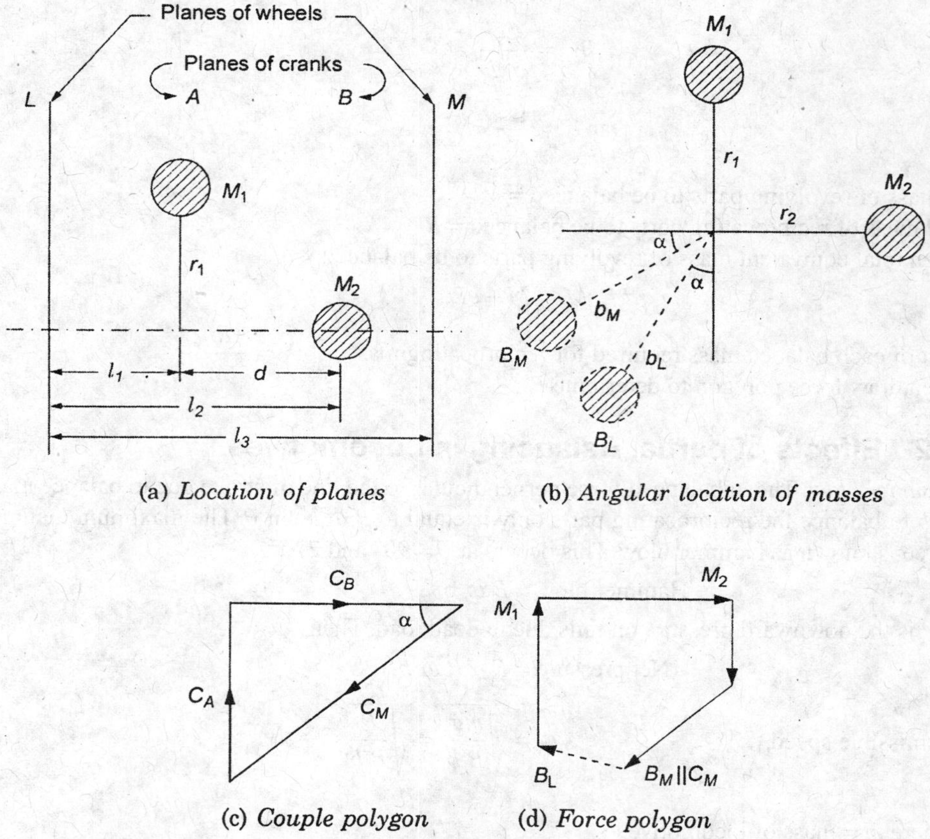

Fig.12.13 Inside cylinders uncoupled locomotive

The location of the cranks and balance masses is shown in Fig.12.13(b) and couple polygen in Fig.12.13(c), and force polygon in Fig.12.13(d). The couple polygon may be drawn by using Table 12.7.

Table 12.7 Partial balancing of uncoupled locomotive

Plane	Mass M	Radius r	$M.r$	Distance from plane L l	Couple Mrl
L	B_L	b_L	$B_L b_L$	0	0
A	M_1	r_1	$M_1 r_1$	l_1	$C_A = M_1 r_1 l_1$
B	M_2	r_2	$M_2 r_2$	l_2	$C_B = M_2 r_2 L_2$
M	B_M	b_M	$B_M b_M$	l_3	$C_M = B_M b_M l_3$

$$C_M = [C_A^2 + C_B^2]^{0.5}$$
$$= B_M b_M l_3$$

or
$$B_M = \frac{C_M}{b_M l_3}$$

$$\tan\alpha = \frac{C_A}{C_B}$$

If mass of revolving parts to be balanced = M
and mass of reciprocating parts to be balanced = R
Then total equivalent mass of revolving parts to be balanced,

$$M_1, M_2 = M + cR$$

Part of each balance mass required for reciprocating masses, $B_r = \dfrac{cRB_M}{M_1}$
Then draw force polygon to determine B_L.

12.4.2 Effects of partial Balancing in Locomotives

(a) *Hammer blow*: The unbalanced force perpendicular to the line of stroke due to balance mass B_r at radius b to balance the reciprocating parts only is equal to $B_r\omega^2 b \sin\theta$. The maximum magnitude of this force is known as hammer blow. This occurs at $\theta = 90°$ and $270°$.

$$\text{Hammer blow} = B_r\omega^2 b \tag{12.17}$$

If P is the downward pressure on rails due to dead load. Then

$$\text{Net pressure} = P - B_r\omega^2 b$$

Permissible speed,
$$\omega = \sqrt{\frac{mg}{B_r b}} = \sqrt{\frac{P}{B_r b}} \tag{12.18}$$

Where m = mass of locomotive.

(b) *Variation of Tractive Effort*

Variation of tractive effort, $F_T = (1 - c)\, R\omega^2 r\, [\cos\theta + \cos(90° + \theta)]$
$= (1 - c)\, R\omega^2 r\, [\cos\theta - \sin\theta]$

For its value to be maximum, $\left(\dfrac{d}{d\theta}\right)(\cos\theta - \sin\theta) = 0$

or $-\sin\theta - \cos\theta = 0$

or $\tan\theta = -1$

or $\theta = -45°$ i.e. $315°$ and $+135°$

At $\theta = 315°$
$$(F_T)_{\max} = (1-c)\, Rr\omega^2\, (\cos 315° - \sin 315°)$$
$$= (1-c)\, Rr\omega^2 \left(\frac{1}{\sqrt{2}} + \frac{1}{\sqrt{2}}\right)$$
$$= +\sqrt{2}(1-c)Rr\omega^2$$

At $\theta = 315°$
$$(F_T)_{\max} = \sqrt{2}(1-c)Rr\omega^2$$

Maximum variation of tractive effort
$$(F_T)_{\max} = \pm\sqrt{2}(1-c)Rr\omega^2 \tag{12.19}$$

(c) *Swaying Couple*: The unbalanced part of the primary disturbing forces cause a horizontal swaying couple to act on the locomotive owing to the distance l between the cylinder centers. Taking moments about the engine center line, the resultant unbalanced couple,

$$T = (1 - c)\,R\omega^2 r \cdot \frac{l}{2} \cdot \left[\cos\theta - \cos(90° + \theta)\right]$$

Swaying couple, $$T = (1 - c)\,R\omega^2 r \cdot \frac{l}{2} \cdot \left[\cos\theta + \sin\theta\right]$$

This is maximum when $\theta = 45°$ and $225°$.

Maximum swaying couple $= \pm\left(\dfrac{1-c}{\sqrt{2}}\right) R\omega^2\, r \cdot l$ (12.20)

Example 12.6

An inside cylinder locomotive has its cylinder centre lines 0.8 m apart and has a stroke of 0.6 m. The rotating masses are equivalent to 150 kg at the crank pin and the reciprocating masses per cylinder are 300 kg. The wheel centre lines are 1.8 m apart. The cranks are at right angles. The whole of the rotating and 2/3rd of the reciprocating masses are to be balanced by masses placed at a radius of 0.5 m. Find (a) the magnitude and direction of the balancing masses, (b) the fluctuation in rail pressure under one wheel, (c) the variation of tractive effort and (d) the magnitude of swaying couple at a crank speed of 300 rpm.

■ **Solution**

Equivalent mass to be balanced $= 150 + 2 \times \dfrac{300}{3} = 350$ kg

Balancing masses

Let M_A and M_D be the balancing masses at angular location θ_A and θ_D respectively. The position of the planes is shown in Fig.12.14(a) and angular position of the masses in Fig.12.14(b). Take A as the reference plane.Table 12.8

Now draw the couple polygon, as shown in Fig.12.14(c), from the data in column 6 of Table 12.8 The closing side *co* represents the balancing couple.

$$0.9\, M_D = \text{vector } c'o' = 3.7 \times 40 = 148$$
$$M_D = 164.4 \text{ kg}$$

Draw o*D* parallel to $c'o'$ in Fig.12.14(b). By measurement, $\theta_D = 250°$.

To find the balancing mass M_A, draw the force polygon, as shown in Fig.12.14(d), from the data in column 4. The vector *do* represents the balancing force.

$$0.5\, M_A = \text{vector } od = 4.1 \times 20 = 82$$

or $$M_A = 164 \text{ kg}$$

To find the angle θ_A, draw *oA* parallel to *od* in Fig.12.14(b). By measurement, $\theta_A = 200°$.

Analytically $$0.9M_D = [(52.5)^2 + (136.5)^2]^{0.5} = 146.248$$

$$M_D = 162.5 \text{ kg}$$

$$\theta_D = 180° + \tan^{-1}\left(\frac{136.5}{52.5}\right)$$

$$= 180° + 68.96° = 248.96° \text{ ccw}$$

$$0.9M_A = \left[(136.5)^2 + (52.5)^2\right]^{0.5} = 146.248$$

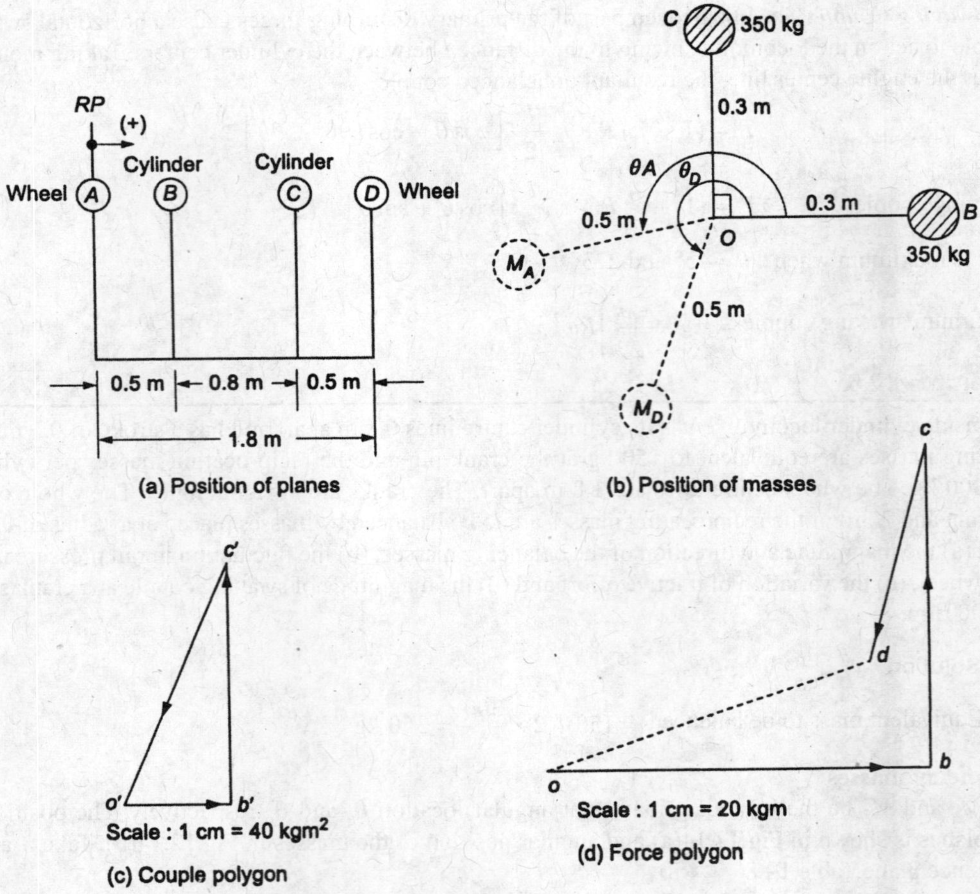

Fig.12.14 Balancing of inside cylinder locomotive

$$M_A = 162.5 \text{ kg}$$

$$\theta_D = 180° + \tan^{-1}\left(\frac{52.5}{136.5}\right)$$

$$= 180° + 21.03° = 201.03° \text{ ccw}$$

Table 12.8

Plane	Mass M	Radius r	$M \cdot r$	Distance from plane A l	Couple Mrl
A	M_A	0.5	$0.5\ M_A$	0	0
B	350	0.3	105	0.5	52.5
C	350	0.3	105	1.3	136.5
D	M_D	0.5	$0.5\ M_D$	1.8	$0.9\ M_D$

(c) Each balance mass = 162.5 kg

Balance mass for the rotating masses $= 150 \times \dfrac{162.5}{300} = 81.25$ kg

Balance mass for the reciprocating masses, $B = \left(\dfrac{2}{3}\right) \cdot \left(\dfrac{300}{350}\right) \times 162.5 = 92.86$ kg

Fluctuation in the rail pressure or hammer blow $= B\omega^2 b = 92.86 \times \left(2\pi \times \dfrac{300}{60}\right)^2 \times 0.5$

$= 45824$ N

(d) Maximum variation of tractive effort $= \pm\sqrt{2}(1-c)R\omega^2 r$

$$= \pm\sqrt{2}\left(1-\frac{2}{3}\right)300 \times \left(2\pi \times \frac{300}{60}\right)^2 \times 0.3$$

$$= \pm 41873 \text{ N}$$

(e) Maximum swaying couple $= \pm\left[\dfrac{1-c}{\sqrt{2}}\right] R\omega^2 r \cdot d$

$$= \pm\left[\frac{1-\frac{2}{3}}{\sqrt{2}}\right] \times 300(10\pi)^2 \times 0.3 \times 0.8$$

$$= \pm 16749 \text{ N m}$$

Example 12.7

The following data refer to an outside cylinder uncoupled locomotive:

Mass of rotating parts per cylinder = 350 kg
Mass of reciprocating parts per cylinder = 300 kg
Angle between cranks = 90°
Crank radius = 0.3 m
Cylinder centers = 1.8 m
Radius of balance masses = 0.8 m
Wheel centers = 1.5 m

If whole of the rotating and 2/3rd of the reciprocating parts are to be balanced in planes of the driving wheels, find (a) magnitude and angular positions of balance masses, (b) speed in km/h at which the wheel will lift off the rails when the load on each driving wheel is 30 kN, and the diameter of tread of driving wheels is 1.8 m, and (c) swaying couple at speed found in (b) above.

■ Solution

(a) Equivalent mass of the rotating parts to be balanced per cylinder,

$$M = 350 + 2 \times \frac{300}{3} = 550 \text{ kg}$$

Let M_B and M_C be the balance masses, and θ_B and θ_c their angular positions. Let B be the reference plane. The position of planes is shown in Fig.12.15(a), and the position of masses in Fig.12.15(b).

Table 12.9

Plane	Mass M	Radius r	$M \cdot r$	Distance from plane B l	Couple Mrl
A	550	0.3	165	–0.15	–24.75
B	M_B	0.8	$0.8\ M_B$	0	0
C	M_c	0.8	$0.8\ M_c$	1.5	$1.2\ M_c$
D	550	0.3	165	1.65	272.25

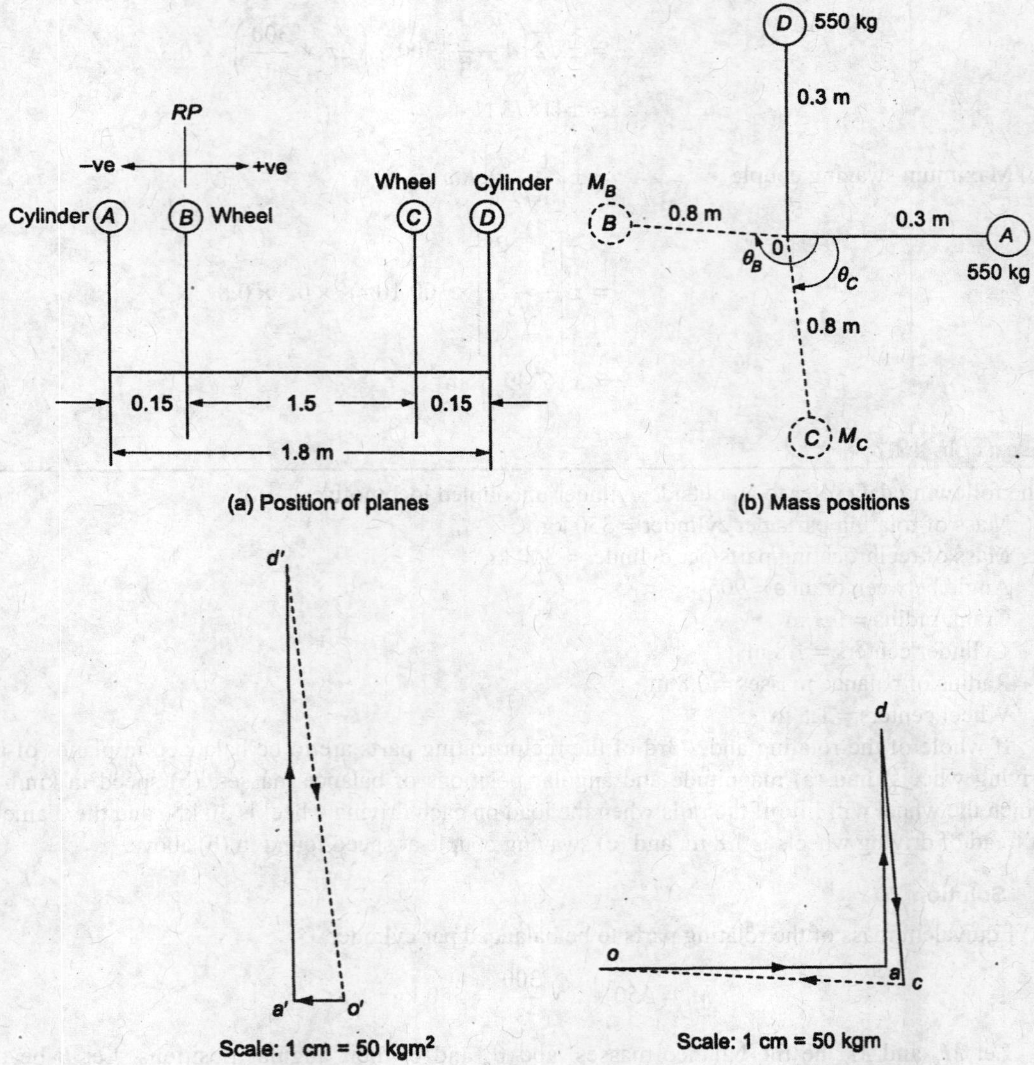

(a) Position of planes (b) Mass positions (c) Couple polygon (d) Force polygon

Fig.12.15 Balancing of outside cylinder uncoupled locomotive

Draw the couple polygon, as shown in Fig.12.14(c), from the data in column 6 of Table 12.9. The closing side *do* represents the balancing couple.

$$1.2\ M_c = \text{vector};\ d'o' = 5.5 \times 50$$

or

$$M_c = 229.17 \text{ kg}$$

Now draw *OC* parallel to *d' o'* in Fig.12.15(b) to find θ_c. By measurement, $\theta_c = 84°$.
To find the balancing mass M_B, draw the force polygon, as shown in Fig.12.15(d), from the data in column 4 of the above table.

$$0.8\ M_B = \text{vector } co = 3.7 \times 50$$

or

$$M_B = 231.25 \text{ kg}$$

Now draw *OB* parallel to coin Fig.12.15(b) to find θ_B. By measurement, $\theta_B = 186°$

Analytically,

$$1.2\ M_c = [(24.75)^2 + (272.25)^2]^{0.5} = 273.37$$

$$M_c = 227.8 \text{ kg}$$

$$\theta_c = \tan^{-1}\left(\frac{272.25}{24.75}\right) = \tan^{-1} 11 = 84.8° \text{ cw}$$

$$M_B = M_C$$

$$\theta_B = 180° + \tan^{-1}\left(\frac{24.75}{272.25}\right)$$

$$= 180° + 5.2° = 185.2°$$

(b) $M_B = M_C = 227.8$ kg

Balancing mass of reciprocating parts, $B = \left(\frac{cR}{M}\right)M_B = \left[\frac{2 \times 300}{3 \times 550}\right] \times 227.8 = 82.83 \text{ kg}$

$$\omega = \left[\frac{P}{B \cdot b}\right]^{0.5} = \left[\frac{30 \times 10^3}{82.83 \times 0.8}\right]^{0.5}$$

$$= 21.28 \text{ rad/s}$$

$$v = \omega \times \frac{D}{2} = 21.28 \times 0.9 \times \frac{3600}{1000} = 68.94 \text{ km/h}$$

(c) Maximum swaying couple

$$= \pm\left[\frac{1-c}{\sqrt{2}}\right] R\omega^2 r \cdot d$$

$$= \pm\left[\frac{1-2/3}{\sqrt{2}}\right] 300\,(21.28)^2\ 0.3 \times 1.8 = 17.293 \text{ k Nm}$$

Example 12.8

The following data rafer to a two cylinder uncoupled locomotive:

Rotating mass per cylinder = 300 kg
Reciprocating mass per cylinder = 330 kg
Distance between the wheels = 1.4 m
Distance between the cylinder centres = 0.6 m
Diameter of treads of the driving wheels = 1.8 m
Crank radius = 0.3 m

Radius of centre of the balance mass = 0.6 m
Speed of the locomotive = 45 km/h
Angle between the cylinder cranks = 90°
Dead load of each wheel = 40 kN

Determine

(a) the balancing mass required in the planes of driving wheels if the complete revolving and 2/3rd of the reciprocating masses are to be balanced;

(b) swaying couple;

(c) variation in tractive effort;

(d) maximum and minimum pressure on rails; and

(e) maximum speed of locomotive without lifting the wheels from the rails.

Solution

(a) Mass to be balanced = 300 = 2 × 300/3 = 520 kg
Taking 1 as the reference plane in Fig.12.16.

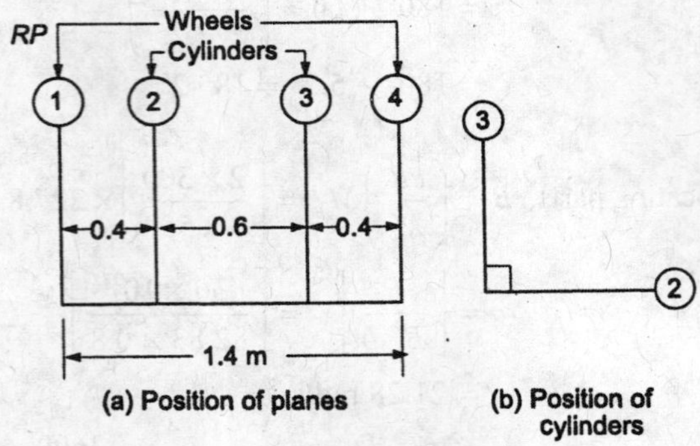

Fig 12.16 Two cylinder uncoupled locomotive

Table 12.10

Plane	M kg	r m	Mr (kg.m)	θ deg	$Mr \times \cos\theta$	$Mr \times \sin\theta$	l m	Mrl (kg.m²)	$Mrl \times \cos\theta$	$Mrl \times \sin\theta$
M_2	520	0.3	156	0	156	0	0.4	62.4	62.4	0
M_3	520	0.3	156	90	0	156	1.0	156	0	156
M_1	M_1	0.6	$0.6\,B_1$	θ_1	$0.6\,M_1 \times \cos\theta_1$	$0.6\,M_1 \times \sin\theta_1$	0	0	0	0
M_4	M_4	0.6	$0.6\,B_4$	θ_4	$0.6\,M_2 \times \cos\theta_4$	$0.8\,M_4 \times \sin\theta_4$	1.4	$0.84\,M_4$	$0.08\,M_4 \times \cos\theta_4$	$0.08\,M_4 \times \sin\theta_4$

From Table 12.10, we have
$$\Sigma M_i r_i \cos\theta_i = 156$$
$$\Sigma M_i r_i \sin\theta_i = 156$$
$$\Sigma M_i r_i l_i \cos\theta_i = 62.4$$
$$\Sigma M_i r_i l_i \sin\theta_i = 156$$
$$\left[\left(\Sigma M_i r_i l_i \cos\theta_i\right)^2 + \left(\Sigma M_i r_i l_i \sin\theta_i\right)^2\right]^{0.5} = M_4 r_4 l_4$$
$$\left[(62.4)^2 + (156)^2\right]^{0.5} = 0.84\, M_4$$
$$M_4 = \frac{168.02}{0.84} = 200 \text{ kg}$$
$$\tan\theta_4 = \frac{-\Sigma M_i r_i l_i \sin\theta_i}{-\Sigma M_i r_i l_i \cos\theta_i} = \frac{-(156)}{-(62.4)} = 2.5$$
$$\theta_4 = 180 + 68.2° = 248.2° \text{ ccw}$$

Since the numerator and denominator are both negative, therefore θ_4 lies in the third quadrant.

$$M_1 r_1 = \left[\left(\Sigma M_i r_i \cos\theta_i + M_4 r_4 \cos\theta_4\right)^2 + \left(\Sigma M_i r_i \sin\theta_i + M_4 r_4 \sin\theta_4\right)^2\right]^{0.5}$$
$$0.6 M_1 = \left[\left(156 + 0.6\times 200\times \cos\ 248.2°\right)^2 + \left(156 + 0.6\times 200\times \sin\ 248.2°\right)^2\right]^{0.5}$$
$$= \left[(111.43)^2 + (44.58)^2\right]^{0.5} = 120.02$$
$$M_1 = 200 \text{ kg}$$
$$\tan\theta_1 = \frac{-\left(\Sigma M_i r_i \sin\theta_i + M_4 r_4 \sin\theta_4\right)}{-\left(\Sigma M_i r_i \cos\theta_i + M_4 r_4 \cos\theta_4\right)} = \frac{-44.58}{-111.43} = 0.4$$
$$\theta_1 = 180° + 21.8° = 201.8°$$

Since the numerator and denominator are both negative, therefore θ_1 lies in the third quadrant.

(b)
$$v = 45\times\frac{1000}{3600} = 12.5 \text{ m/s}$$
$$\omega = \frac{v}{r} = \frac{12.5}{0.9} = 13.889 \text{ rad/s}$$

Swaying couple
$$= \left[\frac{1-c}{\sqrt{2}}\right]\times R r\omega^2 l$$
$$= \left[\frac{1-2/3}{\sqrt{2}}\right]\times 300\times 0.3\times (13.889)^2\times 0.6$$
$$= 2455.3 \text{ Nm}$$

(c) Variation in tractive effort
$$= \pm\sqrt{2}(1-c)R\omega^2 r$$
$$= \pm\sqrt{2}\left(1-\frac{2}{3}\right)\times 300\times (13.889)^2\times 0.3$$
$$= \pm 8184.2 \text{ N}$$

(d) Balance mass for the reciprocating parts only, $R_1 = 200 \times \frac{2}{3} \times \frac{330}{520} = 84.6$ kg

$$\text{Hammer blow} = R_1 b\omega^2 = 84.6 \times 0.6 \times (13.889)^2 = 9791.8 \text{ N}$$
$$\text{Dead weight} = 40 \text{ kN}$$
$$\text{Maximum pressure on rails} = 40{,}000 + 9791.8 = 49791.8 \text{ N}$$
$$\text{Minimum pressure on rails} = 40{,}000 - 9791.8 = 30208.2 \text{ N}$$

(e) Let ω_1 be the speed, then $84.6 \times 0.6 \times \omega_1^2 = 40,000$

$$\omega_1 = 28.07 \text{ rad/s}$$
$$v = 28.07 \times 0.9 \times \frac{3600}{1000} = 90.95 \text{ km/h}$$

12.5 COUPLED LOCOMOTIVES

In a coupled locomotive, the driving wheels are connected to the leading and trailing wheels by an outside coupling rod, as shown in Fig.12.17. By such an arrangement, a greater portion of the engine mass is utilised for tractive purposes. The coupling rod cranks are placed diametrically opposite to the adjacent driving cranks. The coupling rods together with cranks and pins may be treated as rotating masses and completely balanced by masses in the respective wheels. Therefore, in a coupled locomotive, the rotating and reciprocating masses must be treated separately and the balanced masses for the two systems are then suitably combined in the wheel. The hammer blow may also be considerably reduced.

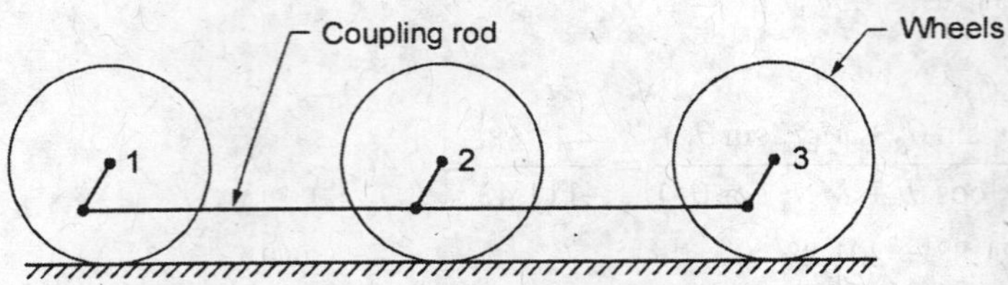

Fig.12.17 Coupled locomotive wheels

Fig.12.18(a) shows the arrangement of coupling rods, wheels, and cylinders of a coupled locomotive, Fig.12.18(b) the angular position of cranks and coupling pin, Fig.12.18(c) the couple polygon when wheel E is driving and Fig.12.18(d) the force polygon when wheel B is the driver.

Example 12.9

The following data refer to a two cylinder locomotive with two coupled wheels on each side:

Length of stroke = 600 mm; Mass of reciprocating parts = 280 kg
Mass of revolving parts =200 kg; Mass of each coupling rod = 240 kg
Radius of centre of coupling rod pin = 250 mm
Distance between cylinders = 0.6 m
Distance between wheels = 1.5 m
Distance between coupling rods = 1.8 m

The main cranks are at right angles and the coupling rod pins are at 180° to their respective main cranks. The balance masses are to be placed in the wheels at a mean radius of 670 mm in order to

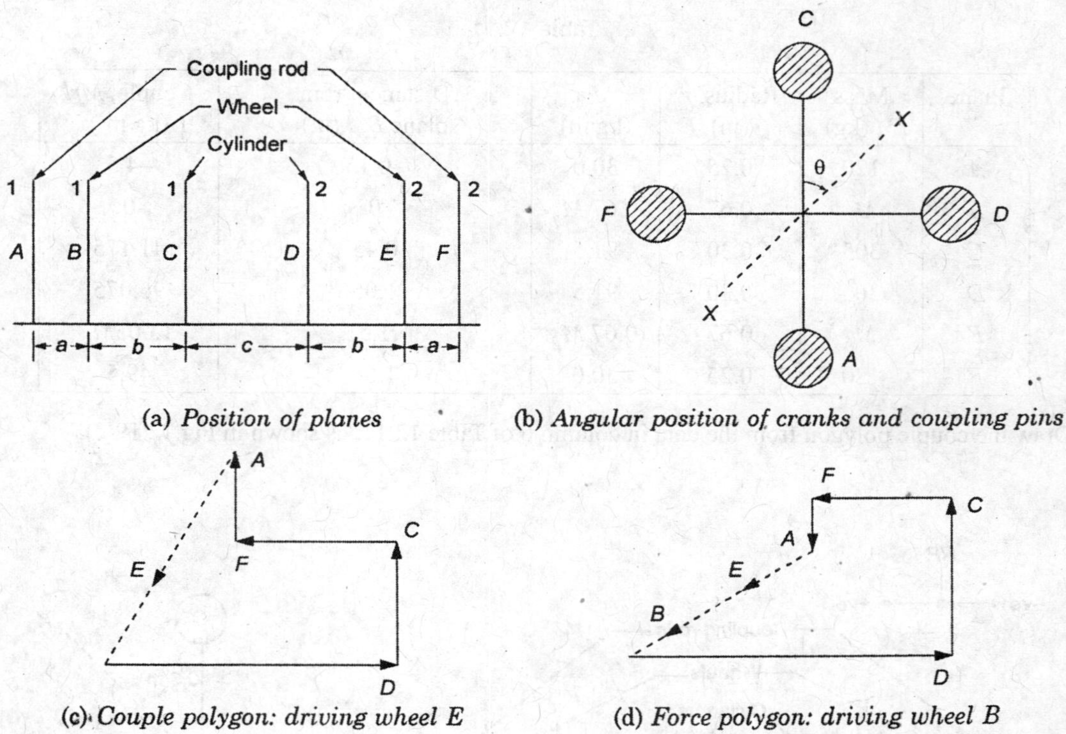

Fig.12.18 Coupled locomotive arrangement

balance the complete revolving and 3/4th of the reciprocating masses. The balance mass for the reciprocating masses is divided equally between the driving wheels and the coupled wheels. Find (a) the magnitude and angular positions of the masses required for the driving and trailing wheels, and (b) the hammer blow at 120 km/h, if the wheels are 1.8 m diameter.

■ Solution

(a) The position of planes for the driving wheels *B* and *E*, cylinders *C* and *D*, and coupling rods *A* and *F*, are shown in Fig.12.19(a). The angular position of cranks *C* and *D* and coupling pins *A* and *F* are shown in Fig.12.19(b).

Mass of the reciprocating parts per cylinder to be balanced $= 3 \times \dfrac{280}{4} = 210$ kg

Mass to be balanced for driving wheels and trailing wheels $= \dfrac{210}{2} = 105$ kg

Masses to be balanced for each driving wheel:

1. Half of the mass of coupling rod $= \dfrac{240}{2} = 120$ kg or $M_A = M_F = 120$ kg
2. Complete the revolving mass (200 kg) and 3/4th the mass of reciprocating parts (105 kg).

or $$M_C = M_D = 200 + 105 = 305 \text{ kg}$$

Driving wheels: Let M_B and M_E be the balance masses placed in the driving wheels *B* and *E*, respectively in plane of *B* as the reference plane.

Table 12.11

Plane	Mass M (kg)	Radius, r (m)	$M \cdot r$ (kg m)	Distance from plane B, l (m)	Couple, Mrl (kg m^2)
A	120	0.25	30.0	–0.15	–4.5
B	M_B	0.67	0.67 M_B	0	0
C	305	0.30	91.5	0.45	41.175
D	305	0.30	91.5	1.05	96.075
E	M_E	0.67	0.67 M_E	1.5	1.005M_E
F	120	0.25	30.0	1.65	49.5

Draw the couple polygon from the data in column 6 of Table 12.11, as shown in Fig.12.19(c).

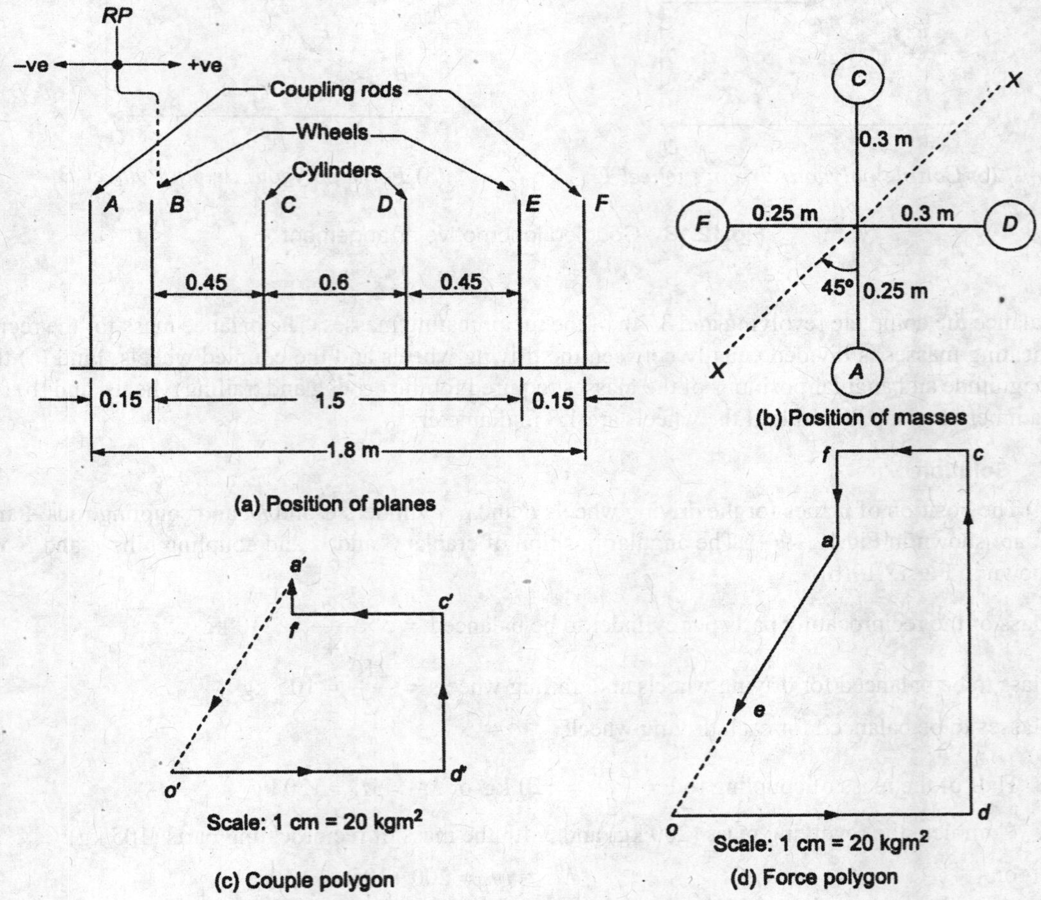

(a) Position of planes

(b) Position of masses

(c) Couple polygon

(d) Force polygon

Fig.12.19 Balancing of two-cylinder coupled wheel locomotive

$$1.005\,M_E = \text{vector } a'o' = 3.3 \times 20$$

or
$$M_E = 65.67 \text{ kg}$$

and
$$\theta_E = 45°$$

Now draw the force polygon from the data in column 4 of the Table 12.11, as shown in Fig.12.19(d).

$$0.67M_B = \text{vector } eo = 2.2 \times 20$$

or
$$M_B = 65.67 \text{ kg}$$

and
$$\theta_B = 45°$$

Trailing wheels: The following masses are to be balanced for each trailing wheel:

1. Half of the mass of the coupling rod. $M_A = M_F = 120$ kg
2. Mass of the reciprocating parts, $M_C = M_D = 105$ kg

Let M_B and M_E be the balanced masses placed in the trailing wheels. We take wheel B as the reference plane.

Table 12.12

Plane	Mass, M (kg)	Radius, r (m)	$M{\cdot}r$ (kg m)	Distance from plane B, l (m)	Couple, Mrl (kgm²)
A	120	0.25	30.0	–0.15	–4.5
B	M_B	0.67	0.67 M_B	0	0
C	105	0.30	31.5	0.45	14.175
D	105	0.30	31.5	1.05	33.075
E	M_E	0.67	0.67 M_E	1.5	1.00 M_E
F	120	0.25	30.0	1.65	49.5

Draw the couple polygon from the data in column 6 from the Table 12.12, as shown in Fig.12.20(a).

$$1.005M_E = \text{vector } a'o' = 2.55 \times 10$$

or
$$M_E = 25.37 \text{ kg}$$

and
$$\theta_E = 41°$$

Now draw the force polygon from the data in column 4 from the above table, as shown in Fig.12.20(b).

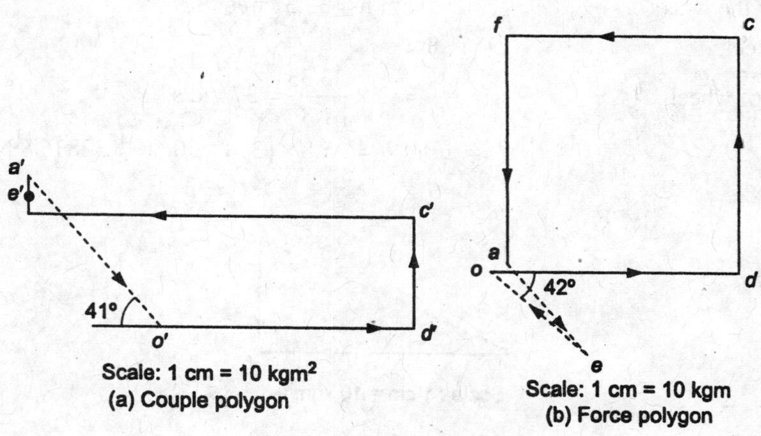

Fig.12.20 Force and couple polygons

$$0.67\,M_B = \text{vector } eo = 1.7 \times 10$$

or $$M_B = 25.37 \text{ kg}$$

and $$\theta_B = 42°$$

The balance masses in all the four wheels are shown in Fig.12.21.

(b) To find the hammer blow, we find the balance mass required for the reciprocating masses only.

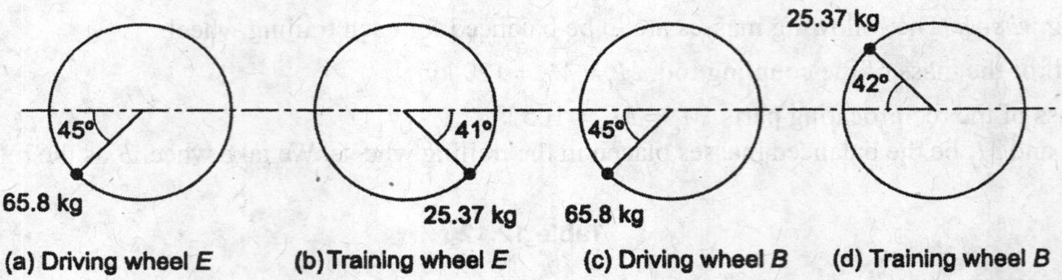

Fig.12.21 Location and magnitude of balance masses

Table 12.13

Plane	Mass M (kg)	Radius r (m)	$M{\cdot}r$ (kg m)	Distance from plane B l (m)	Couple Mrl (kg m²)
B	M_B	0.67	$0.67\,M_B$	0	0
C	105	0.30	31.5	0.45	14.175
D	105	0.30	31.5	1.05	33.075
E	M_E	0.67	0.67	1.5	$1.005\,M_E$

Draw the couple polygon from the data in column 6 of Table 12.13, as shown in Fig.12.22:

$$1.005M_E = \text{vector } c'o' = 3.6 \times 10$$

or $$M_E = 35.8 \text{ kg}$$

Linear speed of the wheel $= 120 \text{ km/h} = 33.33 \text{ m/s}$

Diameter of wheel, $D = 1.8\text{m}$

Angular speed of wheel, $\omega = \dfrac{2v}{D} = 2 \times \dfrac{33.33}{1.8} = 37 \text{ rad/s}$

Hammer blow $= \pm B\omega^2 b = 35.8 \times (37)^2 \times 0.67 = 32836.8 \text{ N}$

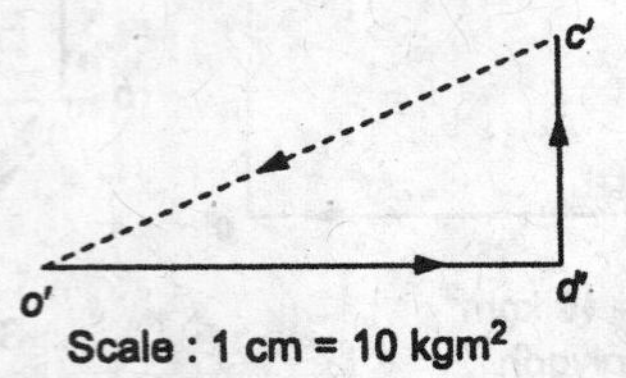

Fig.12.22 Couple polygon

Example 12.10

The following data refer to a four-coupled wheel locomotive with two inside cylinders, as shown in Fig.12.23.

Reciprocating mass per cylinder = 300 kg
Revolving mass per cylinder = 250 kg
Diameter of the driving wheel = 1.9 m
Revolving parts for each coupling rod crank = 120 kg
Engine crank radius = 0.3 m
Coupling rod crank radius = 0.25 m

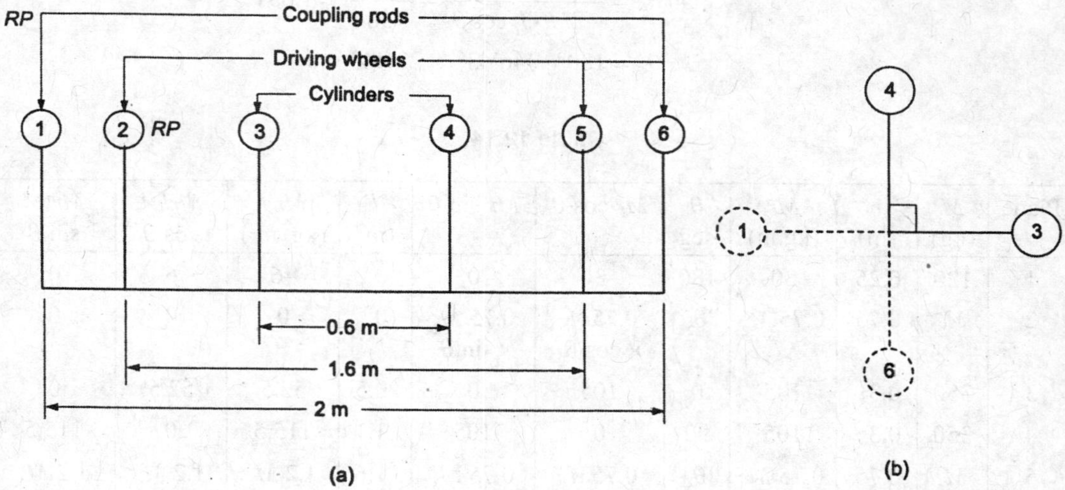

Fig.12.23 Four-coupled wheel locomotive

Distance of centre of balances mass in the planes of the driving wheels from the axle centre = 0.75 m
Angle between the engine cranks = 90°
Angle between the coupling rod crank with adjacent engine crank = 180°
The balance mass required for the reciprocating parts is equally divided between each pair of coupled wheels. Determine

(a) the magnitude and position of the balance mass required to balance 2/3rd of reciprocating and the complete revolving parts.

(b) The hammer blow and

(c) The maximum variation of tractive force when the locomotive speed is 75 km/h.

■ Solution

Leading wheels
Balance mass = 250 + 0.5 × 2/3 × 300 = 350 kg
Take 2 as reference plane with $\theta_3 = 0°$.
From Table 12.14, we have $\Sigma M_i r_i \cos\theta_i = 75$

$$\Sigma M_i r_i \sin\theta_i = 75$$
$$\Sigma M_i r_i l_i \cos\theta_i = 58.5$$
$$\Sigma M_i r_i l_i \sin\theta_i = 61.5$$
$$\left[\left(\Sigma M_i r_i l_i \cos\theta_i\right)^2 + \left(\Sigma M_i r_i l_i \sin\theta_i\right)^2\right]^{0.5} = M_5 r_5 l_5$$
$$\left[(58.5)^2 + (61.5)^2\right]^{0.5} = 1.2\, M_5$$
$$M_5 = \frac{84.88}{1.2} = 70.73 \text{ kg}$$
$$\tan\theta_5 = \frac{-\Sigma M_i r_i l_i \sin\theta_i}{-\Sigma M_i r_i l_i \cos\theta_i} = \frac{-61.5}{-58.5} = 1.0513$$
$$\theta_5 = 180° + 46.43° = 226.43° \text{ ccw}$$

Table 12.14

Plane	M (kg)	r (m)	Mr (kg.m)	θ deg	$Mr\cos\theta$	$Mr\sin\theta$	l (m)	$Mr\,l$ (kg. m²)	$Mrl \times \cos\theta$	$Mrl \times \sin\theta$
1	120	0.25	30	180	–30	0	–0.2	–6	6	0
2	M_2	0.75	$0.75\,M_2$	θ_2	$0.75M_2 \times \cos\theta_2$	$0.75\,M_2 \times \sin\theta_2$	0	0	0	0
3	350	0.3	105	0	105	0	0.5	52.5	52.5	0
4	350	0.3	105	90	0	105	1.1	115.5	0	115.5
5	M_5	0.75	$0.75\,M_5$	θ_5	$0.75M_5 \times \cos\theta_5$	$0.75\,M_5 \times \sin\theta_5$	1.6	$1.2\,M_5$	$1.2\,M_5 \times \cos\theta_5$	$1.2\,M_5 \times \sin\theta_5$
6	120	0.25	30	270	0	–30	1.8	54	0	–54

Since the numerator and denominator are both negative, therefore θ_5 lies in the third quadrant.

$$M_2 r_2 = \left[\left(\Sigma M_i r_i \cos\theta_i + M_5 r_5 \cos\theta_5\right)^2 + \left(\Sigma M_i r_i \sin\theta_i + M_5 r_5 \sin\theta_5\right)^2\right]^{0.5}$$
$$0.75 M_2 = \left[\left(75 + 70.73 \times 0.75 \times \cos 226.43°\right)^2 + \left(75 + 70.73 \times 0.75 \times \sin 226.43°\right)^2\right]^{0.5}$$
$$= \left[(38.43)^2 + (36.56)^2\right]^{0.5} = 53.046$$
$$M_2 = 70.73 \text{ kg}$$
$$\tan\theta_2 = \frac{-\left(\Sigma M_i r_i \sin\theta_i + M_5 r_5 \sin\theta_5\right)}{\left(\Sigma M_i r_i \cos\theta_i + M_5 r_5 \cos\theta_5\right)}$$
$$= \frac{-36.56}{-38.43} = 0.95134$$
$$\theta_2 = 180° + 43.57° = 223.57°$$

Since the numerator and denominator are both negative, therefore θ_2 lies in the third quadrant.

Trailing wheels: Balance mass $= 0.5 \times 2 \times \dfrac{300}{3} = 100$ kg

Take 2 as reference plane with $\theta_3 = 0°$.

Table 12.15

Plane	M (kg)	r (m)	Mr (kg.m)	θ (deg)	$Mr \times \cos\theta$	$Mr \times \sin\theta$	l (m)	$Mr\,l$ (kg.m²)	$Mrl \times \cos\theta$	$Mrl \times \sin\theta$
1	120	0.25	30	180	–30	0	–0.2	–6	6	0
2	M_2	0.75	$0.75\,M_2$	θ_2	$0.75\,M_2 \times \cos\theta_2$	$0.75\,M_2 \times \sin\theta_2$	0	0	0	0
3	100	0.3	30	0	30	0	0.5	15	15	0
4	100	0.3	30	90	0	30	1.1	33	0	33
5	M_5	0.75	$0.75\,M_5$	θ_5	$0.75\,M_5 \times \cos\theta_5$	$0.75\,M_5 \times \sin\theta_5$	1.6	$1.2\,M_5$	$1.2\,M_5 \times \cos\theta_5$	$1.2\,M_5 \times \sin\theta_5$
6	120	0.25	30	270	0	–30	1.8	54	0	–54

From Table 12.15, we have

$$\Sigma M_i r_i \cos\theta_i = 0$$
$$\Sigma M_i r_i \sin\theta_i = 0$$
$$\Sigma M_i r_i l_i \cos\theta_i = 21$$
$$\Sigma M_i r_i l_i \sin\theta_i = -21$$

$$\left[\left(\Sigma M_i r_i l_i \cos\theta_i\right)^2 + \left(\Sigma M_i r_i l_i \sin\theta_i\right)^2\right]^{0.5} = M_5 r_5 l_5$$

$$\left[(21)^2 + (-21)^2\right]^{0.5} = 1.2 M_5$$

$$M_5 = \frac{29.698}{1.2} = 24.75 \text{ kg}$$

$$\tan\theta_5 = \frac{-\Sigma M_i r_i l_i \sin\theta_i}{-\Sigma M_i r_i l_i \cos\theta_i} = \frac{-(-21)}{-(21)} = -1$$

$$\theta_5 = 180° - 45° = 135 \text{ ccw}$$

Since the numerator is positive and denominator is negative, therefore θ_5 lies in the second quadrant.

By symmetry, $M_2 = 24.75$ kg

$$\tan\theta_2 = \frac{-(21)}{-(-21)} = -1$$

$$\theta_2 = 360° - 45° = 315°$$

Since the numerator is negative and denominator is positive, therefore θ_2 lies in the fourth quadrant.

(b)
$$v = 75 \times \frac{1000}{3600} = 20.83 \text{ m/s}$$
$$\omega = v/r = \frac{20.83}{0.95} = 21.93 \text{ rad/s}$$

Neglecting M_1 and M_6, we have
$$1.2M_5 = [15^2 + 33^2]^{0.5} = 36.249$$
$$M_5 = 30.2 \text{ kg}$$

Hammer blow $= M_5 b\omega^2 = 30.2 \times 0.75 \times (21.93)^2 = 10893$ N

(c) Maximum variation in tractive effort $= \pm\sqrt{2}(1 - c)R\omega^2 r$
$$= \pm\sqrt{2}(1 - 2/3) \times 300 \times (21.93)^2 \times 0.3 = \pm 20404 \text{ N}$$

12.6 MULTICYLINDER IN-LINE ENGINES

In a multi cylinder in-line engine, the cylinder centre lines lie in the same plane and on the same side of the crankshaft centre line, as shown in Fig.12.24.

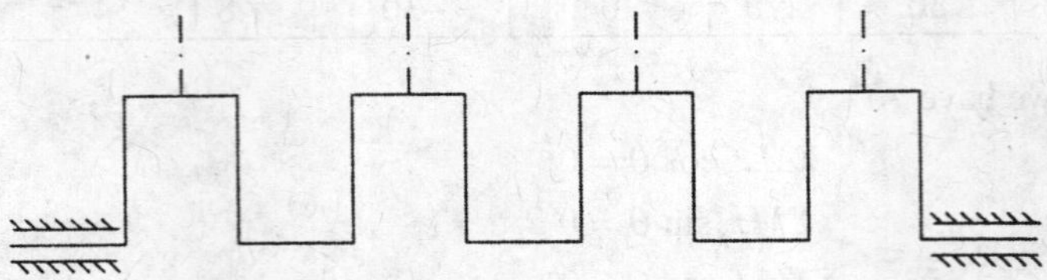

Fig.12.24 Multi-cylinder inline engine

12.6.1 Primary Balancing

The conditions to be satisfied for the primary balancing are:

1. The algebraic sum of the primary forces should be equal to zero, i.e. the primary force polygon must close.
$$\Sigma R\, \omega^2 r \cos\theta = 0 \qquad (12.21a)$$
2. The algebraic sum of the primary couples about any point in the plane of the forces must be equal to zero, i.e. the primary couple polygon must close.
$$\Sigma R\, \omega^2 r\, a \cos\theta = 0 \qquad (12.21b)$$
where a = distance of the plane of rotation of the crank from a parallel reference plane.

Hence, if a system of reciprocating masses is to be in primary balance, the system of reciprocating masses, which is obtained by substituting an equal revolving mass at the crankpin for each reciprocating mass, must be balanced.

The graphical construction for the balancing of primary forces is represented in Fig.12.25.

ef, fg, gh, eh = primary forces

γ = angle turned through by the crankshaft, clockwise.

= angle turned through by the line of stroke, ccw, i.e. PQ goes to PS.

kl, ml, mn, nk_1 = primary forces whose resultant is kk_1.

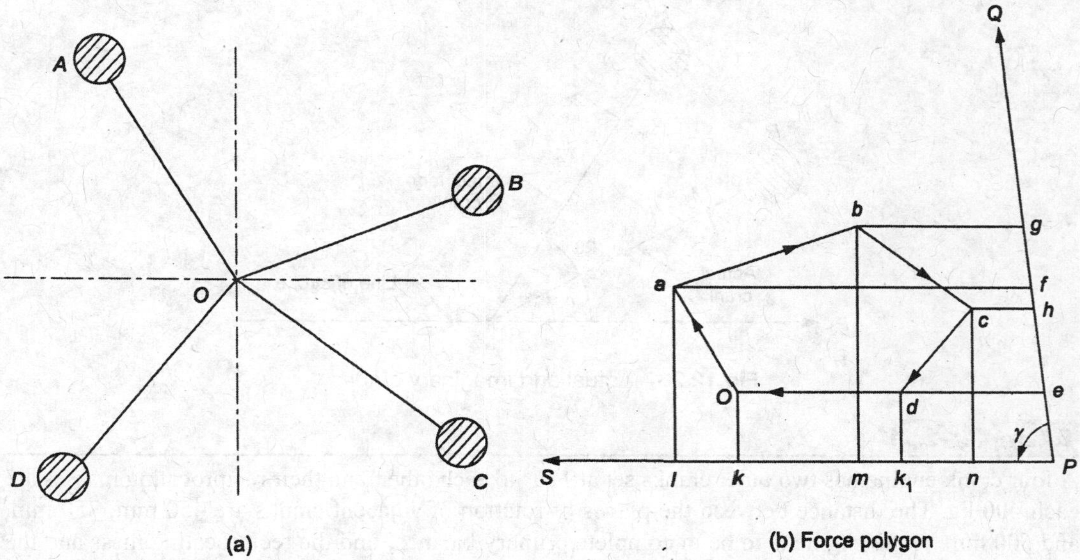

Fig.12.25 Graphical method for primary balancing of multi-cylinder inline engine

For balance of primary forces d must coincide with o. In a similar way, the primary couples can only be balanced if the couple polygon for the corresponding centrifugal forces is closed.

12.6.2 Secondary Balancing

The conditions to be satisfied for the secondary balancing are as follows:

1. The algebraic sum of the secondary forces should be equal to zero, i.e. the secondary force polygon must close.

$$\Sigma R(2\omega)^2\left(\frac{r}{4n}\right)\cos 2\theta = 0 \tag{12.22a}$$

2. The algebraic sum of the secondary couples about any point in the plane of the forces must be equal to zero, i.e. the secondary couple polygon must close.

$$\Sigma R(2\omega)^2\left(\frac{r}{4n}\right)a\cos 2\theta = 0 \tag{12.22b}$$

where a = distance of the plane of rotation of the crank from a parallel reference plane

$$\text{Imaginary crank length} = \frac{r}{4n} \tag{12.23a}$$

$$\text{Speed} = 2\omega \tag{12.23b}$$

Angle made by imaginary secondary crank with inner dead centre = 2θ. The actual and imaginary cranks are shown in Fig.12.26.

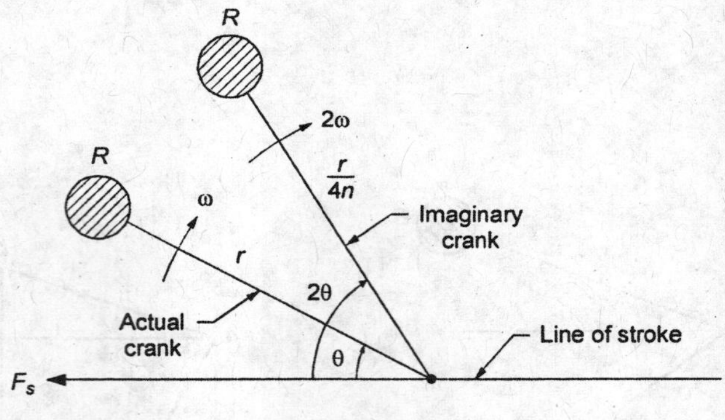

Fig.12.26 Actual and imaginary cranks

Example 12.11

A four crank engine has two outer cranks set at 120° to each other, and their reciprocating masses are each 400 kg. The distance between the planes of rotation of adjacent cranks are 450 mm, 750 mm, and 600 mm. If the engine is to be in complete primary balance, find the reciprocating mass and the relative angular position for each of the inner cranks.

If the length of each cranks is 300 mm, length of each connecting rod is 1.2 m and the speed of rotation is 240 rpm, what is the maximum secondary unbalanced force?

Solution

Reciprocating masses:

Let M_2 and M_3 be the reciprocating masses for inner cranks 2 and 3; θ_2 and θ_3 their angular locations respectively. The position of planes and primary crank positions is shown in Fig.12.27(a) and (b), respectively.

Table 12.16

Plane	Mass, M (kg)	Radius, r (m)	$M \cdot r$ (kg.m)	Distance from plane 2, l (m)	Couple, $M\,r\,l$, (kg.m²)
1	400	0.3	120	–0.45	–54.0
2	M_2	0.3	0.3 M_2	0	0
3	M_3	0.3	0.3 M_3	0.75	0.225 M_3
4	400	0.3	120	1.35	162.0

Since the engine is to be in complete primary balance, therefore, the primary couple and force polygons must close. The primary couple polygon is shown in Fig.12.27(c), drawn from the data in column 6 of Table 12.16.

$$0.225\,M_3 = \text{vector } 0'4' = 4.9 \text{ cm} = 196$$

or $$M_3 = 871 \text{ kg}$$

and $$\theta_3 = 314°$$

The force polygon is drawn in Fig.12.27(d), from the data in column 4.

$$0.3M_2 = \text{vector } 03 = 284$$

or $$M_2 = 947 \text{ kg}$$

and $$\theta_2 = 168°$$

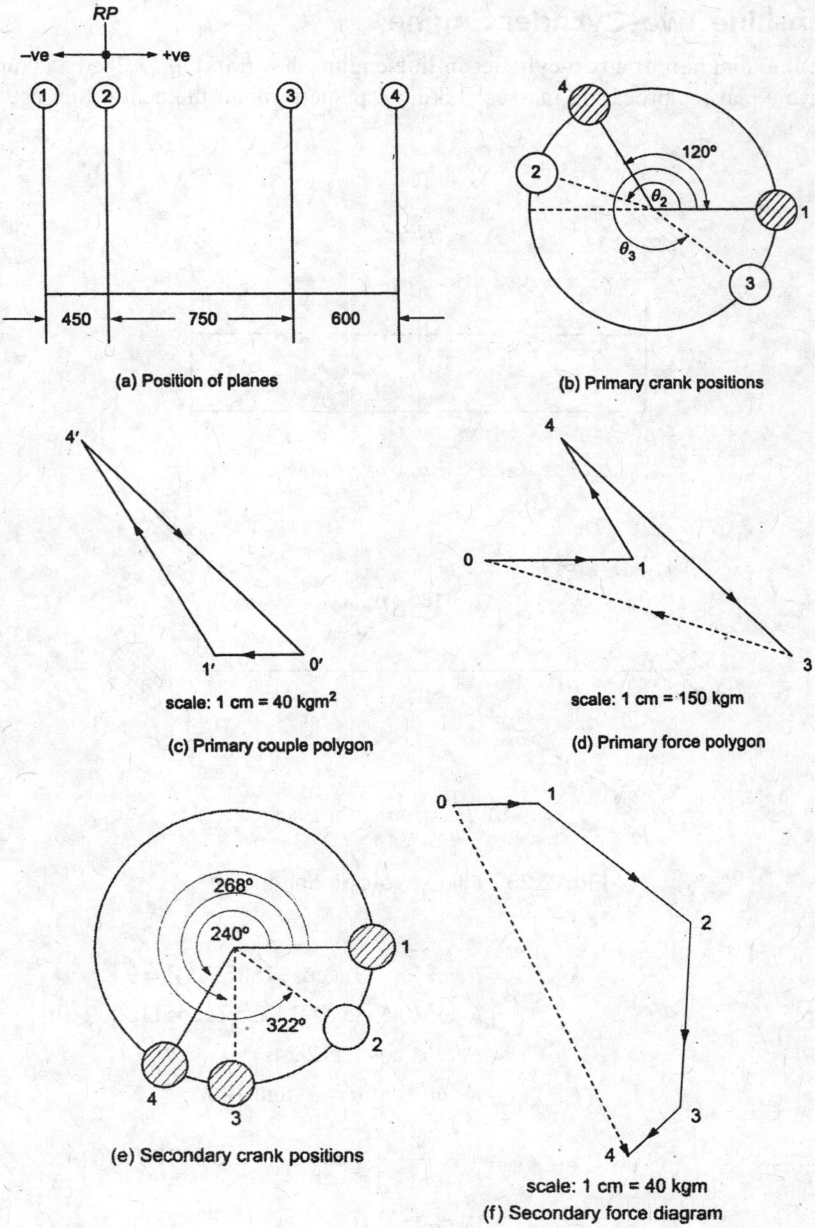

Fig.12.27 Primary and secondary balancing of multi-cylinder inline engine

Secondary unbalanced force: The secondary cranks at twice the angle are shown in Fig.12.27(e). The secondary force polygon is drawn in Fig.12.27(f). The closing side of the polygon gives the unbalanced secondary force.

$$\text{Maximum unbalanced secondary force} = \frac{582\,\omega^2}{n} = 582 \times \frac{\left(2\pi \times \dfrac{240}{60}\right)^2}{1.2/0.3} = 91.96 \text{ KN.}$$

12.6.3 In-Line Two-Cylinder Engine

Consider the line diagram of a two-cylinder in-line engine shown in Fig.12.28. The cranks are 180° apart and have equal reciprocating masses. Taking a plane through the centre line as the reference plane, we have

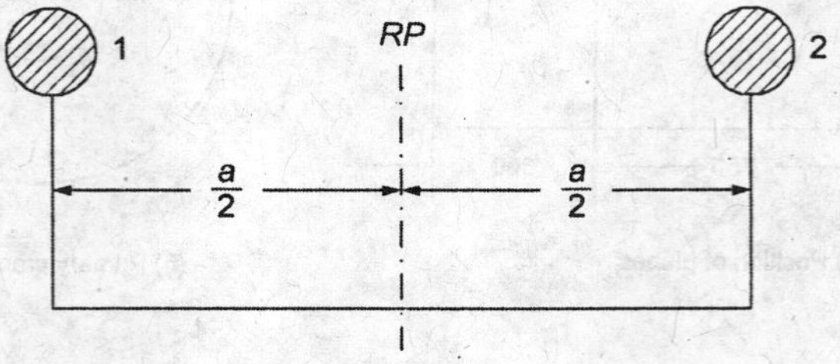

(a) *Position of planes*

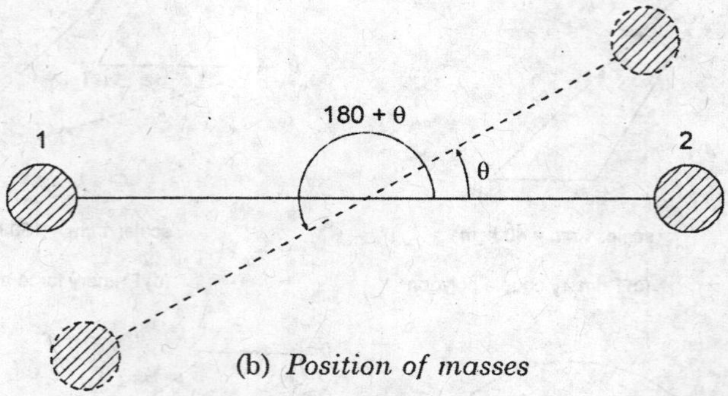

(b) *Position of masses*

Fig.12.28 Inline two-cylinder engine

Primary force, $F_p = Rr\omega^2 [\cos\theta + \cos(180° + \theta)] = 0$

Primary couple, $C_p = Rr\omega^2 [0.5\, \alpha \cos\theta - 0.5\, \alpha \cos(180° + \theta)]$

$= Rr\omega^2 \alpha \cos\theta$ (12.24a)

$$(C_p)_{max} = Rr\, \omega^2\, \alpha \text{ at } \theta = 0° \text{ and } 180°. \quad (12.24b)$$

Secondary force, $F_s = \left(\dfrac{Rr\,\omega^2}{n}\right)\left[\cos 2\theta + \cos 2\,(180° + \theta)\right]$

$$= \frac{Rr\,\omega^2}{n} \cos 2\theta \quad (12.25a)$$

$$(F_s)_{\max} = \frac{2Rr\,\omega^2}{n} \quad \text{at} \quad \theta = 0°, 90°, 180°, 270° \quad (12.25b)$$

Secondary couple, $C_s = \left(\dfrac{2Rr\omega^2}{n}\right)\left[0.5\, a \cos 2\theta - 0.5\, a \cos 2\,(180° + \theta)\right] = 0$

The force and couple polygons for primary and secondary cranks are shown in Fig.12.29 (a) to (d).

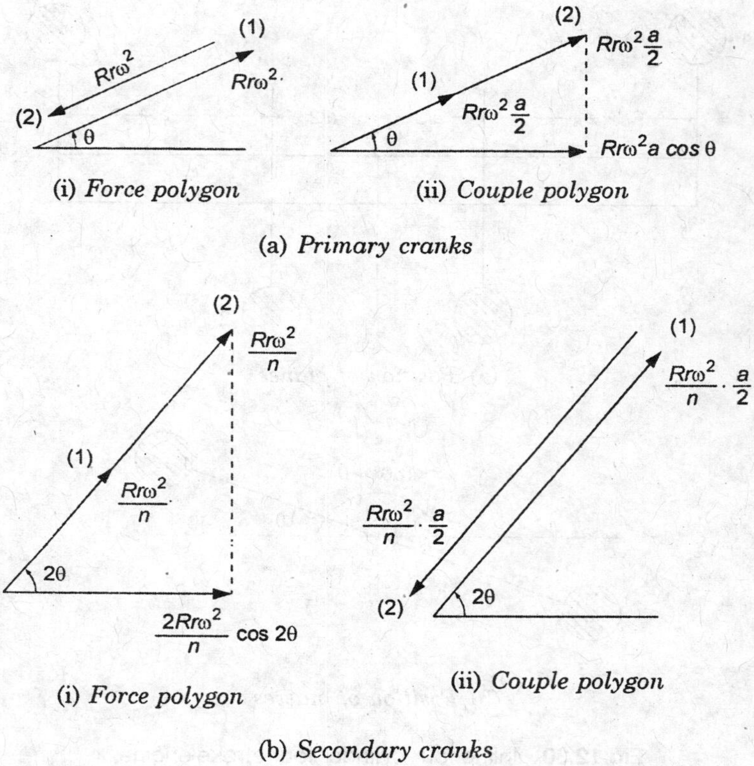

Fig.12.29 Force and couple polygons for inline two cylinder engine

12.6.4 In-line Four-cylinder Four-stroke Engine

A line diagram of a four cylinder engine is shown in Fig.12.30. The forces and couples are as follows:

Primary force, $F_P = Rr\,\omega^2\,[\cos\theta + \cos(180° + \theta) + \cos(180° + \theta) + \cos\theta] = 0$

Primary couple,

$$C_P = Rr\omega^2\,[\,1.5\,\alpha\cos\theta + 0.5\,\alpha\cos(180° + \theta) - 0.5\,\alpha\cos(180° + \theta) - 1.5\,\alpha\cos\theta] = 0$$

Secondary force,

$$F_S = \left(\frac{Rr\,\omega^2}{n}\right)\left[\cos 2\theta + \cos 2(180° + \theta) + \cos 2(180° + \theta) + \cos 2\theta\right]$$

$$= \left(\frac{4Rr\,\omega^2}{n}\right)\cos 2\theta \quad (12.26a)$$

$$(F_S)_{\max} = \left(\frac{4Rr\,\omega^2}{n}\right) \quad \text{at} \quad \theta = 0°, 90°, 180°, 270° \quad (12.26b)$$

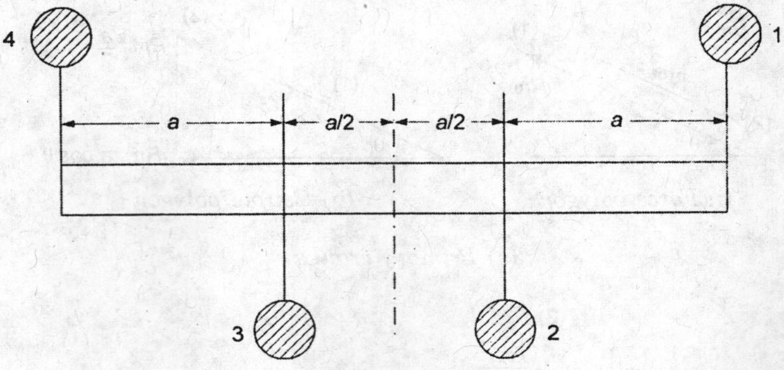

(a) *Position of planes*

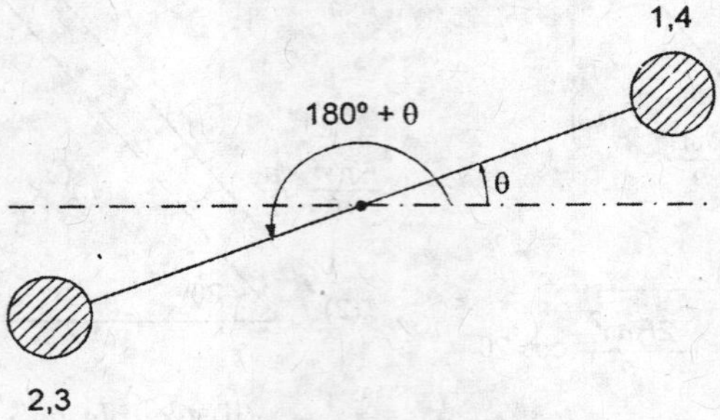

(b) *Position of masses*

Fig.12.30 Inline four-cylinder four-stroke engine.

$$\text{Secondary couple,}\quad C_s = \left(\frac{Rr\,\omega^2}{n}\right)\left[1.5\,a\,\cos 2\theta + 0.5\,a\cos 2\left(180^\circ + \theta\right)\right.$$

$$\left. - 0.5\,a\cos 2\left(180^\circ + \theta\right) - 1.5\,a\cos 2\theta\right]$$

$$= 0$$

The force and couple polygons for primary and secondary cranks are shown in Fig.12.31(a) to (d).

Example 12.12

In a marine oil engine, the cranks of four cylinders are arranged at angular displacements of 90°. The speed of the engine is 105 rpm and the mass of reciprocating parts for each cylinder is 850 kg. Each crank is 0.4 m long. The outer cranks are 3 m apart and the inner cranks are 1.2 m apart and are placed symmetrically between the outer cranks.

Find the firing order of the cylinders for the best primary balancing of reciprocating parts and also the maximum unbalanced primary couple for this arrangement.

■ Solution

$$\omega = 2\pi \times \frac{105}{60} = 10.996 \text{ rad/s}$$

$$r = 0.4 \text{ m}$$

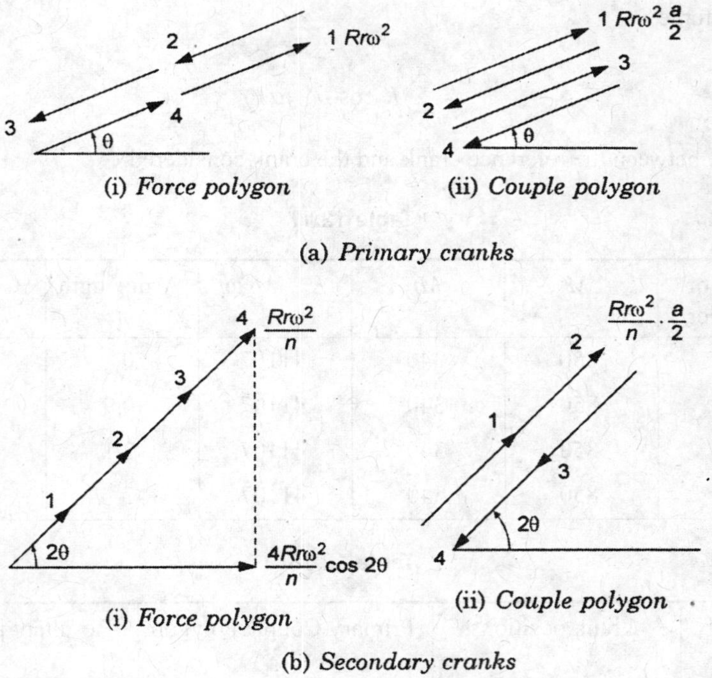

Fig.12.31 Force and couple polygons for four-cylinder four-stroke engine

The primary forces are always balanced as cranks are arranged at an angular displacement of 90° to each other. The primary couples need to be investigated. The position of cranks is as shown in Fig.12.32.

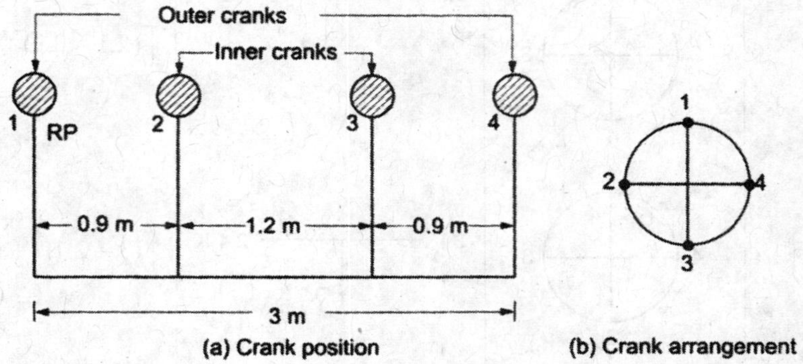

Fig.12.32 Marine oil engine

The possible firing orders are: 1234, 1243, 1423, 1324, 1342, 1432, as shown in Table 12.18.
The disturbing force along the axis of the cylinder = $Mr\omega^2 \cos\theta$
Let $K = Mr\omega^2 = 850 \times 0.4 \times (10.996)^2 = 41107$ N, as shown in Table 12.17.

Total disturbing force

$$= \sum_{I=1}^{4} K\cos(\theta + \alpha I)$$

where α_1 = angle between the reference crank and the crank considered.

Table 12.17

Plane of cylinder	*M*	*Mr*	$K = Mr\omega^2$	Arm length *l*	Couple *Kl*
1	850	340	41107	0	0
2	850	340	41107	0.9	0.9 *K*
3	850	340	41107	2.1	2.1 *K*
4	850	340	41107	3	3 *K*

Table 12.18

Disposition of Cranks	Crank positions	Primary Couple Polygon	Resultant primary Couple
1234	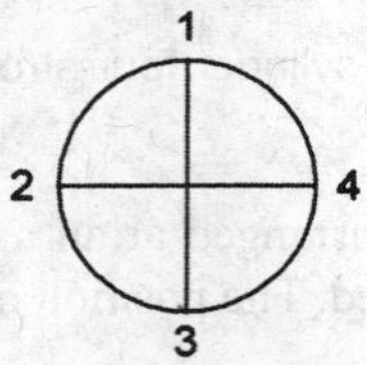	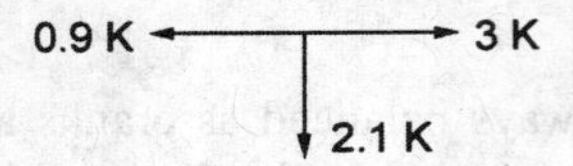	$\left[(3-0.9)^2 + (2.1)^2\right]^{0.5} \times K = 2.97\ K$
1243	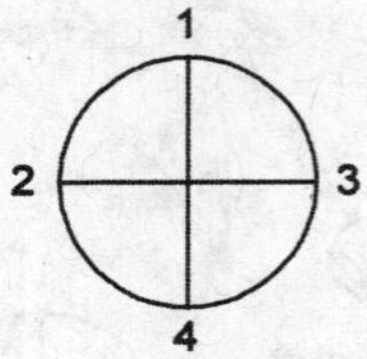	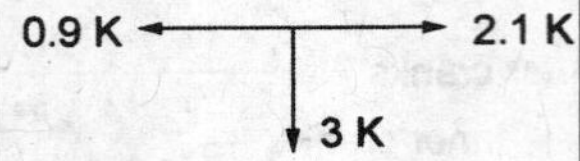	$\left[(2.1-0.9)^2 + (3)^2\right]^{0.5} \times K = 2.97\ K$
1423	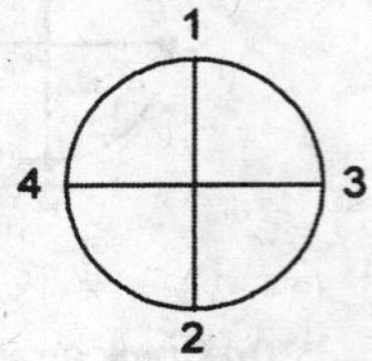	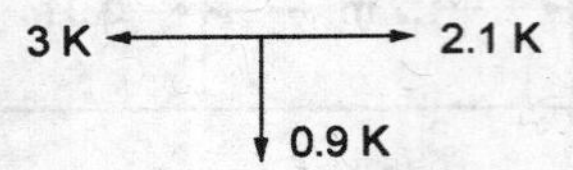	$\left[(3-2.1)^2 + (0.9)^2\right]^{0.5} \times K = 1.273\ K$
1324	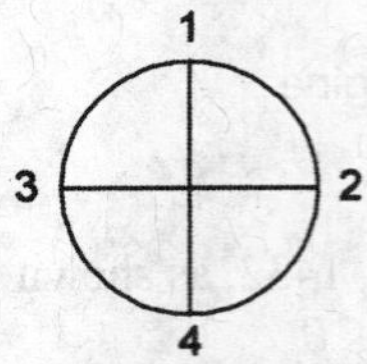	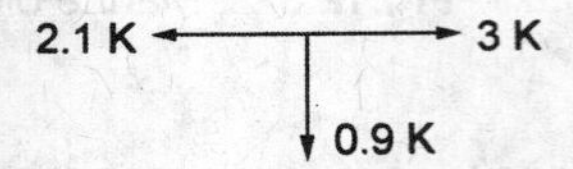	$\left[(3-2.1)^2 + (0.9)^2\right]^{0.5} \times K = 1.273\ K$

Disposition of Cranks	Crank positions	Primary Couple Polygon	Resultant primary Couple
1342	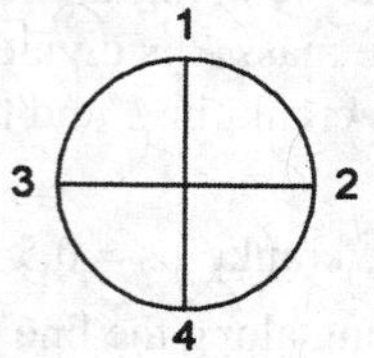	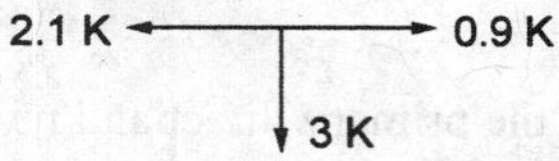	$\left[(2.1-0.9)^2+(3)^2\right]^{0.5}\times K$ $= 3.231$ K
1432	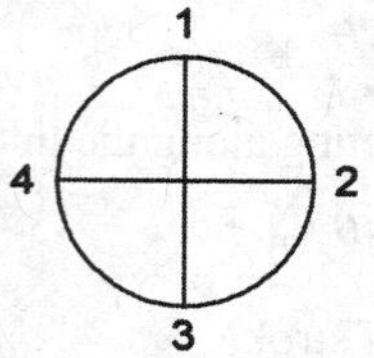	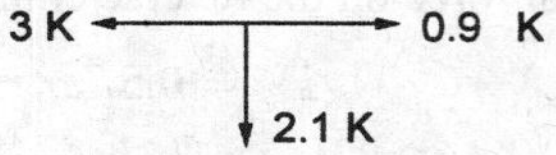	$\left[(3-0.9)^2+(2.1)^2\right]^{0.5}\times K$ $= 2.97$ K

Least value of primary couple = 1.273 K

$= 1.273 \times 41107 = 52329$ N m

Best firing order is 1423 and 1324.

12.7 BALANCING OF RADIAL ENGINES

In the case of radial engines, the cranks are arranged radially and lie in the same plane. We shall study the direct and reverse crank method to balance the radial engines.

12.7.1 Direct and Reverse Cranks Method

This method is used to balance radial or *V*-engines, in which connecting rods are connected to a common crank, as shown in Fig.12.33. Since the plane of various cranks is the same, Therefore, there is no unbalanced primary or secondary couple.

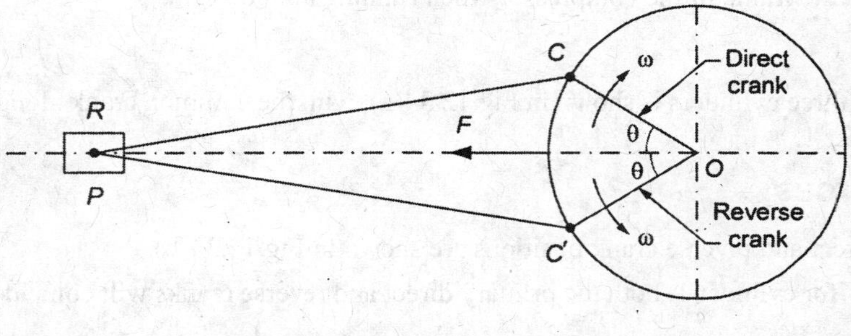

Fig.12.33 Balancing of radial engines

Let the direct crank *OC* rotate uniformly at ω (rad/s) speed in a clockwise direction. Then the reverse crank *OC′* will rotate in the ccw direction. The reverse crank *OC′*s the mirror image of the direct crank *OC*.

1. Primary Forces

Now the primary force, $F_p = R\omega^2 r \cos\theta$

This force is equal to the component of the centrifugal force along the line of stroke, produced by a mass R placed at the crankpin C. Let us suppose that the mass R is divided into two equal parts, each equal to $R/2$. It is assumed that $R/2$ is fixed at the direct crankpin C and the other $R/2$ is fixed at the reverse crankpin C'.

Centrifugal force acting on the primary direct and reverse crankpins = $0.5\ R\omega^2 r$

Component of the centrifugal force on the direct crank acting along the line of stoke from O to P,

$$F_{pd} = 0.5R\omega^2 r \cos\theta$$

Component of the centrifugal force on the reverse crank acting along the line of stroke from O to P

$$F_{pr} = 0.5R\omega^2 r \cos\theta$$

Total component of the centrifugal force along the line of stroke

$$F_p = F_{pd} + P_{pr} = R\omega^2 r \cos\theta \tag{12.27}$$

which is the primary force itself. Hence, for primary force effects, the mass of the reciprocating parts at P may be replaced by two masses at crankpins C and C', each of mass $R/2$ at radii equal to r.

2. Second Forces

The secondary force,

$$F_s = R\omega^2 r \frac{\cos 2\theta}{n} \tag{12.28}$$

In the similar way as discussed for the primary force, the secondary force effect may be taken into account by dividing the mass R into two equal parts and placing it at the imaginary crankpins at radii $r/4n$.

Example 12.13

The three cylinders of an air compressor have their axes 120° to one another and their connecting rods are coupled to a common crank. The stroke is 100 mm and the length of each connecting rod is 150 mm. The mass of the reciprocating parts per cylinder is 2 kg. Find the maximum primary and secondary forces acting on the frame of the compressor when running at 3000 rpm.

■ Solution

The position of three cylinders is shown in Fig.12.34(a), with the common crank along the inner dead centre of cylinder 1.

Primary Forces

The primary direct and reverse crank positions are shown in Fig.12.34(b).

1. Since $\theta = 0°$ for cylinder 1, both the primary direct and reverse cranks will coincide with the common crank.
2. Since $\theta = \pm 120°$ for cylinder 2, the primary direct crank is 120° clockwise and the primary reverse crank is 120° counter-clockwise from the line of stroke of cylinder 2.
3. Since $\theta = \pm 240°$ for cylinder 3, the primary direct crank is 240° clockwise and primary reverse crank at 240° counter-clockwise from the line of stroke of cylinder 3.

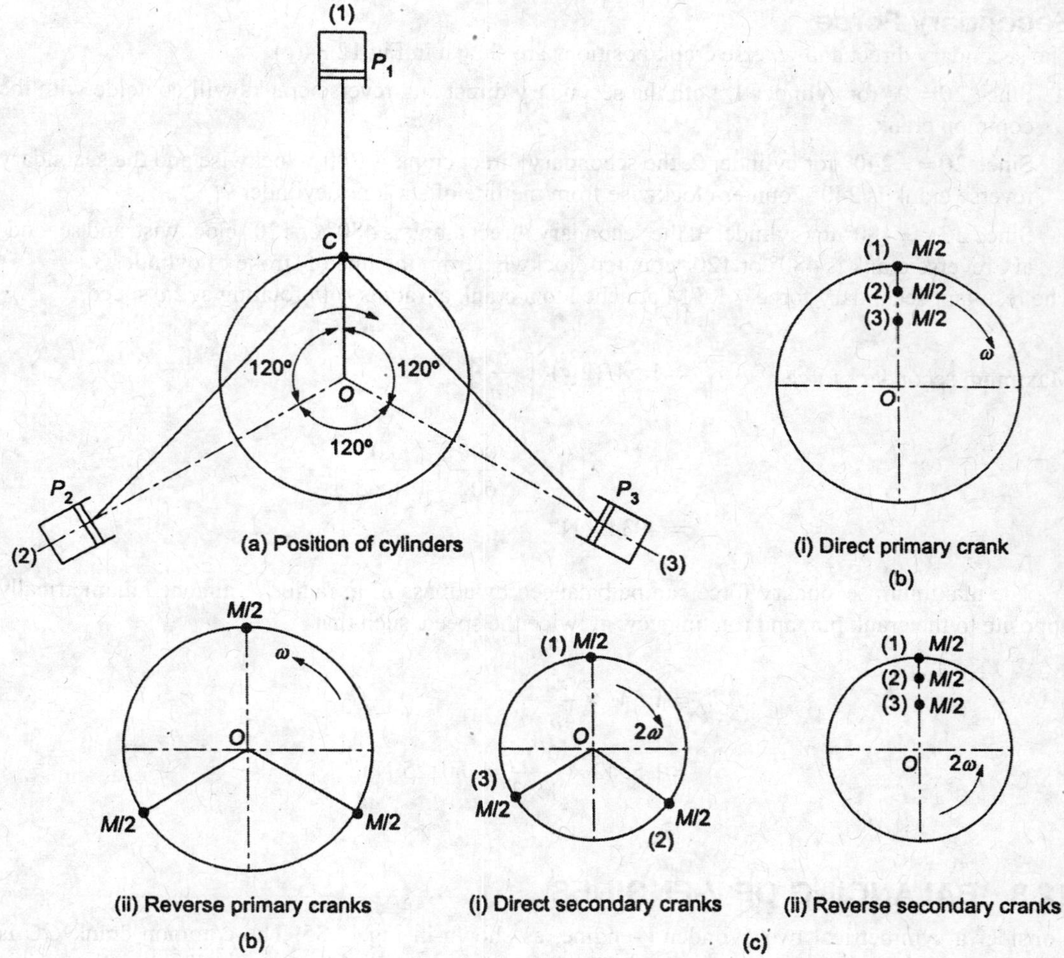

Fig.12.34 Direct and reverse crank method

From Fig.12.34(b-ii), we find that the primary reverse cranks form a balanced system. Therefore, there is no unbalanced primary force due to the reverse cranks. From Fig.12.34(b-i), we find that the resultant primary force is equivalent to the centrifugal force of a mass 1.5 M attached to the end of the crank.

Maximum primary force $= 1.5\, M\omega^2 r$

$$= 1.5 \times 2 \left(2\pi \times \frac{3000}{60} \right)^2 0.05$$

$$= 14804.4 \text{ N}$$

The maximum primary force may be balanced by a mass attached diametrically opposite to the crank pin and rotating with the crank, of magnitude B_1 at radius b_1, such that $B_1 b_1 = 1.5\, Mr = 1.5 \times 2 \times 0.05 = 0.15$ N m

Secondary Force

The secondary direct and reverse crank positions are shown in Fig.12.34(c).

1. Since $2\theta = 0°$ for cylinder 1, both the secondary direct and reverse cranks will coincide with the common crank.
2. Since $2\theta = \pm 240°$ for cylinder 2, the secondary direct crank is 240° clockwise and the secondary reverse crank is 240° counter-clockwise from the line of stroke of cylinder 2.
3. Since $2\theta = \pm 480°$ for cylinder 3, the secondary direct crank is 480° or 120° clockwise and secondary reverse crank is 480° or 120° counter-clockwise from the line of stroke of cylinder 3.

The resultant secondary force = 1.5 M attached to a crank at radius $r/4n$ rotating at 2ω speed.

$$\text{Maximum secondary force} = 1.5M(2w)^2\left(\frac{r}{4n}\right)$$

$$= 1.5 \times 2\left(4\pi \times \frac{3000}{60}\right)^2\left(\frac{0.05}{4 \times 3}\right)$$

$$= 4934.8 \text{ N}$$

The maximum secondary force can be balanced by a mass B_2 at radius b_2 attached diametrically opposite to the crank pin, and rotating ccw at twice the speed, such that

$$B_2 b_2 = 1.5M \times \frac{r}{4n}$$

$$= 1.5 \times 2 \times \frac{0.05}{4 \times 3} = 0.0125 \text{ N m}$$

12.8 BALANCING OF *V*-ENGINES

Consider a symmetrical two cylinder *V*-engine, as shown in Fig.12.35. The common crank *OC* is driven by two connecting rods *PC* and *QC*. The lines of stroke *OP* and *OQ* are inclined to the vertical *OY*, at an angle α.

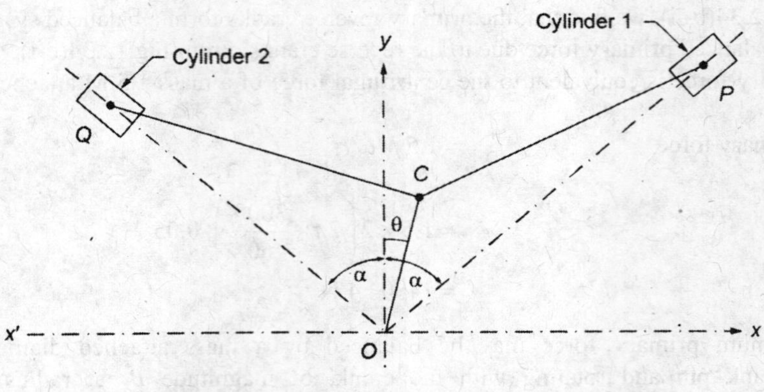

Fig.12.35 Balancing of *V*-engines

Inertia force due to the reciprocating parts of cylinder 1, along the line of stroke

$$= R\omega^2 r\left[\cos(\alpha-\theta)+\frac{\cos 2(\alpha-\theta)}{n}\right]$$

Inertia force due to the reciprocating parts of cylinder 2, along the line of stroke

$$= R\omega^2 r\left[\cos(\alpha+\theta)+\frac{\cos 2(\alpha+\theta)}{n}\right]$$

Since the plane of cranks is the same, therefore, there are no primary or secondary couples.

1. Primary Force

Primary force of cylinder 1 acting along the line of stroke

$$F_{p1} = R\omega^2 r\cos(\alpha-\theta)$$

Component of F_{p1} along the vertical line $OY = F_{p1}\cos\alpha$

$$= R\omega^2 r\cos(\alpha-\theta)\cos\alpha$$

Component of F_{p1} along the horizontal line $OX = F_{p1}\sin\alpha$

$$= R\omega^2 r\cos(\alpha-\theta)\sin\alpha$$

Similarly, for the cylinder 2, we have

$$F_{p2} = R\omega^2 r\cos(\alpha+\theta)$$

Component of $F_p 2$ along the vertical line $OY = F_{p2}\cos\alpha$

$$= R\omega^2 r\cos(\alpha+\theta)\cos\alpha$$

Component of F_{p2} along the horizontal line $OX' = F_{p2}\sin\alpha$

$$= R\omega^2 r\cos(\alpha+\theta)\sin\alpha$$

Total component of primary force along the vertical line OY

$$F_{pv} = R\omega^2 r[\cos(a-\theta)+\cos(a+\theta)]\cos a$$

$$= 2R\omega^2 r\cos^2 a\cos\theta$$

Total component of primary force along the horizontal line OX

$$F_{ph} = R\omega^2 r[\cos(\alpha-\theta)-\cos(\alpha+\theta)]\sin\alpha$$

$$= 2R\omega^2 r\sin^2\alpha\sin\theta$$

Resultant primary force, $F_p = \left[F_{pv}^2 + F_{ph}^2\right]^{0.5}$

$$= 2R\omega^2 r[(\cos^2\alpha\cos\theta)^2 + (\sin^2\alpha\sin\theta)^2]^{0.5}$$

For $\alpha = 45°$, we have

$$F_p = R\omega^2 r \qquad (12.30)$$

2. Secondary Force

Secondary force of cylinder 1 acting along the line of stroke

$$F_{s1} = R\omega^2 r \frac{\cos 2(\alpha - \theta)}{n}$$

Component of F_{s1} along the vertical line $OY = F_{s1} \cos \alpha$

$$= R\omega^2 r \cdot \frac{\cos 2(\alpha - \theta)}{n} \cdot \cos \alpha$$

Component of F_{s1} along the horizontal line $OX = F_{s1} \sin \alpha$

$$= R\omega^2 r \cdot \frac{\cos 2(\alpha - \theta)}{n} \cdot \sin \alpha$$

Similarly, for the cylinder 2, we have

$$F_{s2} = R\omega^2 r \frac{\cos 2(\alpha + \theta)}{n}$$

Component of F_{s2} along the vertical line $OY = F_{s2} \cos \alpha$

$$= R\omega^2 r \cdot \frac{\cos 2(\alpha + \theta)}{n} \cdot \cos \alpha$$

Component of F_{s2} along the horizontal line $OX' = F_{s2} \sin \alpha$

$$= R\omega^2 r \cdot \frac{\cos 2(\alpha + \theta)}{n} \cdot \cos \alpha$$

Total component of secondary force along the vertical line OY

$$F_{sv} = R\omega^2 r [\cos 2(a - \theta) + \cos 2(a + \theta)] \frac{\cos\ a}{n}$$

$$= 2R\omega^2 \left(\frac{r}{n}\right) \cos a \cos 2a \cos 2\theta$$

Total component of secondary force along the horizontal line OX

$$F_{sh} = R\omega^2 r[\cos 2(\alpha - \theta) - \cos 2(\alpha + \theta)] \frac{\sin\ \alpha}{n}$$

$$= 2R\ \omega^2 \left(\frac{r}{n}\right) \sin \alpha \sin 2\alpha \sin 2\theta$$

Resultant secondary force,

$$F_S = [F_{sv}^2 + F_{sh}^2]^{0.5}$$

$$= 2R\omega^2 \left(\frac{r}{n}\right) [(\cos \alpha \cos 2\alpha \cos 2\theta)^2 + (\sin \alpha \sin 2\alpha \sin 2\theta)^2]^{0.5} \qquad (12.31)$$

For $\alpha = 45°$, we have

$$F_s = \left(\frac{\sqrt{3}}{2}\right) R\,\omega^2 \left(\frac{r}{n}\right) \qquad (12.32)$$

Example 12.14

A *V*-twin engine has the cylinder axes at right angles and the connecting rods operate a common crank. The reciprocating mass per cylinder is 10 kg and the crank radius is 80 mm. The length of connecting rod is 0.4 m. Show that the engine may be balanced for primary forces by means of a revolving balance mass.

If the engine speed is 600 rpm, what is the value of maximum resultant secondary force?

Solution

Here $\alpha = 45°, n = \dfrac{0.4}{0.08} = 5$

And $\omega = 2\pi \times \dfrac{600}{60} = 62.86 \text{ rad/s}$

Resultant primary force, $F_p = 2R\omega^2 r[(\cos^2\alpha \cos\theta)^2 + (\sin^2\alpha \sin\theta)^2]^{0.5}$

Since the resultant primary force $R\omega^2 r$ is the centrifugal force of a mass R at the crank pin radius rotating at speed ω, the engine may be balanced by a rotating balance mass.

Maximum resultant secondary force,

$$F_s = 2R\omega^2 \left(\frac{r}{n}\right)[(\cos\alpha \cos 2\alpha \cos 2\theta)^2 + (\sin\alpha \sin 2\alpha \sin 2\theta)^2]^{0.5}$$

For $\alpha = 45°, \quad F_s = \sqrt{2}R\omega^2 \left(\dfrac{r}{n}\right) \sin 2\theta$

For a maximum value $\sin 2\theta = \pm 1$, or $\theta = 45°$ and $135°$.

Maximum resultant secondary force,

$$(F_s)_{\max} = \sqrt{2}\left(R\frac{r}{n}\right)\omega^2 = \sqrt{2}\left(10 \times \frac{0.08}{5}\right)(62.83)^2 = 893.24 \text{ N}$$

Example 12.15

A rotating shaft carries four radial masses A = 8 kg, B, C = 6 kg, D = 5 kg. The mass centres are 30, 40, 40 and 50 mm, respectively, from the axis of the shaft. The axial distance between the planes of rotation of A and B is 400 mm and between B and C is 500 mm. The masses A and C are at right angles to each other. Find for a complete balance, (a) the angle of the masses B and D from mass A, (b) the axial distance between the planes of rotation of C and D, and (c) the magnitude of mass B.

■ Solution

Reference plane B (Fig.12.36).

Table 12.19

Plane	M (kg)	r (m)	Mr (kg.m)	θ (deg)	$Mr\cos\theta$	$Mr\sin\theta$	l (m)	Mrl (kg.m²)	$Mrl \times \cos\theta$	$Mrl \times \sin\theta$
A	8	0.03	0.24	0	0.24	0	–0.4	–0.096	–0.096	0
B	M_2	0.04	$0.04\,M_2$	θ_2	0.04	$M_2\sin\theta_2$	0	0	0	0
C	6	0.04	0.24	90	0	0.24	0.5	0.12	0	0.12
D	5	0.05	0.25	θ_4	$0.25 \times \cos\theta_4$	$0.25 \times \sin\theta_4$	l_4	$0.25l_4$	$0.25\,l_4 \times \cos\theta_4$	$0.25\,l_4 \times \sin\theta_4$

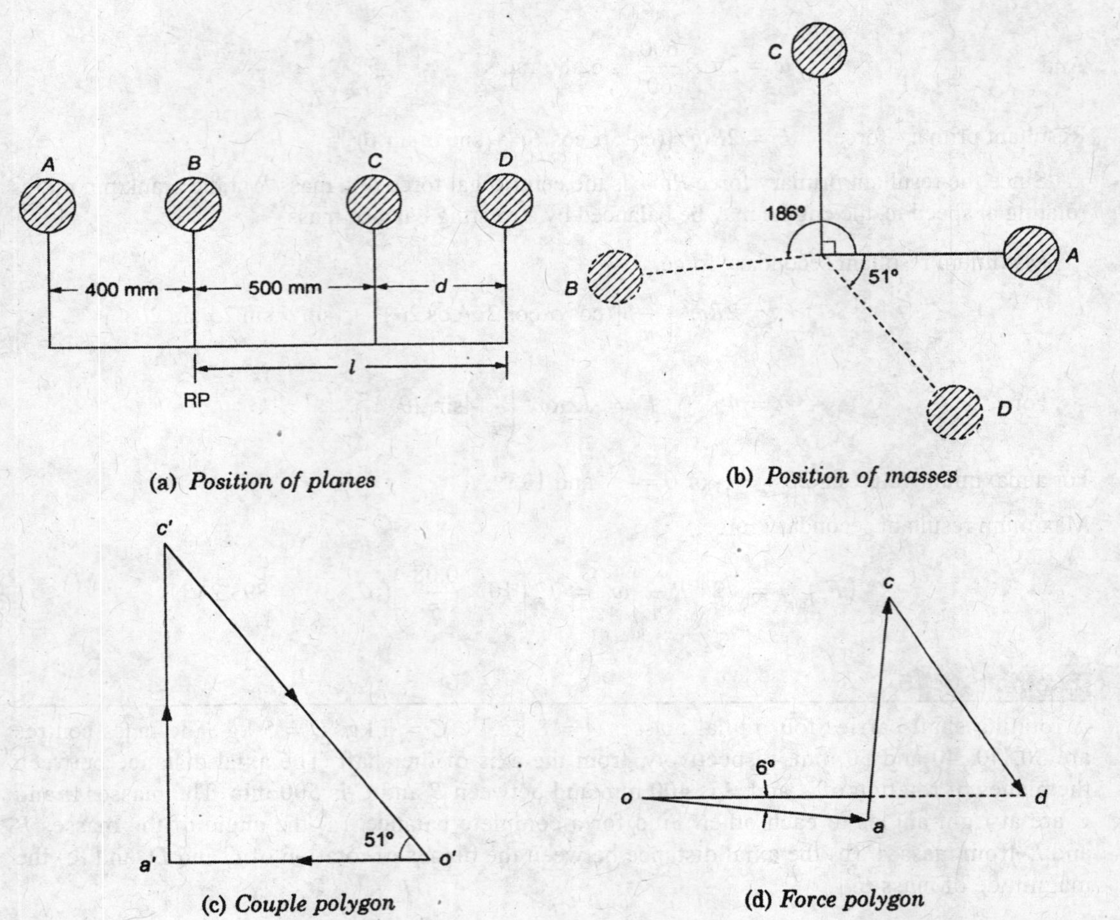

Fig.12.36 Balancing of four masses in different planes.

From Table 12.19, we have

$$\Sigma M_i r_i \cos\theta_i = 0.24 + 0.25\cos\theta_4$$
$$\Sigma M_i r_i \sin\theta_i = 0.24 + 0.25\sin\theta_4$$
$$\Sigma M_i r_i l_i \cos\theta_i = -0.096$$
$$\Sigma M_i r_i l_i \sin\theta_i = 0.12$$

$$\left[(\Sigma M_i r_i l_i \cos\theta_i)^2 + (\Sigma M_i r_i l_i \sin\theta_i)^2\right]^{0.5} = M_2\, r_2\, l_2$$

$$\left[(-0.096)^2 + (0.12)^2\right]^{0.5} = 0.25\, l_4$$

$$l_4 = \frac{0.15367}{0.25} = 0.6147 \text{ m or } 614.7 \text{ mm}$$

$$d = 614.7 - 500 = 114.7 \text{ mm}$$

$$\tan\theta_4 = \frac{-\Sigma M_i r_i l_i \sin\theta_i}{-\Sigma M_i r_i l_i \cos\theta_i}$$
$$= \frac{-(0.12)}{-(-0.096)} = -1.25$$
$$\theta_4 = -51.34° \text{ i.e. } 308.66°$$

Since the numerator is negative and denominator is positive, therefore θ_4 lies in the fourth quadrant.

$$M_2 r_2 = [(\Sigma M_i r_i \cos\theta_i + M_M r_M \cos\theta_M)^2 + (\Sigma M_i r_i \sin\theta_i + M_M r_M \sin\theta_M)^2]^{0.5}$$

$$0.04\, M_2 = \left[(0.25\cos 308.66° + 0.24)^2 + (0.25\sin 308.66° + 0.24)^2\right]^{0.5}$$
$$= \left[(0.3962)^2 + (0.04478)^2\right]^{0.5}$$
$$= 0.39869$$
$$M_2 = 9.967 \text{ kg}$$

$$\tan\theta_2 = \frac{-(\Sigma M_i r_i \sin\theta_i + M_M r_M \sin\theta_M)}{-(\Sigma M_i r_i \cos\theta_i + M_M r_M \cos\theta_M)}$$
$$= \frac{-0.04478}{-0.3962} = 0.11302$$
$$\theta_2 = 6.45° \text{ i.e. } 186.45°$$

Since the numerator and denominator are both negative, therefore, θ_2 lies in the third quadrant.

Graphical Method

From force polygon in Fig.12.36(d), $od = 4.2$ mm, $M_2 = 10.5$ kg, $\theta_2 = 186°$. From couple polygon in Fig.12.36(c), $0.25\, l_4 = o'c' = 7.7$ cm, $l_4 = 616$ mm, $d = 116$ mm, $\theta_d = 309°$.

Example 12.16

A rotating shaft carries four unbalanced masses 20, 15, 18 and 12 kg at radii 50, 60, 70 and 60 mm, respectively. The second, third and fourth masses revolve in planes 100, 150 and 300 mm, respectively, measured from the plane of first mass, and at angular locations of 60°, 120° and 280°, respectively, measured clockwise from the first mass. The shaft is dynamically balanced by two masses, both located at 50 mm radii and revolving in planes midway between those of first and second masses and midway between those of third and fourth masses. Determine graphically the magnitudes of the masses and their angular positions.

■ Solution

Reference plane L (Fig.12.37)

Table 12.20

Plane	M (kg)	r (m)	Mr (kg.m)	θ (deg)	$Mr\cos\theta$	$Mr\sin\theta$	l (m)	Mrl (kg.m^2)	$Mrl \times \cos\theta$	$Mrl\sin\theta$
A	20	0.05	1	0	1	0	–0.05	–0.05	–0.05	0
B	15	0.06	0.9	60	0.45	0.7794	0.05	0.045	0.0225	0.039
C	18	0.07	1.26	120	–0.63	1.0912	0.10	0.126	–0.063	0.1090
D	12	0.06	0.72	280	0.125	–0.7091	0.25	0.18	–03126	–0.1773
L	M_L	0.05	$0.05 M_L$	θ_L	$0.05 M_L \times \cos\theta_L$	$0.05 M_L \times \sin\theta_L$	0	0	0	0
M	M_M	0.05	$0.05 M_M$	θ_M	$0.05 M_M \times \cos\theta_M$	$0.05 M_M \times \sin\theta_M$	0.175	$0.0087 \times M_M \times \cos\theta_M$	$0.0087 \times M_M \times \sin\theta_M$	0.008751

From Table 12.20, we have

$$\Sigma M_i r_i \cos\theta_i = 0.945$$

$$\Sigma M_i r_i \sin\theta_i = 1.1615$$

$$\Sigma M_i r_i l_i \cos\theta_i = 0.0448$$

$$\Sigma M_i r_i l_i \sin\theta_i = -0.2902$$

$$\left[(\Sigma M_i r_i l_i \cos\theta_i)^2 + (\Sigma M_i r_i l_i \sin\theta_i)^2\right]^{0.5} = M_M r_M l_M$$

$$\left[(0.0448)^2 + (-0.0292)^2\right]^{0.5} = 0.00875\, M_M$$

$$M_M = \frac{0.05347}{0.00875} = 6.11 \text{ kg}$$

$$\tan\theta_M = \frac{-\Sigma M_i r_i l_i \sin\theta_i}{-\Sigma M_i r_i l_i \cos\theta_i} = \frac{-(-0.0292)}{-(0.0448)} = -0.65178$$

$$\theta_M = -33° \text{ i.e. } 147°$$

Since the numerator is positive and denominator is negative, therefore θM lies in the second quadrant.

$$M_L r_L = [(\Sigma M_i r_i \cos\theta_i + M_M r_M \cos\theta_M)^2 + (\Sigma M_i r_i \sin\theta_i + M_M r_M \sin\theta_M)^2]^{0.5}$$

$$0.05 M_L = \left[(0.945 + 0.05\times 6.11\times\cos 147°)^2 + (1.1615 + 0.05\times 6.11\times\sin 147°)^2\right]^{0.5}$$

$$= \left[(0.689)^2 + (1.3279)^2\right]^{0.5} = 1.496$$

$$M_L = 29.92 \text{ kg}$$

$$\tan\theta_L = \frac{-(\Sigma M_i r_i \sin\theta_i + M_M r_M \sin\theta_M)}{-(\Sigma M_i r_i \cos\theta_i + M_M r_M \cos\theta_M)}$$

$$= -\frac{-1.3279}{-0.689} = 1.92728$$

$$\theta_L = 62.57° \text{ i.e. } 242.57°$$

Since the numerator and denominator are both negative, therefore θ_L lies in the third quadrant.

Graphical Method

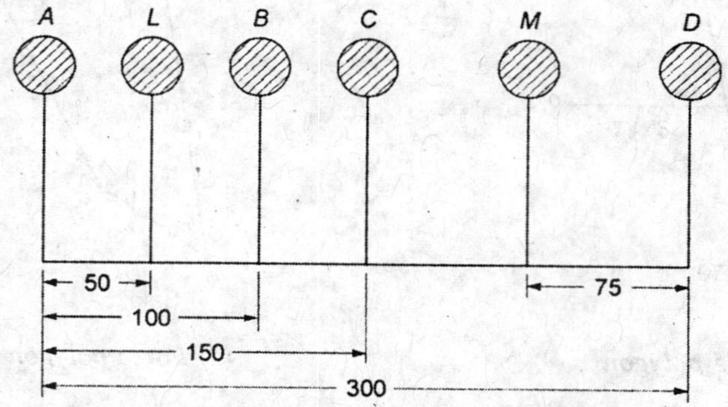

(a) *Position of planes*

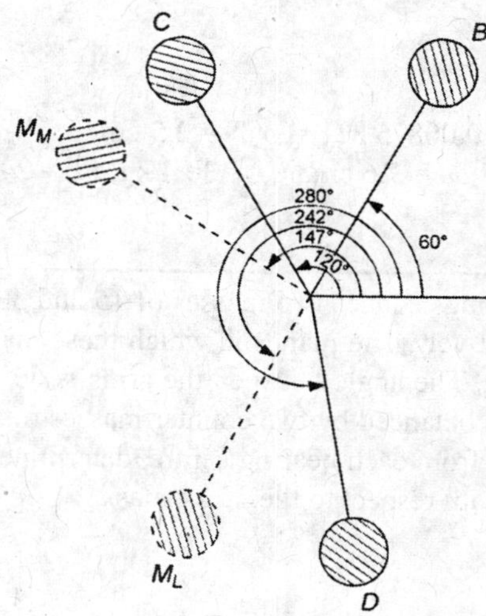

(b) *Position of masses*

Fig.12.37 Graphical method for balancing many masses in different planes

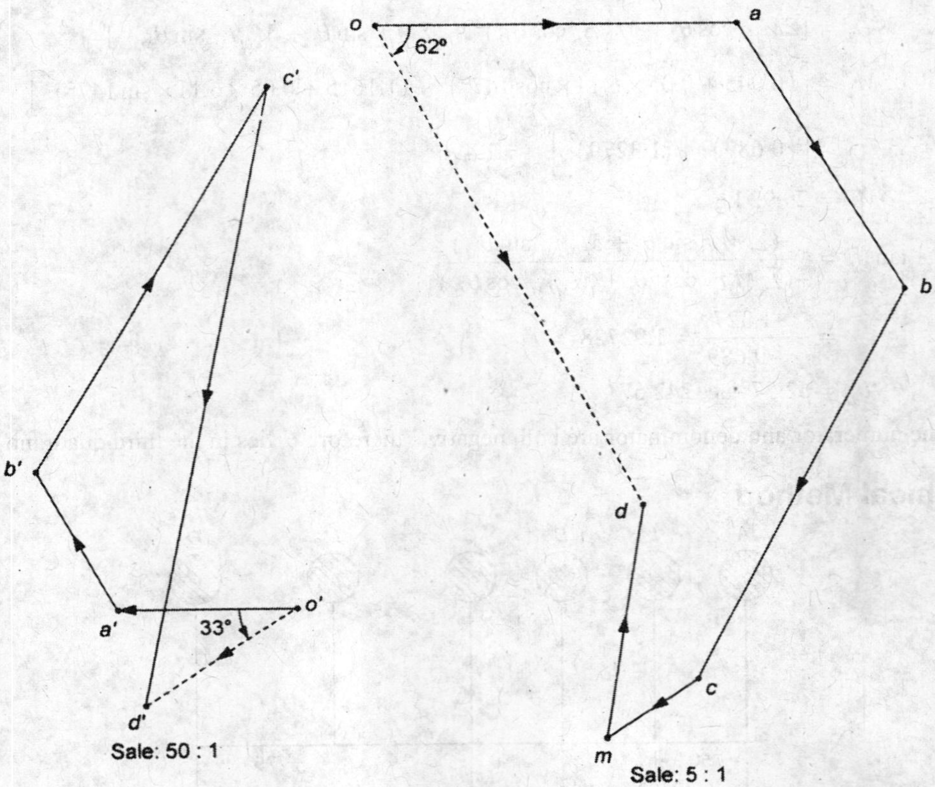

(c) *Couple polygon* (d) *Force polygon*

Fig.12.37 Graphical method for balancing many masses in different planes (*Contd.*)

From couple polygon in Fig.12.37(c), $0.00875\ M_M = o'd' = 2.5$ cm, $M_M = 5.7$ kg, $\theta M = 147°$. From force polygon, Fig.12.37(d), $0.05\ M_L = od = 7.6$ cm, $M_L = 30.4$ kg, $\theta L = 242°$.

Example 12.17

A shaft of span 3 m between two bearings carries two masses of 15 and 30 kg acting at the extremities of the arms 0.5 and 0.6 m, respectively. The planes in which these masses rotate are 1 and 2 m, respectively, from the left end bearing. The angle between the arms is 60°. The speed of rotation of the shaft is 240 rpm. If the masses are balanced by two counter masses rotating with the shaft acting at radii of 0.25 m and placed at 0.3 m from each bearing centre, determine the magnitude of the two balance masses and their orientation with respect to the 15 kg mass.

■ Solution

Reference plane L (Fig.12.38):

Table 12.21

Plane	M (kg)	r (m)	Mr (kg.m)	θ (deg)	$Mr\cos\theta$	$Mr\sin\theta$	l (m)	Mrl (kg.m²)	$Mrl\times\cos\theta$	$Mrl\times\sin\theta$
L	M_L	0.25	$0.25\,M_L$	θ_L	$0.25\,M_L$	$0.25\,M_L\sin\theta_L$	0	0	0	0
A	15	0.5	7.5	0	7.5	0	0.7	5.25	5.25	0
B	30	0.6	18	60	9	15.5884	1.7	30.6	15.3	26.5
M	M_M	0.25	$0.25\,M_M$	θ_M	$0.25\,M_M\times\cos\theta_M$	$0.25\,M_M\times\sin\theta_M$	2.7	$0.675\times M_M$	$0.675\times M_M\cos\theta_M$	$0.675\times M_M\sin\theta_M$

From Table 12.21, we have

$$\Sigma M_i r_i \cos\theta_i = 16.5$$

$$\Sigma M_i r_i \sin\theta_i = 15.5884$$

$$\Sigma M_i r_i l_i \cos\theta_i = 20.55$$

$$\Sigma M_i r_i l_i \sin\theta_i = 26.5$$

$$\left[(\Sigma M_i r_i l_i \cos\theta_i)^2 + (\Sigma M_i r_i l_i \sin\theta_i)^2\right]^{0.5} = M_M r_M l_M$$

$$[(26.5)^2 + (20.55)^2]^{0.5} = 0.675 M_M$$

$$M_M = \frac{34.534}{0.675} = 49.68 \text{ kg}$$

$$\tan\theta_M = \frac{-\Sigma M_i r_i l_i \sin\theta_i}{-\Sigma M_i r_i l_i \cos\theta_i} = \frac{-(26.5)}{-(20.55)} = 1.28954$$

$$\theta_M = 52.2° \text{ i.e. } 232.2°$$

Since the numerator and denominator are both negative therefore θ_M lies the in the third quadrant

$$M_L r_L = [(\Sigma M_i r_i \cos\theta_i + M_M r_M \cos\theta_M)^2 + (\Sigma M_i r_i \sin\theta_i + M_M r_M \sin\theta_M)^2]^{0.5}$$

$$0.25\,M_L = \left[(0.25\times 49.68\cos 232.2° + 16.5)^2 + (0.25\times 49.68\sin 232.2° + 15.5884)^2\right]^{0.5}$$

$$= \left[(8.888)^2 + (5.7746)^2\right]^{0.5} = 10.599$$

$$M_L = 42.4 \text{ kg}$$

$$\tan\theta_L = \frac{-(\Sigma M_i r_i \sin\theta_i + M_M r_M \sin\theta_M)}{-(\Sigma M_i r_i \cos\theta_i + M_M r_M \cos\theta_M)} = \frac{-5.7746}{-8.888} = 0.6497$$

$$\theta_L = 33° \text{ i.e. } 213°$$

Since the numerator and denominator are both negative, therefore θL lies in the third quadrant.

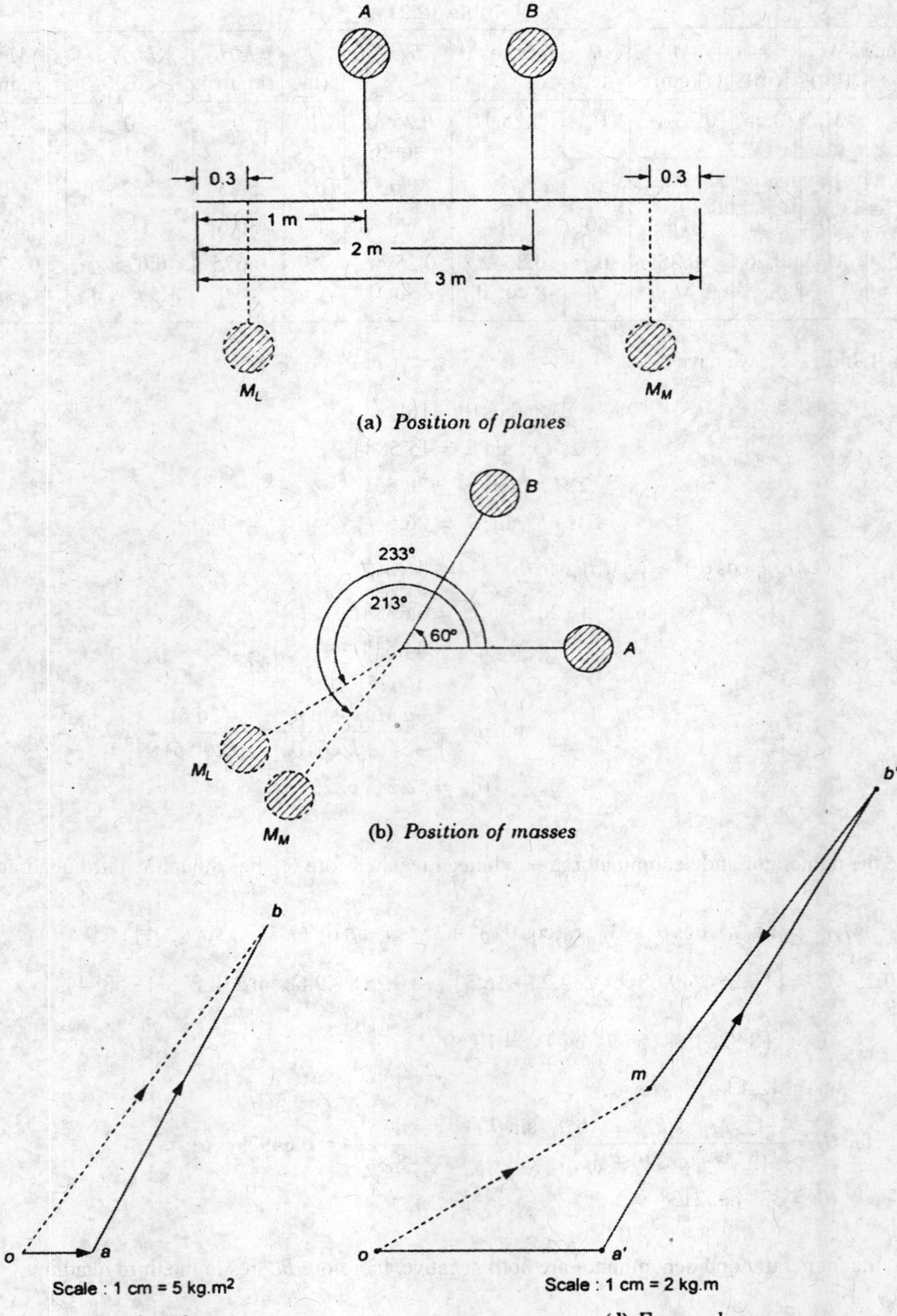

Fig.12.38 Graphical method for balancing of two masses in different planes

Graphical Method:

From couple polygon in Fig.12.38(c), 0.675 $M_M = ob = 6.8$ cm, $M_M = 50.37$ kg, $\theta M = 233°$. From force polygon in Fig.12.38(d), 0.25 $M_L = o'd' = 5.3$ cm, $M_L = 42.4$ kg, $\theta L = 213°$

Example 12.18

A 4 m long shaft carries three pulleys, two at its ends and third at the midpoint. The two end pulleys have mass of 80 and 40 kg and their centre of gravity are 3 and 5 mm, respectively, from the axis of the shaft. The middle pulley mass is 50 kg and its center of gravity is 8 mm from the shaft axis. The pulleys are keyed to the shaft and the assembly is in static balance. The shaft rotates at 300 rpm in two bearings 2.5 m apart with equal overhang on either side. Determine (a) the relative angular positions of the pulleys and (b) dynamic reactions at the two bearings.

■ **Solution**

For static balance, $M_E = 0$, gives (Fig.12.39).

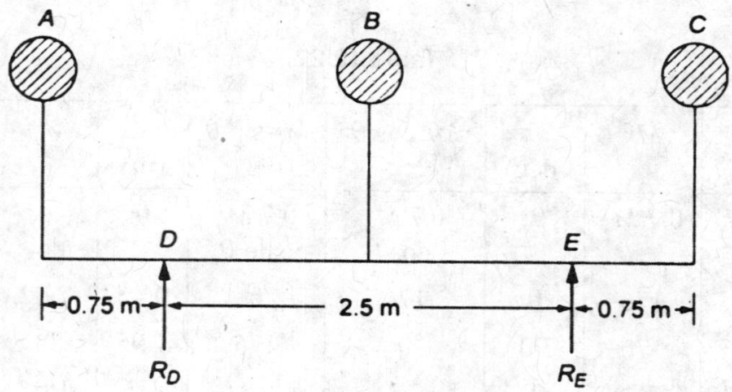

Fig.12.39 Shaft carrying three pulleys

$$R_D \times 2.5 = (80 \times 4.25 + 50 \times 1.25 - 40 \times 0.75) \times 9.81$$

$$R_D = 1147.8 \text{ N}, R_E = 519.9 \text{ N}$$

$$\omega = \frac{2\pi \times 300}{60} = 31.416 \text{ rad/s}$$

$$F_A = 80 \times 0.003 \times (31.416)^2 = 236.87 \text{ N}$$

$$F_B = 50 \times 0.008 \times (31.416)^2 = 394.78 \text{ N}$$

$$F_C = 40 \times 0.005 \times (31.416)^2 = 197.39 \text{ N}$$

$$R_D \times 2.5 = 236.87 \times 4.25 + 394.78 \times 1.25 - 197.39 \times 0.75$$

$$R_D = 446.1N, R_E = 382.94 \text{ N}$$

Taking components of forces in horizontal and vertical directions, we have

$80 + 50 \cos \theta_1 + 40 \cos \theta_2 = 0$

$50 \sin \theta_1 + 40 \sin \theta_2 = 0$

Solving we get,

$$\theta_1 = -24.14°, \theta_2 = 149.25°$$

Example 12.19

A two-cylinder uncoupled locomotive with cranks at 90° has a crank radius of 320 mm. The distance between the centers of driving wheels is 1.5 m. The pitch of cylinders is 0.6 m. The diameter of treads of driving wheels is 1.8 m. The radius of centers of gravity of balance masses is 0.7 m. The pressure due to dead load on each of the wheels is 40 kN. The masses of reciprocating and rotating parts per cylinder are 300 and 350 kg, respectively. The speed of the locomotive is 60 km/h. Find (a) the balancing masses in magnitude and position in the planes of driving wheels to balance whole of the revolving and 2/3rd of the reciprocating parts, (b) the swaying couple, (c) the variation in tractive effort, (d) the maximum and minimum pressure on the rails and (e) the maximum speed at which it is possible to run the locomotive, in order that the wheels are not lifted from the rails.

Solution

Give: $M = 350$ kg, $R = 300$ kg, $r = 0.32$ m, $l = 0.6$ m, $d_w = 1.8$ m, $b = 0.7$ m,

$P = 40$ kN, $v = 60$ km/hr, Equivalent mass $= M + \dfrac{2R}{3} = 350 + \dfrac{2 \times 300}{3} = 550$ kg

Table 12.22

Plane	M (kg)	r (m)	Mr	θ (deg)	$Mr\cos\theta$	$Mr\sin\theta$	l (m)	Mrl	$Mrl \times \cos\theta$	$Mrl \times \sin\theta$
A	M_A	0.7	$0.7\,M_A$	θ_A	$0.7\,M_A \times \cos\theta_A$	$0.7\,M_A \times \sin\theta_A$	0	0	0	0
B	550	0.32	176	0	176	0	0.45	79.2	79.2	0
C	550	0.32	176	90	0	176	1.05	184.8	0	184.8
D	M_D	0.7	$0.7\,M_D$	θ_D	$0.7\,M_D \times \cos\theta_D$	$0.7\,M_D \times \sin\theta_D$	1.5	$1.05 \times M_D$	$1.05\,M_D \times \cos\theta_D$	$1.05\,M_D \times \sin\theta_D$

From Table 12.22, we have

$$\Sigma M_i r_i \cos\theta_i = 176$$
$$\Sigma M_i r_i \sin\theta_i = 176$$
$$\Sigma M_i r_i l_i \cos\theta_i = 79.2$$
$$\Sigma M_i r_i l_i \sin\theta_i = 184.8$$

$$\left[(\Sigma M_i r_i l_i \cos\theta_i)^2 + (\Sigma M_i r_i l_i \sin\theta_i)^2\right]^{0.5} = M_D r_D l_D$$

$$\left[(79.2)^2 + (184.8)^2\right]^{0.5} = 1.05 M_D$$

$$M_D = \frac{201.05}{1.05} = 191.5 \text{ kg}$$

$$\tan\theta_D = \frac{-\Sigma M_i r_i l_i \sin\theta_i}{-\Sigma M_i r_i l_i \cos\theta_i} = \frac{-(184.8)}{-(79.2)} = 2.3333$$

$$\theta_D = 66.8° \text{ i.e. } 246.8°$$

Since the numerator and denominator are both negative, therefore θ_D lies in the third quadrant,

$$M_A r_A = \left[\left(\Sigma M_i r_i \cos\theta_i + M_D r_D \cos\theta_D\right)^2 + \left(\Sigma M_i r_i \sin\theta_i + M_D r_D \sin\theta_D\right)^2\right]^{0.5}$$

$$0.7M_A = \left[\left(0.7\times 191.5\cos 246.8° + 176\right)^2 + \left(0.7\times 191.5\sin 246.8° + 176\right)^2\right]^{0.5}$$

$$= \left[\left(124.192\right)^2 + \left(52.79\right)^2\right]^{0.5} = 134.02$$

$$M_A = 191.5 \text{ kg}$$

$$\tan\theta_A = \frac{-\left(\Sigma M_i r_i \sin\theta_i + M_D r_D \sin\theta_D\right)}{-\left(\Sigma M_i r_i \cos\theta_i + M_D r_D \cos\theta_D\right)}$$

$$= \frac{-52.79}{-124.192} = 0.42852$$

$$\theta_A = 24.2° \text{ i.e. } 204.2°$$

Since the numerator and denominator are both negative, therefore θ_A lies in the third quadrant,

(b)

$$v = \frac{60\times 10^3}{3600} = 16.67 \text{ m/s}$$

$$\omega = \frac{2v}{d_w} = \frac{2\times 16.67}{1.8} = 18.52 \text{ rad/s}$$

Maximum swaying couple

$$= \frac{(1-c)R\omega^2 rl}{\sqrt{2}}$$

$$= \frac{\left(1-\frac{2}{3}\right)\times 300\times(18.52)^2\times 0.32\times 0.6}{\sqrt{2}}$$

$$= 4656.6 \text{ N m}$$

(c) Maximum tractive effort variation $= \pm\sqrt{2}(1-c)R\omega^2 r$

$$= \pm\sqrt{2}\left(1-\frac{2}{3}\right)\times 300\times(18.52)^2\times 0.32$$

$$= \pm 15522 \text{ N}$$

(d) $M_A = M_L = 191.5$ kg

Balance mass for reciprocating parts, $B = \dfrac{191.5\times 2\times 300}{3\times 350} = 69.636$ kg

$$\text{Net pressure on rails} = P \pm B\omega^2 b$$

$$= 40 \pm 69.636\times(18.52)^2\times 0.7$$

$$= 40 \pm 16.719$$

$$= 56.719 \text{ kN, } 24.281 \text{ kN}$$

$$40 \times 10^3 = 69.636 \times \omega^2 \times 0.7$$

$$\omega^2 = 820.59, \ \omega = 28.646 \text{ rad/s}$$

$$v = \frac{28.646 \times 0.9 \times 3600}{1000} = 92.81 \text{ km/h}$$

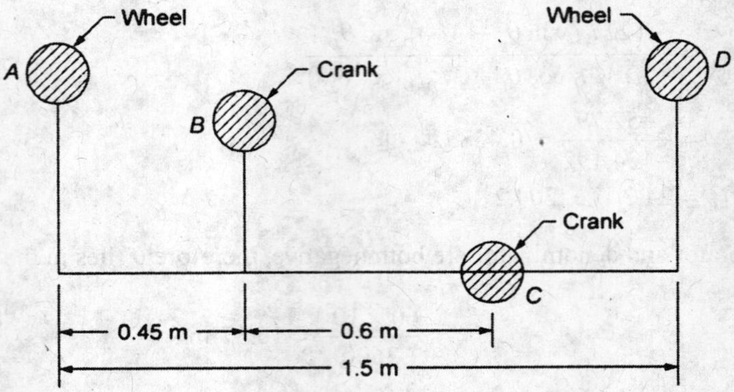

(a) *Position of planes*

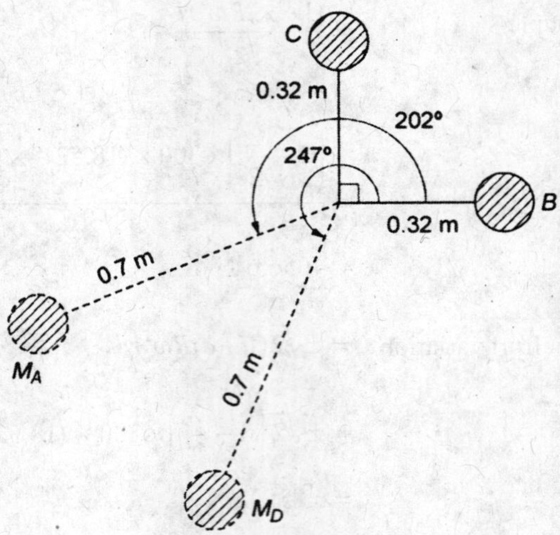

(b) *Position of masses*

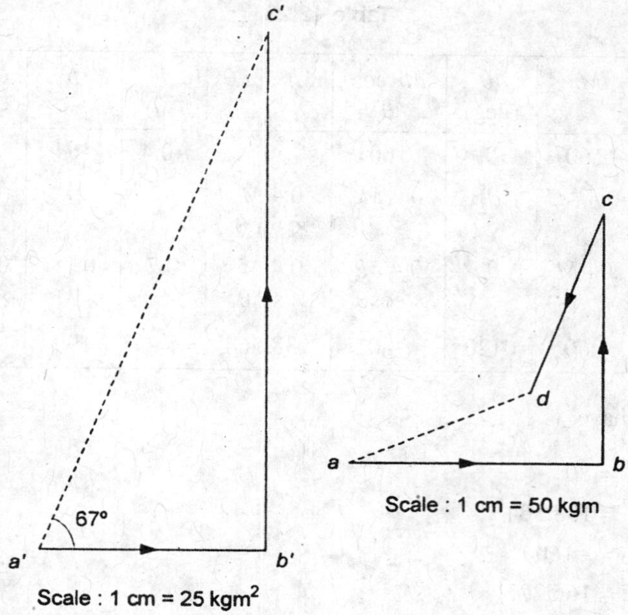

(c) Couple polygon (d) Force polygon

Fig.12.40 Graphical method for balancing of four masses in different planes

Graphical Method

From couple polygon in Fig.12.40(c), 1.05 $M_A = a'c' = 8.05$ cm, $M_A = 191.67$ kg, $\theta_A = 202°$.
From force polygon in Fig.12.40(d), 0.7 $MD = ad = 2.8$ cm, $M_D = 200$ kg, $\theta_D = 247°$.

Example 12.20

A four-cylinder engine has two outer cranks at 120° to each other and their reciprocating masses are each 400 kg. The distance between the planes of rotation of adjacent cranks are 0.4, 0.7, 0.7 and 0.5 m. Find the reciprocating mass and the relative angular position for each of the inner cranks, if the engine is to be in complete primary balance. Also find the maximum secondary force, if the length of each crank is 0.4 m, the length of each connecting rod 1.8 m and the engine speed 480 rpm.

■ **Solution**

(a) Primary cranks

Reference plane 2 (Fig.12.41):

Table 12.23

Plane	M (kg)	r (m)	Mr	θ (deg)	$Mr\cos\theta$	$Mr\sin\theta$	l (m)	Mrl	$Mrl \times \cos\theta$	$Mrl\sin\theta$
1	400	0.4	160	0	160	0	–0.4	–64	–64	0
2	M_2	0.4	0.4 M_2	θ_2	0.4 $M_2 \times \cos\theta_2$	0.4 $M_2 \times \sin\theta_2$	0	0	0	0
3	M_3	0.4	.04 M_3	θ_3	0.4 $M_3 \times \cos\theta_3$	0.4 $M_3 \times \sin\theta_3$	0.7	0.28 $\times M_3$	0.28 $M_3 \times \cos\theta_3$	0.28 $M_3 \times \sin\theta_3$
4	400	0.4	060	0120	–80	138.564	1.2	190	–96	066.277

From Table 12.23, we have

$$\Sigma M_i r_i \cos\theta_i = 80$$
$$\Sigma M_i r_i \sin\theta_i = 138.564$$
$$\Sigma M_i r_i l_i \cos\theta_i = -160$$
$$\Sigma M_i r_i l_i \sin\theta_i = 166.277$$

$$\left[\left(\Sigma M_i r_i l_i \cos\theta_i\right)^2 + \left(\Sigma M_i r_i l_i \sin\theta_i\right)^2\right]^{0.5} = M_3 r_3 l_3$$

$$\left[(-160)^2 + (166.277)^2\right]^{0.5} = 0.28 M_3$$

$$M_3 = \frac{230.755}{0.28} = 824.12 \text{ kg}$$

$$\tan\theta_3 = \frac{-\Sigma M_i r_i l_i \sin\theta_i}{-\Sigma M_i r_i l_i \cos\theta_i} = \frac{-(166.277)}{-(-160)} = -1.03923$$

$$\theta_3 = 46.1° \text{ i.e. } 314.9°$$

Since the numerator is negative and denominator is positive, therefore θ_3 lies in the fourth quadrant.

$$M_2 r_2 = \left[\left(\Sigma M_i r_i \cos\theta_i + M_3 r_3 \cos\theta_3\right)^2 + \left(\Sigma M_i r_i \sin\theta_i + M_3 r_3 \sin\theta_3\right)^2\right]^{0.5}$$

$$0.4\,M_2 = \left[\left(0.4 \times 824.12 \cos 314.9° + 80\right)^2 + \left(0.4 \times 824.12 \sin 314.9° + 138.564\right)^2\right]^{0.5}$$

$$= \left[(308.58)^2 + (-98.964)^2\right]^{0.5}$$

$$= 324.06$$

$$M_2 = 810.15 \text{ kg}$$

$$\tan\theta_2 = \frac{-\left(\Sigma M_i r_i \sin\theta_i + M_3 r_3 \sin\theta_3\right)}{-\left(\Sigma M_i r_i \cos\theta_i + M_3 r_3 \cos\theta_3\right)}$$

$$= \frac{-(-98.964)}{-308.58} = -0.32070$$

$$\theta_2 = -17.78° \text{ i.e. } 162.22°$$

Since the numerator is positive and the denominator is negative, therefore, θ_2 lies in the second quadrant.

(b) Secondary cranks

$$n = \frac{l}{r} = \frac{1.8}{0.4} = 4.5$$

Secondary crank length $= \dfrac{r}{4n} = \dfrac{0.4}{4 \times 4.5} = 0.022$ m

Reference plane 2:

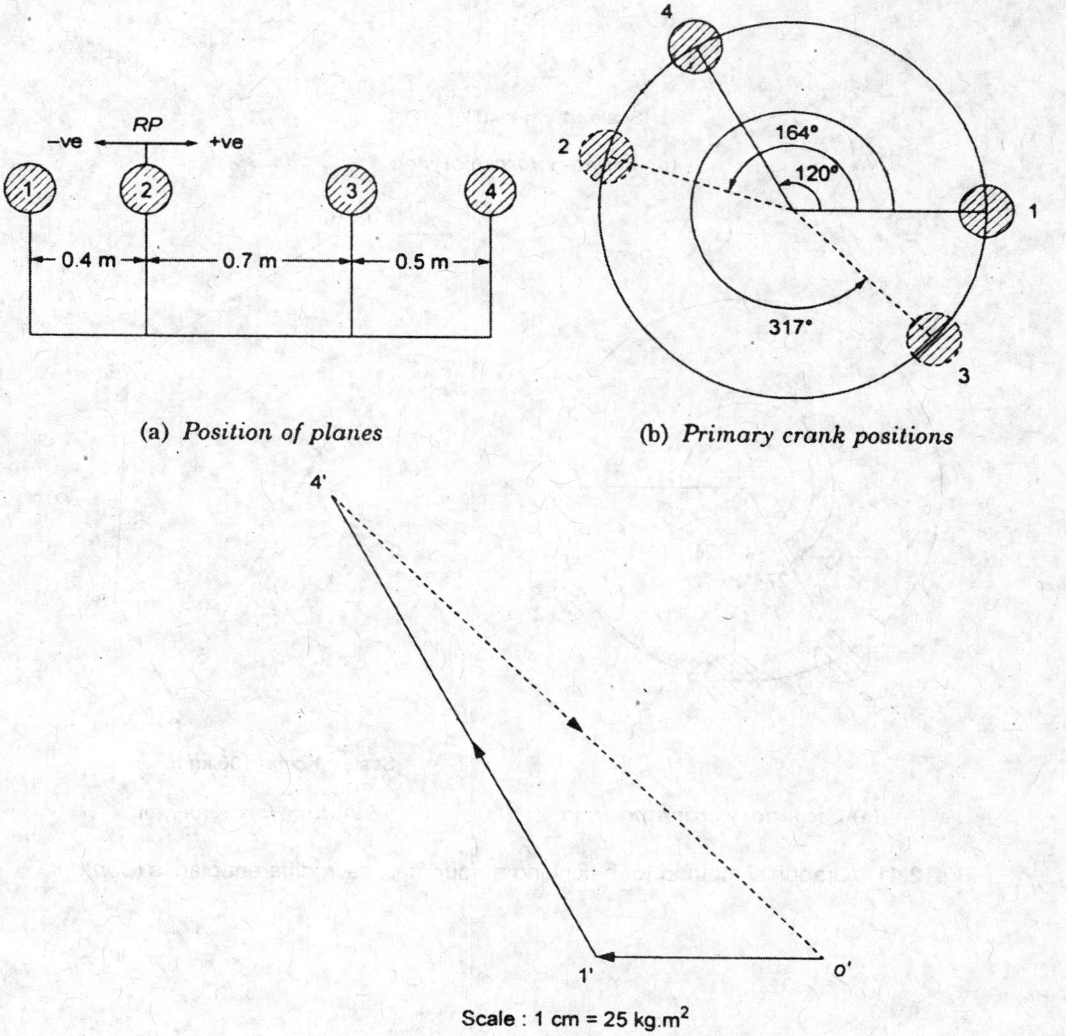

(a) *Position of planes*

(b) *Primary crank positions*

(c) *Primary couple polygon*

Fig.12.41 Graphical method for balancing of four masses in different planes

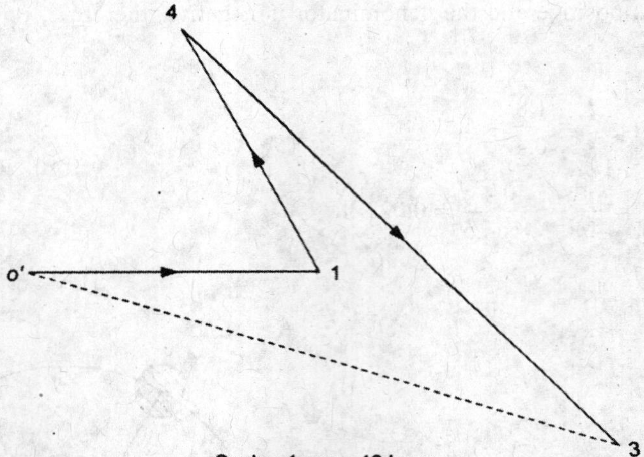

Scale : 1 cm = 40 kg.m

(d) *Primary force polygon*

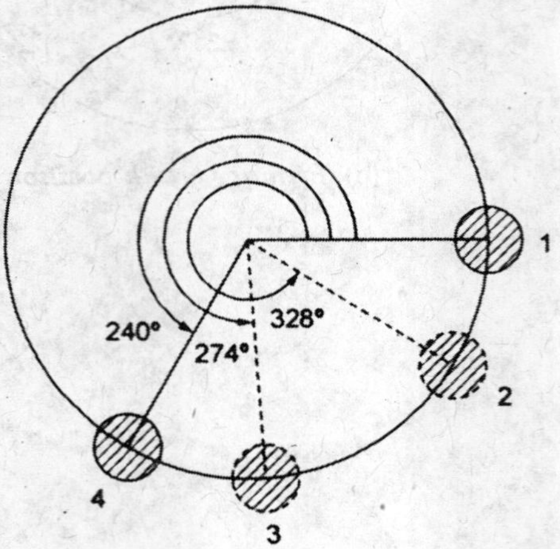

(e) *Secondary crank position*

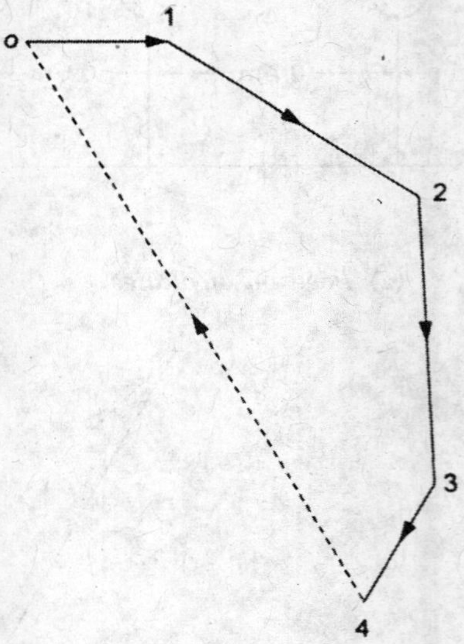

Scale : 1 cm = 100 kg.m

(f) *Secondary force polygon*

Fig.12.41 Graphical method for balancing of four masses in different planes (*Contd.*)

Table 12.24

Plane	M (kg)	r (m)	Mr	θ (deg)	$Mr\cos\theta$	$Mr\sin\theta$	l(m)	Mrl	$Mrl\cos\theta$	$Mrl\sin\theta$
1	400	0.022	8.8	0	8.8	0	–0.4	–4.52	–4.52	0
2	M_2	0.022	0.022 M_2	θ_2	0.022 M_2	0	0	0	0	
3	M_3	0.022	0.022 M_3	θ_3	0.022 M_3	0.022 M_3	0.7	0.0154 M_3	0.015 M_3 × cos θ_3	0.015 M_3 × sinθ_3
4	400	0.022	8.8	240	–4.4	–7.621	1.2	10.56	–5.28	t–9.145

From Table 12.24, we have

$$\Sigma M_i r_i \cos\theta_i = 4.4$$
$$\Sigma M_i r_i \sin\theta_i = -7.621$$
$$\Sigma M_i r_i l_i \cos\theta_i = -8.8$$
$$\Sigma M_i r_i l_i \sin\theta_i = -9.145$$

$$\left[\left(\Sigma M_i r_i l_i \cos\theta_i\right)^2 + \left(\Sigma M_i r_i l_i \sin\theta_i\right)^2\right]^{0.5} = M_3 r_3 l_3$$

$$\left[(-8.8)^2 + (9.145)^2\right]^{0.5} = 0.0154\ M_3$$

$$M_3 = \frac{12.691}{0.0154} = 824.12\ \text{kg}$$

$$\tan\theta_3 = \frac{-\Sigma M_i r_i l_i \sin\theta_i}{-\Sigma M_i r_i l_i \cos\theta_i} = \frac{-(-9.145)}{-(-8.8)} = 1.0392$$

$$\theta_3 = 46.1°$$

Since the numerator and denominator are both positive, therefore θ_3 lies in the first quadrant

$$M_2 r_2 = \left[\left(\Sigma M_i r_i \cos\theta_i + M_3 r_3 \cos\theta_3\right)^2 + \left(\Sigma M_i r_i \sin\theta_i + M_3 r_3 \sin\theta_3\right)^2\right]^{0.5}$$

$$0.022 M_2 = \left[\left(0.022\times 824.12 \cos 46.1° + 4.4\right)^2 + \left(0.022\times 824.12 \sin 46.1° - 7.621\right)^2\right]^{0.5}$$

$$= \left[(16.972)^2 + (5.443)^2\right]^{0.5} = 17.823$$

$$M_2 = 810.15\ \text{kg}$$

$$\tan\ \theta_2 = \frac{-\left(\Sigma M_i r_i \sin\theta_i + M_3 r_3 \sin\theta_3\right)}{-\left(\Sigma M_i r_i \cos\theta_i + M_3 r_3 \cos\theta_3\right)}$$

$$\frac{-5.443}{-16.972} = 0.3207$$

$$\theta_2 = 17.78°\ \text{i.e. } 197.78°$$

Since the numerator and denominator are both negative, therefore, θ_2 lies in the third quadrant.

$$\text{Speed of secondary crank} = 2\pi \times \frac{480}{60} = 50.27 \text{ rad/s}$$

$$\text{Maximum unbalanced secondary force} = \frac{760 \times (50.27)^2}{4.5} = 426795 \text{ N}$$

Graphical Method

From couple polygon for primary crank in Fig.12.41(c), $0.28\ M_3 .\ o'4' = 9.2$ cm, $M_3 = 821.4$ kg, $\theta_3 = 315°$.

From primary force polygon in Fig.12.41(d), $0.4\ M_2 = o3 = 8.2$ cm, $M_2 = 820$ kg, $\theta_2 = 164°$. From secondary crank force polygon in Fig.12.41(f), unbalanced secondary force = $o4$ = 7.6 cm = 760 kg-m.

Example 12.21

In a four-crank symmetrical engine, the reciprocating masses of the two outside cylinders *A* and *D* are each 600 kg and those of the two inside cylinders *B* and *C* are each 900 kg. The distance between the cylinder axes of *A* and *D* is 5 m. Taking the reference line to bisect the angle between the cranks *A* and *D,* and the reference plane to bisect the distance between the cylinder axes of *A* and *D*, find the angles between the cranks and the distance between the cylinder axes of *B* and *C* for complete balance except for secondary couples.

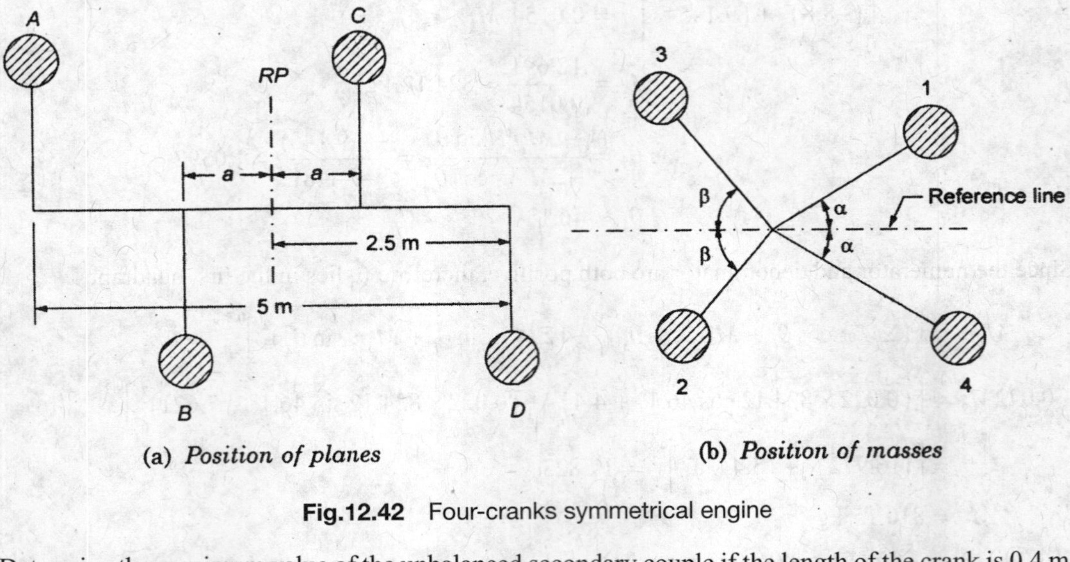

(a) *Position of planes* (b) *Position of masses*

Fig.12.42 Four-cranks symmetrical engine

Determine the maximum value of the unbalanced secondary couple if the length of the crank is 0.4 m, length of connecting rod 1.8 m and speed is 180 rpm.

■ **Solution**

Let the cranks 1 and 4 be inclined at an angle α with the reference line and cranks 3 and 4 at angle β. Also let a = distance of *B* and C from the reference plane, as shown in Fig.12.42(a).

Primary Forces:

Horizontal component,

$$F_{HP} = \omega^2 r\,[M_1 \cos\alpha + M_3 \cos(180° - \beta) + M_2 \cos(180° + \beta) + M_4 \cos(360° - \alpha)]$$
$$= \omega^2 r\,[600 \cos\alpha - 900 \cos\beta - 900 \cos\beta + 600 \cos\alpha]$$
$$= \omega^2 r\,[1200 \cos\alpha - 1800 \cos\beta]$$

Vertical component,

$$F_{VP} = \omega^2 r\,[M_1 \sin\alpha + M_3 \sin(180° - \beta) + M_2 \sin(180° + \beta) + M_4 \sin(360° - \alpha)]$$
$$= \omega^2 r\,[600 \sin\alpha + 900 \sin\beta - 900 \sin\beta - 600 \sin\alpha]$$
$$= 0$$

For primary force balance, $F_{HP} = 0$

$1200 \cos\alpha - 1800 \cos\beta = 0$

$$\frac{\cos\alpha}{\cos\beta} = 1.5 \qquad (12.33)$$

Primary couple about RP:

$$C_{HP} = \omega^2 r\,[M_1 \cos\alpha \times (-2.5) + M_3 \cos(180° - \beta) - \alpha + M_2 \cos(180° + \beta) \times (-\alpha) + M_4 \cos(360° - \alpha) \times 2.5]$$
$$= \omega^2 r\,[-1500 \cos\alpha - 900\,\alpha \cos\beta + 900\,\alpha \cos\beta + 1500 \cos\alpha]$$
$$= 0$$

$$C_{VP} = \omega^2 r\,[M_1 \sin\alpha \times (-2.5) + M_3 \sin(180° - \beta) \times \alpha + M_2 \sin(180° + \beta) \times (-\alpha) + M_4 \sin(360° - \alpha) \times 2.5]$$
$$= \omega^2 r\,[-1500 \sin\alpha + 900\,\alpha \sin\beta + 900\,\alpha \sin\beta - 1500 \sin\alpha]$$
$$\omega^2 r\,[-3000 \sin\alpha + 1800\,\alpha \sin\beta]$$

For complete balance of primary couple, $C_{VP} = 0 - 3000 \sin\alpha + 1800\,\alpha \sin\beta = 0$

$$\frac{\sin\alpha}{\sin\beta} = 0.6\,a \qquad (12.34)$$

From (12.33) and (12.34), we get

$$\frac{\tan\alpha}{\tan\beta} = 0.4\,a \qquad (12.35)$$

Secondary Forces:

Speed $= 2\omega$, Imaginary crank length $= \dfrac{r}{4n} = \dfrac{r^2}{4l}$

$$F_{HS} = \frac{(2\omega)^2 r^2}{4l}[M_1 \cos 2\alpha + M_3 \cos 2(180° - \beta) + M_2 \cos 2(180° + \beta) + M_4 \cos 2(360° - \alpha)]$$

$$= \left(\frac{\omega^2 r^2}{l}\right)[600 \cos 2\alpha + 900 \cos 2\beta + 900 \cos 2\beta + 600 \cos 2\alpha]$$

$$= \left(\frac{\omega^2 r^2}{l}\right)[1200 \cos 2\alpha + 1800 \cos 2\beta]$$

For secondary force to be zero,
$1200 \cos 2\alpha + 1800 \cos 2\beta = 0$

$$\frac{\cos 2\alpha}{\cos 2\beta} = -1.5$$

Vertical component

$$F_{VS} = \left(\frac{\omega^2 r^2}{l}\right)[M_1 \sin 2\alpha + M_3 \sin 2(180° - \beta) + M_2 \sin 2(180° + \beta) + M_4 \sin 2(360° - \alpha)]$$

$$= \left(\frac{\omega^2 r^2}{l}\right)[600 \sin 2\alpha - 900 \sin 2\beta + 900 \sin 2\beta - 600 \sin 2\alpha]$$

$$= 0$$

From Eq. (4), we have

$$2 \cos^2 \alpha - 1 = -1.5 (2 \cos^2 \beta - 1)$$

$$2 \cos^2 \alpha + 3 \cos^2 \beta - 2.5 = 0$$

From Eq. (12.33), we have

$$\cos \beta = \frac{\cos \alpha}{1.5}$$

$$2 \cos^2 \alpha + \frac{3 \cos^2 \alpha}{2.25} - 2.5 = 0$$

$$\cos^2 \alpha\left(2 + \frac{4}{3}\right) - 2.5 = 0$$

$$\cos^2 \alpha = 0.75, \cos a = 0.866, a = 30°, \cos \beta = 0.57735, \beta = 54.73°$$

$$0.6\ a = \frac{\sin 30°}{\sin 54.73°} = 0.6124$$

$$a = 1.02 \text{ m}$$

Secondary Couple

$$C_{HS} = \left(\frac{\omega^2 r^2}{l}\right)[M_1 \cos 2\alpha \times (-2.5) + M_3 \cos 2(180° - \beta) \times a$$

$$+ M_2 \cos 2(180° + \beta) \times (-a) + M_4 \sin 2(360° - \alpha) \times 2.5]$$

$$= \omega^2 r\, [-1500 \cos 2\alpha - 900a \cos 2\beta + 900a \cos 2\beta - 1500 \cos 2\alpha]$$

$$= 0$$

$$C_{VS} = \left(\frac{\omega^2 r^2}{l}\right)[M_1 \sin 2\alpha \times (-2.5) + M_3 \sin 2(180° - \beta) \times a$$

$$+ M_2 \sin 2(180° + \beta) \times (-a) + M_4 \sin 2(360° - \alpha) \times 2.5]$$

$$= \left(\frac{\omega^2 r^2}{l}\right)[-1500 \sin 2\alpha - 900a \sin 2\beta - 900a \sin 2\beta - 1500 \sin 2\alpha]$$

$$= \left(\frac{\omega^2 r^2}{l}\right)[3000 \sin 2\alpha + 1800a \sin 2\beta]$$

$$= -\left[\left(2\pi \times \frac{180}{60}\right)^2 \times \frac{(0.4)^2}{1.8}\right][3000 \sin 60° + 1800 \times 1.02 \sin 109.46°]$$

$$= -136733 \text{ N m}$$

Example 12.22

A three-cylinder radial engine (Fig.12.43) driven by a common crank has the cylinders spaced at 120°. The stroke is 120 mm, length of connecting rod 240 mm and the mass of the reciprocating parts per cylinder is 1 kg and the speed of the crank shaft is 2400 rpm. Determine the magnitude of the primary and secondary forces.

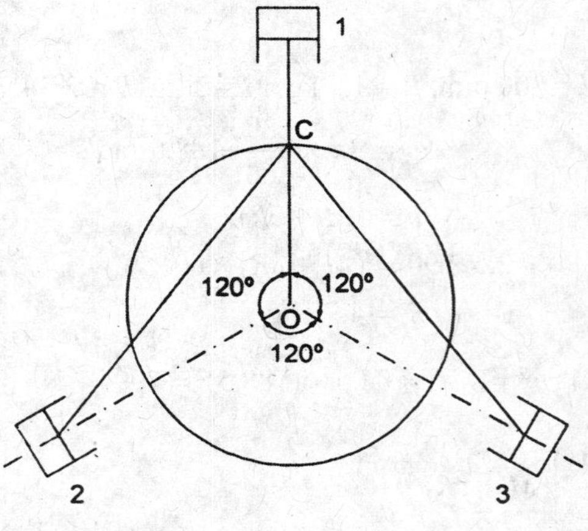

Fig.12.43 Three-cylinder radial engine

Solution

Given: $r = 60$ mm, $l = 240$ mm, $M = 1$ kg, $N = 2400$ rpm.

Maximum primary force

$$= 1.5\, Mr\, \omega^2$$

$$= 1.5 \times 1 \times 0.06 \times \left(2\pi \times \frac{2400}{60}\right)^2$$

$$= 5684.9 \text{ N}$$

$$B1\beta1 = 1.5\, M\rho$$

$$= 1.5 \times 1 \times 0.06$$

$$= 0.09 \text{ N.m}$$

Maximum secondary force $= 1.5M \times (2\omega)^2 \times \dfrac{r}{4n}$

$$= 1.5 \times 1 \times 0.06 \times \frac{\left(4\pi \times \frac{2400}{60}\right)^2}{16} = 1421.2 \text{ N}$$

$$B_2 b_2 = \frac{1.5Mr}{4n} = \frac{1.5 \times 1 \times 0.06}{4 \times 4} = 0.005625 \text{ N m}$$

Example 12.23

A two cylinder V-engine has the cylinders set at an angle of 45°, with both pistons connected to the single crank. The crank radius is 60 mm and the connecting rods are 300 mm long. The reciprocating mass per line is 1.5 kg and the total rotating mass is equivalent to 2 kg at the crank radius. A balance mass fitted opposite to the crank is equivalent to 2.5 kg at a radius of 90 mm. Determine for an engine speed of 1800 rpm, the maximum and minimum values of the primary and secondary forces due to the inertia of reciprocating and rotating masses.

Solution

Given: $\alpha = 45°$ $r = 60$ mm, $l = 300$ mm, $R = 1.5$ kg, $M = 2$ kg, $B = 2.5$ kg, $b = 90$ mm, $N = 1800$ rpm

$$\omega = 2\pi \times \frac{1800}{60} 188.5 \text{ rad/s}$$

$$Bb = (M + cR)r$$

$$2.5 \times 90 = (M + cR) \times 60$$

$$M + cR = 4.75 \text{ kg}$$

Primary force, $F_P = (M + cR)\, r\omega^2 = 4.75 \times 0.06 \times (188.5)^2 = 7994.7$ N

Secondary force, $F_S = \left(\dfrac{\sqrt{3}}{2}\right)(M + cR)\dfrac{\omega^2 r^2}{l}$

$$= \left(\frac{\sqrt{3}}{2}\right)(4.75)\ (188.5)^2\ \frac{(0.06)^2}{0.3} = 1384.7 \text{ N}$$

Example 12.24

A two cylinder locomotive with cranks at 90° has a crank radius of 325 mm. The distance between centres of driving wheels is 1.5 m. The pitch of cylinders is 600 mm. The diameter of treads of driving wheels is 1.8 m. The radius of centres of gravity of balance weights is 650 mm. The pressure due to dead load on each wheel is 40 kN. The weights of reciprocating and rotating parts per cylinder are 4.3 kN and 3 kN respectively. The speed of the locomotive is 60 km/h. Find

(a) the balancing weights both in magnitude and position required to be placed in the planes of driving wheels to balance whole of the revolving and two-third of the reciprocating masses;
(b) the swaying couple;
(c) the variation of tractive effort;
(d) the maximum and minimum pressure on rails; and
(e) what is the maximum speed at which it is possible to run the locomotive, in order that the wheels are not lifted from the rails?

Solution

Given $Mg = 3$ kN, $Rg = 4.3$ kN, $r = 0.325$ m, $l = 0.6$ m, $d_w = 1.8$ m, $b = 0.65$, $P = 40$ kN, $v = 60$ km/h

Reference plane 1 (Fig.12.44):

(a) Mass to be balanced $= \dfrac{\left(3000 + 2 \times \dfrac{4300}{3}\right)}{9.81} = 530$ kg

Table 12.25

Plane	M (kg)	r (m)	Mr	θ (deg)	$Mr\cos\theta$	$Mr\sin\theta$	l (m)	Mrl	$Mrl\cos\theta$	$Mrl\sin\theta$
1	M_1	0.65	$0.65\ M_1$	θ_1	$0.65\ M_1 \times \cos\theta_1$	$0.65\ M_1 \times \sin\theta_1$	0	0	0	0
2	500	0.325	172.25	0	172.25	0	0.45	77.5125	77.5125	0
3	530	0.325	172.25	90	0	172.25	1.05	180.8625	0	180.8625
4	M_4	0.325	M_4	θ_4	$0.65\ M_4 \times \cos\theta_4$	$0.65\ M_4 \times \sin\theta_4$	1.5	$0.97\ M_4$	$0.975\ M_4 \times \cos\theta_4$	$0.957\ M_4 \times \sin\theta_4$

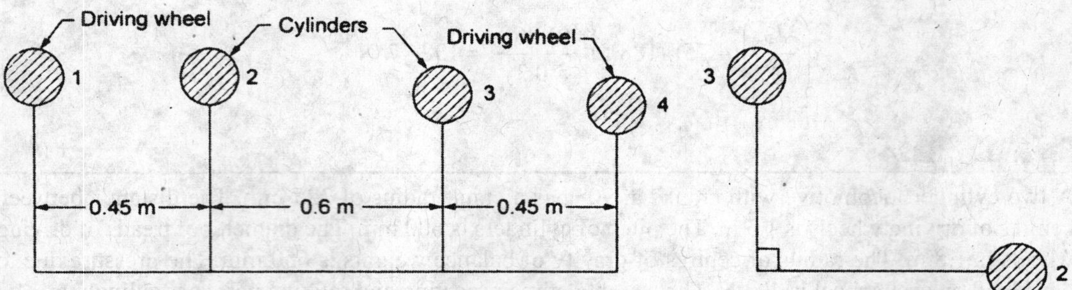

(a) *Position of planes* (b) *Position of masses*

Fig.12.44 Two-cylinder locomotive with orthogonal cranks

From Table 12.25, we have

$$\Sigma M_i r_i \cos\theta_i = 172.25$$

$$\Sigma M_i r_i \sin\theta_i = 172.25$$

$$\Sigma M_i r_i l_i \cos\theta_i = 77.5125$$

$$\Sigma M_i r_i l_i \sin\theta_i = 180.8625$$

$$\left[\left(\Sigma M_i r_i l_i \cos\theta_i\right)^2 + \left(\Sigma M_i r_i l_i \sin\theta_i\right)^2\right]^{0.5} = M_4 r_4 l_4$$

$$\left[(77.5125)^2 + (180.8625)^2\right]^{0.5} = 0.975 M_4$$

$$M_4 = \frac{196.772}{0.975} = 201.81 \text{ kg}$$

$$\tan\theta_4 = \frac{-\Sigma M_i r_i l_i \sin\theta_i}{-\Sigma M_i r_i l_i \cos\theta_i}$$

$$= \frac{-(180.8625)}{-(77.5125)} = 2.3333$$

$$\theta_4 = 66.8° \text{ i.e. } 246.8°$$

Since the numerator and denominator are both negative, therefore, θ_4 lies in the third quadrant.

$$M_1 r_1 = \left[\left(\Sigma M_i r_i \cos\theta_i + M_4 r_4 \cos\theta_4 \right)^2 + \left(\Sigma M_i r_i \sin\theta_i + M_4 r_4 \sin\theta_4 \right)^2 \right]^{0.5}$$

$$0.65 M_1 = \left[\left(0.65 \times 201.81 \cos 246.8 + 172.25 \right)^2 + \left(0.65 \times 201.81 \sin 246.8° + 172.25 \right)^2 \right]^{0.5}$$

$$= [(120.574)^2 + (51.681)^2]^{0.5} = 131.183$$

$$M_1 = 201.81 \text{ kg}$$

$$\tan\theta_1 = \frac{-\left(\Sigma M_i r_i \sin\theta_i + M_3 r_3 \sin\theta_3 \right)}{-\left(\Sigma M_i r_i \cos\theta_i + M_3 r_3 \cos\theta_3 \right)} = \frac{-51.681}{-120.574} = 0.4286$$

$$\theta_1 = 24.2° \text{ i.e. } 204.2°$$

Since the numerator and the denominator are both negative, therefore, θ_1 lies in the third quadrant.

$$\omega = \frac{60 \times 1000 \times 2}{3600 \times 1.8} = 18.52 \text{ rad/s}$$

(b) Swaying couple $= \dfrac{(1-c)Rr\omega^2}{\sqrt{2}}$

$$= \frac{(1 - 2/3) \times 4.3 \times 10^3 \times 0.325 \times (18.52)^2 \times 0.6}{\sqrt{2} \times 9.81}$$

$$= 5303 \text{ N}$$

(c) Variation of tractive effort $= \sqrt{2}(1-c)R\omega^2 r$

$$= \frac{\pm\sqrt{2}(1 - 2/3) \times 4.3 \times 10^3 \times (18.52)^2 \times 0.325}{9.81}$$

$$= \pm 17676.8 \text{ N}$$

(d) Balance mass for reciprocating parts, $B = M_1 \times \left(\dfrac{2}{3} \right) \times \left(\dfrac{R}{M + R} \right)$

$$= 201.81 \times \left(\frac{2}{3} \right) \times \frac{4.3}{5.2} = 111.25 \text{ kg}$$

Hammer blow $= Bb\,\omega^2 = 111.25 \times 0.65 \times (18.52)^2 = 24.8$ kN

Net pressure on rails $= P \pm B\,\omega^2\,b$
$= 40 \pm 24.8$
$= 64.8$ kN, 15.2 kN

(e) $40 \times 10^3 = 111.25 \times \omega^2 \times 0.65$

$$\omega = 73.96 \text{ rad/s}$$

$$v = \frac{73.96 \times 0.9 \times 3600}{1000} = 70.164 \text{ km/h}$$

Example 12.25

Four weights *A*, *B*, *C* and *D* revolve at equal radii and are equally spaced along the shaft. The weights weigh 70 N and the radii of *C* and *D* make angles of 90° and 240° respectively with the radius of *B*. Find the magnitude of the weights *A*, *C* and D and the angular position of *A* so that the system may be completely balanced.

■ Solution

Reference plane B (Fig.12.45)

$$m = \frac{70}{9.81} = 7.136 \text{ kg, let } r = 1 \text{ m.}$$

Table 12.26

Plane	M (kg)	r (m)	Mr	θ (deg)	Mr cos θ	Mr sin θ
A	M_A	1	M_A	θ_A	$M_A \cos \theta_A$	$M_A \sin \theta_A$
B	7.136	1	7.136	0	7.136	0
C	7.136	1	7.136	90	0	7.136
D	7.136	1	7.136	240	– 4.568	– 6180

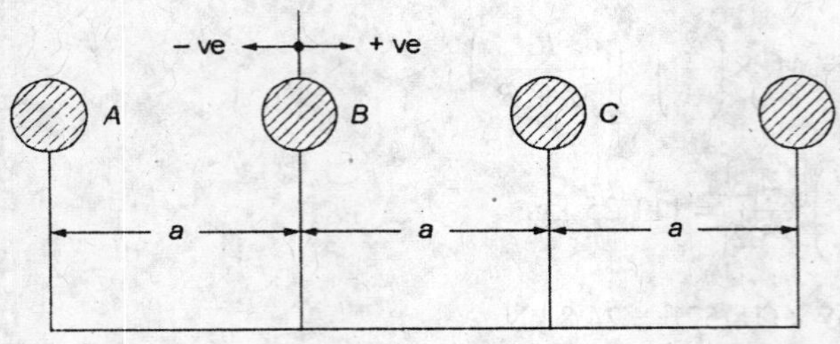

(a) *Position of planes*

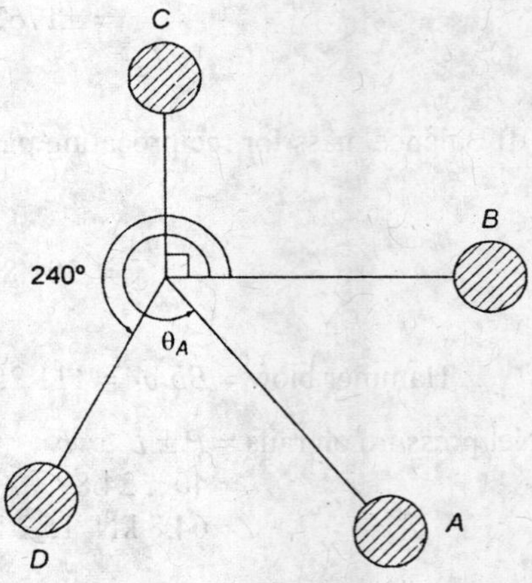

(b) *Position of masses*

Fig.12.45 Balancing of four masses in different planes

From Table 12.26, we have

$$\Sigma M_i r_i \cos\theta_i = 4.568$$

$$\Sigma M_i r_i \sin\theta_i = 0.956$$

$$\Sigma M_i r_i l_i \cos\theta_i = -7.136a$$

$$\Sigma M_i r_i l_i \sin\theta_i = -5.224a$$

$$\left[\left(\Sigma M_i r_i \cos\theta_i\right)^2 + \left(\Sigma M_i r_i \sin\theta_i\right)^2\right]^{0.5} = M_A r_a$$

$$\left[(4.568)^2 + (0.956)^2\right]^{0.5} = M_A$$

$$M_A = 4.64 \text{ kg}$$

$$\tan\theta_4 = \frac{-\Sigma M_i r_i \sin\theta_i}{-\Sigma M_i r_i \cos\theta_i}$$

$$= \frac{-(0.956)}{-(4.568)} = 0.26794$$

$$\theta_A = 15° \text{ i.e. } 195°$$

Since the numerator and denominator are both negative, therefore θ_A lies in the third quadrant.

Example 12.26

A twin cylinder uncoupled locomotive has its cylinders 0.6 m apart and balance weights are 60° apart. The planes are symmetrically placed about the centre line. For each cylinder, the revolving masses are 300 kg at crank pin radius of 320 mm and reciprocating parts 285 kg. All the revolving and 2/3rd of the reciprocating masses are balanced. The driving wheels are 1.8 m diameter. When the engine runs at 60 km/h, find (a) the swaying couple, (b) the variation in tractive effort, and (c) the hammer blow.

■ Solution

Given: $M = 300$ kg, $r = 320$ mm, $R = 285$ kg, $c = \frac{2}{3}$, $l = 0.6$ m, $d_w = 1.8$ m, $v = 60$ km/h

Total mass to be balanced $= M + cR = 300 + 2 \times \frac{285}{3} = 490$ kg

$$v = 60 \times \frac{1000}{3600} = 16.67 \text{ m/s}, \quad \omega = \frac{2v}{d_w} = 2 \times \frac{16.67}{1.8} = 18.52 \text{ rad/s}$$

Reference plane: 1 (Fig.12.46)

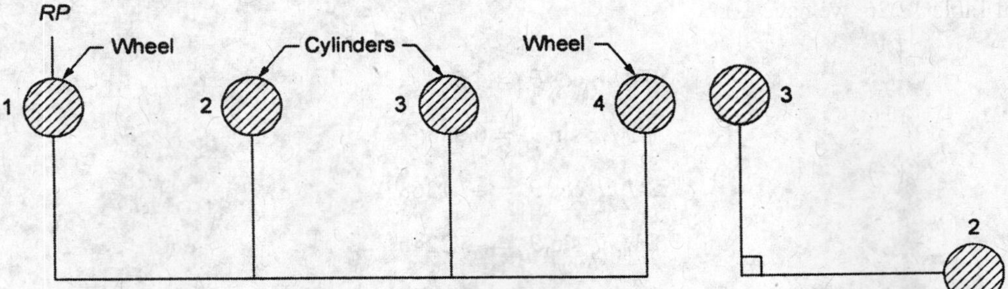

(a) *Position of planes* (b) *Position of masses*

Fig.12.46 Twin-cylinder uncoupled locomotive

Table 12.27

Plane	M (kg)	r (m)	Mr	θ (deg)	$Mr\cos\theta$	$Mr\sin\theta$	l (m)	Mrl	$Mrl\cos\theta$	$Mrl\sin\theta$
1	M_1	0.32	$0.32\,M_1$	θ_1	$0.32\,M_1$	$0.32\,M_1$	0	0	0	0
2	490	0.32	165.8	0	165.8	0	A	$156.8a$	$156.8a$	0
3	490	0.32	156.8	90	0	156.8	$a + 0.6$	$156.8a + 94.08$	0	$156.8a + 94.08$
4	M_4	0.32	$0.32\,M_4$	θ_4	$0.32\,M_4 \times \cos\theta_4$	$0.32\,M_4 \times \sin\theta_4$	$2a + 0.6$	$0.32\,M_4 \times (2a + 0.6)$	$0.32\,M_4 \times (2a + 0.6) \times \cos\theta_4$	$0.32\,M_4 \times (2a + 0.6) \times \sin\theta_4$

From Table 12.27, we have

$$\Sigma M_i r_i \cos\theta_i = 156.8$$

$$\Sigma M_i r_i \sin\theta_i = 156.8$$

$$\Sigma M_i r_i l_i \cos\theta_i = 156.8a$$

$$\Sigma M_i r_i l_i \sin\theta_i = 156.8a + 94.08$$

$$\tan\theta_4 = \frac{-\Sigma M_i r_i l_i \sin\theta_i}{-\Sigma M_i r_i l_i \cos\theta_i}$$

$$= \frac{-156.8a + 94.08}{-156.8a}$$

$$= \frac{a + 0.6}{a}$$

Taking 4 as the reference plane, we get

$$\tan\theta_1 = \frac{a}{a+0.6}$$

$$\tan\theta_1 \times \tan\theta_4 = 1$$

$$\theta_4 = \theta_1 + 60°$$

$$\tan(\theta_1 + 60°) \times \tan\theta_1 = 1$$

$$\tan^2\theta_1 2\sqrt{3}\tan\theta_1 - 1 = 0$$

$$\tan\theta_1 = 0.268, -4.732$$

$$\theta_1 = 195°,\ 285°$$

$$0.268 = \frac{a}{a+0.6}$$

$$a = 0.22 \text{ m}$$

$$M_4 r_4 l_4 = \left[\Sigma M_i r_i l_i \cos\theta_i)^2 + (\Sigma M_i r_i l_i \sin\theta_i)^2\right]^{0.5}$$

$$0.32(2a+0.6)M_4 = \left[(156.8a)^2 + (156.8a + 94.08)^2\right]^{0.5}$$

$$0.32\ (0.44+0.6)M_4 = \left[(156.8\times 0.22)^2 + (156.8\times 0.22 + 94.08)^2\right]^{0.5}$$

$$0.3328\,M_4 = 134.123$$

$$M_4 = 400 \text{ kg}$$

By symmetry, $M_1 = 400$ kg

$$Bb = \frac{400 \times 0.32 \times (213) \times 285}{490} = 49.63 \text{ N m}$$

$$\text{Hammer blow} = Bb\omega^2 = 49.63 \times (18.53)^2 = 17023 \text{ N}$$

$$\text{Swaying couple} = (1-c)\frac{R r \omega^2 l}{\sqrt{2}} = \frac{(1-2/3)\times 285 \times 0.32(18.52)^2 0.6}{\sqrt{2}}$$

$$= 4424.76 \text{ N m}$$

$$\text{Variation of tractive effort} = \pm\sqrt{2}(1-c)R\omega^2 r$$

$$= \pm\sqrt{2}\left(1-\frac{2}{3}\right)\times 285 \times (18.52)^2 \times 0.32$$

$$= \pm 14745.87 \text{ N}$$

Example 12.27

The reciprocating mass per cylinder in a 60° *V*-engine is 1.2 kg. The stroke and the connecting rod length are 100 and 250 mm, respectively. If the engine runs at 2000 rpm, determine the maximum and minimum values of the primary and secondary forces. Also find out the crank positions corresponding to these values.

■ Solution

Given: $R = 1.2$ kg, $2\alpha = 60°$, $r = 100$ mm, $l = 250$ mm, $N = 2000$ rpm, $n = \frac{l}{r} = 2.5$

$$\omega = 2\pi \times \frac{2000}{60} = 209.44 \text{ rad/s}$$

$$F_P = 2\omega^2 r[(\cos^2 \alpha \ \cos\theta)^2 + (\sin^2 \alpha \sin\theta)^2]^{0.5}$$

$$= 2 \times 1.2 \times (209.44)^2 \times 0.1[(\cos^2 30° \cos\theta)^2 + (\sin^2 30° \sin\theta)^2]^{0.5}$$

$$= 10527.6[0.75 \ \cos^2\theta + 0.25 \ \sin^2\theta]^{0.5}$$

For F_P to be maximum or minimum, $\frac{dF_P}{d\theta} = 0$

$-1.5 \cos\theta \sin\theta + 0.5 \sin\theta \cos\theta = 0$

$\sin 2\theta \sin\theta + 0.5 \sin\theta \cos\theta = 0$

$\sin 2\theta = 0$

$\theta = 0°, 90°, 180°$

For $\theta = 0°$, $(F_P)_{max} = 91171.7$ N

For $\theta = 90°$, $(F_P)_{min} = 5264.8$ N

$$F_S = 2R\omega^2 \left(\frac{r}{n}\right)[(\cos\alpha \ \cos 2\alpha \ \cos 2\theta)^2 + (\sin\alpha \ \sin 2\alpha \ \sin 2\theta)^2]^{0.5}$$

For $\theta = 0°$, $F_S = 1824.4$ N

For $\theta = 45°$, $F_S = 1824.4$ N

Example 12.28

The following data refer to a two-cylinder locomotive with cranks at 90°:

Reciprocating mass per cylinder = 300 kg

Crank radius = 300 mm

Diameter of the driving wheels = 1.8 m

Distance between the cylinder centre lines = 0.65 m

Distance between the driving wheel centre planes = 1.55 m

Determine (a) the fraction of the reciprocating masses to be balanced by placing the balancing masses on the driving wheels, if the hammer blow is not to exceed 46 kN at 96.5 km/h, and (b) the variation in tractive effort.

■ Solution

Given: $R = 300$ kg, $r = 300$ mm, $d_w = 1.8$ m, hammer blow = 46 kN, $v = 96.5$ km/h, $l = 0.65$ m,

$$\omega = \frac{96.5 \times 1000}{3600 \times 0.9} = 29.784 \text{ rad/s}$$

Mass to be balanced $M_0 = M + cR = 0 + 300 \times c = 300\, c$ kg

Taking 1 as the reference plane, we have

Table 12.28

Plane	Mass, *M* (kg)	Radius, *r* (m)	*Mr* (kg.m)	θ (deg)	*l* (m)	Couple, *Mrl* (kg.m²)
1	M_1	r_1	M_1r_1	1	0	0
2	300 *c*	0.3	90 *c*	0	0.45	40.5*c*
3	300 *c*	0.3	90 *c*	90	1.10	99.0*c*
4	M_4	r_4	M_4r_4	4	1.55	1.55 M_4r_4

$$M_o \times 0.3 \times 0.45 \cos 0° + M_o \times 0.3 \times 1.1 \cos 90° + M_4 \times r_4 \times 1.55 \cos \theta_4 = 0$$

$$0.135M_o + 1.55\, M_4r_4 \cos \theta_4 = 0 \qquad (12.36)$$

$$M_o \times 0.3 \times 0.45 \sin 0° + M_o \times 0.3 \times 1.1 \sin 90° + M_4\, r_4 \times 1.55 \sin \theta_4 = 0$$

$$0.33M_o + 1.55\, M_4r_4 \sin \theta_4 = 0 \qquad (12.37)$$

From (12.36) and (12.37), we get

$$\tan \theta_4 = \frac{-0.33}{-0.135} = 2.44$$

$$\theta_4 = 247.75°$$

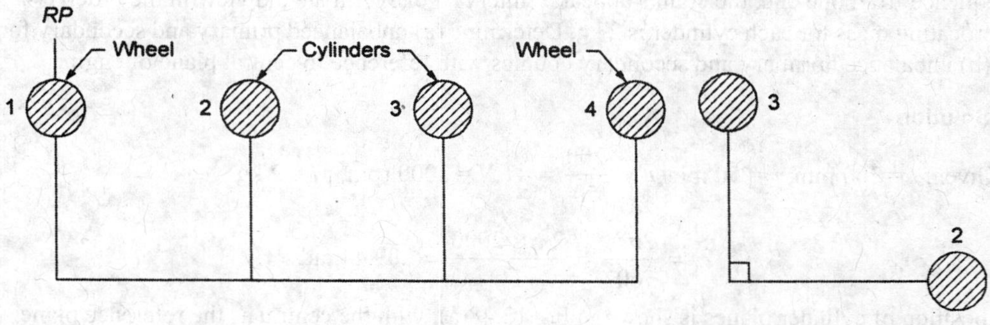

(a) *Position of planes* (b) *Position of masses*

Fig.12.47 Two-cylinder locomotive with orthogonal cranks

$$M_4\, r_4 = \frac{-0.33M_0}{-1.55 \times 0.9255} = 0.23\, M_0$$

By symmetry, $M_1 = M_4$

Taking 4 as the reference plane, we have

$$M_1 r_1 = 0.23\, M_0$$

$$\tan \theta_1 = \frac{-0.135}{-0.33} = 0.409$$

$$\theta_1 = 202.25°$$

Balance mass for reciprocating parts only, $B = \dfrac{0.23\, M_0 \times c \times 300}{b \times 300c}$

$$= \frac{0.23\, M_0}{b}$$

Hammer blow $= B\,\omega^2\, b = 0.23\, M_0 \times (29.784)^2 = 46 \times 10^3$ N m

$$M_0 = 225.46 \text{ kg} = 300c$$

$$c = 0.75 \quad \text{or} \quad \frac{3}{4}$$

Variation of tractive effort $= \sqrt{2}\,(1 - c)\, R\,\omega^2 r$

$$= \pm\sqrt{2}\,(1 - 0.75) \times 300\,(29.784)^2 \times 0.3$$

$$= \pm 28227 \text{ N}$$

Example 12.29

The cranks and connecting rods of a four-cylinder in-line engine running at 2000 rpm are 50 mm and 200 mm each respectively. The cylinders are spaced 0.2 m apart. If the cylinders are numbered 1 to 4 in sequence from one end, the cranks appear at intervals of 90° in an end view in the order 1–4–2–3, reciprocating mass for each cylinder is 2 kg. Determine (a) unbalanced primary and secondary forces, and (b) unbalanced primary and secondary couples with reference to central plane of engine.

Solution

Given: $l = 200$ mm, $r = 50$ mm, $n = \dfrac{200}{50} = 4$, $N = 2000$ rpm, $M = 2$ kg

$$\omega = \frac{2\pi N}{60} = \frac{2\pi \times 2000}{60} = 209.44 \text{ rad/s}$$

The position of cylinder planes is shown in Fig.12.48(a), with the central as the reference plane.

Table 12.29

Plane (1)	Mass, M (kg) (2)	Radius, r (m) (3)	Mr (kg m) (4)	Distance from RP, l (m) (5)	Mrl (kg m^2) (6)
1	2	0.05	0.1	– 0.3	– 0.03
2	2	0.05	0.1	– 0.1	– 0.01
3	2	0.05	0.1	+ 0.1	+ 0.01
4	2	0.05	0.1	+ 0.3	+ 0.03

Primary force polygon: The primary crank position is shown in Fig.12.48(b). The primary force polygon has been drawn in Fig.12.48(c) with the data from column (4). There is no unbalanced primary force as the polygon in closed.

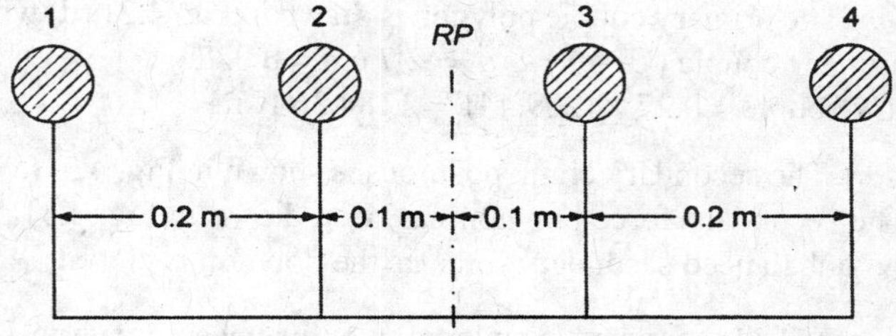

(a) *Cylinder plane position*

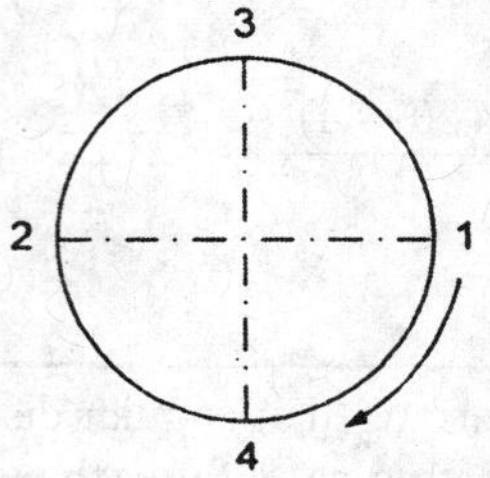

(b) *Primary crank positions*

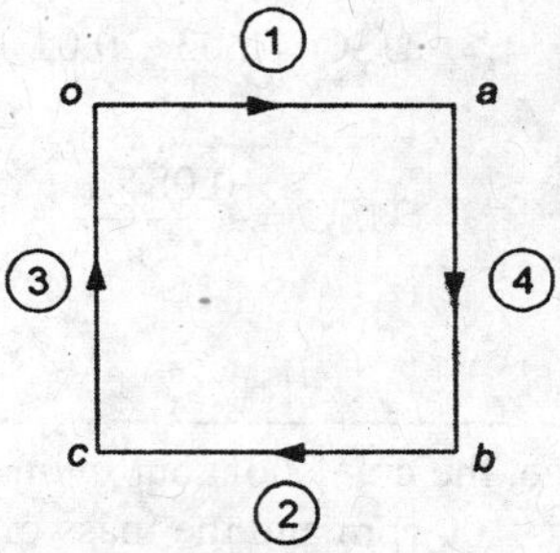

(c) *Primary force polygon*

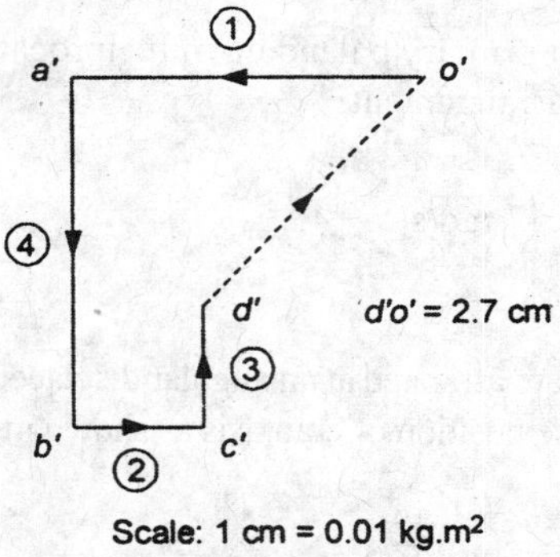

(d) *Primary couple polygon*

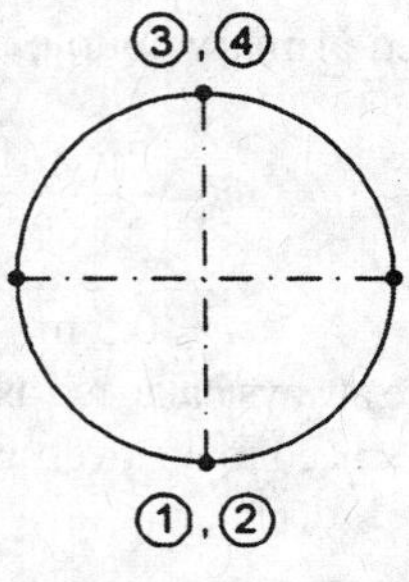

(e) *Secondary crank positions*

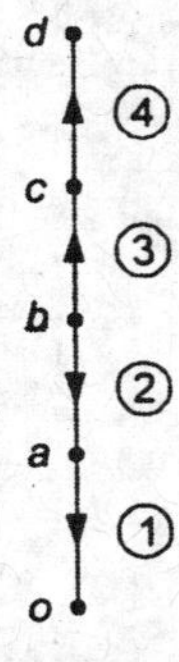

(f) *Secondary force polygon*

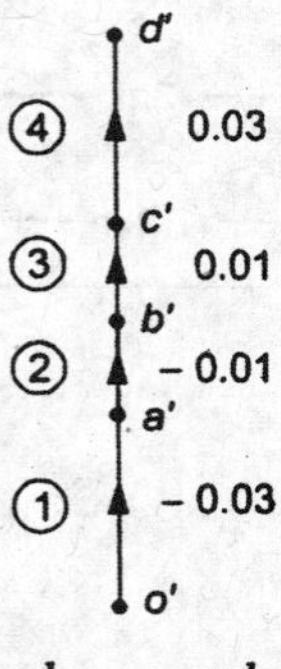

(g) *Secondary couple polygon*

Fig.12.48 Balancing of four-cylinder in-line engine

Primary couple polygon: The primary couple polygon is drawn in Fig. 4.31 (d) with the data in column (6). The unbalanced primary couple (UPC) = $d'o'$ = 2.7 cm = 0.027

Unbalanced primary couple = $0.27 \times (209.44)^2 = 1184.36$ N m

Secondary force polygon: The secondary crank positions is shown in Fig. 4.31(e) by taking crank 3 as the reference crank. The secondary force polygon has been drawn in Fig. 4.31(f) with the data from column (4). There is no unbalanced secondary force as the polygon is closed.

Secondary couple polygon: The secondary couple polygon has been drawn in Fig. 4.31(g). The secondary unbalanced couple (USC) is proportional to,

$$\text{USC} = 0.03 + 0.01 + 0.01 + 0.03 = 0.08 \text{ kg. m}^2$$

or
$$\text{USC} = \frac{0.08 \times \omega^2}{n} = \frac{0.08 \times (209.44)^2}{4} = 877.2 \text{ N m}$$

Example 12.30

In a marine oil engine, the cranks of four cylinders are arranged at angular displacement of 90°. The speed of the engine is 100 rpm and the mass of reciprocating parts for each cylinder is 900 kg. Each crank is 0.5m long. The outer cranks are 3 m apart and the inner cranks are 1.2 m apart and are placed symmetrically between the outer cranks.

Find the firing order of the cylinders for the best primary balancing of reciprocating parts and also the maximum unbalanced primary couple for that arrangement.

■ **Solution**

$$\omega = \frac{2\pi \times 100}{60} = 10.47 \text{ rad/s}$$

$$r = 0.5 \text{ m}$$

The primary forces are always balanced as cranks are arranged at an angular displacement of 90° to each other. The primary couples need to be investigated. The position of cranks is as shown in Fig.12.49.

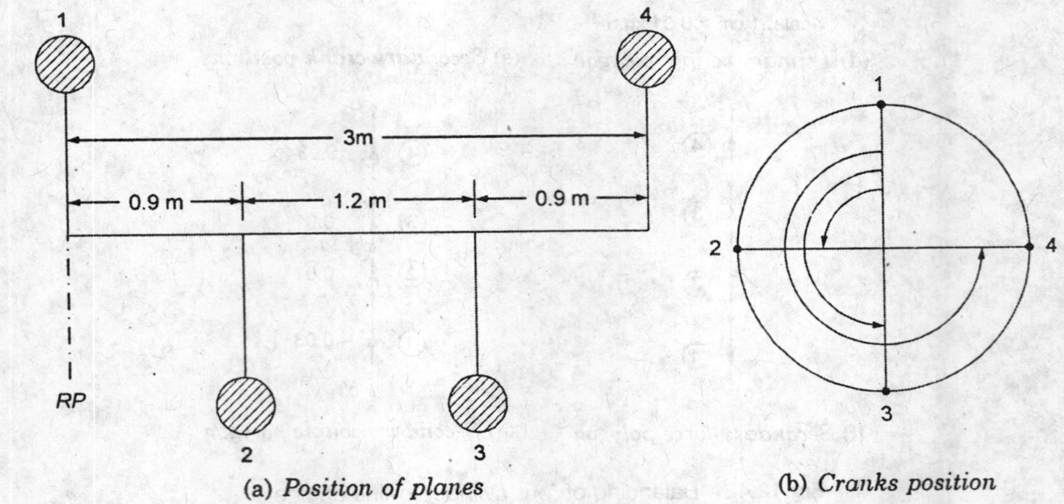

Fig.12.49 Marine oil engine

The possible firing orders are: 1234, 1243, 1423, 1324, 1342, 1432

The disturbing force along the axis of the cylinder $= M\,r\,\omega^2 \cos\theta$

Let $K = M\,r\,\omega^2\ 900 \times 0.5 \times (10.47)^2 = 49348$ N

$$\text{Total disturbing force} = \sum_{i=1}^{4} K\cos(\theta + \dot{\alpha}_i)$$

where α_i = angle between the reference crank and the crank considered.

Table 12.30

Plane of cylinders	M (kg)	r (m)	Mr	$K = Mr\omega^2$	Arm length l(m)	Couple Kl
1	900	0.5	450	49348	0	0
2	900	0.5	450	49348	0.9	0.9 K
3	900	0.5	450	49348	2.1	2.1 K
4	900	0.5	450	49348	3	3 K

Disposition of Cranks	Crank positions	Primary couple Polygon	Resultant primary couple
1234	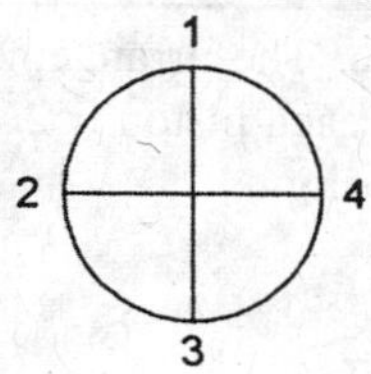	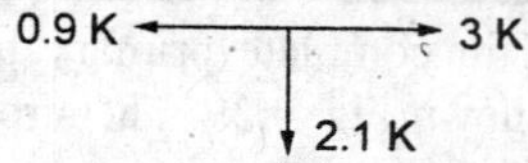	$[(3 - 0.9)^2 + (2.1)^2]^{0.5} \times K = 2.97\,K$
1243	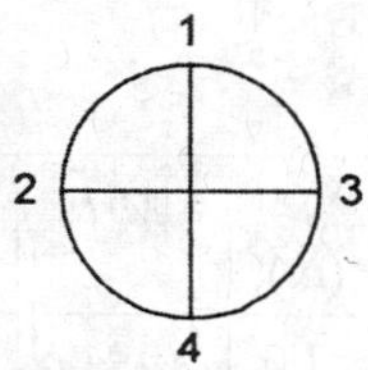	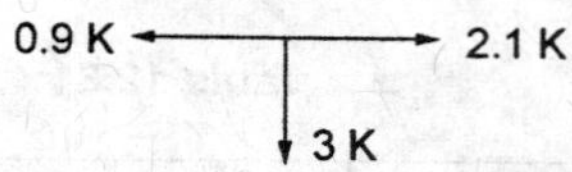	$[(2.1 - 0.9)^2 + (3)^2]^{0.5} \times K = 2.97\,K$
1423	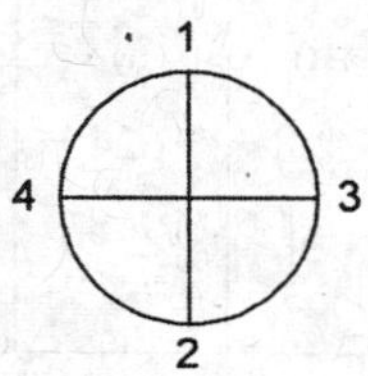	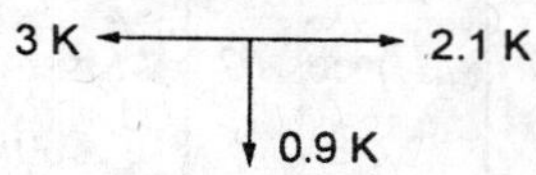	$[(3 - 2.1)^2 + (0.9)^2]^{0.5} \times K = 1.273\,K$
1324	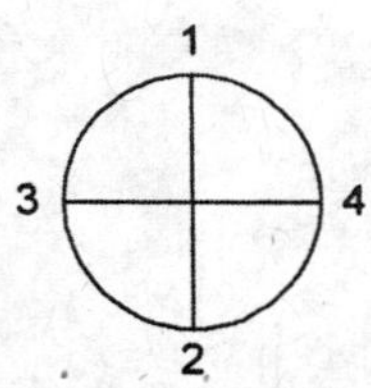	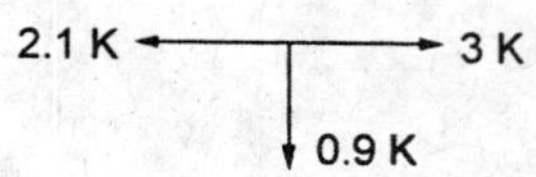	$[(3 - 2.1)^2 + (0.9)^2]^{0.5} \times K = 1.273\,K$

Disposition of Cranks	Crank positions	Primary couple Polygon	Resultant primary couple
1342	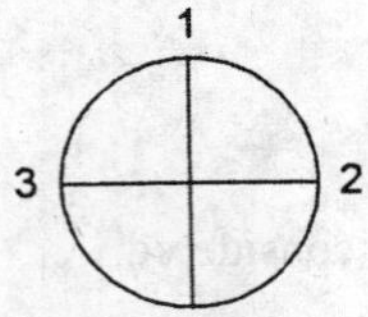	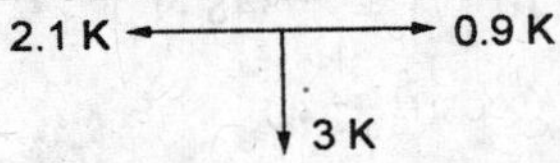	$[(2.1-0.9)^2+(3)^2]^{0.5}\times K=$ 3.231 K
1432	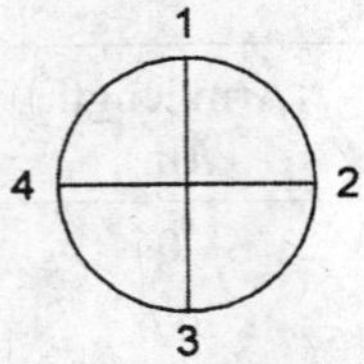	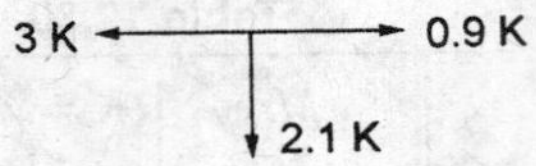	$[(3-0.9)^2+(2.1)^2]^{0.5}\times K=$ 2.97 K

Least value of primary couple = 1.273 K

$= 1.273 \times 49348 = 62820$ N.m

Best firing order is 1423 and 1324.

Example 12.31

A four-cylinder steam engine is in complete primary balance. The arrangement of reciprocating masses in different planes is as shown Fig. 4.39. The stroke of each piston is $2r$ mm. Determine the reciprocating mass of the L.P. cylinder and the relative crank position.

■ Solution

Reference plane 2 (refer to Fig.12.50).

Table 12.31

Plane	M (kg)	r (m)	Mr	θ (deg)	$Mr \times \cos\theta$	$Mr \times \sin\theta$	l (m)	Mrl	$Mrl \times \cos\theta$	$Mrl \times \sin\theta$
1	380	r	$380\,r$	0	$380\,r$	0	–1.3	$-494\,r$	$-494\,r$	0
2	M_2	r	$M_2 r$	$\theta 2$	$M_2 r\cos\theta 2$	$M_2 r\sin\theta 2$	0	0	0	0
3	580	r	$580\,r$	90	0	$580r$	2.8	$1624\,r$	0	$1624\,r$
4	480	r	$480\,r$	θ_4	$480r\cos\theta_4$	$480r\sin\theta_4$	4.1	$1968\,r$	$968\,r \times \cos\theta_4$	$968\,r \times \sin\theta_4$

From Table 12.31, we have

$\Sigma M_i r_i \cos\theta i = 380r + 480r\cos\theta_4$

$\Sigma M_i ri \sin\theta i = 580r + 480r\sin\theta_4$

$\Sigma M_i rili \cos\theta i = -494r$

$\Sigma M_i rili \sin\theta i = 1624r$

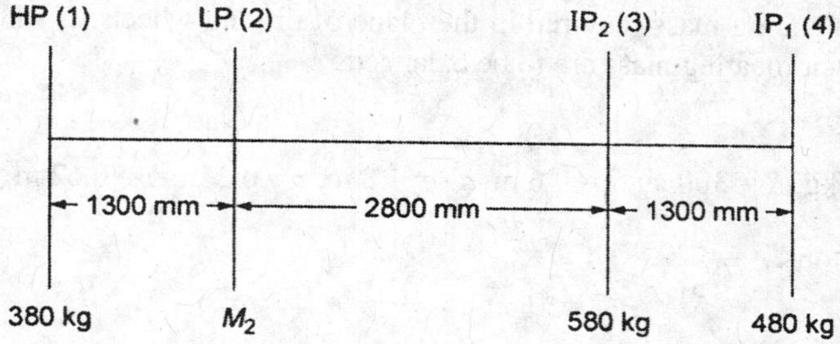

Fig.12.50 Four-cylinder steam engine

$$\tan\theta_4 = \frac{-\Sigma M_i r_i l_i \sin\theta_i}{-\Sigma M_i r_i l_i \cos\theta_i} = \frac{-(1624r)}{-(-494)} = -3.28745$$

θ_4

$$= -73.98 \; i.e. 286.92°$$

Since the numerator is negative and denominator is positive, therefore θ_4 lies in the fourth quadrant.

$$M_2 r_2 = \left[(\Sigma M_i r_i \cos\theta_i + M_M r_M \cos\theta_M)^2 + (\Sigma M_i r_i \sin\theta_i + M_M r_M \sin\theta_M)^2\right]^{0.5}$$

$$M_2 r = [(380r + 480r \cos 286.92°)^2 + (580r + 480r \sin 286.92°)^2]^{0.5}$$

$$= r[(519.7)^2 + (120.78)^2]^{0.5} = 535.55r$$

$$M_2 = 535.55 \text{ kg}$$

$$\tan\theta_2 = \frac{-(\Sigma M_i r_i \sin\theta_i + M_M r_M \sin\theta_M)}{-(\Sigma M_i r_i \cos\theta_i + M_M r_M \cos\theta_M)} = \frac{-120.78}{-519.7} = 0.2324$$

$\theta 2 = 13.08°$, i.e. 193.08°

Since the number and denominator are both negative, therefore θ_2 lies in the third quadrant.

Example 12.32

The following data refer to a two-cylinder uncoupled locomotive:
Rotating mass per cylinder = 280 kg
Reciprocating mass per cylinder = 300 kg
Distance between wheels = 1400 mm
Distance between cylinder centres = 600 mm
Diameter of treads of driving wheels = 1800 mm
Crank radius = 300 mm
Radius of centre of balance mass = 620 mm
Locomotive speed = 50 km/hr
Angle between cylinder cranks = 90°
Dead load on each wheel = 3.5 tons

Determine (i) the balancing mass required in the plane of driving wheels, if whole of the revolving mass and 2/3rd of reciprocating mass are to be balanced.

Solution

Given: $M = 280$ kg, $R = 300$ kg, $l = 0.6$ m, $d_w = 1.8$ m, $r = 0.3$ m, $b = 0.62$ m, $v = 50$ km/hr,

$$P = 3.5 \text{ tons}, c = \frac{2}{3}$$

$$\text{Equivalent mass} = M + \frac{2R}{3} = 280 + \frac{2 \times 300}{3} = 480 \text{ kg}$$

(a) Reference plane A (Refer to Fig.12.40).

Table 12.32

Plane	M (kg)	r (m)	Mr	θ (deg)	$Mr \times \cos\theta$	$Mr \times \sin\theta$	l (m)	Mrl	$Mrl \times \cos\theta$	$Mrl \sin\theta$
A	M_A	0.62	$0.62\, M_A$	θ_A	$0.62\, M_A \times \cos\theta_A$	$0.62\, M_A \times \sin\theta_A$	0	0	0	0
B	480	0.3	144	0	144	0	0.4	57.6	57.6	0
C	480	0.3	144	90	0	144	1.0	144	0	144
D	M_D	0.62	$0.62M_D$	θ_D	$0.62\, M_D \times \cos\theta_D$	$0.62\, M_D \times \sin\theta_D$	1.4	$0.868\, M_D$	$0.868\, M_D \times \cos\theta_D$	$0.868\, M_D \times \sin\theta_D$

From Table 12.32, we have

$$\Sigma M_i r_i \cos\theta i = 144$$

$$\Sigma M_i r_i \sin\theta i = 144$$

$$\Sigma M_i r_i l_i \cos\theta i = 57.6$$

$$\Sigma M_i r_i l_i \sin\theta i = 144$$

$$\left[(\Sigma M_i r_i l_i \cos\theta_i)^2 + (\Sigma M_i r_i l_i \sin\theta_i)^2\right]^{0.5} = M_D r_D l_D$$

$$[(57.6)^2 + (144)^2]^{0.5} = 0.868\, M_D$$

$$M_D = \frac{155.093}{0.868} = 178.68 \text{ kg}$$

$$\tan\theta_D = \frac{-\Sigma M_i r_i l_i \sin\theta_i}{-\Sigma M_i r_i l_i \cos\theta_i}$$

$$= \frac{-(144)}{-(57.6)} = 2.5$$

$$\theta_D = 68.2°, \text{ i.e. } 248.2°.$$

Since the numerator and denominator are both negative, therefore θD lies in third quadrant.

$$M_A r_A = \left[(\Sigma M_i r_i \cos\theta_i + M_D r_D \cos\theta_D)^2 + (\Sigma M_i r_i \sin\theta_i + M_D r_D \sin\theta_D)^2\right]^{0.5}$$

$$0.62 M_A = \left[\left(144 + 0.62 \times 178.68 \cos 248.62°\right)^2 + \left(144 + 0.62 \times 1178.68 \sin 248.2°\right)^2\right]^{0.5}$$

$$= \left[(102.86)^2 + (41.14)^2\right]^{0.5}$$

$$= 110.78$$

$$M_A = 178.68 \text{ kg}$$

$$\tan\theta_A = \frac{(\Sigma M_i r_i \sin\theta_i + M_D r_D \sin\theta_D)}{-(\Sigma M_i r_i \cos\theta_i + M_D r_D \cos\theta_D)} = \frac{-41.14}{-102.86} = 0.4$$

$\theta A = 21.8°$ *i.e.* $201.8°$

Since the numerator and denominator are both negative, therefore θA lies in the third quadrant.

(b) $v = \dfrac{50 \times 10^3}{3600} = 13.89$ m/s

$$\omega = \frac{2v}{d_\omega} = \frac{2 \times 13.89}{1.8} = 15.432 \text{ rad/s}$$

$$\text{Maximum swaying couple} = \frac{(1-c)R\omega^2 rl}{\sqrt{2}} = \frac{\left(\dfrac{1-2}{3}\right) \times 300 \times (15.432)^2 \times 0.3 \times 0.6}{\sqrt{2}}$$

$$= 3031.1 \text{ N m}$$

(c) Maximum tractive effort variation $= \pm\sqrt{2}(1-c)R\omega^2 r$

$$= \pm\sqrt{2}\left(1 - \frac{2}{3}\right) \times 300 \times (15.432)^2 \times 0.3$$

$$= \pm 10103.7 \text{ N}$$

Example 12.33

Four masses *A, B, C* and *D,* as shown below are to be completely balanced:

	A	*B*	*C*	*D*
Mass (kg)	–	30	50	40
Radius (mm)	180	240	120	150

The planes containing masses *B* and *C* are 300 mm apart. The angle between the planes containing *B* and *C* is 90°. *B* and *C* make angles of 210° and 120° respectively with *D* in the same sense. Find (*i*) magnitude and angular position of mass *A* (*ii*) the positions of planes *A* and *D.*

■ Solution

Given: $\angle BOC = 90°$, $\angle BOD = 210°$, $\angle COD = 120°$

Reference plane *B* (Fig. 4.40)

Let M_A = mass at A

a = distance of plane A from B

b = distance of plane D from B

Refer to Fig.12.51.

Table 12.33

Plane	Mass, M (kg)	Radius, r (m)	Centrifugal force, Mr (kg.m)	Distance from RP, l (m)	Couple Mrl (kg.m^2)
A	M_A	0.18	0.18 M_A	$-a$	$-0.18\,M_Aa$
B	30	0.24	7.2	0	0
C	50	0.12	6.0	0.3	1.8
D	40	0.15	6.0	b	6 b

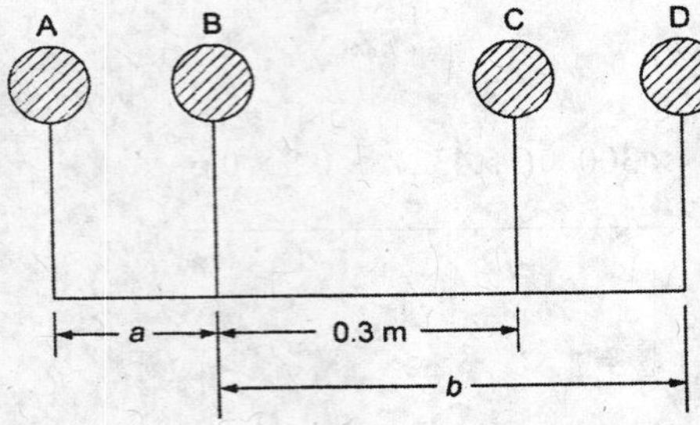

(a) *Position of planes*

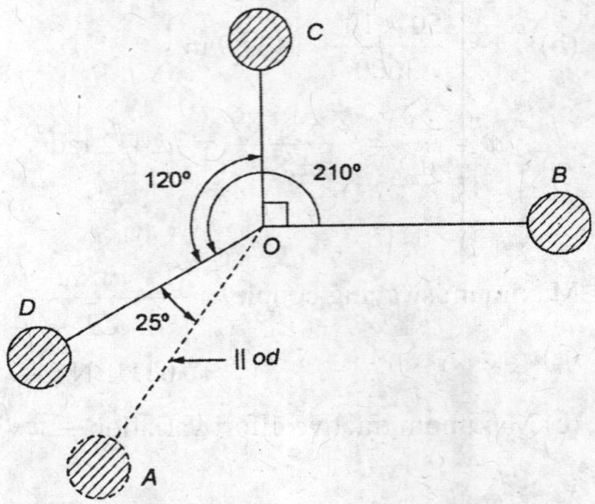

(b) *Angular position of masses.*

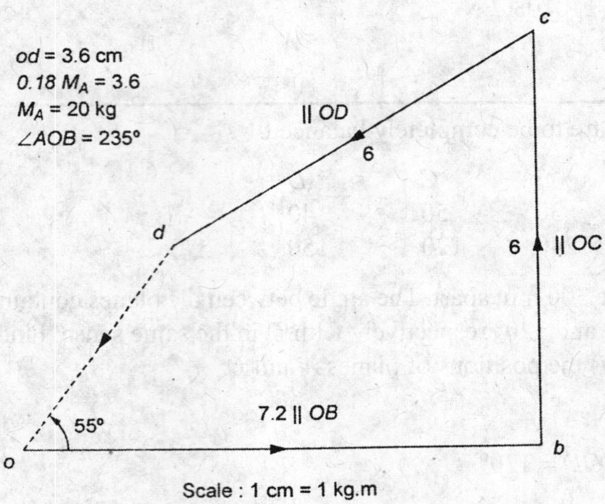

(c) *Force polygon*

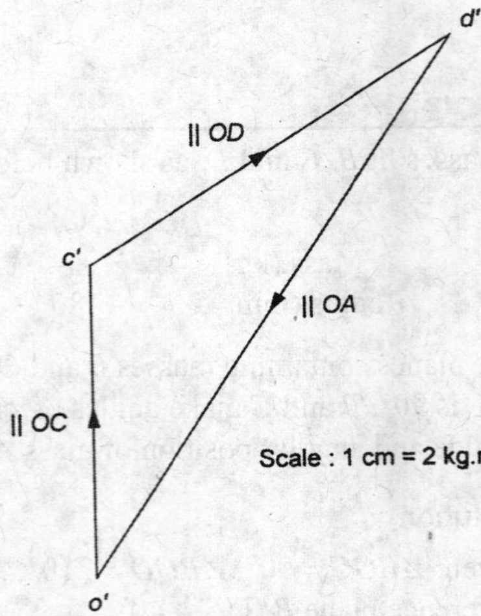

(d) *Couple polygon*

Fig.12.51 Balancing of four masses in different planes

Force Polygon: Draw the force polygon with the data from column 4 as shown in Fig.12.51(c).

1. Draw $ob = 7.2$ units parallel to OB.
2. Draw $bc = 6$ units parallel to OC.
3. Draw $cd = 6$ units parallel to OD.
4. Join od and measure it.

$$od = 3.6 = 0.18\, M_A$$
$$M_A = 20 \text{ kg}$$

To locate the angular position of A, draw OA from O in Fig. 4.40 (b) parallel to od. $\angle AOB = 235°$ and $\angle AOD = 25°$.

Couple polygon: Form the data in column 6 draw the couple polygon as shown in Fig.12.51(d).

1. Draw $o'c'$ parallel to OC and equal to 1.8 units upwards.
2. Draw a line from c' parallel to OD and another line from o' parallel to OA to intersect at d'.
3. $o'd' = 3.8$ units $= -\,0.18\, M_A$

$$a = \frac{-3.8}{0.18 \times 20} = -1.05 \text{ m}$$

Negative sign indicates that plane A is towards right of B instead of the left as assumed.

$$c'd' = 2.4 \text{ units} = 6b$$
$$b = 0.4 \text{ m}$$

We observe that the direction of $c'd'$ is opposite to the direction of mass D. Therefore, the plane of mass D is 0.4 m towards left of plane B and not towards right of plane B as assumed.

Example 12.34

The following data refer to two-cylinder locomotive with cranks at 90°. Reciprocating mass per cylinder = 300 kg, crank radius = 0.3 m, driving wheel diameter = 1.8 m, distance between cylinders centre lines = 0.7 m, distance between the driving wheel central planes = 1.6 m. Determine

(i) The friction of reciprocating masses to be balanced, if the hammer blow is not to exceed 46 kN at 96.5 km/h,
(ii) The variation of tractive effort,
(iii) The maximum swaying couple.

■ **Solution**

Given: $R = 300$ kg, $r = 0.3$ m, $d_w = 1.8$ m, hammer blow = 46 kN, $v = 96.5$ km/h, $l = 0.7$ m

$$\omega = \frac{96.5 \times 1000}{3600 \times 0.9}$$
$$= 29.784 \text{ rad/s}$$

Mass to be balanced $M_o = M + cR$

$$= 0 + 300 \times c = 300c \text{ kg}$$

Reference plane 1 (Fig.12.52)

Table 12.34

Plane	Mass, M (kg)	Radius, r (m)	Mr (kg.m)	θ (deg)	l (m)	Couple, Mrl (kg m²)
1	M_1	r_1	M_1r_1	θ_1	0	0
2	$300c$	0.3	$90c$	0	0.45	$40.5c$
3	$300c$	0.3	$90c$	90	1.15	$103.5c$
4	M_4	r_4	M_4r_4	θ_4	1.6	$1.6M_4r_4$

$$M_o \times 0.3 \times 0.45 \cos 0° + M_o \times 0.3 \times 1.15 \cos 90° + M_4 r_4 \times 1.6 \cos \theta_4 = 0$$

$$0.135\, M_o + 1.6\, M_4 r_4 \cos \theta_4 = 0 \tag{12.38}$$

$$M_o \times 0.3 \times 0.45 \sin 0° + M_o \times 0.3 \times 1.15 \sin 90° + M_4 \times r_4 \times 1.6 \sin \theta_4 = 0$$

$$0.345\, M_o + 1.6\, M_4 r_4 \sin \theta_4 = 0 \tag{12.39}$$

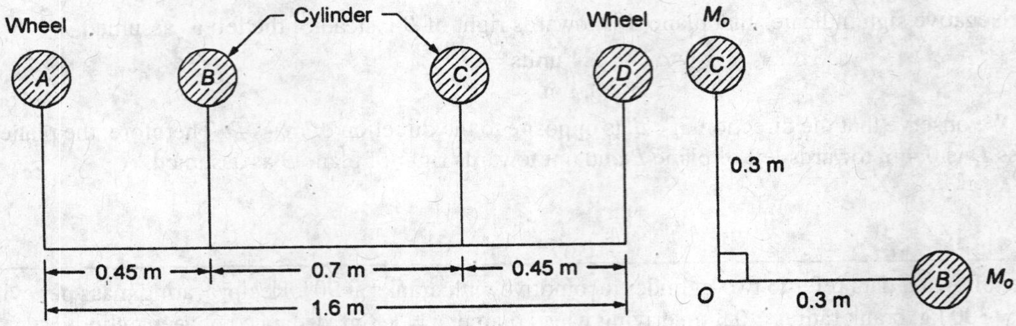

(a) *Position of planes* (b) *Position of cranks*

Fig.12.52 Two-cylinder locomotive with orthogonal cranks

From Eqs. (12.38) and (12.39), we get

$$\tan \theta_4 = \frac{0.345}{-0.135} = 2.5556$$

$$\theta_4 = 248.63°$$

$$M_4 r_4 = \frac{-0.345\, M_o}{-1.6 \times 0.93124} = 0.23154\, M_o$$

By symmetry, $M_1 = M_4$

Taking 4 as the reference plane, we have

$$M_1 r_1 = 0.23154 M_o$$

$$\tan \theta_1 = \frac{-0.135}{-0.345} = 0.39130$$

$$\theta_1 = 201.37°$$

(i) Balance mass for reciprocating parts only, $B = \dfrac{0.23154 M_o \times c \times 300}{b \times 300c}$

$$= \frac{0.23154 M_o}{b}$$

Hammer blow $= B\omega^2 b = 0.23154\ M_o \times (29.784)^2 = 205.4\ M_o$ N m

$= 46 \times 10^3$

$M_o = 224$ kg $= 300\ c$

$$c = \frac{224}{300} = 0.7467$$

(ii) Variation of tractive effort $= \pm\sqrt{2}(1 - c)R\omega^2 r$

$= \pm\sqrt{2}(1 - 0.7467) \times 300 \times (29.784)^2 \times 0.3$

$= 28.6$ kN

(iii) Maximum swaying couple $= \dfrac{(1 - c)R\omega^2 rl}{\sqrt{2}}$

$$= \frac{(1 - 0.7467) \times 300 \times (29.784)^2 \times 0.3 \times 0.7}{\sqrt{2}}$$

$= 10009.83$ N m

Example 12.35

A shaft carries four rotating masses *A, B, C* and *D* in this order along its axis. The mass *A* may be assumed to be concentrated at a radius of 12cm, *B* at 15 cm, *C* at 14 cm and *D* at 18 cm. The masses *A, C* and *D* are 15 kg, 10 kg, and 8 kg respectively. The planes of revolution of *A* and *B* are 15 cm apart and of *B* and *C* are 18 cm apart. The angle between the radii of *A* and *C* is 90°. If the shaft is in complete dynamic balance, determine (*i*) the angles between the radii of *A, B* and *D*, (ii) the distance between the planes of revolution of *C* and *D,* and (*iii*) the mass *B*.

■ Solution

Reference plane *B* (refer to Fig.12.53) Table 12.35

Table 12.35

Plane	M (kg)	r (m)	Mr	θ (deg)	$Mr\cos\theta$	$Mr\sin\theta$	l (m)	Mrl	$Mrl \times \cos\theta$	$Mrl \times \sin\theta$
A	15	0.12	1.8	0	1.8	0	–0.15	–0.27	–0.27	0
B	M_2	0.15	$0.15\ M_2$	θ_2	$0.15\ M_2 \times \cos\theta_2$	$0.15\ M_2 \times \sin\theta_2$	0	0	0	0
C	10	0.14	1.4	90	0	1.4	0.18	0.252	0	0.252
D	8	0.18	1.44	θ_4	$1.44 \times \cos\theta_4$	$1.44 \times \sin\theta_4$	l_4	$1.44\ l_4$	$1.44\ l_4 \times \cos\theta_4$	$1.44\ l_4 \times \cos\theta_4$

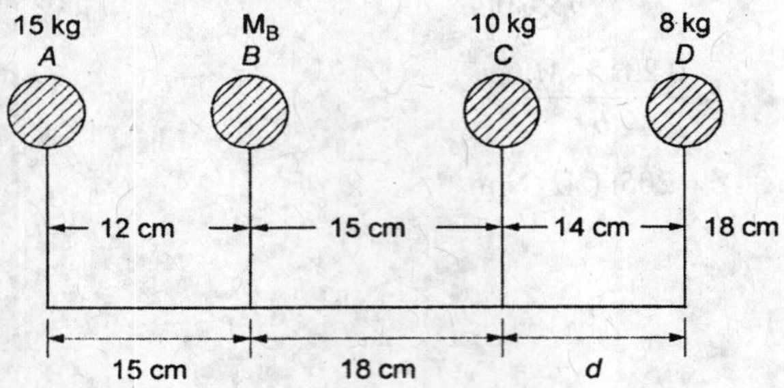

(a) *Position of planes*

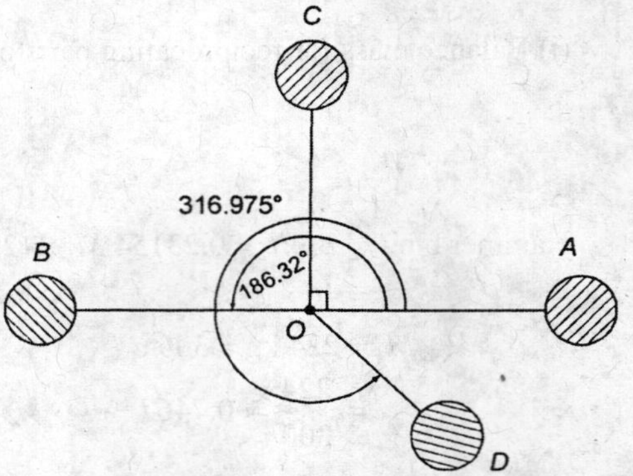

(b) *Position of cranks.*

Fig.12.53 Shaft carrying four masses

From Table 12.35, we have

$$\Sigma M_i r_i \cos\theta_i = 1.8 + 1.44\cos\theta_4$$

$$\Sigma M_i r_i \sin\theta_i = 1.4 + 1.44\sin\theta_4$$

$$\Sigma M_i r_i l_i \cos\theta_i = -0.27$$

$$\Sigma M_i r_i l_i \sin\theta_i = 0.252$$

$$[(\Sigma M_i r_i l_i \cos\theta_i)^2 + (\Sigma M_i r_i l_i \sin\theta_i)^2]^{0.5} = M_2 r_2 l_2$$

$$\left[(-0.27)^2 + (0.252)^2\right]^{0.5} = 1.44\, l_4$$

$$l_4 = \frac{0.36933}{1.44} = 0.2565 \text{ m} \quad \text{or} \quad 25.65 \text{ cm}$$

$$d = 25.65 - 18 = 7.65 \text{ cm}$$

$$\tan\theta_4 = \frac{\Sigma M_i r_i l_i \sin\theta_i}{-\Sigma M_i r_i l_i \cos\theta_i} = \frac{-(0.252)}{-(-0.27)} = 0.93333$$

$$\theta_4 = 43.025° \text{ i.e. } 316.975°$$

Since the numerator is negative and denominator is positive, therefore θ_4 lies in the fourth quadrant.

Now $M_2 r_2 = [(\Sigma M_i r_i \cos\theta_i + M_M r_M \cos\theta_M)^2 + [(\Sigma M_i r_i \sin\theta_i + M_M r_M \sin\theta_M)^2]^{0.5}$

$$0.15M_2 = \left[(1.44\cos 316.975° + 1.8)^2 + (1.44\sin 316.975° + 1.4)^2\right]^{0.5}$$

$$= \left[(2.853)^2 + (0.41746)^2\right]^{0.5}$$

$$= 2.8834$$

$$M_2 = 19.22 \text{ kg}$$

$$\tan\theta_2 = \frac{(\Sigma M_i r_i \sin\theta_i + M_M r_M \sin\theta_M)}{-(\Sigma M_i r_i \cos\theta_i + M_M r_M \cos\theta_M)} = \frac{-041746}{-2.853} = 0.14632$$

$$\theta_2 = 8.32° \text{ i.e. } 188.32°$$

Since the numerator and denominator are both positive, therefore θ_2 lies in the third quadrant.

Example 12.36

Determine the bearing reactions for a system of four unbalance masses, shown in Fig.12.54. The rotor speed is 600 rpm.

Solution

Reference plane 1 (refer to Fig.12.54)

$$\omega = \frac{2\pi N}{60} = \frac{2\pi \times 600}{60} = 62.832 \text{ rad/s}$$

$$\omega^2 = 3947.84 \ (\text{rad/s})^2$$

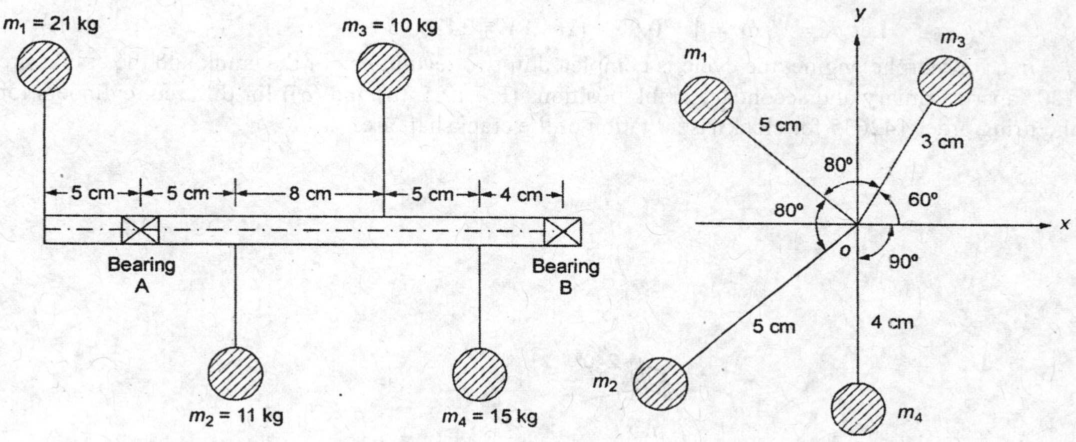

(a) *Position of planes* (b) *Position of masses*

Fig.12.54 Shaft carrying for masses

Table 12.36

Plane	M (kg)	r (cm)	Mr (kg cm)	θ (deg)	Mr cos θ	Mr × sin θ	l (cm)	Mrl	Mrl × cos θ	Mrl × sin θ
1	21	5	105	140	– 107.25	67.49	0	0	0	0
2	11	5	55	220	– 42.13	– 35.35	10	550	– 421.32	– 353.53
3	10	3	30	60	15.0	25.98	18	540	270.00	467.65
4	15	4	60	270	0	– 60	23	1380	0	– 1380

For dynamic balance of the system, taking moments about bearing *B,* we have

$$R_A \times 22 = (105 \times 27 + 55 \times 17 + 30 \times 9 + 60 \times 4) \times 9.81 \times 10^{-2} \times 3947.84$$

$$R_A = 75344.17 \text{ N}$$

$$R_A + R_B = (105 + 55 + 30 + 60) \times 9.81 \times 10^{-2} \times 3947.84 = 96820.77 \text{ N}$$

$$R_B = 21476.6 \text{ N}$$

Example 12.37

The firing order in a six cylinder vertical four-stroke in-line engine is 1– 4 – 2 – 6 – 3 – 5. The piston stroke is 100 mm and the length of each connecting rod is 200 mm. The pitch distances between the cylinder centre lines are 100, 100, 150, 100 and 100 mm, respectively. The reciprocating mass per cylinder is 1 kg and the engine runs at 2500 rpm. Determine the out-of-balance primary and secondary forces and couples on this engine, taking a plane midway between cylinders 3 and 4 as the reference plane.

■ Solution

Given: $L = 100$ mm or $r = 50$ mm, $l = 200$ mm, $M = 1$ kg, $N = 2500$ rpm

$$\omega = \frac{2\pi \times 2500}{60} = 261.8 \text{ rad/s}$$

$$\text{Let } K = Mr\omega^2 = 1 \times 0.05 \times (261.8)^2 = 3426.95 \text{ N}$$

In a four-stroke engine, the cycle is completed in two revolutions of the crank and the cranks are 120° apart. Primary and secondary crank positions (Fig.12.55(*a*) and (*d*)) for different cylinders for the firing order 142635 for clockwise rotation of the crankshaft are:

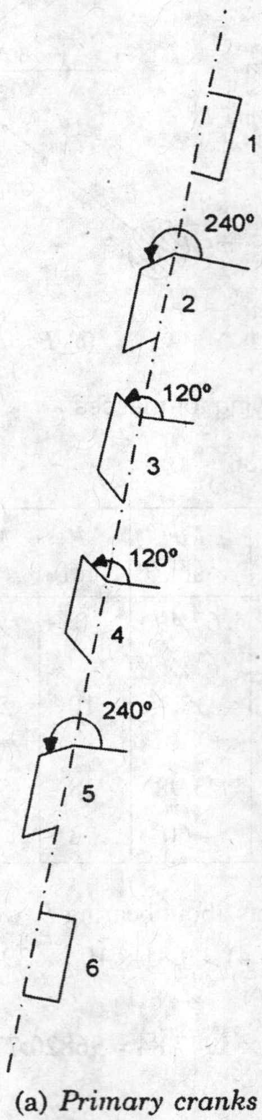

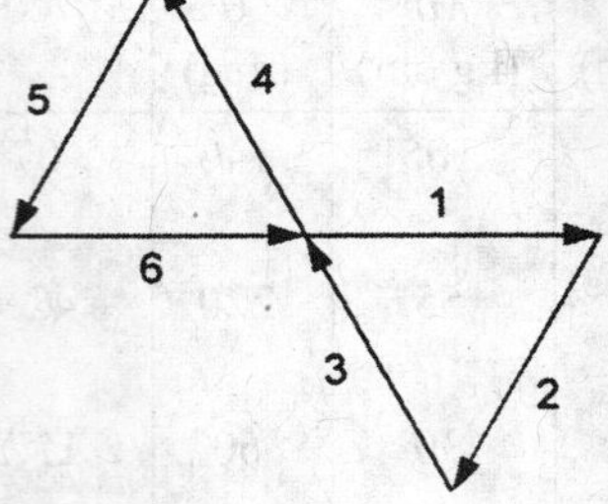

(a) *Primary cranks* (b) *Primary force polygon*

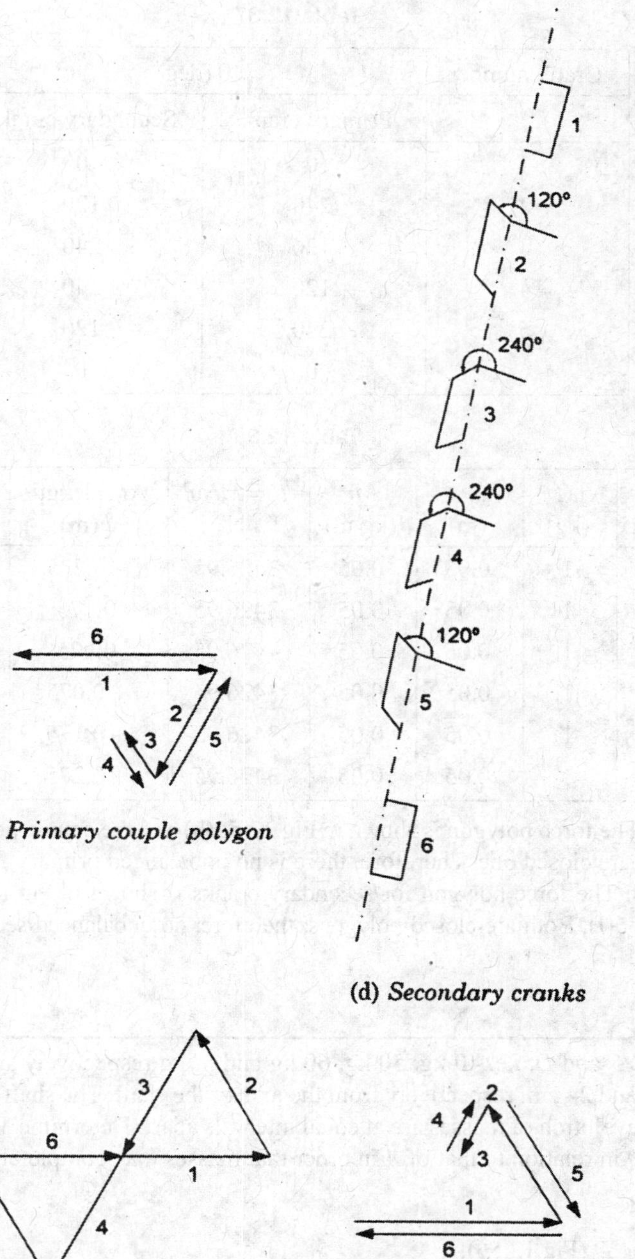

(c) *Primary couple polygon*

(d) *Secondary cranks*

(e) *Secondary force polygon*

(f) *Secondary couple polygon*

Fig 12.55 Firing order 1–4–2–6–3–5 engine

Table 12.37

Crank number	θ (deg)	
	Primary cranks	Secondary cranks
1	0	0
2	240	120
3	120	240
4	120	240
5	240	120
6	0	0

Table 12.38

Plane of cylinders	M (kg)	r (m)	Mr (kg m)	$K = Mr\omega^2$ N	Arm length l (m)	*Couple* Kl (N m)
1	1	0.05	0.05	3426.95	– 0.275	0.275 K
2	1	0.05	0.05	3426.95	0.175	0.175 K
3	1	0.05	0.05	3426.95	0.075	0.075 K
4	1	0.05	0.05	3426.95	+ 0.075	+ 0.075 K
5	1	0.05	0.05	3426.95	+ 0.175	+ 0.175 K
6	1	0.05	0.05	3426.95	+ 0.275	+ 0.275 K

Primary cranks: The force polygon is shown in Fig.12.55(b) and the couple polygon in Fig.12.55(c). Both the polygons are closed one. Therefore, there is no unbalanced primary force and couple.

Secondary cranks: The force polygon for secondary cranks is shown in Fig.12.55(e) and the couple polygon in Fig.12.55(f). Both are closed polygons, therefore, no unbalanced secondary force and couple.

Example 12.38

Four masses, *A, B, C* and *D*, i.e. 40 kg, 50 kg, 60 kg and *M* kg respectively are rigidly connected to shaft at 30, 24, 28 and 24 cm, respectively from the axis of the shaft. The shaft revolves about its axis and the planes of revolution of masses are at equal intervals apart. Determine *M* and the angular positions of *B, C* and *D* in relation to that of *A* in order that masses may completely balance one another.

■ Solution

Reference plane *D* (Fig.12.56)

Table 12.39

Plane	M (kg)	r (m)	Mr (kg m)	l (m)	Mrl (kg m²)
A	40	0.30	12	– 3*d*	–36 *d*
B	50	0.24	12	–2*d*	–24 *d*
C	60	0.28	16.8	–*d*	–16.8 *d*

D	M	0.24	0.24 M	0	0

Couple Polygon: Assume the position of mass A in horizontal direction to the right. Draw the couple polygon shown in Fig.12.56(c) as described below from the data in column 6 of Table 12.39.

1. Draw $o'a' = 36d$ horizontally to the left, being negative, to a scale of 1 cm = $5d$ units.
2. From points o' and a', draw arcs $o'b' = 24d$ and $a'b' = 16.8d$.
3. By measurement, we find that $\theta_B = 238°$ and $\theta c = 33°$

Force polygon: To find the magnitude of mass M at D and its angular location, draw the force polygon as described below from the data in column 4 of Table 12.39.

1. Draw oa = 12 units to the right to a scale of 1 cm = 2 units.
2. Draw ab = 12 units and parallel to $a'b'$.
3. Draw bc = 16.8 units and parallel to $o'b'$.
4. Join oc. Then $oc = 0.24\ M$. By measurement oc = 7.8 cm = 15.6 units

$$0.24\,M = 15.6$$
$$M = 65 \text{ kg}$$
$$\text{and } \theta_D = 200°$$

The angular position of the masses B, C and D in relation to A have been shown in Fig.12.56(b).

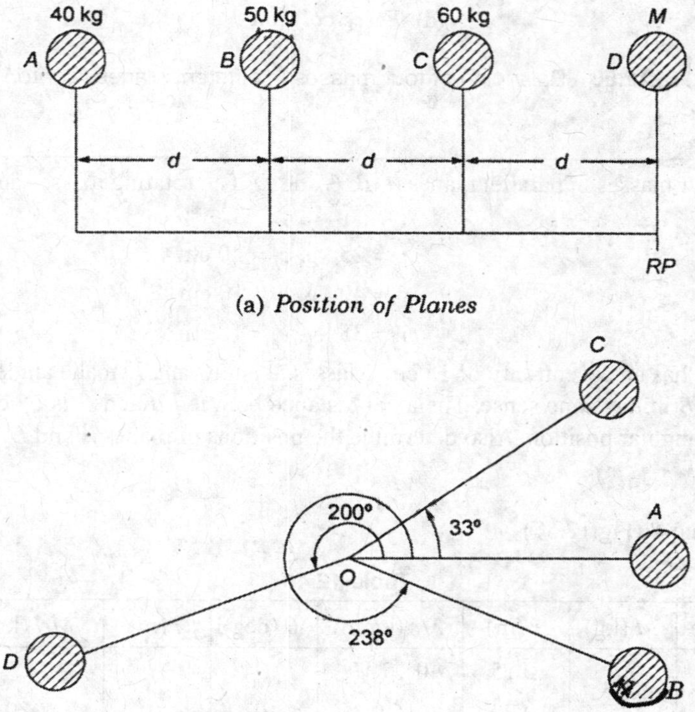

(a) *Position of Planes*

(b) *Angular position of masses*

Fig.12.56 Balancing of four masses in different planes

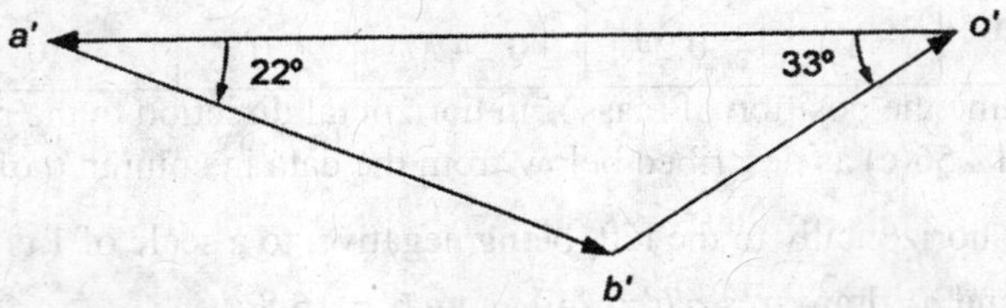

(c) *Couple polygon*

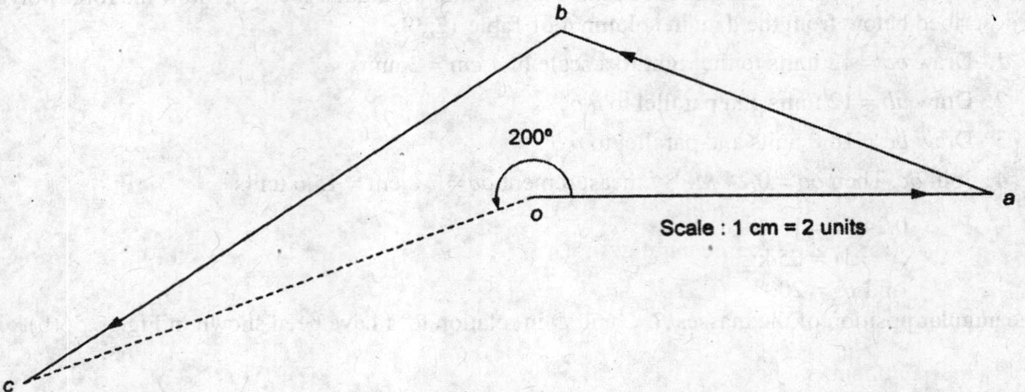

(d) *Force polygon*

Fig.12.56 Balancing of four masses in different planes (*Contd.*)

Example 12.39

A shaft carries four masses in parallel planes *A, B, C* and *D*. The rotating masses and their eccentricities are:

$$m_b = 25 \text{ kg}, e_b = 20 \text{ cm}$$
$$m_c = 40 \text{ kg}, e_c = 10 \text{ cm}$$
$$m_d = 35 \text{ kg}, e_d = 18 \text{ cm}$$

The mass at *A* has an eccentricity of 15 cm. Masses at and *C* and *D* make angles of 90° and 195° respectively with *B* in the same sense. The axial distance between *B* and *C* is 25 cm. Determine the mass at *A* and its angular position. Also determine the positions of planes *A* and *D*.

■ **Solution**

Reference plane *B* (Fig.12.57)

Table 12.40

Plane	*M* (kg)	*r* (m)	*Mr* (kg m)	θ (deg)	*l* (m)	*Mrl* (kg m²)
A	M_1	0.15	$0.15\,M_1$	θ_1	$-l_1$	$-0.15\,M_1l_1$
B	25	0.20	5.0	0	0	0
C	40	0.10	4.0	90	0.25	1.0
D	35	0.18	6.3	195	l_4	$6.3\,l_4$

Force polygon: Draw the force polygon shown in Fig.12.57(c) with the help of data in column 4 of Table 12.40.

1. Draw $ob = 5$ units parallel to OB on a scale of 1 cm = 1 unit.
2. From b draw $bc = 4$ units parallel to OC.
3. From c draw $cd = 6.3$ units parallel to OD.
4. Join d with o. Then od = balanced force. By measurement, we have

$$0.15M_1 = od = 2.5$$
$$M_1 = 16.67 \text{ kg}$$

To locate the angular position of A, draw OA from O in Fig.12.57(b) parallel to od. $\angle DOA = 98°$

Couple polygon: Draw the couple polygon shown in Fig.12.57(d) with the help of data in column 6 of Table 12.40.

1. Draw $o'c' = 1.0$ units parallel to OC.
2. From c' draw a line parallel to OD.
3. From o' draw another line parallel to OA to intersect the above line at d'.

Then $c'd' = 6.3l_4$

$$o'd' = -1.5M_1l_1$$

By measurement, we have

$$-0.15M_1l_1 = 1.0 \text{ units}$$

$$l_1 = \frac{-1.0}{0.15 \times 16.67} = -0.4 \text{ m} \quad \text{or} \quad -40 \text{ cm}$$

–ve sign indicates that the plane of A is towards right of B and to the left as assumed.

$$c'd' = 0.4 \text{ units}$$
$$1.3\, l_4 = 0.4$$
$$l_4 = 0.0635 \text{ m or } 6.35 \text{ cm}$$

Distance between the planes of C and $D = 6.35 - 25 = -18.65$ cm *i.e.* D is to the left of C and not to the right as assumed.

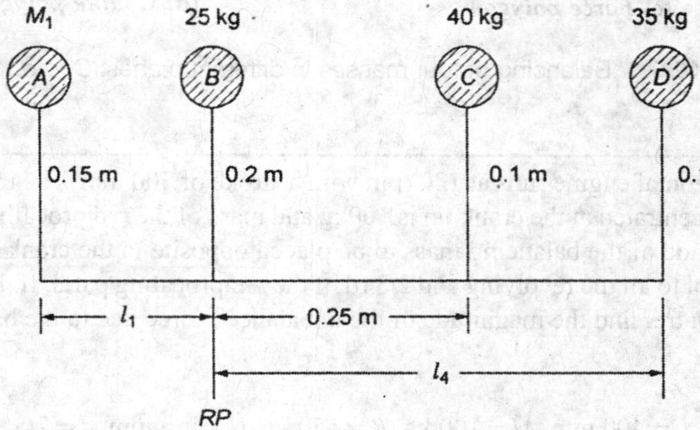

(a) *Position of planes.*

Fig.12.57 Balancing of four masses in different planes

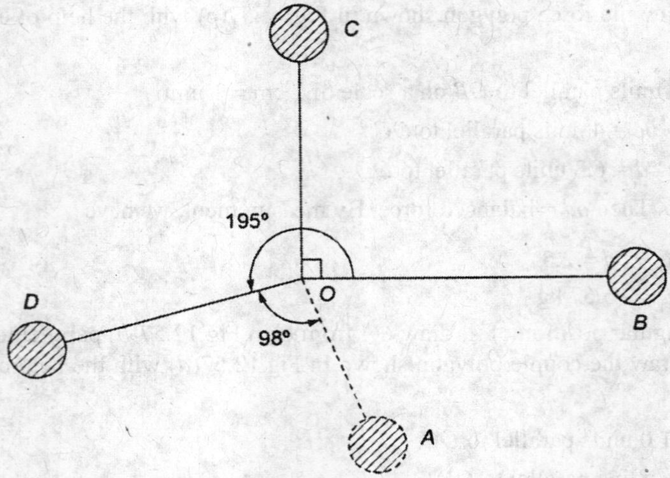

(b) *Angular position of masses.*

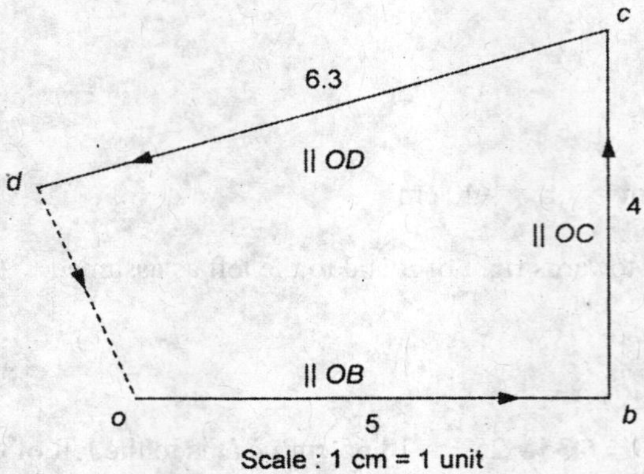

(c) *Force polygon*

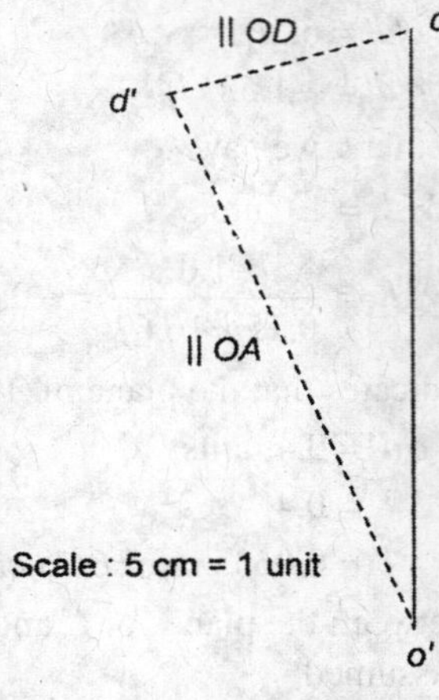

(d) *Couple polygon*

Fig.12.57 Balancing of four masses in different planes (*Contd.*)

Example 12.40

A single cylinder horizontal engine runs at 120 rpm with a stroke of 400 mm. The mass of the revolving parts assumed concentrated at the crankpin is 100kg and mass of the reciprocating parts is 150 kg. Determine the magnitude of the balancing mass to be placed opposite to the crank at a radius of 150 mm which is equivalent to all the revolving and 2/3 rd of the reciprocating parts. If the crank turns 30° from the inner dead centre, find the magnitude of the unbalanced force due to the balancing mass.

Solution

$N = 120$ rpm, $L = 400$ mm, $M = 100$ kg, $R = 150$ kg, $b = 150$ mm, $c = 2/3$, $\theta = 30°$

$Bb = (M + c\,R)\,r$

$0.15\,B = (100 + 2 \times 150/3) \times 0.2$

$B = 266.67$ kg

$\omega = 2 \times 120 / 60 = 12.57$ rad/s

Residual unbalanced force $= R\omega^2 r\,[(1 - c)^2\cos^2\theta + c^2 \sin^2\theta]^{0.5}$

$= 150 \times (12.57)^2 \times 0.2\,[(1 - 2/3)^2 \times \cos^2 30° + 4 \times \sin^2 30°/9]^{0.5}$

$= 2090.2$ N

Example 12.41

A single cylinder engine runs at 240 rpm and has a stroke of 200 mm. The reciprocating parts have a mass of 120 kg and the revolving parts are equivalent to a mass of 80 kg at a radius of 100 mm. A mass is placed opposite to the crank at a radius of 150 mm to balance the whole of the revolving mass and 2/3rd of the reciprocating mass. Determine the magnitude of the balancing mass and the resultant residual unbalance force when the crank has turned 30° from the inner dead centre. Neglect the obliquity of the connecting rod.

■ Solution

$N = 240$ rpm, $L = 200$ mm, $M = 80$ kg, $R = 120$ kg, $b = 150$ mm, $c = 2/3$, $\theta = 30°$

$B\,b = (M + c\,R)\,r$

$0.15\,B = (80 + 2 \times 120 / 3\,) \times 0.1$

$B = 106.67$ kg

$\omega = 2 \times 240 / 60 = 25.13$ rad/s

Residual unbalanced force $= R\omega^2 r\,[(1 - c)^2\cos^2\theta + c^2 \sin^2\theta]^{0.5}$

$= 120 \times (25.13)^2 \times 0.1\,[(1\text{–}2/3)^2 \times \cos^2 30° + 4 \times \sin^2 30°/9]^{0.5}$

$= 3341.7$ N

Example 12.42

A shaft with 3 m span between two bearings carries two weights of 100 N and 200 N acting at the extremities of arms 0.45 m and 0.60 m long, respectively. The planes in which these weights rotate are 1.2 and 2.4 m, respectively from the left end bearing supporting the shaft (Fig.12.58). The angle between these arms is 60°as indicated in the inset (a) of Fig.12.58. The speed of rotation of the shaft is 200 rpm. If the weights are balanced by two counter- weights rotating with the shaft acting at radii of 0.3 m and placed at 0.3 m from each bearing centre, estimate the magnitude of the two balance weights and their orientation with respect to the *x*-axis, i.e. load *A*.

■ Solution

Given: $r_1 = r_4 = 0.3$ m, $r_2 = 0.45$ m, $r_3 = 0.6$ m, $W_2 = 100$ N, $W_3 = 200$ N

$l_2 = 0.9$ m, $l_3 = 2.1$ m, $l_4 = 2.4$ m

Taking 1 as the reference plane, we have

$100 \times 0.45 \times 0.9 \cos 0° + 200 \times 0.6 \times 2.1 \cos 60° + W_4 \times 0.3 \times 2.4 \cos \theta_4 = 0$

$40.5 + 126 + 0.72\,W_4 \cos \theta_4 = 0$

$W_4 \cos \theta_4 = -231.25$

$$100 \times 0.45 \times 0.9 \sin 0° + 200 \times 0.6 \times 2.1 \sin 60° + W_4 \times 0.3 \times 2.4 \sin \theta_4 = 0$$

$W_4 \sin \theta_4 = -303$

$\tan \theta_4 = -303/-231.25 = 1.31$

$\theta_4 = 232.66°$

$W_4 = 381$ N

Taking 4 as the reference plane, we have

$$100 \times 0.45 \times 1.5 \cos 0° + 200 \times 0.6 \times 0.3 \cos 60° + W_1 \times 0.3 \times 2.4 \cos \theta_4 = 0$$

$W_1 \cos \theta_1 = -118.75$

$$100 \times 0.45 \times 1.5 \sin 0° + 200 \times 0.6 \times 0.3 \sin 60° + W_1 \times 0.3 \times 2.4 \sin \theta_4 = 0$$

$W_1 \sin\theta_1 = -43.3$

$\tan \theta_4 = -43.3 / -118.75 = 0.3646$

$\theta_1 = 200°$

$W_1 = 126.6$ N

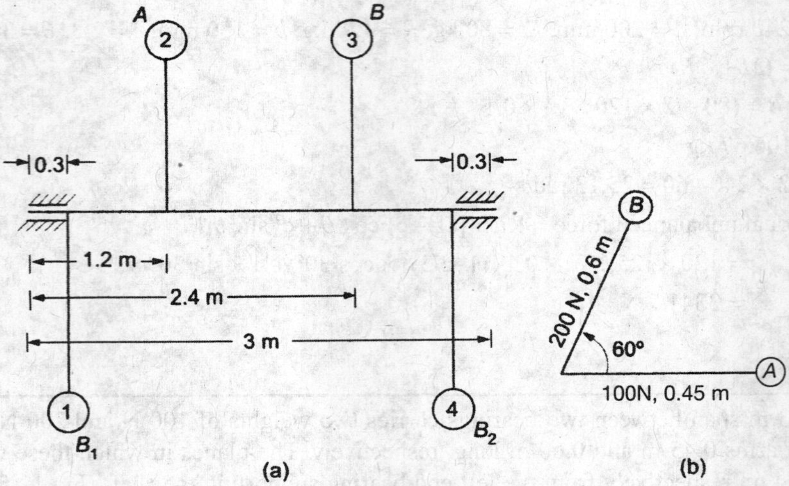

Fig.12.58 Shaft carrying two weights

Example 12.43

A shaft is supported in bearings 2 m apart and projects 0.5 m beyond bearings at each end. The shaft carries three pulleys one at each end and one at the middle of its length. The mass of end pulleys is 50 kg and 20 kg and their centre of gravity are 20 and 15 mm, respectively from the shaft axis. The centre pulley has a mass of 55 kg and its centre of gravity is 15 mm from the shaft axis. If the pulleys are arranged so as to give static balance, determine (*a*) relative angular positions of the pulleys, and (*b*) dynamic forces produced on the bearings when the shaft rotates at 300 rpm.

Solution

Given: $m_A = 50$ kg, $m_B = 55$ kg, $m_C = 20$ kg, $r_A = 20$ mm, $r_B = r_C = 15$ mm, $N = 300$ rpm

The position of shaft and pulleys is shown in Fig.12.59(a).

Let M_L, M_M = mass at the bearings L and M

r_L, r_M = radium of rotation of masses m_L and m_M, respectively.

Table 12.41

Plane (1)	Mass *m (kg)* (2)	Radius *r* (m) (3)	*Mr (kg m)* (4)	*l (m)* (5)	*Mrl (kg m²)* (6)
A	50	0.020	1 0	– 0.5	– 0.5
L(RP)	M_L	r_L	$M_L r_L$	0	0
B	55	0.015	0.825	1.0	0.825
M	M_M	r_M	$M_M r_M$	2.0	$2M_M r_M$
C	20	0.015	0.30	2.5	0.75

$$\omega = \frac{2\pi N}{60} = \frac{2\pi \times 300}{60} = 31.42 \text{ rad/s}$$

Draw the force polygon as shown in Fig.12.59(c) from the data in column (4).

ob = 0.825 kg m is a vertical line. *bc* = 1.0 and *oc* = 0.30.

In Fig.12.59(b), draw *OA* parallel to *BC* and *OC* parallel to *OC*.

$\angle AOB = 165°$, $\angle BOC = 61°$, $\angle AOC = 134°$

Draw couple polygon as shown in Fig.12.59(d).

o'b' = 0.825 kg.m² is vertical, *b'a'* parallel to OA and *a'c'* parallel to *OC*.

$c'o' = 2\, M_M r_M = 9.2$ cm = 1.84 kg. m²

$M_M r_M = 0.92$ kg m

Dynamic force on bearing $M = M_M r_M w^2 = 0.92 \times (31.42)^2 = 908.24$ N

Now draw force polygon as shown in Fig.12.59(e).

Ob = 0.825 kg.m is a vertical line. *bm* || *o'c'* and bm = $M_M r_M$ = 0.92 kg.m, mc || oc, mc = 0.3 kg m, *cd* || *OA* and *cd* = 1.0 kg.m.

$Od = M_L r_L = 4.6$ cm = 0.92 kg m

Dynamic force on bearing $L = M_L r_L w^2 = 0.92 \times (31.42) = 908.24$ N

Example 12.44

A shaft carries four masses in parallel planes *A*, *B*, *C* and *D* in order. The masses at *B* and *C* are 18 kg and 12.5 kg respectively and each has an eccentricity of 6 cm. The masses at *A* and *D* have an eccentricity of 8 cm. The angle between the masses at *B* and *C* is 100° and that between the masses at *B* and *A* is 190° (both angles measured in the same sense). The axial distance between the planes *A* and *B* is 10 cm and between *B* and *C* 20 cm. If the shaft is in complete dynamic balance, determine

(i) the masses at *A* and *D*

(ii) the distance between the planes *C* and *D*

(iii) the angular position of the mass at *D*.

[IAS, 2004]

Solution

Given: $M_B = 18$ kg, $M_C = 12.5$ kg, $r_B = r_C = 60$ mm, $r_A = r_D = 80$ mm,

$\angle BOC = 100°$, $\angle BOA = 190°$

Let M_A, M_D = mass at A and D

L = distance between A and D

Table 12.42

Plane (1)	Mass *m* (kg) (2)	Eccentricity, *r* (m) (3)	*Mr* (kg m) (4)	*L* (m) (5)	*Mrl* (kg m²) (6)
A (RP)	M_A	0.08	$0.08\ M_A$	0	0
B	18	0.06	1.08	0.1	0.108
C	12.5	0.06	0.75	0.3	0.225
D	M_D	0.08	$0.08\ M_D$	*L*	$0.08\ M_D L$

Draw couple polygon as shown in Fig.12.60(c) from the data in column (6) of Table 12.42

$o'c' = 0.08\ M_D L = 5.9$ cm $= 0.236$ kg.m²

In Fig.12.60(b), draw *OD* parallel to *o'c'*. $\angle AOD = 71°$ so that $\angle BOD = 251°$

Now draw the force polygon as shown in Fig.12.60(d) from the data in column (4) of Table 12.42.

1. Draw $ob \parallel OB$, $ob = 1.08$ kg m
2. Draw $bc \parallel OC$, $bc = 0.75$ kg m
3. Draw $cd \parallel OA$ and $od \parallel OD$ to meet at *d*.

$cd = 0.08\ M_A = 3.9$ cm $= 0.78$ kg m, $M_A = 9.75$ kg

$od = 0.08\ M_D = 3.3$ cm $= 0.66$ kg m, $M_D = 8.25$ kg

$0.08 \times 8.25 \times L = 0.236$

$L = 0.3576$ m or 35.76 cm

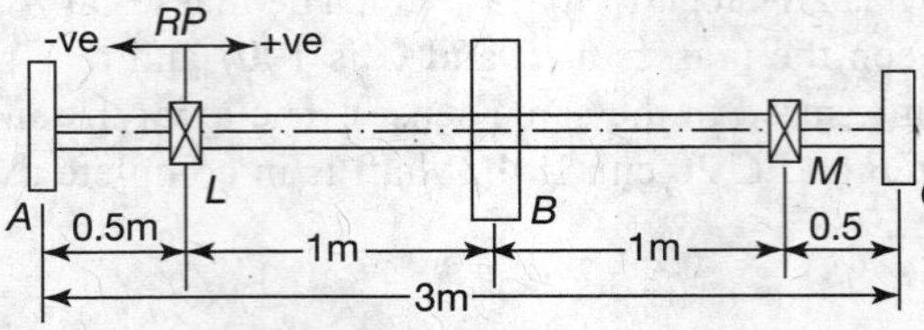

(a) Position of pulleys

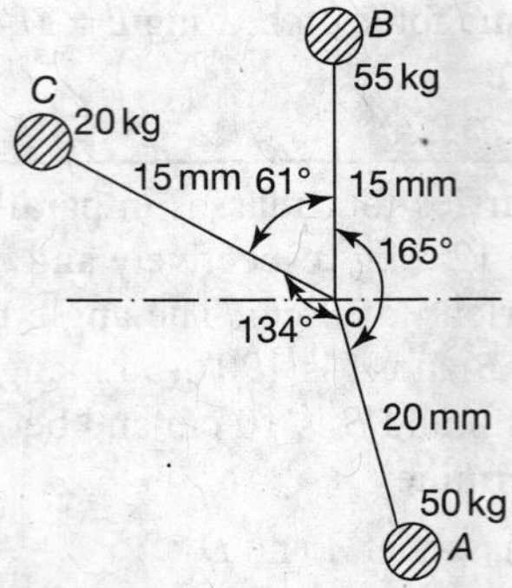

(b) Angular position of pulleys

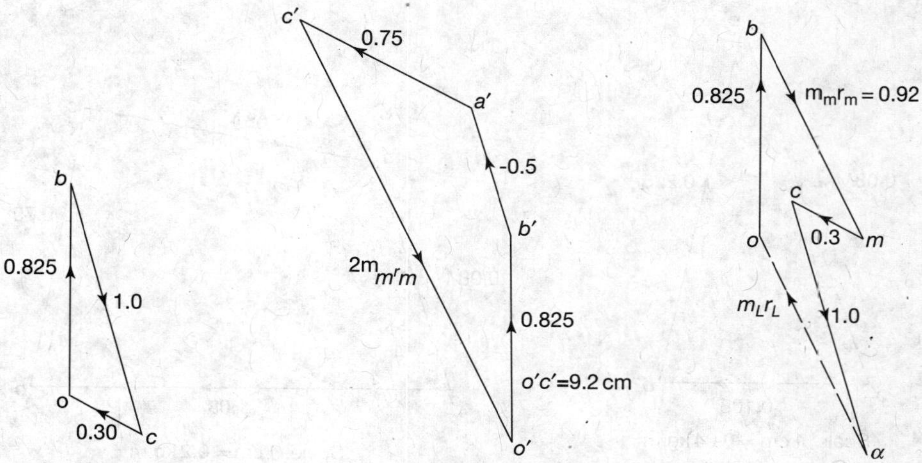

Scale: 1 cm = 0.2 kg.m (c) Force polygon

Scale: 1 cm = 0.2 kg.m² (d) couple polygon

Scale: 1 cm = 0.2 kg.m (e) Force polygon

Fig.12.59 Graphical method for Example 12.43

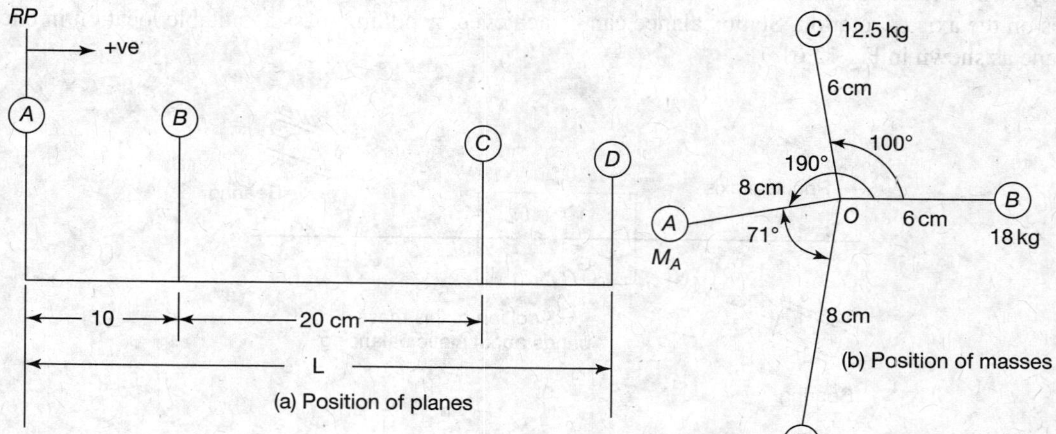

(a) Position of planes

(b) Position of masses

Fig.12.60 Graphical method for Example 12.44

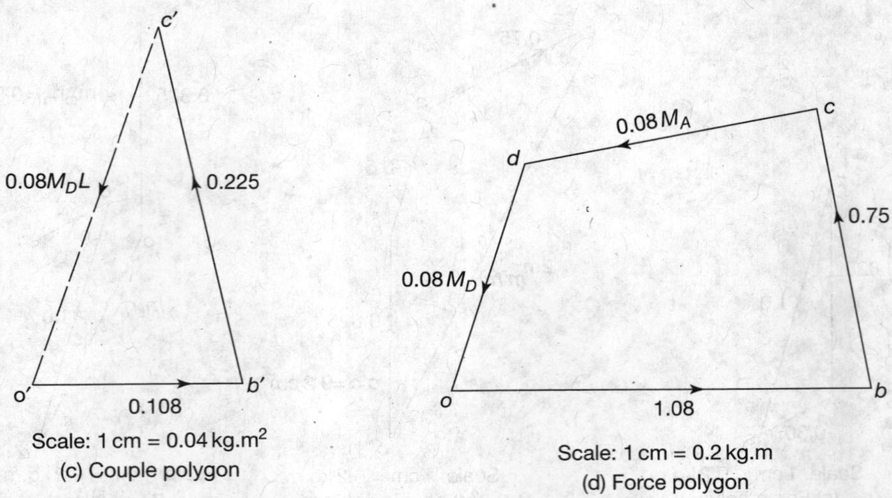

Fig.12.60 Graphical method for Example 12.44 (*Contd.*)

12.9 BALANCING OF ROTORS

A balancing machine is a device to indicate whether a component is in balance or not and if it is not, then to measure the unbalance by indicating its magnitude and location.

12.9.1 Static Balance

A system of rotating masses is said to be in static balance if the combined mass centre of the system lies on the axis of rotation. Static balance can be achieved by adding mass at suitable location in one plane as shown in Fig.12.61.

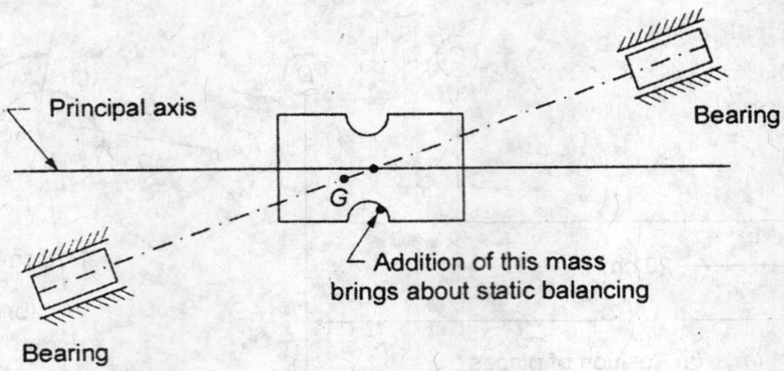

Fig.12.61 Static balancing

12.9.2 Dynamic Balance

A system of rotating masses is said to be in dynamic balance when there does not exist any resultant centrifugal force as well as resultant couple. Dynamic balance is said to have been achieved when the principal axis of the rotor coincides with its axis of rotation. It is implied in the definition of dynamic balance that a dynamically balanced rotor must be statically balanced but a statically balanced rotor may not necessarily be dynamically balanced rotor.

12.9.3 Flexible Rotor Balancing

When the balancing is achieved considering the flexibility of elements like bearings, rotor, lubricant etc., the process is called flexible rotor balancing.

12.9.4 Balancing Machines

Machines that help determine the unbalance and reduce it are called balancing machines.

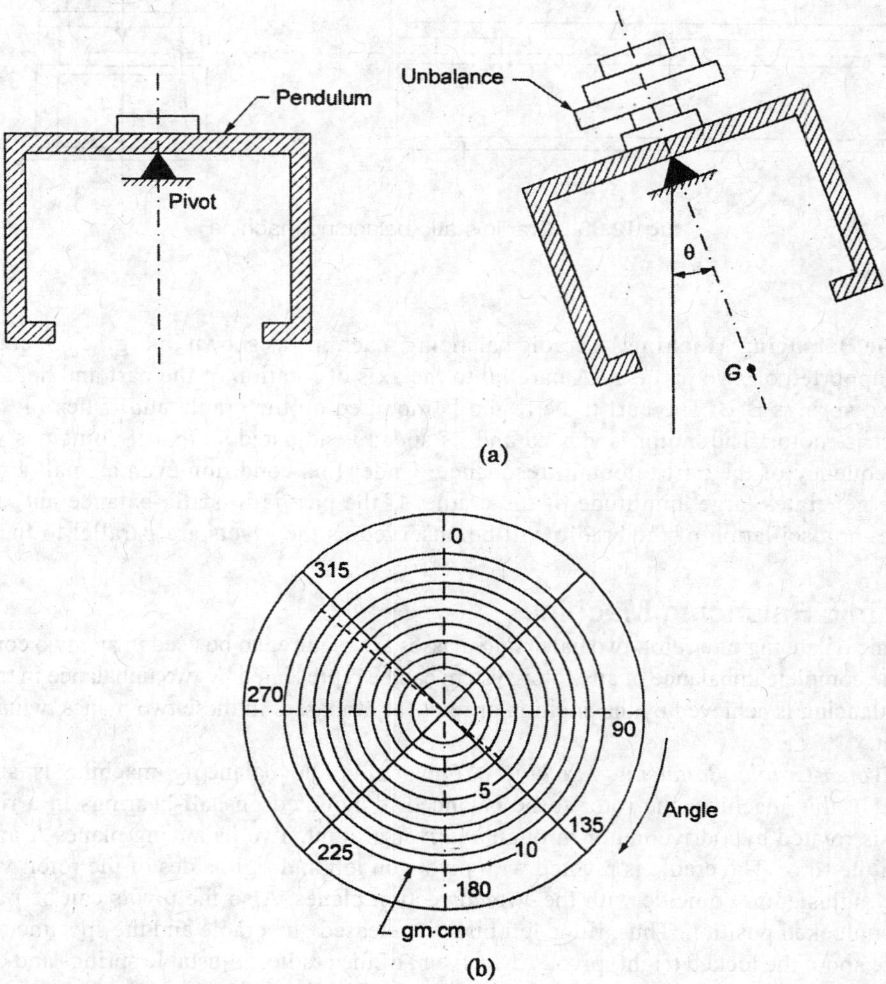

Fig.12.62 Pendulum static balancing machine

1. Static Balancing Machine

(a) *Pendulum Balancing Machine*: A pendulum balancing machine, as shown in Fig.12.62 is a single plane or static balancing machine and is used for small rotors in mass production. In such a case, the reading for the location of the unbalance and its amount could be directly made. Pendulum balancing machines are based on the principle that when an unbalanced rotor is placed on a pendulum, the pendulum will tilt by an amount depending on the extent of unbalance. Usually it is tilted with a universal level indicator which gives the extent and location of unbalance directly by moving the bubble to different locations, as shown in Fig.12.62(b).

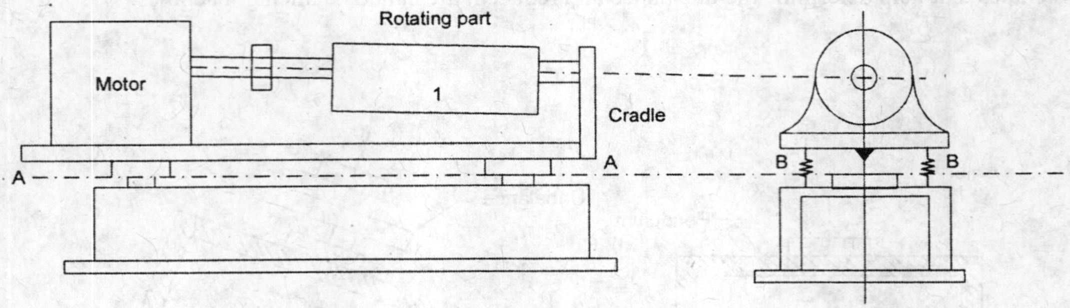

Fig.12.63 Cradle static balancing machine

(b) **Cradle Balancing Machine:** A cradle balancing machine, as shown in Fig.12.63, consists of a cradle supported on two pivots A-A parallel to the axis of rotation of the part and held in position by two springs B-B. The part to be tested is mounted on the cradle and if flexibly coupled to an electric motor. The motor is started and its speed is adjusted so that it coincides with the natural frequency of the part to obtain resonance. Under this condition even a small amount of unbalance generates large amplitude of the cradle. If the part is in static balance and dynamic unbalance, no oscillation of the cradle will be observed as the pivots are parallel to the axis of rotation.

2. Dynamic Balancing Machine

For dynamic balancing of a rotor, two balancing masses are required to be used in any two convenient planes. The complete unbalance of any rotor system can be represented by two unbalance in those two planes. Balancing is achieved by the addition or removal of masses in these two planes, whichever is convenient.

(a) *Pivoted-Cradle Balancing Machine*: A pivoted-cradle balancing machine is shown in Fig.12.64. In this machine, the rotor to be balanced is mounted on half-bearings in a rigid carriage and is rotated by a drive motor through a universal joint. Two balancing planes A and B are chosen on the rotor. The cradle is pivoted with pivots on left and right sides of the rotor which are purpiosely adjusted to coincide with the two correction planes. Also the pivots can be put in the locked or unlocked position. Thus, if the left pivot is released, the cradle and the specimen are free to oscillate about the locked (right) pivot. At each end of the cradle, adjustable springs and dashpots are provides to have a single degree of freedom system. Usually, their natural frequency is tuned to the motor speed.

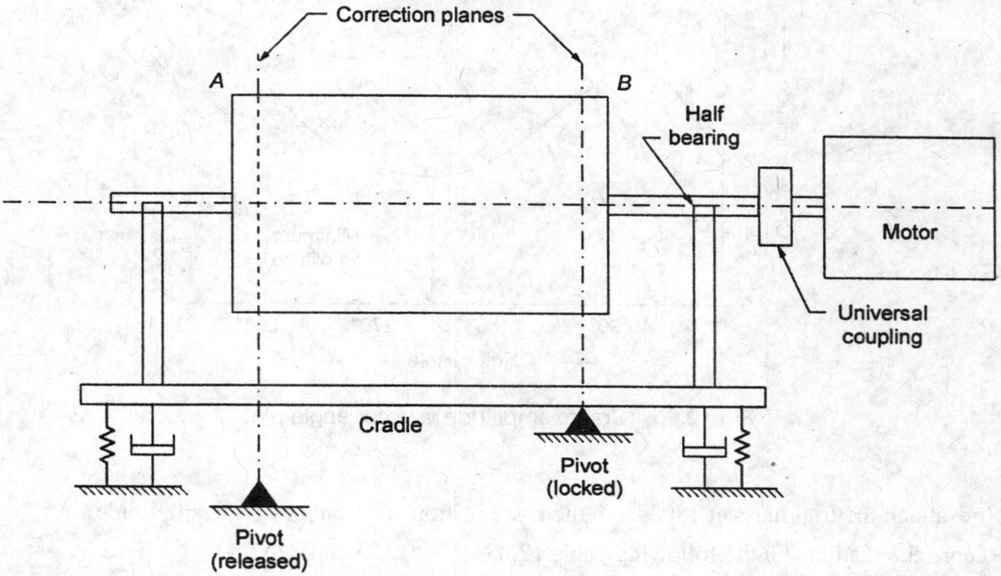

Fig.12.64 Pivoted-cradle dynamic balancing machine

Procedure for Testing

1. Lock the left (say) pivot and take the readings of the amount and angle of location of the correction in the right-hand plane.
2. Attach a trial mass at a known radius to the right hand plane and note the amplitude of oscillations of the cradle.
3. Repeat the procedure at various angular positions with the same trial mass.
4. Plot a graph of amplitude v's angular position of the trial mass to know the optimum angular position for which amplitude is minimum, as shown in Fig.12.65. Then at this position, vary the magnitude of the trial mass and find the exact amount by trial and error which reduces the unbalance to almost zero.
5. Now lock the cradle in the right-hand pivot and release the left hand pivot. Repeat the above procedure to obtain the exact balancing mass required in that plane.
6. This requires a large number of test runs to determine the exact balance mass and is very time consuming.

(b) Balancing by Four Observations
Procedure:

1. Make a test run without attaching any trial mass and note down the amplitude of the cradle vibrations.
2. Attach a trial mass m at some angular position and note down the amplitude of the cradle vibrations by rotating the rotor at the same speed.
3. Detach the trial mass from the present location and again attach it at 90° angular position relative to the first position and at the same radial distance.
4. Rotate the rotor at the same speed and note down the amplitude.

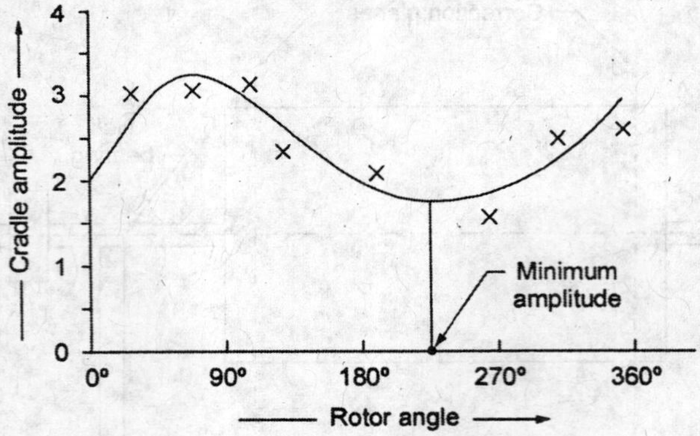

Fig.12.65 Cradle amplitude vs rotor angle plot

5. Now attach the trial mass at 180° from the first position and note down the amplitude.
6. Record the readings in the following Table 12.43.

Table 12.43

Trial mass	*Amplitude*
0	A_1
m at 0°	A_2
m at 90°	A_3
m at 180°	A_4

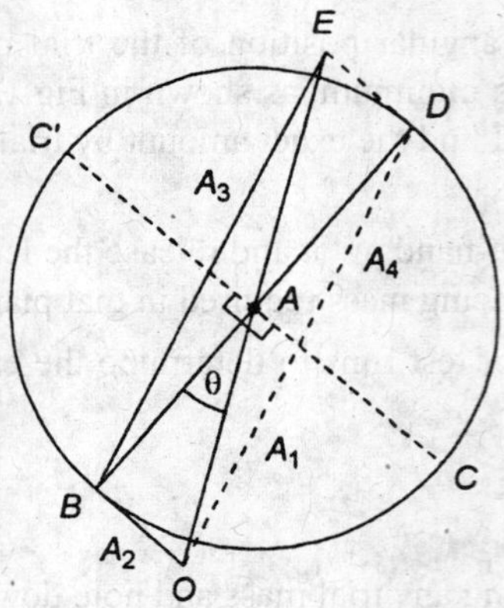

Fig.12.66 Balancing by four observations

Make the following construction, as shown in Fig.12.66.

1. Draw a triangle OBE by taking $OE = 2A_1$, $OB = A_2$ and $BE = A_4$.
2. Mark the mid-point A on OE. Join AB.

3. Now $OB = OA + AB$ (vector sum)
where OB = effect of unbalanced mass + effect of the trial mass m at 0°.
OA = effect of unbalanced mass
Thus AB = effect of the attached mass at 0°
Proof:

2. Extend BA to D such that $AD = AB$.
3. Join OD and DE.
4. When the mass m is attached at 180° at the same radial distance and speed, the effect must be equal and the opposite to the effect at 0°, i.e., if AB represents the effect of the attached mass at 0°, AD represents the effect of the attached mass at 180°.
5. Since $OD = OA + AD$ (vector sum), OD must represent the combined effect of unbalance mass and the trial mass at 180° (A_4).
6. Now, as the diagonals of the quadrilateral $OBED$ bisect each other at A, it is a parallelogram, which means BE is parallel and equal to OD. Thus, BE also represents the combined effect of unbalance mass and trial mass at 180° or A_4, which is true as it is made in the construction.
7. Now, as OA represents the unbalance, the correction has to be equal and opposite of it or OA thus, the correction mass is given by:

$$\frac{m_c}{m} = \frac{OA}{AB}$$

or $$m_c = m \times \frac{OA}{AB} \text{ at an angle } \theta \text{ from the second reading at } 0^\circ$$

For the correction of the unbalance, the mass m_c has to be put in the proper direction relative to AB which may be found by considering the reading A_3.
Draw a circle with A as centre and AB as the radius. As the trial mass as well as the speed of the test run at 90° is the same, the magnitude must be equal to AB or AD, and AC or AC' must represent the effect of the trial mass. If OC represents A_3, then angle measurement is taken in the same direction.

Example 12.45

During the field balancing of a cooling tower fan, the measurement taken are: $A_1 = 0.7$ mm, $A_2 = 1.06$ mm, $A_3 = 1.18$ mm and $A_4 = 0.5$ mm. The trial mass used is 250 g. Determine the necessary balancing mass (to be placed at the same radius as the trial mass) and its angular location with respect to the position of the trial mass during the second run when the vibration amplitude is 1.06 mm.

■ Solution

Refer to Fig.12.61
Draw a triangle OBE by taking $OE = 2A_1$, $OB = A_2$ and $BE = A_4$. Mark the mid-point A of OE. Join AB. Then $AB = 0.46$ mm and $\theta = 135°$.

$$\text{Correction mass, } m_c = \text{m.}\frac{OA}{AB} = 250 \times \frac{0.7}{0.46} = 380.4 \text{ gm.}$$

The balancing mass is attached 135° in the clockwise direction from that of the trial mass during the second run.

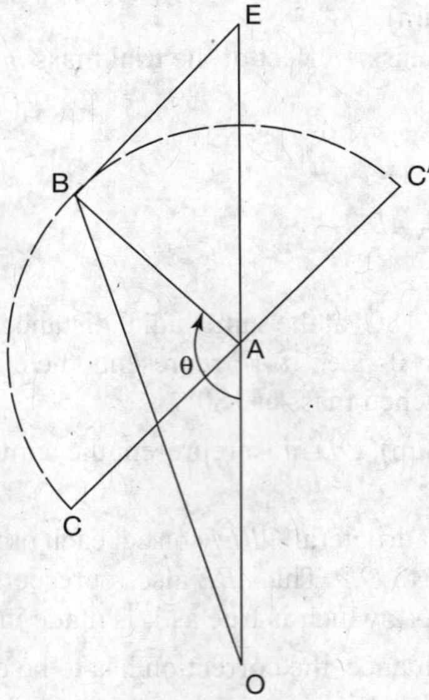

Fig.12.67 Field balancing of a fan

12.9.5 Field Balancing

In heavy machinery like turbines, compressors and generators, it is not possible to balance the motors by mounting them on the balancing machines. In such cases, the balancing of motors has to be done under actual conditions on their own bearings.

Consider two balancing planes *A* and *B* of a motor shown in Fig.12.62(a).

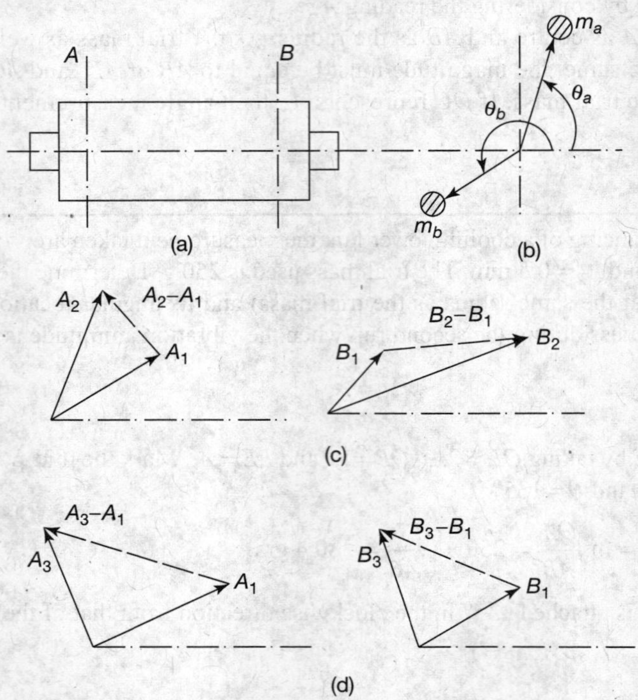

Fig.12.67 Field balancing

The following procedure is adopted to balance the motor:

1. First the motor is rotated at a speed which provides measurable amplitudes at planes A and B. Let the rectors A_1 and B_1 represent the amplitudes due to the unbalance of the motor in planes A and B respectively.
2. Attach a trial mass m_a in plane A at a known radius and angular position. Run the motor at the same speed as in the first case. Measure the amplitudes in the two planes A and B.

Let A_2 and B_2 represent the amplitudes of the motor in planes A and B respectively. Thus, A_2 = effect of unbalance of motor (A_1) + effect of trial mass in plane A.

Thus, effect of trial mass ma in plane $A = A_2 - A_1$, and effect of trial mass m in plane A at $B = B_2 - B_1$, as shown in Fig.12.62(c).

3. Make a third run of the motor by attaching a trial mass m_b in plane B at a known radius and angular position. Run the motor at the same speed as in the first two cases. Measure the amplitudes in the two planes A and B. Let A_3 and B_3 represent the amplitudes of the motor in planes A and B, respectively. Thus B_3 = effect of unbalance of motor B_1 + effect of trial mass in plane B.

Thus, effect of trial mass m_b in plane $B = B_3 - B_1$ and effect of trial mass m_b in plane $A = A_3 - A_1$ as shown in Fig.12.62(d).

Let m_{ca} and m_{cb} be the counter or balancing masses in planes A and B respectively placed at the same radii as the trial masses.

Let $m_{ca} = \alpha_{ma}$ and $m_{cb} = \beta m_b$

where $a = e^{i\theta_a}$ and $\beta = e^{i\theta_b}$

Thus $A_2 - A_1$ = effect of m_{ca} at plane A

$B_2 - B_1$ = effect of m_{ca} at plane B

and $A_3 - A_1$ = effect of m_{cb} at plane A

$B_3 - B_1$ = effect of m_{cb} at plane B

Let A_2 and B_2 represent the amplitudes of the motor in planes A and B respectively. Then effect of trial mass m_a in plane $A = A_2 - A_1$, and effect of trial mass m_a in plane $B = B_2 - B_1$. This is shown vectorially in Fig.12.62 (c)

4. Make a third run of the motor by removing mass m_a and attaching a trial mass m_b in plane B at a known radius and angular position. Run the motor at the same speed as in the first two cases. Measure the amplitudes in the two planes A and B. Let A_3 and B_3 represent the amplitudes of the motor in planes A and B, respectively. Then, effect of trial mass m_b in plane $B = B_3 - B_1$ and effect of trial mass m_b in plane $A = A_3 - A_1$. This is shown vectorially in Fig.12.62(d). Let m_{ca} and m_{cb} be the counter or balancing masses in planes A and B respectively placed at the same radii as the trial masses.

Let $m_{ca} = \alpha m_a$ and $m_{cb} = \beta m_b$

where $a = e^{i\theta_a}$ and $\beta = e^{i\theta_b}$

For complete balancing of the motor, the total effect of m_{ca} and m_{cb} at the plane A should be $-A_1$ and that at the plane B it should be $-B_1$. Thus

$\alpha (A_2 - A_1) + \beta (A_3 - A_1) = -A_1$

$\alpha (B_2 - B_1) + \beta (B_3 - B_1) = -B_1$

Solving for α and β simultaneously, we get

$$\alpha = \frac{B_1(A_3 - A_1) - A_1(B_3 - B_1)}{(A_2 - A_1)(B_3 - B_1) - (A_3 - A_1)(B_2 - B_1)}$$

$$\beta = \frac{A_1(B_2 - B_1) - B_1(A_2 - A_1)}{(A_2 - A_1)(B_3 - B_1) - (A_3 - A_1)(B_2 - B_1)}$$

Example 12.46

For balancing an alternator motor by the field-balancing technique, the experimental results obtained are listed in Table 12.44. Determine the correct balance masses that should be placed in two planes for complete dynamic balancing of the motor.

Table 12.44

Trial number	Trial mass (kg)	Plane *A*		Plane *B*	
		Amplitude cm	Phase angle deg	Amplitude cm	Phase angle deg
1	0	3.5×10^{-4}	15	4.2×10^{-4}	65
2	2 at plane A	4.5×10^{-4}	75	3.1×10^{-4}	81
3	2 at plane B	4.0×10^{-4}	35	2.2×10^{-4}	155

Solution

Fig.12.63(a) shows the vectors A_1, A_2 and A_3 to some scale and Fig.12.63(b) shows B_1, B_2 and B_3 drawn to the same scale.

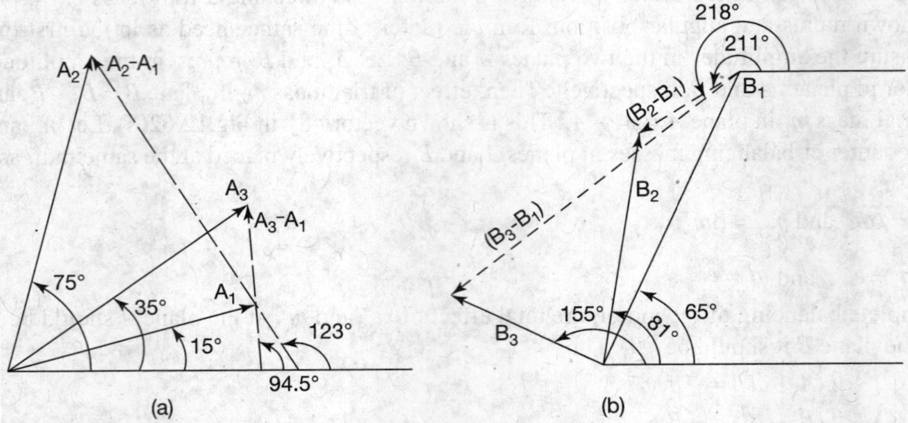

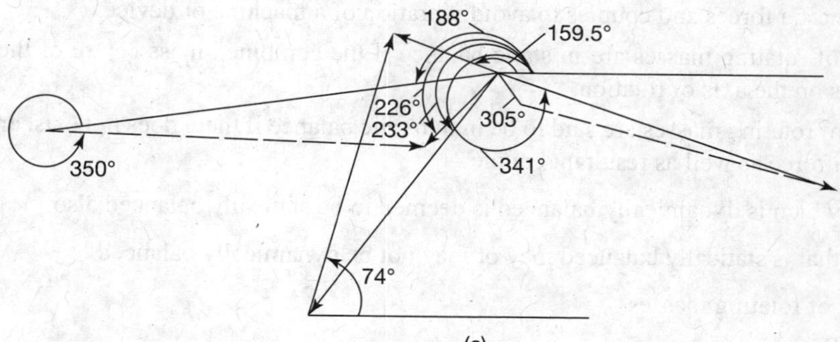

(c)

Fig.12.63 Field balancing of alternator rotor

From Figs. 12.63(a) and (b), we get

$$A_2 - A_1 = 4.1\times 10^{-4}\, e^{-i(123°)},\ A_3 - A_1 = 1.4\times 10^{-4}\, e^{-}$$

$$B_2 - B_1 = 1.5\times 10^{-4}\, e^{-i(211°)},\ B_3 - B_1 = 4.75\times 10^{-4}\, e^{-i(218°)}$$

From the given data

$$A_1 = 3.5\times 10^{-4}\, e^{-i(15°)},\ B_1 = 4.2\times 10^{-4}\, e^{-i(65°)}$$

$$\alpha = \frac{4.2e^{-i(65°)}\times 1.4e^{-i(94.5°)} - 3.5e^{-i(15°)}\times 4.75e^{-i(218°)}}{4.1e^{-i(123°)}\times 4.75e^{-i(218°)} - 1.4e^{-i(94.5°)}\times 1.5e^{-i(211°)}}$$

$$= \frac{5.88e^{-i(159.5°)} - 16.625e^{-i(233°)}}{19.475e^{-i(341°)} - 2.1e^{-i(305.5°)}}$$

$$\beta = \frac{3.5e^{-i(15°)}\times 1.5e^{-i(211°)} - 4.2e^{-i(65°)}\times 4.1e^{-i(123°)}}{19.475e^{-i(341°)}\times 2.1e^{-i(305.5°)}}$$

$$= \frac{5.25e^{-i(226°)} - 17.22e^{-i(188°)}}{19.475e^{-i(341°)} - 2.1e^{-i(138°)}}$$

The valves of α and β are obtained graphically as explained in Fig.12.63(c).

$$\alpha = \frac{1.59e^{-i(74°)}}{1.77e^{-i(345°)}} = 0.902\ e^{-i(-271°)}$$

$$\beta = \frac{1.36e^{-i(350°)}}{1.77e^{-i(345°)}} = 0.768\ e^{-i(5°)}$$

Balance mass at plane $A = 0.902 \times 2 = 1.804$ kg at 271° cw or 89° ccw.
Balance mass at plane $B = 0.768 \times 2 = 1.536$ kg at 5° ccw from the position of trial mess

Summary for Quick Revision

1 By balancing we mean to eliminate either partially or completely the effects due to unbalanced resultant inertia forces and couples to avoid vibration of a machine or device.

2 A system of rotating masses are in static balance if the combined mass centre of the system of masses lies on the axis of rotation.

3 A system of rotating masses are said to be in dynamic balance if there does not exist any resultant centrifugal force as well as resultant couple.

4 A system which is dynamically balanced is deemed to be statically balanced also.

5 A system that is statically balanced may or may not be dynamically balanced.

6 Balancing of rotating masses.

(a) Single Rotating Mass

(i) Balance mass in the same plane as the disturbing mass Place the balancing mass B in the same plane at a radius b and in line with the mass M at 180° such that: $M\,r = B\,b$

(ii) Two balance masses in different planes

1. Balance masses on the same side of the disturbing mass

$$B_1\, b_2 = M\,r\,[l_2 / (l_2 - l_1)]$$

2. Balance masses on the opposite sides of the disturbing mass

$$B_2\, b_2 = M\,r\,[l_1 / (l_1 + l_2)]$$
$$B_1\, b_1 = M\,r\,[l_2 / (l_1 + l_2)]$$

(b) Many masses rotating in the same plane

$$\Sigma\, M_i\, r_i = B\,b$$

The force polygon must close, i.e. the resultant force should be equal to zero.

(c) Many Masses Rotating in Different Planes

The balancing equations are:

$$\Sigma\, M\,r = 0$$
$$\Sigma\, M\,r\,a = 0$$

The force and couple polygons must close, i.e. there should not exist any resultant force and resultant force and resultant couple.

7 Reciprocating masses.

(a) Reciprocating engine.

The full effect on the engine frame of the inertia of the reciprocating mass is equivalent to the force F along the line of stroke at O and the clockwise couple of magnitude S. OP.

$$F = R\,f_c \cos\theta + R\,f_c \cos 2\theta / n$$
$$= F_p + F_s$$

where F_p is the primary force, which represents the inertia force of reciprocating mass having simple harmonic motion, and F_s is the secondary force, which represents the correction required to account for the obliquity of the connecting rod.

The unbalanced force due to the reciprocating mass varies in mágnitude but is constant in direction.

A single revolving mass can neither be used to balance a reciprocating mass, nor vice versa.

(b) Partial primary balance

For complete balancing, $R\,r = B\,b$.

It is usually preferable to make $B\,b = c.\,R\,r$, where $c < 1$.
For unbalanced force to be least, $c = 0.5$
If the balance mass B has to balance the revolving parts M as well as give a partial balance of the reciprocating parts R, then

$$B\,b = M\,r + c\,R\,r = (M + c\,R)\,r$$

In practice, two balance masses, each equal to $B/2$, would be attached to the crank webs.

8 Partial balancing of uncoupled locomotives

(a) Hammer Blow: The maximum magnitude of unbalanced force perpendicular to the line of stroke is known as hammer blow. This occurs at $\theta = 90°$ and $270°$. Hammer blow $= B_r\,\omega^2\,b$
If P is the downward pressure on rails due to dead load. Then
Net pressure $= P \pm B_r\,\omega^2\,b$
Permissible speed, $\omega = [\,P/(B_r\,b)]^{1/2}$
(b) Variation of Tractive Effort $= (1-c)\,R\omega^2\,r\,[\,\cos\theta - \sin\theta\,]$
Maximum variation of tractive effort $= -\sqrt{2}\;(1-c\,)\,R\omega^2$
(c) Swaying couple - The unbalanced part of the primary disturbing forces cause a horizontal swaying couple to act on the locomotive owing to the distance l between the cylinder centers.
Swaying couple $= (1-c)\,R\omega^2\,r\,.\,\ell/2\,.\,[\,\cos\theta + \sin\theta\,]$
Maximum swaying couple $= -\,[(1-c)/\sqrt{2}\,]\,R\omega^2\,\ell$

9 Multiple cylinder in-line engines
In a multi-cylinder in-line engine, the cylinder centre lines lie in the same plane and on the same side of the crankshaft centre line.

(a) Primary Balancing
The conditions to be satisfied for the primary balancing are:
(i) The algebraic sum of the primary forces should be equal to zero, i.e. the primary force polygon must close, i.e. $\Sigma\,R\,\omega^2\,r\cos\theta = 0$
(ii) The algebraic sum of the primary couples about any point in the plane of the forces must be equal to zero, i.e. the primary couple polygon must close, i.e. $\Sigma\,R\,\omega^2\,r\,a\cos\theta = 0$
where a = distance of the plane of rotation of the crank from a parallel reference plane.
(b) Secondary Balancing
The conditions to be satisfied for the secondary balancing are:
(i) The algebraic sum of the secondary forces should be equal to zero, i.e. the secondary force polygon must close, i.e.
$\Sigma\,R\,(2\omega)^2\,[r/(4n)]\cos 2\theta = 0$
(ii) The algebraic sum of the secondary couples about any point in the plane of the forces must be equal to zero, i.e. the secondary couple polygon must close, i.e. $\Sigma\,R\,(2\omega)^2\,[r/(4n)]\,a\cos 2\theta = 0$
where a = distance of the plane of rotation of the crank from a parallel reference plane.
Imaginary crank length $= r/(4n)$, speed $= 2\omega$
Angle made by imaginary secondary crank with inner dead centre $= 2\theta$

10 In-line two-cylinder engine
Primary force, $F_p = R\,r\,\omega^2\,[\,\cos\theta + \cos(180° + \theta)] = 0$
Primary couple, $C_p = R\,r\,\omega^2\,a\cos\theta$
$(C_p)_{max} = R\,r\,\omega^2\,a$
Secondary force, $F_s = (2\,\mathrm{R\,r}\,\omega^2/\mathrm{n})\cos 2\theta$
$(F_s)_{max} = 2\,R\,r\,\omega^2/n$
Secondary couple, $C_s = 0$

11 In-line Four-Cylinder Four – Stroke Engine
Primary force, $F_p = 0$
Primary couple, $C_p = 0$
Secondary force, $F_s = (4\ R\ r\ \omega^2/n) \cos 2\theta$
$(F_s)_{max} = 4\ R\ r\ \omega^2 / n$
Secondary couple, $C_s = 0$

12 Balancing of radial engines

(a) Direct and Reverse Cranks Method

Since the plane of various cranks is the same, therefore, there is no unbalanced primary or secondary couples.

(i) Primary Forces
Component of the centrifugal force on the direct crank acting along the line of stroke from O to P,
$F_{pd} = 0.5\ R\ \omega^2\ r \cos\theta$
Component of the centrifugal force on the reverse crank acting along the line of stroke from O to P,
$F_{pr} = 0.5\ R\ \omega^2\ r \cos\theta$
Total component of the centrifugal force along the line of stroke
$F_p = F_{pd} + F_{pr} = R\ \omega^2\ r \cos\theta$
Which is the primary force itself. Hence, for primary force effect, the mass of the reciprocating parts at P may be replaced by two masses at crankpins C and C', each of mass $R / 2$ at radii equal to r.

(ii) Secondary Forces

The secondary force, $F_s = R\ \omega^2\ r \cos 2\theta/n$

The secondary force effect may be taken into account by dividing the mass R into two equal parts and placing it at the imaginary crankpins at radii $r/4n$.

13 Balancing of V-engines
Resultant primary force, $F_p = 2\ R\ \omega^2\ r\ [\ (\cos^2\alpha \cos\theta)^2 + (\sin^2\alpha \sin\theta)^2]^{0.5}$
Resultant secondary force, $F_s = 2\ \mathrm{R}\ \omega^2\ (r/n)\ [(\cos\alpha \cos 2\alpha \cos 2\theta)^2 + (\sin\alpha \sin\alpha \sin 2\theta)^2]^{0.5}$

Multiple Choice Questions

1 In case of reciprocating engines the ratio of primary to secondary forces is
(a) $\cos\theta/\cos 2\theta$
(b) $\cos\theta/(n \cos 2\theta)$
(c) $n \cos\theta/\cos 2\theta$
(d) $\cos^2\theta/\cos 2\theta$

2 Partial balancing in locomotives results in
(a) hammer blow
(b) variation in tractive effort
(c) swaying couple
(d) all of the above

3 In reciprocating engines, primary forces are
(a) completely balanced
(b) partially balanced
(c) can not be balanced
(d) balanced by secondary forces.

4 In case of locomotives, the effect of hammer blow is counteracted by
(a) flanges of the tyres of the wheels
(b) balancing weights
(c) inside section of the rails
(d) dead weight of the engine.

5 Hammer blow in locomotives results in
(a) pulsating torque
(b) tendency to lift wheels from rails.
(c) uneven speed
(d) variable horizontal force.

6 Swaying couple results due to
(a) primary disturbing force
(b) secondary disturbing force
(c) partial balancing
(d) hammer blow.

7 Inertia force acts
(a) perpendicular to the accelerating force
(b) along the direction of the accelerating force
(c) opposite to the direction of the accelerating force
(d) in any direction with respect to accelerating force.

8 If the balance mass is to be placed in a plane parallel to the plane of the unbalance mass then the minimum number of balance masses required are
(a) one
(b) two
(c) three
(d) four.

9 The frequency of secondary force as compared to that of primary force is
(a) half
(b) twice
(c) four times
(d) sixteen times.

10 If the ratio of the length of connecting rod to crank radius increases, then
(a) primary force increases
(b) primary force decreases
(c) secondary force increases
(d) secondary force decreases.

11 The resultant unbalanced force is minimum in reciprocating engines when the part of the reciprocating mass balanced by rotating masses are
(a) 1/3
(b) 1/2
(c) 2/3
(d) 3/4

12 In partial balancing of locomotives, the maximum variation of tractive effort is
(a) $(2/3)\, M r \omega^2$
(b) $\left(\sqrt{2}/3\right) M r\omega^2$
(c) $\left(3/\sqrt{2}\right) M r\omega^2$
(d) $(3/2)\, M r \omega^2$.

13 Static force balancing involves balancing of
(a) forces
(b) couples
(c) forces as well as couples
(d) masses

14 If a system is dynamically balanced, then it is statically
(a) balanced
(b) unbalanced
(c) partially balanced

Answers

1. (c) 2. (d) 3. (b) 4. (d) 5. (b) 6. (a) 7. (c) 8. (b) 9. (b) 10. (d) 11. (b) 12. (b) 13. (a) 14. (a)

Review Questions

1 What is the necessity of balancing high speed machinery?
2 What do you mean by static and dynamic balance of machinery
3 What do you mean by primary and secondary unbalance in reciprocating engines?
4 What is partial balancing of reciprocating engines?
5 Define hammer blow, tractive effort and swaying couple.
6 What are direct and reverse cranks in radial engines?
7 What is a coupled locomotive?
8 Differentiate between primary and secondary cranks.

Exercises

12.1 A shaft carries four rotating masses *A*, *B*, *C* and *D* in this order along its axis. The mass *A* may be assumed concentrated at a radius of 120 mm, *B* at 150 mm, *C* at 130 mm, and *D* at 180 mm. The masses at *A*, *C* and *D* are 10, 15 and 12 kg, respectively. The planes of rotation of *A* and *B* are 150 mm apart and of *B* and *C* are 200 mm apart. The angle between *A* and *C* is 90°. If the shaft is in complete dynamic balance, determine (a) the angles between the radii of *A*, *B* and *D*, (*b*) the distance between the planes of rotation of *C* and *D*, and (*c*) the mass *B*.

12.2 A shaft rotating at 1000 rpm carries two unbalances of magnitudes 0.2 kg m and 0.1 kg m in planes *A* and *B* respectively. The planes *A* and *B* are 0.5 m apart and the directions of the unbalances are at 90°. A third unbalances has to be attached to the shaft at a location *C* so that the shaft is statically balanced and the magnitude of the bearing reactions is minimum possible. The bearings are 0.6 m from *A* and *B*. Determine (*a*) location of *C* with respect to *A* and *B*, (*b*) the magnitude of unbalance at *C*, (*c*) the angular position of unbalance at *C* with respect to *A* and *B*.

12.3 The cranks and connecting rods of a 4-cylinder in-line engine running at 1800 rpm are 50 mm and 200 mm long respectively. The cylinders are spaced 150 mm apart. If the cylinders are numbered 1 to 4 in sequence from one end, the cranks appear at intervals of 90° in an end view in the order 1–4–2–3. Reciprocating mass corresponding to each cylinder is 1.5 kg. Determine (*a*) unbalanced primary and secondary forces, and (*b*) unbalanced primary and secondary couples with reference to central plane of engine.

12.4 A four-crank engine has the two outer cranks set at 120° to each other, and their reciprocating masses are each 400 kg. The distance between the planes of rotation of adjacent cranks are 0.5, 0.8 and 0.6 m. If the engine is to be in complete primary balance, determine the reciprocating mass and the relative angular position for each of the inner cranks.
If the length of each crank is 0.3 m, the length of each connecting rod 1.2 m and the speed of the engine 240 rpm, what is the maximum secondary unbalanced force?

12.5 A rotating shaft carries four unbalanced masses 20, 15, 18 and 10 kg at radii 50, 60, 70 and 65 mm, respectively. The 2nd, 3rd and 4th masses revolve in planes 80, 150 and 300 mm respectively measured from the plane of the first mass and are angularly located at 60°, 130° and 270° respectively measured anti-clockwise from the first mass looking from this mass end of the shaft. The shaft is dynamically balanced by two masses, both located at 50 mm radii and revolving in planes mid way between those of first and second masses and midway between those of 3rd and 4th masses. Determine the magnitudes of the masses and their respective angular position.

12.6 A single cylinder reciprocating engine has a reciprocating mass of 50 kg. The crank rotates at 80 rpm and the stroke is 300 mm. Mass of revolving parts at 150 mm radius is 40 kg. If 2/3rd of the reciprocating parts and whole of the revolving parts are to be balanced, determine (*a*) the balance mass required at a radius of 340 mm, and (*b*) the unbalanced force when the crank has turned through 40° form *TDC*.

12.7 The reciprocating mass per cylinder in a *V*-twin engine is 1.5 kg. The stroke is 100 mm for each cylinder. If the engine runs at 1800 rpm, determine the maximum and minimum values of the primary forces and the corresponding crank position.

12.8 The cranks of a two-cylinder uncoupled inside cylinder locomotive are at right angles and are 300 mm long. The distance between the centre lines of the cylinders is 650 mm. The wheel centre lines are 1.6 m apart. The reciprocating mass per cylinder is 300 kg. The driving wheel diameter is 1.8 m. If the hammer blow is not to exceed 45 kN at 100 km/h, determine (a) the fraction of reciprocating is 1.8 m. If the hammer blow is not to exceed 45 kN at 100 km/h, determine (a) the fraction of reciprocating masses to be balanced, (b) the variation in tractive effort, and (c) the maximum swaying couple.

12.9 The pistons of a 60° twin V-engine have strokes of 120 mm. The connecting rods driving a common crank have a length of 200 mm. The mass of the reciprocating parts per cylinder is 1 kg and the speed of the crankshaft is 2500 rpm. Determine the magnitude of primary and secondary forces.

12.10 For an inside cylinder locomotive with two cranks at right angles, the reciprocating parts are 300 kg per cylinder. The distance between the cylinder centre lines is 0.6 m and between the plane of rotation of wheels 1.5 m. Each crank is 0.3m long and the driving wheels are 1.8 m diameter. Revolving balance masses are introduced in the planes of wheels partially to balance 2/3 rd of the reciprocating parts. Determine the maximum variative of (*a*) tractive effort, and (*b*) wheel load when the locomotion is running at 90 km/h.

13
GYROSCOPIC AND PRECESSIONAL MOTION

Chapter Outline

13.1 INTRODUCTION

In vehicles having engines with rotating parts of high moment of inertia, gyroscopic forces are in action when the vehicle is changing direction of motion. When automotive vehicles turn with high velocities, gyroscopic forces act on spinning parts such as crankshaft, flywheel, clutch, transmission gears, propeller shaft and wheels. Engine parts as well as the propeller and the gear reduction system of an airplane are under the influence of gyroscopic effects in turns and pullouts. Locomotives and ships are similarly affected. In this chapter, we shall study the gyroscopic and precessional motion of road vehicles, aeroplanes and ships.

13.2 PRECESSIONAL MOTION

Consider a plane disc spinning about the axis OX with angular speed ω, as shown in Fig.13.1(a). After a short interval of time δt, let the disc be spinning with angular speed $\omega + \delta\omega$ about the new axis OX' inclined at a small angle $\delta\theta$ with OX. The angular speed ω is represented by vector OX and $\omega + \delta\omega$ by vector OX' in Fig.13.1(b). The vector XX' represents the change of angular speed in time δt.

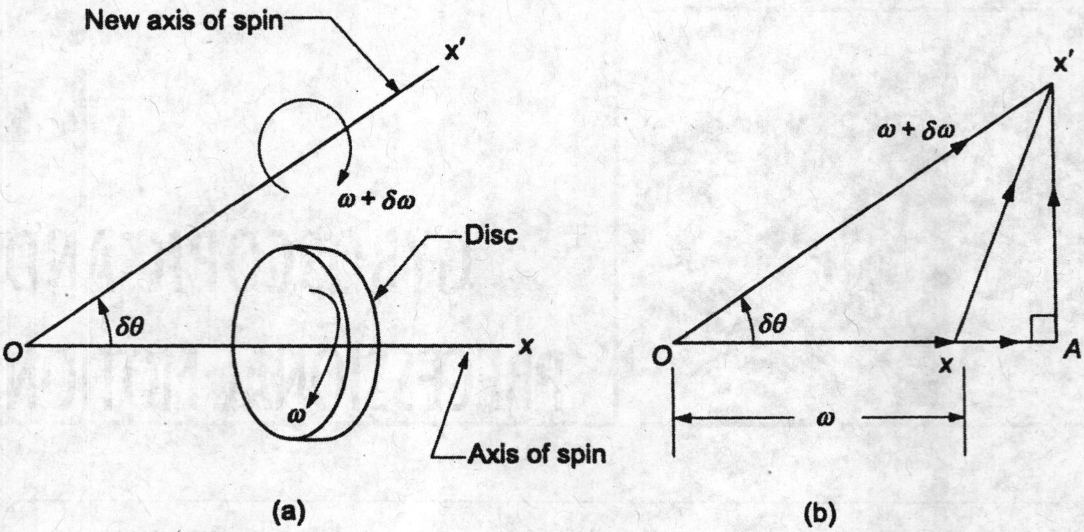

Fig.13.1 Precessional motion

$$\text{Angular acceleration along } OX = \lim_{\delta t \to 0}\left[\frac{(\omega+\delta\omega)\cos\delta\theta-\omega}{\delta t}\right]$$

$$= \lim_{\delta t \to 0}\left[\frac{\omega\cos\delta\theta+\delta\omega\cos\delta\theta-\omega}{\delta t}\right]$$

$$= \lim_{\delta t \to 0}\left[\frac{\omega+\delta\omega-\omega}{\delta t}\right] \qquad \text{(Because } \cos\delta\theta \approx 1\text{)}$$

$$= \lim_{\delta t \to 0}\frac{\delta\omega}{\delta t}$$

$$= \frac{d\omega}{dt} \qquad (13.1)$$

Angular (or gyroscopic) acceleration perpendicular to OX,

$$= \lim_{\delta t \to 0}\left[\frac{(\omega+\delta\omega)\sin\delta\theta}{\delta t}\right]$$

$$= \lim_{\delta t \to 0}\frac{(\omega\cdot\delta\theta+\delta\omega\cdot\delta\theta)}{\delta t}$$

$$= \frac{\omega\cdot d\theta}{dt} \qquad \text{[Neglecting } \delta\omega\cdot\delta\theta\text{, being small]} \qquad (13.2)$$

$$= \omega\cdot\omega_p$$

where ω_p is the precessional angular speed of the spin axis.
Total angular acceleration of disc = XX'

$$= XA \mapsto AX'$$

$$= \frac{d\omega}{dt} \mapsto \omega \cdot \frac{d\theta}{dt}$$

$$= \frac{d\omega}{dt} \mapsto \omega \cdot \omega_p \qquad (13.3)$$

where $\mapsto$ represents vector sum.

13.3 FUNDAMENTALS OF GYROSCOPIC MOTION

Gyroscope: It is a body which while spinning about an axis is free to move in other direction under the action of external forces.

Axis of spin: It is axis about which the body revolves.

Gyroscopic effect: Consider a body spinning about an axis *OX* (Fig.13.2). If a couple represented by a vector *OZ* perpendicular to *OX* is applied, then the body tries to precess about an axis *OY*, which is perpendicular to both *OX* and *OZ*. This combined effect is called gyroscopic or precessional effect. The plane of spin, plane of precession, and plane of gyroscopic couple are mutually perpendicular.

Precession: It means the rotation about the third axis *OY*, which is perpendicular to both the spin axis *OX* and the couple axis *OZ*.

Axis of precession: The third axis *OY* about which a body revolves and is perpendicular to both the spin axis *OX* and couple axis *OZ*, is called the axis of precession.

Mechanical gyroscope: It is a special mechanism generally employed for the control of angular motion of a body. Gyroscope is an instrument, which appears as a device possessing intelligence. Gyroscopic effects can be used in several applications for directional control, e.g. in Gyrocompass used on aeroplanes and ships and in inertia guidance control system for missiles and space craft control.

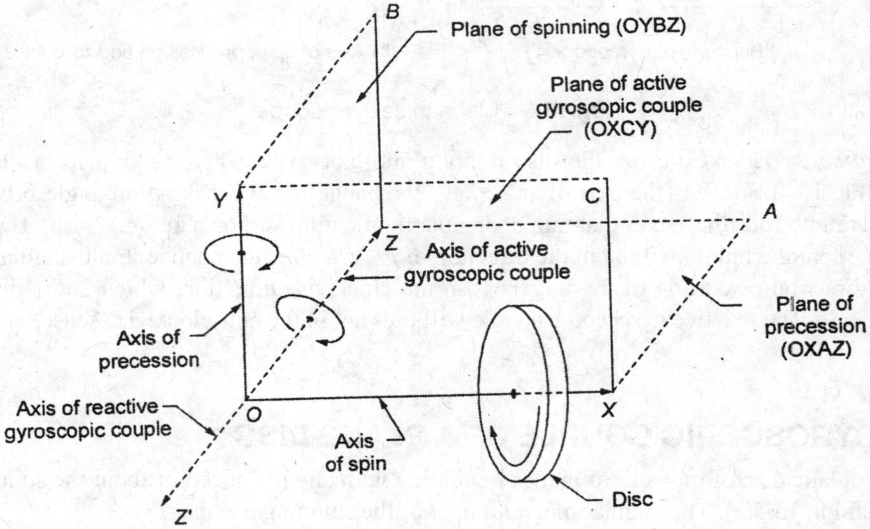

Fig.13.2 Gycoscopic effect

Principle of gyroscope: The principle of a gyroscope can be easily understood from the toy gyroscope shown in Fig.13.3. A rotating disc supported on gimbal rings rotates with an angular velocity, called the velocity of spin, ω. The axis of the rotating shaft rotates about the vertical axis OY with an angular velocity ω_p, called the velocity of precession. A force $F = I\omega\omega_p/l$ will be automatically developed on the bearings of the rotor shaft in the direction indicated. The reaction at the bearings will be $(-F)$. The axis of rotation of the rotating disc would revolve in the horizontal plane XOZ. In case of vehicles, ships and other such devices, the spinning masses may be forced to precess in a desired direction. Thus, couples will be applied to the shaft carrying such spinning masses when the axis of spin is forced to precess in a desired direction.

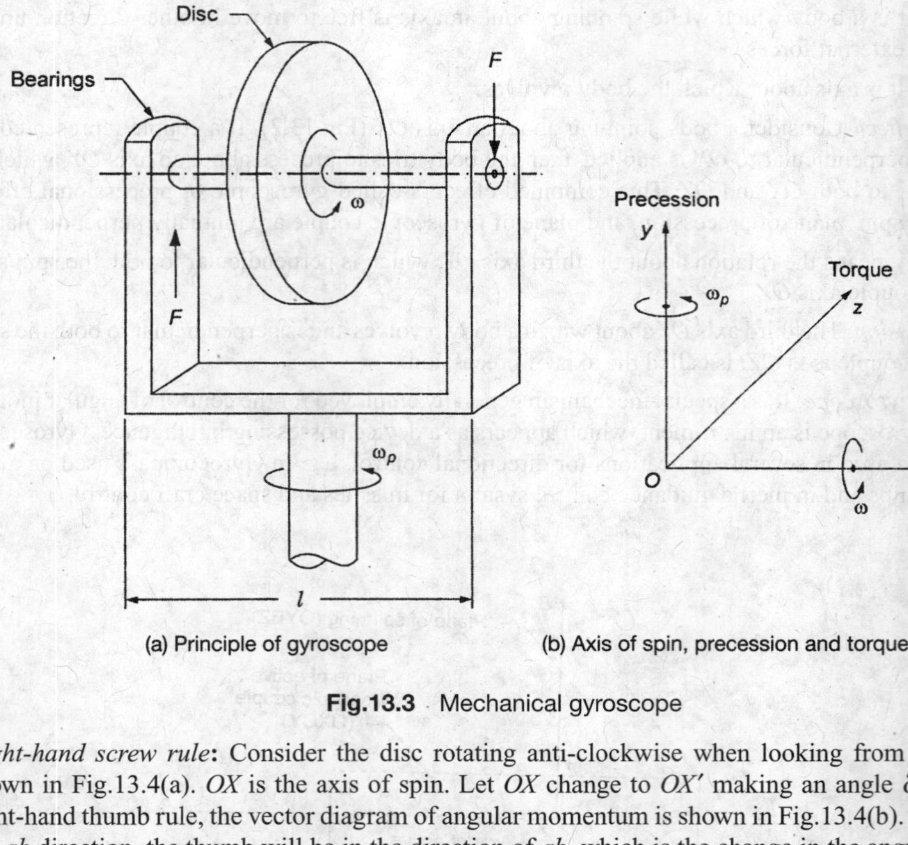

(a) Principle of gyroscope (b) Axis of spin, precession and torque

Fig.13.3 Mechanical gyroscope

Right-hand screw rule: Consider the disc rotating anti-clockwise when looking from the front as shown in Fig.13.4(a). OX is the axis of spin. Let OX change to OX' making an angle $\delta\theta$. Applying right-hand thumb rule, the vector diagram of angular momentum is shown in Fig.13.4(b). If we look in the *ab* direction, the thumb will be in the direction of *ab*, which is the change in the angular momentum. The direction of curling of the fingers is in the clockwise direction, which is the direction of applied couple. The reactive gyroscopic couple will thus act in the anti-clockwise sense.

13.4 GYROSCOPIC COUPLE OF A PLANE DISC

Consider a plane disc of moment of inertia I spinning with angular speed ω about the spinning axis, as shown in Fig.13.5(a). The angular momentum H of the spinning disc is,

$$H = I\omega$$

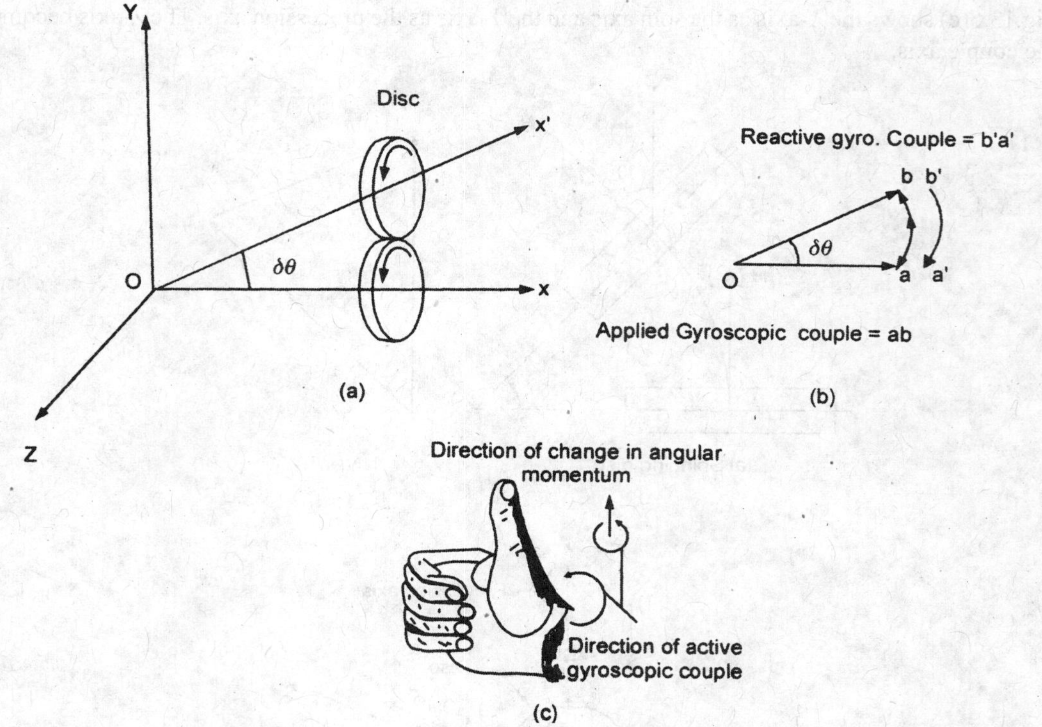

Fig.13.4 Right hand thumb rule

The rate of change of angular momentum with respect to time is proportional to the applied couple C.

$$C = I\alpha = I\frac{d\omega}{dt}$$

$$= \frac{d}{dt}(I\omega) = \frac{dH}{dt} \tag{13.4}$$

where α is the angular acceleration.

If the spin axis is made to change angular position, gyroscopic action results. For constant ω, the magnitude of the angular momentum remains constant for an angular displacement $\delta\theta$ of the spin axis. However, a change in angular momentum δH exists because of the change in direction of the momentum, as shown in Fig.13.5(b).

$$\delta H = (I\omega)\delta\theta$$

$$\lim_{\delta t \to 0}\frac{\delta H}{\delta t} = \lim_{\delta t \to 0}(I\omega)\cdot\left(\frac{\delta\theta}{\delta t}\right)$$

$$\frac{dH}{dt} = I\omega\cdot\frac{d\theta}{dt}$$

$$C = I\cdot\omega\cdot\omega_p \tag{13.5}$$

Fig.13.5(c) shows the X-axis as the spin axis and the Y-axis as the precession axis. The Z-axis becomes the couple axis.

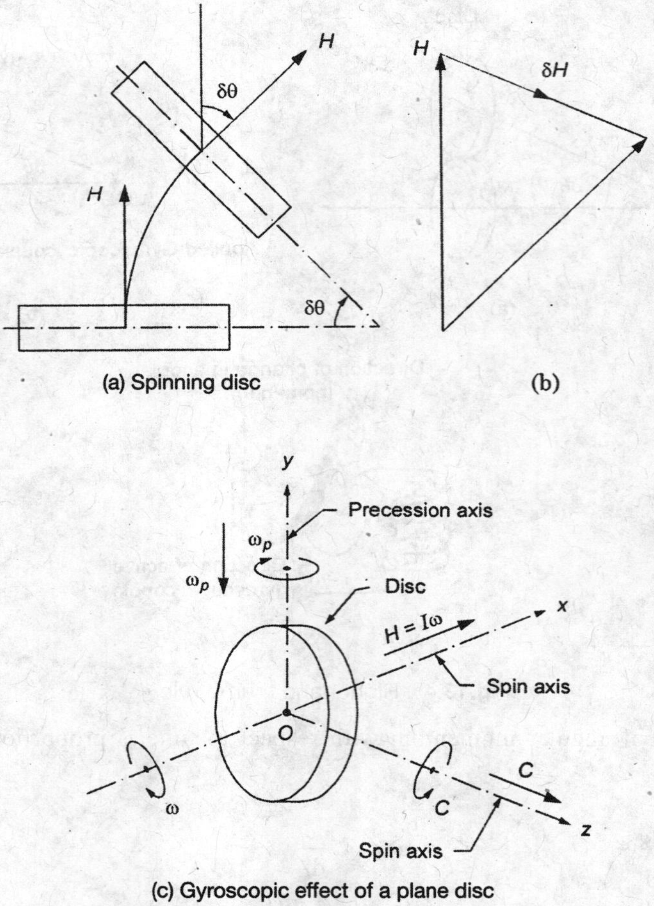

Fig.13.5 Gyroscopic couple

Example 13.1

Determine the gyroscopic couple of a 3 m diameter solid aluminium alloy four-bladed propeller in which each blade has a mass of 20 kg. The test manoeuver of the airplane is a power-on flat spin in which the propeller speed is 1500 rpm and the rotation of the flat spin is 1 rad/s. The radius of gyration of the propeller with respect to the propeller axis is approximately half of the propeller radius.

■ Solution

Radius of gyration, $K = r_m = 0.5 \times 3 = 1.5$ m

Moment of inertia, $I = M K^2$

$$= 20 \times (1.5)^2 = 45 \text{ kg} \cdot \text{m}^2$$

Angular speed, $$\omega = 2\pi \times \frac{1500}{60} = 157.08 \text{ rad/s}$$

Precessional speed, $\omega_p = 1$ rad/s

$$\text{Couple, } C = I\omega \cdot \omega_p$$
$$= 45 \times 157.08 \times 1$$
$$= 7068.6 \text{ Nm}$$

Example 13.2

A boat is propelled by a steam turbine. The moment of inertia of the rotor, shaft and propeller is 60 kg·m². The turbine runs at 3000 rpm in clockwise direction looking from the front. The boat describes a circular path towards the right making one revolution in 10 s. Find the magnitude and direction of the couple acting on the boat hull.

Solution

$$\omega = \frac{2\pi N}{60} = 2\pi \times \frac{3000}{60} = 314.16 \text{ rad/s}$$

Applied couple,

$$\omega_p = \frac{2\pi}{10} = 0.628 \text{ rad/s}$$
$$C = I \cdot \omega \cdot \omega_p$$
$$= 60 \times 314.16 \times 0.628$$
$$= 11837.5 \text{ Nm}$$

The vector diagram of the gyroscopic effect is shown in Fig.13.6. The applied couple *ab* will lower the front and raise the stern. The reaction couple *b′*; *a′* will raise the front and lower the stern of the boat.

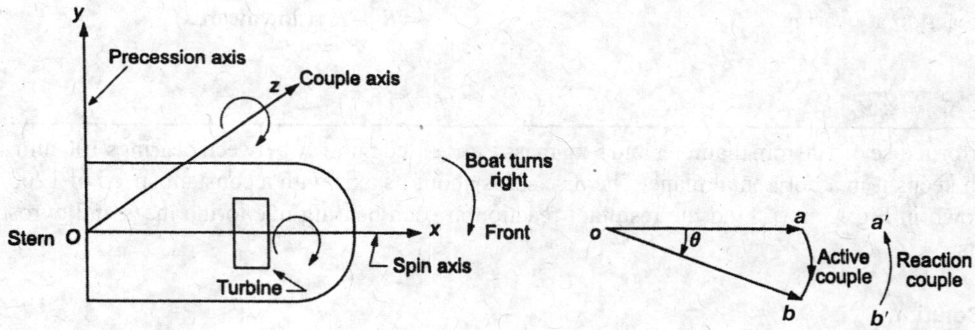

Fig.13.6 Gyroscopic effect on a boat

13.5 EFFECT OF GYROSCOPIC COUPLE ON BEARINGS

Consider a disc of mass m and radius of gyration K mounted centrally on a horizontal axle of length l between the bearings. Let the disc spin with angular speed ω counter-clockwise when viewed from the right-hand side bearing, as shown in Fig.13.7(a). The axle precesses about a vertical axis at ω_p speed in the clock-wise direction when viewed from above. Then

$$I = mK^2$$
$$C = I\,\omega\omega_p$$

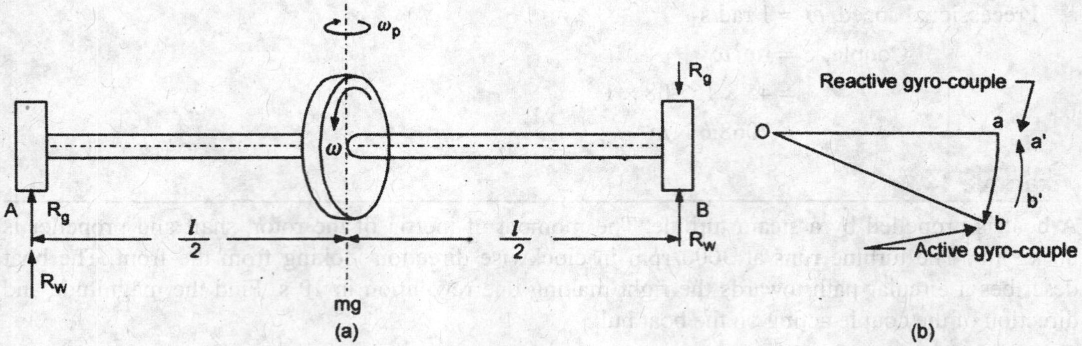

Fig.13.7 Effect of gyro-couple on bearings

The applied (active) and reaction couples are shown in Fig.13.7(b). The reaction couple is clockwise when viewed from front and tends to raise the bearing *A* and lower the bearing *B*.

Force on bearing *A* due to gyro-couple, $R_g = \dfrac{C}{l}$ (upwards)

Force on bearing *B* due to gyro-couple, $R_g = \dfrac{C}{l}$ (downwards)

Force on each bearing due to self weight of disc, $R_w = \dfrac{mg}{2}$ (upwards)

Reaction at bearing *A* $= R_g + R_\omega$ (upwards)

Reaction at bearing *B* $= R_g - R_\omega$ (downwards)

Example 13.3

A uniform disc of 100 mm diameter and 5 kg mass is mounted midway between bearings 100 mm apart, which keeps it in a horizontal plane. The disc spins about its axis with a constant speed of 1200 rpm, as shown in Fig.13.8(a). Find the resultant reaction at each bearing due to the mass and gyroscopic effects.

■ **Solution**

$\omega = 2\pi \times 1200/60 = 125.66$ rad/s

$\omega_p = 2\pi \times 50/60 = 5.236$ rad/s

$I = 0.5\ mr^2 = 0.5 \times 5\ (50 \times 10^{-3})^2 = 0.00625$ kg. $\cdot$ m^2

$C = I\ \omega \cdot \omega_p$

$= 0.00625 \times 125.66 \times 5.236$

$= 4.112$ N m

The direction of reaction gyroscopic couple is shown in Fig.13.8(b).

Bearing reactions:

(a) *Due to self weight of the disc.*

$$R_A = R_B = 5 \times 9.81/2 = 24.525 \text{ N}$$

(b) *Due to reaction gyroscopic couple.*

$$R_A = F = 4.112/0.1 = 41.12 \text{ N}\uparrow$$
$$R_B = F = 41.12\text{N}\downarrow$$

Resultant bearing reactions:

$$R_A = 24.525 + 41.12 = 65.645 \text{ N}\uparrow$$
$$R_B = 24.525 - 41.12 = -16.515 \text{ N}\downarrow$$

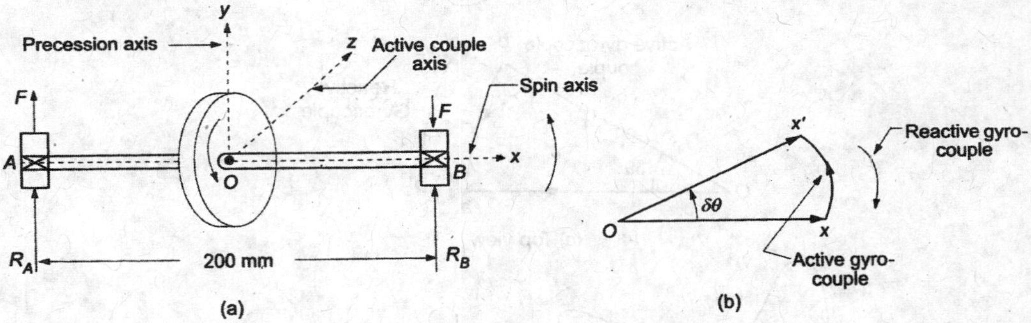

Fig.13.8 Gyro effect on bearing reactions

13.6 GYROSCOPIC COUPLE ON AN AEROPLANE

The top and front views of an aeroplane taking a left turn, when viewed from the rear, are shown in Fig.13.9(a) and (b), respectively. Let the propeller rotate clockwise as seen from the rear (or tail end).

Gyroscopic couple acting on the aeroplane, $C = I\omega \cdot \omega_p$

where I = moment of inertia of the engine and propeller

$= mK^2$

m = mass of engine and propeller

K = radius of gyration

ω = angular speed of engine

ω_p = angular speed of precession = v/R

v = linear velocity of aeroplane

R = radius of curvature.

In Fig.13.9(a), *oa* is the angular momentum vector before turning, *ob* is the momentum vector after turning to left. Vector *ab* is the active gyro-couple and *b'a'* the reactive gyro-couple, which is equal to active gyro-couple but opposite in direction.

The vector *ab* is perpendicular to the plane of applied couple, which is vertical. Its sense is clockwise when seen from the right side view of the plane, as shown in Fig.13.9(b). Using the right hand screw rule, the reactive gyro-couple will be counter-clockwise, which will raise the nose and lower the tail of the aeroplane.

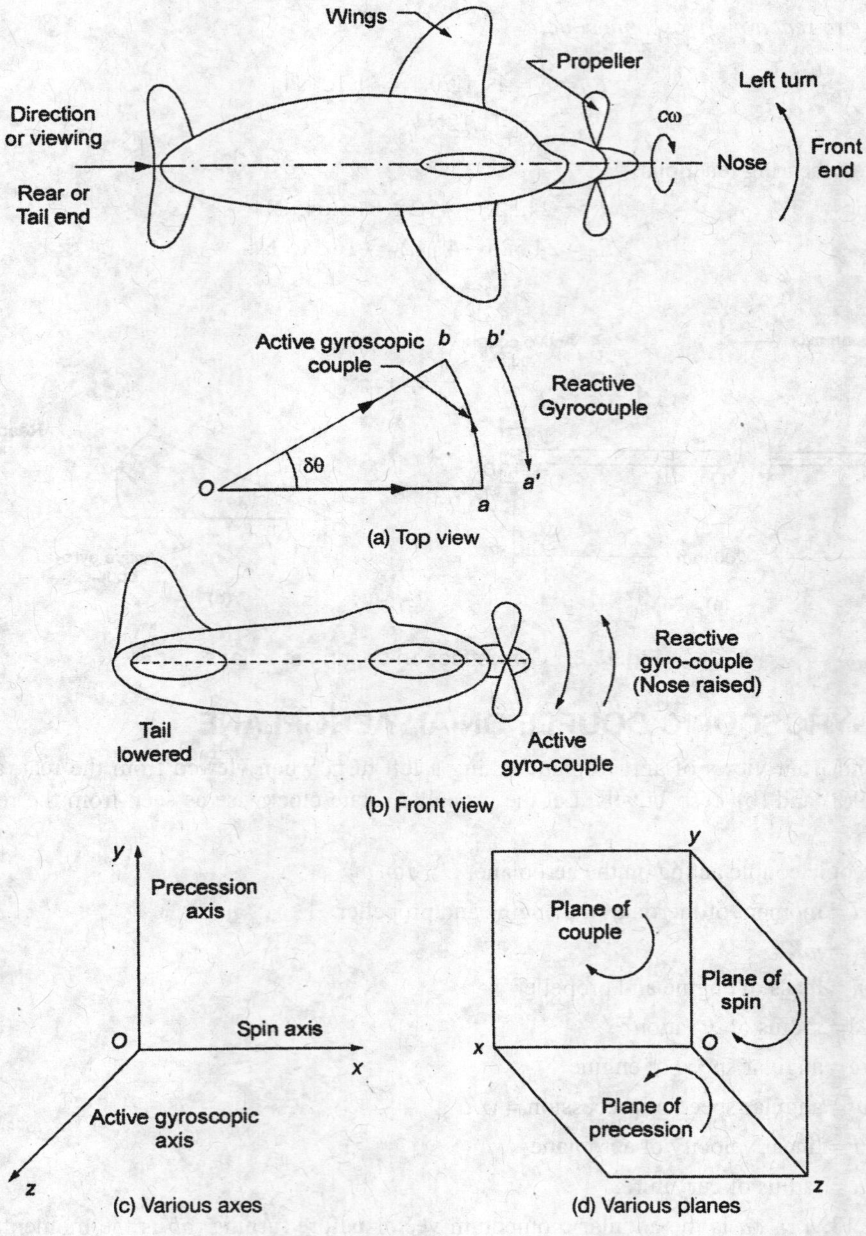

Fig.13.9 Gyroscopic effect on an aeroplane

Example 13.4

An aeroplane makes a complete half circle radius towards left when flying at 210 km/h. The rotary engine and the propeller of the plane is of 50 kg mass having a radius of gyration of 300 mm. The engine rotates at 2400 rpm clockwise as seen from the rear. Find the gyroscopic couple on the aircraft and its effect on the plane.

■ Solution

$$\omega = 2\pi \times 2400/60 = 251.33 \text{ rad/s}$$

$$\omega_p = v/R \times 210 \times 10^3 /(3600 \times 60) = 0.972 \text{ rad/s}$$

$$I = m\,K^2 = 50 \times (0.3)^2 = 4.5 \text{ kg}\cdot\text{m}^2$$

$$C = I\cdot\omega\cdot\omega_p = 4.5 \times 251.33 \times 0.972 = 1099.32 \text{ Nm}$$

The reaction gyro-couple will raise the noise and dip the tail.

13.7 GYROSCOPIC EFFECTS ON A NAVAL SHIP

The following terms for a naval ship in reference to Fig.13.10 are defined:

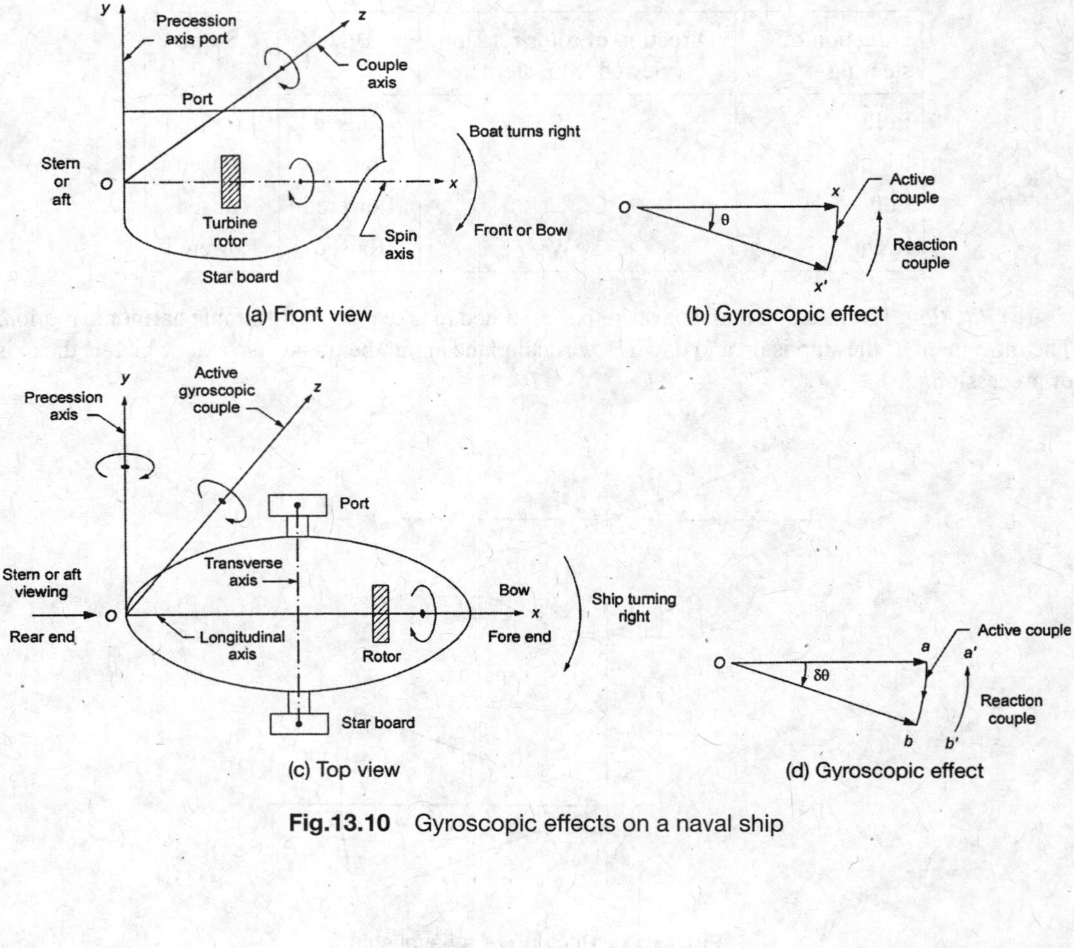

Fig.13.10 Gyroscopic effects on a naval ship

Bow is the fore-end of the ship.

Stern is the rear-end of the ship.

Starboard is the right hand side of the ship while looking in the direction of motion.

Port is the left hand side of the ship while looking in the direction of motion.

Steering is the turning of the ship in a curve while moving forward.

Pitching is the moving of the ship up and down the horizontal position in a vertical plane about transverse axis.

Rolling is the sideway motion of the ship about longitudinal axis.

(a) *Steering*: The gyroscopic effect on a naval ship during steering can be obtained as explained for the aeroplane in Section 13.6. When the ship turns to right, the angular momentum vector changes from *oa* to *ob*, as shown in Fig.13.10(d). The reaction couple $b'a'$ is shown in the reverse direction. If the rotor rotates in clockwise direction looking from the rear end then the reaction couple tends to lower the bow and raise the stern.

Table 13.1 may be used to determine the gyroscopic effects:

Table 13.1 Gyroscopic effects on a naval ship

Direction of steering	Direction of rotor rotation (viewed from stern)	Bow	Stern
Left	CW	Raised	Lowered
Right	CW	Lowered	Raised
Left	CCW	Lowered	Raised
Right	CCW	Raised	Lowered

(b) *Pitching*: The pitching of the naval ship is assumed to take place with simple harmonic motion. The movement of the ship is up and down in vertical plane about the transverse axis, which is the axis of precession.

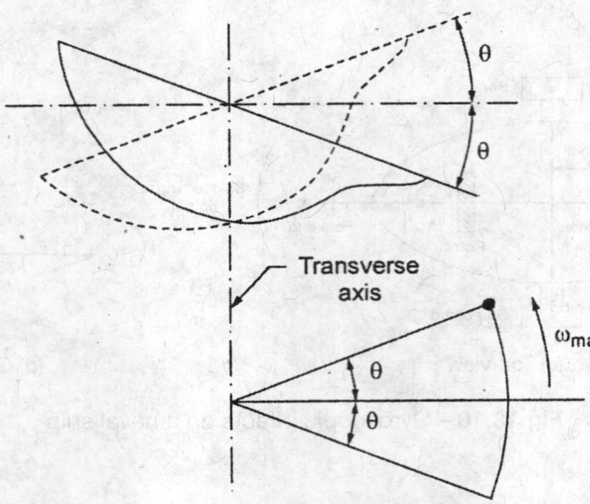

Fig.13.11 Pitching of a naval ship

The pitching angle at time t is (see Fig.13.11).

$$\theta = A \sin \omega_o t$$

where A = amplitude of swing in radians

ω_o = angular velocity of simple harmonic motion

$= \dfrac{2\pi}{t_p}$

t_p = time period of pitching.

Angular velocity of precession,

$$\omega_p = \frac{d\theta}{dt} = A\omega_o \cos \omega_o t$$

$$(\omega_p)_{\max} = A\omega_o = A \times \left(\frac{2\pi}{t_p}\right) \quad \text{for } \cos \omega_o t = 1$$

Maximum gyro-couple, $C_{\max} = I \cdot \omega \cdot (\omega_p)_{\max}$ (13.6)

where I = moment of inertia of turbine rotor and other masses of the naval ship.

ω = angular velocity of rotating masses.

The effects of pitching are as follows:

1. When the pitching is upward, the gyroscopic effect will try to move the ship towards starboard.
2. On the other hand, if the pitching is downward, the gyroscopic effect is to turn the ship towards port side.
3. The pitching of a ship produces forces on the bearings which act horizontally and perpendicular to the motion of the ship.
4. The maximum gyroscopic couple tends to shear the holding down bolts.

Angular acceleration during pitching,

$$\alpha = \frac{d^2\theta}{dt^2} = -A\omega_o^2 \sin \omega_o t$$

$$\alpha_{\max} = A\omega_o^2 = A\left[\frac{2\pi}{t_p}\right]^2 \quad (13.7)$$

(c) *Rolling*: The axis of rolling and that of rotor of turbine are generally same. So, there is no precession of axis of spin and there is no gyroscopic effects during rolling of the naval ship.

13.7.1 Ship Stabilization

A naval ship is normally stable, but it requires stabilization when it faces heavy sea. The ship will either pitch or roll. The amplitude of rolling is much higher than that of pitching. The ship in such a case is stabilized by producing couples in the opposite direction to that of the disturbing couples which are applied by the waves on the ship. The couples in opposite direction are produced by mechanical gyroscopes.

Example 13.5

A ship is propelled by a turbine rotor of mass 500 kg and has a speed of 2400 rpm. The rotor has a radius of gyration of 0.5 m and rotates in clockwise direction when viewed from stern. Find the gyroscopic effects in the following cases:

(a) The ship runs at a speed of 15 knots (1 knot = 1860 m/h). It steers to the left in a curve of 60 m radius.

(b) The spin pitches ± 5° from the horizontal position with the time period of 20 s of simple harmonic motion.

(c) The ship rolls with angular velocity of 0.04 rad/s clockwise when viewed from stern. Also calculate the maximum angular acceleration during pitching.

■ **Solution**

(a) $\omega = 2\pi \times 2400/60 = 251.3$ rad/s

$I = mK^2 = 500 \times (0.5)^2 = 125\ \text{kg}\cdot\text{m}^2$

$\omega_p = 15 \times 1860/(3600 \times 60) = 0.129$ rad/s

$C = I \cdot \omega \cdot \omega_p$

$= 125 \times 251.3 \times 0.129 = 4052.2$ N m

(b) $\omega_o = 2\pi/t_p = 2\pi/20 = 0.314$ rad/s

$(\omega_p)_{max} = A\omega_o = (5\pi/180) \times 0.314 = 0.0274$ rad/s

$C_{max} = I \cdot \omega \cdot (\omega_p)_{max}$

$= 125 \times 251.3 \times 0.0274 = 860.7$ N m

(c) $\omega_p = 0.04$ rad/s

$C = I \cdot \omega \cdot \omega_p$

$= 125 \times 251.3 \times 0.04 = 1256.5$ N m

$\alpha_{max} = A\,\omega_o^2 = (5\pi/180) \times (0.314)^2 = 0.0086\ \text{rad/s}^2$

Example 13.6

An aeroplane makes a complete half circle of 60 m radius towards left when flying at 250 km/h. The rotary engine and the propeller of the plane have a mass of 450 kg with a radius of gyration of 300 mm. The engine runs at 2400 rpm clockwise when viewed from the rear. Find the gyroscopic effect on the aircraft.

■ **Solution**

Given: $R = 60$ m, $v = 250$ km/h, $m = 450$ kg, $K = 300$ mm, $N = 2400$ rpm

$I = mK^2 = 450 \times (0.3)^2 = 40.5\ \text{kg}\cdot\text{m}^2$

$$\omega = \frac{2\pi \times 2400}{60} = 251.33 \text{ rad/s}$$

$$\omega_p = \frac{v}{R} = \frac{250 \times 1000}{3600 \times 60} = 1.157 \text{ rad/s}$$

$C = I\omega\omega_p = 40.5 \times 251.33 \times 1.157 = 11777$ N m

Example 13.7

The rotor of the turbine of a ship makes 1500 rpm clockwise when viewed from stern. The rotor has a mass of 800 kg and its radius of gyration is 300 mm. Find the maximum gyro-couple transmitted to the hull when the ship pitches with maximum angular velocity of 1 rad/s. What is the effect of this couple?

■ **Solution**

Given: $m = 800$ kg, $K = 300$ mm, $N = 1500$ rpm, $\omega_p = 1$ rad/s

$$I = mK^2 = 800 \times (0.3)^2 = 72 \text{ kg}\cdot\text{m}^2$$

$$\omega = \frac{2\pi \times 1500}{60} = 158.08 \text{ rad/s}$$

$$C = I\omega\omega_p = 72 \times 158.08 \times 1 = 11309.7 \text{ Nm}$$

Example 13.8

The mass of a turbine rotor of a ship is 8000 kg and has a radius of gyration of 0.75 m. It rotates at 1800 rpm clockwise when viewed from the stern. Determine the gyroscopic effects in the following cases:

(a) If the ship traveling at 100 km/h steers to the left along a curve of 80 m radius.

(b) If the ship is pitching and the bow is descending with maximum velocity. The pitching is with simple harmonic motion with periodic time of 20 s and the total angular movement between extreme position is 10°.

(c) If the ship is rolling with an angular velocity of 0.03 rad/s clockwise when looking from stern.

In each case, determine the direction in which the ship tends to move.

■ **Solution**

(a) Given: $R = 80$ m, $v = 100$ km/h, $m = 8000$ kg, $K = 0.75$ m, $N = 1800$ rpm

$$I = mK^2 = 8000 \times (0.75)^2 = 4500 \text{ kg}\cdot\text{m}^2$$

$$w = \frac{2p \times 1800}{60} = 188.5 \text{ rad/s}$$

$$w_p = \frac{v}{R} = \frac{100}{3600 \times 60} = 0.347 \text{ rad/s}$$

$$C = Iww_p = 4500 \times 188.5 \times 0.347 = 294531 \text{ Nn}$$

The ship tends to move to the left.

(b) $t_p = 20$ s, $A = 10°$ or $\dfrac{\pi}{18}$ rad

$$(\omega_p)_{\max} = \frac{2\pi A}{t_p} = \frac{2\pi \times \pi}{18 \times 20} = 0.05483 \text{ rad/s}$$

$$C_{\max} = I\omega\,(\omega_p)_{\max} = 4500 \times 188.5 \times 0.05483 = 46510 \text{ Nm}$$

The ship tends to move up.

(c) In case of rolling, $C = 0$ as $\omega_p = 0$.

Example 13.9

A diesel locomotive moving at a speed of 100 km/h turns around a curve of radius 400 m to the right. The driving wheels are 2 m in diameter and along with the axle has a mass of 2000 kg. The radius of gyration of the wheels together with the axle may be taken as 0.6 m. Find the gyro effect on the pair of driving wheels.

■ Solution

Given: $R = 400$ m, $v = 100$ km/h or 28.78 m/s, $m_w = 2000$ kg, $K_w = 0.6$ m, $d_w = 2$ m

$$I_w = m_w K_w^2 = 2000 \times (0.6)^2 = 720 \text{ kg}\cdot\text{m}^2$$

$$\omega = \frac{v}{r_w} = \frac{28.78}{1} = 28.78 \text{ rad/s}$$

$$\omega_p = \frac{v}{R} = \frac{28.78}{400} = 0.06945 \text{ rad/s}$$

$$C = I\omega\omega_p = 720834 \times 28.78 \times 0.06945 = 1389.1 \text{ Nm}$$

Example 13.10

A small high-speed ship is driven by a turbine, the rotor of which is rotating at 12,000 rpm in a clockwise direction, when viewed from the bow. The moment of inertia of the rotor is 15 kg·m² and the ship is travelling at 20 m/s in a curve of 600 m radius, the direction being clockwise viewed from the above.

Determine the gyroscopic couple acting on the ship and its effect on the ship.

■ Solution

Given: $R = 600$m, $v = 20$ m/s, I = 15 kg·m², $N = 12000$ rpm

$$I = mK^2 = 8000 \times (0.75)^2 = 4500 \text{ kg}\cdot\text{m}^2$$

$$\omega = \frac{2\pi \times 12000}{60} = 1256.637 \text{ rad/s}$$

$$\omega_p = \frac{v}{R} = \frac{20}{600} = \frac{1}{30} \text{ rad/s}$$

$$C = Iww_p = 15 \times 1256.637 \times \frac{1}{30} = 628.32 \text{ Nm tending to lift the bow.}$$

Example 13.11

The propeller shaft of an aero-engine is rotating at 1800 rpm. If the distance between the two bearings of the propeller shaft is 1 m, and radius of gyration of propeller is 0.75 m, find the extra pressure on the bearings, when the aero-plane is whirling round in a horizontal circle of 300 m radius at a speed of 300 km/h. The mass of the propeller is 60 kg.

■ Solution

Given: $R = 300$ m, $v = 300$ km/h, $m = 60$ kg, $K = 0.75$ m, $N = 1800$ rpm, $l = 1$ m

$$I = mK^2 = 60 \times (0.75)^2 = 33.75 \text{ kg}\cdot\text{m}^2$$

$$\omega = \frac{2\pi \times 1800}{60} = 188.5 \text{ rad/s}$$

$$\omega_p = \frac{v}{R} = \frac{300 \times 1000}{3600 \times 60} = 0.278 \text{ rad/s}$$

$$C = I\omega\omega_p = 33.75 \times 188.5 \times 0.278 = 1768.2 \text{ Nm}$$

$$\text{Bearing reaction} = \frac{C}{l} = \frac{1768.2}{1} = 1768.2 \text{ N}$$

Example 13.12

A turbine rotor of a ship which rotates clockwise when viewed from aft has a mass of 1500 kg, radius of gyration 0.7 m, and a speed of 2400 rpm. The ship pitches 5° above and below the horizontal position with simple harmonic motion of period 24 s.

Determine the maximum reaction couple exerted by the rotor on the ship and the direction in which the bow will turn when falling.

■ Solution

Given: $m = 1500$ kg, $K = 0.7$ m, $N = 2400$ rpm, $\theta = 5°$, $t_p = 24$ s

$$A = \frac{\pi\theta}{180} = \frac{\pi \times 5}{180} = \frac{\pi}{36} \text{ rad}$$

$$\left(\omega_p\right)_{\max} = \frac{A \times 2\pi}{t_p} = \frac{\pi}{36} \times \frac{2\pi}{24} = 0.0274 \text{ rad/s}$$

$$\omega = \frac{2\pi \times 2400}{60} = 251.32 \text{ rad/s}$$

$$I = mK^2 = 1500 \times (0.7)^2 = 735 \text{ kg}\cdot\text{m}^2$$

$$C_{\max} = I\omega(\omega_p)_{\max} = 735 \times 251.32 \times 0.0274 = 4220.86 \text{ N m}$$

The bow will turn to left (port) (see Table 13.1).

Example 13.13

A turbine rotor of a ship is of 2000 kg mass and has a radius of gyration of 0.8 m. Its speed is 2000 rpm. The ship pitches 5° above and below the mean position. A complete oscillation takes place in 20 s and the motion is simple harmonic. Determine

(a) the maximum couple tending to shear the holding down bolts of the turbine,

(b) the maximum acceleration of the ship during pitching, and

(c) the direction in which the bow will tend to turn while rising, if the rotation of the rotor is clockwise, when looking from aft.

■ Solution

Given: $m = 2000$ kg, $K = 0.8$, $N = 2000$ rpm, $\theta = 5°$, $t_p = 20$ s

$$A = \frac{\pi \times \theta}{180} = \frac{\pi \times 5}{180} \times \frac{\pi}{36} = \text{rad}$$

$$\left(\omega_p\right)_{\max} = \frac{A \times 2\pi}{t_p} = 0.0274 \text{ rad/s}$$

$$\omega = \frac{2\pi \times 2000}{60} = 209.44 \text{ rad/s}$$

$$I = mK^2 = 2000 \times (0.8)^2 = 1280 \text{ kg}\cdot\text{m}^2$$

(a) $C_{\max} = I\,\omega(\omega_p)_{\max} = 1280 \times 209.44 \times 0.0274 = 7345.48$ N m

(b) $a_{\max} = A\left(\frac{2\pi}{t_p}\right)^2 = \frac{\pi}{36}\left(\frac{2\pi}{20}\right)^2 = 8.613 \times 10^{-3}$ rad/s^2

(c) To starboard.

Example 13.14

The turbine rotor of a ship has a mass of 4000 kg. Its radius of gyration is 0.5 m and a speed of 3000 rpm clockwise when looking from stern. Determine the gyroscopic couple and its effect on the ship: (a) when the ship is steering to the left on a curve of 105 m radius at a speed of 36 km/h and (b) when the ship is pitching in a simple harmonic motion, the bow falling with its maximum velocity. The period of pitching is 45 s and the total angular displacement between the two extreme positions of pitching is 15°.

Solution

Given: $m = 4000$ kg, $K = 0.5$ m, $N = 3000$ rpm, $R = 105$ m, $v = 36$ km/h, $t_p = 45$ s, $2A = 15°$

$$v = \frac{36 \times 10^3}{3600} = 10 \text{ m/s}, \ \omega = \frac{2\pi \times 3000}{60} = 314.2 \text{ rad/s}$$

$$I = mK^2 = 4000 \times (0.5)^2 = 1000 \text{ kg}\cdot\text{m}^2$$

$$\omega_p = \frac{v}{R} = \frac{10}{105} = 0.09524 \text{ rad/s}$$

Gyroscopic couple, $C = I\omega\omega_p = 1000 \times 314.2 \times 0.09524 = 29924$ Nm

When the rotor rotates clockwise and ship takes left turn, looking from stern, the effect of reactive gyroscopic couple is to raise the bow and lower the stern.

(b) $$A = 7.5° \text{ or } \frac{\pi}{180} \times 7.5 = 0.131 \text{ rad}$$

Angular velocity of SHM, $\omega_o = \frac{2\pi}{t_p} = \frac{2\pi}{45} = 0.1396$ rad/s

$$(\omega_p)_{max} = A\,\omega_o = 0.131 \times 0.1396 = 0.0183 \text{ rad/s}$$

Gyroscopic couple, $C = I\omega(\omega_p)_{max} = 1000 \times 314.2 \times 0.0183 = 5750$ Nm

When the bow is falling, the effect of gyro-couple is to move the ship towards port side.

Example 13.15

The rolling moment on a ship at a given instant is 12×10^6 Nm clockwise viewed from the rear. The rotor of the stabilizing gyroscope is of 12×10^4 kg mass and spins at 1200 rpm clockwise when viewed from above. If the radius of the wheels about the spin axis is 2 m, determine the angular velocity of the precession to maintain the ship in an upright position.

Solution

Given: $C = 12 \times 10^6$ Nm, $m = 12 \times 10^4$ kg, $N = 1200$ rpm, $d_\omega = 2$ m

$$I = \frac{md^2_w}{4}$$

$$= 12 \times 10^4 \text{ kg} \cdot \text{m}^2$$

$$\omega = \frac{2\pi N}{60} = \frac{2\pi \times 1200}{60}$$

$$= 125.664 \text{ rad/s}$$

$$\omega_p = \frac{C}{I\omega}$$

$$= \frac{12 \times 10^6}{12 \times 10^4 \times 125.664}$$

$$= 0.796 \text{ rad/s}$$

(Counter-clockwise viewed from rear)

Example 13.16

For a single cylinder engine determine the bearing forces caused by the gyroscopic action of the flywheel ($I = 0.32$ kg·m²) as the engine traverses a 305 m radius curve at 96.6 km/h in a turn to the right. The engine speed is 3300 rpm and is turning clockwise when viewed from the front of the engine. The centre distance between the bearings is 152 mm.

■ **Solution**

Given: $I_w = 0.32$ kg·m², $R = 305$ m, $v = 96.6$ km/h or 26.83 m/s, $N_e = 3300$ rpm, $L = 152$ mm

$$\omega_e = \frac{2\pi N_e}{60} = \frac{2\pi \times 3300}{60}$$

$$= 345.6 \text{ rad/s}$$

$$\omega_p = \frac{v}{R}$$

$$= \frac{26.83}{305} = 0.088 \text{ rad/s}$$

$$C = I_w \omega_e \omega_p$$

$$= 0.32 \times 345.6 \times 0.088 = 9.728 \text{ N m}$$

$$\text{Bearing reaction} = \frac{C}{L} = \frac{9.728}{0.152} = 64 \text{ N}$$

13.8 STABILITY OF A FOUR-WHEEL VEHICLE TAKING A TURN

Consider a four-wheel vehicle taking a left turn as shown in Fig.13.12.

Let M = mass of the vehicle
a = track width
l = wheel base
h = height of C.G. of vehicle above the ground
l_1 = horizontal distance of C.G. from the front axle
l_2 = horizontal distance of C.G. from the rear axle

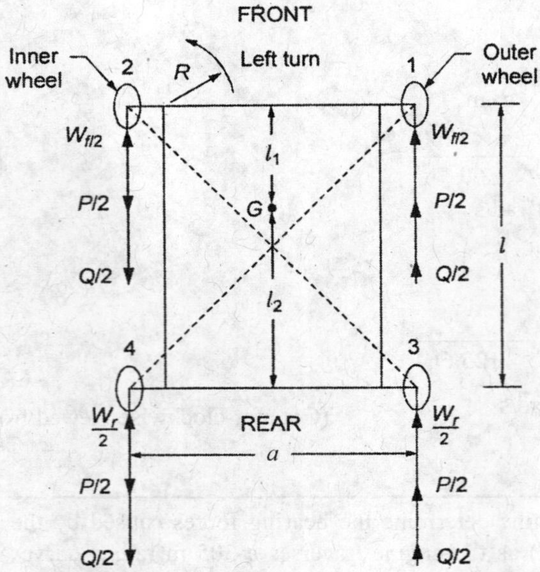

Fig.13.12 Four wheel vehicle taking a turn

M_w = mass of each wheel
K_w = radius of gyration of each wheel
M_e = mass of engine flywheel and transmission etc.
K_e = radius of gyration fo engine flywheel and transmission etc.
i = speed ratio from wheel to engine
R = radius of curve being negotiated
v = liner speed of vehicle

The forces accounting for the stability of the vehicle are:

1. Weight of the vehicle $W = Mg$, giving rise to upward reaction at each wheel.
2. Precession of vehicle.

(a) ***Dead weights***:

Reaction of ground on each front wheel, $\frac{W_f}{2} = \frac{Wl_2}{2l}$ (upwards) (13.8a)

Reaction of ground on each rear wheel, $\frac{W_r}{2} = \frac{Wl_1}{2l}$ (upwards) (13.8b)

(b) ***Centrifugal couple***:

Centrifugal force acting outward at the C.G. of the vehicle, $F_c = \frac{Mv^2}{R}$

Centrifugal couple, $C_c = F_c\, h = \frac{Mv^2 h}{R}$

The couple is balanced by the vertical reactions at the four wheels being +ve at the outer and –ve at the inner wheels.

Reaction of ground on each outer wheel, $\frac{Q}{2} = \frac{C_c}{2a}$ (upwards) (13.9a)

Reaction of ground on each inner wheel, $\frac{Q}{2} = \frac{C_c}{2a}$ (downwards) (13.9b)

(c) *Gyroscopic couple:*

(i) Due to wheels.

Moment of inertia of each wheel, $I_w = M_w K_w^2$

Moment of inertia of engine, flywheel, etc., $I_e = M_e K_e^2$

Velocity of spin of each wheel about its own axis, $\omega_w = \dfrac{v}{r_w}$

Velocity of spin of engine, flywheel etc., $\omega_e = i\omega_w$

Angular velocity of precession of wheels, $\omega_{pw} = \omega_{pe} = \dfrac{v}{R}$

Gyroscopic couple due to four wheels, $C_{gw} = 4I_w \cdot \omega_w \cdot \omega_{pw}$

Reaction of ground on each outer wheel, $\dfrac{P}{2} = \dfrac{C_{gw}}{2a}$ (upwards) (13.10a)

Reaction of ground on each inner wheel, $\dfrac{P}{2} = \dfrac{C_{gw}}{2a}$ (downwards) (13.10b)

(ii) Gyroscopic couple due to other rotating parts of the engine, like flywheel etc.,

$$C_{ge} = I_e \cdot \omega_e \cdot \omega_{pe}$$

Load due to gyro-couple due to engine etc., $\dfrac{F}{2} = \pm \dfrac{C_{ge}}{2l}$

Reaction of ground on each outer wheel, $\dfrac{F}{2} = \pm \dfrac{C_{ge}}{2l}$ (upwards) (13.11a)

Reaction of ground on each inner wheel, $\dfrac{F}{2} = \pm \dfrac{C_{ge}}{2l}$ (downwards) (13.11b)

Take +ve sign when the wheel and engine rotating parts rotate in the same direction, otherwise take –ve sign. Vertical reaction will be produced due to this gyroscopic couple. The reaction will be +ve at the outer wheels and –ve on the inner wheels.

Total vertical reactions at the wheels are given in Table 13.2 with the assumption that the engine and the wheel rotate in the same direction clockwise.

Table 13.2 Vertical reactions in a four wheeler

Wheel (wheel number)	*Vertical reaction*	
	Vehicle taking left turn	*Vehicle taking right turn*
Front outer (1)	$\dfrac{W_f}{2} + \dfrac{P}{2} + \dfrac{Q}{2} + \dfrac{F}{2}$	$\dfrac{W_f}{2} - \dfrac{P}{2} - \dfrac{Q}{2} - \dfrac{F}{2}$
Front inner (2)	$\dfrac{W_f}{2} - \dfrac{P}{2} - \dfrac{Q}{2} - \dfrac{F}{2}$	$\dfrac{W_f}{2} + \dfrac{P}{2} + \dfrac{Q}{2} + \dfrac{F}{2}$
Rear outer (3)	$\dfrac{W_r}{2} + \dfrac{P}{2} + \dfrac{Q}{2} + \dfrac{F}{2}$	$\dfrac{W_r}{2} - \dfrac{P}{2} - \dfrac{Q}{2} - \dfrac{F}{2}$
Rear inner (4)	$\dfrac{W_r}{2} - \dfrac{P}{2} - \dfrac{Q}{2} - \dfrac{F}{2}$	$\dfrac{W_r}{2} + \dfrac{P}{2} + \dfrac{Q}{2} + \dfrac{F}{2}$

The reaction on the rear inner wheel will be minimum when the vehicle is taking a left turn.

$$R_4 = \frac{W_r}{2} - \frac{P}{2} - \frac{Q}{2} - \frac{F}{2} \quad (13.12)$$

When R_4 is zero or negative, the tyres of the vehicle of rear inner wheels will leave the ground tending to overturn the vehicle.

For $$R_4 \leq 0, \frac{W_r}{2} \leq \frac{(P+Q+F)}{2}$$

or $$W_r \leq (P+Q+F) \quad (13.13)$$

Thus, the vehicle may overturn when

1. ω_w is high, i.e., the vehicle is running at a high speed.
2. h is high, i.e., the C.G. of the loaded vehicle is sufficiently high above the ground.
3. R is small, i.e., the vehicle is taking a sharp turn.
4. W is large, i.e., the vehicle is overloaded.

In order to reduce the total gyroscopic couple, the engine must be provided with a heavy flywheel, which should rotate in the opposite direction to that of the wheels.

Example 13.17

A motor car negotiates a curve of 40 m radius at a speed of 60 km/h. Determine the magnitudes of the centrifugal and gyroscopic couples acting on the motor car and state the effect of each of these on the road reactions on the wheels. Assume the following:

(a) Each road wheel has a moment of inertia of 4 kg · m² and an effective wheel radius of 0.5 m.

(b) The rotating parts of the engine and transmission are equivalent to a flywheel of mass 80 kg with a radius of gyration of 0.1 m. The engine turns in a clockwise direction when viewed from the front.

(c) The back axle ratio is 4:1 and the drive through the gear box is direct.

(d) The car weighs 10 kN and has its centre of gravity at 0.6 m above the road level. The car takes a right hand turn.

Solution

Given: $R = 40$ m, $v = 60$ km/h, $I_w = 4$ kg – m², $r_w = 0.5$ m, $M_e = 80$ kg, $K_e = 0.1$ m, $i = 4$, $W = 10$ kN, $h = 0.6$ m

$$v = 60 \times \frac{1000}{3600} = 16.67 \text{ m/s}$$

Angular velocity of wheel,

$$\omega_w = \frac{v}{r} = \frac{16.67}{0.5} = 314.33 \text{ rad/s}$$

Angular velocity of precession of wheels,

$$\omega_{pw} = \frac{v}{R} = \frac{16.67}{40} = 0.417 \text{ rad/s}$$

(a) Gyroscopic couple due to wheels,

$$C_w = 4I_w\ \omega_w\ \omega_{pw}$$
$$= 4 \times 4 \times 314.33 \times 0.417 = 223 \text{ Nm}$$

The reaction gyro-couple due to wheels will tend to lift the inner wheels and depress the outer wheels.

(b) Gyroscopic couple due to engine rotating parts,

$$C_e = I_e\ \omega_e\ \omega_{pe}$$
$$= M_e K_e^2 \times i\omega_w\ \omega_{pe} \qquad [\because\ \omega_{pe} = \omega_{pw}]$$
$$= 80 \times 0.1^2 \times 4 \times 314.33 \times 0.417$$
$$= 44.48\ \text{N m}$$

The reaction gyro-couple due to engine rotating parts will tend to lift the front wheels and depress the rear wheels.

(c) Centrifugal force,

$$F_c = \frac{Wv^2}{gR}$$
$$= 10 \times 1000 \times \frac{(16.67)^2}{9.81 \times 40}$$
$$= 7078.95\ \text{N}$$

Centrifugal couple, $C_c = F_c\, h = 7078.95 \times 0.6 = 4247.4\ \text{N m}$

The reaction centrifugal couple will tend to lift the inner wheels and depress the outer wheels.

Example 13.18

A rear engine automobile is travelling along a track of 100 m mean radius. Each of the four wheels has a moment of inertia of 2 kg·m² and an effective diameter of 0.6 m. The rotating parts of the engine have a moment of inertia of 1.25 kg·m². The engine axis is parallel to the rear axle and the crankshaft rotates in the same direction as the wheels. The gear ratio of engine to back axle is 3:1. The automobile mass is 1500 kg and its centre of gravity is 0.5 m above the road level. The width of track of the vehicle is 1.5 m.

Determine the limiting speed of the vehicle around the curve for all four wheels to maintain contact with the road surface if it is not banked.

■ Solution

Given: $R = 100$ m, $I_w = 2$ kg·m² $d_w = 0.6$ m, $i = 3$, $m = 1500$ kg, $h = 0.5$ m, $a = 1.5$ m

Let v = linear speed of the vehicle, m/s

Angular speed of the vehicle,

$$\omega_w = \frac{2v}{d_w} = \frac{2v}{0.6} = \frac{v}{0.3}\ \text{rad/s}$$

Angular speed of engine,

$$\omega_e = i\omega_\omega = \frac{3v}{0.3} = 10v\ \text{rad/s}$$

Angular velocity of precession,

$$\omega_{pw} = \frac{v}{R} = \frac{v}{100}\ \text{rad/s}$$

Gyroscopic couple of 4 wheels and engine etc., $C_g = 4I_w \cdot \omega_w \cdot \omega_{pw} + I_e \cdot \omega_e \cdot \omega_{pw}$

$$= 4I_w \cdot \frac{v}{0.3} \cdot \frac{v}{100} + I_e \cdot 10v \cdot \frac{v}{100}$$

$$= 4 \times 2 \times \frac{v^2}{30} + 1.25 \times \frac{v^2}{10}$$

$$= (0.2667 + 0.125)\, v^2$$

$$= 0.3917v^2 \text{ N m}$$

Centrifugal force, $F_c = \dfrac{mv^2}{R} = \dfrac{1500v^2}{100} = 15v^2$ N

Centrifugal couple, $C_c = F_c \cdot h$

Maximum lift due to centrifugal couple on one wheel, $Q/2 = \dfrac{C_c}{2a} = F_c \cdot \dfrac{h}{2a}$

$$= 15v^2 \times \frac{0.5}{3} = 2.5v^2 \text{ N}$$

Maximum lift due to gyro-couple,

$$\frac{P}{2} = \frac{C_g}{2a} = 0.3917\frac{v_2}{3} = 0.1306v^2 \text{ N}$$

For safe driving, weight on one wheel should be greater than the maximum lift.

$$\frac{Mg}{4} > \left(\frac{Q}{2} + \frac{P}{2}\right)$$

$$\frac{1500 \times 9.81}{4} > (2.5 + 0.1306)\, v^2$$

or $$v^2 < \frac{1500 \times 9.81}{4 \times 2.6306}$$

or $v < 37.396$ m/s or < 134.62 km/h

Example 13.19

A rail car has a total mass of 4000 kg. The moment of inertia of each wheel together with its gearing is 20 kg·m². The centre distance between the two wheels on an axle is 1.5 m and each wheel is 400 mm radius. Each axle is driven by a motor, the speed ratio between the two being 1:14. Each motor with its gear has a moment of inertia of 15 kg·m² and runs in a direction opposite to that of its axle. The centre of gravity of the car is 1 m above the rails.

Determine the limiting speed for the car when moving on a curve of 250 m radius such that no wheel leaves the rails.

■ Solution

Given: $W = 1000$ N, $I_w = 20$ kg·m², $a = 1.5$ m, $r_w = 0.4$ m, $i = 3$, $I_m = 15$ kg·m², $R = 250$ m, $h = 1$ m.

Let N_w = rpm of wheel, then speed of motor, $N_m = 3N_w$

$$\omega_w = \frac{2\pi \times N_w}{60} = \frac{\pi N_w}{30} \text{ rad/s}, \omega_m = 3\,\omega_w$$

$$v_w = \omega_w r_w = \frac{\pi N_w \times 0.4}{30} = 0.041888\, N_w \text{ m/s}$$

$$\omega_p = \frac{v_w}{R} = \frac{\pi N_w \times 0.4}{30 \times 250} = 1.6755 \times 10^{-4} \times N_w \text{ rad/s}$$

Gyroscopic couple due to four wheels:

$$C_w = 4\, I_w \omega_w \omega_p$$
$$= 4 \times 20\left(\frac{\pi N_w}{30}\right) \times 1.6755 \times 10^{-4} \times N_w$$
$$= 14.037 \times 10^{-4} \times N_w^2 \text{ Nm}$$

Gyroscopic couple due to motor:

$$C_m = I_m \omega_m \omega_p$$
$$= 15 \times \left(\frac{3\pi N_w}{30}\right) \times 1.6755 \times 10^{-4} \times N_w$$
$$= 8.8956 \times 10^{-4} \times N_w^2 \text{ Nm}$$

Total gyroscopic couple:

$$C = C_w - C_m = 6.1414 \times 10^{-4} \times N_w^2 \text{ Nm}$$

Negative sign has been taken because the speed of motor is opposite to that of wheels. Vertical reaction at each of the outer or inner wheels due to gyroscopic effect,

$$\frac{P}{2} = \pm \frac{C}{2a}$$
$$= \frac{\pm 6.1414 \times 10^{-4} \times N_w^2}{2 \times 1.5}$$
$$= \pm 2.04713 \times 10^{-4} \times N_w^2 \text{ N}$$

Centrifugal couple, $C_c = \dfrac{W v_w^2 h}{gR}$

$$= 1000 \times \frac{(0.041888 N_w)^2 \times 1}{9.81 \times 250}$$
$$= 0.7154 \times 10^{-3} \times N_w^2 \text{ Nm}$$

Vertical reaction at each outer or inner wheel due to centrifugal effect,

$$\frac{Q}{2} = \pm \frac{C_c}{2a}$$
$$= \frac{0.7154 \times 10^{-3} \times N_w^2}{3}$$
$$= 2.3848 \times 10^{-4} \times N_w^2 \text{ N}$$

For no wheel to leave the rails,

$$W = 2(P + Q)$$
$$1000 = 4(2.04713 \times 10^{-4} \times N_w^2 + 2.3848 \times 10^{-4} \times N_w^2)$$
$$= 18.72772 \times 10^{-4} \times N_w^2$$
$$N_w = 751 \text{ rpm}$$
$$v_w = 113.26 \text{ km/h}$$

Example 13.20

A four wheel trolley car of total mass 2000 kg running on rails of 1 m gauge, rounds a curve of 30 m radius at 45 km/h. The track is banked at 10°. The wheels have an external diameter of 0.6 m and each pair of an axle has a mass of 250 kg. The radius of gyration of each pair is 250 mm. The height of C.G. of the car above the wheel base is 1 m. Allowing for centrifugal force and gyroscopic couple action, determine the pressure on each rail.

■ Solution

Given: $M = 2000$ kg, $a = 1$ m, $R = 30$ m, $v = 45$ km/h or 12.5 m/s, $\theta = 10°$,
$d_w = 0.6$ m, $M_w = 250$ kg, $K_w = 250$ mm, $h = 1$ m

$$R_A + R_B = \text{Mg}\cos\theta + \frac{Mv^2 \sin\theta}{R}$$

$$= 2000 \times 9.81 \times \cos 10° + \frac{2000 \times (12.5)^2 \times \sin 10°}{30}$$

$$= 21131 \text{ N}$$

Refer to fig. 13.14.
Taking moments about B, we have

$$R_A \times a = \left(Mg\cos\theta + \frac{Mv^2 \sin\theta}{R}\right) \times \frac{a}{2} + \left(Mg\sin\theta - \frac{Mv^2\cos\theta}{R}\right) \times h$$

$$R_A = 3714\ N, \quad R_B = 17417 \text{ N}$$

$$\omega_w = \frac{v}{r_w} = \frac{12.5}{0.3} = 41.67 \text{ rad/s}$$

$$\omega_p = \frac{v}{R} = \frac{12.5}{30} = 0.4167 \text{ rad/s}$$

$$C = I\omega_w \cos\theta\, \omega_p$$

$$= M_w K_w^2 w_w \cos\theta\, \omega_p$$

$$= 272.3 \text{ N m}$$

Force at each pair of wheels on inner wheels on rail due to gyroscopic couple,

$$P = \frac{C}{a} = \frac{272.3}{1} = 272.3 \text{ N}$$

Total pressure on inner wheels, $P_i = R_A - P = 3441.7$ N
Total pressure on outer wheels, $P_o = R_B + P = 17689.3$ N

Example 13.21

A racing car of mass 2000 kg has a wheel base of 2 m and track width of 1 m. The C.G. lies midway between the front and rear axles and is 0.4 m above the ground. The engine of the car has a flywheel rotating in a clockwise direction when seen from the front at 6000 rpm. The moment of inertia of the flywheel is 50 kg·m². If the car takes a curve of 15 m radius towards right, while running at 45 km/h, find the reaction between the wheels and the ground considering the gyroscopic and centrifugal effects of the flywheel and the weight of the car, respectively.

■ **Solution**

Given: $M = 2000$ kg, $a = 1$ m, $l = 2$ m, $I_e = 50$ kg·m², $R = 15$ m, $h = 0.4$ m, $N_e = 6000$ rpm, $v = 45$ km/h.

$$\omega_e = \frac{2\pi N_e}{60} = \frac{2\pi \times 6000}{60} = 628.3 \text{ rad/s}$$

$$\omega_p = \frac{v}{R} = \frac{45 \times 1000}{3600 \times 15} = 0.833 \text{ rad/s}$$

Gyroscopic couple due to engine:

$$C_e = I_e \omega_e \omega_p = 50 \times 628.3 \times 0.833 = 26179 \text{ Nm}$$

$$\frac{P}{2} = \pm\frac{C_e}{2a} = \pm\frac{26179}{2} = 13089.5 \text{ N}$$

Centrifugal couple, $C_c = \dfrac{Mv^2 h}{R} = \dfrac{2000 \times (12.5)^2 \times 0.4}{15} = 8333.3$ Nm

$$\frac{Q}{2} = \pm\frac{C_c}{2a} = \frac{8333.3}{2} = 4166.7 \text{ N}$$

Total pressure on inner wheels, $P_i = \dfrac{Mg}{4} - \dfrac{P}{2} - \dfrac{Q}{2} = -12351.2$ N

Total pressure on outer wheels, $P_o = \dfrac{Mg}{4} + \dfrac{P}{2} + \dfrac{Q}{2} = 22162.2$ N

Example 13.22

A racing car of mass 3000 kg has a wheel base of 2.5 m and track of 1.5 m. The C.G. is located 0.6 m above the ground level and 1.5 m from the rear axle. Each wheel is of 1 m diameter and 0.8 kg·m² moment of inertia. The back axle ratio is 4.5. The drive shaft engine flywheel and transmission are rotating cw when viewed from the front with equivalent mass of 150 kg with radius of gyration 0.2 m. Determine the load distribution on the wheels if the car is rounding a curve of 80 m radius at 120 km/h when (a) taking a right turn, and (b) taking a left turn.

■ **Solution** Given: $M = 3000$ kg, $\ell = 2.5$ m, $a = 1.5$ m, $h = 0.6$ m, $\ell_2 = 1.5$ m, $d_w = 1$m, $I_w = 0.8$ kg·m², $i = 4.5$, $M_c = 150$ kg, $K_e = 0.2$ m, $R = 80$ m, $V = 120$ km/h

Dead weights:

On each rear wheel, $\dfrac{W_r}{2} = \dfrac{M_g \ell_2}{2\ell} = \dfrac{3000 \times 9.81 \times 1}{2.5 \times 2} = 5886$ N

On each front wheel, $\dfrac{W_f}{2} = \dfrac{M_g(\ell - \ell_2)}{2\ell} = \dfrac{3000 \times 9.81 \times 1.5}{2.5 \times 2} = 8829$ N

Centrifugal couple: $C_c = \dfrac{Mv^2 h}{R} = 3000 \times \left(\dfrac{120 \times 1000}{3600}\right)^2 \times \dfrac{0.6}{80} = 25000$ Nm

$\dfrac{Q}{2} = \dfrac{C_c}{2a} = \dfrac{25000}{3} = 8333.3$ N, vertically upwards on outer and downwards on inner wheels.

Gyroscopic couple due to wheels:

$$\omega_w = \frac{v}{r_w} = \frac{120 \times 1000}{3600 \times 0.5} = 66.67 \text{ rad/s}$$

$$\omega_{pw} = \frac{v}{R} = \frac{33.33}{80} = 0.4167 \text{ rad/s}$$

$$C_{gw} = 4I_w\omega_w\omega_{pw} = 4 \times 0.80 \times 66.67 \times 0.4167 = 88.9 \text{ Nm}$$

$$\frac{P}{2} = \frac{C_{gw}}{2a} = \frac{88.9}{3} = 29.63 \text{ N, downwards on inner and upward on outer wheels.}$$

Gyroscopic couple due to engine flywheel and transmission etc.

$$\omega_e = 4.5\ \omega_w \text{ rad/s}$$

$$\omega_{pe} = \omega_{pw},\ I_e = M_eK_e^2 = 150 \times (0.2)^2 = 6 \text{ kg·m}^2$$

$$C_{ge} = I_e\omega_e\omega_{pe} = 6 \times 4.5 \times 66.67 \times 0.4167 = 750 \text{ Nm}$$

$$\frac{F}{2} = \frac{C_{ge}}{2l} = \frac{750}{5} = 150 \text{ N, downwards on front and upwards on rear wheels.}$$

(a) For left turn, the reactions are:

$$R_1 = W_f + Q + P + F = 17341.93 \text{ N}$$
$$R_2 = W_f - Q - P - F = 316.07 \text{ N}$$
$$R_3 = W_r + Q + P + F = 14398.93 \text{ N}$$
$$R_4 = W_r - Q - P - F = -2626.93 \text{ N}$$

(b) For right turn, the reactions are:

$$R_1 = W_f - Q - P - F = 316.07 \text{ N}$$
$$R_2 = W_f + Q + P + F = 17341.93 \text{ N}$$
$$R_3 = W_r - Q - P - F = -2626.93 \text{ N}$$
$$R_4 = W_r + Q + P + F = 14398.93 \text{ N}$$

Example 13.23

A four-wheel vehicle of mass 2500 kg has a wheel base 2.5 m, track width 1.5 m, and height of centre of gravity 0.6 m above the ground level and lies at 1 m from the front axle. Each wheel has an effective diameter of 0.8 m and a moment of inertia of 0.8 kg · m². The drive shaft, engine flywheel and transmission are rotating at four times the speed of road wheels, in clockwise direction when viewed from the front, and is equivalent to a mass of 80 kg having a radius of gyration of 100 mm. If the vehicle is taking a right turn of 60 m radius at 60 km/h, find the load on each wheel.

Solution

Given: $M = 2500$ kg, $\ell = 2.5$ m, $a = 1.5$ m, $h = 0.6$ m, $\ell = 1$ m, $r_w = 0.4$ m, $I_w = 0.8$ kg·m², $i = 4$, $M_e = 80$ kg, $K_e = 100$ mm, $R = 60$ m, $v = 60$ km/h.

Let W_f, W_r = weight on the front and rear wheels, respectively.
Taking moments about the front wheels (Fig.13.13), we have

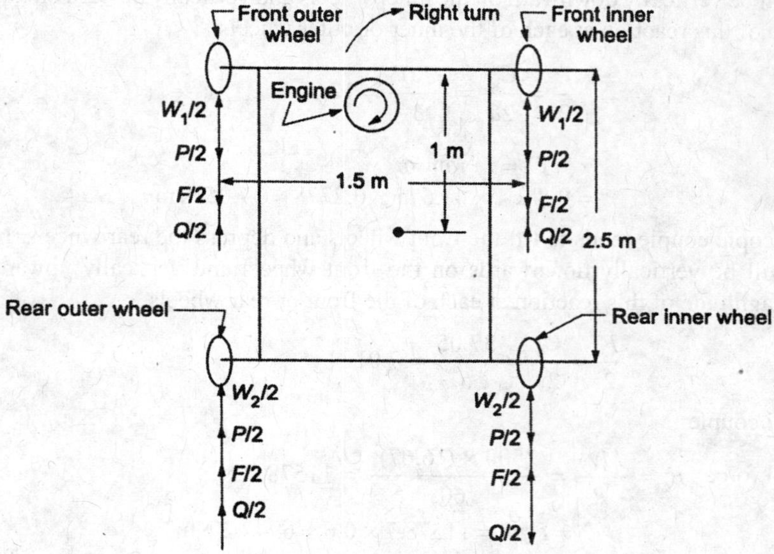

Fig.13.13 Force diagram for a four wheeler taking a turn

$$W_r \times \ell = mg\ell_1$$
$$2.5W_r = 2500 \times 9.81 \times 1$$
$$W_f = 9810 \text{ N}$$
$$W_f = Mg - W_2 = 2500 \times 9.81 - 9810 = 14715 \text{ N}$$

Weight on each of the front wheels $\dfrac{W_f}{2} = 7357.5$ N

Weight on each of the rear wheels $\dfrac{W_r}{2} = 4905$ N

Gyroscopic Effect

$$v = \frac{60 \times 1000}{3600} = 16.67 \text{ m/s}$$
$$I_e = M_e K_e^2 = 80 \times (0.1)^2 = 0.8 \text{ kg} \cdot \text{m}^2$$
$$\omega_w = \frac{v}{r_w} = \frac{16.67}{0.4} = 41.675 \text{ rad/s}$$

$$\omega_p = \frac{v}{R} = \frac{16.67}{60} = 0.2778 \text{ rad/s}$$

$$C_w = 4I_w \cdot \omega_w \cdot \omega_p$$

$$= 4 \times 0.8 \times 41.675 \times 0.2778 = 37.05 \text{ Nm}$$

This gyroscopic couple tends to lift the inner wheels and depress the outer wheels. In other words, the reaction will be vertically downward on the inner wheels and vertically upward on the outer wheels. The magnitude of this reaction at each of the inner or outer wheels,

$$\frac{P}{2} = \frac{C_w}{2a} = \frac{37.05}{3} = 12.35 \text{ N}$$

$$C_e = I_e \cdot i\omega_w \cdot \omega_p$$

$$= 0.8 \times 4 \times 41.675 \times 0.2778 = 37.05 \text{ Nm}$$

This gyroscopic couple tends to lift the front wheels and depress the rear wheels. In other words, the reaction will be vertically downwards on the front wheels and vertically upwards on the rear wheels. The magnitude of this reaction at each of the front or rear wheels,

$$\frac{F}{2} = \frac{C_e}{2\ell} = \frac{37.05}{5} = 7.01 \text{ N}$$

Centrifugal couple

Centrifugal force, $F_c = \frac{Mv^2}{R} = \frac{2500 \times (16.67)^2}{60} = 11578.7 \text{ N}$

$$C_c = F_c \cdot h = 11578.7 \times 0.6 = 6947.2 \text{ Nm}$$

The reactions due to this couple are vertically downwards on the inner wheels and vertically upwards on the outer wheels. The magnitude of this reaction on each of the inner and outer wheels,

$$\frac{Q}{2} = \frac{C_c}{2a} = \frac{6947.2}{3} = 2315.7 \text{ N}$$

Load on the inner front wheel $\frac{W_f}{2} - \frac{P}{2} - \frac{F}{2} - \frac{Q}{2}$

$$= 7357.5 - 12.35 - 7.01 - 2315.7 = 5022.44 \text{ N}$$

Load on the front outer wheel $= \frac{W_f}{2} + \frac{P}{2} - \frac{F}{2} + \frac{Q}{2}$

$$= 7357.5 + 12.35 - 7.01 + 2315.7 = 9678.54 \text{ N}$$

Load on the rear inner wheel $\frac{W_r}{2} - \frac{P}{2} - \frac{F}{2} - \frac{Q}{2}$

$$= 4905 - 12.35 + 7.01 - 2315.7 = 25814.96 \text{ N}$$

Load on the rear outer wheel $= \frac{W_r}{2} + \frac{P}{2} + \frac{F}{2} + \frac{Q}{2}$

$$= 4905 + 12.35 + 7.01 + 2315.7 = 7240.06 \text{ N}$$

Example 13.24

A four-wheel trolley car of total mass 2500 kg running on rails of 1.6 m gauge, negotiates a curve of 40 m radius at 60 km/h. The track is banked at 10°. The wheels have an external diameter of 0.8 m and each pair with axle has a mass of 250 kg. The radius of gyration for each pair is 0.4 m. The height of centre of gravity of the car above the wheel base is 0.9 m. Determine the pressure on each rail, allowing for centrifugal and gyroscopic couple actions.

■ Solution

Here $M = 2500$ kg, $a = 1.6$ m, $R = 40$ m, $v = \dfrac{60 \times 1000}{3600} = 16.67$ m/s, $\theta = 10°$, $r_w = 0.4$ m, $M_w = 250$ kg, $K_w = 0.4$ m, $h = 0.9$ m.

Let R_A and R_B be the reactions at A and B, respectively. The various forces acting on the trolley car are shown in Fig.13.14.

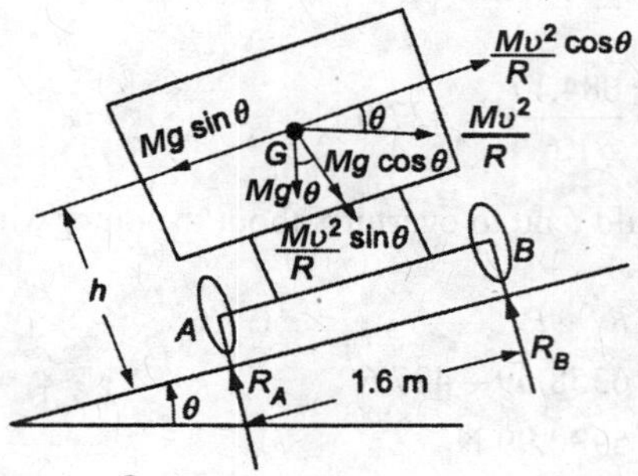

Fig.13.14 Forces acting on a vehicle negotiating a curve

Resolving the forces perpendicular to the track, we have

$$R_A + R_B = Mg\cos\theta + \frac{Mv^2\sin\theta}{R}$$

$$= 2500 \times 9.81 \times \cos 10° + \frac{2500 \times (16.67)^2 \times \sin 10°}{40}$$

$$= 24152.4 + 3015.93$$

$$= 27168.33 \text{ N}$$

Now taking moments about B, we have

$$R_A \times a = \left(Mg\cos\theta + \frac{Mv^2\sin\theta}{R}\right)\cdot\left(\frac{a}{2}\right) + \left(Mg\cos\theta - \frac{Mv^2\cos\theta}{R}\right)\cdot h$$

$$R_A = \left[2500 \times 9.81 \times \cos 10° + \frac{2500 \times (16.67)^2 \times \sin 10°}{40}\right] \times 0.5$$

$$+ \left[2500 \times 9.81 \times \sin 10° - \frac{2500 \times (16.67)^2 \times \cos 10°}{40}\right]\cdot(0.9/1.6)$$

$$= 13584.17 - 7225.58 = 6358.59 \text{ N}$$

$$R_B = 20809.74 \text{ N}$$

$$\omega_w = \frac{v}{r_w} = \frac{16.67}{0.4} = 41.675 \text{ rad/s}$$

$$\omega_p = \frac{v}{R} = \frac{16.67}{40} = 0.41675 \text{ rad/s}$$

$$\begin{aligned} C &= I \cdot \omega \cos\theta \cdot \omega_p \\ &= M_w K_w^2 \cdot \omega_w \cos\theta \cdot \omega_p \\ &= 250 \times (0.4)^2 \times 41.675 \times \cos 10° \times 0.41675 \\ &= 684.17 \text{ Nm} \end{aligned}$$

The force at each pair of wheels on each rail due to the gyroscopic couple,

$$\begin{aligned} P &= \frac{C}{a} \\ &= \frac{684.17}{1.6} = 427.6 \text{ N} \end{aligned}$$

Due to this force the car would tend to overturn about the outer wheels. Total pressure on the inner rail,

$$\begin{aligned} P_i &= R_A - P \\ &= 6358.59 - 427.6 \\ &= 5930.99 \text{ N} \end{aligned}$$

Pressure on the outer rail,

$$\begin{aligned} P_o &= R_B + P \\ &= 20809.74 + 427.6 \\ &= 21237.34 \text{ N} \end{aligned}$$

13.9 STABILITY OF A TWO-WHEEL VEHICLE TAKING A TURN

Consider a two-wheel vehicle taking a right turn as shown in Fig.13.15(a).

Let W = weight of the vehicle and its rider
- h = height of C.G. of the vehicle and the rider
- r_w = wheel radius
- R = track radius
- I_w = moment of inertia of each wheel
- I_e = moment of inertia of engine rotating parts
- ω_w = angular velocity of wheels
- ω_e = angular velocity of engine rotating parts
- $i = \omega_e/\omega_w$ = gear ratio
- v = linear velocity of vehicle = $r_w \omega_w$
- θ = angle of heel or inclination of vehicle to the vertical

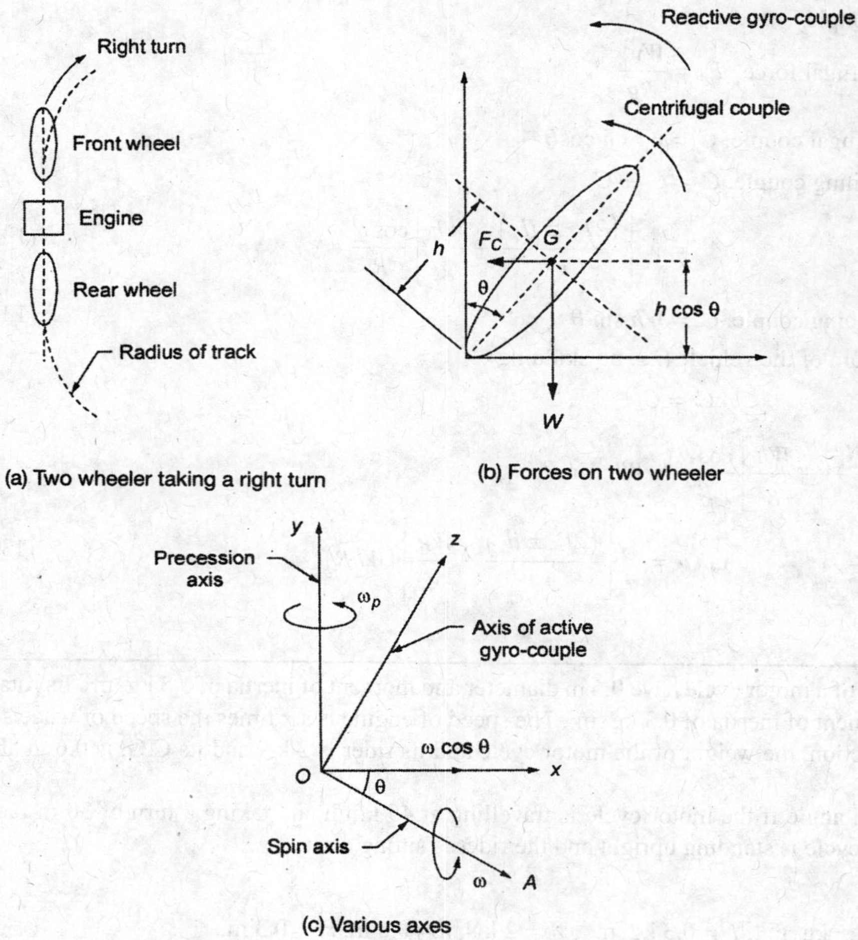

(a) Two wheeler taking a right turn

(b) Forces on two wheeler

(c) Various axes

Fig.13.15 Two wheel vehicle taking a turn

The effects of various forces are as follows:

1. Gyroscopic couple.

$$v = r_w \omega_w$$

or $$\omega_w = \frac{v}{r_w}$$

$$\omega_e = i\omega_w = \frac{iv}{r_w}$$

$$\omega_P = \frac{v}{R}$$

Gyroscopic couple, $C_g = (2I_w\omega_w \pm I_e\omega_e)\omega_p \cos\theta = v^2(2I_w \pm iI_e)\dfrac{\cos\theta}{r_w R}$

Take +ve sign when direction of rotation of wheels and engine is same.

2. Centrifugal couple.

Centrifugal force, $F_c = \dfrac{Wv^2}{Rg}$

Centrifugal couple, $C_c = F_c \cdot h\cos\theta$

Total overturning couple, $C_o = C_g + C_c$

$$= v^2\left[\frac{(2I_w \pm iI_e)}{r_w} + \frac{Wh}{g}\right]\frac{\cos\theta}{R} \quad (13.14)$$

Balancing couple, $C_b = Wh\sin\theta$ (13.15)

For equilibrium of the vehicle (*i.e.* no skidding),

$$C_o = C_b$$

$$v^2\left[\frac{(2I_w \pm iI_e)}{r_w} + \frac{Wh}{g}\right]\frac{\cos\theta}{R} = Wh\sin\theta$$

or

$$\tan\theta = v^2\left[\frac{(2I_w \pm iI_e)}{r_w} + \frac{Wh}{g}\right]/(WhR) \quad (13.16)$$

Example 13.25

The road wheels of a motor cycle have 0.6 m diameter and moment of inertia of 1.5 kg · m². Its rotating parts have a moment of inertia of 0.3 kg · m². The speed of engine is six times the speed of wheels and in the same direction, the weight of the motor cycle and its rider is 2 kN and its C.G. is 0.6 m above the road level.

Find the heel angle if the motor cycle is travelling at 45 km/h and taking a turn of 30 m radius, when the motor cycle is standing upright and the rider is sitting on it.

■ Solution

Given: $I_w = 1.5$ kg · m², $I_e = 0.3$ kg · m², $W = 2$ kN, $h = 0.6$ m, $r_w = 0.3$ m, $v = 45$ km/h, $i = 6$, $R = 30$ m

$\tan\theta = v^2[(2I_w + iI_e)/r_w + Wh/g]/(WhR)$

$= (45\times1000/3600)^2\,[(2\times1.5 + 6\times0.3)/0.3 + 2000\times0.6/9.81]/(2000\times0.6\times30)$

$= 0.60036$

$\theta = 30.98°$

Example 13.26

A thin disc is fixed to a shaft in such a way that it makes an angle of 2° with a plane at right angles to the axis of the shaft. The disc weights 25 N and it has a diameter of 0.5 m. If the shaft rotates at 1000 rpm, find the gyroscopic couple acting on the bearing.

■ Solution

Given: $W = 25$ N, $r = 0.25$ m, $\theta = 2°$, $\omega = 2\pi \times 1000/60 = 104.72$ rad/s

$C_{disc} = W\omega^2 r^2 \sin 2\theta/(8g)$

$= 25 \times (104.72)^2 \times (0.25)^2 \times \sin 4°/(8 \times 9.81)$

$= 184.45$ Nm

Example 13.27

A wheel of a vehicle travelling on a level track at 80 km/h, falls in a spot hole 15 mm deep and rises again in a total time of 0.15 s. The displacement of the wheel of the vehicle takes place with simple harmonic motion. The diameter of wheel is 1.5 m and the distance between the wheel centres is 1.8 m. The wheel pair with axle have a moment of inertia of 500 kg · m². Determine the magnitude and gyroscopic effects produced with this phenomenon.

■ Solution

Given: $v = 80 \times 1000/3600 = 22.22$ m/s

$I_w = 500$ kg · m², $r_w = 0.75$ m, $a = 1.8$ m, $h = 15$ mm, $t = 0.15$ s

Amplitude, $A_o = h/2 = 7.5$ mm

Maximum velocity while falling, $v_{max} = 2\pi A_o/t$

$= 2\pi \times 7.5 \times 10^{-3}/0.15 = 0.314$ m/s

$\omega_p = v_{max}/a = 0.314/1.8 = 0.174$ rad/s

$\omega = v/r_w = 22.22/0.75 = 29.63$ rad/s

$C = I_w \cdot \omega \cdot \omega_p$

$= 500 \times 29.63 \times 0.174 = 2577.8$ N m

As the axle goes down, the effect of this is to tend to turn the vehicle towards left as it moves forward.

Example 13.28

The moment of inertia of each wheel of a motorcycle is 1.5 kg · m². The rotating parts of the engine of the motor cycle have a moment of inertia of 0.28 kg · m². The speed of the engine is six times the speed of the wheels and is in same direction. The mass of the motor cycle is 250 kg and its centre of gravity is 0.6 m above the ground level.

Find the angle of heel if the motor cycle is travelling at 45 km/h and is taking a turn of 30 m radius. The wheel diameter is 0.6 m.

■ Solution

Given $I_w = 1.5$ kg · m², $I_e = 0.28$ kg · m², $M = 250$ kg, $h = 0.6$ m, $v_w = 45$ km/h or 12.5 m/s, $i = 6$, $R = 30$ m, $d_w = 0.6$ m

$$\tan\theta = v_w^2 \left[\frac{\left(\dfrac{2I_w + iI_e}{r_w} + Mh\right)}{WhR}\right]$$

$$= \frac{(12.5)^2\left[\dfrac{2 \times 1.5 + 6 \times 0.28}{0.3} + 250 \times 0.6\right]}{250 \times 9.81 \times 0.6 \times 30}$$

$$= 0.58614$$

$$\theta = 30.376°$$

Example 13.29

A racing motor cycle travels at 150 km/h round a curve of 120 m radius measured horizontally. The motor cycle and rider have a mass of 160 kg and their centre of gravity lies at 0.75 m above the ground level when the motor cycle is vertical. Each wheel is 0.6 m in diameter and has moment of inertia about its axis of rotation 1.5 kg · m². The engine has rotating parts whose moment of inertia about their axis of rotation is 0.3 kg · m² and rotates at five times the wheel speed in the same direction.

Find (a) the correct angle of banking of the track so that there is no tendency to side slip, and (b) the correct angle of inclination of the motor cycle and the rider to the vertical.

■ Solution

Given: $I_w = 1.5\ \text{kg}\cdot\text{m}^2$, $I_e = 0.3\ \text{kg}\cdot\text{m}^2$, $M = 160$ kg, $h = 0.75$ m, $v_w = 150$ km/h or 41.67 m/s, $i = 5$, $R = 120$ m, $d_w = 0.6$ m

$$\tan\theta = v_w^2\left[\frac{\left(\dfrac{2I_w + iI_e}{r_w} + Mh\right)}{WhR}\right]$$

$$= \frac{(41.67)^2\left[\dfrac{2\times 1.5 + 5\times 0.3}{0.3} + 160\times 0.75\right]}{160\times 9.81\times 0.75\times 120}$$

$$= 1.65915$$

$$\theta = 58.922°,\ 31.65915°$$

Example 13.30

The moment of inertia of each wheel of a motor cycle is 2 kg · m². The rotation parts of the engine of the cycle have a moment of inertia of 0.3 kg · m². The speed of the engine is 6 times the speed of the wheels and in the same sense. The weight of the motor cycle together with the rider is 2600 N and its CG is 0.6 m above the ground level when the cycle is standing upright and rider is sitting on it. Find the average angle of inclination with vertical for equilibrium if the cycle is travelling at 60 km/h and taking a turn of 30 m radius. Wheel diameter is 0.6 m.

■ Solution

Given: $I_w = 2\ \text{kg}\cdot\text{m}^2$, $I_e = 0.3\ \text{kg}\cdot\text{m}^2$, $W = 2600$ N, $h = 0.6$ m, $v = 60$ km/h or 16.67 m/s, $R = 30$ m, $d_w = 0.6$ m

$$\omega_w = \frac{v}{r_w} = \frac{16.67}{0.3} = 55.56\ \text{rad/s}$$

$$\omega_e = 6\times 55.56 = 333.33\ \text{rad/s}$$

$$\omega_p = \frac{v}{R} = \frac{16.67}{30} = 0.556\ \text{rad/s}$$

$$C_g = (2I_w\omega_w + I_e\omega_e)\, w_p \cos\theta$$

$$= (12\times 2\times 55.56 + 0.3\times 333.33)\times 0.556\cos\theta$$

$$= 179.165\cos\theta\ \text{N.m.}$$

$$F_c = \frac{Wv^2}{gR},\ C_c = F_c h\ \cos\theta = \frac{2600 \times (16.67)^2 \times 0.6 \times \cos\theta}{9.81 \times 30} = 1473 \cos\theta \text{ Nm}$$

$$C_o = C_g + C_c = (179.165 + 1473)\cos\theta = 1652.165 \cos\theta$$

$$C_b = Wh \sin\theta = 2600 \times 0.6 \sin\theta = 1560 \sin\theta$$

$C_b = C_o$ gives, $1560\ \sin\theta = 1652.165 \cos\theta$

$$\tan\theta = 1.059$$

$$\theta = 46.64^{\circ}$$

13.10 EFFECT OF PRECESSION ON A DISC FIXED RIGIDLY AT A CERTAIN ANGLE TO A ROTATING SHAFT

Consider a disc of radius r fixed rigidly to a rotating shaft (Fig.13.16) at a certain angle such that the polar axis of the disc makes an angle θ with the shaft axis. The shaft revolves with angular speed ω about its axis OX. Let OA be the diametral axis and OP the polar axis of the disc.

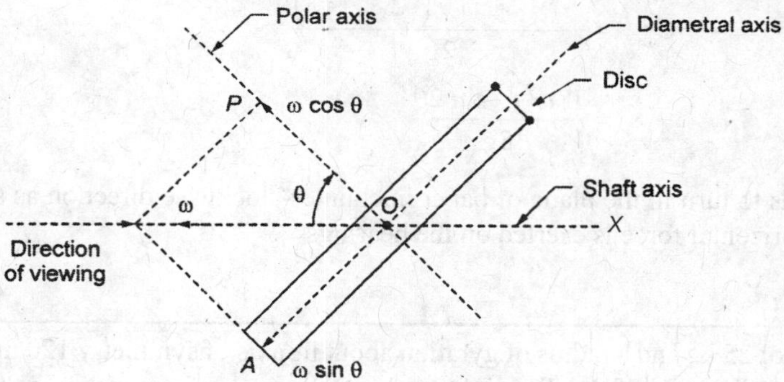

Fig.13.16 Disk fixed rigidly at certain angle to a rotating shaft

Angular velocity of spin of the disc about $OP = \omega \cos\theta$

Angular velocity of precession about $OA = \omega \sin\theta$

Let I_p = polar moment of inertia of the disc about axis $OP = \dfrac{Wr^2}{2g}$

Couple producing the precession, $C_P = I_P\omega \cos\theta\omega \sin\theta$

$$= 0.5\ I_P\omega^2 \sin 2\theta$$

The reaction couple, C_P tends to turn the disc in anti-clockwise direction, when viewed from the top, about an axis through O in the plane of paper.

Now consider the movement of point A about the polar axis OP. In this case, OA is the axis of spin and OP the axis of precession.

Angular velocity of spin about $OA = \omega \sin\theta$

Let I_A = polar moment inertia of the disc about OA

Gyroscopic couple about OA, $C_A = I_A \cdot \omega \sin\theta \cdot \omega \cos\theta$

$$= 0.5\ I_A\omega^2 \sin 2\theta$$

The effect of this couple will be opposite to that of C_P.

Resultant gyroscopic couple acting on the disc,

$$C = C_P - C_A$$
$$= 0.5\ \omega^2 \sin 2\theta\ (I_P - I_A) \quad (13.17)$$

This resultant couple will be acting in anti-clockwise direction, as seen from top.

Now $$I_P = \frac{Wr^2}{2g}, \text{ where } r = \text{radius of disc}$$

$$I_A = \frac{W\left(\frac{b^2}{12} + \frac{r^2}{4}\right)}{g}, \text{ where } b = \text{width of disc}$$

$$\cong \frac{Wr^2}{4g}, \text{ neglecting } l \text{ for a thin disc}$$

The couple exerted by a thin disc on the shaft,

$$C_{\text{disc}} = \frac{W\omega^2 \sin 2\theta \left(\frac{r^2}{2} - \frac{r^2}{4}\right)}{2g}$$

$$= \frac{W\omega^2 r^2 \sin 2\theta}{8g} \quad (13.18)$$

The shaft tends to turn in the plane of paper in counter-clockwise direction as seen from the top. As a result, the horizontal force is exerted on the bearings.

Example 13.31

A disc has a mass of 25 kg and a radius of gyration about its axis of symmetry 120 mm while its radius of gyration about a diameter of the disc right angles to the axis of symmetry is 80 mm. The disc is pressed on to the shaft but due to incorrect boring, the angle between the axis of symmetry and the actual axis of rotation is 0.3°, though both these axes pass through the centre of gravity of the disc. Assuming that the shaft is rigid and is carried between bearings 200 mm apart, determine the bearing forces due to the misalignment at a speed of 4800 rpm.

■ **Solution**

Given: $M = 25$ kg, $K_A = 120$ mm, $K_P = 80$ mm, $\theta = 0.3°$, $N = 4800$ rpm, $l = 200$ mm

$$I_P = MK_P^2 = 25 \times (0.08)^2 = 0.16 \text{ kg} \cdot \text{m}^2$$

$$I_A = MK_A^2 = 25 \times (0.12)^2 = 0.36 \text{ kg} \cdot \text{m}^2$$

$$\omega = \frac{2\pi N}{60} = \frac{2\pi \times 4800}{60} = 502.65 \text{ rad/s}$$

$$C = 0.5\ \omega^2 \sin 2\theta (I_P - I_A)$$

$$= 0.5 \times (1502.65)^2 \times \sin 0.6° \times (0.16 - 0.36) = 264.6 \text{ Nm}$$

$$\text{Bearing force} = \frac{C}{l} = \frac{264.6}{0.2} = 1322.9 \text{ N}$$

Example 13.32

A gyrowheel *D* of mass 0.6 kg and radius of gyration 20 mm is mounted in a pivoted frame *C* as shown in Fig.13.17. The axis *AB* of the pivots passes through the centre of rotation *O* of the wheel, but the centre of gravity *G* of the frame *C* is 10 mm below *O*. The frame has a mass of 0.25 kg and the speed of rotation of the wheel is 3000 rpm in the counter-clockwise direction.

The entire unit is mounted on a vehicle so that the axis *AB* is parallel to the direction of motion of the vehicle. If the vehicle travels at 15 m/s in a curve of 60 m radius, find the inclination of the gyro-wheel from the vertical, when (a) the vehicle moves in the direction of the arrow *X* taking a left hand turn along the curve and (b) the vehicle reverses at the same speed in the direction of arrow *Y* along the same path.

Solution

Here $M_w = 0.6$ kg, $K_w = 0.02$ m, $OG = h = 0.01$ m, $M_f = 0.25$ kg, $N = 3000$ rpm,

$$v = 15 \text{ m/s}, \; R = 60 \text{ m}$$

$$\omega = \frac{2\pi N}{60} = \frac{2\pi \times 3000}{60} \doteq 314.2 \text{ rad/s}$$

$$I_w = M_w K_w^2 = 0.6 \times (0.02)^2 = 0.00024 \text{ kg} \cdot \text{m}^2$$

$$\omega_p = \frac{v}{R} = \frac{15}{60} = 0.25 \text{ rad/s}$$

Let θ = angle of inclination of gyrowheel from the vertical.w

(a) Vehicle moving in the direction of arrow *X* while taking left turn along the curve.
Gyro-couple about *O*, $C_g = I_w \omega \cdot \omega_p \cos\theta = 0.00024 \times 314.2 \times 0.25 \times \cos\theta$

$$= 0.018852 \cos\theta \text{ Nm}$$

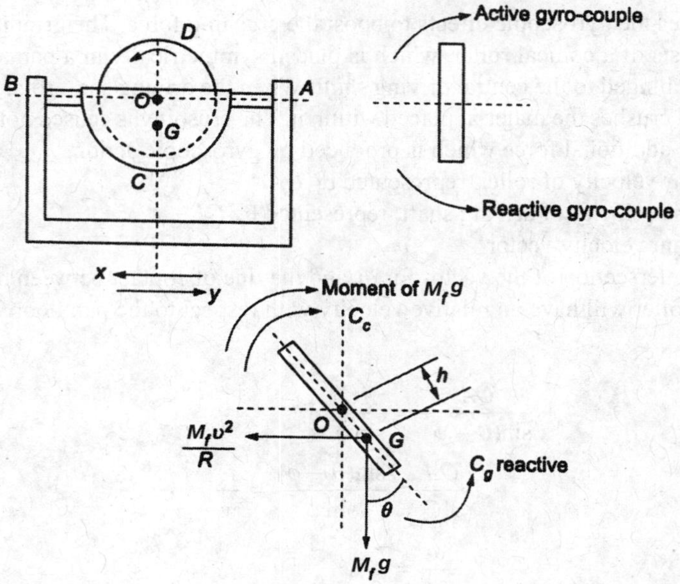

Fig.13.17 Gyrowheel mounted in a pivoted frame

Centrifugal couple about O, $C_c = \left(\frac{M_f v^2}{R}\right) \cdot h \cos\theta = \left(\frac{0.25 \times 15^2}{60}\right) \times 0.01 \times \cos\theta$

$= 0.009375 \cos\theta$ N m

Total overturning couple, $C_o = C_g - C_c = (0.018852 - 0.009375) \cos\theta$

$= 0.009477 \cos\theta$ N m (ccw)

Balancing couple due to the weight of the frame,

$$C_b = M_f gh \sin\theta$$

$$= 0.25 \times 9.81 \times 0.01 \times \sin\theta$$

$$= 0.024525 \sin\theta \text{ N m (cw)}$$

For the equilibrium condition, $C_o = C_b$

$$0.009477 \cos\theta = 0.024525 \sin\theta$$

$$\tan\theta = 0.38642$$

$$\theta = 21.12°$$

(b) Vehicle reverses at the same speed in the direction of arrow Y along the same path.

$$C_o = C_g + C_c = (0.018852 + 0.009375) \cos\theta$$

$$= 0.028227 \cos\theta$$

For the equilibrium condition, $C_o = C_b$

$$0.028227 \cos\theta = 0.024525 \sin\theta$$

$$\tan\theta = 1.15095$$

$$\theta = 49.01°$$

13.11 GYROSCOPIC ANALYSIS OF A GRINDING MILL

A grinding mill uses the gyroscopic effects to boost the crushing force. The grinding mill is shown in Fig.13.18. It consists of a conical roller, which is placed symmetrically in a pan and is free to rotate on a shaft which is hinged to the central driving shaft. When the driving shaft rotates, the roller moves around the pan and crushes the material placed within it. The crushing is caused not only by the weight of the roller but by additional force which is produced by gyroscopic action.

Let ω = angular velocity of roller, represented by OA

ω_1 = angular velocity of driving shaft, represented by OB

ω_r = resultant velocity vector

Point D is the intersection of the vector for ω_r and the line of contact between roller and pan floor. At this point, the roller will have no relative velocity with respect to the pan floor.

In ΔOAC, we have

$$\frac{OA}{\sin(\theta - \phi)} = \frac{AC}{\sin\phi}$$

or

$$\frac{OA}{AC} = \frac{\sin(\theta - \phi)}{\sin\phi} = \frac{\omega}{w_1}$$

Also at point D,

$$\frac{\omega}{\omega_1} = \frac{r_o}{r}$$

where r = radius of the roller at the cross-section containing point D.

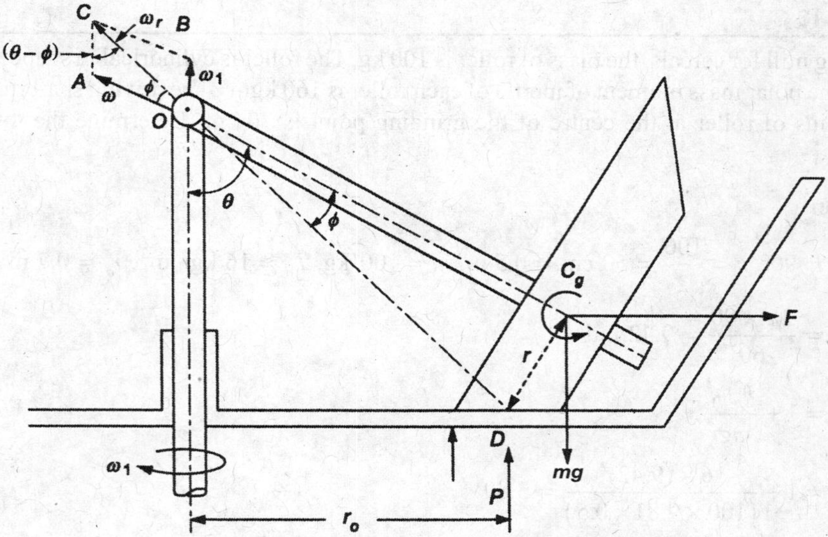

Fig.13.18 Gyroscopic analysis of grinding mill

The gyroscopic couple, $C_g = I_p\omega\omega_1 \sin\theta + \left(I_p - i_e\right)\omega_1^2 \sin\theta\cos\theta$

$$= I_p\omega_1^2\left(\frac{r_o}{r}\right)\sin\theta + \left(I_p - I_e\right)\omega_1^2 \sin\theta\cos\theta$$

$$= \left[(I_p - I_e)\cos\theta + \frac{I_p r_o}{r}\right]\omega_1^2 \sin\theta \quad \text{(a)}$$

By taking moments of all forces, we have

$$C_g = F(r_o + r\cos\theta)\cot\theta - mg(r_o + r\cos\theta) + Pr_o \quad \text{(b)}$$

where P = total crushing force

m = mass of roller

F = centrifugal force

From (a) and (b), we get

$$\frac{P}{mg} = \left[1 + \left(\frac{r}{r_o}\right)\cos\theta\right] + \frac{\omega_1^2\sin\theta}{mgr_o}\left[\left(I_p - I_e\right)\cos\theta + I_p\left(\frac{r_o}{r}\right)\right] - \frac{F\cot\theta}{mg}\left[1 + \left(\frac{r}{r_o}\right)\cos\theta\right] \quad (13.19)$$

For $\theta = 90°$, that is when the roller is circular, we get

$$\frac{P}{mg} = 1 + \frac{I_p\omega_1^2}{mgr}$$

or

$$P = mg + \frac{I_p\omega_1^2}{r} \quad (13.20)$$

Example 13.33

In a crushing mill for cereals, the mass of roller is 100 kg. The roller is cylindrical in shape with 100 cm diameter. The polar mass moment of inertia of each roller is 160 kgm². The driving shaft runs at 90 rpm and the radius of roller at the centre of the grinding point is 70 cm. Determine the total crushing force.

Solution

Here $\theta = 90°$, $r = \dfrac{100}{2} = 50\text{ cm} = 0.5\text{ m}$, $m = 100\text{ kg}$, $I_p = 16\text{ kg}\cdot\text{m}^2$, $r_o = 0.7\text{ m}$

$$\omega_1 = \frac{2\pi \times 90}{60} = 9.42\text{ rad/s}$$

$$\frac{P}{mg} = 1 + \frac{I_p\omega_1^2}{mgr}$$

$$= 1 + \frac{16 \times (9.42)^2}{(100 \times 9.81 \times 0.5)} = 3.89$$

Example 13.34

A thin circular disc is fitted to a shaft as shown in Fig.13.19. Weight of the disc is 500 N and diameter is 1.2 m. Shaft rotates at 300 rpm in anti-clockwise direction when seen from the right side. Find the effect of gyroscopic couple on the shaft and the bearing reactions at A and B taking the effect of weight of the disc.

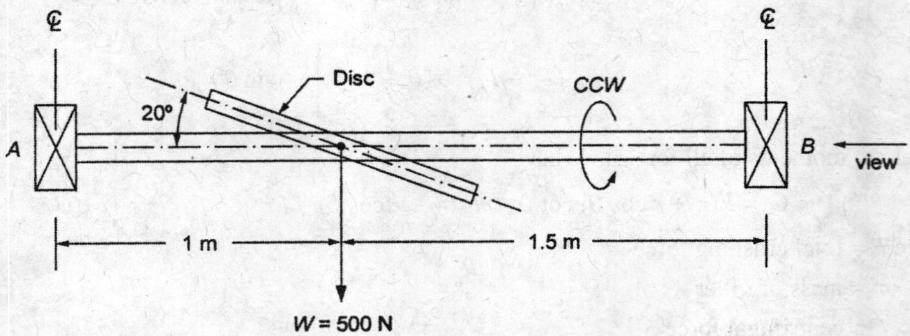

Fig.13.19 Inclined disc mounted on a shaft

Solution

Given: $W = 500$ N, $r = 0.6$ m, $N = 300$ rpm, $\theta = 20°$

$$\omega = \frac{2\pi \times 300}{60} = 31.41\text{ rad/s}$$

$$C = \frac{W\omega^2 r^2 \sin 2\theta}{8g} = \frac{500 \times (31.41)^2 \times (0.6)^2 \times \sin 40°}{8 \times 9.81} = 1454.5\text{ N m}$$

Bearing reactions due to gyroscopic couple, $R'_A = \dfrac{C}{2.5} = 581.8\text{ N} \uparrow$, $R'_B = 581.8\text{ N} \downarrow$,

Bearing reactions due to self weight, $R_A'' = \dfrac{500 \times 1.5}{2.5} = 300 \text{ N} \uparrow, \; R_B'' = 200 \text{ N} \uparrow$

Resultant bearing reactions, $R_A = R_A' + R_A'' = 881.8 \text{ N} \uparrow, \; R_B = R_B' - R_B'' = 318.8 \text{ N} \uparrow$

Example 13.35

The rotor of a turbojet engine has a mass of 200 kg and a radius of gyration 250 mm. The engine rotates at a speed of 1000 rpm in the clockwise direction if viewed from the front of the aeroplane. The aeroplane while flying at 1000 km/h turns with a radius of 2 km to the right. Compute the gyroscopic moment exerted by the rotor on the plane structure. Also determine whether the nose of the plane tends to rise or fall when the plane turns.

■ Solution

Given: $M_e = 200$ kg, $K_e = 250$ mm, $N_e = 1000$ rpm, $v = 1000$ km/h, $R = 2$ km

$$I_e = M_e K_e^2 = 200 \times (0.25)^2 = 12.5 \text{ kg}\cdot\text{m}^2$$

$$\omega_e = \frac{2\pi \times 1000}{60} = 104.72 \text{ rad/s}$$

$$\omega_p = \frac{v}{R} = \frac{1000 \times 1000}{3600 \times 2000} = 0.139 \text{ rad/s}$$

$$C = I_e \omega_e \omega_p = 12.5 \times 104.72 \times 0.139 = 181.95 \text{ Nm}$$

The nose tends to fall.

Example 13.36

A disc of mass 100 kg and radius of gyration 0.5 m is supported as shown in Fig.13.20. When the disc is rotating at 100 rad/s the cord on the right hand side bearing gets broken. Discuss the motion of the disc.

■ Solution

Given: $M = 100$ kg, $K = 0.5$ m, $\omega = 100$ rad/s, $\ell_1 = 0.3$ m, $\ell_2 = 0.7$ m

Applied torque due to disc mass, $C = mg\ell_1 = 100 \times 9.81 \times 0.3 = 294.3$ Nm

Gyroscopic couple, $C_g = mk^2 \omega w_p = 100 \times (0.5)^2 \times 100 \times \omega_p = 2500\, \omega_p$

$C_g = C$ gives, $\omega_p = 0.11772$ rad/s

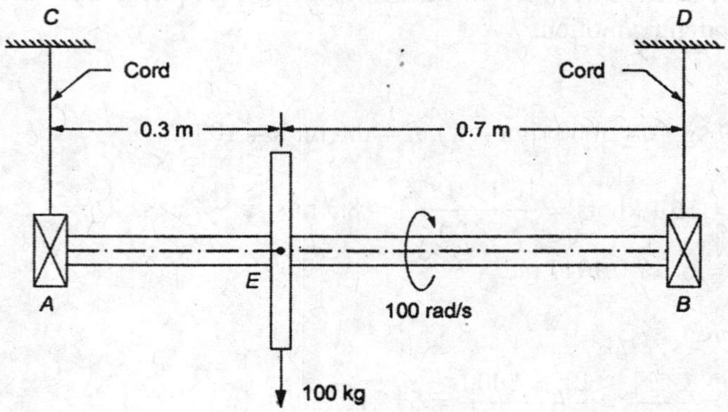

Fig.13.20 Disc mounted on a hanging shaft

Example 13.37

A uniform disc of 250 mm diameter has a mass of 15 kg. It is mounted centrally on a horizontal shaft running bearings 200 mm apart. The disc spins with a uniform speed of 1800 rpm in vertical plane in ccw direction looking from *RHS* bearing. The shaft precesses with a uniform speed of 60 rpm in horizontal plane in ccw direction when looking from top.

Determine the bearing reactions due to the disc mass and gyroscopic effects.

■ Solution

Given: $r = 0.125$ m, $M = 15$ kg, $\ell = 200$ mm, N = 1800 rpm, $N_p = 60$ rpm

$I = 0.5\,mr^2 = 0.5 \times 15 \times (0.125)^2 = 0.1172\ \text{kg}\cdot\text{m}^2$

$$\omega = \frac{2\pi N}{60} = \frac{2\pi \times 1800}{60} = 188.5\ \text{rad/s}$$

$$\omega_p = \frac{2\pi N_p}{60} = \frac{2\pi \times 60}{60} = 6.28\ \text{rad/s}$$

$$C = I\omega w_p 0.1172 \times 188.5 \times 6.28 = 138.81\ \text{Nm (ccw)}$$

Taking moments about bearing *B*, we have

$0.2\,F_A = 138.81,\ F_A = F_B = 694$ N

Reaction force, $F'_A = 694$ N↓, $F'_B = 694$ N↑

Reaction due to disc mass, $R'_A = R'_B = \dfrac{15 \times 9.81 \times 0.1}{0.2} = 73.58$ N ↑

Resultant reactions, $R_A = -620.42$ N ↓, $R_B = 768.58$ N ↑

Example 13.38

The turbine rotor of a ship has a mass of 2000 kg and rotates at a speed of 3000 rpm clockwise when looking from stern. The radius of gyration of the rotor is 0.5 m. Determine the gyroscopic couple and its effects upon the ship when the ship is steering to the right in a curve of 100 m radius at a speed of 16.1 knots (1 knot = 1855 m/hr). Calculate also the torque and its effect when the ship is pitching in simple harmonic motion, the bow falling with its maximum velocity. The period of pitching is 50 and total angular displacement between the two extreme positions of pitching is 12°. Find the maximum acceleration during pitching motion.

■ Solution

Given: $M = 2000$ kg, $N = 3000$ rpm (cw), $K = 0.5$ m, $R = 100$ m,

$$v = 16.1\ \text{knots} = \frac{16.1 \times 1855}{3600} = 8.3\ \text{m/s},\ t_p = 50\ \text{s},\ 2\theta = 12^\circ \text{ or}$$

$$\theta = 6^\circ \text{ or } 0.1047\ \text{rad}$$

(i) *Gyroscopic couple:*

$$\omega = \frac{2\pi \text{N}}{60} = \frac{2\pi \times 3000}{60} = 314.16\ \text{rad/s}$$

$$\omega_p = \frac{v}{R} = \frac{8.3}{100} = 0.083\ \text{rad/s}$$

Moment of inertia, $I = MK^2 = 2000 \times (0.5)^2 = 500$ kg·m²

Gyroscopic couple, $C = I\omega \cdot \omega_p = 500 \times 314.16 \times 0.083 = 13037.6$ Nm

As the rotor rotates clockwise when viewing from the stern and ship steers to the right, therefore, the reaction couple tends to lower the bow and raise the stern.

(ii) *Torque during pitching:*

$$\text{Angular speed of pitching, } \omega_1 = \frac{2\pi}{t_p} = \frac{2\pi}{50} = 0.1257 \text{ rad/s}$$

$$\text{Maximum angular velocity of precession, } (\omega_p)_{max} = \theta\omega_1 = 0.1047 \times 0.1257 = 0.01316 \text{ rad/s}$$

$$\text{Maximum gyro-couple, } C_{max} = I\omega\,(\omega_p)_{max} = 500 \times 314.16 \times 0.01316 = 2066.7 \text{ Nm}$$

$$\text{Maximum angular acceleration, } \alpha_{max} = \theta\omega_1^2 = 0.1047 \times (0.1257)^2 = 1.654 \times 10^{-3} \text{ rad/s}^2$$

Example 13.39

The heavy rotor of a sea vessel rotates at 2000 rpm clockwise looking from the stern, its mass being 750 kg. The vessel pitches with an angular velocity of 1 radian/s.

Determine the gyroscopic couple transmitted to the hull, when bow is rising, if the radius of gyration for the rotor is 250 mm. Also show, in what direction the couple acts on the hull.

Solution

Given: $M = 750$ kg, $N = 2000$ rpm (cw), $K = 0.25$ m, $\omega_p = 1$ rad/s

$$w = \frac{2pN}{60} = \frac{2p \times 2000}{60} = 209.44 \text{ rad/s}$$

$$w_p = \frac{v}{R} = \frac{8.3}{100} = 0.083 \text{ rad/s}$$

Moment of inertia, $I = MK^2 = 750 \times (0.25)^2 = 46.875$ kg·m²

Gyroscopic couple, $C = I\omega \cdot \omega_p = 46.875 \times 209.44 \times 1 = 9817.5$ Nm

When bow is rising the reactive gyro-couple acts in the clockwise direction as viewed from stern and tends to turn the vessel to the right towards star board.

Example 13.40

A four-wheel motor car of mass 2000 kg has a wheel base 2.5 m, track width 1.5 m, and the height of C.G. 500 mm above the ground level and lies at 1 m from the front axle. Each wheel has an effective diameter of 0.8 m and M.I. of 0.8 kg·m². The drive shaft, engine flywheel and transmission are rotating at 4 times the speed of load wheel, in clockwise direction when viewed from the front, and is equivalent to a mass of 75 kg having a radius of gyration of 100 mm. If the car is taking a right turn of 60 m radius at 60 km/h, find the load on each wheels.

Solution

Given: $M = 2000$ kg, $a = 1.5$ m, $l = 2.5$ m, $l_1 = 1$ m, $l_2 = 1.5$ m, $h = 0.5$ m, $d_w = 0.8$ m, $I_w = 0.8$ kg·m², $i = 4$, $M_e = 75$ kg, $K_e = 0.1$ m, $R = 60$ m, $v = 60$ km/h or 16.67 m/s

Refer to Fig.13.21.

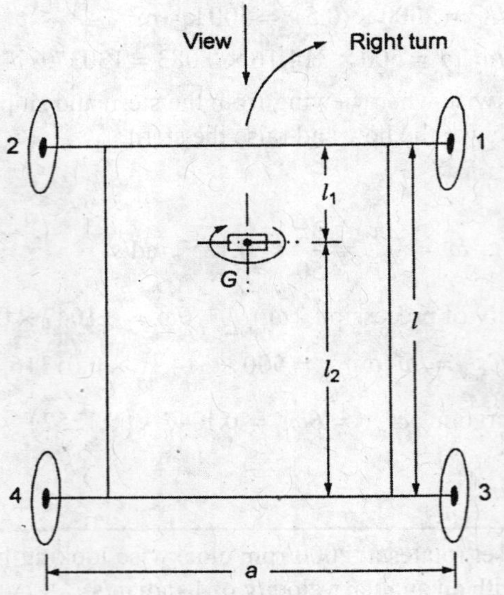

Fig.13.21 Four-wheel vehicle taking a turn

(a) *Dead weights*:

Total weight of motor car, $W = Mg = 2000 \times 9.81 = 19620$ N

On each rear wheel, $\frac{W_r}{2} = \frac{19620 \times 1}{2.5 \times 2} = 3924$ N (upwards)

On each front wheel, $\frac{W_f}{2} = \frac{19620 \times 1.5}{2.5 \times 2} = 5886$ N (upwards)

(b) *Centrifugal couple*:

$$C_c = \frac{Mv^2h}{R} = \frac{2000 \times (16.67)^2 \times 0.5}{60} = 4631.5 \text{ Nm}$$

$$\frac{Q}{2} = \frac{C_c}{2a}$$

$$= \frac{4631.5}{3} = 1543.83 \text{ N}$$, vertically upwards on outer and downwards on inner wheels.

(c) *Gyroscopic couple*:

(i) *Due to wheels*:

$$\omega_w = \frac{v}{r_w} = \frac{16.67}{0.4} = 41.675 \text{ rad/s}$$

$$\omega_{pw} = \frac{v}{R} = \frac{16.67}{60} = 0.278 \text{ rad/s}$$

$$C_{gw} = 4I_w\omega_w\omega_{pw} = 4 \times 0.80 \times 41.675 \times 0.278 = 37.052 \text{ Nm}$$

$$\frac{P}{2} = \frac{C_{gw}}{2a} = \frac{37.052}{3} = 12.35 \text{ N, downwards on inner and upwards on outer wheels.}$$

(ii) *Gyroscopic couple due to engine flywheel and transmission etc*:

$$\omega_e = i \times \omega_w \text{ rad/s}$$

$$\omega_{pe} = \omega_{pw}$$

$$I_e = M_e K_e^2 = 75 \times (0.1)^2 = 0.75 \text{ kg} \cdot \text{m}^2.$$

$$C_{ge} = I_e\omega_e\omega_{pe} = 0.75 \times 4 \times 41.675 \times 0.278 = 34.757 \text{ Nm}$$

$$\frac{F}{2} = \frac{C_{ge}}{2l} = \frac{34.757}{5} = 6.95 \text{ N, downwards on front and upwards on rear wheels.}$$

For motor car taking a right turn, the reactions are:

$$R_1 = \frac{W_f}{2} - \frac{Q}{2} - \frac{P}{2} - \frac{F}{2} = 5886 - 1543.83 - 12.35 - 6.95 = 4322.87 \text{ N}$$

$$R_2 = \frac{W_f}{2} + \frac{Q}{2} + \frac{P}{2} + \frac{F}{2} = 5886 + 1543.83 + 12.35 + 6.95 = 7449.13 \text{ N}$$

$$R_3 = \frac{W_r}{2} - \frac{Q}{2} - \frac{P}{2} - \frac{F}{2} = 3924 - 1543.83 - 12.35 - 6.95 = 2360.87 \text{ N}$$

$$R_4 = \frac{W_r}{2} + \frac{Q}{2} + \frac{P}{2} + \frac{F}{2} = 3924 + 1543.83 + 12.35 + 6.95 = 5487.13 \text{ N}$$

Example 13.41

A motor cycle along with the radius has mass 310 kg and the system centre of gravity is 60 cm above the ground level. Each wheel of the machine has mass 10 kg, radius 30 cm and radius of gyration 25 cm. The rotating parts of the engine have equivalent mass 15 kg and radius of gyration 8 cm and they rotate in the same direction as the road wheels. The gear ratio from wheel to engine is 1.8. Calculate the angel of banking necessary for the machine to ride normal to the banking track on a bend of 80 m radius at a speed of 150 km/h.

Solution

Given: $M = 310$ kg, $h = 0.6$ m, $M_w = 10$ kg, $r_w = 0.3$ m, $K_w = 0.5$ m, $M_e = 15$ kg, $K_e = 0.08$ m, $i = 1.8$, $R = 80$ m, $v = 150$ km/h or 41.67 m/s

Moment of inertia of wheel, $I_w = M_w K_w^2 = 10 \times (0.25)^2 = 0.625 \text{ kg} \cdot \text{m}^2$

Moment of inertia of engine, $I_e = M_e K_e^2 = 15 \times (0.08)^2 = 0.096 \text{ kg} \cdot \text{m}^2$

$$\text{Gyro-couple, } C_g = \left(\frac{v^2}{Rr_w}\right)[2I_w + iI_e]\cos\theta$$

$$= \left[\frac{(41.67)^2}{80 \times 0.3}\right][2 \times 0.625 + 1.8 \times 0.096]\cos\theta$$

$$= 102.939 \cos\theta$$

Centrifugal couple, $C_c = \left(\dfrac{Mv^2h}{R}\right)\cos\theta$

$$= \left[\frac{310\times(41.67)^2\times 0.6}{80}\right]\cos\theta = 4037.1\cos\theta \text{ N m}$$

Balancing couple, $C_b = Mgh \sin\theta = 310\times 9.81\times 0.6\times \sin\theta = 1824.66 \sin\theta$

For equilibrium, $C_b = C_g + C_c$

$$1824.66 \sin\theta = (102.939 + 4037.1)\cos\theta$$

$$\tan\theta = 2.26894$$

Angle of heel, $\theta = 66.21°$

Example 13.42

The engine and the propeller of an aeroplane weights 5 kN and the radius of gyration is 50 cm, The propeller rotates at 3000 rpm in clockwise direction looking from the rear. If the aeroplane makes quarter of a circle turn of radius 100 m towards left hand side while flying at 240 km/hr, what gyroscopic couple will act on the aeroplane frame and what will be its effect?

■ Solution

Given: $W = 5$kN, $K = 0.5$ m, $N = 3000$ rpm (cw), $R = 100$ m, $v = 240$ km/h or 66.67 m/s

$$I = MK^2 = \left(\frac{5000}{9.81}\right)\times(0.5)^2 = 127.421 \text{ kg}\cdot\text{m}^2$$

$$w = \frac{2pN}{60} = \frac{2p\times 3000}{60} = 314.16 \text{ rad/s}$$

$$w_p = \frac{v}{R} = \frac{66.67}{100} = 0.6667 \text{ rad/s}$$

$$C = Iww_p = 127.421\times 314.16\times 0.6667 = 26688 \text{ N m}$$

Effect: Raise the nose and lower the tail of the aeroplane.

Example 13.43

The mass of the turbine rotor of a ship is 15 tonnes and has a radius of gyration of 0.5 m. Its speed is 1800 rpm. The ship pitches 5° above and 5° below the horizontal position. A complete oscillation takes 30 s with SHM. Determine (a) maximum gyroscopic couple, (b) maximum angular acceleration of the ship during pitching, and (c) the direction in which bow will tend to turn when rising, if the rotation of the rotor is clockwise when looking from the aft.

■ Solution

Given: $M = 15{,}000$ kg, $K = 0.5$ m, $N = 1800$ rpm, A = 5°, $t_p = 30$ s

$$\omega = \frac{2\pi N}{60} = \frac{2\pi\times 1800}{60} = 188.5 \text{ rad/s}$$

$$A = \frac{\pi}{180}\times 5 = 0.0873 \text{ rad}$$

$$I = MK^2 = 15{,}000 \times (0.5)^2 = 3750 \text{ kg}\cdot\text{m}^2$$

(a) $$\omega_o = \frac{2\pi}{t_p} = \frac{2\pi}{30} = 0.2094 \text{ rad/s}$$

$$(\omega_p)_{\max} = A\omega_o = 0.0873 \times 0.2094 = 0.0183 \text{ rad/s}$$

$$C_{\max} = I\omega(\omega_p)_{\max} = 3750 \times 188.5 \times 0.0183 = 12936 \text{ Nm}$$

(b) Maximum angular acceleration during pitching $= A\omega_o^2$

$$= 0.0873 \times (0.2094)^2 = 0.00383 \text{ rad/s}^2$$

(c) When the rotation of motor is clockwise, looking from aft while bow is rising, the reactive gyro-couple acts in clockwise direction which tends to turn the bow towards right (Star board)

Example 13.44

Find the angle of inclination with respect to the vertical of a two-wheeler negotiating a turn for the following data:

Combined mass of vehicle with its rider = 300 kg

Moment of inertia of engine flywheel = 0.35 kg · m²

Moment of inertia of each road wheel = 1.10 kg · m²

Speed of engine flywheel = 5 × speed of road wheels in the same direction

Height of centre of gravity of rider with vehicle = 0.6 m

Speed of two-wheeler = 90 km/h

Wheel radius = 0.3 m

Radius of turn = 50 m

■ Solution

Given: $M = 300$ kg, $I_e = 0.35$ kg · m², $I_w = 1.10$ kg · m², $i = 5$, $h = 0.6$ m,

$$v = \frac{90 \times 10^3}{3600} = 25 \text{ m/s}, \; r_\omega = 0.3 \text{ m}, \; R = 50 \text{ m}$$

$$\omega_w = \frac{v}{r_w} = \frac{25}{0.3} = 83.33 \text{ rad/s}$$

$$\omega_e = i\omega_w = 5 \times 83.33 = 416.67 \text{ rad/s}$$

$$\omega_p = \frac{v}{R} = \frac{25}{50} = 0.5 \text{ rad/s}$$

Gyro-couple, $C_g = (2I_w\omega_w + I_e\omega_e)\,\omega_p \cos\theta$

$$= (2 \times 1.10 \times 83.33 + 0.35 \times 416.67) \times 0.5 \cos\theta$$

$$= 164.58 \cos\theta \text{ Nm}$$

Centrifugal force, $$F_c = \frac{Mv^2}{R} = \frac{300 \times 25^2}{50} = 3750 \text{ N}$$

Centrifugal couple, $C_c = F_c h \cos\theta$

$= 3750 \times 0.6 \cos\theta = 2414.58 \cos\theta$ Nm

Balancing couple, $C_b = \text{mgh} \sin\theta = 300 \times 9.81 \times 0.6 \sin\theta = 1765.8 \sin\theta$ Nm

For equilibrium of vehicle, $C_b = C_o$

$$1765.8 \sin\theta = 2414.58 \cos\theta$$

$$\tan\theta = 1.3674$$

$$\theta = 53.82°$$

Example 13.45

A four wheeled vehicle of mass 2200 kg has a wheel base 2.5 m, track width 1.5 m and height of centre of gravity 0.5 m above the ground level and lies at 1 m from the front axle. Each wheel has an effective diameter of 0.8 m and a moment of inertia of 0.85 kg·m². The drive shaft, engine flywheel and transmission are rotating at 4 times the speed of road wheel, in a clockwise direction when viewed from the front, and is equivalent to a mass of 80 kg having a radius of gyration of 0.1 m. If the vehicle is taking a right turn of 60 m radius at 60 km/h, calculate the load on each wheel.

Solution

Given: $M = 2200$ kg, $l = 2.5$ m, $a = 1.5$ m, $h = 0.5$ m, $l_1 = 1$ m, $r_w = 0.4$ m, $I_w = 0.85$ kg·m², $i = 4$,

$M_e = 80$ kg, $K_e = 0.1$ m, $R = 60$ m, $v = \dfrac{60 \times 10^3}{3600} = 16.67$ m/s

$$W = Mg = 2200 \times 9.81 = 21582 \text{ N}$$

$$W_f = \frac{21582 \times 1.5}{2.5} = 12949.2 \text{ N}$$

$$W_r = 21582 - 12949.2 = 8632.8 \text{ N}$$

Weight on each front wheel $= \dfrac{W_f}{2} = \dfrac{12949.2}{2} = 6474.6$ N

Weight on each rear wheel $= \dfrac{W_r}{2} = \dfrac{8632.8}{2} = 4316.4$ N

$$\omega_w = \frac{v}{r_w} = \frac{16.67}{0.4} = 41.675 \text{ rad/s}$$

$$\omega_p = \frac{v}{R} = \frac{16.67}{60} = 0.278 \text{ rad/s}$$

Gyro-couple due to four wheels, $C_w = 4I_w\omega_w\omega_p$

$= 4 \times 0.85 \times 41.675 \times 0.278 = 39.39$ Nm

$$I_e = M_e K_e^2 = 80 \times (0.1)^2 = 0.8 \text{ kg} \cdot \text{m}^2$$

Gyro-couple due to engine etc., $C_e = I_e\omega_e \cdot \omega_p = I_e \times i\omega_w \times \omega_p$

$= 0.8 \times 4 \times 41.675 \times 0.278 = 37.07$ Nm

The gyro-couple due to wheels tends to lift the inner wheels and to press the outer wheels. The reaction will be vertically downwards on the inner wheels and vertically upwards on the outer wheels.

Reaction at each of the inner and outer wheel, $\frac{P}{2} = \frac{C_w}{2a} = \frac{39.39}{2 \times 1.5} = 13.13 \text{ N}$

The gyro-couple due to engine etc. tends to lift the front wheels and to press the rear wheels. The reaction will be vertically downwards on the front wheels and vertically upwards on the rear wheels.

$$\frac{F}{2} = \frac{C_e}{2l} = \frac{37.07}{2 \times 2.5} = 7.41 \text{ N}$$

Centrifugal force, $F_c = \frac{Mv^2}{R} = \frac{2200 \times (16.67)^2}{60} = 10189 \text{ N}$

Centrifugal couple, $C_c = F_c \times h = 10189 \times 0.5 = 5094.6 \text{ Nm}$

The centrifugal couple tends to reduce the pressure on the inner wheels and to increase on the outer wheels. The reactions are vertically downwards on the inner wheels and vertically upwards on the outer wheels.

$$\frac{Q}{2} = \frac{C_c}{2a} = \frac{5094.6}{2 \times 1.5} = 1698.2 \text{ N}$$

The forces acting on the wheels are shown in Fig.13.22.

Load on front wheel $1 = \frac{W_f}{2} - \frac{P}{2} - \frac{F}{2} - \frac{Q}{2}$

$= 6474.6 - 13.13 - 7.41 - 1698.2 = 4755.86 \text{ N}$

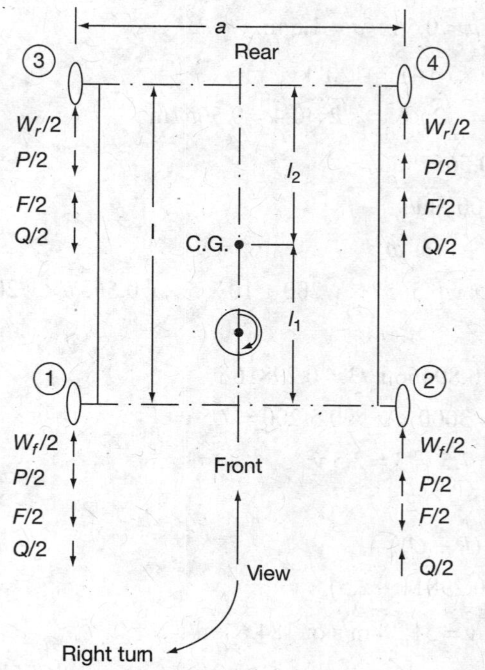

Fig.13.22 Four-wheel vehicle taking a turn

$$\text{Load on front wheel } 2 = \frac{W_f}{2} + \frac{P}{2} - \frac{F}{2} + \frac{Q}{2}$$

$$= 6474.6 + 13.13 - 7.41 + 1698.2 = 8178.52 \text{ N}$$

$$\text{Load on rear wheel } 3 = \frac{W_r}{2} - \frac{P}{2} + \frac{F}{2} - \frac{Q}{2}$$

$$= 4316.4 - 13.13 + 7.41 - 1698.2 = 2612.48 \text{ N}$$

$$\text{Load on rear wheel } 4 = \frac{W_r}{2} + \frac{P}{2} + \frac{F}{2} + \frac{Q}{2}$$

$$= 4316.4 + 13.13 + 7.41 + 1698.2 = 6035.14 \text{ N}$$

Example 13.46

An automobile is travelling along a curved track of 200 m mean radius. Each of the four road wheels have a mass of 80 kg with a radius of gyration of 0.4 m. The rotating parts of the engine have a mass moment of inertia of 10 kg m². The crankshaft rotates in the same direction as the road wheels. The gear ratio of the engine to the back wheels is 5:1. The vehicle has a mass of 3000 kg and its centre of gravity is 0.5 m above the road level. The width of the track of the vehicle is 1.5 m. Calculate the limiting speed of the vehicle around the curve for all four wheels to maintain contact with the road surface.

■ Solution

Given: $R = 200$ m, $m = 80$ kg, $K = 0.4$ m, $\ell_e = 10$ kg · m², $i = 5{:}1$,

$M = 3000$ kg, $h = 0.5$ m, $a = 1.5$ m, $v = ?$

$$I_w = mK^2 = 80 \times (0.4)^2 = 12.8 \text{ kg} \cdot \text{m}^2$$

$$K^2 = 0.5r_w^2, r_w = \sqrt{2} \times K = \sqrt{2} \times 0.4 = 0.5657\text{m}$$

$$\omega_w = v/r_w = v/0.5657 \text{ rad/s,}$$

$$\omega_p = v/R = v/200 \text{ rad/s}$$

$$C = 4\,I_w \omega_w \omega_p + I_e i \omega_w \omega_p$$

$$= 4 \times 12.8 \times v/0.5657 \times v/200 + 10 \times 5 \times v/0.5657 \times v/200$$

$$= 0.89446\, v^2$$

$$P/2 = C/(2a) = 0.89466\, v^2/3 = 0.29816\, v^2$$

$$C_c = Mv^2h/R = 3000 \times v^2 \times 0.5/200 = 7.5\, v^2$$

$$Q/2 = C_c/(2a) = 7.5\, v^2/3 = 2.5\, v^2$$

$$P_i = W/4 - P/2 - Q/2$$

For $P_i \leq 0,\ W \leq 2\,(P + Q)$

$$3000 \times 9.81 = 2 \times 2 \times (0.29816 + 2.5) \times v^2$$

$$v^2 = 2629.41,\ v = 51.28 \text{ m/s or } 184.6 \text{ km/h.}$$

Summary for Quick Review

1 Gyroscope is a body which while spinning about an axis is free to move in other direction under the action of external forces.

2 Axis of spin is the axis about which the body revolves.

3 Gyroscopic effect: Consider a body spinning about an axis *OX*. If a couple represented by a vector *OZ* perpendicular to *OX* is applied, then the body tries to precess about an axis *OY*, which is perpendicular to both *OX* and *OZ*. This combined effect is called gyroscopic or precessional effect.

4 The plane of spin, plane of precession, and plane of gyroscopic couple are mutually perpendicular.

5 Precession: It means the rotation about the third axis *OY*, which is perpendicular to both the spin axis *OX* and couple axis *OZ*.

6 Axis of precession: The third axis *OY* about which a body revolves and is perpendicular to both the spin axis *OX* and couple axis *OZ*, is called the axis of precession.

7 Gyroscopic couple, $C = I\omega \cdot \omega_p$
where ω = angular speed of engine, ω_p = angular speed of precession = v/R, v = linear velocity of boby, R = radius of curvature.

8 Bow is the fore–end of the ship; Stern is the rear–end of the ship, Starboard is the right hand side of the ship while looking in the direction of motion; Port is the left hand side of the ship while looking in the direction of motion.

9 Steering is the turning of the ship in a curve while moving forward. Pitching is the moving of the ship up and down the horizontal position in a vertical plane about transverse axis; Rolling is the sideway motion of the ship about longitudinal axis.

10 Stability of four–wheel vehicle.
Gyroscopic couple due to four wheels, $C_W = 4\, I_w \cdot \omega_W \cdot \omega_p$
Gyroscopic couple due to other rotating parts of the engine, like flywheel etc., $C_e = I_e \,.\, \omega_e\, \omega_p$
Total gyroscopic couple, $C = C_w \pm C_e = 4I_w \cdot \omega_w . \omega_p \pm \mathrm{I}_e \cdot \omega_e \cdot \omega_p = 4I_w \cdot \omega_w \cdot \omega_p \pm I_e \cdot \omega_e \cdot \omega_p$ where $i = \omega_e/\omega_w$ is the gear ratio of engine rotating parts to wheel.
Take +ve sign when the wheel and engine rotating parts rotate in the same direction.
Magnitude of vertical reaction at each of the outer or inner wheels, P/2 = C/(2a)

11 Centrifugal effect.
Couple tending to overturn the wheels, $C_c = Wv^2h/(gR)$
Vertical reaction at each of outer or inner wheel, $Q/2 = C_c/2a$
Total vertical reaction at each inner wheel, $P_i = W/2 - P/2 - Q/2$
Total vertical reaction at each outer wheel, $P_o = W/2 + P/2 + Q/2$
For $P_i \leq 0$, $W/2 \leq (P + Q)/2$
or $W \leq (P + Q)$
Thus, the vehicle may overturn, when

(a) ω_w is high, i.e., the vehicle is running at a high speed.
(b) h is high, i.e., the C.G. of loaded vehicle is sufficiently high above the ground.
(c) R is small, i.e., the vehicle is taking a sharp turn.
(d) W is large, i.e., the vehicle is overloaded.

In order to reduce the total gyroscopic couple, the engine must be provided with a heavy flywheel which should rotate in the opposite direction to that of the wheels.

12 Stability of a two-wheel vehicle

(a) Gyroscopic couple, $C_g = (2I_w\,\omega_w \pm I_e\,\omega_e)\,\omega_p \cos\theta = v^2\,(2\,I_w \pm iI_e)\cos\theta/(r_w R)$

(b) Centrifugal couple, $C_c = F_c \cdot h\cos\theta$

Total overturning couple, $C_o = C_g + C_c = v^2[(2I_w \pm iI_e)/r_w + Wh/g]\cos\theta/R$

Balancing couple, $C_b = W\,h\sin\theta$

For equilibrium of the vehicle (i.e. no skidding), $C_o = C_b$

$$\tan\theta = v^2[\,(2\,I_w \pm iI_e)/r_w + Wh/g]/(W\,h\,R)$$

13 Effect of precession on a disc.

Couple producing the precession, $C_p = 0.5\,I_p\,\omega^2\sin 2\theta$

Gyroscopic couple about OA, $C_A = 0.5\,I_A\,\omega^2\sin 2\theta$

Resultant gyroscopic couple acting on the disc, $C = C_p - C_A = 0.5\,\omega^2\sin 2\theta\,(I_p - I_A)$

The couple exerted by a thin disc on the shaft, $C_{disc} = W\,\omega^2\,r^2\sin 2\theta/(8g)$

14 Grinding mill.

Gyroscopic couple, $C_g = [(I_p - I_e)\cos\theta + I_p\,r_o/r]\,\omega_1^2\sin\theta$

$$= F\,(r_o + r\cos\theta)\cot\theta - mg\,(r_o + r\cos\theta) + Pr_o$$

where P = total crushing force, m = mass of roller, F = centrifugal force

For $\theta = 90°$, i.e. when the roller is circular,

$$P = mg + I_p\omega_1^2/r$$

Multiple Choice Questions

1 When a ship travels in sea, which of the following effects is more dangerous

(a) steering (b) pitching (c) rolling (d) all of the above.

2 The gyroscopic acceleration of a disc rotating at speed w and uniform acceleration is

(a) $d\omega/dt$ (b) $\omega\,d\theta/dt$ (c) $r\omega^2$ (d) $r\,d\omega/dt$

3 The gyroscopic couple acting on a disc of moment of inertia I, rotating with speed w and speed of precession w_p, is given by.

(a) $I\omega^2\,\omega_p$ (b) $I\omega\,\omega_p^2$ (c) $I\omega\,\omega_p$ (d) $I\omega^2\,\omega_p^2$

4 The total reaction of ground on wheels of a vehicle due to gyroscopic couple and centrifugal force while negotiating curve is

(a) increased on inner wheels and decreased on outer wheels

(b) decreased on inner wheels and increased on outer wheels

(c) decreased on all the wheels

(d) increased on all the wheels.

5 The axes of spin, precession and gyroscopic couple are contained in

(a) one plane (b) two planes perpendicular to each other

(c) two parallel planes (d) three planes perpendicular to one another.

6 The gyroscopic couple is introduced in a ship whose spin axis is parallel to starboard, when it is

(a) rolling (b) pitching

(c) pitching or rolling (d) neither pitching nor rolling.

7 The effect of gyroscopic torque on the naval ship when it is rolling and the rotor is spinning about the longitudinal axis is

(a) to raise the bow and lower the stern (b) to lower the bow and raise the stern

(c) to turn the ship to one side (d) no effect.

8 If the propeller of an aeroplane rotates clockwise when viewed from the rear and the aeroplane takes a right turn, the gyroscopic effect will
(a) tend to raise the tail and depress the nose
(b) tend to raise the nose and depress the tail
(c) tilt the aeroplane about spin axis
(d) have no effect.

Answers

1. (b) 2. (b) 3. (c) 4. (b) 5. (b) 6. (b) 7. (d) 8. (a)

Review Questions

1 Define gyroscope and a gyroscopic couple.
2 Define spin and precession.
3 What are gyroscopic planes?
4 How gyroscopic couple affect the motion of an aeroplane while taking a turn.
5 Explain the effect of gyroscopic couple on a naval ship.
6 How a four-wheeled vehicle is affected by gyroscopic couple?
7 Why a two-wheeler rider leans towards the inside while negotiating a turn?
8 Discuss the gyroscopic effect in a grinding mill.

Exercises

13.1 A uniform disc of diameter 250 mm and weighing 4.5 N is mounted at one end of an arm of length 0.5 m. The other end of the arm is free to rotate in a universal bearing. If the disc rotates about the arm with a speed of 240 rpm cw, looking from the front, with what speed will it precess about the vertical axis?

13.2 An aeroplane makes a complete half circle of 60 m radius towards the left when flying at 180 km/h. The rotary engine and the propeller of the plane have a mass of 35 kg with radius of gyration of 0.25 m. The engine runs at 2400 rpm cw, when viewed from the rear. Find the gyroscopic couple on the plane and state its effect on it. What will be the effect if the aeroplane turns to its right instead of the left?

13.3 A motor cycle and its rider together have a mass of 180 kg and their combined centre of gravity is 0.6 m above the ground level when the motor cycle is upright. Each road wheel is of 0.60 m diameter and has a moment of inertia 0.16 kg · m². The engine rotates at 5.5 times the speed of the road wheels and in the same sense. Determine the angle of heel necessary when the motor cycle is rounding a curve of 30 m radius at a speed of 50 km/h.

13.4 A racing car weighs 20 kN. It has a wheel base of 2 m, track width 1 m and height of C.G. 0.3 above the ground level and lies mid-way between the front and rear axles. The engine flywheel rotates at 3000 rpm cw when viewed from the front. The moment of inertia of the flywheel is 4 kg · m² and moment of inertia of each wheel is 3 kg · m². Find the reactions between the wheels and the ground when the car takes a turn on a curve of 15 m radius towards right at 30 km/h. The wheel radius is 0.4 m. Take consideration the gyroscopic and centrifugal effects.

13.5 One of the driving axles of a locomotive with its two wheels has a moment of inertia of 350 kg · m². The wheels are of 1.85 m diameter. The distance between the planes of the wheels is 1.5 m. When travelling at 100 km/h the locomotive passes over a defective rail which causes the right hand wheels to fall 10 mm and rise again in a total time of 0.1 s, the vertical movement of the wheel being S.H.M. Find the maximum gyroscopic couple.

13.6 A ship is pitching through a total angle of 15°, the oscillation may be taken as simple harmonic and the complete time period us 30 s. The turbine rotor mass is 500 kg, its radius of gyration is 0.4 m and it is rotating at 2400 rpm. Calculate the maximum value of gyroscopic couple set up by the rotor and its effect, when the bow is descending and the rotor is rotating clockwise looking from aft. What is the maximum angular acceleration which the ship is subjected to while pitching?

13.7 The propeller of an aircraft weighs 500 N and has radius of gyration of 0.8 m. The propeller shaft rotates at 2000 rpm, cw, as viewed from tail end. The plane turns left, making a U-turn of 120 m radius, at a speed of 350 km/h. Determine the gyroscopic couple and its effect on the aircraft. Also find the extra pressure on bearings if the distance between two bearings of the propeller is 0.80 m.

13.8 A disc with radius of gyration 50 mm and mass of 3 kg is mounted centrally on a horizontal axle of 90 mm length between the bearings. It spins about the axle at 750 rpm ccw when viewed from the right hand side bearing. The axle precesses about a vertical axis at 60 rpm in the ccw direction when viewed from above. Determine the resultant reaction at each bearing due to mass and the gyroscopic effect.

13.9 A two wheeler of 350 mm wheel radius is negotiating a turn of radius 80 m at a speed of 100 km/h. The combines mass of vehicle with its rider is 250 kg. The C.G. of rider is 0.6 m above the ground level. The mass moment of inertia of engine flywheel is 0.3 kg · m² and moment of inertia of each road wheel is 1.0 kg · m². If the speed of the engine is 5 times the speed of the wheel and in the same direction, find angle of heel of vehicle.

13.10 The turbine rotor of a ship has a mass of 2 tonnes and rotates at 1800 rpm clockwise when viewed from the left. The radius of gyration of the rotor is 0.3 m determine the gyroscopic couple and its effect when (a) the ship turns at a radius of 250 m with speed of 30 km/h, (b) the ship pitches with bow rising at an angular velocity of 1 rad/s, and (c) the ship rolls at an angular velocity of 0.1 rad/s.

13.11 Explain what you understand by gyroscopic stabilization. Illustrate with the help of a sketch how this is carried out in ships. Obtain a relation between the gyroscopic torque and the couple applied by the waves for complete stabilization if the waves be sinusoidal.

13.12 For a single cylinder engine determine the bearing forces caused by the gyroscopic action of the flywheel ($I = 0.32$ kg · m²) as the engine traverses a 305 m radius curve at 96.6 km/h in a turn to the right. The engine speed is 3300 rpm and it is turning clockwise when viewed from the front of the engine. The centre distance between the bearings is 152 mm. **[Ans. 64 N]**

13.13 Explain the following:

(a) Gyroscopic stabilization of sea vessels

(b) Effect of gyroscopic couple on the stability of an automobile negotiating a curve

(c) What are the principle of a gyroscope? Discuss the factors that effect the stability of an automobile while negotiating a curve.

(d) How is the magnitude and direction of the gyroscopic couple fixed?

(e) Describe the effect of the gyroscopic couple on pitching, rolling and steering of a ship with neat sketches indicating the direction of couple vector, spin vector and precession vector.

(f) Deduce and expression for the couple that is called into play in the case of a wheel rotating with uniform angular velocity in order to maintain a given rate of precession.

13.14 The rolling moment on a ship at a given instant is 12×10^6 Nm clockwise when viewed from the rear. The rotor of the stabilizing gyroscope is of 12×10^4 kg mass and spins at 1200 rpm clockwise when viewed from above. If the radius of the wheels about the spin axis is 2 m, determine the angular velocity of the precession to maintain the ship in an upright position.

[Ans. 0.796 rad/s (ccw) viewed from rear]Gyroscopic and Precessional Motion

14

GEARS

Chapter Outline

14.1 INTRODUCTION

A gear may be defined as any toothed member designed to transmit or receive motion from another member by successively engaging tooth. The smaller gear is called the *pinion* and the bigger one the *gear wheel*. They are used in metal cutting machine tools, automobiles, tractors, hoisting and transporting machinery, rolling mills, etc.

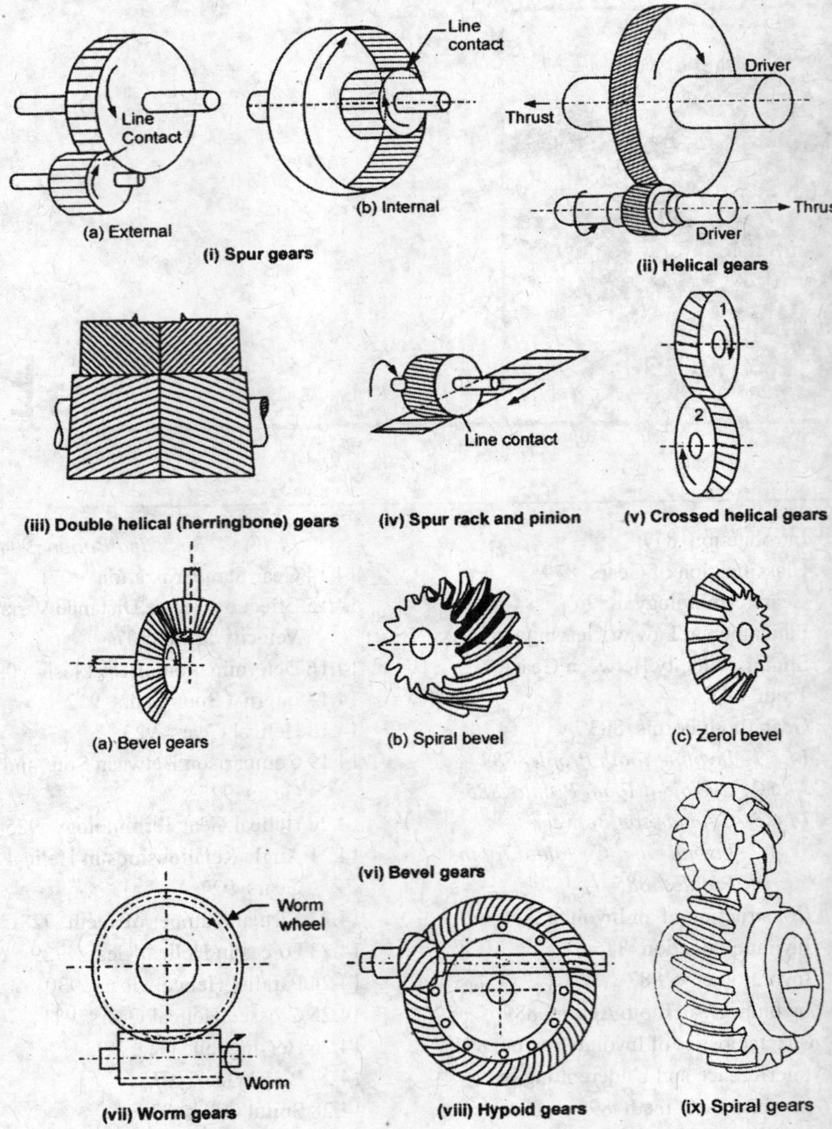

Fig.14.1 Types of gears

The gears provide many advantages over other modes of power transmission likes belts, ropes, and chains etc. Some of their advantages are:

1. They occupy lesser space.
2. There is no slip between the gears in mesh and provide exact speed ratio.
3. They can transmit higher power.
4. Their efficiency is higher.

However, the error in tooth meshing may cause undesirable vibration and noise during operation.

14.2 CLASSIFICATION OF GEARS

The gears may be classified as follows:

1. *Spur gears*: A spur gear is a cylindrical gear whose tooth traces are straight line generators of the reference cylinder [Fig.14.1(i)].
2. *Helical gears*: This is similar to the spur gear in which the tooth traces are helices. [Fig.14.1(ii)].
3. *Double helical (or herringbone) gears*: It is a cylindrical gear in which a part of the face width is right hand and the other left hand, with or without a gap between them [Fig.14.1(iii)].
4. *Spiral gears*: In spiral gears, the tooth traces are curved lines other than helices.[Fig.14.1(ix).
5. *Bevel gears*: The reference surface is a cone in bevel gears. The bevel gears may be straight, spiral, zerol, and face gears. In zerol bevel gears, the teeth are curved in the lengthwise direction and are arranged in such a manner that the effective spiral angle is zero. In face gears, the bevel gear teeth are cut on the flat face of the blank. A crown gear is a bevel gear with a reference cone angle of 90° [Fig 14.1(vi)].
6. *Hypoid gears*: They are similar to the spiral bevel gears with the difference that the axes of the shafts do not intersect. [Fig.14.1(viii)].
7. *Worm gears*: In these gears, there are screw threads on the worm and teeth on the worm wheel. [Fig.14.1(vii)].
8. *Planetary gears*: A gear pair or a gear train one of whose axes, instead of being fixed in position in the mechanism of which the gear pair is a part, moves around the other is called planetary gear train.

Gears may also be classified based on the orientation of the shafts as:

1. Parallel shafts: spur, helical, and double helical gears.
2. Intersecting shafts: straight bevel, spiral bevel, zerol bevel, and face gears.
3. Non-parallel and non-intersecting shafts-spiral, hypoid, and worm gears.

Gears may be of the external, internal, and rack and pinion type. In external gears, the teeth of gears mesh externally, whereas in internal gears the teeth of the two gears mesh internally. A rack is a gear of infinite radius.

14.3 GEAR TERMINOLOGY

A spur gear pair in mesh is shown in Fig.14.2. The various terms releated to gears are defined as follows:

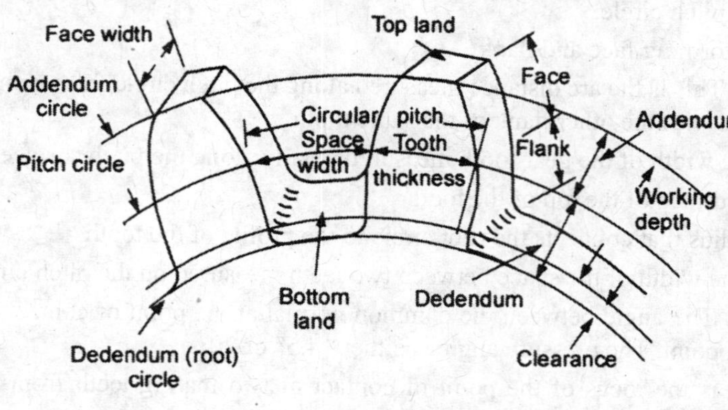

Fig.14.2 Gear terminology

Pitch circle: It is an imaginary circle which by pure rolling action, would give the same motion as the actual gear.

Pitch circle diameter (*d*): It is the diameter of a circle which by pure rolling action would produce the same motion as the toothed gear wheel.

Base circle: It is the circle from which involute form is generated.

Pitch surface: It is the surface of the disc which the toothed gear has replaced at the pitch circle.

Pitch point: It is the pitch of the tangency or the point of contact of the two pitch circles of the mating gears.

Circular pitch (p): It is the distance measured along the circumference of the pitch circle from a point on one tooth to a corresponding point on the adjacent tooth.

$$p = \pi d/z \tag{14.1}$$

where z = number of teeth.

Base pitch (P_b): It is the distance measured along the circumference of the base circle from a point on one tooth to a corresponding point on the adjacent tooth.

$$\text{Base pitch, } p_b = p \times \cos \alpha \tag{14.2}$$

where α = pressure angle of gear tooth profile.

Diametral pitch (*P*): It is expressed as the number of teeth per unit pitch circle diameter.

$$P = z/d \tag{14.3}$$

$$Pp = \pi \tag{14.4}$$

Module (*m*): It is expressed as the length of the pitch circle diameter per unit number of teeth.

$$m = d/z = 1/P \tag{14.5}$$

Addendum (h_a): The radial height of the tooth above pitch circle.

Addendum circle: A circle bounding the top of the teeth.

Dedendum (h_f): The radial depth of a tooth below the pitch circle.

Dedendum circle: A circle passing through the roots of all the teeth.

Clearance (*c*): The radial height difference between addendum and dedendum of a teeth.

Working depth: It is the radial distance of tooth from addendum circle to clearance circle.

Total depth: It is the sum of addendum and dedendum or the radial distance from dedendum circle to addendum circle.

Face: The part of the tooth surface lying below the pitch surface.

Backlash: The minimum distance between the non driving side of a tooth and adjacent side of the mating tooth at the pitch circle.

Profile: The curve forming face and flank.

Tooth thickness (*t*): This is the arc distance measured along the pitch circle from its intercept with one flank to its intercept with the other flank of the same tooth.

Face width (*b*): The width of the gear tooth measured axially along the pitch surface.

Top land: It is the surface of the top of the tooth

Tooth fillet: The radius that connects the root circle to the profile of the tooth.

Tooth space: It is the width of the space between two teeth measured on the pitch circle.

Pressure angle (*a*): The angle between the common normal at the point of contact and the common tangent at the pitch point. The pressure angle is either 14.5° or 20°.

Path of contact: It is the locus of the point of contact of two mating teeth from the beginning of engagement to the end of engagement. It is a straight line.

Path of approach: It is the portion of the path of contact from the beginning of engangement to the pitch point.

Angle of approach: The angle turned by gears during path of approach.

Path of recess: It is the portion of the path of contact from the pitch point to the end of engagement of the two mating teeth.

Angel of recess: The angle turned through during path of recess.

Arc of contact: It is the locus of a point on the pitch circle, from the beginning of engagement to the end of engagement of pair of teeth in mesh.

Involute: The involute of a circle is the curve traced by the end of a thread as it is unwound from a stationary cylinder.

Base circle diameter (d_b): It is the diameter of the base circle.

$$d_b = d \cos \alpha \tag{14.6}$$

Cycloid: It is the locus of a point on the circumference of a circle which rolls without slipping on a fixed straight line.

Centre distance (C): It is the distance between the centres of rotation of the two gears in mesh.

$$\text{C} = (d_1 + d_2)/2 = m(z_1 + z_2)/2 \tag{14.7}$$

14.4 FUNDAMENTAL LAW OF GEARING

Let us consider two curved bodies 1 and 2 rotating about their centers O_1 and O_2 and contacting at point A, as shown in Fig.14.3. A_1 and A_2 are two coincident points, A_1 lying on body 1 and A_2 lyin g on body 2. TT' and NN' represent common tangents and normals, respectively, at point A. Let ω_1 and ω_2 be the angular velocities of body 1 and 2, respectively. Let v_{a1} and v_{a2} be the linear velocities of point A. v_{a1} is perpendicular to O_1A_1 and v_{a2} is perpendicular to O_2A_2. Let the common normal intersect the line of centres at point P. Let O_1G and O_2H be perpendiculars to AP.

If the two bodies are to remain in contact, the component of velocities of A_1 and A_2 along the common normal must be equal.

Therefore
$$v_{a1} \cos \alpha = v_{a2} \cos \beta$$

or

$$\omega_1 \cdot O_1A_1 \cos \alpha = \omega_2 \cdot O_2A_2 \cos\beta$$

$$\omega_1 : O_1A_1 \cdot \frac{A_1F}{A_1C} = \omega_2 \cdot O_2A_2 \cdot \frac{A_2F}{A_2B}$$

The condition for pure rolling is that the point of contact shall lie on the line of centres.

$$\frac{\omega_1}{\omega_2} = \frac{A_1C}{O_1A_1} \cdot \frac{O_2A_2}{A_2B}$$

Triangles A_1CF and $O_1A_1G_1$ are similar. Also triangles A_2FB and O_2A_2H are similar.

Therefore
$$\frac{A_1C}{O_1A_1} = \frac{A_1F}{O_1G}$$

and
$$\frac{A_2B}{O_2A_2} = \frac{A_2F}{O_2H}$$

Hence
$$\frac{\omega_1}{\omega_2} = \frac{A_1F}{O_1G} \cdot \frac{O_2H}{A_2F}$$

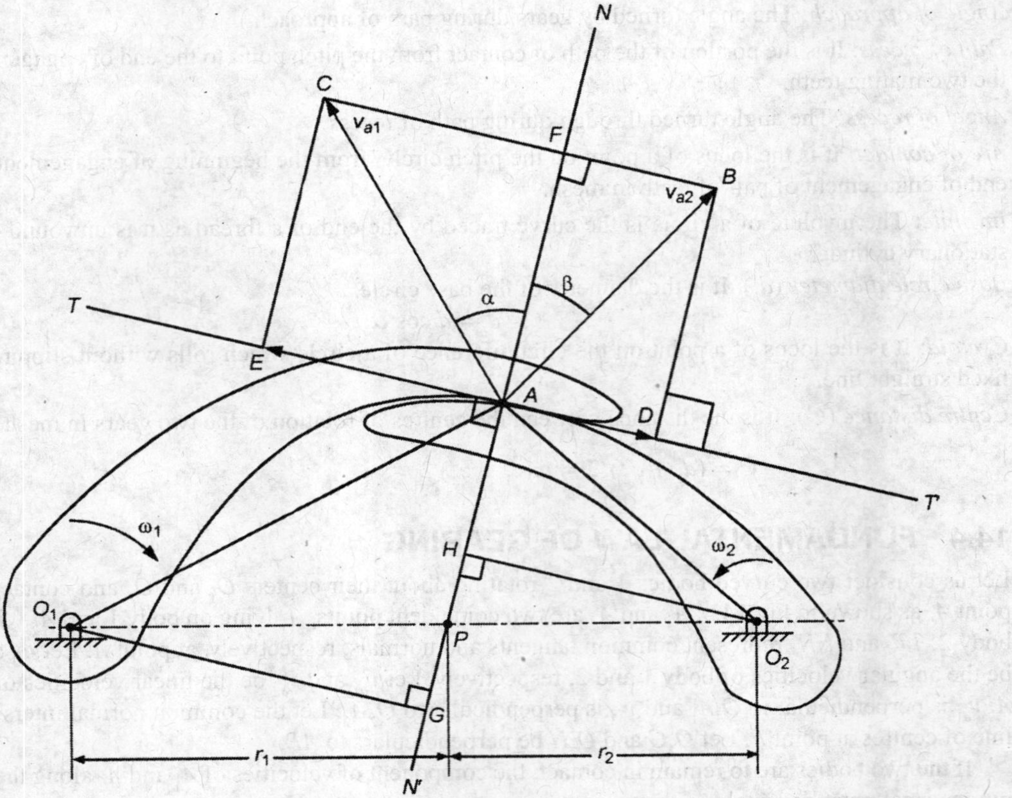

Fig.14.3 Law of gearing

But
$$A_1F = A_2F$$

$$\frac{\omega_1}{\omega_2} = \frac{O_2H}{O_1G}$$

Now triangles O_1PG and O_2PH are similar. Hence

$$\frac{O_2H}{O_1G} = \frac{O_2P}{O_1P}$$

$$\frac{\omega_1}{\omega_2} = \frac{O_2P}{O_1P} \tag{14.8}$$

Therefore, *for constant angular velocity ratio of the two gears in contact the common normal at the point of contact must always intersect the line of centres at a fixed point (pitch point) and divide this line in the inverse ratio of the angular velocities of the two gears.* This is the fundamental law of gearing.

Conjugate action: When the tooth profiles are so shaped so as to produce a constant angular velocity ratio during meshing, then the surfaces are said to be conjugate. Two conjugate surfaces in contact always satisfy the law of gearing.

14.5 SLIDING VELOCITY BETWEEN GEAR TEETH

The relative velocity along the common tangent is called the sliding velocity, v_s. Considering Fig.14.3 again, the sliding velocity v_s along the common tangent,

$$v_s = v_{a1} \sin \alpha + v_{a2} \sin \beta$$

$$= \omega_1 \cdot O_1A_1 \cdot \frac{FC}{A_1C} + \omega_2 \cdot O_2A_2 \cdot \frac{FB}{A_2B}$$

Now $$\frac{A_1C}{O_1A_1} = \frac{A_1F}{O_1G} = \frac{FC}{AG}$$

and $$\frac{A_2B}{O_2A_2} = \frac{A_2F}{O_2H} = \frac{FB}{AH}$$

Hence $$v_s = \omega_1 \cdot AG + \omega_2 \cdot AH$$

$$= \omega_1(AP + PG) + \omega_2(AP - PH)$$

$$= (\omega_1 + \omega_2)AP + \omega_1 \cdot PG - \omega_2 \cdot PH$$

Now $$\frac{\omega_1}{\omega_2} = \frac{O_2P}{O_1P} = \frac{PH}{PG}$$

Hence $$v_s = (\omega_1 + \omega_2)\,AP \tag{14.9}$$

v_s = (sum of the angular velocities) × distance of the point of contact from the pitch point.

Thus, we find that the *velocity of sliding is proportional to the distance of the pitch point from the point of contact.*

14.6 GEAR TOOTH FORMS

There are two types of gear tooth forms: involute and cycloidal. The involute of a circle is the curve traced by the end of a thread as it is unwound from a stationary cylinder. Cycloid is the locus of a point on the circumference of a circle which rolls without slipping on a fixed straight line. If the circle, instead of rolling without slipping on a straight line, rolls on the outside of another circle, the locus of the point on the circumference is called as epicycloids. Conversely, if the circle rolls on the inside of another circle, the corresponding locus of the point on the circumference of the rolling circle is called hypo-cycloid. We shall discuss the involute and cycloidal profiles in brief.

14.6.1 Involute Tooth Profile

Consider two pulleys connected by a crossed wire. The pulleys will rotate in opposite directions with constant angular velocity provided the wire does not slip. Let us assume that one side of the wire is removed and a piece of cardboard is attached to wheel 1, as shown in Fig.14.4(a). Place a pencil at a point Q on the wire and turn wheel 2 counter-clockwise. Point Q will generate an involute on the cardboard relative to wheel 1. If a cardboard is now attached to wheel 2, as shown in Fig.14.4(b), and the process is repeated, an involute is generated on the cardboard of wheel 2. If the cardboards are now cut along the involute, one side of tooth is formed on both the wheels.

The circles that have been used for generating the involutes are known as *base circles*. The angle that is included by a line perpendicular to the line of action through the centre of the base circle and a line from O_1 to O_2 through Q, is known as the *involute pressure angle*, as shown in Fig.14.4(c).

The intersection of the line perpendicular to the base circles and the line of centres has been labelled as P, the pitch point. The circles passing through point P with O_1 and O_2 as centres are called the *pitch circles*, as shown in Fig.14.4(d). At the pitch point, there is pure rolling, and at all other points there is a combination of rolling and sliding.

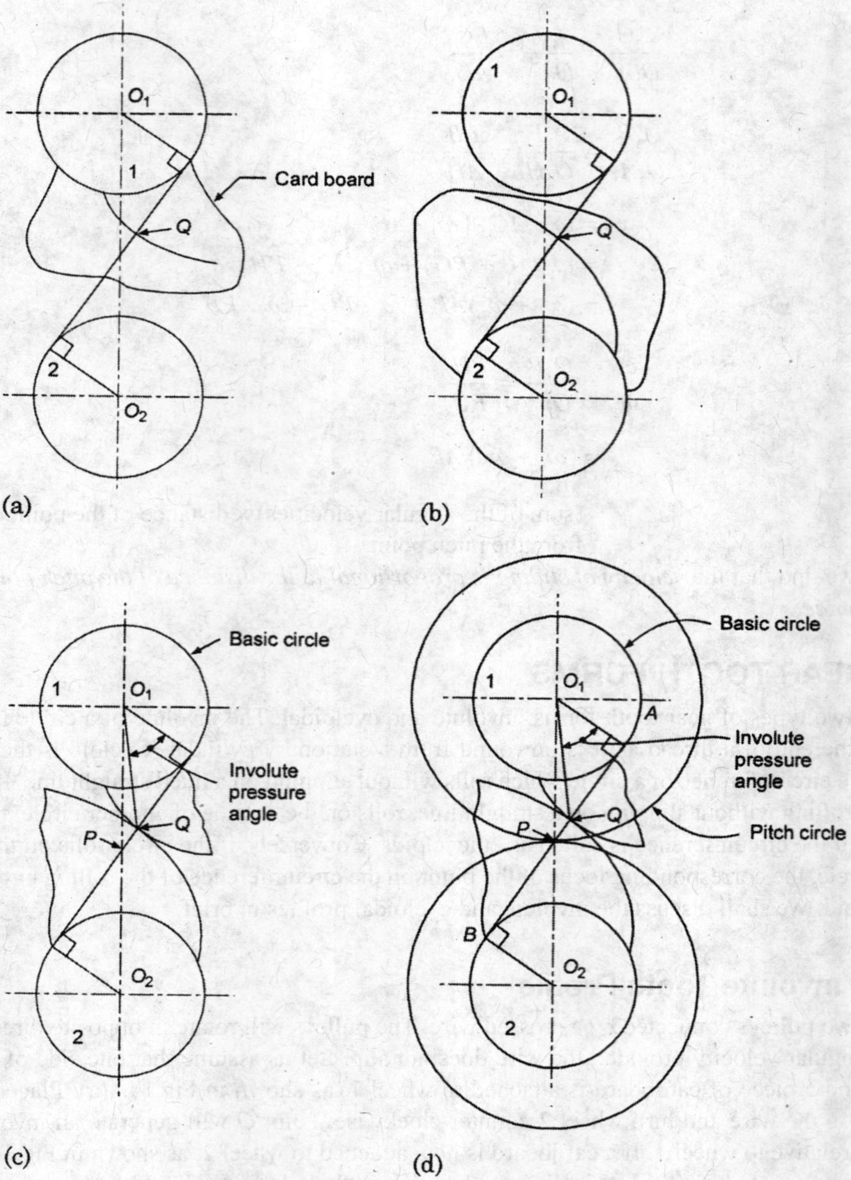

Fig.14.4 Involute tooth profile generation

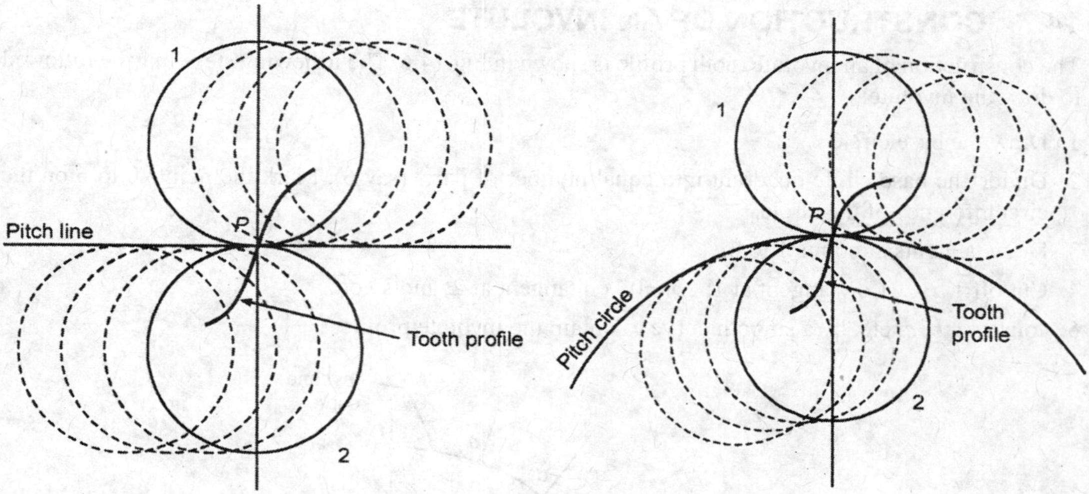

Fig.14.5 Cycloidal tooth profile generation

14.6.2 Cycloidal Tooth Profile

A cycloidal rack tooth profile meshing with a pinion is shown if Fig.14.5(a). The curve generated by a point on a circle 1 by its motion to the right on the straight line, which is the pitch line, is the profile of the face of the cycloidal tooth similarly the curve generated below the pitch line by a point on the rolling circle 2 is the flank of the tooth profile.

Fig.14.5(b) shows the construction of a cycloidal teeth profile of a gear. The circle 1 rolling on the outside of the pitch circle, generates a epicycloid, which is the face portion of the tooth profile. The circle 1 rolls without slipping to the right. The circle 2 rolls without slipping to the left on the inside of the circle generating a hypocycloid, representing the flank profile of the cycloidal tooth.

14.6.3 Comparison between Involute and Cycloidal Tooth Profiles

The comparison of involute and cycloidal tooth profiles is given in Table 14.1. The cycloidal profile is not commonly used for gear tooth, due to the reasons given in Table14.1.

Table 14.1 Comparison between Involute and Cycloidal Tooth Profiles

Characteristic	Involute gears	Cycloidal gears
1. Pressure angle	Constant throughout the engagement	Varies from commencement to end
2. Ease of manufacture	Easy to manufacture	Difficult to manufacture
3. Centre distance	Do not require exact centre distance	Requires exact centre distance
4. Interference	May occur	No interference
5. Strength	Less	More
6. Wear	More wear and tear	Less wear and tear
7. Operation	Smooth	Less smooth

14.7 CONSTRUCTION OF AN INVOLUTE

The construction of an involute tooth profile is shown in Fig.14.6. The following steps may be followed to draw the involute:

1. Draw the base circle.
2. Divide the base circle quadrant into equal number of parts (say 6). Mark the points 0 to 6 on the circumference of the circle.
3. Draw tangents at points 1 to 6.
4. Cut off 1 a = 01 on tangent at 1; $2b$ = 02 on tangent at 2, and so on.
5. Join o, a, b, c, etc. by a smooth curve to obtain the involute profile.

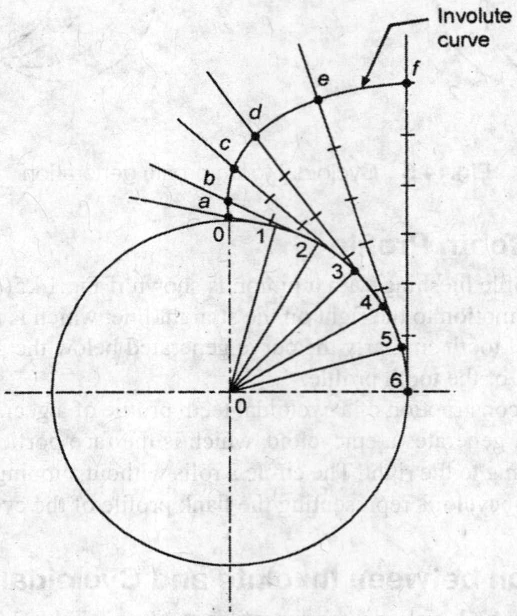

Fig.14.6 Involute profile construction

14.8 INVOLUTE FUNCTION

Consider the involute of a circle shown in Fig.14.7(a).

Let l = length of the thread unwrapped

$$= AB \text{ arc}$$

$$= r_b (\beta + \delta)$$

$$= r_b \tan \alpha$$

Thus $\beta + \delta = \tan \alpha$

where r_b = base radius.

Also from Fig.14.7(b), we have

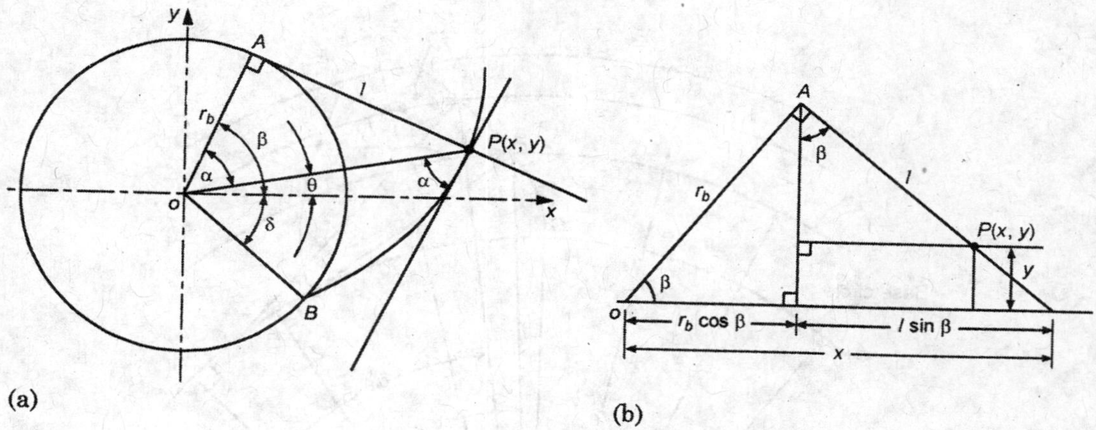

Fig.14.7 Involute function

$$x = r_b \cos\beta + l\sin\beta$$
$$= r_b[\cos\beta + (\beta + \delta)\sin\beta]$$

and
$$y = r_b[\sin\beta - (\beta + \delta)\cos\beta]$$

Also
$$\theta = \alpha = \beta = \tan\alpha - \delta$$

or
$$\theta = \tan\alpha - \alpha - \delta$$
$$= \text{inv}(\alpha) - \delta$$

where
$$\text{inv}(\alpha) = \tan\alpha - \alpha \tag{14.10}$$

Eq. (14.10) represents the involute function. Its values are given in standard tables.

14.9 INVOLUTOMETRY

Fig.14.8 shows an involute which has been generated from a base circle of radius r_b. The involute contains two points A and B with corresponding rodii r_A and r_B and involute pressure angles α_A and α_B.

and
$$r_b = r_A \cos\alpha_A$$
$$r_b = r_B \cos\alpha_B$$

Therefore
$$\cos\alpha_B = \left(\frac{r_A}{r_B}\right)\cos\alpha_A \tag{14.11}$$

It is possible to evaluate the involute pressure angle at any point on the involute profile from Eq. (14.11).

Now
$$\text{arc } DG = \text{length BG}$$
$$\angle DOG = \frac{arc\ DG}{OG} = \frac{BG}{OG}$$
$$\tan\alpha_B = \frac{BG}{OG}$$

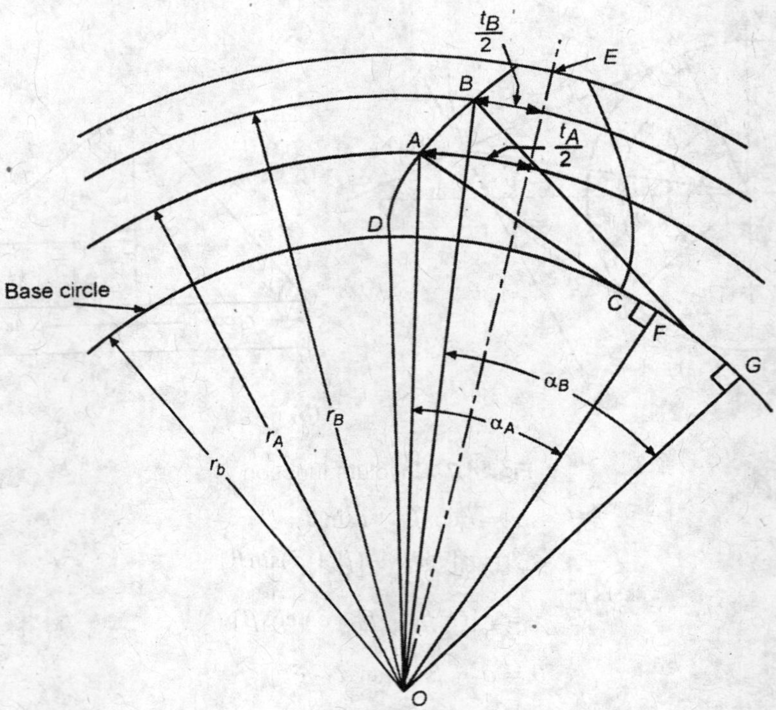

Fig.14.8 Involutometry

Thus
$$\angle DOG = \tan \alpha_B$$

Also
$$\angle DOB = \angle DOG - \alpha_B$$
$$= \tan \alpha_B - \alpha_B$$

and
$$\angle DOA = \tan \alpha_A - \alpha_A$$

Now
$$\angle DOE = \angle DOB + \frac{0.5 t_B}{r_B}$$
$$= \text{inv}\,(\alpha_B) + \frac{t_B}{2r_B}$$

Also
$$\angle DOE = \angle DOA + \frac{0.5 t_A}{r_A}$$
$$= \text{inv}\,(\alpha_A) + \frac{t_A}{2r_A}$$

Thus
$$t_B = 2r_B \left[\frac{t_A}{2r_A} + \text{inv}\,(\alpha_A) - \text{inv}\,(\alpha_B) \right] \tag{14.12}$$

14.10 INVOLUTE GEAR TOOTH ACTION

The gear tooth action between two gears is shown in Fig.14.9. P is the pitch point and line EF is tangent to both the base circles, along which all points of contact of two teeth must lie. Line EF is called the line of action or the pressure line. Line XX' is perpendicular to the line of centres at the pitch point. The angle between XX' and EF is called the pressure angle. If one gear rotates in clockwise direction then the other gear would rotate in the reverse direction of counter clockwise.

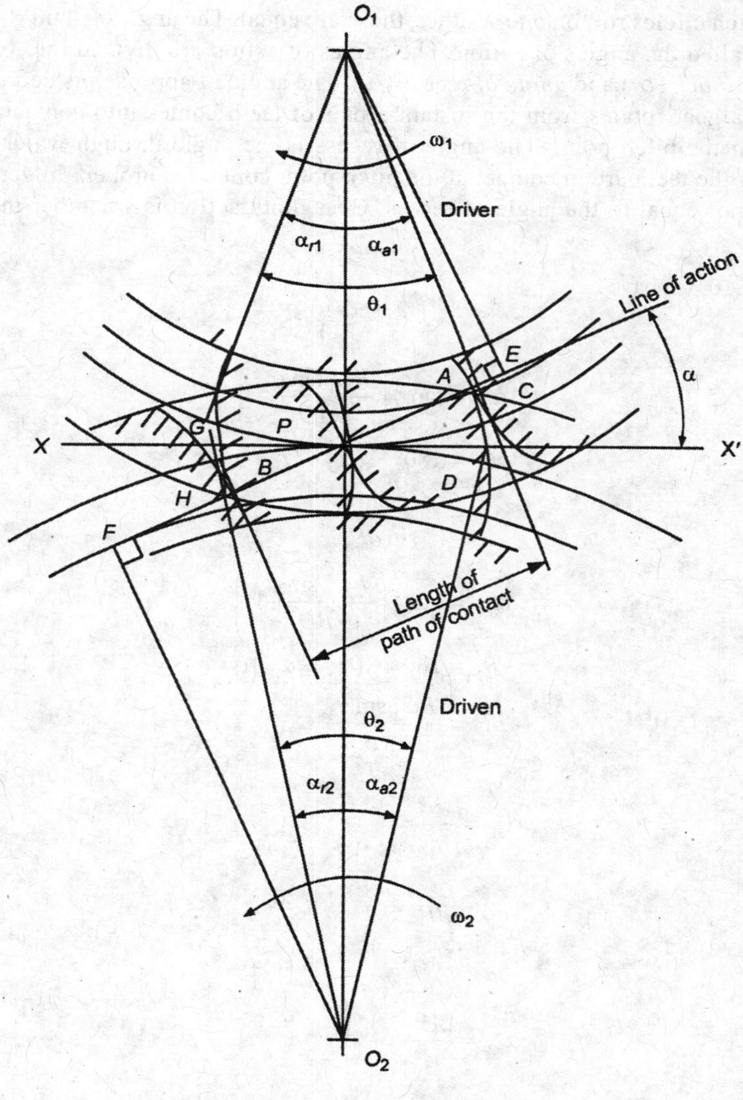

Fig.14.9 Involute gear tooth action

The teeth first come into contact at point A, where the addendum circle of the driven gear cuts the line of action. Contact follows the line of action through point P, and contact ceases at point B, where the addendum circle of the driving gear cuts the line of action. Line AB is called the path of the point of contact, and its length is the length of the path of contact. Point C is the intersection of the tooth profile on gear 2 with its pitch circle when the tooth is at the beginning of contact, and point G is the same point on the profile when the tooth is at the end of contact. Points D and H are the corresponding for gear 1. Arcs CPG and DPH are the arcs on the pitch circles through which the mating tooth profiles move as they pass from the initial to the final point of contact. These arcs action as the *arcs of action*.

Since the pitch circles roll on one another, these are equal. The angles θ_1 and θ_2 which subtend these arcs are called the angles of action. The angles of action are divided into two parts called the *angle of approach* (α_a) and *angle of recess* (α_r). The angle of approach is defined as the angle through which a gear rotates from the instant a pair of teeth comes into contact until the teeth are in contact at the pitch point. The angle of recess is the angle through which a gear rotates from the instant the teeth are in contact at the pitch point contact is broken. In general, the angle of approach is not equal to the angle of recess. Gear tooth action is smoother in recess than in approach.

From Fig.14.10, we have

$$\alpha_{r1} = \sin^{-1}\left(\frac{PO_1 \sin\alpha_a}{AO_1}\right)$$

$$\theta_1 = 180° - (\alpha_a + \alpha_{r1})$$

$$AP = \frac{AO_1 \sin\theta_1}{\sin\alpha_a}$$

$$\alpha_{r2} = \sin^{-1}\left(\frac{PO_2 \sin\alpha_a}{BO_2}\right)$$

$$\theta_2 = 180° - (\alpha_a + \alpha_{r2})$$

$$BP = \frac{BO_2 \sin\theta_2}{\sin\alpha_a}$$

$$\alpha_{a1} = \frac{AP}{r_1} \tag{14.13a}$$

$$\alpha_{r1} = \frac{BP}{r_1} \tag{14.13b}$$

$$\alpha_{a2} = \frac{AP}{r_2} \tag{14.14a}$$

$$\alpha_{r2} = \frac{BP}{r_2} \tag{14.14b}$$

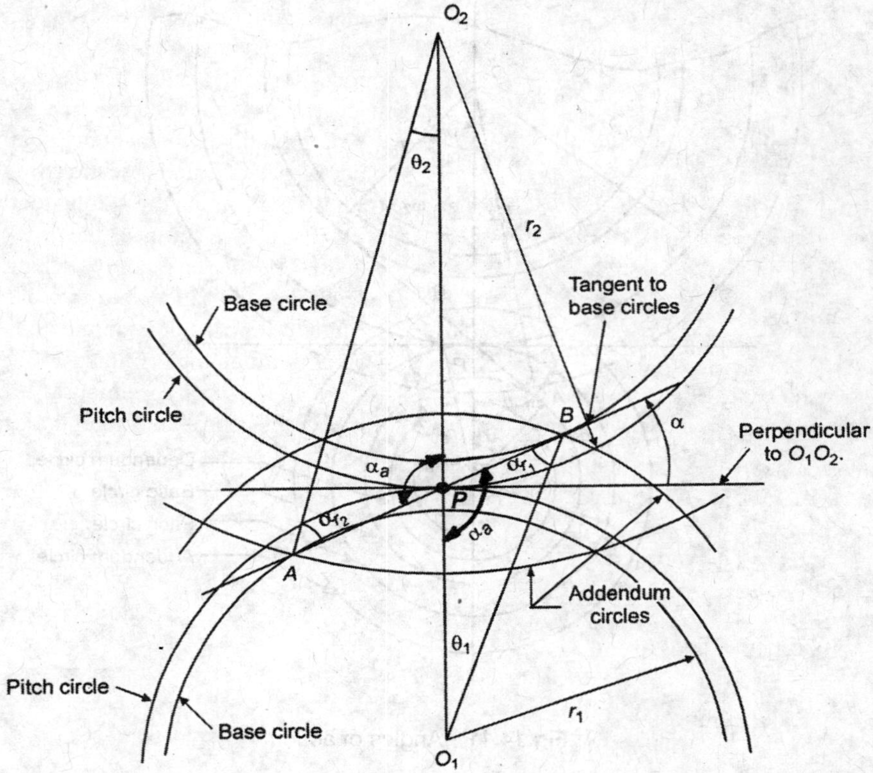

Fig.14.10 Calculating angles of approach and recess

14.11 CHARACTERISTICS OF INVOLUTE ACTION

The characteristics of the involute action are:

1. arc of contact,
2. length of path of contact, and
3. the contact ratio.

As shown in Fig.14.11, the contact of two gear teeth begins at A and ends at B.

Addendum radius of pinion, $r_{a1} = r_1 + h_{a1}$

Base circle radius of pinion, $r_{b1} = r_1 \cos \alpha$

Addendum radius of gear, $r_{b2} = r_{a2} = r_2 + h_{a2}$

Base circle radius of gear $= r_{b2} \cos \alpha$

where

r_1 = pitch circle radius of pinion

r_2 = pitch circle radius of gear

h_{a1} = addendum of pinion

h_{a2} = addendum of gear

r_{b1} = base circle radius of pinion

r_{b2} = base circle radius of gear

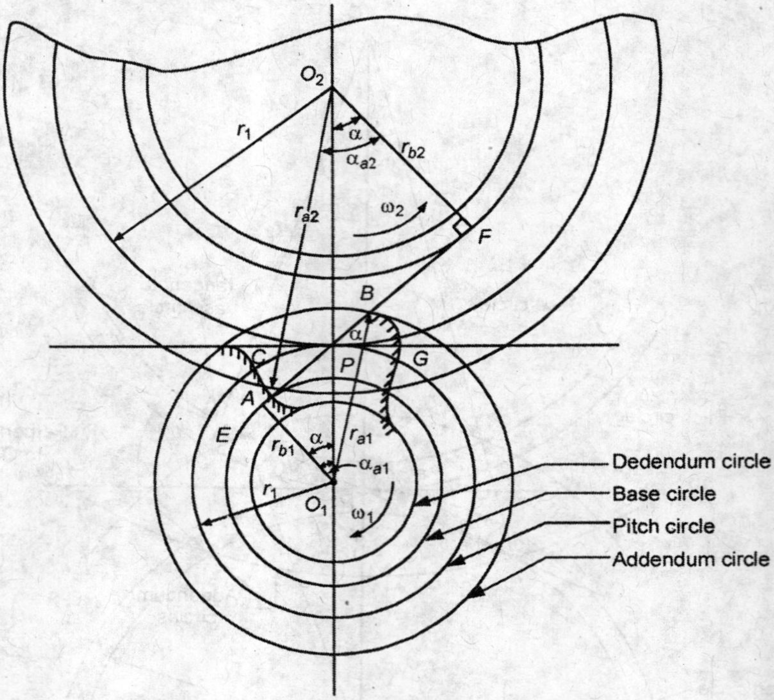

Fig.14.11 Angles of action

Length of path of recess, $\quad L_r = PB = EB - EP$

$$= (r_{a1}^2 - r_{b1}^2)^{0.5} - O_1P \sin\alpha$$

$$= (r_{a1}^2 - r_{a1}^2)^{0.5} - r_1 \sin\alpha \tag{14.15}$$

Length of path of approach, $\quad L_a = AP = AF - EF$

$$= (r_{a2}^2 - r_{b2}^2)^{0.5} - O_2P \sin\alpha$$

$$= (r_{a2}^2 - r_{b2}^2)^{0.5} - r_2 \sin\alpha \tag{14.16}$$

Length of path of contact, $\quad AB = L_p = L_r + L_a$

$$= \left(r_{a1}^2 - r_{b1}^2\right)^{0.5} + \left(r_{a2}^2 - r_{b2}^2\right)^{0.5} - \left(r_1 + r_2\right)\sin\alpha \tag{14.17}$$

Length of arc of contact, $\quad L_c = \text{arc } CG$

$$= \frac{AB}{\cos\alpha} \tag{14.18}$$

Maximum length of path of recess = $r_2 \sin \alpha$ (14.19)
Maximum length of path of approach = $r_1 \sin \alpha$ (14.20)
The *contact ratio* is defined as the average number of pairs of teeth, which are in contact. This can be found by nothing how many times the base pitch fits into the length of the path of contact. The contact ratio (*CR*) can be expressed as:

$$CR = \text{length of path of contact/base pitch}$$

$$= \frac{L_p}{p_b} \quad (14.21)$$

where $p_b = p \cos a = \pi m \cos a$

For a rack and a pinion,

$$L_P = \left(r_{a1}^2 - r_{b1}^2\right)^{0.5} - r_1 \sin \alpha + \frac{a}{\sin \alpha} \quad (14.22)$$

where a = addendum.

Example 14.1

A pinion of 24 teeth drives a gear of 60 teeth at a pressure angle of 20°. The pitch radius of the pinion is 38 mm and the outside radius is 41 mm. The pitch radius of the gear is 95 mm and the outside radius is 98.5 mm. Calculate the length of action and contact ratio.

Solution

Length of path of contact,

$$L_p = (r_{a1}^2 - r_{b1}^2)^{0.5} + (r_{a2}^2 - r_{b2}^2)^{0.5} - (r_1 + r_2)\sin \alpha$$

Here,

$$r_{a1} = 41 \text{ mm}, r_{a2} = 98.5 \text{ mm}, r_1 = 38 \text{ mm}, r_2 = 95 \text{ mm}, \alpha = 20°.$$

$$r_{b1} = r_1 \cos \alpha = 38 \cos 20° = 35.7 \text{ mm}$$

$$r_{b2} = r_2 \cos \alpha = 95 \cos 20° = 811.27 \text{ mm}$$

$$L_p = (41^2 - 35.7^2)^{0.5} + (98.5^2 - 811.27^2)^{0.5} - (38 + 95)\sin 20°$$

$$= 20.16 + 41.63 - 45.49$$

$$= 16.30 \text{ mm}$$

Contact ratio,

$$m_c = \frac{L_p}{P_b}$$

$$P_b = 2\pi \frac{r_{b1}}{z_1}$$

$$= 2\pi \times \frac{35.7}{24} = 11.37 \text{ mm}$$

$$m_c = \frac{16.30}{11.34} = 1.744$$

Example 14.2

Two equal size spur gears in mesh have 36 number of teeth, 20° pressure angle and 6 mm module. If the arc of contact is 1.8 times the circular pitch, find the addendum.

■ Solution

Circular pitch, $p = \pi m = \pi \times 6 = 18.85$ mm

Length of arc of contact, $L_a = 1.8p = 1.8 \times 18.85 = 33.93$ mm

Length of path of contact, $L_p = L_a \cos\alpha$

$= 33.93 \cos 20° = 31.88$ mm

Pitch radii, $r_1 = r_2 = \dfrac{mz}{2} = 6 \times \dfrac{36}{2} = 108$ mm

$$L_p = (r_{a1}^2 - r_{b1}^2)^{0.5} + (r_{a2}^2 - r_{b2}^2)^{0.5} - (r_1 + r_2)\sin\alpha$$

Here $r_{a1} = r_{a2}$ and $r_{b1} = r_{b2} = r\cos\alpha = 108 \cos 20° = 101.49$ mm

$$L_p = 2(r_a^2 - 101.49^2)^{0.5} - 216 \sin 20°$$

$$31.88 = 2(r_a^2 - 101.49^2)^{0.5} - 73.87$$

Addendum, $r_a = 114.44$ mm

$h_a = r_a - r = 114.44 - 108 = 6.44$ mm

Example 14.3

Two 20° involute gears in mesh have a gear ratio of 2 and 20 teeth on the pinion. The module is 5 mm and the pitch line speed is 1.5 m/s. Assuming addendum to be equal to one module, find (a) angle turned through by pinion when one pair of teeth is in mesh, and (b) maximum velocity of sliding.

■ Solution

Given: $i = 2$, $z_1 = 20$, $m = 5$ mm, $v = 1.5$ m/s, $h_{a1} = m = h_{a2}$

(a) $r_1 = \dfrac{mz_1}{2} = 5 \times \dfrac{20}{2} = 50$ mm

$r_2 = ir_1 = 2r_1 = 100$ mm

$r_{a1} = r_1 + h_{a1} = 50 + 5 = 55$ mm

$r_{a2} = r_2 + h_{a2} = 100 + 5 = 105$ mm

$r_{b1} = r_1 \cos 20° = 50 \cos 20° = 46.98$ mm

$r_{b2} = r_2 \cos 20° = 100 \cos 20° = 93.97$ mm

Length of path of approach, $L_a = [r_{a2}^2 - r_{b2}^2]^{0.5} - r_2 \sin\alpha$

$= [105^2 - 93.97^2]^{0.5} - 100 \sin 20°$

$= 12.65$ mm

Length of path of recess, $L_r = \left[r_{a1}^2 - r_{b1}^2\right]^{0.5} - r_1 \sin 20°$

$= [55^2 - 46.98^2]^{0.5} - 50 \sin 20°$

$= 11.50$ mm

Length of path of contact, $L_p = L_a + L_r = 12.65 + 11.50 = 24.15$ mm

Length of arc of contact, $L_c = \dfrac{L_p}{\cos\alpha} = \dfrac{24.15}{\cos 20°} = 25.7$ mm

Angle turned through by the pinion $= L_c \times \dfrac{360}{2\pi r_1}$

$$= 25.7 \times \frac{360}{2\pi \times 50}$$

$$= 211.45°$$

(b) Maximum velocity of sliding, $v_s = (\omega_1 + \omega_2)L_a$

$$\omega_1 = \frac{v}{r_1} = \frac{1.5}{0.05} = 30 \text{ rad/s}$$

$$\omega_2 = \frac{v}{r_2} = \frac{1.5}{0.1} = 15 \text{ rad/s}$$

$$v_s = (30 + 15)12.65 = 5611.25 \text{ mm/s}$$

Example 14.4

The pressure angle of two gears in mesh is 20° and have a module of 10 mm. The number of teeth on pinion are 24 and on gear 60. The addendum of pinion and gear is same and equal to one module. Determine (a) the number of pairs of teeth in contact, (b) the angle of action of pinion and gear, and (c) the ratio of sliding to rolling velocity at the beginning of contact, at pitch point and at the end of contact.

Solution

Given: $\alpha = 20°$, $m = 10$ mm, $z_1 = 24$, $z_2 = 60$, $h_{a1} = h_{a2} = 1$ m $= 10$ mm

$$r_1 = \frac{mz_1}{2} = \frac{10 \times 24}{2} = 120 \text{ mm}, \quad r_2 = \frac{10 \times 60}{2} = 300 \text{ mm}$$

$$r_{a1} = r_1 + h_{a1} = 120 + 10 = 130 \text{ mm}$$

$$r_{a2} = r_2 + h_{a2} = 300 + 10 = 310 \text{ mm}$$

$$r_{b1} = r_1 \cos\alpha = 120 \times \cos 20° = 112.76 \text{ mm}$$

$$r_{b2} = r_2 \cos\alpha = 300 \times \cos 20° = 281.91 \text{ mm}$$

Length of path of recess, $L_r = \left[r_{a1}^2 - r_{b1}^2\right]^{0.5} - r_1 \sin\alpha$

$$= \left[130^2 - (112.76)^2\right]^{0.5} - 120 \sin 20°$$

$$= 23.65 \text{ mm}$$

Length of path of approach, $L_a = \left[r_{a2}^2 - r_{b2}^2\right]^{0.5} - r_2 \sin\alpha$

$$= \left[310^2 - (281.91)^2\right]^{0.5} - 300 \cos 20°$$

$$= 26.33 \text{ mm}$$

Length of path of contact, $L_p = L_a + L_r = 26.33 + 23.65 = 411.98$ mm

(a) Number of pairs of teeth in contact $= \dfrac{L_p}{\pi m \cos\alpha} = \dfrac{411.98}{\pi \times 10 \times \cos 20°}$

$= 1.69$

(b) Angle of action of pinion, $\alpha_{a1} = \text{Arc of contact} \times \dfrac{360}{2\pi r_1}$

$$= \left(\frac{L_p}{\cos\alpha}\right) \times \frac{360}{2\pi r_1}$$

$$= \left(\frac{411.98}{\cos 20°}\right) \times \frac{360}{2\pi \times 120}$$

$$= 25.4°$$

Angle of action of gear, $\alpha_{a2} = \text{Arc of contact} \times \dfrac{360}{2\pi r_2}$

$$= \left(\frac{L_p}{\cos\alpha}\right) \times \frac{360}{2\pi r_2}$$

$$= \left(\frac{411.98}{\cos 20°}\right) \times \frac{360}{2\pi \times 300}$$

$$= 10.16°$$

(c) Ratio of sliding to rolling velocity $= \dfrac{v_s}{v_r}$

$$v_r = r_1\omega_1 = 120\,\omega_1$$

$$\omega_2 = \frac{24\,\omega_1}{60} = 0.4\,\omega_1$$

At the beginning of contact $= \dfrac{\omega_1 + \omega_2}{\dfrac{L_a}{v_r}}$

$$= \frac{(\omega_1 + 0.4\,\omega_1) \times 26.33}{120\,\omega_1}$$

$$= 0.3072$$

At the pitch point, $v_s = 0$. Hence $\dfrac{v_s}{v_r} = 0$

At the end of contact $= \dfrac{\omega_1 + \omega_2}{\dfrac{L_r}{v_r}}$

$$= \frac{(\omega_1 + 0.4\omega_1) \times 23.65}{120\omega_1}$$

$$= 0.276$$

Example 14.5

Two 15 mm module, 20° pressure angle spur gears have addendum equal to one module. The pinion has 25 teeth and the gear 50 teeth. Determine whether interference will occur or not. If it occurs, to what value should the pressure angle be changed to eliminate interference.

Solution

Given: $\alpha = 20°$, $m = 15$ mm, $z_1 = 25$, $z_2 = 50$, $h_{a1} = h_{a2} = 15$ and $m = 10$ mm

Let the pinion be the driver.

$$r_1 = \frac{mz_1}{2} = \frac{15 \times 25}{2} = 187.5 \text{ mm}, \; r_2 = \frac{15 \times 50}{2} = 375 \text{ mm}$$

$$r_{a1} = r_1 + h_{a1} = 187.5 + 15 = 202.5 \text{ mm}$$

$$r_{a2} = r_2 + h_{a2} = 375 + 15 = 390 \text{ mm}$$

$$r_{b2} = r_2 \cos\alpha = 375 \times \cos 20° = 352.4 \text{ mm}$$

Maximum permissible length of path of approach,

$$(L)_{max} = r_1 \sin\alpha$$

$$= 187.5 \sin 20°$$

$$= 64.13 \text{ mm}$$

Length of path of approach,

$$L_a = \left[r_{a2}^2 - r_{b2}^2\right]^{0.5} - r_2 \sin\alpha$$

$$= [390^2 - (352.4)^2]^{0.5} - 375 \sin 20°$$

$$= 38.81 \text{ mm}$$

Since $L_a < (L_{max})$, hence interference will occur.

For $L_a = (L_{max})$, we have

$$r_1 \sin\alpha = \left[r_{a2}^2 - r_{b2}^2\right]^{0.5} - r_2 \sin\alpha$$

$$64.13 = [390^2 - (375\cos\alpha)^2]^{0.5} - 375 \sin\alpha$$

$$(64.13 + 375\sin\alpha)^2 = 390^2 - (375\cos\alpha)^2$$

$$4112.65 + 140625 + 48097.5 \sin\alpha = 152100$$

$$\sin\alpha = 0.15307$$

$$\alpha = 8.8°$$

Example 14.6

For a pair of involute spur gears, $m = 10$ mm, $\alpha = 20°$, $z_1 = 20$, $z_2 = 40$, $n_1 = 60$ rpm. The addendum on each gear is such that the path of approach and the path of recess on each side is 50% of the maximum possible length. Determine the addendum for the pinion and the gear and the length of arc of contact.

Solution

$r_1 = mz_1/2 = 10 \times 20/2 = 100$ mm, $r_2 = mz_2/2 = 10 \times 40/2 = 200$ mm

Let the pinion be the driver.

Maximum possible length of approach $= r_1 \sin\alpha = 100 \sin 20° = 34.2$ mm

$$\text{Actual length of approach} = \left[r_{a2}^2 - r_{b2}^2\right]^{0.5} - r_2 \sin\alpha$$

$$= \left[r_{a2}^2 - (200\cos 20°)^2\right]^{0.5} - 200\sin 20°$$

$$= \left[r_{a2}^2 - 35321\right]^{0.5} - 68.4 = 0.5 \times 34.2 = 17.1$$

or

$$r_{a2}^2 - 35321 = (85.5)^2$$

$$r_{a2} = 206.5 \text{ mm}$$

$$h_{a2} = 206.5 - 200 = 6.05 \text{ mm}$$

$$\text{Maximum possible length of recess} = r_2 \sin\alpha$$

$$= 200 \sin 20° = 68.4 \text{ mm}$$

$$\text{Actual length of recess} = \left[r_{a1}^2 - r_{b1}^2\right]^{0.5} - r_1 \sin\alpha$$

$$= \left[r_{a1}^2 - (100\cos 20°)^2\right]^{0.5} - 100\sin 20°$$

$$= \left[r_{a1}^2 - 8830\right]^{0.5} - 34.2 = 0.5 \times 68.4 = 34.2$$

or

$$r_{a1}^2 - 8830 = (68.4)^2$$

$$r_{a1} = 116.2 \text{ mm}$$

$$h_{a1} = 116.2 - 100 = 16.2 \text{ mm}$$

$$\text{Arc of contact} = \frac{\text{Path of contact}}{\cos\alpha} = 0.5\,(r_1 + r_2)\frac{\sin\alpha}{\cos\alpha}$$

$$= 0.5\,(100 + 200)\tan 20°$$

$$= 54.6 \text{ mm}$$

Example 14.7

Two involute gear wheels having module 3 mm and pressure angle 20° mesh externally to give a velocity ratio of 3. The pinion rotates at 75 rpm and addendum is equal to one module. Determine (a) the number of teeth on each wheel so that interference is just avoided, (b) the length of path and arc of contact, (c) the number of pairs of teeth in contact, and (d) the maximum velocity of sliding between the teeth.

Solution

Given: $m = 3$ mm, $\alpha = 20°$, $i = 3$, $n_1 = 75$ rpm

$$r_{a1} = r_1 + h_{a1} = r_1 + 3,\; r_{a2} = r_2 + h_{a2} = 3r_1 + 3 \quad [\therefore \; r_2 = 3r_1]$$

$$r_{b1} = r_1 \cos 20° = 0.9397 r_1,\; r_{b2} = 3r_1 \cos 20° = 2.819 r_1$$

(a) Let the pinion be the driver.

$$L_a = \left[r_{a2}^2 - r_{b2}^2\right]^{0.5} - r_2 \sin a$$

$$(L_a)_{\max} = r_1 \sin a$$

To avoid interference,

$$L_a = (L_a)_{\max}$$

$$\left[r_{a2}^2 - r_{b2}^2\right]^{0.5} - r_2 \sin a = r_1 \sin a$$

$$r_{a2}^2 - r_{b2}^2 = (r_1 - r_2)^2 \sin^2 a$$

$$(3r_1 + 3)^2 - (2.819)^2 r_1^2 = (4r_1)^2 \sin^2 a$$

or

$$0.8186 r_1^2 - 18 r_1 - 9 = 0$$

$$r_1 = 22.48 \text{ mm}$$

$$r_2 = 67.44 \text{ mm}$$

$$z_1 = 2r_1/m = 2 \times 22.48/3 = 14.98 \cong 15$$

so that $r_1 = 22.5$ mm, and $z_2 = 45$

(b)

$$r_{a1} = 22.5 + 3 = 25.5 \text{ mm}, r_{a2} = 67.5 + 3 = 70.5 \text{ mm}$$

$$r_{b1} = 22.5 \cos 20° = 21.143 \text{ mm}, r_{b2} = 67.5 \cos 20° = 63.429 \text{ mm}$$

$$L_p = \left(r_{a1}^2 - r_{b1}^2\right)^{0.5} + \left(r_{a2}^2 - r_{b2}^2\right)^{0.5} - (r_1 - r_2) \sin \alpha$$

$$= \left[(25.5)^2 - (21.143)^2\right]^{0.5} + \left[(70.5)^2 - (63.429)^2\right]^{0.5}$$

$$-(22.5 + 67.5) \sin 20°$$

$$= 14.247 \text{ mm}$$

$$p = \pi m = \pi \times 3 = 11.425 \text{ mm}$$

Length of arc of contact,

$$L_c = \frac{L_p}{\cos \alpha} = \frac{14.247}{\cos 20°} = 15.16 \text{ mm}$$

(c) Number of pairs of teeth in contact $= \dfrac{L_c}{p} = \dfrac{15.16}{11.425} = 1.6$

(d) Maximum velocity of sliding $= (\omega_1 + \omega_2) r_1 \sin \alpha$

$$= \left(\frac{2\pi}{60}\right)(75 + 25) \times 22.5 \times \sin 20° = 80.58 \text{ mm/s}$$

14.12 INTERFERENCE AND UNDERCUTTING IN INVOLUTE GEAR TEETH

An involute starts at the base circle and is generated outwards. It is therefore impossible to have an involute inside the base circle. The line of action is tangent to the two base circles of a pair of gears in mesh, and these two points represent the extreme limits of the length of action. These

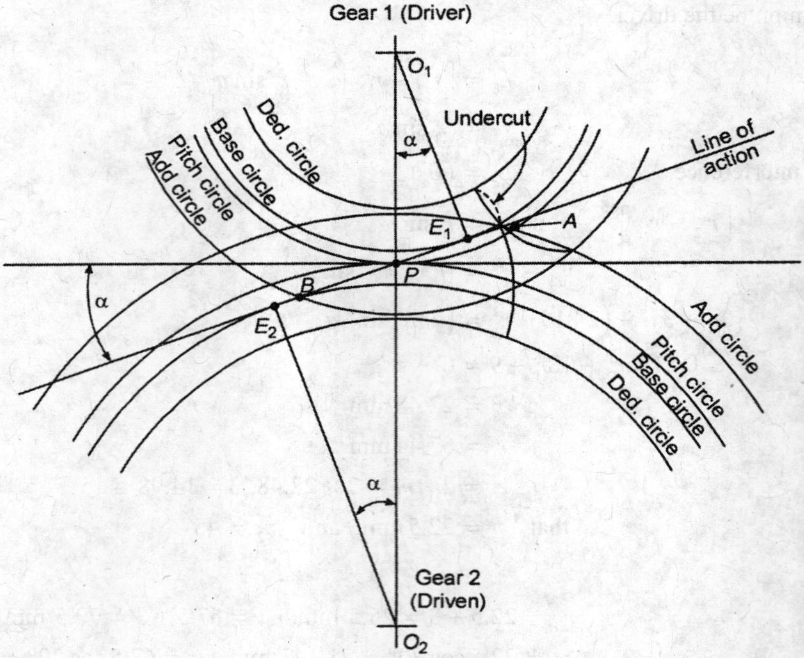

Fig.14.12 Interference in gears

two points are called *interference points*. If the teeth are of such proportion that the beginning of contact occurs before the interference point is met, then the involute portion of the driven gear will mate with a non-involute portion of the driving gear, and *involute interference* is said to occur. This condition is shown in Fig.14.12; E_1 and E_2 show the interference points that should limit the length of action. A shows the beginning of contact, and B shows the end of contact. It can be seen that the beginning of contact occurs before the interference point E_1 is met; therefore, interference is present. The tip of the driven tooth will gauge out or undercut the flank of the driving tooth as shown by the dotted line.

There are several ways of eliminating interference. Interference can be avoided by undercutting, making stub teeth, increasing the pressure angle, and cutting the gears with long and short addendum gear teeth. The method of undercutting is to limit addendum of the driven gear so that it passes through the interference point E_1, thus giving a new beginning of contact. Interference and the resulting undercutting not only weaken the pinion tooth but may also remove and small portion of the involute adjacent to the base circle, which may cause a serious reduction in the length of action.

Fig.14.13 shows a rack and a pinion in mesh. The point of tangency of the line of action and the base circle of the pinion is labelled as the interference point E, which fixes the maximum addendum for the rack. The contact begins as A, and undercutting will occur as shown by the dotted line. If the addendum of the rack extends only to the line that passes through the interference point, E, then the interference point becomes the beginning of contact, and interference is eliminated. If the number of teeth on the pinion is such that it will mesh without interference, it will mesh without interference with any other gear having the same or a larger number of teeth.

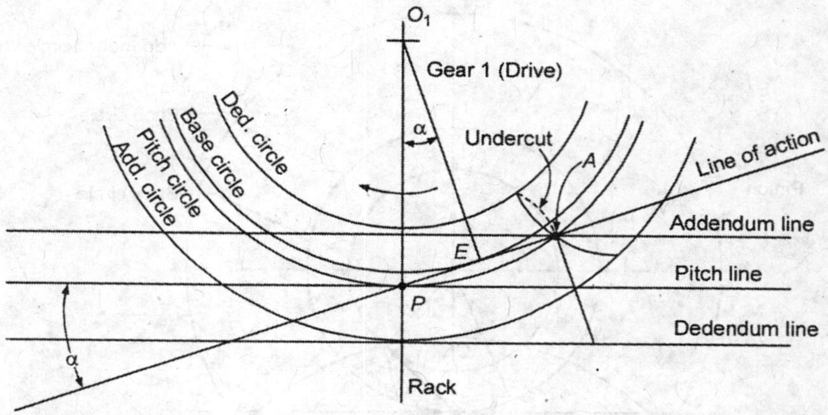

Fig.14.13 Interference in rack and pinion

14.13 MINIMUM NUMBER OF TEETH

14.13.1 Gear Wheel

For a minimum number of teeth to avoid interference, the common tangent to the base circles cuts the addendum circles at A and B, as shown in Fig.14.14.

Let speed ratio, $i = z_2/z_1$

Addendum of pinion, $h_{a1} = a_p m$

Addendum of gear wheel, $h_{a2} = a_\omega m$

where a_p and a_w are the constants by which the module must be multiplied to get the addendum of pinion and gear wheel respectively.

From ΔAO_2P, we have

$$O_2A^2 = O_2P^2 + AP^2 - 2 \cdot O_2P \cdot AP \cdot \cos \angle O_2PA$$

$$= \left(\frac{mz_2}{2}\right)^2 + (O_1P \sin \alpha)^2 - 2 \cdot \frac{mz_2}{2} \cdot O_1P \sin \alpha \cdot \cos\ (90° + \alpha)$$

$$= \left(\frac{mz_2}{2}\right)^2 + \left(\frac{mz_1}{2}\right)^2 \sin^2 \alpha + 2 \cdot \left(\frac{mz_2}{2}\right) \cdot \left(\frac{mz_1}{2}\right) \sin^2 \alpha$$

$$= \left(\frac{mz_2}{2}\right)^2 \left[1 + \left[\frac{1}{i^2}\right] \sin^2 \alpha + \left[\frac{2}{i}\right] \sin^2 \alpha\right]$$

$$O_2A = \left(\frac{mz_2}{2}\right) \left[1 + \left[\frac{1}{i^2}\right] \sin^2 \alpha + \left[\frac{2}{i}\right] \sin^2 \alpha\right]^{0.5} \quad (1)$$

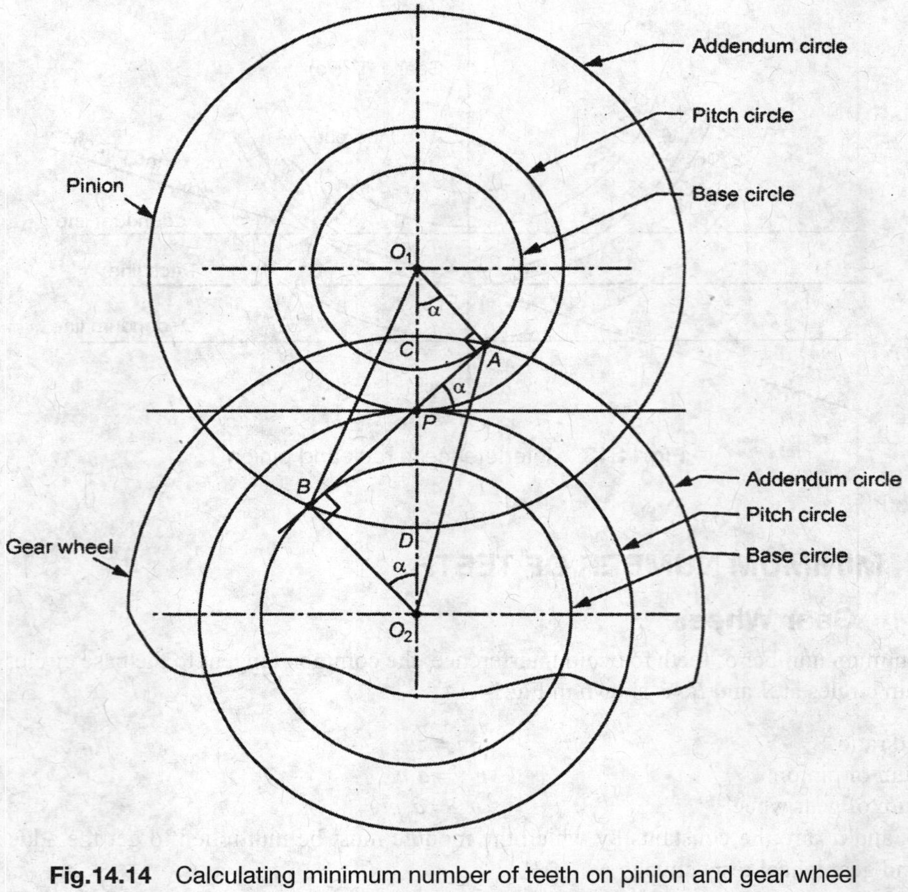

Fig.14.14 Calculating minimum number of teeth on pinion and gear wheel

Also
$$O_2A = O_2P + PC$$
$$= \frac{mz_2}{2} + h_{a2}$$
$$= \frac{mz_2}{2} + a_w m \tag{2}$$

From Eqs. (1) and (2), we get

$$\frac{mz_2}{2} + a_w\ m = \left(\frac{mz_2}{2}\right)\left[1 + \left[\frac{1}{i^2}\right]\sin^2\alpha + \left(\frac{2}{i}\right)\sin^2\alpha\right]^{0.5}$$

or
$$a_w = \left(\frac{z_2}{2}\right)\left[\left\{1 + \left[\frac{1}{i^2}\right]\sin^2\alpha + \frac{2}{i}\sin^2\alpha\right\}^{0.5} - 1\right]$$

or
$$z_2 = \frac{2a_w}{\left[\left\{1 + \left(\frac{1}{i^2}\right)\sin^2\alpha + \left(\frac{2}{i}\right)\sin^2\alpha\right\}^{0.5} - 1\right]} \tag{14.23}$$

or
$$z_2 = \frac{\left[z_1^2 \sin^2\alpha - 4a^2_{\ w}\right]}{\left[4a_w - 2z_1 \sin^2\alpha\right]} \tag{14.24}$$

For
$$a_w = 1,\ z_2 = \frac{\left[z^2_{\ 1} \sin^2\alpha - 4\right]}{\left[4 - 2z_1 \sin^2\alpha\right]} \tag{14.25}$$

For
$$i = 1,\ a_w = a_p \text{ and}$$
$$z_2 = \frac{2a_w}{\left[\left\{1 + 3\sin^2\alpha\right\}^{0.5} - 1\right]} \tag{14.26}$$

14.13.2 Pinion

From ΔBO_1P, we have

$$\begin{aligned}
O_1B^2 &= O_1P^2 + BP^2 - 2\cdot O_1P\cdot BP\cdot\cos\angle O_1PB \\
&= \left(\frac{mz_1}{2}\right)^2 + (O_2P\sin\alpha)^2 - 2\cdot\frac{mz_1}{2}\cdot O_2P\sin\alpha\cdot\cos(90^\circ+\alpha) \\
&= \left(\frac{mz_1}{2}\right)^2 + \left(\frac{mz_2}{2}\right)^2\sin^2\alpha + 2\cdot\left(\frac{mz_1}{2}\right)\cdot\left(\frac{mz_2}{2}\right)\sin^2\alpha \\
&= \left(\frac{mz_1}{2}\right)^2\left[1 + i^2\sin^2\alpha + 2i\sin^2\alpha\right] \\
O_1B &= \left(\frac{mz_1}{2}\right)\left[1 + i^2\sin^2\alpha + 2i\sin^2\alpha\right]^{0.5}
\end{aligned} \tag{1}$$

Also
$$\begin{aligned}
O_1B &= O_1P + PD \qquad (2)\\
&= \frac{mz_1}{2} + h_{a1} \\
&= \frac{mz_1}{2} + a_p m
\end{aligned}$$

From Eqs. (1) and (2), we get

$$\frac{mz_1}{2} + a_p m = \left(\frac{mz_1}{2}\right)\left[1 + i^2\sin^2\alpha + 2i\sin^2\alpha\right]^{0.5}$$

or $$a_p = \left(\frac{z_1}{2}\right)\left[\left\{1 + i^2 \sin^2 \alpha + 2i \sin^2 a\right\}^{0.5} - 1\right]$$

or $$z_1 = \frac{2a_p}{\left[\left\{1 + i^2 \sin^2 \alpha + 2i \sin^2 \alpha\right\}^{0.5} - 1\right]} \tag{14.27}$$

or $$z_1 = \frac{\left[z_2^2 \sin^2 \alpha - 4a_p^2\right]}{\left[4a_p - 2z_2 \sin^2\alpha\right]} \tag{14.28}$$

For $$a_p = 1,\ z_1 = \frac{\left[z_2^2 \sin^2 \alpha - 4\right]}{\left[4 - 2z_2 \sin^2 \alpha\right]} \tag{14.29}$$

For $$\alpha = 14.5°$$

$$(z_1)_{\min} = \left(\frac{z_2^2 - 63.8}{63.8 - 2z_2}\right) \tag{14.30}$$

For $$\alpha = 20°$$

$$(z_1)_{\min} = \left(\frac{z^2{}_2 - 34.2}{34.2 - 2z_2}\right) \tag{14.31}$$

14.13.3 Rack and Pinion

A rack is a gear of infinite pitch radius. Thus its pitch circle is a straight line, called the *pitch line*. The line of action is tangent to the base circle at infinity; hence the involute profile of the rack is straight line and is perpendicular to the line of action. For the rack and pinion shown in Fig.14.15, let

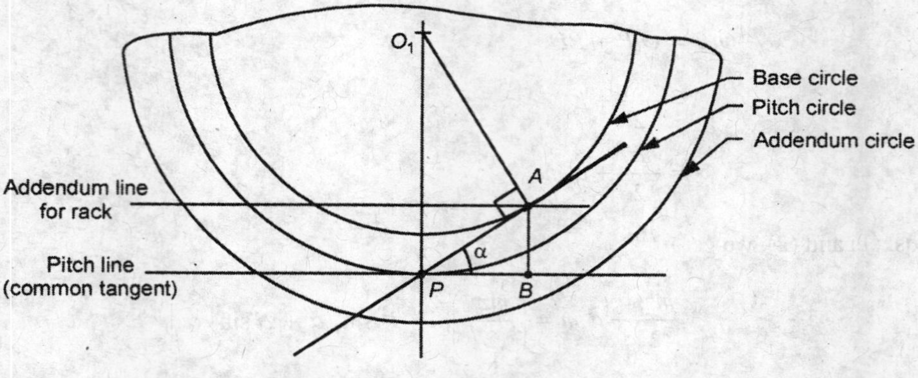

Fig.14.15 Minimum number of teeth on rack and pinion

Addendum of rack, $h_r = a_r m$

$$h_r = AB = AP \sin\alpha$$
$$= O_1P \sin a \sin\alpha$$
$$= O_1P \sin a \sin\alpha$$
$$= r_1 \sin^2\alpha$$

or
$$a_r m = \left(\frac{mz_1}{2}\right)\sin^2\alpha$$
$$a_r = \left(\frac{z_1}{2}\right)\sin^2\alpha$$
$$z_1 = \frac{2a_r}{\sin^2\alpha} \tag{14.32}$$

For $a_r = 1$, the minimum number of teeth on the pinion are given in Table 14.2.

Table 14.2 Minimum Number of Teeth on the Pinion for a Rack

α	14.5°	20°	20° stub	25°
$(z_1)_{min}$	32	18	14	12

Example 14.8

Determine the minimum number of teeth on the 20° pinion in order to avoid interference with a gear to give a gear ratio of 3:1. The addendum on wheel is equal to one module.

■ Solution

$$z_2 = 2a_w/[\{1+(1/i^2)\sin^2\alpha + (2/i)\sin^2\alpha\}^{0.5} - 1]$$
$$z_1 = 2a_w/\text{I}\ [\{1+(1/i^2)\sin^2\alpha + (2/i)\sin^2\alpha\}^{0.5} - 1]$$
$$= 2\times 1/3\left[\{1+(1/9)\sin^2 20° + (2/3)\sin^2 20°\}^{0.5} - 1\right]$$
$$= 14.98 \approx 15$$

Example 14.9

A pinion of 20° involute teeth and 120 mm pitch circle diameter drives a rack. The addendum of both pinion and rack is 6 mm. Determine the least pressure angle which can be used to avoid interference. With the pressure angle find the contact ratio.

■ Solution

Given: $\alpha = 20°$, $d_1 = 120$ mm, $h_a = 6$ mm

$$h_a = r_1 \sin^2 a$$
$$6 = 60 \sin^2 a$$

$$a = 18.435°$$
$$r_{a1} = r_1 + h_a = 60 + 6 = 66 \text{ mm}$$
$$r_{b1} = r_1 \cos a = 60 \cos 18.435° = 56.92 \text{ mm}$$

Length of path of contact, $L_p = \left(r_{a1}^2 - r_{b1}^2 \right)^{0.5}$

$$= \left(66^2 - 56.92^2\right)^{0.5}$$
$$= 33.4 \text{ mm}$$

Base pitch, $p_b = p \cos \alpha = (\pi d_1 / z_1) \cos \alpha$

$$= (\pi \times 120/20) \cos 18.435° = 17.88 \text{ mm}$$

Minimum number of teeth in contact $= L_p / p_b$

$$= 33.4/17.88 = 1.87 \approx 2$$

Example 14.10

Two 3 mm module, 20° pressure angle involute spur gears mesh externally to give a velocity ratio of 4. The addendum is 1.2 times the module. The pinion rotates at 150 rpm. Determine (a) the minimum number of teeth on each gear wheel to avoid interference, and (b) the number of pairs of teeth in contact.

Solution

Given: $m = 3$ mm, $a_w = 1.2$, $h_a = 1.2 \times 3 = 3.6$ mm, $\alpha = 20°$, $i = 4$, $n_1 = 150$ rpm

(a)
$$z_2 = 2 a_w \Big/ \left[\left\{ 1 + \left(1/i^2\right) \sin^2 \alpha + \left(2/i\right) \sin^2 \alpha \right\}^{0.5} - 1 \right]$$
$$= \left(2 \times 1.2\right) / \left[\left\{1 + \left(1/6\right) \sin^2 20° + \left(2/4\right) \sin^2 20°^{0.5} - 1\right]$$
$$= 2.4 \Big/ \left[\left\{ 1 + 7.311 \times 10^{-3} + 0.05849 \right\}^{0.5} - 1 \right]$$
$$= 2.4/\left[1.03237 - 1\right]/$$
$$= 74.13 \cong 76 \text{ so that } z_2 \text{ is divisible by 4.}$$
$$z_1 = 76/4 = 19$$

(b)
$$r_1 = mz_1/2 = 3 \times 19/2 = 28.5 \text{ mm}, \quad r_2 = mz_2/2 = 3 \times 76/2 = 144 \text{ mm}$$
$$r_{a1} = r_1 + h_{a1} = 28.5 + 3.6 = 32.1 \text{ mm}, \quad r_{a2} = r_2 + h_{a2} = 114 + 3.6 = 117.6 \text{ mm}$$
$$r_{b1} = r_1 \cos \alpha = 28.5 \cos 20° = 26.78 \text{ mm}, \quad r_{b2} = r_2 \cos \alpha = 114 \cos 20° = 107.12 \text{ mm}$$
$$L_p = \left(r_{a1}^2 - r_{b1}^2\right)^{0.5} + \left(r_{a1}^2 - r_{b1}^2\right)^{0.5} - \left(r_1 + r_2\right) \sin \alpha$$
$$= \left[\left(32.1\right)^2 - \left(26.78\right)^2\right]^{0.5} + \left[\left(117.6\right)^2 - \left(107.12\right)^2\right]^{0.5} - \left(28.5 + 114\right) \sin 20°$$
$$= 17.49 \text{ mm}$$

Number of pairs of teeth in contact $= L_p/(\pi m \cos a)$

$= 17.49/(\pi \times 3 \times \cos 20°)$

$= 1.975 \simeq 2$

14.14 GEAR STANDARDIZATION

A set of gears is interchangeable when any two gears selected from the set will mesh with each other and satisfy the fundamental law of gearing. For interchangeability, all gears of the set must have the same circular pitch, module, diametral pitch, pressure angle, addendum, and dedendum; and tooth thickness must be one-half of the circular pitch. Standard tooth forms ensure readily availability of gears.

The standard pressure angles are: 14.5° and 20°. The 20° full-depth system has several advantages when compared with the 25° or 30° full-depth system. The lower pressure angle gives a higher contact ratio which results in quieter operation, reduced wear, reduced tooth load, and reduced bearing loads.

The larger pressure angle tooth forms results in broader teeth at the base and hence are stronger in bending. Also fewer teeth may be used on the pinion without undercutting the teeth. The proportions of standard tooth forms are given in Table 14.3.

Table 14.3 Standard Involute Tooth Forms

	14.5° full-depth	20° full-depth	20° stub
Addendum, h_a	m	m	0.800 m
Dedendum, h_f	1.157 m	1.250 m	m
Clearance, c	0.157 m	0.250 m	0.200 m
Fillet radius, r	0.209 m	0.300 m	0.304 m
Tooth thickness, t	1.5708 m	1.5708 m	1.5708 m

14.15 EFFECT OF CENTRE DISTANCE VARIATION ON VELOCITY RATIO

Consider a pair of teeth in contact at L, as shown in Fig.14.16. The angular velocity ratio is,

$$\frac{\omega_1}{\omega_2} = \frac{O_2P}{O_1P}$$

Let the centre distance of rotation of gear 2 be shifted from O_2 to O_2'. As a result of this change, the contact point will shift to L'. Common normal at the point of contact L' is tangent to the base circle, because it is in contact between two involute curves, and they are generated from the base circle. Let the tangent to the base circle $M'N'$ intersect the line joining the centers of rotation O' and O_1 at P'.

Triangles O_1PN and O_2MP are similar. Also triangles $O_1N'P'$ and $O'_2M'P'$ are similar. Therefore,

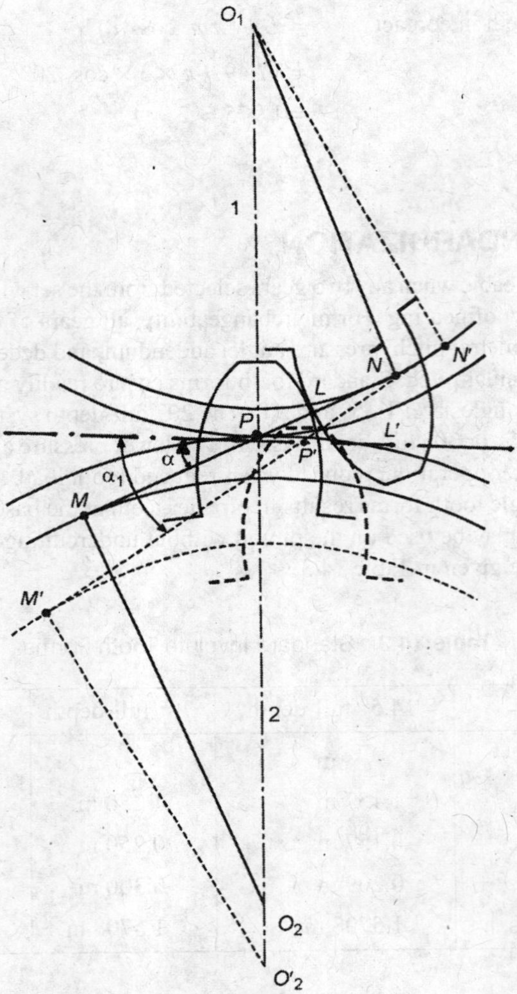

Fig.14.16 Effect of centre distance variation

$$\frac{MO_2'}{N'O_1} = \frac{O_2'P'}{O_1P'}$$

and
$$\frac{MO_2}{NO_1} = \frac{O_2P}{O_1P}$$

But $\quad NO_1 = N'O \quad$ and $\quad O_2M = O_2'M'$

Therefore,
$$\frac{O_2P}{O_1P} = \frac{O_2'P'}{O_1P'} \tag{14.33}$$

Hence, *the variation in the centre distance, within limits, does not affect the angular velocity ratio, But the length of arc of contact is decreased, and the pressure angle is increased.*

14.16 DETERMINATION OF BACKLASH

Two standard gears in mesh are shown in Fig.14.17(a). The standard centre distance with zero backlash is:

$$c = \frac{m}{2}(z_1 + z_2)$$

The cutting pitch circles are known as standard pitch circles. Fig.14.17(b) shows the condition where the two gears have been pulled apart a distance ΔC to give a new centre distance C'. The line of action now crosses the line of centres at a new pitch point P'. The standard pitch radii r_1 and r_2 are now no longer tangent to each other. The pitch point P' divides the centre distance C' into segments which are inversely proportional to the angular velocity ratio. These segments become the radii r_1' and r_2' of new pitch circles that are tangent to each other at point P'. These circles are known as operating pitch circles.

$$\frac{\omega_1}{\omega_2} = \frac{z_2}{z_1} = \frac{r'_2}{r'_1}$$

and

$$C' = r_1' + r_2'$$

to give

$$r_1' = \left[\frac{z_1}{z_1 + z_2}\right] C'$$

and

$$r_2' = \left[\frac{z_2}{z_1 + z_2}\right] C'$$

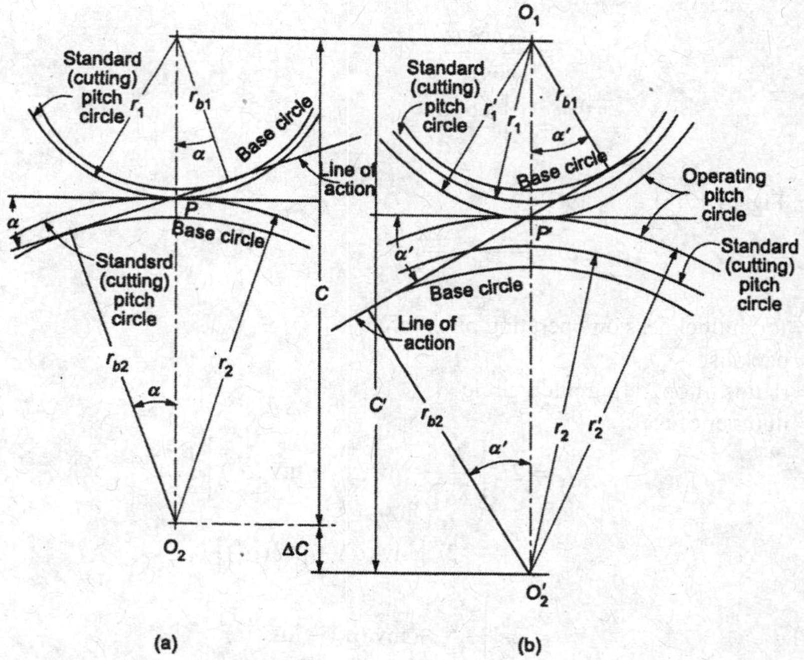

Fig.14.17 Determination of backlash

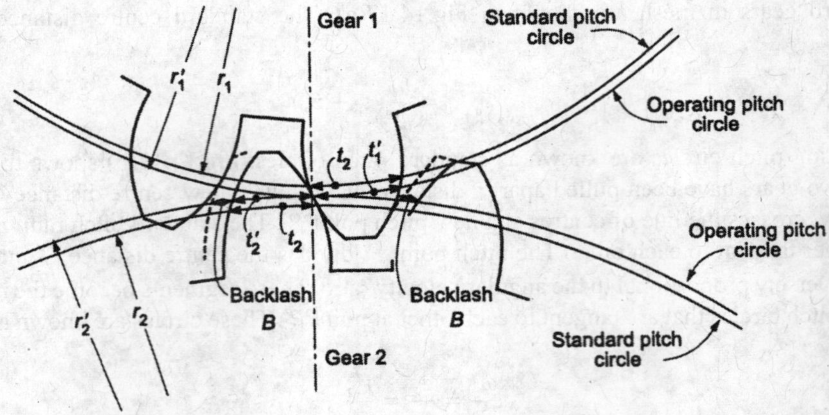

Fig.14.17 Determination of backlash (*Contd.*)

Let α' be the operating pressure angle. Now

$$C' = \frac{r_{b1} + r_{b2}}{\cos\alpha'} = \frac{(r_1 = r_2)\cos\alpha}{\cos\alpha'} = \frac{C\cos\alpha}{\cos\alpha'}$$

or $$\cos\alpha' = \frac{C\cos\alpha}{C'} \tag{14.34}$$

Also
$$\Delta C = C' - C$$
$$= \frac{C\cos\alpha}{\cos\alpha'} - C$$
$$= C\left[\frac{\cos\alpha}{\cos\alpha'} - 1\right] \tag{14.35}$$

Now from Fig.14.17(c), (14.36)

$$t_1' + t_2' + B = \frac{2\pi r_1'}{z_1} = \frac{2\pi r_2'}{z_2} \tag{14.37}$$

where t' = tooth thickness on operating pitch circle
B = backlash
r' = radius of operating pitch circle
z = number of teeth

Now
$$t_1' = 2r_2'\left[\frac{t_1}{2r_1} + \text{inv}(\alpha) - \text{inv}(\alpha')\right]$$
$$= \frac{r_1' t_1}{r_1} - 2r_1'[\text{inv}(\alpha) - \text{inv}(\alpha')] \tag{14.38}$$

$$t_2' = 2r_2'\left[\frac{t_2}{2r_2} + \text{inv}(\alpha) - \text{inv}(\alpha)'\right] \tag{14.39}$$

$$= \frac{r_2' t_2}{r_2} - 2r_2'[\text{inv}(\alpha) - \text{inv}(\alpha)'] \tag{14.40}$$

where t = tooth thickness on standard pitch circle

$$= \frac{p}{2} = \frac{\pi m}{2}$$

r = radius of standard pitch circle

$$= \frac{mz}{2}$$

Also $$= \frac{r_1}{r_1'} = \frac{r_2}{r_2'} = \frac{C}{C'} \tag{14.41}$$

and $$C' = r_1' + r_2' \tag{14.42}$$

Substituting (14.38) to (14.42) in Eq. (14.36), we get

$$B = \left(\frac{C'}{C}\right)\left[\pi m - (t_1 + t_2) + 2C\{\text{inv}(\alpha) - \text{inv}(\alpha')\}\right] \tag{14.43}$$

$$= 2C\left[\text{inv}(\alpha) - \text{inv}(\alpha')\right] \tag{14.44}$$

Example 14.11

A three-module, 20° pinion of 24 teeth drives a gear of 60 teeth. (a) Calculate the length of action and contact ratio, if the gears mesh with zero backlash. (b) If the centre distance is increased 0.5 mm, calculate the radii of the operating pitch circles, the operating pressure angle and the backlash produced.

■ **Solution** Given: $m = 3$ mm, $\alpha = 20°$, $z_1 = 24$, $z_2 = 60$

(a)

$$r_1 = \frac{z_1 m}{2} = 24 \times \frac{3}{2} = 36 \text{ mm}$$

$$r_2 = \frac{z_2 m}{2} = 60 \times \frac{3}{2} = 90 \text{ mm}$$

$$r_{b1} = r_1 \cos a = 36 \times \cos 20° = 33.83 \text{ mm}$$

$$r_{b2} = r_2 \cos a = 90 \times \cos 20° = 84.57 \text{ mm}$$

$$h_{a1} = h_{a2} = m = 3 \text{ mm}$$

$$r_{a1} = r_1 + h_{a1} = 36 = 3 = 39 \text{ mm}$$

$$r_{a2} = r_2 + h_{a2} = 90 + 3 = 93 \text{ mm}$$

$$C = r_1 + r_2 = 36 + 90 = 126 \text{ mm}$$

Length of path of contact, $$AB = \left[r_{a1}^2 - r_{b1}^2\right]^{0.5} + \left[r_{a2}^2 - r_{b2}^2\right]^{0.5} - (r_1 + r_2)\sin\alpha$$

$$= \left(39^2 - 33.83^2\right)^{0.5} + \left(93^2 - 84.57^2\right)^{0.5} - 126 \sin 20°$$

$$= 14.997 \text{ mm}$$

Contact ratio, $$m_c = \frac{AB}{p_b}$$

Base Pitch, $$p_b = 2\pi \times \frac{33.83}{24} = 8.856 \text{ mm}$$

$$m_c = \frac{14.997}{8.856} = 1.693 \simeq 2$$

(b)

$$C' = C + \Delta C = 126 + 0.5 = 126.5 \text{ mm}$$

$$r_1' = \left[\frac{z_1}{z_1 + z_2}\right] C' = \left[\frac{24}{84}\right] 126.5 = 36.143 \text{ mm}$$

$$r_2' = C' - r_1' = 126.5 - 36.143 = 90.357 \text{ mm}$$

$$\cos\alpha' = \frac{C\cos\alpha'}{C'} = \frac{126\cos 20^\circ}{126.5} = 0.93598$$

$$\alpha' = 20.61^\circ$$

Backlash,

$$B = 2C'\,[\text{inv}(\alpha') - \text{inv}(\alpha)]$$

$$= 2 \times 126.5\,[\text{inv}(20.61^\circ) - \text{inv}(20^\circ)]$$

$$= 253\,[0.016362 - 0.014904]$$

$$= 0.3689 \text{ mm}$$

14.17 INTERNAL SPUR GEARS

A pinion in mesh with an internal gear is shown in Fig.14.18. Internal gears have some advantages over the external gears. The most important advantage is the compactness of the drive. Other advantages are : greater length of contact, greater tooth strength, and lower relative sliding velocity between meshing teeth.

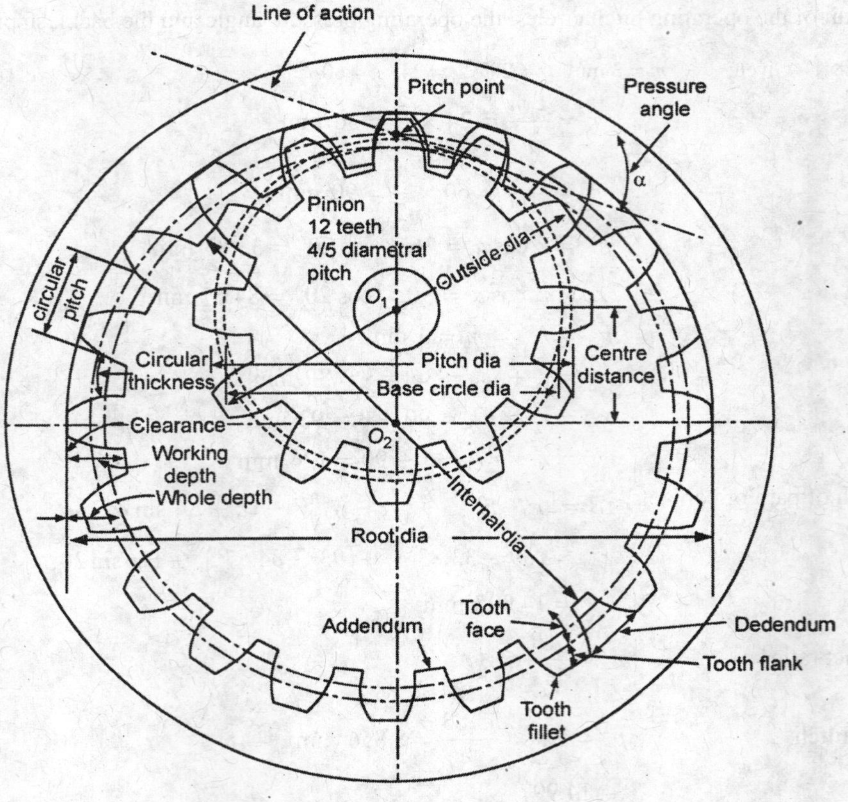

Fig.14.18 Internal spur gears

The tooth profile is concave in internal gears instead of convex as in external gears. Because of this a type of interference called *fouling* may occur in internal gears. Fouling occurs between inactive profiles as the teeth go in and out of the mesh and the there is not sufficient difference between the numbers of teeth on the internal gear and the pinion.

Example 14.12

Two equal spur gears of 48 teeth mesh together with pitch radii of 96 mm and addendum of 4 mm. If the pressure angle is 20°, calculate the length of action and the contact ratio.

Solution

Given: $z_1 = z_2 = 48,\ r = r_1 = r_2 = 96 \text{ mm},\ a = a_w = a_p = 4 \text{ mm},\ \alpha = 20°$

Length of path contact, $L_p = \left(r_{al}^2 - r_{bl}^2\right)^{0.5} + \left(r_{a2}^2 - r_{b2}^2\right)^{0.5} - (r_1 + r_2)\sin\alpha$

$$r_{b1} = r_{b2} = r\cos a = 96\cos 20° = 90.21 \text{ mm}$$

$$r_{a1} = r_{a2} = r + a = 96 + 4 = 100 \text{ mm}$$

$$L_p = \left[(100)^2 - (90.21)^2\right]^{0.5} + \left[(100)^2 - (90.21)^2\right)^{0.5} - (96 + 96)\sin 20°$$

$$= 20.63 \text{ mm}$$

Base pitch, $p_b = \dfrac{2\pi\, r_{b1}}{z_1} = 2\pi \times \dfrac{90.21}{48} = 11.81 \text{ mm}$

Contact ratio, $CR = \dfrac{L_p}{p_b}$

$$= \frac{20.63}{11.81} = 1.747$$

Example 14.13

A pinion with a pitch radius of 40 mm drives a rack. The pressure angle is 20°. Calculate the maximum addendum possible for the rack without having involute interference on the pinion.

Solution

Given: $r_1 = 40 \text{ mm},\ \alpha = 20°$

$$(h_r)_{\max} = r_i\sin^2\alpha = 40\sin^2 20° = 4.68 \text{ mm}$$

Example 14.14

A 0.2-module, 20° pinion of 42 teeth drives a gear of 90 teeth. Calculate the contact ratio. Addendum for pinion and gear is equal to one module.

Solution

Given: $m = 0.2 \text{ mm},\ a = 20°,\ z_1 = 42,\ z_2 = 90$

$$d_1 = mz_1 = 0.2 \times 42 = 8.4 \text{ mm},\ d_2 = mz_2 = 0.2 \times 90 = 18 \text{ mm}$$

$$r_{b1} = r_1\cos a = 4.2\cos 20° = 3.95 \text{ mm},\ r_{b2} = r_2\cos a = 9\cos 20° = 8.46 \text{ mm}$$

$$r_{a1} = r_1 + m = 4.2 + 0.2 = 4.4 \text{ mm},\ r_{a2} = r_2 + m = 9 + 0.2 = 9.2 \text{ mm}$$

Length of path of contact, $L_p = \left(r_{a1}^2 - r_{b1}^2\right)^{0.5} + \left(r_{a2}^2 - r_{b2}^2\right)^{0.5} - (r_1 + r_2)\sin\alpha$

$$= \left[(4.4)^2 - (3.95)^2\right]^{0.5} + \left[(9.2)^2 - (8.46)^2\right]^{0.5} - (4.2 + 9)\sin 20°$$

$$= 1.9384 + 3.6150 - 4.5147 = 1.0387 \text{ mm}$$

Base pitch, $p_b = \dfrac{2\pi r_{b1}}{z_1} = 2\pi \times \dfrac{3.95}{42} = 0.5909$ mm

Contact ratio $= \dfrac{Lp}{p_b} = \dfrac{1.0387}{0.5909} = 1.758 \simeq 2$

Example 14.15

Determine the approximate number of teeth in a 20° involute spur gear so that the base circle diameter will be equal to the dedendum circle diameter.

Solution

Given: Dedendum, $h_f = 1.25$ m, $d_d = d - 2h_f = mz - 1.25$ m

Base circle diameter, $d_b = d\cos\alpha = m z \cos\alpha$

For $d_b = d_d$, $mz\cos\alpha = mz - 1.25$ m

$$z\cos\alpha = z - 1.25$$

$$z\cos 20° = z - 1.25$$

$$0.9397\,z = z - 1.25$$

$$z = 21$$

Example 14.16

A 4-module, 20° pinion with 30 teeth drives a rack. Calculate the length of action and the contact ratio.

Solution

Given: $m = 4$ mm, $\alpha = 20°$, $z_1 = 30$

For a rack, $L_p = \left(r_{a1}^2 - r_{b1}^2\right)^{0.5} - r_1\sin\alpha + \dfrac{h_a}{\sin\alpha}$

$$h_a = m = 4 \text{ mm}, d_1 = mz = 4 \times 30 = 120 \text{ mm}$$

$$r_{a1} = r_1 + h_a = 60 + 4 = 64 \text{ mm}$$

$$r_{b1} = r_1\cos a = 60\cos 20° = 56.38 \text{ mm}$$

$$L_p = \left[(64)^2 - (56.38)^2\right]^{0.5} - 60\sin 20° + \frac{4}{\sin 20°}$$

$$= 21.461 \text{ mm}$$

Base pitch $p_b = \dfrac{2\pi r_{b1}}{z_1} = 2\pi \times \dfrac{56.38}{30} = 11.8082$ mm

Contact ratio $= \dfrac{L_p}{p_b} = \dfrac{21.461}{11.8082} = 1.8174 \simeq 2$

Example 14.17

For a 20° pressure angle, calculate the minimum number of teeth in a pinion to mesh with a rack without involute interference. Also calculate the number of teeth in a pinion to mesh with a gear of equal size without involute interference, The addendum equals the module.

■ **Solution**

Given:
$$z_1 = \frac{2a_r}{\sin^2 \alpha}$$

Here
$$a_r = 1,\ z_1 = \frac{2}{\sin^2 20°} = 17$$

$$z_1 = \frac{2a_p}{\left[\left\{1 + i^2 \sin^2 \alpha + 2i \sin^2 \alpha\right\}^{0.5} - 1\right]}$$

Here
$$i = 1,\ a_p = 1$$

$$z_1 = \frac{2}{\left[\left\{1 + \sin^2 20° + 2 \sin^2 20°\right\}^{0.5} - 1\right]}$$

$$= 12.32 \cong 13$$

Example 14.18

A pair of meshing spur gears has 22 and 38 teeth, a diametral pitch of 0.32, and a pinion running at 1800 rpm. Determine the following: (a) centre distance, (b) pitch diameter, (c) pitch line velocity, and (d) rpm of the gear.

■ **Solution**

Given:
$$z_1 = 22,\ z_2 = 38,\ P = 0.32,\ n_1 = 1800 \text{ rpm}$$

$$m = \frac{1}{P} = \frac{1}{0.32} = 3.125 \text{ mm}$$

(a) Centre distance,
$$C = \frac{m(z_1 + z_2)}{2} = \frac{3.125\ (22 + 38)}{2} = 93.75 \text{ mm}$$

(b) Pitch diameters,
$$d_1 = m z_1 = 3.125 \times 22 = 68.75 \text{ mm},$$
$$d_2 = m z_2 = 3.125 \times 38 = 118.75 \text{ mm}$$

(c) Pitch line velocity,
$$v = \frac{\pi d_1 n_1}{60} = \pi \times 0.06875 \times \frac{1800}{60} = 6.48 \text{ m/s}$$

(d)
$$n_2 = \frac{n_1 z_1}{z_2} = 1800 \times \frac{22}{38} = 1042.1 \text{ rpm}$$

Example 14.19

A pair of spur gears has 16 and 18 teeth, a module of 13 mm, addendum of 13 mm, and pressure angle of 14.5°. Show that the gears have interference. Determine the amount by which the addendum must be reduced to eliminate the interference.

■ Solution

Given: $z_1 = 16, z_2 = 18,\ m = 13$ mm, $h_a = 13$mm, $\alpha = 14.5°$

$$r_1 = \frac{mz_1}{2} = 13 \times \frac{16}{2} = 104 \text{ mm}, r_2 = 13 \times \frac{18}{2} = 117 \text{ mm}$$

$$r_{b1} = r_1 \cos\alpha = 104 \cos 14.5° = 100.7 \text{ mm}, r_{b2} = 117 \cos 14.5° = 113.3 \text{ mm}$$

$$r_{a1} = r_1 + h_a = 104 + 13 = 117 \text{ mm}, r_{a2} = 117 + 13 \text{ mm} = 130 \text{ mm}$$

Let pinion be the driver,

Length of approach, $L_a = \left(r_{a2}^2 - r_{b2}^2\right)^{0.5} - r_2 \sin\alpha$

$$= \left[(130)^2 - (113.3)^2\right]^{0.5} - 117 \sin 14.5°$$

$$= 63.74 - 211.3 = 34.44 \text{ mm}$$

$$(L_a)_{max} = r_1 \sin\alpha = 104 \sin 14.5° = 26.04 \text{ mm.}$$

Since length of approach is more than the maximum length of approach, therefore interference will occur. To eliminate interference, we make $L_a = (L_a)_{max}$

$$26.04 = \left(r_{a2}^2 - r_{b2}^2\right)^{0.5} - r_2 \sin a$$

$$= \left[r_{a2}^2 - (113.3)^2\right]^{0.5} - 117 \sin 14.5°$$

$$= \left[r_{a2}^2 - (113.3)^2\right]^{0.5} - 211.3$$

$$[r_{a2}^2 - (113.3)^2] = (26.04 + 211.3)^2$$

$$r_{a2} = 126.09 \text{ mm}$$

$$h_a' = r_{a2} - r_2 = 126.09 - 117 = 11.09 \text{ mm}$$

Decrease in addendum = $h_a' - h_a = 13 - 11.09 = 3.91$ mm

Example 14.20

An internal spur gear having 200 teeth and 20° pressure angle meshes with a pinion having 40 teeth and a module of 2.5 mm. Determine (a) the velocity ratio if the pinion is the driver, (b) the centre distance, and (c) If the centre distance is increased by 3 mm, find the resulting pressure angle.

■ Solution

Given: $z_1 = 40, z_2 = 200, m = 2.5$ mm

(a) $d_1 = mz_1 = 2.5 \times 40 = 100 \text{ mm}, d_2 = 2.5 \times 200 = 500$ mm

$$i = \frac{z_2}{z_1} = \frac{200}{40} = 5$$

(b) $$C = \frac{d_2 - d_1}{2} = \frac{500 - 100}{2} = 200 \text{ mm}$$

(c) $C' = 200 + 3 = 203$ mm

$$\cos\alpha' = \frac{C\cos\alpha}{C'} = \frac{200\cos 20^\circ}{203} = 0.9258$$

$$\alpha' = 22.2^\circ$$

Example 14.21

Two spur gears of 24 teeth and 36 teeth of 8 mm module and 20° pressure angle are in mesh. Addendum of each gear is 8 mm. The teeth are of involute form and the pinion rotates at 450 rpm. Determine the velocity of sliding when the pinion is at a radius of 102 mm.

Solution

Given: $z_1 = 24, z_2 = 36, m = 8$ mm, $h_a = 8$ mm, $\alpha = 20^\circ$, $n_1 = 450$ rpm, $r = 102$ mm

$$r_1 = \frac{mz_1}{2} = 8\times\frac{24}{2} = 96 \text{ mm}, \; r_2 = 8\times\frac{36}{2} = 144 \text{ mm}$$

$$r_{b1} = r_1\cos\alpha = 96\cos 20^\circ = 90.21 \text{ mm}, \; r_{b2} = 144\cos 20^\circ = 135.31 \text{ mm}$$

$$r_{a1} = r_1 + h_a = 96 + 8 = 104 \text{ mm}, \; r_{a2} = 144 + 8 = 152 \text{ mm}$$

Let pinion be the driver,

Length of recess at 102 mm radius,

$$L_r = (r_{a1}^2 - r_{b1}^2)^{0.5} - r_1\sin a$$

$$= [(102)^2 - (90.21)^2]^{0.5} - 96\sin 20^\circ$$

$$= 47.60 - 32.83 = 14.77 \text{ mm}$$

$$n_2 = \frac{n_1 z_1}{z_2} = 450\times\frac{24}{36} = 300 \text{ rpm}$$

$$\omega_1 = 2\pi\times\frac{450}{60} = 47.124 \text{ rad/s}$$

$$\omega_2 = 2\pi\times\frac{300}{60} = 31.416 \text{ rad/s}$$

Velocity of sliding at 102 mm radius

$$= (\omega_1 + \omega_2)L_r$$

$$= (47.124 + 31.416)\times\frac{14.77}{1000} = 1.16 \text{ m/s}$$

Example 14.22

A pair of spur gears with involute teeth is to give a gear ratio of 3:1.The arc of approach is not to be less than the circular pitch and the pinion is the driver. The pressure angle is 20°. What is the least number of teeth than can be used on each gear?

Solution

Given: For pinion to be the driver, the maximum length of approach $= r_1 \sin \alpha$

$$\text{Maximum length of arch of approach} = \frac{r_1 \sin a}{\cos a} = \pi m$$

$$\left(\frac{m z_1}{2}\right) \tan a = \pi m$$

$$z_1 = \frac{2\pi}{\tan 20°} = 17.26 \cong 18$$

$$z_2 = 3 \times 18 = 54$$

Example 14.23

A pinion with 24 involute teeth of 150 mm pitch circle diameter drives a rack. The addendum of the pinion is 6 mm. Find the least pressure angle which can be used if undercutting of the teeth is to be avoided. Using this pressure angle, find the length of the arc of contact and the minimum number of teeth in contact at one time.

Solution

Given: $z_1 = 24$, $d_1 = 150$ mm, $h = 6$ mm

$$h_{r1} = r_1 \sin^2 \alpha$$

$$6 = 75 \sin^2 \alpha$$

$$\sin^2 \alpha = 0.08$$

$$\sin \alpha = 0.28284$$

$$\alpha = 16.43°$$

$$r_{b1} = r_1 \cos \alpha = 75 \cos 16.43° = 71.94 \text{ mm}$$

$$r_{a1} = r_1 + h_{a1} = 75 + 6 = 81 \text{ mm}, r_{a2} = 144 + 8 = 152 \text{ mm}$$

$$\text{Length of recess, } L_r = L_p = (r_{a1}^2 - r_{b1}^2)^{0.5} - r_1 \sin \alpha$$

$$= [(81)^2 - (71.94)^2]^{0.5} - 7.5 \sin 16.43°$$

$$= 16 \text{ mm}$$

$$\text{Number of teeth in contact} = \frac{L_p}{p_b}$$

$$\frac{16 \times 24}{p \times 150 \times \cos 16.43°} = 0.85 \cong 1$$

Example 14.24

A pair of 20° pressure angle gears in mesh has the following data:

Speed of pinion = 400 rpm

Number of teeth on pinion = 24

Number of teeth on gear = 28

Module = 10 mm

Determine the addendum of the gears if the path of approach and recess is half the maximum value. Determine also the arc of contact and the maximum velocity of sliding between the mating surfaces.

Solution

Given: $z_1 = 24, z_2 = 28, m = 10 \text{ mm}, h_a = ?, a = 20°, n_1 = 400 \text{ rpm}$

$$r_1 = \frac{mz_1}{2} = 10 \times \frac{24}{2} = 120 \text{ mm}, r_2 = 10 \times \frac{28}{2} = 140 \text{ mm}$$

$$r_{b1} = r_1 \cos a = 120 \cos 20° = 112.76 \text{ mm}, r_{b2} = 140 \cos 20° = 131.56 \text{ mm}$$

$$r_{a1} = r_1 + h_{a1} = 120 + h_{a1}, r_{a2} = 140 + h_{a2}$$

Let pinion be the driver,

Length of approach, $L_a = 0.5\,(L_a)_{\max}$

$$(r_{a2}^2 - r_{b2}^2)^{0.5} - r_2 \sin\alpha = 0.5 r_1 \sin\alpha$$

$$[r_{a2}^2 - (131.56)^2]^{0.5} - 140 \sin 20° = 0.5 \times 120 \times \sin 20°$$

$$r_{a2}^2 - (131.56)^2 = (200)^2 \sin^2 20°$$

$$r_{a2}^2 = 21987, r_{a2} = 148.3 \text{ mm}$$

$$h_{a2} = 148.3 - 140 = 8.3 \text{ mm}$$

Length of recess, $L_r = 0.5 \times (L_r)_{\max}$

$$(r_{a1}^2 - r_{b1}^2)^{0.5} - r_1 \sin\alpha = 0.5 r_2 \sin\alpha$$

$$[r_{a1}^2 - (112.76)^2]^{0.5} - 120 \sin 20° = 0.5 \times 140 \times \sin 20°$$

$$r_{a1}^2 - (112.76)^2 = (190)^2 \sin^2 20°$$

$$r_{a1}^2 = 16938, r_{a1} = 130.14 \text{ mm}$$

$$h_{a1} = 130.14 - 120 = 10.14 \text{ mm}$$

Arc of contact

$$= \frac{L_p}{\cos\alpha} = 0.5(r_1 + r_2) \tan\alpha$$

$$= 0.5(120 + 140) \tan 20° = 47.3 \text{ mm}$$

Path of recess,

$$L_r = (r_{a1}^2 - r_{b1}^2)^{0.5} - r_1 \sin a$$

$$= [(130.14)^2 - (112.76)^2]^{0.5} - 120 \sin 20°$$

$$= 23.93 \text{ mm}$$

$$\omega_1 = 2\pi \times \frac{400}{60} = 41.88 \text{ rad/s}$$

$$\omega_2 = 41.88 \times \frac{24}{28} = 35.9 \text{ rad/s}$$

Maximum velocity of sliding $= (\omega_1 + \omega_2) \times L_r$

$$= (41.88 + 35.9) \times \frac{23.91}{1000} = 1.86 \text{ m/s.}$$

Example 14.25

Two gears in mesh have 10 teeth and 40 teeth, respectively. They are full-depth teeth and pressure angle is 20°. The module is 8.5 mm. Determine the (a) reduction in addendum of the gear to avoid interference, and (b) contact ratio.

■ Solution

Given: $z_1 = 10, z_2 = 40, m = 8.5$ mm, $\alpha = 20°$

$$r_1 = mz_1/2 = 8.5 \times \frac{10}{2} = 42.5 \text{ mm}, r_2 = \frac{mz_2}{2} = 8.5 \times \frac{40}{2} = 170 \text{ mm}$$

$$r_{b1} = r_1 \cos a = 42.5 \cos 20° = 311.94 \text{ mm}, r_{b2} = r_2 \cos a = 170 \cos 20° = 1511.75 \text{ mm}$$

$$r_{a2} = r_2 + h_{a2} = 170 + h_{a2}$$

Let pinion be the driver,

Length of approach, $L_a = (L_a)_{max}$

$$(r_{a2}^2 - r_{b2}^2)^{0.5} - r_2 \sin a = r_1 \sin a$$

$$[r_{a2}^2 - (1511.75)^2]^{0.5} - 170 \sin 20° = 42.5 \times \sin 20°$$

$$r_{a2}^2 (1511.75)^2 = (212.5)^2 \sin^2 20°$$

$$r_{a2}^2 = 30802, r_{a2} = 175.5 \text{ mm}$$

$$h_{a2} = 175.5 - 170 = 5.5 \text{ mm}$$

Reduction in addendum = 8.5 – 5.5 = 3 mm

(a) $r_{a1} = 42.5 + 8.5 = 51 \text{ mm}, r_{a2} = 170 + 8.5 = 178.5 \text{ mm}$

$$L_p = (r_{a1}^2 - r_{b1}^2)^{0.5} + (r_{a2}^2 - r_{b2}^2)^{0.5} - (r_1 + r_2) \sin a$$

$$= [(51)^2 - (1511.75)r^2]^{0.5} + [(51)^2 - (311.94)^2)]^{0.5} - (42.5 + 170) \sin 20$$

$$= 38.67 \text{ mm}$$

$$p = 2\pi \frac{r_1}{z_1} = 2\pi \times \frac{42.5}{10} = 26.7 \text{ mm}$$

$$\text{Contact ratio} = \frac{L_p}{p \cos a} = \frac{38.67}{26.7 \cos 20°} = 1.54 \cong 2$$

Example 14.26

A pinion with 24 involute teeth of 150 mm of pitch circle diameter drives a rack. The addendum of the pinion and rack is 6 mm. Find the least pressure angle which can be used if undercutting of the teeth is to be avoided. Using this pressure angle, find the length of arc of contact and the minimum number of teeth in contact at one time?

■ Solution

Given: $z_p = 24, d_p = 150$ mm, $h_p = h_r = 6$ mm

$$h_r = r_p \sin^2 \alpha$$
$$6 = 75 \sin^2 \alpha$$
$$\sin^2 \alpha = \frac{6}{75} = 0.08$$
$$\sin \alpha = 0.28284$$

Least pressure angle, $\alpha = 16.43°$

Addendum radius of pinion, $r_{ap} = r_p + h_p = 75 + 6 = 81$ mm

Base circle radius of pinion, $r_{bp} = r_p \cos \alpha = 75 \cos 16.43° = 71.937$ mm

Maximum length of path of contact, $L_p = (r_{ap}^2 - r_{bp}^2)^{1/2} = \sqrt{81^2 - (71.937)^2}$

$= 37.23$ mm

Maximum length of arc of contact, $L_c = \dfrac{L_p}{\cos \alpha} = \dfrac{37.23}{\cos 16.43°} = 38.814$ mm

Minimum number of teeth in contact, $\dfrac{\pi d_p}{z_p} = \dfrac{\pi \times 150}{24} = 19.635$

Number of pairs of teeth is contact $= \dfrac{L_c}{19.635} = \dfrac{38.814}{19.635} = 1.976 \simeq 2$

Example 14.27

A pinion of 20 involute teeth and 120 mm pitch circle diameter drives a rack. The addendum of both the pinion and rack is 6.00 mm. What is the least pressure angle which can be used to avoid interference? With this pressure angle, find the length of the arc of contact and minimum number of teeth in contact at a time.

■ Solution

Given: $z_p = 20$, $d_p = 120$ mm, $h_p = h_r = 6$ mm,

$$h_r = r_p \sin^2 \alpha$$
$$\sin^2 \alpha = \frac{6}{60} = 0.1$$
$$\sin \alpha = 0.31623$$
$$\alpha = 18.435°$$
$$r_{ap} = r_p + h_p = 60 + 6 = 66 \text{ mm}$$
$$r_{bp} = r_p \cos \alpha = 60 \cos 18.435° = 56.921 \text{ mm}$$
$$(L_p)_{\max} = \sqrt{r_{ap}^2 - r_{ap}^2} = \sqrt{66^2 - (56.921)^2} = 33.4 \text{ mm}$$

Base pitch, $p_b = \dfrac{\pi d_p}{z_p} \cos \alpha = \dfrac{\pi \times 120}{20} \times \cos 18.435° = 17.822$ mm

Minimum number of teeth in contact $= \dfrac{L_p}{p_b} = \dfrac{33.4}{17.882} = 1.868 \simeq 2$

Example 14.28

A pair of involute spur gears with 16° pressure angle and pitch in module 6 mm is in mesh. The number of teeth on pinion is 16 and its speed is 260 rpm. When the gear ratio is 1.8, find in order that the interference is just avoided, (i) the addenda on pinion and the gear wheel, (ii) the length of path of contact (iii) the maximum velocity sliding of teeth on either side of the pitch point.

[PTU, Dec, 2007

Solution

Given: $\alpha = 16°, m = 6 \text{ mm}, z = 16, \text{N} = 260 \text{ rpm},$

$$i = 1.8$$

$$z_2 = 16 \times 1.8 = 28.8 \simeq 29$$

$$d_1 = mz_1 = 6 \times 16 = 96 \text{ mm}, r_1 = 48 \text{ mm}$$

$$d_2 = mz_2 = 6 \times 29 = 174 \text{ mm}, r_2 = 87 \text{ mm}$$

(i) $r_{a1} = r_1 + h_{a1} = 48 + h_{a1}$

$$r_{b1} = r_1 \cos\alpha = 48 \cos 16° = 46.14 \text{ mm}$$

$$r_{a2} = r_2 + h_{a2} = 87 + h_{a2}$$

$$r_{b2} = r_2 \cos\alpha = 87 \cos 16° = 83.63 \text{ mm}$$

Length of path of approach, $L_a = \sqrt{r_{a2}^2 - r_{b2}^2} - r_2 \sin\alpha$

Length of path of recess, $L_r = \sqrt{r_{a1}^2 - r_{b1}^2} - r_1 \sin\alpha$

Maximum length of approach to avoid interference, $(L_a)_{\max} = r_1 \sin\alpha$

Maximum length of recess to avoid interference, $(L_r)_{\max} = r_2 \sin\alpha$

$$\therefore \quad \sqrt{(48 + h_{a1})^2 - (46.14)^2} - 48 \sin 16° = 87 \sin 16°$$

$$(48 + h_{a1})^2 - (46.14)^2 = [(87 + 48) \sin 16°]^2$$

$$= 1384.66$$

$$(48 + h_{a1})^2 = 2513.56$$

$$48 + h_{a1} = 59.27$$

$$h_{a1} = 11.27 \text{ mm}$$

$$\sqrt{(87 + h_{a2})^2 - (83.63)^2} - 87 \sin 16° = 48 \sin 16°$$

$$(87 + h_{a2})^2 - (83.63)^2 = 1384.66$$

$$h_{a2} = 4.53 \text{ mm}$$

(ii) Length of path of contact, $L_p = (r_1 + r_2)\sin\alpha$

$$= (48 + 87)\sin 16° = 37.21 \text{ mm}$$

(iii)
$$\omega_1 = \frac{2\pi N_1}{60} = \frac{2\pi \times 260}{60} = 27.227 \text{ rad/s}$$

$$\omega_2 = \frac{\omega_1}{1.8} = 15.126 \text{ rad/s}$$

Path of recess, $L_r = r_2 \sin\alpha = 87 \sin 16° = 23.98$ mm

Path of approach, $L_a = r_1 \sin\alpha = 48 \sin 16° = 13.23$ mm

Maximum velocity of sliding during approach $= (\omega_1 + \omega_2)L_a$

$$= (27.227 + 15.126) \times 13.23$$

$$= 560.33 \text{ mm/s}$$

Maximum velocity of sliding during recess $= (\omega_1 + \omega_2)L_r$

$$= (27.227 + 15.126) \times 23.98$$

$$= 1015.62 \text{ mm/s}$$

14.18 HELICAL GEARS

A helical gear has teeth in the form of a helix around the gear. The helix may be right handed on one gear and left handed on the other gear. The pitch surfaces are cylindrical like spur gears but the teeth wind around the cylinder helically like screw threads. Helical gears are used to transmit power between parallel shafts.

If a plane is rolled on a base cylinder, a line in the plane parallel to the axis of the cylinder will generate the surface of an involute spur gear tooth. If the generating line is inclined to the axis, the surface of a helical gear tooth will be generated. These two conditions are shown in Fig.14.19.

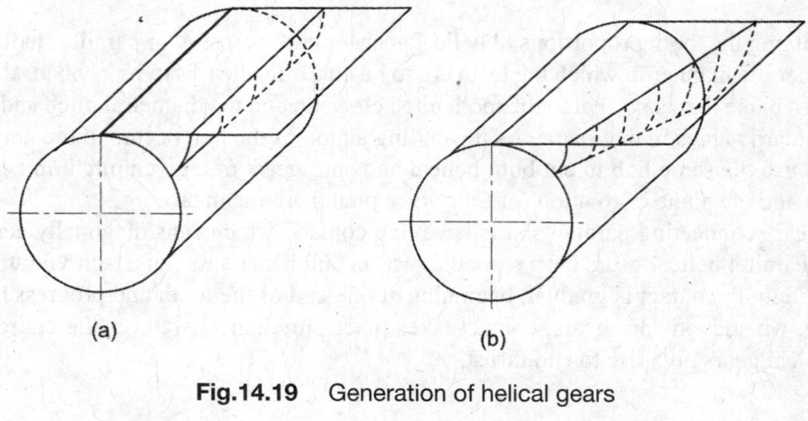

Fig.14.19 Generation of helical gears

Helical gears are used to connect parallel shafts and non-interesecting shafts. The former are known as parallel helical gears and the latter as crossed helical gears, as shown in Fig.14.20.

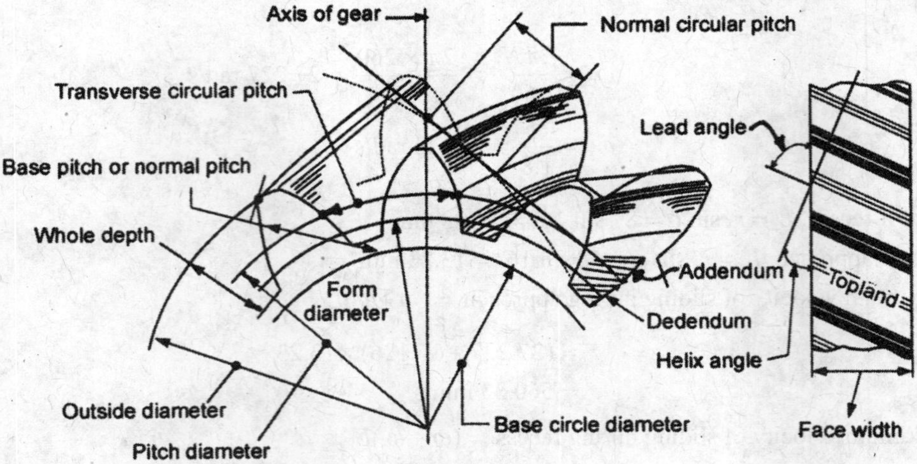

(a) Helical gear

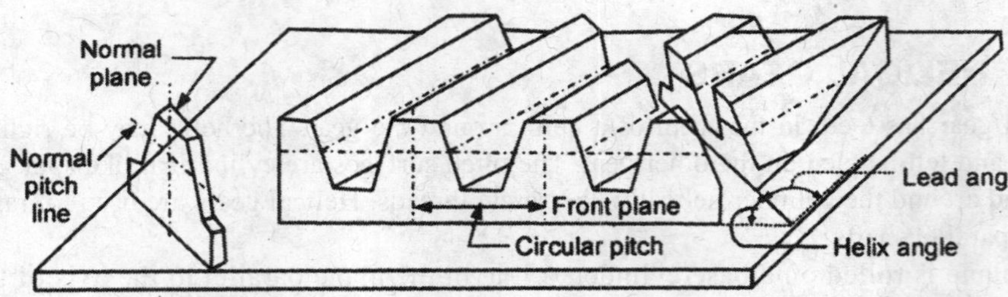

(b) Helical rack

Fig.14.20 Helical gear and rack terminology

In determining the tooth proportions of helical gear for either crossed or parallel shafts, it is necessary to consider the manner in which the teeth are to be cut. If the gear is to be hobbed, all dimensions are figured in a plane that is normal to the tooth pitch element, and the diametral pitch and the pressure angle are standard values in that plane. As the cutting action of the hob occurs in the normal plane, it is possible to use the same hob to cut both helical and spur gears of a given pitch; in a spur gear the normal plane and the plane of rotation (or transverse plane) are identical.

Helical gears connecting parallel shafts have line contact, which runs diagonally across the face of the teeth. Parallel helical gears have smoother action and hence less noise and vibration than spur gears. Also the tooth contact is gradual, beginning at one end of the tooth and progressing across the tooth surface, whereas in spur gears, contact takes place simultaneously over the entire face width. However, helical gears give rise to end thrust.

14.19 COMPARISON BETWEEN SPUR AND HELICAL GEARS

The comparison between spur and helical gears is given is Table 14.4.

Table 14.4 Comparison of Spur and Helical Gears

Spur Gears	Helical Gears
1. Teeth are cut parallel to the axis of the shaft	1. Teeth are cut in the form of a helix on the pitch cylinder between meshing gears.
2. Contact between meshing teeth occurs along the along the entire face width of the tooth.	2. Contact between meshing gears begins with a point on the leading edge of the tooth and gradually extends along the diagonal line across the tooth.
3. Load application is sudden resulting into impact conditions and generating noise in high speed applications.	3. Pick-up of load by the tooth is gradual, resulting in smooth engagement and quiet operation even at high speeds.
4. Used for parallel shafts only.	4. Crossed helical ears are used on shafts with crossed axes.
5. Speed is limited to about 20 m/s.	5. Used in automobiles, turbines and high speed applications upto 50 m/s.
6. Imposes radial load only.	6. Imposes radial and axial thrust loads.
7. Contact ratio is low.	7. Contact ratio is high.

The helical gears may be of single helical type or double helical type (herringbone gears.) The axial thrust present in single helical type is automatically cancelled in double helical type gears.

14.20 HELICAL GEAR TERMINOLOGY

Helix angle (β): It is the angle between a line drawn through one of the teeth and the centre line of the shaft on which the gear is mounted. It varies from 15° to 30°.

Normal circular pitch (p_n): The normal circular pitch is the distance between corresponding points of adjacent teeth as measured in a plane perpendicular (or normal) to the helix. It is the perpendicular distance between two adjacent teeth.

$$p_n = p_t \cos\beta = \pi m \cos\beta \tag{14.45}$$

Normal diametral pitch (P_n): The normal diametral pitch is the diametral pitch measured in the plane normal to the helix. It is equal to the diametral pitch of the *hob.*

$$P_n = \frac{P_t}{\cos\beta} \tag{14.46}$$

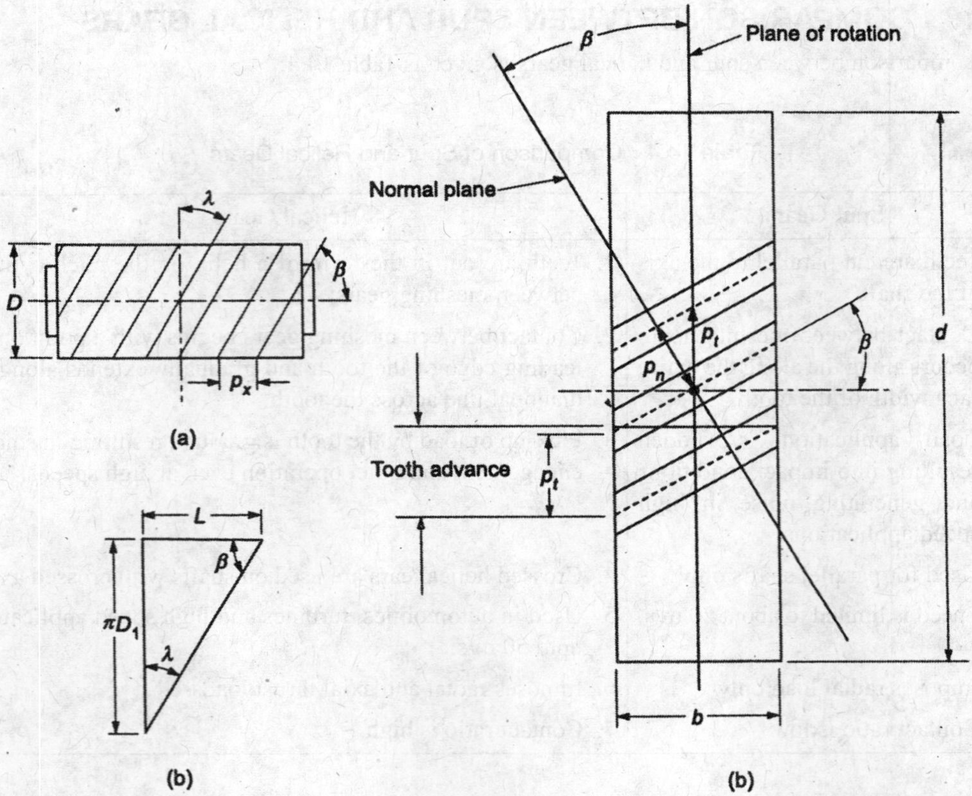

Fig.14.21 Definition of various pitches of helical gears

Transverse circular pitch (p_t): The transverse circular pitch is the distance measured in a plane perpendicular to the shaft axis (or plane of rotation) between corresponding points of adjacent teeth.

$$p_t = \frac{\pi d}{z} = \pi m \tag{14.47}$$

Transverse diametral pitch (P_t): The transverse diametral pitch is the diametral pitch measured in the plane of rotation, that is, transverse to the axis of rotation.

$$P_t = \frac{z}{d} = \frac{1}{m} \tag{14.48}$$

$$P_t p_t = P_n p_n = \pi \tag{14.49}$$

Transverse pressure angle (α_t): It is the pressure angle measured in the transverse plane or plane of rotation.

Normal pressure angel (α_n): It is the pressure angle measured in the normal plane or plane perpendicular to the teeth.

Axial pitch (p_x): The axial pitch is the distance measured in a plane parallel to the shaft axis between corresponding points of adjacent teeth.

$$p_x = p_t \cot \beta \tag{14.50}$$

Lead: The lead is the distance measured parallel to the axis to represent the distance advanced by each tooth per revolution.

Lead angle: The lead angle is the acute angle between the tangent to the helix and a plane perpendicular to the axis of cylinder.

Virtual (*or formative or equivalent*) *number of teeth* (z_v): The number of teeth of the equivalent spur gear in the normal plane are called virtual number of teeth.

$$z_v = \frac{z}{\cos^3 \beta} \tag{14.51}$$

Normal module (m_n): The normal module is the module measured in a plane normal to the helix.

$$\text{Normal module, } m_n = m_t \cos \beta \tag{14.52}$$

Pitch diameter

$$d = \frac{zm_n}{\cos \beta} \tag{14.53}$$

14.21 ANGLE RELATIONSHIPS IN HELICAL GEARS

Consider the cross-sections of the helical gear in the axial plane *x*–*x* and normal plane, *y*–*y*, as shown in Fig.14.22.

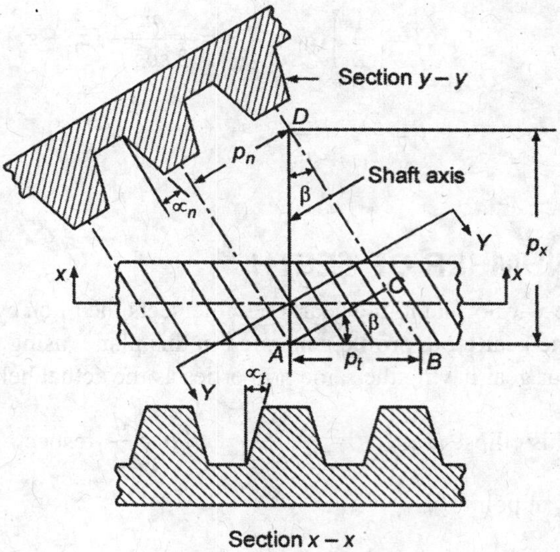

Fig.14.22 Angle relationships for helical gears

In *ABC*, we have

$$\frac{AC}{AB} = \frac{p_n}{p_t} = \cos \beta$$

∴ Normal circular pitch,

$$p_n = p_t \cos \beta \tag{14.54}$$

Transverse circular pitch, $$P_t = \frac{\pi d}{z} \tag{14.55}$$

Transverse diametral pitch, $$P_t = \frac{z}{d}$$

Transverse module, $$m_t = \frac{1}{P_t} = \frac{d}{z}$$

$$p_t P_t = \pi = p_n P_n$$

Normal diametral pitch, $$P_n = \frac{P_t}{\cos\beta} = \frac{z}{d\cos\beta} \tag{14.56}$$

Normal module, $$m_n = m_t \cos\beta \tag{14.57}$$

From Δ *ABD*, Axial pitch, $$p_x = \frac{p_t}{\tan b} = \left(\frac{pd}{z}\right)\cot b \tag{14.58}$$

Also $$\cos\beta = \frac{\tan\alpha_n}{\tan\alpha_t} \tag{14.59}$$

$$d = \frac{zp_t}{\pi} = zm_t = \frac{zm_n}{\cos\beta} \tag{14.60}$$

Centre distance, $$C = \left(\frac{1}{2}\right)(d_1 + d_2) = \frac{m_n}{2\cos\beta}(z_1 + z_2) \tag{14.61}$$

Speed ratio, $$i = \frac{z_2}{z_1} = \frac{n_1}{n_2}$$

14.22 VIRTUAL NUMBER OF TEETH

In helical gears, the plane *x–x* normal to the gear teeth intersects the pitch cylinder to form an ellipse, as shown in Fig.14.23. The gear tooth profile generated in this plane, using the radius of curvature of the ellipse, would be a spur gear having the same properties as the actual helical gear. The semi-major and semi-minor axes of this ellipse are: $a = \frac{d}{2\cos\beta}$ and $b = \frac{d}{2}$ respectively.

The radius of curvature r_c at point *A* is

$$r_c = \frac{a^2}{b} = \frac{d}{2\cos^2\beta}$$

where d = pitch circle diameter.

In the design of helical gears, an imaginary spur gear is considered in the plane *x–x* with a pitch circle radius r_c and module m_n. It is called a "*formative*" or "*virtual*" spur gear.

Pitch circle diameter of virtual gear, $$d_c = \frac{d}{\cos^2\beta}$$

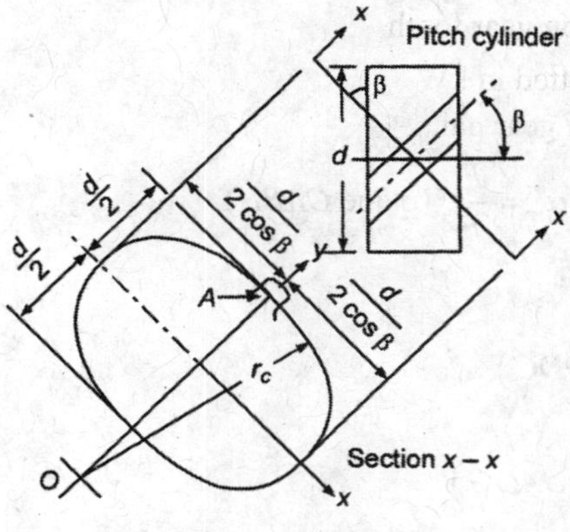

Fig.14.23 Concept of virtual number of teeth on helical gears

The number of teeth of the equivalent (or virtual) spur gear in the normal plane are called the virtual number of teeth, z_v.

$$z_v = \frac{\pi d_c}{p_n} = \frac{\pi d}{\cos^2 \beta} \times \frac{1}{\pi m_n} = \frac{d}{m_n \cos^2 \beta} = \frac{z}{\cos^3 \beta} \tag{14.62}$$

14.23 FORCES IN HELICAL GEARS

Forces in helical gears: The forces in a helical gear are shown in Fig.14.24.

Tangential force,
$$F_t = F_n \cos \alpha_n \cos \beta \tag{14.63}$$
$$= \frac{10^3 P}{v_m} \text{ N}$$

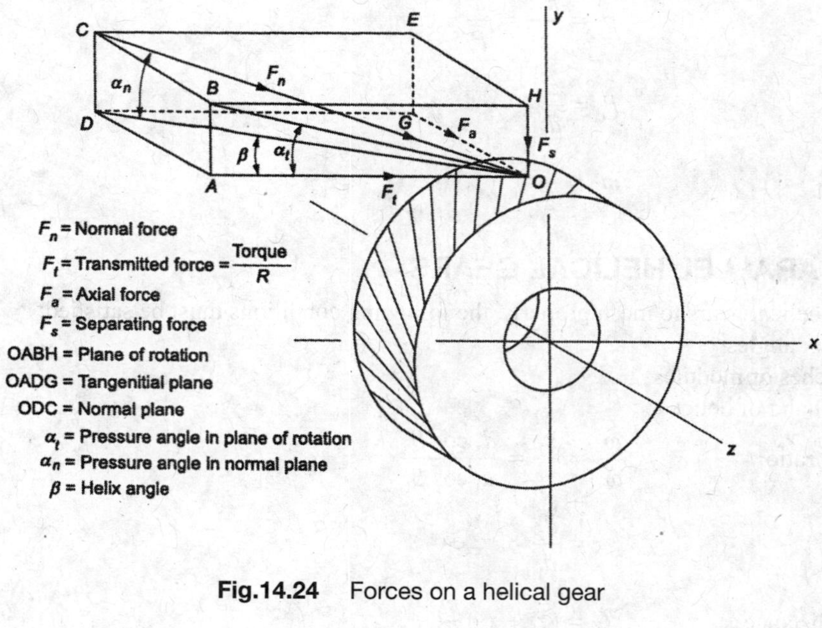

Fig.14.24 Forces on a helical gear

where F_n = normal force on gear tooth
P = power transmitted in kW
v_m = mean speed of gear pair

$$\tan \alpha_t = \frac{F_r}{F_t} \text{ (plane } OABH\text{)}$$

where F_r = radial force

$$\tan \alpha_n = \frac{F_r}{OD} \text{ (plane } ODC\text{)}$$

$$OD = \frac{F_t}{\cos \beta} \text{ (plane } OADG\text{)}$$

Therefore,
$$\tan \alpha_n = \left(\frac{F_r}{F_t}\right) \cos \beta$$

$$\tan \alpha_n = \tan \alpha_t \cos \beta \tag{14.64}$$

Axial force,
$$F_a = F_t \tan \beta \tag{14.65}$$

The minimum number of teeth cut by a *hob*,

$$z_{\min} = \frac{2k \cos \beta}{\sin^2 \alpha_t} \tag{14.66}$$

where addendum = $k \cdot m$, and k is a constant. For full depth teeth, $k = 1.00$; and for the stub system, $k = 0.80$.

If the gear is to be cut by a gear shaping method, the dimensions are considered in the plane of rotation and the diametral pitch and the pressure angle are standard values in that plane. When a helical gear is cut by a gear shaper, the circular pitch, p_t of Fig.14.21 becomes equal to the circular pitch of the cutter so that the following relations apply:

$$p_t = \frac{\pi d}{z} = \frac{\pi}{P_t} = \pi m \tag{14.67}$$

$$P_t = \frac{z}{d} \tag{14.68}$$

$$m_t = \frac{d}{z} \tag{14.69}$$

14.24 PARALLEL HELICAL GEARS

For parallel helical gears to mesh properly, the following conditions must be satisfied:

1. Equal helix angles
2. Equal pitches or modules
3. Opposite hand of helices

Velocity ratio,
$$\frac{\omega_1}{\omega_2} = \frac{z_2}{z_1} = \frac{d_2 \cos \beta_2}{d_1 \cos \beta_1}$$

$$= \frac{d_2}{d_1} \tag{14.70}$$

Centre distance,
$$C = (z_1 + z_2)\frac{m_t}{2} \tag{14.71}$$

14.25 CROSSED HELICAL GEARS

For crossed helical gears to mesh properly, there is only one requirement that they must have common normal pitches or modules. Their pitches in the plane of rotation are not necessarily equal. Their helix angles may or may not be equal and the gears may be of the same or of opposite hand.

$$\text{Velocity ratio,} \quad \frac{\omega_1}{\omega_2} = \frac{z_2}{z_1}$$

$$= \frac{d_2 \cos \beta_2}{d_1 \cos \beta_1} \tag{14.72}$$

The angle between the two shafts,

$$\Sigma = \beta_1 \pm \beta_2 \tag{14.73}$$

The plus and minus signs apply respectively, when the gears have the same or the opposite hand. Fig.14.24 illustrates pairs of crossed helical gears in and out of mesh.

$$\text{Centre distance,} \quad C = \left(\frac{m_n}{2}\right)\left[\frac{z_1}{\cos \beta_1} + \frac{z_2}{\cos \beta_2}\right] \tag{14.74}$$

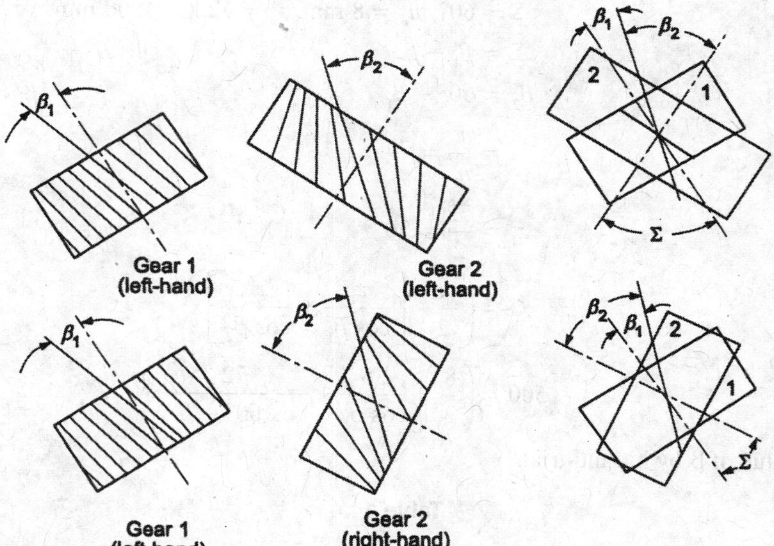

Fig.14.24 Crossed helical gears in and out of mesh

Example 14.29

A pair of crossed helical gears connects two shafts at an angle of 60° with a velocity ratio of 1.5:1. The pinion has a normal diametral pitch of 0.25, a pitch diameter of 200 mm and a helix angle of 35°. Determine the helix angle and the pitch diameter of the gear and the number of teeth on both the pinion and the gear.

Solution

Given: $\Sigma = 60°$ and $\beta_1 = 35°$.

$$\Sigma = \beta_1 + \beta_2$$

Therefore, $\beta_2 = 25°$.

Now
$$\frac{\omega_1}{\omega_2} = \frac{z_2}{z_1} = \frac{d_2 \cos \beta_2}{d_1 \cos \beta_1}$$

$$d_2 = 200 \times \frac{1.5 \cos 35°}{\cos 25°} = 271.15 \text{ mm}$$

$$z_1 = p_n d_1 \cos \beta_1 = 0.25 \times 200 \times \cos 35° = 40.95 \approx 41$$

$$z_2 = 41 \times 1.5 = 61.6 \approx 62$$

Example 14.30

Two left handed helical gears connect two shafts inclined at 60°. The normal module is 8 mm. The larger gear has 72 teeth and the velocity ratio is 1:2. If centre distance is 500 mm, calculate the helix angles of the two gears.

Solution

Given:
$$\Sigma = 60°, m_n = 8 \text{ mm}, z_2 = 72, C = 500 \text{ mm}$$

$$\Sigma = \beta_1 + \beta_2$$

$$\beta_2 = 60° - \beta_1$$

$$i = \frac{n_2}{n_1} = \frac{z_1}{z_2}$$

$$z_1 = \frac{72}{2} = 36$$

$$C = \left(\frac{m_n}{2}\right)\left[\frac{z_1}{\cos \beta_1} + \frac{z_2}{\cos \beta_2}\right]$$

$$500 = \left(\frac{8}{2}\right)\left[\frac{36}{\cos \beta_1} + \frac{72}{\cos(60° - \beta_1)}\right]$$

We find the value of β_1 by hit-and-trial.

Table 14.4

β_1, deg	$\frac{1}{\cos \beta_1}$	$\frac{2}{\cos (60° - \beta_1)}$	$\frac{1}{\cos \beta_1} + \frac{2}{\cos (60° - \beta_1)}$
25	1.1034	2.4415	3.5449
27	1.1223	2.3847	3.5070
28	1.1325	2.3583	3.4908
29	1.1433	2.3332	3.4765
30	1.1547	2.3094	3.4641

or
$$3.472 = \frac{1}{\cos \beta_1} + \frac{2}{\cos(60° - \beta1)}$$

We take $\beta_1 = 29°$, $\beta_2 = 31°$.

Example 14.31

Two standard spur gears are to be replaced by helical gears. The spur gears were cut by a three-module, 20° hob, the velocity ratio is 1.75:1, and the centre distance is 132 mm. The helical gears are to be cut with the same hob and maintain the same centre distance. The helix angle is to be between 15° and 20° and the velocity ratio between 1.70 and 1.75. Find the number of teeth, helix angle, and velocity ratio.

■ **Solution** Given: $m_n = 3\text{mm}$, $i = 1.70 - 1.75$, $C = 132\text{mm}$, $\beta = 15°–20°$,

$$C = m_t (z_1 + z_2)/2$$
$$132 = 1.5 (z_1 + z_2)/\cos 15°$$
$$= z_1 + z_2 = 85$$

Let $i = 1.72 = z_2/z_1$

Thus, $z_v = 31$, $z_2 = 53$

For a helical gear, let $\beta = 15°$, then

$$z_v = z/\cos^3\beta$$
$$z_1 = 31 \cos^3 15° \cong 28, \text{ and } z_2 = 48 \text{ giving } i = 1.71.$$

Hence, $z_1 = 28$, $z_2 = 48$, $\beta = 15°$, $i = 1.71$.

Example 14.32

A pair of helical gears for parallel shafts are to be cut with a three-module hob. The helix angle is to be 20° and the centre distance between 153 and 159 mm. The angular velocity ratio is to approach 2:1 as closely as possible. Calculate the circular pitch and the module in the plane of rotation. Determine the number of teeth, pitch diameters, and the centre distance to satisfy the above conditions.

■ **Solution** Given: $m_n = 3$ mm, $\beta = 20°$, $C = 153 - 159$ mm, $i \approx 2$

$$m_t = m_n/\cos \beta = 3/\cos 20° = 3.19 \text{ mm}$$
$$p_t = \pi m_t = \pi \times 3.19 = 10 \text{ mm}$$

For parallel helical gears, $\beta_1 = \beta_2$

$$z_2/z_1 = 2$$
$$C = 0.5\, m_t (z_1 + z_2) = 0.5\, z_1 m_t (1 + 2) = 1.5 \times 3.19 \times z_1 = 4.875\, z_1$$

For $z_1 = 32$, $C = 153.12$ mm, so that $z_2 = 64$

Example 14.33

Two crossed shafts are connected by helical gears. The velocity ratio is 18:1 and the shaft angle is 45°. If $d_1 = 60$ mm and $d_2 = 95$ mm, calculate the helix angles if both gears have the same hand.

■ **Solution**

Given: $i = 1.8$, $d_1 = 60$ mm, $d_2 = 95$ mm, $\Sigma = 45°$

$$i = d_2 \cos \beta_2/(d_1 \cos \beta_1)$$

$$1.8 = (95/60)\,(\cos\beta_2/\cos\beta_1)$$
$$(\cos\beta_2/\cos\beta_1) = 1.1368$$
$$\Sigma = \beta_1 + \beta_2 = 45°$$
$$\cos(45° - \beta_1) = 1.1368\cos\beta_1$$
$$\tan\beta_1 = 0.60798$$
$$\beta_1 = 31.3°,\ \beta_2 = 13.7°$$

Example 14.34

A pair of helical gears having 30 and 48 teeth and a 23° helix angle transmits power between parallel shafts. The module in the normal plane is 3 mm, and the pressure angle in this plane in 20°. Determine (a) the module in the plane of rotation, (b) pitch diameters, (c) centre distance, (d) circular pitch in the normal plane, and (e) circular pitch in the plane of rotation.

Solution

Given: $z_1 = 30,\ z_2 = 48,\ m_n = 3$ mm, $\alpha_n = 20°,\ \beta = 23°$

(a) $m_t = m_n/\cos\beta = 3/\cos 23° = 3.259$ mm

(b) $d_1 = z_1\, m_n/\cos^3\beta = 30 \times 3/\cos^3 23° = 115.4$ mm

$d_2 = z_2\, m_n/\cos^3\beta = 48 \times 3/\cos^3 23° = 184.64$ mm

(c) $C = 0.5\, m_t(z_1 + z_2) = 0.5 \times 3.259\,(30 + 48) = 127.1$ mm

(d) $p_n = \pi m_n = \pi \times 3 = 11.42$ mm

(e) $p_t = \pi m_t = \pi \times 3.259 = 10.238$ mm

Example 14.35

A pair of crossed helical gears connects shafts making angle of 45°. The right-hand pinion has 36 teeth and a helix angle of 20°. The right-hand gear has 48 teeth and its module in the normal plane is 2.5 mm. Determine (a) The helix angle of the gear, (b) circular pitch in the normal plane, (c) module of the pinion in its plane of rotation, (d) module of the gear in its plane of rotation, and (e) centre distance.

Solution

Given: $z_1 = 36,\ z_2 = 48,\ m_n = 2.5$ mm, $\Sigma = 45°,\ \beta_1 = 20°$

(a) $\Sigma = \beta_1 + \beta_2 = 45°,\ \beta_2 = 45 - 20 = 25°$

(b) $p_n = \pi\, m_n = \pi \times 2.5 = 7.854$ mm

(c) $m_{t1} = m_n/\cos\beta_1 = 2.5/\cos 20° = 2.66$ mm

$m_{t2} = m_n/\cos\beta_2 = 2.5/\cos 25° = 2.758$ mm

(d) $C = 0.5\, m_n\,[z_1/\cos\beta_1 + z_2/\cos\beta_2]$

$= 0.5 \times 2.5\,[\,36/\cos 20° + 36/\cos 25°]$

$= 114.1$ mm

14.26 HERRINGBONE GEARS

It is a gear, half of whose width is cut with a tooth helix in one direction and the other half in the opposite direction. It is in effect a double helical gear cut on a blank. The advantage of herringbone gears is that the end thrust is automatically eliminated.

14.27 BEVEL GEARS

Bevel gears are used to connect shafts whose axes intersect. The shaft angle is defined as the angle between the centre lines which contains the engaging teeth. Figure 14.25 shows the details of a pair of bevel gears.

Pitch cone: The pitch cone is the pitch surface of a bevel gear in a gear pair.

Cone centre: The cone centre is the apex of the pitch cone.

Pitch cone radius (r): The pitch cone radius is the length of the pitch cone element.

Pitch angle (δ): The pitch angle is the angle that the pitch line makes with the axis of the gear.

Reference cone angle: The reference cone angle is the angle between the gear axis and the reference cone generator containing the root cone generator.

Tip (or face) angle (δ_a): The tip angle is the angle between the tip cone generator and the axis of the gear.

Root (or cutting) angle (δ_f): The root angle is angle between the root cone generator and the axis of the gear.

Back cone: The back cone is an imaginary cone the elements of which are perpendicular to the elements of the pitch cone at the larger end of the tooth.

Gear diameter: The gear diameter is the diameter of the largest pitch circle.

Virtual number of teeth: The virtual number of teeth is number of teeth on an imaginary spur gear laid out on a pitch radius equal to the back cone radius.

$$z_v = \frac{z}{\cos \delta} \tag{14.75}$$

Crown gears: The crown gear is a gear pair for which pitch cone angle is 90°.

Miter gears: Mitre gears are two bevel gears of the same size having pitch cone angle 90°.

Angular bevel gears: The shaft angle is greater or less than 90°.

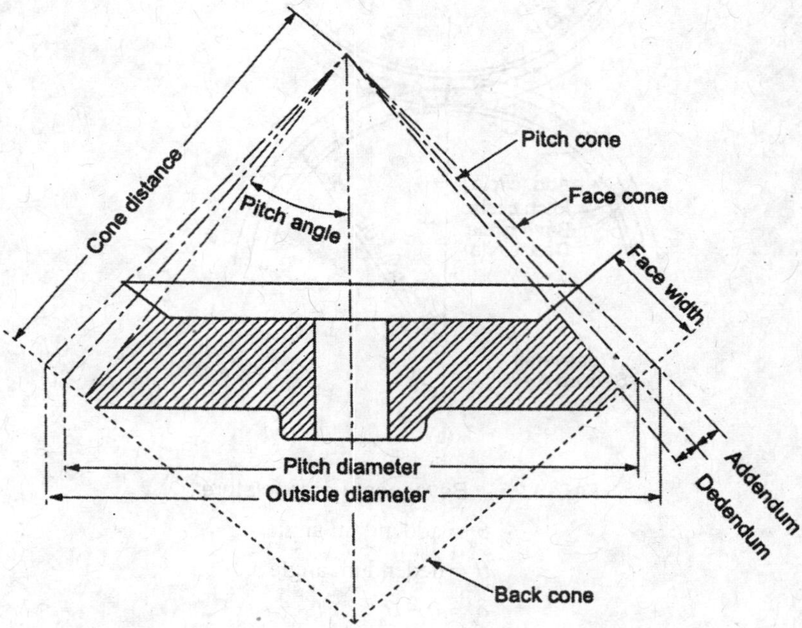

(a) Bevel gear terminology

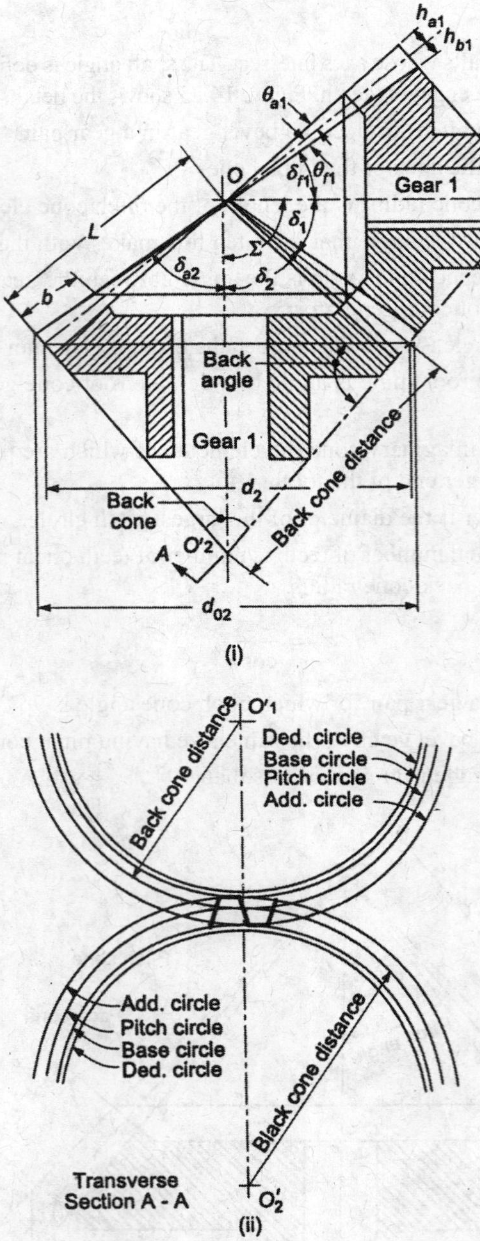

Fig.14.25 Bevel gear nomenclature

Let θ_a = addendum angle

θ_f = dedendum angle

Tip angle, $\delta_a = \delta + \theta_a$

$\delta_f = \delta - \theta_f$

Velocity ratio, $$\frac{\omega_1}{\omega_2} = \frac{d_2}{d_1} = \frac{z_2}{z_1}$$

Module, $$m = \frac{d}{z}$$

Angle relationships:

Let L = length of the pitch cone

$$= \left[\frac{d_1^2 + d_2^2}{4}\right]^{0.5} \tag{14.76}$$

$$\sin\delta_1 = \frac{d_1}{2L} = \sin(\Sigma - \delta_2) = \sin\Sigma\cos\delta_2 - \cos\Sigma\sin\delta_2$$

$$\frac{\sin\delta_1}{\sin\Sigma\sin\delta_2} = \frac{\cos\delta_2}{\sin\delta_2} - \frac{\cos\Sigma}{\sin\Sigma}$$

$$\left(\frac{1}{\sin\Sigma}\right)\left[\frac{\sin\delta_1}{\sin\delta_2} + \cos\Sigma\right] = \frac{1}{\tan\delta_2} \tag{14.77}$$

Also $$\frac{\sin\delta_1}{\sin\delta_2} = \frac{d_1}{d_2}$$

Therefore $$\tan\delta_2 = \frac{\sin\Sigma}{\cos\Sigma + \frac{d_1}{d_2}} = \frac{\sin\Sigma}{\cos\Sigma + \frac{z_1}{z_2}}$$

Similarly $$\tan\delta_1 = \frac{\sin\Sigma}{\cos\Sigma + \frac{d_2}{d_1}}$$

$$= \frac{\sin\Sigma}{\cos\Sigma + \frac{z_2}{z_1}} \tag{14.78}$$

Addendum angle, $$\tan\theta_a = 2h_a \sin\frac{\delta_1}{d_1}$$

$$= 2h_a \sin\frac{\delta_2}{d_2} \tag{14.79}$$

Dedendum angle, $$\tan\theta_f = 2h_f \sin\frac{\delta_1}{d_1}$$

$$= 2h_f \sin\frac{\delta_2}{d_2} \tag{14.80}$$

Outside (or tip) diameter of pinion, $$d_{a1} = d_1 + 2h_a\cos\delta_1 \tag{14.81}$$

Outside diameter of gear, $$d_{a2} = d_2 + 2h_a\cos\delta_2 \tag{14.82}$$

For right angle gears, $\Sigma = 90°$, and

$$\tan \delta_1 = \frac{z_1}{z_2}$$

$$\tan \delta_2 = \frac{z_2}{z_1}$$

For obtuse angle gears, Σ is greater than 90°, and

$$\tan \delta_2 = \frac{\sin (180° - \Sigma)}{\frac{d_1}{d_2} - \cos (180° - \Sigma)}$$

$$= \frac{\sin (180° - \Sigma)}{\frac{z_1}{z_2} - \cos (180° - \Sigma)} \tag{14.83}$$

Similarly

$$\tan \delta_1 = \frac{\sin (180° - \Sigma)}{\frac{d_2}{d_1} - \cos (180° - \Sigma)}$$

$$= \frac{\sin (180° - \Sigma)}{\frac{z_2}{z_1} - \cos (180° - \Sigma)} \tag{14.84}$$

Forces on straight tooth bevel gears are:

Normal force, $$F_n = \frac{F_t}{\cos \alpha_n} \tag{14.85}$$

Radial force, $$F_r = F_t \tan \alpha_n \sin \delta \tag{14.86}$$

Axial force, $$F_a = F_t \tan \alpha_n \cos \delta \tag{14.87}$$

where $$F_t = \frac{10^3 P}{v_m},\ v_m = \frac{\pi d_m n}{10^3 \times 60} \text{ m/s,}$$

and $d_m = d - b \sin \delta$, b = face width of gear tooth.

Example 14.36

A crown bevel gears of 48 teeth and a module of 2 is driven by a 24-tooth pinion. Calculate the pitch angle of the pinion and the shaft angle.

■ **Solution**

Given: $m = 2$ mm, $z_1 = 24$, $z_2 = 48$

$i = z_2/z_1 = 48/24 = 2$

$d_1 = mz_1 = 2 \times 24 = 48$ mm, $d_2 = 2 \times 48 = 96$ mm

$$L = 0.5[d_1^2 + d_2^2]^{0.5} = 0.5\left(48^2 + 96^2\right)^{0.5} = 53.66 \text{ mm}$$

$$\sin\delta_1 = d_1/(2L) = 48/(2 \times 53.66) = 0.4472$$

$$\delta_1 = 26.56°$$
$$\tan \delta_1 = \sin \Sigma/(\cos \delta + i)$$
$$\tan 26.56° = \sin \Sigma/(\cos \Sigma + 2)$$
$$0.5 = \sin \Sigma/(\cos \Sigma + 2)$$
$$0.5 \cos \Sigma + 1 = \sin \Sigma$$

Both sides are equal, if $\Sigma = 90°$.

Hence shaft angle is 90°.

Example 14.37

A 6.35 mm module, straight bevel pinion of 14 teeth drives a gear of 20 teeth. The shaft angle is 90°. Calculate the addendum and dedendum, circular tooth thickness for each gear, and the pitch and base radii of the equivalent spur gear.

Solution

Given

$$m = 6.35 \text{ mm}, z_1 = 14, z_2 = 20, \Sigma = 90°, h_a = h_f = ?$$
$$i = z_2/z_1 = 20/14 = 1.4286$$
$$d_1 = mz_1 = 6.35 \times 14 = 88.9 \text{ mm}, d_2 = 6.25 \times 20 = 127 \text{ mm}$$
$$L = 0.5[d_1^2 + d_2^2]^{0.5} = 0.5\,[(88.9)^2 + (127)^2]^{0.5} = 77.51 \text{ mm}$$
$$\sin \delta_1 = d_1/(2L) = 48/(2 \times 53.66) = 0.4472$$
$$\delta_1 = 26.56°$$
$$\tan \delta_1 = \sin \Sigma/(\cos \Sigma + i)$$
$$= \sin 90°/(\cos 90° + 1.4286) = 0.7$$
$$\delta_1 = 35°, \delta_2 = 90 - 35 = 55°$$
$$h_a = m = 6.35 \text{ mm}. h_f = 1.25 \text{ m} = 1.25 \times 6.35 = 7.94 \text{ mm}$$

For a bevel gear, $z_v = z/\cos \delta$

$$z_{v1} = z_1/\cos \delta_1 = 14/\cos 35° = 17, z_{v2} = 20/\cos 55° = 35$$

For the equivalent spur gear, $d = mz_v$ The pitch diameters are:

$$d_1 = 6.35 \times 17 = 108 \text{ mm}, d_2 = 6.35 \times 35 = 222.25 \text{ mm}$$

Circular tooth thickness $= \pi \times 108/17 = 111.96$ mm

14.28 SPIRAL GEARS

Spiral gears are used to connect non-parallel and non-intersecting shafts, as shown in Fig.14.26. The shaft angle may be less than or greater than 90°, as shown in figure below.

Let

$$\beta_1 = \text{spiral angle of gear 1}$$
$$\beta_2 = \text{spiral angle of gear 2}$$
$$\Sigma = \beta_1 + \beta_2$$

Gear ratio,

$$i = \frac{z_2}{z_1}$$

$$p_1 = \frac{p_n}{\cos \beta_1}, \quad \text{and} \quad p_2 = \frac{p_n}{\cos \beta_2}$$

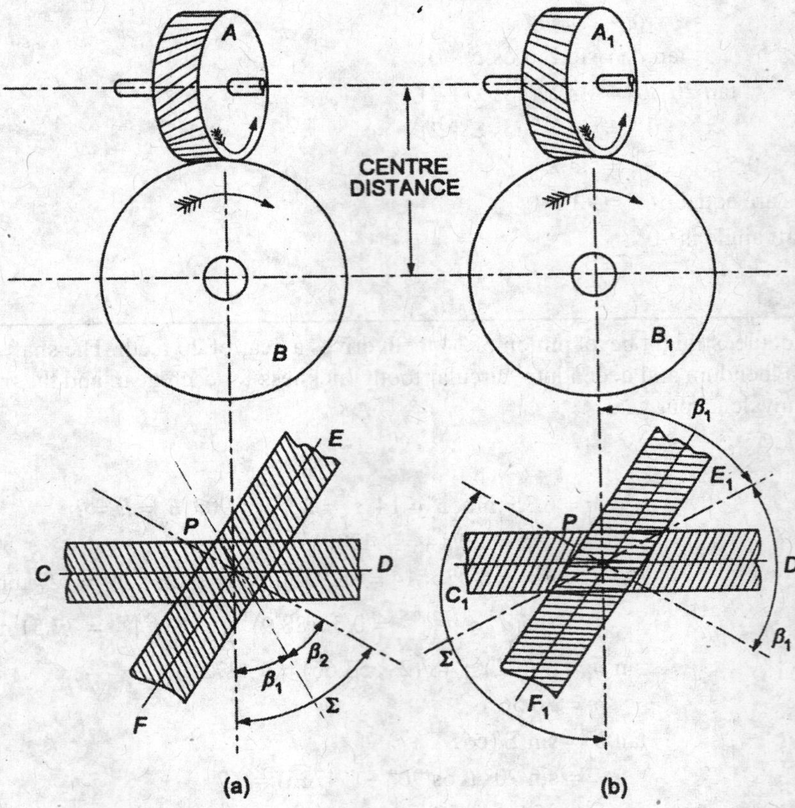

Fig.14.26 Spiral gears in mesh

Also $$d_1 = \frac{z_1 p_1}{\pi} = \frac{z_1 p_n}{\pi \cos \alpha_n}$$

and $$d_2 = \frac{z_2 p_2}{\pi} = \frac{z_2 p_n}{\pi \cos \alpha_n}$$

Centre distance, $$C = \frac{d_1 + d_2}{2}$$

$$= \left(\frac{z_1 p_n}{2\pi}\right)\left[\frac{1}{\cos \beta_1} + \frac{i}{\cos \beta_2}\right] \tag{14.88}$$

Let $d_v = \frac{z_1 p_n}{\pi}$ be the virtual pitch circle diameter of a spur gear with the same number of teeth and normal pitch as the spiral gear 1.

Thus $$C = \left(\frac{d_v}{2}\right)\left[\frac{1}{\cos \beta_1} + \frac{i}{\cos \beta_2}\right] \tag{14.89}$$

14.28.1 Efficiency of Spiral Gears

Two spiral gears in mesh at point P are shown in Fig.14.27. Gear 1 is the driver and 2 is the driven gear.

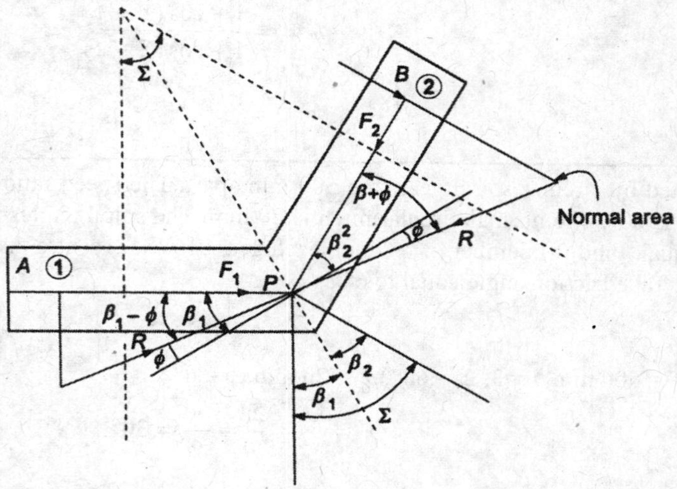

Fig.14.27 Two spiral gears in mesh

Let $\mu = \tan\phi$ be the coefficient of friction between the mating surfaces.

Driving force, $$F_1 = F_2\left[\frac{\cos(\beta_1 - \phi)}{\cos(\beta_2 + \phi)}\right]$$

Without friction, $$F_{10} = \frac{F_2 \cos\beta_1}{\cos\beta_2}$$

Efficiency of the drive, $$\eta = \frac{F_{10}}{F_1} = \frac{\cos\beta_1 \cos(\beta_2 + \phi)}{\cos(\beta_1 - \phi)\cos\beta_2}$$

$$= \frac{\cos\beta_1 \cos(\Sigma - \beta_1 + \phi)}{\cos(\Sigma - \beta_1)\cos(\beta_1 - \phi)}$$

$$= \frac{\cos(\Sigma + \phi) + \cos(2\beta_1 - \Sigma - \phi)}{\cos(\Sigma - \phi) + \cos(\Sigma + \phi - 2\beta_1)}$$

$$= \frac{\cos(\Sigma + \phi) + \cos(2\beta_1 - \Sigma - \phi)}{\cos(\Sigma - \phi) + \cos(2\beta_1 - \Sigma - \phi)}$$

$$= \frac{\cos(\Sigma + \phi) + \cos(\beta_1 - \beta_2 - \phi)}{\cos(\Sigma - \phi) + \cos(\beta_1 - \beta_2 - \phi)} \quad (14.92)$$

where $\Sigma = \beta_1 + \beta_2$.

For efficiency to be maximum,

$$\cos(2\beta_1 - \Sigma - \phi) = 1$$

or $$2\beta_1 - \Sigma - \phi = 0$$

or
$$\beta_1 = \frac{\Sigma + \phi}{2}$$

$$\eta_{\max} = \frac{1 + \cos(\Sigma + \phi)}{1 + \cos(\Sigma - \phi)} \quad (14.93)$$

Example 14.38

Two shafts connected by circular spiral gears are 500 mm apart. The speed ratio is 3 and the angle between the shafts is 60°. The normal circular pitch is 20 mm. The spiral angles for the driving and driven gears are equal. Find (a) number of teeth on each gear, (b) exact centre distance, and (c) efficiency of the drive. Take friction angle equal to 6°.

■ Solution

Given: $C = 500$ mm, $i = 3$, $\Sigma = 60°$, $p_n = 20$ mm, $\phi = 6°$

$$\beta_1 = \beta_2 = \frac{\Sigma}{2} = \frac{60}{2} = 30°$$

Centre distance,
$$C = \left(\frac{p_n z_1}{2\pi}\right)\left[\frac{1}{\cos\beta_1} + \frac{i}{\cos\beta_2}\right]$$

$$= \left(\frac{20 z_1}{2\pi}\right)\left[\frac{1}{\cos 30°} + \frac{3}{\cos 30}\right]$$

$$500 = 14.7 z_1$$

or
$$z_1 = 34$$

$$z_2 = 34 \times 3 = 102$$

Exact centre distance,
$$C = \left(\frac{20 \times 34}{2\pi}\right)\left[\frac{1}{\cos 30°} + \frac{3}{\cos 30°}\right] = 499.87 \text{ mm}$$

Efficiency of the drive,
$$\eta = \frac{\cos(\beta_2 + \phi)\cos\beta_1}{\cos(\beta_1 - \phi)\cos\beta_2}$$

$$= \frac{\cos(30° + 6°)\cos 30°}{\cos(30° - 6°)\cos 30°}$$

$$= 88.56\%$$

Example 14.39

Two spiral gears in mesh have the following data:

Angle of friction = 6°
Normal circular pitch = 20 mm
Shaft angle = 55°
Speed ratio = 3
Approximate centre distance = 400 mm
Spiral angle of pinion = 25°

Determine (a) the exact centre distance, (b) the number of teeth in each wheel, and (c) the efficiency of the drive.

■ Solution

Given: $\phi = 6°, p_n = 20$ mm, $\Sigma = 55°$, $i = 3$, $C \cong 400$ mm, $\beta_1 = 25°$

$$C = \left(\frac{z_1 p_n}{2\pi}\right)\left[\frac{1}{\cos\beta_1} + \frac{i}{\cos\beta_2}\right]$$

$$400 = \left(\frac{z_1 \times 20}{2\pi}\right)\left[\frac{1}{\cos 25°} + \frac{3}{\cos 30°}\right]$$

$$z_1 = 27.51 \approx 28$$

$$z_2 = 84$$

Exact centre distance,

$$C = \left(\frac{z_1 p_n}{2\pi}\right)\left[\frac{1}{\cos\beta_1} + \frac{i}{\cos\beta_2}\right]$$

$$= \left(\frac{28 \times 20}{2\pi}\right)\left[\frac{1}{\cos 25°} + \frac{3}{\cos 30°}\right]$$

$$= 407.08 \text{ mm}$$

Efficiency of drive,

$$= \frac{\cos(\beta_2 + \phi)\cos\beta_1}{\cos(\beta_1 - \phi)\cos\beta_2}$$

$$= \frac{\cos(30° + 6°)\cos 25°}{\cos(25° - 6°)\cos 30°}$$

$$= 81.54\%$$

Example 14.40

Two shafts inclined at an angle of 65° and with a least distance between them of 175 mm are to be connected by spiral gears of normal circular pitch 15 mm to give a reduction ratio 3:1. Find suitable diameters and number of teeth. Determine also the efficiency of the drive if the spiral angles are determined by the condition of maximum efficiency. The angle of firction is 7°.

■ Solution

Given: $\Sigma = 65°$, $C = 175$ mm, $p_n = 15$ mm, $i = 3$, $\phi = 7°$

For maximum efficiency, $\beta_1 = 0.5\,(\Sigma + \phi) = 0.5\,(65 + 7) = 36°$, $\beta_2 = 29°$

$$C = [z_1\, p_n/(2\pi)][1/\cos\beta_1 + i/\cos\beta_2]$$

$$175 = [z_1 \times 15/(2\pi)\ [1/\cos 36° + 3/\cos 29°]$$

$$z_1 = 15.7 \cong 16$$

$$z_2 = 48$$

Maximum efficiency

$$= [1 + \cos(\Sigma + \phi)]/[1 + \cos(\Sigma - \phi)]$$

$$= (1 + \cos 72°)/(1 - \cos 58°) = 85.56\%$$

Example 14.41

A spiral reduction gear of ratio 3:2 is to be used on a machine with the angle between the shafts 80°. The approximate centre distance between the shafts is 125 mm. The normal pitch of the teeth is 10 mm and the gear diameters are equal. Find the number of teeth on each gear, pitch circle diameters and spiral angles. Find the efficiency of the drive if the friction angle is 5°.

Solution

Given: $i = 1.5, \Sigma = 80°, C = 125$ mm, $p_n = 10$ mm, $d_1 = d_2, \phi = 5°$

$$\cos\beta_1/\cos\beta_2 = z_1/z_2 = 2/3$$

$$3\cos\beta_1 = 2\cos(80° - \beta_1)$$

$$= 2(\cos 80° \cos\beta_1 + \sin 80° \sin\beta_1)$$

$$1.5\cos\beta_1 = 0.17365\cos\beta_1 + 0.9898\sin\beta_1$$

$$\tan\beta_1 = 1.34682$$

$$\beta_1 = 53.4°, \beta_2 = 80° - 53.4° = 26.6°$$

$$\eta = [\cos(\Sigma + \phi) + \cos(\beta_1 - \beta_2 - \phi)]/[\cos(\Sigma - \phi) + \cos(\beta_1 - \beta_2 - \phi)]$$

$$= [\cos 85° + \cos 20.8°]/[\cos 75° + \cos 20.8°]$$

$$= 85.62\%$$

$$C = [z_1 p_n/(2\pi)][1/\cos\beta_1 + i/\cos\beta_2]$$

$$125 = [(z_1 \times 10)/(2\pi)] \times (1/\cos 53.4° + 1.5/\cos 26.6°]$$

$$z_1 = 23.41 \cong 24$$

$$z_2 = 24 \times 1.5 = 36$$

$$d_1 = d_2 = z_1 p_n/(\pi \cos\beta_1) = (24 \times 10)/(\pi \times \cos 53.4°) = 128.13 \text{ mm}$$

Example 14.42

The centre distance between two meshing spiral gears is 200 mm and the angle between the shafts 60°. The gear ratio is 2 and normal circular pitch 10 mm. The driven gear has a helix angle of 25°. Determine (a) the number of teeth on each wheel, (b) the exact centre distance, and (c) the efficiency if friction angle is 5°.

Solution

Given: $C = 200$ mm, $i = 2, \Sigma = 60°, p_n = 10$ mm, $\beta_2 = 25°$

(a) $\beta_1 = 60° - 25° = 35°$

$$C = [z_1 p_n/(2\pi)] \times [1/\cos\beta_1 + i/\cos\beta_2]$$

$$200 = [(z_1 \times 10)/(2\pi)] \times [1/\cos 35° + 2/\cos 25°]$$

$$z_1 = 36.66 \cong 37$$

$$z_2 = 37 \times 2 = 74$$

$$d_1 = d_2 = z_1 p_n/(\pi \cos\beta_1) = (24 \times 10)/(\pi \times \cos 53.4°) = 128.13 \text{ mm}$$

(b) Exact centre distance, $C = [(37 \times 10)/(2\pi)] \times [1/\cos 35° + 2/\cos 25°]$

$$= 201.8 \text{ mm}$$

Example 14.43

The angle between two shafts is 90°. They are joined by two spiral gears having a normal circular pitch of 8 mm and gear ratio of 3. If the approximate centre distance between the shafts is 250 mm and friction angle 6°, determine for the maximum efficiency of the drive (a) the number of teeth, (b) the exact centre distance, (c) pitch diameters, and (d) the efficiency.

■ Solution

(a) Given: $\Sigma = 90°, C = 250$ mm, $p_n = 8$ mm, $i = 3, \phi = 6°$

For maximum efficiency, $\beta_1 = 0.5\,(\Sigma + \phi) = 0.5\,(90 + 6) = 48°, \beta_2 = 42°$

$$C = [z_1\, p_n/(2\pi)]\;[1/\cos \beta_1 + i/\cos \beta_2]$$

$$250 = [z_1 \times 8/(2\pi)][1/\cos 48° + 3/\cos 42°]$$

$$z_1 = 35.49 \cong 36$$

$$z_2 = 36 \times 3 = 108$$

(b) Exact centre distance

$$= [36 \times 8/(2\pi)][1/\cos48° + 3/\cos42°]$$

$$= 253.54 \text{ mm}$$

(c)

$$d_1 = z_1\, p_n/(\pi \cos \beta_1) = (36 \times 8)/(\pi \times \cos 48°) = 137 \text{ mm}$$

$$d_2 = z_2\, p_n\, /(\pi \cos \beta_2) = (36 \times 8)/(\pi \times \cos 42°) = 370 \text{ mm}$$

(d)

$$\eta = [1 + \cos\,(\Sigma + \phi)]/[1 + \cos\,(\Sigma - \phi)]$$

$$= [1 + \cos96°]/[1 + \cos84°] = 81\%$$

14.29 WORM GEARS

A worm and worm gear is used to provide a high angular velocity reduction between non-intersecting shafts, which are usually at right angles. The pinion or worm has a small number of teeth (threads), usually one to four. Its mating gear is called the worm wheel. There is a line contact between the worm threads and the worm wheel teeth. Because of this, worm gears can transmit high tooth loads. However, the high sliding velocities give rise to high heating of the worm. The geometry of a worm and worm gear is shown in Fig.14.28 worm and worm gear with shafts at right angles to mesh properly, the following conditions must be satisfied:

1. Lead angle of worm = helix angle of worm gear
2. Axial pitch of worm = circular pitch of worm gear.

Axial diametral pitch (P_x): The axial diametral pitch is the quotient of the number π by the axial pitch.

$$P_x = \frac{\pi}{p_x} \tag{14.92}$$

Diametral quotient (q): The diametral quotient is the ratio of the reference diameter to the axial module,

$$q = \frac{d}{m} \tag{14.93}$$

Axial module (m_x): The axial module is the quotient of the axial pitch by the number π.

$$m_x = \frac{p_x}{\pi} \tag{14.94}$$

Axial circular pitch (p_x): The axial circular pitch is the distance between two consecutive corresponding profiles, measured parallel to the axis of the worm.

Lead (p_z): Lead is the axial distance between two consecutive intersections of a helix and straight generator of the cylinder on which it lies.

$$p_z = p_x z_1 \tag{14.95}$$

where z_1 = number of starts on the worm.

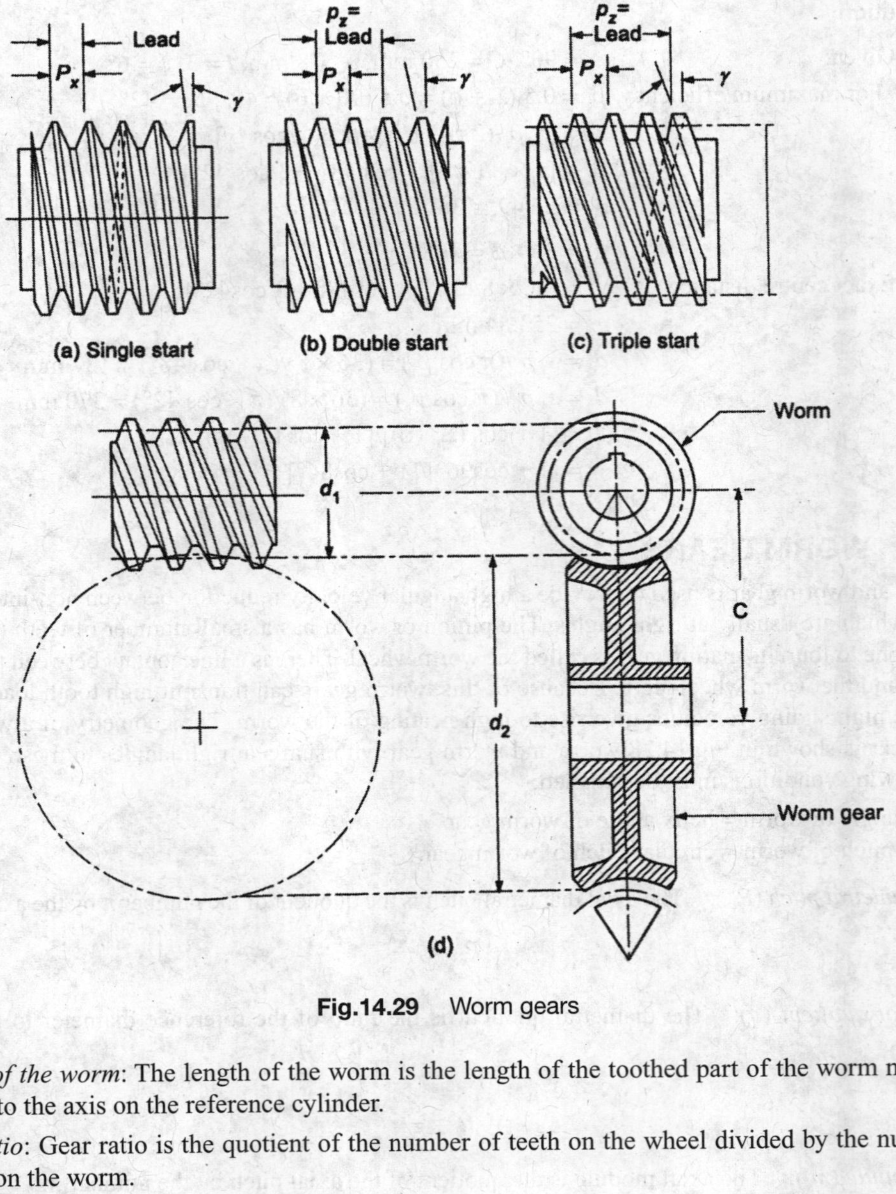

Fig.14.29 Worm gears

Length of the worm: The length of the worm is the length of the toothed part of the worm measured parallel to the axis on the reference cylinder.

Gear ratio: Gear ratio is the quotient of the number of teeth on the wheel divided by the number of threads on the worm.

$$i = \frac{z_2}{z_1} \tag{14.96}$$

Torus: Torus is the surface of revolution generated by the rotation of a circle around an axis external to this circle and situated in its plane.

Gorg: Gorg is part of the tip surface in the form of a portion of a torus with the same middle circle diameter as the reference torus.

Tooth width: Tooth width is the distance between two planes perpendicular to the axis containing the circles of intersection of the reference torus and the lateral faces of the teeth.

Width angle: Width angle is the angle at the centre included between the points of intersection of this circle with the lateral faces of the teeth, in the generating circle of the reference torus.

Lead angle (γ): Lead angle is the angle between a tangent to the pitch helix and the plane of rotation of the worm.

$$\tan\gamma = \frac{p_z}{\pi d_1} = \frac{m_x z_1}{d_1} \tag{14.97}$$

$$\frac{\omega_1}{\omega_2} = \frac{z_2}{z_1} = \frac{\pi d_2}{p_z} \tag{14.98}$$

$$p_n = p_x \cos\gamma \tag{14.99}$$

$$m_x = \frac{m_n}{\tan\gamma} \tag{14.100}$$

Centre distance,
$$C = \frac{d_1 + d_2}{2} = \left(\frac{m_n}{2}\right)\left[\frac{z_1}{\cos\beta_1} + \frac{z_2}{\cos\beta_2}\right]$$

For 90° shafts,
$$\beta_2 = \gamma \text{ and } \beta_1 = 90° - \gamma.$$

Hence
$$C = \left(\frac{m}{2}\right)[z_1 \cot\gamma + z_2]$$

$$= \left(\frac{p_z}{2\pi}\right)[\cos\gamma + i] \tag{14.101}$$

where, velocity ratio,
$$i = \frac{z_2}{z_1} = \frac{\pi d_2}{p_x} = \frac{n_1}{n_2}$$

14.29.1 Efficiency of Worm Gears

The efficiency of the worm gears when worm is the driver,

$$\eta = \frac{\cos\beta_1 \cos(\beta_2 + \phi)}{\cos(\beta_1 - \phi)\cos\beta_2}$$

For $\beta_1 + \beta_2 = 90°$; $\beta_2 = \gamma$ and $\beta_1 = 90° - \gamma$, we get

$$\eta = \frac{\sin\gamma\cos(\gamma + \phi)}{\cos\gamma\sin(\gamma + \phi)}$$

$$= \frac{\tan\gamma}{\tan(\gamma + \phi)} \tag{14.102}$$

When worm wheel is the driver, then

$$\eta = \frac{\tan(\gamma - \phi)}{\tan\gamma} \tag{14.103}$$

A worm and worm gear may be considered self-locking when the lead angle of the worm is less than 5°.

Example 14.44

A triple-threaded worm drives a worm gear of 60 teeth, the shaft angle is 90°. The circular pitch of the worm gear is 30 mm and the pitch diameter of the worm is 95 mm. Determine the lead angle of the worm, the helix angle of the worm gear and the distance between shaft centres.

■ **Solution** Given: $z_1 = 3, z_2 = 60, p_x = 30$ mm, $a_1 = 95$ mm

Lead, $$p_z = p_x z_1 = 30 \times 3 = 90 \text{ mm}$$

$$\tan\gamma = \frac{p_z}{\pi d_1} = \frac{90}{\pi \times 95} = 0.30155$$

$$\gamma = 16.78°$$

Helix angle of worm gear = lead angle of worm

Hence $$\beta_2 = 16.78°$$

$$d_2 = \frac{p_x z_2}{\pi} = \frac{30 \times 60}{\pi} = 572.96 \text{ mm}$$

Centre distance, $$C = \frac{d_1 + d_2}{2}$$

$$= \frac{95 + 572.96}{2} = 333.98 \text{ mm}$$

Example 14.45

A two start worm rotating at 900 rpm drives a 27 tooth worm gear. The worm has a pitch diameter of 60 mm and a pitch of 20 mm. The coefficient of friction is 0.05. Find (a) the helix angle of worm, (b) the speed of gear, (c) centre distance, (d) efficiency and (e) maximum efficiency.

■ **Solution**

Given: $N_1 = 900$ rpm, $z_2 = 27$, $d_1 = 60$ mm, $p_x = 20$ mm, $z_1 = 2$

$$\phi = \tan^{-1} 0.05 = 2.86°, \ p_z = p_x z_1 = 20 \times 2 = 40 \text{ mm}$$

(a) $$\tan\gamma = \frac{p_z}{\pi d_1} = \frac{40}{\pi \times 60} = 0.2122$$

$$\beta_2 = \gamma = 11.98°$$

$$\beta_1 = 90° - \beta_2 = 78.02°$$

(b) $$d_2 = \frac{p_x z_2}{\pi} = \frac{20 \times 27}{\pi} = 171.88 \text{ mm}$$

$$i = \frac{\omega_1}{\omega_2} = \frac{n_1}{n_2} = \frac{\pi d_2}{p_z} = \frac{\pi \times 171.98}{40} = 13.5$$

$$n_2 = \frac{900}{13.5} = 66.67 \text{ rpm}$$

(c) $$C = \left(\frac{p_z}{2\pi}\right)(\cot\gamma + i) = \left(\frac{40}{2\pi}\right)(\cot 11.98° + 13.5) = 115.95 \text{ mm}$$

(d) $$\eta = \frac{\tan\gamma}{\tan(\gamma+\phi)} = \frac{\tan 11.98°}{\tan(11.98° + 2.86°)}$$

$$= 0.8 \quad \text{or} \quad 80\%$$

(e) $$\eta_{max} = \frac{1-\sin\phi}{1+\sin\phi} = \frac{1-\sin 2.86°}{1+\sin 2.86°} = 90.49\%$$

Example 14.46

A double-threaded worm having a lead of 65 mm drives a worm gear with a velocity ratio of 20:1; the angle between the shafts is 90°. If the centre distance is 235 mm, determine the pitch diameter of the worm and worm gear.

Solution

Given: $p_z = 65$ mm, $i = 20$, $\Sigma = 90°$, $C = 235$ mm

$$C = (p_z/2\pi)(\cot\gamma + i)$$

$$235 = (65/2\pi)(\cot\gamma + 20)$$

$$\cot\gamma = 2.71613$$

$$\gamma = 20.21°$$

$$\tan\gamma = p_z/(\pi d_1),\ d_1 = 65/(\pi \times \tan 20.21°) = 56.2 \text{ mm}$$

$$i = \pi d_2/p_z,\ d_2 = 20 \times 65/\pi = 413.8 \text{ mm}$$

$$z_2 = 2 \times 20 = 40$$

Example 14.47

A worm and worm gear with shafts at 90° and a centre distance of 178 mm are to have a velocity ratio of 18:1. If the axial pitch of the worm is to be 26.192 mm, determine the maximum number of teeth in the worm and worm gear that can be used for the drive and their corresponding pitch diameters.

Solution

Given: $p_x = 26.192$ mm, $i = 18$, $\Sigma = 90°$, $C = 178$ mm

$$\text{Let } z_1 = 2,\ p_z = p_x z_1 = 26.192 \times 2 = 52.384 \text{ mm}$$

$$i = \pi d_2/p_z,\ d_2 = 18 \times 52.384/\pi = 300 \text{ mm}$$

$$z_2 = 2 \times 18 = 36$$

$$C = (p_z/2\pi)(\cot\gamma + i)$$

$$178 = (52.384/2\pi)(\cot\gamma + 18)$$

$$\cot\gamma = 3.35$$

$$\gamma = 16.62°$$

$$\tan\gamma = p_z/(\pi d_1)$$

$$d_1 = 52.384/(\pi \times \tan 16.62°) = 55.86 \text{ mm}$$

Example 14.48

A double-threaded worm drives a 31-tooth worm gear with shafts at 90°. If the centre distance is 210 mm and the lead angle of the worm 18.83°, calculate the axial pitch of the worm and the pitch diameters of the two gears.

■ **Solution**

Given:

$$z_1 = 2, z_2 = 31, \Sigma = 90°, C = 210 \text{ mm}, \gamma = 18.83°$$

$$i = z_2/z_1 = 31/2 = 15.5$$
$$C = (p_z/2\pi)(\cot\gamma + i)$$
$$210 = (p_z/2\pi)(\cot 18.83° + 15.5)$$
$$p_z = 71.584$$
$$p_x = p_z/z_1 = 71.584/2 = 35.792 \text{ mm}$$
$$\tan\gamma = p_z/(\pi d_1)$$
$$d_1 = 71.584/(\pi \times \tan 18.83°) = 66.82 \text{ mm}$$
$$d_2 = ip_z/\pi = 15.5 \times 71.584/\pi = 353.2 \text{ mm}$$

Example 14.49

A worm and worm gear with shafts at 90° and a centre distance of 76 mm are to have a velocity ratio of 8:1. Using a lead angle of 28.88°, determine the pitch diameters. Select number of teeth for the gears considering worms with 1 to 10 threads.

■ **Solution**

Given: $\Sigma = 90°, C = 76 \text{ mm}, \gamma = 28.88°, i = 8$

For $z_1 = 2, z_2 = 16$

$$C = (p_z + {}_2\pi)(\cot\gamma + i)$$
$$76 = (p_z/2\pi)(\cot 28.88° + 8)$$
$$p_z = 48.66$$
$$\tan\gamma = p_z/(\pi d_1)$$
$$d_1 = 48.66/(\pi \times \tan 28.88°) = 28.1 \text{ mm}$$
$$d_2 = i\,p_z/\pi = 8 \times 48.66/\pi = 123.9 \text{ mm}$$

Example 14.50

A worm and worm gear have axes at 90° and give a speed reduction of 15 to 1. The triple-thread worm has a lead angle of 20° and an axial pitch of 10 mm. Determine the following for the worm gear : (a) number of teeth, (b) pitch diameter, and (c) helix angle.

■ **Solution**

Given: $i = 15, z_1 = 3, \gamma = 20°, p_x = 10 \text{ mm}$

$$p_z = p_x z_1 = 10 \times 3 = 30 \text{ mm}$$
$$d_2 = ip_z/\pi = 15 \times 30/\pi = 143.24 \text{ mm}$$
$$z_2 = iz_1 = 3 \times 15 = 45$$
$$\beta_2 = \gamma = 20°$$

Summary for Quick Revision

1 A gear may be defined as a toothed member designed to transmit or receive motion from one shaft to another by successively engaging tooth.

2 Gears occupy less space, no slip, transmit higher power, and gives higher efficiency.

3 Gears may be classified as: spur, helical, double helical, spiral, bevel, worm, hypoid and planetary.

4 Pressure angle is the angle between the common normal at the point of contact and the common tangent at the pitch point.

5 Involute of a circle is the curve traced by the end of a thread as it is unwound from a stationary cylinder.

6 Cycloid is the locus of a point on the circumference of a circle which rolls without slipping on a fixed straight line.

7 Basic relations for a spur gear.

Circular pitch, $p = \pi d/z$

where z = number of teeth, d = pitch circle diameter.

Base pitch, $p_b = p \times cos\alpha$

where α = pressure angle of gear tooth profile.

Diametral pitch, $P = z/d$

Relationship between circular and diametral pitch, $P \times p = \pi$

Module, $m = d/z = 1/p$

Base Circle Diameter, $d_b = d \cos\alpha$

Centre distance, $C = (d_1 + d_2)/2 = m\,(z_1 + z_2)/2$

8 Fundamental law of gearing:

For constant angular velocity ratio of the two gears in contact the common normal at the point of contact must always intersect the line of centers at a fixed point (pitch point) and divide this line in the inverse ratio of the angular velocities of the two gears.

9 Relative velocity between gear teeth:

$v_r = (\omega_1 + \omega_2)\,AP$

= (sum of the angular velocities) × distance of the point of contact from the pitch point.

10 Involute function, inv $(\alpha) = \tan\alpha - \alpha$

11 Characteristics of involute action.

Addendum radius of pinion, $r_{a1} = r_1 + h_{a1}$

Base circle radius of pinion, $r_{b1} = r_1 \cos\alpha$

Addendum radius of gear, $r_{a2} = r_2 + h_{a2}$

Base circle radius of gear, $r_{b2} = r_2 \cos\alpha$

Where r_1 = pitch circle radius of pinion

r_2 = pitch circle radius of gear

h_{al} = addendum of pinion

h_{a2} = addendum of gear

r_{b1} = base circle radius of pinion

r_{b2} = base circle radius of gear

Length of path of recess, $L_r = (r_{a1}^2 - r_{b1}^2)^{0.5} - r_1 \sin\alpha$

Length of path of approach, $L_a = (r_{a2}^2 - r_{b2}^2)^{0.5} - r_2 \sin\alpha$

Length of path of contact, $AB = L_p = L_r + L_a = (r_{a1}^2 - r_{b1}^2)^{0.5} + (r_{a2}^2 - r_{b2}^2)^{0.5} - (r_1 + r_2)\sin a$

Length of arc of contact, $L_c = AB/\cos\alpha$

Maximum length of path of recess = $r_2 \sin\alpha$

Maximum length of path of approach = $r_1 \sin\alpha$

Contact ratio, m_c = length of path of contact/base pitch = L_p/p_b
where $p_b = p \cos\alpha = \pi m \cos\alpha$
For a rack and a pinion, $L_p = (r_{a1}^2 - r_{b1}^2)^{0.5} - r_1 \sin\alpha + a/\sin\alpha$
where 'a' = addendum.

12 Interference in gears is a phenomenon in which the tip of the driven gear tooth will dig into the flank of the driving gear tooth.

13 Interference can be avoided by undercutting, making stub tooth, increasing the pressure angle, and cutting the gears with long and short addendum gear teeth.

14 Minimum number of teeth on gear wheel.
Speed ratio, $i = z_2/z_1$, addendum of pinion, $h_{a1} = a_p m$, addendum of gear wheel, $h_{a2} = a_w m$ where a_p and a_w are the constants by which the module must be multiplied to get the addendum of pinion and gear wheel respectively.

$$z_2 = 2a_w \Big/ \left[\left\{1 + (1/i^2)\sin^2 a + (2/i)\sin^2 a\right\}^{0.5} - 1\right]$$

For $a_w = 1$, $z_2 = [z^2_1 \sin^2\alpha - 4]/[4 - 2z_1 \sin^2\alpha]$

For $i = 1$, and $a_w = a_p$

$z_2 = 2a_w/[\{1 + 3\sin^2\alpha\}^{0.5} - 1]$

15 Minimum number of teeth on pinion.

$$z_1 = 2a_p \Big/ \left[\left\{1 + i^2 \sin^2 a + 2i \sin^2 a\right\}^{0.5} - 1\right]$$

For $a_p = 1$, $z_1 = [z^2_2 \sin^2\alpha - 4]/[4 - 2z_2 \sin^2\alpha]$

16 Minimum number of teeth on rack pinion.
$z_1 = 2a_r/\sin^2\alpha$
$z_{min} = 32$ for $\alpha = 14.5°$ and 18 for $\alpha = 20°$

17 Effect of centre distance variation on velocity ratio.
The variation in the centre distance, within limits, does not affect and angular velocity ratio. But the length of arc of contact is decreased, and the pressure angle is increased.

18 Helical gears have teeth in the form of a helix around the gear.

19 Helical gears have line contact which takes place gradually and smoothly, gives less noise as compared to spur gears. However they give rise to end thrust.

20 Basic relations for a helical gear:

Transverse circular pitch, $p_t = \pi d/z = \pi m$
Axial pitch, $p_x = p_t \cot\beta$, where β = helix angle
Virtual number to teeth, $z_v = z/\cos^2\beta$
Normal module, $m_n = m_t \cos\beta$
Normal circular pitch, $p_n = p_t \cos\beta$
Transverse diametral pitch, $P_t = z/d = 1/m$
Normal diametral pitch, $P_n = P_t/\cos\beta$
Speed ratio, $i = z_2/z_1 = n_1/n_2$
Centre distance, $C = (d_1 + d_2)/2 = (m_n/\cos\beta)[(z_1 + z_2)/2]$

21 Crossed helical gears:
Velocity ratio: $i = n_1/n_2 = z_2/z_1 = (d_2 \cos\beta_2)/(d_1 \cos\beta_1)$

Angle between the two shafts, $\Sigma = \beta_1 \pm \beta_2$ (use +ve sign for same hand and –ve sign for opposite hand gears)
Centre distance, $C = (m_n/2)\ [z_1/\cos \beta_1 + z_2/\cos \beta_2]$

22 In herringbone gears, the end thrust is automatically eliminated.

23 Bevel gears are used for shafts whose axes intersect.

24 Crown gears is a gear pair for which pitch cone angle is 90°.

25 Miter gears are two bevel gears of the same size in mesh whose pitch cone angle is 90°.

26 Basic relations for a bevel gear:
Length of the pitch cone, $L = [(d_1^2 + d_2^2)/4]^{1/2}$
Pitch cone angle of pinion, $\delta_1 = \sin^{-1} [d_1/(2L)]$
Angle between the shafts, $\Sigma = \delta_1 + \delta_2$
$\tan \delta_1 = \sin \Sigma/[\cos \Sigma + (z_1 / z_2)]$

27 Spiral gears are used to connect non-parallel and non-intersecting shafts.

28 Basic relations for spiral gears:
Angle between the shafts, $\Sigma = \beta_1 + \beta_2$
Circular pitch, $p = p_n/\cos \beta$
Pitch circle diameter, $d = zp_n/ (\pi \cos \alpha_n)$
Centre distance, $C = [(z_1 p_n)/(2\pi)][1/\cos \beta_1 + 1/\cos \beta_2]$
Efficiency of spiral gears $= [\cos (\Sigma + \phi) + \cos (\beta_1 - \beta_2 - \phi)]/[\cos (\Sigma - \phi) + \cos (\beta_1 - \beta_2 - \phi)]$
Where $\phi = \tan^{-1}\mu$, μ = coefficient of friction
For maximum efficiency, $\beta_1 = (\Sigma + \phi)/2$
Maximum efficiency, $= [1 + \cos (\Sigma + \phi)]/[1 + \cos (\Sigma - \phi)]$

29 Worm gears are used to provide high angular velocity reduction between non-intersecting shafts.

30 Basic relations for worm gears:
Lead angle of worm = Helix angle of worm gear
Axial pitch of worm = Circular pitch of worm gear
Axial diametral pitch, $P_x = \pi/p_x$
Diametral quotient, $q = d/m$
Axial module, $m_x = p_x/\pi$
Lead, $p_z = z_1 p_x$, where, z_1 = number of starts on the worm
Gear ratio, $i = z_2/z_1$
Lead angle, $y = \tan^{-1}[p_z/(\pi d_1)]$
Axial module, $m_x = m_n/\tan y$
Centre distance, $C = (m_n/2)\ [z_1/\cos \beta_1 + z_2/\cos \beta_2]$
Efficiency of worm gears $= [\cos \beta_1 \cos (\beta_2 + \phi)]/[\cos (\beta_1 - \phi) \cos \beta_2]$

Multiple Choice Questions

1 The surface of the gear tooth below the pitch surface is called
(a) addendum portion (b) dendendum portion (c) flank (d) face.

2 The path of contact in involute gears is
(a) a straight line (b) involute path (c) curved path (d) circle.

3 For two meshing gears, their
(a) number of teeth must be same (b) addendum must be same
(c) dedendum must be same (d) module must be same.

4 The size of a gear is usually specified by
(a) circular pitch (b) module (c) pitch circle diameter (d) base diameter.

5 The type of gears used to connect two parallel coplanar shafts are
(a) spur gears (b) bevel gears (c) spiral gears (d) worm gears.

6 The type of gears used to connect two intersecting coplanar shafts are
(a) spur gears (b) straight bevel gears (c) helical gears (d) spiral gears.

7 The type of gears used to connect two non-parallel and non-intersecting shafts are
(a) spur gears (b) bevel gears (c) worm gears (d) spiral gears.

8 The circular pitch of a spur gear is defined as
(a) $\pi d/z$ (b) d/z (c) z/d (d) $z/\pi d$.
where d = pitch circle diameter and z = number of teeth

9 The diametral pitch of a spur gear is defined as
(a) $\pi d/z$ (b) d/z (c) z/d (d) $z/\pi d$.

10 The module of a spur gear is defined as
(a) $\pi d/z$ (b) d/z (c) z/d (d) $z/\pi d$.

11 Choose the correct relationship
(a) $p\,P = \pi$ (b) $p/P = \pi$ (c) $P/p = \pi$ (d) $p\,P = 1/\pi$.

12 Module of a spur gear teeth is
(a) $1/P$ (b) $1/p$ (c) P/π (d) p/π.

13 The range of pressure angle for spur gears is
(a) 10 to 14° (b) 14.5 to 20° (c) 21 to 25° (d) 26 to 30°.

14 Choose the correct statement for involute profile in regard to pressure angle
(a) minimum value when contact begins (b) maximum value when contact ends
(c) remains same for all points of contact (d) interference is zero.

15 In case of cycloidal tooth profile gears
(a) the pressure angle is always constant through the contact
(b) the path of contact is a straight line
(c) the variation in centre distance affects the angular speed ratio
(d) interference is more.

16 The involute function in terms of pressure angle ϕ is
(a) $\tan\phi - \phi$ (b) $\tan^{-1}\phi$ (c) $\tan^{-1}\phi - 1$ (d) $\phi - \tan\phi$.

17 The minimum number of teeth for involute rack of 20° pressure angle is
(a) 17 (b) 24 (c) 32 (d) 34.

18 Path of contact in case of cyclodial tooth profile gears is a
(a) Straight line (b) circle (c) complex curve (d) parabola.

Answers

1. (c) 2. (a) 3. (d) 4. (c) 5. (a) 6. (b) 7. (c) 8. (a) 9. (c) 10. (b)
11. (a) 12. (a) 13. (b) 15. (c) 16. (a) 17. (a) 18. (b)

Review Questions

1 Name the gears for connecting parallel shafts.
2 What are the gears used for intersecting shafts?
3 Which gears are used for non-parallel and non-intersecting shafts?
4 Define pressure angle of a gear.
5 What is the relationship between circular pitch and diameter pitch of a spur gear?
6 State the law of gearing.
7 What is conjugate action in gears?
8 What are the causes for interference in gears?
9 How interference is gears can be minimized?
10 Explain path of recess and path of approach.
11 Compare involute and cycloidel tooth profiles.
12 Define contact ratio.
13 How many are the minimum number of teeth on the pinion for a rack of 20° pressure angle?
14 What is the law for velocity of sliding between a gear pair?
15 What is the effect of centre distance variation on speed ratio in gears?
16 What is a rack?
17 What are the advantages of standard gears?
18 What are the advantages of gear drive?
19 What is backlash in gears?
20 What are the characteristics of involute action?
21 What are herringbone gears? What is their advantages?
22 Explain virtual number of teeth on helical gears.
23 Define axial pitch, normal module and normal pressure angle of helical gears.
24 What are crown and miter bevel gears?
25 Where do we use spiral gears?
26 What are worm gears? Where they are used?

Exercises

(a) Spur gears

14.1 A pair of 20° full involute spur gears having 40 and 60 teeth of module 4 mm are in mesh. The smaller gear rotates at 1440 rpm. Find (a) sliding velocity at engagement and disengagement of the pair of teeth, and (b) contact ratio.

[Ans. 6 m/s, 10.315 m/s; 6]

14.2 Calculate the minimum number of teeth on a pinion to avoid interference to have a speed ratio of 2.5:1. The pressure angle is 20° and addendum of one module of gear may be used.

[Ans. 15]

14.3 A pinion of involute profile has 25 teeth and 150 mm pitch circle diameter. It drives a rack. The addendum of both pinion and rack is 6.25 mm. Calculate the least pressure angle to avoid interference.

[Ans. 16. 78°]

14.4 The following data refer to two meshing involute gears of 20° pressure angle: Number of teeth on pinion = 20, speed ratio = 2, speed of pinion = 250 rpm, module = 12 mm

The addendum of each wheel is such that the path of approach and path of recess on each side are half of the maximum possible length. Calculate (a) addendum for both the wheels, (b)the length of arc of contact, and (c) the maximum possible sliding velocity approach and recess.

[Ans. 19.47 mm, 7.77 mm; 65.51 mm; 0.806 m/s, 1.612 m/s]

14.5 Two spur gear wheels of 80 mm and 120 mm pitch diameters have involute teeth of standard addenda of 3 mm and 20° pressure angle. The module is 1 mm. Determine (a) the length of path of contact, (b) contact ratio, and (c) angle turned through by pinion, while any pair of teeth is in contact.

[Ans. 14.79 mm, 5, 22.49°]

14.6 A pair of involute profile spur gears is to give a speed ratio of 3. The arc of approach is not to be less than the circular pitch. The pressure angle is 20° and pinion is the driver. The module is 4 mm. Calculate (a) minimum number of teeth on gear, and (b) addendum of gear wheel.

[Ans. 57, 5.07 mm]

14.7 A pinion having 30 teeth drives a gear of 80 teeth. The profile of gears is involute with 20° pressure angle, 12 mm module, and 10 mm addendum. Find (a) the length of path of contact, (b) arc of contact, and (c) contact ratio,

[Ans. 52.26 mm, 55.61 mm, 1.47]

14.8 Find the minimum number of teeth on gear wheel to avoid undercutting when the addendum for stub teeth is 0.84 module, if (a) gear ratio is 3:1, and (b) the wheel is used to engage a rack.

[Ans. 16, 18]

14.9 A pair of involute spur gears having 20 and 40 teeth are in mesh, the speed of smaller wheel being 2000 rpm. Calculate the sliding velocity between gear teeth faces (a) at the point of engagement, (b) at the pitch point, and (c) at the point of disengagement, if smaller wheel is the driver. Pressure angle is 20°, addendum = 5 mm, and module = 5 mm.
Also find the angle turned through by the pinion while any one pair of teeth is in contact.

[Ans. 3.97 m/s, 0, 3.61 m/s; 29.43°]

14.10 The thickness of an involute gear tooth is 8 mm at a radius of 90 mm and a pressure angle of 14.5°. Calculate the tooth thickness and radius at a point on the involute which has a pressure angle of 25°. Also calculate the tooth thickness at the base circle.

[Ans. 3.84 mm, 96.14 mm, 8.71 mm]

14.11 A 20° pinion having a module of 2.5 and 40 teeth meshes with a rack with no backlash. If the rack is pulled out 1.25 mm, calculate the backlash error.

[Ans. 1.015 mm]

14.12 A 20° pressure angle pinion with a module of 2 and 18 teeth drives a gear of 54 teeth. If the centre distance at which the gear operates is 75 mm, calculate the operating pressure angle, and backlash produced.

[Ans. 2.595 mm]

(b) Helical gears

14.13 The centre distance between the two shafts connected by two left-handed helical gears is 0.4 m. The shaft angle is 60° and normal module is 6 mm. The gear ratio is 2 and larger gear is having 70 teeth. Find the helix angles of the two gears.

[Ans. 16°, 44°]

14.14 Two left-handed helical gears connect two shafts 60° apart. The normal module is 6 mm. The larger gear has 60 teeth and the velocity ratio is 0.5. The centre distance is 0.3 m. Find the helix angles of the two gears.

[Ans. 25°, 35°]

14.15 Two right-handed helical gears connect two shafts 70° apart. The larger gear has 50 teeth and the smaller 20. The centre distance is 167 mm. Determine the helix angle of the gears. The normal module is 4 mm.

[Ans. 28°, 42°]

(c) Bevel gears

14.16 A pair of bevel gears is mounted on two intersecting shafts at an angle of 72°. The velocity ratio of the gears is 2. Calculate the pitch angles.

[Ans. 22.39°, 49.61°]

14.17 A 6 mm module, straight bevel pinion of 17 teeth drives a gear of 25 teeth. The shaft angle is 90°. Calculate the virtual number of teeth on the pinion and gear.

[Ans. 21, 44]

(d) Spiral gears

14.18 Two spiral gears A and B have 45 and 15 teeth at spiral angles of 20° and 50° respectively. Both gears are of same hand. A is 150 mm in diameter. Find the centre distance between the shafts and the angle between the shafts. If the teeth are of 20° involute form and coefficient of friction is 0.08, find the efficiency if A is the driver.

[Ans. 111.5 mm, 70°, 87.9%]

14.19 Two spiral gears of diameter ratio 1.5 are used on a machine tool. The angle between the shafts is 76° and approximate centre distance is 115 mm. Speed ratio is 1.5 and normal diametral pitch is 10 mm. Calculate the number of teeth on each gear and spiral angles.

[Ans. 23, 35, 38°, 38°]

14.20 A pair of spiral gears is required to connect two shafts, the angle between the non-intersecting axes is 90°. The speed ratio is 3 and pitch circle diameters of the gears are equal. If approximate centre distance is 250 mm, estimate the helix angle of gears and number of teeth on each gear. Take module = 3 mm.

[Ans. 71.56°, 18.44°; 26, 78]

14.21 A drive on a machine tool is to be made by two spiral gears of the same hand and normal pitch of 12.5 mm. The gears are of equal diameters and centre distance between the shafts is approximately 134 mm. The angle between the two shafts is 80° and speed ratio 1.25. Calculate

(a) spiral angle of each wheel, (b) number of teeth on each wheel
(c) efficiency of drive, and (d) maximum efficiency.

The friction angle is 6°.

[Ans. 32.46°, 47; 54°, 30, 24, 82.98%, 83.85%]

(e) Worm gears

14.22 A three-start worm has a pitch diameter of 80 mm and a pitch of 20 mm. It rotates at 600 rpm and drives a 40 tooth worm gear. If coefficient of friction is 0.05, find the

(a) helix angle of worm, (b) speed of the gear,
(c) centre distance, and (d) efficiency and maximum efficiency.

[Ans. 76.56, 45 rpm, 167.3 mm, 81.7% , 90.5%]

15

GEAR TRAINS

Chapter Outline

15.1 INTRODUCTION

A gear train is composed of two or more gears in mesh for the purpose of transmitting motion from one shaft to another. A gear train enables to have larger centre distance between the driving and driven shafts, provides control on the direction of rotation of the driven gear, and facilitates increased transmission ratio with gears of smaller sizes in a lesser space.

15.2 TYPES OF GEAR TRAINS

There are four types of gear trains:

1. Simple gear train
2. Compound gear train
3. Reverted gear train
4. Planetary (or epicyclic) gear train.

1. *Simple Gear Train*: A simple gear train is one in which there is only one gear on each shaft. A simple gear train is often used to change the direction of rotation of a gear without changing its angular velocity. This can be done by placing an idler gear between the driving and driven gears. Consider the simple gear train shown in Fig.15.1

$$\frac{n_1}{n_2} = \frac{z_2}{z_1}$$

$$\frac{n_2}{n_3} = \frac{z_3}{z_2}$$

$$\frac{n_{m-1}}{n_m} = \frac{z_m}{z_{m-1}}$$

Multiplying, we get

$$\frac{n_1}{n_m} = \frac{z_m}{z_1} \tag{15.1}$$

Therefore, the velocity ratio of a simple gear train is the ratio of the angular velocity of the first gear in the train to the angular velocity of the last gear. We find that the intermediate gears do not in any way affect the velocity ratio. These gears are called *idler gears*. If the number of gears in the train are even then the direction of rotation of the last gear is reversed and if the number of gears in the train are odd then the direction of rotation of the last gear remains the same. Idler gears are used for two purposes: to connect gears where a large centre distance is required, and to control the directional relationship between gears.

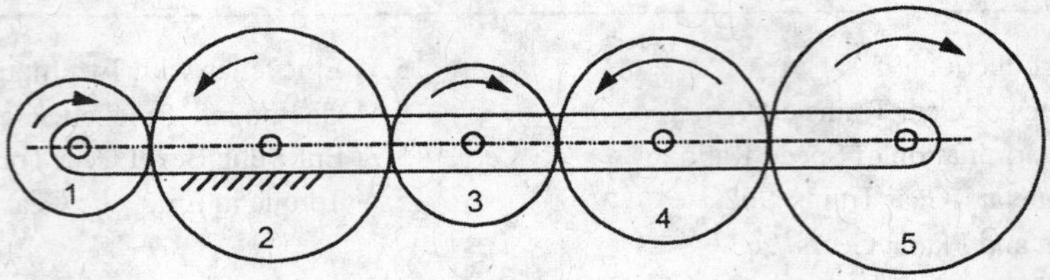

Fig.15.1 Simple gear train

2. *Compound Gear Train*: A pair of gears is compound if they have a common axis and are integral. A compound gear train is a gear train containing compound gears. Consider the compound gear train shown in Fig.15.2. Gears 2 and 3 are on the same shaft.

$$\frac{n_2}{n_1} = \frac{-z_1}{z_2}$$

$$\frac{n_4}{n_3} = \frac{-z_3}{z_4}$$

Multiplying, we get

$$\frac{n_2}{n_1} \times \frac{n_4}{n_3} = \frac{z_1}{z_2} \times \frac{z_3}{z_4}$$

But $n_2 = n_3$

Thus

$$\frac{n_4}{n_1} = \frac{z_1}{z_2} \times \frac{z_3}{z_4}$$

$$= \frac{z_1 z_3}{z_2 z_4}$$

Speed of driven gear/Speed of driving gear

= Product of teeth of driving gears/Product of teeth of driven gears (15.2)

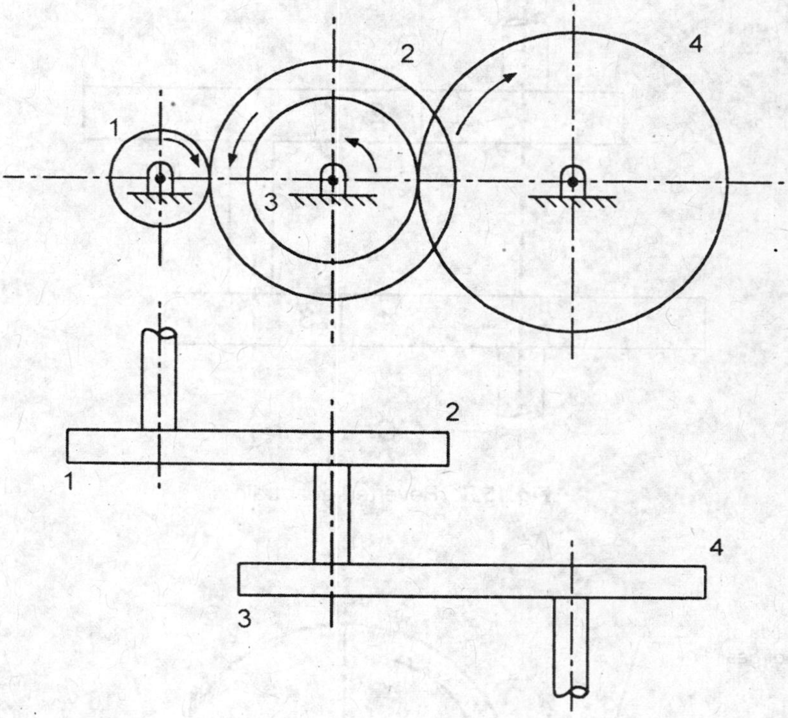

Fig.15.2 Compound gear train

The ratio given by Eq.(15.2) is called the *train value*.

3. *Reverted Gear Train*: In a reverted gear train the first and the last gear are coaxial (same axis). Reverted gear trains are used in automotive transmission, lathe back gears, industrial speed reducers, and in clocks where the minute and hour hand shafts are coaxial. In the reverted gear train shown in Fig.15.3, we have

   Centre distance, $C = r_1 + r_2 = r_3 + r_4$

   For same module ($m = d/z$) of all gears, the pitch radius of gears is proportional to the number of teeth. Hence

$$z_1 + z_2 = z_3 + z_4 \tag{15.3}$$

4. *Planetary Gear Trains*: These are gear trains in which the axis of one or more gears move relative to the frame. The gear at the centre is called the sun, and the gears whose axes move are called the *planets*.

   In Fig.15.4, arm 3 drives gear 1 about gear 2, which is a fixed external gear. Gear 1 rotates about its centre d while this centre rotates about centre O_2 of the fixed gear. As gear 1 rolls on the outside of gear 2, a point on its surface will generate an epicycloid. If gear 2 happens to be an internal gear and gear 1 rolls on the inside of gear 2, then a point on the surface of gear 1 will generate a hypocycloid. Because of the curves generated, a planetary gear train is often called as an *epicyclic*, or *cyclic*, gear train.

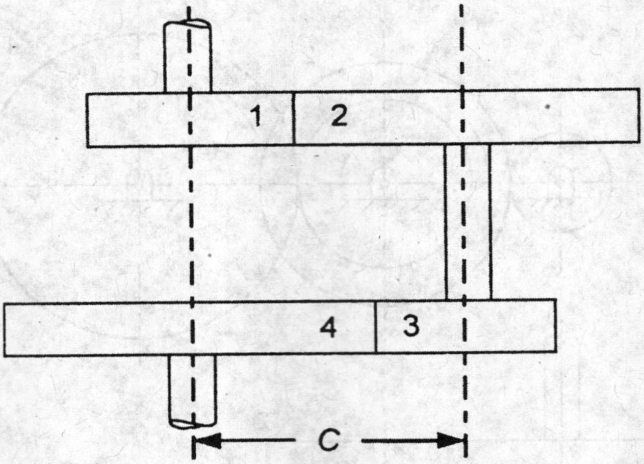

Fig.15.3 Reverted gear train

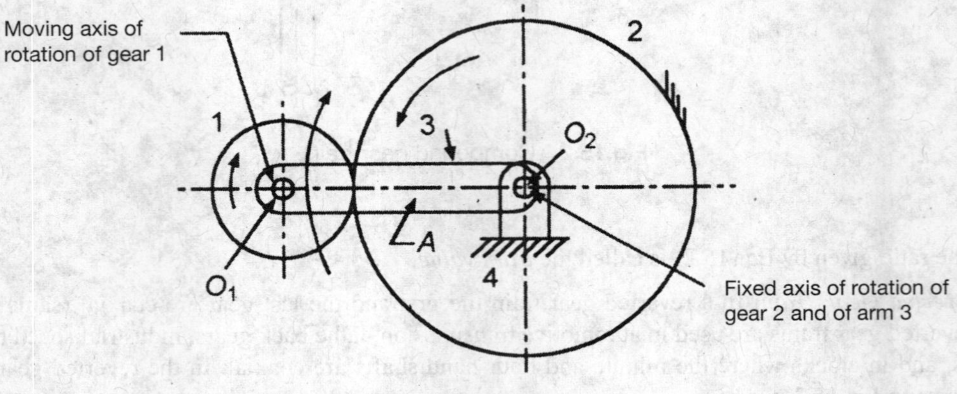

Fig.15.4 Planetary gear train

15.3 DETERMINATION OF SPEED RATIO OF PLANETARY GEAR TRAINS

The speed ratio of a planetary gear train can be determined by the following methods:

1. Relative velocity method
2. Algebraic or tabular method.

1. *Relative Velocity Method*:
   Consider the planetary gear train shown in Fig.15.4.

$$\omega_{13} = \omega_{14} - \omega_{34}$$

$$\omega_{23} = \omega_{24} - \omega_{34}$$

and $$\frac{\omega_{13}}{\omega_{23}} = \frac{\omega_{14} - \omega_{34}}{\omega_{24} - \omega_{34}}$$

Link 3 is the arm, link 1 is the driven gear, link 2 is the driving gear, and link 4 is fixed. If subscript 'a' is used for the arm, then

$$\frac{\omega_{1a}}{\omega_{2a}} = \frac{\omega_1 - \omega_a}{\omega_2 - \omega_a}$$

or $$\frac{n_{1a}}{n_{2a}} = \frac{n_1 - n_a}{n_2 - n_a}$$

Now $$\frac{n_{1a}}{n_{2a}} = \frac{-z_1}{z_2}$$

Hence, $$\frac{n_1 - n_a}{n_2 - n_a} = \frac{-z_1}{z_2}$$

If gear 2 is fixed, then $n_2 = 0$. Thus,

$$\frac{n_1}{n_a} = \left(1 + \frac{z_1}{z_2}\right) \tag{15.4}$$

If arm is fixed, then $n_a = 0$. Thus

$$\frac{n_1}{n_2} = \frac{-z_1}{z_2} \tag{15.5}$$

2. *Tabular Method*: This method is based on the principle of superposition, which states that the resultant revolutions or turns of any gear may be found by taking the number of turns it makes with the arm plus the number of turns it makes relative to the arm. The following steps may be followed for this method:

   1. Assume the arm to be fixed and determine the revolutions of different gears of the train for one revolution of a particular convenient gear.
   2. Multiply all columns in the first row by x and write in the second row.
   3. To account for rotation of arm, add y to the various quantities of second row.
   4. Out of the three quantities involved in the last row, two of them are given. From these the values of x and y can be determined. On substituting in the third, its magnitude can be determined.

   Table 15.1 may be conveniently used for this purpose:

15.4 SUN AND PLANET GEARS

Fig.15.5 shows the sun and planet gears in which gear P is the planet, gear S the sun, arm A, and the annular (or internal) gear 1. The annular gear is fixed and the planet gear rolls over the sun and the annular gear. O_1 is the moving axis of rotation of planet gear P and O_2 is the fixed axis of rotation of sun gear 2 and arm A. The axes O_1 and O_2 are coupled by the arm A. Let z_p, z_s, and z, the number of teeth on the planet, sun, and annular gears respectively. To find the speed of any gear, Table 15.2 may be used to find speed of gears.

Table 15.1 Tabular method for determining speed ratio

Operation	Revolutions of		
	Arm A	Gear 1, z_1	Gear 2, z_2
1. Arm A fixed, +1 revolutions to gear 1, cw	0	+1	$\frac{-z_1}{z_2}$
2. Multiply by x	0	$+x$	$\frac{-xz_1}{z_2}$
3. Add y	$+y$	$x+y$	$\frac{-xz_1}{z_2}+y$

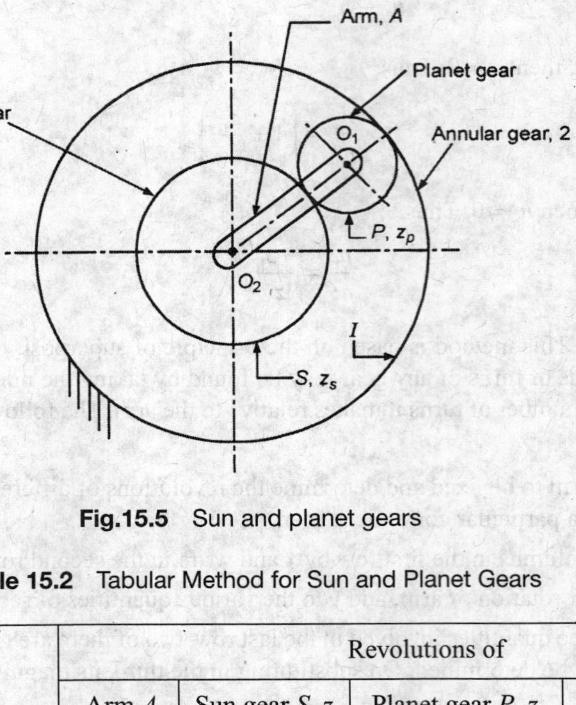

Fig.15.5 Sun and planet gears

Table 15.2 Tabular Method for Sun and Planet Gears

Operation	Revolutions of			
	Arm A	Sun gear S, z_s	Planet gear P, z_p	Annular gear I, z_i
1. Arm A fixed, +1 revolutions to S, cw	0	+ 1	$\frac{-z_s}{z_p}$	$\frac{-z_s}{z_i}$
2. Multiply by x	0	$+x$	$\frac{-z_s x}{z_p}$	$\frac{-z_s x}{z_i}$
3. Add y	y	$x+y$	$\frac{-z_s x}{z_p}+y$	$\frac{-z_s x}{z_i}+y$

Let n_s and n_i be the speed of sun and annular gears respectively. Then

$$x + y = n_s$$

$$-\left(\frac{z_s}{z_i}\right) \cdot x + y = n_i$$

Solving for x and y, we get

$$x = \frac{z_i n_s - z_i n_i}{z_i + z_s}$$

$$y = \frac{z_s n_s + z_i n_i}{z_i + z_s}$$

(a) When sun gear is fixed, then $n_s = 0$. Thus

$$y = \frac{z_i n_i}{z_i + z_s}$$

$$= \frac{n_i}{\left(1 + \frac{z_s}{z_i}\right)} \quad (15.6)$$

(b) When annular gear is fixed, then $n_i = 0$. Thus

$$y = \frac{z_i n_s}{z_i + z_s}$$

$$= \frac{n_s}{\left(1 + \frac{z_i}{z_s}\right)} \quad (15.7)$$

15.5 EPICYCLICS WITH TWO INPUTS

A gear train of this type is shown in Fig.15.6. Let n_1, n_2 and n_0 represent the turns of input 1, input 2 and the output, respectively. By superposition, the number of turns of the output equals the output turns due to input 1 plus the output turns due to input 2. This can be expressed as

$$n_0 = n_1 \left(\frac{n_0}{n_1}\right)_{\text{input 2 held fixed}} + n_2 \left(\frac{n_0}{n_2}\right)_{\text{input 1 held fixed}} \quad (15.8)$$

15.6 COMPOUND EPICYCLIC GEAR TRAIN

A compound epicyclic gear train consists of two or more epicyclic gear trains connected in series. A compound epicyclic gear train is analysed by considering each epicyclic gear train separately. That epicyclic gear train is analysed first where two conditions (or speeds of two elements) are known.

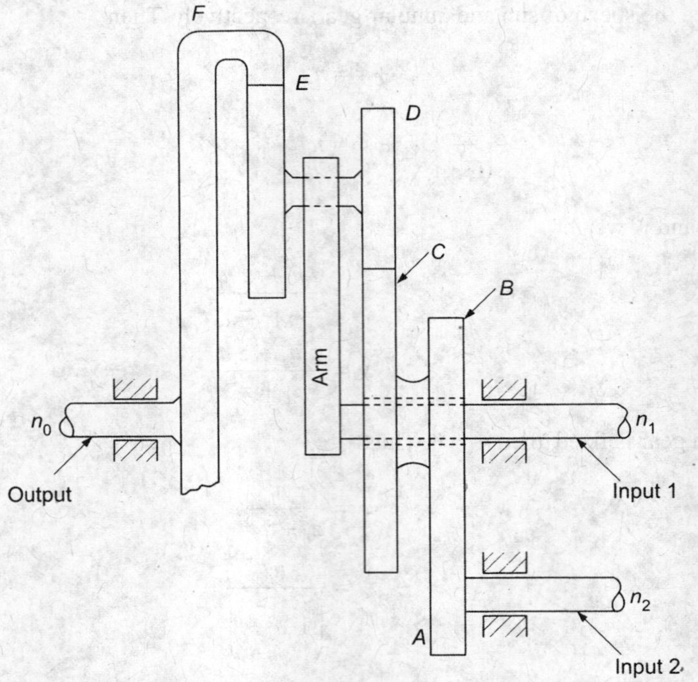

Fig.15.6 Epicyclics with two inputs

15.7 EPICYCLIC BEVEL GEAR TRAINS

Bevel gears can be used to make a more compact epicyclic system and they permit a very high speed reduction with few gears. They find potential applications in speed reduction gears and differential gear of an automobile.

15.8 TORQUE IN EPICYCLIC GEAR TRAINS

Consider the rotating parts of an epicyclic gear train shown in Fig.15.7. if the rotating parts have no angular acceleration, then the gear train is kept in equilibrium by the following three externally applied torques:

1. Input torque (T_1) on the driving member.
2. Output torque (T_2) on the driven member.
3. Braking torque (T_3) on the fixed member.

The net torque applied on the gear train must be zero, i.e.,

$$T_1 + T_2 + T_3 = 0$$

If F_1, F_2, and F_3 are the externally applied forces at radii r_1, r_2, and r_3, then

$$F_1 r_1 + F_2 r_2 + F_3 r_3 = 0$$

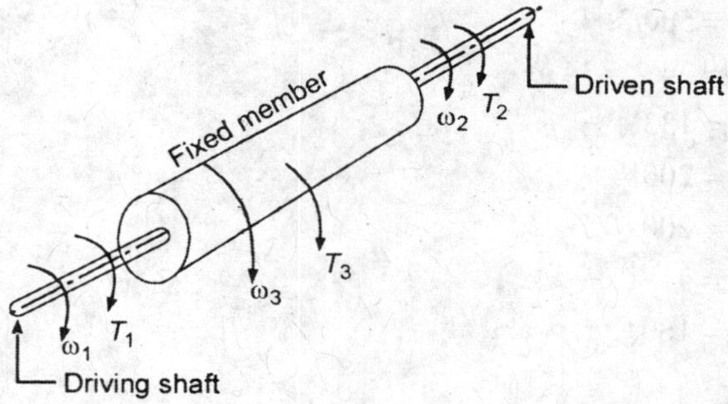

Fig.15.7 Torque in epicyclic gear trains

If ω_1, ω_2, and ω_3 are the angular velocities of the driving, driven, and fixed members respectively, and friction is neglected, then the net kinetic energy dissipated by the gear train must be zero.

$$T_1\omega_1 + T_2\omega_2 + T_3\omega_3 = 0$$

For a fixed member, $\omega_3 = 0$. Therefore

$$T_1\omega_1 + T_2\omega_2 = 0$$

$$T_2 = -\left(\frac{\omega_1}{\omega_2}\right)T_1$$

Holding torque,

$$T_3 = -(T_1 + T_2)$$

$$= T_1\left(\frac{\omega_1}{\omega_2} - 1\right)$$

$$= T_1\left(\frac{n_1}{n_2} - 1\right) \tag{15.9}$$

Example 15.1

The speed ratio of a reverted gear train is to be 15. The module of gears 1 and 2 is 3 mm and that of gears 3 and 4 is 2.5 mm. Calculate the suitable number of teeth for the gears. The centre distance between gear shafts is 250 mm.

■ Solution

Given: $m_1 = m_2 = 3$ mm, $m_3 = m_4 = 2.5$ mm, $c = 250$ mm. Refer to Fig. 15.3.

$$n_1/n_2 = n_3/n_4 = \sqrt{15} = 3.873$$

or

$$z_2/z_1 = z_4/z_3 = 3.873$$

Now

$$r_1 + r_2 = r_3 + r_4 = 250 \text{ mm}$$

or

$$m_1(z_1 + z_2)/2 = m_2(z_3 + z_4)/2 = 250$$

or

$$3(z_1 + z_2) = 2.5(z_3 + z_4) = 500$$

$$z_1 + z_2 = 500/3$$

$$4.873\, z_1 = 500/3$$
or
$$z_1 = 34.2 \approx 34$$
or
$$z_2 = 133$$
$$z_3 + z_4 = 200$$
or
$$4.873\, z_3 = 200$$
or
$$z_3 = 41$$
$$z_4 = 160$$

Example 15.2

In an epicyclic gear train, an arm carries two gears 1 and 2 having 40 and 50 teeth respectively. The arm rotates at 160 rpm ccw about the centre of gear 1, which is fixed. Determine the speed of the gear 2.

■ **Solution**

Given: $z_1 = 40$, $z_2 = 50$, $n_1 = 0$, $n_a = -160$ rpm
Table 15.3 can be used to find the speed of gear 2.

Table 15.3

Operation	Revolutions of		
	Arm A	Gear 1	Gear 2
1. Arm A fixed, +1 revolutions to gear 1, cw.	0	+1	$-z_1/z_2$
2. Multiply by x	0	$+x$	$-x\, z_1/z_2$
3. Add y	$+y$	$x + y$	$-x\, z_1/z_2 + y$

As gear 1 is fixed, therefore
$$x + y = 0$$
or
$$x = -y = -160 \text{ rpm}$$
Speed of gear 2,
$$n_2 = -x\, z_1/z_2 + y$$
$$= 160\,(40/50) + 160$$
$$= 288 \text{ rpm ccw}$$

Example 15.3

In a reverted epicyclic gear train, the arm A carries two gears 1 and 2 at centre of rotation O_1 and a compound gear 3 and 4 at centre of rotation O_2. The gear 1 meshes with gear 4 and the gear 2 meshes with gear 3. The number of teeth are: $z_1 = 75$, $z_2 = 30$, and $z_3 = 90$. Find the speed and direction of gear 2 when gear 1 is fixed and the arm A makes 120 rpm clockwise. Assume all gears to be of the same module.

■ **Solution** Refer to Fig.15.15.

$$r_1 + r_4 = r_2 + r_3$$
or
$$z_1 + z_4 = z_2 + z_3$$
$$z_4 = 30 + 90 - 75 = 45$$

Table 15.4 can be used to find the speed of the gears.

Table 15.4

Operation	Revolutions of			
	Arm A	Compound gear 3, 4	Gear 1	Gear 2
1. Arm A fixed, +1 revolutions to gears 3, 4, ccw	0	+1	$\frac{-z_4}{z_1}$	$\frac{-z_3}{z_2}$
2. Multiply by x	0	$+x$	$\frac{-xz_4}{z_1}$	$\frac{-xz_3}{z_2}$
3. Add y	$+y$	$x+y$	$\frac{-xz_4}{z_1}+y$	$\frac{-xz_3}{z_2}+y$

Since gear 1 is fixed,

$$\frac{-xz_4}{z_1}+y=0$$

or

$$-x\left(\frac{45}{75}\right)+y=0$$

or

$$-0.6+y=0 \tag{1}$$

Arm A makes 120 rpm clockwise, therefore

$$y=-120 \tag{2}$$

From (1) and (2), we get

$$x=-200 \text{ rpm}$$

Now for gear 2, we have

$$n_2=\frac{-xz_3}{z_2}+y$$

$$=200\left(\frac{90}{30}\right)-120$$

$$=480 \text{ rpm ccw}$$

Example 15.4

An epicyclic gear train consists of three gears 1, 2 and 3 as shown in Fig.15.8. The internal gear 1 has 72 teeth and gear 3 has 32 teeth. The gear 2 meshes with both gear 1 and gear 3 and is carried on an arm A which rotates about the centre O_2 at 20 rpm. If the gear 1 is fixed, determine the speed of gears 2 and 3.

Solution

Table 15.5 can be used to find the speed of the gears.

Table 15.5

Operation	Revolutions of			
	Arm A	Gear 3	Gear 1	Gear 2
1. Arm A fixed, +1 revolutions to gear 3, ccw.	0	+1	$\frac{-z_3}{z_1}$	$\frac{-z_3}{z_2}$
2. Multiply by x	0	$+x$	$\frac{-xz_3}{z_1}$	$\frac{-xz_3}{z_2}$
3. Add y	$+y$	$x+y$	$\frac{-xz_3}{z_1}+y$	$\frac{-xz_3}{z_2}+y$

Speed of arm, $y = 20$ rpm

For gear 1 fixed, we have

$$\frac{-xz_3}{z_2 + y} = 0$$

$$-x\left(\frac{32}{72}\right) + 20 = 0$$

$$x = 45$$

Speed of gear 3

$$n_3 = x + y$$

$$= 45 + 20 = 65 \text{ rpm in the direction of arm}$$

Speed of gear 2

$$d_2 + \frac{d_3}{2} = \frac{d_1}{2}$$

or $$2d_2 + d_3 = d_1$$

or $$2z_2 + z_3 = z_1$$

or $$2z_2 + 32 = 72$$

$$z_2 = 20$$

$$n_2 = -x\left(\frac{z_3}{z_2}\right) + y$$

$$= -45\left(\frac{32}{20}\right) + 20$$

$$= -52 \text{ rpm in the opposite direction of arm.}$$

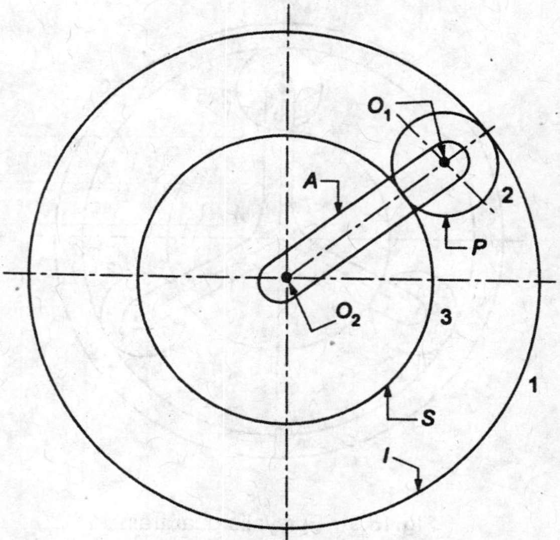

Fig.15.8 Epicyclic gear train

Example 15.5

The pitch circle diameter of the annular gear in the epicyclic gear train shown in Fig.15.9 is 425 mm and the module is 5 mm. When the annular gear 3 is stationary, the spindle *A* makes one revolution in the same sense as the sun gear 1 for every 6 revolutions of the driving spindle carrying the sun gear. All the planet gears are of the same size. Determine the number of teeth on all the gears.

■ Solution

Table 15.6 can be used to find the speed of the gears.

Table 15.6

Operation	Revolutions of			
	Spindle *A*	Gear 1	Gear 2	Gear 3
1. Arm *A* fixed, +1 revolutions to gear 1, ccw.	0	+1	$\frac{-z_1}{z_2}$	$\frac{-z_1}{z_3}$
2. Multiply by *x*	0	$+x$	$-\frac{xz_1}{z_2}$	$-\frac{xz_1}{z_3}$
3. Add *y*	$+y$	$x+y$	$\frac{-xz_1}{z_2}+y$	$\frac{-xz_1}{z_3}+y$

Now $\quad y = +1$

and $\quad x + y = 6$

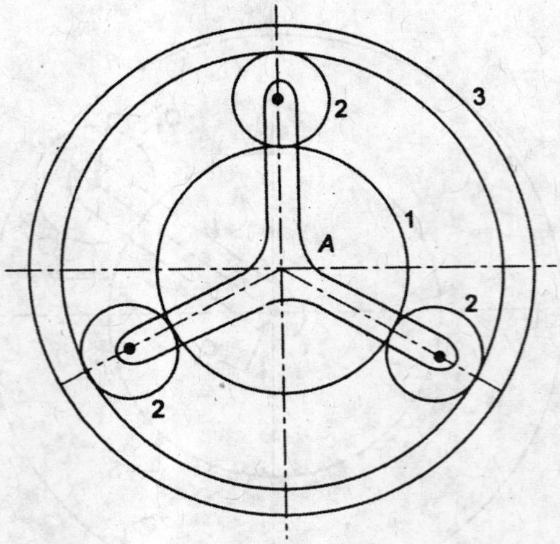

Fig.15.9 Epicyclic gear train

Therefore, $$x = 5$$

Gear 3 is stationary, hence $$\frac{-xz_1}{z_3} + y = 0$$

$$\frac{-5z_1}{z_3} + 1 = 0$$

$$\frac{z_1}{z_3} = \frac{1}{5}$$

Now $$z_3 = \frac{d_3}{m} = \frac{425}{5} = 85$$

$$z_1 = \frac{85}{5} = 17$$

Also $$d_1 + 2d_2 = d_3$$

or $$z_1 + 2z_2 = z_3$$

$$17 + 2z_2 = 85$$

$$z_2 = 34$$

Example 15.6

The Ferguson's paradox epicyclic gear train is shown in Fig.15.10. Gear 1 is fixed to the frame. The arm *A* and gears 2 and 3 are free to rotate on the shaft *S*. Gears 1, 2, and 3 have 100, 101 and 99 teeth respectively. The planet gear has 20 teeth. The pitch circle diameter of all the gears is the same so that the planet gear *P* meshes with all of them. Determine the revolutions of gears 2 and 3 for one revolution of the arm *A*.

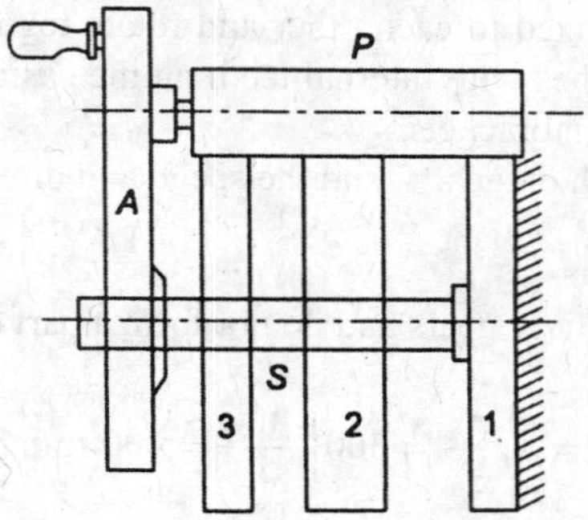

Fig.15.10 Ferguson's paradox epicyclic gear train

Solution

Table 15.7 can be used to find the speed of the gears.

Table 15.7

Operation	Revolutions of			
	Spindle A	Gear 1	Gear 2	Gear 3
1. Arm A fixed, +1 revolutions of gear 1, ccw	0	+1	$\frac{z_1}{z_2}$	$\frac{z_1}{z_3}$
2. Multiply by x	0	$+x$	$\frac{xz_1}{z_2}$	$\frac{xz_1}{z_3}$
3. Add y	$+y$	$x+y$	$\frac{xz_1}{z_2}+y$	$\frac{xz_1}{z_3}+y$

Now

$$y = 1$$

Gear 1 is fixed, therefore $x + y = 0$

or $x = -y = -1$

Also

$$n_2 = \frac{xz_1}{z_2} + y = \frac{-100}{101} + 1 = \frac{1}{101}$$

$$n_3 = \frac{xz_1}{z_3} + y = \frac{-100}{99} + 1 = \frac{-1}{99}$$

Example 15.7

In the gear drive shown in Fig.15.11, the driving shaft A rotates at 300 rpm in the clockwise direction, when seen from the left hand side. The shaft B is the driven shaft. The casing C is held stationary. The wheels E and H are keyed to the central vertical spindle and wheel F can rotate freely on this spindle.

The wheels K and L are rigidly fixed to each other and rotate together freely on a pin fitted on the underside of F. The wheel L meshes with internal teeth on the casing C. The number of teeth on the different gears are indicated within brackets.

Determine the number of teeth on gear C and the speed and direction of rotation of shaft B.

■ Solution

The wheels D and G are auxiliary gears and do not form a part of the epicyclic gear train.

$$n_E = n_A \frac{z_D}{z_E} = 300\left(\frac{40}{30}\right) = 400 \text{ rpm cw}$$

Assuming same module for all the gears,

$$z_C = z_H + z_K + z_L = 40 + 20 + 30 = 90$$

Table 15.8 can be used to find the speed of the gears.

Table 15.8

Operation	Revolutions of			
	Wheel F	Gears E & H	Gear K & L	Gear C
1. Arm A fixed, gear E, H given 1 revolution, cw	0	-1	$\frac{z_H}{z_K}$	$\left(\frac{z_H}{z_K}\right)\left(\frac{z_L}{z_C}\right)$
2. Multiply by x	0	$-x$	$\frac{xz_H}{z_K}$	$x\left(\frac{z_H}{z_K}\right)\left(\frac{z_L}{z_C}\right)$
3. Add $-y$	$-y$	$-x-y$	$\frac{xz_H}{z_K} - y$	$x\left(\frac{z_H}{z_K}\right)\left(\frac{z_L}{z_C}\right) - y$

$$-x - y = -400$$

or $$x + y = 400 \tag{1}$$

Now wheel C is fixed, therefore

$$x\left(\frac{z_H}{z_K}\right)\left(\frac{z_L}{z_C}\right) - y = 0$$

or $$x\left(\frac{40}{20}\right)\left(\frac{30}{90}\right) - y = 0$$

or $$\frac{2x}{3} - y = 0 \tag{2}$$

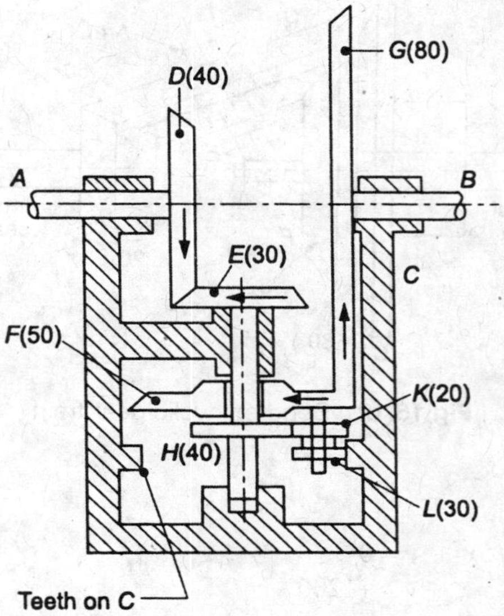

Fig.15.11 Gear drive

From (1) and (2), we have

$$x = 240 \quad \text{and} \quad y = 160$$

Thus $$n_F = -y = -160 \text{ rpm}$$

Speed of wheel G or shaft, $B = -n_F \dfrac{z_F}{z_G}$

$$= -\left(-160 \times \frac{50}{80}\right) = 100 \text{ rpm ccw}$$

Example 15.8

In a gear train, as shown in Fig.15.12, gear B is connected to the input shaft. The arm A carrying the compound wheels D and E, turns freely on the output shaft. If the input speed is 1200 rpm counter-clockwise, when seen from the right, determine the speed of the output shaft under the following conditions: (a) when the gear C is fixed and (b) when gear C rotates at 10 rpm counter-clockwise.

Solution

Table 15.9 can be used to find the speed of the gears.

(a) Gear C is fixed, therefore

$$-x\left(\frac{z_B}{z_C}\right) + y = 0$$

$$-x\left(\frac{20}{80}\right) + y = 0$$

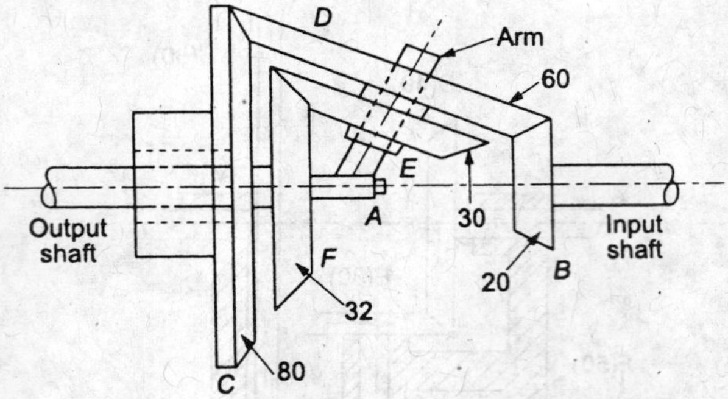

Fig.15.12 Bend epicyclic gear train

or $\quad -0.25x + y = 0 \quad (1)$

Now $\quad x + y = +1200 \quad (2)$

From (1) and (2), we get

$$x = +960 \text{ and } y = +240$$

Table 15.9

Operation	Revolutions of				
	Arm A	Gear B (input shaft)	Compound gears D, E	Gear C	Gear F (output shaft)
1. Arm A fixed, +1 revolution to gear B, ccw.	0	1	$\frac{z_B}{z_D}$	$\frac{-z_B}{z_C}$	$-\left(\frac{z_B}{z_D}\right)\left(\frac{z_E}{z_F}\right)$
2. Multiply by x	0	x	$\frac{xz_B}{z_D}$	$\frac{-xz_B}{z_C}$	$-x\left(\frac{z_B}{z_D}\right)\left(\frac{z_E}{z_F}\right)$
3. Add y	y	$x + y$	$\frac{xz_B}{z_D} + y$	$\frac{-xz_B}{z_C} + y$	$-x\left(\frac{z_B}{z_D}\right)\left(\frac{z_E}{z_F}\right) + y$

$$n_F = -x\left(\frac{z_B}{z_D}\right)\left(\frac{z_E}{z_F}\right) + y$$

$$= -960\left(\frac{20}{80}\right)\left(\frac{30}{32}\right) + 240$$

$$= 15 \text{ rpm (counter-clockwise)}$$

(b) When gear C is rotating at 10 rpm clockwise, we have

$$-x\left(\frac{z_B}{z_C}\right)+y=+10$$

or $$-x\left(\frac{20}{80}\right)+y=10$$

or $$-0.25x+y=10 \tag{3}$$

From (2) and (3), we get

$$x=952 \quad \text{and} \quad y=248$$

$$n_F=-x\left(\frac{z_B}{z_D}\right)\left(\frac{z_E}{z_F}\right)+y$$

$$=-952\left(\frac{20}{80}\right)\left(\frac{30}{32}\right)+248$$

$$= 24.875 \text{ rpm (counter-clockwise)}$$

Example 15.9

The differential gear used in an automobile is shown in Fig.15.13. The pinion A on the propeller shaft has 12 teeth and the crown gear B has 60 teeth. The shafts P and Q form the rear axles to which the road wheels are attached. If the propeller shaft rotates at 1200 rpm and the road wheel attached to axle Q has a speed of 250 rpm while taking a turn, find the speed of road wheel attached to axle P.

Solution

$$n_B=\left(\frac{z_A}{z_B}\right)n_A=\left(\frac{12}{60}\right)1200=240 \text{ rpm}$$

Now $$y=240 \text{ rpm}$$

Also $$-x+y=250$$

or $$x=240-250=-10$$

speed of road wheel attached to axle P = speed of gear C

$$=x+y$$

$$=-10+240=230 \text{ rpm}$$

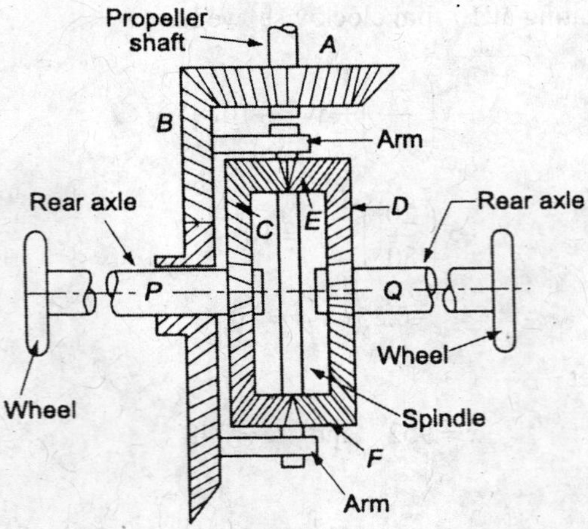

Fig.15.13 Differential gear

Table 15.10 can be used to find the speed of the gears.

Table 15.10

Operation	Revolutions of			
	Gear B	Gear C	Gear E	Gear D
1. Gear B fixed, +1 revolution to gear c, ccw.	0	1	$\frac{z_C}{z_E}$	$-\left(\frac{z_C}{z_E}\right)\left(\frac{z_E}{z_D}\right) = -1$ or $\frac{z_C}{z_D} = 1$
2. Multiply by x	0	x	$x\left(\frac{z_C}{z_E}\right)$	$-x$
3. Add y	y	$x + y$	$x\frac{z_C}{z_E} + y$	$-x + y$

Example 15.10

An epicyclic gear train consists of a sun wheel S, a stationary annular wheel E and three identical planet wheels P carried on a star-shaped carrier C, as shown in Fig.15.14. The size of different toothed wheels is such that the planet carrier C rotates at 1/5 th of the speed of the sun wheel S. The minimum number of teeth on any wheel is 18. The driving torque on the sun wheel is 120 Nm. Determine (a) the number of teeth on different wheels of the train, and (b) the torque necessary to keep the internal gear stationary.

Solution

(a) $$n_C = n_s/5$$

Table 15.11 can be used to find the speed of the gears.

Table 15.11

Operation	Revolutions of			
	Planet carrier C	Sun wheel S	Planet wheel P	Annular wheel E
1. Planet C fixed, +1 revolution to seen wheel S, ccw	0	1	$\frac{-z_S}{z_P}$	$-\left(\frac{z_S}{z_P}\right)\left(\frac{z_P}{z_E}\right) = \frac{-z_S}{z_E}$
2. Multiply by x	0	x	$-x\frac{z_S}{z_P}$	$-x\frac{z_S}{z_E}$
3. Add y	y	$x+y$	$-x\frac{z_C}{z_P}+y$	$y-\frac{xz_C}{z_E}$

Now $$y = 1$$

$$x + y = 5$$

or $$x = 4$$

Gear E is stationary, therefore

$$-x\frac{z_S}{z_E} + y = 0$$

or $$-4\left(\frac{z_S}{z_E}\right) + 1 = 0$$

or $$\frac{z_S}{z_E} = \frac{1}{4}$$

Let $$z_s = 18$$

Then $$z_E = 72$$

Also $$d_s + 2d_p = d_E$$

Assuming same module for all gears, we have

$$z_s + 2z_p = z_E$$

or $$18 + 2z_p = 72$$

or $$z_p = 27$$

(b) $$T_S\omega_S = T_C\omega_C$$

$$120\omega_S = T_C\omega_C$$

or $$T_C = 120 \times 5 = 600 \text{ Nm}$$

Torque required to keep the annular gear stationary = 600 – 120 = 480 Nm

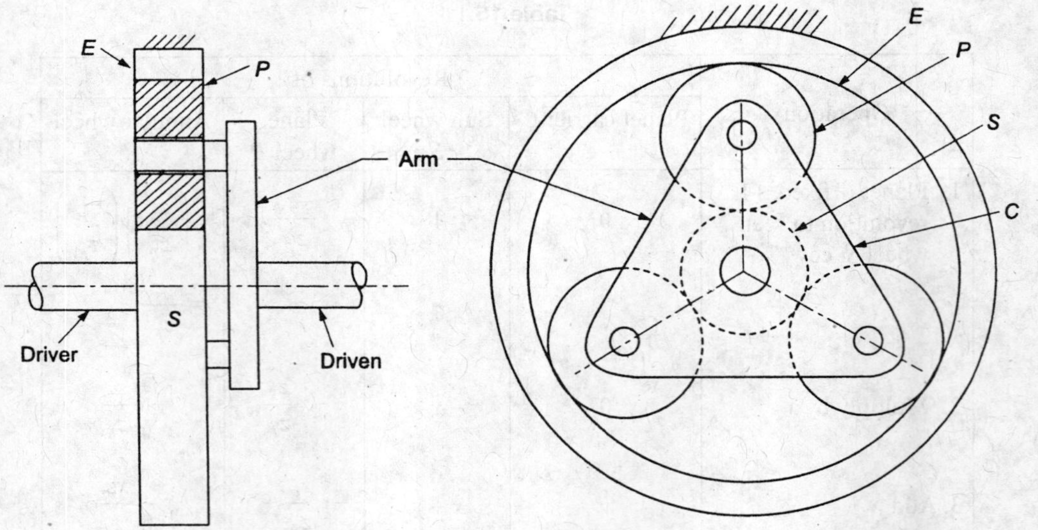

Fig.15.14 Epicyclic gear train

Example 15.11

In a reverted epicyclic gear train shown in Fig.15.15, the arm F carries two wheels A and D and a compound wheel B, C. The wheel A meshes with wheel B and the wheel D meshes with wheel $C.z_A = 80$, $z_D = 48$ and $z_C = 72$. Find the speed and direction of wheel D when wheel A is fixed and arm F makes 240 rpm clockwise.

■ Solution

$$z_A + z_B = z_C + z_D$$
$$z_B = 72 + 48 - 80 = 40$$

Table 15.12 can be used to find the speed of wheel D.

Table 15.12

Operation	Revolutions of			
	Arm F	Wheel A 80	Compound Wheel B, C, 40, 72	Wheel D 48
1. Arm F fixed, +1 revolution to wheel A, ccw.	0	+1	$\frac{-z_A}{z_B} = \frac{-80}{40}$ $= -2$	$\frac{z_A}{z_B} \times \frac{z_C}{z_D}$ $= \frac{80}{40} \times \frac{72}{48}$ $= 3$
2. Multiply by x	0	$+x$	$-2x$	$3x$
3. Add y	y	$y + x$	$y - 2x$	$y + 3x$

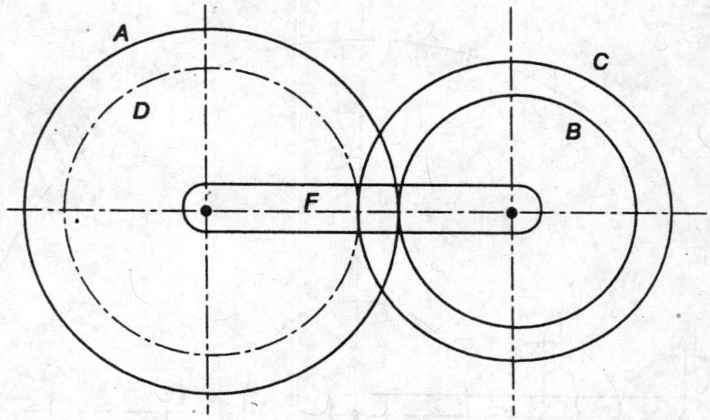

Fig.15.15 Reverted epicyclic gear train

$$y = -240$$
$$x + y = 0$$
$$x = -y = 240 \text{ rpm}$$
$$N_D = y + 3x = -249 + 3 \times 240 = 480 \text{ rpm ccw}$$

Example 15.12

An epicyclic gear train shown in Fig.15.16 is composed of a fixed annular wheel A having 150 teeth. $z_B = 25$, $z_D = 40$ and C is an idle gear. Gear D is concentric with gear A. Wheels B and C are carried on an arm E which revolves clockwise at 120 rpm about the axis of A. Find the number of teeth of gear C and its speed and sense of rotation.

■ **Solution**

$$\frac{d_A}{2} = d_B + d_C + \frac{d_D}{2}$$
$$\frac{z_A}{2} = z_B + z_C + \frac{z_D}{2}$$
$$\frac{150}{2} = 25 + z_C + \frac{40}{2}$$
$$z_C = 75 - 25 - 20 = 30$$

Table 15.13 can be used to find the speed of gear C.

$$y = -120$$
$$x + y = 0; \quad x = 120$$
$$N_C = -5x + y = -5 \times 120 - 120 = -720 \text{ rpm. i.e 720 rpm cw}$$

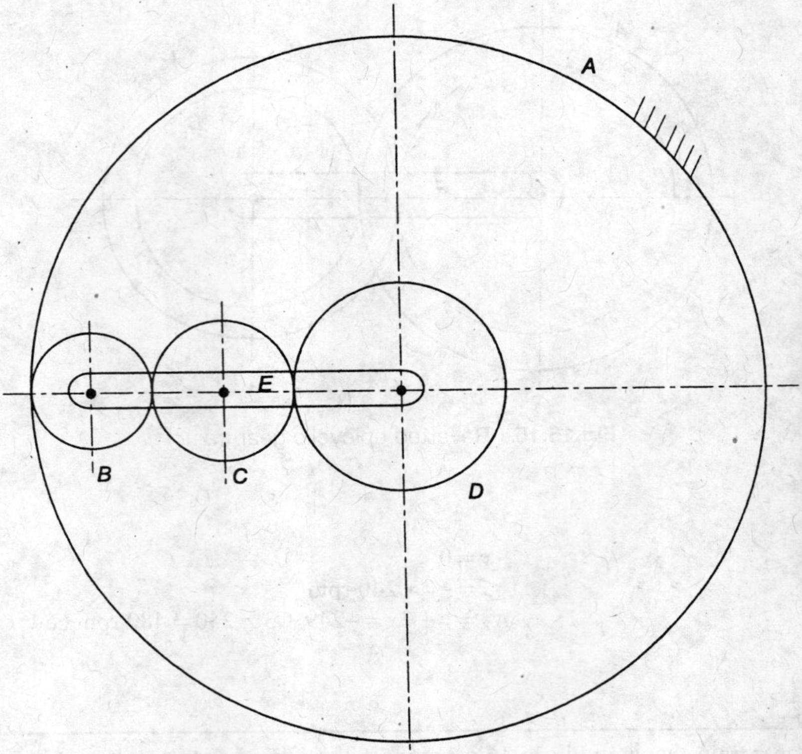

Fig.15.16 Epicyclic gear train

Table 15.13

Operation	Revolutions of				
	Arm E	Gear A, 150	Gear B, 25	Gear C, 30	Gear D, 40
1. Arm E fixed, +1 revolution given to wheel A, ccw.	0	+1	$+\frac{z_A}{z_B}$ $=+\frac{150}{25}$ $=+6$	$-\frac{z_A}{z_c}$ $=-\frac{150}{30}$ $=-5$	$+\frac{z_A}{z_D}$ $=+\frac{150}{40}$ $=+\frac{15}{4}$
2. Multiply by x	0	$+x$	$+6x$	$-5x$	$+15x/4$
3. Add y	y	$y+x$	$y+6x$	$y-5x$	$y+\frac{15x}{4}$

Example 15.13

For the compound epicyclic gear train shown in Fig.15.17, $z_A = 60$, $z_B = 40$, and $z_C = 25$. Find z_D and the speed of shaft connected to arm E,if the speed of shaft connected to sun gear is 120 rpm ccw and gear D is fixed.

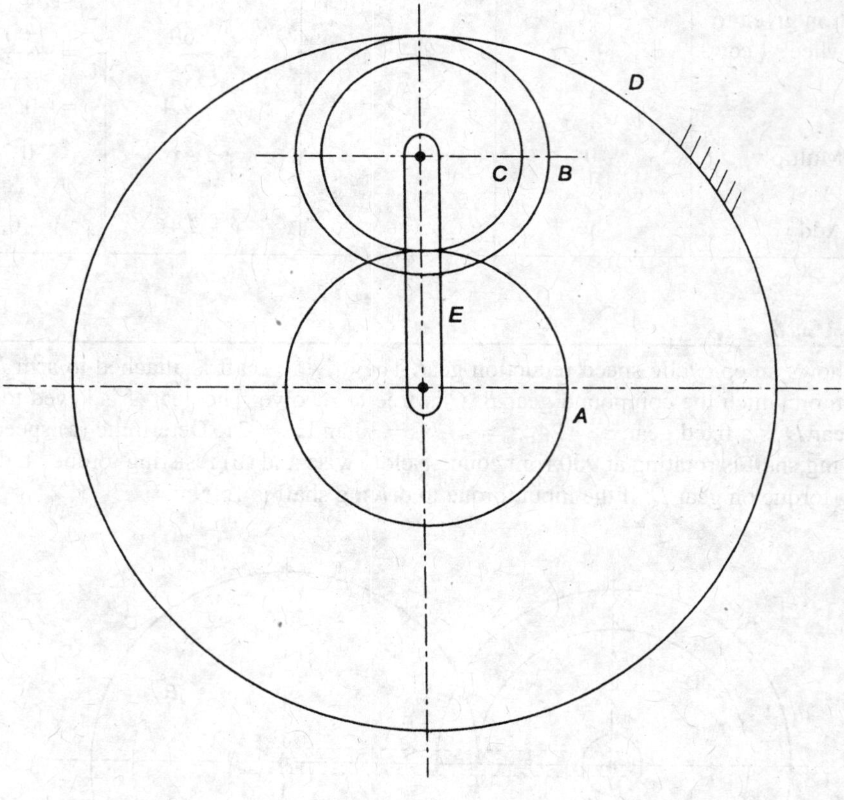

Fig.15.17 Compound epicyclic gear train

Solution

$$d_D = d_A + d_C + d_B$$
$$z_D = z_A + z_C + z_B$$
$$= 60 + 25 + 40 = 125$$

Table 15.14 can be used to find the speed of shaft connected to arm E.

$$y - 0.768x = 0$$
$$x + y = 0$$
$$x = 67.87 \text{ rpm}$$
$$y = 52.13 \text{ rpm ccw}$$

Table 15.14

Operation	Revolutions of			
	Arm E	Gear A, 60	Gear B,C 40, 25	Gear D,40
1. Arm E fixed, +1 revolution given to wheel A ccw.	0	+1	$\frac{-z_A}{z_C}$ $= -\frac{60}{25}$ $= -2.4$	$\frac{-z_A}{z_C} \times \frac{z_B}{z_D}$ $= -\frac{60}{25} \times \frac{40}{125}$ $= -0.768$
2. Multiply by x	0	$+x$	$-2.4x$	$-0.768x$
3. Add y	y	$y+x$	$y-2.4x$	$y-0.768x$

Example 15.14

Fig.15.18 shows an epicyclic speed reduction gear. The driving shaft is attached to arm E. The arm carries a pin on which the compound gear B, C is free to revolve. The gear A is keyed to the driven shaft and gear D is a fixed gear, $z_A = 24$, $z_B = 27$, $z_C = 30$ and $z_D = 21$. Determine (a) speed of driven shaft if driving shaft is rotating at 900 rpm counter-clockwise and (b) resisting torque on driven shaft and holding torque on gear D, if the input torque to driving shaft is 15 Nm.

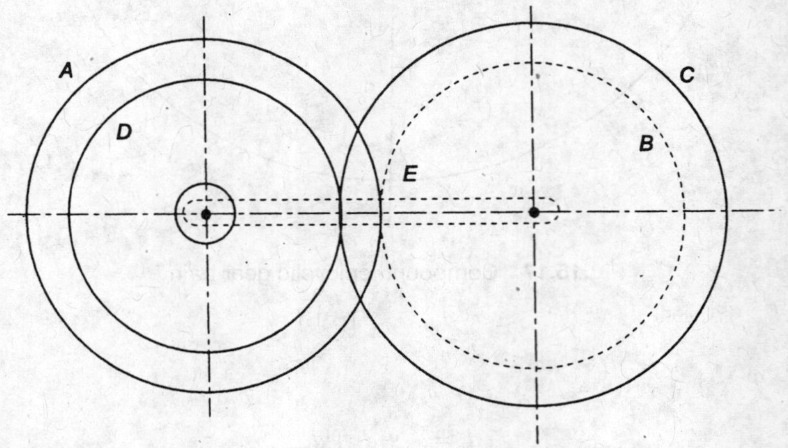

Fig.15.18 Epicyclic speed reduction gear

Solution

Table 15.15 can be used to find the speed of the gears.

(a)

$$N_D = 0,$$

$$y + \frac{80x}{63} = 0$$

$$N_E = y = 900$$

$$900 + \frac{89x}{63} = 0$$

$$x = -708.75$$

$$N_A = x = y = -708.75 + 900 = 191.25 \text{ rpm ccw}$$

(b)

$$M_1 + M_2 + M_3 = 0$$

$$M_1\,\omega_1 + M_2\,\omega_2 + M_3\,\omega_3 = 0$$

For the fixed member,

$$M_3 = 0$$

$$M_1\,\omega_1 + M_2\,\omega_2 = 0$$

Resisting torque,

$$M_2 = \frac{-M_1\omega_1}{\omega_2} = -15 \times \frac{900}{191.25} = -70.59 \text{ Nm}$$

Holding torque,

$$M_3 = -M_1 - M_2 = -15-(-70.59) = 55.59 \text{ Nm}$$

Table 15.15

Operation	Revolutions of			
	Arm E	Gear A, 60	Gears B, C 40, 25	Gear D, 40
1. Arm E fixed, +1 revolution given to wheel A ccw.	0	+1	$\frac{-z_A}{z_B}$ $= -\frac{24}{27}$ $= -\frac{8}{9}$	$\frac{-z_A}{z_B} \times \left(\frac{-z_C}{z_D}\right)$ $= -\left(\frac{8}{9}\right) \times \left(-\frac{30}{21}\right)$ $= \frac{80}{63}$
2. Multiply by x	0	$+x$	$-\frac{8x}{9}$	$\frac{80x}{63}$
3. Add y	y	$Y + x$	$y - \frac{8x}{9}$	$y + \frac{80x}{63}$

Example 15.15

In Fig.15.19, pinion A having 15 teeth is fixed to motor shaft. $z_B = 20$, $z_C = 15$, where B and C are a compound gear wheel. Wheel E is keyed to the machine shaft. Arm F rotates about the same shaft on which A is fixed and carries the compound wheel B, C. If the motor runs at 1200 rpm counter-clockwise, find (a) the speed of the machine shaft, (b) the torque exerted on the machine shaft if the motor develops a troque of 1200 Nm with an efficiency of 95%, and (c) ratio of the reduction gear.

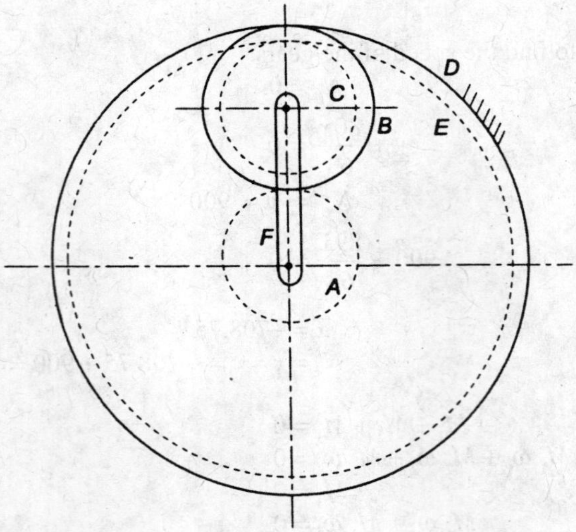

Fig.15.19 Mechanism with a compound gear wheel

Table 15.16

Operation	Revolutions of				
	Arm F	Gear A, 15	Gear B, C 20, 15	Gear E, 50	Gear D, 55
1. Arm F fixed, +1 revolution given to wheel A ccw.	0	+1	$\frac{-z_A}{z_B}$ $= -\frac{15}{20}$ $= -\frac{3}{4}$	$\frac{-z_A}{z_B} \times \left(\frac{z_C}{z_E}\right)$ $= -\left(\frac{3}{4}\right) \times \left(\frac{15}{50}\right)$ $= -\frac{9}{40}$	$\frac{-z_A}{z_B} \times \frac{z_B}{z_D}$ $= \frac{-z_A}{z_D} = -\frac{15}{55}$ $= -\frac{3}{11}$
2. Multiply by x	0	$+x$	$-\frac{3x}{4}$	$-\frac{9x}{40}$	$-\frac{3x}{11}$
3. Add y	y	$y + x$	$y - \frac{3x}{4}$	$y - \frac{9x}{40}$	$y - \frac{3x}{11}$

■ **Solution**

$$d_D = 2d_B + d_A$$
$$z_D = 2z_B + z_A = 2 \times 20 + 15 = 55$$
$$d_E = d_A + d_B + d_C$$
$$z_E = z_A + z_B + z_C = 15 + 20 + 15 = 50$$

Table 15.16 can be used to find the speed of the gears.

(a)

$$N_A = x + y = 1200$$

$$N_D = y - \frac{3x}{11}$$

$$x = 942.86 \text{ rpm}$$

$$y = 257.14 \text{ rpm}$$

$$N_E = y - \frac{9x}{40} = 257.14 - 9 \times \frac{942.86}{40} = 45 \text{ rpm}$$

(b)

$$M_2 = -M_1\omega_1 \frac{\eta}{\omega_2} = -120 \times 1200 \times \frac{0.95}{45} = -3040 \text{ Nm}$$

(c) Ratio of reduction gear

$$= \frac{N_A}{N_E} = \frac{1200}{45} = 26.67$$

Example 15.16

An epicyclic gear train consists of sun wheel *S,* a fixed internal gear *E* and three identical planet wheels *P* carried on a star-shaped planet carrier *C,* as shown in Fig.15.20. The planets rotate at 1/5th of the speed of sun wheel. The minimum number of teeth on any wheel is 18. The driving torque on the sun wheel is 120 Nm. Find (a) number of teeth on different wheels of the train and (b) torque necessary to keep the internal gear stationary.

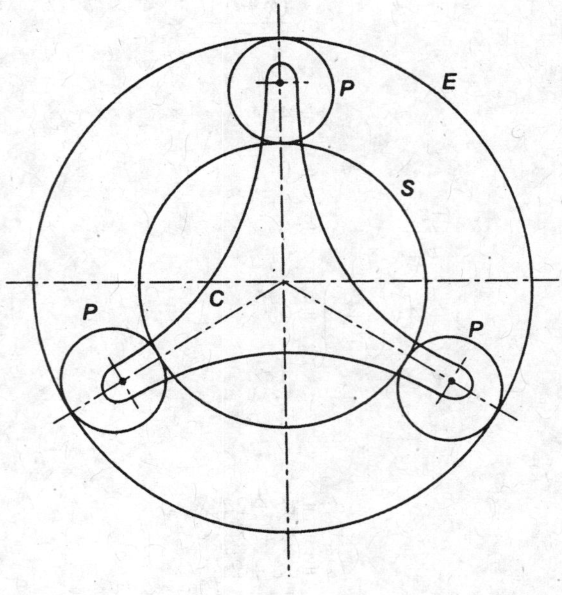

Fig.15.20 Epicyclic gear train

Solution

Table 15.17 can be used to find the speed of the gears.

Table 15.17

Operation	Revolutions of			
	Planet carrier C	Sun gear S	Planet gear P	Annular gear E
1. Planet carrier fixed, +1 revolutions to S, ccw	0	+1	$\frac{-z_s}{z_p}$	$\frac{-z_s}{z_E}$
2. Multiply by x	0	$+x$	$\frac{-z_s}{z_p}\cdot x$	$\frac{-z_s}{z_E}\cdot x$
3. Add y	y	$x+y$	$\frac{-z_s}{z_p}\cdot x+y$	$\frac{-z_s}{z_E}\cdot x+y$

(a)

$$N_C = \frac{N_S}{5}$$

$$y = \frac{(x+y)}{5}$$

or

$$x = 4y$$

$$N_E = 0$$

$$\frac{-z_S}{z_E}\cdot x + y = 0$$

$$y - 4y\left(\frac{z_S}{z_E}\right) = 0$$

$$\frac{z_S}{z_E} = \frac{1}{4}$$

Let $z_S = 18$, then $z_E = 72$

$$d_E = d_s + 2d_P$$

$$z_E = z_s + 2z_P$$

$$72 = 18 + 2z_P$$

$$z_P = 27$$

(b)
$$M_2 = -M_1 \frac{\omega_1}{\omega_2} = -5M_S = -5\times 120 = -600 \text{ Nm}$$
$$M_3 = -M_1 - M_2 = -120 + 600 = 480 \text{ Nm}$$

Example 15.17

Fig.15.21 shows the arrangement of wheels in a compound epicyclic gear train. The sun wheel S_2 is integral with the annular wheel A_1. The two arms are also integral with each other. $z_{s1} = z_{s2} = 24$, $z_{A1} = z_{A2} = 96$.

(a) If the shaft X rotates at 2000 rpm, find the speed of shaft Y, when A_2 is fixed.

(b) At what speed does Y rotate when A_2 rotates at 200 rpm, in the same direction as S_1, which is rotating at 2000 rpm.

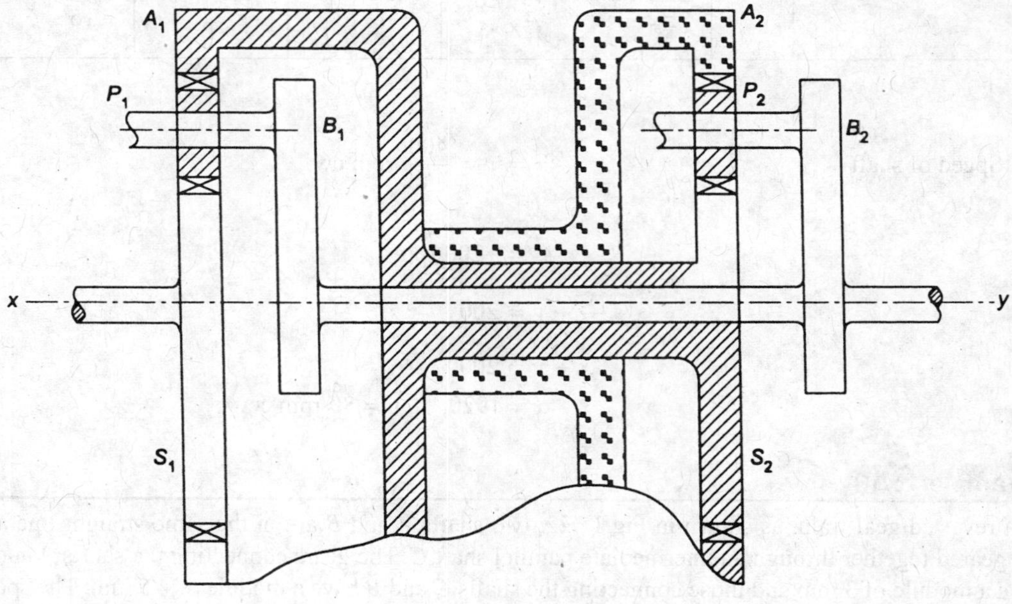

Fig.15.21 Compound epicyclic gear train

■ **Solution**

Table 15.18 can be used to find the speed of the gears.

(a) A_2 fixed,
$$\frac{x}{16} + y = 0$$
$$x = -16y$$
$$x + y = 2000$$
$$y = -\frac{400}{3} \text{ rpm}$$

Table 15.18

Operation	Revolutions of						
	Arm	S_1	P_1	A_1	S_2	P_2	A_2
1. Fix arm B_1, +1 revolution to S_1 ccw	0	+1	$\frac{-24}{z_{P1}}$	$\frac{-24}{96}$ $= -1/4$	$\frac{-24}{96}$ $= -1/4$	$\frac{+24}{96} \times \frac{24}{z_{P2}}$ $= 6/z_{P2}$	$\frac{+24}{96} \times \frac{24}{96}$ $= \frac{+1}{16}$
2. Multiply by x	0	$+x$	$\frac{-24x}{z_{p1}}$	$\frac{-x}{4}$	$\frac{-x}{4}$	$\frac{6x}{z_{p2}}$	$\frac{+x}{16}$
3. Add y	y	$y+x$	$\frac{-24x}{z_{p1}} + y$	$\frac{-x}{4} + y$	$\frac{-x}{4} + y$	$\frac{6x}{z_{p2}} + y$	$\frac{+x}{16} + y$

Speed of shaft $$Y = y = \frac{400}{3} \text{ rpm cw}$$

(b) $$A_2 = +200 \text{ rpm}$$

$$\frac{x}{16} + y = 200$$

$$x + y = 2000$$

$$x = 1920; \quad y = 80 \text{ rpm ccw}$$

Example 15.18

In a reverted gear train, as shown in Fig.15.22, two shafts A and B are in the same straight line and are geared together through an intermediate parallel shaft C. The gears connecting the shafts A and C have a module of 3 mm and those connecting the shafts C and B have a module of 4.5 mm. The speed of shaft A is to be about but greater than 12 times the speed of shaft B. The ratio of each reduction is same. Find suitable number of teeth on all gears. The minimum number of teeth is 18. Also find the exact velocity ratio and the distance of shaft C from A and B.

■ **Solution**

Given: $m_1 = m_2 = 3$ mm, $m_3 = m_4 = 4.5$ mm, $z_{min} = 18$, $n_A > 12n_B$

Now $$\frac{n_1}{n_2} = \frac{n_3}{n_4}$$

But $$n_2 = n_3$$

Therefore, $$n_2^2 = n_1\, n_4 \text{ or } n_2 = \sqrt{n_1 n_4} = \sqrt{12}\, n_4$$

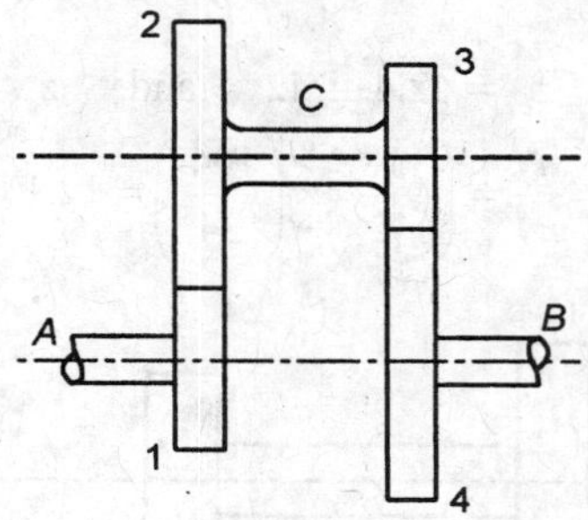

Fig.15.22 Reverted gear train

$$\frac{n_2}{n_4} = \sqrt{12} = 3.464 \cong 3.5$$

For same centre distance,

$$3(z_1 + z_2) = 4.5\,(z_3 + z_4)$$
$$(z_1 + z_2) = 1.5\,(z_3 + z_4)$$
$$z_1\,(1 + 3.5) = 1.5\,z_3\,(1 + 3.5)$$
$$z_1 = 1.5\,z_3$$

Let $z_3 = 20$, then $z_4 = 70$, $z_1 = 30$, $z_2 = 105$

$$\frac{n_A}{n_B} = (3.5)^2 = 12.25$$

Centre distance, $C = 3(30 + 105) = 405$ mm

Example 15.19

The speed ratio of the reverted gear train shown in Fig.15.23 is to be 12. The module of gears *A* and *B* is 3.125 mm and of gears *C* and *D* is 2.5 mm. Calculate the suitable number of teeth for the gears. No gear is to have less than 24 teeth.

■ Solution

Given: $m_a = m_b = 3.125$ mm, $m_c = m_d = 2.5$ mm, $z_{min} = 24$

$$3.125\,(z_a + z_b) = 2.5\,(z_c + z_d) \quad (1)$$
$$(z_c + z_d) = 1.25\,(z_a + z_b)$$
$$3.125\,(z_a + z_b)/2 = 200 \quad (2)$$
$$z_a + z_b = 128$$

Let $z_a = 24$, so that $z_b = 104$

Eq. (1) becomes,

$$z_c + z_d = \frac{200}{1.25} = 160 \quad (3)$$

$$\frac{n_a}{n_d} = 12 = \frac{z_d}{z_c} \times \frac{z_b}{z_a}$$

$$\frac{z_d}{z_c} \times \frac{104}{24} = 12$$

$$\frac{z_d}{z_c} = 2.77$$

From Eq. (3), we get

$$z_c = 42.4 \cong 42 \quad \text{and} \quad z_d = 118$$

$$\frac{n_a}{n_d} = \left(\frac{118}{42}\right) \times \left(\frac{104}{24}\right) = 12.17 \cong 12$$

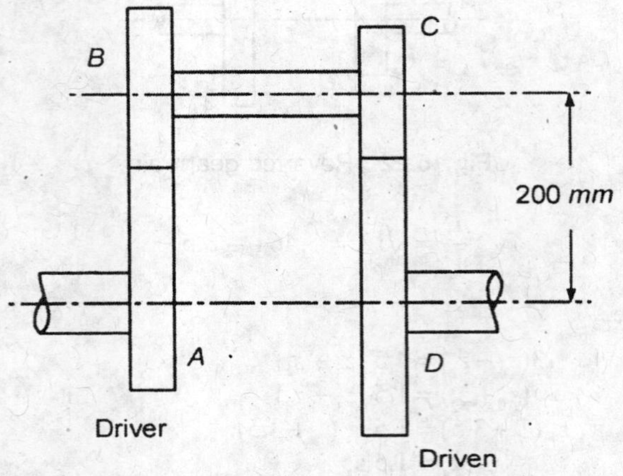

Fig.15.23 Reverted gear train

Example 15.20

Two shafts *x* and *y* are in the same line (axis in one line). They are geared together through an intermediate parallel shaft carrying wheels *B* and *C* which mesh with the wheels on *A* and *D* respectively, as shown in Fig.15.24. Wheels *A* and *B* have a module of 4 mm and the wheels *C* and *D* have a module of 9 mm. The number of teeth on any wheel is not to be less than 15 and the speed of *D* is to be about, but not greater than 1/12 the speed of *A* and the ratio of each reduction is the same. Find (a) suitable number of teeth for the wheels, (b) the actual reduction, and (c) the distance of the intermediate shaft from the axes of the shafts A and *D* (centre distance), (d) Indicate the configuration with a sketch, (e) How is addendum modification related to correction of gears and when are they used in practice? Gear *A* is on shaft *A* and gear *D* on shaft *D*)

■ Solution

Given: $m_a = m_b = 4$ mm, $m_c = m_d = 9$ mm, $z_{\min} = 15$, $\dfrac{n_a}{n_d} = 12$

Now $\dfrac{n_a}{n_b} = \dfrac{n_c}{n_d}$

But $n_b = n_c$

Therefore, $n_b^2 = n_c^2 = n_a n_d \quad \text{or} \quad n_b = n_c = \sqrt{n_a n_d} = \sqrt{12}\, n_d$

$$\frac{n_c}{n_d} = \frac{n_b}{n_d} = \sqrt{12} = 3.464$$

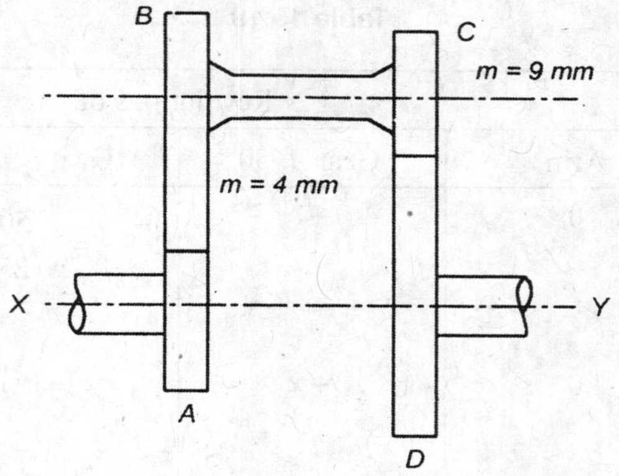

Fig.15.24 Reverted gear train

For same centre distance,

$$4(z_a + z_b) = 9(z_c + z_d)$$
$$(z_a + z_b) = 2.25(z_c + z_d)$$
$$z_a(1 + 3.5) = 2.25z_c(1 + 3.5)$$
$$z_a = 2.25z_c$$

Let $$z_c = 15, \text{ then } z_a = 33.75 \cong 34$$

Now $$\frac{z_d}{z_c} = 3.5, \; z_a = 51.96 \cong 52$$

$$34 + z_b = 2.25(15 + 52)$$
$$z_b = 116.75 = 117$$
$$\frac{n_A}{n_B} = (3.464)^2 = 12, C = 4(34 + 117) = 604 \text{ mm}$$

Example 15.21

In an epicyclic gear train, as shown in Fig.15.25, the number of teeth on wheels *A, B,* and *C* are 50, 25, and 52 respectively. If the arm rotates at 420 rpm cw, find (a) speed of wheel *C* when *A* is fixed, and (b) speed of wheel *A* when *C* is fixed.

■ Solution

Given: $z_A = 50, z_B = 25, z_C = 52$

Table 15.19 may be used to find the speed of gears.

(a) When gear *A* is fixed, $x + y = 0$

Now $y = 420$ rpm

Therefore, $x = -420$ rpm

$$n_c = y + \frac{50x}{52} = 420 - 420 \times \frac{50}{52} = 16.15 \text{ rpm}$$

(b) When *C* is fixed,

Table 15.19

Operation	Revolutions of			
	Arm	Gear A, 50	Gear B, 25	Gear C, 52
1. Arm fixed, +1 revolutions given to gear A, ccw	0	+1	$\frac{-z_a}{z_b} = -\frac{50}{25} = -2$	$+\frac{z_a}{z_c} = +\frac{50}{52}$
2. Multiply by x	0	$+x$	$-2x$	$+\frac{50x}{52}$
3. Add y	y	$y+x$	$y-2x$	$y+\frac{50x}{52}$

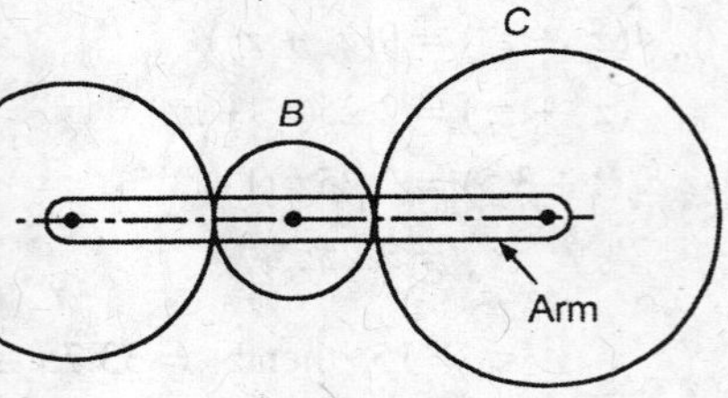

Fig.15.25 Epicyclic gear train

$$n_c = y + x\frac{z_a}{z_c} = 0$$

or $$420 + \frac{50x}{52} = 0$$

or $$x = -436.8 \text{ rpm}$$

$$n_a = x = y = -436.8 + 420 = -16.8 \text{ rpm.}$$

Example 15.22

An epicyclic, gear train, as shown in Fig.15.26 is composed of a fixed annular wheel A having 150 teeth. The wheel A is meshing with wheel B which drives wheel D through an idle wheel C, D being concentric with A. The wheels B and C are carried on an arm which revolves clockwise at 100 rpm about the axis of A and D. If the wheels B and D have 25 teeth and 40 teeth respectively, find the number of teeth on C and the speed and sense of rotation of C.

■ Solution

Given: $z_b = 25,\ z_d = 40$

Table 15.20 is used to find the speed of gears.

For A to be fixed, $x + y = 0,\ y = -100 \text{ rpm},\ x = +100 \text{ rpm}$

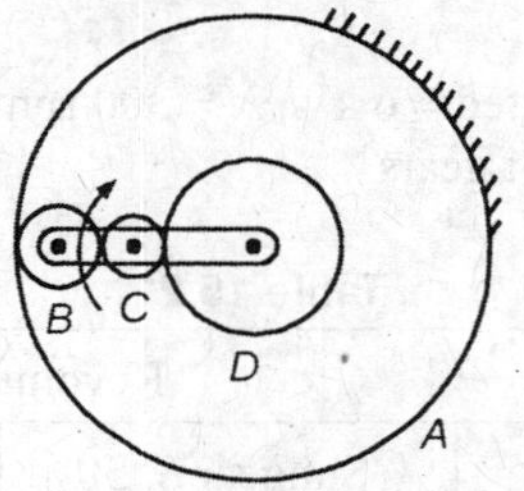

Fig.15.26 Epicyclic gear train

Table 15.20

Operation	Revolutions of				
	Arm	Gear *A*, 150	Gear *B*, 25	Gear *C*	Gear *D*, 40
1. Arm fixed, +1 revolutions given to gear A, ccw	0	+1	$+\frac{z_a}{z_b} = \frac{150}{25} = +6$	$-\frac{z_a}{z_c} = -\frac{150}{z_c}$	$+\frac{z_a}{z_d} = \frac{150}{40} = \frac{+15}{4}$
2. Multiply by *x*	0	$+x$	$+6x$	$\frac{-150x}{z_c}$	$\frac{-15x}{4}$
3. Add *y*	y	$y+x$	$y+6x$	$y-\frac{150x}{z_c}$	$y+\frac{15x}{4}$

Now $$\frac{d_a}{2} = d_b + d_c + \frac{d_d}{2}$$

For same module, $$\frac{z_a}{2} = z_b + z_c + \frac{z_d}{2}$$

$$\frac{150}{2} = 25 + z_c + \frac{40}{2}$$

$$z_c = 30$$

$$n_c = -100 - 150 \times \frac{100}{30} = -600 \text{ rpm, i.e., } 600 \text{ rpm cw.}$$

Example 15.23

An epicyclic gear train, as shown in Fig.15.27, has a sun wheel *S* of 30 teeth and two planet wheels *P* of teeth 50 each. The planet wheels mesh with the teeth of internal gear *A*. The driving shaft carrying the sun wheel transmits 6 kW at 300 rpm. The driven shaft is connected to an arm which carries the planet wheels. Determine the speed of the driven shaft and the torque transmitted, if the overall efficiency is 95%.

■ Solution

Given: $z_s = 30$, $z_p = 50$, power transmitted = 6 kW, $N = 300$ rpm, efficiency = 0.95
Table 15.21 is used to find the speed of gears.

Table 15.21

Operation	Revolutions of			
	Arm	Sungear *S*, 30	Planet *P*, 50	Annular *A*
1. Arm fixed, +1 revolutions given to gear *S*, ccw	0	+1	$\frac{-z_s}{z_p} = \frac{-30}{50} = \frac{-3}{5}$	$\frac{-z_s}{z_a}$
2. Multiply by *x*	0	$+x$	$\frac{-3x}{5}$	$-x\frac{z_s}{z_a}$
3. Add *y*	y	$y+x$	$y-\frac{3x}{5}$	$y-x\frac{z_s}{z_a}$

$$n_s = x + y = 300 \quad (1)$$

Now $\quad d_a = 2d_p + d_s$

For same module, $\quad z_a = 2z_p + z_s$

$$=100 = 30 = 130$$

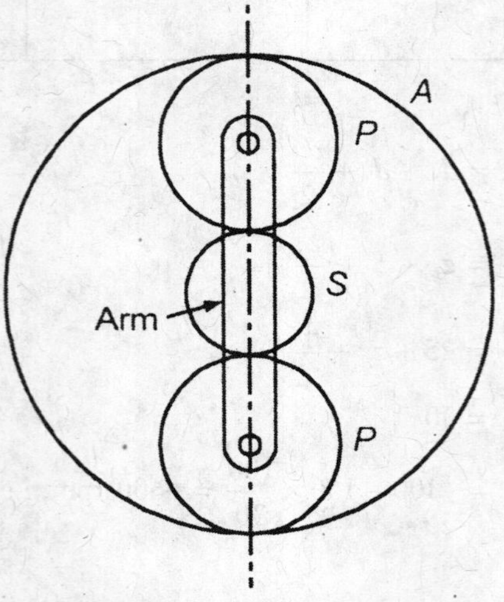

Fig.15.27 Epicyclic gear train

For to be fixed, $\quad \frac{-xz_s}{z_a} + y = 0$

$$= \frac{-3x}{13} + y = 0 \quad (2)$$

From (1) and (2), we get

$$x = \frac{975}{4},\ y = \frac{225}{4}$$

Speed of driven shaft (arm) $= 56.25$ rpm

$$T_s = 6 \times 10^3 \times \frac{60}{(2\pi \times 300)} = 190.986 \text{ N m}$$

$$T_a = -T_s n_s \frac{\eta}{n_a} = 190.986 \times 300 \times \frac{0.95}{56.25} = -967.7 \text{ N m.}$$

Example 15.24

A compound epicyclic gear train is shown in Fig.15.28. The gears A, D and E are free to rotate on the axis P. The compound gear B and C rotate together on the axis Q at the end of arm F. All the gears have equal module. The number of teeth on the gears A, B and C are 18, 45 and 21 respectively. The gears D and E are annular gears. The gear A rotates at 120 rpm ccw and the gear D rotates at 450 rpm cw. Find the speed and direction of the arm and the gear E.

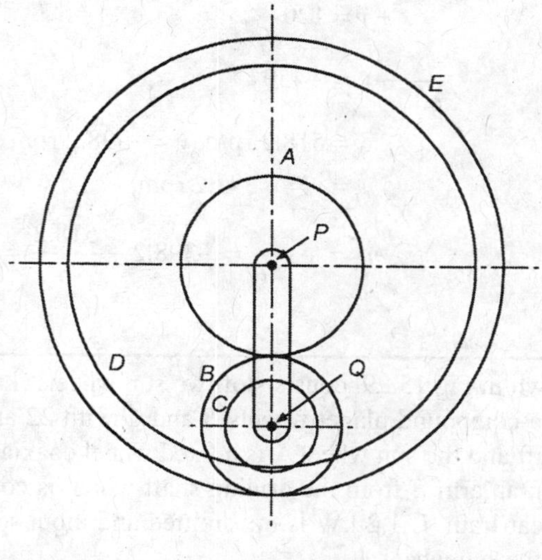

Fig.15.28 Compound epicyclic gear train

■ **Solution**

Given: $z_a = 18$, $z_b = 45$, $z_c = 21$, $n_a = 120$ rpm ccw, $n_d = 450$ rpm cw

$$\frac{z_e}{2} = \frac{z_a}{2} + z_b$$

$$z_e = z_a + 2\,z_b = 18 + 90 = 108$$

$$z_e = z_a + z_b + z_c = 18 + 45 + 21 = 84$$

Table 15.22 is used to find the speed of gears.

Table 15.22

Operation	Revolutions of				
	Arm	Gear *A, 18*	Gear *B, C, 45, 21*	Gear *D, 84*	Gear *E, 108*
1. Arm fixed, + 1 revolutions given to gear *A*, ccw	0	+1	$+\frac{z_a}{z_b}=\frac{-18}{45}=\frac{-2}{5}$	$\frac{-z_a}{z_b}\times\frac{z_c}{z_c}$ $=\frac{-18}{45}\times\frac{21}{84}$ $=\frac{-1}{10}$	$\frac{-z_a}{z_b}\times\frac{z_b}{z_e}$ $=\frac{-18}{45}\times\frac{45}{108}$ $=\frac{-1}{6}$
2. Multiply by *x*	0	$+x$	$=\frac{-2x}{5}$	$\frac{-x}{10}$	$\frac{-x}{6}$
3. Add *y*	y	$y+x$	$y-\frac{2x}{5}$	$y-\frac{x}{10}$	$y-\frac{x}{6}$

$$x+y=120$$

$$\frac{-x}{10}+y=-450$$

$$x=518.2 \text{ rpm}, y=-398.2 \text{ rpm}$$

$$n_f=y=-398.2 \text{ rpm, } i.e., \text{ cw}$$

$$n_e=y-\frac{x}{6}=-398.2-\frac{518.2}{6}=-484.57 \text{ rpm, i.e., cw}$$

Example 15.25

An epicyclic gear train shown in Fig.15.29, consists of two sun wheels *A* and *D* with 28 and 24 teeth respectively, engaged with a compound planet wheels *B* and *C* with 22 and 26 teeth. The sun wheel *D* is keyed to the driven shaft and the sun wheel A is a fixed wheel coaxial with the driven shaft. The planet wheels are carried on an arm *E* from the driving shaft which is coaxial with the driven shaft. Find the velocity ratio of gear train. If 1.2 kW is transmitted and input speed is 120 rpm, determine the torque required to hold the sun wheel *A*.

Solution

Given: $z_a=28, z_d=24, z_b=22, z_c=26$, power transmitted = 1.2 kW, $N=120$ rpm
Table 15.23 is used to find the speed of gears.

$$y=+120 \text{ rpm}$$

$$n_d=y+\frac{91x}{66}=120+\frac{91x}{66}=0$$

$$x=-87 \text{ rpm}$$

$$n_a=x+y=-87=120=33 \text{ rpm}$$

Table 15.23

Operation	Revolutions of			
	Arm, E	Gear A, 28	Gear B, C, 22, 26	Gear D, 24
1. Arm E fixed, +1 revolutions given to gear A, ccw	0	+1	$-\frac{z_a}{z_b} = \frac{-28}{22} = \frac{-14}{11}$	$+\frac{z_a}{z_b} \times \frac{z_c}{z_d} = \frac{28}{22} \times \frac{26}{24} = \frac{91x}{66}$
2. Multiply by x	0	$+x$	$\frac{-14x}{11}$	$\frac{91x}{66}$
3. Add y	y	$y+x$	$y - \frac{14x}{11}$	$y + \frac{91x}{66}$

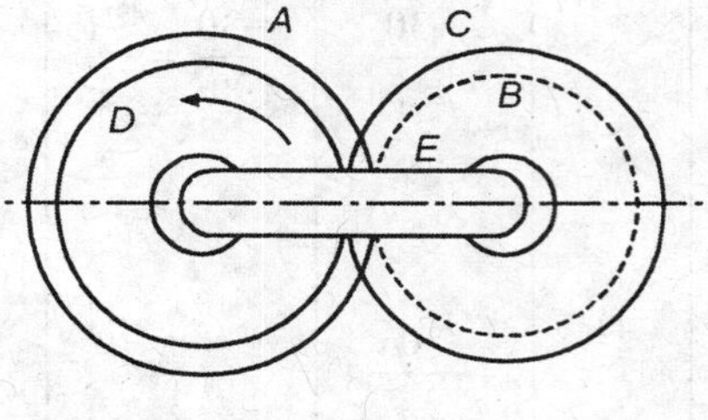

Fig.15.29 Epicyclic gear train

Speed of driven shaft (gear A) = 33 rpm ccw

$$\text{Speed ratio} = \frac{n_d}{n_a} = \frac{87}{33} = 2.637$$

$$M_1 = \frac{1.2 \times 10^3 \times 60}{(2\pi \times 120)} = +95.5 \text{ Nm}$$

$$\omega_1 = \frac{2\pi \times 120}{60} = 12.57 \text{ rad/s,}$$

$$\omega_2 = \frac{2\pi \times 33}{60} = 3.456 \text{ rad/s}$$

$$M_2 = M_1 \times \frac{\omega_1}{\omega_2} = 95.5 \times \frac{12.7}{3.456} = 347.35 \text{ Nm}$$

Holding torque, $M_3 = -M_1 - M_2 = -95.5 - 347.35 = -442.85$ Nm

Example 15.26

An epicyclic gear train is shown in Fig.15.30. The main driving shaft G has a gear S_1 integrally mounted and driving the internal gear A_1 on the casing through two intermediate gears P_1 mounted on either side. The gears P, are free to revolve on arms R, which are integral with gear S_2 which in turn drives the internal gear A_2 on another casing through two gears P_2. The driven shaft H is integral with the casing carrying the internal gear A_1 and arms R_2 on which the gears P_2 are free to rotate. The casing A_2 and gear S_2 are free to rotate on shaft G.

Calculate the speed of shaft H when G rotates at 1000 rpm anticlockwise when (a) A_2 is stationary; (b) A_2 rotates at 500 rpm clockwise.

The number of teeth on gears are: $S_1 = S_2 = 30$, $A_1 = A_2 = 90$.

Solution

Table 15.24 is used to find the speed of gears.

Table 15.24

Operation	Revolutions of						
	Arm, R	S_1, 30	P_1	A_1, 90	S_2, 30	P_2	A_2, 90
1. Arm R_1 fixed, + 1 revolutions given to S_1, ccw	0	+1	$\frac{-30}{z_{p1}}$	$\frac{-30}{90} = \frac{-1}{3}$	$\frac{-1}{3}$	$\frac{-1}{3} \times \frac{30}{z_{p2}} = \frac{-10}{z_{p2}}$	$\frac{1}{3} \times \frac{1}{3} = \frac{1}{9}$
2. Multiply by x	0	$+x$	$\frac{-30x}{z_{p1}}$	$\frac{-x}{3}$	$\frac{-x}{3}$	$\frac{-10x}{z_{p2}}$	$\frac{x}{9}$
3. Add y	y	$y + x$	$y - \frac{30x}{z_{p1}}$	$y - \frac{x}{3}$	$y - \frac{x}{3}$	$y - \frac{10x}{z_{p2}}$	$y + \frac{x}{9}$

(a) $$A_2 \text{ fixed: } \frac{x}{9} + y = 0$$

$$x + y = 1000$$

$$y = -125$$

Speed of shaft $H = y = 125$ rpm cw

(b) $$\frac{x}{9} + y = -500$$

$$\frac{x}{9} + y = -500$$

$$x + y = 1000$$

$$x = 1687.5 \text{ rpm}, y = -687.5 \text{ rpm}$$

Speed of shaft $H = y = 687.5$ rpm cw.

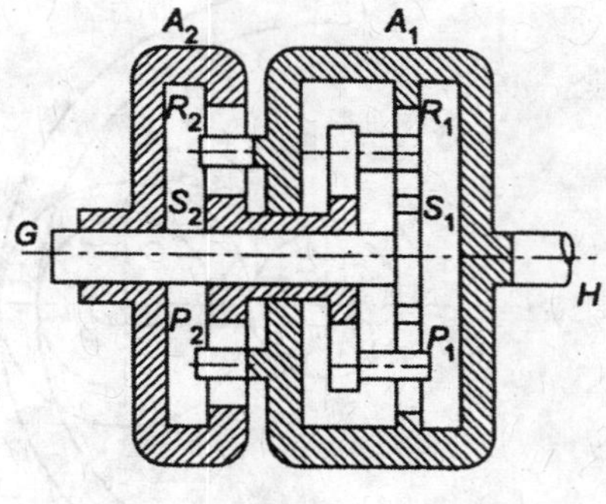

Fig.15.30 Epicyclic gear train

Example 15.27

Fig.15.31 Shows diagrammatically a compound epicyclic gear train. Wheels *A*, *D* and *E* are free to rotate independently on spindle *C*, while *B* and *C* are compound and rotate together on spindle *P*, on the end of arm *OP*. All the teeth on different wheels have the same module. *A* has 12 teeth, *B* has 30 teeth and *C* has 14 teeth cut externally. Find the number of teeth on wheels *D* and *E* which are cut internally.

If the wheel *A* is driven clockwise at 1 rps while *D* is driven clockwise at 5 rps, determine the magnitude and direction of the angular velocities of arm *OP* and wheel *E*.

■ Solution

Given:

$$z_a = 12, z_b = 30, z_c = 14, n_a = 1, n_d = 5$$

$$z_e = z_a + 2z_b = 12 + 60 = 72$$

$$z_d = z_a + z_b + z_c = 12 + 30 + 14 = 56$$

Table 15.25 is used to find the speed of gears

$$x + y = -1$$

$$\frac{-x}{10} + y = -5$$

$$x = \frac{40}{11}, y = -4.636 \text{ rpm}$$

Angular velocity of *OP* $= -2\pi \times \dfrac{4.636}{60} = -0.486$ rad/s cw

Angular velocity of wheel *E*:

$$y - \frac{x}{3} = -4.636 - \frac{3.636}{3} = -5.848 \text{ rpm}$$

$$\omega_e = -2\pi \times \frac{5.848}{60} = -0.612 \text{ rad/s cw}$$

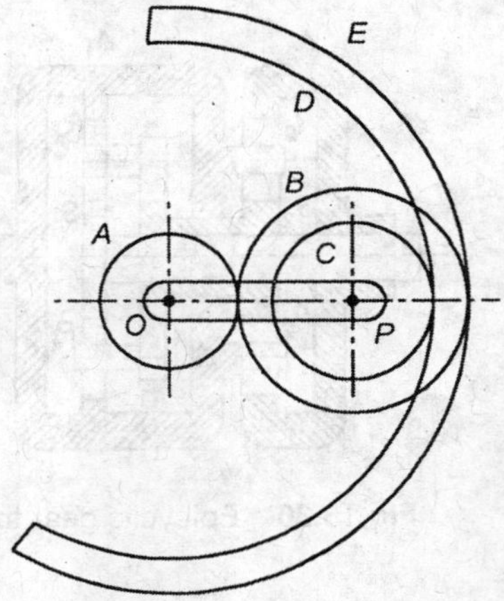

Fig.15.31 Compound epicyclic gear train

Table 15.25

Operation	Revolutions of				
	Arm, OP	Gear A, 12	Gear B, C, 30, 14	Gear D, 56	Gear E, 72
1. Arm fixed, +1 revolutions given to gear A, ccw	0	+1	$-\frac{12}{30}=\frac{2}{5}$	$\frac{-2}{30}\times\frac{14}{56}=\frac{-1}{10}$	$\frac{-24}{30}\times\frac{30}{72}=\frac{-1}{3}$
2. Multiply by x	0	$+x$	$=\frac{-2x}{5}$	$\frac{-x}{10}$	$\frac{-x}{3}$
3. Add y	y	$y+x$	$y-\frac{2x}{5}$	$y-\frac{x}{10}$	$y-\frac{x}{3}$

Example 15.28

A mechanism for recording the distance covered by the bicycle, as shown in Fig.15.32, is as follows:

There is a fixed annular wheel A of 22 teeth and another annular wheel B of 23 teeth, which rotates loosely on the axis of A. An arm driven by the bicycle wheel through gearing not described, also revolves freely on the axis of A and carries on a pin at its extremity two wheels C and D, which are integral with one another. The wheel C has 19 teeth and meshes with A and the wheel D with 20 teeth meshes with B. The diameter of the bicycle wheel is 0.7 m. What must be the velocity ratio between the bicycle wheel and the arm, if B makes one revolution per 1.5 km covered?

Solution

Given: $z_a = 22, z_b = 23; z_c = 19, z_d = 20$

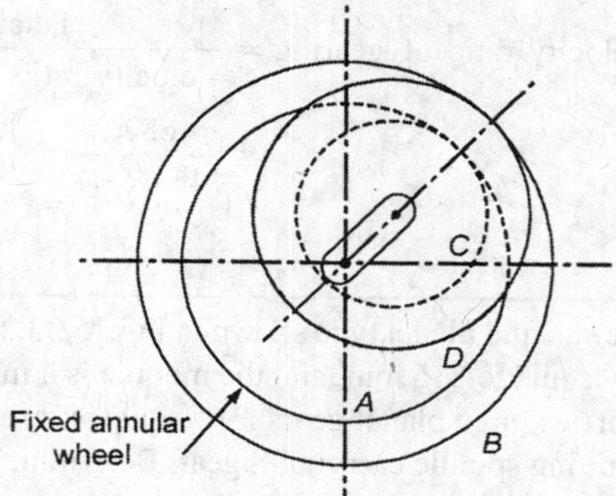

Fig.15.32 Bicycle distance measurment mechanism

Table 15.26 is used to find the speed of gears.

Table 15.26

Operation	Revolutions of			
	Arm, E	Gear A, 22	Gear C, D, 19, 20	Gear B, 23
1. Arm E fixed, +1 revolutions given to gear A, ccw	0	+1	$-\frac{Z_a}{Z_c}=\frac{+22}{19}$	$\frac{+Z_a}{Z_c}\times\frac{Z_d}{Z_b}$ $=\frac{+22}{19}\times\frac{20}{23}$ $=\frac{+440x}{437}$
2. Multiply by x	0	$+x$	$\frac{+22x}{19}$	$\frac{+440x}{437}$
3. Add y	y	$y+x$	$y+\frac{22x}{19}$	$y+\frac{440x}{437}$

A fixed:
$$x+y=0$$

$$\frac{440x}{437}+y=1$$

$$x=\frac{437}{3}=145.67 \text{ revolutions for every 1.5 km}$$

Revolutions made by bicycle wheel per 1.5 km $= 1000\times\frac{1.5}{(\pi\times 1.5)}=682$

$$\text{Velocity ratio of gearing} = \frac{\text{Total velocity ratio}}{\text{Velocity ratio provided by mechanism}}$$

$$= \frac{682}{145.67 \times 1} = 4.682$$

Example 15.29

In an epicyclic gear train of the sun and planet type shown in Fig.15.33, the pitch circle diameter of the annular wheel A is to be nearly equal to 216 mm, and the module is 4 mm. When the annular wheel is stationary, the spider which carries three planet gears P of equal size, has to make one revolution for every five revolutions of the driving spindle carrying S gear. Determine the number of teeth on all the wheels and also the exact pitch circle diameter of A.

Solution

$$d_a = d_s + 2d$$

$$216 = 4(z_s + 2z_p)$$

$$z_s + 2z_p = 54$$

$$d_a = mz_a,\ z_a = \frac{216}{4} = 54 \qquad (1)$$

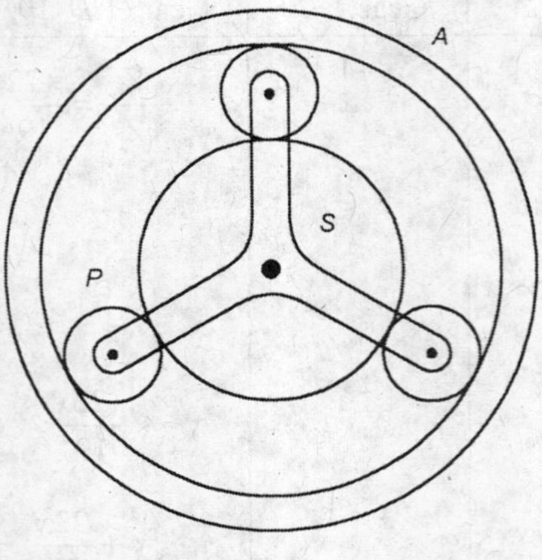

Fig.15.33 Sun and planet gear train

Table 15.27 is used to find the speed of gears.

$$x + y = 5$$

$$y - \frac{xz_s}{z_p} = -1$$

Solving, we get

$$x\left(1 + \frac{z_s}{z_p}\right) = 6 \qquad (2)$$

Also
$$y - \frac{xz_s}{z_a} = 0$$

Table 15.27

Operation	Revolutions of			
	Spider	S	P	A
1. Spider fixed, +1 rev to S ccw	0	+1	$\frac{-z_s}{z_p}$	$\frac{-z_s}{z_a}$
2. Multiply by x	0	$+x$	$\frac{-xz_s}{z_p}$	$\frac{-xz_s}{za}$
3. Add y	y	$y+x$	$y-\frac{xz_s}{z_p}$	$y-\frac{xz_s}{z_a}$

Thus

$$x\left(1+\frac{z_s}{z_a}\right)=5 \tag{3}$$

Solving Eqs. (1), (2), and (3), we get

$$l_c - z_s = 10,\ z_p = 22,\ d_a\ 216 \text{ mm.}$$

Example 15.30

An epicyclic gear train as shown in Fig.15.34, has a sun wheel S of 30 teeth and two planet wheels $P-P$ of 45 teeth. The planet wheels mesh with the internal teeth of a fixed annulus A. The driving shaft carrying the sun wheel transmits 4 kW at 360 rpm. The driven shaft is connected to an arm, which carries the planet wheels. Determine the speed of the driven shaft and the torque transmitted, if the overall efficiency is 95%.

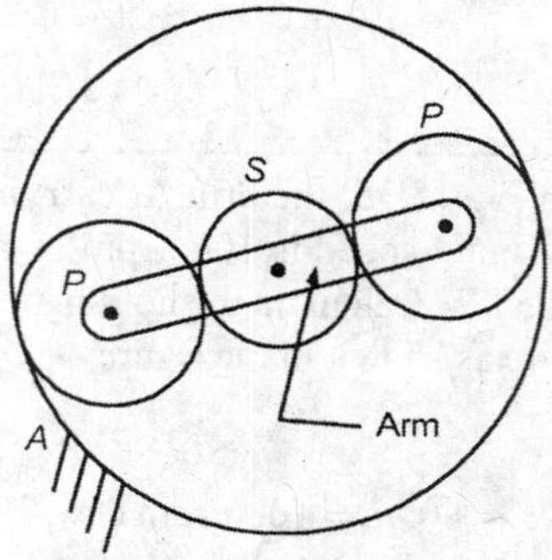

Fig.15.34 Epicyclic gear train

■ **Solution**

Given:

$$z_s = 30,\ z_p = 45$$
$$d_a = d_s + 2\,d_p$$
$$z_a = z_s + 2z_p$$
$$z_a = 30 + 90 = 120$$

Table 15.28 is used to find the speed of gears.

Table 15.28

Operation	Revolutions of			
	Arm	S	P	A
1. Arm fixed, +1 rev to S, ccw	0	+1	$\frac{-z_s}{z_p} = -\frac{30}{45} = -\frac{2}{3}$	$\frac{-z_s}{z_a} = \frac{-30}{120} = -\frac{1}{4}$
2. Multiply by x	0	$+x$	$-\frac{2x}{3}$	$\frac{-x}{4}$
3. Add y	y	$y+x$	$y-\frac{2x}{3}$	$y-\frac{x}{4}$

A fixed:
$$y - \frac{x}{4} = 0$$
$$y + x = 360$$
$$x = 288 \text{ rpm}, y = 72\text{rpm}$$
$$\text{Speed of driven shaft} = 72 \text{ rpm}$$

Input torque,
$$T_1 = \frac{4 \times 10^3 \times 60}{2\pi \times 360} = 106.1 \text{ Nm}$$
$$T_1 n_1 + T_2 n_2 = 0$$
$$T_2 = -106.1 \times \frac{360}{72} = -530.5 \text{ Nm}$$

Example 15.31

In the epicyclic gear train shown in Fig.15.35, the arm A, carrying the compound wheels D and E, turns freely on the output shaft. The input speed is 1000 rpm in counter – clockwise direction when seen from the right. Input power is 7.5 kW. Calculate the holding torque to keep the wheel C fixed. The number of teeth for different gears are as shown in the figure.

Solution

$Z_b = 20$, $z_d = 60$, $z_e = 30$, $z_f = 32$, $z_c = 80$, $n_b = 1000$ rpm ccw, $P_i = 7.5$ kW

Let n_a = rpm of arm. Considering train B, D, C, we have

$(n_b - n_a)/(n_c - n_a) = -z_c/z_b$, $n_a = 0$ being fixed

$(1000 - n_a)/(0 - n_a) = -80/20 = -4$

$n_a = 200$ rpm

Considering train B, D, E, F, we have

$(n_b - n_a)/(n_2 - n_a) = (-z_f/z_e) \times (z_d/z_b)$

$(1000 - 200)/(n_2 - 200) = (-32/30) \times (60/20) = -3.2$

$n_2 = -50$ rpm

$\omega_2 = -2\pi \times 50/60 = -5.236$ rad/s
$T_1 = (7.5 \times 10^3 \times 60)/(2\pi \times 1000) = 71.62$ N.m
$T_1\,\omega_1 + T_2\,\omega_2 = 0$
$T_2 = 7.5 \times 10^3/5.236 = 1432.4$ N.m
Holding torque $T_3 = -T_1 - T_2 = -71.62 - 1432.4 = -1504.02$ N.m

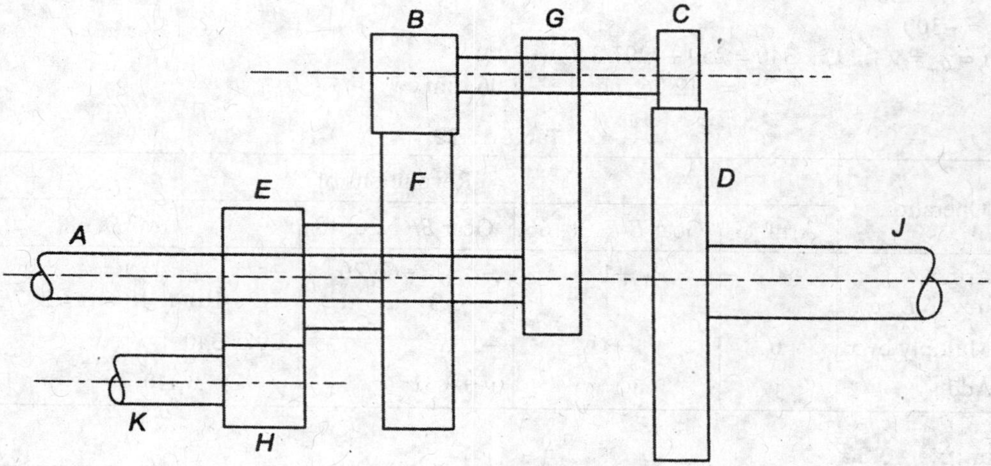

Fig.15.35 Epicyclic gear train

Example 15.32

In the epicyclic gear train shown in Fig.15.36, the compound wheels E and F rotate freely on shaft A which carries the planet carrier G. The planets B and C are compounded gears. The number of teeth on each gear are: $z_e = 30$; $z_b = 20$, $z_c = 18$, $z_d = 68$.

The shafts A and K rotate in the same direction at 250 rpm and 100 rpm respectively. Determine the speed of shaft J.

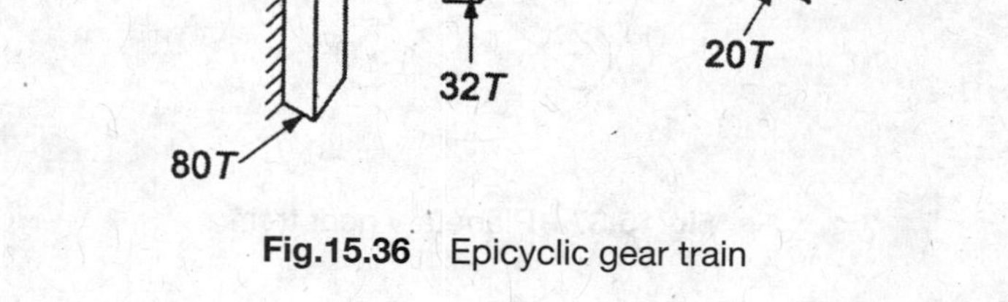

Fig.15.36 Epicyclic gear train

Solution

Table 15.29 is used to find the speed of gears.

$z_f = z_c + z_d - z_b = 18 + 68 - 20 = 66$

$n_a = n_g = y = 250$

$n_e = n_h \times 15/30 = -100 \times 15/30 = -50$

$x + y = -50$

$x = -300$

$n_j = n_d = y + 297\,x/340 = 250 + (297/340)(-300)$

$= -12.06$ rpm, i.e. 12.06 rpm cw

Table 15.29

Operation	Revolutions of			
	Arm, G	Gear E/F, 30, 66	Gear B/C, 20, 18	Gear D, 68
1. Arm fixed, +1 rev to E ccw	0	+1	$-z_f/z_b = -66/20 = -3.3$	$-z_f/z_b \times (-z_c/z_d) = -(33/10)\times(-18/68) = 297/340$
2. Multiply by x	0	$+x$	$-3.3x$	$297x/340$
3. Add y	y	$y + x$	$y - 3.3x$	$y + 297x/340$

Example 15.33

(a) Give a list of the common applications of planetary gear trains. Describe the working of the differential mechanism of a motor car.

(b) In the planetary gear train shown in Fig.15.37, gear 1 has 50 teeth and gear 3 has 90 teeth. Determine the number of equally spaced planets that can be used without overlapping. The gears are standard. The formula used is to be derived, stating the assumptions made.

Solution

$$z_3 = z_1 + 2z_2$$
$$90 = 50 + 2z_2$$
$$z_2 = 20$$

Let n be the number of planets equally spaced. Then

$$n\,d_2 = \pi\,(d_3 - d_2)$$

For same module of all gears,

$$n\,z_2 = \pi\,(z_3 - z_2)$$
$$n = \pi\,(90 - 20)/20 = 11$$

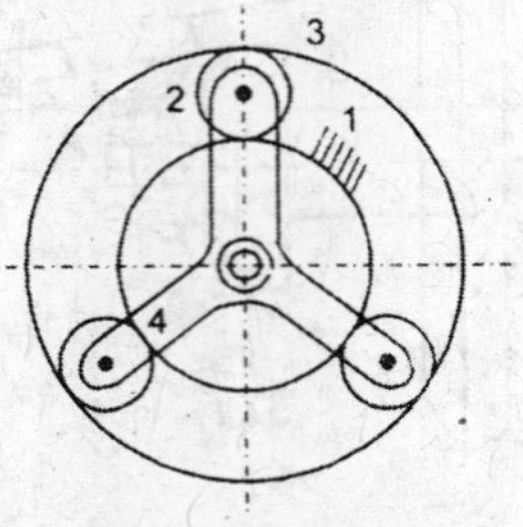

Fig.15.37 Planetary gear train

Example 15.34

Fig.15.38 shows a compound epicyclic gear train in which two sun gears S_1 and S_2 are integral with the input shaft *X*. The arm B_2 is integral with the output shaft *Y*. The planet gear P_2 revolves on a pin attached to arm B_2 and meshes with gear S_2 and annular gear A_2. The annular gear A_2 is coaxial with the input shaft. The planet gear P_1 meshes with the fixed annular gear A_1 and sun gear S_1. The gear P_1 revolves on a pin fixed to gear A_2. The number of teeth on gears are: $S_1 = 26$, $S_2 = 31$, $A_1 = 88$, $A_2 = 83$. The input power on shaft X is 10 kW at 1000 rpm cccw. Find (a) the speed and torque at shaft Y, assuming efficiency of 96 %, and (b) the torque required to hold the gear A_1 stationary.

Solution

(a) Input shaft

For $n_{a1} = 0$

$$y - 26x/88 = 0$$

$$y + x = 1000$$

$$x = 772 \text{ rpm}, y = 228 \text{ rpm}$$

Speed of gear $A_2 = y = 228$ rpm
Speed of arm $B_1 = y = 228$ rpm

(b) Output shaft

$x + y = 1000$

$y - 31x/83 = 228$

$x = 562$ rpm

Speed of arm $B_2 = y = 1000 - 562 = 438$ rpm
Speed of output shaft $= y = 438$ rpm
Input torque, $T_1 = 10 \times 10^3 \times 60/(2\pi \times 1000) = 95.5$ N.m
Output torque, $T_2 = -95.5 \times 1000 \times 0.96/438 = -2011.3$ N.m
Holding torque, $T_3 = -95.5 - (-2011.3) = 113.8$ N.m

Table 15.30

Operation	Revolutions of			
	Arm, B_1	S_1	P_1	A_1
1. Arm B_1 fixed, S_1 given +1 revolutions ccw	0	+1	$-z_{s1}/z_{p1} = -26/z_{p1}$	$-z_{s1}/z_{a1} = -26/88$
2. Multiply by x	0	$+x$	$-26x/z_{p1}$	$-26x/88$
3. Add y	y	$y + x$	$y - 26x/z_{p1}$	$y - 26x/88$

Table 15.31

Operation	Revolutions of			
	Arm B_2	S_2	P_2	A_2
1. Arm B_2 fixed, S_2 given +1 revolutions ccw	0	+1	$-z_{s2}/z_{p2}$	$-z_{s2}/z_{a2} = -31/83$
2. Multiply by x	0	$+x$	$-xz_{s2}/z_{p2}$	$-31x/83$
3. Add y	y	$y + x$	$y - xz_{s2}/z_{p2}$	$y - 31x/83$

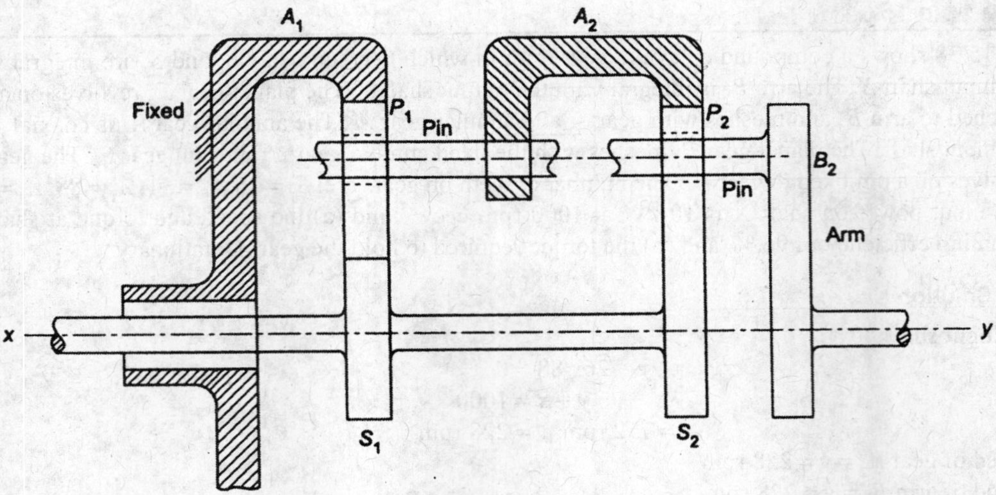

Fig.15.38 Compound epicyclic gear train

Example 15.35

In a reverted epicyclic train (Fig.15.39), the arm A carries two wheels B and C and a compound wheel D-E. The wheel B gears with wheel E and the wheel C gears with wheel D. The number of teeth on wheels B, C and D are 80, 35 and 95 respectively. Find the speed and direction of wheel C wheel B is fixed and the arm A makes 120 rpm clockwise.

■ **Solution**

Given: $z_b = 80, z_c = 35, z_d = 95,$

$$n_b = 0$$

$$n_{arm} = 120 \text{ rpm (cw)}$$

$$r_b + r_e = r_c + r_d$$

or

$$z_b + z_e = z_c + z_d$$

$$z_e = 35 + 95 - 80$$

$$= 50$$

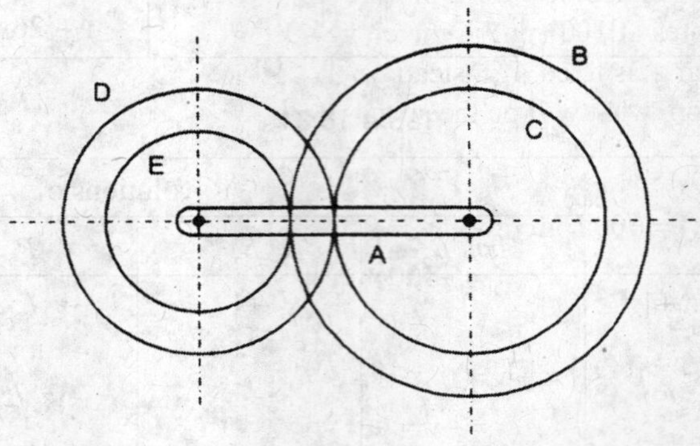

Fig.15.39 Reverted epicyclic gear train

Table 15.29 is used to find the speed of gears.

Table 15.29

Operation	Revolutions of			
	Arm, A	Gears D, E	Gear B	Gear C
1. Arm A fixed, +1 rev to D, E, ccw	0	+1	$-z_e/z_B$	$-z_d/z_C$
2. Multiply by x	0	$+x$	$-x\,z_e/z_B$	$-x\,z_d/z_C$
3. Add y	$+y$	$x+y$	$y-xz_e/z_B$	$y-x\,z_d/z_C$

Since gear B is fixed,

$$y - x\frac{z_e}{z_b} = 0$$

$$y - x \times \frac{50}{80} = 0$$

$$y - \frac{5}{8}x = 0$$

or $$y - 0.625\,x = 0 \tag{1}$$

Arm A makes 120 rpm (cw), therefore

$$y = -120 \tag{2}$$

From eqs (1) and (2), we get

$$-120 - 0.625\,x = 0$$

$$x = -192 \text{ rpm}$$

For gear C, we have

$$n_c = y - x\frac{z_d}{z_e}$$

$$= -120 + 192 \times \frac{95}{35} = 401.143 \text{ rpm (ccw)}$$

Example 15.36

In an epicyclic gear train (Fig.15.40), an arm carries two wheels A and B having 24 and 30 teeth respectively. The arm rotates at 100 rpm in the clockwise direction. Find the speed of the gear B on its own axis, when the gear A is fixed. If instead of being fixed, the wheel A rotates at 200 rpm in the counter clockwise direction, what will be the speed of B?

■ Solution

Given: $z_a = 24$, $z_b = 30$, $n_{\text{arm}} = 100$ rpm (cw), $n_A = 0$ and 200 rpm

Now $$y = +100$$

(i) Gear A is fixed,

$$x + y = 0$$

$$x = -y = -100$$

$$n_b = y - x\,z_a/z_b$$

$$= 100 + 100 \times \frac{24}{30}$$

$$= 180 \text{ rpm (cw)}$$

(ii) Gear A rotates at 200 rpm (ccw)

$$x + y = -200$$

$$x = -200 - y = -200 - 100 = -300$$

$$n_b = 100 + 300 \times \frac{24}{30} = 340 \text{ rpm (cw)}$$

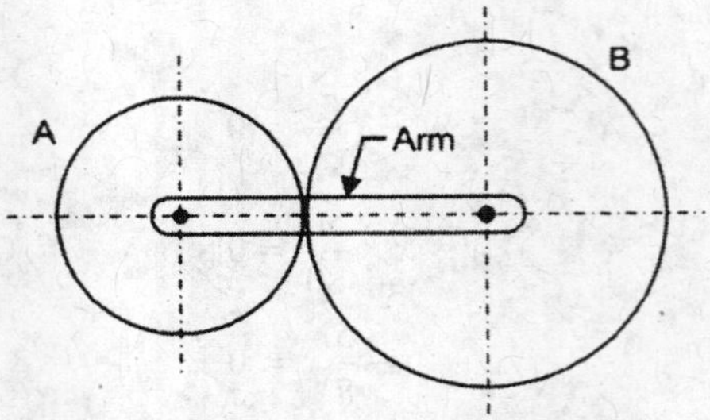

Fig.15.40 Epicyclic gear train

Table 15.30 is used to find speed of gears.

Table 15.30

Operation	Revolutions of		
	Arm	Gear, *A*	Gear, *B*
1. Arm fixed, +1 rev. to A, cw	0	+1	$-z_a/z_b$
2. Multiply by x	0	$+x$	$-xz_a/z_b$
3. Add y	y	$x+y$	$y-x\,z_a/z_b$

Example 15.37

In an epicyclic gear train (Fig.15.41) an annular wheel '*A*' having 54 teeth meshes with a planet wheel '*B*' which gears with a sun wheel '*C*', the wheels '*A*' and '*C*' being co-axial. The wheel *B* is carried on a pin fixed on one end of arm *P* which rotates about the axis of the wheels *A* and *C*. If the wheel *A* makes 20 rpm in a clockwise sense and the arm rotates at 100 rpm in the anticlockwise direction and wheel *C* has 24 teeth, determine rpm and sense of rotation of the wheel *C*.

■ Solution

Given: $z_a = 54$, $z_c = 24$, $n_a = 20$ rpm (cw), $n_p = 100$ rpm (ccw)

Table 15.31 is used to find speed of gears.

(i) $$n_a = -20$$

$$y - x\frac{z_c}{z_a} = -20 \qquad (1)$$

(ii) $$y = 100 \qquad (2)$$

Table 15.31

Operation	Revolutions of			
	Arm, P	Gear, C	Gear, B	Gear A
1. Arm P fixed, +1 rev. to c (ccw)	0	+1	$-z_c/z_b$	$-\frac{z_c}{z_b}\times\frac{z_b}{z_a}=-z_c/z_a$
2. Multiply by x	0	$+x$	$-x\,z_c/z_b$	$-xz_c/z_a$
3. Add y	y	$x+y$	$y-z_c/z_b$	$y-x\,z_c/z_a$

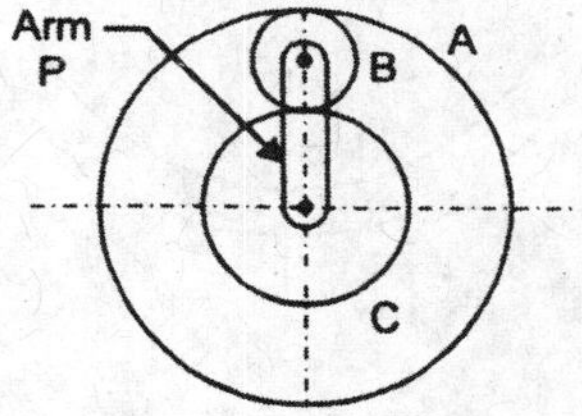

Fig.15.41 Epicyclic gear train

From Eqs. (1) and (2), we get

$$100 - x\times\frac{24}{54} = -20$$

$$x = 120\times\frac{54}{24} = 270$$

$$nc = x + y = 270 + 100 + 370 \text{ rpm (ccw)}$$

Example 15.38

An internal wheel B with 80 teeth is keyed to shaft F (Fig.15.42). A fixed internal wheel C with 82 teeth is concentric with B. A compound wheel D-E gears with two internal wheels; D has 28 teeth and gears with C, while E gears with B. The compound wheel revolves freely on a pin which projects from a disc keyed to a shaft A co-axial with F. If all the wheels have the same pitch and the shaft A makes 800 rpm, what is the speed of F?

■ Solution

Given:

$$z_b = 80,$$
$$z_c = 82,$$
$$z_d = 28,$$
$$n_a = 800 \text{ rpm},$$
$$n_c = 0$$

From the geometry of figure, we have

$$d_e - d_d = d_b - d_e$$

or

$$z_c - z_d = z_b - z_e$$

$$82 - 28 = 80 - z_e$$

$$z_e = 26$$

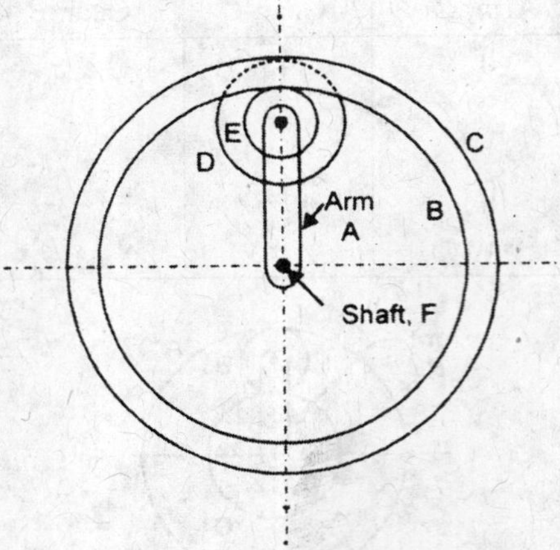

Fig.15.42 Epicyclic gear train

Table 15.32 is used to find speed of gears.

Table 15.32

Operation	Revolutions of			
	Arm, A	Gear, C	Compound gear D, E	Gear B and Shaft F
1. Arm A fixed, +1 revolutions to C, ccw	0	+1	z_c/z_d	$\frac{z_c}{z_d} \times \frac{z_e}{z_b}$
2. Multiply by x	0	x	$x\,z_c/z_d$	$x \times \frac{z_c}{z_d} \times \frac{z_e}{z_b}$
3. Add y	y	$x + y$	$y + x\,z_c/z_d$	$y + x \times \frac{z_c}{z_d} \times \frac{z_e}{z_b}$

c is fixed, $n_c = x + y = 0$

$$n_a = y = 800$$

$$x = -800$$

$$n_f = y + x \times \frac{z_c}{z_d} \times \frac{z_e}{z_b}$$

$$= 800 - 800 \times \frac{82}{28} \times \frac{26}{80} = 38.57 \text{ rpm}$$

Example 15.39

In an epicyclic gear train, the internal wheels *A* and *B* and the compound wheels *C* and *D* rotate independently about axis *O*. The wheels *E* and *F* rotate on pins fixed to the arm *G*. *E* gears with *A* and *C* and *F* gears with *B* and *D*. All wheels have the same module and the number of teeth are:

$$z_c = 28, z_d = 26. z_e = z_f = 18$$

Find (i) The number of teeth on *A* and *B* (ii) if the arm *G* makes 100 rpm clockwise and *A* is fixed, find the speed of *B*.

Solution

Refer to Fig.15.43.

Given:

$$z_c = 28,$$
$$z_d = 26,$$
$$z_e = z_f = 18$$
$$n_g = 100 \text{ rpm (cw)},$$
$$n_a = 0$$

(i) From geometry of figure,

$$d_b = 2d_f + d_D$$
$$d_a = 2d_e + d_C$$

For the same module,

$$z_b = 2z_f + z_D$$
$$= 2 \times 18 + 26$$
$$= 62$$
$$z_a = 2z_e + z_C$$
$$= 2 \times 18 + 28$$
$$= 64$$

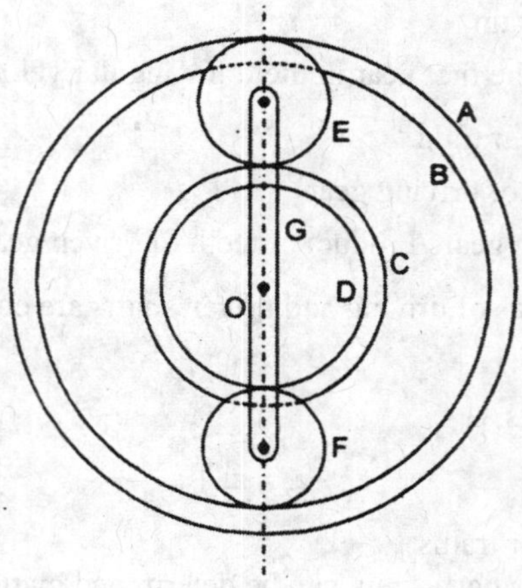

Fig.15.43 Epicyclic gear train

Table 15.33 is used to find speed of gears.

Table 15.33

Operation	Revolutions of					
	Arm G	Gear C, D	Gear E	Gear F	Gear A	Gear B
1. Arm G fixed, +1 rev. to D (ccw)	0	+1	$-z_c/z_e$	$-z_d/z_f$	$-z_c/z_A$	$-z_d/z_B$
2. Multiply by x	0	$+x$	$-x\,z_c/z_e$	$-x\,z_d/z_f$	$-x\,z_c/z_A$	$-x\,z_d/z_B$
3. Add y	y	$x+y$	$y-x\,z_c/z_e$	$y-x\,z_d/z_f$	$y-x\,z_c/z_A$	$y-x\,z_d/z_B$

$$n_g = y = -100$$

$$n_a = y - x\,z_c/z_a = 0$$

or
$$-100 - x \times \frac{28}{64} = 0$$

$$x = -100 \times \frac{64}{28} = -228.57$$

$$n_b = y - x\frac{z_d}{z_b} = -100 + 228.57 \times \frac{26}{62}$$

$$= -4.147 \text{ rpm} \quad \text{or} \quad 4.147 \text{ rpm (cw)}$$

Summary for Quick Revision

1. A gear train is a combination of gears used to transmit motion from one shaft to another.
2. Gear trains may be classified as: simple, compound, reverted and planetary or epicyclic.
3. Speed ratio of simple gear train
   n_1/n_m = angular velocity of the first gear in the train/angular velocity of the last gear
4. Speed ratio of compound gear train
   Speed of driven gear/Speed of driving gear
   = Product of teeth of driving gears/Product of teeth of driven gears
5. Reverted gear train : The axes of driving and driven shafts are coaxial.
   Centre distance, $C = r_1 + r_2 = r_3 + r_4$
   For same module of all gears,
   $$z_1 + z_2 = z_3 + z_4$$
6. Speed ratio of planetary gear trains
   The speed ratio of a planetary gear train can be determined by the following methods:
   1. Relative velocity method
   2. Algebraic or tabular method.

7 Torque on epicyclic gear trains:

$$T_1 + T_2 + T_3 = 0$$

$$T_1\omega_1 + T_2\omega_2 + T_3\omega_3 = 0$$

For a fixed member, $\omega_3 = 0$

Holding torque, $T_3 = T_1[n_1/n_2 - 1]$

Multiple Choice Questions

1 A reverted gear train is one in which the
(a) direction of rotation of first and last gear is the same
(b) direction of rotation of first and last gear is opposite
(c) first and last gear are on the same shaft
(d) first and last gear are essentially on separate but parallel shafts.

2 To connect hour hand to minute hand in a clock mechanism, we use
(a) epicyclic gear train (b) reverted gear train
(c) simple gear train (d) all of the above.

3 In a gear train, where the axes of gears have motions, is called
(a) simple gear train (b) compound gear train
(c) epicyclic gear train (d) reverted gear train.

4 In a simple gear train, if number of idlers is odd, then the direction of rotation of first and last gear shall be
(a) opposite (b) same
(c) depends on type of gears (d) depends on number of teeth on gears.

5 If the axes of first and last gear of a compound gear train are co-axial, the gear train is called
(a) simple (b) compound
(c) reverted (d) epicyclic

6 The train value of a simple gear train having m gears is
(a) $\dfrac{N_1}{N_m}$ (b) $\dfrac{N_m}{N_1}$ (c) $N_1 \times N_m$ (d) $N_m - N_1$
where N = rpm

7 In a compound gear train having m gears, the ratio of speed of last gear to first gear is
(a) $\dfrac{\text{product of teeth of driving gears } (X)}{\text{product of teeth of driven gears } (Y)}$ (b) y/x
(c) $X - Y$ (d) $\left(\dfrac{x}{y}\right)^2$

Answers

1. (c) 2. (b) 3. (c) 4. (b) 5. (c) 6. (a) 7. (a)

Review Questions

1 What is a gear train?

2 What is the difference between a simple and compound gear train?

3 What are the types of gear trains?
4 What is a reverted gear train? Where it is used.
5 What are epicyclic gears? What are their uses?
6 Write the formula for speed ratio of a simple gear train.
7 What is the law for the speed ratio of a compound gear train?
8 What is the fixing torque of a gear in an epicyclic gear train?
9 What is a compound epicyclic gear train?
10 What is a seen and planet gear?

Exercises

15.1 Two coaxial shafts *A* and *B* are geared together through an intermediate parallel shaft *C*. The wheels connecting *A* and *C* having a module of 2 mm and those connecting *C* and *B* have a module of 3.5 mm. Speed of *B* is less than 1/10th that of *A*. If the two pinions have each 20 teeth, find suitable number of teeth for the wheels, the actual velocity ratio, and corresponding centre distance of shafts *C* and *A*.

[Ans. 92, 44; 10.12; 224 mm]

15.2 Two spur gears *A* and *B* of an epicyclic gear train shown in Fig.15.44 have 25 and 35 teeth respectively. The arm *C* rotates at 105 rpm in clockwise direction. Find speed of gear *B* on its own axis, when gear *A* is fixed. If gear *A* rotates at 200 rpm in the counter-clockwise direction, what will be the speed of *B*?

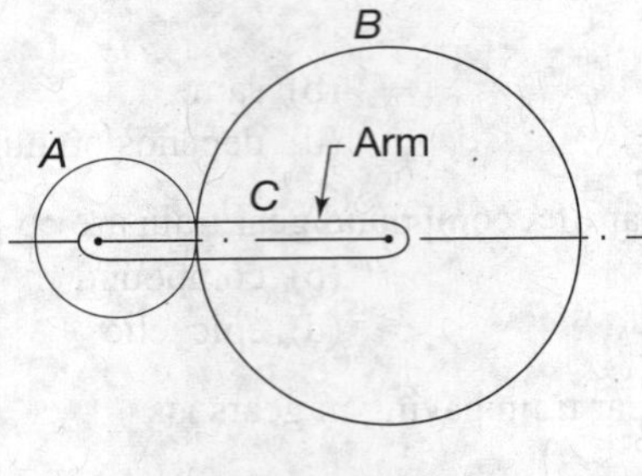

Fig.15.44 Epicyclic gear train

[Ans. 180 rpm cw; 322.86 rpm, cw]

15.3 In an epicyclic gear train shown in Fig.15.45 the number of teeth on gears *A*, *B* and *C* are 50, 25 and 55 respectively. The arm rotates at 450 rpm clockwise. Calculate (a) the speed of gear *C* when *A* is fixed, and (b) speed of gear *A* when *C* is fixed.

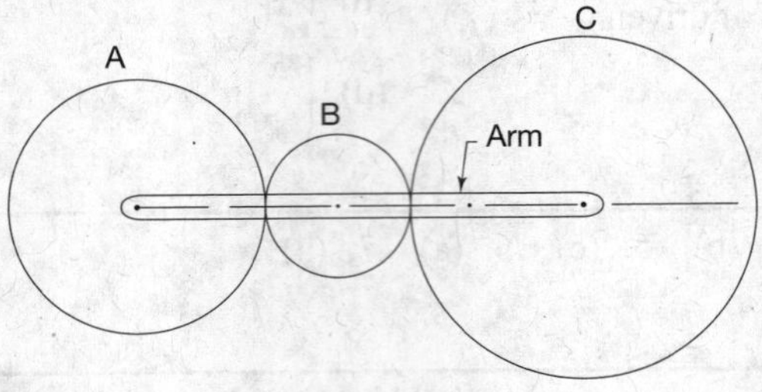

Fig.15.45 Epicyclic gear train

[Ans. 40.9 rpm cw; 45 rpm ccw]

15.4 In an epicyclic gear train shown in Fig 15.46 the arm *A* is fixed to shaft *S*. The wheel *B* having 90 teeth rotates freely on shaft *S* and wheel *F* with 135 teeth is separately driven. If arm *A* runs at 240 rpm clockwise and wheel *F* at 120 rpm in the same direction, find (a) number of teeth on wheel *C*, and (b) speed of wheel *B*.

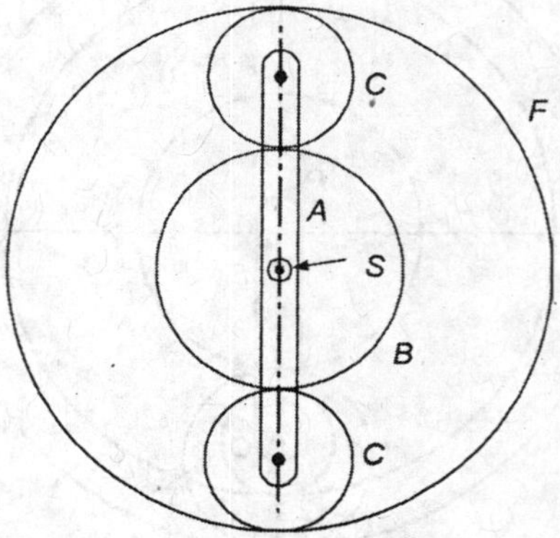

Fig.15.46 Epicyclic gear train

[Ans. 23, 421.3 rpm]

15.5 In an epicyclic gear train shown in Fig.15.47 the internal wheels *A* and *F* and compound wheel *C-D* rotate about the axis *O*. The wheel *B* and *E* rotate on pins fixed to arm *L*. The wheels have the same module and number of teeth are: $Z_b = Z_e = 20$, $Z_c = 30$, and $Z_d = 28$. If arm *L* makes 200 rpm cw, find the speed of *F*, when (a) wheel *A* is fixed, and (b) wheel a makes 20 rpm ccw.

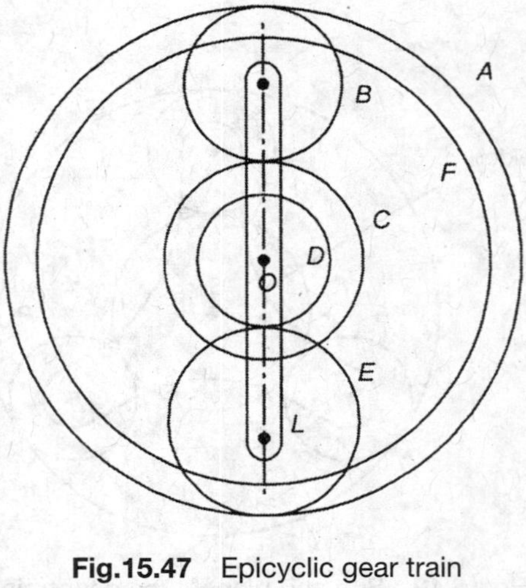

Fig.15.47 Epicyclic gear train

[Ans. 7.84 rpm cw, 11.37 rpm ccw]

15.6 A compound epicyclic gear train is shown in Fig.15.48 The gears *A*, *D* and *E* are free to rotate on axis *P*. The compound gear *B* and *C* rotate together on axis *Q* at the end of arm *F*. All the gears have equal module, $Z_a = 20$, $Z_b = 50$, $Z_c = 25$. Gears D and E are annular gears. Gear A rotates at 100 rpm ccw and gear D rotates at 500 rpm cw. Find speed and direction of arm and gear E.

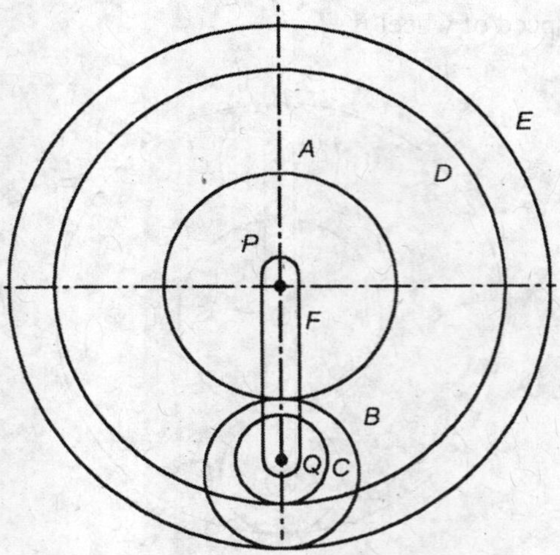

Fig.15.48 Compound epicyclic gear train

[Ans. 328.6 rpm cw, 400 rpm cw]

15.7 In an epicyclic gear of the sun and planet type shown in Fig.15, the annular gear A has 56 teeth and meshes internally. Here planet wheels of equal size mesh with annular gear A and sun wheel S. When gear A is stationary, the spider C which carries the planet wheels is to make one revolution for every 5 rotations of spindle carrying the sun wheel S. Calculate the number of teeth for all the wheels.

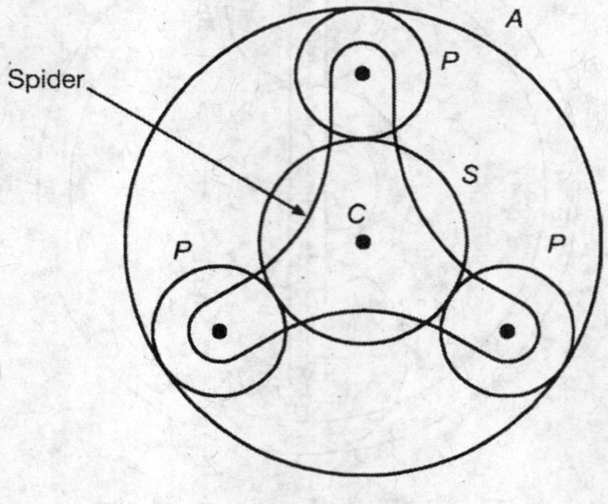

Fig.15.49 Sun and planet type gear train

[Ans. $z_s = 14$, $z_p = 21$]

15.8 The annulus A in the gear train shown in Fig.15.50 rotates at 300 rpm about the axis of the fixed wheel S having 80 teeth. The three arm spider is driven at 180 rpm. Determine the number of teeth required on the wheel P.

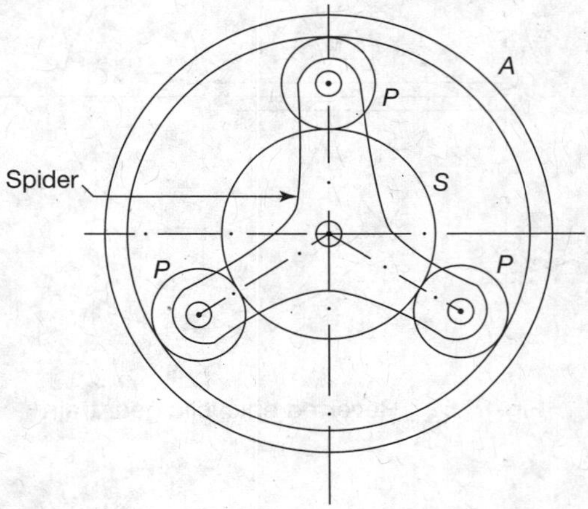

Fig.15.50 Sun and planet type gear train

[Ans. 20]

15.9 Fig.15.51 shows an epicyclic gear train arrangement wheel E is fixed and gears C and D are integrally cost and mounted on one pin. If the arm A makes one revolution per second counter – clockwise, determine the speed and the direction of rotation of wheels B and F.

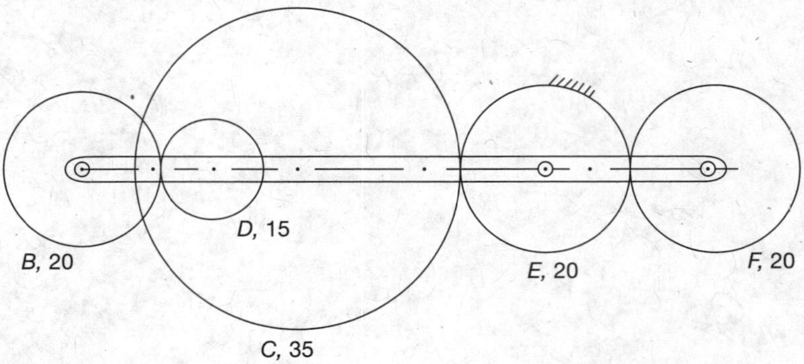

Fig.15.51 Epicyclic gear train

[Ans. 0.428 rps, ccw; 0.952 rps, ccw]

15.10 In a reverted epicclic gear train, as shown in a Fig.15.52 the arm A carries two gears B and C and compound gear D, E. The gear B meshes with gear D. The number of teeth on gears B, C and D are 75, 30 and 90 respectively. Find the speed and direction of gear C when gear B is fixed and the arm A makes 100 rpm clockwise.

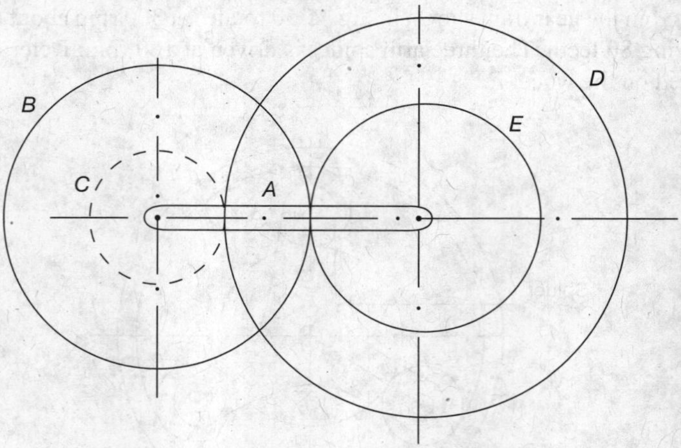

Fig.15.52 Reverted epicyclic gear train

[Ans. 400 rpm, ccw]

16 KINEMATIC SYNTHESIS OF PLANAR MECHANISMS

Chapter Outline

16.1 INTRODUCTION

Kinematic analysis is the process of determination of velocity and acceleration of the various links of an existing mechanism. This was already discussed in Chapters 2 and 3. On the other hand, kinematic synthesis deals with the determination of the lengths and orientation of the various lengths of the links so that a mechanism could be evolved to satisfy certain conditions.

Kinematic synthesis of a mechanism requires the determination of lengths of various links that satisfy the requirements of motion of the mechanism. The usual requirements are related to specified positions of the input and output links. It is easy to design a planar mechanism, when the position of input and output links are known at fewer positions as compared to large number of positions.

16.2 MOVABILITY (OR MOBILITY) OR NUMBER SYNTHESIS

Movability of a mechanism means the number of degrees of freedom, which is equal to the number of independent coordinates required to specify its configurations in order to define its motion. This concept is also known as *number synthesis*.

The Gruebler's (or Kutzbach) criterion for degrees of freedom of planar mechanisms is given by:

$$F = 3\,(n-1) - 2p - h \tag{16.1}$$

where $p = \frac{1}{2}(2n_2 + 3n_3 + 4n_4 + \cdots + in_i)$

= number of simple joints or lower pairs having one degree of freedom

n_2 = number of binary links

n_3 = number of ternary links

n_4 = number of quaternary links, and so on

h = number of higher pairs having two degrees of freedom

$n = n_2 + n_3 + n_4 + \cdots + n_i$ = total number of links.

If $h = 0$, then

$$F = 3\,(n-1) - 2p \tag{16.2}$$

$$\therefore \quad F = 3\,[(n_2 + n_3 + n_4 + \cdots + n_i) - 1] - (2n_2 + 3n_3 + 4n_4 + \cdots + in_i)$$

Simplifying and re-arranging the equation, we get

$$n_2 = (F+3) + [n_4 + 2n_5 + 3n_6 + \cdots + (i-3)\,n_i] \tag{16.3}$$

For a fully constrained mechanism, $F = 1$. Thus

$$n_2 = 4 + [n_4 + \cdots + (i-3)n_i] \tag{16.4}$$

From Eq. (16.4), it is quite evident that the minimum number of binary links is equal to four. Therefore, the four-bar kinematic chain is the simplest mechanism.

For $F = 1,\ n_2 \geq 4$

$F = 2,\ n_2 \geq 5$

$F = 3,\ n_2 \geq 6$

....

$F = k,\ n \geq (k+3)$

For a fully constrained motion, $F = 1$, so that

$$1 = 3\,(n-1) - 2p \tag{16.5}$$

or $$p = \frac{3n}{2} - 2$$

16.3 TRANSMISSION ANGLE

Transmission angle μ is the interior angle between the coupler and output link as shown in Fig.16.1. If link AB is the input link, the force applied to the output link DC is transmitted through the coupler BC. For a particular value of force in the coupler rod, the torque transmitted to the output link (about point D) is maximum when the transmission angle μ is 90°. If links BC and DC become coincident, the transmission angle is zero and the mechanism would lock or jam. If μ deviates significantly from 90°, the

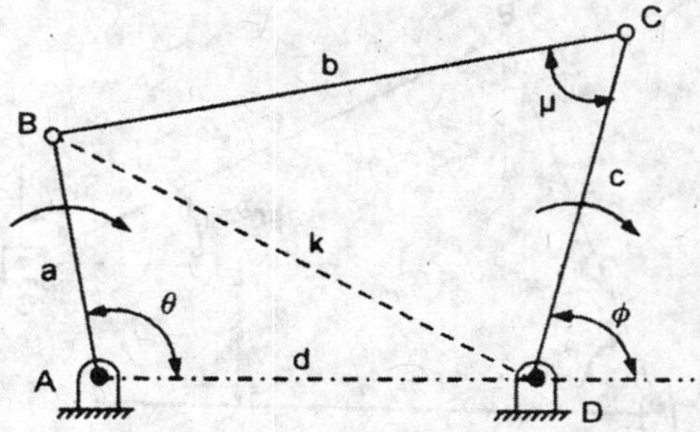

Fig.16.1 Four-bar mechanism

torque on the output link decreases. Sometimes, it may not be sufficient to overcome the friction in the system and the mechanism may be locked or jammed. Hence, μ is generally kept more than 45°. The transmission angle is nearly 90° for the best mechanism. To find the positions where the transmission angle is maximum or minimum, apply cosine law to Δs *ABD* and *BCD*.

$$a^2 + d^2 - 2\,ad\cos\theta = k^2 \tag{1}$$

and
$$b^2 + c^2 - 2\,bc\cos\mu = k^2 \tag{2}$$

From Eqs. (1) and (2), we have

$$a^2 + d^2 - 2\,ad\cos\theta = b^2 + c^2 - 2\,bc\cos\mu$$

or $a^2 + d^2 - b^2 - c^2 - 2\,ad\cos\theta + 2\,bc\cos\mu = 0$

For maximum or minimum values of μ, $\dfrac{d\mu}{d\theta} = 0.$

$$ad\sin\theta - bc\sin\mu \times \frac{d\mu}{d\theta} = 0$$

or
$$\frac{d\mu}{d\theta} = \frac{ad\sin\theta}{bc\sin\mu} = 0$$

or
$$ad\sin\theta = 0$$

Since 'a' and 'd' are not zero, so

$$\sin\theta = 0$$

or
$$\theta = 0° \text{ or } 180°$$

Thus, transmission angle is maximum when $\theta = 180°$, and minimum when $\theta = 0°$.

16.3.1 Transmission Angle in Slider–Crank Mechanism

The slider-crank mechanism *ABC* is shown in Fig.16.2 where μ is the transmission angle. To find the positions of maximum or minimum values of μ, we have

$$BE = BD - ED$$
$$= a\sin\theta - e \tag{1}$$

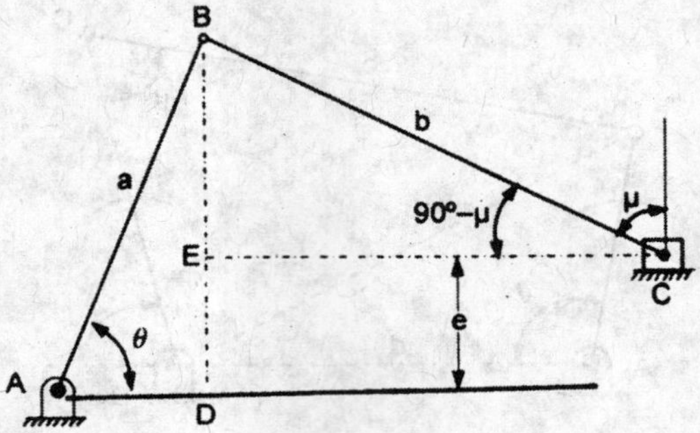

Fig.16.2 Slider-crank mechanism

From Δ *BEC*, we have

$$\frac{BE}{BC} = \sin(90° - \mu) = \cos\mu$$

$$BE = BC\cos\mu = b\cos\mu \tag{2}$$

From Eqs. (1) and (2), we get

$$a\sin\theta - e = b\cos\mu \tag{3}$$

Differentiating Eq. (3) w.r.t. θ and equalizing it to zero, we get

$$a\cos\theta = -b\sin\mu \times \frac{d\mu}{d\theta}$$

or $$\frac{d\mu}{d\theta} = -\frac{a\cos\theta}{b\sin\mu} = 0$$

or $$a\cos\theta = 0$$

As *'a'* is not equal to zero, so $\cos\theta = 0$
Thus $\theta = 90°$ or $270°$ gives the maximum or minimum values of transmission angle μ.

16.4 LIMIT POSITIONS AND DEAD CENTRES OF A FOUR-BAR MECHANISM

For the design of four-link mechanism, limit positions and dead centres are essential.

Limit position: It is the position in which the interior angle between its coupler and input link is either 180° or 360°. In the limit position of the mechanism, the pivot points *A*, *D* and *C* lie on a straight line as shown in Fig.16.3(b). In this case, the angle between the input link *AD* and coupler *DC* is 180°. A four-link mechanism can have maximum two limit positions. The second limit position is shown in Fig.16.3(c) where, the angle between the coupler *DC* and input link *AD* is 360°.

The angle of oscillation of output-link *BC* is given as,

$$\Delta\phi = \phi'' - \phi'$$

where ϕ'' and ϕ' are the two extreme positions of output link, *BC*.

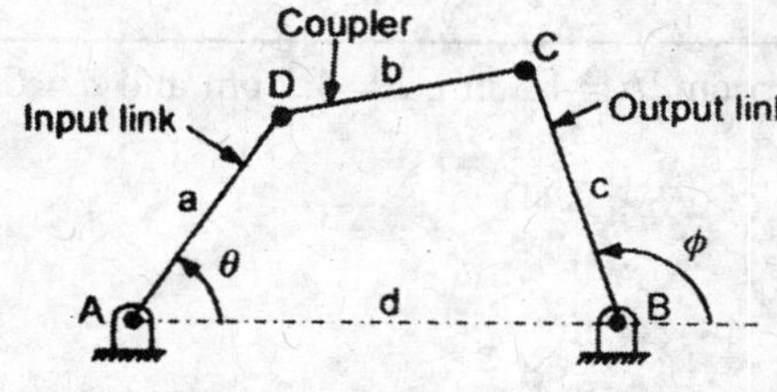

(a) Any Initial position of four-link mechanism

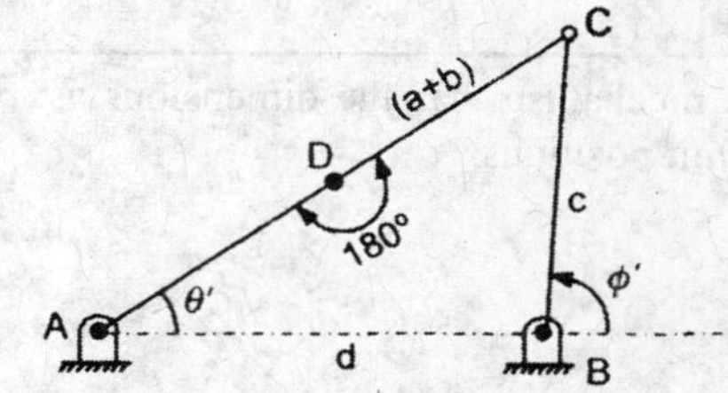

(b) Angle between input link and coupler is 180°

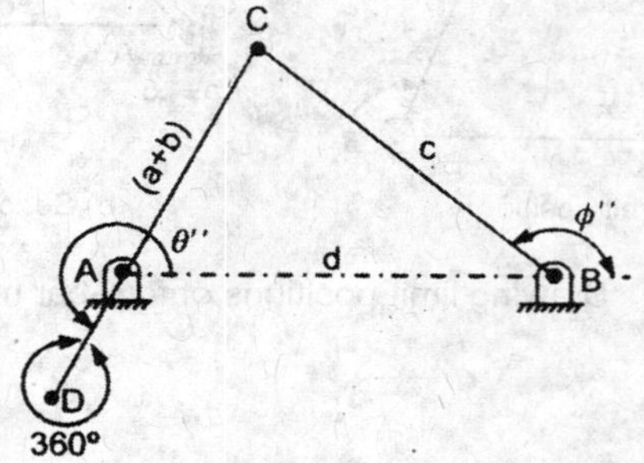

(c) Angle between input link and coupler is 360°

Fig.16.3 Limit positions of four-bar mechanism

Dead centre position: It is the position in which the interior angle between the coupler and the output link (of a four-link mechanism) is either 180° or 360° (*refer to* Fig.16.4). In dead centre positions, the pivot points *B*, *C*, and *D* lie in a straight line. There can be two (maximum) dead centre positions. A crank-crank four-bar link mechanism does not have either a limit position or dead centre positions because both the cranks rotate through 360°.

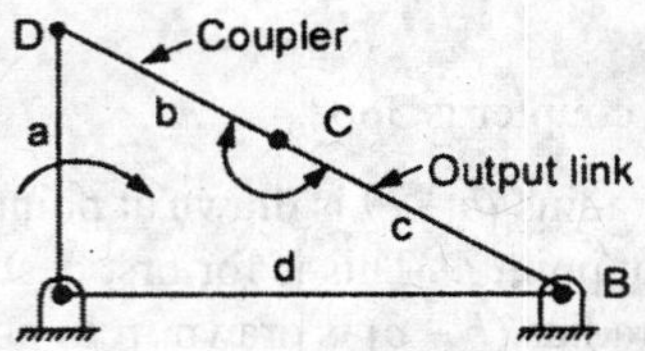

(a) Angle between coupler and output link is 180°

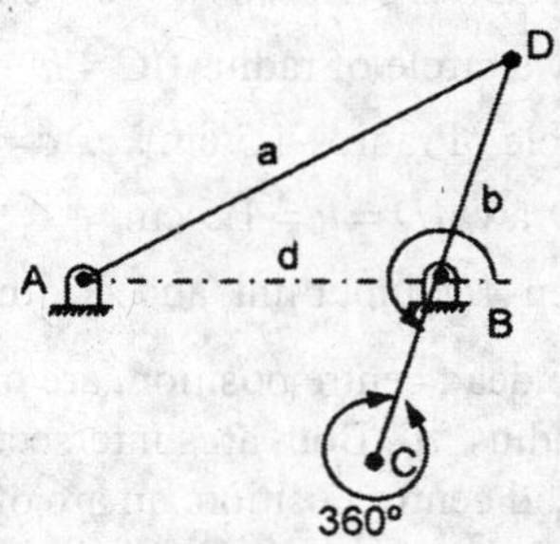

(b) Angle between coupler and output link is 360°

Fig.16.4 Dead centre positions of four-bar mechanism

Example 16.1

A four-link mechanism has the dimensions as: $a = 1.5$ cm, $b = 4.5$ cm, $c = 4.5$ cm and $d = 6.0$ cm. Draw the limit positions.

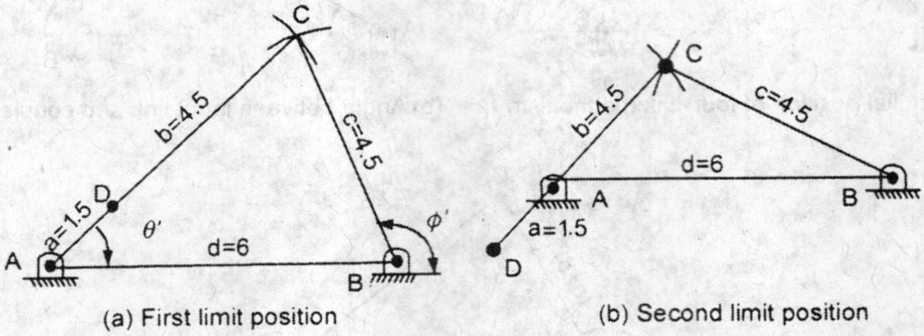

Fig.16.5 Drawing limit positions of four-bar mechanism

■ **Solution**

Procedure:

First limit position

1. Draw points A and B, 6 cm apart representing $d = 6.0$ cm
2. From point A draw an arc of 6 cm ($a + b = 6.0$ cm) and from point B draw an arc of 4.5 cm ($c = 4.5$ cm).
3. These arcs intersect at point C.
4. Locate point D on AC such that $AD = a = 1.5$ cm.
5. AD, DC and BC represent input link, coupler and output link respectively. The angle between AD and DC is 180°. This is the first limit position as shown in Fig.16.5(a).

Second limit position

1. Draw points A and B, 6 cm apart as $d = 6.0$ cm. (Fig.16.5(b)).
2. From point A draw a circle of radius $AC = b - a = 4.5 - 1.5 = 3.0$ cm.
3. From B draw a circle of radius 4.5 cm *i.e.,* $c = 4.5$ cm. These circles (arcs) intersect at point C.
4. Join A, C and B. Draw $AD = a = 1.5$ cm
5. Thus angle between AD, input link and AC, the coupler is 360°.

Note: For drawing the dead centre position, arc of radius $(b + c)$ is drawn at point B and another arc is drawn from A with radius 'a'. Both arcs intersect at point D. This is for first dead centre position.

For the second dead centre position, an arc of radius $(b - c)$ is drawn from point B.

16.5 DIMENSIONAL SYNTHESIS

It deals with the determination of actual dimensions of the mechanism to satisfy the specified motion characteristics. The actual dimensions could be the lengths of the links between adjacent hinge pairs, angles between the arms of a bell-crank lever, or cam contour dimensions, etc.

Synthesis of mechanisms can be carried out by graphical methods or analytical methods.

16.6 GRAPHICAL METHOD

16.6.1 Pole

Consider a four-bar mechanism O_2ABO_4 in two positions $O_2A_1B_1O_2$ and $O_2A_2B_2O_4$ as shown in Fig.16.6. The coupler link AB has moved from the position A_1B_1 to A_2B_2.

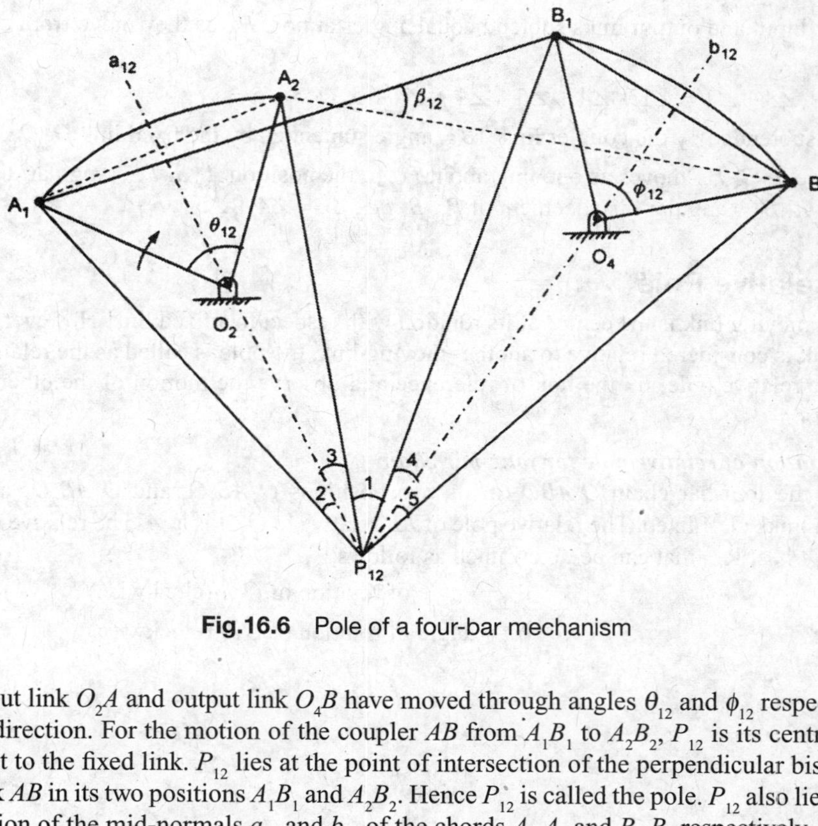

Fig.16.6 Pole of a four-bar mechanism

The input link O_2A and output link O_4B have moved through angles θ_{12} and ϕ_{12} respectively in the clockwise direction. For the motion of the coupler AB from A_1B_1 to A_2B_2, P_{12} is its centre of rotation with respect to the fixed link. P_{12} lies at the point of intersection of the perpendicular bisectors of the coupler link AB in its two positions A_1B_1 and A_2B_2. Hence P_{12} is called the pole. P_{12} also lies at the point of intersection of the mid-normals a_{12} and b_{12} of the chords A_1A_2 and B_1B_2 respectively.

Properties of pole point

1. Since $AB = A_1B_1 = A_2B_2$, the perpendicular bisectors of A_1A_2 and B_1B_2 pass through fixed centres O_2 and O_4.
2. Since coupler AB rotates about P_{12} from position A_1B_1 to A_2B_2, and therefore,

$$\Delta A_1B_1P_{12} = \Delta A_2B_2P_{12},$$

$$\therefore \quad \angle 2 + \angle 3 + \angle 1 = \angle 1 + \angle 4 + \angle 5$$

   or angle subtended by A_1B_1 at P_{12} = angle subtended by A_2B_2 at P_{12}.

3. Since $\angle 2 + \angle 3 + \angle 1 = \angle 1 + \angle 4 + \angle 5$

$$\therefore \quad \angle 2 + \angle 3 = \angle 4 + \angle 5$$

   Thus A_1A_2 and B_1B_2 subtend equal angle at P_{12}.

4. Since P_{12} lies on the perpendicular bisectors of $A_1 A_2$ and $B_1 B_2$,

$\therefore$ $\angle 2 = \angle 3$ and $\angle 4 = \angle 5$

5. Since $\angle 2 + \angle 3 = \angle 4 + \angle 5$

and $\angle 2 = \angle 3$ and $\angle 4 = \angle 5$

$\angle 2 = \angle 4$ and $\angle 3 = \angle 5$

Therefore, input and output links subtend equal angles at pole P_{12} as they move from one position to another.

6. Since $\angle 2 + \angle 3 + \angle 1 = \angle 1 + \angle 4 + \angle 5 = \angle 1 + \angle 4 + \angle 3$ ($\because \angle 5 = 3$)

i.e., angle subtended by the coupler link AB = angle subtended by the fixed link $O_2 O_4$.

7. The triangle $A_1 B_1 P_{12}$ moves as one link about P_{12} to the position $A_2 B_2 P_{12}$. Angular displacement of coupler $A_1 B_1$ = angular displacement of $P_{12} B_1$ *i.e.*, $\beta_{12} = \angle 4 + \angle 5$.

16.6.2 Relative Pole

The pole of a moving link is the centre of its rotation with respect to a fixed link. However, if the rotation of the link is considered relative to another moving link, the pole is called as the relative pole. To determine the relative pole, fix the link of reference and observe the motion of the other link in the reverse direction.

(a) *Determination of relative pole for four-bar chain*

Consider the four-bar chain O_2ABO_4 in its two positions $O_2AB_1O_4$ and $O_2AB_2O_4'$ as shown in Fig.16.7 with links O_2A fixed. The relative pole of AB relative to O_2A is at A. The relative pole of O_4B relative to O_2A is at R_{12}, that can be determined as follows:

Let θ_{12} = angle of rotation of O_2A (clockwise)

ϕ_{12} = angle of rotation of O_4B (clockwise)

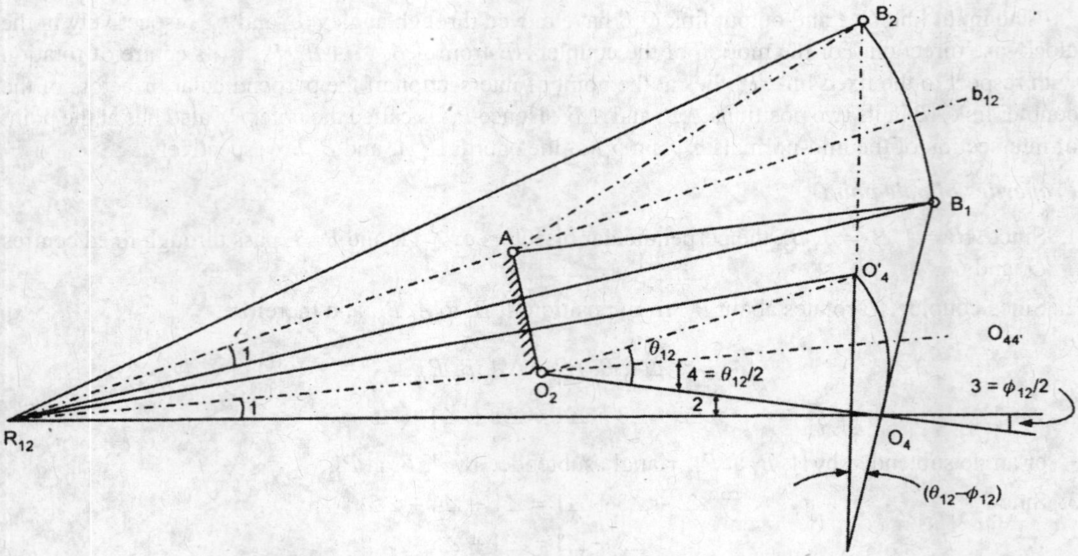

Fig.16.7 Determination of relative pole of four-bar chain

1. Assume O_2 and A as the fixed pivots and rotate O_2O_4 about O_2 through angle θ_{12} in the counter-clockwise direction (opposite to the direction of rotation of O_2A). Let O_4' be the new position after the rotation of O_2O_4'.
2. Locate the point B_2 by drawing arc with centres A and O_4' and radii equal to AB_1 and O_4B_1 respectively. Then $O_2\,AB_2O_4'$ is inversion of $O_2AB_1O_4'$.
3. Draw mid-normals of O_2O_4' and B_1B_2 passing through O_2 and A respectively to intersect at R_{12}, which is the required relative pole.

$\phi_{12} - \theta_{12}$ = angle of rotation of the output link O_4B relative to the input link.

The angle will be negative if $O_4\,B > O_2A$ and positive if $O_4B < O_2A$.

Angular displacement of $R_{12}O_4$ = angular displacement of O_4B_1.

$$\angle O_4\,R_{12}\,O_4' = -(\angle\phi_{12} - \theta_{12}) \qquad \text{(assuming } O_4B > O_2A\text{)}$$

or
$$2\angle 1 = -(\angle\phi_{12} - \angle\theta_{12})$$

or
$$\angle 1 = \frac{1}{2}(\angle\phi_{12} - \angle\theta_{12})$$

In $\Delta O_2\,R_{12}\,O_4$,
$$\angle 4 = \angle 1 + \angle 2$$

or
$$\frac{1}{2}\angle\theta_{12} = -\frac{1}{2}(\angle\phi_{12} - \angle\theta_{12}) + \angle 3 \qquad (\because \angle 2 = \angle 3)$$

$$= \frac{1}{2}\angle\phi_{12} + \frac{1}{2}\angle\theta_{12} + \angle 3$$

$\therefore$
$$\angle 3 = \frac{1}{2}\angle\phi_{12}$$

Procedure:

1. Join O_2O_4 and extend it further. Rotate O_2O_4 about O_2 through an angle $\frac{1}{2}\theta_{12}$ in a direction opposite to that of O_2A.
2. Again rotate O_2O_4 about O_4 through an angle $\frac{1}{2}\phi_{12}$ in a direction opposite to that of O_4B.
3. The point of intersection of these two positions of O_2O_4 after rotation about O_2 and O_4, is the required relative pole R_{12}.

The angles subtended by $O_4\,O_4'$ and $B_1\,B_2$ at R_{12} are the same.

i.e.,
$$\angle O_4\,R_{12}\,O_4' = \angle B_1\,R_{12}\,B_2$$

or
$$2\angle O_4\,R_{12}\,O_2 = 2\angle B_1\,R_{12}\,A$$

or
$$\angle O_4\,R_{12}\,O_2 = \angle B_1\,R_{12}A$$

$\therefore$ Angle subtended by the fixed pivots (O_2 and O_4) at the relative pole = Angle subtended by the coupler AB.

(b) *Determination of relative pole for slider-crank mechanism*

Consider the slider-crank mechanism shown in Fig.16.8. The point B on the slider reciprocates through a horizontal distance x. Its centre of rotation will lie at infinity on a vertical line where the point O_4 can also be assumed to lie. Then $O_2\,O_4$ will also be a vertical line through O_2. Rotate $O_2\,O_4$ about O_4 through 1/2 θ_{12} in the counter-clockwise direction. Rotating O_2O_4 about O_4 through 1/2 θ_{12} would mean a vertical line towards left of O_2, at a distance of x/2. The intersections of the two lines locate R_{12}.

The procedure to locate the relative pole of a slider-crank mechanism is as follows:

1. Draw two parallel lines l_1 and l_2 at a distance equal to the eccentricity e (if there is any eccentricity).
2. Select a line segment $O_2C = \frac{x}{2}$ on line l_1 such that C is measured in a direction opposite to the motion of the slider.
3. Draw perpendicular lines O_2P_1 and $C\,P_2$ to line l_1.
4. Make $\angle\frac{1}{2}\theta_{12}$ at point O_2 with the line O_2P_1 in a direction opposite to the rotation of the input link.
5. The intersection of this line with the line CP_2 extended locates the relative pole R_{12}.

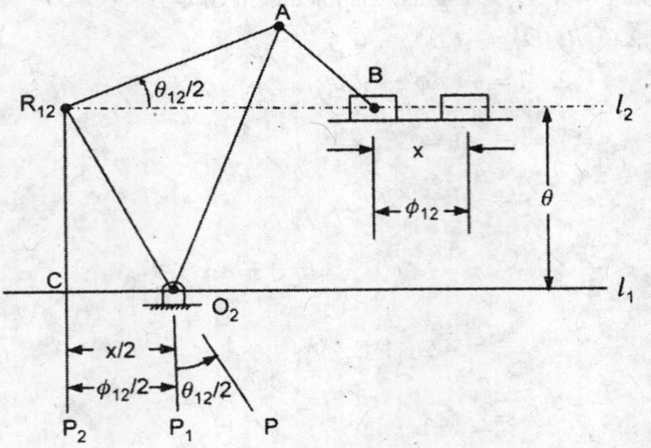

Fig.16.8 Determination of relative pole of slider-crank mechanism

16.7 DESIGN OF MECHANISMS BY RELATIVE POLE METHOD

16.7.1 Four-Bar Mechanism

(a) *Two-position synthesis*

Let for a four-bar mechanism, length of the fixed link O_2O_4 along with the angular displacement θ_{12} (between position 1 and 2) of the input link O_2A and angular displacement ϕ_{12} (between positions 1 and 2) of the output link O_4B, are known.

To design the mechanism (Fig.16.9), first locate the relative pole R_{12} as explained in Section 16.6.2(a).

Now, angle subtended by the coupler AB at R_{12}

$$= \text{angle subtended by the fixed link } O_2O_4 \text{ at } R_{12}$$

$$= \frac{1}{2}\angle\theta_{12} - \frac{1}{2}\angle\phi_{12} \qquad (\text{Assuming } O_4B\ \angle O_2A)$$

$$= \angle\psi_{12}$$

Procedure:

1. Construct an angle ψ_{12} at an arbitrary position R_{12}. Join any two points on the two arms of the angle to obtain the coupler link AB of the mechanism. Join A with O_2 to get the input and output links respectively.

2. Locate point B arbitrarily so that BO_4 is the output link. Construct $\angle BR_{12}Z = \psi_{12}$. Take any suitable point A or $R_{12}Z$. Join AB and O_2A.
3. Instead of locating point B as above, locate point A arbitrarily so that O_2A is the input link. Construct $\angle AR_{12}Y = \psi_{12}$. Select any point B on $R_{12}Y$. Join A and B to get AB. Join B with O_4 to get output link O_4B.

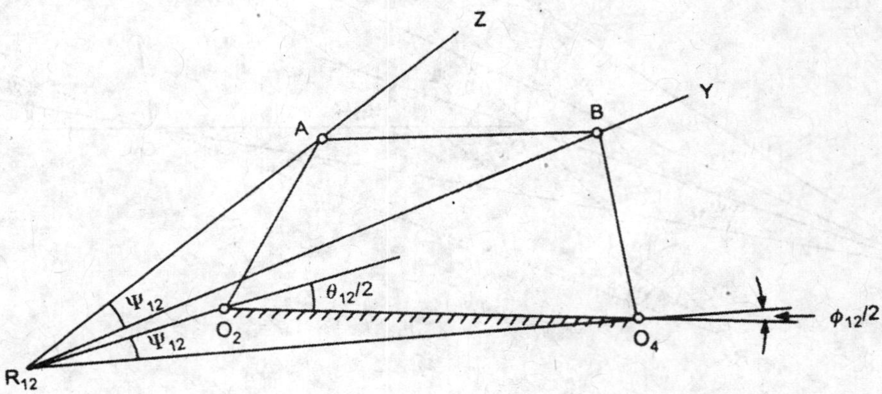

Fig.16.9 Two-position method for four-bar mechanism

(b) ***Three position synthesis***

Let three positions of input link θ_1, θ_2 and θ_3, and three positions of output link ϕ_1, ϕ_2 and ϕ_3 are known. The relative poles R_{12} and R_{13} can be determined considering,

$$\theta_{12} = \theta_2 - \theta_1 \quad \text{and} \quad \phi_{12} = \phi_2 - \phi_1 \text{ to locate } R_{12}$$

and
$$\theta_{13} = \theta_3 - \theta_1 \quad \text{and} \quad \phi_{13} = \phi_3 - \phi_1 \text{ to locate } R_{13}$$

Then angle subtended at R_{12},

$$\psi_{12} = \frac{1}{2}(\theta_{12} - \phi_{12})$$

and at R_{13},

$$\psi_{12} = \frac{1}{2}(\theta_{13} - \phi_{13})$$

Construct the angles ψ_{12} and ψ_{13} at the points R_{12} and R_{13} respectively at arbitrary positions such that the arms of the angles intersect at A and B. Join A and B to get the coupler AB. Join A with O_2 and B with O_4 to get input and output links respectively, as shown in Fig.16.10.

16.7.2 Slider–Crank Mechanism

(a) ***Two position synthesis***

Let θ_{12} = angular displacement of input link O_2A (angle between θ_1 and θ_2) in clockwise direction.

x = linear displacement of slider to the right

e = eccentricity

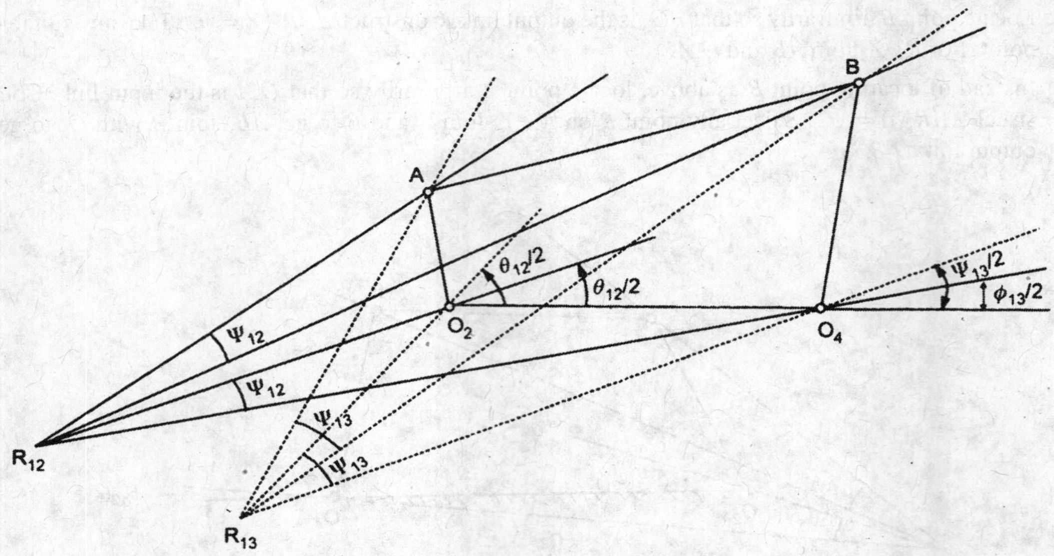

Fig.16.10 Three-position method for four-bar mechanism

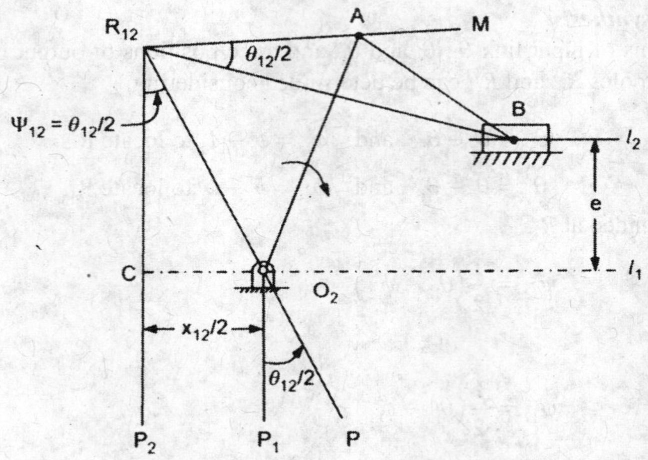

Fig.16.11 Two position method for slider-crank mechanism

Locate the relative pole R_{12} as explained in Section 16.6.2(b) and shown in Fig.16.11.

Procedure:

1. Draw two parallel lines l_1 and l_2 at a distance equal to eccentricity, e.
2. Construct an angle equal to $\theta_{12}/2$ at point R_{12}, chosen arbitrarily and at a convenient position. The intersection of an arm N of this angle with line l_2 gives the position of the slider at point B.
3. Select point A arbitrarily on the other arm M of the angle. Join A with O_2 to get the input link (crank) O_2A. Join A and B to get the coupler (connecting rod).

(b) *Three-position synthesis*

Three positions of input link θ_1, θ_2, and θ_3 are known, along with the corresponding slider positions x_1, x_2 and x_3. Find R_{12} and R_{13} as shown in Fig.16.12.

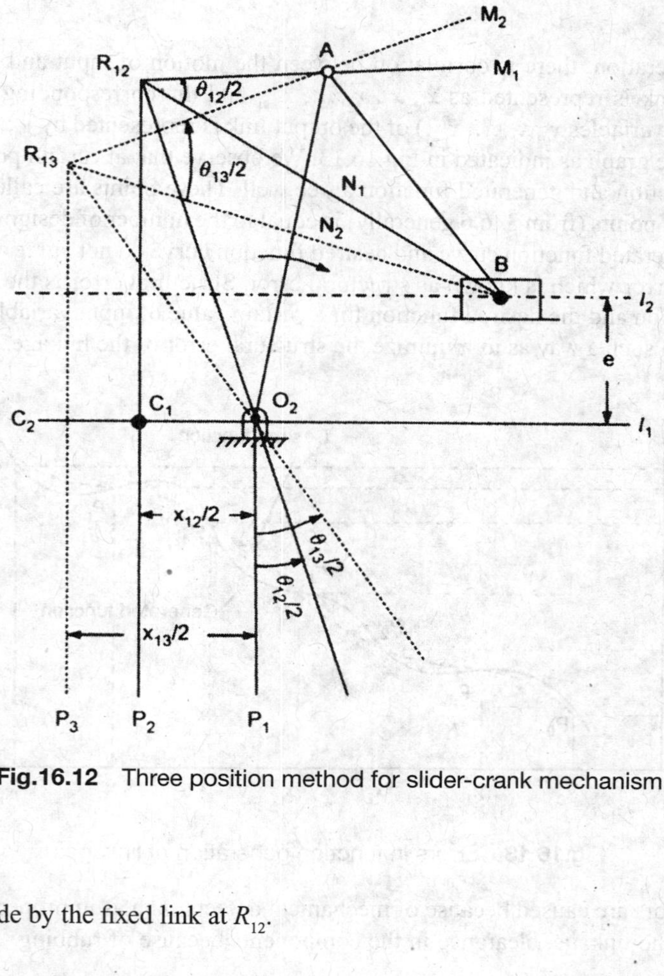

Fig.16.12 Three position method for slider-crank mechanism

Here

$\frac{\theta_{12}}{2}$ = angle made by the fixed link at R_{12}.

$\frac{\theta_{13}}{2}$ = angle made by the fixed link at R_{13}.

Procedure:

1. Construct angle $\frac{\theta_{12}}{2} = \angle M_1 R_{12} N_1$ at R_{12} in an arbitrary position with arm N_1 locating point B.
2. Draw angle $\frac{\theta_{13}}{2} = \angle M_2 R_{13} N_2$ at R_{13} with an arm N_2 along $R_{13}B$.
3. Intersection of the two arms M_1 at R_{12} and M_2 at R_{13} (not through B) of the two angles locates the point A. Join A with B to get the coupler AB.

16.8 ERRORS IN KINEMATIC SYNTHESIS OF MECHANISMS

There are three types of errors present in the design of linkages for function generation. These errors are:

1. Structural,
2. Mechanical, and
3. Graphical.

In function generation, there is correlation between the motion of input and output links. If the motions of input link is represented as $x_0, x_1, x_2, \ldots, x_{n+1}$ and the corresponding motions (which is dependent on input variables $x_0, x_1, \ldots, x_{n+1}$) of the output link is represented by $y_0, y_1, y_2, \ldots, y_{n+1}$, these can be shown on the graph as indicated in Fig.16.13. We observe that at certain points (P_1, P_2, and so on) the desired function and generated function agree well. These points are called precision points. The number of such points (from 3 to 6 generally) is equal to the number of design parameters. Except these points the generated function curve and desired function curve do not agree and are deviating by certain amount of error which is known as structural error. Structural error is the difference between the generated function and the desired function for a certain value of input variable. So, the precision points are spaced in such a way as to minimize the structural error of the linkage.

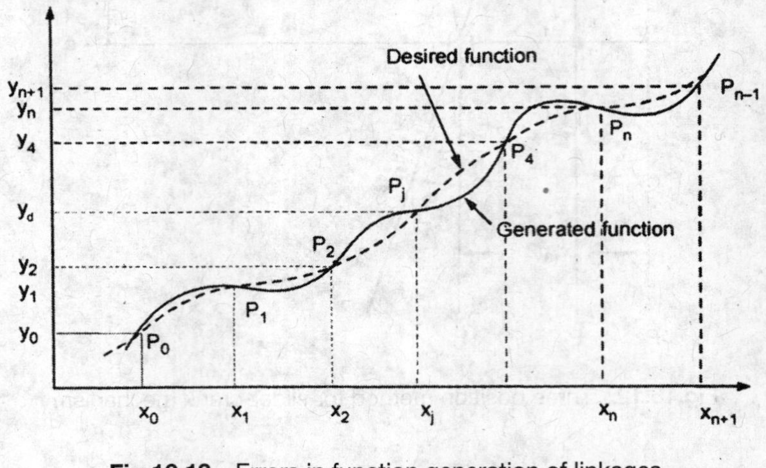

Fig.16.13 Errors in function generation of linkages

Mechanical errors are caused because of mechanical defects such as improper machining, casting of components of the linkage, clearance in the components because of rubbing, overloading of linkages, etc.

Graphical error is caused because of inaccuracy in drawing of perpendicular or parallel lines. It may occur because of wrong graphical construction and wrong choice of scale. Also, there may be human errors in drawing work.

16.9 ANALYTICAL METHOD

16.9.1 Function Generation

In function generation, the motion of input (or driver) link is correlated to the motion of output (or follower) link. Let θ and ϕ be the angles of rotation of input and output links respectively. Let $y = f(x)$ be the function to be generated. The angle of rotation θ of the input link O_2A represents the independent variable x and the angle of rotation ϕ of the output link O_4B represents the dependent variable y, as shown in Fig.16.14. The relation between x and θ and that between y and ϕ is generally assumed to be linear. Let θ_i and ϕ_i be the initial values of θ and ϕ representing x_i and y_i respectively.

Then

$$\frac{\theta - \theta_i}{x - x_i} = r_x = \text{const.} = \frac{\theta_f - \theta_i}{x_f - x_i} \tag{16.6a}$$

and

$$\frac{\phi - \phi_i}{y - y_i} = r_y = \text{const.} = \frac{\phi_f - \phi_i}{y_f - y_i} \tag{16.6b}$$

where the constants r_x and r_y are called scale factors. The subscripts i and f denote the initial and final values.

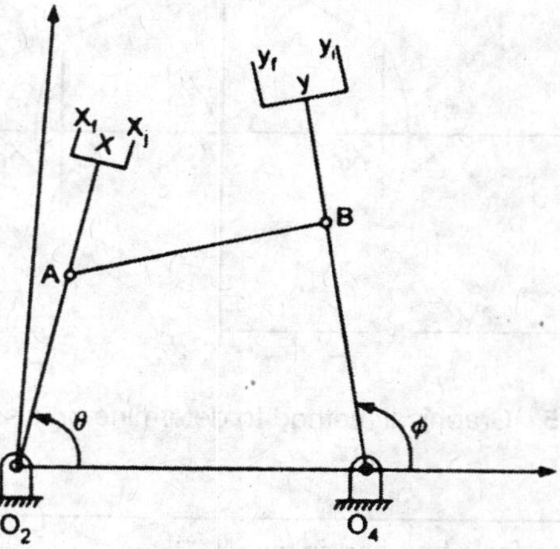

Fig.16.14 Function generation method

16.9.2 Chebyshev's Spacing for Precision Points

Let x_i and x_j be the initial and final values of variable x respectively. A function $f(x)$ is desired to be generated in the interval $x_i \le x \le x_f$. Let the generated function be $F(x, R_1, R_2, \ldots, R_n)$, where $R_1, R_2, \ldots, R_n$ are the design parameters. The difference $E(x)$ between the desired function and generated function can be represented by,

$$E(x) = f(x) - F(x, R_1, R_2, \ldots, R_n) \tag{16.7}$$

At precision points, say for $x = x_1, x_2, \ldots, x_n$, the desired and generated functions agree and $E(x) = 0$. At other points $E(x)$ will have some value, called the structural error. It is desirable that $E(x)$ should be minimum. Therefore, the spacing of precision points is very important. The precision points, according to Chebyshev's spacing, are given by:

$$x_m = a + b\cos\left[\frac{(2m-1)\pi}{2n}\right], m = 1, 2, 3 \tag{16.8}$$

where $a = \dfrac{x_i + x_f}{2}, b = \dfrac{x_f - x_i}{2}$, and

n = number of precision points.

16.9.3 Graphical Method to Locate Precision Points

The Precision points can be obtained by the graphical method from the following steps:

1. Draw a circle of radius '*b*' and centre on the *x*-axis at a distance '*a*' from point *O*.
2. Inscribe a regular polygon of side $2n$ in this circle such that the two sides are perpendicular to the *x*-axis.
3. Determine the locations of *n* accuracy points by projecting the vertices on *x*-axis as shown in Fig.16.15. It is sufficient to draw semi-circles only showing inscribed polygon to get the values of precision points.

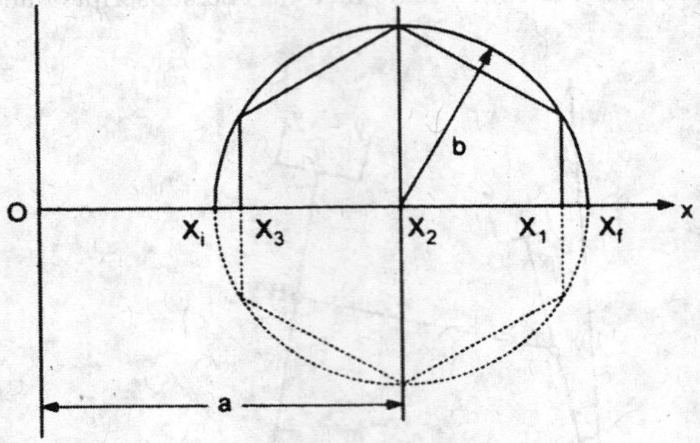

Fig.16.15 Graphical method to determine precision points

Example 16.2

Derive Freudenstein's equation for a four bar linkage.

■ **Solution**

Consider a four-bar mechanism as shown in Fig.16.16 in equilibrium. The magnitudes of the links *AB*, *BC*, *CD* and *DA* are *a*, *b*, *c* and *d* respectively. θ, β and ϕ are the angles of *AB*, *BC* and *DC* respectively with the *x*-axis. *AD* is the fixed link. *AB* is the input link and *DC* the output link.

The displacement along *x*-axis is,

$$a\cos\theta + b\cos\beta = d + c\cos\phi$$

or $$b\cos\beta = c\cos\phi - a\cos\theta + d$$

or $$(b\cos\beta)^2 = (c\cos\phi - a\cos\theta + d)^2$$
$$= c^2\cos^2\phi + a^2\cos^2\theta + d^2 - 2ac\cos\theta\cos\phi - 2ad\cos\theta + 2cd\cos\phi \quad (1)$$

The displacement along *y*-axis is,

$$a\sin\theta + b\sin\beta = c\sin\phi$$

or $$b\sin\beta = c\sin\phi - a\sin\theta$$

or $$(b\sin\beta)^2 = (c\sin\phi - a\sin\theta)^2$$
$$= c^2\sin^2\phi + a^2\sin^2\theta - 2ac\sin\theta\sin\phi \quad (2)$$

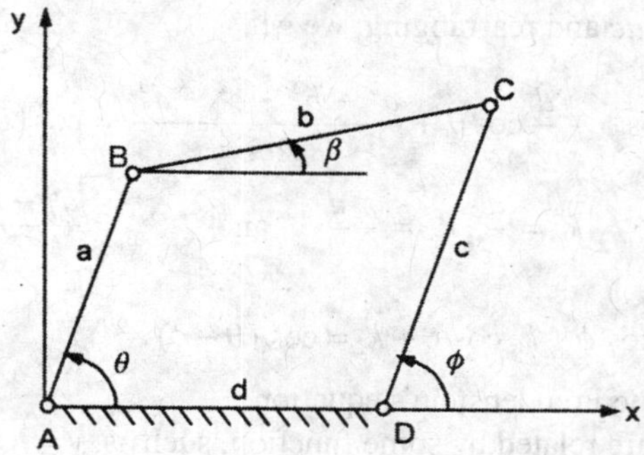

Fig.16.16 Deriving freudenstein's equation for four-bar mechanism

Adding Eqs. (1) and (2), we get

$$b^2 = c^2 + a^2 + d^2 - 2ac\,(\sin\theta\sin\phi + \cos\theta\cos\phi) - 2ad\cos\theta + 2cd\cos\phi$$

or

$$2cd\cos\phi - 2ad\cos\theta + a^2 - b^2 + c^2 + d^2 = 2ac\,(\sin\theta\sin\phi + \cos\theta\cos\phi)$$

Dividing throughout by $2ac$, we get

$$\frac{d}{a}\cos\phi - \frac{d}{c}\cos\theta + \frac{a^2 - b^2 + c^2 + d^2}{2ac} = \cos(\phi - \theta) = \cos(\theta - \phi)$$

or $$k_1\cos\phi + k_2\cos\theta + k_3 = \cos(\theta - \phi) \tag{16.9}$$

where $$k_1 = \frac{d}{a},\ k_2 = \frac{-d}{c},\ \text{and } k_3 = \frac{a^2 - b^2 + c^2 + d^2}{2ac}$$

Eq. (16.9) is known as Freudenstein equation.

16.9.4 Freudenstein's Equation for the Precision Points

Freudenstein's equation helps to determine the length of links of a four-bar mechanism. The displacement equation of a four-bar mechanism, shown in Fig.16.17 is given by:

$$2cd\cos\phi - 2ad\cos\theta + a^2 - b^2 + c^2 + d^2 = 2\,ac\,(\cos\theta\cos\phi + \sin\theta\sin\phi)$$

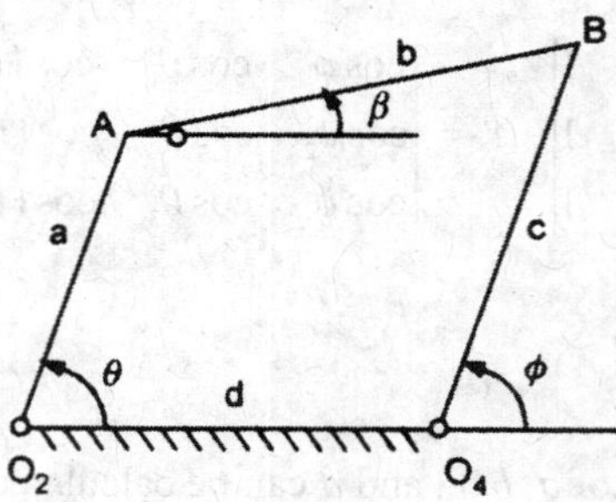

Fig.16.17 Freudenstein's equation for the precision points

Dividing throughout by $2ac$ and rearranging, we get

$$\frac{d}{a}\cos\phi - \frac{d}{c}\cos\theta + \frac{a^2 - b^2 + c^2 + d^2}{2ac} = \cos(\theta - \phi)$$

Let $$k_1 = \frac{d}{a},\ k_2 = -\frac{d}{c} \quad \text{and} \quad k_3 = \frac{a^2 - b^2 + c^2 + d^2}{2ac}$$

Then $$k_1 \cos\phi + k_2 \cos\theta + k_3 = \cos(\theta - \phi) \tag{16.10}$$

Eq. (16.10) is known as the Freudenstein's equation.

Let the input and output are related by some function, such as, $y = f(x)$.

For three specified positions, let

$$\theta_1, \theta_2, \theta_3 = \text{three positions of input link}$$
$$\phi_1, \phi_2, \phi_3 = \text{three positions of output link}$$

Then substituting these values in Eq. (16.10), we get

$$k_1 \cos\phi_1 + k_2 \cos\theta_1 + k_3 = \cos(\theta_1 - \phi_1)$$
$$k_1 \cos\phi_2 + k_2 \cos\theta_2 + k_3 = \cos(\theta_2 - \phi_2)$$
$$k_1 \cos\phi_3 + k_2 \cos\theta_3 + k_3 = \cos(\theta_3 - \phi_3)$$

These equations can be written in the matrix form as:

$$\begin{bmatrix} \cos\phi_1 \ \cos\theta_1 \ 1 \\ \cos\phi_2 \ \cos\theta_2 \ 1 \\ \cos\phi_3 \ \cos\theta_3 \ 1 \end{bmatrix} \begin{Bmatrix} k_1 \\ k_2 \\ k_3 \end{Bmatrix} = \begin{Bmatrix} \cos(\theta_1 - \phi_1) \\ \cos(\theta_2 - \phi_2) \\ \cos(\theta_3 - \phi_3) \end{Bmatrix}$$

These equations can be solved by any numerical technique.

Using Cramer's rule, let

$$A = \begin{vmatrix} \cos\phi_1 & \cos\theta_1 & 1 \\ \cos\phi_2 & \cos\theta_2 & 1 \\ \cos\phi_3 & \cos\theta_3 & 1 \end{vmatrix},\ A_1 = \begin{vmatrix} \cos(\theta_1 - \phi_1) & \cos\theta_1 & 1 \\ \cos(\theta_2 - \phi_2) & \cos\theta_2 & 1 \\ \cos(\theta_3 - \phi_3) & \cos\theta_3 & 1 \end{vmatrix}$$

$$A_2 = \begin{vmatrix} \cos\phi_1 & \cos(\theta_1 - \phi_1) & 1 \\ \cos\theta_2 & \cos(\theta_2 - \phi_2) & 1 \\ \cos\phi_3 & \cos(\theta_3 - \phi_2) & 1 \end{vmatrix},\ A_3 = \begin{vmatrix} \cos\phi_1 & \cos\theta_1 & \cos(\theta_1 - \phi_1) \\ \cos\phi_2 & \cos\theta_2 & \cos(\theta_2 - \phi_2) \\ \cos\phi_3 & \cos\theta_3 & \cos(\theta_3 - \phi_3) \end{vmatrix}$$

Then $k_1 = \dfrac{A_1}{A},\ k_2 = \dfrac{A_2}{A},\ k_3 = \dfrac{A_3}{A}$

Knowing k_1, k_2 and k_3, the values of a, b, c, and d can be calculated. Value of either 'a' or 'd' can be assumed to be unity to obtain the proportionate values of other parameters.

Example 16.3

Design a four-bar mechanism to coordinate three positions of the input and output links given by: $\theta_1 = 25$, $\phi_1 = 30°$; $\theta_2 = 35°$, $\phi_2 = 40°$; $\theta_3 = 50°$, $\phi_3 = 60°$

■ Solution

$$\cos\theta_1 = \cos 25° = 0.9063,\ \cos\theta_2 = \cos 35° = 0.8191,\ \cos\theta_3 = \cos 50° = 0.6428$$

$$\cos\phi_1 = \cos 30° = 0.8660,\ \cos\phi_2 = \cos 40° = 0.7660,\ \cos\phi_3 = \cos 60° = 0.5000$$

$$\cos(\theta_1 - \phi_1) = \cos(25° - 30°) = 0.9962$$

$$\cos(\theta_2 - \phi_2) = \cos(35° - 40°) = 0.9962$$

$$\cos(\theta_2 - \phi_2) = \cos(50° - 60°) = 0.9848$$

$$A = \begin{vmatrix} 0.8660 & 0.9063 & 1 \\ 0.7660 & 0.8191 & 1 \\ 0.5000 & 0.6428 & 1 \end{vmatrix}$$

$$= 0.8660\,(0.8191 - 0.6428) - 0.9063\,(0.7660 - 0.5000) + 1(0.7666 \times 0.6428 - 0.8191 \times 0.5000)$$

$$= 5.5652 \times 10^{-3}$$

$$A_1 = \begin{vmatrix} 0.9962 & 0.9063 & 1 \\ 0.9962 & 0.8191 & 1 \\ 0.9848 & 0.6428 & 1 \end{vmatrix}$$

$$= 0.9962\,(0.8191 - 0.6428) - 0.9063\,(0.9962 - 0.9848) + 1(0.9962 \times 0.6428 - 0.8191 \times 0.9848)$$

$$= -9.9408 \times 10^{-4}$$

$$A_2 = \begin{vmatrix} 0.8660 & 0.9063 & 1 \\ 0.7660 & 0.9962 & 1 \\ 0.5000 & 0.9848 & 1 \end{vmatrix}$$

$$= 0.8660\,(0.9962 - 0.9848) - 0.9962\,(0.7660 - 0.5000) + 1(0.7660 \times 0.9848 - 0.9962) \times 0.5000)$$

$$= -1.14 \times 10^{2}$$

$$A_3 = \begin{vmatrix} 0.8660 & 0.9962 & 0.9962 \\ 0.7660 & 0.8191 & 0.9962 \\ 0.5000 & 0.6428 & 0.9848 \end{vmatrix}$$

$$= 0.8660\,(0.8191 \times 0.9848 - 0.9962 \times 0.6428) - 0.9063\,(0.7660 \times 0.9848 - 0.9962 \times 0.5000)$$

$$= 5.746 \times 10^{-3}$$

$$k_1 = \frac{A_1}{A} = \frac{-9.9408 \times 10^{-4}}{-5.5652 \times 10^{-3}} = 0.1786 = \frac{d}{a}$$

Let $d = 1$ unit, then $a = 5.598$ units

$$k_2 = \frac{A_2}{A} = \frac{1.14 \times 10^{-3}}{-5.5652 \times 10^{-3}} = -0.2048$$

$$= -\frac{d}{c}, \; c = 4.88 \text{ units}$$

$$k_3 = \frac{A_3}{A} = \frac{-5.764 \times 10^{-3}}{-5.5652 \times 10^{-3}} = 1.0272$$

$$= \frac{a^2 - b^2 + c^2 + d^2}{2ac}$$

$$= \frac{(5.598)^2 = b^2 + (4.88)^2}{2 \times 5.598 \times 4.88}, \quad b = 0.176$$

The mechanism is shown in Fig.16.18.

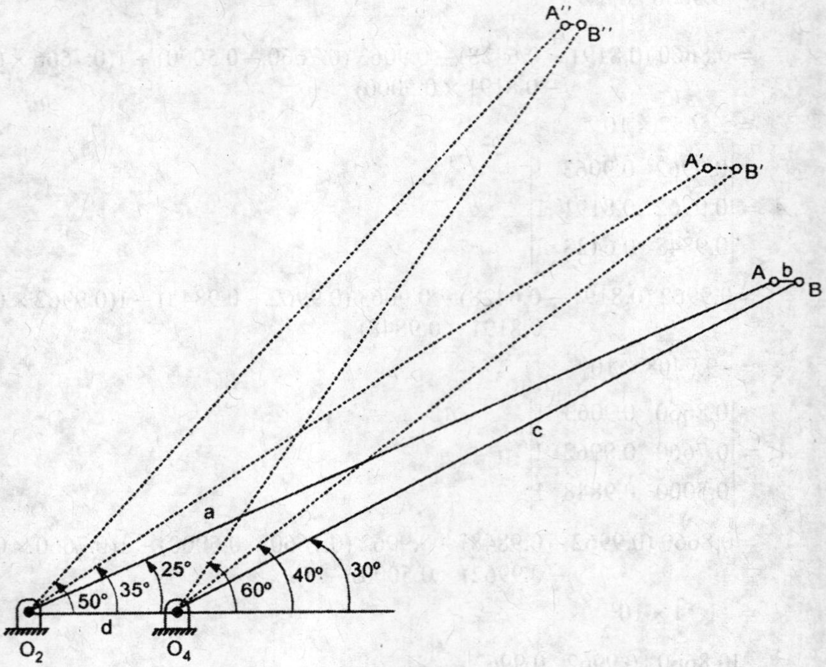

Fig.16.18 Four-bar mechanism developed by three position precision points

Example 16.4

Design a four-bar mechanism when the motions of the input and output links are governed by a function $y = 2x^2$ and x varies 2 to 4 with an interval of 1. Assume θ to vary from 40° to 120° and ϕ from 60° to 132°.

■ Solution

The angular displacement of input link is governed by x whereas that of the output link by y, θ varies from 40° to 120° (*i.e.* through 80°) and ϕ from 60° to 132° (*i.e.*, through 72°). $x = 2, 3, 4$.

The corresponding values of y are: $2 \times 2^2 = 8$, $2 \times 3^2 = 18$, $2 \times 4^2 = 32$.

$$r_x = \frac{\theta_f - \theta_i}{x_f - x_i} = \frac{120° - 40°}{4 - 2} = \frac{80}{2} = 40$$

$$\frac{\theta_2 - \theta_i}{x_2 - x_i} = r_x, \frac{\theta_2 - 40°}{3 - 2} = 40, \theta_2 = 80°$$

$$r_y = \frac{\phi_f - \phi_i}{y_f - y_i} = \frac{132 - 60}{32 - 8} = \frac{72}{24} = 3$$

$$\frac{\phi_2 - \phi_i}{y_f - y_i} = r_y = \frac{\phi_2 - 60}{18 - 8} = 3, \phi_2 = 90°$$

Precision point	x	y	θ, deg	ϕ, deg
1	2	8	40	60
2	3	18	80	90
3	4	32	120	132

$$\cos 40° = 0.7660, \cos 60° = 0.5000, \cos(40° - 60°) = 0.9397$$
$$\cos 80° = 0.1736, \cos 90° = 00, \cos(80° - 90°) = 0.9848$$
$$\cos 120° = -0.5, \cos 132° = 0.6691, \cos(120° - 132°) = 0.9781$$

$$A = \begin{vmatrix} 0.5000 & 0.7660 & 1 \\ 0.0000 & 0.1736 & 1 \\ -0.6691 & -0.5000 & 1 \end{vmatrix} = -0.0596, A_1 = \begin{vmatrix} 0.9397 & 0.7660 & 1 \\ 0.9848 & 0.1736 & 1 \\ 0.9791 & -0.5000 & 1 \end{vmatrix} = -0.03455$$

$$A_2 = \begin{vmatrix} 0.5000 & 0.9397 & 1 \\ 0.0000 & 0.9848 & 1 \\ -0.6691 & 0.9781 & 1 \end{vmatrix} = 0.0335, A_3 = \begin{vmatrix} 0.5000 & 0.7660 & 0.9397 \\ 0.0000 & 0.1736 & 0.9848 \\ -0.6691 & -0.5000 & 0.9781 \end{vmatrix} = -0.0645$$

$$k_1 = \frac{A_1}{A} = 0.5763 = \frac{d}{a}, a = \frac{1}{0.5763} = 1.73$$

$$k_2 = \frac{A_2}{A} = -0.562 = -\frac{d}{c}, c = 1.78$$

$$k_3 = \frac{A_3}{A} = \frac{-0.0645}{-0.0596} = 1.0802$$

$$= \frac{a^2 - b^2 + c^2 + d^2}{2ac} = \frac{(1.73)^2 - b^2 + (1.78)^2 + 1}{2 \times 1.73 \times 1.78}, \quad b = 0.7$$

The mechanism is shown in Fig.16.19.

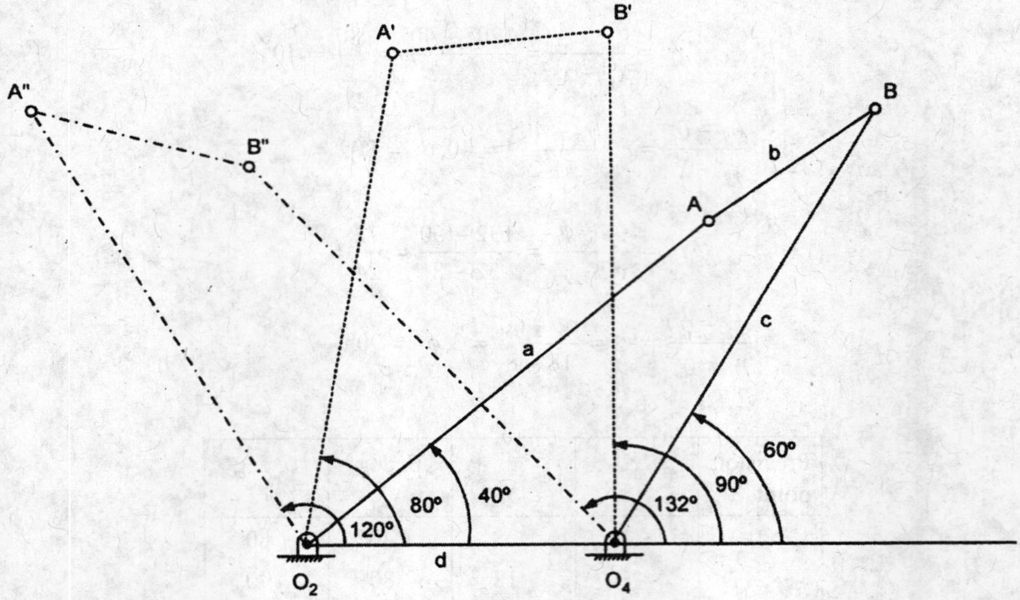

Fig.16.19 Four-bar mechanisms generated by a function $y = 2x^2$

Example 16.5

Determine the lengths of all the four links of a four-bar mechanism to generate $y = \log x$ in the interval $1 \le x \le 11$ for three precision points. The length of the smallest links is 10 cm the range of input angles is $45° \le \theta \le 105°$ and output angles is $135° \le \phi \le 225°$.

■ Solution

$x_i = 1, x_f = 11, n = 3$

Using Chebyshev's precision points,

$$a = \frac{1}{2}\left(x_i + x_f\right) = \frac{1}{2}(1+11) = 6$$

$$b = \frac{1}{2}\left(x_f - x_i\right) = \frac{1}{2}(11-1) = 5$$

$$x_m = a + b\cos\left[\frac{(2m-1)\pi}{2n}\right], m = 1, 2, 3$$

$$x_1 = a + b\cos\left[\frac{(2\times 1-1)\pi}{6}\right] = a + b\cos\frac{\pi}{6} = 6 + 5\ \cos\ \frac{\pi}{6} = 10.33$$

$$x_2 = a + b\cos\frac{\pi}{2} = 6$$

$$x_3 = a + b\cos\left(\frac{5\pi}{6}\right) = 6 + 5\cos\left(\frac{5\pi}{6}\right) = 1.67$$

$$y_1 = \log x_1 = \log 10.33 = 1.014$$
$$y_2 = \log x_2 = \log 6 = 0.778$$
$$y_3 = \log x_3 = \log 1.67 = 0.223$$
$$y_i = \log x_i = \log 1 = 0$$
$$y_f = \log x_f = \log 11 = 1.0414$$

Scale factors are:

$$r_x = \frac{\theta_f - \theta_i}{x_f - x_i} = \frac{105 - 45}{11 - 1} = \frac{60}{10} = 6$$

$$r_y = \frac{\phi_f - \phi_i}{y_f - y_i} = \frac{225 - 135}{1.0414 - 0} = \frac{90}{1.0414} = 86.423$$

$$r_x = \frac{\theta - \theta_i}{x - x_i}, \frac{\theta_1 - \theta_i}{x_1 - x_i} = \frac{\theta_1 - 45°}{10.33 - 1} = 6, \theta_1 = 100.98°$$

$$\frac{\theta_2 - 45°}{6 - 1} = 6, \theta_2 = 75°, \frac{\theta_3 - 45°}{1.67 - 1} = 6, \theta_3 = 49.02°$$

$$r_y = \frac{\phi - \phi_i}{y - y_i}, \frac{\phi_1 - \phi_i}{y_1 - y_i} = \frac{\phi_1 - 135°}{1.014 - 0} = 86.423, \phi_1 = 222.63°$$

$$\frac{\phi_2 - 135°}{0.778 - 0} = 86.423, \phi_2 = 202.24°$$

$$\frac{\phi_3 - 135°}{0.223 - 0} = 86.423, \phi_3 = 154.27°$$

$$d^2 = (0.221 + 1.908 + 1 + 1.538) = b^2$$

$$4.667\, d^2 = b^2$$

$$\frac{d}{b} = 0.463$$

$$\therefore \quad \frac{d}{a} = 2.126, \quad \frac{d}{b} = 0.463, \quad \frac{d}{c} = -0.724$$

$$\therefore \quad \frac{d}{a} > \frac{d}{c} > \frac{d}{b} \text{ or } a \angle d \angle c \angle b$$

$\therefore$ Link a is the smallest. Thus $a = 10$ cm, $d = 21.26$ cm

$$b = \frac{21.26}{0.463} = 45.92 \text{ cm}, \quad c = \frac{-21.26}{0.724} = -29.36 \text{ cm}.$$

–ve sign indicates that the length C is to be drawn in the reverse direction.

Example 16.6

Design a four-bar mechanism so that $\theta_{12} = 70°$ and $\phi_{12} = 50°$. Length of fixed link is 4 cm. Input and output links rotate anti-clockwise.

■ Solution

Refer to Fig.16.20.

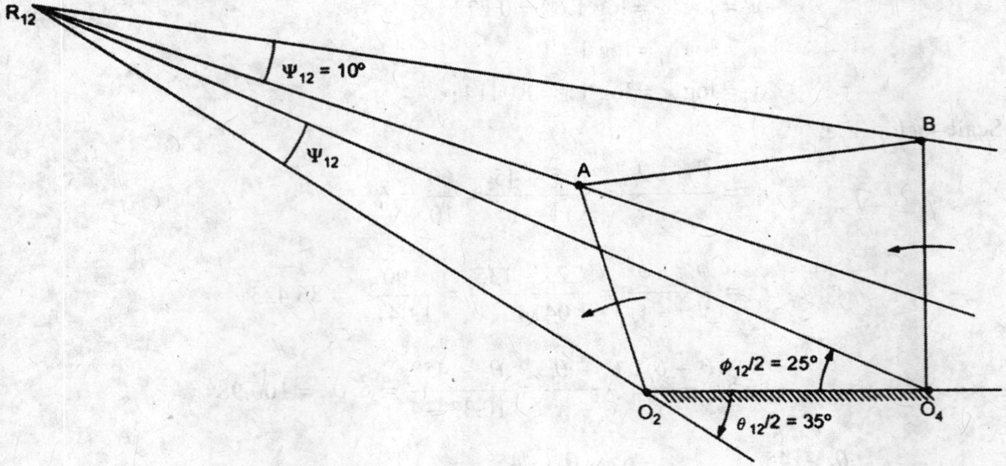

Fig.16.20 Four-bar mechanism generation for $\theta = 70°$ and $\phi_{12} = 50°$

1. Draw $O_2O_4 = 4$ cm. Rotate O_2O_4 about O_2 through $\theta_{12}/2 = 35°$ clockwise.

Precision point	x	y	ϕ, deg	$\cos\phi$	θ, deg	$\cos\theta$	$\theta-\phi$, deg	$\cos(\theta-\phi)$
1	10.33	1.014	222.63	−0.736	100.98	−0.190	−121.65	−0.525
2	6	0.778	202.24	−0.926	75	0.259	−127.24	−0.605
3	1.67	0.223	154.27	−0.901	49.02	0.656	−105.25	−0.263

The Freudenstein's equations becomes

$$\begin{bmatrix} -0.736 & -0.190 & 1 \\ -0.926 & 0.259 & 1 \\ -0.901 & 0.656 & 1 \end{bmatrix} \begin{Bmatrix} k_1 \\ k_2 \\ k_3 \end{Bmatrix} = \begin{Bmatrix} -0.525 \\ -0.605 \\ -0.263 \end{Bmatrix}$$

$$A = \begin{vmatrix} -0.736 & -0.190 & 1 \\ -0.926 & 0.259 & 1 \\ -0.901 & 0.656 & 1 \end{vmatrix} = -0.087$$

$$A_1 = \begin{vmatrix} -0.525 & -0.190 & 1 \\ -0.605 & 0.259 & 1 \\ -0.263 & 0.656 & 1 \end{vmatrix} = -0.185, \; A_2 = \begin{vmatrix} -0.736 & -0.525 & 1 \\ -0.926 & -0.605 & 1 \\ -0.901 & -0.263 & 1 \end{vmatrix} = -0.063$$

$$A_3 = \begin{vmatrix} -0.736 & -0.190 & -0.525 \\ -0.926 & 0.259 & -0.605 \\ -0.901 & 0.656 & -0.263 \end{vmatrix} = -0.103$$

$$k_1 = \frac{A_1}{A} = \frac{-0.185}{-0.087} = 0.126 = \frac{d}{a}, a = \frac{d}{2.126}$$

$$k_2 = \frac{A_2}{A} = \frac{-0.063}{-0.087} = 0.724 = -\frac{d}{c} \text{ or } \frac{d}{c} = -0.724, c = \frac{-d}{0.724}$$

$$k_3 = \frac{A_3}{A} = \frac{-0.103}{-0.087} = 1.184 = \frac{a^2 - b^2 + c^2 + d^2}{2ac}$$

$$\left(\frac{d}{2.126}\right)^2 - b^2 + \left(\frac{-d^2}{0.724}\right) + d^2 = 1.184 \times 2 \times \frac{d}{2.126} \times \left(\frac{-d^2}{0.724}\right)$$

$$\frac{d^2}{4.52} - b^2 + \frac{d^2}{0.524} + d^2 = -1.538\, d^2$$

2. Rotate O_2O_4 about O_4 through $\phi_{12}/2 = 25°$ clockwise. The point of intersection of these two lines locates relative pole R_{12}.
3. Construct an angle $\psi_{12} = \frac{1}{2}(\theta_{12} - \phi_{12}) = \frac{1}{2}(70° - 50°) = 10°$ at R_{12}. Join any two points on the two arms of this angle to obtain the coupler link AB. Join A with O_2 and B with O_4 to get the input and output links.
4. $O_2A\, BO_4$ is the desired mechanism.

Example 16.7

Design a four-bar mechanism such that $\theta_{12} = 120°$, $\theta_{13} = 160°$ and $\phi_{12} = 70°$, $\phi_{13} = 100°$. Input link rotates clockwise and output link also rotates clockwise. A length of fixed link is 5 cm.

Solution

Refer to Fig.16.21.

$$\theta_{12}/2 = \frac{120}{2} = 60°,\ \phi_{12}/2 = \frac{70}{2} = 35°, \Psi_{12} = \frac{1}{2}(\theta_{12} - \phi_{12}) = 25°$$

$$\theta_{13}/2 = \frac{160}{2} = 80°,\ \phi_{13}/2 = \frac{100}{2}, \Psi_{13} = \frac{1}{2}(\theta_{13} - \phi_{12}) = 30°$$

1. Draw $O_2O_4 = 4$ cm. Rotate R_{12} and R_{13}.
2. Construct angles ψ_{13} and ψ_{12} at R_{12} and R_{13} respectively such that the arms of the angles intersect at points A and B. Join AB to get the coupler.
3. Join O_2A and O_4B to get the input and output links respectively.
4. O_2ABO_4 is the required mechanism.

Example 16.8

Design a four-bar mechanism such that $\theta_{12} = 120°$, $\theta_{13} = 160°$ and $\phi_{12} = 70°$, $\phi_{12} = 70°$, $\phi_{13} = 100°$. Input link rotates clockwise and output link rotates anti-clockwise. Length of fixed link is 10 cm.

Solution

Refer to Fig.16.22.

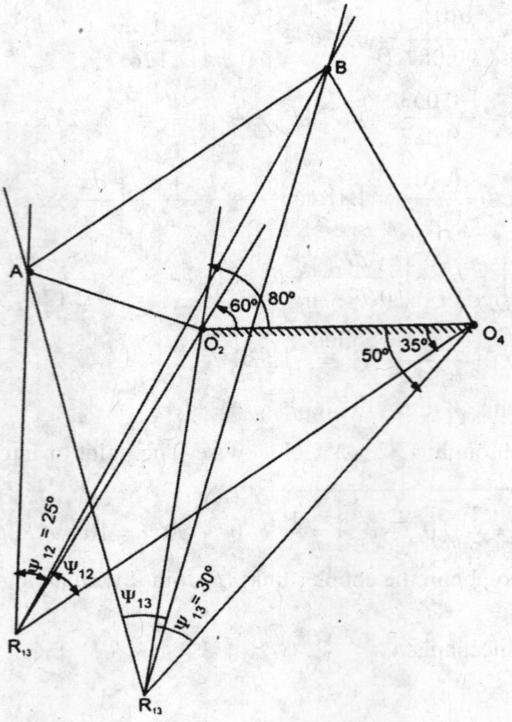

Fig.16.21 Four-bar mechanism generation for Example 16.7

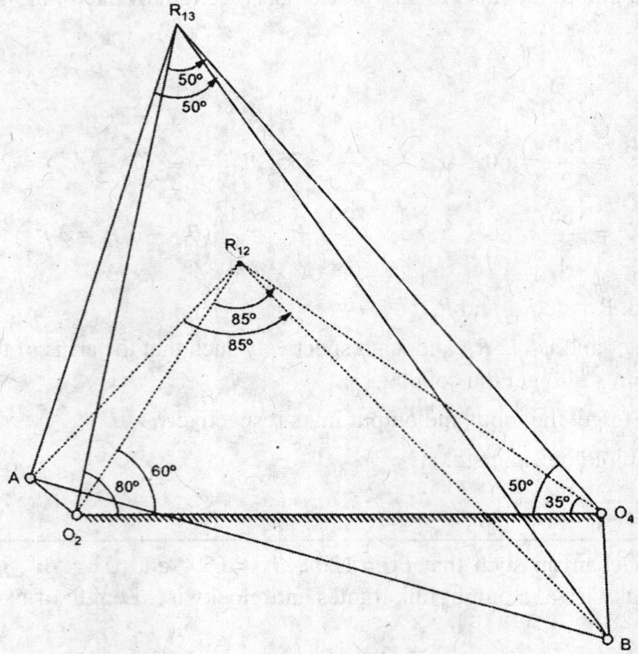

Fig.16.22 Four-bar mechanism generation for Example 16.8

$$\theta_{12}/2 = 60°,\ \phi_{12}/2 = 35°, \Psi_{12} = 180° - \frac{1}{2}(\theta_{12} + \phi_{12}) = 180° - (60 + 35) = 85°$$

$$\theta_{13}/2 = 80°,\ \phi_{13}/2 = 50°, \Psi_{13} = 180° - \frac{1}{2}(\theta_{13} + \phi_{13}) = 180° - (80 + 50) = 50°$$

1. Draw $O_2O_4 = 10$ cm, with O_2 as centre rotate O_2O_4 anti-clockwise through 60°. With O_4 as centre rotate O_2O_4 through 35° clockwise. Intersection of these two arcs give R_{12}.
2. With O_2 as centre rotate O_2O_4 anti-clockwise through 80°. With O_4 as centre rotate O_2O_4 through 50° clockwise. Intersection of these two arcs give R_{13}.
3. Construct angles ψ_{12} and ψ_{13} at the points R_{12} and R_{13} respectively such that the arms of the angles intersect at A and B. Join AB to get the coupler AB.
4. Join A with O_2 and B with O_4 to get the input and output links.

Example 16.9

Synthesize a slider-crank mechanism with eccentricity, $e = 1$ cm for the two input positions of input link, $\theta_{12} = 60°$ and output displacement, $x = 1.5$ cm to the right.

Solution

Refer to Fig.16.23

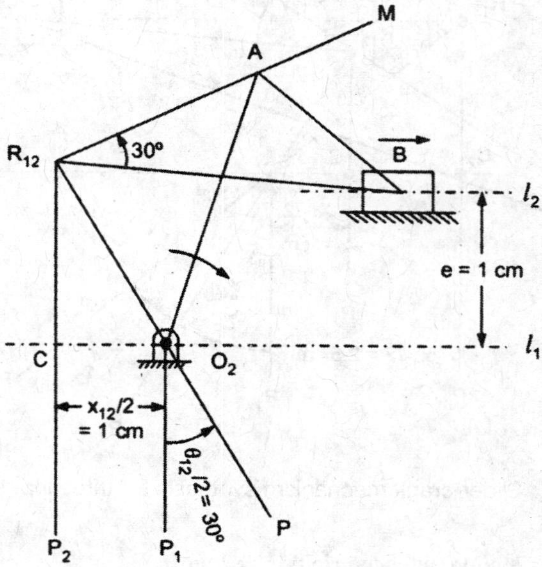

Fig.16.23 Slider-crank mechanism synthesis for two input positions

1. Draw two parallel line l_1 and l_2 at an eccentricity, $e = 1$ cm.
2. Select an arbitrary point O_2 on line l_1. Measure $CO_4 = \frac{x}{2} = \frac{1.5}{2} = 0.75$ cm to the left. Draw CP_2 and O_2P_1 perpendiculars to lines l_1.

3. Make $\angle\theta_{12}/2 = 30° = \angle P_1O_2P$ in the anti-clockwise direction. The intersection of P_1O_2 and P_2C produced locates the relative pole R_{12}. Join R_{12} B.
4. Make $\angle\theta_{12}/2 = 30°$ with BR_{12} in the anti-clockwise direction.
5. Choose an arbitrary point A on the arm of the angle other than the one passing through point B. Join AB to get the coupler. Also join AO_2 to get the input crank.

Example 16.10

Synthesize a slider-crank mechanism for its three positions $\theta_{12} = 60°$ and $\theta_{13} = 100°$ of the input crank and three positions $x_{12} = 2$ cm and $x_{13} = 5$ cm of the output slider block. The eccentricity is 2 cm. The slider is moving outwards.

Solution

Refer to Fig.16.24.

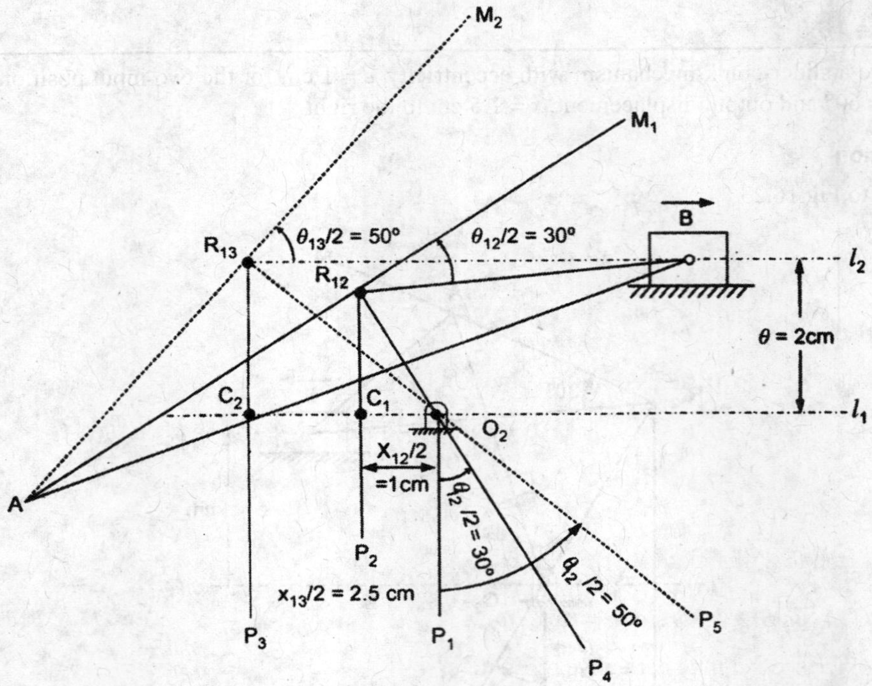

Fig.16.24 Slider-crank mechanism synthesis for three input positions

1. Draw two parallel lines l_1 and l_2 at eccentricity, $e = 2$ cm.
2. Select an arbitrary point O_2 on l_1. Measure $O_2 C_1 = x_{12}/2 = 1$ cm and $O_2 C_2 = x_{13/2} = 2.5$ cm to the left of O_2. Draw O_2P_1, C_1P_2, $C_2 P_3$ perpendiculars on line l_1.
3. Draw $\angle P_1 O_2 P_4 = \theta_{12}/2 = 30°$ and $\angle P_1O_2 P_5 = \theta_{13}/2 = 50°$ in the anti-clockwise direction.
4. Produce $O_2 P_4$ and $O_2 P_5$ lines to meet the perpendiculars at C_1 and C_2 to locate the relative poles R_{12} and R_{13}.
5. Join R_{12} B and R_{13} B.

6. Draw $\angle B R_{12} M_1 = \theta_{12}/2 = 30°$ and $\angle B R_{13} M_2 = \theta_{13}/2 = 50°$ in the anti-clockwise direction. Produce the lines $R_{12} M_1$ and $R_{13} M_2$ backward to meet at A.
7. Join AB and AO_2 to get the coupler and input crank respectively.

Example 16.11

For the four-bar linkage, the following data are given:

$$\theta_2 = 60°,\ \theta_4 = 90°$$
$$\omega_2 = 3 \text{ rad/s},\ \omega_4 = 2 \text{ rad/s}$$
$$\alpha_2 = -1 \text{ rad/s}^2,\ \alpha_4 = 0$$

Determine the link-length ratios.

■ Solution

The Freudenstein's equation for the four-bar mechanism shown in Fig.16.25 is:

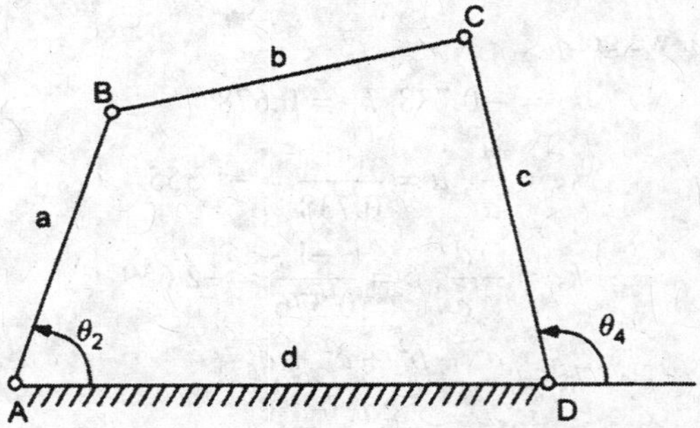

Fig.16.25 Systhesis of four-bar mechanism for Example 16.11

$$k_1 \cos\theta_4 + k_2 \cos\theta_2 + k_3 = \cos(\theta_2 - \theta_4) \quad (1)$$

where
$$k_1 = \frac{d}{a},\ k_2 = \frac{d}{c},\ k_3 = \frac{a^2 - b^2 + c^2 + d^2}{2ac}$$

Taking the first time derivative of Eq. (1), we have

$$k_1\, \omega_4 \sin\theta_4 + k_2\, \omega_2 \sin\theta_2 = (\omega_2 - \omega_4)\sin(\theta_2 - \theta_4) \quad (2)$$

Taking the second time derivative of Eq. (2)

$$k_1\left(\omega_4^2 \cos\theta_4 + \alpha_4 \sin\theta_4\right) + k_2\left(\omega_2^2 \cos\theta_2 + \alpha_2 \sin\theta_2\right)$$
$$= (\alpha_2 - \omega_4)\sin\left(\theta_2 - \theta_4\right) + \left(\omega_2 - \omega_4\right)^2 \cos\left(\theta_2 - \theta_4\right) \quad (3)$$

Substituting the values of various terms in Eqs. (1) and (3), we get

$$k_1 \cos 90° + k_2 \cos 60° + k_3 = \cos(60° - 90°)$$

or
$$0.5\, k_2 + k_3 = \frac{\sqrt{3}}{2}$$

or
$$k_2 + 2k_3 = \sqrt{3} \quad (4)$$

$$k_1 \times 2 \times \sin 90° + k_2 \times 3 \sin 60° = (3-2)\sin(60° - 90°)$$

or $$2k_1 + \frac{3\sqrt{3}}{2}k_2 = -\frac{1}{2}$$

or $$4k_1 + 3\sqrt{3}\,k_2 = -1 \quad (5)$$

$$k_1(4\ \cos 90° + 0 \times \sin 90°) + k_2\,(9 \times \cos 60° - 1 \times \sin 60°)$$
$$= (-1-0) \times \sin\,(60° - 90°) + (3-2)^2 \times \cos(60° - 90°)$$

or $$k_2\left[9 \times \frac{1}{2} - \frac{\sqrt{3}}{2}\right] = (-1) \times \left(-\frac{1}{2}\right) + 1 \times \frac{\sqrt{3}}{2}$$

$$k_2\ (4.5 - 0.866) = \frac{1}{2} + \frac{\sqrt{3}}{2}$$
$$k_2 = 0.376 \quad (6)$$

Solving Eqns. (4) to (6), we get

$$k_1 = -0.738,\ k_3 = 0.678$$
$$k_2 = \frac{d}{a},\ a = \frac{-1}{0.738} = -1.355$$
$$k_2 = -\frac{d}{c},\ c = \frac{-1}{0.376} = -2.659$$
$$k_3 = \frac{a^2 - b^2 + c^2 + d^2}{2ac}$$

$$0.678 \times 2 \times (-1.355) \times (-2.659) = (-1.355)^2 - b^2 + (-2.659)^2 + 1$$
$$b^2 = 5.0207,\ b = 2.24$$

Thus $$d = 1.0,\ a = -1.355,\ b = 2.24,\ c = -2.659.$$

16.10 FREUDENSTEIN'S EQUATION FOR SLIDER-CRANK MECHANISM FOR THREE PRECISION POINTS

The mechanism is shown in Fig.16.26. The displacement of the slider B has to be coordinated with the rotation of the crank OA. Let us assume that the displacement of the slider is proportional to the crank rotation expressed as:

$$s_f - s_i = c(\theta_f - \theta_i)$$

where c = a constant of proportionality

s = distance of slider from the origin O

θ = angle of rotation of crank from the line of stroke

i, f = subscripts for initial and final values respectively.

The coordinates of points A and B are:

$$x_A = l_2 \cos\theta,\ y_A = l_2 \sin\theta$$
$$x_B = s,\ y_B = e$$

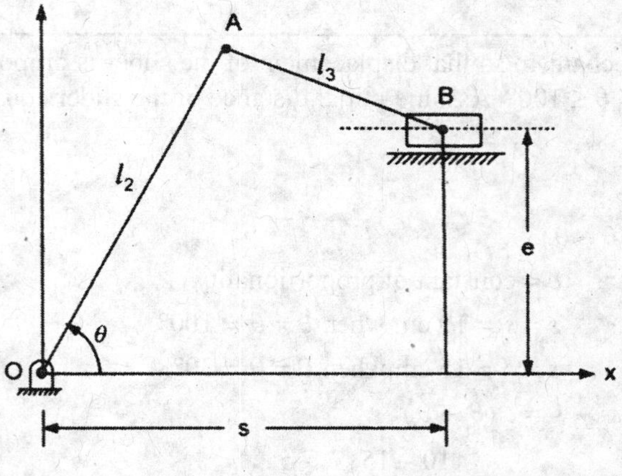

Fig.16.26 Deriving Freudenstein's equation for slider-crank mechanism

Also
$$
\begin{aligned}
AB^2 &= (x_B - x_A)^2 + (y_A - y_B)^2 \\
&= (s - l_2 \cos\theta)^2 + (l_2 \sin\theta - e)^2 \\
&= s^2 + l_2^2 \cos^2\theta - 2\, s\, l_2 \cos\theta + l_2^2 \sin^2\theta + e^2 - 2\, l_2\, e \sin\theta \\
l_3^2 &= s^2 + l_2^2 + e^2 - 2\, s\, l_2 \cos\theta - l_2\, e \sin\theta
\end{aligned}
$$

or
$$s^2 = (l_3^2 - l_2^2 - e^2) + 2\, s\, l_2 \cos\theta + 2l_2\, e \sin\theta$$

Let
$$k_1 = 2l_2, k_2 = 2l_{2e}, \text{ and } k_3 = l_3^2 - l_2^2 - e^2$$

Then
$$s^2 = k_1 s \cos\theta + k_2 \sin\theta + k_3 \tag{16.11}$$

Equation (16.11) is the Freudenstein's equation for the slider-crank mechanism.

Now
$$c = \frac{s_f - s_i}{\theta_f - \theta_i}$$

$\therefore$
$$s - s_i = c(\theta - \theta_i)$$

The precision points according to Chebyshev are given by,

$$\theta_m = a + b\cos\left[\frac{(2m-1)\pi}{2n}\right], m = 1, 2, 3$$

where
$$a = \frac{\theta_i - \theta_f}{2}, b = \frac{\theta_f - \theta_i}{2}, n = 3$$

The following steps are followed to determine lengths.

1. Obtain three accuracy points by using Chebyshev's spacing, *i.e.*, θ_1, θ_2, and θ_3.
2. Calculate values of s_1, s_2, and s_3 from the following equations: $s_m - s_i = c\,(\theta_m - \theta_i)$, $m = 1, 2, 3$.
3. Obtain three equations in k_1, k_2 and k_3.
4. Solve for k_1, k_2 and k_3 using Cramer's rule.
5. Obtain the values of l_2, l_3 and e from the values of k_1, k_2 and k_3.

Example 16.12

Design a slider-crank mechanism so that displacement of the slider is proportional to the crank rotation in the interval $30° \le \theta \le 100°$. Assume initial distance of the slider equal to 15 cm and final distance to be 10 cm.

Solution

Given: $s - s_i \propto (\theta - \theta_i)$

where c = constant of proportionality.

Now $s = s_f = 10$ cm when $\theta = \theta_f = 100°$

and $s = s_i = 15$ cm when $\theta = \theta_i = 30°$.

$$\therefore \quad s_f - s_i = c\,(\theta_f - \theta_i)$$

$$c = \frac{10-15}{100-30} = \frac{-5}{70} = -\frac{1}{14}$$

$$\therefore \quad s - s_i = \left(-\frac{1}{14}\right)(\theta - \theta_i)$$

The Chebyshev's precision spacing is given by:

$$\theta_m = a + b\cos\left[\frac{(2m-1)\pi}{2n}\right], m = 1, 2, 3$$

where $a = \dfrac{\theta_i + \theta_f}{2} = \dfrac{30+100}{2} = 65°$

$$b = \frac{\theta_f - \theta_i}{2} = \frac{1}{2}(100-30) = 35°$$

$$\theta_1 = 65 + 35\cos\frac{\pi}{6} = 95.3°$$

$$\theta_2 = 65 + 35\cos 90° = 65°$$

$$\theta_3 = 65 + 35\cos\frac{5\pi}{6} = 34.69°$$

Now $s_m - s_i = \left(-\dfrac{1}{14}\right)(\theta_m - \theta_i)$

$$\therefore \quad s_1 - 15 = \left(-\frac{1}{14}\right)(98.3 - 30),\ s_1 = 10.336 \text{ cm}$$

$$s_2 - 15 = \left(-\frac{1}{14}\right)(65 - 30),\ s_2 = 12.50 \text{ cm}$$

$$s_3 - 15 = \left(-\frac{1}{14}\right)(34.69 - 30),\ s_3 = 14.665 \text{ cm}$$

Now $k_1 s\cos\theta + k_2 \sin\theta + k_3 = s^2$

$$k_1 \times 10.336 \times \cos 95.3° + k_2 \sin 95.3° + k_3 = (10.336)^2$$

$$-0.95474\,k_1 + 0.99572\,k_2 + k_3 = 106.833 \qquad (1)$$

$$k_1 \times 12.50 \times \cos 65° + k_2 \sin 65° + k_3 = (12.50)^2$$

$$5.28273\ k_1 + 0.90631\ k_2 + k_3 = 156.25 \tag{2}$$

$$k_1 \times 14.665 \times \cos 34.69° + k_2 \sin 34.69° + k_3 = (14.665)^2$$

$$12.0582\ k_1 + 0.56914\ k_2 + k_3 = 215.062 \tag{3}$$

Subtracting Eqs. (1) and (2) from Eq. (3), we get

$$13.013\ k_1 - 0.4266\ k_2 = 108.229 \tag{4}$$

$$6.775\ k_1 - 0.3372\ k_2 = 58.812 \tag{5}$$

$$30.504\ k_1 - k_2 = 253.70 \tag{6}$$

$$20.092\ k_1 - k_2 = 174.41 \tag{7}$$

Subtracting Eq. (7) from Eq. (6), we get

$$10.412\ k_1 = 79.29$$

$\therefore$ $$k_1 = 7.615$$

Then $$k_2 = -21.4$$

$$k_3 = 135.418$$

Now $$k_1 = 2l_2, l_2 = \frac{7.615}{2} = 3.808 \text{ cm}$$

$$k_2 = 2l_2 e, e = \frac{-21.4}{7.615} = -2.81 \text{ cm}$$

$$k_3 = l_3^2 - l_3^2 - e^2$$

$$135.418 = l_3^2 - (3.808)^2 - (2.81)^2$$

$$l_3^2 - 157.815$$

$$l_3 = 12.562 \text{ cm}$$

Thus $$l_2 = 3.808 \text{ cm}, l_3 = 12.562 \text{ cm}, e = -2.81 \text{ cm}$$

16.11 LEAST SQUARE TECHNIQUE

A four-link mechanism can be designed precisely up to five positions of the input and the output links, provided θ and ϕ are measured from some arbitrary reference. In such cases, the synthesis equations become non-linear and cannot be solved by the Cramer's rule. It is also possible to design a mechanism for more than five positions which gives least deviation from the specified positions and provides the average performance. The least-square technique is useful to synthesize such a mechanism.

Consider the Freudenstein's equation,

$$k_1 \cos \phi_i + k_2 \cos \theta_i + k_3 - \cos (\theta_i - \phi_i) = 0$$

Summing up for n values of θ and ϕ and defining,

$$S = \sum_{i=1}^{n} [k_1 \cos \phi_i + k_2 \cos \theta_i + k_3 - \cos (\theta_i - \phi_i)]^2 \tag{16.12}$$

For S to be minimum, the conditions are:

$$\frac{\partial S}{\partial k_1} = 0, \quad \frac{\partial S}{\partial k_2} = 0 \quad \text{and} \quad \frac{\partial S}{\partial k_3} = 0$$

which gives, $\sum_{i=1}^{n} [k_1 \cos\phi_i + k_2 \cos\theta_i + k_3 - \cos(\theta_i - \phi_i)] \cos\phi_i = 0$

or $k_1 \Sigma \cos^2\phi_i + k_2 \Sigma \cos\theta_i \cos\phi_i + k_3 \Sigma \cos\phi_i = \Sigma \cos(\theta_i - \phi_i) \cos\phi_i$ (1)

Similarly,

$k_1 \Sigma \cos\phi_i \cos\theta_i + k_2 \Sigma \cos^2\theta_i + k_3 \Sigma \cos\theta_i = \Sigma \cos(\theta_i - \phi_i) \cos\phi_i$ (2)

and $k_1 \Sigma \cos\phi_i + k_2 \Sigma \cos\theta_i + k_3 \Sigma 1 = \Sigma \cos(\theta_i - \phi_i)$ (3)

Eqs. (1) to (3) are three simultaneous linear, non-homogeneous equations in three unknowns k_1, k_2 and k_3. These can be solved by using the Cramer's rule. These equations can be written in matrix form as:

$$\begin{bmatrix} \Sigma \cos^2\phi_i & \Sigma \cos\theta_i \cos\phi_i & \Sigma \cos\phi_i \\ \Sigma \cos\phi_i \cos\theta_i & \Sigma \cos^2\theta_i & \Sigma \cos\theta_i \\ \Sigma \cos\phi_i & \Sigma \cos\theta_i & 1 \end{bmatrix} \begin{Bmatrix} k_1 \\ k_2 \\ k_3 \end{Bmatrix} = \begin{Bmatrix} \Sigma \cos(\theta_i - \phi_i) \cos\phi_i \\ \Sigma \cos(\theta_i - \phi_i) \cos\theta_i \\ \Sigma \cos(\theta_i - \phi_i) \end{Bmatrix}$$

Example 16.13

A 4-bar mechanism is required such that the input and out angles are coordinated as given below:

Input Crank angle	30°	50°	80°
Output follower angle	0°	30°	50°

Synthesize the four-bar mechanism.

Solution

The Freudenstein's equation for displacement of a four-bar mechanism is,

$$k_i \cos\phi_i + k_2 \cos\theta_i + k_3 = \cos(\theta_i - \phi_i)$$

Position	θ_i deg	$\cos\theta_i$	ϕ_i, deg	$\cos\phi_i$	$\theta_i - \phi_i$, deg	$\cos(\theta_i - \phi_i)$
1	30	0.866	0	1.000	30	0.866
2	50	0.643	30	0.866	20	0.940
3	80	0.174	60	0.500	20	0.940

$$\begin{bmatrix} 1.000 & 0.866 & 1 \\ 0.866 & 0.643 & 1 \\ 0.500 & 0.714 & 1 \end{bmatrix} \begin{Bmatrix} k_1 \\ k_2 \\ k_3 \end{Bmatrix} = \begin{Bmatrix} 0.866 \\ 0.940 \\ 0.940 \end{Bmatrix}$$

Solving by Cramer's rule,

$$\Delta = \begin{bmatrix} 1.000 & 0.866 & 1 \\ 0.866 & 0.643 & 1 \\ 0.500 & 0.174 & 1 \end{bmatrix} = -0.01877$$

$$\Delta_1 = \begin{bmatrix} 0.866 & 0.866 & 1 \\ 0.940 & 0.643 & 1 \\ 0.940 & 0.714 & 1 \end{bmatrix} = -0.0347$$

$$\Delta_2 = \begin{bmatrix} 1.000 & 0.866 & 1 \\ 0.866 & 0.940 & 1 \\ 0.500 & 0.940 & 1 \end{bmatrix} = -0.02708$$

$$\Delta_3 = \begin{bmatrix} 1.000 & 0.866 & 0.866 \\ 0.866 & 0.643 & 0.940 \\ 0.500 & 0.174 & 0.940 \end{bmatrix} = -0.005$$

$$k_1 = \frac{\Delta_1}{\Delta} = \frac{-0.0347}{-0.01877} = 1.8487$$

$$k_2 = \frac{\Delta_2}{\Delta} = \frac{-0.02708}{-0.01877} = -1.4427$$

$$k_3 = \frac{\Delta_3}{\Delta} = \frac{-0.005}{-0.01877} = 0.2664$$

$$k_1 = \frac{d}{a},\ a = \frac{1}{1.8487} = 0.541\text{units for } d = 1 \text{ unit}$$

$$k_2 = -\frac{d}{c},\ c = \frac{-1}{-1.4427} = 0.693 \text{ units}$$

$$k_3 = \frac{a^2 - b^2 + c^2 + d^2}{2ac}$$

$$0.2664 = \frac{(0.541)^2 - b^2 + (0.693)^2 + 1}{2 \times 0.541 \times 0.693}$$

$$b^2 = 1.5732$$

$$b = 1.254 \text{ units}$$

The lengths of various links are:

$a = 0.541$ units, $b = 1.254$ units, $c = 0.693$ units, $d = 1$ unit.

The mechanism is shown in Fig.16.27.

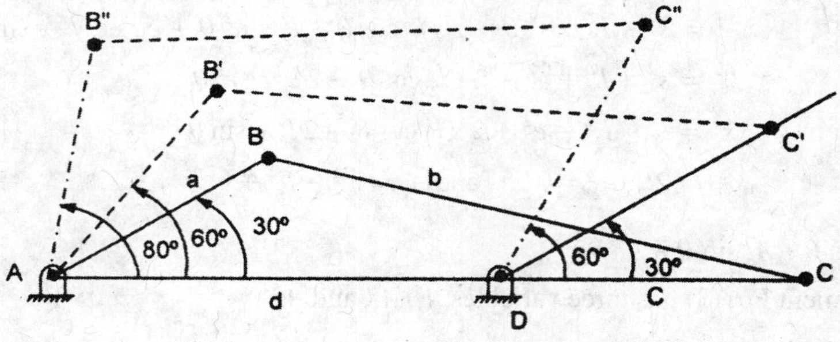

Fig.16.27 Synthesis of four-bar mechanism for Example 16.13

Example 16.14

Design a slider-crank mechanism so that the displacement of the slider is proportional to the square of the crank rotation in the interval $45° \leq \theta \leq 135°$. Use three point Chebyshev spacing.

Solution

Consider the slider-crank mechanism as shown in Fig.16.28.

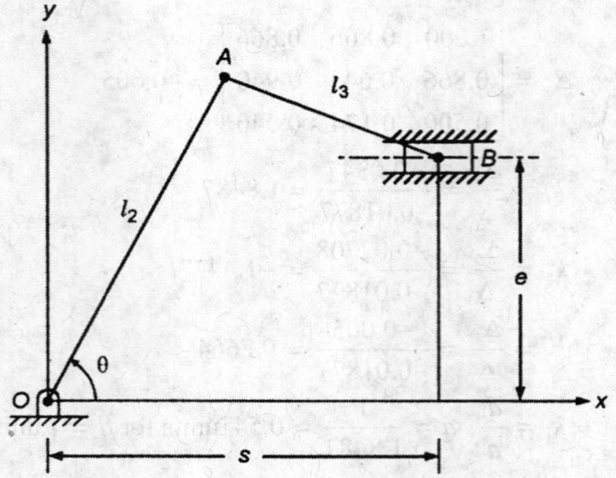

Fig.16.28 Synthesis of slider-crank mechanism using three Chebyshev spacing

Let l_2 = length of crank
l_3 = length of connecting rod
s = distance of slider from the crank shaft
e = eccentricity of slider from crank shaft.

Then
$$\left.\begin{array}{l} x_A = l_2 \cos\theta,\ y_A = l_2 \sin\theta \\ x_B = s,\ y_B = e \end{array}\right\} \quad (1)$$

Now
$$AB^2 = (x_B - x_A)^2 + (y_A - y_B)^2$$
$$= (s - l_2 \cos\theta)^2 + (l_2 \sin\theta - e)^2$$
$$= s^2 + l_2^2 \cos^2\theta - 2\,s\,l_2 \cos\theta + l_2^2 \sin^2\theta + e^2 - 2\,l_2\, e\sin\theta$$
$$l_3^2 = s^2 + l_2^2 + e^2 - 2\,s\,l_2 \cos\theta - 2\,l_2\, e\sin\theta$$

or
$$s^2 = l_3^2 - l_2^2 - e^2 + 2\,s\,l_2 \cos\theta + 2\,l_2\, e\sin\theta$$

Let
$$A_1 = 2l_2,\ A_2 = 2l_2 e,\ \text{and } A_3 = l_2^2 - l_3^2 + e^2 \quad (2)$$

Then $A_1\, s\cos\theta + A_2 \sin\theta - A_3 = s^2$ (3)

The displacement Eq. (3) has three variables A_1, A_2, and A_3.

Let s_i = 10 cm and s_f = 3 cm

The precision points obtained are:

Precision point	s, cm	θ, deg
1	9.97	51.03
2	8.25	90
3	3.91	128.97

Substituting in Eq. (3), we get

$$A_1 \times 9.97 \cos 51.03° + A_2 \sin 51.03° - A_3 = (9.97)^2$$

or
$$6.27\,A_1 + 0.78\,A_2 - A_3 = 99.4 \quad (4)$$

$$A_1 \times 8.25 \cos 90° + A_2 \sin 90° - A_3 = (8.25)^2$$

or
$$A_2 - A_3 = 68.06 \quad (5)$$

$$A_1 \times 3.91 \cos 128.97° + A_2 \sin 128.97° - A_3 = (3.91)^2$$

$$-2.459\,A_1 + 0.78\,A_2 - A_3 = 15.29 \quad (6)$$

We have,

$$\begin{bmatrix} 6.27 & 0.78 & -1 \\ 0 & 1 & -1 \\ -2.459 & 0.78 & -1 \end{bmatrix} \begin{Bmatrix} A_1 \\ A_2 \\ A_3 \end{Bmatrix} = \begin{Bmatrix} 99.4 \\ 68.06 \\ 15.29 \end{Bmatrix}$$

$$[B]\,\{A\} = \{C\}$$

$$|B| = -1.92$$

$$A_1 = \frac{\begin{vmatrix} 99.4 & 0.78 & -1 \\ 68.06 & 1 & -1 \\ 15.29 & 0.78 & -1 \end{vmatrix}}{-1.92} = \frac{-18.5}{-1.92} = 9.64$$

$$A_2 = \frac{\begin{vmatrix} 6.27 & 99.4 & -1 \\ 0 & 68.06 & -1 \\ -2.459 & 15.29 & -1 \end{vmatrix}}{-1.92} = \frac{-253.8}{-1.92} = 132.2$$

$$A_3 = \frac{\begin{vmatrix} 6.27 & 0.78 & 99.4 \\ 0 & 1 & 68.06 \\ -2.459 & 0.78 & 15.29 \end{vmatrix}}{-1.92} = \frac{-123.1}{-1.92} = 64.11$$

$$l_2 = \frac{A_1}{2} = 4.82 \text{ cm}, \; e = \frac{A_2}{2l_2} = \frac{132.2}{2 \times 4.82} = 13.7 \text{ cm}$$

$$A_3 = l_2^2 - l_3^2 + e^2, \; l_3^2 = l_2^2 + e^2 - A_3 = (4.82)^2 + (13.7)^2 - 64.11 = 146.81$$

$$l_3 = 12.12 \text{ cm}$$

where subscripts i and f represent the initial and final values.

Given: $s_f - s_i = c\,(\theta_f - \theta_i)^2$

or $$c = \frac{s_f - s_i}{(\theta_f - \theta_i)^2} = \frac{3-10}{(135-45)^2} = \frac{-7}{(90)^2} = \frac{-7}{8100}$$

where c = constant of proportionality

Using Chebyshev's equation for precision spacing, we have

$$x_m = a - b\cos\left[\frac{(2m-1)\pi}{2n}\right], m = 1, 2, 3$$

where, $a = \dfrac{x_i + x_f}{2}$, $b = \dfrac{x_f - x_i}{2}$, $n = 3$, number of precision points.

Here $x_i = \theta_i$, the angular positions of crank.

$$a = \frac{\theta_i + \theta_f}{2} = \frac{45+135}{2} = 90°$$

$$b = \frac{\theta_f - \theta_i}{2} = \frac{135-45}{2} = 45°$$

$$\theta_1 = 90 - 45\cos\left[\frac{(2\times1-1)\pi}{2\times3}\right] = 90 - 45\cos\frac{\pi}{6} = 51.03°$$

$$\theta_2 = 90 - 45\cos\left[\frac{(2\times2-1)\pi}{6}\right] = 90 - 45\cos\frac{\pi}{2} = 90°$$

$$\theta_3 = 90 - 45\cos\left[\frac{(2\times3-1)\pi}{6}\right] = 90 - 45\cos\frac{5\pi}{6} = 128.97°$$

The given relationship for s and θ is:

$$s - s_i = c\,(\theta - \theta_i)^2$$

$$\therefore \quad s_1 - 10 = \left(-\frac{7}{8100}\right)(51.03-45)^2$$

$$s_1 = 9.97 \text{ cm}$$

$$s_2 - 10 = \left(-\frac{7}{8100}\right)(90-45)^2$$

$$s_2 = 8.25 \text{ cm}$$

$$s_3 - 10 = \left(-\frac{7}{8100}\right)(128.97-45)^2$$

$$s_3 = 3.91 \text{ cm}$$

Example 16.15

Design a four-bar mechanism to coordinate the input and output angles as follows:

Input angles	15°	30°	45°
Output angles	30°	40°	55°

Solution

Freudenstein's equation for a four-bar chain is (Fig.16.29):

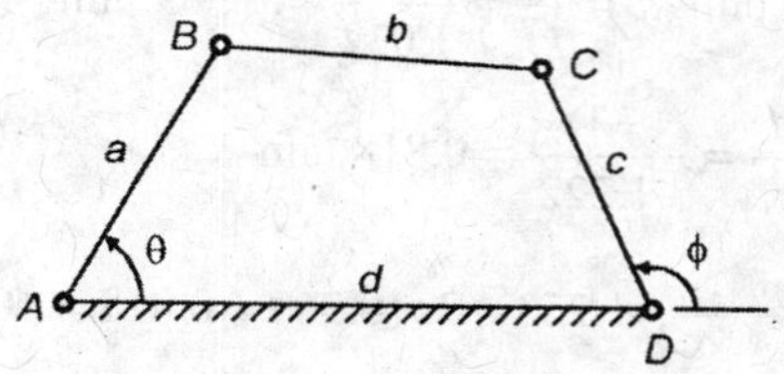

Fig.16.29 Four-bar chain

$$k_1 \cos\phi + k_2 \cos\theta + k_3 = \cos(\theta - \phi)$$

where $$k_1 = \frac{d}{a}, k_2 = -\frac{d}{c} \text{ and } k_3 = \frac{a^2 - b^2 + c^2 + d^2}{2ac}$$

ϕ, deg	$\cos\phi$	θ, deg	$\cos\theta$	$\theta-\phi$, deg	$\cos(\theta-\phi)$
30	0.866	15	0.966	−15	0.966
40	0.766	30	0.866	−10	0.985
55	0.573	45	0.707	−10	0.985

$$\begin{bmatrix} 0.866 & 0.966 & 1 \\ 0.766 & 0.866 & 1 \\ 0.573 & 0.703 & 1 \end{bmatrix} \begin{Bmatrix} k_1 \\ k_2 \\ k_3 \end{Bmatrix} = \begin{Bmatrix} 0.966 \\ 0.985 \\ 0.985 \end{Bmatrix}$$

$$[A]\{k\} = \{B\}$$

$$|A| = \begin{vmatrix} 0.866 & 0.966 & 1 \\ 0.766 & 0.866 & 1 \\ 0.573 & 0.703 & 1 \end{vmatrix} = -3 \times 10^{-3}$$

$$A_1 = \begin{vmatrix} 0.966 & 0.966 & 1 \\ 0.985 & 0.866 & 1 \\ 0.985 & 0.703 & 1 \end{vmatrix} = -3.097 \times 10^{-3}$$

$$A_2 = \begin{vmatrix} 0.866 & 0.966 & 1 \\ 0.766 & 0.985 & 1 \\ 0.573 & 0.985 & 1 \end{vmatrix} = 3.667 \times 10^{-3}$$

$$A_3 = \begin{vmatrix} 0.866 & 0.966 & 0.966 \\ 0.766 & 0.866 & 0.985 \\ 0.573 & 0.703 & 0.985 \end{vmatrix} = -3.758 \times 10^{-3}$$

$$k_1 = \frac{|A_1|}{|A|} = \frac{-3.097}{-3} = +1.032,\ k_2 = \frac{3.667}{-3} = -1.222,\ k_3 = \frac{-3.758}{-3} = 1.253$$

$$k_1 = \frac{d}{a},\ \text{Let } d = 1 \text{ unit},\ a = \frac{d}{k_1} = \frac{1}{1.032} = 0.969 \text{ unit}$$

$$k_2 = -\frac{d}{c},\ c = -\frac{d}{k_2} = \frac{-1}{-1.222} = 0.818 \text{ unit}$$

$$k_3 = \frac{a^2 - b^2 + c^2 + d^2}{2ac},\ 1.253 \times 2 \times 0.969 \times 0.818 = (0.969)^2 - b^2 + (0.818)^2 + 1$$

$$b2 = 0.6217,\ b = 0.788 \text{ unit}$$

The mechanism is shown in Fig.16.30.

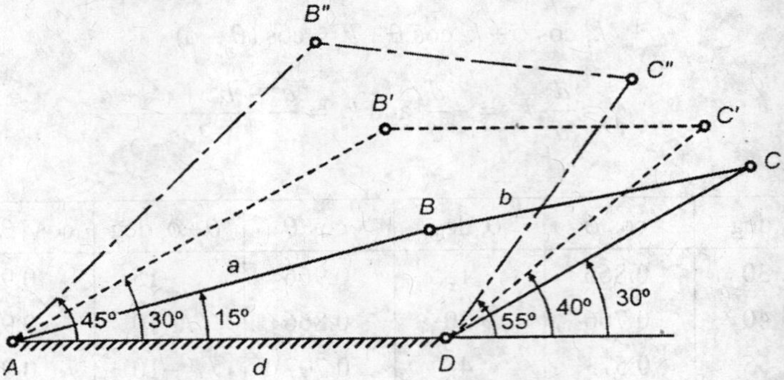

Fig.16.30 Synthesis of four-bar mechanism for Example 16.15

Example 16.16

Synthesize a four-bar function generator to solve the equation: $y = 2x^2 - 1$, $1 < x \le 2$. Use three precision points of Chebyshev spacing. Take $\Delta\phi = 60°$, $\Delta\psi = 90°$, $\phi_o = 30°$ and $\psi_o = 60°$, where $\Delta\phi$, $\Delta\psi$ are ranges of input and output link rotations and ϕ_o, ψ_o are initial angular positions of input and output links respectively.

Solution

Chebyshev's precision points are given by,

$$x_m = a + b \cos\left[\frac{(2m-1)\pi}{2n}\right],\ m = 1, 2, 3$$

$$n = 3,\ x_i = 1,\ x_f = 2$$

$$a = \frac{x_i + x_f}{2} = \frac{1+2}{2} = 1.5,\ b = \frac{x_f - x_i}{2} = \frac{2-1}{2} = 0.5$$

$$\therefore \quad x_i = 1.5 + 0.5 \cos\left[\frac{(2m-1)\pi}{6}\right]$$

$$x_1 = 1.5 + 0.5 \cos \frac{\pi}{6} = 1.933$$

$$x_2 = 1.5 + 0.5 \cos \frac{\pi}{2} = 1.5$$

$$x_3 = 1.5 + 0.5 \cos \frac{5\pi}{2} = 1.067$$

Now $$y = 2x^2 - 1$$

$$y_1 = 2 \times (1.933)^2 - 1 = 6.473$$

$$y_2 = 2 \times (1.5)^2 - 1 = 3.5$$

$$y_3 = 2 \times (1.067)^2 - 1 = 1.277$$

Scale factors are:

$$r_x = \frac{\Delta\phi}{x_f - x_i} = \frac{60}{2 - 1} = 60$$

$$r_x = \frac{\Delta\psi}{y_f - y_i}$$

Now $$y_i = 2 \times 1^2 - 1 = 1$$

$$y_f = 2 \times 2^2 - 1 = 7$$

$$r_y = \frac{90}{7 - 1} = \frac{90}{6} = 15$$

$$r_x = \frac{\phi_m - \phi_i}{x_m - x_i}, \ \phi_i = \phi_0 = 30°, \ x_i = 1$$

$$\frac{\phi_1 - 30}{1.933 - 1} = 60, \ \phi_1 = 85.98°$$

$$\therefore \quad \frac{\phi_2 - 30}{1.5 - 1} = 60, \ \phi_2 = 60°$$

$$\frac{\phi_3 - 30}{1.067 - 1} = 60, \ \phi_3 = 34.02°$$

$$r_y = \frac{\psi_m - \psi_i}{y_m - y_i}, \ \phi_i = \phi_0 = 60°, y_i = 34.02°$$

$$\frac{\psi_1 - 60}{6.473 - 1} = 15, \ \psi_1 = 142.1°$$

$$\frac{\psi_2 - 60}{3.5 - 1} = 15, \ \psi_2 = 97.5°$$

$$\frac{\psi_3 - 60}{1.277 - 1} = 15, \ \psi_3 = 64.155°$$

Precision point	x	y	ψ, deg	$\cos\psi$	ϕ, deg	$\cos\phi$	$(\phi-\psi)$ deg	$\cos(\phi-\psi)$
1	1.933	6.473	142.1	–0.789	85.98	0.070	–56.12	0.557
2	1.5	3.5	97.5	–0.130	60	0.5	–37.5	0.793
3	1.067	1.277	64.155	0.436	34.02	0.829	–30.135	0.865

The Freudenstein's equations become:

$$\begin{bmatrix} -0.789 & 0.070 & 1 \\ -0.130 & 0.5 & 1 \\ 0.436 & 0.829 & 1 \end{bmatrix} \begin{Bmatrix} k_1 \\ k_2 \\ k_3 \end{Bmatrix} = \begin{Bmatrix} 0.557 \\ 0.793 \\ 0.865 \end{Bmatrix}$$

$$[A]\{k\} = \{B\}$$

$$|A| = -0.02657$$

$$|A_1| = \begin{vmatrix} 0.557 & 0.070 & 1 \\ 0.793 & 0.5 & 1 \\ 0.865 & 0.829 & 1 \end{vmatrix} = 0.04668,\ k_1 = \frac{|A_1|}{|A|} = 1.757$$

$$|A_2| = \begin{vmatrix} -0.789 & 0.557 & 1 \\ -0.130 & 0.793 & 1 \\ 0.436 & 0.865 & 1 \end{vmatrix} = -0.57183,\ k_2 = \frac{|A_2|}{|A|} = 21.522$$

$$|A_3| = \begin{vmatrix} -0.789 & 0.070 & 0.557 \\ -0.130 & 0.5 & 0.793 \\ 0.436 & 0.865 & 1 \end{vmatrix} = -0.54377,\ k_3 = \frac{|A_3|}{|A|} = 20.465$$

$$k_1 = \frac{d}{a},\ a = \frac{d}{k_1} = -\frac{1}{1.757} = -0.569$$

$$k_2 = -\frac{d}{c},\ c = -\frac{d}{k_2} = -\frac{1}{21.522} = -0.046$$

$$k_3 = \frac{a^2 - b^2 + c^2 + d^2}{2ac},\ 20.465 \times 2 \times (-0.569) \times (-0.046)^2 = (-0.569)^2 - b^2 + (0.046)^2 + 1$$

$$b^2 = 0.25457,\ b = 0.504$$

$$a = -0.569,\ b = 0.504,\ c = -0.046,\ d = 1$$

The sign indicates the reversed direction of length.

Example 16.17

Synthesize graphically, a four-bar mechanism for three positions of input and output cranks with $\psi_{12} = 35°$, $\phi_{23} = 30°$, $\psi_{12} = 40°$, and $\psi_{23} = 60°$, where ϕ_{ij} and ψ_{ij} are angular distances between position i and j of input and output links respectively.

Solution

Given: $\phi_{12} = 35°,\ \phi_{23} = 30°;\ \psi_{12} = 40°,\ \psi_{23} = 60°$

$\therefore$ $\phi_{13} = 35 + 30 = 65°$ and $\psi_{13} = 100°$

$$\alpha_{12} = \frac{1}{2}(\phi_{12} - \psi_{12}) = \frac{1}{2}(35 - 40) = -2.5°$$

$$\alpha_{13} = \frac{1}{2}(\phi_{13} - \psi_{13}) = \frac{1}{2}(65 - 100) = -17.5°$$

Angle subtended by the coupler at the relative pole = angle subtended by the fixed link at the relative pole.

The four-bar mechanism with three positions may be synthesized as explained below (Fig.16.31):

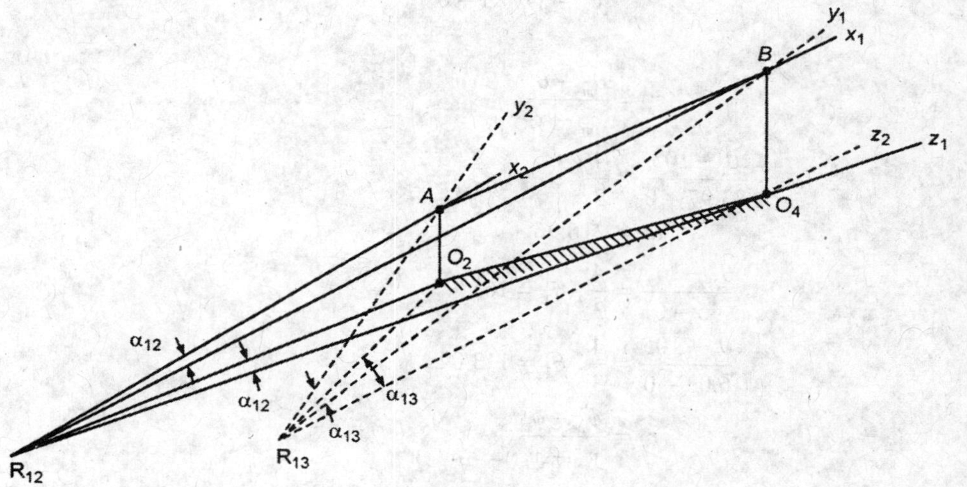

Fig.16.31 Graphical synthesis of four-bar mechanism by three position method

1. Select relative poles R_{12} and R_{13} at positions arbitrarily chosen. Draw $\alpha_{12} = 2.5°$ at R_{12} and $\angle\alpha_{13} = 17.5°$ at R_{13}.
2. Let x_2 and y_2 lines intersect at A and x_1 and y_1 lines at B. Join AB to get the coupler.
3. Select the fixed point O_2 of the fixed link O_2O_4 at an arbitrary position. Join O_2R_{12} and O_2R_{13}.
4. Draw $\angle O_2R_{12}Z_1 = \angle\alpha_{12} = 2.5°$ and $\angle O_2R_{13}Z_2 = \angle\alpha_{13} = 17.5°$. Let $R_{12}Z_1$ and $R_{13}Z_2$ lines intersect at O_4.
5. Join O_2A, O_2O_4 and O_4B. Then O_2ABO_4 is the required four-bar mechanism shown in Fig.16.31.

$$b = \frac{1}{2}(x_f - x_i) = \frac{1}{2}(90 - 0) = 45°$$

$$x_m = 45 + 45\cos\left[\frac{(2m-1)\pi}{6}\right]$$

$$x_1 = 45 + 45\cos\frac{\pi}{6} = 83.97°,\ y_1 = 0.994$$

$$x_2 = 45 + 45\cos\frac{\pi}{2} = 45°,\ y_2 = 0.707$$

$$x_3 = 45 + 45\cos\frac{5\pi}{6} = 6.03°,\ y_3 = 0.105$$

Given $\Delta\theta = 120°$, $\Delta\phi = 60°$

Scale factors are:

$$r_x = \frac{\Delta\theta}{x_f - x_i} = \frac{120}{90 - 0} = \frac{120}{90} = \frac{4}{3}$$

$$y_i = \sin 0° = 0,\ y_f = \sin 90° = 1$$

$$r_y = \frac{\Delta\phi}{y_f - y_i} = \frac{60}{1 - 0} = 60$$

Now $$r_x = \frac{\theta_m - \theta_i}{x_m - x_i}$$

$\therefore$ $$\frac{\theta_1 - 105}{83.97 - 0} = \frac{4}{3}$$

$$\theta_1 = 216.96°$$

$$\frac{\theta_2 - 105}{45 - 0} = \frac{4}{3},\ \theta_2 = 165°$$

$$\frac{\theta_3 - 105}{6.03 - 0} = \frac{4}{3},\ \theta_3 = 113.04°$$

$$r_y = \frac{\phi_m - \phi_i}{y_m - y_i}$$

$$\frac{\phi_1 - 60}{0.994 - 0} = 60,\ \phi_1 = 119.64°$$

$$\frac{\phi_2 - 60}{0.707 - 0} = 60,\ \phi_2 = 102.42°$$

Example 16.18

Synthesize a four-bar mechanism to generate a function $y = \sin x$ for $0 \le x \le 90°$. The range of the output crank may be chosen as 60° while that of input crank be 120°. Assume three precision point obtained form Chebyshev spacing. Assume fixed link to be 50 mm long and $\theta_1 = 105°$ and $\phi_1 = 60°$. Where θ_d and ϕ_d are the angles made by input link at jth position.

Solution

Refer to Fig.16.32.

$y = \sin x$ for $0 \le x \le 90°$

$x_i = 0°, x_f = 90°$

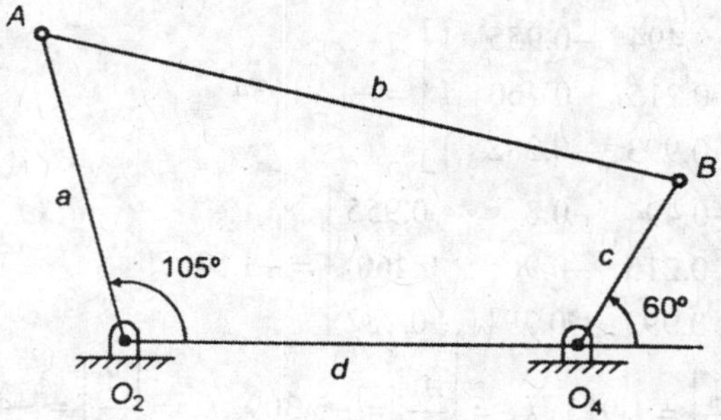

Fig.16.32 Four-bar mechanism configuration

Cheybyshev's precision points are given by:

$$x_m = a + b\cos\left[\frac{(2m-1)\pi}{2n}\right]$$

$$a = \frac{1}{2}(x_i + x_f) = \frac{1}{2}(0+90) = 45°$$

$$\frac{\phi_3 - 60}{0.105 - 0} = 60, \quad \phi_3 = 6.63°$$

Precision point	x deg	y	ϕ deg	$\cos\phi$	θ deg	$\cos\theta$	$\theta-\phi$ deg	$\cos(\theta-\phi)$
1	83.97	0.994	119.64	−0.494	216.96	−0.8	197.32	−0.955
2	45	0.707	102.42	−0.215	165	−0.966	62.58	0.460
3	6.03	0.105	6.63	0.993	113.04	−0.391	−106.41	−0.282

The Freudenstein's equation is

$$k_1 \cos\phi + k_2 \cos\theta + k_3 = \cos(\theta - \phi)$$

where $$k_1 = \frac{d}{a}, k_2 = -\frac{d}{c} \text{ and } k_3 = \frac{a^2 - b^2 + c^2 + d^2}{2ac}$$

Given $d_1 = 50$ mm

Substituting the values, we have

$$\begin{bmatrix} -0.494 & -0.8 & 1 \\ -0.215 & -0.966 & 1 \\ 0.993 & -0.391 & 1 \end{bmatrix} \begin{Bmatrix} k_1 \\ k_2 \\ k_3 \end{Bmatrix} = \begin{Bmatrix} -0.955 \\ 0.460 \\ -0.282 \end{Bmatrix}$$

$$[A]\{k\} = \{B\}$$

$$|A| = 0.36095, \ |A_1| = \begin{vmatrix} -0.955 & -0.8 & 1 \\ 0.460 & -0.966 & 1 \\ -0.282 & -0.391 & 1 \end{vmatrix} = 0.69045$$

$$|A_2| = \begin{bmatrix} -0.494 & -0.955 & 1 \\ -0.215 & 0.460 & 1 \\ 0.993 & -0.282 & 1 \end{bmatrix} = -1.91634$$

$$|A_3| = \begin{bmatrix} -0.494 & -0.8 & -0.955 \\ -0.215 & -0.966 & 0.460 \\ 0.993 & -0.391 & -0.282 \end{bmatrix} = -1.5367$$

$$k_1 = \left|\frac{A_1}{A}\right| = 1.913,\ k_2 = \left|\frac{A_2}{A}\right| = -5.309,\ k_3 = \left|\frac{A_3}{A}\right| = -4.2574$$

$$k_1 = \frac{d}{a},\ a = \frac{d}{k_1} = \frac{50}{1.913} = 26.137 \text{ mm}$$

$$k_2 = -\frac{d}{c},\ c = -\frac{d}{k_2} = \frac{-50}{-5.309} = 9.418 \text{ mm}$$

$$k_3 = \frac{a^2 - b^2 + c^2 + d^2}{2ac}$$

$$-4.2574 \times 2 \times 26.137 \times 9.418 = (26.137)^2 - b^2 + (9.418)^2 + 50^2$$

$$b^2 = 5367.82,\ b = 73.265 \text{ mm}$$

Hence $a = 26.137$ mm, $b = 73.265$ mm, $c = 9.418$ mm, and $d = 50$ mm

Example 16.19

Synthesize a slider-crank mechanism with an eccentricity of 9 mm for the two input positions of input link: $\theta_{12} = 60°$ and output displacement of slider is 16 mm.

■ Solution

Given: $e = 9$ mm, $\theta_{12} = 60°$, $x_{12} = 16$ mm.

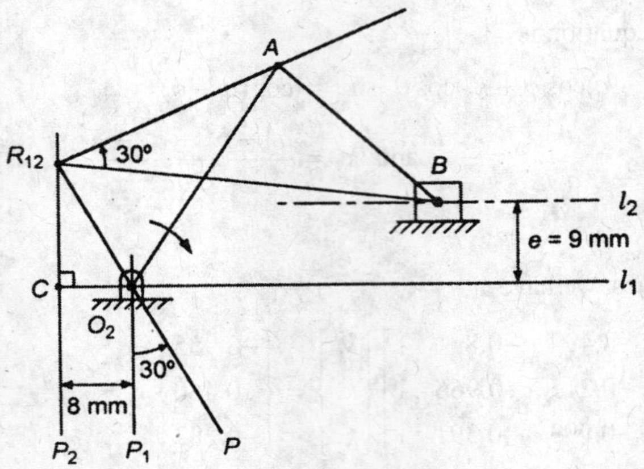

Fig.16.33 Graphical synthesis of slider-crank mechanism for Example 16.19

1. Draw two parallel lines l_1 and l_2 at an eccentricity, $e = 9$ mm (Fig.16.33).
2. Local point O_2 at a convenient position on line l_1. Draw $\angle P_1O_2P = \frac{1}{2}\theta_{12} = 30°$ opposite to the direction of crank rotation, i.e. ccw.

3. Cut off $O_2C = \frac{1}{2} x_{12} = 8$ mm on line l_1 opposite to the direction of displacement of slider. Erect perpendicular P_2C at point C.
4. Extend lines PO_2 and P_2C to intersect at point R_{12}, to give the relative pole.
5. Construct an angle equal to $\frac{1}{2}\theta_{12} = 30°$ at R_{12}. One arm of this angle intersect line l_2 at B to locate the slider B.
6. Select point A arbitrarily and conveniently on the other arm of the above angle.
7. Join AB and AO_2 to give the connecting rod and crank respectively.

Example 16.20

For the four-bar linkage, the following data are given (Fig.16.34):

$$\theta_2 = 60°;\ \theta_4 = 90°$$
$$\omega_2 = 3 \text{ rad/sec};\ \omega_4 = 2 \text{ rad/sec}$$
$$\alpha_2 = -1 \text{ rad/sec}^2;\ \alpha_4 = 0$$

Determine the length of various links.

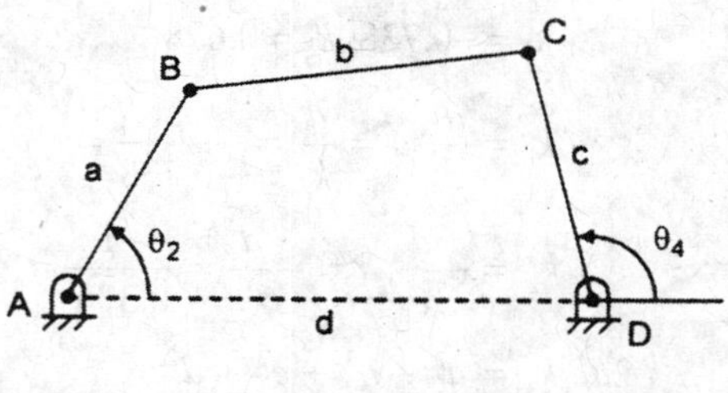

Fig.16.34 Four-bar mechanism configuration

■ **Solution**

The Freudenstein's equation for the displacement of a four-bar linkage is given by:

$$k_1 \cos \theta_4 + k_2 \cos \theta_2 + k_3 = \cos (\theta_2 - \theta_4) \quad (1)$$

where $k_1 = \frac{d}{a}$, $k_2 = -\frac{d}{c}$, $k_3 = \frac{a^2 - b^2 + c^2 + d}{2ac}$

Taking first time derivative of Eq. (1), we have

$$k_1 \omega_4 \sin \theta_4 + k_2 \omega_2 \sin \theta_2 = (\omega_2 - \omega_4) \sin (\theta_2 - \theta_4) \quad (2)$$

Taking the second time derivative of Eq. (2), we have

$$k_1(\omega_4^2 \cos \theta_4 + \alpha_4 \sin \theta_4) + k_2(\omega_2^2 \cos \theta_2 + \alpha_2 \sin \theta_2)$$
$$= (\alpha_2 - \alpha_4) \sin (\theta_2 - \theta_4) + (\omega_2 - \omega_4)^2 \cos (\theta_2 - \theta_4) \quad (3)$$

Substiuting the values of various terms in Eqs. (1) to (3), we get

$$k_1 \cos 90° + k_2 \cos 60° + k_3 = \cos (60° - 90°)$$

or $$0.5\, k_2 + k_3 = \frac{\sqrt{3}}{2}$$

or $$k_2 + 2k_3 = 1.732 \quad (4)$$

$$k_1 \times 2 \times \sin 90° + k_2 \times 3 \times \sin 60° = (3 - 1) \sin (60° - 90°)$$

or $$2k_1 + \frac{3\sqrt{3}}{2} k_2 = -\frac{1}{2}$$

or $$4k_1 + 5.196\, k_2 = -1 \quad (5)$$

$$k_1 (4 \cos 90° + 0 \times \sin 90°) + k_2 (9 \times \cos 60° - 1 \times \sin 60°)$$
$$= (-1 - 0) \sin (60° - 90°) + (3 - 2)^2 \cos (60° - 90°)$$

$$k_2 \left(9 \times \frac{1}{2} - 1 \times \frac{\sqrt{3}}{2} \right) = \frac{1}{2} + 1 + \frac{\sqrt{3}}{2}$$

$$3.634\, k_2 = 1.366$$

$$k_2 = 0.376 \quad (6)$$

Substituting the value of k_2 in Eqs. (5) and (4) gives,

$$k_1 = -0.738,\ k_3 = 0.678$$

Now $$k_1 = \frac{d}{a},\ a = \frac{d}{k_1} = \frac{1}{-0.738} = -1.355$$

$$k_2 = -\frac{d}{c},\ c = -\frac{d}{k_2} = \frac{-1}{0.376} = -2.659$$

$$2ac\, k_3 = a^2 - b^2 + c^2 + d^2$$

$$2 \times (-1.355) \times (-2.659) \times 0.678 = (-1.355)^2 - b^2 + (-2.659)^2 + 1$$

$$b^2 = 5.0207,\ b = 2.24$$

Thus $$a = -1.355,\ b = 2.24,\ c = -2.659,\ d = 1.0$$

Example 16.21

Design graphically a four-bar mechanism such that $\theta_{12} = 120°$, $\theta_{13} = 160°$ and $\phi_{12} = 70°$, $\phi_{13} = 110°$ Input moves anti-clockwise and output moves in clockwise direction.

■ Solution

Given: $\theta_{12} = 120°$, $\phi_{12} = 70°$; $\theta_{13} = 160°$, $\phi_{13} = 110°$

Refer to Fig.16.35.

$$\psi_{12} = \frac{1}{2}(\theta_{12} - \phi_{12}) = \frac{1}{2}(120° - 70°) = 25°$$

$$\psi_{13} = \frac{1}{2}(\theta_{13} - \phi_{13}) = \frac{1}{2}(160° - 110°) = 25°$$

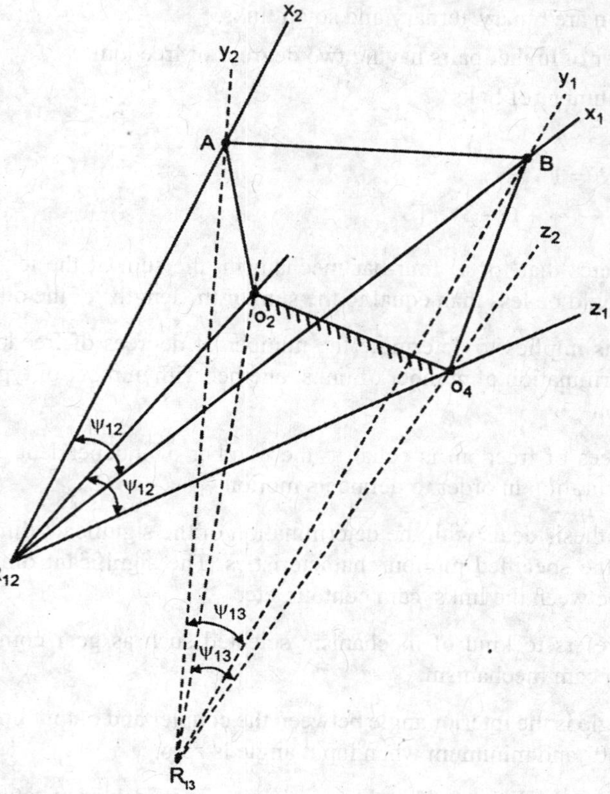

Fig.16.35 Graphical synthesis of four-bar mechanism for Example 16.21

1. Select relative poles R_{12} and R_{13} at arbitrarily chosen positions. Draw $\angle\psi_{12} = \angle x_1R_{12}x_2 = 25°$ at R_{12} and $\angle\psi_{13} = y_1R_{13}y_2 = 25°$ at R_{13}.
2. Let x_2 and y_2 lines intersect at point A and x_1 and y_1 lines at point B. Join AB. Then AB is the coupler.
3. Select fixed point O_2 of fixed link O_2O_4 arbitrarily at a convenient place. Join R_{12} with O_2.
4. Draw $\angle O_2R_{12}Z_1 = \angle\psi_{12} = 25°$.
5. Join R_{13} with O_2 and draw $\angle O_2R_{13}Z_2 = \angle\psi_{13} = 25°$. Let z_1 and z_2 lines intersect at O_4. Join O_2 with O_4. Then O_2O_4 is the fixed link.
6. Join $O_2$with A and O_4 with B. Then O_2ABO_4 is the desired four-bar linkage.

Summary for Quick Revision

1 Gruebler's criterion for degrees of freedom of planar mechanisms

$$F = 3\,(n - 1) - 2p - h$$

where p = number of simple joints or lower pairs having only one degree of freedom

$= (1/2)\,(2n_2 + 3n_3 + \cdots + in_i)$

n_2, n_3, ... and so on are binary, ternary and so on links.

h = number of higher pairs having two degrees of freedom

n = total number of links.

$= n_2 = n_3 + \cdots$

For $h = 0$ and $F = 1$

$n_2 = 4 + [n_4 + \cdots + (i - 3)\, n_i]$

2 Grashof's law states that for a four-bar mechanism, the sum of the lengths of the largest and shortest links should be less than equal to the sum of the lengths of the other two links.

3 Number synthesis implies to determine the number of degrees of freedom of a mechanism. It involves the determination of number of links together with number of type of joints required for a specified motion.

4 Number of degrees of freedom is equal to the number of independent coordinates required to specify its configuration in order to define its motion.

5 Dimensional synthesis deals with the determination of the significant dimensions of the mechanism to satisfy the specified motion characteristics. The significant dimensions could be link lengths, angles between the links, cam contour, etc.

6 Type synthesis refers to kind of mechanism selected such as gear combination, a belt-pulley combination or a cam mechanism.

7 Transmission angle is the interior angle between the coupler and output link. It is maximum when input angle is 180° and minimum when input angle is zero.

8 The pole of the coupler link of a four-bar mechanism is its centre of rotation with respect to the fixed link.

9 The relative pole of the coupler link of a four-bar mechanism is its center of rotation relative to other moving links.

10 Synthesis of four-bar mechanism:

(a) Two-position synthesis: θ_{12} and ϕ_{12} are known.
Angle subtended by coupler at R_{12}, $\angle\psi_{12} = (1/2)(\angle\theta_{12} - \angle\phi_{12})$

(b) Three-position synthesis: $\theta_1, \theta_2, \theta_3$ and ϕ_1, ϕ_2, ϕ_3 are known.

$$\theta_{12} = \theta_2 - \theta_1;\ \theta_{13} = \theta_3 - \theta_1 \text{ and } \phi_{12} = \phi_2 - \phi_1,\ \phi_{13} = \phi_3 - \phi_1$$

Then $\angle\psi_{12} = (1/2)\,(\angle\theta_{12} - \angle\phi_{12})$, $\angle\psi_{13} = (1/2)\,(\angle\theta_{13} - \angle\phi_{13})$

11 Synthesis of slider-crank mechanism:

(a) Two-position synthesis
θ_{12} and x_{12} are known.

(b) Three-position synthesis
θ_{12}, θ_{13} and x_{12}, x_{13} are known.

12 In function generation, the motion of input link is correlated to the motion of output link.

13 In path generation, a point on the coupler link is constrained to describe a path with reference to a fixed frame.

14 Scale factors are:
$r_x = (\theta - \theta_i)/(x - x_i) = (\theta_f - \theta_i)/(x_f - x_i)$
$r_y = (\phi - \phi_i)/(y - y_i) = (\phi_f - \phi_i)/(y_f - y_i)$

15 Chebyshev's spacing for precision points:
$x_m = a + b \cos [(2m - 1)\pi/(2n)]$, $m = 1, 2, 3$
where $a = (x_i + x_f)/2$, $b = (x_f - x_i)/2$ and n = number of precision points.

16 Freudenstein's equation for four-bar mechanism having three precision points:
$k_1 \cos\phi + k_2 \cos\theta + k_3 = \cos(\theta - \phi)$
where $k_1 = d/a$, $k_2 = -d/c$, $k_3 = (a^2 - b^2 + c^2 + d^2)/(2ac)$

17 Freudenstein's equation for single-slider crank mechanism having three precision points:
$s^2 = k_1 s \cos\theta + k_2 \sin\theta + k_3$
where $k_1 = 2\ell_2$, $k_2 = 2\ell_2 e$, $k_3 = \ell_3^2 - \ell_2^2 - e^2$

Multiple Choice Questions

1 Transmission angle is the angle between the
(a) Coupler and driven link (b) Coupler and driving link
(c) Driving link and fixed link (d) Driven link and fixed link.

2 Transmission angle is maximum when input angle is
(a) 0° (b) 90° (c) 180° (d) 45°

3 Transmission angle is minimum when input angle is
(a) 0° (b) 45° (c) 90° (d) 180°

4 The minimum number of links in a kinematic chain are
(a) 3 (b) 4 (c) 5 (d) 6

5 Relative pole of a moving link is its centre of rotation relative to a
(a) Fixed link (b) moving link (c) any link

6 Pole of a moving link is its centre of rotation relative to a
(a) Fixed link (b) moving link (c) any link

7 In function generation, a mechanism is designed to correlate the motion of:
(a) Input and output links (b) Input link and coupler
(c) Output link and coupler (d) Coupler to fixed link.

8 Freudenstein's equation for a four-bar linkage is:
(a) $k_1 \cos\phi + k_2 \cos\theta + k_3 - \cos(\theta - \phi) = 0$
(b) $k_1 \cos\phi + k_2 \cos\theta + k_3 + \cos(\theta - \phi) = 0$
(c) $k_1 \cos\phi + k_2 \cos\theta + k_3 - \cos(\theta - \phi) = 1$
(d) $k_1 \cos\phi + k_2 \cos\theta + k_3 + \cos(\theta - \phi) = 1$

Answers

1. (a) 2. (c) 3. (a) 4. (b) 5. (b) 6. (a) 7. (a) 8. (a)

Review Questions

1 Differentiate between a pole and a relative pole of a coupler link of four-bar mechanism.

2 Define kinematic synthesis.

3 What do you understand by movability of a mechanism?

4 State Grashof's law.

5 When do we get double crank and crank-rocker mechanism?

6 What is a class-II double-rocker mechanism?

7 What are the properties of pole points?

8 What do you mean by function generation?

9 What are Chebyshev's spacing for precision points?

10 Write the Freudenstein's equation of three precision points of a four-bar chain.

Exercises

16.1 Determine the mobility of the linkages shown in Fig.16.36(a) to (c).

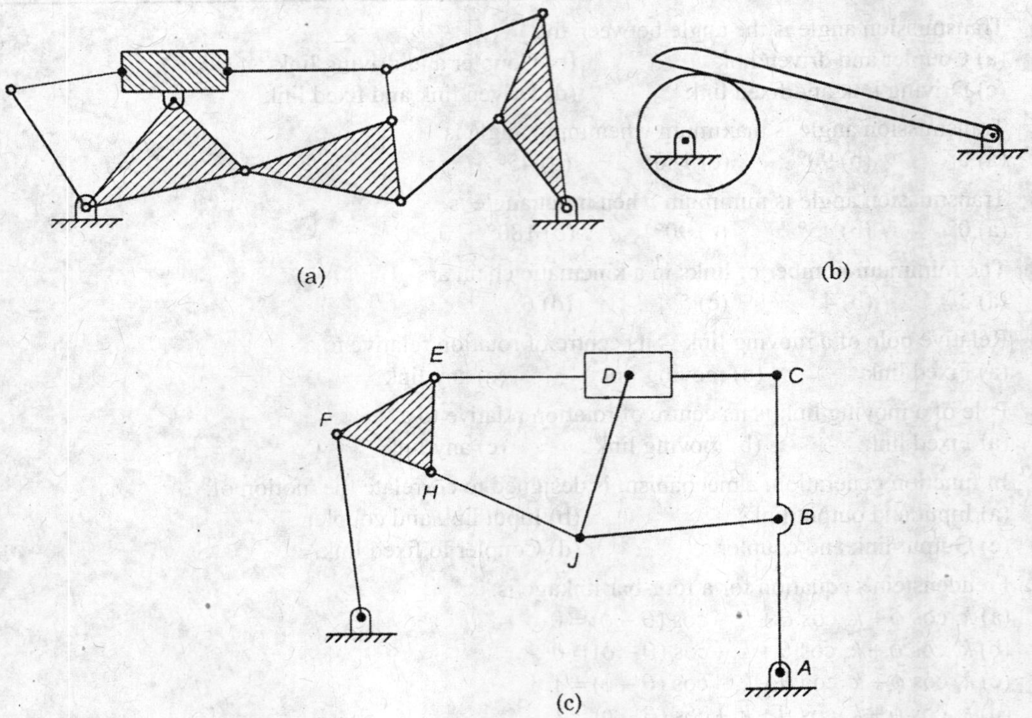

Fig.16.36 Diagram for Exercise 16.1

[Ans. 3, 1, 2]

16.2 Determine the degrees of freedom of the linkage shown in Fig.16.37 (a)

[Ans. 3, 1, 1]

16.3 Fig.16.38 shows a plane mechanism with link lengths given in some unit. If slider A is the driver, will link CG revolve or oscillate? Justify your answer.

[Ans. revolve]

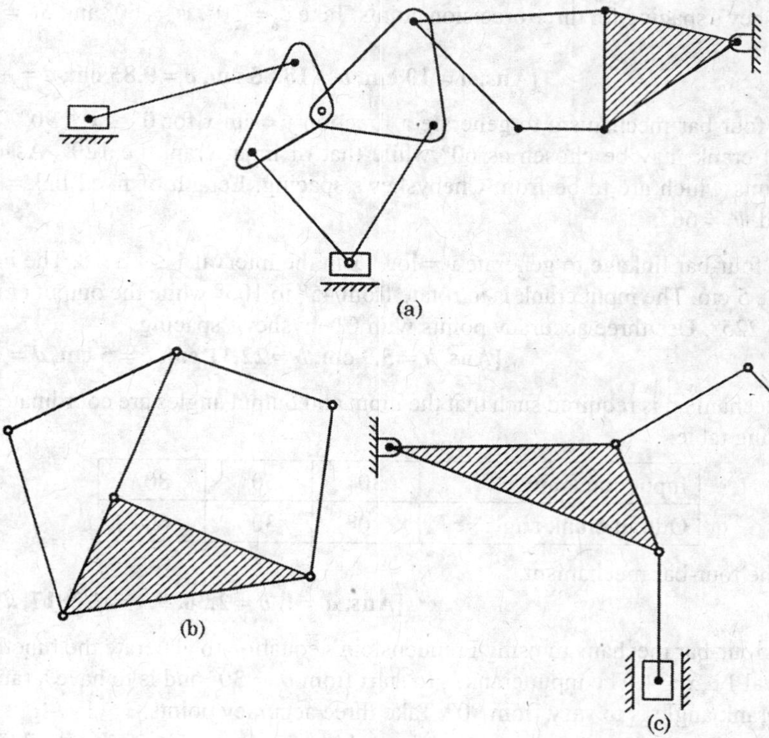

Fig.16.37 Diagram for Exercise 16.2

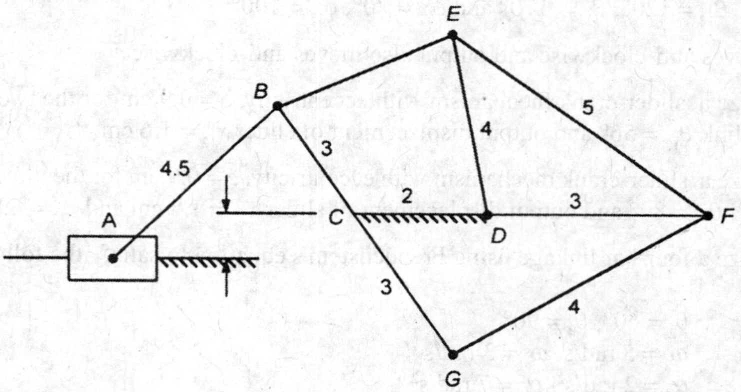

Fig.16.38 Diagram for Exercise 16.3

16.4 Synthesize a four-bar mechanism so that $\theta_{12} = 45°$ and $\phi_{12} = 55°$. Both input and output cranks should move in the ccw direction.

16.5 Synthesis a four-bar linkage that will, in one of its positions, satisfy the following values for the angular velocities and accelerations:

$$y = x^{1.2} \text{ for } 1 \le x \le 5$$

using Chebyshev's spacing for three precision points. Take $\phi_o = 30°$, $\psi_o = 60°$ and $\Delta\phi = \Delta\psi = 90°$ and $a = 10$ cm.

[Ans. $a = 10$ cm, $b = 18.85$ cm, $c = 9.85$ cm, $d = -2.69$ cm]

16.6 synthesize a four-bar mechanism to generate a function $y = \sin x$ for $0° \le x \le 90°$. The range of the output crank may be chosen as 60° while that of input crank be 120°. Assume three precision points which are to be from Chebyshev's spacing. Length of fixed link = 52.5 mm, $\phi_1 = 105°$ and $\psi_1 = 66°$.

16.7 Synthesize a four-bar linkage to generate $y = \log_{10} x$ in the interval $1 \le x \le 10$. The input crank length is to be 5 cm. The input crank is to rotate from 45° to 105° while the output crank moves from 135° to 225°. Use three accuracy points with Chebyshev's spacing.

[Ans. $a = 5.9$ cm, $b = 22.11$ cm, $c = 5$ cm, $d = 10.05$ cm]

16.8 A four-bar mechanism is required such that the input and output angles are coordinated as given in the following table:

Input crank angle	30°	50°	80°
Output crank angle	0°	30°	60°

Synthesize the four-bar mechanism.

[Ans. $a = 1$, $b = 2.3039$, $c = 1.2817$, $d = 1.8321$]

16.9 Synthesize a four-bar mechanism using Freudenstein's equation to generate the function $y = x^{1.5}$ for the interval $1 \le x \le 4$. The input crank is to start from $\theta_2 = 30°$ and is to have a range of 90°. The output crank angle is to vary from 90°. Take three accuracy points.

[Ans. $a = 11.16$, $b = 7.76$, $c = 7.82$, $d = 1.0$]

16.10 Design a four-bar mechanism such that

$$\theta_{12} = 120°,\ \theta_{13} = 170° \text{ and } \phi_{12} = 70°,\ \phi_{13} = 100°$$

Input moves anti-clockwise and output also moves anti-clockwise.

16.11 Synthesize a slider-crank mechanism with eccentricity, $e = 0.9$ cm for the two input positions of input link $\theta_{12} = 56°$ and output displacement of slider $x_{12} = 1.6$ cm.

16.12 Synthesize a slider-crank mechanism with eccentricity, $e = 0.9$ cm for the three input positions $\theta_{12} = 40°$, $\theta_{13} = 80°$ and output displacement of slider $x_{12} = 1.8$ cm and $x_{13} = 4.8$ cm.

16.13 Synthesize a four-bar linkage using Freudenstein's equation to satisfy the following specifications.

$$\theta_2 = 60°,\ \theta_4 = 90°$$
$$\omega_2 = 5 \text{ rad/s},\ \omega_4 = 3 \text{ rad/s}$$
$$\alpha_2 = 2 \text{ rad/s}^2,\ \alpha_4 = 7 \text{ rad/s}^2$$

16.14 Synthesize a four-bar mechanism such that $\theta_{12} = 50°$, $\theta_{23} = 40°$ and $\phi_{12} = 80°$, $\phi_{23} = 50°$.

17 MECHANICAL VIBRATIONS

Chapter Outline

17.1 INTRODUCTION

When a body is displaced from its equilibrium position by the application of an external force and then released, it commences to perform to-and-fro motion. This to-and-fro motion is called vibration.

The mechanism of vibration may be explained as follows:

Work is done by the external forces in producing the initial displacement against the internal elastic forces which resist deformation. This work is stored up as elastic or strain energy in the body. When the external force is removed, the internal elastic forces tend to restore the body to its equilibrium position. Neglecting all resistances offered to the motion of the vibrating body, the whole of elastic or strain energy is converted into kinetic energy at its original equilibrium position. As a result, the motion of the body continues until the whole of the kinetic energy is absorbed in doing work against the internal elastic forces and the energy in the system is once more strain energy. Again the body begins to return to the equilibrium position and the vibration is repeated indefinitely.

17.2 DEFINITIONS

Free (or natural) vibrations: A vibration in which after the initial displacement, no external forces act and the motion is maintained by the internal elastic forces, is termed as free or natural vibration.

Damped vibrations: In practice the energy possessed by a system is gradually dissipated in overcoming internal and external resistances to the motion, and the body finally comes to rest in its original equilibrium position. Such a vibration is said to be damped.

Forced vibration: These type of vibrations are caused when a periodic disturbing force is continuously applied to the body. The vibrations then has the same frequency as the applied force.

Periodic motion: It is a motion which repeats itself after equal intervals of time.

Time period: It is the time taken to complete one cycle.

Frequency: Number of cycles per unit time.

Amplitude: The maximum displacement of a vibrating body from the equilibrium position.

Natural frequency: It is the frequency of free vibrations of a body vibrating of its own without the help of an external agency.

Fundamental (or Principal) modes of vibration: It is the mode of vibration having the lowest natural frequency.

Degrees of freedom: The minimum number of independent coordinates required to specify the motion of a system.

Damping: It is the resistance to the motion of a vibrating body.

Phase difference: It is the angle by which one vibrating system is ahead or behind the other vibrating system.

Resonance: When the frequency of external excitation is equal to the natural frequency of a vibrating body.

Mechanical system: A system consisting of a mass, spring and a damper.

Discrete (or lumped) system: A system with finite number of degrees of freedom.

17.3 TYPES OF FREE VIBRATIONS

Consider the system shown in Fig.17.1, in which a rod, assumed to be weightless and fixed at one end, carries a heavy disc at the free end. This system may be made to vibrate in one of the three ways:

1. ***Longitudinal vibrations***

   If the rod is stretched and compressed so that all particles of the disc vibrate along straight paths parallel to the axis of the rod, then it is termed longitudinal vibrations, as shown in Fig.17.1(a).

2. ***Transverse vibrations***

   If all particles of the disc vibrate nearly along straight paths perpendicular to the axis of the rod so that the rod is bent and subjected alternately to tensile and compressive stresses, then it is termed transverse vibrations as shown in Fig.17.1(b).

3. ***Torsional vibrations***

   When the rod is twisted and untwisted alternately so that torsional shear stresses are induced, the vibrations are termed as torsional. All particles of the disc vibrate along circular arcs whose centres lie on the axis of the rod, as shown in Fig.17.1(c).

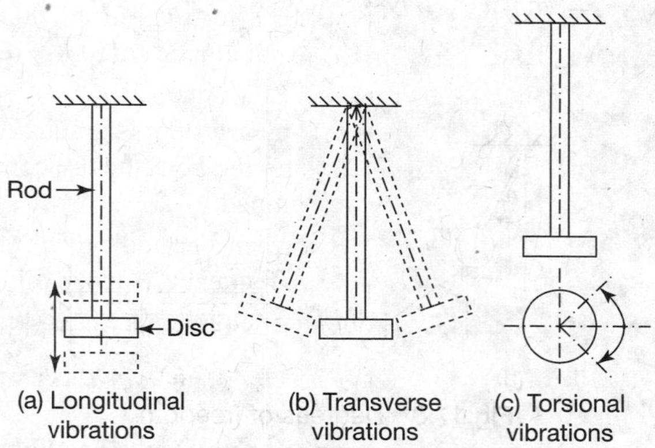

Fig.17.1 Types of free vibrations

17.4 BASIC ELEMENTS OF VIBRATING SYSTEM

The basic elements of an idealized mathematical model for vibrating system are:

1. *Inertial elements*: These are represented by lumped masses (m) for rectilinear motion and lumped moment of inertia (I or J) for angular motion.
2. *Restoring elements*: These elements are represented by massless linear (k) or torsional (k_t) springs for rectilinear and torsional motions respectively.
3. *Damping elements*: These are represented by massless dampers for energy dissipation. The vibration elements are shown in Fig.17.2. They are represented by c for rectilinear motion and c_t for torsional motion.

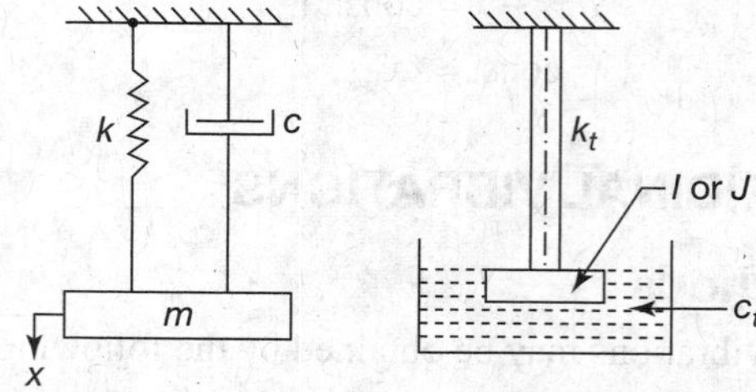

Fig.17.2 Elements of a vibrating system

17.5 DEGREES OF FREEDOM

The number of independent coordinates required to describe the motion of a vibratory system is known as its degrees of freedom. A spring-mass system (Fig.17.3(a)) or a simple pendulum oscillating in a plane (Fig.17.3(b)) are the examples of single degree of freedom systems. A two springs-two mass system (Fig.17.3(c)) and a double pendulum (Fig.17.3(d)) represent two degree of freedom systems. A continuous system in the form of a vibrating beam held between two supports (Fig.17.3(e)) represent an infinite number of degrees.

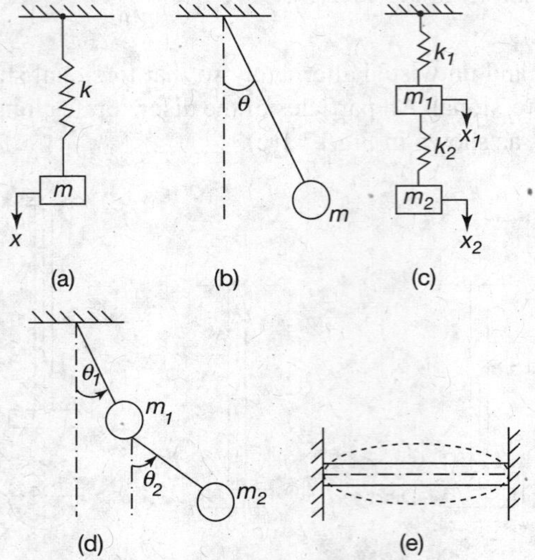

Fig.17.3 Degrees of freedom

17.6 SIMPLE HARMONIC MOTION

When the acceleration of a body is directly proportional to the displacement from the mean position and is always directed towards the mean position, then the motion is said to be simple harmonic. A simple harmonic motion, which is periodic in nature, may be represented by:

$$x = A \sin \omega t$$

Velocity, $$v \text{ or } \dot{x} = \frac{dx}{dt} = A\omega \cos \omega t$$

Acceleration, $$a = \frac{d^2x}{dt^2} \text{ or } \ddot{x} = -A\omega^2 \sin \omega t = -\omega^2 x$$

∴ $$a \propto -x = \text{const.}\,(-x)$$

where $$\text{const} = \omega^2$$

17.7 FREE LONGITUDINAL VIBRATIONS

17.7.1 Solution Methods

The solution of longitudinal vibrations may be obtained by the following methods:

1. Equilibrium method
2. D'Alembect's principle

3. Energy method
4. Rayleigh's method.

We shall illustrate these methods by the spring-mass system shown in Fig.17.4.

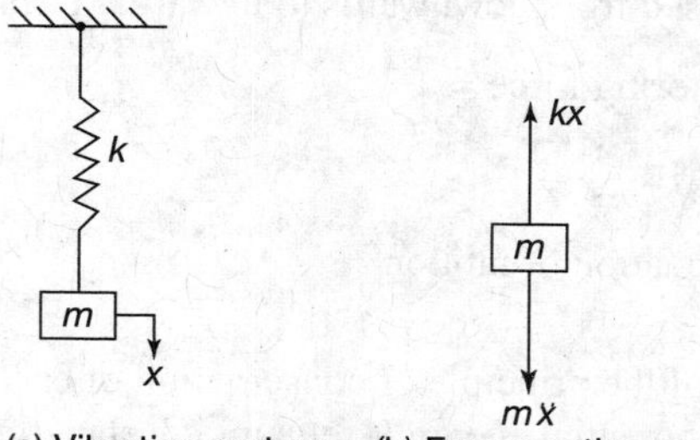

(a) Vibrating system (b) Forces acting on mass m

Fig.17.4 Spring-mass system

1. *Equilibrium method*

This method is based on the Newton's second law of motion, according to which inertia force is equal to the product of mass of the vibrating body and its acceleration in the direction of motion.

Consider the spring-mass system shown in Fig.17.4(a). The forces acting on the mass *m* are shown in Fig.17.4(b). For the equilibrium of the mass *m*, we have

Inertia (or disturbing) force on the mass = Restoring force due to the spring

$$\text{Disturbing force} = \text{mass} \times \text{acceleration}$$

$$= m\ddot{x}$$

$$\text{Restoring force} = -kx$$

–ve sign indicates that the restoring force is opposite to the disturbing force.

$$\therefore \quad m\ddot{x} = -kx$$

or

$$m\ddot{x} + kx = 0 \tag{17.1}$$

$$\ddot{x} + \frac{k}{m}x = 0$$

or

$$\ddot{x} = -\left(\frac{k}{m}\right)x$$

Here the const.

$$= \frac{k}{m} = \omega_n^2$$

$$\therefore \quad \omega_n = \sqrt{\frac{k}{m}} \text{ rad/s} \tag{17.2}$$

Where w_n = natural frequency of shaping–mass system.

Eq. (17.1) is called the equation of motion.

2. *D'Alembert's principle*

This principle states that a dynamic system may be converted into an equivalent static system by adding the inertia force, taken in the reverse direction to the restoring force. To apply the D'Alembert's principle to the spring-mass system shown in Fig.17.4(a), the direction of the inertia force $m\ddot{x}$ on mass m in Fig.17.4(b) has to be reversed from downwards to upwards. Then, we have

Reversed inertia force + restoring force = 0

or $$m\ddot{x} + kx = 0$$

Thus, we obtain the same equation of motion.

3. *Energy method*

The energy method makes use of the principle of conservation of energy. According to this principle, the sum of kinetic energy T and potential energy U remains constant throughout the motion of a vibrating system.

Thus $$T + U = \text{const}$$

or $$\frac{\mathrm{d}}{\mathrm{d}t}(T + U) = 0 \tag{17.3}$$

For the spring-mass system, we have

$$T = \frac{1}{2} m\dot{x}^2$$

$$U = \frac{1}{2} kx^2$$

$\therefore$ $$\frac{\mathrm{d}}{\mathrm{d}t}\left[\frac{1}{2} m\dot{x}^2 + \frac{1}{2} kx^2\right] = 0$$

or $$m\dot{x}\ddot{x} + kx\dot{x} = 0$$

or $$(m\ddot{x} + kx)\,\dot{x} = 0$$

As $\dot{x} \neq 0$, therefore

$$m\ddot{x} + kx = 0$$

which gives the same equation of motion.

4. *Rayleigh's method*

The Rayleigh's method makes use of the fact that the maximum kinetic energy in a vibrating system is equal to the maximum potential energy. For the spring-mass system, the maximum kinetic energy occurs at the mean position and maximum potential energy occurs at the outermost position of oscillations.

$$T_{\max} = U_{\max} \tag{17.4}$$

For $$x = A \sin\omega_n t$$

$$\dot{x} = A\omega_n \cos\omega_n t$$

$$(\dot{x})_{\max} = A\omega_n$$

and $$x_{\max} = A$$

$$T_{max} = \frac{1}{2} m(\dot{x})^2_{max} = \frac{1}{2} m\ A^2 \omega_n^2$$

Now

$$U_{max} = \frac{1}{2} kx^2 = \frac{1}{2} kA^2$$

$\therefore$

$$\frac{1}{2} mA^2\omega_n^2 = \frac{1}{2} kA^2$$

or

$$\omega_n = \sqrt{k/m}$$

which gives the same frequency as obtained in Eq. (17.2).

17.7.2 Single Degree of Freedom System

Consider the single degree of freedom system in the form of a spring-mass shown in Fig.17.5. Before the mass is hung on the spring, the spring in its unstretched position is shown in Fig.17.5(a). When the mass m is hung on the spring, the spring is stretched by an amount δ_{st} = static deflection, from position *A-A* to *B-B*, as shown in Fig.17.5(b). For the static equilibrium of the system, we have,

$$k \cdot \delta_{st} = mg$$

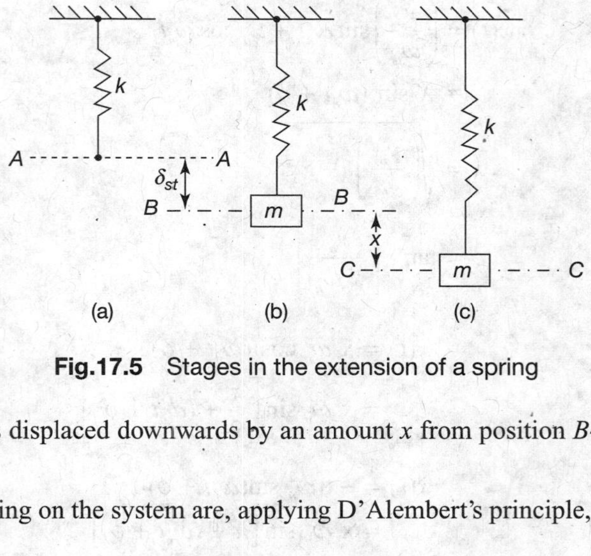

Fig.17.5 Stages in the extension of a spring

Now the mass m is displaced downwards by an amount x from position *B-B* to *C-C*, as shown in Fig.17.5(c).

Now the forces acting on the system are, applying D'Alembert's principle, we have

Upward force $= k(x + \delta_{st})$

Downward force $= -m\ddot{x} + mg$

For the equilibrium of the system, $-m\ddot{x} + mg = k(x + \delta_{st})$

$\therefore$ $$m\ddot{x} + kx = 0 \tag{17.5}$$

Therefore, to write the equation of motion, the forces acting on the vibrating system during static equilibrium position may be ignored.

The solution of Eq. (17.5) is,

$$\omega_n = \sqrt{\frac{k}{m}} = \sqrt{\frac{g}{\delta_{st}}} \text{ rad/s} \tag{17.6}$$

$$\text{Time period, } T = \frac{2\pi}{\omega_n} = 2\pi\sqrt{\frac{m}{k}} \text{ s} \tag{17.7}$$

$$\text{Natural frequency, } f_n = \frac{1}{2\pi}\sqrt{\frac{k}{m}} \text{ Hz (Cycles/s)} \tag{17.8}$$

The solution of Eq. (17.5) is, $x(t) = A\sin\omega_n t + B\cos\omega_n t$

$$\dot{x}(t) = \omega_n\left[A\cos\omega_n t - B\sin\omega_n t\right]$$

Let $x(0) = x_o$ and $\dot{x}(0) = v_o$

Then $x_o = B$

and $v_o = \omega_n A$

or $A = \dfrac{v_o}{\omega_n}$

$$\therefore x(t) = \left(\frac{v_o}{\omega_n}\right)\sin\omega_n t + x_o\cos\omega_n t \tag{17.9}$$

$$= X\sin(\omega_n t + \phi) \tag{17.10}$$

where

$$x = \sqrt{\left(\frac{v_o}{\omega_n}\right)^2 + x_o^2} \tag{17.11}$$

and

$$\phi = \tan^{-1}\left[\frac{v_o}{x_o\omega_n}\right] \tag{17.12}$$

Velocity, $\dot{x}(t) = \text{x}\,\omega_n\cos(\omega_n t + \phi)$

$$= \text{x}\,\omega_n\sin\left[\frac{\pi}{2} + (\omega_n t + \phi)\right]$$

Acceleration, $\ddot{x}(t) = -\text{x}\,\omega_n^2\sin(\omega_n t + \phi)$

$$= \text{x}\,\omega_n^2\sin\left[\pi + (\omega_n t + \phi)\right]$$

We observe that the velocity vector leads the displacement vector by $\pi/2$ and the acceleration vector leads the displacement vector by π.

17.7.3 Effect of the Spring Mass

Consider an element dy of the spring of length l, mass per unit length ρ and stiffness k, as shown in Fig.17.6. We shall use Rayleigh's method to determine the effect of spring-mass on the natural frequency of spring-mass system.

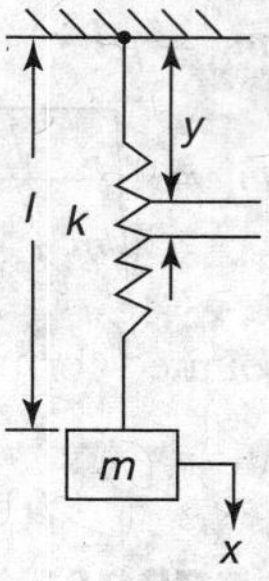

Fig.17.6 Finding effect of spring mass

Maximum kinetic energy of the system,

$$T_{max} = \frac{1}{2} m(\dot{x})^2_{max} + \frac{1}{2}\rho\left(\frac{\dot{x}_{max}}{l}\right)^2 \int_0^{\ell} y^2 dy$$

$$= \frac{1}{2} m\dot{x}^2_{max} + \frac{1}{2}\rho\left(\frac{\dot{x}_{max}}{l}\right)^2 \times \frac{l^3}{3}$$

$$= \frac{1}{2}\left(m + \frac{1}{3}\rho\, l\right)\dot{x}^2_{max}$$

Maximum potential energy,

$$U_{max} = \frac{1}{2} k x^2_{max}$$

Let $\quad x = A \sin \omega_n t$

then $\quad \dot{x} = A\omega_n \cos \omega_n t$

$$\dot{x}_{max} = A\omega_n$$

and $\quad x_{max} = A$

$\therefore \quad T_{max} = \frac{1}{2}(m + \frac{1}{3}\rho l)\, A^2\omega_n^2$

$$U_{max} = \frac{1}{2} k A^2$$

Applying Rayleigh's method, we have

$$T_{max} = U_{max}$$

$\therefore \quad \frac{1}{2}(m + \frac{1}{3}\rho l)\, A^2\omega^2_n = \frac{1}{2} k A^2$

$$\omega_n = \sqrt{\frac{k}{m + \frac{1}{3}\rho l}}$$

Let mass of spring, $m_s = \rho l$

$$\therefore \quad \omega_n = \sqrt{\frac{k}{m + \frac{1}{3}m_s}} \text{ rad/s} \tag{17.13}$$

Therefore, the mass of the spring can be accounted for by adding one-third of its mass to the main mass to calculate the natural frequency.

17.7.4 Equivalent Stiffness of Springs

1. *Springs in series*

Consider two springs in series, as shown in Fig.17.7 of stiffness k_1 and k_2. Let x_1 and x_2 be the extensions of the springs under the force F. Both springs are subjected to the same force, Thus

$$F = k_1x_1 = k_2x_2$$

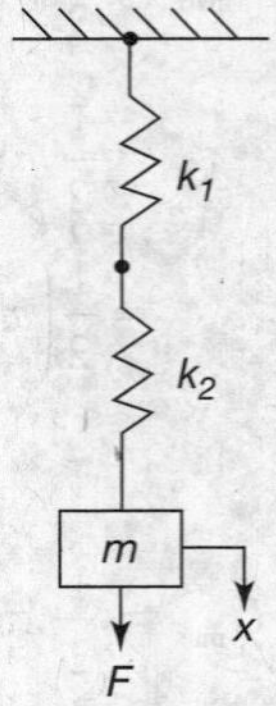

Fig.17.7 Springs in series

Total extension, $x = x_1 + x_2$

$$= F\left(\frac{1}{k_1} + \frac{1}{k_2}\right)$$

Equivalent stiffness, $k_e = \frac{F}{x}$

$$= \frac{1}{\frac{1}{k_1} + \frac{1}{k_2}} = \frac{k_1k_2}{k_1 + k_2}$$

or

$$\frac{1}{k_e} = \frac{1}{k_1} + \frac{1}{k_2} \tag{17.14}$$

In general for n springs in series, we have

$$\frac{1}{k_e} = \sum_{i=1}^{n} \frac{1}{k_i}$$

2. *Springs in parallel*

Consider two springs of stiffness k_1 and k_2 in parallel, as shown in Fig.17.8. The deflection of both the springs is same under the force F, but they experience different forces.

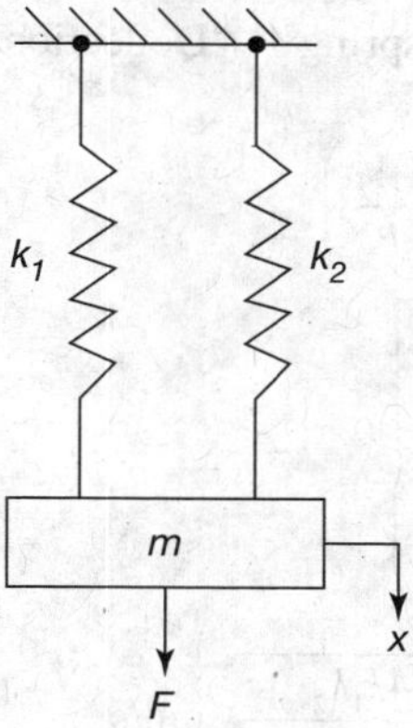

Fig.17.8 Springs in parallel

Let $\quad F_1 = k_1 x \text{ and } F_2 = k_2 x$

Then
$$F = F_1 + F_2$$
$$= (k_1 + k_2)\, x$$

Equivalent stiffness,
$$k_e = \frac{F}{x} = k_1 + k_2 \tag{17.15}$$

In general,
$$k_e = \sum_{i=1}^{n} k_i$$

Example 17.1

Calculate the natural frequency of the systems shown in Fig.17.9(a) and (b).

■ **Solution**

(a) Force in each spring = $2W$

Deflection of weight W, δ_{st} = 2 (deflection of spring 1 + deflection of spring 2)

$$= 2\left(\frac{2W}{k_1} + \frac{2W}{k_2}\right)$$

$$= 4W\left(\frac{k_1 + k_2}{k_1 k_2}\right)$$

Natural frequency,
$$\omega_n = \sqrt{\frac{g}{\delta_{st}}} = \sqrt{\frac{g k_1 k_2}{4W(k_1 + k_2)}}$$

$$= \sqrt{\frac{k_1 k_2}{4m\,(k_1 + k_2)}} \text{ rad/s}$$

(b) Force in spring 1 = W

Force in spring 2 = $\frac{W}{2}$

Deflection of W, δ_{st} = Deflection of spring 1 + Deflection of spring 2

$$= \frac{W}{k_1} + \frac{1}{2}\left(\frac{W}{2} \times \frac{1}{k_2}\right)$$

$$= W\left[\frac{1}{k_1} + \frac{1}{4k_2}\right]$$

$$= W\left(\frac{k_1 + 4k_2}{4k_1 k_2}\right)$$

$$\omega_n = \sqrt{\frac{g}{\delta_{st}}} = \sqrt{\frac{4k_1 k_2}{m(k_1 + 4k_2)}} \text{ rad/s}$$

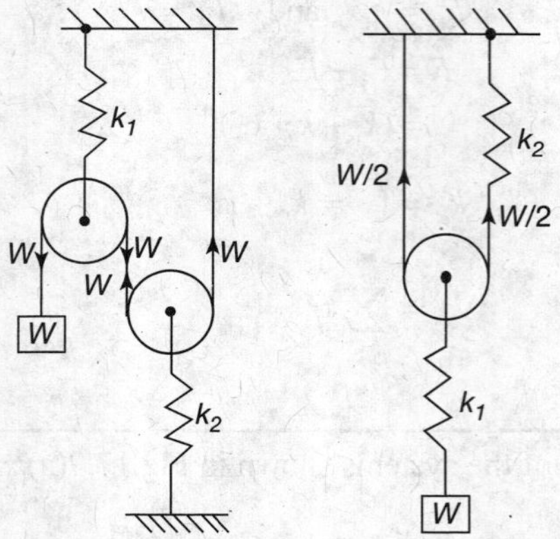

Fig.17.9 Diagram for Example 17.1

Example 17.2

Determine the natural frequency of free vibrations of the system shown in Fig.17.10.

Solution

Force in spring 1, $F_1 = \frac{Wb}{a+b}$

Force in spring 2, $F_2 = \frac{Wa}{a+b}$

Deflection of spring 1, $x_1 = \frac{F_1}{k_1} = \frac{Wb}{k_1(a+b)}$

Deflection of spring 2, $x_2 = \dfrac{F_2}{k_2} = \dfrac{Wa}{k_2(a+b)}$

Deflection of point c, $\delta = x_1 + \left(\dfrac{a}{a+b}\right)(x_2 - x_1)$

$$= \frac{Wb}{k_1(a+b)} + \left(\frac{a}{a+b}\right)\left[\frac{Wa}{k_2(a+b)} - \frac{Wb}{k_1(a+b)}\right]$$

$$= \frac{Wb}{k_1(a+b)} + \frac{Wa}{(a+b)^2}\left[\frac{a}{k_2} - \frac{b}{k_1}\right]$$

$$= \frac{W}{(a+b)^2}\left[\frac{b(a+b)}{k_1} + \frac{a^2}{k_2} - \frac{ab}{k_1}\right]$$

$$= \frac{W}{(a+b)^2}\left[\frac{\{b(a+b) - ab\}k_2 + a^2k_1}{k_1k_2}\right]$$

$$= \frac{W}{(a+b)^2}\left[\frac{b^2}{k_1} + \frac{a^2}{k_2}\right]$$

$$\omega_n = \sqrt{\frac{g}{\delta}} = \sqrt{\frac{(a+b)^2 k_1 k_2}{m(a^2k_1 + b^2k_2)}} \text{ rad/s}$$

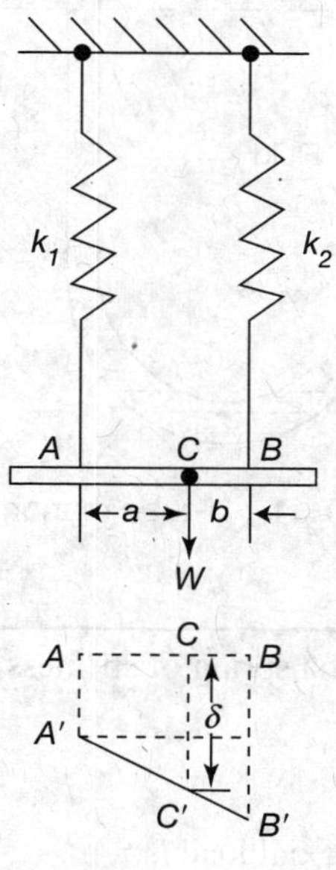

Fig.17.10 Diagram for Example 17.2

Example 17.3

A U-tube manometer, as shown in Fig.17.11, contains a fluid of density ρ. Calculate the frequency of free oscillations of the fluid.

■ Solution

Let l = length of U-tube containing the fluid

A = area of cross-section of the tube

Fluid mass, $m = \rho A l$

x = displacement of fluid column

Weight of fluid responsible for restoring the original fluid level,

$$W = 2\rho A g x$$

The equation of motion becomes,

$$m\ddot{x} + W = 0$$

$$\rho A l \ddot{x} = 2\rho A g x = 0$$

$$\ddot{x} + \frac{2g}{l} x = 0$$

$$\omega_n = \sqrt{\frac{2g}{l}} \text{ rad/s}$$

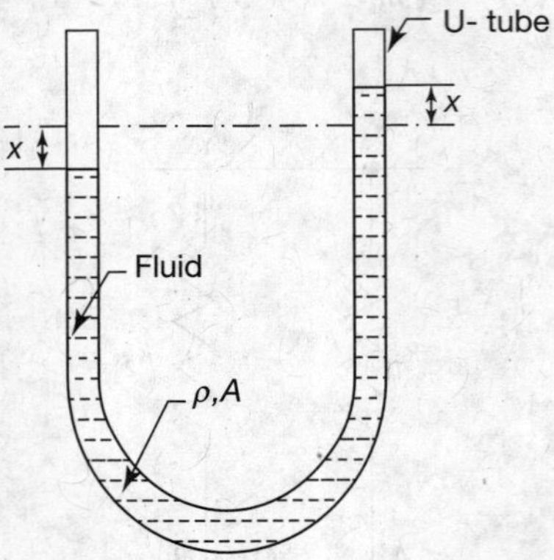

Fig.17.11 U-tube manometer

Example 17.4

A mass of 10 kg is fixed in the middle of a spring of stiffness 5 N/mm, as shown in Fig.17.12. Calculate the natural frequency of the system.

■ Solution

Deflection of a helical spring under axial load is:

$$\delta = \frac{8WD_m^3 n}{Gd^4}$$

$$\therefore \qquad \text{Stiffness, } k = \frac{W}{\delta} = \frac{Gd^4}{8D_m^3 n}$$

or

$$k \propto \frac{1}{n}$$

when the mass m = 10 kg is fixed in the middle of the spring, the number of turns in each portion becomes $n/2$. Thus stiffness of each portion becomes $2k$. Since the two portions of the spring are in parallel, therefore total stiffness becomes $4k$.

Natural frequency,

$$\omega_n = \sqrt{\frac{4k}{m}}$$

$$= \sqrt{\frac{4 \times 5 \times 10^3}{10}}$$

$$= 44.72 \text{ rad/s}$$

or

$$f_n = \frac{\omega_n}{2\pi} = \frac{44.72}{2\pi} = 7.12 \text{ Hz}$$

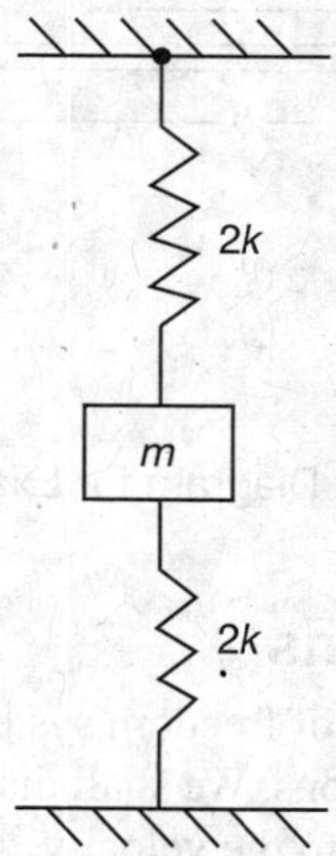

Fig.17.12 Diagram for Example 17.4

Example 17.5

Calculate the natural frequency of the system shown in Fig.17.13.

■ Solution

Force in spring 1, $F_1 = mg$

Force in spring 2, $F_2 = \dfrac{mgb}{a}$

Deflection of mass m,

$$\delta_{st} = \frac{F_1}{k_1} + \frac{F_2}{k_2} \times \frac{b}{a}$$

$$= \frac{mg}{k_1} + \frac{mgb}{k_2 a} \times \frac{b}{a}$$

$$= mg\left[\frac{1}{k_1} + \left(\frac{b}{a}\right)^2 \times \frac{1}{k_2}\right] = mg\left[\frac{k_1 b^2 + k_2 a^2}{k_1 k_2 a^2}\right]$$

Natural frequency,
$$\omega_n = \sqrt{\frac{g}{\delta_{st}}}$$

$$= \sqrt{\frac{k_1 k_2 a^2}{m(k_1 b^2 + k_2 a^2)}} \text{ rad/s}$$

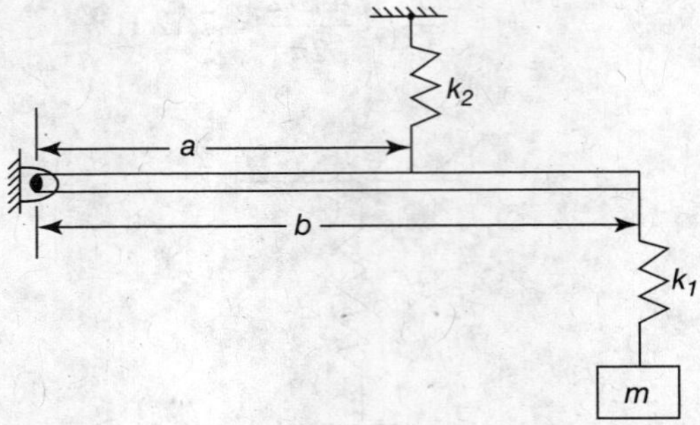

Fig.17.13 Diagram for Example 17.5

17.7.5 Damped Free Vibrations

In damped free vibrations of single degree of freedom systems, a damper is placed in parallel with the spring to decrease the amplitude of vibrations. We shall discuss only viscous type of damper in which the damping force is directly proportional to the velocity.

Consider the spring-mass-damper system as shown in Fig.17.14(a). The forces acting on the mass are shown in Fig.17.14(b). The equation of motion may be written as:

$$m\ddot{x} + c\dot{x} + kx = 0 \qquad (1)$$

where c = damping coefficient, N·s/m

Eq. (1) may be written as:

$$\ddot{x} + \left(\frac{c}{m}\right)\dot{x} + \left(\frac{k}{m}\right)x = 0$$

or
$$\left[D^2 + \left(\frac{c}{m}\right)D + \left(\frac{k}{m}\right)\right]x = 0 \qquad (2)$$

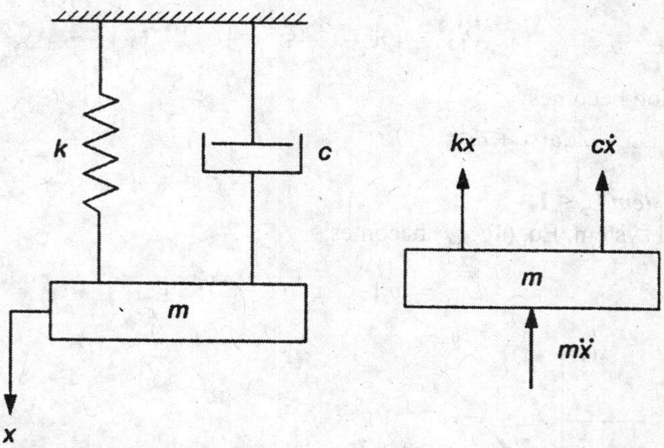

(a) Damped free system (b) Forces acting on mass

Fig.17.14 Damped free vibrations

where $$D = \frac{d}{dt} \quad \text{and} \quad D^2 = \frac{d^2}{dt^2}$$

The characteristic equation is,

$$D^2 + \left(\frac{c}{m}\right)D + \left(\frac{k}{m}\right) = 0 \tag{3}$$

Its roots are,

$$D_{1,2} = -\left(\frac{c}{2m}\right) \pm \left[\left(\frac{c}{2m}\right)^2 - \frac{k}{m}\right]^{1/2} \tag{4}$$

The general solution of Eq. (3) is,

$$x(t) = A \exp(-D_1 t) + B \exp(-D_2 t)$$

where A and B are constants.

For critical damping, the term under the radical sign in Eq. (4) is zero, and the damping coefficient is called the critical damping coefficient, c_c. Thus

$$\left(\frac{c_c}{2m}\right)^2 - \frac{k}{m} = 0$$

or $$\frac{c_c}{2m} = \left(\frac{k}{m}\right)^{1/2} = \omega_n$$

or $$c_c = 2m\omega_n = 2(km)^{1/2} \tag{17.16}$$

We define damping ratio, $$\zeta = \frac{c}{c_c} = \frac{\text{damping coefficient}}{\text{critical damping coefficient}}$$

Now $$\frac{c}{2m} = \left(\frac{c}{c_c}\right)\left(\frac{c_c}{2m}\right) = \zeta\omega_n$$

Thus $$D_{1,2} = \omega_n[-\zeta \pm (\zeta^2 - 1)^{1/2}] \tag{17.17}$$

The equation of motion becomes,

$$\ddot{x} + 2\zeta\omega_n\dot{x} + \omega_n^2 x = 0$$

(a) *Underdamped system* ($\zeta < 1$)

For the underdamped system, Eq. (17.17) becomes,

$$D_{1,2} = \omega_n[-\zeta \pm i\,(1 - \zeta^2)^{1/2}]$$

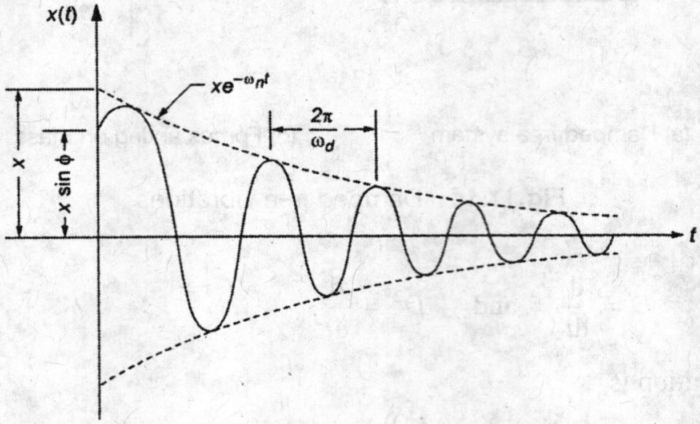

Fig.17.15 Underdamped oscillatory motion ($\zeta < 1$).

Both of these roots are imaginary, Let

$$\omega_d = \omega_n(1 - \zeta^2)^{1/2} = \text{damped natural frequency}$$

The general solution can be written as,

$$x(t) = \exp(-\zeta\omega_n t)[A\cos\omega_d t + B\sin\omega_d t] \tag{17.18}$$
$$= X\exp(-\zeta\omega_n t)\sin(\omega_d t + \phi)$$

Where $x = \sqrt{A^2 + B^2}$ and $\phi = \tan^{-1}(B/A)$

If $x(0) = x_o$ and $\dot{x}(0) = v_o$, then

$$x(t) = \exp(-\zeta\omega_n t)\left[\left(\frac{1}{\omega_d}\right)(v_o + (\zeta\omega_n x_o)\sin\omega_d t + x_o\cos\omega_d t\right] \tag{17.19}$$

The motion represented by Eq. (17.19) is oscillatory, and is shown in Fig.17.15.

(b) *Overdamped system* ($\zeta > 1$)

For the overdamped system, both the roots of Eq. (17.17) are real and negative. The general solution can be written as:

$$x(t) = \exp(-\zeta\omega_n t)\Big[A\exp|\{(\zeta^2 - 1)^{1/2}\omega_n t\}| + B\exp|\{-(\zeta^2 - 1)^{1/2}\omega_n t\}|\Big] \tag{17.20}$$

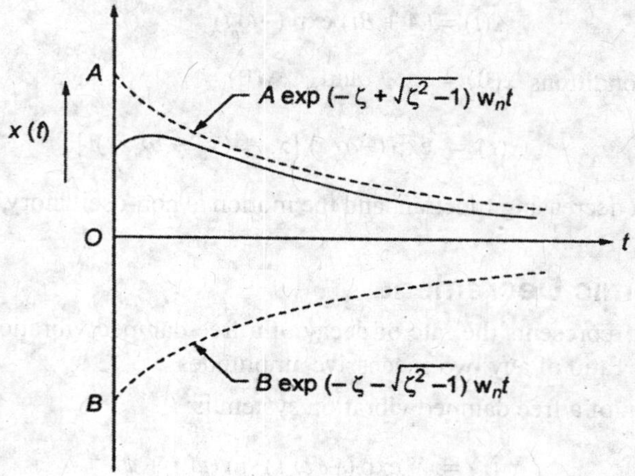

Fig.17.16 Overdamped system ($\zeta > 1$)

if

$$x(0) = x_o \quad \text{and} \quad \dot{x}(0) = v_o, \text{ then}$$

$$A = \left[\frac{v_o + x_o \left\{ \zeta + (\zeta^2 - 1)^{1/2} \right\} \omega_n}{2\omega_n (\zeta^2 - 1)^{1/2}} \right]$$

$$B = \frac{-v_o - x_o \left\{ \zeta + (\zeta^2 - 1)^{1/2} \right\} \omega_n}{2\omega_n (\zeta^2 - 1)^{1/2}}$$

The motion represented by Eq. (17.20) is an exponentially decreasing function as shown in Fig.17.16.

(c) *Criticality damped system* ($\zeta = 1$)

For the critically damped system, both the roots are equal and real, i.e. $D_1 = D_2 = -\omega_n$. The general solution can be written as:

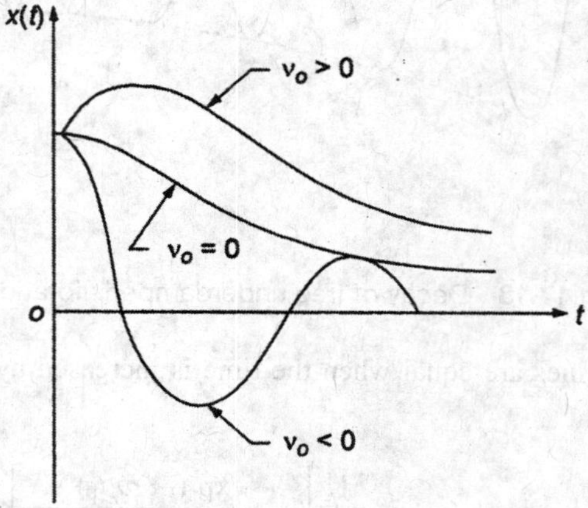

Fig.17.17 Critically damped system ($\zeta = 1$)

$$x(t) = (A + Bt)\exp(-\omega_n t) \qquad (17.21)$$

With initial conditions $x(0) = x_o$ and $\dot{x}(0) = v_o$, we have

$$x(t) = \exp(-\omega_n t)\,[x_o + (v_o + \omega_n x_o)t] \qquad (17.22)$$

Eq. (17.22) represents a decreasing function, and the motion is non-oscillatory, as shown in Fig.17.17.

17.7.6 Logarithmic Decrement

Logarithmic decrement represents the rate of decay of a free damped vibration and is defined as the natural logarithm of the ratio of any two successive amplitudes.

The general solution of a free damped vibration system is:

$$x\,(t) = X\exp(-\zeta\omega_n t)\sin(\omega_d t + \phi)$$

It is shown graphically in Fig.17.18.

Logarithmic decrement,
$$\delta = ln\left(\frac{x_1}{x_2}\right)$$
$$= ln\left[\frac{\exp(-\zeta\omega_n t_1)\sin(\omega_d t_1 + \phi)}{\exp(-\zeta\omega_n(t_1 + T_d))\sin\{\omega_d(t_1 + T_d) + \phi\}}\right]$$

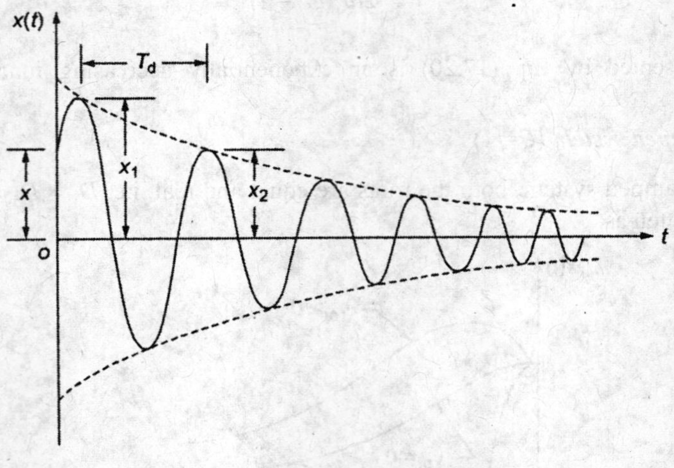

Fig.17.18 Decay of free underdamped vibrations

Since the value of the sines are equal when the time is increased by the damped period T_d, the above relation reduces to,

$$\delta = ln\left[\frac{\exp(-\zeta\omega_n t_1)}{\exp\{-\zeta\omega_n(t_1 + T_d)\}}\right]$$

$$= ln\left[\exp\left(\zeta\omega_n T_d\right)\right]$$
$$= \zeta\omega_n T_d$$

Now $$T_d = \frac{2\pi}{\omega_d} = \frac{2\pi}{\left[\omega_n(1-\zeta^2)^{1/2}\right]}$$

∴ $$\delta = \frac{2\pi\zeta}{(1-\zeta^2)^{1/2}} \tag{17.23}$$

When ζ < 1, then $$\delta \approx 2\pi\zeta \tag{17.24}$$

Now $$\delta = ln\left(\frac{x_o}{x_1}\right) = ln\left(\frac{x_1}{x_2}\right) = \cdots = ln\left(\frac{x_{n-1}}{x_n}\right)$$

Also $$\frac{x_o}{x_n} = \left(\frac{x_o}{x_1}\right)\times\left(\frac{x_1}{x_2}\right)\times\cdots\times\left(\frac{x_{n-1}}{x_n}\right)$$

∴ $$\delta = \left(\frac{1}{n}\right) ln\left(\frac{x_o}{x_n}\right) \tag{17.25}$$

Example 17.6

A mass of 5 kg hangs from a spring and makes damped oscillations. If the time of 50 complete oscillations is found to be 20 s, and the ratio of the first downward displacement to the sixth is found to be 22.5, find the stiffness of the spring and the damping coefficient.

■ Solution

Given: $$m = 5\text{ kg}, f_d = \frac{50}{20} = 2.5\text{ Hz}, \frac{x_1}{x_6} = 22.5$$

Logarithmic decrement, $$\delta = \left(\frac{1}{n}\right) ln\left(\frac{x_o}{x_n}\right) = \left(\frac{1}{5}\right) ln\left(\frac{x_1}{x_6}\right) = \left(\frac{1}{5}\right) ln\ 22.5 = 0.6227$$

$$\delta = \frac{2\pi\ \zeta}{(1-\zeta^2)^{1/2}}$$

$$0.6227 = \frac{2\pi\times\zeta}{(1-\zeta^2)^{1/2}}$$

$$1-\zeta^2 = 101.8\ \zeta^2$$

$$\zeta = 0.0986$$

$$\omega_d = 2\pi f_d = 2\pi\times 2.5 = 15.708\text{ rad/s}$$

$$\omega_n = \frac{\omega_d}{(1-\zeta^2)^{1/2}} = \frac{15.708}{[1-(0.0986)^2]^{1/2}} = 15.785\text{ rad/s}$$

Stiffness of spring, $k = m\omega_n^2 = 5 \times (15.785)^2 = 1245.83$ N/m

Critical damping coefficient, $c_c = 2m\omega_n = 2 \times 5 \times 15.785 = 157.85$ N·s/m

Damping coefficient, $c = \zeta c_c = 0.0986 \times 157.85 = 15.564$ N·s/m

Example 17.7

Determine the undamped and damped natural frequencies of the system shown in Fig.17.19. $k_1 = 2$ kN/m, $k_2 = 3$ kN/m, $c_1 = 100$ N·s/m, $c_2 = 200$ N·s/m, and $m = 15$ kg.

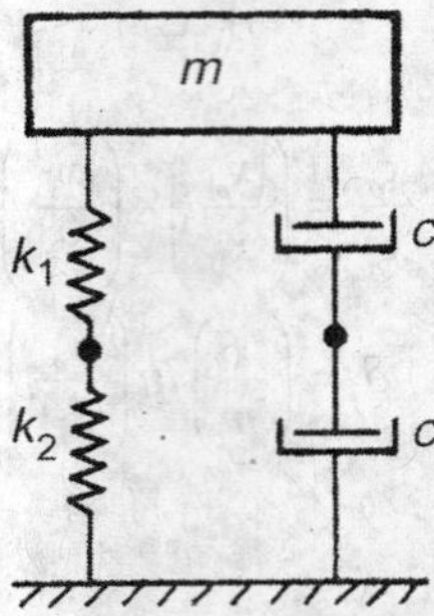

Fig.17.19 Diagram for Example 17.7

Solution

Equivalent stiffness, $k_e = \dfrac{k_1 k_2}{k_1 + k_2}$ for springs in series.

$= \dfrac{2 \times 3}{2 + 3} = 1.2$ kN/m

Equivalent damping coefficient, $c_e = \dfrac{c_1 c_2}{c_1 + c_2}$ for dampers in series.

$= \dfrac{100 \times 200}{100 + 200} = \dfrac{200}{3}$ N·s/m

Undamped natural frequency, $\omega_n = \sqrt{\dfrac{k_e}{m}}$

$= \sqrt{\dfrac{1.2 \times 10^3}{15}} = 8.94$ rad/s

Critical damping coefficient, $c_c = 2m\omega_n = 2 \times 15 \times 8.94 = 268.328$ N·s/m

Damping ratio, $\zeta = \dfrac{c_e}{c_c} = \dfrac{200}{3 \times 268.328} = 0.24845$

Damped natural frequency, $\omega_d = \omega_n\sqrt{1-\zeta^2}$

$$= 8.94\sqrt{1-(0.24845)^2}$$

$$= 8.66 \text{ rad/s}$$

Example 17.8

A spring-mass system consists of a spring of stiffness 350 N/m. The mass is 0.35 kg. The mass is displaced 20 mm beyond the equilibrium position and released. The damping coefficient is 14 N·s/m. Determine (a) critical damping coefficient, (b) damped natural frequency, and (c) logarithmic decrement.

■ Solution

Given : $k = 350$ N/M, m = 0.35 kg, $c = 14$ N.s/m

Undamped natural frequency, $\omega_n = \sqrt{\dfrac{k}{m}}$

$$= \sqrt{\frac{350}{0.35}} = 31.62 \text{ rad/s}$$

(a) Critical damping factor, $c_c = 2m\,\omega_n$

$$= 2 \times 0.35 \times 31.62 = 22.136 \text{ N·s/m}$$

(b) Damping factor, $\zeta = \dfrac{c}{c_c}$

$$= \frac{14}{22.136} = 0.632$$

Damped natural frequency, $\omega_d = \omega_n\sqrt{1-\zeta^2}$

$$= 31.62\sqrt{1-(0.632)^2}$$

$$= 24.5 \text{ rad/s}$$

(c) Logarithmic decrement, $\delta = \dfrac{2\pi\,\zeta}{\sqrt{1-\zeta^2}}$

$$= \frac{2\pi \times 0.632}{\sqrt{1-(0.632)^2}}$$

$$= 5.124$$

Example 17.9

A 25 kg mass is resting on a spring of 5 kN/m stiffness and a dashpot of 150 N·s/m damping coefficient in parallel. If a velocity of 0.1 m/s is applied to the mass at the rest position, what will be its displacement from the equilibrium position at the end of first second?

■ Solution

Given : m = 25 kg, k = 5 kN/M, c = 150 N.s/m, $\dot{x}(o) = 0.1$ m/s

Undamped natural frequency, $\omega_n = \sqrt{\dfrac{k}{m}}$

$$= \sqrt{\frac{5000}{25}} = 14.14 \text{ rad/s}$$

Critical damping coefficient, $c_c = 2m\omega_n$

$= 2 \times 25 \times 14.14 = 707$ N·s/m

Damping factor, $\zeta = \dfrac{c}{c_c} = \dfrac{150}{707} = 0.212$

Damped natural frequency, $\omega_d = \omega_n\sqrt{1-\zeta^2}$

$$= 14.14\sqrt{1-(0.212)^2}$$

$$= 13.82 \text{ rad/s}$$

$$x(t) = e^{-\zeta\omega_n t}[A\sin\omega_d t + B\cos\omega_d t]$$

Now $x(0) = 0$. Hence $B = 0$

$$\dot{x}(t) = e^{-\zeta\omega_n t}[A\omega_d\cos\omega_d t] - \zeta\omega_n e^{-\zeta\omega_n t}A\sin\omega_d t$$

$$0.1 = A\omega_d$$

$$A = \frac{0.1}{13.82} = 0.007236 \text{ m}$$

$$x(t) = 0.007236 \cdot e^{-0.212\times14.14t} \cdot \sin 13.82\,t$$

$$x(1) = 0.343 \text{ mm}$$

Example 17.10

A vibrating system consists of a mass of 30 kg, a spring of stiffness 20 kN/m and a damper of damping factor 0.25. Calculate:

(a) the critical damping coefficient
(b) the natural frequency of damped vibrations
(c) the logarithmic decrement, and
(d) the ratio of two successive amplitudes.

Solution

Given: $m = 30$ kg, $k = 20$ kN/m, $\zeta = 0.25$

(a) $c_c = 2\sqrt{km} = 2\sqrt{20\times10^3\times30} = 1549.2$ N·s/m

(b) $\omega_d = \omega_n\sqrt{1-\zeta^2}$

$$\omega_n = \sqrt{\frac{k}{m}} = \sqrt{\frac{20\times10^3}{30}} = 25.82 \text{ rad/s}$$

$$\omega_d = 25.82\sqrt{1-(0.25)^2} = 25 \text{ rad/s}$$

(c) $\delta = \dfrac{2\pi\zeta}{\sqrt{1-\zeta^2}} = \dfrac{2\pi\times0.25}{\sqrt{1-(0.25)^2}} = 1.622$

(d) $\dfrac{X_n}{X_{n+1}} = e^{\delta} = e^{1.622} = 5.06$

17.7.7 Undamped Forced Vibrations

When a system is subjected to forced harmonic excitation, its vibration response takes place at the same frequency as that of the excitation. Common sources of harmonic excitation are: unbalance in rotating machines, forces produced by reciprocating machines, or the motion of the machine itself.

Consider the spring-mass system shown in Fig.17.20 subjected to harmonic force excitation $F = F_o \sin \omega t$. The equation of motion for this system can be written as:

$$m\ddot{x} + kx = F_o \sin \omega t \tag{1}$$

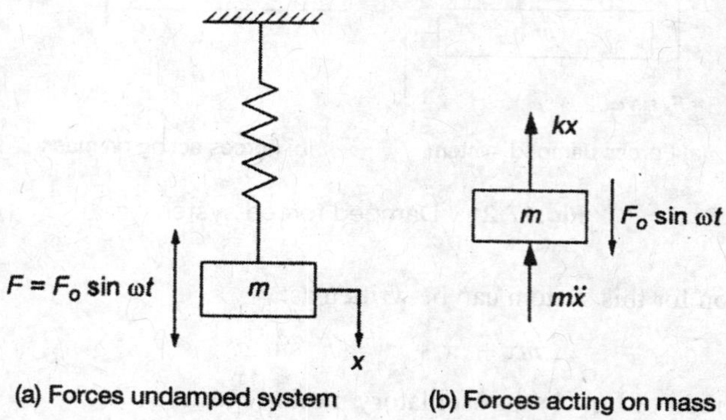

Fig.17.20 Forced underdamped system

The steady state solution of Eq. (1) is,

$$\begin{aligned} x(t) &= \frac{F_o \sin \omega t}{mD^2 + k} \\ &= \frac{F_o \sin \omega t}{-m\omega^2 + k} \\ &= \frac{x_{st} \sin \omega t}{1 - \beta^2} \end{aligned} \tag{17.26}$$

where

$$x_{st} = \frac{F_o}{k} = \text{static deflection}$$

$$\beta = \frac{\omega}{\omega_n} = \text{frequency ratio}$$

$$\omega_n = \left(\frac{k}{m}\right)^{1/2}$$

Amplitude of oscillation,

$$X = \frac{x_{st}}{1 - \beta^2} \tag{17.27}$$

At resonance, $\beta = 1$ and the amplitude tends to infinity.

17.7.8 Damped Forced Vibrations

Consider the damped forced vibration system shown in Fig.17.21 subjected to harmonic excitation.

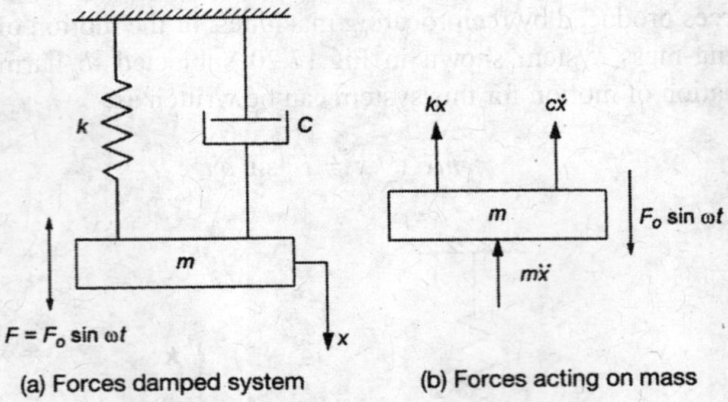

Fig.17.21 Damped forced system

The equation of motion for this system can be written as:

$$m\ddot{x} + c\dot{x} + kx = F_o \sin \omega t \quad (1)$$

The auxiliary solution of Eq. (1) for oscillatory motion ($\zeta < 1$) is:

$$x_a(t) = \exp(-\varsigma\omega_n t)\ [A \cos \omega_d t + B \sin \omega_d t]$$

where $\omega_n = \left(\dfrac{k}{m}\right)^{1/2}$, $\varsigma = \dfrac{c}{c_c}$ and $\omega_d = \omega_n(1 - \varsigma^2)^{1/2}$

The auxiliary solution shall die out in due course of time.

The steady state solution of Eq. (1) can be obtained as:

$$x_s(t) = \frac{F_o \sin \omega t}{mD^2 + cD + k}$$

$$= \frac{F_o \sin \omega t}{-m\omega^2 + cD + k}$$

$$= \frac{\left[(k - m\omega^2) - cD\right] F_o \sin \omega t}{\left[(k - m\omega^2)^2 - c^2D^2\right]}$$

$$= \frac{F_o\left[(k - m\omega^2) \sin \omega t - eco \cos \omega t\right]}{(k - m\omega^2)^2 + c^2\omega^2}$$

Now $\dfrac{c\omega}{k} = \dfrac{c}{c_c} \times 2m\omega_n \times \dfrac{\omega}{k} = 2\zeta\beta$

$$x_s(t) = \frac{x_{st}\left[(1 - \beta^2) \sin \omega t - 2\zeta\beta \cos \omega t\right]}{\left[(1 - \beta^2)^2 + (2\zeta\beta)^2\right]}$$

$$x_s(t) = \frac{x_{st} \sin(\omega t - \phi)}{[(1 - \beta^2)^2 + (2\varsigma\beta)^2]^{1/2}} \quad (17.28)$$

where $$x_{st} = \frac{F_o}{k}$$

The steady state solution shall persist.

Amplitude, $$X = \frac{x_{st}}{\left[(1-\beta^2)^2 + (2\zeta\beta)^2\right]^{1/2}} \tag{17.29}$$

Phase angle, $$\phi = \tan^{-1}\left[\frac{2\zeta\beta}{1-\beta^2}\right] \tag{17.30}$$

Magnification factor, $$M_f = \frac{X}{x_{st}}$$

$$= \frac{1}{[(1-\beta^2)^2 + (2\zeta\beta)^2]^{1/2}} \tag{17.31}$$

For $\beta = 1$, $$M_f = \frac{1}{2\zeta} \tag{17.32}$$

For M_f to be maximum, $\frac{dM_f}{d\beta} = 0$, which gives

$$\beta = (1 - 2\zeta^2)^{1/2} \tag{17.33}$$

$$(M_f)_{max} = \frac{1}{[2\zeta(1-\zeta^2)^{1/2}]} \tag{17.34}$$

Eqs. (17.29) and (17.30) indicate that X and ϕ are functions of β and ζ only. The magnification factor and phase angle are plotted in Figs.17.22 and 17.23 respectively for various values of ζ.

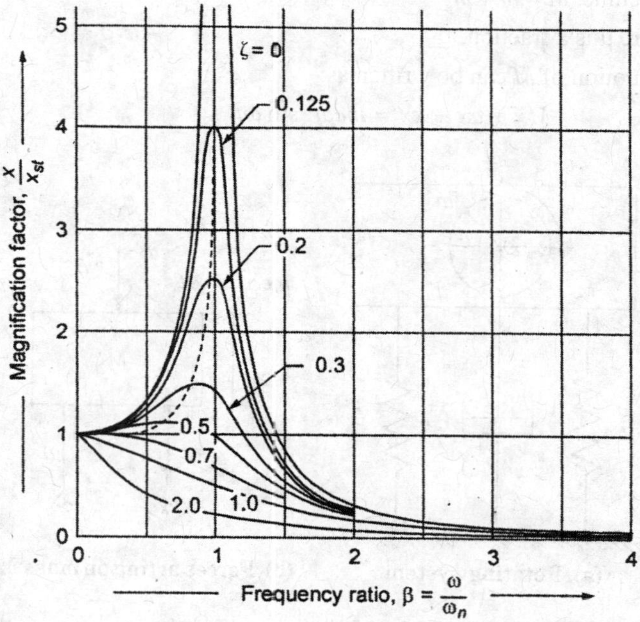

Fig.17.22 Magnification factor X/X_{st} as a function of frequency ratio

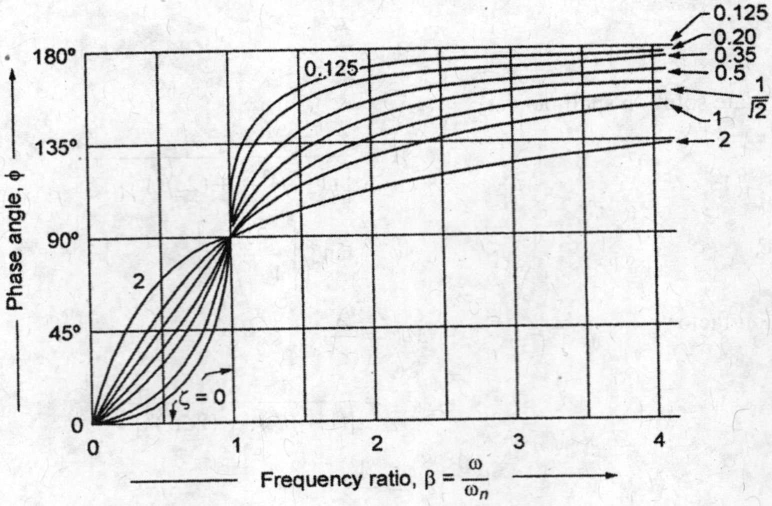

Fig.17.23 Phase angle of direct excitation as a function of frequency

17.7.9 Rotating Unbalance

If the mass centre of a rotor does not coincide with the axis of rotation then there shall be unbalance. This unbalance is a source of vibration excitation. Consider a spring-mass-dashpot system constrained to move in the vertical direction and excited by a rotating unbalances shown in Fig.17.24. Let m be the unbalanced mass of the rotating machine and e its eccentricity rotating at angular speed ω. Let x be the displacement of the non-rotating mass m_o from the static equilibrium position, then

Total mass of machine, $M = m + m_o$

Let $m = \mu M$, where μ is a fraction.

The equation of motion of M can be written as:

$$M\ddot{x} + c\ddot{x} + kx = me\omega^2 \sin \omega t \qquad (1)$$

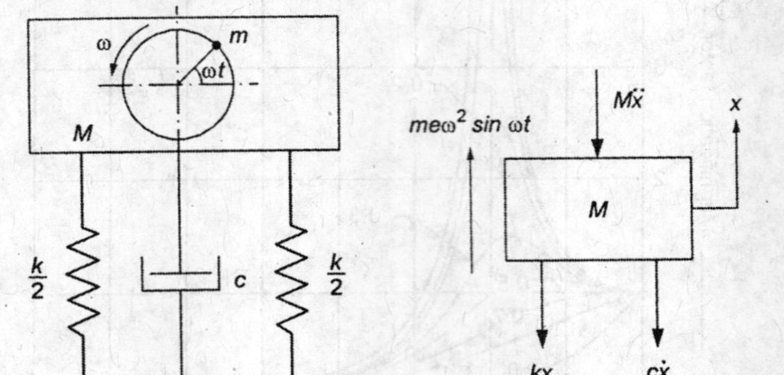

(a) Rotating system (b) Forces acting on mass M

Fig.17.24 Rotating unbalance

Where '*me*' is called the unbalance.
The steady state solution of Eq. (1) is:

$$x = \frac{\mu e \beta^2 \sin(\omega t - \phi)}{\left[(1-\beta^2)^2 + (2\zeta\beta)^2\right]^{1/2}} \tag{17.35}$$

$$\phi = \tan^{-1}\left[\frac{2\zeta\beta}{1-\beta^2}\right] \tag{17.36}$$

Eq. (17.35) is represented graphically in Fig.17.25.

Magnification factor,
$$M_f = \frac{X}{\mu e} = \frac{X}{(me/M)} \tag{17.37}$$

$$= \frac{\beta^2}{\left[(1-\beta^2)^2 + (2\zeta\beta)^2\right]^{1/2}} \tag{17.38}$$

For $\beta = 1, M_f = \dfrac{1}{2\zeta}$ (17.39a)

For $\beta \to \infty, M_f = 1$ (17.39b)

For $\beta = 1, \phi = 90°$, i.e. m_o is out of phase with m. (17.39c)

When $\beta >> 1, \phi = 180$ (17.39d)

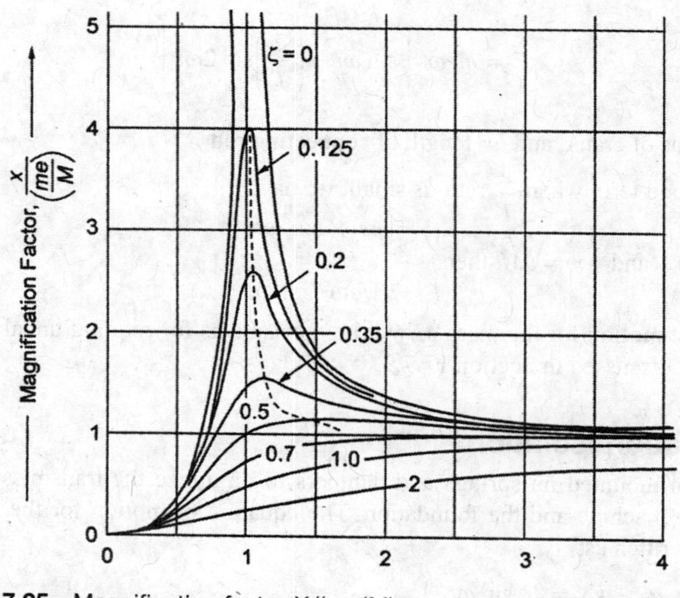

Fig.17.25 Magnification factor $X/(me/M)$ as a function of frequency ratio

17.7.10 Reciprocating Unbalance

A reciprocating system is shown in Fig.17.26. The reciprocating mass m consists of the mass of the piston, the gudgeon pin, and the part of the mass of connecting rod considered reciprocating with the piston. The mass of the rest of the machine is taken as m_o. The exciting force is equal to the inertia force of the reciprocating mass, given by,

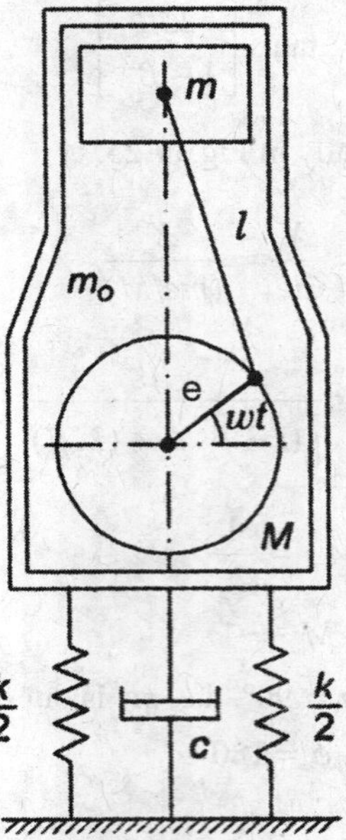

Fig.17.26 Reciprocating unbalance

$$F = me\omega^2\left[\sin\omega t + \left(\frac{e}{l}\right)\sin 2\omega t\right]$$

Where e = radius of crank, and l = length of connecting rod

Neglecting second term when $\frac{e}{l}$ ratio is small, we have

$$F = me\omega^2 \sin\omega t$$

Let $M = m + m_o$ and $m = \mu M$, then

$$F = \mu Me\omega^2 \sin\omega t$$

The equation of motion for the mass M will be the same as for rotating unbalance. The solution will be the same as discussed in Section 17.7.9.

17.7.11 Vibration Isolation

Machines are often mounted on springs and dampers to minimize the transmission of unbalanced forces between the machine and the foundation. The equation of motion for the mass m, shown in Fig.17.27, can be written as:

$$m\ddot{x} + c\dot{x} + kx = F_o \sin\omega t$$

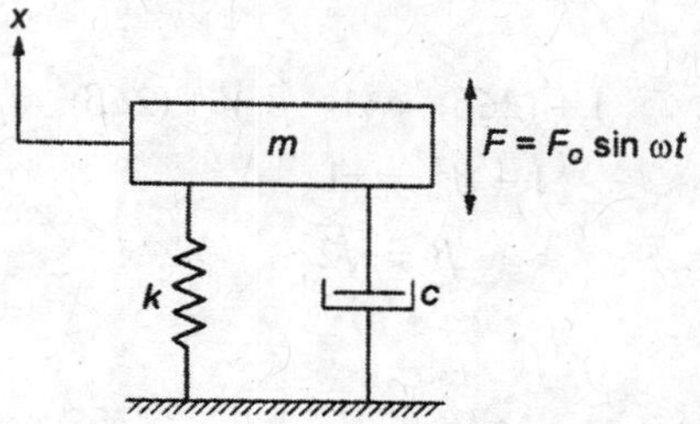

Fig.17.27 Vibration isolation

$$\text{Maximum amplitude, } X = \frac{x_{st}}{[(1-\beta^2)^2 + (2\zeta\beta)^2]^{1/2}} \tag{17.40}$$

Where $x_{st} = \frac{F_o}{k}$ = static deflection

Force transmitted to the foundation,

$$F_{tr} = [(kX)^2 + (c\omega X)^2]^{1/2}$$
$$= X[k^2 + (c\omega)^2]^{1/2}$$
$$= F_o \frac{[1 + (2\zeta\beta)^2]^{1/2}}{[(1-\beta^2)^2 + (2\zeta\beta)^2]^{1/2}} \tag{17.41}$$

$$\text{Force transmissibility, } TR = \frac{F_{tr}}{F_o} = \frac{\left[1 + (2\zeta\beta)^2\right]^{1/2}}{[(1-\beta^2)^2 + (2\zeta\beta)^2]^{1/2}} \tag{17.42}$$

The variation of TR v's β is shown in Fig.17.28. It may be seen that $TR < 1$ when $\beta > \sqrt{2}$.

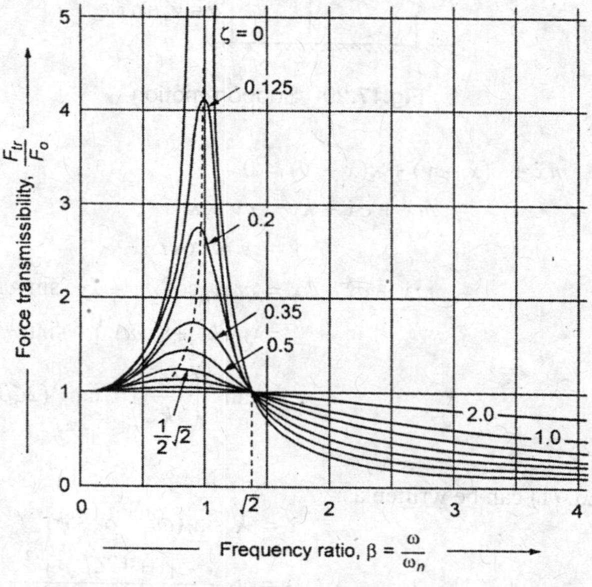

Fig.17.28 Force transmissibility as a function of frequency ratio

(i) For $TR = 1$, we have

$$1 + (2\zeta\beta)^2 = (1 - \beta^2)^2 + (2\zeta\beta)^2$$
$$1 - \beta^2 = \pm 1$$
$$\beta = \sqrt{2} \quad (17.43a)$$

(ii) For $TR > 1$, $\beta < \sqrt{2}$ (17.43b)

(iii) For $TR < 1$, $\beta > \sqrt{2}$ (17.43c)

Phase difference between the transmitted force and the excitation force is given by,

$$\phi - a = \tan^{-1}\left[\frac{2\zeta\beta}{1-\beta^2}\right] - \tan^{-1}(2\zeta\beta) \quad (17.44)$$

17.7.12 Support Motion

(a) ***Absolute amplitude***

Consider the dynamical system excited by the motion of the support, as shown in Fig.17.29. Let $y = y_o \sin \omega t$ be the harmonic displacement of the support and x the absolute displacement of mass m from an inertial reference. The equation of motion for the mass can be written as:

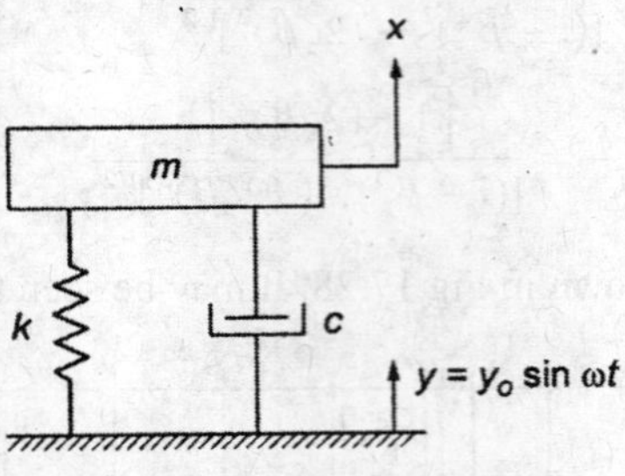

Fig.17.29 Support motion

$$m\ddot{x} + c(\dot{x} - \dot{y}) + k(x - y) = 0$$

or
$$m\ddot{x} + c\dot{x} + kx = c\dot{y} + ky$$

Now
$$\dot{y} = y_o\omega \cos \omega t$$

$\therefore$
$$m\ddot{x} + c\dot{x} + kx = c\omega y_o \cos \omega t + ky_o \sin \omega t$$
$$= y_o[k^2 + (c\omega)^2]^{1/2} \sin(\omega t + a) \quad (1)$$

where
$$\alpha = \tan^{-1}\left(\frac{c\omega}{k}\right) = \tan^{-1}(2\zeta\beta)$$

The solution of Eq. (1) can be written as:

$$x = X \sin(\omega t + \alpha - \phi)$$

where
$$X = y_o \frac{[1 + (2\zeta\beta)^2]^{1/2}}{[(1-\beta^2)^2 + (2\zeta\beta)^2]^{1/2}} \quad (17.45)$$

$$\phi = \tan^{-1}\left[\frac{2\zeta\beta}{1-\beta^2}\right]$$

$$\phi - \alpha = \tan^{-1}\left[\frac{2\zeta\beta}{1-\beta^2}\right] - \tan^{-1}(2\zeta\beta)$$

Displacement Transmissibility, *TR*: It is defined as the ratio of the amplitude of the system to the amplitude of the support.

$$TR = \frac{X}{y_o} = \frac{\left[1+(2\zeta\beta)^2\right]^{1/2}}{\left[(1-\beta^2)^2+(2\zeta\beta)^2\right]^{1/2}} \tag{17.46}$$

For $\beta = 1$,
$$TR = \frac{\left(1+4\zeta^2\right)^{1/2}}{2\zeta} \tag{17.47}$$

(b) *Relative amplitude*

Let $z = x - y$. The equation of motion becomes,

$$m\ddot{z} + c\dot{z} + kz = m\omega^2 y_o \sin \omega t \tag{2}$$

The solution of Eq. (2) can be written as:

$$z = Z\sin(\omega t - \phi)$$

where
$$Z = \frac{\beta^2 y_o}{[1-\beta^2)^2 + (2\zeta\beta)^2]^{1/2}} \tag{17.48}$$

$$\phi = \tan^{-1}\left[\frac{2\zeta\beta}{1-\beta^2}\right] \tag{17.49}$$

Example 17.11

An industrial machine of mass 450 kg is supported on springs with a statistical deflection of 5 mm. If the machine has a rotating unbalance of 0.25 kg · m, determine (a) the force transmitted to the floor at 1200 rpm, and (b) the dynamical amplitude at this speed.

■ Solution

Given: $m = 450$ kg, $\delta_{st} = 5$ mm, $me = 0.25$ kg · m, $N = 1200$ rpm

Angular speed, $\omega = \dfrac{2\pi N}{60} = \dfrac{2\pi \times 1200}{60} = 125.664$ rad/s

Natural frequency, $\omega_n = \left(\dfrac{g}{\delta_{st}}\right)^{1/2} = \left(\dfrac{9.81}{0.005}\right)^{1/2} = 44.29$ rad/s

$$\beta = \frac{\omega}{\omega_n} = \frac{125.664}{44.29} = 2.837$$

$$F_o = (me)\omega^2 = 0.25 \times (125.664)^2 = 3947.84 \text{ N}$$

(a) For $\zeta = 0, F_{tr} = \dfrac{F_o}{\beta^2 - 1} = \dfrac{3947.84}{(2.837)^2 - 1} = 560.1$ N

(b) $\dfrac{MX}{me} = \dfrac{\beta^2}{\beta^2 - 1}$

$$X = 0.25 \times \frac{(2.837)^2}{450\left\{(2.837)^2 - 1\right\}} = 0.6344 \text{ mm}$$

Example 17.12

A weight attached to a spring of stiffness 252 N/m has a viscous damping device. When the weight is displaced and released, the period of vibration is found to be 1.8 s, and the ratio of consecutive amplitudes is 4.2 to 1.0. Determine the amplitude and phase when the force $F = 2 \cos 3t$ N acts on the system.

Solution

Given: $k = 525$ N/m, $T_d = 1.8$ s, $\dfrac{x_n}{x_{n+1}} = 4.2$, $F_o = 2$ N. $\omega = 3$ rad/s

Logarithmic decrement, $\delta = ln\left(\dfrac{x_n}{x_{n+1}}\right) = ln\ 4.2 = 1.435 = \dfrac{2\pi\zeta}{(1-\zeta^2)^{1/2}}$

$$1 - \zeta^2 = 19.1715\ \zeta^2$$

$$\zeta = 0.2226$$

Time period, $T_d = \dfrac{2\pi}{\omega_d} = 1.8$

$$\omega_d = \frac{2\pi}{1.8} = 3.49 \text{ rad/s}$$

$$\omega_n = \frac{\omega_d}{(1-\zeta^2)^{1/2}} = \frac{3.49}{[1-(0.2226)^2]^{1/2}} = 3.58 \text{ rad/s}$$

$$\beta = \frac{\omega}{\omega_n} = \frac{3}{3.58} = 0.838$$

Static deflection, $x_{st} = \dfrac{F_o}{k} = \dfrac{2}{525} = 3.81$ mm

$$X = \frac{x_{st}}{\left[(1-\beta^2)^2 + (2\zeta\beta)^2\right]^{1/2}}$$

$$= \frac{3.81}{\left[\left\{1-(0.838)^2\right\}^2 + (2 \times 0.2226 \times 0.838)^2\right]^{1/2}}$$

$$= 7.982 \text{ mm}$$

Phase difference, $\phi = \tan^{-1}\left[\dfrac{2\zeta\beta}{1-\beta^2}\right]$

$$= \tan^{-1}\left[\frac{2 \times 0.226 \times 0.838}{1-(0.838)^2}\right]$$

$$= 51.41°$$

Example 17.13

A centrifugal fan of mass 5 kg has a rotating unbalance of 0.25 kg·m. When dampers having damping factor 0.2 are used, specify the springs for mounting such that only 10% of the unbalance force is transmitted to the floor and the force transmitted. The fan is running at a constant speed of 1000 rpm.

■ **Solution**

Given: $m = 5$ kg, $me = 0.25$ kg·m, $\zeta = 0.2$, $TR = 0.1$, $N = 1000$ rpm

$$\omega = \frac{2\pi N}{60} = \frac{2\pi \times 1000}{60} = 104.72 \text{ rad/s}$$

$$TR = \frac{\left[1 + (2\zeta\beta)^2\right]^{1/2}}{\left[\left(1-\beta^2\right)^2 + (2\zeta\beta)^2\right]^{1/2}}$$

$$(0.1)^2 = \frac{1 + (2\times 0.2\times \beta)^2}{\left(1-\beta^2\right)^2 + (2\times 0.2\times\beta)^2}$$

$$\beta^4 - 17.84\,\beta^2 - 99 = 0$$

$$\beta^2 = 9.92 \pm [(9.92)^2 + 99]^{1/2} = 22.28$$

$$\beta = 4.72$$

$$\omega_n = \frac{\omega}{\beta} = \frac{104.72}{4.72} = 22.186 \text{ rad/s}$$

$$k = m\omega_n^2 = 5\times(22.186)^2 = 2461.2 \text{ N/m}$$

$$F_o = (me)\,\omega^2 = 0.25\times(104.72)^2 = 2741.57 \text{ N}$$

$$F_{tr} = TR\times F_o = 0.1\times 2741.57 = 274.157 \text{ N}$$

17.8 TRANSVERSE VIBRATIONS

17.8.1 Beam Carrying Single Concentrated Load

Consider a beam carrying a concentrated load at the midspan and vibrating transversally.

Let l = span of beam

I = moment of inertia of beam cross-section

E = Young's modulus of elasticity

W = Central load.

Then $\omega_n = \sqrt{\dfrac{g}{\delta_{st}}}$

where δ_{st} = static deflection

$$= \frac{Wl^3}{48EI} \quad \text{for a simply supported beam}$$

$$\therefore \quad \omega_n = \sqrt{\frac{48EIg}{Wl^3}} \text{ rad/s} \tag{17.50}$$

$$\delta_{st} = \frac{Wl^3}{192\ EI} \quad \text{for a beam fixed at both ends.}$$

and
$$\omega_n = \sqrt{\frac{192EIg}{Wl^3}} \text{ rad/s} \tag{17.51}$$

If the concentrated load is off-midspan, then

$$\delta_{st} = \frac{Wa^2b^2}{3EIl} \quad \text{for a simply supported beam} \tag{17.52}$$

$$= \frac{Wa^3b^3}{3EIl^3} \quad \text{for a beam fixed at both ends} \tag{17.53}$$

For a cantilever carrying end load W,

$$\delta_{st} = \frac{Wl^3}{3EI} \tag{17.54}$$

17.8.2 Beam Carrying Uniformly Distributed Load

Let w = intensity of load per unit length

Then
$$\delta_{st} = \frac{5}{384}\frac{wl^4}{EI} \quad \text{for a simply supported beam}$$

$$\omega_n = \sqrt{\frac{g}{\delta_{st}}} = \sqrt{\frac{384EIg}{5wl^4}} \text{ rad/s} \tag{17.55}$$

and
$$\delta_{st} = \frac{wl^4}{384EI} \quad \text{for a beam fixed at both ends}$$

$$\omega_n = \sqrt{\frac{384EIg}{wl^4}} \text{ rad/s} \tag{17.56}$$

$$\delta_{st} = \frac{wl^4}{8EI} \quad \text{for a cantilever beam}$$

$$\omega_n = \sqrt{\frac{8EIg}{wl^4}} \text{ rad/s} \tag{17.57}$$

17.8.3 Shaft Carrying Several Loads

(a) *Dunkerley's method*

Dunkerley's method suggests the following equation for lower bound on the fundamental frequency:

$$\frac{1}{\omega_n^2} = \frac{1}{\omega_1^2} + \frac{1}{\omega_2^2} + \cdots + \frac{1}{\omega_m^2} + \frac{1}{\omega_s^2} \tag{17.58}$$

where ω_n = fundamental natural frequency of the system

ω_i = natural frequency of the shaft with each lumped mass acting alone at its point of application in the absence of other masses. $i = 1$ to m

ω_s = natural frequency of the shaft alone due to its uniformly distributed mass.

(b) *Rayleigh's method*

Let $W_1, W_2, W_3, \cdots$ = loads on the shaft

$y_1, y_2, y_3, \cdots$ = total deflections under the loads.

Then

$$T_{\max} = \frac{\omega_n^2}{2g}\sum_{i=1}^{m} W_i y_i^2$$

$$U_{\max} = \frac{1}{2}\sum_{i=1}^{m} W_i y_i$$

For

$$T_{\max} = U_{\max}$$

$$\omega_n = \sqrt{\frac{g\,\Sigma W_i y_i}{\Sigma W_i y_i^2}} \quad (17.59)$$

Example 17.14

A shaft of span 1 m and diameter 25 m is simply supported at the ends. It carries a 1.5 kN concentrated load at midspan. If $E = 200$ Gpa, calculate its fundamental frequency.

■ Solution

Given: $l = 1\text{m}$, $d = 25$ mm, $N = 1.5$ kN, $E = 200$ Gpa

$$I = \frac{\pi d^4}{64} = \frac{\pi\left(25\times10^{-3}\right)^4}{64} = 19174.76\times10^{-12}\ \text{m}^4$$

$$\delta_{st} = \frac{Wl^3}{48EI} = \frac{1.5\times10^3\times1}{48\times200\times10^9\times19174.76\times10^{-12}}$$

$$= 8.1487\times10^{-3}\ \text{m}$$

$$\omega_n = \sqrt{\frac{g}{\delta_{st}}} = \sqrt{\frac{9.81}{8.1487\times10^{-3}}} = 34.697\ \text{rad/s}$$

$$f_n = \frac{\omega_n}{2\pi} = 5.52\ \text{Hz}$$

Example 17.15

A shaft 30 mm diameter and 2 m long has a uniformly distributed load of 120 N/m length. It is simply supported at the ends and carries three loads of 1 kN, 1.5 kN and 0.5 kN at 0.6 m, 1 m and 1.5 m respectively from the left end support. Calculate the natural frequency of transverse vibrations. $E = 200$ Gpa.

■ Solution

Given: $d = 30$ mm, $l = 2$ m

$$I = \frac{\pi d^4}{64} = \frac{\pi\left(30\times10^{-3}\right)^4}{64} = 39760.8\times10^{-12}\ \text{m}^4$$

$$\delta_{st} = \frac{Wa^2b^2}{3EIl}$$

$$3EIl = 3 \times 200 \times 10^9 \times 39760.8 \times 10^{-12} \times 2$$

$$= 47712.94 \text{ Nm}^3$$

$$\omega_m = \sqrt{\frac{g}{\delta_{st}}}$$

Table 17.1

Load W, kN	a, m	b, m	Wa^2b^2, N m^4	δ_{st}, m	ω_m, rad/s
1.0	0.6	1.4	705.6	0.0148	25.76
1.5	1.0	1.0	1500	0.0314	17.66
0.5	1.5	0.5	281.25	0.0059	40.79

$$\omega_s = \sqrt{\frac{384EIg}{5\omega l^4}} = \sqrt{\frac{384 \times 200 \times 10^9 \times 39760.8 \times 10^{-12} \times 9.81}{5 \times 120 \times 2^4}}$$

$$= 55.86 \text{ rad/s}$$

Applying Dunkerley's method,

$$\frac{1}{\omega_n^2} = \frac{1}{\omega_1^2} + \frac{1}{\omega_2^2} + \frac{1}{\omega_3^2} + \frac{1}{\omega_s^2}$$

$$= \frac{1}{(25.76)^2} + \frac{1}{(17.66)^2} + \frac{1}{(40.79)^2} + \frac{1}{(55.86)^2}$$

$$\omega_n^2 = 177.46$$

$$\omega_n = 13.32 \text{ rad/s}$$

$$f_n = \frac{\omega_n}{2\pi} = 2.12 \text{ Hz}$$

17.9 CRITICAL SPEED

The critical speed of a rotating shaft is the speed at which the shaft starts to vibrate violently in the transverse direction. Critical speed is also called 'whipping 'or 'whirling' speed. The main reason for the whirling speed is the mass unbalance of the shaft when the mass centre does not coincide with the geometric centre.

17.9.1 Shaft Having a Single Disc

(a) *Without damping*

Consider a light vertical shaft with a disc of mass m at the midspan and rotating with angular speed ω, as shown deflected in Fig.17.30. Let S be the geometric centre of the disc through which the centre line of the shaft passes. Point G is the centre of gravity of the disc where its mass m is assumed to be concentrated. $SG = e$ is the eccentricity due to manufacturing defects to variation in material density of the disc. Point O is the intersection of the bearing centre line with the disc r is the deflection of S from the undefeated position O. Let k be the stiffness of the shaft in the lateral direction. Considering the equilibrium of forces acting on the disc and shaft, we have

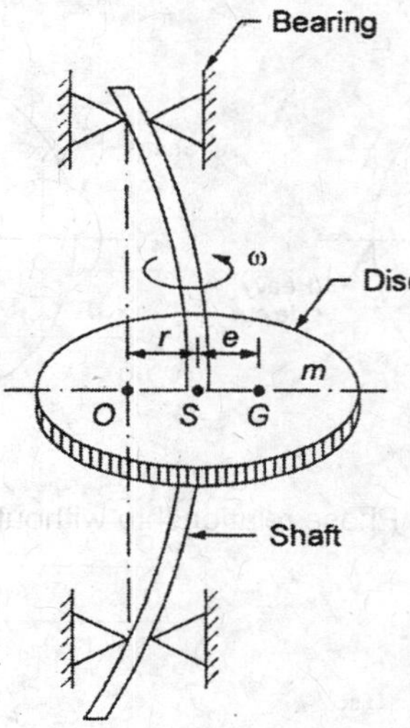

Fig.17.30 Rotating shaft with a disc having eccentric mass

Centrifugal force due to mass m = Restoring force due to lateral stiffness of shaft

$$m(r + e)\,\omega^2 = kr$$

or
$$r = \frac{me\omega^2}{k - m\omega^2}$$
$$= \frac{e\beta^2}{1 - \beta^2} \tag{17.60}$$

where
$$\beta = \frac{\omega}{\omega_m},\ \omega_n = \left(\frac{k}{m}\right)^{\frac{1}{2}}$$

From Eq. (17.60), we find that the deflection of the shaft tends to infinity when $\beta = 1$. Thus, the critical speed of the shaft is equal to the natural frequency of lateral vibrations of the shaft. It may be seen that r is position for $\beta < 1$. i.e. the disc rotates with heavy side outwards, as shown in Fig.17.31(a). For $\beta > 1$, r is negative, i.e. the disc rotates with light side outwards, as shown in Fig.17.31(b). Also $\beta < 1$ corresponds to zero degree phase difference and $\beta > 1$ corresponds to 180° phase difference.

When $\beta >> 1$, $r \to -e$, i.e. the point G approaches O and the disc rotates about its centre of gravity. Therefore, it is always advisable to operate the machine much above its natural frequency.

(b) *With damping*

Consider a shaft with a disc rotating at angular speed ω as shown in Fig.17.32. Let $OG = a$, $OS = r$, $SG = e$, k = stiffness of shaft material, c = damping coefficient of shaft material, $\angle GOS = \alpha$, $\angle GSA = \phi$. The forces acting on the disc are shown in Fig.17.33.

1. Centrifugal force = $ma\omega^2$ acting at G along OG produced.
2. Restoring force = kr at S along SO, and
3. Damping force = $c\omega r$ acting at S perpendicular to OS.

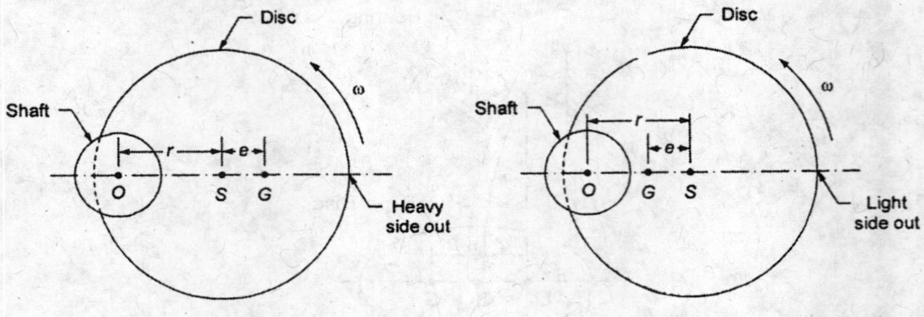

(a) (b)

Fig.17.31 Phase relationship without damping

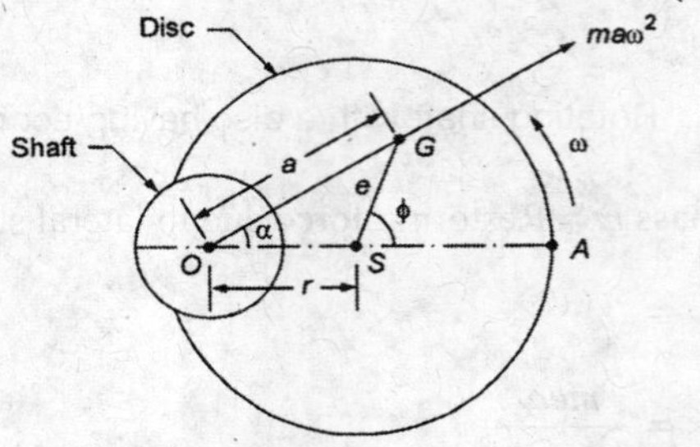

Fig.17.32 Rotating shaft with damping

From the geometry of Fig.17.32, we have

$$a \sin \alpha = e \sin \phi$$

$$a \cos \alpha = r + e \cos \phi$$

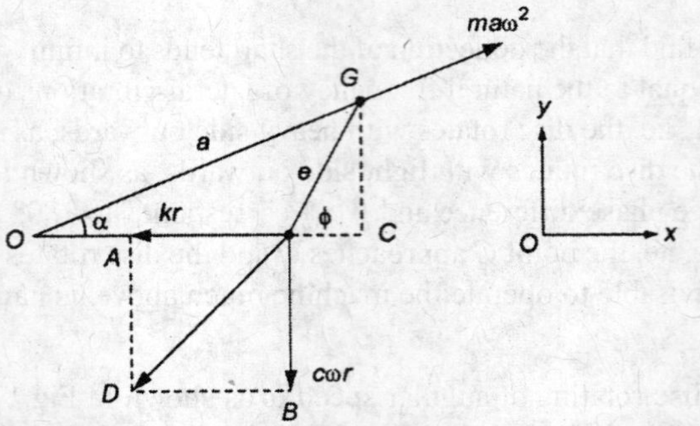

Fig.17.33 Forces acting on the disc

Considering the equilibrium of forces acting on the system along x- and y-directions, we have

$$ma\omega^2 \cos \alpha - kr = 0$$

$$ma\omega^2 \sin \alpha - c\omega r = 0$$

or $$m\omega^2 e \sin \phi - c\omega r = 0$$

or $$me\omega^2 \cos \phi = r\,(k - m\omega^2)$$

$$me\omega^2 \sin \phi = c\omega r$$

Squaring and adding, we get

$$(me\omega^2)^2 = r^2\,[(k - m\omega^2)^2 + (c\omega)^2]$$

or $$\frac{r}{e} = \frac{\omega^2}{\left[\left(k - m\omega^2\right)^2 + \left(c\omega\right)^2\right]^{1/2}}$$

$$= \frac{\beta^2}{\left[\left(1 - \beta^2\right)^2 + \left(2\zeta\beta\right)^2\right]^{1/2}} \tag{17.61}$$

Where $\beta = \dfrac{\omega}{\omega_n}, \zeta = \dfrac{c}{2\left(mk\right)^{1/2}}$

$$\text{Phase difference, } \phi = \tan^{-1}\left(\frac{2\zeta\beta}{1 - \beta^2}\right) \tag{17.62}$$

The following observations may be made from Eq. (17.62):

1. $\phi = 0$ when $\beta << 1$, and heavy side of the disc will be out.
2. $0 < \phi < 90°$ when $\beta < 1$, and heavy side of the disc will be out.
3. $\phi = 90°$ when $\beta = 1$.
4. $90° < \phi < 180°$ when $\beta > 1$, and light side of the disc will be out.
5. $\phi = 180°$ and $r \rightarrow -e$ when $\beta >> 1$, and light side of the disc will be out with the disc rotating about its centre of gravity.

The phase relationships with damping are shown in Fig.17.34.

Example 17.16

A 60 kg compressor rotor is mounted on a shaft of stiffness 15 MN/m. Determine the critical speed of the rotor assuming the bearings to be rigid. If the rotor has an eccentricity of 2 mm and its operating speed is 6500 rpm, determine the unbalance response. The damping factor in the system can be taken as 0.06. If the compressor is started from rest, what will be the maximum whirl amplitude of the rotor before it reaches its full operational speed?

■ Solution

Given: $m = 60$ kg, $k = 15$ MN/m, $e = 2$ mm, $N = 6500$ rpm, $\zeta = 0.06$

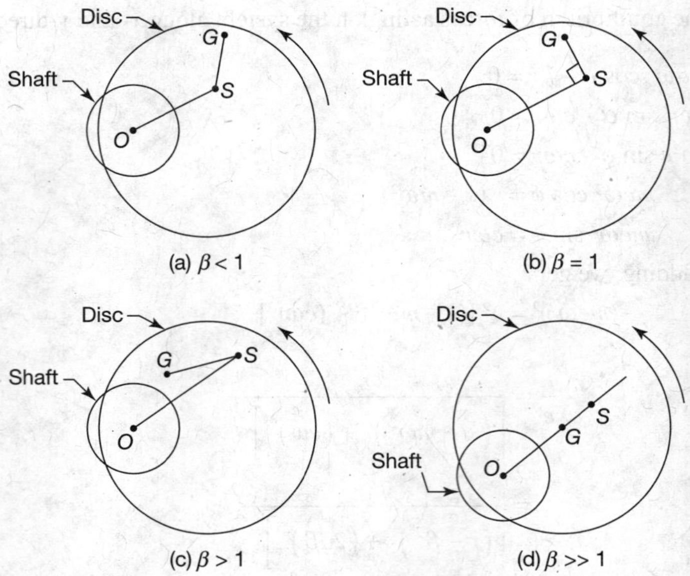

Fig.17.34 Phase relationship with damping

$$\omega_n = \left(\frac{k}{m}\right)^{\frac{1}{2}}$$

$$= \left(\frac{15 \times 10^6}{60}\right)^{\frac{1}{2}} = 500 \text{ rad/s}$$

$$\omega = \frac{2\pi N}{60}$$

$$= \frac{2\pi \times 6500}{60} = 680.678 \text{ rad/s}$$

$$\beta = \frac{\omega}{\omega_n} = \frac{680.678}{500} = 1.36$$

Whirl amplitude, $$r = \frac{e\beta^2}{\left[\left(1-\beta^2\right)^2 + \left(2\zeta\beta\right)^2\right]^{1/2}}$$

$$= \frac{2 \times (1.36)^2}{\left[\left\{1.36)^2\right\}^2 + \left(2 \times 0.06 \times 1.36\right)^2\right]^{1/2}} = 4.276 \text{ mm}$$

Maximum whirl amplitude, $r_{\max} = \dfrac{e}{2\zeta} = \dfrac{2}{2 \times 0.06} = 16.67$ mm

Example 17.17

The rotor of a turbine of mass 15 kg is supported at the midspan of a shaft of span 0.4 m. The rotor has an unbalance of 0.003 kg · m. Determine the force exerted on the bearings at a speed of 6000 rpm. The diameter of steel shaft is 25 mm and $E = 200$ Gpa.

■ **Solution**

$$\omega = \frac{2\pi N}{60} = \frac{2\pi \times 6000}{60} = 628.32 \text{ rad/s}$$

$$I = \frac{\pi d^4}{64} = \frac{\pi\left(25\times10^{-3}\right)^4}{64} = 19174.76\times10^{-12} \text{ m}^4$$

$$k = \frac{48EI}{l^3} = \frac{48\times200\times10^9\times19174.76\times10^{-12}}{(0.4)^3} = 2876214 \text{ N/m}$$

$$\omega_n = \sqrt{\frac{k}{m}} = \sqrt{\frac{2876214}{15}} = 437.89 \text{ rad/s}$$

$$\beta = \frac{\omega}{\omega_n} = \frac{628.32}{437.89} = 1.4349$$

$$r = \frac{-(me)\,\omega^2/k}{\beta^2-1} = \frac{-(0.003)\times(628.32)^2}{2876214\left[(1.4389)^2-1\right]}$$

$$= -0.389 \text{ mm}$$

Total force = $kr = 2876214 \times 0.389 \times 10^{-3} = 1118.5$ N

$$\text{Force on each bearing} = \frac{1118.5}{2} = 559.25 \text{ N}$$

17.10 TORSIONAL VIBRATIONS

17.10.1 Undamped Free Vibration

Consider a circular disc of moment of inertia J about the axis of the shaft, attached to a circular shaft of diameter d and length l, whose other end is fixed, as shown in Fig.17.35.

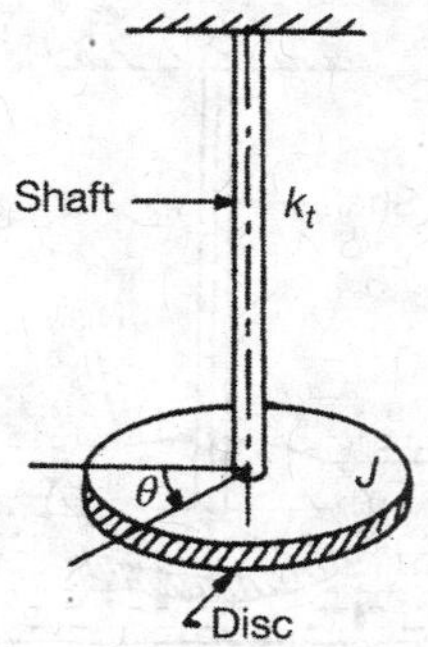

Fig.17.35 Torsional undamped free vibration

Torsional stiffness of shaft,

$$k_t = \pi d^4 G/(32l)$$

where G = modulus of rigidity of the shaft.
For small torsional oscillations of the disc, the equation of motion is,

$$J\ddot{\theta} + k_t\,\theta = 0 \tag{1}$$

Its solution is,

$$\theta(t) = A \sin \omega_n t + B \cos \omega_n t \tag{2}$$

where $$\omega_n = \sqrt{k_t/J} \tag{17.63}$$

and A and B are the constants.
If $\theta(0) = \theta_0$ and $\theta(0) = \omega_0$, then, we get, $A = \omega_0/\omega_n$ and $B = \theta_0$. Eq. (2) becomes,

$$\theta(t) = (\omega_0/\omega_n) \sin \omega_n t + \theta_0 \cos \omega_n t \tag{17.64a}$$

or $$= \left[(\omega_0/\omega_n)^2 + \theta_0^2 \right]^{0.5} \sin(\omega t - \phi) \tag{17.64b}$$

where $\phi = \tan^{-1}\left[\dfrac{\omega_0}{\omega_n \theta_0}\right]$

17.10.2 Damped Free Vibration

Consider the torsional vibration of a viscously damped disc at one end of a circular shaft whose other end is fixed to a rigid support, as shown in Fig.17.36. The equation of motion is,

$$J\ddot{\theta} + c_t\dot{\theta} + k_t\theta = 0 \tag{1}$$

For undamped vibration, the solution of Eq. (1) is,

$$\theta(t) = \exp(-\zeta\omega_n t)\left[A \sin \omega_d t + B \cos \omega_d t\right] \tag{2}$$

where $$\omega_n = (k_t/J)^{0.5} \tag{17.65}$$

$$\zeta = c_t/c_{tc} \tag{17.66}$$

$$c_{tc} = 2J\omega_n = 2(k_t J)^{0.5} \tag{17.67}$$

$$\omega_d = \omega_n (1 - \zeta^2)^{0.5} \tag{17.68}$$

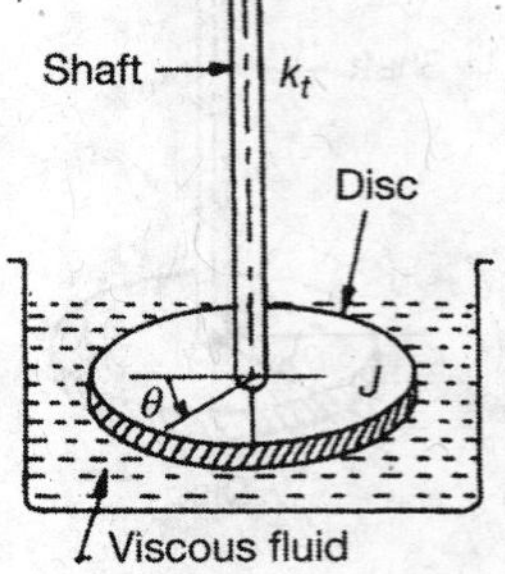

Fig.17.36 Damped free torsional vibration

17.10.3 Damped Forced Vibration

Consider the damped forced vibration of the disc shown in Fig.17.37. The equation of motion is,

$$J\ddot{\theta} + c_t\dot{\theta} + k_t\,\theta = T_0 \sin \omega t$$

The complete solution for under-damped vibration is,

$$\theta(t) = \exp(-\zeta\omega_n t)\sin(\omega_d t + \phi) + \frac{T_o \sin(\omega t - \alpha)}{\sqrt{\left(1-\beta^2\right)^2 + (2\zeta\beta)^2}} \qquad (17.69)$$

where $$\tan\alpha = \frac{2\zeta\beta}{1-\beta^2} \qquad (17.70)$$

$$\beta = \frac{\omega}{\omega_n},\ \omega_n = \sqrt{k_t/J},\ \zeta = \frac{c_t}{c_{tc}},\ c_{tc} = 2J\omega_n = 2\sqrt{k_t J}$$

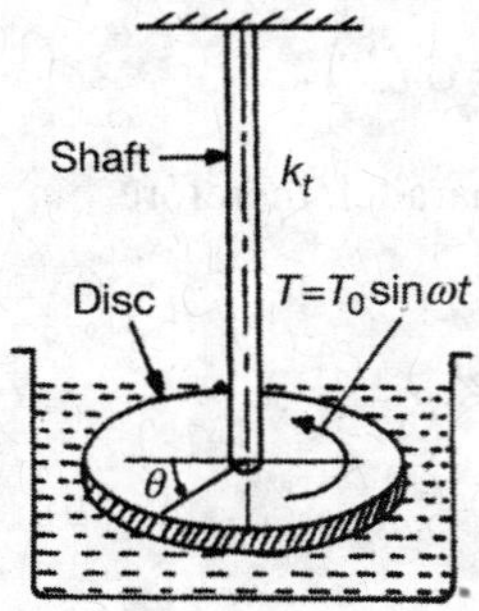

Fig.17.37 Damped forced torsional vibration

17.10.4 Stepped Shaft

Consider a stepped shaft, as shown in Fig.17.38. If a torque is applied to one end of the shaft and the other end is rigidly fixed, then

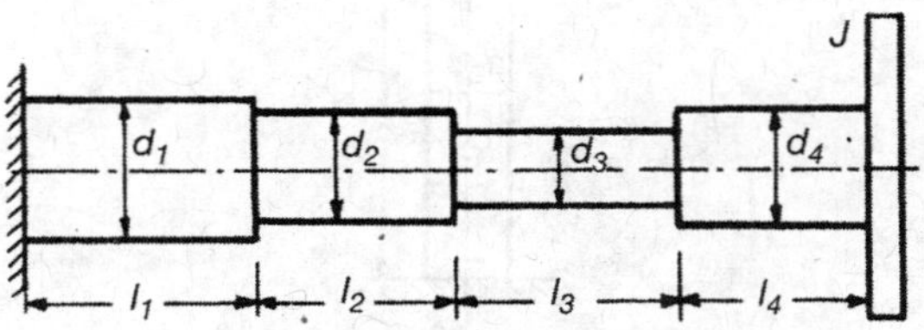

Fig.17.38 Stepped shaft under torsion

$$\theta = \theta_1 + \theta_2 + \theta_3 + \theta_4$$

$$= \frac{32T}{\pi G}\left[\frac{l_1}{d_1^4} + \frac{l_2}{d_2^4} + \frac{l_3}{d_3^4} + \frac{l_4}{d_4^4}\right]$$

$$= \frac{32T}{\pi G d_1^4}\left[l_1 + (d_1/d_2)^4 l_2 + (d_1/d_3)^4 l_3 + (d_1/d_4)^4 l_4\right]$$

The equivalent length l_e of the shaft of diameter d_1 is,

$$l_e = l_1 + (d_1/d_2)^4 l_2 + (d_1/d_3)^4 l_3 + (d_1/d_4)^4 l_4 \tag{17.71}$$

$$\theta = \frac{32\ Tl_e}{\pi G d_1^4}$$

$$k_t = \frac{T}{\theta} = \frac{\pi G d_1^4}{32\ l_e}$$

$$\omega_n = \sqrt{\frac{k_t}{J}}$$

17.10.5 Fixed Shaft with a Rotor

Consider a fixed stepped shaft carrying a rotor, as shown in Fig.17.39.

$$J_1 = \pi d_1^4/32, q_1 = GJ_1/l_1$$

$$J_2 = \pi d_2^4/32, q_2 = GJ_2/l_2$$

The two lengths of the shaft are in parallel, therefore

$$q_e = q_1 + q_2$$

$$= G\,(J_1/l_1 + J_2/l_2)$$

Now $I = mK^2$

Natural frequency, $\omega_n = [q_e/I]^{0.5}$. (17.72)

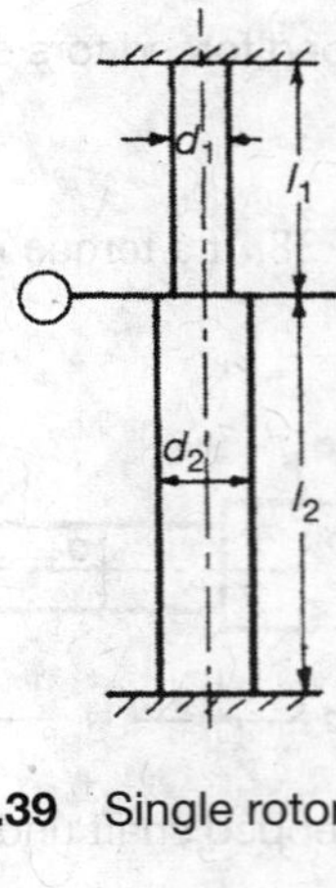

Fig.17.39 Single rotor on a shaft

17.10.6 Two-Degree of Freedom System

Consider the two-degree of freedom rotational system, as shown in Fig.17.40. The equations of motion are,

$$J_1\,\ddot{\theta}_1 + k_t(\theta_1 - \theta_2) = 0$$

$$J_2\,\ddot{\theta}_2 - k_t(\theta_1 - \theta_2) = 0$$

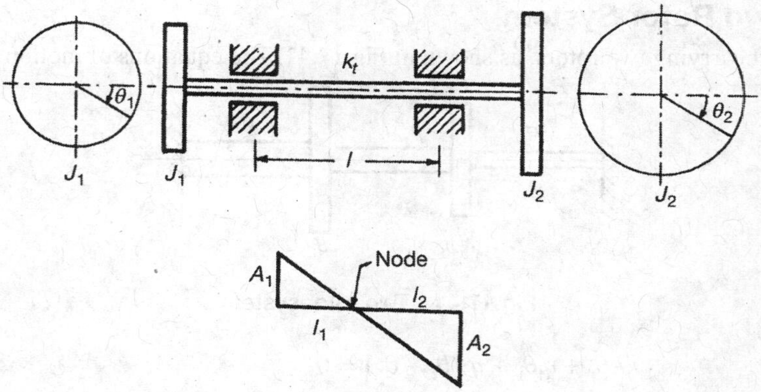

Fig.17.40 Rotational system of two rotor system

Let $$\theta_1 = A_1 \sin \omega_n t$$
$$\theta_2 = A_2 \sin \omega_n t$$

Substituting in equations of motion, we get

$$\left(k_t - J_1\omega_n^2\right) A_1 - k_t A_2 = 0$$
$$-k_t A_1 + \left(k_t - J_2\omega_n^2\right) A_2 = 0$$

For non-trivial solution,

$$\begin{vmatrix} k_t - J_1\omega_n^2 & -kt \\ -k_t & k_t - J_2\omega_n^2 \end{vmatrix} = 0$$

The characteristic equation becomes,

$$\omega_n^2 - \left[(J_1 + J_2)/(J_1 J_2)\right] k_t = 0$$

The natural frequencies are,

$$\omega_{n1} = 0$$
$$\omega_{n2} = [k_t\,(J_1 + J_2)/(J_1\,J_2)]^{0.5} \tag{17.73}$$

The mode shapes are given by,

$$A_1/A_2 = k_t/\left(k_t - J_1\,\omega_n^2\right)$$
$$= +1 \quad \text{for} \quad \omega_{nl} = 0$$
$$= -J_2/J_1 \quad \text{for} \quad \omega n_2 \tag{17.74}$$

To locate the node point, we have

$$l_1/l_2 = A_1/A_2 = J_2/J_1$$

or $$l_2 = J_1\,l_1/J_2$$

and $$l_1 = J_2\,l/(J_1 + J_2) \tag{17.75}$$

17.10.7 Two Rotor System

Consider a shaft carrying two rotors, as shown in Fig.17.41. The equations of motion are,

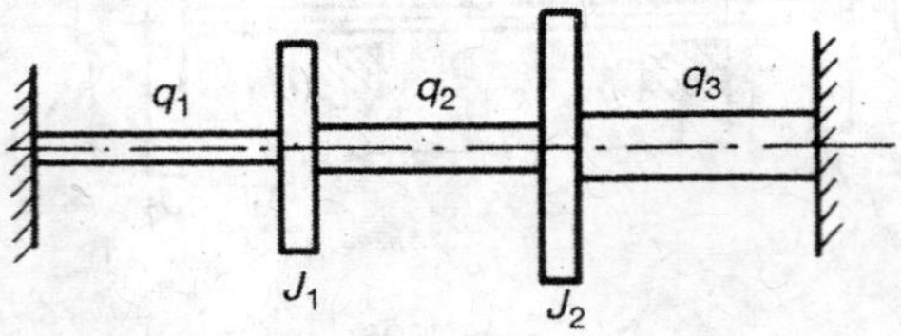

Fig.17.41 Two rotor system

$$J_1\ddot{\theta}_1 + q_1\theta_1 + q_2(\theta_1 - \theta_2) = 0$$

$$J_2\ddot{\theta}_2 - q_2(\theta_1 - \theta_2) + q_3\theta_2 = 0$$

or

$$J_1\ddot{\theta}_1 + (q_1 + q_2)\theta_1 - q_2\theta_2 = 0$$

$$-q_2\theta_1 + J_2\ddot{\theta}_2 + (q_2 + q_3)\theta_2 = 0$$

Let

$$\theta_1 = A_1 \sin \omega_n t$$

$$\theta_2 = A_2 \sin \omega_n t$$

Substituting in equations of motion, for non-trivial solution, we get

$$\begin{vmatrix} q_1 + q_2 - J_1\omega_n^2 & -q_2 \\ -q_2 & q_2 + q_3 - J_2\omega_n^2 \end{vmatrix} = 0$$

The characteristic equation becomes,

$$\omega_n^4 - \left[\frac{q_1 + q_2}{J_1} + \frac{q_2 + q_3}{J_2}\right]\omega_n^2 + \frac{q_1q_2 + q_2q_3 + q_3q_1}{J_1J_2} = 0$$

$$\omega_n^2 = \frac{1}{2}\left[\frac{q_1 + q_2}{J_1} + \frac{q_2 + q_3}{J_2}\right]$$

$$\mp \frac{1}{2}\sqrt{\left[\frac{q_1 + q_2}{J_1} + \frac{q_2 + q_3}{J_2}\right]^2 - 4\frac{(q_1q_2 + q_2q_3 + q_3q_1)}{J_1J_2}} \tag{17.76}$$

$$\frac{A_1}{A_2} = \frac{q_2}{q_1 + q_2 - J_1\omega_n^2} = \frac{q_1 + q_2 - J_2\,\omega_n^2}{q_2} \tag{17.77}$$

17.10.8 Three Rotor System

Consider the three rotor system shown in Fig.17.42. The equations of motion are,

$$J_1\,\ddot{\theta}_1 + q_1(\theta_1 - \theta_2) = 0$$

$$J_2\ddot{\theta}_2 + q_2(\theta_2 - \theta_3) - q_1(\theta_1 - \theta_2) = 0$$

$$J_3\ddot{\theta}_3 - q_2(\theta_2 - \theta_3) = 0$$

Let

$$\theta_1 = A_1 \sin \omega_n t, i = 1 \text{ to } 3$$

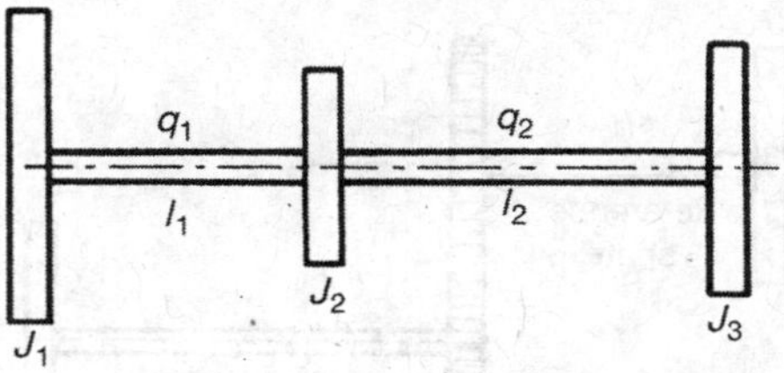

Fig.17.42 Three rotor system

$$\text{Then}\begin{bmatrix} q_1 - J_1\omega_n^2 & -q_1 & 0 \\ -q_1 & q_1 + q_2 - J_2\omega_n^2 & -q_2 \\ 0 & -q_2 & q_2 - J_3\omega_n^2 \end{bmatrix}\begin{Bmatrix} A_1 \\ A_2 \\ A_3 \end{Bmatrix} = \begin{Bmatrix} 0 \\ 0 \\ 0 \end{Bmatrix}$$

The characteristic equation becomes,

$$\omega_n^2\ [J_1J_2J_3\ \omega_n^4 - \{(J_1J_2 + J_1J_3)\, q_2 + (J_2J_3 + J_2J_3)q_1\}\, \omega_n^2 + q_1q_2(J_1 + J_2 + J_3) = 0$$

The roots are $\omega_n^2 = 0$ and

$$\omega_{n1}^2, \omega_{n2}^2 = \frac{1}{2}\left[\frac{q_1}{J_1} + \frac{q_1 + q_2}{J_2} + \frac{q_2}{J_3}\right] \mp \frac{1}{2}\sqrt{\left[\frac{q_1}{J_1} + \frac{q_1 + q_2}{J_2} + \frac{q_2}{J_3}\right]^2 - 4q_1q_2\left(\frac{J_1 + J_2 + J_3}{J_1J_2J_3}\right)} \tag{17.78}$$

$$\left.\begin{aligned} \frac{A_1}{A_2} &= \frac{q_1}{q_1 - J_1\omega_n^2} \\ \frac{A_3}{A_2} &= \frac{q_2}{q_2 - J_3\omega_n^2} \end{aligned}\right\} \tag{17.79}$$

17.11 GEARED SYSTEM

Consider the geared system shown in Fig.17.43. Neglecting the inertia of the gears, let

Gear reduction ratio, $i = n_2/n_1$

Kinetic energy, $T = \frac{1}{2}J_1\ \dot{\theta}_1^2 + \frac{1}{2}J_2\dot{\theta}_2^2$

Potential energy, $U = \frac{1}{2}q_1\ \theta_1^2 + \frac{1}{2}q_2\theta_2^2$

Now $\theta_2 = \theta_1$

Therefore,

$$\begin{aligned} T &= \frac{1}{2}J_1\ \dot{\theta}_1^2 + \frac{1}{2}J_2(i\dot{\theta}_1)^2 \\ &= \frac{1}{2}J_1\ \dot{\theta}_1^2 + \frac{1}{2}(i^2J_2)\dot{\theta}_1^2 \\ &= \frac{1}{2}J_1\ \dot{\theta}_1^2 + \frac{1}{2}J_2'\dot{\theta}_1^2 \end{aligned}$$

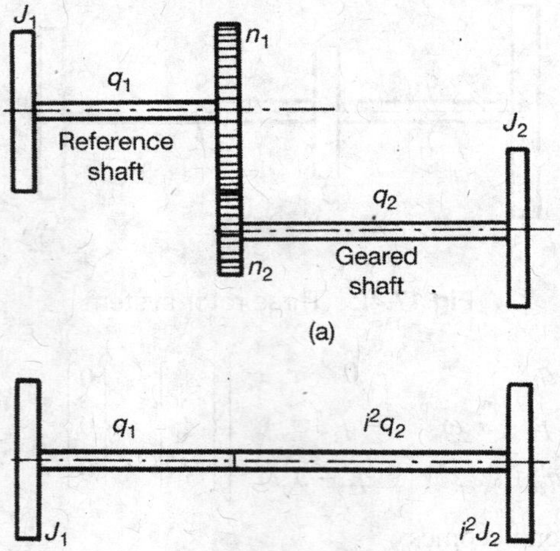

(b) Equivalent two-rotor system

Fig.17.43 Geared system

where $J_2' = i^2 J_2$

$$U = \frac{1}{2} q_1 \theta_1^2 + \frac{1}{2}(i^2 q_2)\theta_1^2$$

$$= \frac{1}{2} q_1 \theta_1^2 + \frac{1}{2} q_2\ \theta_1^2$$

where $q_2' = i^2 q_2$

Equivalent stiffness, $$q_e = \frac{i^2 q_1 q_2}{q_1 + i^2 q_2} \quad (17.80)$$

Natural frequency, $$\omega_n = \sqrt{\frac{q_e(J_1 + J_2')}{J_1 J_2'}} \quad (17.81)$$

The equivalent length of the geared system shown in Fig.17.44 is,

$$l_e = l_1 + l_2(d_1/d_2)^4\ (a_2/a_1)^2 \quad (17.82)$$

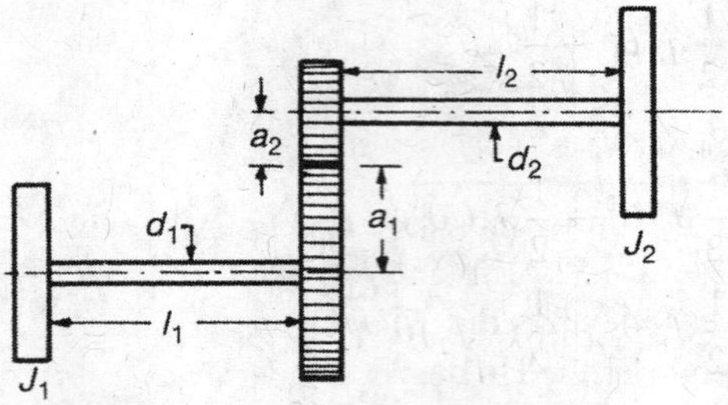

Fig.17.44 Equivalent geared system

The equivalent system for gear system, considering the inertia of the gears, is shown in Fig.17.45. The natural frequency can be determined as explained in Section 17.10.8.

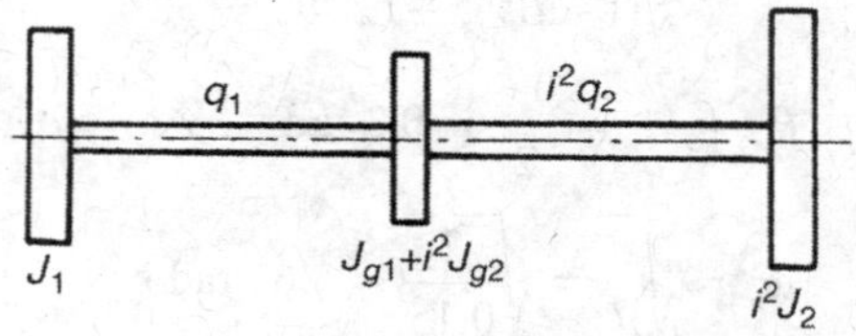

Fig.17.45 Equivalent geared system system considering inertia of gears

Example 17.18

A disc of a torsional pendulum has a moment of inertia of 75×10^{-3} kg·m², and is immersed in a viscous fluid. The brass shaft ($G = 42$ GPa) attached to it is of 10 mm diameter and 0.4 m long. When the pendulum is vibrating, the observed amplitudes on the same side of the rest position for successive cycles are 5°, 3°, and 1.8°. Determine (a) logarithmic decrement, (b) the damping torque at unit speed, and (c) the periodic time of the vibration.

Solution

$J = \pi d^4/32 = \pi \times 10^4/32 = 981.75 \text{ mm}^4$

$q = GJ/l = 42 \times 10^9 \times 981.75 \times 10^{-12}/0.4 = 103.1$ N·m·s/rad

(a) Logarithmic decrement, $\delta = ln\ (5/3) = 0.51$

$$\approx 2\pi\zeta$$

$$\text{or} \quad \zeta = \frac{0.51}{2\pi} = 0.081$$

(b) $\omega_n = \sqrt{q/I} = \sqrt{103.1/75 \times 10^{-3}} = 37.08$ rad/s

$c_{tc} = 2I\omega_n = 2 \times 75 \times 10^{-3} \times 37.08 = 5.56$ N·m·s/rad

$c_t = \zeta c_{tc} = 0.081 \times 5.56 = 0.45$ N·m·s/rad

(c) $\omega_d = \omega_n(1 - \zeta^2)^{0.5}$

$= 37.08\sqrt{1 - (0.081)^2} = 36.96$ rad/s

Periodic time $= 2\pi/\omega_d = 2\pi/36.96 = 0.17$ s

Example 17.19

A periodic torque $T = 0.6 \sin 5t$ N·m is impressed upon a flywheel suspended from a wire. The flywheel has a moment of inertia of 0.15 kg·m² and the wire has a stiffness of 2 N·m/rad. The damping coefficient of viscous damper is 0.4 N·m·s/rad. Find (a) the maximum angular displacement from rest position, (b) the maximum couple applied to dashpot, and (c) the angle by which the angular displacement lags the torque.

■ **Solution**

(a)
$$\theta = \frac{\theta_{st}}{\sqrt{(1-\beta^2)^2 + (2\zeta\beta)^2}}$$

$$\theta_{st} = \frac{T_o}{q} = \frac{0.6}{2} = 0.3 \text{ rad}$$

$$\omega_n = \sqrt{\frac{q}{J}} = \sqrt{\frac{2}{0.15}} = 3.65 \text{ rad/s}$$

$$\beta = \frac{\omega}{\omega_n} = \frac{5}{3.65} = 1.3693$$

$$\zeta = \frac{c_t}{c_{tc}} = \frac{c_t}{2J\omega_n} = \frac{0.4}{2\times 0.15\times 3.65} = 0.3563$$

$$\theta = \frac{0.3}{\sqrt{[1-(1.3693)^2]^2 + (2\times 0.3653\times 1.3693)^2}}$$

$$= \frac{0.3}{1.329} = 0.2257 \text{ rad}$$

(b) Maximum damping couple $= c\omega\theta$

$= 0.4 \times 5 \times 0.2257$

$= 0.4514$ N.m

(c)
$$\phi = \tan^{-1}\left[\frac{2\zeta\beta}{1-\beta^2}\right]$$

$$= \tan^{-1}\left[\frac{2\times 0.3653\times 1.3693}{1-(1.3693)^2}\right] = -48.82° \text{ or } 131.18°$$

Example 17.20

A torsional pendulum has a natural frequency of 200 cycles/min, when vibrating in vacuum. The mass moment of inertia of the disc is 0.025 kg · m². It is then immersed in oil and it is observed that its natural frequency is 180 cycles/min. Determine the damping torque per radian. If the disc is displaced 3° when in oil, find its displacement at the end of the first complete cycle.

■ **Solution**

$$f_d = f_n\sqrt{1-\zeta^2}$$

$$180 = 200\sqrt{1-\zeta^2}$$

$$\zeta = 0.436$$

$$c_{tc} = 2I\omega_n = 4\pi\, If_n$$

$$= 4\pi \times 0.025 \times \frac{200}{60} = 1.048 \text{ N·m·s/rad}$$

$$c_t = c_{tc}\zeta = 1.048\times 0.436 = 0.457 \text{ N·m·s/rad}$$

$$\theta(t) = \exp(-\zeta\omega_n t)(A \sin \omega_d t + B \cos \omega_d t)$$

$$\theta(0) = 3^\circ = \frac{3\pi}{180} = \frac{\pi}{60} \text{ rad, gives}$$

$$B = \frac{\pi}{60}$$

$$\dot{\theta}(t) = \omega_d \exp(-\zeta\omega_n t)[A \cos \omega_d t - B \sin \omega_d t]$$
$$-\zeta\omega_n \exp(-\zeta\ \omega_n\ t)[A \sin \omega_d t + B \cos \omega_d t]$$

$$\dot{\theta}(0) = 0, \text{ gives}$$

$$A = \zeta \frac{\omega_n}{\omega_d} B$$

$$= \frac{0.436 \times 200 \times \pi}{180 \times 60} = 0.02536$$

Time taken to complete one cycle,

$$t = T_d = \frac{2\pi}{\omega_d} = \frac{2\pi \times 60}{2\pi \times 180} = \frac{1}{3} \text{ s}$$

$$\omega_d t = 2\pi$$

$$\theta = \exp\left(\frac{-0.436 \times 2\pi \times 200}{60 \times 3}\right)\left[0.02536 \sin 2\pi + \frac{\pi}{60} \cos 2\pi\right]$$

$$= 0.04765 \times \frac{\pi}{60} = 0.002495 \text{ rad} \quad \text{or} \quad 0.143 \text{ deg}$$

Example 17.21

Calculate the natural frequency of the torsional vibrations of the system shown in Fig.17.46. G = 105 GPa.

Solution

Torsional stiffness, $q = GJ/l$

$$q_1 = 105 \times 10^9 \times \pi\,(25 \times 10^{-3})^4/(32 \times 0.5) = 8053.4 \text{ N·m/rad}$$

$$q_2 = 105 \times 10^9 \times \pi\,(50 \times 10^{-3})^4/(32 \times 1.5) = 92951.5 \text{ N·m/rad}$$

$$q_3 = 105 \times 10^9 \times \pi\,(20 \times 10^{-3})^4/(32 \times 0.3) = 5497.8 \text{ N·m/rad}$$

The stepped shafts are in series. Therefore, the equivalent stiffness q_e is,

$$1/q_e = 1/q_1 + 1/q_2 + 1/q_3$$

$$= 1/8053.4 + 1/92951.5 + 1/5497.8$$

$$q_e = 3036.4 \text{ N·m/rad.}$$

Natural frequency, $\omega_n = \sqrt{q_e/J}$

$$\sqrt{3036.4/20} = 12.32 \text{ rad/s}$$

Example 17.22

The flywheel of an engine dynamo weighs 150 N and has a radius of gyration of 0.25 m. The shaft at the flywheel end has an effective length of 0.20 m and is 50 mm in diameter. The armature weighs 80 N and has a radius of gyration of 0.20 m. The dynamo shaft has a diameter of 40 mm and an effective length of 0.15 m. Neglecting the inertia of the shaft and the coupling, calculate the frequency of torsional vibrations and the position of node.

■ Solution

$$I_1 = \frac{1}{2}\frac{W_1}{g}K_1^2 = \frac{1}{2}\times\frac{150}{9.81}\times(0.25)^2 = 0.4778 \text{ kg·m}^2$$

$$I_2 = \frac{1}{2}\frac{W_2}{g}K_2^2 = \frac{1}{2}\times\frac{80}{9.81}\times(0.20)^2 = 0.163 \text{ kg·m}^2$$

$$q_1 = GJ_1/l_1 = \frac{84\times10^9\times\pi\times(50)^4\times10^{-12}}{0.20\times32} = 257708.8 \text{ N·m/rad}$$

$$q_2 = GJ_2/l_2 = \frac{84\times10\times\pi\times(40)^4\times10^{-12}}{0.15\times32} = 140743.4 \text{ N·m/rad}$$

$$\omega_n = \sqrt{\frac{q_1q_2(I_1+I_2)}{(q_1+q_2)I_1I_2}}$$

$$= \sqrt{\frac{257708.8\times140743.4\ (0.4778+0.163)}{(257708.8+140743.4)\times0.4778\times0.163}}$$

$$= 865.43 \text{ rad/s}$$

$$\text{Position of node from dynamo} = \frac{I_1I_2(q_1+q_2)}{q_1(I_1+I_2)}$$

$$= \frac{0.4778\times0.163\ (257708.8+140743.4)}{257708.8\ (0.4778+0.163)}$$

$$= 0.1579 \text{ m}$$

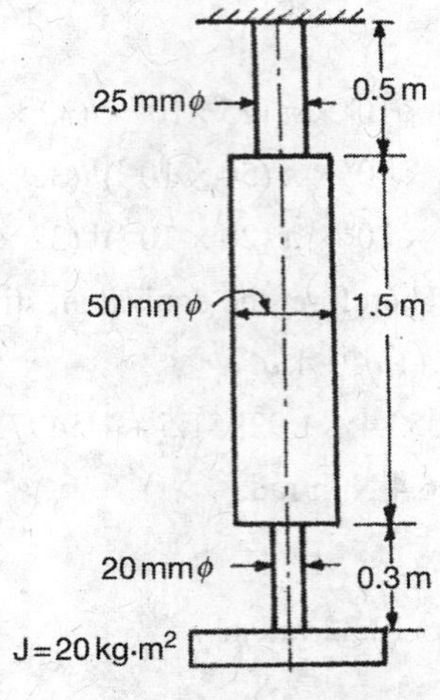

Fig.17.46 Diagram for Example 17.22

Example 17.23

An engine shaft of diameter 50 mm drives a pump shaft of diameter 40 mm through a pair of spur gears, as shown in Fig.17.47, $G = 84$ GPa. Calculate the natural frequencies of the geared system.

Solution

$$q_1 = 84 \times 10^9 \times \pi\,(50 \times 10^{-3})^4/(32 \times 1) = 51542\ \text{N·m/rad}$$

$$q_2 = 84 \times 10^9 \times \pi\,(40 \times 10^{-3})^4/(32 \times 0.3) = 70372\ \text{N·m/rad}$$

$$i = 4$$

$$q'_2 = i^2 q_2 = 16 \times 70372 = 1125946$$

$$q_e = (51542 \times 1125946)/(51542 + 1125946) = 49286\ \text{N·m/rad}$$

$$J_{g1} + i^2 J_{g2} = 40 + 16 \times 5 = 120\ \text{kg}\cdot\text{m}^2$$

$$i^2 J_2 = 16 \times 20 = 320\ \text{kg}\cdot\text{m}^2$$

$$l_e = 1 + 0.3\,(50/40)^4\,(4)^2 = 12.72\ \text{m}$$

$$\omega_n^2 = 0.5\,[51542/800 + 1177488/120 + 1125946/320]$$

$$\pm 0.5\,[(13395.4)^2 - (4 \times 51542 \times 1125946 \times 1240)/(800 \times 120 \times 320)]^{0.5}$$

$$= 177.2,\ 13217.7$$

or $\omega_{n1} = 13.3$ rad/s or 2.12 Hz and $\omega_{n2} = 114.96$ rad/s or 18.3 Hz

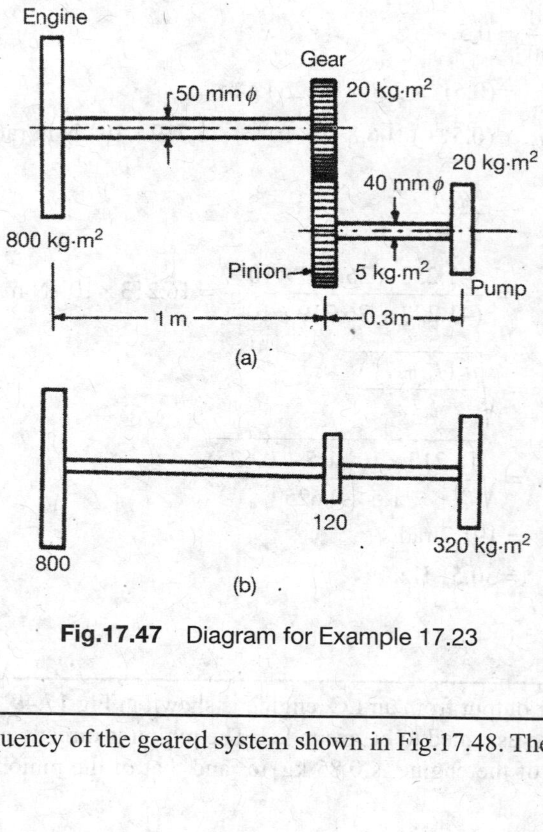

Fig.17.47 Diagram for Example 17.23

Example 17.24

Calculate the natural frequency of the geared system shown in Fig.17.48. The shafts are made of steel for which $G = 84$ GPa.

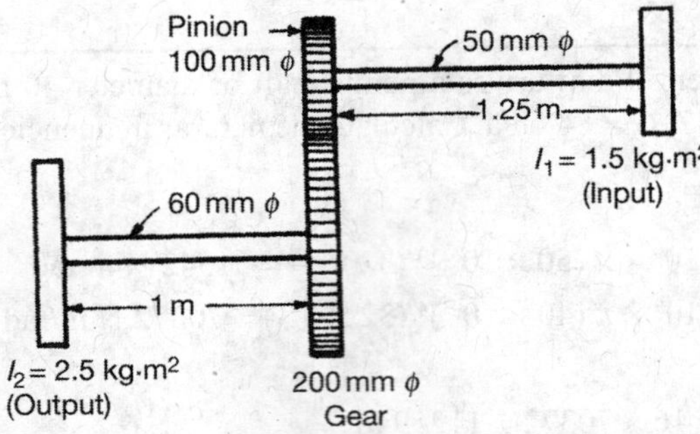

Fig.17.48 Diagram for Example 17.24

Solution

Stiffness of shafts:

$$q_1 = \frac{GJ_1}{l_1} = \frac{84\times10^9\times\pi(50\times10^{-3})^4}{32\times1.25}$$

$$= 41.233\times10^3 \text{ N·m/rad}$$

$$q_2 = \frac{GJ_2}{l_2} = \frac{84\times10^9\times\pi(60\times10^{-3})^4}{32\times1.0} = 106.877\times10^3 \text{ N·m/rad}$$

Speed ratio, $i = \dfrac{100}{200} = 0.5$

$$I_{2'} = i^2 I_2 = (0.5)^2\times2.5 = 0.625 \text{ kg·m}^2$$

$$q_{2'} = i^2 q_2 = (0.5)^2\times106.877\times10^3 = 26.719\times10^3 \text{ N·m/rad}$$

Equivalent stiffness, $q_e = \dfrac{q_1 q_{2'}}{q_1 + q_{2'}}$

$$= \frac{41.233\times26.719\times10^6}{(41.233+26.719\times10^3)} = 16.213\times10^3 \text{ N·m/rad}$$

Natural frequency, $\omega_n = \sqrt{\dfrac{q_e(I_1+I_{2'})}{I_1 I_{2'}}}$

$$= \sqrt{\frac{16.213\times10^3(1.5+0.625)}{1.5\times0.625}}$$

$$= 191.7 \text{ rad/s}$$

$$= 30.51 \text{ Hz}$$

Example 17.25

The mechanism of power output from an I.C. engine is shown in Fig.17.49. The inertia of flywheel is relatively large and can be assumed to be grounded. The pinion is directly coupled to the engine. The mass moment of inertia of the engine is 0.85 kg·m² and that of the pinion and gear, 0.0015 kg·m²

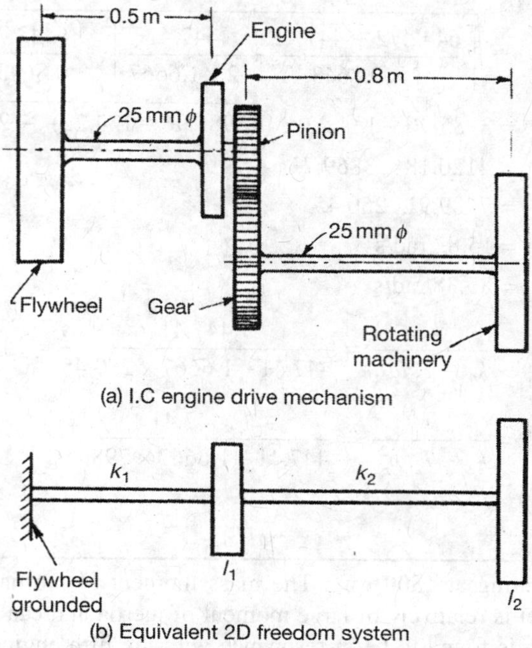

Fig.17.49 Diagram for Example 17.25

and 0.115 kg·m² respectively. The inertia of the rotating machinery is 15 kg·m². The speed reduction from engine to the rotating machinery is 3:1. Determine the natural frequencies and amplitude ratio by reducing the system to a two degree of freedom system. Modulus of rigidity for steel shaft is 84 GPa.

■ **Solution**

The shaft joining the pinion to engine is negligibly small, therefore pinion inertia can be directly added to the engine inertia. As referred to engine speed,

$$I_1 = 0.85 + 0.0015 + \frac{0.115}{3^2} = 0.8643 \text{ kg}\cdot\text{m}^2$$

$$I_2 = \frac{15}{3^2} = 1.6667 \text{ kg}\cdot\text{m}^2$$

$$k_1 = \frac{84\times10^9\times\pi\times(25\times10^{-3})^4}{0.5\times32} = 6442.72 \text{ N·m/rad}$$

$$k_2 = \frac{84\times10^9\times\pi\times(25\times10^{-3})^4}{32\times0.8\times3^2} = 447.41 \text{ N·m/rad}$$

$$\omega_n^2 = \frac{k_1+k_2}{2I_1} + \frac{k_2}{2I_2} \pm \sqrt{\left[\frac{k_1+k_2}{2I_1} + \frac{k_2}{2I_2}\right]^2 - \frac{k_1k_2}{I_1I_2}}$$

$$= \frac{6442.72+447.41}{2\times0.8643} + \frac{447.41}{2\times1.6667}$$

$$\pm\sqrt{\left[\frac{6442.72+447.41}{2\times0.8643}+\frac{447.41}{2\times1.6667}\right]^2-\frac{6442.7\times447.41}{0.8643\times1.6667}}$$

$$= 3985.96 + 134.22 \pm \sqrt{(3985.96+134.22)^2 - 2001027}$$

$$= 4120.18 \pm 3869.73$$

$$= 7989.91,\ 250.45$$

$$\omega_{n1} = 15.82 \text{ rad/s}$$

$$\omega_{n2} = 89.38 \text{ rad/s}$$

$$\left(\frac{\theta_2}{\theta_1}\right)_1 = \frac{k_2}{k_2 - I_2\,\omega_{n1}^2} = \frac{447.41}{447.41 - 1.6667\times250.45} = 14.92$$

$$\left(\frac{\theta_2}{\theta_1}\right)_2 = \frac{k_2}{k_2 - I_2\,\omega_{n2}^2} = \frac{447.41}{447.41 - 1.6667\times7989.91} = -0.03476$$

Example 17.26

An I.C. engine is operating at 1800 rpm. The mass moment of inertia of the engine cylinders is 0.5 kg·m². The flywheel is relatively of large moment of inertia and can be assumed to be grounded (Fig.17.50). The system is found to be in resonance with the fifth engine order excitation torque of 1.25 kN·m amplitude. Design a dynamic torsional vibration absorber so that the resulting two natural frequencies of the system are at least 25% away from the excitation frequency. Also determine the amplitude of the absorber mass. Take $d_2 = 50$ mm, and $G = 84$ GPa.

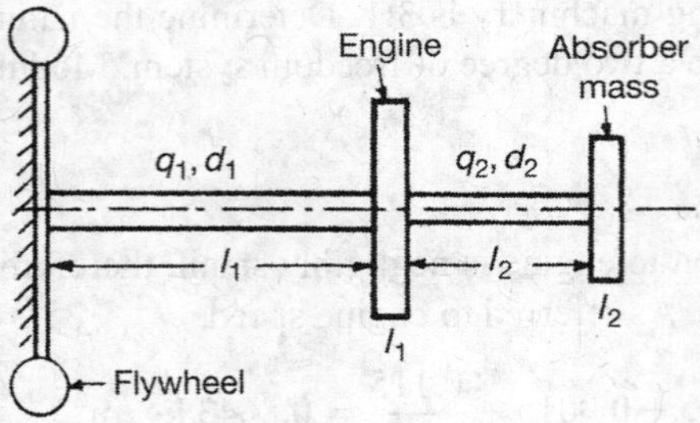

Fig.17.50 Diagram for Example 17.26

■ Solution

For a tuned absorber,

$$\frac{q_2}{q_1} = \frac{I_2}{I_1} = \mu$$

$$\omega = \frac{5\times2\pi\times1800}{60} = 942.5 \text{ rad/s}$$

$$q_1 = \omega^2 I_1 = (942.5)^2\times0.5 = 444132 \text{ N·m/rad}$$

To find the two natural frequencies, we have

$$\left(\frac{\omega}{\omega_{22}}\right)^4 - (2+\mu)\left(\frac{\omega}{\omega_{22}}\right)^2 + 1 = 0$$

$$1.25^4 - (2 + \mu)(1.25)^2 + 1 = 0$$
$$2.441 - 1.5625(2 + \mu) + 1 = 0$$
or
$$2.441 - 3.125 - 1.5625\mu + 1 = 0$$
or
$$\mu = 0.20224$$
and
$$(0.75)^4 - (2 + \mu)(0.75)^2 + 1 = 0$$
$$0.3164 - (2 + \mu) \times 0.5625 + 1 = 0$$
or
$$0.3164 - 1.125 - 0.5625\mu + 1 = 0$$
$$\mu = 0.3403$$

We choose $\mu = 0.3403$ to ensure that both the natural frequencies are at least 25% away from excitation frequency.

$$I_2 = \mu I_1 = 0.3403 \times 0.5 = 0.1701 \text{ kg}\cdot\text{m}^2$$
$$q_2 = \mu q_1 = 0.3403 \times 444132 = 151138 \text{ N·m/rad}$$

Excitation torque, $T_o = 1.25$ kN·m

Amplitude to absorber mass,

$$\theta_2 = \frac{-T_o}{q_2} = -\frac{1.25 \times 10^3}{151138} \times \frac{180}{\pi} = 0.47387$$

Now
$$q_2 = \frac{GJ_2}{l_2}$$
$$l_2 = \frac{84 \times 10^9 \times \pi \times (50 \times 10^{-3})^4}{32 \times 151138} = 0.341 \text{ m}$$

Example 17.27

A stiff rod of mass m with linear and torsional springs is shown in Fig.17.51. Find its natural frequency.

■ Solution

The equation of motion for the system is,

$$\left(\frac{ml^2}{3}\right)\ddot{\theta} + k_t\theta + 2kl^2\theta = 0$$

or
$$\ddot{\theta} + \left[\frac{3(k_t + 2kl^2)}{ml^2}\right]\theta = 0$$

$$\omega_n = \left[\frac{3(k_t + 2kl^2)}{ml^2}\right]^{1/2} \text{ rad/s}$$

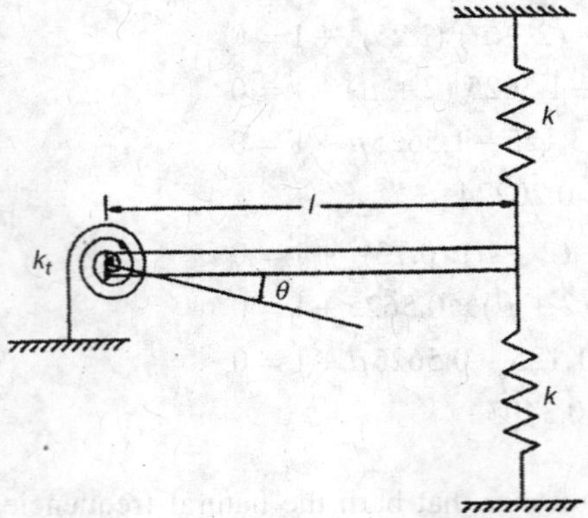

Fig.17.51 Diagram for Example 17.27

Example 17.28

Find the natural frequency of the system shown in Fig.17.52.

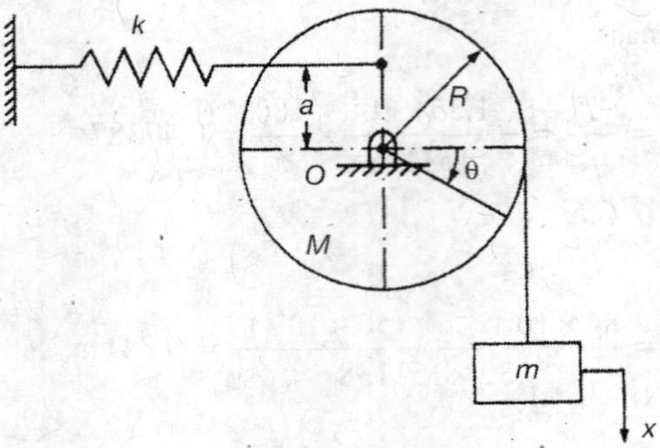

Fig.17.52 Diagram for Example 17.28

■ Solution

The equation of motion is,

$$I_o\ \ddot{\theta} + ka^2\theta + m\ddot{x}R = 0$$

Now $$x = R\theta,\ \ddot{x} = R\ddot{\theta},\ I_o\ \frac{MR^2}{2}$$

$$\therefore \quad \left(\frac{MR^2}{2}\right)\ddot{\theta} + ka^2\theta + mR^2\ddot{\theta} = 0$$

$$\left(m + \frac{M}{2}\right)R^2\ddot{\theta} + ka^2\theta = 0$$

$$\omega_n = \left(\frac{a}{R}\right)\left[\frac{k}{\left(m + \frac{M}{2}\right)}\right]^{1/2} \text{ rad/s}$$

Example 17.29

Determine the damped natural frequency of the system shown in Fig.17.53.

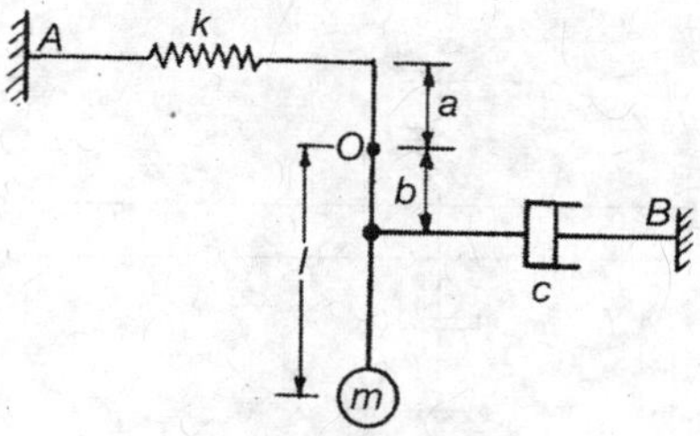

Fig.17.53 Diagram for Example 17.29

Solution

The equation of motion is:

$$ml^2\,\ddot{\theta} + cb^2\,\dot{\theta} + mgl\,\theta + k\,a^2\,\theta = 0$$

$$\ddot{\theta} + \left(\frac{cb^2}{ml^2}\right)\dot{\theta} + \left(\frac{mgl + ka^2}{ml^2}\right)\theta = 0$$

Comparing with the standard equation of motion for the spring dashpot system, we have

$$\ddot{\theta} + 2\zeta\,\omega_n\dot{\theta} + \omega_n^2\,\ddot{\theta} = 0$$

Natural frequency, $$\omega_n = \sqrt{\frac{mgl + ka^2}{ml^2}} \text{ rad/s}$$

$$2\zeta\,\omega_n = \frac{cb^2}{ml^2}$$

Damping factor, $$\zeta = \frac{cb^2}{2}\cdot\sqrt{\frac{1}{ml^2\left(mgl + ka^2\right)}}$$

Damped natural frequency, $$\omega_d = \omega_n\sqrt{1-\zeta^2}$$

$$\omega_d = \sqrt{\frac{mgl + ka^2}{ml^2}} \times \sqrt{1 - \frac{c^2b^4}{4} \times \frac{1}{ml^2\left(mgl + ka^2\right)}}$$

$$= \sqrt{\frac{mgl + ka^2}{ml^2}} \times \sqrt{\frac{4ml^2\left(mgl + ka^2\right) - c^2b^4}{4ml^2\left(mgl + ka^2\right)}}$$

$$= \frac{1}{2ml^2}\sqrt{4ml^2\left(mgl + ka^2\right) - c^2b^4}$$

$$= \sqrt{\frac{mgl + ka^2}{ml^2} - \frac{c^2b^4}{4m^2l^4}} \text{ rad/s}$$

Example 17.30

For the system shown in Fig.17.54, determine the natural frequency of damped vibrations and critical damping coefficient.

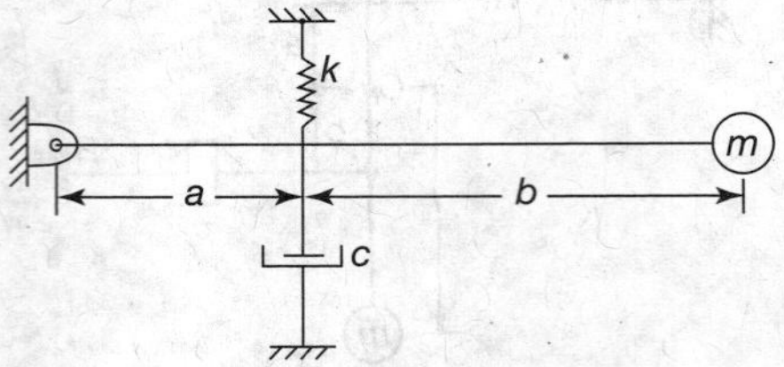

Fig.17.54 Diagram for Example 17.30

Solution

The equation of motion is:

$$m\,(a+b)^2\ddot{\theta} + ca^2\dot{\theta} + ka^2\theta = 0$$

Comparing with the standard equation of motion, we have

$$I = m\,(a+b)^2$$

$$c_{te} = c\,a^2$$

$$q_e = ka^2$$

Undamped natural frequency, $\omega_n = \sqrt{\dfrac{q_e}{I}}$

$$= \sqrt{\frac{ka^2}{m(a+b)^2}}$$

$$= \frac{a}{(a+b)}\sqrt{\frac{k}{m}} \text{ rad/s}$$

Critical damping coefficient, $c_{tc} = 2\,I\,\omega_n$

$$= 2m\,(a+b)^2\left[\left(\frac{a}{a+b}\right)\sqrt{\frac{k}{m}}\right]$$

$$= 2a\,(a+b)\sqrt{km}$$

Damping factor, $\zeta = \dfrac{c_{te}}{c_{tc}}$

$$= \frac{ca^2}{2a\,(a+b)\,\sqrt{km}}$$

$$= \frac{ca}{2\,(a+b)\,\sqrt{km}}$$

$$\text{Damped natural frequency, } \omega_d = \omega_n\sqrt{1-\zeta^2}$$

$$= \left(\frac{a}{a+b}\right)\sqrt{\frac{k}{m}}\sqrt{1-\frac{c^2a^2}{4\,(a+b)^2\,km}}$$

$$= \frac{a}{2\,(a+b)^2\,m}\sqrt{4\,(a+b)^2\,km - c^2a^2}$$

Example 17.31

An integral pulley of mass moment of inertia J about its axis, shown in Fig.17.55, is restrained in its movement about its own axis by a torsional spring of stiffness k_t and a smaller pulley by means of an inextensible string. Determine the natural frequency.

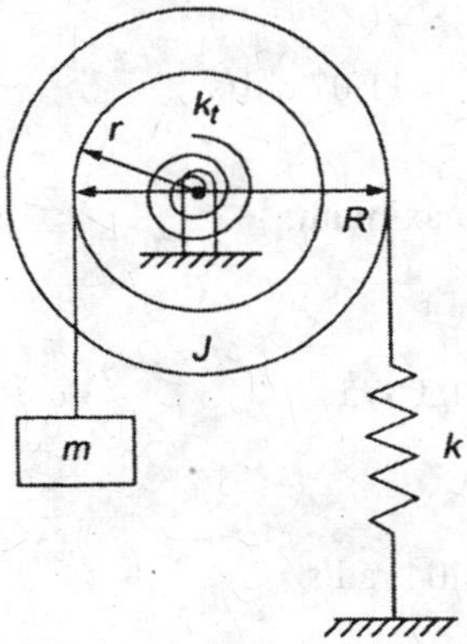

Fig.17.55 Diagram for Example 17.31

■ Solution

The equation of motion is,

$$J\ddot{\theta} + m\ddot{x}r + kyR + k_t\theta = 0$$

Where $x = r\theta$ and $y = R\theta$

$$J\ddot{\theta} + mr^2\ddot{\theta} + (kR^2 + k_t)\theta = 0$$

$$(J + mr^2)\ddot{\theta} + (kR^2 + k_t)\theta = 0$$

or

$$\omega_n = \left(\frac{kR^2 + k_t}{J + mr^2}\right)^{1/2} \text{ rad/s}$$

Example 17.32

The static deflection of an automobile on its springs is 80 mm. Find the critical speed when the automobile is travelling on a road which can be approximated by a sine wave of amplitude 60 mm and a wave length 12 m (Fig.17.56). Assume the damping to be 0.05. Also calculate the amplitude of vibration at 60 km/h.

■ **Solution**

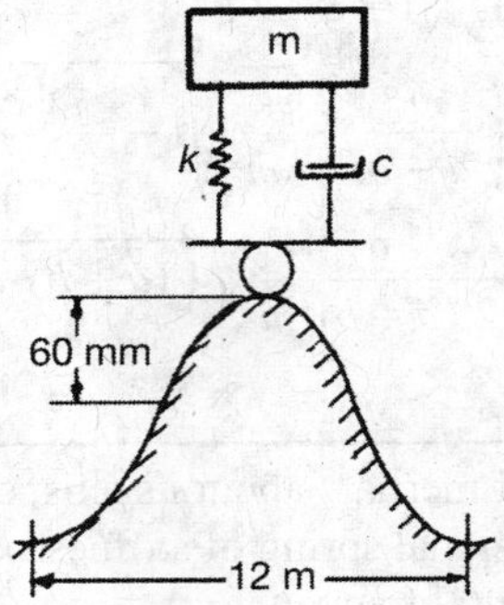

Fig.17.56 Diagram for Example 17.32

$$\omega_n = \sqrt{\frac{g}{\delta_{st}}} = \sqrt{\frac{9.81}{80 \times 10^{-3}}} = 11.07 \text{ rad/s}$$

For dynamic magnification to be maximum,

$$\beta = \sqrt{1 - 2\zeta^2}$$
$$= \sqrt{1 - 2 \times (0.05)^2} = 0.9975$$

Excitation frequency, $\omega = \beta\omega_n$

$$= 0.9975 \times 11.07 = 11.04 \text{ rad/s}$$

Critical velocity of vehicle, $v = f\lambda = \dfrac{\omega\lambda}{2\pi}$

$$\frac{11.04 \times 12}{2\lambda} = 21.09 \text{ m/s}$$
$$= 75.92 \text{ km/h}$$

Excitation frequency, at 60 km/h,

$$\omega = \frac{2\pi \times 60 \times 10^3}{3600 \times 12} = 8.827 \text{ rad/s}$$
$$\beta = \frac{\omega}{\omega_n} = \frac{8.827}{11.07} = 0.7883$$

Amplitude of vehicle, $X = \dfrac{y\sqrt{1 + (2\zeta\beta)^2}}{\sqrt{(1-\beta^2)^2 + (2\zeta\beta)^2}}$

$$= \frac{60\sqrt{1 + (2 \times 0.05 \times 0.7883)^2}}{\sqrt{\{1 - (0.07883)^2\}^2 + (2 \times 0.05 \times 0.7883)^2}} = 0.7883$$
$$= 155.65 \text{ mm}$$

Summary for Quick Revision

1. Vibration is an oscillation where in the quantity is a parameter that defines the motion of a mechanical system.
2. Time period is the time taken to complete one cycle. Frequency is the number of cycles per unit time. Amplitude is the maximum displacement of a vibrating body from the equilibrium position. Natural frequency is the frequency of free vibrations of a body vibrating of its own without the help of an external agency. Fundamental mode of vibration is the mode of vibration having the lowest natural frequency. Degrees of freedom is the minimum number of independent coordinates required to specify the motion of a system. Damping is the resistance to the motion of a vibrating body. Phase difference is the angle by which one vibrating system is ahead or behind the other vibrating system. Resonance is the phenomenon when the frequency of external excitation is equal to the natural frequency of a vibrating body.
3. There are three types of vibrations: longitudinal, transverse, and torsional.
4. There are three elements of a vibrating system: inertial, restoring, and damping.
5. Solution methods for solving vibration problems are: equilibrium method, D'Alembert's principle, energy method, and Rayleigh's method. Equilibrium method is based on the Newton's second law of motion; D'Alembert's method converts the dynamic problem into an equivalent static problem by taking the inertia force in the reverse direction; energy method uses the principle of conservation of energy, and Rayleigh's method equates the maximum kinetic energy to the maximum potential energy.
6. Simple harmonic motion may be defined as a motion in which the acceleration is proportional to the displacement from the mean position and is always directed towards the mean position.
7. Undamped free vibrations are those in which the system vibrates without the help of any external agency. For the spring-mass system,

   Natural frequency, $\omega_n = [\text{k/m}]^{1/2}$ rad/s

   Time period, $T = 2\pi/\omega_n = 2\pi(\text{k/m})^{1/2}$ s

   Natural frequency, $f_n = [1/(2\pi)]\ (\text{k/m})^{1/2}$ Hz
8. In damped free vibrations of single degree of freedom systems, a damper is placed in parallel with the spring to decrease the amplitude of vibrations. For a spring-mass-damper system,

   $c_c = 2m\omega_n = 2\ (\text{k/m})^{1/2}$, $\zeta = c/c_c$ = damping coefficient/critical damping coefficient

   (a) Underdamped system ($\zeta<1$) leads to oscillatory motion.

   $\omega_d = \omega_n(1-\zeta^2)^{1/2}$ = damped natural frequency

   The general solution can be written as,

   $x(t) = X\exp(-\zeta\omega_n t)\sin(\omega_d t + \phi)$

   (b) Overdamped system ($\zeta>1$) leads to exponentially decreasing function.

   (c) Critically damped system ($\zeta = 1$) leads to non-oscillatory motion.
9. Logarithmic decrement is defined as the natural logarithm of the ratio of any two successive amplitudes of vibration.

   Logarithmic decrement,

$$\delta = ln\,(x_n/x_{n+1})$$
$$= 2\pi\zeta/(1-\zeta^2)^{1/2}$$
$$\approx 2\pi\zeta \text{ for } \zeta << 1$$
$$= \left(\frac{1}{n}\right) ln\,(x_o/x_n)$$

10 When a system is subjected to forced harmonic excitation, its vibration response takes place at the same frequency as that of the excitation.

$x(t) = x_{st}\sin\omega t/(1-\beta^2)$

where $x_{st} = F_o/k$ = static deflection, $\beta = \omega/\omega_n$ = frequency ratio, $\omega_n = (k/m)^{1/2}$

Amplitude of oscillation, $X = x_{st}/(1-\beta^2)$

At resonance, $\beta = 1$ and the amplitude tends to infinity.

11 Damped forced vibrations.

The steady state solution is given by,

$$x_s(t) = x_{st}\sin(\omega t-\phi)/[(1-\beta^2)^2 + (2\zeta\beta)^2]^{1/2}$$

where $x_{st} = F_o/k$

Amplitude, $X = x_{st}/[1-\beta^2)^2 + (2\zeta\beta)^2]^{1/2}$

Phase angle, $\phi = \tan^{-1}[2\zeta\beta/(1-\beta^2)]$

Magnification factor, $M_f = X/x_{st} = 1/[(1-\beta^2)^2 + (2\zeta\beta)^2]^{1/2}$

For $\beta = 1,\ M_f = 1/(2\zeta)$

For M_f to be maximum, $\beta = (1-2\zeta^2)^{1/2}$

$$(M_f)_{max} = 1/[2\zeta(1-\zeta^2)^{1/2}]$$

12 Rotating Unbalance

The steady state solution is:

$$x = \mu_e\beta^2\sin(\omega t-\phi)/[1-\beta^2)^2 + (2\zeta\beta)^2]^{1/2}$$
$$\phi = \tan^{-1}[2\zeta\beta/(1-\beta^2)]$$
$$\mu = m/M$$

Magnification factor, $Mf = X/(\mu_e) = X/(me/M) = \beta^2/[1-\beta^2)^2+(2\zeta\beta)^2]^{1/2}$

For $\beta = 1,\ M_f = 1/(2\zeta)$

For $\beta\to\infty,\ M_f = 1$

For $\beta = 1,\ \phi = 90°$, i.e. m_o is out of phase with m.

When $\beta >> 1,\ \phi = 180°$.

13 Reciprocating unbalance

$F = \mu Me\omega^2\sin\omega t$

The equation of motion for the mass M will be the same as for rotating unbalance.

14 Vibration isolation.

Maximum amplitude, $X = x_{st}/[1-\beta^2)^2 + (2\zeta\beta)^2]^{1/2}$

Where $x_{st} = F_o/k$ = static deflection

Force transmitted to the foundation, $F_{tr} = F_o[1 + (2\zeta\beta)^2]^{1/2}/[(1-\beta^2)^2 + (2\zeta\beta)^2]^{1/2}$

Force transmissibility, $TR = F_{tr}/F_o = [1 + (2\zeta\beta)^2]^{1/2}/[(1-\beta^2)^2 + (2\zeta\beta)^2]^{1/2}$

$TR < 1$ when $\beta > \sqrt{2}$, $TR = 1$ for $\beta = \sqrt{2}$, $TR > 1$ for $\beta < \sqrt{2}$, $TR < 1$ for $\beta > \sqrt{2}$

Phase difference between the transmitted force and the excitation force,

$$\phi - \alpha = \tan^{-1}[2\zeta\beta/(1-\beta^2) - \tan^{-1}(2\zeta\beta)$$

15 The critical speed of a rotating shaft is the speed at which the shaft starts to vibrate violently in the transverse direction. Critical speed is also called 'whipping' or 'whirling' speed. The main reason for the whirling speed is the mass unbalance of the shaft when the mass centre does not coincide with the geometric centre.

(a) Single disc without damping

$$r = e\beta^2/(1-\beta^2)$$

where $\beta = \omega/\omega_n$, $\omega_n = (\mathrm{k/m})^{1/2}$

The deflection of the shaft tends to infinity when $\beta = 1$. Thus, the critical speed of the shaft is equal to the natural frequency of lateral vibrations of the shaft. For $\beta < 1$, r is positive, i.e. the disc rotates with heavy side outwards, For $\beta > 1$, r is negative, i.e. the disc rotates with light side outwards, also $\beta < 1$ corresponds to zero degree phase difference and $\beta > 1$ corresponds to 180° phase difference. When $\beta >> 1$, $r \to e$, i.e. the point G approaches O and the disc rotates about its centre of gravity. Therefore, it is always advisable to operate the machine much above its natural frequency.

(b) Single disc with damping

$$r/e = \beta^2/[1-\beta^2]^2 + (2\zeta\beta)^2]^{1/2}$$

Where $\beta = \omega/\omega_n$, $\zeta = c/[2(mk)^{1/2}]$

Phase difference, $\phi = \tan^{-1}[2\zeta\beta/(1-\beta^2)]$

(i) $\phi = 0$ when $\beta << 1$, and heavy side of the disc will be out.

(ii) $0 < \phi < 90°$ when $\beta < 1$, and heavy side of the disc will be out.

(iii) $\phi = 90°$ when $\beta = 1$.

(iv) $90° < \phi < 180°$ when $\beta > 1$, and light side of the disc will be out.

(v) $\phi = 180°$ and $r \to -e$ when $\beta >> 1$, and light side of the disc will be out with the disc rotating about its centre of gravity.

16 Transverse vibrations of a bean:

$$\omega_n = [g/\delta_{st}]^{1/2} \text{ rad/s}$$

Where $\theta_{st} = WL^3/(48EI)$, for a single concentrated load at midspan on a simply supported beam

$= WL^3/(192EI)$, for a single concentrated load at midspan on a beam fixed at both ends

$= 5wL^4/(384EI)$, for a simply supported beam carrying udl on whole span

$= Wa^2b^2/(3EIL)$, for a simply supported beam carrying off centre concentrated load

$= wL^4/(384EI)$, for a beam carrying udl and fixed at both ends

$= WL^3/3EI$, for a cantilever carrying end load

17 Dunkerley's method:

$$1/\omega_n^2 = 1/\omega_1^2 + 1/\omega_2^2 + \cdots + 1/\omega_m^2 + 1\omega_s^2$$

Where ω_i, $i = 1$ to m are the natural frequencies of the shaft with each lumped mass acting alone,

ω_s = natural frequency of the shaft due to its own udl.

ω_n = fundamental natural frequency of the system

Dunkerley's method gives the lower bound on natural frequency of the system.

18 Rayleigh's method:

$$\omega_n = [g\Sigma W_i y_i / \Sigma W_i y_i^2]^{1/2}$$

Rayleigh's method gives the upper bound on the natural frequency of the system.

19 The solution of torsional vibrations is similar to the solution of longitudinal vibrations. The equivalence between these vibrations is:

$$I = m,\ q \text{ or } k_t = k,\ c_t = c,\ T_o = F_o$$

20 Equivalent length of a stepped shaft:

$$l_e = l_1 + (d_1/d_2)^4\, l_2 + (d_1/d_3)^4\, l_3 + \cdots$$

21 For a shaft carrying a single rotor, natural frequency of torsional vibrations,

$$\omega_n = [q/I]^{1/2} \text{ rad/s}$$

22 At the node location, the amplitude of vibration of the shaft is zero.

23 For a shaft carrying two rotors

$$\omega_n = [k_t(J_1 + J_2)/(J_1 J_2)]^{1/2}$$

Position of nodes: $l_1 = J_2/(J_1 + J_2)$

Ratio of amplitudes: $A_1/A_2 = J_2/J_1 = l_1/l_2$

24 For geared shafts with $i = n_1/n_2 = z_1/z_2$,

Equivalent torsional stiffness of driven shaft, $q_2' = i^2 q_2$

Equivalent moment of inertia of rotor on the driven shaft $= J_2' = i^2 J_2$

Equivalent torsional stiffness of the geared shafts, $q_e = i^2 q_1 q_2/(q_1 + i^2 q_2)$

Natural frequency, $\omega_n = [q_e(J_1 + i^2 J_2)/(i^2 J_1 J_2)]^{1/2}$ rad/s

25 For a shaft carrying n rotors, the number of nodes are $(n - 1)$.

Multiple Choice Questions

1 The effect of the spring mass can be accounted for to calculate the natural frequency of a spring-mass system by adding n times the mass of spring to the main mass, where

(a) $n = \frac{1}{2}$ (b) $n = \frac{1}{3}$ (c) $n = \frac{1}{4}$ (d) $n = \frac{3}{4}$.

2 The equivalent stiffness of two springs of equal stiffness in series becomes

(a) $\frac{1}{4}$ (b) $\frac{1}{2}$ (c) $\frac{1}{3}$ (d) 2 times.

3 The equivalent stiffness of two springs of equal stiffness in parallel becomes

(a) Twice (b) One-half (c) One-third (d) One-fourth.

4 Damping ratio (ζ) is defined as:

(a) $\zeta = \frac{c_c}{c}$ (b) $\frac{c}{c_c}$ (c) $c \times c_c$ (d) $\left(\frac{c}{c_c}\right)^2$.

5 For an underdamped system

(a) $\zeta < 1$ (b) $\zeta = 1$ (c) $\zeta > 1$ (d) $\zeta = 0$.

6 For a critically damped system, damping ratio is

(a) 1.0 (b) 0.5 (c) 2 (d) 3.

7 For an underdamped system, motion is

(a) Exponentially decreasing (b) Oscillatory
(c) Non-oscillatory (d) Aperiodic.

8 For a critically damped system, motion is

(a) Non-oscillatory (b) Exponentially decreasing
(c) Oscillatory (d) Aperiodic.

9 Logarithmic decrement (δ) is defined as:

(a) $\delta = ln\left(\dfrac{x_{n+1}}{x_n}\right)$ (b) $\delta = ln\left(\dfrac{x_n}{x_{n+1}}\right)$

(c) $\delta = 2ln\dfrac{x_n}{x_{n+1}}$ (d) $\delta = \dfrac{1}{2}ln\left(\dfrac{x_n}{x_{n+1}}\right)$.

10 The relationship between natural frequency and damped natural frequency is:

(a) $\omega_d = \omega_n\zeta$ (b) $\omega_d = \omega_n\sqrt{1+\zeta^2}$ (c) $\omega_d = \omega_n\sqrt{1-\zeta^2}$ (d) $\omega_d = \omega_n(1-\zeta^2)$.

11 Magnification factor for $\beta = 1$ is

(a) $\dfrac{1}{\zeta}$ (b) $\dfrac{1}{2\zeta}$ (c) $\dfrac{1}{\zeta^2}$ (d) $\dfrac{1}{\zeta^{1/2}}$.

12 Force transmissibility is unity, when

(a) $\beta = 1$ (b) $\beta = \sqrt{2}$ (c) $\beta < \sqrt{2}$ (d) $\beta > \sqrt{2}$.

13 Which of the following methods gives lower bound on the natural frequency?

(a) Dunkerley's method (b) Energy method
(c) Rayleigh's method (d) Equilibrium method.

14 Equivelent length l_e of a stepped shaft is

(a) $l_e = l_1 + \left(\dfrac{d_1}{d_2}\right)^4 l_2 + \left(\dfrac{d_1}{d_3}\right)^4 l_3 + \cdots$ (b) $l_e = l_1 + \left(\dfrac{d_1}{d_2}\right)^3 l_2 + \left(\dfrac{d_1}{d_3}\right)^3 l_3 + \cdots$

(c) $l_e = l_1 + \left(\dfrac{d_1}{d_2}\right)^2 l_2 + \left(\dfrac{d_1}{d_3}\right)^2 l_3 + \cdots$ (d) $l_e = l_1 + \left(\dfrac{d_1}{d_2}\right) l_2 + \left(\dfrac{d_1}{d_3}\right) l_3 + \cdots$

15 A torsional system having m rotors on a vibrating shaft has

(a) m nodes (b) $(m-1)$ nodes
(c) $(m-2)$ nodes (d) $2m$ nodes.

Answers

1. (b) 2. (b) 3. (a) 4. (b) 5. (a) 6. (a) 7. (b) 8. (a) 9. (b) 10. (c) 11. (b) 12. (b)
13. (a) 14. (a) 15. (b)

Review Questions

1 Define vibrations. How they are caused?

2 What are free and damped vibrations?

3 What are forced vibrations?

4 What are the elements of a vibrating system?

5 Define logarithmic decrement. What is its significance?

6 What is the type of motion for underdamped, critically damped, and overdamped system?

7 Define damping coefficient and critical damping coefficient.

8 What is magnification factor?

9 Define the terms vibration isolation and transmissibility.

10 What do you understand by whirling of a shaft?

Exercises

17.1 Determine the frequency of free vibrations of a fluid column of length L in a U-tube if the density of the fluid is ρ and area of cross-section of tube is A.

17.2 A spring-mass system of stiffness k_1 and mass m is suspended at the end of a cantilever of length l_1. The cantilever is supported at a distance l_2 by another spring of stiffness k_2. Assuming the cantilever to be of negligible mass, determine the frequency of natural vibrations of the system.

17.3 A vibrating system consists of a mass of 40 kg and a spring of stiffness 25 N/mm and damper. The damping provided is only 15% of the critical value. Determine (a) the damping factor, (b) critical damping coefficient, (c) damped natural frequency, (d) logarithmic decrement, and (e) ratio of two successive amplitudes.

17.4 In a single-degree damped vibrating system, a suspended mass of 10 kg makes 25 oscillations in 15 s. The amplitude decreases to 1/4th of the initial value after 5 oscillations. Determine (a) stiffness of the spring, (b) logarithmic decrement, (c) damping factor, and (d) damping coefficient.

17.5 A machine part having a mass of 2 kg vibrates in a viscous medium. A harmonic exciting force of 25 N acts on the part and causes a resonant amplitude of 12 mm with a period of 0.2 s. Find the damping coefficient.

If the frequency of the exciting force is changed to 3 Hz, determine the increase in the amplitude of the forced vibrations upon the removal of the damper.

17.6 A single-cylinder vertical diesel engine has a mass of 350 kg and is mounted on a steel frame. The static deflection due to the weight of the frame is 2 mm. The reciprocating masses of the engine amounts to 15 kg and the stroke of the engine is 150 mm. A dashpot with a damping coefficient of 2 N/mm/s is also used to dampen the vibrations. In the steady state of the vibrations, determine (a) amplitude of the vibrations if the driving shaft rotates at 450 rpm, and (b) the speed of the driving at resonance.

17.7 A refrigerator unit having a mass of 40 kg is to be supported on four springs, each having a spring stiffness *k*. The unit operates at 460 rpm. Find the value of stiffness *k* if only 10% of the shaking force is allowed to be transmitted to the supporting structure.

17.8 A rotor has a mass of 10 kg and is mounted midway on a 20 mm diameter horizontal shaft supported at the ends by two bearings 1.2 m apart. The shaft rotates at 2000 rpm. If the centre of rotor of the rotor is 0.10 mm away from the geometric centre of the rotor due to certain manufacturing defect, determine (a) the amplitude of the steady-state vibration, and (b) the dynamic force transmitted to the bearing. For shaft material, $E = 200$ GPa.

17.9 An electric motor running at 450 rpm is supported on a spring and a dashpot. The spring stiffness is 6000 N/m and the dashpot offers resistance of 500 N at 5 m/s. The unbalanced mass 0.5 kg rotates at 6 cm radius and the total mass of vibratory system is 20 kg. Determine (a) damping factor, (b) amplitude of vibration and phase angle, (c) resonant speed and resonant amplitude, and (d) force exerted by the spring and dashpot on the motor.

17.10 A shaft 15 mm diameter and 1 m long is held in long bearings. The weight of the disc at the centre of the shaft is 15 N. The eccentricity of the centre of gravity of the disc from centre of rotor is 0.3 mm. The permissible stress in the shaft material is 65 MPa and its modulus of elasticity is 200 GPa. Determine (a) the critical speed of the shaft, and (b) the range of speed over which it is unsafe to run the shaft.

17.11 For the semi-definite system shown in Fig.17.56, if $J_1 = 1.2$ kg · m², $J_2 = J_3 = 2J_1$, $k_{t1} = 25 \times 10^3$ N m/rad, and $k_{t2} = 2\ k_{t1}$, find the natural frequencies and relative amplitude of the principal modes.

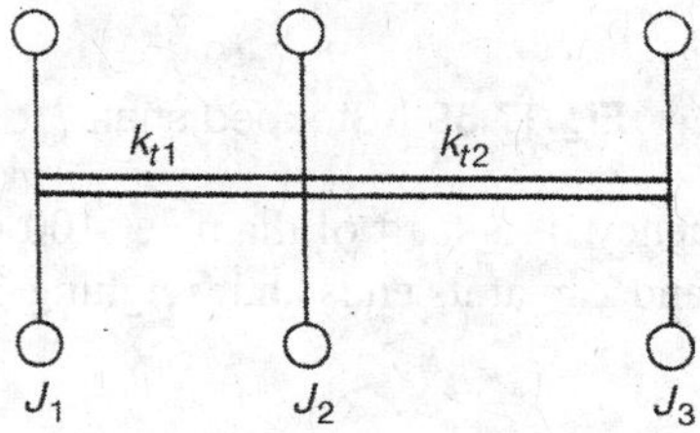

Fig.17.56 Semi-definite system

17.12 Neglecting the inertia effect of the pinion and gear in Fig.17.57, let $J_1 = 0.2$ kg · m², $J_2 = 4J_1$; $k_{t1} = 60 \times 10^3$ N m/rad, $k_{t2} = 7\ k_{t1}$, and the gear ratio 3:1. Find the natural frequencies of the system: (a) referring to shaft 1, and (b) referring to shaft 2.

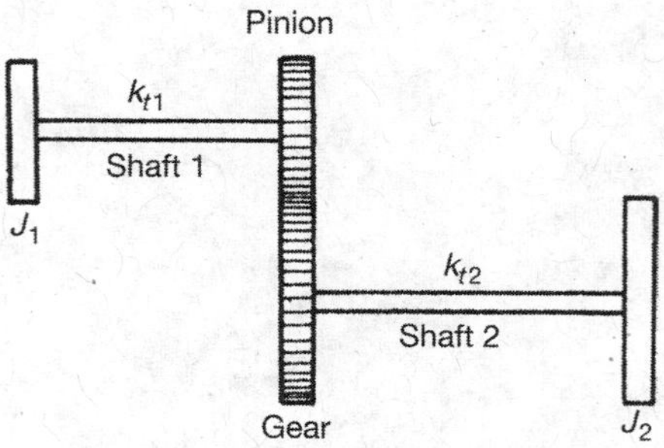

Fig.17.57 Geared system

17.13 A motor shaft of diameter 50 mm drives a pump shaft of diameter 100 mm through a spur gear pair, as shown in Fig.17.58. The motor rotor has a moment of inertia of 500 kg · m² and the pump rotor has 1500 kg · m². The speed ratio is 3:1. G = 84 GPa. Calculate the natural frequency of the gear system.

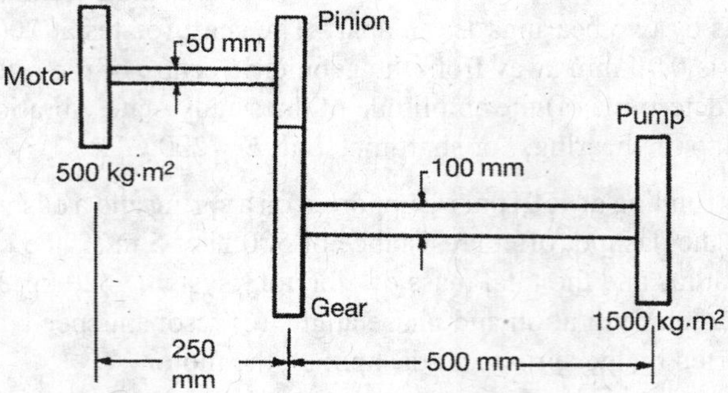

Fig.17.58 Geared system

17.14 Determine the natural frequency and position of the node for the free torsional vibrations of the stepped shaft shown in Fig.17.59. G = 80 GPa.

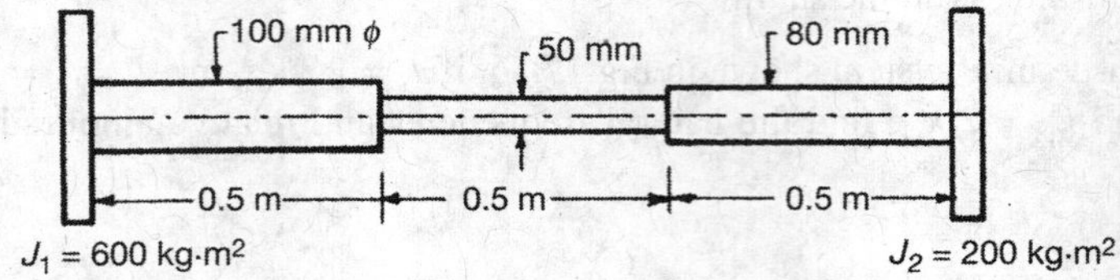

Fig.17.59 Stepped shaft

17.15 Calculate the natural frequency of a shaft of diameter 100 mm and length 3 m carrying two discs of diameters 1.25 m and 2 m at its ends and weighing 500 N and 900 N respectively. For the shaft, G = 84 GPa.

18 AUTOMATIC CONTROL

Chapter Outline

18.1 INTRODUCTION

Mechanisms are required to be adjusted or controlled so that they perform their specified function. This can be done either manually or automatically. Automatic control is desired in order to save the human operator from monotony of work and to bring about efficiency in the system.

Automatic control of mechanisms has been widely used in various devices and machines process, industry, manufacturing systems, and machine tools, etc.

18.2 DEFINITIONS

Automatic controller: It is a mechanism which measures the value of a variable quantity or condition and operates to correct or limit the deviation of the measured value from some selected reference.

Process controller: It is a device which controls a process, for example, a change in pressure, temperature, voltage or speed.

Regulator: It is a device which keeps a quantity at a constant value.

Kinetic control: Which controls the position or velocity or acceleration of a member.

Remote Position control: Which controls only the position of a member.

Servomechanism: when a control system includes a power amplifier also.

Open-loop control: in which the control is achieved essentially by previous design, calibration, and perhaps trial and error.

Closed-loop control: in which the results are continuously monitored and allows correcting action.

Process: includes all functions performed in and by the equipment in which a variable is to be controlled.

Resistance: is opposition to flow. It is measured as potential change required to produce a unit change in flow.

Set-point: is the position in which the control point setting mechanism is set.

Cycling: is a periodic change of the controlled variable from one value to another.

Dead time: is any definite decay /period between two related actions.

Proportional control: in which the controller output is proportional to the input.

Integral control: in which the output of the controller is proportional to the time integral of the activating signal.

Derivative control: in which the output of the controller is proportional to the rate of change of input.

On-off action: in which a final control element is moved from one of two fixed positions to the other.

On-off control: in which the controller operates at either of two levels.

Stable system: is one in which the transient response decays as time increases.

Unstable system: is one in which the transient response increases as time increases.

Overshoot: is defined as the maximum deviation of the response above the steady state value.

Response time: is the time required for the system to reach steady state.

Settling time: is defined as the time required for the output to achieve within 2% of its final value when the system is subjected to a step input.

Delay time: is defined as the time required for the output to reach 50% of its final value in first attempt.

Rise time: is the time required for the output to rise from 10% to 90% of the final value for overdamped system, and zero to 100% of the final value for underdamped system.

Peak time: is the time required for the output to reach the peak of time response or the peak overshoot.

Transducer: is a device, which converts a phenomenon to be measured into a more conveniently measurable quantity, which is directly proportional or analogous to the phenomenon to be measured.

Command: is the result of act of adjustment of a device or mechanism or link.

Response: is the behaviour of the system.

Block diagram: is a pictorial method of portraying the interrelationships among components of a physical system or process.

Transfer function: is the ratio of the Laplace transformation of output to Laplace transformation of input.

18.3 TRANSDUCERS AND SENSORS

Transducers are the basic elements that convert or transform one form of signal to another form which is more convenient to use and measure. A transducer is an essential element of a sensor. A sensor is merely a sophisticated transducer which contains some signal conditioning circuits capable of amplifying and refining the weak and raw signal that is available at the output of the transducer. Some of the commonly used signal conditioning circuits are: amplifiers, filters, Analog to Digital Converter (ADC), etc. Fig.18.1 gives an illustration of a sensor.

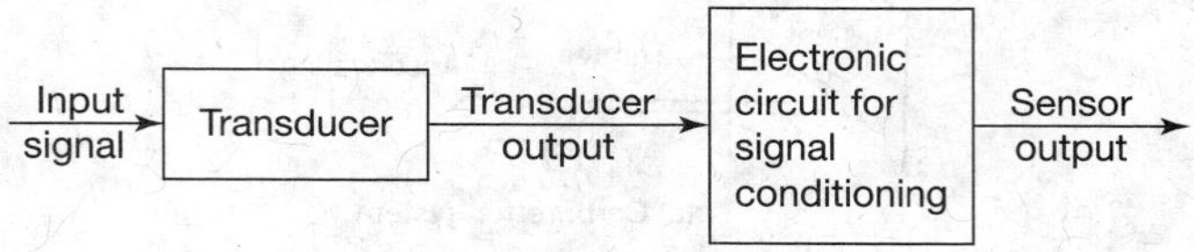

Fig.18.1 Illustration of a sensor

18.3.1 Transducer Types

Transducers are classified based on whether they produce passive or active equivalence. Passive equivalence are: resistance, inductance, and capacitance. On the other hand, active transducers are those, which directly provide electrical signals like voltage, current either in the form of D.C. or A.C. Active and passive type transducers are referred to as the primary and secondary transducers respectively.

Specially designed mechanical structures are also used for the measurement of various physical phenomenon such as movement, proximity, displacement, force, pressure, strain, flow etc.

18.4 ACTUATORS

Actuation is the process of conversion of energy to mechanical form. A device that accomplishes this conversion is termed actuator. Actuators play a very important role while implementing control. The microcontroller provides command signal to the actuator for actuation.

There are many types of actuators, in which energy transformation takes place through multiple forms. The actuators are broadly categorized into following groups:

1. Electromechanical actuators
2. Fluid power actuators
3. Active material based actuators.

Electromechanical actuators are used to efficiently convert electrical energy into mechanical energy. Megnetism is the basis of their principles of operation. Electromechanical actuators are DC, AC, servo, and stepper motors.

Fluid power actuators are of the hydraulic and pneumatic types. Active material based actuators under some sort of transformation through physical interaction such as piezo electric materials, magnetostrictive materials, and electrorheological fluids etc.

18.5 BLOCK DIAGRAMS

A block diagram is a symbolic outline of a system in which various components or operations are represented by rectangles in an ordered sequence. The rectangles are connected by arrows showing the flow of the working medium or of information. The block diagram of an ordinary carburettor depicted in Fig.18.2(a) is shown in Fig.18.2(b).

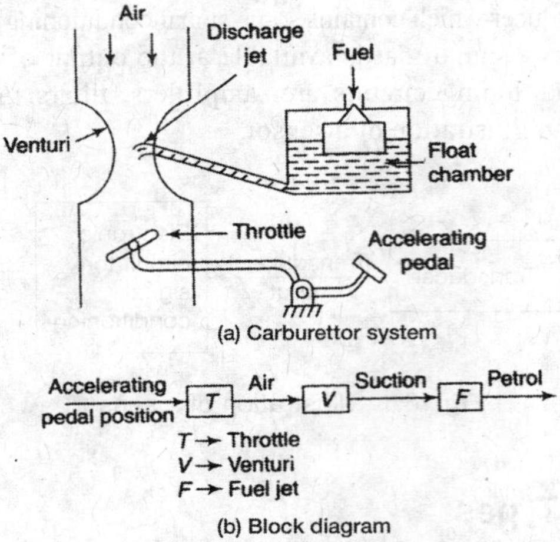

Fig.18.2 Carburettor system block diagram

A mechanical system of the mass-spring-damper system is shown in Fig.18.3(a), whose equivalent block diagram is shown in Fig.18.3(b).

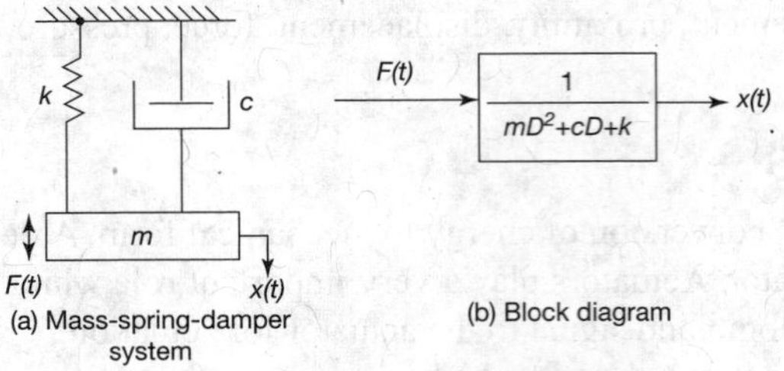

Fig.18.3 Mass-spring-damper system

18.6 SYSTEM MODELING

System modeling is about solving practical problems by creating mathematical models, called model equations. These equations can be manipulated and optimized during the process of system design. The analysis and design procedure can be summarized as follows:

- Identification or description of physical system in terms of basic modeling elements.
- Formulation of a mathematical model.
- Analysis of the model.
- Interpretation, confirmation, and verification.
- Practical behaviour study.

All systems are categorized under four main basic or elemental systems, namely:

- Mechanical system
- Electrical system
- Fluid system
- Thermal system

1. *Mechanical system*

The three basic modeling elements for mechanical systems are: spring, damper, and mass/inertia. The spring element stores potential energy. Spring stiffness is the applied force per unit deflection. The reciprocal of the stiffness is called mechanical capacitance or compliance. The spring can be of the translational or rotational type. The potential energy stored by a spring element is given by:

$$U = \frac{1}{2} k\, x^2 = \frac{1}{2} \frac{F^2}{k}$$

or
$$= \frac{1}{2} k_t \theta^2 = \frac{1}{2} \frac{T^2}{k_t}$$

where k_t = rotational stiffness.

A damper element consumes energy which cannot be recovered. Other name of the damper is dashpot or mechanical resistance. Dampers are of two types: translational and rotational dampers. The loss of energy in a damper is related to the velocity.

Damper resistance, $F = cv$

where c = damping constant

v = velocity

Power loss, $P = Fv = cv^2$

For the rotational damper element,

Torque, $T = c_t \omega$

where c_t = rotational damping constant

ω = angular velocity

Power dissipated, $P = T\omega = c_t \omega^2$

Mass element refers to translational movement where as inertia element corresponds to rotational movement.

Force, $$F = ma = m \frac{dv}{dt}$$

Kinetic energy stored, $$E_k = \frac{1}{2} mv^2$$

For the rotational system,

$$T = J\omega = J\frac{d\theta}{dt}$$

$$E_k = \frac{1}{2}J\omega^2$$

where J = moment of inertia

2. *Electrical system*

The three basic elements of an electrical system are: Inductor, resistor, and capacitor. They are analogous to spring, damper, and mass/inertia.

Stored magnetic energy,

$$E = \frac{1}{2}Li^2$$

Loss dissipated, $P = i^2R$

Stored electrical energy,

$$E_k = \frac{1}{2}Cv^2$$

3. Fluid system

The three basic modeling elements in fluid system are: Inertance, fluid resistance, and fluid capacitance.

Inertance, $\Delta = \Gamma\dot{Q}$

where Δ = pressure momentum

Γ = fluid inertance

Q = flow rate

The element possessing inertance is known as inertor.

Energy stored in inertor, $E = \frac{1}{2}\Gamma Q^2$

An ideal inertor characterizes frictionless, incompressible flow in a uniform passage. For such fluids,

$$\Gamma = \frac{\rho L}{A}$$

where ρ = density of the fluid

L = length of inertor

A = area of passage

A fluid resistor dissipates energy.

Fluid flow rate, $Q = G_f P = \frac{1}{R_f}P$

where P = pressure drop

G_f = fluid conductance (m^3/N.s)

R_f = fluid resistance (N.s/m^3)

Power dissipated, $P_f = PQ = R_f Q^2 = \frac{1}{R_f}P^2$

A fluid capacitor is defined as an element in which the stored energy is a function of fluid pressure. For a fluid capacitor,

Volume, $V = C_f P\ (m^3)$

where C_f = fluid capacitance (m³/N)

P = fluid pressure

$$\frac{dv}{dt} = C_f \frac{dP}{dt} = Q$$

For an ideal fluid capacitor,

$$C_f = \frac{A}{\rho g}$$

where A = area of flow.

4. *Thermal system*

The basic modeling elements of thermal system are: Thermal capacitance and thermal resistance. The thermal capacity of a thermal system is the amount of heat energy it can store.

Heat, $H = C_t T$ (Joules)

where C_t = thermal capacitance (J/K)

T = temperature (k)

Rate of energy storage,

$$q = \frac{dH}{dt} = C_t \frac{dT}{dt}$$

or $$C_t = \frac{q}{\tau}$$

where τ = rate of temperature $= \dfrac{dT}{dt}$

t = time

Thermal resistance, $R_t = \dfrac{T_2 - T_1}{q}\ (K/W)$

Conductance $= \dfrac{1}{R_t}$ (W/K)

18.7 SYSTEM RESPONSE

System response deals with studying the behaviour of the system in which changes occur and in which predictions are desirable. Input-output models form the basis of most classical control systems. There are three main types of behaviour which may be seen at the output. They are:

- Instantaneous response
- Lagging response, and
- Delayed response

Suppose an input u, which is a step signal [Fig.18.4(a)] is given to a system. The typical response shown in Fig.18.4(b) is called the instantaneous response. In this case y responds in a step, but of different amplitude compared to that of u. In the mathematical relationship,

$$y(t) = au(t) + a_0$$

where a = gain

In any system, usually, there is a lag or delay in response due to some inherent causes. In lagging case, as shown in Fig.18.4(c), y starts to change but full/extent of response lags behind the input. After a while, however, y tries to attend the value of the input. In the mathematical relationship,

$$\frac{dy(t)}{dt} = \frac{1}{\tau}[au(t) + a_0 - y(t)]$$

τ = time constant of the system, which determines speed

In the third case, shown in Fig.18.4(d), it is observed that no immediate change in y occurs when u changes. However, after certain time T, y responds to the change in u as in the instantaneous response case. The time T is referred to as a time delay of the system.

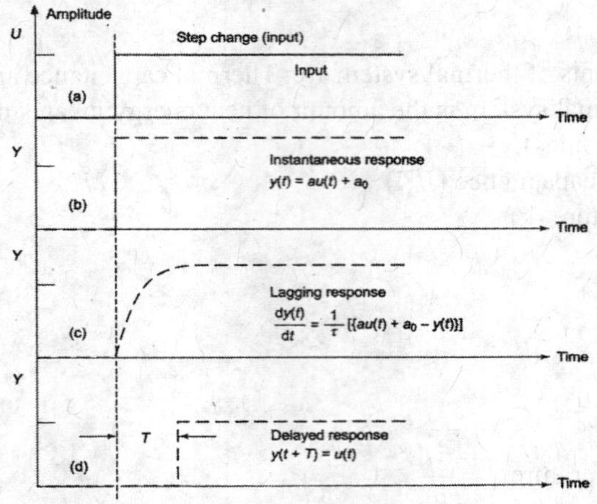

Fig.18.4 Typical behaviour of the system

18.7.1 Transient-Response Specification

If at time $t = 0$, the command signal or control signal is given at the input of a system we expect that the output should respond immediately at the same time. However, in practice the system takes time to produce the output response. This delay and lagging characteristics is due to the presence of energy storing elements within the systems. In effect the system exhibits *transient properties*. The output starts from initial value (the initial value could be zero) and reaches at the *steady-state value* after certain time. How the output reaches at the steady-state value solely depends on the type of the system in hand. (e.g. first order, second order, linear, nonlinear etc.) and the nature of input signal provided. This property of this phenomenon can be specified through *transient-response* specification. This section describes the basics of transient-response specifications. The transient response specifications involves the following terms.

- Peak-time
- Settling-time
- Steady-state value
- Maximum overshoot

The above terminology is based on a step input to the system. Fig.18.5 illustrates the transient response corresponding to a step input signal.

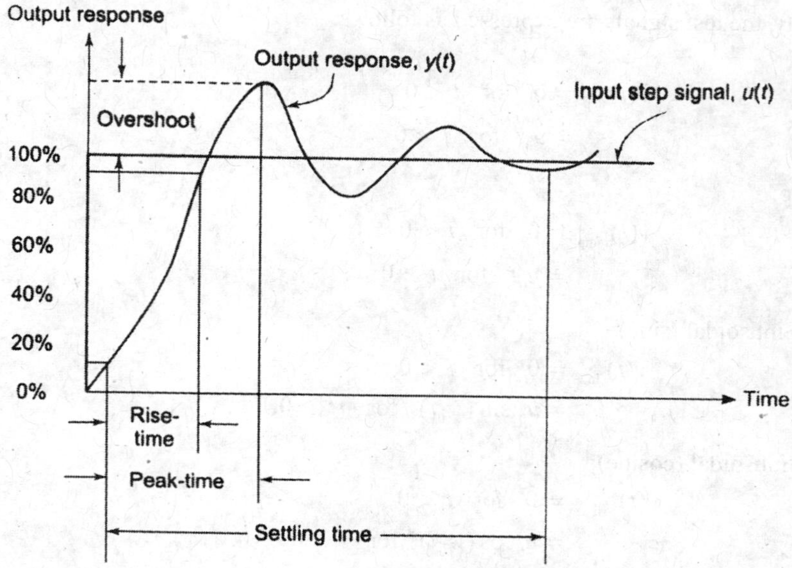

Fig.18.5 Transient-response specifications

Rise-time is the time required to the output to reach 90% from 10% of the input value. *Peak-time* is the time required to reach the peak overshoot value. *Settling-time* is the time at which the response attends within 98% of the final value. An error limiting 2% is called *tolerance*. The output response, settles within this tolerance band only after the settling time is reached and the level of output is referred to as initial steady-state value. Initial steady-state value starts at $t = t_s$. The final steady state value refers to the response value at time $t = \infty$. The maximum overshoot is the difference between the maximum peak and the input value. It is usually expressed in percentage.

18.8 TEST SIGNALS

In order to know the output response of a system, usually some test signals are given at the input and then the outputs are observed. Commonly used test signals are: step, ramp, and sinusoidal signal, as shown in Fig.18.6.

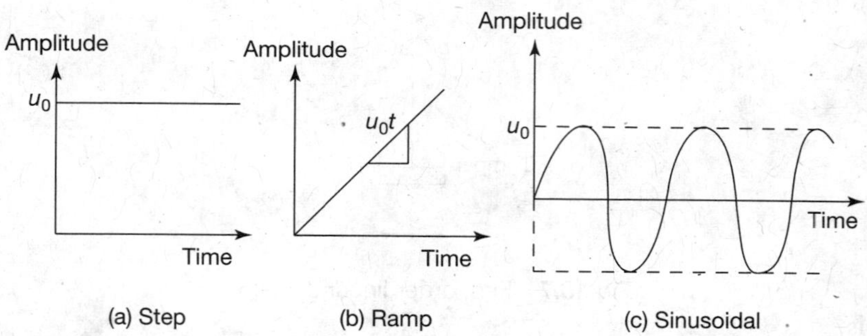

Fig.18.6 Types of test signals

Mathematically, the test signals are expressed as follows:

1. Step:

$$v(t)_{\text{step}} = 0 \quad \text{for} \quad t < 0$$
$$= u_0 \quad \text{for} \quad t \geq 0$$

2. Ramp:

$$v(t)_{\text{ramp}} = 0 \quad \text{for} \quad t < 0$$
$$= u_0 t \quad \text{for} \quad t \geq 0$$

3. (a) Sinusoidal (sine):

$$v(t)_{\sin e} = 0 \quad \text{for} \quad t < 0$$
$$= u_0 \sin(w_0 t) \quad \text{for} \quad t \geq 0$$

(b) Sinusoidal (cosine):

$$v(t)_{\text{cosine}} = 0 \quad \text{for} \quad t < 0$$
$$= u_0 \cos(w_0 t) \quad \text{for} \quad t \geq 0$$

18.9 OUTPUT RESPONSE OF FIRST ORDER SYSTEMS

18.9.1 Linear Systems

(a) *Free response (zero excitation)*

Consider the massless spring of stiffness k and viscous damper of damping coefficient c as shown in Fig.18.7. The equation of motion is:

$$c\dot{y} + ky = 0$$

or

$$\dot{y} + \frac{k}{c} y = 0$$

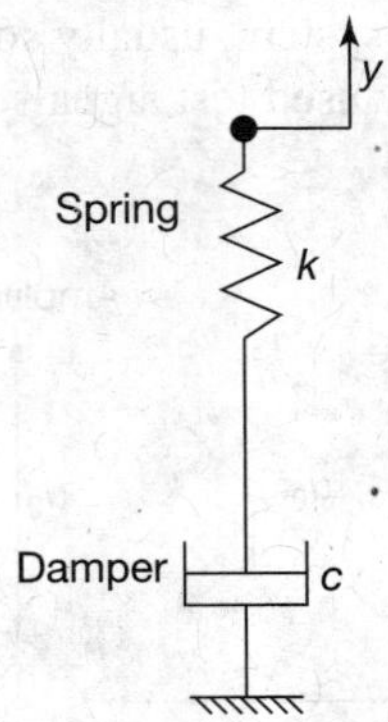

Fig.18.7 First order linear system

Its solutions is,

$$y = A \exp\left(-\frac{k}{c} t\right)$$

Let at $t = 0,\ y = y_0$

$\therefore$ $y_0 = A$

$$y = y_0 \exp\left(\frac{-k}{c}t\right)$$

Let $\tau = \dfrac{c}{k}$, then

$$y = y_0 \exp\left(-\frac{t}{\tau}\right) \tag{18.1}$$

τ is known as the time constant for the system, which determines the speed of the system.

$$\frac{y}{y_0} = \exp\left(-\frac{t}{\tau}\right)$$

It is an exponentially decreasing function. At $t \to \infty$ or when t is very large, the output is approximately zero. Fig.18.8 illustrates the normalized output response $\left(\dfrac{y}{y_0}\right)$ of the system.

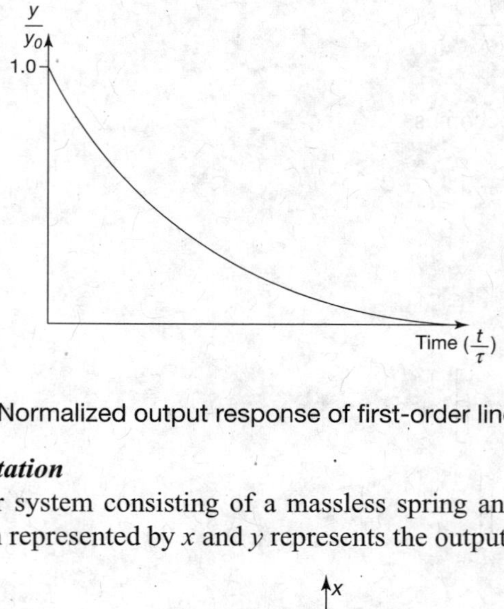

Fig.18.8 Normalized output response of first-order linear system

(b) *Linear system with excitation*

Consider a first order linear system consisting of a massless spring and viscous damper shown in Fig.18.9. A constant input in represented by x and y represents the output of the system.

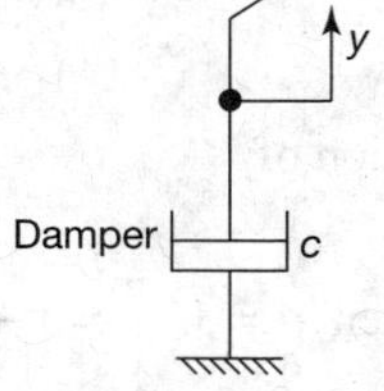

Fig.18.9 Linear system with excitation

The equation of motions is:

$$c\dot{y} = k(x - y)$$

or $$c\dot{y} + ky = kx$$

or $$\dot{y} + \frac{k}{c}y = \frac{k}{c}x$$

This is a first order differential equation.
Complementary function is the solutions of the equation,

$$\dot{y} + \frac{k}{c}y = 0$$

The solution is:

$$y = Ae^{-\frac{k}{c}t}$$

where A is a constant.
The particular integral is:

$$PI = \frac{\frac{k}{c}x}{\Delta + \frac{k}{c}} = \frac{\left(\frac{k}{c}\right)x}{0 + \frac{k}{c}} = x$$

The complete solutions becomes,

$$y = x + A\exp\left(\frac{-k}{c}t\right)$$

when $t = 0, y = 0$

$$\therefore \quad 0 = x + A$$

or $$A = -x$$

$$\therefore \quad y = x\left[1 - \exp\left(\frac{-k}{c}t\right)\right]$$

$$= x\left[1 - \exp\left(\frac{-t}{\tau}\right)\right]$$

Where $\tau = \frac{c}{k}$, is known as time constant for the system

Also $\frac{y}{x} = 1 - \exp\left(\frac{-t}{\tau}\right)$

Fig.18.10 shows graphical representation of $\frac{y}{x}$ v's $\frac{t}{\tau}$. (18.2)

As t increases, y tends to reach x. When $\frac{t}{\tau} = 1$, $\frac{y}{x} = 1 - 0.368 = 0.632$

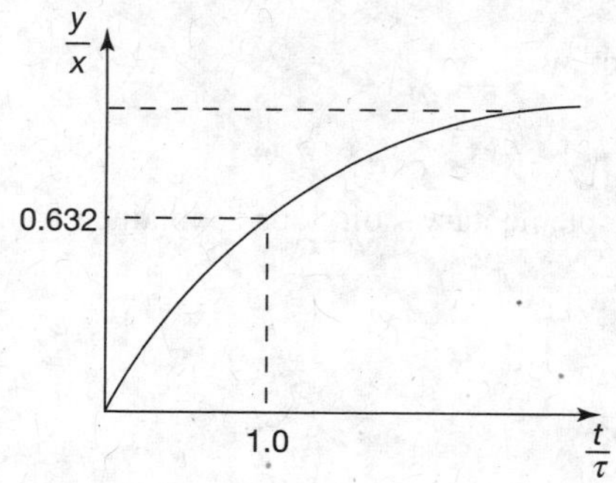

Fig.18.10 $\frac{x}{y}$ v's $\frac{t}{\tau}$ for linear system with excitation

$\text{Exp}\left(-\frac{t}{\tau}\right)$ is known as the dynamic error, which reduces with increase in t and vanishes when t becomes infinitely large. However, one need not wait for an infinitely long time. Instead an accepted value of error is specified and the settling time is obtained when the steady state response enters in a band around the final steady stage value. The usual value of band is taken between 2 to 5 percent.

18.9.2 Step Input

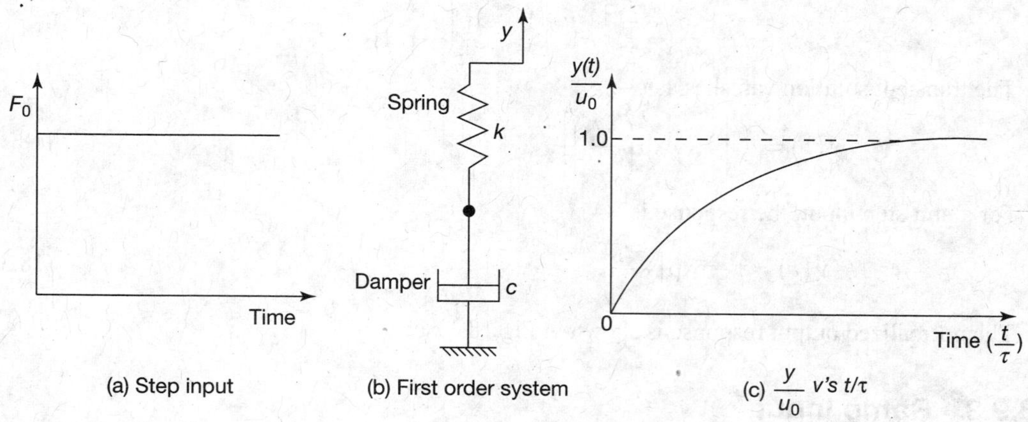

Fig.18.11 Step input applied to first order system

The step input is shown in Fig.18.11(a) and the first order system in Fig.18.11(b). The input-output model equation is:

$$c\dot{y} + ky = F_0$$

or $$\frac{c}{k}\dot{y} + y = \frac{F_0}{k}$$

Let $\tau = \frac{c}{k}$, time constant

and $u_0 = \frac{F_0}{k}$, static deflection.

Then, we have

$$\tau\dot{y} + y = u_0$$

The homogeneous (or complementary) solutions is obtained from the homogeneous equation,

$$\tau\dot{y} + y = 0$$

$$y_c(t) = A\exp\left(\frac{-t}{\tau}\right)$$

The particular solution is,

$$y_p(t) = \frac{u_0}{\tau D + 1} = \frac{u_0}{0+1} = u_0$$

Total solution, $y(t) = y_c(t) + y_p(t)$

$$= A\exp\left(\frac{-t}{\tau}\right) + u_0$$

For, $y(0) = y_0$

$$y_0 = A + u_0$$

or $A = y_0 - u_0$

$\therefore$ $$y(t) = (y_0 - u_0)\exp\left(-\frac{t}{\tau}\right) + u_0$$

$$= y_0\exp\left(-\frac{t}{\tau}\right) + u_0\left[1 - \exp\left(-\frac{t}{\tau}\right)\right]$$

The transient solution vanishes as $t \to \infty$.

$\therefore$ $$y(t) = u_0\left[1 - \exp\left(\frac{-t}{\tau}\right)\right] \tag{18.3a}$$

For a unit step input, the response is

$$y(t) = 1 - \exp\left(\frac{-t}{\tau}\right) \tag{18.3b}$$

The normalized output response is shown in Fig.18.11(c).

18.9.3 Ramp Input

The response equation for the first order system in response to ramp excitation can directly be written as:

$$y(t) = y_0\exp\left(\frac{-t}{\tau}\right) + u_0\left[t - \tau\left\{1 - \exp\left(\frac{-t}{\tau}\right)\right\}\right]$$

where $y(0) = y_0$ and u_0 is the slope of ramp input signal.

If $y(0) = 0$, then

$$y(t) = u_0\left[t - \tau\left\{1 - \exp\left(\frac{-t}{\tau}\right)\right\}\right] \tag{18.4a}$$

If the input function is a unit ramp, then

$$y(t) = t - \tau\left\{1 - \exp\left(\frac{-t}{\tau}\right)\right\} \tag{18.4b}$$

18.9.4 Sinusoidal Excitation

Let the input signal be a cosine function with the frequency ω. Then the governing input-output model equation can be written as:

$$\tau\dot{y} + y = u_0 \cos \omega t$$

Its complementary solution is,

$$y_c(t) = A\exp\left(\frac{-t}{\tau}\right)$$

Particular solutions,

$$\begin{aligned}
y_p(t) &= \frac{u_0 \cos \omega t}{\tau D + 1} \\
&= \frac{(\tau D - 1)\, u_0 \cos \omega t}{\tau^2 D^2 - 1} \\
&= \frac{-u_0 \tau\omega \sin \omega t - u_0 \cos \omega t}{-\tau^2\omega^2 - 1} \\
&= \frac{u_0\left[\omega\tau \sin \omega t + \cos \omega t\right]}{1 + \omega^2\tau^2} \\
&= \frac{u_0\sqrt{1 + \omega^2\tau^2}\,\sin(\omega t + \phi)}{1 + \omega^2\tau^2} \\
&= \frac{u_0}{\sqrt{1 + \omega^2\tau^2}}\cos(\omega t + \phi)
\end{aligned}$$

$$\phi = \tan^{-1}(-\omega\tau)$$

$$\begin{aligned}
y(t) &= y_c(t) + y_p(t) \\
&= A\exp\left[-\frac{t}{\tau}\right] + \frac{u_0}{\Delta}\cos(\omega t + \phi)
\end{aligned}$$

where $\Delta = \sqrt{1 + \omega^2\tau^2}$

For $y(0) = y_0$

$$y_0 = A + \frac{u_0}{\Delta^2}$$

$$A = y_0 - \frac{u_0}{\Delta^2}$$

$$y(t) = y_0 \exp\left(\frac{-t}{\tau}\right) + u_0\left[\frac{1}{\Delta}\cos(\omega t + \phi) - \frac{1}{\Delta^2}\exp\left(\frac{-t}{\tau}\right)\right] \tag{18.5a}$$

If $y(0) = 0$, then

$$y(t) = \frac{1}{\Delta}\left[\cos(\omega t + \phi) - \frac{1}{\Delta}\exp\left(\frac{-t}{\tau}\right)\right] \tag{18.5b}$$

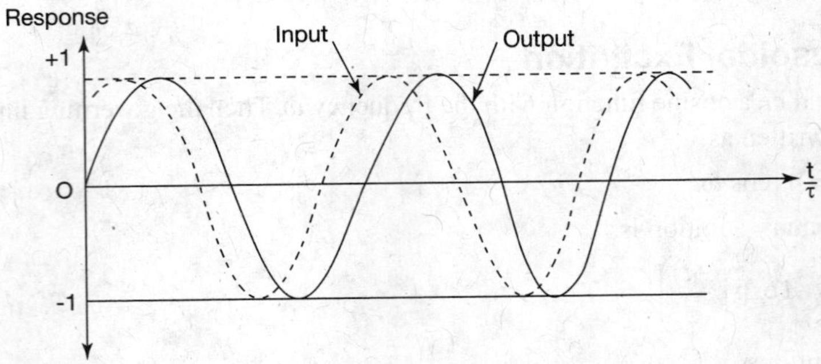

Fig.18.12 Sinusoidal input and normalized output

Fig.18.12 illustrates the input sinusoidal signal as well as corresponding output normalized response. The output lags the input by an amount ϕ.

18.9.5 Torsional System

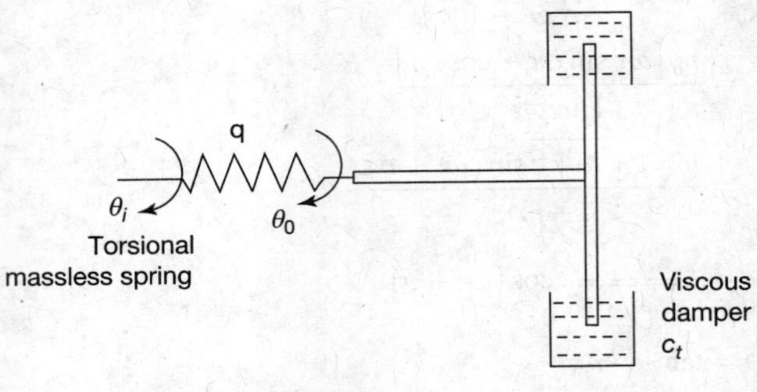

Fig.18.13 Torsional damped system

Fig.18.13 shows a system consisting of a mass less torsional spring of stiffness q and a damper of damping coefficient c_t. First the input signal θ_i is compared with the output signal θ_0. Then the difference $e = \theta_i - \theta_0$ is passed on to the motor which produces an output torque T proportional to e or $= qe$.

The equation of motion is

$$c_t\dot{\theta}_0 = qe = q\left(\dot{\theta}_i - \theta_0\right)$$

or $$c_t\dot{\theta}_0 + q\theta_0 = q\theta_i$$

or $$\dot{\theta}_0 + \frac{q}{c_t}\theta_0 = \frac{q}{c_t}\theta_i$$

It is a first order differential equation.

Complementary function is a solution of the equation

$$\dot{\theta}_0 + \frac{q}{c_t}\theta_0 = 0$$

The Solution is,

$$\theta_0 = A\exp\left(\frac{-q}{c_t}t\right)$$

Particular integral is,

$$PI = \frac{\frac{q}{c_t}\theta_i}{\frac{q}{c_t}D+1} = \theta_i$$

The complete solution becomes,

$$\theta_0(t) = A\exp\left(\frac{-q}{c_t}t\right)+\theta_i$$

For $\theta_0(0) = 0$

$$0 = A+\theta_i$$

$$\text{or } A = -\theta_i$$

$$\theta_0(t) = \theta_i\left[1-\exp\left(\frac{-t}{\tau}\right)\right] \tag{18.6}$$

Where $\tau = c_t/q$ = time constant of the system.

Example 18.1

The time constant of a thermometer is 10 s. Suddenly it is inserted in a bath at temperature 75°C. Calculate the temperature recorded by the thermometer after 5 s.

Solution

$$y = x\left[1-\exp\left(\frac{-t}{\tau}\right)\right]$$

$$= 75\left[1-\exp\left(\frac{-5}{10}\right)\right]$$

$$= 29.5°\text{C}$$

Example 18.2

A scale is fixed to the end of a shaft of torsional stiffness 2.5 N·m/rad. A viscous damping torque of magnitude 1.5 N·m resists the motion of the pointer on a scale at an angular velocity of 2 rad/s. The shaft to which pointer is attached gets the motion from the input shaft through a reduction gear box which has a gear ratio of 8:1. If the input shaft is suddenly rotated through one complete rotation, determine the time taken by the pointer to reach the position within 1% of the final value. Neglect inertia of the rotating system.

Solution

The torsional system is shown in Fig.18.14.

$$\theta_0 = \theta_i\left[1-\exp\left(\frac{-t}{\tau}\right)\right]$$

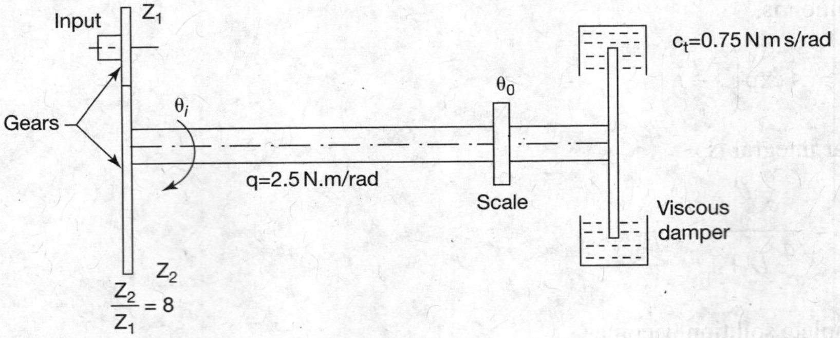

Fig.18.14 Torsional damped system

Rotation of the shaft with pointer, $\theta_i = \frac{2\pi}{8} = \frac{\pi}{4}$ rad

$$c_t = \frac{1.5}{2} = 0.75 \text{ N ms/rad}$$

$$q = 2.5 \text{ N m/rad}$$

$$\tau = \frac{c_t}{q} = \frac{0.75}{2.5} = 0.3\,s$$

$$\theta_0 = \frac{\pi}{4}\left[1 - \exp\left(\frac{-t}{0.3}\right)\right]$$

When $\theta_0 = (1 - 0.01)\,\theta_i = 0.99 \times \frac{\pi}{4} = 0.7775$

$$0.7775 = \frac{\pi}{4}\left[1 - \exp\left(\frac{-t}{0.3}\right)\right]$$

$$0.99 = 1 - \exp\left(\frac{-t}{0.3}\right)$$

$$\exp\left(\frac{-t}{0.3}\right) = 0.01$$

$$-\frac{t}{0.3} = \ln 0.01 = -4.6052$$

$$t = 1.38\,s$$

18.10 OUTPUT RESPONSE OF SECOND ORDER LINEAR SYSTEMS

The governing equation for the second order system is of the form:

$$\ddot{y} + a_1\dot{y} + a_2 y = bu(t)$$

Where the coefficients a_1, a_2, and b are the constants, $u[t]$ is the input and $y(t)$ is the output. $\ddot{y}$ and $\dot{y}$ are the second and first derivatives of $y(t) \cdot u(t)$ could be an arbitrary function or a test signal such as a step function, ramp function, a sinusoidal function and so on. In order to get the output response of the second order system the governing equation has to be solved.

18.10.1 Free Response

The output response is called free response if the forcing function or excitation signal is zero, *i.e.* $u(t) = 0$. Thus

$$\ddot{y} + a_1\dot{y} + a_2 y = 0$$

This equation is analogous to the equation of motion for free vibrations of a mass-spring-dashpot system, where $a_1 = \frac{c}{m}$ and $a_2 = \frac{k}{m}$.

The standard normalized form of this equation is,

$$\ddot{y} + 2\zeta\omega_n\dot{y} + \omega_n^2 y = 0$$

Where $\zeta = \frac{c}{c_c}$ and $\omega_n = \sqrt{\frac{k}{m}}$

The solution of above equation has been discussed in chapter 17.

(a) Underdamped system ($\zeta < 1$)

The solutions of the equation is,

$$y(t) = e^{-\zeta\omega_n t}\left[A\cos\omega_d t + B\sin\omega_d t\right]$$

Where $\omega_d = \omega_n\sqrt{1-\zeta^2}$

For $y(0) = y_0$ and $\dot{y}(0) = \dot{y}_0$, we have

$y_0 = A$

$$\dot{y}(t) = e^{-\zeta\omega_n t}\omega_d\left[-A\sin\omega_d t + B\cos\omega_d t\right] - \zeta\omega_n e^{-\zeta\omega_n t}\left[A\cos\omega_d t + B\sin\omega_d t\right]$$

$$\dot{y}_0 = \omega_d[B] - \zeta\omega_n A = B\omega_d - \zeta\omega_n y_0$$

$$B = \frac{1}{\omega_d}\left[\dot{y}_0 + \zeta\omega_n y_0\right]$$

$$y(t) = e^{-\zeta\omega_n t}\left[y_0\cos\omega_d t + \frac{1}{\omega_d}(\dot{y}_0 + \zeta\omega_n y_0)\sin\omega_d t\right]$$

$$= e^{-\zeta\omega_n t}\left[y_0\left(\cos\omega_d t + \frac{\zeta}{\sqrt{1-\zeta^2}}\sin\omega_d t\right) + \frac{\dot{y}_0}{\omega_d}\sin\omega_d t\right] \quad (18.7a)$$

If $y(0) = y_0$ and $\dot{y}(0) = 0$, then

$$y(t) = e^{-\zeta\omega_n t}\left[y_0\cos\omega_d t + \frac{\zeta}{\sqrt{1-\zeta^2}}\sin\omega_d t\right] \quad (18.7b)$$

(b) Critically damped System ($\zeta = 1$)

The solution of the equation is,

$$y(t) = (A + Bt)e^{-\omega_n t}$$

$$y(0) = y_0 = A$$

$$\dot{y}(t) = e^{-\omega_n t} \cdot B - \omega_n e^{-\omega_n t}(A + Bt)$$

$$\dot{y}_0 = B - \omega_n A$$

$$\therefore \quad B = \dot{y}_0 + \omega_n y_0$$

$$y(t) = e^{-\omega_n t}\left[y_0 + (\dot{y}_0 + \omega_n y_0)t\right] \tag{18.8a}$$

If $y(0) = 1$ and $\dot{y}(0) = 0$, then

$$y(t) = e^{-\omega_n t}[1 + \omega_n t] \tag{18.8b}$$

(c) Over damped system ($\zeta > 1$)

$$y(t) = Ae^{\left(-\zeta + \sqrt{\zeta^2 - 1}\right)\omega_n t} + Be^{\left(-\zeta - \sqrt{\zeta^2 - 1}\right)\omega_n t}$$

$$y(0) = y_0 = A + B$$

$$\dot{y}(t) = A\omega_n\left(-\zeta + \sqrt{\zeta^2 - 1}\right)e^{\left(-\zeta + \sqrt{\zeta^2 - 1}\right)\omega_n t} + B\omega_n\left(-\zeta - \sqrt{\zeta^2 - 1}\right)e^{\left(-\zeta - \sqrt{\zeta^2 - 1}\right)\omega_n t} \tag{18.9}$$

$$\dot{y}(0) = \dot{y}_0 = A\omega_n\left(-\zeta + \sqrt{\zeta^2 - 1}\right) + B\omega_n\left(-\zeta - \sqrt{\zeta^2 - 1}\right)$$

$$= A\omega_n\left(-\zeta + \sqrt{\zeta^2 - 1}\right) + (y_0 - A)\,\omega_n\left(-\zeta - \sqrt{\zeta^2 - 1}\right)$$

$$= 2A\omega_n\sqrt{\zeta^2 - 1} + y_0\omega_n\left(-\zeta - \sqrt{\zeta^2 - 1}\right)$$

$$A = \frac{1}{2\omega_n\sqrt{\zeta^2 - 1}}\left[\dot{y}_0 - y_0\omega_n\left(-\zeta - \sqrt{\zeta^2 - 1}\right)\right]$$

$$B = y_0 - A$$

For $y(0) = 1$ and $\dot{y}(0) = 0$

$$A = \frac{1}{2\omega_n\sqrt{\zeta^2 - 1}}\left[-\omega_n\left(-\zeta - \sqrt{\zeta^2 - 1}\right)\right]$$

$$= \frac{1}{2\sqrt{\zeta^2 - 1}}\left[\zeta + \sqrt{\zeta^2 - 1}\right]$$

$$= \frac{1}{2} + \frac{\zeta}{2\sqrt{\zeta^2 - 1}}$$

$$B = \frac{1}{2} - \frac{\zeta}{2\sqrt{\zeta^2 - 1}}$$

18.10.2 Step Input

The governing equation in standard normalized form for step signal input is:

$$\ddot{y} + 2\zeta\omega_n\dot{y} + \omega_n^2 y = u_0$$

Where $u(t) = u_0$ is a step function.

Let the initial conditions be $y(0) = y_0$ and $\dot{y}(0) = \dot{y}_0$. The characteristic equation for the complementary solution is:

$$\ddot{y} + 2\zeta\omega_n\dot{y} + \omega_n^2 y = 0$$

or $$\left(D^2 + 2\zeta\omega_n D + \omega_n^2\right) y = 0$$

$\therefore$ $$D^2 + 2\zeta\omega_n D + \omega_n^2 = 0$$

The two roots of the characteristic equation are:

$$s_1 = \left(-\zeta + \sqrt{\zeta^2 - 1}\right)\omega_n \text{ and } s_2 = \left(-\zeta - \sqrt{\zeta^2 - 1}\right)\omega_n$$

(i) When the roots are real and distinct ($\zeta > 1$)

$$y(t) = Ae^{s_1 t} + Be^{s_2 t}$$

(ii) When roots are equal ($\zeta = 1$)

$$y(t) = (A + Bt)e^{st}$$

(iii) When roots are imaginary ($\zeta < 1$)

$$y(t) = e^{-\zeta\omega_n t}\left[A\cos\omega_d t + B\sin w_d t\right]$$

where $$\omega_\alpha = \omega_n\sqrt{1 - \zeta^2}$$

(a) Over damped system ($\zeta > 1$)

The particular solution is a constant, since the output at steady state would become close to the input, which is nothing but a constant. Thus

PI of $y(t) = c$

Since $y(t)$ is a constant, $\dot{y} = 0$ and $\ddot{y} = 0$.

$\therefore$ $$\omega_n^2 c = u_0$$

or $$c = \frac{u_0}{\omega_n^2}$$

$\therefore$ $$y(t) = Ae^{s_1 t} + Be^{s_2 t} + \frac{u_0}{\omega_n^2}$$

Applying the initial conditions, we have

$$y(0) = y_0 = A + B + \frac{u_0}{\omega_n^2}$$

$$\dot{y}(0) = \dot{y}_0 = s_1 A + s_2 B$$

Solving for A and B, we have

$$A = \frac{y_0\omega_n^2 s_2 - u_0 s_2 - \dot{y}_0\omega_n^2}{\omega_n^2\left(s_2 - s_1\right)}$$

$$B = \frac{y_0\omega_n^2 s_1 - u_0 s_1 - \dot{y}_0\omega_n^2}{\omega_n^2\left(s_1 - s_2\right)}$$

∴ Complete solution becomes,

$$y(t) = \left[\frac{y_0\omega_n^2 s_2 - u_0 s_2 - \dot{y}_0\omega_n^2}{\omega_n^2\left(s_2 - s_1\right)}\right]e^{s_1 t} + \left[\frac{y_0\omega_n^2 s_1 - u_0 s_1 - \dot{y}_0\omega_n^2}{\omega_n^2\left(s_1 - s_2\right)}\right]e^{s_2 t} + \frac{u_0}{\omega_n^2} \tag{18.10}$$

If $y(0) = y_0 = 0$ and $\dot{y}(0) = \dot{y}_0 = 0$, for unit step, we have

$$A = \frac{s_2}{\omega_n^2\left(s_1 - s_2\right)},\quad B = \frac{-s_1}{\omega_n^2\left(s_1 - s_2\right)}$$

$$y(t) = \frac{1}{\omega_n^2}\left[1 + s_x e^{s_1 t} + s_y e^{s_2 t}\right]$$

where $\quad s_x = \dfrac{s_2}{s_1 - s_2},\quad s_y = \dfrac{s_1}{s_2 - s_1}$

The output response for $\zeta \geq 1$ is shown in Fig.18.15.

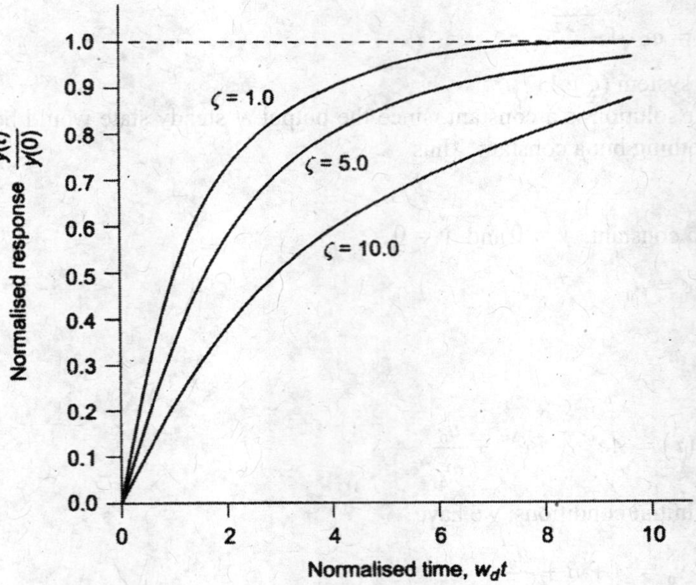

Fig.18.15 Output response of second order system (overdamped situation: ($\zeta > 1$))

(b) Critically damped system. ($\zeta = 1$)

$$s_1 = s_2 = s = -\omega_n$$

$$\therefore \quad y(t) = (A + Bt)e^{st} + \frac{u_0}{\omega_n^2}$$

$$\dot{y}(t) = (A + Bt)se^{st} + Be^{st}$$

$$y(0) = y_0 = A + \frac{u_0}{\omega_n^2} \rightarrow A = y_0 - \frac{u_0}{\omega_n^2} = \frac{y_0\omega_n^2 - u_0}{\omega_n^2}$$

$$\dot{y}(0) = \dot{y}_0 = As + B$$

$$B = \dot{y}_0 - s\left(y_0 - \frac{u_0}{\omega_n^2}\right) = \dot{y}_0 + y_0\omega_n - \frac{u_0}{\omega_n}$$

$$= \frac{\dot{y}_0\omega_n + y_0\omega_n^2 - u_0}{\omega_n}$$

$$y(t) = \left[\left\{\frac{y_0\omega_n^2 - u_0}{\omega_n^2}\right\} + \left\{\frac{\dot{y}_0\omega_n + y_0\omega_n^2 - u_0}{\omega_n}\right\}t\right]e^{-\omega_n t} + \frac{u_0}{\omega_n^2}$$

$$= \frac{1}{\omega_n^2}\left[\left[\left\{y_0\omega_n^2 - u_0\right\} + \left\{\dot{y}_0\omega_n^2 + y_0\omega_n^3 - u_0\omega_n\right\}t\right]e^{-\omega_n t} + u_0\right] \tag{18.11}$$

For $y(0) = 0, \dot{y}(0) = 0$ and $u_0 = 1$

$$y(t) = \frac{1}{\omega_n^2}\left[\left\{-1 - \omega_n t\right\}e^{-\omega_n t} + 1\right]$$

$$= \frac{1}{\omega_n^2}\left[1 - e^{-\omega_n t} - \omega_n t\, e^{-\omega_n t}\right]$$

(c) Underdamped system ($\zeta < 1$)

$$y(t) = e^{-\zeta\omega_n t}\left[A\cos\omega_d t + B\sin\omega_d t\right] + \frac{u_0}{\omega_n^2}$$

$$\dot{y}(t) = e^{-\zeta\omega_n t}.\omega_d\left[-A\sin\omega_d t + B\cos\omega_d t\right] - \zeta\omega_n e^{-\zeta\omega_n t}\left[A\cos\omega_d t + B\sin\omega_d t\right] \tag{18.12}$$

$$y(0) = y_0 = A + \frac{u_0}{\omega_n^2}$$

$$\therefore \quad A = y_0 - \frac{u_0}{\omega_n^2}$$

$$\dot{y}(0) = \dot{y}_0 = \omega_d[\mathrm{B}] - \zeta\omega_n A$$

$$B = \left[\frac{1}{\omega_d}\dot{y}_0 + \zeta\omega_n(y_0 - \frac{u_0)}{\omega_n^2}\right]$$

$$= \frac{1}{\omega_d}\left[\dot{y}_0 + y_0\zeta\omega_n - \frac{u_0\zeta}{\omega_n}\right]$$

$$= \frac{1}{\omega_d}\left[\frac{\dot{y}_0\omega_n + y_0\zeta\omega_n^2 - u_0\zeta}{\omega_n}\right]$$

$$= \frac{\dot{y}_0\omega_n + y_0\zeta\omega_n^2 - u_0\zeta}{\omega_n^2\sqrt{1-\zeta^2}}$$

For $y(0) = \dot{y}(0) = 0$ and $u_o = 1$, we have

$$y(t) = e^{-\zeta\omega_n t}\left[-\frac{1}{\omega_n^2}\cos\omega_d t - \frac{\zeta}{\omega_n^2\sqrt{1-\zeta^2}}\sin\omega_d t\right] + \frac{1}{\omega_n^2}$$

$$= \frac{-e^{-\zeta\omega_n t}}{\omega_n^2}\left[\cos\omega_d t + \frac{\zeta}{\sqrt{1-\zeta^2}}\sin\omega_d t\right] + \frac{1}{\omega_n^2}$$

$$= \frac{1}{\omega_n^2}\left[1 - e^{-\zeta\omega_n t}\left\{\cos\omega_d t + \frac{\zeta}{\sqrt{1-\zeta^2}}\sin\omega_d t\right\}\right]$$

$$= \frac{1}{\omega_n^2}\left[1 - \frac{e^{-\zeta\omega_n t}}{\sqrt{1-\zeta^2}}\cos(\omega_d t - \phi)\right]$$

where $\phi = \tan^{-1}\dfrac{\zeta}{\sqrt{1-\zeta^2}}$

The output response is shown in Fig.18.16.

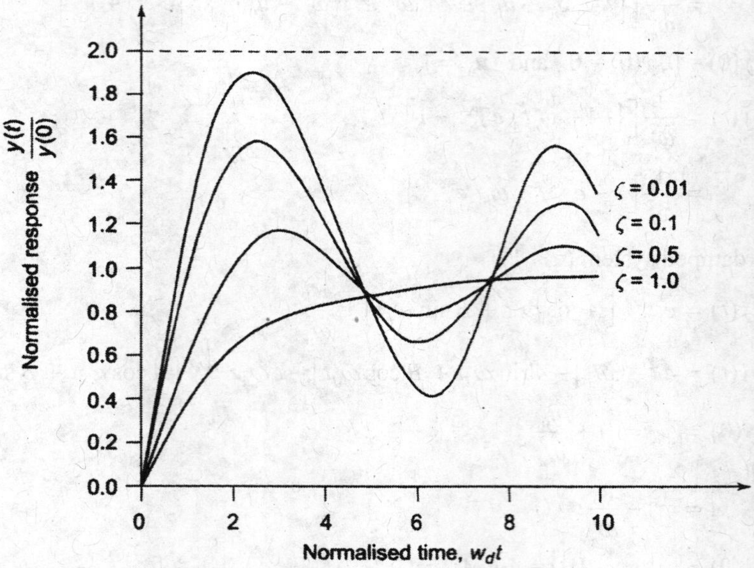

Fig.18.16 Output response of second order system (underdamped situation: (ζ < 1))

18.10.3 Sinusoidal Input

The normalized equation for sinusoidal input to second order system is:

$$\ddot{y} + 2\zeta\omega_n\dot{y} + \omega_n^2 y = u_o\cos\omega t$$

with initial conditions: $y(o) = y_o$ and $\dot{y}(o) = \dot{y}_o$

If the roots of the characteristic equation are distinct and real, then its complete solution is,

$$y(t) = e^{-\zeta\omega_n t}\left[A\cos\omega_d t + B\sin\omega_d t\right] + PI$$

$$PI = \frac{u_0\cos\omega\text{t}}{D^2 + 2\zeta\omega_n D + \omega_n^2}$$

$$= \frac{u_0\cos\omega t}{-\omega^2 + 2\zeta\omega_n D + \omega_n^2}$$

$$= \frac{\left[(\omega_n^2 - \omega^2) - 2\zeta\omega_n\Delta\right]u_0\cos\omega t}{(\omega_n^2 - \omega^2)^2 - (2\zeta\omega_n)^2 D^2}$$

$$= \frac{(\omega_n^2 - \omega^2)u_0\cos\omega t + 2\zeta\omega_n u_0\omega\sin\omega t}{(\omega_n^2 - \omega^2)^2 + (2\zeta\omega_n)^2\omega^2}$$

$$= \frac{(\omega_n^2 - \omega^2)}{(\omega_n^2 - \omega^2)^2 + 4\zeta^2\omega_n^2\omega^2}\cdot u_0\cos\omega t + \frac{2\zeta\omega_n\omega}{(\omega_n^2 - \omega^2)^2 + 4\zeta^2\omega_n^2\omega^2}\cdot u_0\sin\omega t$$

$$y(0) = y_0 = A + \frac{(\omega_n^2 - \omega^2)u_0}{(\omega_n^2 - \omega^2)^2 + 4\zeta^2\omega_n^2\omega^2}$$

$$= A + \frac{(1-\beta^2)u_0}{\omega_n^2\left[(1-\beta^2)^2 + (2\zeta\beta)^2\right]}$$

$$A = y_0 - \frac{(1-\beta^2)u_0}{\omega_n^2 a^2}$$

where $a = \sqrt{(1-\beta^2)^2 + (2\zeta\beta)^2}$

$$\beta = \frac{\omega}{\omega_n}$$

$$\dot{y}(t) = e^{-\zeta\omega_n t}\omega d[-A\sin\omega_d t + B\cos\omega_d t] - \zeta\omega_n e^{-\zeta\omega_n t}[A\cos\omega_d t + B\sin\omega_d t]$$

$$+ \frac{(1-\beta^2)}{\omega_n^2 a^2}(-u_0\omega\sin\omega t) + \frac{2\zeta\omega_n\omega^2(u_0\cos\omega t)}{\omega_n^4 a^2}$$

$$\dot{y}(0) = \dot{y}_0 = \omega_d B - \zeta\omega_n A + \frac{2\zeta\omega_n\omega^2 u_0}{\omega_n^4 a^2}$$

$$= \omega_d B - \zeta\omega_n\left[y_0 - \frac{(1-\beta^2)u_0}{\omega_n^2 a^2}\right] + \frac{2\zeta\omega_n\beta^2 u_0}{\omega_n^2 a^2}$$

$$B = \frac{1}{\omega_d}\left[\dot{y}_0 + \zeta\omega_n\left\{y_0 - \frac{(1-\beta^2)u_0}{\omega_n^2 a^2}\right\} - \frac{2\zeta\omega_n\beta^2 u_0}{\omega_n^2 a^2}\right]$$

$$= \frac{\omega_n^2 a^2(\dot{y}_0 + y_0\omega_n\zeta) - \omega_n\zeta u_0(1+\beta^2)}{\omega_n^2 a^2\omega_d}$$

The complete solution becomes,

$$y(t) = e^{-\zeta\omega_n t}\left[\left\{y_o - \frac{(1-\beta^2)u_o}{\omega_n^2 a^2}\right\}\cos\omega_d t + \left\{\frac{\omega_n^2 a^2(\dot{y}_o + y_o\omega_n\zeta) - \omega_n\zeta u_o(1+\beta)^2}{\omega_n^2 a^2\omega_d}\right\}\sin\omega_d t\right]$$

$$+\frac{u_o}{\omega_n^2}\left[\frac{\cos(\omega t-\phi)}{\sqrt{(1-\beta^2)^2+(2\zeta\beta)^2}}\right] \quad (18.13)$$

where $\phi = \tan^{-1}\left[\frac{2\zeta\beta}{1-\beta^2}\right]$

For $y(o) = \dot{y}(o) = 0$ and $u_o = 1$, we have

$$y(t) = \frac{e^{-\zeta\omega_n t}}{\omega_n^2 a^2}\left[-(1-\beta^2)\cos\omega_d t - \frac{\omega_n\zeta(1+\beta^2)}{\omega_d}\sin\omega_d t\right] + \frac{1}{\omega_n^2}\left[\frac{\cos(\omega t-\phi)}{\sqrt{(1-\beta^2)^2+(2\zeta\beta)^2}}\right]$$

18.11 SECOND ORDER TORSIONAL SYSTEMS

Consider a second order torsional system whose block diagram is shown in Fig.18.17. First the input signed θ_i is compared with the output signal θ_0. Then the difference $e = \theta_i - \theta_0$ is passed on to the motor which produces an output torque T proportional to e (or $= qe$). The system has a viscous resistance with damping coefficient c.

E = error detector, M = motor, T = transducer, L = load, D = damper

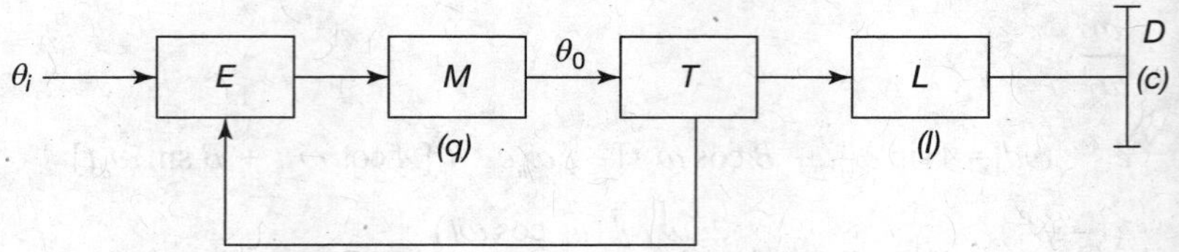

Fig.18.17 Block diagram for second order torsional system

Let I = combined moment of inertia of motor and load
The equation of motion is:

$$I\,\ddot{\theta}_0 + c\,\dot{\theta}_0 + q\theta_0 = q\theta_i$$

or
$$\ddot{\theta}_0 + \frac{c}{I}\dot{\theta}_0 + \frac{q}{I}\theta_0 = \frac{q}{I}\theta_i$$

or
$$\ddot{\theta}_0 + 2\zeta\omega_n\,\dot{\theta}_0 + \omega_n^2\theta_0 = \omega_n^2\theta_i$$

where
$$\zeta = \frac{c}{c_c}, \quad c = 2\zeta I\omega_n$$

The response of the system shall be considered for the following type of input:

(a) Step displacement

(b) Ramp displacement

(c) Harmonic signal

18.11.1 Step Displacement Input

For step input,

$\theta = 0 \quad \text{for} \quad t \le 0$

$\theta = \theta_i \quad \text{for} \quad t > 0$

The complementary equation is,

$$\ddot{\theta}_0 + 2\zeta\omega_n\dot{\theta}_0 + \omega_n^2\theta_0 = 0$$

(a) Under damped system ($\zeta < 1$)

The complementary solution is

$$\left[\theta_0(t)\right]_c = X\, e^{-\zeta\omega_n t} \sin\left(\omega_d t + \phi\right)$$

Particular solution is,

$$PI = \frac{\omega_n^2\, \theta_i}{D^2 + 2\zeta\omega_n\, D + \omega_n^2}$$

As θ_i is constant, $\dot{\theta}_i = 0$

In the steady state, $\dot{\theta}_o = \dot{\theta}_i = 0$

Also $\ddot{\theta}_i = 0$. Hence

$$PI = \frac{\omega_n^2\, \theta_i}{\omega_n^2} = \theta_i$$

The complete solution becomes

$$\theta_0(t) = \theta_i + X\, e^{-\zeta\omega_n t} \sin\left(\omega_d t + \phi\right)$$

Now $\theta_0(0) = 0$

$0 = \theta_i + X \sin\phi$

$$X = \frac{-\theta_i}{\sin\phi}$$

Also $\quad \dot{\theta}_0(0) = 0$

Now $\quad \dot{\theta}_0(t) = \dot{\theta}_i + X\omega_d\, e^{-\zeta\omega_n t} \cos\left(\omega_d t + \phi\right)$

$$-\zeta\omega_n\, X\, e^{-\zeta\omega_n t} \sin\left(\omega_d t + \phi\right)$$

$$0 = 0 + X\,\omega_d \cos\phi - \zeta\omega_n\, X \sin\phi$$

$$\tan\phi = \frac{\omega_d}{\zeta\omega_n}$$

$$\sin\phi = \frac{\omega_d}{\sqrt{\omega_d^2 + \zeta^2\omega_n^2}}$$

$$= \frac{\omega_d}{\sqrt{\omega_n^2\left(1-\zeta^2\right) + \zeta^2\omega_n^2}} = \frac{\omega_d}{\omega_n}$$

$$X = \frac{\theta_i\omega_n}{\omega_d}$$

$$\therefore \quad \theta_0\left(t\right) = \theta_i - \frac{\theta_i\omega_n}{\omega_d}\, e^{-\zeta\omega_n t}\sin\left[\omega_d t + \tan^{-1}\left(\frac{\omega_d}{\zeta\omega_n}\right)\right]$$

$$= \theta_i\left[1 - \frac{1}{\sqrt{1-\zeta^2}}\, e^{-\zeta\omega_n t}\sin\left[\omega_n\sqrt{1-\zeta^2}\,t + \tan^{-1}\left\{\frac{\sqrt{1-\zeta^2}}{\zeta}\right\}\right]\right] \tag{18.14a}$$

For unit step function input, $\theta_i = 1$

$$\therefore \quad \theta_0\left(t\right) = 1 - \frac{1}{\sqrt{1-\zeta^2}}\, e^{-\zeta\omega_n t}\sin\left[\omega_n\sqrt{1-\zeta^2}\,t + \tan^{-1}\left\{\frac{\sqrt{1-\zeta^2}}{\zeta}\right\}\right] \tag{18.14b}$$

The responses for $\zeta = 1$ and $\zeta > 1$ can be obtained in the similar way as explained in previous articles. The plots of θ_0 v's t have been shown in Fig.18.18 for various values of ζ. The salient observations are:

1. If ζ is increased, the frequency of oscillation is reduced.
2. The quickest response with no over shoot is at $\zeta = 1$.
3. For $\zeta \approx 0.6$, the response is even faster with very little over shoot.
4. There is no steady state lag (or error).
5. The steady state is reached quickly if ω_n is large, *i.e.* q is more and I is less.
6. The dynamic error between the response and the input gradually reduces. However, an acceptable error band or a tolerable zone, as shown in Fig.18.19 is prescribed for the system.

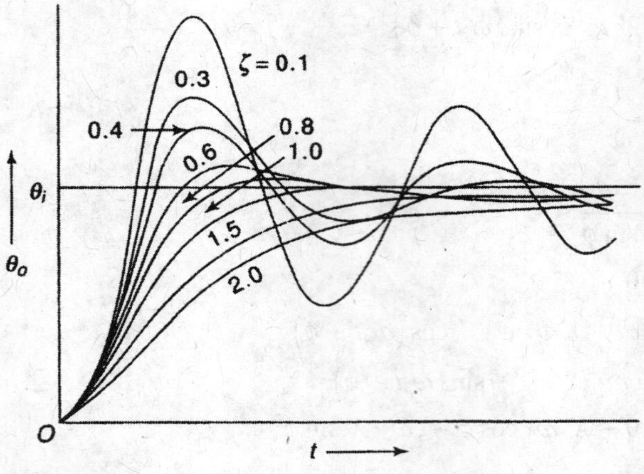

Fig.18.18 Output v's time for second order torsional system with step displacement input

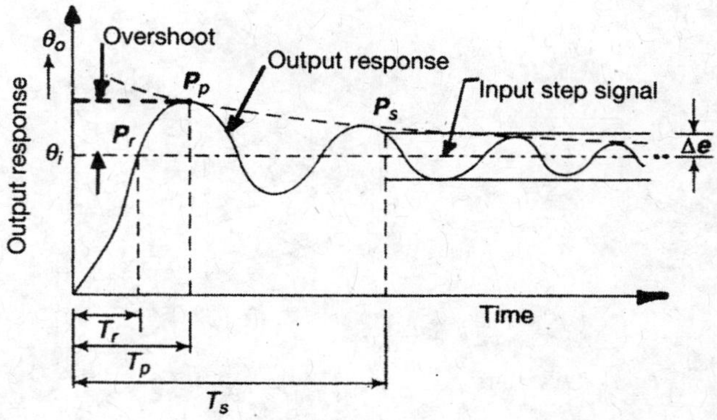

Fig.18.19 Transient response of a torsionel second-order system with step input

Rise Time (T_r)

It is the time taken by the output to be equal to the input for the first time. It corresponds to point P_r in Fig.18.19. For output to be equal to input,

$$1 = 1 - \frac{1}{\sqrt{1-\zeta^2}} e^{-\zeta\omega_n T_r} \sin\left[\sqrt{1-\zeta^2}\,\omega_n\, T_r + \tan^{-1}\frac{\sqrt{1-\zeta^2}}{\zeta}\right]$$

$$\text{or} \quad \frac{1}{\sqrt{1-\zeta^2}} e^{-\zeta\omega_n Tr} \sin\left[\sqrt{1-\zeta^2}\;\omega_n T_r + \tan^{-1}\frac{\sqrt{1-\zeta^2}}{\zeta}\right] = 0$$

This is possible if the term inside the bracket is either zero or π. It cannot be zero because in that case $T_r = 0$, which is not possible.

$$\text{Hence} \quad \sqrt{1-\zeta^2}\;\omega_n T_r + \tan^{-1}\frac{\sqrt{1-\zeta^2}}{\zeta} = \pi$$

$$\text{or} \quad T_r = \frac{\pi - \tan^{-1}\left(\frac{\sqrt{1-\zeta^2}}{\zeta}\right)}{\sqrt{1-\zeta^2}\;\omega_n} \qquad (18.15)$$

Peak Error (Overshoot)

It is the deviation of amplitude of the output from the input. The first peak overshoot occurs at point P_p, as shown in Fig.18.19. The time taken to reach the peek point is given by,

$$\text{Peak time, } T_p = \frac{\pi}{\omega_d} = \frac{\pi}{\omega_n\sqrt{1-\zeta^2}} \qquad (18.16)$$

$$\text{and the peak error, } \Delta e = \frac{1}{\sqrt{1-\zeta^2}} e^{-\zeta\omega_n T_p} \sin\left(\sqrt{1-\zeta^2}\;\omega_n T_p + \phi\right)$$

$$= \frac{1}{\sqrt{1-\zeta^2}} e^{-\zeta\omega_n T_p} \sin(\pi + \phi)$$

$$= \frac{1}{\sqrt{1-\zeta^2}} e^{-\zeta\omega_n T_p} \sqrt{1-\zeta^2}$$

$$= e^{\frac{-\zeta\omega_n \pi}{\omega_n\sqrt{1-\zeta^2}}}$$

$$= e^{-\frac{\zeta\pi}{\sqrt{1-\zeta^2}}}$$

Percentage error $= e^{\frac{-\zeta\pi}{\sqrt{1-\zeta^2}}} \times 100$ (18.17)

Settling Time (T_s):

It is the time taken for the output to reach and remain within a zone of specified percent around the final steady state value. The usual value is within 2.5 percent. If the specified value of error is Δe then after a time T_s the amplitude is to be Δe corresponding to point P_s.

$$\frac{1}{\sqrt{1-\zeta^2}} e^{-\zeta\omega_n T_s} = \Delta e$$

or $$e^{-\zeta\omega_n T_s} = \sqrt{1-\zeta^2} . \Delta e$$

or $$e^{\zeta\omega_n T_s} = \frac{1}{\sqrt{1-\zeta^2 . \Delta e}}$$

$$\zeta\omega_n T_s = ln\left[\frac{1}{\Delta e\sqrt{1-\zeta^2}}\right]$$

$$T_s = \frac{1}{\zeta\omega_n} ln\left[\frac{1}{\Delta e\sqrt{1-\zeta^2}}\right] \quad (18.18)$$

Example 18.3

A measurement system consists of an effective mass of 50 g and a spring constant 2 kN/m. Find (a) the natural frequency of oscillations of the system, (b) the damping constant for critical damping, and (c) the damping constant for 25% overshoot and corresponding period of oscillation.

■ Solution

(a) $\omega_n = \sqrt{\frac{k}{m}} = \sqrt{\frac{2\times 10^3}{50\times 10^{-3}}} = 200$ rad/s

(b) $c_c = 2\sqrt{km} = 2\sqrt{2\times 10^3 \times 50\times 10^{-3}} = 20$ N s/m

(c) For 25 % overshoot,

$$0.25 = \exp\left[\frac{-\pi\zeta}{\sqrt{1-\zeta^2}}\right]$$

or $$\exp\left(\frac{\pi\zeta}{\sqrt{1-\zeta^2}}\right) = 4$$

$$\frac{\pi\zeta}{\sqrt{1-\zeta 2}} = \ell_n 4 = 1.3863$$

$$\pi^2 \zeta^2 = 1.922\,(1 - \zeta^2)$$

$$11.7914\,\zeta^2 = 1.922$$

$$\zeta = 0.404$$

$$c = c_c\,\zeta = 20 \times 0.404 = 8.08 \text{ N.s/m}$$

Period of oscillation, $$T_s = \frac{2\pi}{\omega_n\sqrt{1-\zeta^2}}$$

$$= \frac{2\pi}{200\sqrt{1-(0.404)^2}}$$

$$= 0.034\ s$$

Example 18.4

A second order measurement system has an effective mass of 1.5 kg, the spring stiffness 50 kN/m and damping factor 0.45. Assuming a unit step input, find (a) the rise time, and (b) the settling time for a tolerance band of 2%.

Solution

$$\omega_n = \sqrt{\frac{k}{m}} = \sqrt{\frac{50\times 10^3}{1.5}} = 182.57 \text{ rad/s}$$

(a) Rise time, $$T_r = \frac{\pi - \tan^{-1}\left[\frac{\sqrt{1-\zeta^2}}{\zeta}\right]}{\omega_n\sqrt{1-\zeta^2}}$$

$$= \frac{\pi - \tan^{-1}\left[\frac{\sqrt{1-(0.45)^2}}{0.45}\right]}{182.57\sqrt{1-(0.45)^2}}$$

$$= \frac{\pi - 1.104}{163.04} = 0.0125 \text{ s}$$

(b) Settling time, $$T_s = \frac{1}{\zeta\omega_n}\ln\left[\frac{1}{\Delta e\sqrt{1-\zeta^2}}\right]$$

$$= \frac{1}{0.45\times 182.57}\ln\left[\frac{1}{0.02\sqrt{1-(0.45)^2}}\right]$$

$$= 0.049 \text{ s}$$

18.11.2 Ramp Displacement Input

Now $\theta_i = \omega_i t$

or $\dot{\theta}_i = \omega_i$

The equation of motion becomes:

$$\ddot{\theta}_0 + 2\zeta\omega_n\dot{\theta}_0 + \omega_n^2\,\theta_0 = \omega_n^2\,\omega_i\,t$$

The transient solution is the same as for the step displacement input. The particular solution is

$$q_p(t) = \frac{w_n^2\,w_i\,t}{D^2 + 2zw_n\,D + w_n^2}$$

In the steady state, $\dot{\theta}_o = \dot{\theta}_i = \omega_i$

and $\ddot{\theta}_0 = 0$

$$\therefore \quad q_p(t) = \frac{w_n^2\,w_i\,t}{2zw_n\,D + w_n^2}$$

$$= \frac{w_i\,t}{1 + \dfrac{2zD}{w_n}}$$

$$= \left[1 + \frac{2z}{w_n}D\right]^{-1} w_i t$$

$$= \left[1 - \frac{2z}{w_n}D + \frac{(-1)(-2)}{2}\left(\frac{2z}{w_n}\right)^2 D^2 + \cdots\right] w_i t$$

$$= w_i t - \frac{2z}{w_n}w_i$$

$$= q_i - \frac{2z}{w_n}w_i$$

Thus, the steedy state speeds are equal or there is no error in velocity. But the output θ_0 lags the input by $\dfrac{2\zeta\omega_i}{\omega_n}$.

$$\therefore \quad \theta_i - \theta_0 = \frac{2\zeta\omega_i}{\omega_n} = \frac{2c}{c_c}\frac{\omega_i}{\sqrt{\dfrac{q}{I}}} = \frac{2c\omega_i}{2\sqrt{Iq}\sqrt{\dfrac{q}{I}}} = \frac{c\omega_i}{q}$$

For unit ramp input, $\theta_i - \theta_0 = \dfrac{c}{q}$

$$\text{As } \theta_p = t - \frac{2\zeta}{\omega_n}$$

$$\text{Time lag} = \frac{2\zeta}{\omega_n} \qquad (18.19)$$

For unit ramp input, *i.e.* $\omega_i = 1$

$$\theta_p = t - \frac{2\zeta}{\omega_n} \tag{18.20}$$

The plot of θ_0 v's t is shown in Fig.18.20.

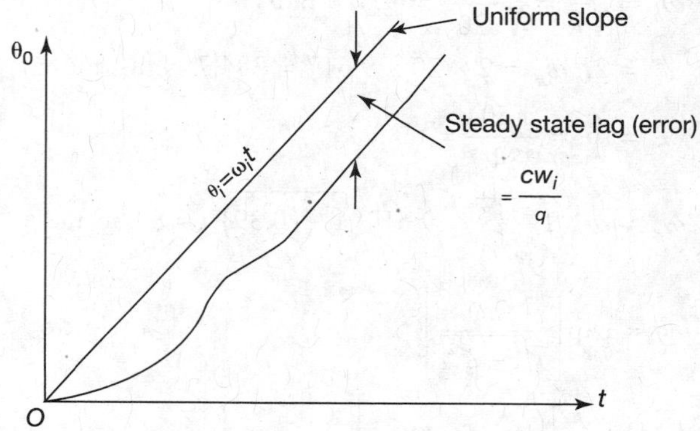

Fig.18.20 θ_0 v'st for ramp displacement input

Example 18.5

A second order measurement system has an effective mass of 1.5 kg and spring stiffness 50 kN/m. Assuming a unit ramp input, find the steady state error and time lag for (a) a critically damped system, and (b) a damping system with $\zeta = 0.45$.

■ Solution

$$\omega_n = \sqrt{\frac{k}{m}} = \sqrt{\frac{50 \times 10^3}{1.5}} = 182.57 \text{ rad/s}$$

(a) For a unit ramp input and critical damping,

Steady state error $= \dfrac{2\zeta}{\omega_n} = \dfrac{2 \times 1}{182.57} = 0.010955$

Now $\omega_i = 1$

Time lag $= 0.010955\ s$

(b) For $\zeta = 0.45$

Steady state error $= \dfrac{2\zeta}{\omega_n} = \dfrac{2 \times 0.45}{182.57} = 0.00493$

Time lag $= 0.00493\ s$

Example 18.6

A servomechanism consists of a system with inertia of 20 kg.m^2 and damping coefficient 150 N.m.s/rad. The torsional stiffness of the controller is 1850 N.m/rad. Determine the damped frequency of oscillation, the peak output and the percentage peak overshoot if the system is suddenly displaced through an angle of 45°.

What will be the steady state position error if the input shaft is rotated at a constant speed of 25 rpm?

■ **Solution**

$$\omega_n = \sqrt{\frac{q}{I}} = \sqrt{\frac{1850}{20}} = 9.617 \text{ rad/s}$$

$$c_c = 2I\,\omega_n = 2 \times 20 \times 9.617 = 384.7 \text{ N.m.s/rad}$$

$$\zeta = \frac{c}{c_c} = \frac{150}{384.7} = 0.39$$

$$\omega_d = \omega_n\sqrt{1-\zeta^2} = 9.617\sqrt{1-(0.39)^2} = 8.856 \text{ rad/s}$$

Peak over shoot $= \exp\left[\dfrac{-\pi\zeta}{\sqrt{1-\zeta^2}}\right]$

$$= \exp\left[\frac{-\pi \times 0.39}{\sqrt{1-(0.39)^2}}\right] = 0.2643 \text{ or } 26.43\%$$

Peak output = Input + peak overshoot
= 45° + 45 × 0.2643 = 56.89°

For $N_i = 25$ rpm,

$$\omega_i = \frac{2\pi \times 25}{60} = 2.618 \text{ rad/s}$$

Steady state position error $= \dfrac{c\omega_i}{q}$

$$= \frac{150 \times 2.618}{1850} = 0.2123 \text{ rad}$$

$$= \frac{0.2123 \times 180}{\pi} = 12.16°$$

18.11.3 Harmonic Input

For a harmonic (or sinusoidal) input, the equation of motion becomes

$$\ddot{\theta}_0 + 2\zeta\omega_n\dot{\theta}_0 + \omega_n^2\theta_0 = \omega_n^2\theta_i \cos\omega_i t$$

Its steady state solution is

$$\frac{\theta_0}{\theta_i} = \frac{1}{\left[\left(1-\beta^2\right)^2 + \left(2\zeta\beta\right)^2\right]^{1/2}} \tag{18.21}$$

Dynamic error, $$\frac{\theta_i - \theta_0}{\theta_i} = 1 - \frac{\theta_0}{\theta_i} \tag{18.22}$$

Phase lag, $$\phi = \frac{2\zeta\beta}{1-\beta^2} \tag{18.23}$$

where $$\beta = \frac{\omega_i}{\omega_n},\ \zeta = \frac{c}{c_c}$$

Example 18.7

For a second order measurement system with a damping factor 0.42, the natural frequency of oscillation is 25 Hz. Determine the dynamic error and time lag at frequency input of 10 Hz. Also find these values at resonance.

■ **Solution**

Here $$\zeta = 0.42,\ \beta = \frac{\omega_i}{\omega_n} = \frac{10}{25} = 0.4$$

$$\frac{\theta_0}{\theta_i} = \frac{1}{\left[\left(1-\beta^2\right)^2 + \left(2\zeta\beta\right)^2\right]^{1/2}}$$

$$= \frac{1}{\left[\left\{1-(0.4)^2\right\}^2 + (2\times 0.42\times 0.4)^2\right]^{1/2}}$$

$$= 1.105$$

Dynamic error $$= 1 - \frac{\theta_0}{\theta_i} = 1 - 1.105 = -0.105$$

$$\phi = \tan^{-1}\left[\frac{2\zeta\beta}{1-\beta^2}\right]$$

$$= \tan^{-1}\left[\frac{2\times 0.42\times 0.4}{1-(0.4)^2}\right] = \tan^{-1} 0.4 = 21.8°$$

Time lag $$= \frac{\phi}{2\pi\omega_i} = \frac{21.8\times\pi}{180}\times\frac{1}{2\pi\times 10} = 0.006056 \text{ s}$$

At resonance, $\beta = 1$

$$\frac{\theta_0}{\theta_i} = \frac{1}{2\zeta} = \frac{1}{2\times 0.42} = 1.190$$

Dynamic error = 1 – 1.190 = – 0.190

$\phi = \tan^{-1}\infty$, $\phi = 90°$

Time lag $= \dfrac{\pi/2}{2\pi \times 25} = 0.01$ s

18.11.4 Step Velocity Input with Error Rate Damping (Derivative Control)

If the controller provides a proportional plus derivative action of the error, then equation of motion becomes

$$I\ddot{\theta}_0 + c\dot{\theta}_0 = qe + q_i\dot{e}$$

$$= q(\theta_i - \theta_0) + q_i(\dot{\theta}_i - \dot{\theta}_0)$$

or $$I\ddot{\theta}_0 + (c + q_i)\dot{\theta}_0 + q\theta_0 = q\theta_i + q_i\dot{\theta}_i$$

or $$\ddot{\theta}_0 + \left(\frac{c + q_i}{I}\right)\dot{\theta}_0 + \frac{q}{I}\theta_0 = \frac{q}{I}\theta_i + \frac{q_i}{I}\dot{\theta}_i$$

Comparing with standard normalized equation,

$$\ddot{\theta}_0 + 2\zeta\omega_n\,\dot{\theta}_0 + \omega_n^2\,\theta_0 = \omega_n^2\,\omega_i\,t$$

we have $$\omega_n = \sqrt{\frac{q}{I}}$$

$$2\zeta\omega_n = \frac{c + q_i}{I}$$

or $$\zeta = \frac{c + q_i}{2\omega_n I} = \frac{c + q_i}{2I\sqrt{\dfrac{q}{I}}} = \frac{c + q_i}{2\sqrt{q\,I}} = \frac{c + q_i}{c_c}$$

For the steady state, $\dot{\theta}_i = \dot{\theta}_0 = \omega_i$ and $\ddot{\theta}_0 = 0$

$\therefore$ $$(\theta_i - \theta_0)\frac{q}{I} = \omega_i\left[\frac{c + q_i - q_i}{I}\right] = \frac{C\omega_i}{I}$$

Steady state error, $$\theta_i - \theta_0 = \frac{c\omega_i}{q} \qquad (18.24)$$

The following observations are made:

1. There is no change in the steady state error by using a derivative control.
2. Damping ratio increases from $\dfrac{c}{c_c}$ to $\dfrac{c + q_i}{c_c}$.

This improves the transient response (overshoot) without the use of excessive damping torque on load.

18.11.5 Step Velocity Input with Integral Control

The equation of motion is:

$$I\ddot{\theta}_0 + c\dot{\theta}_0 = qe + q_i \int_0^t e\mathrm{d}t$$

$$= q(\theta_i - \theta_0) + q_i \left[\int_0^t \theta_i \,\mathrm{d}t - \int_0^t \theta_0 \,\mathrm{d}t\right]$$

Differentiating, we have

$$I\dddot{\theta}_0 + c\ddot{\theta}_0 + q\dot{\theta}_0 + q_i\,\theta_0 = q\,\dot{\theta}_i + q_i\,\theta_i$$

For the steady state,

$$\dot{\theta}_i = \dot{\theta}_0 = \omega_i \text{ and } \ddot{\theta}_0 = \dddot{\theta}_0 = 0$$

$$q\,\omega_i + q_i\,\theta_0 = q\,\theta_i + q_i\,\omega_i$$

or

$$\theta_0 = \theta_i$$

Thus steady state error is zero.

18.12 TRANSFER FUNCTION METHOD

The transfer function is a mathematical model defining the input-output relation of the system. It is a complex function that must have arguments to define itself. The argument is simply frequency. If the system has a single input and a single output, it can be represented by a block diagram, as shown in Fig.18.21. The system response $x(t)$ is caused by an excitation $F(t)$. The transfer function is defined as the ratio of output over the input with all initial conditions equal to zero.

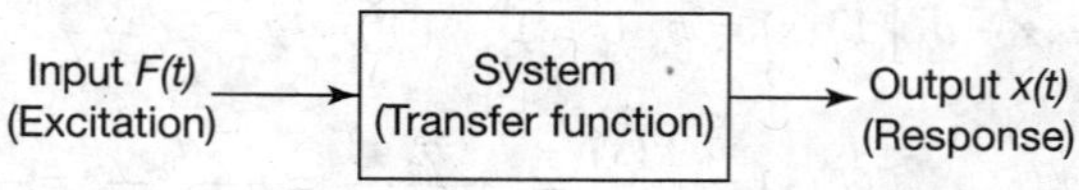

Fig.18.21 Block diagram of transfer function

Thus output $[x(t)]$ = [Transfer function] Input $[F(t)]$

or $$\frac{\text{Output}\,[x(t)]}{\text{Input}\,[F(t)]} = \text{Transfer function}$$

The equation of motion with a single degree freedom system with excitation is,

$$m\ddot{x} + c\dot{x} + kx = F(t)$$

Its transfer function is,

$$\frac{x}{F}(j\omega) = \frac{1}{(k - m\omega^2) + jc\omega t} = G(j\omega)$$

where $j = \sqrt{-1}$ and $G(j\omega)$ = transfer function of the system.

From the engineering point of view it is preferred to define the transfer function as a ratio of Laplace transform of the output to the Laplace transform of the input, assuming initial conditions are zero. Mathematically, it is written as:

$$G(s) = \frac{F(s)}{U(s)} \tag{18.25}$$

Where $G(s)$ = transfer function
$F(s)$ = Laplace transform of output
$U(s)$ = Laplace transform of input

One important point to be noted is that the transfer function does not depend upon the initial conditions.

18.12.1 Transfer Function of First Order Systems

The first order system is defined by,

$$a_1\dot{y} + a_0 y = b_0\, u(t)$$

Where the coefficients a_1, a_0, and b_0 are constants.

$u(t)$ and $y\,(t)$ are input and output respectively. $\dot{y}$ is the first derivative of $y\,(t)$. Let the Laplace transform of output $y\,(t)$ is $Y\,(s)$ and that of the input $u\,(t)$ is $U\,(s)$. Taking the Laplace transform of both sides, we have

$$L\left[a_1\dot{y} + a_0 y\right] = L\left[b_0 u\,(t)\right]$$

$$a_1 s\,Y\,(s) + a_0\,Y\,(s) = b_0\,U\,(s)$$

$$\left[a_1 s + a_0\right] Y\,(s) = b_0\,U\,(s)$$

$$G\,(s) = \frac{Y(s)}{U(s)} = \frac{b_0}{a_1 s + a_0} = \frac{\frac{b_0}{a_0}}{\frac{a_1}{a_0}s + 1} = \frac{G}{\tau\,s + 1} \tag{18.26}$$

$$y\,(t) = L^{-1}\left[\frac{G}{\tau\,s + 1}U\,(s)\right]$$

where $\quad G = \dfrac{b_0}{a_0} \quad$ and $\quad \tau = \dfrac{a_1}{a_0}$

The block diagram of first order system showing transfer function is shown in Fig.18.22.

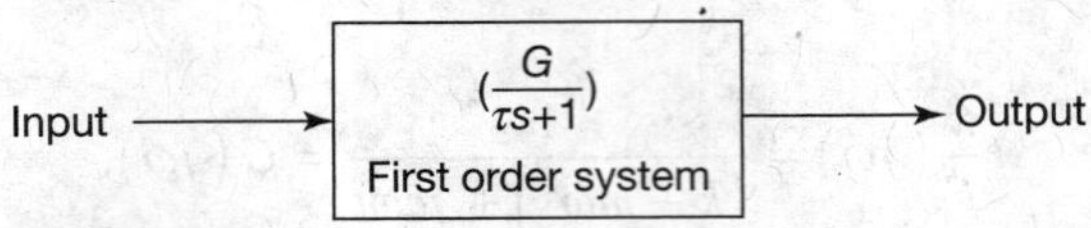

Fig.18.22 Block diagram of first order transfer function

18.12.2 Transfer Function of Second Order Systems

The equation of motion for a second order system is,

$$a_2\,\ddot{y} + a_1\,\dot{y} + a_0 y = b_0 u\,(t)$$

In the standard normalized form, the equation is

$$\ddot{y} + 2\zeta\,\omega_n\,\dot{y} + \omega_n^2 y = G\,u\,(t)$$

Taking Laplace transform of both sides,

$$L\left[\ddot{y} + 2\zeta\,\omega_n\,\dot{y} + \omega_n^2 y\right] = L\left[G\,u\,(t)\right]$$

$$s^2\,Y(s) + 2\zeta\,\omega_n sY(s) + \omega_n^2\,Y(s) = G\,U\,(s)$$

or

$$\left[s^2 + 2\zeta\,\omega_n\,s + \omega_n^2\right]Y(s) = G\,U(s)$$

$$G\,(s) = \frac{Y(s)}{U(s)} = \frac{G}{s^2 + 2\zeta\,\omega_n s + \omega_n^2} \qquad (18.27)$$

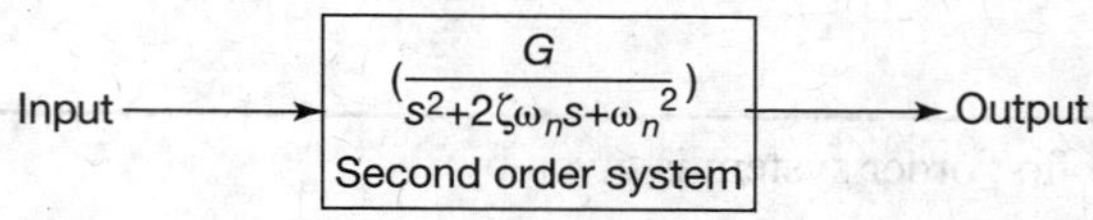

Fig.18.23 Block diagram of second order transfer function

18.12.3 Step Input to First Order System

If a step input u_0 is given to the first order system, the Laplace transform of output is:

$$Y(s) = \frac{Gu_0}{(\tau s + 1)}\left(\frac{1}{s}\right) = U\,\frac{\frac{1}{\tau}}{s\left(s + \frac{1}{\tau}\right)} = U\left[\frac{1}{s} - \frac{1}{s + \frac{1}{\tau}}\right]$$

Now $$L\left(\frac{u_0}{s}\right) = U = G\,u_0$$

Taking the inverse Laplace transform, we have

$$y\,(t) = L^{-1}\left[Y(s)\right] = L^{-1}\left[U\left\{\frac{1}{s} - \frac{1}{s + \frac{1}{\tau}}\right\}\right]$$

$$= U\,(1 - e^{\frac{-t}{\tau}}) \qquad (18.28)$$

18.12.4 Ramp Input to First Order System

If a ramp input u_0 is given to the first order system, the output can be written as,

$$F(s) = G(s)\,U(s)$$

$$Y(s) = \frac{Gu_0}{s^2(\tau s+1)} = U\left[\frac{1}{s^2} - \frac{\tau}{s} + \frac{\tau}{s+\frac{1}{\tau}}\right]$$

Now

$$L\left(\frac{u_0}{s^2}\right) = U = G\,u_0$$

Taking the inverse Laplace transform, we have

$$y(t) = L^{-1}[Y(s)] = L^{-1}\left[U\left\{\frac{1}{s^2} - \frac{\tau}{s} + \frac{\tau}{s+\frac{1}{\tau}}\right\}\right]$$

$$= U(t - \tau - e^{-\frac{t}{\tau}}) \tag{18.29}$$

Example 18.8

The transfer function of a first order system is given by:

$$G(s) = \frac{10}{s+2}$$

Determine the output response by using Laplace transform method if the following signals are applied at the input:

(a) A step signal of 5 units.

(b) A ramp signal with slope 5.

Solution

(a) $$L(5) = \frac{5}{s}$$

Output response, $$y(t) = L^{-1}[G(s)\,U(s)]$$

$$= L^{-1}\left[\frac{10}{s+2} \times \frac{5}{s}\right] = L^{-1}\left[\frac{25}{s} - \frac{25}{s+2}\right]$$

$$= 25 - 25e^{-2t}$$

(b) Laplace transform of ramp input with slope 5 is $\frac{5}{s^2}$.

$$y\ (t) = L^{-1}\left[G\ (s)\ U\ (s)\right] = L^{-1}\left[\frac{10}{s+2}\times\frac{5}{s^2}\right] = L^{-1}\left[\frac{50}{s^2(s+2)}\right]$$

$$= L^{-1}\left[\frac{25}{s^2} - \frac{12.5}{s} + \frac{12.5}{s+2}\right]$$

$$= 25\ t - 12.5 + 12.5\ e^{-2t}$$

18.12.5 Step Input to Second Order System

Now $$Y\ (s) = G\ (s)\ U\ (s) = \frac{G}{s^2 + 2\zeta\cdot\omega_n\ s + \omega_n^2}\left(\frac{u_0}{s}\right)$$

$$= Gu_0\ \frac{1}{s(s^2 + 2\zeta\ \omega_n s + \omega_n^2)}$$

By partial fraction,

$$\frac{1}{s(s^2 + 2\zeta\ \omega_n s + \omega_n^2)} = \frac{1}{s(s+x_1)(s+x_2)} = \frac{A}{s} + \frac{B}{s+x_1} + \frac{C}{s+x_2}$$

Where $$x_1 = \zeta\ \omega_n + \omega_n\sqrt{\zeta^2 - 1}$$

$$x_2 = \zeta\ \omega_n - \omega_n\sqrt{\zeta^2 - 1}$$

A, *B* and *C* are calculated as follows:

$$1 = \mathrm{A}\ (s + \mathrm{x}_1)\ (s + \mathrm{x}_2) + \mathrm{B}\ s\ (s + \mathrm{x}_2) + \mathrm{C}\ s\ (s + \mathrm{x}_1)\ s = 0\ s = -\mathrm{x}_1\ s = -\mathrm{x}_2$$

This gives, $$A = \frac{1}{x_1 x_2},\ B = \frac{1}{x_1(x_1 - x_2)}$$

and $$c = \frac{1}{x_2(x_2 - x_1)}$$

∴ $$Y\ (s) = G\ u_0\left\{\frac{1}{x_1 x_2 s} + \frac{1}{x_1(x_1 - x_2)(s+x_1)} + \frac{1}{x_2(x_2 - x_1)(s+x_2)}\right\}$$

$$= \frac{Gu_0}{x_1 x_2}\left\{\frac{1}{s} - \frac{x_2}{(x_2 - x_1)}\cdot\frac{1}{(s+x_1)} - \frac{x_1}{(x_1 - x_2)}\cdot\frac{1}{(s+x_2)}\right\}$$

Taking the inverse Laplace transform, we have

$$y\ (t) = L^{-1}\left[Y\ (s)\ \right] = \frac{Gu_0}{x_1 x_2}\left[1 - \frac{x_2}{(x_2 - x_1)}e^{-x_1 t} - \frac{x_1}{(x_1 - x_2)}e^{-x_2 t}\right] \qquad (18.30)$$

18.12.6 Ramp Input to Second Order System

If a ramp input u_0 t is given to the second order system, the output response is given by,

$$Y(s) = G(s)\,U(s) = \frac{G}{s^2 + 2\zeta\,\omega_n\,s + \omega_n^2} \cdot \frac{u_0}{s^2}$$

By following the steps of Section 18.12.5, the output response can be derived as:

$$y(t) = L^{-1}\left[Y(s)\right] = Gu_0\left[t - \frac{x_2}{x_1(x_2 - x_1)}e^{-x_1 t} + \frac{x_1}{x_2(x_1 - x_2)}e^{-x_2 t} - \frac{x_1 + x_2}{x_1\,x_2}\right] \quad (18.31)$$

18.13 FREQUENCY RESPONSE OF THE SYSTEM

Frequency response is defined as the ratio of steady state phasor output to the phasor input, where the output and input may either be a voltage or a current signal. Mathematically, the general form of frequency response is written as:

$$H(jw) = \frac{v_{\text{out}}}{v_{\text{in}}}$$

$H(j\omega)$ is referred to as the frequency response of the system and it is a complex function. It can be written as:

$$H(j\omega) = |HC\,(j\omega)|\,e^{\phi(j\omega)}$$

where

$$|H\,(jw)| = \sqrt{[H_R(j\omega)]^2 + [H_I(j\omega)]^2}$$

= magnitude of frequency response

$H_R\,(j\omega)$ = real part

$H_I\,(j\omega)$ = imaginary part

$$\text{Phase angle } f(jw) = \tan^{-1}\left[\frac{H_R\,(jw)}{H_I\,(jw)}\right] \quad (18.32)$$

The relationship between frequency response and transfer function is,

$$H\,(j\omega) = G(s)\Big|_{s\,=\,j\omega} = \frac{Y\,(j\omega)}{U\,(j\omega)} \quad (18.33)$$

18.13.1 Frequency Response of First Order Systems

Consider a first order linear, time-invariant system represented by the transfer function as,

$$G\,(s) = \frac{G}{\tau s + 1}$$

Frequency response, $$H(j\omega) = \frac{G}{\tau(j\omega)+1}$$

$$|H(jw)| = G\sqrt{\left[\frac{1}{1+\omega^2\tau^2}\right]^2 + \left[\frac{\omega\tau}{1+\omega^2\tau^2}\right]^2}$$

$$= \frac{G}{\sqrt{1+\omega^2\tau^2}} \tag{18.34}$$

$$\phi(j\omega) = \tan^{-1}(-\omega\tau) = -\tan^{-1}(\omega\tau)$$

Since the angle is –ve, the output lags behind the input with an angle $\tan^{-1}(\omega\tau)$.

18.13.2 Frequency Response of Second Order Systems

Let $$G(s) = \frac{G}{s^2 + 2\zeta\omega_n s + \omega_n^2}$$

For simplicity, assume $G = \omega_n^2$.

Then $$H(j\omega) = \frac{1}{\left(j\frac{\omega}{\omega_n}\right)^2 + 2\zeta\left(j\frac{\omega}{\omega_n}\right) + 1}$$

$$= \frac{1}{\left[1 - \left(\frac{\omega}{\omega_n}\right)^2\right] + 2\zeta\left(j\frac{\omega}{\omega_n}\right)}$$

$$|H(j\omega)| = \left|\frac{1}{\left(j\frac{\omega}{\omega_n}\right)^2 + 2\zeta\left(j\frac{\omega}{\omega_n}\right) + 1}\right|$$

$$= \frac{1}{\sqrt{\left(1-\frac{\omega}{\omega_n}\right)^2 + \left(2\zeta\frac{\omega}{\omega_n}\right)^2}} \tag{18.35}$$

$$\phi(j\omega) = \tan^{-1}\left[\frac{-2\zeta\frac{\omega}{\omega_n}}{1-\left(\frac{\omega}{\omega_n}\right)^2}\right],$$

Example 18.9

A second-order system has effective mass 20 kg, spring constant 8 kN/m, damping coefficient 130 N·s/m. It is subjected to harmonic input of 24 sin 15 t. Calculate the steady-state response of the system using transfer function approach.

Solution

The transfer function is

$$G(j\omega) = \frac{1}{\left[(k - m\omega^2) + jc\omega\right]}$$

$$= \frac{1}{\left[(8000 - 20 \times 225) + j\,(130 \times 15)\right]}$$

$$= \frac{1}{\left[3500 + j\,1950\right]} = \frac{1}{4006.56\angle 29.12^\circ}$$

$$= 2.5 \times 10^{-4} \angle -29.12^\circ$$

$$\text{Hence} \quad \frac{x}{F}(j\omega) = 2.5 \times 10^{-4} \angle -29.12^\circ$$

$$X = 24\left(2.5 \times 10^{-4}\right) = 0.006\text{m or } 6\text{mm}$$

$$x(t) = x\,(t) = 6 \sin\,(15t - 29.12^\circ)\text{ mm}$$

18.14 CONTROL SYSTEMS

The automatic control systems are of the following two types:

1. ***Open-loop (or unmonitored) system***

   When the input to a system is independent of the output from the system, then the system is called an open loop or unmonitored system. It is also called as a calibrated system. Most measuring instruments are open-loop control systems, where for the same input signal, the readings will depend upon ambient temperature and pressure, etc. Some of the examples of open-loop control systems are:

   (i) A simple Bourdon tube pressure gauge commonly used for measuring pressure.

   (ii) A simple carburettor in which the air-fuel ratio adjusted through venturi remains same irrespective of load conditions.

   (iii) In traffic lights system at road crossings, the timing of lights is preset irrespective of traffic density.

   (iv) Switching off the street lights of a town at a preset time by a time switch irrespective of the setting and rising time of the sun.

2. ***Closed-loop (or monitored) system***

   A closed loop control system uses input as well as some portion of the output to regulate the output. Closed-loop systems are also called feed back control systems. In feedback control the variable required to be controlled is measured. This measurement is compared with a given setpoint. If the error results, the controller takes this error and decides what action should be taken to compensate to remove the error. Errors occur when an operator changes the setpoint intentionally or when a process load changes the process variable accidentally. The error could be positive or negative.

Some examples of closed-loop control system are:

(i) In a traffic control system, if the flow of traffic is measured either by counting the number of vehicles manually or by counting the impulses due to the vehicles passing over a pressure pad and then setting the time of signal lights.

(ii) In a thermostatically controlled water heater, whenever the temperature of water heater rises above the required point, the thermostat senses it and switches the wafer heater off so as to bring the wafer temperature down to the required point. Similarly, when the temperature falls below the required point, the thermostat switches on the water heater to raise the temperature of water to the required point.

18.15 TRANSFER FUNCTION FOR A SYSTEM WITH VISCOUS DAMPED OUTPUT

Consider a shaft which is used to position a load in the form of a pulley or gear as shown in Fig.18.24. The movement of the load is resisted by a viscous damping torque.

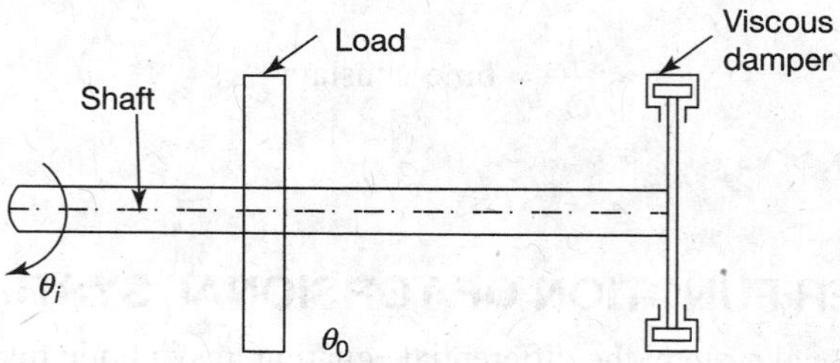

Fig.18.24 Second-order torsional system

Let θ_i = input signal to the shaft

θ_0 = output signal of the shaft

q = torsional stiffness of shaft, N.m/rad

I = moment of inertia of the load, kg.m^2

c_t = viscous torsional damping coefficient of the damper, N.m.s/rad

After a time t,

$$\text{Twist in the shaft} = \theta_i - \theta_o$$

$$\text{Torque transmitted to shaft} = q\,(\theta_i - \theta_o)$$

$$\text{Damping torque} = c_t\omega_o = c_t\,\frac{d\theta_o}{dt}$$

The equation of motion of the system is,

$$I\ddot{\theta}_0 = q(\theta_i - \theta_0) - c_t\dot{\theta}_0$$

$$I\ddot{\theta}_0 + c_t\dot{\theta}_0 + q\theta_0 = q\theta_i$$

$$\ddot{\theta}_0 + \frac{c_t}{I}\dot{\theta}_0 + \frac{q}{I}\theta_0 = \frac{q}{I}\theta_i$$

Now,

$$\omega_n^2 = \frac{q}{I} \text{ and } 2\zeta\omega_n = \frac{c_t}{I} = \frac{c_t}{c_c} \times \frac{2\sqrt{qI}}{I}$$

$$\ddot{\theta}_0 + 2\xi\omega_n\dot{\theta}_0 + \omega_2^2\theta_0 = \omega_n^2\theta_i$$

or

$$\left[D^2 + 2\zeta\omega_n D + \omega_n^2\right]\theta_0 = \omega_n^2\theta_i$$

Transfer function

$$= \frac{\theta_0}{\theta_i} = \frac{\omega_n^2}{D^2 + 2\zeta\ \omega_n^D + \omega_n^2}$$

$$= \frac{1}{\tau^2 D^2 + 2\zeta\ \tau\ D + 1} \quad (18.36)$$

where $\tau = \frac{1}{\omega_n}$ = time constant

18.16 TRANSFER FUNCTION OF TORSIONAL SYSTEM

For a second order torsional system, the differential equation in symbolic form is:

$$\left(D^2 + 2\zeta\omega_n\ D + \omega_n^2\right)\theta_0 = \omega_n^2\ 6$$

Transfer function,

$$\frac{\theta_0}{\theta_i} = \frac{\omega_n^2}{D^2 + 2\zeta\omega_n\ D + \omega_n^2}$$

$$= \frac{1}{\tau^2 D^2 + 2\zeta\ \tau D + 1}$$

where $\tau = \frac{1}{\omega_n}$ is the time constant.

For the first order system,

$$c_i\dot{\theta}_0 + q\theta_0 = q\theta_i$$

or

$$\frac{c_t}{q}\dot{\theta}_0 + \theta_0 = \theta_i$$

$$\frac{c_t}{q}\Delta\theta_0 + \theta_0 = \theta_i$$

$$\frac{\theta_0}{\theta_i} = \frac{1}{1 + \frac{c_t}{q}D} = \frac{1}{1 + \tau D} \qquad (18.37)$$

where $\tau = \frac{c_t}{q}$ is the time constant

18.17 EQUIVALENCE OF TRANSFER FUNCTIONS

1. *Open-loop transfer functions*

If a system with transfer function $G_1(s)$ is connected with another system with transfer function $G_2(s)$, then the overall transfer function of the system is the product of individual transfer functions, as shown in Fig.18.25.

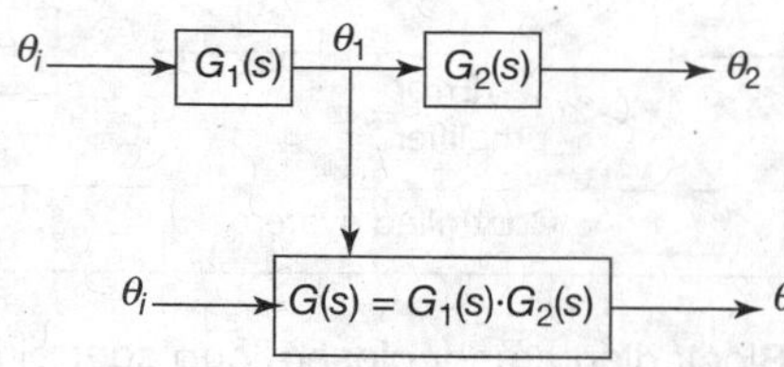

Fig.18.25 Equivalence of open loop transfer functions

In general, in open -loop configuration, the overall transfer function of the composite system is given by the following formula:

$$G(s) = G_1(s) \cdot G_2(s) \cdot G_3(s) \ldots G_n(s) \qquad (18.38)$$

2. *Closed-loop transfer functions*

The closed loop transfer function is defined as the overall transfer function of the entire control system. Consider a closed-loop transfer function consisting of several elements as shown in Fig.18.26.

$$\frac{\theta_0}{\theta_i - \theta_o} = KG\ (s)$$

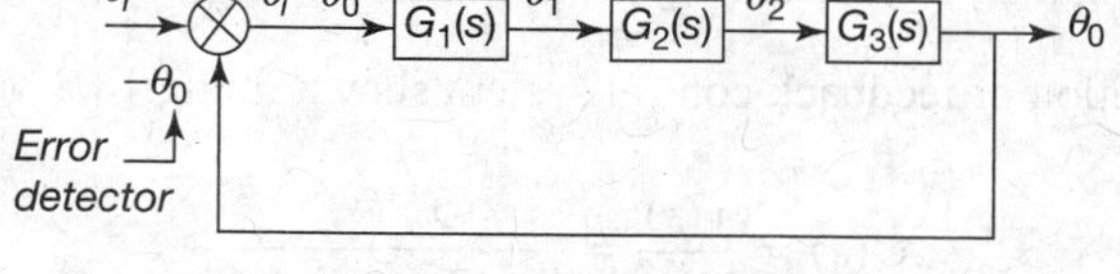

Fig.18.26 Closed-loop transfer functions

$$\theta_0 = K\,G\,(s)\,\theta_i - K\,G\,(s)\,\theta_0$$

or $$[1 + K\,G\,(s)]\theta_0 = K\,G\,(s)\,\theta_i$$

where K = constant representing the overall amplification or gain.

Transfer function, $$TF = \frac{\theta_o}{\theta_i} = \frac{KG(s)}{1 + KG(s)}$$

$$= \frac{\text{open loop } TF}{1 + \text{open loop } TF} \tag{18.39}$$

The equivalence of closed-loop transfer function is shown in Fig.18.27.

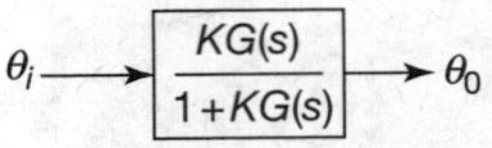

Fig.18.27 Equivalence of closed-loop transfer functions

Now consider the block diagram representing closed-loop control system for a plant as shown in Fig.18.28.

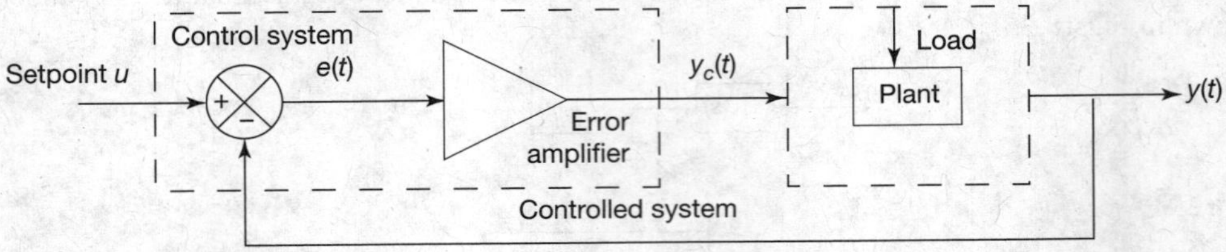

Fig.18.28 Block diagram for closed-loop control for a plant

Fig.18.28 has been represented in another way in terms of block diagram containing the transfer functions of the plant and feedback loop in Fig.18.29(a). Let $G_f\,(s)$ be the transfer function of the feedback control system and $G_p(s)$ that of the plant.

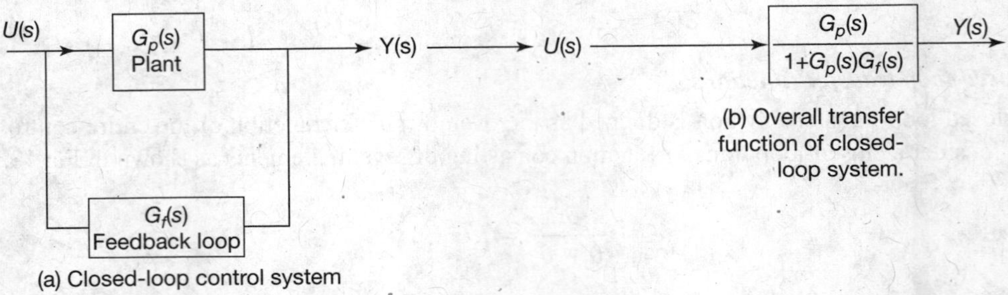

Fig.18.29 Block diagram containing transfer functions of plant and feedback loop

The overall transfer function of feedback control system shown in Fig.18.29(b), can be written as

$$G(s) = \frac{y(s)}{U(s)} = \frac{G_p\,(s)}{1 + G_p(s)\,G_f(s)} \tag{18.40}$$

18.18 THE CONTROLLERS

The controllers can be of the following types:

1. On-off controller
2. Proportional controller
3. Integral controller
4. Derivative controller
5. Proportional-plus-derivative (PD) controller
6. Proportional-plus-integral (PI) controller
7. Proportional-plus-integral-plus-derivative (PID) controller.

The selection of the right controller for an application depends on the following factors:

1. The degree of control required by the application.
2. The individual characteristics of the plant.
3. The desirable performance level including required response, steady state deviation and stability.

1. *On-off controller*

This is the simplest form of control action. The action is simply a switch. The output of the controller has two levels, ON and OFF, *i.e.* the output is either 100% on or 100% off. These two levels are generated based on error signal. If the error signal is greater than zero the ON level is generated and if the error signal is less than zero then the OFF level is generated or vice-versa. Mathematically it can be written as $y_c = c_{ON}$ for $e(t) > 0$ and $y_c = c_{OFF}$ for $e(t) < 0$ or vice-versa. Where, y_c is the controller output, $e(t)$ is the error signal, C_{ON} and C_{OFF} are the two control levels for $e(t) > 0$ and $e(t) < 0$, respectively. Fig.18.30 shows the block diagram of the on-off controller.

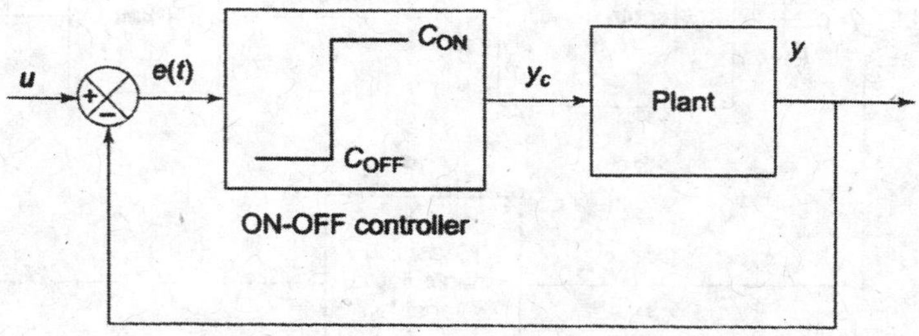

Fig.18.30 Block diagram of the on-off controller

Mostly, thermostat-based heating system uses an ON-OFF type controller. When the output is lower than the setpoint the controller is turned on (*i.e.* provides an ON output), and once the output is more than the setpoint the controller provides OFF output. The turn-ON and turn-OFF in many situation are deliberately made to differ by a small amount, known as the *hysteresis* or *dead-band*, (Fig.18.31) to prevent noise from switching the controller unnecessarily when the output is nearly the setpoint. The hysteresis is designed into the control action between the points at which the control output switches from OFF to ON. This designed in hysteresis prevents the output from switching from OFF to ON too rapidly. If the hysteresis is set too narrow, rapid switching will occur Therefore, the hysteresis should be set so that there is sufficient time delay between the ON and OFF modes of the outputs. The sensitivity of the ON-OFF controller depends on the hysteresis.

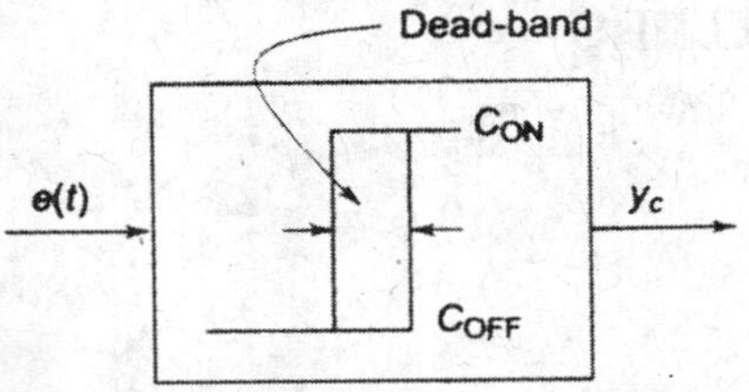

Fig.18.31 On-off controller with hysteresis or dead-band

2. *Proportional controller*

Proportional Control means that the plant input is changed in direct proportion to the error, $e(t)$. This controls the output so that the manipulated variable and the error has a proportional relation. The controller sets the manipulated variable in proportion to the difference between the setpoint and the measured variable (The variable to be controlled, e.g. speed of a motor). The bigger is the difference, the greater is the change in the manipulated variable. The coefficient of deviation is called proportional gain, K_p and is mathematically written as,

$$K_p = G_c(s) = \frac{C_y(s)}{E(s)} \tag{18.41}$$

where, $G_c(s)$, is the transfer function of the proportional controller, $C_y(s)$ is the Laplace transform of the output of the controller, $c_y(t)$ and $E(s)$ is the Laplace transform of the error signal, $e(t)$. A typical proportional controller controlling a plant (e.g. speed of a typical electrical motor) using the feedback hoop is shown in Fig.18.32.

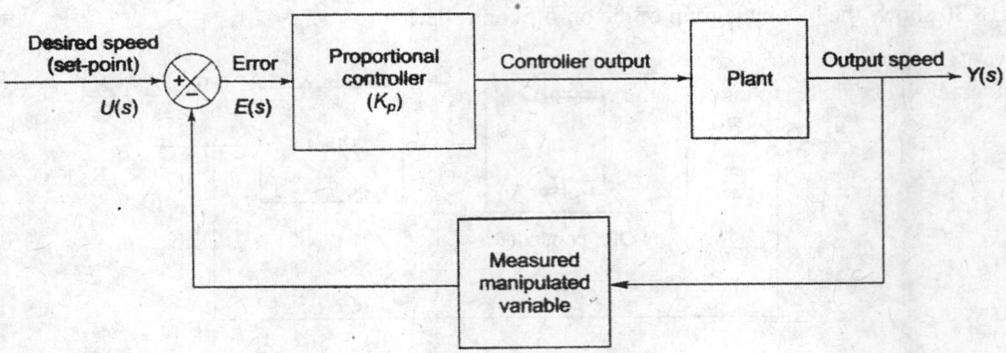

Fig.18.32 Schematic block diagram of a proportional controller

The advantage of proportional controller is that it is relatively easy to implement. However, the disadvantage is that there always involves an *offset* in the output response causing difference between the set-point and the actual output. Other disadvantage of proportional controller is the overshoot problem that arises when a proportional controller is used at high gain.

3. *Integral controller*

The block diagram of the integral controller is shown in the Fig.18.33. The controller controls the output by integrating the error signal. That is,

$$c_y(t) = \frac{1}{T_i}\int_0^{t} e(\tau)d\tau \tag{18.42}$$

Where, $C_y(t)$ is the output of the controller. $e(t)$ is error signal and T_i is called the integral time. The integral correction of output is performed by accumulating the deviation in accordance with time elapsed. Eq. (18.41) can also be written in transfer function form as given below.

$$G_{IC}(s) = \frac{C_y(s)}{E(s)} = \frac{K_i}{s} \tag{18.43}$$

Where, $G_{IC}(s)$ is the transfer function of the integral controller, which is the ratio of the Laplace transform of the output to the Laplace transform of the input (the error signal) of the integral controller. $K_i = 1/T_i$, is called the integral gain which is the reciprocal of the integral time. The offset in the output, and hence the steady-state (Refer Fig.18.33) performance of the system can be improved by employing integral control action. But the integral action may lead to oscillatory output resulting in poor stability.

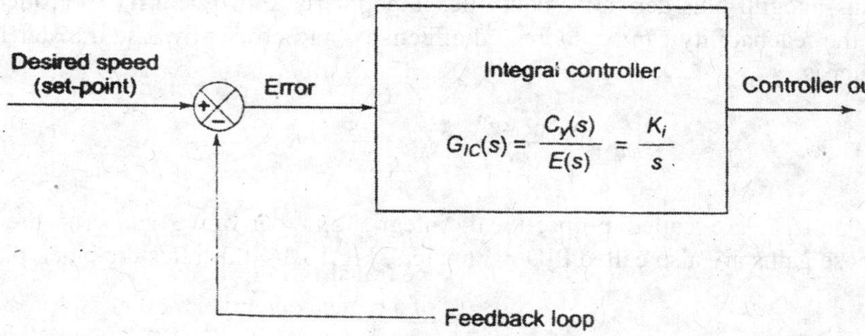

Fig.18.33 Block diagram of an integral controller

4. *Derivative controller*

Derivative controller controls the plant by providing the control signal which is the derivative of the error signal. The transfer function of the controller can be written as,

$$G_D(s) = \frac{C_y(s)}{E(s)} = K_d s \tag{18.44}$$

K_d is the constant of proportionality, usually referred to as derivative time, or simply derivative gain. The derivative action improves the transient performance of the plant. However, derivative control has poor steady-state performance.

5. *Proportional-plus-integral controller*

In short, the proportional-plus-integral controller is referred to as PI controller. The PI controller controls the plant by providing the control signal which is the combination of proportional and integral action over the error signal. As stated earlier, the integral control improves the steady-state performance. On the other hand, the integral action may lead to oscillatory output and hence has poor stability which is not really desirable. Combining proportional and integral action the two constants such as K_p and K_i can be adjusted in order to optimize the system performance or the output response according to the requirement. The transfer function of the PI controller is,

$$G_{PI}(s) = \frac{C_y(s)}{E(s)} = K_P\left(1 + \frac{K_i}{s}\right) = K_P\left(1 + \frac{1}{T_i s}\right) \tag{18.45}$$

6. *Proportional-plus-derivative controller*

Proportional-plus-derivative controller is commonly abbreviated to PD controller. The stability and hence the performance of the system can be improved by employing derivative action along with proportional action into the control system.

Adding a term proportional to the time-derivative of the error signal can take the edge off the overshoot problems that arise when a proportional controller is used at high gain. The transfer function of the PD controller is,

$$G_{PD}(s) = \frac{C_y(s)}{E(s)} = K_p(1 + K_d s) \tag{18.46}$$

7. *Proportional-plus-integral-plus-derivative controller*

Proportional-plus-integral-plus-derivative controller is popularly known as PID controller. This is a method, where the reachability can be addressed effectively and efficiently. The transfer function of the PID controller is,

$$G_{PID}(s) = \frac{C_y(s)}{E(s)} = K_p + \frac{K_i}{s} + K_d s \tag{18.47}$$

where, K_p, K_i, and K_d are called proportional, integral and derivative gains of the controller respectively. These gains are also called PID parameters. Fig.18.34 illustrates the block diagram of a PID controller.

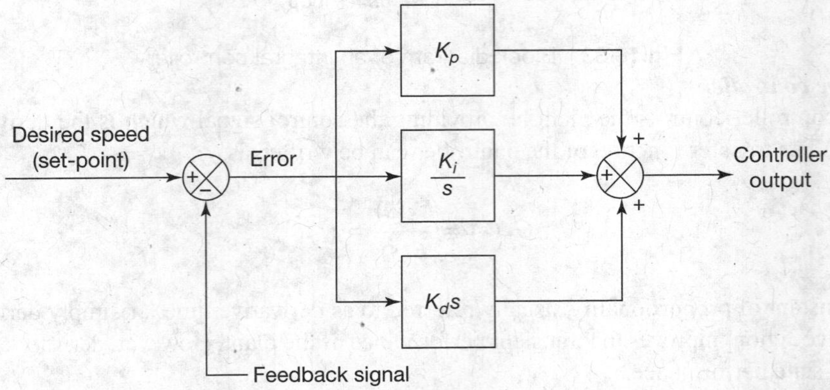

Fig.18.34 Block diagram of a PID controller

Summary for Quick Revision

1. Automatic control is desired to relieve the human operators from monotony of work and bring about efficiency in the system.
2. Set point is the position in which the control point setting mechanism is set.
3. Overshoot is defined as the maximum deviation of the response above the steady state value.
4. Response time is the time required for the system to reach steady state.
5. Settling time is defined as the time required for the output to achieve within 2% of its final value when the system is subjected to a step input.

6 Delay time is defined as the time required for the output to reach 50% of its final value in first attempt.

7 Rise time is the time required to for the output to rise from 10% to 90% of the final value for overdamped system, and zero to 100 % of the final value for underdamped system.

8 Peak time is the time required for the output to reach the peak of time response or the peak overshoot.

9 A transducer is a basic element that converts or transforms one form of signal to another form which is more convenient to use and measure.

10 A sensor is merely a sophisticatedtransducer which contains some signal conditioning circuit capable of amplifying and refining the weak and raw signal.

11 An actuator is a device that converts energy to mechanical form.

12 A block diagram is a symbolic outline of a systemin which various components or operations are represented by rectangles in an ordered sequence. The rectangles are connected by arrows showing the flow of the working medium or of information.

13 System response deals with studying the behavior of the system in which changes occur and in which predictions are desirable.

14 Output response of first order linear system without excitation: $y/y_0 = \exp(-t/\tau)$, where τ = time constant. Output response with excitation, $y/x = 1 - \exp(-t/\tau)$.

15 Transfer function is the ratio of Laplace transform of output response to Laplace transform of input signal.

Multiple Choice Questions

1 A block diagram is represented by
(a) circles (b) triangles (c) rectangles (d) parallelograms

2 The output response of a first order system with excitation is:
(a) $y = x(1 - e^{-t/\tau})$ (b) $y = x\,e^{-t/\tau}$ (c) $y = x(1 + e^{-t/\tau})$ (d) $y = x(1 - e^{t/\tau})$

3 A simple Bourdon tube pressure gauge is a
(a) closed-loop control system (b) open-loop control system
(c) manually operated system (d) feed back control system.

4 The overall transfer function of two blocks $G_1(s)$ and $G_1(s)$ connected in series is
(a) $G_1(s) + G_2(s)$ (b) $G_1(s) \times G_2(s)$ (c) $G_1(s) - G_2(s)$ (d) $G_1(s)/G_2(s)$

5 The equivalent transfer function for a closed-loop control system is
(a) $\dfrac{KG(s)}{1 + KG(s)}$ (b) $\dfrac{KG(s)}{1 - KG(s)}$ (c) $\dfrac{1 + KG(s)}{KG(S)}$ (d) $\dfrac{1 - KG(s)}{KG(s)}$

6 Transfer function is the operational relationship between output and
(a) input (b) error (c) response (d) command

7 Given Fig.18.35 shown a flexible shaft of negligible mass of torsional stiffness K coupled to a viscous damper having a coefficient of viscous damping c. If at any instant the left and right ends of this shaft have angular displacement θ_1 and θ_2 respectively, then the transfer function, θ_2/θ_1 of the system is

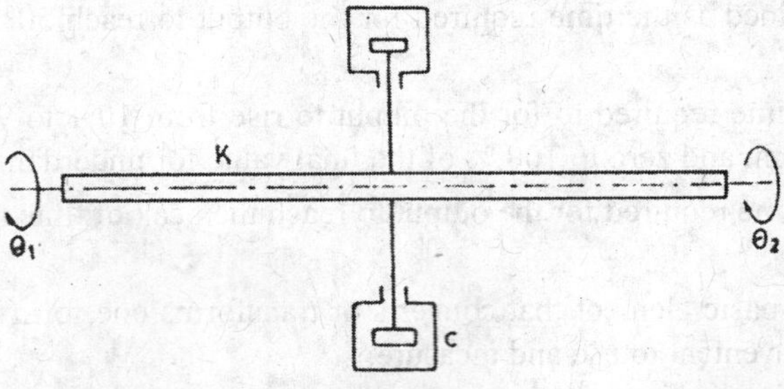

Fig.18.35

(a) $\dfrac{K}{K+c}$ (b) $\dfrac{1}{1+\dfrac{C}{K}s}$ (c) $\dfrac{1}{1+\dfrac{K}{c}s}$ (d) $1+\dfrac{K}{c}s$

8 Consider the following statement in respect of introduction of feedback in a control system:

1. It enhances its gain.
2. It attenuates the unwanted noise.
3. It helps in improving the accuracy of the system

Which of these statements are correct?

(a) 2 and 3 (b) 1, 2 and 3 (c) 1 and 3 (d) 1 and 2

9 Match List-I with List-II and select the correct answer using the codes given below the lists:

List-I	List-II
A. Open loop system	1. Frequency domain analysis
B. Closed loop system	2. More stable
C. Step input	3. Less stable
D. Sinusoidal input	4. Time domain analysis

Codes:

	A	B	C	D
(a)	2	3	4	1
(b)	4	1	2	3
(c)	2	1	4	3
(d)	4	3	2	1

10 Which of the following is a closed-loop control system?

(a) Traffic control on the roads by lights where the timing mechanism is present irrespective of the intensity of traffic

(b) Switching off the street lights of a tower at a predetermined time by a time-switch irrespective of the fact that the sun rises at a different time each day

(c) Switching off an electric heater by a time-switch irrespective of whether the dish has been prepared or not

(d) Human body

11 Match List-I with List-II and select the correct answer using the codes given below the Lists:

List-I (Property)	List-II (System)
A. Resonance	1. Closed-loop control system
B. On-off control	2. Free vibrations
C. Natural frequency	3. Excessively large amplitude
D. Feedback signal	4. Mechanical brake

Codes:

	A	B	C	D
(a)	1	2	4	3
(b)	3	4	2	1
(c)	1	4	2	3
(d)	3	2	4	1

12 A physical system is translated into functional block diagram of the type shown in the Fig.18.36. The command input $r(t)$ and controlled output $c(t)$ of this system are given by

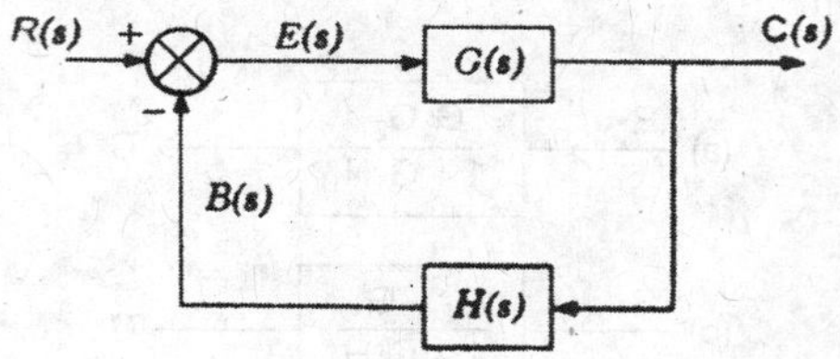

Fig.18.36

(a) $\frac{C(s)}{R(s)} = \frac{G(s)}{1+\frac{G(s)}{H(s)}}$ (b) $\frac{C(s)}{R(s)} = \frac{G(s)}{1+G(s)H(s)}$

(c) $\frac{C(s)}{R(s)} = \frac{G(s)}{1-G(s)H(s)}$ (d) $\frac{C(s)}{R(s)} = \frac{G(s)}{1-G(s)H(s)}$

13 In which one of the following types of control system is the output of the control element proportional to the time rate of change of the input?

(a) Proportional (b) Integral
(c) Proportional and derivative (d) Derivative

14 Traffic control on the roads by lights where the timing mechanism operates irrespective of the intensity of traffic is an example of

(a) Closed loop control (b) Under-damped control
(c) Open loop control (d) Over-damped control

15 What is the value of K for which the relative damping of the closed loop system shown above is equal to 0.5?

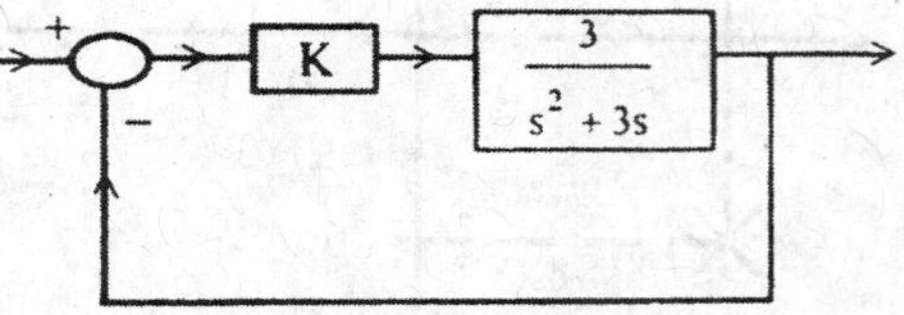

Fig.18.37

(a) 2 (b) 3 (c) 4 (d) 5

16 The block diagram of an automatic control system is shown in the following Fig.18.38.

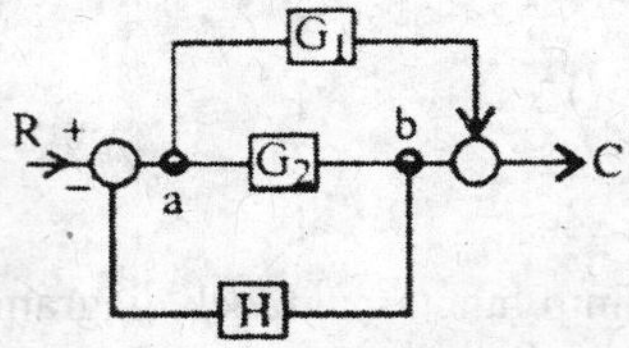

Fig.18.38

Its simplified form will be as in

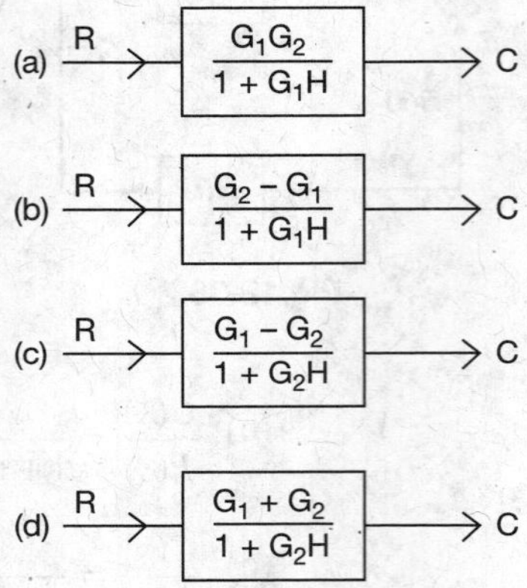

Fig.18.39

17 Given that G = forward path gain and H = feed back path gain, if G and H are functions of frequency, then the feedback would affect gain G of a non-feedback system by a value to

(a) 1 + GH (b) 1 – GH (c) GH (d) 1/GH

18 The Fig.18.40 given below shows the locations of the roots of the characteristic function of a second order, linear, closedloop control system. What is the natural frequency of the system?

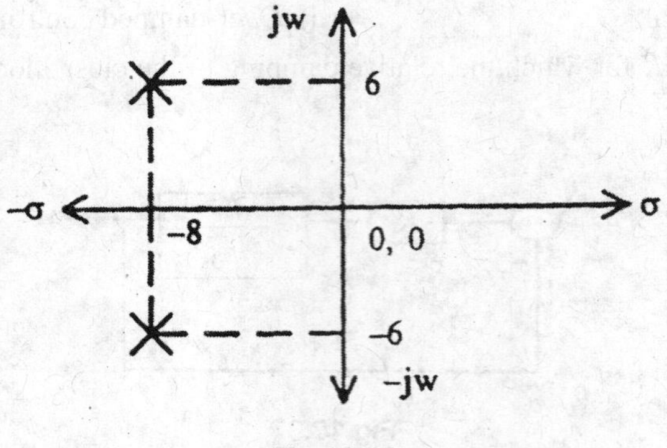

Fig.18.40

(a) 10 rad/s (b) 36 rad/s (c) 48 rad/s (d) 64 rad/s

Answers

1. (c) 2. (a) 3. (b) 4. (b) 5. (a) 6. (a) 7. (b) 8. (c) 9. (a) 10. (d) 11. (b) 12. (d) 13. (c) 14. (c) 15. (b) 16. (a) 17. (a) 18. (a)

Review Questions

1 What is automatic control?

2 Differentiate between an automatic controller and process controller.

3 What is a regulator?

4 Define open-loop control and closed-loop control.

5 What is resistance?

6 What is set point and dead time?

7 Define output response.

8 Differentiate between stable and unstable system.

9 What is overshoot?

10 Explain response time, settling time and delay time.

11 What is a transducer?

12 Define command.

13 What is a transfer function?

14 What are the effects of feed back control.

15 What is a sensor?

16 Differentiate between first-order and second-order systems.

17 Define frequency response of a system.

18 Write the transfer function of first and second order systems.

Exercises

18.1 What is a transducer? Explain the principle of operation of a transducer.

18.2 What is an actuator? List various types of actuators.

18.3 What are the basic modeling elements for a mechanical system? Explain the function of each element.

18.4 Define the following:
(a) Instantaneous response (b) Lagging response
(c) Delayed response.

18.5 What do you mean by rise-time, peak-time and settling time in connection with transient response specification?

18.6 Derive the expression representing the output response of a first order system with input as given below:
(a) Free response (b) Step input
(c) Ramp input

18.7 Derive the expression representing the output response of a second order system with input as given below:

(a) Zero input (b) Step input

(c) Ramp input

18.8 With neat block diagram, explain open-loop and closed-loop control systems.

18.9 State and explain open-loop and closed-loop transfer function.

18.10 The open-loop transfer function of a unity feedback control system is given by:

$$G(s) = \frac{K}{s(1+\tau s)}$$

(a) By what factor the gain be multiplied so that damping ratio is increased from 0.2 to 0.8?

(b) By what factor the constant τ should be multiplied so that the damping ratio is reduced from 0.6 to 0.3?

18.11 A unit step input is applied to a second-order measurement system with effective mass of 40 g and spring constant 2 kN/m. Find the damping constant for 40% overshoot and the corresponding period of oscillation.

[Ans. 5 N.s/m, 0.0293 s]

18.12 A second order measurement system has effective mass of 25 g, spring constant 2500 N/m and damping factor 0.6. Determine the rise time and settling time for a ±2.5% tolerance band.

[Ans. 0.00875s, 0.02062 s]

18.13 A second order measurement system has an effective mass of 60 g and the spring constant 2.25 kN/m. Calculate the steady state error and the time lag for damping factor 0.48 when a unit ramp input is applied.

[Ans. 0.004957, 0.004957 s]

18.14 In a second-order measurement system with a damping ratio of 0.65, the natural frequency is 30 Hz. Calculate the dynamic error and the time lag at frequency input of 10 Hz.

[Ans. –0.0112, 0.00722 s]

18.15 Determine the steady state response of the second – order system shown in Fig.18.41. By using transfer function method.

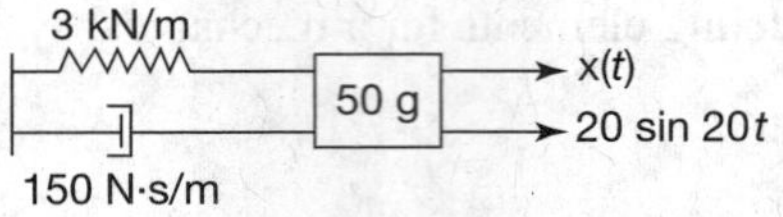

Fig.18.41

A-I

MACHINE THEORY LABORATORY PRACTICE

Experiment 1 To draw the displacement, velocity and acceleration curves for a slider-crank mechanism.

Apparatus Slider-crank apparatus, graph sheet.

Theory The displacement of a slider-crank mechanism, when the crank has rotated by θ from inner dead centre is:

$$x = r\left[(1-\cos\theta)+\left\{n-\left(n^2-\sin^2\theta\right)^{0.5}\right\}\right]$$

Velocity, $$v = \omega r\left[\sin\theta + \frac{\sin 2\theta}{2n}\right]$$

Acceleration, $$f = \omega^2 r\left[\cos\theta + \frac{\cos 2\theta}{n}\right]$$

The slider-crank apparatus shown in Fig.1, consists of the frame F in which the slider S moves in a slot. The graduated wheel W replaces the crank OC. The wheel and the slider are connected by the connecting rod C. When the crank OC is rotated the slider moves to and fro in a linear motion. The motion of the slider can be read on a scale attached to the frame on the side of the slot.

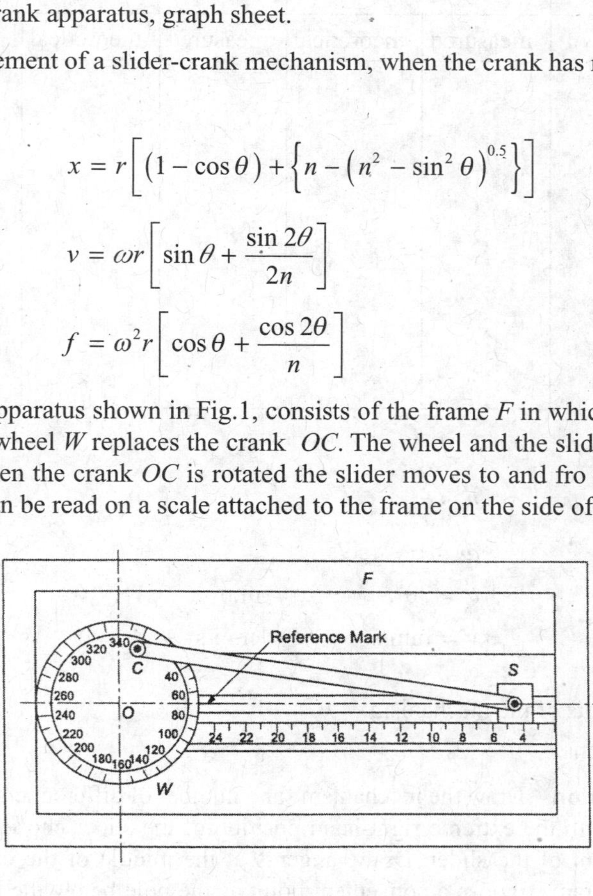

Fig.1 Slider-crank mechanism

Procedure

1. Bring the wheel and the slider to the respective reference marks.
2. For a given angle of rotation of the crank, note down the displacement of the slide.
3. Plot a graph between slider displacement and the crank rotation.
4. Assume that the crank is rotating with a uniform angular speed of 1 rad/s.
5. Convert the crank rotation angle into time and plot the slider displacement v's time.
6. By graphical differentiation, determine the velocity time graph.
7. By differentiation twice, determine the acceleration time graph.
8. Calculate the results with the theoretical values.

Observations

Crank radius, $r =$ mm

Length of connection rod, $\ell =$ mm

Sl. No.	Crank rotation, θ		Time t	Displacement, x mm		Velocity, v mm/s		Acceleration, f mm/s^2	
	deg	rad	s	measured	theoretical	measured	theoretical	measured	theoretical

Calculations

$$n = \frac{\ell}{r}$$

$\omega = 1$ rad/s

$\theta = 30^\circ$; $x =$ mm

$v =$ mm/s: $f =$ mm/s^2

Sources of error

1. Clearance in the joints of the mechanism.
2. Inaccurate graduations.

Graphical differentiation Draw the mechanism for a number of different crank positions by taking 30° crank intervals. With the extreme right-hand position of the slider chosen as the starting point, lay off the displacement of the slider. Draw tangents at the middle of the crank angle interval on the displacement diagram. Choose a convenient point as the pole below the displacement diagram.

From this point, draw lines parallel to the tangents to intersect the vertical axis. Draw horizontal lines from these points. Also project the lines from the points chosen on the displacement diagram to intersect the respective horizontal lines. Join the points of intersection by a smooth curve to get the velocity diagram. Now draw tangents on the velocity diagram and repeat the above procedure to get the acceleration diagram.

Experiment 2 To determine the ratio of times for the crank and slotted lever quick-return mechanism.

Apparatus Crank and slotted lever mechanism, graph sheet.

Theory The crank and slotted lever mechanism is shown in Fig.2. It consists of a graduated disc *A* on which the crank rotation can be measured. The slotted lever *B* is hinged at *O*, and carries a slider *C*. The slotted lever is hinged to an oscillating link *D*, which slides horizontally in link *E* and its other end is attached to the ram *F* on which the cutting tool is mounted.

$$\frac{\text{Time of cutting}}{\text{Time of return}} = \frac{\alpha}{360^\circ - \alpha}$$

where α is the angle of cutting.

Let $OA = d; \quad OB = r; \quad OD = \ell$

$$a = 2\left[90^\circ + \sin^{-1}\left(\frac{r}{d}\right)\right]$$

$$\frac{\text{Maximum velocity during return}}{\text{Maximum velocity during cutting}} = \frac{d+r}{d-r}$$

Procedure

1. Bring the crank and the ram to zero positions.
2. For the given crank angle of rotation, note down the displacement of the ram.
3. Plot the crank rotation v's displacement of the ram.
4. Assume the crank to be rotating at an angular speed of 1 rad/s.
5. Plot the displacement-time graph.
6. By graphical differentiation, determine the velocity-time graph.
7. From the velocity-time graph, determine the maximum velocities during cutting and return.
8. Determine the angle of cutting and angle of return.
9. Determine the ratio of time of cutting and time of return, and the ratio of maximum velocities during return and cutting.
10. Draw the theoretical velocity diagram and calculate the theoretical ratio of velocities. Observations

Length of crank,	$r =$ mm
Length of link *OA*,	$d =$ mm
Length of slotted lever *OD*,	$\ell =$ mm
Angle of forward stroke,	$\alpha =$

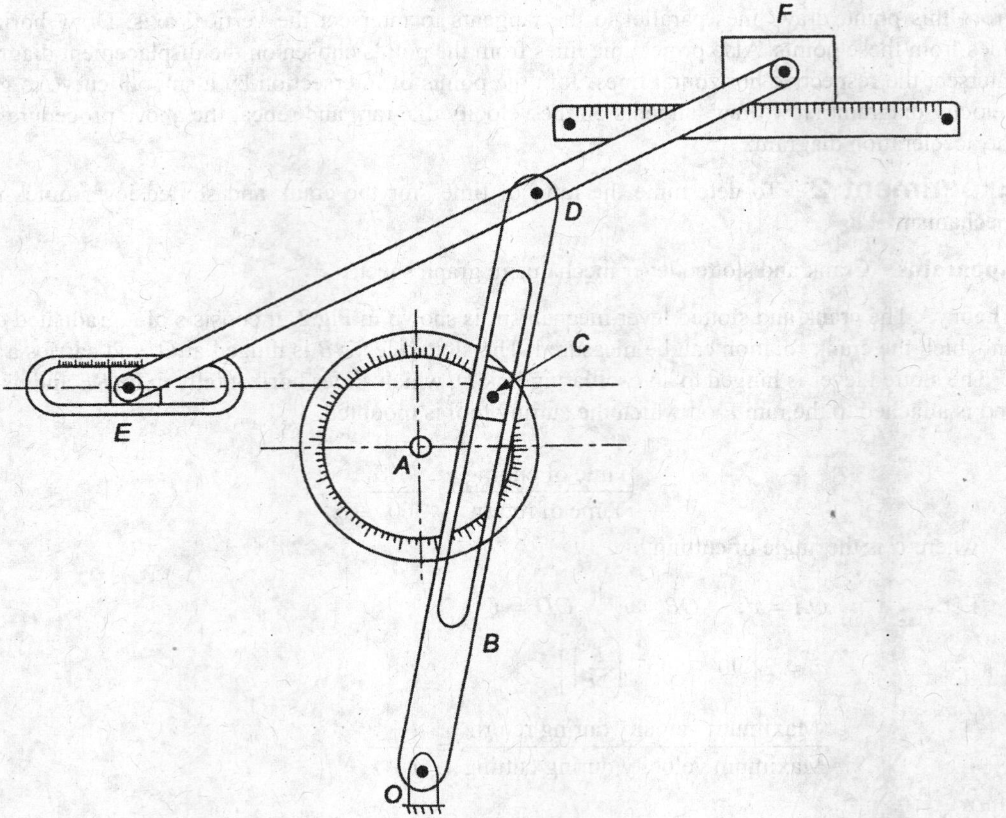

Fig.2 Crank and slotted lever apparatus

Sl. No.	Crank rotation, θ		Time	Ram displacement	Ram velocity
	deg	rad	s	mm	mm/s

Calculations

$$\frac{\text{Time of cutting}}{\text{Time of return}} = \frac{t_c}{t_r} = \frac{a}{360^\circ - a}$$

From graph, $$\frac{t_c}{t_r} =$$

$$\frac{\text{Maximum return velocity}}{\text{Maximum cutting velocity}} = \frac{v_r}{v_c} = \frac{d + r}{d - r}$$

From graph, $$\frac{v_r}{v_c} =$$

Theoretical ratio, $$\frac{v_r}{v_c} =$$

Precautions

1. The slider and slotted lever should be lubricated to decrease friction.
2. Displacement and crank rotation should be measured accurately.

Sources of error

1. Effect of clearances in the joints.
2. Errors during graphical differentiation.

Experiment 3 To determine the ratio of times and tool velocities of Whitworth type quick-return mechanism.

Apparatus Whitworth quick-return mechanism, graph sheets.

Theory The Whitworth quick-return mechanism shown in Fig.3 consists of a graduated disc on which rotation of crank can be measured. The displacement of the tool can be read the scale attached to the ram.

$$\frac{\text{Time of cutting}}{\text{Time of return}} = \frac{\alpha}{360^\circ - \alpha}$$

where a is the angle of cutting.

Procedure

1. Bring the crank and ram to zero positions.
2. For the given crank angle of rotation, note down the displacement of the ram.
3. Plot the crank rotation v's displacement of ram.
4. Assuming the crank to be rotating at 1 rad/s, plot the displacement-time graph.
5. By graphical differentiation, determine the velocity-time graph.
6. Determine the angles of cutting and return strokes.
7. Calculate the ratio of cutting angle and angle of return strokes.
8. Draw the theoretical velocity diagram and calculate the ratio of velocities.

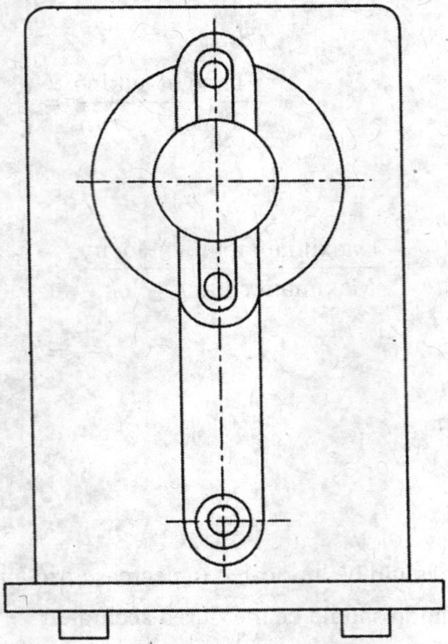

Fig.3 Witworth quick return mechanism apparatus

Observations

Length of crank OA, $r =$ mm

$AB =$ mm

$PB =$ mm

Sl. No.	Crank rotation, θ		Time	Ram displacement	Ram velocity
	deg	rad	s	mm	mm/s

Calculations

Cutting angle, $$\alpha = \frac{\text{Angle of cutting}}{\text{Angle of return}}$$

$$\frac{t_c}{t_r} = \frac{\alpha}{360^\circ - \alpha}$$

From graph, $$\frac{t_c}{t_r} =$$

Theoretical velocity of ram = mm/s

Precautions

1. The slider and slotted lever should be lubricated to decrease friction.
2. Displacement and crank rotation should be measured accurately.

Sources of error

1. Effect of clearances in the joints.
2. Errors during graphical differentiation.1

Experiment 4 To determine the ratio of angular speed of shafts of a Hooke's universal joint.

Apparatus Hooke's joint.

Theory The Hooke joint is shown in Fig.4. It has the provision for measuring the angle of rotation of the driving and the driven shafts. The angle between the driving and driven shafts can also be varied and measured.

Let α = angle between the axes of the two shafts
θ = angle turned through by the driving shaft
ϕ = angle turned through by the driven shaft
ω_1 = angular velocity of the driving shaft
ω_2 = angular velocity of the driven shaft

Then
$$\frac{\omega_1}{\omega_2} = \frac{1 - \cos^2\theta \sin^2\alpha}{\cos\alpha}$$

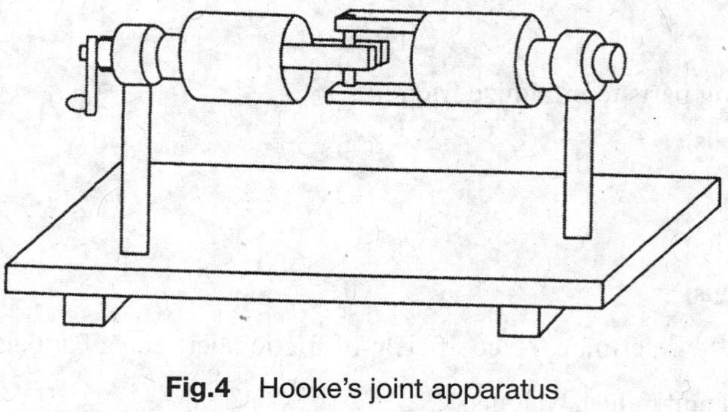

Fig.4 Hooke's joint apparatus

$$\text{Angular acceleration} = \frac{-\omega_1^2 \cos\alpha \sin^2\alpha \sin 2\theta}{\left(1 - \cos^2\theta \sin^2\alpha\right)^2}$$

Procedure

1. Adjust the angle between the shafts to be 15°.
2. Set the angles of driving and driven shafts to be equal to zero degree.
3. Rotate the driving shaft through equal intervals of 30° and note down the corresponding angles of the driven shaft.

4. Change the angle between the shafts to 30°, 45°, 60°, and 90°, etc., and repeat the experiment.
5. Calculate the ratio of incremental angles turned through by the driving and the driven shafts.
6. Calculate the theoretical values and compare.

Observations

Sl. No.	$\alpha = 15°$				$\alpha = 30°$			
	θ	ϕ	ω_1/ω_2		θ	ϕ	ω_1/ω_2	
			Exp.	Theor.			Exp.	Theor.

Calculations

$$\frac{\omega_1}{\omega_2} = \frac{1 - \cos^2\theta \sin^2\alpha}{\cos\alpha} =$$

Also

$$\frac{\omega_1}{\omega_2} = \frac{\Delta\theta}{\Delta\phi} =$$

Precautions

1. Lubricate all moving parts to minimize friction.
2. Errors in graduations.

Sources of error

1. Clearance in the joints.
2. Errors in graduations.

Experiment 5 To determine the coefficient of friction between a flat belt and a pulley.

Apparatus Flat belt and pulley, weights.

Theory The flat belt and pulley system consists of a flat belt and a pulley mounted on bearings. The angle of contact is 180° (Fig.5).

The coefficient of friction is given by,

$$\mu = \left(\frac{1}{\theta}\right) \ln\left(\frac{T_1}{T_2}\right)$$

Where θ = angle of arc of contact in radians
T_1 = tension on the tight side of the belt
T_2 = tension on the slack side of the belt

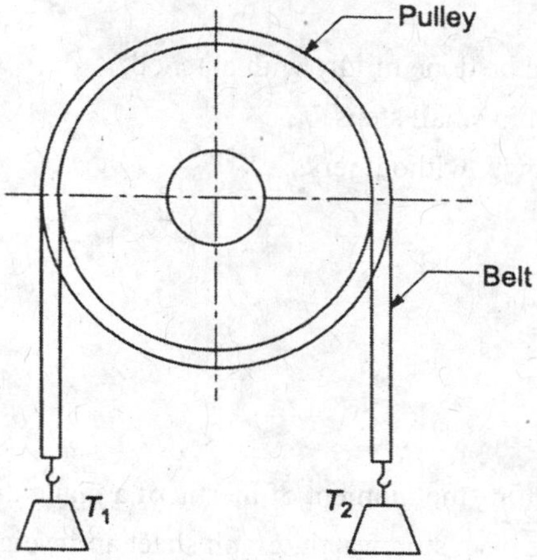

Fig.5 Belt and pulley system

Procedure

1. Note the angle of contact of the belt.
2. Hang some weight on one side of the belt.
3. Put some weight on the other side of the belt. Increase the weight till the belt just starts slipping on tapping the pulley slightly.
4. Note down the values of tight and slack side tensions.
5. Vary the tension on the tight side and repeat the experiment.
6. Calculate the coefficient of friction between the belt and the pulley.

Observations

Material of belt =
Material of pulley =

Sl. No.	T_1	T_2	Angle of contact		Coefficient of friction
	kg	kg	deg	rad	

Calculations

$$\text{Coefficient of friction, } \mu = \left(\frac{1}{\theta}\right)\ln\left(\frac{T_1}{T_2}\right)$$

$$=$$

Precautions

1. Tapping of the pulley should be done mildly with a pencil.
2. Weights should be increased in small steps.
3. Weights should be added slowly without jerks.

Sources of error

1. Worn out old belt.
2. Rusted pulley surface.
3. Friction in pulley bearings.
4. Inaccurate weights.

Experiment 6 To determine the moment of inertia of a plane disc by using a gyroscope.

Apparatus Gyroscope, plane disc, stop watch, graph sheet and weights.

Theory The gyroscope (Fig.6) consists of an electric motor supported within a ring mounted on ball bearings which is carried on a cradle attached to a vertical shaft with ball bearings. A disc is mounted coaxially to the armature. A loading arm carrying a counterpoise and hanger is attached to the ring. The heavy base is of mild steel and has a vertical shaft. It has four levelling screws and a spirit level mounted to the base for levelling. A brass angular scale is fitted to the cradle which enables the angle of the tilt of loading arm to be found when the precession is arrested by stopping the rotation of cradle. Knowing the time for one revolution, the angular velocity of precession can be determined.

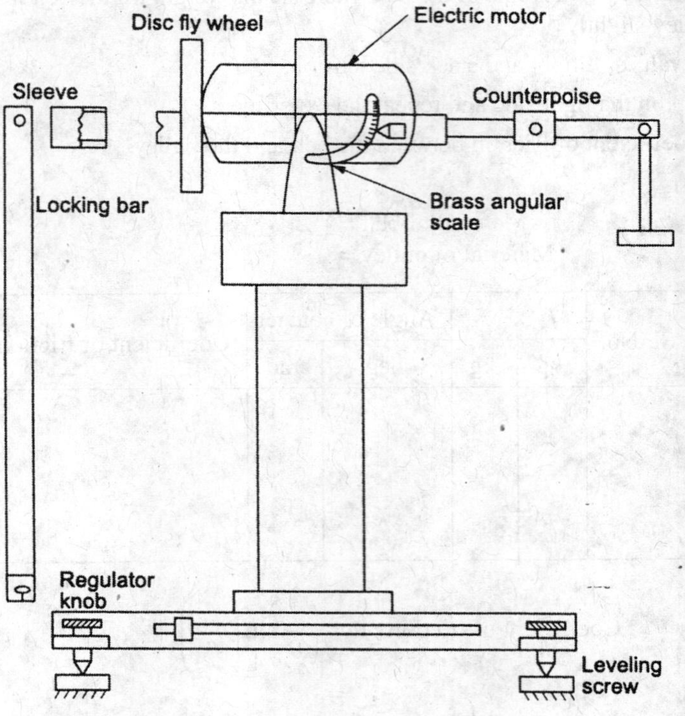

Fig.6 Gyroscope apparatus

The torque, $T = I_p\,\omega\omega_p$

where $\omega = \dfrac{2\pi n}{60}$ rad/s

$$\omega_p = \frac{2\pi}{t_p}$$

t_p = time for one revolution in the horizontal plane

Procedure

1. Set the instrument perfectly horizontal by four levelling screws and the spirit level.
2. Switch on the motor and obtain the desired speed by changing the variable resistance.
3. Determine the motor speed by a tachometer or a strobometer.
4. Move the counterpoise to keep the loading arm horizontal so as to show zero on angular brass scale.
5. Put the hanger with known weight at the end of the loading arm.
6. Note the time for one revolution.
7. Keeping the speed constant, increase the load, thus the torque, to find out corresponding angular speed of precession.
8. Change the motor speed and repeat the experiment.
9. Plot the graph between torque and speed of precession.
10. Calculate the value of moment of inertia of the disc.

Observations

Lever arm = mm

Sl. No.	Motor speed, n_1, rpm				Motor speed, n_2, rpm			
	Weight, W	Torque, T	Time for one revolution T_0	ω_p, $2\pi/t_p$	Weight, W	Torque T	Time for one revolution T_0	ω_p, $2\pi/t_p$
	N	N mm	s	rad/s	N	N mm	S	rad/s

Calculations

Torque, T = N mm

Angular speed, $\omega = \dfrac{2\pi n}{60}$ rad/s =

Speed of precession, $\omega_p = \dfrac{2\pi}{t_p}$

Moment of inertia, $I_p = \dfrac{T}{\omega\omega_p} =$

Precautions

1. The motor speed should be kept constant by a voltage stabilizer.
2. The gyroscope should be leveled properly.
3. The time should be measured accurately.

Sources of error

1. Fluctuations in motor speed.
2. Inaccuracies in measuring time.
3. Personal errors.

Experiment 7 To determine the forces on the spring and stiffness of a Hartnell governor.

Apparatus Hartnell governor, weighing balance, scale and graph sheet.

Theory The Hartnell governor (Fig.7) consists of two bell crank levers hinged in the frame at *A*. The levers carry balls at *B* on the vertical arm and a roller *C* in a fork at the other end. These rollers press against the sleeve *D* which compresses the spring *E* from the bottom. The compression varies with different positions of the sleeve. The initial force in the spring is controlled by the nut *F*. The speed of rotation can be varied by the electric motor and the voltage regulator.

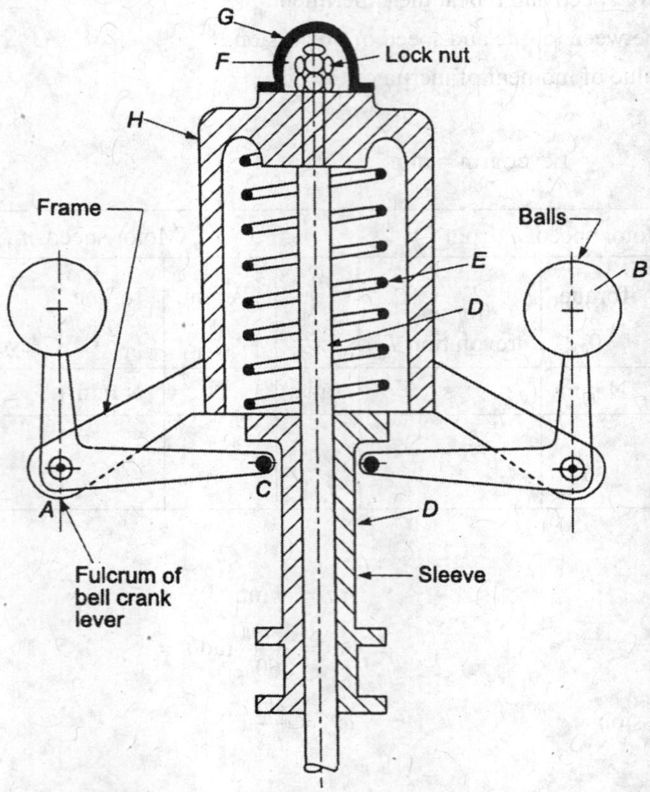

Fig.7 Hartnell governor

The spring stiffness is given by,

$$k = 2\left(\frac{a^2}{b^2}\right)\left[\frac{F_1 - F_2}{r_1 - r_2}\right]$$

where a, b = vertical and horizontal arms of the bell crank lever, respectively.

r_1, r_2 = radii of the balls at the maximum and minimum speeds, respectively.

$F_1 = mr_1\omega_1^2$

$F_2 = mr_2\omega_2^2$

m = mass of the ball

$$\omega = \frac{2\pi n}{60} \text{ rad/s}$$

n = speed in rpm.

The compression of the spring, $\delta = \dfrac{(r_1 - r_2)}{a}$

The radii r_1 and r_2 can be determined by ploting a graph between the displacement of the sleeve from the mean position and the radii of the balls. The motor speed may be measured by a tachometer of a strobometer.

Procedure

1. Plot a graph between the displacement of the sleeve from the mean position and the radii of the balls.
2. Determine the mass of the balls and the length of the arms of the bell crank lever.
3. Start the motor and adjust the speed so that the balls run at the innermost position. Note the sleeve position and from the graph determine the ball radius r_2.
4. Increase the speed and adjust its speed so that the balls run at the outermost position. Again note down the sleeve position and determine the ball radius r_1.
5. Calculate the forces F_1 and F_2.
6. Calculate the spring stiffness k.

Observations

Sleeve position				
Radius of ball, mm				

Mass of ball, m = kg

Lever arm length:

a = mm

b = mm

Sl. No.	Motor speed, n rpm	Ball radius, r mm	$F = mr\omega^2$ N	Spring stiffness k, N/mm

Calculations

Spring force, $F =$

Spring stiffness, $k =$

Precautions

1. Change the speed of the motor slowly.
2. Measure the speed of the motor accurately.
3. Use a constant voltage transformer to keep the speed constant.

Sources of error

1. Friction between the sleeve and the shaft.
2. Friction between the lever roller and the sleeve.
3. Friction at the lever fulcrum.

Experiment 8 To study the motion of the follower for the given cam and to determine the displacement, velocity and acceleration at every point.

Apparatus Cam and follower, graph sheets.

Theory The cam and follower apparatus (Fig.8) consists of a cam with roller follower (or as may be available). The angle of rotation of the cam and follower displacement can be read from the graduations marked on the cam and follower scale.

The cam may be moving with SHM, uniform acceleration and deceleration, or any other type of motion. The various formulae for the displacement, velocity and acceleration may be seen from Chapter 8.

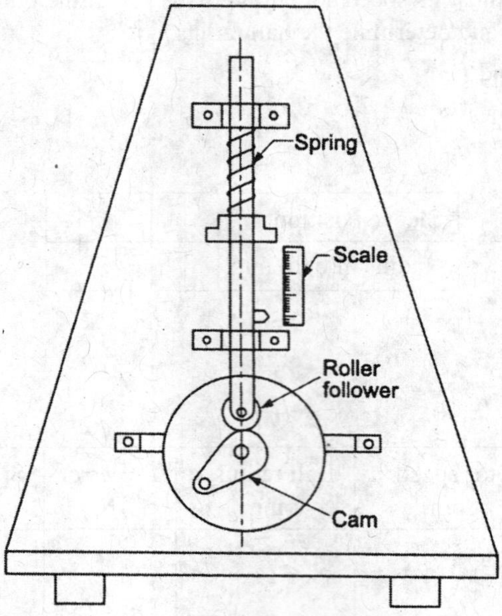

Fig.8 Cam apparatus

Procedure

1. Bring the cam and follower to zero positions.
2. Rotate the cam slowly and note down the angle of rotation of the cam at regular intervals and the corresponding displacement of the follower.
3. Plot a graph between displacement of follower and the angle of rotation of the cam.
4. Plot the velocity and acceleration diagrams by graphical differentiation.
5. Determine the maximum velocity and acceleration during ascent and descent.

Observations

Diameter of roller follower = mm
Base circle diameter of cam = mm
Maximum lift = mm

Sl. No.	Angle of rotation of cam, deg	Displacement of follower, mm

Maximum velocity = mm/s
Maximum acceleration = mm/s^2

Precautions

1. Cam should be rotated slowly and gradually.
2. Cam and roller bearings should be lubricated to reduce friction.

Sources of error

1. Lateral shift in the roller follower and the cam.
2. Effect of clearances in the roller and cam spindles.
3. Effect of elasticity of the links.

Experiment 9 To study the working of Oldham's coupling.

Apparatus Oldham's coupling, graph sheet.

Theory Oldham's coupling apparatus (Fig.9) consists of two shafts having flanges at their ends. The flanges have rectangular slots cut in their middle. An intermediate piece having tongues on both sides perpendicular to each other is used to connect the two flanges. The shafts carrying flanges are mounted on sliding blocks, which enables to change the centre distance between the shafts as desired. The angle of rotation of the flanges and the displacement of the tongue can be measured from the graduated scales. Assuming the speed of rotation of the shafts to be rad/s, we can plot the displacement-time graph. From this graph, we can find the velocity by graphical differentiation.

Maximum sliding speed of each tongue along its slot
= Peripheral velocity of centre of disc along its circular path
= Distance between the axes of the shafts × Angular velocity of each shaft.

Procedure

1. Fix some centre distance between the two shafts.
2. Bring the graduated flange to zero position and note the position of the tongue.

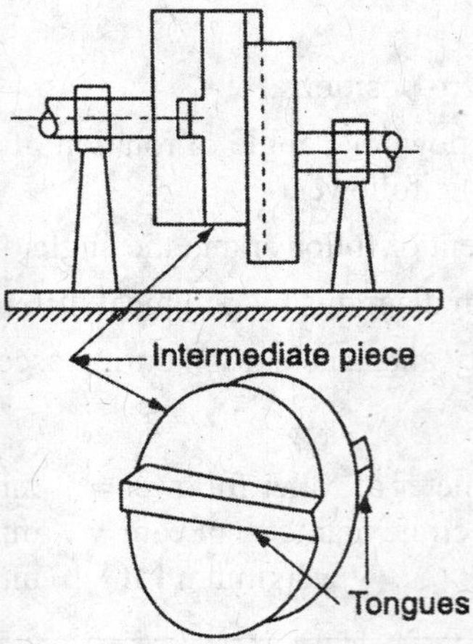

Fig.9 Oldham's coupling

3. Given some known angular rotation to the flange and note the displacement of the tongue.
4. Increase the angular rotation of the flange at regular intervals and note the corresponding displacement of the tongue.
5. Change the centre distance between the shafts and repeat the experiment.
6. Plot displacement of tongue v's angle of rotation of the flange.
7. Assuming angular velocity to be 1 rad/s, plot the displacement-time graph.
8. Determine the velocity-time graph by graphical differentiation.
9. Calculate the theoretical velocity of sliding of the tongue and compare with experimental results.

Observations

Sl. No.	Center distance between shafts, $c =$ mm			
	Angle of rotation of flange, θ, deg	Time = $\pi\theta/180$ s	Displacement of tongue, x, mm	Sliding speed of tongue, v, mm/s

Calculations

Centre distance between shafts, $c =$ mm

Angular speed of shafts = 1 rad/s

Maximum sliding speed of tongue, $v = c$ mm/s

From graph, $v =$ mm/s

Precautions

1. Lubricate the tongue to reduce friction.
2. Fix the shaft bearing firmly after changing the center distance.
3. Measure the angle of flange rotation accurately.

Sources of error

1. Error in the measurement of centre distance.
2. Error in graphical differentiation.

Experiment 10 To determine the speed ratio of a gear train.

Apparatus Spur gear train, string, weights, metre rod, stop watch.

Theory The simple spur gear train in shown in Fig.10. Pulley *D* is mounted on the shaft for gear *A* and pulley *E* on the shaft for gear *C*.

For a simple gear train, the speed ratio is given by,

$$i = \frac{n_i}{n_o} = \frac{z_o}{z_i}$$

where suffixes *i* and *o* represent input and output, respectively.

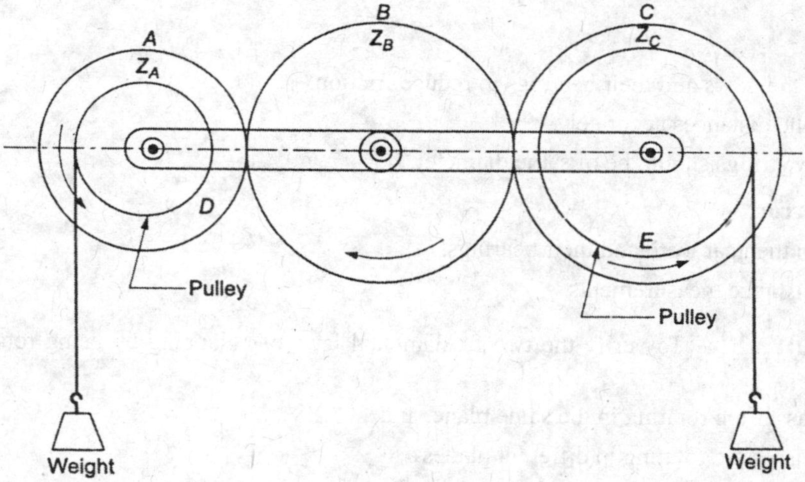

Fig.10 Simple gear train

If the diameters of pulleys *D* and *E* are same, then the speed ratio can be determined by measuring the distances moved by the strings in a given time.

Procedure

1. Put some weights on the string attached to pulley *E*.
2. Add weights on the hanger attached to string passing over pulley *D*.
3. Increase the weight till it starts moving.
4. For the given distance moved by the weight on pulley *E*, determine the distance moved down by the weight on pulley *D* in the same time.
5. Calculate the speed ratio by dividing the distance moved by the weight on pulley *D* to the distance moved by weight on pulley *E*.
6. Calculate the theoretical speed ratio and campare.

Observations

Diameter of pulley $D =$ mm
Diameter of pulley $E =$ mm
Number of teeth, $z_i =$
Number of teeth, $z_o =$

Sl. No.	Distance moved by weight on pulley D, x_1, mm	Distance moved by weight on pulley E, x_0, mm	Speed ratio x_o/x_1

Calculations

$$\text{Theoretical speed ratio} = \frac{z_o}{z_i}$$

Precautions

1. Lubricate the gears and their bearings to reduce friction.
2. Measure the distances accurately.
3. The pulley strings should be of same diameter.

Sources of error

1. Friction in the gear teeth and their bearings.
2. Error in distance measurements.

Experiment 11 To verify the two fundamental laws of balancing by using rotating masses, when

(a) all the masses are rotating in the same plane, and

(b) all the masses are rotating in different planes.

Apparatus Dynamic balancing apparatus for rotating masses.

Theory (a) When all the masses are rotating in the same plane, the following equation results for the equilibrium of the system:

$$\sum_{i=1}^{n} W_i r_i \exp(j\phi_1) + W_b r_b \exp(j\theta) = 0 \tag{1}$$

where W_i = weight of the rotating mass
r_i = radius of ith mass from the centre of rotation
ϕ = angular displacement of the ith mass from the reference axis
W_b = weight of the balancing mass
r_b = radius of the balancing mass from the centre of rotation
θ = angular displacement of the balancing mass from the reference axis.

For a single mass, we have

$$W_i r_i \exp(j\phi_1) = -W_b r_b \exp(j\theta) \tag{2}$$

From (2), we get

$$\cos\phi_1 = -\cos\theta$$

$$\sin\phi_1 = -\sin\theta$$

or
$$\theta = \pi + \phi_1 \tag{3}$$

This implies that weight W_b must be added opposite to the weight W_1. After determining the angle θ, we must determine the product $W_b r_b$ from the following:

$$W_1 r_1 = W_b r_b \tag{4}$$

So, there are two alternatives that we may choose. One of them is to choose the value of r_b and calculate the weight W_b of the balancing mass. The other one is to choose the weight W_b and calculate the distance r_b. The first alternative is preferred because the distances on the disc are fixed. For many rotating masses in the same plane, their resultant has to be determined.

(b). When the masses rotate in different planes, then taking moments about the centre of rotation of balancing mass W_{b2}, we have

$$\sum_{i=1}^{n} W_i r_i \exp(j\phi_i) + \sum_{i=1}^{2} W_{bi} r_{bi} \exp(j\theta_1) = 0 \tag{5}$$

and

$$\sum_{i=1}^{n} W_i r_i l_i \exp(j\phi_i) + W_{b1} r_{b1} l \exp(j\theta_1) = 0 \tag{6}$$

where l_i = axial distance of rotating mass W_i from the balancing mass W_{b2}

l = axial distance of balancing mass W_{b1} from W_{b2}

θ = angle of rotation from first rotating mass

i = angle of rotating mass W_i from W_1

Equations (5) and (6) can be solved simultaneously to determine the unknowns.

Experimental setup The dynamic balancing apparatus for rotating masses is shown in Fig.11. It is intended from primary balancing. It consists of a rectangular steel frame suspended by four springs from a strong steel stand. On rectangular frame, two blocks with ball bearings are mounted which support a steel shaft carrying four balanced discs equally spaced. One of the discs is grooved and this is connected to a balanced 220 volts, A.C. electric motor by V-belt so that the whole system can be rotated. In all the four discs, circumferential slots are provided at four different radii. A number of steel pieces to act as balancing masses are included and these pieces can be attached in the slots of the discs by means of screwed rods and nuts. The discs are marked with radial lines and numbered to read the angular positions of the balancing masses. By attaching the balancing weights to the circumferential slots of the discs at different positions, various combinations of out of balance conditions can be obtained either in one plane or two planes. By switching on the motor and making the system rotate, the out of balance state can be clearly observed due to vibrations and oscillations set up in the system. Now the distance between discs, positions of balancing weights, their magnitude, etc. can be noted and a solution can be obtained analytically or graphically to have balance condition. According to this solution, balancing weights can be attached to the discs at suitable places and the motor can be started. Now the system can be observed, free from oscillations and vibrations and this illustrating the theory of balancing.

Procedure

1. Start the motor and check that the apparatus is completely balanced, i.e. the platform should not oscillate. As the motor is started the platform starts vibrating due to transient vibrations, which die out after some time.

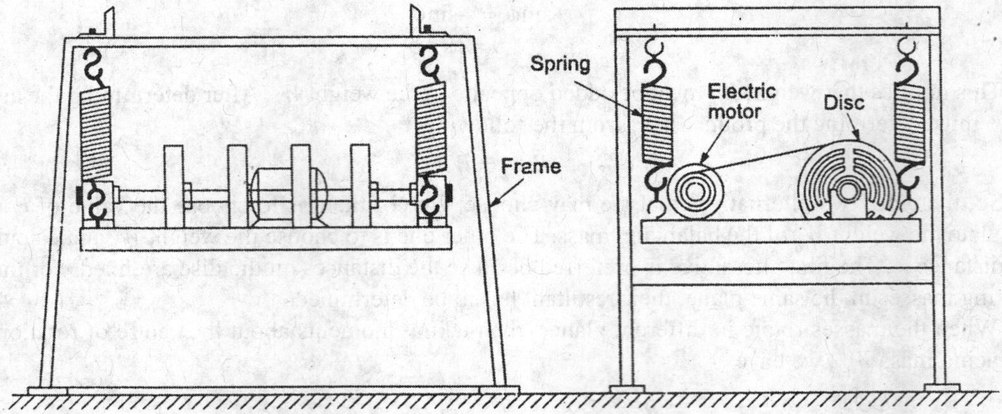

Fig.11 Rotating masses apparatus

2. Fix a known weight in the circumferential slot of any disc, preferably the middle one, at the known distance. Balance the platform by fixing the four different balancing masses at different distances from the centre of rotation of discs.
3. Verify the law $Wr = W_b r_b$ and $\theta = \pi + \phi_1$ by noting the distances and the weight of all the masses.
4. Fix three known weights in the circumferential slots of the same disc at three different or same distance from centre of rotation and balance the platform by fixing the balancing weight. Note down the weight and distances of all the masses. Change to a different balancing weight placed at different distance to balance the platform. Take four different readings. Verify the first law of balancing. Draw the force polygon for any one of the readings.
5. Fix a known weight in the circumferential slot of the same disc at a known distance from the centre of rotation of the disc. Select the two balancing weights and fix them in the slots of two different discs and balance the platform. Note down the axial and radial distances of the weights and value of weight fixed in the slot. Take four different readings for different balancing weights with different axial and radial distances. Verify the second law of balancing.
6. Fix two known weights in the circumferential slots of any of the two different discs at a known axial and radial distances with some angular displacement. Now fix the balancing weights in the remaining two discs. Put the balancing weights in the circumferential slots at a suitable radial distance so that the platform is balanced. Note down the value of weights and the axial and radial distances. Take four readings for different balancing weights at different radial distances. Verify the second law of balancing. Draw the vector diagram.

Precautions

1. Weights should be securely tightened to the discs.

2. Take readings after the system has stabilized.
3. Initial check on the apparatus being completely balanced should be made.

Observations

Sl. No.	Unbalance			Balancing weights		
	Weight	Axial Distance	Radial Distance	Weight	Axial Distance	Radial Distance
	W_1 W_2 W_3			W_{b1} W_{b2} W_{b3}		

Sources of error

1. Error in fixing the weights at an angular position.
2. Error in the value of weights.
3. Error in judging the balanced platform.

A-2

GLOSSARY OF TERMS

Mechanisms

Machine It is a contrivance which transforms energy available in one form or another to do the desired work.

Element It is a part of machine which has been manufactured without the operation of assembling.

Link It is a resistant body or assembly of resistant bodies, which constitute part or parts of a machine, connecting other parts which have motion relative to it.

Rigid link A link is called rigid, when it does not undergo any deformation while transmitting motion, for example a connecting rod, a crank etc.

Flexible link It is a link which while transmitting motion is partly deformed in a manner not to affect transmission of motion, for example, belts and springs.

Floating link It is not connected to the frame.

Fluid link It is formed by having fluid in a receptacle.

Binary link A link having connections at two points.

Ternary link A link having connections at three or more points.

Kinematic pair It is a movable joint of two links which are in contact so that the relative motion between the two links is constrained.

Lower pair It is formed by two links having surface contact while in motion. The relative motion is purely turning or sliding, for example, a universal joint, an automobile steering gear, a shaft revolving in a bearing, a straight line motion mechanism, etc.

Higher pair It is formed by two links having point or line contact while in motion. The relative motion being the combination of sliding and turning, for example a belt, a rope, a chain, gears, cams, ball and roller bearings.

Sliding pair When two links are so connected that one is constrained to have sliding motion relative to another, they form a sliding pair, for example, cross-head and guides.

Turning pair When two links are so connected that one is constrained to turn or revolve about a fixed axis of another link, they form a turning pair, for example a crankshaft turning in a bearing.

Rolling pair When tow links are so connected that one is constrained to work in another link which is fixed, they form a rolling pair, for example ball and roller bearings.

Screw (or helical) pair When one element turns about the other element by means of threads, they form a screw pair. The relative motion is a combination of sliding and turning, for example, a bolt and nut, the lead screw of a lathe.

Spherical pair When one element in the form of a sphere turns about the fixed element, they form a spherical pair, for example a ball and socket joint.

Closed pair When two elements are held together mechanically, forms a closed pair lower pair is a closed pair. Screw pair and spherical pair are closed pairs.

Open (or unclosed) pair When two elements are not held together mechanically, they form an open pair. A cam and follower is an open pair.

Kinematic chain When kinematics pairs are so connected that the last link is joined to the first link to transmit a definite constrained motion, they form a kinematic chain. For a kinematics chain, $L = 2P - 4 = 2(J + 2)/3$, where L = number of links, P = number of pairs and J = number of joints.

Mechanism It is an assemblage of a number of rigid links so formed and connected that they move upon each other with a definite relative motion. A mechanism is formed by fixing one of links of a kinematics chain.

Simple mechanism It is one which has upto four links, for example cams, gears, the beam engine and the elliptical trammel.

Compound mechanism It is one which has more than four links.

Degrees of freedom Degrees of freedom of a mechanism are the number of inputs a mechanism must have in order to fulfill a useful engineering purpose. It may be defined as the number of independent relative motions, a pair can have. $F = 6$ – number of restraints.

Gruebler criterion This criteria states the degrees of freedom of a mechanism, as follows: $F = 3(L - 1) - 2g - h$, where F = degrees of freedom, L = number of links, g = number of lower pairs, h = number of higher pairs.

Structure A mechanism is called a structure if $F = 0$.

Constrained mechanism It is one for which $F = 1$.

Grashof criteria This criteria states that for a mechanism $(l + s) < (a + b)$, where l, s = length of the longest and shortest link respectively and a, b = length of other links.

Plane mechanism It is a mechanism having all the links in the same plane.

Spatial mechanism It is a mechanism having links in different planes.

Complex mechanism It is formed by the inclusion of ternary or higher order floating link to a simple mechanism.

Kinematics of machines It deals with the study of relative motion of parts of which the machines are constituted, neglecting consideration of forces producing it.

Dynamics of machines It deals with the study of motion of a machine under the forces acting on different parts of the machine.

Resistant body It is one which does not suffer appreciable distortion or change in physical form by the forces acting on it. Resistant bodies need not be rigid, such as springs, belts, fluids, etc.

Completely constrained motion It is one in which the motion takes place in a definite direction, for example a rectangular bar moving in a rectangular hole, a shaft with collars at each end rotating in a round hole.

Partially constrained motion It is one in which the constrained motion is not completed by itself but by some other means, for example, a foot step bearing and the rotor of a vertical turbine.

Incomplete constrained motion It is one in which the links are so connected that motion can take place in more than one direction, for example a circular bar moving in a round hole.

Inversion of a mechanism Different mechanisms formed by fixing different links of the same kinematic chain are known as inversions of each other. The inversions of four bar chain are the beam engine, the engine indicator and the coupled wheels of locomotives. The inversions of slider-crank chain are the pendulum pump, the oscillating cylinder engine, the crank and slotted lever type quick-return motion, the Whitworth mechanism and the Gnome engine. Inversions of double slider crank chain are the donkey pump, Oldham's coupling and the elliptical trammel.

Instantaneous centre A link or rigid body as a whole may be considered to be rotating about an imaginary centre or a given centre at a given instant which has zero velocity. Then the link is at rest at this point which is known as the instantaneous centre or centre of rotation. Number of instantaneous centres, $N = n(n-1)/2$, where n = number of links.

Kennedy's theorem of three centres This theorem states that if three bodies have relative motion with respect to each other, their relative instantaneous centres lie on a straight line.

Primary instantaneous centre It is one which is either fixed or permanent.

Secondary instantaneous centre It is one which is neither fixed nor permanent.

Angular velocity ratio theorem This theorem states that the ratio of the angular velocities of any two bodies moving in a constrained system is inversely proportional to the ratio of the distances of their common instantaneous centre from their centre of rotation.

Total acceleration of a point in a rigid link The total acceleration of end B with respect of end A of a rigid link AB is the vector sum of the radial (centripetal) and normal (tangential) accelerations, that is,

$$\overrightarrow{f_{ba}} = w^2 \cdot \overrightarrow{AB} + \alpha \cdot \overrightarrow{AB}$$

Acceleration centre The acceleration centre of a link is one which has zero acceleration.

Klein's construction It is a graphical procedure of drawing the acceleration diagram for a reciprocating engine, that is a slider-crank mechanism.

Coriolis acceleration If the distance between the two points does not remain fixed and the second point slides, the total acceleration will contain an additional component of acceleration, known as coriolis acceleration. Coriolis component of acceleration is equal to $2\upsilon\omega$, where υ is the sliding velocity and ω the angular speed. The direction of Coriolis acceleration is such as to rotate the slider velocity vector in the same sense as the angular velocity of the link.

Motion of Slider-crank Mechanism

Displacement,

$$x = r\cos\theta + l\left[1 - \left(\frac{\sin\theta}{n}\right)^2\right]^{1/2}$$

where $n = l/r$,

l = length of connecting rod,

r = radius of crank,

θ = crank angle,

Velocity of piston,

$$v_p = -r\omega\left[\sin\theta + \frac{\sin 2\theta}{2n\left[1-\left(\frac{\sin\theta}{n}\right)^2\right]^{1/2}}\right] \simeq -r\omega\left[\sin\theta + \frac{\sin 2\theta}{2n}\right]$$

Acceleration,

$$f_p = -r\omega^2\left[\cos\theta + \frac{\cos 2\theta}{n}\right]$$

Lower Pairs

Pantagraph It is a mechanism in which a point describes a path similar to another point. It is used for tracing a curve on a magnified or reduced scale.

Straight line motion mechanisms Peaucellier, Hart and Scott-Russel are for accurate straight line and Grasshopper, Watt and Tchebicheff are for approximate straight line motion.

Engine pressure indicators Simplex, Crosby, Richard, Thompson and Dobbie-McInnes are engine pressure indicators.

Automobile steering gear mechanisms The two steering gears for automobiles are Davis and Ackermann. The later is most commonly used. For correct steering $\cot\phi - \cot\theta = b/l$, where ϕ = outer turning angle and θ = inner turning angle.

The steering mechanism automatically adjusts the values of the inner and outer turning angles. For Davis gear, $\tan\alpha = b/l = 0.4$ to 0.5 and for Ackermann gear, $\tan\alpha = (\sin\phi - \sin\theta)/(\cos\theta + \cos\phi - 2)$ and b/l is nearly 0.455. b = distance between the pivots of front axle and l = wheel base.

Hooke's coupling It is used in the propeller shaft of an automobile. The ratio of the angular speeds of the driven to the driving shafts is $\cos\alpha/(1 - \sin^2\alpha\cos^2\theta)$, where α is the angle between the axes of the two shafts and θ is the angle turned through by the driving shaft. If ϕ is the angle turned through by the driven shaft, then $\tan\theta = \cos\alpha \cdot \tan\phi$. The maximum ratio of angular speeds is $1/\cos\alpha$ at $\phi = 0°$ and $180°$. The minimum ratio is $\cos\alpha$ at $\theta = 90°$ and $270°$.

Angular acceleration of driven shaft $= -\omega_a^2 \cos\alpha \sin^2\alpha \sin 2\theta/(1 - \sin^2\alpha\cos^2\theta)^2$.

For acceleration to be maximum or minimum, $\cos 2\theta \cong 2\sin^2\alpha / (2 - \sin^2 a)$.

Belts Belts are used for power transmission. The two types of belts are flat and V-belts. The velocity ratio is inversely proportional to the pulley diameters. The length of cross-belt is more than the open belt length. The slip between the belt and pulley decreases the speed ratio. The ratio of tight side to slack side tensions is equal to $\exp(\mu\theta)$, where μ is the coefficient of friction between the belt and pulley and θ is the angle of arc of contact. For the v-belt, the virtual coefficient of friction is $\mu/\sin\alpha$, where α is the pulley semi-groove angle. Initial tension is half of the sum of the tight and slack side tensions. The centrifugal tension in the belt decreases the power transmission capacity of the belt.

Speed ratio, $\dfrac{N_2}{N_1} = \dfrac{D_1 + t}{D_2 + t}$ and percentage slip, $s = s_1 + s_2 - 0.01 s_1 s_2$.

Length of open belt, $L_o = \frac{\pi(D_1 + D_2)}{2} + \phi(D_2 - D_1) + 2C\cos\phi$, where $\sin\phi = \frac{D_2 - D_1}{2C}$.

Length of cross belt, $L_c = (\pi + 2\phi)\frac{(D_1 + D_2)}{2} + 2C\cos\phi$, where $\sin\phi = \frac{D_2 - D_1}{2C}$.

Centrifugal tension, $T_c = \frac{wv^2}{g}$, where w = weight of belt per unit length.

For maximum power to be transmitted, $T_c = \frac{T_1}{2}$ or $\frac{T_1 + T_c}{3}$.

Power transmitted, $P = \frac{(T_1 - T_2)v}{1000}$ kW

Pitch surface It is an imaginary surface around the pulley to which the neutral section of the belt is tangential. The radius of this surface is the effective radius of the pulley.

Fast pulley It is one which transmit power.

Loose pulley It is one which does not transmit power.

Idler pulley It is free to rotate on its axis and is used to increase the tension in the belt, taking up stretch in the belt and increasing the angle of contact of the two pulleys.

Crowning of pulleys The convex shape given to the rim of the pulley is called crowning. It prevents the belt from running off the pulley by making the belt run in the centre of the pulley width.

Slip It is the relative motion between the belt and the pulley due to insufficient friction.

Creep It is due to the unequal stretching of the belt on the tight and slack sides. It leads to partial slip and reduced peripheral speed of driven pulley than the driving pulley.

$$\text{Creep of belt} = \frac{T_1 - T_2}{bt\,E}$$

Brakes They are used to decrease the speed of a moving body or to stop it when desired. The brakes are of the following types: block and shoe brake, band brake, band and block brake and internal expanding shoe brake. Railway bogies use block and shoe brake whereas automobiles use internal expanding shoe brake. To stop a moving body, both the translational and rotational kinetic energies have to be absorbed.

Dynamometers It is a device to measure the power being transmitted by a prime mover. Dynamometers are of the following types: Prony (rope) brake, belt transmission and torsion dynamometers.

Governors The function of a governor is to keep the speed of a prime mover constant by adjusting the input. It regulates the speed over a number of cycles of the prime mover. Governors may be classified as follows:

1. Centrifugal governors (a) simple Watt (pendulum type) (b) loaded (i) dead weight type—Porter, Proell (ii) spring controlled type—Hartnell, Wilson-Hartnell, Gravity and spring control, Hartung and Pickering.
2. Intertia governor.

   For simple Watt governor, height of governor, $h = (g/\omega^2)\,[(W + W_1/2)/(w + W_1/3)]$ where W = weight of ball, W_1 = total weight of arm, ω = angular speed.

For Porter governor, $h = (g/\omega^2)\ [(W + w_0)/w_0]$, where W = dead weight, and w_0 = weight of ball.

Controlling force A single force replacing all the forces which tries to pull the ball in a radially inward direction is known as the controlling force.

Quality of a governor It is ascertained by the sensitiveness, staility, effort and power.

Sensitiveness It is defined as the ratio of the range of speed to mean speed.

Stability A governor is said to be stable when for each speed within the working range, there is only one radius of rotation of the governor balls at which the governor is in equilibrium.

Isochronism A governor is termed as isochronous when the equilibrium speed is constant for all radii of rotation of the balls within the working range.

Hunting It is a condition in which the speed of the engine controlled by the governor fluctuates continuously above and below the mean speed.

Effort of governor It is the average force that acts on the sleeve for a given percentage change of speed (generally 1%).

Power of a governor It is defined as the work done on the sleeve for a given percentage change of speed. Thus power is the product of the effort and the displacement of the sleeve.

Cam A cam may be defined as a rotating or a reciprocating element of a mechanism which imparts a rotating. Reciprocating or oscillating motion to another element termed as follower.

Basic circle It is the circle with the least radius of the cam such that the lift of the follower is zero over this circle.

Lift It is the difference between the maximum distance of the lowest point of the follower from the axis of rotation of the cam and the least radius.

Angle of ascent It is the angle moved by the cam, from the instant the follower begins to rise, till it reaches the highest position.

Angle of dwell It is the angle through which the cam rotates during the period in which the follower remains in the highest position.

Angle of descent It is the angle during which the follower returns to its initial position.

Angle of action It is the total angle moved by the cam from the beginning of ascent to the termination of descent.

Pressure angle It is the angle between the line of motion of the follower and a line normal to the cam profile at the point of contact.

The trace point It is a reference point on the follower for the purpose of tracing the cam profile. In the case of a roller follower, it is the centre of the roller and in the case of knife edge follower, the knife edge.

Pitch curve It is the path of the trace point.

The prime circle It is the smallest circle drawn to the pitch curve from the centre of rotation of the cam.

The cam angle It is the angle of rotation of the cam for a definite displacement of the follower.

The pitch point It is the point on the cam pitch curve having the maximum pressure angle.

The pitch circle It is the circle with centre as the centre of the cam axis and radius such that it passes through the pitch point.

The cam profile It is the actual working contour or curve of the cam.

Motion of the follower The motion of the follower may be simple harmonic type, uniform acceleration and deceleration and cycloidal type.

Undercutting It is the condition of the constructed cam profile that has an inadequate curvature to produce correct follower movement. To avoid undercutting in a convex curve, the radius of curvature of pitch curve should be greater than the radius of the roller follower.

Follower motion

1. Simple harmonic motion: $y = 0.5s\,(1 - \cos\theta)$, s = lift, θ = angle turned through.

   $v = 0.5s\omega \sin\theta$, $v_{max} = s\omega/2$; $f = 0.5s\omega^2 \cos\theta$, $f_{max} = s\omega^2/2$.

2. Uniform acceleration and deceleration: $v = ft$, $y = 0.5ft^2$, $v_{max} = 2\omega s/\theta$; $f = 4\omega^2 s/\theta$.

   Cycloidal motion: $y = (s/\pi)\,[\pi\theta/\theta_1 - 0.5\sin(2\pi\theta/\theta_1)]$

   $v = (s\omega/\theta_1)\,[1 - \cos(2\pi\theta/\theta_1)]$, $v_{max} = 2s\omega/\theta_1$

   $f = (2\pi s\omega^2/\theta_1^2)\,[\sin(2\pi\theta/\theta_1)]$

   Maximum pressure angle = 25 to 35°.

Gears

Spur gear It is a cylindrical gear with tooth traces that are straight lines parallel to the gear axis. They are used for connecting shafts whose axes are parallel.

Rack It is a spur gear of infinite diameter.

Helical gear It is a cylindrical gear with teeth that are inclined at an angle to the gear axis.

Herringbone gear It is a gear with half of its width cut with tooth helix in one direction and the other half in the opposite direction.

Straight bevel gear It is a gear with tooth traces that are straight line generators of the cone. They are used for connecting shafts with axes intersecting generally at 90°.

Spiral bevel gears These are gears with tooth traces that are curved and oblique lines.

Hypoid gears They are similar to spiral bevel gears. They are used for connecting shafts whose axes are non-intersecting and non-parallel.

Worm gears They are used for connecting shafts with axes that are perpendicular and non-intersecting. They are used for high speed reductions, of the order of 100:1.

Terminology of Gears

Pitch circle diameter It is the diameter of a circle which would produce the same motion as the toothed gear wheel by pure rolling action.

Base circle It is the circle from which involute form is generated.

Pitch surface It is the surface of the disc which the toothed gear has replaced at the pitch circle.

Pitch point It is the pitch of the tangency or the point of contact of the two pitch circles of the mating gears.

Circular pitch It is the distance measured along the circumference of the pitch circle from a point on one tooth to a corresponding point on the adjacent tooth.

Base pitch It is equal to circular pitch $x \cos \phi$, where ϕ = pressure angle of the gear tooth profile.

Diametral pitch It is expressed as the number of teeth per unit pitch circle diameter.

$$\text{Circular pitch} \times \text{Diametral pitch} = \pi$$

Module It is expressed as the length of the pitch circle diameter per unit number of teeth.

Addendum It is the radial height of the tooth above pitch circle.

Addendum circle It is a circle bounding the top of the teeth.

Dedendum It is the radial depth of a tooth below the pitch circle.

Dedendum circle It is a circle passing through the roots of all the teeth.

Clearance It is the radial height difference between addendum and dedendum of teeth.

Face It is the part of the tooth surface lying below the pitch surface.

Backlash It is the minimum distance between the non-driving side of a tooth and adjacent side of the mating tooth at the pitch circle.

Profile It is the curve forming face and flank.

Tooth thickness It is the arc distance measured along the pitch circle from its intercept with one flank to its intercept with the other flank of the same tooth.

Face width It is the width of the gear tooth measured axially along the pitch surface.

Top land It is the surface of the top of the tooth.

Tooth fillet It is the radius that connects the root circle to the profile of the tooth.

Tooth space It is the width of the space between two teeth measured on the pitch circle.

Pressure angle It is the angle between the common normal at the point of contact and the common tangent at the pitch point. The pressure angle is either 14.5° or 20°.

Path of contact It is the locus of the point of contact of two mating teeth from the beginning of engagement to the end of engagement. It is a straight line.

Path of approach It is the portion of the path of contact from the beginning of engagement to the pitch point.

Angle of approach It is the angle turned by gears during the path of approach.

Path of recess It is the portion of the path of contact from the pitch point to the end of engagement of the two mating teeth.

Angle of recess It is the angle turned through during path of recess.

Arc of contact It is the locus of a point on the pitch circle, from the beginning of engagement to the end of engagement of pair of teeth in mesh.

$$\text{Minimum number of teeth to avoid interference, } z = 2/\sin^2\phi. \text{ For } \phi = 20°, z = 17$$

Law of gearing This law states that the common normal at the point of contact always passes through a fixed point (pitch point) on the line joining the centres of rotation.

For constant angular velocity ratio of gearing, the common normal at the point of contact divides the line joining the centres of rotation in the inverse ratio of the angular velocities.

Velocity of sliding It is equal to the sum of the angular speeds of the driving and driven gears multiplied by the distance of the point of contact from the pitch point.

Involute The involute of a circle is the curve traced by the end of a thread as it is unwound from a stationary cylinder. inv (ϕ) = tan $\phi - \phi$.

Base circle diameter It is equal to pitch circle diameter x cos ϕ.

Cycloid It is the locus of a point on the circumference of a circle which rolls without slipping on a fixed straight line.

Interference It is the portion of a gear tooth below the base circle that cut as a radial line and not an involute curve. Therefore, if contact should occur below the base circle, non-conjugate action would result, leading to interference. Interference can be avoided by undercutting, making stub teeth, increasing the pressure angle and cutting the gears with long and short addendum gear teeth.

Helical Gears

Helix angle It is the angle between a line drawn through one of the teeth and the centre line of the shaft on which the gear is mounted.

Normal circular pitch (p_n) It is the distance between corresponding points of adjacent teeth as measured in a plane perpendicular to the helix. It is the perpendicular distance between two adjacent teeth. $p_n = p_t \cos\beta$, $P_n = P_t/\cos\beta$.

Comparison between involute and cycloidal gears

Characteristic	Involute gears	Cycloidal gears
1. Pressure angle	Constant throughout the engagement	Varies from commencement to end
2. Easie of manufacture	Easy to manufacture	Difficult to manufacture
3. Centre distance	Do not require exact centre distance	Requires exact centre distance
4. Interference	May occur	No interference
5. Strength	Less	More
6. Wear	More	Less
7. Running	Smooth	Less smooth

Transverse circular pitch (P_t) It is the distance measured in a plane perpendicular to the shaft axis between the corresponding points of adjacent teeth. $p_t = \pi d/z$, $P_t = z/d$.

Axial pitch (p_x) It is the distance measured in a plane parallel to the shaft axis between corresponding points of adjacent teeth. $p_x = p_t \cot\beta$.

Lead It is the distance measured parallel to the axis to represent the distance advanced by each tooth per revolution.

Lead angle It is the acute angle between the tangent to the helix and a plane perpendicular to the axis of cylinder.

Virtual (or formative or equivalent) number of teeth (z_v) The number of teeth of the equivalent spur gear in the normal plane is called the virtual number of teeth.

$z_v = z/\cos^3\beta, \beta$ = helix angle; normal module, $m_n = m_t \cos\beta$, $\tan\phi_n = \tan\phi_t \cos\beta \cdot P_t$, $p_t = P_n$

Bevel Gears

Pitch cone It is the pitch surface of a bevel gear in a gear pair.

Cone centre It is the apex of the pitch cone.

Pitch cone radius It is the length of the pitch cone element.

Pitch angle (*d*) It is the angle that the pitch line makes with the axis of the gear.

Reference cone angle It is the angle between the axis and the reference cone generator containing the root cone generator.

Tip (or face) angle It is the angle between the tip cone generator and the axis of the gear.

Root (or cutting) angle It is the angle between the root cone generator and the axis of the gear.

Back cone It is an imaginary cone the elements of which are perpendicular to the elements of the pitch cone at the larger end of the tooth.

Gear diameter It is the diameter of the largest pitch circle.

Virtual number of teeth It is the number of teeth on an imaginary spur gear laid out on a pitch radius equal to the back cone radius. $z_v = z/\cos\delta$.

Crown gears It is a gear pair for which pitch cone angle is 90°.

Miter gears There are two bevel gears of the same size having a pitch cone angle of 90°.

Worm Gears

Axial diametral pitch It is the quotient of the number π by the axial pitch.

Diametral quotient (*q*) It is the ratio of the reference diameter to the axial module, $q = d/m$.

Axial Module (m_x) It is the quotient of the axial pitch by the number π.

Axial circular pitch (p_x) It is the distance, measured parallel to the axis of the worm, between two consecutive corresponding profiles.

Lead (p_z) It is the distance between two consecutive intersections of a helix and a straight generator of the cylinder on which it lies.

Length of the worm It is the length of the toothed part of the worm measured parallel to the axis on the reference cylinder.

Gear ratio It is the quotient of the number of teeth on the wheel divided by the number of threads on the worm.

Torus It is the surface of revolution generated by the rotation of a circle around an axis external to this circle and situated in its plane.

Gorg It is part of the tip surface in the form of a portion of a torus with the same middle circle diameter as the reference torus.

Tooth width It is the distance between two planes perpendicular to the axis containing the circles of intersection of the reference torus and the lateral faces of the teeth.

Width angle In the generating circle of the reference torus, the angle at the centre included between the points of intersection of this circle with the lateral faces of the teeth is called width angle.

Lead angle It is the angle between a tangent to the pitch helix and the plane of rotation of the worm.

$$\tan \gamma = \frac{p_z}{\pi d_w} = \frac{v_g}{v_w}; \; p_z = z_w p_x; \; p_n = p_x \cos \gamma; \; m_x = \frac{m_n}{\tan \gamma}$$

Efficiency, $$\eta = \frac{(\cos \alpha_n - \mu \tan \gamma)}{(\cos \alpha_n + \mu \cot \gamma)}$$

Gear Trains

Simple gear train A simple gear is one in which each shaft carries only one gear. $N_1/N_{n+1} = z_{n+1}/z_1$.

Compound gear train A compound gear train is one in which all the intermediate shafts carry two gears and the first and last shaft carry only one gear.

$$\frac{N_1}{N_4} = \left(\frac{z_2}{z_1}\right) \cdot \left(\frac{z_4}{z_3}\right)$$

Reverted gear train A reverted gear train is one in which the first and the last gears are on the same shaft.

$$d_1 + d_2 = d_3 + d_4; \; \frac{N_1}{N_4} = \left(\frac{z_2}{z_1}\right) \cdot \left(\frac{z_4}{z_3}\right)$$

Epicyclic gear train The axis of rotation of one or more of the gears is carried on an arm which is free to revolve about the axis of rotation of one of the other gears in the train. The speed ratio of these gears trains can be found either by the relative velocity method or by the tabular (or algebraic) method.

Inertia Force in Mechanisms

Dynamical equivalent system Two systems are said to be dynamically equivalent to one another, if by application of equal forces, equal linear and angular accelerations are produced in the two systems. For two masses m_1 and m_2 having accelerations a_1 and a_2 respectively, the conditions for dynamically equivalent system are:

$$m = m_1 + m_2; \quad m_1 a_1 = m_2 a_2; \quad \text{and} \quad mK_G^2 = m_1 a_1^2 + m_2 a_2^2$$

where K_G = radius of gyration about the centre of graving and $K_G^2 = a_1 a_2$.

Dynamics of reciprocating parts Let R = weight of reciprocating parts, then accelerating force for reciprocating parts = $(R/g)\,fp$; $\sin \phi = \sin \theta / n$.

Thrust in connecting rod, $Q = P/\cos \phi$, P = piston effort.

Reaction of guide bar, $S = P \tan \phi$; Crank pin effort, $T = Q \sin (\theta + \phi)$

Force in crank, $W = Q \cos (\theta + \phi)$

Crank effort = $Tr = Pr\,[\sin \theta + \sin 2\theta / \{2(n^2 - \sin^2\theta)^{1/2}\}]$

Flywheel The function of a flywheel is to decrease the variation of speed during one cycle by storing up energy during the working stroke of the engine and releasing it during the idle stroke. Fluctuation of energy = $5.589\times10^{-4}\ WK^2\ (N_1+N_2)(N_1-N_2)$, where W = weight of flywheel and K its radius of gyration.

Gyroscopic and Precessional Motion

Precessional motion It is the motion in which the plane of rotation varies from instant to instant.
Axis of spin It is the axis about which the body revolves.
Gyroscopic effect It is the combined effect of the plane of spin, the plane of precession and the plane of gyroscopic couple. The axis of spin, couple and precession are mutually perpendicular.
Axis of precession It is the third axis about which a body revolves and is perpendicular to both the axis of spin and that of the couple.
Gyroscope It is a body which while spinning about an axis is free to move in other directions under the action of external forces.

Gyroscopic Couple of a plane Disc = Iww_p, where I= moment of inertia of the disc, w = spinning angular velocity, and w_p = angular speed of precession.

Balancing

1. Balancing of a single rotating mass shall require a single mass to balance it rotating in the same plane. $Bb = Mr$
2. Balancing of a single rotating mass by a balanced mass rotating in a different plane parallel to the plane of the unbalanced mass shall require two balancing masses, which can either be arranged in two different planes on the side of the plane of rotation of the unbalanced mass or on the opposite side of the plane of rotation of the unbalanced mass.

$$B_1b_1 = \frac{Mra_2}{d} \quad \text{or} \quad B_2b_2 = \frac{Mra_1}{d}$$

   where d = distance between the planes.
3. For the balancing of several masses rotating in the same plane, the force polygon must close. If the force polygon does not close then the closing side of the polygon taken in the reverse order gives the resultant in magnitude and direction. The balancing mass must be placed at a convenient radius opposite to the resultant force. $\Sigma Mr = 0$.
4. For the balancing of number of masses rotating in different planes, the force polygon and the couple polygon must close.

$$\Sigma Mr = 0 \quad \text{and} \quad \Sigma Mra = 0$$

   a = distance from the reference plane.
5. For the balancing of reciprocating parts, the primary and secondary forces must be balanced. The frequency of the secondary forces is double the frequency of the primary forces. The primary forces are generally balanced partially. For resultant unbalanced primary force to be minimum, the balancing should be 50% but generally 2/3rd of the primary forces are balanced. This gives rise to swaying couple and hammer blow.

$$F_p = R\omega^2 r \cos\theta, F_s = R(2\omega)^2 \left(\frac{r^2}{4l}\right) \cos 2\omega t, c = \frac{Bb}{Rr}$$

6. For the balancing of connecting rod of an engine, 2/3rd of its mass is considered to be rotating at the crank pin and 1/3rd reciprocating along with the gudgeon pin.

Swaying couple It is the couple produced due to unbalanced parts of the primary disturbing forces acting at a distance between the line of stroke of the cylinders.

$$C = (1 - c)R\omega^2 ra\sqrt{2} \text{ at } 45° \quad \text{and} \quad 1350$$

Hammer blow It is the maximum value of the unbalanced vertical force or the balance weights. It is counterbalanced by the self weight of the engine and acts on the rails. Hammer blow = $Bb\omega^2$ at 90° and 270°, where B = balance mass of reciprocating parts alone. To avoid lifting of wheels from the rails, $\omega = [Mg/Bb]^{1/2}$, where Mg = dead load on each wheel. Net pressure on rails = $Mg \pm Bb\omega^2$.

Balancing of In-line Engines

(a) Two-cylinder engines Primary forces are automatically balanced. Primary couples, secondary forces and couples have to be balanced. Equivalent radius of crank for secondary forces is $r^2/4l$ and equivalent frequency is 2ω.

(b) Four-cylinder engine Primary forces and couples are automatically balanced. Secondary force is to be balanced and secondary couple is zero.

Balancing of *V*-Engines Both the resultant primary and secondary forces are to be balanced.

Friction

Flat pivot Frictional moment, $M = (2/3)\,\mu\, WR$ for uniform pressure and $0.5\,\mu\, WR$ for uniform rate of wear. Intensity of pressure = $W/(\pi R^2)$.

Conical pivot $M = (2/3)(\mu/\sin\alpha)\, W R$ for uniform pressure and $(1/2)(\mu/\sin\alpha)\, W R$ for uniform rate of wear, α = semi-cone angle.

Flat collar

$$M = \left(\frac{2}{3}\right)\mu W \frac{(r_1^3 - r_2^3)}{(r_1^2 - r_2^2)} \quad \text{for uniform pressure}$$

$$= \mu W \frac{(r_1 + r_2)}{2} \quad \text{for uniform rate of wear}$$

Intensity of pressure, $$p = \frac{W}{\pi(r_1^2 - r_2^2)}$$

Conical collar

$$M = (2/3)\,(\mu/\sin\alpha)\,W(r_1^3 - r_2^3)/(r_1^2 - r_2^2) \quad \text{for uniform pressure}$$

$$= (1/2)\,(\mu/\sin\alpha)\,W(r_1 + r_2) \quad \text{for uniform rate of wear}$$

$$p = W/\pi\,(r_1^2 - r_2^2)$$

MULTIPLE CHOICE QUESTIONS WITH EXPLANATORY NOTES

A-3

1. Scotch yoke mechanism is used to generate
   (a) sine functions (b) square roots (c) logarithms (d) inversions.

2. The danger of breakage and vibration is maximum
   (a) below critical speed (b) near critical speed
   (c) above critical speed (d) none of the above.

3. In full depth $14\frac{1^\circ}{2}$ degree involute system, the smallest number of teeth in a pinion which meshes with rack with out interference is
   (a) 12 (b) 16 (c) 25 (d) 32.

4. Inversion of a mechanism is
   (a) changing of a higher pair to lower pair
   (b) obtained by fixing different links in a kinematic chain
   (c) turning it upside down
   (d) obtained by reversing the input and output motion.

5. The sense of Coriolis component 2 Vw is the same as that of the relative velocity vector V rotated.
   (a) 45° in the direction of rotation of the link containing the path
   (b) 45° in the direction opposite to the rotation of the link containing the path
   (c) 90° in the direction of rotation of the link containing the path
   (d) 180° in the direction opposite to the rotation of the link containing the path.

6. Under logarithmic decrement, the amplitude of successive vibrations are
   (a) constant (b) in arithmetic progression
   (c) in geometric progression (d) in logarithmic progression.

7. Match the following $14\frac{1^\circ}{2}$ deg composite system of gears

List I	List II
A. Dedendum	1. 2/P
B. Clearance	2. 0.157/P
C. Working depth	3. 1.157/P
D. Addendum	4. 1/P

Codes:

	A	B	C	D
(a)	1	2	3	4
(b)	4	3	2	1
(c)	3	2	1	4
(d)	3	1	2	4

8. A certain machine requires a torque of (500 + 50 sin 2θ) kN · m to drive it, where θ is the angle of rotation of shaft measured from certain datum. The machine is directly coupled to an engine which produces a torque (500 + 50 sin θ) kN · m in a cycle. How many times the value of torque of machine and engine will be identical?

(a) 1 (b) 2 (c) 4 (d) 8.

9. The curve traced by a point on the circumference of a circle which rolls along the inside of a fixed circle, is known as

(a) epicycloid (b) hypocycloid (c) cardiod (d) involute.

10. In Oldham's coupling the condition for maximum speed ratio is

(a) $\frac{\omega_1}{\omega} = \cos a$ (b) $\frac{\omega_1}{\omega} = \sin a$ (c) $\frac{\omega_1}{\omega} = \frac{1}{\cos a}$ (d) $\frac{\omega_1}{\omega} = \frac{1}{\sin a}$.

11. Match the following:

List I (Dynamometer)	List II (Characteristics)
A. Torsion dynamometer	1. High speeds and low power
B. Tesla fluid friction dynamometer	2. Power absorbed independent of size of flywheel.
C. Prony brake	3. Power absorbed available for useful applications
D. Swinging field dynamometer	4. Large powers

Codes:

	A	B	C	D
(a)	4	2	1	3
(b)	2	4	1	3
(c)	3	1	2	4
(d)	4	1	2	3

12. Mitre gears
    (a) spur-gears with gear ratio 1:1
    (b) Skew gears connecting non-parallel and non-intersecting shafts
    (c) Bevel gears transmitting power at more than or less than 90°
    (d) Bevel gears in which the angle between the axis is 90° and the speed ratio of the gears is 1:1.

13. In which of the following case, the turning moment diagram will have least variations:
    (a) Double acting steam engine
    (b) Four stroke single cylinder petrol engine
    (c) 8 cylinder, 4 stroke diesel engine
    (d) Pelton wheel.

14. Which of the following statement is correct:
    1. If a rotor is statically balanced it is always dynamically balanced also
    2. If a rotor is dynamically balanced, it must be statically balanced
    3. If a rotor is dynamically balanced, it may or may not be statically balanced
    4. If a rotor is statically balanced, it may or may not be dynamically balanced

    (a) 1 and 2 only (b) 2 ans 4 only (c) 2 and 3 only (d) 1 and 4 only.

15. Which of the following are inversions of a double slider crank chain?
    1. Whitworth return motion
    2. Scotch Yoke
    3. Oldham's Coupling
    4. Rotary engine

    Select correct answer using the codes given below:

    Codes:

    (a) 1 and 2 (b) 1, 3, and 4 (c) 2 and 3 (d) 2, 3, and 4.

16. Match List I with List II and select the correct answer using the codes given below the lists:

List I	List II
A. Governor	1. Pantograph device
B. Automobile differential	2. Feed-back control
C. Dynamic Absorber	3. Epicylic train
D. Engine Indicator	4. Two-mass oscillator

Codes:

	A	B	C	D
(a)	1	2	3	4
(b)	4	1	2	3
(c)	2	3	4	1
(d)	4	3	2	1

17. Consider the following statements:

Coriolis component of acceleration depends on

1. velocity of slider
2. angular velocity of the link
3. acceleration of slider
4. angular acceleration of link

Of these statements

(a) 1 and 2 are correct (b) 1 and 3 are correct
(c) 2 and 4 are correct (d) 1 and 4 are correct.

18. *ABCD* is a four-bar mechanism in which $AB = 30$ cm and $CD = 45$ cm. *AB* and *CD* are both perpendicular to fixed link *AD*, as shown in the Fig.1. If velocity of *B* at this conditions is *V*, then velocity of C is

(a) V (b) $3/2\ V$ (c) $9/4\ V$ (d) $2/3\ V$

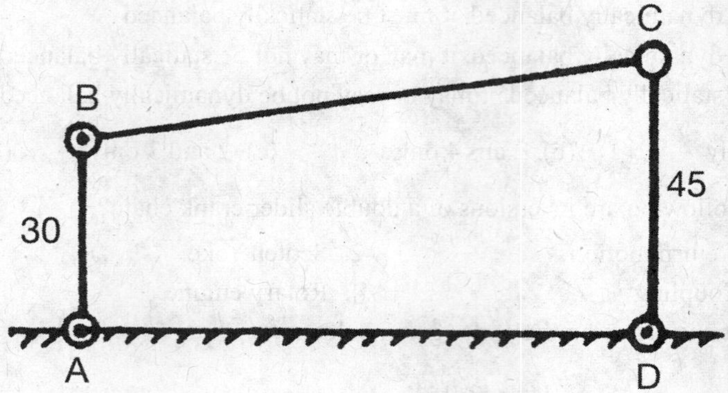

Fig.1

19. Match List I with List II and select the correct answer using the codes given below the lists:

List I (Forces)	List II (Mathematical expressions)
A. Inertia force	1. $c\dfrac{dy}{dt}$
B. Spring force	2. $M\dfrac{d^2y}{dt^2}$
C. Damping force	3. $M\omega^2 R$
D. Centrifugal force	4. ky

Codes:

	A	B	C	D
(a)	1	3	2	4
(b)	2	4	1	3
(c)	2	1	4	3
(d)	1	2	3	4

20. In gears, interference takes place when
   (a) the tip of a tooth of mating gear digs into the portion between base and root circle
   (b) gear do not move smoothly in the absence of lubrication
   (c) pitch of the gear is not same
   (d) gear teeth are undercut.

21. Match List I with List II an select the correct answer using the codes given below the lists:

List I	List II
A. Quadric cycle chain	1. Elliptic trammel
B. Single slider crank chain	2. Rapsons slide
C. Double slider crank chain	3. Ackerman steering
D. Crossed slider crank chain	4. Eccentric mechanism
	5. Pendulum pump

Codes:

	A	B	C	D
(a)	5	4	2	1
(b)	3	1	5	4
(c)	5	3	4	2
(d)	3	5	1	2

22. In a flat collar pivot bearing, the moment due to friction is proportional to (r_1 and r_2 are the outer and inner radii respectively)

   (a) $\frac{r_1^2 - r_2^2}{r_1 - r_2}$ (b) $\frac{r_1^2 - r_2^2}{r_1 + r_2}$ (c) $\frac{r_1^3 - r_2^3}{r_1^2 - r_2^2}$ (d) $\frac{r_1^3 - r_2^3}{r_1 - r_2}$.

23. A friction circle is drawn when a journal rotates in bearing. Its radius depends on the coefficient of friction and the
   (a) magnitudes of the forces on the journal (b) angular velocity of the journal
   (c) clearance between the journal and the bearing (d) radius of the journal.

24. If the rotating mass of a rim type fly wheel is distributed on another rim type flywheel whose mean radius is half mean radius of the former, then energy stored in the latter at the same speed will be
   (a) four times the first one (b) same as the first one
   (c) one-fourth of the first one (d) one and a half times the first one.

25. A flywheel is fitted to the crankshaft of an engine having 'E' amount of indicated work per revolution and permissible limits of co-efficients of fluctuation of energy and speed as K_e and K_s respectively. The kinetic energy of the flywheel is then given by

   (a) $\frac{2K_e E}{K_s}$ (b) $\frac{K_e E}{2K_s}$ (c) $\frac{K_e E}{K_s}$ (d) $\frac{K_e E}{2K_s}$.

26. A Hartnell governor has its controlling force F given by F = p + qr, Where r is the radius of the balls and p and q are constants.
The governor becomes isochronous when
(a) $p = 0$ and q is positive
(b) p is positive and $q = 0$
(c) p is negative and q is positive
(d) p is positive and q is also positive.

27. The plots of controlling force versus radii of rotation of the balls of spring controlled governors are shown in the given Fig.2. A stable governor is characterised by the curve labelled.

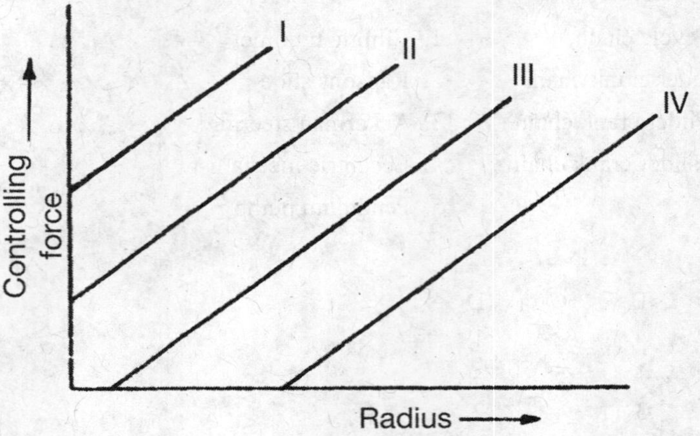

Fig.2

(a) I
(b) II
(c) III
(d) IV.

28. A system in dynamic balance implies that
(a) the system is critically damped
(b) there is on critical speed in the system
(c) the system is also statically balanced
(d) there will be absolutely no wear of bearings.

29. For a twin cylinder V-engine, the crank positions for Primary reverse cranks and Secondary direct cranks are given in the following Fig.3:

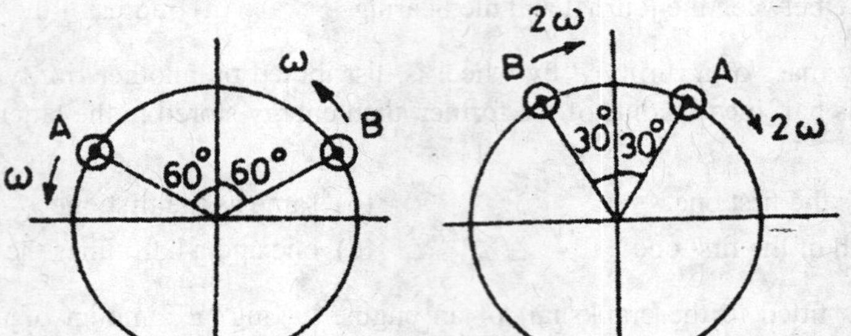

Fig.3

The engine is a
(a) 60° V-engine
(b) 120° V-engine
(c) 30° V-engine
(d) 150° V-engine.

30. Which one of the following can completely balance several masses revolving in different planes on a shaft?

(a) A single mass in one of the planes of the revolving masses
(b) A single mass in a different plane
(c) Two masses in any two planes
(d) Two equal masses in any two planes.

31. With symbols having the usual meanings, the single degree of freedom system, $m\ddot{x} + c\dot{x} + kx = F \sin \omega t$ represents

(a) free vibration with damping
(b) free vibration without damping
(c) forced vibration with damping
(d) forced vibration without damping.

32. In the two-rotor system shown in the given Fig.4, ($I_1 < I_2$), a node of vibration is situated

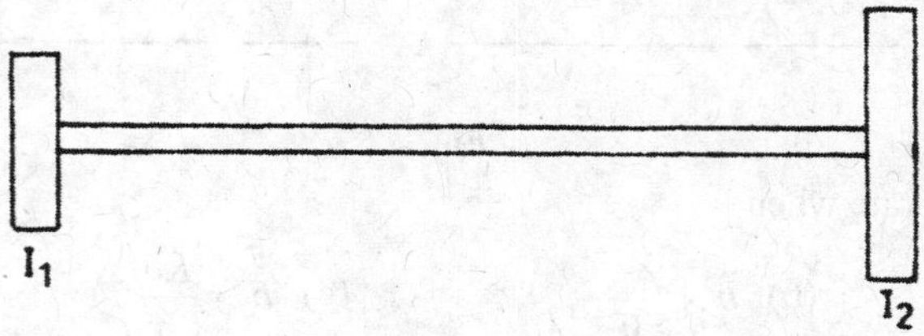

Fig.4

(a) between I_1 and I_2 but nearer to I_1
(b) between I_1 and I_2 but nearer to I_2
(c) exactly in the middle of the shaft
(d) nearer to I_1 but outside.

33. A simple spring mass vibrating system has a natural frequency of N. If the spring stiffness is halfed and the mass is doubled, then the natural frequency will become

(a) $N/2$ (b) $2N$ (c) $4N$ (d) $8N$

34. For the single degree of freedom system shown in the Fig.5, the mass M rolls along an incline of α. The natural frequency of the system will

(a) increases as α increases
(b) decreases as α increases
(c) be independent of α
(d) increase initially as α increases and then decrease with further increase in α.

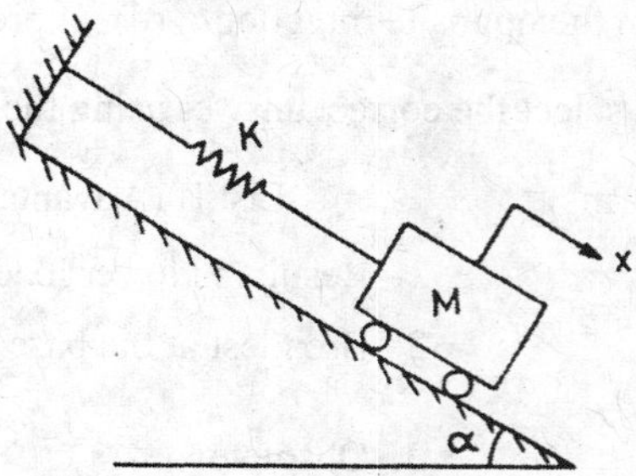

Fig.5

35. For the system shown in the given Fig.6 the moment of inertia of the weight W and the ball about the pivot point is I_o. The natural frequency of the system is given by

$$f_n = \frac{1}{2\pi}\sqrt{\frac{Ka^2 - Wb}{I_o}}$$

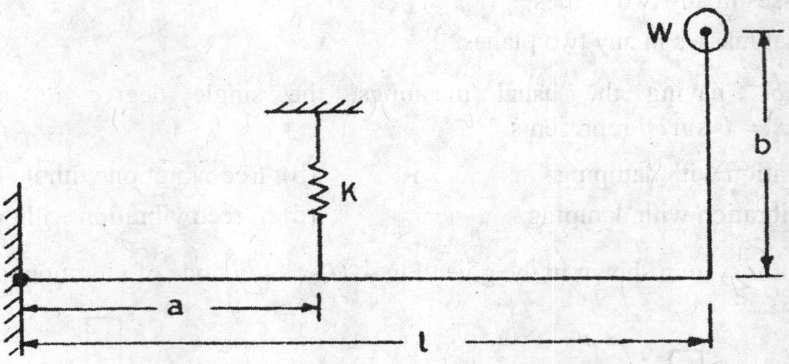

Fig.6

The system will vibrate when

(a) $b < \frac{Ka^2}{W}$ (b) $b = \frac{Ka^2}{W}$ (c) $b > \frac{Ka^2}{W}$ (d) a = 0.

36. Rotating shafts tend to vibrate violently at whirling speeds because
(a) the shafts are rotating at very high speeds
(b) bearing centre line coincides with the shaft axis
(c) the system in unbalanced
(d) resonance is caused due to the heavy weight of the rotor.

37. Critical speed of a shaft with a disc supported in between is equal to the natural frequency of the system in
(a) transverse vibrations (b) torsional vibrations
(c) longitudinal vibrations (d) longitudinal vibrations provided the shaft is vertical.

38. In a automobile service station, an automobile is in a lifted up position by means of a hydraulic jack. A person working in the service station gave a tap to one rear wheel and made it rotate by one revolution. The rotation of another rear wheel is
(a) zero (b) also one revolution in the same direction
(c) also one revolution but in the opposite direction (d) unpredictable.

39. Match List I with List II and select the correct answer using the codes given below the lists:

List I (Standard tooth forms)	List II (Advantage or disadvantages)
A. 20° and 50° system	1. Results in lower loads on bearing
B. $14\frac{1}{2}$ stub-tooth system	2. Broadest at the base and strongest in bending
C. 25° Full depth system	3. Obsolete
D. 20° Full depth system	4. Standards for new applications

Codes:

	A	B	C	D
(a)	4	3	2	1
(b)	3	1	2	4
(c)	3	2	1	4
(d)	4	2	3	1

40. In involute gears the pressure angle is

(a) dependent on the size of teeth (b) dependent on the size of gears

(c) always constant (d) always variable.

41. A thin circular disc is rolling with a uniform linear speed, along a straight path on a plane surface. Consider the following statements in this regard:

1. All points of the disc have the same velocity
2. The centre of the disc has zero acceleration
3. The centre of the disc has centrifugal acceleration
4. The point on the disc making contact with the plane surface has zero acceleration.

Of these statements

(a) 1 and 4 are correct (b) 3 and 4 are correct

(c) 3 alone is correct (d) 2 alone is correct.

42. An elliptic trammel is shown in the given Fig.7. Associated with the motion of the mechanism are fixed and moving centrodes. It can be established analytically on graphically that the moving centrode is a circle with radius and centre respectively of

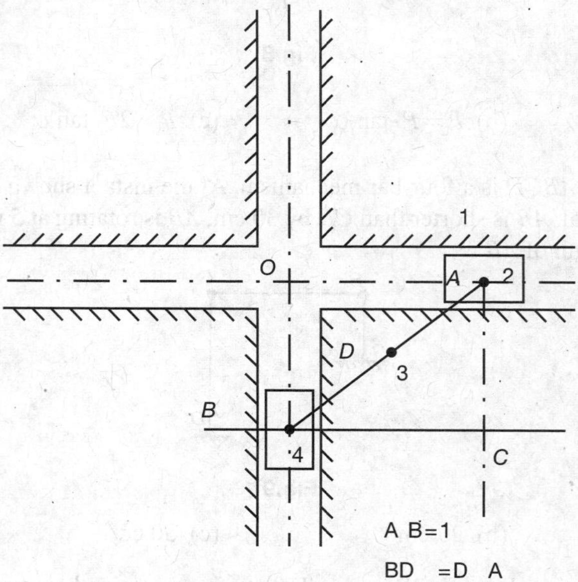

Fig.7

(a) 1 and O (b) 1/2 and B (c) 1/2 and C (d) 1/2 and D.

43. In a circular are cam with roller follower, the acceleration in any position of the lift would depend only upon
   (a) total lift, total angle of lift, minimum radius of cam and cam speed
   (b) radius of circular arc cam, speed, location of centre of circular arc and roller diameter
   (c) weight of cam follower linkage, spring stiffness and cam speed
   (d) total lift, centre of gravity of the cam and cam speed.

44. The Klein's method of constuction for reciprocating engine mechanism
   (a) is a simplified version of instantaneous centre method
   (b) utilises a quadrilateral similar to the diagram of mechanism for reciprocating engine
   (c) enables determination of Corioli's component
   (d) is based on the acceleration diagram.

45. With reference to the mechanism shown in the Fig.8, the relation between F and P is

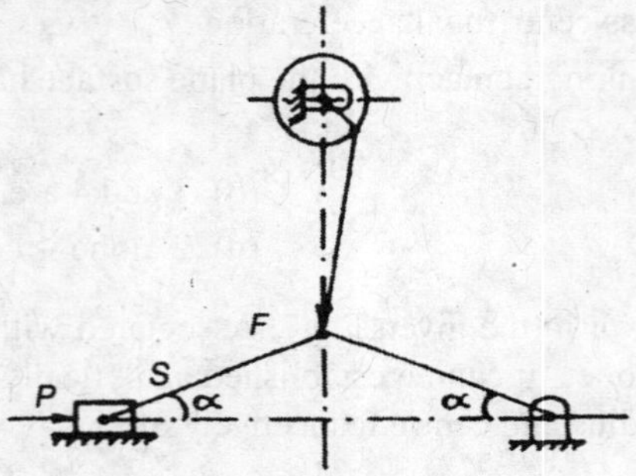

Fig.8

(a) $F = \frac{1}{2} P \cdot \tan a$ (b) $F = P \cdot \tan \alpha$ (c) $P = 2F \cdot \tan \alpha$ (d) $F = 2P \cdot \tan \alpha$.

46. In the given Fig.9, *ABCD* is a four-bar mechanism, At the instant shown *AB* and *CD* are vertical and *BC* is horizontal. *AB* is shorter than *CD* by 30 cm, *AB* is rotating at 5 rad/s and *CD* is rotating at 2 rad/s. The length of *AB* is

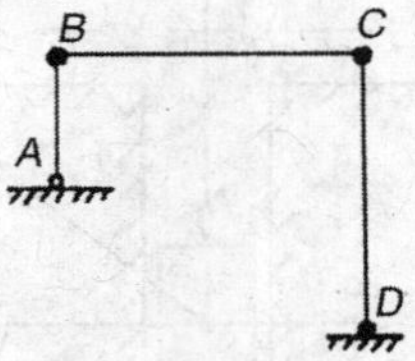

Fig.9

(a) 10 cm (b) 20 cm (c) 30 cm (d) 50 cm.

47. The two-link system, shown in the given Fig.10, is constrained to move with planar motion. It possesses
   (a) 2 – degrees of freedom (b) 3 – degrees of freedom
   (c) 4 – degrees of freedom (d) 6 – degrees of freedom.

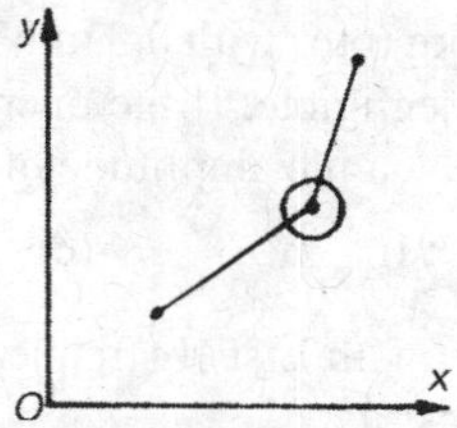

Fig.10

48. Two co-axial rotors having moments of inertia I_1, I_2 and angular speeds ω_1 and ω_2 respectively are engaged together. The loss of energy during engagment is equal to

(a) $\dfrac{I_1 I_2(\omega_1 - \omega_2)^2}{2(I_1 + I_2)}$ (b) $\dfrac{I_1 I_2(\omega_1^2 - \omega_2^2)^2}{2(I_1 + I_2)}$ (c) $\dfrac{2I_1 I_2(\omega_1 - \omega_2)^2}{(I_1 + I_2)}$ (d) $\dfrac{I_1\omega_1^2 + I_2\omega_2^2}{(I_1 + I_2)}$.

49. A spring controlled governor is found unstable. It can be made stable by

(a) increasing the spring stiffness (b) decreasing the spring stiffness
(c) increasing the ball weight (d) decreasing the ball weight.

50. If a number of forces act on a rigid body, each force may be replaced by an equal and parallel force acting through a fixed point, together with a couple. For the rigid body to be in equilibrium,

(a) the resultant force at the fixed point must be zero
(b) the resultant couple on the body must be zero
(c) both resultant force and couple must be zero
(d) none of the above need be zero.

51. A rotor which is balanced statically but not dynamically is supported on two bearings L apart and at high speed of the rotor, reaction of the left bearing is R. The right side of the bearing is shifted to a new position $2L$ apart from the left bearing. At the same rotor speed, dynamic reaction on the left bearing in the new arrangement will

(a) remain same as before (b) become equal to 2 R
(c) become equal to $\dfrac{1}{2}R$ (d) become equal to $\dfrac{1}{4}R$.

52. Consider the following statements regarding a high speed in-line engine with identical reciprocating parts with cranks spaced to give equal firing intervals:

1. All harmonic forces, except those which are multiples of half the number of cylinders, are balanced
2. Couples are balanced if the engine is symmetrical about a plane normal to the axis of the cranks shaft
3. In a four cylinder in-line engine, second and fourth harmonic forces are unbalanced whereas in a six cylinder in a six cylinder in-line engine, second, fourth and sixth harmonic forces are unbalanced

Of these statements

(a) 1, 2 and 3 are correct (b) 1 and 3 are correct
(c) 1 and 3 are correct (d) 2 and 3 are correct.

53. In the statement "an eccentric mass rotating at 3000 rpm will create X times more unbalanced force than 50% of the same mass rotating at 300 rpm," 'X' stands for

(a) 10 (b) 50 (c) 100 (d) 200.

54. A machine of 100 kg mass has a 20 kg rotor with 0.5 mm eccentricity. The mounting springs have stiffness 85 kN/m and damping is negligible. If the operating speed is 20π rad/s and the unit is constrained to move vertically, the dynamic amplitude of the machine will be

(a) 0.470×10^{-4} m (b) 1.000×10^{-4} m (c) 1.270×10^{-4} m (d) 2.540×10^{-4} m.

55. Match List I (force transmissibility) with List II (frequency ratio) and select the correct answer using the codes given below the Lists:

List I	List II
A. 1	1. $\frac{\omega}{\omega_n} > \sqrt{2}$
B. Less than 1	2. $\frac{\omega}{\omega_n} = \sqrt{2}$
C. Greater than 1	3. $\frac{\omega}{\omega_n} >> \sqrt{2}$
D. Tending to zero	4. $\frac{\omega}{\omega_n} \sqrt{2}$

Codes:

	A	B	C	D
(a)	1	2	3	4
(b)	2	1	4	3
(c)	2	1	3	4
(d)	1	2	4	3

56. A mass of 1 kg is attached to the end of a spring with a stiffness 0.7 N/mm. The critical damping coefficient of this system is

(a) 1.40 Ns/m (b) 52.22 Ns/m (c) 52.92 Ns/m (d) 529.20 Ns/m.

57. A system is shown in the following Fig.11. The bar AB is assumed to be rigid and weightless.

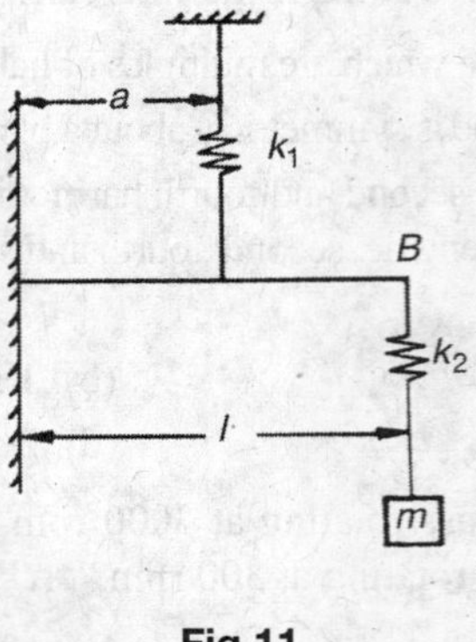

Fig.11

The natural frequency of vibration of the system is given by

(a) $f_n = \dfrac{1}{2\pi}\dfrac{\sqrt{k_1k_2\left(\dfrac{a}{\ell}\right)^2}}{m\left[k_2 + \left(\dfrac{a}{\ell}\right)^2 k_1\right]}$ (b) $f_n = \dfrac{1}{2\pi}\dfrac{\sqrt{k_1k_2}}{m(k_1 + k_2)}$

(c) $f_n = \dfrac{1}{2\pi}\dfrac{\sqrt{k_1}}{mk_2}$ (d) $f_n = \dfrac{1}{2\pi}\dfrac{\sqrt{k_1 + k_2}}{mk_1k_2}$.

58. Two heavy rotating masses are connected by shafts of lengths l_1, l_2 and l_3 and the corresponding diameters are d_1, d_2, and d_3. This system is reduced to a torsionally equivalent system having uniform diameter "d_1" of the shaft. The equivalent length of the shaft is

(a) $\dfrac{l_1 + l_2 + l_3}{3}$ (b) $l_1 + l_2\left(\dfrac{d_2}{d_2}\right)^3 + l_3\left(\dfrac{d_1}{d_3}\right)^3$.

(c) $l_1 + l_2\left(\dfrac{d_1}{d_2}\right)^4 + l_3\left(\dfrac{d_1}{d_3}\right)^4$ (d) $l_1 + l_2 + l_3$.

59. A shaft has two heavy rotors mound on it The transverse natural frequencies, considering each of the rotor separately, are 100 cycles/sec and 200 cycles/sec respectively. The lowest critical speed is

(a) 5367 rpm (b) 6000 rpm (c) 9360 rpm (d) 12000 rpm.

60. A shaft has an attached disc at the centre of its length. The disc has its centre of gravity located at a distance of 2 mm from the axis of the shaft. When the shaft is allowed to vibrate in its natural bow-shaped mode, its has a frequency of vibration of 10 radians/second. When the shaft is rotated a 300 revolutions per minute, it will whirl with a radius of

(a) 2 mm (b) 2.25 mm (c) 2.50 mm (d) 3.00 mm.

61. Let S and G be positions of centre of mass and geometric centre is a disc attached to a rotating disc with axis at O as shown Fig.12. Let the system be resisted by viscous damping. Then at the critical speed, the relative positions of G and S are given by

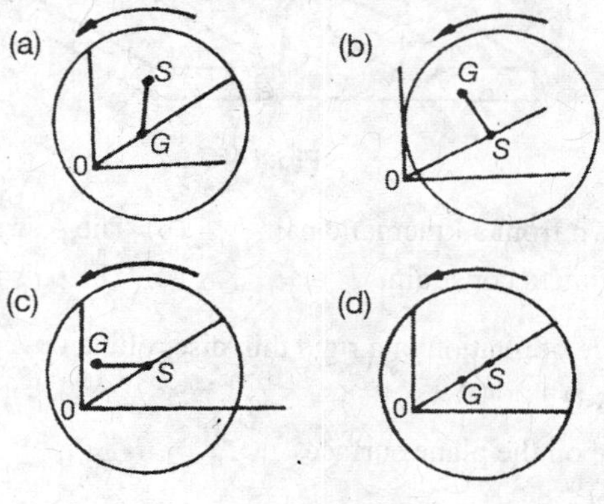

Fig.12

62. Consider the following statements regarding the differential of an automobile
   1. The speed of the crown wheel will always be the mean of the speeds of the two road wheels
   2. The road wheel speeds are independent of the number of teeth on the planets
   3. The difference between the speed of the road wheels depends on the number of teeth on the planets
   4. The ratio of speeds of the road wheels depends upon the number of teeth on the gear wheels attached to them and on the crown wheel

   Of these statements

   (a) 1 and 2 are correct (b) 3 and 4 are correct
   (c) 1 and 3 are correct (d) 2 and 4 are correct.

63. A single epicyclic gear train is shown in the given Fig.13. Wheel *A* is stationary. If the number of teeth on *A* and *B* are 120 and 45 respectively, then when *B* rotates about its own axis at 100 rpm, the speed of *C* would be

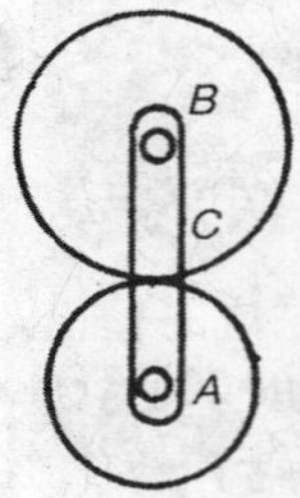

Fig.13

(a) 20 rpm (b) $27\frac{3}{11}$ rpm (c) $19\frac{7}{11}$ rpm (d) 100 rpm.

64. A round bar *A* passes through the cylindrical hole in *B* as shown in the given Fig.14. Which one of the following statements is correct in this regard?

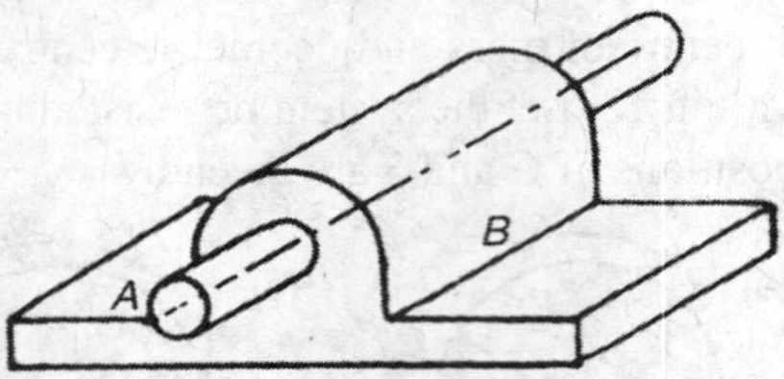

Fig.14

(a) The two links shown from a kinematic pair (b) The pair is completely constrained
(c) The pair has incomplete constraint (d) The pair is successfully constrained.

65. The instantaneous centre of rotation of a rigid thin disc rolling on a plane rigid surface is located at
   (a) the centre of the disc
   (b) an infinite distance on the plane surface
   (c) the point of contact
   (d) the point on the circumference situated vertically opposite to the contact point.

66. The directions of Coriolis component of acceleration, $2\omega V$, of the slider A with respect to the coincident point B is shown in Fig.15. (1 to 4). Directions shown by figures .

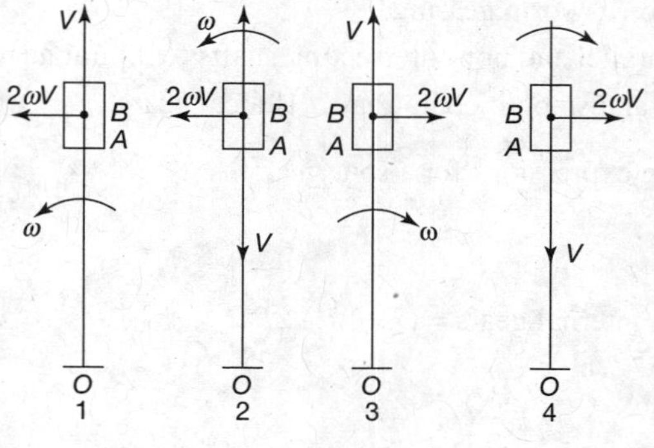

Fig.15

(a) 2 and 4 are wrong
(b) 1 and 2 are wrong
(c) 1 and 3 are wrong
(d) 2 and 3 are wrong.

67. Klein's construction for determining the acceleration of piston P is shown in the given Fig.16. When N coincides with O,

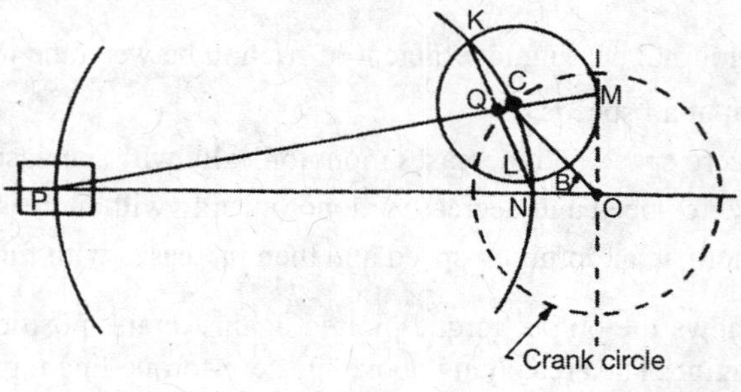

Fig.16

(a) acceleration of piston is zero and its velocity is zero
(b) acceleration is maximum and velocity is maximum
(c) acceleration is maximum and velocity is zero
(d) acceleration is zero and velocity is maximum.

68. A torsional system with discs of moment of inertia I_1 and I_2, shown in the given Fig.17, is gear driven such that the ratio of the speed of shaft B to shaft A is 'n'. Neglecting the inertia of gears, the equivalent inertia of disc 2 at the speed of shaft A is equal to

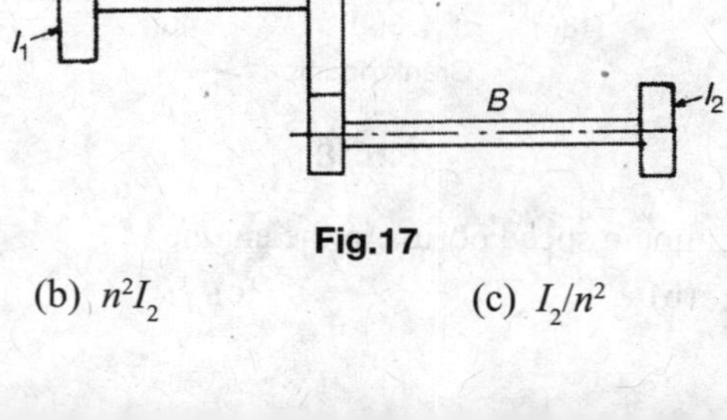

Fig.17

(a) nI_2
(b) n^2I_2
(c) I_2/n^2
(d) I_2/n

69. Which one of the following pairs is not correctly matched?
   (a) Positive drive ... Belt drive
   (b) High velocity ratio ... Worm gearing
   (c) To connect non-parallel and non-intersecting shafts ... Spiral gearing
   (d) Diminished noise and smooth operation ... Helical gears.

70. Consider the following expressions for a spur gear:
   1. Addendum = $1/p_d$
   2. Clearance = $p_c/20$
   3. Centre distance for internal gears = $(T_1 + T_2)/2P_d$
   4. Dendendum = $1.157\,7\, p_c/\pi$

   Of these expressions
   (a) 1, 2, 3, and 4 are correct
   (b) 1 and 2 are correct
   (c) 1, 2, and 3 are correct
   (d) 1, 2, and 4 are correct.

71. Babbit lining is used on brass/bronze bearing to
   (a) increases bearing resistance
   (b) increase compressive strength
   (c) provide antifriction properties
   (d) increase wear resistance.

72. In an oil-lubricated journal bearing, coefficient of friction between the journal and the bearing.
   (a) remains constant at all speeds
   (b) is minimum at zero speed and increases monotonically with increases in speed
   (c) is maximum at zero speed and decreases monotonically with increase in speed
   (d) becomes minimum at an optimum speed and then increases with further increase in speed.

73. The given Fig.18 shows the output torque plotted against crank positions for a single cylinder four-stroke-cycle engine. The areas lying above the zero-torque line represent positive work and the areas below represent negative work. The engine drives a machine which offers a resisting torque equal to the average torque. The relative magnitudes of the hatched areas given by the numbers (in the areas) as shown:

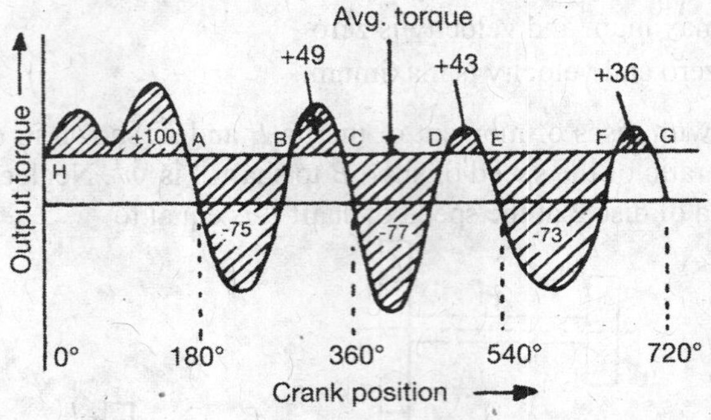

Fig.18

During the cycle, the minimum speed occurs in the engine at
   (a) B
   (b) D
   (c) H
   (d) F

74. For a spring controlled governor to be stable, the controlling force (F) is related to the radius (r) by the equation.

(a) $F = ar - b$ (b) $F = ar + b$ (c) $F = ar$ (d) $F = a/r + b$

75. A rotor supported at A and B, carries two masses as shown in the given Fig.19. The rotor is

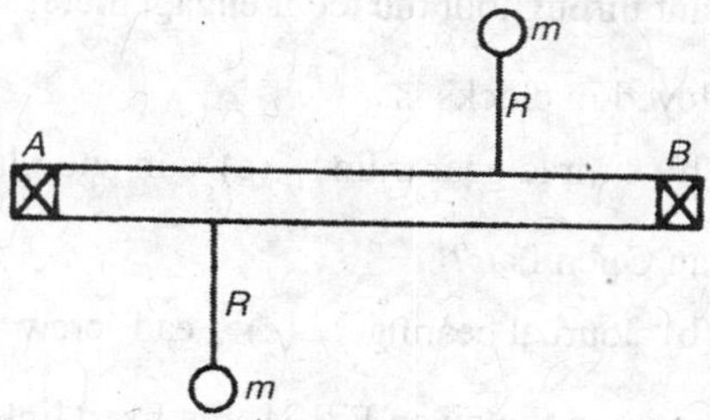

Fig.19

(a) dynamically balanced (b) statically balanced
(c) statically and dynamically balanced (d) not balanced.

76. The equation of free vibrations of a system is $\ddot{X} + 36\pi^2 X = 0$. Its natural frequency is

(a) 46 Hz (b) 3π Hz (c) 3 Hz (d) 6π Hz.

77. Which of the following methods can be used to determine the damping of machine element?

1. Logarithmic method 2. Band-width method 3. Rayleigh method 4. Holzer method.

Select the correct answer using the codes given below:

Codes:

(a) 1 and 3 (b) 1 and 2 (c) 3 and 4 (d) 1, 3, and 4.

78. If $\omega/\omega_n = \sqrt{2}$, where ω is the frequency of excitation and ω_n is the natural frequency of vibrations, then the transmissibility of vibrations will be

(a) 0.5 (b) 1.0 (c) 1.5 (d) 2.0

79. A slender shaft supported on two bearing at its ends carries a disc with an eccentricity e from the axis of rotation. The critical speed of the shaft is N. If the disc is replaced by a second one of same weight but mounted with an eccentricity $2e$, critical speed of the shaft in the second case is

(a) $1/2\ N$ (b) $1/\sqrt{2}\ N$ (c) N (d) $2\ N$

80. For the spring-mass system shown in the Fig.20(a), the frequency of vibration is N. What will be the frequency when one more similar spring is added in series, as shown in Fig.20(b)?

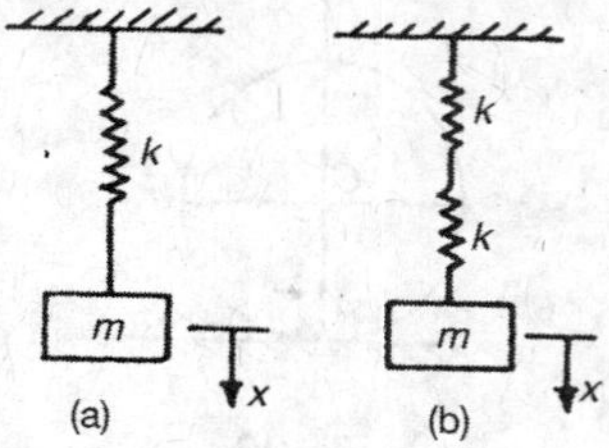

Fig.20

(a) $N/2$ (b) $N/\sqrt{2}$ (c) $\sqrt{2}/N$ (d) $2N$.

81. Which one of the following is true for involute gears?
   (a) Interference is inherently absent
   (b) Variation in centre distance of shafts increases radial force
   (c) A convex flank is always in contact with concave flank
   (d) Pressure angle is constant throughout the teeth engagement.

82. The gear train usually employed in clocks is a
   (a) reverted gear train (b) simple gear train (c) sun and planet gear (d) differential gear.

83. Which one of following is an Open Pair?
   (a) Ball and socket joint (b) Journal bearing (c) Lead screw and nut (d) Cam and follower.

84. In the mechanism *ABCD* shown in the given Fig.21, the fixed link is denoted as (1), Crank *AB* as (2), rocker *BD* (3), Swivel trunnion at *C* as (4) The instantaneous centre I_4 is at

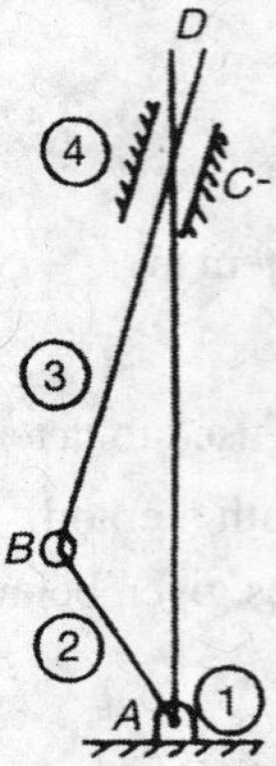

Fig.21

   (a) the centre of swivel trunnion
   (b) the intersection of line *AB* and a perpendicular to *BD*
   (c) infinity along *AC*
   (d) infinity perpendicular to *BD*.

85. The instantaneous centre of motion of rigid-thin disc-wheel rolling on plane rigid surface shown in the Fig.22, is located at the point.

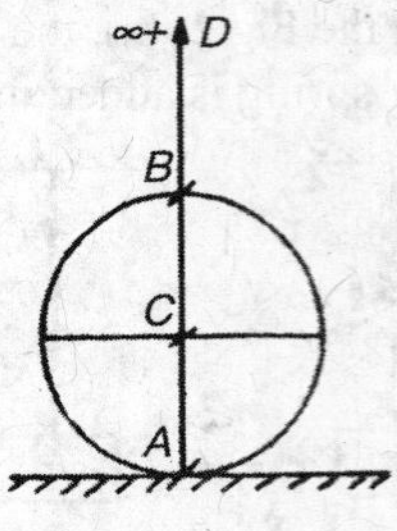

Fig.22

   (a) *A* (b) *B* (c) *C* (d) *D*.

86. In a cam drive with uniform velocity follower, the slope of the displacement must be as shown in Fig.23(a). But in actual practice it is as shown in Fig.23(b) (i.e. rounded at the corners). This is because of

(a) the difficulty in manufacturing cam profile
(b) loose contact of follower with cam surface
(c) The acceleration in the beginning and retardation at the end of stroke would require to be infinitely high
(d) uniform velocity motion is a partial parabolic motion.

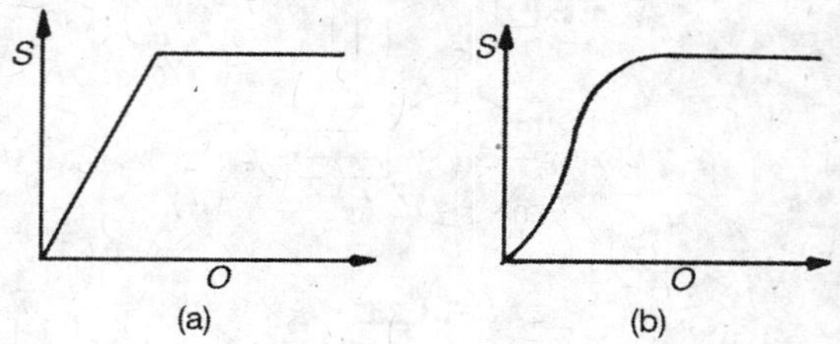

Fig.23

87. In a 4-stroke I.C engine, the turning moment during the compression stroke is

(a) positive throughout
(b) negative throughout
(c) positive during major portion of the stroke
(d) negative during major portion of the stroke.

88. With reference to the engine mechanism shown in the given Fig.24, match List I with List II and select the correct answer

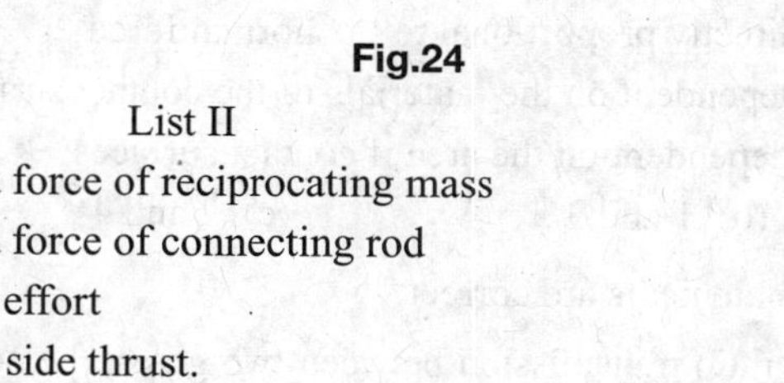

Fig.24

List I	List II
A. F_Q	1. Inertia force of reciprocating mass
B. F_R	2. Inertia force of connecting rod
C. F_W	3. Crank effort
D. F_C	4. Piston side thrust.

Codes:

	A	B	C	D
(a)	1	2	4	3
(b)	1	2	3	4
(c)	4	1	2	3
(d)	4	1	3	2

89. A compound train consisting of spur, bevel and spiral gears is shown in the given Fig.25 along with the teeth numbers marked against the wheels. Over-all speed ratio of the train is

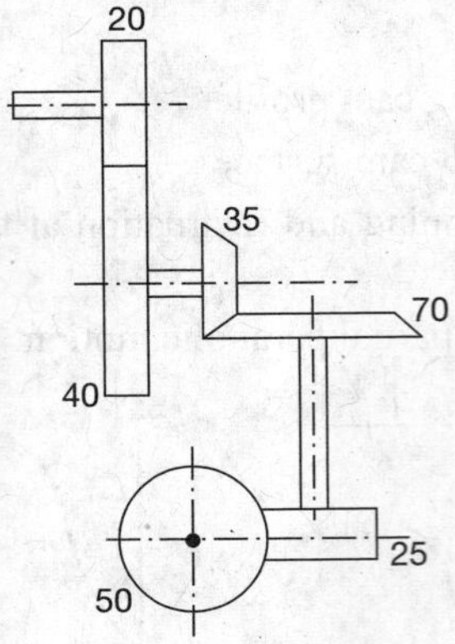

Fig.25

(a) 8 (b) 2 (c) $\frac{1}{2}$ (d) $\frac{1}{8}$.

90. Which of the following statements hold good for a multi-collar thrust bearing carrying an axial thrust of W units?

1. Friction moment is independent of the number of collars
2. The intensity of pressure is affected by the number of collars
3. Co-efficient of friction of the bearing surface is affected by the number of collars

(a) 1 and 2 (b) 1 and 3 (c) 2 and 3 (d) 1, 2, and 3.

91. The centre of gravity of the coupler link in a 4-bar mechanism would experience

(a) no acceleration (b) only linear acceleration
(c) only angular acceleration (d) both linear and angular accelerations.

92. Which of the following statements regarding laws governing the friction between dry surfaces are correct?

1. The friction force is dependent on the velocity of sliding
2. The friction force is directly proportional to the normal force
3. The friction force is dependent on the materials of the contact surfaces
4. The Friction force is dependent on the area if contact surfaces

(a) 2, 3, and 4 (b) 1 and 3 (c) 2 and 4 (d) 1, 2, 3, and 4.

93. Which of the following statements are correct?

1. For constant velocity ratio transmission between two gears, the common normal at the point of contact must always pass through a fixed point on the line joining the centres of rotation of the gears
2. For involute gears the pressure angle changes with change in centre distance between gears
3. The velocity ratio of compound gear train depends upon the number of teeth of the input and output gears only
4. Epicyclic gear trains involve rotation of at least one gear axis about some other gear axis

(a) 1, 2, and 3 (b) 1 and 3 (c) 1, 2, and 4 (d) 2, 3, and 4.

94. Which one of the following equations is valid with reference to the given Fig.26.

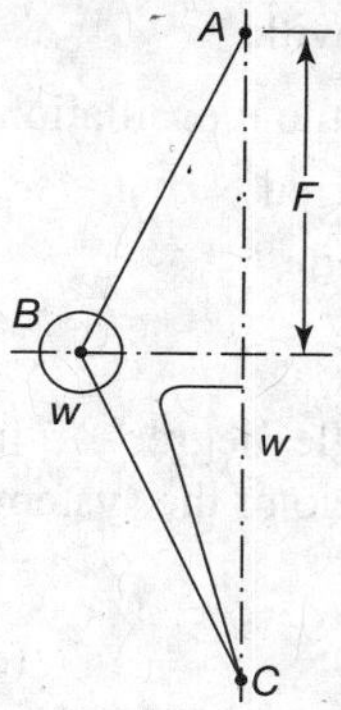

Fig.26

(a) $\omega^2 = \left(\frac{W}{w}\right)\left(\frac{g}{h}\right)$ (b) $\omega^2 = \left(\frac{W+w}{w}\right)\left(\frac{g}{h}\right)^{1/2}$

(c) $\omega^2 = \left(\frac{w}{W+w}\right)\left(\frac{h}{g}\right)^{1/2}$ (d) $\omega^2 = \left(\frac{W+w}{w}\right)\left(\frac{g}{h}\right)$.

95. Match list I with List II and select the correct answer

List I	List II
A. Hunting	1. One radius of rotation for each speed
B. Isochronism	2. Too sensitive
C. Stability	3. Mean force exerted at the sleeve during change of speed
D. Effort	4. Constant equilibrium speed for all radii of rotation

Codes:

	A	B	C	D
(a)	2	4	1	3
(b)	3	1	4	3
(c)	2	1	4	3
(d)	1	2	3	4

96. A system of masses rotating in different parallel planes is in dynamic balance if the resultant.
   (a) force is equal to zero
   (b) couple is equal to zero
   (c) force and the resultant couple are both equal to zero
   (d) force is numerically equal to the resultant couple, but neither of them need necessarily be zero.

97. When shaking force is transmitted through the spring, damping becomes detrimental when the ratio of its frequency to the natural frequency its greater than
   (a) 0.25 (b) 0.50 (c) 1.00 (d) sqrt 2.

98. When the mass of a critically damped single degree of freedom system is deflected from its equilibrium position and released, it will

(a) return to equilibrium position without oscillation
(b) oscillate with increasing time period
(c) oscillate with decreasing amplitude
(d) oscillate with constant amplitude.

99. The equation of motion for a single degree of freedom system with viscous damping is $4\ddot{x} + 9\dot{x} + 16x = 0$. The damping ratio of the system is

(a) $\frac{9}{128}$ (b) $\frac{9}{16}$ (c) $\frac{9}{8\sqrt{2}}$ (d) $\frac{9}{8}$.

100. For the spring-mass system shown in the given Fig.27, the frequency of oscillations of the block along the axis of the spring is

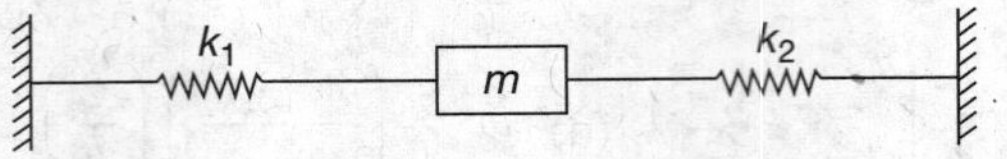

Fig.27

(a) $\frac{1}{2\pi}\sqrt{\frac{k_1 - k_2}{m}}$ (b) $\frac{1}{2\pi}\frac{\sqrt{k_1 k_2}}{[(k_1 + k_2)m]^{1/2}}$

(c) $\frac{1}{2\pi}\sqrt{\frac{k_1 + k_2}{m}}$ (d) $\frac{1}{2\pi}\sqrt{\frac{m}{k_1 + k_2}}$.

101. The critical speed of a rotating shaft depends upon

(a) mass (b) stiffness
(c) mass and stiffness (d) mass, stiffness and eccentricity.

102. A fixed gear having 200 teeth is in mesh with another gear having 50 teeth. The two gears are connected by an arm. The number of turns made by the smaller gear for one revolution of arm about the centre of the bigger gear is

(a) $\frac{2}{4}$ (b) 3 (c) 4 (d) 5.

103. An involute pinion and gear are in mesh. If both have the same size of addendum, then there will be an interference between the

(a) tip of the gear tooth and flank of pinion
(b) tip of the pinion and flank of gear
(c) flanks of both gear and pinion
(d) tips of both gear and opinion.

104. Match List I with List II and select the answer using the codes given below the Lists.

List I	List II
A. Helical gears	1. Non-interchangeable
B. Herringbone gears	2. Zero axial thrust
C. Worm gears	3. Quiet motion
D. Hypoid Gears	4. Extreme speed reduction

Codes:

	A	B	C	D
(a)	1	2	3	4
(b)	3	2	1	4
(c)	3	1	4	2
(d)	3	2	4	1

105. Given θ = angle through which the axis of the outer forward wheel turns
ϕ = angle through which the axis of the inner forward wheel turns
a = distance between the pivots of front axle, and
b = wheel base.

For correct steering, centre lines of the axes of four wheels of an automobile should meet at a common point. This condition will be satisfied if

(a) $\cos\theta - \cos\phi = a/b$ (b) $\cot\theta - \cot\phi = a/b$
(c) $\cos\theta + \cos\phi = a/b$ (d) $\tan\theta + \tan\phi = b/a$.

106. If air resistance is neglected, while it is executing small oscillations the acceleration of the bob of a simple pendulum at the mid-point of its swing will be

(a) zero
(b) a minimum but not equal to zero
(c) a maximum
(d) not determinable unless the length of the pendulum and the mass of the bob are known.

107. In the Fig.28 shown crank *AB* is 15 cm long and is rotating at 10 rad/s. *C* is vertically above *A*. *CA* equals 24 cm. *C* is swivel trunnion through which *BD* (40 cm) slides. If *ABCD* becomes a vertical line during its motion, the angular velocity of the swivel trunnion at that instant will be

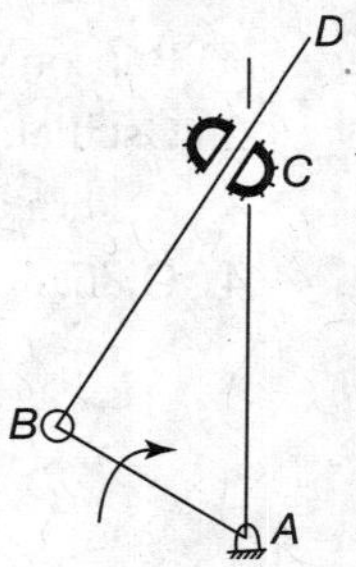

Fig.28

(a) zero (b) (100/25) rad/s (c) (100/15) rad/s (d) (100/10) rad/s.

108. In order to draw the acceleration diagram, it is necessary to determine the Coriolis component of acceleration in the case of

(a) crank and slotted lever quick return mechanism
(b) slider-crank mechanism
(c) four bar mechanism
(d) pantograph.

109. What is the correct sequence of the following steps in engine analysis?

1. Vibration analysis
2. Inertia force analysis
3. Balancing analysis
4. Velocity and Acceleration analysis

Select the correct answer using the codes given below:

(a) 2, 4, 1, 3 (b) 2, 4, 3, 1 (c) 4, 2, 1, 3 (d) 4, 2, 3, 1

110. If μ is the actual coefficient of friction in a belt moving in grooved pulley, the groove angle being 2 α, the virtual coefficient of friction will be

(a) $\mu/\sin\alpha$ (b) $\mu/\cos\alpha$ (c) $\mu\sin\alpha$ (d) $\mu\cos\alpha$

111. Match List I (Positioning of two shafts) with List II (Possible connection) and select the correct answer using the codes given below the Lists:

List I	List II
A. Parallel shaft with slight offset	1. Hookes joint
B. Parallel shafts at a reasonable distance	2. Worm and wheel
C. Perpendicular shafts	3. Oldham coupling
D. Intersecting shafts	4. Belt and pulley

Codes:

	A	B	C	D
(a)	4	3	2	1
(b)	4	3	1	2
(c)	3	4	1	2
(d)	3	4	2	1

112. Match List I with List II and select the correct answer using the codes given below the Lists:

List I	List II
A. Quadric cycle chain	1. Rapson's slide
B. Single slider crank chain	2. Oscillating cylinder engine mechanism
C. Double slider crank chain	3. Ackermann steering mechanism
D. Crossed slider crank chain	4. Oldham coupling

Codes:

	A	B	C	D
(a)	1	2	4	3
(b)	4	3	2	1
(c)	3	4	1	2
(d)	3	2	4	1

113. Match List I with List II and select the correct answer using the codes given below the Lists:

List I (Bearing)	List II (Purpose)
A. Ball bearing	1. Heavy loads with oscillatory motion
B. Tapered Roller bearings	2. Light loads
C. Spherical Roller bearings	3. Carrying both radial and thrust loads
D. Needle roller bearings	4. Self-aligning property

Codes:

	A	B	C	D
(a)	4	1	3	2
(b)	2	1	4	3
(c)	2	3	1	4
(d)	2	3	4	1

114. In a journal bearings, the radius of the friction circle increases with the increase in

(a) load (b) radius of the journal
(c) speed of the journal (d) viscosity of the lubricant.

115. Match List I with List II and select the correct answer using the codes given below the Lists:

List I	List II
A. Flywheel	1. Dunkerley Method
B. Governor	2. Turning Moment
C. Critical speed	3. D' Alembert's Principle
D. Inertia force	4. Speed control on par with load

Codes:

	A	B	C	D
(a)	4	2	3	1
(b)	4	2	1	3
(c)	2	4	3	1
(d)	2	4	1	3

116. The sensitivity of an isochronous governor is

(a) zero (b) one (c) two (d) infinity.

117. When the primary direct crank of a reciprocating engine is positioned at 30° clockwise, the secondary reverse crank for balancing will be at

(a) 30° anti-clockwise (b) 60° anti-clockwise
(c) 30° clockwise (d) 60° clockwise.

118. A statically-balanced system is shown in the given Fig.29. Two equal weights W, each with an eccentricity, 'e' are placed on opposite sides of the axis in the same axial plane. The axial distance between them is 'a'. The total dynamic reactions at the supports will be

(a) zero (b) $\frac{W}{g}\omega^2 e\frac{a}{L}$ (c) $\frac{W}{g}\omega^2 e\frac{a}{L}$ (d) $\frac{W}{g}\omega^2 e\frac{L}{a}$.

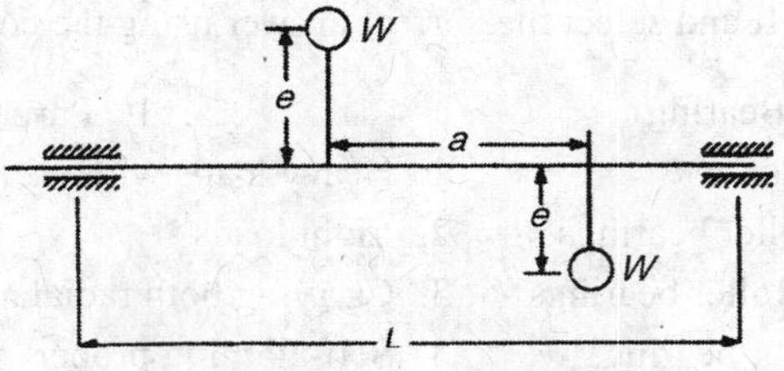

Fig.29

119. A damped free vibration is expressed by the general equation $x = Xe^{-\zeta\omega_n t}\sin\left(\sqrt{1-\zeta^2}\ \omega_n t + \phi\right)$ which is shown in Fig.30:

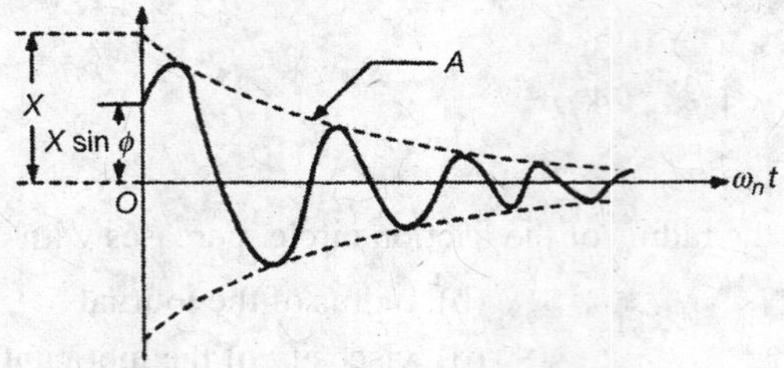

Fig.30

The envelope A has the equation:

(a) Xe^{-1} (b) $X\sin\left(\sqrt{1-\zeta^2}\right)\omega_n t$ (c) $e^{-\zeta\omega_n t}$ (d) $Xe^{-\zeta\omega_n t}$.

120. What is the equivalent stiffness (i.e. spring constant) of the system shown in the given Fig.31.

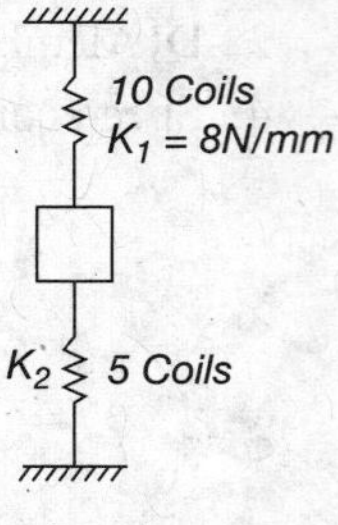

Fig.31

(a) 24 N/mm (b) 16 N/mm (c) 4 N/mm (d) 5.3 N/mm.

121. The given Fig.32 depicts a vector diagram of forces and displacements in the case of Forced Damped Vibration. If vector A represents the forcing function $P = P_o \sin \omega t$, vector B the displacement $y = Y \sin \omega t$, and ϕ the phase angle between them, then the vectors C and D represent respectively.

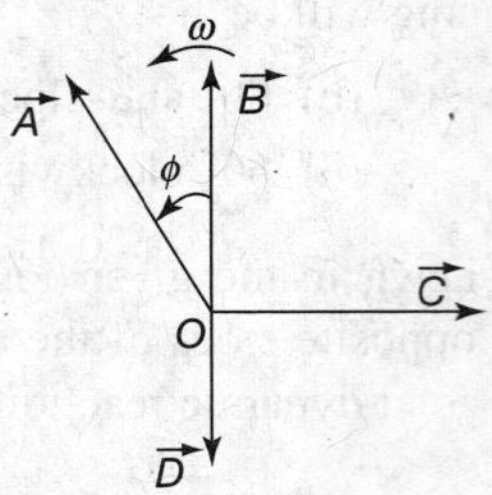

Fig.32

(a) the force of inertia and the force of damping
(b) the elastic force and the damping force
(c) the damping force and the inertia force
(d) the damping force and the elastic force.

122. Two heavy rotating masses are connected by shafts of length l_1, l_2, and l_3 and the corresponding diameters and d_1, d_2, and d_3. This system is reduced to a torsionally equivalent system having uniform diameter d_1 of the shaft. The equivalent length of the shaft is equal to

(a) $l_1 + l_2 + l_3$
(b) $\dfrac{l_1 + l_2 + l_3}{3}$
(c) $l_1 + l_2\left(\dfrac{d_1}{d_2}\right)^3 + l_3\left(\dfrac{d_1}{d_3}\right)^3$
(d) $l_1 + l_2\left(\dfrac{d_1}{d_2}\right)^4 + l_3\left(\dfrac{d_1}{d_3}\right)^4$.

123. An axial flow fan balanced at one speed often exhibits substantial vibrational effects when operated at other speeds, mainly due to

(a) primary critical speed effect
(b) secondary critical speed effect
(c) unbalanced parts of the fan
(d) aerodynamic unbalance.

124. A reverted gear train is one in which the output shaft and input shaft?

(a) rotate in opposite directions
(b) are co-axial
(c) are at right angles to each other
(d) are at an angle to each other.

125. In the case of an involute toothed gear, involute starts from

(a) addendum circle
(b) dedundum circle
(c) pitch circle
(d) base circle.

126. In the epicyclic gear train shown in the given Fig.33, A is fixed. A has 100 teeth and B has 20 teeth. If the arm C makes three revolutions, the number of revolutions made by B will be

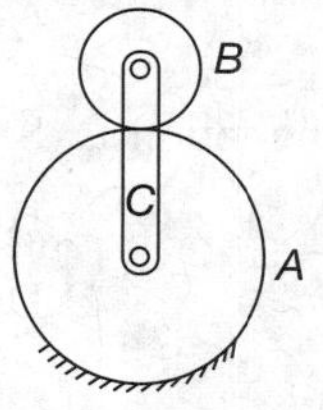

Fig.33

(a) 12
(b) 15
(c) 18
(d) 24.

127. Given that G = forward path given and H = feed back path gain, if G and H are functions of frequency, then the feedback would affect gain G of a non-feedback system by a value to

(a) $1 + GH$
(b) $1 - GH$
(c) GH
(d) $1/GH$.

128. Two geared shafts A and B having moments of inertia I_a and I_b and angular acceleration α_a and α_b respectively are meshed together. B rotates at G times the speed of A. If the gearing efficiency of the two shafts in η, then in order to accelerate B, the torque which must be applied to A will be

(a) $I_a\alpha_a + G^2 I_b\alpha/\eta$
(b) $G^2 I_a \alpha_a/\eta$
(c) $G^2 I_a \alpha_a/\eta$
(d) $G^2 I_b \alpha_b/\eta$.

129. Which of the following pair (*s*) is/are correctly matched?

I. Four bar chain	Oscillating-oscillating converter
II. Inertia governor	Rate of change of engine speed
III. Hammer blow	Reciprocating unbalance.

Select the correct answer using the codes given below:

(a) I alone (b) I, II, and III (c) II and III (d) I and III.

130. Which of the following are examples of kinematic chain?

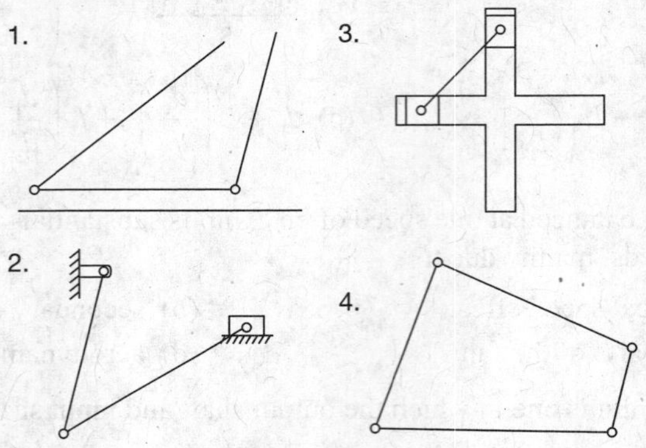

Fig.34

Select the correct answer using the codes given below:

(a) 1, 3, and 4 (b) 2 and 4 (c) 1, 2, and 3 (d) 1, 2, 3, and 4.

131. Which of the following pairs are correctly matched? Select the correct answer using the codes given below the pairs.

Mechanism	Chain from which derived
1. Whitworth quick return motion	Single slider crank chain
2. Oldham's coupling	Four bar chain
3. Scotch Yoke	Double slider crank chain

Codes:

(a) 1 and 2 (b) 1, 2, and 3 (c) 1 and 3 (d) 2 and 3.

132. In S.H.M. with respect to the displacement vector, the positions of Velocity vector and Acceleration vectors will be respectively

(a) 180° and 90° (b) 90° and 180° (c) 0° and 90° (d) 90° and 0°.

133. When a slider moves with a velocity '*V*' on a link rotating at an angular speed of ω, the Corioli's component of acceleration is given by

(a) $\sqrt{2}V\omega$ (b) $V\omega$ (c) $\frac{V\omega}{2}$ (d) $2\,V\omega$.

134. The total number of instantaneous centres for a mechanism consisting of 'n' links is

(a) $n/2$ (b) n (c) $\frac{n-1}{2}$ (d) $\frac{n(n-1)}{2}$.

135. Two links OA and OB are connected by a pin joint at 'O'. The link OA turns with angular velocity ω_1 rad/s in the clockwise direction and the link OB turns with angular velocity ω_2 rad/s in the anticlockwise direction. If the radius of the pin at 'O' is 'r', then the rubbing velocity at the pin joint 'O' will be

(a) $\omega_1 \cdot \omega_2 \cdot r$ (b) $(\omega_1 - \omega_2)\, r$ (c) $(\omega_1 + \omega_2)\, r$ (d) $(\omega_1 + \omega_2)\, 2r$.

136. In a cam drive, it is essential to off-set the axis of a follower to

(a) Decrease the side thrust between the follower and guide
(b) Decrease the wear between follower and cam surface
(c) Take care of space limitation
(d) Reduce the cost.

137. The working surface above the pitch surface of the gear tooth is termed as

(a) Addendum (b) Dedendum (c) Flank (d) Face.

138. In the case of a flywheel, the maximum fluctuation of energy is the

(a) Sum of maximum and minimum energies
(b) Difference between the maximum and minimum energies
(c) Ratio of the maximum, and minimum energy
(d) Ratio of the minimum and maximum energy.

139. Which one of the following figures respresenting Hooke's jointed inclined shaft system will result in a velocity ratio of unity?

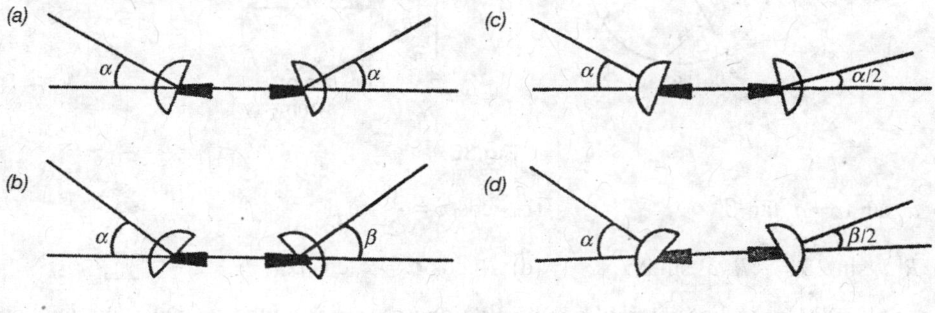

Fig.35

140. Given that:

m = mass of the ball of the governor,
ω = angular velocity of the governor, and
g = accelerate due to gravity,
The height of Watt's governor is given by

(a) $\frac{g}{2\omega^2}$ (b) $\frac{g}{\omega^2}$ (c) $\frac{\sqrt{2g}}{\omega^2}$ (d) $\frac{2g}{\omega^2}$.

141. Match List I with List II and select the correct answer using the codes given below the lists:

List I	List II
A. End thrust	1. Plain bearing
B. No cage	2. Ball bearing
C. More accurate centering	3. Needle bearing
D. Can be overloaded	4. Tapered roller bearing.

Codes:

	A	B	C	D
(a)	3	4	2	1
(b)	4	3	1	2
(c)	3	4	1	2
(d)	4	3	2	1

142. For a given fractional change of speed, if the displacement of the sleeve is high, then the governor is said to be

(a) Hunting (b) Isochronous (c) Sensitive (d) Stable.

143. A four-cylinder symmetrical in-line engine is shown in the given Fig.36. Reciprocating weights per cylinder are R_1 and R_2 and the corresponding angular disposition of the crank are α and β Which one of the following equations should be satisfied for its primary force balance?

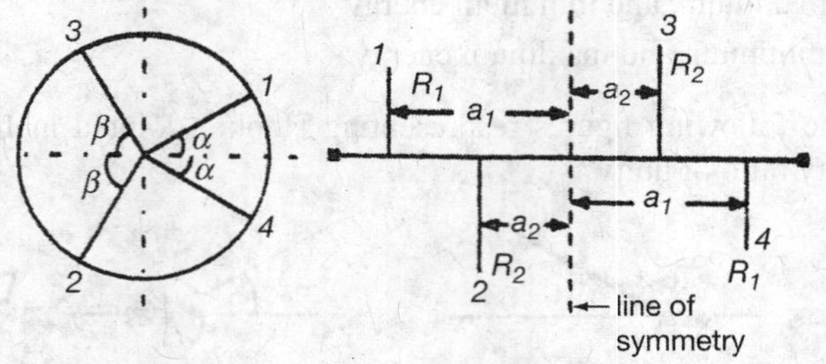

Fig.36

(a) $a_1 \tan \alpha = a_2 \tan \beta$ (b) $\cos\alpha = \frac{1}{2} \sec \beta$

(c) $R_1 a_1 \sin 2\alpha = -R_2 a_2 \sin 2\beta$ (d) $R_1 \cos \alpha = R_2 \cos \beta$.

144. In a multicylinder in-line internal combustion engine, even number of cylinders is chosen so that

(a) Uniform firing order is obtained (b) The couples are balanced

(c) Primary forces are balanced (d) Secondary forces are balanced.

145. The amplitude versus time curve of a damped-free vibration is shown in the below Fig.37. Curve labelled '*A*' is

(a) A logarithmic decrement curve (b) An exponentially decreasing curve

(c) A hyperbolic curve (d) A linear curve.

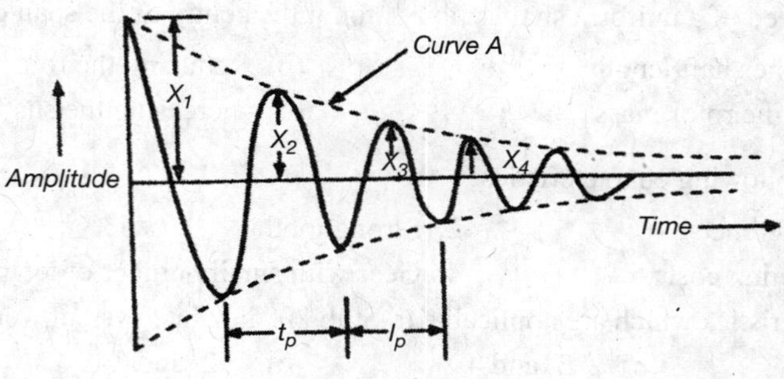

Fig.37

146. If a mass 'm' oscillates on a spring having a mass m_s and stiffness 'k', then the natural frequency of the system is given by

(a) $\sqrt{\dfrac{k}{m+\dfrac{m_s}{3}}}$ (b) $\sqrt{\dfrac{k}{\dfrac{m}{3}+m}}$ (c) $\sqrt{\dfrac{3k}{m+m_s}}$ (d) $\sqrt{\dfrac{k}{m+m_s}}$

147. Match List I with List II and select the correct answer using the codes given below the lists:

List I	List II
A. Node and mode	1. Geared vibration
B. Equivalent inertia	2. Damped-free vibration
C. Log decrement	3. Forced vibration
D. Resonance	4. Multi-rotor vibration

Codes:

	A	B	C	D
(a)	1	4	3	2
(b)	4	1	2	3
(c)	1	4	2	3
(d)	4	1	3	2

148. Two shafts are shown in the below Fig.38. These two shafts will be torsionally equivalent to each other if their

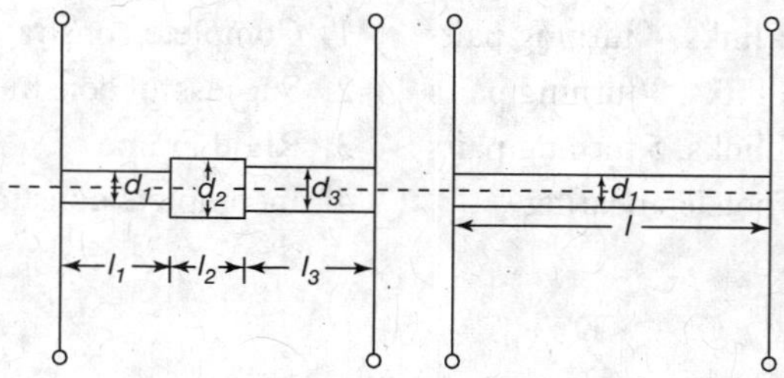

Fig.38

(a) Polar moment of inertias are the same (b) Total angle of twists are the same
(c) Length are the same (d) Strain energies are the same.

149. The critical speed of a uniform shaft with a rotor at the centre of the span can be reduced by

(a) Reducing the shaft length (b) Reducing the rotor mass
(c) Increasing the rotor mass (d) Increasing the shaft diameter.

150. Consider the following characteristics:

1. Small interference 2. Strong tooth
3. Low production cost 4. Gear with small number of teeth

Those characteristics which are applicable to Stub 20° involute system would include

(a) 1 alone (b) 2, 3, and 4 (c) 1, 2, and 3 (d) 1, 2, 3, and 4.

151. A physical system is translated into functional block diagram of the type shown in the Fig.39. The command input *r(t)* and controlled output *c(t)* of this system are given by

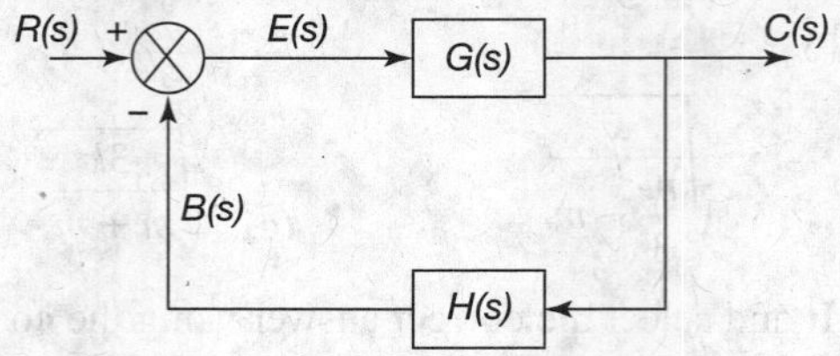

Fig.39

(a) $\dfrac{C(s)}{R(s)} = \dfrac{G(s)}{1 + \dfrac{G(s)}{H(s)}}$ (b) $\dfrac{C(s)}{R(s)} = \dfrac{H(s)}{1 + G(s)H(s)}$

(c) $\dfrac{C(s)}{R(s)} = \dfrac{G(s)}{1 - G(s)H(s)}$ (d) $\dfrac{C(s)}{R(s)} = \dfrac{G(s)}{1 - G(s)H(s)}$.

152. If a compression coil spring is cut into two equal parts and the parts are then used in parallel, the ratio of the spring rate to its initial value will be

(a) 1 (b) 2
(c) 4 (d) Indeterminable for want of sufficient data.

153. Match List I with List II and select the correct answer using the codes given below the Lists:

List I	List II
A. 4 links, 4 turning pairs	1. Complete constraint
B. 3 links, 3 turning pairs	2. Successful constraint
C. 5 links, 5 turning pairs	3. Rigid Frame
D. Footstep bearing	4. Incomplete constraint

Codes:

	A	B	C	D
(a)	3	1	4	2
(b)	1	3	2	4
(c)	3	1	2	4
(d)	1	3	4	2

154. The relative acceleration of two points which are at variable distance apart on a moving link can be determined by using the

(a) Three centres in line theorem (b) Instantaneous centre of rotation method
(c) Coriolis component of acceleration method (d) Klein's construction.

155. Consider a four-bar mechanism shown in Fig.40.

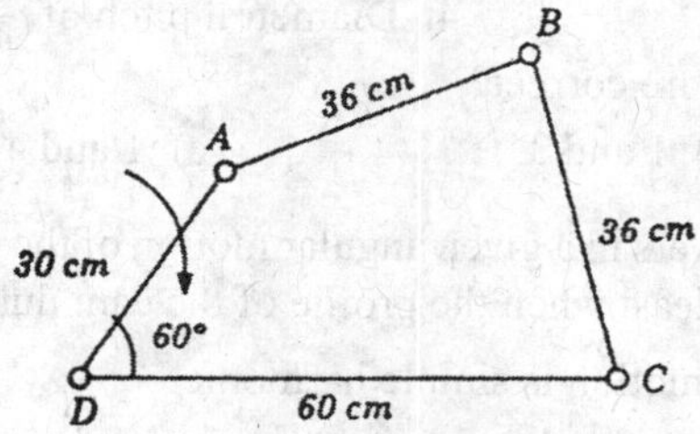

Fig.40

The driving link DA is rotating uniformly at a speed of 100 rpm. Clockwise.
The velocity of A will be

(a) 300 cm/s (b) 314 cm/s (c) 325 cm/s (d) 400 cm/s.

156. Which one of the following pairs is correctly matched?

(a) Governors ... Interference (b) Gears ... Hunting
(c) Klein's construction ... Acceleration of piston (d) Cam ... Pinion.

157. The primary disturbing force due to inertia of reciprocating parts of mass m at radius r moving with an angular velocity ω is given by

(a) $m\omega^2 r\sin\theta$ (b) $m\omega^2 r\cos\theta$

(c) $m\omega^2 r\sin\left(\frac{2\theta}{n}\right)$ (d) $m\omega^2 r\cos\left(\frac{2\theta}{n}\right)$.

158. A link AB is subjected to a force $F(\rightarrow)$ at a point P perpendicular to the link at a distance 'a' from the CG as shown in Fig.41.

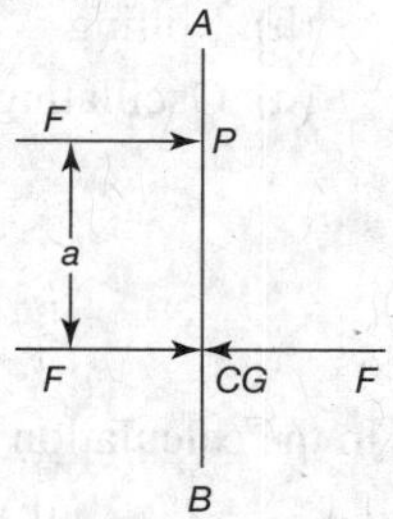

Fig.41

This will result in

(a) An inertia force $F(\rightarrow)$ through the CG and no inertia torque
(b) An inertia torque $F \cdot a$ (↺) and no inertia force
(c) Both inertia force $F(\rightarrow)$ through the CG and inertia torque $F \cdot a$(↺)
(d) Both inertia force $F(\leftarrow)$ through the CG and inertia torque $F \cdot a$(↻).

159. Consider the following statements:

A pinion of $14\frac{1^\circ}{2}$ pressure angle and 48 involute teeth has a pitch circle diameter of 28.8 cm. It has

1. Module of 6 mm
2. Circular pitch of 18 mm
3. Addendum of 6 mm
4. Diametral pitch of $\frac{11}{113}$

Which of these statements are correct?

(a) 2 and 3 (b) 1 and 3 (c) 1 and 4 (d) 2 and 4.

160. For a given lift of the followers in a given angular motion of the cam, the acceleration/retardation of the follower will be the least when the profile of the cam during the rise portion is

(a) Such that the follower motion is simple harmonic
(b) Such that the follower motion has a constant velocity from start to end
(c) A straight line, it being a tangent cam
(d) Such that the follower velocity increases linearly for half the rise portion and then decrease linearly for the remaining half of the rise portion.

161. Consider the following statements regarding the choice of conjugate teeth for the profile of mating gears:

1. They will transmit the desired motion
2. They are difficult to manufacture
3. Standardisation is not possible
4. The cost of production is low

Which of these statements are correct?

(a) 1, 2, and 3 (b) 1, 2, and 4 (c) 2, 3, and 4 (d) 1, 3, and 4.

162. The motion transmitted between the teeth of two spur gears in mesh is generally

(a) Sliding
(b) Rolling
(c) Rotary
(d) Partly sliding and partly rolling.

163. In a single slide four-bar linkage, when the slider is fixed, it forms a mechanism of

(a) Hand pump
(b) Rolling?
(c) Quick return
(d) Oscillating cylinder.

164. Consider the following parameters:

1. Limit of peripheral speed
2. Limit of centrifugal stress
3. Coefficient of fluctuation of speed
4. Weight of the rim

Which of these parameters are used in the calculation of the diameter of flywheel rim?

(a) 1, 3, and 4 (b) 2, 3, and 4 (c) 1, 2, and 3 (d) 1, 2, and 4.

165. Consider the following speed governors:

1. Porter governor
2. Hartnell governor
3. Watt governor
4. Proell governor

The correct sequence of development of these governor is

(a) 1, 3, 2, 4 (b) 3, 1, 4, 2 (c) 3, 1, 2, 4 (d) 1, 3, 4, 2.

166. If a two-mass system is dynamically equivalent to a rigid body, then the system will not satisfy the condition that the
    (a) Sum of the two masses must be equal to that of the rigid body
    (b) Polar moment of inertia of the system should be equal to that of the rigid body
    (c) Centre of gravity (*c.g.*) of the system should coincide with that of the rigid body
    (d) Total moment of inertia about the axis through *c.g.* must be equal to that of the rigid body.

167. A rigid shaft when laid on horizontal parallel ways will not roll if the
    (a) Centre of gravity falls parallel
    (b) Centre of gravity lies on the shaft axis
    (c) Horizontal moments are large
    (d) Vertical moments are large.

168. If the ratio of the length of connecting rod to the crank radius increases, then
    (a) Primary unbalanced forces will increase
    (b) Primary unbalanced forces will decrease
    (c) Secondary unbalanced forces will increase
    (d) Secondary unbalanced forces will decrease.

169. If a spring-mass-dashpot system is subjected to excitation by a constant harmonic force, then at resonance, its amplitude of vibration will be
    (a) Infinity
    (b) Inversely proportional to damping
    (c) Directly proportional to damping
    (d) Decreasing exponentially with time.

170. In a forced vibration with viscous damping, maximum amplitude occurs when forced frequency is
    (a) Equal to natural frequency
    (b) Slightly less than natural frequency
    (c) Slightly greater than natural frequency
    (d) Zero.

171. The value of the natural frequency obtained by Rayleigh's method
    (a) Is always greater than the actual fundamental frequency
    (b) Is always less than the actual fundamental frequency
    (c) Depends upon the initial deflection curve chosen and may be greater than or less than the actual fundamental frequency
    (d) Is independent of the initial deflection curve chosen.

172. In a multi-rotor system of torsional vibration maximum number of nodes that can occur is
    (a) Two
    (b) Equal to the number of rotor plus one
    (c) Equal to the number of rotors
    (d) Equal to the number of rotors minus one.

173. A rotating shaft carries a flywheel which overhangs on the bearing as a cantilever. If this flywheel weight is reduced to half of its original weight, the whirling speed will
    (a) Be double
    (b) Increase by $\sqrt{2}$ times
    (c) Decrease by $\sqrt{2}$ times
    (d) Be half.

174. Consider the gear train shown in the given Fig.42 and table of gears and their number of teeth.

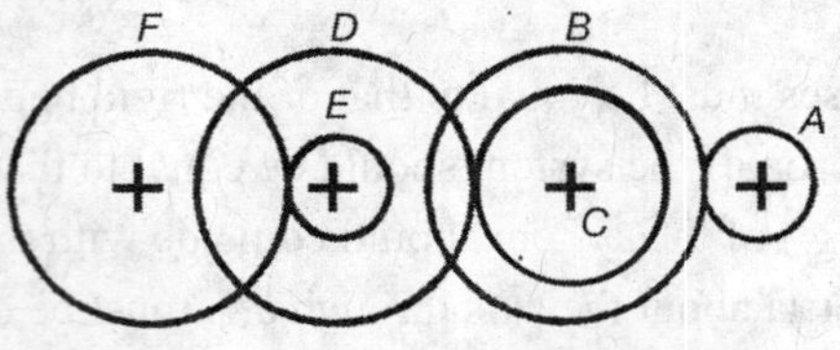

Fig.42

Gear:	A	B	C	D	E	F
No. of teeth:	20	50	25	75	26	65

Gears *BC* and *DE* are moulded on parallel shaft rotating together.
If the speed of *A* is 975 rpm, the speed of *F* will be

(a) 39 rpm (b) 52 rpm (c) 75 rpm (d) 80 rpm.

175. Consider the following statements in respect of introduction of feedback in a control system:
1. It enhances its gain
2. It attenuates the unwanted noise
3. It helps in improving the accuracy of the system

Which of these statements are correct?

(a) 2 and 3 (b) 1, 2, and 3 (c) 1 and 3 (d) 1 and 2.

176. The kinematic chain shown in the given Fig.43 is a

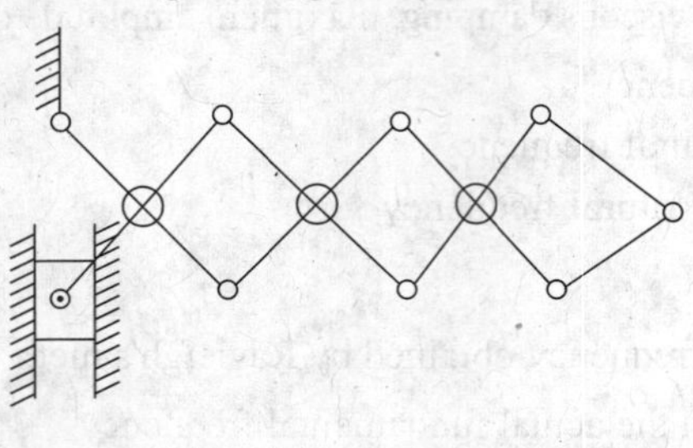

Fig.43

(a) structure
(b) mechanism with one degree of freedom
(c) mechanism with two degrees of freedom
(d) mechanism with more than two degrees of freedom.

177. A point on a link connecting a double slider crank chain will trace a

(a) straight line (b) circle (c) parabola (d) ellipse.

178. A wheel is rolling on a straight level track with a uniform velocity '*v*'. The instantaneous velocity of a point on the wheel lying at the mid-point of a radius.

(a). varies between $3\frac{v}{2}$ and $-\frac{v}{2}$ (b) varies between $\frac{v}{2}$ and $-\frac{v}{2}$

(c) varies between $3\frac{v}{2}$ and $\frac{v}{2}$ (d) does not vary and is equal to *v*.

179. A four-bar chain has
 (a) all turning pairs
 (b) one turning pair and the others are sliding pairs
 (c) one sliding pair and the others are turning pairs
 (d) all sliding pairs.

180 Sensitiveness of a governor is defined as
 (a) $\frac{\text{Range of speed}}{2 \times \text{Mean speed}}$ (b) $\frac{2 \times \text{Mean speed}}{\text{Range of speed}}$
 (c) Mean speed × Range of speed (d) $\frac{\text{Range of speed}}{\text{Mean speed}}$.

181. Masses B_1, B_2 and 9 kg are attached to a shaft in parallel planes as shown in the given Fig.44. If the shaft is rotating at 100 rpm, the mass B_2 is

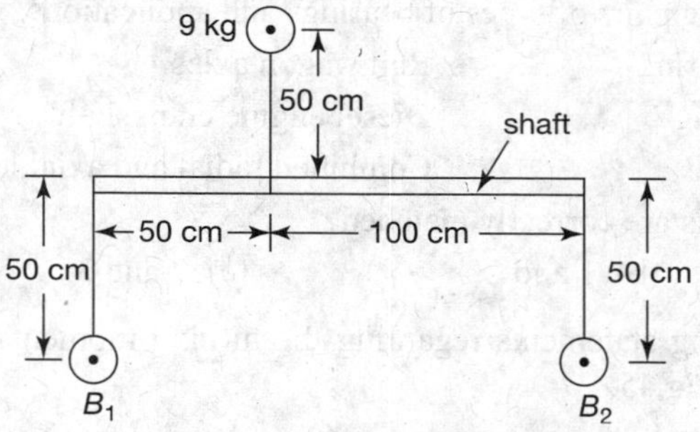

Fig.44

 (a) 3 kg (b) 6 kg (c) 9 kg (d) 27 kg.

182. The equation of motion for a damped viscous vibration is
 $3\ddot{x} + 9\dot{x} + 27x = 0$
 The damping factor is
 (a) 0.25 (b) 0.50 (c) 0.75 (d) 1.00.

183. A mass is suspended at the bottom of two springs in series having stiffness 10 N/mm and 5 N/mm. The equivalent spring stiffness of the two springs is nearly.
 (a) 0.3 N/mm (b) 3.3 N/mm (c) 5 N/mm (d) 15 N/mm.

184. The velocity ratio in the case of compound train of wheels is equal to
 (a) $\frac{\text{No. of teeth on first driver}}{\text{No. of teeth on last follower}}$ (b) $\frac{\text{No. of teeth on last follower}}{\text{No. of teeth on first driver}}$
 (c) $\frac{\text{Product of teeth on the drivers}}{\text{Product of teeth on the followers}}$ (d) $\frac{\text{Product of teeth on the followers}}{\text{Product of teeth on the drivers}}$.

185. Consider the following pairs of parts:
 1. Pair of gear in mesh 2. Belt and pulley
 3. Cylinder and piston 4. Cam and follower

Among these, the higher pairs are

(a) 1 and 4 (b) 2 and 4 (c) 1, 2, and 3 (d) 1, 2, and 4.

186. Which one of the following sets of accelerations is involved in the motion of the piston inside the cylinder of a uniformly rotating cylinder mechanism?

(a) Coriolis and radial acceleration (b) Radial and tangential acceleration

(c) Coriolis and gyroscopic acceleration (d) Gyroscopic and tangential acceleration.

187. Consider the following statements:

1. Round bar in a round hole forms a turning pair
2. A square bar in a square hole forms a sliding pair
3. A vertical shaft in a foot-step bearing forms a successful constraint

Which of these statements are correct?

(a) 1 and 3 (b) 1 and 2 (c) 2 and 3 (d) 1, 2, and 3.

188. Consider the following pairs of types of bearings and applications:

1. Partial Journal bearing ... Rail wagon axles
2. Full Journal bearing ... Diesel engine crankshaft
3. Radial bearing ... Combined radial and axial loads

Which of these pairs is/are correctly matched?

(a) 1 alone (b) 1 and 2 (c) 2 and 3 (d) 1, 2, and 3.

189. Consider the following statements regarding the turning moment diagram of a reciprocating engine shown in the Fig.45:

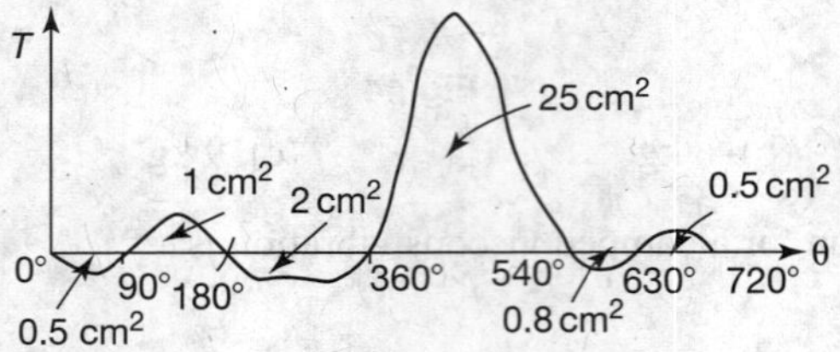

Fig.45

(Scale 1 cm² = 100 Nm)

1. It is a four stroke IC engine
2. The compression stroke is 0° to 180°
3. Mean turning moment $T_m = \dfrac{580}{\pi}$ Nm
4. It is a multi-cylinder engine

Which of these statements are correct?

(a) 1, 2, and 3 (b) 1, 2, and 4 (c) 2, 3, and 4 (d) 1, 3, and 4.

190. The pitching of a ship in the ocean is an oscillatory periodic motion. A ship is pitching 6° above and 6° below with a period of 20s from its horizontal plane. Consider the following statements in this regard:

1. The motion has a frequency of oscillation (*i.e.* pitching) of 3 cycles/minute
2. The motion has an angular frequency of 3.14 rad/s

3. The angular velocity of prescission of ship's rotor is $\frac{\pi^2}{300}$ rad/s

4. The amplitude of pitching is $\frac{\pi}{30}$ rad

Which of these statements are correct?

(a) 1 and 2 (b) 1, 2, and 4 (c) 2, 3, and 4 (d) 1, 3, and 4.

191. The critical speed of a shaft is affected by the
   (a) diameter and the eccentricity of the shaft
   (b) span and the eccentricity of the shaft
   (c) diameter and the span of the shaft
   (d) span of the shaft.

192. Match List I with List II and select the correct answer using the codes given below the lists:

List I	List II
A. Compound train	1. Hart mechanism
B. Quick return Mechanism	2. Coriolis force
C. Exact straight line motion	3. Transmission of motion around bends and corners
D. Approximate straight line motion	4. Watt mechanism

Codes:

A	B	C	D	
(a)	1	2	3	4
(b)	3	2	1	4
(c)	3	4	1	2
(d)	1	4	3	2

193. Match List I (Kinematic inversions) with List II (Applications) and select the correct answer using the codes given below the lists:

List I	List II
A. 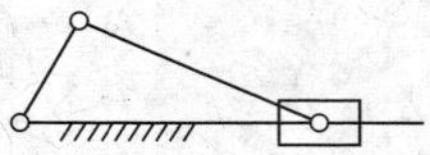	1. Hand pump
B. 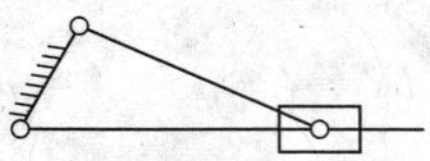	2. Compressor
C. 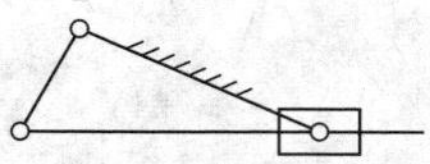	3. Whitworth quick return mechanism
D. 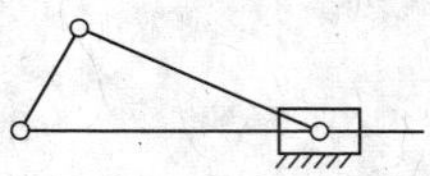	4. Oscillating Cylinder Engine

Codes:

	A	B	C	D
(a)	1	3	4	2
(b)	2	4	3	1
(c)	2	3	4	1
(d)	1	4	3	2

194. Match List I (Applications) with List II (Features of vibration) and select the correct answer using the codes given below the lists:

List I	List II
A. Vibration damper	1. Frequency of free vibration
B. Shock absorber	2. Forced vibration
C. Frahm Tachometer	3. Damping of vibration
D. Oscillator	4. Transverse vibration
	5. Absorption of vibration

Codes:

	A	B	C	D
(a)	5	3	2	1
(b)	3	1	4	2
(c)	5	3	4	1
(d)	3	4	2	5

195. When the intensity of pressure is uniform in a flat pivot bearing of radius r, the friction force is assumed to act at

(a) r (b) $\frac{r}{2}$ (c) $\frac{2r}{3}$ (d) $\frac{r}{3}$.

196. Consider a harmonic motion $x = 1.25 \sin (5t - \pi/6)$ cm. Match List I with List II and select the correct answer using the codes given below the Lists:

List I	List II
A. Amplitude (cm)	1. $\frac{5}{2\pi}$
B. Frequency (cycle/s)	2. 1.25
C. Phase angle (rad)	3. $\frac{1}{5}$
D. Time period (s)	4. $\frac{\pi}{6}$

Codes:

	A	B	C	D
(a)	4	1	2	3
(b)	2	3	4	1
(c)	4	3	2	1
(d)	2	1	4	3

197. Which of the flowing pairs of devices and their functions are correctly matched?

1. Flywheel ... For storing kinetic energy
2. Governors ... For controlling speeds
3. Lead screw in lathe ... For providing feed to the slides
4. Fixtures ... For locating workpiece and guiding tools

Select the correct answer using the codes given below:

Codes:

(a) 1, 3, and 4 (b) 2 and 3 (c) 1 and 2 (d) 2 and 4.

198. Match list I with list II and select the correct answer using the codes given below the lists. (Notations have their usual meanings):

List I	List II
A. Law of correct steering	1. $f = 3(n-1) - 2j$
B. Displacement relation of Hooke's joint	2. $x = R\left[(1-\cos\theta) + \dfrac{\sin^2\theta}{2n}\right]$
C. Relation between kinematic pairs and links	3. $\cot\theta - \cot\phi = \dfrac{a}{b}$
D. Displacement equation of reciprocating engine piston	4. $\tan\theta - \tan\phi = \cos a$

Codes:

	A	B	C	D
(a)	1	4	3	2
(b)	1	2	3	4
(c)	3	4	1	2
(d)	3	2	1	4

199. Force required to accelerate a cylindrical body which rolls without slipping on a horizontal plane (mass of cylindrical body is m, radius of the cylindrical surface in contact with plane is r, radius of gyration of body is k and acceleration of the body is a) is

(a) $m(k^2/r^2 + 1)\cdot a$ (b) $(mk^2/r^2)\cdot a$ (c) $mk^2\cdot a$ (d) $(mk^2/r + 1)\cdot a$.

200. Consider the following statements regarding motions in machines:

1. Tangential acceleration is a function of angular velocity and the radial acceleration is a function of angular acceleration.
2. The resultant acceleration of a point A with respect to a point B on a rotating link is perpendicular to AB.
3. The direction of the relative velocity of a point A with respect to a point B on a rotating link is perpendicular to AB.

Which of these statements is/are correct?

(a) 1 alone (b) 2 and 3 (c) 1 and 2 (d) 3 alone.

201. Consider the following statements:

In petrol engine mechanism, the piston is at its dead centre position when piston

1. acceleration is zero
2. acceleration is maximum
3. velocity is zero
4. velocity is infinity.

Which of these statements are correct?
(a) 1 and 4 (b) 1 and 3 (c) 2 and 3 (d) 2 and 4.

202. The speed of driving shaft of a Hooke's joint of angle 19.5° (given sin 19.5° = .33, cos 19.5°= .94) is 500 rpm. The maximum speed of the driven shaft is nearly
(a) 168 rpm (b) 444 rpm (c) 471 rpm (d) 531 rpm.

203. The given Fig.46 shows the Klein's construction for acceleration of the slider-crank mechanism. Which one of the following quadrilaterals represents the required acceleration diagram?

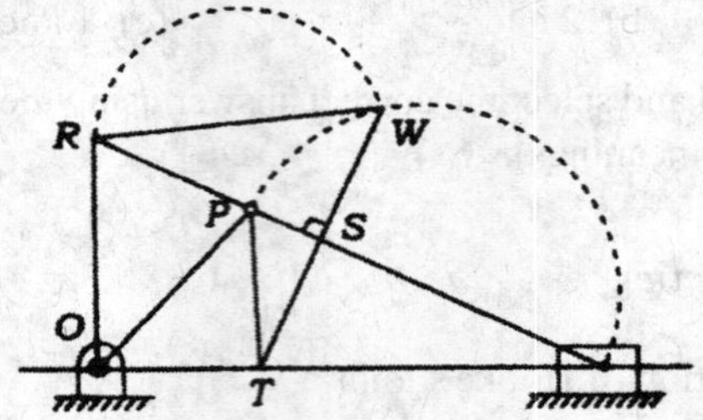

Fig.46

(a) *ORST* (b) *OPST* (c) *ORWT* (d) *ORPT*

204. A servomotor is connected through a gear ratio of 10 (*i.e.*, motor speed: load side speed = 10:1) to a load having moment of inertia *J*. The equivalent parameter referred to motor shaft side is
(a) J_{eq} = 0.01 J (b) J_{eq} = 10 J (c) J_{eq} = 0.1 J (d) J_{eq} = 100 J

205. Match List I with List II and select the correct answer using the codes given below the Lists:

List I	List II
A. Cam and follower	1. Grubler's rule
B. Screw pair	2. Grashof's linkage
C. 4-bar mechanism	3. Pressure angle
D. Degree of freedom of planar mechanism	4. Single degree of freedom

Codes:

	A	B	C	D
(a)	3	4	2	1
(b)	1	2	4	3
(c)	1	4	2	3
(d)	3	2	4	1

206. Consider the following statements:
When two gears are meshing, the clearance is given by the
1. difference between dedendum of one gear and addendum of the mating gear
2. difference between total and the working depth of a gear tooth
3. distance between the bottom land of one gear and the top land of the mating gear
4. difference between the radii of the base circle and the dedendum circle

Which of these statements are correct?
(a) 1, 2, and 3 (b) 2, 3, and 4 (c) 1, 3, and 4 (d) 1, 2, and 4.

207. A body of mass m and radius of gyration k is to be replaced by two masses m_1 and m_2 located at distances h_1 and h_2 from the CG of the original body. An equivalent dynamic system will result, if

(a) $h_1 + h_2 = k$ (b) $h_1^2 + h_2^2 = k^2$ (c) $h_1 + h_2 = k^2$ (d) $\sqrt{h_1 h_2} = k^2$.

208. Match List I and List II and select the correct answer using the code given below the Lists:

List I	List II
A. Undercutting	1. Beam strength
B. Addendum	2. Interference
C. Lewis equation	3. Large speed reduction
D. Worm and wheel	4. Intersecting axes
	5. Module

Codes:

	A	B	C	D
(a)	2	5	1	3
(b)	1	5	4	3
(c)	1	3	4	5
(d)	2	3	1	5

209. The natural frequency of transverse vibration of a massless beam of length L having a mass m attached at its midspan is given by (EI is the flexural rigidity of the beam)

(a) $\left(\dfrac{mL^3}{48EI}\right)^{\frac{1}{2}}$ rad/s (b) $\left(\dfrac{48mL^3}{EI}\right)^{\frac{1}{2}}$ rad/s (c) $\left(\dfrac{48EI}{mL^3}\right)^{\frac{1}{2}}$ rad/s (d) $\left(\dfrac{3EI}{mL^3}\right)^{\frac{1}{2}}$ rad/s.

210. Match List I with List II and select the correct answer using the codes given below the Lists:

List I	List II
A. 6 d.o.f. system	1. Vibrating beam
B. 1 d.o.f. system	2. Vibrating absorber
C. 2 d.o.f. system	3. A rigid body in space
D. Multi d.o.f. system	4. Pure rolling of a cylinder

Codes:

	A	B	C	D
(a)	1	2	4	3
(b)	1	4	2	3
(c)	3	2	4	1
(d)	3	4	2	1

211. A shaft carries a weigh W at the centre. The CG of the weight is displaced by an amount e from the axis of the rotation. If y is the additional displacement of the CG from the axis of rotation due to the centrifugal force, then the ratio of y to e (where ω_c is the critical speed of shaft and ω is the angular speed of shaft) is given by

(a) $\dfrac{1}{\left[\dfrac{\omega_c}{\omega}\right]^2 + 1}$ (b) $\dfrac{1}{\left[\dfrac{\omega_c}{\omega}\right]^2 - 1}$ (c) $\left[\dfrac{\omega_c}{\omega}\right]^2 + 1$ (d) $\dfrac{\omega}{\left[\dfrac{\omega_c}{\omega}\right]^2 - 1}$.

212. In a simple gear train, if the number of idler gears is odd, then the direction of motion of driven gear will

(a) be same as that of the driving gear
(b) be opposite to that of the driving gear
(c) depend upon the number of teeth on the driving gear
(d) depend upon the total number of teeth on all gears of the train.

213. When a vehicle travels on a rough road whose undulations can be assumed to be sinusoidal, the resonant conditions of the base-excited vibrations, are determined by the

(a) mass of the vehicle, stiffness of the suspension spring, speed of the vehicle, wavelength of the roughness curve
(b) speed of the vehicle only
(c) speed of the vehicle and the stiffness of the suspension spring
(d) amplitude of the undulations.

214. During torsional vibration of a shaft, the node is characterized by the

(a) maximum angular velocity
(b) maximum angular displacement
(c) maximum angular acceleration
(d) zero angular displacement.

215. In a slider-crank mechanism, the maximum acceleration of slider is obtained when the crank is

(a) at the inner dead centre position
(b) at the outer dead centre position
(c) exactly midway position between the two dead centres
(d) slightly in advance of the midway position between the two dead centres.

216. Consider the following statements for completely balancing a single rotating mass:

1. Another rotating mass placed diametrically opposite in the same plane balances the unbalanced mass.
2. Another rotating mass placed diametrically opposite in a parallel plane balances the unbalanced mass.
3. Two masses placed in two different parallel planes balance the unbalanced mass.

Which of the above statements is/are correct?

(a) 1 only (b) 1 and 2 (c) 2 and 3 (d) 1 and 3.

217. Consider the following statements in case of reverted gear train:

1. The direction of rotation of the first and the last gear is the same
2. The direction of rotation of the first and the last gear is opposite
3. The first and the last gears are on the same shaft
4. The first and the last gears are on separate but co-axial shafts.

Which of these statements is/are correct?

(a) 1 and 3 (b) 2 and 3 (c) 2 and 4 (d) 4 alone.

218. The instantaneous centre of rotation of a rigid thin disc rolling without slip on a plane rigid surface is located at

(a) the centre of the disc
(b) an infinite distance perpendicular to the plane surface
(c) the point of contact
(d) the point on the circumference situated vertically opposite to the contact point.

219. Match List I (Kinematic pairs) with List II (Practical example) and select the correct answer using the codes given below the lists:

List I (Kinematic pairs)	List II (Practical example)
A. Sliding pair	1. A road roller rolling over the ground
B. Revolute pair	2. Crank shaft in a journal bearing in an engine
C. Rolling pair	3. Ball and socket joint
D. Spherical pair	4. Piston and cylinder
	5. Nut and screw

Codes:

	A	B	C	D
(a)	5	2	4	3
(b)	4	3	1	2
(c)	5	3	4	2
(d)	4	2	1	3

220. The choice of displacement diagram during rise or return of a follower of a cam-follower mechanism is based on dynamic considerations. For high speed cam follower mechanism, the most suitable displacement for the follower is

(a) cycloidal motion
(b) simple harmonic motion
(c) parabolic or uniform acceleration motion
(d) uniform motion or constant velocity motion.

221. A linkage is shown below in the Fig.47 in which links *ABC* and *DEF* are ternary links whereas *AF*, *BE* and *CD* binary links.

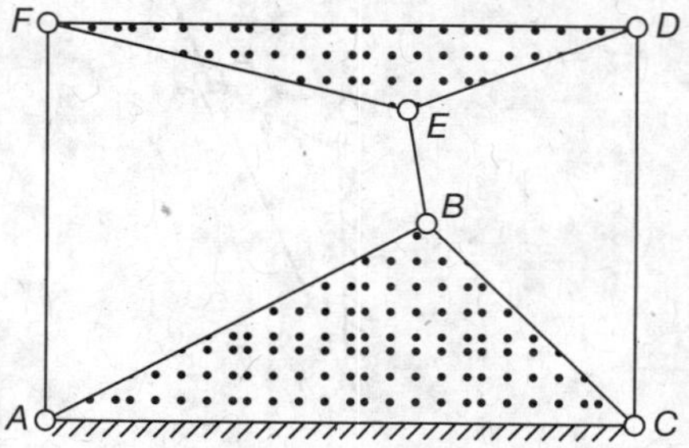

Fig.47

The degrees of freedom of the linkage when link *ABC* is fixed are

(a) 0 (b) 1 (c) 2 (d) 3.

222. Match List I (Mechanism) with List II (Motion) and select the correct answer using the codes given below the lists:

List I (Mechanism)	List II (Motion)
A. Hart mechanism	1. Quick return motion
B. Pantograph	2. Copying mechanism
C. Whitworth mechanism	3. Exact straight line motion
D. Scotch yoke	4. Simple harmonic motion
	5. Approximate straight line motion

Codes:

	A	B	C	D
(a)	5	1	2	3
(b)	3	2	1	4
(c)	5	2	1	3
(d)	3	1	2	4

223. Match List (Connecting shafts) with List II (Couplings) and select the correct answer using the codes given below the lists:

List I (Connecting shaft)	List II (Couplings)
A. In perfect alignment	1. Oldham coupling
B. With angular misalignment of 10°	2. Rigid coupling
C. Shafts with parallel misalignment	3. Universal joint
D. Where one of the shafts may undergo more deflection with respect to the other	4. Pin type flexible coupling

Codes:

	A	B	C	D
(a)	2	1	3	4
(b)	4	3	1	2
(c)	2	3	1	4
(d)	4	1	3	2

224. The crank and slotted lever quick-return motion mechanism is shown in Fig.48. The length of links O_1O_2, O_1C and O_2A are 10 cm, 20 cm and 5 cm respectively.

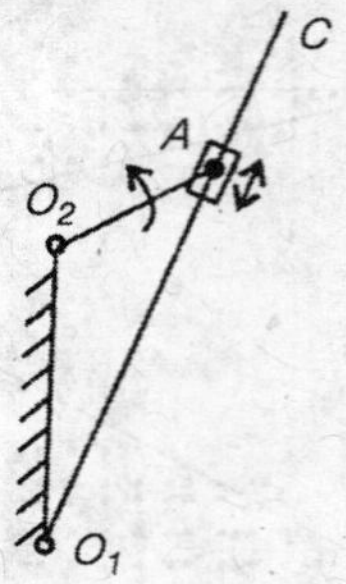

Fig.48

The quick return ratio of the mechanism is

(a) 3.0 (b) 2.75 (c) 2.0 (d) 0.

225. If the rotating mass of a rim type flywheel is distributed on another rim type fly wheel whose mean radius is half the mean radius of the former, then energy stored in the latter at the same speed will be

(a) four times the first one (b) same as the first one
(c) one-fourth of the first one (d) two times the first one

226. The nature of the governors is shown by the graph between radius (r) of rotation and controlling force (F). Which of the following is an isochronous governor?

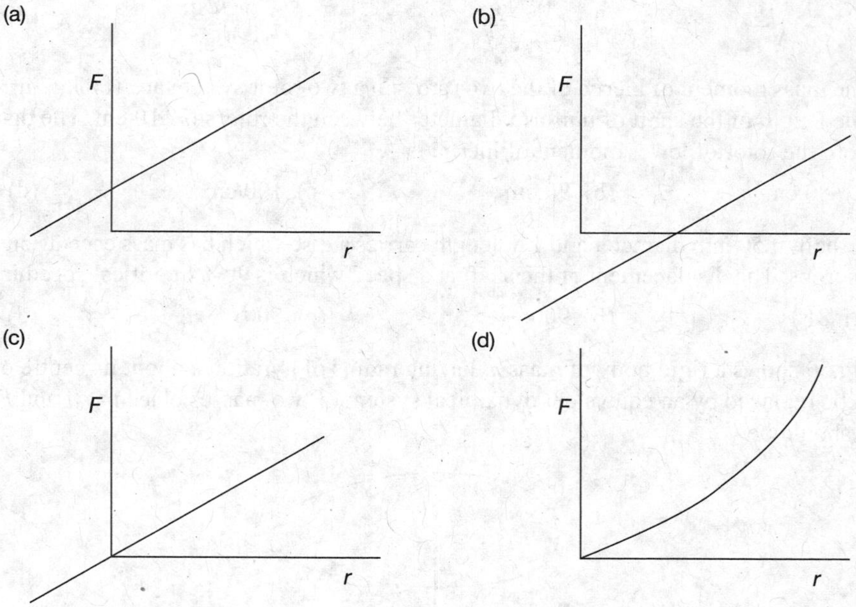

Fig.49

227. The radius of the friction circle in a journal bearing is dependent on coefficient of friction and the

(a) angular velocity of the journal
(b) radius of the journal
(c) magnitudes of the forces on the journal
(d) journal and bearing clearance.

228. In a collar thrust bearing, the number of collars have been doubled while maintaining coefficient of friction and axial thrust same. It will result in

(a) same friction torque and same bearing pressure
(b) double friction torque and half bearing pressure
(c) double friction torque and same bearing pressure
(d) same friction torque and half bearing pressure.

229. In a Hartnell governor, the mass of each ball is 2.5 kg. Maximum and minimum speeds of rotation are 10 rad/s and 8 rad/s respectively. Maximum and minimum radii of rotation are 20 cm and 14 cm respectively. The lengths of horizontal and vertical arms of bell crank levers are 10 cm and 20 cm respectively. Neglecting obliquity and gravitational effects, the lift of the sleeve is

(a) 1.5 cm (b) 3.0 cm (c) 6.0 cm (d) 12.0 cm.

230. A rod of uniform diameter is suspended from one of its ends in vertical plane. The mass of the rod is 'm' and length 'l', the natural frequency of this rod in Hz for small amplitude is

(a) $\frac{1}{2\pi}\sqrt{\frac{g}{l}}$ (b) $\frac{1}{2\pi}\sqrt{\frac{g}{3l}}$ (c) $\frac{1}{2\pi}\sqrt{\frac{2g}{3l}}$ (d) $\frac{1}{2\pi}\sqrt{\frac{3g}{3l}}$.

231. The mass moment of inertia of the two rotors in a two rotor system are 100 kg · m² and 10kg · m². The length of the shaft of uniform diameter between the rotors is 110 cm. The distance of node from the rotor of lower moment of inertia is

(a) 80 cm (b) 90 cm (c) 100 cm (d) 110 cm.

232. A shaft of 50 mm diameter and 1 m length carries a disc which has mass eccentricity equal to 190 microns. The displacement of the shaft at a speed which is 90% of critical speed in microns is

(a) 810 (b) 900 (c) 800 (d) 820.

233. Fig.50 shows a rigid body of mass m having radius of gyration k about its centre of gravity. It is to be replaced by an equivalent dynamical system of two masses placed at A and B. The mass at A should be

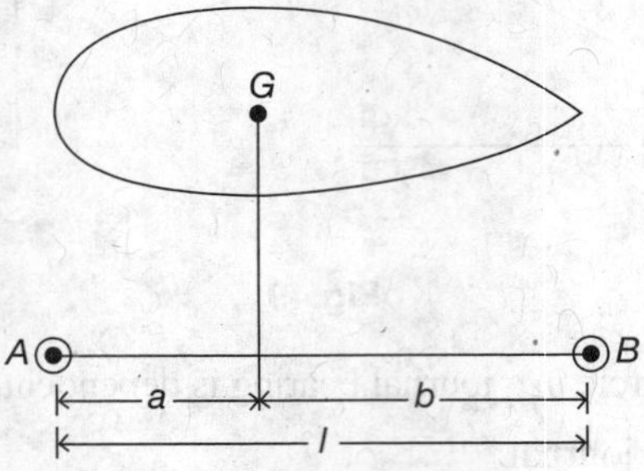

Fig.50

(a) $\frac{a \times m}{a+b}$ (b) $\frac{b \times m}{a+b}$ (c) $\frac{m}{3} \times \frac{a}{b}$ (d) $\frac{m}{2} \times \frac{b}{a}$.

234. $f = 3(n-1) - 2j$. In the Grubler's equation for planar mechanisms given, j is the

(a) Number of mobile links (b) Number of links
(c) Number of lower pairs (d) Length of the longest link.

235. Which of the following are examples of forced closed kinematic pairs?

1. Cam and roller mechanism 2. Door closing mechanism
3. Slider-crank mechanism 4. Automotive clutch operating mechanism

Select the correct answer using the codes given below:

Codes:

(a) 1, 2, and 4 (b) 1 and 3 (c) 2, 3, and 4 (d) 1, 2, 3, and 4.

236. Which of the mechanisms shown in Fig.51 do/does not have single degree of freedom?

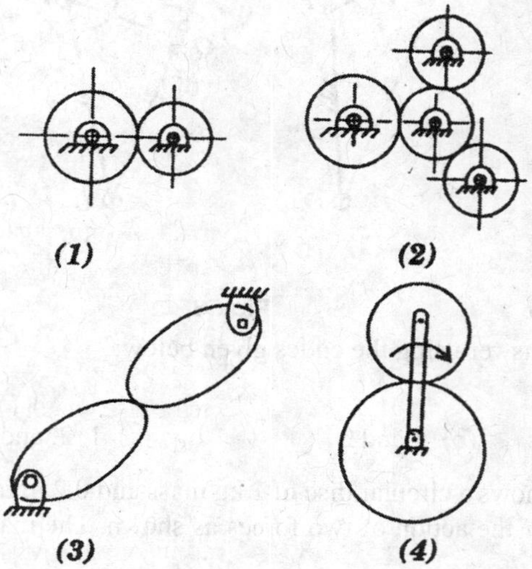

Fig.51

(a) 3 and 4 (b) 2 and 3 (c) 3 only (d) 4 only.

237. Two points, *A* and *B* located along the radius of a wheel, as shown in the Fig.52, have velocities of 80 and 140 m/s, respectively. The distance between points *A* and *B* is 300 mm. The radius of wheel is

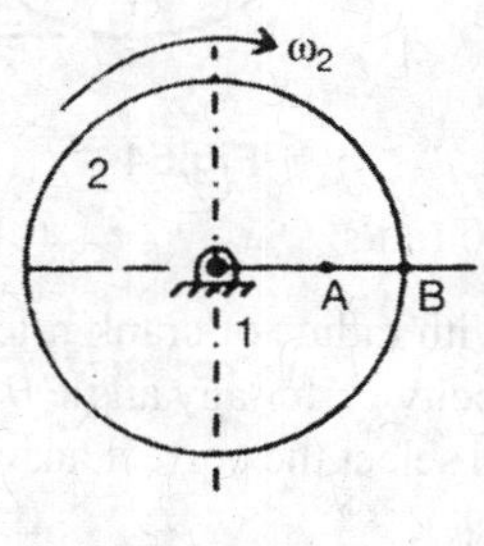

Fig.52

(a) 400 mm (b) 500 mm (c) 600 mm (d) 700 mm.

238. In a slider-crank mechanism, the velocity of piston becomes maximum when

(a) Crank and connecting rod are in line with each other

(b) Crank is perpendicular to the line of stroke of the piston

(c) Crank and connecting rod are mutually perpendicular

(d) Crank is 120° with the line of stroke.

239. Three positions of the quick-return mechanism are shown in Fig.53. In which of the cases does the Coriolis component of acceleration exist?

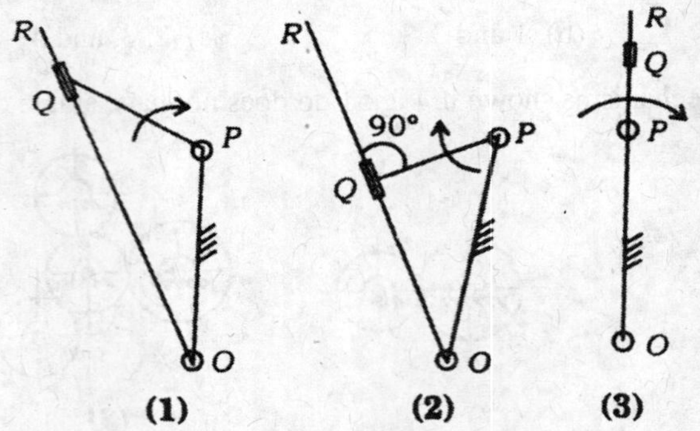

Fig.53

Select the correct answer using the codes given below:

Codes:

(a) 1 only (b) 1 and 2 (c) 1, 2, and 3 (d) 2 and 3.

240. The below Fig.54 shows a circular disc of 1 kg mass and 0.2 m radius undergoing unconstrained planar motion under the action of two forces as shown. The magnitude of angular acceleration α of the disc is

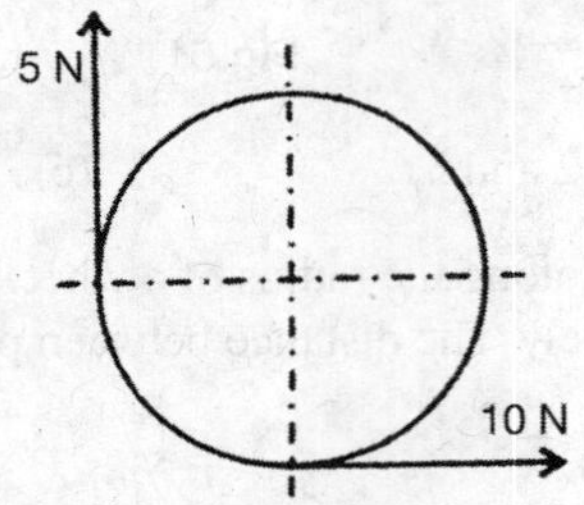

Fig.54

(a) 50 rad/s^2 (b) 100 rad/s^2 (c) 25 rad/s^2 (d) 20 rad/s^2.

241. For a slider-crank mechanism with radius of crank r, length of connecting rod l, obliquity ratio n, crank rotating at an angular velocity ω; for any angle θ of the crank match, List I (Kinematic Variable) with List II (Equation) and select the correct answer using the codes given below the Lists:

List I (Kinematic Variable)	List II (Equation)
A. Velocity of piston	1. $\frac{\omega}{n} \cdot \cos\theta$
B. Acceleration of piston	2. $\omega^2 r \cdot \left(\cos\theta + \frac{\cos 2\theta}{n}\right)$
C. Angular velocity of connecting rod	3. $-\frac{\omega^2}{n} \cdot \sin\theta$
D. Angular acceleration of connecting rod	4. $\omega r \cdot \left(\sin\theta + \frac{\sin 2\theta}{2n}\right)$

Codes:

	A	B	C	D
(a)	4	2	3	1
(b)	2	4	3	1
(c)	4	2	1	3
(d)	2	4	1	3

242. Fig.55 shows Klien's construction for slider-crank mechanism *OCP* drawn to full scale. What velocity does *CD* represent?

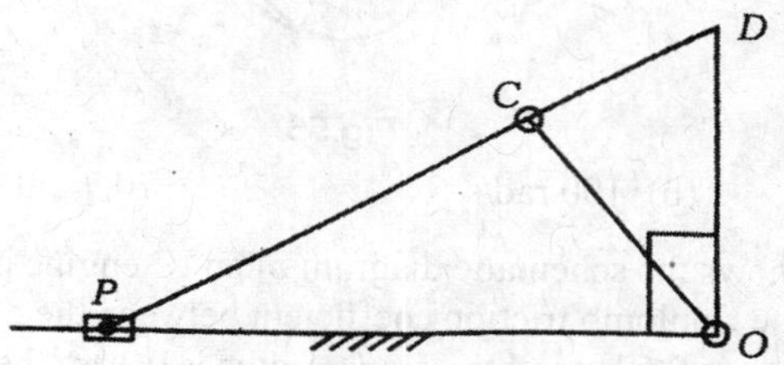

Fig.55

(a) Velocity of the crank pin
(b) Velocity of the piston
(c) Velocity of the piston with respect to crank pin
(d) Angular velocity of the connecting rod.

243. The maximum fluctuation of energy E_f, during a cycle for a flywheel is

(a) $I(\omega^2_{max} - \omega^2_{min})$

(b) $\frac{1}{2} \cdot I \cdot \omega_{av} \cdot (\omega_{max} - \omega_{min})$

(c) $\frac{1}{2} \cdot I \cdot K_{es} \cdot \omega_{av}^2$

(d) $I \cdot \omega_{av}^2 \cdot K_{es}$

(where I = Mass moment of inertia of the flywheel
ω_{av} = Average rotational speed
K_{es} = Coefficient of fluctuation of speed)

244. For minimizing speed fluctuations of an engine as a prime mover, it must have
(a) Only a flywheel fitted to the crankshaft
(b) A governor provided in the system
(c) Both a flywheel and a governor provided in the system
(d) Neither a flywheel nor a governor.

245. Effect of friction, at the sleeve of a centrifugal governor is to make it
(a) More sensitive
(b) More stable
(c) Insensitive over a small range of speed
(d) Unstable.

246. A circular disc having a mass of 30 kg is moved asymmetrically between two bearings A and B as shown in the Fig.56. It is used as an eccentric cam with an eccentricity of 0.01 m. If the shaking force on each of the bearings is not to exceed 1500 N, the speed of rotation of the cam should not exceed

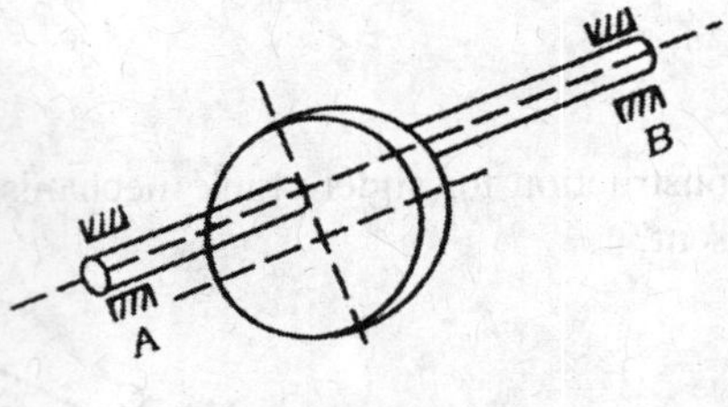

Fig.56

(a) 10 rad/s (b) 100 rad/s (c) 70.7 rad/s (d) 140 rad/s

247. The below Fig.57 shows the schematic diagram of an IC engine producing a torque $T = 41$Nm at the given instant. The Coulomb friction coefficient between the cylinder and the piston is 0.08. If the mass of the piston is 0.5 kg and the crank radius is 0.1 m, the Coulomb friction force occurring at the piston cylinder interface is

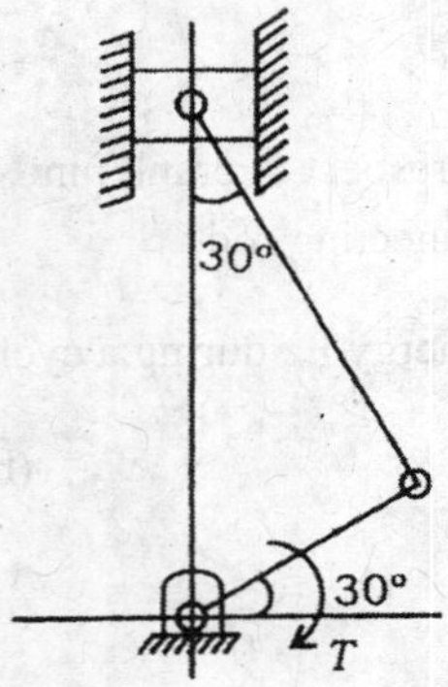

Fig.57

(a) 16 N (b) 0.4 N (c) 4 N (d) 16.4 N.

248. Consider the following, modifications regarding avoiding the interference between gears:

1. The centre distance between meshing gears be increased
2. Addendum of the gear be modified
3. Teeth should be undercut slightly at the root
4. Pressure angle should be increased
5. Circular pitch be increased

Which of these are effective in avoiding interference?

(a) 1, 2, and 3 (b) 2, 3, 4, and 5 (c) 1, 4, and 5 (d) 3, 4, and 5.

249. In a reverted gear train, two gears P and Q are meshing, Q–R is a compound gear, and R and S are meshing. The modules of P and R are 4 mm and 5 mm respectively. The number of teeth in P, Q and R are 20, 40 and 25 respectively. The number of teeth in S is

(a) 23 (b) 35 (c) 50 (d) 53.

250. When two spur gears having involute profiles on their teeth engage, the line of action is tangential to the

(a) Pitch circles (b) Dedendum circles (c) Addendum circles (d) Base circles.

251. If the annular wheel of an epicyclic gear train has 100 teeth and the planet wheel has 20 teeth, the number of teeth on the sun wheel is

(a) 80 (b) 60 (c) 40 (d) 20.

252. The double slider-crank chain is shown below in the Fig.58 in its three possible inversions. The link shown hatched is the fixed link:

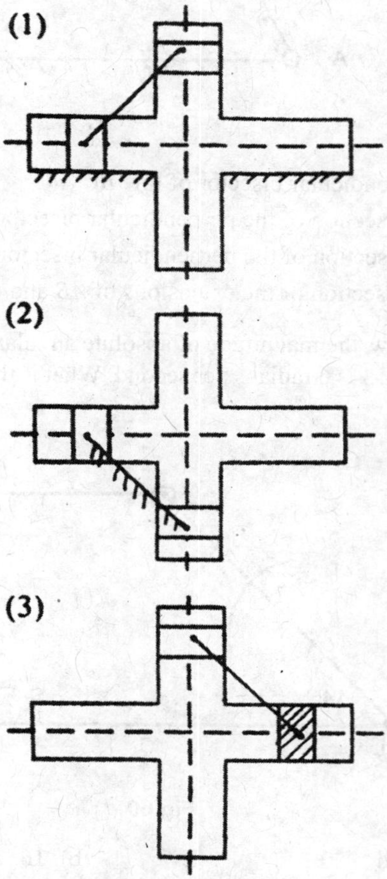

Fig.58

Which one of the following statements is correct?

(a) Inversion (1) is for ellipse trammel and inversion (2) is for Oldham coupling

(b) Inversion (1) is for ellipse trammel and inversion (3) is for Oldham coupling

(c) Inversion (2) is for ellipse trammel and inversion (3) is for Oldham coupling

(d) Inversion (3) is for ellipse trammel and inversion (2) is for Oldham coupling.

253. *ABCD* is a four bar mechanism, in which *AD* is the fixed link, and link *BC*, is in the form of a circular disc with centre *P*. In which one of the following cases *P* will be the instantaneous centre of the disc?

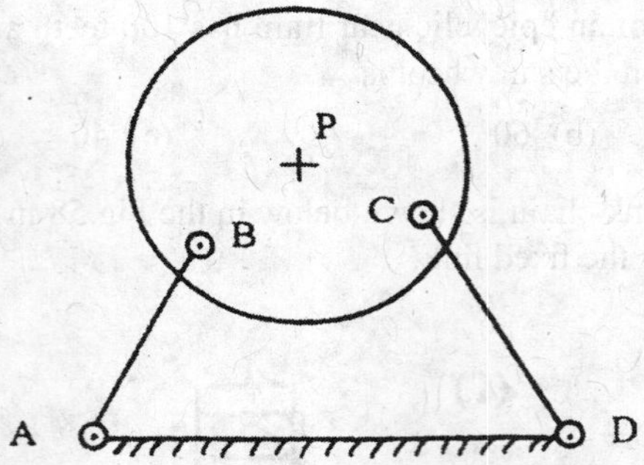

Fig.59

(a) If it lies on the perpendicular bisector of line *BC*
(b) If it lies on the intersection of the perpendicular bisectors of *BC* and *AD*
(c) If it lies on the intersection of the perpendicular bisectors of *AB* and *CD*
(d) If it lies on the intersection of the extensions of *AB* and *CD*.

254. In the Fig.60 given below, the magnitude of absolute angular velocity of link 2 is 10 radians per second while that of link 3 is 6 radians per second. What is the angular velocity of link 3 relative to 2?

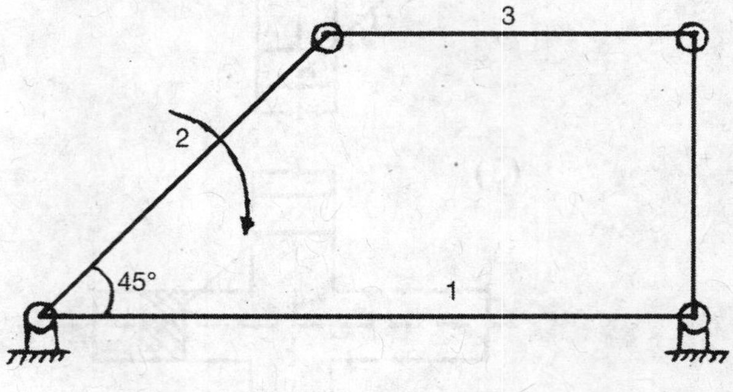

Fig.60

(a) 6 radians per second
(b) 16 radians per second
(c) 4 radians per second
(d) 14 radians per second.

255. *ABCD* is a mechanism with link lengths *AB* = 200, *BC* = 300, *CD* = 400 and *DA* = 350. Which one of the following links should be fixed for the resulting mechanism to be a double crank mechanism? (All lengths are in mm)

(a) *AB* (b) *BC* (c) *CD* (d) *DA*.

256. The crank of the mechanism shown below in the Fig.61 rotates at a uniform angular velocity ω.

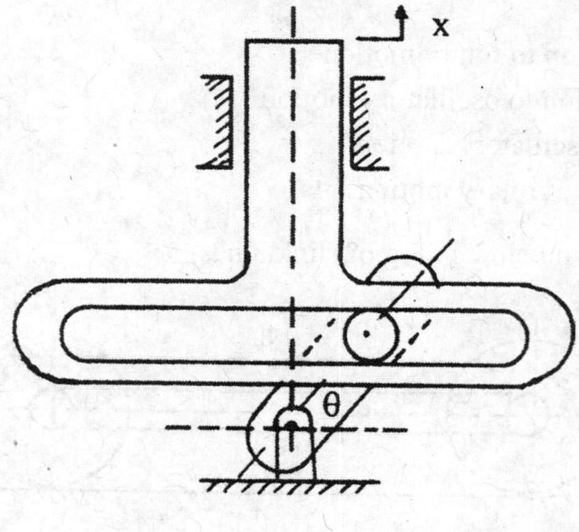

Fig.61

Which one of the following diagrams shows the velocity of slider x with respect to the crank angle?

(a)

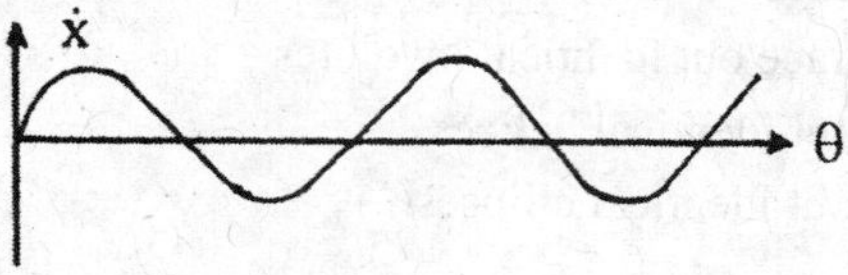

(c)

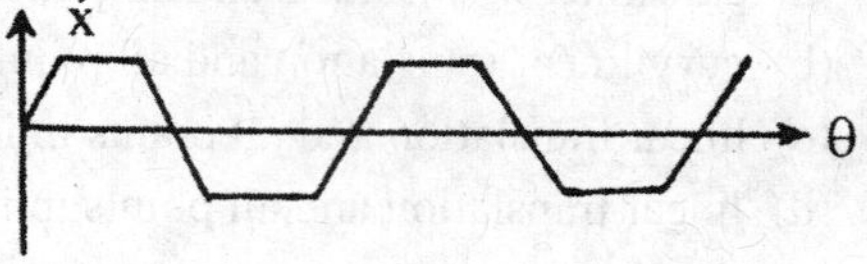

(b)

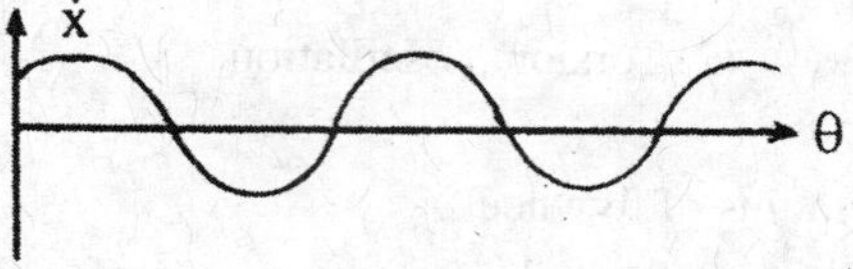

(d)

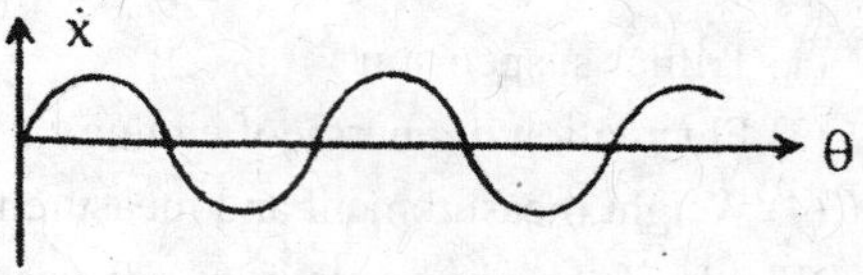

257. In the plate cam mechanism with reciprocating roller follower, in which one of the following cases the follower has constant acceleration?

(a) Cycloidal motion
(b) Simple harmonic motion
(c) Parabolic motion
(d) 3-4-5 polynimial motion.

258. Which one of the following statements is correct?

In a petrol engine mechanism the velocity of the piston is maximum when the crank is

(a) at the dead centres
(b) at right angles to the line of stroke
(c) slightly less than 90° to line of stroke
(d) slightly above 90° to line of stroke.

259. In a differential mechanism, two equal sized bevel wheels A and B are keyed to the two halves of the rear axle of a motor car. The car follows a curved path. Which one of the following statements is correct?

The wheels A and B will resolve at different speeds and the casing will revolve at a speed which is equal to the

(a) difference of speeds of A and B
(b) arithmetic mean of the speeds A and B
(c) geometric mean of the speeds of A and B
(d) harmonic mean of the speeds A and B

260. Which one of the following conversions is used by a lawn-sprinkler which is a four bar mechanisms?
(a) Reciprocating motion to rotary motion
(b) Reciprocating motion to oscillatory motion
(c) Rotary motion to oscillatory motion
(d) Oscillatory motion to rotary motion.

261. In the mechanism shown below in Fig.62, link 3 has

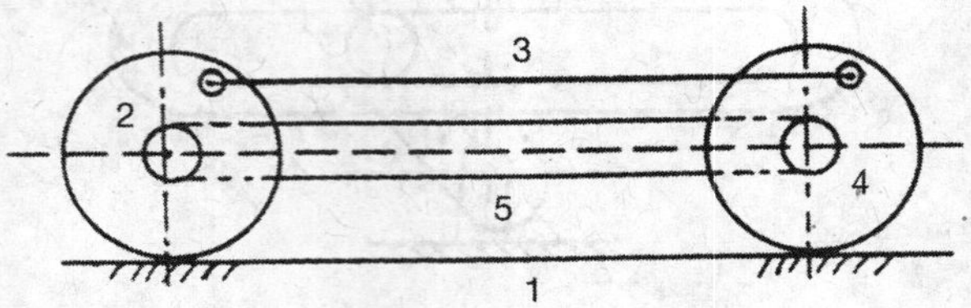

Fig.62

(a) curvilinear translation and all points in it trace out identical cycloids
(b) curvilinear translation and all points in it trace out identical involutes
(c) linear translation and all points in it trace out identical helices
(d) linear translation and all points in it trace out identical ellipses.

262. Consider the following methods:
(1) Trifilar suspension (2) Torsional oscillation
(3) Fluctuation of energy of engine
(4) Weight measurement and measurement of radius of flywheel
Which of the above methods are used to determine the polar mass moment of inertia of an engine flywheel with arms?
(a) 1 and 4 (b) 2 and 3 (c) 1, 2, and 3 (d) 1, 2, and 4.

263. A connecting rod has a mass of 0.5 kg. The radius of gyration through its centre of gravity is 5 cm and its acceleration is 2×10^4 rad/s^2. The equivalent two-mass system for the connecting rod has a radius of gyration 6 cm. What is the correction couple of the equivalent system?
(a) 11 Nm (b) 9 Nm (c) 6 Nm (d) 2 Nm.

264. Which one of the following statements is correct?
A governor will be stable if the radius of rotation of the balls
(a) increases as the equilibrium speed decreases
(b) decreases as the equilibrium speed increases
(c) increases as the equilibrium speed increases
(d) remains unaltered with the change in equilibrium speed.

265. Which one of the following statements in the context of balancing in engines is correct?
(a) Magnitude of the primary unbalancing force is less than the secondary unbalancing force
(b) The primary unbalancing force attains its maximum value twice in one revolution of the crank

(c) The hammer blow in the locomotive engines occurs due to unbalanced force along the line of stroke of the piston

(d) The unbalanced force due to reciprocating masses varies in magnitude and direction.

266. A four-cylinder in-line reciprocating engine is shown in the Fig.63 given below. The cylinders are numbered 1 to 4 and the firing order is 1-4-2-3:

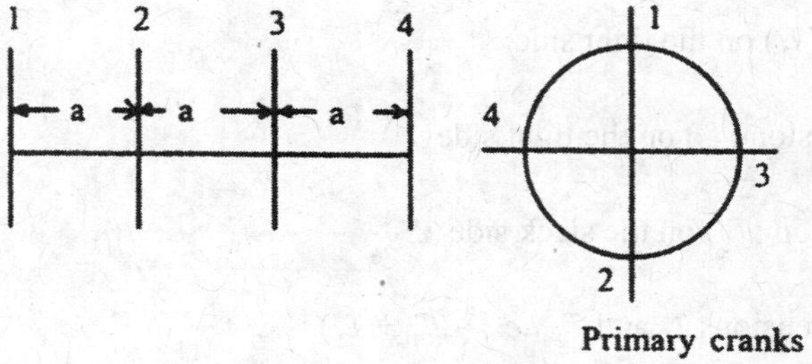

Fig.63

Which one of the following statements is correct?

(a) Both primary and secondary forces are balanced

(b) Only primary force is balanced

(c) Only secondary force is balanced

(d) Both primary and secondary forces are unbalanced.

267. Match List I with List II and select the correct answer using the codes given below the lists:

List I	List II
A. Open loop system	1. Frequency domain analysis
B. Closed loop system	2. More stable
C. Step input	3. Less stable
D. Sinusoidal input	4. Time domain analysis

Codes:

	A	B	C	D
(a)	2	3	4	1
(b)	4	1	2	3
(c)	2	1	4	3
(d)	4	3	2	1

268. An epicyclic gear train has 3 shafts *A*, *B* and *C*. *A* is an input shaft running at 100 rpm clockwise. *B* is an output shaft running at 250 rpm clockwise. Torque on *A* is 50 kNm (clockwise). *C* is a fixed shaft. The torque to fix *C*

(a) is 20 kNm anticlockwise
(b) is 30 kNm anticlockwise
(c) is 30 kNm clockwise
(d) cannot be determined as the data is insufficient.

269. Which of the following is a closed-loop control system?

(a) Traffic control on the roads by lights where the timing mechanism is present irrespective of the intensity of traffic

(b) Switching off the street lights of a tower at a predetermined time by a time-switch irrespective of the fact that the sun rises at a different time each day

(c) Switching off an electric heater by a time-switch irrespective of whether the dish has been prepared or not
(d) Human body.

270. The power transmitted by a belt is dependent on the centrifugal effect in the belt. The maximum power can be transmitted when the centrifugal tension is

(a) $\frac{1}{3}$ of tension (T_1) on the tight side
(b) $\frac{1}{3}$ of total tension (T_t) on the tight side
(c) $\frac{1}{3}$ of the tension (T_2) on the slack side
(d) $\frac{1}{3}$ of sum of tensions T_1 and T_2 *i.e.* $\frac{1}{3}(T_1 + T_2)$.

271. The length of the belt in the case of a cross-belt drive is given in terms of centre distance between pulleys (C), diameters of the pulleys D and d as

(a) $2C + \frac{\pi}{2}(D + d) + \frac{(D + d)^2}{4C}$ (b) $2C + \frac{\pi}{2}(D - d) + \frac{(D + d)^2}{4C}$
(c) $2C + \frac{\pi}{2}(D + d) + \frac{(D - d)^2}{4C}$ (d) $2C + \frac{\pi}{2}(D - d) + \frac{(D - d)^2}{4C}$

272. Consider the following statements:
1. Coriolis acceleration component in a slotted bar mechanism is always perpendicular to the direction of the slotted bar
2. In a 4-link mechanism, the instantaneous centre of rotation of the input link and output link always lies on a straight line along the coupler

Which of the statements given above is/are correct?
(a) 1 only (b) 2 only (c) Both 1 and 2 (d) Neither 1 nor 2.

273. Which one of the following is the correct value of the natural frequency (ω_n) of the system given above?

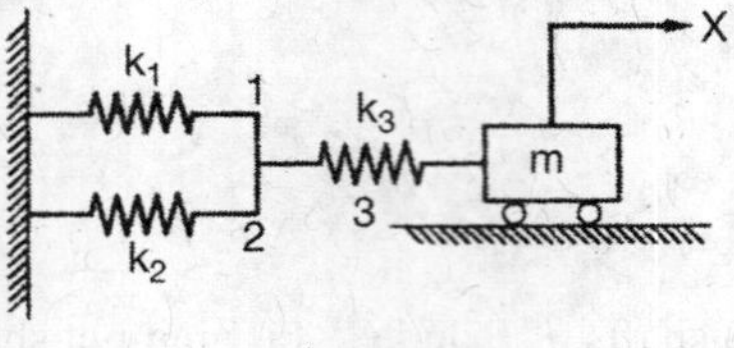

Fig.64

(a) $\left[\frac{1}{\left\{\frac{1}{k_1 + k_2} + \frac{1}{k_3}\right\} m}\right]^{1/2}$ (b) $\left(\frac{3k}{m}\right)^{1/2}$ (c) $\left(\frac{k}{3m}\right)^{1/2}$ (d) $\left[\frac{k_3 + \left(\frac{1}{\frac{1}{k_1} + \frac{1}{k_2}}\right)}{m}\right]^{1/2}$

274. Consider the following statements concerning centrifugal governors:
   1. The slope of the controlling force curve should be less than that of the straight line representing the centripetal force at the speed considered for the stability of a centrifugal governor
   2. Isochronism for a centrifugal governor can be achieved only at the expense of stability
   3. When sleeve of a centrifugal governor reaches its topmost position, the engine should develop maximum power

   Which of the statements given above is/are correct?
   (a) 1 and 2 (b) 2 and 3 (c) 2 only (d) 3 only.

275. Consider the following statements for a 4-cylinder in-line engine whose cranks are arranged at regular intervals of 90°.
   1. There are 8 possible firing orders for the engine
   2. Primary force will remain unbalanced for some firing orders

   Which of the statements given above is/are correct?
   (a) 1 only (b) 2 only (c) Both 1 and 2 (d) Neither 1 nor 2.

276. Spiral gears are used to connect
   (a) two parallel shafts.
   (b) two intersecting shafts.
   (c) two non-parallel and non-intersecting shafts.
   (d) None of the above.

277. In the below Fig.65 shown, if the speed of the input shaft of the spur gear train is 2400 rpm and the speed of the output shaft is 100 rpm, what is the module of the gear 4?

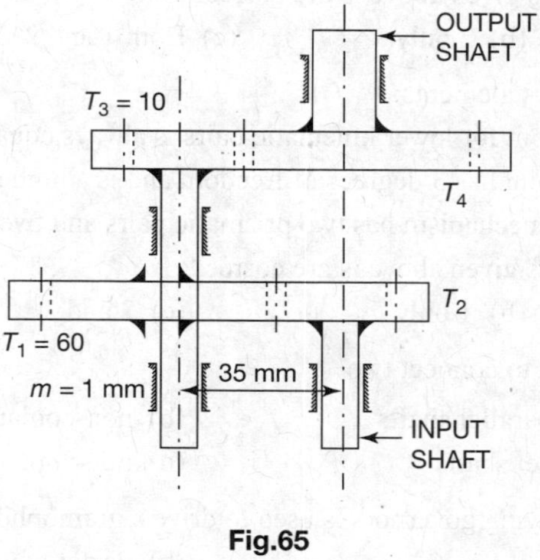

Fig.65

   (a) 1.2 (b) 1.4 (c) 2 (d) 2.5.

278. The crank of slider-crank punching press has a mass moment of inertia of 1 kgm^2. The below Fig.66 shows the torque demand per revolution for a punching operation. If the speed of the crank is found to drop from 30 rad/s to 20 rad/s during punching, what is the maximum torque demand during the punching operation?

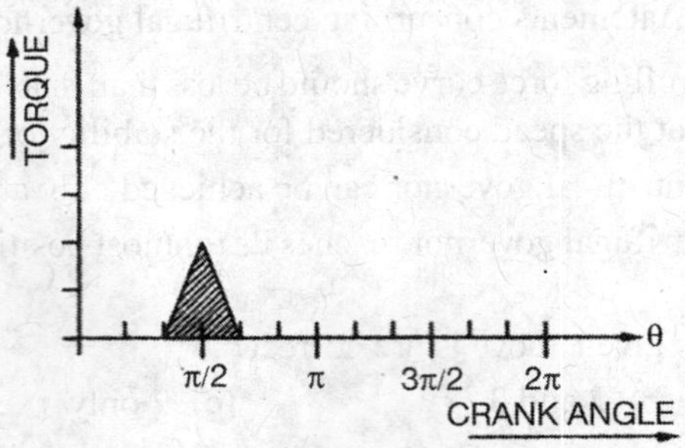

Fig.66

(a) 95.4 Nm (b) 104.7 Nm (c) 477.2 Nm (d) 523.8 Nm.

279. A single cylinder, four-stroke I.C. engine rotating at 900 rpm has a crank length of 50 mm and a connecting rod length of 200 mm. If the effective reciprocating mass of the engine is 1.2 kg, what is the approximate magnitude of the maximum 'shaking force' created by the engine?
(a) 533 N (b) 666 N (c) 133 N (d) None of the above.

280. In a kinematic chain, a quaternary joint is equivalent to
(a) one binary joint (b) two binary joints (c) three binary joints (d) four binary joints.

281. Consider the following statements:
1. The effect of gyroscopic couple on a car while negotiating a curve is that its outer wheels tend to get lifted from the ground
2. If spin vector is rotated about the precession vector axis in a direction opposite to that of precession through 90°, the new position of the spin vector indicates the direction of the torque vector
Which of the statements given above is/are correct?
(a) 1 only (b) 2 only (c) Both 1 and 2 (d) Neither 1 nor 2.

282. Consider the following statements:
1. The degree of freedom for lower kinematic pairs is always equal to one
2. A ball-and-socket joint has 3 degrees of freedom and is a higher kinematic pair
3. Oldham's coupling mechanism has two prismatic pairs and two revolute pairs
Which of the statements given above is/are correct?
(a) 1, 2 and 3 (b) 1 only (c) 2 and 3 (d) 3 only.

283. A Hooke's joint is used to connect two
(a) coplanar and non-parallel shafts (b) non-coplanar and non-parallel shafts
(c) coplanar and parallel shafts (d) non-coplanar and parallel shafts.

284. Which one of the following governors is used to drive a gramophone?
(a) Watt governor (b) Porter governor
(c) Pickering governor (d) Hartnell governor.

285. In the Fig.67 given below, the link 2 rotates at an angular velocity of 2 rad/s. What is the magnitude of Coriolis acceleration experienced by the link 4?
(a) 0 (b) 0.8 m/s^2 (c) 0.24 m/s^2 (d) 0.32 m/s^2.

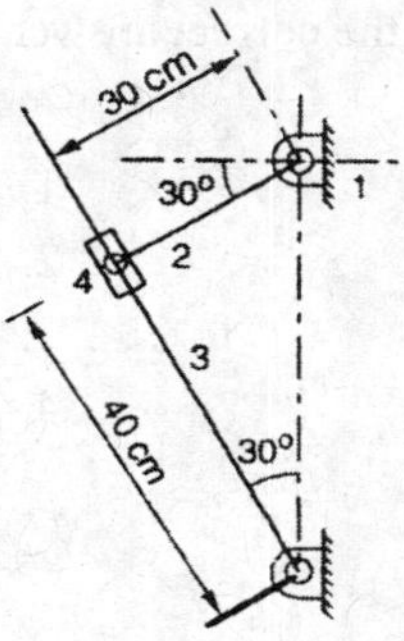

Fig.67

286. The below Fig.67 shows a cam with a circular profile, rotating with a uniform angular velocity of ω rad/s. What is the nature of displacement of the follower?

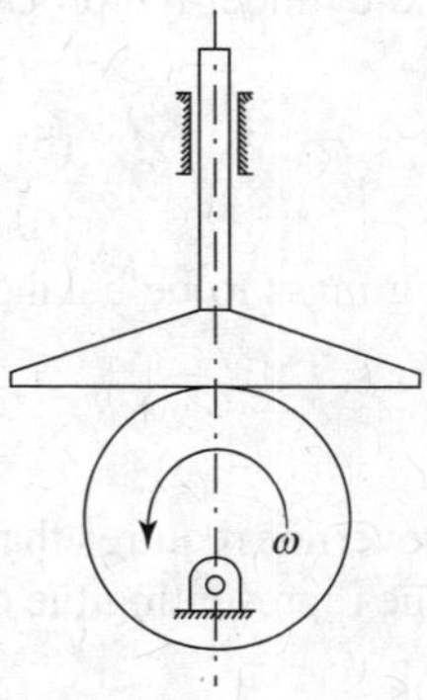

Fig.68

(a) Uniform (b) Parabolic (c) Simple harmonic (d) Cycloidal.

287. Which one of the following can completely balance several masses revolving in different planes on a shaft?

(a) A single mass in one of the planes of the revolving mass

(b) A single mass in any one plane

(c) Two masses in any two planes

(d) Two equal masses in any two planes.

288. Which one of the following expresses the sensitiveness of a governor?

(a) $\dfrac{N_1 + N_2}{2N_1N_2}$ (b) $\dfrac{N_1 - N_2}{2N_1N_2}$ (c) $\dfrac{2(N_1 + N_2)}{N_1 - N_2}$ (d) $\dfrac{2(N_1 - N_2)}{N_1 + N_2}$

N_1 = Maximum equilibrium speed
N_2 = Minimum equilibrium speed.

289. What is the number of nodes in a shaft carrying three rotors?

(a) Zero (b) 2 (c) 3 (d) 4.

290. Match List I with List II and select the correct answer using the codes given below the Lists:

List I (Property)	List II (System)
A. Resonsnce	1. Closed-loop control system
B. On-off control	2. Free vibrations
C. Natural frequency	3. Excessively large amplitude
D. Feedback signal	4. Mechanical brake

Codes:

	A	B	C	D
(a)	1	2	4	3
(b)	3	4	2	1
(c)	1	4	2	3
(d)	3	2	4	1

291. In case of partial balancing of single-cylinder reciprocating engine, what is the primary disturbing force along the line of stroke?

(a) $cmr\,\omega^2 \cos\theta$ (b) $(1 - c^2)\,mr\,\omega^2 \cos\theta$

(c) $(1 - c)\,m\omega^2 r \cos 2\theta$ (d) $(1 - c)\,m\omega^2 r \cos 2\theta$

where, c = Fraction of reciprocating mass to be balanced; ω = Angular velocity of crankshaft; θ = Crank angle.

292. Consider the following statements:

1. The condition of stability of a governor requires that the slope of the controlling force curve should be less than that of the line representing the centripetal force at the equilibrium speed under consideration
2. For a centrifugal governor when the load on the prime mover drops suddenly, the sleeve should at once reach the lower-most position

Which of the statements given above is/are correct?

(a) Only 1 (b) Only 2 (c) Both 1 and 2 (d) Neither 1 nor 2.

293. What is the number of instantaneous centres of rotation for a 6-link mechanism?

(a) 4 (b) 6 (c) 12 (d) 15.

294. In which one of the following is a flywheel generally employed?

(a) Lathe (b) Electric motor (c) Punching machine (d) Gearbox.

295. What is the value of pressure angle generally used for involute gears?

(a) 35° (b) 30° (c) 25° (d) 20°.

296. Consider the following statements:

1. A stub tooth has a working depth larger than that of a full-depth tooth.
2. The path of contact for involute gears in an arc of a circle.

Which of the statements given above is/are correct?

(a) Only 1 (b) Only 2 (c) Both 1 and 2 (d) Neither 1 nor 2.

297. Consider the following statements:

Cam followers are generally classified according to

1. the nature of its motion
2. the nature of its surface in contact with the cam
3. the speed of the cam

Which of the statements given above are correct?

(a) 1, 2, and 3 (b) Only 1 and 2 (c) Only 2 and 3 (d) Only 1 and 3.

298. Consider the following statements:

Radius of friction circle for a journal bearing depends upon

1. coefficient of friction
2. radius of the journal
3. angular speed of rotation of the shaft

Which of the statements given above are correct?

(a) 1, 2 and 3 (b) Only 1 and 2 (c) Only 2 and 3 (d) Only 1 and 3.

299. What is the maximum acceleration of a cam follower undergoing simple harmonic motion?

(a) $\frac{h}{2}\left(\frac{\pi\omega}{\phi}\right)^2$ (b) $4h\left(\frac{\omega^2}{\phi^2}\right)$ (c) $4h\left(\frac{\omega^2}{\phi}\right)$ (d) $\frac{2h\pi\omega^2}{\phi^2}$

where, h = Stroke of the follower;

ω = Angular velocity of the cam;

ϕ = Cam rotation angle for the maximum follower displacement.

300. Consider the following statements:

1. Lower pairs are more resistant than the higher pairs in a plane mechanism.
2. In a 4-bar mechanism (with 4 turning pairs), when the link opposite to the shortest link is fixed a double rocker mechanism results.

Which of the statements given above is/are correct?

(a) Only 1 (b) Only 2 (c) Both 1 and 2 (d) Neither 1 nor 2.

301. Consider the following follower motions in respect of a given lift, speed of rotation and angle of stroke of a cam:

1. Cycloidal motion
2. Simple harmonic motion
3. Uniform velocity motion

Which one of the following is the correct sequence of the above in the descending order of maximum velocity?

(a) 3 – 2 – 1 (b) 1 – 2 – 3 (c) 2 – 3 – 1 (d) 3 – 1 – 2.

302. If α = helix angle, and p_c = circular pitch; then which one of the following correctly expresses the axial pitch of a helical gear?

(a) $p_c \cos\alpha$ (b) $\frac{p_c}{\cos\alpha}$ (c) $\frac{p_c}{\tan\alpha}$ (d) $p_c \sin\alpha$.

303. In a slider-bar mechanism, when does the connecting rod have zero angular velocity?
(a) When crank angle = 0° (b) When crank angle = 90°
(c) When crank angle = 45° (d) Never.

304. The turning moment diagram for a single cylinder double acting steam engine consists of +ve and –ve loops above and below the average torque line. For the +ve loop, the ratio of the speeds of the flywheel at the beginning and the end is which one of the following?
(a) Less than unity (b) Equal to unity (c) Greater than unity (d) Zero.

305. Which one of the following is the correct statement?
In meshing gears with involute gear teeth, the contact begins at the intersection of the
(a) line of action and the addendum circle of the driven gear
(b) line of action and the pitch circle of the driven gear
(c) dedendum circle of the driver gear and the addendum circle of the driven gear
(d) addendum circle of the driver gear and the pitch circle of the driven gear.

306. Interference between the teeth of two meshing involute gears can be reduced or eliminated by
1. Increasing the addendum of the gear teeth and correspondingly reducing the addendum of the pinion
2. Reducing the pressure angle of the teeth of the meshing gears
3. Increasing the centre distance
Which of the statements given above is/are correct?
(a) 1 and 2 (b) 2 and 3 (c) 1 only (d) 3 only.

307. What is the direction of the Coriolis components of acceleration in a slotted lever-crank mechanism?
(a) Along the sliding velocity vector
(b) Along the direction of the crank
(c) Along a line rotated 90° from the sliding velocity vector in a direction opposite to the angular velocity of the slotted lever
(d) Along a line rotated 90° from the sliding velocity vector in a direction same as that of the angular velocity of the slotted lever.

308. The controlling force curves for a spring-controlled governor are shown in the below Fig.69. Which curve represents a stable governor?

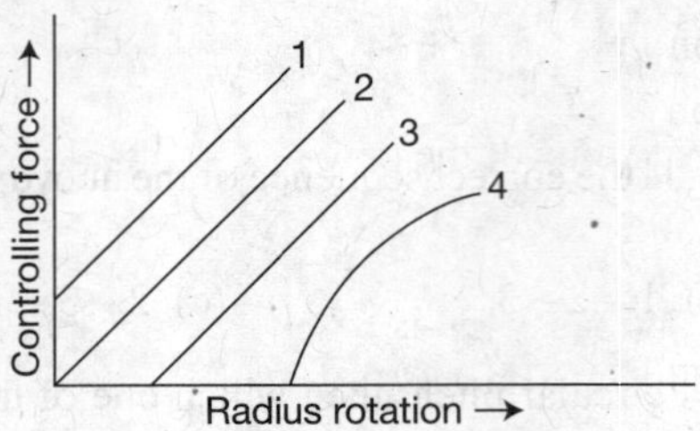

Fig.69

(a) 1 (b) 2 (c) 3 (d) 4.

309. For a governor running at constant speed, what is the value of the force acting on the sleeve?

(a) Zero (b) Variable depending upon the load

(c) Maximum (d) Minimum.

310. What is the condition for dynamic balancing of a shaft-rotor system?

(a) $\Sigma M = 0$ and $\Sigma F = 0$ (b) $\Sigma M = 0$ (c) $\Sigma F = 0$ (d) $\Sigma M + \Sigma F = 0$.

311. (W = Weight of reciprocating parts per cylinder).

For a three-cylinder radial engine, the primary and direct reverse cranks are as shown in the below Fig.70.

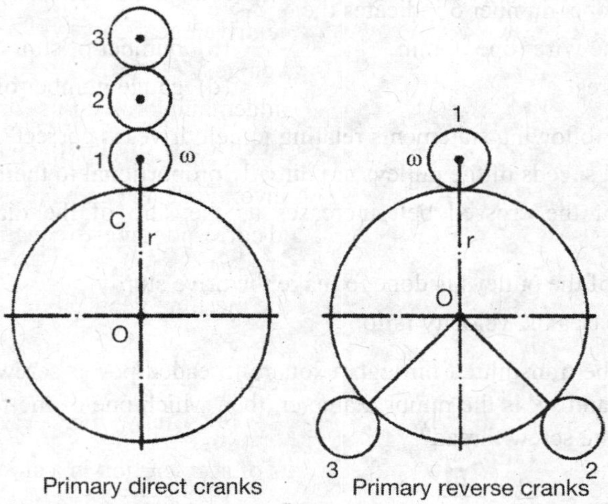

Fig.70

Which one of the following pairs is not correctly matched in this regard?

(a) Primary direct force ... $\frac{3W}{2g} \omega^2 \cdot r$ (b) Primary reverse force ... Zero

(c) Primary direct crank speed ... ω (d) Primary reverse crank speed ... 2ω.

312. A uniform bar, fixed at one end carries a heavy concentrated mass at the other end. The system is executing longitudinal vibrations. The inertia of the bar may be taken into account by which one of the following portions of the mass of the bar at the free end?

(a) $\frac{5}{384}$ (b) $\frac{1}{48}$ (c) $\frac{33}{140}$ (d) $\frac{1}{3}$.

313. A motion is aperiodic at what value of the damping factor?

(a) 1.0 or above (b) 0.5 (c) 0.3 (d) 0.866.

314. A rolling disc of radius 'r' and mass 'm' is connected to one end of a linear spring of stiffness 'k', as shown in the below Fig.71. The natural frequency of oscillation is given by which one of the following?

(a) $\omega = \sqrt{\frac{2}{3}\frac{k}{m}}$ (b) $\omega = \sqrt{\frac{k}{m}}$ (c) $\omega = \sqrt{\frac{k}{2m}}$ (d) $\omega = \sqrt{\frac{2k}{m}}$.

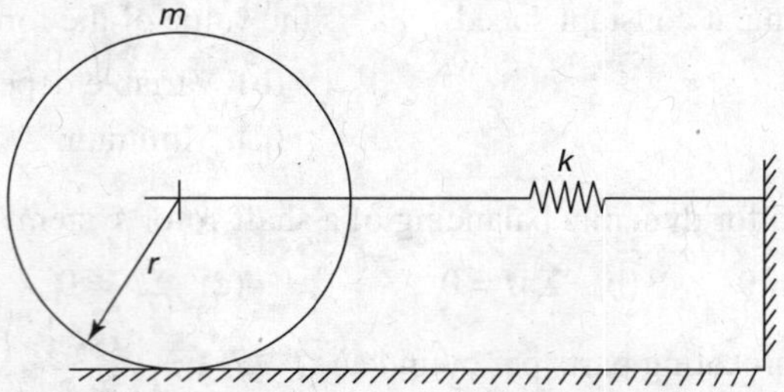

Fig.71

315. In a 6 × 20 wire rope, number 6 indicates the

(a) diameter of the wire rope in mm
(b) number of strands in the wire rope
(c) number of wires
(d) gauge number of the wire.

316. Which one of the following statements relating to belt drives is correct?

(a) The rotational speeds of the pulleys are directly proportional to their diameters
(b) The length of the crossed belt increases as the sum of the diameters of the pulleys increases
(c) The crowing of the pulleys is done to make the drive sturdy
(d) The slip increases the velocity ratio.

317. A force '*F*' is to be transmitted through a square threaded power screw into a nut. If '*t*' is the height of the nut and '*d*' is the minor diameter, then which one of the following is the average shear stress over the screw thread?

(a) $\frac{2F}{\pi dt}$
(b) $\frac{F}{\pi dt}$
(c) $\frac{F}{2\pi dt}$
(d) $\frac{4F}{\pi dt}$.

318. In the compound gear train shown in the below Fig.72, gears *A* and *C* have equal numbers of teeth and gears *B* and *D* have equal numbers of teeth. When *A* rotates at 800 rpm, *D* rotates at 200 rpm. The rotational speed of compound gears *BC* would then be

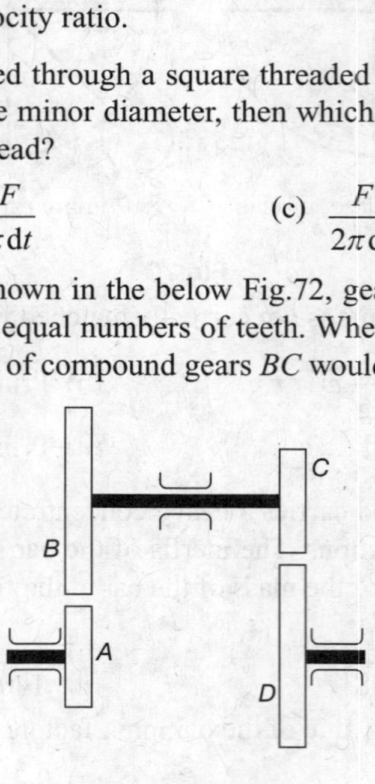

Fig.72

(a) 300 rpm
(b) 400 rpm
(c) 500 rpm
(d) 600 rpm.

319. Under service conditions involving jurking, vibration and pulsation of the working load, the bolt of choice would be a

(a) short bolt with high rigidity
(b) long bolt with increased elasticity
(c) bolt with a dished washer
(d) bolt with castle nut.

320. Gearing contact is which one of the following?

(a) Sliding contact
(b) Sliding contact, only rolling at pitch point
(c) Rolling contact
(d) Rolling and sliding at each point of contact.

321. Maximum angular velocity of the connecting rod with a crank to connecting rod ratio 1:5 for crank speed of 3000 rpm is around:

(a) 300 rad/s
(b) 60 rad/s
(c) 30 rad/s
(d) 3000 rad/s.

322. Consider the following statements:

1. One way of improving vibration isolation is to decrease the mass of the vibrating object
2. For effective isolation, the natural frequency of the system should be far less than the exciting frequency

Which of the statements given above is/are correct?

(a) 1 only
(b) 2 only
(c) Both 1 and 2
(d) Neither 1 nor 2.

323. Common contact ratio of a pair of spur pinion and gear is

(a) less than 1.0
(b) equal to 1
(c) between 2 and 3
(d) greater than 3.

324. In case of a multiple disc clutch, if n_1 is the number of discs on the driving shaft and n_2 is the number of discs on the driven shaft, then what is the number of pairs of contact surfaces?

(a) $n_1 + n_2$
(b) $n_1 + n_2 - 1$
(c) $n_1 + n_2 + 1$
(d) $n_1 + 2\,n_2$.

325. For roller chain drive with sprocket having 10 teeth, the velocity of the driven shaft with respect to that of drive will be approximately

(a) same
(b) 5% above
(c) 5% below
(d) 5% above to 5% below.

326. Which of the following in-line engines working on four-stroke cycle is completely balanced inherently?

(a) 2 cylinder engine
(b) 3 cylinder engine
(c) 4 cylinder engine
(d) 6 cylinder engine.

327. Match List I with List II and select the correct answer using the code given below the lists:

List I (Principle/method)	List II (Corresponding Application)
A. Klein's construction	1. Instantaneous centres in linkages
B. Kennedy's theorem	2. Relative acceleration of linkages
C. D'Alembert's principle	3. Mobility of linkages
D. Grubler's rule	4. Dynamic forces in linkages

Codes:

	A	B	C	D
(a)	4	1	2	3
(b)	2	3	4	1
(c)	4	3	2	1
(d)	2	1	4	3

328. Which mechanism produces intermittent rotary motion from continuous rotary motion?

(a) Whitworth mechanism (b) Scotch Yoke mechanism
(c) Geneva mechanism (d) Elliptical trammel.

329. The Fig.73 below shows the schematic of an automobile having a mass of 900 kg and the suspension spring constant of 81×40^4 N/m. If it travels at a speed of 72 km/hr on a rough road with periodic waviness as shown, what is the forcing frequency of the road on the wheel?

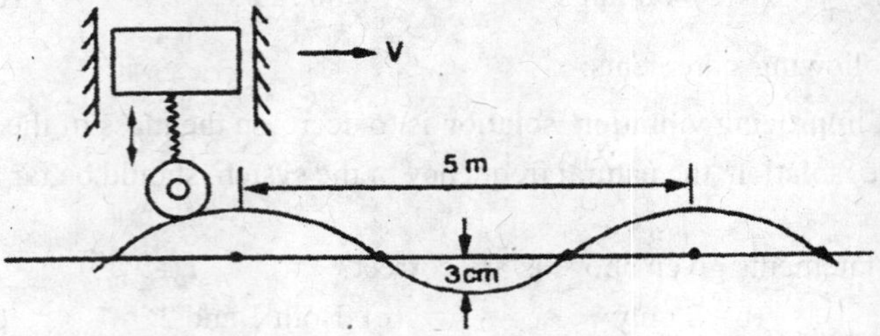

Fig.73

(a) 10 HZ (b) 4 HZ (c) 1.5 Hz (d) 20 HZ.

330. Which one of the following mechanisms represents an inversion of the single slider crank chain?

(a) Elliptical trammel (b) Oldham's coupling
(c) Whitworth quick return mechanism (d) Pantograph mechanism.

331. At a given instant, a disc is spinning with angular velocity ω in a plane at right angles to the paper (see the Fig.74) and after a short interval of time δt, it is spinning with angular velocity $\omega + \delta\omega$ and the axis of spin has changed direction by the amount $\delta\theta$.

In this situation what is the component of acceleration parallel to OA?

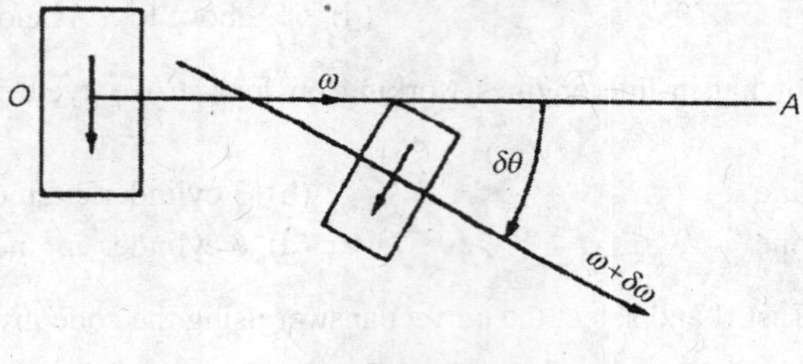

Fig.74

(a) $d\theta/dt$ (b) $\omega\,(d\theta/dt)$ (c) $d\omega/dt$ (d) $d\theta/d\omega$

332. The given Fig.75 shows a slider crank mechanism in which link 1 is fixed. The number of instantaneous centres would be

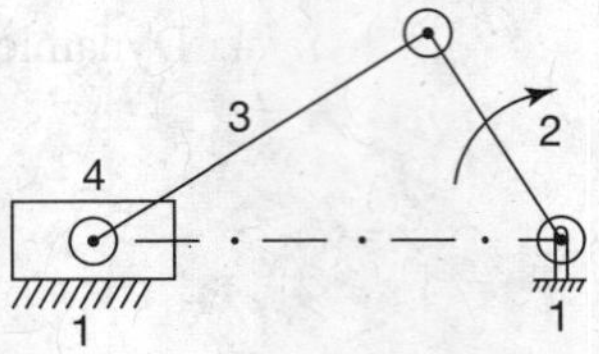

Fig.75

(a) 4 (b) 5 (c) 6 (d) 12.

333. Consider the following statements:

The Klein's construction for slider crank mechanism with crank rotating at constant angular velocity provides values of

1. piston velocity
2. piston acceleration
3. normal acceleration of crank pin
4. angular acceleration of the connecting rod

Of these statements

(a) 1 and 2 are correct
(b) 1, 2, 3, and 4 are correct
(c) 1, 2, and 4 are correct
(d) 3 and 4 are correct.

334. If two parallel shafts are to be connected and the distance between the axes of shafts is small and variable, then one would need to use

(a) a clutch (b) a universal joint (c) an Oldham's coupling (d) a knuckle joint

335. Consider the following statements relating to the curve for the inertia torque *v/s* crank angle for a horizontal, single cylinder petrol engine shown in the given Fig.76:

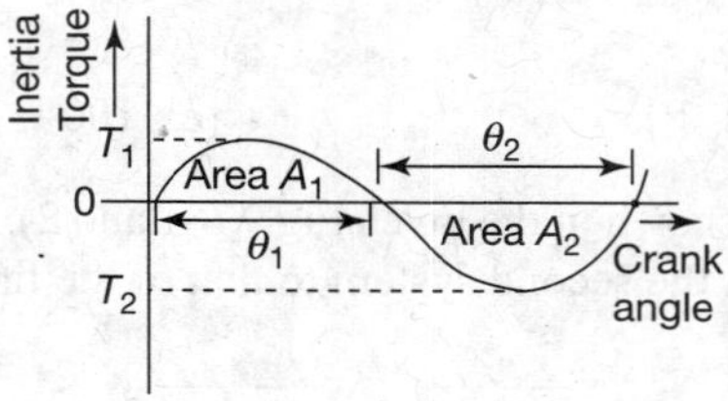

Fig.76

1. $\theta_1 + \theta_2 = 180°$
2. $T_1 = T_2$
3. $\theta_1 \neq \theta_2$
4. $A_1 = A_2$

Of these statements

(a) 1 and 3 are correct
(b) 2 and 3 are correct
(c) 1, 2 and 4 are correct
(d) 1, 3 and 4 are correct.

336. The height h of Porter governor with equal arms pivoted at equal distance from axis of rotation is expressed as (where m = mass of balls of the governor, M = mass of sleeve of the governor and N = rpm)

(a) $h = 91.2\left[\dfrac{m+M}{m}\right]\dfrac{g}{N^2}$
(b) $h = 91.2\left[\dfrac{mg - Mg}{mg}\right]\dfrac{g}{N^2}$

(c) $h = 91.2\left[\dfrac{m}{mM}\right]\dfrac{g}{N^2}$
(d) $h = 91.2\left[\dfrac{M}{m}\right]\dfrac{g}{N^2}$

337. Match List I (Type of Governor) with List II (Characteristics) and select the correct answer using the codes given below the lists:

List I	List II
A. Isochronous governor	1. Continuously fluctuates above and below mean speed
B. Sensitive governor	2. For each given speed there is only one radius of rotation
C. Hunting governor	3. Higher displacement of sleeve for fractional change of speed
D. Stable governor	4. Equilibrium speed is constant for all radii of rotation

Codes:

	A	B	C	D
(a)	4	3	2	1
(b)	2	4	1	3
(c)	2	4	3	1
(d)	4	3	1	2

338. Consider the following statements:
An in-line four-cylinder four-stroke engine is completely balanced for
1. primary forces
2. secondary forces
3. primary couples
4. secondary couples

Of these statements
(a) 1, 3, and 4 are correct (b) 1, 2, and 4 are correct
(c) 1 and 3 are correct (d) 2 and 4 are correct.

339. A uniform cantilever beam undergoes transverse vibrations. The number of natural frequencies associated with the beam is
(a) 1 (b) 10 (c) 100 (d) infinite.

340. Two vibratory systems are shown in the given Fig.77 (1 and 2). The ratio of the natural frequency of longitudinal vibration of the second system to that of the first is

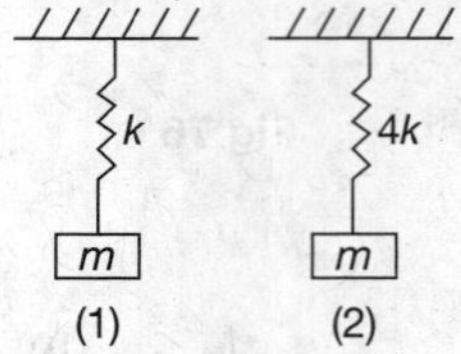

Fig.77

(a) 4 (b) 2 (c) 0.5 (d) 0.25.

341. A shaft, supported on two bearings at its ends, carries two flywheels '*L*' apart. Mass moment of inertia of the two flywheels are I_a and I_b, I being the polar moment of inertia of cross-sectional area of the shaft. Distance l_a of the mode of torsional vibration of the shaft from the flywheel I_a is given by

(a) $l_a = \dfrac{LI_b}{I_a + I_b}$ (b) $l_a = \dfrac{LI_a}{I_a + I_b}$ (c) $l_a = \dfrac{LI_b}{I_a + I_b - I}$ (d) $l_a = \dfrac{LI_a}{I_a + I_b - I}$

342. A rack is a gear of
(a) infinite diameter (b) infinite module (c) zero pressure angle (d) large pitch.

343. The maximum efficiency for spiral gears in mesh is given by
(where θ = shaft angle and ϕ = friction angle)

(a) $\dfrac{1+\cos(\theta-\phi)}{1+\cos(\theta+\phi)}$ (b) $\dfrac{1+\cos(\theta+\phi)}{1+\cos(\theta-\phi)}$

(c) $\dfrac{1-\cos(\theta-\phi)}{1+\cos(\theta+\phi)}$ (d) $\dfrac{1-\cos(\theta+\phi)}{1+\cos(\theta-\phi)}$.

344. The block diagram of an automatic control system is shown in the following Fig.78.

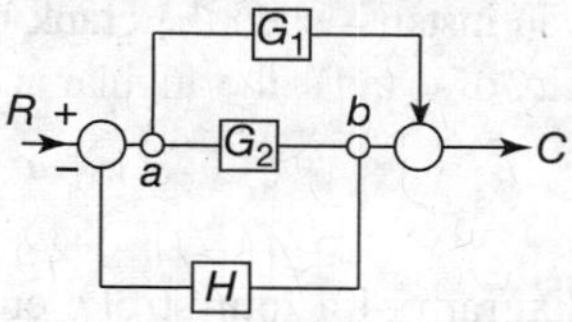

Fig.78

Its simplified form will be as in

(a) $R \rightarrow \boxed{\dfrac{G_1G_2}{1+G_1H}} \rightarrow C$ (b) $R \rightarrow \boxed{\dfrac{G_2-G_1}{1+G_2H}} \rightarrow C$

(c) $R \rightarrow \boxed{\dfrac{G_1-G_2}{1+G_2H}} \rightarrow C$ (d) $R \rightarrow \boxed{\dfrac{G_1+G_2}{1+G_2H}} \rightarrow C$

345. Consider the following statements in respect of a body executing simple harmonic motion:

1. Periodic time is the time for one complete revolution
2. The acceleration is directed towards the centre of suspension
3. The acceleration is proportional to distance from mean position
4. The velocity will be maximum when it passes through mean position

Of these statements

(a) 1, 2, and 3 are correct (b) 2, 3, and 4 are correct

(c) 1, 3, and 4 are correct (d) 1, 2, and 4 are correct.

346. A four-bar mechanism *ABCD* is shown in the given Fig.79. If the linear velocity 'V_B' of the point '*B*' is 0.5 m/s, then the linear velocity 'V_c' of point '*C*' will be

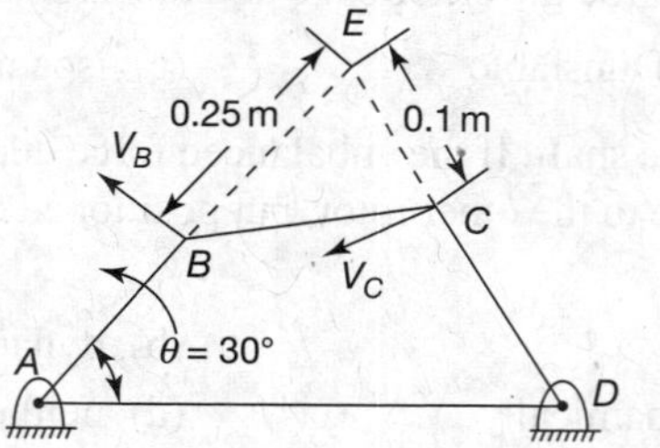

Fig.79

(a) 1.25 m/s (b) 0.5 m/s (c) 0.4 m/s (d) 0.2 ms.

347. If reduction ratio of about 50 is required in a gear drive, then the most appropriate gearing would be

(a) spur gears (b) bevel gears (c) double helical gears (d) worm and worm wheel.

348. A spring of stiffness '*k*' extended from a displacement x_1 to a displacement x_2. The work done by the spring is

(a) $\frac{1}{2}kx_1^2 - \frac{1}{2}kx_2^2$ (b) $\frac{1}{2}k(x_1 - x_2)^2$ (c) $\frac{1}{2}k(x_1 + x_2)^2$ (d) $k\left[\frac{x_1 + x_2}{2}\right]^2$.

349. In a reciprocating engine mechanism, the crank and connecting rod of same length r metres are at right angles to each-other at a given instant, when the crank makes an angle of 45° with IDC. If the crank rotates with a uniform velocity of ω rad/s, the angular acceleration of the connecting rod will be

(a) $2\omega^2 r$ (b) $\omega^2 r$ (c) $\frac{\omega^2}{r}$ (d) zero.

350. A simplified turning moment diagram of a four-stroke engine is shown in the given Fig.80. If the mean torque 'T_m' is 10 Nm, the estimated peak torque 'T_p' will be (assuming negative torque demand is negligible)

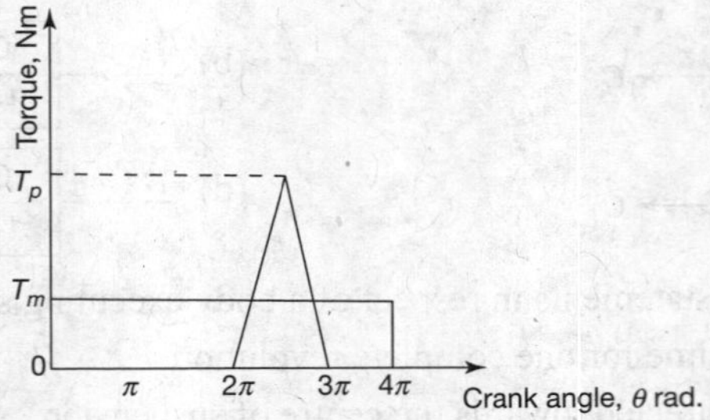

Fig.80

(a) 80 Nm (b) 120 Nm (c) 60 Nm (d) 40 Nm.

351. The height of a simple Watt governor running at a speed 'N' is proportional to

(a) N (b) $\frac{1}{N}$ (c) N^2 (d) $\frac{1}{N^2}$.

352. The controlling force curve of spring-loaded governor is given by the equation $F = ar - c$, (where r is the radius of rotation of the governor balls and a, c are constants). The governor is

(a) stable (b) unstable (c) isochronous (d) insensitive.

353. Two rotors are mounted on a shaft. If the unbalanced force due to one rotor is equal in magnitude to the unbalanced force due to the other rotor, but positioned exactly 180° apart, then the system will be balanced

(a) statically (b) dynamically
(c) statically as well as dynamically (d) neither statically nor dynamically.

354. The primary direct crank of a reciprocating engine is located at an angle θ clockwise. The secondary direct crank will be located at an angle

(a) 2 θ clockwise (b) 2 θ anti-clockwise (c) θ clock-wise (d) θ anticlockwise.

355. The given Fig.81 shows vibrations of a mass 'M' isolated by means of springs and a damper. If an external force 'F' (= $A \sin \omega t$) acts on the mass and the damper is not used, then

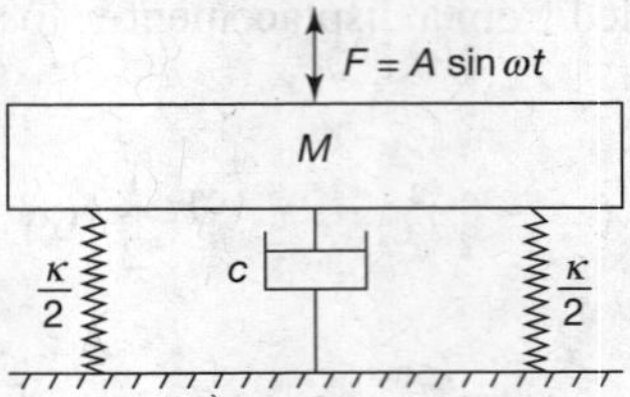

Fig.81

(a) $\omega = \sqrt{\frac{k}{M}}$ (b) $\omega = \frac{1}{2}\sqrt{\frac{k}{M}}$ (c) $\omega = 2\sqrt{\frac{k}{M}}$ (d) $\omega = \sqrt{\frac{k}{2M}}$

356. The transmitted force through a mass-spring damper system will be greater than that transmitted through rigid supports for all values of damping factors, if the frequency ratio $\left(\frac{\omega}{\omega_n}\right)$ is

(a) more than $\sqrt{2}$ (b) less than $\sqrt{2}$ (c) equal to one (d) less than one.

357. In a forced vibrations with viscous damping, maximum amplitude occurs when the forced frequency is

(a) equal to natural frequency (b) slightly less than natural frequency
(c) slightly greater than natural frequency (d) zero.

358. The rotor of a turbine is generally rotated at

(a) the critical speed (b) a speed much below the critical speed
(c) a speed much above the critical speed (d) a speed having no relation to critical speed.

359. The characteristic equation of a closed-loop automatic control system in time domain is given by

$$D^5 + 2D^4 + 4D^2 + D + 1 = 0$$

Consider the following statements in this regard:

1. The system is linear 2. The system is non-linear
3. The system is stable 4. The system is unstable

Of these statements

(a) 1 and 3 are correct (b) 2 and 3 are correct
(c) 1 and 4 are correct (d) 2 and 4 are correct.

360. The given Fig.82 shows a/an

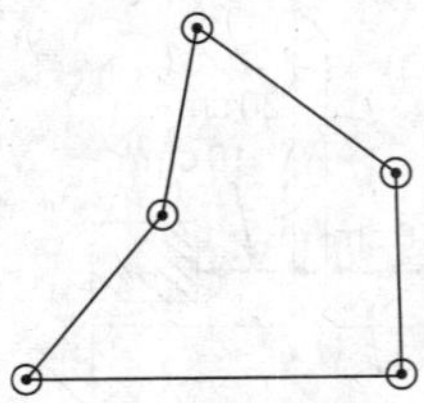

Fig.82

(a) locked chain (b) constrained kinematic chain
(c) unconstrained kinematic chain (d) mechanism.

361. In a four-link kinematic chain, the relation between the number of links (L) and number of pairs (J) is

(a) $L = 2J + 4$ (b) $L = 2J - 4$ (C) $L = 4J + 2$ (d) $L = 4J - 2$.

362. An imaginary circle which by pure rolling action, gives the same motion as the actual gear, is called

(a) addendum circle (b) pitch circle (c) dedendum circle (d) base circle.

363. The pressure angle of a spur gear normally varies from

(a) 14° to 20° (b) 20° to 25° (c) 30° to 36° (d) 40° to 50°.

364. If the number of teeth on the wheel rotating at 300 rpm is 90, then the number of teeth on the mating pinion rotating at 1500 rpm is

(a) 15 (b) 18 (c) 20 (d) 60.

365. Which one the following is the lubrication regime during normal operation of a rolling element bearing?

(a) Hydrodynamic lubrication (b) Hydrostatic lubrication
(c) Elastohydrodynamic lubrication (d) Boundary lubrication.

366. To carry a large axial load in a flat collar bearing, a number of collars is provided to

(a) reduce frictional torque (b) increase frictional torque
(c) decrease intensity of pressure (d) increase intensity of pressure.

367. The amount of energy absorbed by a flywheel is determined from the

(a) torque-crank angle diagram (b) acceleration-crank angle diagram
(c) speed-space diagram (d) speed-energy diagram.

368. Sensitiveness of a governor is defined as the ratio of the

(a) maximum equilibrium speed to the minimum equilibrium speed
(b) difference between maximum and minimum equilibrium speeds to the mean equilibrium speed
(c) difference between maximum and minimum equilibrium speeds to the maximum equilibrium speed
(d) maximum difference in speeds to the minimum equilibrium speed.

369. A rigid rotor consists of a system of two masses located as shown in the given Fig.83.

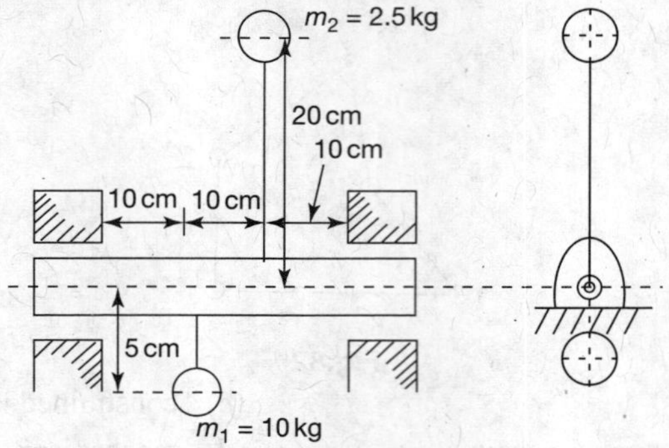

Fig.83

The system is

(a) statically balanced (b) dynamically balanced
(c) statically unbalanced (d) both statically and dynamically unbalanced.

370. A viscous damping system with free vibrations will be critically damped if the damping factor is

(a) zero (b) less than one (c) equal to one (d) greater than one.

371. In a simple spring mass vibrating system, the natural frequency ω_n of the system is (k is spring stiffness, m is mass and m_s is spring mass)

(a) $\sqrt{\dfrac{k}{m - \dfrac{m_s}{3}}}$ (b) $\sqrt{\dfrac{k}{m + \dfrac{m_s}{3}}}$ (c) $\sqrt{\dfrac{k}{m + 3m_s}}$ (d) $\sqrt{\dfrac{k}{m - 3m_s}}$.

372. In which one of the following types of control system is the output of the control element proportional to the time rate of change of the input?

(a) Proportional (b) Integral
(c) Proportional and derivative (d) Derivative.

373. A pulley and belt in a belt drive form a

(a) cylindrical pair (b) turning pair (c) rolling pair (d) sliding pair

374. Match List I (Terms) with List II (Definitions) and select the correct answer using the codes given below the Lists:

List I	List II
A. Module	1. Radial distance of a tooth from the pitch circle to the top of the tooth
B. Addendum	2. Radial distance of a tooth from the pitch circle to the bottom of the tooth
C. Circular pitch	3. Distance on the circumference of the pitch circle from a point of one tooth to the corresponding point on the next tooth
	4. Ratio of pitch circle diameter in mm to the number of teeth

Codes:

	A	B	C
(a)	4	1	3
(b)	4	2	3
(c)	3	1	2
(d)	3	2	4

375. The stiffness of spring k used in the Hartnell governor as shown in the given Fig.84 (F_1 and F_2 are centrifugal forces at maximum and minimum radii of rotation r_1 and r_2 respectively) is

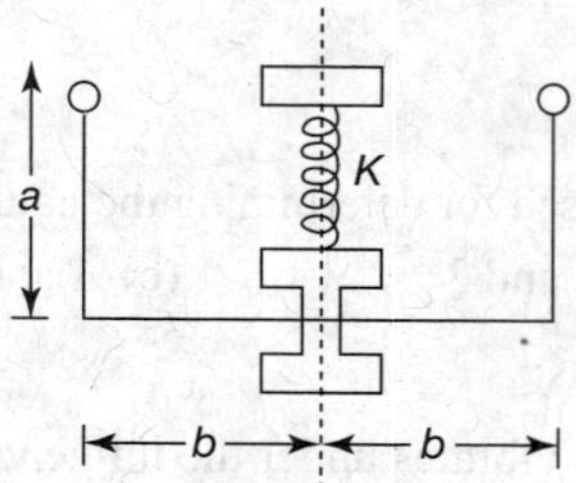

Fig.84

(a) $2\left(\dfrac{b}{a}\right)^2\left(\dfrac{F_1 - F_2}{r_1 - r_2}\right)$ (b) $2\left(\dfrac{a}{b}\right)^2\left(\dfrac{F_1 - F_2}{r_1 - r_2}\right)$

(c) $2\left(\dfrac{b}{a}\right)^2\left(\dfrac{F_1 + F_2}{r_1 - r_2}\right)$ (d) $2\left(\dfrac{a}{b}\right)^2\left(\dfrac{F_1 + F_2}{r_1 - r_2}\right)$.

376. The balancing weights are introduced in planes parallel to the plane of rotation of the disturbing mass. To obtain complete dynamic balance, the minimum number of balancing weights to be introduced in different planes is

(a) 1 (b) 2 (c) 3 (d) 4.

377. Consider the following statements:

The unbalanced force in a single-cylinder reciprocating engine is

1. equal to inertia force of the reciprocating masses
2. equal to gas force
3. always fully balanced

Which of the statements(s) is/are correct?

(a) 1 alone (b) 2 alone (c) 1 and 3 (d) 2 and 3.

378. The equivalent spring stiffness for the system shown in the given Fig.85 (S is the spring stiffness of each of the three springs) is

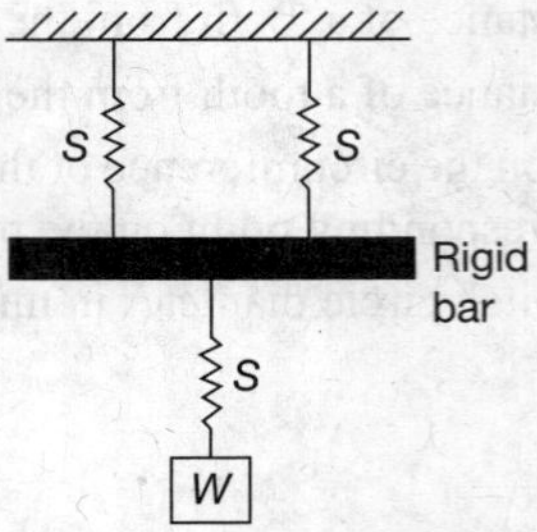

Fig.85

(a) $\frac{S}{2}$ (b) $\frac{S}{3}$ (c) $\frac{2S}{2}$ (d) S.

379. Consider the following methods:

1. Energy method
2. Equilibrium method
3. Rayleigh's method

Which of these methods can be used for determining the natural frequency of the free vibrations?

(a) 1 and 2 (b) 1, 2, and 3 (c) 1 and 3 (d) 2 and 3.

380. Consider the following statements:

1. In forced vibrations, the body vibrates under the influence of an applied force
2. In damped vibrations, amplitude reduces over every cycle of vibration
3. In torsional vibrations, the disc moves parallel to the axis of shaft
4. In transvers vibrations, the particles of the shaft moves approximately perpendicular to the axis of the shaft

Which of these statements are correct?

(a) 1, 2, and 3 (b) 1, 3, and 4 (c) 2, 3, and 4 (d) 1, 2, and 4.

381. Consider the following specifications of gears A. B. C and D:

Gears	A	B	C	D
Numbers of teeth	20	60	20	60
Pressure angle	14 ½°	14 ½°	20°	14 ½°
Module	1	3	3	1
Material	Steel	Brass	Brass	Steel

Which of these gears form the pair of spur gears to achieve a gear ratio of 3?
(a) A and B (b) A and D (c) B and C (d) C and D.

382. Minimum number of teeth for involute rack and pinion arrangement for pressure angle of 20° is
(a) 18 (b) 20 (c) 30 (d) 34.

383. Rope brake dynamometer uses
(a) water as lubricant (b) oil as lubricant (c) grease as lubricant (d) no lubricant.

384. Consider the following statements:

If the fluctuation of speed during a cycle is ±5% of mean speed of a flywheel, the coefficient of fluctuation of speed will

1. increase with increase of mean speed of prime mover
2. decrease with increase of mean speed of prime mover
3. remain same with increase of mean speed of prime mover

Which of the statements(s) is/are correct?
(a) 1 and 3 (b) 1 and 2 (c) 3 alone (d) 2 alone.

385. Which one of the following "Kinematic pairs" has 3 degrees of freedom between the pairing elements?

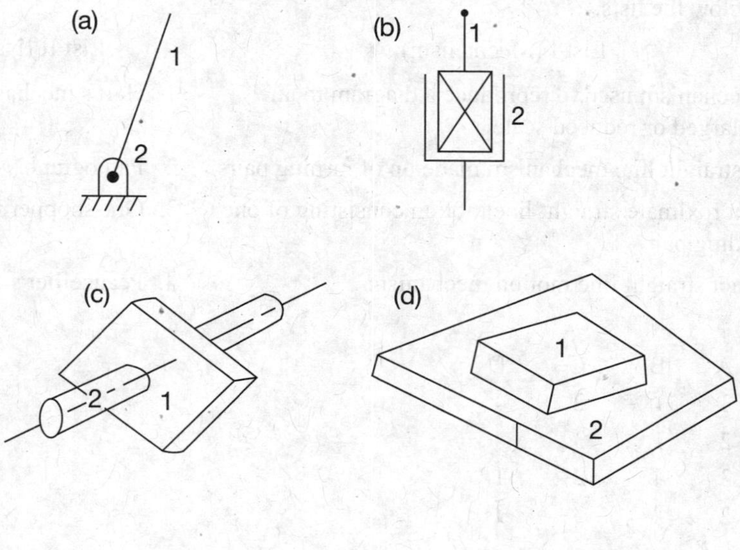

386. The below Fig.86 shows a four bar mechanism. If the radial acceleration of the point C is 5cm/s^2, the length of the link CD is

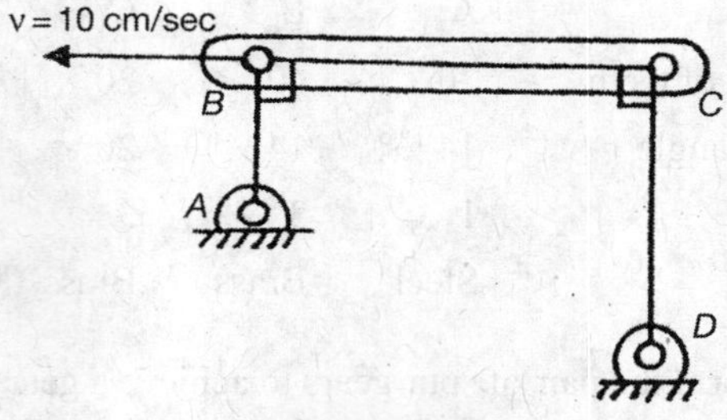

Fig.86

(a) 2 cm (b) 10 cm (c) 20 cm (d) 100 cm.

387. A slider sliding at 10 cm/s on a link which is rotating at 60 rpm. is subjected to Corioli's acceleration of magnitude

(a) 40 π^2 cm/s^2 (b) 0.4 π^2 cm/s^2 (c) 40 π cm/s^2 (d) 40 π cm/s^2.

388. Consider the following necessary and sufficient conditions for replacing a rigid body by a dynamical equivalent system of two masses:

1. Total mass must be equal to that of the rigid body
2. Sum of the squares of radii of gyration of two masses about the c.g. of the rigid body must be equal to square of its radius of gyration about the same point
3. The c.g. of two masses must coincide with that of the rigid body
4. The total moment of inertia of two masses about an axis through the c.g. must be equal to that of the rigid body

Which of the above conditions are correct?

(a) 1, 2, and 3 (b) 1, 3, and 4 (c) 2, 3, and 4 (d) 1, 2, and 4.

389. Match List I (Mechanism) with List II (Name) and select the correct answer using the codes given below the lists:

List I (Mechanism)	List II (Name)
A. Mechanism used to reproduce a diagram to an enlarged or reduced scale	1. Hart's mechanism
B. A straight line mechanism made up of turning pairs	2. Pantograph
C. Approximate straight line motion consisting of one sliding pair	3. Grasshopper mechanism
D. Exact straight line motion mechanism	4. Peaucellier's mechanism

Codes:

	A	B	C	D
(a)	3	1	2	4
(b)	2	1	3	4
(c)	3	4	2	1
(d)	2	4	3	1

390. Consider the following mechanisms:

1. Oscillating cylinder engine mechanism
2. Toggle mechanism
3. Radial cylinder engine mechanism
4. Quick Return Mechanism

Which of the above are inversions of slider-crank mechanism?

(a) 1, 2, and 4 (b) 2, 3, and 4 (c) 1, 2, and 3 (d) 1, 3, and 4.

391. The frictional torque transmitted in a flat pivot bearing, assuming uniform wear, is

(a) μWR (b) $\frac{3}{4}\mu WR$ (c) $\frac{2}{3}\mu WR$ (d) $\frac{1}{2}\mu WR$.

(where μ = Coefficient of friction
W = Load over the bearing
R = Radius of bearing)

392. With usual notations for different parameters involved, the maximum fluctuation of energy for a flywheel is given by

(a) $2\,EC_s$ (b) $\frac{EC_s}{2}$ (c) $2\,EC_s^2$ (d) $2\,E^2C_s$.

393. Hammer blow

(a) is the maximum horizontal unbalanced force caused by the mass provided to balance the reciprocating masses
(b) is the maximum vertical unbalanced force caused by the mass added to balance the reciprocating masses
(c) varies as the square root of the speed
(d) varies inversely with the square of the speed.

394. Whirling speed of shaft is the speed at which

(a) shaft tends to vibrate in longitudinal direction
(b) torsional vibrations occur
(c) shaft tends to vibrate vigorously in transverse direction
(d) combination of tranverse and longitudinal vibration occurs.

395 The velocity of sliding of meshing gear teeth is

(a) $(\omega_1 \times \omega_2)\,x$ (b) $\frac{\omega_1}{\omega_2}x$ (c) $(\omega_1 \times \omega_2)\,x$ (d) $(\omega_1 + \omega_2)\,x$

(where ω_1 and ω_2 = angular velocities of meshing gears
x = distance between point of contact and the pitch point).

396. A speed reducer unit consists of a double-threaded worm of pitch = 11 mm and a worm wheel of pitch diameter = 84 mm. The ratio of the output torque to the input torque is

(a) 7.6 (b) 12 (c) 24 (d) 42.

397. When the system is given a constant angular velocity rather than an angular displacement, it is known as

(a) step function input (b) harmonic input
(c) unit step displacement (d) variable input.

398. Consider the following statements:

1. When frequency ratio is < 2, the force transmitted to the foundations is more than the exicting force
2. When frequency ratio is > 2, the force transmitted to the foundations increases as the damping is decreased
3. The analysis of base-excited vibrations is similar to that of forced vibrations

Which of these statements are correct?

(a) 1 and 2 (b) 2 and 3 (c) 1 and 3 (d) 1, 2, and 3.

399. Consider the following statements:

1. Critical or whirling speed of the shaft is the speed at which it tends to vibrate violently in the transverse direction
2. To find the natural frequency of a shaft carrying several loads, the energy method gives approximate results
3. Dunkerley's method gives accurate results of the natural frequency of a shaft carrying several loads

Which of these statements is/are correct?

(a) 1 only (b) 2 and 3 (c) 1 and 3 (d) 1, 2, and 3.

400. Consider the following statements:

Coriolis acceleration component appears in the acceleration analysis of the following planar mechanisms:

1. Whitworth quick-return mechanism
2. Slider-crank mechanism
3. Scotch-Yoke mechanism

Which of these statements is/are correct?

(a) 1, 2, and 3 (b) 1 and 2 (c) 2 and 3 (d) 1 only.

401. Which one of the following is an exact straight line mechanism using lower pairs?

(a) Watt's mechanism (b) Grasshopper mechanism
(c) Robert's mechanism (d) Peaucellier's mechanism.

402. Consider the following statements in respect of four-bar mechanisms:

1. It is possible to have the length of one link greater than the sum of lengths of the other three links
2. If the sum of the lengths of the shortest and the longest links is less than that the sum of lengths of the other two, it is known as Grashoff's linkage
3. It is possible to have the sum of the lengths of the shortest and the longest links greater than that of the remaining two links

Which of these statements is/are correct?

(a) 1, 2 and 3 (b) 2 and 3 (c) 2 only (d) 3 only.

403. The height of Watt's governor is

(a) directly proportional to the speed (b) directly proportional to the $(\text{speed})^2$
(c) inversely proportional to the speed (d) inversely proportional to the $(\text{speed})^2$.

404. The Fig.87 shows a critically damped spring-mass system undergoing single degree of freedom vibrations. If $m = 5$ kg and $k = 20$ N/m, the value of viscous damping coefficient is

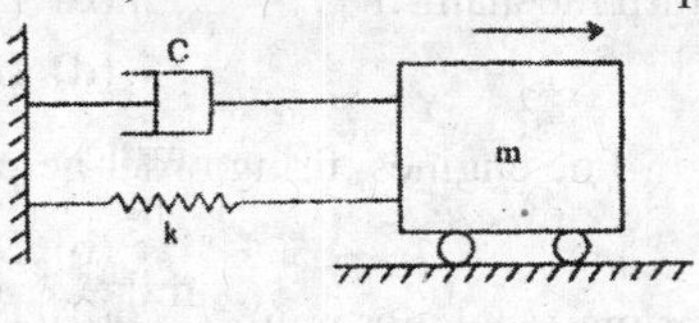

Fig.87

(a) 10 Ns/m (b) 20 Ns/m (c) 4 Ns/m (d) 8 Ns/m.

405. In a system subjected to damped forced vibrations, the ratio of maximum displacement to the static deflection is known as

(a) Critical damping ratio (b) Damping factor
(c) Logarithmic decrement (d) Magnification factor.

406. In the epicyclic gear train shown in the Fig.88, $T_A = 40$, $T_B = 20$. For three revolutions of the arm, the gear *B* will rotate through

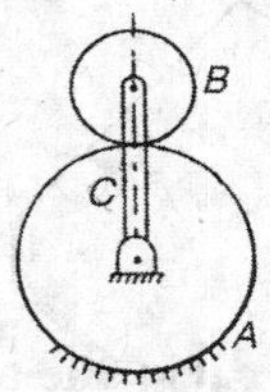

Fig.88

(a) 6 revolutions (b) 2.5 revolutions (c) 3 revolutions (d) 9 revolutions.

407. Which one of the following statements in respect of involute profiles for gear teeth is *not* correct?

(a) Interference occurs in involute profiles
(b) Involute tooth form is sensitive to change in centre distance between the base circles
(c) Basic rack for involute profile has straight line form
(d) Pitch circle diameters of two mating involute gears are directly proportional to the base circle diameters.

408. Traffic control on the roads by lights where the timing mechanism operates irrespective of the intensity of traffic is an example of

(a) Closed loop control (b) Under-damped control
(c) Open loop control (d) Over-damped control.

409. In the given configuration of the mechanism as shown in the Fig.89, $V_A = 40$ m/s and $V_B = 30$ m/s. The magnitude of velocity of slider *B* relative to the slider *A* is

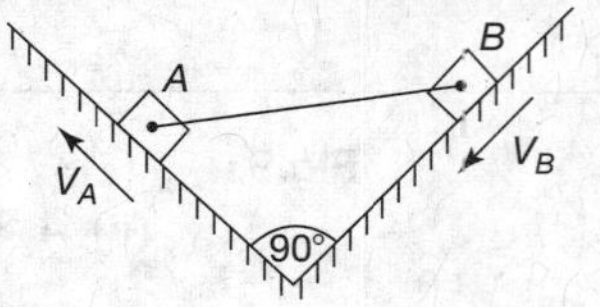

Fig.89

(a) 30 m/s (b) 40 m/s (c) 50 m/s (d) 30.5 m/s

410. Oldham's coupling is an inversion of the kinematic chain used in

(a) Whitworth quick-return mechanism (b) Elliptical trammel
(c) Rotary engine (d) Universal joint.

411. In balancing of 4-stroke in-line engines, firing order helps to control the magnitude of

(a) Primary forces only (b) Secondary forces only
(c) Primary forces and primary couples only (d) Primary and secondary couples only.

412. The method of direct and reverse cranks is used in engines for

(a) the control of speed fluctuations (b) balancing of forces and couples
(c) kinematic analysis (d) vibration analysis.

413. The Fig.90 shows a rigid body oscillating about the pivot A. If J is mass moment of inertia of the body about the axis of rotation, its natural frequency for small oscillations is proportional to

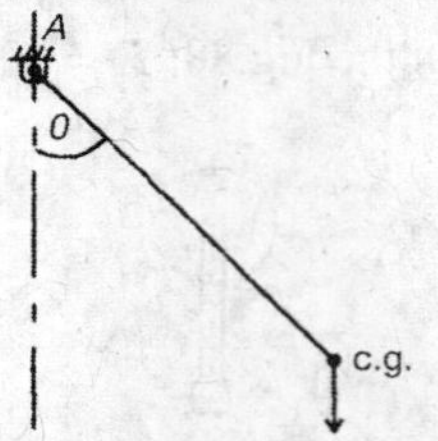

Fig.90

(a) J (b) J^2 (c) $\frac{1}{J}$ (d) $\frac{1}{\sqrt{J}}$.

414. Consider the following statements:
Two rotors mounted on a single shaft can be considered to be equivalent to a geared-shaft system having two rotors provided

1. the kinetic energy of the equivalent system is equal to that of the original system
2. the strain energy of the equivalent system is equal to that of the original system
3. the shaft diameters of the two systems are equal

Which of these statements are correct?

(a) 1, 2, and 3 (b) 1 and 2 (c) 2 and 3 (d) 1 and 3.

415. What is the value of K for which the relative damping of the closed loop system as shown in Fig.91 below is equal to 0.5?

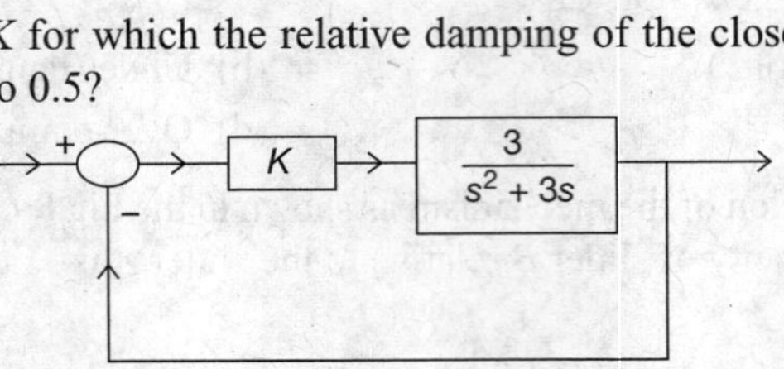

Fig.91

(a) 2 (b) 3 (c) 4 (d) 5.

416. What is the number of instantaneous centres for an eight link mechanism?

(a) 15 (b) 28 (c) 30 (d) 8.

417. For the rotor system as shown in Fig.92, the mass required for its complete balancing is

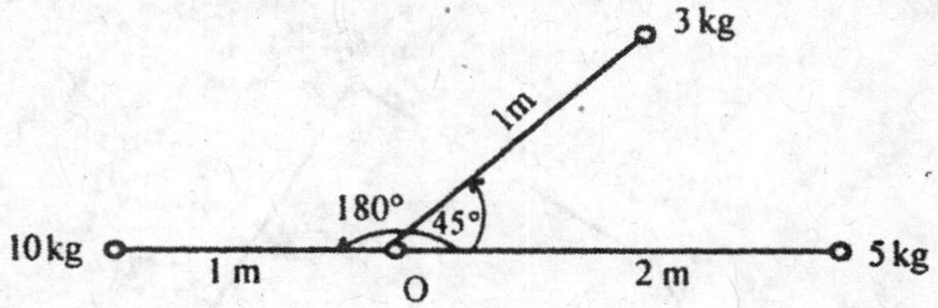

Fig.92

(a) 1.5 kg at 2 m radius and at 225° from reference
(b) 3 kg at 1 m radius and at 45° from reference
(c) 8 kg at 1 m radius and at 225° from reference
(d) 4 kg at 2 m radius and at 45° from reference.

418. The Fig.93 given below show different schemes suggested to transmit continuous rotary motion from axis *A* to axis *B*. Which of these schemes are *not* dynamically balanced?

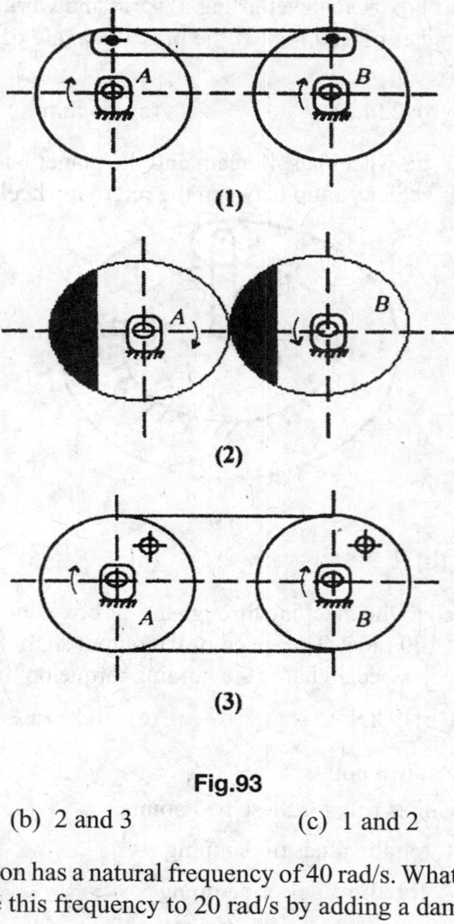

Fig.93

(a) 1 and 3 (b) 2 and 3 (c) 1 and 2 (d) 1, 2, and 3.

419. A spring-mass suspension has a natural frequency of 40 rad/s. What is the damping ratio required if it is desired to reduce this frequency to 20 rad/s by adding a damper to it?

(a) $\frac{\sqrt{3}}{2}$ (b) $\frac{1}{2}$ (c) $\frac{1}{\sqrt{2}}$ (d) $\frac{1}{4}$.

420. The four bar mechanism shown in the Fig.94 (Given: $OA = 3$ cm, $AB = 5$ cm, $BC = 6$ cm, $OC = 7$ cm) is a

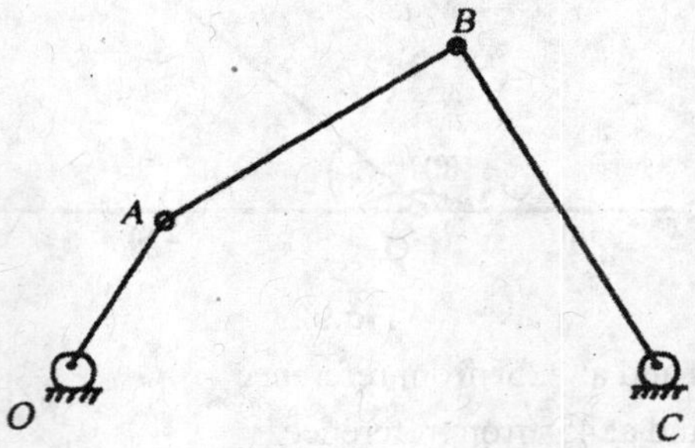

Fig.94

(a) Double crank mechanism (b) Double rocker mechanism
(c) Crank rocker mechanism (d) Single slider mechanism.

421. Speed reduction in a gear box is achieved using a worm and worm wheel. The worm wheel has 30 teeth and a pitch diameter of 210 mm. If the pressure angle of the worm is 20°, what is the axial pitch of the worm?

(a) 7 mm (b) 22 mm (c) 14 mm (d) 63 mm.

422. In the Fig.95 shown, the sun wheel has 48 teeth and the planet has 24 teeth. If the sun wheel is fixed, what is the angular velocity ratio between the internal wheel and arm?

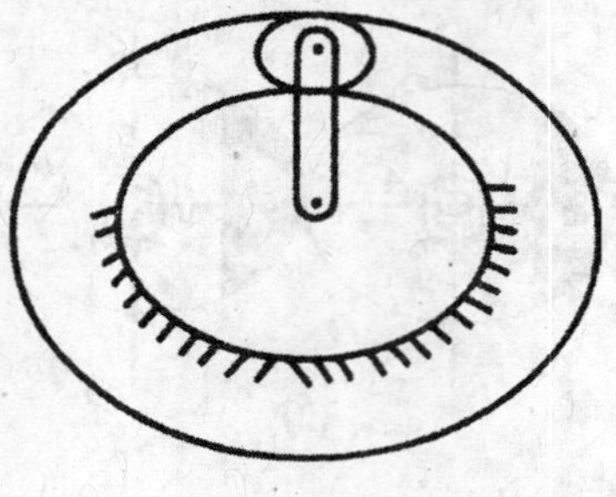

Fig.95

(a) 3.0 (b) 1.5 (c) 2.0 (d) 4.0.

423. 100 kW power is supplied to the machine through a gear box which uses an epicyclic gear train. The power is supplied at 100 rad/s. The speed of the output shaft of the gear box is 10 rad/s in a sense opposite to the input speed. What is the holding torque on the fixed gear of the train?

(a) 8 kNm (b) 9 kNm (c) 10 kNm (d) 11 kNm.

424. Transmissibility is unity at two points.
Which one of the following is true for these two points?

(a) ω/ω_n is zero and $\sqrt{3}$ for all values of damping
(b) ω/ω_n is zero and $\sqrt{2}$ for all values of damping
(c) ω/ω_n is unity and 2 for all values of damping
(d) ω/ω_n is unity and $\sqrt{3}$ for all values of damping.

425. Examine the Fig.96 shown below wherein the numbers indicate the links:

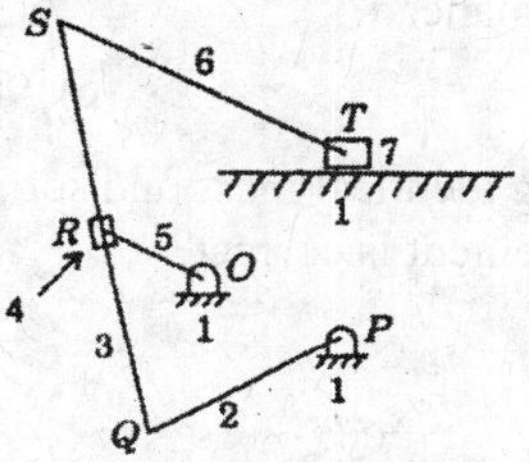

Fig.96

Which of the statements given below are correct?

1. I_{34} is at ∞, perpendicular to QS
2. I_{45} is at ∞, perpendicular to QS
3. I_{71} is at T
4. I_{45} is at R.

Select the correct answer from the code given below :

(a) 1 and 2 (b) 1 and 4 (c) 2 and 3 (d) 1, 3, and 4.

426. Match List I (Gear Train) with List II (Application) and select the correct answer using the code given below the lists:

List I (Gear Train)	List II (Application)
A. Compound gear train	1. Automobile gear box
B. Eipcyclic spur gear train with brake bands	2. Automatic transmission of automobile
C. Worm and worm-wheel gear train	3. Speed reducers for lifts
D. Epicyclic bevel gear train	4. Automobile differential

Code:

	A	B	C	D
(a)	1	2	3	4
(b)	3	4	1	2
(c)	1	4	3	2
(d)	3	2	1	4

427. Cycloidal tooth profile of a cycloidal gear tooth is a combination of

(a) Hypocycloid and involute of a circle
(b) Hypocycloid and epicycloid
(c) Epicycloid and involute of a circle
(d) Straight line and epicycloid.

428. Which one of the following statements is not correct?

(a) Response of an inertia governor is faster than that of a centrifugal governor
(b) An I.C. engine prime-mover always requires both governor and the flywheel
(c) Spring loaded centrifugal governors are effective over a wide range of operating speeds
(d) Flywheel is not necessary in case of electric motor driven punch press.

429. The problem of hunting of a centrifugal governor becomes very acute when the governor becomes

(a) Less sensitive (b) Highly sensitive (c) Highly stable (d) Less stable.

430. What should be the angle between the cylinder axes if the primary forces of a 2-cylinder V-engine are to be completely balanced?

(a) 45° (b) 60° (c) 90° (d) 120°.

431. Systems A and B (Fig.97) having identical mass and springs are set in simple harmonic motion. Which one of the following statement is correct?

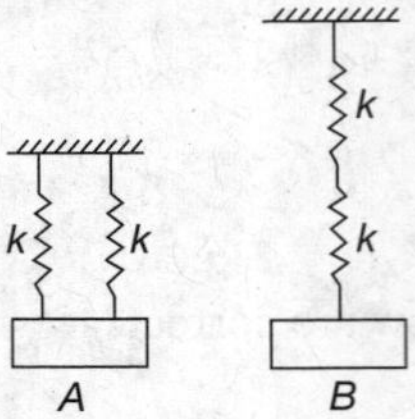

Fig.97

(a) For systems A and B, the period of vibration is same
(b) System B has a period of vibration twice that of System A
(c) System B has a period of vibration half that of system A
(d) System B has a period of vibration $\sqrt{2}$ times that of system A.

432. What is the polar for the function $\frac{1}{(1+j\tau\omega)}$, where ω stands for the frequency (angular), τ the time constant and $j = \sqrt{-1}$?

(a) Semi circle (b) Parabola (c) Ellipse (d) Circle.

433. Suppose two rotors with inertias I_1 and I_2 on shafts of lengths l_1 and l_2 respectively, are connected by gears such that the speed of the I_2 rotor is always G times that of the I_1 rotor. This system may, for vibration analysis, be treated as being on one shaft (integral with l_1) if

(a) I_1 is changed to $G^2 I_1$ (b) I_2 is changed to $G^2 I_2$
(c) I_1 is changed to I_1/G^2 (d) I_2 is changed to I_2/G^2.

434. Two rotors A and B are connected to the two ends of a shaft of uniform diameter. The mass moment of inertia of rotor A about the axis of the shaft is four times that of B. If the length of the shafts is 1 m and C is the position of node for torsional vibrations, then what is the length of AC?

(a) 1/5 m (b) 4/5 m (c) 1/25 m (d) 16/25 m.

435. If the frequency of fluctuations in engine speed coincides with the natural frequency of oscillations of the governor, then, due to resonance, the amplitude of oscillations becomes very high. Consequently the governor tends to intensify the speed variations. What is such a situation?

(a) Sensitiveness (b) Stability (c) Isochronism (d) Hunting.

436. What is the minimum number of arbitrarily chosen parallel planes in which the balancing mass/masses may be placed for complete dynamic balance of a system of unbalanced revolving masses in different transverse planes of a rotating shaft?

(a) 1 (b) 2 (c) 3 (d) 4.

437. Which one of the following is realized by hydraulic dash-pot shock absorbers?

(a) Viscous damping (b) Structural damping
(c) Coulomb damping (d) Spring damping.

438. Which one of the following is the correct statement?

Transmissibility is defined as the ratio of the

(a) force applied to the machine to the force transmitted to the foundation
(b) force transmitted to the foundation to the force applied to the machine
(c) force applied to the machine to the vector sum of spring forces
(d) damping forces to the spring forces.

439. The criterion of constraint of a chain connecting the number of binary joints (J), number of higher pairs (H) and the number of links (L) is:

$$J + \frac{H}{2} = \frac{3L}{2} - 2.$$

When is the chain locked?

(a) L.H.S. = R.H.S. (b) L.H.S. > R.H.S.
(c) L.H.S. < R.H.S. (d) The chain will never get locked.

440. Which one of the following statements is correct?

Transmission angle is the angle between

(a) the output link and the frame (b) the output link and the coupler
(c) the input link and the coupler (d) the input link and the frame.

441. In typical power transmission with reduction from an induction motor of speed 1450 rpm to a speed as low as 1 rpm, which one of the following order of reduction is desirable?

(a) Worm drive-spur drive-belt drive (b) Belt drive-spur drive-worm drive
(c) Worm drive-belt drive-spur drive (d) Spur drive-worm drive-belt drive.

442. Which one of the following is the correct statement?

The relative velocity of sliding in the teeth of gears in mesh is zero at

(a) the point of engagement
(b) the point of disengagement
(c) the pitch point
(d) the point between point of engagement and pitch point.

443. Which one of the following is the correct statement?

The consequence of a slight increase in the centre distance between two mating involute gears is that

(a) The law of gearing is not satisfied perfectly
(b) interference occurs
(c) pressure angle increases
(d) pressure angle decreases.

444. Which one of the following is correct for a shaft carrying two rotors at its ends?

(a) It has no node (b) It has one node
(c) It has two nodes (d) It has three nodes.

445. The Fig.98 given below shows the locations of the roots of the characteristic function of a second order, linear, closedloop control system. What is the natural frequency of the system?

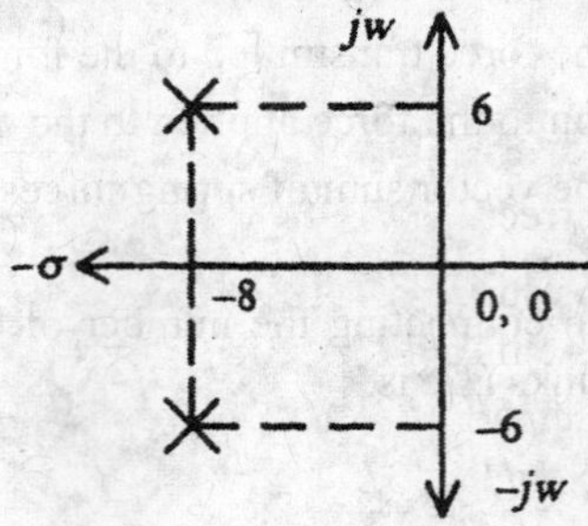

Fig.98

(a) 10 rad/s (b) 36 rad/s (c) 48 rad/s (d) 64 rad/s.

446. A type-2 system has its transfer function for open loop represented by

$$G(s)\ H(s) = \frac{4(1+s)}{s^2(1+0.1s)} \text{ in the } s \text{ - plane.}$$

Nyquist plot shows that there are no poles within the path of values of s. It further shows that there is no encirclement of $-1 + j \cdot 0$. Then what is the number of zeros of the said plot?

(a) 1 (b) 0 (c) 2 (d) 3.

447. For a spring-loaded roller-follower driven with a disc cam,

(a) the pressure angle should be larger during rise than that during return for ease of transmitting motion
(b) the pressure angle should be smaller during rise than that during return for ease of transmitting motion
(c) the pressure angle should be large during rise as well as during return for ease of transmitting motion
(d) the pressure angle does not affect the ease of transmitting motion.

448. For the planar mechanism shown Fig.99, select the most appropriate choice for the motion of link 2 when link 4 is moved upwards.

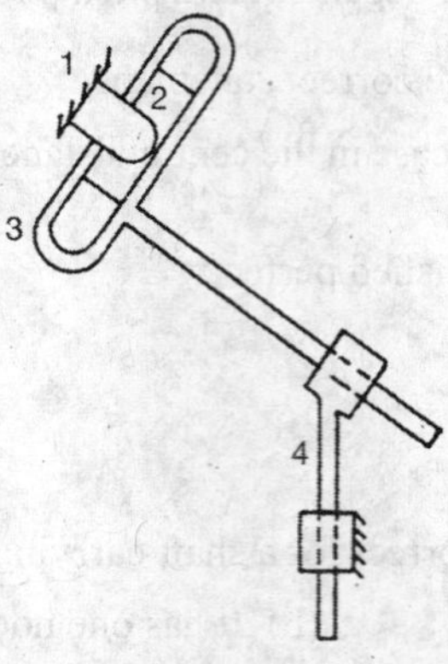

Fig.99

(a) Link 2 rotates clockwise
(b) Link 2 rotates counter-clockwise
(c) Link 2 does not move
(d) Link 2 motion cannot be determined.

449. Which of the following statement is correct?
(a) Flywheel reduces speed fluctuations during a cycle for a constant load, but flywheel does not control the mean speed of the engine if the load changes
(b) Flywheel does not reduce speed fluctuations during a cycle for a constant load, but flywheel does control the mean speed of the engine if the load changes
(c) Governor controls speed fluctuations during a cycle for a constant load, but governor does not control the mean speed of the engine if the load changes
(d) Governor controls speed fluctuations during a cycle for a constant load, and governor also controls the mean speed of the engine if the load changes.

450. In spur gears, the circle on which the involute is generated is called the
(a) pitch circle (b) clearance circle (c) base circle (d) addendum circle.

451. The ratio of tension on the tight side to that on the slack side in a flat belt drive is
(a) proportional to the product of coefficient of friction and lap angle
(b) an exponential function of the product of coefficient of friction and lap angle
(c) proportional to the lap angle
(d) proportional to the coefficient of friction.

452. To make a worm drive reversible, it is necessary to increase
(a) centre distance (b) worm diameter factor
(c) number of starts (d) reduction ratio.

453. Instantaneous centre of a body rolling with sliding on a stationary curved surface lies
(a) at the point of contact
(b) on the common normal at the point of contact
(c) on the common tangent at the point of contact
(d) at the centre of curvature of the stationary surface.

454. The number of degrees of freedom of a five link plane mechanism with five revolute pairs as shown in the Fig.100 is

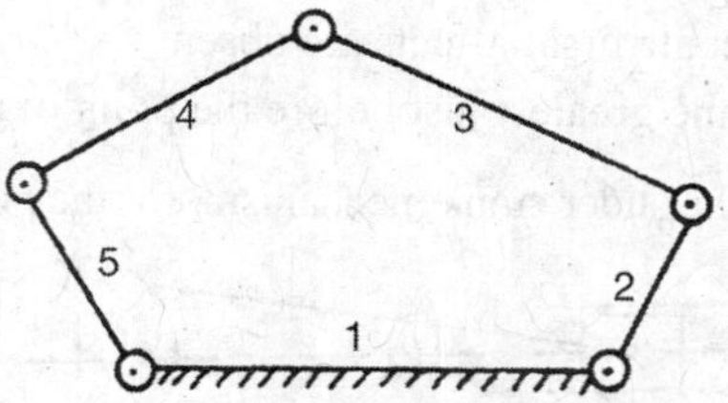

Fig.100

(a) 3 (b) 4 (c) 2 (d) 1.

455. In a plate cam mechanism with reciprocating roller follower, the follower has a constant acceleration in the case of

(a) cycloidial motion (b) simple harmonic motion
(c) parabolic motion (d) polynomial motion.

456. The sun gear in the Fig.101 is driven clockwise at 100 rpm. The ring gear is held stationary. For the number of teeth shown on the gears, the arm rotates at

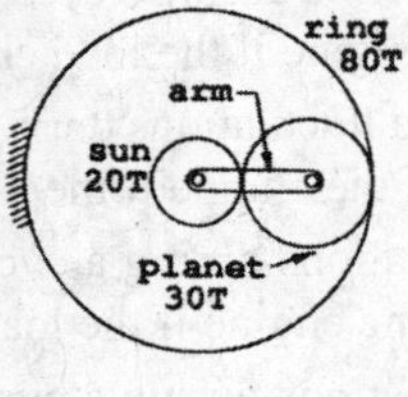

Fig.101

(a) 0 rpm (b) 20 rpm (c) 33.33 rpm (d) 66.67 rpm.

457. For the audio cassette mechanism shown Fig.102 below, where is the instantaneous centre of rotation (point) of the two spools?

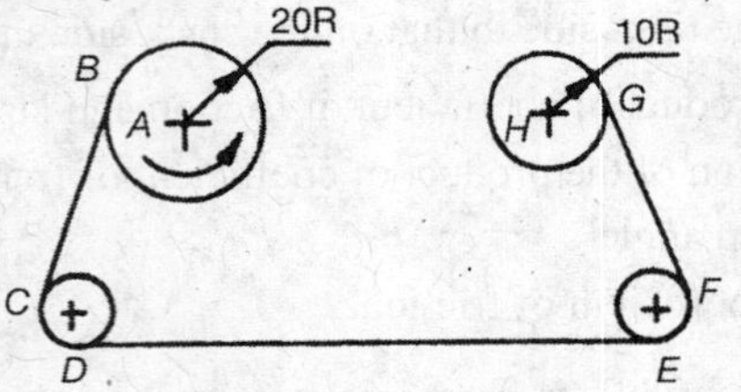

Fig.102

(a) Point P lies to the left of both the spools but at infinity along the line joining A and H
(b) Point P lies in between the two spools on the line joining A and H, such that $\overline{PH} = 2\overline{AP}$
(c) Point P lies to the right of both the spools on the line joining A and H, such that $\overline{AH} = \overline{HP}$
(d) Point P lies at the intersection of the line joining B and C and the line joining G and F.

458. With regard to belt drives with given pulley diameters, centre distance and coefficient of friction between the pulley and the belt materials, which of the statements below are FALSE?

(a) A crossed flat belt configuration can transmit more power than an open flat belt configuration
(b) A "V" belt has greater power transmission capacity than an open flat belt
(c) Power transmission is greater when belt tension is higher due to centrifugal effects than the same belt drive when centrifugal affects are absent
(d) Power transmission is the greatest just before the point of slipping is reached.

459. The cross head velocity in the slider crank mechanism, for the position shown in Fig.103 below is:

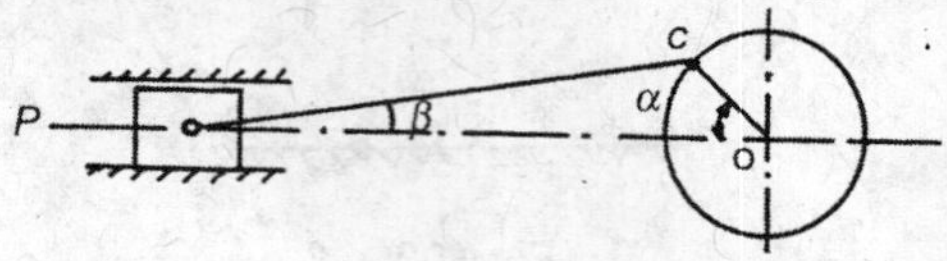

Fig.103

(a) $v_C = \cos\left(90 - \overline{\alpha + \beta}\right)\cos\beta$ (b) $v_C = \cos\left(90 - \overline{\alpha + \beta}\right)\sec\beta$

(c) $v_C = \cos\left(90 - \overline{\alpha - \beta}\right)\cos$ (d) $v_C = \cos\left(90 - \overline{\alpha - \beta}\right)\sec\beta$.

Where v_C is the linear velocity of the crank pin.

460. An automobile of weight *W* is shown in Fig.104. *A* pull '*P*' is applied as shown. The reaction at the front wheels (location *A*) is

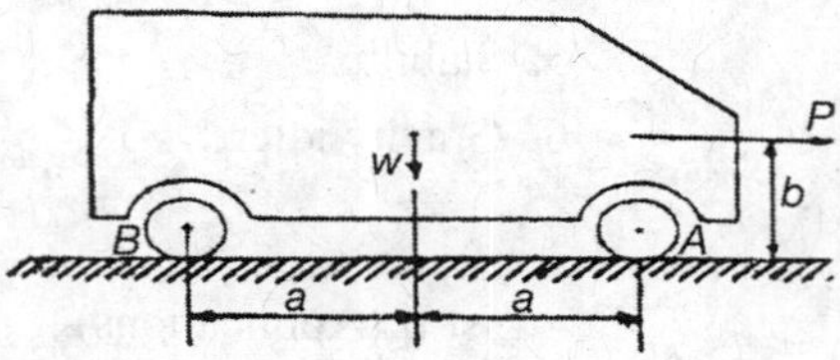

Fig.104

(a) $W/2 - Pb/2a$ (b) $W/2 + Pb/2a$ (c) $W/2 - Pa/2b$ (d) $W/2$.

461. The percentage improvement in power capacity of a flat belt drive, when the wrap angle at the driving pulley is increased from 150° to 210° by an idler arrangement for a friction coefficient of 0.3, is

(a) 22.61 (b) 33.92 (c) 40.17 (d) 67.85.

462. In an experiment to find the velocity and acceleration of a particular cam rotating at 10 rad/s, the values of displacements and velocities are recorded. The slope of displacement curve at an angle of 'θ' is 1.5 m/s and the slope of velocity curve at the same angle is –0.05 m/s². The velocity and acceleration of the cam at the instant are respectively

(a) 15 m/s and –5 m/s² (b) 15 m/s and 5 m/s²

(c) 1.2 m/s and 0.5 m/s² (d) 1.2 m/s and –0.5 m/s².

463. Consider the triangle formed by the connecting rod and the crank of an IC engine as the two sides of the triangle. If the maximum area of this triangle occurs when the crank angle is 75° the ratio of connecting rod length to crank radius is

(a) 5 (b) 4 (c) 3.73 (d) 3.

464. The difference between tensions on the tight and slack sides of a belt drive is 3000 N. If the belt speed is 15 m/s, the transmitted power in kW is

(a) 45 (b) 22.5 (c) 90 (d) 100.

465. The profile of a cam in a particular zone is given by $x = \sqrt{3}\cos\theta$ and $y = \sin\theta$. The normal to the cam profile at $\theta = \frac{\pi}{4}$ is at an angle (with respect to *x* axis)

(a) $\frac{\pi}{4}$ (b) $\frac{\pi}{2}$ (c) $\frac{\pi}{3}$ (d) 0.

466. A flywheel of moment of inertia 9.8 kg.m² fluctuates by 30 rpm for a fluctuation in energy of 1936 Joules. The mean speed of the flywheel is (in rpm)

(a) 600 (b) 900 (c) 968 (d) 2940.

467. Math List I with List II

List I	List II
(A) Collision of bodies	1. Kinetics
(B) Minimum potential energy	2. Reciprocating unbalance
(C) Degree of freedom	3. Dynamics
(D) Prony brake	4. Coefficient of restitution
(E) Hammer blow	5. Stability
(F) Ellipse trammel	6. Gravity idler

468. Math List I with List II

List I (Gear types)	List II (Applications)
(a) Worm gears	1. Parallel shafts
(b) Cross helical gears	2. Non-parallel, intersecting shafts
(c) Bevel gears	3. Non-parallel, non-intersecting shafts
(d) Spur gears	4. Large speed ratios

469. Fig.105 shows a quick return mechanism. The crank OA rotates clockwise uniformly. $OA = 2$ cm, $OO' = 4$ cm. The ratio of time for forward motion to that for return motion is

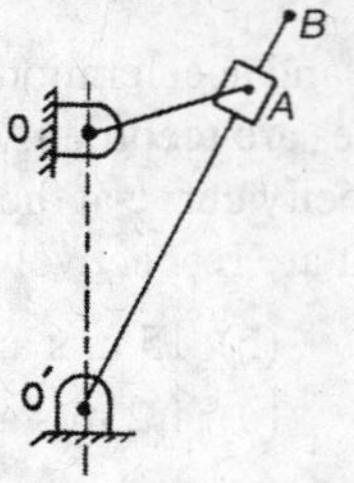

Fig.105

(a) 0.5 (b) 2.0 (c) $\sqrt{12}$ (d) 1.

470. The arm OA of an epicyclic gear train shown in Fig.106 below revolves counter-clockwise about O with an angular velocity of 4 rad/s. Both gears are of same size. The angular velocity of gear C, if the sun gear B is fixed, is

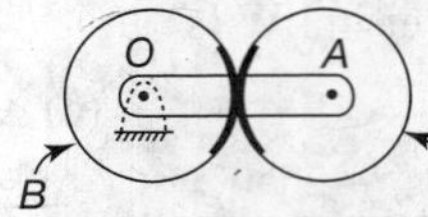

Fig.106

(a) 4 rad/s (b) 8 rad/s (c) 10 rad/s (d) 12 rad/s.

471. The mechanism used in a shaping machine is

(a) a closed 4-bar chain having 4 revolute pairs
(b) a closed 6-bar chain having 6 revolute pairs
(c) a closed 4-bar chain having 4 revolute and 2 sliding pairs
(d) an inversion of the single slider-crank chain.

472. The lengths of the links of a 4-bar linkage with revolute pairs only are p, q, r and s units. Given that $p < q < r < s$. Which of these links should be the fixed one, for obtaining a "double crank" mechanism?

(a) link of length p (b) link of length q (c) link of length r (d) link of length s.

473. For a certain engine having an average speed of 1200 rpm, a flywheel approximated as a solid disc, is required for keeping the fluctuation of speed within 2% above the average speed. The fluctuation of kinetic energy per cycle is found to be 2 kg.m. What is the least possible mass of the flywheel if its diameter is not to exceed 1 m?

(a) 40 kg (b) 51 kg (c) 62 kg (d) 73 kg

474. In a band brake the ratio of tight side band tension to the tension on the slack side is 3. If the angle of overlap of band on the drum is 180°, the coefficient of friction required between drum and the band is

(a) 0.20 (b) 0.25 (c) 0.30 (d) 0.35.

475. Two mating spur gears have 40 and 120 teeth respectively. The pinion rotates at 1200 rpm and transmits a torque of 20 N m. The torque transmitted by the gear is

(a) 6.6 Nm (b) 20 Nm (c) 40 Nm (d) 60 Nm.

476. For a mechanism shown in Fig.107 below, the mechanical advantage for the given configuration is

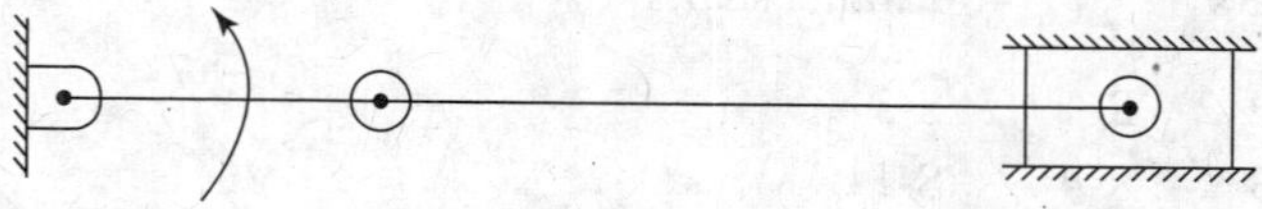

Fig.107

(a) 0 (b) 0.5 (c) 1.0 (d) ∞.

477. In the Fig.108 shown, the relative velocity of link 1 with respect to link 2 is 12 m/sec. Link 2 rotates at a constant speed of 120 rpm. The magnitude of Coriolis component of acceleration of link 1 is

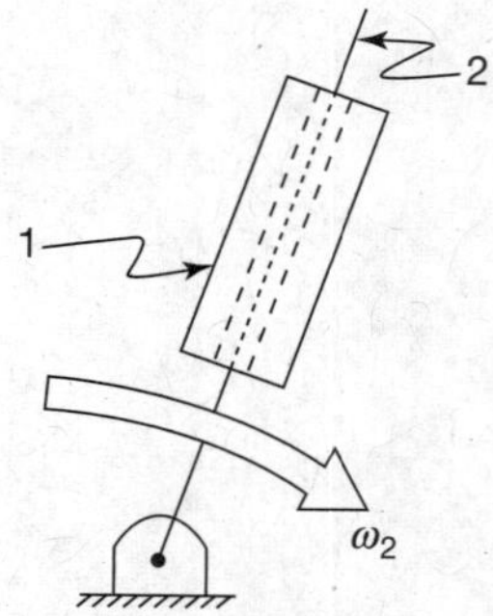

Fig.108

(a) 302 m/s² (b) 604 m/s² (c) 906 m/s² (d) 1208 m/s²

478. The Fig.109 below shows a planar mechanism with single degree of freedom. The instant aneous centre 24 for the given configuration is located at a position

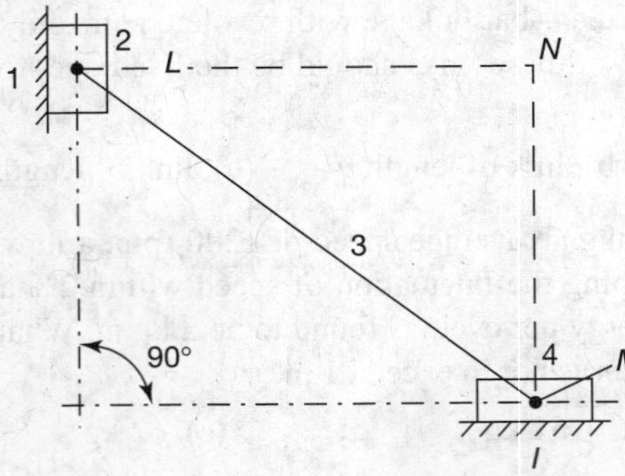

Fig.109

(a) L (b) M (c) N (d) ∞.

479. Match the following:

Type of gears	Arrangement of shafts
P. Bevel gears	1. Non-parallel off-set shafts
Q. Worm gears	2. Non-parallel intersecting shafts
R. Herringbone gears	3. Non-parallel, non-intersecting shafts
S. Hypoid gears	4. Parallel shafts

(a) P-4 Q-2 R-1 S-3
(b) P-2 Q-3 R-4 S-1
(c) P-3 Q-2 R-1 S-4
(d) P-1 Q-3 R-4 S-2

480. Match the following with respect to spatial mechanisms.

Type of Joint	Degrees of constraint
P-Revolute	1. Three
Q-Cylindrical	2. Five
R-Spherical	3. Four
	4. Two
	5. Zero

(a) P-1 Q-3 R-3
(b) P-5 Q-4 R-3
(c) P-2 Q-3 R-1
(d) P-4 Q-5 R-3

Common Data for Questions 481 to 483

An instantaneous configuration of a four-bar mechanism, whose plane is horizontal, is shown in the Fig.110 below. At this instant, the angular velocity and angular acceleration of link O_2A are $\omega = 8$ rad/s and $\alpha = 0$, respectively, and the driving torque (T) is zero. The link O_2A is balanced so that its centre of mass falls at O_2.

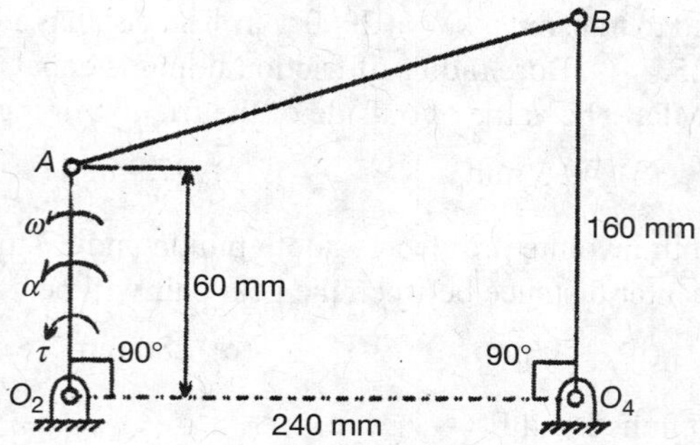

Fig.110

481. Which kind of 4-bar mechanism is O_2ABO_4?

(a) Double-crank mechanism (b) Crank-rocker mechanism
(c) Double-rocker mechanism (d) Parallelogram mechanism.

482. At the instant considered, what is the magnitude of the angular velocity of O_4B?

(a) 1 rad/s (b) 3 rad/s (c) 8 rad/s (d) $\frac{64}{3}$ rad/s

483. At the same instant, if the component of the force at joint A along AB is 30 N, then the magnitude of the joint raction at O_2

(a) is zero (b) is 30 N
(c) is 78 N (d) cannot be determined from the given data

Statement for Linked Answer Questions 484 and 485

A band brake consists of a lever attached to one end of the band. The other end of the band is fixed to the ground. The wheel has a radius of 200 mm and the wrap angle of the band is 270°. The braking force applied to the lever is limited to 100 N, and the coefficient of friction between the band and the wheel is 0.5. No other information is given Fig.111.

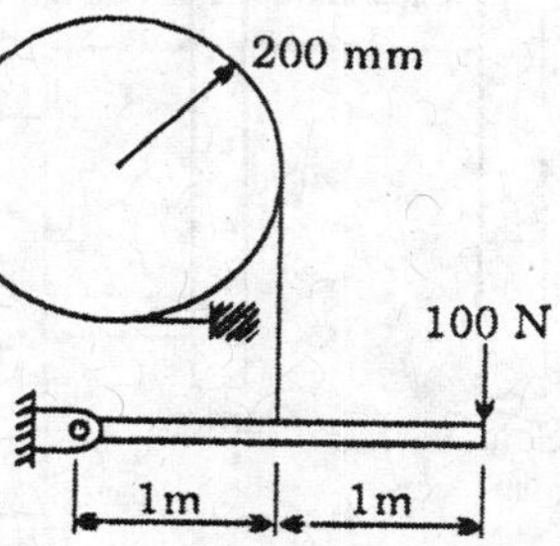

Fig.111

484. The maximum tension that can be generated in the band during braking is

(a) 1200 N (b) 2110 N (c) 3224 N (d) 4420 N.

485. The maximum wheel torque that can be completely braked is

(a) 200 Nm (b) 382 Nm (c) 604 Nm (d) 844 Nm.

486. A disk clutch is required to transmit 5 kW at 2000 rpm. The disk has a friction lining with coeffient of friction equal to 0.25. Bore radius of friction lining is equal to 25 mm. Assume uniform contact pressure of 1 MPa. The value of outside of the friction lining is

(a) 39.4 mm (b) 49.5 mm (c) 97.9 mm (d) 142.9 mm.

487. Twenty degree full depth involute profiled 19-tooth pinion and 37-tooth gear are in mesh. If the module is 5 mm, the center distance between the gear pair will be

(a) 140 mm (b) 150 mm (c) 280 mm (d) 300 mm.

488. Match the items in columns I and II

Column I	Column II
P. Higher kinematic pair	1. Grubler's equation
Q. Lower kinematic pair	2. Line contact
R. Quick return mechanism	3. Euler's equation
S. Mobility of a linkage	4. Planer
	5. Shaper
	6. Surface contact

(a) P–2, Q–6, R–4, S–3 (b) P–6, Q–2, R–4, S–1
(c) P–6, Q–2, R–5, S–3 (d) P–2, Q–6, R–5, S–1

Common Data for Questions 489 and 490

A planetary gear train has four gears and one carrier. Angular velocities of the gears and ω_1, ω_2, ω_3 and ω_4, respectively. The carrier rotates with angular velocity ω_5.

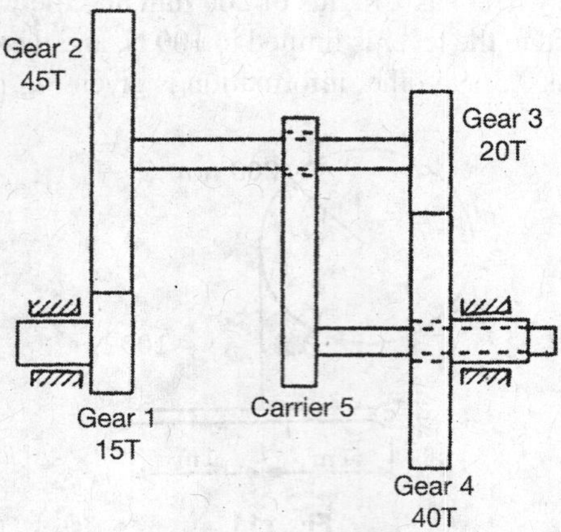

Fig.112

489. What is the relation between the angular velocites of Gear 1 and Gear 4?

(a) $\dfrac{\omega_2 - \omega_5}{\omega_4 - \omega_5} = 6$ (b) $\dfrac{\omega_4 - \omega_5}{\omega_1 - \omega_5} = 6$

(c) $\dfrac{\omega_1 - \omega_2}{\omega_4 - \omega_3} = -\left(\dfrac{2}{3}\right)$ (d) $\dfrac{\omega_2 - \omega_5}{\omega_4 - \omega_5} = \dfrac{8}{9}$

490. For $\omega_1 = 60$ rpm clockwise (cw) when looked from the left, what is the angular velocity of the carrier and its direction so that Gear 4 rotates in counterclockwise (ccw) direction at twice the angular velocity of Gear 1 when looked from the left?

(a) 130 rpm, cw (b) 223 rpm, ccw (c) 256 rpm, cw (d) 156 rpm, ccw

Answers

1.(a)	2.(b)	3.(d)	4.(b)	5.(c)	6.(d)	7.(c)	8.(c)	9.(b)	10.(c)
11.(d)	12.(d)	13.(d)	14.(b)	15.(c)	16.(c)	17.(a)	18.(b)	19.(b)	20.(a)
21.(d)	22.(c)	23.(d)	24.(c)	25.(b)	26.(a)	27.(d)	28.(c)	29.(a)	30.(c)
31.(c)	32.(b)	33.(a)	34.(c)	35.(a)	36.(d)	37.(a)	38.(c)	39.(d)	40.(c)
41.(d)	42.(d)	43.(b)	44.(b)	45.(d)	46.(b)	47.(c)	48.(a)	49.(b)	50.(c)
51.(a)	52.(a)	53.(d)	54.(c)	55.(b)	56.(c)	57.(a)	58.(c)	59.(a)	60.(b)
61.(b)	62.(d)	63.(b)	64.(c)	65.(c)	66.(a)	67.(d)	68.(b)	69.(a)	70.(d)
71.(c)	72.(d)	73.(d)	74.(a)	75.(b)	76.(c)	77.(b)	78.(b)	79.(c)	80.(b)
81.(d)	82.(a)	83.(d)	84.(d)	85.(a)	86.(c)	87.(b)	88.(c)	89.(d)	90.(a)
91.(d)	92.(a)	93.(a)	94.(d)	95.(a)	96.(c)	97.(d)	98.(a)	99.(b)	100.(c)
101.(c)	102.(d)	103.(a)	104.(d)	105.(b)	106.(a)	107.(a)	108.(a)	109.(c)	110.(a)
111.(d)	112.(d)	113.(d)	114.(b)	115.(d)	116.(d)	117.(d)	118.(c)	119.(d)	120.(a)
121.(c)	122.(c)	123.(d)	124.(b)	125.(d)	126.(c)	127.(a)	128.(a)	129.(c)	130.(d)
131.(c)	132.(b)	133.(d)	134.(d)	135.(c)	136.(b)	137.(c)	138.(b)	139.(a)	140.(b)
141.(d)	142.(c)	143.(d)	144.(a)	145.(b)	146.(a)	147.(b)	148.(b)	149.(c)	150.(d)
151.(d)	152.(c)	153.(d)	154.(c)	155.(b)	156.(c)	157.(b)	158.(c)	159.(b)	160.(b)
161.(a)	162.(d)	163.(a)	164.(a)	165.(b)	166.(d)	167.(b)	168.(d)	169.(a)	170.(a)
171.(a)	172.(d)	173.(b)	174.(b)	175.(c)	176.(c)	177.(d)	178.(b)	179.(a)	180.(d)
181.(a)	182.(b)	183.(b)	184.(c)	185.(d)	186.(a)	187.(d)	188.(b)	189.(a)	190.(d)
191.(c)	192.(b)	193.(c)	194.(c)	195.(c)	196.(d)	197.(b)	198.(c)	199.(a)	200.(d)
201.(c)	202.(d)	203.(c)	204.(a)	205.(d)	206.(a)	207.(c)	208.(a)	209.(c)	210.(a)
211.(b)	212.(a)	213.(a)	214.(d)	215.(b)	216.(a)	217.(d)	218.(c)	219.(d)	220.(a)
221.(a)	222.(b)	223.(c)	224.(c)	225.(c)	226.(c)	227.(b)	228.(b)	229.(b)	230.(d)
231.(c)	232.(a)	233.(b)	234.(c)	235.(a)	236.(c)	237.(d)	238.(b)	239.(c)	240.(a)
241.(c)	242.(c)	243.(d)	244.(c)	245.(c)	246.(b)	247.(a)	248.(b)	249.(a)	250.(d)
251.(b)	252.(a)	253.(d)	254.(c)	255.(a)	256.(d)	257.(c)	258.(b)	259.(b)	260.(a)
261.(a)	262.(c)	263.(a)	264.(c)	265.(b)	266.(a)	267.(a)	268.(b)	269.(d)	270.(b)
271.(a)	272.(a)	273.(a)	274.(b)	275.(a)	276.(c)	277.(a)	278.(c)	279.(b)	280.(c)
281.(c)	282.(a)	283.(c)	284.(c)	285.(a)	286.(b)	287.(c)	288.(d)	289.(b)	290.(b)
291.(c)	292.(c)	293.(d)	294.(c)	295.(d)	296.(d)	297.(b)	298.(b)	299.(a)	300.(c)
301.(b)	302.(c)	303.(b)	304.(c)	305.(a)	306.(c)	307.(d)	308.(c)	309.(a)	310.(a)
311.(d)	312.(d)	313.(a)	314.(a)	315.(b)	316.(b)	317.(a)	318.(b)	319.(d)	320.(d)
321.(b)	322.(a)	323.(c)	324.(b)	325.(c)	326.(d)	327.(d)	328.(c)	329.(b)	330.(c)
331.(b)	332.(c)	333.(b)	334.(c)	335.(a)	336.(a)	337.(d)	338.(a)	339.(d)	340.(b)

341.(a)	342.(a)	343.(b)	344.(d)	345.(c)	346.(d)	347.(d)	348.(b)	349.(d)	350.(a)
351.(d)	352.(a)	353.(a)	354.(a)	355.(a)	356.(b)	357.(a)	358.(c)	359.(b)	360.(c)
361.(b)	362.(b)	363.(a)	364.(b)	365.(d)	366.(c)	367.(a)	368.(b)	369.(d)	370.(c)
371.(b)	372.(d)	373.(c)	374.(b)	375.(b)	376.(b)	377.(a)	378.(c)	379.(b)	380.(d)
381.(b)	382.(a)	383.(a)	384.(d)	385.(d)	386.(c)	387.(c)	388.(b)	389.(b)	390.(a)
391.(d)	392.(a)	393.(b)	394.(c)	395.(d)	396.(b)	397.(b)	398.(c)	399.(c)	400.(d)
401.(d)	402.(b)	403.(d)	404.(b)	405.(d)	406.(d)	407.(c)	408.(a)	409.(c)	410.(b)
411.(c)	412.(b)	413.(d)	414.(d)	415.(b)	416.(b)	417.(a)	418.(c)	419.(a)	420.(c)
421.(b)	422.(b)	423.(d)	424.(b)	425.(a)	426.(a)	427.(a)	428.(d)	429.(b)	430.(a)
431.(b)	432.(b)	433.(b)	434.(a)	435.(d)	436.(b)	437.(a)	438.(b)	439.(b)	440.(b)
441.(a)	442.(c)	443.(c)	444.(b)	445.(a)	446.	447.(c)	448.(b)	449.(b)	450.(c)
451.(b)	452.(c)	453.(d)	454.(c)	455.(c)	456.(b)	457.(d)	458.(c)	459.(a)	460.(b)
461.(a)	462.(a)	463.(c)	464.(a)	465.(c)	466.(a)	467.(x)*	468.(y)*	469.(b)	470.(b)
471.(c)	472.(a)	473.(b)	474.(d)	475.(d)	476.(a)	477.(a)	478.(c)	479.(b)	480.(c)
481.(b)	482.(b)	483.(b)	484.(b)	485.(b)	486.(a)	487.(a)	488.(d)	489.(a)	490.(d)

x^* = 467: A–4, B-5, E–2, F–3
y^* = 468: A–4, B-3, C–2, D–1

Explanatory Notes

1. (a) Scotch yoke is used to generate sine functions.
2. (b) The danger of breakage and vibration is maximum near critical speed.
3. (d) $z_{\min} = \dfrac{2}{\sin^2 14.5°} = 32$
4. (d) Inversion of a mechanism is obtained by fixing different links in a kinematic chain.
5. (c)
6. (d) Logarithmic decrement, $\delta = \ln\left[\dfrac{x_n}{x_{n+1}}\right]$

   $\dfrac{x_n}{x_{n+1}} = e^{\delta}$
7. (c) A–3, B–2, C–1, D–4
8. (c) $500 + 50 \sin 2\theta = 500 + 50 \sin \theta$

   $\sin 2\theta - \sin\theta = 0$

   $2\sin\theta\cos\theta - \sin\theta = 0$

   $\sin\theta(2\cos\theta - 1) = 0$

   Either $\sin\theta = 0$, giving $\theta = 0°, 180°, 360°$

   or $\cos\theta = \dfrac{1}{2}$, giving $\theta = 60°, 300°$

   $\therefore \theta = 0°, 60°, 180°, 300°$, *i.e.* 4 times.
9. (b)
10. (c)
11. (d) A–4, B–1, C–2, D–3
12. (d)
13. (d)
14. (b)
15. (c)
16. (c) A–2, B–3, C–4, D–1
17. (a) $f^{cr} = 2v\omega$
18. (b) $V_C = \dfrac{CD}{AB} \times V_B = \dfrac{45}{30} \times V = \dfrac{3}{2} V$
19. (b) A–2, B–4, C–1, D–2
20. (a)
21. (d) A–3, B–5, C–1, D–2
22. (c)
23. (d)
24. (c) Energy stored, $E \alpha I$ and $I \alpha R^2$

    $\dfrac{E_2}{E_1} = \left(\dfrac{R_2}{R_1}\right)^2 = \left(\dfrac{1}{2}\right)^2 = \dfrac{1}{4}$
25. (b) K.E. of flywheel $= \dfrac{K_e E}{2K_s}$
26. (a) For an isochronous governor, $F = qr$.
27. (d) For a stable governor, the controlling force must increase as the radius of rotation increases. Therefore, the controlling force line when produced must intersect the controlling force axis below the origin, *i.e.* line IV.
28. (c)
29. (a)
30. (c)
31. (c)
32. (b) $\dfrac{l_1}{l_2} = \dfrac{I_2}{I_1}$

    As $I_1 < I_2$, therefore $l_1 > l_2$

    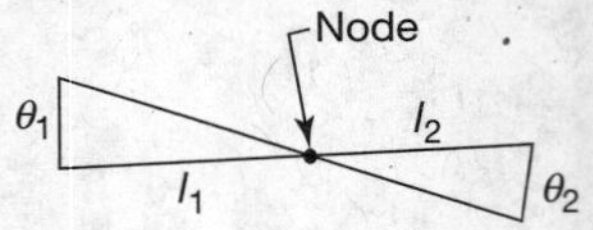

33. (a) $\omega_n = \sqrt{\dfrac{k}{m}}$

    $\dfrac{\omega_{n2}}{\omega_{n1}} = \dfrac{\sqrt{(k/2)/(2m)}}{\sqrt{(k/m)}} = \sqrt{\dfrac{1}{4}} = \dfrac{1}{2}$

    $N_2 = N/2$
34. (c)
35. (a) For the system to vibrate,

    $(ka^2 - wb) > 0$

    or $ka^2 < wb$

    or $b < \dfrac{ka^2}{w}$
36. (d)
37. (a)
38. (c)
39. (d) A–4, B–2, C–3, D–1

40. (c)
41. (d)
42. (d)
43. (b)
44. (b)
45. (d) $F = 2P \tan \alpha$
46. (b) $\omega_1 \times AB = \omega_2 \times CD = \omega_2 (AB + 30)$

$$5 \times AB = 2 (AB + 30)$$
$$3AB = 60$$
$$AB = 20 \text{ cm}$$

47. (c) $n = 2, p = 1, h = 0$

No link is fixed in the two-link system.

$\therefore F = 3n - 2p = 3 \times 2 - 2 \times 1 = 4$

48. (a) Loss of energy,

$$\Delta\text{kE} = \frac{1}{2}(I_1\omega_1^{\,2} + I_2\omega_1^{\,2}) - \frac{1}{2}(I_1\omega_1'^2 + I_2\omega_2'^2)$$

For momentum balance,

$$I_1\omega_1 + I_2\omega_2 = I_1\omega_1' + I_2\omega_2'$$

$$\omega_1' = \omega_2' = \frac{I_1\omega_1 + I_2\omega_2}{I_1 + I_2}$$

$$\Delta\text{kE} = \frac{1}{2}(I_1\omega_1^2 + I_2\omega_2^2) - \frac{1}{2}(I_1 + I_2)\left[\frac{I_1\omega_1 + I_2\omega_2}{I_1 + I_2}\right]^2$$

$$= \left[\frac{1}{2}(I_1 + I_2)^2(I_1\omega_1^2 + I_2\omega_2^2) - \frac{1}{2}(I_1 + I_2)(I_1\omega_1 + I_2\omega_2)^2\right] \times \frac{1}{(I_1 + I_2)^2}$$

$$= \frac{1}{2}\Big[(I_1^3\omega_1^2 + I_1^2 I_2\omega_2^2 + I_1 I_2^2\omega_1^2 + I_2^3\omega_2^2 + 2I_1^2 I_2\omega_1^2 + 2I_1 I_2^2\omega_2^2)$$
$$- (I_1^3\omega_1^2 + I_1 I_2^2\omega_2^2 + I_1^2 I_2\omega_1^2 + I_2^3\omega_2^2 + 2I_1^2 I_2\omega_1\omega_2 + 2I_1 I_2^2\omega_1\omega_2)\Big] \times \frac{1}{(I_1 + I_2)^2}$$

$$= \frac{1}{2(I_1 + I_2)^2}\left[I_1^2 I_2\omega_2^2 + I_1 I_2^2\omega_1^2 + I_1^2 I_2\omega_1^2 + I_1 I_2^2\omega_2^2 - 2I_1 I_2^2\omega_1\omega_2 - 2I_1^2 I_2\omega_1\omega_2\right]$$

$$= \frac{I_1 I_2}{2(I_1 + I_2)^2}\left[I_1\omega_2^2 + I_2\omega_1^2 + I_1\omega_1^2 + I_2\omega_2^2 - 2I_2\omega_1\omega_2 - 2I_1\omega_1\omega_2\right]$$

$$= \frac{I_1 I_2}{2(I_1 + I_2)^2}\left[\omega_1^2(I_1 + I_2) + \omega_2^2(I_1 + I_2) - 2\omega_1\omega_2(I_1 + I_2)\right]$$

$$= \frac{I_1 I_2}{2(I_1 + I_2)}(\omega_1 - \omega_2)^2$$

49. (b) For a stable governor, $F_c = ar - b$

For an unstable governor, $F_c = ar + b$

$+b$ can be changed to $-b$ by decreasing the spring stiffness.

50. (c)
51. (a) $R_1 = \dfrac{W}{2}$ or $W = 2R_1$

$$R_2 \times 2L = W \times L = 2R_1 \times L$$
$$R_2 = R_1$$

52. (a)
53. (d) $m_1 N_1^2 = (m_2 N_2^2) \times X$

$$m_1 \times 3000^2 = 0.5\text{m} \times 300^2 \times X$$
$$X = 2 \times 100 = 200$$

54. (c)

$$\omega_n = \sqrt{\frac{k}{M}} = \sqrt{\frac{85 \times 10^3}{100}} = 29.155 \text{ rad/s}$$

$$\beta = \frac{\omega}{\omega_n} = \frac{20\,\pi}{29.155} = 2.155$$

$$X = \frac{\left(\frac{me}{M}\right)\beta^2}{\pm(1-\beta^2)} = \frac{20 \times 0.5}{100} \times \frac{(2.155)^2 \times 10^{-3}}{\left[(2.155)^2 - 1\right]}$$

$$= 1.270 \times 10^{-4} \text{ m}$$

55. (b) A–2, B–1, C–4, D–3

56. (c)

$$c_C = 2\sqrt{mk} = 2\sqrt{1 \times 0.7 \times 10^3} = 52.92 \text{ N·s/m}$$

57. (a) Force in spring 2 = mg

Force in spring $1 = \frac{mgl}{a}$

Deflection of mass m, δ_{st} = deflection of spring 2 + deflection of spring $1 \times \frac{l}{a}$

$$= \frac{mg}{k_2} + \frac{mgl}{a} \times \frac{l}{a} \times \frac{1}{k_1}$$

$$= mg\left[\frac{1}{k_2} + \frac{l^2}{k_1 a^2}\right]$$

Natural frequency, $f_n = \frac{1}{2\pi}\sqrt{\frac{g}{\delta_{st}}}$

$$= \frac{1}{2\pi}\frac{1}{\sqrt{m\left[\frac{1}{k_2} + \frac{l^2}{a^2} + \frac{1}{k_1}\right]}}$$

$$= \frac{1}{2\pi}\sqrt{\frac{k_1 k_2 a^2}{m(a^2 k_1 + k_2 l^2)}}$$

$$= \frac{1}{2\pi}\left[\frac{k_1 k_2}{m(k_1 + \left(\frac{l}{a}\right)^2 k_2}\right]^{1/2}$$

$$\text{or } f_n = \frac{1}{2\pi}\sqrt{\frac{k_1 k_2 \left(\frac{a}{l}\right)^2}{m\left[k_2 + \left(\frac{a}{l}\right)^2 k_1\right]}}\ Hz$$

58. (c)

59. (a) Applying Dunkerly's method,

$$\frac{1}{f_n^2} = \frac{1}{f_1^2} + \frac{1}{f_2^2}$$

$$= \frac{1}{100^2} + \frac{1}{200^2} = 1.25 \times 10^{-4}$$

$$f_n = 89.44 \text{ Hz} \quad \text{or} \quad 5367 \text{ rpm}$$

60. (b) $\omega = \frac{2\pi N}{60} = \frac{2\pi \times 300}{60} = 10\pi$ rads/s

$$\beta = \frac{\omega}{\omega_n} = \frac{10\pi}{10} = \pi$$

$$r = \frac{e\beta^2}{\beta^{2-1}} = \frac{2 \times \pi^2}{\pi^{2-1}} = 2.25 \text{ mm}$$

61. (b) When $\beta = 1, \phi = 90°$

62. (d)

63. (b)

Operation	Revolutions of		
	Arm, C	Gear A $z_a = 120$	Gear B $z_b = 45$
1. Arm C is fixed, +1 revolutions given to gear *A*, ccw	0	+1	$\frac{-z_a}{z_b} = -\frac{120}{45} = -\frac{8}{3}$
2. Multiply by *x*	0	x	$\frac{-8x}{3}$
3. Add *y*	y	$x + y$	$\frac{-8x}{3} + y$

$$x + y = 0$$

$$\frac{-8x}{3} + y = 100$$

or $\frac{-8x}{3} - x = 100, x = \frac{-300}{11} = -27\frac{3}{11}$

or $y = 27\frac{3}{11}$ rpm

64. (c)
65. (c)
66. (a) The direction of the coriolis acceleration is such as to rotate the sliding velocity vector in the same sense as the angular velocity of the link.
67. (d) Acceleration of piston $P = \omega^2 \times ON$ When N coincides with O, acceleration of piston is zero and velocity is maximum.
68. (b)
69. (a)
70. (d)
71. (c)
72. (d)
73. (d)

$$T_A = +100,\ T_B = 100 - 75 = 25, T_C$$
$$= 25 + 49 = 74,$$
$$T_D = 74 - 77 = -3, T_E = -3 + 43 = 40,$$
$$T_F = 40 - 73 = -33, T_G = -33 + 36 = +3$$

Minimum torque occurs at F.

74. (a) For a stable governor, $F=ar-b$
75. (b) The rotor is statically balanced not dynamically balanced.
76. (c) $f_n = \frac{1}{2\pi}\sqrt{36\pi^2} = \frac{6\pi}{2\pi} = 3$ HZ
77. (b)
78. (b) For $\beta = \sqrt{2}, TR = 1$ as $TR = \frac{1}{\beta^2 - 1}$
79. (c) critical speed does not depend on eccentricity.
80. (b) $\omega_{n1} = \sqrt{\frac{k}{m}},\ ke = \frac{k}{2}, \omega_{n2} = \sqrt{\frac{k}{2m}}$

$$\omega_{n2} = \frac{N}{\sqrt{2}}$$

81. (d)
82. (a)
83. (d)
84. (d)
85. (a)
86. (c)
87. (b)
88. (c) A–4, B–1, C–2, D–3
89. (d) Overall speed ratio $= \frac{20}{40} \times \frac{35}{70} \times \frac{25}{50} = \frac{1}{8}$
90. (a)
91. (d)
92. (a)
93. (a)
94. (d)
95. (a) A–2, B–4, C–1, D–3
96. (c)
97. (d) At resonance, $\beta^2 - 1 = 1$ or $\beta = \sqrt{2}$
98. (a) For critical damping, $\zeta = 1$ and amplitude, $x(t) = (A + Bt)\exp(-\omega_n t)$ Thus, there are no oscillations.
99. (b) $m = 4kg,\ c = 9$ N.s/m, $k = 16$ N/m

$$c_c = 2\sqrt{mk} = 2\sqrt{4 \times 16} = 16 \text{ N.s/m}$$
$$\zeta = \frac{c}{c_c} = \frac{9}{16}$$

100. (c) The springs are in parallel, equivalent stiffness, $k_e = k_1 + k_2$

$$f_n = \frac{1}{2\pi}\sqrt{\frac{k_e}{m}} = \frac{1}{2\pi}\sqrt{\frac{k_1 + k_2}{m}}$$

101. (c)

102. (d)

Operation	Revolutions of		
	Arm, C	Gear A, $z_A = 200$	Gear B, $z_B = 50$
1. Arm C fixed, +1 revolution given to gear A, ccw	0	+1	$\frac{-z_A}{z_B} = -\frac{200}{50}$ $= -4$
2. Multiply by x	0	x	$-4x$
3. Add y	y	$x + y$	$-4x + y$

$x + y = 0$, as gear A is fixed

For $\quad y = 1, x = -1$

$N_B = -4x + y = -4x(-1) + 1 = 5$

103. (a)

104. (d) A–3, B–2, C–4, D–1

105. (b) For correct steering,

$$\cot\theta - \cot\phi = \frac{a}{b}$$

106. (a) For a simple pendulum,

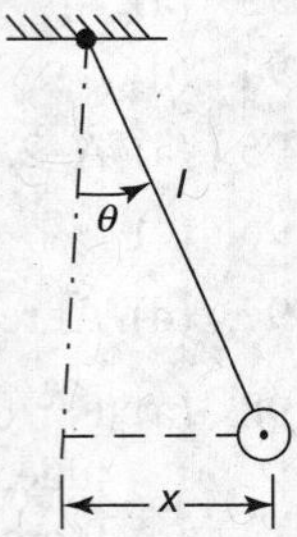

$$x = l\sin\theta$$

$$\dot{x} = l\cos\theta.\frac{d\theta}{dt} = l\omega\cos\theta$$

$$\ddot{x} = -l\omega^2\sin\theta = -\omega^2 x$$

At $x = 0$, $\quad \ddot{x} = 0$

107. (a)

108. (a)

109. (c)

110. (a)

111. (d) A–3, B–4, C–2, D–1

112. (d) A–3, B–2, C–4, D–1

113. (d) A–2, B–3, C–4, D–1

114. (b)

115. (d) A–2, B–4, C–1, D–3

116. (d) Sensitivity of a governor

$$= \frac{N_1 + N_2}{2(N_1 - N_2)}$$

$\therefore$ For an isochronous governor, $N_1 = N_2$

Sensitivity = infinity.

117. (d) $\theta_{sr} = 2\theta_{pd}$ in the same direction

$= 2 \times 30° = 60°$ cw

118. (c)

$$\text{Dynamic force} = \frac{W}{g}e\omega^2$$

$$\text{Couple} = \frac{W}{g}e\omega^2 a$$

$$\text{Reaction at the bearings} = \frac{W}{g}e\omega^2\frac{a}{L}$$

119. (d) The equation of envelope is: $Xe^{-\zeta\omega_n t}$

120. (a) $k_1 = 8$ N/mm, $n_1 = 10$, $n_2 = 5$

$k\,a\,\frac{1}{n}$, $\quad \therefore k_2 = 16$ N/mm

k_1 and k_2 are in parallel.

$\therefore k_e = 8 + 16 = 24$ N/mm

121. (c)

$$y = Y\sin\omega t$$

$$\dot{y} = Y\omega\cos\omega t = Y\omega\sin\left(\frac{\pi}{2} + \omega t\right)$$

$$\text{Damping force, } F_a = c\dot{y} = cY\omega\sin\left(\frac{\pi}{2} + \omega t\right)$$

$$\text{Inertia force, } F_i = m\ddot{y} = -mY\omega^2\sin\omega t$$

$$= mY\omega^2\sin(\pi + \omega t)$$

$\therefore \vec{C}$: damping force

$\vec{D}$: inertia force

122. (c)

123. (d)
124. (b)
125. (d)
126. (c)

Operation	Revolutions of		
	Arm C	Gear A $z_A = 100$	Gear B $z_B = 20$
1. Arm C fixed, +1 revolutions given to gear A, ccw	0	+1	$-\frac{z_A}{z_B} = -\frac{100}{20}$ $= -5$
2. Multiply by x	0	x	$-5x$
3. Add y	y	$x + y$	$-5x + y$

$x + y = 0$ or $x = -y$

For $y = 3, x = -3$

$N_B = -5 \times (-3) + 3 = 18$ rpm

127. (a)
128. (a) $T_A = I_a \alpha_a + \dfrac{G^2 I_b \alpha_b}{\eta}$
129. (c)
130. (d)
131. (c)
132. (b)
133. (d)
134. (d)
135. (c)
136. (b)
137. (c)
138. (b)
139. (a) The velocity ratio will be unity when the driving and driven shafts make equal angles with the intermediate shaft and forks of intermediate shaft should be in the same plane. This happens in case (a).
140. (b) Height of Watt's governor $= \dfrac{g}{\omega^2}$
141. (d) A–4, B–3, C–2, D–1
142. (c)
143. (d)
144. (a)
145. (b)
146. (a)
147. (b) A–4, B–1, C–2, D–3
148. (b)
149. (c) $\omega_c = \omega_n = \sqrt{\dfrac{48EIg}{Wl^3}} = \sqrt{\dfrac{48EI}{ml^3}}$

ω_c reduces if m increases

150. (d)
151. (d)
152. (c) Let k = initial spring rate.

Now $k \alpha \dfrac{1}{n}$

When the spring is cut into two parts, then stiffness of each part becomes $2k$. When these two parts are put in parallel, $k_e = 4k$.

153. (d) A–1, B–3, C–4, D–2
154. (c)
155. (b)

$$v_A = \omega \times DA = \frac{2\pi \times 100}{60} \times 30 = 314 \text{ cm/s}$$

156. (c)
157. (b)
158. (c)
159. (b)

$a = 14.5°, z = 48, d = 28.8$ cm

Module, $m = \dfrac{d}{z} = \dfrac{28.8 \times 10}{48} = 6$ mm

Circular pitch, $p = \dfrac{\pi d}{z} = \pi m = 18.84$ mm

Addendum, $h_a = m = 6$ mm

Diametral pitch, $P = \dfrac{1}{m} = \dfrac{1}{6}$

160. (b)
161. (a)
162. (d)
163. (a)
164. (a)
165. (b)

166. (d)
167. (b)
168. (d) Secondary force, $F_s \alpha \frac{1}{n} \alpha \frac{r}{l}$

If n increases, F_s decreases.

169. (a)
170. (a)
171. (a)
172. (d) Number of nodes = Number of rotors–1
173. (b) Stiffness of cantilever with end load,

$$k = \frac{3EI}{L^3}, \omega_n = \sqrt{\frac{kg}{w}} \alpha \sqrt{\frac{1}{w}}$$

$$\frac{\omega_{n2}}{\omega_{n1}} = \sqrt{2}$$

174. (b) $\frac{N_F}{N_A} = \frac{z_A \times z_C \times z_E}{z_B \times z_D \times z_F}$

$$= \frac{20 \times 25 \times 26}{50 \times 75 \times 65} = \frac{4}{75}$$

$$N_F = 975 \times \frac{4}{75} = 52 \text{ rpm}$$

175. (c)
176. (c) $n = 10, p = 12, h = 1$

$$F = 3(n-1) - 2p - h$$
$$= 3(10-1) - 2 \times 12 - 1$$
$$= 27 - 24 - 1 = 2$$

177. (d)
178. (b) $v \alpha r, v_A = -\frac{v}{2}, v_B = +\frac{v}{2}$
179. (a)
180. (d)
181. (a) $B_2 \times 50 \times 150 = 9 \times 50 \times 50$

$$B_2 = 3 \text{ kg}$$

182. (b)

$$m = 3\text{kg}, \quad c = 9\text{N.s/m}, \ k = 27\text{N/m}$$
$$c_c = 2\sqrt{mk} = 2\sqrt{3 \times 27} = 18\text{N.s/m}$$
$$\zeta = \frac{c}{c_c} = \frac{9}{18} = 0.5$$

183. (b)

$$k_e = \frac{k_1 k_2}{k_1 + k_2} = \frac{10 \times 5}{10 + 5} = \frac{50}{15} = 3.3 \text{ N/mm}$$

184. (c)
185. (d)
186. (a)
187. (d)
188. (b)
189. (a)

$$T_m = (-0.5 + 1 - 2 + 25 - 0.8 + 0.5) \times \frac{100}{4\pi}$$
$$= \frac{580}{\pi} \text{N.m}$$

190. (d)

$$\text{Frequency of oscillations} = \frac{60}{20} = 3 \text{ cpm}$$
$$\text{Angular velocity of precession} = \frac{2\pi}{20} \times \frac{\pi}{30}$$
$$= \frac{\pi^2}{300} \text{ rad/s}$$
$$\text{Amplitude of pitching} = \frac{6\pi}{180} = \frac{\pi}{30} \text{ rad}$$

191. (c)
192. (b) A–3, B–2, C–1, D–4
193. (c) A–2, B–3, C–4, D–1
194. (c) A–5, B–3, C–2, D–1
195. (c)
196. (d)

$$X = 1.25; \omega = 5, f = \frac{5}{2\pi}; T = \frac{1}{5}, \phi = \frac{\pi}{6}$$

197. (b)
198. (c) A–3, B–4, C–1, D–4
199. (a)
200. (d)
201. (c)
202. (d) For maximum speed, $\frac{\omega_b}{\omega_a} = \frac{1}{\cos \alpha}$

$$\omega_b = \frac{500}{\cos 19.5°} = 531 \text{ rpm}$$

203. (c)

204. (a) $J_{eq} = \frac{J}{i^2} = \frac{J}{10^2} = 0.01\,J$

205. (d) A–3, B–2, C–4, D–1

206. (a)

207. (c)

208. (a) A–2, B–5, C–1, D–3

209. (c)

$$\delta_{st} = \frac{mgL^3}{48\,EI}, \omega_n = \sqrt{\frac{g}{\delta_{st}}} = \sqrt{\frac{48\,EI}{mL^3}} \text{ rad/s}$$

210. (a) A–1, B–2, C–4, D–3

211. (b)

212. (a)

213. (a)

214. (d)

215. (b)

216. (a)

217. (d)

218. (c)

219. (d) A–4, B–2, C–1, D–3

220. (a)

221. (a) $F = 3(n-1) - 2p - h$

$n_2 = 3,\ n_3 = 2$

$p = \frac{1}{2}(3 \times 2 + 2 \times 3) = 6$

$n = 5, h = 0$

$F = 3(5-1) - 2 \times 6 = 0$

222. (b) A–3, B–2, C–1, D–4.

223. (c) A–2, B–3, C–1, D–4.

224. (c) $O_1O_2 = 10$ cm, $O_1C = 20$ cm, $O_2A = 5$ cm

$$\cos\frac{\beta}{2} = \frac{O_2A}{O_1O_2} = \frac{5}{10} = \frac{1}{2}$$

$$\frac{\beta}{2} = 60°$$

$$\beta = 120°$$

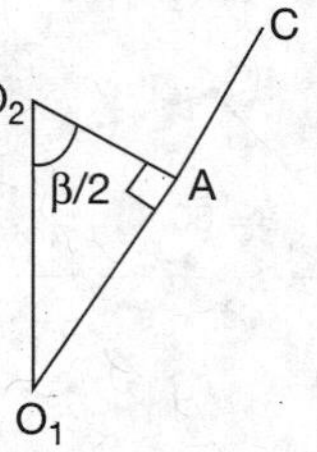

Quick return ratio =

$$\frac{360° - \beta}{\beta} = \frac{360 - 120}{120} = \frac{240}{120} = 2$$

225. (c) Energy stored in a flywheel rim $\alpha\, K^2\ \alpha\, r^2$ where K = radius of gyration.

$$\frac{E_2}{E_1} = \frac{(r/2)^2}{r^2} = \frac{1}{4}$$

226. (c) For an isochronous governor, $F = ar$, *i.e* F v's r curve passes through the origin.

227. (b) Radius of the friction circle = μr where r = journal radius

228. (b) In a collar bearing,

$$p\ \alpha\ \frac{1}{n} \quad \text{and} \quad T_f\ \alpha\ n$$

where n = number of collars.
Therefore, pressure will be halved and friction torque doubled.

229. (b) Lift of sleeve, $h = \frac{b}{a}(r_1 - r_2)$

$$= \frac{10}{20}(20 - 14) = 3 \text{ cm}$$

230. (d) Equation of motion is:

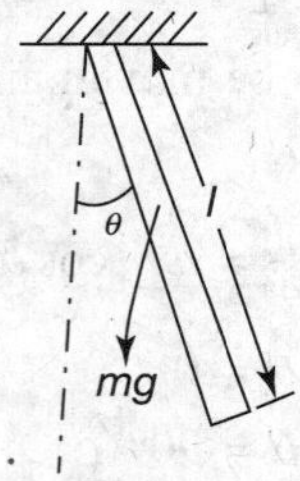

$$\left(\frac{ml^2}{3}\right)\ddot{\theta} + \left(\frac{mgl}{2}\right)\theta = 0$$

or $$\ddot{\theta} + \frac{3}{2}\left(\frac{g}{l}\right)\theta = 0$$

$$\omega_n = \sqrt{\frac{3g}{2l}}$$

$$f_n = \frac{1}{2\pi}\sqrt{\frac{3g}{2l}}$$

231. (c) $J_1 l_1 = J_2 l_2$

$$l_2 = \frac{J_1 l_1}{J_2} = \frac{J_1 l}{J_1 + J_2} = \frac{100 \times 110}{100 + 10} = 100 \text{ cm}$$

232. (a)

$$r = \frac{e\beta^2}{1 - \beta^2} = \frac{190 \times (0.9)^2}{1 - (0.9)^2} = 810 \text{ microns.}$$

233. (b) $m_A = \dfrac{mb}{a + b}$

234. (c)

235. (a)

236. (c)

237. (d) $\dfrac{v_A}{v_B} = \dfrac{r_A}{r_B} = \dfrac{80}{140} = \dfrac{4}{7}$

$$r_B = r_A + 300$$
$$7r_A = 4(r_A + 300)$$
$$3r_A = 1200$$
$$r_A = 400 \text{ mm}$$
$$r_B = 700 \text{ mm}$$

238. (b) $v_p = \omega r\left[\sin\theta + \dfrac{\sin 2\theta}{2n}\right]$

For v_p to be maximum, neglecting second term,

$$\frac{dv_p}{d\theta} = \omega r\left[\cos\theta\right] = 0$$

or $\cos\theta = 0$

$\therefore \quad \theta = 90°$

239. (c)

240. (a) $T = I\alpha$

$$I = \frac{1}{2}mr^2 = \frac{1}{2} \times 1 \times (0.2)^2 = 0.02 \text{ kg.m}^2$$

$$T = (F_1 - F_2)\, r = (10 - 5) \times 0.2 = 1 \text{ N.m}$$

$$\alpha = \frac{T}{I} = \frac{1}{0.02} = 50 \text{ rad/s}^2$$

241. (c) A–4, B–2, C–1, D–3

242. (c)

243. (d) $E_f = \dfrac{1}{2} I\left(\omega_{max}^2 - \omega_{min}^2\right)$

$$= \frac{1}{2} I\left(\omega_{max} + \omega_{min}\right)\left(\omega_{max} - \omega_{min}\right)$$
$$= I\omega_{mean}^2 \times \left[\frac{\omega_{max} - \omega_{min}}{\omega_{mean}}\right]$$
$$= I\omega_{mean}^2 \cdot K_s$$

244. (c)

245. (c)

246. (b) $R_A = R_B = 1500$ N

$$Fc = mr\,\omega^2$$
$$= 30 \times 0.01 \times \omega^2$$
$$= 0.3\,\omega^2$$
$$R_A + R_B = F_c = 300$$
$$0.3\,\omega^2 = 300$$
$$\omega = 100 \text{ rad/s}$$

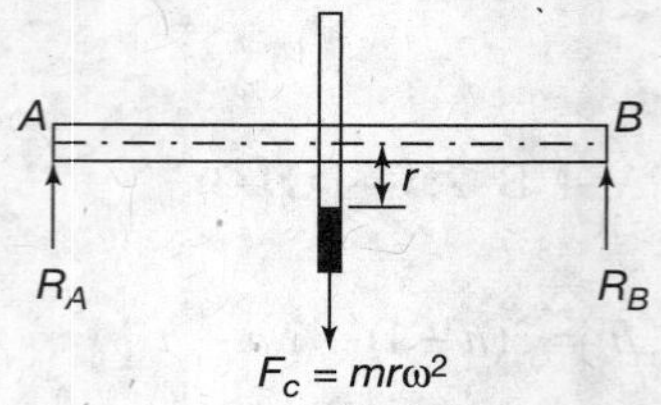

247. (a) Torque,

$$T = F_p r \frac{\sin(\theta + \phi)}{\cos\phi}$$

$$F_n = F_p \tan\phi = \frac{T \sin\phi}{r \sin(\theta + \phi)}$$

$$= \frac{40 \sin 30°}{0.1 \sin(60° + 30°)} = 200 \text{ N}$$

Coulomb friction force $= \mu F_n = 0.08 \times 200 = 16$ N

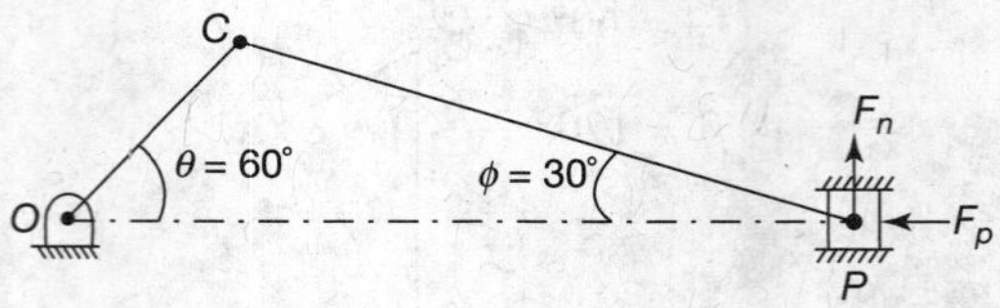

248. (b)

249. (a) $C = \frac{1}{2} m_p \left(z_p + z_q\right) = \frac{1}{2} m_r \left(z_r + z_s\right)$

$$4(20 + 40) = 5(25 + z_s)$$

$$z_s = \frac{4 \times 60}{5} - 25 = 23$$

250. (d)

251. (b) $\frac{2d_p + d_s}{2} = \frac{d_A}{2}$

$$2d_p + d_s = d_A$$

or $2z_P + z_s = z_A$

$$2 \times 20 + z_s = 100$$

$$z_s = 60$$

252. (a)

253. (d)

254. (c) $\omega_{32} = \omega_2 - \omega_3 = 10 - 6 = 4$ rad/s

255. (a)

$CD = l = 400$ mm

$AB = s = 200$ mm

$l + s = 400 + 200 = 600$ mm

$p + q = 300 + 350 = 650$ mm

$(l + s) < (p + q)$

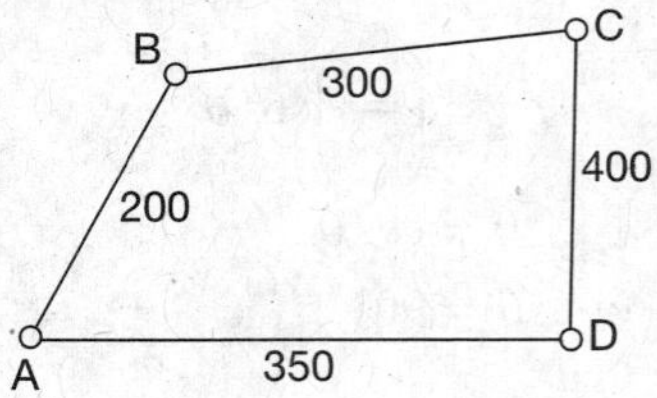

Grashof's criteria is satisfied,

To obtain double crank mechanism, the shortest link *AB* is to be fixed.

256. (d) $x = r(1 - \cos\theta) = r(1 - \cos\omega t)$

$\dot{x} = r\omega \sin\omega t$

It is a sine curve as depicted by (d)

257. (c)

258. (b)

259. (b)

260. (a)

261. (a)

262. (c)

263. (a) $m = 0.5$ kg, $K_G = 5$ cm, $K_e = 6$ cm, $\alpha = 2 \times 10^4$ rad/s^2

Correction couple, $T_C = m\left(K_e^2 - K_G^2\right)\alpha$

$= 0.5\,(36 - 25) \times 10^{-4} \times 2 \times 10^4$

$= 11$ Nm

264. (c)

265. (b) Primary unbalance force, $F_p\ \alpha \cos\theta$, and it is maximum when $\theta = 0°$ and 180°.

266. (a)

267. (a) A–2, B–3, C–4, D–1

268. (b) $N_A = 100$ rpm, *cw* ; $NB = 250$ rpm, $c\omega$;

$T_A = 50$ kN.m, *cw* ; $T_C = 0$

$T_A N_A + T_B N_B + T_C N_C = 0$

$50 \times 100 + T_B \times 250 = 0$

$$T_B = -\frac{50 \times 100}{250} = -20 \text{ kN.B}$$

$T_C = -T_A - T_B = -50 - (-20) = -30$ kN.m

l.e 30 kN.m ccw

269. (d)

270. (b)

271. (a)

272. (a)

273. (a)

$$\frac{1}{k_e} = \frac{1}{k_1 + k_2} + \frac{1}{k_3}$$

$$\omega_n = \sqrt{\frac{k_e}{m}} = \sqrt{\frac{1}{\frac{m}{k_e}}}$$

$$= \left[\frac{1}{m\left[\frac{1}{k_1 + k_2} + \frac{1}{k_3}\right]}\right]^{1/2} \text{ rad/s}$$

274. (b)

275. (a)

276. (c)

277. (a)

$$C = \frac{m}{2}(z_1 + z_2)$$

$$35 = \frac{1}{2}(60 + z_2)$$

$$z_2 = 10$$

$$N_1 = \frac{2400 \times z_2}{z_1} = \frac{2400 \times 10}{60} = 400 \text{ rpm}$$

$$Z_4 = \frac{N_3 z_3}{N_4} = \frac{400 \times 10}{100} = 40$$

$$35 = \frac{m}{2}(z_3 + z_4) = \frac{m}{2}(10 + 40)$$

$$m = \frac{70}{50} = 1.2 \text{ mm}$$

278. (c)

$$\frac{1}{2} I\left(\omega_{\max}^2 - \omega_{\min}^2\right) = \frac{1}{2} T_{\max} \cdot \theta$$

$$1\left(30^2 - 20^2\right) = T_{\max} \times \frac{\pi}{180} \times 60$$

$$T_{\max} = \frac{500 \times 180}{\pi \times 60} = 477.46 \text{ Nm}$$

279. (b)

$$n = \frac{l}{r} = \frac{200}{50} = 4$$

$$\omega = \frac{2\pi N}{60} = \frac{2\pi \times 900}{60} = 94.28 \text{ rad/s}$$

$$f_p = \omega^2 r\left[\cos\theta + \frac{\cos 2\theta}{n}\right]$$

$$= (94.28)^2 \times 50 \times 10^{-3}\left[1 + \frac{1}{4}\right] \text{ for } \theta = 0°$$

$$= 555.16 \text{ m/s}^2$$

Shaking force $= Rf_p = 1.2 \times 555.16$
$= 666$ N

280. (c)
281. (c)
282. (a)
283. (c)
284. (c)
285. (a) For the configuration shown, linear velocity of slotted lever = 0. Hence, angular velocity is also zero.
Coriolis acceleration, $f^{cr} = 2v\omega = 0$
286. (b)
287. (c)
288. (d)
289. (b)
290. (b) A–3, B–4, C–2, D–1
291. (c)
292. (c)
293. (d) $N = \frac{n(n-1)}{2} = \frac{6 \times 5}{2} = 15$
294. (c)
295. (d)
296. (d)
297. (b)
298. (b)
299. (a)
300. (c)
301. (b)
302. (c) $p_x = \frac{p_t}{\tan\alpha}$
303. (b) $\omega_c = \left(\frac{\omega}{n}\right)\cos\theta$

$$0 = \cos\theta$$

or $\theta = 90°$

304. (c) The speed of flywheel increases due to surplus energy in the positive loop. Thus, speed ratio will be greater than unity.
305. (a)
306. (c)
307. (d)
308. (c)
309. (a)
310. (a)
311. (d)
312. (d) Let w = weight of bar per unit length.

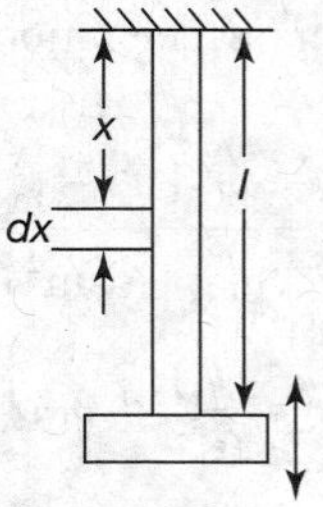

v = velocity of free end of bar

l = length of bar

Velocity of the element dx $= \dfrac{vx}{l}$

K.E. of element dx $= \dfrac{1}{2}\dfrac{wdx}{g}\left(\dfrac{vx}{l}\right)^2$

$$= \frac{wv}{2gl^2}\cdot x^2 \text{dx}$$

Total K.E. of bar

$$= \frac{wv^2}{2gl^2}\cdot\int_o^e x^2 dx = \left(\frac{wl}{3}\right)\left(\frac{v^2}{2g}\right)$$

313. (a)

314. (a)

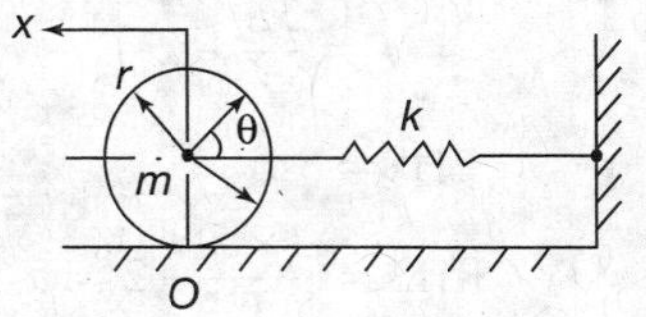

$J_0 = \dfrac{1}{2}\text{m}r^2$

$x = r\,\theta$

$\ddot{x} = r\,\ddot{\theta}$

Equation of motion:

$$J_0\ddot{\theta} + m\ddot{x}\,r + k\,x\,r = 0$$

$$\frac{1}{2}mr^2\ddot{\theta} + mr^2\ddot{\theta} + kr^2\theta = 0$$

$$\frac{3}{2}mr^2\ddot{\theta} + k\,r^2\theta = 0$$

or

$$\frac{3}{2}m\,\ddot{\theta} + k\,\theta = 0$$

$$\omega_n = \sqrt{\frac{2k}{3\text{m}}}\ \text{rad/s}$$

315. (b)

316. (b)

317. (a) Thread thickness, $t_0 = \dfrac{p}{2}$

Number of threads, $n = \dfrac{\text{Length of nut}}{\text{pitch}}$

$$= \frac{t}{p}$$

Shearing area, $A_s = \pi dt_0 n = \dfrac{\pi dt}{2}$

Average shear stress $= \dfrac{F}{A_s} = \dfrac{2F}{\pi dt}$

318. (b) $z_A = z_C$, $z_B = z_D$, $N_A = 800$ rpm, $N_D = 200$ rpm

$$\frac{N_A}{N_B} = \frac{N_C}{N_D} = \sqrt{\frac{800}{200}} = 2$$

$$N_B = \frac{N_A}{2} = \frac{800}{2} = 400 \text{ rpm}$$

$$N_C = 2N_D = 2\times 200 = 400 \text{ rpm}$$

319. (d)

320. (d)

321. (b) Angular velocity of connecting rod,

$$\omega_c = \frac{\omega\cos\theta}{\left(n^2 - \sin^2\theta\right)^{1/2}}$$

$$n = 5,\ \omega = \frac{2\pi\times 3000}{60} = 314.16 \text{ rad/s}$$

$$(\omega_c)_{\max} = \frac{\omega}{n} = \frac{314.16}{5} = 62.83 \text{ rad/s}$$

$$\simeq 60 \text{ rad/s}$$

322. (a)

323. (c)

324. (b)

325. (c)

326. (d)

327. (d) A–2, B–1, C–4, D–3

328. (c)

329. (b) $y_0 = 3$ cm, $\lambda = 5$ m, $m = 900$ kg,

$k = 81\times 10^4$ N/m, $v = \dfrac{72\times 10^3}{3600} = 20$ m/s

Excitation frequency of road on wheel,

$$f = \frac{v}{\lambda} = \frac{20}{5} = 4 \text{ Hz}$$

330. (c)
331. (b)
332. (c) $n = 4, N = \frac{n(n-1)}{2} = \frac{4 \times 3}{2} = 6$
333. (b)
334. (c)
335. (a) $\theta_1 + \theta_2 = 180°, \theta_1 \neq \theta_2$
336. (a)
337. (d) A–4, B–3, C–1, D–2
338. (a)
339. (d)
340. (b) $\omega_{n1} = \sqrt{\frac{k}{m}}, \omega_{n2} = \sqrt{\frac{4k}{m}}$

$$\frac{\omega_{n2}}{\omega_{n1}} = \sqrt{4} = 2$$

341. (a) $I_a l_a = I_b l_b$

$$\frac{l_a}{l_b} = \frac{I_b}{I_a}$$

$$\frac{l_a}{l_a + l_b} = \frac{I_b}{I_a + I_b}$$

$$l_a = \frac{I_b L}{I_a + I_b}$$

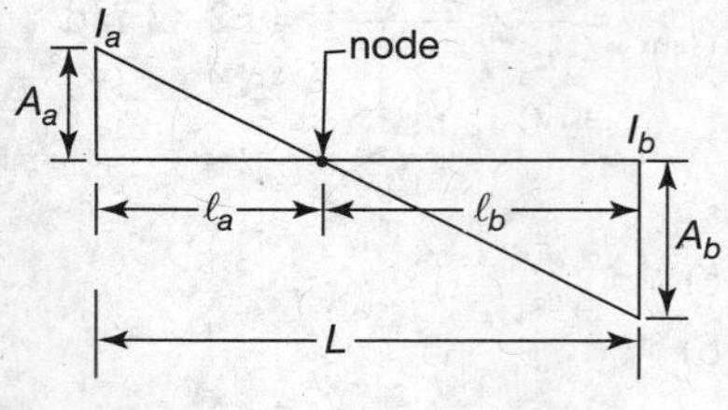

342. (a)
343. (b)
344.
345. (c)
346. (d) $\frac{v_c}{v_b} = \frac{EC}{EB}$

$$v_c = 0.5 \times \frac{0.1}{0.25} = 0.2 \text{ m/s}$$

347. (d)
348. (b)
349. (d) Angular acceleration of connecting rod,

$$\alpha_c = -\frac{\omega^2 (n^2 - 1)\sin\theta}{(n^2 - \sin^2\theta)^{3/2}}$$

For $n = \frac{l}{r} = 1.0, \alpha_c = 0$

350. (a) $\frac{1}{2} T_p \times \pi = T_m \times 4\pi$

$$T_p = 8\ T_m = 8 \times 10 = 80 \text{ Nm}$$

351. (d) $h = \frac{g}{\omega^2} \alpha \frac{1}{N^2}$
352. (a)
353. (a)
354. (a)
355. (a)
356. (b)TR>1 if $\beta < \sqrt{2}$
357. (a)
358. (c)
359. (b)
360. (c) $L = 5, J = 5$

$$J = \frac{3L}{2} - 2$$

LHS = 5

$$\text{RHS} = \frac{3 \times 5}{2} - 2 = 5.5$$

LHS < RHS

The chain is unconstrained

361. (b)
362. (b)
363. (a)
364. (b) $\frac{z_1}{z_2} = \frac{N_2}{N_1}$

$$z_2 = \frac{90 \times 300}{1500} = 18$$

365. (d)
366. (c)
367. (a)
368. (b)

369. (d) $m_1 r_1 = 10 \times 5 = 50$ kg·cm
$m_2 r_2 = 2.5 \times 20 = 50$ kg·cm
Therefore, the rotor system is dynamically balanced.

370. (c) For a critically damped system, the damping factor is equal to one.

371. (b)

372. (d)

373. (c)

374. (b) A–4, B–2, C–3

375. (b)

376. (b)

377. (a)

378. (c) $\frac{1}{S_e} = \frac{1}{2S} + \frac{1}{S}$

$$S_e = \frac{2S^2}{3S} = \frac{2}{3}S$$

379. (b)

380. (d)

381. (b) $\frac{z_D}{z_A} = \frac{60}{20} = 3$

Gears C and D and B and C cannot mesh as they have different pressure angles. Only gears A and D can mesh as $\alpha_A = \alpha_D$ and $m_A = m_D$.

382. (a) $z_{\min} = \frac{2}{\sin^2 20°} = 18$

383. (a)

384. (d) $N_{\max} - N_{\min} = 0.1$ Nm

$$K_s = \frac{N_{\max} - N_{\min}}{N_m}$$

As N_m increases, K_s decreases.

385. (d)

386. (c) $f_{cd}^n = \frac{v_{cd}^2}{CD}$

$$CD = \frac{v_{cd}^2}{f_{cd}^n} = \frac{10^2}{5} = 20 \text{ cm}$$

387. (c)

$$f^{cr} = 2v\omega = 2 \times 10 \times \frac{2\pi \times 60}{60} = 40\pi \text{ cm/s}^2$$

388. (b)

389. (b) A–2, B–1, C–3, D–4

390. (a)

391. (d)

392. (a)

393. (b)

394. (c)

395. (d)

396. (b) Lead, $p_z = 2 \times \text{pitch} = 2 \times 11 = 22$ mm

$d_g = 84$ mm

$$\frac{\omega_1}{\omega_2} = \frac{\pi \text{dg}}{p_z} = \frac{\pi \times 84}{22} = 12$$

397. (b)

398. (c)

399. (c)

400. (d)

401. (d)

402. (b)

403. (d)

404. (b) $c_c = 2\sqrt{mk} = 2\sqrt{5 \times 20} = 20$ N.s/m

405. (d)

406. (d)

Operation	Revolutions of		
	Arm C	Gear A $z_A = 40$	Gear B $z_B = 20$
1. Arm C fixed, +1 revolutions to geal A, ccw	0	+1	$\frac{-40}{20} = -2$
2. Multiply by x	0	x	$-2x$
3. Add y	y	$x + y$	$-2x + y$

$$x + y = 0$$

$$x = -y$$

$$y = 3, x = -3$$

$$N_B = -2x + y = -2\,(-3) + 3 = 9 \text{ revolutions}$$

407. (c)

408. (a)

409. (c)

$$v_{ba} = \sqrt{v_a^2 + v_b^2} = \sqrt{40^2 + 30^2} = 50 \text{ m/s}$$

410. (b)

411. (c)

412. (b)

413. (d) $\omega_n \, \alpha \sqrt{\frac{1}{J}}$

414. (d)

415. (b)

416. (b) $N = \frac{n(n-1)}{2} = \frac{8 \times 7}{2} = 28$

417. (a)

M kg	r m	Mr kg·m	θ deg	H = Mr × cos θ Kg·m	V = Mr × sin θ kg·m
5	2	10	0	10	0
10	1	10	180°	–10	0
3	1	3	45°	2.12	2.12

ΣH = 2.12 kg.m, Σv = 2.12 kg.m

$$R = \sqrt{(2.12)^2 + (2.12)^2} = 3 \text{ kg.m}$$

Let B = mass required for balancing at 2 m radius.

Then $2B = 3$

$B = 1.5$ kg

$$\tan\theta = \frac{2.12}{2.12} 1, \theta = 45°$$

Angle of balance mass from 5 kg mass = 180 + 45 = 225°

418. (c)

419. (a) $\omega_d = \omega_n\sqrt{1-\zeta^2}$

$$20 = 40\sqrt{1-\zeta^2}$$

$$1-\zeta^2 = 0.25$$

$$\zeta^2 = 0.75$$

$$\zeta = \frac{\sqrt{3}}{2}$$

420. (c) l = 7cm, $s = 3, p = s, q = 6$

$L + s = 7 + 3 = 10$

$P + q = 5 + 6 = 11$

$(l + s) < (p + q)$

∴ Grashof's criteria is satisfied.

The adjacent link OC to the shortest link OA is fixed. Therefore, it is a crank–rocker mechanism.

421. (b) $Z_g = 30$, $d_g = 210$ mm,

Axiel pitch of worm = circular pitch of worm gear

$$= \frac{\pi dg}{Zg} = \frac{\pi \times 210}{30} = 22 \text{ mm}$$

422. (b)

Operation	Revolutions of			
	Arm A	Sungear S $z_s = 48$	Planet P $z_p = 24$	Angular gear I
1. Arm fixed, +1 revolution to s, ccw	0	+1	$-\frac{z_s}{z_p}$	$-\frac{z_s}{z_I}$
2. Multiply by x	0	x	$\frac{-xz_s}{z_p}$	$\frac{-xz_s}{z_I}$
3. Add y	y	$x + y$	$\frac{-xz_s}{z_p} + y$	$-xz_s/z_1 + y$

Sun gear is fixed, $x + y = 0, x = -y$

$2z_p + z_s = z_I$

$z_I = 2 \times 24 + 48 = 96$

$$N_I = -x\frac{z_s}{z_I} + y = y\left(\frac{z_s}{z_I} + 1\right)$$

$$= y = \left(\frac{48}{96} + 1\right) = \frac{3y}{2}$$

$$\frac{N_I}{N_A} = \frac{3}{2} = 1.5$$

423. (d) $T_1\, \omega_1 + T_2\, \omega_2 = 0$
$P = 100$ kW, $\omega_1 = 100$ rad/s

$$P = \frac{T_1\omega_1}{10^3},\ T_1 = \frac{100 \times 10^3}{100} = 1000 \text{ Nm}$$

$\omega_2 = -10$ rad/s

$$\text{T}_2 = -\frac{T_1\omega_1}{\omega_2} = -\left(\frac{1000 \times 100}{-10}\right) = 10^4 \text{ Nm}$$

Holding torque, $\text{T}_3 = -(\text{T}_1 + \text{T}_2)$
$= -(1000 + 10^4)$
$= -11$ kNm

424. (b) $TR = 1$ at $\beta = 0$ and $\beta = \sqrt{2}$
425. (a)
426. (a) A–1, B–2, C–3, D–4
427. (a)
428. (d)
429. (b)
430. (a)

431. (b) System A: $k_e = 2k,\ \omega_n = \sqrt{\dfrac{k_e}{m}} = \sqrt{\dfrac{2k}{m}}$

$$T_A = \frac{2\pi}{\omega_n} = 2\pi\sqrt{\frac{m}{2k}} \times \sqrt{\frac{1}{2k}}$$

System B: $k_e\ \dfrac{k_1 k_2}{k_1 + k_2} = \dfrac{k^2}{2k} = \dfrac{k}{2}$

$$T_B \propto \sqrt{\frac{2}{k}}$$

$$\frac{T_B}{T_A} = \sqrt{\frac{2}{k}}\Big/\sqrt{\frac{1}{2k}} = 2$$

432. (b)
433. (b) $I_2 = G^2\, I_2$

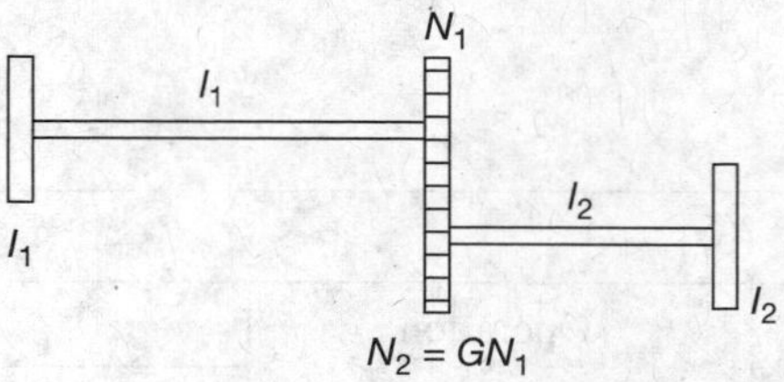

434. (a) $I_A = 4\, I_B$

$$AC = \frac{I_B \times AB}{I_A + I_B}$$

$$= \frac{I_B \times 1}{5\, I_B} = 0.2 \text{ m}$$

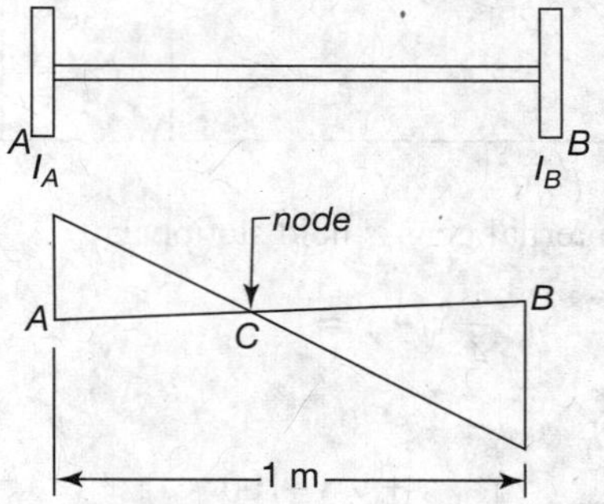

435. (d)
436. (b)
437. (a)
438. (b)
439. (b)
440. (b)
441. (a)
442. (c)
443. (c)
444. (b)
445. (a) $\omega_n = \sqrt{8^2 + 6^2} = 10$ rad/s
446. (c)
447. (c)
448. (b)
449. (b)
450. (c)

451. (b)

452. (c)

453. (d)

454. (c) $n = 5, p = 5$

$$F = 3(n-1) - 2p$$
$$= 3 \times 4 - 2 \times 5 = 2$$

455. (c)

456. (b)

Operation	Revolutions of			
	Arm A	Sun gear S, 20	Planet P, 30	Internal gear I, 80
1. Arm fixed, +1 revolution to S, ccw	0	+1	$\frac{-20}{30} = \frac{-2}{3}$	$\frac{-20}{80} = \frac{-1}{4}$
2. Multiply by x	0	x	$\frac{-2}{3}x$	$\frac{-1}{4}x$
3. Add y	y	$x + y$	$\frac{-2}{3}x + y$	$\frac{-1}{4}x + y$

Internal gear is held stationary.

$$\therefore \quad -\frac{1}{4}x + y = 0$$

or $\quad -x + 4y = 0 \qquad (1)$

$$x + y = -100 \qquad (2)$$

Adding Eqs. (1) and (2),

$$5y = -100$$
$$y = -20$$

or $\quad N_A = 20$ rpm cw

457. (d)

458. (c) and (d)

459. (a)

$$v = v_c \cos[90° - (\alpha + \beta)]$$
$$v_p = v \cos\beta$$
$$= v_c \cos[90° - (\alpha + \beta)] \cos\beta$$

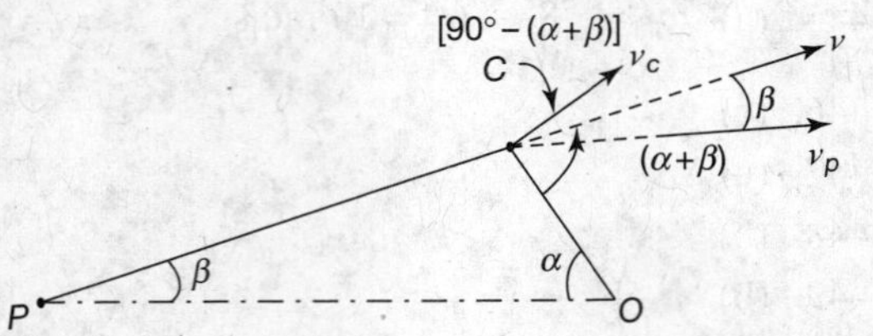

460. $\Sigma M_B = 0$ gives

(b) $R_A \times 2a = W \times a + Pb$

$$R_A = \frac{W}{2} + \frac{Pb}{2a}$$

461. (a) $P = (T_1 - T_2)v$

1. $\frac{T_1}{T_2} = e^{\mu\theta} = e^{0.3 \times \frac{\pi}{180} \times 150} = 2.193$

$T_2 = 0.456\, T_1$

$P = T_1(1 - 0.456)v = 0.544\, T_1 v$

2. $\frac{T_1}{T_2} = e^{0.3 \times \frac{\pi}{180} \times 210} = 3.003$

$T_2 = 0.333\, T_1$

$P = T_1 (1 - 0.333)\, v = 0.667\, T_1 v$

Improvement in power capacity of belt

462. (a)

$$= \left(\frac{0.667 - 0.544}{0.544}\right) \times 100 = 22.6\%$$

$\omega = 10$ rad/s

Velocity = $\omega \times$ slope of displacement curve

$$\frac{\omega\, ds}{d\theta} = 10 \times 1.5 = 15 \text{ m/s}$$

Acceleration = $\omega^2 \times$ slope of velocity curve

$$= 100 \times (-0.05) = -5 \text{ m/s}$$

463. (c) Maximum area of ΔAOB will occur when

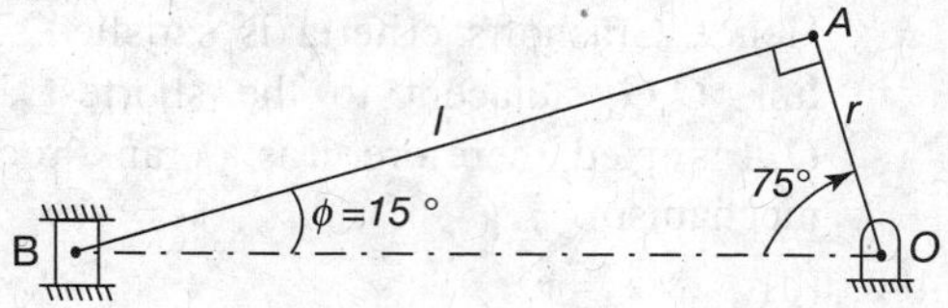

$\angle OAB = 90°$.

$$\tan\phi = \frac{r}{l}$$

or $$\frac{l}{r} = \cot 15° = 3.732$$

464. (a) $T_1 - T_2 = 3000$ N, $v = 15$ m/s

$$\text{Power transmitted} = \frac{(T_1 - T_2)v}{10^3}$$

$$= \frac{3000 \times 15}{10^3} = 45 \text{ kW}$$

465. (c) $x = \sqrt{3}\cos\theta$

$y = \sin\theta$

$\sin^2\theta + \cos^2\theta = 1$

$$y^2 + \frac{x^2}{3} = 1$$

or $$x^2 + 3y^2 = 3$$

At $$\theta = \frac{\pi}{4},\ x = \sqrt{3} \times \frac{1}{\sqrt{2}} = \sqrt{\frac{3}{2}}$$

$$y = \sin\frac{\pi}{4} = \frac{1}{\sqrt{2}}$$

$$\frac{d}{dx}\left(x^2 + 3y^2\right) = 2x + 6y\frac{dy}{dx} = 0$$

Slope of tangent,

$$\frac{dy}{dx} = -\frac{2x}{6y} = -\frac{x}{3y} = -\sqrt{\frac{3}{2}} \times \frac{\sqrt{2}}{3} = -\frac{1}{\sqrt{3}}$$

Slope of normal $$= -\frac{dx}{dy} = \sqrt{3}$$

$$\therefore \quad \theta = \frac{\pi}{3}$$

466. (a) $$E_f = \frac{1}{2} I (\omega_1^2 - \omega_2^2)$$

$$1936 = \frac{1}{2} \times 9.8 (\omega_1 + \omega_2)(\omega_1 - \omega_2)$$

$$\omega_1 - \omega_2 = 2\pi \times \frac{30}{60} = 3.14 \text{ rad/s}$$

$$\omega_1 + \omega_2 = 2\,\omega_m$$

$$1936 = 9.8 \times 3.14 \times co_m$$

$$\omega_m = 62.914 \text{ rad/s}$$

$$\frac{2\pi N_m}{60} = 62.914$$

$$N_m = 600 \text{ rpm}$$

467. A–4, B–5, E–2, F–3

468. A–4, B–3, C–2, D–1

469. (b) At extreme positions,

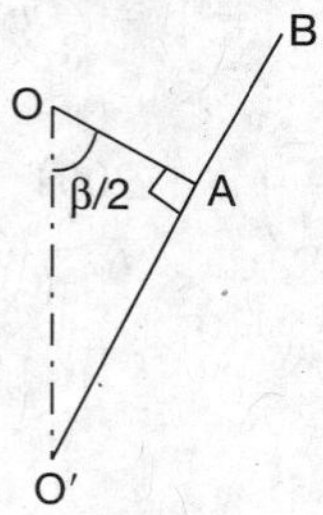

$$\cos\frac{\beta}{2} = \frac{OA}{OO'} = \frac{2}{4} = \frac{1}{2}$$

$$\frac{\beta}{2} = 60°$$

$\beta = 120°$

$\alpha = 360° - 120° = 240°$

$$\text{Ratio} = \frac{240}{120} = 2$$

470. (b) $z_B = z_C$

Operation	Angular speed of		
	Arm *A*	Gear *B*	Gear *C*
1. Arm *A* fixed, +1 revolution to gear *B*, ccw	0	+1	–1
2. Multiply by *x*	0	x	–*x*
3. Add *y*	*y*	$x+y$	$-x+y$

$x + y = 0, x = -y, y = 4$ rad/s, $x = -4$ rad/s

$\omega_c = -(-4) + 4 = 8$ rad/s

471. (c)

472. (a) Longest link should be fixed to obtain double crank mechanism.

473. (b)

$$E_f = \frac{1}{2} I\left(\omega_1^2 - \omega_2^2\right)$$

$$= \frac{1}{2} MK^2 (\omega_1 + \omega_2)(\omega_1 - \omega_2)$$

$$= MK^2\omega_m^2 ks = \left(\frac{2\pi}{60}\right)^2 N_m^2 mk^2 ks$$

$$K^2 = \frac{1}{2}r^2 = \frac{1}{2}\left(\frac{1}{2}\right)^2 = \frac{1}{8} m^2$$

$$2\times10^3 = \left(\frac{2\pi}{60}\right)^2 \times (1200)^2 \times M \times \frac{1}{8} \times 0.02$$

$M = 50.66 – 51$ kg

474. (d) $\dfrac{T_1}{T_2} = e^{\mu\theta}$

$3 = e^{\pi\mu}$

$\pi\mu = \ln 3 = 1.0986$

$\mu = 0.35$

475. (d) $N_g = 1200 \times \dfrac{40}{120} = 400$ rpm

$T_g N_g = T_p N_p$

$$T_g = \frac{20\times1200}{400} = 60 \text{ N m}$$

476. (a)

477. (a)

$$f^{cr} = 2v\omega = 2\times12\times\left(\frac{2\pi\times120}{60}\right) = 302 \text{ m/s}^2$$

478. (c)

479. (b) P–2, Q–3, R–4, S–1

480. (c) P–2, Q–3, R–1

481. (b) $l = AB = \sqrt{240^2 + 100^2} = 260$ mm

$s = O_2A = 60$ mm, $p = 240$ mm, $q = 160$ mm

$l + s = 260 + 60 = 320$ mm

$p + q = 240 + 160 = 400$ mm

$(l + s) < (p + q)$

Hence Grashof's criteria is satisfied. The link O_2O_4 adjacent to the shortest link O_2A is fixed, therefore, it is a crank–rocker mechanism.

482. (b) $v_a = v_b$

$8 \times 60 = \omega_b \times 160$

$\omega_b = 3$ rad/s

483. (b) $\tan\alpha = \dfrac{100}{240} = 0.4167$

$\alpha = 22.62°$

$F_{AB} = 30$ N

$F_1 = F_{AB} \cos\alpha$

$= 30 \cos 22.62° = 27.7$ N

$F_2 = F_{AB} \sin\alpha$

$= 30 \sin 22.62° = 11.54$ N

$R = \sqrt{F_1^2 + F_2^2} = 30$ N

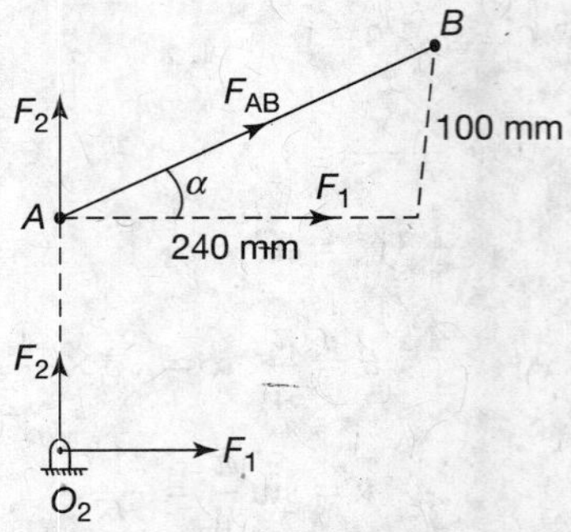

484. (b) $\frac{T_1}{T_2} = e^{\mu\theta} = e^{0.5\times\frac{3\pi}{2}}$

$= 10.55$

ΣM_o gives:

$100 \times 2 - T_2 \times 1 = 0$

$T_2 = 200$ N

$T_1 = 10.55 \times 200 = 2110$ N

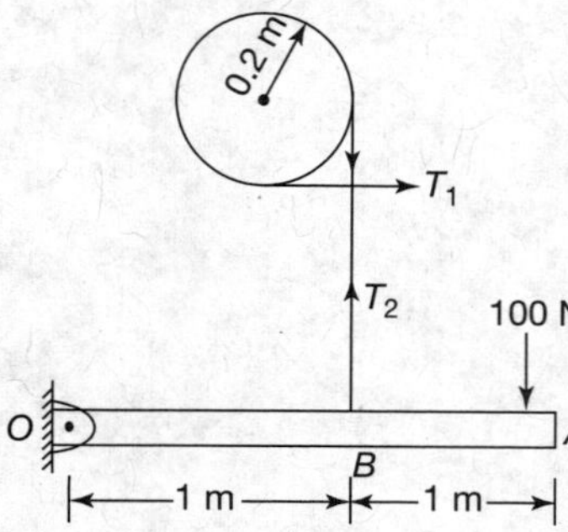

485. (b) Torque $= (T_1 - T_2)\, r$

$= (2110 - 200) \times 0.2$

$= 382$ Nm

486. (a) $P = 5$ kw, $N = 2000$ rpm, $\mu = 0.25$

$r_1 = 25$mm, $p = 2$ MPa

$$P = \frac{2\pi NT}{60 \times 10^3}$$

$$T = \frac{5 \times 60 \times 10^3}{2\pi \times 2000} = 23.873 \text{ N.m}$$

$$p = \frac{F}{\pi\left(r_2^2 - r_1^2\right)}$$

$$F = 1 \times \pi\left(r_2^2 - 25^2\right) = \pi\left(r_2^2 - 625\right) N$$

$$T = \frac{2}{3}\mu F\left[\frac{r_2^3 - r_1^3}{r_2^2 - r_1^2}\right]$$

$$23.873 \times 10^3 = \frac{2}{3} \times 0.25 \times \pi\left(r_2^2 - 625\right)\left[\frac{r_2^3 - 25^3}{r_2^2 - 625}\right]$$

$$45594 = r_2^3 - 25^3$$

$$r_2^3 = 61219$$

$$r_2 = 39.4 \text{ mm}$$

487. (a) $\alpha = 20°$, $z_1 = 19$, $z_2 = 37$, $m = 5$ mm,

$d_1 = \text{m}z_1 = 5 \times 19 = 95$ mm

$d_2 = \text{m}z_2 = 5 \times 37 = 185$ mm

$$C = \frac{1}{2}\left(d_1 + d_2\right) = \frac{1}{2}\left(95 + 185\right) = 140 \text{ mm}$$

488. (d) P–2, Q–6, R–5, S–1

A-4

LAPLACE TRANSFORMS

The Laplace transform $F(s)$ of a function $f(t)$ is defined as follows:

$$L[f(t)] = \int_0^\infty f(t)e^{-st}\,dt$$

where 's' is a complex variable called the Laplace operator and is equal to $j\omega\left(s = \sigma + j\omega\right)$. e^{-st} is called the kernel of the Laplace transformation.

1. Step input

$$f(t) = 0 \text{ for } t < 0$$
$$= u_0 \text{ for } t \geq 0$$
$$F(s) = L[f(t)] = \frac{u_0}{s}$$

For a unit step, $$F(s) = \frac{1}{s}$$

2. Pulse input

$$f(t) = 0 \text{ for } t < 0 \quad \text{and} \quad t > t_0$$
$$= u_0 \text{ for } 0 < t < t_0$$
$$F(s) = \frac{u_0}{s}(1 - e^{-st})$$

3. Exponentially decaying function

$$f(t) = e^{-at}$$
$$F(s) = \frac{1}{s + a}$$

4. Ramp function

$$f(t) = u_0 t \text{ for } t \geq 0$$
$$F(s) = \frac{u_0}{s^2}$$

5. Sinusoidal input

$$f(t) = u_0 \sin \omega t$$

$$F(s) = \frac{u_0 \omega}{s^2 + \omega^2}$$

For $$f(t) = u_0 \cos \omega t$$

$$F(s) = \frac{u_0 s}{s^2 + \omega^2}$$

6. Derivatives

$$L[\dot{f}(t)] = sF(s) - f(0)$$
$$L[\ddot{f}(t)] = s^2 F(s) - sf(0) - \dot{f}(0)$$

7. Inverse Laplace Transform

$$f(t) = L^{-1}[F(s)]$$

Sl. No.	$F(s)$	$f(t)$
1.	$\frac{1}{(s+a)(s+b)}$	$\frac{e^{-at} - e^{-bt}}{b-a}$
2.	$\frac{a}{(s+b)^2}$	ate^{-bt}
3.	$\frac{1}{s(as+1)}$	$1 - e^{-t/a}$
4.	$\frac{s}{(s^2+\omega^2)^2}$	$\frac{1}{2\omega} t \sin \omega t$

INDEX

C

D